Mastering Quantum Mechanics

Mastering Quantum Mechanics

Essentials, Theory, and Applications

Barton Zwiebach

The MIT Press
Cambridge, Massachusetts
London, England

The MIT Press would like to thank the anonymous peer reviewers who provided comments on drafts of this book. The generous work of academic experts is essential for establishing the authority and quality of our publications. We acknowledge with gratitude the contributions of these otherwise uncredited readers.

This book was set in Stone Serif and Stone Sans by Westchester Publishing Services. Printed and bound in the United States of America.

Library of Congress Cataloging-in-Publication Data.

Names: Zwiebach, Barton, 1954– author.
Title: Mastering quantum mechanics : essentials, theory, and applications / Barton Zwiebach, Massachusetts Institute of Technology, Cambridge, MA.
Description: Cambridge, MA : The MIT Press, [2022] | Includes bibliographical references and index.
Identifiers: LCCN 2021016609 | ISBN 9780262046138 (hardcover)
Subjects: LCSH: Quantum theory.
Classification: LCC QC174.12 .Z85 2022 | DDC 530.12—dc23
LC record available at https://lccn.loc.gov/2021016609

10 9 8 7 6 5 4 3 2 1

To my wife, Gaby, with love

Contents

Preface

Mastering Quantum Mechanics offers students a complete overview of quantum mechanics, beginning with an introduction to the essential concepts and results, followed by the theoretical foundations that provide the conceptual framework of the subject, and closing with the tools and applications they will need for advanced studies and research. The book emerged naturally from my many years of teaching quantum mechanics to both Massachusetts Institute of Technology (MIT) undergraduates and students of all ages and backgrounds worldwide. In fact, while teaching MIT's cycle of quantum mechanics courses, I began working with the university's online initiatives, edX and MITx, to offer the courses online at the *same* high academic level that my colleagues and I teach on campus, a feat possible thanks to the sophisticated software of the online platforms. As I developed and taught these courses over the last ten years, I collected and refined a large amount of lecture notes and received comments and input from hundreds of learners. These materials form the backbone of this book. It is my hope that students will use this book in the manner best suited for them, whether to gain a solid understanding of the key components of quantum mechanics or to prepare themselves for graduate work in physics. Most importantly, I hope that my book will guide students at all levels through the complex and fascinating world of quantum mechanics in a clear, precise, accessible, and useful manner. I hope to convey the beauty and elegance of the subject and my enthusiasm for it.

Mastering Quantum Mechanics aims to present quantum mechanics in a modern and approachable way while still including traditional material that remains essential to a well-rounded understanding of the subject. The book is accessible to anyone with a working knowledge of the standard topics typically covered during the first three semesters of undergraduate physics (mechanics, electromagnetism, and waves). It assumes some fluency in mathematics, specifically with multivariable calculus and ordinary differential equations. Familiarity with linear algebra is helpful but not strictly required. I do not assume any previous knowledge of quantum mechanics. By working through this textbook, students will master the essential tools and internalize the main concepts of quantum mechanics. They will gain the strong foundation in quantum theory that is required for graduate work in physics, whether in quantum field theory, string theory, condensed matter physics, atomic physics, quantum computation, or any other subject that uses quantum mechanics. Put differently, *Mastering Quantum Mechanics* need not be a student's final stop in the study of quantum mechanics. On the contrary, the knowledge gained here can be a springboard for future studies, whether a graduate-level course in quantum mechanics or specialized courses in the above-mentioned fields.

When writing this book, I tried to find the most efficient way to make quantum mechanics comprehensible and help the student internalize it. Below are a few ways in which this book attempts to smooth the learning process, in terms of pedagogy.

1. There is a gradual increase in difficulty as the book progresses, closely matching students' increasingly sophisticated understanding of the material. I aim to keep pushing learners at the right pace throughout, challenging them at a constant level.
2. I give a fairly in-depth discussion of the mathematical background required to understand the ideas of quantum mechanics clearly and precisely. This material, dealing with complex vector spaces and linear operators, is developed in two chapters at the beginning of part II. The concepts are illustrated at every step with examples from the quantum systems covered in part I. This approach helps students remember and internalize what they have already learned, as well as understand the math. With this mathematical background in place, students can focus on comprehending the physics at hand.
3. The exposition style is explicit and deliberate, aiming to help students' thought processes by anticipating their potential questions. I also attempt, when possible, to look at facts from various angles and perspectives so that students pause to appreciate the nuances.
4. I have addressed a number of topics at various points at different levels of detail. Readers will encounter Hermitian operators, uncertainty, the harmonic oscillator, angular momentum, and central potentials both in parts I and II. The hydrogen atom appears in all three parts! It would take too long to do full justice to any of these subjects in part I, and students of part I are not yet prepared to absorb all of their important implications. Returning to these topics in parts II and III, with more experience, is in fact helpful to the learning process.
5. When useful, I quote and use important results that will be proven later in the book. For example, while perturbation theory is seen in full detail in part III, some of its results are cited and used in parts I and II. This is a win-win strategy: the student who will not get to part III will gain some exposure to the subject, and the student who gets to part III approaches the subject with some familiarity.
6. I discuss multiparticle states and tensor products ahead of turning to the subject of addition of angular momentum. This sequencing greatly facilitates understanding the latter, which is often a stumbling block in the learning of quantum mechanics.

7. Undergraduate textbooks generally devote limited resources to the advanced material in part III. I have endeavored, at various points, to present the more extended analysis needed to properly grasp this intrinsically subtle material. I do that for density matrices, degenerate perturbation theory, semiclassical approximation, and the adiabatic theorem.
8. Readers can support their studies by viewing videos, recorded in a classroom setting, in which I explain about 90 percent of the material in this book. These videos are available for free on MIT OpenCourseWare (OCW). There are also three online courses (8.04x, 8.05x, and 8.06x) that run at various times and include computer-graded exercises and problems.

Mastering Quantum Mechanics assumes no previous knowledge of quantum mechanics. It is an undergraduate textbook suitable for physics sophomores or juniors. While three semesters are needed to cover most of the material, the book can be used in a number of different ways: for self-learning, for a one-semester or a two-semester course, or to supplement a graduate-level course. All chapters contain exercises and problems. The exercises, inserted at various points throughout the text, are relatively straightforward and can be used by learners to assess their comprehension. The problems at the ends of the chapters are more challenging and sometimes develop new ideas. Mastery of the material requires solving all the exercises and a good fraction of the problems. I have starred the sections that one can safely skip in a first reading of this book.

The book is organized into three parts: part I, "Essentials," presents the basic knowledge of the subject. Part II, "Theory," provides the conceptual framework of quantum mechanics, solidifying the understanding of part I and extending it in various new directions. Finally, part III, "Applications," allows students to internalize all they have learned as valuable techniques are introduced. For the benefit of interested readers and instructors, I sketch below the main features and contents of each part.

I: Essentials This material aims to give the student taking just one semester of quantum mechanics a sound introduction to the subject. Such an introduction, I believe, must include exposure to the following subjects: states and probability amplitudes, the Schrödinger equation, energy eigenstates of particles in potentials, the harmonic oscillator, angular momentum, the hydrogen atom, and spin one-half.

Chapter 1 gives a preview of many of the key ideas that will be developed in the book. Chapter 2 begins with the very surprising possibility of interaction-free measurements in the context of photons on a Mach-Zehnder interferometer, a kind of two-state system illustrating the concept of probability amplitudes. We then trace the ideas that led to the development of quantum theory, beginning with the photoelectric effect and continuing with Compton scattering and de Broglie waves. I explain the stationary phase principle, a tool used throughout this book. Then comes the Schrödinger equation in chapter 3, at which point I introduce position and momentum operators. While deriving the probability current, Hermitian operators appear for the first time. In chapter 4 we use wave packets to motivate Heisenberg's uncertainty principle, and we examine how they evolve in time. Through Fourier transformation and Plancherel's theorem, I introduce the idea of momentum space. Expectation values of operators, and their time dependence, are explored in chapter 5. With the introduction of an inner product, the notation is streamlined to discuss Hermitian operators and their spectrum. We take a preliminary look at the axioms of quantum mechanics. We learn about the uncertainty of an operator in a state, showing that this uncertainty vanishes if and only if the state is an eigenstate of the operator.

With chapter 6 we begin a three-chapter sequence on energy eigenstates. We begin with instructive examples: the particle on a circle, the infinite square well, the finite square well, the delta function potential, and the linear potential, a system naturally solved in momentum space. Chapter 7 focuses on general features. We explore the properties of bound states in one dimension and develop the semiclassical approximation to understand qualitatively the behavior of energy eigenstates. We continue with the node theorem and explore the numerical calculation of energy eigenstates with the shooting method. Here, Students learn how to "remove" the units from the Schrödinger equation, making numerical work efficient. We continue with the virial theorem, the variational principle, and the Hellman-Feynman lemma. The third chapter in the sequence, chapter 8, deals with energy eigenstates that are part of the continuum–scattering states. We use the stationary phase principle to analyze the time-dependent process in which a packet hitting a barrier gives rise to a reflected packet and a transmitted packet.

In chapter 9 we study the harmonic oscillator, finding the energy eigenstates by solving the Schrödinger differential equation, as well as using algebraic methods, with creation and annihilation operators. Our first look into angular momentum and central potentials comes in chapter 10, where we do just enough to be able to understand the hydrogen atom properly. We find the angular momentum commutator algebra and show that we can find simultaneous eigenstates of $\hat{\mathbf{L}}^2$ and $\hat{L}_z$—these are spherical harmonics. For central potentials we discuss the nature of the spectrum of bound states and the boundary conditions at $r=0$. In chapter 11 we study the hydrogen atom, going first through the reduction of the two-body problem into a trivial center-of-mass motion and a relative motion in a Coulomb potential. We derive the energy spectrum from an analysis of the differential equation, discuss the degeneracies, and consider Rydberg atoms. Part I concludes with chapter 12, where we consider the "simplest" quantum system and discover that the natural operators in this system describe intrinsic angular momentum. We discuss the Stern-Gerlach experiment and see how to construct general states of spin one-half particles.

Part I does not use Dirac's bra-ket notation except for a preview in chapter 12. A proper discussion of bra-kets is lengthy, as one needs to think of bras as elements of dual vector spaces. This subject is taken up in part II, chapter 14. Instead of bra-kets, part I uses inner products. This mathematical structure, well worth learning about, makes the definition of Hermitian operators very simple.

II: Theory Part II aims to develop the theory required to understand the foundations of quantum mechanics, to better understand the quantum systems of part I, and to extend the scope of quantum systems under study. It begins with mathematical tools and then turns to the pictures of quantum mechanics and the axioms of quantum mechanics. We learn about entanglement while introducing tensor products, and we deepen our understanding of angular momentum by studying the addition of angular momentum. Identical particles complete part II.

Chapters 13 and 14 give the reader the requisite foundations in complex vector spaces and linear operators. At every step, physical examples illustrate the mathematical concepts. These examples are from systems already encountered in part I, thus serving both as a review of concepts as well as a way to deepen the understanding of these systems. These chapters also present important computational tools: index manipulation, Pauli matrix identities, matrix representations, matrix exponentials, commutator identities, including Hadamard's lemma, and simple cases of the Campbell-Baker-Hausdorff formula. We discuss orthogonal projectors and the construction of rotation operators for spin states. Finally, after touching upon it in part I, I present the bra-ket notation of Dirac, showing how it relates to inner products and to dual spaces.

In chapter 15 we review the concept of uncertainty, giving it a geometric interpretation and then proving the uncertainty inequality. We prove the spectral theorem for finite dimensional vector spaces, leading to the concept of a complete orthonormal set of projectors. I demonstrate how to build a complete set of commuting observables. Chapter 16 discusses unitary time evolution, showing how it implies the Schrödinger equation. The chapter also presents the Heisenberg picture of quantum mechanics. Finally, having all the mathematical and physics background ready, we discuss the axioms of quantum mechanics. Chapter 17 studies the dynamics of quantum systems. We explore two different, important systems: coherent states of the harmonic oscillator and nuclear magnetic resonance. We also develop the factorization, or "supersymmetric" method, useful for finding algebraically the spectrum of one-dimensional potentials. Multiparticle states and tensor products of vector spaces appear in chapter 18. We consider entangled states, the questions raised by Albert Einstein, Boris Podolsky, and Nathan Rosen, and the remarkable resolution provided by the work of John Bell. Teleportation and no-cloning are also presented. Chapter 19 offers our second look at angular momentum and central potentials. This time we use the algebra of angular momentum to work out the finite-dimensional representations. We study spherical free-particle solutions and derive Rayleigh's formula, useful for scattering. In chapter 20 we focus on addition of angular momentum, giving a preview of perturbation theory that allows us to study this subject in the context of the hydrogen atom. We present a derivation of the spectrum of the hydrogen atom using the simultaneous conservation of angular momentum and of the Runge-Lenz vector. The last chapter of part II, chapter 21, deals with identical particles. We emphasize the action of the permutation group on tensor products, giving a clear motivation for the symmetrization postulate and the existence of bosons and fermions. We briefly explain why exotic statistics are allowed when space is two-dimensional.

III: Applications The material in part III helps the student master the key theoretical concepts of part II while developing a wider perspective and a host of tools valuable for research and advanced work. Apart from the important subjects of density matrices and particles on electromagnetic fields, most of the work deals with approximation methods.

Chapter 22 introduces mixed states and density matrices. I discuss bipartite systems and give a brief introduction to open systems, decoherence, and the Lindblad equation. We also discuss measurements in quantum mechanics, touching on the possible relevance of decoherence. We turn to quantum computation in chapter 23, showing how quantum superpositions and interference allow for a surprising speedup of certain computations. In chapter 24 we study the coupling of charged particles to electromagnetic fields, whose gauge transformations are supplemented by transformations of the wave function. We derive the Landau levels of a charged particle inside a uniform constant magnetic field.

We then begin a sequence of chapters on approximation methods: for small, time-independent perturbations; for potentials that vary slowly with position; for small, time-dependent perturbations; and, finally, for perturbations that vary slowly in time. Chapter 25 deals with time-independent perturbation theory. We pay particular attention to the perturbation of degenerate states and to the possibility that degeneracies are not lifted to first order in the perturbation. The fine structure of the hydrogen atom and the Zeeman effect are worked out to illustrate the theory. In chapter 26 the students learn about the WKB approximation, useful for slowly varying potentials. We introduce the so-called connection formulae, show how to use them, and derive them using complex-analytic methods. We apply the WKB method to tunneling and to the delicate calculation of level splitting in double-well potentials.

Chapter 27 is devoted to time-dependent perturbation theory. We obtain a number of results for first-order transitions and give a derivation of Fermi's golden rule, both for time-independent and time-dependent transitions. As a classic application of this theory, we discuss the interaction of light and atoms. We also model and solve for the time dependence of a system where a discrete state is coupled to a continuum. The subject of chapter 28 is the adiabatic approximation, where we consider perturbations that vary slowly in time. We state and prove the quantum adiabatic theorem and go on to discuss Landau-Zener transitions, Berry's phase, and the Born-Oppenheimer approximation.

The last two chapters of part III, chapters 29 and 30, deal with scattering. The first of these examines scattering in one dimension, or, more precisely, scattering on the half line. This is a good setup to learn many of the ideas relevant to three-dimensional scattering. We discuss Levinson's theorem as well as resonances. Turning to three-dimensional scattering in the last chapter, we discuss cross sections, phase shifts, and partial waves. We conclude by setting up integral methods that lead to the Born approximation.

A note on units Unless explicitly noted, I use Gaussian-cgs units. This is particularly convenient when dealing with the hydrogen atom, magnetic moments, and particles in electromagnetic fields. With these units, the Coulomb potential energy of hydrogen, for example, is $V = -e^2/r$. Numerical estimates are easily done by using the following values for the fine-structure constant α and for the product $\hbar c$:

$$\alpha = \frac{e^2}{\hbar c} \simeq \frac{1}{137}, \quad \hbar c \simeq 197.3\,\text{MeV} \cdot \text{fm}.$$

When dealing with magnetic fields and magnetic moments, I often also give values in SI units (from the French Système International). Thus, magnetic fields are often expressed in gauss as well as in tesla. Recall that tesla $= (10^4/c)$ gauss, with the factor of c needed because magnetic fields have different units in the Gaussian and SI systems. This issue first comes up in section 12.2.

How to Use This Textbook

This book emerged from a three-semester structure but is easily adaptable to other settings. Instructors can pick and choose, as there is lots of material for extra learning. Here are some ideas on how to deal with one-semester or two-semester courses and how to use the book to supplement a graduate-level course.

A one-semester course Such a course would be naturally built from part I, which contains most of what could be considered basic material in quantum mechanics. Some instructors wishing to supplement this content could do it as follows. Discuss the basics of time-independent perturbation theory, as summarized in section 20.3, after the study of the Hellmann-Feynman lemma in section 7.9. Give the algebraic derivation of angular momentum multiplets in section 19.3, after studying chapter 10. Supplement the discussion of bra-ket notation in chapter 12 by including the material in section 14.10.

A two-semester course Parts I and II are the natural foundation for a one-year undergraduate course. By skipping some of the part II material (such as the factorization method, Rayleigh's formula, Runge-Lenz vector, and hidden symmetry in the hydrogen atom, among others) the instructor can make some room to include part III material. Prime candidates for inclusion are chapter 22 on density matrices and decoherence, and chapter 24 on charged particles in electromagnetic fields.

Supplementing a graduate-level course Some of the part II and much of the part III material could be used to supplement a graduate course suitable for students who have had just two semesters of undergraduate quantum mechanics. Such a course, using part II material, could start with the axioms of quantum mechanics, as discussed in section 16.6, including background content from the spectral theorem. This would be followed by chapter 18 on multiparticle states and by chapter 21 on identical particles. From part III, many choices are possible, as the chapters are largely independent. Density matrices, charged particles on electromagnetic fields, semiclassical approximation, and adiabatic approximation would be natural choices for students who have already learned time-independent and time-dependent perturbation theory.

Updated information about this book, including errata and additional materials, can be found at https://mitpress.mit.edu/books/mastering-quantum-mechanics.

Acknowledgments

This book grew out of teaching, over a period of ten years, several iterations on the three-course sequence of undergraduate quantum mechanics at MIT. I am grateful to Robert Jaffe, Richard Milner, John Negele, and Krishna Rajagopal for sharing their course notes. I want to thank those colleagues I taught with for sharing their expertise and insights with me. They are Allan Adams, Ray Ashoori, Will Detmold, Edward Farhi, Aram Harrow, Roman Jackiw, Wolfgang Ketterle, Maxim Metlitski, Washington Taylor, Jesse Thaler, and Vladan Vuletic. I also want to thank Ashoke Sen for years of fruitful discussions and productive collaboration. Thank you to Matthew Headrick for the helpful advice and detailed comments. I am particularly grateful to Maxim Metlitski, who shared his insights on a number of subjects, including hydrogen ionization, exotic statistics, double-well potentials, and the adiabatic theorem. A few sections in this book rely heavily on explanations and notes he shared with me.

As I began teaching the quantum sequence, I inherited a good collection of problems that had been previously used for problem sets and for exams. A fraction of those problems have been incorporated into this book, suitably modified and rewritten. I have tried, with mixed success, to locate the original authors of the problems. Some may have originated from other textbooks. If known, the name of the author appears at the end of the problem title. A few problems were inspired by problems or material in other textbooks; in those cases I identify the textbook explicitly.

Starting in 2014, I began working with MITx to construct the online versions of the three-course quantum sequence at MIT. This five-year effort resulted in three online courses, 8.04x, 8.05x, and 8.06x, which have been offered worldwide largely free of charge through edX and are being used at MIT as equivalent online versions of the traditional courses. I wish to thank Physics Department head Peter Fisher, former associate head (now dean of Science) Nergis Mavalvala, and dean for Digital Learning Krishna Rajagopal for supporting this project and for their encouragement in creating this book. The material in this book was tested and improved while running these online courses, where we received input and comments from hundreds of learners, as well as hundreds of comments from learner Johnny Gleeson. I am grateful to Saif Rayyan for his enthusiasm and his help in launching 8.05x and then 8.04x, work that was continued with dedication and care by Jolyon Bloomfield, who helped improve 8.04x and launch 8.06x.

Running the edX courses, we were lucky to have the help of Mark Weitzman, who began by assisting the learners with their questions and soon turned his expertise to helping us proof the material. In preparing this book, I relied on Mark, who read and criticized various versions of the drafts, proposed multiple improvements, and worked through all the problems. I am truly indebted to him for his wisdom and advice.

I wish to thank graduate students Sarah Geller, Andrew Turner, and Gherardo Vita for their help typesetting course notes. I am grateful to Steven Drasco for a careful reading of the 8.04 notes and Luen Malshi for her questions. When the notes started to take final form, former MIT undergraduate Billy Woltz read the full document and suggested a stream of improvements and clarifications. I want also to thank MIT undergraduate Faisal Alsallom, who reviewed much of the draft and asked a number of perceptive questions that helped improve the exposition. I am grateful to Jermey Matthews, editor at MIT Press, for inviting me to publish with them, for his advice, and for his careful work throughout the publishing process. Haley Biermann, also at MIT Press, supported this effort effectively.

I would like to use this opportunity to thank my parents, Betty and Oscar Zwiebach. They followed each step of this project with enthusiasm. Their questions and interest were a constant source of encouragement and motivation. I also want to thank my family for their cheer, my daughter Cecile for timely proofreading, and my wife, Gaby, for her constant support on a project that took years to complete and essentially full-time dedication toward the end. She believed in my ability to explain. To her this book is dedicated.

Barton Zwiebach

Newton, Massachusetts, February 2021

I Essentials

1 Key Features of Quantum Mechanics

After one hundred years of quantum mechanics, we are still discovering some of its surprising features, and the theory remains the subject of much investigation and speculation. The framework of quantum mechanics is a rich and elegant extension of the framework of classical physics. It is also counterintuitive and seems, at first, almost paradoxical. Quantum mechanics has a few key features. The theory is linear, a sign of profound simplicity. Complex numbers are essential to its formulation. The theory is not deterministic: the results of experiments cannot be predicted with certainty; only probabilities can be assessed. General quantum states are represented by superpositions, and when dealing with more than one particle, there is the possibility of entanglement. Quantum mechanics is needed to understand atoms, which are impossible in classical mechanics.

Quantum physics has replaced classical physics as the correct fundamental description of our physical universe. It is used routinely to describe most phenomena that occur at short distances. It also describes some unusual states of macroscopic systems, such as superconductivity and superfluidity. Quantum physics is the result of applying the framework of quantum mechanics to different physical phenomena. We thus have quantum electrodynamics when quantum mechanics is applied to electromagnetism, quantum optics when it is applied to light and optical devices, or quantum gravity when it is applied to gravitation. Quantum mechanics provides a remarkably coherent and elegant framework. The era of quantum physics began in earnest in 1925 with the discoveries of Erwin Schrödinger and Werner Heisenberg. The seeds for these discoveries were planted by Max Planck, Albert Einstein, Niels Bohr, Louis de Broglie, and others. It is a tribute to the human imagination that we have been able to discover the counterintuitive and abstract set of rules that define quantum mechanics. In this chapter we aim to explain and provide some perspective on the main features of this framework.

We will begin by discussing the property of linearity, which quantum mechanics shares with electromagnetic theory. This property tells us what kind of theory quantum mechanics is and why, it could be argued, it is simpler than classical mechanics. We explain that quantum mechanics is the first theory in physics in which complex numbers play an essential role. We then turn to photons, the particles of light. We use photons and polarizers to explain why quantum physics is not deterministic and why, in contrast with classical physics, the results of some experiments simply cannot be predicted. Quantum mechanics is a framework in which we can only predict the *probabilities* for the various outcomes of any given experiment. We continue with quantum superpositions, in which a quantum object somehow appears to exist simultaneously in two mutually incompatible states. A quantum light bulb, for example, could be in a state in which, seemingly, it is both on and off at the same time! We look at the entangled states of two quantum particles, a situation in which the particles, even when far away from each other, can be connected in a subtle and fascinating way. We conclude with a discussion of atoms, explaining why they are impossible in classical physics and how the uncertainty principle of quantum mechanics resolves this predicament.

1.1 Linearity of the Equations of Motion

In physics a theory is usually described by a set of equations for some quantities called the *dynamical variables* of the theory. In classical mechanics, the position and the velocity of a particle are suitable dynamical variables. After writing a theory, the most important task is to find solutions of the equations. A solution of the equations describes a possible reality, according to the theory. Because an expanding universe is a solution of Albert Einstein's gravitational equations, for example, it follows that an expanding universe is possible, according to this theory. A single theory may have a number of solutions, each describing a possible reality.

There are linear theories and nonlinear theories. Nonlinear theories are more complicated than linear theories. Remarkably, if you add two solutions of a linear theory you obtain a third, equally valid solution of the theory. An example of a beautiful linear theory is James Clerk Maxwell's theory of electromagnetism, which governs the behavior of electric and magnetic fields, the dynamical variables in this theory. A field, as you probably know, is a quantity whose values can depend on position and on time. A simple solution of this theory describes an electromagnetic wave propagating in a given direction. Another simple solution could describe an electromagnetic wave propagating in a different direction. Because the theory is linear, a new and consistent solution, the sum solution, has the two waves propagating simultaneously without affecting each other. The electric field in the sum solution is the sum of the electric field in the first solution plus the electric field in the second solution. The same goes for the magnetic field: the magnetic field in the sum solution is the sum of the magnetic field in the first solution plus the magnetic field in the second solution. In fact, you can add any number of solutions and still find a solution. Even if this sounds esoteric, you are totally familiar with it. The air around you is full of electromagnetic waves, each one propagating oblivious to the others. There are waves from thousands of cell phones, waves carrying hundreds of wireless internet messages, waves from a plethora of radio and TV stations, and many, many more. Today, a single transatlantic cable can simultaneously carry millions of telephone calls, together with huge amounts of video and internet data. All of that is courtesy of linearity.

More concretely, we say that Maxwell's equations are **linear** equations. A solution of Maxwell's equations is described by an electric field $\mathbf{E}$, a magnetic field $\mathbf{B}$, a charge density ρ, and a current density $\mathbf{J}$, all collectively denoted as $(\mathbf{E}, \mathbf{B}, \rho, \mathbf{J})$. This collection of fields and sources satisfies Maxwell's equations. Linearity implies that if $(\mathbf{E}, \mathbf{B}, \rho, \mathbf{J})$ is a solution, so is $(\alpha\mathbf{E}, \alpha\mathbf{B}, \alpha\rho, \alpha\mathbf{J})$, where all fields and sources have been multiplied by the constant α. Given two solutions

$$(\mathbf{E}_1, \mathbf{B}_1, \rho_1, \mathbf{J}_1) \quad \text{and} \quad (\mathbf{E}_2, \mathbf{B}_2, \rho_2, \mathbf{J}_2), \tag{1.1.1}$$

linearity also implies that we can obtain a new solution by adding them:

$$(\mathbf{E}_1 + \mathbf{E}_2,\ \mathbf{B}_1 + \mathbf{B}_2,\ \rho_1 + \rho_2,\ \mathbf{J}_1 + \mathbf{J}_2). \tag{1.1.2}$$

The new solution may be called the *superposition* of the two original solutions.

We will consider equations for which the solutions satisfy the following conditions: (i) given a solution, any multiple of the solution is also a solution, and (ii) given two solutions, the sum of the solutions is also a solution. We will call these **linear equations**, and they take the general form

$$L u = 0, \tag{1.1.3}$$

where u denotes the unknown. The unknown may be a number, a function of time, a function of space, a function of spacetime—essentially, anything unknown! In fact, u could represent a collection $(u_1, u_2, \ldots)$ of unknowns. The symbol L denotes a **linear operator**, an object that acts on u and satisfies the following two properties:

$$L(a u) = aLu, \qquad L(u_1 + u_2) = Lu_1 + Lu_2, \tag{1.1.4}$$

where a is a number. When L is a linear operator, the equation $L u = 0$ is a linear equation that satisfies the two conditions (i) and (ii) listed above. Indeed, $Lu = 0$ implies $L(au) = 0$ using the first equation above, and $Lu_1 = Lu_2 = 0$ implies $L(u_1 + u_2) = 0$ using the second equation above. Note that, with α and β two numbers, the linearity of L implies that

$$L(\alpha u_1 + \beta u_2) = \alpha Lu_1 + \beta Lu_2, \tag{1.1.5}$$

showing that if u_1 is a solution ($Lu_1 = 0$) and u_2 is a solution ($Lu_2 = 0$) then $\alpha u_1 + \beta u_2$ is also a solution. We call $\alpha u_1 + \beta u_2$ the **general superposition** of the solutions u_1 and u_2.

Example 1.1. *Linearity of a differential equation.*

Consider the equation

$$\frac{du}{dt} + \frac{1}{\tau} u = 0, \tag{1.1.6}$$

where τ is a constant with units of time. This differential equation can be cast in the form $L u = 0$ if we define the operator L to act as follows:

$$L u \equiv \frac{du}{dt} + \frac{1}{\tau} u. \tag{1.1.7}$$

One sometimes writes the operator L as

$$L \equiv \frac{d}{dt} + \frac{1}{\tau}, \tag{1.1.8}$$

where it is implicit that the derivative acts on whatever is to the right of L, and the constant term $1/\tau$ acts by multiplication. We can quickly verify that L is a linear operator. For any arbitrary constant a we have

$$L(au) = \frac{d(au)}{dt} + \frac{1}{\tau} au = a\Big(\frac{du}{dt} + \frac{1}{\tau} u\Big) = a\,Lu. \tag{1.1.9}$$

Moreover, for two functions u_1 and u_2 we also find that

$$\begin{aligned} L(u_1+u_2) &= \frac{d(u_1+u_2)}{dt} + \frac{1}{\tau}(u_1+u_2) = \frac{du_1}{dt} + \frac{du_2}{dt} + \frac{1}{\tau}u_1 + \frac{1}{\tau}u_2 \\ &= \frac{du_1}{dt} + \frac{1}{\tau}u_1 + \frac{du_2}{dt} + \frac{1}{\tau}u_2 = Lu_1 + Lu_2. \end{aligned} \tag{1.1.10}$$

This shows that the differential equation is actually linear. With a little practice, linear equations are easily recognized by inspection. □

Exercise 1.1. *Let L_1 and L_2 be two linear operators. We define the operator L_1+L_2 by $(L_1+L_2)u \equiv L_1u + L_2u$ and the operator L_1L_2 by $(L_1L_2)u \equiv L_1(L_2u)$. Is L_1+L_2 a linear operator? Is L_1L_2 a linear operator?*

Einstein's theory of general relativity is a nonlinear theory whose dynamical variable is the gravitational field, which describes, for example, how planets move around a star. Being a nonlinear theory, two gravitational wave solutions cannot be added to form a consistent, new gravitational wave solution. This makes Einstein's theory rather complicated; by all accounts, much more complicated than Maxwell's theory. In fact, classical mechanics, as invented mostly by Isaac Newton, is also a nonlinear theory! In classical mechanics the dynamical variables are positions and velocities of particles, acted on by forces. There is no general way to use two solutions to build a third.

Indeed, consider the equation of motion for a particle on a line under the influence of a time-independent potential $V(x)$, which is in general an arbitrary function of x. The dynamical variable in this problem is $x(t)$, the position as a function of time. Letting V' denote the derivative of V with respect to its argument, Newton's second law takes the form

$$m\frac{d^2x(t)}{dt^2} = -V'(x(t)). \tag{1.1.11}$$

The left-hand side is the mass times acceleration, and the right-hand side is the force experienced by the particle in the potential. It is probably worth emphasizing that $V'(x(t))$ is the function $V'(x)$ evaluated for x set equal to $x(t)$:

$$V'(x(t)) \equiv \frac{\partial V(x)}{\partial x}\Big|_{x=x(t)}. \tag{1.1.12}$$

While we could have used an ordinary derivative here, we wrote a partial derivative as is commonly done for the general case of time-dependent potentials. The reason Newton's second law (1.1.11) is not a linear equation for $x(t)$ is because the right-hand side fails to be a linear function of $x(t)$. The left-hand side is linear in $x(t)$, but the function $V'(x(t))$ is usually not linear in $x(t)$. For arbitrary potentials $V(x)$, we expect that

$$V'(ax) \neq aV'(x), \quad \text{and} \quad V'(x_1+x_2) \neq V'(x_1) + V'(x_2). \tag{1.1.13}$$

As a result, given a solution $x(t)$, the scaled solution $\alpha x(t)$ is not expected to be a solution. Given two solutions $x_1(t)$ and $x_2(t)$, $x_1(t)+x_2(t)$ is not expected to be a solution either.

Exercise 1.2. *What is the most general potential $V(x)$ for which the equation of motion for $x(t)$ is linear?*

Quantum mechanics is a linear theory. The signature equation in this theory, the so-called Schrödinger equation, is a linear equation that determines the time evolution of a quantity called the **wave function.** The wave function is the dynamical variable in quantum mechanics, but curiously, its physical interpretation was not clear to Erwin Schrödinger when he wrote his equation in 1925. It was Max Born who months later suggested that the wave function encodes probabilities. This was the correct physical interpretation, but it was thoroughly disliked by many, including Schrödinger, who remained unhappy about it for the rest of his life. The linearity of quantum mechanics implies a profound simplicity. In some sense quantum mechanics is simpler than classical mechanics, as solutions can be added to form new solutions.

The wave function Ψ, read as *psi*, depends on time and often depends on space and other variables. The Schrödinger equation is a partial differential equation that takes the form

$$i\hbar \frac{\partial \Psi}{\partial t} = \hat{H}\Psi, \tag{1.1.14}$$

where i is the imaginary unit ($i=\sqrt{-1}$), and $\hbar$ (pronounced "h-bar") is the (reduced) Planck's constant $\hbar = \frac{h}{2\pi}$, with h as the standard Planck's constant. On the right-hand side of the equation, $\hat{H}$ is the Hamiltonian, also known as the energy operator. The wave function is a complex-valued function, as we will elaborate below. The Hamiltonian $\hat{H}$ is a linear operator that acts on wave functions:

$$\hat{H}(\alpha\Psi) = \alpha\hat{H}\,\Psi, \qquad \hat{H}(\Psi_1 + \Psi_2) = \hat{H}(\Psi_1) + \hat{H}(\Psi_2), \tag{1.1.15}$$

with α an arbitrary complex number. Of course, $\hat{H}$ itself does not depend on the wave function! To check that the Schrödinger equation is linear, we cast it in the form $L\Psi = 0$ with L defined as

$$L\Psi \equiv i\hbar \frac{\partial \Psi}{\partial t} - \hat{H}\Psi. \tag{1.1.16}$$

It is now a simple matter to verify that L is a linear operator. Physically, this means that if Ψ_1 and Ψ_2 are solutions to the Schrödinger equation, then so is the superposition $\alpha\Psi_1 + \beta\Psi_2$, where α and β are both complex numbers.

Exercise 1.3. *As we will soon see, $\hbar$ has units of energy · time. Based on the consistency of the Schrödinger equation, what are the units of the Hamiltonian operator $\hat{H}$?*

1.2 Complex Numbers Are Essential

Quantum mechanics is the first physics theory that truly requires the use of *complex* numbers. The numbers most of us use for daily life (integers, fractions, decimals) are *real* numbers. The set of complex numbers is denoted by $\mathbb{C}$, and the set of real numbers is denoted by $\mathbb{R}$. Complex numbers appear when we combine real numbers with the imaginary unit i, defined to be equal to the square root of minus one: $i \equiv \sqrt{-1}$. This means that i squared must give minus one: $i^2 = -1$. Complex numbers are fundamental in mathematics. An equation like $x^2 = -4$, for an unknown x, cannot be solved if x has to be real. No real number squared gives you minus four. But if we allow for complex numbers, we have the solutions $x = \pm 2i$. You know that quadratic equations often have complex solutions. Mathematicians have shown that all polynomial equations of degree one or greater can be solved in terms of complex numbers. The formula to solve a general quadratic equation tells you that this is true for polynomial equations of degree two.

A complex number z, in all generality, is a number of the form

$$z = a + ib \in \mathbb{C}, \quad a, b \in \mathbb{R}. \tag{1.2.1}$$

Here a and b are real numbers, and ib denotes the product of i and b. The symbol $\in$ means "belongs to," in this case saying that z belongs to the complex numbers. The real number a is called the real part of z, and the real number b is called the imaginary part of z:

$$\text{Re}\, z = a, \qquad \text{Im}\, z = b. \tag{1.2.2}$$

The complex conjugate of z, denoted z^*, is defined by

$$z^* \equiv a - ib. \tag{1.2.3}$$

You can quickly verify that a complex number z is real if $z^* = z$, and it is purely imaginary if $z^* = -z$. The number 0 is both real and imaginary. For any complex number $z = a + ib$, one can define the *norm* of z, denoted $|z|$, to be the *nonnegative, real* number given by

$$|z| \equiv \sqrt{a^2 + b^2}. \tag{1.2.4}$$

You can quickly check that

$$|z|^2 = zz^*. \tag{1.2.5}$$

Complex numbers are represented as vectors in a two-dimensional complex plane. The real part of the complex number is the x component of the vector, and the imaginary part of the complex number is the y component (see figure 1.1). The norm of a complex number, as defined above, is the length of the corresponding vector in the complex plane. Consider the unit-length vector in the complex plane making an angle θ with the x axis: it has x component $\cos\theta$ and y component $\sin\theta$. The vector can therefore be represented by the complex number $\cos\theta + i\sin\theta$. Euler's identity relates this complex number to the exponential of $i\theta$:

$$e^{i\theta} = \cos\theta + i\sin\theta. \tag{1.2.6}$$

A complex number of the form $e^{i\chi}$, with χ real, is called a *pure phase*. Any complex number $z = a + ib$ can be written in *polar form*: $z = re^{i\theta}$, with $r = |z|$ and $\theta = \arctan(b/a)$.

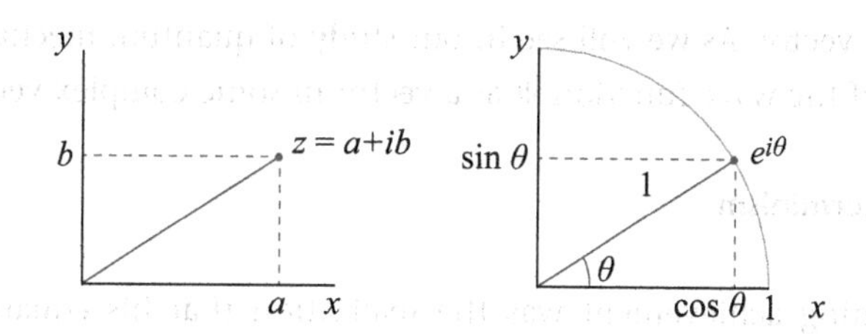

Figure 1.1
Left: The representation of the complex number $z=a+ib$ in the complex plane. *Right*: The unit-norm complex number $e^{i\theta}$ has real part $\cos\theta$ and imaginary part $\sin\theta$.

Exercise 1.4. *Write i as a pure phase. What are the two square roots of i?*

While complex numbers are sometimes useful in classical mechanics or Maxwell's theory, they are not strictly needed. None of their dynamical variables, which correspond to measurable quantities, is a complex number. In fact, all measurements in physics result in real numbers. In quantum mechanics, however, complex numbers are fundamental. The Schrödinger equation (1.1.14) explicitly involves the imaginary unit i on the left-hand side, making the existence of real solutions impossible. Indeed, the wave function, the dynamical variable of quantum mechanics, is itself a complex-valued function. This means that if, for example, the wave function depends on the position $\mathbf{x}$ of a particle and the time t, we have

$$\Psi(\mathbf{x},t)\in\mathbb{C},\quad \text{for all values of } \mathbf{x} \text{ and } t. \tag{1.2.7}$$

At each point in space and at any time, the wave function evaluates to a complex number. As we will learn, the Schrödinger equation admits solutions where Ψ is a wave. Standard wave equations involve second-order partial derivatives with respect to time. The Schrödinger equation, however, only involves a first-order partial derivative with respect to time. The factor of i multiplying this time derivative in fact allows the existence of waves.

Since complex numbers cannot be directly measured, the relation between the wave function and a measurable quantity must be somewhat indirect. Born's natural proposal was to identify probabilities, which are always nonnegative real numbers, with the square of the norm of the wave function. If we write the wave function of our quantum system as Ψ, the probabilities for possible events are computed from $|\Psi|^2$. For this reason, the wave function Ψ is often called the *probability amplitude*. Quantum mechanics gives a surprising way to calculate probabilities. Underlying the probabilities in the theory are complex probability amplitudes. True probabilities arise by taking the norm-squared of the probability amplitude. The equation of the theory, the Schrödinger equation, is not an equation for probabilities but for probability amplitudes.

The mathematical framework required to express the laws of quantum mechanics consists of complex vector spaces. In any vector space, we have objects called vectors that can be added together. In a complex vector space, a vector multiplied by a complex number is still a vector. As we will see in our study of quantum mechanics, it is often useful to think of the wave function Ψ as a vector in some complex vector space.

1.3 Loss of Determinism

Maxwell's crowning achievement was the realization that his equations of electromagnetism allowed for the existence of propagating waves. In particular, in 1865 he conjectured that light was an electromagnetic wave, a propagating fluctuation of electric and magnetic fields. He was proven right in subsequent experiments. Toward the end of the nineteenth century, physicists were convinced that light was a wave. The certainty, however, did not last too long. Experiments on blackbody radiation and on the photoemission of electrons suggested that the behavior of light had to be more complicated than that of a simple wave. Max Planck and Albert Einstein were the most prominent contributors to the resolution of the puzzles those experiments raised.

In order to explain the features of the photoelectric effect, Einstein postulated (1905) that the energy in a light beam comes in quanta so that, in fact, the beam is composed of discrete packets of energy. Einstein essentially implied that light is made up of particles, each carrying a fixed amount of energy. He himself found this idea disturbing, convinced like most other contemporaries that, as Maxwell had shown, light is a wave. He anticipated that a physical entity, like light, that could behave both as a particle and as a wave could bring about the demise of classical physics and would require a completely new physical theory. He was in fact right. Though he never quite liked quantum mechanics, his ideas about particles of light, later given the name *photons*, helped construct this theory.

It took physicists until 1925 to accept that light could behave as a particle. The experiments of Arthur Compton (1923) eventually convinced most skeptics. Nowadays, particles of light, or photons, are routinely manipulated in laboratories around the world. Even if mysterious, we have grown accustomed to them. Each photon of visible light carries very little energy—a small laser pulse can contain many billions of photons. Our eye, however, is a very good photon detector: in total darkness we are able to see light when as few as ten photons hit upon our retina. When we say that light behaves like a particle, we mean a quantum mechanical particle: a packet of energy and momentum that is not composed of smaller packets. We *do not* mean a Newtonian corpuscle, also known as a classical point particle, which is a zero-size object with a definite position and velocity.

As it turns out, the energy of a photon depends only on the color of the light. As Einstein discovered, the energy E and frequency ν for a photon are related by

$$E = h\nu, \tag{1.3.1}$$

where $h \simeq 6.626 \times 10^{-34}\,\mathrm{J \cdot s}$ is Planck's constant. Here J stands for joule and s for seconds, so Planck's constant has units of *energy · time*. In the Schrödinger equation (1.1.14), we find the "reduced" Planck constant $\hbar$, defined by

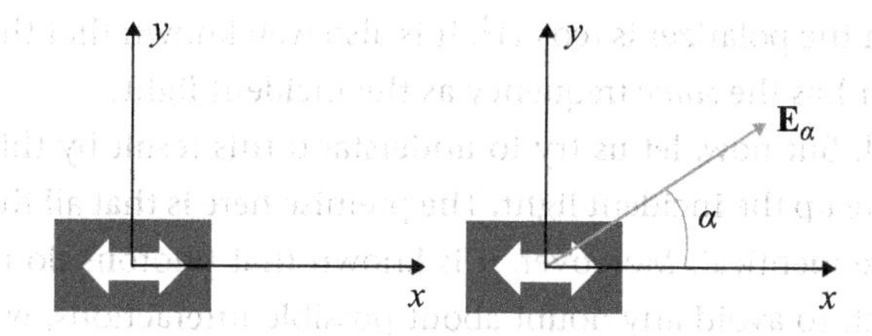

Figure 1.2
Left: A polarizer that transmits light linearly polarized along the $\hat{\mathbf{x}}$ direction. *Right*: Light linearly polarized at an angle α relative to the x axis, hitting the polarizer. The large double-sided arrow denotes the preferential direction of the polarizer, in this case along the x direction.

$$\hbar = \frac{h}{2\pi} \simeq 1.055 \times 10^{-34}\mathrm{J}\cdot\mathrm{s} = 6.582 \times 10^{-16}\ \mathrm{eV}\cdot\mathrm{s}, \tag{1.3.2}$$

where eV stands for electron volt, a unit of energy. The frequency of a photon determines the wavelength λ of the light through the relation $\nu\lambda = c$, where $c \simeq 2.998 \times 10^8$m/s is the speed of light in vacuum. All photons with a wavelength of 520 nanometers are green and have the same energy. To increase the energy in a green light beam, one simply needs more photons.

As we now explain, the existence of photons implies that quantum mechanics is not deterministic. By this we mean that the result of an experiment cannot be determined, as it would be in classical physics, by the conditions that are under the control of the experimenter.

Consider a polarizer whose preferential direction is aligned along the $\hat{\mathbf{x}}$ direction, as shown in figure 1.2, *left*. Imagine sending a beam of light traveling in the z direction, orthogonal to the plane of the figure. If the light is linearly polarized along the $\hat{\mathbf{x}}$ direction, meaning that the electric field lies along the x axis while oscillating back and forth, the light goes through the polarizer. If the polarization of the incident light is orthogonal to the $\hat{\mathbf{x}}$ direction, the light will not go through at all. Thus, light linearly polarized in the $\hat{\mathbf{y}}$ direction will be totally absorbed by the polarizer. Now consider sending light polarized along a direction forming an angle α with the x axis, as shown to the right of the figure. What will happen?

Thinking of the light as a propagating wave, the incident electric field $\mathbf{E}_\alpha$ makes an angle α with the x axis and therefore takes the form

$$\mathbf{E}_\alpha = E_0 \cos\alpha\ \hat{\mathbf{x}} + E_0 \sin\alpha\ \hat{\mathbf{y}}. \tag{1.3.3}$$

This is an electric field of amplitude E_0. Here we are ignoring the time and space dependence of the wave; they are not relevant to our discussion. When this electric field hits the polarizer, the component along $\hat{\mathbf{x}}$ goes through, and the component along $\hat{\mathbf{y}}$ is absorbed. Thus, beyond the polarizer the electric field $\mathbf{E}$ is

$$\mathbf{E} = E_0 \cos\alpha\ \hat{\mathbf{x}}. \tag{1.3.4}$$

You may recall that the energy in an electromagnetic wave is proportional to the square of the amplitude of the electric field. This means that the fraction of the beam's energy that goes through the polarizer is $(\cos\alpha)^2$. It is also well known that the light emerging from the polarizer has the *same* frequency as the incident light.

So far, so good. But now, let us try to understand this result by thinking about the photons that make up the incident light. The premise here is that all the photons in the incident beam are identical. Moreover, it is known that photons do not interact with each other. In fact, to avoid any doubt about possible interactions, we could send the whole energy of the incident light beam one photon at a time. Since all the light that emerges from the polarizer has the same frequency as the incident light and thus the same energy, we must conclude that each individual photon either goes through or is absorbed. If a fraction of a photon were to go through, it would emerge as a photon of lower energy and thus lower frequency, which is something that does not happen.

But now we have a problem. As we know from our wave analysis, a fraction $(\cos\alpha)^2$ of the photons must go through since it is the fraction of the energy that is transmitted. Consequently, a fraction $1-(\cos\alpha)^2$ of the photons must be absorbed. But if all the photons are identical, why is it that what happens to one photon does not happen to all of them?

The answer in quantum mechanics is that there is indeed a loss of determinism. No one can predict whether a photon will go through or will get absorbed. The best anyone can do is to predict probabilities. In this case there would be a probability $(\cos\alpha)^2$ of going through and a probability $1-(\cos\alpha)^2$ of failing to go through.

Two escape routes suggest themselves. Perhaps the polarizer is not really a homogeneous object, and depending exactly on where the photon hits, it either gets absorbed or goes through. Experiments show this is not the case. A more intriguing possibility was suggested by Einstein and others. A possible way out, they claimed, was the existence of *hidden variables*. The photons, while apparently identical, must have other *hidden* properties, not currently understood, that determine with certainty which photon goes through and which photon gets absorbed. Hidden variable theories might seem to be untestable, but in fact they can be tested. Through the work of John Bell and others, physicists have devised clever experiments that rule out most versions of hidden-variable theories. No one has figured out how to restore determinism to quantum mechanics. It seems to be an impossible task.

When we try to describe photons quantum mechanically, we can use wave functions or, equivalently, the language of states. A wave function and a state are essentially equivalent ways of representing the same object. A photon polarized along the x axis is *not* represented using an electric field but rather a *state*:

$$|\text{photon}; x\rangle. \tag{1.3.5}$$

We will learn the rules needed to manipulate such objects, but for the time being, you can think of it as a vector in a space of general photon states. Another state of a photon is

$$|\text{photon}; y\rangle, \tag{1.3.6}$$

representing a photon polarized along the y axis, another vector in the space of photon states. We now claim that each photon in the beam that is polarized along the direction

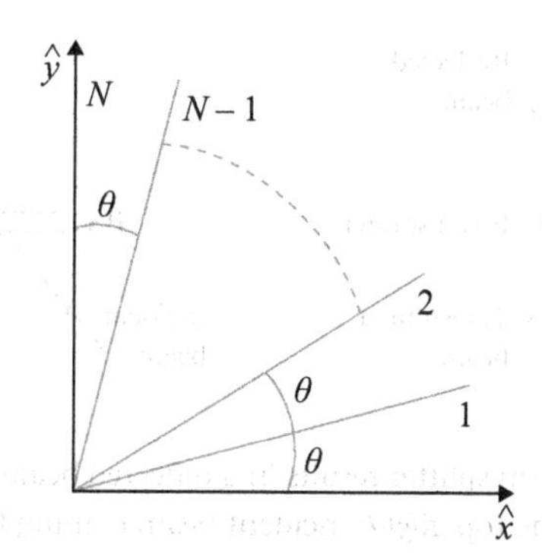

Figure 1.3
N polarizers whose preferential directions are tilted relative to each other.

α is in a state $|\text{photon}; \alpha\rangle$ that can be written as a superposition of the above two states:

$$|\text{photon}; \alpha\rangle = |\text{photon}; x\rangle \cos\alpha + |\text{photon}; y\rangle \sin\alpha. \tag{1.3.7}$$

This equation should be compared with the decomposition (1.3.3) of the electric field. While there are some similarities, both are superpositions, one refers to electric fields, and the other refers to states of a single photon. Any photon that emerges beyond the polarizer will necessarily be polarized along the x axis and therefore will be in the state we defined in (1.3.5):

$$\text{Beyond the polarizer:} \quad |\text{photon}; x\rangle. \tag{1.3.8}$$

This can be compared with the electric field (1.3.4) beyond the polarizer. That field, with the factor $\cos\alpha$, carries information about the amplitude of the wave. Here, for a single photon, there is no room for such a factor.

At the famous Fifth Solvay International Conference of 1927, the world's most notable physicists gathered to discuss the newly formulated quantum theory. Seventeen out of the twenty-nine attendees were or became Nobel Prize winners. Einstein, unhappy with the uncertainty in quantum mechanics, stated the now famous quote "God does not play dice," to which Niels Bohr is said to have answered, "Einstein, stop telling God what to do." Bohr was willing to accept the loss of determinism; Einstein was not.

Exercise 1.5. *Light polarized along the x axis hits two polarizers, one after the other. The first polarizer has preferential direction at 45°, in between the x and y axes. The second polarizer has preferential direction along the y axis. What fraction of the incident photons emerge from the second polarizer?*

Exercise 1.6. *Suppose we now put sequentially N polarizers ($N \geq 2$) where each polarizer's preferential direction is rotated counterclockwise by the same amount θ with respect to the previous one. The first polarizer is at an angle θ relative to the x axis, and the last polarizer is aligned along the y axis (see figure 1.3). Assume light polarized along the x axis hits the system. Find the fraction of photons expected to emerge beyond the last polarizer. Suppose you*

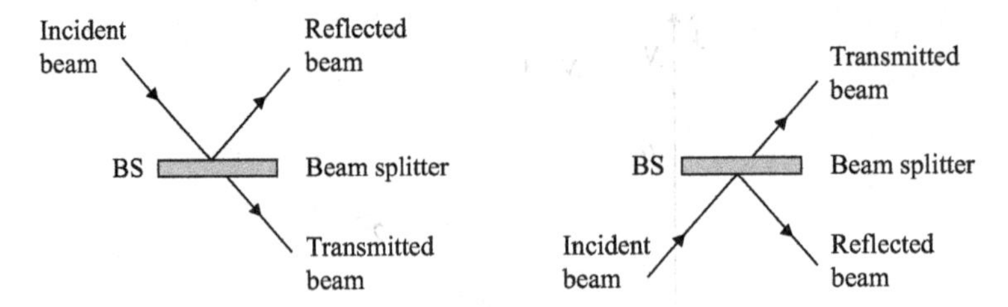

Figure 1.4
An incident beam hitting a beam splitter results in a reflected beam and a transmitted beam. *Left*: incident beam coming from the top. *Right*: incident beam coming from the bottom.

use N = 500 polarizers and send in 1,000 photons. How many photons are expected to emerge beyond the last polarizer? (Answer: 995.)

1.4 Quantum Superpositions

We have already discussed the concept of linearity, the idea that the sum of two solutions representing physical realities represents a new, allowed physical reality. This superposition of solutions has a straightforward meaning in classical physics. In the case of electromagnetism, for example, if we have two solutions, each with its own electric and magnetic field, the sum solution is simply understood: its electric field is the sum of the electric fields of the two solutions, and its magnetic field is the sum of the magnetic fields of the two solutions. In quantum mechanics, as we have explained, linearity holds. The interpretation of a superposition, however, is very surprising.

One interesting example is provided by a Mach-Zehnder interferometer, an arrangement of beam splitters, mirrors, and detectors devised by Ludwig Zehnder (1891) and improved by Ludwig Mach (1892) to study interference between two beams of light.

A beam splitter (BS), as its name indicates, splits an incident beam into two beams: one that is reflected from the splitter and one that goes through the splitter. Our beam splitters will often be balanced, meaning they split a given beam into two beams of equal intensity (figure 1.4). The light that bounces off is called the reflected beam, and the light that goes through is called the transmitted beam. The incident beam can hit the beam splitter from the top or from the bottom.

The Mach-Zehnder configuration, shown in figure 1.5, has a left beam splitter (BS1) and a right beam splitter (BS2). In between we have two perfect mirrors: M1 on the top and M2 on the bottom. An incident beam from the left is split by BS1 into two beams, each of which hits a mirror, and is then sent into BS2. At BS2 the beams are recombined and sent into two outgoing beams that go into photon detectors D0 and D1.

It is relatively simple to arrange the beam splitters so that the incident beam, upon splitting at BS1 and recombining at BS2, emerges in the top beam that goes into D0. In this arrangement, no light at all goes into D1. This requires a precise interference effect at BS2. Note that we have two beams incident upon BS2; the top beam is called a, and

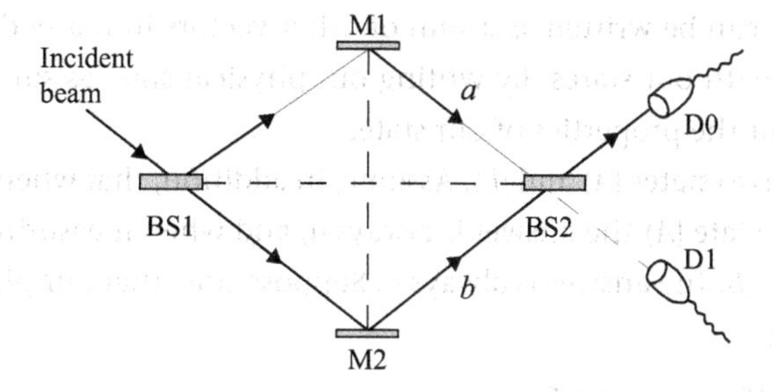

Figure 1.5
A Mach-Zehnder interferometer consists of two beam splitters BS1 and BS2, two mirrors M1 and M2, and two detectors D0 and D1. An incident beam will be split into two beams by BS1. One beam, beam a, goes through the upper branch, which contains M1. The other beam, beam b, goes through the lower branch, which contains M2. The beams on the two branches recombine at BS2 and are then sent into the detectors. The configuration is prepared to produce an interference so that all incident photons end at the detector D0, with none at D1.

the lower beam is called b. Two contributions go toward D0: the reflection of a at BS2 and the transmission from b at BS2. These two contributions interfere constructively to give a beam going into D0. Two contributions also go toward D1: the transmission from a at BS2 and the reflection from b at BS2. It turns out that these two can be arranged to interfere destructively to give no beam going into D1.

It is instructive to think of the incident beam as a sequence of photons that we send into the interferometer, one photon at a time. This shows that at the level of photons the interference is not that of one photon with another photon. Each photon must interfere with *itself* to give the result. Indeed, interference between two photons is not possible: destructive interference, for example, would require that two photons end up giving no photon, which would violate energy conservation. Energy conservation holds in quantum mechanics just as it does in classical mechanics.

Therefore, each photon does the very strange thing of going through both branches of the interferometer! Each photon is in a superposition of two states: a state in which the photon is in the top beam or upper branch added to a state in which the photon is in the bottom beam or lower branch. Thus, the photon in the interferometer is in a funny state in which the photon seems to be doing two incompatible things at the same time.

Equation (1.3.7) for a photon state is another example of a quantum superposition. The photon state has a component along an x-polarized photon and a component along a y-polarized photon.

As we mentioned before, we speak of wave functions and states as equivalent descriptions of a quantum system. We also sometimes refer to states as vectors. A quantum state may not be a vector like the familiar vectors in three-dimensional space, but it is a vector nonetheless because we can do with states what we do with ordinary vectors: we can add states, and we can multiply states by numbers. Linearity in quantum mechanics guarantees that adding wave functions (or states, or vectors) is a sensible thing to do.

Just as any vector can be written as a sum of other vectors in many different ways, we will do the same with our states. By writing our physical state as sums of other states, we can learn about the properties of our state.

Consider now two states $|A\rangle$ and $|B\rangle$. Assume, in addition, that when measuring some property Q in the state $|A\rangle$ the answer is always a, and when measuring the same property Q in the state $|B\rangle$ the answer is always b. Suppose now that our physical state $|\Psi\rangle$ is the superposition

$$|\Psi\rangle = \alpha|A\rangle + \beta|B\rangle, \qquad \alpha, \beta \in \mathbb{C}. \tag{1.4.1}$$

What happens now if we measure property Q in the system described by the state $|\Psi\rangle$? It may seem reasonable that one gets some intermediate value between a and b. This is *not* what happens. A measurement of Q will yield either a or b. There is no certain answer; classical determinism is lost, but the answer is always one of these two values and not an intermediate one. The coefficients α and β in the above superposition affect the probabilities with which we may obtain the two possible values. In fact, the probability of obtaining a is proportional to $|\alpha|^2$, and the probability of obtaining b is proportional to $|\beta|^2$:

$$\text{Probability}(a) \sim |\alpha|^2, \quad \text{Probability}(b) \sim |\beta|^2. \tag{1.4.2}$$

Since the only two possibilities are to measure a or b, the actual probabilities must sum to one, and therefore they are given by:

$$\text{Probability}(a) = \frac{|\alpha|^2}{|\alpha|^2 + |\beta|^2}, \quad \text{Probability}(b) = \frac{|\beta|^2}{|\alpha|^2 + |\beta|^2}. \tag{1.4.3}$$

If we measure the value a, immediate repeated measurements are known to give a with complete certainty, so the state after the measurement must be $|A\rangle$. The same happens for b, so we have the following:

$$\begin{aligned} &\text{After measuring } a, \text{ the state becomes } |\Psi\rangle = |A\rangle\,; \\ &\text{After measuring } b, \text{ the state becomes } |\Psi\rangle = |B\rangle. \end{aligned} \tag{1.4.4}$$

Exercise 1.7. *Assume that the measurement of the energy in a state $|A\rangle$ always yields the value E_a, and the measurement of the energy in a state $|B\rangle$ always yields the value E_b. Now consider a quantum system in the superposition state $|\Psi\rangle = (1+2i)|A\rangle + (1-i)|B\rangle$. What are the probabilities P_a and P_b of measuring energy E_a and E_b, respectively? (Partial answer: $P_aP_b =$ 10/49). The experiment is repeated a very large number of times using many identical copies of the system, and the results for the measured energies are tabulated. What is the expected "average" value of the list consistent with the calculated probabilities?*

In quantum mechanics one makes the following assumption: *Superposing a state with itself doesn't change the physics* or change the state in a nontrivial way. Since superimposing a state with itself simply changes the overall number multiplying it, we have Ψ and $\alpha\Psi$ representing the same physics for any nonzero complex number α. Thus, letting $\cong$ represent physical equivalence:

$$|A\rangle \cong 2|A\rangle \cong i|A\rangle \cong -|A\rangle. \tag{1.4.5}$$

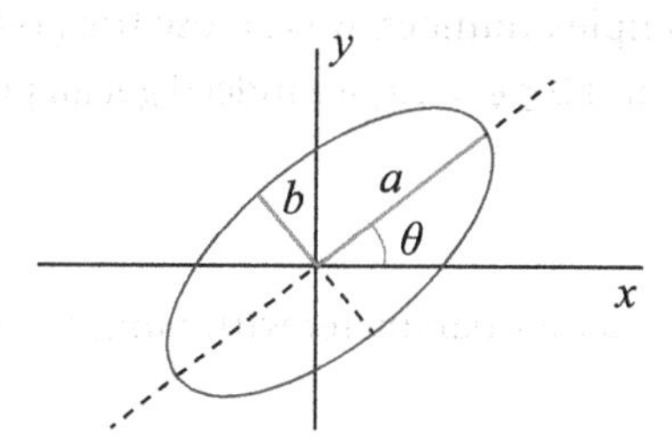

Figure 1.6
An elliptically polarized plane wave is specified by the parameters θ and a/b.

This assumption is necessary to verify that the polarization of a photon state has the expected number of degrees of freedom. The polarization of a plane wave, as one studies in electromagnetism, is described by two real numbers. A wave of general polarization is an elliptically polarized wave, as shown in figure 1.6. At any given point, as time changes, the electric field vector traces an ellipse. The shape of the ellipse is encoded by the ratio a/b of the semimajor to semiminor axes (the first real parameter), and a tilt encoded by the angle θ (the second real parameter). Consider now a general photon state $|\Psi\rangle$ formed by the superposition of the two independent polarization states $|\text{photon}; x\rangle$ and $|\text{photon}; y\rangle$:

$$|\Psi\rangle = \alpha|\text{photon}; x\rangle + \beta|\text{photon}; y\rangle, \quad \alpha, \beta \in \mathbb{C}. \tag{1.4.6}$$

At first sight it looks as if the superposition is determined by two nonzero complex parameters α and β or, equivalently, four real parameters. But since the overall factor does not matter, we can multiply this state by $1/\alpha$ to get the equivalent state that encodes all the physics:

$$|\Psi\rangle \simeq |\text{photon}; x\rangle + \tfrac{\beta}{\alpha}\,|\text{photon}; y\rangle, \tag{1.4.7}$$

showing that we really have one complex parameter, the ratio β/α. This is equivalent to two real parameters, as expected.

Exercise 1.8. *Let $|A\rangle$ and $|B\rangle$ be two nonequivalent quantum states. One of the following four states*

$$(1+i)|A\rangle + (1-i)|B\rangle, \quad i|A\rangle + |B\rangle, \quad (i-1)|A\rangle + (i+1)|B\rangle, \quad |A\rangle + i|B\rangle,$$

is not equivalent to the other three. Which one is it?

Example 1.2. *The space of nonequivalent quantum states built by superposition.*
Our discussion of photon polarization showed that the nonequivalent states obtained by the superposition of two different states are described by two real parameters. Here we will examine this question again, finding a suggestive description of the real parameters. We considered in (1.4.1) the states $|\Psi\rangle$ constructed by the superposition of two states $|A\rangle$ and $|B\rangle$. Since scaling gives an equivalent state, consider dividing the state by α, assumed nonzero, to find

$$|\Psi\rangle = |A\rangle + \tfrac{\beta}{\alpha}|B\rangle, \quad \alpha, \beta \in \mathbb{C}. \tag{1.4.8}$$

Since β/α is an arbitrary complex number, we can use the polar representation to write it as $\frac{\beta}{\alpha} = re^{i\phi}$. If we take $r \geq 0$ and $0 \leq \phi < 2\pi$, we indeed get all possible complex numbers. Thus, we have

$$|\Psi\rangle = |A\rangle + re^{i\phi}|B\rangle. \tag{1.4.9}$$

We can write $r = \tan \nu$ with ν a new parameter with range $0 \leq \nu < \frac{\pi}{2}$, which results in all $r \geq 0$:

$$|\Psi\rangle = |A\rangle + \tan \nu\, e^{i\phi}|B\rangle, \quad 0 \leq \nu < \tfrac{\pi}{2}, \quad 0 \leq \phi < 2\pi. \tag{1.4.10}$$

In this convention, which chooses $|A\rangle$ to appear with a unit coefficient, no further rescalings are possible, and the coefficient of $|B\rangle$ determines the state. Since the coefficient of B changes when ν and ϕ change, the states obtained as we sweep the ranges of ν and ϕ are nonequivalent. A more geometric description of the states arises if we use a new parameter $\theta = 2\nu$ with range $0 \leq \theta < \pi$:

$$|\Psi\rangle = |A\rangle + \tan \tfrac{\theta}{2}\, e^{i\phi}|B\rangle, \quad 0 \leq \theta < \pi, \quad 0 \leq \phi < 2\pi. \tag{1.4.11}$$

We can make the treatment of the states $|A\rangle$ and $|B\rangle$ more symmetric by multiplying $|\Psi\rangle$ by $\cos \frac{\theta}{2}$, giving us our final *canonical* form:

$$\boxed{|\Psi\rangle = \cos \tfrac{\theta}{2}|A\rangle + \sin \tfrac{\theta}{2}\, e^{i\phi}|B\rangle, \quad 0 \leq \theta < \pi, \quad 0 \leq \phi < 2\pi.} \tag{1.4.12}$$

The result is particularly nice because parameters θ, ϕ with the above ranges can play the role of polar and azimuthal angles in spherical coordinates (the point with $\theta = \pi$ can be included if we treat "infinity" as an ordinary number). Thus, if we are considering superpositions of two physically nonequivalent states, the most general state can be specified by a direction in a three-dimensional space! We will learn in section 12.3 that this representation is precisely the one required to describe the general states of a spin one-half particle! □

Let us do a further example of superposition, this time using electrons. Electrons are particles with spin angular momentum. Classically, we imagine them as tiny balls spinning around an axis that goes through the particle itself. The true quantum electron is much stranger. First, it is a point particle, without apparent size. Second, once an oriented axis is fixed for measurement, the electron's rotation may be clockwise or counterclockwise, but in both cases the magnitude of the angular momentum is the *same*. These opposite ways of spinning are called *spin up* and *spin down* along the axis (see figure 1.7). Up and down refer to the sign of the angular momentum measured: up for positive along the axis and down for negative along the axis. According to quantum mechanics and as verified by multiple experiments, spin up or spin down are the only possibilities that arise *whatever* axis we use to measure the spin of the electron. The electron is said to be a particle of spin one-half, as the magnitude of the angular momentum measured turns out to be one-half of the number $\hbar$.

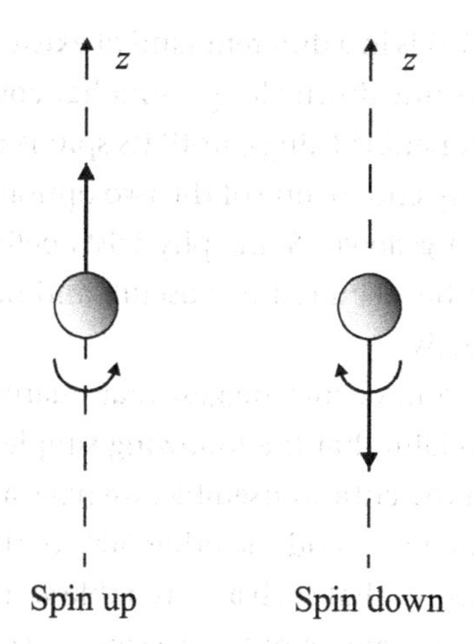

Figure 1.7
An electron with spin along the z axis. *Left*: the electron is said to have spin up along z. *Right*: the electron is said to have spin down along z. The up and down arrows represent the direction of the angular momentum of the spinning electron.

Physicists usually set up coordinate systems in space by choosing three orthogonal directions, those of the x, y, and z axes. Let us choose to describe a spinning electron using the z axis. Then, one possible state of an electron is to spin up along the z axis. Such a state is written as $|\uparrow;z\rangle$, with an arrow pointing up and the label z indicating that the spin component is along the increasing z direction. Another possible state of the electron is to spin down along the z axis. Such a state is written as $|\downarrow;z\rangle$, with an arrow pointing down, meaning this time the spin component is along the decreasing z direction. In fact, these states are a bit mysterious. Suppose you have an electron in the $|\uparrow;z\rangle$ state, and you measure its spin along the x axis. The result is not zero; you either find it up along x or down along x. So the picture of the state $|\uparrow;z\rangle$ is not a classical one in which the angular momentum points along z, for this would imply zero angular momentum components along x and along y.

If the two states $|\uparrow;z\rangle$ and $|\downarrow;z\rangle$ are possible, so is the state $|\Psi\rangle$ represented by the sum

$$|\Psi\rangle = |\uparrow;z\rangle + |\downarrow;z\rangle. \tag{1.4.13}$$

What kind of physics does this sum $|\Psi\rangle$ represent? It represents a state in which a measurement of the spin along the z axis would result in two possible outcomes with equal probabilities: an electron with spin up or an electron with spin down. Since we can only speak of probabilities, any experiment must be repeated multiple times until probabilities can be estimated. Suppose we have a large ensemble of such electrons, all in the above state $|\Psi\rangle$. As we measure their spin along z, one at a time, we find about half of them spinning up along z and the other half spinning down along z. There is no way to predict which option will be realized for a particular electron until we measure it. It is not easy to imagine the state of the electron in superposition, but one may try as follows.

An electron in the state (1.4.13) is in a different kind of existence in which it is able to be spinning up along z and spinning down along z simultaneously! It is in a ghostly, eerie state, seemingly doing incompatible things, until its spin is measured. Once measured, the electron must immediately choose one of the two options; we always find electrons either spinning up or spinning down. Some physicists believe a physical picture of a state in superposition is neither required nor useful, and one only needs the rules to manipulate such states correctly.

A critic of quantum mechanics could suggest that quantum superpositions are not needed. Such a critic would claim that the following simpler ensemble results in identical experimental results. In the critic's ensemble, we have a large number of electrons, with 50% of them in the state $|\uparrow; z\rangle$ and the other 50% of them in the state $|\downarrow; z\rangle$. The critic would then assert, correctly, that such an ensemble would yield the same measurements of spins along z as the ensemble of those esoteric $|\Psi\rangle$ states. The new ensemble could provide a simpler explanation of the result without having to invoke quantum superpositions.

Quantum mechanics, however, allows for further experiments that can distinguish between the ensemble of our friendly critic and the ensemble of $|\Psi\rangle$ states. While it would take us too far afield to explain this now, if we measured the spin of the electrons in the x direction, instead of the z direction, the results would be *different* in the two ensembles. In the ensemble of our critic, we would find 50% of the electrons up along x and 50% of the electrons down along x. In our ensemble of $|\Psi\rangle$ states, however, one can show that measurement yields a surprisingly simple result: all states pointing up along x. The critic's ensemble is therefore not equivalent to our quantum mechanical ensemble. The critic is wrong in claiming that quantum mechanical superpositions are not required.

A strange situation, imagined by Schrödinger, postulated an apparatus that could put a cat in a superposition of an "alive cat" state and a "dead cat" state. In fact, Schrödinger considered this example to suggest, incorrectly in our present view, that the rules of quantum mechanics could lead to ridiculous results. Most physicists do not believe such states of a real cat exist, but the reasons are largely technical: quantum superpositions are extremely fragile, and any interaction with the environment would destroy them. While we can control and suppress the interactions of an electron with the environment, it is impossible to do that with a cat. Thus, a real cat is either alive or dead. Nevertheless, stimulated by this speculation, physicists have managed to build quantum superpositions of many-particle states. One such example follows the proposal of Anthony Legget to use SQUIDs (superconducting quantum interference devices) to construct quantum states of many particles, whimsically called *cat states* in honor of Schrödinger. SQUIDs are used in practice as sensitive magnetometers, capable of measuring magnetic fields as small as 10^{-18} tesla. In a superconductor, the *supercurrent* is a current that flows without dissipation.

A SQUID, illustrated schematically in figure 1.8, is a superconducting ring (*shaded*) that contains a Josephson junction (*black*). The junction is a thin interruption of the

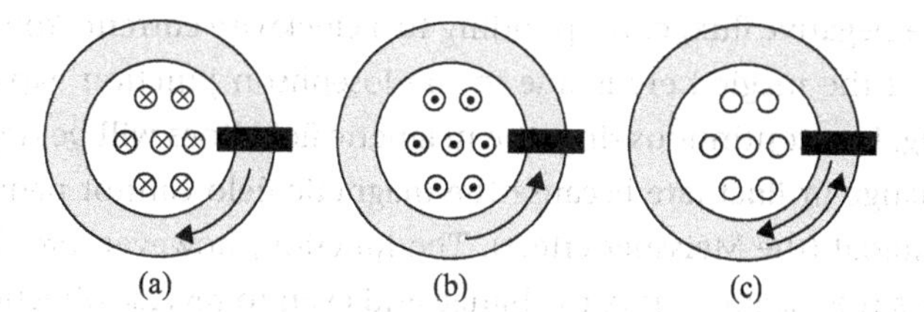

Figure 1.8
Three SQUIDs, each with a superconducting ring (*shaded*) and a Josephson junction (*dark black*). (a) SQUID in a state |clockwise current⟩ and a magnetic field going into the page. (b) SQUID in a state |counterclockwise current⟩ and a magnetic field going out the page. (c) SQUID in a quantum superposition of clockwise and counterclockwise states, making the supercurrent and the magnetic field undetermined. The ring links magnetic field lines and thus, magnetic flux.

ring by a small slice of insulating material. While it is an insulating barrier, the Josephson junction does not interrupt the flow of supercurrents in the ring. The supercurrents actually tunnel through the energy barrier created by the junction. Tunneling is a quantum mechanical process—forbidden in classical physics—in which a particle can cross through a region that is energetically banned. A supercurrent in the SQUID is a flow of paired electrons—in fact, a flow of a large number of paired electrons. The supercurrent will create a magnetic field with field lines linked by the ring. In particular, the magnetic field at the center of the ring is perpendicular to the ring. The direction of the magnetic field depends on the direction of the circulating supercurrent. A typical supercurrent is of the order of a microampere, which can be realized by having about a billion electrons going around the ring.

In a SQUID we can have two types of current states: a state |clockwise current⟩ and a state |counterclockwise current⟩, representing clockwise and counterclockwise supercurrents in the ring, respectively. They are also called *flux states*, since each one is associated with a flux of the magnetic field linked by the loop, in one case negative and in the other case positive. Each one is a macroscopic state, involving the order of a billion circulating electrons. These are the cat states of the SQUID, and they are shown in parts (a) and (b) of figure 1.8.

We can then consider a superposition of such cat states, defining a state $|1\rangle$ by

$$|1\rangle \equiv |\text{clockwise current}\rangle + |\text{counterclockwise current}\rangle. \tag{1.4.14}$$

This is a state of indefinite current and indefinite magnetic field (see figure 1.8c). It is *not* a state of zero current and zero magnetic field, as you would get classically by superposing two oppositely directed identical currents. The state $|1\rangle$ is remarkable in that the two summands are macroscopic states. In this superposition we have billions of electrons in an eerie state in which they are supposedly rotating in both directions simultaneously. This state is definitely not classical!

Suppose you have many SQUIDs, all in the state $|1\rangle$, and you measure the magnetic flux through the ring, which is tantamount to measuring the current. You will sometimes get a positive flux, corresponding to a counterclockwise current, and you will sometimes get a negative flux, corresponding to a clockwise current. You will never get zero flux. Part of the magic here is due to the Josephson junction gap on the superconducting ring. In a continuous ring, the magnetic field lines will be trapped, and the flux cannot change or fluctuate because the magnetic field cannot penetrate a superconducting material (the Meissner effect). The junction, however, provides an escape route and allows the magnetic flux to change and even to reverse direction. Indeed, in the state $|1\rangle$ the flux is indefinite—one could say fluctuating—and, when measured, it is sometimes found to be positive and sometimes found to be negative.

A skeptic could claim that the results that follow from measuring the flux in the state $|1\rangle$ would be achieved with an ensemble of SQUIDs in which half are in a clockwise current state, and the other half are in a counterclockwise current state. This is true, but as we now describe, there are experiments that show a surprising phenomenon explained naturally by the quantum superposition.

To begin, note that by superposing our flux (or current) states differently, we can get the state $|2\rangle$:

$$|2\rangle \equiv |\text{clockwise current}\rangle - |\text{counterclockwise current}\rangle. \tag{1.4.15}$$

The sign difference relative to $|1\rangle$ matters, although it does not affect results if you are just measuring flux. Using the above equations, we can solve for the flux states in terms of the $|1\rangle$ and $|2\rangle$ states:

$$\begin{aligned} |\text{clockwise current}\rangle &= \tfrac{1}{2}(|1\rangle + |2\rangle), \\ |\text{counterclockwise current}\rangle &= \tfrac{1}{2}(|1\rangle - |2\rangle). \end{aligned} \tag{1.4.16}$$

It turns out that the states $|1\rangle$ and $|2\rangle$ have slightly different energies. The rules of quantum mechanics imply that states with different energies evolve differently in time. If we start with the state $|\text{clockwise current}\rangle$ at some time, the states $|1\rangle$ and $|2\rangle$ on the right-hand side evolve differently. In fact, after some time that depends on their energy difference, the initial linear combination with a plus sign becomes proportional to the second linear combination of $|1\rangle$ and $|2\rangle$ with a minus sign (see problem 1.4). At that point the current that started as clockwise has become counterclockwise, and the flux has flipped. At all times in between, the state was a superposition of clockwise and counterclockwise states. These oscillations have been observed experimentally, confirming the above picture of superpositions.

1.5 Entanglement

When we consider the superposition of states of *two* particles, we can get the remarkable phenomenon called *quantum mechanical entanglement*. Entangled states of two particles are those in which we can't speak separately of the state of each particle. The particles are in a common state in which they are somehow linked.

Let us consider two *noninteracting* particles. Particle A could be in any of the states

$$\{|A_1\rangle, |A_2\rangle, \cdots\}, \tag{1.5.1}$$

while particle B could be in any of the states

$$\{|B_1\rangle, |B_2\rangle, \cdots\}. \tag{1.5.2}$$

It may seem reasonable to conclude that the state of the full system, including particle A and particle B, would be specified by stating the state of particle A and the state of particle B. If that were the case, the possible states would be written as

$$|A_i\rangle \otimes |B_j\rangle, \tag{1.5.3}$$

for some specific choice of i and j that specifies the state of A and B, respectively. Here we have used the symbol $\otimes$, which means *tensor* product, to combine the two states into a single state for the two-particle system. We will study $\otimes$ later, but for the time being, we can think of it as a kind of product that distributes over addition and obeys simple rules, such as

$$\begin{aligned}(\alpha_1|A_1\rangle + \alpha_2|A_2\rangle) \otimes (\beta_1|B_1\rangle + \beta_2|B_2\rangle) = \alpha_1\beta_1|A_1\rangle \otimes |B_1\rangle + \alpha_1\beta_2|A_1\rangle \otimes |B_2\rangle \\ + \alpha_2\beta_1|A_2\rangle \otimes |B_1\rangle + \alpha_2\beta_2|A_2\rangle \otimes |B_2\rangle.\end{aligned} \tag{1.5.4}$$

The numbers can be moved across the $\otimes$, but the order of the states must be preserved. The state on the left-hand side—shown expanded out on the right-hand side—is still one in which we combine a state $(\alpha_1|A_1\rangle + \alpha_2|A_2\rangle)$ of particle A with a state $(\beta_1|B_1\rangle + \beta_2|B_2\rangle)$ of particle B. Just like any one of the states listed in (1.5.3), this state is not entangled.

Using the states in (1.5.3), however, we can construct some intriguing superpositions. Superpositions must be allowed, as we have explained before. Consider the following one:

$$|A_1\rangle \otimes |B_1\rangle + |A_2\rangle \otimes |B_2\rangle. \tag{1.5.5}$$

A state of two particles is said to be **entangled** if it cannot be written in the factorized form $(\cdots) \otimes (\cdots)$. Recall that this factorized form allows us to describe the overall state by simply stating the state of each particle. We can prove that the state (1.5.5) cannot be factorized. If it could, it would be possible to write it as the product shown on the left-hand side of (1.5.4), with suitable constants $\alpha_1, \alpha_2, \beta_1, \beta_2$. To determine these constants, we compare the right-hand side of (1.5.4) with our state (1.5.5) and conclude that we need

$$\alpha_1\beta_1 = 1, \quad \alpha_1\beta_2 = 0, \quad \alpha_2\beta_1 = 0, \quad \alpha_2\beta_2 = 1. \tag{1.5.6}$$

It is clear that there is no solution here. The second equation, for example, requires either α_1 or β_2 to be zero. Having $\alpha_1 = 0$ contradicts the first equation, and having $\beta_2 = 0$ contradicts the last equation. This confirms that our state (1.5.5) is indeed an entangled state. There is no way to describe the state by specifying a state for each of the particles.

Let us illustrate the above discussion using electrons and their spin states. Consider a state of two electrons denoted as $|\uparrow\rangle \otimes |\downarrow\rangle$, omitting the label z on the states for brevity. As the notation indicates, the first electron, described by the state to the left of $\otimes$, is up along z. The second electron, described by the state to the right of $\otimes$, is down along z. This is not an entangled state. Another possible state is one in which they do exactly the opposite: in $|\downarrow\rangle \otimes |\uparrow\rangle$ the first electron is down, and the second is up. This state is not entangled either. However, the superposition of these two states,

$$|\uparrow\rangle \otimes |\downarrow\rangle + |\downarrow\rangle \otimes |\uparrow\rangle, \tag{1.5.7}$$

is an entangled state of the pair of electrons.

Exercise 1.9. *Show that the above state cannot be factorized and is therefore entangled.*

In the state (1.5.7), the first electron is up along z if the second electron is down along z (first term), or the first electron is down along z if the second electron is up along z (second term). There is a correlation between the spins of the two particles; they always point in opposite directions. Imagine that the two entangled electrons are very far away from each other: Alice has one electron of the pair on planet Earth and Bob has the other electron on the moon. Nothing we know is connecting these particles, but nevertheless the states of the electrons are linked, meaning that the measurements we do on the separate particles exhibit correlations. Suppose Alice measures the spin of the electron on Earth. If she finds it up along z, it means that the first term in the above superposition is realized because in that term the first particle is up. As discussed before, the state of the two particles immediately becomes that of the first term. The state of Bob's electron will be spin down along z. Bob could measure and reach this conclusion before a message, carried with the speed of light, could reach the moon telling him that a measurement has been taken by Alice on Earth, and the result was spin up. Of course, experiments must be done with an ensemble that contains many pairs of particles, each pair in the same entangled state above. Half of the times the electron on Earth will be found up, with the electron on the moon down, and the other half of the times the electron on Earth will be found down, with the electron on the moon up.

Our friendly critic could now say, correctly, that such correlations between the measurements of spins along z could have been produced by preparing a *conventional* ensemble in which 50% of the pairs are in the state $|\uparrow\rangle \otimes |\downarrow\rangle$, and the other 50% of the pairs are in the state $|\downarrow\rangle \otimes |\uparrow\rangle$. Such objections were dealt with conclusively in 1964 by John Bell, who showed that if Alice and Bob are able to measure spin in *three* arbitrary directions, the correlations predicted by the quantum entangled state differ from the classical correlations of *any* conceivable conventional ensemble. Quantum correlations in entangled states are very subtle, and it takes sophisticated experiments to show they are not reproducible as classical correlations. Indeed, experiments with entangled states have confirmed the existence of quantum correlations. The surprising properties of well-separated entangled particles do not lead to paradoxes or, as it may seem, to contradictions with the ideas of special relativity. You cannot use quantum mechanical entangled states to send information faster than the speed of light.

One can also consider the entangled states of a large number of spin one-half particles. As in the case of two particles, such entangled states cannot be described by just giving the state of each of the particles. With a large number of particles, the entangled state is a delicate macroscopic quantum state that is easily destroyed by interactions with an environment. Nevertheless, if such states could be constructed and manipulated, they would allow for *quantum computation*, a new way of doing what classical computers do with some new and intriguing capabilities. Some computations that are hard with classical computers may become easy with a quantum computer. We will consider this subject in chapter 23.

1.6 Making Atoms Possible

The relevance of quantum mechanics to the understanding of our physical world is illustrated most dramatically by the following fact: atoms are simply impossible in classical physics! This can be seen by simple arguments. Consider an electron orbiting a proton, the classical model of the hydrogen atom. Since the proton is much more massive than the electron, we can assume that the electron is going in circles around a stationary proton. The proton has charge e, and the electron has charge $-e$. The Coulomb potential energy of the electron-proton system is $V(r) = -e^2/r$, where r is the distance between the particles. Associated with this potential is a force of attraction of magnitude $|F| = e^2/r^2$. For a circular orbit, we equate the force to the electron mass m_e times the centripetal acceleration to find that

$$m_e \frac{v^2}{r} = \frac{e^2}{r^2} \Rightarrow r = \frac{e^2}{m_e v^2}. \tag{1.6.1}$$

This shows that the size of the electron orbit, which we can take as an estimate of the size of the hydrogen atom, is not fixed by the equation of motion and can take any value, depending on the speed v. We can attempt to fix the atomic size by trying to minimize the total energy of the electron. This classical energy E_{cl} is the sum of the electron kinetic energy and the potential energy $V(r)$ discussed above:

$$E_{\text{cl}} = \frac{1}{2} m_e v^2 - \frac{e^2}{r}. \tag{1.6.2}$$

The left equation in (1.6.1) multiplied by $r/2$ tells us that $\frac{1}{2} m_e v^2 = \frac{e^2}{2r}$, and we can use this result to find the total energy as a function of the radius only, assuming circular motion:

$$E_{\text{cl}}(r) = -\frac{e^2}{2r}. \tag{1.6.3}$$

This energy is shown in figure 1.9, *left*. The energy is unbounded from below, and therefore energy minimization does not help fix the atomic size; the smaller r is, the smaller the energy. One could perhaps think that the atomic size is fixed somehow, just like the radii of the planets orbiting the sun, and that the orbits are stable. But there is an additional problem. Charged particles in accelerated motion radiate. Circular motion is accelerated motion, and the electron would radiate electromagnetic waves as it orbits the proton. The energy carried away by these waves is energy loss for the electron, which begins to spiral into the proton. As you will examine in problem 1.3, this happens very quickly: it only takes about 10^{-11} seconds for the atom to collapse. Clearly, atoms are not possible in classical physics.

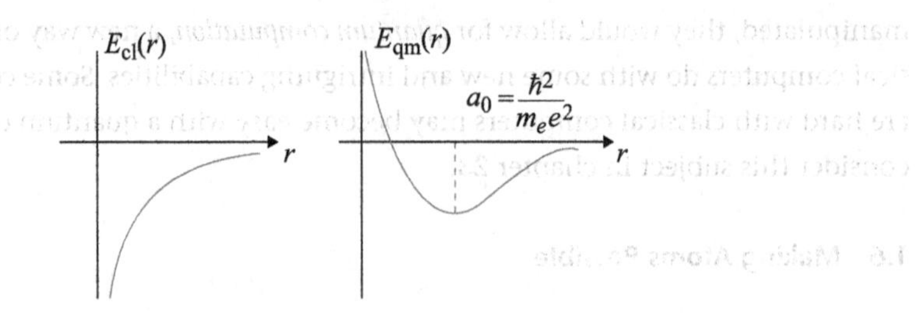

Figure 1.9
Left: Classical energy of an electron on a circular orbit of radius r around the proton. *Right*: The quantum-mechanical energy of an electron on a circular orbit of radius r around the proton. The classical energy is unbounded below, but the quantum-mechanical energy has a minimum at a radius equal to a_0, the Bohr radius.

How does quantum mechanics allow for atoms? It does on account of the uncertainty principle, a key constraint on observables in quantum mechanics. This principle changes the balance of energies as a function of radius. What follows is an argument that, while nonrigorous, captures well the spirit of the quantum solution of this classical conundrum.

Applied to position and momentum, the uncertainty principle states that associated with an uncertainty Δx in the position of a particle there is *necessarily* an uncertainty Δp in the momentum of the particle, and this uncertainty is at least of order

$$\Delta p \sim \frac{\hbar}{\Delta x}. \tag{1.6.4}$$

The smaller the position uncertainty, the larger the momentum uncertainty. Recall that $p = m_e v$, and that kinetic energy is written in terms of momentum as $p^2/(2m_e)$.

If the quantum electron is localized within a distance r from the proton, we can estimate its potential energy as $\bar{V} \simeq -e^2/r$. Moreover, we can think of r as the uncertainty in the position of the electron, thus $\Delta x \sim r$. But then the momentum uncertainty is $\Delta p \simeq \hbar/r$. Even if the average momentum of the electron is zero, there is kinetic energy $\bar{K}$ due to Δp, which we estimate as

$$\bar{K} \simeq \frac{(\Delta p)^2}{2m_e} \simeq \frac{\hbar^2}{2m_e r^2}. \tag{1.6.5}$$

As the value of the radius goes down, lowering the potential energy, this time the kinetic energy goes up like $1/r^2$, not like $1/r$, as it did classically. The total energy now, called $E_{qm}(r)$, for quantum mechanical, is estimated as

$$E_{qm}(r) \simeq \bar{K} + \bar{V} = \frac{\hbar^2}{2m_e r^2} - \frac{e^2}{r}. \tag{1.6.6}$$

Viewed as a function of r, the energy now has a minimum, as shown in figure 1.9, *right*. Indeed, for very small r the kinetic energy dominates, and it is large and positive. For very large r, the second term dominates, and the total energy is negative but increasing as $r \to \infty$. A minimum must therefore exist. We find it quickly by differentiation:

$$\frac{dE_{qm}}{dr} \simeq -\frac{\hbar^2}{m_e r^3} + \frac{e^2}{r^2} = 0. \tag{1.6.7}$$

Solving this equation, we find the atomic radius selected by quantum mechanics:

$$r \simeq \frac{\hbar^2}{m_e e^2} \equiv a_0. \tag{1.6.8}$$

This is in fact the Bohr radius a_0 of the hydrogen atom. In the quantum-mechanical picture of the atom, we do not have an electron going around in circles. In the ground state of the hydrogen atom, the electron has a spherically symmetric probability distribution, and the most probable radius is in fact $r = a_0$. In this state the electron has zero angular momentum! The electron is just a spherical "cloud." It does not radiate; it is in a quantum stationary state, in which it can stay forever. Quantum mechanics indeed selects a size for atoms, something classical mechanics cannot do. Moreover, it eliminates the classical instability due to radiation.

Problems

Problem 1.1. *Complex number practice.*

A complex number z can be written in either Cartesian or polar form:

$$z = a + ib = re^{i\theta}, \quad |z| = \sqrt{a^2 + b^2}.$$

The real numbers a and b are, respectively, the real and imaginary parts of z. The real numbers r and θ are, respectively, the magnitude and argument of z. We call $|z|$ the norm of z. Use this definition for z in the following:

1. Use Taylor expansions to derive the Euler formula:

   $$e^{i\theta} = \cos\theta + i\sin\theta.$$

   In other words, write out the Taylor expansion of the left-hand side and manipulate it to look like the Taylor expansion of the right-hand side.
2. Express a and b in terms of r and θ and vice versa (assume that $a > 0$).
3. Complex numbers are often viewed as vectors in a two-dimensional space known as the complex plane. Multiplication of a complex number by a pure phase (a complex number of unit magnitude) is equivalent to a *rotation* in the complex plane.

   Consider a complex number $z = re^{i\theta}$. What modulus and argument does $ze^{i\phi}$ have? What angle does z rotate through if multiplied by i?
4. The complex conjugate of a complex number $z = a + ib$, denoted z^*, is $z^* = a - ib$. A complex number is real if $z = z^*$, meaning that its imaginary part is zero. A complex number z is imaginary if $z = -z^*$, which implies that its real part is zero.
   - What number is both real and purely imaginary? What is $(z^*)^*$ in terms of z?
   - Write z^* in modulus and argument form, given $z = re^{i\theta}$.
   - Write the real and imaginary parts of z in terms of z and z^*.
   - Show that zz^* is real.
5. Use the Euler formula to derive formulae for $\cos(\alpha + \beta)$ and $\sin(\alpha + \beta)$. Write your answers in terms of $\sin\alpha$, $\cos\alpha$, $\sin\beta$, and $\cos\beta$ only.

6. Calculate the product $(2+i)(3+i)$, and use the answer to prove Euler's relation

   $\frac{\pi}{4} = \arctan\frac{1}{2} + \arctan\frac{1}{3}$.

   Using the Taylor expansion of arctan, this kind of formula is useful for the numerical evaluation of π. Consider the product $(5+i)^4(239-i)$ (expanded using your favorite software) to find Machin's expression for the calculation of $\pi/4$. As of January 2020, π has been calculated to fifty trillion digits.

Problem 1.2. *Quantized energies in the Bohr model.*

Consider an electron in circular motion around a fixed (heavy) proton as a model for the hydrogen atom. Let $V = -e^2/r$ denote the potential energy of the electron and $F = e^2/r^2$ be the magnitude of the electrostatic force acting on the electron.

1. Find expressions for the kinetic energy K of the electron and for the potential energy V of the electron in terms of the total energy E of the electron.
2. Assume that the magnitude L of the electron's angular momentum is quantized and equal to $n\hbar$, where n is a positive integer. Find the quantized values for the total energy E_n and the associated orbit radii r_n.

   Write your answers in terms of n, the rest energy $E_e = m_e c^2$ of the electron, the reduced Compton wavelength $\lambda\!\!\!^{-} = \frac{\hbar}{m_e c}$, and the fine structure constant $\alpha = \frac{e^2}{\hbar c}$.

Problem 1.3. *Radiative collapse of a classical atom.*

In classical physics, a hydrogen atom would be built by placing an electron in a circular orbit around a proton. We know, however, that a nonrelativistic, accelerating electron radiates energy at a rate given by the Larmor formula:

$$\frac{dE}{dt} = -\frac{2}{3}\frac{e^2 a^2}{c^3}.$$

Here E is the energy of the particle, e is the electron charge, a is the magnitude of the electron acceleration, and c is the speed of light.

Because of this energy loss, the classical atom has a stability problem. We want to figure out how big this effect is. The electron potential energy in the presence of the proton is $V = -e^2/r$, and the magnitude of the force of attraction is e^2/r^2. The total energy E of the electron is the sum of its kinetic energy K and its potential energy V.

1. Assuming a circular orbit, show that for a nonrelativistic electron the energy $|\Delta E|$ lost per revolution is small compared to the electron's kinetic energy K. Do this by computing the ratio $|\Delta E|/K$ and expressing it in terms of the speed v of the electron and c.

   Your answer should lead you to conclude that for a nonrelativistic electron spiraling into the proton, we can indeed regard the orbit as circular at any instant.
2. A good estimate for the size of the hydrogen atom is 50 pm (picometers), and a good estimate for the size of the nucleus is 1 fm (femtometer). Compare the classically calculated velocity of the electron to the velocity of light at an orbital radius of 50 pm, 1 pm, and 1 fm. Express your answers as a fraction of the speed of light.

Hint: To deal with e^2, use the fine structure constant $\alpha = e^2/\hbar c \simeq 1/137$. It's also helpful to know $m_e c^2 \simeq 0.511\text{MeV}$ and $\hbar c \simeq 197.3\,\text{MeV.fm}$. This set of numbers is quite useful, so we suggest writing it down, if not memorizing it!

3. Find an expression for the time $t(r_i \to r_f)$ it takes for an electron to spiral in from an initial radius r_i to a final radius r_f. Give your answer in terms of r_i, r_f, and constants such as m_e, e, and c. Calculate how long (in seconds) it would take for the electron to spiral from 50 pm to 1 pm. Is it justified to ignore relativistic corrections in this answer?
4. In this nonrelativistic analysis, what happens to the total energy E of the electron as it approaches the proton? Is there a minimum value for the total energy of the electron?

Problem 1.4. *Quantum oscillations in SQUIDs.*

To illustrate the oscillations between flux states in a SQUID, consider the states

$$\begin{aligned} |\text{clockwise current}\rangle &= \tfrac{1}{2}\big(|1\rangle + |2\rangle\big), \\ |\text{counterclockwise current}\rangle &= \tfrac{1}{2}\big(|1\rangle - |2\rangle\big). \end{aligned}$$

States $|1\rangle$ and $|2\rangle$ are (time-independent) states of definite energy E_1 and E_2, respectively. As we will learn later, states of fixed energy evolve in time as follows:

$$|1, t\rangle = |1\rangle e^{-iE_1 t/\hbar}, \quad |2, t\rangle = |2\rangle e^{-iE_2 t/\hbar}.$$

Implicit in this notation is that $|1, t=0\rangle = |1\rangle$, and $|2, t=0\rangle = |2\rangle$. This implies that a state $|A\rangle = \frac{1}{2}\big(|1\rangle + |2\rangle\big)$, at time equal zero, evolves in time as

$$|A, t\rangle = \tfrac{1}{2}\big(|1, t\rangle + |2, t\rangle\big).$$

Let $\Delta = E_2 - E_1 \neq 0$ and assume that at time equal to zero the state of a SQUID is $|\text{clockwise current}\rangle$. Find the shortest time $T > 0$ it takes the state to become $|\text{counterclockwise current}\rangle$. Your answer will depend on Δ and $\hbar$ only.

2 Light, Particles, and Waves

We use a quantum-mechanical wave function to discuss the behavior of a photon in a Mach-Zehnder interferometer. Quantum interference gives the surprising possibility of an interaction-free measurement, which we illustrate in the context of Elitzur-Vaidman bombs. We then retrace the developments that led to quantum mechanics. The photoelectric effect gave evidence for the existence of photons, quanta of light that carry fixed amounts of energy and momentum. Further evidence came from Compton scattering, where a quantum of light collides with an electron. The wave-particle duality of light states that the full description of light requires both wave and particle features. De Broglie conjectured that wave-particle duality applies to all matter particles, with wavelength inversely proportional to the momentum of the particle. The frequency of a de Broglie wave is proportional to the energy of the particle, and this relationship implies the correct group velocity for the motion of wave packets.

2.1 Mach-Zehnder Interferometer

We have previously discussed the Mach-Zehnder interferometer, which we show again in figure 2.1. It contains two beam splitters, BS1 and BS2, and two mirrors, both denoted by M. Inside the interferometer we have two beams, one going along the upper branch and one going along the lower branch. This extends beyond BS2: the upper branch continues to the detector D0 while the lower branch continues to the detector D1.

Imagine vertical cuts in figure 2.1, lying between BS1 and BS2. Any such cut intersects the two beams, and we can ask what the probability is of finding a photon in each of the two beams at that cut. For this we need two probability *amplitudes*, which we recall are two complex numbers whose norm squared give probabilities. We can encode this information in a two-component vector:

$$\begin{pmatrix} \alpha \\ \beta \end{pmatrix}. \tag{2.1.1}$$

Here α is the probability amplitude of being in the upper beam, and β the probability amplitude of being in the lower beam. Therefore, $|\alpha|^2$ is the probability of finding the photon in the upper beam, and $|\beta|^2$ is the probability of finding the photon in the

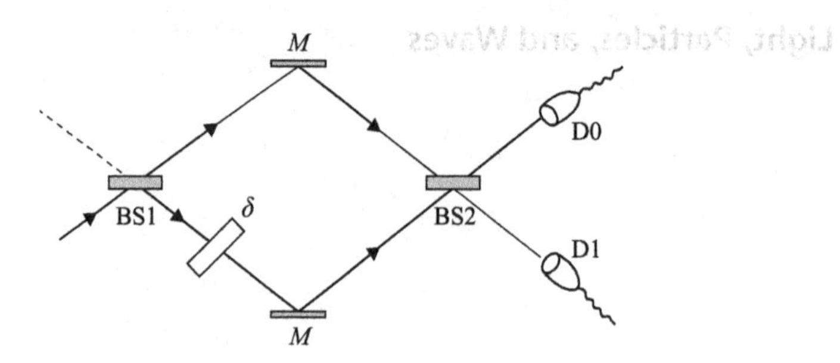

Figure 2.1
The Mach-Zehnder interferometer.

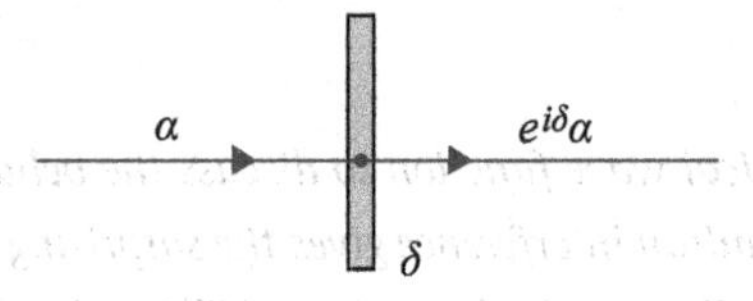

Figure 2.2
A phase shifter of phase δ multiplies the amplitude by the phase factor $e^{i\delta}$.

lower beam. Since the photon must be found in either one of the beams, we must have that

$$|\alpha|^2 + |\beta|^2 = 1. \tag{2.1.2}$$

Following this notation, the states in which the photon is definitely in one or the other beam are

$$\text{photon in upper beam: } \begin{pmatrix} 1 \\ 0 \end{pmatrix}, \text{ photon in bottom beam: } \begin{pmatrix} 0 \\ 1 \end{pmatrix}. \tag{2.1.3}$$

We can view the state (2.1.1) as a superposition of these two simpler states using the rules of vector addition and multiplication:

$$\begin{pmatrix} \alpha \\ \beta \end{pmatrix} = \begin{pmatrix} \alpha \\ 0 \end{pmatrix} + \begin{pmatrix} 0 \\ \beta \end{pmatrix} = \alpha \begin{pmatrix} 1 \\ 0 \end{pmatrix} + \beta \begin{pmatrix} 0 \\ 1 \end{pmatrix}. \tag{2.1.4}$$

For convenience, when writing in the text, a state $\begin{pmatrix} \alpha \\ \beta \end{pmatrix}$ will be displayed as (α, β).

A photon state or photon wave function (α, β) is said to be **normalized** if $|\alpha|^2 + |\beta|^2 = 1$. A normalized state can be used to compute probabilities consistently. If we are tracking a photon as it goes through an interferometer, the state must be normalized at all times.

The lower branch of the interferometer shown in figure 2.1 includes a *phase shifter*, a piece of material whose only effect is to multiply the probability *amplitude* by a fixed phase factor $e^{i\delta}$, with $\delta \in \mathbb{R}$. As shown in figure 2.2, a probability amplitude α to the left of the phase shifter becomes a probability amplitude $e^{i\delta}\alpha$ to the right of the phase shifter. Since the norm of a pure phase factor is one, the phase shifter does not change

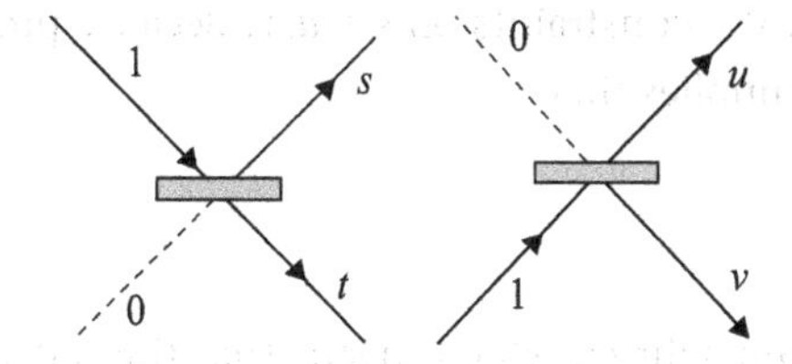

Figure 2.3
Left: A photon $(1,0)$ incident from the top; s and t are the reflected and transmitted amplitudes, respectively. *Right*: A photon $(0,1)$ incident from the bottom; v and u are the reflected and transmitted amplitudes, respectively.

the probability of finding the photon in the beam. When the $\delta = \pi$, the effect of the phase shifter is to change the sign of the amplitude since $e^{i\pi} = -1$.

Let us now consider the effect of beam splitters in detail. If the incident photon hits the left beam splitter from the top, we consider this photon as belonging to the upper branch and represent it by $(1,0)$. If the incident photon hits the beam splitter from the bottom, we consider this photon as belonging to the lower branch and represent it by $(0,1)$. We show the two cases in figure 2.3. The effect of the beam splitter is to give an output wave function for each of the two cases:

$$\text{left BS:} \begin{pmatrix} 1 \\ 0 \end{pmatrix} \Rightarrow \begin{pmatrix} s \\ t \end{pmatrix}, \quad \text{right BS:} \begin{pmatrix} 0 \\ 1 \end{pmatrix} \Rightarrow \begin{pmatrix} u \\ v \end{pmatrix}, \tag{2.1.5}$$

with s, t, u, and v some complex numbers. We will show that these numbers are not completely arbitrary.

As you can see from the diagram, for the photon hitting from above, s can be viewed as a reflection amplitude and t as a transmission amplitude. Similarly, for the photon hitting from below, v can be viewed as a reflection amplitude and u as a transmission amplitude. The four numbers s, t, u, v, by linearity, completely characterize the beam splitter. They can be used to predict the output given any incident photon, which may have amplitudes to hit both from above and from below. Indeed, an incident photon state (α, β) would give

$$\begin{pmatrix} \alpha \\ \beta \end{pmatrix} = \alpha \begin{pmatrix} 1 \\ 0 \end{pmatrix} + \beta \begin{pmatrix} 0 \\ 1 \end{pmatrix} \Rightarrow \alpha \begin{pmatrix} s \\ t \end{pmatrix} + \beta \begin{pmatrix} u \\ v \end{pmatrix} = \begin{pmatrix} \alpha s + \beta u \\ \alpha t + \beta v \end{pmatrix} = \begin{pmatrix} s & u \\ t & v \end{pmatrix} \begin{pmatrix} \alpha \\ \beta \end{pmatrix}. \tag{2.1.6}$$

In summary, we see that the beam splitter produces the following effect:

$$\begin{pmatrix} \alpha \\ \beta \end{pmatrix} \Rightarrow \begin{pmatrix} s & u \\ t & v \end{pmatrix} \begin{pmatrix} \alpha \\ \beta \end{pmatrix}. \tag{2.1.7}$$

We can therefore represent the action of the beam splitter on wave functions as multiplication by the two-by-two matrix:

$$\begin{pmatrix} s & u \\ t & v \end{pmatrix}. \tag{2.1.8}$$

We must now figure out the constraints on s, t, u, v. Because probabilities must add up to one, equation (2.1.5) implies that

$$|s|^2 + |t|^2 = 1, \tag{2.1.9}$$

$$|u|^2 + |v|^2 = 1\,. \tag{2.1.10}$$

Consider a balanced beam splitter, which means that the reflection and transmission probabilities are the same. So all four constants must have equal norm-squared values:

$$|s|^2 = |t|^2 = |u|^2 = |v|^2 = \tfrac{1}{2}. \tag{2.1.11}$$

This means all coefficients are determined up to phases. Let's try a guess for the values. Could we have

$$\begin{pmatrix} s & u \\ t & v \end{pmatrix} = \begin{pmatrix} \frac{1}{\sqrt{2}} & \frac{1}{\sqrt{2}} \\ \frac{1}{\sqrt{2}} & \frac{1}{\sqrt{2}} \end{pmatrix}? \tag{2.1.12}$$

To decide, we note the key consistency check: acting on a normalized wave function, the action of the beam splitter must still give a normalized wave function. Clearly, if the total probability of finding the photon before it hits the beam splitter is one, that probability should still be one after emerging from the beam splitter since a beam splitter does not absorb photons! A beam splitter must conserve probability. So we try with a couple of wave functions:

$$\begin{pmatrix} \frac{1}{\sqrt{2}} & \frac{1}{\sqrt{2}} \\ \frac{1}{\sqrt{2}} & \frac{1}{\sqrt{2}} \end{pmatrix}\begin{pmatrix} 1 \\ 0 \end{pmatrix} = \begin{pmatrix} \frac{1}{\sqrt{2}} \\ \frac{1}{\sqrt{2}} \end{pmatrix}, \quad \begin{pmatrix} \frac{1}{\sqrt{2}} & \frac{1}{\sqrt{2}} \\ \frac{1}{\sqrt{2}} & \frac{1}{\sqrt{2}} \end{pmatrix}\begin{pmatrix} \frac{1}{\sqrt{2}} \\ \frac{1}{\sqrt{2}} \end{pmatrix} = \begin{pmatrix} 1 \\ 1 \end{pmatrix}. \tag{2.1.13}$$

While the first example works out, the second one does not, as $|1|^2 + |1|^2 = 2 \neq 1$. An easy fix is achieved by changing the sign of v:

$$\begin{pmatrix} s & u \\ t & v \end{pmatrix} = \begin{pmatrix} \frac{1}{\sqrt{2}} & \frac{1}{\sqrt{2}} \\ \frac{1}{\sqrt{2}} & -\frac{1}{\sqrt{2}} \end{pmatrix} = \frac{1}{\sqrt{2}}\begin{pmatrix} 1 & 1 \\ 1 & -1 \end{pmatrix}. \tag{2.1.14}$$

Let's check that this matrix works in general. Consider acting on a normalized state (α, β) so that $|\alpha|^2 + |\beta|^2 = 1$. We then find

$$\frac{1}{\sqrt{2}}\begin{pmatrix} 1 & 1 \\ 1 & -1 \end{pmatrix}\begin{pmatrix} \alpha \\ \beta \end{pmatrix} = \frac{1}{\sqrt{2}}\begin{pmatrix} \alpha + \beta \\ \alpha - \beta \end{pmatrix}. \tag{2.1.15}$$

Indeed, the resulting state is normalized:

$$\begin{aligned} \tfrac{1}{2}|\alpha + \beta|^2 + \tfrac{1}{2}|\alpha - \beta|^2 &= \tfrac{1}{2}(|\alpha|^2 + |\beta|^2 + \alpha\beta^* + \alpha^*\beta) + \tfrac{1}{2}(|\alpha|^2 + |\beta|^2 - \alpha\beta^* - \alpha^*\beta) \\ &= |\alpha|^2 + |\beta|^2 \;=\; 1. \end{aligned}$$

As you can quickly verify, the minus sign in the bottom-right entry of (2.1.14) means that a photon incident from below, as it is reflected, will have its amplitude changed by a sign or, equivalently, a phase shift by π.

This effect is in fact familiar in electromagnetic theory. A typical beam splitter consists of a glass plate with a reflective dielectric coating on one side. The refractive index of the coating is chosen to be bigger than that of air, and smaller than that of glass.

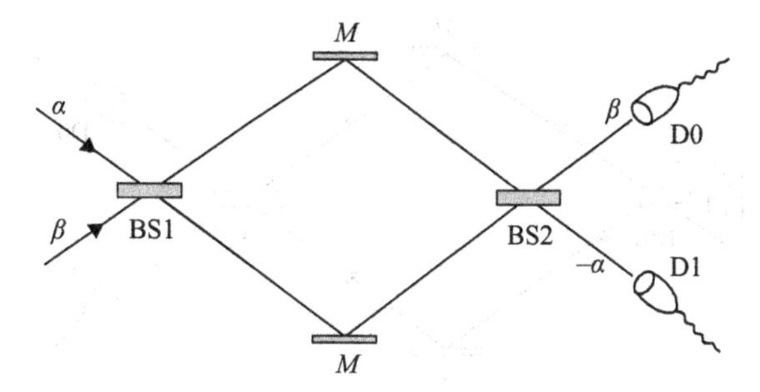

Figure 2.4
The interferometer with input wave function (α, β) has an output wave function $(\beta, -\alpha)$.

When an electromagnetic wave in a medium is incident on a medium of higher refractive index, the reflected wave is phase shifted by π. This is the case as a wave hits the coating from the air but not when the wave hits the coating from glass. Thus, the beam splitter represented by (2.1.14) would have its coating on the bottom side. Transmitted waves have no phase shift. The phase shift in the electromagnetic wave picture is in fact the phase shift for the quantum amplitudes.

Another possibility for a beam splitter matrix is

$$\frac{1}{\sqrt{2}}\begin{pmatrix} -1 & 1 \\ 1 & 1 \end{pmatrix}, \tag{2.1.16}$$

which would be realized by a dielectric coating on the top side. You can quickly check that this beam splitter also conserves probability. With these two options, beam splitter one (BS1) and beam splitter two (BS2) will be associated with matrices U_1 and U_2, respectively:

$$\text{BS1: } U_1 = \frac{1}{\sqrt{2}}\begin{pmatrix} -1 & 1 \\ 1 & 1 \end{pmatrix}, \quad \text{BS2: } U_2 = \frac{1}{\sqrt{2}}\begin{pmatrix} 1 & 1 \\ 1 & -1 \end{pmatrix}. \tag{2.1.17}$$

These two beam splitters are part of the interferometer shown in figure 2.4, with BS1 on the left and BS2 on the right. In addition to the beam splitters, we have two mirrors, one for each of the two possible paths. A mirror acts like a $\delta = \pi$ phase shifter, changing the relevant amplitude by a sign. The two mirrors change both the upper and lower amplitudes of the photon by a minus sign, thus changing the photon state by an overall minus sign. Since an overall sign on the output state does not matter, we will ignore the effect of the mirrors in the following analysis.

If we now assume that the interferometer has an incident photon wave function (α, β), the output wave function that goes into the detectors is obtained by acting first with the BS1 matrix U_1 and then with the BS2 matrix U_2:

$$\begin{aligned} \text{output} &= U_2 \cdot U_1 \begin{pmatrix} \alpha \\ \beta \end{pmatrix} = \frac{1}{\sqrt{2}}\begin{pmatrix} 1 & 1 \\ 1 & -1 \end{pmatrix} \frac{1}{\sqrt{2}}\begin{pmatrix} -1 & 1 \\ 1 & 1 \end{pmatrix}\begin{pmatrix} \alpha \\ \beta \end{pmatrix} \\ &= \frac{1}{2}\begin{pmatrix} 0 & 2 \\ -2 & 0 \end{pmatrix}\begin{pmatrix} \alpha \\ \beta \end{pmatrix} = \begin{pmatrix} \beta \\ -\alpha \end{pmatrix}. \end{aligned} \tag{2.1.18}$$

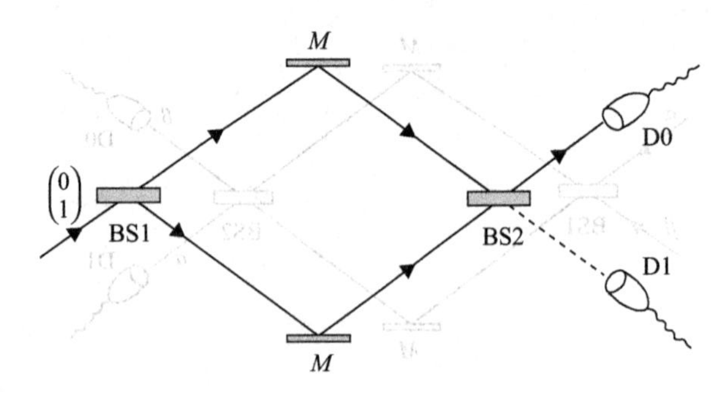

Figure 2.5
Incident photon from below will go into D0.

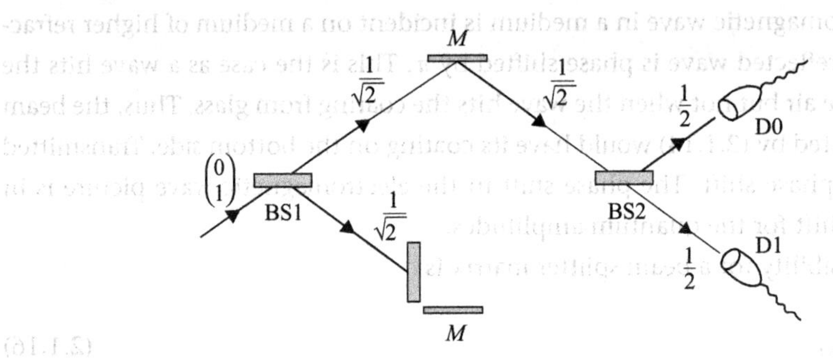

Figure 2.6
The probability of detecting the photon at D1 is increased by blocking one of the paths.

The labels on figure 2.4 show the input and output amplitudes. With the help of this result, for any input photon state we can immediately write the output photon state that goes into the detectors.

If the incident photon state is $(0, 1)$, the output from the interferometer is $(1, 0)$, and therefore the photon will be detected at D0. This is shown in figure 2.5. We can make a very simple table with the possible outcomes and their respective probabilities p:

All pathways open: (2.1.19)

Outcome	p
Photon at D0	1
Photon at D1	0

Let us now block the lower path, as indicated in figure 2.6. What happens then? It is best to track things down systematically. The incident photon acted on by BS1 gives

$$U_1\begin{pmatrix}0\\1\end{pmatrix}=\frac{1}{\sqrt{2}}\begin{pmatrix}-1&1\\1&1\end{pmatrix}\begin{pmatrix}0\\1\end{pmatrix}=\frac{1}{\sqrt{2}}\begin{pmatrix}1\\1\end{pmatrix}. \tag{2.1.20}$$

This is indicated in the figure, to the right of BS1. Then the lower branch is stopped while the photon continues on the upper branch. The upper beam reaches BS2, and here the input is $(\frac{1}{\sqrt{2}}, 0)$ because nothing is coming from the lower branch of the interferometer. We therefore get an output

$$U_2 \begin{pmatrix} \frac{1}{\sqrt{2}} \\ 0 \end{pmatrix} = \frac{1}{\sqrt{2}} \begin{pmatrix} 1 & 1 \\ 1 & -1 \end{pmatrix} \begin{pmatrix} \frac{1}{\sqrt{2}} \\ 0 \end{pmatrix} = \begin{pmatrix} \frac{1}{2} \\ \frac{1}{2} \end{pmatrix}. \tag{2.1.21}$$

In this experiment there are three possible outcomes: the photon can be absorbed by the block or can go into either of the two detectors. Looking at the diagram we can easily read the probabilities:

Blocked lower pathway: (2.1.22)

Outcome	p
Photon at block	$\frac{1}{2}$
Photon at D0	$\frac{1}{4}$
Photon at D1	$\frac{1}{4}$

Before blocking the lower path, we could not get a photon to D1. The probability of reaching D1 is now 1/4 and was in fact *increased* by blocking a path.

2.2 Elitzur-Vaidman Bombs

To see that allowing the photon to reach D1 by blocking a path is very strange, we consider an imaginary situation proposed by physicists Avshalom Elitzur and Lev Vaidman, from Tel Aviv University, in Israel. They imagined bombs with a special type of trigger: a photon detector. A narrow tube goes across each bomb, and in the middle of the tube is a photon detector. To detonate the bomb, one sends a photon into the tube. The photon is then detected by the photon detector and the bomb explodes. If the photon detector is defective, however, the photon is not detected at all. It propagates *freely through the tube* and comes out of the bomb. The bomb, in this case, does not explode.

Here is the situation we want to address. Suppose we have a number of Elitzur-Vaidman (EV) bombs, and we suspect some have become defective. How can we tell if a bomb is operational without detonating it? Assume, for the sake of the problem, that we are unable to examine the detector without destroying the bomb.

We seem to be facing an impossible situation. If we send a photon into the detector tube and nothing happens, we know the bomb is defective, but if the bomb is operational, it will simply explode. It seems impossible to confirm that the photon detector in the bomb is working without testing it. Indeed, it is impossible in classical physics. It is not impossible in quantum mechanics, however. As we will see, we can perform what can be called an interaction-free measurement!

We now place an EV bomb on the lower path of the interferometer, with the detector tube properly aligned. Suppose we send in a photon as pictured in figure 2.7. If the bomb is defective, it is as if there is no detector; the lower branch of the interferometer is free, and all the photons that we send in will end up in D0, just as they did in figure 2.5.

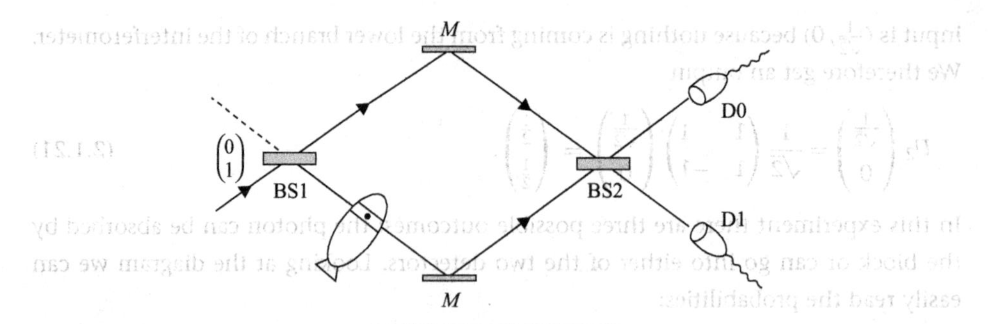

Figure 2.7
A Mach-Zehnder interferometer with an Elitzur-Vaidman bomb inserted on the lower branch, with the detector tube properly aligned. If the bomb is defective, all incident photons will end up at D0. If a photon ends up at D1, we know that the bomb is operational, even though the photon never went into the bomb detector!

Defective bomb: (2.2.1)

Outcome	p
Photon at D0 (no explosion)	1
Photon at D1 (no explosion)	0
Bomb explodes	0

If the bomb is working, on the other hand, we have the situation we had in figure 2.6, in which we placed a block in the lower branch of the interferometer:

Operational bomb: (2.2.2)

Outcome	p
Bomb explodes	$\frac{1}{2}$
Photon at D0 (no explosion)	$\frac{1}{4}$
Photon at D1 (no explosion)	$\frac{1}{4}$

Assume the bomb is working. Then 50% of the time the photon will hit it and it will explode; 25% of the time the photon will end in D0, and we will not be able to tell if it is defective or not. But 25% of the time the photon will end in D1, and since this would be impossible for a defective bomb, we will learn that the bomb is operational!

This is shocking and counterintuitive. We learned that a bomb works even though the photon *never made it* into the bomb; it ended on D1. How could we have learned the bomb detector is working without the photon ever reaching it? However strange it may seem, this kind of interaction-free measurement is indeed possible. Experiments, thankfully without using bombs, have confirmed the correctness of the above discussion.

Our setup with the Mach-Zehnder interferometer makes the point clear, but it is not optimal if we are trying to determine if bombs are operational. Indeed, 50% of the operational bombs will explode at the first run of the experiment. You may think that perhaps this is the best we can do. It turns out, however, that it is possible to do better. We will consider next a setup in which during the process of certifying an operational bomb, the probability of exploding it can be made arbitrarily small!

Exercise 2.1. *While you do not know it, you have been handed a functioning EV bomb. You are allowed to use a Mach-Zehnder interferometer to test this bomb two times. Your testing protocol is such that if the first test is inconclusive, you will test it a second time. What is the probability of an inconclusive result? What is the probability that you certify that the bomb is operational without detonating it? What is the probability that the bomb will explode when subject to the protocol? (Partial answer: the product of the three probabilities is 25/2048).*

2.3 Toward Perfect Bomb Detection

As a matter of principle, we wish to demonstrate that this kind of interaction-free measurement can be optimized, at the cost of using more refined beam splitters and a more intricate testing protocol. We modify the Mach-Zehnder setup of the previous section in order to increase toward 100% the fraction of EV bombs that we can verify as working without detonating them.

To improve detection we use a high reflectivity beam splitter, represented by a two-by-two matrix U that depends on a fixed, positive integer N:

$$U = \begin{pmatrix} \cos\frac{\pi}{2N} & i\sin\frac{\pi}{2N} \\ i\sin\frac{\pi}{2N} & \cos\frac{\pi}{2N} \end{pmatrix}. \tag{2.3.1}$$

Exercise 2.2. *Confirm that the beam splitter based on U conserves probability.*

The matrix U above is actually a unitary matrix, meaning that it satisfies $U^\dagger U = \mathbb{1}$ where $U^\dagger$ is obtained from U by complex conjugation and transposition, and $\mathbb{1}$ is the two-by-two identity matrix. In problem 2.3 you will show that any two-by-two matrix conserving probability is unitary. We assume that for any two-by-two matrix that conserves probability, such as U, the corresponding beam splitter can be built.

For an input photon $(1, 0)$, the output photon would be

$$U\begin{pmatrix} 1 \\ 0 \end{pmatrix} = \begin{pmatrix} \cos\frac{\pi}{2N} \\ i\sin\frac{\pi}{2N} \end{pmatrix}. \tag{2.3.2}$$

It follows that the probability of reflection, called the reflectivity R, and the probability of transmission, called the transmissivity T, are given by

$$R = \left(\cos\frac{\pi}{2N}\right)^2, \qquad T = \left(\sin\frac{\pi}{2N}\right)^2, \quad R + T = 1.$$

For large N, which is the case of interest, R is close to one, and T is close to zero. This is indeed a high-reflectivity beam splitter.

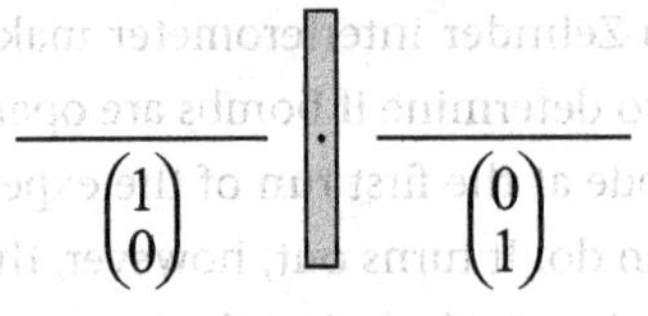

Figure 2.8
(1, 0) and (0, 1) are states to the left and to the right of the beam splitter, respectively.

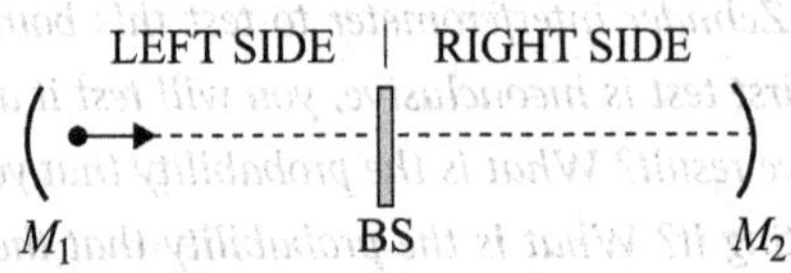

Figure 2.9
A cavity with the beam splitter positioned vertically and two perfectly reflecting vertical mirrors equidistant to the beam splitter. The initial photon state is on the left and moving toward the beam splitter.

If you think of our photons in the previous arrangements with the Mach-Zehnder interferometer, we have really been describing a system with two states or, more precisely, two basis states. The first basis state, (1, 0), depending on which section of the apparatus we were talking about, represented a photon incident from the top, a photon in the upper branch of the interferometer, or a photon traveling toward D0. The second basis state, (0, 1), represented a photon incident from the bottom, a photon in the lower branch of the interferometer, or a photon going into D1. This was our choice of states. They are basis states because any arbitrary state of a photon can be built by the superposition of these two states.

We now think of the beam splitter placed vertically. Again we have a two-state system. It is then natural to think of photons to the left of the beam splitter, whether they are moving away or toward it, as represented by (1, 0). A state in which the photon is to the right of the beam splitter, moving away or toward it, is represented by (0, 1). This is shown in figure 2.8.

We now construct a cavity in which the beam splitter (BS) is placed in between perfectly reflecting mirrors M_1 and M_2, which are placed at equal distances to the left and to the right of the beam splitter, respectively. A photon is sent in from the left, as shown in figure 2.9. The photon will hit BS and split, the reflected and transmitted components will bounce off the mirrors and hit BS a second time, and so on and so forth.

After the photon coming from the left hits the beam splitter, the state of the photon becomes the vector s_1, given by

$$s_1 = U\begin{pmatrix}1\\0\end{pmatrix}. \tag{2.3.3}$$

The top component of s_1 is the amplitude of having a photon to the left of the beam splitter and moving away from it, and the bottom component of s_1 is the amplitude of having a photon to the right of the beam splitter, moving away from it. The photon state described by s_1 is reflected by the mirrors and remains the state s_1, now incident upon the beam splitter. After hitting the beam splitter, the state s_1 becomes s_2 with

$$s_2 = Us_1 = U^2 \begin{pmatrix} 1 \\ 0 \end{pmatrix}. \tag{2.3.4}$$

After k hits into the beam splitter, the original photon state $(1, 0)$ becomes s_k, given by

$$s_k = U^k \begin{pmatrix} 1 \\ 0 \end{pmatrix}. \tag{2.3.5}$$

To calculate the s_k state, we need to raise U to the k-th power. This is simple after we learn the following result:

$$\begin{pmatrix} \cos\alpha & i\sin\alpha \\ i\sin\alpha & \cos\alpha \end{pmatrix} \begin{pmatrix} \cos\beta & i\sin\beta \\ i\sin\beta & \cos\beta \end{pmatrix} = \begin{pmatrix} \cos(\alpha+\beta) & i\sin(\alpha+\beta) \\ i\sin(\alpha+\beta) & \cos(\alpha+\beta) \end{pmatrix}. \tag{2.3.6}$$

Exercise 2.3. *Prove (2.3.6).*

It now follows that

$$U^k = \begin{pmatrix} \cos\frac{\pi}{2N} & i\sin\frac{\pi}{2N} \\ i\sin\frac{\pi}{2N} & \cos\frac{\pi}{2N} \end{pmatrix}^k = \begin{pmatrix} \cos\frac{k\pi}{2N} & i\sin\frac{k\pi}{2N} \\ i\sin\frac{k\pi}{2N} & \cos\frac{k\pi}{2N} \end{pmatrix}. \tag{2.3.7}$$

As a result, the state s_k is given by

$$s_k = U^k \begin{pmatrix} 1 \\ 0 \end{pmatrix} = \begin{pmatrix} \cos\frac{k\pi}{2N} \\ i\sin\frac{k\pi}{2N} \end{pmatrix}. \tag{2.3.8}$$

If $k \ll N$, the amplitude to be to the left of the beam splitter is much greater than the amplitude to be to the right of the beam splitter, which is consistent with the high reflectivity of the beam splitter. Since s_k represents the state of the photon after k hits to the beam splitter, the probability $p_k(L)$ that the photon will be found on the left side of the cavity and the probability $p_k(R)$ that it will be found on the right side of the cavity are

$$p_k(L) = \cos^2\frac{k\pi}{2N}, \quad p_k(R) = \sin^2\frac{k\pi}{2N}. \tag{2.3.9}$$

While the probability of remaining to the left of the beam splitter is high for $k \ll N$, if we wait for N hits, we have

$$p_N(L) = \cos^2\tfrac{\pi}{2} = 0, \quad p_N(R) = \sin^2\tfrac{\pi}{2} = 1. \tag{2.3.10}$$

At this point, despite the high reflectivity, the photon is certain to be found to the right of the beam splitter! If we wait for $2N$ hits, the photon will again be found on the left side of the cavity, as it was at the beginning. This is happening for the empty cavity. A cavity in which we place a defective EV bomb to the right of the beam splitter is effectively empty, as the photon detector on the bomb is not operational and lets the

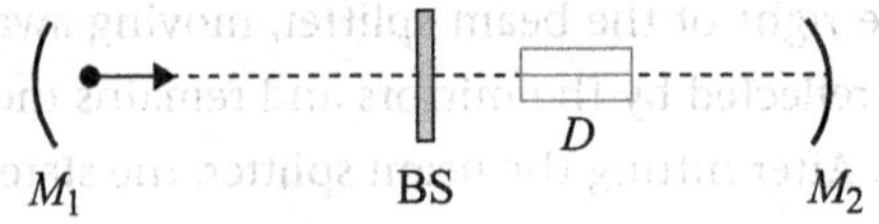

Figure 2.10
The cavity with a photon detector D to the right of the beam splitter. The initial photon state is on the left and moving toward the beam splitter.

photons go through. So after N hits, the photon would be expected to be on the right side of the cavity.

A photon detector D is now inserted on the right side of the cavity so that any photon reaching the right side will be detected and absorbed (figure 2.10). To distinguish this case from the previous ones, the probabilities will be written with an uppercase P. As before, a photon is sent in from the left. After the first hit—call it the $k=1$ hit—the photon is either detected or, if not, will certainly be on the left side of the cavity. The probability $P_{k=1}(L)$ of being on the left and the probability $P_{k=1}(D)$ of having been detected add up to one. They are

$$P_{k=1}(L) = \left(\cos\frac{\pi}{2N}\right)^2, \quad P_{k=1}(D) = 1 - \left(\cos\frac{\pi}{2N}\right)^2. \tag{2.3.11}$$

If the photon is on the left, it will bounce off the mirror and hit the beam splitter again. Once more it will either bounce back to the left or be detected. We then have

$$P_{k=2}(L) = \left(\cos\frac{\pi}{2N}\right)^4, \quad P_{k=2}(D) = 1 - \left(\cos\frac{\pi}{2N}\right)^4. \tag{2.3.12}$$

Perhaps it helps to think in terms of time. After we wait for the time t_2 that would take a free photon to make two hits, the probability that the photon is to the left is $P_{k=2}(L)$, and the probability that the photon was detected at some point is $P_{k=2}(D)$. Of course, the time between hits depends on the length of the cavity.

Exercise 2.4. *Confirm that $P_{k=2}(D)$ equals the probability of detection after the first hit plus the probability of detection after the second hit.*

We can now readily generalize. After waiting for the time t_k that a free photon would take to hit the beam splitter k times, our photon will either be to the left with probability $P_k(L)$ or will have been detected at some point with probability $P_k(D)$:

$$P_k(L) = \left(\cos\frac{\pi}{2N}\right)^{2k}, \quad P_k(D) = 1 - \left(\cos\frac{\pi}{2N}\right)^{2k}. \tag{2.3.13}$$

Applied to the case $k=N$, we find that

$$P_N(L) = \left(\cos\frac{\pi}{2N}\right)^{2N}, \quad P_N(D) = 1 - \left(\cos\frac{\pi}{2N}\right)^{2N}. \tag{2.3.14}$$

All our calculations are now in place, and we can describe our *testing protocol*: We place an EV bomb in the cavity, to the right of the beam splitter, just like we placed the detector. We insert two detectors, D_L and D_R, that are turned on by a signal and are initially off, letting photons go through freely. We send in a photon from the left and

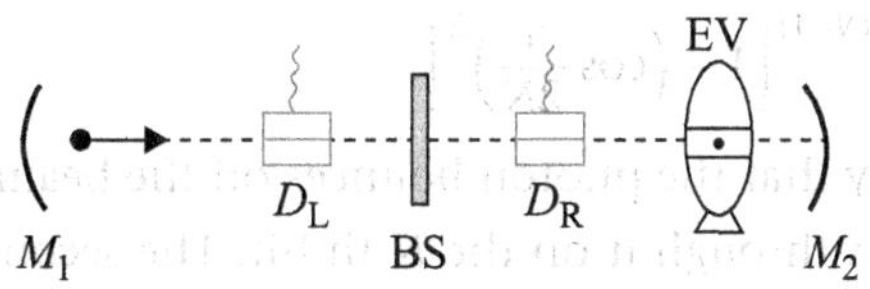

Figure 2.11
The configuration to test an Elitzur-Vaidman (EV) bomb located to the right of the beam splitter. A photon is sent in from the left. Immediately after time t_N, the photon detectors D_L and D_R are activated to determine where the photon is.

wait for the time t_N that it would take a free photon to hit the beam splitter N times. Assuming the bomb has not exploded, at that point we activate the detectors D_L and D_R to see where the photon is (figure 2.11).

If the EV bomb is *not* operational, it is as if we had inserted nothing, and at time t_N the photon will be found on the right side of the cavity, as shown in (2.3.10). If the EV bomb *is* operational, it is effectively a working detector. In that case the probability of detection by time t_N is $P_N(D)$, and it is in fact the probability P_{explode} that the bomb will explode in the laboratory:

$$P_{\text{explode}} = P_D(N) = 1 - \left(\cos\frac{\pi}{2N}\right)^{2N}. \tag{2.3.15}$$

This probability is very small for large N. Indeed, for small ϵ we have $\cos\epsilon \simeq 1 - \frac{1}{2}\epsilon^2 + \mathcal{O}(\epsilon^4)$, and therefore

$$P_{\text{explode}} = P_D(N) \simeq 1 - \left(1 - \frac{\pi^2}{8N^2}\right)^{2N} \simeq 1 - \left(1 - 2N\frac{\pi^2}{8N^2}\right), \tag{2.3.16}$$

where we used $(1+\epsilon)^k \simeq 1 + k\epsilon$, valid when $k\epsilon$ is small. This therefore gives

$$P_{\text{explode}} \simeq \frac{\pi^2}{4N}. \tag{2.3.17}$$

As we hoped for, we can make the probability of explosion arbitrarily small by letting N be large. For $N = 250$, for example, $P_{\text{explode}} \simeq \pi^2/1000 \simeq 0.0099$, less than 1 percent!

Suppose now that we do not know if the EV bomb is operational or is defective. Of course, it is either one or the other. Imagine we detect the photon at time t_N to the *left* of the beam splitter. Then we are *certain* the bomb is operational. Why? Because if it were not, the photon would have been found on the right side of the beam splitter.

Imagine now that we detect the photon at time t_N to the *right* of the beam splitter. We will argue that the bomb is very likely defective, but this is not completely certain. Thus, the photon detection to the right of the beam splitter is a bit inconclusive. Certainly, a defective bomb leads to this detection, but we must ask: What is the probability $p(R|\text{op})$ that the photon is to the right of the beam splitter when the bomb is actually operational? For this to happen, our detection at time t_N must have prevented an explosion that was about to occur as the photon was finally rushing into the bomb. The probability $p(R|\text{op})$ is therefore given by

$$p(R|\text{op}) = \left(\cos\frac{\pi}{2N}\right)^{2(N-1)}\left[1-\left(\cos\frac{\pi}{2N}\right)^2\right], \tag{2.3.18}$$

which is the probability that the photon bounces off the beam splitter on each of the first $N-1$ hits and goes through it on the N-th hit. The second factor above is a sine squared, and for large N, the first factor is roughly one. Therefore,

$$p(R|\text{op}) \simeq \left(\sin\frac{\pi}{2N}\right)^2 \simeq \frac{\pi^2}{4N^2}. \tag{2.3.19}$$

This is a very tiny probability for an inconclusive test of an operational bomb.

Exercise 2.5. *Let $N = 250$ and imagine testing 25,000 operational bombs with the protocol, one at a time. Confirm that we would expect to certify without doubt that 24,752 bombs are operational. We would also expect 247 bombs to explode and one bomb to test inconclusively.*

2.4 Photoelectric Effect

In this and in the following section, we will discuss some foundational experiments relating to photons. The first is the photoelectric effect. The second is Compton scattering. Together, these two experiments convinced physicists that photons are quanta of light.

The photoelectric effect was first observed by Heinrich Hertz in 1887. He found that when polished metal plates are irradiated with light, they can emit electrons, nowadays called *photoelectrons*, as we now know they arise due to the impact of photons. The emitted electrons, if they are large in number, produce a detectable *photoelectric current*. The key observations of Hertz included the following:

- There is a threshold frequency ν_0 for the incident light: only for frequencies $\nu > \nu_0$ is there a photoelectric current. The frequency ν_0 depends on the metal and is affected by nonhomogeneities on the metal surface.
- The magnitude of the photoelectric current is proportional to the intensity of the light source.
- The energy of the photoelectrons is *independent* of the intensity of the light source.

A natural explanation for the features of this effect didn't come until 1905, when Einstein explained the above observations by postulating that the energy in light is carried by discrete quanta, later called photons, with energy $h\nu$. In Einstein's postulate h was Planck's constant. Previously, Planck had found the need to introduce the constant h to describe the dependence of the energy density in blackbody radiation as a function of frequency.

A given material has a characteristic energy W, called the *work function*, which is the minimum energy required to eject an electron. The energy W is on the order of a few electron volts, but it is not easily calculated because it is the result of an interaction of many electrons with a background of atoms. It is easily measured, however. When the surface of the material is irradiated, electrons in the material absorb the energy of the incident photons. If the energy imparted on an electron by the absorption of a single

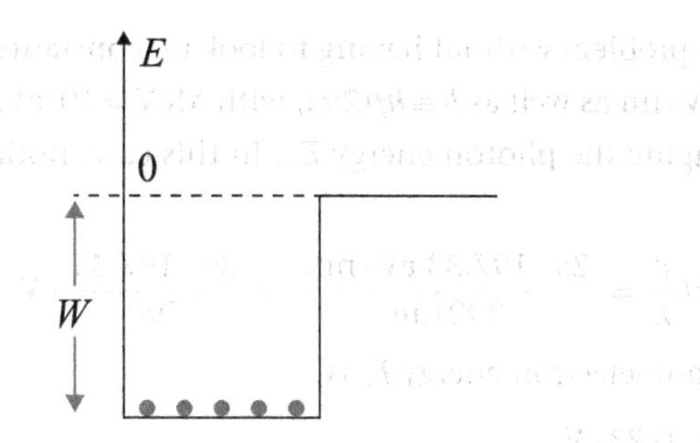

Figure 2.12
Electrons in a metal are bound. A photon with energy greater than the work function W can eject an electron.

photon is greater than the work function W, then the electron is ejected with kinetic energy E_e equal to the difference of the photon energy $E_\gamma = h\nu$ and the work function (figure 2.12):

$$E_e = \tfrac{1}{2}m_e v^2 = E_\gamma - W = h\nu - W. \tag{2.4.1}$$

Here v is the velocity of the ejected photoelectron. This equation, written by Einstein, explains the experimental features noted above if we assume that the quanta act on individual electrons to eject them. Since their kinetic energy must be positive, getting photoelectrons requires $h\nu > W$. The threshold frequency ν_0 is therefore given by

$$h\nu_0 = W, \tag{2.4.2}$$

as it leads to a photoelectron with zero kinetic energy. For $\nu > \nu_0$, the electrons will be ejected. We then have

$$E_e = h(\nu - \nu_0). \tag{2.4.3}$$

Increasing the intensity of the light source means increasing the rate of incident photons. This will increase the number of photoelectrons but will not change the *energy* of the photoelectrons because it does not change the energy of each incident light quantum.

In accordance with (2.4.3), Einstein made a prediction: the kinetic energy of the photoelectrons increases linearly with the frequency of light, or strictly speaking, it is linear in $\nu - \nu_0$. Einstein's prediction was confirmed experimentally in 1915 by Millikan, who carefully measured the photoelectron energies and observed the expected linear dependence. Millikan's meticulous work allowed him to determine the value of Planck's constant to better than 1% accuracy! Still, skepticism remained and physicists were not yet convinced about the particle nature of these light quanta. Maxwell's equations had been just too successful, and those equations treated light as a wave.

Example 2.1. *Photoelectron energy and speed.*

Consider ultraviolet (UV) light with a wavelength $\lambda = 290\,\text{nm}$ (nm$= 10^{-9}$m) incident on a metal with work function $W = 4.05\,\text{eV}$. We want to determine the photoelectron energy in electron volts and its speed in kilometers per second (km/s).

To solve this problem without having to look up constants, recall the useful relation $\hbar c \simeq 197.33\,\text{MeV}\cdot\text{fm}$ as well as $\hbar \equiv h/(2\pi)$, with $\text{MeV} = 10^6\text{eV}$, and $\text{fm} = 10^{-15}\text{m}$. Let us use this to compute the photon energy E_γ. In this case, noting that MeV·fm = eV·nm, we have

$$E_\gamma = h\nu = 2\pi\hbar\frac{c}{\lambda} = \frac{2\pi \cdot 197.33\,\text{eV}\cdot\text{nm}}{290\,\text{nm}} = \frac{2\pi \cdot 197.33}{290}\,\text{eV} \simeq 4.28\,\text{eV}, \tag{2.4.4}$$

and thus the photoelectron energy E_e is

$$E_e = E_\gamma - W = 0.23\,\text{eV}. \tag{2.4.5}$$

To compute the velocity of the electron, we set

$$0.23\,\text{eV} = \tfrac{1}{2}m_e v^2 = \tfrac{1}{2}(m_e c^2)\left(\tfrac{v}{c}\right)^2. \tag{2.4.6}$$

Recalling that $m_e c^2 \simeq 511{,}000$ eV, one finds that

$$\tfrac{0.46}{511\,000} = \left(\tfrac{v}{c}\right)^2 \quad\Rightarrow\quad \tfrac{v}{c} = 0.0009488. \tag{2.4.7}$$

Using $c \simeq 300{,}000$ km/s, we finally get $v \simeq 284.6$ km/s. □

Exercise 2.6. *Estimate the wavelength λ_γ (in nanometers) of the minimum-energy photon required to ionize hydrogen (bound-state energy 13.6 eV). What's the ratio of that wavelength to the approximate size 50 pm of the atom?*

This is a good time to consider the units of h. We ask: Is there a physical quantity that has the units of h? The answer is yes. From the equation $E = h\nu$, we have

$$[h] = \left[\frac{E}{\nu}\right] = \frac{ML^2/T^2}{1/T} = L\cdot M\frac{L}{T}, \tag{2.4.8}$$

where the brackets $[\cdot]$ give the units of a quantity, and M, L, T are units of mass, length, and time, respectively. We have written the rightmost expression as a product of a length and a momentum. Therefore,

$$[h] = [\mathbf{r} \times \mathbf{p}] = [\mathbf{L}]. \tag{2.4.9}$$

We see that h has units of angular momentum! As we will learn in chapter 12, the magnitude of the spin angular momentum for a spin one-half particle is in fact $\frac{1}{2}\hbar$. The constant h also has units of "action"—that is, units of energy E times time T. Physicists sometimes call $\hbar$ the quantum of action. For a dynamical system, the action is obtained by integrating over time a certain combination of energies. In the Lagrangian formulation of classical physics, the dynamical equations of motion arise from the principle of least action.

With $[h] = [r][p]$, we also find a canonical way to associate a *length* to any particle of a given mass m. Indeed, using the speed of light we can construct the momentum $p = mc$, and then the length is given by the ratio h/p. This actually is the **Compton wavelength** λ_C of a particle:

$$\lambda_C = \frac{h}{mc}. \tag{2.4.10}$$

The Compton wavelength is *independent* of the velocity of the particle. The de Broglie wavelength of a particle, to be discussed later, is a length scale obtained using the *true* momentum of the particle. Compton and de Broglie wavelengths must not be confused! Sometimes the *reduced* Compton wavelength $\lambda\!\!\!^{-}_C$ is useful:

$$\lambda\!\!\!^{-}_C = \frac{\hbar}{mc}. \tag{2.4.11}$$

This expression uses the (reduced) Planck constant $\hbar$, instead of h.

It is possible to get some physical intuition for the Compton wavelength λ_C of a particle. We claim that *λ_C is the wavelength of a photon whose energy is equal to the rest energy of the particle.* Indeed, we would have

$$mc^2 = h\nu = h\frac{c}{\lambda} \quad \Rightarrow \quad \lambda = \frac{h}{mc} = \lambda_C, \tag{2.4.12}$$

confirming the claim. Suppose you are trying to localize a point particle of mass m. If you use light to do so, the possible accuracy in the position of the particle is roughly the wavelength of the light. Once we use light with $\lambda < \lambda_C$, the photons carry more energy than the rest energy of the particle. It is possible then that the energy of the photons goes into creating more particles of mass m, making it difficult, if not impossible, to localize the particle. The Compton wavelength is the length scale at which we need *relativistic quantum field theory* to take into account the possible processes of particle creation and annihilation.

Let us calculate the Compton wavelength of the electron:

$$\lambda_C(e) = \frac{h}{m_e c} = \frac{2\pi\hbar c}{m_e c^2} \simeq \frac{2\pi \cdot 197.3\,\text{MeV} \cdot \text{fm}}{0.511\,\text{MeV}} \simeq 2{,}426\,\text{fm} = 2.426\,\text{pm}. \tag{2.4.13}$$

This length is roughly twenty times smaller than the Bohr radius (53 pm) and about two thousand times larger than the size of a proton ($\sim$1 fm). As we will see in the next section, the Compton wavelength of the electron appears in the formula for the change of photon wavelength in the process called Compton scattering.

Exercise 2.7. *The proton rest energy is 937 MeV, and its size l_p is approximately 0.8 fm. Find the approximate ratio of the Compton wavelength of the proton to its size.*

2.5 Compton Scattering

Originally, Einstein did not make it clear that a light quantum meant a particle of light. In 1916, however, he posited that the quantum would carry momentum as well as energy, making the case for a particle much clearer. In relativity, the energy, momentum, and rest mass of a particle are related by

$$E^2 - p^2c^2 = m^2c^4. \tag{2.5.1}$$

Of course, one can also express the energy and momentum of the particle in terms of the velocity:

$$E=\frac{mc^2}{\sqrt{1-\frac{v^2}{c^2}}},\quad \mathbf{p}=\frac{m\mathbf{v}}{\sqrt{1-\frac{v^2}{c^2}}}. \tag{2.5.2}$$

Here $|\mathbf{p}|=p$. You should use these expressions to confirm that (2.5.1) holds. The nonrelativistic analogs of the above equations are $E=\frac{1}{2}mv^2$, and $\mathbf{p}=m\mathbf{v}$, giving the energy-momentum relation $E=\mathbf{p}^2/(2m)$. A particle that moves with the speed of light, like the photon, must have zero rest mass, otherwise its energy and momentum would be infinite due to the vanishing denominators. With the rest mass set to zero, equation (2.5.1) gives the relation between the photon energy E_γ and the photon momentum p_γ:

$$E_\gamma=p_\gamma c. \tag{2.5.3}$$

Then, recalling that $E_\gamma=h\nu$ and $\lambda\nu=c$, we reach

$$p_\gamma=\frac{E_\gamma}{c}=\frac{h\nu}{c}=\frac{h}{\lambda}. \tag{2.5.4}$$

We will see this relation again when we discuss matter waves.

Compton's famous experiments (1923–1924), scattered X-rays off a carbon target. X-ray photons have energies that range from 100 eV to 100 keV (keV $=10^3$eV). The goal was to scatter X-ray photons off of free electrons. While the electrons in carbon are bound, when bombarded with X-ray photons whose energy is much larger than the binding energy, they can behave as free electrons.

The classical counterpart of the Compton experiment is *Thomson scattering*, the scattering of low-energy electromagnetic waves off of free electrons. The mechanism is as follows. An electromagnetic wave is incident on an electron, and the electric field of the wave shakes the electron, making it oscillate with the frequency of the incident field. The electron oscillation produces a radiated field of the *same frequency* as that of the incident radiation. For unpolarized light incident upon the electron, the Thomson differential scattering cross section $d\sigma$ is given by

$$d\sigma=\left(\frac{e^2}{m_ec^2}\right)^2\frac{1}{2}\left(1+\cos^2\theta\right)d\Omega, \tag{2.5.5}$$

where θ is the angle between the incident and scattered electromagnetic waves, and $d\Omega$ is a small solid angle around the scattering direction. This is shown in figure 2.13. The cross section $d\sigma$ has units of length squared, or area, as it should. It represents the area that would extract from the *incident* plane wave the amount of energy scattered by the electron into the narrow cone defined by $d\Omega$. The quantity $e^2/(m_ec^2)$, appearing squared on the above right-hand side, is called the *classical electron radius*, and it is about 2.8 fm, not much bigger than a proton, which is about 1 fm! The above formula gives the angular dependence of the distribution of scattered waves, with peaks in the forward and backward directions but nonzero in all directions. Calculating the value of the classical electron radius is instructive, as it teaches us how to deal with e^2. In the Gaussian-cgs units we use, the unit-free quantity that represents e^2 is the **fine-structure constant** α defined by

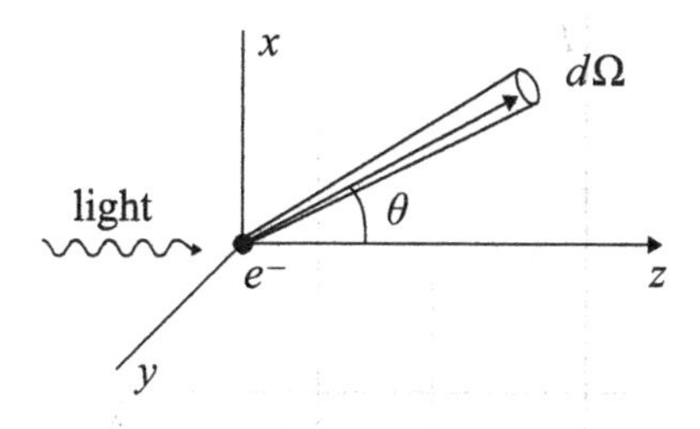

Figure 2.13
Unpolarized light incident on an electron scatters at an angle θ. Classically, this is described by Thomson scattering. The light does not change frequency in this process.

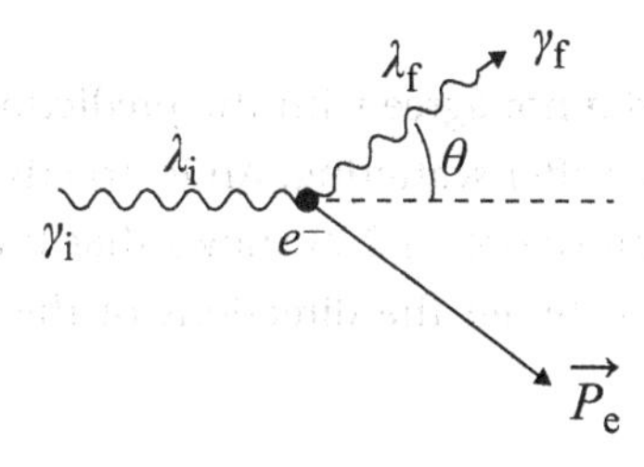

Figure 2.14
Compton scattering: An incident photon hits an electron at rest. The photon kicks the electron, giving it some energy, and is deflected at some angle θ relative to the incident direction.

$$\alpha \equiv \frac{e^2}{\hbar c} \simeq \frac{1}{137}. \tag{2.5.6}$$

The small value of this constant tells us of the relative weakness of the electromagnetic force. The fine structure constant is useful for a number of evaluations. For the classical electron radius r_0 defined above, we have

$$r_0 = \frac{e^2}{m_e c^2} = \frac{e^2}{\hbar c} \frac{\hbar c}{m_e c^2} \simeq \frac{1}{137} \frac{197.33\text{MeV} \cdot \text{fm}}{0.511\,\text{MeV}} \simeq 2.82\,\text{fm}. \tag{2.5.7}$$

In Compton scattering we treat the light as photons. As seen in the laboratory, the elementary process going on is a collision between an incident photon and a stationary electron. As shown in figure 2.14, an incident photon of wavelength λ_i hits a stationary electron e^-. After the collision we find a photon emerging at an angle θ relative to the incident direction, with a final wavelength λ_f. The electron is also scattered. Two facts can be quickly demonstrated:

- The incident photon cannot just be absorbed by the electron. It is inconsistent with energy and momentum conservation (see Problem 2.6).
- The photon must lose some energy, and thus the final photon wavelength λ_f must be larger than the initial photon wavelength λ_i. This is clear in the laboratory frame, where the initially stationary electron must recoil and thus acquire some kinetic energy.

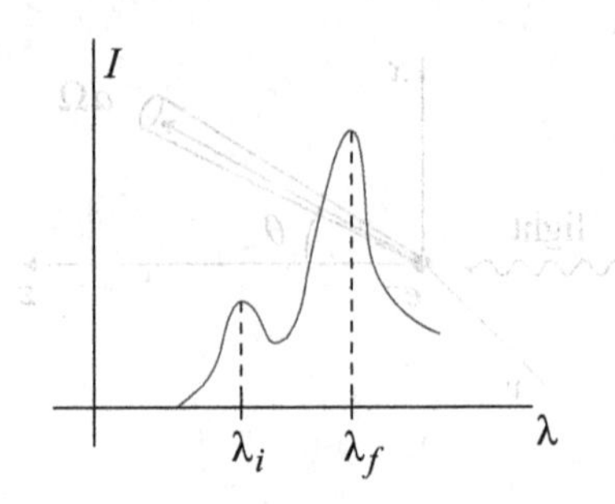

Figure 2.15
Compton's experiment: The intensity I of photons scattered at $\theta = 90°$, as a function of wavelength λ. There is a small peak at the incident photon wavelength λ_i and a large peak at $\lambda_f \simeq \lambda_i + \lambda_C$.

Compton's observations did not agree with the predictions of Thomson scattering: the X-rays changed frequency after scattering. An instructive calculation using energy and momentum conservation (Problem 2.7) shows that the change of wavelength is correlated with the angle θ between the directions of the scattered photon and the original photon:

$$\lambda_f = \lambda_i + \frac{h}{m_e c}(1 - \cos\theta) \;=\; \lambda_i + \lambda_C\,(1 - \cos\theta). \tag{2.5.8}$$

Note the appearance of the Compton wavelength of the electron, the particle the photon scatters from. The maximum energy loss for the photon occurs when it bounces back so that $\theta = \pi$:

$$\lambda_f(\theta = 180°) = \lambda_i + 2\lambda_C. \tag{2.5.9}$$

The maximum possible change in wavelength is $2\lambda_C$. For $\theta = \frac{\pi}{2}$, the change of wavelength is exactly λ_C:

$$\lambda_f(\theta = 90°) = \lambda_i + \lambda_C. \tag{2.5.10}$$

Compton's experiment used a carbon target and molybdenum X-rays with energy and wavelength

$$E_\gamma \approx 17.5\,\text{keV}, \qquad \lambda_i = 0.0709\,\text{nm}. \tag{2.5.11}$$

With the photon detector placed at an angle $\theta = 90°$, the plot of the intensity (or number of photons scattered) as a function of wavelength is shown in figure 2.15. One finds a large peak for $\lambda_f = 0.0731$ nm but also a small second peak at the original wavelength $\lambda_i = 0.0709$ nm.

The peak at λ_f is the expected one: $\lambda_f - \lambda_i \simeq 2.2\,\text{pm}$, which is about the electron Compton wavelength of 2.4 pm. Given that the photons have energies of 17 keV and the bound-state energy of carbon is about 300 eV, the expected peak represents instances when the atom is ionized by the collision, and it is a fine approximation to consider the collision as one between a photon and an electron that is free.

The peak at λ_i represents a process in which an electron is kicked by the photon but still remains bound. This is not that unlikely: the typical momentum of a bound electron is actually comparable to the momentum of the photon (see the exercises below). In this case the photon scatters off, and the recoil momentum is carried by the whole atom. The relevant Compton wavelength is therefore that of the atom. Since the mass of the carbon atom is several thousand times larger than the mass of the electron, the Compton wavelength of the atom is much smaller than the electron Compton wavelength, and there is no detectable change in the wavelength of the photon. Effectively, the photon is scattering off the whole atom. The photon is deflected, but its energy is essentially unchanged. This is analogous to classical elastic scattering of an object off a hard target, like in the bouncing of a ball off a hard wall.

Exercise 2.8. *In the hydrogen atom, the Bohr radius a_0 represents the size of the atom, and the Rydberg $Ry = \frac{e^2}{2a_0} \simeq 13.6\,\text{eV}$ is its binding energy. In this atom, $p_e = \frac{\hbar}{a_0}$ represents the typical electron momentum. What is the energy of a photon with momentum equal to p_e? (Answer: $274\,Ry \simeq 3.7\,\text{keV}$).*

Exercise 2.9. *For an atom of atomic number Z, the size is of order a_0/Z, and the typical electron momentum is of magnitude $Z\hbar/a_0$. What is the energy of a photon with momentum the size of the typical electron momentum in carbon ($Z = 6$)? (Answer: $22.4\,\text{keV}$).*

Exercise 2.10. *A photon of wavelength λ_i equal to the Compton wavelength of an electron hits an electron and is scattered exactly backward. Find the velocity parameter $\beta = v/c$ of the recoiling electron.*

2.6 Matter Waves

As we have seen, light behaves as both a particle and a wave. This kind of behavior is usually said to be a **duality**: the complete reality of an object is captured using *both* the wave and particle features of the object. The photon is a particle of energy E_γ but has frequency ν, which is a wave attribute, with $E_\gamma = h\nu$. It is a particle with momentum $\mathbf{p}_\gamma$ ($|\mathbf{p}_\gamma| = p_\gamma$), but it also has a wavelength λ, a wave attribute, given by

$$\lambda = \frac{h}{p_\gamma}. \tag{2.6.1}$$

In 1924 Louis de Broglie proposed that the wave-particle duality of the photon was universal and thus valid for matter particles too. In this way he conjectured the *wave nature of matter*. Inspired by (2.6.1), de Broglie postulated that associated with a matter particle with momentum $\mathbf{p}$, there is a *plane wave* of wavelength λ given by

$$\lambda = \frac{h}{p}, \tag{2.6.2}$$

where $p \equiv |\mathbf{p}|$. This is a quantum property, as it involves Planck's h; classical particles have no wave properties. It is natural that the formula uses the *magnitude* of the momentum since a wavelength is, by definition, a positive number. We call λ the **de Broglie wavelength** of the particle.

Left unsaid in this proposal is the physical interpretation of the matter wave. What is waving, and what does it represent? This was not answered for a while because the

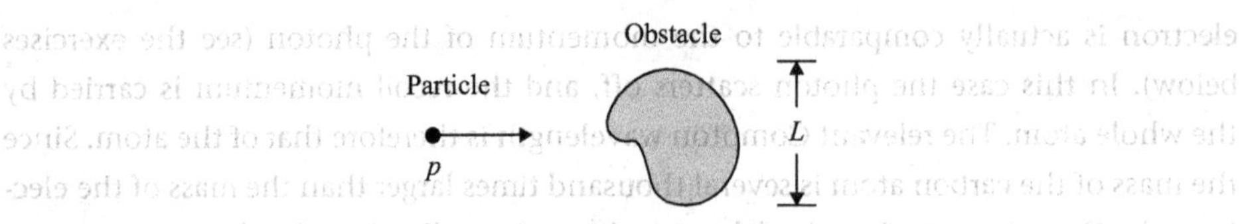

Figure 2.16
A particle of momentum p, incident on an obstacle of size L. Quantum effects are small when $\lambda = h/p \ll L$.

answer was unexpected: what is oscillating is a complex probability amplitude, a quantity whose norm squared is related to probabilities. Even for photons, the quantum wave is *not* an electromagnetic field but rather a probability amplitude wave. This is indeed the way we discussed photons in the Mach-Zehnder interferometer.

If matter particles have wave properties, matter particles can diffract or interfere! In the famous Davisson-Germer experiment (1927), electrons strike a nickel target, and the angular distribution of reflected electrons shows a diffraction pattern similar to that obtained for X-rays. The peaks showed the effect of constructive interference from scattering off the lattice of atoms in the metal, demonstrating the wave nature of the incident electrons.

One can also perform a two-slit interference experiment with electrons. The incident beam of electrons produces an interference pattern on the target screen. The experiment can also be done shooting one electron at a time, even when waiting a while in between electrons. The interference pattern shows up anyway as one records the impact points of a very large number of electrons. This demonstrates that each electron interferes with itself! An experiment by Eibenberger et al. (2013) reports interference using molecules with 810 atoms and mass exceeding 10,000 amu (amu stands for "atomic mass unit," with one amu as the approximate mass of the hydrogen atom). That's a molecule whose mass is about twenty million times the mass of the electron!

The de Broglie wavelength is often calculated to estimate if quantum effects are important. Consider for this purpose a particle of mass m and momentum p incident upon an object of size L that is at rest (figure 2.16). Let $\lambda = h/p$ denote the de Broglie wavelength of the particle. The wave nature of the particle is not important if λ is much smaller than L. Thus, the classical approximation, in which wave effects are negligible, requires that

$$\lambda \ll L. \tag{2.6.3}$$

The above is a guideline; classical behavior is sometimes obtained as a subtle limit of quantum mechanics. A classical electromagnetic field requires a large number of photons. But this is not sufficient: a state with an exact, *fixed* number of photons, even if large, is not classical. Classical electromagnetic states are so-called coherent states, in which the number of photons fluctuates (section 17.4). The guideline provided by the de Broglie wavelength, however, works quite well to understand the phenomenon of Bose-Einstein condensation, as we discuss in example 2.2 of the following section.

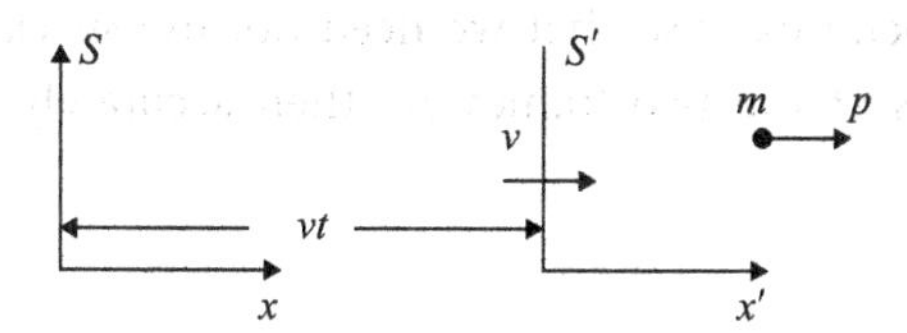

Figure 2.17
The S' frame moving with velocity $v > 0$ along the x-direction of the S frame. As seen in the S frame, a particle of mass m is moving with velocity $\widetilde{v}$ and momentum $p = m\widetilde{v}$.

2.7 De Broglie Wavelength and Galilean Transformations

We have seen that with any particle with momentum $\mathbf{p}$, we can associate a plane wave, or a "matter wave," with de Broglie wavelength

$$\lambda = \frac{h}{p}, \quad \text{where } p = |\mathbf{p}|. \tag{2.7.1}$$

This relation between the particle momentum and the wavelength of the associated wave can also be written as

$$p = \frac{h}{\lambda} = \frac{h}{2\pi}\frac{2\pi}{\lambda} = \hbar k, \tag{2.7.2}$$

where we recall that $\hbar = h/(2\pi)$, and

$$k \equiv \frac{2\pi}{\lambda}. \tag{2.7.3}$$

We call k the *wave number*, and it is a simple function of the wavelength. Some results, like $\lambda = h/p$, are best written in terms of h, while some other results, like $p = \hbar k$, are best written in terms of $\hbar$. We will simply call both h and $\hbar$ the Planck constant.

A de Broglie wave is an example of a *wave function*. Wave functions, as we will see, are governed by the Schrödinger equation and encode probability amplitudes. The purpose of the following discussion is to gain some intuition about de Broglie waves.

We can ask an obvious question: Do de Broglie waves have polarization properties like those of electromagnetic waves? Yes, de Broglie waves for particles with spin have polarization! We will not discuss spin now since the effects of spin are often small. We will focus on the simplest case where there is no spin, and the de Broglie wave Ψ is just a complex-valued function, a complex number that depends on space and time:

$$\Psi(\mathbf{x}, t) \in \mathbb{C}. \tag{2.7.4}$$

We say that $\Psi(\mathbf{x}, t)$ is a **wave function**, a *wave* described by a *function* of a set of coordinates. Additional questions come to mind. Is the wave function measurable? What kind of object is it?

In order to make progress, let us consider how observers in different frames of reference determine the de Broglie wavelength of a particle. Assume we have two frames, S and S', sketched in figure 2.17, with the x and x' axes aligned. We assume S is at rest, with S' moving to the right along the $+x$ direction of S with constant velocity $v > 0$. At time equal zero, the two reference frames coincide.

We assume the velocity $v \ll c$ so that we need not use special relativity. The time and spatial coordinates of the two frames are then accurately related by a *Galilean transformation*:

$$x' = x - vt, \quad t' = t. \tag{2.7.5}$$

Time runs at the same speed in all Galilean frames, and the relation between x and x' is manifest from the arrangement shown in figure 2.17.

Now assume both observers focus on a particle of mass m that moves with nonrelativistic speed along the common direction of x and x'. Call the velocity and momentum of the particle in the S frame $\tilde{v}$ and $p = m\tilde{v}$, respectively. Both are assumed to be positive so that $\lambda = h/p$:

$$\tilde{v} > 0, \qquad p = m\tilde{v} > 0, \qquad \lambda = \frac{h}{p}. \tag{2.7.6}$$

It follows by differentiation with respect to $t = t'$ of the first equation in (2.7.5) that

$$\frac{dx'}{dt'} = \frac{dx}{dt} - v, \tag{2.7.7}$$

which means that the particle velocity $\tilde{v}'$ in the S' frame is given by

$$\tilde{v}' = \tilde{v} - v. \tag{2.7.8}$$

Multiplying by the mass m we find the relation between the momenta in the two frames

$$p' = p - mv. \tag{2.7.9}$$

The momentum p' in the S' frame can be appreciably different from the momentum p in the S frame. Thus, the observers in S' and in S will obtain rather different de Broglie wavelengths λ' and λ! Indeed,

$$\lambda' = \frac{h}{|p'|} = \frac{h}{|p - mv|} \neq \lambda = \frac{h}{p}. \tag{2.7.10}$$

This is actually very strange! As we review now, for familiar waves like sound waves or water waves, Galilean observers will record different frequencies but the same wavelength. This is intuitively clear: to find the wavelength, one need only take a picture of the wave at some given time, and both observers looking at the picture will agree on the value of the wavelength. On the other hand, to measure frequency each observer must wait some time to see a full period of the wave go through. This will take different amount of time for the different observers.

Let us demonstrate these claims quantitatively. We begin with the statement that the phase $\phi = kx - \omega t$ of such a wave is a Galilean invariant. The wave itself may be $\cos\phi$, $\sin\phi$, $e^{i\phi}$, or some other function, but if the wave is observable, its value at any point must be agreed upon by the two observers. In a water wave, for example, the height of the water displacement at any point and at any time is agreed upon by all observers. Since all the features of a wave (peaks, zeroes, and so on) are controlled by the phase, the two observers must agree on the value of the phase.

In the S frame, the phase can be written as follows:

$$\phi = kx - \omega t = k(x - \tfrac{\omega}{k}t) = \frac{2\pi}{\lambda}(x - Vt) = \frac{2\pi x}{\lambda} - \frac{2\pi V}{\lambda}t, \tag{2.7.11}$$

where $V = \frac{\omega}{k}$ is the velocity of the wave. Note that the wavelength can be read from the coefficient k of x, while the angular frequency ω appears as minus the coefficient of t. The two observers should agree on the value of ϕ. That is, we should have

$$\phi'(x', t') = \phi(x, t), \tag{2.7.12}$$

where the coordinates and times are related by a Galilean transformation. Therefore,

$$\phi'(x', t') = \frac{2\pi}{\lambda}(x - Vt) = \frac{2\pi}{\lambda}(x' + vt' - Vt') = \frac{2\pi}{\lambda}x' - \frac{2\pi(V - v)}{\lambda}t'. \tag{2.7.13}$$

Since the right-hand side is expressed in terms of primed variables, we can read λ' from the coefficient of x' and ω' as minus the coefficient of t':

$$\lambda' = \lambda, \tag{2.7.14}$$

$$\omega' = \frac{2\pi}{\lambda}(V - v) = \frac{2\pi V}{\lambda}\left(1 - \frac{v}{V}\right) = \omega\left(1 - \frac{v}{V}\right). \tag{2.7.15}$$

This confirms that, as claimed, the wavelength is a Galilean invariant and the frequency transforms.

Exercise 2.11. *Let T denote the period of an ordinary wave in the frame S and T' denote the period of the same wave in the frame S'. Show that*

$$\frac{1}{T} - \frac{1}{T'} = \frac{v}{\lambda}.$$

So what does it mean, then, that the wavelength of de Broglie waves Ψ changes under a Galilean transformation? It means that the value of Ψ does not correspond to a measurable quantity for which all Galilean observers must agree. Thus, the wave function need not be invariant under Galilean transformations:

$$\Psi(x, t) \neq \Psi'(x', t'), \tag{2.7.16}$$

where (x, t) and (x', t') are related by Galilean transformations and thus represent the same point and time. You will figure out the correct relation between $\Psi(x, t)$ and $\Psi'(x', t')$ in problem 3.4, which you will be able to solve after learning about the Schrödinger equation.

We have seen that for a de Broglie wave, the relation

$$p = \hbar k \tag{2.7.17}$$

fixes the wavelength in terms of the magnitude p of the particle momentum. What fixes the angular frequency ω of the de Broglie wave? As also postulated by de Broglie, the angular frequency ω of the wave is determined by the relation

$$E = \hbar\omega, \tag{2.7.18}$$

where E is the energy of the particle. Recall that the angular frequency ω is related to the period T of the wave by the relation $\omega = \frac{2\pi}{T}$. Note that $\hbar\omega = h\nu$, with ν the frequency, so this postulated for particles what had been known for photons. For nonrelativistic particles, the energy E is determined by the momentum through the relation

$$E = \frac{p^2}{2m}. \tag{2.7.19}$$

Thus, given the momentum of the particle, we can find the energy, and therefore the frequency, of the de Broglie wave.

We give three pieces of evidence that the energy/frequency relation (2.7.18) is reasonable:

1. A single de Broglie plane wave extends all over space and actually represents a delocalized particle. To represent a more conventional, localized particle, it turns out we must superpose de Broglie waves to form a wave packet. The wave packet will move with the so-called group velocity v_g, which, as we will soon review, is found by differentiation of ω with respect to k:

$$v_g = \frac{d\omega}{dk}. \tag{2.7.20}$$

   We now confirm that the group velocity of the wave packet, as it should, coincides with the velocity of the particle:

$$v_g = \frac{d\omega}{dk} = \frac{d\,\hbar\omega}{d\,\hbar k} = \frac{dE}{dp} = \frac{d}{dp}\left(\frac{p^2}{2m}\right) = \frac{p}{m} = v. \tag{2.7.21}$$

   In fact, the phase velocity $v_p = \frac{\omega}{k}$ is not equal to the velocity of the particle: $v_p = \frac{\omega}{k} = \frac{E}{p} = \frac{p}{2m} = \frac{v}{2}$. This is not a problem because the phase velocity is not the velocity of the wave packet.

2. The relation is also suggested by special relativity. The energy and the momentum components of a particle form a four-vector:

$$\left(\frac{E}{c}, \mathbf{p}\right). \tag{2.7.22}$$

   Similarly, for waves whose phases are relativistically invariant, we have another four-vector:

$$\left(\frac{\omega}{c}, \mathbf{k}\right). \tag{2.7.23}$$

   Setting two four-vectors equal to each other is a consistent choice: it would be valid in all Lorentz frames. As you can see, both de Broglie relations follow from the equality

$$\left(\frac{E}{c}, \mathbf{p}\right) = \hbar\left(\frac{\omega}{c}, \mathbf{k}\right). \tag{2.7.24}$$

   The de Broglie relation

$$\mathbf{p} = \hbar\mathbf{k} \tag{2.7.25}$$

is the three-dimensional extension of our earlier expression $p=\hbar k$. A wave of the form $\exp(i\mathbf{k}\cdot\mathbf{x})$, for example, is a wave with wave number $\mathbf{k}$. For this wave we write $k=|\mathbf{k}|$, and since $p=|\mathbf{p}|$, the norm of (2.7.25) gives $p=\hbar k$. Moreover, $k=2\pi/\lambda$.

You could wonder, with de Broglie relations compatible with special relativity, why we focus on low-energy nonrelativistic physics. Why not work with wave functions that could be Lorentz invariant? There are two parts to the answer. First, nonrelativistic quantum mechanics is all you need in a large number of applications. Second, relativistic quantum mechanics, to be fully consistent, requires particle creation and annihilation. The right framework to handle this is *quantum field theory*.

3. As mentioned before, the energy/frequency relation (2.7.18) is consistent with Einstein's proposal for the energy of photons:

$$E = h\nu = \hbar\, 2\pi\nu = \hbar\omega\,. \tag{2.7.26}$$

In summary we have

$$\boxed{\mathbf{p}=\hbar\,\mathbf{k}, \quad E=\hbar\omega.} \tag{2.7.27}$$

These are called the *de Broglie relations*, and they are valid for all particles.

Example 2.2. *Bose-Einstein condensation.*

Consider a gas of N particles at temperature T in a box of volume V. Assume the particles are all identical bosons of mass m. As we will learn later, bosons are particles that can all be in the same quantum state. The same is not true for fermions: two or more fermions are never found in the same quantum state. For our bosons in the box, they could all be in the ground state of the box. This is a state in which the wave of each particle is extended over the whole box, and the particles essentially lose their individuality, each one doing exactly the same thing. When this happens we have a Bose-Einstein condensate, and the N particles in the box form a remarkable macroscopic quantum state. This possibility was predicted by Einstein in 1925, following the earlier work of Bose. Bose-Einstein condensates were realized experimentally in 1995, by Eric Cornell and Carl Wieman at the University of Colorado, Boulder, and by Wolfgang Ketterle at MIT.

With N particles in a box of volume V, there is a length scale ℓ, viewed as the typical separation between the particles. It is easily estimated by imagining that N cubes of side length ℓ fill the box: $N\ell^3=V$. Therefore,

$$\ell^3 \sim 1/n, \tag{2.7.28}$$

where $n=N/V$ is the number of particles per unit volume. When the temperature is high, the thermal energy of the particles is large, and so is their momentum, resulting in a very short de Broglie wavelength λ. We can imagine each individual particle as a tiny wave packet of de Broglie waves, a packet the size of the order λ. As the temperature is decreased, λ grows. In fact, the condensate forms at a critical temperature for which the de Broglie wavelength and the length scale ℓ are comparable! The particle's individual wave packets begin to overlap and then merge to form a large delocalized wave.

We can estimate the critical temperature as follows. The thermal kinetic energy of a particle is taken to be $\frac{3}{2}k_B T$, as for a monoatomic gas, where k_B is the Boltzmann constant. This gives us an estimate for the velocity and thus for the momentum:

$$\tfrac{3}{2}k_B T = \tfrac{1}{2}mv^2 \quad \Rightarrow \quad v \sim \sqrt{\frac{k_B T}{m}} \quad \Rightarrow \quad p \sim \sqrt{mk_B T}. \tag{2.7.29}$$

The de Broglie wavelength, estimated with $\hbar$ rather than h, is set equal to the scale ℓ:

$$\frac{\hbar}{\sqrt{mk_B T}} \sim n^{-1/3} \quad \Rightarrow \quad T \sim \frac{\hbar^2 n^{2/3}}{mk_B}. \tag{2.7.30}$$

This correctly reproduces the dependence of T on the density and mass of the particles. The exact result of an intricate calculation fixes the numerical coefficient in front of the right-hand side, giving $T \simeq 3.31 \frac{\hbar^2 n^{2/3}}{mk_B}$.

2.8 Stationary Phase and Group Velocity

A single wave of the form $f(kx - \omega t)$, where f is some arbitrary function, has a well-defined *phase velocity*. This velocity is the velocity with which any peak, any zero, or in fact any point in the wave moves. It is computed by finding the relation between x and t that makes the phase $kx - \omega t$ of the wave constant. Indeed, one has that

$$kdx - \omega dt = 0 \quad \rightarrow \quad v_p = \frac{dx}{dt} = \frac{\omega}{k}. \tag{2.8.1}$$

To understand group velocity, we form wave packets and investigate how fast they move (for a simpler, less general derivation of the group velocity, see problem 2.10). For generality, we will assume that $\omega(k)$ is some arbitrary function of k. Consider a superposition of plane waves $e^{i(kx-\omega(k)t)}$ obtained by adding contributions with different values of k weighted with a function $\Phi(k)$:

$$\psi(x,t) = \int dk\, \Phi(k) e^{i(kx-\omega(k)t)}. \tag{2.8.2}$$

We assume that the function $\Phi(k)$ is real, continuous, and smooth and has a sharp peak at some wave number $k = k_0$, as shown in figure 2.18. This means that the nonvanishing contributions to the integral occur only for values of k near k_0.

Since $\Phi(k)$ is real, the phase φ of the *integrand* arises solely from the exponential:

$$\varphi(k) = kx - \omega(k)t. \tag{2.8.3}$$

We wish to understand the values of x and t for which the quantity $|\psi(x,t)|$, representing the packet, takes large values. If the packet describes a particle, those values of x and t where $|\psi(x,t)|$ is large define the location x of the particle at time t.

For this purpose we use the *stationary phase principle*:

Stationary phase principle: Since only for $k \sim k_0$ does the integral over k have a chance to give a large contribution, the phase factor must be *stationary* at $k = k_0$—that is, the derivative of the phase must vanish at $k = k_0$.

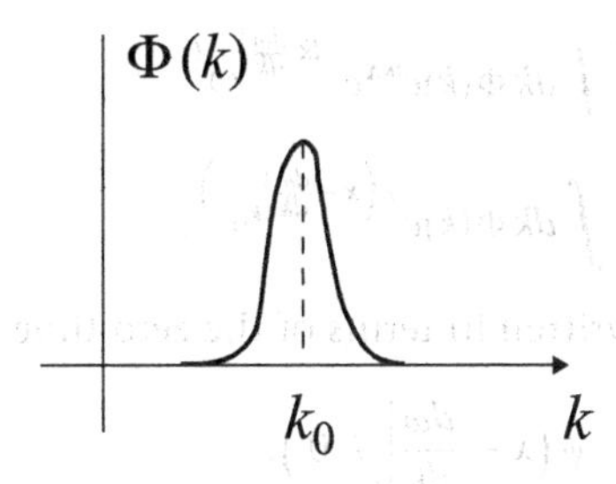

Figure 2.18
The function $\Phi(k)$ is assumed to peak around $k=k_0$.

The idea is intuitively plausible: if a slowly varying function of k is multiplied by a phase that is a rapidly varying function of k, the integral over k largely cancels out; contributions can only accumulate if the phase slows down. Certainly, $\Phi(k)$ is slowly varying near k_0; in fact, it peaks at $k=k_0$. The application of the stationary phase principle to our integral requires that we find the derivative of the phase $\varphi(k)$ and set it equal to zero at k_0:

$$\frac{d\varphi}{dk}\bigg|_{k_0} = x - \frac{d\omega}{dk}\bigg|_{k_0} t = 0. \tag{2.8.4}$$

This means that $|\psi(x,t)|$ is appreciable when x and t are related by

$$x = \frac{d\omega}{dk}\bigg|_{k_0} t, \tag{2.8.5}$$

showing that the packet moves with *group velocity*

$$v_g = \frac{d\omega}{dk}\bigg|_{k_0}. \tag{2.8.6}$$

Exercise 2.12. *If $\Phi(k)$ is not real, write $\Phi(k)=|\Phi(k)|e^{i\phi(k)}$. Find the new version of (2.8.5) and show that the velocity of the wave is not changed.*

Let us now do a more detailed calculation that confirms the above analysis and gives some extra insight. Notice first that at time zero the wave packet (2.8.2) takes the form

$$\psi(x,0) = \int dk\, \Phi(k) e^{ikx}. \tag{2.8.7}$$

We expand $\omega(k)$ in a Taylor expansion around $k=k_0$:

$$\omega(k) = \omega(k_0) + (k-k_0)\frac{d\omega}{dk}\bigg|_{k_0} + \mathcal{O}\left((k-k_0)^2\right). \tag{2.8.8}$$

Back into $\psi(x,t)$ in (2.8.2), and neglecting the $\mathcal{O}((k-k_0)^2)$ terms in the exponent, we find that

$$\psi(x,t) = \int dk\, \Phi(k)\, e^{ikx}\, e^{-i\omega(k_0)t} e^{-i(k-k_0)\frac{d\omega}{dk}\big|_{k_0} t}. \tag{2.8.9}$$

It is convenient to take out of the integral all the factors that do not depend on k:

$$\psi(x,t) = e^{-i\omega(k_0)t + ik_0 \frac{d\omega}{dk}\big|_{k_0} t} \int dk\, \Phi(k) e^{ikx} e^{-ik \frac{d\omega}{dk}\big|_{k_0} t}$$
$$= e^{-i\omega(k_0)t + ik_0 \frac{d\omega}{dk}\big|_{k_0} t} \int dk\, \Phi(k) e^{ik\left(x - \frac{d\omega}{dk}\big|_{k_0} t\right)}. \tag{2.8.10}$$

The above integral can be written in terms of the zero-time wave packet in (2.8.7):

$$\psi(x,t) = e^{-i\omega(k_0)t + ik_0 \frac{d\omega}{dk}\big|_{k_0} t}\, \psi\left(x - \frac{d\omega}{dk}\Big|_{k_0} t, 0\right). \tag{2.8.11}$$

The phase factors in front of the expression are not important in tracking where the wave packet is. In particular we can take the norm of both sides of the equation to find that

$$|\psi(x,t)| = \left|\psi\left(x - \frac{d\omega}{dk}\Big|_{k_0} t, 0\right)\right|. \tag{2.8.12}$$

If $|\psi(x,0)|$ peaks at some value x_0, it is clear from the above equation that $|\psi(x,t)|$ peaks for

$$x - \frac{d\omega}{dk}\Big|_{k_0} t = x_0 \quad \Rightarrow \quad x = x_0 + \frac{d\omega}{dk}\Big|_{k_0} t, \tag{2.8.13}$$

showing that the peak of the packet moves with velocity $v_g = \frac{d\omega}{dk}\big|_{k_0}$. This is what we wanted to confirm.

Exercise 2.13. *Consider a wave in which ω and k are related by $c^2k^2 = \omega^2 - \omega_0^2$. Here c is the speed of light, and ω_0 is a constant. Calculate the product of the group and phase velocities.*

The stationary phase principle will help us understand the behavior of superpositions of de Broglie waves that describe the motion of particles. Such superpositions will be solutions of the Schrödinger equation, thus ensuring they describe allowed quantum behavior. We will begin our study of Schrödinger's equation in chapter 3.

Problems

Problem 2.1. *Mach-Zehnder interferometer with phase shifters.*

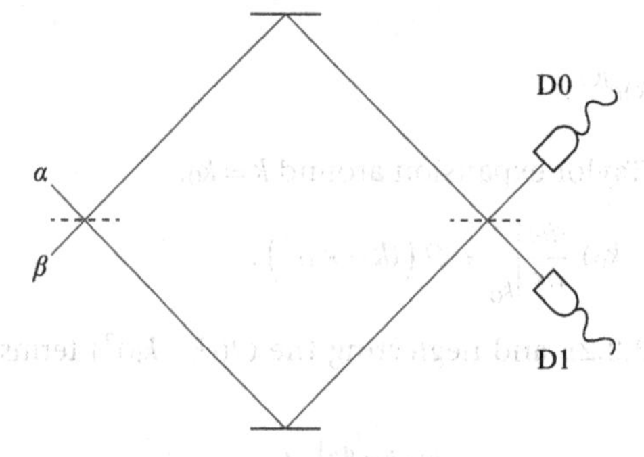

Consider the above interferometer and assume an input beam of the form (α, β), where α and β are complex amplitudes satisfying $|\alpha|^2 + |\beta|^2 = 1$. Call P_0 and P_1 the detection probabilities at D0 and D1, respectively. The left and right beam splitters BS1 and BS2 act via the matrices U_1 and U_2 defined in (2.1.17).

1. Calculate P_0 and P_1, assuming we insert a phase shifter with phase δ_l on the *lower* branch of the interferometer. Write your answers for P_0 and P_1 in the form

$$(\cdots)|\alpha|^2 + (\cdots)|\beta|^2 + (\cdots)\mathrm{Im}(\alpha\beta^*),$$

where you must determine the δ_l-dependent factors in parentheses.

2. Calculate P_0 and P_1, assuming we insert a phase shifter with phase δ_u on the *upper* branch of the interferometer. Again, write your answer in the form specified in part 1. [Hint: Try to manipulate the phase shifter matrix to take advantage of your computation from part 1.]

3. Calculate P_0 and P_1, assuming we insert the two phase shifters simultaneously—the one on the upper branch with phase δ_u and the one on the lower branch with phase δ_l. You should find that your answer depends only on $\delta_u - \delta_l$, so express your answer in terms of $\Delta = (\delta_u - \delta_l)/2$.

Problem 2.2. *Testing Elitzur-Vaidman bombs.*

1. Suppose you decide to test Elitzur-Vaidman bombs with a Mach-Zehnder interferometer repeatedly, until the status of any given bomb is certain beyond reasonable doubt. What fraction of the *working* bombs are certified without detonation?

2. Suppose 80% of the bombs in your possession are defective. You choose one at random and test it with a Mach-Zehnder interferometer by sending in one photon. You detect the photon at D0. What is the probability that the bomb is defective? [Hint: You may want to use Bayes' theorem.]

Problem 2.3. *Two-by-two matrices and linear devices.*

Consider a Mach-Zehnder interferometer and a photon represented by the two-component column vector u:

$$u = \begin{pmatrix} u_1 \\ u_2 \end{pmatrix}, \quad \text{with} \quad |u_1|^2 + |u_2|^2 = 1\,.$$

Any *linear* optical element in the interferometer can be represented by a two-by-two matrix R, such that with input beam u the output is a beam u' given by matrix multiplication as

$$u' = \begin{pmatrix} u_1' \\ u_2' \end{pmatrix} = R\,u\,.$$

Our goal is to show that conservation of probability for *arbitrary* u requires R to be a unitary matrix. A (finite size) matrix R is said to be unitary if $R^\dagger R = \mathbb{1}$, where the dagger denotes the operation of transposition and complex conjugation, and $\mathbb{1}$ is the identity matrix.

1. Let us begin by computing $u^\dagger u$. The answer should be a pure number. By conservation of probability, what is $(u')^\dagger u'$?

2. Write $(u')^\dagger u'$ in terms of u, $u^\dagger$, R, and $R^\dagger$. From this result, we see that if $R^\dagger R = \mathbb{1}$, then $(u')^\dagger u' = u^\dagger u$, as we require to conserve probability. So R being a unitary matrix is a *sufficient* condition. We still need to show that it is also a *necessary* condition.
3. The quantity $R^\dagger R$ is a two-by-two matrix. So let us represent it by

$$R^\dagger R = \begin{pmatrix} a & b \\ c & d \end{pmatrix}.$$

   We require $(u')^\dagger u' = 1$ for *any* valid u. Writing the left-hand side in terms of u_1, u_2, a, b, c, and d, what consistency condition do you obtain?
4. Your answer to part (3) has to be true for any valid u. Try the following u vectors and obtain all the information you can about a, b, c, and d:

$$u = \begin{pmatrix} 1 \\ 0 \end{pmatrix}, \quad u = \begin{pmatrix} 0 \\ 1 \end{pmatrix}, \quad u = \frac{1}{\sqrt{2}}\begin{pmatrix} 1 \\ 1 \end{pmatrix}, \quad u = \frac{1}{\sqrt{2}}\begin{pmatrix} 1 \\ i \end{pmatrix}.$$

   Think about the implications of your answer.

Problem 2.4. *Improving on bomb detection.*

We modify the Mach-Zehnder interferometer to increase the percentage of Elitzur-Vaidman bombs that can be confirmed to work without our detonating them. For this purpose, we build a beam splitter with reflectivity R and transmissivity T. A photon incident (from either port) has a probability R to be reflected and a probability T to be transmitted ($R + T = 1$). Let r and t denote the *positive* square roots:

$$r \equiv \sqrt{R}, \quad t \equiv \sqrt{T}.$$

1. Build the two-by-two matrix U that represents the beam splitter. To do so, consider what happens when a photon hits the beam splitter from the top side with input $(1, 0)$ and when it hits it from the bottom side with input $(0, 1)$. To fix conventions, U will have all entries positive (and real) except for the bottom rightmost element (the $(2, 2)$ element). Is U unitary?
2. The interferometer with detectors D0 and D1 (shown below) uses two identical copies of the beam splitter (indicated by dashed lines). The incident photon arrives from the top side.

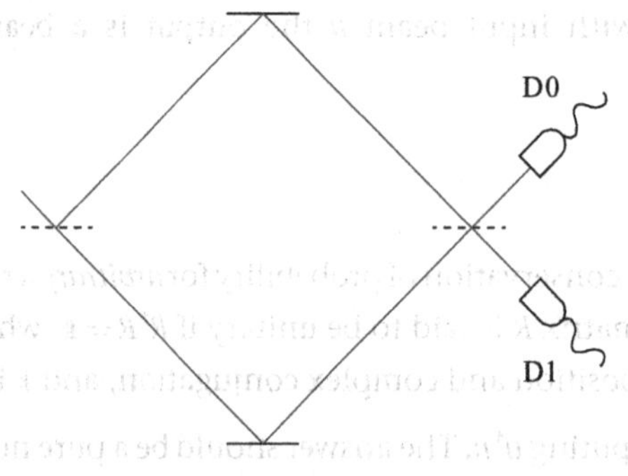

   A defective bomb is inserted on the lower branch of the interferometer. What are the detection probabilities P_0 and P_1 at D0 and D1, respectively?

A functioning bomb is inserted on the lower branch of the interferometer. What is the detonation probability P_{boom} and the detection probabilities P_0 and P_1?

3. You test bombs until you are reasonably sure that either they malfunction or are operational. What fraction f of the operational bombs can be certified to be good without detonating them?

   The ratio f cannot be arbitrarily large, given the allowed values of R. Consider the inequality $f \leq f_0$ and determine the lowest possible value of the positive constant f_0. Can bomb testing work with $f = f_0$?

Problem 2.5. *De Broglie relations and the scale of quantum effects.*

Matter particles as waves. If a wavelength can be associated with every moving particle, why are we not aware of this property in many situations? To get an idea of why, calculate the de Broglie wavelength $\lambda = h/p$ ($h = 6.63 \times 10^{-34}\,\text{J}\cdot\text{s}$) of each of the following particles:

1. An automobile of mass 2,000 kg traveling at a speed of 50 mph (=22.4 m/s).
2. A marble of mass 10 g moving with a speed of 10 cm/s.
3. A smoke particle of diameter 100 nm and a mass of 1 fg (f is for femto, implying a factor of 10^{-15}) being jostled about by air molecules at room temperature ($T = 300\,\text{K}$). Assume that the particle has the same translational kinetic energy $\frac{3}{2}k_B T$ as the air molecules ($k_B = 1.38 \times 10^{-23}\,\text{J/K}$).
4. An ^{87}Rb atom that has been laser cooled to a temperature of $T = 100\,\mu\text{K}$. Again, assume the kinetic energy is $\frac{3}{2}k_B T$.

Photons. Light of frequency ν can be regarded as consisting of photons of energy $E = h\nu$. Recall that $\hbar c \simeq 197.33\,\text{eV}\cdot\text{nm}$.

5. Visible light has a wavelength in the range 400–700 nm. Calculate the energy and frequency for both a photon of wavelength 400 nm and a photon of wavelength 700 nm. Give your energies in eV and frequencies in Hz (hertz).
6. A small microwave oven operates at roughly 2.5 GHz (gigahertz) at a maximum power of 300 W (watts). How many photons per second can it emit?
7. How many such microwave photons does it take to warm a 200 mL (milliliter) glass of water by 10°C? (The heat capacity of water is roughly 4.2 J/g/K, and the density of water is 1 g/mL.)
8. At a given power of an electromagnetic wave, do you expect a classical wave description to work better for radio frequencies or for X-rays? Why?

Problem 2.6. *Can a free electron absorb a photon?*

We would like to know if a free electron can completely absorb a photon. For maximum learning, we suggest that you stop reading now and investigate the question before following the steps below. You may also want to consider how this process would look in the center of momentum frame.

Consider now a laboratory frame where the electron is initially at rest, and the photon is moving toward the electron. After the collision, if the process is possible, we would just have an electron moving at some speed v.

1. By comparing the momentum of the system before and after the collision, construct a relationship between $h\nu$ and the final speed of the electron. Note that you will need to use relativistic momentum!
2. By computing the relativistic energy of the system before and after the collision and using conservation of energy, find a relationship between $h\nu$ and variables describing the electron.
3. We now have two expressions for $h\nu$, one from conservation of momentum and one from conservation of energy. By setting these two expressions equal to each other, you should find that two possible speeds v can result from such an interaction. What are these speeds, as fractions of c?
4. Only one of the velocities that you found is actually physically attainable. For that velocity, what must the corresponding energy of the photon be?
5. Is it possible for a free electron to absorb a photon?

Problem 2.7. *Deriving the Compton scattering wavelength shift.*

We aim here to derive (2.5.8), relating the shift $\lambda_f - \lambda_i$ in the photon wavelength to the deflection angle θ of the photon:

$$\lambda_f - \lambda_i = \frac{h}{m_e c}(1 - \cos\theta). \tag{1}$$

See figure 2.14. Recall that for a photon, the energy, momentum, and wavelength are related by $E = pc = h\nu$ with $\nu\lambda = c$. For an electron the energy and momentum are related by $E^2 - p^2c^2 = m_e^2c^4$. Call λ_i and ν_i the wavelength and frequency of the incident photon, and call λ_f and ν_f the wavelength and frequency of the final scattered photon.

1. Use energy conservation to show that

$$p_e^2c^2 = (h\nu_i - h\nu_f + m_ec^2)^2 - m_e^2c^4.$$

2. Consider the vector equation for momentum conservation and square it to show that

$$p_e^2c^2 = h^2(\nu_i^2 + \nu_f^2 - 2\nu_i\nu_f\cos\theta).$$

3. Set the right-hand sides from the last two parts equal and show that (1) follows.

Problem 2.8. Bohr radius, Compton wavelength of the electron, and classical electron radius.

1. The Bohr radius a_0 is the length scale that can be constructed from e^2, $\hbar$, and m_e and no extra numerical constants. Find the formula for the Bohr radius by dimensional analysis. Evaluate this length in units of pm $= 10^{-12}$m.
2. Let $\bar{\lambda}_C$ denote the reduced Compton wavelength of the elecron and $r_0 = \frac{e^2}{m_ec^2}$ denote the classical electron radius. Compute the constants c_1 and c_2 in the following

equation:

$$a_0 : \lambda\!\!\!^{-}_C : r_0 = 1 : c_1 : c_2.$$

Note: the relation $a : b : c = a' : b' : c'$ implies the equalities $b/a = b'/a'$ and $c/a = c'/a'$. Your answers for c_1 and c_2 should be in terms of the fine structure constant α.

Use these ratios to compute the values for $\lambda\!\!\!^{-}_C$ and r_0 in femtometers (fm).

Problem 2.9. *De Broglie wavelength of nonrelativistic and relativistic particles.*

1. The de Broglie wavelength $\lambda_{\rm nr}$ of a *nonrelativistic* electron with kinetic energy $E_{\rm kin}$ can be written as

$$\lambda_{\rm nr} = \frac{\delta}{\sqrt{E_{\rm kin}/{\rm eV}}}\ \text{Å}.$$

In this formula δ is a unit-free constant, and the answer comes out in angstroms ($\text{Å} = 10^{-10}$m). Give the value of δ.

2. The de Broglie wavelength $\lambda_{\rm r}$ of a *relativistic* electron with energy E can be calculated in terms of the γ factor of the electron: $E = \gamma m_e c^2$. One finds

$$\lambda_{\rm r} = \frac{\ell}{\sqrt{\gamma^2 - 1}}.$$

What is the value of ℓ in pm $= 10^{-12}$m? Is this a well-known length?

3. Consider an electron moving with speed v. Find the ratio $\lambda_{\rm r}/\lambda_{\rm nr}$ of the relativistic and nonrelativistic values of the de Broglie wavelength. What can you say about the ratio for any $0 < v < c$?

4. What is the velocity parameter $\beta = v/c$ of a massive particle whose de Broglie wavelength is equal to its Compton wavelength? Is that particle relativistic—that is, is $\beta > 0.2$?

5. The de Broglie wavelength of a particle gives you a rough idea of the distance scale it can explore in a collision experiment. The International Linear Collider, which may be built in the future, is expected to accelerate electrons to 1 TeV = 1,000 GeV. What is the de Broglie wavelength of such electrons?

   Find the de Broglie wavelength of 7 TeV protons at the LHC (Large Hadron Collider) in Geneva, Switzerland.

Problem 2.10. *Simple derivation of phase and group velocities.*

Consider the following superposition of two waves:

$$\psi(x,t) = \cos(k_1 x - \omega_1 t) + \cos(k_2 x - \omega_2 t),$$

where k_2 differs slightly from k_1, and ω_2 differs slightly from ω_1:

$$k_2 - k_1 \ll k_1, k_2, \quad \omega_2 - \omega_1 \ll \omega_1, \omega_2.$$

Moreover, we assume $k_2 > k_1$, and $\omega_2 > \omega_1$. Show that you can rewrite the wave as

$$\psi(x,t) = 2\cos(kx - \omega t)\cos(\Delta k x - \Delta\omega t), \quad \text{with} \quad \Delta k = \tfrac{1}{2}(k_2 - k_1), \quad \Delta\omega = \tfrac{1}{2}(\omega_2 - \omega_1).$$

Sketch $\psi(x,0)$, assuming $\Delta k \ll k$. The result for $\psi(x,t)$ is a product of two factors, each a wave. Identify the short-wavelength factor as that moving with the phase velocity ω/k and the long-wavelength factor as that moving with the group velocity $\frac{\Delta\omega}{\Delta k}$. Note that $\frac{\Delta\omega}{\Delta k}$ is a good proxy for the derivative of ω with respect to k.

Problem 2.11. *Tests of the stationary phase approximation.*

Consider integrals of the form

$$\Psi(x) = \int_{-\infty}^{\infty} dk\, \Phi(k) e^{ikx},$$

where $\Phi(k)$ is a function that is sharply localized around $k = k_0$. In each of the two following cases, use the stationary phase argument to predict the location of the peak of $|\Psi(x)|$. Then compute the integral exactly to find $\Psi(x)$ and $|\Psi(x)|$ and to confirm your prediction.

1. $\Phi(k) = Ce^{-L^2(k-k_0)^2}$.
2. $\Phi(k) = Ce^{-L^2(k-k_0)^2} e^{-ikx_0}$.

In the above expressions, L and x_0 are constants with units of length. The constant C, which you need not calculate, allows for proper normalization of the wave functions.

Useful integral: For complex constants a and b, with $\text{Re}(a) > 0$: $\int_{-\infty}^{\infty} e^{-ax^2+bx} dx = \sqrt{\frac{\pi}{a}} \exp(\frac{b^2}{4a})$.

Problem 2.12. *A wave packet and stationary phase.*

We send in a particle of mass m from $x = \infty$ toward $x = 0$. This is described by the localized incident wave packet $\Psi_{\text{inc}}(x,t)$ given by

$$\Psi_{\text{inc}}(x,t) = \int_0^{\infty} dk\, f(k)\, e^{-ikx} e^{-iE(k)t/\hbar}, \qquad \text{for } x > R,$$

with $f(k)$ a real function that peaks sharply for $k = k_0 > 0$. As usual $E = \frac{p^2}{2m} = \frac{\hbar^2 k^2}{2m}$. This incident wave encounters some kind of obstacle in the region $x \in [0, R]$, as well as an impenetrable barrier at $x = 0$. A reflected, outgoing packet is produced of the form

$$\Psi_{\text{out}}(x,t) = -\int_0^{\infty} dk\, f(k) e^{2i\delta(k)} e^{ikx} e^{-iE(k)t/\hbar}, \qquad \text{for } x > R,$$

where $\delta(k)$ is a calculable function called a phase shift. Assume $\delta(k)$ is known.

1. Use the stationary phase approximation to find the relation between x and t that describes the motion of the "peak" of the incident packet $\Psi_{\text{inc}}(x,t)$. What can you tell about $\Psi_{\text{inc}}(x,t)$ in the region $x > R$ for large positive t?
2. Find the relation between x and t that describes the motion of the outgoing packet $\Psi_{\text{out}}(x,t)$. What can you tell about $\Psi_{\text{out}}(x,t)$ in the region $x > R$ for large negative t?
3. Imagine a situation in which the incident particle, coming in with the relevant group velocity, simply bounces off from $x = 0$ at $t = 0$. Compared to this picture, the outgoing particle represented by $\Psi_{\text{out}}(x,t)$ is actually delayed. Calculate the delay time τ as a function of $\delta(k)$, its derivative, and constants.

3 Schrödinger Equation

We use the free-particle wave function to introduce a momentum operator and an energy operator. We then derive the differential equation—Schrödinger equation—satisfied by the free-particle wave function. Extending the definition of the energy operator, we are led to the general Schrödinger equation for the motion of a nonrelativistic particle in a potential. We define the position operator and calculate the commutator of position and momentum operators. We explore the probabilistic interpretation of the wave function, noting that a normalized wave function must remain normalized under time evolution. We show that this will happen if the energy operator, or Hamiltonian, is a Hermitian operator. Our strategy to prove Hermiticity is to show that the probability density is accompanied by a probability current, and together they satisfy a conservation equation. Wave functions remain normalized in time because the flow of probability current vanishes at infinity.

3.1 The Wave Function for a Free Particle

We have discussed de Broglie matter waves for particles, but we have not yet written down an expression representing such waves, not even for the simplest case of a free particle. So we ask: What is the mathematical form of the de Broglie wave, or wave function, associated with a particle with energy E and momentum p? We know that ω and k are determined from $E = \hbar\omega$ and $p = \hbar k$. Let's assume our wave is propagating in the $+x$ direction. All the following are examples of waves that could be candidates for the particle wave function:

1. $\sin(kx - \omega t)$,
2. $\cos(kx - \omega t)$,
3. $e^{i(kx-\omega t)} = e^{ikx}e^{-i\omega t}$ (fixed time dependence $\propto e^{-i\omega t}$),
4. $e^{-i(kx-\omega t)} = e^{-ikx}e^{i\omega t}$ (fixed time dependence $\propto e^{+i\omega t}$).

The third and fourth options use time dependences with opposite signs. We will use superposition to decide which of these four candidates is the right one. Let's take them one by one:

(i) Starting from (1), we build a superposition in which the particle has an equal probability of being found moving in the $+x$ and the $-x$ directions. Such a state must exist, and we build it by adding two waves of type (1):

$$\Psi(x,t) = \sin(kx - \omega t) + \sin(kx + \omega t). \tag{3.1.1}$$

Expanding the trigonometric functions, this can be simplified to

$$\Psi(x,t) = 2 \sin kx \cos \omega t. \tag{3.1.2}$$

But this result is not sensible. The wave function vanishes identically for all x at the special times $\omega t = \frac{\pi}{2}, \frac{3\pi}{2}, \frac{5\pi}{2}, \ldots$. A wave function that is zero all over space at some particular time cannot represent a particle. Thus, (1) is not a good candidate.

(ii) Constructing a wave function from (2) with a superposition of left-moving and right-moving waves, we encounter a similar problem:

$$\Psi(x,t) = \cos(kx - \omega t) + \cos(kx + \omega t) = 2 \cos kx \cos \omega t. \tag{3.1.3}$$

This choice is no good; it also vanishes identically when $\omega t = \frac{\pi}{2}, \frac{3\pi}{2}, \ldots$.

(iii) Let's try a similar superposition of exponentials from (3), with both having the same time dependence:

$$\begin{aligned} \Psi(x,t) &= e^{i(kx-\omega t)} + e^{i(-kx-\omega t)} \\ &= (e^{ikx} + e^{-ikx})\, e^{-i\omega t} \\ &= 2 \cos kx\, e^{-i\omega t}. \end{aligned} \tag{3.1.4}$$

This wave function meets our criteria! It is never zero for all values of x because $e^{-i\omega t}$ is never zero.

(iv) A superposition of exponentials from (4) also meets our criteria

$$\begin{aligned} \Psi(x,t) &= e^{-i(kx-\omega t)} + e^{-i(-kx-\omega t)} \\ &= (e^{ikx} + e^{-ikx})\, e^{i\omega t} \\ &= 2 \cos kx\, e^{i\omega t}. \end{aligned} \tag{3.1.5}$$

This, too, is never zero for all values of x.

Since both options (3) and (4) seem to work, we ask: Can we use *both* (3) and (4) to represent a particle moving in the $+x$ direction? Let's assume that we can. Then, since adding a state to itself should not change the state, we could represent the right-moving particle by using the sum of (3) and (4):

$$\Psi(x,t) = e^{i(kx-\omega t)} + e^{-i(kx-\omega t)} = 2 \cos(kx - \omega t). \tag{3.1.6}$$

This, however, is the same as (2), which we already showed leads to difficulties. Therefore, we must choose between (3) and (4). The choice is a matter of convention, and happily, most physicists use the same convention. We take the free-particle wave function to be the following:

$$\boxed{\text{free-particle wave function:} \quad \Psi(x,t) = e^{i(kx-\omega t)},} \tag{3.1.7}$$

representing a particle with momentum p and energy E given by

$$p = \hbar k, \quad \text{and} \quad E = \hbar\omega. \tag{3.1.8}$$

An alternative argument leading to this same result is explored in problem 3.1. In three dimensions the corresponding wave function would be the following:

$$\boxed{\text{free-particle wave function:} \quad \Psi(\mathbf{x}, t) = e^{i(\mathbf{k}\cdot\mathbf{x}-\omega t)},} \tag{3.1.9}$$

representing a particle with

$$\mathbf{p} = \hbar\mathbf{k}, \quad \text{and} \quad E = \hbar\omega. \tag{3.1.10}$$

Exercise 3.1. *For the free-particle wave function in one dimension describing a particle with momentum p and energy E, we have*

$$\Psi(x + x_0, t) = A\,\Psi(x, t), \qquad \Psi(x, t + t_0) = B\,\Psi(x, t). \tag{3.1.11}$$

Determine A and B in terms of the momentum p, the energy E, $\hbar$, x_0, and t_0. Note that A and B are constant phases, independent of x and t. As such, these factors do not change the physics of the wave function.

3.2 Equations for a Wave Function

We determined that the wave function or de Broglie wave for a free particle moving in one dimension with momentum p and energy E is given by (3.1.7):

$$\Psi_p(x, t) = e^{i(kx-\omega t)}. \tag{3.2.1}$$

We added the subscript p to Ψ to indicate the value of the momentum. The constants k and ω are determined from

$$p = \hbar k, \quad E = \hbar\omega = \frac{p^2}{2m}. \tag{3.2.2}$$

The wave function Ψ_p represents a state of definite momentum. We now look for an *operator* that extracts the value of the momentum from the wave function. The operator, to be called the *momentum operator*, must involve a derivative with respect to x, as this would bring down a factor of k from the exponential. More precisely, we can see that

$$\frac{\hbar}{i}\frac{\partial}{\partial x}\Psi_p(x, t) = \frac{\hbar}{i}(ik)\Psi_p(x, t) = \hbar k\Psi_p(x, t) = p\,\Psi_p(x, t), \tag{3.2.3}$$

where the p factor in the last right-hand side is in fact the momentum of the state. We thus identify the operator $\frac{\hbar}{i}\frac{\partial}{\partial x}$ as the *momentum operator* $\hat{p}$:

$$\boxed{\hat{p} \equiv \frac{\hbar}{i}\frac{\partial}{\partial x}.} \tag{3.2.4}$$

We have verified that acting on the wave function $\Psi_p(x, t)$ describing a particle of momentum p, the operator $\hat{p}$ gives precisely the number p times the wave function:

$$\hat{p}\,\Psi_p = p\,\Psi_p. \tag{3.2.5}$$

The hat in $\hat{p}$ tells us that this is the momentum *operator*, not the value of the momentum. Acting on general wave functions—that is, on general functions of x and t, the momentum operator gives us another function of x and t. But acting on the special wave function Ψ_p, it gives the number p times Ψ_p. We say that Ψ_p is an **eigenstate** of $\hat{p}$ with eigenvalue p. We also say that Ψ_p is a state of **definite momentum**. The implication of this statement is that an ideal measurement of the momentum in the state Ψ_p will always give the value p.

This terminology is analogous to that used in matrix algebra. Matrices act by multiplication on column vectors to give new column vectors. An eigenvector of a matrix is a special column vector. The matrix acting on an eigenvector gives a number, called the eigenvalue, times the eigenvector. Operators like $\hat{p}$ are analogous to matrices, and wave functions, or states, are analogous to column vectors. The special wave functions that are eigenstates are the analogs of the special column vectors that are eigenvectors. We thus have the following correspondences:

$$\begin{aligned} \text{operators} &\leftrightarrow \text{matrices} \\ \text{wave functions} &\leftrightarrow \text{column vectors} \\ \text{eigenstates} &\leftrightarrow \text{eigenvectors} \end{aligned} \tag{3.2.6}$$

Let us now consider defining an energy operator $\hat{E}$ for the free particle. Since the energy of a free particle is given in terms of its momentum, we will find that the energy operator can be written in terms of the momentum operator. While the energy of the free-particle wave function can be extracted using a time derivative, we do not follow that approach to define the energy operator. Instead, as we will see below, we use time derivatives to find a differential equation for the wave function.

Begin with the product of the energy times the wave function Ψ_p:

$$E\,\Psi_p = \frac{p^2}{2m}\Psi_p = \frac{p}{2m}(p\Psi_p) = \frac{p}{2m}\frac{\hbar}{i}\frac{\partial}{\partial x}\Psi_p, \tag{3.2.7}$$

where we used equation (3.2.5) to write $p\Psi_p$ as the momentum operator acting on Ψ_p. Since the remaining p on the last expression is a constant, we can move it across the derivative until it is close to the wave function, at which point we replace it by the momentum operator:

$$E\,\Psi_p = \frac{1}{2m}\frac{\hbar}{i}\frac{\partial}{\partial x}(p\Psi_p) = \frac{1}{2m}\frac{\hbar}{i}\frac{\partial}{\partial x}\Big(\frac{\hbar}{i}\frac{\partial}{\partial x}\Psi_p\Big). \tag{3.2.8}$$

This can be written as

$$E\,\Psi_p = \frac{1}{2m}\hat{p}\,\hat{p}\,\Psi_p = \frac{\hat{p}^2}{2m}\Psi_p, \tag{3.2.9}$$

which suggests the following definition of the *energy operator* $\hat{E}$:

$$\hat{E} \equiv \frac{\hat{p}^2}{2m} = -\frac{\hbar^2}{2m}\frac{\partial^2}{\partial x^2}. \tag{3.2.10}$$

Indeed, with this definition, (3.2.9) shows that

$$\hat{E}\,\Psi_p = E\,\Psi_p, \tag{3.2.11}$$

as befits the energy operator. The state Ψ_p is a state of definite energy E, the eigenvalue of the energy operator.

Let us now find a differential equation for which our de Broglie wave function is a solution. First, we note that a suitable time derivative also extracts the energy eigenvalue from Ψ_p. Using (3.2.1), we find that

$$i\hbar\frac{\partial}{\partial t}\Psi_p(x,t) \;=\; i\hbar(-i\omega)\Psi_p(x,t) \;=\; \hbar\omega\,\Psi_p(x,t) \;=\; E\,\Psi_p(x,t). \tag{3.2.12}$$

We can now replace the final right-hand side $E\Psi_p$ by $\hat{E}\,\Psi_p$, giving us a differential equation satisfied by Ψ_p:

$$i\hbar\,\frac{\partial}{\partial t}\Psi_p(x,t) \;=\; -\frac{\hbar^2}{2m}\frac{\partial^2}{\partial x^2}\Psi_p(x,t). \tag{3.2.13}$$

Since the momentum p appears only in the wave function, we are inspired by this differential equation to think of it more generally as an equation for general wave functions $\Psi(x,t)$ of a free particle:

$$\boxed{i\hbar\,\frac{\partial}{\partial t}\Psi(x,t) \;=\; -\frac{\hbar^2}{2m}\frac{\partial^2}{\partial x^2}\Psi(x,t).} \tag{3.2.14}$$

This inspired guess is the **free-particle Schrödinger equation**. Schematically, using the energy operator, it can be written as

$$\boxed{i\hbar\,\frac{\partial}{\partial t}\Psi(x,t) \;=\; \hat{E}\,\Psi(x,t).} \tag{3.2.15}$$

It is worth rechecking that our de Broglie wave function satisfies the Schrödinger equation (3.2.14). Indeed, for $\Psi_p = e^{i(kx-\omega t)}$ we find that

$$i\hbar(-i\omega)\Psi_p \;=\; -\frac{\hbar^2}{2m}(ik)^2\Psi_p. \tag{3.2.16}$$

This is a solution since the Ψ_p factors cancel, and all that is needed is the equality

$$\hbar\omega = \frac{\hbar^2k^2}{2m}, \tag{3.2.17}$$

which is recognized as the familiar relation $E = \frac{p^2}{2m}$, assumed to hold when writing Ψ_p.

Note that the Schrödinger equation admits solutions that are more general than the de Broglie wave function for a particle of definite momentum and energy. This often happens in physics. Limited evidence leads to some equation that happens to contain much more physics than the unsuspecting inventor ever imagined. Since the equation we wrote is linear, any superposition of plane wave solutions with different values of k is a solution. Take, for example,

$$\Psi(x,t) = a_1\, e^{i(k_1 x-\omega_1 t)} + a_2\, e^{i(k_2 x-\omega_2 t)}, \tag{3.2.18}$$

where a_1 and a_2 are arbitrary complex numbers, and $k_1 \neq k_2$. This is a solution provided the pairs (k_1,ω_1) and (k_2,ω_2) each satisfy (3.2.17). While each summand is a state of definite momentum, the total solution is not a state of definite momentum. Indeed,

$$\hat{p}\,\Psi(x,t) = a_1\, \hbar k_1\, e^{i(k_1 x-\omega_1 t)} + a_2\, \hbar k_2\, e^{i(k_2 x-\omega_2 t)}, \tag{3.2.19}$$

and the right-hand side cannot be written as a number times $\Psi(x,t)$. The full state is not a state of definite energy either. The general solution of the free Schrödinger equation is the most general superposition of plane waves:

$$\Psi(x,t) = \int_{-\infty}^{\infty} dk\, \Phi(k)\, e^{i(kx-\omega(k)t)}, \tag{3.2.20}$$

where $\Phi(k)$ is an *arbitrary* function of k that controls the superposition, and we have written $\omega(k)$ to emphasize that ω is a function of the momentum, as in (3.2.17). The momentum eigenstate plane waves Ψ_p are not localized in space; they are nowhere vanishing. General superpositions, as in the above expression, can be made to represent localized solutions by choosing a suitable $\Phi(k)$. In that case they are said to describe *wave packets*.

Exercise 3.2. *Verify that the wave packet* $\Psi(x,t)$ *in (3.2.20) solves the free Schrödinger equation.*

Exercise 3.3. *A free nonrelativistic particle has a wave function*

$$\Psi(x,t) = e^{i\left(\frac{x}{L}-\frac{t}{\tau}\right)}, \tag{3.2.21}$$

where $L>0$ *and* $\tau>0$ *are constants with units of length and time, respectively. Use the free Schrödinger equation to determine the mass* m *of the particle.*

We now have the tools to evolve in time any initial state of a free particle: given the initial wave function $\Psi(x,0)$ at time zero, we can obtain $\Psi(x,t)$. (This is a preview of a more detailed discussion to come in section 4.3.) Indeed, if we know $\Psi(x,0)$, we can write:

$$\Psi(x,0) = \int_{-\infty}^{\infty} dk\, \Phi(k)\, e^{ikx}, \tag{3.2.22}$$

where $\Phi(k)$ is, by definition, the Fourier transform of $\Psi(x,0)$. The Fourier transform $\Phi(k)$ can be calculated in terms of $\Psi(x,0)$. Once we know $\Phi(k)$, the time evolution simply requires the replacement

$$e^{ikx} \;\to\; e^{ikx}e^{-i\omega(k)t} \tag{3.2.23}$$

in the above integrand so that the answer for the time evolved $\Psi(x,t)$ is in fact given by (3.2.20).

As we have discussed before, the velocity of a wave packet described by (3.2.20) is given by the group velocity (2.8.6) evaluated for the dominant value of k. We confirm that this is indeed reasonable:

$$v_g \equiv \frac{d\omega}{dk} = \frac{d\hbar\omega}{d\hbar k} = \frac{dE}{dp} = \frac{d}{dp}\left(\frac{p^2}{2m}\right) = \frac{p}{m}, \tag{3.2.24}$$

which is the expected velocity for a nonrelativistic particle with momentum p and mass m.

The Schrödinger equation (3.2.14) has an explicit i on the left-hand side. This i shows that it is impossible to find a solution for real Ψ. If Ψ were real, the right-hand side of the equation would be real, but the left-hand side would be imaginary. Thus, the Schrödinger equation makes it compulsory to work with complex wave functions.

Note also that the Schrödinger equation does not take the form of an *ordinary* wave equation. Such a wave equation for a function $\phi(x,t)$ takes the form:

$$\frac{\partial^2\phi}{\partial x^2} - \frac{1}{v_0^2}\frac{\partial^2\phi}{\partial t^2} = 0, \tag{3.2.25}$$

where v_0 is a constant velocity. The general solutions of this linear equation are traveling waves $f_\pm(x \pm v_0 t)$, with f an arbitrary function. This equation certainly allows for real solutions, which are not acceptable in quantum theory. The Schrödinger equation has no second-order time derivatives; it is first order in time! In going from the wave equation above to the Schrödinger equation, we essentially trade a time derivative for a factor of i! The result is an equation that also has wave solutions, but they are complex valued.

Diffusion equations are also linear differential equations with first-order time derivatives and second-order spatial derivatives. As opposed to the Schrödinger equation, however, there is no i in the diffusion equation, and the solutions are real.

3.3 Schrödinger Equation for a Particle in a Potential

Suppose now that our quantum particle is not free but rather is moving in some external potential $V(x,t)$. Here $V(x,t)$ is a real quantity with units of energy. In this case the total energy of the particle is the sum of kinetic and potential energies:

$$E = \frac{p^2}{2m} + V(x,t). \tag{3.3.1}$$

This naturally suggests that the energy operator should take the form

$$\hat{E} = \frac{\hat{p}^2}{2m} + V(x,t). \tag{3.3.2}$$

The first term, as we already know, involves second derivatives with respect to x. The second term acts multiplicatively: acting on any wave function $\Psi(x,t)$, it simply multiplies it by $V(x,t)$. This term thus represents a simple kind of operator, one for which we do not generally include a hat on top. We now postulate that the Schrödinger equation for a particle in a potential takes the earlier form (3.2.15) valid for free particles with $\hat{E}$ replaced by the above energy operator:

$$i\hbar\frac{\partial}{\partial t}\Psi(x,t)=\Big(-\frac{\hbar^2}{2m}\frac{\partial^2}{\partial x^2}+V(x,t)\Big)\Psi(x,t). \tag{3.3.3}$$

The energy operator $\hat{E}$ is usually called the **Hamiltonian** operator $\hat{H}$, so one has

$$\hat{H}\equiv -\frac{\hbar^2}{2m}\frac{\partial^2}{\partial x^2}+V(x,t), \tag{3.3.4}$$

and the Schrödinger equation takes the form

$$i\hbar\frac{\partial}{\partial t}\Psi(x,t) = \hat{H}\,\Psi(x,t). \tag{3.3.5}$$

Let's immediately point out two basic properties of the Schrödinger equation:

1. The differential equation is first order in time. This means that as an initial condition it suffices to know the wave function completely at some initial time t_0, and the Schrödinger equation then determines the wave function for all times. This can be understood very explicitly. If we know $\Psi(x,t_0)$ for all x, then the right-hand side of the Schrödinger equation (3.3.3), which just involves x derivatives and multiplication, can be evaluated at any point x. This means that at any point x we know the time derivative of the wave function (left-hand side of the Schrödinger equation), and this allows us to calculate the wave function a little later.
2. Linearity and superposition. The Schrödinger equation is a linear equation for complex wave functions. Therefore, given two solutions Ψ_1 and Ψ_2 we can form new solutions as linear combinations $\alpha\Psi_1+\beta\Psi_2$ with complex coefficients α and β.

Let us reconsider the way in which the potential $V(x,t)$ is an operator. We can do this by introducing a **position** operator $\hat{x}$ that acting on functions of x gives another function of x as follows:

$$\hat{x}f(x) \equiv xf(x). \tag{3.3.6}$$

It is worth noting that this equation is on a *very* different footing from $\hat{p}\,\Psi_p=p\,\Psi_p$, which states that Ψ_p is a $\hat{p}$ eigenstate with eigenvalue p, a number. In writing $\hat{x}f(x)=xf(x)$, the function $f(x)$ is not an eigenstate of $\hat{x}$ because x is not an eigenvalue; x is a function, not a number.

It follows from (3.3.6), and successive applications of it, that for any positive integer k,

$$\hat{x}^kf(x) = x^kf(x). \tag{3.3.7}$$

If the potential $V(x,t)$ can be written as some series expansion in terms of x, it then follows that

$$V(\hat{x},t)\Psi(x,t) = V(x,t)\Psi(x,t). \tag{3.3.8}$$

With this notation we can write the Hamiltonian in a way that makes manifest its operator nature:

$$\boxed{\hat{H} \equiv \frac{\hat{p}^2}{2m} + V(\hat{x}, t),} \tag{3.3.9}$$

The operators we are dealing with (momentum, position, Hamiltonian) are all declared to be linear operators. A **linear operator** $\hat{A}$, as we stated early on in (1.1.4), satisfies

$$\hat{A}(a\phi) = a\hat{A}\phi, \quad \hat{A}(\phi_1 + \phi_2) = \hat{A}\phi_1 + \hat{A}\phi_2, \tag{3.3.10}$$

where a is an arbitrary constant. Two linear operators $\hat{A}$ and $\hat{B}$ that act on the same set of objects can always be added: $(\hat{A} + \hat{B})\phi \equiv \hat{A}\phi + \hat{B}\phi$. They can also be multiplied, and the product $\hat{A}\hat{B}$ is a linear operator defined by $\hat{A}\hat{B}\phi \equiv \hat{A}(\hat{B}\phi)$, meaning that you act first with $\hat{B}$, which is closest to ϕ, and then act on the result with $\hat{A}$. The order of multiplication matters, and thus $\hat{A}\hat{B}$ and $\hat{B}\hat{A}$ may not be the same operator. To quantify this possible difference, one introduces the **commutator** $[\hat{A}, \hat{B}]$ of two operators, defined to be the linear operator

$$\boxed{[\hat{A}, \hat{B}] \equiv \hat{A}\hat{B} - \hat{B}\hat{A}.} \tag{3.3.11}$$

If the commutator vanishes, the two operators are said to commute. It is also clear that

$$[\hat{A}, \hat{A}] = 0, \quad \text{for any operator } \hat{A}. \tag{3.3.12}$$

We have operators $\hat{x}$ and $\hat{p}$ that are clearly somewhat related. We would like to know their commutator $[\hat{x}, \hat{p}]$. For this we let $[\hat{x}, \hat{p}]$ act on some arbitrary function $\phi(x)$ and then attempt simplification. Let's do it. We begin by writing out the commutator:

$$[\hat{x}, \hat{p}]\phi(x) = (\hat{x}\hat{p} - \hat{p}\hat{x})\phi(x) = \hat{x}\hat{p}\,\phi(x) - \hat{p}\hat{x}\,\phi(x) = \hat{x}(\hat{p}\phi(x)) - \hat{p}(\hat{x}\phi(x)). \tag{3.3.13}$$

We can now let $\hat{p}$ act on the first term and $\hat{x}$ act on the second term to find

$$[\hat{x}, \hat{p}]\phi(x) = \hat{x}\Big(\frac{\hbar}{i}\frac{\partial\phi(x)}{\partial x}\Big) - \hat{p}(x\phi(x)). \tag{3.3.14}$$

At this point the $\hat{x}$ in the first term acts by multiplication because it is applied to a function of x. In the second term, we can also use the derivative representation of $\hat{p}$ since it is also acting on a function of x. We thus have

$$\begin{aligned} [\hat{x}, \hat{p}]\phi(x) &= x\frac{\hbar}{i}\frac{\partial\phi(x)}{\partial x} - \frac{\hbar}{i}\frac{\partial}{\partial x}(x\phi(x)) \\ &= \frac{\hbar}{i}x\frac{\partial\phi(x)}{\partial x} - \frac{\hbar}{i}x\frac{\partial\phi(x)}{\partial x} - \frac{\hbar}{i}\phi(x) \\ &= -\frac{\hbar}{i}\phi(x) = i\hbar\,\phi(x) \end{aligned} \tag{3.3.15}$$

so that, all in all, we have shown that for arbitrary $\phi(x)$ one has

$$[\hat{x}, \hat{p}]\phi(x) = i\hbar\,\phi(x). \tag{3.3.16}$$

Since this equation holds for any ϕ, it really represents the equality of two operators. Indeed, whenever we have operators $\hat{A}$ and $\hat{B}$ for which $\hat{A}\phi = \hat{B}\phi$ for arbitrary ϕ, we simply say that $\hat{A} = \hat{B}$; the operators are the same because they give the same result acting on every possible state! We can therefore peel off the ϕ from (3.3.16) to obtain the most fundamental commutation relation in quantum mechanics:

$$\boxed{[\hat{x}, \hat{p}] = i\hbar.} \tag{3.3.17}$$

The right-hand side is a number but should be viewed as an operator: acting on any function, it multiplies the function by the number. We will see later that this commutation relation is the key ingredient in the proof of Heisenberg's uncertainty principle. This principle implies that in any quantum state the product of the position uncertainty and the momentum uncertainty must be greater or equal to $\hbar/2$.

Exercise 3.4. *Show that $[\hat{x}, \hat{p}^2] = 2i\hbar\hat{p}$. Do it by moving the $\hat{x}$ operator across $\hat{p}^2 = \hat{p}\hat{p}$ by repeated use of the commutator (3.3.17).*

The idea that operators can fail to commute may remind you of matrix multiplication, which is also noncommutative. This is in accord with the correspondences in (3.2.6). In fact, in the matrix formulation of quantum mechanics these correspondences are actually concrete and workable. Matrix mechanics was worked out in 1925 by Werner Heisenberg and clarified by Max Born and Pascual Jordan.

As an example of useful matrices that do not commute, consider the *Pauli matrices*, three two-by-two matrices $\sigma_1, \sigma_2, \sigma_3$ given by:

$$\sigma_1 = \begin{pmatrix} 0 & 1 \\ 1 & 0 \end{pmatrix}, \quad \sigma_2 = \begin{pmatrix} 0 & -i \\ i & 0 \end{pmatrix}, \quad \sigma_3 = \begin{pmatrix} 1 & 0 \\ 0 & -1 \end{pmatrix}. \tag{3.3.18}$$

Actually, these matrices are essentially angular momentum operators for spin one-half particles. The so-called spin operator $\hat{\mathbf{S}}$, to be discussed in chapter 12, has three components $(\hat{S}_1, \hat{S}_2, \hat{S}_3)$ that are given by $\hat{S}_i = \frac{\hbar}{2}\sigma_i$, for $i = 1, 2, 3$. Let us now see if σ_1 and σ_2 commute:

$$\begin{aligned} \sigma_1\sigma_2 &= \begin{pmatrix} 0 & 1 \\ 1 & 0 \end{pmatrix}\begin{pmatrix} 0 & -i \\ i & 0 \end{pmatrix} = \begin{pmatrix} i & 0 \\ 0 & -i \end{pmatrix}, \\ \sigma_2\sigma_1 &= \begin{pmatrix} 0 & -i \\ i & 0 \end{pmatrix}\begin{pmatrix} 0 & 1 \\ 1 & 0 \end{pmatrix} = \begin{pmatrix} -i & 0 \\ 0 & i \end{pmatrix}. \end{aligned} \tag{3.3.19}$$

We then see that

$$[\sigma_1, \sigma_2] = \begin{pmatrix} 2i & 0 \\ 0 & -2i \end{pmatrix} = 2i\begin{pmatrix} 1 & 0 \\ 0 & -1 \end{pmatrix} = 2i\sigma_3. \tag{3.3.20}$$

In fact, one has

$$\begin{aligned} [\sigma_1, \sigma_2] &= 2i\,\sigma_3, \\ [\sigma_2, \sigma_3] &= 2i\,\sigma_1, \\ [\sigma_3, \sigma_1] &= 2i\,\sigma_2, \end{aligned} \tag{3.3.21}$$

a nice set of commutation relations. Since matrices are ultimately operators, it is of interest to find their eigenstates (or eigenvectors) and their eigenvalues, just as we learned that plane waves are momentum eigenstates.

Exercise 3.5. *Show that the three Pauli matrices have eigenvalues* ± 1. *Find the eigenvectors of* σ_3 *and* σ_1.

If we were to write the $\hat{x}$ and $\hat{p}$ operators in matrix form, they would require infinite dimensional matrices. One can prove that there are no finite-size matrices that commute to give a number times the identity matrix, as is required from (3.3.17). This shouldn't be a surprise: on the real line, there are an infinite number of linearly independent wave functions, and in view of the correspondences in (3.2.6), it would suggest an infinite number of basis vectors. The relevant matrices must therefore be infinite dimensional. You will construct matrices for $\hat{x}$ and $\hat{p}$ after learning about the harmonic oscillator (problem 9.10).

We have written the Schrödinger equation for a particle in a one-dimensional potential. How about for the case of a particle in a three-dimensional potential? As we will see now, this is easily done once we realize that in three dimensions the position and momentum operators have several components. Recall that the de Broglie wave function

$$\Psi(\mathbf{x}, t) = e^{i(\mathbf{k}\cdot\mathbf{x}-\omega t)} = e^{i(k_x x + k_y y + k_z z - \omega t)} \tag{3.3.22}$$

corresponds to a particle carrying momentum $\mathbf{p} = \hbar\mathbf{k}$, with $\mathbf{k} = (k_x, k_y, k_z)$. Just as we did in (3.2.3), we can try to extract the vector momentum by using a differential operator. The relevant operator is the gradient:

$$\nabla = \left(\frac{\partial}{\partial x}, \frac{\partial}{\partial y}, \frac{\partial}{\partial z}\right), \tag{3.3.23}$$

with which we try:

$$\frac{\hbar}{i}\nabla\Psi(x,t) = \frac{\hbar}{i}\left(ik_x, ik_y, ik_z\right)\Psi(x,t) = \hbar\mathbf{k}\,\Psi(x,t) = \mathbf{p}\,\Psi(x,t). \tag{3.3.24}$$

We therefore define the momentum operator $\hat{\mathbf{p}}$ as follows:

$$\hat{\mathbf{p}} \equiv \frac{\hbar}{i}\nabla. \tag{3.3.25}$$

If we define $(\hat{p}_1, \hat{p}_2, \hat{p}_3) \equiv (\hat{p}_x, \hat{p}_y, \hat{p}_z)$ and $(x_1, x_2, x_3) \equiv (x, y, z)$, then we have the components of the above equation as

$$\hat{p}_k = \frac{\hbar}{i}\frac{\partial}{\partial x_k}, \quad k = 1, 2, 3. \tag{3.3.26}$$

Just like when we defined a position operator $\hat{x}$, we now have three position operators $(\hat{x}_1, \hat{x}_2, \hat{x}_3)$ making up $\hat{\mathbf{x}}$. With three position and three momentum operators, we now should state the nine possible commutation relations. If you recall our derivation of $[\hat{x}, \hat{p}] = i\hbar$, you will note that the commutator vanishes unless the subscripts on $\hat{x}$ and $\hat{p}$ are the same. This means that the general commutator takes the form

$$[\hat{x}_i, \hat{p}_j] = i\hbar\,\delta_{ij}. \tag{3.3.27}$$

Here, the Kronecker delta is defined by

$$\delta_{ij} = \begin{cases} 1 & \text{if } i = j, \\ 0 & \text{if } i \neq j. \end{cases} \tag{3.3.28}$$

The Kronecker delta vanishes unless the two indices take the same value, in which case the Kronecker delta evaluates to one. In order to write the general Schrödinger equation, we need to consider the energy operator, or Hamiltonian:

$$\hat{H} = \frac{\hat{\mathbf{p}}^2}{2m} + V(\mathbf{x}, t). \tag{3.3.29}$$

This time we have that

$$\hat{\mathbf{p}}^2 = \hat{\mathbf{p}} \cdot \hat{\mathbf{p}} = \frac{\hbar}{i}\nabla \cdot \frac{\hbar}{i}\nabla = -\hbar^2\nabla^2, \tag{3.3.30}$$

where ∇^2 is the Laplacian operator:

$$\nabla^2 \equiv \frac{\partial^2}{\partial x^2} + \frac{\partial^2}{\partial y^2} + \frac{\partial^2}{\partial z^2}. \tag{3.3.31}$$

The Schrödinger equation finally takes the following form:

$$\boxed{i\hbar\frac{\partial}{\partial t}\Psi(\mathbf{x}, t) = \Big(-\frac{\hbar^2}{2m}\nabla^2 + V(\mathbf{x}, t)\Big)\Psi(\mathbf{x}, t).} \tag{3.3.32}$$

3.4 Interpreting the Wave Function

Schrödinger thought the wave function Ψ represents a particle that could spread out and disintegrate. He suggested that the *fraction* of the particle to be found at x would be proportional to $|\Psi|^2$. This was problematic, as noted by Max Born (1882–1970). Born solved the Schrödinger equation for the scattering of a particle off a potential, finding a wave function that fell off like $1/r$, with r the distance to the scattering center. But Born also noticed in the experiment that one does not find fractions of particles going in multiple directions; rather, particles remain whole. Born suggested a probabilistic interpretation:

> Max Born: The wave function $\Psi(\mathbf{x}, t)$ doesn't tell us how much of the particle is at position $\mathbf{x}$ at time t, but rather it determines the probability that upon measurement taken at time t we would find the particle at position $\mathbf{x}$.

To make this precise, we use an infinitesimal volume element $d^3\mathbf{x}$ centered around some arbitrary point $\mathbf{x}$. According to Born, the probability dP of finding the particle within the volume element $d^3\mathbf{x}$ at time t is

$$dP = |\Psi(\mathbf{x}, t)|^2\, d^3\mathbf{x}. \tag{3.4.1}$$

Note that the probability is proportional to the norm *squared* of the wave function. Born originally suggested that the probability would be proportional to the norm itself but then changed his mind; the latter proposal was more parallel to the situation in classical physics in which the energy density of a wave is typically proportional to the square of its amplitude. Consistency requires that the total probability of finding the particle *somewhere* in the whole space is unity. Thus, the integral of dP over all of space must give one

$$\int_{\text{all space}} d^3\mathbf{x}\, |\Psi(\mathbf{x},t)|^2 = 1. \tag{3.4.2}$$

This is the *normalization condition* on the wave function. The Schrödinger equation is linear, and therefore it does not fix the scale of the wave function: given a solution, you can multiply it by any number, and you still have a solution. On the other hand, the normalization condition fixes the scale of the wave function; if a wave function Ψ satisfies the condition, 2Ψ, for example, will not.

Since the normalization condition must hold for all time, we will have to explore the consistency of this constraint with time evolution. We will see there is no problem and while doing so learn more about the Schrödinger equation. Born's proposal is presently considered an axiom of quantum mechanics. We will have a first look at the axioms of quantum mechanics in section 5.3 and a more detailed discussion of them in section 16.6.

3.5 Normalization and Time Evolution

We have seen that the wave function $\Psi(x,t)$ that describes the quantum mechanics of a particle of mass m moving in a one-dimensional potential $V(x,t)$ satisfies the Schrödinger equation:

$$i\hbar\frac{\partial\Psi(x,t)}{\partial t} = \Big(-\frac{\hbar^2}{2m}\frac{\partial^2}{\partial x^2} + V(x,t)\Big)\Psi(x,t). \tag{3.5.1}$$

The operator acting on the wave function on the right-hand side is called the Hamiltonian $\hat{H}$, and thus the Schrödinger equation is more briefly written as

$$i\hbar\,\frac{\partial\Psi(x,t)}{\partial t} = \hat{H}\,\Psi(x,t). \tag{3.5.2}$$

The quantity $dP = |\Psi(x,t)|^2 dx$ is the probability of finding the particle in the interval dx centered on x at time t; this is the one-dimensional version of Born's rule (3.4.1). The probabilities of finding the particle at all possible points must add up to one. Therefore, the one-dimensional version of (3.4.2) is

$$\int_{-\infty}^{\infty} |\Psi(x,t)|^2\, dx = 1. \tag{3.5.3}$$

Any Ψ for which the above left-hand side is a finite number is said to be square integrable. We will try to understand how this equation is compatible with the time evolution prescribed by the Schrödinger equation. First, however, let us examine what kind of conditions are required from wave functions in order for equation (3.5.3) to hold.

Suppose the wave function Ψ has well-defined limits as $x \to \pm\infty$. If those limits are different from zero, the limits of $|\Psi|^2$ as $x \to \pm\infty$ are also different from zero, and the integral around infinity would produce an infinite result, inconsistent with the claim that the total integral is one. Therefore, the limits should be zero:

$$\boxed{\lim_{x\to\pm\infty} \Psi(x,t) = 0.} \tag{3.5.4}$$

It is in principle possible to have square-integrable functions that have no well-defined limit at infinity, but such cases do not seem to appear in practice, so we will assume that (3.5.4) holds. It would also be reasonable to assume that the spatial derivative of Ψ vanishes as $x \to \pm\infty$, but as we will see soon, it suffices to assume that the limit of the spatial derivative of Ψ is bounded:

$$\boxed{\lim_{x\to\pm\infty} \left|\frac{\partial\Psi(x,t)}{\partial x}\right| < \infty.} \tag{3.5.5}$$

The absolute value symbol is needed to prevent the derivative from becoming infinitely large and negative. With this constraint, we are stating that the limit of the derivative is just some (finite) number. Generally, the number is zero.

We have emphasized before that the overall numerical factor multiplying the wave function is not physical. But equation (3.5.3) seems to be in conflict with this: if a given Ψ satisfies it, the presumed equivalent 2Ψ will not! To make precise sense of probabilities, it is *convenient* to work with normalized wave functions. This is not strictly necessary, however, as we explain now. Since time plays no role in this argument, assume in all that follows that the equations refer to some fixed but arbitrary time t_0. Suppose you have a wave function Ψ such that

$$\int_{-\infty}^{\infty} dx\,|\Psi|^2 = \mathcal{N} \neq 1. \tag{3.5.6}$$

We now claim that the probability dP of finding the particle in the interval dx about x is given by

$$dP = \frac{1}{\mathcal{N}}|\Psi|^2\,dx. \tag{3.5.7}$$

This is consistent because, independent of the value of $\mathcal{N}$, the probabilities are correctly normalized:

$$\int dP = \frac{1}{\mathcal{N}}\int_{-\infty}^{\infty} dx\,|\Psi|^2 = \frac{1}{\mathcal{N}}\mathcal{N} = 1. \tag{3.5.8}$$

Note that dP in (3.5.7) does not change when Ψ is multiplied by any number c because $\Psi(x) \to c\Psi(x)$ results in

$$|\Psi(x)|^2 \to |c|^2|\Psi(x)|^2 \quad \text{and} \quad \mathcal{N} \to |c|^2\mathcal{N}. \tag{3.5.9}$$

This makes it clear that the overall scale of Ψ contains no physics. As long as $\int|\Psi|^2dx < \infty$, the wave function is said to be **normalizable**, or **square integrable**. By adjusting the overall coefficient of Ψ, we can then make it **normalized**. Indeed, again assuming (3.5.6), the new wave function Ψ' defined by

$$\Psi' = \frac{1}{\sqrt{\mathcal{N}}}\Psi \tag{3.5.10}$$

is properly normalized. We can verify this fact directly:

$$\int_{-\infty}^{\infty} dx|\Psi'|^2 = \frac{1}{\mathcal{N}}\int_{-\infty}^{\infty}|\Psi|^2 dx = 1. \tag{3.5.11}$$

We sometimes work with wave functions for which the normalization integral (3.5.3) is infinite and thus *cannot* be normalized. Such wave functions can be very useful. In fact, the de Broglie plane wave $\Psi = \exp(ikx - i\omega t)$ for a free particle cannot be normalized since $|\Psi|^2 = 1$. This means that $\exp(ikx - i\omega t)$, holding for all $x \in (-\infty, \infty)$, does not truly represent a particle. To construct a square-integrable wave function, we can use a superposition of plane waves. It is indeed a pleasant surprise that the superposition of *infinitely* many non–square integrable waves is often square integrable!

Exercise 3.6. *Explain why a normalized wave function Ψ remains normalized after multiplication by a constant phase factor $e^{i\alpha}$, with α a real constant.*

3.6 The Wave Function as a Probability Amplitude

Let's begin with a normalized wave function $\Psi(x, t_0)$ at some initial time t_0:

$$\int_{-\infty}^{\infty} \Psi^*(x, t_0)\Psi(x, t_0)\, dx = 1. \tag{3.6.1}$$

The wave function also satisfies the conditions (3.5.4) and (3.5.5) we have imposed at infinity but is otherwise completely arbitrary. Since the initial value $\Psi(x, t_0)$ and the Schrödinger equation determine $\Psi(x, t)$ for all times, we can ask:

$$\text{Does} \int_{-\infty}^{\infty} \Psi^*(x, t)\Psi(x, t)dx = 1 \text{ hold for all } t\,? \tag{3.6.2}$$

We define the **probability density** $\rho(x, t)$ from the relation $dP = \rho(x, t)dx$:

$$\rho(x, t) \equiv \Psi^*(x, t)\Psi(x, t) = |\Psi(x, t)|^2. \tag{3.6.3}$$

We are trying to evaluate the quantity $\mathcal{N}(t)$, defined as the integral of the probability density throughout space:

$$\mathcal{N}(t) \equiv \int_{-\infty}^{\infty} \rho(x, t)dx. \tag{3.6.4}$$

The normalization statement in (3.6.1) is an initial condition on the value of $\mathcal{N}$:

$$\mathcal{N}(t_0) = 1. \tag{3.6.5}$$

The condition that it remain normalized for all later times is $\mathcal{N}(t) = 1$. This would be guaranteed if we showed that for all times

$$\frac{d\mathcal{N}(t)}{dt} = 0. \tag{3.6.6}$$

We call this *conservation* of probability: the wave function can be used at all times to compute probabilities that, consistently, add up to one. Our goal is to understand how the Schrödinger equation ensures conservation of probability. For this we begin by computing the time derivative of $\mathcal{N}$. We assume that for the functions and integrals we work with we may freely exchange the order of differentiation and integration, making it possible to move time derivatives across the integral sign. We thus have

$$\frac{d\mathcal{N}(t)}{dt} = \int_{-\infty}^{\infty} \frac{\partial\rho(x,t)}{\partial t} dx = \int_{-\infty}^{\infty} \left(\frac{\partial\Psi^*}{\partial t} \Psi(x,t) + \Psi^*(x,t) \frac{\partial\Psi}{\partial t} \right) dx. \tag{3.6.7}$$

The derivatives of the wave function can be evaluated using the Schrödinger equation:

$$i\hbar \frac{\partial\Psi}{\partial t} = \hat{H}\Psi \implies \frac{\partial\Psi}{\partial t} = -\frac{i}{\hbar}\hat{H}\Psi \tag{3.6.8}$$

and its complex conjugate:

$$\frac{\partial\Psi^*}{\partial t} = \frac{i}{\hbar}(\hat{H}\Psi)^*. \tag{3.6.9}$$

To conjugate the left-hand side, we noted that the complex conjugate of the time derivative of Ψ is just the time derivative of the complex conjugate of Ψ. To conjugate the right-hand side, we simply added the star symbol to the product $\hat{H}\Psi$, without attempting to conjugate separately the operator $\hat{H}$ and Ψ. We now use these derivatives in (3.6.7) to find that

$$\frac{d\mathcal{N}(t)}{dt} = \int_{-\infty}^{\infty} \frac{\partial\rho(x,t)}{\partial t} dx = \frac{i}{\hbar}\Big(\int_{-\infty}^{\infty} (\hat{H}\Psi)^* \,\Psi dx - \int_{-\infty}^{\infty} \Psi^*(\hat{H}\Psi) dx \Big). \tag{3.6.10}$$

To show that the time derivative of $\mathcal{N}(t)$ vanishes, it suffices to show that

$$\boxed{\int_{-\infty}^{\infty} dx\, (\hat{H}\Psi)^* \,\Psi = \int_{-\infty}^{\infty} dx\, \Psi^*(\hat{H}\Psi).} \tag{3.6.11}$$

This equality should hold for arbitrary Ψ that satisfies conditions (3.5.4) and (3.5.5) at infinity. If $\hat{H}$ satisfies (3.6.11), probability is conserved. In fact, if $\hat{H}$ is a Hermitian operator, the condition above will be satisfied. An operator $\hat{M}$ is said to be a **Hermitian** operator if it satisfies the following:

$$\boxed{\text{Hermitian operator } \hat{M}: \quad \int_{-\infty}^{\infty} dx\, (\hat{M}\Psi_1)^* \,\Psi_2 = \int_{-\infty}^{\infty} dx\, \Psi_1^*(\hat{M}\Psi_2)} \tag{3.6.12}$$

for any two arbitrary wave functions Ψ_1 and Ψ_2 that satisfy conditions (3.5.4) and (3.5.5) at infinity. As you can see, a Hermitian operator can be switched from acting on the first wave function, under the conjugation, to acting on the second wave function. Of course, for a Hermitian operator this equation also holds when the two wave functions are the same. It follows that for $\hat{H}$ Hermitian, condition (3.6.11) holds. Hermitian operators are special; not all operators are Hermitian.

It is worth closing this circle of ideas by defining a new linear operator. Given a linear operator $\hat{T}$, we define its **Hermitian conjugate** $\hat{T}^\dagger$ via the following relation:

$$\int_{-\infty}^{\infty} dx\,(\hat{T}^\dagger\Psi_1)^*\,\Psi_2 = \int_{-\infty}^{\infty} dx\,\Psi_1^*\,(\hat{T}\Psi_2). \tag{3.6.13}$$

The wave functions Ψ_1 and Ψ_2 here are arbitrary, apart from satisfying our conditions at infinity. Since $\hat{T}$ is presumed known, the operator $\hat{T}^\dagger$ is calculated by starting from the right-hand side and trying to recast the expression with no operator acting on Ψ_2, at which point it becomes possible to identify $\hat{T}^\dagger$. Note that an operator $\hat{T}$ is Hermitian if it is equal to its Hermitian conjugate:

$$\hat{T} \text{ is Hermitian if } \quad \hat{T}^\dagger = \hat{T}. \tag{3.6.14}$$

Hermitian operators are very important in quantum mechanics. They can be shown to have real eigenvalues and provide orthonormal eigenstates that are a basis for the state space. It turns out that observables in quantum mechanics are represented by Hermitian operators, and the possible measured values of those observables are the eigenvalues. At this point, however, our quest to show that the normalization of the wave function is preserved under time evolution has come down to showing that the Hamiltonian operator is Hermitian.

Exercise 3.7. *Consider the operator $\hat{T} = ia$, which multiplies any function by the constant ia, with a assumed real. Use (3.6.13) to show that $\hat{T}^\dagger = -ia$.*

Exercise 3.8. *Consider the operator $\hat{T} = \frac{\partial}{\partial x}$, which acting on a function takes its x derivative. Use (3.6.13) with wave functions that vanish at $x = \pm\infty$ to show that $\hat{T}^\dagger = -\hat{T}$.*

3.7 The Probability Current

We will now demonstrate the existence of a probability current intimately related to the previously defined probability density. The properties of this current at infinity guarantee the Hermiticity of the Hamiltonian. For this purpose let's take a closer look at the integrand of equation (3.6.10). Using the explicit expression for the Hamiltonian and recalling that the potential $V(x,t)$ is real, we find that

$$\begin{aligned}\frac{\partial\rho}{\partial t} &= \frac{i}{\hbar}\big((\hat{H}\Psi)^*\,\Psi - \Psi^*(\hat{H}\Psi)\big)\\ &= \frac{i}{\hbar}\Big[\Big(-\frac{\hbar^2}{2m}\frac{\partial^2\Psi}{\partial x^2} + V(x,t)\Psi\Big)^*\,\Psi - \Psi^*\Big(-\frac{\hbar^2}{2m}\frac{\partial^2\Psi}{\partial x^2} + V(x,t)\Psi\Big)\Big]\\ &= \frac{i}{\hbar}\Big[-\frac{\hbar^2}{2m}\Big(\frac{\partial^2\Psi^*}{\partial x^2}\Psi - \Psi^*\frac{\partial^2\Psi}{\partial x^2}\Big) + V(x,t)\Psi^*\Psi - \Psi^* V(x,t)\Psi\Big].\end{aligned} \tag{3.7.1}$$

The terms involving the potential cancel out and we get

$$\frac{\partial\rho}{\partial t} = \frac{\hbar}{2im}\Big(\frac{\partial^2\Psi^*}{\partial x^2}\Psi - \Psi^*\frac{\partial^2\Psi}{\partial x^2}\Big). \tag{3.7.2}$$

The Hermiticity of $\hat{H}$ requires the integral of the right-hand side to be zero. This can be guaranteed if the right-hand side is a total derivative. Indeed it is, as you can confirm immediately by evaluation:

$$\frac{\partial \rho}{\partial t} = \frac{\partial}{\partial x}\left[\frac{\hbar}{2im}\left(\frac{\partial \Psi^*}{\partial x}\Psi - \Psi^*\frac{\partial \Psi}{\partial x}\right)\right]. \tag{3.7.3}$$

We now rewrite this equation as follows:

$$\begin{aligned}\frac{\partial \rho}{\partial t} &= -\frac{\partial}{\partial x}\left[\frac{\hbar}{2im}\left(\Psi^*\frac{\partial \Psi}{\partial x} - \frac{\partial \Psi^*}{\partial x}\Psi\right)\right] = -\frac{\partial}{\partial x}\left[\frac{\hbar}{2im}\,2i\,\mathrm{Im}\left(\Psi^*\frac{\partial \Psi}{\partial x}\right)\right] \\ &= -\frac{\partial}{\partial x}\left[\frac{\hbar}{m}\mathrm{Im}\left(\Psi^*\frac{\partial \Psi}{\partial x}\right)\right],\end{aligned} \tag{3.7.4}$$

where we used that $z - z^* = 2i\,\mathrm{Im}(z)$. The result obtained so far implies that

$$\frac{\partial \rho}{\partial t} + \frac{\partial}{\partial x}\left[\frac{\hbar}{m}\mathrm{Im}\left(\Psi^*\frac{\partial \Psi}{\partial x}\right)\right] = 0. \tag{3.7.5}$$

As we will discuss further below, this equation actually encodes the statement of probability conservation. The equation is of the type known as a conservation equation:

$$\frac{\partial \rho}{\partial t} + \frac{\partial J}{\partial x} = 0, \tag{3.7.6}$$

where $J(x,t)$ is the current associated with the density ρ. Since our ρ is a probability density, we have therefore identified a probability current:

$$\boxed{J(x,t) \equiv \frac{\hbar}{m}\mathrm{Im}\left(\Psi^*\frac{\partial \Psi}{\partial x}\right) = \frac{\hbar}{2im}\left(\Psi^*\frac{\partial \Psi}{\partial x} - \Psi\frac{\partial \Psi^*}{\partial x}\right).} \tag{3.7.7}$$

There is just one component for this current since the particle moves in one dimension. The units of J are one over time, or probability per unit time, as we now verify.

For one spatial dimension, $[\Psi] = L^{-1/2}$, with L denoting units of length. This comes from the requirement that the spatial integral $\int dx|\Psi|^2$ is unit-free. (When working in d spatial dimensions, the wave function will have units of $L^{-d/2}$.) We then have

$$\left[\Psi^*\frac{\partial \Psi}{\partial x}\right] = \frac{1}{\sqrt{L}}\frac{1}{L}\frac{1}{\sqrt{L}} = \frac{1}{L^2}, \quad [\hbar] = \frac{ML^2}{T}, \quad \left[\frac{\hbar}{m}\right] = \frac{L^2}{T}. \tag{3.7.8}$$

Combining the first and third equalities, we have $[J] = 1/T$, as claimed.

We can finally show very easily that the time derivative of $\mathcal{N}$ is zero. This was, after all, our goal from the beginning. Indeed, using (3.7.6) we have

$$\frac{d\mathcal{N}}{dt} = \int_{-\infty}^{\infty} dx\,\frac{\partial \rho}{\partial t} = -\int_{-\infty}^{\infty} \frac{\partial J}{\partial x}dx = -(J(\infty,t) - J(-\infty,t)). \tag{3.7.9}$$

The derivative of $\mathcal{N}$ vanishes if the probability current vanishes at infinity. Recalling the explicit formula (3.7.7) for the current, we see that it vanishes because we restrict ourselves to wave functions that vanish and spatial derivatives that remain bounded as $x \to \pm\infty$. We therefore have, as we wanted to show,

$$\boxed{\frac{d\mathcal{N}}{dt}=0.} \tag{3.7.10}$$

Summarizing, we went from probability conservation to the condition of having a Hermitian $\hat{H}$. The Hermiticity of $\hat{H}$ was guaranteed because we could define a probability current associated with the probability density, and this current vanishes at infinity. The vanishing of the current at infinity is in fact the main justification for conditions (3.5.4) and (3.5.5). Note that our result (3.7.10), together with (3.6.10), indeed implies the Hermiticity of $\hat{H}$.

To illustrate how the probability conservation equation (3.7.6) works more generally in one dimension, focus on a segment $x \in [a, b]$. Then the probability P_{ab} to find the particle in the segment $[a, b]$ is given by

$$P_{ab}=\int_a^b \rho(x,t)\,dx. \tag{3.7.11}$$

Taking a time derivative and, as before, using current conservation, we get

$$\frac{dP_{ab}}{dt}=-\int_a^b \frac{\partial J(x,t)}{\partial x}dx=-J(b,t)+J(a,t). \tag{3.7.12}$$

This is the expected result. If the amount of probability in the region $[a, b]$ changes in time, it must be due to the probability current flowing in or out at the edges of the interval. Assuming the currents at $x=b$ and at $x=a$ are positive, we note that probability is flowing out at $x=b$ and is coming in at $x=a$. The signs on the above right-hand side correctly reflect the effect of these flows on the rate of change of the total probability inside the segment.

3.8 Probability Current in Three Dimensions and Current Conservation

The determination of the probability current $\mathbf{J}$ for a particle moving in three dimensions follows the route taken before, but we use the three-dimensional version of the Schrödinger equation. After some work (problem 3.6), the probability density and the current are determined to be

$$\boxed{\rho(\mathbf{x},t) \;=\; |\Psi(\mathbf{x},t)|^2, \qquad \mathbf{J}(x,t) \;=\; \frac{\hbar}{m}\,\mathrm{Im}\,(\Psi^*\nabla\Psi).} \tag{3.8.1}$$

This is natural: the partial derivative in the one-dimensional current becomes the gradient in the three-dimensional current. The above quantities satisfy the conservation equation

$$\frac{\partial\rho}{\partial t}+\nabla\cdot\mathbf{J} \;=\; 0. \tag{3.8.2}$$

In three spatial dimensions, the wave function has units $[\Psi]=L^{-\frac{3}{2}}$, and the units of $\mathbf{J}$ are quickly determined to be probability per unit time and per unit area:

$$[\Psi^* \nabla \Psi] = \frac{1}{L^4}, \quad \left[\frac{\hbar}{m}\right] = \frac{L^2}{T} \quad \Rightarrow \quad [\mathbf{J}] = \frac{1}{TL^2}. \tag{3.8.3}$$

The conservation equation (3.8.2) is particularly clear in integral language. Consider a fixed region V of space and the probability $Q_V(t)$ of finding the particle inside the region:

$$Q_V(t) = \int_V \rho(\mathbf{x}, t)\, d^3\mathbf{x}. \tag{3.8.4}$$

The time derivative of the probability is then calculated using the conservation equation

$$\frac{dQ_V}{dt} = \int_V \frac{\partial \rho}{\partial t} d^3\mathbf{x} = -\int_V \nabla \cdot \mathbf{J}\, d^3\mathbf{x}. \tag{3.8.5}$$

Finally, using Gauss' divergence theorem we find that

$$\frac{dQ_V}{dt} = -\int_S \mathbf{J} \cdot d\mathbf{a}, \tag{3.8.6}$$

where S is the boundary of the volume V, and $d\mathbf{a}$ is the infinitesimal area element with outwardly oriented normal. The interpretation here is clear: the probability of finding the particle inside V changes in time only if there is flux of the probability current across the boundary of the region. Note that in the one-dimensional analog (3.7.12) the flux involves no integral!

When the volume extends throughout space, the boundary is at infinity, which in spherical coordinates is placed at $r \to \infty$. If the current flux at infinity is zero, the Hamiltonian is Hermitian, probability is conserved, and all is good. While zero wave function and bounded derivatives at infinity imply zero current and zero flux at infinity, all we actually *need* is a current that vanishes sufficiently fast as $r \to \infty$ so that its flux across ever-growing spheres vanishes in the limit $r \to \infty$. A precise analysis is sometimes needed, and we will do it when required.

Our probability density, probability current, and probability conservation are, for probability, the exact analogs of the familiar electric charge density, electric current density, and charge conservation. In electromagnetism charges flow; in quantum mechanics probability flows. The terms of the correspondence are summarized by the following table.

	Electromagnetism	Quantum mechanics
ρ	Charge density	Probability density
Q_V	Charge in a volume V	Probability of finding particle in V
$\mathbf{J}$	Current density	Probability current density

Exercise 3.9. *Consider the plane wave $\Psi(\mathbf{x}, t) = Ae^{i\mathbf{k}\cdot\mathbf{x}}e^{-i\omega t}$, where A is a complex number. Show that the probability current is $\mathbf{J} = \frac{\hbar \mathbf{k}}{m}|A|^2$.*

Problems

Problem 3.1. *Plane wave for matter particles.*

This problem gives an alternative understanding of the construction of plane waves in section 3.1. Assume we want to represent the wave for a matter particle moving in the x-direction with momentum $p = \hbar k$. A reasonable guess for such a wave is

$$\Psi(x,t) = \cos(kx - \omega t) + \gamma \sin(kx - \omega t),$$

where γ is a constant. A physical requirement is that an arbitrary displacement of x or an arbitrary shift of t should not alter the character of the wave. Note that the effect of both of these shifts is to simply alter the phase of each of the sinusoids by the same amount. Therefore, under a change of phase by some finite θ (representing a shift in space and/or time), we require that

$$\cos(kx - \omega t + \theta) + \gamma \sin(kx - \omega t + \theta) = a\,[\cos(kx - \omega t) + \gamma \sin(kx - \omega t)],$$

for some constant a that may depend on θ. By solving the equations that follow from this requirement, find the two possible solutions for γ, one with a positive imaginary part and the other with a negative imaginary part. For each solution give γ and a and write Ψ, thus constructing two candidate solutions for a matter wave.

Problem 3.2. *Normalizability of wave functions.*

Which of the following wave functions are normalizable (explain briefly)?

1. $\Psi(x,t) = e^{i(kx-\omega t)}$, for all x and t.
2. $\Psi(x,t) = \cos kx\, e^{-i\omega t}$, for all x and t.
3. $\Psi(x,t) = \sin(\frac{\pi x}{L})e^{-i\omega t}$ for all t and $|x| < L$, and $\Psi(x,t) = 0$ for $|x| > L$.
4. $\Psi(x,t) = \left(1 - \exp(-\frac{|x|}{L})\right)e^{-i\omega t}$, for all x and t.
5. $\Psi(x,t) = \exp(-\frac{x^2}{L^2})\exp(-i\omega t)$, for all x and t.

Problem 3.3. *Electron wave function for the ground state of hydrogen.*

The wave function of an electron in the ground state of hydrogen at $t = 0$ is known to be

$$\Psi(\mathbf{x}, 0) = N e^{-r/a_0},$$

where $r = |\mathbf{x}|$ is the three-dimensional distance to the origin, where the proton is, and a_0 is the Bohr radius (about 52 pm).

1. Determine N (assumed real and positive) so that the integral of $|\Psi|^2$ over all space is equal to one, as required. [Useful integral: $\int_0^\infty dx\, e^{-x} x^n = n!$ (for $n \geq 0$). Recall that $0! = 1! = 1$.]
2. Find the approximate probability P_{b_0} that the electron is found inside a tiny sphere centered at the origin with radius $b_0 \ll a_0$. [Hint: the exact calculation is much harder than the approximate estimate.] What is the value of P_{b_0} for $b_0 = 1$ fm, the approximate size of the proton?

Problem 3.4. *Galilean invariance of the free Schrödinger equation.*

Show that the free-particle one-dimensional Schrödinger equation for the wave function $\Psi(x, t)$,

$$i\hbar\frac{\partial\Psi}{\partial t} = -\frac{\hbar^2}{2m}\frac{\partial^2\Psi}{\partial x^2},$$

is invariant under the Galilean transformations

$$x' = x - vt, \quad t' = t.$$

By this we mean that there is a $\Psi'(x', t')$ of the form

$$\Psi'(x', t') = f(x, t)\Psi(x, t),$$

where

1. $f(x, t)$ may involve x, t, $\hbar$, m, and v,
2. $f(x, t)$ does not depend on $\Psi(x, t)$,
3. $|f(x, t)| = 1$ so that $|\Psi'(x', t')| = |\Psi(x, t)|$,
4. $f(0, 0) = 1$ to fix an overall phase ambiguity,

and such that Ψ' satisfies the corresponding Schrödinger equation in primed variables:

$$i\hbar\frac{\partial\Psi'}{\partial t'} = -\frac{\hbar^2}{2m}\frac{\partial^2\Psi'}{\partial x'^2}.$$

Why is $|f(x, t)| = 1$ required?

1. We would like to find the function $f(x, t)$. Before we do so, we need to understand how derivatives in the primed coordinate system relate to derivatives in the unprimed coordinate system. Using the chain rule, write $\frac{\partial}{\partial x'}$ and $\frac{\partial}{\partial t'}$ in terms of $\frac{\partial}{\partial x}$ and $\frac{\partial}{\partial t}$.
2. To find the function $f(x, t)$, we demand that Ψ' obeys the Schrödinger equation. Recalling that Ψ also obeys the Schrödinger equation, we can obtain an equation in the following form:

$$\Big(\cdots\cdots\Big)\Psi + \Big(\cdots\cdots\Big)\frac{\partial\Psi}{\partial x} = 0.$$

   Find the missing terms above.
3. Derive an expression for $f(x, t)$ and write Ψ' in terms of Ψ. [Hint: The function $f(x, t)$ cannot depend on any observable of Ψ; it is a universal function that is used to transform *any* Ψ. Thus, if Ψ is a (single) plane wave, f cannot depend on its momentum or its energy.]
4. Consider the plane wave solution

$$\Psi(x, t) = Ae^{i(kx-\omega t)} = Ae^{\frac{i}{\hbar}\left(px - \frac{p^2}{2m}t\right)},$$

   where we can use $p = \hbar k$, and $E = \hbar\omega = \frac{p^2}{2m}$, as we are describing a nonrelativistic particle. Compute the transformation of Ψ to the S' frame and write out Ψ' as a function of x' and t'. What are the energy and momentum of the particle in the S' frame?

Problem 3.5. *Probability current in one dimension.*

1. Calculate the probability current $J(x)$ for $\Psi(x,0)=Ae^{\gamma x}$, where A is a complex constant, and γ is a real constant.
2. Calculate the probability current $J(x)$ for $\Psi(x,0)=N(x)e^{iS(x)/\hbar}$, where $N(x)$ and $S(x)$ are real.
3. Calculate the probability current $J(x)$ for $\Psi(x,0)=Ae^{ikx}+Be^{-ikx}$, where A, B are complex constants, and k is a real constant.

Problem 3.6. *Current conservation in three dimensions.*

We already derived the expression for the one-dimensional probability current $J(x,t)$ starting from $\rho(x,t)=|\Psi(x,t)|^2$ and using the one-dimensional Schrödinger equation to write $\partial_t\rho+\partial_x J=0$. Repeat the same steps starting from $\rho(\mathbf{x},t)=|\Psi(\mathbf{x},t)|^2$ and using the three-dimensional Schrödinger equation to derive the form of the probability current $\mathbf{J}(\mathbf{x},t)$ that should appear in the conservation equation $\partial_t\rho+\nabla\cdot\mathbf{J}=0$.

1. Use the Schrödinger equation and its complex conjugate to compute $\partial\rho/\partial t$.
2. Use the identity $\nabla\cdot(f\mathbf{A})=(\nabla f)\cdot\mathbf{A}+f(\nabla\cdot\mathbf{A})$, with f a scalar function and $\mathbf{A}$ a vector field, to compute $\nabla\cdot(f\nabla g-g\nabla f)$.
3. Use (1) and (2) to write $\partial\rho/\partial t$ as the divergence of a quantity.
4. Compare your result with the relation $\frac{\partial\rho}{\partial t}=-\nabla\cdot\mathbf{J}$ to identify the probability current $\mathbf{J}$.

Problem 3.7. *Time evolution of an overlap between two states (Merzbacher).*

Consider two *normalizable* wave functions $\Psi_1(x,t)$ and $\Psi_2(x,t)$, each describing the time evolution of a wave packet moving in an arbitrary potential. Both wave functions are solutions of the same Schrödinger equation. We define the overlap integral $\gamma(t)$ as

$$\gamma(t)=\int_{-\infty}^{\infty}\Psi_1^*(x,t)\Psi_2(x,t)dx.$$

At $t=0$ the value of $|\gamma(0)|$ is small. As the packets evolve and spread, we want to find out what will happen to $|\gamma(t)|$.

1. Let $\Psi\equiv\Psi_1+\Psi_2$ and compute $\int_{-\infty}^{\infty}|\Psi(x,t)|^2dx$. Now, let $\tilde{\Psi}\equiv\Psi_1+i\Psi_2$ and compute $\int_{-\infty}^{\infty}|\tilde{\Psi}(x,t)|^2dx$. Knowing that Ψ_1 and Ψ_2 are normalizable, what do you learn about the real and imaginary parts of $\gamma(t)$?
2. Compute the time rate of change of $\gamma(t)$ directly, using the Schrödinger equation for Ψ_1 and Ψ_2 to evaluate appropriate time derivatives.

Problem 3.8. *Probability density and current for an approximate solution.*

For a particle of mass m and *total* energy E moving in a potential $V(x)$, a solution of the Schrödinger equation takes the form

$$\Psi(x,t) = \psi(x)\exp\Big(-\frac{iEt}{\hbar}\Big). \tag{1}$$

1. In the classical picture, the value $p(x)$ of the momentum the particle would have at x is obtained from energy conservation:

$$\frac{(p(x))^2}{2m}+V(x)=E.$$

The Schrödinger equation for $\Psi(x,t)$ implies an equation for $\psi(x)$. Find this equation and write it in the form

$$\hat{p}^2\,\psi(x)\;=\;(\cdots)\psi(x),\tag{2}$$

where $\hat{p}$ is the momentum operator, and the expression represented in dots, that you should calculate, can be written in terms of $p(x)$ and possibly some constants.

2. When the potential $V(x)$ is a slowly varying function of position, one can show that an approximate solution of equation (2) takes the form

$$\psi(x)\simeq\frac{A}{\sqrt{p(x)}}\exp\Big(\frac{i}{\hbar}\int_{x_0}^{x}p(x')dx'\Big),\tag{3}$$

where A is a (possibly complex) normalization constant, x_0 is a constant of integration, and we assume $p(x)>0$. Calculate the probability density $\rho(x,t)$ associated with this approximate solution. Simplify your answer as much as possible. (Do not attempt to normalize the solution.)

3. Calculate the probability current $J(x,t)$ associated with this solution. The final answer is quite simple. [Hint: you may find it useful to calculate x derivatives in the form $\psi'=(\cdots)\psi$.]

4. Explain how the differential statement for probability conservation is indeed satisfied.

Problem 3.9. *Probability currents in three-dimensional scattering.*

In the elastic scattering of particles in three-dimensional space off a spherically symmetric target at the origin, the wave function for large values r of the radial distance takes the approximate form

$$\Psi(\mathbf{x})=\Psi_1(\mathbf{x})+\Psi_2(\mathbf{x}),\;\;\text{with}\;\;\Psi_1(\mathbf{x})=e^{ikz},\;\;\text{and}\;\;\Psi_2(\mathbf{x})=\frac{f_k(\theta)}{r}e^{ikr}.$$

The time dependence $e^{-iEt/\hbar}$, with E the energy $\hbar^2k^2/(2m)$ of the particles, has been suppressed because it plays no role here. The term Ψ_1 represents the incoming particles, moving in the $+z$ direction. The Ψ_2 term represents the amplitude for particles moving radially out—the scattered particles. This amplitude depends on the polar angle θ but is ϕ independent; $f_k(\theta)$ is a complex function of θ that carries the information about the scattering. Here, $z=r\cos\theta$.

Recall that the probability current associated with a wave function is $\mathbf{J}=\frac{\hbar}{m}\text{Im}\,(\Psi^*\nabla\Psi)$. For our Ψ, this gives $\mathbf{J}=\mathbf{J}_1+\mathbf{J}_2+\mathbf{J}_{12}$, where

$$\mathbf{J}_1=\frac{\hbar}{m}\text{Im}\,(\Psi_1^*\nabla\Psi_1),\;\mathbf{J}_2=\frac{\hbar}{m}\text{Im}\,(\Psi_2^*\nabla\Psi_2),\;\mathbf{J}_{12}=\frac{\hbar}{m}\text{Im}\,(\Psi_1^*\nabla\Psi_2+\Psi_2^*\nabla\Psi_1)$$

represent the contribution to the probability current from the plane wave ($\mathbf{J}_1$), the contribution from the spherical waves ($\mathbf{J}_2$), and an interference term ($\mathbf{J}_{12}$).

1. Calculate the probability current $\mathbf{J}_1$ and the total flux Φ_1 of this current over a large sphere of radius R centered at the origin $r=0$.
2. Calculate the radial component $\hat{\mathbf{r}} \cdot \mathbf{J}_2$ of the probability current $\mathbf{J}_2$. Here $\hat{\mathbf{r}}$ is the radial unit vector. Use this to calculate the flux Φ_2 of this current over a radius R sphere centered at the origin in the limit as $R \to \infty$. Your answer should be left as an integral of the form $\int_0^\pi d\theta \cdots$.
3. In preparation for the last part of this problem, compute $\hat{\mathbf{r}} \cdot \nabla \left(e^{ikz}\right)$, writing your answer in terms of r, k, and θ (and not z).
4. Calculate the radial component $\hat{\mathbf{r}} \cdot \mathbf{J}_{12}$ of the interference term, keeping only the leading part in $1/r$ (i.e., ignore $1/r^2$ terms and higher). Show that

$$\hat{\mathbf{r}} \cdot \mathbf{J}_{12} \simeq \frac{\hbar k}{2mr}(1+\cos\theta)\left[f_k(\theta)e^{ikr(1-\cos\theta)} + f_k^*(\theta)e^{-ikr(1-\cos\theta)}\right]. \tag{1}$$

Calculating the flux of $\mathbf{J}_{12}$ over the large sphere is delicate. We will leave that work for problem 30.2. The end result is the so-called *optical theorem.*

4 Wave Packets, Uncertainty, and Momentum Space

We use the Fourier theorem to study wave packets and derive the relation $\Delta x \Delta k \sim 1$ with Δx, the width of the packet and Δk the width of the wave number distribution. With the identification of momentum with $\hbar k$ and the interpretation of widths as uncertainties, we arrive at the Heisenberg uncertainty product $\Delta x \Delta p \approx \hbar$. We discuss the time evolution of free wave packets and estimate the effect of shape distortion. Fourier's theorem furnishes a momentum-dependent function $\Phi(p)$, an alternative but equivalent representation of the position-dependent wave function $\Psi(x)$. Together with Plancherel's identity, one is led to a probabilistic interpretation for $\Phi(p)$ and the idea of momentum space.

4.1 Wave Packets and Uncertainty

A wave packet can be built as a superposition of plane waves e^{ikx} with various values of the wave number k. This means various values of the wavelength $2\pi/k$. In quantum mechanics this also means various values of the momentum $\hbar k$. A general wave packet at $t=0$ is of the form

$$\Psi(x,0) = \frac{1}{\sqrt{2\pi}} \int_{-\infty}^{\infty} \Phi(k) e^{ikx} dk. \tag{4.1.1}$$

This is the Fourier representation of the wave packet. It gives $\Psi(x,0)$ as a superposition of plane waves with different momentum, weighted by the **Fourier transform** $\Phi(k)$. If $\Phi(k)$ is known, $\Psi(x,0)$ is calculated by doing the above integral. If we know $\Psi(x,0)$, then $\Phi(k)$ is calculable using the inverse Fourier transform:

$$\Phi(k) = \frac{1}{\sqrt{2\pi}} \int_{-\infty}^{\infty} \Psi(x,0) e^{-ikx} dx. \tag{4.1.2}$$

The two equations above are structurally similar. Our goal here is to understand how the uncertainties associated with $\Psi(x,0)$ and $\Phi(k)$ are related. As we will make more precise below, the uncertainty associated with $\Psi(x,0)$ arises because $|\Psi(x,0)|^2$ is nonvanishing for extended domains on the x-axis, and thus the position of the quantum particle is uncertain. Since $\Phi(k)$ is nonvanishing for a range of k, the wave packet receives contributions from a range of momenta, making the momentum of the particle uncertain.

As a preparatory step, let us consider the question of reality. We ask: What is the condition on $\Phi(k)$ that guarantees the reality of $\Psi(x,t)$? The answer is quite simple:

$$\boxed{\Psi(x,0) \text{ is real if and only if } \Phi^*(-k)=\Phi(k).} \tag{4.1.3}$$

To prove this, begin by complex conjugating the expression (4.1.1) for $\Psi(x,0)$:

$$\Psi^*(x,0)=\frac{1}{\sqrt{2\pi}}\int_{-\infty}^{\infty}\Phi^*(k)e^{-ikx}dk=\frac{1}{\sqrt{2\pi}}\int_{-\infty}^{\infty}\Phi^*(-k)e^{ikx}dk. \tag{4.1.4}$$

In the second step, we let $k\to -k$ in the integral, which is allowed because we are integrating over *all* k. The two sign flips, one from the measure dk and one from switching the limits of integration, cancel each other out. If $\Phi^*(-k)=\Phi(k)$, then

$$\Psi^*(x,0)=\frac{1}{\sqrt{2\pi}}\int_{-\infty}^{\infty}\Phi(k)e^{ikx}dk=\Psi(x,0), \tag{4.1.5}$$

as we wanted to check. If, on the other hand, we know that $\Psi(x,0)$ is real, then the equality of Ψ^* and Ψ gives:

$$\frac{1}{\sqrt{2\pi}}\int_{-\infty}^{\infty}\Phi^*(-k)e^{ikx}dk=\frac{1}{\sqrt{2\pi}}\int_{-\infty}^{\infty}\Phi(k)e^{ikx}dk. \tag{4.1.6}$$

This is equivalent to

$$\frac{1}{\sqrt{2\pi}}\int_{-\infty}^{\infty}[\Phi^*(-k)-\Phi(k)]e^{ikx}dk=0. \tag{4.1.7}$$

This equation actually means that the object in between the brackets must vanish. Indeed, the integral is computing a Fourier transform of that object, and it tells us that it is zero. But a function with a zero Fourier transform must be zero itself, by the Fourier inversion formula. Therefore, reality implies $\Phi^*(-k)=\Phi(k)$, as we wanted to show.

Condition (4.1.3) is simple to understand. Consider two exponentials e^{ikx} and e^{-ikx} and their superposition with coefficients a and b:

$$a\,e^{ikx}+b\,e^{-ikx}. \tag{4.1.8}$$

This sum is real if $b=a^*$:

$$a\,e^{ikx}+a^*\,e^{-ikx}, \tag{4.1.9}$$

since the second term is then the complex conjugate of the first. We therefore see that, for reality, the coefficient multiplying the exponential with wave number k and the coefficient multiplying the exponential with wave number $(-k)$ are complex conjugates of each other. This is the content of (4.1.3). If we start with a real wave function $\Psi(x,0)$, the Fourier transform $\Phi(k)$ satisfies the stated condition. If, however, we start with an arbitrary $\Phi(k)$, the wave function $\Psi(x,0)$ need not be real.

Let us now begin our analysis of uncertainty. We start, however, without making reference to quantum mechanics; we have a function and its Fourier transform, and our

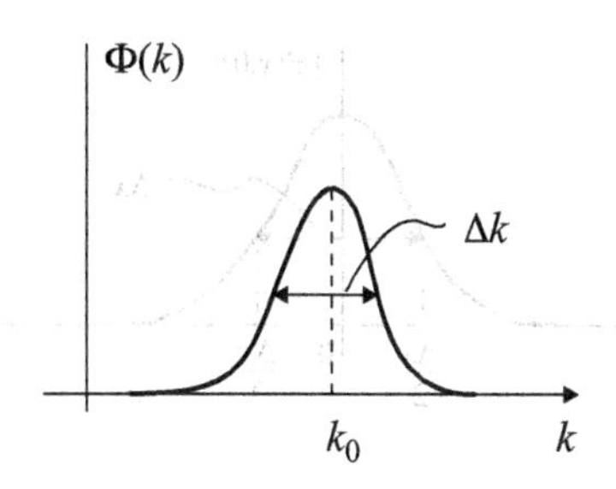

Figure 4.1
A $\Phi(k)$ that is centered about $k=k_0$ and has width Δk.

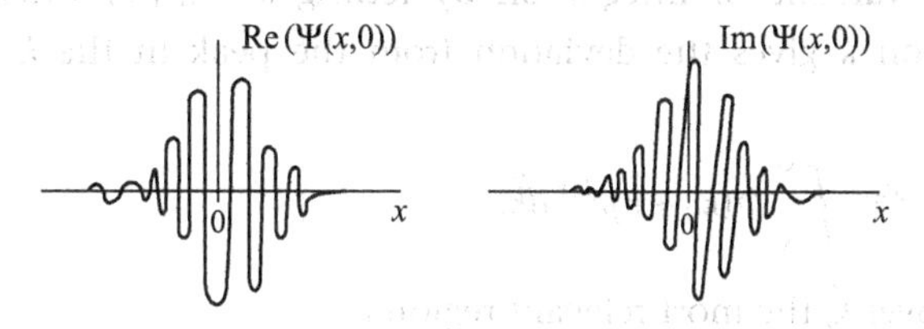

Figure 4.2
The real and imaginary parts of $\Psi(x,0)$.

first objective is to relate the relevant *widths* in these objects. Let us therefore consider a function $\Phi(k)$ that we will assume real, symmetric about a maximum at $k=k_0$, and of width Δk, as shown in figure 4.1. The name width for Δk is appropriate, as this represents the range of wave numbers that contribute to the wave packet. We will not attempt to define the width precisely here. We could take Δk to be defined, for example, as the distance between the points where $\Phi(k)$ is half of the peak value. A precise choice will not be required.

With $\Phi(k)$ so defined, the resulting wave function $\Psi(x,0)$ is centered around $x=0$. This follows directly from the stationary phase argument of section 2.8 applied to (4.1.1), as you should confirm. The wave function $\Psi(x,0)$ is not real because the condition $\Phi^*(-k)=\Phi(k)$ is not satisfied. This condition states that whenever Φ is nonzero for some k, it must also be nonzero for $-k$. This is not true for our chosen $\Phi(k)$: there is a bump around k_0, but there is no corresponding bump around $-k_0$. Therefore, $\Psi(x,0)$ will have both a real and an imaginary part. Since $\Psi(x,0)$ peaks at $x=0$, we can expect that, typically, both real and imaginary parts are centered and peak around $x=0$, as shown in figure 4.2. At any rate, $|\Psi(x,0)|$ should peak at $x=0$, with some width Δx, as shown in figure 4.3. Again, we do not define this width precisely. Our goal is to find a relation between the widths Δx and Δk in the spatial and k distributions, respectively.

To find such relation, consider again the integral representation for $\Psi(x,0)$:

$$\Psi(x,0)=\frac{1}{\sqrt{2\pi}}\int_{-\infty}^{\infty}\Phi(k)e^{ikx}dk, \tag{4.1.10}$$

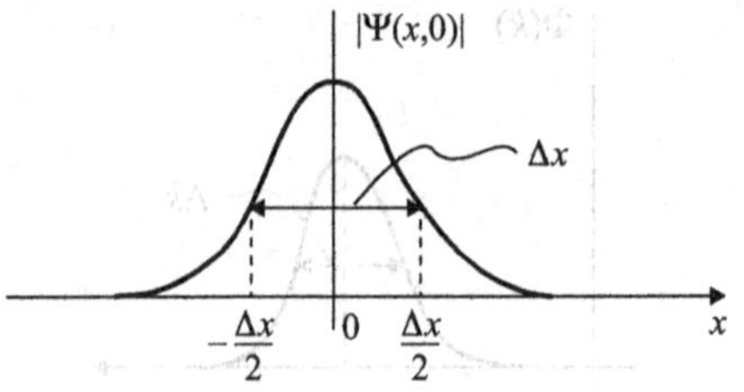

Figure 4.3
The $|\Psi(x,0)|$ associated with $\Phi(k)$ in figure 4.1. $|\Psi(x,0)|$ peaks around $x=0$ with width Δx.

and change the variable of integration by letting $k=k_0+\tilde{k}$, where the new variable of integration $\tilde{k}$ gives the deviation from the peak in the k distribution. We then have

$$\Psi(x,0)=\frac{1}{\sqrt{2\pi}}e^{ik_0x}\int_{-\infty}^{\infty}\Phi(k_0+\tilde{k})e^{i\tilde{k}x}\,d\tilde{k}. \tag{4.1.11}$$

As we integrate over $\tilde{k}$, the most relevant region is

$$\tilde{k}\in\left[-\tfrac{\Delta k}{2},\,\tfrac{\Delta k}{2}\right] \tag{4.1.12}$$

because this is where $\Phi(k_0+\tilde{k})$ is large. As we sweep this region, the phase $\tilde{k}x$ in the exponential varies in the interval

$$\tilde{k}x\in\left[-\tfrac{\Delta k}{2}|x|,\,\tfrac{\Delta k}{2}|x|\right]. \tag{4.1.13}$$

The absolute value of x is there in order to have a proper interval in which the first entry is smaller than the second entry for any value of x. The total change in the phase of $e^{i\tilde{k}x}$ as $\tilde{k}$ varies over the relevant region is therefore $\Delta k\,|x|$. We will get a substantial contribution to the integral if the total phase change is small; if the phase change is large, the integral will get washed out. This is the same argument we used to motivate the stationary phase approximation: we are integrating the slowly varying $\Phi(k)$ against a phase factor, and we require the phase not to vary much over the relevant region of integration to get a contribution. A variation of order π is reasonably modest, and therefore, we get a significant contribution to the integral for values of x such that $\Delta k|x|\lesssim\pi$:

$$|\Psi(x,0)| \text{ is substantial only for } \Delta k\,|x|\lesssim\pi. \tag{4.1.14}$$

Since, by definition, $\Psi(x,0)$ is large only for $x\in(-\frac{1}{2}\Delta x,\,\frac{1}{2}\Delta x)$, and the largest $|x|$ in this range is $\frac{1}{2}\Delta x$, we must have $\Delta k\frac{1}{2}\Delta x\approx\pi$. Factors of two, π, and so on are clearly unreliable in this argument, but nevertheless we record

$$\boxed{\Delta x\,\Delta k\approx 2\pi.} \tag{4.1.15}$$

This is what we wanted to show: the product of the width in the k distribution and the width in the spatial distribution is a constant of order one. This relation between the widths is not quantum mechanical; as you have seen, it follows from the properties of Fourier transforms.

Exercise 4.1. *Evaluate the norm*

$$\left|\int_{\alpha}^{\alpha+\Delta} e^{i\theta}\, d\theta\right|, \tag{4.1.16}$$

showing that the result only depends on Δ. Confirm that the integral of a pure phase over an excursion $\Delta > 0$ first peaks for $\Delta = \pi$. This is support for our using π as a reasonable phase excursion for a finite contribution in an integral.

The quantum mechanical version of the identity (4.1.15) requires two ingredients. First, we identify $\hbar k$ as the momentum p. Second, we think in terms of uncertainties rather than widths. Quantum mechanically, the particle has some probability to be found wherever $|\Psi(x,0)|^2$ is finite. The position where it will be found cannot be predicted, but the range of possible positions can be predicted and corresponds to Δx, which is better thought of as a position uncertainty. Similarly, we have built the wave packet as a superposition of waves with different momenta. If we measured the momentum a number of times, a range of values would be obtained. That range is $\hbar\Delta k$ and is identified with the momentum uncertainty:

$$\Delta p = \hbar\Delta k. \tag{4.1.17}$$

As a result, we can multiply the $\Delta x \Delta k \approx 1$ relation by $\hbar$ to find

$$\Delta x\, \Delta p \approx \hbar. \tag{4.1.18}$$

This is the rough version of the Heisenberg uncertainty product. The precise version requires defining the uncertainties Δx and Δp precisely for arbitrary wave packets, as we will do in section 5.5. Equipped with precise definitions, one can then show (section 15.2) the

$$\boxed{\text{Heisenberg uncertainty product:} \quad \Delta x\, \Delta p \geq \tfrac{\hbar}{2}.} \tag{4.1.19}$$

The product of uncertainties has in fact a lower bound. This inequality is sometimes called Heisenberg's *uncertainty principle*. In section 1.6 we used the uncertainty principle to explain how quantum mechanics resolves the difficulties of classical models of the atom.

Example 4.1. *A finite-extent step in k-space.*

Let $\Phi(k)$ be a finite step of width Δk and height $1/\sqrt{\Delta k}$, as shown in figure 4.4. We wish to find $\Psi(x,0)$ and estimate the value of Δx. Note that $\Psi(x,0)$ should be real because $\Phi^*(-k) = \Phi(k)$. From the integral representation,

$$\begin{aligned}\Psi(x,0) &= \frac{1}{\sqrt{2\pi}} \int_{-\frac{\Delta k}{2}}^{\frac{\Delta k}{2}} \frac{1}{\sqrt{\Delta k}} e^{ikx} dk = \frac{1}{\sqrt{2\pi\,\Delta k}}\, \frac{e^{ikx}}{ix}\bigg|_{-\frac{\Delta k}{2}}^{\frac{\Delta k}{2}} \\ &= \frac{1}{\sqrt{2\pi\,\Delta k}}\, \tfrac{2}{x} \sin \tfrac{\Delta k x}{2} = \sqrt{\frac{\Delta k}{2\pi}}\, \frac{\sin \frac{\Delta k x}{2}}{\frac{\Delta k x}{2}}.\end{aligned} \tag{4.1.20}$$

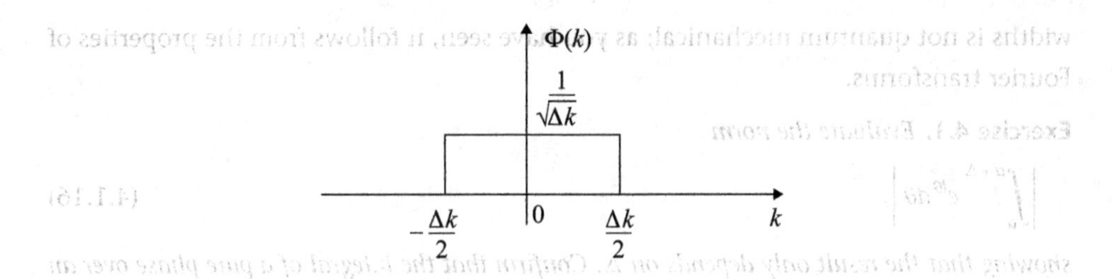

Figure 4.4
A momentum distribution $\Phi(k)$ that is a finite-extent step.

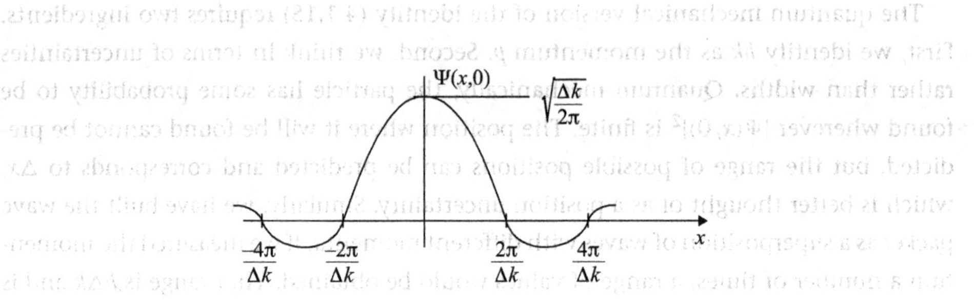

Figure 4.5
The $\Psi(x, 0)$ associated with the finite-step momentum distribution $\Phi(k)$ in figure 4.4.

Figure 4.5 shows $\Psi(x, 0)$, and we estimate $\Delta x \approx 4\pi/\Delta k$ as the width of the central lobe. This immediately gives $\Delta x \Delta k \approx 4\pi$, consistent with (4.1.15). □

4.2 Wave Packet Shape Changes

Back in section 2.8, we looked at the general solution of the Schrödinger equation in order to appreciate general features of the motion of a wave packet:

$$\Psi(x, t) = \frac{1}{\sqrt{2\pi}} \int_{-\infty}^{\infty} \Phi(k) e^{i(kx - \omega(k)t)} dk. \tag{4.2.1}$$

Under the assumption that $\Phi(k)$ peaks around some value $k = k_0$, we expanded the frequency $\omega(k)$ in a Taylor expansion around $k = k_0$. Keeping terms up to and including $(k - k_0)^2$, we have

$$\omega(k) = \omega(k_0) + (k - k_0) \left.\frac{d\omega}{dk}\right|_{k_0} + \frac{1}{2}(k - k_0)^2 \left.\frac{d^2\omega}{dk^2}\right|_{k_0}. \tag{4.2.2}$$

The second term on the right-hand side played a role in the determination of the group velocity. We now claim that the next term, with the second derivative of ω, is responsible for the shape distortion that occurs as time goes by. The intuition is clear: with $\frac{d\omega}{dk}$ as the group velocity, $\frac{d^2\omega}{dk^2}$ is the derivative of the group velocity. A derivative in the group velocity means that, in the wave packet, waves with different wave numbers move with different velocity. Thus, we get distortion.

To estimate the effect of distortion, we begin by evaluating the derivatives in (4.2.2):

$$\frac{d\omega}{dk} = \frac{dE}{dp} = \frac{p}{m} = \frac{\hbar k}{m}, \qquad \frac{d^2\omega}{dk^2} = \frac{\hbar}{m}. \tag{4.2.3}$$

Since all higher derivatives vanish, the expansion in (4.2.2) is actually exact as written. What kind of phase contribution are we neglecting when we ignore the last term in (4.2.2)? The frequency $\omega(k)$ enters as part of the integrand in (4.2.1) as follows:

$$e^{-i\omega(k)t} = e^{-i\omega(k_0)t}\, e^{-i(k-k_0)\left.\frac{d\omega}{dk}\right|_{k_0} t}\, e^{-i\frac{1}{2}(k-k_0)^2\frac{\hbar}{m}t}. \tag{4.2.4}$$

The contribution in question is the last exponential on the right-hand side above. Assume we start with the packet at $t=0$ and evolve in time to $t>0$. This exponential is essentially equal to one as long as the magnitude of its phase is significantly less than one:

$$(k-k_0)^2\frac{\hbar}{m}t \ll 1. \tag{4.2.5}$$

We can estimate $(k-k_0)^2 \approx (\Delta k)^2$ since the relevant k values must be within the width of the k distribution. Moreover, since $\Delta p = \hbar\Delta k$, we get

$$\frac{(\Delta p)^2 t}{m\hbar} \ll 1. \tag{4.2.6}$$

Thus, the condition for minimal shape change is

$$\boxed{t \ll \frac{m\hbar}{(\Delta p)^2}.} \tag{4.2.7}$$

We can express the inequality in terms of position uncertainty using $\Delta x \Delta p \geq \hbar/2$. This means that

$$\Delta p \geq \frac{\hbar}{2\Delta x} \quad\to\quad \frac{1}{\Delta p} \leq \frac{2\Delta x}{\hbar} \quad\to\quad \frac{1}{(\Delta p)^2} \leq \frac{4(\Delta x)^2}{\hbar^2}. \tag{4.2.8}$$

We then get, ignoring small numerical factors,

$$\boxed{t \ll \frac{m}{\hbar}(\Delta x)^2.} \tag{4.2.9}$$

Also from (4.2.7), we can write

$$\frac{\Delta p\, t}{m} \ll \frac{\hbar}{\Delta p}, \tag{4.2.10}$$

which gives

$$\boxed{\frac{\Delta p}{m}t \ll \Delta x.} \tag{4.2.11}$$

This inequality has a simple interpretation. First note that $\Delta p/m$ represents the uncertainty in the velocity of the packet. There will be shape change when sufficient time elapses for this velocity uncertainty to produce position uncertainties comparable to the width Δx of the wave packet.

In all of the above inequalities, we use $\ll$ and this gives the condition for *negligible* change of shape. If we replace $\ll$ by $\approx$, we are setting an estimate for a *measurable* change of shape.

Example 4.2. *A localized electron.*
Assume we have localized an electron down to $\Delta x = 10^{-10}$m at time equal zero. The electron is then let free. We can estimate the maximum time t that it may remain localized to that precision. Using (4.2.9), we have

$$t \approx \frac{m(\Delta x)^2}{\hbar} = \frac{mc^2(\Delta x)^2}{\hbar c \cdot c} = \frac{0.5\,\text{MeV} \cdot 10^{-20}\text{m}^2}{200\,\text{MeV} \cdot \text{fm} \cdot 3 \times 10^8\text{m/s}} \approx 10^{-16}\,\text{s}. \tag{4.2.12}$$

If we originally had localized the electron to $\Delta x = 10^{-2}$m, we would have gotten $t \approx 1$s! This is a pretty large time in particle physics. □

4.3 Time Evolution of a Free Wave Packet

Suppose you know the wave function $\Psi(x, 0)$ at time equal zero, and your goal is to find $\Psi(x, t)$ for all times. This is accomplished in a few simple steps.

1. Use $\Psi(x, 0)$ and (4.1.2) to compute $\Phi(k)$:

$$\Phi(k) = \frac{1}{\sqrt{2\pi}} \int_{-\infty}^{\infty} dx\, \Psi(x, 0) e^{-ikx}. \tag{4.3.1}$$

2. Use $\Phi(k)$ and (4.1.1) to rewrite $\Psi(x, 0)$ as a superposition of plane waves:

$$\Psi(x, 0) = \frac{1}{\sqrt{2\pi}} \int_{-\infty}^{\infty} \Phi(k) e^{ikx} dk. \tag{4.3.2}$$

   This is useful because we know how plane waves evolve in time. The above is in fact the Fourier representation of $\Psi(x, 0)$.

3. A plane wave e^{ikx} evolves in time into $e^{i(kx-\omega(k)t)}$ with $\hbar\omega(k) = \frac{\hbar^2 k^2}{2m}$. Using superposition, we have that (4.3.2) evolves into

$$\Psi(x, t) = \frac{1}{\sqrt{2\pi}} \int_{-\infty}^{\infty} \Phi(k) e^{i(kx-\omega(k)t)} dk. \tag{4.3.3}$$

   This is in fact the answer for $\Psi(x, t)$. One can easily confirm that this is the solution because (i) it solves the Schrödinger equation (check that!), and (ii) setting $t = 0$ in $\Psi(x, t)$ gives us the initial wave function (4.3.2) that represented the initial condition.

4. If possible, do the integral (4.3.3) over k to find a closed-form expression for $\Psi(x, t)$. If it is too hard, the integral can always be done numerically.

Exercise 4.2. *Let $\Psi(x, 0) = \sin kx$ be a free-particle wave function at $t = 0$. Find $\Psi(x, t)$.*

Example 4.3. *Evolution of a free Gaussian wave packet.* Let $\Psi_a(x, 0)$ be a Gaussian wave packet at $t = 0$:

$$\Psi_a(x, 0) = \frac{1}{(2\pi)^{1/4}\sqrt{a}} e^{-\frac{x^2}{4a^2}}. \tag{4.3.4}$$

The constant a has units of length and represents the width of the Gaussian. The state Ψ_a is properly normalized, as you can check that $\int dx|\Psi_a(x,0)|^2 = 1$.

We will not do the calculations here (see problem 4.3), but we can imagine that this packet will change shape as time evolves. What is the timescale τ for shape changes? Equation (4.2.9) gives us a clue. The right-hand side represents a time scale for change of shape. So we must have $\tau \simeq ma^2/\hbar$. This is in fact right. You will discover when evolving the Gaussian that a convenient time constant is actually twice the above time:

$$\tau \equiv \frac{2ma^2}{\hbar}. \tag{4.3.5}$$

Consider the norm squared of the wave function:

$$|\Psi_a(x,0)|^2 = \frac{1}{\sqrt{2\pi}}\frac{1}{a}e^{-\frac{x^2}{2a^2}}. \tag{4.3.6}$$

It turns out that after time evolution the norm squared of the wave function remains a Gaussian:

$$|\Psi_a(x,t)|^2 = \frac{1}{\sqrt{2\pi}}\frac{1}{a(t)}e^{-\frac{x^2}{2a^2(t)}}. \tag{4.3.7}$$

Here, $a(t)$ is a time-dependent width. The goal of your calculation will be to determine the function $a(t)$ and its dependence on the time scale τ. □

4.4 Uncovering Momentum Space

We now begin a series of developments that lead to the idea of momentum space as a counterpoint or dual version of position space. In this section the time dependence of wave functions will play no role. Therefore, we will simply suppress time dependence. You can imagine all wave functions evaluated at time equal zero or at some arbitrary time t_0.

We begin by recalling the Fourier theorem identities given in (4.1.1) and (4.1.2):

$$\begin{aligned}\Psi(x) &= \frac{1}{\sqrt{2\pi}}\int_{-\infty}^{\infty}\Phi(k)e^{ikx}dk,\\ \Phi(k) &= \frac{1}{\sqrt{2\pi}}\int_{-\infty}^{\infty}\Psi(x)e^{-ikx}dx.\end{aligned} \tag{4.4.1}$$

The Fourier transform $\Phi(k)$ has all the information carried by the wave function $\Psi(x)$. This is clear because knowing $\Phi(k)$ means knowing $\Psi(x)$. The function $\Phi(k)$ is also the weight with which we add plane waves with momentum $\hbar k$ to form $\Psi(x)$.

We will now see that the consistency of the above equations can be used to derive an integral representation for a delta function. Such a representation is needed for our upcoming discussion. The idea is to replace $\Phi(k)$ in the first equation by the value given in the second equation. In order to keep the notation clear, we must use x' as a dummy variable of integration in the second equation:

$$\begin{aligned}\Psi(x) &= \frac{1}{\sqrt{2\pi}}\int_{-\infty}^{\infty}dk\,e^{ikx}\frac{1}{\sqrt{2\pi}}\int_{-\infty}^{\infty}dx'\,e^{-ikx'}\Psi(x')\\ &= \int_{-\infty}^{\infty}dx'\,\Psi(x')\left\{\frac{1}{2\pi}\int_{-\infty}^{\infty}dk\,e^{ik(x-x')}\right\},\end{aligned} \tag{4.4.2}$$

where we rearranged the order of integration to arrive at the final expression. Something curious is happening here: the integral shown within braces is a function of $x-x'$ and is acting to reduce the overall x' integral to an evaluation of Ψ at x. We know that a delta function can do exactly that (for delta function basics, see problem 4.2). More precisely, $\delta(x'-x)$ is the function that for general $f(x)$,

$$\int_{-\infty}^{\infty} dx'\, f(x')\,\delta(x'-x) = f(x). \tag{4.4.3}$$

We thus conclude that the integral shown in braces is a delta function:

$$\delta(x'-x) = \frac{1}{2\pi}\int_{-\infty}^{\infty} dk\, e^{ik(x-x')}. \tag{4.4.4}$$

In this integral one can let $k\to -k$, and since $\int dk$ is left invariant under this replacement, we find that $\delta(x'-x)=\delta(x-x')$ or, more plainly, $\delta(x)=\delta(-x)$. We will record the integral representation of the delta function using the other sign:

$$\boxed{\delta(x-x') = \frac{1}{2\pi}\int_{-\infty}^{\infty} dk\, e^{ik(x-x')}.} \tag{4.4.5}$$

A useful property of delta functions can be derived from the above representation. One finds that for arbitrary real $a\neq 0$,

$$\delta(ax) = \frac{1}{|a|}\delta(x). \tag{4.4.6}$$

Exercise 4.3. *Evaluate the integrals I_1 and I_2:*

$$I_1 = \int_{-\infty}^{\infty} dx f(x)\delta(-2x), \qquad I_2 = \int_{-\infty}^{\infty} dx f(x)\frac{d}{dx}\delta(x-x_0).$$

Assume $f(x)$ is continuous and has a continuous first derivative.

Exercise 4.4. *Consider the function $H(x)$ defined by integrating a delta function as follows: $H(x)=\int_{-\infty}^{x}\delta(x')dx'$. Explain why $H(x)$ is a step function: it vanishes for $x<0$ and takes value 1 for $x>0$ (the value at zero is ambiguous). Conclude that the derivative of a step function is the delta function: $H'(x)=\delta(x)$.*

At this point we may ask: How does the normalization condition for $\Psi(x)$ look in terms of $\Phi(k)$? We find out by calculation. Using the Fourier representation of $\Psi(x)$ and its complex conjugate, we have

$$\int_{-\infty}^{\infty} dx\,\Psi^*(x)\Psi(x) = \int_{-\infty}^{\infty} dx \frac{1}{\sqrt{2\pi}}\int_{-\infty}^{\infty}\Phi^*(k)e^{-ikx}dk\,\frac{1}{\sqrt{2\pi}}\int_{-\infty}^{\infty}\Phi(k')e^{ik'x}dk'. \tag{4.4.7}$$

We rearrange the integrals to do the x integration first:

$$\int_{-\infty}^{\infty} dx\,\Psi^*(x)\Psi(x) = \int_{-\infty}^{\infty} dk\,\Phi^*(k)\int_{-\infty}^{\infty} dk'\,\Phi(k')\frac{1}{2\pi}\int_{-\infty}^{\infty} dx\, e^{i(k'-k)x}. \tag{4.4.8}$$

The x integral, with the $1/(2\pi)$ prefactor, is precisely a delta function, and it makes the k' integration immediate:

$$\begin{aligned}\int_{-\infty}^{\infty} dx\, \Psi^*(x)\Psi(x) &= \int_{-\infty}^{\infty} dk\, \Phi^*(k) \int_{-\infty}^{\infty} dk'\, \Phi(k')\, \delta(k'-k) \\ &= \int_{-\infty}^{\infty} dk\, \Phi^*(k)\Phi(k).\end{aligned} \tag{4.4.9}$$

Our final result is therefore

$$\boxed{\int_{-\infty}^{\infty} dx\, |\Psi(x)|^2 = \int_{-\infty}^{\infty} dk\, |\Phi(k)|^2.} \tag{4.4.10}$$

This is known as Plancherel's theorem (or sometimes as Parseval-Plancherel's theorem). This equation relates the $\Psi(x)$ normalization to a perfectly analogous normalization for $\Phi(k)$. This is a hint that, just as for $|\Psi(x)|^2$, we have a probability interpretation for $|\Phi(k)|^2$.

Exercise 4.5. *Consider the normalized Gaussian wave function*

$$\Psi(x,0) = N \exp\Big(-\frac{x^2}{4\Delta^2}\Big), \quad \textit{with} \quad N^2 = \frac{1}{\sqrt{2\pi}\,\Delta},$$

where Δ happens to be the rigorously defined position uncertainty of the wave function. Show that the Fourier transform of $\Psi(x,0)$ is

$$\Phi(k) = \tilde{N} e^{-k^2\Delta^2}$$

with a normalization constant $\tilde{N}$ you can determine using Plancherel's theorem.

Since our plane waves are momentum eigenstates, let us rewrite Plancherel's theorem using momentum $p=\hbar k$. Instead of integrals over k, we will have integrals over p, and we note that $dk = dp/\hbar$. Letting $\tilde{\Phi}(p) \equiv \Phi(k)$, when $p=\hbar k$, equations (4.4.1) become

$$\begin{aligned}\Psi(x) &= \frac{1}{\sqrt{2\pi}\,\hbar} \int_{-\infty}^{\infty} \tilde{\Phi}(p) e^{ipx/\hbar} dp, \\ \tilde{\Phi}(p) &= \frac{1}{\sqrt{2\pi}} \int_{-\infty}^{\infty} \Psi(x) e^{-ipx/\hbar} dx.\end{aligned} \tag{4.4.11}$$

For a more symmetric pair of equations, we redefine the function $\tilde{\Phi}(p)$. We will let $\tilde{\Phi}(p) \to \Phi(p)\sqrt{\hbar}$ in equations (4.4.11). We then obtain our final form for Fourier's relations in terms of momentum:

$$\boxed{\begin{aligned}\Psi(x) &= \frac{1}{\sqrt{2\pi\hbar}} \int_{-\infty}^{\infty} \Phi(p)\, e^{ipx/\hbar}\, dp, \\ \Phi(p) &= \frac{1}{\sqrt{2\pi\hbar}} \int_{-\infty}^{\infty} \Psi(x) e^{-ipx/\hbar} dx.\end{aligned}} \tag{4.4.12}$$

Similarly, Plancherel's theorem (4.4.10) becomes

$$\boxed{\int_{-\infty}^{\infty} dx\, |\Psi(x)|^2 = \int_{-\infty}^{\infty} dp\, |\Phi(p)|^2.} \tag{4.4.13}$$

Exercise 4.6. *Verify that the redefinitions we did to arrive at (4.4.12) indeed yield (4.4.13) when starting from (4.4.10).*

In the top equation in (4.4.12), $\Phi(p)$ denotes the weight with which we add the momentum state $e^{ipx/\hbar}$ in the superposition that represents $\Psi(x)$. This state $e^{ipx/\hbar}$ is a state with momentum p. This suggests that we can roughly view $\Phi(p)$ as the amplitude for the particle to have momentum p and $|\Phi(p)|^2$ as roughly the probability of the particle having momentum p. Just as we say that $\Psi(x)$ is the wave function in position space x, we can think of $\Phi(p)$ as the **wave function in momentum space** p. Plancherel's identity (4.4.13) suggests that viewing $\Phi(p)$ as a probability amplitude is in fact consistent. Given that a properly normalized $\Psi(x)$ leads to a $\Phi(p)$ that satisfies $\int dp|\Phi(p)|^2 = 1$, we postulate that

$$\boxed{\begin{array}{c}|\Phi(p)|^2 dp \text{ is the probability of finding the particle with} \\ \text{momentum in the range } (p, p+dp).\end{array}} \tag{4.4.14}$$

This postulate makes the analogy between position and momentum space compelling. In the same way we defined the action of the operator $\hat{x}$ on position-space wave functions, we postulate the following action of the momentum operator on momentum-space wave functions:

$$\hat{p}\,\Phi(p) \equiv p\,\Phi(p). \tag{4.4.15}$$

This is not an eigenvalue-type equation because p here is a function, not a number. Just as in position space we represent the momentum operator as a derivative with respect to position, in momentum space we can represent the position as a derivative with respect to momentum. In fact we claim that, in momentum space,

$$\hat{x} = i\hbar\frac{\partial}{\partial p}. \tag{4.4.16}$$

Most importantly, this is consistent with the commutation relation $[\hat{x}, \hat{p}] = i\hbar$. We test this claim acting with this commutator on a function $\Phi(p)$, just as we did in position space in (3.3.15). This time we find

$$\begin{aligned}[\hat{x},\,\hat{p}\,]\Phi(p) &= (\hat{x}\hat{p} - \hat{p}\hat{x})\Phi(p) = \hat{x}(p\Phi(p)) - \hat{p}\left(i\hbar\frac{\partial\Phi}{\partial p}\right) \\ &= i\hbar\frac{\partial}{\partial p}(p\Phi(p)) - i\hbar p\frac{\partial\Phi}{\partial p} = i\hbar\,\Phi(p),\end{aligned} \tag{4.4.17}$$

confirming the consistency of the momentum-space representation $\hat{x} = i\hbar\frac{\partial}{\partial p}$.

We conclude this discussion of momentum space by looking at the Schrödinger equation itself in momentum space. In preparation for this, we look at the Fourier transforms that allow us to bring the potential to momentum space. Let us first calculate the Fourier transform of $x^n\Psi(x)$, assuming we know the Fourier transform $\Phi(p)$ of $\Psi(x)$. To do this we write

$$x^n\Psi(x) = x^n\frac{1}{\sqrt{2\pi\hbar}}\int_{-\infty}^{\infty} dp\,\Phi(p)\,e^{ipx/\hbar} = \frac{1}{\sqrt{2\pi\hbar}}\int_{-\infty}^{\infty} dp\,\Phi(p)\left(-i\hbar\frac{\partial}{\partial p}\right)^n e^{ipx/\hbar}, \tag{4.4.18}$$

since the action of the n iterated derivatives on the exponential would precisely produce the x^n factor that can exit the integral. We now integrate by parts the n derivatives, getting a minus sign for each one so that

$$x^n \Psi(x) = \frac{1}{\sqrt{2\pi\hbar}} \int_{-\infty}^{\infty} dp \left[\left(i\hbar \frac{\partial}{\partial p} \right)^n \Phi(p) \right] e^{ipx/\hbar}. \tag{4.4.19}$$

The expression in brackets is the desired Fourier transform. With this result, we can consider now the Fourier transform of a potential function $V(x)$ assumed to have a regular series expansion around $x = 0$:

$$V(x) = \sum_{n=0}^{\infty} v_n x^n. \tag{4.4.20}$$

Here, the v_n are expansion coefficients. It then follows from (4.4.19) and linearity that

$$\begin{aligned} V(x)\Psi(x) &= \frac{1}{\sqrt{2\pi\hbar}} \int_{-\infty}^{\infty} dp \left[\sum_{n=0}^{\infty} v_n \left(i\hbar \frac{\partial}{\partial p} \right)^n \Phi(p) \right] e^{ipx/\hbar} \\ &= \frac{1}{\sqrt{2\pi\hbar}} \int_{-\infty}^{\infty} dp \left[V\left(i\hbar \frac{\partial}{\partial p} \right) \Phi(p) \right] e^{ipx/\hbar}. \end{aligned} \tag{4.4.21}$$

The Fourier transform of $V(x)\Psi(x)$, shown in brackets, is the differential operator $V(i\hbar \frac{\partial}{\partial p})$ acting on $\Phi(p)$.

We are now ready to derive the form of the Schrödinger equation in momentum space. Including the time dependence in both the position-space and momentum-space wave functions, we write, following (4.4.12),

$$\Psi(x,t) = \frac{1}{\sqrt{2\pi\hbar}} \int_{-\infty}^{\infty} dp\, \Phi(p,t)\, e^{ipx/\hbar}. \tag{4.4.22}$$

Starting with the Schrödinger equation (3.3.3), we then have

$$i\hbar \frac{\partial}{\partial t} \int_{-\infty}^{\infty} dp\, \Phi(p,t)\, e^{ipx/\hbar} = \left(-\frac{\hbar^2}{2m} \frac{\partial^2}{\partial x^2} + V(x) \right) \int_{-\infty}^{\infty} dp\, \Phi(p,t)\, e^{ipx/\hbar}, \tag{4.4.23}$$

where we canceled the common constant prefactor in the Fourier transform. On the left-hand side, the time derivative simply goes into the integral and acts on $\Phi(p,t)$. On the right-hand side, the x partial derivatives act on the exponential inside the integral, and for $V(x)$ we use the formula just derived. We thus have

$$\int_{-\infty}^{\infty} dp \left[i\hbar \frac{\partial}{\partial t} \Phi(p,t) \right] e^{ipx/\hbar} = \int_{-\infty}^{\infty} dp \left[\frac{p^2}{2m} \Phi(p,t) + V\left(i\hbar \frac{\partial}{\partial p} \right) \Phi(p,t) \right] e^{ipx/\hbar}. \tag{4.4.24}$$

The above equality implies the equality of the objects inside brackets on each side of the equation. This is the Schrödinger equation in momentum space:

$$\boxed{ i\hbar \frac{\partial}{\partial t} \Phi(p,t) = \left[\frac{p^2}{2m} + V\left(i\hbar \frac{\partial}{\partial p} \right) \right] \Phi(p,t). } \tag{4.4.25}$$

The equation is still first order in time derivatives. The spatial derivatives in the kinetic energy have been traded by the simple multiplicative factor $p^2/(2m)$. The potential is more complicated in this picture: each power of x in $V(x)$ turns into a derivative with respect to momentum. This equation is particularly nice for a potential that is linear in x, as we will discuss in section 6.7.

Let's consider the generalization of the Fourier transforms to three dimensions. Fourier's theorem in momentum-space language (namely, using $\mathbf{p}$ as opposed to $\mathbf{k}$) takes the form

$$\boxed{\begin{aligned}\Psi(\mathbf{x}) &= \frac{1}{(2\pi\hbar)^{3/2}}\int d^3\mathbf{p}\,\Phi(\mathbf{p})e^{i\mathbf{p}\cdot\mathbf{x}/\hbar},\\ \Phi(\mathbf{p}) &= \frac{1}{(2\pi\hbar)^{3/2}}\int d^3\mathbf{x}\,\Psi(\mathbf{x})e^{-i\mathbf{p}\cdot\mathbf{x}/\hbar}.\end{aligned}} \tag{4.4.26}$$

Just as we did in the one-dimensional case, if we insert the Fourier transform into the expression for $\Psi(\mathbf{x})$ we find an integral representation for the three-dimensional δ function:

$$\begin{aligned}\Psi(\mathbf{x}) &= \frac{1}{(2\pi\hbar)^3}\int d^3\mathbf{p}\,e^{i\mathbf{p}\cdot\mathbf{x}/\hbar}\int d^3\mathbf{x}'\,\Psi(\mathbf{x}')e^{-i\mathbf{p}\cdot\mathbf{x}'/\hbar}\\ &= \int d^3\mathbf{x}'\,\Psi(\mathbf{x}')\frac{1}{(2\pi\hbar)^3}\int d^3\mathbf{p}\,e^{i\mathbf{p}\cdot(\mathbf{x}-\mathbf{x}')/\hbar}\\ &= \int d^3\mathbf{x}'\,\Psi(\mathbf{x}')\frac{1}{(2\pi)^3}\int d^3\mathbf{k}\,e^{i\mathbf{k}\cdot(\mathbf{x}-\mathbf{x}')},\end{aligned} \tag{4.4.27}$$

which leads to the identification

$$\delta^3(\mathbf{x}-\mathbf{x}') = \frac{1}{(2\pi)^3}\int d^3\mathbf{k}\,e^{i\mathbf{k}\cdot(\mathbf{x}-\mathbf{x}')}. \tag{4.4.28}$$

It is then straightforward to derive Plancherel's identity:

$$\int_{-\infty}^{\infty} d^3\mathbf{x}\,|\Psi(\mathbf{x})|^2 = \int d^3\mathbf{p}\,|\Phi(\mathbf{p})|^2. \tag{4.4.29}$$

In three-dimensional momentum space, we use the same probability interpretation: we declare $|\Phi(\mathbf{p})|^2 d^3\mathbf{p}$ to be the probability of finding the particle with momentum in the range $d^3\mathbf{p}$ centered on $\mathbf{p}$.

In all of the above Fourier pairs, one can introduce the time dependence without effort. Since time plays no role in the integrals, one simply replaces $\Phi(x)$ by $\Phi(x,t)$ and $\Phi(p)$ by $\Phi(p,t)$. Similar replacements can also be done for the three-dimensional Fourier pairs. In all of these equations, time is just an extra parameter that goes along for the ride.

Exercise 4.7. *What is the Fourier transform $\Phi(\mathbf{p})$ of the three-dimensional delta function $\Psi(\mathbf{x}) = \delta(\mathbf{x}-\mathbf{x}')$? Is $\Phi(\mathbf{p})$ localized in momentum space?*

Problems

Problem 4.1. *Exercises on packets changing shape.*

1. A free proton is localized within $\Delta x = 10^{-10}$m. Estimate the time t_s it takes the packet to spread appreciably. Repeat the calculation for a proton localized within 1 cm.

2. Consider a wave packet that satisfies the relation $\Delta x \Delta p \sim \hbar$. Find the ratio of the time t_{cross} it takes for this packet to cross a point to the time t_{spread} it takes to spread appreciably, in terms of the momentum p of the wave packet and the uncertainty Δp in the momentum.

Problem 4.2. *Delta function definition and properties.*

A one-dimensional delta function can be represented with the help of the following function:

$$D_\alpha(x) = \frac{1}{\sqrt{\pi}|\alpha|} \exp\Big(-\frac{x^2}{\alpha^2}\Big), \quad \alpha \in \mathbb{R}.$$

The delta function $\delta(x)$ is sometimes inaccurately viewed as $\lim_{\alpha\to 0} D_\alpha$; this limit does not exist. It is not a function. The correct way to take the limit is to consider integrals. Then, one claims,

$$\lim_{\alpha\to 0} \int_{-\infty}^{\infty} f(x) D_\alpha(x) = f(0). \tag{1}$$

This result is the basis for writing, less precisely, that $\int f(x)\delta(x) = f(0)$.

1. Confirm that $\int_{-\infty}^{\infty} D_\alpha(x) = 1$. Sketch $D_\alpha(x)$ as a function of x for small α. How do the features of this function depend on α?
2. Assume that $f(x)$ is a function that for all x can be represented by a Taylor expansion about $x=0$. It follows that $f(x) = f(0) + \sum_{k=1}^{\infty} f^{(k)}(0)\frac{x^k}{k!}$. Evaluate the integral in equation (1) and show that the answer holds.
3. Show that $\lim_{\alpha\to 0} \int f(x) D_\alpha(x-b)dx = f(b)$, which is written as $\int f(x)\delta(x-b)dx = f(b)$.
4. Show that for any real $a \neq 0$,

$$D_\alpha(ax) = \frac{1}{|a|} D_{\alpha/|a|}(x)$$

and that as a result, $\lim_{\alpha\to 0} \int f(x) D_\alpha(ax) = \frac{1}{|a|} f(0)$. This is written as $\delta(ax) = \frac{1}{|a|}\delta(x)$.
5. Argue that for very small α, one has the approximate relation

$$D_\alpha(x^2 - b^2) \simeq \frac{1}{2|b|}\big(D_{\frac{\alpha}{2|b|}}(x-b) + D_{\frac{\alpha}{2|b|}}(x+b)\big).$$

This is expressed as $\int \delta(x^2 - b^2) f(x) = \frac{1}{2|b|}(f(b) + f(-b))$.

Problem 4.3. *Evolving the Gaussian wave packet.*

Consider the normalized wave packet representing the state of a particle of mass m at $t=0$:

$$\Psi_a(x, 0) = \frac{1}{(2\pi)^{1/4}\sqrt{a}} \exp\left(-\frac{x^2}{4a^2}\right).$$

Here a is a length parameter that represents the width of the packet at $t=0$. You should confirm that $\Psi_a(x, 0)$ is properly normalized. The following integral will be useful for this problem:

$$\int_{-\infty}^{\infty} dx\, e^{-\alpha x^2+\beta x} = \sqrt{\frac{\pi}{\alpha}}\exp\left(\frac{\beta^2}{4\alpha}\right), \quad \alpha,\beta,\in\mathbb{C}, \;\; \text{Re}(\alpha)>0. \tag{1}$$

See example 4.3 for some background discussion.

1. Find the Fourier representation of $\Psi_a(x,0)$. Namely, determine the function $\Phi_a(k)$ such that

$$\Psi_a(x,0)=\frac{1}{\sqrt{2\pi}}\int_{-\infty}^{\infty}\Phi_a(k)e^{ikx}dk.$$

2. Assume the particle is free and find the wave function $\Psi_a(x,t)$ for arbitrary $t>0$. The answer is a bit messy but can be written more clearly using the time constant $\tau\equiv 2ma^2/\hbar$ built from the constants in the problem.
3. At time zero the probability density is

$$|\Psi_a(x,0)|^2=\frac{1}{\sqrt{2\pi}a}\exp\left(-\frac{x^2}{2a^2}\right)\equiv G(x;a)\,,$$

where we defined the Gaussian $G(x;a)$ with width parameter a. What is the probability density $|\Psi_a(x,t)|^2$ for $t>0$? $|\Psi_a(x,t)|^2$ can be written in terms of the Gaussian G with a time-dependent width parameter $a(t)$. What is $a(t)$?

Problem 4.4. *Time evolution of a wave as seen in momentum and position space.*

At $t=0$ the momentum space wave function $\Phi(p,t)$ for a particle of mass m is a step function stretching from $-p_0$ to p_0, where $p_0>0$ is a constant momentum:

$$\Phi(p,0)=\begin{cases}\frac{1}{\sqrt{2p_0}}, & -p_0<p<p_0,\\ 0, & \text{otherwise.}\end{cases}$$

1. Use the momentum space Schrödinger equation to determine $\Phi(p,t)$.
2. Use the Fourier transformation to calculate $\Psi(x,t)$. Explicitly note that the integral you write using $\Phi(p,t)$ is consistent with the prescription discussed in section 4.3 to evolve wave packets. Manipulate the answer to show that it can be put in the following form:

$$\sqrt{\pi x_0}\,\Psi(x,t)=\int_0^1 e^{-iu^2\tilde{t}}\cos u\tilde{x}\,du \;\equiv\; F(\tilde{x},\tilde{t}),$$

where $\tilde{x}$ and $\tilde{t}$ are unit-free versions of x and t, respectively:

$$\tilde{x}=x/x_0,\quad \tilde{t}=t/t_0,\quad x_0=\hbar/p_0,\quad t_0=2m\hbar/p_0^2.$$

Here x_0 and t_0 are, respectively, a length scale and a timescale built using the constants $\hbar$, m, and p_0 available in the problem. Is $\Psi(x,t)$ a symmetric function of x for all times?
3. We can evaluate $F(\tilde{x},\tilde{t})\sim\Psi(x,t)$ numerically. Plot the absolute value $|F(\tilde{x},\tilde{t})|$ as a function of $\tilde{x}>0$ for various values of $\tilde{t}=0,1,2,3,4$. For $t=0$ the magnitude of the wave function vanishes at $x=n\pi$, with $n=1,2,\cdots$; does it vanish for any x for the t

values you examined? Confirm that $|F(0,4)| = 0.4638$. Explore numerically the time dependence of $|F(0,t)|$.

Problem 4.5. *Overlap between moving Gaussians.*

We learned in problem 3.7 that the overlap $\int dx \Psi_1^* \Psi_2$ between two different wave packets $\Psi_1(x,t)$ and $\Psi_2(x,t)$ is, on general grounds, time independent. The following example seems to be in conflict with this fact: consider two Gaussians that coincide at $t=0$ but are moving in opposite directions with momenta large compared to their momenta uncertainties. Now we make two claims:

(i) At $t=0$ the overlap is big.

(ii) Once the centers of the packets are separated by a small multiple of the position uncertainty, the overlap is small.

The purpose of this problem is to find a flaw in the above claims, since together they contradict the time independence of the overlap. Consider a *free particle* and a normalized Gaussian wave packet $\hat{\Psi}$ at zero time:

$$\hat{\Psi}(x,0) = \frac{1}{(2\pi)^{1/4}\sqrt{a}} \exp\Big(-\frac{x^2}{4a^2}\Big).$$

From this we create two wave packets Ψ_1 and Ψ_2:

$$\Psi_1(x,0) = e^{iqx/\hbar}\,\hat{\Psi}(x,0), \qquad \Psi_2(x,0) = e^{-iqx/\hbar}\,\hat{\Psi}(x,0).$$

Here q is a real quantity with units of momentum.

1. What is $\langle \hat{p} \rangle$ for $\hat{\Psi}(x,0)$? Will this expectation value change in time? Explain.
2. What is $\langle \hat{p} \rangle$ for $\Psi_1(x,0)$? What is $\langle \hat{p} \rangle$ for $\Psi_2(x,0)$? Do these change in time?
3. Compute the zero-time overlap $\gamma(0) = \int \Psi_1^*(x,0)\Psi_2(x,0)dx$ of the two packets. The integral (1) in problem 4.3 may be useful.
4. Write an inequality that expresses the fact that the momentum of the wave packets is *large* compared to the momentum uncertainty.
5. What was wrong with the claims (i) and (ii)?

Problem 4.6. *Plancherel's identity and momentum-space wave function of hydrogen.*

1. Consider the Fourier pair $(\Psi(\mathbf{x}), \Phi(\mathbf{p}))$ of three-dimensional (3D) wave functions. Use the Fourier relations and the integral form for the delta function to prove the 3D version of Plancherel's identity:

$$\int d^3x\, |\Psi(\mathbf{x})|^2 = \int d^3p\, |\Phi(\mathbf{p})|^2.$$

2. In the hydrogen atom, the ground-state wave function takes the form $\Psi(\mathbf{x}) = Ne^{-r/a_0}$, where $r = |\mathbf{x}|$, a_0 is the Bohr radius, and N is a real, positive normalization constant. Find N.

3. The Fourier transform $\Phi(\mathbf{p})$ of the ground-state wave function, which you need not derive, takes the form

$$\Phi(\mathbf{p}) = N'\left(1 + \frac{a_0^2 p^2}{\hbar^2}\right)^{-2},$$

for some constant N' and with $p = |\mathbf{p}|$. Find $|N'|^2$.

4. Calculate the probability that the electron may be found with a momentum whose magnitude exceeds $\hbar/a_0$. Write your integrals explicitly, but you may evaluate them with a computer. (The momentum distribution was measured by ionization of atomic hydrogen by a high-energy electron beam. See B. Lohan, and E. Weigold "Direct measurement of the electron momentum probability distribution in atomic hydrogen," Phys. Lett. **86A**, 139–141 (1981).)

5 Expectation Values and Hermitian Operators

We define the expectation value of an operator in a quantum state and derive a simple expression for its time derivative. We introduce an inner product that helps manipulate Hermitian operators and write succinct normalization conditions and expectation values. Observables in quantum mechanics are represented by Hermitian operators. We discuss the key properties of Hermitian operators, introduce the measurement axiom, and briefly discuss the other axioms in the formulation of quantum mechanics. Finally, we give the precise definition of the uncertainty $\Delta\hat{Q}$ *of an operator* $\hat{Q}$ *in a state* Ψ *and show that such uncertainty vanishes if and only if* Ψ *is an eigenstate of* $\hat{Q}$*.*

5.1 Expectation Values of Operators

In this section we learn how to define the expectation value of a quantum mechanical **observable**. An observable is a quantity that can be measured in an experiment. In quantum mechanics observables are operators that happen to be Hermitian. We will find a suitable definition for the expectation value of an operator by analogy with random variables and by exploring the particular case of the momentum operator acting on momentum-space wave functions.

Consider a random variable Q. This variable takes values in the set $\{Q_1, \ldots, Q_n\}$ and does so randomly with respective, nonzero probabilities $\{p_1, \ldots, p_n\}$ adding to one. The *expectation value* $\langle Q \rangle$, or the expected value of Q, is defined to be

$$\langle Q \rangle \equiv \sum_{i=1}^{n} Q_i p_i \,. \tag{5.1.1}$$

The expected value can be thought of heuristically as a *long-run mean*: as more and more values of the random variable are collected, the mean of that set approaches the expected value.

Since quantum mechanics gives us probabilities, we can use the above definition in simple cases. In the example that follows, the random variable takes continuous values, and the above definition of the mean must be modified by turning the sum into an integral.

As we have seen, in a quantum system the probability of a particle to be found in the interval $[x, x+dx]$ at time t is given by

$$\Psi^*(x,t)\Psi(x,t)\,dx, \tag{5.1.2}$$

where $\Psi(x,t)$ is a normalized wave function. The expected value of the position of the particle, denoted as $\langle \hat{x} \rangle$, is therefore

$$\langle \hat{x} \rangle \equiv \int_{-\infty}^{\infty} x\,\Psi^*(x,t)\Psi(x,t)dx. \tag{5.1.3}$$

Note that this expected value depends on t. What does $\langle \hat{x} \rangle$ correspond to physically? If we consider many exact copies of the physical system and measure the position x at a time t in each of them, the average value recorded will approach the number $\langle \hat{x} \rangle$ as the number of measurements approaches infinity.

Let's discuss the expectation value for the momentum. Since we learned in (4.4.14) that

$$\Phi^*(p,t)\Phi(p,t)\,dp \tag{5.1.4}$$

is the probability of finding the particle with momentum in the range $[p, p+dp]$ at time t, we define the expectation $\langle \hat{p} \rangle$ of the momentum operator as

$$\langle \hat{p} \rangle \equiv \int_{-\infty}^{\infty} p\,\Phi^*(p,t)\Phi(p,t)dp. \tag{5.1.5}$$

We will now manipulate this expression to find what form it takes in position space. Using the second equation in (4.4.12) and its complex conjugate version, with time labels attached, we have

$$\begin{aligned}\langle \hat{p} \rangle &= \int_{-\infty}^{\infty} p\,\Phi^*(p,t)\Phi(p,t)\,dp \\ &= \int_{-\infty}^{\infty} dp\,p \int_{-\infty}^{\infty} \frac{dx}{\sqrt{2\pi\hbar}} e^{ipx/\hbar}\,\Psi^*(x,t) \int_{-\infty}^{\infty} \frac{dx'}{\sqrt{2\pi\hbar}} e^{-ipx'/\hbar}\Psi(x',t).\end{aligned} \tag{5.1.6}$$

We now change the order of integration so that the p integral is done first:

$$\langle \hat{p} \rangle = \int_{-\infty}^{\infty} dx\,\Psi^*(x,t) \int_{-\infty}^{\infty} dx'\,\Psi(x',t)\,\frac{1}{2\pi\hbar}\int_{-\infty}^{\infty} dp\,p\,e^{ipx/\hbar}e^{-ipx'/\hbar}. \tag{5.1.7}$$

The factor of p in the last integrand can be replaced in favor of a $\partial/\partial x$ derivative that we choose to place as far to the left as possible:

$$\langle \hat{p} \rangle = \int_{-\infty}^{\infty} dx\,\Psi^*(x,t)\left(\frac{\hbar}{i}\frac{\partial}{\partial x}\right)\int_{-\infty}^{\infty} dx'\,\Psi(x',t)\,\frac{1}{2\pi\hbar}\int_{-\infty}^{\infty} dp\,e^{ipx/\hbar}e^{-ipx'/\hbar}. \tag{5.1.8}$$

Note that this derivative acts on $e^{ipx/\hbar}$, the only term with x dependence to the right of the derivative, and produces the desired factor of p. This trick is useful to perform the last integral, where we let $p = \hbar k$:

$$\frac{1}{2\pi\hbar}\int_{-\infty}^{\infty} dp\, e^{ipx/\hbar}e^{-ipx'/\hbar} = \frac{1}{2\pi}\int_{-\infty}^{\infty} dk\, e^{ik(x-x')} = \delta(x-x'), \tag{5.1.9}$$

recalling the integral representation (4.4.5). As a result, we have

$$\langle\hat{p}\rangle = \int_{-\infty}^{\infty} dx\, \Psi^*(x,t)\left(\frac{\hbar}{i}\frac{\partial}{\partial x}\right)\int_{-\infty}^{\infty} dx'\, \Psi(x',t)\delta(x'-x), \tag{5.1.10}$$

given that $\delta(x-x') = \delta(x'-x)$. The x' integral is now easily done, and we find

$$\langle\hat{p}\rangle = \int_{-\infty}^{\infty} dx\, \Psi^*(x,t)\left(\frac{\hbar}{i}\frac{\partial}{\partial x}\right)\Psi(x,t). \tag{5.1.11}$$

We have thus shown that

$$\boxed{\langle\hat{p}\rangle = \int_{-\infty}^{\infty} dx\, \Psi^*(x,t)\,\hat{p}\,\Psi(x,t), \quad \hat{p} = \frac{\hbar}{i}\frac{\partial}{\partial x}.} \tag{5.1.12}$$

This equation confirms the correctness of our earlier arguments that led to the identification of the differential operator $\frac{\hbar}{i}\frac{\partial}{\partial x}$ as the momentum operator $\hat{p}$ in position space.

Notice the position of the $\hat{p}$ operator in the expectation value: it acts on $\Psi(x)$. This motivates the following definition for the expectation value $\langle\hat{Q}\rangle$ of *any* operator $\hat{Q}$:

$$\boxed{\langle\hat{Q}\rangle \equiv \int_{-\infty}^{\infty} dx\, \Psi^*(x,t)\,\hat{Q}\Psi(x,t).} \tag{5.1.13}$$

The expectation value depends on the state of the system–that is, it depends on the wave function $\Psi(x,t)$. Given that we integrate over x, the expectation value of $\hat{Q}$ is ultimately a function of time. It must be emphasized that our definitions assume that $\Psi(x,t)$ is a *normalized* wave function. It should also be clear that the expectation value is not affected by the phase factor ambiguity of the normalized wave function.

Example 5.1. *The kinetic energy operator is positive.*

Consider the kinetic energy operator $\hat{T}$ for a particle moving in one dimension:

$$\hat{T} = \frac{\hat{p}^2}{2m} = -\frac{\hbar^2}{2m}\frac{\partial^2}{\partial x^2}. \tag{5.1.14}$$

The definition gives

$$\langle\hat{T}\rangle = -\frac{\hbar^2}{2m}\int_{-\infty}^{\infty} dx\, \Psi^*(x,t)\frac{\partial^2}{\partial x^2}\Psi(x,t). \tag{5.1.15}$$

The kinetic energy is said to be a positive operator because it is proportional to the square of another operator, in this case the momentum operator. We expect therefore that the expectation value of the kinetic operator evaluated in arbitrary normalized states is positive. We now make this positivity manifest, using integration by parts of one of the x derivatives. Note that

$$\int_{-\infty}^{\infty} dx\, \Psi^* \frac{\partial^2}{\partial x^2} \Psi = \int_{-\infty}^{\infty} dx \left[\frac{\partial}{\partial x} \left(\Psi^* \frac{\partial}{\partial x} \Psi \right) - \frac{\partial \Psi^*}{\partial x} \frac{\partial \Psi}{\partial x} \right]$$
$$= \left(\Psi^* \frac{\partial}{\partial x} \Psi \right) \Bigg|_{\infty}^{\infty} - \int_{-\infty}^{\infty} dx \frac{\partial \Psi^*}{\partial x} \frac{\partial \Psi}{\partial x}. \tag{5.1.16}$$

The boundary terms at plus and minus infinity vanish as a result of our familiar conditions on normalizable states. Moreover, noting that $\frac{\partial \Psi^*}{\partial x} = (\frac{\partial \Psi}{\partial x})^*$, we find that

$$\langle \hat{T} \rangle = \frac{\hbar^2}{2m} \int_{-\infty}^{\infty} dx \left| \frac{\partial \Psi(x,t)}{\partial x} \right|^2. \tag{5.1.17}$$

As desired, this right-hand side is manifestly positive. The expectation value of $\hat{T}$ can also be computed in momentum space using the probabilistic interpretation that led to (5.1.5):

$$\langle \hat{T} \rangle = \int dp \frac{p^2}{2m} |\Phi(p,t)|^2. \tag{5.1.18}$$

This is also manifestly positive. □

Other examples of operators whose expectation values we can now compute include the momentum operator $\hat{\mathbf{p}} \to \frac{\hbar}{i} \nabla$ in three dimensions, the potential energy operator $V(\hat{\mathbf{x}})$, and the angular momentum operator $\hat{\mathbf{L}} = \hat{\mathbf{r}} \times \hat{\mathbf{p}}$, for which

$$\begin{aligned} \hat{L}_x &= \hat{y}\hat{p}_z - \hat{z}\hat{p}_y = \frac{\hbar}{i}\left(y\frac{\partial}{\partial z} - z\frac{\partial}{\partial y} \right), \\ \hat{L}_y &= \hat{z}\hat{p}_x - \hat{x}\hat{p}_z = \frac{\hbar}{i}\left(z\frac{\partial}{\partial x} - x\frac{\partial}{\partial z} \right), \\ \hat{L}_z &= \hat{x}\hat{p}_y - \hat{y}\hat{p}_x = \frac{\hbar}{i}\left(x\frac{\partial}{\partial y} - y\frac{\partial}{\partial x} \right). \end{aligned} \tag{5.1.19}$$

Exercise 5.1. *Find the expectation value $\langle \hat{x} \rangle$ and $\langle \hat{x}^2 \rangle$ on the $t = 0$ normalized Gaussian wave function*

$$\Psi(x, t=0) = \frac{1}{(2\pi\Delta^2)^{1/4}} e^{-\frac{x^2}{4\Delta^2}}. \tag{5.1.20}$$

Exercise 5.2. *Show that $\hat{L}_x, \hat{L}_y$, and $\hat{L}_z$ all vanish acting on a wave function of the form $\Psi(\mathbf{x}, t) = f(r)h(t)$, with $r = |\mathbf{x}|$, $f(r)$ a function that is differentiable for all r, and $h(t)$ general. Conclude that the expectation values of the angular momentum operators vanish in a* rotationally invariant *state Ψ. We say Ψ is rotationally invariant because $\Psi(\mathbf{x}', t) = \Psi(\mathbf{x}, t)$ when $|\mathbf{x}'| = |\mathbf{x}|$, which happens when $\mathbf{x}$ and $\mathbf{x}'$ can be mapped onto each other by a rotation about the origin.*

5.2 Time Dependence of Expectation Values

The expectation values of operators are in general time dependent because the wave functions representing the states are time dependent. We will consider here operators $\hat{Q}$

that do not have *explicit* time dependence—that is, operators in which the time variable t does not appear, and time derivatives do not appear either. To investigate the time dependence of expectation values, we calculate their time derivative. Using the time independence of $\hat{Q}$, we find that the time derivatives only act on the wave functions:

$$\begin{aligned} i\hbar\frac{d}{dt}\langle\hat{Q}\rangle &= i\hbar\frac{d}{dt}\int_{-\infty}^{\infty} d^3x\,\Psi^*(x,t)\hat{Q}\Psi(x,t) \\ &= \int_{-\infty}^{\infty} d^3x\left(i\hbar\frac{\partial\Psi^*}{\partial t}\hat{Q}\Psi + \Psi^*\hat{Q}\,i\hbar\frac{\partial\Psi}{\partial t}\right). \end{aligned} \tag{5.2.1}$$

We use the Schrödinger equation to evaluate the time derivatives:

$$i\hbar\frac{d}{dt}\langle\hat{Q}\rangle = \int_{-\infty}^{\infty} d^3x\left(-(\hat{H}\Psi)^*\hat{Q}\Psi + \Psi^*\hat{Q}\hat{H}\Psi\right). \tag{5.2.2}$$

We now recall the Hermiticity of $\hat{H}$, which implies that

$$\int_{-\infty}^{\infty} dx\,(\hat{H}\Psi_1)^*\Psi_2 = \int_{-\infty}^{\infty} dx\,\Psi_1^*\hat{H}\Psi_2. \tag{5.2.3}$$

This can be used to move $\hat{H}$ into the other wave function in the first term on the right-hand side of (5.2.2) so that

$$\begin{aligned} i\hbar\frac{d}{dt}\langle Q\rangle &= \int_{-\infty}^{\infty} d^3x\left(-\Psi^*\hat{H}\hat{Q}\Psi + \Psi^*\hat{Q}\hat{H}\Psi\right) = \int_{-\infty}^{\infty} d^3x\,\Psi^*(\hat{Q}\hat{H} - \hat{H}\hat{Q})\Psi \\ &= \int_{-\infty}^{\infty} d^3x\,\Psi^*\left[\hat{Q},\hat{H}\right]\Psi, \end{aligned} \tag{5.2.4}$$

where we note the appearance of the commutator. All in all, we have proven that for operators $\hat{Q}$ that do not explicitly depend on time,

$$\boxed{i\hbar\frac{d}{dt}\langle\hat{Q}\rangle = \left\langle[\hat{Q},\hat{H}]\right\rangle.} \tag{5.2.5}$$

In deriving this equation, we did not have to assume that $\hat{Q}$ is a Hermitian operator: the equation holds for *any* time-independent operator.

On the right hand-side, we have the expectation value of the commutator of $\hat{Q}$ with the Hamiltonian. In general that commutator must be simplified, and there are a number of commutator identities that can be of help. We will discuss them briefly now.

First, note the antisymmetry of the commutator, which follows directly from the definition $[A,B] = AB - BA$:

$$[A,B] = -[B,A] \quad \text{and} \quad [A,A] = 0. \tag{5.2.6}$$

We also have a distributive property:

$$\begin{aligned} [A, B+C] &= [A,B] + [A,C]\,, \\ [A+B, C] &= [A,C] + [B,C]\,. \end{aligned} \tag{5.2.7}$$

You should find the above identities rather easy to prove. With a bit more effort (problem 5.1), one can also prove the so-called derivation properties:

$$[A, BC] = [A, B]\, C + B\, [A, C],$$
$$[AB, C] = A\, [B, C] + [A, C]\, B. \tag{5.2.8}$$

Finally, we have the Jacobi identity:

$$0 = [A, [B, C]] + [B, [C, A]] + [C, [A, B]]\,. \tag{5.2.9}$$

This is proven by expanding the products implicit in all the commutators and noting that the twelve terms that arise cancel in pairs.

Let us consider a set of commutators that appear often in applications. Begin with $[\hat{p}, \hat{x}] = -i\hbar$, and let us examine $[\hat{p}, \hat{x}^2]$. Using the derivation property above, we see that

$$[\hat{p}, \hat{x}^2] = [\hat{p}, \hat{x}]\hat{x} + \hat{x}[\hat{p}, \hat{x}] = -i\hbar\hat{x} + \hat{x}(-i\hbar) = -i\hbar\, 2\hat{x}. \tag{5.2.10}$$

This result leads us to claim that

$$\boxed{[\hat{p}, \hat{x}^n] = -i\hbar\, n\hat{x}^{n-1}.} \tag{5.2.11}$$

We can easily prove this by induction. We have seen that it holds for $n=1$ and $n=2$. Assume it holds as written above and now consider

$$\begin{aligned}[\hat{p}, \hat{x}^{n+1}] &= [\hat{p}, \hat{x}^n\, \hat{x}] = [\hat{p}, \hat{x}^n]\hat{x} + \hat{x}^n[\hat{p}, \hat{x}] \\ &= -i\hbar\, n\hat{x}^{n-1}\hat{x} + \hat{x}^n(-i\hbar) = -i\hbar\,(n+1)\hat{x}^n.\end{aligned} \tag{5.2.12}$$

This shows that (5.2.11) holds for $\hat{x}^{n+1}$, thus completing the induction argument. We see that, apart from the $(-i\hbar)$ factor, the $\hat{p}$ commutator takes the $\hat{x}$ derivative of $\hat{x}^n$. We can extend this to an arbitrary polynomial as

$$[\hat{p}, a_0 + a_1\hat{x} + a_2\hat{x}^2 + a_3\hat{x}^3 + \cdots] = -i\hbar\,(a_1 + 2a_2\hat{x} + 3a_3\hat{x}^2 + \cdots). \tag{5.2.13}$$

We deduce that for any function $f(x)$ that has a Taylor expansion about $x=0$ we have

$$\boxed{[\hat{p}, f(\hat{x})] = \frac{\hbar}{i}\frac{df(\hat{x})}{d\hat{x}}.} \tag{5.2.14}$$

This result is consistent with our early identification $\hat{p} = \frac{\hbar}{i}\frac{\partial}{\partial x}$ in (3.2.4). We have now shown how this works for commutators of $\hat{p}$ with functions of the position operator.

Exercise 5.3. *Use induction to show that*

$$[\hat{x}, \hat{p}^n] = i\hbar\, n\hat{p}^{n-1}. \tag{5.2.15}$$

This means that, effectively, $\hat{x} = i\hbar\frac{\partial}{\partial \hat{p}}$, *as we claimed already in (4.4.16).*

5.3 Hermitian Operators and Axioms of Quantum Mechanics

In this section we introduce an efficient notation for integrals, expectation values, normalization condition, and Hermitian operators. We state and discuss four important properties of Hermitian operators. We conclude with a discussion of the measurement axiom as well as a brief mention of the other axioms of quantum mechanics.

It is convenient to use a practical notation for the integrals of pairs of functions. Given two (possibly complex) functions $f(x)$ and $g(x)$, we define

$$\boxed{\langle f\,,\,g\rangle \equiv \int_{-\infty}^{\infty} dx\, f^*(x)\,g(x).} \tag{5.3.1}$$

We call the object $\langle\cdot\,,\,\cdot\rangle$ an **inner product** on the space of functions. Note that given two functions f and g, the inner product $\langle f, g\rangle$ is a number. For any constant a, we have that

$$\langle af, g\rangle = a^*\langle f, g\rangle, \quad \langle f, ag\rangle = a\langle f, g\rangle. \tag{5.3.2}$$

The inner product does not treat the first and second functions symmetrically: in the integrand the first function is complex conjugated, while the second function is not. In fact, as you can quickly check, complex conjugation exchanges the order of the inputs:

$$\langle f, g\rangle^* = \langle g, f\rangle. \tag{5.3.3}$$

For functions g_i and f_i and constants c_i, you should check that the inner product satisfies

$$\begin{aligned}\Big\langle f\,,\sum_i c_i g_i\Big\rangle &= \sum_i c_i\langle f\,,\,g_i\rangle,\\ \Big\langle\sum_i c_i f_i\,,\,g\Big\rangle &= \sum_i c_i^*\langle f_i\,,\,g\rangle.\end{aligned} \tag{5.3.4}$$

If two functions have zero inner product, we say that the functions are orthogonal. We define the **norm** $\|f\| \geq 0$ of the function f by the equation

$$\|f\|^2 \equiv \langle f, f\rangle = \int_{-\infty}^{\infty} dx\, |f(x)|^2 \geq 0. \tag{5.3.5}$$

This equation defines a real norm $\|f\| \geq 0$ because its square $\|f\|^2$ is nonnegative. Recall that the norm $|a|$ of a constant is defined by $|a|^2 = a^*a$.

Exercise 5.4. *Show that $\|f\| = 0$ implies $f(x) = 0$ for all x. Assume $f(x)$ is continuous.*

Exercise 5.5. *Show that $\|af\| = |a|\,\|f\|$, where f is a function, and $a \in \mathbb{C}$ is a constant.*

A normalized wave function Ψ satisfies

$$\|\Psi\|^2 = \langle\Psi, \Psi\rangle = 1. \tag{5.3.6}$$

Such a wave function has a unit norm $\|\Psi\| = 1$.

Recall now the definition of the Hermitian conjugate $T^\dagger$ of an operator T. We did this in (3.6.13), where we wrote

$$\int_{-\infty}^{\infty} dx\,(T^\dagger\Psi_1)^*\,\Psi_2 \;=\; \int_{-\infty}^{\infty} dx\,\Psi_1^*\,(T\Psi_2). \tag{5.3.7}$$

With the present notation, this reads as

$$\big(T^\dagger\Psi_1\,,\Psi_2\big) = \big(\Psi_1\,,T\Psi_2\big). \tag{5.3.8}$$

$T^\dagger$ is the operator that acting on Ψ_1 on the left-hand side reproduces the result of the right-hand side. An operator $\hat{Q}$ is Hermitian if $\hat{Q}^\dagger = \hat{Q}$, and therefore for the class of wave functions Ψ we work with,

$$\hat{Q}\ \text{ is Hermitian:}\quad \big(\hat{Q}\Psi_1,\Psi_2\big) \;=\; \big(\Psi_1,\hat{Q}\Psi_2\big). \tag{5.3.9}$$

The expectation value of $\hat{Q}$ was defined by $\langle\,\hat{Q}\,\rangle_\Psi = \int dx\,\Psi^*\,\hat{Q}\Psi$. Now we write

$$\langle\,\hat{Q}\,\rangle_\Psi \;=\; \big(\Psi\,,\,\hat{Q}\Psi\big). \tag{5.3.10}$$

Under $\Psi \to a\Psi$ with constant a, we have

$$\big(\Psi\,,\,\hat{Q}\Psi\big) \;\to\; |a|^2\big(\Psi\,,\,\hat{Q}\Psi\big). \tag{5.3.11}$$

For the formula for $\langle\,\hat{Q}\,\rangle_\Psi$ to be meaningful, the state Ψ must be normalized. Indeed, a normalized wave function is only ambiguous up to a phase factor. A phase factor has unit norm, and the equation above shows that this ambiguity does not affect the expectation value.

Exercise 5.6. *Let $\hat{Q}_1$ and $\hat{Q}_2$ be two Hermitian operators. Show that*

- *the sum $\alpha\hat{Q}_1 + \beta\hat{Q}_2$, with $\alpha, \beta \in \mathbb{R}$ is a Hermitian operator;*
- *the product $\hat{Q}_1\hat{Q}_2$ is not Hermitian unless the operators commute.*

We now consider four key properties of Hermitian operators. The first two we prove; the last two are also true but we only prove them partially.

Claim 1. *The expectation value of a Hermitian operator is real.* To prove this we complex conjugate the above definition (5.3.10) of $\langle\hat{Q}\rangle_\Psi$ and check that the result is the expectation value itself:

$$\big(\langle Q\rangle_\Psi\big)^* \;=\; \big(\Psi,\,\hat{Q}\Psi\big)^* \;=\; \big(\hat{Q}\Psi,\Psi\,\big) \;=\; \big(\Psi,\hat{Q}\Psi\,\big) \;=\; \langle Q\rangle_\Psi. \tag{5.3.12}$$

For the second equality, we used the conjugation property (5.3.3) of the inner product. For the third equality, we used the Hermiticity of $\hat{Q}$. The expectation value is thus real. □

Claim 2. *The eigenvalues of a Hermitian operator are real.* Assume the operator $\hat{Q}$ has an eigenvalue q_1 associated with a normalized eigenfunction $\psi_1(x)$:

$$\hat{Q}\psi_1(x) \;=\; q_1\psi_1(x). \tag{5.3.13}$$

We now see that the expectation value of $\hat{Q}$ in the state ψ_1 is in fact the eigenvalue q_1:

$$\langle\hat{Q}\rangle_{\psi_1} = \langle\psi_1\,,\hat{Q}\psi_1\rangle = \langle\psi_1\,,q_1\,\psi_1\rangle \;=\; q_1\langle\psi_1\,,\psi_1\rangle = q_1. \tag{5.3.14}$$

By claim 1, the expectation value of $\hat{Q}$ is real, and this proves the reality of the eigenvalue q_1, as we wanted to show. □

Exercise 5.7. *Let Ψ be a normalized $\hat{Q}$ eigenstate. Show that*

$$\hat{Q}\Psi = \langle\hat{Q}\rangle_\Psi \Psi. \tag{5.3.15}$$

Exercise 5.8. *An operator $\hat{A}$ is said to be antihermitian if $\hat{A}^\dagger = -\hat{A}$. Let Ψ be a normalized eigenstate of $\hat{A}$ with eigenvalue λ. Show that $\lambda^* = -\lambda$, which means λ is purely imaginary. Note that $\lambda = 0$ is also a possibility: 0 is in fact both real and imaginary.*

Consider now the collection $\{\psi_i(x), q_i\}$ of eigenfunctions and eigenvalues of the Hermitian operator $\hat{Q}$:

$$\hat{Q}\,\psi_i(x) = q_i\psi_i(x). \tag{5.3.16}$$

The list may be finite $(i=1,\ldots,N)$ or infinite $(i=1,\ldots,\infty)$. We are assuming for simplicity that the eigenvectors are countable.

Claim 3. *The eigenfunctions of a Hermitian operator $\hat{Q}$ can be chosen to be orthonormal:*

$$\langle\psi_i, \psi_j\rangle = \int dx\,\psi_i^*(x)\psi_j(x) = \delta_{ij}. \tag{5.3.17}$$

For $i=j$, this is just a matter of properly normalizing each eigenfunction, which we can easily do. The equation also states that different eigenfunctions are orthogonal, or have zero inner product. We now explain why this is necessarily so for two eigenfunctions ψ_i and ψ_j such that $q_i \neq q_j$. For this we evaluate $\langle\psi_i, \hat{Q}\psi_j\rangle$ in two different ways. First,

$$\langle\psi_i, \hat{Q}\psi_j\rangle = \langle\psi_i, q_j\psi_j\rangle = q_j\langle\psi_i, \psi_j\rangle. \tag{5.3.18}$$

Second, using the Hermiticity of $\hat{Q}$ and the reality of the eigenvalues,

$$\langle\psi_i, \hat{Q}\psi_j\rangle = \langle\hat{Q}\psi_i, \psi_j\rangle = \langle q_i\psi_i, \psi_j\rangle = q_i\langle\psi_i, \psi_j\rangle. \tag{5.3.19}$$

Equating the final right-hand sides in the two evaluations, we get

$$(q_j - q_i)\,\langle\psi_i\,, \psi_j\rangle = 0. \tag{5.3.20}$$

Since the eigenvalues were assumed to be different, the first factor cannot vanish, proving that $\langle\psi_i, \psi_j\rangle = 0$, as claimed. This is not yet a full proof of (5.3.17) because it is possible to have **degenerate** eigenfunctions—namely, different eigenfunctions with the *same* eigenvalue. In that case the above argument does not work. One must then show that it is possible to *choose* linear combinations of the degenerate eigenfunctions that are mutually orthogonal. This choice of orthonormal degenerate eigenfunctions is guaranteed to be possible by the Gram-Schmidt procedure discussed in section 14.2. The degenerate eigenfunctions are automatically orthogonal to the eigenfunctions outside the degenerate subspace because the eigenvalues are in this case different.

Claim 4. *The orthonormal eigenfunctions of a Hermitian operator $\hat{Q}$ form a complete set of basis functions: any reasonable Ψ can be written as a superposition of $\hat{Q}$ eigenfunctions.* This is the so-called spectral theorem, a deep result in the theory of complex vector spaces that will not be proven here but will be established for finite-dimensional vector spaces in section 15.6. This means that for any wave function at any fixed time we can write

$$\Psi(x) = \alpha_1\psi_1(x) + \alpha_2\psi_2(x) + \cdots = \sum_i \alpha_i\psi_i(x), \tag{5.3.21}$$

with calculable coefficients α_i given by the inner product of ψ_i and Ψ:

$$\alpha_i = \langle\psi_i, \Psi\rangle. \tag{5.3.22}$$

We can quickly confirm that this value of α_i arises from (5.3.21) using the linearity property (5.3.4) and the orthonormality of the eigenfunctions:

$$\langle\psi_i, \Psi\rangle = \Big\langle\psi_i, \sum_j \alpha_j\psi_j\Big\rangle = \sum_j \alpha_j\langle\psi_i, \psi_j\rangle = \sum_j \alpha_j\delta_{ij} = \alpha_i. \tag{5.3.23}$$

The condition that Ψ is normalized implies a condition on the coefficients α_i. We have

$$\langle\Psi\,,\,\Psi\rangle = \Big\langle\sum_i \alpha_i\psi_i\,,\,\sum_j \alpha_j\psi_j\Big\rangle = \sum_{i,j} \alpha_i^*\alpha_j\langle\psi_i\,,\,\psi_j\rangle = \sum_{i,j} \alpha_i^*\alpha_j\delta_{ij} = \sum_i \alpha_i^*\alpha_i. \tag{5.3.24}$$

The wave function Ψ is normalized if

$$\sum_i |\alpha_i|^2 = 1. \tag{5.3.25}$$

We are finally in the position to state the measurement axiom of quantum mechanics. The measurement axiom follows the *Copenhagen interpretation* of quantum mechanics. With a little abuse of language, we refer to the measurement of an observable on a state as the measurement of the corresponding Hermitian operator on the state. So measuring the momentum of the state Ψ is stated as measuring $\hat{p}$ on Ψ.

Measurement axiom. *If we measure the Hermitian operator $\hat{Q}$ on the (normalized) state Ψ, the possible outcomes for the measured values are the eigenvalues $q_1, q_2, \ldots$ of $\hat{Q}$ associated with the orthonormal eigenvectors $\psi_1, \psi_2, \ldots$. With the state written as $\Psi = \sum_i \alpha_i\psi_i$, the probability p_i of measuring q_i is given by*

$$p_i = |\alpha_i|^2. \tag{5.3.26}$$

After the outcome q_i, the state of the system becomes

$$\Psi = \psi_i. \tag{5.3.27}$$

This is called the collapse of the wave function. If the spectrum of $\hat{Q}$ is degenerate after measuring a degenerate eigenvalue, the wave function collapses in the associated degenerate subspace (see remark 4 below).

Remarks:

1. As required, the probabilities p_i of measuring the various eigenvalues add to one:

$$\sum_i p_i = \sum_i |\alpha_i|^2 = 1, \tag{5.3.28}$$

by the normalization condition for Ψ given in (5.3.25).

2. If the state of the particle is a $\hat{Q}$ eigenstate ψ_i, the axiom implies that the measurement of $\hat{Q}$ yields q_i with a probability equal to one—there is no uncertainty in the measured value. The state remains ψ_i right after the measurement.
3. The collapse of the wave function is an instantaneous, discontinuous change in the wave function. Immediately after the measurement that yields q_i, the state becomes ψ_i, implying that an immediate repeated measurement of $\hat{Q}$ will yield q_i with no uncertainty. In between measurements the wave function evolves in time continuously as dictated by the Schrödinger equation.
4. Suppose $\hat{Q}$ has a degenerate eigenvalue q_k with a number $p \geq 2$ of orthonormal eigenstates $\psi_k^{(1)}, \cdots, \psi_k^{(p)}$. Assume the wave function Ψ takes the form

$$\Psi = \alpha_k^{(1)}\psi_k^{(1)} + \cdots + \alpha_k^{(p)}\psi_k^{(p)} + \sum_{i \neq k} \alpha_i \psi_i. \tag{5.3.29}$$

The measurement of $\hat{Q}$ yields q_k with probability p_k, given by

$$p_k = |\alpha_k^{(1)}|^2 + \cdots + |\alpha_k^{(p)}|^2. \tag{5.3.30}$$

The state after this measurement is

$$\Psi = \frac{\alpha_k^{(1)}\psi_k^{(1)} + \cdots + \alpha_k^{(p)}\psi_k^{(p)}}{\sqrt{|\alpha_k^{(1)}|^2 + \cdots + |\alpha_k^{(p)}|^2}} = \frac{\alpha_k^{(1)}\psi_k^{(1)} + \cdots + \alpha_k^{(p)}\psi_k^{(p)}}{\sqrt{p_k}}. \tag{5.3.31}$$

The square root denominator provides the proper normalization to Ψ, as you should check. The wave function collapses within the degenerate subspace.
5. Recall that any state can be written as a sum of different states in a number of ways. If we are to measure $\hat{Q}_1$, we expand the state in $\hat{Q}_1$ eigenstates; if we are to measure $\hat{Q}_2$, we expand the state in $\hat{Q}_2$ eigenstates, and so on and so forth. Each decomposition is suitable for a particular measurement. Each decomposition reveals the various probabilities for the outcomes of the specific observable.

Example 5.2. *Expectation value of a Hermitian operator.*
We wish to calculate the expectation value $\langle\hat{Q}\rangle$ on the state $\Psi = \sum_i \alpha_i \psi_i$. This is a matter of the computation

$$\begin{aligned}\langle\hat{Q}\rangle &= (\Psi, \hat{Q}\Psi) = \sum_i \sum_j \alpha_i^* \alpha_j \, (\psi_i, \hat{Q}\psi_j) = \sum_i \sum_j \alpha_i^* \alpha_j \, (\psi_i, q_j \psi_j) \\ &= \sum_i \sum_j \alpha_i^* \alpha_j \, q_j \delta_{ij} = \sum_i \alpha_i^* \alpha_i q_i = \sum_i |\alpha_i|^2 q_i = \sum_i p_i q_i.\end{aligned} \tag{5.3.32}$$

This is a good consistency check: the expectation value of $\hat{Q}$, as anticipated, is the sum of the possible outcomes q_i multiplied by the corresponding probabilities p_i. □

Exercise 5.9. *Convince yourself that any nontrivial superposition of $\hat{Q}$ eigenfunctions with different eigenvalues is* not *a $\hat{Q}$ eigenfunction. On the other hand, show that any superposition of degenerate $\hat{Q}$ eigenfunctions is a $\hat{Q}$ eigenfunction.*

Axioms of quantum mechanics We have seen enough of quantum mechanics to have a first look at the ideas that show us how to apply quantum mechanics to physical systems. These are called the *axioms* of quantum mechanics. Indeed, above we discussed the measurement axiom, a key part of the axiomatic structure of quantum mechanics. Let's list the axioms and then discuss them briefly.

A1. The wave function is the complete description of a quantum system.

A2. Hermitian operators are observables.

A3. The measurement axiom.

A4. Time evolution via the Schrödinger equation.

Axiom A1 states that the wave function represents the most that can be known about the quantum system. The wave function can be thought of as a vector in a complex vector space equipped with an inner product. We have not yet defined such objects in generality, but in the cases we have already looked at, the vector space was the space of complex functions, and the corresponding inner product $\langle\cdot,\cdot\rangle$ was defined in (5.3.1).

Axiom A2 states that Hermitian operators are observables. Hermitian operators are those that can be moved freely from the first to the second input in the inner product. We have shown above that the eigenvalues of such operators are real and have argued that the eigenfunctions provide a complete, orthonormal set of functions. These properties allow Hermitian operators to be measured, as explained in the measurement axiom.

Axiom A3 is the measurement axiom. It was stated above (see (5.3.26)) in the context of the measurement of a Hermitian operator. This is the axiom where probabilities make their appearance. In general the result of a measurement cannot be predicted with certainty.

Axiom A4 is the last of the axioms. It stipulates that the wave function evolves deterministically according to the Schrödinger equation. The Schrödinger equation is fully specified once the Hamiltonian $\hat{H}$ is known. Note that in a measurement the state changes nondeterministically. The Schrödinger equation does not govern the change in the wave function during a measurement. This fact has been quite puzzling to a number of physicists, who have wondered are measurements special. Despite such concerns, taken as axioms, A3 and A4 seem completely consistent with all observations.

We will revisit the above axioms in chapter 16, after we have learned more about quantum systems, complex vector spaces, and the structure of Hermitian operators.

Example 5.3. *Incompatible observables.* Consider two observables $\hat{Q}_1$ and $\hat{Q}_2$. Suppose there is a state ψ that is an eigenstate of both operators:

$$\hat{Q}_1\psi = \lambda_1\psi\,, \quad \hat{Q}_2\psi = \lambda_2\psi. \tag{5.3.33}$$

If your quantum system is in state ψ, a $\hat{Q}_1$ measurement would yield λ_1, and a $\hat{Q}_2$ measurement would yield λ_2. Both are valid simultaneously, meaning that the state can be characterized by the values of λ_1 and λ_2. Indeed, if you measure $\hat{Q}_1$ on the state, you get λ_1, with the state, remaining ψ immediately after the measurement (remark 2 below the measurement axiom). Measuring $\hat{Q}_2$ at this point would therefore yield λ_2.

Two observables $\hat{A}$ and $\hat{B}$ are said to be *incompatible* if there is no state ψ that is a simultaneous eigenstate of the operators. There are no definite measured values of both $\hat{A}$ and $\hat{B}$ on any state. It is sometimes easy to tell that two observables are incompatible. Suppose the commutator of $\hat{A}$ and $\hat{B}$ is a number,

$$[\hat{A}, \hat{B}] = ic, \tag{5.3.34}$$

where i is the imaginary unit, and c is a *nonzero* real number. We now show that there cannot be a simultaneous eigenstate of both operators. Assume one such state ψ exists so that $\hat{A}\psi = \lambda_A \psi$, and $\hat{B}\psi = \lambda_B \psi$. Now let the commutator of $\hat{A}$ and $\hat{B}$ act on ψ:

$$[\hat{A}, \hat{B}]\psi = \hat{A}(\hat{B}\psi) - \hat{B}(\hat{A}\psi) = \lambda_B \hat{A}\psi - \lambda_A \hat{B}\psi = (\lambda_B \lambda_A - \lambda_A \lambda_B)\psi = 0. \tag{5.3.35}$$

We get zero. On the other hand, the right-hand side of (5.3.34) acting on the state simply yields $ic\psi$, which is different from zero. This contradiction shows that no simultaneous eigenstate can exist.

The classic examples of incompatible operators are the position and momentum operators. Indeed, their commutator is a constant: $[\hat{x}, \hat{p}] = i\hbar$. Their incompatibility means that, given a particle, you cannot simultaneously know both its position and its momentum. Of course, the possibility of such knowledge is the basis of classical mechanics. □

5.4 Free Particle on a Circle—a First Look

Consider now the problem of a particle of mass m confined to a circle of circumference L. The coordinate along the circle is called x, and we can view the circle as the interval $x \in [0, L]$ with the end points $x = 0$ and $x = L$ identified or declared to be the same point. Indeed, if you have a finite piece of thread and you tie one end point to the other, the thread forms a closed curve that is topologically a circle. It is clearer, however, to think of the circle as the full real line $x \in (-\infty, \infty)$ with the identification

$$x \sim x + L, \tag{5.4.1}$$

which means that two points whose coordinates are related in this way are to be considered *the same point*. Given this identification, we can indeed choose to view the circle as the set of points $x \in [0, L]$ with $x = 0$ and $x = L$ identified (see figure 5.1). The identification (5.4.1) suggests that the wave function must satisfy the periodicity condition

$$\Psi(x + L, t) = \Psi(x, t). \tag{5.4.2}$$

We will declare this equation to hold for all x. From this it follows that not only Ψ is periodic; all of its derivatives with respect to x are periodic too.

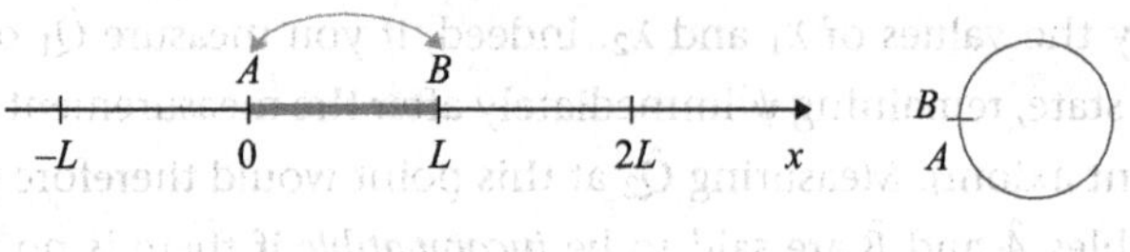

Figure 5.1
A circle of circumference L presented as the real line x with the identification $x \sim x+L$. After the identification, all points on the line are represented by those on $[0, L]$, with point A at $x=0$ declared identical to point B at $x=L$. The result is the circle shown to the right.

Since the particle is free, the Hamiltonian of this system is just the kinetic energy operator:

$$\hat{H} = \frac{\hat{p}^2}{2m} = -\frac{\hbar^2}{2m}\frac{\partial^2}{\partial x^2}. \tag{5.4.3}$$

Assume that at some fixed time, say, $t=0$, we have the wave function

$$\Psi(x, 0) = \sqrt{\frac{2}{L}}\Big(\frac{1}{\sqrt{3}}\sin\frac{2\pi x}{L} + \sqrt{\frac{2}{3}}\cos\frac{6\pi x}{L}\Big). \tag{5.4.4}$$

This wave function satisfies the periodicity condition, as you should check. It is also properly normalized, as you will verify later. We want to know the possible values of the momentum and their corresponding probabilities.

Given the measurement axiom, we must find the set of momentum eigenstates and rewrite the wave function as a superposition of such states. We recall that in the full line an exponential e^{ikx} is a momentum eigenstate with momentum $\hbar k$. Two things happen on the circle that do not happen on the full line. First, the momentum will be quantized as a consequence of the periodicity condition (5.4.2). Second, since the circle has finite length the momentum eigenfunctions will be normalizable! Consider first the periodicity condition as applied to e^{ikx}. We require that

$$e^{ikx} = e^{ik(x+L)} \quad\Rightarrow\quad e^{ikL} = 1 \quad\Rightarrow\quad kL = 2\pi n,\ \ n \in \mathbb{Z}. \tag{5.4.5}$$

Note that n can be any integer: positive, negative, or zero. We thus write for the momentum eigenstates, labeled by n:

$$\psi_n(x) = N_n e^{\frac{2\pi i n x}{L}}, \tag{5.4.6}$$

with N_n a real normalization constant. The normalization condition gives

$$1 = \int_0^L |\psi_n(x)|^2 dx = N_n^2 \int_0^L dx = N_n^2 L \quad\Rightarrow\quad N_n = \frac{1}{\sqrt{L}}. \tag{5.4.7}$$

Therefore, our normalized momentum eigenstates are

$$\psi_n(x) = \frac{1}{\sqrt{L}} e^{\frac{2\pi i n x}{L}}, \tag{5.4.8}$$

and these are states with momentum p_n, which is calculated as follows:

$$\hat{p}\,\psi_n = \frac{\hbar}{i}\frac{\partial}{\partial x}\psi_n = \frac{2\pi\, n\hbar}{L}\psi_n \quad\Rightarrow\quad p_n = \frac{2\pi\, n\hbar}{L}. \tag{5.4.9}$$

Now that we are equipped with the momentum eigenstates we must rewrite the wave function (5.4.4) as a superposition of such states. Rewriting the sin and cos functions as sums of exponentials,

$$\Psi(x,0) = \sqrt{\frac{2}{3}}\frac{1}{2i}\frac{1}{\sqrt{L}}\left(e^{\frac{2\pi ix}{L}} - e^{-\frac{2\pi ix}{L}}\right) + \frac{1}{\sqrt{3}}\frac{2}{2}\frac{1}{\sqrt{L}}\left(e^{\frac{6\pi ix}{L}} + e^{-\frac{6\pi ix}{L}}\right). \tag{5.4.10}$$

We then recognize that we have

$$\Psi(x,0) = \sqrt{\frac{2}{3}}\frac{1}{2i}\,\psi_1(x) - \sqrt{\frac{2}{3}}\frac{1}{2i}\,\psi_{-1}(x) + \frac{1}{\sqrt{3}}\,\psi_3(x) + \frac{1}{\sqrt{3}}\,\psi_{-3}(x). \tag{5.4.11}$$

This is what we wanted: the original wave function written as a superposition of momentum eigenstates $\psi_m(x)$. We were lucky that we did not have to do any integration to calculate the coefficients of this expansion.

Exercise 5.10. *Verify that the above wave function is correctly normalized.*

Using the measurement axiom, we can now give the possible values p_n of the momentum and their corresponding probabilities P_n:

$$\begin{aligned}
p_1 &= \frac{2\pi\hbar}{L}, \quad P_1 = \left|\sqrt{\frac{2}{3}}\frac{1}{2i}\right|^2 = \frac{1}{6},\\
p_{-1} &= -\frac{2\pi\hbar}{L}, \quad P_{-1} = \left|-\sqrt{\frac{2}{3}}\frac{1}{2i}\right|^2 = \frac{1}{6},\\
p_3 &= \frac{6\pi\hbar}{L}, \quad P_3 = \left|\sqrt{\frac{1}{3}}\right|^2 = \frac{1}{3},\\
p_{-3} &= -\frac{6\pi\hbar}{L}, \quad P_{-3} = \left|\sqrt{\frac{1}{3}}\right|^2 = \frac{1}{3}.
\end{aligned} \tag{5.4.12}$$

Exercise 5.11. *Consider again $\Psi(x,0)$ as given in (5.4.4). Each term here is an eigenstate of the energy operator $\hat{H}$. Imagine measuring the energy. Find the possible values E_i of the measured energy and their probabilities $p(E_i)$.*

Exercise 5.12. *For the particle on a circle, show that momentum eigenstates are energy eigenstates. What is the energy of $\psi_n(x)$? Are ψ_n and ψ_{-n} degenerate energy eigenstates? Are there degenerate momentum eigenstates?*

5.5 Uncertainty

For random variables, the uncertainty is the *standard deviation*: the square root of the expected value of the square of deviations from the average value. Let Q be a random variable that takes on values $Q_1, \ldots, Q_n$ with probabilities $p_1, \ldots, p_n$, respectively. As usual, $0 < p_i \leq 1$ for all i, and $\sum_i p_i = 1$. The expected value $\langle Q\rangle$ of the random variable Q is given by

$$\langle Q\rangle = \sum_i p_i Q_i. \tag{5.5.1}$$

The variance, which is the square of the standard deviation $\Delta Q \geq 0$, is given by

$$\text{var}\, Q = (\Delta Q)^2 \equiv \sum_i p_i (Q_i - \langle Q \rangle)^2. \tag{5.5.2}$$

The right-hand side is greater than or equal to zero; in fact, it is a sum of contributions, all of which are greater than or equal to zero. This definition makes it clear that if $\Delta Q = 0$, the random variable has just one possible value: each term in the above sum must vanish, making $Q_i = \langle Q \rangle$ for all i.

We find another useful expression for the variance by expanding the right-hand side in the above definition:

$$\begin{aligned}(\Delta Q)^2 &= \sum_i p_i Q_i^2 - 2\sum_i p_i Q_i \langle Q \rangle + \sum_i p_i \langle Q \rangle^2 \\ &= \sum_i p_i Q_i^2 - 2\langle Q \rangle \sum_i p_i Q_i + \langle Q \rangle^2 \sum_i p_i \\ &= \langle Q^2 \rangle - 2\langle Q \rangle \langle Q \rangle + \langle Q \rangle^2 \\ &= \langle Q^2 \rangle - \langle Q \rangle^2,\end{aligned} \tag{5.5.3}$$

where we use $\sum_i p_i = 1$. Therefore,

$$(\Delta Q)^2 = \langle Q^2 \rangle - \langle Q \rangle^2. \tag{5.5.4}$$

Since by definition $(\Delta Q)^2 \geq 0$, we have an interesting inequality:

$$\langle Q^2 \rangle \geq \langle Q \rangle^2. \tag{5.5.5}$$

Exercise 5.13. *Let Q be a random variable with two possible values: $Q = +1$ with probability p_1, and $Q = -1$ with probability p_2, with $p_1 + p_2 = 1$. Calculate the expectation value $\langle Q \rangle$ and the uncertainty ΔQ. For what values of p_1 and p_2 is the uncertainty a maximum?*

Exercise 5.14. *A random variable Q has a uniform probability distribution $\mathcal{P}(Q)$ over a limited range:*

$$\mathcal{P}(Q) = \begin{cases} \frac{1}{2L}, & |Q| < L, \\ 0, & \text{otherwise.} \end{cases} \tag{5.5.6}$$

As required, $\int_{-\infty}^{\infty} \mathcal{P}(Q) dQ = 1$. Find the expectation value $\langle Q \rangle$ and the uncertainty ΔQ.

Now let us consider the quantum mechanical case. We have already defined expectation values of Hermitian operators, so we can now mirror (5.5.4) and declare that the uncertainty $\Delta\hat{Q}(\Psi)$ of an operator in a state Ψ is a real number greater than or equal to zero whose square is given by

$$(\Delta\hat{Q}(\Psi))^2 \equiv \langle \hat{Q}^2 \rangle_\Psi - (\langle \hat{Q} \rangle_\Psi)^2. \tag{5.5.7}$$

For brevity, we often omit the state Ψ on the symbols, writing the above as

$$\boxed{(\Delta\hat{Q})^2 \equiv \langle \hat{Q}^2 \rangle - \langle \hat{Q} \rangle^2.} \tag{5.5.8}$$

It is with this definition that one can precisely formulate the Heisenberg uncertainty principle (4.1.19): for any wave packet, the uncertainties Δx and Δp satisfy $\Delta x \Delta p \geq \frac{\hbar}{2}$. This inequality will be proven in section 15.2.

We will now prove two claims that allow us to write the uncertainty in useful ways.

Claim 1. The uncertainty can be written as the expectation value of the square of the difference between the operator and its expectation value:

$$(\Delta \hat{Q})^2 = \left\langle (\hat{Q} - \langle \hat{Q} \rangle)^2 \right\rangle. \tag{5.5.9}$$

Indeed, by expanding the square on the right-hand side, recalling that $\langle \hat{Q} \rangle$ is a number and thus commutes with $\hat{Q}$, and noting that the expectation value of a sum of operators is the sum of expectation values, we show that

$$\left\langle (\hat{Q} - \langle \hat{Q} \rangle)^2 \right\rangle = \langle \hat{Q}^2 - 2\langle \hat{Q} \rangle \hat{Q} + \langle \hat{Q} \rangle^2 \rangle = \langle \hat{Q}^2 \rangle - 2\left\langle \langle \hat{Q} \rangle \hat{Q} \right\rangle + \left\langle \langle \hat{Q} \rangle^2 \right\rangle. \tag{5.5.10}$$

For any number c, we have $\langle c\hat{Q} \rangle = c\langle \hat{Q} \rangle$. Since $\langle \hat{Q} \rangle$ is a number, this helps simplify the second term on the final right-hand side above. In addition, $\langle c \rangle = c$, and this helps simplify the last term since $\langle \hat{Q} \rangle^2$ is also a number. Therefore,

$$\left\langle (\hat{Q} - \langle \hat{Q} \rangle)^2 \right\rangle = \langle \hat{Q}^2 \rangle - 2\langle \hat{Q} \rangle \langle \hat{Q} \rangle + \langle \hat{Q} \rangle^2 = \langle \hat{Q}^2 \rangle - \langle \hat{Q} \rangle^2 = (\Delta \hat{Q})^2, \tag{5.5.11}$$

completing the proof of claim 1. □

Claim 2. The uncertainty $\Delta \hat{Q}$ is in fact the norm of wave function $(\hat{Q} - \langle \hat{Q} \rangle)\Psi$:

$$\Delta \hat{Q} = \| (\hat{Q} - \langle \hat{Q} \rangle)\Psi \|. \tag{5.5.12}$$

To prove this we begin with the claim 1 expression (5.5.9) for the squared uncertainty. By definition of the expectation value, we find that

$$(\Delta \hat{Q})^2 = \left\langle (\hat{Q} - \langle \hat{Q} \rangle)(\hat{Q} - \langle \hat{Q} \rangle) \right\rangle = \left(\Psi, (\hat{Q} - \langle \hat{Q} \rangle)(\hat{Q} - \langle \hat{Q} \rangle)\Psi \right). \tag{5.5.13}$$

The operator $\hat{Q} - \langle \hat{Q} \rangle$ is Hermitian because $\hat{Q}$ is Hermitian and $\langle \hat{Q} \rangle$ is real. We can therefore move the leftmost $\hat{Q} - \langle \hat{Q} \rangle$ factor into the first entry of the inner product:

$$(\Delta \hat{Q})^2 = \left((\hat{Q} - \langle \hat{Q} \rangle)\Psi, (\hat{Q} - \langle \hat{Q} \rangle)\Psi \right) = \left\| (\hat{Q} - \langle \hat{Q} \rangle)\Psi \right\|^2, \tag{5.5.14}$$

recalling the definition (5.3.5) of the norm. This completes the proof of claim 2. □

Exercise 5.15. *Show that for a Hermitian operator $\hat{Q}$ we have*

$$\langle \hat{Q}^2 \rangle_\Psi = \| \hat{Q}\Psi \|^2. \tag{5.5.15}$$

This identity is the essence of the proof of claim 2, which has $\hat{Q}$ in this identity replaced by the operator $\hat{Q} - \langle \hat{Q} \rangle$. Having shown that

$$\left\langle (\hat{Q} - \langle \hat{Q} \rangle)^2 \right\rangle = \langle \hat{Q}^2 \rangle - \langle \hat{Q} \rangle^2, \tag{5.5.16}$$

the now manifest positivity of the left-hand side implies that

$$\langle \hat{Q}^2 \rangle \geq \langle \hat{Q} \rangle^2, \tag{5.5.17}$$

just as was the case for random variables.

It is useful to have a simple characterization of states with zero uncertainty for a given Hermitian operator $\hat{Q}$. Happily, such characterization exists. If $\Delta\hat{Q} = 0$, by claim 2 the norm of $(\hat{Q} - \langle\hat{Q}\rangle)\Psi$ vanishes. But a state of zero norm must vanish (exercise 5.4), and therefore:

$$(\hat{Q} - \langle\hat{Q}\rangle)\Psi = 0, \quad \rightarrow \quad \hat{Q}\,\Psi = \langle\hat{Q}\rangle\,\Psi. \tag{5.5.18}$$

Since $\langle\hat{Q}\rangle$ is a number, we see that Ψ is an eigenstate of $\hat{Q}$. So zero $\hat{Q}$ uncertainty means Ψ is a $\hat{Q}$ eigenstate. On the other hand, the converse is also true. If Ψ is a $\hat{Q}$ eigenstate, then $\hat{Q}\Psi = \langle\hat{Q}\rangle\Psi$ (exercise 5.7). It follows that $(\hat{Q} - \langle\hat{Q}\rangle)\Psi = 0$ and the uncertainty vanishes. All in all, we have established that the uncertainty of $\hat{Q}$ vanishes in a state Ψ *if and only if* Ψ is a $\hat{Q}$ eigenstate:

$$\boxed{\Delta\hat{Q}(\Psi) = 0 \iff \Psi \text{ is an eigenstate of } \hat{Q}.} \tag{5.5.19}$$

A nonvanishing uncertainty $\Delta\hat{Q}(\Psi)$ signals the failure of Ψ to be a $\hat{Q}$ eigenstate. We will explore this statement further in section 15.1.

Exercise 5.16. *Consider again the mass m particle on the circle $x \in [0, L]$, with $t = 0$ wave function*

$$\Psi(x, 0) = \sqrt{\frac{2}{3}}\frac{1}{2i}\psi_1(x) - \sqrt{\frac{2}{3}}\frac{1}{2i}\psi_{-1}(x) + \frac{1}{\sqrt{3}}\psi_3(x) + \frac{1}{\sqrt{3}}\psi_{-3}(x).$$

Here ψ_n denotes the normalized momentum eigenstate of momentum $2\pi n\hbar/L$. Calculate the expected value $\langle\hat{p}\rangle$ of the momentum and the uncertainty $\Delta\hat{p}$ in the state Ψ.

Problems

Problem 5.1. *Exercises with commutators.*

Let A, B, and C be linear operators.

1. Compute $[A, BC]$ in terms of $[A, B]$ and $[A, C]$.
   Compute $[AB, C]$ in terms of $[A, C]$ and $[B, C]$.
2. Verify that $[A, [B, C]] + [B, [C, A]] + [C, [A, B]]$ vanishes.
3. Compute $[AB, CD]$ in terms of $[A, C]$, $[A, D]$, $[B, C]$, and $[B, D]$.
4. Calculate $[\hat{x}\hat{p}, \hat{x}^2]$ and $[\hat{x}\hat{p}, \hat{p}^2]$.

Problem 5.2. *Expectation value of the momentum.*

Consider a wave function that at a given time is real and vanishes as $x \to \pm\infty$. Use the position-space representation to show that $\langle\hat{p}\rangle$ is zero. Show clearly how you reach the same conclusion using the momentum space representation of $\langle\hat{p}\rangle$. (Hint: recall that the reality of the position-space wave function implies a certain property for the momentum-space wave function.)

Problem 5.3. *Free particle evolution.*

Consider the state of a *free* particle of mass m that at $t=0$ is represented by the wave function

$$\Psi(x,0) = \sin k_0 x\,, \quad k_0 \in \mathbb{R}.$$

1. Find the probability current $J(x,0)$.
2. If we measure the momentum of the particle (at $t=0$), what are the possible values that we may obtain?
3. Calculate $\Psi(x,t)$.

Problem 5.4. *Ehrenfest's theorem.*

Consider a particle moving in one dimension with the Hamiltonian $\hat{H}=\frac{\hat{p}^2}{2m}+V(\hat{x})$. Compute $\frac{d}{dt}\langle\hat{x}\rangle$ and $\frac{d}{dt}\langle\hat{p}\rangle$. The resulting quantum equations, analogous to those of classical mechanics, are the contents of Ehrenfest's theorem.

Problem 5.5. *Phase of the wave function.*

For q, a constant with units of momentum, define the boost operator $\hat{B}_q$ as the operator that acts on arbitrary functions of x by multiplication by a q-dependent phase:

$$\hat{B}_q f(x) = e^{iq\hat{x}/\hbar} f(x) = e^{iqx/\hbar} f(x).$$

Suppose $\psi_0(x)$ is a properly normalized wave function with $\langle\hat{x}\rangle_{\psi_0}=x_0$ and $\langle\hat{p}\rangle_{\psi_0}=p_0$, where x_0 and p_0 are constants. Now consider a new wave function obtained by boosting ψ_0:

$$\psi_{\text{new}}(x) = \hat{B}_q \psi_0(x).$$

1. What is the expectation value $\langle\hat{x}\rangle_{\psi_{\text{new}}}$ in the state ψ_{new}?
2. What is the expectation value $\langle\hat{p}\rangle_{\psi_{\text{new}}}$ in the state ψ_{new}?
3. What is the physical effect of adding an overall factor $e^{iqx/\hbar}$ to a wave function representing a momentum eigenstate?
4. Compute $[\hat{p}\,,\,\hat{B}_q]$ and $[\hat{x}\,,\,\hat{B}_q]$. Your answers should be independent of $\hat{x}$ and $\hat{p}$.

Problem 5.6. *Exercises with a particle in a box.*

Consider a particle of mass m that is free to move in the interval $x\in[0,a]$. The potential $V(x)$ is zero in this interval and infinite elsewhere. For this system consider a solution of the Schrödinger equation of the form

$$\Psi_n(x,t) = N_n \sin\left(\frac{n\pi}{a}x\right) e^{-i\phi_n(t)}, \quad x\in[0,a],$$

where $\Psi_n(x,t)=0$ for $x<0$, and $x>a$. Here $n\geq 1$ is an integer.

1. Find the expression for the (real) phase $\phi_n(t)$ so that the above wave function solves the Schrödinger equation. You may ignore a constant offset that arises as a constant of integration. Find the normalization constant N_n, assumed to be real and positive.
2. Use $\Psi_n(x,0)$ to calculate $\langle\hat{x}\rangle$, $\langle\hat{x}^2\rangle$, and Δx.
3. Use $\Psi_n(x,0)$ to calculate $\langle\hat{p}\rangle$, $\langle\hat{p}^2\rangle$, and Δp.

4. In this example, is the Heisenberg uncertainty inequality satisfied? Is it saturated?
5. What answers in parts (2) and (3) change for $\Psi_n(x,t)$?

Problem 5.7. *Momentum uncertainty for a free wave packet.*

Consider a wave packet moving freely—that is, in the absence of any potential. Compute $\frac{d}{dt}\langle\hat{p}\rangle$ and $\frac{d}{dt}\langle\hat{p}^2\rangle$. Using your answers, compute the rate of change of the momentum uncertainty $\frac{d}{dt}(\Delta p)^2$. What can you tell about the time dependence, if any, of the uncertainty Δp of a free wave packet?

Problem 5.8. *Gaussians and uncertainty product saturation.*

Consider the Gaussian wave function

$$\psi(x) = N\exp\left(-\frac{1}{4}\frac{x^2}{a^2}\right),$$

where $N > 0$ is real, and a is a real positive constant with units of length. Below are two useful integrals:

$$\int_{-\infty}^{\infty} dx\, e^{-\alpha x^2+\beta x} = \sqrt{\frac{\pi}{\alpha}}\exp\left(\frac{\beta^2}{4\alpha}\right), \quad \int_{-\infty}^{\infty} dx\, x^2 e^{-\alpha x^2} = \frac{1}{2\alpha}\int_{-\infty}^{\infty} dx\, e^{-\alpha x^2}.$$

Both are valid when $\text{Re}(\alpha) > 0$.

1. Use the position-space wave function $\psi(x)$ to calculate the uncertainties Δx and Δp. [Hint: These calculations are quite brief if done right! Using the second of the above integrals, you can avoid a calculation of N.]

   Compute $\Delta x\Delta p$, and compare to the Heisenberg uncertainty product $\Delta x\Delta p \geq \frac{\hbar}{2}$. How does your computation compare to this bound?
2. Calculate the Fourier transform $\phi(p)$ of $\psi(x)$. The answer should be proportional to N.
3. Compute $\int_{-\infty}^{\infty} |\phi(p)|^2\, dp$ in terms of a and N. Recalling that $\psi(x)$ is properly normalized and Plancherel's theorem, what is N^2?

Problem 5.9. *Complex Gaussians and the uncertainty product.*

Consider the Gaussian wave function

$$\psi(x) = N\exp\left(-\frac{1}{4}\frac{x^2}{\Delta^2}\right), \quad \Delta \in \mathbb{C},$$

where N is a real and positive normalization constant, and Δ is a complex constant with $\text{Re}(\Delta^2) > 0$. The integrals in problem 5.8 will be useful once more.

1. Compute $\text{Re}(1/z)$ in terms of $\text{Re}(z)$ and $|z|$ for any $z \neq 0$. Use the position-space representation $\psi(x)$ of the wave function to calculate the uncertainties Δx and Δp. Leave your answer in terms of $|\Delta|$ and $\text{Re}(\Delta^2)$.
2. Calculate the Fourier transform $\phi(p)$ of $\psi(x)$. Your answer should be proportional to N. Use Plancherel's theorem to test your answer for $\phi(p)$ by checking that the determination of N^2 from momentum- and position-space wave functions agree. Then recalculate Δp using momentum space.
3. We call ϕ_Δ the argument of Δ so that we have $\Delta = |\Delta|e^{i\phi_\Delta}$. Calculate the product $\Delta x\Delta p$, writing your answer in terms of a trigonometric function of ϕ_Δ and noting that

$|\Delta|$ drops out of the result. Consider the following two cases: $\phi_\Delta = 0$ and $\phi_\Delta = \pi/4$. In each case state if Δ^2 is real, imaginary, or complex, if the Heisenberg uncertainty product is saturated, and if the wave functions are normalizable.

4. We have previously seen (problem 4.3) that a real Gaussian wave packet

$$\Psi(x, t=0) = \left(\frac{1}{2\pi a^2}\right)^{1/4} \exp\left(-\frac{x^2}{4a^2}\right)$$

evolves in time into

$$\Psi(x,t) = \left(\frac{1}{2\pi a^2(1+it/\tau)^2}\right)^{1/4} \exp\left(-\frac{x^2}{4a^2(1+it/\tau)}\right),$$

where $\tau = 2ma^2/\hbar$. Evidently, the width of the Gaussian grows in time. From this wave function and your previous results, what is $\Delta x(t)$? Using this wave function and your previous results, what is Δp? Note that although Δx grows in time, Δp is *independent* of time.

Problem 5.10. *General time evolution of Δx for a free particle.*

Consider a free-particle Hamiltonian $\hat{H} = \frac{\hat{p}^2}{2m}$ for a particle of mass m. In problem 5.7 you must have concluded that $\langle \hat{p} \rangle$, $\langle \hat{p}^2 \rangle$, and Δp are all time-independent constants.

1. Consider the evaluation of

$$\frac{d\langle \hat{x} \rangle}{dt} = \ldots . \tag{1}$$

Find the right-hand side, and solve the resulting differential equation for $\langle \hat{x} \rangle$ as an explicit function of t. Your answer will feature the constant $\langle \hat{p} \rangle$ and the expectation value $\langle \hat{x} \rangle_0$ of $\hat{x}$ at time equals zero.

2. We now want to investigate the time dependence of $(\Delta x)^2$. For this we begin with the time dependence of $\langle \hat{x}^2 \rangle$. Calculate its time rate of change and put it in the form

$$\frac{d\langle \hat{x}^2 \rangle}{dt} = \frac{1}{m} \langle \hat{R} \rangle, \tag{2}$$

where $\hat{R}$ is a Hermitian operator constructed from $\hat{x}$ and $\hat{p}$ that you should determine.

3. Calculate a second derivative of $\langle \hat{x}^2 \rangle$, and set it in the form

$$\frac{d^2\langle \hat{x}^2 \rangle}{dt^2} = \frac{2}{m^2} \langle \cdots \rangle. \tag{3}$$

The expectation value on the right-hand side (if calculated correctly) is time independent.

4. A function of time like $\langle \hat{x}^2 \rangle$ with a constant second time derivative is necessarily of the form

$$\langle \hat{x}^2 \rangle = A\,t^2 + B\,t + C,$$

where A, B, and C are constants. For example, it is clear that $C = \langle \hat{x}^2 \rangle_0$, the expectation value of $\hat{x}^2$ at zero time. Find the constants A, B and write out $\langle \hat{x}^2 \rangle$.

5. Use the solution for $\langle \hat{x} \rangle$ in part (1) to finally calculate the function $(\Delta x)^2$.

6 Stationary States I: Special Potentials

Stationary states are separable solutions of the Schrödinger equation that have a simple time dependence and represent energy eigenstates obtained by solving the time-independent equation $\hat{H}\psi = E\psi$. The expectation value of any time-independent Hermitian operator on a stationary state is time independent. For stationary states of a particle in a potential, the wave function and its spatial derivative are continuous unless the potential has delta functions or hard walls, in which case the derivative is discontinuous. We study the energy eigenstates of a free particle in a circle and those of a particle in an infinite square well. We then turn to the finite square well and find the spectrum of bound states, normalizable energy eigenstates, with energies fixed by transcendental equations. Next, we consider an attractive delta function potential, which is shown to admit a single bound state. We conclude with an analysis of the linear potential, solving for the spectrum using the momentum-space version of the Schrödinger equation.

6.1 Stationary States

Stationary states are a class of simple and useful solutions of the Schrödinger equation. They give us intuition and help us build up general solutions of this equation. Stationary states have time dependence, but this dependence is so simple that in such states observables are in fact time independent. For the case of a particle moving in a potential, stationary states exist if the potential is time independent.

Let us therefore consider the Schrödinger equation for the wave function $\Psi(x,t)$, assuming that the potential energy is time independent and thus written as $V(x)$:

$$i\hbar\frac{\partial\Psi}{\partial t} = \hat{H}\Psi(x,t) = \left(-\frac{\hbar^2}{2m}\frac{\partial^2}{\partial x^2} + V(x)\right)\Psi(x,t). \tag{6.1.1}$$

The signature property of a stationary state is that the position and the time dependence of the wave function factorize, which means that

$$\Psi(x,t) = g(t)\,\psi(x) \tag{6.1.2}$$

for some function g depending only on time and a function ψ depending only on position. To see if this ansatz for Ψ is possible, we substitute it in the Schrödinger equation.

We then find

$$\left(i\hbar\frac{dg(t)}{dt}\right)\psi(x)=g(t)\,\hat{H}\psi(x) \tag{6.1.3}$$

because $g(t)$ can be moved across $\hat{H}$. We can then divide this equation by $\Psi(x,t)=g(t)\psi(x)$ to obtain

$$i\hbar\frac{1}{g(t)}\frac{dg(t)}{dt}=\frac{1}{\psi(x)}\hat{H}\psi(x). \tag{6.1.4}$$

The left-hand side is manifestly a function of t only, while the right-hand side is a function of x only because $\hat{H}$ contains no time-dependent potential. Moreover, each side has units of energy: the left-hand side because it has units of $\hbar$ over time and the right-hand side because it has the units of $\hat{H}$. The only way the two sides can equal each other for all values of t and x is for both sides to be equal to a *constant* E with units of energy. We therefore get two separate equations. The first equation, from the left-hand side, reads

$$i\hbar\frac{dg}{dt}=Eg. \tag{6.1.5}$$

This is solved by

$$g(t)=e^{-iEt/\hbar}, \tag{6.1.6}$$

and the most general solution is simply a constant times the above right-hand side. The second equation, from the x-dependent side of the equality, is

$$\boxed{\hat{H}\psi(x)\;=\;E\psi(x).} \tag{6.1.7}$$

This equation is an eigenvalue equation for the Hermitian operator $\hat{H}$. A $\psi(x)$ solving this equation is called an **energy eigenstate**; it is an eigenstate of the energy operator $\hat{H}$. We showed that the eigenvalues of Hermitian operators must be real, thus the constant E must be real. The equation above is called the **time-independent Schrödinger equation**. More explicitly, it reads:

$$\boxed{\left(-\frac{\hbar^2}{2m}\frac{d^2}{dx^2}+V(x)\right)\psi(x)=E\psi(x).} \tag{6.1.8}$$

Note that this equation does not determine the overall normalization of the energy eigenstate ψ: if ψ is a solution, so is $a\psi$, with a a constant. The full solution of the Schrödinger equation associated with $\psi(x)$, called a **stationary state**, is obtained from (6.1.2) using the $g(t)$ obtained above:

$$\boxed{\text{stationary state:}\quad \Psi(x,t)=e^{-iEt/\hbar}\,\psi(x),\quad \text{with}\quad E\in\mathbb{R}\ \text{and}\ \hat{H}\psi=E\psi.} \tag{6.1.9}$$

A stationary state is the time-dependent solution associated with an energy eigenstate. It is worth noting that not only is $\psi(x)$ an $\hat{H}$ eigenstate, the associated stationary state is also an $\hat{H}$ eigenstate,

$$\hat{H}\Psi(x,t) = E\,\Psi(x,t), \tag{6.1.10}$$

since the time-dependent function in Ψ cancels out.

We have noted that the energy E must be real. If it were not, we would not be able to normalize the stationary state consistently. The normalization condition for Ψ, if E is not real, would give

$$\begin{aligned}1 &= \int_{-\infty}^{\infty} dx\,\Psi^*(x,t)\Psi(x,t) = \int_{-\infty}^{\infty} dx\, e^{iE^*t/\hbar}e^{-iEt/\hbar}\psi^*(x)\psi(x)\\ &= e^{i(E^*-E)t/\hbar}\int_{-\infty}^{\infty} dx\,\psi^*(x)\psi(x) = e^{2\,\mathrm{Im}(E)t/\hbar}\int_{-\infty}^{\infty} dx\,\psi^*(x)\psi(x).\end{aligned} \tag{6.1.11}$$

The final expression has a time dependence due to the exponential. On the other hand, the normalization condition states that this expression must be equal to one. It follows that the exponent must be zero; that is E is real. Given this, we also see that the normalization condition yields

$$\boxed{(\Psi,\Psi) = 1 \;\Rightarrow\; (\psi,\psi) = \int_{-\infty}^{\infty} dx\,\psi^*(x)\psi(x) = 1.} \tag{6.1.12}$$

The energy eigenvalue E is in fact the expectation value of $\hat{H}$ on the state Ψ:

$$\begin{aligned}\langle\hat{H}\rangle_\Psi &= (\Psi,\hat{H}\Psi) = \int_{-\infty}^{\infty} dx\,\Psi^*(x,t)\hat{H}\Psi(x,t)\\ &= \int_{-\infty}^{\infty} dx\Psi^*(x,t)E\Psi(x,t) = E\int_{-\infty}^{\infty} dx\Psi^*(x,t)\Psi(x,t) = E.\end{aligned} \tag{6.1.13}$$

Since the stationary state is an eigenstate of $\hat{H}$, the uncertainty $\Delta\hat{H}$ of the Hamiltonian in a stationary state is zero (recall (5.5.19)). If we measure the energy of a stationary state, the result is always the energy eigenvalue E. Stationary states are energy eigenstates.

Suppose we have a system with a time-independent $\hat{H}$, and we have determined the full set of energy eigenstates $\psi_1, \psi_2, \ldots$ and their energies $E_1, E_2, \ldots$. Since the energy eigenstates are eigenfunctions of the Hermitian operator $\hat{H}$, any reasonable wave function can be written as a superposition of energy eigenstates (claim 4, section 5.3). Thus, at $t=0$ we have

$$\Psi(x,t=0) = \sum_n c_n\,\psi_n(x), \tag{6.1.14}$$

with c_n some complex constants. Since each $\psi(x)$ evolves in time as shown in (6.1.9), the above wave function evolves to become

$$\Psi(x,t) = \sum_n c_n\, e^{-iE_nt/\hbar}\psi_n(x). \tag{6.1.15}$$

Exercise 6.1. *Confirm explicitly that the wave function in (6.1.15) solves the time-dependent Schrödinger equation.*

The general strategy above is one reason why energy eigenstates are so useful: the time evolution of any wave function is easily done once the energy eigenstates are known. One simply expands the initial wave function in terms of the energy eigenstates and then evolves by adding the corresponding time-dependent phase factors. We followed this strategy for the particular case of plane waves in section 4.3.

There are two important observations on stationary states:

1. The expectation value of any time-independent operator $\hat{Q}$ on a stationary state Ψ is time independent:

$$\langle\hat{Q}\rangle_{\Psi(x,t)} = \int_{-\infty}^{\infty} dx\, \Psi^*(x,t)\,\hat{Q}\Psi(x,t) = \int_{-\infty}^{\infty} dx\, e^{iEt/\hbar}\psi^*(x)\hat{Q}e^{-iEt/\hbar}\psi(x). \tag{6.1.16}$$

Since any time-dependent function can be moved across $\hat{Q}$, the two time-dependent exponentials now cancel and we find that

$$\langle\hat{Q}\rangle_{\Psi(x,t)} = \int_{-\infty}^{\infty} dx\, \psi^*(x)\hat{Q}\psi(x) = \langle\psi, \hat{Q}\psi\rangle = \langle\hat{Q}\rangle_{\psi(x)}, \tag{6.1.17}$$

which is manifestly time independent.

2. The superposition of stationary states with different energies is not stationary. This is clear because a stationary state is a product of a function of time and a function of position: two such states with different energies have different functions of time, and the sum cannot be written as a product of a function of time and a function of position. We now show that a time-independent observable $\hat{Q}$ may have a time-dependent expectation value in such a state. Consider a superposition

$$\Psi(x,t) = c_1 e^{-iE_1t/\hbar}\psi_1(x) + c_2 e^{-iE_2t/\hbar}\psi_2(x), \tag{6.1.18}$$

where ψ_1 and ψ_2 are $\hat{H}$ eigenstates with energies E_1 and E_2, respectively. Consider a Hermitian operator $\hat{Q}$. With the system in the above state Ψ, its expectation value (omitting the $-\infty$ and ∞ limits on the integrals, for brevity) is

$$\begin{aligned}\langle\hat{Q}\rangle_\Psi &= \int dx\, \Psi^*(x,t)\hat{Q}\Psi(x,t)\\ &= \int dx\, \big(c_1^* e^{iE_1t/\hbar}\psi_1^*(x) + c_2^* e^{iE_2t/\hbar}\psi_2^*(x)\big)\\ &\qquad \big(c_1 e^{-iE_1t/\hbar}\hat{Q}\psi_1(x) + c_2 e^{-iE_2t/\hbar}\hat{Q}\psi_2(x)\big)\\ &= \int dx\, \Big(|c_1|^2\psi_1^*\hat{Q}\psi_1 + |c_2|^2\psi_2^*\hat{Q}\psi_2\\ &\qquad + c_1^*c_2 e^{i(E_1-E_2)t/\hbar}\psi_1^*\hat{Q}\psi_2 + c_2^*c_1 e^{-i(E_1-E_2)t/\hbar}\psi_2^*\hat{Q}\psi_1\Big).\end{aligned} \tag{6.1.19}$$

The first two terms after the last equal sign are time-independent expectation values; the time dependence arises from the last two terms. Using the hermiticity of $\hat{Q}$ in the last term, we then get

$$\begin{aligned}\langle \hat{Q}\rangle_\Psi = & |c_1|^2\langle \hat{Q}\rangle_{\psi_1} + |c_2|^2\langle \hat{Q}\rangle_{\psi_2} \\ & + c_1^* c_2\, e^{i(E_1-E_2)t/\hbar}\int dx\, \psi_1^* \hat{Q}\psi_2 \\ & + c_1 c_2^*\, e^{-i(E_1-E_2)t/\hbar}\int dx\, \psi_1 (\hat{Q}\psi_2)^*. \end{aligned} \tag{6.1.20}$$

The last two terms are complex conjugates of each other, and therefore

$$\langle Q\rangle_\Psi = |c_1|^2\langle \hat{Q}\rangle_{\psi_1} + |c_2|^2\langle \hat{Q}\rangle_{\psi_2} + 2\,\text{Re}\Big[c_1^* c_2 e^{i(E_1-E_2)t/\hbar}\langle \psi_1, \hat{Q}\psi_2\rangle\Big]. \tag{6.1.21}$$

This expectation value is indeed time dependent if $E_1 \neq E_2$ and $\langle \psi_1, \hat{Q}\psi_2\rangle$ is nonzero. The expectation value $\langle \hat{Q}\rangle_\Psi$ is real, as it must be for any Hermitian operator.

While a stationary state wave function $\Psi(x,t) = e^{-iEt/\hbar}\psi(x)$ depends on time, it is physically time independent. This is, in fact, the content of observation (1) above; no expectation value shows time dependence. We can see this time independence more conceptually as follows. Consider the stationary state at time t and at time $t+t_0$, with t_0 some arbitrary constant time. We see that

$$\Psi(x, t+t_0) = e^{-iE(t+t_0)/\hbar}\psi(x) = e^{-iEt_0/\hbar}\,\Psi(x,t). \tag{6.1.22}$$

Since the stationary-state wave functions at t and at $t+t_0$ differ by an overall *constant* phase, they are physically equivalent, they are the *same* state. The phase is a constant because it has no t or x dependence. We have emphasized from the beginning that overall constants multiplying the wave function do not change the state (see (1.4.5)).

Exercise 6.2. *Let $\psi_1(x)$ and $\psi_2(x)$ be two normalized stationary states with energies E_1 and E_2, respectively, with $E_2 > E_1$. At $t=0$, the state Ψ of our system is*

$$\Psi(x,0) = \frac{1}{\sqrt{2}}\left(\psi_1(x) + \psi_2(x)\right).$$

Determine the smallest time $t_0 > 0$ for which $\Psi(x, t_0) \propto \psi_1(x) - \psi_2(x)$.

6.2 Solving for Energy Eigenstates

The time dependence of a stationary state is always a simple phase factor $e^{-iEt/\hbar}$, where E is the value of the energy. The space dependence is harder to obtain, and it is determined by the time-independent Schrödinger equation:

$$\hat{H}\psi(x) = E\,\psi(x). \tag{6.2.1}$$

For a given Hamiltonian $\hat{H}$, we are interested in finding the eigenstates ψ and the eigenvalues E, which are corresponding energies. Perhaps the most interesting feature of the above equation is that it determines the allowed values of E. Just as finite-size matrices have a set of eigenvalues, the above time-independent Schrödinger equation may have a discrete set of possible energies. A continuous set of possible energies is often allowed as well. Indeed, for any given potential there are many solutions of the above equation.

The set of energies that are solutions of the time-independent Schrödinger equation is called the **energy spectrum** of the theory. The name "spectrum" is appropriate, as an energy eigenstate is associated to a time-dependent exponential with frequency $\omega = E/\hbar$. (More generally, one speaks of the spectrum of any Hermitian operator as the set of its eigenvalues.) Let us assume, for simplicity, that the eigenstates and their energies can be counted so we can write the lists

$$\begin{array}{ll} \psi_1(x), & E_1, \\ \psi_2(x), & E_2, \\ \vdots & \vdots \end{array} \tag{6.2.2}$$

Our discussion of Hermitian operators in section 5.3 applies here because $\hat{H}$ is Hermitian. In particular, the energy eigenstates can be organized to form a *complete set of orthonormal functions*:

$$\int_{-\infty}^{\infty} dx\, \psi_i^*(x)\psi_j(x) = \delta_{ij}. \tag{6.2.3}$$

We now wish to understand some general features of the spatial dependence $\psi(x)$ of energy eigenstates. For this consider the time-independent Schrödinger equation (6.2.1) written as

$$\frac{d^2\psi}{dx^2} = -\frac{2m}{\hbar^2}(E - V(x))\,\psi. \tag{6.2.4}$$

The solutions $\psi(x)$ depend on the properties of the potential $V(x)$. It is difficult to make general statements about the wave function unless we restrict the types of potentials. We will certainly consider continuous potentials. We will also consider potentials that are not continuous but are piecewise continuous; that is, they have a number of discontinuities. Our potentials can easily fail to be bounded, as is the case for the harmonic oscillator potential $V(x) \sim x^2$, which becomes infinite as $|x| \to \infty$. We allow delta functions in one-dimensional potentials but do not consider powers or derivatives of delta functions. We allow for potentials that become infinite beyond certain points. These points represent hard walls because no finite amount of kinetic energy allows a particle to enter a region of infinite potential energy.

To make our discussion a bit more precise, we define finite discontinuities. A function $f(x)$ has a finite discontinuity $\Delta_a f(x)$ at $x = a$ if

$$\Delta_a f(x) \equiv \lim_{\epsilon \to 0} \Big(f(a+\epsilon) - f(a-\epsilon) \Big) \tag{6.2.5}$$

is different from zero and is finite. This is intuitively clear: a function is discontinuous at $x = a$ if it reaches different values as the point $x = a$ is approached from the right and from the left. It is also true that the products of continuous functions are continuous. The product of a continuous function and a finitely discontinuous function is a finitely discontinuous function.

Claim. *The wave function must be continuous at all points.* Assume ψ fails to be continuous by having a set of finite discontinuities. Since the derivative of a step function is a delta function (exercise 4.4), we conclude that the derivative ψ' would contain delta functions, and ψ'', on the left-hand side of (6.2.4), would contain derivatives of delta functions. This would require the right-hand side to have derivatives of delta functions, and those would have to appear in the potential since ψ only has finite discontinuities. Having declared that our potentials contain no derivatives of delta functions, we conclude that ψ cannot contain finite discontinuities. Worse discontinuities would lead to similar problems. We must indeed have a continuous ψ. The precise mathematical definition of continuity implies that a function fails to be continuous at points where it takes infinite value. In fact, we find it physically consistent to assume that *our wave functions take finite values at all points.*

Consider now four possibilities concerning the potential:

1. $V(x)$ is continuous. In this case the continuity of $\psi(x)$ implies that ψ'', given by the right-hand side of (6.2.4), is also continuous. If ψ'' is continuous, ψ' must also be continuous.
2. $V(x)$ has finite discontinuities. In this case ψ'' has finite discontinuities because according to (6.2.4) it includes the product of a continuous ψ against a finitely discontinuous V. But then $\psi' \sim \int \psi''$ must be continuous because the integral of a function with finite discontinuities is continuous.
3. $V(x)$ contains delta functions. In this case ψ'' also contains delta functions as it includes the product of a continuous ψ and a delta function in V. With ψ'' having delta functions, $\psi' \sim \int \psi''$ must have finite discontinuities because the integral of a function that includes delta functions has finite discontinuities.
4. $V(x)$ contains a hard wall. A potential that is finite immediately to the left of $x = a$ and becomes infinite for $x > a$ is said to have a hard wall for $x \geq a$. In such a case, the wave function will vanish for $x \geq a$. The slope ψ' will be finite as $x \to a$ from the left and will vanish for $x > a$. Thus, ψ' is discontinuous at the wall. This point will be explained in more detail in the context of the infinite square well (section 6.4).

In the first two cases, ψ' is continuous, and in the second two cases, it can have a finite discontinuity. In conclusion, we have obtained the following conditions,

Both ψ and ψ' are continuous unless the potential has delta functions or hard walls, in which case ψ' may have finite discontinuities.	(6.2.6)

Exercise 6.3. *A particle of mass m moves in a potential $V(x)$. Suppose an energy eigenstate with energy $E_0 = \frac{1}{2}\hbar\omega$ has a wave function*

$$\psi(x) = N \exp\left(-\frac{1}{2}\frac{m\omega}{\hbar}x^2\right),$$

where N is a normalization constant. What is the potential $V(x)$?

6.3 Free Particle on a Circle—a Second Look

Consider again the problem of a particle confined to a circle of circumference L, which we examined in section 5.4. The coordinate along the circle is called x, and as discussed before, we can view the circle as the full real line x with the identification

$$x \sim x + L. \tag{6.3.1}$$

This means that two points whose coordinates are related in this way are to be considered *the same point*. All wave functions and, in particular, all energy eigenstates must satisfy the periodicity condition

$$\psi(x+L) = \psi(x). \tag{6.3.2}$$

From this it follows that $\psi(x)$ is not only periodic but all of its derivatives are also periodic. The periodicity of an energy eigenstate $\psi(x)$ implies that its time-dependent version, obtained by multiplication by $\exp(-iEt/\hbar)$, with E its energy, is also periodic in x.

Let us consider the case in which there is no potential on the circle, $V(x) = 0$, and the particle is thus free. The time-independent Schrödinger equation is then

$$-\frac{\hbar^2}{2m}\frac{d^2\psi}{dx^2} = E\psi(x). \tag{6.3.3}$$

We want to find the allowed values of E and the corresponding wave functions. The ground state energy is the lowest value of E for which we can find a solution. Let us first demonstrate, without solving the equation, that any solution must have $E \geq 0$. For this multiply the above equation by $\psi^*(x)$ and integrate over the circle $x \in [0, L)$. Since ψ is normalized, we get

$$-\frac{\hbar^2}{2m}\int_0^L \psi^*(x)\frac{d^2\psi}{dx^2}\,dx = E\int_0^L \psi^*(x)\psi(x)dx = E. \tag{6.3.4}$$

A rewriting of the left-hand side gives the equation

$$-\frac{\hbar^2}{2m}\int_0^L \Big[\frac{d}{dx}\Big(\psi^*\frac{d\psi}{dx}\Big) - \frac{d\psi^*}{dx}\frac{d\psi}{dx}\Big]dx = E. \tag{6.3.5}$$

The total derivative can be integrated, and we find

$$-\frac{\hbar^2}{2m}\Big[\Big(\psi^*\frac{d\psi}{dx}\Big)\Big|_{x=L} - \Big(\psi^*\frac{d\psi}{dx}\Big)\Big|_{x=0}\Big] + \frac{\hbar^2}{2m}\int_0^L \Big|\frac{d\psi}{dx}\Big|^2 dx = E. \tag{6.3.6}$$

Since $\psi(x)$ and its derivatives are periodic, the contributions from $x = L$ and $x = 0$ cancel out, and we are left with

$$E = \frac{\hbar^2}{2m}\int_0^L \Big|\frac{d\psi}{dx}\Big|^2 dx \geq 0, \tag{6.3.7}$$

which establishes our claim. We also see that the lowest possible energy is $E = 0$ and requires a vanishing integrand and therefore a wave function with a zero derivative all along the circle. Such wave function is a constant that cannot be zero because a zero solution implies the particle is nowhere to be found and is thus inconsistent. Thus, a constant nonzero wave function represents the ground state of the particle on the circle!

Having shown that all solutions must have $E \geq 0$, let us go back to the Schrödinger equation, which can be rewritten as

$$\frac{d^2\psi}{dx^2} = -\frac{2mE}{\hbar^2}\psi. \tag{6.3.8}$$

We can then introduce k, defined by

$$k^2 \equiv \frac{2mE}{\hbar^2} \geq 0. \tag{6.3.9}$$

Since $E \geq 0$, the constant k is real and $k \in (-\infty, \infty)$. Note that this definition is very natural since it makes

$$E = \frac{\hbar^2 k^2}{2m}, \tag{6.3.10}$$

which means that, as usual, $p = \hbar k$. Using k instead of E, the differential equation takes the familiar form

$$\frac{d^2\psi}{dx^2} = -k^2\psi. \tag{6.3.11}$$

Each solution of this equation is an energy eigenstate. We can write the general solution in terms of sines and cosines of kx, or complex exponentials of kx. Let's consider a solution defined by a complex exponential:

$$\psi(x) \sim e^{ikx}. \tag{6.3.12}$$

This solves the differential equation, and therefore it is an energy eigenstate, but it is also a momentum eigenstate! In fact, it has momentum $\hbar k$. Since (6.3.11) is a second-order differential equation, we have two solutions, e^{ikx} and e^{-ikx}. It is convenient, however, to consider them separately since each is a momentum eigenstate but any nontrivial superposition of them is not.

As we saw in section 5.4, the periodicity condition (6.3.2) applied to the above $\psi(x)$ requires that

$$e^{ik(x+L)} = e^{ikx} \quad \Rightarrow \quad e^{ikL} = 1 \quad \Rightarrow \quad kL = 2\pi n, \quad n \in \mathbb{Z}. \tag{6.3.13}$$

The wave number is quantized, and therefore momentum is quantized! The wave number has discrete values

$$k_n \equiv \frac{2\pi n}{L}, \qquad n \in \mathbb{Z}. \tag{6.3.14}$$

All integers, positive, negative, and zero, are allowed and are in fact necessary because they all correspond to *different* values of the momentum $p_n = \hbar k_n$. The solutions to the Schrödinger equation can then be indexed by the integer n:

$$\psi_n(x) = N_n e^{ik_n x}, \tag{6.3.15}$$

where N_n is a real normalization constant whose value is determined from

$$1 = \int_0^L \psi_n^*(x)\psi_n(x)dx = \int_0^L N_n^2 dx = N_n^2 L \quad \Rightarrow \quad N_n = \frac{1}{\sqrt{L}}. \tag{6.3.16}$$

Our normalized energy *and* momentum eigenstates thus take the form

$$\boxed{\psi_n(x) = \frac{1}{\sqrt{L}} e^{ik_n x} = \frac{1}{\sqrt{L}} e^{\frac{2\pi i n x}{L}}, \quad n \in \mathbb{Z}.} \tag{6.3.17}$$

The associated energies and momenta are

$$\begin{aligned} E_n &= \frac{\hbar^2 k_n^2}{2m} = \frac{\hbar^2 4\pi^2 n^2}{2mL^2} = 2\pi^2 n^2 \frac{\hbar^2}{mL^2}, \\ p_n &= \hbar k_n = 2\pi n \frac{\hbar}{L}. \end{aligned} \tag{6.3.18}$$

The ground state is obtained for $n=0$, and as anticipated, $\psi_0 = 1/\sqrt{L}$ is just a constant. The ground state has zero energy and zero momentum. There are infinitely many energy eigenstates. The energy spectrum has degeneracies because $E_n \sim n^2$, and thus both ψ_n and ψ_{-n} have energy E_n. The only nondegenerate eigenstate is the ground state ψ_0.

Whenever we find degenerate energy eigenstates, we must wonder what makes those states different, given that they have the same energy. To answer this, one must find an observable that takes different values on the states. Happily, in our case we know the answer. Our degenerate energy eigenstates can be distinguished by their momentum: ψ_n has momentum $2\pi n\hbar/L$, and ψ_{-n} has momentum $(-2\pi n\hbar/L)$. Note that the spectrum of the Hermitian operator $\hat{p}$ is nondegenerate: there is one and only one state for each possible momentum eigenvalue.

We noted when discussing stationary states that the superposition of two energy eigenstates with different energies is not an energy eigenstate. The situation changes if we have degenerate eigenstates. Given two degenerate energy eigenstates, *any* linear combination of these states is an eigenstate with the same energy. Indeed, if ψ_1 and ψ_2 are degenerate,

$$\hat{H}\psi_1 = E\psi_1, \quad \hat{H}\psi_2 = E\psi_2, \tag{6.3.19}$$

then $a\psi_1 + b\psi_2$, with a and b arbitrary complex constants, is also a state of energy E:

$$\hat{H}(a\psi_1 + b\psi_2) = a\hat{H}\psi_1 + b\hat{H}\psi_2 = aE\psi_1 + bE\psi_2 = E(a\psi_1 + b\psi_2). \tag{6.3.20}$$

We can therefore form two linear combinations of the degenerate eigenstates ψ_n and ψ_{-n} to obtain another description of the normalized energy eigenstates for $n \neq 0$:

$$\frac{1}{\sqrt{2}}(\psi_n + \psi_{-n}) = \sqrt{\frac{2}{L}}\cos(k_n x), \qquad \frac{1}{i\sqrt{2}}(\psi_n - \psi_{-n}) = \sqrt{\frac{2}{L}}\sin(k_n x). \tag{6.3.21}$$

While these energy eigenstates are real functions on the circle, they are not momentum eigenstates. Only our exponentials are simultaneous eigenstates of both $\hat{H}$ and $\hat{p}$.

Exercise 6.4. *Consider the states proportional to* $\cos(k_n x)$. *To obtain the full set of inequivalent states in this class, what values do we need to allow n to take? Consider the same question for the states* $\sin(k_n x)$.

We have learned that eigenstates of Hermitian operators with different eigenvalues are orthogonal (claim 3, section 5.3). The energy eigenstates ψ_n are *orthonormal* since they are all normalized, and they are eigenstates of the Hermitian operator $\hat{p}$ with no degeneracies. Indeed, we easily verify that

$$\int_0^L \psi_n^*(x)\psi_m(x)dx = \frac{1}{L}\int_0^L e^{\frac{2\pi i(m-n)x}{L}}dx = \delta_{mn}. \tag{6.3.22}$$

The energy eigenstates are also complete: We can construct a general wave function on the circle as a superposition that is in fact a Fourier series! For any $\Psi(x,0)$ that satisfies the periodicity condition, we can write

$$\Psi(x,0) = \sum_{n\in\mathbb{Z}} a_n\,\psi_n(x), \tag{6.3.23}$$

where, as you should check, the coefficients a_n are determined by the integrals

$$a_n = \int_0^L dx\,\psi_n^*(x)\,\Psi(x,0). \tag{6.3.24}$$

The initial state $\Psi(x,0)$ is then easily evolved in time:

$$\Psi(x,t) = \sum_{n\in\mathbb{Z}} a_n\,\psi_n(x)e^{-\frac{iE_n t}{\hbar}}. \tag{6.3.25}$$

Let us discuss an interesting subtlety of the problem of a particle on a circle. We have dealt with the momentum operator $\hat{p}$ and its eigenstates. How about the position operator $\hat{x}$? Perhaps surprisingly, the position operator is not well defined! The problem arises because of the identification $x \sim x+L$ that defines the circle. The coordinate of each point on the circle is ambiguous up to additive multiples of L. If we try to fix this ambiguity by declaring that the coordinate on the circle is $x \in [0, L)$, then the result is a discontinuity in the coordinate as we approach $x = L$, which is equivalent to $x = 0$. Alternatively, if we tried to define the operator $\hat{x}$ via

$$\hat{x}\,\psi(x) \stackrel{?}{\equiv} x\psi(x), \tag{6.3.26}$$

we would still have complications: while $\psi(x)$ is assumed to be periodic in order to be well defined on the circle, $x\psi(x)$ is not periodic. The $\hat{x}$ operator acting on any legal state does not give a legal state. All is not lost, however. We can define a set of operators $\hat{Q}_n$ as exponentials of $\hat{x}$:

$$\hat{Q}_n \equiv \exp\Big(\frac{2\pi in\,\hat{x}}{L}\Big). \tag{6.3.27}$$

The $\hat{Q}_n$ are well defined because they are unchanged under the replacement $\hat{x} \to \hat{x} + L$. Acting on wave functions, we have

$$\hat{Q}_n\,\psi(x) = e^{2\pi inx/L}\psi(x), \tag{6.3.28}$$

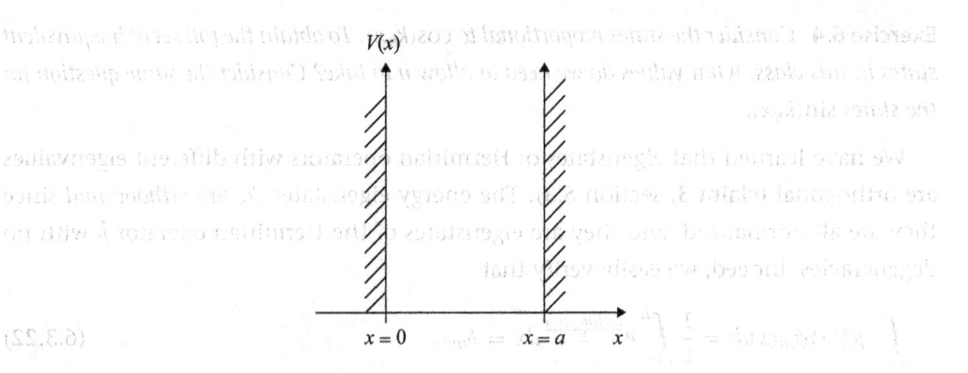

Figure 6.1
The infinite square well potential.

in the same way as we defined $V(\hat{x})\psi(x) = V(x)\psi(x)$ for potentials. The action of $\hat{Q}_n$ is consistent since it does not spoil the periodicity condition. The famous commutator $[\hat{x}, \hat{p}] = i\hbar$, valid on the full real line, does not hold here because $\hat{x}$ is not a well-defined operator on the circle.

Exercise 6.5. *A particle on a circle $x \in [0, L]$ at $t = 0$ has a wave function $\Psi(x, 0)$ that to a good approximation is constant for $x \in [\frac{L}{4}, \frac{3L}{4}]$ and is zero elsewhere. Write the normalized wave function and find the probability that a momentum measurement yields $p = 2\pi\hbar/L$.*

6.4 The Infinite Square Well

We have already studied the quantum states of a particle moving freely in a circle. We now introduce another instructive one-dimensional problem, the case of a particle moving on an **infinite square well**. A suitable potential forces a particle to live on an interval $x \in [0, a]$ of the real line. At the boundaries $x = 0$ and $x = a$ of the interval are hard walls that prevent the particle from going into $x < 0$ and $x > a$, respectively. One wall extends for all $x \geq a$; the other for all $x \leq 0$. The potential is zero for $0 < x < a$ and is infinite beyond the ends of the interval (figure 6.1):

$$V(x) = \begin{cases} 0, & 0 < x < a, \\ \infty, & x \leq 0, x \geq a. \end{cases} \tag{6.4.1}$$

It is reasonable to assume that the wave function must vanish in the region where the potential is infinite. Classically, any region where the potential exceeds the energy of the particle is forbidden. This is not so in quantum mechanics. But even in quantum mechanics, a particle can't be in a region of *infinite* potential. We will be able to justify these claims by studying the more complicated *finite* square well in the limit as the height of the potential goes to infinity. But for the meantime, we simply state the fact that

$$\psi(x) = 0 \quad \text{for } x < 0 \text{ and for } x > a. \tag{6.4.2}$$

Since the wave function must be continuous, it should vanish at $x=0$ and at $x=a$:

1. $\psi(x=0)=0$.
2. $\psi(x=a)=0$.

These are our boundary conditions. We claimed in chapter 5 that at hard walls the derivative ψ' of the wave function is discontinuous. Let us discuss this point now. Since the wave function vanishes outside the interval, the derivative vanishes outside of the interval as well. If the derivative ψ' were continuous at the wall, ψ' would vanish at 0 and at a. But this is impossible. A solution of the time-independent Schrödinger's equation (a second-order differential equation) for which *both* the wave function and its derivative vanish at a point is identically zero, as will be elaborated in section 7.3. A zero solution does not describe a particle. If a solution exists, we must accept that ψ' can have discontinuities at an infinite wall. Therefore, we do not require ψ' to vanish at the boundaries; in fact, we impose no condition on ψ' at the boundaries. The two conditions above on ψ will suffice to find a solution. In that solution ψ' is discontinuous at the boundaries.

In the region $x \in [0, a]$, the potential vanishes, and the time-independent Schrödinger equation takes the form

$$\frac{d^2\psi}{dx^2} = -\frac{2mE}{\hbar^2}\psi. \tag{6.4.3}$$

As we did for the case of a particle on a circle, we can show that the energy E must be positive (do it!). This allows us to define a real quantity k such that

$$k^2 \equiv \frac{2mE}{\hbar^2} \quad \Rightarrow \quad E = \frac{\hbar^2 k^2}{2m}. \tag{6.4.4}$$

The differential equation is then

$$\frac{d^2\psi}{dx^2} = -k^2\psi. \tag{6.4.5}$$

The general solution can be written as

$$\psi(x) = c_1 \cos kx + c_2 \sin kx, \tag{6.4.6}$$

with constants c_1 and c_2 to be determined. For this we use our boundary conditions.

The condition $\psi(x=0)=0$ implies that c_1 in equation (6.4.6) must be zero. The coefficient c_2 of $\sin kx$ need not vanish since this function vanishes automatically for $x=0$. Therefore, the solution so far reads

$$\psi(x) = c_2 \sin kx. \tag{6.4.7}$$

Note that if we demanded continuity of ψ', we would have to ask for $\psi'(x=0)=0$, and that would require c_2 to be equal to zero and thus ψ to be identically zero. That is not a physical solution. There is no particle if $\psi=0$.

Having dealt with the point $x=0$, we must now impose the vanishing of ψ at $x=a$:

$$c_2 \sin ka = 0 \quad \Rightarrow \quad ka = n\pi \quad \Rightarrow \quad k_n = \frac{n\pi}{a}. \tag{6.4.8}$$

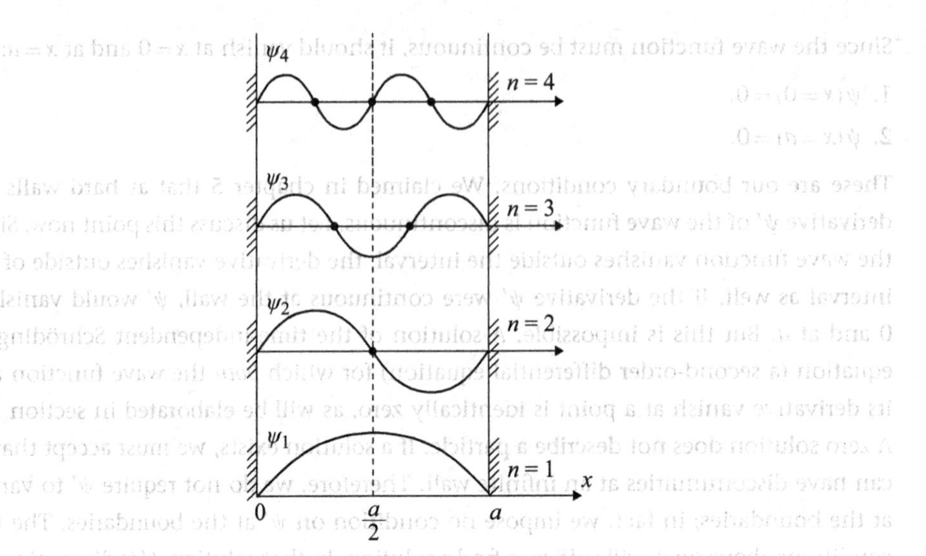

Figure 6.2
The four lowest-energy eigenstates for the infinite square well potential. The wave function ψ_n has $n-1$ nodes, indicated by the heavy dots. The solutions alternate as symmetric and antisymmetric about the midpoint $x=a/2$.

Here n must be an integer, and the solution would be

$$\psi_n(x) = N_n \sin\Big(\frac{n\pi x}{a}\Big), \tag{6.4.9}$$

with N_n a normalization constant. Which integers n are acceptable here? Having $n=0$ is not acceptable because it would make the wave function zero. Moreover, n and $-n$ give the same wave function up to a sign. Since the sign of a wave function is irrelevant, it would be double counting to include both positive and negative n's. We restrict ourselves to n being a positive integer. To solve for the coefficient N_n, we demand that $\psi_n(x)$ be normalized:

$$1 = N_n^2 \int_0^a \sin^2\Big(\frac{n\pi x}{a}\Big)dx = N_n^2\,\tfrac{1}{2}\,a \quad\Rightarrow\quad N_n = \sqrt{\frac{2}{a}}, \tag{6.4.10}$$

noting that the average value of $\sin^2 x$ or $\cos^2 x$ over an integer number of half periods is $\frac{1}{2}$. Therefore, the normalized energy eigenstates are

$$\boxed{\psi_n = \sqrt{\frac{2}{a}}\sin\Big(\frac{n\pi x}{a}\Big), \quad E_n = \frac{\hbar^2 k_n^2}{2m} = \frac{\hbar^2\pi^2 n^2}{2ma^2}, \quad n=1,2,\cdots.} \tag{6.4.11}$$

Each value of n gives a different energy, implying that in the one-dimensional infinite square well there are no degeneracies in the energy spectrum. The ground state—the lowest energy state—corresponds to $n=1$ and has nonzero energy. Figure 6.2 shows the first four energy eigenstates of the infinite square well, labeled from $n=1$ to $n=4$.

Exercise 6.6. *A particle of mass m is placed on an infinite square well of width equal to its Compton wavelength. What would be its ground state energy? Your result should confirm that attempting to confine a particle within its Compton wavelength, as discussed at the end of section 2.4, is made difficult by possible particle creation.*

Comments:

1. The ground state ψ_1 has no nodes. A **node** of a wave function is a point where the wave function changes sign. Clearly, the wave function vanishes at a node, but a point with vanishing wave function is not necessarily a node. The zeroes of ψ_1 at $x=0$ and $x=a$ are not nodes. Since $\psi_1(x)$ does not vanish anywhere in the interior of $[0,a]$, it has no nodes. It is in fact true that any normalizable ground state wave function of a one-dimensional potential does not have nodes.
2. The first excited state ψ_2 has one node at $x=a/2$, the midpoint of the interval. The second excited state ψ_3 has two nodes. The pattern in fact continues: the ***n*th excited state ψ_{n+1} has *n* nodes**. This pattern of nodes is an illustration of the so-called node theorem, which we will discuss at length in section 7.4.
3. In the figure the dotted vertical line marks the interval midpoint $x=a/2$. We note that the ground state is symmetric under reflection about $x=a/2$. The first excited state is antisymmetric; indeed its node is at $x=a/2$. The second excited state is again symmetric. Symmetry and antisymmetry alternate forever.
4. We will later prove that the normalizable energy eigenstates of a one-dimensional even potential are either even or odd (section 7.1, corollary 2). A potential $V(x)$ is said to be even if $V(-x)=V(x)$. Our potential (6.4.1) is not even, but this could have been easily fixed with no physical effect. All that really matters in the definition of a well is the length a of the interval, so letting the well extend over $x\in[-a/2,a/2]$ makes the potential even (figure 6.3). The symmetry or antisymmetry of the wave functions about $x=0$ is then manifest and guaranteed by the general result. We did not choose to present the infinite well in this form because the expressions for the wave functions are a bit more complicated. Instead of all being given by the sine function, they are given by either sine or cosine, depending on the number of nodes.
5. The most general allowed wave function in the infinite square well is a superposition of energy eigenstates $\psi_n(x)$ with $n=1,2,\ldots$ and vanishes at the end points. The energy eigenstates thus form a basis for the set of allowed wave functions. A wave function that does not vanish at the end points is not an allowed state in the space of states.

We pointed out that for the free particle on a circle there is no well-defined position operator $\hat{x}$. For the particle in the infinite square well, there is also a subtlety. An operator is well defined if acting on the space of allowed states gives an allowed state. The momentum operator $\hat{p}$ is in fact problematic because when acting on allowed wave functions—those that vanish at the end points—it gives us wave functions that are not allowed. Indeed, acting on the energy eigenstate ψ_n, we have $\hat{p}\psi_n(x)\sim\cos(n\pi x/a)$,

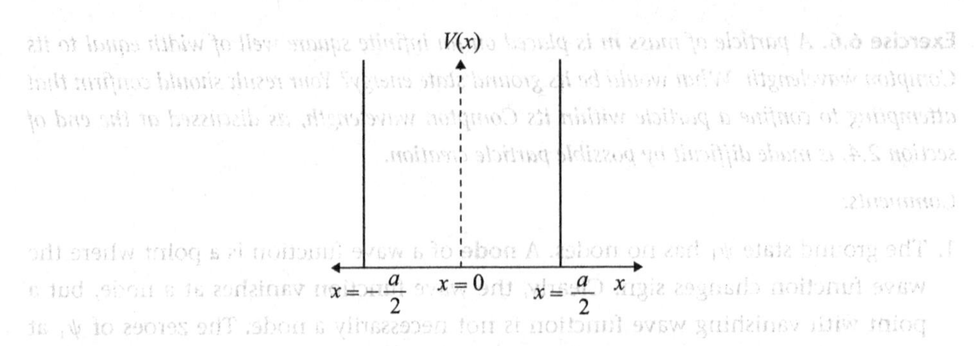

Figure 6.3
The infinite square well of width a centered at $x=0$ defines a symmetric $V(x)$.

which does not vanish at the end points. Even more, momentum eigenstates, which are necessarily of the form $e^{ipx/\hbar}$, are also not allowed wave functions. While the action of $\hat{p}$ is not well defined, the action of $\hat{p}^2$ is well defined: acting on the eigenstates, we have $\hat{p}^2\psi_n \sim \psi_n'' \sim \psi_n$. It is important that $\hat{p}^2$ is well defined since the Hamiltonian $\hat{H}$ contains this operator.

Exercise 6.7. *Consider a particle of mass m on an infinite square well that extends from $x=-a$ to $x=a$. Find the ground state wave function $\psi(x)$ and its energy E_g.*

Exercise 6.8. *To establish the symmetry or antisymmetry of the energy eigenstates in the infinite square well $x\in[0,a]$, we need an identity comparing $\psi_n(a-x)$ to $\psi_n(x)$ since x and $a-x$ are points related by reflection about the midpoint $\frac{a}{2}$ of the well. Use a suitable trigonometric identity to find the requisite sign factor σ_n in the following equation:*

$$\sin\left(\frac{n\pi(a-x)}{a}\right)=\sigma_n\cdot\sin\left(\frac{n\pi x}{a}\right).$$

Use your result to identify the symmetry or antisymmetry of the energy eigenstates ψ_n.

6.5 The Finite Square Well

We now examine the finite square well, shown in figure 6.4 and defined as follows:

$$V(x)=\begin{cases}-V_0, & \text{for } |x|\leq a,\\ 0, & \text{for } |x|\geq a.\end{cases}\quad \text{with } V_0>0. \tag{6.5.1}$$

Note that the potential energy is zero for $|x|>a$. The potential energy is negative and equal to $-V_0$ in the well because we defined V_0 to be a positive number. The width of the well has been chosen to be $2a$, twice as wide as our choice for the infinite square well. Note also that we have placed the bottom of the well differently. The bottom of the infinite square well was at zero potential energy. Now the top of the finite square well is at zero potential energy. To obtain the infinite square well as a limit of the finite square well, we have to take V_0 to infinity, but care is needed to compare energies. Those in the infinite square well are measured with respect to a bottom of the potential

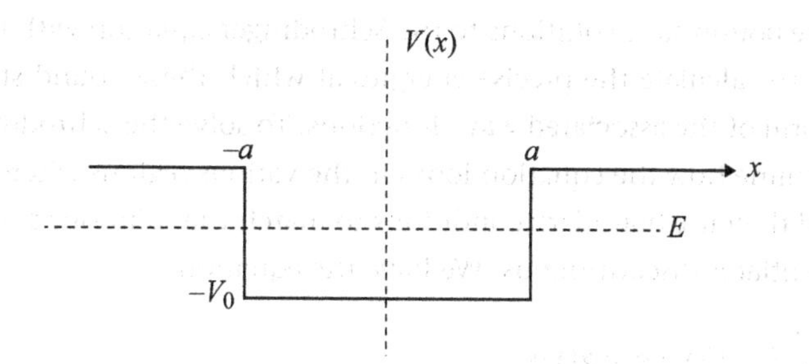

Figure 6.4
The finite square well potential.

at zero energy. They can be compared with the finite square well energies measured relative to the bottom of the potential at $-V_0$.

We will be interested in **bound states**: energy eigenstates that are normalizable. Normalizability makes bound states localized states: they must vanish as $|x| \to \infty$. The energy E of a finite well bound state must be *negative*. This is readily understood. If $E > 0$, in the region $x > a$ where the potential vanishes the wave function must satisfy the equation

$$-\frac{\hbar^2}{2m}\frac{d^2\psi}{dx^2} = E\psi, \tag{6.5.2}$$

which is equivalent to

$$\frac{d^2\psi}{dx^2} = -k^2\psi, \quad \text{with} \quad E = \frac{\hbar^2 k^2}{2m}. \tag{6.5.3}$$

Any solution $c_1 e^{ikx} + c_2 e^{-ikx}$ of this equation is not normalizable given that it holds for all $x > a$. Thus, positive energies cannot lead to bound states. The energy E of a bound state is shown as a dashed line in the figure, and we have the range

$$-V_0 < E < 0. \tag{6.5.4}$$

Since E is negative, $E = -|E|$. For a bound state of energy E, the energy $\tilde{E}$ measured with respect to the bottom of the potential is

$$\tilde{E} \equiv E - (-V_0) = V_0 - |E| > 0. \tag{6.5.5}$$

Those $\tilde{E}$ energies are those that can be compared with the energies of the infinite square well in the limit as $V_0 \to \infty$.

For our bound states, the region $|x| < a$ is the *classically allowed* region: the energy E is larger than the value $-V_0$ of the potential, and the kinetic energy is thus positive. On the other hand, the region $|x| > a$ is the *classically forbidden* region: its total energy E is below the potential energy $V = 0$, implying a manifestly unphysical negative kinetic energy. In quantum mechanics such a particle *can be* in the classically forbidden region; the wave function need not vanish in this region but tends to be small. The probability of finding the particle quickly falls off and goes to zero as $|x| \to \infty$.

What are the bound state solutions to the Schrödinger equation with this potential? We would like to calculate the precise energies at which these bound states exist and visualize the form of the associated wave functions. To solve the Schrödinger equation, we have to examine how the equation looks in the various regions where the potential is constant and then use boundary conditions to match the solutions across the points where the potential is discontinuous. We have the equation

$$\frac{d^2\psi}{dx^2} = -\frac{2m}{\hbar^2}(E - V(x))\,\psi = \alpha(x)\psi, \tag{6.5.6}$$

where we have defined the factor $\alpha(x)$ that multiplies the wave function on the right-hand side. We then consider the solutions of this equation in two regions:

- $|x| < a$: This is the classically allowed region. Here $\alpha(x)$ is a *negative constant*, and the wave function is constructed with sines and cosines.
- $|x| > a$: This is the classically forbidden region. Here $\alpha(x)$ is a *positive constant*, and the wave function is constructed with real exponentials.

The potential $V(x)$ for the finite square well is an even function of x: $V(-x) = V(x)$. We can therefore use the theorem cited earlier (section 7.1, corollary 2) that for an even potential the bound states are either symmetric or antisymmetric. We begin by looking for even bound states: wave functions ψ for which $\psi(-x) = \psi(x)$.

Even bound states Since the potential is piecewise continuous, we must study the differential equation in two regions:

- $|x| < a$. This is the region inside the well, where $V(x) = -V_0$. Equation (6.5.6) becomes

$$\frac{d^2\psi}{dx^2} = -\frac{2m}{\hbar^2}(E - (-V_0))\psi = -\frac{2m}{\hbar^2}(V_0 - |E|)\psi. \tag{6.5.7}$$

Since $V_0 - |E|$ is a positive constant, we can define a real $k > 0$ by

$$\boxed{k^2 \equiv \frac{2m}{\hbar^2}(V_0 - |E|) > 0, \quad k > 0.} \tag{6.5.8}$$

It is interesting to note that this equation is analogous to the equation $k^2 = 2mE/\hbar^2$ for a free particle with kinetic energy E. Indeed, $V_0 - |E|$ is the kinetic energy of the particle when inside the well. The differential equation to be solved now reads

$$\psi'' = -k^2\psi, \tag{6.5.9}$$

for which the only possible even solution is

$$\boxed{\psi(x) = \cos kx, \quad |x| < a.} \tag{6.5.10}$$

We are not including a normalization constant because we do not aim for normalized eigenstates. Our eigenstates, being bound states, will be *normalizable*. Of course, without a normalization factor the above wave function does not have the required units, but this will not matter in what follows.

- $|x| > a$. This is the region outside the well, where $V(x) = 0$. Equation (6.5.6) then becomes

$$\psi'' = -\frac{2m}{\hbar^2}(E - 0)\,\psi = \frac{2m|E|}{\hbar^2}\,\psi. \tag{6.5.11}$$

This time we define a real positive constant κ by the relation

$$\boxed{\kappa^2 \equiv \frac{2m|E|}{\hbar^2}, \quad \kappa > 0.} \tag{6.5.12}$$

The differential equation to be solved now reads

$$\psi'' = \kappa^2 \psi, \tag{6.5.13}$$

and the solutions are exponentials. In fact, we need exponentials that decay as $x \to \pm\infty$; otherwise, the wave function will not be normalizable. This should be physically intuitive. In a classically forbidden region, the probability of being far away from the well must be vanishingly small. For $x > a$, we choose the decaying exponential

$$\boxed{\psi(x) = A\,e^{-\kappa x}, \quad x > a,} \tag{6.5.14}$$

where A is a constant to be determined by the boundary conditions. More generally, given that the solution is even, we have

$$\psi(x) = A\,e^{-\kappa|x|}, \quad |x| > a. \tag{6.5.15}$$

The deeper the energy E of the bound state, the larger the value of κ and the faster the wave function decays in the forbidden region.

All in all, we have for an

$$\text{even solution:} \qquad \psi(x) = \begin{cases} \cos(kx), & |x| < a, \\ Ae^{-\kappa|x|}, & |x| > a. \end{cases} \tag{6.5.16}$$

It is now useful to note that κ^2 and k^2 satisfy a simple relation. Using their definitions above, we see that the energy $|E|$ drops out of their sum, and we have

$$k^2 + \kappa^2 = \frac{2mV_0}{\hbar^2}. \tag{6.5.17}$$

At this point we can improve the notation by introducing the *unit-free* constants η, ξ, and z_0 as follows:

$$\eta \equiv ka > 0, \quad \xi \equiv \kappa a > 0, \quad z_0^2 \equiv \frac{2mV_0 a^2}{\hbar^2}. \tag{6.5.18}$$

Clearly, η is a proxy for k, and ξ is a proxy for κ. Both depend on the energy of the bound state. The parameter z_0, unit-free, depends on the depth and width of the potential and the mass of the particle. If you are given a potential, you know the number z_0. A very deep and/or wide potential has very large z_0, while a very shallow and/or narrow potential has small z_0. As we will see, the value of z_0 tells us how many bound states the square well has.

Multiplying (6.5.17) by a^2 and using our definitions above, we get

$$\boxed{\eta^2 + \xi^2 = z_0^2.} \tag{6.5.19}$$

Let us make clear that solving for ξ is actually like solving for the bound state energy. From equation (6.5.12), we can see that

$$\xi^2 = \kappa^2 a^2 = \frac{2m|E|a^2}{\hbar^2} = \frac{2mV_0 a^2}{\hbar^2}\frac{|E|}{V_0} = z_0^2 \frac{|E|}{V_0}, \tag{6.5.20}$$

and from this we get

$$\frac{|E|}{V_0} = \left(\frac{\xi}{z_0}\right)^2. \tag{6.5.21}$$

This is a nice equation. The left-hand side gives the energy as a fraction of the depth V_0 of the well, and the right-hand side involves ξ and the constant z_0 of the potential. The quantity η also encodes the energy in a slightly different way. From (6.5.8) we show that

$$\eta^2 = k^2 a^2 \equiv \frac{2ma^2}{\hbar^2}(V_0 - |E|) = z_0^2\left(1 - \frac{|E|}{V_0}\right). \tag{6.5.22}$$

Using (6.5.5), we see that this provides the energy $\tilde{E}$, measured relative to the bottom of the potential

$$\frac{\tilde{E}}{V_0} = 1 - \frac{|E|}{V_0} = \left(\frac{\eta}{z_0}\right)^2. \tag{6.5.23}$$

This formula is convenient to understand how the infinite square well energy levels appear in the limit as the depth of the finite well goes to infinity. Note the similarity of equations (6.5.21) and (6.5.23).

Let us finally complete the computation. We must impose the continuity of ψ and the continuity of ψ' at $x = a$. We need not worry about the continuity of ψ and ψ' at $x = -a$ since ψ for $x < 0$ is determined from ψ for $x > 0$ by the condition that ψ is even; if ψ and ψ' are continuous at $x = a$, they will also be continuous at $x = -a$. Using the expression for ψ in (6.5.16), the continuity conditions give

$$\begin{aligned} \psi \text{ continuous at } x = a: &\quad \cos(ka) = Ae^{-\kappa a}, \\ \psi' \text{ continuous at } x = a: &\quad -k\sin(ka) = -\kappa A e^{-\kappa a}. \end{aligned} \tag{6.5.24}$$

Dividing the second equation by the first, we eliminate the constant A and find a second relation between k and κ! This is exactly what is needed. The result is

$$k \tan ka = \kappa \quad \Rightarrow \quad ka \tan ka = \kappa a \quad \Rightarrow \quad \xi = \eta \tan \eta. \tag{6.5.25}$$

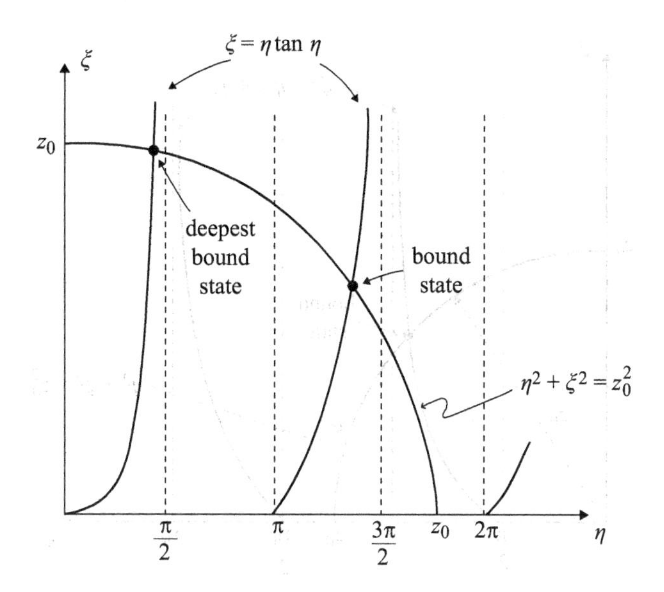

Figure 6.5
Graphic representation of the simultaneous equations (6.5.26) for even bound states. The intersections of the circle with the $\eta \tan \eta$ function represent even bound states in the finite square well potential. The deepest bound state is the one with lowest η or with highest ξ.

Our task of finding the even bound states is now reduced to finding solutions to the simultaneous equations:

$$\boxed{\text{even solutions: } \eta^2 + \xi^2 = z_0^2, \quad \xi = \eta \tan \eta, \quad \xi, \eta > 0.} \tag{6.5.26}$$

These equations are transcendental, and a general algebraic solution is not possible. In special limits, however, the equations simplify and can give us important qualitative insights. The equations can also be solved numerically to find all the solutions that exist for a given fixed value of z_0. Each solution represents a bound state. We can understand the solution space by plotting these two equations in the *first quadrant* of an (η, ξ) plane, as shown in figure 6.5.

The first equation in (6.5.26) is a piece of a circle of radius z_0. The second equation, $\xi = \eta \tan \eta$, gives infinitely many curves as η grows from zero to infinity. The value of ξ goes to infinity when η approaches each odd multiple of $\pi/2$. The bound states are represented by the intersections of the circle with the curves.

In figure 6.5 we see two intersections, which means two even bound states. The first intersection takes place near $\eta = \pi/2$ with large $\xi \sim z_0$. This is the ground state, or the most deeply bound bound state. This can be seen from (6.5.21), noting that this is the solution with largest ξ. Alternatively, it can be seen from equation (6.5.23), noting that this is the solution with smallest η. The second solution occurs for η near $3\pi/2$. As the radius of the circle becomes bigger, we get more and more intersections; z_0 controls the number of even bound states. Finally, note that there is always an even solution, no

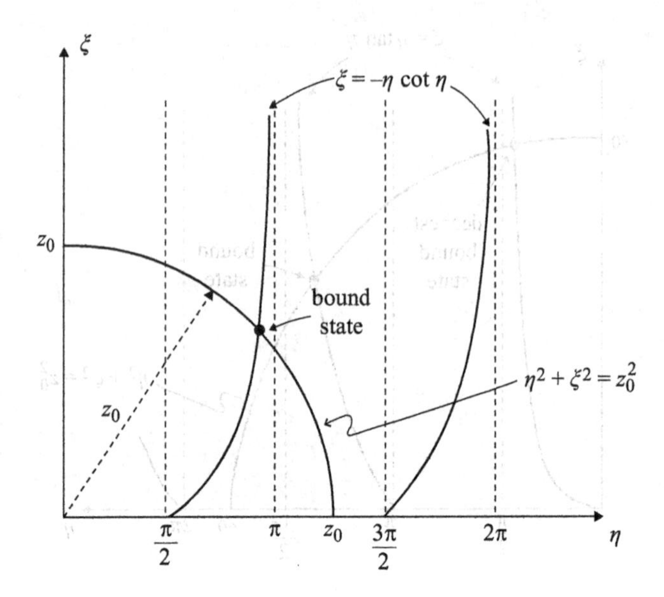

Figure 6.6
Graphic representation of (6.5.28) for odd bound states. The intersections of the circle with the curves $\xi = -\eta \cot \eta$ are odd bound state solutions in the finite square well potential. In the case displayed, there is just one bound state.

matter how small z_0 is, because the arc of the circle will always intersect the first curve of the $\xi = \eta \tan \eta$ plot. Thus, at least one bound state exists regardless of how shallow or narrow the finite well is.

Odd bound states For odd solutions all of our definitions $(k, \kappa, z_0, \eta, \xi)$ remain the same. The wave function now is of the form

$$\psi(x) = \begin{cases} A\, e^{-\kappa x}, & x > a, \\ \sin kx, & -a < x < a, \\ -A\, e^{\kappa x}, & x < -a. \end{cases} \tag{6.5.27}$$

Matching ψ and ψ' at $x = a$ now gives $\xi = -\eta \cot \eta$ (do it!). As a result, the relevant simultaneous equations are now

$$\boxed{\text{odd solutions:} \quad \eta^2 + \xi^2 = z_0^2, \quad \xi = -\eta \cot \eta, \quad \xi, \eta > 0.} \tag{6.5.28}$$

In figure 6.6 the curve $\xi = -\eta \cot \eta$ does not appear for $\eta < \pi/2$ because ξ is then negative. For $z_0 < \pi/2$, there are no odd bound state solutions, but we still have the even bound state. Comparing the figures for even and for odd bound states, one reaches the conclusion that as a function of the bound state energy the even and odd bound states alternate.

Exercise 6.9. *Consider a finite square well with $z_0 = (2.001)\pi$. How many even bound states does this potential have? How many odd bound states does this potential have? How would you qualitatively describe the energy of the even bound state with the highest energy?*

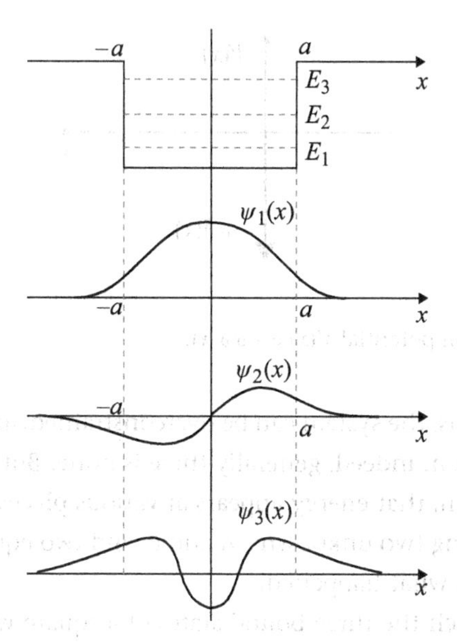

Figure 6.7
Sketching eigenstates of a finite square well potential. The energies are $E_1 < E_2 < E_3$, and the corresponding eigenstates are ψ_1, ψ_2, and ψ_3, with ψ_1 the ground state.

We can actually give a formula for the *total* number n_b of bound states of the finite square well as a function of z_0. This formula counts both even and odd states. First, note that however small z_0 may be, there is always one bound state. Second, looking at figures 6.5 and 6.6, we see that there are curves to be intersected by the z_0-radius circle in each interval of size $\pi/2$ along the horizontal axis. Therefore, a new state appears each time z_0 increases by $\pi/2$. To write the answer for n_b, we use the "floor" function floor(x), defined as the greatest integer less than or equal to x. Consistent with these observations, we have

$$n_b(z_0) = 1 + \text{floor}\left(\frac{2z_0}{\pi}\right). \tag{6.5.29}$$

The argument of the floor function here is z_0 divided by $\pi/2$. Note that n_b is a discontinuous function of z_0; the number of states increases by one each time z_0 hits a multiple of $\pi/2$. Moreover, if z_0 is an exact multiple of $\pi/2$, the formula is counting the zero-energy bound state (usually called a *threshold* bound state). Make sure you test the above formula with your result for exercise 6.9.

We could have anticipated the quantization of the energy by the following heuristic argument. Suppose you try to find energy eigenstates that, as far as solving the Schrödinger equation, are determined up to an overall normalization. Assume you don't know the energy is quantized, and you fix some arbitrary fixed energy. As we did in both the even and the odd case, we can set the coefficient of the $\cos kx$ or $\sin kx$ function inside the well equal to one. The coefficient of the decaying exponential outside the well was undetermined; we called it A. Therefore we have just one unknown, A. But we have two constraints: the continuity of ψ and of ψ' at $x = a$. With one unknown and

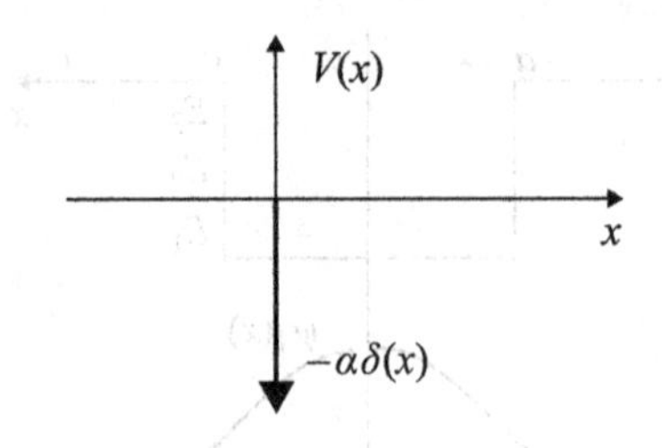

Figure 6.8
An attractive delta function potential $V(x) = -\alpha\,\delta(x)$.

two constraint equations, the system can be overconstrained, and we have no reason to believe there is a solution. Indeed, generally there is none. But then, if we think of the energy E as an unknown, that energy appears at various places in the constraint equations (in k and κ). Having two unknowns, A and E, and two equations, we can expect a solution! This is indeed what happened.

In figure 6.7 we sketch the three bound states of a square well potential. The states ψ_1, ψ_2, and ψ_3 have energies E_1, E_2, and E_3, respectively, with $E_1 < E_2 < E_3 < 0$. The state ψ_1 is the ground state, ψ_2 is the first excited state, and ψ_3 is the second excited state. A few features of the wave functions are manifest. They alternate in parity: ψ_1 is even, ψ_2 is odd, and ψ_3 is even. The number of nodes increases as the energy increases: ψ_1 has no nodes, ψ_2 has one node, and ψ_3 has two nodes. The exponential decay in the region $|x| > a$ is fastest for the ground state and slowest for the least bound state. This is as expected: the deeper the bound state energy, the faster the wave function must decay in the forbidden region.

6.6 The Delta Function Potential

Consider a particle of mass m moving in a rather singular one-dimensional potential $V(x)$ that vanishes everywhere except at $x = 0$, where it has infinite value. More precisely, the potential is a delta function localized at $x = 0$ and is written as

$$V(x) = -\alpha\,\delta(x), \quad \alpha > 0. \tag{6.6.1}$$

Here α is a constant chosen to be positive. Because of the explicit minus sign, the potential is infinitely negative at $x = 0$; this means that the potential is attractive. The potential is shown in figure 6.8, where we represent the delta function by an arrow pointing downward.

We want to know if this potential admits bound states. For a bound state, the energy E must be negative: this ensures that all of $x \neq 0$ is classically forbidden, and the wave function will decay rapidly, allowing a normalized solution.

A bit of intuition comes by thinking of the delta function as a finite square well in the limit as the width of the well goes to zero, and the depth goes to infinity in such a way that the product, representing the "area," is finite. This must be so because the

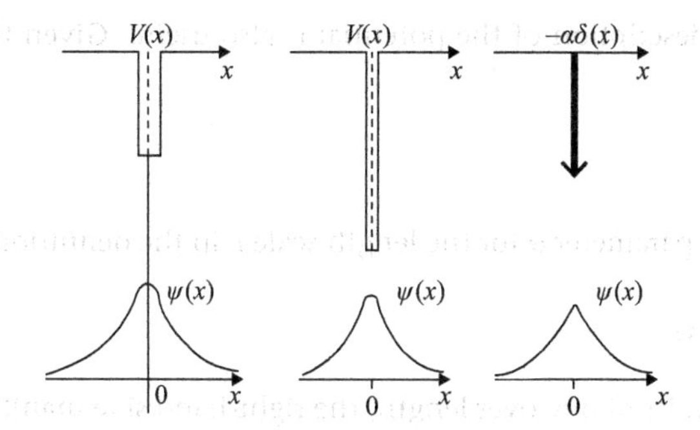

Figure 6.9
The delta function potential as the limit where the finite square well becomes narrower and deeper simultaneously. We expect to get a wave function with a discontinuous derivative.

delta function is a function with unit area, as is clear from its integral. In figure 6.9 we show two finite-well representations and sketch the ground state wave function. We can see that the middle region provides the curving of the wave function needed to have a smooth derivative. In the limit as the width of the well goes to zero we would expect, if there is a bound state, to have a *discontinuous* derivative.

We can get further insight by considering units. The dimensional constants in the problem are $\alpha, m,$ and $\hbar$. We want to learn how to construct a length scale L and an energy scale E in terms of $\alpha, m,$ and $\hbar$. Since a $\delta(x)$ has units of one over length, the constant α must have the units of energy times length for the potential to have units of energy. Thus, we find that the scales are related by

$$E = \frac{\alpha}{L}, \tag{6.6.2}$$

but, as usual, we can construct an energy scale with

$$E = \frac{\hbar^2}{mL^2}. \tag{6.6.3}$$

From these two equations, we find that

$$L = \frac{\hbar^2}{m\alpha} \quad \Rightarrow \quad E = \frac{m\alpha^2}{\hbar^2}. \tag{6.6.4}$$

The units of energy must be carried by the above combination of the constants of the problem. Therefore, the energy E_b of any bound state must be a number times that combination:

$$E_b = -\# \frac{m\alpha^2}{\hbar^2}, \tag{6.6.5}$$

where # is a unit-free positive number that we aim to determine. It is good to see α appearing in the numerator. This means that as the strength of the delta function increases, the depth of the bound state also increases, as we would naturally expect.

An alternative description of the potential is also useful. Given that we saw above that

$$\alpha = \frac{\hbar^2}{mL}, \tag{6.6.6}$$

we could trade the parameter α for the length scale L in the definition of the potential:

$$V(x) = -\frac{\hbar^2}{mL}\,\delta(x). \tag{6.6.7}$$

With $\delta(x)$ having units of one over length, the right-hand side manifestly has the right units. In this picture the strength of the delta function potential is characterized by the value of L. The larger the value of L, the weaker the strength of the potential.

Let us now turn to the relevant equations. We want to find all possible $E<0$ states. The wave function is constrained by the time-independent Schrödinger equation

$$-\frac{\hbar^2}{2m}\frac{d^2\psi}{dx^2} = (E - V(x))\psi. \tag{6.6.8}$$

For $x \neq 0$, we have $V(x) = 0$, so this becomes

$$\frac{d^2\psi}{dx^2} = \left(-\frac{2mE}{\hbar^2}\right)\psi = \kappa^2\psi, \qquad \kappa^2 \equiv -\frac{2mE}{\hbar^2} > 0. \tag{6.6.9}$$

The solutions to this differential equation are linear combinations of the functions

$$e^{\kappa x}, \quad e^{-\kappa x}, \quad \kappa > 0. \tag{6.6.10}$$

Since the delta function effectively separates the $x>0$ and $x<0$ regions, we need solutions in both regions, and they must match appropriately at $x=0$ to form a complete solution.

Before setting up the relevant equations, we explain why we expect a single bound state. The potential is even: $\delta(-x) = \delta(x)$, just like the finite square well potential. In the finite square well, the ground state is even and has no nodes, and the first excited state is odd and has one node. Trusting that the delta function potential is the limit of a square well, we assert that if the delta function potential has a ground state it must be even and without nodes. Odd states must have a node at $x=0$, for otherwise the wave function would be discontinuous at $x=0$. But no such odd solution can be built from the functions above: for $x>0$, we must take $e^{-\kappa x}$ for the wave function to be normalizable, but then this function does not vanish at $x=0$. Therefore, there are no odd solutions. The even solution is expected to be unique because of the node theorem applied to the delta function potential: excited bound state solutions must have more and more nodes, and neither the function $e^{-\kappa x}$ for $x>0$ nor the function $e^{\kappa x}$ for $x<0$ vanishes anywhere. Intuitively, the even excited states of the square well do not help because as the well goes to zero size all the oscillatory part of the solution shrinks away, and only the condition of an even wave function remains. The delta function potential must have just one bound state, an even one.

Let us use the above solutions to build the ground state wave function. Again, for $x > 0$ we must discard the solution $e^{\kappa x}$ because it diverges as $x \to \infty$. Similarly, we must discard $e^{-\kappa x}$ for $x < 0$. Since the wave function must be continuous at $x = 0$, the solution must be of the form

$$\psi(x) = \begin{cases} A\, e^{-\kappa x}, & x > 0, \\ A\, e^{\kappa x}, & x < 0. \end{cases} \tag{6.6.11}$$

Alternatively, we can write

$$\psi(x) \;=\; A\, e^{-\kappa |x|}. \tag{6.6.12}$$

Is any value of κ allowed for this solution? No, we will get another constraint by considering the derivative of the wave function and learning that, as anticipated, it is discontinuous. Indeed, the Schrödinger equation gives us a constraint for this discontinuity. Starting with

$$-\frac{\hbar^2}{2m}\frac{d^2\psi}{dx^2} + V(x)\psi = E\psi, \tag{6.6.13}$$

we integrate this equation from $x = -\epsilon$ to $x = \epsilon$, with $0 < \epsilon \ll 1$, a range that includes the position of the delta function. This gives

$$-\frac{\hbar^2}{2m}\left(\frac{d\psi}{dx}\bigg|_{\epsilon} - \frac{d\psi}{dx}\bigg|_{-\epsilon}\right) + \int_{-\epsilon}^{\epsilon} dx\,(-\alpha\delta(x))\psi(x) = E\int_{-\epsilon}^{\epsilon} dx\,\psi(x). \tag{6.6.14}$$

The integral on the left-hand side returns a finite value due to the delta function. In the limit as $\epsilon \to 0$ the integral on the right-hand side vanishes because $\psi(x)$ is finite for all x, while the region of integration is squeezed to zero size. Integrating the delta function yields

$$-\frac{\hbar^2}{2m}\lim_{\epsilon\to 0}\left(\frac{d\psi}{dx}\bigg|_{\epsilon} - \frac{d\psi}{dx}\bigg|_{-\epsilon}\right) - \alpha\psi(0) = 0. \tag{6.6.15}$$

Following (6.2.5), we define the discontinuity Δ_0 of ψ' at $x = 0$ by

$$\Delta_0\left(\frac{d\psi}{dx}\right) \;\equiv\; \lim_{\epsilon\to 0}\left(\frac{d\psi}{dx}\bigg|_{\epsilon} - \frac{d\psi}{dx}\bigg|_{-\epsilon}\right). \tag{6.6.16}$$

We have therefore learned that

$$\boxed{\Delta_0\left(\frac{d\psi}{dx}\right) \;=\; -\frac{2m\alpha}{\hbar^2}\,\psi(0).} \tag{6.6.17}$$

Note that the discontinuity in ψ' at the position of the delta function is proportional to the value of the wave function at that point. This has a simple consequence. Suppose a bound state ψ of some potential has a node. You can modify the potential by adding a delta function at the position of the node, and your bound state ψ would still be valid because the constraint (6.6.17) is automatically satisfied given that ψ' is continuous at the node.

Applying the discontinuity equation to our solution (6.6.11), we have

$$\lim_{\epsilon\to 0}\left(\frac{d\psi}{dx}\bigg|_{\epsilon}-\frac{d\psi}{dx}\bigg|_{-\epsilon}\right)=\lim_{\epsilon\to 0}\left(-\kappa A e^{-\kappa\epsilon}-\kappa A e^{-\kappa\epsilon}\right)=-2\kappa A=-\frac{2m\alpha}{\hbar^2}A. \quad (6.6.18)$$

This relation fixes the value of κ as

$$\kappa=\frac{m\alpha}{\hbar^2}, \quad (6.6.19)$$

and therefore the value E_b of the bound state energy:

$$E_b=-\frac{\hbar^2\kappa^2}{2m}=-\frac{1}{2}\frac{m\alpha^2}{\hbar^2}. \quad (6.6.20)$$

As we anticipated with the unit analysis, the answer takes the required form (6.6.5). The undetermined constant # takes the value 1/2.

In terms of the length scale $L=\hbar^2/(m\alpha)$ for which the potential reads $V=-\frac{\hbar^2}{mL}\delta(x)$, the above results are more transparent. The constant κ becomes

$$\kappa=\frac{1}{L}. \quad (6.6.21)$$

With $\psi\sim e^{-\kappa|x|}=e^{-|x|/L}$, we interpret L as the penetration length into the forbidden region. The length scale L is roughly the uncertainty in the position of the particle. The bound state energy in this language reads

$$E_b=-\frac{1}{2m}\left(\frac{\hbar}{L}\right)^2, \quad (6.6.22)$$

an energy associated with a momentum $\hbar/L$. Finally, the discontinuity condition (6.6.17) takes the simpler form

$$\Delta_0\left(\frac{d\psi}{dx}\right)=-\frac{2}{L}\psi(0), \quad (6.6.23)$$

where the units are manifestly consistent.

Exercise 6.10. *Show that the uncertainty in the position of the particle in the ground state of the delta function potential is* $\Delta x=L/\sqrt{2}$.

While a delta function potential in one dimension is a nice and tractable system, delta functions in two- and three-dimensional potentials are too singular to give simple bound states. We can anticipate such complications from the analysis of units.

Exercise 6.11. *Consider an attractive delta function potential for a particle of mass m moving in two dimensions:* $V(x,y)=-\alpha\delta(x)\delta(y)$, *with* $\alpha>0$. *Show that there is no way to construct a quantity with units of energy from the constants* α, m, *and* $\hbar$ *of the problem.*

Exercise 6.12. *Consider an attractive delta function potential for a particle of mass m moving in three dimensions:* $V(x,y,z)=-\alpha\delta(x)\delta(y)\delta(z)$, *with* $\alpha>0$. *Build a quantity with units of energy from the constants* α, m, *and* $\hbar$ *of the problem. Does this quantity have a reasonable dependence on* α?

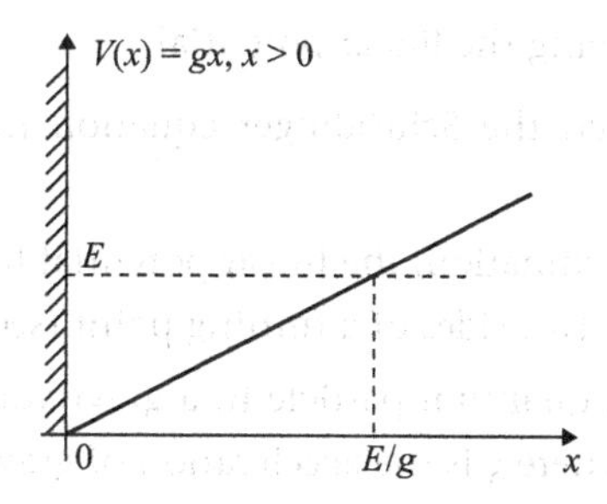

Figure 6.10
A linear potential with $V(x)=gx$ for $x>0$ and a hard wall at $x=0$. For a state of energy E, there is a turning point at $x=E/g$.

6.7 The Linear Potential

Consider a particle of mass m in a potential that is linearly increasing for $x>0$:

$$V(x)=\begin{cases} \frac{\hbar^2}{2mL^2}\frac{x}{L}, & \text{for } x>0, \\ \infty, & \text{for } x<0. \end{cases} \tag{6.7.1}$$

With L a positive-length parameter, this expression makes manifest the energy units of the potential. There is also an infinite wall at $x\leq 0$ so that the motion of the particle is constrained to $x>0$ (figure 6.10). The smaller the value of L, the steeper the potential. It is useful to define the positive constant g by

$$g\equiv\frac{\hbar^2}{2mL^3}. \tag{6.7.2}$$

The potential, for positive x, can then be written more briefly as

$$V(x)=gx, \quad x>0. \tag{6.7.3}$$

The wall plus the linear potential will display an infinite set of bound states. Just like the infinite square well, for any value of the energy, assumed positive, the particle encounters classically forbidden regions to the left and to the right of the classically allowed region. For energy E, the classically forbidden region extends over $x<0$ and $x>E/g$ (figure 6.10). For an energy eigenstate, the wave function would decay quickly as x grows beyond E/g, making it normalizable.

In order to study the particle in the potential $V(x)$, with its wall, it is actually simpler to consider the motion in a potential $\tilde{V}(x)$ that is linear for *all* values of x, positive and negative:

$$\tilde{V}(x)=\frac{\hbar^2}{2mL^2}\frac{x}{L}, \quad -\infty<x<\infty. \tag{6.7.4}$$

Once we understand the behavior of wave functions in this potential, we will find a simple way to determine the bound states of the potential V with the wall.

We have reasons for studying the linear potential:

1. It is a potential for which the Schrödinger equation in momentum space leads quickly to a solution.
2. In the semiclassical approximation, the linear potential helps relate the behavior of the wave function on the two sides of a turning point (section 26.5).
3. The potential describes a quantum particle in a gravitational field. In this case the potential is $V(x) = mgx$, where g is the acceleration of gravity.

Recall the momentum-space version (4.4.25) of the Schrödinger equation, which for a potential $\tilde{V}(x)$ reads

$$i\hbar\frac{\partial}{\partial t}\Phi(p,t) = \left[\frac{p^2}{2m} + \tilde{V}\left(i\hbar\frac{\partial}{\partial p}\right)\right]\Phi(p,t). \tag{6.7.5}$$

For an energy eigenstate, we write $\Phi(p,t) = \phi(p)e^{-iEt/\hbar}$. The above equation, after canceling the common time dependence, gives us

$$\left[\frac{p^2}{2m} + \tilde{V}\left(i\hbar\frac{d}{dp}\right)\right]\phi(p) = E\phi(p), \tag{6.7.6}$$

where we can use ordinary derivatives with respect to p. For the linear potential in (6.7.4), we have

$$\frac{p^2}{2m}\phi + g\,i\hbar\frac{d\phi}{dp} = E\phi(p). \tag{6.7.7}$$

The Schrödinger equation, which in position space always involves second-order spatial derivatives, has become a *first-order* differential equation in momentum space! Hence the simplicity of the approach. Rearranging terms,

$$\frac{d\phi}{dp} = \frac{i}{\hbar g}\left(\frac{p^2}{2m} - E\right)\phi \quad\Rightarrow\quad \frac{d\phi}{\phi} = \frac{i}{\hbar g}\left(\frac{p^2}{2m} - E\right)dp. \tag{6.7.8}$$

This is quickly integrated:

$$\phi(p) = C\exp\left(\frac{i}{\hbar g}\frac{p^3}{6m} - \frac{i}{\hbar}\frac{E}{g}p\right), \tag{6.7.9}$$

with C an arbitrary constant of proportionality. Since the constant at the end would be fixed by normalization, we will simply ignore it in the following steps. The position-space wave function $\psi(x)$ is obtained by Fourier transformation:

$$\psi(x) \propto \int_{-\infty}^{\infty} dp\,\exp\left(\frac{i}{\hbar g}\frac{p^3}{6m} - \frac{i}{\hbar}\frac{E}{g}p + \frac{ixp}{\hbar}\right). \tag{6.7.10}$$

The integral cannot be done in terms of simple functions, but we can appreciate its features by passing to a dimensionless variable of integration. We will trade the dimensional p for a unit-free w via

$$p = \frac{\hbar}{L}w. \tag{6.7.11}$$

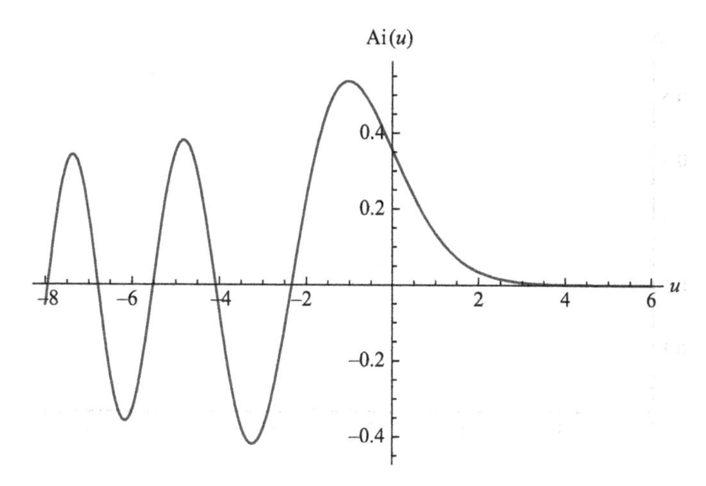

Figure 6.11
A plot of the Airy function Ai(u).

The integral now becomes

$$\psi(x) \propto \int_{-\infty}^{\infty} dw \exp\Big(i\frac{\hbar^2}{2mg}\frac{1}{L^3}\frac{w^3}{3} - i\frac{E}{gL}w + i\frac{x}{L}w\Big). \tag{6.7.12}$$

Using the relation between g and L, we have

$$\psi(x) \propto \int_{-\infty}^{\infty} dw \exp\Big(i\frac{w^3}{3} + i\frac{1}{L}\big(x - \tfrac{E}{g}\big)w\Big). \tag{6.7.13}$$

We can expand the exponential into a cosine term and a sine term. Since the sine term is odd under $w \to -w$, it provides no contribution to the integral. The cosine term is even under $w \to -w$, and the integral can be restricted to run from zero to infinity. We then write

$$\psi(x) \propto \frac{1}{\pi}\int_{0}^{\infty} dw \cos\Big(\tfrac{1}{3}w^3 + \tfrac{1}{L}\big(x - \tfrac{E}{g}\big)w\Big). \tag{6.7.14}$$

This integral, not calculable in terms of simple functions, defines the Airy function "Ai":

$$\psi(x) \propto \mathrm{Ai}\big(\tfrac{1}{L}\big(x - \tfrac{E}{g}\big)\big), \quad \text{with} \quad \mathrm{Ai}(u) \equiv \frac{1}{\pi}\int_{0}^{\infty} dw \cos\big(\tfrac{1}{3}w^3 + uw\big). \tag{6.7.15}$$

The Airy function can be calculated numerically using the above integral. A plot of Ai(u) as a function of u is shown in figure 6.11. The function decays quickly as u turns positive, with an inflection point at $u = 0$. For negative u the function oscillates ever more rapidly and with slowly decreasing amplitude. Airy functions will be considered in more detail in section 26.4. These functions are named after George Biddell Airy (1801–1892), England's seventh Astronomer Royal, who worked at Greenwich during the period 1835–1881. The astronomical data collected by Airy led to the selection of the meridian at Greenwich as the prime meridian. His studies of optics led to the functions that bear his name.

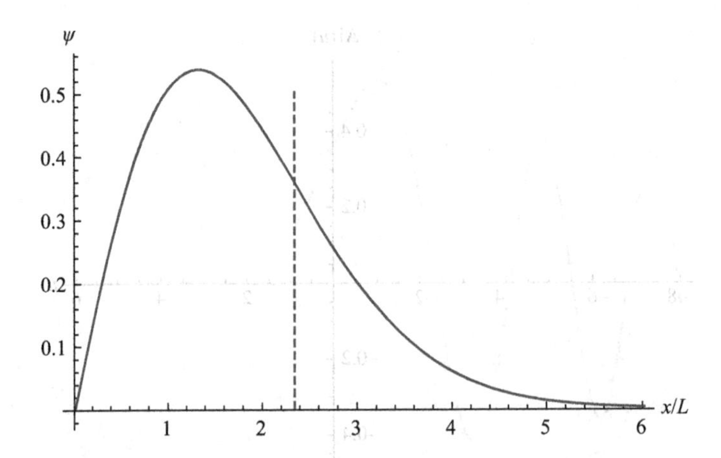

Figure 6.12
The ground state wave function $\psi(x) = \text{Ai}(\frac{x}{L} + a_1)$ on a linear potential. The vertical dashed line separates the classically allowed and classically forbidden regions.

The wave function we obtained, $\psi(x) \propto \text{Ai}(\frac{1}{L}(x - \frac{E}{g}))$, is the wave function for a linear potential that extends for all values of x. We do not expect bound states for this potential since for any value of E the region to the left of $x = E/g$ is allowed, and the wave function will not be normalizable. Note that at the classical turning point $x = E/g$, the wave function evaluates to a number proportional to Ai(0), and $u = 0$ is indeed the inflection point of the Airy function Ai(u).

To use our results for the case of a potential with a wall at $x = 0$, the strategy is clear: the solution we got, restricted to $x > 0$, is a valid solution for the problem with a wall *if* the solution vanishes at the wall. The part of the solution for $x < 0$ is then simply discarded. Thus, we need

$$\psi(0) \propto \text{Ai}(-\tfrac{E}{gL}) = 0. \tag{6.7.16}$$

The bound state energies should be positive, since the potential is always positive. Happily, this works out since all the zeroes of Ai(u) are for negative u. Ordered by increasing absolute value, the zeroes are listed as $a_1, a_2, \ldots$. The energies E_n are therefore

$$E_n = -gL\, a_n = \frac{\hbar^2}{2mL^2}\,|a_n| = \left(\frac{\hbar^2 g^2}{2m}\right)^{1/3} |a_n| > 0, \quad n = 1, 2, \ldots, \tag{6.7.17}$$

recalling from (6.7.2) that g and L are not independent constants. For reference, $a_1 \simeq -2.338107$ and $a_2 \simeq -4.087949$ are the first two zeroes of the Airy function Ai, obtained numerically. The ground state wave function in the linear potential is shown in figure 6.12. The wave function, as a function of x/L, is the Airy function displaced by $|a_1| \simeq 2.34$ to the right (compare with figure 6.11). Higher-energy eigenstates of the linear potential are found by displacing the Airy function to the right by $|a_n|$ with $n \geq 2$.

Problems

Problem 6.1. *A hard wall.*

A particle of mass m is moving in one dimension, subject to the potential $V(x)$:

$$V(x) = \begin{cases} 0, & \text{for } x > 0, \\ \infty, & \text{for } x \leq 0. \end{cases}$$

The potential vanishes for $x > 0$, but there is a wall at $x = 0$.

1. Find the energy eigenstates $\psi(x)$ and their energies E by solving the Schrödinger equation. You will need a continuous parameter to describe your states. Write your answers in terms of k where $k^2 = \frac{2mE}{\hbar^2}$. These states are scattering states: they are not normalizable, but the value $|\psi(x)|$ of the wave function remains bounded for all x. Choose a coefficient for your wave function such that $\psi'(0) = k$.
2. What values of k give the full set of energy eigenstates without overcounting? What energies E are allowed?

Problem 6.2. *A step up on the infinite line.*

A particle of mass m is moving in one dimension, subject to the potential $V(x)$:

$$V(x) = \begin{cases} V_0, & \text{for } x > 0, \quad V_0 > 0, \\ 0, & \text{for } x \leq 0. \end{cases}$$

Find the energy eigenstates $\psi(x)$ that exist for energies $0 < E < V_0$ by solving the Schrödinger equation. You will use a couple of continuous parameters to describe your states. Write your answers in terms of k and κ where

$$k^2 \equiv \frac{2mE}{\hbar^2} \quad \text{and} \quad \kappa^2 \equiv \frac{2m(V_0 - E)}{\hbar^2}.$$

The energy eigenstates are not normalizable. Choose a coefficient for your wave function such that $\psi(0) = -k/\kappa$. The wave function must be described by giving $\psi(x)$ for $x \leq 0$ and $\psi(x)$ for $x > 0$.

Problem 6.3. *Mimicking hydrogen with a one-dimensional square well.*

For the hydrogen atom, the Bohr radius a_0 and the ground state energy E_0 are given by

$$a_0 = \frac{\hbar^2}{me^2} \simeq 0.529\,\text{Å}, \quad E_0 = -\frac{e^2}{2a_0} \simeq -13.6\,\text{eV}.$$

The ground state is a bound state, and the potential goes to zero at infinity. We want to design a one-dimensional finite square well that simulates the hydrogen atom:

$$V(x) = \begin{cases} -V_0, & \text{for } |x| < a_0, \quad (V_0 > 0), \\ 0, & \text{for } |x| \geq a_0. \end{cases}$$

Calculate V_0 in eV so that the ground state of the well is at the right depth E_0. Does the square well that you have constructed have a second bound state?

Problem 6.4. *Particle in a square well (Ohanian).*

A particle of mass m moves in an infinite square well $x \in [0, a]$. Its $t = 0$ wave function is

$$\Psi(x, 0) = \sqrt{\frac{3}{4}}\sqrt{\frac{2}{a}} \sin\frac{2\pi x}{a} + \frac{1}{2}\sqrt{\frac{2}{a}} \sin\frac{3\pi x}{a}.$$

1. Is Ψ an energy eigenstate? Find $\Psi(x, t)$.
2. Suppose we measure the energy of the state. What are the possible values of the result, and what are the corresponding probabilities? Do these answers depend on time?
3. Calculate the time-dependent expectation value $\langle \hat{x} \rangle$. Carefully explain how this result can be used to immediately obtain the time-dependent expectation value $\langle \hat{p} \rangle$.

Problem 6.5. *Infinite rectangular well in the plane.*

Consider a particle of mass m moving in the (x, y) plane with a potential that is zero inside the rectangular box comprising all points (x, y) for which

$$0 \leq x \leq L_x, \quad 0 \leq y \leq L_y$$

and is infinite elsewhere.

1. Use the two-dimensional Schrödinger equation and separation of variables to find the *normalized* energy eigenstates $\psi(x, y)$. [For separation of variables, write $\psi(x, y) = \psi(x)\phi(y)$.] Note that you will need two integer parameters: n_x and n_y, associated with quantization in the x and y directions, respectively. What are the energies of the energy eigenstates?
2. Consider the case $L_x = L_y = L$. You will see that there are degeneracies in the energy spectrum. Some degeneracies have a simple symmetry explanation. If (n'_x, n'_y) and (n_x, n_y) are degenerate because of this symmetry, what is the relationship between them?

   Other symmetries are completely accidental and are seemingly random. Give an example of such a symmetry.
3. Suppose that $(L_x/L_y)^2$ is irrational. Are degeneracies possible?
4. Suppose that $L_x/L_y = p/q$, where p and q are integers so that the ratio is rational. Are there any degeneracies in this case?

Problem 6.6. *Infinite square well with an extra dimension.*

A particle in a one-dimensional infinite square well of width a is a particle forced to move on a line *segment* $x \in [0, a]$. Consider a particle moving on a *cylinder* of length a. The cylinder has circumference L, and it can be represented as a rectangular region in the (x, y) plane, with the y coordinate along the circumference of the cylinder and the horizontal lines with arrows identified, as shown below.

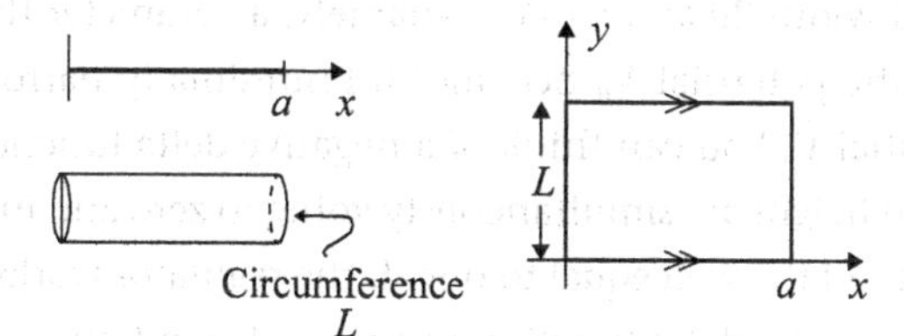

The system is described by the two-dimensional Schrödinger equation with a potential that vanishes in the rectangle $\{(x, y): 0 \leq x \leq a,\ 0 \leq y \leq L\}$ and is infinite on the vertical edges at $x = 0$ and $x = a$.

1. Perform separation of variables in the Schrödinger equation, and give the two equations that help determine the energy eigenstates. State the boundary conditions that apply.
2. Solve for the energies $E_{n\ell}$ and the normalized eigenstates $\psi_{n\ell}(x, y)$, where n and ℓ are quantum numbers for the x and y dependence, respectively. State the ranges n and ℓ run over. What is the ground state energy of the particle?
3. Assume henceforth that a and L are such that no accidental degeneracies occur (accidental degeneracies require special relations between a and L). List the energy eigenvalues for the particle in the cylinder that coincide with those for the one-dimensional segment $x \in [0, a]$.
4. What are (or is) the lowest energy levels that exist on the cylinder but do not exist in the segment?
5. The y dimension that turns the segment into a cylinder may be considered as a yet undetected small extra dimension. Suppose the size L of the extra dimension is about one thousand times smaller than the size a of a small interval where an experimenter has localized a particle. Assume also that the length a and the particle mass m are such that $\frac{\hbar^2}{2ma^2} = 1$ eV. Estimate the minimum energy that the experimenter needs to find evidence for the extra dimension.

Problem 6.7. *Bound state energy for shallow well.*

Consider the familiar finite square well potential $V(x) = -V_0$ for $-a \leq x \leq a$, and $V(x) = 0$ for $|x| > a$. Assume that the potential is very shallow and/or narrow so that $z_0^2 = 2mV_0a^2/\hbar^2$ is a very small number, and as a result, there is just one bound state. Estimate the energy E of this state in terms of V_0 and z_0 (i.e., find the leading term of the energy in the expansion in terms of z_0, as $z_0 \to 0$).

Problem 6.8. *From square well to delta function.*

In this problem we ask you to derive the bound state energy of a particle of mass m on the delta function potential $\hat{V}(x) = -\alpha\,\delta(x)$, $\alpha > 0$, starting from the problem of a particle of mass m on a finite square well potential. For this purpose consider a square well with potential $V_a(x)$:

$$V_a(x) = \begin{cases} -V_0, & \text{for } |x| < a, \quad V_0 > 0, \\ 0, & \text{for } |x| > a. \end{cases}$$

We think of the total width $2a$ as a *regulator*–namely, a parameter that will be taken to zero in the limit as the potential V_a becomes infinitesimally narrow to represent the delta function potential $\hat{V}$. You can think of a negative delta function as the limit of a well whose width and height are simultaneously going to zero and infinity, respectively, while keeping the area of the well equal to one. If the regulator works properly, the final answer for the energy of the delta function must not depend on a.

1. For a given value of a, fix the value of V_0 so that in the limit $a \to 0$ the potential V_a represents $\hat{V}$ correctly. Give your answer in terms of α and a.
2. What is the value of z_0^2 for the well V_a? As $a \to 0$, what happens to z_0? Explain why such behavior is reasonable.
3. Work with a very small but nonzero and calculate *leading* approximations for η and ξ in terms of z_0.
4. Determine the bound state energy of the delta function potential from your analysis.

Problem 6.9. *Finite square well turning into the infinite square well.*

Consider the standard square well potential $V(x) = -V_0$ for $|x| \le a$ and zero elsewhere, with $V_0 > 0$. The wave function for an even state is determined by its values for $x > 0$, which are:

$$\psi(x) = \begin{cases} \frac{1}{\sqrt{a}} \cos kx, & \text{for } 0 \le x \le a, \\ \frac{A}{\sqrt{a}} e^{-\kappa(x-a)}, & \text{for } x > a. \end{cases}$$

We have included the $\frac{1}{\sqrt{a}}$ prefactor to have consistent units for ψ, and A is a constant required by continuity at $x = a$. Recall that the bound state constraints are $\xi^2 + \eta^2 = z_0^2$, and $\xi = \eta \tan \eta$, with

$$\eta = ka, \qquad \xi = \kappa a, \qquad k^2 = \frac{2m(V_0 - |E|)}{\hbar^2}, \quad \text{and} \quad \kappa^2 = \frac{2m|E|}{\hbar^2}.$$

We want to explore the limit $V_0 \to \infty$ and show that the discontinuity of ψ' in the infinite well does not cause trouble. Keeping m and a constant as we let V_0 grow large is the same as letting z_0 grow large, where $z_0^2 = 2ma^2 V_0/\hbar^2$.

1. Consider the ground state of the potential. Compute η and ξ to leading order in z_0 in the limit of large z_0. Your answers should have no trigonometric functions in them. (Hint: Let the graphic solution guide you in your approximations.) Show that, consistent with your leading solutions for η and ξ, we have $A = \frac{c}{z_0}$, with c a constant you must determine. Use the solution for $x > a$ to write $\psi(x)$ and determine $\psi(a)$ and $\psi'(a)$ in this large z_0 limit.
2. We want to see if the expectation value of the Hamiltonian receives a singular contribution from the classically forbidden region $|x| > a$. Since the potential $V(x)$ vanishes there, we only have the contribution from the kinetic energy operator $\hat{K} = \frac{\hat{p}^2}{2m}$. Calculate the contribution to the expectation of $\hat{K}$ from the forbidden region $x > a$:

$$\langle \hat{K} \rangle \Big|_{x>a} = \int_a^\infty dx\, \psi^*(x) \hat{K} \psi(x).$$

The answer should have a z_0 dependence. The above computation is meaningful because $\psi(x)$ is accurately normalized. Indeed, as $z_0 \to \infty$, one can check that $\psi(x)$ takes the expected form for $|x| < a$ and vanishes for $|x| > a$.

Problem 6.10. *Infinite square well with an attractive delta function.*

Consider a particle of mass m in an infinite square well $x \in [0, a]$ with an attractive delta function at the midpoint $x = a/2$. The potential is then

$$V(x) = \begin{cases} \infty, & \text{for } x < 0 \text{ and for } x > a, \\ -\frac{\hbar^2}{mL}\, \delta(x - \frac{a}{2}), & \text{for } 0 < x < a. \end{cases} \tag{1}$$

The strength of the delta function is defined by a length scale L, which is a known parameter of the potential. Assume the strength of the delta function is sufficiently large to allow for a ground state with negative energy $E < 0$.

We will define $\xi \geq 0$ and $\kappa \geq 0$ from the equations $\xi^2 \equiv \kappa^2 a^2 \equiv \frac{2m|E|a^2}{\hbar^2}$. We also remind you that for the above delta function potential in the *absence* of a square well, the bound state energy E_0 is given by $E_0 = -\frac{\hbar^2}{2mL^2}$. All the work in this problem will focus on the *ground state* of the potential $V(x)$.

1. Write an ansatz for the wave function (your equation should have an expression for $0 < x < a/2$ and an expression for $a/2 < x < a$). Find the appropriate boundary conditions at $x = a/2$ and derive the transcendental equation that determines the energy. Your answer should take the form

   $$\frac{L}{a} = h(\xi),$$

   with $h(\xi)$ a function that you should determine. Sketch a plot of $h(\xi)$ for $\xi \in [0, \infty)$.
2. As a check on your answer, consider the appropriate limit of L/a for which the ground state energy E approaches the value E_0 quoted above. Show that your equation indeed gives $E \to E_0$ in this limit.
3. A physicist has designed an apparatus to study the delta function potential. In the experiment the length a represents the size of a real box in which the delta function potential is placed. The physicist wants the box to have a small effect on the bound state energy.

   The first nontrivial correction to the energy of part (2) takes the form $E = E_0(1 + \beta)$, where β is a unit-free quantity that is a function of the ratio a/L. Calculate β.

   Determine the value of a/L so that the ground state energy is that of the pure delta function potential with an error of at most 1%. Your answer can be given in the form $a/L = \ln(\#)$, where # is a number you should determine.
4. What is the ratio a/L for which the ground state energy E becomes zero? Find the normalized ground state wave function in this case.

7 Stationary States II: General Features

Here, we study the general properties of energy eigenstates for a particle in a one-dimensional potential. The spectrum of bound states is always nondegenerate, and the bound states of even potentials are either even or odd. We introduce the semiclassical approximation that applies to particles moving in slowly varying potentials and gives information about the amplitude and oscillations of energy eigenstates. We discuss and motivate the node theorem, which prescribes the number of nodes in bound state wave functions. We show how to set up a numerical calculation of bound states for a large class of potentials using the shooting method. We discuss the virial theorem that relates expectation values of different operators evaluated on stationary states. We develop the variational method for the calculation of upper bounds on the energies of stationary states. Finally, we present the Hellmann-Feynman lemma, which gives a simple formula for the rate of change of the energy of stationary states when the Hamiltonian changes.

7.1 General Properties

We have already learned quite a bit about the stationary states of a particle moving in one-dimensional potentials. These energy eigenstates are of two types. The normalizable ones are called bound states, and for them the wave function and its derivative vanish as $|x| \to \infty$. Nonnormalizable energy eigenstates are also of interest. The simplest example of these are the momentum eigenstates of a free particle. For nonnormalizable energy eigenstates, we demand that the value of $|\psi(x)|$ is bounded.

In chapter 6 we solved for the energy eigenstates of interesting one-dimensional problems: the particle on a circle, a particle in an infinite square well, a particle in a finite square well, a particle in a delta function potential, and a particle in a linear potential plus a wall. In the first two examples and in the last one, all the energy eigenstates are normalizable. For the finite square well and the delta function potential, there are nonnormalizable energy eigenstates, but we did not consider them. We only discussed their bound states.

It is now a good time to examine some general features of energy eigenstates, some of which were anticipated in these examples. These eigenstates are solutions of the

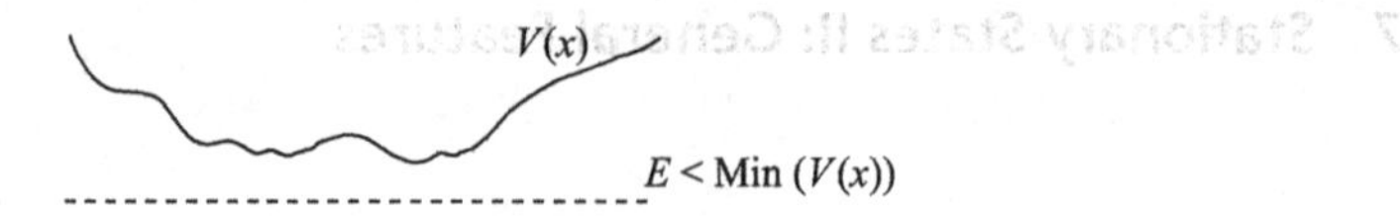

Figure 7.1
There are no energy eigenstates with energy E less than the minimum of the potential.

time-independent Schrödinger equation:

$$-\frac{\hbar^2}{2m}\frac{d^2\psi}{dx^2} + (V(x) - E)\psi = 0. \tag{7.1.1}$$

Let us first state two important properties that you will prove in the end-of-chapter problems:

1. There are no energy eigenstates with energy $E < \min_x V(x)$ (problem 7.1). In other words, the situation indicated in figure 7.1 cannot occur.

   A key idea in the proof is that any solution $\psi(x)$ with energy less than the minimum of the potential would have $|\psi(x)|$ growing without bound. As discussed above this is not allowed for any energy eigenstate, even nonnormalizable ones.
2. A one-dimensional potential on the open line $x \in (-\infty, \infty)$ has no degenerate *bound* states.

   The proof (problem 7.2) begins by considering two degenerate energy eigenstates and goes on to show that the states must be proportional to each other and are in fact the same state. The result *does not* apply to nonnormalizable energy eigenstates, which can be degenerate. The result does not apply to a particle on a circle either.

Let us now prove explicitly two additional results. We first show that the reality of $V(x)$ allows us to work with real wave functions $\psi(x)$. Even though complex solutions exist, we can choose to work with real ones without any loss of generality. This is perhaps not so surprising since there is no i in the time-independent Schrödinger equation.

Theorem 7.1.1. *The energy eigenstates $\psi(x)$ can be chosen to be real.*

Proof. Consider the time-independent Schrödinger equation satisfied by the complex wave function $\psi(x)$:

$$\psi'' + \frac{2m}{\hbar^2}(E - V(x))\psi = 0. \tag{7.1.2}$$

Since $(\psi'')^* = (\psi^*)''$ and $V(x)$ is real, the complex conjugation of the above equation gives

$$(\psi^*)'' + \frac{2m}{\hbar^2}(E - V(x))\psi^* = 0. \tag{7.1.3}$$

We see that $\psi^*(x)$ is another solution of the Schrödinger equation with the *same energy*. The solution $\psi^*(x)$ is said to be different from $\psi(x)$ if they are not proportional to each other—that is, if there is no constant c such that $\psi^* = c\psi$. In that case ψ^* and ψ represent two degenerate solutions, and by superposition, we can obtain two *real* degenerate solutions:

$$\psi_r \equiv \frac{1}{2}(\psi + \psi^*), \quad \psi_{im} \equiv \frac{1}{2i}(\psi - \psi^*). \tag{7.1.4}$$

These are, of course, the real and imaginary parts of ψ. These are the real solutions we can choose to work with, as claimed.

If ψ and ψ^* are the same solution, then $\psi^* = c\psi$ with some constant c, and the real and imaginary parts above yield

$$\psi_r \equiv \frac{1}{2}(1 + c)\psi, \quad \psi_{im} \equiv \frac{1}{2i}(1 - c)\psi. \tag{7.1.5}$$

Since these are, by construction, real and both are proportional to ψ, these are the same real solution. In either case we can work with a real solution. □

Exercise 7.1. *Consider the condition $\psi^* = c\psi$ with c a constant. What is the most general possibility for c?*

Exercise 7.2. *Use the condition $\psi^* = c\psi$ and the solution to exercise 7.1 to show directly that ψ_r and ψ_{im} in (7.1.5) are real.*

If we are dealing with bound states of one-dimensional potentials, more can be said: It is not that we can choose to work with real solutions but rather that any solution *is* real. More precisely, we have the following:

Corollary 7.1. *Any bound state $\psi(x)$ of a one-dimensional potential is real, up to an overall constant phase factor.*

Proof. Since one-dimensional potentials have no degenerate bound states, the two real solutions ψ_r and ψ_{im} considered above must be equal up to a constant c' that can only be real:

$$\psi_{im}(x) = c' \, \psi_r(x), \quad \text{with } c' \in \mathbb{R}. \tag{7.1.6}$$

It then follows that $\psi = \psi_r + i\psi_{im} = (1 + ic')\psi_r$. Writing $1 + ic' = |1 + ic'|\, e^{i\beta}$ with real β shows that ψ is equal to a real solution, up to a constant phase factor $e^{i\beta}$. Since an overall phase is unobservable, any bound state solution is represented by a real function and can thus be assumed to be real. This is useful to keep in mind! □

The second result shows that for a potential that is an even function of x, we can work with energy eigenstates that are either symmetric or antisymmetric functions of x.

Theorem 7.1.2. *If $V(-x) = V(x)$, the energy eigenstates can be chosen to be symmetric or antisymmetric under $x \to -x$.*

Proof. Consider again the Schrödinger equation

$$\psi''(x) + \frac{2m}{\hbar^2}(E - V(x))\psi = 0. \tag{7.1.7}$$

Recall that primes here denote a derivative with respect to the argument, so $\psi''(x)$ means the function "second derivative of ψ" evaluated at x. Similarly, $\psi''(-x)$ means the function "second derivative of ψ" evaluated at $-x$. Thus, we can change x for $-x$ with impunity in the above equation, getting

$$\psi''(-x) + \frac{2m}{\hbar^2}(E - V(x))\psi(-x) = 0, \tag{7.1.8}$$

where we recalled that V is even. We now want to make clear that the above equation implies that $\psi(-x)$ is another solution of the Schrödinger equation with the same energy, a fact that seems obvious except for the term with two derivatives. For this let us define a function $\varphi(x)$ and take two derivatives of it:

$$\varphi(x) \equiv \psi(-x) \quad \Rightarrow \quad \frac{d}{dx}\varphi(x) = \psi'(-x)\cdot(-1), \qquad \frac{d^2}{dx^2}\varphi(x) = \psi''(-x). \tag{7.1.9}$$

Using the last equation, (7.1.8) becomes

$$\frac{d^2}{dx^2}\varphi(x) + \frac{2m}{\hbar^2}(E - V(x))\varphi(x) = 0, \tag{7.1.10}$$

showing that $\varphi(x) = \psi(-x)$ provides a degenerate solution to the Schrödinger equation. Equipped with the degenerate solutions $\psi(x)$ and $\psi(-x)$, we can now form symmetric (s) and antisymmetric (a) combinations that are, respectively, even and odd under $x \to -x$:

$$\psi_s(x) \equiv \tfrac{1}{2}(\psi(x) + \psi(-x)), \quad \psi_a(x) \equiv \tfrac{1}{2}(\psi(x) - \psi(-x)). \tag{7.1.11}$$

These are the solutions claimed to exist. □

Note that the above proof does not work for odd potentials $V(-x) = -V(x)$. No general statements seem possible in that case.

Again, if we focus on *bound states* of one-dimensional even potentials, the absence of degeneracy has a stronger implication: the solutions are *automatically* even or odd. There are no solutions that are neither odd nor even.

Corollary 7.2. *Any bound state of a one-dimensional even potential is either even or odd.*

Proof: The absence of degeneracy implies that the solutions $\psi(x)$ and $\psi(-x)$ must be the same solution. Because of corollary 1, we can choose $\psi(x)$ to be real, and thus we must have

$$\psi(-x) = c\,\psi(x), \quad \text{with} \quad c \in \mathbb{R}. \tag{7.1.12}$$

Letting $x \to -x$ in the above equation, we get $\psi(x) = c\psi(-x) = c^2\psi(x)$, from which we learn that $c^2 = 1$. The only possibilities are $c = \pm 1$. So $\psi(x)$ is either even or odd under $x \to -x$. □

We used corollary 7.2 to find the bound states of the finite square well. Since that potential is even, we could restrict our work to search for even bound states and odd bound states. The potential does not have bound states that are neither even nor odd.

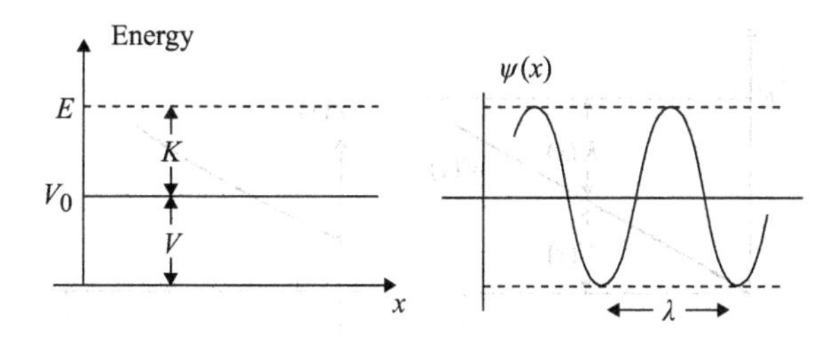

Figure 7.2
Left: A constant potential $V = V_0 < E$. The kinetic energy K is constant, and $E = K + V_0$ is the total energy. *Right*: A sketch of the wave function for a particle in this potential. Constant K implies constant momentum and therefore constant de Broglie wavelength.

Exercise 7.3. *A potential $V(x)$ defined for all x is even about the point $x = a$. What condition on $V(x)$ expresses this fact? Are all energy eigenstates of this potential necessarily even or odd about $x = a$?*

7.2 Bound States in Slowly Varying Potentials

We will now consider some insights that classical physics can give about the behavior of energy eigenstates of a particle in a potential. Indeed, classical ideas sometimes help us obtain an approximate description of the quantum states. This is called the **semi-classical approximation**. We will give a more complete and systematic analysis of the semiclassical approximation in chapter 26.

To begin, recall that the total energy E of a particle is the sum of a potential energy V and a kinetic energy K. In classical physics, when V is a function of position, the kinetic energy K must also be a function of position in order for the sum E to remain constant. Since the kinetic energy cannot be negative, the total energy must always be bigger than the potential energy. In our first example, shown in figure 7.2, the potential V is constant, and the energy E is greater than V. With a constant V, the particle experiences no force, and it will have a constant kinetic energy K and thus a constant momentum $|p| = \sqrt{2mK}$. We now claim that the wave representing the quantum particle has a de Broglie wavelength λ equal to Planck's constant h divided by the magnitude of the *classical* momentum.

Indeed, from the Schrödinger equation,

$$\psi'' = -\frac{2m}{\hbar^2}(E - V)\psi = -\frac{2mK}{\hbar^2}\psi = -\frac{p^2}{\hbar^2}\psi, \tag{7.2.1}$$

leading to solutions of the form

$$\psi = c_1 \cos\left(\frac{p}{\hbar}x\right) + c_2 \sin\left(\frac{p}{\hbar}x\right) = c_1 \cos\left[\frac{2\pi}{(h/p)}x\right] + c_2 \sin\left[\frac{2\pi}{(h/p)}x\right], \tag{7.2.2}$$

with c_1 and c_2 constants. As claimed, the wavelength of ψ is the de Broglie wavelength of a particle with momentum p. We are particularly interested in real solutions, as we have

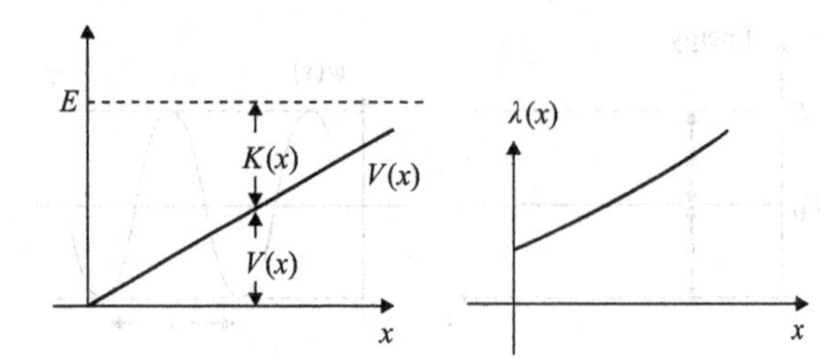

Figure 7.3
A linearly varying potential $V(x)$. The energy $E=K(x)+V(x)$ is fixed. As x increases, $K(x)$ decreases, $p(x)$ decreases, and $\lambda(x)$ increases, resulting in a wave function of increasing wavelength.

shown that energy eigenstates can be assumed to be real without loss of generality. If the above solution is real (setting c_1 and c_2 real), it cannot be a momentum eigenstate. It is a superposition of a state with momentum p and a state of momentum $-p$. In classical mechanics, the kinetic energy K only determines p^2 and not the sign of p.

Consider now the situation depicted in figure 7.3, where a classical particle is moving in a linearly increasing potential $V(x)$. This time the kinetic energy $K(x)$ is also position dependent. As a result, the magnitude of the momentum of the classical particle $p(x)=\sqrt{2mK(x)}$ is also position dependent. We can now define a position-dependent de Broglie wavelength $\lambda(x)$ given by

$$\lambda(x) \equiv \frac{h}{p(x)}. \tag{7.2.3}$$

This *local* de Broglie wavelength is particularly interesting when the particle moves in a potential that is a slowly varying function of position. By slowly varying we mean that the change in the potential over a distance comparable to the local de Broglie wavelength is very small compared to the potential:

$$\lambda(x)\left|\frac{dV}{dx}\right| \ll |V(x)|. \tag{7.2.4}$$

Indeed, the left-hand side of the inequality is an estimate for the change in V over a distance $\lambda(x)$. In this situation the semiclassical approximation applies, and to a good approximation, *the wave function will have a slowly varying wavelength equal to the local de Broglie wavelength* $\lambda(x)$.

As we see in figure 7.3, because $V(x)$ is linearly growing with x, a particle with total energy E will have decreasing kinetic energy $K(x)$ as x increases. Thus, the classical momentum will decrease, and we anticipate that the wave function will have an increasing local de Broglie wavelength $\lambda(x)$ as x increases. We will show the wave function after we learn what happens to the amplitude of the wave function.

Exercise 7.4. *Explain why the statement that the de Broglie wavelength varies little compared to itself over a distance equal to the de Broglie wavelength is expressed by the relation*

$$\left|\frac{d\lambda(x)}{dx}\right| \ll 1. \tag{7.2.5}$$

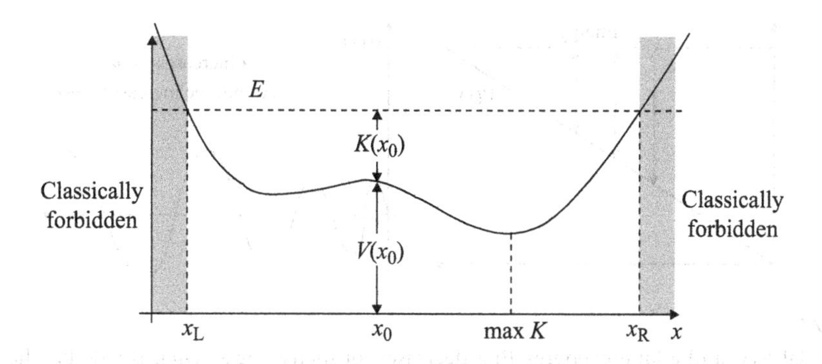

Figure 7.4
An arbitrary potential with turning points x_L and x_R for a motion with energy E. The regions to the right of x_R and to the left of x_L are classically forbidden. At any point $x \in [x_L, x_R]$, the sum of the potential energy $V(x)$ and the kinetic energy $K(x)$ is equal to the total energy E. The kinetic energy K is maximum where the potential is minimum.

Exercise 7.5. *Write an expression for a wave function $\psi(x)$ for which the action of the momentum operator upon it gives the local momentum $p(x)$—namely, $\hat{p}\psi(x) = p(x)\psi(x)$.*

In figure 7.4 we show an arbitrary potential $V(x)$ and consider the classical motion of a particle of total energy E. For any point x_0, we have $V(x_0) + K(x_0) = E$. The maximum kinetic energy occurs for the minimum value of the potential. Classically, a particle cannot have negative kinetic energy; thus, the particle cannot be found at points where $V(x)$ is greater than the energy E. In the figure this happens for $x > x_R$ and for $x < x_L$, and these regions are called classically forbidden regions. A particle of energy E will oscillate from x_L to x_R and back. As it moves, its velocity $v(x)$ is a function of position. The points x_L and x_R are called **turning points**, points where $E - V(x)$ changes sign. Therefore, at a turning point: (i) $E - V(x)$ vanishes, (ii) the particle motion reverses direction, and (iii) the point separates a classically allowed from a classically forbidden region.

It turns out that the classical behavior of the bouncing particle with energy E tells us plenty about the quantum wave function for an energy eigenstate of energy E. This is actually surprising, given that the energy eigenstate does not represent a bouncing quantum particle. As we have learned, expectation values of (time-independent) operators are constant on energy eigenstates. Thus, $\langle \hat{x} \rangle$, for example, which would vary in time if the quantum particle were bouncing, is in fact constant.

The oscillating classical particle spends more time in the regions where its velocity is small and less time in the regions where its velocity is large. As a result, in the semiclassical approximation the *amplitude* of the oscillatory wave function is bigger at the places the particle spends more time and smaller at places it spends less time. The particle is more likely to be found in those places where it spends more time. This can be quantified. Consider the probability $|\psi(x)|^2 dx$ of finding the particle within an infinitesimal interval dx centered about x. This quantity is set proportional to the

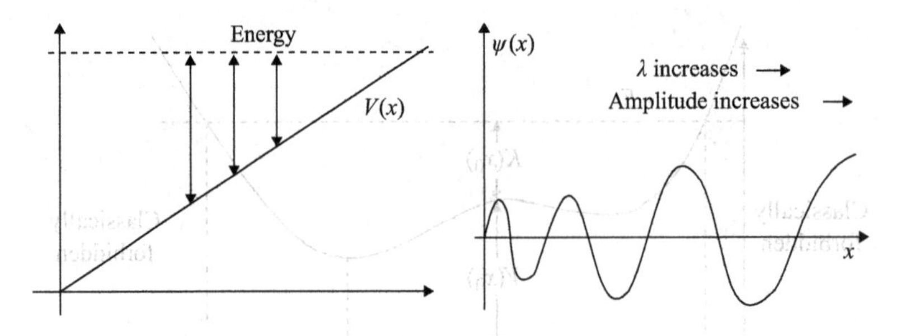

Figure 7.5
A potential $V(x)$ and a kinetic energy that decreases for increasing x. Then the de Broglie wavelength increases with x. Equation (7.2.8) then implies that the amplitude of $\psi(x)$ also increases with x.

fraction of time the particle spends at dx:

$$|\psi(x)|^2 dx \simeq \frac{dt}{T_{1/2}}. \tag{7.2.6}$$

Here dt is the time required to traverse dx, and $T_{1/2}$ is the half period of oscillation, the time the particle takes to go from one turning point to the other. With $v(x) = p(x)/m$ the local velocity of the particle and $dt = dx/v(x)$, we find that

$$|\psi(x)|^2 dx \simeq \frac{dx}{v(x)T_{1/2}} = \frac{m}{T_{1/2}} \frac{1}{p(x)} dx \quad \Rightarrow \quad |\psi(x)|^2 \sim \frac{1}{p(x)}. \tag{7.2.7}$$

This final relation has to be interpreted with some care. Recall that $\psi(x)$ has a very small wavelength for large $p(x)$. So $|\psi(x)|^2$ oscillates between zero and some peak value over very short distances. On the other hand, the momentum $p(x)$ appearing to the right has no such oscillations. Therefore, by $|\psi(x)|^2$ in the above relation we really mean the average of $|\psi(x)|^2$ over a few oscillations near x. For an oscillatory function, this average is proportional to the square of the *amplitude* of the wave $\psi(x)$, defined as the local maximum of $|\psi(x)|$. Writing the amplitude as $\text{Amp}(\psi(x))$, we have

$$\boxed{\text{Amp}(\psi(x)) \sim \frac{1}{\sqrt{p(x)}} \sim \sqrt{\lambda(x)}.} \tag{7.2.8}$$

The amplitude of the wave function is proportional to the square root of the de Broglie wavelength. In figure 7.5 as x increases the momentum $p(x)$ of the particle decreases and $\lambda(x)$ increases. Therefore, the amplitude of the wave increases with x, as sketched to the right. The result (7.2.8) is accurate for eigenstates of large energy, where $\psi(x)$ oscillates rapidly over the region in between turning points. It is not to be applied to ground states or low-energy eigenstates.

To illustrate the need to average $|\psi|^2$, consider a state of large energy in the infinite well potential, as shown in figure 7.6. Within the box, the classical particle "bounces" between the walls with constant speed. As a result, the particle spends the same amount of time in every equal size interval dx within the box. The classical probability density

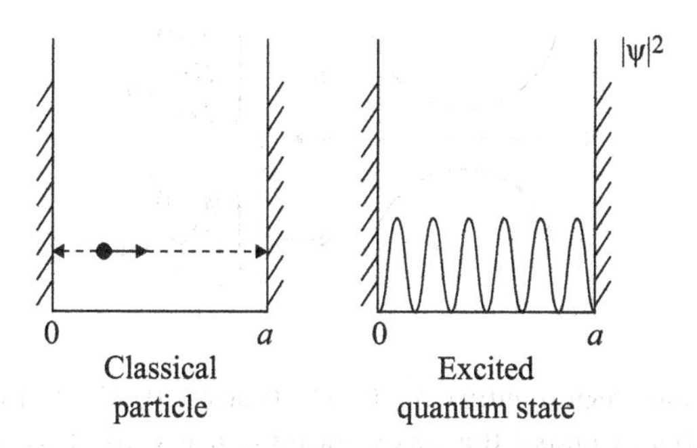

Figure 7.6
Left: Classical particle in a one-dimensional box bouncing with constant speed. *Right*: Probability density for a highly excited eigenstate.

$\rho_{cl}(x)$ within the box is uniform:

$$\rho_{cl}(x)dx = \frac{1}{a}dx \quad \rightarrow \quad \rho_{cl}(x) = \frac{1}{a}, \tag{7.2.9}$$

since the integration over the box $x \in [0, a]$ gives $\int_0^a \rho_{cl}(x)dx = 1$. Now consider an energy eigenstate $\psi_n(x)$ with large n (see (6.4.11)):

$$\psi_n(x) = \sqrt{\frac{2}{a}} \sin\left(\frac{n\pi x}{a}\right). \tag{7.2.10}$$

The associated quantum-mechanical probability density ρ_n is

$$\rho_n(x) = |\psi_n(x)|^2 = \frac{2}{a} \sin^2\left(\frac{n\pi x}{a}\right). \tag{7.2.11}$$

For large n, this is a rapidly oscillating function, seemingly quite different from the classical probability density $\rho_{cl} = 1/a$ (figure 7.6). While the classical probability density never vanishes, the quantum probability density has many zeroes! Still, over arbitrary distances larger than a/n (presumed small since n is very large) the average of the quantum probability density approaches the classical probability density. Recall that the average of $\sin^2 x$ over any integer number of oscillations is $1/2$ so that its average over *any* interval that includes a large but not necessarily integer number of oscillations is approximately equal to $1/2$. We then have, as claimed:

$$\text{Average}_x(\rho_n(x)) \simeq \frac{2}{a} \cdot \frac{1}{2} = \frac{1}{a} = \rho_{cl}(x). \tag{7.2.12}$$

The semiclassical approximation is an explicit realization of what is sometimes called the *correspondence principle*, the idea that there exist quantum states whose properties can be understood by an analysis of the corresponding classical system. Our earlier comments apply to the square well; an energy eigenstate of high quantum number is *not* a close quantum analog of the bouncing particle. The best quantum analog would be a narrow wave packet, built as a superposition of energy eigenstates. That wave packet, at least for times that are small enough, would also bounce in between the walls.

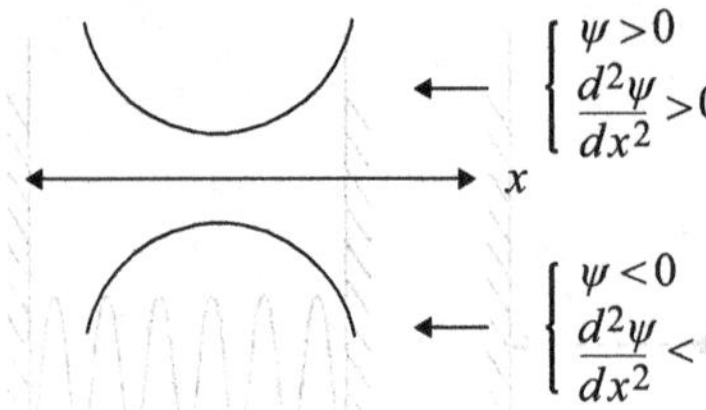

Figure 7.7
In the classically forbidden region, either $\psi > 0$, $\psi'' > 0$ or $\psi < 0$, $\psi'' < 0$. In both cases the wave function is said to be *convex toward* the x-axis, meaning it is convex when positive and is the reflection of a convex function when negative.

7.3 Sketching Wave Function Behavior

Let's examine the behavior of an energy eigenstate for the case of a general potential $V(x)$. We rewrite the time-independent Schrödinger equation (7.1.1) by dividing by ψ to get

$$\frac{\psi''(x)}{\psi(x)} = -\frac{2m}{\hbar^2}(E - V(x)), \tag{7.3.1}$$

which is convenient because there is no wave function on the right-hand side. We will consider the equation in two regions and at some special points.

- $E - V(x) < 0$. This condition defines the *classically forbidden region* because the total energy is smaller than the potential. In this case the right-hand side of equation (7.3.1) is positive. This means that the wave function and its second derivative have the same sign: both ψ and ψ'' are positive, or both ψ and ψ'' are negative. These possibilities are shown in figure 7.7. Such possible behaviors are summarized by saying that the wave function is **convex toward the axis.** This is not standard mathematical terminology, but it is used in physics. In mathematics a function of a single variable is said to be convex if, given any two points on its graph, the portion of the graph between the points lies below the segment that joins the points. Therefore, in figure 7.7 the graph of $\psi(x)$ is convex when $\psi > 0$ and not convex when $\psi < 0$. The reflection of the $\psi < 0$ graph about the x-axis, however, is convex. This is what is meant by convex *toward* the axis. Equivalently, and more concisely, a function of a single variable is convex toward the axis if given any two points on its graph, the portion of the graph between the two points lies closer to the axis than the segment joining the two points. When classically forbidden regions reach to $x = -\infty$ or $x = \infty$, the behavior is of the type shown in figure 7.8.
- $E - V(x) > 0$. This condition defines the *classically allowed region* because the energy is larger than the potential. In this case the right-hand side of equation (7.3.1) is negative. This means that the wave function and its second derivative have opposite signs: either $\psi > 0$ and $\psi'' < 0$, or $\psi < 0$ and $\psi'' > 0$. Both options are shown in figure 7.9

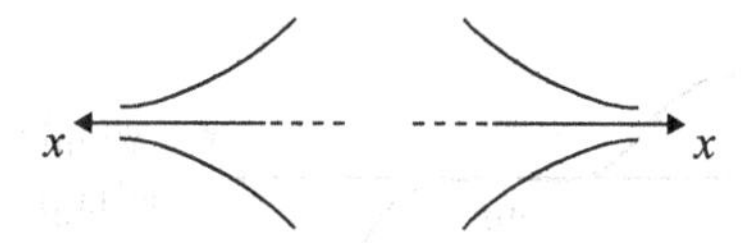

Figure 7.8
Wave functions approaching $x = -\infty$ (*left*) or $x = \infty$ (*right*) within classically forbidden regions. The graphs are all convex toward the axis.

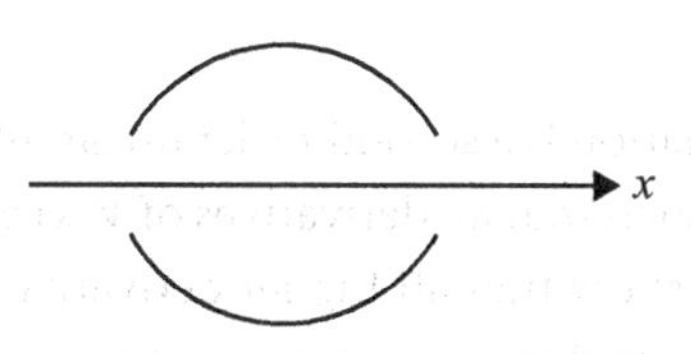

Figure 7.9
In a classically allowed region, we have $\psi > 0,\ \psi'' < 0$ or $\psi < 0,\ \psi'' > 0$. The wave function is *concave toward* the x-axis: it is concave when positive and the reflection of a concave function about the x-axis when negative.

and are summarized by saying that the wave function is **concave toward the axis.** Again, that means that $\psi(x)$ is concave when positive and the reflection of a concave function about the x-axis when negative. A concave function is one in which the graph of the function lies above the segment joining any two points in the graph. A function of a single variable is concave toward the axis if given any two points on its graph, the portion of the graph between the two points lies further from the axis than the segment joining the two points. Note that the behaviors allowed for in figure 7.9 occur both for $\sin x$ and $\cos x$. These are indeed functions for which the function and its second derivative have opposite signs.

- $V(x_0) = E$. At x_0 the potential energy $V(x_0)$ is equal to the total energy E, making the kinetic energy zero. If $E - V(x)$ changes sign at x_0 (as is usually the case), then x_0 is, by definition, a turning point. A turning point, we claim, is an **inflection point** in the graph of $\psi(x)$, a point where the second derivative of ψ changes sign. To see this recall that

$$\psi''(x) = -\frac{2m}{\hbar^2}(E - V(x))\psi(x). \tag{7.3.2}$$

Since $E - V(x)$ changes sign at x_0, while ψ does not, ψ'' changes sign at x_0. Not all inflection points are turning points. In fact, nodes of the wave function are inflection points. This is also clear from the above equation. Since, by definition, ψ changes sign at a node, while $E - V(x)$ does not, the sign of ψ'' changes at a node.

It is important to emphasize that while there are points where the wave function vanishes and there are points where its derivative vanishes, having *both* ψ and ψ' vanishing at any point in the domain of the wave function is not allowed (figure 7.10). This is

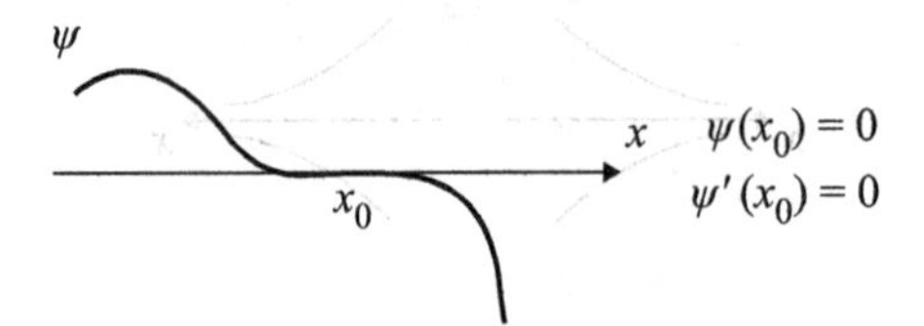

Figure 7.10
It is impossible to have both ψ and ψ' vanish at any point x_0; the Schrödinger equation would then force $\psi(x)$ to vanish identically.

because the Schrödinger equation is a second-order linear differential equation. Assume $\psi = \psi' = 0$ at some point x_0 such that all derivatives of $V(x)$ exist at x_0. One can then use the differential equation to show that all higher derivatives of ψ vanish at x_0. Assuming the wave function has a Taylor expansion around x_0, we conclude that the wave function must vanish identically in some neighborhood of x_0. In fact, mathematicians show that for $V(x)$ continuous in some interval I containing x_0 there is a unique solution in some interval I' containing x_0 once we specify $\psi(x_0)$ and $\psi'(x_0)$. Specifying $\psi(x_0) = \psi'(x_0) = 0$ is consistent with the solution $\psi(x) = 0$, which, by uniqueness, is the only solution. Since the zero solution is not physical, we cannot allow both ψ and ψ' to vanish at any point x_0. Hard walls are not an exception to this claim, as we have learned that $\psi' \neq 0$ at such points. One conclusion of this result is that at a node the derivative of the wave function must not vanish. For a node at $x = 0$, one must have $\psi(x) \sim x$, near $x = 0$. A wave function $\psi \sim x^2$ does not describe a node at $x = 0$ since the wave function does not change sign at zero. While $\psi \sim x^3$ changes sign at zero, this is not allowed because here $\psi(0) = \psi'(0) = 0$.

Exercise 7.6. *Use the time-independent Schrödinger equation to show that* $\psi(x_0) = \psi'(x_0) = 0$ *implies* $\psi''(x_0) = 0$ *and* $\psi'''(x_0) = 0$.

We conclude this section by explaining intuitively the quantization of the energy levels for bound states of an even potential. As shown in figure 7.11, *top*, we have an even potential, and we have marked four energies $E_1 < E_2 < E_3 < E_4$. *Not all* correspond to energy eigenstates. We imagine integrating numerically the Schrödinger equation from $x = \infty$ down to $x = 0$. This could be done starting with some very large x where we set ψ to vanish and ψ' to be a small positive value. Since any exact solution ψ does not strictly vanish for any large x, this introduces a tiny error, which is made smaller and smaller by taking larger and larger values of x for the initial point.

As shown in the inset slightly to the right and below the potential, we assume $\psi > 0$ as $x \to \infty$. Since we have shown that any bound state of an even potential is automatically even or odd, the picture for large negative x must be either the one with $\psi > 0$ (the even extension) or the one with $\psi < 0$ (the odd extension). To have a solution, the picture from the right must match properly at $x = 0$ with one of the two possible extensions at $x < 0$.

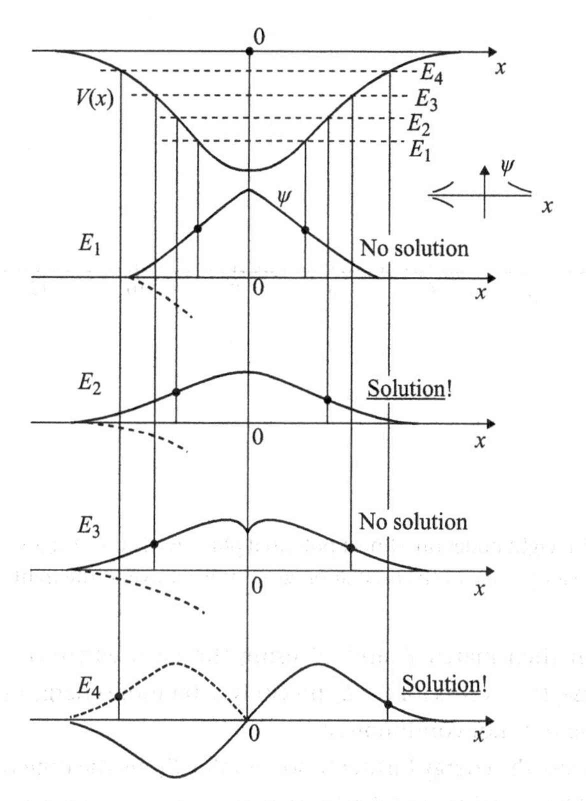

Figure 7.11
An even potential (top) and the result of integrating the Schrödinger equation from $x=\infty$ toward zero for values $E_1 < E_2 < E_3 < E_4$ of the energy. We get a solution when $\psi(x)$ for $x \geq 0$ can be matched, with continuous ψ and ψ', to an even or odd extension valid for $x<0$. There is no solution for E_1 and E_3. The ground state arises for E_2, and the first excited state arises for E_4.

Consider, for example, the integration of the Schrödinger equation for $E=E_1$, the lowest of our collection of energy choices. Coming in from $x=\infty$, the solution has an inflection point at the end of the classically forbidden region and becomes concave toward the axis. Beyond this inflection point, the first derivative ψ' decreases. Still, as shown in the figure, ψ' does not vanish at $x=0$; it is positive at this point. As a result, the even extension for $x<0$ does not match properly with the $x>0$ solution at $x=0$: ψ' fails to be continuous. Matching with the odd extension is not possible: ψ would fail to be continuous at $x=0$. Thus, E_1 is not an allowed energy.

As we increase the energy to E_3, the positive turning point moves to larger x, and ψ' is negative by the time we get to $x=0$, while ψ is still positive. The solution can't be matched with the even extension because of the discontinuous derivative. This is not an allowed energy either.

For $E=E_1$, the derivative $\psi'(0)$ is positive, while for $E=E_3$, the derivative $\psi'(0)$ is negative. It follows that there must be an energy $E=E_2$, with $E_1 < E_2 < E_3$ for which

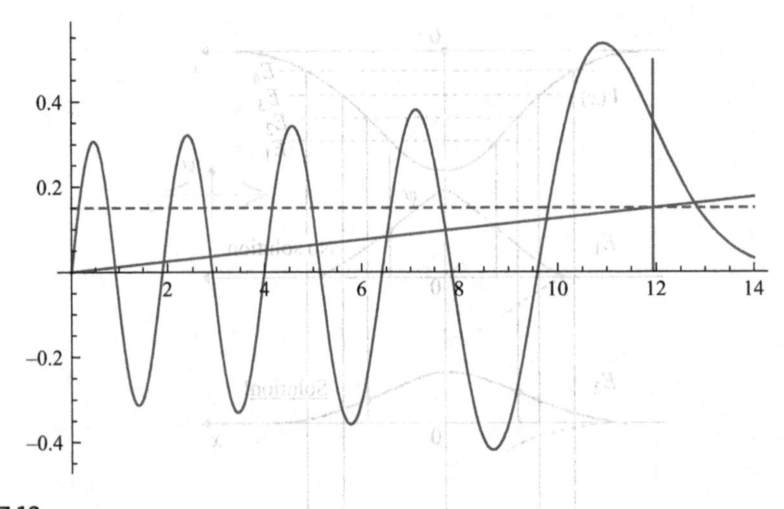

Figure 7.12
A wave function with eight nodes on a linear potential plus a wall at the origin. Both the de Broglie wavelength and the amplitude of the oscillation grow as we move to the right.

$\psi'(0)=0$. We can then match ψ and ψ' using the even extension, and we have a solution. Of course, this occurs for $E=E_2$ precisely; a bit more energy or a bit less energy and the derivative ψ' is not continuous at $x=0$.

If we now increase the energy further to some value E_4, by the time we reach $x=0$ the derivative ψ' will be negative, and ψ will be exactly zero. As you can see in the figure, we can get a good match with the *odd* extension to $x<0$. This is the first excited state. It is an odd wave function with a single node at $x=0$.

Here is a checklist you can use when trying to sketch the behavior of a bound state $\psi(x)$ of a one-dimensional potential:

1. An nth *excited state* $\psi(x)$ must have n nodes. Recall that a zero at a hard wall is not a node, nor do we count as nodes $\pm\infty$ when $\lim_{x\to\pm\infty}\psi(x)=0$.
2. Include inflection points at turning points and at the nodes.
3. In the classically forbidden region, the wave function must be convex toward the axis. The rate of decay of the wave function as we move into the forbidden region correlates with $V(x)-E$: the larger this value, the faster the decay.
4. In the classically allowed region, the wave function is oscillatory and concave toward the axis. For slowly varying potentials, or high-energy bound states, the de Broglie wavelength $\lambda(x)\sim 1/p(x)$, with $p(x)$ the value of the local momentum. The amplitude of the wave function is proportional to $\sqrt{\lambda(x)}$.
5. For even potentials, the eigenstates are even if the number of nodes is even or odd if the number of nodes is odd.

An example illustrating some of these facts is shown in figure 7.12, which plots the exact wave function on a linear potential (shown with a heavy line), for a value of the energy indicated by the horizontal dashed line. There is also a wall at $x=0$. Since the wave function has eight nodes, it is the eighth excited state (see (1)). There are inflection points at the nodes and at the turning point of the potential, indicated by the vertical line (see (2)). The wave function is convex toward the axis beyond the turning point (see (3)). The de Broglie wavelength grows as we move to the right and so does the amplitude of the wave (see (4)). The linear potential was studied in section 6.7.

7.4 The Node Theorem

The node theorem is a statement about the number of nodes in bound states of one-dimensional potentials. A node is a point where the wave function changes sign—a crossing of the zero value. We have also seen that at a node the wave function vanishes, but its derivative does not. While the wave function vanishes at nodes, not all points of vanishing wave function are nodes. The points at infinity where the wave function may vanish or approach zero are not nodes. Nor are nodes the points where we have hard walls, and the wave function vanishes automatically.

We motivated the node theorem with the infinite square well potential:

$$V(x) = \begin{cases} 0, & 0 < x < a, \\ \infty, & \text{elsewhere.} \end{cases} \tag{7.4.1}$$

The bound states take the form

$$\psi_n(x) = \sqrt{\frac{2}{a}} \sin\left(\frac{n\pi x}{a}\right), \quad n = 1, 2, \ldots, \infty, \tag{7.4.2}$$

with ever-increasing energies $E_n \sim n^2$. The eigenstate ψ_{n+1} has n nodes. The ψ_1 state, for example, only vanishes at the hard walls $x = 0$ and $x = a$, so the state has no nodes.

This leads us to the *node theorem*. Consider a potential $V(x)$ that is continuous and satisfies $V(x) \to \infty$ as $|x| \to \infty$. This potential has an infinite number of bound states (energy eigenstates that satisfy $\psi \to 0$ as $|x| \to \infty$), which we label as $\psi_1, \psi_2, \psi_3, \ldots$ as we order them by energy

$$E_1 < E_2 < E_3 < \cdots. \tag{7.4.3}$$

We have strict inequalities here because there are no degenerate bound states in one dimension. The node theorem states that for any integer $n \geq 0$, the eigenstate ψ_{n+1} has n nodes. We will give an intuitive argument explaining this phenomenon. For a less powerful but rigorous result, see problem 7.6.

For the argument that follows, we will use a result discussed in the previous section: one cannot have a vanishing derivative at a zero of the wave function. That applies to nodes or finite end points where we have hard walls. First, we examine the potential $V(x)$ and, for convenience, fix the location of $x = 0$ at any local minimum, as shown in figure 7.13. We then define the *screened* potentials $V_a(x)$ as follows:

$$V_a(x) = \begin{cases} V(x), & |x| < a, \\ \infty, & |x| > a. \end{cases} \tag{7.4.4}$$

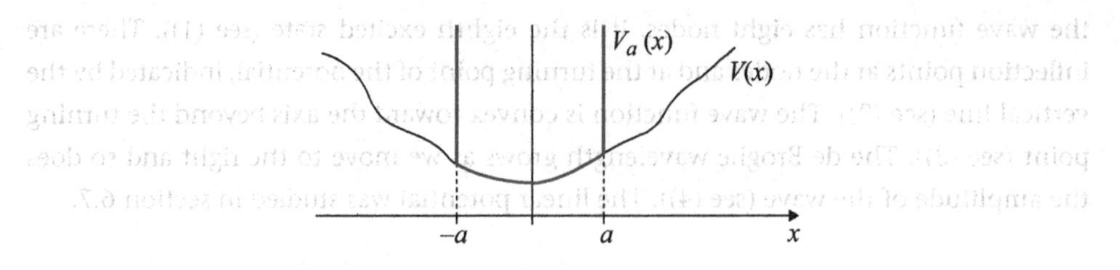

Figure 7.13
The potential $V(x)$ and the screened potential $V_a(x)$.

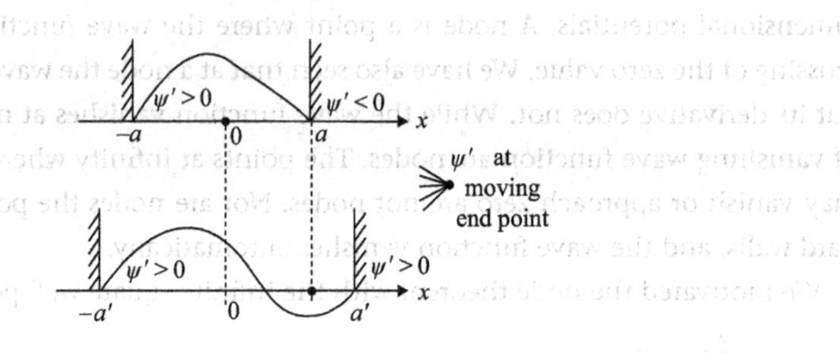

Figure 7.14
Introducing a single node requires changing the sign of the derivative at the right end point: $\psi'(a) < 0$ but $\psi'(a') > 0$. At some intermediate screen size, the value of ψ' at the right end point must become zero (see inset to the right, in between the figures). But this is impossible.

As shown in the figure, the screened potential $V_a(x)$ is an infinite well of width $2a$ whose bottom is that of $V(x)$. The argument below is based on two plausible assumptions. First, as $a \to \infty$, the bound states of $V_a(x)$ become the bound states of $V(x)$. Second, as a is increased, the wave function and its derivative are continuously stretched and deformed. Moreover, the energies change continuously. If this is the case, no new states can suddenly appear as we vary the value of a of the screened potential.

When a is very small, $V_a(x)$ is approximately a very narrow infinite well with a flat bottom—an infinite *square* well. This is because we chose $x = 0$ to be a minimum, and any minimum is locally flat. On this infinite square well, the node theorem holds. The ground state, for example, will vanish at the end points and will have no nodes. We will now argue that as the screen is enlarged we can't generate new nodes. This applies to the ground state, as we explicitly discuss below, and to all other states as well. If we can't generate nodes by screen enlargement, the node theorem applies to $V(x)$. Note that the ordering of the states by energy must also be preserved because the energies vary continuously, and they cannot ever coincide by nondegeneracy of the bound state spectrum.

Consider how we might develop an additional node while stretching the screen. To start, consider the ground state at the top of figure 7.14. There is no node at this value of the screen, and we have $\psi'(-a) > 0$ (left wall) and $\psi'(a) < 0$ (right wall). Suppose that as we increase a we produce a node, shown for the larger screen a' below. For this to

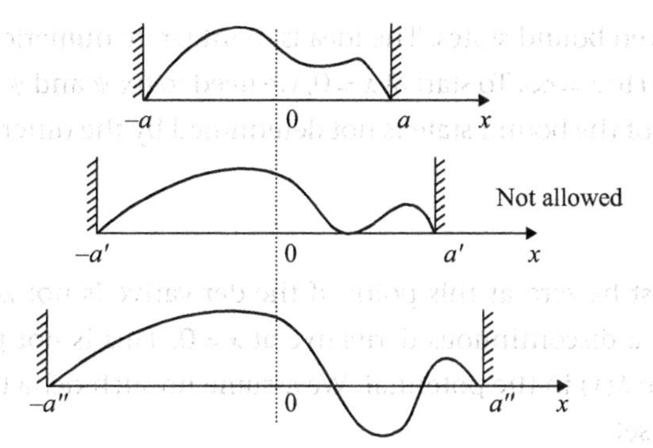

Figure 7.15
Introducing two nodes by having the wave function cross the x-axis in between the two boundaries (compare top and bottom). This is not possible, as it would require an intermediate screen (middle) in which $\psi = \psi' = 0$ at some point.

happen, the sign of ψ' at one of the end points must change. In the case shown in the figure, it is the right end point that experiences a change in the sign of ψ'. With the assumption of continuity, there would have to be some intermediate screen size at which $\psi' = 0$ at the right end point. But in that case, $\psi = \psi' = 0$ at this end point, which is impossible.

It is possible to introduce nodes without changing the sign of ψ' at either end point. In this process, shown in figure 7.15, the wave function dips and produces two new nodes. This process can't take place, however. Indeed, for some intermediate screen the wave function must be tangential to the x-axis, and at this point we will have $\psi = \psi' = 0$, which is impossible.

We conclude that we cannot change the number of nodes of *any* wave function as we stretch the screen and take $a \to \infty$. The nth excited state of the tiny infinite square well with n nodes will turn into the nth excited state of $V(x)$ with n nodes. In the tiny infinite square well, the energy levels are ordered in increasing energy by the number of nodes. The same is true at all stages of the stretching screen and therefore true for $V(x)$.

7.5 Shooting Method

The *shooting method* is an efficient numerical method to find bound state solutions to the Schrödinger equation

$$\frac{d^2\psi}{dx^2} + \frac{2m}{\hbar^2}(E - V(x))\psi = 0. \tag{7.5.1}$$

The method is easily implemented for even potentials $V(x)$, in which case we can separately search for even and for odd bound states. The method gives us both the energies and the associated normalizable wave functions.

Consider first even bound states. The idea is to integrate numerically the differential equation from $x=0$ to $x=\infty$. To start at $x=0$, we need to fix ψ and ψ' at this point. Since the normalization of the bound state is not determined by the differential equation, we can choose

$$\psi(x=0) = 1. \tag{7.5.2}$$

The derivative must be zero at this point: if the derivative is not zero, the even wave function will have a discontinuous derivative at $x=0$. This is not possible unless one has a delta function $\delta(x)$ in the potential. We assume no such delta function exists, and therefore we must set

$$\psi'(x=0) = 0. \tag{7.5.3}$$

To integrate the differential equation, we now need to guess some energy. Pick some arbitrary value E_1 very slightly above the minimum of the potential, since the energy of a bound state is bounded below by this minimum. We know that arbitrary values of the energy do not yield bound states, so what goes wrong if we now integrate the differential equation? You get a solution that cannot be normalized. As you work with your computer, you notice that beyond a certain point in the x-axis, $|\psi(x)|$ begins to grow without bound.

Assume that for the chosen value E_1 of the energy ψ eventually grows positive and without bound. Since E_1 was chosen essentially at the bottom of the potential, the lowest bound state must have energy higher than E_1. You must therefore change the value of the energy until you find some value E_2 for which the solution eventually grows negative and without bound as x is increased. This is a signal that there is some energy in between E_1 and E_2 for which a normalizable solution exists. You then try to narrow the interval. Since $E_1 < E_2$, you can do that by finding values larger than E_1 for which the wave function still grows positive without bound and values lower than E_2 for which the wave function still grows negative without bound. One can always try the value $(E_1+E_2)/2$ first. As you narrow the interval, you will still find this growth in $|\psi(x)|$, but it occurs for larger and larger values of x, which is progress. Indeed, you are getting a better and better approximation of the bound state energy (see figure 7.16).

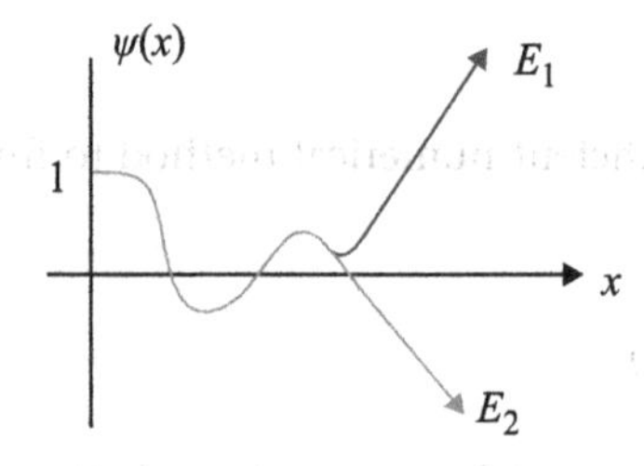

Figure 7.16
Numerical integration of the Schrödinger equation with a guess for a bound state energy leads to solutions in which $|\psi(x)|$ grows without bound beyond some value of x. Bound states are found in between a value E_1 of the energy for which $\psi(x)$ grows positive without bound and another value E_2 for which $\psi(x)$ grows negative without bound.

For odd solutions, the procedure is the same, but the boundary conditions at $x=0$ are

$$\begin{aligned} \psi(x=0) &= 0, \\ \psi'(x=0) &= 1. \end{aligned} \tag{7.5.4}$$

The first is needed for a continuous odd function. The second is arbitrary, up to normalization. If a potential is not even but has a hard wall, this provides a good starting point for integration of the Schrödinger equation. At the wall the wave function is set equal to zero, and the derivative can be set equal to one. If the potential has two hard walls, we can integrate starting from one wall and then demand that the solution reach exactly zero by the time we get to the second wall.

A practical note: To carry out the above numerical integrations clearly, it is first necessary to remove the units from the Schrödinger equation. We show below how this is done.

7.6 Removing Units from the Schrödinger Equation

To explain how to remove units from the time-independent Schrödinger equation, let us examine a specific example (for further practice, see problems 7.4 and 7.5). Consider a particle of mass m moving in the quartic potential

$$V(x)=\alpha x^4, \quad \alpha>0, \tag{7.6.1}$$

with α a constant. The time-independent Schrödinger equation for energy eigenstates $\psi(x)$ with energy E then reads

$$-\frac{\hbar^2}{2m}\frac{d^2\psi}{dx^2}+\alpha x^4\,\psi=E\psi. \tag{7.6.2}$$

As usual, we do not know $\psi(x)$ or the allowed values for the energy E. The equation has three dimensional constants $(\hbar, m, \alpha)$, which can be considered parameters of the system, and the energy E. We want to remove these dimensional parameters and replace the energy eigenvalue E with a unit-free eigenvalue. Indeed, if you wished to put the equation in a computer, it would be cumbersome to deal with $\hbar$, m, and α. By cleaning up the equation, we learn that all these constants really play *no role* in the search for the energies and wave functions. The way these constants appear in a solution is governed by dimensional analysis, not by the differential equation.

Note that each term in the time-independent Schrödinger equation (7.6.2) must have units of energy times units of ψ because the right-hand side of the equation has those units. In fact, the units of ψ are not relevant to the consistency of the equation because ψ appears on each term. The units of energy on the left-hand side arise on the first term by a combination of constants and derivatives and in the second term by a combination of α and the quartic power of x.

The first step in this process is to replace x for a unit-free variable u. The relation between them is of the form $x=Lu$, where L is a quantity with units of length that will be built from the parameters $(\hbar, m, \alpha)$ in the problem. Once this change is done, the terms on the left-hand side of the Schrödinger equation will have ψ acted on by unit-free u derivatives or u factors and multiplied by an energy-like quantity E_0 built from the parameters of the theory. In the resulting differential equation, the energy E of the bound state is encoded by the dimensionless quantity $\mathcal{E} = E/E_0$.

Let us do this now, aiming to construct L and E_0 from $\hbar, m, \alpha$. Since the potential has units of energy, the constant α that multiplies x^4 can be written in terms of the energy E_0 and the length parameter L:

$$\alpha = \frac{E_0}{L^4}, \qquad E_0 = \frac{\hbar^2}{mL^2}, \tag{7.6.3}$$

where we noted that, additionally, the energy E_0 can also be written in terms of L, m, and $\hbar$. As we did in writing these relations, is it customary not to include additional numerical constants, although doing so can sometimes be convenient. Combining these two, we can indeed express L and E_0 in terms of the parameters of the system:

$$L^6 = \frac{\hbar^2}{\alpha m}, \qquad E_0^3 = \frac{\alpha \hbar^4}{m^2}. \tag{7.6.4}$$

The quantities L and E_0 are the natural length scale and energy scale, respectively, in this quantum system.

Now we let $x = Lu$ in the differential equation (7.6.2), finding that

$$-\frac{\hbar^2}{2mL^2}\frac{d^2\psi}{du^2} + \alpha L^4 u^4 \psi = E\psi. \tag{7.6.5}$$

In here we used

$$\frac{d}{dx} = \frac{du}{dx}\frac{d}{du} = \frac{1}{L}\frac{d}{du} \quad \Rightarrow \quad \frac{d^2}{dx^2} = \frac{d}{dx}\frac{d}{dx} = \frac{1}{L^2}\frac{d^2}{du^2}. \tag{7.6.6}$$

As expected, using (7.6.3) we find multiplicative factors of E_0 on the left-hand side:

$$-\frac{1}{2}E_0\frac{d^2\psi}{du^2} + E_0 u^4 \psi = E\psi \quad \Rightarrow \quad -\frac{1}{2}\frac{d^2\psi}{du^2} + u^4 \psi = \frac{E}{E_0}\psi. \tag{7.6.7}$$

Letting

$$\mathcal{E} \equiv \frac{E}{E_0} \tag{7.6.8}$$

denote the dimensionless energy eigenvalue, we finally get the desired unit-free version of the Schrödinger equation:

$$-\frac{1}{2}\frac{d^2\psi}{du^2} + u^4 \psi = \mathcal{E}\,\psi. \tag{7.6.9}$$

If we study this equation numerically, we find unit-free energy eigenvalues $\mathcal{E}_k$, with $k = 1, 2, \ldots$, for which the associated energies E_k are simply

$$E_k = \mathcal{E}_k E_0 = \mathcal{E}_k\Big(\frac{\alpha\hbar^4}{m^2}\Big)^{1/3}. \tag{7.6.10}$$

This analysis shows that the problem is solved once and for all, independent of the values of α and m, by finding the unit-free energy eigenvalues $\mathcal{E}$ and eigenfunctions $\psi(u)$ of the unit-free equation (7.6.9). To find the physical position dependence of the eigenfunctions, we simply replace $u = x/L$ in $\psi(u)$. This removal of units from the Schrödinger equation is actually the first step even in analytic work, as we will see when we study the harmonic oscillator in chapter 9 and the hydrogen atom in chapter 11.

7.7 Virial Theorem

In quantum mechanics the virial theorem is a particular case of a general property of stationary states. Since we focus on energy eigenstates, we are considering time-independent Hamiltonians. We will write the stationary states as follows:

$$\Psi(\mathbf{x}, t) = e^{-iEt/\hbar}\,\psi(\mathbf{x}). \tag{7.7.1}$$

Consider now the main equation (5.2.5) governing the time dependence of the expectation value of a time-independent operator $\hat{Q}$:

$$i\hbar\frac{d}{dt}\langle\hat{Q}\rangle = \Big\langle[\hat{Q}, \hat{H}]\Big\rangle. \tag{7.7.2}$$

If the expectation value is taken on a stationary state Ψ, then both sides of the equation vanish. The left-hand side vanishes because, as we showed in section 6.1, the expectation value of any time-independent operator on a stationary state is time independent. Therefore, the time derivative kills $\langle\hat{Q}\rangle$. The vanishing of the right-hand side is also simple to see:

$$\Big\langle[\hat{Q}, \hat{H}]\Big\rangle = \langle\Psi, [\hat{Q}, \hat{H}]\Psi\rangle = \langle e^{-iEt/\hbar}\psi, [\hat{Q}, \hat{H}]e^{-iEt/\hbar}\psi\rangle. \tag{7.7.3}$$

Since the operators $\hat{Q}$ and $\hat{H}$ do not involve time derivatives, the time-dependent phases can go out of the inner product and cancel each other:

$$\Big\langle[\hat{Q}, \hat{H}]\Big\rangle = \langle\psi, [\hat{Q}, \hat{H}]\psi\rangle. \tag{7.7.4}$$

Finally, expanding out the commutator using the Hermiticity of $\hat{H}$ and the relation $\hat{H}\psi = E\psi$, we have

$$\Big\langle[\hat{Q}, \hat{H}]\Big\rangle = \langle\psi, \hat{Q}\hat{H}\psi\rangle - \langle\hat{H}\psi, \hat{Q}\psi\rangle = E\langle\psi, \hat{Q}\psi\rangle - E\langle\psi, \hat{Q}\psi\rangle = 0. \tag{7.7.5}$$

We have thus confirmed the following:

Lemma. *For any time-independent* $\hat{Q}$*, a time-independent Hamiltonian* $\hat{H}$*,* and an energy eigenstate $\Psi = \exp(-iEt/\hbar)\psi$, we have $\langle[\hat{Q}, \hat{H}]\rangle_\Psi = \langle[\hat{Q}, \hat{H}]\rangle_\psi = 0$.

You may have noticed that requiring $\hat{Q}$ to be time independent is not really needed for the above lemma; all that is required is that $\hat{Q}$ contain no time derivatives. Still, all familiar applications use time-independent $\hat{Q}$ operators.

This lemma is useful because, using suitably chosen $\hat{Q}$ operators, we can often learn important facts about the expectation values of operators on energy eigenstates.

The vanishing of $\langle[\hat{Q}, \hat{H}]\rangle_\psi$ has interesting consequences. The most familiar application deals with the Hamiltonian for a point particle moving in a potential. If the problem is one-dimensional, the Hamiltonian is the sum of the kinetic plus potential energies:

$$\hat{H} = \frac{\hat{p}^2}{2m} + V(\hat{x}). \tag{7.7.6}$$

We then choose $\hat{Q} = \hat{x}\hat{p}$ and compute

$$[\hat{x}\hat{p}, \hat{H}] = \Big[\hat{x}\hat{p}, \frac{\hat{p}^2}{2m}\Big] + [\hat{x}\hat{p}, V(x)]. \tag{7.7.7}$$

The first commutator is computed using the derivation property ((5.2.8), second equation), as well as (5.2.15):

$$\Big[\hat{x}\hat{p}, \frac{\hat{p}^2}{2m}\Big] = \frac{1}{2m}[\hat{x}, \hat{p}^2]\hat{p} = \frac{1}{2m}2i\hbar\hat{p}\,\hat{p} = 2i\hbar\frac{\hat{p}^2}{2m}. \tag{7.7.8}$$

The second commutator gives us

$$[\hat{x}\hat{p}, V(\hat{x})] = \hat{x}[\hat{p}, V(\hat{x})] = \frac{\hbar}{i}\hat{x}\frac{dV}{d\hat{x}}, \tag{7.7.9}$$

using the result (5.2.14) for the commutator of $\hat{p}$ with an arbitrary function of $\hat{x}$. Having computed the two contributions to $[\hat{x}\hat{p}, \hat{H}]$, the vanishing expectation value on a stationary state gives

$$\Big\langle [\hat{x}\hat{p}, \hat{H}] \Big\rangle = 2i\hbar\Big\langle\frac{\hat{p}^2}{2m}\Big\rangle - i\hbar\Big\langle\hat{x}\frac{dV}{d\hat{x}}\Big\rangle = 0. \tag{7.7.10}$$

We have therefore shown that

$$\boxed{\Big\langle\frac{\hat{p}^2}{2m}\Big\rangle = \frac{1}{2}\Big\langle\hat{x}\frac{dV}{d\hat{x}}\Big\rangle.} \tag{7.7.11}$$

This is the **virial theorem** for one-dimensional potentials: it relates the expectation value of the kinetic energy, the left-hand side, to the expectation value of a simple function of the potential, both evaluated on the same arbitrary energy eigenstate. Note that for a quadratic potential $V(\hat{x}) = \gamma\hat{x}^2$, with γ a constant,

$$\tfrac{1}{2}\hat{x}\frac{d}{d\hat{x}}V(\hat{x}) = \tfrac{1}{2}\hat{x}\,2\gamma\hat{x} = \gamma\hat{x}^2 = V(\hat{x}), \tag{7.7.12}$$

which means that for a quadratic potential the expectation values of the kinetic and the potential energies are the same!

The three-dimensional version of the virial theorem gives a useful equality when we consider a particle moving in a central potential, meaning that $V(\mathbf{x})$ is in fact a function $V(r)$ of the length r of the position vector $\mathbf{x}$. The Hamiltonian is then

$$\hat{H} = \frac{\hat{\mathbf{p}}^2}{2m} + V(r). \tag{7.7.13}$$

This time we consider the commutator of $\hat{H}$ with $\hat{\mathbf{x}} \cdot \hat{\mathbf{p}}$, the analog of the one-dimensional $\hat{x}\hat{p}$:

$$\hat{\mathbf{x}} \cdot \hat{\mathbf{p}} = \sum_{k=1}^{3} \hat{x}_k \hat{p}_k = \hat{x}_1 \hat{p}_1 + \hat{x}_2 \hat{p}_2 + \hat{x}_3 \hat{p}_3. \tag{7.7.14}$$

We thus have

$$[\hat{\mathbf{x}} \cdot \hat{\mathbf{p}}, \hat{H}] = \Big[\hat{\mathbf{x}} \cdot \hat{\mathbf{p}}, \frac{\hat{\mathbf{p}}^2}{2m} + V(r)\Big] = \frac{1}{2m}[\hat{\mathbf{x}} \cdot \hat{\mathbf{p}}, \hat{\mathbf{p}}^2] + \sum_{k=1}^{3} \hat{x}_k [\hat{p}_k, V(r)], \tag{7.7.15}$$

where we noted that $\hat{x}_k$ commutes with $V(r)$ because all $\hat{x}$ operators commute with each other, and ultimately, r is a function of the $\hat{x}_k$'s. Let us look at the first commutator:

$$\begin{aligned} [\hat{\mathbf{x}} \cdot \hat{\mathbf{p}}, \hat{\mathbf{p}}^2] &= [\hat{x}_1 \hat{p}_1 + \hat{x}_2 \hat{p}_2 + \hat{x}_3 \hat{p}_3\,,\ \hat{p}_1^2 + \hat{p}_2^2 + \hat{p}_3^2] \\ &= [\hat{x}_1 \hat{p}_1\,, \hat{p}_1 \hat{p}_1] + [\hat{x}_2 \hat{p}_2\,, \hat{p}_2 \hat{p}_2] + [\hat{x}_3 \hat{p}_3\,, \hat{p}_3 \hat{p}_3], \end{aligned} \tag{7.7.16}$$

because the only nontrivial commutators happen between $\hat{x}$'s and $\hat{p}$'s with the same index. The above commutators are of the same form as in the one-dimensional case. You can quickly check that

$$[\hat{\mathbf{x}} \cdot \hat{\mathbf{p}}\,, \hat{\mathbf{p}}^2] = 2i\hbar\, \hat{\mathbf{p}}^2. \tag{7.7.17}$$

The second commutator in (7.7.15) involves the commutator $[\hat{p}_k, V(r)]$. Recall that in one dimension we showed that $[\hat{p}, V(\hat{x})] = \frac{\hbar}{i}\frac{dV}{dx}$. Since, as an operator in position space, $\hat{p}_k = \frac{\hbar}{i}\frac{\partial}{\partial x_k}$ (see (3.3.26)), this time we have $[\hat{p}_k, V(\mathbf{x})] = \frac{\hbar}{i}\frac{\partial V}{\partial x_k}$. Therefore,

$$\sum_{k=1}^{3} \hat{x}_k [\hat{p}_k, V(r)] = \sum_{k=1}^{3} \hat{x}_k \frac{\hbar}{i}\frac{\partial V(r)}{\partial x_k} = -i\hbar \sum_{k=1}^{3} \hat{x}_k \frac{\partial V(r)}{\partial r}\frac{\partial r}{\partial \hat{x}_k}. \tag{7.7.18}$$

Using $r^2 = \hat{x}_1^2 + \hat{x}_2^2 + \hat{x}_3^2$, we can evaluate the partial derivative of r with respect to any coordinate:

$$\frac{\partial r}{\partial \hat{x}_k} = \frac{1}{2r}\frac{\partial r^2}{\partial \hat{x}_k} = \frac{1}{2r}(2\hat{x}_k) = \frac{\hat{x}_k}{r}. \tag{7.7.19}$$

Therefore, the commutator gives

$$\sum_{k=1}^{3} \hat{x}_k [\hat{p}_k, V(r)] = -i\hbar \sum_{k=1}^{3} \frac{\hat{x}_k \hat{x}_k}{r}\frac{\partial V(r)}{\partial r} = -i\hbar\, r \frac{\partial V(r)}{\partial r}. \tag{7.7.20}$$

All in all, we have found that

$$[\hat{\mathbf{x}} \cdot \hat{\mathbf{p}}\,, \hat{H}] = 2i\hbar \frac{\hat{\mathbf{p}}^2}{2m} - i\hbar\, r \frac{\partial V(r)}{\partial r}. \tag{7.7.21}$$

The vanishing expectation value of the commutator on any energy eigenstate now gives the relation

$$\left\langle \frac{\hat{\mathbf{p}}^2}{2m} \right\rangle = \frac{1}{2}\left\langle r\frac{\partial V}{\partial r}\right\rangle. \tag{7.7.22}$$

This is the three-dimensional version of the virial theorem in quantum mechanics.

The first formulation of the virial theorem, due to Rudolf Clausius (1870), was in the context of classical mechanics. The analog of expectation value in quantum mechanics is time average. Similarly, the analog of a stationary state is a system exhibiting periodic motion or, if not periodic, at least motion in which positions and momenta are bounded in time. For a quantity $\mathcal{O}$ built from position and momenta, the classical average $\langle \cdots \rangle_{\text{cl}}$ over a time period τ is defined by

$$\langle \mathcal{O} \rangle_{\text{cl}} \equiv \frac{1}{\tau}\int_0^{\tau} \mathcal{O}(t)dt. \tag{7.7.23}$$

For a periodic system with period T, the average over T of the time derivative of $\mathcal{O}$ vanishes:

$$\left\langle \frac{d\mathcal{O}}{dt} \right\rangle_{\text{cl}} = \frac{1}{T}\int_0^T \frac{d\mathcal{O}}{dt}(t)dt = \mathcal{O}(T) - \mathcal{O}(0) = 0, \tag{7.7.24}$$

by the periodicity condition. This is the classical analog of the vanishing of the time derivative of the expectation value of a quantum operator $\hat{Q}$. As you will verify in problem 7.13, the classical form of the virial theorem follows by choosing $\mathcal{O} = \sum_\alpha \mathbf{r}_\alpha \cdot \mathbf{p}_\alpha$, where the sum of the products of position and momenta is over all the particles of the system. The resulting form of the classical virial theorem is completely analogous to (7.7.22).

The virial theorem was applied by Fritz Zwicky in 1933 to the Coma cluster of galaxies, a cluster containing over one thousand galaxies and located about 320 million light-years away from us. Estimates of average kinetic energies obtained by Doppler shift measurements, together with the virial theorem, led to an estimate for the total mass of the cluster. From an analysis of luminosity, it became clear that the matter in stars and in hot interstellar gas cannot account, by a large factor, for the total mass of the cluster. This led Zwicky to postulate the existence of *dark matter*. With much better measurements since, it seems that about 90% of the mass of the Coma cluster is in the form of dark matter.

7.8 Variational Principle

Sometimes we have quantum systems whose stationary states can take time and effort to find. Perhaps an analytic solution is just not possible. It is then useful to have tools that can give us some partial information about the energy spectrum. The variational principle is one such tool. It provides upper bounds on the energy of certain states. It is practical and, in fact, fun to use.

Consider a system with Hamiltonian $\hat{H}$ and focus on the time-independent Schrödinger equation:

$$\hat{H}\psi_E = E\psi_E. \tag{7.8.1}$$

Here ψ_E is an energy eigenstate of energy E. Let us assume that the system has a collection of normalizable energy eigenstates, including a ground state with ground state energy E_{gs}. We leave the coordinate dependence of the wave function implicit; the discussion applies to quantum systems in any number of spatial dimensions and to particles with other degrees of freedom that we will study later, such as spin. Our first goal is to learn something about the ground state energy *without* solving the Schrödinger equation or trying to figure out the ground state wave function.

For this purpose, consider an *arbitrary*, normalized wave function ψ:

$$\langle \psi, \psi \rangle = 1. \tag{7.8.2}$$

By arbitrary we mean a wave function that need not satisfy the time-independent Schrödinger equation or, equivalently, a wave function that need not be an energy eigenstate. This is called a **trial wave function**; it is a wave function that we simply choose as we wish. The **variational principle** states that the ground state energy E_{gs} of the Hamiltonian is smaller than or equal to the expectation value of $\hat{H}$ in this arbitrary, normalized trial wave function ψ, namely,

$$\boxed{E_{gs} \le \langle \psi, \hat{H}\psi \rangle, \quad \text{normalized } \psi.} \tag{7.8.3}$$

When the right-hand side of the above inequality is evaluated, we get an energy and learn that the ground state energy must be smaller than or equal to this value. Thus, *any* trial wave function provides an *upper bound* for the ground state energy. Better and better trial wave functions will produce lower and lower upper bounds, converging toward the ground state energy. Note that if the trial wave function is set equal to the (unknown) ground state wave function, the expectation value of $\hat{H}$ becomes exactly E_{gs}, and the inequality is saturated.

Proof of (7.8.3). For simplicity, we will consider here the case where the energy eigenstates ψ_n of $\hat{H}$ are denumerable, and their corresponding energies E_n are ordered as

$$E_{gs} = E_1 \le E_2 \le E_3 \le \cdots, \tag{7.8.4}$$

with the $\le$ signs allowing for degeneracies. Of course, $\hat{H}\psi_n = E_n\psi_n$. Since the energy eigenstates can be chosen to be a complete orthonormal set (claim 4, section 5.3), any trial wave function can be expanded as

$$\psi = \sum_{n=1}^{\infty} b_n \psi_n, \tag{7.8.5}$$

with b_n some constants. The normalization condition (7.8.2) gives:

$$1 = \langle \psi, \psi \rangle = \sum_{m,n=1}^{\infty} \langle b_m\psi_m, b_n\psi_n \rangle = \sum_{m,n=1}^{\infty} b_m^* b_n \langle \psi_m, \psi_n \rangle = \sum_{m,n=1}^{\infty} b_m^* b_n \delta_{m,n} \tag{7.8.6}$$

so that we get

$$\sum_{n=1}^{\infty} |b_n|^2 = 1. \tag{7.8.7}$$

The evaluation of the right-hand side in (7.8.3) is also quickly done:

$$\langle \psi, \hat{H}\psi \rangle = \sum_{m,n=1}^{\infty} b_m^* b_n \langle \psi_m, \hat{H}\psi_n \rangle = \sum_{m,n=1}^{\infty} b_m^* b_n E_n \delta_{m,n} = \sum_{n=1}^{\infty} |b_n|^2 E_n. \tag{7.8.8}$$

Since $E_n \geq E_1$ for all n, we can replace the E_n on the above right-hand side for E_1, getting a smaller or equal value:

$$\langle \psi, \hat{H}\psi \rangle = \sum_{n=1}^{\infty} |b_n|^2 E_n \geq \sum_{n=1}^{\infty} |b_n|^2 E_1 = E_1 \sum_{n=1}^{\infty} |b_n|^2 = E_1 = E_{gs}, \tag{7.8.9}$$

where we used the normalization constraint (7.8.7). This is in fact the claim (7.8.3) of the variational principle. □

It is generally more convenient not to worry about the normalization of the trial wave functions. Given a trial wave function ψ that is not normalized, the wave function

$$\frac{\psi}{\sqrt{\langle \psi, \psi \rangle}} \tag{7.8.10}$$

is normalized and can be used in (7.8.3). Since $\langle \psi, \psi \rangle$ is a number, it comes out of expectation values, and we therefore find that

$$\boxed{E_{gs} \leq \frac{\langle \psi, \hat{H}\psi \rangle}{\langle \psi, \psi \rangle} \equiv \mathcal{F}[\psi].} \tag{7.8.11}$$

This formula can be used for trial wave functions that are not normalized. We also introduced the definition of the functional $\mathcal{F}[\psi]$. A functional is a machine that given a function, in this case the wave function ψ, gives us a number. Our result states that the ground state energy arises as the minimum value that the functional $\mathcal{F}$ can produce. The inequality in (7.8.11) is our main result from the variational principle.

The above inequality can be used to find good upper bounds for the ground state energy of quantum systems that are not exactly solvable. For this purpose it is useful to construct trial wave functions

$$\psi(\beta_1, \ldots, \beta_m)$$

that depend on a set of parameters $\beta_1, \ldots, \beta_m$. One then computes $\mathcal{F}[\psi]$, which, of course, is a function of the parameters. Any random values for the parameters will give an upper bound for the ground state energy, but by minimizing $\mathcal{F}[\psi]$ over the parameter space, we get the lowest possible upper bound consistent with the chosen form for the trial wave function.

Example 7.1. *Trial wave functions for the one-dimensional delta function potential.* Consider a one-dimensional problem with the delta function potential:

$$V(x) = -\alpha\,\delta(x), \quad \alpha > 0. \tag{7.8.12}$$

The ground state energy was obtained in (6.6.20) and is given by

$$E_{gs} = -\frac{m\alpha^2}{2\hbar^2}. \tag{7.8.13}$$

We use the variational principle to find an upper bound for the ground state energy. As a trial wave function consider an unnormalized Gaussian, with a real parameter $\beta > 0$:

$$\psi(x) = e^{-\frac{1}{2}\beta^2 x^2}, \quad \langle\psi,\psi\rangle = \int_{-\infty}^{\infty} dx\, \psi^2(x) = \frac{\sqrt{\pi}}{\beta}, \quad \beta > 0. \tag{7.8.14}$$

The functional $\mathcal{F}$ in (7.8.11) is then

$$\frac{\langle\psi,\hat{H}\psi\rangle}{\langle\psi,\psi\rangle} = \frac{\beta}{\sqrt{\pi}}\int_{-\infty}^{\infty} dx\, e^{-\frac{1}{2}\beta^2x^2}\Big(-\frac{\hbar^2}{2m}\frac{d^2}{dx^2} - \alpha\delta(x)\Big)e^{-\frac{1}{2}\beta^2x^2}. \tag{7.8.15}$$

It helps the computation to integrate by parts the second derivatives:

$$\begin{aligned}\frac{\langle\psi,\hat{H}\psi\rangle}{\langle\psi,\psi\rangle} &= \frac{\beta}{\sqrt{\pi}}\frac{\hbar^2}{2m}\int_{-\infty}^{\infty} dx\Big[\frac{d}{dx}e^{-\frac{1}{2}\beta^2x^2}\Big]^2 - \frac{\beta}{\sqrt{\pi}}\alpha\\ &= \frac{\beta}{\sqrt{\pi}}\frac{\hbar^2}{2m}\frac{\beta\sqrt{\pi}}{2} - \frac{\beta}{\sqrt{\pi}}\alpha\\ &= \frac{\beta^2\hbar^2}{4m} - \frac{\beta}{\sqrt{\pi}}\alpha.\end{aligned} \tag{7.8.16}$$

The first term on the last right-hand side is the kinetic energy expectation value, and the second term is the potential energy. The right-hand side is a function of β with a linear term with a negative coefficient and a quadratic term with a positive coefficient. This defines a function with a minimum. For any value of β, the right-hand side above provides an upper bound for the ground state energy, and the best upper bound, to be called E_0, is the lowest one. Therefore, the ground state energy E_{gs} satisfies

$$E_{gs} \le \text{Min}_\beta\Big(\frac{\beta^2\hbar^2}{4m} - \frac{\beta}{\sqrt{\pi}}\alpha\Big) \equiv E_0. \tag{7.8.17}$$

The minimum is easily found:

$$\beta = \frac{2m\alpha}{\hbar^2\sqrt{\pi}} \quad \Rightarrow \quad E_0 = \frac{2}{\pi}\Big(-\frac{m\alpha^2}{2\hbar^2}\Big). \tag{7.8.18}$$

Comparing it with (7.8.13), we see that $E_0 = \frac{2}{\pi}E_{gs} \simeq 0.64 E_{gs}$ so that E_{gs} is, correctly, more negative than E_0. The trial wave function brought us to about 64% of the correct value. □

In the end-of-chapter problems, you will develop the following results:

1. Consider an attractive one-dimensional potential defined as a nowhere positive potential that approaches zero at infinity. This potential has a state with energy less than zero, a bound state (problem 7.15).
2. Restricted to trial wave functions orthogonal to the ground state, the functional $\mathcal{F}$ gives upper bounds for the energy of the first excited state (problem 7.16).
3. We have shown that the functional $\mathcal{F}[\psi]$ has a minimum when ψ is the ground state wave function. A minimum is a stationary point, but not all stationary points of functions (or functionals) are minima; some can also be maxima or, more generally, saddle points. It turns out that $\mathcal{F}[\psi]$ is in fact stationary at each and every energy eigenstate. For eigenstates of energies higher than the ground state, $\mathcal{F}$ has a saddle point (problem 7.16).

7.9 Hellmann-Feynman Lemma

It is sometimes useful to see how the spectrum of a theory changes when the Hamiltonian is changed. A simple and powerful result was found independently by several physicists, including Hans Hellmann in 1937 and Richard Feynman in 1939. They imagined a situation in which the Hamiltonian $\hat{H}(\lambda)$ depends on a parameter λ. As a result, the normalized energy eigenstates $\psi_n(\lambda)$, with $n = 1, 2, \ldots$ depend on λ, and their energies $E_n(\lambda)$ also depend on λ. Motivated by the possibility that λ could change, one speaks of $\psi_n(\lambda)$ as the instantaneous energy eigenstates and of $E_n(\lambda)$ as the instantaneous energies.

In the **Hellmann-Feynman lemma**, one finds a simple expression for the rate of change of the energies with respect to λ:

$$\boxed{\frac{dE_n(\lambda)}{d\lambda} = \Big\langle \psi_n(\lambda), \frac{d\hat{H}(\lambda)}{d\lambda}\psi_n(\lambda)\Big\rangle.} \tag{7.9.1}$$

The lemma states that the rate of change of the instantaneous energies is given by the expectation value of the operator "rate of change of the Hamiltonian" on the instantaneous eigenstates.

Proof. Since $\psi_n(\lambda)$ is an energy eigenstate of $\hat{H}(\lambda)$ with eigenvalue $E_n(\lambda)$:

$$\hat{H}(\lambda)\,\psi_n(\lambda) = E_n(\lambda)\psi_n(\lambda). \tag{7.9.2}$$

Forming the inner product with $\psi_n(\lambda)$, we have

$$\langle \psi_n(\lambda),\ \hat{H}(\lambda)\,\psi_n(\lambda)\rangle = E_n(\lambda)\big\langle \psi_n(\lambda),\ \psi_n(\lambda)\big\rangle. \tag{7.9.3}$$

Recalling that the state is normalized, the above equation gives

$$E_n(\lambda) = \langle \psi_n(\lambda), \hat{H}(\lambda)\psi_n(\lambda)\rangle. \tag{7.9.4}$$

Now we differentiate this relation with respect to λ. Note that on the right-hand side we must differentiate the wave functions as well as the Hamiltonian:

$$\begin{aligned}\frac{dE_n}{d\lambda}(\lambda) &= \Big\langle \frac{d}{d\lambda}\psi_n(\lambda), \hat{H}(\lambda)\psi_n(\lambda)\Big\rangle + \Big\langle \psi_n(\lambda), \hat{H}(\lambda)\frac{d}{d\lambda}\psi_n(\lambda)\Big\rangle \\ &\quad + \Big\langle \psi_n(\lambda), \frac{d\hat{H}(\lambda)}{d\lambda}\psi_n(\lambda)\Big\rangle.\end{aligned} \tag{7.9.5}$$

We now use the eigenstate equation for the first two terms, directly for the first, and use the Hermiticity of $\hat{H}(\lambda)$ for the second:

$$\begin{aligned}\frac{dE_n}{d\lambda}(\lambda) &= E_n(\lambda)\Big\langle \frac{d}{d\lambda}\psi_n(\lambda), \psi_n(\lambda)\Big\rangle + E_n(\lambda)\Big\langle \psi_n(\lambda), \frac{d}{d\lambda}\psi_n(\lambda)\Big\rangle \\ &\quad + \Big\langle \psi_n(\lambda), \frac{d\hat{H}(\lambda)}{d\lambda}\psi_n(\lambda)\Big\rangle.\end{aligned} \tag{7.9.6}$$

The first two terms combine into a single derivative and therefore:

$$\frac{dE_n}{d\lambda}(\lambda) = E_n(\lambda)\frac{d}{d\lambda}\langle \psi_n(\lambda), \psi_n(\lambda)\rangle + \Big\langle \psi_n(\lambda), \frac{d\hat{H}(\lambda)}{d\lambda}\psi_n(\lambda)\Big\rangle. \tag{7.9.7}$$

Since the state is normalized for all λ, the first term vanishes, and we are left with the result we wanted to prove. Note that the Hellmann-Feynman lemma is an exact result. □

Remark: In proving this lemma, we assumed we could deal with a single state $\psi_n(\lambda)$. This is easily done when the state we are considering is not degenerate and remains nondegenerate as λ changes. If the state is degenerate, energy eigenstates are not uniquely defined, and the lemma does not apply.

Example 7.2. *Infinite square well with a delta function perturbation.*

As an application of the Hellmann-Feynman lemma, we consider a particle of mass m in an infinite square well $x \in [0, a]$. The Hamiltonian $\hat{H}$ is now supplemented by a delta function perturbation located at the midpoint $x = \frac{a}{2}$. As a result, the total Hamiltonian $\hat{H}(\lambda)$ is given by

$$\hat{H}(\lambda) = \hat{H} + \lambda\, a\delta\big(x - \tfrac{a}{2}\big). \tag{7.9.8}$$

Here λ is a parameter with units of energy. We wish to find the approximate energy $E(\lambda)$ of the ground state for small values of λ.

Our strategy will be as follows. The lemma gives us derivatives of the energy by calculation of an expectation value on instantaneous eigenstates, presumed known. Since we only know the energy eigenstates for $\lambda = 0$, we can determine the derivative of the energy at $\lambda = 0$. But this, via Taylor expansion, gives us an approximate value of the energy:

$$E(\lambda) = E(0) + \lambda \frac{dE(\lambda)}{d\lambda}\Big|_{\lambda=0} + \mathcal{O}(\lambda^2). \tag{7.9.9}$$

For the infinite square well, the normalized ground state wave function $\psi(0)$ and the energy $E(0)$, with the 0 referring to $\lambda=0$, are read from (6.4.11):

$$\psi(0)=\sqrt{\frac{2}{a}}\sin\frac{\pi x}{a}, \quad E(0)=\frac{\pi^2\hbar^2}{2ma^2}. \tag{7.9.10}$$

By the lemma,

$$\frac{dE(\lambda)}{d\lambda}\bigg|_{\lambda=0}=\Big(\psi(0), \frac{d\hat{H}}{d\lambda}\bigg|_{\lambda=0}\psi(0)\Big)=\big(\psi(0), a\delta\big(x-\tfrac{a}{2}\big)\,\psi(0)\big), \tag{7.9.11}$$

where the derivative of $\hat{H}(\lambda)$ was evaluated using (7.9.8). Using the form of the wave function, we have

$$\frac{dE(\lambda)}{d\lambda}\bigg|_{\lambda=0}=\int_0^a dx\,\frac{2}{a}\Big(\sin\frac{\pi x}{a}\Big)^2 a\delta\big(x-\tfrac{a}{2}\big)=2. \tag{7.9.12}$$

Back in (7.9.9) we learned that

$$E(\lambda)=\frac{\pi^2\hbar^2}{2ma^2}+2\lambda\ +\mathcal{O}(\lambda^2) \tag{7.9.13}$$

is the approximate value of the ground state energy when λ is small, meaning small compared with the first term, which fixes the energy scale. □

Example 7.3. *Nondegenerate perturbation from Hellmann-Feynman.*

The previous example suggests a general approach. Consider the Hamiltonian

$$\hat{H}(\lambda)=\hat{H}^{(0)}+\lambda\,\delta H, \tag{7.9.14}$$

where we assume that $\hat{H}^{(0)}$ is a familiar Hamiltonian for which energy eigenstates $\psi_k^{(0)}$ and energies $E_k^{(0)}$ are known:

$$\hat{H}^{(0)}\psi_k^{(0)}=E_k^{(0)}\psi_k^{(0)}. \tag{7.9.15}$$

The extra term in the Hamiltonian is a perturbation proportional to the operator δH. Again, as in the previous example, calling $E_k(\lambda)$ the energy of the state $\psi_k^{(0)}$ once λ becomes nonzero,

$$\begin{aligned} E_k(\lambda)&=E_k^{(0)}+\lambda\frac{dE_k(\lambda)}{d\lambda}\bigg|_{\lambda=0}+\mathcal{O}(\lambda^2)\\ &=E_k^{(0)}+\lambda\Big(\psi_k^{(0)}, \frac{d\hat{H}(\lambda)}{d\lambda}\bigg|_{\lambda=0}\psi_k^{(0)}\Big)+\mathcal{O}(\lambda^2). \end{aligned} \tag{7.9.16}$$

The derivative of the Hamiltonian is δH, and bringing the λ parameter inside the expectation value, we find that

$$E_k(\lambda)=E_k^{(0)}+\big(\psi_k^{(0)},\ \lambda\delta H\,\psi_k^{(0)}\big)+\mathcal{O}(\lambda^2). \tag{7.9.17}$$

This is a very nice result: the first, order correction to the energy of the state, given by the second term in the above equation's right-hand side, is equal to the expectation value of the full perturbation $\lambda\delta H$ on the unperturbed state! There is no need to discover how the states change in order to find out how the energy changes to first order. This is sometimes said to be the most important result in perturbation theory! □

The Hellmann-Feynman lemma is also useful for computing the expectation values of operators on known energy eigenstates. Suppose a Hamiltonian is a function of a collection of parameters $\beta_1, \dots, \beta_m$, denoted collectively as β. If we know the energy $E(\beta)$ and the wave function $\psi(\beta)$ of an energy eigenstate, by the lemma,

$$\frac{\partial E}{\partial \beta_i} = \left\langle \frac{\partial \hat{H}}{\partial \beta_i} \right\rangle_{\psi(\beta)}, \quad i = 1, \dots, m. \tag{7.9.18}$$

This follows from the statement of the Hellmann-Feynman lemma by treating each β_i as the parameter λ. In the above relation, the right-hand side is the expectation value of some operator evaluated on the state $\psi(\beta)$, and the left-hand side is determined by the known energy $E(\beta)$. This form of the Hellmann-Feynman lemma is useful for computing expectation values of some operators in hydrogen atom eigenstates (problem 11.5).

Problems

Problem 7.1. *No energy eigenstates with E less than V(x).*

Consider a normalized energy eigenstate $\psi(x)$ in a potential $V(x)$ with energy E:

$$-\frac{\hbar^2}{2m}\psi''(x) + [V(x) - E]\psi(x) = 0.$$

Let V_{min} denote the minimum value of the potential $V(x)$, over all x.

1. Prove that $E > V_{\text{min}}$ by noting that on an energy eigenstate $E = \langle \hat{H} \rangle$.
2. Try to understand intuitively where this result comes from. Assume that $\psi(x)$ is real. If $E < V_{min}$, what do you know about the signs of $\psi''(x)$ and $\psi(x)$ at any point x? Try to sketch a normalizable wave function over all $x \in (-\infty, \infty)$ that satisfies this sign condition.

Problem 7.2. *Nondegeneracy of bound states in one dimension.*

In this problem you will prove that there are no degenerate (same-energy) bound states in one-dimensional potentials $V(x)$ with $x \in (-\infty, \infty)$. Recall that for bound states we assume that the wave function and its derivative vanish as $|x| \to \infty$. Assume the particle has mass m and is moving in a potential $V(x)$.

1. Begin by assuming that we have two (different) bound energy eigenstates $\psi_1(x)$ and $\psi_2(x)$, both with energy E. Without loss of generality, we assume both states are real. Use the Schrödinger equation to prove that for all x,

$$\psi_2 \frac{d\psi_1}{dx} - \psi_1 \frac{d\psi_2}{dx} = \text{constant}$$

2. What is the value of the constant in the equation above? Find a relationship between $\psi_1(x)$ and $\psi_2(x)$ and use it to conclude that degeneracy is not possible.

3. Can there be degenerate normalizable states in a one-dimensional potential for which x is restricted to a circle of finite circumference?
4. Consider a one-dimensional potential $V(x)$ with hard walls at $x < a$ and $x > b$, with $a < b$. Does the nondegeneracy proof apply?

Problem 7.3. *An infinite square well with a step.*

A particle of mass m is moving in one dimension subject to the potential $V(x)$:

$$V(x) = \begin{cases} \infty, & \text{for } x < 0, \\ 0, & \text{for } 0 < x < a, \\ V_0, & \text{for } a < x < 2a, \quad (V_0 > 0), \\ \infty, & \text{for } x > 2a. \end{cases}$$

Sketch the potential.

1. Find the equations that determine the eigenstates with energies $0 < E < V_0$. For this define, as usual,

$$k^2 = \frac{2mE}{\hbar^2}, \quad \kappa^2 = \frac{2m(V_0 - E)}{\hbar^2}, \quad z_0^2 = \frac{2ma^2 V_0}{\hbar^2}, \quad \eta = ka, \quad \xi = \kappa a,$$

   where we use k for classically allowed regions and κ for classically forbidden regions. You should find two equations constraining ξ and η.
2. As a numerical application, consider $z_0 = 2\pi$. How many states do you get with $E < V_0$? For these states find the numerical value of the ratio E/V_0, to at least four significant digits.

Problem 7.4. *Removing dimensional constants from the Schrödinger equation.*

In the hydrogen atom, the length scale is the Bohr radius $a_0 = \frac{\hbar^2}{me^2}$. Consider now the "radial equation" for the radial part $\psi(r)$ of the wave function:

$$\left(-\frac{\hbar^2}{2m}\frac{d^2}{dr^2} + \frac{\hbar^2 \ell(\ell+1)}{2mr^2} - \frac{e^2}{r} \right) \psi(r) = E\psi(r).$$

Here ℓ is a nonnegative integer. Clean up the equation by defining a unit-free coordinate u and a unit-free energy $\mathcal{E}$ so that the equation will take the form

$$\left(-\frac{d^2}{du^2} + \ldots \right) \psi(u) = \mathcal{E}\, \psi(u).$$

How are r and u related? How are $\mathcal{E}$ and E related? Complete the above equation.

Problem 7.5. *Particle in a linear potential.*

A particle of mass m moves in a potential of the form

$$V(x) = \begin{cases} \infty, & \text{for } x \le 0, \\ \frac{1}{2} g_0 x, & \text{for } x > 0. \end{cases}$$

Note the hard wall at $x = 0$ and the linearly growing potential for $x > 0$. The constant $g_0 > 0$ is assumed known. This problem asks for a coordinate space analysis of the linear potential considered in section 6.7.

1. Use g_0, m, and $\hbar$ to construct a length scale L_0 and an energy scale E_0 (your expressions should not include unnecessary numerical constants).
2. Use a unit-free length coordinate u and a unit-free energy $\mathcal{E}$ defined by

$$u \equiv \frac{x}{L_0}, \quad \mathcal{E} \equiv \frac{2E}{E_0}$$

to show that a unit-free (time-independent) Schrödinger equation for an energy eigenstate $\psi(u)$ reads

$$\psi''(u) = (u - \mathcal{E})\psi(u).$$

3. Use the differential equation to obtain the leading behavior of $\psi(u)$ in the large u limit. Try an ansatz of the form

$$\psi(u) \simeq \exp(\alpha\, u^{\beta}), \quad \beta > 0,$$

with α and β constants that you must fix, and select values that are consistent with a normalizable solution.
4. Do a qualitative sketch of the ground state wave function and of the seventh excited state.

Problem 7.6. *Nodes in wave functions.*

Consider the Schrödinger equation in the form

$$\psi'' + \frac{2m}{\hbar^2}(E - V(x))\psi = 0.$$

Let ψ_k and ψ_{k+1} be, respectively, energy eigenstates with energies E_k and E_{k+1}, with $E_k < E_{k+1}$.

1. Show that

$$\left(\psi_{k+1}\psi_k' - \psi_k\psi_{k+1}'\right)\Big|_a^b = \frac{2m}{\hbar^2}(E_{k+1} - E_k)\int_a^b dx\, \psi_k\psi_{k+1}.$$

2. Let now a, b with $a < b$ be two successive zeroes of $\psi_k(x)$ and assume for convenience that $\psi_k(x) > 0$ for $a < x < b$. By making use of (1), show that ψ_{k+1} must change sign in the interval (a, b). That is, ψ_{k+1} must have at least one zero in between each pair of zeroes of ψ_k. Hint: consider the sign of each side of the equation in (1) under the assumption that ψ_{k+1} does not change sign in (a, b).
3. Explain why this result implies that ψ_{k+1} must have at least one node more than ψ_k.

Problem 7.7. *Shooting method and application.*

For a particle of mass m in a quartic potential $V(x) = \alpha x^4$, after rescaling x into a unit-free variable u, the Schrödinger equation takes the form (7.6.9)

$$-\frac{1}{2}\frac{d^2\psi}{du^2} + \left(u^4 - \mathcal{E}\right)\psi = 0,$$

where $\mathcal{E}$ is a unit-free measure of the energy eigenvalue.

The Mathematica instructions that allow you to find the values of $\mathcal{E}$ (written as e) for the even solutions of this potential are given below. These instructions produce a plot for the solution $\psi(u)$, for $u \in [0, 3.5]$ with some suitable initial conditions, and for the chosen value of the energy $\mathcal{E}$.

```
Clear[e, psi]
v[x_]:= x^4
e=0.65;
psi = psi/. NDSolve[{-(1/2)psi''[u] + (v[u]-e)psi[u]==0,
  psi[0]==1, psi'[0]==0}, psi, {u, 0, 3.5}][[1]];
Plot[psi[u], {u, 0, 3.5}]
```

After executing these instructions, if you write `psi[0.5]`, for example, the program will return the value of ψ at $u = 0.5$. Play with this to familiarize yourself. The initial value of $\mathcal{E}$ set above is 0.65, but the ground state energy is a little bit higher. Compute the ground state energy $\mathcal{E}$ to six significant digits (a zero followed by six correct digits).

We now revisit problem 7.3, setting $z_0 = 2\pi$. Use $x = au$, with $u \in [0, 2]$ unit-free, and write $V = V_0 f(u)$ for a suitably defined $f(u)$ to obtain a differential equation for the energy eigenstates in which no units appear and the energy eigenvalue is encoded by the pure number $e = E/V_0$. For the lowest two bound states in this system, you found $E_1 = 0.\#\#436\, V_0$ and $E_2 = 0.\#\#747\, V_0$. Test your differential equation using the shooting method to recover the above values of E_1 and E_2 and then find the next two energy levels E_3 and E_4 to four significant digits.

Problem 7.8. *Wavelength-amplitude relation in a slowly varying potential.*

We have stated that for slowly varying potentials the amplitude of the wave function is roughly proportional to the square root of the "local" de Broglie wavelength (7.2.8). We want to test the accuracy of this claim for a potential $V(x)$, which we can solve numerically:

$$V(x) = \begin{cases} \infty, & x < 0 \text{ and for } x \geq 3a, \\ 0, & 0 \leq x < a, \\ \frac{x-a}{a}\, V_0, & a \leq x < 2a, \\ V_0, & 2a \leq x < 3a. \end{cases}$$

Sketch the potential. As before, let $z_0^2 \equiv 2ma^2 V_0/\hbar^2$, and $e = E/V_0$, with $E > 0$ the energy of the energy eigenstate. We will fix $z_0 = 2\pi$. Code the differential equation for the eigenstates and use the shooting method to construct the energy eigenstate with ten nodes. For a state with many nodes, one can view the potential as slowly varying. Let A_L and A_R denote the amplitudes of your wave function on the left ($0 < x < a$) and right ($2a < x < 3a$) sides of the square well.

1. For the ten-node eigenstate, give e and A_L/A_R (with five significant digits).
2. Find a formula for the de Broglie wavelength λ of an energy E eigenstate in a region where the potential is a constant equal to V (answer in terms of $E, V, m, \hbar$). Use this

result to write λ/a in terms of z_0, e, V, and V_0. Determine λ_L/a and λ_R/a for the left and right sides of the potential for the ten-node state under consideration. Does the approximate theoretical prediction $\frac{A_L}{A_R} \simeq \sqrt{\frac{\lambda_L}{\lambda_R}}$ provide an accurate estimate of how the amplitude changes over the slowly varying potential?

Problem 7.9. *Hydrogen ion using the square well model.*

In problem 6.3 you modeled the hydrogen atom size a_0 and ground state energy E_0 using the square well potential of depth $-V_0$ and width $2a_0$. You previously found that this well has $z_0 = 1.3192$ and $V_0 = z_0^2|E_0| = 1.7403|E_0|$. The ground state energy of the square well potential equals the ground state energy of the hydrogen atom.

To simulate the *molecular hydrogen ion* H_2^+ (having two protons and one electron), we will construct an *even* potential with two identical square well models of hydrogen separated by a small distance $2\gamma a_0$, where γ is a small, positive, unit-free constant. The potential $V(x)$ is therefore

$$V(x) = \begin{cases} 0, & \text{for } |x| < \gamma a_0, \\ -V_0, & \text{for } \gamma a_0 < |x| < (2+\gamma)a_0, \quad (V_0 > 0), \\ 0, & \text{for } |x| > (2+\gamma)a_0. \end{cases}$$

Sketch the potential. For definiteness, work with $\gamma = 0.2$. In order to make your differential equation dimensionless, use $x = a_0 u$ with u unit-free, write $V = V_0 f(u)$ with suitably defined $f(u)$, and let $\mathcal{E} = E/V_0$. Since we are looking for bound states, we have $\mathcal{E} < 0$.

1. Use the shooting method to find the energy of the lowest-energy eigenstate—namely, the bound state energy of an electron shared by two protons. Find the value of $\mathcal{E}$ for the state and also express the ground state E energy in eV.
2. The *binding* energy of the ion is estimated by adding to the ground state energy above the positive energy due to the repulsion of the two protons. Assume that the protons are located at the centers of the wells. What binding energy do you get? How does this compare with the experimental value?

Problem 7.10. *Two delta functions: a first look.*

Consider a particle of mass m moving in a one-dimensional double well potential:

$$V(x) = -g\delta(x-a) - g\delta(x+a), \quad g > 0.$$

This is an attractive potential with δ-function dips at $x = \pm a$.

1. Find transcendental equations for the bound state energy eigenvalues of the system. In solving the Schrödinger equation, define

$$\kappa \equiv \sqrt{-\frac{2mE}{\hbar^2}}, \qquad \xi \equiv \kappa a, \qquad \lambda \equiv \frac{mag}{\hbar^2}.$$

Both ξ and λ are dimensionless. Write the transcendental equations in terms of λ and ξ alone and solve for λ. You should find that even and odd wave functions satisfy different equations, which you can write as $\lambda_{\text{even}} = \cdots$ and $\lambda_{\text{odd}} = \cdots$, where the dots are (different) functions of ξ. Is λ_{even} a monotonically increasing function of ξ for $\xi > 0$? How about λ_{odd}?

2. There is a critical value λ_{crit} such that $\lambda > \lambda_{\text{crit}}$ is required for an odd solution to exist. Find λ_{crit}. If we define $E_0 \equiv \frac{\hbar^2}{2ma^2}$, what is the relationship between the energy E, ξ, and E_0?
3. Discuss the behavior of the number N of bound states in this potential as a function of $g > 0$. What are the minimum and the maximum possible values for N?
4. Using your favorite mathematical software, plot $\lambda(\xi)$ for both odd and even states. From your plot, find approximate ξ values for $\lambda = 1$ and $\lambda = 2$ and compute the energies of the even and odd states at these values of λ. Check your approximate answers by solving the transcendental equation numerically.
5. In the limit of large λ, find an approximate formula for the energy difference $\Delta E > 0$ between the first excited state and the ground state. (Hint: start by finding approximate expressions for ξ_{even} and ξ_{odd} in the limit of large λ.)

Problem 7.11. *Two delta functions: a second look.*

Consider again the situation of problem 7.10, where a particle of mass m is moving in a one-dimensional double delta function potential $V(x) = -g\delta(x-a) - g\delta(x+a)$ with $g > 0$. You found the value of the bound state energy E for the even state in terms of the energy $E_0 = \hbar^2/(2ma^2)$. You had

$$\frac{E}{E_0} = -\xi^2, \quad \text{where} \quad \frac{\xi}{1+e^{-2\xi}} = \lambda, \quad \text{and} \quad \lambda = \frac{mag}{\hbar^2},$$

with λ, unit-free, encoding the intensity g of the delta functions (while keeping a constant) or the separation of the delta functions (while keeping g constant). We can thus write

$$\lambda = \frac{a}{a_0}, \quad a_0 = \frac{\hbar^2}{mg},$$

with a_0 a natural length scale in the problem once g is fixed. Also define the energy $E_\infty \equiv \frac{mg^2}{2\hbar^2}$. Here $(-E_\infty)$ is the bound state energy of a particle in the single delta function potential $V(x) = -g\delta(x)$.

1. Write E in terms of E_∞, ξ, and λ. What are the values of E for $a \to 0$ and for $a \to \infty$?
2. We would like to plot E/E_∞ as a function of $\lambda = a/a_0$ in order to understand how the ground state energy varies as a function of the separation between the δ function wells. Lacking a closed-form solution for $E(\lambda)$, construct this plot numerically. Your plot should be consistent with the limits found in the previous question. Compute E/E_∞ at $\lambda = 1$ to five significant digits.

Assume now that this is a model for a diatomic molecule with interatomic distance $2a$. The bound state electron helps overcome the repulsive energy between the ions. With x the distance between the atoms, let the repulsive potential energy $V_r(x)$ be given by

$$V_r(x) = \frac{\beta g}{x}, \quad \beta > 0,$$

where β is a small number. The total potential energy V_{tot} of the configuration is estimated as the sum of the negative energy E of the bound state and the positive repulsive energy:

$$V_{\text{tot}} = E + V_r(2a).$$

3. Write $V_r(2a)$ in terms of E_∞, β, and λ. Now consider the total potential energy V_{tot} and plot it as a function of $a/a_0 = \lambda$ for various values of β. You should find a critical stable point for the potential for sufficiently small β.
4. Compute to two significant digits the value of a/a_0 at the critical point of the potential for $\beta = 0.31$.

Problem 7.12. *Virial theorem in classical mechanics.*

Consider the condition

$$\left\langle \frac{d\mathcal{O}}{dt} \right\rangle_{\text{cl}} = 0, \quad \mathcal{O} \equiv \sum_\alpha \mathbf{r}_\alpha \cdot \mathbf{p}_\alpha.$$

Here, $\mathbf{r}_\alpha$ and $\mathbf{p}_\alpha$ are, respectively, the position and momenta of a set of particles labeled by α, and $\langle \cdots \rangle_{\text{cl}}$ denotes the classical average. Show that

$$2\langle T \rangle_{\text{cl}} = \sum_\alpha \left\langle \mathbf{r}_\alpha \cdot \frac{\partial V}{\partial \mathbf{r}_\alpha} \right\rangle_{\text{cl}},$$

with T the kinetic energy of the system of particles and $V(\mathbf{r}_1, \mathbf{r}_2, \cdots)$ the potential energy of a general configuration.

Problem 7.13. *Virial theorem for the linear potential.*

Consider a particle of mass m in a linear potential $V = \frac{\hbar^2}{2mL^3}x$ (with $L > 0$), plus a wall at $x = 0$, as discussed in section 6.7. Focus on the energy eigenstate of energy E_n, as given in (6.7.17). Use the virial theorem to show that in this state the expectation value of the position is $\langle x \rangle = \frac{2}{3}L|a_n|$.

Problem 7.14. *Variational principle applied to the infinite square well.*

1. Consider a particle of mass m in a box of size L so the wave function vanishes at $x = \pm L/2$. Find an *upper* bound on the ground state energy using a simple trial wave function: a quadratic function of x that vanishes at $\pm L/2$. Compare with the exact ground state energy, noting that your result implies the inequality $\pi^2 < 10$.
2. Work with a unit-free coordinate $u \in [-1, 1]$ with $u = 2x/L$. Show that in this case the Schrödinger equation becomes $-4\frac{d^2\psi}{du^2} = e\psi$, with e a unit-free energy that you should relate to the usual energy E and constants $\hbar$, m, and L. Confirm that for the ground state $e = \pi^2$.
3. Now try an ansatz of the form $\psi = (u^2 - 1)(u^2 + B)$. Use the variational principle (and an algebraic manipulator) to show that

$$e \leq \frac{66+84B+210B^2}{1+6B+21B^2}.$$

Confirm that the minimum value of the fraction above is $56-4\sqrt{133}$. Check that this number is remarkably close to π^2.

Problem 7.15. *Variational principle comparison method.*

Consider two Hamiltonians for a point particle of mass m. The Hamiltonians have the same kinetic term but different potentials. The Hamiltonian $\hat{H}_1$ has a potential $V_1(\mathbf{x})$, and the Hamiltonian $\hat{H}_2$ has a potential $V_2(\mathbf{x})$. Additionally, assume that for all points in space the potential energy V_1 is less than or equal to V_2:

$$V_1(\mathbf{x}) \leq V_2(\mathbf{x}), \quad \text{for all } \mathbf{x}.$$

Prove that the ground state energies $E_{gs}^{(1)}$ and $E_{gs}^{(2)}$ of the two systems satisfy

$$E_{gs}^{(1)} \leq E_{gs}^{(2)}.$$

Use this result to argue that an attractive potential, defined as a nowhere positive potential that approaches zero at infinity, has a state with energy less than zero. For this, compare the attractive potential to a suitable square well potential.

Problem 7.16. *Developing the variational principle.*

1. Consider normalized trial wave functions ψ that are orthogonal to the ground state wave function ψ_1: $\langle\psi_1, \psi\rangle = 0$. Show that the first excited energy E_2 is bounded as

   $$E_2 \leq \langle\psi, \hat{H}\psi\rangle.$$

   This result has a clear generalization (that you need not prove): trial wave functions orthogonal to the lowest n energy eigenstates give an upper bound for the energy of the $(n+1)$th state.

2. Assume we can use real wave functions and consider the variational functional

   $$\mathcal{F}(\psi) = \frac{\langle\psi, \hat{H}\psi\rangle}{\langle\psi, \psi\rangle}.$$

   This functional has a remarkable property: it is stationary at the energy eigenstates! You will do a computation that confirms this for a special case while giving you insight into the nature of the critical point. Let us take

   $$\psi = \psi_2 + \sum_{n=1} \epsilon_n \psi_n. \tag{7.9.19}$$

   This is the first excited state perturbed by small additions of all energy eigenstates: the ϵ's are all taken to be small. Evaluate the functional $\mathcal{F}$ for this wave function, including terms quadratic in the ϵ's but ignoring terms that are cubic or higher order. Confirm that all linear terms in ϵ's cancel, showing that the functional is indeed stationary at $\psi_2(x)$. Does any ϵ drop out to quadratic order? Discuss the nature of the critical point (maximum, minimum, flat directions, saddle).

Problem 7.17. *Variational analysis of the potential $V(x)=\alpha x^4$.*

As shown in section 7.6 for the potential $V(x)=\alpha x^4$, the unit-free version of the Schrödinger equation takes the form

$$-\frac{1}{2}\frac{d^2\psi}{du^2}+(u^4-\mathcal{E})\,\psi=0.$$

1. Use an algebraic manipulator and the shooting method to determine the value of $\mathcal{E}$ for the ground state energy to six digits of accuracy ($\mathcal{E}\simeq 0.67$). (The Mathematica instructions for this calculation were used in problem 7.7). Determine the value of $\mathcal{E}$ for the first excited state to four digits of accuracy.
2. Use a one-parameter Gaussian as a trial wave function to determine, using the variational principle, an upper bound for the ground state energy. Write a one-parameter trial wave function to determine an upper bound for the first excited state energy. Compare your bounds with the accurate results obtained in (1).

Problem 7.18. *Square well with repulsive delta function.*

Consider the one-dimensional infinite square well $0\le x\le a$. We add a repulsive delta function at the middle of the well

$$V(x) \;=\; V_0 a\,\delta\!\left(x-\tfrac{a}{2}\right), \quad V_0>0,$$

with V_0 a large value with units of energy. In fact, V_0 is large compared to the natural energy scale of the well:

$$\frac{V_0}{\left(\frac{\hbar^2}{ma^2}\right)} \equiv \gamma \gg 1.$$

The dimensionless number γ is taken to be large. The delta function is creating a barrier between the left side and the right side of the well. Calculate the ground state energy, including corrections of order $1/\gamma$ but ignoring higher-order ones. Compare with the energy of the first excited state. What is happening to the energy difference between these two levels?

Problem 7.19. *A three-delta-function potential.*

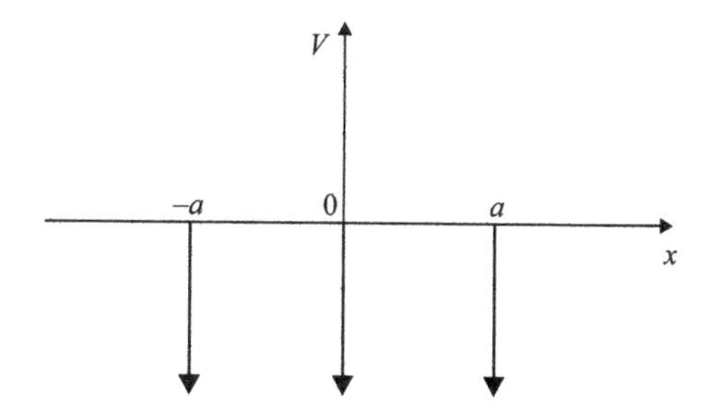

A particle of mass m moves in one dimension in a potential $V(x)$ that is the sum of three evenly spaced attractive delta functions:

$$V(x) = -V_0\, a \sum_{n=-1}^{1} \delta(x - na)\,, \text{ where } V_0 > 0,\ a > 0 \text{ are constants.}$$

1. Calculate the discontinuity in the first derivative of the wave function at $x = -a$, 0, and a.
2. Consider the possible number and locations of nodes in bound state wave functions.
   (i) How many nodes are possible in the region $x > a$?
   (ii) How many nodes are possible in the region $0 < x < a$?
   (iii) Can there be a node at $x = a$?
   (iv) Can there be a node at $x = 0$?
3. For arbitrarily large V_0, how many bound states are there? Sketch them qualitatively.
4. Consider the lowest-energy *antisymmetric* bound state and introduce the definitions:

$$z_0^2 \equiv \frac{2mV_0a^2}{\hbar^2}\,, \quad \kappa \equiv \sqrt{\frac{2m|E|}{\hbar^2}}\,, \quad \xi \equiv \kappa a.$$

   Here z_0 and ξ are proxies for V_0 and the energy E of the bound state, respectively. The condition for the state to exist allows one to write $z_0^2 = \cdots$, where the dots represent a function of ξ that you should determine. Find the minimum value of z_0 for such a bound state to exist.

8 Stationary States III: Scattering

We study nonnormalizable energy eigenstates of one-dimensional potentials. We examine in detail the finite step potential, finding energy eigenstates that represent waves that are incident onto the step and give rise to reflected and transmitted waves. We build normalizable wave packets by superposition of energy eigenstates and use them to represent the physical process in which a particle scatters off the potential. We find that for energies less than the height of the potential the reflected wave packet experiences a time delay. Applied to a finite square well, we encounter the phenomenon of resonant transmission, explaining the curious results found by Ramsauer and Townsend in the scattering of electrons by rare gases.

8.1 The Step Potential

We have studied in some detail bound states—that is, energy eigenstates that are normalizable. We now begin our study of nonnormalizable energy eigenstates. For a nonnormalizable state $\psi(x)$, the integral over space of $|\psi(x)|^2$ diverges. The most familiar such states are the momentum eigenstates of a free particle, which are indeed energy eigenstates. We call nonnormalizable energy eigenstates **scattering states** because they are used to describe the motion of free particles as they encounter obstacles or potentials and scatter off of them. While scattering states cannot be normalized, we still find that for any such state $|\psi(x)|$ is bounded; in other words, it never becomes infinite. Strictly speaking, a nonnormalizable energy eigenstate is not the state of a particle; one must superpose nonnormalizable states to produce *normalizable* states that can represent a particle undergoing scattering in some potential.

We will now study the energy eigenstates of the step potential shown in figure 8.1. The potential is simple: it is zero for all negative x, and it is equal to the positive constant V_0 for all positive x. We thus have

$$V(x) = \begin{cases} 0, & x < 0, \\ V_0, & x \geq 0. \end{cases} \tag{8.1.1}$$

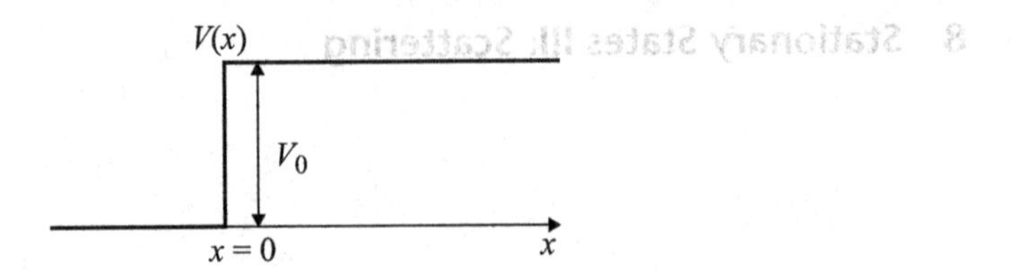

Figure 8.1
The step potential.

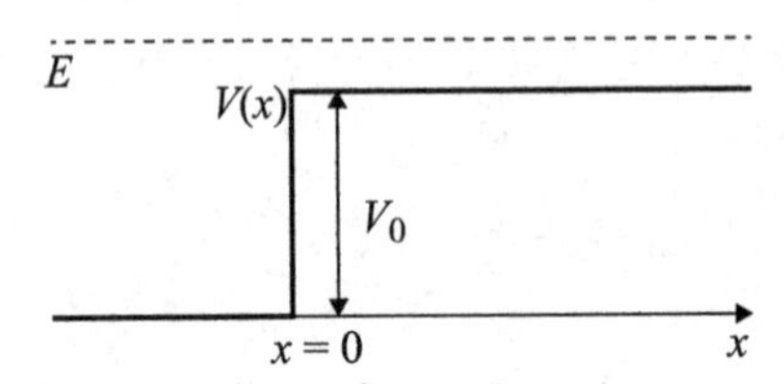

Figure 8.2
The full x-axis is classically allowed when the energy E of the state is greater than V_0.

The potential is discontinuous at $x = 0$. From the viewpoint of a particle incident from the left, the potential represents a finite height barrier. This potential has no bound states. Indeed, all energy eigenstates must have positive energy since the energy cannot be less than the minimum of the potential—in this case, zero. With positive energy the solution cannot be localized for $x < 0$, where the potential vanishes, and cannot be normalized. Our solutions to the Schrödinger equation with this potential will thus be scattering states of definite energy $E > 0$. There is no quantization of the energy. We will find eigenstates for every value of the energy greater than zero. We can consider two cases: $E > V_0$, or energy above the barrier, and $E < V_0$, or energy below the barrier. In both cases eigenstates extend infinitely to the left and are nonnormalizable. Let us begin with the case $E > V_0$, shown in figure 8.2.

Energy eigenstates with $E > V_0$.

A stationary state with energy E is of the form

$$\Psi(x,t) = \psi(x)e^{-iEt/\hbar}, \tag{8.1.2}$$

and we aim to determine the unknown $\psi(x)$, which will exist for all $E > V_0$. Note that the probability density is time independent: $\rho(x,t) = \Psi^*(x,t)\Psi(x,t) = \psi^*(x)\psi(x)$. There are, in fact, two energy eigenstates for each value of the energy, just as with zero potential we have two degenerate states $e^{\pm ikx}$ corresponding to waves moving in opposite directions. Intuitively, two eigenstates exist for any energy that makes the motion of the particle classically allowed both to the far left and to the far right of the potential. With $E > V_0$, one energy eigenstate in this barrier problem can be associated with a wave incident from the left, and the other can be associated with a wave incident from the right, facing a step down.

Let us select, for definiteness, the case of a wave incident from the left. With such a wave traveling in the direction of increasing x, we expect to get a reflected wave and a transmitted wave. The reflected wave, moving in the direction of decreasing x, would exist for $x < 0$. The transmitted wave, moving in the direction of increasing x, would exist for $x > 0$. Since complex exponentials are solutions of the time-independent Schrödinger equation for particles moving in constant potentials, we guess that the energy eigenstate contains the following three pieces:

$$\psi(x) = \begin{cases} Ae^{ikx} + Be^{-ikx}, & x < 0, \\ Ce^{i\bar{k}x}, & x > 0. \end{cases} \tag{8.1.3}$$

Recall that e^{ikx}, with $k > 0$, represents a wave moving in the direction of increasing x, given the universal time dependence in (8.1.2). Therefore, A is the coefficient of the incident wave, B is the coefficient of the reflected wave, and C is the coefficient of the transmitted wave. The waves for $x < 0$ have wave number $k > 0$, and the wave for $x > 0$ has wave number $\bar{k} > 0$. The wave number $\bar{k}$ is *not* the complex conjugate of k; it is just another name for a wave number (we generally use $*$ for complex conjugation). These wave numbers are fixed by the time-independent Schrödinger equation and are given by

$$k^2 = \frac{2mE}{\hbar^2}, \qquad \bar{k}^2 = \frac{2m(E - V_0)}{\hbar^2}. \tag{8.1.4}$$

Two conditions constrain our coefficients A, B, and C: both the wave function and its derivative must be continuous at $x = 0$. With these conditions we can solve for B and C in terms of A. This is all we can expect to do: because of linearity the overall scale of these three coefficients must remain undetermined. In fact, if we think of A as the amplitude of a given incoming wave, the amplitudes B and C of the reflected and the transmitted waves, respectively, must be proportional to A. Let us begin:

- $\psi(x)$ continuous at $x = 0$ requires

$$A + B = C. \tag{8.1.5}$$

- $\psi'(x)$ continuous at $x = 0$ requires

$$ikA - ikB = i\bar{k}C \quad \rightarrow \quad A - B = \frac{\bar{k}}{k}C. \tag{8.1.6}$$

Solving for B and C in terms of A, we get

$$\boxed{\frac{B}{A} = \frac{k - \bar{k}}{k + \bar{k}}, \qquad \frac{C}{A} = \frac{2k}{k + \bar{k}}.} \tag{8.1.7}$$

We see that if A is real, B and C are also real.

We obtain further insight into the solution by evaluating the probability current to the left and to the right of the $x = 0$ step. Using the form (3.7.7) of the probability current for a wave function $\Psi(x, t) = \psi(x)e^{-iEt/\hbar}$ gives

$$J(x) = \frac{\hbar}{m}\text{Im}\left(\psi^* \frac{\partial \psi}{\partial x}\right), \tag{8.1.8}$$

without time dependence. A short calculation shows that the current J_L to the left of the step is

$$J_L = \frac{\hbar k}{m}(|A|^2 - |B|^2) = J_A - J_B, \quad J_A = \frac{\hbar k}{m}|A|^2, \quad J_B = \frac{\hbar k}{m}|B|^2. \tag{8.1.9}$$

The incident and reflected waves turn out to contribute separately to the current: the total current to the left of the step is simply the current J_A associated with the incident wave minus the current J_B associated with the reflected wave. Note that this is not a priori obvious since the current is a quadratic function of the wave function. Note also that there is no x dependence for the current to the left of the step. The current J_R to the right of the step is

$$J_R = \frac{\hbar \bar{k}}{m}|C|^2 = J_C. \tag{8.1.10}$$

Again, J_R has neither time nor space dependence. In any stationary state, there cannot be an accumulation of probability at any region of space because the probability density ρ is time independent everywhere. This is clear from $\rho = \Psi^*\Psi = \psi^*\psi$. While probability is continuously flowing in scattering solutions, it must be conserved. From the conservation equation

$$\partial_x J + \partial_t \rho = 0, \tag{8.1.11}$$

the time independence of ρ implies that the current J must be x independent. We have already confirmed the space independence of J for $x < 0$ and for $x > 0$. But more is needed; the current must be the same to the left and to the right of $x = 0$. Our solution (8.1.7) must imply that $J_L = J_R$. Let us verify this:

$$\begin{aligned} J_L = & \frac{\hbar k}{m}(|A|^2 - |B|^2) = \frac{\hbar k}{m}\left(1 - \left(\frac{k-\bar{k}}{k+\bar{k}}\right)^2\right)|A|^2 \\ = & \frac{\hbar k}{m}\left(\frac{4k\bar{k}}{(k+\bar{k})^2}\right)|A|^2 = \frac{\hbar \bar{k}}{m}\frac{4k^2}{(k+\bar{k})^2}|A|^2 = \frac{\hbar \bar{k}}{m}|C|^2 = J_R, \end{aligned} \tag{8.1.12}$$

as expected. The equality of J_L and J_R implies that

$$J_A - J_B = J_C \quad \Rightarrow \quad J_A = J_B + J_C \quad \Rightarrow \quad 1 = \frac{J_B}{J_A} + \frac{J_C}{J_A}. \tag{8.1.13}$$

We now define the *reflection coefficient* R as the ratio of the probability current in the reflected wave to the probability current in the incoming wave:

$$\boxed{R \equiv \frac{J_B}{J_A} = \frac{|B|^2}{|A|^2} = \left(\frac{k-\bar{k}}{k+\bar{k}}\right)^2 \leq 1.} \tag{8.1.14}$$

This ratio happens to be the norm squared of the ratio B/A, and it is manifestly less than one, as it should be. We also define the *transmission coefficient* T as the ratio of the *probability* current in the transmitted wave to the probability current in the incoming wave:

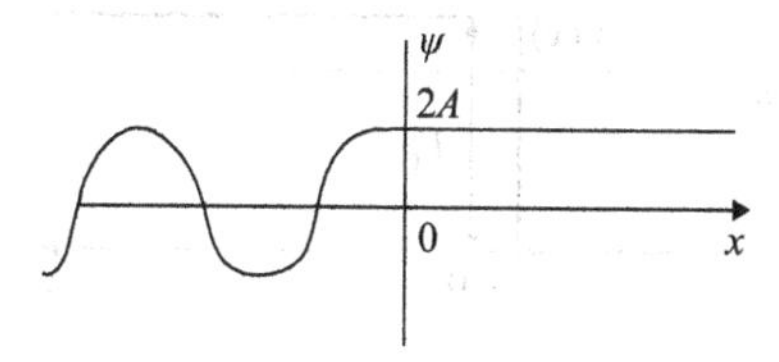

Figure 8.3
Energy eigenstate of the step potential for $E = V_0$.

$$T \equiv \frac{J_C}{J_A} = \frac{\bar{k}}{k}\frac{|C|^2}{|A|^2} = \frac{\bar{k}}{k}\frac{4k^2}{(k+\bar{k})^2} = \frac{4k\bar{k}}{(k+\bar{k})^2}. \tag{8.1.15}$$

The above definitions are sensible because R and T, given in terms of current ratios, add up to one:

$$R + T = 1, \tag{8.1.16}$$

as follows by inspection of (8.1.13). Note that $T \neq |C|^2/|A|^2$ because the wave number $\bar{k}$ to the right of the step and the wave number k to the left of the step are not equal. The reflection coefficient R turned out to be equal to the ratio $|B|^2/|A|^2$ of the wave amplitudes because both the incoming and reflected waves have the same wave number k.

When the energy is exactly equal to the height of the barrier, $E = V_0$, we have $\bar{k} = 0$, and equations (8.1.7) give $B = A$ and $C = 2A$. Therefore, for $E = V_0$ the energy eigenstate is

$$E = V_0: \qquad \psi(x) = \begin{cases} 2A\cos kx, & x < 0, \\ 2A, & x > 0. \end{cases} \tag{8.1.17}$$

The wave function, for A real and positive, is shown in figure 8.3. While constant for $x > 0$, note the spatial dependence of $\psi(x)$ for $x < 0$. This means that the probability density ρ also has spatial dependence for $x < 0$. This is in fact a general feature for arbitrary energies E and arises due to the interference between incoming and reflected waves to the left of the step.

When $E = V_0$, we find $R = 1$ (see (8.1.14)), and thus $T = 0$, consistent with a vanishing probability current for the *constant* wave function that exists for $x > 0$. We can also understand the vanishing of T for $E = V_0$ from continuity. The coefficients R and T must be continuous functions of the energy E. For $E < V_0$, as we will see next, we have an exponentially decaying wave function for $x > 0$. Such a wave function carries zero probability current, and therefore we have $T = 0$ for *any* $E < V_0$. For continuity, T must still be zero for $E = V_0$.

Exercise 8.1. *Assume the energy E is very large so that $V_0/E \ll 1$. Take $A = 1$ and find the coefficient B to leading order in the small parameter V_0/E.*

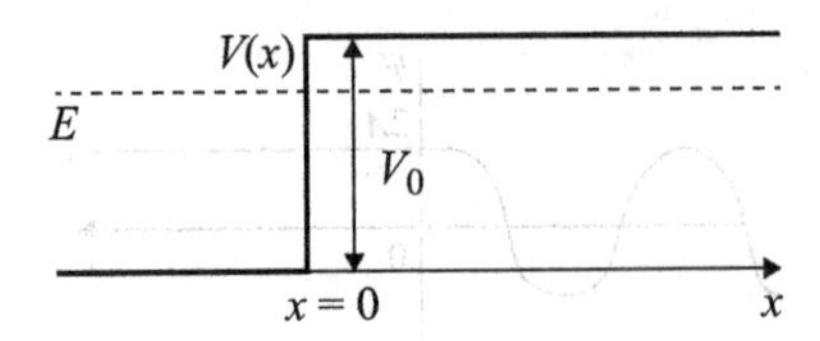

Figure 8.4
When $E < V_0$, only the region to the left of the barrier is classically allowed.

Energy eigenstates with $E < V_0$.

When $E < V_0$, the region $x > 0$ is classically forbidden (see figure 8.4). For each value of $E < V_0$, there is just one energy eigenstate; if the barrier were infinite, it would be the state $\psi = \sin kx$ for $x < 0$ and $\psi = 0$ for $x > 0$. For the present problem, we have an incident wave, a reflected wave, and some decaying solution for $x > 0$. Let us try to solve for the energy eigenstate without redoing all the work involved in solving for B and C in terms of A. For this purpose we first note that the ansatz (8.1.3) for $x < 0$ can be left unchanged. On the other hand, for $x > 0$ the earlier ansatz

$$\psi(x) = Ce^{i\bar{k}x}, \quad \bar{k}^2 = \frac{2m(E - V_0)}{\hbar^2}, \tag{8.1.18}$$

should be changed to a decaying exponential:

$$\psi(x) = Ce^{-\kappa x}, \quad \kappa^2 = \frac{2m(V_0 - E)}{\hbar^2}, \tag{8.1.19}$$

with $\kappa \geq 0$. We note that the former becomes the latter upon the replacement

$$\bar{k} \to i\kappa. \tag{8.1.20}$$

This means that we can simply perform this replacement in our earlier expressions for B/A and C/A, and we obtain the new expressions. In particular, from (8.1.7) we get

$$\frac{B}{A} = \frac{k - i\kappa}{k + i\kappa}. \tag{8.1.21}$$

It is convenient to rewrite B/A so that its limit as the energy E goes to zero is manifest. In that limit $k \to 0$, while κ approaches a fixed value, thus giving us $B/A \to -1$. This is reasonable since for small energy the step looks like a gigantic barrier, and one would expect the wave function to be very small at $x = 0$. Indeed, when $B/A = -1$ we get $\psi(0) = A + B = 0$. Multiplying the numerator and the denominator of the above fraction by i, we find that

$$\frac{B}{A} = -\frac{\kappa + ik}{\kappa - ik}. \tag{8.1.22}$$

Apart from the minus sign, the ratio has a complex number in the numerator and its complex conjugate in the denominator. Thus, the ratio is a complex number of unit norm, a pure phase factor. This implies that the magnitude of A is equal to the magnitude of B. As a result, $J_A = J_B$, implying $R = 1$ and $J_C = 0$, consistent with $T = 0$. There is

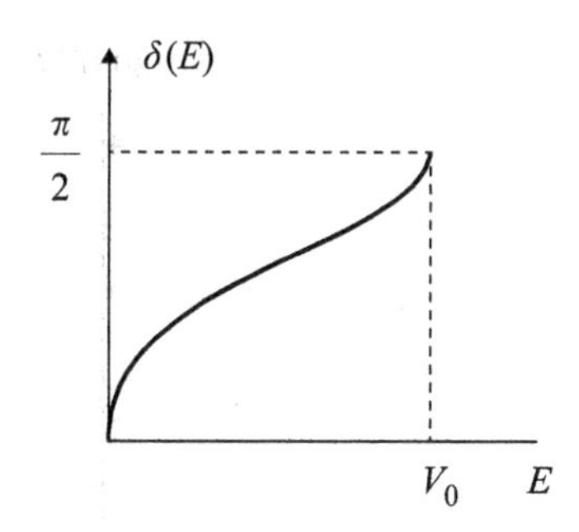

Figure 8.5
The phase shift $\delta(E)$ as a function of energy for $E < V_0$.

zero probability of transmission and complete reflection of the wave. The vanishing of J_C is manifest from $\psi(x) = Ce^{-\kappa x}$, for any value of the complex constant C.

It is useful to write the ratio B/A in terms of a phase factor or phase shift $\delta(E)$:

$$\kappa + ik = |\kappa + ik|\, e^{i\delta(E)}, \tag{8.1.23}$$

where $\delta(E)$, defined by this equation, is the argument of $\kappa + ik$. Since both κ and k are defined to be nonnegative, $\delta(E) \in [0, \frac{\pi}{2}]$. The energy dependence of $\delta(E)$ arises because both k and κ are energy dependent. In fact, we can write

$$\boxed{\delta(E) = \tan^{-1}\left(\frac{k}{\kappa}\right) = \tan^{-1}\sqrt{\frac{E}{V_0 - E}} \quad \in [0, \tfrac{\pi}{2}].} \tag{8.1.24}$$

It also follows from the above that

$$\sin^2 \delta(E) = \frac{E}{V_0}. \tag{8.1.25}$$

Complex conjugating (8.1.23), we also have

$$\kappa - ik = |\kappa + ik| e^{-i\delta(E)}, \tag{8.1.26}$$

recalling that $|z| = |z^*|$ for arbitrary complex z. It follows that

$$\frac{B}{A} = -\frac{\kappa + ik}{\kappa - ik} = -\, e^{2i\delta(E)}. \tag{8.1.27}$$

For zero energy the phase shift $\delta(E)$ vanishes, consistent with our earlier discussion. This can be seen from (8.1.24):

$$\delta(E) \to 0, \quad \text{as} \quad E \to 0. \tag{8.1.28}$$

A sketch of $\delta(E)$ as a function of the energy is given in figure 8.5.

Exercise 8.2. *Show that*

$$\frac{C}{A} = -2i\sqrt{\frac{E}{V_0}}\, e^{i\delta(E)}. \tag{8.1.29}$$

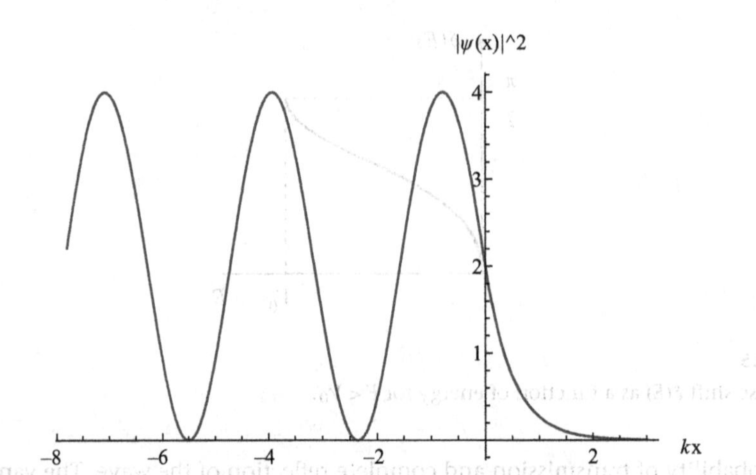

Figure 8.6
Probability density $|\psi(x)|^2$ as a function of position for an energy eigenstate with $E = \frac{1}{2}V_0$, giving $\delta(E) = \frac{\pi}{4}$ and $k = \kappa$, and setting $A = 1$ in (8.1.32). The value of $|\psi(0)|$ (equal to $\sqrt{2}$ here) is nonzero for arbitrary phase shift $\delta(E)$. The probability density decays exponentially for $x > 0$.

When $E < V_0$, the properties of the eigenstate are captured by the phase shift $\delta(E)$. We will see in the following section that for $E < V_0$, reflected wave packets emerge with a time delay that is controlled by $\delta(E)$. Now, however, we conclude by showing that the phase shift controls the value of the wave function at the barrier. Using the expression for the ratio B/A, the wave function $\psi(x)$ for $x < 0$ takes the form

$$\begin{aligned}\psi(x) &= Ae^{ikx} + (-Ae^{2i\delta(E)})e^{-ikx} \\ &= Ae^{i\delta(E)}\left(e^{-i\delta(E)}e^{ikx} - e^{i\delta(E)}e^{-ikx}\right) \\ &= 2iAe^{i\delta(E)}\sin(kx - \delta(E)).\end{aligned} \tag{8.1.30}$$

This means that the probability density is

$$|\psi(x)|^2 = 4A^2\sin^2(kx - \delta(E)), \quad x < 0. \tag{8.1.31}$$

The probability density is positive and exhibits spatial oscillations for $x < 0$, while it is a decaying exponential for $x > 0$ (figure 8.6). The wave function does not vanish at $x = 0$, where we have

$$|\psi(0)| = 2|A|\sin\delta(E) = 2|A|\sqrt{\frac{E}{V_0}}. \tag{8.1.32}$$

For a vanishing phase shift, we have perfect reflection, and the wave function vanishes at the barrier. For $E = V_0$, we get the maximal possible phase shift: $\delta = \pi/2$. Then we have $|\psi(0)| = 2|A|$, as noted in figure 8.3.

8.2 Wave Packets in the Step Potential

Now we examine the more physical scenario. As we've seen with the free particle, momentum eigenstates are not normalizable. Physical particles are therefore represented by normalizable wave packets built with an infinite superposition of momentum eigenstates. We can do similarly with the energy eigenstates in the step potential. We will consider superpositions of energy eigenstates with $E > V_0$ or, equivalently, eigenstates with $k > \hat{k}$, where

$$k = \sqrt{\frac{2mE}{\hbar^2}} > \hat{k} \equiv \sqrt{\frac{2mV_0}{\hbar^2}}. \tag{8.2.1}$$

We are using the energy eigenstates we discussed for $E > V_0$—namely, those associated with waves incoming from the left. To begin, let us write the energy eigenstates following (8.1.3) but including the time dependence. We also set $A = 1$ and use the values for B/A and C/A in (8.1.7). This gives an energy eigenstate $\Psi_k(x,t)$ that we label with the wave number k:

$$\Psi_k(x,t) = \begin{cases} \left(e^{ikx} + \frac{k-\bar{k}}{k+\bar{k}}\, e^{-ikx}\right) e^{-iE(k)t/\hbar}, & x < 0, \\ \frac{2k}{k+\bar{k}}\, e^{i\bar{k}x}\, e^{-iE(k)t/\hbar}, & x > 0. \end{cases} \tag{8.2.2}$$

We can form a superposition of these solutions by multiplying by an arbitrary function $f(k)$ and integrating over k:

$$\Psi(x,t) = \int_{\hat{k}}^{\infty} dk f(k) \Psi_k(x,t). \tag{8.2.3}$$

We have only included eigenstates with energy greater than V_0 by having the integral's lower limit set equal to $\hat{k}$. The above $\Psi(x,t)$ is guaranteed to be a solution of the full Schrödinger equation for arbitrary $f(k)$ because it is a superposition of energy eigenstates. More explicitly, in terms of the parts valid for $x < 0$ and $x > 0$ the solution $\Psi(x,t)$ is

$$\Psi(x,t) = \begin{cases} \int_{\hat{k}}^{\infty} dk f(k) \left(e^{ikx} + \frac{k-\bar{k}}{k+\bar{k}}\, e^{-ikx}\right) e^{-iE(k)t/\hbar}, & x < 0, \\ \int_{\hat{k}}^{\infty} dk f(k)\, \frac{2k}{k+\bar{k}}\, e^{i\bar{k}x}\, e^{-iE(k)t/\hbar}, & x > 0. \end{cases} \tag{8.2.4}$$

Note that the same function $f(k)$ controls both the $x < 0$ and the $x > 0$ expressions.

We can split the solution $\Psi(x,t)$ into incident, reflected, and transmitted waves as follows:

$$\Psi(x,t) = \begin{cases} \Psi_{\text{inc}}(x,t) + \Psi_{\text{ref}}(x,t), & x < 0, \\ \Psi_{\text{tr}}(x,t), & x > 0. \end{cases} \tag{8.2.5}$$

Naturally, both $\Psi_{\text{inc}}(x,t)$ and $\Psi_{\text{ref}}(x,t)$ exist for $x < 0$, and $\Psi_{\text{tr}}(x,t)$ exists for $x > 0$. We then have, explicitly:

$$\begin{aligned} \Psi_{\text{inc}}(x,t) &= \int_{\hat{k}}^{\infty} dk f(k) e^{ikx} e^{-iE(k)t/\hbar}, & x < 0, \\ \Psi_{\text{ref}}(x,t) &= \int_{\hat{k}}^{\infty} dk f(k) \left(\frac{k-\bar{k}}{k+\bar{k}}\right) e^{-ikx} e^{-iE(k)t/\hbar}, & x < 0, \\ \Psi_{\text{tr}}(x,t) &= \int_{\hat{k}}^{\infty} dk f(k) \left(\frac{2k}{k+\bar{k}}\right) e^{i\bar{k}x} e^{-iE(k)t/\hbar}, & x > 0. \end{aligned} \tag{8.2.6}$$

The incident and reflected waves are not defined for $x > 0$, and the transmitted wave is not defined for $x < 0$. For the transmitted wave, we view $\bar{k}$ as a function of k using (8.1.4). The above is a complete solution: for a given choice of $f(k)$, all integrals can be evaluated, numerically if necessary, to find the wave function $\Psi(x, t)$ at any point x and at any time t.

To better understand the solution, we let $f(k)$ be a real function of k that is vanishingly small except for a narrow peak about a wave number $k_0 > \hat{k}$. This makes the incident, reflected, and transmitted waves into wave packets that are localized in space at any instant of time.

At a fixed time t, the peak of $|\Psi_{\rm inc}(x, t)|$ tells us where we are most likely to find the particle within the incident wave. That peak is also the "center" of the wave packet associated with $\Psi_{\rm inc}$. We now ask: How does the peak of $|\Psi_{\rm inc}(x, t)|$ move? For this we recall that the main contribution to the associated integral occurs when the phase in the integrand is stationary for $k = k_0$. We therefore require that

$$\frac{d}{dk}\left(kx - \frac{\hbar^2 k^2}{2m}\frac{t}{\hbar}\right)\bigg|_{k_0} = 0 \quad \Rightarrow \quad x - \frac{\hbar k_0}{m} t = 0. \tag{8.2.7}$$

This gives the

$$\text{incident wave peak:} \quad x = \frac{\hbar k_0}{m} t. \tag{8.2.8}$$

This is the relation between time t and the x position of the peak of $|\Psi_{\rm inc}|$. It describes a peak moving with constant velocity $\hbar k_0/m > 0$. Since $\Psi_{\rm inc}(x, t)$ only exists for $x < 0$, the above condition shows that we get the peak only for $t < 0$. The peak of the packet gets to the origin $x = 0$ at $t = 0$.

For $t > 0$, $\Psi_{\rm inc}(x, t)$ is not zero, but it must be rather small since the (x, t) relation above from the stationary phase condition cannot be satisfied for any x in the domain $x < 0$. For small positive time, $\Psi_{\rm inc}$ describes the lagging tail of the incident wave packet that is still traveling toward the origin.

Now consider the motion of the peak in the reflected wave $\Psi_{\rm ref}(x, t)$. This time the stationary phase condition is

$$\frac{d}{dk}\left(-kx - \frac{\hbar^2 k^2}{2m}\frac{t}{\hbar}\right)\bigg|_{k_0} = 0 \quad \Rightarrow \quad x + \frac{\hbar k_0}{m} t = 0. \tag{8.2.9}$$

This gives the

$$\text{reflected wave peak:} \quad x = -\frac{\hbar k_0}{m} t. \tag{8.2.10}$$

This relation represents a peak moving with constant negative velocity $-\hbar k_0/m$. Since $\Psi_{\rm ref}(x, t)$ only exists for $x < 0$, the above condition shows that we get the peak only for $t > 0$, as befits a reflected wave arising from an incident wave that hits the step at $t = 0$.

For $t < 0$, $\Psi_{\rm ref}(x, t)$ is not zero, but it must be rather small since the (x, t) relation above from the stationary phase condition cannot be satisfied for any x in the domain $x < 0$.

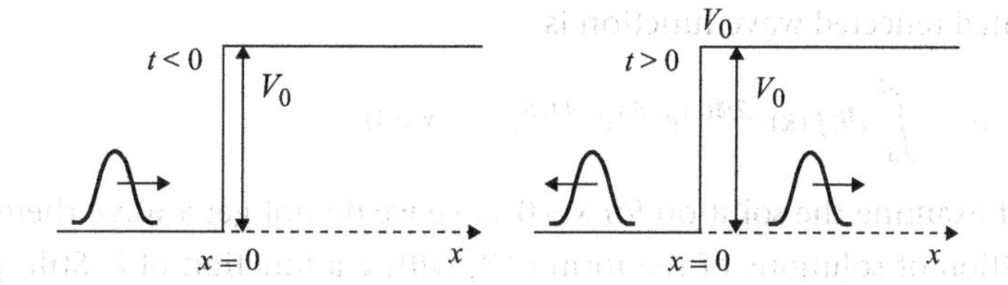

Figure 8.7
Left: At large negative times, an incoming wave packet is traveling in the $+x$ direction. The plot is that of $|\Psi(x,t)|$ for $t \ll 0$. *Right*: At large positive times, we have a reflected wave packet traveling in the $-x$ direction and a transmitted wave packet traveling in the $+x$ direction. The plot is that of $|\Psi(x,t)|$ for $t \gg 0$.

For small negative time, Ψ_{ref} describes the reflection of the leading tail of the incident wave packet that reaches the origin before time zero.

Finally, let us consider the motion of the peak in the transmitted wave $\Psi_{\text{tr}}(x,t)$. The stationary phase condition reads

$$\frac{d}{dk}\left(\bar{k}x - \frac{\hbar^2 k^2}{2m}\frac{t}{\hbar}\right)\bigg|_{k_0} = 0 \quad\Rightarrow\quad \frac{d\bar{k}}{dk}\bigg|_{k_0} x - \frac{\hbar k_0}{m} t = 0. \tag{8.2.11}$$

Using the relation (8.1.4) between k and $\bar{k}$, we have

$$\bar{k}^2 = k^2 - \frac{2mV_0}{\hbar^2} \quad\Rightarrow\quad \frac{d\bar{k}}{dk} = \frac{k}{\bar{k}} \quad\Rightarrow\quad \frac{d\bar{k}}{dk}\bigg|_{k_0} = \frac{k_0}{\bar{k}(k_0)}, \tag{8.2.12}$$

where $\bar{k}(k_0)$ is the value of $\bar{k}$ for $k = k_0$. Back to (8.2.11), we immediately find the

$$\text{transmitted wave peak:} \quad x = \frac{\hbar\bar{k}(k_0)}{m} t. \tag{8.2.13}$$

Since $x > 0$ is the domain of Ψ_{tr}, this describes a peak moving to the right with velocity $\hbar\bar{k}/m$ for $t > 0$. For $t < 0$, $\Psi_{\text{tr}}(x,t)$ is not zero, but it must be rather small since the (x,t) relation above cannot be satisfied for any x in the domain $x > 0$. For small negative time, Ψ_{tr} describes the transmission of the leading edge of the incident wave packet that reaches the origin before time zero.

In summary, for large negative time Ψ_{inc} dominates, and both Ψ_{ref} and Ψ_{tr} are very small. For large positive time, both Ψ_{ref} and Ψ_{tr} dominate, and Ψ_{inc} becomes very small. These situations are sketched in figure 8.7. Of course for small times, positive or negative, all three waves exist, and together they describe the complex process of collision with the step, in which a reflected and a transmitted wave are generated from the incident wave.

Let us now examine a wave packet built with energies $E < V_0$. Recall that in this situation $B/A = -e^{2i\delta(E)}$. Therefore, for an incident wave, all of whose momentum components have energy less than V_0,

$$\Psi_{\text{inc}}(x,t) = \int_0^{\hat{k}} dk\, f(k)\, e^{ikx} e^{-iEt/\hbar}, \qquad x < 0. \tag{8.2.14}$$

The associated reflected wave function is

$$\Psi_{\text{ref}}(x,t) = -\int_0^{\hat{k}} dk\, f(k)\, e^{2i\delta(E)} e^{-ikx} e^{-iEt/\hbar}, \qquad x<0. \tag{8.2.15}$$

We will not examine the solution for $x>0$ since we do not get a wave there but rather a superposition of solutions of the form $e^{-\kappa x}$, with κ a function of k. Still, you should write the solution out.

Exercise 8.3. *Use the result of exercise 8.2 to write the transmitted solution* $\Psi_{\text{tr}}(x,t)$.

The peak in the incident wave moves as determined earlier. Using the stationary phase condition again to find the motion of the peak in $|\Psi_{\text{ref}}|$, this time we find

$$\frac{d}{dk}\left(2\delta(E) - kx - \frac{Et}{\hbar}\right)\Bigg|_{k_0} = 0 \;\Rightarrow\; 2\delta'(E)\frac{\hbar^2 k_0}{m} - x - \frac{\hbar k_0 t}{m} = 0. \tag{8.2.16}$$

Here $\delta'(E)$ denotes the derivative of δ with respect to its argument E. From this we quickly find the

$$\text{reflected wave peak:}\quad x = -\frac{\hbar k_0}{m}\left(t - 2\hbar\,\delta'(E)\right), \quad (E<V_0), \tag{8.2.17}$$

where the derivative is evaluated at $E(k_0)$. The reflected wave packet is moving toward more negative x as time grows positive. This is as it should be. But there is a time delay associated with the reflected packet that is evident when we compare the above equation with $x = -\frac{\hbar k_0}{m}t$. In the latter the packet is emerging from $x=0$ at $t=0$. In the former the packet emerges from $x=0$ at $t = 2\hbar\,\delta'(E)$. The time delay Δt is then

$$\boxed{\Delta t = 2\hbar\,\delta'(E).} \tag{8.2.18}$$

Making use of the expression (8.1.24) for $\delta(E)$, the derivative with respect to the energy is readily computed:

$$\delta'(E) \equiv \frac{d\delta(E)}{dE} = \frac{1}{2}\sqrt{\frac{1}{E(V_0-E)}}. \tag{8.2.19}$$

The derivative is positive and becomes infinite both for $E\to 0$ and for $E\to V_0$ (figure 8.8). The delay can be large for wave packets of little energy or those with energies just below V_0.

We conclude this section with some observations about particles in the forbidden region. For energy eigenstates with $E<V_0$ and for the wave packets built as a superposition of such eigenstates, there is some nonvanishing probability density in the forbidden region $x>0$. Such nontrivial probability density implies some likelihood of detecting the particle in the forbidden region. If so, it would seem that such a particle would have an unphysical negative kinetic energy. This conclusion does not follow in quantum mechanics, however. Once we measure the particle position, the wave function collapses to become narrowly localized within the forbidden region, in a region the

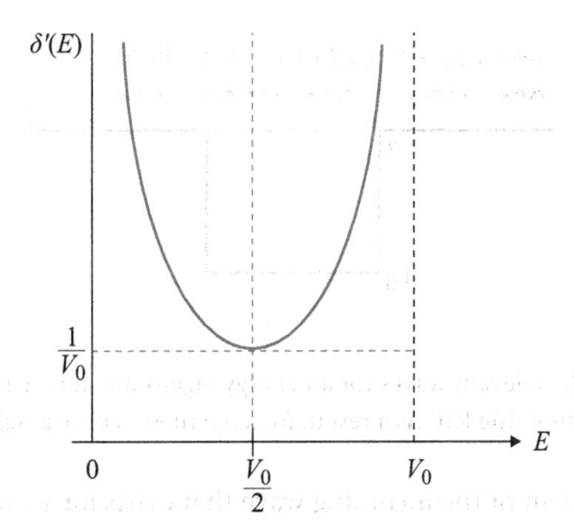

Figure 8.8
The derivative $\delta'(E)$ of the phase shift for a step potential of height V_0.

size of the detector resolution. This means that the collapsed wave function is *no longer* an energy eigenstate. The measured value of the energy becomes ambiguous. For any localized wave function, moreover, the expected value $\langle \frac{\hat{p}^2}{2m} \rangle$ of the kinetic energy is always positive.

Exercise 8.4. *Suppose the state of a particle of mass m is a Gaussian with position uncertainty* $\Delta x \equiv \Delta$. *The wave function* $\psi(x)$ *then takes the form* $\psi(x) = N \exp\left(-x^2/(4\Delta^2)\right)$. *The uncertainty in momentum is then* $\Delta p = \frac{\hbar}{2\Delta}$. *What is the expectation value* $\langle \frac{\hat{p}^2}{2m} \rangle$ *of the kinetic energy?*

8.3 Resonant Transmission in a Square Well

Consider once again the attractive finite square well with $V_0 > 0$:

$$V(x) = \begin{cases} 0, & \text{for } |x| > a, \\ -V_0, & \text{for } |x| < a. \end{cases} \tag{8.3.1}$$

We have already studied the bound states of this potential. We now consider energy eigenstates with $E > 0$; that is, nonnormalizable states that can be used to build normalizable wave packets approaching the well and scattering off of it. For any $E > 0$, there are two degenerate eigenstates. We can choose one to represent a wave incoming from the left and another to represent a wave incoming from the right. We now write an ansatz for the eigenstate that represents a wave incident from the left:

$$\psi(x) = \begin{cases} Ae^{ikx} + Be^{-ikx}, & x < -a, \\ Ce^{ik_2x} + De^{-ik_2x}, & |x| < a, \\ Fe^{ikx}, & x > a. \end{cases} \tag{8.3.2}$$

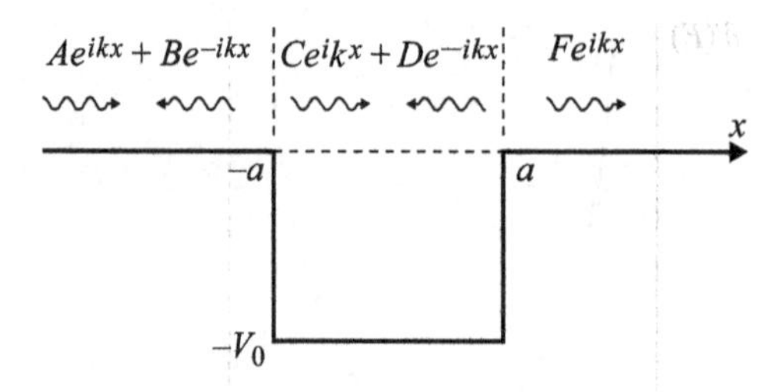

Figure 8.9
The square well with the relevant waves for an energy eigenstate in which we have a wave, with coefficient A, incident from the left. As a result, for $x > a$ there is only a right-moving wave.

Here, A is the coefficient of the incoming wave that exists for $x < a$, and B is the coefficient of the reflected wave that exists in that same region (see figure 8.9). Both of these waves have wave number k. In the well region $|x| < a$, we have a wave moving to the right with coefficient C and a wave moving to the left with coefficient D. The wave number in this region is called k_2. Consistent with an incoming wave from the left, to the right of the well we have just one wave moving to the right. Its coefficient is F and its wave number k. Even though the potential is an even function of x, a nonnormalizable energy eigenstate need not be even or odd. In our ansatz (8.3.2), the symmetry is broken by the condition that our wave is incident from the left. The values of k and k_2, both positive, are determined by the Schrödinger equation and are

$$k = \sqrt{\frac{2mE}{\hbar^2}}, \quad k_2 = \sqrt{\frac{2m(E+V_0)}{\hbar^2}}. \tag{8.3.3}$$

There are four boundary conditions: the continuity of ψ and ψ' at $x = -a$ and at $x = a$. These four equations can be used to fix the coefficients B, C, D, and F in terms of A. We define reflection and transmission coefficients R and T as

$$R \equiv \frac{|B|^2}{|A|^2}, \quad T \equiv \frac{|F|^2}{|A|^2}. \tag{8.3.4}$$

From probability current conservation, we know that the currents to the left and to the right of the well must be equal so that

$$|A|^2 - |B|^2 = |F|^2. \tag{8.3.5}$$

This is not an independent equation; it must follow from the boundary conditions. It implies that

$$R + T = \frac{|B|^2}{|A|^2} + \frac{|F|^2}{|A|^2} = 1, \tag{8.3.6}$$

showing that our definition of R and T makes sense.

Solving for R and T is straightforward but laborious. Let us just quote the answer one gets. The transmission coefficient T is the following function of the energy E of the eigenstate:

$$\frac{1}{T} = 1 + \frac{1}{4}\frac{V_0^2}{E(E+V_0)}\sin^2(2k_2a). \tag{8.3.7}$$

Since the second term on the right-hand side is manifestly positive, we have $T \leq 1$. As $E \to 0$, and assuming $\sin^2(2k_2a) \neq 0$ (more on this below), we have $\frac{1}{T} \to 1 + \infty$, which means $T \to 0$. As $E \to \infty$, we have $T \to 1$.

We can remove all units from this result by defining

$$e \equiv \frac{E}{V_0}, \quad z_0^2 \equiv \frac{2ma^2 V_0}{\hbar^2}, \tag{8.3.8}$$

where z_0 is the familiar parameter that characterizes finite square wells. Then,

$$(k_2a)^2 = \frac{2ma^2(E+V_0)}{\hbar^2} = \frac{2ma^2 V_0}{\hbar^2}(1+e) \quad \Rightarrow \quad 2k_2a = 2z_0\sqrt{1+e}, \tag{8.3.9}$$

so we have

$$\frac{1}{T} = 1 + \frac{1}{4e(1+e)} \sin^2(2z_0\sqrt{1+e}). \tag{8.3.10}$$

The well becomes transparent, making $T=1$ for certain values of the energy. All we need is for the argument of the sine function to be a multiple of π:

$$2z_0\sqrt{1+e} = n\pi, \quad n \in \mathbb{Z}. \tag{8.3.11}$$

Not all integers are allowed. Because $e > 0$, the left-hand side is bigger than or equal to $2z_0$, and therefore

$$n \geq \frac{2z_0}{\pi}. \tag{8.3.12}$$

We call E_n the energies associated with $T=1$ and write $E_n = e_n V_0$. Then from (8.3.11) we have

$$e_n + 1 = \frac{n^2\pi^2}{4z_0^2} = \frac{n^2\pi^2\hbar^2}{2m(2a)^2 V_0} \tag{8.3.13}$$

so that

$$E_n + V_0 = \frac{n^2\pi^2\hbar^2}{2m(2a)^2}. \tag{8.3.14}$$

Note that $E_n + V_0$ is the energy of the scattering state measured with respect to the bottom of the square well. The right-hand side is the energy of the nth bound state of the *infinite* square well of width $2a$ (see (6.4.11)). We thus have a rather surprising result: *we get full transmission for those positive energies E_n such that $E_n + V_0$ are in the spectrum of the infinite square well extension of our finite square well.* The infinite square well extension of the finite square is shown in figure 8.10. The inequality $n \geq \frac{2z_0}{\pi}$ guarantees that $E_n > 0$. Since infinite square well bound states are characterized by fitting an integer number of half wavelengths, we have a resonance-type situation in which perfect transmission is happening when the scattering waves of wave number k_2 fit perfectly inside the *finite* square well. The phenomenon we observe is called *resonant transmission.*

The fitting of an exact number of half wavelengths can also be seen directly from the vanishing of the sine function in (8.3.7), giving $k_2(2a) = n\pi$. Writing $k_2 = 2\pi/\lambda_2$, where

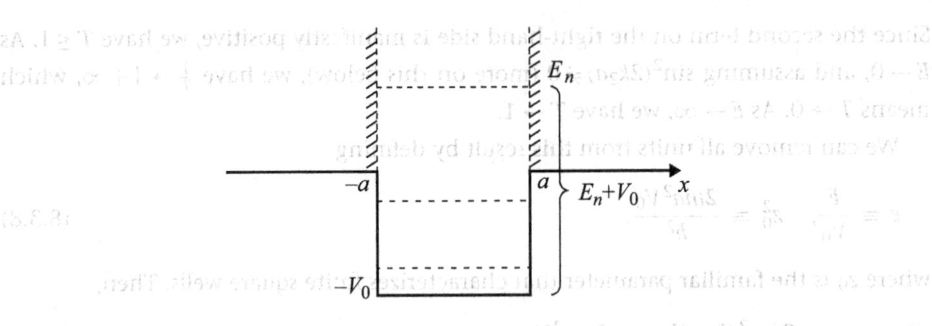

Figure 8.10
We get resonant transmission across the finite well at the positive bound state energies of a would-be infinite well whose bottom coincides with that of the finite square well.

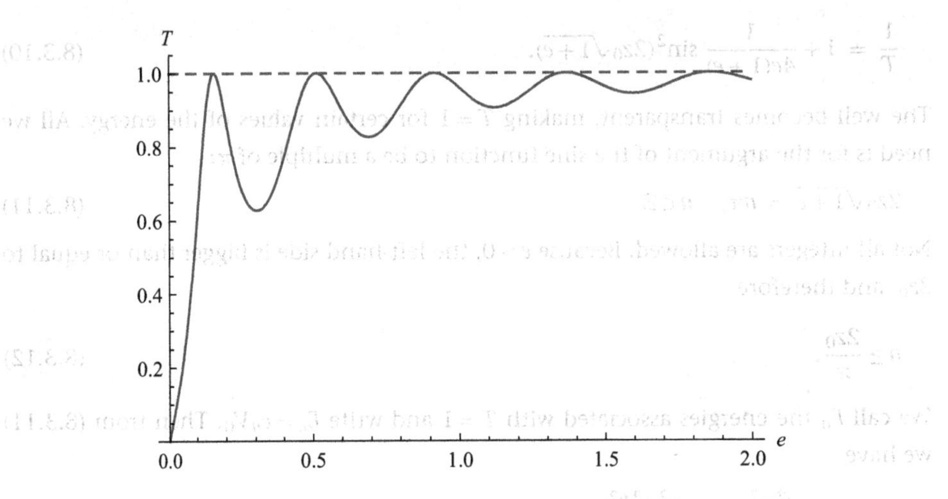

Figure 8.11
The transmission coefficient T as a function of $e = E/V_0$ for a square well with $z_0 = 13\pi/4$. At the energies for which $T = 1$, we have resonant transmission. There are three resonances for $0 < E < V_0$; for $E_7 \simeq 0.16V_0$, for $E_8 \simeq 0.51V_0$, and for $E_9 \simeq 0.92V_0$. Note that the spacing between the $T = 1$ points grows with e.

λ_2 is the de Broglie wavelength on the well, we find that

$$\frac{2\pi}{\lambda_2}(2a) = n\pi \quad \Rightarrow \quad 2a = n\frac{\lambda_2}{2}. \tag{8.3.15}$$

We show in figure 8.11 the transmission coefficient T as a function of $e = E/V_0$ for a square well with $z_0 = 13\pi/4$. In this case we must have $n \geq \frac{13}{2}$ or $n \geq 7$.

The Ramsauer-Townsend effect Carl Ramsauer and, separately, John Sealy Townsend did seminal work in the early 1920s when they examined the elastic scattering of low-energy electrons by noble gas atoms. These atoms have fully filled electronic shells, are very unreactive, and have high ionization energies. The potential created by the nucleus becomes visible as the incident electron penetrates the electron cloud. This potential is a spherically symmetric, attractive potential for the electrons—some kind

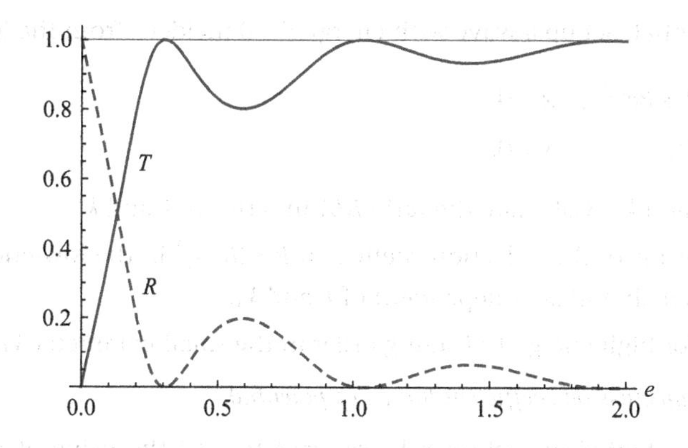

Figure 8.12
A sketch of reflection and transmission coefficients R (*dashed line*) and T (*continuous line*), respectively, as a function of energy for the Ramsauer-Townsend effect. Note that $R+T=1$. For $R=0$, there is no scattering; the electrons go straight through the noble gas atoms via resonant transmission. This first occurs for energies of about 1 eV. In this plot (done with $z_0=7\pi/4$) the lowest energy for which $R=0$ is about $0.3V_0$.

of finite spherical well. In the experiment, some electrons collide with the atoms and scatter, mostly bouncing back. With the benefit of hindsight, we can view their results as information on the reflection coefficient R for electrons scattering off a potential well!

Ramsauer and Townsend reported a very unusual phenomenon, separately publishing the results of their investigations in 1921. At very low energies, the scattering cross section was high. But the energy dependence was surprising. As the energy was increased, the scattering went down toward zero, to go up again as the energy was increased further. Such mysterious behavior had no sensible classical explanation. What one has at play is quantum-mechanical *resonant scattering*. The scattering going to zero means the reflection coefficient going to zero and the transmission coefficient going to one! The first resonant transmission happens for electrons of around one electron volt (such electrons have a speed of about 600 km/s). Figure 8.12 provides a sketch of both R and T as a function of energy. Our one-dimensional square well potential does not provide a good quantitative match for the data, but it illustrates the physical phenomenon. A three-dimensional spherical well is needed for a quantitative analysis.

Problems

Problem 8.1. *Step-down potential.*

In this problem we show that in quantum mechanics a particle can bounce off a step down. For this, consider the $V_0>0$ step-down potential:

$$V(x)=\begin{cases} 0, & x<0, \\ -V_0, & x>0. \end{cases}$$

Sketch the potential. Set up a wave with energy $E > 0$ incident from the left with

$$\psi(x) = \begin{cases} Ae^{ikx} + Be^{-ikx}, & x < 0, \\ Ce^{i\bar{k}x}, & x > 0. \end{cases}$$

1. What are k^2 and $\bar{k}^2$? Calculate the ratio B/A in terms of k and $\bar{k}$.
2. What is the value of the reflection coefficient $R = |B/A|^2$ in the low-energy limit $k \to 0$? Your answer should be independent of E and V_0.
3. Calculate R for high energy to leading order in the small parameter V_0/E.

Problem 8.2. *Transmission coefficient for a step potential.*

In the step-up potential of section 8.1, we want to find the value of $E/V_0 > 1$ for a given value of the transmission coefficient T or, alternatively, the reflection coefficient $R = 1 - T$.

1. Let $x \equiv \bar{k}/k < 1$ and show that the reflection coefficient $R = 1 - T$ is related to x as follows:

$$R = \left(\frac{1-x}{1+x}\right)^2.$$

2. Use this to show that

$$\frac{E}{V_0} = \frac{(1+\sqrt{R})^2}{4\sqrt{R}}.$$

3. Find E/V_0 for $T = R = 1/2$ (try to guess the answer before computing). What is E/V_0 in order to have a 99% probability of transmission?

Problem 8.3. *Reflection of a wave packet off a step potential.*

Consider a step potential with step height V_0:

$$V(x) = \begin{cases} V_0, & x > 0, \\ 0, & x < 0. \end{cases}$$

We send in from $x = -\infty$ a wave packet, all of whose momentum components have energies less than the energy V_0 of the step so that

$$k \leq \hat{k}, \quad \hat{k}^2 = \frac{2mV_0}{\hbar^2}.$$

We will then write the incident wave packet as

$$\Psi_{\text{inc}}(x, t) = \sqrt{a} \int_0^{\hat{k}} dk\, \Phi(k) e^{ikx} e^{-iE(k)t/\hbar}, \quad x < 0.$$

Here a is the constant with units of length uniquely determined by the constants m, V_0, and $\hbar$ in this problem, and $\Phi(k)$ is a real, unit-free function that has peaked at some $k_0 < \hat{k}$:

$$a = \frac{\hbar}{\sqrt{mV_0}}, \quad \Phi(k) = e^{-\beta^2 a^2 (k-k_0)^2}.$$

The real constant β, to be fixed below, controls the width of the momentum distribution. The units of Ψ_{inc} above are $L^{-1/2}$, thus the $\sqrt{a}$ prefactor. Recall that dk has units of L^{-1}.

1. Write the reflected wave function $\Psi_{\text{ref}}(x,t)$ (valid for $x<0$) as an integral, similar to Ψ_{inc}. This integral involves the known phase shift $\delta(E)$ for the potential.
2. Introduce a unit-free version K of the wave number k, a unit-free version u of the coordinate x, and a unit-free version τ of the time t as follows:

$$k=\frac{K}{a},\quad x=au,\quad t=\frac{\hbar}{V_0}\tau.$$

Naturally, we will write $k_0=K_0/a$ and $\hat{k}=\hat{K}/a$. Note that $kx=Ku$.

Show that the group velocity $\frac{dx}{dt}$ (measuring the speed of the center of the wave packet) and the uncertainty relation $\Delta x\Delta p\geq \hbar/2$ for the incoming packet can be written in the form

$$\frac{du}{d\tau}=AK_0,\quad \Delta u\,\Delta K\geq B,$$

where A and B represent *numerical* constants that you should find.

Determine ΔK in terms of β, using the approximation that instead of truncating Φ at $k=\hat{k}$, Φ is actually the full Gaussian. (It may help to know that the Gaussian wave function with position uncertainty Δ takes the form $\psi(x)\sim\exp(-x^2/(4\Delta^2))$.)

Given your expression for ΔK, what is the minimum possible uncertainty Δu_{min} (using the same assumption about Φ being the full Gaussian)?

3. Complete the following equations by fixing the constants A, B, C, D, and F:

$$E(k)=AV_0K^2,$$

$$e^{2i\delta(E)}=B+CK^2+iK\sqrt{D+FK^2}\equiv\omega(K).$$

(Note that these constants are different from those in (2).)

4. Show that the delay $\Delta t=2\hbar\delta'(E(k_0))$ experienced by the reflected wave implies a dimensionless time delay $\Delta\tau$ given by

$$\Delta\tau=\frac{2}{K_0\sqrt{B+CK_0^2}},$$

where $K_0=ak_0$. Find the constants B and C (again, different from those in (2) and (3)).

5. Prove that the complete wave function $\Psi(x,t)$ that is valid for $x<0$ and all times, which we now view as $\Psi(u,\tau)$ valid for $u<0$ and all τ, takes the form

$$a^{1/2}\Psi(u,\tau)=\int_0^A dK\,e^{-\beta^2(K-K_0)^2}e^{-iBK^2\tau}\left(e^{iKu}-e^{-iKu}\omega(K)\right)$$

and determine the two missing constants A and B.

6. Set $\beta = 4$ and $K_0 = 1$. What are the values of ΔK and $\Delta u_{\min}$? What is the predicted time delay $\Delta\tau$? Estimate how rapidly the wave packet will spread out: How long in τ would it take for its position uncertainty to double?

   You have the integral form for $\Psi(u, \tau)$, but the integrals are too difficult to evaluate. Write a program to plot $a|\Psi|^2$ at different times (using the chosen values of β and K_0). Observe what happens as you step forward in time from around $\tau = -20$ to $\tau = 20$.

7. Numerically determine the time delay $\Delta\tau$ by tracking the position of the peak of the packet from $\tau = -20$ to $\tau = 20$.

Problem 8.4. *Scattering off a rectangular barrier.*

Consider a rectangular barrier

$$V(x) = \begin{cases} V_0, & \text{for } |x| \leq a, \quad (V_0 > 0), \\ 0, & \text{for } |x| > a. \end{cases}$$

1. A wave of energy $E = V_0$ is incident from the left. Compute the transmission coefficient T. Recalling that $z_0^2 = 2mV_0 a^2/\hbar^2$, write your answer in terms of z_0.
2. When $E > V_0$, the transmission coefficient T is given by

$$\frac{1}{T} = 1 + \frac{V_0^2}{4E(E - V_0)} \sin^2\left(\frac{2a}{\hbar}\sqrt{2m(E - V_0)}\right).$$

   What is the condition on E for *resonant transmission*—that is, $T = 1$?

Problem 8.5. *A numerical test of stationary phase.*

We have used the stationary phase condition to figure out the time dependence of the position of peaks in wave packets constructed from integral representations. More generally, the stationary phase approximation can help find the value of the integral itself.

Consider the integral of a Gaussian, peaked at $x = 2$, against a phase factor with phase $\phi(\lambda, x)$:

$$f(\lambda) = \int_{-\infty}^{\infty} dx\, e^{-100(x-2)^2} e^{i\phi(\lambda,x)}, \quad \phi(\lambda, x) = 50\left(x - \tfrac{1}{32}\lambda x^4\right), \quad \lambda \in \mathbb{R}.$$

1. What is the width Δ at half maximum for the Gaussian? In other words, what is the largest Δ for which for all x in $|x - 2| \leq \frac{1}{2}\Delta$ the Gaussian is larger than half maximum? If you had to determine the integral numerically, would it be safe to integrate from 1 to 3?
2. Use the stationary phase method to find the critical value λ_* of λ for which $f(\lambda)$ would have the largest magnitude. Write $\phi(\lambda_*, x)$ as a Taylor expansion around $x = 2$ up to and including terms that are quadratic in $(x - 2)$.
3. What is the excursion $\Delta\phi$ of the phase $\phi(\lambda_*, x)$ for $|x - 2| < \frac{1}{2}\Delta$? That is, in the interval $x = 2 \pm \Delta/2$, what is the maximum value of ϕ minus the minimum value of ϕ? The excursion gives you an idea of how much ϕ is varying over the important range of the

integral. Make sure to use the exact formula for $\phi(\lambda_*, x)$, rather than the approximate version found in (2).

Your result, expressed in units of π, should imply that it is a decent approximation to ignore the phase *variation* at the critical λ. By assuming that the phase is constant in x at $\lambda = \lambda_*$, perform the resulting integral analytically. The answer $f(\lambda_*)$ is a complex number. Write your answer as a magnitude times a complex phase in the range $[0, 2\pi]$.

4. Perform the integral analytically using the quadratic approximation for the phase. Write your answer in terms of a magnitude times a complex phase (in the range $[0, 2\pi]$) with four significant digits of accuracy.
5. Write a program to evaluate numerically $f(\lambda)$ for arbitrary λ.

   Find the value of $f(\lambda_*)$, where λ_* is the value you estimated previously. Write your answer in terms of a magnitude times a complex phase (in the range $[0, 2\pi]$) with four significant digits of accuracy. How does your numerical result compare with your quadratic estimate?

   Plot $|f(\lambda)|$ and find numerically, to four significant digits, the value of λ that actually maximizes $|f(\lambda)|$.

Problem 8.6. *Resonant transmission across two delta functions.*

Consider a potential with two positive-strength delta functions, one at $x = -a$ and another at $x = a$:

$$V(x) = g\delta(x+a) + g\delta(x-a), \quad g > 0.$$

Note the unit-free combination $\lambda = \frac{mag}{\hbar^2} \geq 0$ that represents the effective strength of the potential. In solving the general scattering problem of a particle incident from the left, one sets up a wave function

$$\psi(x) = \begin{cases} Ae^{ikx} + Be^{-ikx}, & x < -a, \\ Ce^{ikx} + De^{-ikx}, & |x| < a, \\ Fe^{ikx}, & x > a. \end{cases}$$

Here A, B, C, D, F are complex constants that must be adjusted for this to be a solution of the time-independent Schrödinger equation. We are interested in finding the energies for which there is resonant transmission; namely, the energies that make the transmission coefficient equal to one.

1. Which of the complex constants in the above ansatz for ψ should vanish for resonant transmission? Explain briefly.
2. Assume this constant vanishes, and find the four equations that implement the boundary conditions. Clean them up and put them in the form

$$C + De^{\cdots} = \cdots$$

$$C + De^{\cdots} = \cdots$$

$$C - De^{\cdots} = \cdots$$

$$C - De^{\cdots} = \cdots$$

The expressions indicated by dots should be written in terms of ka, λ, constants in the ansatz for ψ, and numerical constants.

3. Show that the existence of a solution for the above equations requires

$$\xi \cot \xi = -2\lambda, \quad \text{with} \quad \xi = 2ka. \tag{1}$$

Sketch a plot of $\xi \cot \xi$ for $\xi \in [0, 3\pi]$. Show the -2λ line in the plot for both very small and very large λ. For $\lambda \ll 1$, what are the approximate values of ka for perfect transmission? For $\lambda \gg 1$, what are the approximate values of ka for perfect transmission?

4. Under condition (1), one can prove that

$$\frac{C}{D} = -\frac{1}{\cos(2ka)}, \quad C = \left(1 + \frac{\lambda}{ika}\right)A.$$

Consider the case of $\lambda \gg 1$ and the first resonant transmission. Find an approximate formula for ψ in the region $|x| < a$. Setting $A = 1$, do a rough plot of $|\psi(x)|^2$ for all x. Comment on the features of your plot.

9 Harmonic Oscillator

We use the classical harmonic oscillator as inspiration to define the quantum harmonic oscillator. We search for energy eigenstates by looking into the time-independent Schrödinger equation, a second-order differential equation solved by a series expansion after isolating the large-argument behavior of the solution. Normalized wave functions exist for quantized values of the energy and are written in terms of Hermite polynomials. We examine a factorization of the Hamiltonian that yields the spectrum by considering a first-order differential equation. We introduce creation and annihilation operators, which allow for a compact presentation of the energy eigenstates.

9.1 Harmonic Oscillator

The classical harmonic oscillator is a rich and interesting dynamic system. It governs many kinds of oscillations in complex systems. The simplest example is that of a particle of mass m moving in one dimension under the action of a restoring force $F=-kx$, with $k>0$. In this case the total energy E of the particle is given by

$$E = \tfrac{1}{2}mv^2 + \tfrac{1}{2}kx^2. \tag{9.1.1}$$

The first term is the kinetic energy, and the second term is the potential energy:

$$V(x) = \tfrac{1}{2}kx^2. \tag{9.1.2}$$

The potential is quadratic in x. In such a system, the particle performs oscillatory motion with angular frequency ω given by

$$\omega = \sqrt{\frac{k}{m}} \quad \Rightarrow \quad k = m\omega^2. \tag{9.1.3}$$

Trading k for ω and using the momentum to express the kinetic energy, we can rewrite the energy E of the classical oscillator as follows:

$$E = \frac{p^2}{2m} + \tfrac{1}{2}m\omega^2 x^2. \tag{9.1.4}$$

The quadratic potential is ubiquitous in physics, as it generally arises to first approximation when we expand an *arbitrary* potential around a minimum. To show this consider an arbitrary potential $V(x)$ with a minimum at x_0. For x near x_0, we can use a Taylor expansion to write

$$V(x) = V(x_0) + (x - x_0)V'(x_0) + \tfrac{1}{2}(x - x_0)^2 V''(x_0) + \mathcal{O}((x - x_0)^3). \tag{9.1.5}$$

Since x_0 is a critical point, $V'(x_0) = 0$. Dropping the higher-order terms, we find that the potential is approximately quadratic:

$$V(x) \simeq V(x_0) + \tfrac{1}{2}V''(x_0)(x - x_0)^2. \tag{9.1.6}$$

This is a good approximation for x close to x_0. Since x_0 is a minimum, $V''(x_0) > 0$, and this is a harmonic oscillator centered at x_0 and with $k = V''(x_0)$. The additive constant $V(x_0)$ has no effect on the dynamics.

Faced with the question of defining a *quantum* harmonic oscillator, we are inspired by expression (9.1.4) for the energy. We *declare* that $\hat{x}$ and $\hat{p}$ will be operators with the commutation relation $[\hat{x}, \hat{p}] = i\hbar$ and that the Hamiltonian $\hat{H}$ will be given by

$$\hat{H} = \frac{\hat{p}^2}{2m} + \tfrac{1}{2}m\omega^2\hat{x}^2, \qquad [\hat{x}, \hat{p}] = i\hbar. \tag{9.1.7}$$

The harmonic oscillator potential in here is

$$V(x) = \tfrac{1}{2}\, m\omega^2 x^2. \tag{9.1.8}$$

The quantum harmonic oscillator is a natural extension of the classical oscillator. As in the classical theory, ω has units of frequency: $[\omega] = 1/T$. We can use ω and $\hbar$ to construct a

$$\text{characteristic energy: } \hbar\omega. \tag{9.1.9}$$

This is in fact the unique way we construct an energy with the physical constants $(\hbar, m, \omega)$ available in the harmonic oscillator.

Given this quantum system, we begin by finding the energy eigenstates. As usual, we set

$$\Psi(x, t) = \varphi(x)\, e^{-iEt/\hbar}, \tag{9.1.10}$$

where E is the energy of the eigenstate, and we use $\varphi(x)$ instead of $\psi(x)$ to denote the spatial wave functions of the oscillator. The time-independent Schrödinger equation then reads

$$-\frac{\hbar^2}{2m}\frac{d^2\varphi(x)}{dx^2} + \frac{1}{2}m\omega^2 x^2 \varphi(x) = E\varphi(x). \tag{9.1.11}$$

Here both E and $\varphi(x)$ are unknown. We expect that energy eigenstates only exist for certain quantized values of E. This is a differential equation we must solve.

As a first step, we will clean the equation of dimensional constants since this helps us appreciate its structure. As we discussed in section 7.6, if you aim to put the equation into a computer it would be cumbersome to deal with $\hbar, m$, and ω. By cleaning up the equation, we demonstrate that the dependence of observables on these constants is governed by dimensional analysis. Moreover, our analytical work will be much simpler.

The key, as indicated in section 7.6, is to work with a unit-free coordinate u instead of x. In that case the units of energy are produced by the constants in the problem, and as we have seen, the only possibility is $\hbar\omega$. A common $\hbar\omega$ factor will then allow us to replace E for a unit-free energy. We therefore begin by introducing a unit-free coordinate u to replace the conventional coordinate x. We set

$$u \equiv \frac{x}{L_0}, \qquad u \text{ unit-free,} \tag{9.1.12}$$

where L_0 must be a constant with units of length. To construct L_0 in terms of $\hbar, m$, and ω, we equate a characteristic kinetic energy to a characteristic potential energy:

$$\frac{\hbar^2}{mL_0^2} = m\omega^2 L_0^2. \tag{9.1.13}$$

This fixes L_0 to be the

$$\text{oscillator size } L_0: \quad L_0 = \sqrt{\frac{\hbar}{m\omega}}. \tag{9.1.14}$$

The length scale L_0, to be called the **oscillator size**, is as relevant to the harmonic oscillator as the Bohr radius a_0 is to the hydrogen atom. As we will see below, the ground state energy of the quantum oscillator is $\hbar\omega/2$, and as it turns out, L_0 is precisely the classical oscillation amplitude of an oscillator with that energy.

Exercise 9.1. *Show that a classical oscillator with oscillation amplitude L_0 has energy $\hbar\omega/2$.*

Plugging $x = L_0 u$ into the time-independent Schrödinger equation yields

$$-\frac{\hbar^2}{2mL_0^2}\frac{d^2\varphi(u)}{du^2} + \frac{1}{2}m\omega^2 L_0^2 u^2 \varphi(u) = E\varphi(u). \tag{9.1.15}$$

Now note that

$$\frac{\hbar^2}{mL_0^2} = m\omega^2 L_0^2 = \hbar\omega \tag{9.1.16}$$

so that the differential equation becomes

$$-\tfrac{1}{2}\hbar\omega\frac{d^2\varphi(u)}{du^2} + \tfrac{1}{2}\hbar\omega u^2\varphi(u) = E\varphi(u). \tag{9.1.17}$$

Multiplying by $\frac{2}{\hbar\omega}$, we find that

$$-\frac{d^2\varphi(u)}{du^2} + u^2\varphi(u) = \mathcal{E}\varphi(u), \tag{9.1.18}$$

where we have defined a unit-free energy $\mathcal{E}$:

$$\mathcal{E} \equiv \frac{2E}{\hbar\omega} \quad \Rightarrow \quad E = \tfrac{1}{2}\hbar\omega\,\mathcal{E}. \tag{9.1.19}$$

If we know the pure number $\mathcal{E}$, we know the energy E. Rearranging, we reach the cleaned-up, unit-free version of the time-independent Schrödinger equation:

$$\boxed{\frac{d^2\varphi}{du^2} = (u^2 - \mathcal{E})\,\varphi\,.} \tag{9.1.20}$$

It is clearly less cluttered than (9.1.11), and all dimensional constants have disappeared. Of course, if we solve the equation to find $\varphi(u)$, then to go to physical variables we use

$$u = \frac{x}{L_0} = x\sqrt{\frac{m\omega}{\hbar}}\,. \tag{9.1.21}$$

This is how the dimensionful quantities appear in the solution. They also appear in the energy E expressed in terms of the dimensionless energy parameter $\mathcal{E}$.

Exercise 9.2. *Consider (9.1.20) and examine a solution of the form $\varphi = \exp(\alpha u^2/2)$, with constant α. What values of α provide a solution to the differential equation? What is the dimensionless energy $\mathcal{E}$ of the solution that has a normalizable wave function?*

9.2 Solving the Harmonic Oscillator Differential Equation

The boxed differential equation (9.1.20) has solutions for all values of the energy parameter $\mathcal{E}$; after all you could integrate it on a computer! Quantization arises because solutions are not normalizable except for special values of $\mathcal{E}$. To understand this issue, we examine solutions for large values of $|u|$, particularly those for which $u^2 \gg \mathcal{E}$. In this limit, $\mathcal{E}$ can be ignored, and we have the approximate equation

$$\varphi''(u) \simeq u^2\varphi(u). \tag{9.2.1}$$

This equation cannot be solved by any polynomial (recall that any polynomial has a maximal degree). If φ is a polynomial of degree n, the degree of the left-hand side would be $n-2$ and that of the right-hand side would be $n+2$. This shows that a polynomial does not give a solution; the solution must be nonpolynomial. Let us therefore try a solution of the form

$$\varphi(u) = u^k e^{\alpha u^2/2}, \tag{9.2.2}$$

where k is an integer, α a constant, and both are unknown. The exponential factor here is nonpolynomial. Since u is large, the leading term in $\varphi''(u)$ arises when we differentiate the exponential, as this multiplies the function by a factor of u:

$$\varphi''(u) \simeq \alpha^2 u^2 \varphi(u) \quad \text{as } |u| \to \infty. \tag{9.2.3}$$

The subleading terms would be of order $u\varphi(u)$ and can be ignored. Comparing with (9.2.1), we have solutions for $\alpha = \pm 1$. We therefore have

$$\varphi(u) \simeq Au^k e^{-u^2/2} + Bu^k e^{u^2/2} \quad \text{as } |u| \to \infty. \tag{9.2.4}$$

The solution with coefficient B would not yield an energy eigenstate because it diverges as $|u| \to \infty$ and would not be normalizable. Note that the u^k factor in (9.2.2) played no role in the analysis. This factor, however, suggests that a *polynomial* multiplying $e^{-u^2/2}$ could be a solution of the differential equation. In other words, the solution could take the form

$$\varphi(u) = h(u)e^{-u^2/2}, \tag{9.2.5}$$

where $h(u)$ is a polynomial in u. This will happen if the exponential takes care of all the nonpolynomiality in the solution. We will find below that this is the case if the energy parameter $\mathcal{E}$ takes special values. Note that there is no assumption or loss of generality in writing the above expression: any function $\varphi(u)$ can be written as some other function times $e^{-u^2/2}$, as is made immediately clear by writing $(\varphi(u)e^{u^2/2})e^{-u^2/2}$. In setting up (9.2.5), we are only hoping that the differential equation for $h(u)$ admits polynomial solutions. Clearly, if we find $h(u)$, we have found $\varphi(u)$.

Plugging (9.2.5) into (9.1.20) and simplifying, we find a second-order linear differential equation for $h(u)$. This takes a bit of straightforward work that you should consider doing. The answer is

$$\boxed{\frac{d^2h}{du^2} - 2u\frac{dh}{du} + (\mathcal{E} - 1)h = 0.} \tag{9.2.6}$$

Exercise 9.3. *Consider the above differential equation for $h(u)$. Recall that $E = \hbar\omega\mathcal{E}/2$. Examine a solution with constant $h(u) = c_0$. What is its energy E? Examine a solution that is a polynomial of degree one in u: $h(u) = u + c_0$, for some constant c_0. Find c_0 and the energy E.*

Before doing a systematic analysis of equation (9.2.6), we show that getting a polynomial solution requires the quantization of $\mathcal{E}$. Indeed, assume that $h(u)$ is a polynomial of degree j:

$$h(u) \;=\; u^j + a_1 u^{j-1} + a_2 u^{j-2} + \ldots . \tag{9.2.7}$$

There is no loss of generality in assuming the leading term has coefficient one because the leading term cannot vanish and rescaling $h(u)$ does not change the fact that it is a solution. In equation (9.2.6) the first term is then a polynomial of degree $j-2$. Each of the other two terms is a polynomial of degree j. For the equation to hold, the contributions to the coefficient of u^j and the coefficient of u^{j-1} must vanish. Plugging the expansion of $h(u)$ into the left-hand side of (9.2.6), one quickly finds the

$$\text{coefficient of } u^j\text{:} \;\; -2j + \mathcal{E} - 1 = 0 \;\;\Rightarrow\;\; \mathcal{E} = 2j + 1. \tag{9.2.8}$$

This *is* the quantization of energy: a polynomial solution $h(u)$ of degree j requires $\mathcal{E} = 2j + 1$. You may wonder about the subleading term of degree $j-1$ whose coefficient must also vanish:

$$\text{coefficient of } u^{j-1}\text{:} \;\; (-2(j-1) + \mathcal{E} - 1)a_1 = 0. \tag{9.2.9}$$

With the value of energy $\mathcal{E}$ already determined, this condition requires that

$$2a_1 = 0. \tag{9.2.10}$$

We must therefore set $a_1 = 0$. Thus, the polynomial is actually of the form

$$h(u) = u^j + a_2 u^{j-2} + \cdots. \tag{9.2.11}$$

The vanishing of a_1 could have been anticipated. Since the harmonic oscillator potential is even, bound states *must be* either even or odd. Because $e^{-u^2/2}$ is even, the solution $\varphi(u)$ will be either even or odd if $h(u)$ is even or odd. If a_1 had not vanished, $h(u)$ would have two consecutive powers of u and would fail to be either even or odd.

We can analyze the equation more systematically using a series expansion with undetermined coefficients a_k:

$$h(u) = \sum_{k=0}^{\infty} a_k u^k. \tag{9.2.12}$$

At this point the conventional route is to simply plug this full expansion into (9.2.6) and manipulate it into a single expression by shifting sums appropriately. A simpler way to find how the differential equation (9.2.6) constrains the coefficients a_k is to select from each term in the differential equation the contribution to the coefficient of u^j, with j some arbitrary positive integer. For this we focus on three consecutive terms within the expansion of $h(u)$:

$$h(u) = \ldots + a_j u^j + a_{j+1} u^{j+1} + a_{j+2} u^{j+2} + \ldots \tag{9.2.13}$$

and write the contribution of each term in the differential equation to the coefficient of u^j:

$$\begin{aligned}
&\text{Contribution from} \quad \frac{d^2h}{du^2} \; : \; (j+2)(j+1)a_{j+2},\\
&\text{contribution from} \quad -2u\frac{dh}{du} \; : \; -2ja_j,\\
&\text{contribution from} \quad (\mathcal{E}-1)h \; : \; (\mathcal{E}-1)a_j.
\end{aligned} \tag{9.2.14}$$

You should convince yourself that no further terms in (9.2.13) could have contributed to the coefficient of u^j in the differential equation. The total coefficient of u^j on the left-hand side of the differential equation must be set to zero, for all values of j, for the differential equation to be satisfied. Therefore,

$$(j+2)(j+1)a_{j+2} - 2ja_j + (\mathcal{E}-1)a_j = 0, \qquad j = 0, 1, 2, \ldots \tag{9.2.15}$$

This can be rewritten as

$$a_{j+2} = \frac{2j+1-\mathcal{E}}{(j+2)(j+1)}\, a_j. \tag{9.2.16}$$

This type of equation is called a *recursion relation*; if you know a_j and $\mathcal{E}$, it determines a_{j+2} for you. If you choose $a_0 = 1$, you can construct a solution that contains only even index coefficients, $a_2, a_4, \cdots$, as determined recursively by the above relation. That solution, of the form

$$h_e(u) = 1 + a_2 u^2 + a_4 u^4 + \cdots, \tag{9.2.17}$$

would be even. Another solution is constructed by choosing $a_1 = 1$ and then using the above recursion to find $a_3, a_5, \ldots$. That solution, of the form

$$h_o(u) = u + a_3 u^3 + a_5 u^5 + \cdots, \tag{9.2.18}$$

would be odd. For arbitrary values of $\mathcal{E}$, both even and odd solutions exist. The general solution $h(u)$ to (9.2.6) with arbitrary $\mathcal{E}$ is a superposition of the even and odd solutions with arbitrary coefficients β_0 and β_1:

$$\begin{aligned} h(u) &= \beta_0 h_e(u) + \beta_1 h_o(u) \\ &= \beta_0(1 + a_2 u^2 + a_4 u^4 + \cdots) + \beta_1(u + a_3 u^3 + a_5 u^5 + \cdots). \end{aligned} \tag{9.2.19}$$

This makes sense because $\beta_0 = h(0)$ and $\beta_1 = h'(0)$, and the solution of a second-order differential equation is determined by knowing the function and its derivative at a point. The problem is that neither $h_e(u)$ nor $h_o(u)$ are expected to be polynomial for arbitrary $\mathcal{E}$ because the recursion relation may not terminate. Neither one nor any linear combination of them would be expected to be a good energy eigenstate for arbitrary $\mathcal{E}$.

Let us now demonstrate that if the series for $h(u)$ does not terminate, the corresponding $\varphi(u)$ is not an acceptable energy eigenstate. Let us see what the large u behavior of $h(u)$ would be if it does not terminate. For large j the recursion relation (9.2.16) gives

$$\frac{a_{j+2}}{a_j} \simeq \frac{2}{j}. \tag{9.2.20}$$

What kind of function of u has coefficients that grow this way? Note that

$$e^{u^2} = \sum_{n=0}^{\infty} \frac{1}{n!}\left(u^2\right)^n = \sum_{j \in \text{ even}} \frac{1}{(j/2)!} u^j. \tag{9.2.21}$$

This series has coefficients $c_j = \frac{1}{(j/2)!}$ for even j, and so we see that

$$\frac{c_{j+2}}{c_j} = \frac{(j/2)!}{((j+2)/2)!} = \frac{1}{\frac{j}{2}+1} = \frac{2}{j+2} \simeq \frac{2}{j} \tag{9.2.22}$$

for large j. This is just the behavior noted in (9.2.20) for the $h(u)$ coefficients, showing that $h(u) \sim e^{u^2}$ for large u. Therefore, if the series for $h(u)$ does not terminate, the wave function behaves like

$$\varphi(u) = h(u) e^{-u^2/2} \simeq e^{u^2} e^{-u^2/2} \simeq e^{u^2/2}, \tag{9.2.23}$$

which is in fact the bad solution identified in (9.2.4). This proves that $h(u)$ must be a polynomial, and the recursion relation must terminate for us to get an energy eigenstate!

Now we discuss how to get a polynomial $h(u)$, although the main conclusion was anticipated earlier in (9.2.8). If $h(u)$ is to be of degree j, it must have nonvanishing a_j and *vanishing* a_{j+2}, as determined from the recursion relation (9.2.16). The numerator in this recursion relation must vanish, and we must choose $\mathcal{E}$ such that

$$2j+1-\mathcal{E}=0 \quad \rightarrow \quad \mathcal{E}=2j+1, \quad j=0,1,2,\ldots. \tag{9.2.24}$$

The solution will then take the form

$$\begin{aligned} h(u) &= a_j u^j + a_{j-2}u^{j-2} + \cdots + a_0, \quad \text{if } j \text{ is even,} \\ h(u) &= a_j u^j + a_{j-2}u^{j-2} + \cdots + a_1 u, \quad \text{if } j \text{ is odd,} \end{aligned} \tag{9.2.25}$$

with powers decreasing in steps of two because this is what the recursion relation demands for having a solution. The solution will therefore be automatically even (if j is even) or odd (if j is odd). Say j is even, and the solution is even with energy parameter $\mathcal{E}=2j+1$ as required. The second solution of the differential equation for that value of the energy would be odd, but the energy parameter $\mathcal{E}=2j+1$ that made the even solution terminate will not make the odd solution terminate. This means that the second solution of the differential equation is not an energy eigenstate.

Exercise 9.4. *When $\mathcal{E}=1$, the differential equation for $h(u)$ becomes $h''(u)-2uh'(u)=0$. Consider the polynomial ansatz $h(u)=1+a_1u+a_2u^2$. What values do a_1 and a_2 need to take? Being a second-order differential equation, there is a second solution, which in this case is non-polynomial. This solution has a power series expansion that begins with $h=u+\cdots$. Find the next three nonzero terms in the power series expansion.*

Physicists usually use the letter n (instead of j) to denote the degree of the solution. Therefore, we say that

$$\mathcal{E}=2n+1, \quad n=0,1,2,\ldots \tag{9.2.26}$$

corresponds to the polynomial solution

$$h_n(u) = a_n u^n + a_{n-2}u^{n-2} + \cdots, \quad n=0,1,2,\ldots \tag{9.2.27}$$

that defines the energy eigenstate as

$$\varphi_n(u)=h_n(u)e^{-u^2/2}. \tag{9.2.28}$$

The corresponding physical energies E_n are given by (9.1.19) and are therefore

$$E_n = \frac{\hbar\omega}{2}\mathcal{E}=\frac{\hbar\omega}{2}(2n+1). \tag{9.2.29}$$

This is a key result, and we write it as

$$\boxed{E_n=\hbar\omega(n+\tfrac{1}{2}), \quad n=0,1,2,\ldots.} \tag{9.2.30}$$

We see that the energies are quantized, and the energy levels are evenly spaced with spacing $\hbar\omega$. The ground state energy E_0 is not zero; we have $E_0=\hbar\omega/2$. We have one energy eigenstate for each value of the energy. The spectrum is nondegenerate, as it should be since all energy eigenstates of the harmonic oscillator are bound states of a one-dimensional potential.

The polynomial solutions $h_n(u)$ that we have obtained are in fact Hermite polynomials, usually denoted as $H_n(u)$ and normalized with a specific choice of the leading coefficient:

$$H_n(u) = 2^n u^n \pm \cdots. \tag{9.2.31}$$

The Hermite polynomials are solutions of (9.2.6) with $\mathcal{E} = 2n+1$; therefore, they satisfy the differential equation

$$\frac{d^2H_n}{du^2} - 2u\frac{dH_n}{du} + 2nH_n = 0. \tag{9.2.32}$$

The first few Hermite polynomials are

$$\begin{aligned} H_0(u) &= 1, \\ H_1(u) &= 2u, \\ H_2(u) &= 4u^2 - 2, \\ H_3(u) &= 8u^3 - 12u. \end{aligned} \tag{9.2.33}$$

There exists a generating function for Hermite polynomials $H_n(u)$. It is a function of u and a formal expansion parameter z:

$$e^{-z^2+2zu} = \sum_{n=0}^{\infty} \frac{z^n}{n!} H_n(u). \tag{9.2.34}$$

We call z a formal parameter because it is included just to organize information efficiently; z is not related here to any physical quantity. The above relation says that if we expand the left-hand side as a power series in the parameter z, the expansion coefficients are functions of u that are in fact the Hermite polynomials. It is not hard to show (problem 9.4) that the polynomials defined by this expansion satisfy the requisite differential equation (9.2.32) and are normalized as claimed in (9.2.31).

Exercise 9.5. *Use the recursion relation (9.2.16) to determine the next term in the solution (9.2.31). That is, find a_{n-2} in the expression $H_n(u) = 2^n u^n + a_{n-2}u^{n-2} + \cdots$.*

The energy eigenstates $\varphi_n(u)$ in (9.2.28) are now written in terms of the Hermite polynomials as

$$\varphi_n(u) = N_n H_n(u) e^{-u^2/2}, \tag{9.2.35}$$

where N_n is a normalization constant. We can write this eigenstate as a function of the physical coordinate x. Recalling that $u = x/L_0$, where $L_0 = \sqrt{\hbar/(m\omega)}$, we have

$$\varphi_n(x) = N_n H_n\left(x\sqrt{\tfrac{m\omega}{\hbar}}\right) e^{-\frac{m\omega}{2\hbar}x^2}, \quad n = 0, 1, 2, \ldots. \tag{9.2.36}$$

Note that the energy eigenstates are real functions of x. The normalization constant N_n is determined from the condition

$$\langle \varphi_n, \varphi_n \rangle = \int_{-\infty}^{\infty} dx\, \varphi_n(x)\varphi_n(x) = 1. \tag{9.2.37}$$

Once this holds, we have full orthonormality of the energy eigenstates:

$$\langle \varphi_n, \varphi_m \rangle = \delta_{mn}. \tag{9.2.38}$$

Indeed, the orthogonality of different φ's follows because they are eigenstates of the Hermitian operator $\hat{H}$ with different eigenvalues. The ground state is obtained for $n=0$:

$$\varphi_0(x) = N_0 H_0 (x/L_0)\, e^{-\frac{x^2}{2L_0^2}}. \tag{9.2.39}$$

Using $H_0(u) = 1$ and calculating N_0 for proper normalization, we get

$$\varphi_0(x) = N_0\, e^{-\frac{x^2}{2L_0^2}}, \quad \text{with } N_0^2 = \sqrt{\frac{m\omega}{\pi\hbar}} = \frac{1}{\sqrt{\pi}L_0}. \tag{9.2.40}$$

The ground state is a pure Gaussian in x. We also see that the excited states are polynomials times that *same* Gaussian.

Exercise 9.6. *Explain why* $(\varphi_0(x))^2$ *is in fact a representation of the delta function* $\delta(x)$ *as* $L_0 \to 0$.

9.3 Algebraic Solution for the Spectrum

We have already seen how to calculate energy eigenstates for the simple harmonic oscillator by solving a second-order differential equation, the time-independent Schrödinger equation. We now try a method that gives a solution for the eigenstates by solving a *first-order* differential equation. This is, of course, much easier than solving a second-order differential equation.

In this method the first step is to attempt a "factorization" of the harmonic oscillator Hamiltonian. By this we mean, roughly, writing the Hamiltonian as the product $V^\dagger V$, with V an operator and $V^\dagger$ its Hermitian conjugate. As a first step, we rewrite the Hamiltonian (9.1.7) as a factor times a sum of squares:

$$\hat{H} = \tfrac{1}{2} m\omega^2 \Big(\hat{x}^2 + \frac{\hat{p}^2}{m^2\omega^2} \Big). \tag{9.3.1}$$

Now recall that for real numbers a and b, the quadratic form $a^2 + b^2$ can be factorized over the complex numbers. We have $a^2 + b^2 = (a - ib)(a + ib)$, an identity that holds because $i(ab - ba) = 0$. If a and b were operators, this cancellation might not happen since operators can fail to commute. Inspired by this, we examine whether the above expression in parentheses can be written as a product:

$$\left(\hat{x}-\frac{i\hat{p}}{m\omega}\right)\left(\hat{x}+\frac{i\hat{p}}{m\omega}\right) = \hat{x}^2+\frac{\hat{p}^2}{m^2\omega^2}+\frac{i}{m\omega}(\hat{x}\hat{p}-\hat{p}\hat{x}), \\ = \hat{x}^2+\frac{\hat{p}^2}{m^2\omega^2}-\frac{\hbar}{m\omega}, \tag{9.3.2}$$

where the right-hand side shows the desired term plus a constant that arises from $[\hat{x},\hat{p}]= i\hbar$. While the factorization did not work exactly as expected, it is good enough for our purposes. We now define the rightmost factor in the above left-hand side to be $\hat{V}$:

$$\hat{V} \equiv \hat{x}+\frac{i\hat{p}}{m\omega}. \tag{9.3.3}$$

Since $\hat{x}$ and $\hat{p}$ are Hermitian operators, we then have:

$$\hat{V}^\dagger = \hat{x}-\frac{i\hat{p}}{m\omega}, \tag{9.3.4}$$

and this is, in fact, the leftmost factor in the product! We can therefore rewrite (9.3.2) as

$$\hat{x}^2+\frac{\hat{p}^2}{m^2\omega^2} = \hat{V}^\dagger\hat{V}+\frac{\hbar}{m\omega}, \tag{9.3.5}$$

and the Hamiltonian becomes

$$\hat{H} = \tfrac{1}{2}m\omega^2\,\hat{V}^\dagger\hat{V} + \tfrac{1}{2}\hbar\omega. \tag{9.3.6}$$

This is a factorized form of the Hamiltonian: up to the additive constant $\frac{1}{2}\hbar\omega$, $\hat{H}$ is the product of a positive constant times the operator product $\hat{V}^\dagger\hat{V}$. The commutator of $\hat{V}$ and $\hat{V}^\dagger$ is simple:

$$[\,\hat{V},\hat{V}^\dagger\,] = \left[\hat{x}+\frac{i\hat{p}}{m\omega},\hat{x}-\frac{i\hat{p}}{m\omega}\right] = -\frac{i}{m\omega}[\hat{x},\hat{p}]+\frac{i}{m\omega}[\hat{p},\hat{x}] = \frac{2\hbar}{m\omega}. \tag{9.3.7}$$

This implies that

$$\left[\sqrt{\frac{m\omega}{2\hbar}}\,\hat{V}\,,\,\sqrt{\frac{m\omega}{2\hbar}}\,\hat{V}^\dagger\right]=1 \tag{9.3.8}$$

and suggests the definition of unit-free operators $\hat{a}$ and $\hat{a}^\dagger$,

$$\hat{a} \equiv \sqrt{\frac{m\omega}{2\hbar}}\,\hat{V}, \\ \hat{a}^\dagger \equiv \sqrt{\frac{m\omega}{2\hbar}}\,\hat{V}^\dagger, \tag{9.3.9}$$

that satisfy the simple commutator

$$\boxed{[\hat{a},\hat{a}^\dagger\,]=1.} \tag{9.3.10}$$

The operator $\hat{a}$ is called the *annihilation* operator, and $\hat{a}^\dagger$ is called the *creation* operator. The justification for these names will be seen below. Using the definitions of $\hat{V}$ and $\hat{V}^\dagger$, we obtain expressions for $\hat{a}$ and $\hat{a}^\dagger$ in terms of $\hat{x}$ and $\hat{p}$:

$$\hat{a} = \sqrt{\frac{m\omega}{2\hbar}}\left(\hat{x} + \frac{i\hat{p}}{m\omega}\right),$$
$$\hat{a}^\dagger = \sqrt{\frac{m\omega}{2\hbar}}\left(\hat{x} - \frac{i\hat{p}}{m\omega}\right). \tag{9.3.11}$$

The inverse relations are useful many times as well:

$$\hat{x} = \sqrt{\frac{\hbar}{2m\omega}}(\hat{a}+\hat{a}^\dagger) = \frac{1}{\sqrt{2}} L_0 (\hat{a}+\hat{a}^\dagger),$$
$$\hat{p} = i\sqrt{\frac{m\omega\hbar}{2}}(\hat{a}^\dagger - \hat{a}) = \frac{i}{\sqrt{2}}\frac{\hbar}{L_0}(\hat{a}^\dagger - \hat{a}). \tag{9.3.12}$$

While neither $\hat{a}$ nor $\hat{a}^\dagger$ is Hermitian, they are Hermitian conjugates of each other, and the above equations are consistent with the Hermiticity of $\hat{x}$ and $\hat{p}$, as you can check. We can also write the Hamiltonian in terms of the $\hat{a}$ and $\hat{a}^\dagger$ operators. Using (9.3.9), we find that

$$\hat{V}^\dagger \hat{V} = \frac{2\hbar}{m\omega}\hat{a}^\dagger \hat{a}, \tag{9.3.13}$$

and therefore back in (9.3.6) we get

$$\hat{H} = \hbar\omega(\hat{a}^\dagger \hat{a} + \tfrac{1}{2}) = \hbar\omega(\hat{N} + \tfrac{1}{2}), \quad \hat{N} \equiv \hat{a}^\dagger \hat{a}. \tag{9.3.14}$$

The above form of the Hamiltonian is factorized: up to an additive constant, $\hat{H}$ is the product of a positive constant times the operator product $\hat{a}^\dagger \hat{a}$. This product of operators is called the **number operator** $\hat{N}$. By construction, $\hat{N}$ is a Hermitian operator equal to the Hamiltonian, up to a scale and an additive constant. An eigenstate of $\hat{H}$ is also an eigenstate of $\hat{N}$, and vice versa. It follows from the above relation that the eigenvalues E and N of $\hat{H}$ and $\hat{N}$, respectively, are related by

$$E = \hbar\omega\left(N + \tfrac{1}{2}\right). \tag{9.3.15}$$

Let us now show the powerful conclusions that arise from the factorized Hamiltonian. Consider the expectation value of $\hat{H}$ on an arbitrary normalized state ψ:

$$\langle \hat{H} \rangle_\psi = \langle \psi, \hat{H}\psi \rangle = \hbar\omega \langle \psi, \hat{a}^\dagger \hat{a}\psi \rangle + \tfrac{1}{2}\hbar\omega \langle \psi, \psi \rangle. \tag{9.3.16}$$

Moving the $\hat{a}^\dagger$ to the first input in the inner product, we get

$$\langle \hat{H} \rangle_\psi = \hbar\omega \langle \hat{a}\psi, \hat{a}\psi \rangle + \tfrac{1}{2}\hbar\omega \geq \tfrac{1}{2}\hbar\omega \tag{9.3.17}$$

because any expression of the form $\langle \phi, \phi \rangle$ is greater than or equal to zero. For any energy eigenstate with energy E: $\hat{H}\psi = E\psi$, we have $\langle \hat{H} \rangle_\psi = E$, and therefore all energy eigenstates satisfy

$$E \geq \tfrac{1}{2}\hbar\omega. \tag{9.3.18}$$

This important result about the spectrum followed directly from the factorization of the Hamiltonian, without solving any differential equation! We also get the information required to find the ground state wave function. The minimum possible energy $\frac{1}{2}\hbar\omega$ will be realized for a state ψ if the term $\langle \hat{a}\psi, \hat{a}\psi\rangle$ in (9.3.17) vanishes. For this to vanish, $\hat{a}\psi$ must vanish. Therefore, the ground state wave function φ_0 must satisfy

$$\hat{a}\,\varphi_0 = 0. \tag{9.3.19}$$

The operator $\hat{a}$ annihilates the ground state, and this why $\hat{a}$ is called the annihilation operator. Using the definition of $\hat{a}$ in (9.3.11) and the position space representation of $\hat{p}$, this becomes the condition

$$\left(x + \frac{i}{m\omega}\frac{\hbar}{i}\frac{d}{dx}\right)\varphi_0(x) = 0 \quad \Rightarrow \quad \left(x + L_0^2\,\frac{d}{dx}\right)\varphi_0(x) = 0. \tag{9.3.20}$$

Remarkably, this is a **first-order** differential equation for the ground state, not a second-order equation, like the Schrödinger equation that determines the general energy eigenstates. This is a dramatic simplification afforded by the factorization of the Hamiltonian into a product of first-order differential operators. The above equation is rearranged as

$$\frac{d\varphi_0}{dx} = -\frac{1}{L_0^2}\,x\varphi_0 \quad \Rightarrow \quad \frac{d\varphi_0}{\varphi_0} = -\frac{1}{L_0^2}\,x dx. \tag{9.3.21}$$

Integration now yields

$$\varphi_0(x) \;=\; N_0\, e^{-\frac{x^2}{2L_0^2}}, \tag{9.3.22}$$

recovering the result in (9.2.40) and including the normalization factor that makes $\langle \varphi_0, \varphi_0\rangle = 1$. Note that φ_0 is indeed an energy eigenstate with energy E_0:

$$\hat{H}\varphi_0 = \hbar\omega(\hat{a}^\dagger\hat{a} + \tfrac{1}{2})\varphi_0 = \tfrac{1}{2}\hbar\omega\varphi_0 \quad \Rightarrow \quad E_0 \;=\; \tfrac{1}{2}\,\hbar\omega. \tag{9.3.23}$$

Exercise 9.7. *Try to find a state $\psi(x)$ that is annihilated by the creation operator; that is, $\hat{a}^\dagger\psi = 0$. Show that you can find a solution, but $\psi(x)$ is not normalizable and grows without bound as $x \to \pm\infty$.*

Before proceeding with the analysis of excited states, let us consider the consequences of factorization more generally. Factorizing a Hamiltonian means finding an operator $\hat{A}$ such that we can rewrite the Hamiltonian as $\hat{A}^\dagger\hat{A}$ up to an additive constant E_0 with units of energy:

$$\boxed{\hat{H} \;=\; \hat{A}^\dagger\hat{A} \;+\; E_0.} \tag{9.3.24}$$

The constant E_0 does not complicate the task of finding energy eigenstates: any eigenstate of $\hat{A}^\dagger\hat{A}$ with eigenvalue λ is an eigenstate of $\hat{H}$ with eigenvalue $\lambda + E_0$. Two key properties follow from the factorization (9.3.24):

1. Any energy eigenstate must have energy greater than or equal to E_0. First, note that for an *arbitrary* normalized $\psi(x)$ we have

$$\langle\psi,\hat{H}\psi\rangle = (\psi,\hat{A}^\dagger\hat{A}\,\psi) + E_0\langle\psi,\psi\rangle = \langle\hat{A}\psi,\hat{A}\psi\rangle + E_0. \tag{9.3.25}$$

   Since $\langle\hat{A}\psi,\hat{A}\psi\rangle \geq 0$, we have shown that

$$\boxed{\langle\psi,\hat{H}\psi\rangle \geq E_0.} \tag{9.3.26}$$

   If we take ψ to be an energy eigenstate of energy E: $\hat{H}\psi = E\psi$, the above relation gives

$$E \geq E_0. \tag{9.3.27}$$

   As claimed, all possible energies are greater than or equal to E_0.

2. A wave function ψ_0 that satisfies

$$\hat{A}\,\psi_0 = 0 \tag{9.3.28}$$

   is an energy eigenstate that *saturates* the inequality (9.3.27). Indeed,

$$\hat{H}\psi_0 = \hat{A}^\dagger\hat{A}\,\psi_0 + E_0\psi_0 = \hat{A}^\dagger(\hat{A}\,\psi_0) + E_0\psi_0 = E_0\psi_0. \tag{9.3.29}$$

   The state ψ_0 satisfying $\hat{A}\,\psi_0 = 0$ is the ground state. For conventional Hamiltonians this is a first-order differential equation for ψ_0 and much easier to solve than the Schrödinger equation.

9.4 Excited States of the Oscillator

We have seen that all energy eigenstates are eigenstates of the Hermitian number operator $\hat{N} = \hat{a}^\dagger\hat{a}$. This is because $\hat{H} = \hbar\omega(\hat{N} + \frac{1}{2})$. Note that since $\hat{a}\varphi_0 = 0$ we also have

$$\hat{N}\varphi_0 = 0. \tag{9.4.1}$$

We say that the ground state has "number" equal to zero. To understand further the significance of the number operator, we compute commutators. We can quickly check that

$$\begin{aligned}
[\hat{N},\hat{a}] &= [\hat{a}^\dagger\hat{a},\hat{a}] = [\hat{a}^\dagger,\hat{a}]\,\hat{a} = -\hat{a},\\
[\hat{N},\hat{a}^\dagger] &= [\hat{a}^\dagger\hat{a},\hat{a}^\dagger] = \hat{a}^\dagger[\hat{a},\hat{a}^\dagger] = \hat{a}^\dagger,
\end{aligned} \tag{9.4.2}$$

which we summarize as

$$\begin{aligned}
[\hat{N},\,\hat{a}\,] &= -\,\hat{a},\\
[\hat{N},\,\hat{a}^\dagger\,] &= \hat{a}^\dagger\,.
\end{aligned} \tag{9.4.3}$$

Using these identities, you should be able to show by induction that

$$\begin{aligned}[\hat{N}, (\hat{a})^k] &= -k\,(\hat{a})^k, \\ [\hat{N}, (\hat{a}^\dagger)^k] &= k\,(\hat{a}^\dagger)^k.\end{aligned} \tag{9.4.4}$$

These relations hint at the reason why $\hat{N}$ is called the number operator. Acting on powers of creation or annihilation operators by commutation, $\hat{N}$ gives the same object multiplied by (plus or minus) the *number* k of creation or annihilation operators. Closely related commutators are also useful:

$$\begin{aligned}[\hat{a}^\dagger, (\hat{a})^k] &= -k\,(\hat{a})^{k-1}, \\ [\hat{a}, (\hat{a}^\dagger)^k] &= k\,(\hat{a}^\dagger)^{k-1}.\end{aligned} \tag{9.4.5}$$

A quick heuristic reproduces the above commutators. Identify $\hat{a} = \frac{d}{d\hat{a}^\dagger}$, which is consistent with the commutator $[\hat{a}, \hat{a}^\dagger] = 1$. Then, the commutator on the second line follows. For the commutator on the first line, identify $\hat{a}^\dagger = -\frac{d}{d\hat{a}}$, which is also consistent with $[\hat{a}, \hat{a}^\dagger] = 1$. This is analogous to our thinking of $\hat{p}$ as $\frac{\hbar}{i}\frac{\partial}{\partial x}$ in position space and of $\hat{x}$ as $i\hbar\frac{\partial}{\partial p}$ in momentum space.

In the computations that follow we will also make use of a simple *commutator lemma* that applies when we have an operator $\hat{A}$ that kills an arbitrary state ψ, and we aim to simplify $\hat{A}\hat{B}\psi$, where $\hat{B}$ is another operator. Here is the result:

$$\boxed{\text{Commutator lemma: If } \hat{A}\,\psi = 0, \text{ then } \hat{A}\,\hat{B}\,\psi = [\hat{A}, \hat{B}]\psi.} \tag{9.4.6}$$

Acting on a state ψ killed by $\hat{A}$, we can replace the product $\hat{A}\hat{B}$ by the commutator $[\hat{A}, \hat{B}]$. This is easily proved. First note that

$$\hat{A}\hat{B} = [\hat{A}, \hat{B}] + \hat{B}\hat{A}, \tag{9.4.7}$$

as can be quickly checked by expanding the right-hand side. It then follows that

$$\hat{A}\,\hat{B}\,\psi = ([\hat{A}, \hat{B}] + \hat{B}\hat{A})\psi = [\hat{A}, \hat{B}]\psi \tag{9.4.8}$$

because $\hat{B}\hat{A}\,\psi = \hat{B}(\hat{A}\psi) = 0$. This is what we wanted to show. With these preparatory results, we can now proceed to construct the states of the harmonic oscillator.

Since $\hat{a}$ annihilates φ_0, consider acting on the ground state with $\hat{a}^\dagger$. It is clear that $\hat{a}^\dagger$ cannot also annihilate φ_0. If it did, acting with both sides of the commutator identity $[\hat{a}, \hat{a}^\dagger] = 1$ on φ_0 would lead to a contradiction: the left-hand side would vanish, but the right-hand side would not. Thus, consider the wave function

$$\varphi_1 \equiv \hat{a}^\dagger \varphi_0. \tag{9.4.9}$$

Let us show that φ_1 is an energy eigenstate. For this purpose we act on it with the number operator:

$$\hat{N}\varphi_1 = \hat{N}\hat{a}^\dagger\varphi_0 = [\hat{N}, \hat{a}^\dagger]\varphi_0, \tag{9.4.10}$$

where we note that $\hat{N}\varphi_0 = 0$ and use the commutator lemma. Given that $[\hat{N}, \hat{a}^\dagger] = \hat{a}^\dagger$, we get

$$\hat{N}\varphi_1 = \hat{a}^\dagger\varphi_0 = \varphi_1. \tag{9.4.11}$$

Thus, φ_1 is an $\hat{N}$ eigenstate with eigenvalue one. Since φ_0 has an $\hat{N}$ eigenvalue of zero, the effect of acting on φ_0 with $\hat{a}^\dagger$ was to increase the eigenvalue of the number operator by one unit. The operator $\hat{a}^\dagger$ is called the *creation* operator because it creates an excited state out of the ground state. Alternatively, it is also called the *raising* operator because it raises by one unit the eigenvalue of $\hat{N}$. Since $N=1$ for φ_1, it follows that φ_1 is an energy eigenstate with energy E_1, given by

$$E_1 = \hbar\omega(1+\tfrac{1}{2}) = \tfrac{3}{2}\hbar\omega. \tag{9.4.12}$$

It also turns out that φ_1 is properly normalized:

$$\langle\varphi_1, \varphi_1\rangle = \langle\hat{a}^\dagger\varphi_0, \hat{a}^\dagger\varphi_0\rangle = \langle\varphi_0, \hat{a}\hat{a}^\dagger\varphi_0\rangle, \tag{9.4.13}$$

where we moved the $\hat{a}^\dagger$ acting on the left input into the right input, where it goes as $(\hat{a}^\dagger)^\dagger = \hat{a}$. We then have

$$\langle\varphi_1, \varphi_1\rangle = \langle\varphi_0, \hat{a}\hat{a}^\dagger\varphi_0\rangle = \langle\varphi_0, [\hat{a}, \hat{a}^\dagger]\varphi_0\rangle = \langle\varphi_0, \varphi_0\rangle = 1, \tag{9.4.14}$$

where we used the commutator lemma in the evaluation of $\hat{a}\hat{a}^\dagger\varphi_0$. Indeed the state φ_1 is correctly normalized. Next consider the state

$$\varphi_2' \equiv \hat{a}^\dagger\hat{a}^\dagger\varphi_0. \tag{9.4.15}$$

For this state,

$$\hat{N}\varphi_2' = \hat{N}\hat{a}^\dagger\hat{a}^\dagger\varphi_0 = \left[\hat{N}, \hat{a}^\dagger\hat{a}^\dagger\right]\varphi_0 = 2\hat{a}^\dagger\hat{a}^\dagger\varphi_0 = 2\varphi_2', \tag{9.4.16}$$

where we used the second commutator in (9.4.4) for the case $k=2$. The result implies that φ_2' is a state with number $N=2$ and energy $E_2 = \frac{5}{2}\hbar\omega$. Is it properly normalized? To find out we calculate its norm squared:

$$\begin{aligned}(\varphi_2', \varphi_2') &= \langle\hat{a}^\dagger\hat{a}^\dagger\varphi_0, \hat{a}^\dagger\hat{a}^\dagger\varphi_0\rangle = \langle\varphi_0, \hat{a}\hat{a}\hat{a}^\dagger\hat{a}^\dagger\varphi_0\rangle = \langle\varphi_0, \hat{a}[\hat{a}, \hat{a}^\dagger\hat{a}^\dagger]\varphi_0\rangle \\ &= \langle\varphi_0, 2\hat{a}\hat{a}^\dagger\varphi_0\rangle = 2\langle\varphi_0, \varphi_0\rangle = 2.\end{aligned} \tag{9.4.17}$$

The properly normalized wave function is therefore

$$\varphi_2 \equiv \frac{1}{\sqrt{2}}\hat{a}^\dagger\hat{a}^\dagger\varphi_0. \tag{9.4.18}$$

It is time to generalize. We now claim that the normalized nth excited state of the simple harmonic oscillator is

$$\boxed{\varphi_n \equiv \frac{1}{\sqrt{n!}}\underbrace{\hat{a}^\dagger\cdots\hat{a}^\dagger}_{n}\varphi_0 = \frac{1}{\sqrt{n!}}(\hat{a}^\dagger)^n\varphi_0.} \tag{9.4.19}$$

Exercise 9.8. *Verify that the state φ_n has $\hat{N}$ eigenvalue n:*

$$\hat{N}\varphi_n = n\,\varphi_n. \tag{9.4.20}$$

This means that $\hat{N}$ *counts* the number of creation operators acting on the ground state. Since the $\hat{N}$ eigenvalue of φ_n is n, its energy E_n is given by

$$E_n = \hbar\omega(n + \tfrac{1}{2}). \tag{9.4.21}$$

Exercise 9.9. *Verify that the state φ_n is correctly normalized.*

As we pointed out earlier, the various states φ_n are eigenstates of a Hermitian operator (the Hamiltonian $\hat{H}$) with different eigenvalues, so they form an orthonormal set:

$$\langle \varphi_n, \varphi_m \rangle = \delta_{m,n}. \tag{9.4.22}$$

Let us investigate what happens when we act with a destruction operator or a creation operator on a general energy eigenstate. Begin with a destruction operator acting on φ_n. First note that $\hat{a}\varphi_n$ is a state with $n-1$ creation operators acting on φ_0 because the $\hat{a}$ eliminates one of the creation operators. Thus, we expect $\hat{a}\varphi_n \sim \varphi_{n-1}$. We can make this precise, as follows:

$$\hat{a}\,\varphi_n = \hat{a}\frac{1}{\sqrt{n!}}\,(\hat{a}^\dagger)^n\varphi_0 = \frac{1}{\sqrt{n!}}\,[\hat{a},\,(\hat{a}^\dagger)^n]\varphi_0 = \frac{n}{\sqrt{n!}}\,(\hat{a}^\dagger)^{n-1}\varphi_0. \tag{9.4.23}$$

At this point we use (9.4.19) with n set equal to $n-1$. The result is

$$\hat{a}\,\varphi_n = \frac{n}{\sqrt{n!}}\sqrt{(n-1)!}\,\varphi_{n-1} = \sqrt{n}\,\varphi_{n-1}. \tag{9.4.24}$$

Similarly, by the action of $\hat{a}^\dagger$ on φ_n we get

$$\hat{a}^\dagger\varphi_n = \frac{1}{\sqrt{n!}}(\hat{a}^\dagger)^{n+1}\varphi_0 = \frac{1}{\sqrt{n!}}\sqrt{(n+1)!}\,\varphi_{n+1} = \sqrt{n+1}\,\varphi_{n+1}. \tag{9.4.25}$$

Collecting the results, we have

$$\boxed{\begin{aligned} \hat{a}\,\varphi_n &= \sqrt{n}\,\varphi_{n-1}, \\ \hat{a}^\dagger\varphi_n &= \sqrt{n+1}\,\varphi_{n+1}. \end{aligned}} \tag{9.4.26}$$

These relations make it clear that $\hat{a}$ lowers the number of any energy eigenstate by one unit, except for the vacuum φ_0, which it kills. The raising operator $\hat{a}^\dagger$ increases the number of any eigenstate by one unit.

Example 9.1. Position uncertainty Δx in the nth energy eigenstate.
We determine this uncertainty by calculation. From the definition of uncertainty,

$$(\Delta x)_n^2 = \langle \hat{x}^2\rangle_{\varphi_n} - \langle \hat{x}\rangle_{\varphi_n}^2. \tag{9.4.27}$$

The expectation value $\langle \hat{x}\rangle$ vanishes for any energy eigenstate since we are integrating x, which is odd, against $|\varphi_n(x)|^2$, which is always even. Still, it is instructive to see how this happens explicitly:

$$\langle \hat{x}\rangle_{\varphi_n} = \langle \varphi_n, \hat{x}\varphi_n\rangle = \frac{L_0}{\sqrt{2}}\big\langle \varphi_n,\, (\hat{a} + \hat{a}^\dagger)\varphi_n\big\rangle, \tag{9.4.28}$$

using the formula for $\hat{x}$ in terms of $\hat{a}$ and $\hat{a}^\dagger$. This is a good strategy in computing the expectation values of $\hat{x}$ and $\hat{p}$ dependent operators on energy eigenstates: expand $\hat{x}$ and $\hat{p}$ in terms of $\hat{a}$ and $\hat{a}^\dagger$ operators. The above overlap vanishes because $\hat{a}\varphi_n \sim \varphi_{n-1}$, $\hat{a}^\dagger\varphi_n \sim \varphi_{n+1}$, and both φ_{n-1} and φ_{n+1} are orthogonal to φ_n. Now we compute the expectation value of $\hat{x}^2$:

$$\begin{aligned}\langle \hat{x}^2 \rangle_{\varphi_n} &= \langle \varphi_n, \hat{x}^2 \varphi_n \rangle = \frac{L_0^2}{2} \langle \varphi_n, (\hat{a}+\hat{a}^\dagger)(\hat{a}+\hat{a}^\dagger)\varphi_n \rangle \\ &= \frac{L_0^2}{2} \langle \varphi_n, (\hat{a}\hat{a} + \hat{a}\hat{a}^\dagger + \hat{a}^\dagger\hat{a} + \hat{a}^\dagger\hat{a}^\dagger)\varphi_n \rangle.\end{aligned} \tag{9.4.29}$$

Since $\hat{a}\hat{a}\varphi_n \sim \varphi_{n-2}$ and $\hat{a}^\dagger\hat{a}^\dagger\varphi_n \sim \varphi_{n+2}$ and both φ_{n-2} and φ_{n+2} are orthogonal to φ_n, the $\hat{a}\hat{a}$ and $\hat{a}^\dagger\hat{a}^\dagger$ terms do not contribute. We are left with

$$\langle \hat{x}^2 \rangle_{\varphi_n} = \frac{L_0^2}{2} \langle \varphi_n, (\hat{a}\hat{a}^\dagger + \hat{a}^\dagger\hat{a})\varphi_n \rangle. \tag{9.4.30}$$

At this point we recognize that $\hat{a}^\dagger\hat{a} = \hat{N}$ and that $\hat{a}\hat{a}^\dagger = [\hat{a}, \hat{a}^\dagger] + \hat{a}^\dagger\hat{a} = 1 + \hat{N}$. As a result,

$$\langle \hat{x}^2 \rangle_{\varphi_n} = \frac{L_0^2}{2} \langle \varphi_n, (1+2\hat{N})\varphi_n \rangle = \frac{L_0^2}{2}(1+2n). \tag{9.4.31}$$

We therefore have

$$(\Delta x)_n^2 = L_0^2 \left(n + \tfrac{1}{2}\right). \tag{9.4.32}$$

The square of the position uncertainty grows linearly with the number. For large n we have $(\Delta x)_n \sim L_0\sqrt{n}$. This is roughly comparable to the amplitude of the classical oscillation of a particle with energy $\hbar\omega n$. □

Exercise 9.10. *Verify that the expectation value of $\hat{p}$ in an energy eigenstate vanishes, and demonstrate that*

$$(\Delta p)_n^2 = \langle \hat{p}^2 \rangle_{\varphi_n} = \left(\tfrac{\hbar}{L_0}\right)^2 \left(n + \tfrac{1}{2}\right). \tag{9.4.33}$$

It follows from the position and momentum uncertainties obtained above that

$$(\Delta x)_n (\Delta p)_n = \hbar\left(n + \tfrac{1}{2}\right), \quad \text{on the state } \varphi_n. \tag{9.4.34}$$

Only for the ground state φ_0 does the product of uncertainties saturate the lower bound given by the Heisenberg uncertainty inequality (4.1.19): $\Delta x \Delta p \geq \frac{\hbar}{2}$.

Exercise 9.11. *Select from the following five states those that are not $\hat{N}$ eigenstates:*

$$\varphi_0, \qquad \hat{a}^\dagger\hat{a}^\dagger\varphi_0, \qquad \varphi_0 + \hat{a}^\dagger\varphi_0, \qquad \hat{a}\hat{a}^\dagger\hat{a}^\dagger\varphi_0, \qquad (\hat{a}+\hat{a}^\dagger)^2\varphi_0.$$

Exercise 9.12. *Calculate the commutator $[\hat{N}, \hat{x}]$ and leave the answer in terms of $\hat{x}$ and $\hat{p}$, as needed. Use this result to determine the Hermitian operator $\hat{O}$ that appears in the commutator*

$$[\hat{N}, \hat{x}^2] = -\frac{i}{m\omega}\hat{O}.$$

Problems

Problem 9.1. *Harmonic oscillators beyond the turning points.*

1. Consider the simple harmonic oscillator energy eigenstate with $n=0$. Calculate the probability $P(\text{out}|n=0)$ that the distance $|x|$ to the origin takes a value greater than the amplitude of a classical oscillator of the same energy. The particle is thus "out" of the classically allowed region. Write your answer in terms of an integral $\frac{2}{\sqrt{\pi}}\int_1^\infty(\cdots)du$, where u is the dummy variable of integration. This integral cannot be performed analytically and defines a special function related to the error function $\text{erf}(x)$. Evaluate the integral numerically to obtain a numerical probability for $P(\text{out}|n=0)$.
2. Repeat the calculation for the energy eigenstates with $n=1$ and $n=2$. Find the probabilities $P(\text{out}|n=1)$ and $P(\text{out}|n=2)$.

Problem 9.2. *Harmonic oscillator with a wall.*

Consider a *half* harmonic oscillator described by the potential

$$V(x)=\begin{cases}\frac{1}{2}m\omega^2x^2, & \text{for } x>0,\\ \infty, & \text{for } x<0.\end{cases}$$

This system represents, for example, a spring that can be stretched but not compressed. Carefully explain how the eigenstates of this system and their energies are related to the eigenstates and energies of a (full) harmonic oscillator.

Problem 9.3. *Harmonic oscillator expectation value.*

Calculate the expectation value of the operator $\hat{x}^4$ on the energy eigenstate with number n. Show that

$$\langle\varphi_n, \hat{x}^4\varphi_n\rangle = \tfrac{1}{4}L_0^4\,(6n^2+6n+3).$$

Problem 9.4. *Generating function for Hermite polynomials.*

1. Consider the polynomials $H_n(\xi)$ defined by the generating function $e^{-s^2+2s\xi}$ as follows:

$$e^{-s^2+2s\xi}=\sum_{n=0}^{\infty}H_n(\xi)\frac{s^n}{n!}. \tag{1}$$

   By comparing powers of s in the power series expansion of the generating function with the infinite sum over $H_n(\xi)$, compute the leading-order term of $H_n(\xi)$:

$$H_n(\xi)=\cdots+\text{terms of order } \xi^{n-2} \text{ and lower.}$$

2. Find an operation that you can apply to both sides of (1) to generate the following sum:

$$\sum_{n=0}^{\infty}nH_n(\xi)\frac{s^n}{n!}.$$

Compute the sum. Also compute the following two sums:

$$\sum_{n=0}^{\infty} H_n'(\xi)\frac{s^n}{n!}, \qquad \sum_{n=0}^{\infty} H_n''(\xi)\frac{s^n}{n!}.$$

Use your results to compute the following sum:

$$\sum_{n=0}^{\infty} (H_n''(\xi) - 2\xi H_n'(\xi) + 2nH_n)\frac{s^n}{n!}.$$

Identify the differential equation satisfied by the Hermite polynomials.

Problem 9.5. *A particular Hermitian operator.*

Consider the operator $\hat{O} \equiv \hat{x}\hat{p} + \hat{p}\hat{x}$.

1. Is $\hat{O}$ Hermitian?
2. Suppose you have a real normalizable wave function $\psi(x)$ that vanishes at $x = \pm\infty$. What is the expectation value of $\hat{O}$?
3. Assume your quantum system is a harmonic oscillator. Write $\hat{O}$ in terms of $\hat{a}$ and $\hat{a}^\dagger$.
4. Calculate the expectation value of $\hat{O}$ in an arbitrary eigenstate φ_n of the harmonic oscillator.
5. Show that any operator of the form $[\hat{N}, \hat{K}]$, with arbitrary $\hat{K}$, has zero expectation value in a harmonic oscillator eigenstate. Does the calculation in exercise 9.12 then help you understand your answer to the previous question?

Problem 9.6. *Harmonic trap for Bose-Einstein condensation.*

For all the estimates required below, ignore factors of 2, π, 1/3, and so on. Give your answers in terms of the variables indicated, with no additional numerical constants that in any case cannot be estimated reliably. This problem is a variation on the estimates obtained in example 2.2. Identical bosons of mass m are put in a harmonic trap at temperature T. We wish to estimate the (minimum) number N of particles required to form a Bose-Einstein condensate at temperature T.

1. The particles are put in a (spherical) harmonic trap, which means that *each* particle a distance r from the center of the trap has a potential energy $U(r) = \frac{1}{2}m\omega^2 r^2$. Assuming the total energy of each particle is the thermal energy, estimate the volume V they occupy on the trap in terms of $k_BT, m,$ and ω. Assuming there are N particles on the trap, estimate the mean separation ℓ between the particles in terms of $k_BT, m, \omega,$ and N.
2. Estimate the de Broglie wavelength λ of a free particle of mass m at temperature T and use the condition for Bose-Einstein condensation to show that

$$N \simeq \left(\frac{k_BT}{\hbar\omega}\right)^3.$$

   Note that the mass of the particle does not feature in the answer.
3. The exact answer of a complete analysis is $N = 1.202\left(\frac{k_BT}{\hbar\omega}\right)^3$. Let the temperature be $T = 2\mu$K and consider a trap of resonant frequency $\nu \simeq 500$ Hz. Show that $N \simeq$ 630,000.

Problem 9.7. *Harmonic oscillators oscillating!*

At time equal zero, a particle of mass m in a harmonic oscillator with frequency ω has a wave function

$$\Psi(x,0)=\frac{1}{\sqrt{2}}(\varphi_0(x)+\varphi_1(x)),$$

where $\varphi_0(x)$ and $\varphi_1(x)$ are (real) normalized eigenstates of the Hamiltonian with number eigenvalue zero and one, respectively.

1. Write down $\Psi(x,t)$ and $|\Psi(x,t)|^2$. Leave your expressions in terms of $\varphi_0(x)$ and $\varphi_1(x)$.
2. Find $\langle\hat{x}\rangle$ as a function of time.
3. Find $\langle\hat{p}\rangle$ as a function of time.
4. For an *arbitrary* harmonic oscillator state, find the smallest time T for which

   $$|\Psi(x,t)|^2=|\Psi(x,t+T)|^2.$$

   Write $\Psi(x,t+T)$ in terms of $\Psi(x,t)$.

Problem 9.8. *Ground state and first excited state of a potential.*

Consider the even potential sketched below, with height V_0 except for a dip around $x=0$, near which the potential is accurately described by a quadratic function:

$$V(x)\simeq\tfrac{1}{2}\alpha x^2,\quad x\text{ near }0,$$

where $\alpha>0$ is a constant.

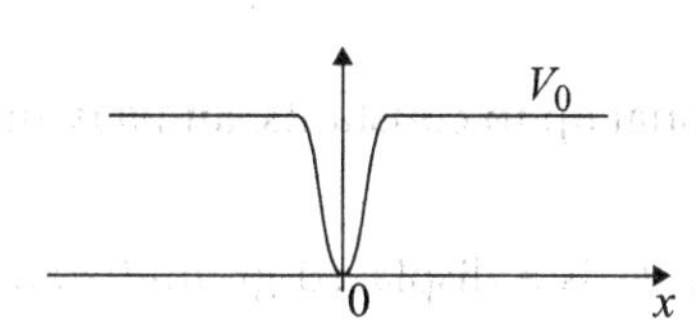

1. Use the harmonic oscillator to give an estimate for the ground state energy. Find an inequality satisfied by α, V_0, m, and $\hbar$, which is required for the accuracy of your result.
2. Now consider the even potential built by using two copies of the potential above with well-separated centers at $\pm x_0$, as shown in the figure below.

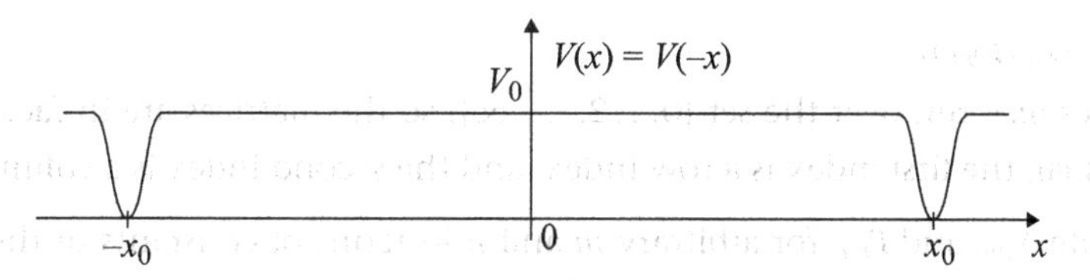

   Give the approximate energies of the ground state and the first excited state of this potential and sketch the associated wave functions. Write an inequality involving $x_0, \alpha, \hbar$, and m, required for the accuracy of your result.

3. Combine your two inequalities into the form $\cdots \ll \alpha \ll \cdots$, where the dots represent quantities you found.

Problem 9.9. *Coherent states as displaced ground states.*

Consider the state ψ_λ defined by

$$\psi_\lambda \equiv N e^{\lambda \hat{a}^\dagger} \varphi_0,$$

with $\lambda \in \mathbb{C}$ a complex number and φ_0 the normalized ground state wave function. This is known as a *coherent state*. For (1) and (2), you will find it useful to Taylor expand the exponential.

1. Find the constant N (real and positive) needed for the state ψ_λ to be normalized.
2. Compute the effect of the annihilation operator $\hat{a}$ on the state ψ_λ, writing $\hat{a}\psi_\lambda$ in terms of ψ_λ and no annihilation operator.
3. Find the expectation value of the harmonic oscillator Hamiltonian in the state ψ_λ.
4. Find the uncertainty ΔE_λ in the energy of the state ψ_λ.
5. Use your result from (2) to find the differential equation that $\psi_\lambda(x)$ must satisfy. Express your answer in the form

$$\frac{d\psi_\lambda(u)}{du} = \ldots$$

using the dimensionless variable $u = x/L_0$. Solve this differential equation to obtain $\psi_\lambda(u)$. Do not worry about normalizing the wave function.
6. Write the ground state wave function φ_0 as a function of u. Then show that

$$\psi_\lambda(u) \sim \varphi_0(u - b),$$

where $\sim$ means proportional up to constants, and b is an λ-dependent constant you should determine.
7. When λ is real, the state ψ_λ is a displaced ground state. What is the value of the displacement? If λ is complex, the real part determines the displacement. What is the physical import of the imaginary part of λ?

Problem 9.10. *Matrices for $\hat{x}$ and $\hat{p}$ in the harmonic oscillator.*

We can define matrices X_{mn} and P_{mn} associated with the $\hat{x}$ and $\hat{p}$ operators using the basis $\{\varphi_0, \varphi_1, \cdots\}$ of energy eigenstates of the harmonic oscillator:

$$X_{mn} \equiv \langle \varphi_m, \hat{x}\, \varphi_n \rangle,$$

$$P_{mn} \equiv \langle \varphi_m, \hat{p}\, \varphi_n \rangle.$$

The indices m, n run over the set $\{0, 1, 2, \ldots, \infty\}$, so the matrices are in fact of infinite size. As usual, the first index is a row index, and the second index is a column index.

1. Calculate X_{mn} and P_{mn} for arbitrary m and n in terms of constants of the oscillator $(\hbar, m, \omega)$, the integers m, n, Kronecker deltas $(\delta_{p,q})$, and numerical constants.
2. Use the results you obtained in (1) to write the explicit 3-by-3 matrices X and P that result upon truncation to $m, n \in \{0, 1, 2\}$. State if the matrices are Hermitian, symmetric, or antisymmetric.
3. Calculate the commutator $[X, P]$ of your 3-by-3 matrices. Comment on your answer.

10 Angular Momentum and Central Potentials

The study of central potentials leads to the introduction of angular momentum operators $\hat{L}_x$, $\hat{L}_y$, and $\hat{L}_z$. These operators form an algebra under commutation, and the total angular momentum operator squared $\hat{\mathbf{L}}^2$ commutes with all of them. We search for simultaneous eigenstates of $\hat{L}_z$ and $\hat{\mathbf{L}}^2$ and find the quantization of angular momentum eigenvalues. The $\hat{\mathbf{L}}^2$ eigenvalues are parameterized by a nonnegative integer ℓ, and the $\hat{L}_z$ eigenvalues are parametrized by an integer m with $|m| \leq \ell$. The simultaneous eigenfunctions are the spherical harmonics, which are functions of θ and ϕ. With the angular dependence determined, the Schrödinger equation in three dimensions reduces to a one-dimensional radial equation that fixes the r dependence of the wave function.

10.1 Angular Momentum in Quantum Mechanics

We have so far considered a number of Hermitian operators: the position operator, the momentum operator, and the energy operator, or Hamiltonian. These operators are observables, and their eigenvalues are the possible results of measuring them on states. We will be discussing another operator: angular momentum. It is a vector operator, just like the momentum operator. It will lead to three components, $\hat{L}_x$, $\hat{L}_y$, and $\hat{L}_z$, each of which is a Hermitian operator, and thus a measurable quantity. The definition of the angular momentum operators, as you will see, arises from their familiar classical mechanics counterparts. Several properties of the operators, however, will be rather new and surprising.

In quantum mechanics, angular momentum can be of two types. One can have *orbital* angular momentum and *spin* angular momentum. To appreciate their distinction, it is useful to consider how these two types share a common origin in classical mechanics. For this purpose we will examine a collection of moving particles, each with a mass m_i and a time-dependent position $\mathbf{R}_i(t)$ measured relative to some chosen origin for the coordinate system. Here the index i labels the particles. Each particle has a velocity $\mathbf{V}_i$, and the total mass of the system is called M:

$$\mathbf{V}_i(t) = \dot{\mathbf{R}}_i(t), \qquad M = \sum_i m_i. \tag{10.1.1}$$

Here a dot denotes a time derivative. For such particles we can define the center of mass (CM) position $\mathbf{R}(t)$ and the center of mass velocity $\mathbf{V}(t)$:

$$\mathbf{R} = \frac{1}{M}\sum_i m_i \mathbf{R}_i, \quad \mathbf{V} = \dot{\mathbf{R}} = \frac{1}{M}\sum_i m_i \mathbf{V}_i. \tag{10.1.2}$$

Finally, for each particle we introduce positions $\mathbf{r}_i$ and velocities $\mathbf{v}_i$ *relative* to the CM:

$$\mathbf{R}_i = \mathbf{R} + \mathbf{r}_i, \quad \mathbf{V}_i = \mathbf{V} + \mathbf{v}_i, \tag{10.1.3}$$

where the second relation follows by taking time derivatives of the first. It is worth showing that the following sum vanishes:

$$\sum_i m_i \mathbf{r}_i = \sum_i m_i(\mathbf{R}_i - \mathbf{R}) = M\mathbf{R} - \mathbf{R}\sum_i m_i = 0. \tag{10.1.4}$$

We therefore have

$$\sum_i m_i \mathbf{r}_i = 0, \quad \sum_i m_i \mathbf{v}_i = 0, \tag{10.1.5}$$

with the second relation following by taking time derivatives of the first. Let us now compute the total angular momentum $\mathbf{J}$ of the full set of particles, relative to the origin. Adding the angular momentum of each of the particles, we find that

$$\mathbf{J} = \sum_i \mathbf{R}_i \times (m_i \mathbf{V}_i) = \sum_i (\mathbf{R} + \mathbf{r}_i) \times (m_i(\mathbf{V} + \mathbf{v}_i)), \tag{10.1.6}$$

writing the positions and velocities in terms of those relative to the CM. Expanding the cross product, we have

$$\mathbf{J} = \mathbf{R} \times \mathbf{V}\sum_i m_i + \mathbf{R} \times \sum_i m_i \mathbf{v}_i + \Big(\sum_i m_i \mathbf{r}_i\Big) \times \mathbf{V} + \sum_i \mathbf{r}_i \times (m_i \mathbf{v}_i). \tag{10.1.7}$$

Because of the sum rules (10.1.5), the second and third terms on the right-hand side vanish and we find

$$\mathbf{J} = \mathbf{R} \times (M\mathbf{V}) + \sum_i \mathbf{r}_i \times (m_i \mathbf{v}_i). \tag{10.1.8}$$

This is the expression we were after: it expresses the total angular momentum relative to the origin as the sum of the angular momentum of the CM relative to the origin (first term), plus an angular momentum due to motion relative to the center of mass (second term). If we were to replace sums with integrals, this result would apply equally well for a continuous mass distribution. If we have a finite-size object moving and rotating, the above formula gives the total angular momentum as a sum of an *orbital* part, representing the contribution to the angular momentum due to the CM motion about the origin, plus what we could call a *spin* part, representing the angular momentum due to the rotation of the object relative to its own CM. Writing $\mathbf{L}$ for the former and $\mathbf{S}$ for the latter, we have

$$\mathbf{J} = \mathbf{L} + \mathbf{S}, \quad \mathbf{L} = \mathbf{R} \times (M\mathbf{V}), \quad \mathbf{S} = \sum_i \mathbf{r}_i \times (m_i \mathbf{v}_i). \tag{10.1.9}$$

This is all classical. It would apply to a finite-size particle moving and spinning about an axis through its center of mass. Applied to an ordinary classical *point* particle, however, the total angular momentum is just orbital. The **S** contribution vanishes because the center of mass coordinate is the position of the particle, and since the particle has no size, all the vectors $\mathbf{r}_i$ relative to the CM vanish.

In quantum mechanics, however, it is common for a point particle to have spin angular momentum, an *intrinsic* angular momentum that is always present, even when the particle is not moving. One can wonder: Is it possible to imagine the quantum point particle as a tiny rotating ball in the limit as the size is very small? The answer is no. All attempts to model intrinsic angular momentum in this way have failed; the surface of the rotating ball would have to move faster than the speed of light. Intrinsic angular momentum in quantum mechanics has no origin as a physical rotation of material degrees of freedom. It has a more abstract description that we will explore in chapter 12. Still, intrinsic angular momentum in the quantum theory is angular momentum, and it is often added to the orbital angular momentum to find the total angular momentum. The quantum mechanical addition of angular momentum will be considered in chapter 20. Our focus in this chapter is on orbital angular momentum.

We have learned that operators act on states, and we have in fact defined operators through that action. It turns out that *functions* of the operators can act on states in ways that are quite intuitive and physical, thus allowing us to understand the operators better.

Consider, for example, the momentum operator $\hat{p}$ defined on wave functions $\psi(x)$ on the real line. With the identification $\hat{p}=\frac{\hbar}{i}\frac{d}{dx}$, we have that

$$\hat{p}\,\psi(x)=\frac{\hbar}{i}\frac{d\psi}{dx}. \tag{10.1.10}$$

In calculus you learned that Taylor expansions allow you to calculate the value of a function $\psi(x)$ near $x=x_0$ as follows:

$$\psi(x_0-a)=\psi(x_0)-a\frac{d\psi}{dx}\bigg|_{x_0}+\frac{a^2}{2!}\frac{d^2\psi}{dx^2}\bigg|_{x_0}-\frac{a^3}{3!}\frac{d^3\psi}{dx^3}\bigg|_{x_0}+\cdots. \tag{10.1.11}$$

The derivatives here are evaluated at x_0, and a is a small deviation from x_0. If such an expansion is valid for all values of x_0, we can write

$$\psi(x-a)=\psi(x)-a\frac{d\psi}{dx}+\frac{a^2}{2!}\frac{d^2\psi}{dx^2}-\frac{a^3}{3!}\frac{d^3\psi}{dx^3}+\cdots. \tag{10.1.12}$$

All derivatives here are evaluated at x, so there is no need for the evaluation symbol. Here, $\psi(x-a)$ is the function that arises by the translation of $\psi(x)$ by a distance a. Indeed, if $\psi(x)$ peaks at $x=0$, then it is clear that $\psi(x-a)$ peaks at $x=a$. The goal is now to recreate the right-hand side above via the action of a function of the momentum operator. The right function, we claim, is an operator U_a obtained as an exponential of the momentum:

$$U_a=\exp\Big(-\frac{ia\hat{p}}{\hbar}\Big)=\exp\Big(-a\frac{d}{dx}\Big). \tag{10.1.13}$$

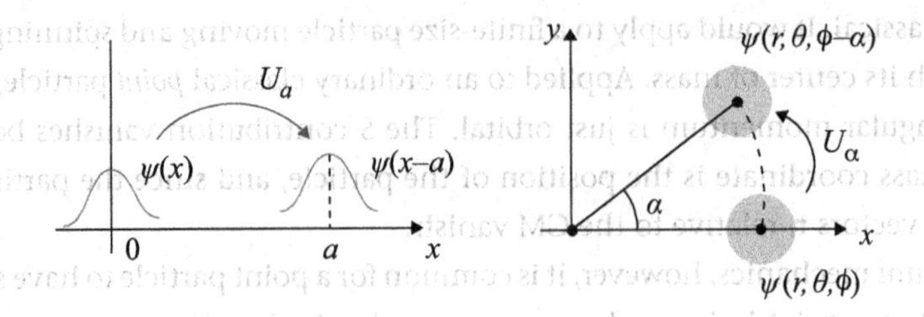

Figure 10.1
Left: The operator U_a translates the state $\psi(x)$ by a distance a. *Right*: The operator U_α rotates the state $\psi(r,\theta,\phi)$ about the z-axis by an angle α.

Letting this operator act on a wave function $\psi(x)$, expanding the exponential, and recalling that a is a constant gives

$$\begin{aligned} U_a\psi(x) &= \exp\left(-a\frac{d}{dx}\right)\psi(x) \\ &= \left(1 - a\frac{d}{dx} + \frac{a^2}{2!}\frac{d^2}{dx^2} - \frac{a^3}{3!}\frac{d^3}{dx^3} + \dots\right)\psi(x) \\ &= \psi(x) - a\frac{d\psi}{dx} + \frac{a^2}{2!}\frac{d^2\psi}{dx^2} - \frac{a^3}{3!}\frac{d^3\psi}{dx^3} + \cdots . \end{aligned} \tag{10.1.14}$$

Comparing the right-hand side with the expansion in (10.1.12), we have shown that

$$U_a\psi(x) = \psi(x-a) \tag{10.1.15}$$

and have demonstrated that the operator U_a generates a translation by a (figure 10.1, *left side*). Since U_a is an exponential of the momentum operator, we say that the momentum operator generates translations. This really means it generates translations when multiplied by the imaginary unit i, a distance parameter, divided by $\hbar$, and exponentiated. Since $\hat{p}$ is a Hermitian operator and a is real, U_a is a unitary operator, as you will now verify.

Exercise 10.1. *Use the series expansion of U_a in terms of $\hat{p}$ to show that $U_a^\dagger = U_{-a}$ and that U_a is unitary: $U_a^\dagger U_a = \mathbb{1}$. Here, $\mathbb{1}$ is the identity operator: $\mathbb{1}\psi = \psi$ for all ψ. You may assume that $e^M e^{-M} = \mathbb{1}$ for arbitrary operator M (exercise 13.13).*

Exercise 10.2. *Let M be a Hermitian operator. Show that* $\exp(iM)$ *is unitary.*

A similar story happens for angular momentum. We will see in section 10.4 that *angular momentum operators generate rotations about the origin*. In this situation we consider wave functions $\psi(\mathbf{x})$ in three dimensions and properly exponentiated angular momentum operators. For example, using the angular momentum operator $\hat{L}_z$, we build the operator U_α, defined as follows:

$$U_\alpha \equiv \exp\left(-\frac{i\alpha\hat{L}_z}{\hbar}\right). \tag{10.1.16}$$

Here α is an angle and therefore unit-free, and the argument of the exponential is unit-free because $\hbar$ has units of angular momentum. Since $\hat{L}_z$ is Hermitian, U_α is unitary, by the result of exercise 10.2 with $\hat{M} = -\alpha\hat{L}_z/\hbar$. Acting on a wave function, we claim, the operator U_α performs a rotation about the z-axis by an angle α. In spherical coordinates (r, θ, ϕ), this amounts to changing ϕ by the addition of α. Thus, acting on a wave function $\psi(r, \theta, \phi)$, the rotation operator must give

$$U_\alpha\, \psi(r,\theta,\phi) = \psi(r,\theta,\phi-\alpha). \tag{10.1.17}$$

Indeed, if a particular feature of $\phi(r,\theta,\phi)$ appears for a value ϕ_0 of the angle ϕ, that same feature appears at $\phi_0 + \alpha$ on the right-hand side. To the right of figure 10.1, we show a wave function $\psi(r,\theta,\phi)$, supported around some point on the x-axis, and a wave function $\psi(r,\theta,\phi-\alpha)$, obtained after a rotation by an angle α about the z-axis. The latter wave function is obtained from the former by the action of the operator U_α.

10.2 Schrödinger Equation in Three Dimensions and Angular Momentum

In this section we sketch the work to be done in the rest of this chapter and at the beginning of chapter 11. We begin by considering a particle moving in three spatial dimensions under the influence of a potential.

We recall that in three dimensions the position and momentum operators are triplets of operators, often referred to as *vector operators*:

$$\hat{\mathbf{x}} = (\hat{x}, \hat{y}, \hat{z}), \quad \hat{\mathbf{p}} = (\hat{p}_x, \hat{p}_y, \hat{p}_z) = \frac{\hbar}{i}\nabla = \frac{\hbar}{i}\Big(\frac{\partial}{\partial x}, \frac{\partial}{\partial y}, \frac{\partial}{\partial z}\Big). \tag{10.2.1}$$

The commutation relations are as follows:

$$[\,\hat{x}, \hat{p}_x\,] = i\hbar, \quad [\,\hat{y}, \hat{p}_y\,] = i\hbar, \quad [\,\hat{z}, \hat{p}_z\,] = i\hbar, \tag{10.2.2}$$

with all other commutators involving the three coordinates and the three momenta equal to zero. The state of a particle is represented by a three-dimensional wave function $\psi(\mathbf{x}) = \psi(x,y,z)$. The particle is moving in a three-dimensional potential $V(\mathbf{r}) = V(x,y,z)$. The Hamiltonian $\hat{H}$ is

$$\hat{H} = -\frac{\hbar^2}{2m}\nabla^2 + V(\mathbf{r}), \tag{10.2.3}$$

and as a result, the time-independent Schrödinger equation for energy eigenstates takes the form

$$-\frac{\hbar^2}{2m}\nabla^2\psi(\mathbf{r}) + V(\mathbf{r})\psi(\mathbf{r}) = E\psi(\mathbf{r}). \tag{10.2.4}$$

We have a **central potential** if $V(\mathbf{r}) = V(r)$. A central potential has no angular dependence. The value of the potential depends only on the distance r from the origin. A central potential is spherically symmetric; the surfaces of constant potential are spheres centered at the origin. For this reason the function $V(r)$ is said to be rotationally invariant: it has the same value at any two points $\mathbf{x}$ and $\mathbf{x}'$ that are related by a rotation about the origin. This is clear because such points have the same distance to the origin. For a central potential, the equation above reads

$$-\frac{\hbar^2}{2m}\nabla^2\psi(\mathbf{r}) + V(r)\psi(\mathbf{r}) = E\psi(\mathbf{r}). \tag{10.2.5}$$

This equation will be the main object of our study. Note that because the wave function is a general function of $\mathbf{r}$, it will only be rotational invariant for the simplest kinds of solutions. Given the rotational symmetry of the potential, we will express the Schrödinger equation and energy eigenfunctions using spherical coordinates.

In spherical coordinates the Laplacian is known to be given by

$$\nabla^2\psi = (\nabla\cdot\nabla)\psi = \frac{1}{r}\frac{\partial^2}{\partial r^2}(r\psi) + \frac{1}{r^2}\left(\frac{1}{\sin\theta}\frac{\partial}{\partial\theta}\sin\theta\frac{\partial}{\partial\theta} + \frac{1}{\sin^2\theta}\frac{\partial^2}{\partial\phi^2}\right)\psi. \tag{10.2.6}$$

Therefore, the Schrödinger equation for a particle in a central potential becomes

$$-\frac{\hbar^2}{2m}\left[\frac{1}{r}\frac{\partial^2}{\partial r^2}r + \frac{1}{r^2}\left(\frac{1}{\sin\theta}\frac{\partial}{\partial\theta}\sin\theta\frac{\partial}{\partial\theta} + \frac{1}{\sin^2\theta}\frac{\partial^2}{\partial\phi^2}\right)\right]\psi + V(r)\psi = E\psi. \tag{10.2.7}$$

We will aim to establish two facts:

1. The angular dependent piece of the ∇^2 operator can be identified as the magnitude squared of the angular momentum operator, with a minus sign:

$$\frac{1}{\sin\theta}\frac{\partial}{\partial\theta}\sin\theta\frac{\partial}{\partial\theta} + \frac{1}{\sin^2\theta}\frac{\partial^2}{\partial\phi^2} = -\frac{\hat{\mathbf{L}}^2}{\hbar^2}, \tag{10.2.8}$$

where

$$\hat{\mathbf{L}}^2 = \hat{L}_x\hat{L}_x + \hat{L}_y\hat{L}_y + \hat{L}_z\hat{L}_z. \tag{10.2.9}$$

It follows that the Laplacian operator in (10.2.6) is written as

$$\nabla^2 = \frac{1}{r}\frac{\partial^2}{\partial r^2}r - \frac{1}{r^2}\frac{\hat{\mathbf{L}}^2}{\hbar^2}. \tag{10.2.10}$$

Note that in the second term of this equation there is no concern about the order in which $\hat{\mathbf{L}}^2$ and $1/r^2$ act: $\hat{\mathbf{L}}^2$ is purely angular, and therefore it commutes with any function of r. The Hamiltonian now reads

$$H = -\frac{\hbar^2}{2m}\frac{1}{r}\frac{\partial^2}{\partial r^2}r + \frac{1}{2mr^2}\hat{\mathbf{L}}^2 + V(r). \tag{10.2.11}$$

This will imply that the Schrödinger equation (10.2.7) becomes

$$-\frac{\hbar^2}{2m}\left[\frac{1}{r}\frac{\partial^2}{\partial r^2}r - \frac{1}{r^2}\frac{\hat{\mathbf{L}}^2}{\hbar^2}\right]\psi + V(r)\psi = E\psi \tag{10.2.12}$$

or, expanding out,

$$\boxed{-\frac{\hbar^2}{2m}\frac{1}{r}\frac{\partial^2}{\partial r^2}(r\psi) + \frac{\hat{\mathbf{L}}^2}{2mr^2}\psi + V(r)\psi = E\psi.} \tag{10.2.13}$$

2. The central-potential Schrödinger equation (10.2.5) describes the nontrivial quantum dynamics for a *two-body* problem when the two-body potential is just a function of the distance between the particles:

$$V(\mathbf{r}_1, \mathbf{r}_2) = V(|\mathbf{r}_1 - \mathbf{r}_2|). \tag{10.2.14}$$

This is true for the electrostatic potential energy between the proton and the electron forming a hydrogen atom. Therefore, we will be able to treat the hydrogen atom as a central potential problem.

In the rest of this chapter, we will develop the theory of angular momentum, establish the first fact above, and obtain the so-called radial equation that governs the radial dependence of the three-dimensional wave functions. The discussion of the two-body problem is in chapter 11.

10.3 The Angular Momentum Operator

The angular momentum of classical mechanics is defined as the cross product of $\mathbf{r}$ and $\mathbf{p}$: $\mathbf{L}=\mathbf{r}\times\mathbf{p}$. Writing out the components of the vectors,

$$\mathbf{r}=(x,y,z),\quad \mathbf{p}=(p_x,p_y,p_z),\quad \text{and}\quad \mathbf{L}=(L_x,L_y,L_z), \tag{10.3.1}$$

the cross product gives

$$\begin{aligned} L_x &= yp_z - zp_y, \\ L_y &= zp_x - xp_z, \\ L_z &= xp_y - yp_x. \end{aligned} \tag{10.3.2}$$

To define the quantum *orbital* angular momentum operators $(\hat{L}_x,\hat{L}_y,\hat{L}_z)=\hat{\mathbf{L}}$, we literally copy the above relations with (x,y,z) replaced by the operators $(\hat{x},\hat{y},\hat{z})$ and (p_x,p_y,p_z) replaced by the operators $(\hat{p}_x,\hat{p}_y,\hat{p}_z)$:

$$\begin{aligned} \hat{L}_x &\equiv \hat{y}\hat{p}_z - \hat{z}\hat{p}_y, \\ \hat{L}_y &\equiv \hat{z}\hat{p}_x - \hat{x}\hat{p}_z, \\ \hat{L}_z &\equiv \hat{x}\hat{p}_y - \hat{y}\hat{p}_x. \end{aligned} \tag{10.3.3}$$

In passing from the classical to the quantum formula, we do not encounter ordering ambiguities. The classical expressions for angular momentum are not changed if the variables that are multiplied are written in reverse order. Happily, the quantum expressions are not changed either. For example, had we written $L_x=p_zy-p_yz$, clearly the same as the above L_x, and then defined $\hat{L}_x=\hat{p}_z\hat{y}-\hat{p}_y\hat{z}$, this new $\hat{L}_x$ would in fact be equal to the above $\hat{L}_x$ since $\hat{p}_z$ and $\hat{y}$ commute and so do $\hat{p}_y$ and $\hat{z}$. Indeed, in the quantum expressions the operator products involve a coordinate and a momentum along different axes, so they commute and their order does not matter. The operators are called orbital angular momentum operators because they are defined in analogy to the classical counterparts that capture the contribution to the angular momentum due to orbital motion.

It is simple to check that the Hermiticity of the angular momentum operators follows from the Hermiticity of $\hat{\mathbf{x}}$ and $\hat{\mathbf{p}}$. Take $\hat{L}_x$, for example. Recalling that $(AB)^\dagger=B^\dagger A^\dagger$ for any two operators A and B, we find that

$$(\hat{L}_x)^\dagger = (\hat{y}\hat{p}_z-\hat{z}\hat{p}_y)^\dagger = (\hat{y}\hat{p}_z)^\dagger-(\hat{z}\hat{p}_y)^\dagger = \hat{p}_z^\dagger\hat{y}^\dagger-\hat{p}_y^\dagger\hat{z}^\dagger. \tag{10.3.4}$$

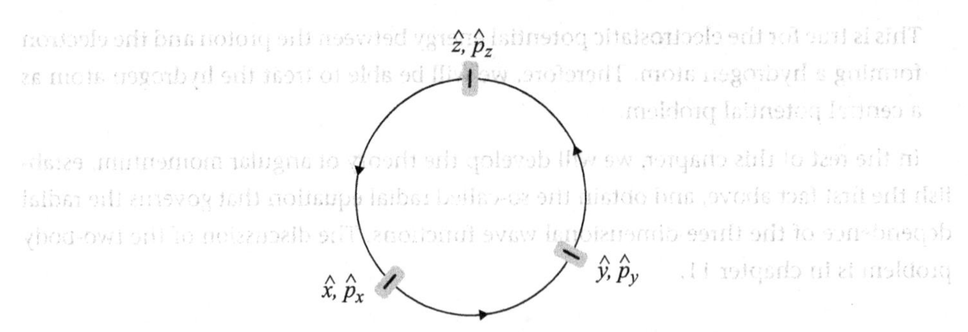

Figure 10.2
The commutation relations for angular momentum satisfy cyclicity.

Since all position and momenta are Hermitian operators, we have

$$(\hat{L}_x)^\dagger = \hat{p}_z\hat{y} - \hat{p}_y\hat{z} = \hat{y}\hat{p}_z - \hat{z}\hat{p}_y = \hat{L}_x, \tag{10.3.5}$$

where we moved the momenta to the right of the coordinates by virtue of vanishing commutators. The other two angular momentum operators are also Hermitian, so we have

$$\hat{L}_x^\dagger = \hat{L}_x, \quad \hat{L}_y^\dagger = \hat{L}_y, \quad \hat{L}_z^\dagger = \hat{L}_z. \tag{10.3.6}$$

All the angular momentum operators are Hermitian, and therefore they are all observables.

Given a set of Hermitian operators, it is natural to seek to determine their commutators. This computation enables us to see if we can have simultaneous eigenstates of the operators, as we will discuss in the following section. Let us calculate the commutator of $\hat{L}_x$ with $\hat{L}_y$:

$$[\hat{L}_x, \hat{L}_y] = [\hat{y}\hat{p}_z - \hat{z}\hat{p}_y, \hat{z}\hat{p}_x - \hat{x}\hat{p}_z]. \tag{10.3.7}$$

We now see that these terms fail to commute only because $\hat{z}$ and $\hat{p}_z$ fail to commute. In fact, the first term of $\hat{L}_x$ only fails to commute with the first term of $\hat{L}_y$. Similarly, the second term of $\hat{L}_x$ only fails to commute with the second term of $\hat{L}_y$. Therefore,

$$\begin{aligned} [\hat{L}_x, \hat{L}_y] &= [\hat{y}\hat{p}_z, \hat{z}\hat{p}_x] + [\hat{z}\hat{p}_y, \hat{x}\hat{p}_z] \\ &= [\hat{y}\hat{p}_z, \hat{z}]\hat{p}_x + \hat{x}[\hat{z}\hat{p}_y, \hat{p}_z] \\ &= \hat{y}[\hat{p}_z, \hat{z}]\hat{p}_x + \hat{x}[\hat{z}, \hat{p}_z]\hat{p}_y \\ &= \hat{y}(-i\hbar)\hat{p}_x + \hat{x}(i\hbar)\hat{p}_y \\ &= i\hbar(\hat{x}\hat{p}_y - \hat{y}\hat{p}_x). \end{aligned} \tag{10.3.8}$$

We now recognize that the operator on the final right-hand side is $\hat{L}_z$, so we have shown that

$$[\hat{L}_x, \hat{L}_y] = i\hbar\hat{L}_z. \tag{10.3.9}$$

The $\hat{x}$ and $\hat{p}$ commutation relations are completely cyclic, as illustrated in figure 10.2. In any $\hat{x}, \hat{p}$ commutation relation, if we cycle the position operator as in $\hat{x} \to \hat{y} \to \hat{z} \to \hat{x}$ and, simultaneously, the momentum operator as in $\hat{p}_x \to \hat{p}_y \to \hat{p}_z \to \hat{p}_x$, we will obtain another consistent commutation relation. You can also see, by looking at (10.3.3), that such cycling takes $\hat{L}_x \to \hat{L}_y \to \hat{L}_z \to \hat{L}_x$. We therefore claim that we do not have to calculate additional angular momentum commutators, and (10.3.9) leads to

$$\boxed{\begin{aligned}[\hat{L}_x, \hat{L}_y] &= i\hbar \hat{L}_z,\\ [\hat{L}_y, \hat{L}_z] &= i\hbar \hat{L}_x,\\ [\hat{L}_z, \hat{L}_x] &= i\hbar \hat{L}_y.\end{aligned}} \tag{10.3.10}$$

This is the full set of commutators of orbital angular momentum operators. The set of commutators defines the **algebra of angular momentum**. Notice that while the operators $\hat{L}_x, \hat{L}_y$, and $\hat{L}_z$ were defined in terms of coordinates and momenta, the final answers for the commutators involve neither coordinates nor momenta: commutators of angular momenta give angular momenta! The algebra of angular momentum is a structure realized by the particular construction we have given, but other possible constructions are also physically relevant. Any three operators that satisfy the above commutation relations form an algebra of angular momentum. An algebra where all products of operators appear in the form of commutators is called a *Lie algebra*. General Lie algebras can involve any number of operators.

A little notation allows us to write the algebra more briefly. Defining

$$\hat{L}_1 \equiv \hat{L}_x, \quad \hat{L}_2 \equiv \hat{L}_y, \quad \hat{L}_3 \equiv \hat{L}_z, \tag{10.3.11}$$

we can use index notation for $\hat{L}_i$ operators with $i = 1, 2, 3$. The above commutators are in fact summarized by the single equation

$$\boxed{[\hat{L}_i, \hat{L}_j] = i\hbar\, \epsilon_{ijk}\, \hat{L}_k.} \tag{10.3.12}$$

In here ϵ_{ijk} is the totally antisymmetric symbol defined by the relation

$$\epsilon_{123} = +1. \tag{10.3.13}$$

Totally antisymmetric means the object is antisymmetric under the exchange of *any* two of its indices:

$$\epsilon_{ijk} = -\epsilon_{jik} = -\epsilon_{kji} = -\epsilon_{ikj}. \tag{10.3.14}$$

Any epsilon with a repeated index vanishes, by the required antisymmetry, thus $\epsilon_{133} = 0$, for example. The ϵ_{ijk} symbol vanishes unless the indices are some permutation of $1, 2, 3$. For example, $\epsilon_{321} = -\epsilon_{123} = -1$, by exchange of the first and third index. Moreover, the repeated index k in (10.3.12) is summed over all three possible values 1, 2, 3. Therefore, we have

$$[\hat{L}_1, \hat{L}_2] = i\hbar\epsilon_{12k}\hat{L}_k \equiv i\hbar \sum_{k=1}^{3} \epsilon_{12k}\hat{L}_k = i\hbar\epsilon_{123}\hat{L}_3 = i\hbar\hat{L}_3, \tag{10.3.15}$$

since $\epsilon_{121} = \epsilon_{122} = 0$, and only the term with $k = 3$ can contribute. The result is the first commutator in (10.3.10).

We stated in section 10.1 that in quantum mechanics one naturally has *intrinsic* or spin angular momentum. The spin angular momentum operators $\hat{S}_x, \hat{S}_y$, and $\hat{S}_z$ cannot be written in terms of coordinates and momenta, making them quite different from the orbital operators. They are more abstract entities. In fact, their simplest construction is as 2-by-2 matrices, as we will learn in chapter 12. Still, they are angular momentum operators, so they satisfy exactly the same algebra as their orbital cousins:

$$\begin{aligned}
&[\hat{S}_x, \hat{S}_y] = i\hbar\,\hat{S}_z,\\
&[\hat{S}_y, \hat{S}_z] = i\hbar\,\hat{S}_x \quad \Rightarrow \quad [\hat{S}_i, \hat{S}_j] = i\hbar\,\epsilon_{ijk}\hat{S}_k,\\
&[\hat{S}_z, \hat{S}_x] = i\hbar\,\hat{S}_y.
\end{aligned} \tag{10.3.16}$$

Sometimes physicists speak of angular momentum without assuming that it is orbital, spin, or any combination thereof. In that case the operators are often written as $\hat{\mathbf{J}} = (\hat{J}_x, \hat{J}_y, \hat{J}_z)$, and their algebra is

$$\begin{aligned}
&[\hat{J}_x, \hat{J}_y] = i\hbar\,\hat{J}_z\\
&[\hat{J}_y, \hat{J}_z] = i\hbar\,\hat{J}_x \quad \Rightarrow \quad [\hat{J}_i, \hat{J}_j] = i\hbar\,\epsilon_{ijk}\hat{J}_k.\\
&[\hat{J}_z, \hat{J}_x] = i\hbar\,\hat{J}_y
\end{aligned} \tag{10.3.17}$$

10.4 Commuting Operators and Rotations

We have seen that the nonvanishing commutator $[\hat{x}, \hat{p}] = i\hbar$ implies that we cannot have simultaneous eigenstates of position and of momentum (example 5.3). Let us now see what the commutators of the angular momentum operators tell us. In particular we begin by asking: Can we have simultaneous eigenstates of $\hat{L}_x$ and $\hat{L}_y$ with nonzero eigenvalues? As it turns out, the answer is no. We demonstrate this as follows. Let's assume there exists a wave function ϕ_0 that is simultaneously an eigenstate of $\hat{L}_x$ and $\hat{L}_y$:

$$\begin{aligned}
&\hat{L}_x\,\phi_0 = l_x\phi_0,\\
&\hat{L}_y\,\phi_0 = l_y\phi_0.
\end{aligned} \tag{10.4.1}$$

Here l_x and l_y denote the corresponding eigenvalues. Letting the first commutator identity of (10.3.10) act on ϕ_0, we have

$$[\hat{L}_x, \hat{L}_y]\phi_0 = i\hbar\,\hat{L}_z\phi_0. \tag{10.4.2}$$

The left-hand side is quickly evaluated and found to vanish:

$$[\hat{L}_x, \hat{L}_y]\phi_0 = \hat{L}_x\hat{L}_y\phi_0 - \hat{L}_y\hat{L}_x\phi_0 = \hat{L}_x l_y\phi_0 - \hat{L}_y l_x\phi_0 = (l_x l_y - l_y l_x)\phi_0 = 0. \tag{10.4.3}$$

It follows now that the right-hand side of the commutator identity must vanish as well:

$$\hat{L}_z\phi_0 = 0. \tag{10.4.4}$$

The eigenstate in question must be annihilated by $\hat{L}_z$. Consider now the other two commutator identities acting on ϕ_0:

$$\begin{aligned}
&[\hat{L}_y, \hat{L}_z]\phi_0 = i\hbar\,\hat{L}_x\phi_0,\\
&[\hat{L}_z, \hat{L}_x]\phi_0 = i\hbar\,\hat{L}_y\phi_0.
\end{aligned} \tag{10.4.5}$$

With $\hat{L}_z\phi_0 = 0$, we confirm that the left-hand sides above vanish:

$$\begin{aligned}[\hat{L}_y, \hat{L}_z]\phi_0 &= \hat{L}_y\hat{L}_z\phi_0 - \hat{L}_z\hat{L}_y\phi_0 = \hat{L}_y \cdot 0 - l_y\hat{L}_z\phi_0 = 0,\\ [\hat{L}_z, \hat{L}_x]\phi_0 &= \hat{L}_z\hat{L}_x\phi_0 - \hat{L}_x\hat{L}_z\phi_0 = l_x\hat{L}_z\phi_0 - \hat{L}_x \cdot 0 = 0.\end{aligned} \tag{10.4.6}$$

Back in (10.4.5), the first relation implies that $\hat{L}_x\phi_0 = 0$, and the second relation implies that $\hat{L}_y\phi_0 = 0$. These mean that $l_x = l_y = 0$. All in all, assuming that ϕ_0 is a simultaneous eigenstate of $\hat{L}_x$ and $\hat{L}_y$ has led to

$$\hat{L}_x\phi_0 = \hat{L}_y\phi_0 = \hat{L}_z\phi_0 = 0. \tag{10.4.7}$$

The state is annihilated by *all* the angular momentum operators. We have learned that it is impossible to find states that are simultaneous eigenstates of any two angular momentum operators with nonzero eigenvalues. Rotationally invariant wave functions depend only on r and are, in fact, annihilated by all the angular momentum operators.

Exercise 10.3. *Show that $\hat{L}_i f(r) = 0$ for $i = 1, 2, 3$ and for arbitrary f by representing the $\hat{L}_i$ as differential operators on account of $\hat{p}_k = \frac{\hbar}{i}\frac{\partial}{\partial x_k}$.*

For Hermitian operators that commute, there is no obstacle to finding simultaneous eigenstates. This is helpful because the states can then be characterized by the eigenvalues of the operators. In fact, we will show in chapter 15 that commuting Hermitian operators always have a *complete* set of simultaneous eigenstates. This claim, for the particular case in which at least one of the two operators has no degeneracies, is simple to establish (problem 10.3). Suppose we select $\hat{L}_z$ as one of the operators we want to measure. Can we now find a second Hermitian operator that commutes with it? Neither $\hat{L}_x$ nor $\hat{L}_y$ will do, but there is an operator that works. As it turns out, $\hat{\mathbf{L}}^2$, defined in (10.2.9), commutes with $\hat{L}_z$ and is an interesting choice for a second operator. Indeed, we quickly check that

$$\begin{aligned}[\hat{L}_z, \hat{\mathbf{L}}^2] &= [\hat{L}_z, \hat{L}_x\hat{L}_x] + [\hat{L}_z, \hat{L}_y\hat{L}_y]\\ &= [\hat{L}_z, \hat{L}_x]\hat{L}_x + \hat{L}_x[\hat{L}_z, \hat{L}_x] + [\hat{L}_z, \hat{L}_y]\hat{L}_y + \hat{L}_y[\hat{L}_z, \hat{L}_y]\\ &= i\hbar\hat{L}_y\hat{L}_x + i\hbar\hat{L}_x\hat{L}_y - i\hbar\hat{L}_x\hat{L}_y - i\hbar\hat{L}_y\hat{L}_x = 0.\end{aligned} \tag{10.4.8}$$

So we should be able to find simultaneous eigenstates of both $\hat{L}_z$ and $\hat{L}^2$. We will do this in the following section. The operator $\hat{\mathbf{L}}^2$ is a *Casimir* operator for the algebra of angular momentum, which means it commutes with all angular momentum operators. Just as $\hat{\mathbf{L}}^2$ commutes with $\hat{L}_z$, it also commutes with $\hat{L}_x$ and with $\hat{L}_y$:

$$\boxed{[\hat{\mathbf{L}}^2, \hat{L}_i] = 0, \quad i = 1, 2, 3.} \tag{10.4.9}$$

To understand the angular momentum operators a little better, let us write them in spherical coordinates. For this we need the relation between (r, θ, ϕ) and the Cartesian coordinates (x, y, z):

$$\begin{aligned} x &= r\sin\theta\cos\phi, & r &= \sqrt{x^2 + y^2 + z^2},\\ y &= r\sin\theta\sin\phi, & \theta &= \cos^{-1}\left(\tfrac{z}{r}\right),\\ z &= r\cos\theta, & \phi &= \tan^{-1}\left(\tfrac{y}{x}\right).\end{aligned} \tag{10.4.10}$$

In spherical coordinates, rotations about the z-axis are the simplest: they change ϕ but leave θ invariant. Both rotations about the x- and y-axes change θ *and* ϕ. We can therefore hope that $\hat{L}_z$ is simple in spherical coordinates and expressible in terms of $\frac{\partial}{\partial \phi}$. To check this, we first write $\partial/\partial\phi$ in terms of derivatives of Cartesian coordinates:

$$\frac{\partial}{\partial \phi}=\frac{\partial y}{\partial \phi}\frac{\partial}{\partial y}+\frac{\partial x}{\partial \phi}\frac{\partial}{\partial x}+\frac{\partial z}{\partial \phi}\frac{\partial}{\partial z}=x\frac{\partial}{\partial y}-y\frac{\partial}{\partial x}, \tag{10.4.11}$$

where we used (10.4.10) to evaluate the partial derivatives and noted that z is ϕ independent. On the other hand, using the definition $\hat{L}_z=\hat{x}\hat{p}_y-\hat{y}\hat{p}_x$ gives

$$\hat{L}_z=\frac{\hbar}{i}\Big(x\frac{\partial}{\partial y}-y\frac{\partial}{\partial x}\Big). \tag{10.4.12}$$

We therefore conclude that

$$\boxed{\hat{L}_z=\frac{\hbar}{i}\frac{\partial}{\partial \phi}.} \tag{10.4.13}$$

This is a very simple and useful representation. Since ϕ is a unit-free angle, the units of angular momentum are carried by $\hbar$. The formula, with a derivative along ϕ, shows that $\hat{L}_z$ is some kind of momentum along an angle direction. Since ϕ and $\phi+2\pi$ represent the same point in spherical coordinates, we can say that ϕ lives in the circle $\phi\sim\phi+2\pi$. This implies that $\hat{L}_z$ eigenvalues are quantized, just like momentum is quantized on a circle. The quantization of $\hat{L}_z$ will be confirmed in the following section.

While the $\hat{L}_z$ operator takes a simple form in spherical coordinates, the $\hat{L}_x$ and $\hat{L}_y$ operators are more complicated. Rather than calculating these two separately, it is more useful to consider the complex linear combinations $\hat{L}_\pm\equiv\hat{L}_x\pm i\hat{L}_y$. As you will confirm in problem 10.2,

$$\hat{L}_\pm\equiv\hat{L}_x\pm i\hat{L}_y=\pm\hbar e^{\pm i\phi}\Big(\frac{\partial}{\partial\theta}\pm i\cot\theta\frac{\partial}{\partial\phi}\Big). \tag{10.4.14}$$

Additional calculation in that problem shows that, as suggested earlier, the $\hat{\mathbf{L}}^2$ operator encodes the angular part of the Laplacian:

$$\boxed{-\frac{\hat{\mathbf{L}}^2}{\hbar^2}=\frac{1}{\sin\theta}\frac{\partial}{\partial\theta}\Big(\sin\theta\frac{\partial}{\partial\theta}\Big)+\frac{1}{\sin^2\theta}\frac{\partial^2}{\partial\phi^2}.} \tag{10.4.15}$$

Let us close this section by showing that, as claimed in section 10.1, angular momentum operators generate rotations. We focus on the z component $\hat{L}_z$ and consider the operator $U_\alpha=\exp(-i\alpha\hat{L}_z/\hbar)$. With the above representation of $\hat{L}_z$, we find that

$$U_\alpha=\exp\Big(-\alpha\frac{\partial}{\partial\phi}\Big). \tag{10.4.16}$$

Just as we did for the case of the momentum operator in (10.1.14) and (10.1.15), we now see that, acting on a wave function expressed in spherical coordinates $\psi(r,\theta,\phi)$,

$$U_\alpha \psi(r,\theta,\phi) = \exp\Big(-\alpha \frac{\partial}{\partial\phi}\Big)\psi(r,\theta,\phi) = \psi(r,\theta,\phi-\alpha). \tag{10.4.17}$$

A feature of the original wave function at $\phi = 0$ now appears on the resulting wave function at $\phi = \alpha$, with the same values of r and θ. This means the state has been rotated by an angle α about the z-axis. This is what we wanted to show. Rotations about general directions take more effort to implement. They will be studied in section 14.7.

10.5 Eigenstates of Angular Momentum

We have already observed that the Hermitian operators $\hat{L}_z$ and $\hat{\mathbf{L}}^2$ commute. We now aim to construct the simultaneous eigenfunctions of these operators. They will be functions of θ and ϕ, and we will call them $\psi_{\ell m}(\theta,\phi)$. The conditions they must meet to be eigenfunctions are

$$\begin{aligned} \hat{L}_z\,\psi_{\ell m} &= \hbar m\,\psi_{\ell m}, & m\in\mathbb{R}, \\ \hat{\mathbf{L}}^2\,\psi_{\ell m} &= \hbar^2\,\ell(\ell+1)\,\psi_{\ell m}, & \ell\in\mathbb{R}. \end{aligned} \tag{10.5.1}$$

As befits Hermitian operators, the eigenvalues are real; both m and l are unit-free real numbers. There is an $\hbar$ in the $\hat{L}_z$ eigenvalue and an $\hbar^2$ in the $\hat{\mathbf{L}}^2$ eigenvalue because angular momentum has units of $\hbar$. Note that we have written the eigenvalue of $\hat{\mathbf{L}}^2$ as $\ell(\ell+1)$ instead of a simpler ℓ, a notation that will prove convenient soon. Note that $\ell(\ell+1)$ grows monotonically from zero to infinity as ℓ ranges from zero to infinity. It follows that a positive value of $\ell(\ell+1)$ uniquely fixes a positive value of ℓ.

We first show that the eigenvalues of $\hat{\mathbf{L}}^2$ can't be negative. This follows if we can demonstrate that

$$\langle\psi,\,\hat{\mathbf{L}}^2\psi\rangle \geq 0. \tag{10.5.2}$$

If this holds, taking ψ to be a normalized eigenfunction with $\hat{\mathbf{L}}^2$ eigenvalue λ we immediately find $\langle\psi,\,\lambda\psi\rangle = \lambda \geq 0$, as desired. To prove the above inequality, we simply expand and use Hermiticity of the angular momentum operators:

$$\begin{aligned} \langle\psi,\,\hat{\mathbf{L}}^2\psi\rangle &= \langle\psi,\,\hat{L}_x^2\psi\rangle + \langle\psi,\,\hat{L}_y^2\psi\rangle + \langle\psi,\,\hat{L}_z^2\psi\rangle \\ &= \langle\hat{L}_x\psi,\,\hat{L}_x\psi\rangle + \langle\hat{L}_y\psi,\,\hat{L}_y\psi\rangle + \langle\hat{L}_z\psi,\,\hat{L}_z\psi\rangle \geq 0. \end{aligned} \tag{10.5.3}$$

The last inequality holds because each of the three summands is greater than or equal to zero. Having shown that the eigenvalues of $\hat{\mathbf{L}}^2$ are nonnegative, we can indeed restrict ourselves to $\ell \geq 0$ in the definition (10.5.1).

Let us now solve the first eigenvalue equation in (10.5.1) using the position representation (10.4.13) for the $\hat{L}_z$ operator:

$$\frac{\hbar}{i}\frac{\partial\psi_{\ell m}}{\partial\phi} = \hbar m \psi_{\ell m} \quad\Rightarrow\quad \frac{\partial\psi_{\ell m}}{\partial\phi} = im\,\psi_{\ell m}. \tag{10.5.4}$$

This determines the ϕ dependence of the solution and we write

$$\psi_{\ell m}(\theta,\phi) = e^{im\phi}\,P_\ell^m(\theta), \tag{10.5.5}$$

where the function $P_\ell^m(\theta)$ captures the still undetermined θ dependence of the eigenfunction $\psi_{\ell m}$. We will require that $\psi_{\ell m}$ be uniquely defined as a function of the angles, and this requires that

$$\psi_{\ell m}(\theta, \phi + 2\pi) = \psi_{\ell m}(\theta, \phi). \tag{10.5.6}$$

There is no similar condition for θ, whose range is from 0 to π. The above condition requires that

$$e^{im(\phi+2\pi)} = e^{im\phi} \quad \Rightarrow \quad e^{2\pi i m} = 1. \tag{10.5.7}$$

This equation implies that m must be an integer:

$$\boxed{m \in \mathbb{Z}.} \tag{10.5.8}$$

The eigenvalues $\hbar m$ of $\hat{L}_z$ are indeed quantized. This completes our analysis of the first eigenvalue equation in (10.5.1).

Let us now turn to the second eigenvalue equation. Using our expression (10.4.15) for $\hat{\mathbf{L}}^2$ gives

$$-\hbar^2\left(\frac{1}{\sin\theta}\frac{\partial}{\partial\theta}\left(\sin\theta\frac{\partial}{\partial\theta}\right) + \frac{1}{\sin^2\theta}\frac{\partial^2}{\partial\phi^2}\right)\psi_{\ell m} = \hbar^2\ell(\ell+1)\psi_{\ell m}. \tag{10.5.9}$$

We multiply through by $\sin^2\theta$ and cancel the $\hbar^2$ to get

$$\left(\sin\theta\frac{\partial}{\partial\theta}\sin\theta\frac{\partial}{\partial\theta} + \frac{\partial^2}{\partial\phi^2}\right)\psi_{\ell m} = -\ell(\ell+1)\sin^2\theta\,\psi_{\ell m}. \tag{10.5.10}$$

Using $\psi_{\ell m} = e^{im\phi}P_\ell^m(\theta)$, we can evaluate the action of $\frac{\partial^2}{\partial\phi^2}$ on $\psi_{\ell m}$ and then cancel the overall $e^{im\phi}$ to arrive at the differential equation for the functions P_ℓ^m:

$$\sin\theta\frac{d}{d\theta}\left(\sin\theta\frac{dP_\ell^m}{d\theta}\right) + \left(\ell(\ell+1)\sin^2\theta - m^2\right)P_\ell^m = 0. \tag{10.5.11}$$

We now want to make clear that we can view P_ℓ^m as a function of $\cos\theta$ by writing the differential equation in terms of $x = \cos\theta$, with $x \in [-1, 1]$. Indeed, this gives

$$\frac{d}{d\theta} = \frac{dx}{d\theta}\frac{d}{dx} = -\sin\theta\frac{d}{dx} \quad \Rightarrow \quad \sin\theta\frac{d}{d\theta} = -(1-x^2)\frac{d}{dx}. \tag{10.5.12}$$

The differential equation becomes

$$(1-x^2)\frac{d}{dx}\left[(1-x^2)\frac{dP_\ell^m}{dx}\right] + [\ell(\ell+1)(1-x^2) - m^2]P_\ell^m(x) = 0. \tag{10.5.13}$$

Dividing by $1-x^2$, we get the final form

$$\boxed{\frac{d}{dx}\left[(1-x^2)\frac{dP_\ell^m}{dx}\right] + \left[\ell(\ell+1) - \frac{m^2}{1-x^2}\right]P_\ell^m(x) = 0.} \tag{10.5.14}$$

The $P_\ell^m(x)$ are called the *associated Legendre functions*. As we will see, the $P_\ell^m(x)$ are not polynomials in x unless $m=0$. All we know at this point is that m is an integer, and ℓ is not negative. We will soon discover that ℓ is a nonnegative *integer* and that for a given value of ℓ there is a range of possible values of m.

To find the conditions on ℓ, we consider the above equation for $m=0$. In that case we define $P_\ell(x) \equiv P_\ell^0(x)$, and the $P_\ell(x)$ must satisfy

$$\frac{d}{dx}\left[(1-x^2)\frac{dP_\ell}{dx}\right] + \ell(\ell+1)P_\ell(x) = 0. \tag{10.5.15}$$

This is the well-studied Legendre differential equation. We try finding a series solution by writing

$$P_\ell(x) = \sum_{k=0}^{\infty} a_k x^k, \tag{10.5.16}$$

assuming that $P_\ell(x)$ is regular at $x=0$. Just as we did for the harmonic oscillator differential equation in section 9.2, we consider a few terms in the series expansion

$$P_\ell(x) = \cdots + a_k x^k + a_{k+1}x^{k+1} + a_{k+2}x^{k+2} + \cdots \tag{10.5.17}$$

and examine how they contribute to the total coefficient of x^k when substituted into the left-hand side of the differential equation (10.5.15). This coefficient must vanish for all values of k for the differential equation to hold. A short calculation gives the condition:

$$(k+1)(k+2)a_{k+2} + [\ell(\ell+1) - k(k+1)]a_k = 0. \tag{10.5.18}$$

Equivalently, we have that

$$\frac{a_{k+2}}{a_k} = \frac{k(k+1) - \ell(\ell+1)}{(k+1)(k+2)}. \tag{10.5.19}$$

The large k behavior of the coefficients is such that unless the series terminates, P_ℓ diverges at $x=\pm 1$ (problem 10.4). Since $x=\cos\theta$, this corresponds to $\theta=\{0,\pi\}$, the north and south poles of the sphere, respectively. In order for the series to terminate, we must have

$$\ell(\ell+1) = k(k+1), \tag{10.5.20}$$

for some integer $k \geq 0$. The utility of picking $\hbar^2\ell(\ell+1)$ as the eigenvalue of $\hat{\mathbf{L}}^2$ now becomes clear. We can simply pick $\ell=k$ so that $a_{k+2}=0$, making $P_k(x)$ a degree k polynomial. The other solution of the quadratic equation (10.5.20) gives a negative ℓ, which we already know is not allowed. We have thus learned that the possible values of ℓ are

$$\boxed{\ell = 0, 1, 2, 3, \ldots.} \tag{10.5.21}$$

This is the quantization of ℓ. Just like the m values, the ℓ values are also quantized. We will not discuss here in detail the construction of the Legendre polynomials $P_\ell(x)$.

We just note that they are given by the Rodriguez formula,

$$P_\ell(x) = \frac{1}{2^\ell \ell!}\left(\frac{d}{dx}\right)^\ell \left(x^2-1\right)^\ell, \tag{10.5.22}$$

and that they have a simple generating function:

$$\frac{1}{\sqrt{1-2xs+s^2}} = \sum_{\ell=0}^{\infty} P_\ell(x) s^\ell. \tag{10.5.23}$$

The first three Legendre polynomials are

$$\begin{aligned} P_0(x) &= 1, \\ P_1(x) &= x, \\ P_2(x) &= \tfrac{1}{2}\left(3x^2-1\right). \end{aligned} \tag{10.5.24}$$

$P_\ell(x)$ is a degree ℓ polynomial of definite parity. For even ℓ the polynomials are even polynomials, and for odd ℓ the polynomials are odd polynomials.

Exercise 10.4. *Use the generating function (10.5.23) to show that*

$$P_\ell(1) = 1, \quad \textit{for all } \ell. \tag{10.5.25}$$

Exercise 10.5. *Use the Rodriguez formula to show that the leading term in the Legendre polynomial P_ℓ is*

$$P_\ell(x) = \frac{(2\ell)!}{2^\ell(\ell!)^2}\, x^\ell + \cdots \tag{10.5.26}$$

Having solved the $m=0$ equation, we now have to discuss the general equation for $P_\ell^m(x)$. The differential equation (10.5.14) involves m^2 and not m, so we can take the solutions for m and $-m$ to be the same. Happily, the $m \neq 0$ functions can be obtained easily from the Legendre polynomials. One can show that taking $|m|$ derivatives of the Legendre polynomials and multiplying by a prefactor gives a solution for $P_\ell^m(x)$:

$$P_\ell^m(x) = (1-x^2)^{|m|/2}\left(\frac{d}{dx}\right)^{|m|} P_\ell(x). \tag{10.5.27}$$

Since P_ℓ is a polynomial of degree ℓ, the above gives a nonzero answer only for $|m| \leq \ell$. We thus have solutions for

$$\boxed{-\ell \leq m \leq \ell.} \tag{10.5.28}$$

These solutions are regular; they do not diverge anywhere in the interval $x \in [-1, 1]$. It is possible to prove that no other regular solutions exist.

At this point we have solved completely for the associated Legendre functions. Therefore, we have obtained all $\hat{L}_z$ and $\hat{\mathbf{L}}^2$ eigenfunctions $\psi_{\ell m} = e^{im\phi} P_\ell^m(\cos\theta)$. These are first determined by the integer ℓ associated with the magnitude of the angular momentum and, for a fixed ℓ, by the choice of m. There are $2\ell+1$ choices of m: $-\ell, -\ell+1, \ldots, \ell$.

For $\ell = 0$ we just have $m = 0$, for $\ell = 1$ we have $m = -1, 0, 1$, and so on. The eigenstates $\psi_{\ell m}$ with a fixed value of ℓ and all allowed values of m form an ℓ *multiplet*, also called a multiplet of angular momentum ℓ. Here are the first few multiplets:

$$\begin{aligned} \ell = 0: \quad & m = \ \ 0, \\ \ell = 1: \quad & m = -1, 0, 1, \\ \ell = 2: \quad & m = -2, -1, 0, 1, 2, \\ \ell = 3: \quad & m = -3, -2, -1, 0, 1, 2, 3. \end{aligned} \tag{10.5.29}$$

All states in a multiplet have angular momenta of the same magnitude.

Exercise 10.6. *Compute the associated Legendre function $P_2^1(x)$.*

Exercise 10.7. *Compute the associated Legendre function $P_\ell^\ell(x)$ for arbitrary $\ell \geq 0$.*

Our $\psi_{\ell m}$ eigenfunctions, with suitable normalization, are called the **spherical harmonics** $Y_{\ell m}(\theta, \phi)$. The properly normalized spherical harmonics for $m \geq 0$ are

$$Y_{\ell m}(\theta, \phi) \equiv \sqrt{\frac{2\ell + 1}{4\pi} \frac{(\ell - m)!}{(\ell + m)!}} \, (-1)^m \, e^{im\phi} \, P_\ell^m(\cos\theta). \tag{10.5.30}$$

For $m < 0$, we use

$$Y_{\ell m}(\theta, \phi) = (-1)^m \left[Y_{\ell, -m}(\theta, \phi) \right]^* . \tag{10.5.31}$$

We thus have, from (10.5.1) and the lessons we have learned,

$$\boxed{\begin{aligned} \hat{\mathbf{L}}^2 \, Y_{\ell m} &= \hbar^2 \, \ell(\ell + 1) \, Y_{\ell m}, \quad \ell = 0, 1, \ldots \\ \hat{L}_z \, Y_{\ell m} &= \hbar m \, Y_{\ell m}, \qquad\quad m = -\ell, \ldots, \ell. \end{aligned}} \tag{10.5.32}$$

The first few spherical harmonics are

$$\begin{aligned} Y_{00}(\theta, \phi) &= \frac{1}{\sqrt{4\pi}}, \\ Y_{1,\pm 1}(\theta, \phi) &= \mp \sqrt{\frac{3}{8\pi}} e^{\pm i\phi} \sin\theta = \mp \sqrt{\frac{3}{8\pi}} \frac{x \pm iy}{r}, \\ Y_{10}(\theta, \phi) &= \sqrt{\frac{3}{4\pi}} \cos\theta = \sqrt{\frac{3}{4\pi}} \frac{z}{r}. \end{aligned} \tag{10.5.33}$$

Note also that $Y_{\ell 0}$ is, up to normalization, the Legendre polynomial $P_\ell(\cos\theta)$:

$$Y_{\ell 0}(\theta) = \sqrt{\frac{2\ell + 1}{4\pi}} P_\ell(\cos\theta). \tag{10.5.34}$$

This follows directly from (10.5.30), recalling that $P_\ell(x) = P_\ell^0(x)$.

Exercise 10.8. *Show that*

$$Y_{\ell\ell} = \frac{1}{2^\ell \ell!} \sqrt{\frac{(2\ell + 1)!}{4\pi}} (-1)^\ell e^{i\ell\phi} (\sin\theta)^\ell. \tag{10.5.35}$$

Note that $Y_{\ell\ell} \sim (x + iy)^\ell / r^\ell$.

Being eigenstates of the Hermitian operators $\hat{L}_z$ and $\hat{\mathbf{L}}^2$ with different eigenvalues, spherical harmonics with different ℓ and m subscripts are automatically orthogonal. The Hermiticity of these operators is relative to the inner product that is defined by integration over all of space. In spherical coordinates this involves integration over r, θ, and ϕ, as follows:

$$\langle \psi_1, \psi_2 \rangle = \int d^3\mathbf{x}\, \psi_1^* \psi_2 = \int_0^\infty r^2 dr \int_0^\pi \sin\theta\, d\theta \int_0^{2\pi} d\phi\, \psi_1^*\, \psi_2. \tag{10.5.36}$$

The angular integration is over solid angle $d\Omega$ and is written as

$$\langle \psi_1, \psi_2 \rangle = \int_0^\infty r^2 dr \int d\Omega\, \psi_1^*\, \psi_2, \quad \text{with} \quad \int d\Omega \equiv \int_0^{2\pi} d\phi \int_{-1}^{1} d(\cos\theta). \tag{10.5.37}$$

If we examine the Hermiticity of $\hat{L}_z$ and $\hat{\mathbf{L}}^2$ in spherical coordinates, the integral over r plays no role because these operators are purely angular. Thus, the Hermiticity of $\hat{L}_z$ and $\hat{\mathbf{L}}^2$ also holds in an *angular* inner product $\langle \cdot, \cdot \rangle_a$ defined as

$$\langle \psi_1, \psi_2 \rangle_a \equiv \int d\Omega\, \psi_1^*\, \psi_2 \tag{10.5.38}$$

for wave functions with angular dependence.

Exercise 10.9. *Confirm explicitly that the operators $\hat{L}_z$ and $\hat{\mathbf{L}}^2$ are Hermitian relative to the angular inner product:*

$$\langle \psi_1, \hat{L}_z \psi_2 \rangle_a = \langle \hat{L}_z \psi_1, \psi_2 \rangle_a, \quad \langle \psi_1, \hat{\mathbf{L}}^2 \psi_2 \rangle_a = \langle \hat{\mathbf{L}}^2 \psi_1, \psi_2 \rangle_a. \tag{10.5.39}$$

The statement that the spherical harmonics form an orthonormal set of functions is taken to mean that, using the angular inner product, we find

$$\big\langle Y_{\ell' m'}, Y_{\ell m} \big\rangle_a = \delta_{\ell' \ell}\, \delta_{m' m}. \tag{10.5.40}$$

More explicitly,

$$\int d\Omega\, Y^*_{\ell' m'}(\theta, \phi)\, Y_{\ell m}(\theta, \phi) = \delta_{\ell \ell'} \delta_{m m'}. \tag{10.5.41}$$

The complicated normalization factor in the explicit form (10.5.30) of the spherical harmonics is needed for unit normalization.

10.6 The Radial Equation

It is time to go back to the problem that motivated our discussion of angular momentum: the Schrödinger equation for a particle moving in a central potential. We can use what we have learned about angular dependence to find an equation for the radial dependence of the wave function.

Our ansatz for the solution of the Schrödinger equation is the product of a purely radial function $R_{E\ell}(r)$ and a spherical harmonic:

$$\psi_{E\ell m}(r, \theta, \phi) = R_{E\ell}(r) Y_{\ell m}(\theta, \phi). \tag{10.6.1}$$

We have put subscripts E and ℓ on the radial function. We did not include m because, as we will see, the equation for the radial function does not depend on m. The label E is for the energy of the stationary state we are trying to find. We can now insert this into the Schrödinger equation (10.2.13) to find that

$$-\frac{\hbar^2}{2m}\frac{1}{r}\frac{\partial^2}{\partial r^2}(rR_{E\ell}Y_{\ell m}) + \frac{\hat{\mathbf{L}}^2}{2mr^2}R_{E\ell}Y_{\ell m} + V(r)R_{E\ell}Y_{\ell m} = ER_{E\ell}Y_{\ell m}. \tag{10.6.2}$$

Since the spherical harmonics are $\hat{\mathbf{L}}^2$ eigenstates, we can simplify the equation:

$$-\frac{\hbar^2}{2m}\frac{1}{r}\frac{d^2(rR_{E\ell})}{dr^2}Y_{\ell m} + \frac{\hbar^2\ell(\ell+1)}{2mr^2}R_{E\ell}Y_{\ell m} + V(r)R_{E\ell}Y_{\ell m} = ER_{E\ell}Y_{\ell m}. \tag{10.6.3}$$

Canceling the common spherical harmonic and multiplying by r, we get a purely radial equation for the quantity $(rR_{E\ell})$:

$$-\frac{\hbar^2}{2m}\frac{d^2(rR_{E\ell})}{dr^2} + \frac{\hbar^2\ell(\ell+1)}{2mr^2}(rR_{E\ell}) + V(r)(rR_{E\ell}) = E\,(rR_{E\ell}). \tag{10.6.4}$$

It is now convenient to define

$$u_{E\ell}(r) \equiv rR_{E\ell}(r). \tag{10.6.5}$$

This allows us to rewrite the entire differential equation as

$$\boxed{-\frac{\hbar^2}{2m}\frac{d^2u_{E\ell}}{dr^2} + \left(V(r) + \frac{\hbar^2\ell(\ell+1)}{2mr^2}\right)u_{E\ell} = Eu_{E\ell}.} \tag{10.6.6}$$

This is called the **radial equation.** It looks like the familiar time-independent Schrödinger equation in one dimension but with an effective potential

$$V_{\text{eff}}(r) = V(r) + \frac{\hbar^2\ell(\ell+1)}{2mr^2} \tag{10.6.7}$$

that features the original potential $V(r)$ supplemented by a *centrifugal term,* a repulsive potential proportional to $\ell(\ell+1)$. Because of this term, the radial equation is a little different for each value of ℓ. As anticipated, the quantum number m does not appear in the radial equation. The same radial solution $u_{E\ell}(r)$ must be used for *all* allowed values of m. Once we have solved for $u_{E\ell}$, the full energy eigenstate $\psi_{E\ell m}$ is

$$\boxed{\psi_{E\ell m}(r,\theta,\phi) = R_{E\ell}(r)Y_{\ell,m}(\theta,\phi) = \frac{u_{E\ell}(r)}{r}Y_{\ell m}(\theta,\phi).} \tag{10.6.8}$$

The normalization condition requires

$$1 = \int d^3x\,|\psi_{E\ell m}|^2 = \int r^2 dr d\Omega\,\frac{|u_{E\ell}|^2}{r^2}Y^*_{\ell m}Y_{\ell m}. \tag{10.6.9}$$

The angular integral gives one, the explicit factors of r cancel, and we get the suggestive

$$\int_0^\infty dr\,|u_{E\ell}|^2 = 1. \tag{10.6.10}$$

Indeed, $u_{E\ell}(r)$ plays the role of a one-dimensional wave function for a particle moving in the effective potential $V_{\text{eff}}(r)$.

There is one peculiarity, however. Unlike the x coordinate of one-dimensional potentials, which ranges from minus infinity to plus infinity, the radial coordinate only exists for $r \geq 0$. In Cartesian coordinates the point $r=0$ is the origin $\mathbf{x}=0$, a completely regular point, special only because the potential is defined in reference to $\mathbf{x}=0$. For the radial equation, however, the point $r=0$ is a point where space comes to an end. It is a boundary on the half line $r \geq 0$, as far as the radial wave function $u(r)$ is concerned. In the radial equation, we often imagine some kind of hard wall at $r=0$, preventing the particle from going beyond this point.

We claim that all radial solutions must vanish at $r=0$:

$$\lim_{r \to 0} u_{E\ell}(r) = 0. \tag{10.6.11}$$

This requirement does *not* arise from normalization: As you can see in (10.6.10), a finite $u_{E\ell}$ at $r=0$ would cause no trouble. To see the necessity of the above condition and to refine it a bit further, let us consider first the case $\ell=0$.

Imagine a solution $u_{E,0}(r)$ with $\ell=0$ that approaches a constant as $r \to 0$:

$$\lim_{r \to 0} u_{E,0}(r) = c \neq 0. \tag{10.6.12}$$

The full solution $\psi(\mathbf{x})$ near the origin would then take the form

$$\psi(\mathbf{x}) \simeq \frac{c}{r} Y_{00} = \frac{c'}{r}, \tag{10.6.13}$$

since Y_{00} is simply a constant. The problem with this wave function is that it simply *does not solve* the Schrödinger equation! You may remember from electromagnetism that the Laplacian of $1/r$ is a delta function at the origin so that, as a result,

$$\nabla^2 \psi(\mathbf{x}) = -4\pi c' \delta(\mathbf{x}). \tag{10.6.14}$$

Since the Laplacian is part of the Hamiltonian, this delta function must be canceled by some other contribution, but there is none, since the potential $V(r)$ does not have delta functions. In fact, delta function potentials in more than one dimension are singular and require a regulator parameter to make sense (see exercises 6.11 and 6.12, as well as problem 19.6). This confirms that the radial wave function $u_{E,0}$ cannot approach a constant at the origin. It must vanish.

We can learn about the behavior of the radial solution at the origin under the reasonable assumption that the *centrifugal barrier dominates the potential as* $r \to 0$. In this case the most singular terms of the radial differential equation must cancel each other out, leaving less singular terms that we can ignore in this leading-order calculation. So we set

$$-\frac{\hbar^2}{2m}\frac{d^2 u_{E\ell}}{dr^2} + \frac{\hbar^2}{2mr^2}\ell(\ell+1) u_{E\ell} = 0, \quad \text{as } r \to 0. \tag{10.6.15}$$

Canceling out constants, we get

$$\frac{d^2 u_{E\ell}}{dr^2} = \frac{\ell(\ell+1)}{r^2} u_{E\ell}. \tag{10.6.16}$$

The solutions can be taken to be of the form $u_{E\ell} = r^s$ with s a constant to be determined. We then find a quadratic equation with two solutions:

$$s(s-1) = \ell(\ell+1) \quad \rightarrow \quad s = \ell+1,\ s = -\ell. \tag{10.6.17}$$

This leads to two possible behaviors near $r=0$:

$$u_{E\ell} \sim r^{\ell+1}, \qquad u_{E\ell} \sim \frac{1}{r^{\ell}}. \tag{10.6.18}$$

For $\ell=0$, the second behavior was shown to be inconsistent with the Schrödinger equation at $r=0$. For $\ell>0$, the second behavior is not consistent with normalization. Therefore, we have established that

$$\boxed{u_{E\ell}(r) \sim r^{\ell+1}, \quad \text{as } r \to 0.} \tag{10.6.19}$$

Note that $u_{E\ell}$ vanishes at $r=0$ for all $\ell \geq 0$. The radial dependence of the wave function is obtained by dividing $u_{E\ell}$ by r:

$$R_{E\ell}(r) \sim r^{\ell}. \tag{10.6.20}$$

This allows for a constant nonzero wave function at the origin only when $\ell=0$. Only for $\ell=0$ solutions can a particle be found at the origin. For $\ell \neq 0$, the angular momentum barrier prevents the particle from reaching the origin.

Example 10.1. *Free particle in spherical coordinates.*

A free particle means vanishing potential: $V(r)=0$. We actually know a complete description of the energy eigenstates of a free particle. In Cartesian coordinates they are plane waves, momentum eigenstates with all possible values of the momentum. The solutions can be labeled by the three components of the momentum, which determine the energy. The momentum is $\hbar\mathbf{k}$ with the magnitude $k=|\mathbf{k}|$ fixed by the energy:

$$k = \sqrt{\frac{2mE}{\hbar^2}}. \tag{10.6.21}$$

It may sound surprising that we want to reconsider this seemingly trivial problem in spherical coordinates. It is, however, both interesting and useful to find *radial* solutions that represent energy eigenstates of a free particle. These radial solutions will be labeled by the energy E and (ℓ, m), the two integers that describe the angular dependence of the solutions (of course, ℓ also affects the radial dependence). Ultimately, a radial solution (E, ℓ, m) is a superposition of infinitely many Cartesian solutions $(E, \mathbf{k})$ with the same energy.

This analysis of the free particle is useful for studying spherical potentials, like the infinite square well, that are zero in some spherical region and nonzero elsewhere. It is also useful for scattering problems to understand the behavior of the solutions far away from the scattering center, where they are approximately outgoing spherical waves. Our work here will be a first look into this subject. We will reconsider the subject in more detail in chapter 19.

To find the energy eigenstates, we try to solve the radial equation (10.6.6), which for zero potential $V(r)$ reads

$$-\frac{\hbar^2}{2m}\frac{d^2u_{E\ell}}{dr^2}+\frac{\hbar^2}{2m}\frac{\ell(\ell+1)}{r^2}u_{E\ell} = Eu_{E\ell}. \tag{10.6.22}$$

Expressing E in terms of k gives

$$-\frac{d^2u_{E\ell}}{dr^2}+\frac{\ell(\ell+1)}{r^2}u_{E\ell} = k^2u_{E\ell}, \qquad E=\frac{\hbar^2k^2}{2m}. \tag{10.6.23}$$

We do not expect the energy to be quantized; after all a free particle can have any nonnegative value of the energy. We can make this clear by redefining the radial coordinate in a way that the energy parameter k simply disappears from the equation. Letting

$$\rho=kr, \tag{10.6.24}$$

the radial equation becomes

$$-\frac{d^2u_{E\ell}}{d\rho^2}+\frac{\ell(\ell+1)}{\rho^2}u_{E\ell} = u_{E\ell}. \tag{10.6.25}$$

This second-order linear differential equation, which is rather nontrivial, must be solved for all values of $\ell=0,1,\ldots$. To gain some intuition, let us consider the $\ell=0$ and $\ell=1$ solutions that have the requisite behavior $u_{E\ell}\sim r^{\ell+1}$ as $r\to 0$.

Case $\ell=0$: the radial equation becomes

$$-\frac{d^2u_{E,0}}{d\rho^2}=u_{E,0}. \tag{10.6.26}$$

The solutions for $u_{E,0}$ are arbitrary linear combinations of $\sin\rho$ and $\cos\rho$. Since an $\ell=0$ solution must behave as $u_{E,0}\sim r$ for $r\to 0$, we must have $u_{E,0}\sim\rho$ for $\rho\to 0$. The $\cos\rho$ solution is thus discarded, and we choose

$$u_{E,0}=\sin\rho. \tag{10.6.27}$$

Since $\ell=0$, we must also have $m=0$. The complete wave function is obtained from (10.6.8) and is then

$$\psi_{E,0,0}(\mathbf{x}) = \frac{\sin kr}{kr}Y_{00}=\frac{\sin kr}{kr}\frac{1}{\sqrt{4\pi}}. \tag{10.6.28}$$

This wave function exists for all values of $k\geq 0$. It is not normalizable, just as the Cartesian-coordinate plane waves are not:

$$\int_0^\infty dr|u_{E,0}|^2=\int_0^\infty dr\sin^2 kr=\frac{1}{k}\int_0^\infty d\rho\sin^2\rho=\infty. \tag{10.6.29}$$

Case $\ell=1$: the radial equation becomes

$$-\frac{d^2u_{E,1}}{d\rho^2}+\frac{2}{\rho^2}u_{E,1} = u_{E,1}. \tag{10.6.30}$$

A solution with proper behavior near $\rho = 0$ takes some effort to find, but here it is:

$$u_{E,1}(\rho) = \frac{\sin\rho}{\rho} - \cos\rho. \tag{10.6.31}$$

An expansion around $\rho = 0$ gives $u_{E,1} \simeq \frac{1}{3}\rho^2 + \mathcal{O}(\rho^4)$, consistent with the required $u_{E,1} \sim r^2$. The complete wave functions are therefore

$$\psi_{E,1m}(\mathbf{x}) = \frac{1}{kr}\Big(\frac{\sin kr}{kr} - \cos kr\Big)\, Y_{1,m}, \quad m = 0, \pm 1. \tag{10.6.32}$$

These nonnormalizable eigenstates exist for all values of $k \geq 0$. □

The above solutions can be used to find the energy eigenstates of an infinite spherical well (problem 10.7). The general solutions of (10.6.25) for arbitrary ℓ can be written in terms of the spherical Bessel functions $j_\ell(\rho)$ and $n_\ell(\rho)$, the latter singular as $\rho \to 0$. In fact, the eigenstates considered above arise from $u_{E\ell} = \rho j_\ell(\rho)$. These matters will be discussed in detail in chapter 19.

Problems

Problem 10.1. *A few commutators and a few expectation values.*

1. Compute the following commutators: $[\hat{L}_z, \hat{x}]$, $[\hat{L}_z, \hat{y}]$, and $[\hat{L}_z, \hat{z}]$.
2. Calculate the following commutators: $[\hat{L}_z, \hat{p}_x]$, $[\hat{L}_z, \hat{p}_y]$, and $[\hat{L}_z, \hat{p}_z]$.

Assume ψ_m is an $\hat{L}_z$ eigenfunction: $\hat{L}_z\psi_m = \hbar m\psi_m$ with m an integer.

3. Compute the expectation values for $\hat{p}_x$ and $\hat{p}_y$ on the state ψ_m.
4. Compute the expectation values for $\hat{x}$ and $\hat{y}$ on the state ψ_m.

Problem 10.2. *Angular momentum in spherical coordinates.*

Spherical polar coordinates (r, θ, ϕ) are related to Cartesian coordinates as written in (10.4.10).

1. Calculate the nine partial derivatives of the spherical coordinates (r, θ, ϕ) with respect to the Cartesian coordinates (x, y, z), expressing your answers in terms of the spherical coordinates.
2. Use your results from (1) to calculate $L_\pm = \hat{L}_x \pm i\hat{L}_y$, and confirm it is as given in (10.4.14).
3. Find a relation of the form $\hat{\mathbf{L}}^2 = \hat{L}_+\hat{L}_- + \cdots$ where the dots involve only the operator $\hat{L}_z$. Use this relation and the result of the previous part to show that $\hat{\mathbf{L}}^2$ is the differential operator given in (10.4.15).

Problem 10.3. *Simultaneous eigenstates.*

Let $\hat{A}$ and $\hat{B}$ be two Hermitian operators that *commute*. Assume that $\hat{A}$ has *no degeneracies* in its spectrum. That is, if $\hat{A}\psi_a = a\psi_a$ *and* $\hat{A}\psi_b = a\psi_b$ for two states ψ_a and ψ_b, then ψ_a and ψ_b are the same state, up to a multiplicative constant. Given that there are no degeneracies, we can let the eigenstates of $\hat{A}$ be labeled by their eigenvalues so that the eigenstate ψ_a has eigenvalue a.

1. If $\hat{A}\chi = a\chi$, what is the most general expression for the state χ?
2. Simplify $\hat{A}\hat{B}\psi_a$, writing it in terms of an expression without the $\hat{A}$ operator.
3. Combining your results of (1) and (2), what can you conclude about $\hat{B}\psi_a$? Are $\hat{A}$ eigenstates also $\hat{B}$ eigenstates? Are $\hat{B}$ eigenstates also $\hat{A}$ eigenstates?

Problem 10.4. *Legendre differential equation.*

In setting up a series solution for the Legendre equation, we found a two-step recursion relation (10.5.19):

$$\frac{a_{k+2}}{a_k} = \frac{k(k+1) - \ell(\ell+1)}{(k+1)(k+2)}.$$

1. Examine the above ratio for large k and verify that we have $a_k \simeq C/k$, with C an arbitrary constant.
2. Consider both even and odd solutions that do not truncate: $\sum_{k\,\text{even}} a_k x^k$ and $\sum_{k\,\text{odd}} a_k x^k$. Both of these series diverge as $x \to \pm 1$. Use the series expansion

$$\ln(1-x) = -\Big(x + \frac{x^2}{2} + \frac{x^3}{3} + \cdots\Big)$$

to argue that even and odd solutions diverge as $\ln(1-x^2)$ and $\ln(\frac{1+x}{1-x})$, respectively.

Problem 10.5. *A wave function in a central potential.*

Consider the following normalized wave function:

$$\psi(r,\theta,\phi) = \frac{u_1(r)}{r} Y_{\ell_1,m_1}(\theta,\phi) + \frac{u_2(r)}{r} Y_{\ell_2,m_2}(\theta,\phi).$$

Assume that $\ell_1 \neq \ell_2$ and $m_1 \neq m_2$.

1. Derive the condition satisfied by u_1 and u_2 that implies that ψ is normalized.
2. Define the positive constants α_1 and α_2:

$$\alpha_1 = \int_0^\infty dr|u_1|^2, \qquad \alpha_2 = \int_0^\infty dr|u_2|^2.$$

What is the expectation value of $\hat{L}_z$ in the state ψ? Give your answer in terms of $\alpha_1, \alpha_2, m_1, m_2$, and $\hbar$.
3. Calculate the uncertainty ΔL_z. For a nice, simple answer, you will need to use what (1) tells you about the constants α_1, α_2.

Problem 10.6. *Isotropic harmonic oscillator.*

Consider a particle of mass m moving in a spherically symmetric harmonic oscillator potential in three dimensions:

$$V(r) = \tfrac{1}{2} m\omega^2 r^2.$$

1. Write the relevant radial equation for the radial wave function $u(r)$ of an energy eigenstate with energy E. Remove units from the differential equation by setting $r = ax$, where x is unit-free, and a carries units of length. What is a^2 in terms of $\hbar, m,$ and ω (as usual, a is solely built from these quantities without additional numerical factors)?

   Show that the final differential equation takes the form

$$-\frac{d^2u}{dx^2} + \left(x^2 + \frac{\ell(\ell+1)}{x^2}\right)u = \mathcal{E}\, u$$

   and determine how the unit-free energy parameter $\mathcal{E}$ is related to the energy E and $\hbar\omega$.

2. Consider both the $x \to 0$ and $x \to \infty$ behavior of the differential equation for $u(x)$ to write an ansatz of the form

$$u(x) = x^{\#}\, w(x) e^{-\# x^{\#}}$$

   and determine the value of the constants represented by # (not all the same).

3. Show that the resulting differential equation for $w(x)$ takes the form

$$w'' + 2\left(\frac{\ell+1}{x} - x\right) w' + (\mathcal{E} - 2\ell - 3)w = 0.$$

4. We now write an ansatz

$$w(x) = \sum_{k=0}^{\infty} a_k\, x^k.$$

   Show that the differential equation for w leads to the recursion of the form

$$\frac{a_{k+2}}{a_k} = \frac{\cdots\cdots}{\cdots\cdots}$$

   and determine completely the right-hand side, which is a function of k, ℓ, and $\mathcal{E}$.

5. It can be shown that $w(x)$ must be a polynomial for the wave functions to be normalizable. Assume $w(x)$ is a polynomial of degree $N \geq 0$ and determine the energy $\mathcal{E}$ of the solution. Explain carefully why consistency with the ansatz for $u(x)$ in (2) requires that N be an *even* number.

6. Represent the energy levels in a two-dimensional diagram where the values of ℓ are indicated on the horizontal axis and the values of $\mathcal{E}$ are indicated on the vertical axis. Denote each state by a short segment and label it with the value of N. Include all states with energies $\mathcal{E} \leq 11$. Including the degeneracies associated with the m quantum number, how many states of the oscillator have energy $\mathcal{E} = 9$?

Problem 10.7. *Spherical wells.*

Consider a particle of mass m moving in the *infinite* spherical well

$$V(r) = \begin{cases} 0, & \text{if } r < a, \\ \infty, & \text{if } r > a. \end{cases}$$

1. Solve for the radial wave function $u(r)$ for $\ell = 0$ and find the possible energy levels.
2. The spectrum you have found should coincide with the spectrum of a one-dimensional infinite square well. What is the width of that well in order for the spectra to coincide?
3. Find the lowest three energies for $\ell = 1$ eigenstates (see example 10.1).

Now consider states of a particle of mass m moving in a *finite* spherical well with $V_0 > 0$:

$$V(r) = \begin{cases} -V_0, & \text{if } r < a, \\ 0, & \text{if } r > a. \end{cases}$$

4. What is the smallest value of V_0 for the potential to have a bound state?

11 Hydrogen Atom

The hydrogen atom is a two-body problem involving an electron and a proton. Passing to center-of-mass (CM) and relative coordinates and momenta, the problem reduces to that of the motion of a single particle in a Coulomb potential. The resulting radial equation can be solved in terms of a differential equation with a one-step recursion relation. Quantization emerges because polynomial solutions are required for normalizable solutions. A principal quantum number $n = 1, 2, \ldots$ *determines the energy. For each* n *there are degenerate states with angular momentum values* $\ell = 0, 1, \ldots, n-1$ *and with* $m \in (-\ell, \ell)$*. So-called Rydberg atoms have large* n *and have remarkable semiclassical properties.*

In this chapter we begin our study of the hydrogen atom. This is a classic example, of much physical interest, that demonstrates the power of quantum theory. It is also the basis for all of atomic physics. We will obtain the bound state spectrum of the hydrogen atom: the set of bound states of an electron and a proton. We make some simplifying assumptions about the hydrogen atom. The most important is that we will ignore the effects of the electron spin. To first approximation, the electron spin simply doubles the number of energy eigenstates of the hydrogen atom: any state we find here exists with the electron spin "up" and with the electron spin "down." But there are also more subtle effects due to spin, which affect the spectrum slightly but measurably. These effects, together with other relativistic effects, will be studied in a second iteration (chapter 25). Our treatment here will ignore both spin and relativistic effects. Still, a good number of physical properties of the atom will emerge from the analysis.

11.1 The Two-Body Problem

Our goal here is to show that the two-body quantum mechanical problem of the hydrogen atom can be recast as one in which we have center mass (CM) degrees of freedom that behave like a free particle and relative-motion degrees of freedom for which the dynamics is controlled by a central potential.

The hydrogen atom consists of a proton and an electron moving in three dimensions. We label the position and momentum operators of the proton as $\hat{\mathbf{x}}_p, \hat{\mathbf{p}}_p$ and those of the electron as $\hat{\mathbf{x}}_e, \hat{\mathbf{p}}_e$. These are canonical variables, meaning they satisfy the canonical commutation relations:

$$\left[\hat{x}_{p,i}\,,\, \hat{p}_{p,j}\right] = i\hbar\delta_{ij}, \quad \left[\hat{x}_{e,i}\,,\, \hat{p}_{e,j}\right] = i\hbar\delta_{ij}. \tag{11.1.1}$$

The subscripts $i, j = 1, 2, 3$ denote the various components of the vector operators. Furthermore, the proton variables *commute* with the electron variables. We have two pairs of *independent* canonical variables.

The wave function for the system is a function of the positions of *both* particles:

$$\psi(\mathbf{x}_p, \mathbf{x}_e). \tag{11.1.2}$$

It is important to emphasize that we do not have two wave functions, one for the electron and one for the proton. Any quantum system, however many particles it describes, has a *single* wave function satisfying the Schrödinger equation. That wave function depends on the positions of each of the particles. A solution of the form $\psi(\mathbf{x}_p, \mathbf{x}_e) = \psi_p(\mathbf{x}_p)\psi_e(\mathbf{x}_e)$ would be woefully inadequate to describe the system, as it would miss the interplay between the two particles. The Hamiltonian couples the degrees of freedom of the two particles via the potential.

The norm squared of the wave function (11.1.2) is a joint probability distribution so that the quantity

$$|\psi(\mathbf{x}_p, \mathbf{x}_e)|^2\, d^3\mathbf{x}_p\, d^3\mathbf{x}_e \tag{11.1.3}$$

is the probability of finding the proton within a window $d^3\mathbf{x}_p$ of $\mathbf{x}_p$ and the electron within a window $d^3\mathbf{x}_e$ of $\mathbf{x}_e$. The normalization condition is

$$\int |\psi(\mathbf{x}_p, \mathbf{x}_e)|^2\, d^3\mathbf{x}_p\, d^3\mathbf{x}_e = 1. \tag{11.1.4}$$

The Hamiltonian of the system is given by

$$\hat{H} = \frac{\hat{\mathbf{p}}_p^2}{2m_p} + \frac{\hat{\mathbf{p}}_e^2}{2m_e} + V(|\mathbf{x}_e - \mathbf{x}_p|). \tag{11.1.5}$$

Here m_p and m_e denote, respectively, the masses of the proton and of the electron. Note that the kinetic energy is simply the sum of the kinetic energy of the proton and the kinetic energy of the electron. The potential only depends on the magnitude of the separation between the two particles, not on their individual positions.

In order to simplify the problem, we will introduce two new pairs of independent canonical variables. The first pair is associated with the CM motion. We introduce the total momentum operator $\hat{\mathbf{P}}$ and the CM position operator $\hat{\mathbf{X}}$ as follows:

$$\hat{\mathbf{P}} = \hat{\mathbf{p}}_p + \hat{\mathbf{p}}_e\,, \quad \hat{\mathbf{X}} = \frac{m_e\hat{\mathbf{x}}_e + m_p\hat{\mathbf{x}}_p}{m_e + m_p}. \tag{11.1.6}$$

The operator $\hat{\mathbf{X}}$ is given by the familiar expression for the center of mass of the system but with the positions replaced by position operators. Using the commutation relations

(11.1.1), we can show that $\hat{\mathbf{X}}$ and $\hat{\mathbf{P}}$ are canonical conjugates:

$$\begin{aligned}\left[\hat{\mathbf{X}}_i, \hat{\mathbf{P}}_j\right] &= \left[\frac{m_e\hat{\mathbf{x}}_{e,i} + m_p\,\hat{\mathbf{x}}_{p,i}}{m_e + m_p}, \hat{\mathbf{p}}_{p,j} + \hat{\mathbf{p}}_{e,j}\right] \\ &= \frac{m_e}{m_e + m_p}\left[\hat{\mathbf{x}}_{e,i}, \hat{\mathbf{p}}_{e,j}\right] + \frac{m_p}{m_e + m_p}\left[\hat{\mathbf{x}}_{p,i}, \hat{\mathbf{p}}_{p,j}\right] \\ &= \frac{m_e}{m_e + m_p}\, i\hbar\delta_{ij} + \frac{m_p}{m_e + m_p}\, i\hbar\delta_{ij},\end{aligned} \tag{11.1.7}$$

resulting in the expected

$$\left[\hat{\mathbf{X}}_i, \hat{\mathbf{P}}_j\right] = i\hbar\delta_{ij}. \tag{11.1.8}$$

For the second pair of canonical variables, we will define relative position and momentum operators. The relative position operator $\hat{\mathbf{x}}$ is the natural variable implied by the form of the potential:

$$\hat{\mathbf{x}} = \hat{\mathbf{x}}_e - \hat{\mathbf{x}}_p. \tag{11.1.9}$$

Since the second pair of canonical variables must commute with the first pair, we must check that $\hat{\mathbf{x}}$, defined above, commutes with $\hat{\mathbf{X}}$ and with $\hat{\mathbf{P}}$. The commutation with $\hat{\mathbf{X}}$ is automatic, and the commutation with $\hat{\mathbf{P}}$ works thanks to the minus sign in the above definition. We must now construct a relative momentum operator $\hat{\mathbf{p}}$ that is canonically conjugate to $\hat{\mathbf{x}}$. It must be built from the momentum operators of the two particles, so we write

$$\hat{\mathbf{p}} = \alpha\,\hat{\mathbf{p}}_e - \beta\,\hat{\mathbf{p}}_p, \tag{11.1.10}$$

with the α and β coefficients to be determined. To be canonically conjugate, the relative operators must satisfy

$$\left[\hat{\mathbf{x}}_i, \hat{\mathbf{p}}_j\right] = i\hbar\delta_{ij} \quad\Rightarrow\quad \alpha + \beta = 1, \tag{11.1.11}$$

using the above definitions of $\hat{\mathbf{x}}$ and $\hat{\mathbf{p}}$ and the proton and electron commutators. Finally, the relative momentum must commute with the CM coordinate:

$$[\hat{\mathbf{X}}_i, \hat{\mathbf{p}}_j] = 0 \quad\Rightarrow\quad m_e\alpha - m_p\beta = 0. \tag{11.1.12}$$

The two equations for α and β can be solved to find that

$$\alpha = \frac{m_p}{m_e + m_p}, \quad \beta = \frac{m_e}{m_e + m_p}. \tag{11.1.13}$$

We define the total mass M and the **reduced mass** μ as follows:

$$M = m_e + m_p, \quad \mu = \frac{m_e m_p}{m_e + m_p}. \tag{11.1.14}$$

The reduced mass of a pair of particles with very different masses is approximately equal to the mass of the lower-mass particle. Using these definitions, we have

$$\alpha = \frac{\mu}{m_e}, \quad \beta = \frac{\mu}{m_p}. \tag{11.1.15}$$

The relative variables are therefore

$$\hat{\mathbf{p}} = \mu\left(\frac{\hat{\mathbf{p}}_e}{m_e} - \frac{\hat{\mathbf{p}}_p}{m_p}\right) = \frac{m_p}{M}\,\hat{\mathbf{p}}_e - \frac{m_e}{M}\,\hat{\mathbf{p}}_p\,, \qquad \hat{\mathbf{x}} = \hat{\mathbf{x}}_e - \hat{\mathbf{x}}_p. \tag{11.1.16}$$

The relative momentum $\hat{\mathbf{p}}$ can be written in terms of velocities as follows: $\hat{\mathbf{p}} = \mu(\hat{\mathbf{v}}_e - \hat{\mathbf{v}}_p)$. The relative momentum vanishes if the motion is only CM motion, in which case $\hat{\mathbf{x}}$ is constant, and the velocities of the two particles are the same.

We can now rewrite the Hamiltonian in terms of the new variables. Solving for the electron and proton momentum operators in terms of $\hat{\mathbf{P}}$ and $\hat{\mathbf{p}}$, we find

$$\hat{\mathbf{p}}_p = \frac{m_p}{M}\,\hat{\mathbf{P}} - \hat{\mathbf{p}}, \qquad \hat{\mathbf{p}}_e = \frac{m_e}{M}\,\hat{\mathbf{P}} + \hat{\mathbf{p}}. \tag{11.1.17}$$

We can then rewrite the kinetic terms of the Hamiltonian in the form

$$\begin{aligned}\frac{\hat{\mathbf{p}}_p^2}{2m_p} + \frac{\hat{\mathbf{p}}_e^2}{2m_e} &= \frac{1}{2m_p}\left(\frac{m_p^2}{M^2}\hat{\mathbf{P}}^2 - \frac{2m_p}{M}\,\hat{\mathbf{P}}\cdot\hat{\mathbf{p}} + \hat{\mathbf{p}}^2\right)\\ &\quad + \frac{1}{2m_e}\left(\frac{m_e^2}{M^2}\hat{\mathbf{P}}^2 + \frac{2m_e}{M}\,\hat{\mathbf{P}}\cdot\hat{\mathbf{p}} + \hat{\mathbf{p}}^2\right)\\ &= \frac{\hat{\mathbf{P}}^2}{2M} + \frac{\hat{\mathbf{p}}^2}{2\mu}.\end{aligned} \tag{11.1.18}$$

Happily, the term coupling the two momenta vanishes. Thus, the CM degrees of freedom and the relative degrees of freedom give independent contributions to the kinetic energy. The Hamiltonian can then be written as

$$\hat{H} = \frac{\hat{\mathbf{P}}^2}{2M} + \frac{\hat{\mathbf{p}}^2}{2\mu} + V(|\hat{\mathbf{x}}|). \tag{11.1.19}$$

In position space, the total and relative momentum operators can be expressed as gradients:

$$\hat{\mathbf{P}} \to \frac{\hbar}{i}\nabla_{\mathbf{X}}, \qquad \hat{\mathbf{p}} \to \frac{\hbar}{i}\nabla_{\mathbf{x}}. \tag{11.1.20}$$

Each ∇ has a subscript indicating the type of coordinate we use to take the derivatives. We introduced first a wave function $\psi(\mathbf{x}_e, \mathbf{x}_p)$, but the new canonical variables require that we now think of the wave function as a function $\psi(\mathbf{X}, \mathbf{x})$ of the new coordinates. In problem 11.1 you will confirm that, consistent with (11.1.4), this wave function is normalized as follows:

$$\int |\psi(\mathbf{X}, \mathbf{x})|^2\, d^3\mathbf{X}\, d^3\mathbf{x} = 1. \tag{11.1.21}$$

We solve the time-independent Schrödinger equation by using separation of variables:

$$\psi(\mathbf{X}, \mathbf{x}) = \psi_{\mathrm{CM}}(\mathbf{X})\,\psi_{\mathrm{rel}}(\mathbf{x}). \tag{11.1.22}$$

Since the Hamiltonian does not couple CM and relative degrees of freedom, this strategy is well motivated. General solutions are obtained by superposition of these factorized solutions. Plugging the ansatz into the time-independent Schrödinger equation $\hat{H}\psi = E\psi$, we find

$$\begin{aligned}&\left[\frac{\hat{\mathbf{p}}^2}{2M}\psi_{\rm CM}(\mathbf{X})\right]\psi_{\rm rel}(\mathbf{x}) + \left[\frac{\hat{\mathbf{p}}^2}{2\mu}\psi_{\rm rel}(\mathbf{x}) + V(|\mathbf{x}|)\psi_{\rm rel}(\mathbf{x})\right]\psi_{\rm CM}(\mathbf{X})\\ &\quad = E\psi_{\rm CM}(\mathbf{X})\psi_{\rm rel}(\mathbf{x}).\end{aligned} \tag{11.1.23}$$

Dividing by the total wave function $\psi_{\rm CM}(\mathbf{X})\psi_{\rm rel}(\mathbf{x})$, this becomes

$$\frac{1}{\psi_{\rm CM}(\mathbf{X})}\left[\frac{\hat{\mathbf{P}}^2}{2M}\psi_{\rm CM}(\mathbf{X})\right] + \frac{1}{\psi_{\rm rel}(\mathbf{x})}\left[\frac{\hat{\mathbf{p}}^2}{2\mu} + V(|\mathbf{x}|)\right]\psi_{\rm rel}(\mathbf{x}) = E. \tag{11.1.24}$$

The first term on the left-hand side is a function of $\mathbf{X}$ only, and the second term on the left-hand side is a function of $\mathbf{x}$ only. Their sum is equal to the constant E, and since $\mathbf{x}$ and $\mathbf{X}$ are independent variables, each term must individually be constant. We thus set the first term equal to the constant $E_{\rm CM}$ and the second term equal to the constant $E_{\rm rel}$, resulting in the following equations:

$$\frac{\hat{\mathbf{P}}^2}{2M}\psi_{\rm CM}(\mathbf{X}) = E_{\rm CM}\psi_{\rm CM}(\mathbf{X}), \tag{11.1.25}$$

$$\left[\frac{\hat{\mathbf{p}}^2}{2\mu} + V(|\mathbf{x}|)\right]\psi_{\rm rel}(\mathbf{x}) = E_{\rm rel}\psi_{\rm rel}(\mathbf{x}), \tag{11.1.26}$$

$$E = E_{\rm CM} + E_{\rm rel}. \tag{11.1.27}$$

We get two Schrödinger equations. The first equation tells us that the center of mass moves as a free particle of mass M. As a result, the CM energy is not quantized, and we get plane wave solutions. The second equation is for the relative motion, and as we wanted to show, it describes motion in a central potential. The third equation tells us that the total energy is the sum of the CM energy and the energy from the relative motion. Of course, there is still just one wave function $\psi = \psi_{\rm CM}\psi_{\rm rel}$.

The work we have done applies to *any* system of two particles with a potential that is a function of the distance between the particles. The Hamiltonian need only take the general form in (11.1.5), with arbitrary values of the masses and with arbitrary V.

11.2 Hydrogen Atom: Potential and Scales

We now have the tools to study the hydrogen atom, for which the potential V gives the Coulomb energy associated with the electron and proton. In Gaussian-cgs units the potential energy is given by the product of the charges divided by the distance between them:

$$V(\mathbf{x}_p, \mathbf{x}_e) = -\frac{e^2}{|\mathbf{x}_p - \mathbf{x}_e|}. \tag{11.2.1}$$

The minus sign is there because the electron and proton have opposite charges, and thus the product of charges is $-e^2$. The two particles attract one another, and the potential energy is lowered as the particles approach each other. In terms of the relative coordinate $\mathbf{x} = \mathbf{x}_e - \mathbf{x}_p$, the potential energy is written as

$$V(\mathbf{x}) = -\frac{e^2}{|\mathbf{x}|}. \tag{11.2.2}$$

Letting $r = |\mathbf{x}|$, this is our radial potential:

$$V(r) = -\frac{e^2}{r}. \tag{11.2.3}$$

Since the potential energy vanishes when the charges are infinitely separated and is negative otherwise, bound states will be states of negative energy.

We will also identify important physical constants and scales:

- The fine structure constant α defined in (2.5.6):

$$\alpha = \frac{e^2}{\hbar c} \simeq \frac{1}{137}. \tag{11.2.4}$$

  This constant encodes the value of e^2. Its small value reflects the weakness of the electrostatic interaction.

- The *Bohr radius* a_0. This is the characteristic length scale in the hydrogen atom. It can be calculated by equating kinetic and potential energies expressed in terms of a_0 and ignoring all numerical constants:

$$\frac{\hbar^2}{m_e a_0^2} = \frac{e^2}{a_0} \quad \Rightarrow \quad a_0 = \frac{\hbar^2}{m_e e^2}. \tag{11.2.5}$$

  Note that the Bohr radius does not depend on the speed of light; our treatment is nonrelativistic, and the speed of light does not appear. Here the mass should be the reduced mass of the proton-electron pair, which in this case can be taken rather accurately to be the mass m_e of the electron. At any rate, from now on we will simply write m for the mass. We then have, explicitly,

$$\begin{aligned} a_0 &= \frac{\hbar^2}{me^2} = \frac{\hbar^2 c^2}{mc^2 e^2} = \frac{\hbar c}{(mc^2)\frac{e^2}{\hbar c}} = \frac{\hbar c}{mc^2 \alpha} \\ &= \frac{197.3\,\text{MeV}\cdot\text{fm}}{0.51\times 10^6\,\text{eV}(\frac{1}{137})} = \frac{1973\,\text{eV}\cdot\text{angstrom}}{0.51\times 10^6\,\text{eV}} \times 137 \\ &= 0.529\,\text{angstroms} \simeq 53\,\text{pm}. \end{aligned} \tag{11.2.6}$$

  The length a_0 is also about one-twentieth of a nanometer.

- The energy scale Ry of the hydrogen atom, called the Rydberg. Since the potential $V(r)$ has units of energy, we get energy from the ratio e^2/a_0. The definition of the Rydberg includes an extra factor of one-half:

$$\text{Ry} \equiv \frac{e^2}{2a_0} = e^2\left(\frac{me^2}{2\hbar^2}\right) = \left(\frac{e^4}{\hbar^2 c^2}\right)\frac{1}{2}mc^2 = \alpha^2\,\frac{1}{2}mc^2. \tag{11.2.7}$$

  The speed of light helps the numerical evaluation even though it cancels out. It also helps conceptually: we see above that the energy scale Ry, up to a factor of one-half, is the rest energy of the electron suppressed by two powers of the fine structure constant. We evaluate Ry as follows:

$$\text{Ry} = \frac{1}{(137)^2} \times \tfrac{1}{2}(511\,000\,\text{eV}) \simeq 13.6 \text{ eV}. \tag{11.2.8}$$

The Rydberg happens to be the magnitude of the bound state energy for the ground state of the hydrogen atom, as we will soon see.

11.3 Hydrogen Atom: Bound State Spectrum

Our task is now to solve the Schrödinger equation for the hydrogen atom and find the spectrum of bound states. From our analysis of the two-body problem, we now have to solve equation (11.1.26), the equation relevant to the relative motion:

$$\Big(\frac{\hat{\mathbf{p}}^2}{2\mu} + V(|\mathbf{x}|)\Big)\psi_{\text{rel}}(\mathbf{x}) = E_{\text{rel}}\psi_{\text{rel}}(\mathbf{x}). \tag{11.3.1}$$

For simplicity we omit the "rel" subscripts and also set $\mu = m$:

$$\Big(\frac{\hat{\mathbf{p}}^2}{2m} + V(r)\Big)\psi(\mathbf{x}) = E\,\psi(\mathbf{x}). \tag{11.3.2}$$

We showed in section 10.6 that this Schrödinger equation has solutions of the form

$$\psi(\mathbf{x}) = \frac{u_{E\ell}(r)}{r} Y_{\ell m}(\theta, \phi), \tag{11.3.3}$$

where E is the energy, and $\ell \geq 0$ is the angular momentum. Moreover, the radial function $u_{E\ell}$ satisfies the radial equation (10.6.6). In our analysis below, we will use a more general potential, relevant to the study of an electron around a nucleus of arbitrary charge. If the nucleus has Z protons, the nuclear charge is Ze, and the potential becomes

$$V(r) = -\frac{Ze^2}{r}. \tag{11.3.4}$$

For hydrogen we simply take $Z = 1$. With this potential, bound states with angular momentum ℓ correspond to $E < 0$ solutions of the equation

$$\Big(-\frac{\hbar^2}{2m}\frac{d^2}{dr^2} + \frac{\hbar^2\ell(\ell+1)}{2mr^2} - \frac{Ze^2}{r}\Big)u_{E\ell} = E\,u_{E\ell}. \tag{11.3.5}$$

These states are bound states, and thus normalizable, because for any $E < 0$ there is a classically forbidden region that extends beyond some value of r. For a fixed ℓ, the possible values E of the energy are to be determined, along with the eigenstates $u_{E\ell}$. Positive energy solutions represent electrons that are not bound to the proton. These are interesting eigenstates, but we will not consider them here.

As usual, we like to work with a unit-free coordinate. This could be achieved by writing $r = a_0 x$, with x unit-free and a_0 carrying the length units of r. It will be more convenient to use a slight variation to eliminate both Z and some factors of two from the equation. We will take the new unit-free coordinate x to be defined from

$$r \equiv \frac{a_0}{2Z}\, x, \tag{11.3.6}$$

which implies that $\frac{d}{dr} = \frac{2Z}{a_0}\frac{d}{dx}$ so that the Schrödinger equation then becomes

$$\left(\frac{2\hbar^2 Z^2}{m a_0^2}\left(-\frac{d^2}{dx^2} + \frac{l(l+1)}{x^2}\right) - \frac{2Z^2 e^2}{a_0}\frac{1}{x}\right) u_{E\ell} = E\, u_{E\ell}. \tag{11.3.7}$$

Note that

$$\frac{2\hbar^2 Z^2}{m a_0^2} = \frac{2\hbar^2 Z^2}{m a_0}\frac{m e^2}{\hbar^2} = \frac{2Z^2 e^2}{a_0} = 4Z^2\text{Ry}. \tag{11.3.8}$$

The differential equation therefore reduces to

$$\left(-\frac{d^2}{dx^2} + \frac{\ell(\ell+1)}{x^2} - \frac{1}{x}\right) u_{E\ell} = \frac{E}{4Z^2\text{Ry}}\, u_{E\ell}. \tag{11.3.9}$$

We now define the unit-free parameter κ that encodes the energy:

$$\kappa^2 = -\frac{E}{4Z^2\text{Ry}} > 0. \tag{11.3.10}$$

The differential equation is then

$$\left(-\frac{d^2}{dx^2} + \frac{\ell(\ell+1)}{x^2} - \frac{1}{x}\right) u_{E\ell} = -\kappa^2 u_{E\ell}. \tag{11.3.11}$$

Since κ is unit-free, we can pass from the unit-free coordinate x to a unit-free coordinate ρ:

$$\rho \equiv \kappa\, x = \frac{2\kappa Z}{a_0}\, r. \tag{11.3.12}$$

The differential equation for $u_{E\ell}$ then simplifies to its final form:

$$\left(-\frac{d^2}{d\rho^2} + \frac{\ell(\ell+1)}{\rho^2} - \frac{1}{\kappa\,\rho}\right) u_{E\ell} = -u_{E\ell}. \tag{11.3.13}$$

We did not get κ to disappear from the equation. This is good news: the equation should fix the possible values of κ and thus the possible energies. The equation above is not ready for a series solution; you should be able to see that it gives a complicated three-term recursion relation. To make progress we discuss the behavior for small and large ρ.

For $\rho \to \infty$, the above differential equation reduces to

$$\frac{d^2}{d\rho^2} u_{E\ell} = u_{E\ell} \quad \Rightarrow \quad u_{E\ell} = A\, e^{\pm\rho}. \tag{11.3.14}$$

Of course, we hope for $u = Ae^{-\rho}$ for normalizability. As we have discussed previously, for $\rho \to 0$ the radial solution must be of the form $u_{E\ell} \sim r^{(l+1)}$. This information about the behavior for small and for large ρ suggests a good ansatz for $u_{E\ell}(\rho)$:

$$u_{E\ell}(\rho) = \rho^{\ell+1}\, W_{E\ell}(\rho)\, e^{-\rho}, \tag{11.3.15}$$

where $W_{E\ell}(\rho)$ is a yet to be determined function that we hope satisfies a simpler differential equation. To derive this differential equation for $W_{E\ell}(\rho)$, we plug our ansatz into (11.3.13). For a little help with the calculation, we give an intermediate result:

$$-u''_{E\ell} + \frac{\ell(\ell+1)}{\rho^2}u_{E\ell} + u_{E\ell} = \left(-W''_{E\ell} - \frac{2(\ell+1)}{\rho}W'_{E\ell} + \frac{2(\ell+1)}{\rho}W_{E\ell} + 2W'_{E\ell}\right)\rho^{\ell+1}e^{-\rho}.$$

With a little more work, we finally get the differential equation for $W_{E\ell}(\rho)$:

$$\rho\frac{d^2W_{E\ell}}{d\rho^2} + 2(\ell+1-\rho)\frac{dW_{E\ell}}{d\rho} + \left[\frac{1}{\kappa} - 2(\ell+1)\right]W_{E\ell} = 0. \tag{11.3.16}$$

This looks more complicated than the differential equation we started with, but it leads to a very nice one-step recursion relation. As usual, we write $W_{E\ell}$ as a series expansion:

$$W_{E\ell} = \sum_{m=0}^{\infty} c_m\rho^m. \tag{11.3.17}$$

We focus on the coefficient of ρ^k in the differential equation (11.3.16), when the series expansion is introduced. One can see that only two terms in the expansion can contribute:

$$W_{E\ell} = \cdots + c_k\rho^k + c_{k+1}\rho^{k+1} + \cdots. \tag{11.3.18}$$

The full coefficient of ρ^k must vanish and we get

$$c_{k+1}k(k+1) + 2(\ell+1)(k+1)c_{k+1} - 2kc_k + \left[\frac{1}{\kappa} - 2(\ell+1)\right]c_k = 0. \tag{11.3.19}$$

This gives

$$\frac{c_{k+1}}{c_k} = \frac{2(k+\ell+1) - \frac{1}{\kappa}}{(k+1)(k+2\ell+2)}. \tag{11.3.20}$$

For normalizable wave functions, the series must terminate. To see this we examine the large k behavior of the above ratio:

$$\frac{c_{k+1}}{c_k} \simeq \frac{2k}{k^2} = \frac{2}{k}. \tag{11.3.21}$$

To the approximation we are working, we can replace the last fraction by the more easily tractable $\frac{2}{k+1}$. Note that $\frac{2}{k+1} < \frac{2}{k}$, and therefore, if the ratio $\frac{2}{k+1}$ leads to a divergence so will the ratio $\frac{2}{k}$. Taking

$$\frac{c_{k+1}}{c_k} = \frac{2}{k+1} \quad \Rightarrow \quad c_{k+1} = \frac{2}{k+1}c_k, \tag{11.3.22}$$

and this is solved exactly by

$$c_k = \frac{2^k}{k!}c_0, \tag{11.3.23}$$

which should be an accurate representation of the coefficients for large k. Therefore, the sum

$$W_{E\ell} = \sum_{k=0}^{\infty} c_k \rho^k \simeq c_0 \sum_{k=0}^{\infty} \frac{2^k \rho^k}{k!} = c_0 e^{2\rho}. \tag{11.3.24}$$

This behavior indeed makes the ansatz in (11.3.15) nonnormalizable, showing that the series expansion of W must terminate.

Suppose $W_{E\ell}$ is a polynomial of degree $N \geq 0$ so the coefficients satisfy

$$c_N \neq 0 \quad \text{and} \quad c_{N+1} = 0. \tag{11.3.25}$$

From equation (11.3.20) this implies

$$\frac{1}{2\kappa} = N + \ell + 1. \tag{11.3.26}$$

Quantization has happened! The energy-encoding parameter κ is now related to integers! Note that ℓ can take values $\ell = 0, 1, 2, \ldots$ as befits an angular momentum quantum number. Moreover, N can take values $N = 0, 1, 2, \ldots$ since a polynomial of degree zero exists, being equal to a constant. Define the **principal quantum number** n as follows:

$$n \equiv N + \ell + 1 = \frac{1}{2\kappa}, \quad \text{with } \ell \geq 0,\ N \geq 0, \text{ and } n \geq 1. \tag{11.3.27}$$

The polynomial $W_{E\ell}$ so determined is called $W_{n\ell}$, trading the energy label for the quantum number n, which also determines the energy, as it fixes κ. The polynomial $W_{n\ell}$ is of degree $N = n - \ell - 1$. It is a polynomial with a nonvanishing constant term and, in fact, nonvanishing coefficients all the way to ρ^N. The associated radial solution is called $u_{n\ell}$ and is defined from (11.3.15):

$$u_{n\ell}(\rho) = \rho^{\ell+1} W_{n\ell}(\rho)\, e^{-\rho}. \tag{11.3.28}$$

The nonvanishing constant term in $W_{n\ell}$ implies that, indeed, the leading behavior of $u_{n\ell}$ near $\rho = 0$ is $\rho^{\ell+1}$. It follows from (11.3.27) that for a fixed n we must have that

$$0 \leq \ell \leq n - 1, \quad \text{and} \quad 0 \leq N \leq n - 1. \tag{11.3.29}$$

If n and ℓ are known, N is determined; n and ℓ are independent quantum numbers. Remarkably, the energies depend only on n, since κ depends only on n. There is no ℓ dependence of the energies. Equation (11.3.10) gives the energy dependence on κ:

$$E = -4Z^2 \text{Ry}\, \kappa^2, \tag{11.3.30}$$

and using $\kappa = \frac{1}{2n}$, we get the well-known energy dependence on the principal quantum number:

$$E_n = -Z^2\, \text{Ry}\, \frac{1}{n^2} = -\frac{Z^2 e^2}{2a_0} \frac{1}{n^2}, \quad n \geq 1. \tag{11.3.31}$$

We write E_n for the energy of the states with principal quantum number n. These are the energy levels of the hydrogen atom! Since at any fixed value of $n \geq 2$ there are various possible ℓ values, the spectrum is highly degenerate. Moreover, each value of ℓ amounts

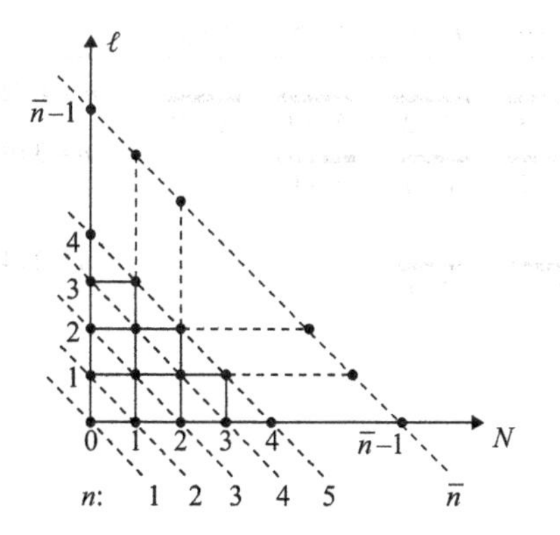

Figure 11.1
All points with integer $N, \ell \geq 0$ represent hydrogen atom states. For any $n \geq 1$, the allowed N and ℓ satisfy $n = N + \ell + 1$. The dotted diagonal lines include the states for fixed values of n, shown for $n = 1, 2, 3, 4, 5$, and $\bar{n}$.

to $2\ell + 1$ degenerate states, given the possible values of m. One way to visualize the spectrum is shown in figure 11.1. Each integer point in the (N, ℓ) positive quadrant represents an ℓ multiplet. The states with a common value of n lie on the dashed diagonal lines. A state in the hydrogen atom spectrum is uniquely specified by three

$$\text{hydrogen quantum numbers: } (n, \ell, m). \tag{11.3.32}$$

If we know n and ℓ, we also know N. Each quantum number has a very important physical meaning: n fixes the energy eigenvalue, $\hbar^2 \ell(\ell + 1)$ is the eigenvalue of the square of angular momentum, and $\hbar m$ is the eigenvalue of the z component of angular momentum.

Exercise 11.1. *Take $n = 2$ and set $a_0 = 1$. Use the recursion relationship (11.3.20) to find the polynomials $W_{2\ell}$ for the allowed values of ℓ.*

It is interesting to know the degeneracy of the energy level with principal quantum number n. Recall that for each n, ℓ can take values from $0, \ldots, n-1$, and for each value of ℓ, m takes values from $-\ell$ to ℓ. The following table counts the states for the first few values of the principal quantum number n:

n value	ℓ values	m values	Total states
$n=1$	$\ell=0$	$m=0$	1 state
$n=2$	$\ell=0$	$m=0$	1
	$\ell=1$	$m=-1,0,1$	+3
			$=4$ states
$n=3$	$\ell=0$	$m=0$	1
	$\ell=1$	$m=-1,0,1$	+3
	$\ell=2$	$m=-2,\ldots,2$	+5
			$=9$ states

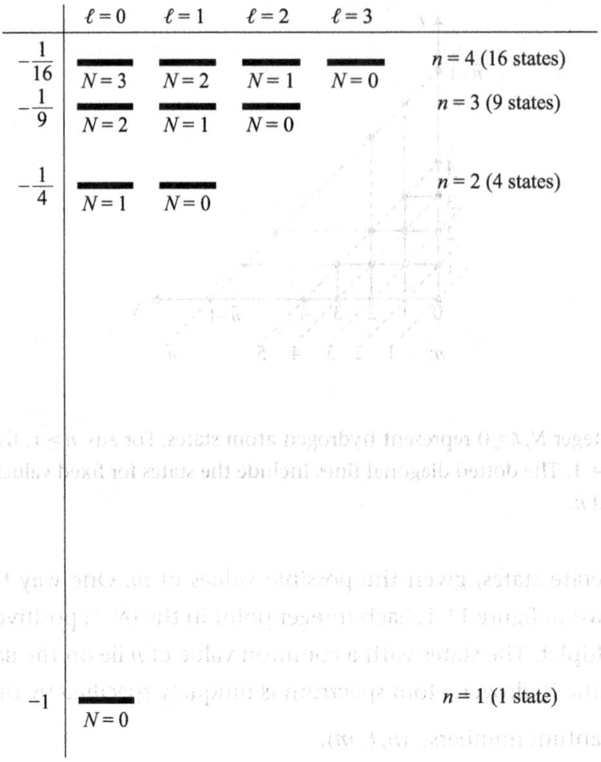

$\ell=0$ $\ell=1$ $\ell=2$ $\ell=3$

$-\frac{1}{16}$ $N=3$ $N=2$ $N=1$ $N=0$ $n=4$ (16 states)

$-\frac{1}{9}$ $N=2$ $N=1$ $N=0$ $n=3$ (9 states)

$-\frac{1}{4}$ $N=1$ $N=0$ $n=2$ (4 states)

-1 $N=0$ $n=1$ (1 state)

Figure 11.2
The bound state spectrum of the hydrogen atom, including levels with $n=1,2,3,4$. Each bar is an angular momentum multiplet, and $n=N+\ell+1$. The vertical line gives the value of $-\frac{1}{n^2}$, which sets the scale for the bound state energies $E=-Z^2\text{Ry}/n^2$ ($Z=1$ for hydrogen).

The calculation for arbitrary principal quantum number n is not difficult:

$$\text{the number of states for } n = \sum_{\ell=0}^{n-1}(2\ell+1) = \frac{2(n-1)n}{2} + n = n^2 - n + n = n^2. \quad (11.3.33)$$

The energy level n is n^2-fold degenerate (or $2n^2$ degenerate if we include the electron spin). The sum of consecutive odd integers is indeed a perfect square.

A more familiar representation of the states of hydrogen is given in the energy diagram of figure 11.2. The different columns indicate the different values of ℓ. We also indicate in the figure the values of N, the degree of the polynomial entering the radial solution. Note that for a given ℓ—that is, for a fixed radial equation, the value of N starts at zero and increases as we go up the column. The number N corresponds to the number of nodes in the radial solution.

Recall that we defined $\rho = 2\kappa Zr/a_0$ in (11.3.12). Together with $\kappa = \frac{1}{2n}$, this gives the final form of the relation between ρ and r:

$$\rho = \frac{Z}{na_0} r. \quad (11.3.34)$$

The eigenstates are labeled by the quantum numbers (n, ℓ, m), and the wave functions are

$$\psi_{n\ell m}(r,\theta,\phi) = N' \frac{u_{n\ell}(r)}{r} Y_{\ell m}(\theta,\phi) = N\rho^{\ell} W_{n\ell}(\rho)e^{-\rho} Y_{\ell m}(\theta,\phi), \tag{11.3.35}$$

where we used (11.3.28) to write $u_{n\ell}$ in terms of $W_{n\ell}$ and other factors. The constants N' and N are included for normalization. Using the expression for ρ and absorbing all constants into a normalization constant that we call $\mathcal{N}$, we have

$$\boxed{\psi_{n\ell m}(r,\theta,\phi) = \mathcal{N}\left(\frac{r}{a_0}\right)^{\ell}\begin{pmatrix}\text{polynomial in } \frac{r}{a_0} \\ \text{of degree } N=n-(\ell+1)\end{pmatrix} e^{-\frac{Zr}{na_0}}\, Y_{\ell m}(\theta,\phi).} \tag{11.3.36}$$

For the ground state of hydrogen, we have $Z=1$ and $(n,\ell,m)=(1,0,0)$. Having zero angular momentum, the associated wave function has no angular dependence. Moreover, $N=0$ so the polynomial in r/a_0 is just a constant. As a result, $\psi_{100} \sim e^{-r/a_0}$. The *normalized* wave function is

$$\psi_{100}(r,\theta,\phi) = \frac{1}{\sqrt{\pi a_0^3}}\, e^{-r/a_0}. \tag{11.3.37}$$

The associated u_{10} for this solution, in the notation $\psi = \frac{u_{E\ell}}{r} Y_{\ell m}$ of (11.3.3), is then

$$u_{10}(r) = 2a_0^{-3/2}\, re^{-r/a_0}. \tag{11.3.38}$$

It is simple to find the normalized ground state for $Z \neq 1$. From our general expression (11.3.36), we see that the exponential dependence must be changed from e^{-r/a_0} to e^{-Zr/a_0}. In essence, this changes $a_0 \to a_0/Z$, making the effective Bohr radius smaller. Since the normalization of (11.3.37) works for any value of a_0, it works for a_0/Z, and the desired $Z \neq 1$ ground state is simply

$$\psi_{100}(r,\theta,\phi) = \sqrt{\frac{Z^3}{\pi a_0^3}}\, e^{-Zr/a_0}. \tag{11.3.39}$$

Exercise 11.2. *The following is an unnormalized radial wave function for hydrogen:*

$$u(r) \sim (r^5 + \alpha_1 r^6 + \cdots)\exp\left(-\frac{r}{17a_0}\right). \tag{11.3.40}$$

What is the value of the principal quantum number n? What is the value of the angular momentum quantum number ℓ? How many nodes does this wave function have?

Exercise 11.3. *How many hydrogen atom eigenstates have energy* $E = -\frac{e^2}{450\, a_0}$*, as well as a radial wave function containing eight nodes?*

It is instructive to find simple explanations for the features of the solution (11.3.36). Much of it is clear. The spherical harmonic in that solution is needed because we are describing a solution with angular momentum ℓ, or, more precisely, $\hat{\mathbf{L}}^2$ eigenvalue $\hbar^2\ell(\ell+1)$, as well as with z component of angular momentum $\hbar m$. The leading power r^{ℓ} is required as the expected behavior near $r \to 0$ of a solution with angular momentum ℓ.

The polynomial $W_{n\ell}$ in the solution is consistent with the node theorem applied to the solution $u_{n\ell}=\rho^{\ell+1}W_{n\ell}e^{-\rho}$ of the radial equation. Note that the product $\rho^{\ell+1}e^{-\rho}$ generates no nodes, so all nodes must arise from $W_{n\ell}$. A state with angular momentum ℓ can have the principal quantum number n taking values

$$n=\ell+1,\ \ell+2,\ \ldots \tag{11.3.41}$$

all the way to infinity. These correspond to the solutions of the radial equation with ℓ fixed, increasing energy and increasing number of nodes. The solution with $n=\ell+1$ is the ground state solution of the radial equation. The solution with $n=\ell+N+1$ is the Nth excited state of the radial equation and should have $N=n-\ell-1$ nodes. Indeed, the polynomial in (11.3.36) has degree N consistent with having that number of zeroes.

The value of the principal quantum number n can be read from the exponential behavior of the solution for large r. That behavior can be quickly obtained from the original differential equation (11.3.5). In the limit $r\to\infty$, we may ignore the effective potential, and the radial equation becomes

$$-\frac{\hbar^2}{2m}u''_{E\ell}(r)\simeq E\,u_{E\ell}(r)\quad\Rightarrow\quad u''_{E\ell}(r)\simeq-\frac{2mE}{\hbar^2}\,u_{E\ell}(r). \tag{11.3.42}$$

When the principal quantum number is n, the energy E_n is given in (11.3.31), and we get

$$-\frac{2mE}{\hbar^2}=\frac{2m}{\hbar^2}\frac{Z^2e^2}{2a_0}\frac{1}{n^2}=\frac{m}{\hbar^2}\frac{Z^2}{a_0}\frac{\hbar^2}{ma_0}\frac{1}{n^2}=\Big(\frac{Z}{na_0}\Big)^2. \tag{11.3.43}$$

Back into the differential equation, we have

$$u''_{E\ell}(r)\simeq\Big(\frac{Z}{na_0}\Big)^2u_{E\ell}(r)\quad\Rightarrow\quad u_{E\ell}(r)\simeq\exp\Big(-\frac{Zr}{na_0}\Big), \tag{11.3.44}$$

explaining the leading large r behavior of (11.3.36).

We have obtained the spectrum of the hydrogen atom by solving the Schrödinger equation. Experimentally, much about the spectrum was known well before the discovery of quantum mechanics in 1925. The knowledge came from the spectrum of photons emitted in transitions between two different levels of the hydrogen atom. Clues from the spectrum were key to the development of Bohr's model of the atom in 1913.

Given principal quantum numbers n and n' with $n>n'\geq 1$, the transition from the less bound level n to the more bound level n' is accompanied by the emission of a photon that, by energy conservation, carries the energy difference between the two levels. The energy E of a photon determines its wavelength λ via $E=hc/\lambda$, and therefore we have

$$\frac{hc}{\lambda}=\text{Ry}\Big(\frac{1}{n'^2}-\frac{1}{n^2}\Big). \tag{11.3.45}$$

The Lyman series was discovered by Theodore Lyman starting around 1906. Here $n'=1$, so we have transitions to the ground state. Since these involve large energies, the photons turn out to be in the ultraviolet spectrum. For $n=2$ we get $\lambda\simeq 122\,\text{nm}$, and as n approaches infinity λ approaches 91 nm. The Balmer series, in fact, was seen first. Johann Balmer discovered in 1885 an empirical equation, equivalent to (11.3.45), based on the measurements of hydrogen lines by Anders Jonas Angström. The Balmer series corresponds to $n'=2$; that is, transitions to the first excited state. The first four lines in this series are in the visible spectrum, defined as wavelengths ranging from 700 to 400 nm. In fact, for $n=3$ one gets $\lambda \simeq 656$ nm, and as n approaches infinity, the wavelength approaches 365 nm. The next series corresponds to $n'=3$ and was observed by Friedrich Paschen in 1908. The transitions here correspond to the infrared spectrum. For $n=4$ one gets $\lambda \simeq 1{,}875$ nm, and as n approaches infinity the wavelength approaches 820 nm.

11.4 Rydberg Atoms

A Rydberg atom is an atom in which the outermost electron is in a high principal quantum number n. Assume the atom has Z protons in the nucleus. Being neutral, it will have Z electrons in a cloud about the nucleus. The outermost electron feels the effect of Z protons and $Z-1$ electrons. As a whole, it feels a net charge of $+1$, as if there were just one proton. It follows that if the outermost electron is far from the nucleus and the other electrons, its physics will be that of a hydrogen atom!

To understand Rydberg atoms, we need to calculate the size of the atom. This will be, roughly, the expected value of r in a state with principal quantum number n. A naive use of (11.3.36) would read from the exponential a relevant length scale of na_0. This is wrong: the size of the atom is larger. We will later explain how to derive the correct answer from (11.3.36), but first we will do it from the virial theorem, which we discussed in section 7.7. For a three-dimensional radial potential, the result in (7.7.22) implies that on energy eigenstates we have the following equality:

$$\langle \hat{T} \rangle = \frac{1}{2}\Big\langle r\frac{\partial V}{\partial r}\Big\rangle. \tag{11.4.1}$$

Here $\hat{T}$ is the kinetic energy operator, and V is the potential energy. In our case $V=-e^2/r$, and therefore,

$$r\frac{\partial V}{\partial r} = -e^2 r\frac{\partial}{\partial r}\frac{1}{r} = \frac{e^2}{r} = -V. \tag{11.4.2}$$

We thus find that

$$\langle \hat{T} \rangle = -\tfrac{1}{2}\langle V \rangle. \tag{11.4.3}$$

This relation implies that the bound state energy E, which is the expectation value of the Hamiltonian in the energy eigenstate, is given by

$$E = \langle \hat{T} + V \rangle = \langle \hat{T} \rangle + \langle V \rangle = \tfrac{1}{2}\langle V \rangle = -\langle \hat{T} \rangle. \tag{11.4.4}$$

It is noteworthy that the magnitude $|E|$ of the bound state energy is exactly equal to the expectation value of the kinetic energy. The bound state energy and the expectation

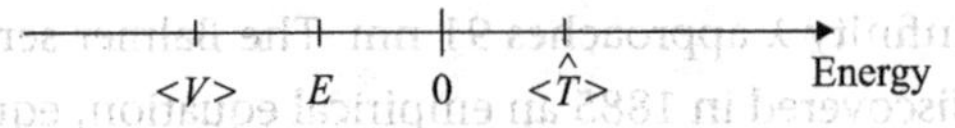

Figure 11.3
For any hydrogen atom bound state, the virial theorem shows that the expectation values of $\hat{T}$ and V are simply related to the bound state energy E: $\langle \hat{T} \rangle = |E|$ and $\langle V \rangle = 2E$.

values of $\hat{T}$ and V are shown on a horizontal energy plot in figure 11.3. Note how $\langle V \rangle, E, 0,$ and $\langle \hat{T} \rangle$ are equally separated.

Let us now consider the relation

$$\langle V \rangle = 2E. \tag{11.4.5}$$

Substituting the form of the potential and the value of the bound state energy E for a state with principal quantum number n, we find that

$$\Big\langle -\frac{e^2}{r} \Big\rangle = 2\Big(-\frac{e^2}{2a_0}\frac{1}{n^2}\Big). \tag{11.4.6}$$

Canceling common factors, this immediately gives

$$\Big\langle \frac{1}{r} \Big\rangle = \frac{1}{n^2 a_0}. \tag{11.4.7}$$

This nice result is ℓ independent! It hints that $\langle r \rangle \sim n^2 a_0$, and this is indeed approximately true. The exact value of $\langle r \rangle$ takes more effort to compute, and we will not do it here. The result, however, is

$$\langle r \rangle = n^2 a_0 \Big[1 + \frac{1}{2}\Big(1 - \frac{\ell(\ell+1)}{n^2}\Big)\Big] = \tfrac{1}{2}\, a_0 \big(3n^2 - \ell(\ell+1)\big). \tag{11.4.8}$$

There is some ℓ dependence in $\langle r \rangle$, with its value decreasing as ℓ increases. For $\ell = 0$,

$$\langle r \rangle = \tfrac{3}{2}\, n^2 a_0, \qquad \ell = 0. \tag{11.4.9}$$

For the largest possible value of ℓ—namely, $\ell = n - 1$, the above gives

$$\langle r \rangle = n^2 a_0 \Big[1 + \frac{1}{2n}\Big], \qquad \ell = n - 1. \tag{11.4.10}$$

For practical purposes we can assume that the size of the Rydberg atom with principal quantum number n is roughly $n^2 a_0$. This, of course, grows very fast with n.

Let us now show how one can obtain this estimate starting from the general form of the solution (11.3.36), written as

$$\psi_{n\ell m}(r, \theta, \phi) = R_{n\ell}(r) Y_{\ell m}(\theta, \phi), \tag{11.4.11}$$

with $R_{n\ell}$ taking the specific form

$$R_{n\ell} \sim r^{\ell} (\text{polynomial of degree } n - (\ell + 1))\, e^{-\frac{r}{n a_0}}. \tag{11.4.12}$$

It is useful to define a radial probability density associated with the above wave function. Let $p(r)dr$ be the probability of finding the electron in a spherical shell of radius r and thickness dr. This probability is obtained by integrating $|\psi|^2$ over that shell, so we have

$$p(r)dr = r^2 dr \int d\Omega \, |\psi_{n\ell m}|^2 = r^2 |R_{n\ell}(r)|^2 \, dr \int d\Omega \, Y^*_{\ell m} Y_{\ell m}. \tag{11.4.13}$$

The integral gives one, and we obtain the **radial probability distribution**:

$$\boxed{p(r)dr = r^2 |R_{n\ell}(r)|^2 \, dr.} \tag{11.4.14}$$

We can now return to our case of interest. We estimate the expectation value of r as the value for which the above distribution peaks. We can approximately determine this by selecting from $R_{n\ell}(r)$ the highest power of r instead of the full polynomial:

$$R_{n\ell}(r) \sim r^\ell r^{n-\ell-1} e^{-\frac{r}{na_0}} = r^{n-1} e^{-\frac{r}{na_0}}. \tag{11.4.15}$$

For this $R_{n\ell}$, the radial probability distribution $p(r)$ is

$$p(r) \sim r^2 r^{2n-2} e^{-\frac{2r}{na_0}} = r^{2n} e^{-\frac{2r}{na_0}}. \tag{11.4.16}$$

Here is the crux of the matter: while the exponential in $p(r)$ suggests a length scale na_0, the power of r is crucial for the determination of the maximum. The function $p(r)$ vanishes for $r=0$ due to the power of r while it also vanishes for $r \to \infty$ due to the exponential. The peak is a compromise, found by differentiation:

$$p'(r) = \left(\frac{2n}{r} - \frac{2}{na_0}\right) r^{2n} e^{-\frac{2r}{na_0}} = 0. \tag{11.4.17}$$

This gives, consistent with our earlier estimates,

$$r = n^2 a_0. \tag{11.4.18}$$

Since the polynomial in $R_{n\ell}$ is of the form $r^\ell + \cdots + r^{n-1}$, we should not be surprised that this estimate is accurate when $\ell = n-1$, and the polynomial indeed becomes equal to the single term we kept in our analysis above. In fact, compare with the exact answer quoted in (11.4.10). A plot of $p(r)$ for $\ell = n-1$ is shown in figure 11.4.

Rydberg atoms were first discovered in interstellar gas in 1965 by B. Höglund and Peter G. Metzler, who saw hydrogen atom transitions occurring at around $n = 100$ levels.

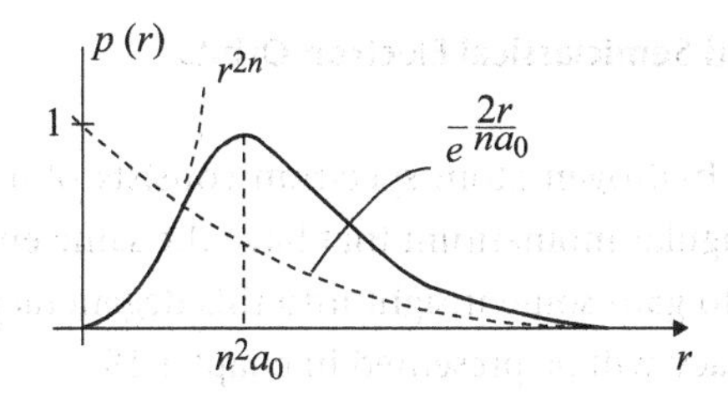

Figure 11.4

The radial probability distribution $p(r)$ for an energy eigenstate with principal quantum number n and $\ell = n-1$ vanishes for $r=0$ and approaches zero as $r \to \infty$. For the intermediate value $r = n^2 a_0$, it reaches its peak value.

Those atoms were created at recombination, when the universe cooled enough for protons to be able to bind electrons. Rydberg atoms with n as large as 350 have been detected in outer space. Their size is

$$r \simeq 0.53 \times 10^{-10}\text{m} \times (350)^2 \simeq 6.5 \times 10^{-6}\text{m}. \quad \rightarrow \quad r \simeq 6\,\text{microns!} \tag{11.4.19}$$

This is a gigantic atom. The diameter of a red blood cell is about 8 microns. The diameter of a human hair is about 50 microns. The Kleppner laboratory at MIT has created Rydberg atoms with $n = 60$ using three lasers acting on alkali metal atoms. In the laboratory, Rydberg atoms are detected because they can easily be ionized with electric fields. This is not the case for ordinary atoms, for which ionization requires electric fields in the order of 10^8 V/cm.

Exercise 11.4. *Calculate the magnitude of the energy difference Δ between the $n+1$ and n levels of a hydrogen atom in the approximation that n is very large. Show that the wavelength of a photon emitted in the transition between these two levels is $\lambda \simeq 2\pi a_0 n^3/\alpha$, with α the fine structure constant.*

For future reference let us collect a number of expectation values of powers of r on hydrogen atom eigenstates $\psi_{n\ell m}$. We have

$$\begin{aligned}
\langle r \rangle &= \tfrac{1}{2}\, a_0\big(3n^2 - \ell(\ell+1)\big),\\
\left\langle \frac{1}{r} \right\rangle &= \frac{1}{a_0 n^2},\\
\left\langle \frac{1}{r^2} \right\rangle &= \frac{1}{a_0^2 n^3 (\ell + \frac{1}{2})},\\
\left\langle \frac{1}{r^3} \right\rangle &= \frac{1}{a_0^3 n^3 \ell\, (\ell + \frac{1}{2})\,(\ell+1)}.
\end{aligned} \tag{11.4.20}$$

We derived the expectation value of $1/r$ from the virial theorem, and the expectation value of $1/r^2$ can be obtained from the Hellmann-Feynman lemma (problem 11.5). The other two are usually obtained from the so-called Kramers relation, an equation that follows from the radial equation for hydrogen and relates $\langle r^k \rangle$, $\langle r^{k-1} \rangle$, and $\langle r^{k-2} \rangle$ for arbitrary integer k.

11.5 Degeneracies and Semiclassical Electron Orbits

The degeneracy of the hydrogen atom spectrum consists of a number of states with various values of the angular momentum that have the same energy. Below we will use a semiclassical picture to gain some insight into this degeneracy. An algebraic analysis explaining the degeneracy will be presented in chapter 19.

To understand the physics of the bound states of hydrogen, it is useful to consider the effective potential V_{eff} entering the radial equation:

$$V_{\text{eff}}(r) = -\frac{e^2}{r} + \frac{\hbar^2 \ell(\ell+1)}{2mr^2}. \tag{11.5.1}$$

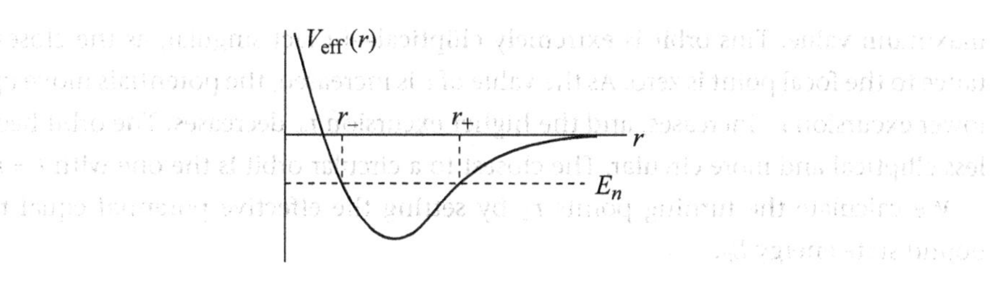

Figure 11.5
The effective potential V_{eff} in the hydrogen atom radial equation. In the classical limit, for an electron of energy E_n its motion would be an ellipse in which the maximal and minimal distance to the focal point are the values of r_+ and r_-, respectively.

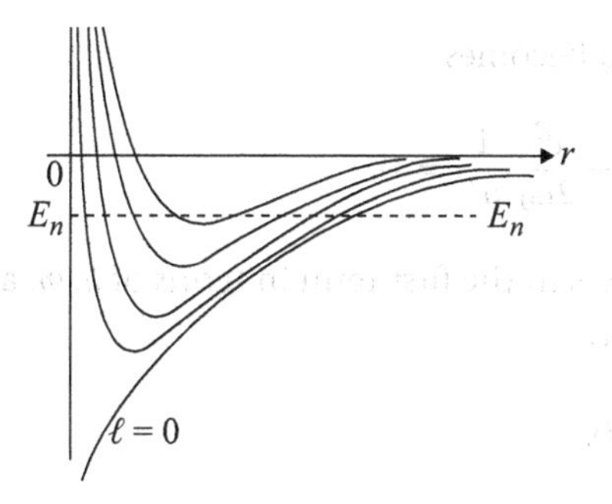

Figure 11.6
The effective potential V_{eff} plotted for various values of ℓ, beginning with $\ell = 0$, for which there is no centrifugal barrier. The picture makes clear that for a fixed energy E_n the classical elliptical orbits become more circular as the value of ℓ increases.

As $r \to 0$, the centrifugal barrier due to angular momentum dominates over the Coulomb potential because $1/r^2$ diverges faster than $1/r$. As a result, the effective potential goes to plus infinity as $r \to 0$. For sufficiently large r, the Coulomb potential dominates. Thus, for sufficiently large and increasing r, the effective potential is negative and increasing, approaching zero as $r \to \infty$. It follows that V_{eff} has a minimum for some value of r where it takes a negative value, as shown in figure 11.5.

We also show in the figure a value E_n of the total energy. In the classical picture, the orbit is elliptical, and the turning points r_- and r_+ are, respectively, the smallest and largest distance from the particle to the focus of the ellipse. As we will see below, the semiclassical approximation is valid when the principal quantum number n is large. The quantum particle then detects the potential as slowly varying.

We have seen that for a given principal quantum number n, the angular momentum ℓ can take values $0, 1, \ldots, n-1$, and all those states are degenerate. If we think of classical orbits, which are highly elliptical and which are almost circular? This can be answered without computation by looking at figure 11.6, where we show the effective potential V_{eff}, which is a function of ℓ, for various values of ℓ, including $\ell = 0$. The intersections with the line at energy $E_n < 0$ determine the mimimum and maximum excursions of the radius for each value of ℓ. For $\ell = 0$, we see that the excursion is from zero to some large maximum value. This orbit is extremely elliptical—in fact singular, as the closest distance to the focal point is zero. As the value of ℓ is increased, the potentials move up, the lower excursion r_- increases, and the higher excursion r_+ decreases. The orbit becomes less elliptical and more circular. The closest to a circular orbit is the one with $\ell = n-1$.

We calculate the turning points $r_\pm$ by setting the effective potential equal to the bound state energy E_n:

$$\frac{\hbar^2 \ell(\ell+1)}{2mr^2} - \frac{e^2}{r} = -\frac{e^2}{2a_0}\frac{1}{n^2}. \tag{11.5.2}$$

To remove the units, we introduce x as a unit-free version of r:

$$r = a_0 x. \tag{11.5.3}$$

Then the above equation becomes

$$\frac{\hbar^2}{2ma_0^2}\frac{\ell(\ell+1)}{x^2} - \frac{e^2}{a_0}\frac{1}{x} + \frac{e^2}{2a_0}\frac{1}{n^2} = 0. \tag{11.5.4}$$

Writing one of the two a_0's in the first term in terms of $\hbar$, m, and e, we quickly find that the equation simplifies to

$$\frac{\ell(\ell+1)}{x^2} - \frac{2}{x} + \frac{1}{n^2} = 0. \tag{11.5.5}$$

This is a quadratic equation for x^{-1} and has solutions

$$\frac{1}{x} = \frac{1}{\ell(\ell+1)}\left(1 \pm \sqrt{1 - \frac{\ell(\ell+1)}{n^2}}\right). \tag{11.5.6}$$

Inverting and eliminating roots from the denominator, one finds

$$x_\pm = n^2\left(1 \pm \sqrt{1 - \frac{\ell(\ell+1)}{n^2}}\right). \tag{11.5.7}$$

Therefore, the turning points $r_\pm$ for the classical orbit associated with principal quantum number n and orbital angular momentum ℓ are given by

$$r_\pm = n^2 a_0\left(1 \pm \sqrt{1 - \frac{\ell(\ell+1)}{n^2}}\right), \quad \ell = 0, 1, \cdots, n-1. \tag{11.5.8}$$

Since r_- and r_+ are, respectively, the shortest and longest distances to the focal point, their sum $r_+ + r_-$ is the length of the major axis of the ellipse. In fact, for all values of ℓ that sum takes the same value! Using (11.5.8), one immediately finds

$$\tfrac{1}{2}(r_+ + r_-) = n^2 a_0, \tag{11.5.9}$$

which tells us that the length of the semimajor axis is $n^2 a_0$. We see that orbits with the same energy (fixed n) have the *same semimajor axis but different amounts of eccentricity*, as controlled by ℓ (see figure 11.7). This is a familiar fact in planetary elliptic orbits where the period and the energy of an orbit depend only on the value of the semimajor axis. The eccentricity ϵ of the ellipse is defined by

$$\epsilon \equiv \frac{r_+ - r_-}{r_+ + r_-} = \sqrt{1 - \frac{\ell(\ell+1)}{n^2}}. \tag{11.5.10}$$

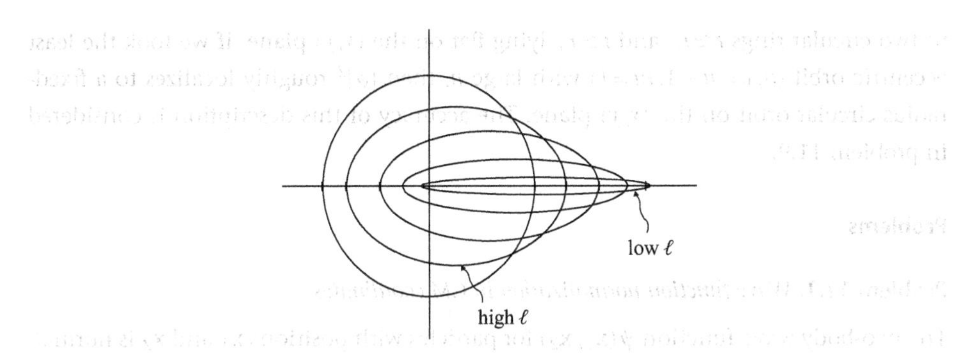

Figure 11.7
All the semiclassical orbits associated with a fixed principal quantum number n are ellipses with the same semimajor axis. The orbits with low ℓ are highly elliptical; those with ℓ below but near n are almost circular. In this figure the focal point of all ellipses is at the origin.

A circle is an ellipse of zero eccentricity: $r_+ = r_-$. You can verify that no allowed value of ℓ gives us zero eccentricity. As the eccentricity grows, the ellipse becomes less and less circular. The largest possible eccentricity is $\epsilon = 1$, which we find for all $\ell = 0$ states with different values of n. This corresponds to a degenerate ellipse with $r_- = 0$ (see figure 11.6). For a fixed n, the eccentricity is minimized for the largest value of ℓ—namely, $n-1$. From (11.5.10), we find

$$\epsilon = \frac{1}{\sqrt{n}}, \quad \text{for} \quad \ell = n-1. \tag{11.5.11}$$

The above information about orbits is relevant quantum mechanically if the semiclassical approximation holds. We now claim that this approximation holds for large n Rydberg atoms. More precisely, we want N and ℓ large for a striking semiclassical picture. Of course, with $n = N + \ell + 1$, this implies large n.

Consider, for example, the case $n = 100$ and $\ell = 60$. Using (11.5.8) gives us

$$r_- \simeq 2{,}038\, a_0, \quad r_+ \simeq 17{,}962\, a_0. \tag{11.5.12}$$

Such an ellipse has $e \simeq 0.796$. For this energy eigenstate, we have $N = n - (\ell + 1) = 39$, so the radial wave function is oscillatory with thirty-nine nodes. A rapidly oscillating wave function has a short de Broglie wavelength that effectively makes the potential slowly varying. Following the semiclassical intuition developed in section 7.2, the amplitude of the wave function will be largest near the turning points, small in between them, and vanishingly small beyond them.

Moreover, if the quantum number m takes the largest possible value $m = \ell$, the wave function is actually localized at the equatorial plane $\theta = \pi/2$. Indeed, we see from (10.5.35) that the highest m spherical harmonic $Y_{\ell\ell}$ satisfies $|Y_{\ell\ell}|^2 \sim (\sin\theta)^{2\ell}$ without any ϕ dependence. Since $\sin\theta$ vanishes at $\theta = 0$ and π and peaks with value one at $\theta = \pi/2$, a high-power $(\sin\theta)^{2\ell}$ very narrowly peaks about $\theta = \pi/2$, where it attains a value equal to one. Thus, large $\ell = m$ localizes $|\psi|^2$ to the (x, y) plane, without any ϕ dependence! It follows that for the $(n, \ell, m) = (100, 60, 60)$ state $|\psi|^2$ roughly localizes to two circular rings $r \simeq r_-$ and $r \simeq r_+$ lying flat on the (x, y) plane. If we took the least eccentric orbit $(n, \ell = n-1, m = \ell)$ with large n, then $|\psi|^2$ roughly localizes to a fixed-radius circular orbit on the (x, y) plane. The accuracy of this description is considered in problem 11.9.

Problems

Problem 11.1. *Wave function normalization in CM coordinates.*

The two-body wave function $\psi(\mathbf{x}_1, \mathbf{x}_2)$ for particles with positions $\mathbf{x}_1$ and $\mathbf{x}_2$ is normalized when

$$1 = \iint d^3\mathbf{x}_1\, d^3\mathbf{x}_2\, |\psi(\mathbf{x}_1, \mathbf{x}_2)|^2. \tag{1}$$

Consider now CM and relative coordinates $\mathbf{X}$ and $\mathbf{x}$, respectively:

$$\mathbf{X} = \frac{m_1\mathbf{x}_1 + m_2\mathbf{x}_2}{m_1 + m_2}, \quad \mathbf{x} = \mathbf{x}_2 - \mathbf{x}_1.$$

In the new coordinates, the new wave function ψ' is simply equal to the old wave function, expressed in terms of $\mathbf{X}$ and $\mathbf{x}$:

$$\psi'(\mathbf{X}, \mathbf{x}) = \psi(\mathbf{x}_1(\mathbf{X}, \mathbf{x}), \mathbf{x}_2(\mathbf{X}, \mathbf{x})).$$

We wish to show that (1) implies that the normalization condition for ψ' reads

$$1 = \iint d^3\mathbf{X}\, d^3\mathbf{x}\, |\psi'(\mathbf{X}, \mathbf{x})|^2. \tag{2}$$

To see this, consider the transformation of the measure of integration:

$$d^3\mathbf{X}\, d^3\mathbf{x} = |J|\, d^3\mathbf{x}_1\, d^3\mathbf{x}_2.$$

Compute the Jacobian determinant $|J|$ of the transformation, the determinant of a 6×6 matrix. Show that $|J| = 1$ and explain why it implies that (2) holds on account of (1). (Hint: recall that one can add to any column (or row) an arbitrary multiple of another column (or row) without changing the determinant.)

Problem 11.2. *Hydrogen atom with total momentum.*

When the motion of the nucleus is taken into account, the state of the hydrogen atom can be represented by a wave function $\psi(\mathbf{X}, \mathbf{x})$, with $\mathbf{X}$ the CM coordinate and $\mathbf{x} = \mathbf{x}_e - \mathbf{x}_p$ the relative coordinate pointing from the proton to the electron.

Suppose that the atom is in such a state that the *total* momentum has equal probabilities for the values $\mathbf{p}_0$ and $-\mathbf{p}_0$. Moreover, the internal states are $\phi_{1,0,0}(\mathbf{x})$ or $\phi_{2,1,1}(\mathbf{x})$ with probabilities 1/4 and 3/4, respectively (we use the notation $\phi_{n,\ell,m}$). These probabilities are not correlated with the total momentum.

1. What is $\psi(\mathbf{X}, \mathbf{x})$? For concreteness, make all coefficients in your expression positive real numbers. Assemble a solution consistent with the previous description of the state. In order to normalize the wave function, assume that the CM

coordinate $\mathbf{X}$ is restricted to a large three-dimensional box all of whose sides are of length L.

Consistent with the statement about various probabilities and ignoring an overall phase, how many independent phases can be introduced into the above expression for $\psi(\mathbf{X}, \mathbf{x})$?

2. What is the expectation value for the total energy of the state in (1)?

   If we introduced complex phases for different terms in ψ, would the expectation value for the total energy change?

Problem 11.3. *Hydrogen atom in motion.*

Consider the hydrogen atom wave function

$$\psi(\mathbf{X}, \mathbf{x}, t) = \int d^3\mathbf{K}\, e^{i\mathbf{K}\cdot\mathbf{X}}\, g(\mathbf{K})\, \psi_0(\mathbf{x})\, e^{-iEt/\hbar},$$

where $\mathbf{X}$ and $\mathbf{x}$ are, respectively, CM and relative coordinates. Here, $\psi_0(\mathbf{x})$ is an eigenstate of the relative motion Hamiltonian with energy E_0. In this wave packet, $g(\mathbf{K})$ is a real function that narrowly peaks about $\mathbf{K} = \mathbf{K}_0$. Let M denote the total mass of the atom and μ denote the reduced mass. The following questions refer to the above wave function ψ.

1. What is E as a function of $\mathbf{K}$?
2. What is the approximate expected value $\langle E \rangle$ of the total energy?
3. What is the approximate expected value of the total momentum $\langle \mathbf{P} \rangle$ of the atom?
4. What is the approximate expected location of this atom at time t? (Hint: use the stationary phase principle.)

Problem 11.4. *Velocity parameter in hydrogen atom eigenstates.*

For any hydrogen atom eigenstate, write $\langle \hat{T} \rangle = \frac{1}{2} m \langle v^2 \rangle$, where, to a very good approximation, m is the mass of the electron. Express the ratio $\sqrt{\langle v^2 \rangle}/c$ in terms of the fine structure constant $\alpha = \frac{e^2}{\hbar c} \approx \frac{1}{137}$ and the principal quantum number n. Is the electron relativistic? What is the ratio $\sqrt{\langle v^2 \rangle}/c$ when the nucleus has Z protons?

Problem 11.5. *Hellmann-Feynman for hydrogen atom expectation values.*

The Hellmann-Feynman lemma (section 7.9) considers Hamiltonians $\hat{H}(\lambda)$ that depend on a parameter λ and gives a simple formula for the rate of change of the energy of an eigenstate with respect to λ. We will use this result to compute expectation values in the hydrogen atom eigenstates $\psi_{n\ell m}$. The Hamiltonian $\hat{H}$ for radial wave functions in the hydrogen atom is given by

$$\hat{H} = -\frac{\hbar^2}{2m}\frac{d^2}{dr^2} + \frac{\hbar^2}{2m}\frac{\ell(\ell+1)}{r^2} - \frac{e^2}{r}.$$

The hydrogen atom energies are $E_n = -\frac{e^2}{2a_0}\frac{1}{n^2}$ with $a_0 = \frac{\hbar^2}{me^2}$. In solving the radial equation, one sets $n = N + \ell + 1$, where N is the degree of the radial polynomial.

1. Use the lemma with parameter $\lambda = e^2$ to calculate $\langle 1/r \rangle$ on a hydrogen eigenstate $\psi_{n\ell m}$.
2. Use the lemma with parameter $\lambda = \ell$ to calculate $\langle 1/r^2 \rangle$ on the eigenstate $\psi_{n\ell m}$. Assume, as it turns out correctly, that ℓ can be treated as a continuous variable and show that

$$\left\langle \frac{1}{r^2} \right\rangle = \frac{1}{a_0^2 n^3 (\ell + \frac{1}{2})}.$$

Problem 11.6. *Electron beyond the classical orbit.*

Consider a hydrogen atom in the ground state, the state with energy $E_0 = -\frac{e^2}{2a_0}$.

1. Find the orbital radius of an electron with energy E_0 in *classical* mechanics.
2. Calculate the probability of finding the quantum ground state electron beyond the classical orbit-radius.

Problem 11.7. *Exercises on hydrogen atom and positronium.*

1. Find $\langle r \rangle$ and $\langle r^2 \rangle$ for the ground state of hydrogen. What is the most probable value of r in the ground state? Give your answers in terms of a_0.
2. Assume that the nucleus of hydrogen has a radius of one femtometer. Calculate the probability that the ground state electron is found inside the nucleus. Make approximations to simplify your work and still get a very accurate answer.
3. Positronium is a bound state of an electron and a positron, which has charge $+e$ and the same mass as an electron. Positronium behaves in some respects like the hydrogen atom. What is the ratio of the energy in the nth state of positronium to the energy in the nth state of hydrogen? How does the size of positronium compare with the size of a hydrogen atom?

Problem 11.8. *Instantaneous decay and electron wave function.*

Consider an electron in the ground state of tritium, an unstable $Z=1$ isotope of hydrogen with a nucleus composed of a proton and two neutrons. A process of beta decay turns a neutron into a proton instantaneously (with an electron and an antineutrino flying out) so that we get helium-3, a $Z=2$ ion. Assume that the electron wave function does not change during this fast decay. What is the probability that it will be found in the ground state of helium-3, right after the decay?

Problem 11.9. *Electron orbit in the hydrogen atom.*

In this problem we consider a hydrogen atom with a fixed principal quantum number n, with $\ell = n-1$, and $m = n-1$. The value n is arbitrary and possibly large.

1. Write the wave function $\psi_{n,\ell,m}(r, \theta, \phi)$ in terms of the relevant spherical harmonic and a radial factor fully determined except for an overall unit-free normalization constant N.
2. Give, up to normalization, the radial probability density $p(r)$ for which $p(r)dr$ is the probability of finding the electron in the interval $(r, r+dr)$. For what value of r is $p(r)$

maximum? For large n this is actually a rather sharp maximum. Estimate the large n dependence of the width of the peak and show that it behaves like $a_0\sqrt{n}$.

3. Recall that up to normalization $|Y_{\ell,\ell}(\theta,\phi)|^2 \simeq (\sin\theta)^{2\ell}$. Sketch $|Y_{\ell,\ell}|^2$ as a function of $\theta \in [0,\pi]$ when ℓ is a large integer. Estimate the large ℓ dependence of the width of the peak and show that it behaves like $1/\sqrt{\ell}$.

   Describe in words, or with a picture, the locus where the electron is likely to be found for large n and $\ell = m = n-1$.

12 The Simplest Quantum System: Spin One-Half

We discuss the simplest possible quantum system, one in which the state space is spanned by two states. By considering Hermitian operators in this system, we discover the 2×2 matrices $\hat{S}_x, \hat{S}_y$, and $\hat{S}_z$, satisfying an algebra of angular momentum with eigenvalues that indicate the existence of states with angular momentum one-half, in units of $\hbar$. The Stern-Gerlach experiment demonstrated that the magnetic moment of the electron takes quantized values. By inference, one is led to conclude that the electron is a spin one-half particle, thus realizing the simplest quantum system. We explicitly construct spin operators pointing in arbitrary directions and the associated spin states. We conclude by showing how the quantum states of a two-state system can be used to build a protocol for quantum key distribution.

12.1 A System with Two States

We have considered in some detail the quantum mechanics of particles moving in one and three dimensions, solving the Schrödinger equation for states that have nontrivial spatial dependence. We just studied the hydrogen atom, a quantum system that is the cornerstone of any study of atomic physics. By all accounts, this is an intricate quantum theory, with *infinitely many* states, each with elaborate spatial dependence.

We want to return to simpler quantum systems, such as when we examined photons in a Mach-Zehnder interferometer. There we described the physics using a two-component column vector. The photon could be in the up branch or in the down branch of the interferometer, and the entries in the column vector, two complex *numbers*, were the probability amplitudes for these two options. This column vector played the role of a wave function, and there was no need to include spatial dependence. As the photon moved in the interferometer, the change in this column vector was interpreted as due to the passage of time. The lesson here is that we can do interesting quantum mechanics with a system that has two states and thus requires two time-dependent numbers to describe it.

We aim to better appreciate the essence of quantum mechanics by stripping away non essential complications. We will describe a two-state system. If you thought of a particle, we want to go from a wave function $\Psi(x,t)$ defined all over the real line for all times to a wave function that only exists at two points. Such a wave function has two degrees of freedom: the probability amplitude $\Psi_1(t)$ for the particle to be at x_1 and the probability amplitude $\Psi_2(t)$ for the particle to be at x_2. We assemble the wave function into a column vector:

$$\Psi(t) = \begin{pmatrix} \Psi_1(t) \\ \Psi_2(t) \end{pmatrix}, \quad \Psi_1(t), \Psi_2(t) \in \mathbb{C}. \tag{12.1.1}$$

In this interpretation, the complex numbers $\Psi_1(t)$ and $\Psi_2(t)$ represent, respectively, the values of the wave function at the two points x_1 and x_2 at time t. We require that

$$|\Psi_1(t)|^2 + |\Psi_2(t)|^2 = 1 \tag{12.1.2}$$

so that the probabilities $|\Psi_1(t)|^2$ and $|\Psi_2(t)|^2$ for the particle to be at x_1 and x_2, respectively, add up to one at all times.

The physics of a two-state system can appear in many forms; we do not have to limit ourselves to a particle that can only be at two points. The system in question could be a box with a partition that divides it into a left side L and a right side R, with Ψ_1 and Ψ_2 the amplitudes for the particle to be on the L and R sides, respectively. Or we could have a molecule in which the bond structure can exist in two possible configurations. The system could also be a photon that could be polarized along x or along y. Or perhaps, more intriguingly, the system could be a particle with an intrinsic attribute with two values called $\pm$, for lack of a better name. Then Ψ_1 and Ψ_2 would be the amplitudes for the particle to be in the $+$ and $-$ states, respectively.

Writing the two-state wave function (12.1.1) as

$$\Psi(t) = \Psi_1(t) \begin{pmatrix} 1 \\ 0 \end{pmatrix} + \Psi_2(t) \begin{pmatrix} 0 \\ 1 \end{pmatrix}, \tag{12.1.3}$$

we see that *the* two states in the two-state system are

$$\begin{pmatrix} 1 \\ 0 \end{pmatrix} \text{ and } \begin{pmatrix} 0 \\ 1 \end{pmatrix}. \tag{12.1.4}$$

Indeed, the general wave function is a superposition of these two *basis vectors*: we multiply the first basis vector by a number Ψ_1 and the second basis vector by Ψ_2 and add them to get the general vector. This makes it clear that the name *two-state system* should not be taken literally. It is a system with two *basis states*. The number of possible states is infinite, as there are infinitely many constants Ψ_1 and Ψ_2 satisfying the normalization condition.

What we really have here is a vector space with two basis vectors, and the wave function is a vector in this vector space. We say this is a two-dimensional *complex vector space*: *two-dimensional* because there are two basis vectors and *complex vector space* because the basis vectors are multiplied by complex numbers and added to form a general vector. A vector in a two-dimensional complex vector space is also known as a *spinor*. If the basis vectors can only be multiplied by real numbers, we have a *real* vector space. In quantum mechanics, we need complex vector spaces.

In this situation the Schrödinger equation reads

$$i\hbar\frac{\partial\Psi(t)}{\partial t}=\hat{H}\Psi(t), \tag{12.1.5}$$

with $\Psi(t)$ the time-dependent wave function and $\hat{H}$ a Hamiltonian to be specified. When we first considered the Schrödinger equation, we learned that the consistency of the probability interpretation of the wave function requires the Hermiticity of the Hamiltonian. But Hermiticity is only defined when we have an inner product. So we have to understand how to define an inner product in the two-state system. The clue comes from wave functions on the line, where the inner product between two wave functions $\psi(x)$ and $\phi(x)$ is given by

$$\langle\psi,\phi\rangle\equiv\int dx\,\psi^*(x)\phi(x). \tag{12.1.6}$$

If space were discrete with equally spaced points x_i with $i=1,2,\ldots$, the integral on the right would be proportional to

$$\int dx\,\psi^*(x)\phi(x)\sim\sum_i\psi^*(x_i)\phi(x_i), \tag{12.1.7}$$

and if we had just two points, the sum on the right would be

$$\psi^*(x_1)\phi(x_1)+\psi^*(x_2)\phi(x_2). \tag{12.1.8}$$

The top and bottom components of our states can be thought of as wave functions at two different points. It is therefore natural for us, given two states

$$\Psi=\begin{pmatrix}\Psi_1\\ \Psi_2\end{pmatrix},\quad \Phi=\begin{pmatrix}\Phi_1\\ \Phi_2\end{pmatrix}, \tag{12.1.9}$$

to define the inner product mimicking (12.1.8):

$$\boxed{\langle\Psi,\Phi\rangle\;\equiv\;\Psi_1^*\Phi_1+\Psi_2^*\Phi_2,\quad \Psi=\begin{pmatrix}\Psi_1\\ \Psi_2\end{pmatrix},\quad \Phi=\begin{pmatrix}\Phi_1\\ \Phi_2\end{pmatrix}.} \tag{12.1.10}$$

This definition is good in that the normalization condition (12.1.2) can now be written in the familiar form:

$$\langle\Psi(t),\Psi(t)\rangle=1. \tag{12.1.11}$$

As usual, we define the norm $\|\Psi\|$ of a state from the relation:

$$\|\Psi\|^2=\langle\Psi,\Psi\rangle. \tag{12.1.12}$$

With an inner product, we can discuss Hermiticity. As stated in (5.3.8), the Hermitian conjugate $T^\dagger$ of a linear operator T is defined via the relation

$$\langle T^\dagger u,v\rangle=\langle u,Tv\rangle, \tag{12.1.13}$$

with u and v arbitrary vectors:

$$u = \begin{pmatrix} u_1 \\ u_2 \end{pmatrix}, \quad v = \begin{pmatrix} v_1 \\ v_2 \end{pmatrix}. \tag{12.1.14}$$

A linear operator acting on two-component vectors is a 2×2 matrix whose entries are just complex numbers. Thus, we can write

$$T = \begin{pmatrix} T_{11} & T_{12} \\ T_{21} & T_{22} \end{pmatrix}. \tag{12.1.15}$$

It is now possible to compute the Hermitian conjugate of T:

Exercise 12.1. *Use (12.1.13) to show that the matrix for* $T^\dagger$ *is given by*

$$T^\dagger = \begin{pmatrix} T_{11}^* & T_{21}^* \\ T_{12}^* & T_{22}^* \end{pmatrix}, \tag{12.1.16}$$

where the star denotes complex conjugation. This means that $T^\dagger = (T^t)^*$*—namely,* $T^\dagger$ *is obtained by transposing* T *to give* T^t *and then complex conjugating all of its elements.*

With this insight we can now find the most general Hamiltonian $\hat{H}$ for a two-state system. The only condition $\hat{H}$ must satisfy is Hermiticity. Therefore, let us consider a 2×2 matrix $\hat{H}$ and the condition that it is equal to its Hermitian conjugate:

$$\hat{H} = \begin{pmatrix} h_{11} & h_{12} \\ h_{21} & h_{22} \end{pmatrix} = \begin{pmatrix} h_{11}^* & h_{21}^* \\ h_{12}^* & h_{22}^* \end{pmatrix} = \hat{H}^\dagger. \tag{12.1.17}$$

The equality gives three conditions: h_{11} is real, h_{22} is real, and $h_{21} = h_{12}^*$. We can represent the most general solution to these conditions using four *real* constants $h_0, h_1, h_2,$ and h_3:

$$\hat{H} = \begin{pmatrix} h_0 + h_3 & h_1 - ih_2 \\ h_1 + ih_2 & h_0 - h_3 \end{pmatrix}. \tag{12.1.18}$$

Note that we have expressed the diagonal elements h_{11} and h_{22}, which must be real, as the sum and the difference of two real numbers h_0 and h_3. This was done for convenience. Moreover, the real constants h_1 and h_2 are the real and imaginary parts of the complex number h_{21}. It is clear by inspection that the above $\hat{H}$ is Hermitian, in fact the most general Hermitian 2×2 matrix. If the real numbers h_0, h_1, h_2, h_3 are all time independent, the above is the most general time-independent Hamiltonian of a two-state system. If h_0, h_1, h_2, h_3 are real functions of time, the above is the most general time-dependent Hamiltonian of a two-state system.

Let us investigate this general $\hat{H}$ a bit more. Since we have four independent real numbers fixing $\hat{H}$, we can make this fact manifest by rewriting $\hat{H}$ as follows:

$$\hat{H} = h_0 \begin{pmatrix} 1 & 0 \\ 0 & 1 \end{pmatrix} + h_1 \begin{pmatrix} 0 & 1 \\ 1 & 0 \end{pmatrix} + h_2 \begin{pmatrix} 0 & -i \\ i & 0 \end{pmatrix} + h_3 \begin{pmatrix} 1 & 0 \\ 0 & -1 \end{pmatrix}. \tag{12.1.19}$$

This rewriting associates a Hermitian matrix to each of the four constants. This expansion shows that these four matrices form a basis for the *real* vector space of Hermitian 2×2 matrices: the most general Hermitian 2×2 matrix is obtained by multiplying the basis matrices by real numbers and adding them. These basis matrices are quite important, and they have been given names. The first is the identity matrix $\mathbb{1}$, and the other three are the Pauli matrices we encountered earlier in (3.3.18) when talking about matrix mechanics:

$$\sigma_1 = \begin{pmatrix} 0 & 1 \\ 1 & 0 \end{pmatrix}, \quad \sigma_2 = \begin{pmatrix} 0 & -i \\ i & 0 \end{pmatrix}, \quad \sigma_3 = \begin{pmatrix} 1 & 0 \\ 0 & -1 \end{pmatrix}. \tag{12.1.20}$$

We also calculated the commutators:

$$[\sigma_1, \sigma_2] = 2i\sigma_3, \quad [\sigma_2, \sigma_3] = 2i\sigma_1, \quad [\sigma_3, \sigma_1] = 2i\sigma_2. \tag{12.1.21}$$

This list is reminiscent of the commutation relation of angular momentum operators. We can make the connection to angular momentum manifest by defining three operators proportional to the Pauli matrices:

$$\hat{S}_x = \tfrac{\hbar}{2}\sigma_1, \quad \hat{S}_y = \tfrac{\hbar}{2}\sigma_2, \quad \hat{S}_z = \tfrac{\hbar}{2}\sigma_3. \tag{12.1.22}$$

More explicitly,

$$\hat{S}_x = \frac{\hbar}{2}\begin{pmatrix} 0 & 1 \\ 1 & 0 \end{pmatrix}, \quad \hat{S}_y = \frac{\hbar}{2}\begin{pmatrix} 0 & -i \\ i & 0 \end{pmatrix}, \quad \hat{S}_z = \frac{\hbar}{2}\begin{pmatrix} 1 & 0 \\ 0 & -1 \end{pmatrix}. \tag{12.1.23}$$

As you can easily check, the commutators of the Pauli matrices imply the following commutators for the $\hat{S}$ operators:

$$\begin{aligned} [\hat{S}_x, \hat{S}_y] &= i\hbar\hat{S}_z, \\ [\hat{S}_y, \hat{S}_z] &= i\hbar\hat{S}_x, \\ [\hat{S}_z, \hat{S}_x] &= i\hbar\hat{S}_y. \end{aligned} \tag{12.1.24}$$

This is a remarkable result. These operators satisfy the algebra of angular momentum! The above commutators are exactly those in (10.3.10), with the $\hat{L}$s replaced by $\hat{S}$s. In fact, we anticipated in (10.3.16) that angular momentum-like operators could arise from 2×2 constant matrices rather than from $\hat{\mathbf{x}}$ and $\hat{\mathbf{p}}$ operators. The operators $\hat{S}$ have the units of angular momentum since they have the units of $\hbar$. Note that the Pauli matrices have no units. Note also that the common $\frac{\hbar}{2}$ factor in the definition (12.1.22) is *necessary* for the $\hat{S}$ operators to satisfy precisely the commutation relations of angular momentum. It is surprising that we have found angular momentum operators in the two-state system, and one would imagine, correctly, that there is a deep significance to this. We will find that this angular momentum is the angular momentum of a particle with spin one-half. Spin is an intrinsic angular momentum, a property of particles that cannot be described with the orbital angular momentum operators $\hat{L}$. Of course, two-state systems can sometimes describe other physics, and the angular momentum operators then play a different role.

The angular momentum operators we have uncovered are Hermitian operators and can therefore be measured on states. In fact, the basis vectors in (12.1.4) are eigenstates of $\hat{S}_z$ with eigenvalues $\pm\hbar/2$:

$$\hat{S}_z \begin{pmatrix} 1 \\ 0 \end{pmatrix} = \frac{\hbar}{2}\begin{pmatrix} 1 & 0 \\ 0 & -1 \end{pmatrix}\begin{pmatrix} 1 \\ 0 \end{pmatrix} = \frac{\hbar}{2}\begin{pmatrix} 1 \\ 0 \end{pmatrix},$$
$$\hat{S}_z \begin{pmatrix} 0 \\ 1 \end{pmatrix} = \frac{\hbar}{2}\begin{pmatrix} 1 & 0 \\ 0 & -1 \end{pmatrix}\begin{pmatrix} 0 \\ 1 \end{pmatrix} = -\frac{\hbar}{2}\begin{pmatrix} 0 \\ 1 \end{pmatrix}. \tag{12.1.25}$$

These are therefore states that have a z-component of angular momentum equal to $\hbar/2$ and $-\hbar/2$, respectively. We can call the basis states with names that emphasize their properties relative to $\hat{S}_z$:

$$|z;+\rangle \equiv \begin{pmatrix} 1 \\ 0 \end{pmatrix}, \quad |z;-\rangle \equiv \begin{pmatrix} 0 \\ 1 \end{pmatrix}. \tag{12.1.26}$$

You can quickly verify using the inner product (12.1.10) that these are orthonormal basis vectors. The objects $|\cdots\rangle$ are **kets** of the Dirac bra-ket notation. A ket is just a vector, or a state in a quantum theory. The labels represented by the dots help characterize the state. The plus and minus are, respectively, for positive and negative values of the angular momentum in the z-direction, the direction that the z label defines. The eigenstate/eigenvalue properties are then summarized by the relation:

$$\hat{S}_z|z;\pm\rangle = \pm\tfrac{\hbar}{2}\,|z;\pm\rangle. \tag{12.1.27}$$

There is something seemingly unusual here. The basis states are eigenstates of $\hat{S}_z$, but what will happen if we try to measure $\hat{S}_x$ or $\hat{S}_y$ on them? We can begin with a simpler question: How do we build eigenstates of $\hat{S}_x$? Certainly, the basis states themselves are not $\hat{S}_x$ eigenstates:

$$\hat{S}_x|z;+\rangle = \frac{\hbar}{2}\begin{pmatrix} 0 & 1 \\ 1 & 0 \end{pmatrix}\begin{pmatrix} 1 \\ 0 \end{pmatrix} = \frac{\hbar}{2}\begin{pmatrix} 0 \\ 1 \end{pmatrix} = \frac{\hbar}{2}|z;-\rangle,$$
$$\hat{S}_x|z;-\rangle = \frac{\hbar}{2}\begin{pmatrix} 0 & 1 \\ 1 & 0 \end{pmatrix}\begin{pmatrix} 0 \\ 1 \end{pmatrix} = \frac{\hbar}{2}\begin{pmatrix} 1 \\ 0 \end{pmatrix} = \frac{\hbar}{2}|z;+\rangle. \tag{12.1.28}$$

The operator $\hat{S}_x$ in fact turns a plus state into a minus state and vice versa. It is therefore simple to see that the following linear combinations of $\hat{S}_z$ eigenstates,

$$|x;+\rangle \equiv \frac{1}{\sqrt{2}}\Big(|z;+\rangle + |z;-\rangle\Big) = \frac{1}{\sqrt{2}}\begin{pmatrix} 1 \\ 1 \end{pmatrix},$$
$$|x;-\rangle \equiv \frac{1}{\sqrt{2}}\Big(|z;+\rangle - |z;-\rangle\Big) = \frac{1}{\sqrt{2}}\begin{pmatrix} 1 \\ -1 \end{pmatrix} \tag{12.1.29}$$

are in fact $\hat{S}_x$ eigenstates,

$$\hat{S}_x\,|x;\pm\rangle = \pm\tfrac{\hbar}{2}\,|x;\pm\rangle. \tag{12.1.30}$$

The $1/\sqrt{2}$ factors in the superposition are required to have normalized $\hat{S}_x$ eigenstates, as you can easily verify using the rightmost expressions in (12.1.29). Having found the $\hat{S}_x$ eigenstates, we can now answer the question: What happens when we measure $\hat{S}_x$ on the state $|z;+\rangle$? For this, we use (12.1.29) to solve for $|z;+\rangle$ in terms of $\hat{S}_x$ eigenstates:

$$|z;+\rangle = \frac{1}{\sqrt{2}}|x;+\rangle + \frac{1}{\sqrt{2}}|x;-\rangle. \tag{12.1.31}$$

We can see that the probability amplitude for $|z;+\rangle$ to be in the state $|x;+\rangle$ is $\frac{1}{\sqrt{2}}$ and equal to the probability amplitude for $|z;+\rangle$ to be in the state $|x;-\rangle$. So if we measure $\hat{S}_x$ on $|z;+\rangle$, we have probability $\frac{1}{2}$ to land on the state $|x;+\rangle$ and probability $\frac{1}{2}$ to land on the state $|x;-\rangle$. You can quickly verify that you also get equal probabilities if you measure $\hat{S}_x$ on $|z;-\rangle$.

For completeness let us consider the eigenstates of $\hat{S}_y$. We get, again, two eigenstates, $|y;\pm\rangle$ with eigenvalue $\pm\hbar/2$:

$$\hat{S}_y|y;\pm\rangle = \pm\tfrac{\hbar}{2}\,|y;\pm\rangle. \tag{12.1.32}$$

The states are given by

$$\begin{aligned} |y;+\rangle &= \frac{1}{\sqrt{2}}\Big(|z;+\rangle + i|z;-\rangle\Big) = \frac{1}{\sqrt{2}}\begin{pmatrix}1\\ i\end{pmatrix}, \\ |y;-\rangle &= \frac{1}{\sqrt{2}}\Big(|z;+\rangle - i|z;-\rangle\Big) = \frac{1}{\sqrt{2}}\begin{pmatrix}1\\ -i\end{pmatrix}. \end{aligned} \tag{12.1.33}$$

Note that this time the superposition of $|z;\pm\rangle$ states involves complex numbers!

Exercise 12.2. *Use the matrix expression for $\hat{S}_y$ to confirm that the above are indeed eigenstates with the indicated eigenvalues.*

Exercise 12.3. *Verify that $|x;+\rangle$ is neither orthogonal to $|y;+\rangle$ nor orthogonal to $|y;-\rangle$.*

Again we can write $|z;+\rangle$ as a linear superposition of $\hat{S}_y$ eigenstates and confirm that if we measure $\hat{S}_y$ on $|z;+\rangle$ we have probability $\frac{1}{2}$ of landing on the state $|y;+\rangle$ and probability $\frac{1}{2}$ of landing on the state $|y;-\rangle$. The same holds if we measure $\hat{S}_y$ on the state $|z;-\rangle$. So we conclude that the states $|z;\pm\rangle$ have definite values of $\hat{S}_z$ but indefinite values (with zero average) of $\hat{S}_x$ and $\hat{S}_y$. We think of $|z;\pm\rangle$ as states of angular momentum $\pm\frac{\hbar}{2}$ in the z-direction. We say $|z;+\rangle$ has angular momentum up along z, and $|z;-\rangle$ has angular momentum down along z. This is *not* to say that these states, when measured, have zero angular momentum along x or y.

In order to consider dynamics, we must choose a Hamiltonian. From the general $\hat{H}$ in (12.1.19), we know that

$$\hat{H} = h_0\mathbf{1} + h_1\sigma_1 + h_2\sigma_2 + h_3\sigma_3. \tag{12.1.34}$$

For simplicity, let us choose all h equal to zero except for h_1, which we set equal to $\frac{\hbar}{2}\omega$, with ω some fixed angular frequency:

$$\hat{H} = \tfrac{\hbar}{2}\,\omega\sigma_1 = \omega\hat{S}_x. \tag{12.1.35}$$

This has the right units of energy: $\hat{S}_x$ has units of $\hbar$ and ω units of inverse time. Suppose now that at time $t=0$ the state of the system is $|z;+\rangle$:

$$\Psi(t=0) = |z;+\rangle = \frac{1}{\sqrt{2}}\Big(|x;+\rangle + |x;-\rangle\Big), \tag{12.1.36}$$

which we expanded in terms of $\hat{S}_x$ eigenstates $|x;\pm\rangle$ because these are in fact $\hat{H}$ eigenstates with energies $E_\pm = \pm\frac{1}{2}\hbar\omega$. Since an energy eigenstate of energy E evolves with an $e^{-iEt/\hbar}$ phase factor, the time evolution of $\Psi(t=0)$ gives

$$\Psi(t) = \frac{1}{\sqrt{2}}\Big(e^{-iE_+t/\hbar}|x;+\rangle + e^{-iE_-t/\hbar}|x;-\rangle\Big). \tag{12.1.37}$$

Substituting the values of the energies, we get

$$\Psi(t) = \frac{1}{\sqrt{2}}\Big(e^{-i\omega t/2}|x;+\rangle + e^{i\omega t/2}|x;-\rangle\Big). \tag{12.1.38}$$

We wish to understand the time evolution in terms of the original basis $|z;\pm\rangle$, so we write

$$\begin{aligned}\Psi(t) &= \frac{1}{\sqrt{2}}\Big(e^{-i\omega t/2}\frac{1}{\sqrt{2}}\big(|z;+\rangle + |z;-\rangle\big) + e^{i\omega t/2}\frac{1}{\sqrt{2}}\big(|z;+\rangle - |z;-\rangle\big)\Big)\\ &= \frac{1}{2}\Big(|z;+\rangle\big(e^{-i\omega t/2} + e^{i\omega t/2}\big) - |z;-\rangle\big(e^{i\omega t/2} - e^{-i\omega t/2}\big)\Big)\\ &= |z;+\rangle\cos\tfrac{\omega t}{2} - i|z;-\rangle\sin\tfrac{\omega t}{2}.\end{aligned} \tag{12.1.39}$$

We thus see oscillations! At $t=0$ the state has angular momentum up along z, but by the time $t=\pi/\omega$, the angular momentum is down along z. By time $t=2\pi/\omega$, the state has returned to its original direction up along z. There is indeed interesting dynamics in two-state systems.

12.2 The Stern-Gerlach Experiment

In 1922, at the University of Frankfurt in Germany, Otto Stern and Walther Gerlach conducted fundamental experiments to measure the deflection of beams of silver atoms as they were sent through nonhomogeneous magnetic fields. As we will explain below, the silver atom has a magnetic dipole moment, and magnetic dipoles get deflected when exposed to spatially varying magnetic fields. Rather than finding the expected continuous range of deflections, however, the incoming beam was split into two beams, as deduced from the two separate spots on the target screen. These experiments demonstrated that these silver atoms have quantized magnetic moments that can have one of two values.

A little knowledge of chemistry shows that a silver atom in the Stern-Gerlach experiment acts like a heavy electron. The silver atom has forty-seven electrons; twenty-eight of them completely fill the $n=1, 2$, and 3 shells, and eighteen more fill the s, p, and d orbitals of the $n=4$ shell. Finally, there is just one electron with $n=5$ and $\ell=0$: the zero angular momentum 5s state. Since this is the only unpaired electron, the magnetic dipole moment of the silver atom is to a good approximation just the magnetic dipole moment of the 5s electron; the nucleus has a much smaller dipole moment and can be ignored. For a charged particle like the electron, a magnetic moment arises because the particle has spin.

Although the quantized magnetic moments found by Stern and Gerlach were consistent with the idea that the electron had spin, this suggestion took some time to develop. Pauli introduced a "two-valued" degree of freedom for electrons, without suggesting a physical interpretation. Ralph Kronig suggested in 1925 that this degree of freedom originated from the self-rotation of the electron. This idea was severely criticized by Pauli, and Kronig did not publish it. George Uhlenbeck and Samuel Goudsmit had a similar idea, and Paul Ehrenfest encouraged them to publish it. They did so in 1925 and are now credited with discovering that the electron has an intrinsic spin with value "one-half." Much of the mathematics of spin one-half was developed by Pauli himself in 1927 and goes along the lines we followed in the previous section. In fact, it took until 1927 for anyone to realize that the Stern-Gerlach experiment measured the magnetic moment of the electron.

Let us begin by recalling what a magnetic dipole moment is and how it is related to angular momentum. A current on a closed loop induces a magnetic dipole moment. The magnetic moment vector $\boldsymbol{\mu}$ is proportional to the current I on the loop and the area vector $\mathbf{A}$ of the loop:

$$\boldsymbol{\mu} = \frac{I}{c}\mathbf{A}. \tag{12.2.1}$$

Here c is the speed of light, and it features in the above formula when we use Gaussian units. For a planar loop, the vector area is a vector normal to the loop with length equal to the value of the area. The direction of the normal is determined from the direction of the current and the right-hand rule. If a magnetic dipole is placed in a uniform magnetic field $\mathbf{B}$, the energy E of the system is

$$E = -\boldsymbol{\mu} \cdot \mathbf{B}. \tag{12.2.2}$$

The dipole tends to align with the magnetic field. We see that the product μB of the magnitude μ of the magnetic moment times the magnitude B of the magnetic field has units of energy, so the units of μ are

$$[\mu] = \frac{\text{erg}}{\text{gauss}}. \tag{12.2.3}$$

In SI units $[\mu] = \text{joule/tesla}$, where

$$\text{tesla} = \frac{10^4}{c}\ \text{gauss}. \tag{12.2.4}$$

One says that a tesla (T) is 10,000 gauss, but this hides the factor of c needed to relate formulae in the International System of Units (SI) to formulae in Gaussian units. Since

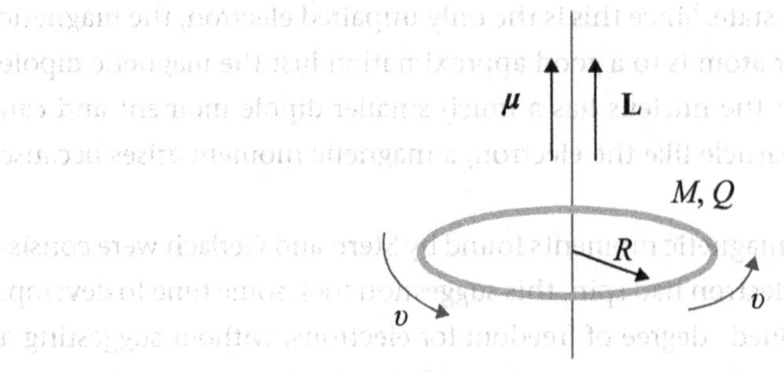

Figure 12.1
A rotating ring of mass M and charge Q gives rise to both a magnetic moment μ and an angular momentum **L**. These are parallel vectors, and the ratio of their magnitudes is independent of the radius R of the ring and the rotation velocity v.

you probably want the freedom to work with both kinds of units, we will give key results in both systems.

A rotating charge distribution results in a magnetic moment and, if the distribution has mass, an angular momentum. The magnetic moment and the angular momentum turn out to be proportional to each other, and the constant of proportionality is universal. To see this consider a ring of charge with radius R that has a uniform charge distribution and total charge Q (figure 12.1). Assume that the ring is rotating about an axis perpendicular to the plane of the ring and going through its center. Let the tangential velocity at the ring be v. The current at the loop is equal to the linear charge density λ times the velocity:

$$I = \lambda v = \frac{Q}{2\pi R} v. \tag{12.2.5}$$

It follows that the magnitude μ of the dipole moment of the loop is

$$\mu = \frac{I}{c} A = \frac{Q}{2\pi Rc} v\pi R^2 = \frac{Q}{2c} Rv. \tag{12.2.6}$$

Let the mass of the ring be M, also uniformly distributed. The magnitude L of the angular momentum of the ring is then $L = R(Mv)$. As a result,

$$\mu = \frac{Q}{2Mc} RMv = \frac{Q}{2Mc} L, \tag{12.2.7}$$

leading to the notable ratio

$$\frac{\mu}{L} = \frac{Q}{2Mc}. \tag{12.2.8}$$

The ratio does not depend on the radius of the ring nor on its velocity, only on the ratio Q/M. By superposition, any rotating object with axially symmetric, uniform mass and charge density distributions will have a ratio μ/L as above, with Q the total charge and M the total mass. This formula certainly applies to a uniformly charged rotating sphere. Noting that the direction of the magnetic moment coincides with that of the angular momentum, the above relation is rewritten as

$$\mu = \frac{Q}{2Mc}\mathbf{L}. \tag{12.2.9}$$

Does the electron have an intrinsic magnetic moment μ because it is a tiny spinning ball? Not really. The electron has an intrinsic μ, but it cannot really be viewed as a rotating little ball of charge; this was part of Pauli's objection to the original idea of spin. We currently view the electron as an elementary particle with zero size, so the idea that it rotates is just not sensible. The classical relation above, however, points to the correct quantum result. Even if it has no size, the electron has intrinsic spin angular momentum and thus, quantum mechanically, a spin angular momemtum operator $\hat{\mathbf{S}}$. One could guess that the quantum analog of (12.2.9) is

$$\hat{\mu} \stackrel{?}{=} -\frac{e}{2m_e c}\hat{\mathbf{S}}. \tag{12.2.10}$$

We included a minus sign because the charge of the electron is $-e$, with $e > 0$. Note also that this relation can only mean that in quantum mechanics the dipole moment is itself an operator. Since angular momentum and spin have the same units, we can write this as

$$\hat{\mu} \stackrel{?}{=} -\frac{e\hbar}{2m_e c}\frac{\hat{\mathbf{S}}}{\hbar}. \tag{12.2.11}$$

This is not exactly right, however. For electrons the magnetic moment is actually twice as large. One uses a constant *g-factor* to describe this effect:

$$\hat{\mu} = -g\frac{e\hbar}{2m_e c}\frac{\hat{\mathbf{S}}}{\hbar}, \quad g = 2 \text{ for an electron.} \tag{12.2.12}$$

This factor of two is in fact predicted by the nonrelativistic Pauli equation for the electron (see section 24.6), as well as by the relativistic Dirac equation for the electron. It has also been verified experimentally. That is to say, for this value of $\hat{\mu}$, the quantum Hamiltonian coupling the electron to the magnetic field is given by

$$\hat{H} = -\hat{\mu}\cdot\mathbf{B}, \tag{12.2.13}$$

as suggested by the classical energy relation (12.2.2). This formula is valid for any particle with magnetic dipole moment $\hat{\mu}$.

To write the value of the magnetic moment more briefly, one introduces a canonical value μ_B of the dipole moment called the **Bohr magneton:**

$$\mu_B = \frac{e\hbar}{2m_e c} \simeq 9.274 \times 10^{-21}\,\frac{\text{erg}}{\text{gauss}} = 5.788 \times 10^{-9}\,\frac{\text{eV}}{\text{gauss}}. \tag{12.2.14}$$

On the other hand, in SI units one has

$$\mu_B \equiv \frac{e\hbar}{2m_e} = 9.274 \times 10^{-24}\,\frac{\text{joule}}{\text{tesla}} = 5.788 \times 10^{-5}\,\frac{\text{eV}}{\text{tesla}}, \tag{12.2.15}$$

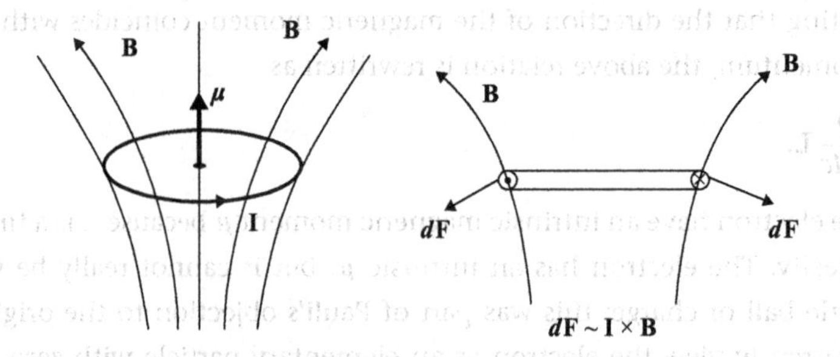

Figure 12.2
A magnetic dipole in a nonuniform magnetic field will experience a force. The force points in the direction for which $\boldsymbol{\mu}\cdot\mathbf{B}$ grows the fastest. In this case the force is downward.

consistent with the above Gaussian values and the relation (12.2.4) between a tesla and a gauss. Using the Gaussian formula for μ_B, for an

$$\boxed{\text{electron:}\quad \hat{\boldsymbol{\mu}} = -2\mu_B\frac{\hat{\mathbf{S}}}{\hbar} = -\mu_B\boldsymbol{\sigma} = -\frac{e\hbar}{2m_ec}\boldsymbol{\sigma},} \tag{12.2.16}$$

where we recalled that $\hat{\mathbf{S}} = \frac{\hbar}{2}\boldsymbol{\sigma}$. It follows from the above relation and (12.2.13) that the Hamiltonian coupling a magnetic field to an electron is

$$\hat{H} = \frac{2\mu_B}{\hbar}\mathbf{B}\cdot\hat{\mathbf{S}} = \mu_B\,\mathbf{B}\cdot\boldsymbol{\sigma}\quad\text{(electron)}. \tag{12.2.17}$$

Another feature of magnetic dipoles is needed for our discussion: a dipole placed in a nonuniform magnetic field experiences a force. An illustration is given in figure 12.2, where to the left we show a current ring whose associated dipole moment $\boldsymbol{\mu}$ points upward. The magnetic field lines diverge as we move up, so the magnetic field is stronger as we move down. This dipole will experience a force pointing down, which can be deduced as follows. On a small piece of wire, the force $d\mathbf{F}$ is proportional to $\mathbf{I}\times\mathbf{B}$. The vectors $d\mathbf{F}$ are sketched on the right part of the figure. Their horizontal components cancel out, but the result is a net force downward.

In classical electromagnetism the force $\mathbf{F}$ on a dipole $\boldsymbol{\mu}$ in a magnetic field $\mathbf{B}$ is given by

$$\mathbf{F} = \nabla(\boldsymbol{\mu}\cdot\mathbf{B}). \tag{12.2.18}$$

Note that the force points in the direction for which $\boldsymbol{\mu}\cdot\mathbf{B}$ increases the fastest. Given that in our situation $\boldsymbol{\mu}$ and $\mathbf{B}$ are parallel, this direction is that in which the magnitude of $\mathbf{B}$ increases the fastest.

In the Stern-Gerlach experiment, silver is vaporized in an oven, and with the help of a collimating slit, a narrow beam of silver atoms is sent down to a magnet configuration. In the situation described by figure 12.3, the magnetic field points mostly in the positive z-direction, and the gradient is also in the positive z-direction. As a result, the above

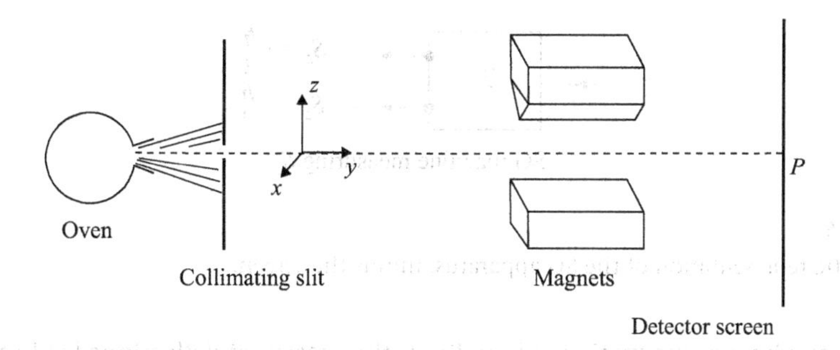

Figure 12.3
A sketch of the SG apparatus. An oven and a collimating slit produce a narrow beam of silver atoms. The beam goes through a region with a strong magnetic field and a strong gradient, both in the z-direction. The screen, *right*, acts as a detector.

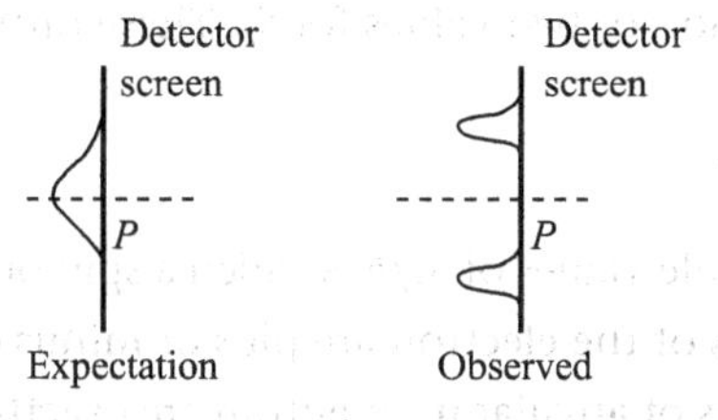

Figure 12.4
Left: The pattern on the detector screen that would be expected from classical physics. *Right*: The observed pattern, showing two separated peaks corresponding to positive and negative magnetic moment μ_z.

equation gives

$$\mathbf{F} \simeq \nabla(\mu_z B_z) = \mu_z \nabla B_z \simeq \mu_z \frac{\partial B_z}{\partial z} \mathbf{e}_z, \tag{12.2.19}$$

and thus, classically, the atoms experience a force and deflection in the z-direction proportional to the z-component of their magnetic moment. Undeflected atoms would hit the detector screen at point P. Atoms with positive μ_z should be deflected upward, and atoms with negative μ_z should be deflected downward.

The oven produces atoms with magnetic moments pointing in random directions, and thus the expectation was that the z-component of the magnetic moment would have a smooth probability distribution leading to a detection that would be roughly like that indicated on the left side of figure 12.4. Surprisingly, the observed result was two separate peaks, as if all atoms had either a fixed positive μ_z or a fixed negative μ_z. This is shown on the right side of the figure. The fact that the peaks are spatially separated led to the original misleading name of *space quantization*. The Stern-Gerlach experiment demonstrates the quantization of the dipole moment, and by theoretical inference from (12.2.16), the quantization of the spin (or intrinsic) angular momentum.

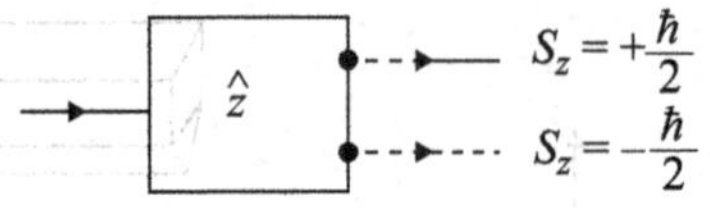

$S_z = +\frac{\hbar}{2}$

$\hat{z}$

$S_z = -\frac{\hbar}{2}$

SG machine measuring S_z

Figure 12.5
A schematic representation of the SG apparatus, minus the screen.

Given the known magnetic field gradient, the measured deflections lead to a determination of the two values of μ_z. Since (12.2.16) relates magnetic moments to spin angular momentum as

$$\mu_z = -2\mu_B \frac{S_z}{\hbar}, \tag{12.2.20}$$

this leads to a determination of two values for S_z. These turn out to be

$$S_z = \pm\tfrac{\hbar}{2}, \text{ or } \tfrac{S_z}{\hbar} = \pm\tfrac{1}{2}. \tag{12.2.21}$$

A particle with such possible values of $S_z/\hbar$ is called a spin one-half particle. The values of the magnetic moments of the electron are plus or minus one Bohr magneton. Note that such quantized values of angular momentum are exactly those we obtained in our discussion of the simplest quantum system.

With the magnetic field and its gradient along the z-direction, the Stern-Gerlach (SG) apparatus can help measure the component of the spin in the z-direction. To streamline our pictures, we will denote such an apparatus as a box with a $\hat{z}$ label, as in figure 12.5. The input beam comes in from the left, and the box lets out two beams from the right. If we placed a detector to the right, the top beam would be identified as having atoms with $S_z = \hbar/2$, and the bottom beam would be identified as having atoms with $S_z = -\hbar/2$. In the quantum mechanical view of the experiment, a single atom can be in both output beams, with different amplitudes. Only the act of measurement, realized by the placement of the detector screen, forces the atom to decide into which beam it goes.

Let us now consider thought experiments in which we put a few SG apparatus in series. In the first configuration, shown at the top of figure 12.6, the first box is a $\hat{z}$-type SG machine, where we block the $S_z = -\hbar/2$ output beam and let only the $S_z = \hbar/2$ beam enter the next machine. This machine acts as a filter. The second SG apparatus is also a $\hat{z}$-type machine. Since all ingoing particles have $S_z = \hbar/2$, the second machine releases them from the top output, and nothing exits the bottom output. The quantum mechanical lesson here is that $S_z = \hbar/2$ states have no component or amplitude along $S_z = -\hbar/2$ states. $S_z = \hbar/2$ and $S_z = -\hbar/2$ are thus said to be orthogonal states. This is also in accord with our previous analysis: the basis states $|z;\pm\rangle$ in (12.1.26) carry precisely the correct amount of angular momentum and are indeed orthogonal states. Note the deep geometric difference between spin states and vectors. When ordinary vectors in

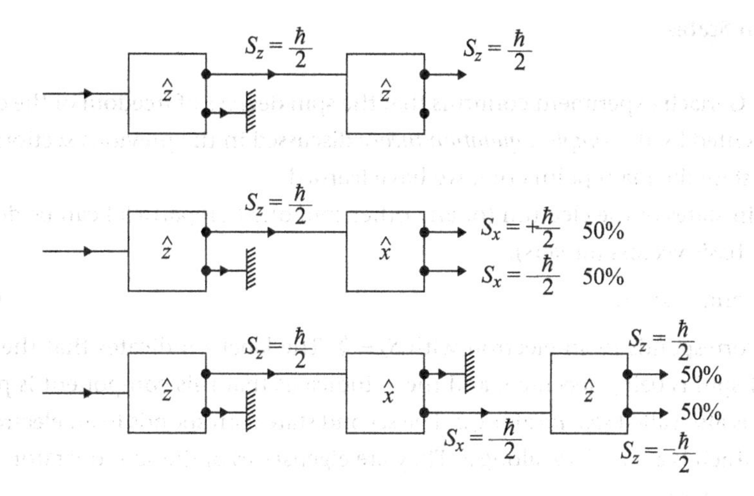

Figure 12.6
Three configurations of SG boxes.

three dimensions are orthogonal, we mean they make right angles with each other; they are perpendicular to each other. While the states $|z;+\rangle$ and $|z;-\rangle$ have zero inner product and are thus orthogonal, the directions the spins point to, $+z$ and $-z$, are antiparallel, not perpendicular.

The second configuration in the figure shows the outgoing $S_z = \hbar/2$ beam from the first machine entering an $\hat{x}$-type machine. The outputs of this machine are—in analogy to the $\hat{z}$-type machine—$S_x = \hbar/2$ and $S_x = -\hbar/2$. Classically, an object with angular momentum along the z-axis has no component of angular momentum along the x-axis; these are orthogonal directions. But the result of the experiment indicates that this is not true for quantum spins. About half of the $S_z = \hbar/2$ atoms exit through the top $S_x = \hbar/2$ output, and the other half exit through the bottom $S_x = -\hbar/2$ output. Quantum mechanically, a state with a definite value of S_z has an amplitude along the state $S_x = \hbar/2$ as well as an amplitude along the state $S_x = -\hbar/2$. This is another feature that is properly accounted for in our two-state model of spin. We showed that if we measured $\hat{S}_x$ on an $\hat{S}_z$ eigenstate, both answers $\hat{S}_x = \pm\hbar/2$ were equally probable.

In the third and bottom configuration, the $S_z = \hbar/2$ beam from the first machine goes into the $\hat{x}$-type machine, and the top output is blocked so we only have an $S_x = -\hbar/2$ output. That beam is fed into a $\hat{z}$-type machine. One could speculate that the beam entering the third machine has both $S_x = -\hbar/2$ *and* $S_z = \hbar/2$, as it is composed of silver atoms that made it through both machines. If that were the case, the third machine would release all atoms from the top output. This speculation is falsified by the result. There is no memory of the first filter: the particles escaping the second machine do not have $S_z = \hbar/2$ anymore. We find that half of the particles make it out of the third machine with $S_z = \hbar/2$ and the other half with $S_z = -\hbar/2$. The two-state model of spin is also consistent with this experiment. No states are simultaneously $\hat{S}_z$ and $\hat{S}_x$ eigenstates. We verified this explicitly: the $\hat{S}_z$ eigenstates are not $\hat{S}_x$ eigenstates and vice versa.

12.3 Spin States

The Stern-Gerlach experiment confirms that the spin degrees of freedom of the electron are represented by the *simplest quantum theory* discussed in the previous section. Let us therefore state the main points that we have learned.

The spin states of the electron (or any other spin one-half particle) can be described using two basis vectors (or kets):

$$|z;+\rangle \quad \text{and} \quad |z;-\rangle. \tag{12.3.1}$$

The first corresponds to an electron with $S_z = \frac{\hbar}{2}$. The label z indicates that the z component of spin is being specified, and the $+$ indicates that this component is positive. This state is also called *spin up* along z. The second state corresponds to an electron with $S_z = -\frac{\hbar}{2}$, which is a *spin down* along z. They are eigenstates of the spin operator $\hat{S}_z = \frac{\hbar}{2}\sigma_3$ with eigenvalues:

$$\hat{S}_z|z;\pm\rangle = \pm\frac{\hbar}{2}\,|z;\pm\rangle. \tag{12.3.2}$$

With two basis states, the state space of electron spin is a two-dimensional *complex vector space*. Each vector in this vector space represents a possible state of the electron spin. Note that we are not discussing other degrees of freedom of the electron, such as its position or momentum, just the spin. The general vector in the two-dimensional space is an arbitrary linear combination of the basis states:

$$\Psi = c_1|z;+\rangle + c_2|z;-\rangle, \quad \text{with } c_1, c_2 \in \mathbb{C}. \tag{12.3.3}$$

It is customary to call the state $|z;+\rangle$ the *first* basis state and denote it by $|1\rangle$. The state $|z;-\rangle$ is called the *second* basis state and is denoted by $|2\rangle$. In a two-dimensional vector space, a vector is explicitly *represented* as a column vector with two components:

$$|z;+\rangle = |1\rangle = \begin{pmatrix}1\\0\end{pmatrix}, \quad |z;-\rangle = |2\rangle = \begin{pmatrix}0\\1\end{pmatrix}. \tag{12.3.4}$$

Using these options the state in (12.3.3) takes the possible forms

$$\Psi = c_1|z;+\rangle + c_2|z;-\rangle = c_1|1\rangle + c_2|2\rangle = c_1\begin{pmatrix}1\\0\end{pmatrix} + c_2\begin{pmatrix}0\\1\end{pmatrix} = \begin{pmatrix}c_1\\c_2\end{pmatrix}. \tag{12.3.5}$$

The inner product was given in (12.1.10):

$$\langle\Psi,\Phi\rangle = c_1^* d_1 + c_2^* d_2, \quad \text{for} \quad \Psi = \begin{pmatrix}c_1\\c_2\end{pmatrix}, \quad \Phi = \begin{pmatrix}d_1\\d_2\end{pmatrix}. \tag{12.3.6}$$

In the bra-ket notation of Dirac, the states Ψ and Φ are displayed as kets $|\Psi\rangle$ and $|\Phi\rangle$, with the inner product $\langle\Psi,\Phi\rangle$ written as

$$\langle\Psi|\Phi\rangle \equiv \langle\Psi,\Phi\rangle \tag{12.3.7}$$

so that the equation above reads

$$\langle\Psi|\Phi\rangle = c_1^* d_1 + c_2^* d_2, \quad \text{for} \quad |\Psi\rangle = \begin{pmatrix}c_1\\c_2\end{pmatrix}, \quad |\Phi\rangle = \begin{pmatrix}d_1\\d_2\end{pmatrix}. \tag{12.3.8}$$

In this language,

$$\langle z;-|z;+\rangle = 0, \quad \langle z;+|z;+\rangle = 1,$$
$$\langle z;+|z;-\rangle = 0, \quad \langle z;-|z;-\rangle = 1. \tag{12.3.9}$$

Labeling the basis states as $|1\rangle$ and $|2\rangle$, we have a simple form that summarizes the four equations above:

$$\langle i|j\rangle = \delta_{ij}, \quad i,j=1,2. \tag{12.3.10}$$

We have the operators $\hat{S}_x = \frac{\hbar}{2}\sigma_1$, $\hat{S}_y = \frac{\hbar}{2}\sigma_2$, and $\hat{S}_z = \frac{\hbar}{2}\sigma_3$ given in (12.1.23) and repeated here for convenience:

$$\hat{S}_x = \frac{\hbar}{2}\begin{pmatrix} 0 & 1 \\ 1 & 0 \end{pmatrix}, \quad \hat{S}_y = \frac{\hbar}{2}\begin{pmatrix} 0 & -i \\ i & 0 \end{pmatrix}, \quad \hat{S}_z = \frac{\hbar}{2}\begin{pmatrix} 1 & 0 \\ 0 & -1 \end{pmatrix}. \tag{12.3.11}$$

These operators satisfy the angular momentum commutation relations:

$$[\hat{S}_x, \hat{S}_y] = i\hbar\hat{S}_z, \quad [\hat{S}_y, \hat{S}_z] = i\hbar\hat{S}_x, \quad [\hat{S}_z, \hat{S}_x] = i\hbar\hat{S}_y. \tag{12.3.12}$$

Using numerical subscripts for the components $(\hat{S}_1 = \hat{S}_x, \hat{S}_2 = S_y, \hat{S}_3 = S_z)$, the above commutators are summarized by

$$[\hat{S}_i, \hat{S}_j] = i\hbar\,\epsilon_{ijk}\,\hat{S}_k. \tag{12.3.13}$$

The epsilon symbol ϵ_{ijk} was defined following equation (10.3.12).

Exercise 12.4. *Check that the set of commutation relations of the spin operators are in fact preserved when we replace $\hat{S}_x \to -\hat{S}_y$ and $\hat{S}_y \to \hat{S}_x$.*

Eigenstates for $\hat{S}_x$ satisfying $\hat{S}_x|x;\pm\rangle = \pm\frac{\hbar}{2}|x;\pm\rangle$ and eigenstates for $\hat{S}_y$ satisfying $\hat{S}_y|y;\pm\rangle = \pm\frac{\hbar}{2}|y;\pm\rangle$ were found in section 12.1 and are given by

$$|x;+\rangle = \frac{1}{\sqrt{2}}|z;+\rangle + \frac{1}{\sqrt{2}}|z;-\rangle, \qquad |y;+\rangle = \frac{1}{\sqrt{2}}|z;+\rangle + \frac{i}{\sqrt{2}}|z;-\rangle,$$
$$|x;-\rangle = \frac{1}{\sqrt{2}}|z;+\rangle - \frac{1}{\sqrt{2}}|z;-\rangle, \qquad |y;-\rangle = \frac{1}{\sqrt{2}}|z;+\rangle - \frac{i}{\sqrt{2}}|z;-\rangle. \tag{12.3.14}$$

Since $\hat{S}_x$, $\hat{S}_y$, and $\hat{S}_z$ are the spin operators in the x-, y-, and z-directions, respectively, we can ask: What would we mean by a spin operator in some arbitrary fixed direction? To define such an operator, we specify a direction using a unit vector $\mathbf{n}$ with Cartesian components (n_x, n_y, n_z) and pointing in the direction specified by the polar and azimuthal angles (θ, ϕ), of spherical coordinates:

$$\mathbf{n} = (n_x, n_y, n_z) = (\sin\theta\cos\phi,\ \sin\theta\sin\phi,\ \cos\theta)\,. \tag{12.3.15}$$

We can also think of the spin operators as a triplet $\hat{\mathbf{S}}$:

$$\hat{\mathbf{S}} = (\hat{S}_x,\ \hat{S}_y,\ \hat{S}_z) = \tfrac{\hbar}{2}(\sigma_1,\ \sigma_2,\ \sigma_3) = \tfrac{\hbar}{2}\boldsymbol{\sigma}. \tag{12.3.16}$$

We sometimes refer to such a triplet as a *vector operator*. The spin operator $\hat{S}_{\mathbf{n}}$ in the direction $\mathbf{n}$ is defined to be

$$\hat{S}_{\mathbf{n}} \equiv \mathbf{n}\cdot\hat{\mathbf{S}} \equiv n_x\hat{S}_x + n_y\hat{S}_y + n_z\hat{S}_z = \tfrac{\hbar}{2}\,\mathbf{n}\cdot\boldsymbol{\sigma}. \tag{12.3.17}$$

This definition gives the expected operator when the unit vector is along one of the coordinate axes. For example, when $\mathbf{n}$ points along z, we have $(n_x, n_y, n_z) = (0, 0, 1)$ and $\hat{S}_{\mathbf{n}} = \hat{S}_z$. The same holds, of course, for unit vectors pointing in x- and y-directions. Note that $\hat{S}_{\mathbf{n}}$ is an operator, a Hermitian 2×2 matrix, in fact. To see the explicit form of $\hat{S}_{\mathbf{n}}$, we use the Pauli matrices to expand the last right-hand side in (12.3.17):

$$\hat{S}_{\mathbf{n}} = \frac{\hbar}{2}\left[n_x\begin{pmatrix}0 & 1\\ 1 & 0\end{pmatrix} + n_y\begin{pmatrix}0 & -i\\ i & 0\end{pmatrix} + n_z\begin{pmatrix}1 & 0\\ 0 & -1\end{pmatrix}\right] = \frac{\hbar}{2}\begin{pmatrix}n_z & n_x - in_y\\ n_x + in_y & -n_z\end{pmatrix}. \tag{12.3.18}$$

Using the values of the components of $\mathbf{n}$ in spherical coordinates, we find that

$$\hat{S}_{\mathbf{n}} = \frac{\hbar}{2}\begin{pmatrix}\cos\theta & \sin\theta e^{-i\phi}\\ \sin\theta e^{i\phi} & -\cos\theta\end{pmatrix}. \tag{12.3.19}$$

Since the z-axis could have been chosen in any arbitrary direction, just like $\hat{S}_z$, we want $\hat{S}_{\mathbf{n}}$ to have eigenvalues $\pm\hbar/2$ for arbitrary $\mathbf{n}$. This is quickly confirmed:

Exercise 12.5. *Confirm by direct computation that the eigenvalues of $\hat{S}_{\mathbf{n}}$ are $\pm\hbar/2$.*

Associated to the eigenvalues $\pm\hbar/2$ of $\hat{S}_{\mathbf{n}}$, we have eigenstates $|\mathbf{n};\pm\rangle$:

$$\hat{S}_{\mathbf{n}}|\mathbf{n};\pm\rangle = \pm\tfrac{\hbar}{2}|\mathbf{n};\pm\rangle. \tag{12.3.20}$$

We say that $|\mathbf{n};\pm\rangle$ are states with angular momentum up or down along $\mathbf{n}$. Let us first find the eigenvector $|\mathbf{n};+\rangle$. To do so we write an ansatz

$$|\mathbf{n};+\rangle = c_1|+\rangle + c_2|-\rangle = \begin{pmatrix}c_1\\ c_2\end{pmatrix}, \qquad |\pm\rangle \equiv |z;\pm\rangle, \tag{12.3.21}$$

where, for notational simplicity, we have introduced the abbreviations $|\pm\rangle$ for $|z;\pm\rangle$, and the complex constants $c_1, c_2 \in \mathbb{C}$ are to be determined. The eigenvector equation (12.3.20) can be written as $(\hat{S}_{\mathbf{n}} - \frac{\hbar}{2}\mathbb{1})|\mathbf{n};+\rangle = 0$ and reads

$$\frac{\hbar}{2}\begin{pmatrix}\cos\theta - 1 & \sin\theta e^{-i\phi}\\ \sin\theta e^{i\phi} & -\cos\theta - 1\end{pmatrix}\begin{pmatrix}c_1\\ c_2\end{pmatrix} = 0. \tag{12.3.22}$$

There are two equations here. Either one gives the same relation between c_1 and c_2. The top equation, for example, gives

$$c_2 = e^{i\phi}\,\frac{1-\cos\theta}{\sin\theta}\,c_1 = e^{i\phi}\,\frac{\sin\frac{\theta}{2}}{\cos\frac{\theta}{2}}\,c_1. \tag{12.3.23}$$

Exercise 12.6. *Check that the bottom equation in (12.3.22) gives the same relation.*

We want normalized states, and therefore

$$|c_1|^2 + |c_2|^2 = 1 \quad\Rightarrow\quad |c_1|^2\left[1 + \frac{\sin^2\frac{\theta}{2}}{\cos^2\frac{\theta}{2}}\right] = 1 \quad\Rightarrow\quad |c_1|^2 = \cos^2\tfrac{\theta}{2}. \tag{12.3.24}$$

Since the overall phase of the state is not observable, we take the simplest option for c_1:

$$c_1 = \cos\tfrac{\theta}{2}, \qquad c_2 = \sin\tfrac{\theta}{2}\exp(i\phi), \tag{12.3.25}$$

leading to

$$|\mathbf{n};+\rangle = \cos\tfrac{\theta}{2}|+\rangle + \sin\tfrac{\theta}{2}e^{i\phi}|-\rangle. \tag{12.3.26}$$

As a quick check, we see that for $\theta=0$, which corresponds to a unit vector $\mathbf{n}=\mathbf{e}_3$ along the plus z-direction, we get $|\mathbf{e}_3;+\rangle=|+\rangle$. Note that even though the coordinate ϕ is ambiguous when $\theta=0$ (the north pole of the sphere), this does not affect our answer, since the term with ϕ dependence vanishes. In the same way, one can obtain the normalized eigenstate $|\mathbf{n};-\rangle$ corresponding to $-\hbar/2$. A simple phase choice gives

$$|\mathbf{n};-\rangle = \sin\tfrac{\theta}{2}|+\rangle - \cos\tfrac{\theta}{2}e^{i\phi}|-\rangle. \tag{12.3.27}$$

If we again consider the $\theta=0$ direction, this time the ambiguity of ϕ remains in the term that contains the $|-\rangle$ state. It is convenient to multiply this state by the phase $-e^{-i\phi}$. Doing this, the pair of eigenstates reads

$$\boxed{\begin{aligned} |\mathbf{n};+\rangle &= \cos\tfrac{\theta}{2}\,|+\rangle + \sin\tfrac{\theta}{2}e^{i\phi}|-\rangle, \\ |\mathbf{n};-\rangle &= -\sin\tfrac{\theta}{2}e^{-i\phi}|+\rangle + \cos\tfrac{\theta}{2}|-\rangle. \end{aligned}} \tag{12.3.28}$$

The above states are normalized. Furthermore, they are orthogonal:

$$\langle\mathbf{n};-|\mathbf{n};+\rangle = -\sin\tfrac{\theta}{2}e^{i\phi}\cos\tfrac{\theta}{2} + \cos\tfrac{\theta}{2}\sin\tfrac{\theta}{2}e^{i\phi} = 0. \tag{12.3.29}$$

Therefore, $|\mathbf{n};+\rangle$ and $|\mathbf{n};-\rangle$ are an orthonormal pair of states. We view $\mathbf{n}$ as the *spin arrow* for the spin state $|\mathbf{n};+\rangle$. It is the direction the spin state points in. The formulae in (12.3.28) work nicely at the north pole ($\theta=0$), giving the states $|\pm\rangle$. At the south pole $\theta=\pi$, however, the ϕ ambiguity shows up again. If one works near the south pole, multiplying the results in (12.3.28) by suitable phases will do the job. The fact that no formula works unambiguously for the full span of spherical angles is a topological property of spin states! Note also that the expression for $|\mathbf{n};+\rangle$ was anticipated in chapter 1, equation (1.4.12), as the natural way to express a general superposition of two basis states.

Let us verify that the $|\mathbf{n};\pm\rangle$ reduce to the known results as $\mathbf{n}$ points along the z-, x-, and y-axes. Again, if $\mathbf{n}=(0,0,1)=\mathbf{e}_3$, we have $\theta=0$, and hence

$$|\mathbf{e}_3;+\rangle=|+\rangle, \quad |\mathbf{e}_3;-\rangle=|-\rangle, \tag{12.3.30}$$

which are, as expected, the familiar eigenstates of $\hat{S}_z$. If we point along the x-axis, $\mathbf{n}=(1,0,0)=\mathbf{e}_1$, which corresponds to $\theta=\pi/2$, $\phi=0$. Hence,

$$\begin{aligned} |\mathbf{e}_1;+\rangle &= \frac{1}{\sqrt{2}}(|+\rangle+|-\rangle) = |x;+\rangle, \\ |\mathbf{e}_1;-\rangle &= \frac{1}{\sqrt{2}}(-|+\rangle+|-\rangle) = -|x;-\rangle, \end{aligned} \tag{12.3.31}$$

when we compared with (12.3.14). The overall minus sign in the second state is physically irrelevant, so we indeed recovered the eigenvectors of $\hat{S}_x$. Finally, if

$\mathbf{n} = (0, 1, 0) = \mathbf{e}_2$, we have $\theta = \pi/2$, $\phi = \pi/2$ and therefore, with $e^{\pm i\phi} = \pm i$, we find that

$$\begin{aligned} |\mathbf{e}_2; +\rangle &= \frac{1}{\sqrt{2}}(|+\rangle + i|-\rangle) = |y; +\rangle, \\ |\mathbf{e}_2; -\rangle &= \frac{1}{\sqrt{2}}(i|+\rangle + |-\rangle) = i\frac{1}{\sqrt{2}}(|+\rangle - i|-\rangle) = i|y; -\rangle, \end{aligned} \tag{12.3.32}$$

which are, up to an immaterial phase for the second state, the eigenvectors of $\hat{S}_y$.

As a matter of notation, when we only deal with $|\mathbf{n}; +\rangle$ we will just write it as $|\mathbf{n}\rangle$:

$$|\mathbf{n}\rangle \equiv |\mathbf{n}; +\rangle. \tag{12.3.33}$$

The plus and minus labels are useful if we have to deal with both states. We call $|\mathbf{n}\rangle$ the spin state pointing in the $\mathbf{n}$ direction, or the spin state corresponding to the spin arrow $\mathbf{n}$.

Exercise 12.7. *Show that, up to a phase, the state $|\mathbf{n}; -\rangle$ in (12.3.28) can be obtained from $|\mathbf{n}; +\rangle$ by changing $\mathbf{n} \to -\mathbf{n}$, thus changing θ and ϕ appropriately. This confirms that $|\mathbf{n}; -\rangle$ is the spin state pointing along $-\mathbf{n}$.*

Exercise 12.8. *Consider the state $|\Psi(t)\rangle = \cos\frac{\omega t}{2}|z; +\rangle - i\sin\frac{\omega t}{2}|z; -\rangle$, obtained in (12.1.39) by time evolution of the $t = 0$ state $|z; +\rangle$. For times $0 < \omega t < 2\pi$, find the time-dependent values of θ and ϕ that define the spin state. Describe the motion of the spin arrow.*

12.4 Quantum Key Distribution

The quantum states of a two-state system can be used for a remarkable application: a secure method of secret communication that cannot be broken by an eavesdropper that is able to intercept the secret message. The procedure is called quantum key distribution (QKD), and it is a method of quantum cryptography.

At issue is the transmission of a secret key between Alice and Bob, who are physically separate. Alice and Bob require the key to both encrypt and decrypt their messages. The method can be simple. All messages are sent in binary code and are composed of fixed-size strings of zeroes and ones. Each string represents a number in binary code. The key can be another string of zeroes and ones, another binary number. Encryption can be done by adding the key to each string in a message and decryption by subtracting the key from each string. The goal of QKD is to allow Alice and Bob to establish a key remotely while any attempt by Eve, the eavesdropper, to intercept the key will be revealed to Alice and Bob.

The procedure we will describe was proposed by Charles Bennett and Gilles Brassard in 1984. As a result, the protocol is referred to briefly as BB84. QKD has been tested experimentally repeatedly and has been developed commercially.

The protocol uses photons whose polarization states are described by a two-state system. We saw this when we discussed linearly polarized photons and measurements by polarizers in section 1.3. The two basis states chosen can be a photon polarized along the x-direction and a photon polarized along the y-direction. Since photon polarization is associated with the direction of an electric field and the electric field oscillates in time, there is no distinction between polarization along x or $-x$. We write the basis states as

$$|\text{photon; x}\rangle, \quad |\text{photon; y}\rangle. \tag{12.4.1}$$

These states are orthonormal: orthogonal and normalized. Any pair of orthogonal polarization states provides an alternative basis. Let $|\text{photon}; \alpha\rangle$ denote a photon polarized along a line obtained from the x-axis by a counterclockwise rotation with angle α. Then the following could form a pair of basis states:

$$|\text{photon}; \alpha\rangle, \quad |\text{photon}; \alpha + 90^\circ\rangle. \tag{12.4.2}$$

These two basis states are orthogonal for any choice of α; with photon states we do not have the subtlety we noticed for spin states, where a state with spin along z is not orthogonal to a state with spin along x. For photons, if the polarization directions are orthogonal, the states themselves are orthogonal.

The protocol requires choosing *two* sets of basis states. The first set is that in (12.4.1), with a little change of terminology. We will write $|\text{H}\rangle = |\text{photon; x}\rangle$, the H for horizontal, and $|\text{V}\rangle = |\text{photon; y}\rangle$, the V for vertical. The second pair is from (12.4.2), and for this we choose $\alpha = -45^\circ$. This gives us two states, which, omitting the word *photon*, will be denoted as $|-45^\circ\rangle$ and $|+45^\circ\rangle$ (note that $|-45^\circ\rangle \neq -|45^\circ\rangle$; the minus inside the ket is a label). Thus, our two sets of basis states are as follows:

$$\begin{aligned} &\text{Basis set 1:} \quad |\text{H}\rangle, \quad |\text{V}\rangle, \\ &\text{Basis set 2:} \quad |-45^\circ\rangle, \quad |+45^\circ\rangle. \end{aligned} \tag{12.4.3}$$

Using polarizers, Alice is able to produce photons in any of these four states. There are simple linear relations between the two sets of basis states. They are obtained by thinking of the above states as ordinary vectors in a two-dimensional plane. We then have

$$\begin{aligned} |\text{H}\rangle &= \tfrac{1}{\sqrt{2}}\left(\ |-45^\circ\rangle + |+45^\circ\rangle\right), \\ |\text{V}\rangle &= \tfrac{1}{\sqrt{2}}\left(-|-45^\circ\rangle + |+45^\circ\rangle\right). \end{aligned} \tag{12.4.4}$$

Indeed, the top equation holds because a horizontal vector of length one arises by adding to vectors of length one at $\pm 45^\circ$ and scaling with a factor of $1/\sqrt{2}$. Similarly, we have the inverse relations

$$\begin{aligned} |-45^\circ\rangle &= \tfrac{1}{\sqrt{2}}\left(|\text{H}\rangle - |\text{V}\rangle\right), \\ |+45^\circ\rangle &= \tfrac{1}{\sqrt{2}}\left(|\text{H}\rangle + |\text{V}\rangle\right). \end{aligned} \tag{12.4.5}$$

These relations tell us what happens when we measure photon states. We can do two types of measurements. One will be denoted as $+$, and the other will be denoted as $\times$. In a $+$ measurement, we force the photon to decide if it is along $|\text{H}\rangle$ or along $|\text{V}\rangle$; the outgoing photon is one of the two. In a $\times$ measurement, we force the photon to decide if it is along $|-45°\rangle$ or along $|+45°\rangle$; the outgoing photon is one of the two. Such measurements can be done with birefringent calcite crystals.

If a photon is in either $|H\rangle$ or $|V\rangle$ and we do a + measurement, the photon is unaffected; it remains in its original state. Similarly, if a photon is in either $|-45°\rangle$ or $|+45°\rangle$ and we do a × measurement, the photon remains in its original state. In both cases we are measuring along a basis *suitable* for the states. If one is measuring along a suitable basis, the result of the measurement tells us the state of the photon *before* measurement.

Now imagine a photon is in $|H\rangle$ or in $|V\rangle$, and we do a × measurement. In either case, using (12.4.4) we see that the photon will have a probability $\frac{1}{2}$ of emerging as $|-45°\rangle$ and a probability $\frac{1}{2}$ of emerging as $|+45°\rangle$. Similarly, imagine a photon is in $|-45°\rangle$ or in $|+45°\rangle$, and we do a + measurement. In either case (see (12.4.5)) the photon will have a probability $\frac{1}{2}$ of emerging as $|H\rangle$ and a probability $\frac{1}{2}$ of emerging as $|V\rangle$. In both cases we are measuring along an *unsuitable* basis. If one is measuring along an unsuitable basis, the result of the measurement does not tell us what the state was before measurement.

Photons are easy to create, polarize, transmit, and detect. We could have in principle used spin one-half states of electrons, with basis one comprising $|z;\pm\rangle$ and basis two comprising $|x;\pm\rangle$. The facts explained above would hold equally well. But it is harder to manipulate and transmit electrons. Still, the point is that any two-state system can be used.

Alice and Bob must agree on one more thing: the way to assign a 0 or a 1 to the basis states. After all, the states that are sent and received must be turned into a binary sequence for the key. So let us say they agree to assign a 0 to $|H\rangle$ and to assign a 1 to $|V\rangle$. Moreover, $|-45°\rangle$ is also assigned 0 and $|+45°\rangle$ is also assigned 1. Thus, the assignment *rules* are

$$\begin{aligned} |H\rangle &\to 0, \quad |V\rangle \to 1, \\ |-45°\rangle &\to 0, \quad |+45°\rangle \to 1. \end{aligned} \tag{12.4.6}$$

We have dealt with all the preliminaries. Now we can explain what Alice and Bob do to establish a key they can share. Here are the steps:

1. Alice produces a series of photons, chosen at random. Each photon is in one of the four states $|H\rangle$, $|V\rangle$, $|-45°\rangle$, or $|+45°\rangle$. She sends the photons to Bob over a nonsecure channel.
2. On each photon Bob performs a measurement. He selects a random sequence of measurements, each being either a + or a × measurement. Bob carefully records the measurement he takes on each photon as well as the result of the measurement.
3. Over a (nonsecure) phone line, Bob tells Alice which measurement type he conducted on each of the photons she sent him. He does not tell her the result of each measurement. For each measurement Alice, who has kept a record of her photons, tells him "Yes" if he used a suitable basis to measure with or "No" if the basis he measured with was unsuitable.

4. For each yes, both Alice and Bob know the photon state with full certainty, and both assign to it a 0 or a 1, following the rules. From the list of yes's, they create a list of binary digits that is the key. This is the key they now share and can use to encrypt and decrypt messages.

We can illustrate this with a table. The first line contains the sequence of photons Alice sends. The second line shows Bob's choices of measurements. The third line has Bob's results. The fourth line is the list of yes's and no's arising from suitable and unsuitable bases, respectively. The last line contains the key, 1101, found using the rules on the yes answers.

If Alice sends a large number N of photons polarized randomly among the four possible states, we would expect the key to be roughly of length $N/2$. Indeed, for any photon, and with two choices of measurement type, Bob has a probability $\frac{1}{2}$ of choosing the suitable basis, which will lead to a yes and one verified binary digit for the key.

With the procedure clear, let us see what happens if Eve eavesdrops on this communication (figure 12.7). Since the channel is not secure, she can check out the photons that Alice is sending Bob and listen in to the phone conversation to try to deduce the key. But Eve faces a challenging task. She must measure the photons, or at least many of them; otherwise, she is gaining zero information, and listening to the phone conversation will not help her. If Eve measures, however, there is a question and then a dilemma. Which basis to measure along, $+$ or $\times$? And once she decides and measures, the state is likely to be altered, and she must send a replacement photon, or just by counting photons, Bob and Alice will notice her actions. But what photon should she send? She does not really know the state of the photon *before* measurement because she does not know if she used a suitable or an unsuitable basis. Eve does not have the option of producing

Table 12.1
Table of photon states sent by Alice.

Alice sends	V	H	H	+45°	V	V	−45°	+45°	H	+45°
Bob uses	+	×	×	+	+	×	×	+	×	×
Bob finds	V	+45°	+45°	H	V	−45°	−45°	H	+45°	+45°
Alice says	Yes	No	No	No	Yes	No	Yes	No	No	Yes
Key	1	–	–	–	1	–	0	–	–	1

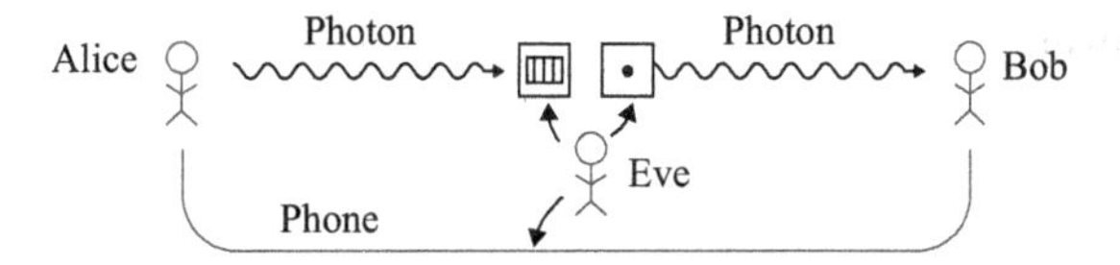

Figure 12.7
Alice sends a sequence of photon states to Bob. These photons are intercepted by Eve, who measures them with her left box and then must generate replacement photons for Bob with her right box. Eve can also eavesdrop on Alice and Bob's phone conversation.

a copy of the photon state before measuring it because cloning quantum states is not possible, as will be discussed in section 18.9.

To understand the effect of Eve's mischief, let us assume she eavesdrops on every photon Alice sends, and as a result, she has to send in a replacement photon each time. Assume also that, each time, Eve randomly picks a basis to measure and then simply sends along the photon state she just measured. Under these conditions, consider what happens when Alice sends a photon in one of the two states of a given basis. Assuming that this photon will end up contributing to the key, what is the probability p_{agree} that both Alice and Bob will agree on its value in the key?

A photon can end up contributing or not contributing to the key. Of course, the photon can end up contributing to the key with Alice and Bob agreeing or disagreeing on its value. Contributing to the key, with agreement, can happen in two ways: (i) with Eve measuring in the right basis and Bob measuring in the right basis, with probability $\frac{1}{2}\cdot\frac{1}{2}=\frac{1}{4}$ and (ii) with Eve measuring in the wrong basis, Bob measuring in the right basis, and Bob getting the right state, with probability $\frac{1}{2}\cdot\frac{1}{2}\cdot\frac{1}{2}=\frac{1}{8}$. So the total probability of contributing to the key with agreement is $\frac{3}{8}$. Contributing to the key with disagreement will happen with Eve measuring in the wrong basis, Bob measuring in the right basis, and Bob getting the wrong state, with probability $\frac{1}{2}\cdot\frac{1}{2}\cdot\frac{1}{2}=\frac{1}{8}$. As a result, if the photon ends up contributing to the key, the probability p_{agree} that Alice and Bob record the same value is

$$p_{\text{agree}} = \tfrac{3}{8}/(\tfrac{3}{8}+\tfrac{1}{8}) = \tfrac{3}{4}. \tag{12.4.7}$$

Despite mischief, three out of four times a digit in the key is the same for Alice and Bob. It follows that, given mischief, the probability p_n that Alice's and Bob's length-n keys are identical is equal to p_{agree} raised to the nth power:

$$p_n = (\tfrac{3}{4})^n. \tag{12.4.8}$$

For $n=72$, we get $p_{72} \simeq 1.0\times 10^{-9}$, meaning that if Alice and Bob test and agree on a *full* 72-digit key, the chance that Eve has been reading all their photons is about one in a billion! In practical terms, Alice and Bob could aim for a 70-digit key, but rather than sending about 140 photons for this purpose, they could send 280, leading roughly to a 140-digit key. They can compare aloud over the phone the first 70 digits of their keys, and if there is perfect agreement, almost surely Eve has not been tinkering with the photons. They can then confidently use the second set of 70 digits as a key, without need of comparison.

Problems

Problem 12.1. *Invariances in the algebra of angular momentum.*

The spin operators $\hat{S}_x$, $\hat{S}_y$, and $\hat{S}_z$ are said to satisfy the algebra of angular momentum because

$$[\hat{S}_x, \hat{S}_y] = i\hbar \hat{S}_z, \quad [\hat{S}_y, \hat{S}_z] = i\hbar \hat{S}_x, \quad [\hat{S}_z, \hat{S}_x] = i\hbar \hat{S}_y.$$

We now consider defining new operators $\hat{S}'_x, \hat{S}'_y$, and $\hat{S}'_z$ in terms of the unprimed $\hat{S}$s. For each of the following transformations, check whether the primed objects satisfy the algebra of angular momentum.

1. $(\hat{S}'_x, \hat{S}'_y, \hat{S}'_z) = (\hat{S}_y, \hat{S}_z, \hat{S}_x)$. This is a cyclic permutation.
2. $(\hat{S}'_x, \hat{S}'_y, \hat{S}'_z) = (\hat{S}_x, \hat{S}_y, -\hat{S}_z)$. This is changing the sign of one of the operators.
3. $(\hat{S}'_x, \hat{S}'_y, \hat{S}'_z) = (\hat{S}_x, -\hat{S}_y, -\hat{S}_z)$. This is changing the sign of two of the operators.
4. $(\hat{S}'_x, \hat{S}'_y, \hat{S}'_z) = (-\hat{S}_x, -\hat{S}_y, -\hat{S}_z)$. This is changing the sign of all the operators.
5. $(\hat{S}'_x, \hat{S}'_y, \hat{S}'_z) = (\hat{S}_y, -\hat{S}_x, \hat{S}_z)$. This can be viewed as a rotation by 90° about the z-axis.

If the primed operators satisfy the algebra of angular momentum, we say that the transformation is an *automorphism* of the algebra.

Problem 12.2. *Experiments with SG devices.*

An experimentalist has prepared a beam of spin one-half atoms, all in the state $|+\rangle$ that points in the positive z-direction.

- These atoms are passed through a first SG device designed to measure $\hat{S}_{\mathbf{n}}$, with $\mathbf{n}$ a unit vector in the (x, z) plane specified by $\phi = 0$, $\theta \neq 0$. Only those atoms that have eigenvalue $\hbar/2$ are kept.
- The remaining atoms go through a second SG device, oriented in the z-direction.

Of all the atoms in the original beam, what fraction are found after the second SG device to be in the state $|+\rangle$, and what fraction are found to be in the state $|-\rangle$? What fraction never made it to the second SG device? The answers are functions of the angle θ. Argue that your answers make sense for $\theta = 0, \pi/2$, and π.

Problem 12.3. *Measurements on a beam of spin one-half particles.*

A student of quantum mechanics has prepared a beam of electrons, all of which are in the same spin state. The following is known:

- If the beam is passed through an SG device in the z-direction, half the electrons have spin $\frac{\hbar}{2}$, and half have spin $-\frac{\hbar}{2}$.
- If the beam is passed through an SG device in the x-direction, the mean value of the spin is $\langle \hat{S}_x \rangle = \frac{\hbar}{2} g$, where $|g| \leq 1$.
- If the beam is passed through an SG device in the y-direction, $\langle \hat{S}_y \rangle \geq 0$.

Determine the spin state for the beam. Find, as a function of g, the probability $p_y(+)$ that measurement of the spin along the y-axis will yield $\frac{\hbar}{2}$.

Problem 12.4. *Spin one-half state along a specific direction.*

An (unnormalized) spin state is given by

$$(1+i)|+\rangle - (1+i\sqrt{3})|-\rangle.$$

What direction does this spin state point to?

Problem 12.5. *Spin in a magnetic field along the z-direction.*

At $t=0$ an electron has its spin pointing in the $+x$-direction: $\Psi(t=0)=|x;+\rangle$. The dynamics of this electron is governed by the Hamiltonian $\hat{H}$ given by

$$\hat{H}=\omega\hat{S}_z,$$

where $\omega>0$ is a constant with the units of angular frequency. Find the state $\Psi(t)$ for all later times $t>0$. Describe the time evolution as that of a rotating spin state. The Hamiltonian above describes the time evolution of the spin of the electron subject to a uniform constant magnetic field of magnitude B pointing in the positive z-direction, where $\omega=\frac{eB}{m_e c}$.

Problem 12.6. *Spin in a magnetic field along the y-direction.*

At $t=0$ an electron has its spin pointing in the direction defined by the angles $\theta=\theta_0$ and $\phi=0$. A magnetic field along the y-direction interacts with the magnetic dipole of the spin, resulting in a Hamiltonian

$$\hat{H}=-\omega\hat{S}_y,$$

where $\omega>0$ is a constant with units of angular frequency. Calculate the state $\Psi(t)$. Write your answer in terms of the basis states $|+\rangle$ and $|-\rangle$ along the z-direction. Describe the time-dependent spin orientation in terms of functions $\theta(t)$ and $\phi(t)$.

Problem 12.7. *Expectation values of spin operators on general spin states.*

Consider a general spin state $|\mathbf{n}\rangle$, with $\mathbf{n}$ the unit vector defined by angles θ and ϕ. Show that the *expectation values* of the spin operators on the state $|\mathbf{n}\rangle$ are given by

$$\Big(\langle\mathbf{n}|\hat{S}_x|\mathbf{n}\rangle,\ \langle\mathbf{n}|\hat{S}_y|\mathbf{n}\rangle,\ \langle\mathbf{n}|\hat{S}_z|\mathbf{n}\rangle\Big)=\frac{\hbar}{2}(n_x,n_y,n_z),$$

with n_x, n_y, and n_z the Cartesian components of $\mathbf{n}$. With the notation $\hat{\mathbf{S}}=(\hat{S}_x,\hat{S}_y,\hat{S}_z)$, this result is often written as $\langle\mathbf{n}|\hat{\mathbf{S}}|\mathbf{n}\rangle=\frac{\hbar}{2}\mathbf{n}$.

Problem 12.8. *Six-state protocol for QKD.*

Alice and Bob decide to use three basis pairs, X, Y, and Z, for a modified QKD protocol:

Basis set X: $|x;+\rangle$, $|x;-\rangle$,

Basis set Y: $|y;+\rangle$, $|y;-\rangle$,

Basis set Z: $|z;+\rangle$, $|z;-\rangle$.

As before Alice will send Bob at random a sequence of spins in any of the six states above. Bob measures at random in any of three bases X, Y, or Z. To produce a key, they agree that a (+)-type state is assigned a 0, and a (−)-type state is assigned a 1.

1. Write a table with twelve random spins sent by Alice (analogous to table 12.1) that result in the key 0101.

2. If Alice sends N spins, with large N, how long is the key expected to be?
3. Suppose Eve eavesdrops on every spin state sent by Alice, measures randomly among the three basis sets, and replaces each spin state by the state she measured. Find the probability $\bar{p}_n$ that a key of length n is identical for Alice and Bob despite Eve's mischief.
4. For BB84 and $n=72$, we found a 10^{-9} chance that the key is consistent despite eavesdropping. What do you get for this six-state protocol?

II Theory

13 Vector Spaces and Operators

We define real and complex vector spaces. Using subspaces, we introduce the concept of the direct sum representation of a vector space. We define spanning sets, linearly independent sets, basis vectors, and the concept of dimensionality. Linear operators are linear functions from a vector space to itself. The set of linear operators on a vector space itself forms a vector space, with a product structure that arises by composition. While the product is associative, it is not in general commutative. We discuss when operators are injective, surjective, and invertible. We explain how operators can be represented by matrices once we have made a choice of basis vectors on the vector space. We derive the relation between matrix representations in different bases and show that traces and determinants are basis independent. Finally, we discuss eigenvectors and eigenvalues of operators, connecting them to the idea of invariant subspaces.

13.1 Vector Spaces

In quantum mechanics the state of a physical system is a *vector* in a *complex vector space*. Observables are linear operators—in fact, Hermitian operators acting on this complex vector space. The purpose of this chapter is to learn the basics of vector spaces and the operators that act on them.

Complex vector spaces are somewhat different from the more familiar real vector spaces. They have more powerful properties. In order to understand complex vector spaces, it is useful to compare them often to their real-dimensional friends. In a vector space, one has vectors and numbers, also sometimes referred to as scalars. We can add vectors to get vectors, and we can multiply vectors by numbers to get vectors. If the numbers we use are real, we have a real vector space. If the numbers we use are complex, we have a complex vector space. More generally, the numbers we use belong to what is called a *field*, denoted by the letter $\mathbb{F}$. We will discuss just two cases, $\mathbb{F}=\mathbb{R}$, meaning the numbers are real, and $\mathbb{F}=\mathbb{C}$, meaning the numbers are complex.

The definition of a vector space is the same for any field $\mathbb{F}$. A vector space V is a set of vectors with an operation of **addition** $(+)$ that assigns an element $u+v \in V$ to each $u, v \in V$. Because the operation of addition takes two elements of V to another element of V, we say that V is *closed* under addition. There is also a **scalar multiplication** by elements of $\mathbb{F}$, with $av \in V$ for any $a \in \mathbb{F}$ and $v \in V$. This means the space V is also closed under multiplication by numbers. These operations must satisfy the following additional properties:

1. $u+v=v+u \in V$ for all $u, v \in V$ (addition is commutative).
2. $u+(v+w)=(u+v)+w$, and $(ab)u=a(bu)$ for any $u, v, w \in V$ and $a, b \in \mathbb{F}$ (associativity of addition and scalar multiplication).
3. There is a vector $0 \in V$ such that $0+u=u$ for all $u \in V$ (additive identity).
4. For each $v \in V$, there is a $u \in V$ such that $v+u=0$ (additive inverse).
5. The element $1 \in \mathbb{F}$ satisfies $1v=v$ for all $v \in V$ (multiplicative identity).
6. $a(u+v)=au+av$, and $(a+b)v=av+bv$ for every $u, v \in V$ and $a, b \in \mathbb{F}$ (distributive property).

This definition is very efficient. Several familiar properties follow from it by short proofs that we will not give but that are not complicated and you may try to produce. For example:

- The additive identity is unique: any vector $0'$ that acts like 0 is actually equal to 0.
- $0v=0$, for any $v \in V$, where the first zero is a number, and the second one is a vector. This means that the number zero acts as expected when multiplying a vector.
- $a0=0$, for any $a \in \mathbb{F}$. Here both zeroes are vectors. This means that the zero vector multiplied by any number is still the zero vector.
- The additive inverse of any vector $v \in V$ is unique. It is denoted by $-v$ and in fact $-v=(-1)v$.

We must stress that while the numbers in $\mathbb{F}$ are sometimes real or complex, we *do not* speak of the vectors themselves as real or complex. A vector multiplied by a complex number, for example, is not said to be a complex vector. The vectors in a real vector space are not said to be real, and the vectors in a complex vector space are not said to be complex.

The definition of a vector space does *not* introduce a multiplication of vectors. Only in very special cases is there a natural way to multiply vectors to give vectors. One such example is the cross product in three spatial dimensions.

As is customary in the mathematics literature, vectors are denoted by symbols such as v, u, w, without arrows added on top or the use of boldface. In physics we sometimes use arrows or boldface. We hope the various notations will not cause confusion.

A guide to the examples In this and the following chapter, as we introduce the mathematical concepts we illustrate their use with examples from quantum systems we have examined before. This should help the reader appreciate the material. Moreover, in this way the reader will review those important quantum systems. For the case of spin one-half systems, the examples develop considerably the theory discussed in chapter 12. The examples also discuss the harmonic oscillator. Finally, the examples include mathematical techniques used in computations, such as index notation and commutator algebra. Below we list the examples so the reader can access them quickly. Examples of a purely mathematical nature are not included in the lists below.

Examples and developments for spin one-half systems:

- Example 13.2. Vector space $\mathbb{C}^2$ for spin one-half.
- Example 13.7. The real vector space of 2×2 Hermitian matrices.
- Example 13.11. Working with Pauli matrices.
- Example 13.12. Is there a linear operator that reverses the direction of all spin states?
- Example 13.14. Null space and range of spin operators on $\mathbb{C}^2$.
- Example 13.22. Exponentials of linear combinations of Pauli matrices.
- Example 14.1. Inner product in $\mathbb{C}^2$ for spin one-half.
- Example 14.2. Inner product of spin states.
- Example 14.4. An orthonormal basis of Hermitian matrices in two dimensions.
- Example 14.7. Orthogonal projector onto a spin state $|\mathbf{n}\rangle$.
- Section 14.7. Rotation operators for spin states.
- Example 14.13. Orthogonal projector onto a spin state $|\mathbf{n}\rangle$ revisited.

Examples and developments for the simple harmonic oscillator:

- Example 13.4. Fock states of the one-dimensional harmonic oscillator.
- Example 13.8. State space $\mathcal{H}$ of the simple harmonic oscillator.
- Example 13.16. Right inverse for the annihilation operator $\hat{a}$.
- Example 13.18. Matrix representation for the harmonic oscillator $\hat{a}$ and $\hat{a}^\dagger$ operators.
- Example 13.21. Eigenvectors of the annihilation operator $\hat{a}$ of the harmonic oscillator.
- Example 14.16. Harmonic oscillator in bra-ket notation.

Examples for computational techniques:

- Example 13.10. Review of index manipulations.
- Section 13.7. Functions of linear operators and key identities.
- Example 13.23. Five ways to do a computation.
- Example 14.3. Inner product for operators.
- Example 14.12. Trace of the operator $|u\rangle\langle w|$.
- Example 14.14. Adjoint in bra-ket description of operators.

Additional examples:

- Example 13.3. Vector space $\mathbb{C}^{2\ell+1}$ for orbital angular momentum ℓ.
- Example 13.5. The state space of the infinite square well.
- Example 13.9. State space of hydrogen atom bound states.
- Example 14.11. From Hermitian to unitary operators.

Example 13.1. *A few vector spaces.*

1. The vector space $\mathbb{C}^N$ is the set of N-component vectors:

$$\begin{pmatrix} a_1 \\ \vdots \\ a_N \end{pmatrix}, \quad a_i \in \mathbb{C}, \quad i = 1, 2, \ldots, N. \tag{13.1.1}$$

We add two vectors by adding the corresponding components:

$$\begin{pmatrix} a_1 \\ \vdots \\ a_N \end{pmatrix} + \begin{pmatrix} b_1 \\ \vdots \\ b_N \end{pmatrix} = \begin{pmatrix} a_1 + b_1 \\ \vdots \\ a_N + b_N \end{pmatrix}. \tag{13.1.2}$$

Multiplication by a number $\alpha \in \mathbb{C}$ is defined by multiplying each component by α:

$$\alpha \begin{pmatrix} a_1 \\ \vdots \\ a_N \end{pmatrix} = \begin{pmatrix} \alpha\, a_1 \\ \vdots \\ \alpha\, a_N \end{pmatrix}. \tag{13.1.3}$$

The zero vector is the vector with the number zero for each entry. You should verify that all the axioms are then satisfied. This is the classic example of a complex vector space. If all entries and numbers are real, this would be a real vector space.

2. Here is a slightly more unusual example in which matrices are the vectors of the vector space. Consider the set of $M \times N$ matrices with complex entries:

$$\begin{pmatrix} a_{11} & \ldots & a_{1N} \\ \vdots & & \vdots \\ a_{M1} & \ldots & a_{MN} \end{pmatrix}, \quad a_{ij} \in \mathbb{C}. \tag{13.1.4}$$

We define addition as follows:

$$\begin{pmatrix} a_{11} & \ldots & a_{1N} \\ \vdots & & \vdots \\ a_{M1} & \ldots & a_{MN} \end{pmatrix} + \begin{pmatrix} b_{11} & \ldots & b_{1N} \\ \vdots & & \vdots \\ b_{M1} & \ldots & b_{MN} \end{pmatrix} = \begin{pmatrix} a_{11} + b_{11} & \ldots & a_{1N} + b_{1N} \\ \vdots & & \vdots \\ a_{M1} + b_{M1} & \ldots & a_{MN} + b_{MN} \end{pmatrix}. \tag{13.1.5}$$

Multiplication by a constant $f \in \mathbb{C}$ ends up multiplying all entries:

$$f \begin{pmatrix} a_{11} & \ldots & a_{1N} \\ \vdots & & \vdots \\ a_{M1} & \ldots & a_{MN} \end{pmatrix} = \begin{pmatrix} fa_{11} & \ldots & fa_{1N} \\ \vdots & & \vdots \\ fa_{M1} & \ldots & fa_{MN} \end{pmatrix}. \tag{13.1.6}$$

The zero matrix is defined as the matrix in which all entries are the number zero. With these definitions the set of $M \times N$ matrices forms a complex vector space.

3. Consider the set of $N \times N$ Hermitian matrices. These are matrices with complex entries that are left invariant by the successive operations of transposition and complex conjugation. Curiously, the set of $N \times N$ Hermitian matrices form a *real* vector space. This is because multiplication by real numbers preserves the property

of Hermiticity, while multiplication by complex numbers does not. This illustrates the earlier claim that we should not use the labels *real* or *complex* for the vectors themselves.

4. Here is another slightly unusual example in which polynomials are the vectors. Consider the set $\mathcal{P}(\mathbb{F})$ of polynomials. A polynomial $p \in \mathcal{P}(\mathbb{F})$ is a function from $\mathbb{F}$ to $\mathbb{F}$: acting on the variable $z \in \mathbb{F}$, it gives a value $p(z) \in \mathbb{F}$. Each nonzero polynomial p is defined by coefficients $a_0, a_1, \ldots a_n \in \mathbb{F}$, with n a finite, nonnegative integer called the degree of the polynomial:

$$p(z) = a_0 + a_1 z + a_2 z^2 + \cdots + a_n z^n, \quad a_n \neq 0. \tag{13.1.7}$$

The zero polynomial $p(z) = 0$ has $n = 0$ and $a_0 = 0$. The addition of polynomials works as expected. If $p_1, p_2 \in \mathcal{P}(\mathbb{F})$, then $p_1 + p_2 \in \mathcal{P}(\mathbb{F})$ is defined by

$$(p_1 + p_2)(z) = p_1(z) + p_2(z), \tag{13.1.8}$$

and multiplication works as $(ap)(z) = ap(z)$, for $a \in \mathbb{F}$ and $p \in \mathcal{P}(\mathbb{F})$. The zero vector is the zero polynomial. The space $\mathcal{P}(\mathbb{F})$ of all polynomials forms a vector space over $\mathbb{F}$.

This vector space has a simple generalization when z is just a formal variable not valued in $\mathbb{F}$, while the coefficients a_k are still valued in $\mathbb{F}$. This vector space actually represents an interesting subspace of states of the quantum harmonic oscillator (example 13.4).

5. Consider the set $\mathbb{F}^\infty$ of infinite sequences $(x_1, x_2, \ldots)$ of elements $x_i \in \mathbb{F}$. Here

$$\begin{aligned} (x_1, x_2, \ldots) + (y_1, y_2, \ldots) &= (x_1 + y_1, x_2 + y_2, \ldots), \\ a(x_1, x_2, \ldots) &= (ax_1, ax_2, \ldots), \quad a \in \mathbb{F}. \end{aligned} \tag{13.1.9}$$

This is a vector space over $\mathbb{F}$.

6. The set of complex functions $f(x)$ on an interval $x \in [0, L]$ form a vector space over $\mathbb{C}$. Here the functions are the vectors of the vector space. The required definitions are

$$(f_1 + f_2)(x) = f_1(x) + f_2(x), \quad (af)(x) = af(x) \tag{13.1.10}$$

with $f_1(x)$ and $f_2(x)$ complex valued functions on the interval and a a complex number. This vector space contains, for example, the wave functions of a particle in a one-dimensional potential that confines it to the interval $x \in [0, L]$. □

Let us now see in more detail how some of the above vector spaces are suitable to the description of familiar quantum systems.

Example 13.2. *Vector space $\mathbb{C}^2$ for spin one-half.*

We had our first look at spin one-half in chapter 12. The state space there was that of two-component complex vectors (as in item 1 of example 13.1). The quantum states Ψ of the spin one-half particle take the form

$$\Psi = \begin{pmatrix} c_1 \\ c_2 \end{pmatrix}, \quad \text{with } c_1, c_2 \in \mathbb{C}. \tag{13.1.11}$$

It takes just two complex numbers to specify completely the spin state of the particle. Note how this differs from the data required to specify the position state of the particle: a whole wave function $\psi(x)$ worth of data. Using the rule for multiplying vectors by complex constants, we can rewrite Ψ as

$$\Psi = c_1 \begin{pmatrix} 1 \\ 0 \end{pmatrix} + c_2 \begin{pmatrix} 0 \\ 1 \end{pmatrix}. \tag{13.1.12}$$

The column vectors with a single 1 and a single 0 have been given names:

$$|z;+\rangle = |1\rangle = \begin{pmatrix} 1 \\ 0 \end{pmatrix}, \quad |z;-\rangle = |2\rangle = \begin{pmatrix} 0 \\ 1 \end{pmatrix}. \tag{13.1.13}$$

We identified $|z;+\rangle$ as the state of a particle with its spin pointing in the positive z-direction and $|z;-\rangle$ as the state of a particle with its spin pointing in the negative z-direction. With this notation the state in (13.1.11) is $\Psi = c_1|z;+\rangle + c_2|z;-\rangle$. □

Example 13.3. *Vector space $\mathbb{C}^{2\ell+1}$ for orbital angular momentum ℓ.*
We learned in chapter 10 that in a central potential energy eigenstates are described with wave functions of the form $\psi(r,\theta,\phi) = R(r)F(\theta,\phi)$. The angular dependence $F(\theta,\phi)$ determines the angular momentum of the state. Restricting ourselves to quantum states of angular momentum ℓ means that we pick wave functions $F_\ell(\theta,\phi)$ that are $\hat{\mathbf{L}}^2$ eigenstates:

$$\hat{\mathbf{L}}^2 F_\ell(\theta,\phi) = \hbar^2 \ell(\ell+1) F_\ell(\theta,\phi). \tag{13.1.14}$$

In fact, we found that the wave functions $Y_{\ell m}(\theta,\phi)$ with $m = -\ell,\ldots,\ell$ all have angular momentum ℓ. Indeed, from (10.5.32) we have that

$$\begin{aligned} \hat{\mathbf{L}}^2\, Y_{\ell m} &= \hbar^2\, \ell(\ell+1)\, Y_{\ell m}, \\ \hat{L}_z\, Y_{\ell m} &= \hbar m\, Y_{\ell m}, \qquad m = -\ell,\ldots,\ell. \end{aligned} \tag{13.1.15}$$

Any angular wave function with angular momentum ℓ must be a linear superposition of the $Y_{\ell m}$'s with various values of m. It is thus natural to define the vector space $\mathbb{C}^{2\ell+1}$ of angular wave functions by encoding general wave functions of angular momentum ℓ into a column vector of size $2\ell+1$ with arbitrary complex entries:

$$F_\ell(\theta,\phi) = c_1 Y_{\ell\ell} + c_2 Y_{\ell,\ell-1} + \cdots c_{2\ell+1} Y_{\ell,-\ell} \iff \begin{pmatrix} c_1 \\ c_2 \\ \vdots \\ c_{2\ell+1} \end{pmatrix}, \quad c_i \in \mathbb{C}. \tag{13.1.16}$$

The addition of vectors is natural, and so is multiplication by a constant. The spherical harmonics themselves can be thought of as simple vectors with a single nonvanishing entry equal to one:

$$Y_{\ell\ell} = e_1 = \begin{pmatrix} 1 \\ 0 \\ \vdots \\ 0 \end{pmatrix}, \quad \cdots \quad, Y_{\ell,-\ell} = e_{2\ell+1} = \begin{pmatrix} 0 \\ \vdots \\ 0 \\ 1 \end{pmatrix}. \tag{13.1.17}$$

For each possible value of $\ell = 0, 1, \ldots$, there is a state space $\mathbb{C}^{2\ell+1}$ of states. For $\ell = 0$, the space is just the space of complex constants $\mathbb{C}$. □

Example 13.4. *Fock states of the one-dimensional harmonic oscillator.*

There exist interesting states, called *Fock states*, that consist of *finite* linear superpositions of energy eigenstate states $\varphi_n \sim (\hat{a}^\dagger)^n \varphi_0$, with φ_0 the ground state (see chapter 9). A Fock state ψ_k takes the form

$$\psi_k = (\gamma_0 + \gamma_1 \hat{a}^\dagger + \gamma_2 (\hat{a}^\dagger)^2 + \cdots + \gamma_k (\hat{a}^\dagger)^k)\varphi_0, \quad \gamma_i \in \mathbb{C}. \tag{13.1.18}$$

Fock states are in one-to-one correspondence with the vector space of *formal* polynomials discussed in example 13.1, item 4. We consider the space $\mathcal{P}(\mathbb{C})$ of polynomials with complex coefficients and a formal variable z identified with $\hat{a}^\dagger$. In this correspondence, the above state ψ_k is unambiguously associated with the polynomial $p_k(z)$ below:

$$p_k(z) = \gamma_0 + \gamma_1 z + \gamma_2 z^2 + \cdots + \gamma_k z^k. \tag{13.1.19}$$

Fock states are manifestly normalizable; they are, after all, a finite sum of energy eigenstates.

Not all states in the state space of the harmonic oscillator are Fock states. We will find normalizable coherent states and squeezed states that are not polynomials in the creation operator acting on the ground state. They are, instead, exponentials of linear and quadratic functions of the creation operator. The normalizability of non-Fock states must be checked since they are built as *infinite* linear superpositions of energy eigenstates. □

Example 13.5. *The state space of the infinite square well.*

The energy eigenstates of the infinite square well $x \in [0, a]$ (section 6.4) are represented by orthonormal wave functions:

$$\psi_n(x) = \sqrt{\frac{2}{a}} \sin\left(\frac{n\pi x}{a}\right) \quad \text{with} \quad n = 1, 2, \cdots. \tag{13.1.20}$$

The specification of a state of the square well is tantamount to the specification of an infinite sequence $(c_1, c_2, c_3, \ldots)$ of complex numbers that result in the state $\psi(x)$:

$$\psi(x) = \sum_{n=1}^{\infty} c_n \psi_n(x) = c_1 \psi_1(x) + c_2 \psi_2(x) + \cdots. \tag{13.1.21}$$

This is in fact the space $\mathbb{C}^\infty$ of infinite sequences presented in example 13.1, item 5. The states of the square well, however, must be normalizable so that $\|\psi\|^2 = \sum_{n=1}^{\infty} |c_n|^2 < \infty$. The infinite sequences in $\mathbb{C}^\infty$ satisfying this constraint still form a vector space. Item 6 of that same example examines the space of complex functions on an interval. This is the space we are considering when restricted to functions that vanish at the end points and are square integrable. □

13.2 Subspaces, Direct Sums, and Dimensionality

To better understand a vector space, one can try to figure out its possible subspaces. A **subspace** of a vector space V is a subset of V that is also a vector space. To verify that a subset U of V is a subspace, you must check that U contains the vector 0 and that U is closed under addition and scalar multiplication. All other properties required by the axioms for U to be a vector space are automatically satisfied because U is contained in V (think about this!).

Example 13.6. *Subspaces of* $\mathbb{R}_2$.

Let $V = \mathbb{R}^2$ so that elements of V are pairs (v_1, v_2) with $v_1, v_2 \in \mathbb{R}$. Now introduce the subsets W_r defined by a real number r:

$$W_r \equiv \{(v_1, v_2) | \; 3v_1 + 4v_2 = r, \text{ with } r \in \mathbb{R}\}. \tag{13.2.1}$$

When is W_r a subspace of $\mathbb{R}$? Since we need the zero vector $(0, 0)$ to be contained, this requires $3 \cdot 0 + 4 \cdot 0 = r$ or $r = 0$. Indeed, one can readily verify that W_0 is closed under addition and scalar multiplication and is therefore a subspace of V.

It is possible to visualize all nontrivial subspaces of $\mathbb{R}^2$. These are the lines that go through the origin. Each line is a vector space: it contains the zero vector (the origin), and all vectors defined by points on the line can be added or multiplied to find vectors on the same line. □

Exercise 13.1. *Let U_1 and U_2 be two subspaces of V. Is $U_1 \cap U_2$ a subspace of V?*

To understand a complicated vector space, it is useful to consider subspaces that together build up the space. Let $U_1, \ldots, U_m$ be a collection of subspaces of V. We say that the space V is the **direct sum of the subspaces** $U_1, \ldots, U_m$ and we write

$$V = U_1 \oplus \cdots \oplus U_m \tag{13.2.2}$$

if any vector in V can be written *uniquely* as the sum

$$u_1 + \cdots + u_m, \quad \text{where } u_i \in U_i. \tag{13.2.3}$$

This can be viewed as a decomposition of any vector into a sum of vectors, one in each of the subspaces. Part of the intuition here is that while the set of all subspaces fills the whole space, the various subspaces cannot overlap. More precisely, their only common element is zero: $U_i \cap U_j = \{0\}$ for $i \neq j$. If this is violated, the decomposition of vectors in V would not be unique. Indeed, for if some vector $v \in U_i \cap U_j$ $(i \neq j)$ then also $-v \in U_i \cap U_j$ (why?), and therefore letting $u_i \to u_i + v$ and $u_j \to u_j - v$ would leave the total sum unchanged, making the decomposition nonunique. The condition of zero mutual overlaps is necessary for the uniqueness of the decomposition, but it is not in general sufficient. It suffices, however, when we have two summands: to show that $V = U \oplus W$, one must prove that any vector can be written as $u + w$ with $u \in U$ and $w \in W$ and that $U \cap W = 0$. In general, uniqueness of the sum in (13.2.3) follows if the only way to write 0 as a sum $u_1 + \cdots + u_m$ with $u_i \in U_i$ is by taking all u_i's equal to zero. Direct sum decompositions appear rather naturally when we consider the addition of angular momentum.

Given a vector space, we can produce lists of vectors. A **list** $(v_1, \ldots, v_n)$ of vectors in V contains, by definition, a *finite* number of vectors. The number of vectors in a list is the length of the list. The **span** of a list of vectors $(v_1, \ldots, v_n)$ in V, denoted as $\text{span}(v_1, \ldots, v_n)$, is the set of all linear combinations of these vectors:

$$a_1 v_1 + \cdots + a_n v_n, \qquad a_i \in \mathbb{F}. \tag{13.2.4}$$

A vector space V is spanned by a list $(v_1, \ldots, v_n)$ if $V = \text{span}(v_1, \ldots, v_n)$.

Now comes a very natural definition: A vector space V is said to be **finite-dimensional** if it is spanned by some list of vectors in V. If V is not finite-dimensional, it is **infinite-dimensional**. In such a case, no list of vectors from V can span V. Note that by definition, any finite-dimensional vector space has a spanning list.

Let us explain why the vector space of all polynomials $p(z)$ in example 13.1, item 4, is an infinite-dimensional vector space. Indeed, consider any list of polynomials. Since a list is always of finite length, there is a polynomial of maximum degree in the list. Thus, polynomials of higher degree are not in the span of the list. Since no list can span the space, it is infinite-dimensional.

For example 13.1, item 1, consider the list of vectors $(e_1, \ldots, e_N)$ with

$$e_1 = \begin{pmatrix} 1 \\ 0 \\ \vdots \\ 0 \end{pmatrix}, \quad e_2 = \begin{pmatrix} 0 \\ 1 \\ \vdots \\ 0 \end{pmatrix}, \quad \ldots \quad e_N = \begin{pmatrix} 0 \\ 0 \\ \vdots \\ 1 \end{pmatrix}. \tag{13.2.5}$$

This list spans the space: the general vector displayed in (13.1.1) is $a_1 e_1 + \cdots + a_N e_N$. This vector space is therefore finite-dimensional.

To make further progress, we need the concept of linear independence. A list of vectors $(v_1, \ldots, v_n)$ with $v_i \in V$ is said to be **linearly independent** if the equation

$$a_1 v_1 + \cdots + a_n v_n = 0 \tag{13.2.6}$$

only has the solution $a_1 = \cdots = a_n = 0$. One can prove a key result: *the length of any linearly independent list is less than or equal to the length of any spanning list.* This is reasonable, as we discuss now. Spanning lists can be enlarged as much as desired because adding vectors to a spanning list still gives a spanning list. They cannot be reduced arbitrarily, however, because at some point the remaining vectors will fail to span. For linearly independent lists, the situation is exactly reversed: they can be easily shortened because dropping vectors will not disturb the linear independence but cannot be enlarged arbitrarily because at some point the new vectors can be expressed in terms of those already in the list. As it happens, in a finite vector space the length of the longest list of linearly independent vectors is the same as the length of the shortest list of spanning vectors. This leads to the concept of dimensionality, as we will see below.

We can now explain what a basis for a vector space is. A **basis** of V is a list of vectors in V that both spans V and is linearly independent. It is not hard to prove that any finite-dimensional vector space has a basis. While bases are not unique, all bases of a finite-dimensional vector space have the same length. The **dimension** of a finite-dimensional vector space is equal to the length of any list of basis vectors. If V is a space of dimension n, we write $\dim V = n$. It is also true that for a finite-dimensional vector space a list of vectors of length $\dim V$ is a basis if it is a linearly independent list or if it is a spanning list.

The list $(e_1, \ldots, e_N)$ in (13.2.5) is not only a spanning list but a linearly independent list (prove it!). Thus, the dimensionality of the space is N.

Exercise 13.2. *Explain why the vector space in example 13.1, item 2, has dimension $M \cdot N$.*

The vector space $\mathbb{F}^\infty$ of infinite sequences in example 13.1, item 5, is infinite-dimensional, as we now justify. Assume $\mathbb{F}^\infty$ is finite-dimensional, in which case it has a spanning list of some length n. Define s_k as the element in $\mathbb{F}^\infty$ with a one in the kth position and zero elsewhere. The list $(s_1, \ldots, s_m)$ is clearly a linearly independent list of length m, with m arbitrary. Choosing $m > n$, we have a linearly independent list longer than a spanning list. This is a contradiction, and therefore $\mathbb{F}^\infty$ cannot be finite-dimensional. Recall this is the state space of the square well. The space of complex functions on the interval $[0, L]$ (example 13.1, item 5) is also infinite-dimensional.

Equipped with the concept of dimensionality, there is a simple way to see if we have a direct sum decomposition of a vector space. In fact, we have $V = U_1 \oplus \cdots \oplus U_m$ if any vector in V can be written as $u_1 + \cdots + u_m$, with $u_i \in U_i$ and if $\dim U_1 + \cdots + \dim U_m = \dim V$. The proof of this result is not complicated.

Example 13.7. *The real vector space of 2 × 2 Hermitian matrices.*

Consider example 13.1, item 3, and focus on the case of the vector space of 2 × 2 Hermitian matrices. Recall that the most general Hermitian 2 × 2 matrix takes the form

$$\begin{pmatrix} a_0 + a_3 & a_1 - ia_2 \\ a_1 + ia_2 & a_0 - a_3 \end{pmatrix}, \qquad a_0, a_1, a_2, a_3 \in \mathbb{R}. \tag{13.2.7}$$

Now consider the following list of four "vectors," $(\mathbb{1}, \sigma_1, \sigma_2, \sigma_3)$, with σ_i the Pauli matrices (12.1.20) and $\mathbb{1}$ the 2 × 2 identity matrix. All entries in this list are Hermitian matrices, so this is a list of vectors in the space. Moreover, the list spans the space since the general Hermitian matrix shown above is $a_0\mathbb{1} + a_1\sigma_1 + a_2\sigma_2 + a_3\sigma_3$. The list is linearly independent since

$$a_0\mathbb{1} + a_1\sigma_1 + a_2\sigma_2 + a_3\sigma_3 = 0 \quad \Rightarrow \quad \begin{pmatrix} a_0 + a_3 & a_1 - ia_2 \\ a_1 + ia_2 & a_0 - a_3 \end{pmatrix} = \begin{pmatrix} 0 & 0 \\ 0 & 0 \end{pmatrix}, \tag{13.2.8}$$

and you can quickly see that this implies that a_0, a_1, a_2, and a_3 are all zero. So the list is a basis, and the space in question is a four-dimensional real vector space. □

Example 13.8. *State space $\mathcal{H}$ of the simple harmonic oscillator.*

The energy eigenstates of the harmonic oscillator can be used to give a direct sum representation of the state space $\mathcal{H}$ for the harmonic oscillator. Let U_n be the one-dimensional subspace that is the span of the energy eigenstate φ_n of the oscillator (see chapter 9).

The state φ_n is an eigenstate of the number operator $\hat{N}$ with eigenvalue n:

$$U_n \equiv \{\alpha\varphi_n,\ \alpha \in \mathbb{C},\ \hat{N}\varphi_n = n\varphi_n\}. \tag{13.2.9}$$

The space U_n is an $\hat{N}$-invariant subspace of $\mathcal{H}$. Since any state of the oscillator can be written uniquely as a sum of energy eigenstates, we have the direct sum decomposition:

$$\mathcal{H} = \bigoplus_{n=1}^{\infty} U_n = U_0 \oplus U_1 \oplus U_2 \oplus \cdots . \tag{13.2.10}$$

The space $\mathcal{H}$, of course, is infinite-dimensional, the direct sum of an infinite countable set of one-dimensional subspaces. A little care is needed here to describe $\mathcal{H}$ precisely. While it is clear that any harmonic oscillator state can be written uniquely as a sum of energy eigenstates, not all sums of energy eigenstates correspond to physical states of the harmonic oscillator. Finite linear combinations always do; these are the Fock states considered in example 13.4. Some but not all infinite linear combinations also give physical states. Physical states must be normalizable. □

Example 13.9. *State space of hydrogen atom bound states.*

The bound states of the hydrogen atom span an important subspace $\mathcal{H}$ of this quantum system. The bound state spectrum was determined in section 11.3, with energy levels indexed by the principal quantum number $n = 1, 2, \ldots$. Calling $\mathcal{H}_n$ the vector subspace spanned by the degenerate energy eigenstates at principal quantum number n, we see that

$$\mathcal{H} = \bigoplus_{n=1}^{\infty} \mathcal{H}_n = \mathcal{H}_1 \oplus \mathcal{H}_2 \oplus \cdots . \tag{13.2.11}$$

Ignoring the spin of the electron, $\dim \mathcal{H}_n = n^2$. We can refine the description by giving a direct sum decomposition of $\mathcal{H}_n$. In fact, for any fixed n the orbital angular momentum runs from $\ell = 0$ to $\ell = n-1$, a total of n angular momentum multiplets. We write this as

$$\mathcal{H}_n = \bigoplus_{\ell=0}^{n-1} \mathcal{H}_{n,\ell} = \mathcal{H}_{n,0} \oplus \cdots \oplus \mathcal{H}_{n,n-1}. \tag{13.2.12}$$

The space $\mathcal{H}_{n,\ell}$ is simply a vector space of states with angular momentum ℓ and principal quantum number n. It has dimension $2\ell+1$, and on account of example 13.3, $\mathcal{H}_{n,\ell} = \mathbb{C}^{2\ell+1}$. The hydrogen atom spectrum is special in that it has a large amount of degeneracies: multiplets with different values of ℓ but the same value of n are degenerate.

For a general central potential, the only degeneracies are those within ℓ multiplets. The spectrum of bound states can *always* be organized by angular momentum, and we find that

$$\mathcal{H} = \bigoplus_{\ell=0}^{\infty} \hat{\mathcal{H}}_\ell. \tag{13.2.13}$$

Each $\hat{\mathcal{H}}_\ell$ is a collection of many, perhaps infinitely many, multiplets of angular momentum ℓ. For the hydrogen atom, for example, $\hat{\mathcal{H}}_{\ell=0}$ contains states with $n=1,2,\ldots$. More generally, in hydrogen $\hat{\mathcal{H}}_\ell$ has states with principal quantum number $n>\ell$, for all such values of n. □

Example 13.10. *Review of index manipulations.*

Let us review and summarize the basic elements of index manipulation. For any vector $\mathbf{a}=(a_1,a_2,a_3)$, we denote the components with an index so that we have components a_i with index i running over the set $i=1,2,3$. With another vector $\mathbf{b}=(b_1,b_2,b_3)$, the dot product is written as

$$\mathbf{a}\cdot\mathbf{b}=a_1b_1+a_2b_2+a_3b_3=\sum_{i=1}^{3}a_ib_i. \tag{13.2.14}$$

It is a useful convention that repeated indices are summed over the values they run over. Thus, in the above, since we have the repeated i, we simply write

$$\mathbf{a}\cdot\mathbf{b}=a_i\,b_i. \tag{13.2.15}$$

A repeated index is sometimes called a *dummy* index, and the particular letter we use for it is immaterial: $a_ib_i=a_kb_k$, for example. In general, there should be no more than two indices with the same label in any expression. A useful symbol is the Kronecker delta δ_{ij}, symmetric in i and j and defined as

$$\delta_{ij}=\begin{cases}0, \text{ if } i\neq j,\\ 1, \text{ if } i=j.\end{cases} \tag{13.2.16}$$

Note that by the summation convention $\delta_{ii}=\delta_{11}+\delta_{22}+\delta_{33}=1+1+1=3$. Moreover, one often has to simplify $\delta_{ij}B_j$. In fact,

$$\delta_{ij}B_j=B_i. \tag{13.2.17}$$

This holds because as we sum over j, the Kronecker delta vanishes unless j is equal to i, in which case it equals one.

The other important object is the three-index Levi-Civita symbol ϵ_{ijk}. We encountered this object in describing the commutator of angular momentum operators: $[\hat{L}_i,\hat{L}_j]=i\hbar\,\epsilon_{ijk}\,\hat{L}_k$. Since each index can run over three values, in principle this object has $3\times3\times3=27$ values to be specified. But the ϵ symbol is defined as being totally antisymmetric, meaning antisymmetric under the exchange of any pair of indices: $\epsilon_{ijk}=-\epsilon_{jik}=-\epsilon_{ikj}=-\epsilon_{kji}$. This implies that in order for ϵ_{ijk} to be nonvanishing no two indices can have the same value, and therefore i,j,k must be some permutation of $1,2,3$. We declare that

$$\epsilon_{123}=1, \tag{13.2.18}$$

and this determines all other cases, such as, for example, $\epsilon_{312}=-\epsilon_{132}=+\epsilon_{123}=+1$. The ϵ symbol is often used to write cross products. As you know, the cross product has components

$$\mathbf{a}\times\mathbf{b} = \big(\, a_2b_3 - a_3b_2,\ a_3b_1 - a_1b_3,\ a_1b_2 - a_2b_1\big), \tag{13.2.19}$$

where the three objects in parentheses are the three components of $(\mathbf{a}\times\mathbf{b})$. We claim that

$$(\mathbf{a}\times\mathbf{b})_i = \epsilon_{ijk}\, a_j b_k. \tag{13.2.20}$$

We check one case (you do the others!). Take the first component:

$$(\mathbf{a}\times\mathbf{b})_1 = \epsilon_{1jk}a_jb_k = \epsilon_{123}a_2b_3 + \epsilon_{132}a_3b_2 = a_2b_3 - a_3b_2. \tag{13.2.21}$$

The product of two epsilon symbols with one common index satisfies a useful identity:

$$\epsilon_{ijk}\epsilon_{ipq} = \delta_{jp}\delta_{kq} - \delta_{jq}\delta_{kp}. \tag{13.2.22}$$

As a simple consistency check, you should verify that the right-hand side, just like the left-hand side, is antisymmetric under the exchange of j and k as well as under the exchange of p and q. A consequence of this identity is a formula for the product of two symbols with two summed indices. For this we set $p = j$, finding that

$$\epsilon_{ijk}\epsilon_{ijq} = \delta_{jj}\delta_{kq} - \delta_{jq}\delta_{kj} = 3\delta_{kq} - \delta_{kq} \quad\Rightarrow\quad \epsilon_{ijk}\epsilon_{ijq} \;=\; 2\delta_{kq}. \tag{13.2.23}$$

The classic application of the double epsilon identity is to the simplification of the double cross product $\mathbf{a}\times(\mathbf{b}\times\mathbf{c})$. To do this we calculate its ith component:

$$\begin{aligned}(\mathbf{a}\times(\mathbf{b}\times\mathbf{c}))_i &= \epsilon_{ijk}a_j(\mathbf{b}\times\mathbf{c})_k\\ &= \epsilon_{ijk}a_j\, \epsilon_{kpq}b_pc_q\\ &= \epsilon_{kij}\epsilon_{kpq}\, a_j\, b_p\, c_q\\ &= (\delta_{ip}\delta_{jq} - \delta_{iq}\delta_{jp})\, a_j\, b_p\, c_q\\ &= a_jb_ic_j - a_jb_jc_i = b_i(\mathbf{a}\cdot\mathbf{c}) - (\mathbf{a}\cdot\mathbf{b})c_i.\end{aligned} \tag{13.2.24}$$

From this it follows that

$$\mathbf{a}\times(\mathbf{b}\times\mathbf{c}) = \mathbf{b}(\mathbf{a}\cdot\mathbf{c}) - (\mathbf{a}\cdot\mathbf{b})\mathbf{c}. \tag{13.2.25}$$

Most identities of vector algebra can be derived using the above methods. □

Exercise 13.3. *Write* $(\mathbf{a}\times\mathbf{b})\cdot(\mathbf{c}\times\mathbf{d})$ *in terms of dot products only.*

13.3 Linear Operators

A linear map is a particular kind of function from one vector space V to another vector space W. When the linear map takes the vector space V to itself, we call the linear map a linear operator. We will focus our attention on these operators. In quantum mechanics linear operators produce the time evolution of states. Moreover, physical observables are associated with linear operators.

A **linear operator** T on a vector space V is a function that takes V to V with the following properties:

1. $T(u+v) = Tu + Tv$, for all $u, v \in V$.
2. $T(au) = aTu$, for all $a \in \mathbb{F}$ and $u \in V$.

In the above notation, Tu, for example, means the result of the action of the operator T on the vector u. It could also be written as $T(u)$, but it is simpler to write it as Tu, in a way that makes the action of T on u look "multiplicative."

A simple consequence of the axioms is that the action of a linear operator on the zero vector is the zero vector:

$$T0 = 0. \tag{13.3.1}$$

This follows from $Tu = T(u+0) = Tu + T0$ and canceling the common Tu term.

Let us consider a few examples of linear operators;

1. Let $V = \mathcal{P}[x]$ denote the space of real polynomials $p(x)$ of a real variable x with real coefficients. Here are two linear operators T and S on V:
   - Let T denote differentiation: $Tp = p'$ where $p' \equiv \frac{dp}{dx}$. This operator is linear because
   $$\begin{aligned} T(p_1 + p_2) = &\ (p_1 + p_2)' = p_1' + p_2' = Tp_1 + Tp_2, \\ T(ap) = &\ (ap)' = ap' = a\,Tp. \end{aligned} \tag{13.3.2}$$
   - Let S denote multiplication by x: $Sp = xp$. S is also a linear operator.
2. In the space $\mathbb{F}^\infty$ of infinite sequences, define the **left-shift** operator L by
   $$L(x_1, x_2, x_3, \ldots) = (x_2, x_3, \ldots). \tag{13.3.3}$$
   By shifting to the left, we lose the information about the first entry, but that is perfectly consistent with linearity. We also have the **right-shift** operator R that acts by shifting to the right and creating a new first entry as follows:
   $$R(x_1, x_2, \ldots) = (0, x_1, x_2, \ldots). \tag{13.3.4}$$
   The first entry after the action of R is zero. It could not be any other number because the zero element (a sequence of all zeroes) should be mapped to itself (by linearity).
3. For any vector space V, we define the **zero operator** 0 that, acting on any vector in V, maps it to the zero vector: $0v = 0$ for all $v \in V$. This map is very simple, almost trivial, but certainly linear. Note that now we have the zero number, the zero vector, and the zero operator, all denoted by the symbol 0.
4. For any vector space V, we define the **identity operator** $\mathbb{1}$ that leaves all vectors in V invariant: $\mathbb{1}v = v$ for all $v \in V$.

On any vector space V, there are many linear operators. We call $\mathcal{L}(V)$ the set of all linear operators on V. Since operators on V can be added and can also be multiplied by numbers, the set **$\mathcal{L}(V)$ is itself a vector space,** where the vectors are the operators. Indeed, for any two operators $S, T \in \mathcal{L}(V)$ we have the natural definition

$$\begin{aligned} (S+T)v &= Sv + Tv, \\ (aS)v &= a(Sv). \end{aligned} \tag{13.3.5}$$

A vector space must have an additive identity. Here it is an operator that can be added to other operators with no effect. The additive identity in the vector space $\mathcal{L}(V)$ is the zero operator on V, considered in (3) above.

In the vector space $\mathcal{L}(V)$, there is a surprising new structure: the vectors (the operators!) can be naturally multiplied. There is a **multiplication of linear operators** that gives a linear operator: we just let one operator act first and the other next! So given $S, T \in \mathcal{L}(V)$, we define the operator ST as

$$(ST)v \equiv S(Tv). \tag{13.3.6}$$

We easily verify linearity:

$$(ST)(u+v) = S(T(u+v)) = S(Tu+Tv) = S(Tu) + S(Tv) = (ST)(u) + (ST)(v), \tag{13.3.7}$$

and you can also verify that $(ST)(av) = a(ST)(v)$.

The product just introduced in the space of linear operators is **associative**. This is a fundamental property of operators and means that for S, T, U, linear operators

$$S(TU) = (ST)U. \tag{13.3.8}$$

This equality holds because acting on any vector v both the left-hand side and the right-hand side give $S(T(U(v)))$. The product has an identity element: the identity operator $\mathbb{1}$ of (4). If we have a product, we can ask if the elements (the operators) have inverses. As we will see later, some operators have inverses and some do not.

Finally, and crucially, this product is in general **noncommutative**. We can check this using the two operators T and S of (1), acting on the polynomial $p = x^n$. Since T differentiates and S multiplies by x, we get

$$\begin{aligned}(TS)x^n &= T(Sx^n) = T(x^{n+1}) = (n+1)x^n,\\ (ST)x^n &= S(Tx^n) = S(nx^{n-1}) = nx^n.\end{aligned} \tag{13.3.9}$$

We quantify the failure of commutativity by the difference $TS - ST$, which is itself a linear operator:

$$(TS - ST)x^n = (n+1)x^n - nx^n = x^n = \mathbb{1}\, x^n, \tag{13.3.10}$$

where we inserted the identity operator at the last step. Since this relation is true acting on x^n, for any $n \geq 0$, it holds by linearity acting on any polynomial—namely, on any element of the vector space. So we can simply write

$$[T, S] = \mathbb{1}, \tag{13.3.11}$$

where we introduced the **commutator** $[\cdot, \cdot]$ of two linear operators X, Y, defined by

$$[X, Y] \equiv XY - YX. \tag{13.3.12}$$

Exercise 13.4. *Calculate the commutator $[L, R]$ of the left-shift and right-shift operators. Express your answer using the identity operator and the operator P_1 defined by $P_1(x_1, x_2, \ldots) = (x_1, 0, 0, \ldots)$.*

Example 13.11. *Working with Pauli matrices.*

The Pauli matrices σ_i, with $i = 1, 2, 3$, or the associated spin operators $S_i = \frac{\hbar}{2}\sigma_i$ are indeed operators on $\mathbb{C}^2$, the vector space of spin states (see example 13.2). We should be able to manipulate these 2×2 matrices efficiently. We first recall their explicit form

$$\sigma_1 = \begin{pmatrix} 0 & 1 \\ 1 & 0 \end{pmatrix}, \quad \sigma_2 = \begin{pmatrix} 0 & -i \\ i & 0 \end{pmatrix}, \quad \sigma_3 = \begin{pmatrix} 1 & 0 \\ 0 & -1 \end{pmatrix}. \tag{13.3.13}$$

These matrices are Hermitian; in fact, together with the identity matrix they span the real vector space of 2 × 2 Hermitian matrices (example 13.7). They are also traceless:

$$\text{tr}\,\sigma_i = 0, \quad i = 1, 2, 3. \tag{13.3.14}$$

The Pauli matrices square to the identity matrix, as one can check explicitly:

$$(\sigma_1)^2 = (\sigma_2)^2 = (\sigma_3)^2 = \mathbb{1}. \tag{13.3.15}$$

This property implies that the eigenvalues of each of the Pauli matrices can only be plus or minus one. Indeed, the eigenvalues of a matrix satisfy the algebraic equation that the matrix satisfies. Thus, the eigenvalues must satisfy $\lambda^2 = 1$, showing that $\lambda = \pm 1$ are the only options. Since the sum of eigenvalues equals the trace, which is vanishing, each Pauli matrix has an eigenvalue $+1$ and an eigenvalue -1.

The commutation relations for the spin operators $[\hat{S}_i, \hat{S}_j] = i\hbar\,\epsilon_{ijk}\hat{S}_k$ together with $\hat{S}_i = \frac{\hbar}{2}\sigma_i$ imply that

$$[\sigma_i, \sigma_j] = 2i\,\epsilon_{ijk}\sigma_k. \tag{13.3.16}$$

Make sure never to confuse the imaginary number i with the index i. If you compute a commutator of Pauli matrices by hand, you might notice a curious property. Take the commutator $[\sigma_1, \sigma_2] = 2i\sigma_3$. If you do the matrix multiplications, you find that $\sigma_1\sigma_2 = i\sigma_3$ while $\sigma_2\sigma_1 = -i\sigma_3$. These two products differ by a sign:

$$\sigma_1\sigma_2 = -\sigma_2\sigma_1. \tag{13.3.17}$$

We say that σ_1 and σ_2 *anticommute*: they can be moved across each other at the cost of a sign. Just as we define the commutator of two operators X, Y by $[X, Y] \equiv XY - YX$, we define the **anticommutator**, denoted by curly brackets, by the following:

$$\text{anticommutator:} \quad \{X, Y\} \equiv XY + YX. \tag{13.3.18}$$

In this language we have checked that $\{\sigma_1, \sigma_2\} = 0$, and the property $\sigma_1^2 = \mathbb{1}$, for example, can be rewritten as $\{\sigma_1, \sigma_1\} = 2 \cdot \mathbb{1}$. In fact, you can check (by examining the two remaining cases) that any two different Pauli matrices anticommute:

$$\{\sigma_i, \sigma_j\} = 0, \quad \text{for } i \neq j. \tag{13.3.19}$$

We can improve this equation to make it also work when i is equal to j. We claim that

$$\{\sigma_i, \sigma_j\} = 2\delta_{ij}\,\mathbb{1}. \tag{13.3.20}$$

Indeed, when $i \neq j$ the right-hand side vanishes, as needed, and when i is equal to j, the right-hand side gives $2 \cdot \mathbb{1}$, which is also needed since the Pauli matrices square to the identity.

The commutator and anticommutator identities for the Pauli matrices can be summarized in a single equation. This is possible because for any two operators X, Y we have

$$XY = \tfrac{1}{2}\{X, Y\} + \tfrac{1}{2}[X, Y], \tag{13.3.21}$$

as you should confirm by expansion. Applied to the product of two Pauli matrices and using our expressions for the commutator and anticommutator, we get

$$\boxed{\sigma_i\sigma_j \;=\; \delta_{ij}\,\mathbb{1} \;+\; i\,\epsilon_{ijk}\,\sigma_k.} \tag{13.3.22}$$

Note that $\sigma_k\sigma_{k+1} = i\sigma_{k+2}$, where we use arithmetic modulo 3 in the subscripts ($4 \equiv 1$, $5 \equiv 2$). The equation for $\sigma_i\sigma_j$ can be recast in vector notation if we introduce the "vector" triplet of Pauli matrices:

$$\boldsymbol{\sigma} \;\equiv\; (\sigma_1, \sigma_2, \sigma_3). \tag{13.3.23}$$

We can construct a matrix by the dot product of a vector $\mathbf{a} = (a_1, a_2, a_3)$ with the "vector" $\boldsymbol{\sigma}$. Here the components a_i of $\mathbf{a}$ are assumed to be numbers. We define

$$\mathbf{a}\cdot\boldsymbol{\sigma} \;\equiv\; a_1\sigma_1 + a_2\sigma_2 + a_3\sigma_3 \;=\; a_i\sigma_i. \tag{13.3.24}$$

Note that $\mathbf{a}\cdot\boldsymbol{\sigma}$ is just a single 2×2 matrix. The components of $\mathbf{a}$, being numbers, commute with matrices, and this dot product is commutative: $\mathbf{a}\cdot\boldsymbol{\sigma} = \boldsymbol{\sigma}\cdot\mathbf{a}$. To rewrite (13.3.22) we multiply this equation by a_ib_j to get

$$\begin{aligned} a_i\sigma_i\, b_j\sigma_j \;&=\; a_ib_j\delta_{ij}\,\mathbb{1} \;+\; i\,(a_ib_j\epsilon_{ijk})\,\sigma_k \\ &=\; (\mathbf{a}\cdot\mathbf{b})\,\mathbb{1} \;+\; i\,(\mathbf{a}\times\mathbf{b})_k\,\sigma_k \end{aligned} \tag{13.3.25}$$

so that, finally, we get the matrix equation

$$(\mathbf{a}\cdot\boldsymbol{\sigma})(\mathbf{b}\cdot\boldsymbol{\sigma}) = (\mathbf{a}\cdot\mathbf{b})\,\mathbb{1} + i\,(\mathbf{a}\times\mathbf{b})\cdot\boldsymbol{\sigma}. \tag{13.3.26}$$

This equation holds even if the components of $\mathbf{a}$ and $\mathbf{b}$ are operators, provided the operators commute with the Pauli matrices, as is often the case in applications. Indeed, in deriving the above equation we never had to move any a_i across any b_j. As a simple application, we take $\mathbf{b} = \mathbf{a}$, with components of ordinary numbers. We then have $\mathbf{a}\cdot\mathbf{a} = |\mathbf{a}|^2$ as well as $\mathbf{a}\times\mathbf{a} = 0$, an equation that can fail when $\mathbf{a}$ has operator components. The above equation then gives

$$(\mathbf{a}\cdot\boldsymbol{\sigma})^2 = |\mathbf{a}|^2\,\mathbb{1}. \tag{13.3.27}$$

When $\mathbf{a}$ is a unit vector $\mathbf{n}$, this becomes

$$(\mathbf{n}\cdot\boldsymbol{\sigma})^2 = \mathbb{1}, \quad \mathbf{n}\cdot\mathbf{n} = 1. \tag{13.3.28}$$

Since $\mathbf{n}\cdot\boldsymbol{\sigma}$ is Hermitian (being a superposition of Pauli matrices with real coefficients) and traceless, it follows that $\mathbf{n}\cdot\boldsymbol{\sigma}$, just like any Pauli matrix, has eigenvalues ± 1. It thus follows that the spin operator $\hat{S}_\mathbf{n} = \frac{\hbar}{2}\mathbf{n}\cdot\boldsymbol{\sigma}$ has eigenvalues $\pm\frac{\hbar}{2}$. This was the reason we could think of $\hat{S}_\mathbf{n}$ as a spin operator in the direction of $\mathbf{n}$. □

Example 13.12. *Is there a linear operator that reverses the direction of all spin states?*

A simple way to define a linear operator on a vector space V is to define its action on a set of *basis* vectors of V. Once you know how the operator acts on the basis vectors, you know by linearity how it acts on arbitrary vectors. It is far more delicate to define a linear operator by stating how it acts on *every* vector in V. In that case one must check the consistency of the definition with linearity.

We ask if there is a linear operator T that reverses the direction of all spin states in $\mathbb{C}^2$. If it existed, it must take an arbitrary spin state $|\mathbf{n};+\rangle$ into the state $|\mathbf{n};-\rangle$, up to a constant. Let us test whether this is possible. If T reverses every spin state, it must send $|+\rangle$ to $|-\rangle$ and vice versa. Of course, in general, it can do this up to nonvanishing constants α and β to be determined:

$$T|+\rangle=\alpha|-\rangle, \quad T|-\rangle=\beta|+\rangle, \quad \alpha,\beta\in\mathbb{C}. \tag{13.3.29}$$

Let us test this on the spin states along the x-axis (see (12.3.14)): $|x;\pm\rangle=\frac{1}{\sqrt{2}}(|+\rangle\pm|-\rangle)$. Acting on the plus state with T,

$$T\,|x;+\rangle=\tfrac{1}{\sqrt{2}}(\alpha|-\rangle+\beta|+\rangle)=\tfrac{\beta}{\sqrt{2}}(|+\rangle+\tfrac{\alpha}{\beta}|-\rangle). \tag{13.3.30}$$

For the result to point along $|x;-\rangle$, we need $\alpha/\beta=-1$. Now consider spin states along the y-axis (see (12.3.14)): $|y;\pm\rangle=\frac{1}{\sqrt{2}}(|+\rangle\pm i|-\rangle)$. Acting on the plus state with T, we get

$$T\,|y;+\rangle=\tfrac{1}{\sqrt{2}}(\alpha|-\rangle+i\beta|+\rangle)=\tfrac{i\beta}{\sqrt{2}}(|+\rangle-i\tfrac{\alpha}{\beta}|-\rangle). \tag{13.3.31}$$

For the result to point along $|y;-\rangle$ this time, we need $\alpha/\beta=+1$. The inconsistent constraints on α/β demonstrate that we *cannot* build a linear operator T that reverses the directions of all spin states. There is a basic reason why this operator does not exist. As we will learn later, on a complex vector space any linear operator has at least one eigenvalue and one eigenvector. The eigenvalue cannot be zero since by definition T does not kill spin states. But a nonzero eigenvalue implies an eigenvector, thus a vector that acted by T is just multiplied by the eigenvalue. Such vector a does not change direction, showing no linear operator T can reverse all spin states. There is a map that flips all spin states, but it is not a linear operator. □

13.4 Null Space, Range, and Inverses of Operators

When we encounter a linear operator on a vector space, there are two questions we can ask to determine the most basic properties of the operator: What vectors are mapped to zero by the operator? What vectors in V are obtained from the action of T on V? The first question leads to the concept of null space, the second to the concept of range.

The **null space** or **kernel** of $T\in\mathcal{L}(V)$ is the subset of vectors in V that are mapped to zero by T:

$$\text{null}\,T \;=\; \{v\in V;\; Tv=0\}. \tag{13.4.1}$$

Actually, null T is a *subspace* of V. Indeed, the null space contains the zero vector and is clearly a closed set under addition and scalar multiplication.

A linear operator $T:V\to V$ is said to be **injective** if $Tu=Tv$, with $u,v\in V$, implies $u=v$. An injective map is called a *one-to-one* map because two different elements cannot be mapped to the same one (physicist Sean Carroll has suggested that a better name would be *two-to-two*, as injectivity really means that two different elements are mapped by T to two different elements). In fact, an operator is injective if and only if its null space vanishes:

$$T \text{ injective} \iff \text{null}\, T = 0. \tag{13.4.2}$$

To show this equivalence, we first prove that injectivity implies zero null space. Indeed, if $v \in \text{null}\, T$ then $Tv = T0$ (both sides are zero), and injectivity shows that $v = 0$, proving that $\text{null}\, T = 0$. In the other direction, zero null space means that $T(u - v) = 0$ implies $u - v = 0$ or, equivalently, that $Tu = Tv$ implies $u = v$. This is injectivity.

As mentioned above, it is also of interest to consider the elements of V of the form Tv. We define the **range** of T as the image of V under the map T:

$$\text{range}\, T = \{Tv;\ v \in V\}. \tag{13.4.3}$$

Actually, $\text{range}\, T$ is a *subspace* of V (try proving it!). A linear operator T is said to be **surjective** if $\text{range}\, T = V$. That is, for a surjective T the image of V under T is the complete V.

Example 13.13. *Left- and right-shift operators.*
Recall the action of the left- and right-shift operators on infinite sequences:

$$L(x_1, x_2, x_3, \ldots) = (x_2, x_3, \ldots), \qquad R(x_1, x_2, \ldots) = (0, x_1, x_2, \ldots). \tag{13.4.4}$$

We can immediately see that $\text{null}\, L = (x_1, 0, 0, \ldots)$. Being different from zero, L is not injective. But L is surjective because any sequence can be obtained by the action of L: $(x_1, x_2, \ldots) = L(x_0, x_1, x_2, \ldots)$ for arbitrary x_0. The null space of R is zero, and thus R is injective. R is not surjective: we cannot get any element whose first entry is nonzero. In summary:

$$\begin{aligned} &L: \ \text{not injective, surjective,} \\ &R: \ \text{injective,} \quad \text{not surjective.} \end{aligned} \tag{13.4.5}$$

We will consider further properties of these operators in example 13.15. □

Let us now consider finite-dimensional vector spaces. Since both the null space and the range of a linear operator $T : V \to V$ are themselves vector spaces, one can calculate their dimensions. The larger the null space of an operator, the more vectors are mapped to zero, and one would expect the range to be reduced accordingly. The smaller the null space, the larger we expect the range to be. This intuition is made precise by the *rank-nullity* theorem. For **any** linear operator on V, the sum of the dimensions of its null space and its range is equal to the dimension of the vector space V:

$$\boxed{\dim(\text{null}\, T) + \dim(\text{range}\, T) = \dim V.} \tag{13.4.6}$$

The dimension of the range of T is called the **rank** of T, thus the name rank-nullity theorem. We only sketch the main steps. Let $(e_1, \ldots, e_m)$ with $m = \dim(\text{null}\, T)$ be a basis for $\text{null}(T)$. This basis can be extended to a basis $(e_1, \ldots, e_m, f_1, \ldots, f_n)$ of the full vector space V, where $m + n = \dim V$. The final step consists in showing that the Tf_i form a basis for the range of T. This is done in two steps:

Exercise 13.5. *Show that the vectors $(Tf_1, \ldots, Tf_n)$ (i) span the range of T and (ii) are linearly independent.*

Remark: The rank-nullity theorem in its general form applies to linear maps that relate spaces of different dimensionality. If $T: V \to W$ is a linear map from a vector space V to a vector space W, then the range of T is a subspace of W, and the null space of T is a subspace of V. Nevertheless, one still has exactly (13.4.6). The result does not involve the dimensionality of the space W.

Example 13.14. *Null space and range of spin operators on $\mathbb{C}^2$.*
Let us consider curious linear combinations $\hat{S}_\pm$ of spin operators of a spin one-half particle:

$$\hat{S}_\pm = \hat{S}_x \pm i\hat{S}_y. \tag{13.4.7}$$

While both $\hat{S}_x$ and $\hat{S}_y$ are Hermitian, using i to form a linear combination does not preserve Hermiticity. In fact, $\hat{S}_\pm^\dagger = \hat{S}_\mp$; the operators are Hermitian conjugates of each other. It is instructive to find their matrix representatives

$$\hat{S}_\pm = \tfrac{\hbar}{2}(\sigma_x \pm i\sigma_y) = \tfrac{\hbar}{2}\left[\begin{pmatrix} 0 & 1 \\ 1 & 0 \end{pmatrix} \pm i \begin{pmatrix} 0 & -i \\ i & 0 \end{pmatrix}\right] = \tfrac{\hbar}{2}\begin{pmatrix} 0 & 1\pm 1 \\ 1\mp 1 & 0 \end{pmatrix}. \tag{13.4.8}$$

We thus have the two matrices

$$\hat{S}_+ = \hbar \begin{pmatrix} 0 & 1 \\ 0 & 0 \end{pmatrix}, \qquad \hat{S}_- = \hbar \begin{pmatrix} 0 & 0 \\ 1 & 0 \end{pmatrix}. \tag{13.4.9}$$

Let us focus on $\hat{S}_+$; the case of $\hat{S}_-$ is completely analogous. Consider the null space of $\hat{S}_+$ first:

$$\hbar \begin{pmatrix} 0 & 1 \\ 0 & 0 \end{pmatrix} \begin{pmatrix} c_1 \\ c_2 \end{pmatrix} = \begin{pmatrix} 0 \\ 0 \end{pmatrix} \quad \rightarrow \quad c_2 = 0 \tag{13.4.10}$$

so that the general vector in the null space is $\begin{pmatrix} c_1 \\ 0 \end{pmatrix}$ with $c_1 \in \mathbb{C}$. This is the state we call e_1, or $|+\rangle$, a spin state pointing in the positive z-direction. Therefore,

$$\text{null}\, \hat{S}_+ = \text{span} \begin{pmatrix} 1 \\ 0 \end{pmatrix} = \text{span}\, e_1 = \text{span}\, |+\rangle. \tag{13.4.11}$$

Following the notation used to sketch the derivation of the rank-nullity theorem, we have a basis (e_1, f_1) for $\mathbb{C}^2$ with $f_1 = \begin{pmatrix} 0 \\ 1 \end{pmatrix} = |-\rangle$. To find the range of $\hat{S}_+$, we let it act on a general vector:

$$\hbar \begin{pmatrix} 0 & 1 \\ 0 & 0 \end{pmatrix} \begin{pmatrix} c_1 \\ c_2 \end{pmatrix} = \hbar \begin{pmatrix} c_2 \\ 0 \end{pmatrix}, \tag{13.4.12}$$

and we conclude that

$$\text{range}\, \hat{S}_+ = \text{span}\, e_1 = \text{span}\, |+\rangle. \tag{13.4.13}$$

This may seem unusual as the range and null spaces are the same. But all is fine; both are one-dimensional, adding to total dimension two, as required by the rank-nullity theorem. The picture in terms of spin states is simple: acting on $|-\rangle$, the operator $\hat{S}_+$ gives $|+\rangle$, raising the $\hat{S}_z$ eigenvalue of the state. Acting on $|+\rangle$, the operator $\hat{S}_+$ gives zero: the eigenvalue of $\hat{S}_z$ cannot be raised anymore. Clearly, the range of $\hat{S}_+$ and its null space coincide: they are both equal to the span of $|+\rangle$.

Additionally, since $e_1 = \hat{S}_+ f_1$ we actually have range $\hat{S}_+ = \text{span}(\hat{S}_+ f_1)$, in accordance with the sketch of the argument that leads to the rank-nullity theorem. You can quickly confirm that the operator $\hat{S}_-$ lowers the value of $\hat{S}_z$ acting on $|+\rangle$ and kills $|-\rangle$. The range and null space of $\hat{S}_-$ are both the span of $|-\rangle$. The operators $\hat{S}_\pm$ are raising and lowering operators for spin angular momentum. They will be studied further in chapter 19. □

Since linear operators can be multiplied, given an operator we can ask if it has an inverse. It is interesting to consider the question in some detail, as there are some subtleties. To do so we have to discuss left inverses and right inverses.

Let $T \in \mathcal{L}(V)$ be a linear operator. The linear operator S is a **left inverse** for T if

$$ST = \mathbb{1}. \tag{13.4.14}$$

Namely, the product of S and T with S to the left of T is equal to the identity matrix. Analogously, the linear operator S' is a **right inverse** for T if

$$TS' = \mathbb{1}. \tag{13.4.15}$$

Namely, the product of T and S' with S' to the right of T is equal to the identity matrix. If both inverses exist, then they are actually equal. This is easily proven using the above defining relations and associativity of the product:

$$S' = \mathbb{1}S' = (ST)S' = S(TS') = S\mathbb{1} = S. \tag{13.4.16}$$

If both left and right inverses of T exist, then T is said to be **invertible.**

The left and right inverses are relevant for systems of linear equations. Assume we have an operator $T \in \mathcal{L}(V)$, a known vector $c \in V$, and an unknown vector $x \in V$ to be determined from the equation:

$$Tx = c. \tag{13.4.17}$$

Suppose all you have is a left inverse S for T. Then acting with S on the equation gives you $STx = Sc$ and therefore $x = Sc$. Have you solved the equation? Not quite. If you try to check that this is a solution, you fail! Indeed, inserting the value $x = Sc$ on the left-hand side of the equation gives TSc, which may not equal c because S is not known to be a right inverse. All we can say is that $x = Sc$ is the *only possible solution,* given that it follows from the equation but cannot be verified without further analysis. If all you have is a right inverse S', you can now check that $x = S'c$ does solve the equation! This time, however, there is no guarantee that the solution is unique. Indeed, if T has a null space, the solution is clearly not unique since any vector in the null space can be added to the solution to give another solution. Only if both left and right inverses exist are we guaranteed that a unique solution exists!

It is reasonable to ask when a linear operator $T \in \mathcal{L}(V)$ has a left inverse. Think of two pictures of V and T mapping elements from the first picture to elements in the second picture. A left inverse should map each element in the second picture back to the element it came from in the first picture. If T is not injective, two different elements in the first picture are sometimes mapped to the same element in the second picture. The inverse operator can at best map back to one element so it fails to act as an inverse for the other element. This complication is genuine. A left inverse for T exists if and only if T is injective:

$$T \text{ has a left inverse} \iff T \text{ is injective.} \tag{13.4.18}$$

The proof from left to right is easy. Assume $Tv_1 = Tv_2$. Then multiply from the left by the left inverse S, finding $v_1 = v_2$, which proves injectivity. To prove that injectivity implies a left inverse, we begin by considering a basis of V denoted by the collection of vectors $\{v_i\}$. We do not list the vectors because the space V could be infinite-dimensional. Then define

$$Tv_i = w_i \tag{13.4.19}$$

for all v_i's. One can use injectivity to show that the w_i's are linearly independent. Since the map T may not be surjective, the $\{w_i\}$ may not be a basis. They can be completed with a collection $\{y_k\}$ of vectors to form a basis for V. Then we define the action of S by stating how it acts on this basis:

$$\begin{aligned} Sw_i &= v_i, \\ Sy_k &= 0. \end{aligned} \tag{13.4.20}$$

We then verify that

$$ST\Big(\sum_i a_i v_i\Big) = S\Big(\sum_i a_i w_i\Big) = \sum_i a_i v_i, \tag{13.4.21}$$

showing that $ST = \mathbb{1}$ when acting on any element of V. Setting $Sy_k = 0$ is a natural option to define S fully, but the final verification did not make use of that choice.

For the existence of a right inverse S' of T, we need the operator T to be surjective:

$$T \text{ has a right inverse} \iff T \text{ is surjective.} \tag{13.4.22}$$

The necessity of surjectivity is quickly understood: if we have a right inverse, we have $TS'(v) = v$, or, equivalently, $T(S'v) = v$ for all $v \in V$. This says that any $v \in V$ is in the range of T. This is surjectivity of T. A more extended argument is needed to show that surjectivity implies the existence of a right inverse.

Since an operator is invertible if it has both a left and a right inverse, the two boxed results above imply that

$$T \in \mathcal{L}(V): \quad T \text{ is invertible} \iff T \text{ is injective and surjective.} \tag{13.4.23}$$

This is a completely general result, valid for infinite- and finite-dimensional vector spaces.

Example 13.15. *Inverses for the left- and right-shift operators.*

Recall the properties (13.4.5) of the left- and right-shift operators L and R. Since L is surjective, it must have a right inverse. Since R is injective, it must have a left inverse. The right inverse of L is actually R, and the left inverse of R is actually L. These two facts are encoded by the single equation

$$LR = \mathbb{1}, \tag{13.4.24}$$

which is easily confirmed:

$$LR(x_1, x_2, \ldots,) = L(0, x_1, x_2, \ldots) = (x_1, x_2, \ldots). \tag{13.4.25}$$

Neither L nor R is invertible. □

Example 13.16. *Right inverse for the annihilation operator $\hat{a}$.*

We described in example 13.8 the state space of the harmonic oscillator as a direct sum of one-dimensional spaces U_n spanned by the energy eigenstates φ_n. The operator $\hat{a}$ maps U_n to U_{n-1}, as it is after all the lowering operator. It does so via the relation $\hat{a}\varphi_n = \sqrt{n}\varphi_{n-1}$. It is clear that the operator $\hat{a}$ is not injective: its null space is U_0, the space spanned by the ground state. The operator $\hat{a}$, however, is surjective: the full state space is in the range of $\hat{a}$. This means that $\hat{a}$ has a right inverse S'. This inverse must increase the number of the state by one unit, but, as we will see, it is not the $\hat{a}^\dagger$ operator.

To find the right inverse S' of $\hat{a}$, we write $S'\varphi_n = s_n\varphi_{n+1}$ with s_n a constant to be determined. Then we demand that $\hat{a}S' = \mathbb{1}$ when acting on any U_n:

$$\hat{a}\, S'\varphi_n = s_n\hat{a}\varphi_{n+1} = s_n\sqrt{n+1}\,\varphi_n \quad \rightarrow \quad s_n = \frac{1}{\sqrt{n+1}}. \tag{13.4.26}$$

The right inverse S' of $\hat{a}$ is therefore

$$S'\varphi_n = \frac{1}{\sqrt{n+1}}\varphi_{n+1}, \quad \hat{a}\, S' = \mathbb{1}. \tag{13.4.27}$$

Recall that $\hat{a}^\dagger$ satisfies $\hat{a}^\dagger\varphi_n = c_n\varphi_{n+1}$, with $c_n = \sqrt{n+1}$. Therefore, S' is not equal to $\hat{a}^\dagger$. It is not even proportional to $\hat{a}^\dagger$ because s_n/c_n is n dependent. S' acts just like $\hat{a}^\dagger$ on the ground state but differs from the $\hat{a}^\dagger$ action by n-dependent constants on $U_{n>0}$. In fact, $S'\varphi_n = \frac{1}{n+1}\hat{a}^\dagger\varphi_n$. □

Exercise 13.6. *Explain why the creation operator $\hat{a}^\dagger$ on the state space of the harmonic oscillator is injective, and determine its left inverse S.*

If the vector space V is finite-dimensional, the results are simpler. Any injective operator is surjective, and any surjective operator is injective. Therefore, any injective operator or any surjective operator is also invertible. The following three properties are therefore completely equivalent for operators on finite-dimensional vector spaces:

$$\dim V = \text{finite:} \quad \boxed{T \text{ is invertible} \iff T \text{ is injective} \iff T \text{ is surjective.}} \tag{13.4.28}$$

Proving these results is accomplished with simple exercises.

13.5 Matrix Representation of Operators

To get an extra handle on linear operators, we sometimes represent them as matrices. This is actually a completely general statement: after choosing a basis on the vector space V, *any* linear operator $T \in \mathcal{L}(V)$ can be represented by a particular matrix. This representation carries *all* the information about the linear operator. The matrix form of the operator can be very useful for explicit computations. The only downside of matrix representations is that they depend on the chosen basis. On the upside, a clever choice of basis may result in a matrix representation of unusual simplicity, which can be quite valuable. Additionally, some quantities computed easily from the matrix representation of an operator do not depend on the choice of basis.

The **matrix representation** of a linear operator $T \in \mathcal{L}(V)$ is a matrix whose components $T_{ij}(\{v\})$ are read from the action of the operator T on each of the elements of a basis $(v_1, \ldots, v_n)$ of V. The notation $T_{ij}(\{v\})$ reflects the fact that the matrix components depend on the choice of basis. If the choice of basis is clear by the context, we simply write the matrix components as T_{ij}. The rule that defines the matrix is simple:

Rule: The jth column of the matrix T is the list of components of Tv_j when expanded along the basis.

$$\begin{pmatrix} \cdots & \cdots & T_{1j} & \cdots \\ \cdots & \cdots & T_{2j} & \cdots \\ \vdots & \vdots & \vdots & \vdots \\ \vdots & \vdots & T_{nj} & \vdots \end{pmatrix}, \qquad Tv_j = T_{1j}\, v_1 + T_{2j}\, v_2 + \cdots + T_{nj}\, v_n. \tag{13.5.1}$$

Example 13.17. *Matrix representation for an operator in* $\mathbb{R}^3$.
The action of T on some basis vectors (v_1, v_2, v_3) of $\mathbb{R}^3$ is given by

$$\begin{aligned} Tv_1 &= -v_1 + 7v_3, \\ Tv_2 &= 2v_1 + v_2 + 3v_3, \\ Tv_3 &= 6v_1 - 5v_2 + 8v_3. \end{aligned} \tag{13.5.2}$$

The matrix representation of T is then

$$\begin{pmatrix} -1 & 2 & 6 \\ 0 & 1 & -5 \\ 7 & 3 & 8 \end{pmatrix}. \tag{13.5.3}$$

This follows by direct application of the rule. From the action of T on v_1, for example, we have

$$Tv_1 = -v_1 + 7v_3 = -v_1 + 0 \cdot v_2 + 7v_3 = T_{11}v_1 + T_{21}v_2 + T_{31}v_3, \tag{13.5.4}$$

allowing us to read the first column of the matrix. The other columns follow similarly. □

The equation in (13.5.1) can be written more briefly as

$$\boxed{Tv_j = \sum_{i=1}^{n} T_{ij}\, v_i.} \tag{13.5.5}$$

We write the summation sign explicitly for clarity; it is sometimes omitted on account of the summation convention that states that repeated indices are understood to be summed over. While operators are represented by matrices, vectors in V are represented by **column vectors**: the entries on the column vector are the components of the vector along the basis vectors. For a vector $a \in V$,

$$a = a_1 v_1 + \cdots + a_n v_n \quad \longleftrightarrow \quad a = \begin{pmatrix} a_1 \\ \vdots \\ a_n \end{pmatrix}. \tag{13.5.6}$$

It is a simple consequence that the basis vector v_k is represented by a column vector of zeroes, with a one on the kth entry:

$$v_k = \begin{pmatrix} 0 \\ \vdots \\ 1 \\ \vdots \\ 0 \end{pmatrix} \leftarrow k\text{th}. \tag{13.5.7}$$

With this you can now see why our key definition $Tv_j = \sum_{i=1}^{n} T_{ij} v_i$ is consistent with the familiar rule for multiplication of a matrix times a vector:

$$Tv_j = \begin{pmatrix} T_{11} & \cdots & T_{1j} & \cdots & T_{1n} \\ T_{21} & \cdots & T_{2j} & \cdots & T_{2n} \\ \vdots & \vdots & \vdots & \vdots & \vdots \\ T_{n1} & \cdots & T_{nj} & \cdots & T_{nn} \end{pmatrix} \begin{pmatrix} 0 \\ \vdots \\ 1 \\ \vdots \\ 0 \end{pmatrix} j\text{th} = \begin{pmatrix} T_{1j} \\ T_{2j} \\ \vdots \\ T_{nj} \end{pmatrix} \tag{13.5.8}$$

$$= T_{1j} \begin{pmatrix} 1 \\ 0 \\ \vdots \\ 0 \end{pmatrix} + T_{2j} \begin{pmatrix} 0 \\ 1 \\ \vdots \\ 0 \end{pmatrix} + \cdots + T_{nj} \begin{pmatrix} 0 \\ 0 \\ \vdots \\ 1 \end{pmatrix} = T_{1j} v_1 + T_{2j} v_2 + \ldots + T_{nj} v_n.$$

Exercise 13.7. *Verify that in any basis the matrix representation of the identity operator is a diagonal matrix with an entry of one at each element of the diagonal and zero elsewhere.*

The rules for representations are not only consistent with the formula for multiplication of matrices times vectors; they actually *imply* this familiar formula. In fact, they also imply the famous rule for matrix multiplication. We discuss both of these now.

Consider vectors a, b that, expanded along the basis $(v_1, \ldots, v_n)$, read

$$a = a_1 v_1 + \cdots + a_n v_n,$$
$$b = b_1 v_1 + \cdots + b_n v_n. \tag{13.5.9}$$

Assume the vectors are related by the equation

$$b = Ta. \tag{13.5.10}$$

We want to see how this looks in terms of the representations of T, a, and b. We have

$$b = Ta = T\sum_j a_j v_j = \sum_j a_j Tv_j = \sum_{i,j} a_j T_{ij} v_i = \sum_i \Big(\sum_j T_{ij} a_j\Big) v_i. \tag{13.5.11}$$

The object in parentheses is the ith component of b:

$$b_i = \sum_j T_{ij} a_j. \tag{13.5.12}$$

This is how $b = Ta$ is represented; we see on the right-hand side the familiar product of the matrix for T and the column vector for a.

Let us now examine the product of two operators and their matrix representation. Consider the operator TS acting on v_j:

$$(TS)v_j = T(Sv_j) = T\sum_p S_{pj} v_p = \sum_p S_{pj}\, Tv_p = \sum_p S_{pj} \sum_i T_{ip} v_i \tag{13.5.13}$$

so that changing the order of the sums we find that

$$(TS)v_j = \sum_i \Big(\sum_p T_{ip} S_{pj}\Big) v_i. \tag{13.5.14}$$

Using the identification implicit in (13.5.5), we see that the object in parentheses is $(TS)_{ij}$, the i, j element of the matrix that represents TS. Therefore, we find that

$$(TS)_{ij} = \sum_p T_{ip} S_{pj}, \tag{13.5.15}$$

which is precisely the familiar formula for matrix multiplication. The matrix that represents TS is the product of the matrix that represents T and the matrix that represents S, in that order.

Changing basis and its effect on matrix representations While matrix representations are very useful for concrete visualization, they are basis dependent. It is a good idea to try to determine if there are quantities that can be calculated using a matrix representation that are, nevertheless, guaranteed to be basis independent. One such quantity is the **trace** of the matrix representation of a linear operator. The trace is the sum of the matrix elements on the diagonal. Remarkably, that sum is the same independent of the basis used. This allows us to speak of the trace of an *operator*. The **determinant** of a matrix representation is also basis independent. We can therefore speak of the determinant of an operator. We will prove the basis independence of the trace and the determinant once we learn how to relate matrix representations in different bases.

Let us then consider the effect of a change of basis on the matrix representation of an operator. Consider a vector space V and two sets of basis vectors: $(v_1, \ldots, v_n)$ and

$(u_1, \ldots, u_n)$. Consider then two linear operators $A, B \in \mathcal{L}(V)$ such that for any $i = 1, \ldots, n$, A acting on v_i gives u_i, and B acting on u_i gives v_i:

$$A: \begin{array}{ccc} v_1 & \ldots & v_n \\ \downarrow & \ldots & \downarrow, \\ u_1 & \ldots & u_n \end{array} \qquad B: \begin{array}{ccc} v_1 & \ldots & v_n \\ \uparrow & \ldots & \uparrow. \\ u_1 & \ldots & u_n \end{array} \tag{13.5.16}$$

This can also be written as

$$Av_k = u_k, \qquad Bu_k = v_k, \quad \text{for all } k \in \{1, \ldots, n\}. \tag{13.5.17}$$

These relations define the operators A and B completely: we have stated how they act on basis sets. We now verify the obvious: A and B are inverses of each other. Indeed,

$$\begin{aligned} BAv_k &= B(Av_k) = Bu_k = v_k, \\ ABu_k &= A(Bu_k) = Av_k = u_k, \end{aligned} \tag{13.5.18}$$

being valid for all k, shows that

$$BA = \mathbb{1} \quad \text{and} \quad AB = \mathbb{1}. \tag{13.5.19}$$

Thus, B is the inverse of A, and A is the inverse of B.

Operators like A or B that map one basis into another, vector by vector, have a remarkable property: *their matrix representations are the same in each of the bases they relate.* Let us prove this for A. By definition of matrix representations, we have

$$A\, v_k = \sum_i A_{ik}(\{v\}) v_i, \qquad \text{and} \qquad A\, u_k = \sum_i A_{ik}(\{u\}) u_i. \tag{13.5.20}$$

Since $u_k = Av_k$, we then have, acting with another A,

$$Au_k = A(Av_k) = A \sum_i A_{ik}(\{v\}) v_i = \sum_i A_{ik}(\{v\})\, Av_i = \sum_i A_{ik}(\{v\})\, u_i. \tag{13.5.21}$$

Comparison with the second equation immediately above yields the claimed

$$A_{ik}(\{u\}) = A_{ik}(\{v\}). \tag{13.5.22}$$

The same holds for the B operator. We can simply call A_{ij} and B_{ij} the matrices that represent A and B because these matrices are the same in the $\{v\}$ and $\{u\}$ bases, and these are the only bases at play here. On account of (13.5.19), these are matrix inverses:

$$B_{ij} A_{jk} = \delta_{ik} \quad \text{and} \quad A_{ij} B_{jk} = \delta_{ik}, \tag{13.5.23}$$

and we can write $B_{ij} = (A^{-1})_{ij}$. At this point we will use the convention that repeated indices are summed over to avoid clutter.

We can now apply these preparatory results to the matrix representations of the operator T. We have, by definition,

$$Tv_k = T_{ik}(\{v\})\, v_i. \tag{13.5.24}$$

We need to calculate Tu_k in order to read the matrix representation of T on the u basis:

$$Tu_k = T_{ik}(\{u\})\, u_i. \tag{13.5.25}$$

Computing the left-hand side, using the linearity of the operator T, we have

$$Tu_k = T(Av_k) = T(A_{jk}v_j) = A_{jk}Tv_j = A_{jk}\, T_{pj}(\{v\})\, v_p. \tag{13.5.26}$$

We need to express the rightmost v_p in terms of u vectors. For this,

$$v_p = Bu_p = B_{ip}\, u_i = (A^{-1})_{ip}\, u_i \tag{13.5.27}$$

so that

$$Tu_k = A_{jk}\, T_{pj}(\{v\})\, (A^{-1})_{ip}\, u_i \;=\; (A^{-1})_{ip}\, T_{pj}(\{v\})\, A_{jk}\, u_i, \tag{13.5.28}$$

where we reordered the matrix elements to clarify the matrix products. All in all,

$$Tu_k \;=\; \big(A^{-1}T(\{v\})A\big)_{ik}\, u_i, \tag{13.5.29}$$

which, comparing with (13.5.25), allows us to read

$$T_{ij}(\{u\}) \;=\; \big(A^{-1}T(\{v\})A\big)_{ij}. \tag{13.5.30}$$

Omitting the indices, this matrix relation is written as

$$\boxed{T(\{u\}) \;=\; A^{-1}T(\{v\})\,A, \quad \text{when} \quad u_i = A\,v_i.} \tag{13.5.31}$$

Note that in the first relation A stands for a matrix, but in $u_i = Av_i$ it stands for an operator. This is the result we wanted to obtain. In general, if two matrices R, S are related by $S = M^{-1}RM$ for some matrix M, we say that S is obtained from R by a similarity transformation generated by M. In this language the matrix representation $T(\{u\})$ is obtained from the matrix representation $T(\{v\})$ by a similarity transformation generated by the matrix A that represents the operator that changes the basis from $\{v\}$ to $\{u\}$.

The trace of a matrix is equal to the sum of its diagonal entries. Thus, the trace of T is given by T_{ii}, with the sum over i understood. The trace is cyclic when acting on the product of various matrices:

$$\text{tr}(S_1S_2\ldots S_{k-1}S_k) = \text{tr}(S_kS_1S_2\ldots S_{k-1}). \tag{13.5.32}$$

This result is a simple consequence of $\text{tr}(S_1S_2) = \text{tr}(S_2S_1)$, which is easily verified by writing out the explicit products and taking the traces. With the help of (13.5.31) and the cyclicity of the trace, the basis independence of the trace follows quickly:

$$\text{tr}(T(\{u\})) \;=\; \text{tr}(A^{-1}T(\{v\})A) \;=\; \text{tr}(AA^{-1}T(\{v\})) \;=\; \text{tr}(T(\{v\})). \tag{13.5.33}$$

For the determinant we recall that

$$\det(S_1S_2) = (\det S_1)(\det S_2). \tag{13.5.34}$$

This means that $\det(S)\det(S^{-1}) = 1$ and that the determinant of the product of multiple matrices is also the product of determinants. From (13.5.31) we then find that

$$\det T(\{u\}) \;=\; \det(A^{-1}T(\{v\})A) \;=\; \det(A^{-1})\det T(\{v\})\det A \;=\; \det T(\{v\}), \tag{13.5.35}$$

showing that the determinant of the matrix that represents a linear operator is independent of the chosen basis.

Example 13.18. *Matrix representation for the harmonic oscillator $\hat{a}$ and $\hat{a}^\dagger$ operators.*

The one-dimensional simple harmonic oscillator state space is infinite-dimensional, and an *orthonormal* basis $\{e_1, e_2, \ldots\}$ is provided by the infinite set of nondegenerate energy eigenstates:

$$\{e_1, e_2, e_3, \ldots\} = \{\varphi_0, \varphi_1, \varphi_2, \ldots\}, \tag{13.5.36}$$

where, as explained in section 9.4,

$$\varphi_n = \frac{1}{\sqrt{n!}}(\hat{a}^\dagger)^n \varphi_0, \tag{13.5.37}$$

and φ_0 is the ground state wave function. Moreover, recalling the basic commutation relation $[\hat{a}, \hat{a}^\dagger] = 1$, you can quickly recheck the result (9.4.26):

$$\begin{aligned} \hat{a}\varphi_n &= \sqrt{n}\,\varphi_{n-1}, \\ \hat{a}^\dagger \varphi_n &= \sqrt{n+1}\,\varphi_{n+1}. \end{aligned} \tag{13.5.38}$$

Given that $e_n = \varphi_{n-1}$ for $n = 1, \ldots$, the above relations become

$$\begin{aligned} \hat{a}\, e_n &= \sqrt{n-1}\, e_{n-1}, \\ \hat{a}^\dagger e_n &= \sqrt{n}\, e_{n+1}. \end{aligned} \tag{13.5.39}$$

The matrix representations now follow from the definition (13.5.5). Letting $\{\hat{a}\}_{mn}$ and $\{\hat{a}^\dagger\}_{mn}$ denote, respectively, the matrix elements of $\hat{a}$ and $\hat{a}^\dagger$, we have

$$\begin{aligned} \hat{a}\, e_n &= \sum_m \{\hat{a}\}_{mn} e_m = \sqrt{n-1}\, e_{n-1} \quad \Rightarrow \quad \{\hat{a}\}_{mn} = \sqrt{m}\,\delta_{m,n-1}, \\ \hat{a}^\dagger\, e_n &= \sum_m \{\hat{a}^\dagger\}_{mn}\, e_m = \sqrt{n}\; e_{n+1} \quad \Rightarrow \quad \{\hat{a}^\dagger\}_{mn} = \sqrt{n}\,\delta_{m,n+1}. \end{aligned} \tag{13.5.40}$$

The matrix $\{\hat{a}\}_{mn}$ is upper diagonal: the Kronecker delta $\delta_{m,n-1}$ tells us that for any row index m, the value of the column index n must be one unit higher. Similarly, the matrix $\{\hat{a}^\dagger\}_{mn}$ is lower diagonal. We have, explicitly,

$$\{\hat{a}\} = \begin{pmatrix} 0 & 1 & 0 & 0 & \cdots \\ 0 & 0 & \sqrt{2} & 0 & \cdots \\ 0 & 0 & 0 & \sqrt{3} & \cdots \\ 0 & 0 & 0 & 0 & \cdots \\ \vdots & \vdots & \vdots & \vdots & \vdots \end{pmatrix}, \quad \{\hat{a}^\dagger\} = \begin{pmatrix} 0 & 0 & 0 & 0 & \cdots \\ 1 & 0 & 0 & 0 & \cdots \\ 0 & \sqrt{2} & 0 & 0 & \cdots \\ 0 & 0 & \sqrt{3} & 0 & \cdots \\ \vdots & \vdots & \vdots & \vdots & \vdots \end{pmatrix}. \tag{13.5.41}$$

The operators $\hat{a}$ and $\hat{a}^\dagger$ are Hermitian conjugates of each other. Their matrices are also Hermitian conjugates of each other: they are related by transposition and complex conjugation. In fact, being real, the matrices are just related by transposition. A Hermitian matrix is one left invariant by transposition and complex conjugation. As we will show in section 14.4, the matrix representation of a Hermitian operator is Hermitian when we use an orthonormal basis. The above representations imply that the number operator $\hat{N} = \hat{a}^\dagger \hat{a}$ is represented as the infinite diagonal matrix $\{\hat{N}\} = \text{diag}(0, 1, 2, 3, \cdots)$. This is consistent with $\hat{N}\varphi_n = n\varphi_n$. □

Exercise 13.8. *In example 13.16 and exercise 13.6, you found the right inverse for $\hat{a}$ and the left inverse for $\hat{a}^\dagger$, respectively. Write down the corresponding matrices as in (13.5.41).*

13.6 Eigenvalues and Eigenvectors

In quantum mechanics we need to consider the eigenvalues and eigenstates of Hermitian operators acting on complex vector spaces. These operators are called observables, and their eigenvalues represent possible results of a measurement. In order to acquire a better perspective on these matters, we consider the eigenvalue/eigenvector problem more generally.

One way to understand the action of an operator $T \in \mathcal{L}(V)$ on a vector space V is to describe how it acts on subspaces of V. Let U denote a subspace of V. In general, the action of T may take elements of U outside U. We have a noteworthy situation if T acting on any element of U gives an element of U. In this case U is said to be **invariant under** T, and T is then a well-defined linear operator on U. A very interesting situation arises if a suitable list of invariant subspaces builds up the space V as a direct sum.

Of all subspaces, one-dimensional subspaces are the simplest. Given some nonzero vector $u \in V$, one can consider the one-dimensional subspace U spanned by u:

$$U = \{cu\colon\ c \in \mathbb{F}\}. \tag{13.6.1}$$

We can ask if the one-dimensional subspace U is invariant under T. For this Tu must be equal to a number times u, as this guarantees that $Tu \in U$. Calling the number λ, we write

$$T u = \lambda u. \tag{13.6.2}$$

This equation is so ubiquitous that names have been invented to label the objects involved. The number $\lambda \in \mathbb{F}$ is called an **eigenvalue** of the linear operator T *if there is a nonzero vector* $u \in V$ such that the equation above is satisfied. It is convenient to call any vector that satisfies (13.6.2) for a given λ an **eigenvector** of T corresponding to λ. In doing so we are including the zero vector as a solution and thus as an eigenvector. Note that with these definitions, having an eigenvalue *means* having associated eigenvectors.

Suppose we find for some specific λ a nonzero vector u satisfying (13.6.2). Then it follows that cu, for any $c \in \mathbb{F}$, also satisfies the equation so that the solution space of the equation includes the subspace U spanned by u, which is an invariant subspace under T.

It can often happen that for a given λ there are several linearly independent eigenvectors. In this case we say that the eigenvalue λ is **degenerate.** The full invariant subspace associated with a degenerate eigenvalue is higher dimensional, and it is spanned by a maximal set of linearly independent eigenvectors; the dimension of this space is called the **geometric multiplicity** of the eigenvalue. The set of eigenvalues of T is called the **spectrum** of T.

Equation (13.6.2) is equivalent to

$$(T - \lambda \mathbb{1})\, u = 0, \tag{13.6.3}$$

for some nonzero u, so that $T - \lambda \mathbb{1}$ has a nonzero null space and is therefore not injective and not invertible:

$$\boxed{\lambda \text{ is an eigenvalue} \iff (T - \lambda \mathbb{1}) \text{ is not injective nor invertible.}} \tag{13.6.4}$$

We also note that

$$\boxed{\text{eigenvectors of } T \text{ with eigenvalue } \lambda = \text{null}\,(T - \lambda \mathbb{1}).} \tag{13.6.5}$$

The null space of T is simply the subspace of eigenvectors of T with eigenvalue zero.

It should be noted that the eigenvalues of T and the associated invariant subspaces of eigenvectors are basis independent objects: nowhere in our discussion did we have to invoke the use of a basis. Below, we will review the familiar calculation of eigenvalues and eigenvectors using a matrix representation of the operator T.

Example 13.19. *Rotation operator in three dimensions.*
Take a real three-dimensional vector space V (our space to great accuracy!). Consider the rotation operator T that rotates all vectors by a fixed angle about the z-axis. To find eigenvalues and eigenvectors, we just think of the invariant subspaces. We must ask: Which vectors do not change their direction as a result of this rotation? Only the vectors along the z-direction satisfy this condition. So the vector space spanned by $\mathbf{e}_z$ is the invariant subspace, or the space of eigenvectors. The eigenvectors are associated with the eigenvalue $\lambda = 1$ since the vectors are not scaled by the rotation. □

Example 13.20. *Rotation operator in two dimensions.*
Now consider the case where T is a rotation by ninety degrees on a two-dimensional *real* vector space V. Are there one-dimensional subspaces invariant under T? No, *all* vectors are rotated; none is left invariant. Thus, there are *no eigenvalues* or, of course, eigenvectors. If you tried calculating the eigenvalues by the usual recipe, you would find complex numbers. A complex eigenvalue is meaningless in a real vector space. □

Although we will not prove the following result, it follows from the facts we have introduced and no extra machinery. It is of interest, being completely general and valid for both real and complex vector spaces:

Theorem 13.6.1. *Let $T \in \mathcal{L}(V)$, and assume $\lambda_1, \ldots, \lambda_n$ are distinct eigenvalues of T and $u_1, \ldots, u_n$ are corresponding nonzero eigenvectors. Then $(u_1, \ldots, u_n)$ are linearly independent.*

Comments: Note that we cannot ask whether the eigenvectors are orthogonal to each other as we have not yet introduced an inner product on the vector space V. There may be more than one linearly independent eigenvector associated with some eigenvalues. In that case any one eigenvector will do. Since an n-dimensional vector space V does not have more than n linearly independent vectors, the theorem implies that no linear operator on V can have more than n distinct eigenvalues.

We saw that some linear operators in real vector spaces can fail to have eigenvalues. Complex vector spaces are nicer:

Theorem 13.6.2. *Every linear operator on a finite-dimensional complex vector space has at least one eigenvalue.*

Proof. This claim means there is at least a one-dimensional invariant subspace spanned by a nonzero eigenvector. The above theorem is a fundamental result, provable with simple tools. The key idea is to consider an arbitrary nonzero vector v in the n-dimensional vector space V and to build the list $(v, Tv, T^2v, \ldots, T^nv)$ of $n+1$ vectors. If some entry different from the first vanishes, it would mean that zero is an eigenvalue of T and that the claim holds. Indeed, if $T^kv=0$ is the first term in the list that vanishes, it means $T^{k-1}v$ is a T eigenvector with eigenvalue zero. Assume now that none of the vectors in the list vanishes. In that case, with the list longer than the dimension of V, the vectors are not linearly dependent, and there is some linear relation of the form

$$a_0v+a_1Tv+\cdots+a_nT^nv=0, \tag{13.6.6}$$

which is satisfied with some set of coefficients $a_i \in \mathbb{C}$, not all of them zero. Let m be the highest value for which $a_m \neq 0$ in this relation. It follows that

$$a_0v+a_1Tv+\cdots+a_mT^mv=0, \tag{13.6.7}$$

with $m \leq n$ but also $m \geq 1$ since $v \neq 0$. This equation can be written as

$$(a_0\mathbb{1}+a_1T+\cdots+a_mT^m)v=0. \tag{13.6.8}$$

Inspired by this expression, consider the polynomial $p(z)$ defined by

$$p(z)=a_0+a_1z+\cdots a_mz^m. \tag{13.6.9}$$

By the fundamental theorem of algebra, any degree m polynomial over the complex numbers can be written as the product of m linear factors:

$$p(z)=a_0+a_1z+\cdots a_mz^m=a_m(z-\lambda_1)\cdots(z-\lambda_m), \tag{13.6.10}$$

where the various λ_i may include repeated values. It follows that the above factorization applies to (13.6.8) and allows us to conclude that

$$a_m(T-\lambda_1\mathbb{1})\cdots(T-\lambda_m\mathbb{1})\,v=0. \tag{13.6.11}$$

If all the $(T-\lambda_i\mathbb{1})$ factors above were injective, the equality could not hold: the left-hand side would be nonzero. Therefore, at least one factor $(T-\lambda_k\mathbb{1})$ must fail to be injective. This shows λ_k is an eigenvalue of T. □

In order to efficiently find eigenvalues and eigenvectors, one usually considers determinants. When λ is an eigenvalue, $T-\lambda\mathbb{1}$ is not invertible, and in any basis, the matrix representative of $T-\lambda\mathbb{1}$ is noninvertible. A matrix is noninvertible if and only if it has zero determinant, therefore,

$$\boxed{\lambda \text{ is an eigenvalue} \iff \det(T-\lambda\mathbb{1})=0.} \tag{13.6.12}$$

In an n-dimensional vector space, the condition for λ to be an eigenvalue of T is

$$\det\begin{pmatrix} T_{11}-\lambda & T_{12} & \ldots & T_{1n} \\ T_{21} & T_{22}-\lambda & \ldots & T_{2n} \\ \vdots & \vdots & \vdots & \vdots \\ T_{n1} & T_{n2} & \ldots & T_{nn}-\lambda \end{pmatrix}=0. \tag{13.6.13}$$

The left-hand side, when computed and expanded out, is a polynomial $f(\lambda)$ in λ of degree n called the *characteristic polynomial*:

$$f(\lambda) = \det(T - \lambda \mathbb{1}) = (-\lambda)^n + b_{n-1}\lambda^{n-1} + \cdots + b_1\lambda + b_0, \tag{13.6.14}$$

where the b_i are constants calculable in terms of the T_{ij}'s. The equation

$$f(\lambda) = 0 \tag{13.6.15}$$

determines all eigenvalues. Over the complex numbers, the characteristic polynomial can be factorized, again, by the fundamental theorem of algebra,

$$f(\lambda) = (-1)^n(\lambda - \lambda_1)(\lambda - \lambda_2)\ldots(\lambda - \lambda_n). \tag{13.6.16}$$

The notation does not preclude the possibility that some of the λ_i's may be equal. The λ_i's are the eigenvalues, since they lead to $f(\lambda) = 0$ for $\lambda = \lambda_i$. Even when all factors in the characteristic polynomial are identical, we still have one eigenvalue. For any eigenvalue λ, the operator $(T - \lambda \mathbb{1})$ is not injective and thus has a nonvanishing null space. Any vector in this null space is an eigenvector with eigenvalue λ.

If all eigenvalues of T are different, the spectrum of T is said to be **nondegenerate.** If an eigenvalue λ_i appears k times, the characteristic polynomial includes the factor $(\lambda - \lambda_i)^k$, and λ_i is said to be a degenerate eigenvalue with **algebraic multiplicity** k. In general, the geometric multiplicity of an eigenvalue is less than or equal to the algebraic multiplicity, implying that the number of linearly independent eigenvectors with eigenvalue λ_i can be less than or equal to k. For Hermitian operators in complex vector spaces, however, the two multiplicities are the same.

Exercise 13.9. *Show that the matrix* $\begin{pmatrix} 1 & 0 \\ -1 & 1 \end{pmatrix}$ *has an eigenvalue with algebraic multiplicity two and geometric multiplicity one.*

Example 13.21. *Eigenvectors of the annihilation operator $\hat{a}$ of the harmonic oscillator.*

Assume ψ_λ is an eigenstate of $\hat{a}$ with eigenvalue λ:

$$\hat{a}\,\psi_\lambda = \lambda \psi_\lambda. \tag{13.6.17}$$

We wish to find the allowed values of λ and the corresponding expressions for the eigenvectors ψ_λ. We write a general ansatz for the state ψ_λ:

$$\psi_\lambda = \varphi_0 + \sum_{n=1}^{\infty} c_n \varphi_n. \tag{13.6.18}$$

In here we used the normalization ambiguity to scale the coefficient of the ground state to one. The ground state must be present—if not, the leading term of the state would be some φ_k with $k \geq 1$, and the action of $\hat{a}$ on the state would create a term $\sim \varphi_{k-1}$ not present in ψ_λ, making it impossible to satisfy the eigenvalue equation. Acting with $\hat{a}$ on the state, we have

$$\hat{a}\psi_\lambda = c_1\varphi_0 + \sum_{n=2}^{\infty} c_n\sqrt{n}\varphi_{n-1} = c_1\Big(\varphi_0 + \sum_{n=1}^{\infty} \frac{c_{n+1}}{c_1}\sqrt{n+1}\varphi_n\Big). \tag{13.6.19}$$

For the right-hand side to be equal to $\lambda\psi_\lambda$, we must have $c_1=\lambda$. This suggests that, in fact, we may get a solution of the eigenvalue equation for arbitrary $\lambda\in\mathbb{C}$. Writing $c_1=\lambda$, the above equation becomes

$$\hat{a}\psi_\lambda=\lambda\Big(\varphi_0+\sum_{n=1}^{\infty}\frac{c_{n+1}}{\lambda}\sqrt{n+1}\,\varphi_n\Big). \tag{13.6.20}$$

For the state inside the parentheses to be ψ_λ, as in (13.6.18), we must have

$$\frac{c_{n+1}}{\lambda}\sqrt{n+1}=c_n \quad\Rightarrow\quad c_{n+1}=\frac{\lambda}{\sqrt{n+1}}\,c_n, \quad n=1,2,\ldots. \tag{13.6.21}$$

With $c_1=\lambda$, the solution for any $n\geq 1$ is quickly checked to be

$$c_n=\frac{\lambda^n}{\sqrt{n!}}, \quad n\geq 1. \tag{13.6.22}$$

We can then rewrite the state ψ_λ in (13.6.18):

$$\psi_\lambda=\varphi_0+\sum_{n=1}^{\infty}\frac{\lambda^n}{\sqrt{n!}}\frac{(\hat{a}^\dagger)^n}{\sqrt{n!}}\varphi_0=\Big(1+\sum_{n=1}^{\infty}\frac{(\lambda\hat{a}^\dagger)^n}{n!}\Big)\varphi_0. \tag{13.6.23}$$

The factor in parentheses is in fact the expansion of an exponential. We can then write

$$\hat{a}\,\psi_\lambda=\lambda\psi_\lambda, \qquad \psi_\lambda=\exp(\lambda\,\hat{a}^\dagger)\,\varphi_0, \quad \lambda\in\mathbb{C}. \tag{13.6.24}$$

There is no quantization of the spectrum: all complex values of λ are allowed. The eigenvalues can be complex because $\hat{a}$ is not Hermitian. The states ψ_λ, called coherent states, were considered in problem 9.9 and will be studied in more detail in chapter 17. The ground state φ_0 is a coherent state ψ_0 with $\lambda=0$. □

Theorem 13.6.2, establishing that any linear operator on a finite-dimensional complex vector space has an eigenvalue, does not apply for $\hat{a}$ because the relevant vector space is infinite-dimensional. We were not guaranteed to find an eigenvalue, but in fact, we found an infinite number of them. This need not always be the case:

Exercise 13.10. *Explain why the creation operator $\hat{a}^\dagger$ has no eigenvalues in the harmonic oscillator state space.*

13.7 Functions of Linear Operators and Key Identities

We often find it necessary in quantum mechanics to construct linear operators starting from some other linear operators. This can be done using some ordinary functions. Consider, for example, the exponential function e^x of a single variable x. If we had a linear operator $M\in\mathcal{L}(V)$ in some vector space V, how could we define the exponential e^M? One simple way to do this is to consider the Taylor expansion of e^x about $x=0$,

$$e^x=\sum_{n=0}^{\infty}\frac{x^n}{n!}=1+x+\tfrac{1}{2!}x^2+\tfrac{1}{3!}x^3+\cdots, \tag{13.7.1}$$

and attempt to use this formula with x replaced by M. This leaves the question of what to do with the leading term in the above expansion, the number 1. The number 1 should be replaced by a linear operator that, just like 1, is a multiplicative identity. The clear choice is to replace 1 by the identity operator $\mathbb{1} \in \mathcal{L}(V)$. Thus, we set

$$e^M \equiv \mathbb{1} + M + \tfrac{1}{2!}M^2 + \tfrac{1}{3!}M^3 + \cdots = \sum_{n=0}^{\infty} \frac{M^n}{n!}, \tag{13.7.2}$$

with the last sum assuming that the zeroth power of M is the identity operator: $M^0 = \mathbb{1}$. In this definition $e^M \in \mathcal{L}(V)$ because each term in the sum that defines the exponential is a linear operator. If we have a set of basis vectors $\{e_i\}$ for V, we can speak of the matrix M_{ij} representing the linear operator M. The matrix $(e^M)_{ij}$ associated with the linear operator e^M is precisely given by the above definition, with M on the right-hand side set equal to the matrix for M, and $\mathbb{1}$ the identity matrix. Just as the Taylor expansion of e^x converges and is well defined for all x in the complex plane, the definition of e^M is well defined in that the right-hand side converges for any linear operator M in a finite-dimensional vector space.

Exercise 13.11. *Convince yourself that for a diagonal matrix $M = diag(m_1, \cdots, m_n)$, we have $e^M = diag(e^{m_1}, \cdots, e^{m_n})$.*

Properties that hold for functions $f(x)$ sometimes hold for $f(M)$. Consider, for example, the trivial commutativity property $xe^x = e^x x$. Since linear operators do not necessarily commute, one must ask if M and e^M commute. They do, as we can check using the definition above. We form the two products,

$$M\left(\mathbb{1} + M + \tfrac{1}{2!}M^2 + \tfrac{1}{3!}M^3 + \cdots\right) = \left(\mathbb{1} + M + \tfrac{1}{2!}M^2 + \tfrac{1}{3!}M^3 + \cdots\right)M, \tag{13.7.3}$$

and immediately see that multiplying out we get exactly the same terms. The general lesson here is simple: since M commutes with itself, a function $f(M)$ that involves only M and no other operator except the identity will commute with M.

We know that $e^x e^{-x} = 1$, and a clumsy but direct way to check it is by multiplying the series expansions of each:

$$e^x e^{-x} = \left(1 + x + \tfrac{1}{2!}x^2 + \tfrac{1}{3!}x^3 + \cdots\right)\left(1 - x + \tfrac{1}{2!}x^2 - \tfrac{1}{3!}x^3 + \cdots\right) = 1. \tag{13.7.4}$$

You can check a few terms to see that this holds or prove it in all generality using the sums that summarize the expansions, but you know this works. If this works for x, it also works when x is replaced by M, and therefore

$$e^M e^{-M} = \mathbb{1}. \tag{13.7.5}$$

This means e^M is an invertible linear operator, and its inverse is e^{-M}. This is not a trivial result because, in particular, one property of the exponential function that does not hold is the following. Given two arbitrary linear operators M_1 and M_2,

$$e^{M_1+M_2} \neq e^{M_1}\, e^{M_2} \neq e^{M_2}\, e^{M_1}. \tag{13.7.6}$$

The equalities hold if M_1 and M_2 commute: $[M_1, M_2]=0$. We will further discuss this point below.

Another useful property involves derivatives. We know that $\frac{d}{dt}e^{tx}=xe^{tx}$. The analogous property holds for matrices, as you should check:

Exercise 13.12. *Prove that for $M \in \mathcal{L}(V)$ we have*

$$\frac{d}{dt}e^{tM} = Me^{tM} = e^{tM}M. \tag{13.7.7}$$

Exercise 13.13. *Define $h(t)=e^{tM}e^{-tM}$, with $M \in \mathcal{L}(V)$ and t a real parameter taking values from zero to one. Calculate $\frac{dh}{dt}$ and use your result to show that $e^M e^{-M}=\mathbb{1}$.*

The Euler identity

$$e^{i\theta} = \cos\theta + i\sin\theta, \tag{13.7.8}$$

with θ real, is an inspiration for a set of useful operator identities. In particular, consider the linear operator $e^{iM\theta}$. We then have $e^{iM\theta}=\cos(\theta M)+i\sin(\theta M)$, where we would define the trigonometric functions via their Taylor expansions around zero argument. This is not very useful, however. Matters simplify greatly if the linear operator M satisfies $M^2=\mathbb{1}$. In fact, we claim that

$$\boxed{e^{iM\theta} = \mathbb{1}\cos\theta + iM\sin\theta, \quad \text{when } M^2=\mathbb{1}.} \tag{13.7.9}$$

You should prove this relation using Taylor expansions, but it is worthwhile noticing that the identity is quite plausible. Think of replacing the complex number i by a primed version $i'=iM$, an operator, so the object to be evaluated is $e^{i'\theta}$. Note that $(i')^2=-\mathbb{1}$, exactly analogous to $i^2=-1$. This suggests $e^{i'\theta}=\mathbb{1}\cos\theta+i'\sin\theta$, which is what we claim to have. Let us give a proof of (13.7.9) using a technique that is helpful in many other contexts. Define the operator

$$g(\theta) \equiv e^{iM\theta}, \tag{13.7.10}$$

viewed as a function of θ, and attempt to find a differential equation for it. Taking one derivative with respect to θ, denoted by a prime, we have

$$g'(\theta) = \frac{d}{d\theta}e^{iM\theta} = iMg(\theta). \tag{13.7.11}$$

Taking a second derivative, we have

$$g''(\theta) = iMg'(\theta) = (iM)(iM)g(\theta) = -M^2g(\theta) = -g(\theta), \tag{13.7.12}$$

using the assumed $M^2=\mathbb{1}$. The differential equation $g''(\theta)=-g(\theta)$ is the familiar equation with the solution formed by the arbitrary superposition of $\cos\theta$ and $\sin\theta$. Here, with g an operator, we have the solution

$$g(\theta) = A\cos\theta + B\sin\theta, \tag{13.7.13}$$

with A, B constant *operators* on V, meaning θ-independent operators. It follows from the definition (13.7.10) that $g(0)=\mathbb{1}$, implying that $A=\mathbb{1}$. Moreover, it follows from (13.7.11) that $g'(0) = iM\mathbb{1} = iM$, implying that $B = iM$. We then have $g(\theta) = \mathbb{1}\cos\theta + iM\sin\theta$, which is what we wanted to prove.

An important application of (13.7.9) allows us to exponentiate arbitrary linear combinations of Pauli matrices. We explore this in the following example.

Example 13.22. *Exponentials of linear combinations of Pauli matrices.*
Examples of matrices that square to the identity were considered in exercise 13.11, where we showed that the linear combination $\mathbf{n}\cdot\boldsymbol{\sigma}$ of Pauli matrices with $\mathbf{n}$ a unit vector satisfies $(\mathbf{n}\cdot\boldsymbol{\sigma})^2 = \mathbb{1}$ (see (13.3.28)). It thus follows that

$$e^{i\mathbf{n}\cdot\boldsymbol{\sigma}\,\theta} = \mathbb{1}\cos\theta + i\mathbf{n}\cdot\boldsymbol{\sigma}\sin\theta. \tag{13.7.14}$$

A more general version of the identity follows from (13.3.27): $(\mathbf{a}\cdot\boldsymbol{\sigma})^2 = a^2\,\mathbb{1}$, with $a \equiv |\mathbf{a}|$ the norm of the vector $\mathbf{a}$. We can then consider the exponential

$$e^{i\,\mathbf{a}\cdot\boldsymbol{\sigma}} = e^{i\,\hat{\mathbf{a}}\cdot\boldsymbol{\sigma}\,a}, \tag{13.7.15}$$

where we wrote $\mathbf{a} = \hat{\mathbf{a}}a$ with $\hat{\mathbf{a}}$ the unit vector along $\mathbf{a}$. Since $\hat{\mathbf{a}}\cdot\boldsymbol{\sigma}$ squares to the identity, we have

$$\boxed{e^{i\,\mathbf{a}\cdot\boldsymbol{\sigma}} = \mathbb{1}\cos a + i\,\hat{\mathbf{a}}\cdot\boldsymbol{\sigma}\,\sin a, \quad \mathbf{a} = \hat{\mathbf{a}}\,a.} \tag{13.7.16}$$

This kind of exponential will help us construct unitary rotation operators acting on spin states in section 14.7. □

Invertible operators act nicely on operators when they do so by *similarity*. Let A be an invertible operator. A similarity transformation $s_A\colon \mathcal{L}(V) \to \mathcal{L}(V)$ generated by A maps an arbitrary operator B as follows:

$$s_A : B \to s_A(B) = A^{-1}BA. \tag{13.7.17}$$

Similarity preserves a number of properties of the operator. It maps the identity to itself, and it maps a product of operators to the product of the images:

$$s_A(B_1B_2) = s_A(B_1)s_A(B_2), \tag{13.7.18}$$

as you can easily check. Moreover, matrix representations of an operator in different bases are related by similarity; see (13.5.31).

Exercise 13.14. *Let $|b\rangle$ be an eigenvector of B with eigenvalue b. Use this to find the associated eigenvector of $A^{-1}BA$. What is the corresponding eigenvalue?*

A particularly interesting similarity transformation is generated by the exponential e^{-A} of an arbitrary linear operator A. Recall that this matrix is invertible, with inverse e^{A}, so the following is a similarity transformation of the linear operator B:

$$e^{A}\,B\,e^{-A}. \tag{13.7.19}$$

The remarkable fact is that this operator can be simplified and written in terms of B and commutators of B with A. Indeed, the leading terms are easily seen by expanding the exponentials and multiplying out:

$$e^{A}\,B\,e^{-A} = (\mathbb{1} + A + \mathcal{O}(A^2))B(\mathbb{1} - A + \mathcal{O}(A^2)) = B + [A, B] + \mathcal{O}(A^2). \tag{13.7.20}$$

You will show in problem 13.4 that the complete answer, known as *Hadamard's lemma*, takes the form

$$e^A B e^{-A} = B + [A, B] + \frac{1}{2!}[A, [A, B]] + \frac{1}{3!}[A, [A, [A, B]]] + \cdots \tag{13.7.21}$$

where the coefficients are those of the exponential function: a term with k appearances of A has a coefficient $1/k!$. To write this equation more neatly and in closed form, it is convenient to define an operator acting on operators to give operators. We define the "adjoint action" operator ad_A that acts on operators X via the commutator:

$$\text{ad}_A(X) \equiv [A, X]. \tag{13.7.22}$$

With this notation, you will show that the complete version of (13.7.21) becomes

$$e^A B e^{-A} = e^{\text{ad}_A}(B). \tag{13.7.23}$$

Another useful commutator identity is the following:

$$[A, e^B] = [A, B]\, e^B, \quad \text{when} \quad [[A, B], B] = 0. \tag{13.7.24}$$

This identity is valid when the commutator $[A, B]$ is an operator that commutes with B. Often, $[A, B]$ is just a multiple of the identity. To prove this identity, consider

$$[A, e^B]e^{-B} = A - e^B A e^{-B} = A - e^{\text{ad}_B} A = A - (A + [B, A]) = [A, B], \tag{13.7.25}$$

since $[A, B]$ commuting with B implies that the expansion of $e^{\text{ad}_B} A$ truncates after two terms. Multiplying the above relation by e^B from the right gives the desired result. Note, finally, that the factor $[A, B]$ on the right-hand side of (13.7.24) can be moved to the right of e^B without changing the result. As an example of a computation that uses this result, consider the commutator $[\hat{a}, e^{\beta\hat{a}^\dagger\hat{a}^\dagger}]$. Since $[\hat{a}, \beta\hat{a}^\dagger\hat{a}^\dagger] = 2\beta\hat{a}^\dagger$, the commutator itself commutes with the exponent $\beta\hat{a}^\dagger\hat{a}^\dagger$, and the conditions for the identity hold. Thus, we have

$$[\hat{a}, e^{\beta\hat{a}^\dagger\hat{a}^\dagger}] = 2\beta\hat{a}^\dagger\, e^{\beta\hat{a}^\dagger\hat{a}^\dagger}. \tag{13.7.26}$$

We remarked before that the exponentials for general linear operators A and B do not multiply in a simple way: $e^A e^B \neq e^{A+B}$. The Baker-Campbell-Hausdorff (BCH) formula gives us a way to write $e^A e^B$ using only one exponential. It states that

$$e^A e^B = e^C, \quad \text{with } C = A + B + \tfrac{1}{2}[A, B] + \tfrac{1}{12}\big([A, [A, B]] + [B, [B, A]]\big) + \cdots. \tag{13.7.27}$$

The dots represent terms, all of which can be written as nested commutators. This is in fact a nontrivial result, something that is far from obvious if one computes the product of the expansions of e^A and e^B. Baker, Campbell, and Hausdorff realized this, but the first determination of all the terms in the series for C is due to Eugene Dynkin in 1947 (it is not a simple formula!). If $[A, B] = 0$, then we have the familiar result:

$$e^A e^B = e^{A+B}, \quad \text{if } [A, B] = 0. \tag{13.7.28}$$

If the commutator $[A,B]$ is nonzero but commutes with *both* A and B, then we have the simple result

$$e^A e^B = e^{A+B+\frac{1}{2}[A,B]}, \quad \text{if } [A,[A,B]] = [B,[A,B]] = 0. \tag{13.7.29}$$

Since $[A,B]$ commutes with $A+B$, the use of (13.7.28) implies that the exponent can be separated:

$$\boxed{e^A e^B = e^{A+B}\, e^{\frac{1}{2}[A,B]}, \quad \text{if } [A,[A,B]] = [B,[A,B]] = 0.} \tag{13.7.30}$$

This is a very useful formula, as it allows us to combine products of exponentials. You will derive it in problem 13.5. With $[A,B]$ typically equal to a number times the identity operator, the result is particularly simple. Note that $e^{c\mathbb{1}} = e^c\,\mathbb{1}$, so when $[A,B] = c\mathbb{1}$, the factor $e^{\frac{1}{2}[A,B]}$ above can be replaced by the number $e^{c/2}$. Moving the last exponential to the left-hand side, one can also view this formula as giving the expansion of the exponential of a sum of operators:

$$e^{A+B} = e^A e^B e^{-\frac{1}{2}[A,B]}, \quad \text{if } [A,[A,B]] = [B,[A,B]] = 0. \tag{13.7.31}$$

The position of the third exponential relative to $e^A e^B$ does not matter because, by assumption, it commutes with A and B and thus with e^A and e^B.

Example 13.23. *Five ways to do a computation.*

The purpose of this example is to show how a standard computation can be done in different ways and in this way illustrate the methods that are generically useful. The goal is simplification of the expression

$$e^{i\frac{\theta}{2}\sigma_3}\,\sigma_1\, e^{-i\frac{\theta}{2}\sigma_3}. \tag{13.7.32}$$

We will assume that you are familiar with the basic manipulations of Pauli matrices, as reviewed in examples 13.11 and 13.22.

1. Differential equation: Define the expression to be calculated as a function of θ, and find a differential equation for it by differentiation:

$$f(\theta) = e^{i\frac{\theta}{2}\sigma_3}\,\sigma_1\, e^{-i\frac{\theta}{2}\sigma_3}, \quad f(0) = \sigma_1. \tag{13.7.33}$$

   Differentiation with respect to θ, denoted by a prime, encounters two objects: the first exponential and the second exponential. It is a simple but crucial fact that because of the opposite signs in their exponents the derivatives end up producing a commutator with the operator in between:

$$f'(\theta) = e^{i\frac{\theta}{2}\sigma_3}\,[\tfrac{i}{2}\sigma_3, \sigma_1]\, e^{-i\frac{\theta}{2}\sigma_3}. \tag{13.7.34}$$

   In obtaining this, the derivative of the first exponential was written to the right of the exponential, and the derivative of the second exponential was written to the left of the exponential, using (13.7.7). The commutator is quickly evaluated: $[\frac{i}{2}\sigma_3, \sigma_1] = \frac{i}{2}\cdot 2i\sigma_2 = -\sigma_2$, and therefore we find that

$$f'(\theta) = -e^{i\frac{\theta}{2}\sigma_3}\,\sigma_2\, e^{-i\frac{\theta}{2}\sigma_3}, \quad f'(0) = -\sigma_2. \tag{13.7.35}$$

Taking another derivative, we get

$$f''(\theta) = -e^{i\frac{\theta}{2}\sigma_3}\,[\tfrac{i}{2}\sigma_3, \sigma_2]\,e^{-i\frac{\theta}{2}\sigma_3} = -e^{i\frac{\theta}{2}\sigma_3}\,\sigma_1\,e^{-i\frac{\theta}{2}\sigma_3} = -f(\theta). \tag{13.7.36}$$

The differential equation $f'' = -f$ has a solution $f = A\cos\theta + B\sin\theta$, with A and B constant operators. With the $\theta = 0$ initial conditions obtained above, we find $f = \sigma_1\cos\theta - \sigma_2\sin\theta$. In summary,

$$e^{i\frac{\theta}{2}\sigma_3}\,\sigma_1\,e^{-i\frac{\theta}{2}\sigma_3} = \sigma_1\cos\theta - \sigma_2\sin\theta. \tag{13.7.37}$$

2. Direct expansion of the exponentials. Each exponential can be expanded out using the Euler formula for matrices. In fact, we can use (13.7.14) with $\mathbf{n} = (0,0,1)$ and $\theta \to \theta/2$ to write

$$e^{i\frac{\theta}{2}\sigma_3}\,\sigma_1\,e^{-i\frac{\theta}{2}\sigma_3} = (\mathbb{1}\cos\tfrac{\theta}{2} + i\sigma_3\sin\tfrac{\theta}{2})\sigma_1\,(\mathbb{1}\cos\tfrac{\theta}{2} - i\sigma_3\sin\tfrac{\theta}{2}). \tag{13.7.38}$$

Multiplying the first two factors and recalling that $\sigma_3\sigma_1 = i\sigma_2$, we find that

$$e^{i\frac{\theta}{2}\sigma_3}\,\sigma_1\,e^{-i\frac{\theta}{2}\sigma_3} = (\sigma_1\cos\tfrac{\theta}{2} - \sigma_2\sin\tfrac{\theta}{2})(\mathbb{1}\cos\tfrac{\theta}{2} - i\sigma_3\sin\tfrac{\theta}{2}). \tag{13.7.39}$$

Multiplying out, you get

$$\begin{aligned} e^{i\frac{\theta}{2}\sigma_3}\,\sigma_1\,e^{-i\frac{\theta}{2}\sigma_3} &= \sigma_1(\cos^2\tfrac{\theta}{2} - \sin^2\tfrac{\theta}{2}) - \sigma_2\,2\sin\tfrac{\theta}{2}\cos\tfrac{\theta}{2} \\ &= \sigma_1\cos\theta - \sigma_2\sin\theta. \end{aligned} \tag{13.7.40}$$

3. Using explicit matrices. Since the usual representation of σ_3 is diagonal, the exponentials in question are immediately turned into matrices using exercise 13.11:

$$e^{i\frac{\theta}{2}\sigma_3} = \exp\begin{pmatrix} i\frac{\theta}{2} & 0 \\ 0 & -i\frac{\theta}{2}\end{pmatrix} = \begin{pmatrix} e^{i\frac{\theta}{2}} & 0 \\ 0 & e^{-i\frac{\theta}{2}}\end{pmatrix}. \tag{13.7.41}$$

We thus have

$$e^{i\frac{\theta}{2}\sigma_3}\,\sigma_1\,e^{-i\frac{\theta}{2}\sigma_3} = \begin{pmatrix} e^{i\frac{\theta}{2}} & 0 \\ 0 & e^{-i\frac{\theta}{2}}\end{pmatrix}\begin{pmatrix} 0 & 1 \\ 1 & 0\end{pmatrix}\begin{pmatrix} e^{-i\frac{\theta}{2}} & 0 \\ 0 & e^{i\frac{\theta}{2}}\end{pmatrix} = \begin{pmatrix} 0 & e^{i\theta} \\ e^{-i\theta} & 0\end{pmatrix}. \tag{13.7.42}$$

Using $e^{\pm i\theta} = \cos\theta \pm i\sin\theta$, the final matrix is recognized as $\sigma_1\cos\theta - \sigma_2\sin\theta$.

4. Hadamard's lemma: In the notation of this lemma, we are trying to compute

$$e^{i\frac{\theta}{2}\sigma_3}\,\sigma_1\,e^{-i\frac{\theta}{2}\sigma_3} = e^{\text{ad}_A}\sigma_1, \quad \text{with } A = i\tfrac{\theta}{2}\sigma_3. \tag{13.7.43}$$

Therefore, we need to compute the repeated action of ad_A on σ_1. We begin with

$$\begin{aligned} \text{ad}_A(\sigma_1) &= [i\tfrac{\theta}{2}\sigma_3, \sigma_1] = -\theta\sigma_2, \\ (\text{ad}_A)^2(\sigma_1) &= [i\tfrac{\theta}{2}\sigma_3, -\theta\sigma_2] = -\theta^2\sigma_1. \end{aligned} \tag{13.7.44}$$

The second equation can be used to immediately write the action of $(\text{ad}_A)^{2k}$ on σ_1:

$$\begin{aligned} (\text{ad}_A)^{2k}(\sigma_1) &= (-1)^k\theta^{2k}\,\sigma_1, \\ (\text{ad}_A)^{2k+1}(\sigma_1) &= -(-1)^k\theta^{2k+1}\sigma_2, \end{aligned} \tag{13.7.45}$$

with the second equation following from the first by one additional action of ad_A. Expanding the exponential e^{ad_A}, splitting the full sum into sums over even and odd terms, and using the above results, we get

$$
\begin{aligned}
e^{\mathrm{ad}_A}\sigma_1 &= \sum_{k=0}^{\infty}\frac{(\mathrm{ad}_A)^{2k}}{(2k)!}\sigma_1 + \sum_{k=0}^{\infty}\frac{(\mathrm{ad}_A)^{2k+1}}{(2k+1)!}\sigma_1 \\
&= \sum_{k=0}^{\infty}(-1)^k\frac{\theta^{2k}}{(2k)!}\sigma_1 - \sum_{k=0}^{\infty}(-1)^k\frac{\theta^{2k+1}}{(2k+1)!}\sigma_2 \\
&= \sigma_1\cos\theta_1 - \sigma_2\sin\theta_1,
\end{aligned}
\tag{13.7.46}
$$

recognizing the Taylor expansions of sine and cosine. This is indeed the answer.

5. Partial guesswork: Expand to first order in θ:

$$
\begin{aligned}
e^{i\frac{\theta}{2}\sigma_3}\,\sigma_1\,e^{-i\frac{\theta}{2}\sigma_3} &= \Big(1 + i\tfrac{\theta}{2}\sigma_3 + \mathcal{O}(\theta^2)\Big)\sigma_1\Big(1 - i\tfrac{\theta}{2}\sigma_3 + \mathcal{O}(\theta^2)\Big) \\
&= \sigma_1 + i\tfrac{\theta}{2}[\sigma_3,\sigma_1] + \mathcal{O}(\theta^2) \\
&= \sigma_1 - \theta\sigma_2 + \mathcal{O}(\theta^2).
\end{aligned}
\tag{13.7.47}
$$

At this point, and admittedly with some hindsight, one identifies θ as the small-angle expansion of $\sin\theta$ and the unit factor accompanying σ_1 as the small-angle expansion of $\cos\theta$, giving

$$
e^{i\frac{\theta}{2}\sigma_3}\,\sigma_1\,e^{-i\frac{\theta}{2}\sigma_3} = \sigma_1\cos\theta - \sigma_2\sin\theta. \tag{13.7.48}
$$

A consistency check is reassuring: the square of the left-hand side is the identity matrix since $\sigma_1^2 = \mathbb{1}$. You can also confirm that the right-hand side squared is the identity, recalling that the anticommutator of two different Pauli matrices vanishes. □

Problems

Problem 13.1. *Direct sum of vector subspaces.*

Consider the following statement: U_1, U_2, and W are subspaces of V and the following holds:

$$V = U_1 \oplus W \quad \text{and} \quad V = U_2 \oplus W.$$

Can you conclude that $U_1 = U_2$? Namely, are they the same subspace? If yes, prove it. If no, give a counterexample.

Problem 13.2. *Identities for commutators.*

1. A function $f(x)$ can be expanded in a power series in x as $f(x) = \sum_{n=0}^{\infty} f_n x^n$. Show that

$$[\hat{p}, f(\hat{x})] = \tfrac{\hbar}{i} f'(\hat{x}).$$

2. On the space of x-dependent functions $\psi(x)$, the operator $f(\hat{x})$ acts multiplicatively: $f(\hat{x})\psi(x) = f(x)\psi(x)$ and $\hat{p}$ acts as $\hat{p}\psi(x) = \frac{\hbar}{i}\frac{\partial\psi}{\partial x}$. Calculate $[\hat{p}, f(\hat{x})]$ by letting this commutator act on an arbitrary wave function.

Problem 13.3. *Useful operator identity and translations.*

Suppose that A and B are two operators that do not commute; $[A,B]\neq 0$. Assume, however, that $[A,B]$ commutes with A.

1. Prove that

$$e^A B e^{-A} = B + [A,B], \quad \text{if } [A,[A,B]] = 0.$$

   For this, define an operator-valued function $F(t)\equiv e^{tA}Be^{-tA}$. What is $F(0)$? Derive a differential equation for $F(t)$ and integrate it.

2. Let a be a real number and $\hat{p}$ be the momentum operator. Show that the unitary translation operator T_a defined by

$$T_a \equiv e^{-ia\hat{p}/\hbar},$$

   translates the position operator:

$$T_a^\dagger \hat{x}\, T_a = \hat{x} + a.$$

Problem 13.4. *Proof of the Hadamard lemma.*

The goal here is to prove the Hadamard lemma (13.7.21):

$$e^A B e^{-A} = B + [A,B] + \frac{1}{2!}[A,[A,B]] + \frac{1}{3!}[A,[A,[A,B]]] + \cdots. \quad (1)$$

1. Define $f(t)\equiv e^{tA}Be^{-tA}$, and calculate the first few derivatives of $f(t)$ evaluated at $t=0$. Then use Taylor expansions. Calculating explicitly the first three derivatives suffices to obtain (1).

2. Do this to all orders by finding the form of the $(n+1)$th term on the right-hand side of (1). To write the answer in a neat form, define the operator ad_A that acts on operators X to give operators via the commutator

$$\text{ad}_A(X) \;\equiv\; [A,X].$$

   Confirm that with this notation, the complete version of equation (1) becomes

$$e^A B e^{-A} = e^{\text{ad}_A}(B).$$

Problem 13.5. *Special case of the BCH formula.*

Consider two operators A and B, such that $[A,B]$ commutes both with A and with B. You will prove here the BCH identity (13.7.31):

$$e^{A+B} = e^B e^A e^{\frac{1}{2}[A,B]} = e^A e^B e^{-\frac{1}{2}[A,B]}. \quad (1)$$

For this purpose we will consider the operator-valued function of t:

$$G(t) \equiv e^{t(A+B)}e^{-tA}.$$

1. Show that

$$G^{-1}\frac{d}{dt}G(t) = B + [A,B]t.$$

2. Solve this differential equation for G. Verify that the answer can be put in the form

$$G(t) \;=\; G(0)\, e^{tB}\, e^{\frac{1}{2}[A,B]t^2}.$$

Consider $G(1)$ to obtain the first equality in equation (1). Relabel the operators to obtain the other equality.

Problem 13.6. *Translation operators.*

Consider the coordinate-space and momentum-space translation operators T_x and $\tilde{T}_p$, respectively:

$$T_x = \exp\Big(-\frac{i\hat{p}x}{\hbar}\Big), \quad \tilde{T}_p = \exp\Big(\frac{ip\hat{x}}{\hbar}\Big).$$

1. Verify that the above are translation operators by calculation of

$$T_x^\dagger \, \hat{x} \, T_x \quad \text{and} \quad \tilde{T}_p^\dagger \, \hat{p} \, T_p.$$

2. Since $\hat{x}$ and $\hat{p}$ do not commute, the translation operators T_x and $\tilde{T}_p$ do not generally commute. But they sometimes do! Compute the commutator

$$[\,T_x, \tilde{T}_p\,] = \ldots$$

You should find the CBH formula useful. What is the condition satisfied by x and p that guarantees that T_x and $\tilde{T}_p$ commute?

Problem 13.7. *Rotation of spin states.*

The operator $\hat{R}_{\mathbf{n}}(\alpha)$, with α real and $\mathbf{n}$ a unit vector, is defined by

$$\hat{R}_{\mathbf{n}}(\alpha) \equiv \exp\Big(-\frac{i\alpha\hat{S}_{\mathbf{n}}}{\hbar}\Big) = \exp\Big(-i\,\frac{\alpha}{2}\,\mathbf{n}\cdot\boldsymbol{\sigma}\Big),$$

where we noted that $\hat{S}_{\mathbf{n}} = \frac{\hbar}{2}\,\mathbf{n}\cdot\boldsymbol{\sigma}$. Using the exponential function identities, we have

$$\hat{R}_{\mathbf{n}}(\alpha) = \mathbb{1}\cos\tfrac{\alpha}{2} - i\,\boldsymbol{\sigma}\cdot\mathbf{n}\,\sin\tfrac{\alpha}{2}.$$

1. Verify by direct computation that $\hat{R}_{\mathbf{n}}(\alpha)$ is unitary: $\hat{R}_{\mathbf{n}}^\dagger(\alpha)\hat{R}_{\mathbf{n}}(\alpha) = \mathbb{1}$.
2. For brevity we write $\hat{R}_y(\alpha)$ for $\hat{R}_{\hat{\mathbf{e}}_y}(\alpha)$. Evaluate the operator

$$\hat{R}_y^\dagger(\alpha)\,\hat{S}_z\,\hat{R}_y(\alpha)$$

in terms of $\hat{S}_x$, $\hat{S}_y$, and $\hat{S}_z$.

3. Find the state obtained by acting with $\hat{R}_y(\alpha)$ on $|+\rangle$. For what operator is the resulting state an eigenstate with eigenvalue $\hbar/2$? Explain why we can think of $\hat{R}_y(\alpha)$ as a rotation operator.

We will prove in section 14.7 that for arbitrary $\mathbf{n}$ and α the operator $\hat{R}_{\mathbf{n}}(\alpha)$ rotates spin states by an angle α about an axis pointing along $\mathbf{n}$.

14 Inner Products, Adjoints, and Bra-kets

An inner product is an extra structure that can be added to a vector space. With an inner product, the Schwarz inequality holds, we are able to build orthonormal bases, and we can introduce orthogonal projectors. The inner product allows the definition of a new linear operator: the adjoint $T^\dagger$ of a linear operator T. In an orthonormal basis, the matrix for $T^\dagger$ is found by complex conjugation and transposition of the matrix for T. An operator is Hermitian if it is equal to its adjoint. Unitary operators are invertible operators that preserve the norm of vectors. A unitary operator U satisfies $UU^\dagger = U^\dagger U = \mathbb{1}$. We build unitary operators that rotate spin states. We introduce the bra-ket notation of Dirac, where kets represent states and bras are linear functionals on vectors. The notation affords some flexibility and simplifies a number of manipulations.

14.1 Inner Products

We have been able to go a long way without introducing any additional structure on vector spaces. We have been able to consider linear operators, matrix representations, traces, invariant subspaces, eigenvalues, and eigenvectors. It is now time to put some additional structure on the vector spaces. In this section we consider a function called an *inner product* that allows us to construct numbers from vectors. With inner products we can introduce orthonormal bases for vector spaces and the concept of orthogonal projectors. Inner products will also allow us to define the *adjoint* of an operator. With adjoints available, we can define self-adjoint operators, usually called Hermitian operators in physics. We can also define unitary operators. Some of these ideas were first presented in part I of this book, in less generality and depth.

An **inner product** on a vector space V over a field $\mathbb{F}$ ($\mathbb{R}$ or $\mathbb{C}$) is a machine that takes an *ordered* pair of elements of V—that is, two vectors—and yields a number in $\mathbb{F}$. A vector space with an inner product is called an *inner-product space*. In order to motivate the definition of an inner product, we first discuss real vector spaces and begin by recalling the way in which we associate a length to a vector in $\mathbb{R}^n$.

The length, or **norm** of a vector, is a real nonnegative number equal to zero if the vector is the zero vector. A vector $a=(a_1,\dots,a_n)$ in $\mathbb{R}^n$ has norm $\|a\|$ defined by

$$\|a\| = \sqrt{a_1^2+\cdots+a_n^2}. \tag{14.1.1}$$

Squaring this, we view $\|a\|^2$ as the *dot product* of a with a:

$$\|a\|^2 = a\cdot a = a_1^2+\cdots+a_n^2. \tag{14.1.2}$$

This suggests that the dot product of any two vectors $a, b\in\mathbb{R}^n$ is defined by

$$a\cdot b \equiv a_1b_1+\cdots+a_nb_n. \tag{14.1.3}$$

We now generalize the dot product on $\mathbb{R}^n$ to an inner product, also denoted with a dot, on real vector spaces. This inner product is required to satisfy the following properties:

1. $a\cdot a\geq 0$, for all vectors a.
2. $a\cdot a=0$ if and only if $a=0$.
3. $a\cdot(b_1+b_2) = a\cdot b_1 + a\cdot b_2$.
4. $a\cdot(\alpha\, b) = \alpha\, a\cdot b$, with $\alpha\in\mathbb{R}$ a number.
5. $a\cdot b = b\cdot a$.

Along with these axioms, the length or norm $\|a\|$ of a vector a is the positive or zero number defined by relation

$$\|a\|^2 = a\cdot a. \tag{14.1.4}$$

The third property above is additivity on the second entry. Because of the fifth property, commutativity, additivity also holds for the first entry.

These axioms are satisfied by the definition (14.1.3) but do not require it. A new dot product defined by $a\cdot b = c_1a_1b_1+\cdots+c_na_nb_n$, with $c_1,\dots,c_n$ positive constants, would do equally well! (Which axiom goes wrong if we take some c_n equal to zero?) It follows that any result we can prove with these axioms holds true not only for the conventional dot product but for many others as well.

Exercise 14.1. *For the standard, consistent inner product in three dimensions, we have $a\cdot b = \|a\|\,\|b\|\cos\theta_{ab}$ with θ_{ab} the angle between the two vectors a and b. Suppose you try to define a new inner product $*$ using half the angle in between the vectors, as in $a*b = \|a\|\,\|b\|\cos\frac{1}{2}\theta_{ab}$. Find some reason why this is inconsistent with the axioms (a possible answer will be discussed in example 14.2).*

The above axioms guarantee a fundamental result, the **Schwarz inequality**:

$$|a\cdot b| \leq \|a\|\,\|b\|. \tag{14.1.5}$$

On the left-hand side, the bars denote absolute value. If any of the two vectors is zero, the inequality is trivially satisfied. To prove the inequality, consider two nonzero vectors a and b and then examine the shortest vector joining a point on the line defined by the direction of b to the end of a (figure 14.1). This is the vector $a_\perp$, given by

$$a_\perp \equiv a - \frac{a\cdot b}{b\cdot b}\, b. \tag{14.1.6}$$

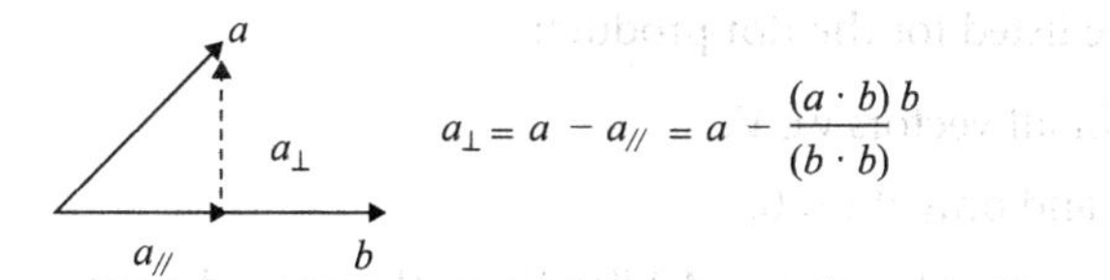

Figure 14.1
The Schwarz inequality follows from the statement that the vector $a_\perp$ must have a nonnegative norm.

The subscript $\perp$ indicates that the vector is perpendicular to b: $a_\perp \cdot b = 0$, as you can quickly see. To write the above vector, we subtracted from a the component of a parallel to b. Note that the vector $a_\perp$ is not changed as $b \to cb$ with c a constant; it does not depend on the overall length of b. Moreover, as it should, the vector $a_\perp$ is zero if and only if the vectors a and b are parallel.

The Schwarz inequality follows from $a_\perp \cdot a_\perp \geq 0$ as required by axiom (1). Using the explicit expression for $a_\perp$, a short computation gives

$$a_\perp \cdot a_\perp = a \cdot a - \frac{(a \cdot b)^2}{b \cdot b} \geq 0. \tag{14.1.7}$$

Since b is not the zero vector, we then have

$$(a \cdot b)^2 \leq (a \cdot a)(b \cdot b). \tag{14.1.8}$$

Taking the square root of this relation, we obtain the Schwarz inequality (14.1.5). The inequality becomes an equality only if $a_\perp = 0$ or, as discussed above, when $a = cb$ with c a real constant.

For complex vector spaces, some modifications are necessary. Recall that the length $|\gamma|$ of a complex number γ is given by $|\gamma| = \sqrt{\gamma^* \gamma}$, where the asterisk denotes complex conjugation. It is not hard to generalize this a bit. Let $z = (z_1, \ldots, z_n)$ be a vector in $\mathbb{C}^n$. Then the norm $\|z\|$ of z is a real number greater than or equal to zero defined by

$$\|z\|^2 \equiv z_1^* z_1 + \cdots + z_n^* z_n. \tag{14.1.9}$$

We must use complex conjugates to produce a real number greater than or equal to zero. The above suggests that for vectors $w = (w_1, \ldots, w_n)$ and $z = (z_1, \ldots, z_n)$ in $\mathbb{C}^n$, an inner product, denoted by $\langle \cdot, \cdot \rangle$, could be given by

$$\langle w, z \rangle = w_1^* z_1 + \cdots + w_n^* z_n. \tag{14.1.10}$$

Note that we are not treating the two vectors in a symmetric way. There is the first vector, in this case w, whose components are conjugated and a second vector z whose components are not conjugated. If the order of vectors is reversed, the new inner product is the complex conjugate of the original. The order of vectors matters for the inner product in complex vector spaces. We can, however, define an inner product in a way that applies both to complex and real vector spaces. Let us do this now.

An **inner product** on a vector space V over a field $\mathbb{F}$ is a map from an ordered pair (u, v) of vectors in V to a number $\langle u, v \rangle$ in the field. The axioms for $\langle u, v \rangle$ are inspired by the axioms we listed for the dot product:

1. $\langle v, v\rangle \geq 0$, for all vectors $v \in V$.
2. $\langle v, v\rangle = 0$ if and only if $v = 0$.
3. $\langle u, v_1 + v_2\rangle = \langle u, v_1\rangle + \langle u, v_2\rangle$. Additivity in the second entry.
4. $\langle u, \alpha v\rangle = \alpha \langle u, v\rangle$, with $\alpha \in \mathbb{F}$. Homogeneity in the second entry.
5. $\langle u, v\rangle = \langle v, u\rangle^*$. Conjugate exchange symmetry.

The **norm** $\|v\|$ of a vector $v \in V$ is defined by relation

$$\|v\|^2 = \langle v, v\rangle. \tag{14.1.11}$$

Comparing with the dot product axioms for real vector spaces, the key difference is in (5): the inner product in complex vector spaces is not symmetric. For the above axioms to apply to vector spaces over $\mathbb{R}$, we just define the obvious: complex conjugation of a real number is the same real number. On a real vector space, complex conjugation has no effect, and the inner product is strictly symmetric.

Let us make a few remarks. One can use axiom (3) with $v_2 = 0$ to show that for all $u \in V$,

$$\langle u, 0\rangle = 0 \quad \Rightarrow \quad \langle 0, u\rangle = 0, \tag{14.1.12}$$

where the second equation follows by axiom (5). Axioms (3) and (4) amount to full linearity in the second entry. It is important to note that additivity holds for the first entry as well:

$$\begin{aligned} \langle u_1 + u_2, v\rangle &= \langle v, u_1 + u_2\rangle^* \\ &= (\langle v, u_1\rangle + \langle v, u_2\rangle)^* \\ &= \langle v, u_1\rangle^* + \langle v, u_2\rangle^* \\ &= \langle u_1, v\rangle + \langle u_2, v\rangle. \end{aligned} \tag{14.1.13}$$

Homogeneity works differently on the first entry, however:

$$\langle \alpha u, v\rangle = \langle v, \alpha u\rangle^* = (\alpha\langle v, u\rangle)^* = \alpha^* \langle u, v\rangle. \tag{14.1.14}$$

In summary, we get linearity and *conjugate homogeneity* on the first entry:

$$\boxed{\begin{aligned} \langle u_1 + u_2, v\rangle &= \langle u_1, v\rangle + \langle u_2, v\rangle, \\ \langle \alpha u, v\rangle &= \alpha^* \langle u, v\rangle. \end{aligned}} \tag{14.1.15}$$

For a real vector space, conjugate homogeneity is just plain homogeneity.

Two vectors $u, v \in V$ are said to be **orthogonal** if $\langle u, v\rangle = 0$. This, of course, means that $\langle v, u\rangle = 0$ as well. The zero vector is orthogonal to all vectors, including itself.

The inner product we have defined is **nondegenerate:** any vector orthogonal to all vectors in the vector space must be equal to zero. Indeed, if $x \in V$ is such that $\langle x, v\rangle = 0$ for all v, pick $v = x$ so that $\langle x, x\rangle = 0$ implies $x = 0$ by axiom (2).

The **Pythagorean** identity holds for the norm squared of orthogonal vectors in an inner-product vector space. As you can quickly verify,

$$\|u+v\|^2 = \|u\|^2 + \|v\|^2, \quad \text{for } u, v \in V, \text{ orthogonal vectors: } \langle u, v\rangle = 0. \tag{14.1.16}$$

The **Schwarz inequality** can be proven by an argument fairly analogous to the one we gave above for real vector spaces. The result now reads

$$\boxed{\text{Schwarz inequality:} \quad |\langle u, v\rangle| \leq \|u\| \, \|v\| \,.} \tag{14.1.17}$$

The inequality is saturated if and only if one vector is a multiple of the other. You will prove this identity in a slightly different way in problem 14.1. You will also consider there the **triangle inequality**:

$$\|u+v\| \leq \|u\| + \|v\|, \tag{14.1.18}$$

which is saturated when $u = cv$ for c, a real, positive constant. Our definition (14.1.11) of a norm on a vector space V is mathematically sound: a norm is required to satisfy the triangle inequality. Other properties are required: (i) $\|v\| \geq 0$ for all v, (ii) $\|v\| = 0$ if and only if $v = 0$, and (iii) $\|cv\| = |c| \, \|v\|$ for any constant c. Our norm satisfies all of them.

A finite-dimensional complex vector space with an inner product is a **Hilbert space**. Our study of quantum mechanics will often involve *infinite-dimensional* vector spaces. An infinite-dimensional complex vector space with an inner product is a Hilbert space if an additional *completeness* property holds: all Cauchy sequences of vectors must converge to vectors in the space. An infinite sequence of vectors v_i, with $i = 1, 2, \ldots, \infty$ is a Cauchy sequence if for any $\epsilon > 0$ there is an N such that $\|v_n - v_m\| < \epsilon$ whenever $n, m > N$. For the infinite-dimensional vector spaces we have to deal with, the extra condition holds, and we will not have to concern ourselves with it. We often denote Hilbert spaces with the symbol $\mathcal{H}$.

Example 14.1. *Inner product in $\mathbb{C}^2$ for spin one-half.*

We had a first look at spin one-half states and their inner product in sections 12.1 and 12.3. The inner product is described by

$$\langle \Psi, \Phi \rangle = \psi_1^* \phi_1 + \psi_2^* \phi_2, \quad \Psi = \begin{pmatrix} \psi_1 \\ \psi_2 \end{pmatrix}, \quad \Phi = \begin{pmatrix} \phi_1 \\ \phi_2 \end{pmatrix}, \quad \psi_1, \psi_2, \phi_2, \phi_2 \in \mathbb{C}. \tag{14.1.19}$$

This definition obeys all five axioms of the inner product, as you can quickly verify. In bra-ket notation the inner product is described as

$$\langle \Psi | \Phi \rangle = \psi_1^* \phi_1 + \psi_2^* \phi_2, \quad \text{for} \quad |\Psi\rangle = \begin{pmatrix} \psi_1 \\ \psi_2 \end{pmatrix}, \quad |\Phi\rangle = \begin{pmatrix} \phi_1 \\ \phi_2 \end{pmatrix}. \tag{14.1.20}$$

Recall that we defined $\langle \Psi | \Phi \rangle \equiv \langle \Psi, \Phi \rangle$. While the above inner product is natural and obvious for a complex vector space, the intuition behind it is surprising when using the vectors that represent the direction of spin states, which we consider next. □

Example 14.2. *Inner product of spin states.*
We have defined the spin state $|\mathbf{n}\rangle \in \mathbb{C}^2$ as the normalized eigenstate of $\hat{\mathbf{S}} \cdot \mathbf{n}$ with eigenvalue $\hbar/2$. We think of this spin state as the state of a particle whose spin vector points in the direction of the unit vector $\mathbf{n}$ (we are using the notation $|\mathbf{n}\rangle = |\mathbf{n}; +\rangle$ and $|{-\mathbf{n}}\rangle = |\mathbf{n}; -\rangle$). In fact, any spin state in $\mathbb{C}^2$ is, up to a multiplicative constant, a state $|\mathbf{n}\rangle$ for some direction $\mathbf{n}$. This notation associates to any vector in spin space $\mathbb{C}^2$ a vector in ordinary space $\mathbb{R}^3$. Some confusion can arise because the natural inner products on the two spaces are related in a surprising way. In particular, given two spin states $|\mathbf{n}\rangle$ and $|\mathbf{n}'\rangle$, their inner product in $\mathbb{C}^2$ is *not* given by the inner (dot) product $\mathbf{n} \cdot \mathbf{n}'$ in $\mathbb{R}^3$.

For example, the states $|\mathbf{n}\rangle$ and $|{-\mathbf{n}}\rangle$ are orthogonal spin states, just like the states $|+\rangle$ and $|-\rangle$. Nevertheless, in $\mathbb{R}^3$ the vectors $\mathbf{n}$ and $-\mathbf{n}$ are antiparallel, and their inner product is nonzero: $\mathbf{n} \cdot (-\mathbf{n}) = -1$. Similarly, spin states along the z-axis and along the x-axis are not orthogonal, while vectors along those directions are.

We can do a simple computation of an inner product between the state $|+\rangle$ whose spin points along the positive z-axis and a state whose vector $\mathbf{n}$ lies on the (x, z) plane and is defined by spherical angles θ and $\phi = 0$. Using the expression for spin states in (12.3.28), we get

$$|\mathbf{n}\rangle = \cos\tfrac{\theta}{2}\,|+\rangle + \sin\tfrac{\theta}{2}\,|-\rangle. \tag{14.1.21}$$

It follows that

$$|\langle +|\mathbf{n}\rangle|^2 = \cos^2\tfrac{\theta}{2}. \tag{14.1.22}$$

This formula is instructive, as the inner product of the two states is related to the cosine of *half* the angle between the spin vectors in $\mathbb{R}^3$. As you will show in problem 14.2, this result holds for arbitrary spin states:

$$|\langle \mathbf{n}'|\mathbf{n}\rangle|^2 = \frac{1 + \mathbf{n} \cdot \mathbf{n}'}{2} = \cos^2\tfrac{\gamma}{2}, \tag{14.1.23}$$

where γ is the angle between the two unit vectors in $\mathbb{R}^3$: $\cos\gamma = \mathbf{n} \cdot \mathbf{n}'$. The above formula demonstrates that the dot product in $\mathbb{R}^3$ can be used to compute the *absolute* value of inner products in spin space, but, of course, it does not determine the possibly complex inner products themselves. We asked in exercise 14.1 if one could define a dot product $*$ in $\mathbb{R}^3$ by using the cosine of half the angle formed by the vectors, as is tempting from the above result. This is not possible: since a and $-a$ are vectors that form an angle of π and $\cos\frac{\pi}{2} = 0$, we would have $a * (-a) = 0$, in contradiction with $a * (-a) = -(a * a) < 0$ for any nonzero vector a. □

Example 14.3. *Inner product for operators.*
We have considered the vector space $\mathcal{L}(V)$ formed by the linear operators that act on a vector space V. In this vector space, the operators are now the "vectors." Given two operators $A, B \in \mathcal{L}(V)$, we now define an inner product $\langle A, B\rangle \in \mathbb{C}$ satisfying all the desired

axioms. In order to do so, we think of the operators as matrices in some arbitrary basis. Inspiration for a definition comes from an attempt to define the norm squared of a single operator A:

$$\langle A, A\rangle = \|A\|^2 . \tag{14.1.24}$$

Imagine a matrix A with all of its entries. We want the norm squared to be such that when it vanishes it sets to zero every single entry—this is what is needed for the matrix to be set to zero, as required by the axioms. There is a simple way to do this. We declare the norm proportional to the sum of squares of the absolute values of each of the entries A_{ij}:

$$\langle A, A\rangle = \tfrac{1}{2}\sum_{i,j}|A_{ij}|^2 = \tfrac{1}{2}\sum_{i,j}A_{ij}^*A_{ij}. \tag{14.1.25}$$

This definition certainly does what we wanted. If $\langle A, A\rangle = 0$, each of the entries of the matrix A must vanish. A clearer description of the norm requires a little manipulation in terms of the Hermitian conjugate of the matrix A. Recalling that $A_{ij}^* = (A^\dagger)_{ji}$, we have

$$\langle A, A\rangle = \tfrac{1}{2}\sum_{i,j}(A^\dagger)_{ji}A_{ij} = \tfrac{1}{2}\sum_{j}(A^\dagger A)_{jj} = \tfrac{1}{2}\text{tr}\,(A^\dagger A). \tag{14.1.26}$$

This now suggests the general definition for the inner product of two operators:

$$\langle A, B\rangle \equiv \tfrac{1}{2}\,\text{tr}\,(A^\dagger B). \tag{14.1.27}$$

The appearance of the trace is reassuring: we learned that the trace of an operator is basis independent, and therefore, as defined above, the inner product is basis independent. This is as it should be. An example illustrating the above definition will be given in the following section. □

Exercise 14.2. *Show that the inner product (14.1.27) satisfies $\langle A, B\rangle = \langle B, A\rangle^*$. Show that for Hermitian matrices the inner product is always a real number.*

14.2 Orthonormal Bases

In an inner-product space, we can demand that basis vectors have special properties. A list of vectors is said to be **orthonormal** if all vectors have norm one and are pairwise orthogonal. If $(e_1, \ldots, e_n)$ is a list of orthonormal vectors in V, then

$$\langle e_i, e_j\rangle = \delta_{ij}, \;\forall\, i, j = 1, \ldots, n. \tag{14.2.1}$$

We also have a simple expression for the norm of $a_1e_1 + \cdots + a_ne_n$, with a_i constants in the relevant field:

$$\begin{aligned}\|a_1e_1 + \cdots + a_ne_n\|^2 &= \big\langle a_1e_1 + \cdots + a_ne_n,\; a_1e_1 + \cdots + a_ne_n\big\rangle \\ &= \langle a_1e_1, a_1e_1\rangle + \cdots + \langle a_ne_n, a_ne_n\rangle \\ &= |a_1|^2 + \cdots + |a_n|^2.\end{aligned} \tag{14.2.2}$$

This result implies the somewhat nontrivial fact that *the vectors in any orthonormal list are linearly independent*. Indeed, if $a_1e_1+\cdots+a_ne_n=0$, then its norm squared is zero and so is $|a_1|^2+\cdots+|a_n|^2$. This implies all $a_i=0$, thus proving the claim.

An **orthonormal basis** of V is a list of orthonormal vectors that is also a basis for V. Let $(e_1,\ldots,e_n)$ denote an orthonormal basis. Then any vector v can be written as

$$v = a_1e_1+\cdots+a_ne_n, \tag{14.2.3}$$

for some constants a_i that can be calculated as follows:

$$a_i = \langle e_i, v\rangle. \tag{14.2.4}$$

Indeed,

$$\langle e_i, v\rangle=\sum_j\langle e_i, a_je_j\rangle=\sum_j a_j\langle e_i, e_j\rangle=\sum_j a_j\delta_{ij}=a_i. \tag{14.2.5}$$

Therefore, any vector v can be written as

$$v = \langle e_1, v\rangle\, e_1+\cdots+\langle e_n, v\rangle\, e_n = \sum_i\langle e_i, v\rangle\, e_i. \tag{14.2.6}$$

To find an orthonormal basis on an inner-product space V, we can start with any basis and follow a procedure that yields the desired orthonormal basis. A little more generally, the Gram-Schmidt procedure achieves the following:

Gram-Schmidt: Given a list $(v_1,\ldots,v_n)$ of linearly independent vectors in V, one can construct a list $(e_1,\ldots,e_n)$ of orthonormal vectors such that both lists span the same subspace of V.

The Gram-Schmidt procedure goes as follows. You take e_1 to be v_1, scaled to have unit norm:

$$e_1=\frac{v_1}{\|v_1\|}. \tag{14.2.7}$$

Clearly, $\langle e_1, e_1\rangle=1$. Then take

$$f_2\equiv v_2+\alpha\, e_1 \tag{14.2.8}$$

and fix the constant α to make f_2 orthogonal to e_1: $\langle e_1, f_2\rangle=0$. The answer, as you can check, is

$$f_2 = v_2-\langle e_1, v_2\rangle e_1. \tag{14.2.9}$$

This vector, divided by its norm, is set equal to e_2, the second vector in our orthonormal list:

$$e_2 = \frac{v_2-\langle e_1, v_2\rangle e_1}{\|v_2-\langle e_1, v_2\rangle e_1\|}. \tag{14.2.10}$$

In fact, we can write the general vector in a recursive fashion. If we have orthonormal $e_1, e_2,\ldots, e_{j-1}$, we can write the next orthonormal vector e_j as follows:

$$e_j = \frac{v_j-\langle e_1, v_j\rangle e_1-\cdots-\langle e_{j-1}, v_j\rangle e_{j-1}}{\left\|v_j-\langle e_1, v_j\rangle e_1-\cdots-\langle e_{j-1}, v_j\rangle e_{j-1}\right\|}. \tag{14.2.11}$$

It should be clear to you by inspection that this vector, as required, satisfies $\langle e_i, e_j\rangle = 0$ for all $i < j$ and that it has unit norm. The Gram-Schmidt procedure is quite practical.

If we have an orthonormal basis $(e_1, \ldots, e_n)$ for a vector space V, there is a simple formula for the matrix elements of any operator $T \in \mathcal{L}(V)$. Consider the inner product

$$\langle e_i, Te_j\rangle = \langle e_i, \sum_k T_{kj} e_k\rangle = \sum_k T_{kj}\langle e_i, e_k\rangle = \sum_k T_{kj}\delta_{ik} = T_{ij}. \tag{14.2.12}$$

We thus have

$$\boxed{T_{ij} = \langle e_i, Te_j\rangle \text{ in an orthonormal basis.}} \tag{14.2.13}$$

This formula is so familiar that one could be led to believe that an inner product is required to define the matrix elements of an operator. We know better: a basis suffices. If the basis is orthonormal, the simple formula above is available.

Example 14.4. *An orthonormal basis of Hermitian matrices in two dimensions.*

Two-by-two Hermitian matrices are important in quantum mechanics, mostly because they define the most general Hamiltonian for a quantum system with two basis states. In fact, we showed in example 13.7 that 2×2 Hermitian matrices form a real vector space of dimension four, with basis vectors $(\mathbb{1}, \sigma_1, \sigma_2, \sigma_3)$. Now we can demonstrate that with the inner product in (14.1.27) this is in fact an orthonormal basis of operators. As noted in exercise 14.2, for Hermitian matrices the inner product is real and thus suitable for a real vector space.

To manipulate the operators, it is convenient to use an index μ that runs over four values, zero to three, so that we can use the value zero for the identity matrix:

$$\sigma_\mu = \{\sigma_0, \sigma_1, \sigma_2, \sigma_3\}, \quad \sigma_0 = \mathbb{1}, \quad \mu = 0, 1, 2, 3. \tag{14.2.14}$$

With the inner product introduced before, the basis vectors are orthonormal:

$$\langle \sigma_\mu, \sigma_\nu\rangle = \tfrac{1}{2}\text{tr}(\sigma_\mu^\dagger \sigma_\nu) = \tfrac{1}{2}\text{tr}(\sigma_\mu\sigma_\nu) = \delta_{\mu\nu}. \tag{14.2.15}$$

Here we used the Hermiticity of σ_μ, and the last step, giving us the Kronecker delta, is verified by explicit computation:

$$\begin{aligned} \langle \sigma_0, \sigma_0\rangle &= \tfrac{1}{2}\text{tr}(\mathbb{1}\mathbb{1}) = \tfrac{1}{2}\text{tr}\mathbb{1} = \tfrac{1}{2}\cdot 2 = 1, \\ \langle \sigma_0, \sigma_i\rangle &= \tfrac{1}{2}\text{tr}(\mathbb{1}\sigma_i) = \tfrac{1}{2}\text{tr}\sigma_i = 0, \\ \langle \sigma_i, \sigma_j\rangle &= \tfrac{1}{2}\text{tr}(\sigma_i\sigma_j) = \tfrac{1}{2}\text{tr}(\delta_{ij}\mathbb{1} + i\epsilon_{ijk}\sigma_k) = \tfrac{1}{2}\delta_{ij}\text{tr}\mathbb{1} = \delta_{ij}. \end{aligned} \tag{14.2.16}$$

On account of (14.2.3) and (14.2.4), we know that any Hermitian matrix $M = M^\dagger$ can be written as

$$M = m_0\mathbb{1} + m_1\sigma_1 + m_2\sigma_2 + m_3\sigma_3, \quad \text{with} \quad m_\mu = \langle \sigma_\mu, M\rangle = \tfrac{1}{2}\text{tr}(\sigma_\mu M). \tag{14.2.17}$$

As a result, we have

$$M = \tfrac{1}{2}(\text{tr}M)\,\mathbb{1} + \tfrac{1}{2}\sum_{i=1}^{3}\text{tr}(\sigma_i M)\,\sigma_i, \tag{14.2.18}$$

showing how to write any Hermitian matrix as a superposition of $\mathbb{1}$ and Pauli matrices. □

14.3 Orthogonal Projectors

We now turn to the definition and construction of *orthogonal projectors*. In general, projectors in some vector space V are linear operators that acting on arbitrary vectors give us a vector in some subspace U of V. In other words, the range of the operator is U. Orthogonal projectors do this but also more: they give zero acting on vectors that are orthogonal to any vector on U. We will discuss this in detail now. A particularly simple characterization of an orthogonal projector will wait for section 14.5, after we discuss the adjoint of an operator. We will show that orthogonal projectors are in fact Hermitian operators (theorem 14.5.4).

We begin our work by noting that an inner product can help us construct interesting subspaces of a vector space V. Consider any *subset* U of vectors in V. Then we can define a *subspace* $U^{\perp}$, called the **orthogonal complement** of U as the set of all vectors orthogonal to the vectors in U:

$$U^{\perp} = \{v \in V \,|\, \langle v, u\rangle = 0, \text{ for all } u \in U\}. \tag{14.3.1}$$

This is clearly a subspace of V. Something remarkable happens when the set U, rather than being just a subset, is itself a *subspace*. In that case, U and $U^{\perp}$ actually give a direct sum decomposition of the full space:

Theorem 14.3.1. *If U is a subspace of V, then $V = U \oplus U^{\perp}$.*

Proof. This is a fundamental result and is actually not hard to prove. Let $(e_1, \ldots, e_n)$ be an orthonormal basis for the subspace U. We can then easily write any vector v in V as a sum of a vector in U and a vector in $U^{\perp}$:

$$v = \underbrace{(\langle e_1, v\rangle e_1 + \cdots + \langle e_n, v\rangle e_n)}_{\in U} + \underbrace{(v - \langle e_1, v\rangle e_1 - \cdots - \langle e_n, v\rangle e_n)}_{\in U^{\perp}}. \tag{14.3.2}$$

On the right-hand side, the first vector in parenthese is clearly in U as it is written as a linear combination of U basis vectors. The second vector is clearly in $U^{\perp}$: it is orthogonal to all the basis vectors of U and thus orthogonal to any vector in U. To complete the proof, one must show that there is no vector except the zero vector in the intersection $U \cap U^{\perp}$ (recall the comments below (13.2.3)). Let $v \in U \cap U^{\perp}$. Then v is in U and in $U^{\perp}$, so it should satisfy $\langle v, v\rangle = 0$. But then $v = 0$, completing the proof. □

Given $V = U \oplus U^{\perp}$, any vector $v \in V$ can be written *uniquely* as

$$v = u + w, \quad \text{with } u \in U \text{ and } w \in U^{\perp}. \tag{14.3.3}$$

One can define a linear operator P_U, called the **orthogonal projection** of V onto U, that acting on v above gives the vector u: $P_U v = u$.

It follows from this definition that

$$\begin{aligned} P_U u &= u, \quad \text{for } u \in U, \\ P_U w &= 0, \quad \text{for } w \in U^{\perp}. \end{aligned} \tag{14.3.4}$$

Indeed, for the first we write $u = u + 0$ with $0 \in U^{\perp}$, and for the second we write $w = 0 + w$ with $0 \in U$. In addition, we have the other following properties:

1. range $P_U = U$.

   The definition of P_U implies that range $P_U \subset U$. The first line in (14.3.4) shows that range $P_U \supset U$. These two together prove the claim. If U is a proper subspace of V (that is, $U \neq V$), P_U is not surjective.

2. null $P_U = U_\perp$.

   The second line in (14.3.4) shows that null $P_U \supset U^\perp$. On the other hand, if $\mu \in$ null P_U, we have $P_U\mu = 0$, and by (14.3.3), $\mu = 0 + \mu$, with $0 \in U$, and $\mu \in U^\perp$. This means null $P_U \subset U^\perp$, thus establishing the claim. If U is a proper subspace of V, P_U is not invertible.

3. $P_U^2 = P_U$.

   The first line in (14.3.4) tells us that P_U acting on U is the identity operator. Thus, if we act twice with P_U on any vector, the second action has no effect as it is acting on a vector in U. More explicitly, for $v = u + w$ with $u \in U$ and $w \in U^\perp$, we find that

$$P_U P_U v = P_U(P_U v) = P_U\, u = u = P_U v. \tag{14.3.5}$$

   Since v is arbitrary, this establishes the claim.

4. With $(e_1, \ldots, e_n)$ an orthonormal basis for U, the action of the projector is explicitly obtained from (14.3.2):

$$P_U v = \langle e_1, v\rangle e_1 + \cdots + \langle e_n, v\rangle e_n. \tag{14.3.6}$$

5. $\|P_U v\| \leq \|v\|$.

   The action of P_U cannot increase the length of a vector. Using the decomposition (14.3.3) and the Pythagorean theorem, we find that

$$\|v\|^2 = \|u + w\|^2 = \|u\|^2 + \|w\|^2 \geq \|u\|^2 = \|P_U v\|^2\,. \tag{14.3.7}$$

   The claim follows by taking the square root.

Given the results in (1) and (2), the relation $V = U \oplus U^\perp$ becomes

$$V = \text{range}\, P_U \oplus \text{null}\, P_U, \tag{14.3.8}$$

and for all $u \in$ range P_U and all $w \in$ null P_U, we have $\langle w, u\rangle = 0$.

Remarks: For any linear operator $T \in \mathcal{L}(V)$, the dimensions of null T and range T add up to the dimension of V (see (13.4.6)). But for an arbitrary T, as opposed to an orthogonal projector, the spaces null T and range T typically have a nonzero intersection and thus do not provide a direct sum decomposition of V. Orthogonal projectors do. The $\hat{S}_\pm$ operators in example 13.14 are operators for which the null space and the range are in fact the same!

The eigenvalues and eigenvectors of P_U are easy to describe. Since all vectors in U are left invariant by the action of P_U, an orthonormal basis for U provides a set of orthonormal eigenvectors of P all with eigenvalue one. If we choose an orthonormal basis for $U^\perp$, that basis provides orthonormal eigenvectors of P all with eigenvalue zero.

In fact, $P_U P_U = P_U$ implies that the eigenvalues of P_U can only be zero or one. Recall that the eigenvalues of an operator satisfy whatever equation the operator satisfies. Therefore, $\lambda^2 = \lambda$ holds, and $\lambda = 0, 1$, are the only possible eigenvalues.

A matrix representation of orthogonal projectors helps us to understand how the projector encodes the dimension of the space U it projects onto. Consider a vector space $V = U \oplus U^\perp$ that is $(n+k)$-dimensional, where U is n-dimensional, and $U^\perp$ is k-dimensional. Let $(e_1, \ldots, e_n)$ be an orthonormal basis for U and $(f_1, \ldots, f_k)$ an orthonormal basis for $U^\perp$. We then see that the list of vectors

$$(e_1, \ldots, e_n, f_1, \ldots, f_k) \text{ is an orthonormal basis for } V. \tag{14.3.9}$$

Since $P_U e_i = e_i$, for $i = 1, \ldots, n$, and $P_U f_j = 0$ for $j = 1, \ldots, k$, these orthonormal basis vectors are in fact eigenvectors of P_U. It follows that in this basis the projector operator is represented by the diagonal matrix:

$$P_U = \text{diag}(\underbrace{1, \ldots, 1}_{n \text{ entries}}, \underbrace{0, \ldots, 0}_{k \text{ entries}}). \tag{14.3.10}$$

As expected from its noninvertibility, $\det P_U = 0$. More interestingly, the trace of the matrix P_U is n. Therefore,

$$\boxed{\text{tr}\, P_U = \dim U.} \tag{14.3.11}$$

The dimension of U is the dimension of the range of P_U and thus gives the rank of the projector P_U. Rank-one projectors are the most common projectors. They project to one-dimensional subspaces of the vector space.

Projection operators are useful in quantum mechanics in two ways. First, the act of measuring an observable projects the physical state vector instantaneously to some invariant subspace of the observable; the projection can be seen to be done by an orthogonal projector. Second, orthogonal projectors themselves can be considered observables. These ideas will become clearer once we show that orthogonal projectors are in fact Hermitian operators.

Example 14.5. *Orthogonal projector onto a two-dimensional subspace in* $\mathbb{R}^3$.

Two-dimensional *vector subspaces* of $\mathbb{R}^3$ are planes going through the origin. We can specify planes $\mathcal{C}_\mathbf{n}$ using a unit vector $\mathbf{n} = (n_1, n_2, n_3)$ orthogonal to the plane. Then points $\mathbf{x} = (x_1, x_2, x_3)$ on the plane satisfy the constraint

$$\mathbf{n} \cdot \mathbf{x} = n_1 x_1 + n_2 x_2 + n_3 x_3 = 0, \quad \mathbf{x} \in \mathcal{C}_\mathbf{n}. \tag{14.3.12}$$

How do we build the orthogonal projector $P_\mathbf{n}$ onto $\mathcal{C}_\mathbf{n}$? The projector can be thought of as a 3×3 matrix that acting on arbitrary vectors gives the appropriate vector on the plane. At this point it may be a good idea if you try to determine the answer yourself. Then keep reading.

Given any vector $\mathbf{x}$, we rewrite it as follows:

$$\mathbf{x} = \underbrace{\mathbf{x} - (\mathbf{n} \cdot \mathbf{x})\mathbf{n}}_{\mathcal{C}_\mathbf{n}} + \underbrace{(\mathbf{n} \cdot \mathbf{x})\mathbf{n}}_{\mathcal{C}_\mathbf{n}^\perp}. \tag{14.3.13}$$

The first vector on the right-hand side is in $\mathcal{C}_\mathbf{n}$ because $\mathbf{n} \cdot (\mathbf{x} - (\mathbf{n} \cdot \mathbf{x})\mathbf{n}) = 0$, recalling that $\mathbf{n} \cdot \mathbf{n} = 1$. The second vector is along $\mathbf{n}$ and thus orthogonal to the plane $\mathcal{C}_\mathbf{n}$. The above formula indeed implements the decomposition $\mathbb{R}^3 = \mathcal{C}_\mathbf{n} \oplus \mathcal{C}_\mathbf{n}^\perp$, and the projector $P_\mathbf{n}$ acts as

$$P_\mathbf{n}\,\mathbf{x} \equiv \mathbf{x} - (\mathbf{n}\cdot\mathbf{x})\mathbf{n}. \tag{14.3.14}$$

We can describe this in matrix notation by looking at the ith component of both sides. With repeated indices summed over,

$$(P_\mathbf{n})_{ij}x_j = x_i - (n_j x_j)n_i = (\delta_{ij} - n_i n_j)x_j. \tag{14.3.15}$$

From this we read that

$$(P_\mathbf{n})_{ij} = \delta_{ij} - n_i n_j = \begin{pmatrix} 1-n_1^2 & -n_1n_2 & -n_1n_3 \\ -n_2n_1 & 1-n_2^2 & -n_2n_3 \\ -n_3n_1 & -n_3n_2 & 1-n_3^2 \end{pmatrix}. \tag{14.3.16}$$

The matrix for $P_\mathbf{n}$ is symmetric. Moreover, $P_\mathbf{n}$ is a rank-two projector since $\mathcal{C}_\mathbf{n}$ is of dimension two. Consistent with this, $\text{tr}P_\mathbf{n} = 3 - (n_1^2+n_2^2+n_3^2) = 3-1=2$. □

Example 14.6. *Nonorthogonal projector in* $\mathbb{R}^2$.

To better appreciate what an orthogonal projector is, we consider a simple example of a nonorthogonal one. We define a projector P_α taking $\mathbb{R}^2$ to the x_1-axis. Here α is a fixed angle in the range $\alpha \in (0, \frac{\pi}{2})$. For any point (x_1, x_2), draw a line through the point that makes an angle α with the x_1-axis (figure 14.2, *left*). Let x_* denote the coordinate value where the line intersects the x_1-axis. We set

$$P_\alpha \begin{pmatrix} x_1 \\ x_2 \end{pmatrix} = \begin{pmatrix} x_* \\ 0 \end{pmatrix}. \tag{14.3.17}$$

A look at the figure shows that

$$x_* = x_1 - \frac{\cos\alpha}{\sin\alpha}\,x_2. \tag{14.3.18}$$

This implies that

$$P_\alpha = \begin{pmatrix} 1 & -\frac{\cos\alpha}{\sin\alpha} \\ 0 & 0 \end{pmatrix}. \tag{14.3.19}$$

It is simple to see that $P_\alpha P_\alpha = P_\alpha$, as it befits a projector. The null space of P_α is the set of points on the line going through the origin with angle α: $(t\cos\alpha, t\sin\alpha)$, $t \in (-\infty, \infty)$. The trace of the projector matrix is one, equal to the dimensionality of its range. One sign that the projector is not orthogonal is that its matrix is not symmetric—as we already said (but did not show), orthogonal projectors are Hermitian operators. The other sign is that the projector P_α sometimes increases the length of a vector. In figure 14.2, right, the line OR, making an angle of 2α with the x_1-axis, contains the set of points that are projected by P_α with the length unchanged. For points to the "left" of the line OR and above the real line, the projection increases the length. □

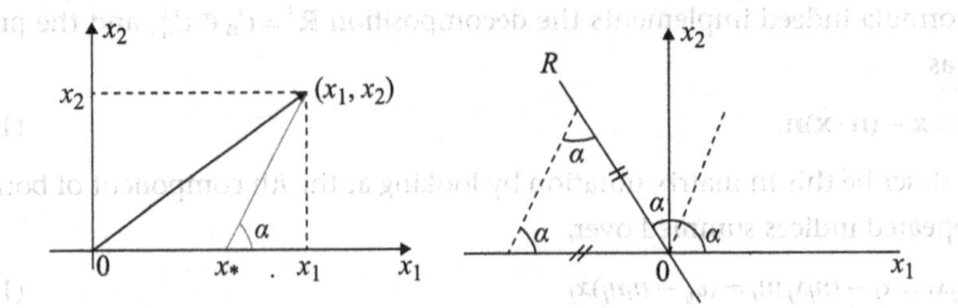

Figure 14.2
Left: The projector P_α takes the point (x_1, x_2) to the point $(x_*, 0)$ on the x_1-axis. *Right*: The projector P_α preserves the length of vectors on the *OR* line, which makes an angle 2α with the x_1-axis. Acting on vectors making an angle greater than 2α and less than π with the x_1-axis, the projector increases their length.

Example 14.7. *Orthogonal projector onto a spin state* $|\mathbf{n}\rangle$.
Let $P_{\mathbf{n}}$ project to the one-dimensional space generated by $e_1 = |\mathbf{n}\rangle$. This projector, acting on an arbitrary spin state v, is $P_{\mathbf{n}}v = e_1\langle e_1, v\rangle$. In the notation we have been using for spin states, v is the ket $|v\rangle$ and then $\langle e_1, v\rangle = \langle \mathbf{n}|v\rangle$. As a result, we have

$$P_{\mathbf{n}}|v\rangle = |\mathbf{n}\rangle\langle \mathbf{n}|v\rangle. \tag{14.3.20}$$

Using two-component vectors, we have

$$|\mathbf{n}\rangle = \begin{pmatrix} \cos\frac{\theta}{2} \\ \sin\frac{\theta}{2}e^{i\phi} \end{pmatrix}, \quad \langle \mathbf{n}| = \left(\cos\tfrac{\theta}{2},\, \sin\tfrac{\theta}{2}e^{-i\phi}\right), \quad |v\rangle = \begin{pmatrix} v_1 \\ v_2 \end{pmatrix}. \tag{14.3.21}$$

The above expression for $P_{\mathbf{n}}$ acting on a state can therefore be rewritten as follows:

$$P_{\mathbf{n}}|v\rangle = \begin{pmatrix} \cos\frac{\theta}{2} \\ \sin\frac{\theta}{2}e^{i\phi} \end{pmatrix} \left(\cos\tfrac{\theta}{2},\, \sin\tfrac{\theta}{2}e^{-i\phi}\right) \begin{pmatrix} v_1 \\ v_2 \end{pmatrix}. \tag{14.3.22}$$

While conventionally we would multiply the last two factors to form an inner product, it is best to multiply the first two factors, a column vector times a row vector, to find the matrix that represents the action of $P_{\mathbf{n}}$:

$$P_{\mathbf{n}}|v\rangle = \begin{pmatrix} \cos^2\frac{\theta}{2} & \cos\frac{\theta}{2}\sin\frac{\theta}{2}e^{-i\phi} \\ \sin\frac{\theta}{2}\cos\frac{\theta}{2}e^{i\phi} & \sin^2\frac{\theta}{2} \end{pmatrix} \begin{pmatrix} v_1 \\ v_2 \end{pmatrix}. \tag{14.3.23}$$

Since the state $|v\rangle$ is arbitrary, the matrix acting on it is the matrix representation of the projector $P_{\mathbf{n}}$. Using double-angle identities, we identify

$$P_{\mathbf{n}} = \tfrac{1}{2}\begin{pmatrix} 1+\cos\theta & \sin\theta e^{-i\phi} \\ \sin\theta e^{i\phi} & 1-\cos\theta \end{pmatrix} = \tfrac{1}{2}\mathbb{1} + \tfrac{1}{2}\begin{pmatrix} n_3 & n_1 - in_2 \\ n_1 + in_2 & -n_3 \end{pmatrix}, \tag{14.3.24}$$

recalling that in spherical coordinates $\mathbf{n} = (n_1, n_2, n_3) = (\sin\theta\cos\phi, \sin\theta\sin\phi, \cos\theta)$. Finally, we recognize the dot product of $\mathbf{n}$ against the Pauli matrices triplet, so the final result reads

$$P_{\mathbf{n}} = \tfrac{1}{2}(\mathbb{1} + \mathbf{n}\cdot\boldsymbol{\sigma}). \tag{14.3.25}$$

This is the orthogonal projector we were looking for. It is a rank-one projector. □

14.4 Linear Functionals and Adjoint Operators

When we consider a linear operator T on a vector space V equipped with an inner product, we can construct a related but generally different linear operator $T^\dagger$ on V called the **adjoint** of T. When the adjoint $T^\dagger$ happens to be equal to T, the operator is said to be Hermitian. To understand adjoints, we first need to develop the concept of a linear functional.

A **linear functional** ϕ on the vector space V is a linear map from V to the numbers $\mathbb{F}$: for $v \in V$, $\phi(v) \in \mathbb{F}$ (as usual, $\mathbb{F}$ is either $\mathbb{R}$ or $\mathbb{C}$). A linear functional has the following two properties:

1. $\phi(v_1 + v_2) = \phi(v_1) + \phi(v_2)$, with $v_1, v_2 \in V$.
2. $\phi(av) = a\phi(v)$ for $v \in V$ and $a \in \mathbb{F}$.

Example 14.8. *Linear functional on $\mathbb{R}^3$.*
Consider the real vector space $\mathbb{R}^3$ with the inner product equal to the familiar dot product. For any $v = (v_1, v_2, v_3) \in \mathbb{R}^3$, we define the functional ϕ as follows:

$$\phi(v) = 3v_1 + 2v_2 - 4v_3. \tag{14.4.1}$$

Linearity follows because the components v_1, v_2, and v_3 appear linearly on the right-hand side. We can actually use the vector $u = (3, 2, -4)$ to write the linear functional as an inner product. Indeed, one can readily see that for any $v \in \mathbb{R}^3$,

$$\phi(v) = \langle u, v\rangle = \langle (3, 2, -4), (v_1, v_2, v_3)\rangle = 3v_1 + 2v_2 - 4v_3. \tag{14.4.2}$$

This is no accident; we will now prove that any linear functional $\phi(v)$ on a vector space V admits such representation with some suitable choice of vector u. □

Theorem 14.4.1. *Let ϕ be a linear functional on V. There is a unique vector $u \in V$ such that $\phi(v) = \langle u, v\rangle$ for all $v \in V$.*

Proof. Consider an orthonormal basis $(e_1, \ldots, e_n)$, and write the vector v as

$$v = \langle e_1, v\rangle e_1 + \cdots + \langle e_n, v\rangle e_n. \tag{14.4.3}$$

When ϕ acts on v, we find

$$\begin{aligned}\phi(v) &= \phi\big(\langle e_1, v\rangle e_1 + \cdots + \langle e_n, v\rangle e_n\big)\\ &= \langle e_1, v\rangle \phi(e_1) + \cdots + \langle e_n, v\rangle \phi(e_n)\\ &= \langle \phi(e_1)^* e_1, v\rangle + \cdots + \langle \phi(e_n)^* e_n, v\rangle\\ &= \langle \phi(e_1)^* e_1 + \cdots + \phi(e_n)^* e_n, v\rangle,\end{aligned} \tag{14.4.4}$$

where we first used linearity and then conjugate homogeneity to bring the constants $\phi(e_i)$ inside the inner products. We have thus shown that, as claimed,

$$\phi(v) = \langle u, v\rangle \quad \text{with} \quad u = \phi(e_1)^* e_1 + \cdots + \phi(e_n)^* e_n. \tag{14.4.5}$$

Next, we prove that this u is unique. If there exists another vector u' that also gives the correct result for all v, then $\langle u', v\rangle = \langle u, v\rangle$, which implies $\langle u - u', v\rangle = 0$ for all v. Taking

$v = u' - u$, we see that this implies $u' - u = 0$, or $u' = u$, proving uniqueness. This proof applies for finite-dimensional vector spaces. The result, however, is true for infinite-dimensional vector spaces when ϕ is what is called a *continuous* linear functional. □

We can now address the construction of the adjoint. Consider a linear operator T and a functional $\phi(v)$ defined as follows:

$$\phi(v) = \langle u, Tv \rangle. \tag{14.4.6}$$

This is clearly a linear functional, whatever the operator T is. Since any linear functional can be written as $\langle w, v \rangle$, with some suitable vector w, we write

$$\langle u, Tv \rangle = \langle \#, v \rangle. \tag{14.4.7}$$

Of course, the vector # must depend on the vector u that appears on the left-hand side. Moreover, it must have something to do with the operator T, which does not appear on the right-hand side. We can think of # as a function of the vector u and thus write $\# = T^\dagger u$, where $T^\dagger$ denotes a function, not obviously linear, from V to V. So we think of $T^\dagger u$ as the vector obtained by acting with some function $T^\dagger$ on u. The above equation is written as

$$\langle u, Tv \rangle = \langle T^\dagger u, v \rangle. \tag{14.4.8}$$

Our next step is to show that, in fact, $T^\dagger$ is a linear operator on V.

Theorem 14.4.2. $T^\dagger \in \mathcal{L}(V)$.

Proof. For this purpose, consider

$$\langle u_1 + u_2, Tv \rangle = \langle T^\dagger (u_1 + u_2), v \rangle \tag{14.4.9}$$

with u_1, u_2, and v arbitrary vectors. Expand the left-hand side to get

$$\begin{aligned} \langle u_1 + u_2, Tv \rangle &= \langle u_1, Tv \rangle + \langle u_2, Tv \rangle \\ &= \langle T^\dagger u_1, v \rangle + \langle T^\dagger u_2, v \rangle \\ &= \big\langle T^\dagger u_1 + T^\dagger u_2, v \big\rangle. \end{aligned} \tag{14.4.10}$$

Comparing the right-hand sides of the last two equations, we get the desired

$$T^\dagger (u_1 + u_2) = T^\dagger u_1 + T^\dagger u_2. \tag{14.4.11}$$

Having established linearity, we now establish homogeneity. Consider

$$\langle au, Tv \rangle = \langle T^\dagger (au), v \rangle, \tag{14.4.12}$$

with $a \in \mathbb{C}$, and u and v arbitrary vectors. The left-hand side is

$$\langle au, Tv \rangle = a^* \langle u, Tv \rangle = a^* \langle T^\dagger u, v \rangle = \langle a T^\dagger u, v \rangle. \tag{14.4.13}$$

This time we conclude that

$$T^\dagger (au) = a T^\dagger u. \tag{14.4.14}$$

This completes the proof that $T^\dagger$, so defined, is a linear operator on V. □

The operator $T^\dagger \in \mathcal{L}(V)$ is called the **adjoint** of T. Its operational definition is the relation

$$\boxed{\langle u, Tv\rangle = \langle T^\dagger u, v\rangle .} \tag{14.4.15}$$

A couple of important properties are readily proven. The first is

$$(ST)^\dagger = T^\dagger S^\dagger . \tag{14.4.16}$$

To see this, apply the operational definition twice: $\langle u, STv\rangle = \langle S^\dagger u, Tv\rangle = \langle T^\dagger S^\dagger u, v\rangle$. By definition, $\langle u, STv\rangle = \langle (ST)^\dagger u, v\rangle$. Comparison leads to the claimed relation. The second property states that the adjoint of the adjoint of an operator is the original operator:

$$(S^\dagger)^\dagger = S. \tag{14.4.17}$$

To see this, first consider $\langle u, S^\dagger v\rangle = \langle (S^\dagger)^\dagger u, v\rangle$. But $\langle u, S^\dagger v\rangle = \langle S^\dagger v, u\rangle^* = \langle v, Su\rangle^* = \langle Su, v\rangle$. Comparing with the first result, we have shown that $(S^\dagger)^\dagger u = Su$, for any u. This is the content of the claimed relation. It is a direct result of $(S^\dagger)^\dagger = S$ that $\langle Su, v\rangle = \langle u, S^\dagger v\rangle$, as you can see by reading this equation from right to left. In summary, an operator can be moved from the left input of the inner product to the right input, and vice versa, by adding a dagger.

Example 14.9. *Adjoint operator on* $\mathbb{C}^3$.
Let $v = (v_1, v_2, v_3)$, with $v_i \in \mathbb{C}$, denote a vector in the three-dimensional complex vector space $\mathbb{C}^3$. Define a linear operator T that acts on v as follows:

$$T(v_1, v_2, v_3) = (\,2v_2 + iv_3,\; v_1 - iv_2,\; 3iv_1 + v_2 + 7v_3\,). \tag{14.4.18}$$

Assume the inner product is the standard one on $\mathbb{C}^3$: $\langle u, v\rangle = u_1^* v_1 + u_2^* v_2 + u_3^* v_3$. We want to find the action of $T^\dagger$ on a vector and wish to determine the matrix representations of T and $T^\dagger$ using the orthonormal basis $e_1 = (1,0,0), e_2 = (0,1,0), e_3 = (0,0,1)$.

We introduce the vector $u = (u_1, u_2, u_3)$ and use the basic identity $\langle Tu, v\rangle = \langle u, T^\dagger v\rangle$. The left-hand side of the identity gives

$$\langle\, Tu, v\,\rangle = (2u_2 + iu_3)^* v_1 \;+\; (u_1 - iu_2)^* v_2 \;+\; (3iu_1 + u_2 + 7u_3)^* v_3 . \tag{14.4.19}$$

To identify this with $\langle u, T^\dagger v\rangle$, we rewrite the right-hand side, factoring the various u_i^*'s:

$$\langle u, T^\dagger v\rangle = u_1^*(v_2 - 3iv_3) + u_2^*(2v_1 + iv_2 + v_3) + u_3^*(-iv_1 + 7v_3). \tag{14.4.20}$$

We can therefore read the desired action of $T^\dagger$:

$$T^\dagger(v_1, v_2, v_3) = (\,v_2 - 3iv_3,\; 2v_1 + iv_2 + v_3,\; -iv_1 + 7v_3\,). \tag{14.4.21}$$

To find the matrix representations, think of (14.4.18) as follows:

$$\begin{pmatrix} * & * & * \\ * & * & * \\ * & * & * \end{pmatrix}\begin{pmatrix} v_1 \\ v_2 \\ v_3 \end{pmatrix} = \begin{pmatrix} 2v_2 + iv_3 \\ v_1 - iv_2 \\ 3iv_1 + v_2 + 7v_3 \end{pmatrix}. \tag{14.4.22}$$

The matrix on the left-hand side is the representation of T. We immediately see that

$$T = \begin{pmatrix} 0 & 2 & i \\ 1 & -i & 0 \\ 3i & 1 & 7 \end{pmatrix}, \qquad T^\dagger = \begin{pmatrix} 0 & 1 & -3i \\ 2 & i & 1 \\ -i & 0 & 7 \end{pmatrix}, \tag{14.4.23}$$

using (14.4.21) to write the matrix for $T^\dagger$. These matrices are related: one is the transpose and complex conjugate of the other. This is not an accident, as we will see now. □

Let us calculate adjoints using matrix notation. Let $u=e_i$ and $v=e_j$ where e_i and e_j are *orthonormal* basis vectors. Then the definition $\langle T^\dagger u, v\rangle = \langle u, Tv\rangle$ can be written (with repeated indices summed) as

$$\begin{aligned} \langle T^\dagger e_i, e_j\rangle &= \langle e_i, Te_j\rangle, \\ \langle (T^\dagger)_{ki}e_k, e_j\rangle &= \langle e_i, T_{kj}e_k\rangle, \\ ((T^\dagger)_{ki})^*\,\delta_{kj} &= T_{kj}\,\delta_{ik}, \\ ((T^\dagger)_{ji})^* &= T_{ij}. \end{aligned} \tag{14.4.24}$$

Relabeling i and j and taking the complex conjugate, we find the familiar relation between a matrix and its adjoint:

$$\text{In an orthonormal basis,} \quad (T^\dagger)_{ij} = (T_{ji})^*. \tag{14.4.25}$$

The adjoint matrix is the transpose and complex conjugate matrix *as long* as we use an orthonormal basis. If we did not, in the equation above, $\langle e_i, e_j\rangle = \delta_{ij}$ would be replaced by $\langle e_i, e_j\rangle = g_{ij}$, where g_{ij} is some constant matrix that would appear in the rule for the construction of the adjoint matrix.

Exercise 14.3. *Let* $(e_1, \cdots, e_n)$ *be a basis with inner product* $\langle e_i, e_j\rangle = m_i\delta_{ij}$*(i not summed), where* $m_i > 0$ *for all i. Show that*

$$(T^\dagger)_{ij} = \frac{m_j}{m_i}\, T^*_{ji}. \tag{14.4.26}$$

14.5 Hermitian and Unitary Operators

Before we begin looking at special kinds of operators, let us consider a very surprising fact about operators on complex vector spaces. Suppose we have an operator T such that for any vector $v \in V$ the following inner product vanishes:

$$\langle v, Tv\rangle = 0 \quad \text{for all } v \in V. \tag{14.5.1}$$

What can we say about the operator T? The condition states that T is an operator that acting on any vector v gives a vector orthogonal to v. In a two-dimensional *real* vector space, this could be the operator that rotates any vector by ninety degrees, a nontrivial operator. It is quite surprising and important that for *complex* vector spaces any such operator *necessarily* vanishes. This is a theorem:

Theorem 14.5.1. *Let T be a linear operator in a complex vector space V:*

$$\textit{If } \langle v, Tv\rangle = 0 \textit{ for all } v \in V, \textit{ then } T = 0. \tag{14.5.2}$$

Proof. Any proof must be such that it fails to work for a real vector space. Note that the result follows if we can prove that $\langle u, Tv\rangle = 0$, for all $u, v \in V$. Indeed, if this holds then take $u = Tv$. Then $\langle Tv, Tv\rangle = 0$ for all v implies that $Tv = 0$ for all v, and therefore $T = 0$.

We will thus try to show that $\langle u, Tv\rangle = 0$ for all $u, v \in V$. All we know is that objects of the form $\langle \#, T\#\rangle$ vanish, whatever # is. So we must aim to form linear combinations of such terms in order to reproduce $\langle u, Tv\rangle$. We begin by trying the following:

$$\langle u+v, T(u+v)\rangle \; - \; \langle u-v, T(u-v)\rangle \; = \; 2\langle u, Tv\rangle \; + \; 2\langle v, Tu\rangle. \tag{14.5.3}$$

We see that the "diagonal" terms vanish, but instead of getting just $\langle u, Tv\rangle$, we also get $\langle v, Tu\rangle$. Here is where complex numbers help. We can get the same two terms but with opposite signs as follows:

$$\langle u+iv, T(u+iv)\rangle \; - \; \langle u-iv, T(u-iv)\rangle \; = \; 2i\langle u, Tv\rangle \; - \; 2i\,\langle v, Tu\rangle. \tag{14.5.4}$$

In checking this don't forget that, for example, $\langle iv, u\rangle = -i\langle v, u\rangle$. It follows from the last two relations that

$$\begin{aligned}\langle u, Tv\rangle \; = \; \tfrac{1}{4}\Big(& \langle u+v, T(u+v)\rangle - \langle u-v, T(u-v)\rangle \\ & + \tfrac{1}{i}\langle u+iv, T(u+iv)\rangle \; - \; \tfrac{1}{i}\langle u-iv, T(u-iv)\rangle\Big).\end{aligned} \tag{14.5.5}$$

The condition $\langle v, Tv\rangle = 0$ for all v implies that each term on the above right-hand side vanishes, thus showing that $\langle u, Tv\rangle = 0$ for all $u, v \in V$. As explained above this proves the result. □

Exercise 14.4. *If a nonvanishing operator T existed that acting on any vector v on a complex vector space gave a vector orthogonal to v, it would contradict the above theorem. Give an independent reason why such an operator cannot exist (Hint: if nothing comes to mind, reread example 13.12).*

An operator T is said to be **Hermitian** if $T^\dagger = T$. Hermitian operators are pervasive in quantum mechanics. The above theorem in fact helps us discover Hermitian operators. You recall that the expectation value of a Hermitian operator, on any state, is real. It is also true, however, that any operator whose expectation value is real for all states must be Hermitian:

$$\boxed{T = T^\dagger \;\text{ if and only if }\; \langle v, Tv\rangle \in \mathbb{R} \text{ for all } v.} \tag{14.5.6}$$

To prove this we first go from left to right. If $T = T^\dagger$, then

$$\langle v, Tv\rangle \; = \; \langle T^\dagger v, v\rangle \; = \; \langle Tv, v\rangle = \langle v, Tv\rangle^*, \tag{14.5.7}$$

showing that $\langle v, Tv\rangle$ is real. To go from right to left, first note that the reality condition means that

$$\langle v, Tv\rangle \; = \; \langle Tv, v\rangle \; = \; \langle v, T^\dagger v\rangle. \tag{14.5.8}$$

Now the leftmost and rightmost terms can be combined to give $\langle v, (T - T^\dagger)v\rangle = 0$, which holding for all v implies, by the theorem, that $T = T^\dagger$.

We have shown (theorem 13.6.2) that on a complex vector space any linear operator has at least one eigenvalue. Below we show that the eigenvalues of a Hermitian operator are real numbers. Moreover, while eigenvectors corresponding to different eigenvalues are in general linearly independent, for Hermitian operators they are guaranteed to be orthogonal. These results were established in section 5.3 for state spaces of wave functions with the obvious inner product arising from integration. Here the inner product is completely general, and so is the state space.

Theorem 14.5.2. *The eigenvalues of Hermitian operators are real.*

Proof. Let v be a nonzero eigenvector of the Hermitian operator T with eigenvalue λ: $Tv = \lambda v$. Taking the inner product with v, we find that

$$\langle v, Tv \rangle = \langle v, \lambda v \rangle = \lambda \langle v, v \rangle. \tag{14.5.9}$$

Since T is Hermitian, we can also evaluate $\langle v, Tv \rangle$ as follows:

$$\langle v, Tv \rangle = \langle Tv, v \rangle = \langle \lambda v, v \rangle = \lambda^* \langle v, v \rangle. \tag{14.5.10}$$

The above equations give $(\lambda - \lambda^*)\langle v, v \rangle = 0$, and since v is not the zero vector, we conclude that $\lambda^* = \lambda$, showing that λ is real. □

Theorem 14.5.3. *Eigenvectors of a Hermitian operator associated with different eigenvalues are orthogonal.*

Proof. Let v_1 and v_2 be eigenvectors of the operator T:

$$Tv_1 = \lambda_1 v_1, \qquad Tv_2 = \lambda_2 v_2, \tag{14.5.11}$$

with λ_1 and λ_2 real, by Theorem 14.5.2, and different from each other. Consider the inner product $\langle v_2, Tv_1 \rangle$, and evaluate it in two different ways by following the direction of the arrows emanating from the central term:

$$\lambda_2 \langle v_2, v_1 \rangle = \langle \lambda_2 v_2, v_1 \rangle = \langle Tv_2, v_1 \rangle \underset{\leftarrow}{=} \langle v_2, Tv_1 \rangle \underset{\rightarrow}{=} \langle v_2, \lambda_1 v_1 \rangle = \lambda_1 \langle v_2, v_1 \rangle. \tag{14.5.12}$$

Going left, we used the Hermiticity of T. Equating the leftmost and rightmost terms, we find that

$$(\lambda_1 - \lambda_2)\langle v_1, v_2 \rangle = 0. \tag{14.5.13}$$

The assumption $\lambda_1 \neq \lambda_2$ leads to the claimed orthogonality: $\langle v_1, v_2 \rangle = 0$. □

We can finally give a rather simple characterization of orthogonal projectors.

Theorem 14.5.4. *An operator P such that $P^2 = P$ and $P^\dagger = P$ is an orthogonal projector.*

Proof. For any $v \in V$, we can write

$$v = (v - Pv) + Pv, \quad v - Pv \in \text{null}\, P, \quad \text{and} \quad Pv \in \text{range}\, P. \tag{14.5.14}$$

Any vector both in the null space of P and in the range of P must be zero. Indeed, let $x \in \text{range}\, P$ be nonzero. This means there is a $y \in V$, nonzero, such that $x = Py$. From this it follows that $Px = P^2 y = Py = x$, demonstrating that x cannot be in the null space of P. All of this shows that $P^2 = P$ implies that

$$V = \text{range}\, P \oplus \text{null}\, P. \tag{14.5.15}$$

Now it remains to be shown that the range and null spaces above are orthogonal. Let $u \in \text{range}\, P$ and $w \in \text{null}\, P$. Since P leaves u invariant, is Hermitian, and kills w, we have

$$\langle u, w \rangle = \langle Pu, w \rangle = \langle u, Pw \rangle = 0, \tag{14.5.16}$$

proving the desired orthonormality. □

Let us now consider another important class of linear operators on a complex vector space, the so-called unitary operators. An operator $U \in \mathcal{L}(V)$ in a complex vector space V is said to be a **unitary operator** if it is surjective and does not change the norm of the vector it acts upon:

$$\|Uv\| = \|v\|\,, \text{ for all } v \in V. \tag{14.5.17}$$

Note that U can only kill vectors of zero length, and since the only such vector is the zero vector, $\text{null}\, U = 0$ and U is injective. By including the condition of surjectivity, we tailored the definition to be useful even for infinite-dimensional spaces. In finite-dimensional vector spaces, injectivity implies surjectivity and invertibility. In infinite-dimensional vector spaces, however, invertibility requires both injectivity and surjectivity. Since U is also assumed to be surjective, a unitary operator U is always invertible.

Example 14.10. *A unitary operator proportional to the identity.*
The operator $\lambda \mathbb{1}$ with λ a complex number of unit norm is unitary. Indeed, with $|\lambda| = 1$ we have $\|\lambda \mathbb{1} v\| = \|\lambda v\| = |\lambda|\, \|v\| = \|v\|$ for all $v \in V$, showing the operator preserves the length of all vectors. Moreover, the operator is clearly surjective since for any $v \in V$ we have $v = (\lambda \mathbb{1}) \frac{1}{\lambda} v$. □

For another useful characterization of unitary operators, we begin by squaring the invariance of the norm condition (14.5.17):

$$\langle Uu, Uu \rangle = \langle u, u \rangle. \tag{14.5.18}$$

By the definition of adjoint,

$$\langle u, U^\dagger U\, u \rangle = \langle u, u \rangle \;\Rightarrow\; \langle u,\, (U^\dagger U - \mathbb{1})u \rangle \,=\, 0 \;\text{ for all }\; u. \tag{14.5.19}$$

Theorem 14.5.1 then implies $U^\dagger U = \mathbb{1}$. Since U is invertible, $U^\dagger$ is the inverse of U, and we also have $UU^\dagger = \mathbb{1}$:

$$\boxed{U \text{ is unitary} \iff U^\dagger U = UU^\dagger = \mathbb{1}.} \tag{14.5.20}$$

The right-to-left arrow holds because any operator U that obeys these identities is unitary (invertible and norm preserving). Unitary operators also preserve *inner products*:

$$\langle\, Uu,\; Uv \rangle \;=\; \langle u, v \rangle. \tag{14.5.21}$$

This follows immediately by moving the second U to act on the first input and using $U^\dagger U = \mathbb{1}$.

Example 14.11. *From Hermitian to unitary operators.*

We noted earlier that the exponentiation of certain Hermitian operators gives interesting operators. In section 10.1, for example, we considered the momentum operator, which is Hermitian, multiplied it by i as well as some real constants, and exponentiated it to obtain a translation operator. We did similarly with the angular momentum operator $\hat{L}_z$, a Hermitian operator that multiplied by i and a real constant is exponentiated to give a rotation operator. We said that the momentum operator generates translations, and the angular momentum operator generates rotations. The general rule was noted in exercise 10.2:

$$\boxed{\hat{M} \text{ Hermitian} \Rightarrow e^{i\hat{M}} \text{ unitary.}} \tag{14.5.22}$$

This is an important result, so let us make sure it is completely clear. Since $(\hat{A}_1 + \hat{A}_2)^\dagger = \hat{A}_1^\dagger + \hat{A}_2^\dagger$ and $\mathbb{1}^\dagger = \mathbb{1}$, the series expansion of the exponential implies that $(e^{\hat{A}})^\dagger = e^{\hat{A}^\dagger}$. It is then clear that for a Hermitian $\hat{M}$,

$$\left(e^{i\hat{M}}\right)^\dagger = e^{(i\hat{M})^\dagger} = e^{-i\hat{M}}. \tag{14.5.23}$$

It then follows that

$$\left(e^{i\hat{M}}\right)^\dagger e^{i\hat{M}} = e^{-i\hat{M}} e^{i\hat{M}} = \mathbb{1}, \tag{14.5.24}$$

confirming that $e^{i\hat{M}}$ is a unitary operator, as its Hermitian conjugate is its inverse. In quantum mechanics, unitary operators act naturally on states, as they preserve their norm.

A Hermitian operator $\hat{Q}$ that commutes with the Hamiltonian defines a conserved quantity: as you have learned, the expectation value of $\hat{Q}$ is constant in time. We also say that the operator generates a symmetry transformation. Indeed, the operator $e^{i\alpha\hat{Q}}$, with α a real constant that makes the exponent unit-free, is a unitary operator that commutes with the Hamiltonian. As a result, when acting on a nondegenerate energy eigenstate, it preserves the state, and when acting on a degenerate subspace of eigenvectors, it preserves the subspace. □

Assume the vector space V is finite-dimensional and has an orthonormal basis $(e_1, \ldots, e_n)$. Consider another set of vectors $(f_1, \ldots, f_n)$ where the f's are obtained from the e's by the action of a unitary operator U:

$$f_i = U e_i. \tag{14.5.25}$$

This also means that $e_i = U^\dagger f_i$. The new vectors are also orthonormal:

$$\langle f_i, f_j \rangle = \langle U e_i, U e_j \rangle = \langle e_i, e_j \rangle = \delta_{ij}. \tag{14.5.26}$$

They are linearly independent because any list of orthonormal vectors is linearly independent. They span V because dim V linearly independent vectors span V. Thus, the f_i's are an orthonormal *basis*. This is an important result: *the action of a unitary operator on an orthonormal basis gives us another orthonormal basis.*

Let the matrix elements of U in the e-basis be denoted as $U_{ki} = \langle e_k, Ue_i \rangle$. The matrix elements U'_{ki} of U in the f-basis are in fact the same:

$$U'_{ki} = \langle f_k, Uf_i \rangle = \langle Ue_k, Uf_i \rangle = \langle e_k, f_i \rangle = \langle e_k, Ue_i \rangle = U_{ki}. \tag{14.5.27}$$

We first saw this equality when we studied general changes of bases (section 13.5).

14.6 Remarks on Complex Vector Spaces

In this as well as in chapter 13, we examined in some detail vector spaces and the structures that can be defined on them. We began our study of quantum mechanics in chapter 1, noting how complex numbers were essential to the formulation of the theory and why the wave function was necessarily complex valued. Equipped with the requisite mathematical tools, it is now evident that one must view the wave function as a vector in some complex vector space. This can be considered an axiom in quantum mechanics.

Given that we are familiar with both real and complex vector spaces, it is now a good time to summarize some of the key differences between these two types of spaces—differences that make complex vector space the natural arena for quantum mechanics.

1. Operators on complex vector spaces have at least one eigenvalue (theorem 13.6.2). In real vector spaces, operators can fail to have eigenvalues. In quantum mechanics Hermitian operators are observables, and their eigenvalues are the possible values of the observables. As we will show in section 15.6, Hermitian operators not only have eigenvalues; the associated eigenvectors can be arranged to provide an orthonormal basis for the state space.
2. For *any* operator in a complex vector space, there is a basis for which the operator is represented by an upper-triangular matrix: a matrix in which all entries below the diagonal are zero. This presentation is generally not possible in a real vector space. In this presentation the diagonal elements are in fact the eigenvalues of the operator. If an operator is diagonalizable, there is a basis where the operator is a diagonal matrix. For nondiagonalizable operators on complex vector spaces, the Jordan form is always available. The *Jordan form* represents the operator as an upper-triangular matrix in which all elements are zero except on the diagonal and on the line immediately above the diagonal.
3. In complex vector spaces, the trace of *any* operator, defined as the sum of the diagonal elements in any matrix representation, is equal to the sum of its eigenvalues. This follows from (2) above. It does not matter if the operator is not diagonalizable. In real vector spaces, we have no such result.
4. In a complex vector space, an operator T for which $\langle v, Tv \rangle = 0$ for all $v \in V$ must vanish (theorem 14.5.1). This result does not hold for real vector spaces, as elaborated in problem 14.4. The result is important because it helps characterize Hermitian operators as those for which any expectation value is real and unitary operators as operators U that by virtue of preserving the norm of vectors satisfy $U^\dagger U = \mathbb{1}$.

Since stronger results hold, complex vector spaces are arguably more tractable than real vector spaces. In fact, a number of results for real vector spaces follow by considering the *complexification* of the vector space, a complex vector space where the real vector space is embedded naturally.

14.7 Rotation Operators for Spin States

We have seen examples of orbital angular momentum operators generating particular rotations (section 10.1). The operators generating rotations are unitary operators. We now wish to construct general rotation operators. For this purpose we will focus here on spin angular momentum. As we will explain, the main results apply for general angular momentum, such as orbital angular momentum.

We will show how unitary operators constructed from the spin operators $\hat{\mathbf{S}}$ act on the spin operators themselves, and from this, we will learn how they rotate spin states. A few relations are useful in our analysis, and we discuss them now. The first one gives the expectation value of $\hat{\mathbf{S}}$ in a spin state $|\mathbf{n}\rangle$. We claim that

$$\langle\mathbf{n}|\hat{\mathbf{S}}|\mathbf{n}\rangle = \tfrac{\hbar}{2}\,\mathbf{n}. \tag{14.7.1}$$

This is, in fact, a way to measure the direction $\mathbf{n}$ of a spin state: the components of $\mathbf{n}$ are proportional to the expectation values of the components of the spin operator. This is quickly checked explicitly, recalling that with $\mathbf{n}=(n_x,n_y,n_z)=(\sin\theta\cos\phi,\sin\theta\sin\phi,\cos\theta)$ we have $|\mathbf{n}\rangle=\cos\frac{\theta}{2}|+\rangle+\sin\frac{\theta}{2}e^{i\phi}|-\rangle$. Thus, for example,

$$\begin{aligned}\langle\mathbf{n}|\hat{S}_x|\mathbf{n}\rangle &= \tfrac{\hbar}{2}\Big(\cos\tfrac{\theta}{2}\langle+|+\sin\tfrac{\theta}{2}e^{-i\phi}\langle-|\Big)\sigma_x\Big(\cos\tfrac{\theta}{2}|+\rangle+\sin\tfrac{\theta}{2}e^{i\phi}|-\rangle\Big)\\ &= \tfrac{\hbar}{2}\Big(\cos\tfrac{\theta}{2}\langle+|+\sin\tfrac{\theta}{2}e^{-i\phi}\langle-|\Big)\Big(\cos\tfrac{\theta}{2}|-\rangle+\sin\tfrac{\theta}{2}e^{i\phi}|+\rangle\Big)\\ &= \tfrac{\hbar}{2}\cos\tfrac{\theta}{2}\sin\tfrac{\theta}{2}(e^{i\phi}+e^{-i\phi})=\tfrac{\hbar}{2}\sin\theta\cos\phi=\tfrac{\hbar}{2}n_x.\end{aligned} \tag{14.7.2}$$

Exercise 14.5. *Check that* $\langle\mathbf{n}|\hat{S}_y|\mathbf{n}\rangle=\frac{\hbar}{2}n_y$, *and* $\langle\mathbf{n}|\hat{S}_z|\mathbf{n}\rangle=\frac{\hbar}{2}n_z$, *thus completing the verification of (14.7.1).*

The second property we need is just a simple commutator. For an arbitrary vector $\mathbf{a}$, we have

$$[\mathbf{a}\cdot\hat{\mathbf{S}},\hat{\mathbf{S}}\,] = -i\hbar\,\mathbf{a}\times\hat{\mathbf{S}}. \tag{14.7.3}$$

This is quickly checked by explicit computation:

Exercise 14.6. *Verify that* $[\mathbf{a}\cdot\hat{\mathbf{S}},\,\hat{S}_k]=-i\hbar\,(\mathbf{a}\times\hat{\mathbf{S}})_k$, *thus proving (14.7.3).*

The third and final set of needed results involves rotations. It is known in mechanics that a vector $\mathbf{v}$ in three-dimensional space rotating around the directed axis $\mathbf{n}$ with angular velocity ω satisfies the differential equation

$$\frac{d\mathbf{v}}{dt}=\boldsymbol{\omega}\times\mathbf{v},\quad\text{with}\quad \boldsymbol{\omega}\equiv\omega\mathbf{n}. \tag{14.7.4}$$

At any instant of time, the vector $\mathbf{v}(t)$ is obtained from the time equal zero vector $\mathbf{v}(0)$ by a rotation of angle ωt about the axis $\mathbf{n}$. This rotation can be viewed as a rotation matrix $\mathcal{R}_{\mathbf{n}}(\omega t)$ acting on the time equal zero vector:

$$\mathbf{v}(t) = \mathcal{R}_{\mathbf{n}}(\omega t)\, \mathbf{v}(0). \tag{14.7.5}$$

The subscript in the rotation matrix gives the unit vector specifying the directed axis of rotation, and the argument is the angle of rotation. Interestingly (problem 14.6), the explicit time evolution of the vector can be written neatly because the rotation matrix takes a simple form. For a rotation angle α about the directed axis $\mathbf{n}$, we find that the action on an arbitrary vector $\mathbf{u}$ gives

$$\mathcal{R}_{\mathbf{n}}(\alpha)\, \mathbf{u} = (1 - \cos\alpha)\, (\mathbf{n} \cdot \mathbf{u})\, \mathbf{n} + (\cos\alpha)\, \mathbf{u} + (\sin\alpha)\, (\mathbf{n} \times \mathbf{u}). \tag{14.7.6}$$

Note that $\mathcal{R}_{\mathbf{n}}(0)\mathbf{u} = \mathbf{u}$, as expected: the zero angle rotation is the identity matrix. The rotation matrix is a periodic function of α with period 2π, also as expected. Acting on a vector, $\mathcal{R}_{\mathbf{n}}(\alpha)$ rotates it by an angle α about the axis defined by $\mathbf{n}$. A couple of extra properties of rotations are worth discussing (their verification is in problem 14.6). If we perform rotations about the *same* axis successively, the composition is simply a rotation about that axis with the total angle given by the sum of the angles:

$$\mathcal{R}_{\mathbf{n}}(\beta)\, \mathcal{R}_{\mathbf{n}}(\alpha) = \mathcal{R}_{\mathbf{n}}(\alpha + \beta). \tag{14.7.7}$$

This shows that $\mathcal{R}_{\mathbf{n}}(-\alpha)$ is the inverse of $\mathcal{R}_{\mathbf{n}}(\alpha)$, as is also clear from the definition. Another property is interesting. In $\mathbb{R}^3$ the dot product is invariant under rotations. This means that the inner product of two vectors is unchanged when both vectors are subject to the *same rotation*. Thus, for vectors $\mathbf{u}$ and $\mathbf{v}$ we have

$$\mathbf{u} \cdot \mathbf{v} = \big(\mathcal{R}_{\mathbf{n}}(\alpha)\, \mathbf{u}\big) \cdot \big(\mathcal{R}_{\mathbf{n}}(\alpha)\, \mathbf{v}\big). \tag{14.7.8}$$

We can now begin the detailed analysis of spin rotations. For this we define the *unitary* operator $\hat{R}_{\mathbf{n}}(\alpha)$:

$$\hat{R}_{\mathbf{n}}(\alpha) = e^{-\frac{i}{\hbar}\alpha \hat{S}_{\mathbf{n}}} = e^{-\frac{i\alpha}{\hbar}\mathbf{n}\cdot\hat{\mathbf{S}}} = e^{-\frac{i\alpha}{2}\mathbf{n}\cdot\boldsymbol{\sigma}}. \tag{14.7.9}$$

We now claim that the action of $\hat{R}_{\mathbf{n}}(\alpha)$ on the vector operator, or operator triplet $\hat{\mathbf{S}}$, rotates the components as if they were the components of an ordinary vector in three dimensions and that its action on a spin state $|\mathbf{n}'\rangle$ gives a spin state pointing at the direction defined by the rotation of $\mathbf{n}'$. This is the content of the following theorem:

Theorem 14.7.1. *With $\hat{R}_{\mathbf{n}}(\alpha) = \exp(-\frac{i}{\hbar}\alpha\hat{S}_{\mathbf{n}})$, we find that*

$$\hat{R}_{\mathbf{n}}^{\dagger}(\alpha)\, \hat{\mathbf{S}}\, \hat{R}_{\mathbf{n}}(\alpha) \;=\; \mathcal{R}_{\mathbf{n}}(\alpha)\, \hat{\mathbf{S}}, \tag{14.7.10}$$

as well as

$$\hat{R}_{\mathbf{n}}(\alpha)|\mathbf{n}'\rangle = |\mathbf{n}''\rangle, \quad \textit{with} \;\; \mathbf{n}'' = \mathcal{R}_{\mathbf{n}}(\alpha)\, \mathbf{n}', \tag{14.7.11}$$

with the spin states defined up to phases.

Comments: Note that the first equation above, (14.7.10), relates triplets: on the left-hand side the unitary operator acts on each component of $\hat{\mathbf{S}}$ by similarity, and on the right-hand side $\mathcal{R}_{\mathbf{n}}(\alpha)\, \hat{\mathbf{S}}$ is defined as in (14.7.6). We emphasize that $\mathcal{R}_{\mathbf{n}}(\alpha)$ is defined to act on vectors, or triplets, while $\hat{R}_{\mathbf{n}}(\alpha)$ is a unitary operator on spin space $\mathbb{C}^2$. The second equation, (14.7.11), tells us how $\hat{R}_{\mathbf{n}}(\alpha)$ rotates spin states.

Proof. We define the vector function $\mathbf{G}(\alpha)$ as the quantity we wish to evaluate:

$$\mathbf{G}(\alpha) \equiv \hat{R}_{\mathbf{n}}^{\dagger}(\alpha)\, \hat{\mathbf{S}}\, \hat{R}_{\mathbf{n}}(\alpha) = e^{\frac{i\alpha}{\hbar}\mathbf{n}\cdot\hat{\mathbf{S}}}\, \hat{\mathbf{S}}\, e^{-\frac{i\alpha}{\hbar}\mathbf{n}\cdot\hat{\mathbf{S}}}. \tag{14.7.12}$$

Note that $\mathbf{G}(0) = \hat{\mathbf{S}}$. Now differentiate with respect to α to obtain

$$\frac{d\mathbf{G}(\alpha)}{d\alpha} = \frac{i}{\hbar}\, e^{\frac{i\alpha}{\hbar}\mathbf{n}\cdot\hat{\mathbf{S}}}\, [\mathbf{n}\cdot\hat{\mathbf{S}}, \hat{\mathbf{S}}]\, e^{-\frac{i\alpha}{\hbar}\mathbf{n}\cdot\hat{\mathbf{S}}}. \tag{14.7.13}$$

Using the commutator identity (14.7.3), we now get

$$\frac{d\mathbf{G}(\alpha)}{d\alpha} = e^{\frac{i\alpha}{\hbar}\mathbf{n}\cdot\hat{\mathbf{S}}}\, \mathbf{n}\times\hat{\mathbf{S}}\, e^{-\frac{i\alpha}{\hbar}\mathbf{n}\cdot\hat{\mathbf{S}}} = \mathbf{n}\times\left(e^{\frac{i\alpha}{\hbar}\mathbf{n}\cdot\hat{\mathbf{S}}}\, \hat{\mathbf{S}}\, e^{-\frac{i\alpha}{\hbar}\mathbf{n}\cdot\hat{\mathbf{S}}}\right), \tag{14.7.14}$$

showing that we have the differential equation:

$$\frac{d\mathbf{G}(\alpha)}{d\alpha} = \mathbf{n}\times\mathbf{G}(\alpha). \tag{14.7.15}$$

We recognize this as the equation (14.7.4) for a rotating operator, with the role of time played by α and with angular speed $\omega = 1$. The solution (14.7.5) applies; after all, the equation is a linear matrix equation, and it makes no difference whether the unknowns are operators or commuting objects. We therefore find that

$$\mathbf{G}(\alpha) = \mathcal{R}_{\mathbf{n}}(\alpha)\, \mathbf{G}(0) = \mathcal{R}_{\mathbf{n}}(\alpha)\, \hat{\mathbf{S}}. \tag{14.7.16}$$

This completes the proof of (14.7.10). To prove the second claim, giving the specification of $\mathbf{n}''$, we evaluate the expectation value of $\hat{\mathbf{S}}$ on the state $|\mathbf{n}''\rangle$:

$$\langle\mathbf{n}''|\hat{\mathbf{S}}\,|\mathbf{n}''\rangle = \langle\mathbf{n}'|\hat{R}_{\mathbf{n}}^{\dagger}(\alpha)\, \hat{\mathbf{S}}\, \hat{R}_{\mathbf{n}}(\alpha)|\mathbf{n}'\rangle = \langle\mathbf{n}'|\mathcal{R}_{\mathbf{n}}(\alpha)\hat{\mathbf{S}}\,|\mathbf{n}'\rangle, \tag{14.7.17}$$

where the rotation matrix is acting on $\hat{\mathbf{S}}$. Equivalently, it is acting on the vector defined by the expectation value:

$$\langle\mathbf{n}''|\hat{\mathbf{S}}\,|\mathbf{n}''\rangle = \mathcal{R}_{\mathbf{n}}(\alpha)\langle\mathbf{n}'|\hat{\mathbf{S}}\,|\mathbf{n}'\rangle. \tag{14.7.18}$$

Recalling the expectation value in (14.7.1), we get the relation we wished to prove:

$$\tfrac{\hbar}{2}\,\mathbf{n}'' = \tfrac{\hbar}{2}\,\mathcal{R}_{\mathbf{n}}(\alpha)\mathbf{n}' \quad\Rightarrow\quad \mathbf{n}'' = \mathcal{R}_{\mathbf{n}}(\alpha)\mathbf{n}'. \tag{14.7.19}$$

The operator $\hat{R}_{\mathbf{n}}(\alpha)$ rotates a spin state by rotating the vector that defines it with the associated matrix operator $\mathcal{R}_{\mathbf{n}}(\alpha)$. □

It is useful to elucidate how the property $\hat{R}_{\mathbf{n}}^{\dagger}(\alpha)\, \hat{\mathbf{S}}\, \hat{R}_{\mathbf{n}}(\alpha) = \mathcal{R}_{\mathbf{n}}(\alpha)\, \hat{\mathbf{S}}$ leads to the action of the rotation operators on a spin operator $\hat{S}_{\mathbf{n}'}$. For this we take the dot product of this equation with the unit vector $\mathbf{n}'$:

$$\hat{R}_{\mathbf{n}}^{\dagger}(\alpha)\, \mathbf{n}'\cdot\hat{\mathbf{S}}\, \hat{R}_{\mathbf{n}}(\alpha) = \mathbf{n}'\cdot\mathcal{R}_{\mathbf{n}}(\alpha)\, \hat{\mathbf{S}}. \tag{14.7.20}$$

On the left-hand side, the dot product gives the spin operator $\hat{S}_{\mathbf{n}'}$. On the right-hand side, we use the invariance (14.7.8) of the dot product to act with $\mathcal{R}_{\mathbf{n}}(-\alpha)$ on both vectors:

$$\hat{R}_{\mathbf{n}}^{\dagger}(\alpha)\, \hat{S}_{\mathbf{n}'}\, \hat{R}_{\mathbf{n}}(\alpha) = (\mathcal{R}_{\mathbf{n}}(-\alpha)\mathbf{n}')\cdot\hat{\mathbf{S}}. \tag{14.7.21}$$

The right-hand side is now a spin operator, and we find that

$$\hat{R}_{\mathbf{n}}^{\dagger}(\alpha)\, \hat{S}_{\mathbf{n}'}\, \hat{R}_{\mathbf{n}}(\alpha) = \hat{S}_{\tilde{\mathbf{n}}'} \quad\text{with}\quad \tilde{\mathbf{n}}' = \mathcal{R}_{\mathbf{n}}(-\alpha)\mathbf{n}'. \tag{14.7.22}$$

The vector $\mathbf{n}'$ that defines the spin operator becomes $\tilde{\mathbf{n}}'$ using the inverse of the rotation that transformed the states in (14.7.11). Letting $\alpha \to -\alpha$ in the above relation and recalling that the rotation operators are unitary, we find that

$$\boxed{\hat{R}_{\mathbf{n}}(\alpha)\,\hat{S}_{\mathbf{n}'}\,\hat{R}^{\dagger}_{\mathbf{n}}(\alpha) = \hat{S}_{\mathbf{n}''} \quad \text{with} \quad \mathbf{n}'' = \mathcal{R}_{\mathbf{n}}(\alpha)\mathbf{n}'.} \tag{14.7.23}$$

You can quickly see that an inner product $\langle \mathbf{n}_1 | \hat{S}_{\mathbf{n}_2} | \mathbf{n}_3 \rangle$ is invariant under the simultaneous transformations:

$$|\mathbf{n}_1\rangle \to \hat{R}_{\mathbf{n}}(\alpha)|\mathbf{n}_1\rangle, \quad |\mathbf{n}_3\rangle \to \hat{R}_{\mathbf{n}}(\alpha)|\mathbf{n}_3\rangle, \quad \hat{S}_{\mathbf{n}_2} \to \hat{R}_{\mathbf{n}}(\alpha)\,\hat{S}_{\mathbf{n}_2}\,\hat{R}^{\dagger}_{\mathbf{n}}(\alpha). \tag{14.7.24}$$

This invariance is the statement that the theory has rotational symmetry: nothing changes when states and operators are simultaneously rotated.

If you go over the above derivations, the key result, $\hat{R}^{\dagger}_{\mathbf{n}}(\alpha)\,\hat{\mathbf{S}}\,\hat{R}_{\mathbf{n}}(\alpha) = \mathcal{R}_{\mathbf{n}}(\alpha)\,\hat{\mathbf{S}}$, was obtained using *only* the commutator algebra of the spin angular momentum operators. Since orbital angular momentum operators obey the same commutator algebra, this result also holds for orbital angular momentum. More generally, letting $\hat{\mathbf{J}} = (\hat{J}_1, \hat{J}_2, \hat{J}_3)$ be angular momentum operators $[\hat{J}_i, \hat{J}_j] = i\hbar\epsilon_{ijk}\hat{J}_k$, we have

$$\boxed{\hat{R}^{\dagger}_{\mathbf{n}}(\alpha)\,\hat{\mathbf{J}}\,\hat{R}_{\mathbf{n}}(\alpha) = \mathcal{R}_{\mathbf{n}}(\alpha)\,\hat{\mathbf{J}}, \quad \hat{R}_{\mathbf{n}}(\alpha) = e^{-i\frac{\alpha}{\hbar}\mathbf{n}\cdot\hat{\mathbf{J}}}.} \tag{14.7.25}$$

The rotation matrix $\mathcal{R}_{\mathbf{n}}(\alpha)$ is the same one we used above.

14.8 From Inner Products to Bra-kets

Paul Dirac invented an alternative notation for inner products that leads to the concepts of *bras* and *kets*. Dirac's notation is sometimes more efficient than the conventional mathematical notation we have developed. It is also widely used.

In this and the following sections, we discuss Dirac's notation and spend some time rewriting some of our results using bras and kets. This will afford you the opportunity to appreciate the identities by looking at them in a different notation. Operators can also be written in terms of bras and kets, using their matrix representation. We are providing here a detailed explanation for some of the rules we followed when we began writing spin states as kets in chapter 12.

A classic application of the bra-ket notation is to systems with a nondenumerable basis of states, such as position states $|x\rangle$ and momentum states $|p\rangle$ of a particle moving in one dimension. We will consider these in section 14.10.

The construction begins by writing the inner product differently. The first step in the Dirac notation is to define the so called bra-ket pair $\langle u|v\rangle$ of two vectors u and v using inner products:

$$\langle u|\, v\rangle \equiv \langle\, u, v\,\rangle. \tag{14.8.1}$$

It is as if the inner-product comma is replaced by a vertical bar! Since things look a bit different in this notation, let us rewrite a few of the properties of inner products in bra-ket notation.

We now say, for example, that $\langle v|v\rangle \geq 0$ for all v, while $\langle v|v\rangle = 0$ if and only if $v = 0$. The conjugate exchange symmetry becomes $\langle u|v\rangle = \langle v|u\rangle^*$. Additivity and homogeneity on the second entry is written as

$$\langle u|c_1v_1 + c_2v_2\rangle = c_1\langle u|v_1\rangle + c_2\langle u|v_2\rangle, \quad c_1, c_2 \in \mathbb{C}, \tag{14.8.2}$$

while conjugate homogeneity (14.1.15) on the first entry is summarized by

$$\langle c_1u_1 + c_2u_2|v\rangle = c_1^*\langle u_1|v\rangle + c_2^*\langle u_2|v\rangle. \tag{14.8.3}$$

Two vectors u and v are orthogonal if $\langle u|v\rangle = 0$, and the norm $\|v\|$ of a vector is $\|v\|^2 = \langle v|v\rangle$. The Schwarz inequality for any pair of vectors u and v, reads $|\langle u|v\rangle| \leq \|u\|\,\|v\|$.

A set of basis vectors $\{e_i\}$ with $i = 1, \ldots, n$ is said to be orthonormal if

$$\langle e_i|e_j\rangle = \delta_{ij}. \tag{14.8.4}$$

An arbitrary vector can be written as a linear superposition of basis states:

$$v = \sum_i \alpha_i e_i. \tag{14.8.5}$$

We then see that the coefficients are determined by the inner product

$$\langle e_k|v\rangle = \Big\langle e_k\Big|\sum_i \alpha_i e_i\Big\rangle = \sum_i \alpha_i\langle e_k|e_i\rangle = \alpha_k. \tag{14.8.6}$$

We can therefore write, just as we did in (14.2.6),

$$v = \sum_i e_i\,\langle e_i|v\rangle. \tag{14.8.7}$$

The next step is to isolate bras and kets from the bra-ket. To do this we reinterpret the bra-ket form of the inner product. We want to "split" the bra-ket into two ingredients, a bra and a ket:

$$\langle u|v\rangle \;\Rightarrow\; \langle u|\;\;|v\rangle. \tag{14.8.8}$$

Here the symbol $|v\rangle$ is called a **ket**, and the symbol $\langle u|$ is called a **bra**. The bra-ket is recovered when the space between the bra and the ket collapses.

We will view the ket $|v\rangle$ as another way to write the vector v. There is a bit of redundancy in this notation that may be confusing: both $v \in V$ and $|v\rangle \in V$. Both are vectors in V, but sometimes the ket $|v\rangle$ is called a *state* in V. The enclosing symbol $|\;\rangle$ is a decoration added to the vector v without changing its meaning, perhaps like the familiar arrows added above a symbol to denote a vector. In this case the label in the ket is a vector, and the ket itself is that vector!

When the label of the ket is a vector, the bra-ket notation is a direct rewriting of the mathematical notation. Sometimes, however, the label of the ket is not a vector. The label could be the value of some quantity that characterizes the state. In such cases the notation affords some extra flexibility. We used such labeling, for example, when we wrote $|+\rangle$ and $|-\rangle$ for the spin states that point along the positive z-direction and along the negative z-direction, respectively. We will encounter similar situations in this chapter.

$$\boxed{\begin{array}{l}\text{Sometimes the label inside a ket is the vector itself;}\\ \text{other times it is an object that characterizes the vector.}\end{array}} \tag{14.8.9}$$

Let T be an operator in a vector space V. We wrote Tv as the vector obtained by the action of T on the vector v. Now the same action would be written as $T|v\rangle$. With kets labeled by vectors, we can simply identify

$$|Tv\rangle \equiv T|v\rangle. \tag{14.8.10}$$

When kets are labeled by vectors, operators go in or out of the ket without change. If the ket labels are not vectors, the above identification is not possible. Imagine a non-degenerate system where we label the states by their energies, as in $|E_i\rangle$, where E_i is the value of the energy for the ith state. Acting with the momentum operator $\hat{p}$ on the state is denoted as $\hat{p}\,|E_i\rangle$. It would be confusing, however, to rewrite this as $|\hat{p}\,E_i\rangle$ since E_i is not a vector. It is an energy, and $\hat{p}$ does not act on energies.

Bras are rather different from kets, although we also label them by vectors. Bras are linear functionals on the vector space V. We defined linear functionals in section 14.4: they are linear maps ϕ from V to the numbers: $\phi(v) \in \mathbb{F}$. The set of *all* linear functionals on V is in fact a new vector space over $\mathbb{F}$, the vector space V^* **dual to** V. The vector space structure of V^* follows from the natural definitions of sums of linear functionals and the multiplication of linear functionals by numbers:

1. For $\phi_1, \phi_2 \in V^*$, we define the sum $\phi_1 + \phi_2 \in V^*$ by

$$(\phi_1 + \phi_2)v \equiv \phi_1(v) + \phi_2(v). \tag{14.8.11}$$

2. For $\phi \in V^*$ and $a \in \mathbb{F}$, we define $a\phi \in V^*$ by

$$(a\phi)(v) = a\phi(v). \tag{14.8.12}$$

We proved before that for any linear functional $\phi \in V^*$ there is a unique vector $u \in V$ such that $\phi(v) = \langle u, v\rangle$. We can make this more explicit by labeling the linear functional by u and thus writing

$$\phi_u(v) = \langle u, v\rangle. \tag{14.8.13}$$

Since the elements of V^* are characterized uniquely by vectors, the vector space V^* has the same dimensionality as V.

A bra is also labeled by a vector. The bra $\langle u|$ can also be viewed as a linear functional because it has a natural action on vectors. The bra $\langle u|$, acting on the vector $|v\rangle$, is defined to give the bra-ket number $\langle u|v\rangle$:

$$\langle u| \,:\, |v\rangle \;\rightarrow\; \langle u|v\rangle. \tag{14.8.14}$$

Compare this with

$$\phi_u \,:\, v \;\rightarrow\; \langle u, v\rangle. \tag{14.8.15}$$

Since $\langle u, v\rangle = \langle u|v\rangle$, the last two equations mean that we can identify

$$\boxed{\phi_u \Longleftrightarrow \langle u|.} \tag{14.8.16}$$

This identification will allow us to work out how to manipulate bras.

Once we choose a basis, a vector can be represented by a column vector, as discussed in (13.5.6). If kets are viewed as column vectors, then *bras should be viewed as row vectors.* In this way a bra to the left of a ket in the bra-ket makes sense: matrix multiplication of a row vector times a column vector gives a number. Indeed, for vectors

$$u = \begin{pmatrix} u_1 \\ \vdots \\ u_n \end{pmatrix}, \quad v = \begin{pmatrix} v_1 \\ \vdots \\ v_n \end{pmatrix}, \tag{14.8.17}$$

the canonical inner product gives

$$\langle u|v\rangle = u_1^* v_1 + \cdots + u_n^* v_n. \tag{14.8.18}$$

If we think of this as having a bra and a ket,

$$\langle u| = (u_1^*, \ldots, u_n^*), \quad |v\rangle = \begin{pmatrix} v_1 \\ \vdots \\ v_n \end{pmatrix}, \tag{14.8.19}$$

then matrix multiplication gives us the desired bra-ket:

$$\langle u|v\rangle = (u_1^*, \ldots, u_n^*) \cdot \begin{pmatrix} v_1 \\ \vdots \\ v_n \end{pmatrix} = u_1^* v_1 + \cdots + u_n^* v_n. \tag{14.8.20}$$

The row representative of the bra $\langle u|$ was obtained by the transposition and complex conjugation of the column vector representative of $|u\rangle$.

The key properties needed to manipulate bras follow from the properties of linear functionals and identification (14.8.16). For our linear functionals, you can quickly verify that for $u_1, u_2 \in V$ and $a \in \mathbb{F}$,

$$\begin{aligned} \phi_{u_1+u_2} &= \phi_{u_1} + \phi_{u_2}, \\ \phi_{au} &= a^* \phi_u. \end{aligned} \tag{14.8.21}$$

With the noted identification with bras, these become

$$\boxed{\begin{aligned} \langle u_1 + u_2| &= \langle u_1| + \langle u_2|, \\ \langle au| &= a^* \langle u|. \end{aligned}} \tag{14.8.22}$$

If $\phi_u = \phi_{u'}$, then $u = u'$. Thus, we conclude that $\langle u| = \langle u'|$ implies $u = u'$.

A rule to pass from general kets to general bras is useful. We can obtain such a rule by considering the ket

$$|v\rangle = |\alpha_1 u_1 + \alpha_2 u_2\rangle = \alpha_1 |u_1\rangle + \alpha_2 |u_2\rangle. \tag{14.8.23}$$

Then

$$\langle v| \;=\; \langle \alpha_1 u_1 + \alpha_2 u_2| \;=\; \alpha_1^*\langle u_1| + \alpha_2^*\langle u_2|, \tag{14.8.24}$$

using the relations in (14.8.22). We have thus shown that the rule to pass from kets to bras, and *vice versa*, is

$$\boxed{|v\rangle = \alpha_1|u_1\rangle + \alpha_2|u_2\rangle \quad\Longleftrightarrow\quad \langle v| = \alpha_1^*\,\langle u_1| + \alpha_2^*\,\langle u_2|.} \tag{14.8.25}$$

As we mentioned earlier, we sometimes write kets with labels other than vectors. Let us reconsider the basis vectors $\{e_i\}$ discussed in (14.8.4). The ket $|e_i\rangle$ is simply called $|i\rangle$, and the orthonormal condition reads

$$\langle i\,|j\rangle \;=\; \delta_{ij}. \tag{14.8.26}$$

The expansion (14.8.5) of a vector now reads

$$|v\rangle \;=\; \sum_i \,|\,i\rangle\,\alpha_i. \tag{14.8.27}$$

As in (14.8.6), the expansion coefficients are $\alpha_i \;=\; \langle i\,|\,v\rangle$ so that

$$\boxed{|v\rangle \;=\; \sum_i |i\rangle\langle i\,|\,v\rangle.} \tag{14.8.28}$$

We placed the numerical component $\langle i\,|\,v\rangle$ to the right of the ket $|\,i\rangle$. This is useful because we will soon discover that the sum $\sum_i |\,i\rangle\langle\, i|$ has a special meaning.

14.9 Operators, Projectors, and Adjoints

Let T be an operator in a vector space V. This means that it acts on kets to give kets. As we explained before, we denote by $T|v\rangle$ the vector obtained by acting with T on the vector v. Our identification of vectors with kets implies that

$$|Tv\rangle \;\equiv\; T|v\rangle. \tag{14.9.1}$$

Note that given a linear operator T on V, we can define an associated linear operator $\mathcal{O}_T$ on the dual space V^*:

$$\mathcal{O}_T :\; V^* \to V^*. \tag{14.9.2}$$

We write this as

$$\mathcal{O}_T :\; \langle u| \;\to\; \langle u|T \in V^*. \tag{14.9.3}$$

Here, $\langle u|T$ is *defined* as the bra (or linear functional) that acting on the ket $|v\rangle$ gives the number $\langle u|T|v\rangle$. Since any linear functional is represented by a vector, we can ask: What is the vector that represents $\langle u|T$? We can answer this quickly by using both bra-kets and inner products:

$$\langle u|T|v\rangle = \langle u|Tv\rangle = \langle u, Tv\rangle = \langle T^\dagger u, v\rangle = \langle T^\dagger u|v\rangle, \tag{14.9.4}$$

which holding for all v shows that

$$\boxed{\langle u|T = \langle T^\dagger u|.} \tag{14.9.5}$$

Recalling that $(T^\dagger)^\dagger = T$, we also find that

$$\boxed{\langle u|T^\dagger = \langle Tu|.} \tag{14.9.6}$$

An operator enters or leaves the bra by turning into its adjoint! Since $\langle Tu|$ is the bra associated with $|Tu\rangle$, the above relation says that

$$\boxed{\langle u|T^\dagger \text{ is the bra associated with } T|u\rangle.} \tag{14.9.7}$$

We can actually write operators using bras and kets, written in a suitable order. As an example, consider a bra $\langle u|$ and a ket $|w\rangle$, with $u, w \in V$. We claim that S, defined by

$$S = |u\rangle\langle w|, \tag{14.9.8}$$

is naturally viewed as a linear operator on V. While surprising, this could have been anticipated: the matrix product of a column vector (the ket) times a row vector (the bra), in that order, gives a matrix. Indeed, letting S act on a vector as the bra-ket notation suggests, we find that

$$S|v\rangle \equiv |u\rangle\langle w|v\rangle \propto |u\rangle, \text{ since } \langle w|v\rangle \text{ is a number.} \tag{14.9.9}$$

Acting on a bra, it gives a bra:

$$\langle v|S \equiv \langle v|u\rangle\langle w| \propto \langle w|, \text{ since } \langle v|u\rangle \text{ is a number.} \tag{14.9.10}$$

Let us now review the description of operators as matrices. For simplicity we will usually consider orthonormal bases, and we recall that in this case matrix elements of an operator are easily read from an inner product, as given in (14.2.13):

$$T_{ij} = \langle e_i, Te_j\rangle. \tag{14.9.11}$$

In bra-ket notation this reads

$$T_{ij} = \langle e_i|Te_j\rangle = \langle e_i|T|e_j\rangle. \tag{14.9.12}$$

Labeling the kets with the subscripts on the basis vectors, we write

$$\boxed{T_{ij} = \langle i|T|j\rangle.} \tag{14.9.13}$$

There is one additional claim: the operator T itself can be written in terms of the matrix elements and basis bras and kets. We claim that

$$\boxed{T = \sum_{i,j} |i\rangle\, T_{ij}\, \langle j|.} \tag{14.9.14}$$

We can verify that this is correct by computing the matrix elements of this T:

$$\begin{aligned}\langle i'|T|j'\rangle &= \sum_{i,j} \langle i'|(|i\rangle\, T_{ij}\, \langle j|)|j'\rangle = \sum_{i,j} \langle i'|\, i\rangle\, T_{ij}\, \langle j|j'\rangle \\ &= \sum_{i,j} \delta_{ii'}\, T_{ij}\, \delta_{jj'} = T_{i'j'},\end{aligned} \tag{14.9.15}$$

consistent with (14.9.13). The above boxed result means that $|i\rangle\langle j|$ is represented by a matrix with all zeroes except for a one at the (i,j) position. Since the trace of an operator is obtained by summing the diagonal elements of its matrix representation, $\mathrm{tr}\, T = \sum_i T_{ii}$, in an orthonormal basis we have

$$\mathrm{tr}\, T = \sum_i \langle i|T|i\rangle. \tag{14.9.16}$$

Let us now reconsider projector operators. Choose one element $|m\rangle$ from the orthonormal basis to form an operator P_m defined by

$$P_m \equiv |m\rangle\langle m|. \tag{14.9.17}$$

This operator maps any vector $|v\rangle \in V$ to a vector along $|m\rangle$. Indeed, acting on $|v\rangle$ it gives

$$P_m|v\rangle = |m\rangle\langle m|v\rangle \sim |m\rangle. \tag{14.9.18}$$

It follows that P_m is a projector to the one-dimensional subspace spanned by $|m\rangle$. It is in fact manifestly an orthogonal projector because any vector $|v\rangle$ killed by P must satisfy $\langle m|v\rangle = 0$ and is therefore orthogonal to $|m\rangle$. In the chosen basis, P_m is represented by a matrix in which all elements are zero except for the element $(P_m)_{mm}$, which is one:

$$P_m = \mathrm{diag}\,(0, \ldots, 1, \ldots, 0). \tag{14.9.19}$$

As befits a projector, $P_m P_m = P_m$:

$$P_m P_m = \big(|m\rangle\langle m|\big)\big(|m\rangle\langle m|\big) = |m\rangle\, \langle m|m\rangle\, \langle m| = |m\rangle\langle m|, \tag{14.9.20}$$

since $\langle m|m\rangle = 1$. The operator P_m is a *rank-one* projection operator since it projects to a one-dimensional subspace of V, the subspace generated by $|m\rangle$. The rank of the projector is also equal to the trace of its matrix representation, which for P_m is equal to one.

Using the basis vectors $|m\rangle$ and $|n\rangle$ with $m \neq n$, we can define

$$P_{m,n} \equiv |m\rangle\langle m| + |n\rangle\langle n|. \tag{14.9.21}$$

It should be clear to you that this is a projector to the two-dimensional subspace spanned by $|m\rangle$ and $|n\rangle$. It should also be clear that it is an orthogonal projector of rank two. You should also verify that $P_{m,n}P_{m,n} = P_{m,n}$. Similarly, we can construct a rank-three projector by adding to $P_{m,n}$ an extra term $|k\rangle\langle k|$ with $k \neq m, n$. If we include *all* basis vectors of a vector space of dimension N, we will have the operator

$$P_{1,\ldots,N} \equiv |1\rangle\langle 1| + \cdots + |N\rangle\langle N|. \tag{14.9.22}$$

As a matrix, $P_{1,\ldots,N}$ has a one on every element of the diagonal and a zero everywhere else. This is therefore the unit matrix, which represents the identity operator $\mathbb{1}$.

We thus have the surprising relation

$$\mathbb{1} = \sum_i |i\rangle\langle i|. \tag{14.9.23}$$

This equation is sometimes called a *resolution* of the identity: it decomposes the identity operator as a sum of projectors to one-dimensional orthogonal subspaces. The fact that the above right-hand side is the identity was hinted at in (14.8.28):

$$|v\rangle = \sum_i |i\rangle\langle i\,|\,v\rangle. \tag{14.9.24}$$

This expansion of any vector v along a basis follows from $|v\rangle = \mathbb{1}|v\rangle$ and the use of (14.9.23). We can view equation (14.9.23) as a *completeness relation* for the chosen orthonormal basis.

For a spin one-half system, the unit operator can be written as a sum of two terms since the vector space is two-dimensional. Using the orthonormal basis vectors $|+\rangle$ and $|-\rangle$ for spins along the positive and negative z-directions, respectively, we have

$$\mathbb{1} = |+\rangle\langle +| + |-\rangle\langle -|. \tag{14.9.25}$$

We can use the completeness relation to show that our formula (14.9.13) for matrix elements is consistent with matrix multiplication. Indeed, for the product T_1T_2 of two operators, we write

$$\begin{aligned}(T_1T_2)_{mn} = \langle m|T_1T_2|n\rangle &= \langle m|T_1\,\mathbb{1}\,T_2|n\rangle = \langle m|T_1\Big(\sum_{k=1}^{N}|k\rangle\langle k|\Big)T_2|n\rangle \\ &= \sum_{k=1}^{N}\langle m|T_1|k\rangle\langle k|T_2|n\rangle = \sum_{k=1}^{N}(T_1)_{mk}(T_2)_{kn}.\end{aligned} \tag{14.9.26}$$

This is the expected result for the product of T_1 and T_2.

Example 14.12. *Trace of the operator* $|u\rangle\langle w|$.

Consider two states $|u\rangle, |w\rangle \in V$, and use them to define the operator $|u\rangle\langle w|$. We claim that the trace of this operator is simply the overlap between the two states:

$$\text{tr}\,(|u\rangle\langle w|) = \langle w|u\rangle. \tag{14.9.27}$$

We can easily prove this. Assume for this purpose that $|i\rangle$ with $i = 1, \cdots, N$ forms an orthonormal basis. Recalling that $\text{tr}M = \sum_i \langle i|M|i\rangle$, for any operator M, we see that

$$\text{tr}\big(|u\rangle\langle w|\big) = \sum_{i=1}^{N}\langle i|\big(|u\rangle\langle w|\big)|i\rangle = \sum_{i=1}^{N}\langle i|u\rangle\langle w|i\rangle. \tag{14.9.28}$$

With a reordering of the multiplicative factors in the last expression, we encounter a resolution of the identity:

$$\text{tr}\big(|u\rangle\langle w|\big) = \sum_{i=1}^{N}\langle w|i\rangle\langle i|u\rangle = \langle w|\Big(\sum_{i=1}^{N}|i\rangle\langle i|\Big)|u\rangle = \langle w|u\rangle, \tag{14.9.29}$$

giving us the claimed result. The result can also be viewed as a corollary of the *generalized cyclicity* property of the trace: $\text{tr}(AB) = \text{tr}(BA)$ holds even when A and B are not square matrices. All that is needed is that they must be compatible for multiplication and the product must be a square matrix: thus, A can be of size $m \times p$ and B of type $p \times m$, with p and m arbitrary. In the trace of $|u\rangle\langle w|$, we view $|u\rangle$ as a $n \times 1$ matrix and $\langle w|$ as a $1 \times n$ matrix. The answer then follows from cyclicity. □

In preparation for the example that follows, recall that for a spin state $|\mathbf{n}\rangle$ pointing along the unit vector $\mathbf{n}$ we have $\langle \mathbf{n}|\hat{\mathbf{S}}|\mathbf{n}\rangle = \frac{\hbar}{2}\mathbf{n}$ (see (14.7.1)). Since $\hat{\mathbf{S}} = \frac{\hbar}{2}\boldsymbol{\sigma}$, we know that

$$\langle \mathbf{n}|\boldsymbol{\sigma}|\mathbf{n}\rangle = \mathbf{n}. \tag{14.9.30}$$

Example 14.13. *Orthogonal projector onto a spin state $|\mathbf{n}\rangle$ revisited.*
We constructed in example 14.7 the projector $P_{\mathbf{n}}$ onto the span of $|\mathbf{n}\rangle$. We now reconsider the construction. Following the discussion above, we immediately identify $P_{\mathbf{n}} = |\mathbf{n}\rangle\langle \mathbf{n}|$ and attempt to write this operator as a sum of Hermitian matrices.

The way to do this was worked out in example 14.4, where we derived (14.2.18), enabling us to write an arbitrary Hermitian matrix M in terms of Pauli matrices and the identity matrix. Using this formula for the present case,

$$|\mathbf{n}\rangle\langle \mathbf{n}| = \tfrac{1}{2}\,\text{tr}(|\mathbf{n}\rangle\langle \mathbf{n}|)\,\mathbb{1} + \tfrac{1}{2}\sum_{i=1}^{3}\text{tr}(\sigma_i|\mathbf{n}\rangle\langle \mathbf{n}|)\,\sigma_i. \tag{14.9.31}$$

Since $\text{tr}(|\mathbf{n}\rangle\langle \mathbf{n}|) = \langle \mathbf{n}|\mathbf{n}\rangle = 1$ and $\text{tr}(\sigma_i|\mathbf{n}\rangle\langle \mathbf{n}|) = \langle \mathbf{n}|\sigma_i|\mathbf{n}\rangle = n_i$, the above equation gives

$$|\mathbf{n}\rangle\langle \mathbf{n}| = \tfrac{1}{2}\,\mathbb{1} + \tfrac{1}{2}\sum_{i=1}^{3} n_i\,\sigma_i. \tag{14.9.32}$$

Therefore, we have shown that the projector $P_{\mathbf{n}}$ is in fact given by

$$\boxed{P_{\mathbf{n}} = |\mathbf{n}\rangle\langle \mathbf{n}| = \tfrac{1}{2}(\mathbb{1} + \mathbf{n}\cdot\boldsymbol{\sigma}).} \tag{14.9.33}$$

You can see from this formula that the trace of the projector is indeed one. □

Let us finish this section with elaborations on adjoints, Hermitian operators, and unitary operators. Letting the relation $\langle u|T^\dagger = \langle Tu|$ derived earlier act on the ket $|v\rangle$, we get

$$\boxed{\langle u|T^\dagger|v\rangle = \langle v|T|u\rangle^*, \quad \forall u, v \in V.} \tag{14.9.34}$$

This equation is useful to compute arbitrary matrix elements of $T^\dagger$ in terms of those of T. Taking u and v to be orthonormal basis vectors, we find the familiar matrix representation of the adjoint operator: $\langle i|T^\dagger|j\rangle = \langle j|T|i\rangle^*$, leading to $(T^\dagger)_{ij} = (T_{ji})^*$.

Exercise 14.7. *Show that $(T_1T_2)^\dagger = T_2^\dagger T_1^\dagger$ by taking matrix elements.*

Example 14.14. *Adjoint in bra-ket description of operators.*
Consider arbitrary vectors u, w and define the operator T by the ket-bra expression $T = |u\rangle\langle w|$. We wish to find the ket-bra expression for $T^\dagger$. We claim that

$$\boxed{T = |u\rangle\langle w| \;\Rightarrow\; T^\dagger = |w\rangle\langle u|.} \tag{14.9.35}$$

This equation allows us to compute in bra-ket language the adjoint of any operator. To establish this claim, we act with T on an arbitrary $|v\rangle$ and then pass to the associated bras:

$$T|v\rangle = |u\rangle\langle w|v\rangle \;\Rightarrow\; \langle v|T^\dagger = \langle v|w\rangle\langle u|. \tag{14.9.36}$$

Since this equation is valid for any bra $\langle v|$, we read that indeed $T^\dagger = |w\rangle\langle u|$. □

Recall that a linear operator T is said to be Hermitian if $T^\dagger = T$. An operator A is said to be *anti*-Hermitian if $A^\dagger = -A$.

Exercise 14.8. *Show that the commutator* $[T_1, T_2]$ *of two Hermitian operators* T_1 *and* T_2 *is anti-Hermitian. Show that the commutator* $[A_1, A_2]$ *of two anti-Hermitian operators* A_1 *and* A_2 *is also anti-Hermitian. Show that the commutator* $[T, A]$ *of a Hermitian operator* T *and an anti-Hermitian operator* A *is Hermitian.*

It should be clear from our earlier work that for $T = T^\dagger$,

$$\langle Tu|v\rangle = \langle u|Tv\rangle, \quad \text{and} \quad \langle v|T|u\rangle^* = \langle u|T|v\rangle, \;\; \forall u, v, \;\; T \text{ Hermitian}. \tag{14.9.37}$$

Inside a bra-ket, a Hermitian operator moves freely from the bra to the ket and vice versa.

A unitary operator U satisfies $U^\dagger U = UU^\dagger = \mathbb{1}$ and, as a consequence, preserves inner products:

$$\langle Uu|Uw\rangle = \langle u|U^\dagger|Uw\rangle = \langle u|U^\dagger U|w\rangle = \langle u|w\rangle. \tag{14.9.38}$$

Unitary operators acting on an orthonormal basis give another orthonormal basis. If states $|e_i\rangle$ form an orthonormal basis, then $|f_i\rangle$ defined by

$$|f_i\rangle \equiv U|e_i\rangle, \quad \forall i \tag{14.9.39}$$

form another orthonormal basis, as you can quickly check. In fact, we can write the unitary operator explicitly in terms of the two sets of basis vectors:

$$U = \sum_k |f_k\rangle\langle e_k|. \tag{14.9.40}$$

It is clear that this sum acting on $|e_i\rangle$ gives $|f_i\rangle$, as desired. Using (14.9.35) to compute the adjoint, we have

$$U^\dagger = \sum_k |e_k\rangle\langle f_k|, \tag{14.9.41}$$

and now you can quickly verify that $U^\dagger U = UU^\dagger = \mathbb{1}$. This construction shows that given any two orthonormal bases on the state space there is a unitary operator that relates them.

Exercise 14.9. *Prove that* $\langle e_i|U|e_j\rangle = \langle f_i|U|f_j\rangle$. *This is the by-now-familiar property that the matrix elements of a basis-changing operator are the same in both bases. This is almost obvious in the new notation!*

14.10 Nondenumerable Basis States

In this section we describe the use of bras and kets for the position and momentum states of a particle moving on the real line $x \in \mathbb{R}$. Let us begin with position. We will introduce position states $|x\rangle$ where the label x in the ket is the value of the position. Roughly, $|x\rangle$ represents the state of the system where the particle is at position x. The full state space requires position states $|x\rangle$ for all values of x. Physically, we consider all of these states to be linearly independent: the state of a particle at some point x_0 can't be built by the superposition of states where the particle is elsewhere. Since x is a continuous variable, the basis states form a nondenumerable infinite set:

$$\text{basis states: } |x\rangle, \quad \forall x \in \mathbb{R}. \tag{14.10.1}$$

Since we have an infinite number of basis vectors, this state space is an infinite-dimensional complex vector space. This should not surprise you. The states of a particle on the real line can be represented by wave functions, and the set of possible wave functions form an infinite-dimensional complex vector space.

Note here that the label in the ket is not a vector; it is the position on a line. If we did not have the decoration provided by the ket, it would be hard to recognize that the object is a state in an infinite-dimensional complex vector space. Therefore, the following should be noted:

$$\begin{aligned} |ax\rangle &\neq a|x\rangle, && \text{for any real } a \neq 1, \\ |-x\rangle &\neq -|x\rangle, && \text{unless } x = 0, \\ |x_1 + x_2\rangle &\neq |x_1\rangle + |x_2\rangle. \end{aligned} \tag{14.10.2}$$

All these equations would hold if the labels inside the kets were vectors. In the first line, roughly, the left-hand side is a state with a particle at ax, while the right-hand side is a state with a particle at x. Analogous remarks hold for the other lines. Note also that $|0\rangle$ represents a particle at $x = 0$, not the zero vector on the state space, for which we would probably have to use the symbol 0.

For the quantum mechanics of a particle moving in three spatial dimensions, we would have position states $|\mathbf{x}\rangle$. Here the label is a vector in a three-dimensional real vector space, while the ket is a vector in the infinite-dimensional complex vector space of the theory. Again, the decoration enclosing the vector label plays a crucial role: it reminds us that the state lives in an infinite-dimensional complex vector space.

Let us go back to our position basis states for the one-dimensional problem. The inner product must be defined, so we will take

$$\langle x|y\rangle \equiv \delta(x - y). \tag{14.10.3}$$

It follows that position states with different positions are orthogonal to each other. The norm of a position state is infinite: $\langle x|x\rangle = \delta(0) = \infty$, so these are not allowed states of particles. We visualize the state $|x\rangle$ as the state of a particle perfectly localized at x, but this is an idealization. We can easily construct normalizable states using the superpositions of position states. We also have a completeness relation:

$$\mathbb{1} = \int_{-\infty}^{\infty} dx\, |x\rangle\langle x|. \tag{14.10.4}$$

This is consistent with our inner product above. Letting the above equation act on $|y\rangle$, we find an equality:

$$|y\rangle = \int dx\, |x\rangle\langle x|y\rangle = \int dx\, |x\rangle\, \delta(x-y) = |y\rangle. \tag{14.10.5}$$

All integrals are now assumed to run from $-\infty$ to $+\infty$. The position operator $\hat{x}$ is defined by its action on the position states. Not surprisingly, we define

$$\hat{x}\,|x\rangle \equiv x\,|x\rangle, \tag{14.10.6}$$

thus declaring that $|x\rangle$ are $\hat{x}$ eigenstates with eigenvalue equal to the position x. We can also show that $\hat{x}$ is a Hermitian operator by checking that $\hat{x}^\dagger$ and $\hat{x}$ have the same matrix elements:

$$\langle x_1|\hat{x}^\dagger|x_2\rangle = \langle x_2|\hat{x}|x_1\rangle^* = [x_1\delta(x_1-x_2)]^* = x_2\delta(x_1-x_2) = \langle x_1|\hat{x}|x_2\rangle, \tag{14.10.7}$$

using the reality of x_1 and $\delta(x_1-x_2)$ and the symmetry of the delta function to change x_1 into x_2. We thus conclude that $\hat{x}^\dagger = \hat{x}$. As a result, the bra associated with (14.10.6) is

$$\langle x|\hat{x} = x\langle x|. \tag{14.10.8}$$

The wave function associated with a state is formed by taking the inner product of a position state with the given state. Given the state $|\psi\rangle$ of a particle, we define the associated position state wave function $\psi(x)$ by

$$\psi(x) \equiv \langle x|\psi\rangle \;\in \mathbb{C}. \tag{14.10.9}$$

This is sensible: $\langle x|\psi\rangle$ is a number that depends on the value of x and is thus a function of x. We can now do a number of basic computations. First, we write any state as a superposition of position eigenstates by inserting $\mathbb{1}$, as in the completeness relation:

$$|\psi\rangle \;=\; \mathbb{1}|\psi\rangle \;=\; \int dx\, |x\rangle\, \langle x|\psi\rangle \;=\; \int dx\, |x\rangle\, \psi(x). \tag{14.10.10}$$

As expected, $\psi(x)$ is the component of $|\psi\rangle$ along the state $|x\rangle$. The overlap of states can also be written in position space:

$$\langle\phi|\psi\rangle \;=\; \langle\phi|\mathbb{1}|\psi\rangle \;=\; \int dx\, \langle\phi|x\rangle\, \langle x|\psi\rangle \;=\; \int dx\, \phi^*(x)\psi(x). \tag{14.10.11}$$

Matrix elements involving $\hat{x}$ are also easily evaluated:

$$\begin{aligned}\langle\phi|\hat{x}|\psi\rangle &= \langle\phi|\hat{x}\mathbb{1}|\psi\rangle = \int dx\, \langle\phi|\hat{x}|x\rangle\, \langle x|\psi\rangle \\ &= \int dx\, \langle\phi|x\rangle\, x\, \langle x|\psi\rangle = \int dx\, \phi^*(x)\, x\, \psi(x).\end{aligned} \tag{14.10.12}$$

We now introduce momentum states $|p\rangle$ that are eigenstates of the momentum operator $\hat{p}$, in complete analogy to the position states:

$$\begin{aligned}
&\text{Basis states: } |p\rangle, \quad \forall p \in \mathbb{R}. \\
&\langle p'|p\rangle = \delta(p-p'), \\
&\quad \mathbb{1} = \int dp\, |p\rangle\langle p|, \\
&\hat{p}\,|p\rangle = p\,|p\rangle.
\end{aligned} \tag{14.10.13}$$

Just as for position space, we also find that

$$\hat{p}^\dagger = \hat{p}, \quad \text{and} \quad \langle p|\hat{p} = p\langle p|. \tag{14.10.14}$$

In order to relate the two bases, we need the value of the overlap $\langle x|p\rangle$. Since $|p\rangle$ is the state of a particle with momentum p, we must interpret $\langle x|p\rangle$ as the wave function for a particle with momentum p:

$$\langle x|p\rangle = \frac{e^{ipx/\hbar}}{\sqrt{2\pi\hbar}}, \tag{14.10.15}$$

where the normalization was adjusted to be compatible with the completeness relations. Indeed, consider the $\langle p'|p\rangle$ overlap and use completeness in x to evaluate it:

$$\begin{aligned}
\langle p'|p\rangle &= \int dx \langle p'|x\rangle\langle x|p\rangle = \frac{1}{2\pi\hbar}\int dx e^{i(p-p')x/\hbar} \\
&= \frac{1}{2\pi}\int du\, e^{i(p-p')u} = \delta(p-p'),
\end{aligned} \tag{14.10.16}$$

where we let $u = x/\hbar$ and used the integral representation of the delta function obtained from Fourier's theorem in (4.4.5).

We can now ask: What is $\langle p|\psi\rangle$? The answer is quickly obtained by computation:

$$\langle p|\psi\rangle = \int dx \langle p|x\rangle\langle x|\psi\rangle = \frac{1}{\sqrt{2\pi\hbar}}\int dx e^{-ipx/\hbar}\psi(x) = \tilde{\psi}(p), \tag{14.10.17}$$

which is the Fourier transform of $\psi(x)$, as defined in (4.4.12). Thus, the *Fourier transform of $\psi(x)$ is the wave function in the momentum representation.*

It is often necessary to evaluate $\langle x|\hat{p}|\psi\rangle$. This is, by definition, the wave function of the state $\hat{p}|\psi\rangle$. We would expect it to equal the familiar action of the momentum operator on the wave function for $|\psi\rangle$. There is no need to speculate, because we can calculate this matrix element with the rules defined so far. We do so by inserting a complete set of momentum states:

$$\langle x|\,\hat{p}\,|\psi\rangle = \int dp\, \langle x|p\rangle\langle p|\hat{p}\,|\psi\rangle = \int dp\, (p\langle x|p\rangle)\langle p|\psi\rangle. \tag{14.10.18}$$

Now we notice that

$$p\langle x|p\rangle = \frac{\hbar}{i}\frac{d}{dx}\langle x|p\rangle, \tag{14.10.19}$$

and therefore,

$$\langle x|\,\hat{p}\,|\psi\rangle = \int dp \left(\frac{\hbar}{i}\frac{d}{dx}\langle x|p\rangle\right)\langle p|\psi\rangle. \tag{14.10.20}$$

The derivative can be moved out of the integral since no other part of the integrand depends on x:

$$\langle x|\hat{p}\,|\psi\rangle = \frac{\hbar}{i}\frac{d}{dx}\int dp\,\langle x|p\rangle\langle p|\psi\rangle. \tag{14.10.21}$$

The completeness sum is now trivial and can be discarded to obtain, as anticipated,

$$\boxed{\langle x|\hat{p}\,|\psi\rangle = \frac{\hbar}{i}\frac{d}{dx}\langle x|\psi\rangle = \frac{\hbar}{i}\frac{d}{dx}\psi(x).} \tag{14.10.22}$$

Exercise 14.10. *Show that*

$$\langle x|\hat{p}^n\,|\psi\rangle = \Big(\frac{\hbar}{i}\frac{d}{dx}\Big)^n\psi(x). \tag{14.10.23}$$

Exercise 14.11. *Show that*

$$\langle p|\,\hat{x}\,|\psi\rangle = i\hbar\frac{d}{dp}\tilde{\psi}(p). \tag{14.10.24}$$

Example 14.15. *Ket version of the Schrödinger equation.*

Given a state $|\psi\rangle$, we defined the wave function $\psi(x) = \langle x|\psi\rangle$. For the time-dependent state $|\Psi, t\rangle$, we define the Schrödinger wave function $\Psi(x,t)$ similarly:

$$\Psi(x,t) \equiv \langle x|\Psi,t\rangle. \tag{14.10.25}$$

In here, the time dependence simply goes along for the ride. Consider the familiar form of the Schrödinger equation for a particle of mass m moving in a one-dimensional potential $V(x,t)$:

$$i\hbar\frac{\partial\Psi(x,t)}{\partial t} = \Big(-\frac{\hbar^2}{2m}\frac{\partial^2}{\partial x^2} + V(x,t)\Big)\Psi(x,t). \tag{14.10.26}$$

Since the bra $\langle x|$ is time independent, (14.10.25) implies that the left-hand side of the above equation can be written as follows:

$$i\hbar\frac{\partial\Psi(x,t)}{\partial t} = \langle x|i\hbar\frac{\partial}{\partial t}|\Psi,t\rangle. \tag{14.10.27}$$

Similarly, using (14.10.23) for $n = 2$, we find that

$$-\frac{\hbar^2}{2m}\frac{\partial^2}{\partial x^2}\Psi(x,t) = \langle x|\frac{\hat{p}^2}{2m}|\Psi,t\rangle. \tag{14.10.28}$$

Finally,

$$V(x,t)\Psi(x,t) = \langle x|V(\hat{x},t)|\Psi,t\rangle. \tag{14.10.29}$$

It follows from the last three equations that the Schrödinger equation (14.10.26) can be rewritten as

$$\langle x|\,i\hbar\frac{\partial}{\partial t}|\Psi,t\rangle = \langle x|\Big(\frac{\hat{p}^2}{2m} + V(\hat{x},t)\Big)|\Psi,t\rangle. \tag{14.10.30}$$

Since this holds for arbitrary bra $\langle x|$, it follows that we have the ket equality:

$$i\hbar\frac{\partial}{\partial t}|\Psi,t\rangle = \Big(\frac{\hat{p}^2}{2m} + V(\hat{x},t)\Big)|\Psi,t\rangle. \tag{14.10.31}$$

Identifying the operator on the right-hand side as the Hamiltonian $\hat{H}$, we get

$$i\hbar\frac{\partial}{\partial t}|\Psi, t\rangle = \hat{H}\,|\Psi, t\rangle, \tag{14.10.32}$$

which is the "ket" version of the Schrödinger equation. □

Example 14.16. *Harmonic oscillator in bra-ket notation.*

The harmonic oscillator is a quantum system with a countable set of basis states: the energy eigenstates $\varphi_n(x)$ with $n = 0, 1, \ldots$. Since these wave functions have x dependence, it is natural to define kets $|n\rangle$ representing energy eigenstates and identify the wave functions as overlaps with the position states:

$$\varphi_n(x) = \langle x|n\rangle. \tag{14.10.33}$$

In particular, the ground state is now called $|0\rangle$, and we have

$$\varphi_0(x) = \langle x|0\rangle. \tag{14.10.34}$$

Do not confuse the oscillator ground state $|0\rangle$ with the zero vector or with a state of zero energy! The wave function $\varphi_0(x)$ arises from the condition $\hat{a}|0\rangle = 0$. To get a differential equation for φ_0, we act on $\hat{a}|0\rangle = 0$ with the position bra $\langle x|$:

$$\langle x|\hat{a}|0\rangle = 0 \quad\Rightarrow\quad \langle x|\Big(\hat{x} + \frac{i\hat{p}}{m\omega}\Big)\,|0\rangle = 0. \tag{14.10.35}$$

Using the identity (14.10.22) to turn $\hat{p}$ into a differential operator, we find that

$$\Big(x + \frac{i}{m\omega}\frac{\hbar}{i}\frac{d}{dx}\Big)\,\varphi_0(x) = 0 \Rightarrow \Big(\frac{d}{dx} + \frac{x}{L_0^2}\Big)\,\varphi_0 = 0, \tag{14.10.36}$$

where L_0 is the familiar oscillator size. The solution is indeed $\varphi_0(x) = N_0\,\exp(-\frac{x^2}{2L_0^2})$, with $N_0^2 = \frac{1}{\sqrt{\pi}}\frac{1}{L_0}$ for unit normalization. In the new notation, the formula $\varphi_n = \frac{1}{\sqrt{n!}}(\hat{a}^\dagger)^n\varphi_0$, derived in section 9.4, takes the form

$$|n\rangle \;=\; \frac{1}{\sqrt{n!}}(a^\dagger)^n|0\rangle. \tag{14.10.37}$$

These states are eigenstates of the number operator with eigenvalue n: $\hat{N}|n\rangle = n|n\rangle$. Recall also that the Hamiltonian is $\hat{H} = \hbar\omega(\hat{N} + \frac{1}{2})$. The action of creation and annihilation operators on the energy eigenstates was determined in (9.4.26). In the new notation,

$$\begin{aligned}\hat{a}^\dagger|n\rangle &= \sqrt{n+1}\,|n+1\rangle\\ \hat{a}\,|n\rangle &= \sqrt{n}\;|n-1\rangle.\end{aligned} \tag{14.10.38}$$

The matrix elements of an operator $\mathcal{O}$ are rewritten as $\langle\varphi_m, \mathcal{O}\varphi_n\rangle = \langle m|\mathcal{O}|n\rangle$. □

Problems

Problem 14.1. *Schwarz inequality and triangle inequality.*

1. For real vector spaces, the dot product satisfies the Schwarz inequality $(\mathbf{a}\cdot\mathbf{b})^2 \leq (\mathbf{a}\cdot\mathbf{a})(\mathbf{b}\cdot\mathbf{b})$. Prove this inequality as follows. Consider the vector $\mathbf{a} - \lambda\mathbf{b}$, with λ a

real constant. Note that

$$f(\lambda) \equiv (\mathbf{a} - \lambda \mathbf{b}) \cdot (\mathbf{a} - \lambda \mathbf{b}) \geq 0 \tag{14.10.39}$$

for all λ, and therefore the minimum over λ is still nonnegative: $\min_\lambda f(\lambda) \geq 0$. When is the Schwarz inequality saturated?

2. For a complex vector space, the Schwarz inequality reads $|\langle a, b\rangle| \leq \|a\| \, \|b\|$, with the norm defined by $\|a\|^2 = \langle a, a\rangle$. Prove this inequality using the vector $v(\lambda) \equiv a - \lambda b$, with λ a *complex* constant and noting that

$$f(\lambda) \equiv \langle v(\lambda), v(\lambda)\rangle \geq 0, \tag{14.10.40}$$

for all λ so that the minimum of $f(\lambda)$ over λ is nonnegative. When is the Schwarz inequality saturated? [Hint: To minimize over a complex variable (such as λ), one must vary the real and imaginary parts. Equivalently, show that you can treat λ and λ^* as if they were independent variables in the sense of partial derivatives. Confirm that since $f(\lambda)$ is real, the stationary condition for λ is equivalent to the stationary condition for λ^*.]

3. For a complex vector space, one has the *triangle inequality*

$$\|a + b\| \leq \|a\| + \|b\| \, . \tag{14.10.41}$$

Prove this inequality starting from the expansion of $\|a + b\|^2$. You will have to use the property $|\mathrm{Re}(z)| \leq |z|$, which holds for any complex number z, as well as the Schwarz inequality. Show that the triangle inequality is saturated if and only if $a = cb$ for c, a *real* positive constant.

Problem 14.2. *Overlap of two spin one-half states.*

Consider a spin state $|\mathbf{n}\rangle$ where $\mathbf{n}$ is the unit vector defined by the polar and azimuthal angles θ and ϕ and the spin state $|\mathbf{n}'\rangle$ where $\mathbf{n}'$ is the unit vector defined by the polar and azimuthal angles θ' and ϕ'. Let γ denote the angle between the vectors $\mathbf{n}$ and $\mathbf{n}'$: $\mathbf{n} \cdot \mathbf{n}' = \cos\gamma$. Show by direct computation that the overlap of the associated spin states is controlled by *half* the angle between the unit vectors:

$$|\langle \mathbf{n}'|\mathbf{n}\rangle|^2 = \cos^2 \tfrac{\gamma}{2} = \tfrac{1}{2}(1 + \mathbf{n} \cdot \mathbf{n}'). \tag{14.10.42}$$

Problem 14.3. *Orthogonal projections and approximations (Axler).*

Consider a vector space V with an inner product and a subspace U of V. The question is: Given a vector $v \in V$ that is not in U, what is the vector in U that best approximates v? As we also have a norm, we can ask a more precise question: What is the vector $u \in U$ for which $|v - u|$ is smallest? The answer is nice and simple: the vector u is given by $P_U v$, the orthogonal projection of v to U!

1. Prove the above claim by showing that for any $u \in U$ one has

$$|v - u| \geq |v - P_U v|. \tag{14.10.43}$$

As an application consider the infinite-dimensional vector space of real functions in the interval $x \in [-1, 1]$. The inner product of two functions f and g on this interval is taken to be

$$\langle f,g\rangle = \int_{-1}^{1} f(x)g(x)dx. \tag{14.10.44}$$

Take U to be the six-dimensional subspace of functions spanned by $\{1,x,x^2,x^3,x^4,x^5\}$. (For this problem please use an algebraic manipulator that does integrals.)

2. Use the Gram-Schmidt algorithm to find an orthonormal basis $(e_1,\ldots,e_6)$ for U.
3. Consider approximating the functions $\sin\pi x$ and $\cos\pi x$ with the best possible representatives from U. Calculate exactly these two representatives and write them as polynomials in x with coefficients that depend on powers of π and other constants. Also write the polynomials using numerical coefficients with six significant digits.
4. Do a plot for each of the functions ($\sin\pi x$ and $\cos\pi x$) where you show the original function, its best approximation in U calculated above, and the approximation in U that corresponds to the truncated Taylor expansion about $x=0$.

The polynomials you found in (2) are in fact proportional to the Legendre polynomials, which are orthogonal but normalized differently.

Problem 14.4. *Elaborations on a theorem.*

Theorem 14.5.1 states that for complex vector spaces, the condition $\langle v,Tv\rangle=0$ for all $v\in V$ implies that $T=0$. The result does not hold for real vector spaces. To distinguish the two cases, we consider separate conditions:

$$\text{Real case: } \langle u,Su\rangle=0, \text{ for all } u, \quad \text{complex case: } \langle v,Tv\rangle=0, \text{ for all } v. \tag{14.10.45}$$

We first examine the case of dimension two to see why the theorem is true and why it fails for real vector spaces. Then we extend to higher dimensions.

1. Let S be represented by a real 2×2 matrix S_{ij} and u by two real components u_i, with $i,j=1,2$. Similarly, let T be represented by a 2×2 matrix T_{ij} with complex entries and v by two complex components v_i, with $i,j=1,2$. Write out the quadratic forms and then apply the conditions under which they vanish for all u and v, respectively. Show that $T_{ij}=0$. For what kind of matrices S does the vanishing of $\langle u,Su\rangle$ imply the vanishing of S?
2. Extend your argument to arbitrary size matrices, showing that $T_{ij}=0$ and stating for what kind of matrices S the theorem holds in the real case.
3. Consider a complex vector space and an arbitrary linear operator. A basis can be shown to exist for which the matrix representing the operator has an upper-triangular form (the elements below the diagonal vanish). In light of the above analysis, explain why the same does not hold for arbitrary linear operators on real vector spaces.

Problem 14.5. *Another characterization of orthogonal projectors.*

Consider a vector space V and a linear operator P that satisfies the equation $P^2=P$. Theorem 14.5.4 demonstrates that this implies $V=\text{null}\,P\oplus\text{range}\,P$. This is not enough, however, to show that P is an orthogonal projector. Show that orthogonality is guaranteed if

$$|Pv| \leq |v| \quad \text{for all } v \in V. \tag{14.10.46}$$

You may find it useful to prove first the following characterization of orthogonal vectors: Let $u, v \in V$. Then $\langle u, v\rangle = 0$ if and only if $|u| \leq |u + av|$ for any constant a.

Problem 14.6. *Rotation matrix for vectors.*

Given a vector $\mathbf{u}$ in $\mathbb{R}^3$, the rotation matrix $\mathcal{R}_{\mathbf{n}}(\alpha)$ acting on the vector is supposed to give us the result of rotating $\mathbf{u}$ by an angle α about an axis oriented along $\mathbf{n}$. We want to show that, as indicated in (14.7.6),

$$\mathcal{R}_{\mathbf{n}}(\alpha)\,\mathbf{u} = (1 - \cos\alpha)\,(\mathbf{n}\cdot\mathbf{u})\,\mathbf{n} + (\cos\alpha)\,\mathbf{u} + (\sin\alpha)\,(\mathbf{n}\times\mathbf{u}). \tag{1}$$

We can denote the rotated vector by $\mathbf{u}(\alpha) = \mathcal{R}_{\mathbf{n}}(\alpha)\,\mathbf{u}$, with $\mathbf{u} = \mathbf{u}(0)$. As discussed, arising by rotation means that it satisfies the differential equation

$$\frac{d\mathbf{u}(\alpha)}{d\alpha} = \mathbf{n}\times\mathbf{u}(\alpha). \tag{2}$$

1. Verify the correctness of (1) by showing that it satisfies the differential equation (2).
2. Construct the solution (1) directly by inspection of the rotation geometry. For this write the $\alpha = 0$ vector $\mathbf{u}$ as follows:

   $$\mathbf{u} = (\mathbf{u}\cdot\mathbf{n})\mathbf{n} + \mathbf{u}_\perp, \quad \mathbf{u}_\perp = \mathbf{u} - (\mathbf{u}\cdot\mathbf{n})\,\mathbf{n}.$$

   Note that as α starts to differ from zero it is $\mathbf{u}_\perp$ that begins changing in time by rotating in a plane spanned by $\mathbf{u}_\perp$ and $\mathbf{n}\times\mathbf{u}_\perp$.
3. Use index notation to describe the rotation as $u_i(\alpha) = \mathcal{R}_{\mathbf{n}}(\alpha)_{ij}u_j$ with

   $$\mathcal{R}_{\mathbf{n}}(\alpha)_{ij} = (1 - \cos\alpha)n_i n_j + \cos\alpha\,\delta_{ij} + \sin\alpha\,\epsilon_{ikj}n_k.$$

   Use this to prove the rotational invariance (14.7.8) of the dot product.
4. Confirm explicitly the composition rule (14.7.7) for rotations.

Problem 14.7. *Sum rules and the quantum virial theorem.*

Consider the Hamiltonian $\hat{H} = \frac{\hat{p}^2}{2m} + V(\hat{x})$ for a one-dimensional quantum system. Assume $\hat{H}$ has a discrete set of eigenfunctions: $\hat{H}|a\rangle = E_a|a\rangle$, with a running over some set of values.

1. Prove the Thomas-Reiche-Kuhn sum rule: $\sum_{a'} \left|\langle a|\hat{x}|a'\rangle\right|^2 (E_{a'} - E_a) = \frac{\hbar^2}{2m}$. (Hint: consider $[[\hat{x}, \hat{H}], \hat{x}]$.)
2. Show that $\langle a|\hat{p}|a'\rangle = \frac{im}{\hbar}(E_a - E_{a'})\langle a|\hat{x}|a'\rangle$. (Hint: consider $[\hat{H}, \hat{x}]$.) Use this result to prove the energy-weighted sum rule:

   $$\sum_{a'} \left|\langle a|\hat{x}|a'\rangle\right|^2 (E_a - E_{a'})^2 = \frac{\hbar^2}{m^2}\langle a|\hat{p}^2|a\rangle.$$

3. Consider the commutator $[\hat{x}\hat{p}, \hat{H}]$ to show that

   $$2\langle a|\frac{\hat{p}^2}{2m}|a\rangle = \langle a|\hat{x}\partial_{\hat{x}}V(\hat{x})|a\rangle.$$

This is, in fact, the quantum-mechanical virial theorem, usually stated as $2\langle\hat{T}\rangle = \langle x\frac{dV}{dx}\rangle$, where $\hat{T}$ denotes kinetic energy, and the expectation values are for a stationary state. Write the resulting relation between expectation values of the kinetic and potential energy when $V(x) = \alpha x^n$.

Problem 14.8. *Exercises on the one-dimensional harmonic oscillator.*

1. Show that a state of the oscillator with a negative number has a negative norm squared. This means such a state is inconsistent.
2. We showed that the ground energy eigenstate $|0\rangle$ is the unique state annihilated by the lowering operator $\hat{a}$.
   - Show algebraically that the excited states of the oscillator are nondegenerate by showing that a degeneracy would imply a degeneracy of the ground state.
   - Show that the existence of a state with a positive but fractional number implies the existence of states of negative norm squared.
3. At $t=0$ a particle in the harmonic oscillator is in the superposition $|\psi(0)\rangle = \frac{1}{\sqrt{2}}\big(|0\rangle - |1\rangle\big)$. Find the time-dependent expectation values $\langle\hat{x}\rangle(t)$ and $\langle\hat{p}\rangle(t)$.
4. Consider a normalized state $|\lambda\rangle$ in the harmonic oscillator satisfying

$$\hat{a}|\lambda\rangle = \lambda\,|\lambda\rangle, \quad \lambda \in \mathbb{C}, \tag{1}$$

   where $\hat{a}$ is the annihilation operator, and λ is a complex constant. Note that the states above are coherent states, as discussed in example 13.21, with $\psi_\lambda(x) = \langle x|\lambda\rangle$. Calculate both the expectation value $\langle\hat{H}\rangle$ of the harmonic oscillator Hamiltonian and the energy uncertainty ΔH in the $|\lambda\rangle$ state. Equation (1) is the only property of the states needed for these computations.

Problem 14.9. *Parity operator and oscillator states.*

Let P denote a parity operator defined by its action on position eigenstates:

$$P: |x\rangle \to |-x\rangle, \quad \text{for all } x \in \mathbb{R}.$$

1. Given a state $|\psi\rangle$ with position space wave function $\psi(x)$, what is the wave function associated with the state $P|\psi\rangle$? What does P give when acting on the ground state of the harmonic oscillator?
2. Show that $P\hat{x} = -\hat{x}P$ and $P\hat{p} = -\hat{p}P$, where $\hat{x}$ and $\hat{p}$ are the position and momentum operators, respectively. Conclude that $P\hat{a}^\dagger = -\hat{a}^\dagger P$, where $\hat{a}^\dagger$ is the harmonic oscillator creation operator.
3. Show that the energy eigenstate $|n\rangle$ of the harmonic oscillator satisfies $P|n\rangle = (-1)^n|n\rangle$. What does this imply for the associated wave function $\varphi_n(x) = \langle x|n\rangle$?

15 Uncertainty Principle and Compatible Operators

The uncertainty of a Hermitian operator on a state of a quantum system vanishes if and only if the state is an eigenstate of the operator. We derive the Heisenberg uncertainty principle, which gives a lower bound for the product of uncertainties of two Hermitian operators. When one of the operators is the Hamiltonian, we are led to energy-time uncertainty relations. The uncertainty inequality is used to derive rigorous lower bounds for the energies of certain ground states. We discuss the diagonalization of operators and prove the spectral theorem, which states that any Hermitian operator or, more generally, any normal operator provides an orthonormal basis of eigenvectors for the state space. We prove that commuting Hermitian operators can be simultaneously diagonalized, addressing the issue of degeneracy. Finally, we discuss complete sets of commuting observables.

15.1 Uncertainty Defined

In quantum mechanics, observables are Hermitian operators. Given one such operator, $\hat{Q}$, we can measure it on a quantum system represented by a state Ψ. If the state Ψ is a $\hat{Q}$ eigenstate, there is no uncertainty in the value of the observable, which coincides with the eigenvalue of $\hat{Q}$ on Ψ. We only have uncertainty in the measured value of $\hat{Q}$ if Ψ is not a $\hat{Q}$ eigenstate but rather a superposition of $\hat{Q}$ eigenstates with different eigenvalues. We gave the definition of uncertainty and its basic properties in section 5.5. In this section we begin with a brief review of the main facts about uncertainty. Then we discuss a geometric interpretation of the uncertainty in terms of an orthogonal projection.

We call $\Delta Q(\Psi)$ the uncertainty of the Hermitian operator $\hat{Q}$ on the state Ψ. The uncertainty is a nonnegative real number and should vanish if and only if the state is an eigenstate of $\hat{Q}$. Recall that on normalized states Ψ, the expectation value of $\hat{Q}$ is given by

$$\langle \hat{Q} \rangle = \langle \Psi, \hat{Q}\Psi \rangle. \tag{15.1.1}$$

We could write $\langle \hat{Q} \rangle_\Psi$ to emphasize that the expectation value depends on Ψ, but this is usually not done to avoid cluttering the notation. The expectation $\langle \hat{Q} \rangle$ is guaranteed to be a real number since $\hat{Q}$ is Hermitian. We then define the uncertainty $\Delta Q(\Psi)$ as the norm of the vector obtained by acting with $(\hat{Q} - \langle\hat{Q}\rangle\mathbb{1})$ on the state Ψ:

$$\boxed{\Delta Q(\Psi) \equiv \|(\hat{Q} - \langle\hat{Q}\rangle\mathbb{1})\Psi\|.} \tag{15.1.2}$$

In the above, $\mathbb{1}$ is the identity operator. Defined as a norm, the uncertainty is manifestly nonnegative. We can quickly see that zero uncertainty means the state is an eigenstate of $\hat{Q}$. Indeed, a state of zero norm must be the zero state, and therefore

$$\Delta Q(\Psi) = 0 \;\Rightarrow\; (\hat{Q} - \langle\hat{Q}\rangle\mathbb{1})\Psi = 0 \;\Rightarrow\; \hat{Q}\,\Psi = \langle\hat{Q}\rangle\Psi. \tag{15.1.3}$$

Since $\langle\hat{Q}\rangle$ is a number, the last equation shows that Ψ is an eigenstate of $\hat{Q}$. You should also note that $\langle\hat{Q}\rangle$ is, on general grounds, the eigenvalue. Taking the eigenvalue equation $\hat{Q}\Psi = \lambda\Psi$ and forming the inner product with another Ψ, we get

$$(\Psi, \hat{Q}\Psi) = \lambda(\Psi, \Psi) = \lambda \quad\Rightarrow\quad \lambda = \langle\hat{Q}\rangle. \tag{15.1.4}$$

Alternatively, if the state Ψ is a $\hat{Q}$ eigenstate, we now know that the eigenvalue is $\langle\hat{Q}\rangle$, and therefore the state $(\hat{Q} - \langle\hat{Q}\rangle\mathbb{1})\Psi$ vanishes, and its norm is zero. We have therefore reconfirmed that the uncertainty $\Delta Q(\Psi)$ vanishes if and only if Ψ is a $\hat{Q}$ eigenstate.

To compute the uncertainty, one usually squares the expression in (15.1.2) so that

$$(\Delta Q(\Psi))^2 = \big((\hat{Q} - \langle\hat{Q}\rangle\mathbb{1})\Psi, (\hat{Q} - \langle\hat{Q}\rangle\mathbb{1})\Psi\big). \tag{15.1.5}$$

One then uses the Hermiticity of the operator $\hat{Q} - \langle\hat{Q}\rangle\mathbb{1}$ to move it from the left entry to the right entry, multiplies out, and evaluates.

Exercise 15.1. *Show that the above steps give the useful expression*

$$(\Delta Q(\Psi))^2 = \langle\hat{Q}^2\rangle - \langle\hat{Q}\rangle^2, \tag{15.1.6}$$

and conclude that $\langle\hat{Q}^2\rangle \geq \langle\hat{Q}\rangle^2$.

Geometric interpretation A geometric interpretation of the uncertainty is obtained by thinking about the relation between the states Ψ and $\hat{Q}\Psi$. Let U_Ψ be the one-dimensional vector subspace generated by Ψ. Now consider the state $\hat{Q}\Psi$. If Ψ is not an eigenstate of $\hat{Q}$, the state $\hat{Q}\Psi$ does not lie on U_Ψ. We now claim that

1. the orthogonal projection of $\hat{Q}\Psi$ to U_Ψ is in fact $\langle\hat{Q}\rangle\Psi$,
2. the component of $\hat{Q}\Psi$ in the orthogonal subspace $U_\Psi^\perp$ has a norm equal to ΔQ.

Figure 15.1 shows the various vectors in these claims. To prove (1) and (2), we consider the orthogonal projector P_{U_Ψ} to U_Ψ defined by its action on an arbitrary state η:

$$P_{U_\Psi}\eta = \Psi\,(\Psi, \eta). \tag{15.1.7}$$

As it should, the projector leaves invariant any state $c\Psi$ for arbitrary $c \in \mathbb{C}$ and kills any state orthogonal to Ψ. Claim (1) is quickly verified by calculating the projection of $\hat{Q}\Psi$ to U_Ψ:

$$P_{U_\Psi}\hat{Q}\Psi = \Psi\,(\Psi, \hat{Q}\Psi) = \Psi\,\langle\hat{Q}\rangle. \tag{15.1.8}$$

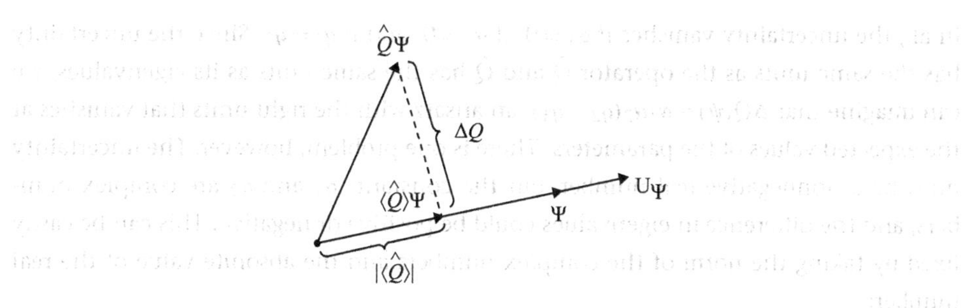

Figure 15.1
A state Ψ and the one-dimensional subspace U_Ψ generated by it. The projection of $\hat{Q}\Psi$ to U_Ψ is $\langle\hat{Q}\rangle\Psi$. The norm of the orthogonal component $(\hat{Q}\Psi)_\perp$ is the uncertainty $\Delta\hat{Q}$.

Moreover, the vector $\hat{Q}\Psi$ minus its projection to U_Ψ must be a vector $(\hat{Q}\Psi)_\perp$ orthogonal to Ψ:

$$\hat{Q}\Psi - \langle\hat{Q}\rangle\Psi = (\hat{Q}\Psi)_\perp, \tag{15.1.9}$$

as is easily confirmed by taking the inner product with Ψ. Since the norm of the above left-hand side is the uncertainty, we confirm that

$$\Delta Q = \|(\hat{Q}\Psi)_\perp\|, \tag{15.1.10}$$

as claimed in (2). These results are illustrated in figure 15.1. The uncertainty is the shortest distance from $\hat{Q}\Psi$ to the subspace U_Ψ. The magnitude of the expectation value of $\hat{Q}$ is the length of the projection of $\hat{Q}\Psi$ to U_Ψ. Letting χ denote the unit-norm state in the direction of $(\hat{Q}\Psi)_\perp$, we can also write

$$\hat{Q}\Psi = \langle\hat{Q}\rangle\Psi + \Delta Q\,\chi, \quad \langle\chi,\chi\rangle = 1, \quad \langle\chi,\Psi\rangle = 0. \tag{15.1.11}$$

Figure 15.1 makes explicit the Pythagorean relationship:

$$\langle\hat{Q}\rangle^2 + (\Delta Q)^2 = \|\hat{Q}\Psi\|^2, \tag{15.1.12}$$

where the right-hand side is $\langle\hat{Q}\Psi, \hat{Q}\Psi\rangle = \langle\hat{Q}^2\rangle$. This is just (15.1.6).

Example 15.1. *Uncertainty for a linear combination of two different eigenstates.*
Consider two normalized, orthogonal eigenstates ψ_1 and ψ_2 of a linear operator $\hat{Q}$, with eigenvalues q_1 and q_2, respectively:

$$\hat{Q}\psi_1 = q_1\psi_1, \quad \hat{Q}\psi_2 = q_2\psi_2. \tag{15.1.13}$$

Define a normalized state ψ formed as a superposition of the two eigenstates:

$$\psi = \alpha_1\psi_1 + \alpha_2\psi_2, \qquad \alpha_1, \alpha_2 \in \mathbb{C}. \tag{15.1.14}$$

We wish to calculate the uncertainty $\Delta Q(\psi)$ of the operator $\hat{Q}$ in the state ψ.

It is useful to anticipate the answer based on what we know about uncertainty. It is clear that if α_1 vanishes or α_2 vanishes then ψ becomes a $\hat{Q}$ eigenstate, and the uncertainty must vanish. Moreover, if $q_1 = q_2$ the states ψ_1 and ψ_2 are degenerate, and any superposition is an eigenstate of $\hat{Q}$. The uncertainty must vanish in this case as well. All in all, the uncertainty vanishes if $\alpha_1 = 0$, if $\alpha_2 = 0$, and if $q_1 = q_2$. Since the uncertainty has the same units as the operator $\hat{Q}$ and $\hat{Q}$ has the same units as its eigenvalues, we can imagine that $\Delta Q(\psi) \sim \alpha_1 \alpha_2 (q_2 - q_1)$, an ansatz with the right units that vanishes at the expected values of the parameters. There is one problem, however. The uncertainty must be a nonnegative real number, but the constants α_1 and α_2 are complex numbers, and the difference in eigenvalues could be positive or negative. This can be easily fixed by taking the norm of the complex numbers and the absolute value of the real number:

$$\Delta Q(\psi) \sim |\alpha_1|\,|\alpha_2|\,|q_2 - q_1|. \tag{15.1.15}$$

Let us assume that the above is correct and try to fix the overall constant. For this imagine $\alpha_1 = \alpha_2 = 1/\sqrt{2}$ so that the state $\psi = (\psi_1 + \psi_2)/\sqrt{2}$ is equally likely to be found in the first or in the second state. It follows that the expectation value of $\hat{Q}$ here must be the average of the eigenvalues $(q_1 + q_2)/2$. In the statistical interpretation, the square of the uncertainty is the variance, which is the average of the square of the deviations relative to the mean. But in this case since the mean lies at the midpoint between the possible observed values q_1 and q_2, the absolute value of all deviations is the same and equal to $|q_1 - q_2|/2$. It follows that the variance is this quantity squared, and the standard deviation or uncertainty ΔQ is the quantity $|q_1 - q_2|/2$ itself. This value for ΔQ is predicted by the above formula without any additional multiplicative constant. Thus, the evidence is that

$$\Delta Q(\psi) = |\alpha_1|\,|\alpha_2|\,|q_2 - q_1|. \tag{15.1.16}$$

Let us now confirm this by explicit calculation. Using the orthonormality of ψ_1 and ψ_2 and recalling that ψ is normalized, we have

$$\begin{aligned}
\langle \psi, \psi \rangle &= |\alpha_1|^2 + |\alpha_2|^2 = 1, \\
\langle \psi, \hat{Q}\psi \rangle &= |\alpha_1|^2 q_1 + |\alpha_2|^2 q_2, \\
\langle \psi, \hat{Q}^2\psi \rangle &= |\alpha_1|^2 q_1^2 + |\alpha_2|^2 q_2^2.
\end{aligned} \tag{15.1.17}$$

It then follows that

$$\begin{aligned}
(\Delta Q(\psi))^2 &= \langle \psi, \hat{Q}^2\psi \rangle - \langle \psi, \hat{Q}\psi \rangle^2 \\
&= |\alpha_1|^2 q_1^2 + |\alpha_2|^2 q_2^2 - \left(|\alpha_1|^4 q_1^2 + |\alpha_2|^4 q_2^2 + 2|\alpha_1|^2|\alpha_2|^2 q_1 q_2\right) \\
&= q_1^2|\alpha_1|^2(1 - |\alpha_1|^2) + q_2^2|\alpha_2|^2(1 - |\alpha_2|^2) - 2q_1q_2|\alpha_1|^2|\alpha_2|^2 \\
&= (q_1^2 - 2q_1q_2 + q_2^2)|\alpha_1|^2|\alpha_2|^2 \\
&= |\alpha_1|^2|\alpha_2|^2(q_1 - q_2)^2.
\end{aligned} \tag{15.1.18}$$

Taking the square root of this final result, we obtain the claimed (15.1.16). For the generalization of this result to the case of a superposition of multiple orthonormal eigenstates, see problem 15.3. □

15.2 The Uncertainty Principle

The uncertainty principle is an inequality satisfied by the product of the uncertainties of two Hermitian operators that fail to commute. Since the uncertainty of an operator on any given physical state is a number greater than or equal to zero, the product of uncertainties is also a real number greater than or equal to zero. The uncertainty inequality gives us a *lower bound* for the product of uncertainties. When the two operators in question commute, the uncertainty inequality gives no information: it states that the product of uncertainties must be greater than or equal to zero, which we already know.

Let us state the uncertainty inequality. Consider two Hermitian operators $\hat{A}$ and $\hat{B}$ and a physical state Ψ of the quantum system. Let ΔA and ΔB denote the uncertainties of $\hat{A}$ and $\hat{B}$, respectively, in the state Ψ. Then, the uncertainty inequality states that

$$(\Delta A)^2(\Delta B)^2 \geq \left(\langle\Psi|\tfrac{1}{2i}[\hat{A},\hat{B}]|\Psi\rangle\right)^2. \tag{15.2.1}$$

The left-hand side is a real, nonnegative number. For this to be a sensible inequality, the right-hand side must also be a real, nonnegative number. Since the square of a complex number would be either complex or negative, the object within parentheses must be real:

$$\langle\Psi|\tfrac{1}{2i}[\hat{A},\hat{B}]|\Psi\rangle \in \mathbb{R}. \tag{15.2.2}$$

This holds because the operator $\frac{1}{2i}[\hat{A},\hat{B}]$ is in fact Hermitian. To see this, first note that the commutator of two Hermitian operators is *anti*-Hermitian:

$$[\hat{A},\hat{B}]^\dagger = (\hat{A}\hat{B})^\dagger - (\hat{B}\hat{A})^\dagger = \hat{B}^\dagger\hat{A}^\dagger - \hat{A}^\dagger\hat{B}^\dagger = \hat{B}\hat{A} - \hat{A}\hat{B} = -[\hat{A},\hat{B}]. \tag{15.2.3}$$

The presence of the i then makes the operator $\frac{1}{2i}[\hat{A},\hat{B}]$ Hermitian, as claimed.

Taking square roots, the uncertainty inequality can also be written as

$$\boxed{\Delta A\,\Delta B \geq \left|\langle\Psi|\tfrac{1}{2i}[\hat{A},\hat{B}]|\Psi\rangle\right|.} \tag{15.2.4}$$

The bars on the right-hand side denote absolute value. The right-hand side is just the norm of the expectation value of $\frac{1}{2i}[\hat{A},\hat{B}]$ on the state Ψ. In general, all the quantities in the uncertainty inequality can be time dependent, and the inequality must hold for all times. The time dependence arises because Ψ is time dependent, and if it is not an energy eigenstate, expectation values and uncertainties can be time dependent. Of course, if Ψ is an energy eigenstate and $\hat{A}$ and $\hat{B}$ are time-independent operators, all quantities in the uncertainty inequality are time independent.

Before we prove the uncertainty inequality, let's do the canonical example!

Example 15.2. *Position-momentum uncertainty inequality.*

Taking $\hat{A}=\hat{x}$ and $\hat{B}=\hat{p}$ gives the position-momentum uncertainty relation:

$$\Delta x \Delta p \geq |\langle\Psi|\tfrac{1}{2i}[\hat{x},\hat{p}]|\Psi\rangle|. \tag{15.2.5}$$

Since $[\hat{x},\hat{p}]/(2i) = \hbar/2$ and Ψ is normalized, we get

$$\boxed{\Delta x\, \Delta p \geq \tfrac{\hbar}{2}.} \tag{15.2.6}$$

This is the most famous application of the uncertainty inequality. It is sometimes called Heisenberg's uncertainty principle. □

We now go over the proof of the uncertainty inequality (15.2.4). We do this not only because you should know how such an important result is derived. The derivation is needed to learn under what conditions, and for what kinds of states, the uncertainty inequality is saturated. Since for an arbitrary state the uncertainties can easily multiply to a value larger than the lower bound in the uncertainty inequality, states that saturate the inequality are in some sense minimum uncertainty states.

Proof of (15.2.4). We define the following two states:

$$\begin{aligned} |f_A\rangle &\equiv (\hat{A} - \langle\hat{A}\rangle\mathbb{1})|\Psi\rangle, \\ |f_B\rangle &\equiv (\hat{B} - \langle\hat{B}\rangle\mathbb{1})|\Psi\rangle. \end{aligned} \tag{15.2.7}$$

Note that by the definition (15.1.2) of uncertainty,

$$\begin{aligned} \langle f_A|f_A\rangle &= (\Delta A)^2, \\ \langle f_B|f_B\rangle &= (\Delta B)^2. \end{aligned} \tag{15.2.8}$$

The Schwarz inequality immediately furnishes an inequality involving the uncertainties:

$$\langle f_A|f_A\rangle\langle f_B|f_B\rangle \geq |\langle f_A|f_B\rangle|^2, \tag{15.2.9}$$

and therefore,

$$(\Delta A)^2(\Delta B)^2 \geq |\langle f_A|f_B\rangle|^2 = (\mathrm{Re}\langle f_A|f_B\rangle)^2 + (\mathrm{Im}\langle f_A|f_B\rangle)^2. \tag{15.2.10}$$

Our task is now to compute each of the two terms on the above right-hand side. For convenience we introduce *checked* operators

$$\begin{aligned} \check{A} &\equiv \hat{A} - \langle\hat{A}\rangle\mathbb{1}, \\ \check{B} &\equiv \hat{B} - \langle\hat{B}\rangle\mathbb{1}. \end{aligned} \tag{15.2.11}$$

We now compute

$$\begin{aligned} \langle f_A|f_B\rangle &= \langle\Psi|\check{A}\check{B}|\Psi\rangle = \langle\Psi|(\hat{A} - \langle\hat{A}\rangle\mathbb{1})(\hat{B} - \langle\hat{B}\rangle\mathbb{1})|\Psi\rangle \\ &= \langle\Psi|\hat{A}\hat{B}|\Psi\rangle - 2\langle\hat{A}\rangle\langle\hat{B}\rangle + \langle\hat{A}\rangle\langle\hat{B}\rangle \end{aligned} \tag{15.2.12}$$

so that, simplifying, we have

$$\begin{aligned} \langle f_A|f_B\rangle &= \langle\Psi|\check{A}\check{B}|\Psi\rangle = \langle\Psi|\hat{A}\hat{B}|\Psi\rangle - \langle\hat{A}\rangle\langle\hat{B}\rangle, \\ \langle f_B|f_A\rangle &= \langle\Psi|\check{B}\check{A}|\Psi\rangle = \langle\Psi|\hat{B}\hat{A}|\Psi\rangle - \langle\hat{B}\rangle\langle\hat{A}\rangle, \end{aligned} \tag{15.2.13}$$

where the second equation follows because $|f_A\rangle$ and $|f_B\rangle$ go into each other as we exchange $\hat{A}$ and $\hat{B}$. We now use this to find a nice expression for the imaginary part of $\langle f_A|f_B\rangle$:

$$\text{Im}\langle f_A|f_B\rangle = \tfrac{1}{2i}(\langle f_A|f_B\rangle - \langle f_B|f_A\rangle) = \tfrac{1}{2i}\langle\Psi|[\hat{A},\hat{B}]|\Psi\rangle. \tag{15.2.14}$$

For the real part, the expression is not that simple because the product of expectation values does not cancel. It is best to write the real part as the anticommutator of the checked operators:

$$\text{Re}\langle f_A|f_B\rangle = \tfrac{1}{2}(\langle f_A|f_B\rangle + \langle f_B|f_A\rangle) = \tfrac{1}{2}\langle\Psi|\{\check{A},\check{B}\}|\Psi\rangle. \tag{15.2.15}$$

Back in (15.2.10) we get

$$(\Delta A)^2(\Delta B)^2 \geq \left(\langle\Psi|\tfrac{1}{2i}[\hat{A},\hat{B}]|\Psi\rangle\right)^2 + \left(\langle\Psi|\tfrac{1}{2}\{\check{A},\check{B}\}|\Psi\rangle\right)^2. \tag{15.2.16}$$

This can be viewed as the most complete form of the uncertainty inequality. It turns out, however, that the second term on the right-hand side is seldom simple enough to be of use, and often it can be made equal to zero for certain states. At any rate, the term is positive or zero so it can be dropped while preserving the inequality. This is usually done, giving the familiar forms (15.2.1) and (15.2.4) that we have now established. □

What are the conditions for the uncertainty inequality to be saturated and thus achieve the minimum possible product of uncertainties? As the proof above shows, saturation is achieved under two conditions:

1. The Schwarz inequality is saturated. For this we need $|f_B\rangle = \beta|f_A\rangle$ where $\beta \in \mathbb{C}$.
2. $\text{Re}(\langle f_A|f_B\rangle) = 0$ so that the last term in (15.2.16) vanishes. This means that $\langle f_A|f_B\rangle + \langle f_B|f_A\rangle = 0$.

Using $|f_B\rangle = \beta|f_A\rangle$ in condition (2), we get

$$\langle f_A|f_B\rangle + \langle f_B|f_A\rangle = \beta\langle f_A|f_A\rangle + \beta^*\langle f_A|f_A\rangle = (\beta+\beta^*)\langle f_A|f_A\rangle = 0, \tag{15.2.17}$$

which requires that $\beta + \beta^* = 0$ or equivalently that the real part of β vanishes. It follows that β must be purely imaginary. So $\beta = i\lambda$, with λ real. Therefore, the uncertainty inequality will be saturated if and only if

$$|f_B\rangle = i\lambda|f_A\rangle, \quad \lambda \in \mathbb{R}. \tag{15.2.18}$$

More explicitly, this requires the

$$\boxed{\text{saturation condition:} \quad (\hat{B} - \langle\hat{B}\rangle\mathbb{1})|\Psi\rangle = i\lambda(\hat{A} - \langle\hat{A}\rangle\mathbb{1})|\Psi\rangle, \quad \lambda \in \mathbb{R}.} \tag{15.2.19}$$

This must be viewed as an equation for Ψ, given any two operators $\hat{A}$ and $\hat{B}$. Moreover, $\langle\hat{A}\rangle$ and $\langle\hat{B}\rangle$ are constants that happen to be Ψ dependent. What is λ, physically? The absolute value of λ is in fact fixed by the equation. Taking the norm of both sides, we get

$$\Delta B = |\lambda|\Delta A \quad \Rightarrow \quad |\lambda| = \frac{\Delta B}{\Delta A}. \tag{15.2.20}$$

The value of λ fixes the uncertainties ΔA and ΔB since the product $\Delta A \Delta B$ is, at saturation, equal to the right-hand side in the inequality and thus fixed once $|\Psi\rangle$ is known.

The saturation condition (15.2.19) can be written as an eigenvalue equation. By moving the $\hat{A}$ operator to the left-hand side and $\langle\hat{B}\rangle$ to the right-hand side, we have

$$(\hat{B}-i\lambda\hat{A})|\Psi\rangle = (\langle\hat{B}\rangle - i\lambda\langle\hat{A}\rangle)|\Psi\rangle. \tag{15.2.21}$$

Since the factor multiplying $|\Psi\rangle$ on the right-hand side is a number, this is indeed an eigenvalue equation. It is a curious one, however, since the operator $\hat{B}-i\lambda\hat{A}$ is *not* Hermitian. The eigenvalues are not real, and they are not expected to be quantized either. The presence of expectation values on the right-hand side may obscure the fact that this *is* a standard eigenvalue problem. Indeed, consider the related equation

$$(\hat{B}-i\lambda\hat{A})|\Psi\rangle = (b-i\lambda a))|\Psi\rangle, \tag{15.2.22}$$

with b and a *real* constants to be determined. This is the familiar eigenvalue problem. Taking expectation value by applying $\langle\Psi|$ to both sides of the equation, we get

$$\langle\hat{B}\rangle - i\lambda\langle\hat{A}\rangle = b-i\lambda a. \tag{15.2.23}$$

Since $a, b, \langle\hat{A}\rangle, \langle\hat{B}\rangle$, and λ are real, this equation implies $b=\langle\hat{B}\rangle$ and $a=\langle\hat{A}\rangle$. Therefore, when solving (15.2.21) you can simply take $\langle\hat{A}\rangle, \langle\hat{B}\rangle$ to be numbers, and if you get a solution, those numbers will indeed be the expectation values of the operators in your solution!

The classic illustration of this saturation condition is that for the $\hat{x}, \hat{p}$ uncertainty inequality $\Delta x\Delta p \geq \hbar/2$. Let us consider this next.

Example 15.3. *Saturating the position-momentum uncertainty inequality.*

We are looking here for wave functions $\psi(x)$ for which the uncertainties Δx and Δp multiply exactly to $\hbar/2$, thus saturating the inequality (15.2.6). Such wave functions are sometimes called *minimum uncertainty* states. More precisely, since the product $\Delta x\Delta p$ is fixed, for a fixed momentum uncertainty the states have the minimum position uncertainty, and for a fixed position uncertainty, the states have a minimum momentum uncertainty. The states here, having uncertainties in x and p, are neither position nor momentum eigenstates.

Using equation (15.2.19) with $\hat{A}=\hat{x}$ and $\hat{B}=\hat{p}$, the condition for saturation is

$$(\hat{p}-\langle\hat{p}\rangle\mathbb{1})|\psi\rangle = i\lambda\,(\hat{x}-\langle\hat{x}\rangle\mathbb{1})|\psi\rangle. \tag{15.2.24}$$

As discussed above, in searching for ψ we can treat $\langle\hat{p}\rangle$ and $\langle\hat{x}\rangle$ as constants, which we will call p_0 and x_0, to have more suggestive notation. If we solve the equation, these constants will in fact be, respectively, the $\hat{p}$ and $\hat{x}$ expectation values on ψ. Using the coordinate space description of ψ, the equation reads

$$\Big(\frac{\hbar}{i}\frac{d}{dx}-p_0\Big)\psi(x) = i\lambda\,(x-x_0)\psi(x). \tag{15.2.25}$$

The wave function $\psi(x)$ is in fact a function of x_0, p_0 and λ. With a little rearrangement, the differential equation becomes

$$\frac{d\psi}{dx} = \Big(\frac{ip_0}{\hbar}-\frac{\lambda}{\hbar}(x-x_0)\Big)\psi, \tag{15.2.26}$$

and integration gives

$$\ln \psi(x) = \frac{ip_0 x}{\hbar} - \frac{\lambda(x - x_0)^2}{2\hbar} + C', \tag{15.2.27}$$

with C' a constant of integration. Calling $e^{C'} = C$ the wave function takes the form

$$\psi(x) = C \exp\Big(-\frac{\lambda(x - x_0)^2}{2\hbar} + \frac{ip_0 x}{\hbar}\Big), \quad \lambda \in \mathbb{R}. \tag{15.2.28}$$

Now we see that λ, known to be real, should be positive in order for $\psi(x)$ to be normalizable. The wave function is a Gaussian centered at x_0, with a plane-wave modulating factor $\exp(ip_0 x/\hbar)$. This is the minimum uncertainty wave packet. On account of (15.2.20), we now have

$$\lambda = \frac{\Delta p}{\Delta x}. \tag{15.2.29}$$

As $\lambda \to 0$, we have $\Delta p \to 0$, and the wave function becomes a plane wave with momentum p_0. As $\lambda \to \infty$, the wave function localizes at $x = x_0$, and it approaches a position eigenstate. We can write λ in terms of the more physical position uncertainty. Recalling that saturation implies $\Delta x \Delta p = \hbar/2$, we find that

$$\lambda = \frac{\hbar}{2} \frac{1}{(\Delta x)^2}. \tag{15.2.30}$$

We can then rewrite the wave function as follows:

$$\psi(x) = \frac{1}{[2\pi(\Delta x)^2]^{1/4}} \exp\Big(-\frac{(x - x_0)^2}{4(\Delta x)^2} + \frac{ip_0 x}{\hbar}\Big), \quad x_0 = \langle \hat{x} \rangle, \; p_0 = \langle \hat{p} \rangle. \tag{15.2.31}$$

We have included in this formula the normalization factor. This is our final form for the general Gaussian minimum uncertainty packet. Rewriting equation (15.2.24) with all operators on the left-hand side,

$$(\hat{p} - i\lambda \hat{x})|\psi\rangle = (\langle \hat{p} \rangle - i\lambda \langle \hat{x} \rangle)|\psi\rangle, \tag{15.2.32}$$

it becomes clear that the minimum uncertainty packet is in fact an eigenstate of the *non-Hermitian* operator $\hat{p} - i\lambda \hat{x}$ with complex eigenvalue $p_0 - i\lambda x_0$. For a fixed $\lambda \geq 0$, eigenstates exist for any complex eigenvalue, since p_0 and x_0 can be fixed to any value. The minimum uncertainty states are some kind of generalized coherent states. Coherent states of the harmonic oscillator, to be studied in chapter 17, are eigenstates of the non-Hermitian annihilation operator $\hat{a} \sim \hat{p} - im\omega \hat{x}$, with arbitrary complex eigenvalues! These are minimum uncertainty wave functions with fixed $\lambda = m\omega$, and thus, from (15.2.30), $\Delta x = \sqrt{\frac{\hbar}{2m\omega}}$. □

15.3 Energy-Time Uncertainty

A more subtle form of the uncertainty relation deals with energy and time. The inequality is sometimes stated vaguely in the form $\Delta E \Delta t \gtrsim \hbar$. In here, there is no problem in defining ΔE precisely; after all, we have the Hamiltonian operator, and its uncertainty

ΔH is a perfect candidate for the energy uncertainty. The problem is time. Time is not an operator in quantum mechanics; it is a parameter, a real number used to describe the way systems change. Unless we define an uncertainty Δt precisely, we cannot hope for a well-defined uncertainty relation.

We describe a familiar setting in order to illustrate the spirit of the inequality $\Delta E \Delta t \gtrsim \hbar$. Consider a photon that is detected at some point in space, as a passing oscillatory wave of exact duration Δt. Without any quantum mechanical considerations, we can ask the observer for the value of the angular frequency ω of the pulse. In order to answer our question, the observer will attempt to count the number N of complete oscillations of the waveform that went through. Of course, this number N is given by Δt divided by the period $2\pi/\omega$ of the wave:

$$N = \frac{\omega}{2\pi}\Delta t. \tag{15.3.1}$$

The observer, however, will typically fail to count full waves because as the pulse gets started from zero and later dies off, the waveform may cease to follow the sinusoidal pattern. Thus, we expect an uncertainty $\Delta N \gtrsim 1$. Given the above relation, with Δt known exactly, this implies an uncertainty $\Delta\omega$ in the value of the angular frequency:

$$\Delta\omega\,\Delta t \gtrsim 2\pi. \tag{15.3.2}$$

This is all still classical; the above identity is routinely used by electrical engineers. It represents a limit on the ability to ascertain the frequency of a wave that is observed for a limited amount of time. It becomes quantum mechanical if we speak of a single photon whose energy is $E = \hbar\omega$. Then $\Delta E = \hbar\Delta\omega$ so that multiplying the above inequality by $\hbar$ we get

$$\Delta E\,\Delta t \gtrsim h. \tag{15.3.3}$$

In this uncertainty inequality, Δt is the duration of the pulse. It is a reasonable relation, but the presence of $\gtrsim$ betrays its lack of precision.

Russian physicists Leonid Mandelstam and Igor Tamm found a way out shortly after the formulation of the uncertainty principle. Consider a time-independent Hermitian operator $\hat{Q}$ that measures "Q-ness." We can find a precise energy-Q uncertainty inequality by applying the uncertainty inequality to the Hamiltonian $\hat{H}$ and $\hat{Q}$, finding

$$\Delta H\,\Delta Q \geq \left|\langle\Psi|\frac{1}{2i}[\hat{H},\hat{Q}]|\Psi\rangle\right|. \tag{15.3.4}$$

The right-hand side is interesting because the expectation value of the commutator $[\hat{H},\hat{Q}]$ controls the time dependence of the expectation value of $\hat{Q}$. Since $\hat{Q}$ is time independent, the time dependence of the expectation value originates from the time dependence of the states. The relevant equation was derived in section 5.2 and reads

$$\frac{\hbar}{i}\frac{d}{dt}\langle\hat{Q}\rangle = \langle[\hat{H},\hat{Q}]\rangle \quad \text{for time-independent } \hat{Q}. \tag{15.3.5}$$

This equation reminds us that an operator that commutes with $\hat{H}$ is conserved: its *expectation value* is time independent. With this result, the inequality (15.3.4) can be simplified:

$$\Delta H \Delta Q \geq \left| \left\langle \frac{1}{2i}[\hat{H}, \hat{Q}] \right\rangle \right| = \left| \frac{1}{2i} \frac{\hbar}{i} \frac{d\langle \hat{Q} \rangle}{dt} \right| = \frac{\hbar}{2} \left| \frac{d\langle \hat{Q} \rangle}{dt} \right|. \tag{15.3.6}$$

Therefore, we find that

$$\boxed{\Delta H \, \Delta Q \geq \frac{\hbar}{2} \left| \frac{d\langle \hat{Q} \rangle}{dt} \right|, \quad \text{for time-independent } \hat{Q}.} \tag{15.3.7}$$

This is a precise uncertainty inequality. It also suggests a definition of a time Δt_Q:

$$\Delta t_Q \equiv \frac{\Delta Q}{\left| \frac{d\langle \hat{Q} \rangle}{dt} \right|}. \tag{15.3.8}$$

This quantity has units of time. It is the time it would take $\langle \hat{Q} \rangle$ to change by ΔQ if both ΔQ and the velocity $\frac{d\langle \hat{Q} \rangle}{dt}$ were time independent. When $\langle \hat{Q} \rangle$ and ΔQ are roughly of the same size, we can view Δt_Q as the time for "appreciable" change in $\langle \hat{Q} \rangle$. In terms of Δt_Q, the uncertainty inequality reads

$$\boxed{\Delta H \Delta t_Q \geq \tfrac{\hbar}{2}.} \tag{15.3.9}$$

This is a precise inequality because Δt_Q has a precise definition.

The uncertainty relation involves ΔH. It is natural to ask if this quantity is time dependent. We will now show that the uncertainty ΔH, evaluated for any state of a quantum system, is time independent if the Hamiltonian is also time independent. Indeed, if $\hat{H}$ is time independent, we can use $\hat{H}$ and $\hat{H}^2$ for $\hat{Q}$ in (15.3.5) so that

$$\begin{aligned} \frac{d}{dt}\langle \hat{H} \rangle &= \frac{i}{\hbar}\big\langle [\hat{H}, \hat{H}] \big\rangle = 0, \\ \frac{d}{dt}\langle \hat{H}^2 \rangle &= \frac{i}{\hbar}\big\langle [\hat{H}, \hat{H}^2] \big\rangle = 0. \end{aligned} \tag{15.3.10}$$

It then follows that the energy uncertainty is conserved:

$$\frac{d}{dt}(\Delta H)^2 = \frac{d}{dt}\big(\langle \hat{H}^2 \rangle - \langle \hat{H} \rangle^2\big) = 0, \tag{15.3.11}$$

showing that ΔH is a constant. So we have shown that

$$\boxed{\text{if } \hat{H} \text{ is time independent, the uncertainty } \Delta H \text{ is constant in time.}} \tag{15.3.12}$$

The conservation of energy uncertainty can be used to understand some aspects of atomic decays. Consider the hyperfine transition in the hydrogen atom. Since both the proton and the electron are spin one-half particles, the ground state of hydrogen is fourfold degenerate, corresponding to the four possible combinations of spins (up-up, up-down, down-up, down-down). The magnetic interaction between the spins actually breaks this degeneracy and produces the so-called hyperfine splitting. This is a very tiny split: 5.88×10^{-6}eV, much smaller than the 13.6 eV binding energy of the ground state. For a hyperfine atomic transition, the emitted photon carries the energy difference: $E_\gamma = 5.88 \times 10^{-6}$eV, resulting in a wavelength of 21.1 cm and a frequency of $\nu = 1420.405751786(30)$ MHz. The eleven significant digits of this frequency attest to the sharpness of the emission line.

The issue of uncertainty arises because the excited state of the hyperfine splitting has a lifetime τ_H for decay to the ground state and emission of a photon. This lifetime is extremely long—in fact, $\tau_H \sim 11$ million years ($= 3.4 \times 10^{14}$ s (second), recalling that a year is about $\pi \times 10^7$s, accurate to better than 1%). This lifetime can be viewed as the time it takes some observable of the electron-proton system to change significantly (its total spin angular momentum, perhaps), so by the uncertainty principle, it must be related to some energy uncertainty $\Delta E \sim \hbar/\tau_H \simeq 2 \times 10^{-30}$ eV of the original excited state of the hydrogen atom. Once the decay takes place, the atom goes to the fully stable ground state, without any possible energy uncertainty. By the conservation of energy uncertainty, the photon must carry the uncertainty ΔE. But $\Delta E/E_\gamma \sim 3 \times 10^{-25}$, an absolutely infinitesimal effect on the photon, resulting in no broadening of the 21 cm line! That's one reason it is so useful in astronomy. For decays with much shorter lifetimes, there can be an observable broadening of an emission line due to the energy-time uncertainty principle.

The energy-Q uncertainty relation (15.3.7) can be used to derive an upper bound for the rate of change of the overlap $|\langle \Psi(0)|\Psi(t)\rangle|$ between a state and its initial value at time equal zero (problem 15.10). Sometimes a state can evolve in such a way that the above overlap vanishes for some value of time, the state turning orthogonal to the time-equal-zero state. The time $\Delta t_\perp$ that it takes to do so is in fact bounded below:

$$\Delta H\, \Delta t_\perp \geq \frac{h}{4}. \tag{15.3.13}$$

Note the h, not $\hbar$, to the right of the inequality. Moreover, the constant ΔH can be evaluated on the initial state. A state cannot evolve into an orthogonal state arbitrarily fast; there is a lower bound for the time it takes to do so, and this bound is inversely proportional to the energy uncertainty. Indeed, if the energy uncertainty is zero, the state is in fact an energy eigenstate and evolves only by a phase: $|\Psi(t)\rangle = \exp(-iEt/\hbar)|\Psi(0)\rangle$, with E the energy. In that case the overlap $|\langle \Psi(0)|\Psi(t)\rangle|$ remains one for all times $t > 0$. The bound may play a role in limiting the maximum possible speed of a quantum computer.

15.4 Lower Bounds for Ground State Energies

We have used the variational principle to find upper bounds on ground state energies. The uncertainty principle can be used to find *lower* bounds for the ground state energy of certain systems. These two approaches work together nicely in some cases. Below we use the uncertainty principle in the form $\Delta x \Delta p \geq \hbar/2$ to find rigorous lower bounds for the ground state energy of one-dimensional Hamiltonians.

This is best illustrated by example. Consider the Hamiltonian $\hat{H}$ for a particle in a one-dimensional quartic potential:

$$\hat{H} = \frac{\hat{p}^2}{2m} + \alpha\, \hat{x}^4, \quad \alpha > 0. \tag{15.4.1}$$

You did a variational analysis of this potential in problem 7.17, and the relevant energy scales were discussed in section 7.6. Our goal is to find a *lower bound* for the ground state energy $\langle \hat{H} \rangle_{\rm gs}$. Taking the ground state expectation value of the Hamiltonian, we find that

$$\langle \hat{H} \rangle_{\rm gs} = \frac{\langle \hat{p}^2 \rangle_{\rm gs}}{2m} + \alpha\, \langle \hat{x}^4 \rangle_{\rm gs}. \tag{15.4.2}$$

For the ground state, or in fact any bound state, the expectation value of $\hat{p}$ vanishes. Therefore, $\langle \hat{p} \rangle_{\rm gs} = 0$, and

$$\langle \hat{p}^2 \rangle_{\rm gs} = (\Delta p)^2_{\rm gs}. \tag{15.4.3}$$

From the inequality $\langle \hat{Q}^2 \rangle \geq \langle \hat{Q} \rangle^2$, we find that

$$\langle \hat{x}^4 \rangle \geq \langle \hat{x}^2 \rangle^2. \tag{15.4.4}$$

Moreover, $(\Delta x)^2 = \langle \hat{x}^2 \rangle - \langle \hat{x} \rangle^2$ leads to $\langle \hat{x}^2 \rangle \geq (\Delta x)^2$ so that, on arbitrary states,

$$\langle \hat{x}^4 \rangle \geq (\Delta x)^4. \tag{15.4.5}$$

Therefore,

$$\langle \hat{H} \rangle_{\rm gs} = \frac{\langle \hat{p}^2 \rangle_{\rm gs}}{2m} + \alpha\, \langle \hat{x}^4 \rangle_{\rm gs} \geq \frac{(\Delta p_{\rm gs})^2}{2m} + \alpha\, (\Delta x_{\rm gs})^4. \tag{15.4.6}$$

From the uncertainty principle,

$$\Delta x_{\rm gs}\, \Delta p_{\rm gs} \geq \frac{\hbar}{2} \quad \Rightarrow \quad \Delta p_{\rm gs} \geq \frac{\hbar}{2 \Delta x_{\rm gs}}. \tag{15.4.7}$$

Back to the value of $\langle \hat{H} \rangle_{\rm gs}$, we get

$$\langle \hat{H} \rangle_{\rm gs} \geq \frac{\hbar^2}{8m(\Delta x_{\rm gs})^2} + \alpha\, (\Delta x_{\rm gs})^4 \equiv f(\Delta x_{\rm gs}). \tag{15.4.8}$$

The quantity to the right of the inequality defines the function $f(\Delta x_{\rm gs})$. Figure 15.2 shows a plot of $f(\Delta x)$.

If we knew the value of $\Delta x_{\rm gs}$, we could immediately use $\langle \hat{H} \rangle_{\rm gs} \geq f(\Delta x_{\rm gs})$. Since we don't know the value of $\Delta x_{\rm gs}$, however, the only thing we can say for sure is that $\langle \hat{H} \rangle_{\rm gs}$ is bigger than the *minimum* value that can be taken by $f(\Delta x_{\rm gs})$ as we vary $\Delta x_{\rm gs}$:

$$\langle \hat{H} \rangle_{\rm gs} \geq {\rm Min}_{\Delta x}\Big(\frac{\hbar^2}{8m(\Delta x)^2} + \alpha\, (\Delta x)^4 \Big). \tag{15.4.9}$$

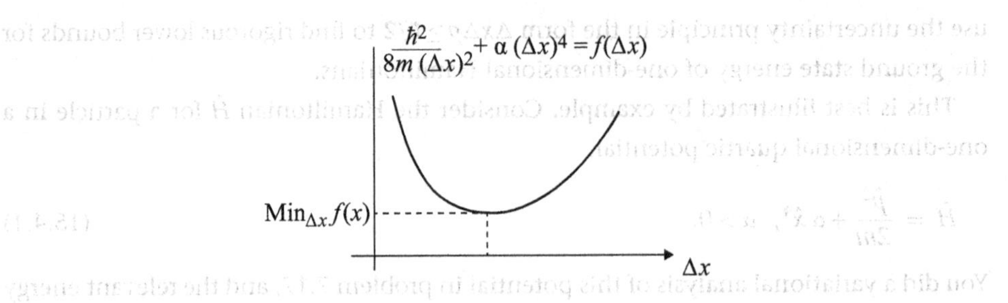

Figure 15.2
We know that $\langle \hat{H}_{gs} \rangle \geq f(\Delta x_{gs})$, but we don't know the value of Δx_{gs}. As a result, we can only be certain that $\langle \hat{H}_{gs} \rangle$ is greater than or equal to the *minimum* value the function $f(\Delta x_{gs})$ can take.

The minimization problem is straightforward. In fact, you can check that

$$\mathrm{Min}_x \Big(\frac{A}{x^2} + Bx^4 \Big) = 2^{\frac{1}{3}} \tfrac{3}{2} (A^2 B)^{\frac{1}{3}}. \tag{15.4.10}$$

Applied to (15.4.9), we obtain

$$\langle \hat{H} \rangle_{gs} \geq 2^{\frac{1}{3}} \frac{3}{8} \Big(\frac{\hbar^2 \sqrt{\alpha}}{m} \Big)^{\frac{2}{3}} \simeq 0.4724 \Big(\frac{\hbar^2 \sqrt{\alpha}}{m} \Big)^{\frac{2}{3}}. \tag{15.4.11}$$

This is the final lower bound for the ground state energy. It is actually not too bad: for the exact ground state, instead of 0.4724 we would have 0.668 (obtained numerically by the shooting method in problem 7.7).

15.5 Diagonalization of Operators

When we have operators we wish to understand, it can be useful to find a basis on the vector space for which the operators are represented by matrices that take a simple form. Diagonal matrices are those where all nondiagonal entries vanish. If we can find a set of basis vectors for which the matrix representing an operator is diagonal, the operator is said to be **diagonalizable.**

If an operator T is diagonal in some basis $(u_1, \ldots, u_n)$ of the vector space V, its matrix takes the form diag $(\lambda_1, \ldots, \lambda_n)$, with constants λ_i, and we have

$$Tu_1 = \lambda_1 u_1, \ \ldots, \ \ Tu_n = \lambda_n u_n. \tag{15.5.1}$$

The basis vectors are thus eigenvectors with eigenvalues given by the diagonal elements. It follows that *a matrix is diagonalizable if and only if it possesses a set of eigenvectors that span the vector space.*

Recall that all operators T on finite-dimensional complex vector spaces have at least one eigenvalue and thus at least one eigenvector. But even in complex vector spaces, not all operators have enough eigenvectors to span the space. Those operators cannot be diagonalized. The simplest example of such an operator is provided by the 2×2 matrix

$$\begin{pmatrix} 0 & 1 \\ 0 & 0 \end{pmatrix}. \tag{15.5.2}$$

The only eigenvalue of this matrix is $\lambda = 0$, and the associated eigenvector is $(1, 0)$, or any multiple thereof. One basis vector cannot span a two-dimensional vector space. As a result, this matrix cannot be diagonalized.

Suppose we have a vector space V, and we have chosen a basis $(v_1, \ldots, v_n)$ such that a linear operator has a matrix representation $T_{ij}(\{v\})$ that is not diagonal. We can change the basis to a new one $(u_1, \ldots, u_n)$ using a linear operator A that acts as follows:

$$u_k = A\, v_k. \tag{15.5.3}$$

As we learned in section 13.5, the matrix representation $T_{ij}(\{u\})$ of the operator in the new basis takes the form

$$T(\{u\}) = A^{-1} T(\{v\}) A \quad \text{or} \quad T_{ij}(\{u\}) = (A^{-1})_{ik}\, T_{kp}(\{v\})\, A_{pj}, \tag{15.5.4}$$

where the matrix A_{ij} is the representation of A in either the original v-basis or the new u-basis. T is diagonalizable if there is an operator A such that $T_{ij}(\{u\})$ is diagonal. The matrices $T(\{u\})$ and $T(\{v\})$ are related by similarity.

There are two equivalent ways of thinking about the diagonalization of T:

1. The matrix representation of T is diagonal when using the u-basis obtained by acting with A on the original v-basis. Thus, the u_i are the eigenvectors of T.
2. The operator $A^{-1}TA$ is diagonal in the *original* v-basis.

The second viewpoint requires justification. Since T is diagonal in the u-basis, $Tu_i = \lambda_i u_i$ (i is not summed). This implies that $TA\, v_i = \lambda_i A v_i$. Acting with A^{-1} we find $(A^{-1}TA)\, v_i = \lambda_i v_i$, which confirms that $A^{-1}TA$ is represented by a diagonal matrix in the original v-basis. Both viewpoints are valuable.

Using the second viewpoint, we write the following matrix equation in the *original basis*:

$$\boxed{A^{-1}TA = D_T.} \tag{15.5.5}$$

Here D_T is a diagonal matrix, and we say that A is a matrix that diagonalizes T by similarity. It is useful to note the following:

The columns of the matrix A are the eigenvectors of T. (15.5.6)

We see this as follows. Recall that the eigenvectors of T are the u_k and therefore,

$$u_k = Av_k = \sum_i A_{ik} v_i = \begin{pmatrix} A_{1k} \\ \vdots \\ A_{nk} \end{pmatrix}. \tag{15.5.7}$$

In the last step, we noted that the basis vector v_i is represented by a column vector of zeroes with a single unit entry at the ith position. This confirms that the kth column of A is the kth eigenvector of T.

While not all operators on complex vector spaces can be diagonalized, the situation is much improved for Hermitian operators. Hermitian operators can be diagonalized and so can unitary operators. But even more is true: these operators take the diagonal form in an orthonormal basis!

An operator M is said to be **unitarily diagonalizable** if there is an *orthonormal* basis $\{\tilde{e}_i\}$ in which its matrix representation is a diagonal matrix. The basis $\{\tilde{e}_i\}$ is therefore an *orthonormal basis of eigenvectors*. This uses the first viewpoint on diagonalization.

Alternatively, start with an arbitrary orthonormal basis $(e_1, \ldots, e_n)$, where the matrix representation of M is written simply as the matrix M. We saw at the end of section 14.9 that orthonormal bases are mapped onto each other by unitary operators. It follows that there is a unitary operator U that maps the $\{e_i\}$ basis vectors to the orthonormal basis $\{\tilde{e}_i\}$ of eigenvectors. With $U^{-1} = U^\dagger$ and using the $\{e_i\}$ basis, we have the following matrix equation, with D_M a diagonal matrix:

$$U^\dagger M U = D_M. \tag{15.5.8}$$

This equation holds for the same reason that equation (15.5.5) holds. We say that the matrix M has been diagonalized by a unitary transformation, thus the terminology *unitarily diagonalizable.*

Exercise 15.2. *Consider the operators T_1, T_2, T_3, and define $\tilde{T}_i = A^{-1}T_iA,\ i = 1, 2, 3$ obtained by similarity transformation with A. Show that $T_1T_2 = T_3$ implies $\tilde{T}_1\tilde{T}_2 = \tilde{T}_3$.*

15.6 The Spectral Theorem

While we could prove that Hermitian operators are unitarily diagonalizable, this result holds for the more general class of *normal* operators. The proof in the more general case is not harder than the one for Hermitian operators. An operator M is said to be **normal** if it commutes with its adjoint:

$$M \text{ is normal}: \quad [M^\dagger, M] = 0. \tag{15.6.1}$$

Hermitian operators are clearly normal. So are anti-Hermitian operators ($M^\dagger = -M$ means M is anti-Hermitian). Unitary operators U are normal because $U^\dagger U = UU^\dagger = \mathbb{1}$, showing that U and $U^\dagger$ commute. If an operator is normal, a similarity transformation with a unitary operator gives another normal operator:

Exercise 15.3. *If M is normal, show that $V^\dagger MV$, where V is a unitary operator, is also normal.*

It is a useful fact that a normal operator T and its adjoint $T^\dagger$ share the same set of eigenvectors:

Lemma. *Let w be an eigenvector of the normal operator M: $Mw = \lambda w$. Then w is also an eigenvector of $M^\dagger$ with a complex conjugate eigenvalue:*

$$M^\dagger w = \lambda^* w. \tag{15.6.2}$$

Proof. Define $u = (M^\dagger - \lambda^* \mathbb{1})w$. The result holds if u is the zero vector. To show this we compute the norm squared of u:

$$\langle u, u\rangle = \langle (M^\dagger - \lambda^* \mathbb{1})w, (M^\dagger - \lambda^* \mathbb{1})w\rangle. \tag{15.6.3}$$

Using the adjoint property to move the operator in the first entry to the second entry,

$$\langle u, u\rangle = \langle w, (M - \lambda \mathbb{1})(M^\dagger - \lambda^* \mathbb{1})w\rangle. \tag{15.6.4}$$

Since M and $M^\dagger$ commute, so do the two factors in parentheses, and therefore,

$$\langle u, u\rangle = \langle w, (M^\dagger - \lambda^* \mathbb{1})(M - \lambda \mathbb{1})w\rangle = 0, \tag{15.6.5}$$

since $(M - \lambda \mathbb{1})$ kills w. It follows that $u = 0$ and therefore (15.6.2) holds. □

We can now state the main result: the **spectral theorem**. It states that a matrix is unitarily diagonalizable if and only if it is normal. We will prove this result for finite-dimensional matrices.

Theorem 15.6.1. Spectral theorem. *Let M be an operator in a finite-dimensional complex vector space. The vector space has an orthonormal basis of M eigenvectors if and only if M is normal.*

Proof. It is easy to show that if M is unitarily diagonalizable, it is normal. Indeed, from (15.5.8), when M is unitarily diagonalizable, there is a unitary U such that

$$M = UD_M U^\dagger \quad \text{and therefore} \quad M^\dagger = UD_M^\dagger U^\dagger.$$

We then get

$$M^\dagger M = UD_M^\dagger D_M U^\dagger \quad \text{and} \quad MM^\dagger = UD_M D_M^\dagger U^\dagger,$$

so that

$$[M^\dagger, M] = U(D_M^\dagger D_M - D_M D_M^\dagger)U^\dagger = 0$$

because any two diagonal matrices commute.

We must now prove that for any normal M, viewed as a matrix on an arbitrary orthonormal basis, there is a unitary matrix U such that $U^\dagger M U$ is diagonal. By our general discussion, this implies that the eigenvectors of M are an orthonormal basis. We will prove this by induction in the dimension of the vector space where the result holds.

The result is clearly true for $\dim V = 1$: any 1×1 matrix is normal and automatically diagonal. We now assume that a normal matrix in an $(n-1)$-dimensional vector space is unitarily diagonalizable and try to prove the same is true for a normal matrix in an n-dimensional space.

Let M be an $n \times n$ normal matrix referred to the orthonormal basis $(|1\rangle, \ldots, |n\rangle)$ of V so that $M_{ij} = \langle i|M|j\rangle$. We know there is at least one eigenvalue λ_1 of M with a nonzero eigenvector $|x_1\rangle$ of unit norm:

$$M|x_1\rangle = \lambda_1 |x_1\rangle \quad \text{and} \quad M^\dagger |x_1\rangle = \lambda_1^* |x_1\rangle, \tag{15.6.6}$$

in view of the lemma. We claim now that there is a unitary matrix U_1 such that

$$|x_1\rangle = U_1|1\rangle \quad \Rightarrow \quad U_1^\dagger|x_1\rangle = |1\rangle. \tag{15.6.7}$$

U_1 is not unique and can be constructed as follows: Take $|x_1\rangle$ and $n-1$ additional vectors that together span V and use the Gram-Schmidt procedure to construct an orthonormal basis $|x_1\rangle, \ldots, |x_n\rangle$. Then write $U_1 = \sum_i |x_i\rangle\langle i|$ that, as required, maps $|1\rangle$ to $|x_1\rangle$.

Now define

$$M_1 \equiv U_1^\dagger M U_1. \tag{15.6.8}$$

M_1 is also normal, and $M_1|1\rangle = U_1^\dagger M U_1|1\rangle = U_1^\dagger M|x_1\rangle = \lambda_1 U_1^\dagger|x_1\rangle = \lambda_1|1\rangle$ so that

$$M_1|1\rangle = \lambda_1|1\rangle, \tag{15.6.9}$$

which says that the first column of M_1 has zeroes in all entries except the first. Indeed,

$$\langle j|M_1|1\rangle = \lambda_1\langle j|1\rangle = \lambda_1\delta_{1,j}. \tag{15.6.10}$$

The normality of M_1 implies that the first row of M_1 is also zero except for the first element. Indeed,

$$\langle 1|M_1|j\rangle = (\langle j|M_1^\dagger|1\rangle)^* = (\lambda_1^*\langle j|1\rangle)^* = \lambda_1\langle 1|j\rangle = \lambda_1\delta_{1,j}, \tag{15.6.11}$$

where we used $M_1^\dagger|1\rangle = \lambda_1^*|1\rangle$, which follows from the lemma. It follows from the last two equations that M_1, in the original basis, takes the form

$$M_1 = \left(\begin{array}{c|ccc} \lambda_1 & 0 & \ldots & 0 \\ \hline 0 & & & \\ \vdots & & M' & \\ 0 & & & \end{array}\right),$$

where M' is an $(n-1)$ by $(n-1)$ matrix. Since M_1 is normal and matrices multiply in blocks, one can quickly see that M' is also normal. By the induction hypothesis, M' can be unitarily diagonalized, so there exists an $(n-1)$ by $(n-1)$ unitary matrix U' such that $U'^\dagger M' U'$ is diagonal:

$$U'^\dagger M' U' = D_{M'}, \quad \text{with } D_{M'} \text{ diagonal.} \tag{15.6.12}$$

The matrix U' can be extended to an n by n unitary matrix $\hat{U}$ as follows:

$$\hat{U} = \left(\begin{array}{c|ccc} 1 & 0 & \ldots & 0 \\ \hline 0 & & & \\ \vdots & & U' & \\ 0 & & & \end{array}\right). \tag{15.6.13}$$

We can now confirm that $\hat{U}$ diagonalizes M_1—that is, $\hat{U}^\dagger M_1 \hat{U}$ is a diagonal matrix D_M:

$$\hat{U}^\dagger M_1 \hat{U} = \begin{pmatrix} 1 & 0 & \dots & 0 \\ 0 & & & \\ \vdots & & U'^\dagger & \\ 0 & & & \end{pmatrix} \begin{pmatrix} \lambda_1 & 0 & \dots & 0 \\ 0 & & & \\ \vdots & & M' & \\ 0 & & & \end{pmatrix} \begin{pmatrix} 1 & 0 & \dots & 0 \\ 0 & & & \\ \vdots & & U' & \\ 0 & & & \end{pmatrix}$$

$$= \begin{pmatrix} \lambda_1 & 0 & \dots & 0 \\ 0 & & & \\ \vdots & & U'^\dagger M' U' & \\ 0 & & & \end{pmatrix} = \begin{pmatrix} \lambda_1 & 0 & \dots & 0 \\ 0 & & & \\ \vdots & & D_{M'} & \\ 0 & & & \end{pmatrix} = D_M. \tag{15.6.14}$$

But then, using the definition of M_1,

$$D_M = \hat{U}^\dagger M_1 \hat{U} = \hat{U}^\dagger U_1^\dagger M U_1 \hat{U} = (U_1\hat{U})^\dagger M (U_1\hat{U}). \tag{15.6.15}$$

Since the product of unitary matrices is unitary, $\tilde{U} \equiv U_1\hat{U}$ is unitary, and we have shown that $\tilde{U}^\dagger M \tilde{U}$ is diagonal. This is the desired result. We have used the induction hypothesis to prove that an n by n normal matrix M is unitarily diagonalizable. This completes the induction argument and thus the proof. □

This theorem implies that Hermitian and unitary operators are unitarily diagonalizable: their eigenvectors can be chosen to form an orthonormal basis. The proof did not require a separate discussion of degeneracies. If an eigenvalue of M is degenerate and appears k times, then k orthonormal eigenvectors are associated with the corresponding k-dimensional, M-invariant subspace of the vector space.

Let us now describe the general situation we encounter when diagonalizing a normal operator T on a vector space V. In general, we expect degeneracies in the eigenvalues so that each eigenvalue λ_k is repeated $d_k \geq 1$ times. An eigenvalue λ_k is degenerate if $d_k > 1$. It follows that V has T-invariant subspaces of different dimensionalities. Let U_k denote the T-invariant subspace of dimension $d_k \geq 1$ spanned by eigenvectors with eigenvalue λ_k:

$$U_k \equiv \{v \in V \mid Tv = \lambda_k v\}, \quad \dim U_k = d_k. \tag{15.6.16}$$

By the spectral theorem, U_k has a basis comprised by d_k orthonormal eigenvectors:

$$(u_1^{(k)}, \ \dots \ , u_{d_k}^{(k)}).$$

The full space V is decomposed as the direct sum of the invariant subspaces of T:

$$V = U_1 \oplus \cdots \oplus U_m, \quad \dim V = \sum_{i=1}^{m} d_i, \ m \geq 1. \tag{15.6.17}$$

All U_i subspaces are guaranteed to be orthogonal to each other. In fact, the full list of eigenvectors is a list of orthonormal vectors that form a basis for V and is conveniently ordered as follows:

$$(u_1^{(1)}, \dots, u_{d_1}^{(1)}, \ \dots, \ u_1^{(m)}, \dots, u_{d_m}^{(m)}). \tag{15.6.18}$$

The matrix T is manifestly diagonal in this basis because each vector above is an eigenvector of T. The matrix representation of T reads

$$T = \text{diag}\,(\underbrace{\lambda_1, \ldots, \lambda_1}_{d_1 \text{ times}}, \ldots, \underbrace{\lambda_m, \ldots, \lambda_m}_{d_m \text{ times}}). \tag{15.6.19}$$

This is clear because the first d_1 vectors in the list are in U_1, the second d_2 vectors are in U_2, and so on until the last d_m vectors are in U_m.

Let us now consider the uniqueness of the basis (15.6.18). In other words, we ask how much we can change the basis vectors without changing the matrix representation of T. If we have no degeneracies in the spectrum of T ($d_i = 1$, for all i), each basis vector can at most be multiplied by a phase. On the other hand, with degeneracies the list can be changed considerably without changing the matrix representation of T. Let V_k be a unitary operator on U_k—namely, $V_k : U_k \to U_k$ for each $k = 1, \ldots, m$. We claim that the following basis of eigenvectors leads to the same matrix T:

$$\big(V_1 u_1^{(1)}, \ldots, V_1 u_{d_1}^{(1)}, \ldots\ldots, V_m u_1^{(m)}, \ldots, V_m u_{d_m}^{(m)} \big). \tag{15.6.20}$$

This is still a collection of orthonormal T eigenvectors because the first d_1 vectors are still orthonormal eigenvectors in U_1, the second d_2 vectors are still orthonormal eigenvectors in U_2, and so on. More explicitly, we can calculate the matrix elements of T within U_k in the new basis:

$$\big\langle V_k u_i^{(k)}, T(V_k u_j^{(k)}) \big\rangle = \lambda_k \langle V_k u_i^{(k)}, V_k u_j^{(k)} \rangle = \lambda_k \langle u_i^{(k)}, u_j^{(k)} \rangle = \lambda_k \delta_{ij}. \tag{15.6.21}$$

In the first step, we noted that any vector in U_k has T eigenvalue λ_k. In the second step, we used the unitarity of V_k. This shows that in the U_k subspace the matrix for T is still diagonal with all entries equal to λ_k.

The spectral theorem affords us a simple way to write a normal operator. For this, consider the basis of orthonormal eigenvectors for the T-invariant subspace U_k of dimension d_k. Let P_k denote the orthogonal projector to this subspace. The projector P_k has rank d_k, and its action on an arbitrary vector v can be written as

$$P_k v = \sum_{i=1}^{d_k} u_i^{(k)} \langle u_i^{(k)}, v \rangle, \tag{15.6.22}$$

following the prescription in (14.3.6). In bra-ket notation we have

$$P_k = \sum_{i=1}^{d_k} |u_i^{(k)}\rangle\langle u_i^{(k)}|. \tag{15.6.23}$$

By construction, the basis vectors of U_k are left invariant by P_k, and the basis vectors of any U_q, with $q \neq k$, are killed by P_k:

$$P_k u_i^{(q)} = u_i^{(q)} \delta_{kq}. \tag{15.6.24}$$

The various P_k satisfy the following properties:

$$P_k^\dagger = P_k, \quad P_k P_l = \delta_{kl} P_l, \quad \sum_k P_k = \mathbb{1}. \tag{15.6.25}$$

The first relation is manifest from (15.6.23). The second follows because each P_k is a projector, and U_k and U_l are orthogonal subspaces when $k \neq l$. The third expression is also clear from (15.6.23) because by the time we sum over the allowed values of $k = 1, \ldots, m$, we are summing the ket-bra combinations for all the basis vectors of the state space. This last property is the counterpart of $V = U_1 \oplus \cdots \oplus U_m$. A set of projectors P_k satisfying (15.6.25) is called a **complete set of orthonormal projectors.** We have thus seen that any normal operator gives rise to one such complete set. Finally, we now claim that the operator T itself can be written as follows:

$$T = \sum_k \lambda_k P_k. \tag{15.6.26}$$

This equation, in the form of a sum of terms that are each an eigenvalue times its associated projector, is called the **spectral decomposition** of T. The proof of this formula is simple; we just need to check that the right-hand side acts on the vectors of the orthonormal basis exactly as T does. Acting on $u_i^{(q)}$ and using (15.6.24), we see that

$$\sum_k \lambda_k P_k u_i^{(q)} = \sum_k \lambda_k u_i^{(q)} \delta_{kq} = \lambda_q u_i^{(q)}, \tag{15.6.27}$$

consistent with $Tu_i^{(q)} = \lambda_q u_i^{(q)}$. This proves the decomposition (15.6.26).

15.7 Simultaneous Diagonalization of Hermitian Operators

We say that two operators S and T in a vector space V can be **simultaneously diagonalized** if there is some basis of V in which both the matrix representation of S and the matrix representation of T are diagonal. It then follows that each vector in this basis is an eigenvector of S *and* an eigenvector of T. There is therefore an entire basis of *shared* eigenvectors.

A necessary condition for simultaneous diagonalization is that the operators S and T commute. Indeed, if they can be simultaneously diagonalized, there is a basis where both are diagonal, and they manifestly commute. If the operators don't commute, this is a basis-independent statement, and therefore a simultaneous diagonal presentation cannot exist. Since arbitrary linear operators S and T on a complex vector space cannot be diagonalized, the vanishing of $[S, T]$ does not guarantee simultaneous diagonalization. But if the operators are Hermitian, it does, as we show now.

Theorem 15.7.1. *If S and T are commuting Hermitian operators, they can be simultaneously diagonalized.*

Proof. The main complication is that degeneracies in the spectrum require some discussion. Either both operators have degeneracies or at least one has no degeneracies. Without loss of generality, we can assume there are two cases to consider:

1. There is no degeneracy in the spectrum of T, or
2. both T and S have degeneracies in their spectrum.

Case (1). Since T is nondegenerate, there is a basis $(u_1, \ldots, u_n)$ of eigenvectors of T with different eigenvalues:

$$Tu_i = \lambda_i u_i, \quad i \text{ not summed}, \quad \lambda_i \neq \lambda_j \text{ for } i \neq j. \tag{15.7.1}$$

We now want to understand what kind of vector Su_i is. For this we act with T on it:

$$T(Su_i) = S(Tu_i) = S(\lambda_i u_i) = \lambda_i(S u_i). \tag{15.7.2}$$

It follows that Su_i is also an eigenvector of T with eigenvalue λ_i, thus it must equal u_i, up to scale:

$$Su_i = \omega_i u_i, \tag{15.7.3}$$

showing that u_i is also an eigenvector of S, this time with eigenvalue ω_i. Thus, any eigenvector of T is also an eigenvector of S, showing that these operators are simultaneously diagonalizable.

Case (2). Since T has degeneracies, as explained in the previous section, we have a decomposition of V in T-invariant subspaces U_k spanned by eigenvectors:

$$\begin{gathered} U_k \equiv \{u \,|\, Tu = \lambda_k u\}, \quad \dim U_k = d_k, \qquad V = U_1 \oplus \cdots U_m, \\ \text{orthonormal basis for } V: \quad (u_1^{(1)}, \ldots, u_{d_1}^{(1)}, \;\ldots\;, u_1^{(m)}, \ldots, u_{d_m}^{(m)}). \\ T = \text{diag}\,\big(\underbrace{\lambda_1, \ldots, \lambda_1}_{d_1 \text{ times}}, \;\ldots\;, \underbrace{\lambda_m, \ldots, \lambda_m}_{d_m \text{ times}}\big) \text{ in this basis.} \end{gathered} \tag{15.7.4}$$

We also explained that the alternative orthonormal basis of V given by

$$(V_1 u_1^{(1)}, \ldots, V_1 u_{d_1}^{(1)}, \;\ldots\;, V_m u_1^{(m)}, \;\ldots\;, V_m u_{d_m}^{(m)}) \tag{15.7.5}$$

leads to the same matrix for T when each V_k is a unitary operator on U_k.

We now claim that the U_k are also S-invariant subspaces! To show this, let $u_k \in U_k$, and examine the vector Su_k. We have

$$T(Su_k) = S(Tu_k) = \lambda_k Su_k \quad \Rightarrow \quad Su_k \in U_k. \tag{15.7.6}$$

It follows that in the basis (15.7.4) the matrix for S takes *block-diagonal* form, with blocks on each of the U_k subspaces. We cannot guarantee, however, that S is diagonal within each square block; we only know that $Su_i^{(k)} \in U_k$.

Since S, restricted to each S-invariant subspace U_k, is Hermitian, we can find an orthonormal basis of U_k in which the matrix S is diagonal. This new basis is unitarily related to the original basis $(u_1^{(k)}, \ldots, u_{d_k}^{(k)})$ and thus takes the form $(V_k u_1^{(k)}, \ldots, V_k u_{d_k}^{(k)})$ with V_k a unitary operator in U_k. Note that the eigenvalues of S in this block need not be degenerate. Doing this for each block, we find a basis of the form (15.7.5) in which S is diagonal. But T is still diagonal and unchanged in this new basis, so both S and T have been simultaneously diagonalized. □

Remarks:

1. The above proof gives an algorithmic way to produce the common list of eigenvectors. If one of the operators is nondegenerate, we are in case (1), and its eigenvectors will do the job even if the other operator is degenerate. If both operators are degenerate, we are in case (2). One diagonalizes one of the operators and constructs the second matrix in the basis of eigenvectors of the first. This second matrix is block diagonal, where the blocks are organized by the degeneracies in the spectrum of the first matrix. One must then diagonalize within the blocks of the second matrix to select new eigenvectors in each degenerate subspace. It is guaranteed that the new basis that works for the second matrix also works for the first.
2. If we had to simultaneously diagonalize three different commuting Hermitian operators S_1, S_2, and S_3, all of which have degenerate spectra, we would proceed as follows. We would diagonalize S_1 and fix a basis in which S_1 is diagonal. In this basis we must find that S_2 and S_3 have exactly the same block structure. The corresponding block matrices are simply the matrix representations of S_2 and S_3 in each of the invariant spaces U_k appearing in the diagonalization of S_1. Since S_2 and S_3 commute, their restrictions to U_k commute. These restrictions can be diagonalized simultaneously, as guaranteed by our theorem that works for two matrices. The new basis in U_k that makes the restriction of both S_2 and S_3 diagonal will not disturb the diagonal form of S_1 in this block. This would be repeated for each block, until we got a common basis of eigenvectors.
3. An inductive argument is now apparent. If we know how to simultaneously diagonalize n commuting Hermitian operators, we can diagonalize $n+1$ of them, $S_1, \ldots, S_{n+1}$, as follows. We diagonalize S_1 and then consider the remaining n operators in the basis that makes S_1 diagonal. We are guaranteed a common block structure for the n operators. The problem becomes one of simultaneous diagonalization of n commuting Hermitian block matrices, which is possible by the induction assumption. We have thus proven the corollary below.

Corollary. If $\{S_1, \ldots, S_n\}$ is a set of mutually commuting Hermitian operators, they can all be simultaneously diagonalized.

Example 15.4. *Simultaneous diagonalization of two matrices.*
Consider two Hermitian matrices A_1 and A_2 that commute:

$$A_1 = \begin{pmatrix} 1 & 0 & 1 \\ 0 & 0 & 0 \\ 1 & 0 & 1 \end{pmatrix}, \quad A_2 = \begin{pmatrix} \frac{5}{4} & \frac{1}{2\sqrt{2}} & -\frac{1}{4} \\ \frac{1}{2\sqrt{2}} & \frac{3}{2} & -\frac{1}{2\sqrt{2}} \\ -\frac{1}{4} & -\frac{1}{2\sqrt{2}} & \frac{5}{4} \end{pmatrix}. \tag{15.7.7}$$

We wish to find simultaneous eigenvectors for the two matrices. This takes some effort because both matrices have a degenerate spectrum: the A_1 eigenvalues are $(2,0,0)$, and the A_2 eigenvalues are $(2,1,1)$. So we are not in the simple situation in which one of the matrices is nondegenerate, and its eigenvectors automatically work for the other matrix. We must then follow the route summarized in remark (1) above. We pick the matrix A_1, which looks simpler, and we quickly find its eigenvalues and orthonormal eigenvectors:

$$\lambda_1 = 2,\ u_1 = \frac{1}{\sqrt{2}}\begin{pmatrix}1\\0\\1\end{pmatrix},\quad \lambda_2 = 0,\ u_2 = \frac{1}{\sqrt{2}}\begin{pmatrix}-1\\0\\1\end{pmatrix},\quad \lambda_3 = 0,\ u_3 = \begin{pmatrix}0\\1\\0\end{pmatrix}. \tag{15.7.8}$$

Since $\lambda_1 = 2$ is nondegenerate, its eigenvector u_1 must be an eigenvector of A_2. One quickly finds that $A_2 u_1 = u_1$, so it has eigenvalue one. We now build the matrix for the operator A_2 in the basis of A_1 eigenvectors u_1, u_2, u_3. Because u_1 is an A_2 eigenvector of eigenvalue one, the first column in the matrix, which we call A_2', is $(1,0,0)$. Thus, we have

$$A_2' = \begin{pmatrix}1 & 0 & 0\\ 0 & a'_{22} & a'_{23}\\ 0 & a'_{32} & a'_{33}\end{pmatrix}, \tag{15.7.9}$$

since the matrix must be Hermitian. Now we need to determine the rest of the entries. A short calculation gives

$$\begin{aligned} A_2 u_2 &= \ \ \tfrac{3}{2}u_2 - \tfrac{1}{2}u_3,\\ A_2 u_3 &= -\tfrac{1}{2}u_2 + \tfrac{3}{2}u_3. \end{aligned} \tag{15.7.10}$$

The left-hand sides have the matrix A_2 in (15.7.7) acting on the second and third eigenvectors in (15.7.8). The results are easily written as superpositions of u_2 and u_3 by inspection. More systematically, you could write the u_i's in terms of the original basis vectors e_1, e_2, e_3 used to describe A_1 and A_2 and then solve for the e_i in terms of the u_i, which allows you to write any column vector in terms of u_i's. At any rate the above relations imply that the full matrix A_2' is

$$A_2' = \begin{pmatrix}1 & 0 & 0\\ 0 & \frac{3}{2} & -\frac{1}{2}\\ 0 & -\frac{1}{2} & \frac{3}{2}\end{pmatrix}. \tag{15.7.11}$$

As a good check on our arithmetic, we confirm the invariance of the trace: $\text{tr}A_2 = \text{tr}A_2' = 4$. We now must diagonalize the 2×2 block that, as expected, is associated with the degenerate subspace of A_1 spanned by the eigenvectors u_2 and u_3. Calling the eigenvalues t_2, t_3 and the associated eigenvectors w_2, w_3, we have

$$\begin{pmatrix} \frac{3}{2} & -\frac{1}{2} \\ -\frac{1}{2} & \frac{3}{2} \end{pmatrix}, \quad t_2=2, \; w_2=\frac{1}{\sqrt{2}}\begin{pmatrix} -1 \\ 1 \end{pmatrix}, \quad t_3=1, \; w_3=\frac{1}{\sqrt{2}}\begin{pmatrix} 1 \\ 1 \end{pmatrix}. \tag{15.7.12}$$

Since this matrix is relative to the $\{u_2, u_3\}$ basis, the eigenvectors are in fact

$$w_2=\frac{1}{\sqrt{2}}(-u_2+u_3), \quad w_3=\frac{1}{\sqrt{2}}(u_2+u_3). \tag{15.7.13}$$

These are the missing common eigenvectors of the two matrices. All in all, the three simultaneous eigenvectors of A_1 and A_2 are u_1, w_2, w_3:

$$u_1=\frac{1}{\sqrt{2}}\begin{pmatrix} 1 \\ 0 \\ 1 \end{pmatrix}, \; (2,1); \quad w_2=\frac{1}{2}\begin{pmatrix} 1 \\ \sqrt{2} \\ -1 \end{pmatrix}, \; (0,2); \quad w_3=\frac{1}{2}\begin{pmatrix} -1 \\ \sqrt{2} \\ 1 \end{pmatrix}, \; (0,1). \tag{15.7.14}$$

Following each eigenvector, $(\#_1, \#_2)$ denote the A_1 and A_2 eigenvalues, respectively. □

15.8 Complete Set of Commuting Observables

We have discussed the problem of finding the eigenstates and eigenvalues of a Hermitian operator S. In a quantum system, S is a quantum mechanical observable. The eigenstates of S are states of the system in which the observable S can be measured with certainty. The result of the measurement is the eigenvalue associated with the eigenstate. Moreover, the eigenstates of S form a basis for the state space.

If the Hermitian operator S has a nondegenerate spectrum, all eigenvalues are different, and we have a rather nice situation in which each eigenstate can be uniquely labeled with the corresponding eigenvalue of S. The physical value of the observable distinguishes the various eigenstates. In this case the operator S provides a *complete set of commuting observables* or a CSCO. The set here has just one observable, the operator S.

The situation is more nontrivial if the Hermitian operator S has a degenerate spectrum. This means that V has an S-invariant subspace of dimension $d>1$ spanned by orthonormal eigenstates $(u_1, \ldots, u_d)$, all of which have a common S eigenvalue. This time, the eigenvalue of S does not allow us to uniquely label the basis eigenstates of the invariant subspace. Physically, this is an unsatisfactory situation as we have different basis states—the various u_i's—that we cannot tell apart by the measurement of S. This time S does not provide a CSCO.

We are thus physically motivated to find another Hermitian operator T that is compatible with S. Two Hermitian operators are said to be **compatible observables** if they commute since then we can find a basis of V comprised by simultaneous eigenvectors of the operators. These states can be labeled by two observables—namely, the two eigenvalues. If we are lucky, the basis eigenstates in each of the S-invariant subspaces of dimension higher than one can be organized into T eigenstates of different eigenvalues. In this case T breaks the spectral degeneracy of S, and using T eigenvalues, as well as S eigenvalues, we can uniquely label a basis of orthonormal states of V. In this case we say that S and T form a CSCO.

We are now ready for a definition of a complete set of commuting observables. Consider a set of commuting observables—namely, a set $\{S_1, \ldots, S_k\}$ of Hermitian operators, all of which commute with each other. The operators act on a complex vector space V that represents the physical state-space of some quantum system. By the corollary in the previous section, we can find an orthonormal basis of vectors in V such that each vector is an eigenstate of every operator in the set. Let each eigenstate in the basis be labeled by the ordered list of eigenvalues of the S_i operators, with $i = 1, \ldots, k$. The set $\{S_1, \ldots, S_k\}$ is said to be a CSCO if no two eigenstates have the same labels.

This idea can be expressed with equations by writing the spectral decomposition for each of the commuting observables $S_1, \ldots, S_k$. Assume $\dim V = N$, and in writing the spectral decompositions, we use the rank-one projectors $P_1, \ldots, P_N$. Each projector is written as $P_i = |u_i\rangle\langle u_i|$, with $|u_i\rangle, i = 1, \ldots, N$, the common basis of eigenvectors. We then have

$$
\begin{aligned}
S_1 &= \lambda_1^{(1)} P_1 + \cdots + \lambda_N^{(1)} P_N, \\
&\vdots \qquad \vdots \qquad\quad \vdots \qquad \vdots \\
S_k &= \lambda_1^{(k)} P_1 + \cdots + \lambda_N^{(k)} P_N.
\end{aligned}
\tag{15.8.1}
$$

In this presentation the operators S_i all manifestly commute. In each row a number of eigenvalues are repeated, as each operator has a degenerate spectrum (if one operator has no degeneracies, it could serve as a complete set just by itself!). We have a complete set of commuting observables if the "columns" of eigenvalues associated to a single projector are all different. As ordered sets, we must have

$$
\{\lambda_i^{(1)}, \ldots, \lambda_i^{(k)}\} \neq \{\lambda_j^{(1)}, \ldots, \lambda_j^{(k)}\}, \text{ for } i \neq j. \tag{15.8.2}
$$

This is in fact the condition that each eigenstate u_i is uniquely determined by the list of eigenvalues provided by the complete set of observables.

It is a physically motivated assumption that for any quantum system there is a complete set of commuting observables, for otherwise there is no physical way to distinguish the various states that span the vector space. It would mean having states that are manifestly different but have no single observable property in which they differ! In any physical problem, we are urged to find commuting observables, and we must add observables until all degeneracies are resolved. A CSCO need not be unique. Once we have a CSCO, adding another observable causes no harm, although it is not necessary. Also, if the pair (S_1, S_2) form a CSCO, so will the pair $(S_1 + S_2, S_1 - S_2)$, for example. It is often useful to have CSCOs with the smallest possible number of operators.

The first operator that is usually included in a CSCO is the Hamiltonian $\hat{H}$. For bound state problems in one dimension, energy eigenstates are nondegenerate, and the energy can be used to uniquely label the $\hat{H}$ eigenstates. A simple example is the infinite square well. Another example is the one-dimensional harmonic oscillator. In such cases $\hat{H}$ forms the CSCO. If we have, however, a two-dimensional isotropic harmonic oscillator in the (x, y) plane, the Hamiltonian has degeneracies. At the first excited level, we can have the first excited state of the x harmonic oscillator or, at the same energy, the first excited state of the y harmonic oscillator. We thus need another observable that can be used to distinguish these states. There are several options (problem 15.12).

Example 15.5. *CSCO for bound states of hydrogen.*

The hydrogen atom bound states, obtained in section 11.3, are represented by wave functions $\psi_{n\ell m}(r,\theta,\phi)$ taking the form

$$\psi_{n\ell m}(r,\theta,\phi)=\frac{u_{n\ell}(r)}{r}\,Y_{\ell m}(\theta,\phi). \tag{15.8.3}$$

The labels on the wave function suffice to distinguish all states and tell us about observables. Here $n\geq 1$ is the principal quantum number, and it, alone, determines the value of the energy E_n of the state. With Hamiltonian $\hat{H}$ we have

$$\hat{H}\psi_{n\ell m}=E_n\psi_{n\ell m},\quad E_n=-\frac{e^2}{2a_0}\frac{1}{n^2}. \tag{15.8.4}$$

We include $\hat{H}$ in the set of commuting observables. Since for each n we have $\ell=0,\ldots,n-1$ and for each value of ℓ we have $(2\ell+1)$ values of m, there is plenty of degeneracy and clearly the label n does not suffice. The other two labels, ℓ and m, arise because the spherical harmonics are eigenstates of $\hat{\mathbf{L}}^2$ and $\hat{L}_z$, as we showed in section 10.5:

$$\begin{aligned}\hat{\mathbf{L}}^2\,Y_{\ell m} &= \hbar^2\,\ell(\ell+1)\,Y_{\ell m},\\ \hat{L}_z\,Y_{\ell m} &= \hbar m\,Y_{\ell m}.\end{aligned} \tag{15.8.5}$$

The angular momentum operators only involve angular derivatives and are independent of r. As a result, they ignore the radial dependence in the wave functions $\psi_{n\ell m}$, and we have

$$\begin{aligned}\hat{\mathbf{L}}^2\,\psi_{n\ell m} &= \hbar^2\,\ell(\ell+1)\,\psi_{n\ell m},\\ \hat{L}_z\,\psi_{n\ell m} &= \hbar m\,\psi_{n\ell m}.\end{aligned} \tag{15.8.6}$$

The label ℓ encodes the eigenvalue of $\hat{\mathbf{L}}^2$, and the label m encodes the eigenvalue of $\hat{L}_z$. We thus choose the set of observables to be

$$\{\hat{H},\,\hat{\mathbf{L}}^2,\,\hat{L}_z\}. \tag{15.8.7}$$

We have seen that their labels uniquely specify the states. The only question that remains is if all commutators vanish. We checked in section 10.4 that $\hat{\mathbf{L}}^2$ and $\hat{L}_z$ commute. Both of these operators, we claim, commute with the Hamiltonian, which, as we showed in (10.2.11), takes the form

$$\hat{H}=-\frac{\hbar^2}{2m}\frac{1}{r}\frac{\partial^2}{\partial r^2}r+\frac{1}{2mr^2}\hat{\mathbf{L}}^2+V(r). \tag{15.8.8}$$

The $[H,\hat{\mathbf{L}}^2]=0$ claim is clear: $\hat{\mathbf{L}}^2$ commutes with itself and ignores all radial dependence—it is a purely angular operator. The $[H,\hat{L}_z]=0$ claim is also clear: $\hat{L}_z$ also ignores all radial dependence and commutes with $\hat{\mathbf{L}}^2$. This confirms that the set in (15.8.7) is indeed a CSCO for the bound state spectrum of hydrogen. □

Problems

Problem 15.1. *Uncertainty in a spin one-half state.*

Consider the unnormalized state of a spin one-half particle

$$|\psi\rangle = |-\rangle + \epsilon|+\rangle, \quad \epsilon \in \mathbb{C}.$$

Calculate the uncertainty $\Delta S_z(\psi)$. Make sure your answer is correct for $\epsilon = 1$. For $|\epsilon| \ll 1$, what is this uncertainty to leading order in ϵ?

Problem 15.2. *Spin states and uncertainty inequalities.*

1. Derive an uncertainty relation for the $\hat{S}_x$ and $\hat{S}_y$ operators. That is, given a state $|\psi\rangle$, derive an inequality of the form $(\Delta S_x)(\Delta S_y) \geq \ldots$, where the right-hand side is a function of $|\psi\rangle$.
2. What is the minimum possible value of the product $\Delta S_x \Delta S_y$? For which states $|\psi\rangle$ is this minimum achieved?
3. What is the maximum possible value of ΔS_i, for any $i = 1, 2, 3$? Explain.
4. Consider a general spin state $|\mathbf{n}\rangle$ described in terms of the angles θ and ϕ that define the unit vector $\mathbf{n}$. Calculate ΔS_x, ΔS_y, and ΔS_z. Based on your result, what states maximize the product $\Delta S_x \Delta S_y$? Explain.

Problem 15.3. *Uncertainty in a generic linear combination of eigenstates.*

Consider an orthonormal set of $N \geq 2$ states $\psi_i, i = 1, \ldots N$, all of which are eigenstates of the Hermitian operator $\hat{Q}$: $\hat{Q}\psi_i = q_i\psi_i$. Now form a *normalized* linear superposition ψ:

$$\psi = \sum_{i=1}^{N} \alpha_i \psi_i, \quad \alpha_i \in \mathbb{C}, \quad i = 1, \ldots, N.$$

Show that the uncertainty $\Delta Q(\psi)$ of $\hat{Q}$ on the state ψ is given by

$$(\Delta Q(\psi))^2 = \sum_{1 \leq i < j \leq N} |\alpha_i|^2 |\alpha_j|^2 (q_i - q_j)^2.$$

Problem 15.4. *Maximum position uncertainty in an infinite square well.*

Consider a particle in an infinite square well $x \in [-\frac{L}{2}, \frac{L}{2}]$. The potential is zero within this length L interval and is infinite outside this interval. The position uncertainty Δx depends on the wave function $\psi(x)$ in the well. Prove that Δx is bounded above as follows:

$$\Delta x \leq \tfrac{L}{2}.$$

The proof should explain why the value $\Delta x = \frac{L}{2}$ is only reached for a singular wave function.

Problem 15.5. *Confirming expectation values and uncertainties on a Gaussian.*

Consider the Gaussian states in (15.2.31) saturating the position-momentum uncertainty inequality:

$$\psi(x) = N \exp\Big(-\frac{(x-x_0)^2}{4\Delta^2} + \frac{ip_0 x}{\hbar}\Big), \quad N^2 = \frac{1}{\sqrt{2\pi}\,\Delta}.$$

Confirm by direct computation the following facts: (i) the wave function is normalized, (ii) $\langle \hat{x} \rangle = x_0$, (iii) $\langle \hat{p} \rangle = p_0$, (iv) $\Delta x = \Delta$, and (iv) $\Delta p = \hbar/(2\Delta x)$, proving saturation of the uncertainty inequality. All the integrals can be computed by hand efficiently with a little thought. In particular, all one needs is the familiar formula $\int_{-\infty}^{\infty} du e^{-u^2} = \sqrt{\pi}$ and $\int_{-\infty}^{\infty} du u^2 e^{-u^2}$, which is quickly derived from the first.

Problem 15.6. *Saturating uncertainty relations with spin states.*

As in problem 15.2, consider the uncertainty inequality $\Delta S_x \, \Delta S_y \geq \cdots$ and rewrite it in terms of uncertainties for Pauli operators:

$$\Delta\sigma_x \, \Delta\sigma_y \geq \ldots.$$

1. Use equation (15.2.19) to find the spin states that saturate the above uncertainty inequality. Solve explicitly for the states when $|\lambda| < 1$ and describe them: Where do they point to? What are the associated values of $\Delta\sigma_x$ and $\Delta\sigma_y$? Trace the evolution of the states as λ varies from -1 to 1.
2. You may be able to infer, without computation, where saturating spin states point to when $|\lambda| > 1$. Confirm your guess by calculation, and again, trace the evolution of the states as λ goes from -1 to $-\infty$ and when λ goes from 1 to ∞.

Problem 15.7. *Upper and lower bounds for ground state energy.*

Consider the harmonic oscillator Hamiltonian

$$\hat{H} = \frac{\hat{p}^2}{2m} + \tfrac{1}{2} m\omega^2 x^2.$$

Use a Gaussian trial wave function and the variational principle to find an upper bound for the ground state energy. Use the uncertainty principle to derive a lower bound for that same ground state energy. Show that your two bounds determine the ground state energy.

Problem 15.8. *A test of the time-energy uncertainty inequality.*

Consider a state that at $t = 0$ is a superposition, with equal amplitudes, of two orthonormal stationary states ψ_1 and ψ_2 of some time-independent Hamiltonian:

$$\Psi(x, 0) = \tfrac{1}{\sqrt{2}}(\psi_1 + \psi_2).$$

Let E_1 and E_2 denote the energies of ψ_1 and ψ_2, respectively. Assume that $E_1 < E_2$.

1. Calculate the energy uncertainty ΔH in the $t=0$ state. Is this uncertainty preserved under time evolution?
2. Calculate the smallest time $\Delta t_\perp > 0$ required for the state Ψ to become a state orthogonal to $\Psi(x,0)$. Confirm that the bound $\Delta H \Delta t_\perp \geq \frac{h}{4}$ is in fact saturated.

Problem 15.9. *Exploring the time evolution of an overlap.*

Consider a physical system governed by a *time-independent* Hamiltonian $\hat{H}$. Let $|\Psi(0)\rangle$ denote the state of the system at $t=0$ and $|\Psi(t)\rangle$ the state of the system at time $t \geq 0$. The state of the system satisfies the Schrödinger equation and is taken to be normalized. Now consider the overlap of the time-evolved state with the initial state, squared:

$$|\langle\Psi(0)|\Psi(t)\rangle|^2.$$

1. At time equal zero, the above equals one. Explain why it cannot ever exceed one. What is the value of the overlap if $|\Psi(0)\rangle$ is an energy eigenstate?
2. Calculate the overlap in a power series expansion valid for small t, neglecting terms cubic and higher in t; namely, determine the terms represented by the dots in the equation

$$|\langle\Psi(0)|\Psi(t)\rangle|^2 = \cdots + \mathcal{O}(t^3).$$

Your answer will depend only on t, $\hbar$, and the uncertainty ΔH of the Hamiltonian!

Problem 15.10. *Exact inequalities for the time evolution of an overlap.*

We will explore inequalities for the overlap $|\langle\Psi(0)|\Psi(t)\rangle|^2$. For this purpose and for ease of manipulation, we will write

$$\cos^2\phi(t) \equiv |\langle\Psi(0)|\Psi(t)\rangle|^2,$$

which makes clear that the right-hand side is never larger than one. Knowing the phase $\phi(t)$ is equivalent to knowing the overlap. At $t=0$ we take $\phi(0)=0$. As time evolves and the right-hand side becomes smaller than one, we take ϕ to be in the interval:

$$0 \leq \phi(t) \leq \tfrac{\pi}{2}.$$

This suffices, as it allows us to consider the possibility that the overlap becomes zero when ϕ reaches the upper bound. We denote the state of the system by $|\Psi(t)\rangle$ and define $\hat{Q}$ to be the projector to the state at $t=0$:

$$\hat{Q} \equiv |\Psi(0)\rangle\langle\Psi(0)|.$$

1. Use the energy-Q uncertainty inequality to establish a limit on the rate of change of the phase ϕ:

$$\left|\frac{d\phi}{dt}\right| \leq \frac{\Delta H}{\hbar}. \tag{1}$$

This is a very simple bound: the velocity of ϕ is limited by the energy uncertainty! Conclude that in a system governed by a time-independent Hamiltonian (so that ΔH is time independent), the minimum time $\Delta t_\perp$ needed for any state with energy uncertainty ΔH to evolve into an orthogonal state satisfies the constraint

$$\Delta H \Delta t_{\perp} \geq \frac{h}{4}.$$

2. Show that as a consequence of equation (1) we have

$$|\langle\Psi(0)|\Psi(t)\rangle|^2 \geq \cos^2\Big(\frac{\Delta H t}{\hbar}\Big), \text{ for } t \leq \frac{\pi\hbar}{2\Delta H}.$$

Problem 15.11. *Simultaneous diagonalization of two Hermitian matrices.*

Consider the Hermitian matrices A_1 and A_2:

$$A_1 = \begin{pmatrix} 1 & 1 & 0 & -1 \\ 1 & 1 & -1 & 0 \\ 0 & -1 & 1 & 1 \\ -1 & 0 & 1 & 1 \end{pmatrix}, \quad A_2 = \begin{pmatrix} 1 & \frac{1}{2} & -1 & \frac{1}{2} \\ \frac{1}{2} & 1 & \frac{1}{2} & -1 \\ -1 & \frac{1}{2} & 1 & \frac{1}{2} \\ \frac{1}{2} & -1 & \frac{1}{2} & 1 \end{pmatrix}.$$

These matrices commute so they can be simultaneously diagonalized:

$$U^{\dagger} A_1 U = D_1, \quad U^{\dagger} A_2 U = D_2,$$

where D_1 and D_2 are two diagonal matrices and U is unitary. Determine the matrices U, D_1, and D_2. Find the common eigenvectors of the two matrices and label them as u_{a_1,a_2} where a_1 and a_2 are the eigenvalues of A_1 and A_2, respectively. (You are urged to use a mathematical manipulator to avoid tedious arithmetic!)

Problem 15.12. *Two-dimensional oscillator.*

Suppose a particle of mass m is free to move on the (x, y) plane subject to a harmonic potential centered at the origin. Assume the restoring forces in the x and y directions are different. The Hamiltonian for this system is

$$\hat{H} = \frac{1}{2m}\hat{p}_x^2 + \frac{1}{2m}\hat{p}_y^2 + \tfrac{1}{2}m\omega_x^2\hat{x}^2 + \tfrac{1}{2}m\omega_y^2\hat{y}^2,$$

where $[\hat{x}, \hat{p}_x] = [\hat{y}, \hat{p}_y] = i\hbar$, and all other commutators between $\hat{x}$, $\hat{y}$, $\hat{p}_x$, and $\hat{p}_y$ vanish.

1. Introduce lowering and raising operators $\hat{a}_x$, $\hat{a}_y$, $\hat{a}_x^{\dagger}$, and $\hat{a}_y^{\dagger}$ as well as number operators $\hat{N}_x = \hat{a}_x^{\dagger}\hat{a}_x$ and $\hat{N}_y = \hat{a}_y^{\dagger}\hat{a}_y$, with eigenvalues n_x and n_y, respectively. What is $\hat{H}$ in terms of these operators? Find expressions for the energy eigenstates and the energy eigenvalues.

2. Plot an energy-level diagram for this system. Assume, for a clearer diagram, that $\omega_x \approx \omega_y$, and $\omega_x > \omega_y$. Include at least the first three groups of states. Indicate the values of n_x and n_y.

   Now define new operators $\hat{N}$ and $\hat{n}$ as follows:

   $$\hat{N} = \hat{N}_x + \hat{N}_y, \quad \hat{n} = \hat{N}_x - \hat{N}_y,$$

   and notice that they commute with $\hat{H}$. The energy eigenstates can therefore be labeled by n and N, the eigenvalues of $\hat{n}$ and $\hat{N}$, respectively.

3. What is $E_{N,n}$? Redraw the energy-level diagram, and label the states with the quantum numbers n and N. Use your pictures to decide which of the following are CSCOs: $\{\hat{N}\}$, $\{\hat{N}, \hat{n}\}$, $\{\hat{N}_x, \hat{N}_y\}$, and $\{\hat{H}\}$. How do your answers change if you take $\omega_x = \omega_y$? How do your answers change if ω_x/ω_y is equal to a rational number?

   From now on let $\omega_x = \omega_y = \omega$, and define the angular momentum operator $\hat{\ell}$ as follows:

$$\hat{\ell} = \hat{x}\hat{p}_y - \hat{y}\hat{p}_x.$$

4. Write $\hat{\ell}$ in terms of the operators $\hat{a}_x$, $\hat{a}_y$, $\hat{a}_x^\dagger$, and $\hat{a}_y^\dagger$. Show that $\hat{\ell}$ commutes with $\hat{H}$, implying that they can be simultaneously diagonalized.
5. Consider the degenerate subspace consisting of all the energy eigenstates that have $\hat{N}$ eigenvalue equal to N. Find a basis for this subspace such that the basis vectors are eigenstates of $\hat{\ell}$. Classify these basis states by their angular momentum eigenvalues, and show that $\hat{H}$ and $\hat{\ell}$ together constitute a CSCO for the entire state space.

   For this purpose define

$$\hat{a}_L = \frac{1}{\sqrt{2}}(\hat{a}_x + i\hat{a}_y), \quad \hat{a}_R = \frac{1}{\sqrt{2}}(\hat{a}_x - i\hat{a}_y), \quad \hat{N}_L = \hat{a}_L^\dagger \hat{a}_L, \quad \hat{N}_R = \hat{a}_R^\dagger \hat{a}_R,$$

   and express $\hat{H}$ and $\hat{\ell}$ in terms of $\hat{N}_L$ and $\hat{N}_R$.

16 Pictures of Quantum Mechanics

We explain the postulate of unitary time evolution and use it to derive the Schrödinger equation. In the Schrödinger picture of quantum mechanics, physical states evolve in time, and we have a set of fundamental time-independent operators. We introduce the Heisenberg picture of quantum mechanics in which we define time-dependent Heisenberg operators associated to Schrödinger operators. In this picture the dynamics of Schrödinger states is captured by the time-dependent operators, and Heisenberg states are taken to be time independent. We then turn to the axioms of quantum mechanics, defining the states of the system, observables, measurement, and dynamics. Two additional postulates explain how to build composite systems and how to treat identical particles.

16.1 Schrödinger Picture and Unitary Time Evolution

The state space of quantum mechanics—the Hilbert space $\mathcal{H}$ of states—is best thought of as a space with time-independent basis vectors. There is no role for time in the definition of the state space $\mathcal{H}$. In the Schrödinger "picture" of the dynamics, the state that represents a quantum system depends on time. Time is viewed as a parameter: at different times the state of the system is represented by different states in the Hilbert space. We write the state vector as

$$|\Psi, t\rangle, \tag{16.1.1}$$

and its components along the basis vectors of $\mathcal{H}$ are time dependent. If we call those basis vectors $|u_i\rangle$, we write

$$|\Psi, t\rangle = \sum_i |u_i\rangle c_i(t), \tag{16.1.2}$$

where the $c_i(t)$ are some functions of time. Assuming the state is normalized, which is almost always convenient, we can imagine $|\Psi, t\rangle$ as a unit vector whose tip, as a function of time, sweeps a trajectory in $\mathcal{H}$. We will discuss the postulate of unitary time evolution in this section and then show that the Schrödinger equation follows from this postulate.

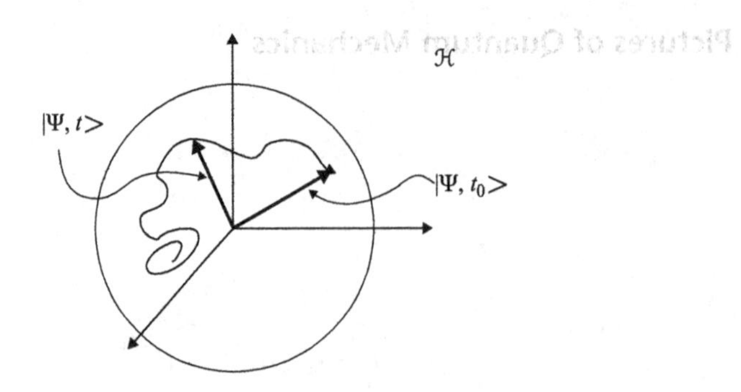

Figure 16.1
The initial state $|\Psi, t_0\rangle$ can be viewed as a vector in the complex vector space $\mathcal{H}$. As time goes by the vector moves, evolving by unitary transformations, so that its norm is preserved.

We declare that for any quantum system there is a *unitary* operator $\mathcal{U}(t, t_0)$ such that for *any* state $|\Psi, t_0\rangle$ of the system at time t_0 the state at time t is

$$|\Psi, t\rangle = \mathcal{U}(t, t_0)|\Psi, t_0\rangle, \quad \forall\, t, t_0. \tag{16.1.3}$$

It must be emphasized that the operator $\mathcal{U}$ generates time evolution for *any* possible state at time t_0—it does *not* depend on the chosen state at time t_0. A physical system has a single operator $\mathcal{U}$ that generates the time evolution of all possible states. The above equation is valid for all times t, so t can be greater than, equal to, or less than t_0. As defined, the operator $\mathcal{U}$ is unique: if there is another operator $\mathcal{U}'$ that generates exactly the same evolution, then $(\mathcal{U} - \mathcal{U}')|\Psi, t_0\rangle = 0$, and since the state $|\Psi, t_0\rangle$ is arbitrary, the operator $\mathcal{U} - \mathcal{U}'$ vanishes, showing that $\mathcal{U} = \mathcal{U}'$.

The unitary property of $\mathcal{U}$ means that its Hermitian conjugate is its inverse:

$$(\mathcal{U}(t, t_0))^\dagger \mathcal{U}(t, t_0) = \mathbb{1}. \tag{16.1.4}$$

In order to avoid extra parentheses, we will write

$$\mathcal{U}^\dagger(t, t_0) \equiv (\mathcal{U}(t, t_0))^\dagger \tag{16.1.5}$$

so the unitarity property reads

$$\mathcal{U}^\dagger(t, t_0)\mathcal{U}(t, t_0) = \mathbb{1}. \tag{16.1.6}$$

The unitarity of $\mathcal{U}$ also implies that the norm of the state is conserved by time evolution:

$$\langle \Psi, t\,|\Psi, t\rangle = \langle \Psi, t_0\,|\mathcal{U}^\dagger(t, t_0)\mathcal{U}(t, t_0)|\Psi, t_0\rangle = \langle \Psi, t_0|\,\Psi, t_0\rangle. \tag{16.1.7}$$

This is illustrated in figure 16.1.

Remarks:

1. For time $t = t_0$, equation (16.1.3) gives no time evolution:

$$|\Psi, t_0\rangle = \mathcal{U}(t_0, t_0)|\Psi, t_0\rangle. \tag{16.1.8}$$

Since this equality holds for *any* possible state at $t = t_0$, the unitary evolution operator with equal time arguments must be the identity operator:

$$\mathcal{U}(t_0, t_0) = \mathbb{1}, \quad \forall t_0. \tag{16.1.9}$$

2. Composition: Consider the evolution from t_0 to t_2 as a two-step procedure, from t_0 to t_1 first, followed by evolution from t_1 to t_2:

$$|\Psi, t_2\rangle = \mathcal{U}(t_2, t_1)|\Psi, t_1\rangle = \mathcal{U}(t_2, t_1)\mathcal{U}(t_1, t_0)|\Psi, t_0\rangle. \tag{16.1.10}$$

This equation and $|\Psi, t_2\rangle = \mathcal{U}(t_2, t_0)|\Psi, t_0\rangle$ imply that $\mathcal{U}$ composes as follows:

$$\mathcal{U}(t_2, t_0) = \mathcal{U}(t_2, t_1)\mathcal{U}(t_1, t_0). \tag{16.1.11}$$

3. Inverses: Consider the above composition law (16.1.11), and set $t_2 = t_0$ and $t_1 = t$. Then using (16.1.9), we get

$$\mathbb{1} = \mathcal{U}(t_0, t)\mathcal{U}(t, t_0). \tag{16.1.12}$$

We then have

$$\mathcal{U}(t_0, t) = (\mathcal{U}(t, t_0))^{-1} = (\mathcal{U}(t, t_0))^\dagger, \tag{16.1.13}$$

where the first relation follows from (16.1.12) and the second by unitarity. Again, declining to use parentheses that are not really needed, we write

$$\boxed{\mathcal{U}(t_0, t) = \mathcal{U}^{-1}(t, t_0) = \mathcal{U}^\dagger(t, t_0).} \tag{16.1.14}$$

Simply said, inverses or the Hermitian conjugation of $\mathcal{U}$ reverses the order of the time arguments.

16.2 Deriving the Schrödinger Equation

The time evolution of states has been specified in terms of a unitary operator $\mathcal{U}$ assumed known. We now ask a "reverse engineering" question. What kind of differential equation do states satisfy for which the solution is unitary time evolution? The answer is simple and satisfying: the states satisfy the Schrödinger equation. The discussion that follows can be called a derivation of the Schrödinger equation.

We begin by taking the time derivative of the time evolution postulate (16.1.3) to find that

$$\frac{\partial}{\partial t}|\Psi, t\rangle = \frac{\partial \mathcal{U}(t, t_0)}{\partial t}|\Psi, t_0\rangle. \tag{16.2.1}$$

We want the right-hand side to involve the ket $|\Psi, t\rangle$ so we write

$$\frac{\partial}{\partial t}|\Psi, t\rangle = \frac{\partial \mathcal{U}(t, t_0)}{\partial t}\,\mathcal{U}(t_0, t)|\Psi, t\rangle. \tag{16.2.2}$$

This now looks like a differential equation for the state $|\Psi, t\rangle$. Let us introduce a name for the operator appearing on the right-hand side:

$$\frac{\partial}{\partial t}|\Psi, t\rangle = \Lambda(t, t_0)|\Psi, t\rangle, \quad \text{with} \quad \Lambda(t, t_0) \equiv \frac{\partial \mathcal{U}(t, t_0)}{\partial t}\,\mathcal{U}(t_0, t). \tag{16.2.3}$$

The operator Λ has units of inverse time. Note also that

$$\Lambda^\dagger(t, t_0) = \mathcal{U}(t, t_0)\frac{\partial \mathcal{U}(t_0, t)}{\partial t}, \tag{16.2.4}$$

since the adjoint operation reverses the order of the operators, changes the order of time arguments in each $\mathcal{U}$, and does not interfere with the time derivative.

We now want to prove two important facts about Λ:

1. $\Lambda(t, t_0)$ is anti-Hermitian. To prove this begin with $\mathcal{U}(t, t_0)\mathcal{U}(t_0, t) = \mathbb{1}$, and take a derivative with respect to time to find

$$\frac{\partial \mathcal{U}(t, t_0)}{\partial t}\mathcal{U}(t_0, t) + \mathcal{U}(t, t_0)\frac{\partial \mathcal{U}(t_0, t)}{\partial t} = 0. \tag{16.2.5}$$

Glancing at (16.2.3) and (16.2.4), we see that we got

$$\Lambda(t, t_0) + \Lambda^\dagger(t, t_0) = 0, \tag{16.2.6}$$

proving that $\Lambda(t, t_0)$ is indeed anti-Hermitian.

2. $\Lambda(t, t_0)$ is actually independent of t_0. This is important because in the differential equation (16.2.3) t_0 appears nowhere except in Λ. We will show that $\Lambda(t, t_0)$ is actually equal to $\Lambda(t, t_1)$ for any other time t_1 different from t_0. To prove this we begin with the expression for $\Lambda(t, t_0)$ in (16.2.3) and insert a suitable form of the unit operator in between the two factors:

$$\Lambda(t, t_0) = \frac{\partial \mathcal{U}(t, t_0)}{\partial t}\big(\mathcal{U}(t_0, t_1)\mathcal{U}(t_1, t_0)\big)\mathcal{U}(t_0, t). \tag{16.2.7}$$

Since the time derivative ignores both t_0 and t_1, we have, as claimed,

$$\begin{aligned}\Lambda(t, t_0) &= \frac{\partial}{\partial t}\Big(\mathcal{U}(t, t_0)\mathcal{U}(t_0, t_1)\Big)\Big(\mathcal{U}(t_1, t_0)\mathcal{U}(t_0, t)\Big)\\ &= \frac{\partial \mathcal{U}(t, t_1)}{\partial t}\mathcal{U}(t_1, t) = \Lambda(t, t_1).\end{aligned} \tag{16.2.8}$$

It follows that we can write $\Lambda(t) \equiv \Lambda(t, t_0)$, and thus equation (16.2.3) becomes

$$\frac{\partial}{\partial t}|\Psi, t\rangle = \Lambda(t)|\Psi, t\rangle. \tag{16.2.9}$$

We define an operator $\hat{H}(t)$ by multiplication of Λ by $i\hbar$:

$$\boxed{\hat{H}(t) \equiv i\hbar\Lambda(t) = i\hbar\frac{\partial \mathcal{U}(t, t_0)}{\partial t}\mathcal{U}(t_0, t).} \tag{16.2.10}$$

Since Λ is anti-Hermitian and has units of inverse time, $\hat{H}(t)$ is a *Hermitian* operator with units of energy. $\hat{H}(t)$ is called the (quantum) Hamiltonian of the quantum system. Multiplying (16.2.9) by $i\hbar$, we find the

$$\boxed{\text{Schrödinger equation:}\quad i\hbar\frac{\partial}{\partial t}|\Psi, t\rangle = \hat{H}(t)\,|\Psi, t\rangle.} \tag{16.2.11}$$

This is our main result. Unitary time evolution implies this equation. In this derivation the Hamiltonian $\hat{H}(t)$ follows from the knowledge of $\mathcal{U}$, as shown in (16.2.10). If you are handed the unitary operator that generates time evolution, you can quickly reconstruct the Hamiltonian. Most often, however, we know the Hamiltonian and wish to calculate the time evolution operator $\mathcal{U}$.

The above equation for $|\Psi, t\rangle$ is the ket form of the Schrödinger equation. As discussed in example 14.15, the ordinary form of the Schrödinger equation for a wave function with coordinate dependence arises by writing $\Psi(x,t) \equiv \langle x|\Psi, t\rangle$.

There are basically two reasons why the quantity $\hat{H}(t)$ appearing in the above Schrödinger equation is called the Hamiltonian, or energy operator. First, in quantum mechanics the momentum operator is given by $\hbar/i$ times the derivative with respect to a spatial coordinate. In special relativity, energy corresponds to the time component of the momentum four-vector, and it is reasonable to view the energy operator as an operator proportional to a time derivative. The second argument is based on analogy to classical mechanics, as we explain now.

We have used the Schrödinger equation (16.2.11) to derive an equation for the time evolution of expectation values of observables. For a time-independent observable $\hat{Q}$, this took the form (15.3.5), which we rewrite as

$$\frac{d\langle\hat{Q}\rangle}{dt} = \Big\langle \tfrac{1}{i\hbar}[\,\hat{Q}, \hat{H}\,]\Big\rangle. \tag{16.2.12}$$

This equation is a natural generalization of the Hamiltonian equations in classical mechanics, and $\hat{H}$ plays a role analogous to that of the classical Hamiltonian. Indeed, in classical mechanics one has Poisson brackets $\{\cdot,\cdot\}_{\text{pb}}$ defined for functions of x and p by

$$\{A, B\}_{\text{pb}} = \frac{\partial A}{\partial x}\frac{\partial B}{\partial p} - \frac{\partial A}{\partial p}\frac{\partial B}{\partial x}. \tag{16.2.13}$$

Note that $\{x, p\}_{\text{pb}} = 1$. Moreover, just like commutators, Poisson brackets are antisymmetric: $\{A, B\}_{\text{pb}} = -\{B, A\}_{\text{pb}}$. For any function $Q(x, p)$ without explicit time dependence, its time derivative is given by taking the Poisson bracket of Q with the classical Hamiltonian H. To see this begin with

$$\frac{dQ}{dt} = \frac{\partial Q}{\partial x}\dot{x} + \frac{\partial Q}{\partial p}\dot{p}, \tag{16.2.14}$$

where we used the chain rule for derivatives, and dots mean time derivatives. Hamilton's equations of motion state that

$$\dot{x} = \frac{\partial H}{\partial p}, \qquad \dot{p} = -\frac{\partial H}{\partial x}. \tag{16.2.15}$$

If you have not seen this before, you can at least quickly check that for a classical Hamiltonian $H(x,p) = \frac{p^2}{2m} + V(x)$ the equations above give exactly what you would expect. Using the values for $\dot{x}$ and $\dot{p}$ in (16.2.14), we now have

$$\frac{dQ}{dt} = \frac{\partial Q}{\partial x}\frac{\partial H}{\partial p} - \frac{\partial Q}{\partial p}\frac{\partial H}{\partial x}. \tag{16.2.16}$$

The right-hand side is indeed the Poisson bracket of Q with the Hamiltonian, so we have that

$$\frac{dQ}{dt} = \{Q, H\}_{\text{pb}}. \tag{16.2.17}$$

The similarity to the time derivative of the quantum expectation values (16.2.12) is quite striking and suggests that $\hat{H}$ is indeed a quantum version of the classical Hamiltonian H. Note also that this comparison suggests that quantum commutators behave as $i\hbar$ times Poisson brackets:

$$[\hat{A}, \hat{B}] \iff i\hbar\,\{A, B\}_{\text{pb}}. \tag{16.2.18}$$

This correspondence is sometimes an equality: $\{x, p\}_{pb} = 1$ while $[\hat{x}, \hat{p}] = i\hbar$. But for general functions $A(x,p)$ and $B(x,p)$, there are ordering ambiguities in the quantum analogs $\hat{A}(\hat{x}, \hat{p})$ and $\hat{B}(\hat{x}, \hat{p})$ and in passing from the result of the Poisson bracket to its quantum analog. This is illustrated in the following exercise:

Exercise 16.1. *Show that* $\{x^2, p^2\}_{\text{pb}} = 4xp$, *while* $[\hat{x}^2, \hat{p}^2] = i\hbar(2\hat{x}\hat{p} + 2\hat{p}\hat{x}) = i\hbar(4\hat{x}\hat{p} - 2i\hbar)$.

While the reasons discussed above justify our calling $\hat{H}$ the Hamiltonian, ultimately any Hermitian operator with units of energy has the right to be called a Hamiltonian regardless of any connection to a classical theory. The value of the classical theory is that it suggests potentially interesting quantum Hamiltonians, as we saw, for example, in setting up the quantum harmonic oscillator.

16.3 Calculating the Time Evolution Operator

Typically, the Hamiltonian $\hat{H}(t)$ is known, and we wish to calculate the unitary operator $\mathcal{U}$ that implements time evolution. For this it is useful to find an equation for $\mathcal{U}$. Multiplying equation (16.2.10) from the right by $\mathcal{U}(t, t_0)$ gives

$$i\hbar\frac{\partial \mathcal{U}(t, t_0)}{\partial t} = \hat{H}(t)\,\mathcal{U}(t, t_0). \tag{16.3.1}$$

This is indeed a differential equation for the *operator* $\mathcal{U}$. Note also that letting both sides of this equation act on $|\Psi, t_0\rangle$ gives us back the Schrödinger equation.

Since there is no possible confusion with the time derivatives, we do not need to write them as partial derivatives. Then the above equation takes the form

$$\frac{d\mathcal{U}}{dt} = -\frac{i}{\hbar}\hat{H}(t)\,\mathcal{U}(t). \tag{16.3.2}$$

If we view operators as matrices, this is a differential equation for the *matrix* $\mathcal{U}$, involving the matrix $\hat{H}$. Solving this equation is in general quite difficult. We will consider three cases of increasing complexity.

Case 1. $\hat{H}$ is time independent. In this case, equation (16.3.2) is structurally of the form

$$\frac{d\mathcal{U}}{dt} = \hat{K}\,\mathcal{U}(t), \quad \text{with } \hat{K} = -\frac{i}{\hbar}\hat{H}. \tag{16.3.3}$$

Here, $\hat{K}$ is a time-independent matrix. If the matrices were 1×1, this would reduce to the plain differential equation

$$\frac{du}{dt} = ku(t) \quad \Rightarrow \quad u(t) = e^{kt}u(0). \tag{16.3.4}$$

For the matrix case (16.3.3), we claim that

$$\mathcal{U}(t) = e^{t\hat{K}}\,\mathcal{U}(0). \tag{16.3.5}$$

Here, the exponential of $t\hat{K}$ is multiplied from the right by the matrix $\mathcal{U}(0)$. The ansatz clearly works at time equal zero. The exponential of a matrix, as usual, is defined by the Taylor series of the exponential function (section 13.7). With $\hat{K}$ time independent, we have the derivative

$$\frac{d}{dt}e^{t\hat{K}} = \hat{K}e^{t\hat{K}} = e^{t\hat{K}}\hat{K}. \tag{16.3.6}$$

With this result we readily verify that (16.3.5) solves (16.3.3):

$$\frac{d\mathcal{U}}{dt} = \frac{d}{dt}\left(e^{t\hat{K}}\mathcal{U}(0)\right) = \hat{K}e^{t\hat{K}}\mathcal{U}(0) = \hat{K}\,\mathcal{U}(t). \tag{16.3.7}$$

Using the explicit form of the matrix $\hat{K}$, the solution (16.3.5) is therefore

$$\mathcal{U}(t, t_0) = e^{-\frac{i}{\hbar}\hat{H}t}\,\mathcal{U}_0, \tag{16.3.8}$$

where $\mathcal{U}_0$ is a constant matrix. Recalling that $\mathcal{U}(t_0, t_0) = \mathbb{1}$, we have $\mathbb{1} = e^{-\frac{i}{\hbar}\hat{H}t_0}\,\mathcal{U}_0$, and therefore $\mathcal{U}_0 = e^{\frac{i}{\hbar}\hat{H}t_0}$. The two exponentials can be combined into a single one, and the full solution becomes

$$\boxed{\mathcal{U}(t, t_0) = \exp\Big[-\frac{i}{\hbar}\hat{H}(t - t_0)\Big], \quad \text{time-independent } \hat{H}.} \tag{16.3.9}$$

Exercise 16.2. *Verify that the ansatz $\mathcal{U}(t) = \mathcal{U}(0)e^{tK}$, consistent for $t = 0$, would* not *have provided a solution of (16.3.3).*

Case 2. $[\hat{H}(t_1), \hat{H}(t_2)\,] = 0$ for all t_1, t_2. Here the Hamiltonian is time dependent, but, despite this, the Hamiltonians at different times commute. Of course, they trivially commute when both are evaluated at the same time. One example is provided by the Hamiltonian for a spin in a magnetic field of time-dependent magnitude but constant direction. We claim that the time evolution operator is now given by

$$\boxed{\mathcal{U}(t, t_0) = \exp\Big[-\frac{i}{\hbar}\int_{t_0}^{t} dt'\hat{H}(t')\Big], \quad \hat{H} \text{ at different times commute.}} \tag{16.3.10}$$

If the Hamiltonian is time independent, the integral in the exponent is easily done, and the above solution reduces correctly to (16.3.9). To prove that (16.3.10) solves the differential equation (16.3.2), we streamline notation by writing

$$\hat{R}(t) \equiv -\frac{i}{\hbar}\int_{t_0}^{t} dt' \hat{H}(t') \;\Rightarrow\; \hat{R}' = -\frac{i}{\hbar}\hat{H}(t), \tag{16.3.11}$$

where primes denote time derivatives. We claim that $\hat{R}'(t)$ and $\hat{R}(t)$ commute. Indeed

$$[\hat{R}'(t), \hat{R}(t)] = -\frac{1}{\hbar^2}\Big[\hat{H}(t), \int_{t_0}^{t} dt'\hat{H}(t')\Big] = -\frac{1}{\hbar^2}\int_{t_0}^{t} dt' \,[\hat{H}(t), \hat{H}(t')] = 0, \tag{16.3.12}$$

recalling that Hamiltonians at different times commute. The claimed solution is

$$\mathcal{U} = \exp \hat{R}(t). \tag{16.3.13}$$

Since $\hat{R}$ and $\hat{R}'$ commute, it follows that for any $n \geq 1$,

$$(\hat{R}^n)' = n\hat{R}'\hat{R}^{n-1} \tag{16.3.14}$$

because in all n terms produced by the derivative the $\hat{R}'$ factor can be moved to the left without impediment. Since first-order derivatives thus work as usual, we find that

$$\frac{d\mathcal{U}}{dt} = \frac{d}{dt}\exp\hat{R} = \hat{R}' \exp\hat{R} = -\frac{i}{\hbar}\hat{H}(t)\,\mathcal{U}, \tag{16.3.15}$$

which is exactly what we wanted to show.

Case 3. $[\hat{H}(t_1), \hat{H}(t_2)] \neq 0$. This is the most general situation, and there is only a series solution. The solution for $\mathcal{U}$ is given by the so-called time-ordered exponential, denoted by the symbol T in front of an exponential:

$$\begin{aligned} \mathcal{U}(t, t_0) &= \mathrm{T}\exp\Big[-\frac{i}{\hbar}\int_{t_0}^{t} dt'\hat{H}(t')\Big] = \mathrm{T}\sum_{n=0}^{\infty}\frac{1}{n!}\Big(-\frac{i}{\hbar}\Big)\int_{t_0}^{t} dt_1\hat{H}(t_1)\cdots\int_{t_0}^{t} dt_n\hat{H}(t_n) \\ &\equiv \mathbb{1} + \Big(-\frac{i}{\hbar}\Big)\int_{t_0}^{t} dt_1\hat{H}(t_1) \\ &\quad + \Big(-\frac{i}{\hbar}\Big)^2\int_{t_0}^{t} dt_1\hat{H}(t_1)\int_{t_0}^{t_1} dt_2\hat{H}(t_2) \\ &\quad + \Big(-\frac{i}{\hbar}\Big)^3\int_{t_0}^{t} dt_1\hat{H}(t_1)\int_{t_0}^{t_1} dt_2\hat{H}(t_2)\int_{t_0}^{t_2} dt_3\hat{H}(t_3) \\ &\qquad\vdots \qquad\qquad \vdots \\ &\quad + \Big(-\frac{i}{\hbar}\Big)^n\int_{t_0}^{t} dt_1\hat{H}(t_1)\int_{t_0}^{t_1} dt_2\hat{H}(t_2)\cdots\int_{t_0}^{t_{n-1}} dt_n\hat{H}(t_n) \\ &\qquad\vdots \qquad\qquad \vdots \qquad\qquad \vdots \end{aligned} \tag{16.3.16}$$

The action of time ordering T on each term of the exponential is defined by the above expression. The time ordering refers to the fact that in the nth term of the series we have a product $\hat{H}(t_1)\hat{H}(t_2)\ldots\hat{H}(t_n)$ of *noncommuting* operators with integration ranges that force ordered times $t_1 \geq t_2 \geq t_3 \cdots \geq t_n$. Note also that the nth term in the time-ordered exponential does not have the $1/n!$ combinatorial factor of the exponential.

In the exponential, all $t_1, \ldots, t_n$ are integrated from t_0 to t, and the full region of integration splits over $n!$ subregions, each with a different ordering of the times. Effectively, the T operator reorders the operators within each region, giving the same integral for each. Thus, the $1/n!$ is canceled by the $n!$ contributions. Some aspects of this construction are explored in problem 16.4.

Exercise 16.3. *Prove that (16.3.16) indeed solves the equation* $\frac{d\mathcal{U}}{dt} = -\frac{i}{\hbar}\hat{H}\,\mathcal{U}$.

16.4 Heisenberg Picture

The idea here is to confine the dynamical evolution to the operators. We will "fold" the time dependence of the states into the operators. Since the objects we usually calculate are time dependent expectation values of operators, this approach turns out to be quite effective. The result is a new "picture" of quantum mechanics, the *Heisenberg picture* in which the operators carry the time dependence and the states do not evolve.

We will define time-dependent Heisenberg operators starting from Schrödinger operators. Schrödinger operators are in fact the operators we have been using all along, such as $\hat{x}, \hat{p}, \hat{L}_i, \hat{S}_i$, and others, as well as operators constructed from them, possibly including time. Schrödinger operators come in two types: time independent, like $\hat{x}$ or $\hat{p}$, and time dependent, like Hamiltonians with time-dependent potentials. For each Schrödinger operator, we will associate a Heisenberg operator.

Let us consider a Schrödinger operator $\hat{A}_S$, with the subscript S for Schrödinger. This operator may or may not have time dependence. We now examine a matrix element of $\hat{A}_S$ in between time-dependent states $|\alpha, t\rangle$ and $|\beta, t\rangle$ and use the time-evolution operator to convert the states to time zero:

$$\langle \alpha, t|\hat{A}_S|\beta, t\rangle = \langle \alpha, 0|\,\mathcal{U}^\dagger(t,0)\,\hat{A}_S\,\mathcal{U}(t,0)\,|\beta, 0\rangle. \tag{16.4.1}$$

We simply define the Heisenberg operator $\hat{A}_H(t)$ associated with $\hat{A}_S$ as the object in between the time equal zero states:

$$\boxed{\hat{A}_H(t) \equiv \mathcal{U}^\dagger(t,0)\,\hat{A}_S\,\mathcal{U}(t,0).} \tag{16.4.2}$$

The Heisenberg operator is obtained from the Schrödinger operator by a similarity transformation generated by the time-evolution operator $\mathcal{U}(t,0)$.

Let us consider a number of important consequences of this definition.

1. At $t = 0$, the Heisenberg operator becomes equal to the Schrödinger operator:

$$\hat{A}_H(0) = \hat{A}_S. \tag{16.4.3}$$

The Heisenberg operator associated with the identity operator is the identity operator:

$$\mathbb{1}_H = \mathcal{U}^\dagger(t,0)\,\mathbb{1}\,\mathcal{U}(t,0) = \mathbb{1}. \tag{16.4.4}$$

2. The Heisenberg operator associated with the product of Schrödinger operators is equal to the product of the corresponding Heisenberg operators:

$$\hat{C}_S = \hat{A}_S\hat{B}_S \quad \Rightarrow \quad \hat{C}_H(t) = \hat{A}_H(t)\hat{B}_H(t). \tag{16.4.5}$$

Indeed,

$$\begin{aligned} \hat{C}_H(t) &= \mathcal{U}^\dagger(t,0)\,\hat{C}_S\,\mathcal{U}(t,0) = \mathcal{U}^\dagger(t,0)\,\hat{A}_S\hat{B}_S\,\mathcal{U}(t,0) \\ &= \hat{\mathcal{U}}^\dagger(t,0)\,\hat{A}_S\,\mathcal{U}(t,0)\mathcal{U}^\dagger(t,0)\,\hat{B}_S\,\mathcal{U}(t,0) = \hat{A}_H(t)\hat{B}_H(t). \end{aligned} \tag{16.4.6}$$

3. It also follows from (16.4.5) that if we have a commutator of Schrödinger operators the corresponding Heisenberg operators satisfy the same commutation relations:

$$[\hat{A}_S, \hat{B}_S] = \hat{C}_S \quad \Rightarrow \quad [\hat{A}_H(t), \hat{B}_H(t)] = \hat{C}_H(t). \tag{16.4.7}$$

Since $\mathbb{1}_H = \mathbb{1}$, equation (16.4.7) implies that, for example,

$$[\hat{x}, \hat{p}] = i\hbar\mathbb{1} \quad \Rightarrow \quad [\hat{x}_H(t), \hat{p}_H(t)] = i\hbar\mathbb{1}. \tag{16.4.8}$$

4. Schrödinger and Heisenberg Hamiltonians: Consider a Schrödinger Hamiltonian H_S that depends on some Schrödinger momenta and position operators $\hat{p}$ and $\hat{x}$:

$$\hat{H}_S(\hat{x}, \hat{p}; t). \tag{16.4.9}$$

Since the $\hat{x}$ and $\hat{p}$ operators in $\hat{H}_S$ appear in products, property (2) implies that the associated Heisenberg Hamiltonian $\hat{H}_H$ takes the same form but with $\hat{x}$ and $\hat{p}$ replaced by their Heisenberg counterparts:

$$\hat{H}_H(t) = \hat{H}_S(\hat{x}_H(t), \hat{p}_H(t); t). \tag{16.4.10}$$

5. Equality of Hamiltonians: When $[\hat{H}_S(t), \hat{H}_S(t')] = 0$, for all t, t', the Heisenberg Hamiltonian is in fact equal to the Schrödinger Hamiltonian. To see this, recall that for this type of Hamiltonian the time evolution operator is

$$\mathcal{U}(t,0) = \exp\Big[-\frac{i}{\hbar}\int_0^t dt'\hat{H}_S(t')\Big]. \tag{16.4.11}$$

Moreover, by definition,

$$\hat{H}_H(t) = \mathcal{U}^\dagger(t,0)\,\hat{H}_S(t)\mathcal{U}(t,0). \tag{16.4.12}$$

Since the $\hat{H}_S$ commute at different times, $\hat{H}_S(t)$ commutes both with $\mathcal{U}(t,0)$ and $\mathcal{U}^\dagger(t,0)$. Therefore, the $\hat{H}_S(t)$ in (16.4.12) can be moved, say, to the right, giving us

$$\boxed{\hat{H}_H(t) = \hat{H}_S(t), \quad \text{when } [\hat{H}_S(t), \hat{H}_S(t')] = 0.} \tag{16.4.13}$$

Clearly, this equality holds for time-independent Schrödinger Hamiltonians. The meaning of this equality becomes clearer when we use (16.4.10) and (16.4.9) to write

$$\hat{H}_S(\hat{x}_H(t), \hat{p}_H(t); t) = \hat{H}_S(\hat{x}, \hat{p}; t). \tag{16.4.14}$$

Operationally, this means that if we take $\hat{x}_H(t)$ and $\hat{p}_H(t)$ and plug them into the Schrödinger Hamiltonian (left-hand side), the result is as if we had simply plugged $\hat{x}$ and $\hat{p}$. We will confirm this for the case of the simple harmonic oscillator.

6. Equality of operators: If a Schrödinger operator $\hat{A}_S$ commutes with the Hamiltonian $\hat{H}_S(t)$ for all times, then $\hat{A}_S$ commutes with $\mathcal{U}(t,0)$ since this operator (even in the most complicated of cases) is built using $\hat{H}_S(t)$. It follows that $\hat{A}_H(t)=\hat{A}_S$; the Heisenberg operator is equal to the Schrödinger operator. In summary,

$$[A_S, \hat{H}_S(t)]=0, \quad \forall\, t \;\Rightarrow\; \hat{A}_H(t)=\hat{A}_S. \tag{16.4.15}$$

7. Expectation values: Consider (16.4.1) and let $|\alpha,t\rangle=|\beta,t\rangle=|\Psi,t\rangle$. The matrix element now becomes an expectation value and we have

$$\boxed{\langle\Psi,t|\hat{A}_S|\Psi,t\rangle \;=\; \langle\Psi,0|\,\hat{A}_H(t)\,|\Psi,0\rangle.} \tag{16.4.16}$$

With a little abuse of notation, we simply write this equation as

$$\boxed{\langle\hat{A}_S\rangle \;=\; \langle\hat{A}_H(t)\rangle.} \tag{16.4.17}$$

When writing such an equation, you should realize that on the left-hand side you compute the expectation value using the time-dependent state, while on the right-hand side you compute the expectation value using the state at time equal zero.

16.5 Heisenberg Equations of Motion

We can calculate the Heisenberg operator associated with a Schrödinger one using the definition (16.4.2). Alternatively, Heisenberg operators satisfy a differential equation: the Heisenberg equation of motion. This equation looks very much like the equations of motion of classical dynamical variables—so much so that people trying to invent quantum theories sometimes begin with the equations of motion of some classical system and postulate the existence of Heisenberg operators that satisfy similar equations. In that case they must also find a Heisenberg Hamiltonian and show that the equations of motion indeed arise in the quantum theory.

To determine the equation of motion of Heisenberg operators, we will simply take time derivatives of the definition (16.4.2). For this purpose we recall (16.3.1), which we copy here using the subscript S for the Hamiltonian:

$$i\hbar\,\frac{\partial\mathcal{U}(t,t_0)}{\partial t} \;=\; \hat{H}_S(t)\,\mathcal{U}(t,t_0). \tag{16.5.1}$$

Taking the adjoint of this equation, we find that

$$i\hbar\,\frac{\partial\mathcal{U}^\dagger(t,t_0)}{\partial t} \;=\; -\,\mathcal{U}^\dagger(t,t_0)\,\hat{H}_S(t). \tag{16.5.2}$$

We can now calculate. Using (16.4.2) we find

$$\begin{aligned} i\hbar\frac{d}{dt}\hat{A}_H(t) &= \Big(i\hbar\frac{\partial\mathcal{U}^\dagger}{\partial t}(t,0)\Big)\hat{A}_S(t)\,\mathcal{U}(t,0) + \mathcal{U}^\dagger(t,0)\,\hat{A}_S(t)\Big(i\hbar\frac{\partial\mathcal{U}}{\partial t}(t,0)\Big) \\ &\quad + \mathcal{U}^\dagger(t,0)\,i\hbar\frac{\partial\hat{A}_S(t)}{\partial t}\,\mathcal{U}(t,0). \end{aligned} \tag{16.5.3}$$

Using (16.5.1) and (16.5.2), we have

$$\begin{aligned} i\hbar\frac{d}{dt}\hat{A}_H(t) &= -\mathcal{U}^\dagger(t,0)\,\hat{H}_S(t)\,\hat{A}_S(t)\,\mathcal{U}(t,0) + \mathcal{U}^\dagger(t,0)\,\hat{A}_S(t)\,\hat{H}_S(t)\,\mathcal{U}(t,0) \\ &\quad + \mathcal{U}^\dagger(t,0)\,i\hbar\frac{\partial\hat{A}_S(t)}{\partial t}\,\mathcal{U}(t,0). \end{aligned} \tag{16.5.4}$$

We now use (16.4.5) for the top line and recognize that in the bottom line we have the Heisenberg operator associated with the time derivative of $\hat{A}_S$:

$$\begin{aligned} i\hbar\frac{d}{dt}\hat{A}_H(t) &= -\hat{H}_H(t)\,\hat{A}_H(t) + \hat{A}_H(t)\,H_H(t) + i\hbar\Big(\frac{\partial\hat{A}_S(t)}{\partial t}\Big)_H \\ &= [\hat{A}_H(t),\,\hat{H}_H(t)] + i\hbar\Big(\frac{\partial\hat{A}_S(t)}{\partial t}\Big)_H. \end{aligned} \tag{16.5.5}$$

Properly understood, in the last term, the operations of taking a time derivative and then going into the Heisenberg picture commute. To make the point clearly, note that in general $\hat{A}_S$ is built from some time-independent operators that we can denote collectively by $\hat{\mathcal{O}}$ and, separately, has some explicit time dependence. To reflect this, the operator is written as $\hat{A}_S(\hat{\mathcal{O}};t)$. It then follows that

$$\begin{aligned} \Big(\frac{\partial\hat{A}_S}{\partial t}\Big)_H &= \Big(\lim_{\epsilon\to0}\frac{1}{\epsilon}\big(\hat{A}_S(\hat{\mathcal{O}};t+\epsilon)-\hat{A}_S(\hat{\mathcal{O}};t)\big)\Big)_H \\ &= \lim_{\epsilon\to0}\frac{1}{\epsilon}\big(\hat{A}_S(\hat{\mathcal{O}};t+\epsilon)-\hat{A}_S(\hat{\mathcal{O}};t)\big)_H \\ &= \lim_{\epsilon\to0}\frac{1}{\epsilon}\big(\hat{A}_S(\hat{\mathcal{O}}_H(t);t+\epsilon)-\hat{A}_S(\hat{\mathcal{O}}_H(t);t)\big) \\ &= \frac{\partial}{\partial t}\hat{A}_S(\hat{\mathcal{O}}_H(t);t) = \frac{\partial\hat{A}_H}{\partial t}, \end{aligned} \tag{16.5.6}$$

where we note that $\hat{A}_H = \hat{A}_S(\hat{\mathcal{O}}_H(t);t)$, since making $\hat{A}_S$ into a Heisenberg operator just means turning $\hat{\mathcal{O}}$ into a Heisenberg operator. We emphasize that the partial time derivative does *not* act on the Heisenberg operator $\hat{\mathcal{O}}_H(t)$. With this understanding, we write (16.5.5) in its final form:

$$\boxed{i\hbar\,\frac{d\hat{A}_H(t)}{dt} = [\hat{A}_H(t),\,\hat{H}_H(t)] + i\hbar\,\frac{\partial\hat{A}_H(t)}{\partial t}.} \tag{16.5.7}$$

A few comments are in order.

1. Schrödinger operators without time dependence: if the operator $\hat{A}_S$ has no explicit time dependence, the last term in (16.5.7) vanishes, and we have the simpler

$$i\hbar \frac{d\hat{A}_H(t)}{dt} = [\hat{A}_H(t), \hat{H}_H(t)]. \tag{16.5.8}$$

2. Time dependence of expectation values: Let $\hat{A}_S$ be a Schrödinger operator without time dependence. Let us now take the time derivative of the expectation value relation in (16.4.16):

$$\begin{aligned} i\hbar\frac{d}{dt}\langle\Psi,t|\hat{A}_S|\Psi,t\rangle &= i\hbar\frac{d}{dt}\langle\Psi,0|\,\hat{A}_H(t)\,|\Psi,0\rangle = \langle\Psi,0|\,i\hbar\frac{d\hat{A}_H(t)}{dt}\,|\Psi,0\rangle \\ &= \langle\Psi,0|\,[\hat{A}_H(t), \hat{H}_H(t)]\,|\Psi,0\rangle. \end{aligned} \tag{16.5.9}$$

We write this as

$$\boxed{i\hbar\frac{d}{dt}\langle\hat{A}_H(t)\rangle = \big\langle[\hat{A}_H(t), \hat{H}_H(t)]\big\rangle.} \tag{16.5.10}$$

Notice that this equation takes exactly the same form in the Schrödinger picture (recall the comments below (16.4.17)):

$$\boxed{i\hbar\frac{d}{dt}\langle\hat{A}_S\rangle = \big\langle[\hat{A}_S, \hat{H}_S]\big\rangle.} \tag{16.5.11}$$

3. A time-independent operator $\hat{A}_S$ is said to be **conserved** if it commutes with the Hamiltonian:

$$\text{conserved operator } \hat{A}_S\text{:}\quad [\hat{A}_S, \hat{H}_S] = 0. \tag{16.5.12}$$

It follows that $[\hat{A}_H(t), \hat{H}_H(t)] = 0$, and using (16.5.8), we find that

$$\frac{d\hat{A}_H(t)}{dt} = 0. \tag{16.5.13}$$

The Heisenberg operator is constant. Thus, the expectation value of the operator is also constant. This is consistent with (6) in the previous section: $\hat{A}_H$ is in fact equal to $\hat{A}_S$!

Example 16.1. *Heisenberg operators for the harmonic oscillator.*
Our main goal here is to obtain the explicit form of the Heisenberg operators $\hat{x}_H(t)$ and $\hat{p}_H(t)$ for the simple harmonic oscillator. For this we will have to solve their Heisenberg equations of motion.

The Schrödinger Hamiltonian $\hat{H}_S$ for the oscillator takes the form

$$\hat{H}_S = \frac{\hat{p}^2}{2m} + \tfrac{1}{2}m\omega^2\hat{x}^2. \tag{16.5.14}$$

Using (16.4.10), the associated Heisenberg Hamiltonian $\hat{H}_H(t)$ is obtained by replacing $\hat{x}$ and $\hat{p}$ above by their Heisenberg counterparts:

$$\hat{H}_H(t) = \frac{\hat{p}_H^2(t)}{2m} + \tfrac{1}{2}\,m\omega^2\hat{x}_H^2(t). \tag{16.5.15}$$

The Heisenberg equation of motion for $\hat{x}_H(t)$ is

$$\begin{aligned}\frac{d}{dt}\hat{x}_H(t) &= \frac{1}{i\hbar}[\hat{x}_H(t), \hat{H}_H(t)] = \frac{1}{i\hbar}\Big[\hat{x}_H(t), \frac{\hat{p}_H^2(t)}{2m}\Big] \\ &= \frac{1}{i\hbar}\,2\,\frac{\hat{p}_H(t)}{2m}\Big[\hat{x}_H(t), \hat{p}_H(t)\Big] = \frac{1}{i\hbar}\frac{\hat{p}_H(t)}{m}\,i\hbar = \frac{\hat{p}_H(t)}{m}\end{aligned} \tag{16.5.16}$$

so that our first equation is

$$\frac{d}{dt}\hat{x}_H(t) = \frac{\hat{p}_H(t)}{m}. \tag{16.5.17}$$

For the momentum operator, we get

$$\begin{aligned}\frac{d}{dt}\hat{p}_H(t) &= \frac{1}{i\hbar}[\hat{p}_H(t), \hat{H}_H(t)] = \frac{1}{i\hbar}\Big[\hat{p}_H(t), \frac{1}{2}m\omega^2\hat{x}_H^2(t)\Big] \\ &= \frac{1}{i\hbar}\frac{1}{2}m\omega^2\cdot 2(-i\hbar)\hat{x}_H(t) = -m\omega^2\,\hat{x}_H(t),\end{aligned} \tag{16.5.18}$$

so our second equation is

$$\frac{d}{dt}\hat{p}_H(t) = -m\omega^2\,\hat{x}_H(t). \tag{16.5.19}$$

Taking another time derivative of our first equation and using the second one, we get

$$\frac{d^2}{dt^2}\hat{x}_H(t) = -\omega^2\,\hat{x}_H(t). \tag{16.5.20}$$

The solution of this differential equation takes the form

$$\hat{x}_H(t) = \hat{A}\,\cos\omega t + \hat{B}\,\sin\omega t, \tag{16.5.21}$$

where $\hat{A}$ and $\hat{B}$ are time-independent operators to be determined by initial conditions. Using (16.5.17), we can calculate the associated momentum operator:

$$\hat{p}_H(t) = m\frac{d}{dt}\hat{x}_H(t) = -m\omega\hat{A}\sin\omega t + m\omega\hat{B}\cos\omega t. \tag{16.5.22}$$

At zero time the Heisenberg operators must equal the Schrödinger ones, so

$$\hat{x}_H(0) = \hat{A} = \hat{x}, \quad \hat{p}_H(0) = m\omega\hat{B} = \hat{p}. \tag{16.5.23}$$

We thus find that

$$\hat{A} = \hat{x}, \quad \hat{B} = \frac{1}{m\omega}\hat{p}. \tag{16.5.24}$$

Finally, back in (16.5.21) and (16.5.22) we have the full solution for the simple harmonic oscillator Heisenberg operators:

$$\boxed{\begin{aligned}\hat{x}_H(t) &= \hat{x}\,\cos\omega t + \frac{1}{m\omega}\hat{p}\,\sin\omega t, \\ \hat{p}_H(t) &= \hat{p}\cos\omega t - m\omega\,\hat{x}\,\sin\omega t.\end{aligned}} \tag{16.5.25}$$

Let us confirm that the Heisenberg Hamiltonian is time independent and in fact equal to the Schrödinger Hamiltonian. This must hold because the Schrödinger Hamiltonian is time independent. Starting with (16.5.15) and using (16.5.25), we show that

$$\begin{aligned}
\hat{H}_H(t) &= \frac{\hat{p}_H^2(t)}{2m} + \frac{1}{2}m\omega^2\hat{x}_H^2(t) \\
&= \frac{1}{2m}(\hat{p}\cos\omega t - m\omega\hat{x}\sin\omega t)^2 + \frac{1}{2}m\omega^2\Big(\hat{x}\cos\omega t + \frac{1}{m\omega}\hat{p}\sin\omega t\Big)^2 \\
&= \frac{\cos^2\omega t}{2m}\hat{p}^2 + \frac{m^2\omega^2\sin^2\omega t}{2m}\hat{x}^2 - \frac{\omega}{2}\sin\omega t\cos\omega t(\hat{p}\hat{x} + \hat{x}\hat{p}) \\
&\quad + \frac{\sin^2\omega t}{2m}\hat{p}^2 + \frac{m\omega^2\cos^2\omega t}{2}\hat{x}^2 + \frac{\omega}{2}\cos\omega t\sin\omega t\,(\hat{x}\hat{p} + \hat{p}\hat{x}) \\
&= \frac{\hat{p}^2}{2m} + \frac{1}{2}m\omega^2\hat{x}^2,
\end{aligned} \tag{16.5.26}$$

confirming that $\hat{H}_H(t) = \hat{H}_S$. □

Example 16.2. *Heisenberg creation and annihilation operators.*

Let us determine the Heisenberg operators corresponding to the simple harmonic oscillator creation and annihilation operators. For simplicity, the Heisenberg operator associated with $\hat{a}$ will be denoted by $\hat{a}(t)$. Since the harmonic oscillator Hamiltonian is time independent, $\mathcal{U} = e^{-i\hat{H}t/\hbar}$ and we have

$$\hat{a}(t) \equiv e^{i\hat{H}t/\hbar}\,\hat{a}\,e^{-i\hat{H}t/\hbar} = e^{i\omega t\hat{N}}\,a\,e^{-i\omega t\hat{N}}, \tag{16.5.27}$$

where we wrote $\hat{H} = \hbar\omega(\hat{N} + \frac{1}{2})$ and noted that the additive constant has no effect on the result. A simple way to evaluate $\hat{a}(t)$ goes through a differential equation. We take the time derivative of the above to find

$$\frac{d}{dt}\hat{a}(t) = i\omega\,e^{i\omega t\hat{N}}\,[\hat{N}, \hat{a}]e^{-i\omega t\hat{N}} = -i\omega\,e^{i\omega t\hat{N}}\,\hat{a}\,e^{-i\omega t\hat{N}}. \tag{16.5.28}$$

We recognize on the final right-hand side the operator $\hat{a}(t)$, so we have obtained the differential equation

$$\frac{d}{dt}\hat{a}(t) = -i\omega\,\hat{a}(t). \tag{16.5.29}$$

Since $\hat{a}(t=0) = \hat{a}$, the solution is

$$\hat{a}(t) = e^{-i\omega t}\,\hat{a}. \tag{16.5.30}$$

Together with the adjoint of this formula, we have:

$$\boxed{\begin{aligned}\hat{a}(t) &= e^{-i\omega t}\,\hat{a}, \\ \hat{a}^\dagger(t) &= e^{i\omega t}\,\hat{a}^\dagger.\end{aligned}} \tag{16.5.31}$$

Clearly, $[\hat{a}(t), \hat{a}^\dagger(t)] = 1$, as expected. □

Exercise 16.4. *Starting with* $\hat{x} = \sqrt{\frac{\hbar}{2m\omega}}\,(\hat{a} + \hat{a}^\dagger)$, *find* $x_H(t)$ *using the just-obtained Heisenberg versions of* $\hat{a}$ *and* $\hat{a}^\dagger$. *Confirm that you get the result in (16.5.25).*

16.6 Axioms of Quantum Mechanics

We built the mathematical foundations of quantum mechanics by taking a detailed look at complex vector spaces in chapters 13 and 14. We then turned to the spectral theorem and the spectral representation of operators in chapter 15, learning about complete sets of orthogonal projectors. In this chapter we discussed unitary time evolution and how it implies a Schrödinger equation. Having built, additionally, some experience in quantum systems, it is a good time to scrutinize the axioms of quantum mechanics.

We encountered these axioms before, in section 5.3, but we did not aim to phrase them in all generality. Sometimes these axioms are called *"postulates"* of quantum mechanics. The words *axioms* and *postulates* are largely considered synonymous: they are statements that cannot be proven and are treated as self-evident truths upon which a theory is built. Still, there is a sense in which axioms are used for universal truths, while postulates are used for specific theories or applications. In this spirit we will consider four axioms and two postulates. The axioms apply to any isolated quantum system. A system that is not isolated is called an *open* system, and it interacts with other systems. If all the systems together compose a whole system that is isolated, the axioms apply to the whole system. Open systems will be explored when we discuss density matrices in chapter 22.

Let us now state the axioms, denoted as A1, A2, A3, and A4, valid for any *isolated* quantum system:

A1. States of the system The complete description of a quantum system is given by a *ray* in a Hilbert space $\mathcal{H}$.

Remarks:

- A ray in a vector space is a nonzero vector $|\Psi\rangle$ with the equivalence relation $|\Psi\rangle \simeq c|\Psi\rangle$ for any nonzero $c \in \mathbb{C}$. Any vector in this ray is a representative of the state of the system. The vector $|\Psi\rangle$ is also called a wave function.
- Affirming that the state gives a complete description of the system, the axiom implies that the state describes the *most* that can be known about the system.
- A finite-dimensional Hilbert space is a complex vector space with an inner product $\langle \cdot, \cdot \rangle$ satisfying the axioms listed in section 14.1. When the Hilbert space is infinite-dimensional, the space must also be complete in the norm: all Cauchy sequences of vectors must converge to vectors in the space, as explained above example 14.1. The Hilbert space $\mathcal{H}$ is the state space of the system.
- The state $|\Psi\rangle$ of the system has a representative with unit norm. This representative is a normalized state or wave function.
- However complicated the quantum system and however many particles it contains, just *one* state, one wave function, represents the full quantum state of the system.

A2. Observables Hermitian operators on the state space $\mathcal{H}$ are observables.

Remarks:

- An observable of a system is a property of the system that can be measured. This postulate says that such properties arise from Hermitian operators.
- The spectral theorem (section 15.6) implies that any observable $\hat{A} = \hat{A}^\dagger$ can be written as the sum

$$\hat{A} = \sum_k a_k P_k, \tag{16.6.1}$$

where the sum runs over all the *different* eigenvalues a_k of $\hat{A}$, and the P_k are a complete set of orthogonal projectors into the corresponding eigenspaces:

$$P_k^\dagger = P_k, \quad P_k P_l = \delta_{kl} P_k, \quad \sum_k P_k = \mathbb{1}. \tag{16.6.2}$$

If an eigenvalue is nondegenerate, the associated projector is rank one. If an eigenvalue has a multiplicity $l > 1$, the associated projector is rank l and projects into an l-dimensional eigenspace.

A3. Measurement Let P_k, with $k = 1, \ldots,$ denote a complete set of orthogonal projectors, and let $\mathcal{H}_k$ denote the subspace P_k projects into. Measurement along this set of projectors is a process in which the state Ψ is projected to $\mathcal{H}_k$ with probability $p(k)$ given by

$$p(k) = \langle \Psi | P_k | \Psi \rangle = \| P_k | \Psi \rangle \|^2. \tag{16.6.3}$$

The normalized state after measurement is

$$\frac{P_k |\Psi\rangle}{\| P_k |\Psi\rangle \|}. \tag{16.6.4}$$

Measuring an observable $\hat{A}$ is measuring along the complete set of orthogonal projectors associated with its spectral decomposition (16.6.1). The probability $p(k)$ for the state to be projected to $\mathcal{H}_k$ is the probability of measuring a_k.

Remarks:

- Measurement of an observable is a nondeterministic physical process. We cannot in general predict the result of the measurement, just the probabilities for the various possible results.
- The measurement axiom does not give any prescription for measuring the state $|\Psi\rangle$ itself. The state is the full description of the system, but it cannot be directly measured. We can only measure observables, and such measurements give us some information about the state.
- The probabilities $p(k)$ add up to one, as they should. Using the completeness of the set of projectors, we indeed find that

$$\sum_k p(k) = \sum_k \langle \Psi | P_k | \Psi \rangle = \langle \Psi | \Big(\sum_k P_k \Big) | \Psi \rangle = \langle \Psi | \Psi \rangle = 1. \tag{16.6.5}$$

- When measuring $\hat{A}$, if the eigenvalue a_k is nondegenerate with eigenvector $|k\rangle$, then $P_k = |k\rangle\langle k|$, and the probability $p(k)$ is

$$p(k) = \langle\Psi|k\rangle\langle k|\Psi\rangle = |\langle k|\Psi\rangle|^2. \tag{16.6.6}$$

- When measuring $\hat{A}$, if the eigenvalue a_k is degenerate with multiplicity l, the associated eigenspace is spanned by l orthonormal eigenvectors $|k;1\rangle, \ldots, |k;l\rangle$, and

$$P_k = \sum_{i=1}^{l} |k;i\rangle\langle k;i| \tag{16.6.7}$$

so that we have

$$p(k) = \langle\Psi|\sum_{i=1}^{l} |k;i\rangle\langle k;i|\Psi\rangle = \sum_{i=1}^{l} |\langle k;i|\Psi\rangle|^2. \tag{16.6.8}$$

- Measurement along an orthonormal basis $\{|i\rangle\}$ means measuring along the complete set of rank-one orthogonal projectors $P_i = |i\rangle\langle i|$. The probability $p(i)$ of being found in the state $|i\rangle$ arising by projection via P_i is

$$p(i) = \langle\Psi|P_i|\Psi\rangle = |\langle i|\Psi\rangle|^2. \tag{16.6.9}$$

- When we say we are measuring an orthogonal projector P, we mean treating P as a Hermitian operator, with eigenvalues one and zero.
- The axiom applies (at least formally) to the traditional measurement of the particle position on the real line for a wave function $\psi(x)$. Consider for this purpose the following operator defined for real a and b with $a < b$:

$$P_{a,b} \equiv \int_a^b dx |x\rangle\langle x|. \tag{16.6.10}$$

It is clear that this $P_{a,b}$ is Hermitian, and a quick calculation shows that it satisfies $P_{a,b}P_{a,b} = P_{a,b}$, making it an orthogonal projector. You can also check that any $|x'\rangle$ with $x' \in (a,b)$ is an eigenstate of $P_{a,b}$ with eigenvalue one. The projector $P_{a,b}$ thus projects to the subspace $\mathcal{H}_{ab}$ of states spanned by *all* the kets $|x'\rangle$ with $x' \in (a,b)$.

The projector $P_{a,b}$, together with the projector $\widetilde{P}_{a,b} = \mathbb{1} - P_{a,b}$, forms a complete set of orthogonal projectors. If we measure $|\psi\rangle$ along this set of projectors, the probability $p(a,b)$ of being found in $\mathcal{H}_{ab}$ is

$$\begin{aligned} p(a,b) &= \langle\psi|P_{a,b}|\psi\rangle = \int_a^b dx \langle\psi|x\rangle\langle x|\psi\rangle \\ &= \int_a^b dx\, \psi^*(x)\psi(x) = \int_a^b dx |\psi(x)|^2. \end{aligned} \tag{16.6.11}$$

This is the probability of finding the particle in the range $x \in [a,b]$.

A4. Dynamics Time evolution is unitary: given any state $|\Psi, t_0\rangle$ of the system at time t_0, the state $|\Psi, t_1\rangle$ at time t_1 is obtained by the action of a unitary operator $\mathcal{U}(t_1, t_0)$:

$$|\Psi, t_1\rangle = \mathcal{U}(t_1, t_0)|\Psi, t_0\rangle. \tag{16.6.12}$$

Remarks:

- Time evolution is deterministic: if the state is known exactly at some time it is known exactly at a later time.
- The same operator $\mathcal{U}(t_1, t_0)$ evolves any possible state of the system at time t_0.
- We showed in section 16.2 that unitary time evolution means the state satisfies the Schrödinger equation $i\hbar\partial_t|\Psi\rangle = \hat{H}|\Psi\rangle$ with $\hat{H}$ as the Hamiltonian, a Hermitian operator with units of energy. Thus, axiom A4 implies that any quantum system has a Schrödinger equation that controls the time evolution of the wave function.

It is worth noting how surprising the measurement axiom is. While axiom A4 states that the time evolution of the state is generated by a unitary operator, when we measure, the "evolution" of the state is nondeterministic and happens from the action of one out of several projectors. Projectors, moreover, are not unitary operators: they kill some states, while unitary operators kill none. This means that measurement is not time evolution in the sense of A4. This has mystified many physicists who argue that measurement devices are physical systems governed by quantum laws and wonder why and how unitary time evolution fails to hold. We will discuss this issue further, without resolving it, in section 22.7.

The axiomatic formulation described above follows the *Copenhagen interpretation* of quantum mechanics developed by Bohr, Heisenberg, and others. Perhaps one day it will be improved and replaced with a less mysterious one, but to date, this formulation is consistent with all known facts about quantum mechanics.

In setting up certain quantum mechanical systems, the use of some guiding principles that appear not to follow from the above axioms seem needed. One such principle is relevant to the construction of composite systems, and the other is relevant to systems with identical particles. We will call both of them postulates, as they apply in specific circumstances and lack the generality of the four axioms stated above. Since we have not yet studied composite systems or systems with identical particles, the statements below will not be explained at this point; they are included for completeness. The reader may return to this part after studying the relevant chapters.

P1. Composite system postulate Assume system A has a state space $\mathcal{H}_A$, and system B has a state space $\mathcal{H}_B$. The state space of the composite system AB is the tensor product $\mathcal{H}_A \otimes \mathcal{H}_B$. If system A is prepared in $|\Psi_A\rangle$ and system B is prepared in $|\Psi_B\rangle$, the state of AB is $|\Psi_A\rangle \otimes |\Psi_B\rangle$.

P2. Symmetrization postulate In a system with N identical particles, the states that are physically realized are either totally symmetric under the exchange of the particles, in which case the particles are said to be bosons, or they are totally antisymmetric under the exchange, in which case they are said to be fermions.

The composite system postulate, more than an axiom, seems like a prescription for building or defining a composite system in which we can implement axioms A1 to A4. The material relevant to this postulate is discussed in chapter 18. The symmetrization postulate, stated in chapter 21, is the known way to resolve the problem of *exchange degeneracy*. In fact, this postulate is proven in relativistic quantum field theory under some weak set of assumptions. Moreover, in quantum systems with two spatial dimensions, particles that are neither bosons nor fermions can exist (section 21.4).

Problems

Problem 16.1. *Spin in a time-varying magnetic field.*

A spin is placed in a uniform but oscillating magnetic field $\mathbf{B} = B_0 \hat{\mathbf{z}} \cos \omega t$. The spin is initially an eigenstate of $\hat{S}_x$ with eigenvalue $\hbar/2$.

1. Find the unitary operator $\mathcal{U}(t)$ that generates time evolution.
2. Calculate the time evolution of the state and describe it by giving the time-dependent angles $\theta(t)$ and $\phi(t)$ that define the direction of the spin.
3. Find the time-dependent probability of finding the spin with $S_x = -\hbar/2$. The spin is said to flip if, at some time, it is an $\hat{S}_x$ eigenstate with eigenvalue $-\hbar/2$. What is the largest value of ω that allows a flip?

Problem 16.2. *Heisenberg operators for spin.*

Consider the time-independent Schrödinger Hamiltonian $\hat{H}$ for a spin in a uniform and constant magnetic field of magnitude B along the z-direction:

$$\hat{H} = -\lambda B \hat{S}_z.$$

Here λ is the (real) constant that relates the dipole moment to the spin. Find the explicit time evolution for the Heisenberg operators $\hat{S}_x(t), \hat{S}_y(t)$, and $\hat{S}_z(t)$ associated with the Schrödinger operators $\hat{S}_x, \hat{S}_y$, and $\hat{S}_z$.

Problem 16.3. *The Heisenberg picture and Newton's laws.*

1. Consider the Hamiltonian $\hat{H} = \hat{p}^2/(2m) + V(\hat{x})$, and derive the Heisenberg equations of motion for $\hat{x}_H(t)$ and $\hat{p}_H(t)$. Use your results to obtain Ehrenfest's theorem

$$\frac{d}{dt}\langle\hat{x}\rangle = \frac{\langle\hat{p}\rangle}{m}, \qquad \frac{d}{dt}\langle\hat{p}\rangle = -\langle V'(\hat{x})\rangle,$$

   where $\langle\hat{x}\rangle = \langle\psi, 0|\hat{x}_H(t)|\psi, 0\rangle = \langle\psi, t|\hat{x}|\psi, t\rangle$, and so on. Combine them to derive an equation for $\frac{d^2}{dt^2}\langle\hat{x}\rangle$. Explain the conditions on the potential such that this equation reduces to the classical Newton's law.

2. Consider a free particle in a normalized state whose expected values of position and momentum at $t=0$ are x_0 and p_0, respectively. Use Ehrenfest's theorem to determine $\langle\hat{x}\rangle$ as a function of time.
3. Now imagine that this particle has a charge q, and consider applying an electric field that varies with time, so $V(\hat{x}; t) = qE_0\, \hat{x} \sin(\omega t)$. Demonstrate that now $[\hat{H}(t_1), \hat{H}(t_2)] \neq 0$ for $t_1 \neq t_2$. Look back at the steps underlying the derivation of Ehrenfest's theorem in part (1), and explain why it still holds.
4. Find $\langle\hat{x}\rangle$ as a function of time for the situation in part (3).

Problem 16.4. *Time ordering and Heisenberg Hamiltonian.*

The time ordering operator T is defined to act on a product of two time-dependent operators as follows:

$$\mathrm{T}(\hat{H}(t_1)\hat{H}(t_2)) = \begin{cases} \hat{H}(t_1)\hat{H}(t_2), & \text{if } t_1 > t_2, \\ \hat{H}(t_2)\hat{H}(t_1), & \text{if } t_2 > t_1. \end{cases}$$

1. Show that

$$\mathrm{T}\int_{t_0}^{t} dt_1\hat{H}(t_1)\int_{t_0}^{t} dt_2\hat{H}(t_2) = 2\int_{t_0}^{t} dt_1\hat{H}(t_1)\int_{t_0}^{t_1} dt_2\hat{H}(t_2).$$

   This result explains, to quadratic order, the action of T on the exponential (16.3.16).

2. Prove that the Heisenberg Hamiltonian $\hat{H}_H(t)$ associated with a general Schrödinger Hamiltonian $\hat{H}_S(t)$ is given to cubic order in $\hat{H}_S(t)$ by

$$\begin{aligned} \hat{H}_H(t) = {} & \hat{H}_S(t) + \Big(-\frac{i}{\hbar}\Big)\int_{t_0}^{t} dt_1 [\hat{H}_S(t), \hat{H}_S(t_1)] \\ & + \Big(-\frac{i}{\hbar}\Big)^2 \int_{t_0}^{t} dt_1 \int_{t_0}^{t_1} dt_2 \Big[[\hat{H}_S(t), \hat{H}_S(t_1)], \hat{H}_S(t_2)\Big] + \mathcal{O}(\hat{H}_S^4). \end{aligned}$$

Problem 16.5. *Virial theorem revisited.*

Consider a Hamiltonian for a particle in three dimensions under the influence of a central potential:

$$\hat{H} = \frac{\hat{\mathbf{p}}^2}{2m} + V(r),$$

as well as the Schrödinger operator $\Omega \equiv \hat{\mathbf{r}}\cdot\hat{\mathbf{p}}$. We let $\Omega_H(t)$ denote the associated Heisenberg operator.

1. Use the Heisenberg equation of motion to calculate the time rate of change $\frac{d}{dt}\Omega_H(t)$. Your answer for the right-hand side should be in terms of the Heisenberg operators $\hat{\mathbf{p}}_H^2$, $\hat{\mathbf{r}}_H$, derivatives of $V(r_H)$, and constants.

2. Consider a stationary state $|\Psi, t\rangle$ and *any* Heisenberg operator $\mathcal{O}_H(t)$ arising from a time-independent Schrödinger operator. Explain carefully why

$$\langle\Psi, 0|\frac{d}{dt}\mathcal{O}_H(t)|\Psi, 0\rangle = 0.$$

3. Use your results from (1) and (2) to show that for a potential $V(r) = c/r^k$, with c constant and k a positive integer,

$$\langle\hat{T}\rangle = -\frac{k}{2}\langle V\rangle.$$

   Here the expectation value is taken on a stationary state, $\hat{T}$ denotes the kinetic energy operator $\frac{\hat{\mathbf{p}}^2}{2m}$, and V denotes the potential.

Problem 16.6. *Time evolution in the Heisenberg picture.*

In this problem we will study the time evolution of a wave packet acted upon by a constant force. This is a case where the Schrödinger equation is hard to solve, but the Heisenberg equations of motion for the time dependence of operators can be solved easily, and quite a bit can be learned about the motion.

Suppose a quantum particle is described by a Hamiltonian $\hat{H}$ taking the form

$$\hat{H} = \frac{\hat{p}^2}{2m} + g\,\hat{x}.$$

This corresponds to a particle subject to a constant force $F = -\frac{dV}{dx} = -g$.

1. Use the Heisenberg equations of motion to show that the Heisenberg operators $\hat{x}_H(t)$ and $\hat{p}_H(t)$ obey an analog of Newton's law $F = ma$. Integrate the Heisenberg equations of motion to obtain $\hat{x}_H(t)$ in terms of $\hat{x}_H(0) = \hat{x}$ and $\hat{p}_H(0) = \hat{p}$.
2. Suppose that at $t = 0$ a particle has coordinate space wave function,

   $$\langle x|\psi\rangle = \psi(x) = N e^{-\frac{x^2}{4\Delta^2}},$$

   where N is a constant that normalizes ψ to unity. Compute $\langle\psi|\hat{x}_H(t)|\psi\rangle$ and show that it behaves classically.
3. Compute $(\Delta x(t))^2 = \langle\hat{x}_H^2(t)\rangle - \langle\hat{x}_H(t)\rangle^2$. Show that $(\Delta x(t))^2$ grows quadratically with time,

   $$(\Delta x(t))^2 = (\Delta x(0))^2 + \lambda t^2,$$

   and find the value of the coefficient λ. How does the spreading of the wave packet depend on the value of g?

17 Dynamics of Quantum Systems

To illustrate the methods used to determine the dynamics of quantum systems, we focus on two broad classes of examples. In the first class, we consider coherent states of the harmonic oscillator. In their simplest form, they arise by translation of the ground state. We use the Heisenberg picture to show that, as opposed to energy eigenstates, coherent states have classical behavior. Looking at the dynamics of electromagnetic fields in a cavity, we are led to a harmonic oscillator description of photon states as well as field operators for electric and magnetic fields. For the second class, we consider two-state systems, quantum systems with two basis states. The general Hamiltonian of such a system is a 2 × 2 Hermitian matrix. We discuss Larmor precession of spin one-half states in a magnetic field and nuclear magnetic resonance, in which one follows the evolution of spin states in a magnetic field with a large longitudinal component and a rotating radio-frequency component. Finally, we discuss the factorization, or supersymmetric method, that gives an algebraic solution for the spectrum of one-dimensional potentials.

17.1 Basics of Coherent States

Coherent states are quantum states that exhibit classical behavior. We will introduce them here and explore their properties. In preparation for this, we first examine translation operators. We took a brief look at translation operators in section 10.1.

Let us construct a unitary **translation** operator T_{x_0} that acting on states moves them, or translates them, by a distance x_0, where x_0 is a constant with units of length. We then claim as the

$$\text{translation operator:} \quad T_{x_0} \equiv e^{-\frac{i}{\hbar}\hat{p}x_0}. \tag{17.1.1}$$

This operator is unitary because it is the exponential of an anti-Hermitian operator (see example 14.11). The multiplication of two such operators is simple:

$$T_{x_0}T_{y_0} = e^{-\frac{i}{\hbar}\hat{p}x_0}e^{-\frac{i}{\hbar}\hat{p}y_0} = e^{-\frac{i}{\hbar}\hat{p}(x_0+y_0)}, \tag{17.1.2}$$

since the exponents commute, and $e^A e^B = e^{A+B}$ if $[A,B]=0$. As a result,

$$T_{x_0} T_{y_0} = T_{x_0+y_0}. \tag{17.1.3}$$

The translation operators form a group: the product of two translations is a translation, there is a unit element $T_0 = \mathbb{1}$, corresponding to $x_0 = 0$, and each element T_{x_0} has an inverse T_{-x_0}. The group multiplication rule is commutative. It follows from the explicit definition of the translation operator that

$$(T_{x_0})^\dagger = e^{\frac{i}{\hbar}\hat{p}x_0} = e^{-\frac{i}{\hbar}\hat{p}(-x_0)} = T_{-x_0} = (T_{x_0})^{-1}, \tag{17.1.4}$$

confirming that the operator is unitary. In the following we write $(T_{x_0})^\dagger$ simply as $T^\dagger_{x_0}$. We say that T_{x_0} translates by x_0 because its action on the operator $\hat{x}$ is as follows:

$$T^\dagger_{x_0}\,\hat{x}\,T_{x_0} = e^{\frac{i}{\hbar}\hat{p}x_0}\,\hat{x}\,e^{-\frac{i}{\hbar}\hat{p}x_0} = \hat{x} + \tfrac{i}{\hbar}\,[\hat{p},\hat{x}]x_0 = \hat{x} + x_0, \tag{17.1.5}$$

where we used $e^A B e^{-A} = B + [A,B]$, valid when $[A,B]$ commutes with A. Further motivation for the identification of the T operators as translations is obtained by considering a normalized state $|\psi\rangle$ and the expectation value of $\hat{x}$ on this state:

$$\langle\hat{x}\rangle_\psi = \langle\psi|\,\hat{x}\,|\psi\rangle. \tag{17.1.6}$$

Now we ask: What is the expectation value of $\hat{x}$ on the state $T_{x_0}|\psi\rangle$? We find

$$\langle\hat{x}\rangle_{T_{x_0}\psi} = \langle\psi|T^\dagger_{x_0}\,\hat{x}\,T_{x_0}|\psi\rangle = \langle\psi|(\hat{x}+x_0)|\psi\rangle = \langle\hat{x}\rangle_\psi + x_0. \tag{17.1.7}$$

The expectation value of $\hat{x}$ on the displaced state is indeed equal to the expectation value of $\hat{x}$ in the original state plus x_0, confirming that *we should view $T_{x_0}|\psi\rangle$ as the state $|\psi\rangle$ displaced a distance x_0.*

We claim that when acting on position states the translation operator T_{x_0} does what we would expect; it displaces the state by x_0:

$$T_{x_0}|x_1\rangle = |x_1 + x_0\rangle. \tag{17.1.8}$$

We can prove this by acting on the above left-hand side with an arbitrary momentum bra $\langle p|$:

$$\langle p|T_{x_0}|x_1\rangle = \langle p|e^{-\frac{i}{\hbar}\hat{p}x_0}|x_1\rangle = e^{-\frac{i}{\hbar}px_0}\langle p|x_1\rangle. \tag{17.1.9}$$

Recalling the value $\langle x|p\rangle = e^{ipx}/\sqrt{2\pi\hbar}$ of the overlap (14.10.15), we get

$$\langle p|T_{x_0}|x_1\rangle = e^{-\frac{i}{\hbar}px_0}\frac{e^{-\frac{i}{\hbar}px_1}}{\sqrt{2\pi\hbar}} = \frac{e^{-\frac{i}{\hbar}p(x_1+x_0)}}{\sqrt{2\pi\hbar}} = \langle p|x_1 + x_0\rangle, \tag{17.1.10}$$

proving the desired result, given that $\langle p|$ is arbitrary. It also follows from unitarity and (17.1.8) that

$$T^\dagger_{x_0}|x_1\rangle = T_{-x_0}|x_1\rangle = |x_1 - x_0\rangle. \tag{17.1.11}$$

Passing to bras, the previous result gives

$$\langle x_1|T_{x_0} = \langle x_1 - x_0|. \tag{17.1.12}$$

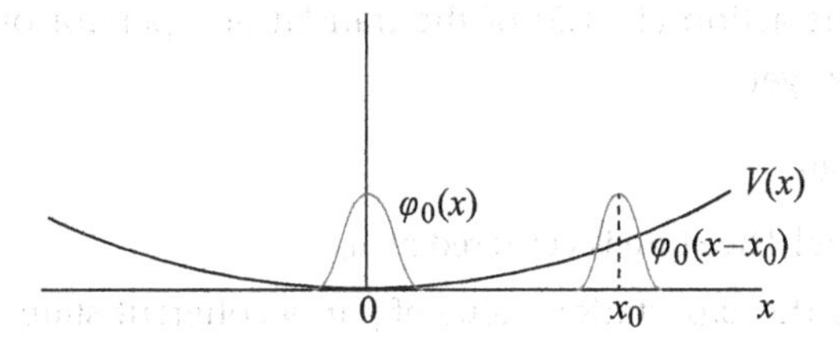

Figure 17.1
The ground state wave function $\varphi_0(x)$ displaced to the right a distance x_0 is the wave function $\varphi_0(x-x_0)$. The corresponding state, denoted as $|x_0\rangle_c$, is the simplest example of a coherent state.

We can also discuss the action of the translation operator in terms of arbitrary states $|\psi\rangle$ and their wave functions $\psi(x)=\langle x|\psi\rangle$. Then the "translated" state $T_{x_0}|\psi\rangle$ has a wave function

$$\langle x|T_{x_0}|\psi\rangle \;=\; \langle x-x_0|\psi\rangle \;=\; \psi(x-x_0). \tag{17.1.13}$$

Indeed, $\psi(x-x_0)$ is the function $\psi(x)$ translated by the distance $+x_0$. For example, the value that $\psi(x)$ takes at $x=0$ is taken by the function $\psi(x-x_0)$ at $x=x_0$.

We are finally ready to define a coherent state $|x_0\rangle_c$ of the simple harmonic oscillator. The state is labeled by x_0, and the c subscript on the ket is there to remind you that it is a coherent state, *not* a position state. Here we define the

$$\boxed{\text{coherent state:}\quad |x_0\rangle_c \;\equiv\; T_{x_0}|0\rangle \;=e^{-\frac{i}{\hbar}\hat{p}x_0}|0\rangle,} \tag{17.1.14}$$

where $|0\rangle$ denotes the ground state of the oscillator. The coherent state is simply the translation of the ground state by a distance x_0. This state has no time dependence displayed, so it may be thought of as the state of the system at $t=0$. As t increases, the state will evolve according to the Schrödinger equation, and we will later consider this evolution. Note that the coherent state is normalized:

$${}_c\langle x_0|x_0\rangle_c \;=\langle 0|T_{x_0}^\dagger T_{x_0}|0\rangle \;=\; \langle 0|0\rangle \;=1. \tag{17.1.15}$$

This had to be: it is the action of the unitary operator T_{x_0} on the normalized ground state.

The wave function ψ_{x_0} associated to the coherent state is easily obtained:

$$\psi_{x_0}(x) \;\equiv\; \langle x|x_0\rangle_c \;=\; \langle x|T_{x_0}|0\rangle \;=\; \langle x-x_0|0\rangle \;=\; \varphi_0(x-x_0), \tag{17.1.16}$$

where we used (17.1.12), and $\langle x|0\rangle=\varphi_0(x)$ is the ground state wave function. As expected the wave function for the coherent state is just the ground state wave function displaced a distance x_0 to the right. This is illustrated in figure 17.1.

Let us now do a few sample calculations to better understand these states.

1. We first find the expectation value of $\hat{x}$ in a coherent state. This is quickly done:

$${}_c\langle x_0|\hat{x}\,|x_0\rangle_c \;=\; \langle 0|T_{x_0}^\dagger\,\hat{x}\,T_{x_0}|0\rangle \;=\; \langle 0|(\hat{x}+x_0)|0\rangle, \tag{17.1.17}$$

where we used the action (17.1.5) of the translation operator on $\hat{x}$. Recalling now that $\langle 0|\hat{x}|0\rangle = 0$, we get

$$ {}_c\langle x_0|\hat{x}\,|x_0\rangle_c = x_0. \tag{17.1.18}$$

Not that surprising! The state is centered at x_0.

2. Now we calculate the expectation value of $\hat{p}$ in a coherent state. Since $\hat{p}$ commutes with T_{x_0}, we find that

$$ {}_c\langle x_0|\hat{p}\,|x_0\rangle_c = \langle 0|T^{\dagger}_{x_0}\hat{p}\,T_{x_0}|0\rangle = \langle 0|\hat{p}\,T^{\dagger}_{x_0}\,T_{x_0}|0\rangle = \langle 0|\,\hat{p}\,|0\rangle = 0, \tag{17.1.19}$$

recalling that the expectation value of $\hat{p}$ in any harmonic oscillator energy eigenstate vanishes. The coherent state we built has no momentum. We will later consider more general coherent states that have momentum.

3. Finally, we calculate the expectation value of the energy in a coherent state. Note that the coherent state is not an energy eigenstate. It is also neither a position eigenstate nor a momentum eigenstate! With $\hat{H}$ the Hamiltonian, we have

$$ {}_c\langle x_0|\hat{H}|x_0\rangle_c = \langle 0|T^{\dagger}_{x_0}\hat{H}T_{x_0}|0\rangle. \tag{17.1.20}$$

We now compute

$$\begin{aligned} T^{\dagger}_{x_0}\hat{H}T_{x_0} &= T^{\dagger}_{x_0}\Big(\frac{\hat{p}^2}{2m}+\tfrac{1}{2}m\omega^2\hat{x}^2\Big)T_{x_0} = \frac{\hat{p}^2}{2m}+\tfrac{1}{2}m\omega^2(\hat{x}+x_0)^2 \\ &= \hat{H}+m\omega^2x_0\hat{x}+\tfrac{1}{2}m\omega^2x_0^2. \end{aligned} \tag{17.1.21}$$

Back in (17.1.20),

$$ {}_c\langle x_0|\hat{H}|x_0\rangle_c = \langle 0|\hat{H}|0\rangle + m\omega^2x_0\langle 0|\hat{x}|0\rangle + \tfrac{1}{2}m\omega^2x_0^2. \tag{17.1.22}$$

Recalling that the ground state energy is $\hbar\omega/2$ and that in the ground state $\hat{x}$ has no expectation value, we finally get

$$ {}_c\langle x_0|\hat{H}|x_0\rangle_c = \tfrac{1}{2}\hbar\omega + \tfrac{1}{2}m\omega^2x_0^2. \tag{17.1.23}$$

This is reasonable: the total energy is the zero-point energy plus the potential energy of a particle at x_0. The coherent state $|x_0\rangle_c$ is the quantum version of a point particle on a spring held stretched at $x = x_0$.

17.2 Heisenberg Picture for Coherent States

We will later discuss the explicit time evolution of coherent states. In the meantime we can study the time evolution of expectation values quite efficiently using the Heisenberg picture since we already calculated in (16.5.25) the time-dependent Heisenberg operators $\hat{x}_H(t)$ and $\hat{p}_H(t)$.

If at time equal zero we have the coherent state $|x_0\rangle_c$, at time t we write the time-evolved state as $|x_0, t\rangle_c$. We now ask what the (time-dependent) expectation value of $\hat{x}$ is on this state:

$$\langle \hat{x} \rangle_{x_0}(t) = {}_c\langle x_0, t | \hat{x} | x_0, t \rangle_c = {}_c\langle x_0 | \hat{x}_H(t) | x_0 \rangle_c, \tag{17.2.1}$$

the last equality being the definition of the Heisenberg operator itself. Using the explicit form of $\hat{x}_H(t)$, we get

$$\langle \hat{x} \rangle_{x_0}(t) = {}_c\langle x_0 | \left(\hat{x} \cos\omega t + \frac{1}{m\omega} \hat{p} \sin\omega t \right) | x_0 \rangle_c. \tag{17.2.2}$$

Finally, using (17.1.18) and (17.1.19), we find that

$$\langle \hat{x} \rangle_{x_0}(t) = x_0 \cos\omega t. \tag{17.2.3}$$

The expectation value of $\hat{x}$ is performing oscillatory motion! This confirms the classical interpretation of the coherent state. For the momentum the calculation is quite similar:

$$\langle \hat{p} \rangle_{x_0}(t) = {}_c\langle x_0 | \hat{p}_H(t) | x_0 \rangle_c = {}_c\langle x_0 | \left(\hat{p} \cos\omega t - m\omega \hat{x} \sin\omega t \right) | x_0 \rangle_c, \tag{17.2.4}$$

resulting in

$$\langle \hat{p} \rangle_{x_0}(t) = -m\omega x_0 \sin\omega t, \tag{17.2.5}$$

which is the expected result, as it is equal to $m\frac{d}{dt}\langle \hat{x} \rangle_{x_0}(t)$.

We know that the harmonic oscillator ground state is a minimum uncertainty state. We will now discuss the extension of this fact to coherent states. We begin by calculating the uncertainties Δx and Δp in a coherent state at $t=0$. We will see that, just like the ground state, the coherent state minimizes the product of uncertainties. Then we will calculate uncertainties of the coherent state as a function of time!

Let us begin with the position uncertainty Δx. We find that

$${}_c\langle x_0 | \hat{x}^2 | x_0 \rangle_c = \langle 0 | T_{x_0}^\dagger \hat{x}^2 T_{x_0} | 0 \rangle = \langle 0 | (\hat{x} + x_0)^2 | 0 \rangle = \langle 0 | \hat{x}^2 | 0 \rangle + x_0^2. \tag{17.2.6}$$

The first term on the right-hand side was calculated in (9.4.31) and results in

$${}_c\langle x_0 | \hat{x}^2 | x_0 \rangle_c = \tfrac{1}{2} L_0^2 + x_0^2. \tag{17.2.7}$$

Since ${}_c\langle x_0 | \hat{x} | x_0 \rangle_c = x_0$, the position uncertainty is given by

$$(\Delta x)^2 = \tfrac{1}{2} L_0^2, \quad \text{on the state } |x_0\rangle_c. \tag{17.2.8}$$

For the momentum uncertainty, the computation is quite analogous:

$${}_c\langle x_0 | \hat{p}^2 | x_0 \rangle_c = \langle 0 | T_{x_0}^\dagger \hat{p}^2 T_{x_0} | 0 \rangle = \langle 0 | \hat{p}^2 | 0 \rangle = \tfrac{1}{2} \left(\tfrac{\hbar}{L_0} \right)^2, \tag{17.2.9}$$

where the last equality follows from (9.4.33). Recalling that ${}_c\langle x_0 | \hat{p} | x_0 \rangle_c = 0$, we have

$$(\Delta p)^2 = \tfrac{1}{2} \left(\tfrac{\hbar}{L_0} \right)^2, \quad \text{on the state } |x_0\rangle_c. \tag{17.2.10}$$

As a result,

$$\Delta x \Delta p = \tfrac{\hbar}{2}, \quad \text{on the state } |x_0\rangle_c. \tag{17.2.11}$$

The coherent state has minimum $\Delta x \Delta p$ at time equal zero, as befits a state that is just a displaced ground state.

To find the time-dependent uncertainties, our expectation values must be evaluated on the time-dependent state $|x_0, t\rangle_c$. We begin with the position operator:

$$
\begin{aligned}
(\Delta x)^2(t) &= {}_c\langle x_0, t|\hat{x}^2|x_0, t\rangle_c - {}_c\langle x_0, t|\hat{x}|x_0, t\rangle_c^2 \\
&= {}_c\langle x_0|\hat{x}_H^2(t)|x_0\rangle_c - {}_c\langle x_0|\hat{x}_H(t)|x_0\rangle_c^2 \qquad (17.2.12) \\
&= {}_c\langle x_0|\hat{x}_H^2(t)|x_0\rangle_c - x_0^2 \cos^2 \omega t,
\end{aligned}
$$

using (17.2.3). The computation of the first term, $I = {}_c\langle x_0|\hat{x}_H^2(t)|x_0\rangle_c$, takes a few steps:

$$
\begin{aligned}
I &= {}_c\langle x_0|\Big(\hat{x}\cos\omega t + \frac{1}{m\omega}\hat{p}\sin\omega t\Big)^2|x_0\rangle_c \\
&= {}_c\langle x_0|\hat{x}^2|x_0\rangle_c \cos^2\omega t + {}_c\langle x_0|\hat{p}^2|x_0\rangle_c\Big(\frac{\sin\omega t}{m\omega}\Big)^2 + \frac{\cos\omega t\sin\omega t}{m\omega}\,{}_c\langle x_0|(\hat{x}\hat{p}+\hat{p}\hat{x})|x_0\rangle_c \\
&= \Big(\tfrac{1}{2}L_0^2 + x_0^2\Big)\cos^2\omega t + \frac{m\hbar\omega}{2}\Big(\frac{\sin\omega t}{m\omega}\Big)^2 + \frac{\cos\omega t\sin\omega t}{m\omega}\,{}_c\langle x_0|(\hat{x}\hat{p}+\hat{p}\hat{x})|x_0\rangle_c.
\end{aligned}
$$

The last expectation value vanishes, as you should prove:

Exercise 17.1. *Show that* ${}_c\langle x_0|(\hat{x}\hat{p}+\hat{p}\hat{x})|x_0\rangle_c = \langle 0|(\hat{x}\hat{p}+\hat{p}\hat{x})|0\rangle = 0$. *(The vanishing of* $\langle 0|(\hat{x}\hat{p}+\hat{p}\hat{x})|0\rangle$ *was also discussed in problem 9.5.)*

Returning to our computation, the expectation value I of $x_H^2(t)$ in the coherent state is

$$
I = \Big(\tfrac{1}{2}L_0^2 + x_0^2\Big)\cos^2\omega t + \frac{m\hbar\omega}{2}\Big(\frac{\sin\omega t}{m\omega}\Big)^2 = \tfrac{1}{2}L_0^2 + x_0^2\cos^2\omega t. \qquad (17.2.13)
$$

Therefore, finally, going back to (17.2.12) we get

$$
(\Delta x)^2(t) = \tfrac{1}{2}L_0^2. \qquad (17.2.14)
$$

The uncertainty Δx does not change in time as the state evolves! This suggests, but does not yet prove, that the state does not change shape as it moves. Not changing shape means that at different times $|\psi(x,t)|^2$ and $|\psi(x,t')|^2$ differ only by an overall displacement in x. To settle the issue of shape change, it is useful to calculate the time-dependent uncertainty in the momentum:

$$
\begin{aligned}
(\Delta p)^2(t) &= {}_c\langle x_0, t|\hat{p}^2|x_0, t\rangle_c - {}_c\langle x_0, t|\hat{p}|x_0, t\rangle_c^2 \\
&= {}_c\langle x_0|\hat{p}_H^2(t)|x_0\rangle_c - {}_c\langle x_0|\hat{p}_H(t)|x_0\rangle_c^2 \qquad (17.2.15) \\
&= {}_c\langle x_0|\hat{p}_H^2(t)|x_0\rangle_c - m^2\omega^2 x_0^2\sin^2\omega t,
\end{aligned}
$$

where we used (17.2.5). The rest of the computation is recommended:

Exercise 17.2. *Show that*

$$
{}_c\langle x_0|\hat{p}_H^2(t)|x_0\rangle_c = \tfrac{1}{2}\Big(\frac{\hbar}{L_0}\Big)^2 + m^2\omega^2 x_0^2\sin^2\omega t. \qquad (17.2.16)
$$

This result then implies that

$$
(\Delta p)^2(t) = \tfrac{1}{2}\Big(\frac{\hbar}{L_0}\Big)^2. \qquad (17.2.17)
$$

This, together with (17.2.14), gives

$$\Delta x(t)\Delta p(t) = \tfrac{\hbar}{2}, \quad \text{on the state } |x_0, t\rangle_c. \tag{17.2.18}$$

The coherent state remains a minimum uncertainty packet for all times. Since only Gaussians have such minimum uncertainty (example 15.3), the state remains a Gaussian for all times. Since Δx is constant, the Gaussian does not change shape, thus the name *coherent state*. The state does not spread out in time, changing shape; it just moves "coherently."

Note that

$$\Delta x(t) = \frac{L_0}{\sqrt{2}}, \quad \text{and} \quad \Delta p(t) = \frac{1}{\sqrt{2}}\frac{\hbar}{L_0}. \tag{17.2.19}$$

The quantum oscillator size L_0 is very small for a macroscopic oscillator. A coherent state with a large $x_0 \gg L_0$ is classical in the sense that the typical excursion x_0 is much larger than the position uncertainty $\sim L_0$. Similarly, the typical momentum $m\omega x_0 \sim \frac{\hbar}{L_0}\frac{x_0}{L_0}$ is much larger than the momentum uncertainty, by just the same factor $\sim x_0/L_0$.

Exercise 17.3. *Prove that on the coherent state $|x_0\rangle_c$ we have*

$$\frac{\Delta p(t)}{\sqrt{\overline{\langle \hat{p}^2\rangle(t)}}} = \frac{\Delta x(t)}{\sqrt{\overline{\langle \hat{x}^2\rangle(t)}}} = \frac{1}{\sqrt{1+\frac{x_0^2}{L_0^2}}}, \tag{17.2.20}$$

where the overlines on the expectation values denote time average.

Coherent states in the energy basis We can get an interesting expression for the coherent state $|x_0\rangle_c$ by rewriting the momentum operator in terms of creation and annihilation operators. From (9.3.12) we have that

$$\hat{p} = \frac{i}{\sqrt{2}}\frac{\hbar}{L_0}(\hat{a}^\dagger - \hat{a}). \tag{17.2.21}$$

It follows that the coherent state (17.1.14) is given by

$$|x_0\rangle_c = \exp\Big(-\frac{i}{\hbar}\hat{p}\,x_0\Big)|0\rangle = \exp\Big(\frac{1}{\sqrt{2}}\frac{x_0}{L_0}(\hat{a}^\dagger - \hat{a})\Big)|0\rangle. \tag{17.2.22}$$

Since $\hat{a}|0\rangle = 0$, the above formula admits simplification: we should be able to get rid of all the $\hat{a}$'s! We could do this if we could split the exponential into two exponentials, one with the $\hat{a}^\dagger$'s to the *left* of another one with the $\hat{a}$'s. The exponential with the $\hat{a}$'s would stand near the vacuum and give no contribution, as we will see below. For this purpose we recall the commutator identity (13.7.31):

$$e^{A+B} = e^A e^B e^{-\frac{1}{2}[A,B]}, \quad \text{if } [A,B] \text{ commutes with } A \text{ and with } B. \tag{17.2.23}$$

Consider the exponential in (17.2.22), and write it as e^{A+B}, identifying A and B as follows:

$$A = \frac{1}{\sqrt{2}}\frac{x_0}{L_0}\hat{a}^\dagger, \quad B = -\frac{1}{\sqrt{2}}\frac{x_0}{L_0}\hat{a}. \tag{17.2.24}$$

Then $[A,B] = x_0^2/(2L_0^2)$ and we find

$$\exp\Big(\frac{x_0}{\sqrt{2}L_0}\,\hat{a}^\dagger - \frac{x_0}{\sqrt{2}L_0}\,\hat{a}\Big) = \exp\Big(\frac{x_0}{\sqrt{2}L_0}\,\hat{a}^\dagger\Big)\exp\Big(-\frac{x_0}{\sqrt{2}L_0}\,\hat{a}\Big)\exp\Big(-\frac{1}{4}\frac{x_0^2}{L_0^2}\Big). \tag{17.2.25}$$

Since the last exponential is just a number and the second exponential acts like the identity on the vacuum, the coherent state in (17.2.22) becomes

$$|x_0\rangle_c = \exp\Big(-\frac{i\hat{p}\,x_0}{\hbar}\Big)|0\rangle = \exp\Big(-\frac{1}{4}\frac{x_0^2}{L_0^2}\Big)\exp\Big(\frac{x_0}{\sqrt{2}L_0}\,\hat{a}^\dagger\Big)|0\rangle. \tag{17.2.26}$$

While this form is poised to produce an expansion in energy eigenstates, the unit normalization of the state is no longer manifest. Expanding the exponential with creation operators, we get

$$\begin{aligned}|x_0\rangle_c &= \sum_{n=0}^{\infty}\exp\Big(-\frac{1}{4}\frac{x_0^2}{L_0^2}\Big)\cdot\frac{1}{n!}\Big(\frac{x_0}{\sqrt{2}L_0}\Big)^n(\hat{a}^\dagger)^n|0\rangle\\ &= \sum_{n=0}^{\infty}\exp\Big(-\frac{1}{4}\frac{x_0^2}{L_0^2}\Big)\cdot\frac{1}{\sqrt{n!}}\Big(\frac{x_0}{\sqrt{2}L_0}\Big)^n|n\rangle.\end{aligned} \tag{17.2.27}$$

We thus have the desired expansion of the coherent state as a linear superposition of all energy eigenstates:

$$|x_0\rangle_c = \sum_{n=0}^{\infty}c_n|n\rangle, \quad \text{with } c_n = \exp\Big(-\frac{1}{4}\frac{x_0^2}{L_0^2}\Big)\cdot\frac{1}{\sqrt{n!}}\Big(\frac{x_0}{\sqrt{2}L_0}\Big)^n. \tag{17.2.28}$$

Since the probability P_n of finding the energy $E_n = \hbar\omega(n+\frac{1}{2})$ is equal to $|c_n|^2 = c_n^2$, we have

$$P_n = c_n^2 = \exp\Big(-\frac{x_0^2}{2L_0^2}\Big)\cdot\frac{1}{n!}\Big(\frac{x_0^2}{2L_0^2}\Big)^n. \tag{17.2.29}$$

Introducing a dimensionless λ, the probability can be written neatly as follows:

$$P_n = \frac{\lambda^n}{n!}e^{-\lambda}, \quad \text{with } \lambda \equiv \frac{x_0^2}{2L_0^2}. \tag{17.2.30}$$

The values of P_n must define a probability distribution for all integers $n \geq 0$, parameterized by λ. This is in fact the familiar *Poisson distribution*. It is straightforward to verify that, as required,

$$\sum_{n=0}^{\infty}P_n = e^{-\lambda}\sum_{n=0}^{\infty}\frac{\lambda^n}{n!} = e^{-\lambda}e^{\lambda} = 1. \tag{17.2.31}$$

The physical interpretation of λ can be obtained by computing the expectation value of n in this probability distribution:

$$\langle n\rangle \equiv \sum_{n=0}^{\infty}n\,P_n = e^{-\lambda}\sum_{n=0}^{\infty}n\frac{\lambda^n}{n!} = e^{-\lambda}\sum_{n=0}^{\infty}\lambda\frac{d}{d\lambda}\frac{\lambda^n}{n!} = e^{-\lambda}\lambda\frac{d}{d\lambda}e^{\lambda} = \lambda. \tag{17.2.32}$$

Therefore, λ is equal to the expected value $\langle n\rangle$—that is, the expected value of the number operator $\hat{N}$ on the coherent state.

Exercise 17.4. *Show that*

$$\langle n^2\rangle \equiv \sum_{n=0}^{\infty} n^2 P_n = \lambda^2 + \lambda. \tag{17.2.33}$$

It now follows that

$$(\Delta n)^2 = \langle n^2\rangle - \langle n\rangle^2 = \lambda \quad \Rightarrow \quad \Delta n = \sqrt{\lambda}. \tag{17.2.34}$$

In terms of energy, we have $E = \hbar\omega(n + \frac{1}{2})$, and therefore

$$\langle E\rangle = \hbar\omega(\langle n\rangle + \tfrac{1}{2}) = \hbar\omega(\lambda + \tfrac{1}{2}). \tag{17.2.35}$$

Moreover, $E = \hbar\omega(n + \frac{1}{2})$ also implies that $\Delta E = \hbar\omega\Delta n$; the shift by $\hbar/2$ is immaterial to the uncertainty, and the scaling by $\hbar\omega$ just goes through. Therefore,

$$\Delta E = \hbar\omega\sqrt{\lambda} = \hbar\omega\,\frac{x_0}{\sqrt{2}L_0}. \tag{17.2.36}$$

Note now that for large λ the average energy is much larger than its uncertainty:

$$\frac{\langle E\rangle}{\Delta E} = \sqrt{\lambda} + \frac{1}{2\sqrt{\lambda}} \simeq \sqrt{\lambda}. \tag{17.2.37}$$

All in all, for large λ,

$$\boxed{\sqrt{\lambda} = \frac{\Delta E}{\hbar\omega} \simeq \frac{\langle E\rangle}{\Delta E}.} \tag{17.2.38}$$

We see that the uncertainty ΔE is much larger than the separation between energy levels: it is big enough to contain about $\sqrt{\lambda}$ levels. At the same time, ΔE is much smaller than the expected value $\langle E\rangle$ of the energy, the latter a factor $\sqrt{\lambda}$ larger than the former. Of course, large $\sqrt{\lambda}$ means $x_0/L_0 \gg 1$.

17.3 General Coherent States

Consider again the coherent state $|x_0\rangle_c$, written in terms of creation and annihilation operators. As we had in (17.2.22),

$$|x_0\rangle_c = e^{\alpha\,(\hat{a}^\dagger - \hat{a})}|0\rangle, \quad \text{with } \alpha = \frac{1}{\sqrt{2}}\frac{x_0}{L_0}. \tag{17.3.1}$$

An obvious generalization is to let α be a complex number: $\alpha \in \mathbb{C}$. This must be done with care since the operator in the exponential (17.3.1) must be anti-Hermitian, making the exponential unitary. We must therefore replace $\alpha\,(\hat{a}^\dagger - \hat{a})$ with $\alpha\hat{a}^\dagger - \alpha^*\hat{a}$, which is anti-Hermitian when α is complex. We thus define

$$\boxed{|\alpha\rangle \equiv D(\alpha)|0\rangle \equiv \exp(\alpha\hat{a}^\dagger - \alpha^*\hat{a})|0\rangle, \quad \text{with } \alpha \in \mathbb{C}.} \tag{17.3.2}$$

The coherent state, now denoted by $|\alpha\rangle$, is obtained by acting on the vacuum with the unitary *displacement* operator

$$D(\alpha) \equiv \exp(\alpha \hat{a}^\dagger - \alpha^* \hat{a}). \tag{17.3.3}$$

Since $D(\alpha)$ is unitary, it is clear that $\langle\alpha|\alpha\rangle = 1$. In the present notation, the coherent state is recognized by using a Greek letter as the label of the ket.

The action of the annihilation operator on the states $|\alpha\rangle$ is quite interesting:

$$\begin{aligned} \hat{a}|\alpha\rangle &= \hat{a}\, e^{\alpha\hat{a}^\dagger - \alpha^*\hat{a}}|0\rangle = [\hat{a},\, e^{\alpha\hat{a}^\dagger - \alpha^*\hat{a}}]|0\rangle \\ &= [\hat{a},\, \alpha\hat{a}^\dagger - \alpha^*\hat{a}]e^{\alpha\hat{a}^\dagger - \alpha^*\hat{a}}|0\rangle = \alpha e^{\alpha\hat{a}^\dagger - \alpha^*\hat{a}}|0\rangle, \end{aligned} \tag{17.3.4}$$

where the commutator was evaluated using (13.7.24). We conclude that

$$\boxed{\hat{a}\,|\alpha\rangle = \alpha\,|\alpha\rangle.} \tag{17.3.5}$$

This result is a bit shocking: we have found eigenstates of the *non-Hermitian* operator $\hat{a}$. Because $\hat{a}$ is not Hermitian, our theorems about eigenstates and eigenvectors of Hermitian operators do not apply. For example, the eigenvalues need not be real ($\alpha \in \mathbb{C}$), two eigenvectors with different eigenvalues need not be orthogonal (they are not!), and the set of eigenvectors need not form a complete basis (coherent states actually give an overcomplete basis!). We actually determined explicitly the eigenstates of the annihilation operator in example 13.21.

Exercise 17.5. *Ordering the exponential in the state $|\alpha\rangle$ in (17.3.2), show that*

$$|\alpha\rangle = e^{-\frac{1}{2}|\alpha|^2} e^{\alpha\hat{a}^\dagger}|0\rangle. \tag{17.3.6}$$

Exercise 17.6. *Show that*

$$\langle\beta|\alpha\rangle = \exp\Big(-\tfrac{1}{2}(|\alpha|^2 + |\beta|^2) + \beta^*\alpha\Big). \tag{17.3.7}$$

Hint: You may find it helpful to evaluate $e^{\beta^*\hat{a} + \alpha\hat{a}^\dagger}$ in two different ways using (17.2.23).

Exercise 17.7. *The above formula for the overlap $\langle\beta|\alpha\rangle$ does not make $\langle\alpha|\alpha\rangle = 1$ manifest. Show that the above can be rewritten as*

$$\langle\beta|\alpha\rangle = e^{-\frac{1}{2}|\alpha-\beta|^2} e^{i\,\mathrm{Im}(\beta^*\alpha)}. \tag{17.3.8}$$

When $\beta = \alpha$, both exponents vanish manifestly, and thus the overlap is clearly equal to one. Note that the result implies that $|\langle\beta|\alpha\rangle|^2 = e^{|\alpha-\beta|^2}$.

To find the physical interpretation of the complex number α, we first note that when real, as in (17.3.1), α encodes the initial position x_0 of the coherent state. More precisely, it encodes the expectation value of $\hat{x}$ in the state at $t = 0$. For complex α, its real part is still related to the initial position:

$$\langle\alpha|\hat{x}|\alpha\rangle = \frac{L_0}{\sqrt{2}}\langle\alpha|(\hat{a} + \hat{a}^\dagger)|\alpha\rangle = \frac{L_0}{\sqrt{2}}(\alpha + \alpha^*) = L_0\sqrt{2}\,\mathrm{Re}(\alpha), \tag{17.3.9}$$

where we used (17.3.5), both on bras and on kets. We have thus learned that

$$\text{Re}(\alpha) = \frac{1}{\sqrt{2}}\frac{\langle\hat{x}\rangle}{L_0}. \tag{17.3.10}$$

It is natural to conjecture that the imaginary part of α is related to the momentum expectation value on the initial state. So we explore

$$\langle\alpha|\hat{p}|\alpha\rangle = \frac{i}{\sqrt{2}}\frac{\hbar}{L_0}\langle\alpha|(\hat{a}^\dagger - \hat{a})|\alpha\rangle = -\frac{i}{\sqrt{2}}\frac{\hbar}{L_0}(\alpha - \alpha^*) = \sqrt{2}\frac{\hbar}{L_0}\,\text{Im}(\alpha) \tag{17.3.11}$$

and learn that

$$\text{Im}(\alpha) = \frac{1}{\sqrt{2}}\frac{L_0}{\hbar}\langle\hat{p}\rangle. \tag{17.3.12}$$

The identification of α in terms of expectation values of $\hat{x}$ and $\hat{p}$ is now complete:

$$\boxed{\alpha = \frac{1}{\sqrt{2}}\Big(\frac{\langle\hat{x}\rangle}{L_0} + i\frac{L_0\langle\hat{p}\rangle}{\hbar}\Big).} \tag{17.3.13}$$

Exercise 17.8. *Show that the anti-Hermitian operator in the exponent of $D(\alpha)$ can be written as follows:*

$$\alpha\,\hat{a}^\dagger - \alpha^*\hat{a} = -\tfrac{i}{\hbar}(\hat{p}\,\langle x\rangle - \langle\hat{p}\rangle\,\hat{x}). \tag{17.3.14}$$

Assuming α is defined as in (17.3.13), this result allows us to rewrite the general coherent state (17.3.2) as follows:

$$\boxed{|\alpha\rangle = \exp\Big(-\frac{i\hat{p}\,\langle\hat{x}\rangle}{\hbar} + \frac{i\langle\hat{p}\rangle\,\hat{x}}{\hbar}\Big)|0\rangle.} \tag{17.3.15}$$

In order to find the time dependence of the general coherent state $|\alpha\rangle$, we use the time-evolution operator:

$$|\alpha, t\rangle \equiv e^{-\frac{i\hat{H}t}{\hbar}}|\alpha\rangle = \Big(e^{-i\frac{Ht}{\hbar}}\exp(\alpha\hat{a}^\dagger - \alpha^*\hat{a})\,e^{i\frac{Ht}{\hbar}}\Big)e^{-i\frac{Ht}{\hbar}}|0\rangle. \tag{17.3.16}$$

To proceed, we can use the Heisenberg picture. For a time-independent Hamiltonian (as that of the simple harmonic oscillator) and a Schrödinger operator $\mathcal{O}$, we have $\mathcal{O}_H(t) = e^{iHt/\hbar}\mathcal{O}\,e^{-iHt/\hbar}$. With the opposite signs for the exponentials, we get $e^{-iHt/\hbar}\mathcal{O}\,e^{iHt/\hbar} = \mathcal{O}_H(-t)$. Such a relation is also valid for any function of an operator: $e^{-iHt/\hbar}F(\mathcal{O})e^{iHt/\hbar} = F(\mathcal{O}_H(-t))$, as you can convince yourself whenever $F(x)$ has a convergent Taylor expansion in powers of x. It then follows that back in (17.3.16) we have:

$$|\alpha, t\rangle = \exp\Big(\alpha\hat{a}^\dagger(-t) - \alpha^*\hat{a}(-t)\Big)e^{-i\omega t/2}|0\rangle. \tag{17.3.17}$$

Recalling from (16.5.31) that $\hat{a}(t) = e^{-i\omega t}\,\hat{a}$ and $\hat{a}^\dagger(t) = e^{i\omega t}\,\hat{a}^\dagger$, we find that

$$|\alpha, t\rangle = e^{-i\omega t/2}\exp\Big(\alpha e^{-i\omega t}\hat{a}^\dagger - \alpha^* e^{i\omega t}\hat{a}\Big)|0\rangle. \tag{17.3.18}$$

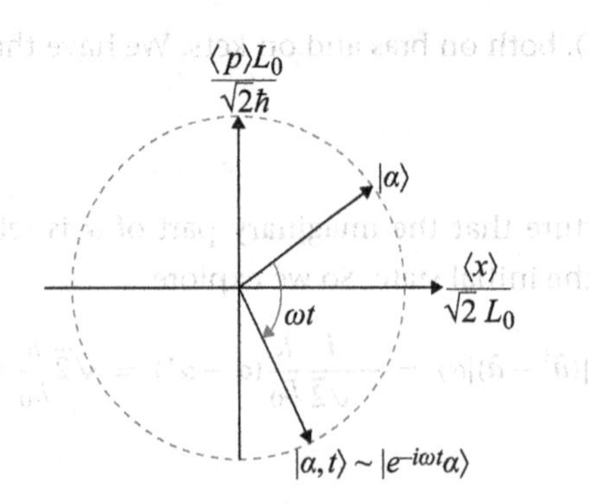

Figure 17.2
Time evolution of the coherent state $|\alpha\rangle$. The real and imaginary parts of α determine, respectively, the expectation values $\langle\hat{x}\rangle$ and $\langle\hat{p}\rangle$. As time goes by, the α parameter of the coherent state rotates clockwise with angular velocity ω.

Looking at the exponential, we see that it is in fact the displacement operator with α replaced by $\alpha e^{-i\omega t}$. As a result, we have shown that

$$\boxed{|\alpha, t\rangle = e^{-i\omega t/2}|e^{-i\omega t}\alpha\rangle.} \tag{17.3.19}$$

This is how a coherent state $|\alpha\rangle$ evolves in time: up to an irrelevant phase, the state remains a coherent state with a time-varying parameter $e^{-i\omega t}\alpha$. In the complex α plane, the state is represented by a vector that rotates clockwise with angular velocity ω. The α plane can be viewed as having a real axis that gives $\langle\hat{x}\rangle$, up to a proportionality constant, and an imaginary axis that gives $\langle\hat{p}\rangle$, up to a proportionality constant. The evolution of any state is represented by a circle. This is illustrated in figure 17.2.

17.4 Photon States

As an application of the harmonic oscillator and coherent states, we will now consider electromagnetic oscillations in a cavity. As it turns out, their quantum description is through a harmonic oscillator whose states are photon states. Moreover, the coherent states of this oscillator turn out to represent approximate classical states of the electromagnetic field.

The energy E in a classical electromagnetic field is obtained by adding the contributions of the electric and magnetic fields $\mathbf{E}$ and $\mathbf{B}$:

$$E = \int d^3x \, \frac{1}{8\pi}\Big[\mathbf{E}^2(\mathbf{x}, t) + \mathbf{B}^2(\mathbf{x}, t)\Big]. \tag{17.4.1}$$

Now consider a rectangular cavity of volume V with a single mode of the electromagnetic field—namely, a single frequency ω, with corresponding wave number $k = \omega/c$ and a single polarization state. The electromagnetic fields form a standing wave in which the electric and magnetic fields are out of phase. They can take the form

$$E_x(z,t) = \sqrt{\frac{8\pi}{V}}\,\omega\, q(t)\sin kz, \quad B_y(z,t) = \sqrt{\frac{8\pi}{V}}\, p(t)\,\cos kz. \tag{17.4.2}$$

Here, $q(t)$ and $p(t)$ are classical time-dependent functions. As we will see below, in the quantum theory they become Heisenberg operators $\hat{q}(t)$ and $\hat{p}(t)$ satisfying $[\hat{q}(t), \hat{p}(t)] = i\hbar$.

The energy (17.4.1) associated with the above fields is quickly calculated, recalling that with periodic boundary conditions on the fields, the average of $(\sin kz)^2$ or $(\cos kz)^2$ over the volume V is 1/2. We then find that

$$E = \tfrac{1}{2}\big(p^2(t) + \omega^2 q^2(t)\big). \tag{17.4.3}$$

There is some funny business here with units. The variables $q(t)$ and $p(t)$ do not have their familiar units, as you can see from the expression for the energy. We are missing a quantity with units of mass that divides the p^2 contribution and multiplies the q^2 contribution. Here p has units of $\sqrt{E}$, and q has units of $T\sqrt{E}$. Still, the product of q and p has units of $\hbar$, which is useful. Since photons are massless particles, there is no quantity with units of mass that we can use. Note that the dynamical variable $q(t)$ is not a position; it is essentially the electric field. The dynamical variable $p(t)$ is not a momentum; it is essentially the magnetic field.

The quantum theory of this electromagnetic field uses the structure implied by the classical results above. From the energy above, we *postulate* a Hamiltonian $\hat{H}$ of the form

$$\hat{H} = \tfrac{1}{2}(\hat{p}^2 + \omega^2 \hat{q}^2), \tag{17.4.4}$$

with Schrödinger operators $\hat{q}$ and $\hat{p}$ that satisfy $[\hat{q}, \hat{p}] = i\hbar$ and associated Heisenberg operators $\hat{q}(t)$ and $\hat{p}(t)$ with the same commutator. As soon as we declare that the classical variables $q(t)$ and $p(t)$ are to become operators, the electric and magnetic fields in (17.4.2) become *field operators*, space and time-dependent operators. This oscillator is our familiar oscillator but with m set equal to one, which is allowed given the unusual units of $\hat{q}$ and $\hat{p}$. With the familiar (9.3.11) and $m = 1$, we have

$$\hat{a} = \frac{1}{\sqrt{2\hbar\omega}}(\omega\hat{q} + i\hat{p}) \quad \hat{a}^\dagger = \frac{1}{\sqrt{2\hbar\omega}}(\omega\hat{q} - i\hat{p}), \quad [\hat{a}, \hat{a}^\dagger] = 1. \tag{17.4.5}$$

It follows that

$$\hbar\omega\,\hat{a}^\dagger\hat{a} = \tfrac{1}{2}(\omega\hat{q} - i\hat{p})(\omega\hat{q} + i\hat{p}) = \tfrac{1}{2}(\hat{p}^2 + \omega^2\hat{q}^2 + i\omega[\hat{q}, \hat{p}]) = \tfrac{1}{2}(\hat{p}^2 + \omega^2\hat{q}^2 - \hbar\omega), \tag{17.4.6}$$

and comparing with (17.4.4), we can rewrite the Hamiltonian in terms of $\hat{a}$ and $\hat{a}^\dagger$:

$$H = \hbar\omega\Big(\hat{a}^\dagger\hat{a} + \tfrac{1}{2}\Big) = \hbar\omega\big(\hat{N} + \tfrac{1}{2}\big). \tag{17.4.7}$$

This was the expected answer: this formula does not depend on m, and our setting $m = 1$ had no import. At this point we got photons: We *interpret* the state $|n\rangle$ of the above harmonic oscillator as the state with *n photons*. This state has energy $\hbar\omega(n + \frac{1}{2})$, which is, up to the zero-point energy $\hbar\omega/2$, the energy of n photons of energy $\hbar\omega$ each. A photon is the basic quantum of the electromagnetic field.

For more intuition we consider the electric field operator. For this we first note that

$$\hat{q} = \sqrt{\frac{\hbar}{2\omega}}\,(\hat{a}+\hat{a}^\dagger), \tag{17.4.8}$$

and the corresponding Heisenberg operator is, using (16.5.31),

$$\hat{q}(t) = \sqrt{\frac{\hbar}{2\omega}}\,(\hat{a}e^{-i\omega t}+\hat{a}^\dagger e^{i\omega t}). \tag{17.4.9}$$

In quantum field theory—which is what we are doing here—the electric field is a Hermitian operator. Its form is obtained by substituting (17.4.9) into (17.4.2):

$$\boxed{\hat{E}_x(z,t) = \mathcal{E}_0\,(\hat{a}e^{-i\omega t}+\hat{a}^\dagger e^{i\omega t})\sin kz, \quad \mathcal{E}_0 = \sqrt{\frac{4\pi\hbar\omega}{V}}.} \tag{17.4.10}$$

This is a *field operator*: an operator that depends on position, in this case z, as well as on time. The coordinates x, y, or z are *not* operators in this analysis. The constant $\mathcal{E}_0$ is sometimes called the electric field of a photon.

A classical electric field can be identified as the expectation value of the electric field operator in the given photon state. We immediately see that in the n photon state $|n\rangle$ the expectation value of $\hat{E}_x$ vanishes! Indeed,

$$\langle\hat{E}_x(z,t)\rangle = \mathcal{E}_0\,\big(\langle n|\hat{a}|n\rangle e^{-i\omega t}+\langle n|\hat{a}^\dagger|n\rangle e^{i\omega t}\big)\sin kz \;=\; 0, \tag{17.4.11}$$

since the matrix elements of $\hat{a}$ and $\hat{a}^\dagger$ vanish. Energy eigenstates of the photon field do not correspond to classical electromagnetic fields. Now consider the expectation value of the field in a coherent state $|\alpha\rangle$, with $\alpha\in\mathbb{C}$. This time, we get

$$\langle\hat{E}_x(z,t)\rangle = \mathcal{E}_0\,\big(\langle\alpha|\hat{a}|\alpha\rangle e^{-i\omega t}+\langle\alpha|\hat{a}^\dagger|\alpha\rangle e^{i\omega t}\big)\sin kz. \tag{17.4.12}$$

Recalling that $\hat{a}|\alpha\rangle = \alpha|\alpha\rangle$,

$$\langle\hat{E}_x(z,t)\rangle = \mathcal{E}_0\,(\alpha\, e^{-i\omega t}+\alpha^*\, e^{i\omega t})\sin kz. \tag{17.4.13}$$

This is a standing wave. To make this clear, we write $\alpha = |\alpha|e^{i\theta}$ and find that

$$\langle\hat{E}_x(z,t)\rangle = 2\mathcal{E}_0\,\mathrm{Re}(\alpha e^{-i\omega t})\sin kz \;=\; 2\mathcal{E}_0\,|\alpha|\,\cos(\omega t-\theta)\sin kz. \tag{17.4.14}$$

Coherent photon states with large $|\alpha|$ give rise to classical electric fields! In the state $|\alpha\rangle$, the expectation value of the number operator $\hat{N}$ is $|\alpha|^2$. Thus, the above electric field is the classical field associated with a quantum state with about $|\alpha|^2$ photons.

17.5 Spin Precession in a Magnetic Field

In this section we begin our study of a second, broad class of dynamical systems. This is the class of *two-state* systems. A two-state system does not just have two states! It has two *basis* states; the state space is the two-dimensional complex vector space $\mathbb{C}^2$. For such a state space, the Hamiltonian can be viewed as the most general Hermitian 2×2 matrix. When the Hamiltonian is time independent, this Hermitian matrix is characterized by four real numbers.

Two-state systems are useful idealizations when other degrees of freedom can be ignored. A spin one-half particle is a two-state system with regard to spin. Being a particle, however, it may move and thus has position or momentum degrees of freedom that imply a much larger, higher-dimensional state space. Only when we ignore these degrees of freedom—perhaps because the particle is at rest—can we speak of a two-state system.

In this section our two-state system will be a spin one-half particle. This is to prepare for the more complex example of nuclear magnetic resonance to be considered in the following section. The mathematics of two-state systems is always the same, making it possible to visualize any two-state system as a spin system (section 17.7). In the problems at the end of the chapter, we will consider other examples, including the ammonia molecule, which exhibits curious oscillations between two states.

To examine spin precession, let us first recall our earlier discussion of magnetic dipole moments of particles from section 12.2, where we described the Stern-Gerlach experiment. Classically, we had the relation (12.2.9), valid for a spinning charged particle with charge q and mass m uniformly distributed:

$$\boldsymbol{\mu} = \frac{q}{2mc}\mathbf{L}. \tag{17.5.1}$$

Here $\boldsymbol{\mu}$ is the magnetic dipole moment, and $\mathbf{L}$ is its angular momentum. In the quantum world, particles have spin angular momentum operators $\hat{\mathbf{S}}$ and magnetic moment operators $\hat{\boldsymbol{\mu}}$, and the above relation gets modified by the inclusion of a unit-free constant factor g, which differs for each particle:

$$\hat{\boldsymbol{\mu}} = g\frac{q}{2mc}\hat{\mathbf{S}} = g\frac{q\hbar}{2mc}\frac{\hat{\mathbf{S}}}{\hbar}. \tag{17.5.2}$$

If we consider electrons and protons with masses m_e and m_p, respectively, the Bohr magneton μ_B and the nuclear magneton μ_N are defined as follows:

$$\mu_B = \frac{e\hbar}{2m_ec} = 5.788\times 10^{-9}\,\frac{\text{eV}}{\text{gauss}}, \quad \mu_N = \frac{e\hbar}{2m_pc} = 3.152\times 10^{-12}\,\frac{\text{eV}}{\text{gauss}}. \tag{17.5.3}$$

The corresponding expressions in SI units are

$$\mu_B = \frac{e\hbar}{2m_e} = 5.788\times 10^{-5}\,\frac{\text{eV}}{\text{tesla}}, \quad \mu_N = \frac{e\hbar}{2m_p} = 3.152\times 10^{-8}\,\frac{\text{eV}}{\text{tesla}}. \tag{17.5.4}$$

Note that the nuclear magneton is about two thousand times smaller than the Bohr magneton. As a result, nuclear magnetic dipole moments are much smaller than the electron dipole moment. For an electron, $g_e = 2$, and since the electron charge is negative, we get

$$\hat{\boldsymbol{\mu}}_e = -2\mu_B\,\tfrac{1}{\hbar}\,\hat{\mathbf{S}}. \tag{17.5.5}$$

The dipole moment and the angular momentum are antiparallel. For a proton, the experimental result is

$$\hat{\boldsymbol{\mu}}_p = 5.586\,\mu_N\,\tfrac{1}{\hbar}\,\hat{\mathbf{S}}. \tag{17.5.6}$$

The neutron is neutral, so one may expect no magnetic dipole moment. But the neutron, just like the proton, is not an elementary particle: it is made of gluons, electrically charged valence quarks, and virtual quarks. A dipole moment is thus possible, depending on the way quarks are distributed. Indeed, experimentally,

$$\hat{\mu}_n = -3.826\,\mu_N\,\frac{1}{\hbar}\,\hat{\mathbf{S}}. \tag{17.5.7}$$

Somehow, the negative charge contributes more than the positive charge to the magnetic dipole of the neutron. The compositeness of the proton also accounts for the unusual value of its magnetic dipole moment.

For notational convenience, we introduce the *gyromagnetic ratio* γ that summarizes the relation between the magnetic dipole moment and the spin as follows:

$$\boxed{\hat{\mu} = \gamma\,\hat{\mathbf{S}}.} \tag{17.5.8}$$

The values of γ for the electron, the proton, and the neutron can be read directly from the three previous equations. In general, for any particle or nuclei, equation (17.5.2) gives $\gamma = \frac{gq}{2mc}$, where g, q, and m are, respectively, the g factor, the charge, and the mass of the particle. The $g_e = 2$ factor of the electron has additional quantum corrections that can be evaluated using quantum field theory. These corrections are in agreement with a remarkably precise measurement that gives $g_e = 2.002\,319\,304\,3617(15)$. Also with great accuracy is the Bohr magneton given by $\mu_B = 5.788\,381\,8012(26) \times 10^{-5}\,\text{eV}{\cdot}\text{T}^{-1}$. Just for fun, we also know that $\hbar = 6.582\,119\,569 \times 10^{-16}\text{eV}{\cdot}\text{s}$.

Exercise 17.9. *Show that the SI gyromagnetic ratios γ_e and γ_p for the electron and the proton are*

$$\gamma_e = -1.760\,8596 \times 10^{11}\,\text{s}^{-1}\text{T}^{-1}, \quad \gamma_p = 2.675 \times 10^{8}\,\text{s}^{-1}\text{T}^{-1}. \tag{17.5.9}$$

This gives $|\gamma_e/\gamma_p| \sim 660$.

If we insert the particle in a magnetic field $\mathbf{B}$, the Hamiltonian $\hat{H}_S$ for the spin system is

$$\hat{H}_S = -\hat{\mu}\cdot\mathbf{B} = -\gamma\,\mathbf{B}\cdot\hat{\mathbf{S}} = -\gamma\,(B_x\hat{S}_x + B_y\hat{S}_y + B_z\hat{S}_z). \tag{17.5.10}$$

For a magnetic field $\mathbf{B} = B\hat{\mathbf{z}}$ along the z-axis, for example, we get

$$\hat{H}_S = -\gamma B\hat{S}_z, \tag{17.5.11}$$

and the associated time-evolution unitary operator is given by

$$\mathcal{U}(t,0) = \exp\Big(-\frac{i\hat{H}_S t}{\hbar}\Big) = \exp\Big(-\frac{i(-\gamma Bt)\hat{S}_z}{\hbar}\Big). \tag{17.5.12}$$

In section 14.7 we discussed in great detail the unitary rotation operator $\hat{R}_{\mathbf{n}}(\alpha)$ defined by a unit vector $\mathbf{n}$ and an angle α:

$$\hat{R}_{\mathbf{n}}(\alpha) = \exp\Big(-\frac{i\alpha\hat{S}_{\mathbf{n}}}{\hbar}\Big), \quad \text{with } \hat{S}_{\mathbf{n}} \equiv \mathbf{n}\cdot\hat{\mathbf{S}}. \tag{17.5.13}$$

We showed that, acting on a spin state, the operator rotates it by an angle α about the direction defined by the vector $\mathbf{n}$. Comparing (17.5.13) and (17.5.12), we conclude that $\mathcal{U}(t,0)$ generates a rotation by the angle $(-\gamma Bt)$ about the z-axis. We now confirm this explicitly.

Consider a spin state that at $t=0$ points along the direction specified by the angles (θ_0, ϕ_0):

$$|\Psi,0\rangle = \cos\tfrac{\theta_0}{2}|+\rangle + \sin\tfrac{\theta_0}{2}e^{i\phi_0}|-\rangle. \tag{17.5.14}$$

Given the Hamiltonian $\hat{H}_S = -\gamma B\hat{S}_z$ in (17.5.11), we find that

$$\hat{H}_S|\pm\rangle = \mp\tfrac{1}{2}\gamma B\hbar|\pm\rangle. \tag{17.5.15}$$

Therefore, the time-evolved state is given by

$$\begin{aligned}|\Psi,t\rangle &= e^{-i\hat{H}_S t/\hbar}|\Psi,0\rangle = e^{-i\hat{H}_S t/\hbar}\Big(\cos\tfrac{\theta_0}{2}|+\rangle + \sin\tfrac{\theta_0}{2}e^{i\phi_0}|-\rangle\Big)\\ &= \cos\tfrac{\theta_0}{2}e^{+i\gamma Bt/2}|+\rangle + \sin\tfrac{\theta_0}{2}e^{i\phi_0}e^{-i\gamma Bt/2}|-\rangle.\end{aligned} \tag{17.5.16}$$

To identify the direction of the resulting state, we factor out the phase that multiplies the $|+\rangle$ state:

$$|\Psi,t\rangle = e^{+i\gamma Bt/2}\Big(\cos\tfrac{\theta_0}{2}|+\rangle + \sin\tfrac{\theta_0}{2}e^{i(\phi_0-\gamma Bt)}|-\rangle\Big). \tag{17.5.17}$$

We now recognize that the spin state points along the direction defined by angles

$$\begin{aligned}\theta(t) &= \theta_0,\\ \phi(t) &= \phi_0 - \gamma Bt.\end{aligned} \tag{17.5.18}$$

Keeping θ constant while changing ϕ indeed corresponds to a rotation about the z-axis and, after time t, the spin has rotated an angle $(-\gamma Bt)$ as claimed above. It rotated with an angular speed of magnitude $\omega = \gamma B$, assuming $\gamma, B > 0$. Indeed, with $\gamma, B > 0$, the spin direction rotated with angular velocity pointing along the $-z$-direction, so we had $\boldsymbol{\omega} = -\gamma\mathbf{B}$.

We will now show that the above result is general: in a time-independent magnetic field, spin states precess with **Larmor angular frequency** ω_L given by

$$\boxed{\boldsymbol{\omega}_L = -\gamma\,\mathbf{B}.} \tag{17.5.19}$$

To see this, note that the Hamiltonian of a spin in a magnetic field (17.5.10) becomes

$$\boxed{\hat{H}_S = -\hat{\boldsymbol{\mu}}\cdot\mathbf{B} = -\gamma\mathbf{B}\cdot\hat{\mathbf{S}} = \boldsymbol{\omega}_L\cdot\hat{\mathbf{S}}.} \tag{17.5.20}$$

With the magnetic field assumed time independent, $\boldsymbol{\omega}_L$ is also time independent, and the evolution operator is simply

$$\mathcal{U}(t,0) = \exp(-i\hat{H}_S t/\hbar) = \exp\Big(-i\frac{\boldsymbol{\omega}_L\cdot\hat{\mathbf{S}}}{\hbar}t\Big). \tag{17.5.21}$$

Letting $\mathbf{n}$ denote the direction of ω_L, we write

$$\omega_L = \omega_L\,\mathbf{n}, \quad \mathbf{n}\cdot\mathbf{n}=1, \quad \omega_L \geq 0. \tag{17.5.22}$$

In this notation, the time-evolution operator becomes

$$\mathcal{U}(t,0) = \exp\Big(-i\frac{\omega_L t\,\hat{S}_{\mathbf{n}}}{\hbar}\Big) = \hat{R}_{\mathbf{n}}(\omega_L t), \tag{17.5.23}$$

when comparing with (17.5.13). The time-evolution operator $\mathcal{U}(t,0)$ rotates the spin states by the angle $\omega_L t$ about the $\mathbf{n}$-axis. In other words, we have shown that

$$\boxed{\text{with } \hat{H}_S = \omega_L\cdot\hat{\mathbf{S}}, \text{ spin states precess with angular velocity } \omega_L.} \tag{17.5.24}$$

The precession of a spin state means the precession of the unit vector $\mathbf{n}(t)$ that fixes the direction of the spin state $|\mathbf{n}(t)\rangle$. As the above equation shows, the angular frequency of precession is easily read directly from the Hamiltonian.

17.6 Nuclear Magnetic Resonance

In nuclear magnetic resonance, spins are subject to a time-dependent magnetic field. This magnetic field has a time-independent z-component and a circularly polarized component representing a magnetic field rotating on the (x, y) plane. More concretely, we have

$$\mathbf{B}(t) = B_0\,\mathbf{z} + B_1(\mathbf{x}\cos\omega t - \mathbf{y}\sin\omega t). \tag{17.6.1}$$

Generally, the constant z-component $B_0 > 0$ is larger than B_1, the magnitude of the radio-frequency (RF) signal. Associated to the longitudinal magnetic field B_0, the Larmor angular frequency ω_0 is defined by

$$\omega_0 \equiv \gamma B_0. \tag{17.6.2}$$

The time-dependent part of the field points along the x-axis at $t=0$ and is rotating with angular velocity $\omega > 0$ in the clockwise direction of the (x, y) plane. This corresponds to *negative* angular velocity about the positive z-axis. The spin Hamiltonian is

$$\hat{H}_S(t) = -\gamma\,\mathbf{B}(t)\cdot\hat{\mathbf{S}} = -\gamma\,[B_0\hat{S}_z + B_1(\cos\omega t\,\hat{S}_x - \sin\omega t\hat{S}_y)]. \tag{17.6.3}$$

This Hamiltonian is not only time dependent; the Hamiltonians at different times do not commute. This is therefore a nontrivial time-evolution problem.

We attempt to simplify the problem by considering a frame of reference that rotates just like the RF signal in the above magnetic field. To analyze this first imagine the case where the full magnetic field vanishes, and therefore $\hat{H}_S = 0$. With no magnetic field, spin states would simply be static; they do not precess. What should the Hamiltonian be in the frame rotating about the z-axis like the RF signal, with negative angular velocity of magnitude ω? It cannot be zero because in this frame the spin states are rotating with *positive* angular velocity ω about the z-direction. There must be a Hamiltonian that has that effect. The unitary operator $\mathcal{U}_\omega$ that generates this rotation is

$$\mathcal{U}_\omega(t) = \exp\Big(-\frac{i\omega t\hat{S}_z}{\hbar}\Big) \quad\Rightarrow\quad \hat{H}_{\mathcal{U}_\omega} = \omega\hat{S}_z, \tag{17.6.4}$$

with $\hat{H}_{\mathcal{U}_\omega}$ the associated time-independent Hamiltonian. The Hamiltonian $\hat{H}_{\mathcal{U}_\omega}$ governs the time evolution of the "rotating-frame" state $|\Psi_R, t\rangle$ when $\hat{H}_S = 0$.

To implement this concretely, we postulate that even when $\hat{H}_S \neq 0$ the state $|\Psi_R, t\rangle$ and the laboratory-frame state $|\Psi, t\rangle$ are related as follows:

$$\boxed{|\Psi_R, t\rangle \equiv \mathcal{U}_\omega(t)|\Psi, t\rangle.} \tag{17.6.5}$$

This relation is what we want when $\hat{H}_S = 0$: the state $|\Psi, t\rangle$ is then time independent, and $\mathcal{U}_\omega(t)$ alone generates the time evolution. Note that if we can find $|\Psi_R, t\rangle$, this relation determines $|\Psi, t\rangle$ since $\mathcal{U}_\omega$ is known. Moreover, at $t = 0$ the two states agree: $|\Psi_R, 0\rangle = |\Psi, 0\rangle$.

We now use (17.6.5) to find the full rotating-frame Hamiltonian $\hat{H}_R$ when $\hat{H}_S \neq 0$. Indeed, this relation is rewritten as

$$|\Psi_R, t\rangle = \mathcal{U}_\omega(t)|\Psi, t\rangle = \mathcal{U}_\omega(t)\,\mathcal{U}_S(t)|\Psi, 0\rangle = \mathcal{U}_\omega(t)\,\mathcal{U}_S(t)|\Psi_R, 0\rangle, \tag{17.6.6}$$

with $\mathcal{U}_S(t)$ the unitary operator generating the time evolution associated with $\hat{H}_S(t)$. Since the Hamiltonian associated to an arbitrary unitary time-evolution operator $\mathcal{U}$ is $i\hbar(\partial_t\mathcal{U})\mathcal{U}^\dagger$ (see (16.2.10)), we have

$$\hat{H}_R(t) = i\hbar\,\partial_t(\mathcal{U}_\omega\mathcal{U}_S)\,\mathcal{U}_S^\dagger\,\mathcal{U}_\omega^\dagger = i\hbar\,(\partial_t\mathcal{U}_\omega)\mathcal{U}_\omega^\dagger + \mathcal{U}_\omega\, i\hbar\,(\partial_t\mathcal{U}_S)\mathcal{U}_S^\dagger\,\mathcal{U}_\omega^\dagger. \tag{17.6.7}$$

This means we have shown that

$$\hat{H}_R = \hat{H}_{\mathcal{U}_\omega} + \mathcal{U}_\omega\,\hat{H}_S\,\mathcal{U}_\omega^\dagger. \tag{17.6.8}$$

This is a nice result: when $\hat{H}_S = 0$, it gives the expected $\hat{H}_{\mathcal{U}_\omega}$, and when $\hat{H}_S \neq 0$, the rotating-frame Hamiltonian receives an extra contribution.

We now check that $\hat{H}_R$ is much simpler than $\hat{H}_S$; it is in fact time independent. Using the expressions for $\hat{H}_{\mathcal{U}_\omega}$ and $\mathcal{U}_\omega$ from (17.6.4) and the formula for $\hat{H}_S$, we find that

$$\begin{aligned} \hat{H}_R &= \omega\hat{S}_z - \gamma\, e^{-\frac{i\omega t\hat{S}_z}{\hbar}}\Big(B_0\hat{S}_z + B_1\big(\cos\omega t\,\hat{S}_x - \sin\omega t\hat{S}_y\big)\Big)e^{\frac{i\omega t\hat{S}_z}{\hbar}} \\ &= (-\gamma B_0 + \omega)\hat{S}_z - \gamma B_1\hat{M}(t), \end{aligned} \tag{17.6.9}$$

where we defined

$$\hat{M}(t) \equiv e^{-\frac{i\omega t\hat{S}_z}{\hbar}}\big(\cos\omega t\,\hat{S}_x - \sin\omega t\hat{S}_y\big)e^{\frac{i\omega t\hat{S}_z}{\hbar}}. \tag{17.6.10}$$

We show that $\hat{M}(t)$ is time independent by calculating the time derivative of $\hat{M}$:

$$\partial_t\hat{M} = e^{-\frac{i\omega t\hat{S}_z}{\hbar}}\Big(-\tfrac{i\omega}{\hbar}\,[\hat{S}_z, \cos\omega t\,\hat{S}_x - \sin\omega t\,\hat{S}_y] + (-\omega\sin\omega t\,\hat{S}_x - \omega\cos\omega t\,\hat{S}_y)\Big)e^{\frac{i\omega t\hat{S}_z}{\hbar}}.$$

The commutator arises from differentiation of the exponentials in $\hat{M}$ and the other terms from differentiation of the expression within the exponentials in $\hat{M}$. Evaluating the commutators, we find a complete cancellation: $\partial_t \hat{M} = 0$. Since $\hat{M}$ is time independent, we can evaluate it at any time. The simplest time is $t = 0$, giving

$$\hat{M}(t) = \hat{S}_x. \tag{17.6.11}$$

As a result, the rotating-frame Hamiltonian $\hat{H}_R$ of (17.6.9) becomes

$$\hat{H}_R = (-\gamma B_0 + \omega)\hat{S}_z - \gamma B_1 \hat{S}_x. \tag{17.6.12}$$

The time independence of $\hat{H}_R$ means that the time evolution of $|\Psi_R, t\rangle$ is easily calculated. A little rewriting allows us to read an effective magnetic field $\mathbf{B}_R$ associated with $\hat{H}_R$:

$$\hat{H}_R = -\gamma\Big[B_1\hat{S}_x + \Big(B_0 - \frac{\omega}{\gamma}\Big)\hat{S}_z\Big] = -\gamma\Big[B_1\hat{S}_x + B_0\Big(1 - \frac{\omega}{\omega_0}\Big)\hat{S}_z\Big], \tag{17.6.13}$$

using $\omega_0 = \gamma B_0$ for the Larmor frequency associated with the constant component of the field. We thus have

$$\boxed{\hat{H}_R = -\gamma\, \mathbf{B}_R \cdot \hat{\mathbf{S}} \quad \Rightarrow \quad \mathbf{B}_R = B_1 \mathbf{x} + B_0\Big(1 - \frac{\omega}{\omega_0}\Big)\mathbf{z}.} \tag{17.6.14}$$

Note that the RF signal contributes to the effective magnetic field a component B_1 pointing along the x-axis. The longitudinal effective magnetic field is also changed: the initial value B_0 is now multiplied by an ω-dependent factor.

The full solution for the state is obtained beginning with (17.6.5) and (17.6.4):

$$|\Psi, t\rangle = \mathcal{U}_\omega^\dagger(t)|\Psi_R, t\rangle = \exp\Big[\frac{i\omega t\hat{S}_z}{\hbar}\Big]|\Psi_R, t\rangle. \tag{17.6.15}$$

Since $\hat{H}_R$ is time independent, the time evolution of $|\Psi_R, t\rangle$ is easily taken into account:

$$|\Psi, t\rangle = \exp\Big[\frac{i\omega t\hat{S}_z}{\hbar}\Big]\exp\Big[-i\frac{(-\gamma \mathbf{B}_R \cdot \hat{\mathbf{S}})\,t}{\hbar}\Big]|\Psi_R, 0\rangle. \tag{17.6.16}$$

Recalling that $|\Psi_R, 0\rangle = |\Psi, 0\rangle$, we finally get

$$\boxed{|\Psi, t\rangle = \exp\Big[\frac{i\omega t\hat{S}_z}{\hbar}\Big]\exp\Big[i\frac{\gamma \mathbf{B}_R \cdot \hat{\mathbf{S}}\,t}{\hbar}\Big]|\Psi, 0\rangle.} \tag{17.6.17}$$

This is the complete solution to the time evolution of an arbitrary spin state in a longitudinal plus transverse RF magnetic field.

Exercise 17.10. *Verify that for $B_1 = 0$ the above solution reduces to the one describing precession about the z-axis.*

In the applications to be discussed below, we always find that the magnitude of the RF signal is far smaller than the magnitude of the longitudinal magnetic field:

$$B_1 \ll B_0. \tag{17.6.18}$$

Now consider the evolution of a spin that initially points in the positive z-direction. We look at two cases:

1. $\omega \ll \omega_0$. In this case the effective magnetic field (17.6.14) in the rotating frame is approximately given by

$$\mathbf{B}_R \simeq B_0 \mathbf{z} + B_1 \mathbf{x}. \tag{17.6.19}$$

This is a field mostly along the z-axis but tipped a little toward the x-axis. The right most exponential in (17.6.17) makes the spin precess rapidly about the direction of $\mathbf{B}_R$. Since $|\mathbf{B}_R| \sim B_0$, the angular rate of precession is pretty much ω_0. The next exponential in (17.6.17) induces a rotation about the z-axis with smaller angular velocity ω.

2. $\omega = \omega_0$. This is a resonance condition, with the RF frequency set equal to the longitudinal Larmor frequency. This condition makes the longitudinal component of the effective magnetic field in the rotating frame vanish. Indeed, equation (17.6.14) gives

$$\mathbf{B}_R = B_1 \mathbf{x}. \tag{17.6.20}$$

In this case the rightmost exponential in (17.6.17) makes the spin precess about the x-axis. As a result, with $\gamma B_1 > 0$, the spin that points initially along the z-axis will rotate toward the positive y-axis with angular velocity $\omega_1 = \gamma B_1$. If we set the RF signal to last a time T such that

$$\omega_1 T = \tfrac{\pi}{2}, \tag{17.6.21}$$

the state $|\Psi_R, T\rangle$ will point along the y-axis. The effect of the other exponential in (17.6.17) is just to rotate the spin about the z-axis. We then have

$$|\Psi, t\rangle = \exp\Big[\frac{i\omega_0 t\hat{S}_z}{\hbar}\Big]|\Psi_R, t\rangle, \quad t < T, \tag{17.6.22}$$

and if the RF pulse turns off after time T,

$$|\Psi, t\rangle = \exp\Big[\frac{i\omega_0 t\hat{S}_z}{\hbar}\Big]|\Psi_R, T\rangle, \quad t > T. \tag{17.6.23}$$

The state $|\Psi, t\rangle$ can be visualized as a spin that is slowly rotating with angular velocity ω_1 from the z-axis toward the y-axis while rapidly rotating around the z-axis with angular velocity ω_0. As a result the tip of the spin vector is performing a spiral motion on the surface of a hemisphere. By the time the polar angle reaches $\pi/2$, the RF signal turns off, and the spin now just rotates on the (x, y) plane. This is called a 90° pulse. The motion of the tip of the spin state is sketched in figure 17.3.

The value of the longitudinal magnetic field B_0 in experimental setups is of the order of a few tesla. We have defined the Larmor angular frequency $\omega_0 = \gamma B_0$. It follows that the Larmor *frequency* f_0 is given by $f_0 = \frac{\omega_0}{2\pi} = \frac{\gamma}{2\pi} B_0$. From the results in exercise 17.9, we find that

$$\Big|\frac{\gamma_e}{2\pi}\Big| = 28.0\,\text{GHz/T}, \quad \Big|\frac{\gamma_p}{2\pi}\Big| = 42.6\,\text{MHz/T}. \tag{17.6.24}$$

For a magnetic field of two tesla, the proton precesses with a frequency of about 85 MHz.

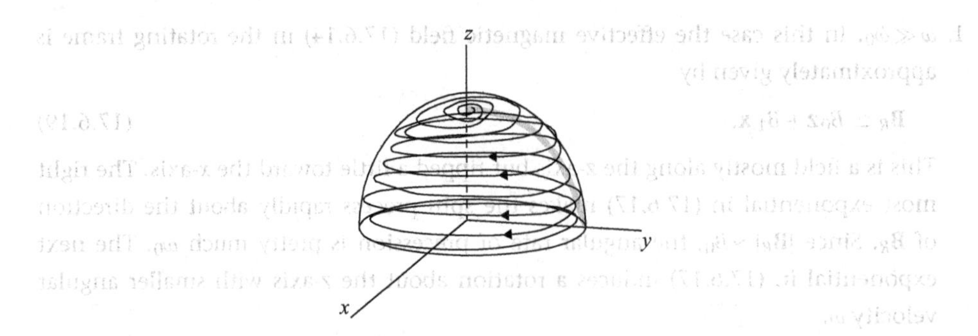

Figure 17.3
The time evolution of a spin state, initially pointing along the positive z-axis at $t = 0$ and subject to a 90° RF pulse. The tip of the vector representing the spin follows a spiral trajectory resulting from the composition of a slow rotation about the x-axis and a fast rotation about the z-axis.

Magnetic resonance imaging (MRI) This technology was developed in the late 1970s using earlier research on nuclear magnetic resonance by Felix Bloch and by Edward Purcell, working at MIT's Radiation Laboratory. In 1952 Bloch and Purcell received the Nobel Prize in Physics for this work. Magnetic resonance imaging has advantages over X-rays: it allows one to distinguish various soft tissues and does not involve radiation.

The human body is mostly composed of water molecules (H_2O). We thus have many hydrogen atoms, whose nuclei are protons and are the main players through their magnetic dipole moments. The MRI machine produces a large and constant magnetic field B_0 along the axis of the machine, a direction we choose to call the z-direction. Despite the disordering effects of body temperature, there is a net alignment of nuclear spins along B_0. This *longitudinal magnetization* puts a large number of spins in play.

We apply a 90° pulse so we get the spins to rotate with Larmor frequency ω_0 in the (x, y) plane. These rotating dipoles produce an oscillating magnetic field, and this signal is picked up by a receiver. The magnitude of the signal is proportional to the proton density. This is the first piece of information and allows differentiation of tissues.

The above signal from the rotation of the spins decays with a time constant T_2 that is typically much smaller than a second. This decay is attributed to interactions between the spins that quickly dampen their rotation. A T_2-weighted image allows doctors to detect the abnormal accumulation of fluids (edema).

There is another time constant T_1, of order one second, that controls the time needed to regain the longitudinal magnetization. This effect is due to the spins interacting with the rest of the lattice of atoms. White matter, gray matter, and cerebrospinal fluids have about the same proton density but are distinguished by different T_1 constants. The constants T_1 and T_2 are in fact associated with processes of decoherence. Such processes can be studied phenomenologically using the Lindblad equation, as we will do in section 22.6.

MRIs commonly include the use of contrast agents, which are substances that shorten the time constant T_1 and are usually administered by injection into the bloodstream. The contrast agent (gadolinium) can accumulate at organs or locations where information is valuable. For a number of substances, one can use the MRI apparatus to determine their (T_1, T_2) constants and build a table of data. This table can then be used as an aid to evaluating the results of other MRIs.

The typical MRI machine has a B_0 of about two tesla or twenty thousand gauss. This requires a superconducting magnet with liquid helium cooling. For people with claustrophobia there are "open" MRI scanners that work with lower magnetic fields. In addition, the machines are equipped with a number of *gradient* magnets, each of about two hundred gauss. They change locally the value of B_0 and provide spatial resolution by making the Larmor frequency spatially dependent. One can then attain spatial resolutions of about half a millimeter! MRIs are considered safe, as there is no evidence of biological harm caused by very large static magnetic fields.

17.7 Two-State System Viewed as a Spin System

The most general time-independent Hamiltonian for a two-state system is a Hermitian operator, and it can be represented, using an orthonomal basis $\{|1\rangle, |2\rangle\}$, by the most general Hermitian 2×2 matrix $\hat{H}$. That matrix, as discussed in detail in example 13.7, can be characterized by four real constants $h_0, h_1, h_2, h_3 \in \mathbb{R}$ as follows:

$$\hat{H} = \begin{pmatrix} h_0 + h_3 & h_1 - ih_2 \\ h_1 + ih_2 & h_0 - h_3 \end{pmatrix} = h_0 \mathbb{1} + h_1\sigma_1 + h_2\sigma_2 + h_3\sigma_3. \tag{17.7.1}$$

On the right-hand side, we wrote $\hat{H}$ as a sum of matrices, where the identity $\mathbb{1}$ and the Pauli matrices σ_i, with $i = 1, 2, 3$, are all Hermitian. We write $\mathbf{h} = (h_1, h_2, h_3)$ and then define

$$\mathbf{h} \cdot \boldsymbol{\sigma} \equiv h_1\sigma_1 + h_2\sigma_2 + h_3\sigma_3. \tag{17.7.2}$$

In this notation,

$$\boxed{\hat{H} = h_0 \mathbb{1} + \mathbf{h} \cdot \boldsymbol{\sigma}.} \tag{17.7.3}$$

It is again convenient to introduce the magnitude h and the direction $\mathbf{n}$ of $\mathbf{h}$:

$$\mathbf{h} = h\mathbf{n}, \quad \mathbf{n} \cdot \mathbf{n} = 1, \quad h = \sqrt{h_1^2 + h_2^2 + h_3^2} \geq 0. \tag{17.7.4}$$

Do not confuse h here with Planck's constant. Now the Hamiltonian reads

$$\hat{H} = h_0 \mathbb{1} + h\,\mathbf{n} \cdot \boldsymbol{\sigma}. \tag{17.7.5}$$

Recall now that the spin states $|\mathbf{n}; \pm\rangle$ are eigenstates of $\mathbf{n} \cdot \boldsymbol{\sigma}$:

$$\mathbf{n} \cdot \boldsymbol{\sigma}\, |\mathbf{n}; \pm\rangle = \pm |\mathbf{n}; \pm\rangle. \tag{17.7.6}$$

In writing the spin states, however, you must recall that what we call the z-up and z-down states are just the first and second basis states: $|+\rangle = |1\rangle$ and $|-\rangle = |2\rangle$. With this noted, the spin states $|\mathbf{n};\pm\rangle$ are indeed $\hat{H}$ eigenstates since using the last two equations above we have

$$\hat{H}|\mathbf{n};\pm\rangle = (h_0 \pm h)|\mathbf{n};\pm\rangle. \tag{17.7.7}$$

This also shows that the energy eigenvalues are $h_0 \pm h$. In summary we have the

$$\begin{aligned} \text{spectrum:} \quad & |\mathbf{n};+\rangle \text{ with energy } h_0 + h, \\ & |\mathbf{n};-\rangle \text{ with energy } h_0 - h. \end{aligned} \tag{17.7.8}$$

Here, $|+\rangle = |1\rangle$ and $|-\rangle = |2\rangle$. The $|1\rangle$ and $|2\rangle$ states, in general, have no relation to spin states. They are simply the basis states of any two-state system in which the Hamiltonian takes the matrix form (17.7.1).

To understand the time evolution of states with the Hamiltonian $\hat{H}$, we rewrite $\hat{H}$ in terms of the spin operators, recalling that $\hat{\mathbf{S}} = \frac{\hbar}{2}\boldsymbol{\sigma}$. Using (17.7.3), we find

$$\hat{H} = h_0 \mathbb{1} + \frac{2}{\hbar}\mathbf{h}\cdot\hat{\mathbf{S}}. \tag{17.7.9}$$

Comparison with the spin Hamiltonian $\hat{H}_S = \boldsymbol{\omega}_L \cdot \hat{\mathbf{S}}$ shows that in the system described by $\hat{H}$ the states precess with an angular velocity $\boldsymbol{\omega}$ given by

$$\boxed{\boldsymbol{\omega} = \frac{2}{\hbar}\mathbf{h}.} \tag{17.7.10}$$

The part $h_0\mathbb{1}$ of the Hamiltonian $\hat{H}$ does not rotate states during time evolution; it simply multiplies all the states by an overall time-dependent phase $\exp(-ih_0 t/\hbar)$.

If the Hamiltonian $\hat{H}$ is known, the vector $\boldsymbol{\omega}$ above is immediately calculable. Suppose we are given a normalized state $|\Psi, 0\rangle$ of the system:

$$|\Psi, 0\rangle = \alpha|1\rangle + \beta|2\rangle, \qquad \alpha, \beta \in \mathbb{C}. \tag{17.7.11}$$

To find its time evolution, we first introduce the spin state $|\mathbf{n}\rangle$ defined by

$$|\mathbf{n}\rangle \equiv \alpha|+\rangle + \beta|-\rangle. \tag{17.7.12}$$

Since the constants α, β are known, the direction $\mathbf{n}$ is determined. Time evolution gives the state $|\mathbf{n}(t)\rangle$, with $\mathbf{n}(t)$ obtained by the rotation of $\mathbf{n}$ about the axis and the angular velocity specified by $\boldsymbol{\omega}$. If you want an explicit formula for $\mathbf{n}(t)$, you can use the rotation matrix $\mathcal{R}$ given in (14.7.6). The time-evolved state $|\Psi, t\rangle$ is simply $|\mathbf{n}(t)\rangle$, with the replacements $|+\rangle \to |1\rangle$ and $|-\rangle \to |2\rangle$.

17.8 The Factorization Method

We determined the spectrum of the harmonic oscillator in two ways: first, by simply solving the time-independent Schrödinger equation and, second, by factorizing the Hamiltonian. Indeed, in this second approach the Hamiltonian was written as $\hat{H} = \hbar\omega(\hat{a}^\dagger\hat{a} + \frac{1}{2})$, showing that, up to additive and multiplicative constants, we have the factorization $\hat{H} \sim \hat{a}^\dagger\hat{a}$ into the product of two operators, $\hat{a}$ and its Hermitian conjugate $\hat{a}^\dagger$.

As explained in section 9.3, this factorization brings a number of simplifications: one can show that the spectrum is bounded below, the ground state can be found by solving a first-order differential equation, and excited states are easily constructed.

In this section we study a generalization of the above factorization. First, we will consider general operators $\hat{A}$ and $\hat{A}^\dagger$. More importantly, we will consider *two* Hamiltonians, one associated with the product $\hat{A}^\dagger\hat{A}$ and the other with the product $\hat{A}\hat{A}^\dagger$. These two Hamiltonians have closely related spectra, and their interplay helps understand both systems. We will illustrate the factorization method by solving for the energy eigenstates of a free particle in spherical coordinates. The factorization method is also called the supersymmetric method, as is appropriate when the two Hamiltonians are imagined to describe two different kinds of particles. In conventional supersymmetric theories, those would be bosons and fermions.

We will work with Hamiltonians after they have been written, for convenience, in terms of a unit-free coordinate that we will call x. For one-dimensional problems, $x \in (-\infty, \infty)$. In three-dimensional problems, x is the radial variable and $x \in [0, \infty)$. In terms of unit-free coordinates, the annihilation operator $\hat{a}$ in the harmonic oscillator takes the form $\hat{a} \sim \frac{d}{dx} + x$. In generalizing this to an operator $\hat{A}$, we will posit that

$$\hat{A} \equiv \frac{d}{dx} + W(x), \tag{17.8.1}$$

where $W(x)$ is some real function of x to be chosen at our convenience. The function $W(x)$ is often called the *superpotential*, a terminology from supersymmetry. With our conventional inner product on spatial wave functions, the Hermitian conjugate of $\hat{A}$, called $\hat{A}^\dagger$, is given by

$$\hat{A}^\dagger \equiv -\frac{d}{dx} + W(x). \tag{17.8.2}$$

Let us now define the two Hamiltonians associated to products of $\hat{A}$ and $\hat{A}^\dagger$:

$$\begin{aligned} \hat{H}^{(1)} &\equiv \hat{A}^\dagger\hat{A} = -\frac{d^2}{dx^2} + W^2(x) - W'(x), \\ \hat{H}^{(2)} &\equiv \hat{A}\hat{A}^\dagger = -\frac{d^2}{dx^2} + W^2(x) + W'(x), \end{aligned} \tag{17.8.3}$$

where primes denote derivatives with respect to the argument. The calculation of the operator products that give the final expressions to the right is done by letting them act on an arbitrary function $f(x)$, as follows:

$$\begin{aligned} \hat{A}^\dagger\hat{A} f &= \Big(-\frac{d}{dx} + W(x)\Big)\Big(\frac{df}{dx} + Wf\Big) \\ &= -\frac{d^2 f}{dx^2} - \frac{dW}{dx}f - W\frac{df}{dx} + W\frac{df}{dx} + W^2 f \\ &= \Big(-\frac{d^2}{dx^2} + W^2 - \frac{dW}{dx}\Big)f, \end{aligned} \tag{17.8.4}$$

thus confirming the result for $\hat{H}^{(1)}$. The calculation for $\hat{H}^{(2)}$ is almost identical. The two Hamiltonians define two potentials $V^{(1)}$ and $V^{(2)}$. These are given by

$$V^{(1)}(x) \equiv W^2(x) - W'(x), \quad V^{(2)}(x) \equiv W^2(x) + W'(x). \tag{17.8.5}$$

The potentials are determined by the superpotential $W(x)$. In general, these potentials can be rather different. This makes the factorization analysis interesting, showing that Hamiltonians with different potentials can have closely related physics. Let us introduce some notation for the energy eigenstates ϕ_n of the Hamiltonians. We will write

$$\begin{aligned} \hat{H}^{(1)}\phi_n^{(1)} &= E_n^{(1)}\phi_n^{(1)}, \quad n=0,1,\ldots, \\ \hat{H}^{(2)}\phi_n^{(2)} &= E_n^{(2)}\phi_n^{(2)}, \quad n=0,1,\ldots. \end{aligned} \tag{17.8.6}$$

We have in mind normalizable states, in which case we are dealing with bound states. Moreover, given that we are considering one-dimensional potentials, the bound state spectrum of the Hamiltonians is nondegenerate. In some examples we can deal with nonnormalizable states, and many of the results hold. While the above eigenstates can include those with zero energy, for clarity we will denote zero-energy eigenstates with the symbol χ.

Here is a set of observations that elucidate the properties of the Hamiltonians:

1. The energies $E_n^{(1)}$ and $E_n^{(2)}$ are all nonnegative. This is easily shown:

$$\begin{aligned} E_n^{(1)} &= \langle \phi_n^{(1)}, \hat{H}^{(1)}\phi_n^{(1)} \rangle = \langle \phi_n^{(1)}, \hat{A}^\dagger\hat{A}\phi_n^{(1)} \rangle = \langle \hat{A}\phi_n^{(1)}, \hat{A}\phi_n^{(1)} \rangle \geq 0, \\ E_n^{(2)} &= \langle \phi_n^{(2)}, \hat{H}^{(2)}\phi_n^{(2)} \rangle = \langle \phi_n^{(2)}, \hat{A}\hat{A}^\dagger\phi_n^{(2)} \rangle = \langle \hat{A}^\dagger\phi_n^{(2)}, \hat{A}^\dagger\phi_n^{(2)} \rangle \geq 0. \end{aligned} \tag{17.8.7}$$

2. A zero-energy eigenstate $\chi^{(1)}$ of $\hat{H}^{(1)}$ satisfies

$$\hat{A}\,\chi^{(1)} = 0. \tag{17.8.8}$$

   This follows from item 1 above: if the energy of $\chi^{(1)}$ is zero, then $0 = \langle \hat{A}\chi^{(1)}, \hat{A}\chi^{(1)} \rangle$, which implies that the state in the inner product vanishes. Similarly, a zero-energy eigenstate $\chi^{(2)}$ of $\hat{H}^{(2)}$ satisfies

$$\hat{A}^\dagger\,\chi^{(2)} = 0. \tag{17.8.9}$$

3. The spectrum of $\hat{H}^{(1)}$ and $\hat{H}^{(2)}$ match for nonzero energies.

   This is a consequence of the following two properties: (i) if $\phi_n^{(1)}$ is an $\hat{H}^{(1)}$ eigenstate with nonzero energy, then $\hat{A}\phi_n^{(1)}$ is an $\hat{H}^{(2)}$ eigenstate with the *same* energy, and (ii) if $\phi_n^{(2)}$ is an $\hat{H}^{(2)}$ eigenstate with nonzero energy, then $\hat{A}^\dagger\phi_n^{(2)}$ is an $\hat{H}^{(1)}$ eigenstate with the *same* energy.

   These are easily proven. Indeed, for an $\hat{H}^{(1)}$ eigenstate we have

$$\hat{A}^\dagger\hat{A}\phi_n^{(1)} = E_n^{(1)}\phi_n^{(1)}. \tag{17.8.10}$$

   Multiplying by $\hat{A}$, we find that

$$\hat{A}\hat{A}^\dagger(\hat{A}\phi_n^{(1)}) = E_n^{(1)}(\hat{A}\phi_n^{(1)}) \quad \Rightarrow \quad \hat{H}^{(2)}\,(\hat{A}\phi_n^{(1)}) = E_n^{(1)}(\hat{A}\phi_n^{(1)}), \tag{17.8.11}$$

   as claimed. Similarly, for an $\hat{H}^{(2)}$ eigenstate we have

$$\hat{A}\hat{A}^\dagger\phi_n^{(2)} = E_n^{(2)}\phi_n^{(2)}. \tag{17.8.12}$$

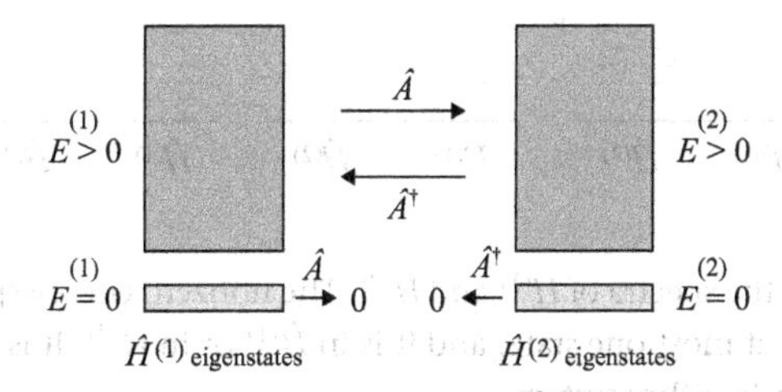

Figure 17.4
The operator $\hat{A}$ maps $\hat{H}^{(1)}$ eigenstates of nonzero energy to $\hat{H}^{(2)}$ eigenstates of the same energy while killing zero-energy eigenstates. The operator $\hat{A}^\dagger$ maps $\hat{H}^{(2)}$ eigenstates of nonzero energy to $\hat{H}^{(1)}$ eigenstates of the same energy while killing zero-energy eigenstates. The nonzero energy eigenstates of $\hat{H}^{(1)}$ and $\hat{H}^{(2)}$ are equal in number and paired.

Multiplying by $\hat{A}^\dagger$, we find that

$$\hat{A}^\dagger\hat{A}(\hat{A}^\dagger\phi_n^{(2)}) = E_n^{(2)}(\hat{A}^\dagger\phi_n^{(1)}) \quad \Rightarrow \quad \hat{H}^{(1)}(\hat{A}^\dagger\phi_n^{(2)}) = E_n^{(2)}(\hat{A}^\dagger\phi_n^{(2)}). \tag{17.8.13}$$

We have that $\hat{A}$ maps $\hat{H}^{(1)}$ eigenstates to $\hat{H}^{(2)}$ eigenstates, and $\hat{A}^\dagger$ maps $\hat{H}^{(2)}$ eigenstates to $\hat{H}^{(1)}$ eigenstates. These maps are, up to scale, inverses of each other. Indeed, acting on an $\hat{H}^{(1)}$ eigenstate with $\hat{A}$ first and with $\hat{A}^\dagger$ after gives the state back: $\hat{A}^\dagger\hat{A}\phi_n^{(1)} = E_n^{(1)}\phi_n^{(1)}$. If the spectrum is nondegenerate (if we are dealing with bound states), this suffices to claim that the nonzero energy spectrum of $\hat{H}^{(1)}$ and $\hat{H}^{(2)}$ agree.

But more is true. The map $\hat{A}$ acting on the nonzero energy spectrum of $\hat{H}^{(1)}$, even if degenerate, is injective. Suppose we had two degenerate $\hat{H}^{(1)}$ eigenstates ψ_1, ψ_2 of nonzero energy such that $\hat{A}\psi_1 = \hat{A}\psi_2$, implying noninjectivity. Then we would have $\hat{A}(\psi_1 - \psi_2) = 0$. Since the only states annihilated by $\hat{A}$ are zero-energy eigenstates χ of $\hat{H}^{(1)}$, we would have $\psi_1 = \psi_2 + c\chi$, for nonzero constant c, which is not consistent with both ψ_1 and ψ_2 being eigenstates of the same nonzero energy. Moreover, the map $\hat{A}$ is surjective: any state $\phi_n^{(2)}$ is proportional to $\hat{A}\hat{A}^\dagger\phi_n^{(2)} = \hat{A}(\hat{A}^\dagger\phi_n^{(2)})$ and is thus in the image of $\hat{A}$. The injective and surjective map $\hat{A}$ establishes that the nonzero energy spectra of $\hat{H}^{(1)}$ and $\hat{H}^{(2)}$ are the same.

Note that the maps fail to do anything interesting for zero-energy states. While $\hat{A}$ turns $\hat{H}^{(1)}$ eigenstates of nonzero energy into $\hat{H}^{(2)}$ eigenstates, zero-energy states of $\hat{H}^{(1)}$ are killed by $\hat{A}$: $\hat{A}\chi^{(1)} = 0$, so we get no state. Similarly, $\hat{A}^\dagger$ kills zero-energy $\hat{H}^{(2)}$ eigenstates. Thus, zero-energy states need not be matched. In fact, as we see next, they are not. The action of the $\hat{A}$ and $\hat{A}^\dagger$ maps on the spectrum is shown in figure 17.4.

4. The zero-energy eigenstate $\chi^{(1)}$ of $\hat{H}^{(1)}$ is obtained by solving the differential equation $\hat{A}\chi^{(1)} = 0$, or equivalently,

$$\frac{d\chi^{(1)}}{dx} + W(x)\chi^{(1)} = 0 \quad \Rightarrow \quad \chi^{(1)} = C_1 \exp\Big(-\int_0^x W(x')dx'\Big). \tag{17.8.14}$$

For the zero-energy eigenstate $\chi^{(2)}$ of $\hat{H}^{(2)}$, the differential equation $\hat{A}^\dagger\chi^{(2)} = 0$ gives

$$-\frac{d\chi^{(2)}}{dx} + W(x)\chi^{(2)} = 0 \quad \Rightarrow \quad \chi^{(2)} = C_2 \exp\Big(\int_0^x W(x')dx'\Big). \tag{17.8.15}$$

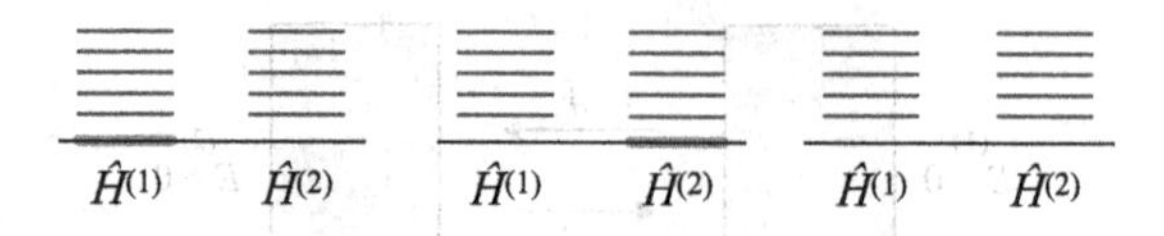

Figure 17.5
The possible options for the spectra of $\hat{H}^{(1)}$ and $\hat{H}^{(2)}$. The nonzero energy spectra agree and match. For zero energy, there is at most one state, and it is in $\hat{H}^{(1)}$ or in $\hat{H}^{(2)}$. It is also possible that there are no zero-energy states in either system.

The two solutions have the same exponential but with opposite signs. If one solution is normalizable, it is because its exponential goes to zero as $|x| \to \infty$. But then the other solution cannot be normalized. There are three possibilities: (i) $\hat{H}^{(1)}$ has a zero-energy state but $\hat{H}^{(2)}$ does not, (ii) $\hat{H}^{(2)}$ has a zero-energy state but $\hat{H}^{(1)}$ does not, or (iii) neither $\hat{H}^{(1)}$ nor $\hat{H}^{(2)}$ have zero-energy states. The three possibilities are shown in figure 17.5.

Exercise 17.11. *Let $W(x) = gx^2$ with $g > 0$ and $x \in (-\infty, \infty)$. Show that neither $\hat{H}^{(1)}$ nor $\hat{H}^{(2)}$ has a zero-energy state. Plot the potentials and explain physically why the nonzero energy spectra of the Hamiltonians agree.*

The system composed of the two Hamiltonians is said to be an example of supersymmetric quantum mechanics. For this, one assembles the two Hamiltonians into a single $\hat{H}$ and defines a supersymmetry generator $\hat{Q}$ as follows:

$$\hat{H} \equiv \begin{pmatrix} \hat{H}^{(1)} & 0 \\ 0 & \hat{H}^{(2)} \end{pmatrix}, \quad \hat{Q} \equiv \begin{pmatrix} 0 & \hat{A}^\dagger \\ -\hat{A} & 0 \end{pmatrix}. \tag{17.8.16}$$

The supersymmetry generator $\hat{Q}$ is anti-Hermitian, $\hat{Q}^\dagger = -\hat{Q}$, and satisfies

$$\hat{Q}^2 = \hat{Q}\hat{Q} = -\hat{H}. \tag{17.8.17}$$

In a supersymmetric theory, the Hamiltonian is indeed obtained by squaring supersymmetry generators. Eigenstates of $\hat{H}$ come in two types, the $\hat{H}^{(1)}$ eigenstates, which we call *bosons*, and the $\hat{H}^{(2)}$ eigenstates, which we call *fermions*:

$$\text{boson:} \begin{pmatrix} \phi_n^{(1)} \\ 0 \end{pmatrix}, \quad \text{fermion:} \begin{pmatrix} 0 \\ \phi_n^{(2)} \end{pmatrix}. \tag{17.8.18}$$

Bosons are σ_3 eigenstates with eigenvalue one, and fermions are σ_3 eigenstates with eigenvalue minus one. Note that $\hat{Q}$ maps bosons into fermions. This is consistent with the property $\sigma_3\hat{Q} = -\hat{Q}\sigma_3$, which you can check. In fact, $\hat{Q}$ precisely implements the maps that show the equivalence of the nonzero energy spectrum. For each nonzero energy boson, there is a nonzero energy fermion, and vice versa.

A supersymmetric ground state is a state annihilated by $\hat{Q}$. You can verify that a zero-energy state $\chi^{(1)}$ of $\hat{H}^{(1)}$ is a supersymmetric bosonic ground state. Similarly, a zero-energy state $\chi^{(2)}$ of $\hat{H}^{(2)}$ is a supersymmetric fermionic state. Whenever a supersymmetric ground state exists, we say that supersymmetry is unbroken. If no supersymmetric ground state exists, we say supersymmetry is spontaneously broken. Thus, supersymmetry is unbroken in the first two cases shown in figure 17.5, but it is spontaneously broken in the third case. A signal of supersymmetry breaking is indeed a nonzero ground state energy. Supersymmetry transformations are realized by the unitary operator $\exp(\epsilon\hat{Q})$ with ϵ real. Had we defined $\hat{Q}$ as Hermitian by having no minus sign acting on $\hat{A}$, the unitary operator would have required an explicit factor of i in the exponent. This would be inconvenient when dealing with real wave functions.

Example 17.1. *Radial solutions for a free particle in three dimensions.*

When studying the radial equation in section 10.6, we considered a free particle in spherical coordinates (example 10.1). For a particle with wave number k and energy $E=\frac{\hbar^2k^2}{2m}$ and using a unit-free coordinate $\rho=kr$, one finds that the radial equation for u_ℓ takes the form given in (10.6.25):

$$-\frac{d^2u_\ell}{d\rho^2}+\frac{\ell(\ell+1)}{\rho^2}u_\ell = u_\ell. \tag{17.8.19}$$

As noted then, this equation does not imply quantization of the energy; the energy in fact does not appear explicitly in the equation. Defining the Hamiltonian $\hat{H}_\ell$ as follows,

$$\hat{H}_\ell \equiv -\frac{d^2}{d\rho^2}+\frac{\ell(\ell+1)}{\rho^2}, \tag{17.8.20}$$

the equation we want to solve is

$$\hat{H}_\ell\, u_\ell = u_\ell. \tag{17.8.21}$$

Since this is a second-order linear differential equation, we expect two solutions. However, we want to solve this equation for *all* values of ℓ—namely, $\ell=0,1,2,\ldots$. The factorization method here will help us find solutions for higher ℓ in terms of the $\ell=0$ solutions.

To use the method, with ρ playing the role of x, we write a $W(\rho)$ so that $\hat{H}_\ell$ is the Hamiltonian $\hat{H}_\ell^{(1)}$ defined in (17.8.3). We choose

$$W(\rho)=-\frac{\ell+1}{\rho}, \quad W'(\rho)=\frac{\ell+1}{\rho^2} \tag{17.8.22}$$

so that

$$\hat{H}_\ell^{(1)}=-\frac{d^2}{d\rho^2}+W^2-W'=-\frac{d^2}{d\rho^2}+\frac{\ell(\ell+1)}{\rho^2}=\hat{H}_\ell, \tag{17.8.23}$$

just as we wanted. Let us now find $\hat{H}_\ell^{(2)}$:

$$\hat{H}_\ell^{(2)}=-\frac{d^2}{d\rho^2}+\frac{(\ell+1)^2}{\rho^2}+\frac{\ell+1}{\rho^2}=-\frac{d^2}{d\rho^2}+\frac{(\ell+1)(\ell+2)}{\rho^2}=\hat{H}_{\ell+1}. \tag{17.8.24}$$

We have shown that $\hat{H}_\ell^{(2)}=\hat{H}_{\ell+1}$, and this relation will allow us to change ℓ. The logic goes as follows. Assume we have a solution of

$$\hat{H}_\ell u_\ell = u_\ell \;\Rightarrow\; \hat{H}_\ell^{(1)}u_\ell = u_\ell, \tag{17.8.25}$$

since $\hat{H}_\ell^{(1)} = \hat{H}_\ell$. But then, by the general result, the operator $\hat{A}_\ell$ maps u_ℓ to an $\hat{H}_\ell^{(2)}$ eigenstate of the same eigenvalue:

$$\hat{H}_\ell^{(2)} \hat{A}_\ell u_\ell = \hat{A}_\ell u_\ell. \tag{17.8.26}$$

This means that $\hat{H}_{\ell+1} \hat{A}_\ell u_\ell = \hat{A}_\ell u_\ell$, and therefore $\hat{A}_\ell u_\ell$ is the desired eigenstate $u_{\ell+1}$:

$$u_{\ell+1} = \hat{A}_\ell \, u_\ell. \tag{17.8.27}$$

Using this relation iteratively, we find that

$$u_\ell = \hat{A}_{\ell-1} \hat{A}_{\ell-2} \cdots \hat{A}_0 \, u_0. \tag{17.8.28}$$

This is the complete solution, with

$$\hat{A}_\ell = \frac{d}{d\rho} + W_\ell = \frac{d}{d\rho} - \frac{\ell+1}{\rho} = \rho^{\ell+1} \frac{d}{d\rho} \frac{1}{\rho^{\ell+1}}, \tag{17.8.29}$$

where the last equality holds when $\frac{d}{d\rho}$ acts on everything to the right of it. This rewriting helps us simplify the product of ℓ different $\hat{A}$ operators appearing in (17.8.28). We have

$$\begin{aligned} u_\ell &= \Big(\rho^\ell \frac{d}{d\rho} \frac{1}{\rho^\ell}\Big)\Big(\rho^{\ell-1} \frac{d}{d\rho} \frac{1}{\rho^{\ell-1}}\Big) \cdots \Big(\rho^2 \frac{d}{d\rho} \frac{1}{\rho^2}\Big)\Big(\rho \frac{d}{d\rho} \frac{1}{\rho}\Big) u_0 \\ &= \rho^{\ell+1} \Big(\frac{1}{\rho}\frac{d}{d\rho}\Big)\Big(\frac{1}{\rho}\frac{d}{d\rho}\Big) \cdots \Big(\frac{1}{\rho}\frac{d}{d\rho}\Big)\Big(\frac{1}{\rho}\frac{d}{d\rho}\Big) \frac{u_0}{\rho} \\ &= \rho \cdot \rho^\ell \Big(\frac{1}{\rho}\frac{d}{d\rho}\Big)^\ell \frac{u_0}{\rho}. \end{aligned} \tag{17.8.30}$$

The solutions for u_0 are found by solving the differential equation $\hat{H}_0 u_0 = u_0$, which reads

$$-\frac{d^2 u_0}{d\rho^2} = u_0 \;\Rightarrow\; u_0 = c_1 \sin\rho + c_2 \cos\rho. \tag{17.8.31}$$

We have two choices for u_0, the sine and cosine functions. The sine function is consistent with the expectation that $u_\ell \sim \rho^{\ell+1}$ and leads to a regular wave function at $r=0$. The cosine function does not, but it is still of interest for solutions holding away from the origin. Based on the final form in equation (17.8.30), one conventionally defines the spherical Bessel functions $j_\ell(\rho)$ and $n_\ell(\rho)$ using the sine and cosine u_0 solutions as follows:

$$\begin{aligned} j_\ell(\rho) &\equiv (-1)^\ell \rho^\ell \Big(\frac{1}{\rho}\frac{d}{d\rho}\Big)^\ell \frac{\sin\rho}{\rho}, \\ n_\ell(\rho) &\equiv (-1)^{\ell+1} \rho^\ell \Big(\frac{1}{\rho}\frac{d}{d\rho}\Big)^\ell \frac{\cos\rho}{\rho}. \end{aligned} \tag{17.8.32}$$

In particular, this gives

$$\begin{aligned} j_0(\rho) &= \frac{\sin\rho}{\rho}, & n_0(\rho) &= -\frac{\cos\rho}{\rho}, \\ j_1(\rho) &= \frac{\sin\rho}{\rho^2} - \frac{\cos\rho}{\rho}, & n_1(\rho) &= -\frac{\cos\rho}{\rho^2} - \frac{\sin\rho}{\rho}. \end{aligned} \tag{17.8.33}$$

The general solution for u_ℓ is therefore

$$u_\ell(\rho) = c_1\,\rho j_\ell(\rho) + c_2\,\rho\, n_\ell(\rho), \tag{17.8.34}$$

with c_1 and c_2 arbitrary constants. These solutions, for all values of ℓ, are required for solving central potential problems (section 19.5) and for the derivation of Rayleigh's formula (section 19.6), a key tool in scattering theory.

Problems

Problem 17.1. *Shifted harmonic oscillator.*

A quantum harmonic oscillator perturbed by a constant force of magnitude F in the positive x-direction is described by the Hamiltonian

$$\hat{H} = \frac{\hat{p}^2}{2m} + \frac{1}{2}m\omega^2\hat{x}^2 - F\hat{x}.$$

Note that given $[\hat{x}, \hat{p}] = i\hbar$ we also have $[\hat{x} - x_0, \hat{p}] = i\hbar$, for any constant x_0, demonstrating that $\hat{y} \equiv \hat{x} - x_0$ and $\hat{p}$ form a pair of conjugate variables.

1. Find the ground state energy of $\hat{H}$. What is $\langle \hat{x} \rangle$ in the ground state?
2. The ground state $|0'\rangle$ of $\hat{H}$ can be written as $|0'\rangle = Ne^{\alpha\hat{a}^\dagger}|0\rangle$, where $\hat{a}^\dagger$ and $|0\rangle$ are, respectively, the raising operator and ground state of the *unperturbed* $F = 0$ Hamiltonian. Find the real number α. [Hint: Consider operators $\hat{a}_y$ and $\hat{a}_y^\dagger$ based on $\hat{y}$ and $\hat{p}$.]

Problem 17.2. *Wave function for a coherent state.*

Consider the normalized coherent state $|\alpha\rangle = e^{\alpha\hat{a}^\dagger - \alpha^*\hat{a}}|0\rangle$, with

$$\alpha = \frac{x_0}{\sqrt{2}L_0} + i\frac{p_0 L_0}{\sqrt{2}\hbar}, \quad L_0 = \sqrt{\frac{\hbar}{m\omega}}, \quad x_0, p_0 \in \mathbb{R}.$$

Calculate the wave function $\psi_\alpha(x) = \langle x|\alpha\rangle$. Your answer for this wave function should come out manifestly normalized and can be written in terms of the ground state wave function.

Problem 17.3. *Coherent states of the harmonic oscillator.*

This problem is a review of coherent states. Most of the questions are answered in the previous pages. To maximize your learning, try solving the problem without looking back!

For arbitrary $\alpha \in \mathbb{C}$, consider the coherent state $|\alpha\rangle \equiv e^{\alpha\hat{a}^\dagger - \alpha^* a}|0\rangle$.

1. Show that $\hat{a}|\alpha\rangle = \alpha|\alpha\rangle$.
2. Write the state $|\alpha\rangle$ as a superposition of energy eigenstates $|n\rangle$. What is the probability that a particle in the state $|\alpha\rangle$ has energy E_n?
3. Show that $|\alpha\rangle$ and $|\beta\rangle$, with α, β arbitrary, are *not* orthogonal by calculating $\langle\beta|\alpha\rangle$.
4. Evaluate $\Delta H/\langle H\rangle$ on the state $|\alpha\rangle$ and show that it decreases as $|\alpha|$ increases.
5. Evaluate $\langle\alpha|\hat{x}|\alpha\rangle$ and $\langle\alpha|\hat{p}|\alpha\rangle$. Evaluate Δx and Δp for the state $|\alpha\rangle$ and show that this is a minimum uncertainty state.

6. Expand the $t=0$ state $|\alpha\rangle$ as a superposition of energy eigenstates to write an expression for the state at all subsequent time $t>0$. Show that the state continues to be a coherent state for some value of $\alpha(t)$ that you must determine.
7. Compute the time-dependent expectation values $\langle\hat{x}\rangle$, $\langle\hat{p}\rangle$, $\langle\hat{H}\rangle$, Δx, and Δp, assuming that at $t=0$ we have a coherent state $|\alpha_0\rangle$ with α_0 real.

Problem 17.4. *Squeezed states from an instantaneous change of the Hamiltonian.*

Consider a harmonic oscillator Hamiltonian $\hat{H}_1$ with parameters m_1, ω_1 leading to an oscillator size $L_0^{(1)}$. This Hamiltonian governs the dynamics of a particle for $t<0$. The Hamiltonian changes discontinuously at $t=0$ such that for all $t>0$ the new Hamiltonian $\hat{H}_2$ has parameters m_2, ω_2 leading to an oscillator size $L_0^{(2)}$. Assume that at $t=0^-$, the state of the system is the ground state $|0\rangle_1$ of $\hat{H}_1$. Since the Hamiltonian changes instantaneously, the state of the system at $t=0^+$ is still $|0\rangle_1$. This is not, however, the ground state of $\hat{H}_2$. We will see that $|0\rangle_1$ can be viewed as a squeezed state of $\hat{H}_2$. For this we will express $|0\rangle_1$ in terms of the ground state $|0\rangle_2$ and creation operators of $\hat{H}_2$.

1. The $\hat{x}$ and $\hat{p}$ operators are not changed in the transition. Use this to show that the creation and annihilation operators $(\hat{a}_1^\dagger, \hat{a}_1)$ of $\hat{H}_1$ are related to the creation and annihilation operators $(\hat{a}_2^\dagger, \hat{a}_2)$ of $\hat{H}_2$ as follows:

$$\begin{aligned} \hat{a}_1 &= \hat{a}_2 \cosh\gamma \, + \, \hat{a}_2^\dagger \sinh\gamma, \\ \hat{a}_1^\dagger &= \hat{a}_2 \sinh\gamma \, + \, \hat{a}_2^\dagger \cosh\gamma, \quad \text{with } e^\gamma = \frac{L_0^{(2)}}{L_0^{(1)}} = \sqrt{\frac{m_1\omega_1}{m_2\omega_2}}. \end{aligned}$$

2. We write an ansatz for $|0\rangle_1$ in terms of $|0\rangle_2$. Since both states have wave functions that are even under $x \to -x$, we expect $|0\rangle_1$ to involve $|0\rangle_2$ acted upon by even numbers of creation operators. We thus write

$$|0\rangle_1 = \mathcal{N}(\gamma)\exp\left(-\tfrac{1}{2} f(\gamma)\,\hat{a}_2^\dagger\hat{a}_2^\dagger\right)|0\rangle_2,$$

with $f(\gamma)$ and $\mathcal{N}(\gamma)$ unknown. Use the condition $\hat{a}_1|0\rangle_1=0$ to show that $f(\gamma)=\tanh\gamma$.

3. Calculate $\mathcal{N}(\gamma)$ by first noting that $\mathcal{N}(\gamma)={}_2\langle 0|0\rangle_1$ and then introducing a complete set of states $|x\rangle$ in between the two vacua. Show that $\mathcal{N}(\gamma)=1/\sqrt{\cosh\gamma}$. All in all,

$$|0\rangle_1 = \frac{1}{\sqrt{\cosh\gamma}}\exp\left(-\tfrac{1}{2}\tanh\gamma\,\hat{a}_2^\dagger\hat{a}_2^\dagger\right)|0\rangle_2. \tag{1}$$

The state on the right-hand side is a squeezed state of the theory with $\hat{H}_2$ Hamiltonian.

4. Verify that the position uncertainty Δx of the squeezed state (1) can be written as

$$\Delta x = e^{-\gamma}\frac{L_0^{(2)}}{\sqrt{2}}, \tag{2}$$

implying that, on the squeezed state, the position uncertainty is that of the $\hat{H}_2$ ground state multiplied by $e^{-\gamma}$. For $\gamma>0$, the state is literally squeezed.

5. More generally, dropping the reference to systems one and two, we define the squeezed vacuum $|0_\gamma\rangle$ of the harmonic oscillator with size L_0 and operators $(\hat{a}^\dagger, \hat{a})$:

$$|0_\gamma\rangle \equiv \frac{1}{\sqrt{\cosh\gamma}} \exp(-\tfrac{1}{2} \tanh\gamma\, \hat{a}^\dagger\hat{a}^\dagger)\,|0\rangle. \tag{3}$$

The earlier results should make clear that $\Delta x = e^{-\gamma}\frac{L_0}{\sqrt{2}}$ and that

$$(\hat{a}\cosh\gamma + \hat{a}^\dagger \sinh\gamma)|0_\gamma\rangle = 0. \tag{4}$$

Now we claim that $|0_\gamma\rangle$ can be built by acting with a unitary operator $S(\gamma)$ on the vacuum:

$$|0_\gamma\rangle = S(\gamma)\,|0\rangle, \quad \text{with} \quad S(\gamma) = \exp(-\tfrac{\gamma}{2}(\hat{a}^\dagger\hat{a}^\dagger - \hat{a}\hat{a})), \quad \gamma \in \mathbb{R}. \tag{5}$$

To prove this relation, consider the following steps.

a. Calculate the operator $\hat{a}(\gamma)$ defined by

$$\hat{a}(\gamma) = S^\dagger(\gamma)\,\hat{a}\,S(\gamma).$$

To do this, derive a second-order differential equation for $\hat{a}(\gamma)$. Also write the corresponding expression for $\hat{a}^\dagger(\gamma) \equiv (\hat{a}(\gamma))^\dagger$, and check the value of the commutator $[\hat{a}(\gamma),\, \hat{a}^\dagger(\gamma)]$.

b. Show that $S(\gamma)|0\rangle$ is annihilated by $\hat{a}\cosh\gamma + \hat{a}^\dagger \sinh\gamma$, as it should be, on account of (4) (note that $\hat{a}(\gamma)$ does not kill the squeezed vacuum $S(\gamma)|0\rangle$).

c. Explain why any state annihilated by $\hat{a}\cosh\gamma + \hat{a}^\dagger \sinh\gamma$ is unique up to normalization. Conclude that $|0_\gamma\rangle$ and $S(\gamma)|0\rangle$ must be the same. In summary,

$$|0_\gamma\rangle = \frac{1}{\sqrt{\cosh\gamma}} \exp(-\tfrac{1}{2} \tanh\gamma\, \hat{a}^\dagger\hat{a}^\dagger)\,|0\rangle = \exp(-\tfrac{\gamma}{2}(\hat{a}^\dagger\hat{a}^\dagger - \hat{a}\hat{a}))|0\rangle. \tag{6}$$

Problem 17.5. *More general squeezed states.*

In problem 17.4 we considered the squeezed vacuum state $|0_\gamma\rangle = S(\gamma)\,|0\rangle$, with $\gamma \in \mathbb{R}$, and $S(\gamma) = \exp(-\frac{\gamma}{2}(\hat{a}^\dagger\hat{a}^\dagger - \hat{a}\hat{a}))$. For coherent states $|\alpha\rangle = D(\alpha)|0\rangle$, $\alpha \in \mathbb{C}$ and $D(\alpha) = \exp(\alpha\,\hat{a}^\dagger - \alpha^*\hat{a})$. More general squeezed states $|\alpha, \gamma\rangle$ arise by first squeezing and then translating

$$|\alpha, \gamma\rangle \;\equiv\; D(\alpha)S(\gamma)|0\rangle.$$

Note that $|0, \gamma\rangle = |0_\gamma\rangle$, and $|\alpha, 0\rangle = |\alpha\rangle$. In problem 17.4 you also calculated $\hat{a}(\gamma) = S^\dagger(\gamma)\,\hat{a}\,S(\gamma)$.

1. Calculate the expectation value $\langle\hat{N}\rangle$ of the number operator $\hat{N} = \hat{a}^\dagger\hat{a}$ and its uncertainty ΔN on the squeezed vacuum state $|0_\gamma\rangle$. The answers should just be functions of γ. Find the ratio $\Delta N/\langle N\rangle$, and plot it as a function of γ. Can this ratio be made small?
2. Calculate the expectation value $\langle\hat{N}\rangle$ on the generalized state $|\alpha, \gamma\rangle$.
3. For a single-mode electromagnetic field of frequency ω, the time-dependent part of the electric field operator in (17.4.10) takes the form $\widehat{E}(t) = \mathcal{E}_0(\hat{a}e^{-i\omega t} + \hat{a}^\dagger e^{i\omega t})$, with $\mathcal{E}_0$ a real constant. Calculate the expectation value $\langle\widehat{E}(t)\rangle$ and the uncertainty $\Delta E(t)$ on the state $|\alpha, \gamma\rangle$.

Problem 17.6. *Position and momentum eigenstates in the harmonic oscillator.*

A position state $|x\rangle$ in the harmonic oscillator can be viewed as a generalized squeezed state: we first squeeze the ground state to zero width and then displace it to x.

1. Show that for $\gamma \to \infty$ the squeezed vacuum $|0_\infty\rangle \sim \exp(-\frac{1}{2}\hat{a}^\dagger\hat{a}^\dagger)|0\rangle$ is annihilated by $\hat{x}$, suggesting its wave function is proportional to $\delta(x)$.
2. We claim that the position state $|x\rangle$ is given by

$$|x\rangle = \varphi_0(x)\exp\Big(\sqrt{2}\frac{x}{L_0}\hat{a}^\dagger - \frac{1}{2}\hat{a}^\dagger\hat{a}^\dagger\Big)|0\rangle.$$

   Here $\varphi_0(x)$ is the ground state wave function. Confirm the claim by showing that $\hat{x}|x_0\rangle = x_0|x_0\rangle$ and that the normalization is correct because $\langle 0|x\rangle$ takes the expected value.
3. Use the expression for $|x\rangle$ to calculate the wave function for the excited state $|2\rangle = \frac{1}{\sqrt{2}}a^\dagger a^\dagger|0\rangle$.
4. Show that for $\gamma \to -\infty$ the squeezed vacuum $|0_{-\infty}\rangle \sim \exp(\frac{1}{2}\hat{a}^\dagger\hat{a}^\dagger)|0\rangle$ is annihilated by $\hat{p}$, suggesting its momentum space wave function is proportional to $\delta(p)$.
5. Explicitly construct the properly normalized *momentum* eigenstates $|p\rangle$ of the harmonic oscillator. These states satisfy the familiar property $\hat{p}|p\rangle = p|p\rangle$.

Problem 17.7. *Maxwell's equations and photon states.*

In Gaussian units and in vacuum, Maxwell's equations read

$$\nabla\cdot\mathbf{E} = 0, \quad \nabla\cdot\mathbf{B} = 0, \quad \nabla\times\mathbf{E} = -\frac{1}{c}\frac{\partial\mathbf{B}}{\partial t}, \quad \nabla\times\mathbf{B} = \frac{1}{c}\frac{\partial\mathbf{E}}{\partial t}.$$

1. Examine the cavity electromagnetic fields in (17.4.2), and derive the conditions on $q(t)$ and $p(t)$ implied by Maxwell's equations.
2. Derive the Heisenberg equations of motion for $\hat{q}(t)$ and $\hat{p}(t)$ using the Hamiltonian $\hat{H}$ in (17.4.4). Show that they coincide with the equations obtained in (1). Clearly, $\hat{H}$ provides a good quantum description of the classical electromagnetic mode in the cavity.

Problem 17.8. *Altering an oscillation.*

Consider a two-state system with basis states $|1\rangle$ and $|2\rangle$ and a Hamiltonian $\hat{H}$ given by

$$\hat{H} = \begin{pmatrix} 0 & -\Delta \\ -\Delta & 0 \end{pmatrix} = -\Delta\,\sigma_1, \text{ with } \Delta > 0.$$

If the system is initially in state $|1\rangle$, you can quickly confirm that the probability of the state remaining in $|1\rangle$ varies periodically between one and zero as a function of time.

Add to $\hat{H}$ a time-independent term along σ_3 so that instead the probability of the system being in state $|1\rangle$ varies periodically between one and some minimum value $p_{min} > 0$. Write the new term in $\hat{H}$ in terms of Δ and p_{min}. [Hint: it might be useful to write $\hat{H}$ as the dot product of a vector with the spin operator and to think about time evolution as precession.]

Problem 17.9. *The ammonia molecule.*

The ammonia molecule NH_3 has the shape of a flattened tetrahedron, with three hydrogen atoms at the base and a nitrogen atom on top. This molecule is viewed as a two-state system: in the first basis state, called $|\Uparrow\rangle$, the nitrogen is up relative to the hydrogen base; in the second basis state, called $|\Downarrow\rangle$, the nitrogen is down relative to the hydrogen base. To a first approximation, these two states are degenerate eigenstates of energy E_0. Choosing basis states $|1\rangle = |\Uparrow\rangle$ and $|2\rangle = |\Downarrow\rangle$, the Hamiltonian is

$$\hat{H}^{(1)} = \begin{pmatrix} E_0 & 0 \\ 0 & E_0 \end{pmatrix}.$$

The nitrogen atom is in fact subject to a potential with critical points corresponding to the up and down positions and a finite barrier in between. To a second approximation, we incorporate the barrier by allowing tunneling between the up and down states:

$$\hat{H}^{(2)} = \begin{pmatrix} E_0 & -\Delta \\ -\Delta & E_0 \end{pmatrix}, \quad \Delta > 0, \tag{1}$$

where Δ, with units of energy, was chosen to be positive for convenience.

1. Determine the ground state $|G\rangle$ and the excited state $|E\rangle$ of $\hat{H}^{(2)}$. What are their energies? Let z be the axis perpendicular to the plane of the hydrogen atoms, and let $\psi_\Uparrow(z) = \langle z|\Uparrow\rangle$ be a real wave function peaked at some positive z_0, describing nitrogen as up. Similarly, let $\psi_\Downarrow(z) = \langle z|\Downarrow\rangle$ be a real wave function peaked at $-z_0$, describing nitrogen as down. Sketch these wave functions as well as the wave functions corresponding to the ground and excited states.
2. Assume at $t=0$ the state of the nitrogen is $|\Uparrow\rangle$. Calculate the time-dependent probabilities $P_\Uparrow(t)$ and $P_\Downarrow(t)$ of finding the nitrogen up and down, respectively.

   A photon emitted in a transition between the two levels of the ammonia molecule has wavelength $\lambda \simeq 1.256\,\text{cm}$. What is its frequency in GHz? What is the energy difference between the levels? Starting up at $t=0$, how long does it take for the nitrogen atom to be again up with probability one?

Problem 17.10. *The ammonia molecule in an external electric field.*

Let us now consider the electrostatic properties of the ammonia molecule discussed in problem 17.9. The electrons tend to cluster toward the nitrogen, leaving the nitrogen vertex slightly negative and the hydrogen plane slightly positive. As a result, we get an electric dipole moment that points down when the nitrogen is up and, conversely, points up when the nitrogen is down.

The energy E of a dipole in an external electric field gets a negative contribution when the dipole and the field are aligned and a positive contribution when they are anti-aligned. The contribution is given by the product of the magnitudes of the dipole moment and the electric field. With an electric field of magnitude ε along the positive z-axis and a dipole of magnitude μ, the state $|\Uparrow\rangle$ gets an extra positive contribution $\mu\varepsilon$ to the energy, while the state $|\Downarrow\rangle$ gets an extra negative contribution $-\mu\varepsilon$ to the energy.

The new Hamiltonian, including the effects of the electric field, follows from that in problem 17.9, equation (1). Since it does not affect the dynamics, we set the constant E_0 to zero and find that

$$\hat{H} = \begin{pmatrix} \mu\varepsilon & -\Delta \\ -\Delta & -\mu\varepsilon \end{pmatrix}.$$

Recall that $\Delta > 0$ and that we are using the basis $|1\rangle = |\Uparrow\rangle$, $|2\rangle = |\Downarrow\rangle$.

1. Show that the Hamiltonian may be written as

   $$\hat{H} = -K\mathbf{n}\cdot\boldsymbol{\sigma}$$

   for some constant $K > 0$ and some unit vector $\mathbf{n}$ defined by polar angles (θ, ϕ) that you should specify in terms of variables in the original Hamiltonian.
2. Find the energy eigenvalues and eigenstates. Write the eigenstates using the $|\Uparrow\rangle$, $|\Downarrow\rangle$ basis and the angles θ and ϕ.
3. Consider the cases $\mu\varepsilon \ll \Delta$, or effectively $\mu\varepsilon = 0$, and $\mu\varepsilon \gg \Delta$, or effectively $\Delta = 0$. In each case determine the values of θ, the energy eigenvalues, and the associated eigenstates. Discuss your results.
4. Make a plot of the two energy eigenvalues as a function of ε. Include on your plot the limits you considered above.
5. Initially, the electric field is turned off, $\varepsilon = 0$, and the system starts out in the state $|\psi(0)\rangle = |\Downarrow\rangle$. Describe the evolution of the state viewed as a spin state rotating in time. At some later time, we turn on a very strong electric field so that $\mu\varepsilon \gg \Delta$. What is the further evolution of the state? What does turning on this strong electric field do to the probability of finding the system in the state $|\Uparrow\rangle$?

Electric fields with gradients can be used to isolate ammonia molecules in the excited state $|E\rangle$. If these molecules enter and exit a cavity tuned to 23.9 GHz, with the right velocity, the molecules will transition to the ground state $|G\rangle$, and a coherent electromagnetic field will build up in the cavity. This is the operating principle of the ammonia *maser*.

Problem 17.11. *Time evolution in a three-state system.*

Carbon dioxide is a linear molecule (OCO) that can pick up an extra electron and become a negatively charged ion. Suppose the electron would have energy E_O if it were attached to either oxygen atom or energy E_C if it were attached to the carbon atom in the middle. Call these states $|L\rangle$, $|C\rangle$, and $|R\rangle$, for left oxygen, carbon, and right oxygen. The energy eigenstates need not, however, have either energy E_O or E_C because there is some probability that the electron may hop between an oxygen atom and the carbon atom. Assume that the probability of jumping directly from oxygen to oxygen can be neglected.

1. Write down a model Hamiltonian to describe the physics of the electron outlined above. You will need to introduce a parameter Δ, with units of energy, that characterizes the tunneling between C and O. Find the energy eigenvalues of this three-state system in terms of E_C, E_O, and Δ.

2. In this part and the next, assume $E_C = E_O$. Find the energy eigenstates.
3. Assume that at time $t=0$ the electron is in state $|L\rangle$; that is, it is localized on the left oxygen atom. What is the probability that at some later time t the electron will be in state $|L\rangle$? In state $|C\rangle$? In state $|R\rangle$? Plot these three probabilities as functions of time.

Problem 17.12. *The evolution of the spin in nuclear magnetic resonance.*

A spin one-half particle with magnetic moment $\hat{\boldsymbol{\mu}} = \gamma \hat{\mathbf{S}}$ is placed in the time-dependent magnetic field of (17.6.1), tuned for resonance:

$$\mathbf{B}(t) = B_0\mathbf{z} + B_1(\mathbf{x}\cos\omega_0 t - \mathbf{y}\sin\omega_0 t), \quad \omega_0 = \gamma B_0.$$

Use the general time-dependent solution (17.6.17) to answer the following questions.

1. Assuming that, initially, the spin points in the positive z-direction, $|\Psi, 0\rangle = |+\rangle$, give $|\Psi, t\rangle$, and describe it as a spin pointing in a time-dependent direction $\mathbf{n}(t)$ specified by the time-dependent angles $\theta(t)$ and $\phi(t)$ that you determine.
2. Use your answer above to eliminate t, and write ϕ as a function of θ. Where does the spin point to as it lies for the first time along the (x, y) plane.

Problem 17.13. *Factorization method and Pöschl-Teller Hamiltonians.*

With a unit-free coordinate x, the Pöschl-Teller Hamiltonians are defined to be

$$\hat{H}_n = -\frac{d^2}{dx^2} - n(n+1)\,(\text{sech}\, x)^2, \quad n = 0, 1, 2, \ldots.$$

Recall that $\text{sech}\, x = 1/\cosh x$. Sketch the potential $V_n(x)$ associated to $\hat{H}_n$. We wish to find the bound state spectrum of $\hat{H}_n$ for all integers $n \geq 0$.

1. Consider the functions $W_n = n \tanh x$. Construct the associated Hamiltonians $\hat{H}_n^{(1)}$ and $\hat{H}_n^{(2)}$, and verify that for $n = 0, 1, \ldots$ we have
$$\hat{H}_n^{(1)} = \hat{H}_{n+1}^{(2)} - (2n+1). \qquad (1)$$
Relate $\hat{H}_n^{(1)}$ and $\hat{H}_n^{(2)}$ to the Pöschl-Teller Hamiltonians $\hat{H}_n$ and $\hat{H}_{n-1}$, respectively.
2. Show that $\hat{H}_n^{(1)}$, for $n \geq 1$, has a zero-energy bound state. Find its wave function $\psi_{0,n}$ (up to normalization).
3. Use the existence of the $\hat{H}_1^{(1)}$ ground state and equation (1) to conclude that $\hat{H}_2^{(1)}$ must have a bound state with a particular positive energy. What is that energy?
4. Explain why all eigenstates of $\hat{H}_0^{(1)}$ have energy $E \geq 0$, and none is a bound state. Show this means that $\hat{H}_1^{(1)}$ cannot have a second bound state (whose energy would have to be between 0 and 1).
5. Use the fact that $\hat{H}_1^{(1)}$ has only one bound state to show that $\hat{H}_n^{(1)}$, with $n \geq 1$, has precisely n bound states, which we can label as $\psi_{k,n}$ with $k = 0, \ldots, n-1$. Find the energy eigenvalues $E_{k,n}^{(1)}$. Write an expression for the wave function $\psi_{k,n}$ in the form of differential operator(s) acting on ground state wave functions. [Hint: It may help to figure out explicitly the bound states of $\hat{H}_2^{(1)}, \hat{H}_3^{(1)}$, and $\hat{H}_4^{(1)}$ so that the pattern becomes clear.]
6. What are all the bound state energies of the Pöschl-Teller Hamiltonians $\hat{H}_n$?

18 Multiparticle States and Tensor Products

We introduce the tensor product $V \otimes W$ of two vector spaces V and W as the state space of two particles, that separately have state spaces V and W, respectively. Operators and inner products on $V \otimes W$ arise from operators and inner products on V and W. Entangled states in $V \otimes W$ are states that can't be factorized into a state in V and a state in W. We construct an entangled state of two spin one-half particles that has zero total angular momentum. We introduce Bell states, which are four entangled states that form a basis for the state space of two spin one-half particles. We explain how quantum teleportation of a spin state works with the help of an entangled pair. We discuss the local realism claims of Einstein, Podolsky, and Rosen (EPR) and the surprising discovery of Bell inequalities, which made the claims of EPR testable and experimentally refuted. Finally, we show it is impossible to construct a machine capable of cloning arbitrary quantum states of any system.

18.1 Introduction to the Tensor Product

In this section we develop the tools needed to describe a system that contains more than one particle. Most of the new ideas appear when we consider systems with two particles. We will assume the particles are distinguishable. For indistinguishable particles quantum mechanics imposes some additional constraints on the allowed set of states. We will study those constraints in chapter 21. The same tools apply to the case of one particle, if such particle has a number of independent degrees of freedom. The material we are about to develop will be needed to understand the addition of angular momenta. In that problem one may be adding the angular momenta of two or more particles in a system or, alternatively, adding the independent angular momenta of a single particle.

Consider two particles. Below we list the state space and the operators associated with each particle:

- Particle 1: Its states are elements of a complex vector space V. In this space we have operators $T_1, T_2, \ldots$.

- Particle 2: Its states are elements of a complex vector space W. In this space we have operators $S_1, S_2, \ldots$.

This list of operators for each particle may include some or many of the operators you are already familiar with: position, momentum, spin, Hamiltonians, projectors, and so on.

Once we have two particles, the two of them *together* form our system. We are after the description of quantum states of this two-particle system. On first thought, we may believe that any state of this system should be described by giving the state $v \in V$ of the first particle and the state $w \in W$ of the second particle. This information could be represented by the ordered list (v, w), where the first item is the state of the first particle and the second item the state of the second particle. This is *a* state of the two-particle system, but it is far from being the general state of the two-particle system. It misses remarkable new possibilities, as we shall soon see.

We thus introduce a new notation. Instead of representing the state of the two-particle system with particle 1 in v and particle 2 in w as (v, w), we will represent it as $v \otimes w$. This element $v \otimes w$ will be viewed as a vector in a new vector space $V \otimes W$ that contains the quantum states of the two-particle system. This $\otimes$ operation is called the **tensor product**. In this case we have two complex vector spaces, and the tensor product $V \otimes W$ is a new complex vector space:

$$v \otimes w \in V \otimes W \quad \text{when} \quad v \in V,\ w \in W. \tag{18.1.1}$$

In the tensor product $v \otimes w$, there is no multiplication to be carried out; we are just placing one vector to the left of $\otimes$ and another to the right of $\otimes$.

We have only described some elements of $V \otimes W$, not quite given its definition yet. We now explain two physically motivated rules that define the tensor product completely.

1. If the vector representing the state of the first particle is scaled by a complex number, this is equivalent to scaling the state of the two particles. The same holds true for the second particle. So we declare

$$\boxed{(av) \otimes w = v \otimes (aw) = a\,(v \otimes w), \qquad a \in \mathbb{C}.} \tag{18.1.2}$$

2. If the state of the first particle is a superposition of two states, the state of the two-particle system is also a superposition. We thus demand distributive properties for the tensor product:

$$\boxed{\begin{aligned} (v_1 + v_2) \otimes w &= v_1 \otimes w + v_2 \otimes w, \\ v \otimes (w_1 + w_2) &= v \otimes w_1 + v \otimes w_2. \end{aligned}} \tag{18.1.3}$$

The tensor product $V \otimes W$ is thus defined to be the vector space whose elements are (complex) linear combinations of elements of the form $v \otimes w$, with $v \in V, w \in W$, with the above rules for manipulation. The tensor product $V \otimes W$ is the complex vector space of states of the two-particle system!

Remarks:

1. The vector $0 \in V \otimes W$ is equal to $0 \otimes w$ or $v \otimes 0$. Indeed, with $a = 0$ we have $av = 0$, and the equality of the first and last term in (18.1.2) gives $0 \otimes w = 0(v \otimes w) = 0 \in V \otimes W$, since in any vector space the product of the number zero times any vector is the zero vector.
2. Let $v_1, v_2 \in V$ and $w_1, w_2 \in W$. Consider a vector in $V \otimes W$ built by superposition:

$$\alpha_1(v_1 \otimes w_1) + \alpha_2(v_2 \otimes w_2) \in V \otimes W, \quad \alpha_1, \alpha_2 \in \mathbb{C}. \tag{18.1.4}$$

This is a state of the two-particle system that, with α_1 and α_2 both nonzero, cannot be described by giving the state of the first particle and the state of the second particle. The above superpositions give rise to entangled states. An entangled state of the two particles is one that, roughly, cannot be described by giving the state of the first particle and the state of the second particle. We will make this precise soon.

If $(e_1, \ldots, e_m)$ is a basis of V and $(f_1, \ldots, f_n)$ is a basis of W, then the set of elements $e_i \otimes f_j$ where $i = 1, \ldots, m$ and $j = 1, \ldots, n$ forms a basis for $V \otimes W$:

$$\{\, e_i \otimes f_j\,;\ i = 1, \ldots, m,\ \ j = 1, \ldots, n.\} \text{ is a basis for } V \otimes W. \tag{18.1.5}$$

It is simple to see these span $V \otimes W$. First note that for any $v \otimes w$ we have $v = \sum_i v_i e_i$, and $w = \sum_j w_j f_j$ so that

$$v \otimes w = \Big(\sum_i v_i e_i\Big) \otimes \Big(\sum_j w_j f_j\Big) = \sum_{i,j} v_i w_j\, e_i \otimes f_j. \tag{18.1.6}$$

Since the basis spans elements of the form $v \otimes w$ for all v, w, it will span all linear superpositions of such elements, which is to say, it will span $V \otimes W$. With $n \cdot m$ basis vectors, the dimensionality of $V \otimes W$ is equal to the *product* of the dimensionalities of V and W:

$$\dim(V \otimes W) = \dim(V) \times \dim(W). \tag{18.1.7}$$

Dimensions are multiplied, not added, in a tensor product. The most general state in $V \otimes W$ takes the form

$$\sum_{i=1}^{m} \sum_{j=1}^{n} c_{ij}\, e_i \otimes f_j, \tag{18.1.8}$$

with $c_{ij} \in \mathbb{C}$ arbitrary numbers.

Example 18.1. *State space of two spin one-half particles.*
Consider two spin one-half particles. For the first particle, we have the state space V_1 with basis states $|+\rangle_1$ and $|-\rangle_1$. For the second particle, we have the state space V_2 with basis states $|+\rangle_2$ and $|-\rangle_2$. The tensor product $V_1 \otimes V_2$ has four basis vectors:

$$|+\rangle_1 \otimes |+\rangle_2; \quad |+\rangle_1 \otimes |-\rangle_2; \quad |-\rangle_1 \otimes |+\rangle_2; \quad |-\rangle_1 \otimes |-\rangle_2. \tag{18.1.9}$$

The most general state of the two-particle system is a linear superposition of the four basis states:

$$|\Psi\rangle = \alpha_1|+\rangle_1 \otimes |+\rangle_2 + \alpha_2|+\rangle_1 \otimes |-\rangle_2 + \alpha_3|-\rangle_1 \otimes |+\rangle_2 + \alpha_4|-\rangle_1 \otimes |-\rangle_2. \tag{18.1.10}$$

Here the α_i with $i=1,2,3,4$ are complex constants. If we follow the convention that the first ket corresponds to particle 1 and the second ket corresponds to particle 2, we need not write the subscripts, and the notation is simpler. The above state would read

$$|\Psi\rangle = \alpha_1|+\rangle \otimes |+\rangle + \alpha_2|+\rangle \otimes |-\rangle + \alpha_3|-\rangle \otimes |+\rangle + \alpha_4|-\rangle \otimes |-\rangle. \tag{18.1.11}$$

Using the properties of the tensor product, the state can be written in the form

$$|\Psi\rangle = |+\rangle \otimes (\alpha_1|+\rangle + \alpha_2|-\rangle) + |-\rangle \otimes (\alpha_3|+\rangle + \alpha_4|-\rangle). \tag{18.1.12}$$

Expanding out the products, we recover the original form. □

18.2 Operators on the Tensor Product Space

How do we construct operators that act in the vector space $V \otimes W$? Let T be an operator in V and S be an operator in W. In other words, $T \in \mathcal{L}(V)$, and $S \in \mathcal{L}(W)$. We can then construct an operator $T \otimes S$ acting on the tensor product:

$$T \otimes S \in \mathcal{L}(V \otimes W). \tag{18.2.1}$$

The operator is defined to act as follows. For any $v \in V$ and $w \in W$,

$$T \otimes S\,(v \otimes w) \equiv Tv \otimes Sw. \tag{18.2.2}$$

This is the only "natural" option: we let T act on the vector it knows how to act on and S act on the vector it knows how to act on. The identity operator in the tensor product is $\mathbb{1} \otimes \mathbb{1}$, the tensor product of the respective identity operators: the $\mathbb{1}$ to the left is the identity operator on V, and the $\mathbb{1}$ to the right is the identity operator on W. A general operator on $V \otimes W$ is a sum $\sum_i T_i \otimes S_i$ with $T_i \in \mathcal{L}(V)$, and $S_i \in \mathcal{L}(W)$. We will elaborate on this idea at the end of this section.

Suppose that we want the operator $T \in \mathcal{L}(V)$ that acts on the first particle to act on the tensor product $V \otimes W$, even though we have not supplied an operator S to act on the W part. The idea is to choose $S = \mathbb{1}$—namely, the identity operator. In this way we "upgrade" the operator T that acts on a single vector space to $T \otimes \mathbb{1}$ that acts on the tensor product:

$$T \in \mathcal{L}(V) \quad \Rightarrow \quad T \otimes \mathbb{1} \in \mathcal{L}(V \otimes W), \qquad T \otimes \mathbb{1}\,(v \otimes w) \equiv Tv \otimes w. \tag{18.2.3}$$

Similarly, an operator S belonging to $\mathcal{L}(W)$ is upgraded to $\mathbb{1} \otimes S$ to act on the tensor product. It is useful to realize that upgraded operators of the first particle *commute* with upgraded operators of the second particle. Indeed,

$$\begin{aligned} (T \otimes \mathbb{1}) \cdot (\mathbb{1} \otimes S)\,(v \otimes w) &= (T \otimes \mathbb{1})(v \otimes Sw) = Tv \otimes Sw, \\ (\mathbb{1} \otimes S) \cdot (T \otimes \mathbb{1})\,(v \otimes w) &= (\mathbb{1} \otimes S)\,(Tv \otimes w) = Tv \otimes Sw, \end{aligned} \tag{18.2.4}$$

and therefore for any S, T we have

$$[T\otimes \mathbb{1},\ \mathbb{1}\otimes S] = 0. \tag{18.2.5}$$

Given a system of two particles, we can construct a simple total Hamiltonian $\hat{H}_T$ describing no interactions by upgrading the single-particle Hamiltonians $\hat{H}_1$ and $\hat{H}_2$ and then adding them:

$$\hat{H}_T \equiv \hat{H}_1\otimes \mathbb{1} + \mathbb{1}\otimes \hat{H}_2. \tag{18.2.6}$$

The two terms in $\hat{H}_T$ commute with each other.

Exercise 18.1. *Convince yourself that for an arbitrary operator $\hat{A}$ we have*

$$\exp(\hat{A}\otimes \mathbb{1}) = (\exp\hat{A})\otimes \mathbb{1}, \quad \text{and} \quad \exp(\mathbb{1}\otimes\hat{A}) = \mathbb{1}\otimes(\exp\hat{A}). \tag{18.2.7}$$

Exercise 18.2. *Assume $\hat{H}_1$ and $\hat{H}_2$ are time independent. Convince yourself that the time-evolution operator for the two-particle Hamiltonian $\hat{H}_T$ above takes the product form*

$$\exp\Big(-\frac{i\hat{H}_T t}{\hbar}\Big) = \exp\Big(-\frac{i\hat{H}_1 t}{\hbar}\Big)\otimes \exp\Big(-\frac{i\hat{H}_2 t}{\hbar}\Big). \tag{18.2.8}$$

Example 18.2. *Spin angular momentum of a state of two spin one-half particles.*

Let us now find out how the total angular momentum operator acts on a state of two spin one-half particles. Consider, therefore, a general state $|\Psi\rangle$ of the two particles:

$$|\Psi\rangle = \alpha_1|+\rangle\otimes|+\rangle + \alpha_2|+\rangle\otimes|-\rangle + \alpha_3|-\rangle\otimes|+\rangle + \alpha_4|-\rangle\otimes|-\rangle, \tag{18.2.9}$$

with α_i, $i=1,\ldots,4$, complex constants. Recall that in each term on the above right-hand side the first ket corresponds to the first particle, and the second ket corresponds to the second particle. Consider now the *total* z-component of spin angular momentum. Roughly, the total angular momentum in the z-direction would be the sum of the z-components of each individual particle. However, we know better at this point—summing the two angular momenta really means constructing a new operator in the tensor product vector space:

$$\hat{S}_z^{\text{tot}} \equiv \hat{S}_z^{(1)}\otimes \mathbb{1} + \mathbb{1}\otimes\hat{S}_z^{(2)}. \tag{18.2.10}$$

We act with $\hat{S}_z^{\text{tot}}$ on the state $|\Psi\rangle$. The contributions from the two operators on the above right-hand side are

$$\begin{aligned}
(\hat{S}_z^{(1)}\otimes\mathbb{1})|\Psi\rangle &= \alpha_1\hat{S}_z|+\rangle\otimes|+\rangle + \alpha_2\hat{S}_z|+\rangle\otimes|-\rangle + \alpha_3\hat{S}_z|-\rangle\otimes|+\rangle + \alpha_4\hat{S}_z|-\rangle\otimes|-\rangle \\
&= \tfrac{\hbar}{2}\big(\alpha_1|+\rangle\otimes|+\rangle + \alpha_2|+\rangle\otimes|-\rangle - \alpha_3|-\rangle\otimes|+\rangle - \alpha_4|-\rangle\otimes|-\rangle\big), \\
(\mathbb{1}\otimes\hat{S}_z^{(2)})|\Psi\rangle &= \alpha_1|+\rangle\otimes\hat{S}_z|+\rangle + \alpha_2|+\rangle\otimes\hat{S}_z|-\rangle + \alpha_3|-\rangle\otimes\hat{S}_z|+\rangle + \alpha_4|-\rangle\otimes\hat{S}_z|-\rangle \\
&= \tfrac{\hbar}{2}\big(\alpha_1|+\rangle\otimes|+\rangle - \alpha_2|+\rangle\otimes|-\rangle + \alpha_3|-\rangle\otimes|+\rangle - \alpha_4|-\rangle\otimes|-\rangle\big).
\end{aligned}$$

Adding these together, we have

$$\hat{S}_z^{\text{tot}}|\Psi\rangle = \hbar\big(\alpha_1|+\rangle_1\otimes|+\rangle_2 - \alpha_4|-\rangle_1\otimes|-\rangle_2\big). \tag{18.2.11}$$

One can derive this result quickly by noting that since $\hat{S}_z^{(1)}$ is diagonal in the basis for V_1 and $\hat{S}_z^{(2)}$ is diagonal in the basis for V_2, the total $\hat{S}_z^{\text{tot}}$ is diagonal in the tensor product basis. As a result, its eigenvalues on these basis states are the sum of the $\hat{S}_z$ eigenvalues for particle 1 and particle 2. Thus,

$$\begin{aligned}
\hat{S}_z^{\text{tot}}|+\rangle\otimes|+\rangle &= \left(\tfrac{\hbar}{2}+\tfrac{\hbar}{2}\right)|+\rangle\otimes|+\rangle = \hbar\,|+\rangle\otimes|+\rangle,\\
\hat{S}_z^{\text{tot}}|+\rangle\otimes|-\rangle &= \left(\tfrac{\hbar}{2}-\tfrac{\hbar}{2}\right)|+\rangle\otimes|-\rangle = 0,\\
\hat{S}_z^{\text{tot}}|-\rangle\otimes|+\rangle &= \left(-\tfrac{\hbar}{2}+\tfrac{\hbar}{2}\right)|-\rangle\otimes|+\rangle = 0,\\
\hat{S}_z^{\text{tot}}|-\rangle\otimes|-\rangle &= \left(-\tfrac{\hbar}{2}-\tfrac{\hbar}{2}\right)|-\rangle\otimes|-\rangle = -\hbar\,|-\rangle\otimes|-\rangle.
\end{aligned} \tag{18.2.12}$$

The result in (18.2.11) follows quickly from the four relations above. Suppose we are only interested in states in the tensor product that have zero $\hat{S}_z^{\text{tot}}$ or, equivalently, states $|\Psi\rangle$ that satisfy $\hat{S}_z^{\text{tot}}|\Psi\rangle=0$. This requires that

$$\alpha_1=\alpha_4=0 \quad\Rightarrow\quad |\Psi\rangle=\alpha_2|+\rangle\otimes|-\rangle+\alpha_3|-\rangle\otimes|+\rangle. \tag{18.2.13}$$

This is the *most general* state in the tensor product of two spin one-half particles that has zero total spin angular momentum in the z-direction. □

Example 18.3. *State of two spin one-half particles with zero total spin.*

We found in the previous example that the general state with zero total S_z^{tot} is

$$|\Psi\rangle=\alpha_2|+\rangle\otimes|-\rangle+\alpha_3|-\rangle\otimes|+\rangle, \tag{18.2.14}$$

with α_2 and α_3 arbitrary complex constants. We now calculate the total x-component $\hat{S}_x^{\text{tot}}$ of spin angular momentum on the above states. For this we recall that $\hat{S}_x|\pm\rangle=\frac{\hbar}{2}|\mp\rangle$, and we write

$$\hat{S}_x^{\text{tot}}=\hat{S}_x\otimes\mathbb{1}+\mathbb{1}\otimes\hat{S}_x. \tag{18.2.15}$$

The calculation proceeds as follows:

$$\begin{aligned}
\hat{S}_x^{\text{tot}}|+\rangle\otimes|-\rangle &= \hat{S}_x|+\rangle\otimes|-\rangle + |+\rangle\otimes\hat{S}_x|-\rangle = \tfrac{\hbar}{2}\big(|-\rangle\otimes|-\rangle + |+\rangle\otimes|+\rangle\big),\\
\hat{S}_x^{\text{tot}}|-\rangle\otimes|+\rangle &= \hat{S}_x|-\rangle\otimes|+\rangle + |-\rangle\otimes\hat{S}_x|+\rangle = \tfrac{\hbar}{2}\big(|+\rangle\otimes|+\rangle + |-\rangle\otimes|-\rangle\big).
\end{aligned} \tag{18.2.16}$$

Therefore,

$$\begin{aligned}
\hat{S}_x^{\text{tot}}|\Psi\rangle &= \alpha_2\,\tfrac{\hbar}{2}\big(|-\rangle\otimes|-\rangle+|+\rangle\otimes|+\rangle\big)+\alpha_3\,\tfrac{\hbar}{2}\big(|+\rangle\otimes|+\rangle+|-\rangle\otimes|-\rangle\big)\\
&= \tfrac{\hbar}{2}\,(\alpha_2+\alpha_3)\big(|+\rangle\otimes|+\rangle+|-\rangle\otimes|-\rangle\big).
\end{aligned} \tag{18.2.17}$$

If we demand that $\hat{S}_x^{\text{tot}}$ is also zero acting on the state $|\Psi\rangle$ we must have $\alpha_2=-\alpha_3$. Thus, the following state is the unique state with zero $\hat{S}_x^{\text{tot}}$ and $\hat{S}_z^{\text{tot}}$:

$$|\Psi\rangle = \alpha\Big(|+\rangle\otimes|-\rangle - |-\rangle\otimes|+\rangle\Big). \tag{18.2.18}$$

As it turns out, this state also has zero $\hat{S}_y^{\text{tot}}$ (exercise below). Since all three operators $\hat{S}_z^{\text{tot}}, \hat{S}_x^{\text{tot}}$, and $\hat{S}_y^{\text{tot}}$ annihilate $|\Psi\rangle$, the state has zero total spin angular momentum. It is a state of two spin one-half particles in which the spin angular momenta add up to zero. □

Exercise 18.3. *Verify that the state (18.2.18) satisfies $\hat{S}_y^{\text{tot}}|\Psi\rangle=0$.*

Let us discuss further the structure of linear operators on the tensor product $V\otimes W$. The space of those linear operators is called $\mathcal{L}(V\otimes W)$. We claimed that the most general element in this space was constructed as the sum $\sum_i T_i\otimes S_i$ with $T_i\in\mathcal{L}(V)$ and $S_i\in\mathcal{L}(W)$. In fact, the precise statement is that

$$\mathcal{L}(V\otimes W)=\mathcal{L}(V)\otimes\mathcal{L}(W). \tag{18.2.19}$$

We can explain why this holds using basis states. Suppose we have basis states $|e_i^V\rangle$ with $i=1,\ldots,m$ for the space V and basis states $|e_a^W\rangle$ with $a=1,\ldots,n$ for the space W. Let us now list basis vectors for the other relevant spaces:

$$\begin{aligned}&\text{basis vectors for } \mathcal{L}(V)=\{\,|e_i^V\rangle\langle e_j^V|,\ i,j=1,\ldots,m\},\\ &\text{basis vectors for } \mathcal{L}(W)=\{\,|e_a^W\rangle\langle e_b^W|,a,b=1,\ldots,n\},\end{aligned} \tag{18.2.20}$$

since the general operator in a vector space is a linear superposition of basis ket-bra operators. For the tensor product we then have

$$\text{basis vectors for } \mathcal{L}(V)\otimes\mathcal{L}(W)=\left\{\,|e_i^V\rangle\langle e_j^V|\otimes|e_a^W\rangle\langle e_b^W|,\ \begin{matrix}i,j=1,\ldots,m\\ a,b=1,\ldots,n\end{matrix}\right\}, \tag{18.2.21}$$

since here a basis vector is a basis vector in the first factor $\mathcal{L}(V)$ tensored with a basis vector in the second factor $\mathcal{L}(W)$. Finally, we also have

$$\text{basis vectors for } \mathcal{L}(V\otimes W)=\left\{\,|e_i^V\rangle\otimes|e_a^W\rangle\langle e_j^V|\otimes\langle e_b^W|,\ \begin{matrix}i,j=1,\ldots,m\\ a,b=1,\ldots,n\end{matrix}\right\}, \tag{18.2.22}$$

constructed as ket-bra operators of $V\otimes W$. We can now discuss the claimed relation (18.2.19). We assert that the basis vectors of $\mathcal{L}(V\otimes W)$ and the basis vectors of $\mathcal{L}(V)\otimes\mathcal{L}(W)$ in fact represent the same operator. This follows from the way these basis vectors are defined to act. Letting similarly labeled basis vectors act on $|v\rangle\otimes|w\rangle$, we have

$$\begin{aligned}&\text{from } \mathcal{L}(V\otimes W):\ |e_i^V\rangle\otimes|e_a^W\rangle\langle e_j^V|\otimes\langle e_b^W|\ \ |v\rangle\otimes|w\rangle\ =\ |e_i^V\rangle\otimes|e_a^W\rangle\langle e_j^V|v\rangle\langle e_b^W|w\rangle,\\ &\text{from } \mathcal{L}(V)\otimes\mathcal{L}(W):\ |e_i^V\rangle\langle e_j^V|\ \otimes\ |e_a^W\rangle\langle e_b^W|\ \ |v\rangle\otimes|w\rangle\ =\ |e_i^V\rangle\langle e_j^V|v\rangle\otimes|e_a^W\rangle\langle e_b^W|w\rangle.\end{aligned}$$

We can see that the results in both lines are the same. This shows that the spaces of linear operators on the left-hand side and the right-hand side of (18.2.19) agree.

It is interesting to consider the case when the spaces V and W are the same space V, and we have a "swap" operator $S\in\mathcal{L}(V\otimes V)$ that acts as follows:

$$S(v\otimes\tilde{v})=\tilde{v}\otimes v,\quad \text{for all } v,\tilde{v}\in V. \tag{18.2.23}$$

It may seem puzzling that S can be constructed from sums of products of operators that act separately on the two vector spaces. But, in fact, one can easily build this operator. With basis vectors $|e_i\rangle, i=1,\ldots,n$ for V, the swap operator $S\in\mathcal{L}(V\otimes V)$ is given by

$$S=\sum_{i,j=1}^{n}|e_i\rangle\langle e_j|\otimes|e_j\rangle\langle e_i|. \tag{18.2.24}$$

Exercise 18.4. *Show that this operator satisfies the requisite action (18.2.23).*

18.3 Inner Products for Tensor Spaces

We now consider the definition of an *inner product* in $V \otimes W$. This product can be defined naturally if we have inner products on V and on W. In this case, with vectors $v, \tilde{v} \in V$ and $w, \tilde{w} \in W$, we simply let

$$\langle v \otimes w, \tilde{v} \otimes \tilde{w} \rangle \equiv \langle v, \tilde{v} \rangle \langle w, \tilde{w} \rangle, \tag{18.3.1}$$

where the inner products on the right-hand side are those in V and in W. To find the inner product of general vectors in $V \otimes W$, we must declare that with vectors $X, Y, Z \in V \otimes W$ the following distributive properties hold:

$$\begin{aligned} \langle X + Y, Z \rangle &= \langle X, Z \rangle + \langle Y, Z \rangle, \\ \langle X, Y + Z \rangle &= \langle X, Y \rangle + \langle X, Z \rangle. \end{aligned} \tag{18.3.2}$$

Note that (18.3.1) implies that for a complex constant a,

$$\begin{aligned} \langle v \otimes w, a(\tilde{v} \otimes \tilde{w}) \rangle &= \langle v \otimes w, (a\tilde{v}) \otimes \tilde{w} \rangle = \langle v, a\tilde{v} \rangle \langle w, \tilde{w} \rangle = a \langle v \otimes w, \tilde{v} \otimes \tilde{w} \rangle, \\ \langle a(v \otimes w), \tilde{v} \otimes \tilde{w} \rangle &= \langle (av) \otimes w, \tilde{v} \otimes \tilde{w} \rangle = \langle av, \tilde{v} \rangle \langle w, \tilde{w} \rangle = a^* \langle v \otimes w, \tilde{v} \otimes \tilde{w} \rangle, \end{aligned} \tag{18.3.3}$$

using the properties of the inner product in V. The distributive properties then show that, in general, the inner product in $V \otimes W$ satisfies the expected

$$\begin{aligned} \langle X, aY \rangle &= a \langle X, Y \rangle, \\ \langle aX, Y \rangle &= a^* \langle X, Y \rangle. \end{aligned} \tag{18.3.4}$$

It is useful to describe the inner product in a basis $\{e_i \otimes f_j; i = 1, \ldots, m, \ j = 1, \ldots, n\}$ for the tensor product, with $\{e_i; i = 1, \ldots, m\}$ and $\{f_j; j = 1, \ldots, m\}$ *orthonormal* bases for V and W. Using (18.3.1), we immediately find

$$\langle e_i \otimes f_j, e_p \otimes f_q \rangle = \langle e_i, e_p \rangle \langle f_j, f_q \rangle = \delta_{ip} \delta_{jq}. \tag{18.3.5}$$

This makes the tensor product basis vectors $e_i \otimes f_j$ orthonormal. The verification that the inner product on $V \otimes W$ satisfies the remaining axioms of an inner product (section 14.1) is addressed in the exercises below. For both exercises, assume that $X, Y \in V \otimes W$, and write the most general such vectors as $X = \sum_{ij} x_{ij} e_i \otimes f_j$, and $Y = \sum_{ij} y_{ij} e_i \otimes f_j$. Then proceed using (18.3.2), (18.3.4), and (18.3.5).

Exercise 18.5. *Show that* $\langle X, X \rangle \geq 0$, *and* $\langle X, X \rangle = 0$ *if and only if* $X = 0$.

Exercise 18.6. *Show that* $\langle X, Y \rangle = \langle Y, X \rangle^*$.

It is often convenient to use bra-ket notation for inner products in the tensor product. Kets and bras in the tensor product are often written as follows:

$$\begin{aligned} |v \otimes w\rangle &= |v\rangle_1 \otimes |w\rangle_2, \\ \langle v \otimes w| &= {}_1\langle v| \otimes {}_2\langle w|. \end{aligned} \tag{18.3.6}$$

Notice that for both bras and kets we write the state of particle 1 to the left of the state of particle 2. We then write (18.3.1) as

$$\langle v \otimes w | \tilde{v} \otimes \tilde{w} \rangle = \big({}_1\langle v| \otimes {}_2\langle w| \big) \big(|\tilde{v}\rangle_1 \otimes |\tilde{w}\rangle_2 \big) = \langle v|\tilde{v}\rangle \langle w|\tilde{w}\rangle. \tag{18.3.7}$$

Going back to example 18.3, we can now normalize the state we built with two spin one-half particles and zero total spin angular momentum. Our four basis vectors $|+\rangle_1 \otimes |+\rangle_2$, $|+\rangle_1 \otimes |-\rangle_2$, $|-\rangle_1 \otimes |+\rangle_2$, and $|-\rangle_1 \otimes |-\rangle_2$ are orthonormal. We had the unnormalized state in (18.2.18) given by

$$|\Psi\rangle = \alpha \Big(|+\rangle_1 \otimes |-\rangle_2 \; - \; |-\rangle_1 \otimes |+\rangle_2 \Big). \tag{18.3.8}$$

The associated bra is then

$$\langle\Psi| = \alpha^* \Big({}_1\langle +| \otimes {}_2\langle -| \; - \; {}_1\langle -| \otimes {}_2\langle +| \Big). \tag{18.3.9}$$

We then have

$$\begin{aligned} \langle\Psi|\Psi\rangle &= \alpha\alpha^* \Big({}_1\langle +| \otimes {}_2\langle -| \; - \; {}_1\langle -| \otimes {}_2\langle +| \Big)\Big(|+\rangle_1 \otimes |-\rangle_2 \; - \; |-\rangle_1 \otimes |+\rangle_2 \Big) \\ &= \alpha\alpha^* \Big({}_1\langle +| \otimes {}_2\langle -||+\rangle_1 \otimes |-\rangle_2 \; + \; {}_1\langle -| \otimes {}_2\langle +||-\rangle_1 \otimes |+\rangle_2 \Big) \end{aligned} \tag{18.3.10}$$

since only terms where the spin states are the same for the first particle and for the second particle survive. We thus have, for normalization,

$$\langle\Psi|\Psi\rangle \;=\; |\alpha|^2(1+1) = 2|\alpha|^2 = 1 \quad \Rightarrow \quad \alpha = \frac{1}{\sqrt{2}}. \tag{18.3.11}$$

The normalized state with zero total spin angular momentum is then

$$\boxed{|\Psi\rangle = \frac{1}{\sqrt{2}} \Big(|+\rangle_1 \otimes |-\rangle_2 - |-\rangle_1 \otimes |+\rangle_2 \Big).} \tag{18.3.12}$$

This is a rather interesting state, sometimes called the spin singlet state. We will see that it is an entangled state of the two particles. Finally, consistent with having zero total spin, it is a rotationally invariant state (problem 18.1).

Exercise 18.7. *Show that $(S\otimes T)^\dagger = S^\dagger \otimes T^\dagger$. For this it suffices to show that $\langle (S\otimes T)u, v\rangle = \langle u, (S^\dagger \otimes T^\dagger)v\rangle$, where u, v are basis vectors of the type $e_i \otimes f_j$.*

18.4 Matrix Representations and Traces

We have considered linear operators $\mathcal{L}(V\otimes W)$ on the tensor product $V\otimes W$, with V and W complex vector spaces of dimensions m and n, respectively. In general such operators are sums of operators that take the form $A\otimes B$, where $A \in \mathcal{L}(V)$ and $B \in \mathcal{L}(W)$. We learned in section 13.5 how to represent an operator on a vector space as a matrix. Thus, imagine we have the $m\times m$ matrix representing A and the $n \times n$ matrix representing B. How do we then build the matrix representation of the tensor product operator $A\otimes B$? The answer is not unique; it depends on a choice of an ordered list of basis vectors for $V\otimes W$. Rather than focus on this choice, we give first a natural and simple possibility that works and then discuss the corresponding choice of basis vectors. The matrix takes the form

$$A\otimes B = \begin{pmatrix} A_{11}B & \cdots & A_{1m}B \\ \vdots & \vdots & \vdots \\ A_{m1}B & \dots & A_{mm}B \end{pmatrix}. \tag{18.4.1}$$

Here the entries $A_{ij}B$ are $n \times n$ blocks obtained by multiplying the $n \times n$ matrix B against the number A_{ij}. We see that, as it should be, $A \otimes B$ is an mn by mn matrix. The matrix is formed by inserting the B matrix as a block at the various i, j positions of the A matrix and multiplying by the corresponding A_{ij}.

We first do a consistency check. Given operators $A, C \in \mathcal{L}(V)$ and $B, D \in \mathcal{L}(W)$, the matrix representations should then implement the product relation

$$(A \otimes B) \cdot (C \otimes D) = (AC) \otimes (BD). \tag{18.4.2}$$

Since matrices that can be partitioned in equal-size blocks can be multiplied in blocks, the matrices for $A \otimes B$ and $C \otimes D$ multiply properly:

$$\begin{pmatrix} A_{11}B & \cdots & A_{1m}B \\ \vdots & \vdots & \vdots \\ A_{m1}B & \ldots & A_{mm}B \end{pmatrix} \begin{pmatrix} C_{11}D & \cdots & C_{1m}D \\ \vdots & \vdots & \vdots \\ C_{m1}D & \ldots & C_{mm}D \end{pmatrix} = \begin{pmatrix} (AC)_{11}BD & \cdots & (AC)_{1m}BD \\ \vdots & \vdots & \vdots \\ (AC)_{m1}BD & \ldots & (AC)_{mm}BD \end{pmatrix},$$

and we recognize on the right-hand side the matrix representation for $AC \otimes BD$.

Exercise 18.8. *Convince yourself that a representation of $A \otimes B$ as blocks of A multiplying the elements of B would also give an mn by mn matrix that satisfies the composition rule (18.4.2).*

Exercise 18.9. *Write down the matrix $M_{xy} = \sigma_x \otimes \sigma_y$, and the matrix $M_{yx} = \sigma_y \otimes \sigma_x$. Confirm that $M_{xy} \neq M_{yx}$. Show that $M_{xy}M_{yx} = M_{yx}M_{xy} = \sigma_z \otimes \sigma_z$.*

To justify the claimed representation (18.4.1), we must display the list of basis vectors in $V \otimes W$ for which the matrix is in fact the representation. Recall that for basis vectors v_i and an operator T, the matrix representation is defined by $Tv_i = \sum_k T_{ki}v_k$. This means that the ith column of the matrix T is filled with the action of T on the ith basis vector. In the present case with V being m-dimensional and W being n-dimensional, we write the basis vectors as follows:

$$\begin{aligned} &\text{orthonormal basis for } V: e_1, \ldots, e_m, \\ &\text{orthonormal basis for } W: f_1, \ldots, f_n. \end{aligned} \tag{18.4.3}$$

It is natural to expect that the basis vectors we are using for $V \otimes W$ are $e_i \otimes f_j$ for $i = 1, \ldots, m$ and $j = 1, \ldots, n$. If this is so, how are they ordered? For this let $e_i \otimes f_j$ be the kth basis vector, where $k = k(i, j)$ is some function of i, j and possibly m, n that we wish to determine, as it would fix the ordering of the basis vectors. We now compute

$$(A \otimes B)e_i \otimes f_j = Ae_i \otimes Bf_j = \Big(\sum_p A_{pi}e_p\Big) \otimes \Big(\sum_q B_{qj}f_q\Big) = \sum_{p,q} A_{pi}B_{qj}\, e_p \otimes f_q. \tag{18.4.4}$$

By our earlier remarks, we now know that $A_{pi}B_{qj}$ must belong to the kth column of the matrix $A \otimes B$, for all values of p and q. The element A_{pi} appears on the ith column of the *block* matrix (18.4.1). Within that block, $A_{pi}B_{qj}$ appears on the jth column. As a result, we have

$$k = (i-1)n + j, \tag{18.4.5}$$

since there are $i-1$ size $n\times n$ blocks before we reach the ith block, and then, within that block, we move by j further to the right. The number k is the index number for the vector $e_i\otimes f_j$ in our ordered list of basis vectors. We have thus learned that

$$e_i\otimes f_j \text{ is the } [(i-1)n+j]\text{th basis vector on the list.} \tag{18.4.6}$$

The first n vectors arise from $i=1$ and $j=1,\dots,n$. The full list goes as follows:

$$e_1\otimes f_1,\cdots,e_1\otimes f_n,\; e_2\otimes f_1,\cdots,e_2\otimes f_n,\;\dots,\; e_m\otimes f_1,\cdots,e_m\otimes f_n. \tag{18.4.7}$$

Exercise 18.10. *It is clear from (18.4.5) that i and j determine k. Explain why this same relation implies that k, in the range* $1,\dots,mn$, *determines i and j uniquely.*

The relation $Tv_i=\sum_k T_{ki}v_k$ also shows that the element T_{ki} multiplying the kth vector v_k appears on the kth row of the matrix representation. Therefore, in the relation

$$(A\otimes B)e_i\otimes f_j=\sum_{p,q}A_{pi}B_{qj}\,e_p\otimes f_q, \tag{18.4.8}$$

obtained above, we must verify that the row where $A_{pi}B_{qj}$ appears in $A\otimes B$ equals the index number for the basis vector $e_p\otimes f_q$.

Exercise 18.11. *Explain why* $A_{pi}B_{qj}$ *indeed appears in the* $[(p-1)n+q]$*th row of* $A\otimes B$.

Exercise 18.12. *List explicitly the ordered list of basis vectors for the representation you built in exercise 18.9.*

Let us conclude this section with a discussion of traces. For an operator A, the trace is defined as $\text{tr}A=\sum_i A_{ii}$, the sum of the diagonal matrix elements in any matrix representation. If we use our matrix representation (18.4.1), we see that the diagonal elements are all products of diagonal elements of A and diagonal elements of B. In fact,

$$\text{tr}(A\otimes B)=A_{11}\text{tr}B+A_{22}\text{tr}B+\cdots+A_{mm}\text{tr}B, \tag{18.4.9}$$

and therefore we recognize that

$$\boxed{\text{tr}(A\otimes B)=\text{tr}A\cdot\text{tr}B,\quad A\in\mathcal{L}(V),\; B\in\mathcal{L}(W).} \tag{18.4.10}$$

It should be emphasized that for a general operator in $\mathcal{L}(V\otimes W)$ there is no such formula. Indeed, by linearity of the trace,

$$\begin{aligned}\text{tr}(A_1\otimes B_1+A_2\otimes B_2)&=\text{tr}(A_1\otimes B_1)+\text{tr}(A_2\otimes B_2)\\&=\text{tr}A_1\cdot\text{tr}B_1+\text{tr}A_2\cdot\text{tr}B_2,\end{aligned} \tag{18.4.11}$$

and no further simplification is possible.

In $\mathcal{L}(V\otimes W)$, in addition to a standard trace, there exist **partial traces**: a trace tr_V where we trace over the first vector space and a trace tr_W where we trace over the second vector space. In this notation, the full trace of the operator is written as $\text{tr}_{V\otimes W}$. It is simplest to describe these traces using the orthonormal basis vectors e_i and f_j of V and W, respectively, as well as bra-ket notation. For a general operator $\mathcal{O}\in\mathcal{L}(V\otimes W)$, we have the conventional definition, following (14.9.16):

$$\text{tr}_{V\otimes W}\,\mathcal{O}=\sum_{k,l}\langle e_k|\langle f_l|\mathcal{O}|e_k\rangle|f_l\rangle. \tag{18.4.12}$$

For the partial traces, we define

$$\text{tr}_V\, \mathcal{O} \equiv \sum_k \langle e_k|\mathcal{O}|e_k\rangle \in \mathcal{L}(W),$$
$$\text{tr}_W\, \mathcal{O} \equiv \sum_l \langle f_l|\mathcal{O}|f_l\rangle \in \mathcal{L}(V). \tag{18.4.13}$$

As noted above, the partial traces are still operators: if we trace $\mathcal{O}$ over V, the result is an operator on W, and if we trace $\mathcal{O}$ over W, the result is an operator on V. Further traces of those operators, a tr_W of $\text{tr}_V\,\mathcal{O}$ or a tr_V of $\text{tr}_W\mathcal{O}$, would give us numbers—in fact the same number, which is also equal to the full trace $\text{tr}_{V\otimes W}$ on the tensor product:

$$\boxed{\text{tr}_{V\otimes W}\mathcal{O} = \text{tr}_W\, \text{tr}_V\, \mathcal{O} = \text{tr}_V\, \text{tr}_W\mathcal{O}.} \tag{18.4.14}$$

To prove this we can consider a general operator $\mathcal{O} \in \mathcal{L}(V \otimes W) = \mathcal{L}(V) \otimes \mathcal{L}(W)$ written

$$\mathcal{O} = \sum_{i,j;\,p,q} \mathcal{O}_{ij,pq}\, |e_i\rangle\langle e_j| \otimes |f_p\rangle\langle f_q|. \tag{18.4.15}$$

We then have

$$\text{tr}_V\, \mathcal{O} = \sum_k \langle e_k|\mathcal{O}|e_k\rangle = \sum_k \sum_{i,j;\,p,q} \mathcal{O}_{ij,pq}\, \langle e_k|e_i\rangle\langle e_j|e_k\rangle\; |f_p\rangle\langle f_q|. \tag{18.4.16}$$

The overlaps set both i and j equal to k, and therefore,

$$\text{tr}_V\, \mathcal{O} = \sum_k \sum_{p,q} \mathcal{O}_{kk,pq}\, |f_p\rangle\langle f_q|,$$
$$\text{tr}_W\, \mathcal{O} = \sum_l \sum_{i,j} \mathcal{O}_{ij,ll}\, |e_i\rangle\langle e_j|, \tag{18.4.17}$$

including the analogous result for $\text{tr}_W\, \mathcal{O}$. It is now clear that additional traces give

$$\text{tr}_W\, \text{tr}_V\, \mathcal{O} = \sum_{k,l} \mathcal{O}_{kk,ll} = \text{tr}_V\, \text{tr}_W\, \mathcal{O}. \tag{18.4.18}$$

You can quickly confirm that for the full trace we also have $\text{tr}_{V\otimes W}\mathcal{O} = \sum_{k,l} \mathcal{O}_{kk,ll}$, thus completing the proof of (18.4.14). These results on traces will be useful when we consider density matrices for quantum systems composed of two subsystems (chapter 22).

18.5 Entangled States

You have learned that $V \otimes W$ includes states

$$\Psi = \sum_i \alpha_i\, v_i \otimes w_i, \tag{18.5.1}$$

obtained by the linear superposition of simpler states of the form $v_i \otimes w_i$. If handed such a Ψ, you might want to know whether you can write it as a single term $v_* \otimes w_*$ for some $v_* \in V$ and $w_* \in W$:

$$\text{Can one write the state as } \Psi = v_* \otimes w_*\ ? \tag{18.5.2}$$

If no such v_* and w_* exist, we say that Ψ is an **entangled state** of the two particles. If, on the other hand, v_* and w_* exist, then you are able to describe the state of the particles in Ψ independently: particle 1 is in state v_* and particle 2 in state w_*, and we say that Ψ is not an entangled state. Schrödinger called entanglement the essential feature of quantum mechanics.

Entanglement is a basis-independent property. Indeed, if the state can be factorized into $v_* \otimes w_*$ for some basis choice in V and W, it can be factorized for any other basis choice by simply rewriting v_* and w_* in the new basis. If the state cannot be factorized into $v_* \otimes w_*$ for some basis choice in V and U, it cannot be factorized for any other basis choice because factorization with another basis choice would then imply factorization in the original basis choice. The *tensor product* basis vectors can be chosen to not be entangled, as we did for the orthonormal basis $e_i \otimes f_j$ in (18.1.5), or chosen to be entangled, as we will do for the Bell basis relevant to a pair of spin one-half particles.

In the tensor product of two two-dimensional complex vector spaces, it is not hard to decide when a state is entangled. Let V have a basis e_1, e_2 and W have a basis f_1, f_2. Then the most general state in $V \otimes W$ is

$$\Psi_A = A_{11}\, e_1 \otimes f_1 + A_{12}\, e_1 \otimes f_2 + A_{21}\, e_2 \otimes f_1 + A_{22}\, e_2 \otimes f_2, \tag{18.5.3}$$

with coefficients A_{ij} that can be encoded by a matrix A:

$$A = \begin{pmatrix} A_{11} & A_{12} \\ A_{21} & A_{22} \end{pmatrix}. \tag{18.5.4}$$

The state is *not* entangled if there exist constants a_1, a_2, b_1, b_2 such that

$$\begin{aligned} &A_{11}\, e_1 \otimes f_1 + A_{12}\, e_1 \otimes f_2 + A_{21}\, e_2 \otimes f_1 + A_{22}\, e_2 \otimes f_2 \\ &= (a_1 e_1 + a_2 e_2) \otimes (b_1 f_1 + b_2 f_2). \end{aligned} \tag{18.5.5}$$

Note that these four unknown constants are not uniquely determined: we can, for example, multiply a_1 and a_2 by some constant $c \neq 0$ and divide b_1 and b_2 by c to obtain a different solution. Indeed $v \otimes w = (cv) \otimes (w/c)$ for any $c \neq 0$. Using the distributive laws for $\otimes$ to expand the right-hand side of (18.5.5) and recalling that $e_i \otimes f_j$ are basis vectors in the tensor product, we see that the equality requires the following four relations:

$$A_{11} = a_1 b_1, \quad A_{12} = a_1 b_2, \quad A_{21} = a_2 b_1, \quad A_{22} = a_2 b_2. \tag{18.5.6}$$

Combining these four expressions gives us a consistency condition:

$$\det A = A_{11} A_{22} - A_{12} A_{21} = a_1 b_1 a_2 b_2 - a_1 b_2 a_2 b_1 = 0. \tag{18.5.7}$$

In other words, if Ψ_A is *not* entangled the determinant of the matrix A must be zero. We can in fact show that $\det A = 0$ implies that Ψ_A is not entangled. To do this we simply have to present a solution for the equations above under the condition $\det A = 0$.

Assume first that $A_{11} = 0$. Then $\det A = 0$ implies $A_{12} A_{21} = 0$. If $A_{12} = 0$, then

$$\Psi_A = A_{21} e_2 \otimes f_1 + A_{22} e_2 \otimes f_2 = e_2 \otimes (A_{21} f_1 + A_{22} f_2), \tag{18.5.8}$$

and the state is indeed not entangled. If $A_{21} = 0$, then

$$\Psi_A = A_{12} e_1 \otimes f_2 + A_{22} e_2 \otimes f_2 = (A_{12} e_1 + A_{22} e_2) \otimes f_2, \tag{18.5.9}$$

and again, the state is not entangled. Thus, we can solve all equations when $A_{11}=0$. Now assuming $A_{11}\neq 0$, we can readily find a factorization that works:

$$\Psi_A = \left(\sqrt{A_{11}}e_1 + \frac{A_{21}}{\sqrt{A_{11}}}e_2\right)\otimes\left(\sqrt{A_{11}}f_1 + \frac{A_{12}}{\sqrt{A_{11}}}f_2\right), \tag{18.5.10}$$

noting that the $\det A=0$ condition means that $A_{22}=A_{12}A_{21}/A_{11}$. We have thus proved that

$$\boxed{\Psi_A = \sum_{i,j=1}^{2} A_{ij}\, e_i\otimes f_j \text{ is entangled if and only if } \det A\neq 0.}$$

Example 18.4. *Entangled state of two spin one-half particles.*
Consider the state $|\Psi\rangle = \frac{1}{\sqrt{2}}(|+\rangle_1\otimes|-\rangle_2 - |-\rangle_1\otimes|+\rangle_2)$ of zero total spin angular momentum (18.3.12). With basis vectors $e_1=|+\rangle_1, e_2=|-\rangle_1$ and $f_1=|+\rangle_2,\ f_2=|-\rangle_2$, the state is

$$|\Psi\rangle = \tfrac{1}{\sqrt{2}}\, e_1\otimes f_2 - \tfrac{1}{\sqrt{2}}\, e_2\otimes f_1 \;\Rightarrow\; A = \begin{pmatrix} 0 & \frac{1}{\sqrt{2}} \\ -\frac{1}{\sqrt{2}} & 0 \end{pmatrix}, \tag{18.5.11}$$

reading the associated A matrix. Since $\det A = \frac{1}{2}\neq 0$, the state is entangled. □

Exercise 18.13. *Consider the operator $S\otimes T$ on $\mathcal{L}(V\otimes W)$, with $S\in\mathcal{L}(V)$ and $T\in\mathcal{L}(W)$. Explain why the action of $S\otimes T$ on a nonentangled state leaves it nonentangled.*

Exercise 18.14. *Consider the operator $S\otimes T$ on $\mathcal{L}(V\otimes W)$, with $S\in\mathcal{L}(V)$ and $T\in\mathcal{L}(W)$. Does the action of $S\otimes T$ on an entangled state give an entangled state? If yes, prove it. If no, give an example.*

18.6 Bell Basis States

Bell states are a set of four *entangled, orthonormal basis vectors* in the state space of two spin one-half particles. To describe this basis consider the tensor product $V_1\otimes V_2$, with V_1 and V_2 both the two-dimensional complex vector space appropriate to spin one-half particles. For brevity of notation, we will leave out the 1 and 2 subscripts on the states as well as the $\otimes$ in between the states. It is always understood that in $V_1\otimes V_2$ the state in V_1 appears to the left of the state of V_2. Consider now the state

$$|\Phi_0\rangle \equiv \tfrac{1}{\sqrt{2}}\big(|+\rangle|+\rangle + |-\rangle|-\rangle\big). \tag{18.6.1}$$

This is clearly an entangled state: its associated matrix is diagonal with equal entries of $1/\sqrt{2}$ and thus a nonzero determinant. Moreover, this state is unit normalized:

$$\langle\Phi_0|\Phi_0\rangle = 1. \tag{18.6.2}$$

We use this state as the first of our basis vectors for $V_1\otimes V_2$. Since this tensor product is four-dimensional, we need three more entangled basis states. Here they are

$$|\Phi_i\rangle \equiv (\mathbb{1}\otimes\sigma_i)|\Phi_0\rangle,\; i=1,2,3. \tag{18.6.3}$$

It is clear that these states are entangled. If $|\Phi_i\rangle$ were not entangled, it would follow that $(\mathbb{1}\otimes\sigma_i)|\Phi_i\rangle$ (i not summed) is also not entangled (exercise 18.13). But using $\sigma_i^2=\mathbb{1}$, we see that this last state is in fact $|\Phi_0\rangle$, which is entangled. This contradiction shows that $|\Phi_i\rangle$ must be entangled. Since the operator $\mathbb{1}\otimes\sigma_i$ is unitary, it follows from the definition that all $|\Phi_i\rangle$ are unit normalized.

Let us look at the form of $|\Phi_1\rangle$:

$$|\Phi_1\rangle = (\mathbb{1}\otimes\sigma_1)\tfrac{1}{\sqrt{2}}\big(|+\rangle|+\rangle + |-\rangle|-\rangle\big) = \tfrac{1}{\sqrt{2}}\big(|+\rangle|-\rangle + |-\rangle|+\rangle\big). \tag{18.6.4}$$

By analogous calculations we obtain the full list of **Bell states**:

$$\begin{aligned}|\Phi_0\rangle &= \mathbb{1}\otimes\mathbb{1}\,|\Phi_0\rangle = \tfrac{1}{\sqrt{2}}\big(|+\rangle|+\rangle+|-\rangle|-\rangle\big),\\ |\Phi_1\rangle &= \mathbb{1}\otimes\sigma_1|\Phi_0\rangle = \tfrac{1}{\sqrt{2}}\big(|+\rangle|-\rangle+|-\rangle|+\rangle\big),\\ |\Phi_2\rangle &= \mathbb{1}\otimes\sigma_2|\Phi_0\rangle = \tfrac{i}{\sqrt{2}}\big(|+\rangle|-\rangle-|-\rangle|+\rangle\big),\\ |\Phi_3\rangle &= \mathbb{1}\otimes\sigma_3|\Phi_0\rangle = \tfrac{1}{\sqrt{2}}\big(|+\rangle|+\rangle-|-\rangle|-\rangle\big).\end{aligned} \tag{18.6.5}$$

Note that $|\Phi_2\rangle$ is the spin singlet state (18.3.12). We can confirm by inspection that Φ_0 is orthogonal to the other three: $\langle\Phi_0|\Phi_i\rangle=0$. It is not much work either to see that the basis is in fact orthonormal. But the calculation is kind of fun. Since $(S\otimes T)^\dagger=S^\dagger\otimes T^\dagger$ and $\sigma_i^\dagger=\sigma_i$, we find that

$$\langle\Phi_i| = \langle\Phi_0|(\mathbb{1}\otimes\sigma_i). \tag{18.6.6}$$

We can then compute

$$\begin{aligned}\langle\Phi_i|\Phi_j\rangle &= \langle\Phi_0|(\mathbb{1}\otimes\sigma_i)(\mathbb{1}\otimes\sigma_j)|\Phi_0\rangle\\ &= \langle\Phi_0|\mathbb{1}\otimes\sigma_i\sigma_j|\Phi_0\rangle\\ &= \langle\Phi_0|\mathbb{1}\otimes\big(\mathbb{1}\delta_{ij}+i\epsilon_{ijk}\sigma_k\big)|\Phi_0\rangle\\ &= \delta_{ij}\langle\Phi_0|\mathbb{1}\otimes\mathbb{1}|\Phi_0\rangle + i\epsilon_{ijk}\langle\Phi_0|\mathbb{1}\otimes\sigma_k|\Phi_0\rangle\\ &= \delta_{ij}\langle\Phi_0|\Phi_0\rangle + i\epsilon_{ijk}\langle\Phi_0|\Phi_k\rangle = \delta_{ij}.\end{aligned} \tag{18.6.7}$$

Indeed, we have an orthonormal basis of entangled states.

We can solve for the nonentangled basis states in terms of the Bell states. We quickly find from (18.6.5) that

$$\begin{aligned}|+\rangle|+\rangle &= \tfrac{1}{\sqrt{2}}\big(|\Phi_0\rangle+|\Phi_3\rangle\big),\\ |-\rangle|-\rangle &= \tfrac{1}{\sqrt{2}}\big(|\Phi_0\rangle-|\Phi_3\rangle\big),\\ |+\rangle|-\rangle &= \tfrac{1}{\sqrt{2}}\big(|\Phi_1\rangle-i|\Phi_2\rangle\big),\\ |-\rangle|+\rangle &= \tfrac{1}{\sqrt{2}}\big(|\Phi_1\rangle+i|\Phi_2\rangle\big).\end{aligned} \tag{18.6.8}$$

Introducing labels A and B for the two spaces in a tensor product $V_A\otimes V_B$, we can rewrite the above equations as

$$
\begin{aligned}
|+\rangle_A|+\rangle_B &= \tfrac{1}{\sqrt{2}}\big(|\Phi_0\rangle_{AB}+|\Phi_3\rangle_{AB}\big),\\
|-\rangle_A|-\rangle_B &= \tfrac{1}{\sqrt{2}}\big(|\Phi_0\rangle_{AB}-|\Phi_3\rangle_{AB}\big),\\
|+\rangle_A|-\rangle_B &= \tfrac{1}{\sqrt{2}}\big(|\Phi_1\rangle_{AB}-i|\Phi_2\rangle_{AB}\big),\\
|-\rangle_A|+\rangle_B &= \tfrac{1}{\sqrt{2}}\big(|\Phi_1\rangle_{AB}+i|\Phi_2\rangle_{AB}\big),
\end{aligned} \tag{18.6.9}
$$

where $|\Phi_i\rangle_{AB}$ are the Bell states we defined above, with the first state in V_A and the second state in V_B.

Let us now discuss measurements that can be done on an entangled pair of particles. Recall that given an orthonormal basis $|e_1\rangle, \ldots, |e_n\rangle$ we can measure a state $|\Psi\rangle$ along this basis (see axiom A3 and equation (16.6.9)). We have the probability $p(i) = |\langle e_i|\Psi\rangle|^2$ of being found in the state $|i\rangle$. After measurement, the state will be in one of the basis states $|e_i\rangle$.

For a state of two spin one-half particles A, B, we may choose the four Bell states as our orthonormal basis for measurement. If so, after measurement the state will be in one of the Bell states $|\Phi_i\rangle_{AB}$, with probability $|_{AB}\langle\Phi_i|\Psi\rangle|^2$.

If Alice and Bob each has one of the particles in an entangled pair, more sophisticated measurements are possible. We examine those now.

Partial measurement Suppose we have a general entangled state $|\Psi\rangle \in V\otimes W$ of two particles. Alice has access to the first particle and decides to measure along the basis $|e_1\rangle, \ldots, |e_n\rangle$ of V. This is analyzed with measurement axiom A3, using a complete set of mutually orthogonal projectors M_i:

$$
M_i \equiv |e_i\rangle\langle e_i| \otimes \mathbb{1}, \quad i=1,\ldots,n. \tag{18.6.10}
$$

The projectors act trivially on the state space of the second particle and act on the state space of the first particle as expected. Clearly, $M_i^\dagger = M_i$, $M_iM_j = M_i\delta_{ij}$, and $\sum_{i=1}^n M_i = \mathbb{1}\otimes\mathbb{1}$, which is the identity in the tensor product. To simplify the writing of probabilities, consider the measurement of a general state $|\Psi\rangle$ written with the help of the basis vectors $|e_i\rangle$ as

$$
|\Psi\rangle = \sum_i |e_i\rangle \otimes |w_i\rangle. \tag{18.6.11}
$$

Here, the $|w_i\rangle \in W$ are some calculable vectors that in general are neither normalized nor orthogonal. Such a writing of $|\Psi\rangle$ is always possible. From axiom A3, the probability $p(i)$ that Alice will find the first particle to be in the state $|i\rangle$ is

$$
p(i) = \langle\Psi|M_i|\Psi\rangle = \sum_{p,q}\langle e_p|\otimes\langle w_p|\big(|e_i\rangle\langle e_i|\otimes\mathbb{1}\big)\,|e_q\rangle\otimes|w_q\rangle = \sum_{p,q}\langle e_p|e_i\rangle\langle e_i|e_q\rangle\,\langle w_p|w_q\rangle.
$$

Using the orthonormality of the basis,

$$
p(i) = \langle w_i|w_i\rangle. \tag{18.6.12}
$$

If Alice finds her particle in $|e_i\rangle$, the state of the system after measurement is $M_i|\Psi\rangle$, suitably normalized:

$$M_i|\Psi\rangle = |e_i\rangle\langle e_i| \otimes \mathbb{1} \sum_p |e_p\rangle \otimes |w_p\rangle = |e_i\rangle \otimes |w_i\rangle. \tag{18.6.13}$$

Normalizing, we see that after the measurement the state of the system will be

$$|e_i\rangle \otimes \frac{|w_i\rangle}{\sqrt{\langle w_i|w_i\rangle}}, \quad \text{for some value of } i. \tag{18.6.14}$$

Exercise 18.15. *Show that one also has $p(i) = \big\| \langle e_i|\Psi\rangle \big\|^2$, an expression formally analogous to the familiar rule for measuring along a basis. Note that the norm is needed because $\langle e_i|\Psi\rangle \in W$.*

Exercise 18.16. *Alice and Bob measure the entangled state $|\Psi\rangle$ along the basis states $|e_i\rangle \otimes |f_j\rangle$ of $V \otimes W$. Show that the probability of finding the state in $|e_i\rangle \otimes |f_j\rangle$ is $p(i,j) = |\langle e_i| \otimes \langle f_j|\Psi\rangle|^2$. Show, additionally, that $p(i) = \sum_j p(i,j)$.*

As a simple illustration of partial measurement, consider the entangled spin single state:

$$|\Psi\rangle = \tfrac{1}{\sqrt{2}}\big(|+\rangle_1 \otimes |-\rangle_2 - |-\rangle_1 \otimes |+\rangle_2\big). \tag{18.6.15}$$

If we are to measure the first particle along the $|+\rangle, |-\rangle$ basis, we rewrite the state in the form (18.6.11):

$$|\Psi\rangle = |+\rangle_1 \otimes \tfrac{1}{\sqrt{2}}|-\rangle_2 + |-\rangle_1 \otimes \big(-\tfrac{1}{\sqrt{2}}|+\rangle_2\big). \tag{18.6.16}$$

On account of (18.6.12), the probabilities $p(+)$ and $p(-)$ are then given by

$$\begin{aligned} p(+) &= \tfrac{1}{\sqrt{2}}\tfrac{1}{\sqrt{2}}\langle -|-\rangle = \tfrac{1}{2}; \ \text{ state after measurement: } |+\rangle_1 \otimes |-\rangle_2, \\ p(-) &= \big(-\tfrac{1}{\sqrt{2}}\big)^2\langle +|+\rangle = \tfrac{1}{2}; \ \text{ state after measurement: } |-\rangle_1 \otimes |+\rangle_2, \end{aligned} \tag{18.6.17}$$

the states after measurement given in the form (18.6.14). After the measurement of the first particle, a measurement of the second particle will show that its spin is always opposite to the spin of the first particle.

As a more nontrivial example, consider a state of three particles A, B, C. Such a state lives in $V_A \otimes V_B \otimes V_C$. To analyze what happens if Alice decides to do a Bell measurement of the pair AB, the state Ψ of the system must be written in the form

$$|\Psi\rangle = |\Phi_0\rangle_{AB} \otimes |u_0\rangle_C + \sum_{i=1}^{3} |\Phi_i\rangle_{AB} \otimes |u_i\rangle_C. \tag{18.6.18}$$

In general, the states $|u_\mu\rangle$ with $\mu = 0, 1, 2, 3$ are neither normalized nor orthogonal to each other. After measurement, the state of the particles AB will be one of the Bell states $|\Phi_\mu\rangle_{AB}$. The probability $p_{AB}(\Phi_\mu)$ that the AB particles are in the state $|\Phi_\mu\rangle$ is

$$p_{AB}(\Phi_\mu) = \langle u_\mu|u_\mu\rangle. \tag{18.6.19}$$

Moreover, the state after measurement is

$$|\Phi_\mu\rangle_{AB} \otimes \frac{|u_\mu\rangle_C}{\sqrt{\langle u_\mu|u_\mu\rangle}}, \quad \text{for some } \mu \in \{0, 1, 2, 3\}. \tag{18.6.20}$$

18.7 Quantum Teleportation

Teleportation is impossible in classical physics: there is no basis for dematerializing an object and recreating it somewhere else. In 1993, Bennet, Brassand, Crépeau, Jozsa, Peres, and Wooters discovered that, surprisingly, it *is* possible to teleport a quantum state.

Imagine that Alice is handed a spin one-half particle in some quantum state $|\Psi\rangle$. She does not know what the state is, but of course, it can be written as

$$|\Psi\rangle = \alpha|+\rangle + \beta|-\rangle, \tag{18.7.1}$$

where $\alpha, \beta \in \mathbb{C}$ are constants. We will call particle C the particle imprinted with this state and will write the state of the particle as $|\Psi\rangle_C$:

$$|\Psi\rangle_C = \alpha|+\rangle_C + \beta|-\rangle_C. \tag{18.7.2}$$

Alice's goal is to teleport the state of the particle—called a *quantum bit*, or *qubit* in the language of quantum computation—to Bob, who is far away. Particle C itself does not change position. If Alice and Bob share an entangled pair of spin one-half particles, teleportation will imprint the quantum state $|\Psi\rangle$ on the spin one-half particle available to Bob.

Teleporting is a practical solution to the problem of making the state of particle C available to Bob quickly and efficiently. Alternatively, Alice could perhaps isolate particle C in a safe box and send the box to Bob through the mail. One thing she can't do is clone the state of C and send the copy to Bob. The quantum *no-cloning* principle, to be discussed in section 18.9, prevents Alice from creating a copy of a state that is unknown to her. Indeed, when Alice is handed particle C, she has no way of finding out what α and β are. Measuring the state with some Stern-Gerlach apparatus will not help; the spin will just point up or point down. What has she learned? Almost nothing. Only with many copies of the state would she be able to learn about the values of α and β. Having just one particle, she is unable to measure α and β and send those values to Bob.

A diagram showing the teleportation setup is shown in figure 18.1. The key tool Alice and Bob use is an entangled state of two particles A, B in which Alice has access to particle A, and Bob has access to particle B. One pair AB of entangled particles will allow Alice to teleport the state of particle C. The state of C will be imprinted on particle B. Teleporting quantum states is by now routinely done.

Alice has a console with four lights labeled $\mu = 0, 1, 2, 3$. She will do a Bell measurement on AC, the pair containing particle A of the shared, entangled pair and particle C, whose state is to be teleported. When she does, one of her four lights will blink: if it is the μth light, it is because she ended up with the Bell state $|\Phi_\mu\rangle_{AC}$. Bob, who is in possession of particle B, has a console with four boxes that generate unitary transformations. The first box, labeled $\mu = 0$, does nothing to the state. The ith box (with $i = 1, 2, 3$) applies the operator σ_i. Alice communicates to Bob that the μth light blinked. Then Bob

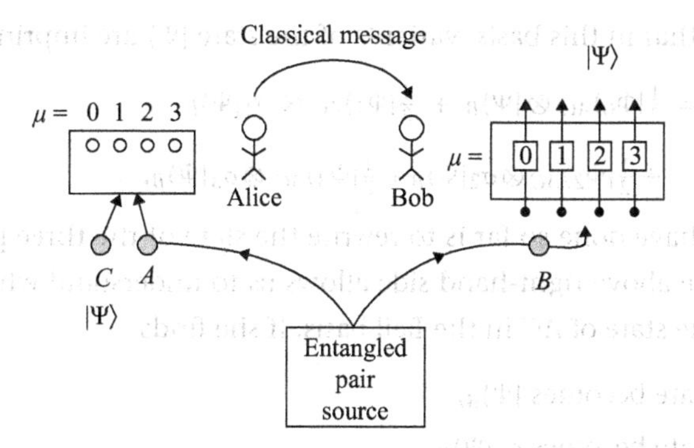

Figure 18.1
Alice has particle C in the state $|\Psi\rangle$, to be teleported, as well as particle A. Particle A is entangled with particle B, which is in Bob's possession. Alice performs a Bell measurement on particles A and C. After she measures, Bob's particle will carry the state $|\Psi\rangle$, up to a simple unitary transformation.

submits particle B to the μth box and out comes, we claim, the state $|\Psi\rangle$ imprinted on particle B.

To prove this, let the entangled, shared pair be the first Bell basis state:

$$|\Phi_0\rangle_{AB} = \frac{1}{\sqrt{2}}\left(|+\rangle_A|+\rangle_B + |-\rangle_A|-\rangle_B\right). \tag{18.7.3}$$

The total state of our three particles A, B, C is therefore

$$\begin{aligned} |\Phi_0\rangle_{AB} \otimes |\Psi\rangle_C &= |\Phi_0\rangle_{AB} \otimes (\alpha|+\rangle_C + \beta|-\rangle_C) \\ &= \frac{1}{\sqrt{2}}\left(|+\rangle_A|+\rangle_B + |-\rangle_A|-\rangle_B\right) \otimes \left(\alpha|+\rangle_C + \beta|-\rangle_C\right). \end{aligned} \tag{18.7.4}$$

Expanding out and reordering the states to have A followed by C and then by B, we have

$$\begin{aligned} |\Phi_0\rangle_{AB} \otimes |\Psi\rangle_C = \frac{1}{\sqrt{2}}\Big(& \alpha \underbrace{|+\rangle_A|+\rangle_C}|+\rangle_B + \beta \underbrace{|+\rangle_A|-\rangle_C}|+\rangle_B \\ & + \alpha \underbrace{|-\rangle_A|+\rangle_C}|-\rangle_B + \beta \underbrace{|-\rangle_A|-\rangle_C}|-\rangle_B \Big). \end{aligned} \tag{18.7.5}$$

Note that as long as we label the states, the order in which we write them does not matter. We now write these basis states with braces in the Bell basis using (18.6.9). We find that

$$\begin{aligned} |\Phi_0\rangle_{AB} \otimes |\Psi\rangle_C = & \tfrac{1}{2}\Big(|\Phi_0\rangle_{AC} + |\Phi_3\rangle_{AC}\Big)\, \alpha|+\rangle_B + \tfrac{1}{2}\Big(|\Phi_1\rangle_{AC} - i|\Phi_2\rangle_{AC}\Big)\, \beta|+\rangle_B \\ & + \tfrac{1}{2}\Big(|\Phi_1\rangle_{AC} + i|\Phi_2\rangle_{AC}\Big)\, \alpha|-\rangle_B + \tfrac{1}{2}\Big(|\Phi_0\rangle_{AC} - |\Phi_3\rangle_{AC}\Big)\, \beta|-\rangle_B. \end{aligned} \tag{18.7.6}$$

Collecting the Bell states,

$$\begin{aligned} |\Phi_0\rangle_{AB} \otimes |\Psi\rangle_C = & \tfrac{1}{2}|\Phi_0\rangle_{AC}\left(\alpha|+\rangle_B + \beta|-\rangle_B\right) + \tfrac{1}{2}|\Phi_1\rangle_{AC}\left(\alpha|-\rangle_B + \beta|+\rangle_B\right) \\ & + \tfrac{1}{2}|\Phi_2\rangle_{AC}\left(i\alpha|-\rangle_B - i\beta|+\rangle_B\right) + \tfrac{1}{2}|\Phi_3\rangle_{AC}\left(\alpha|+\rangle_B - \beta|-\rangle_B\right). \end{aligned} \tag{18.7.7}$$

We can then see that in this basis, variants of the state $|\Psi\rangle$ are imprinted on particle B:

$$\begin{aligned}|\Phi_0\rangle_{AB}\otimes|\Psi\rangle_C = \tfrac{1}{2}|\Phi_0\rangle_{AC}\otimes|\Psi\rangle_B + \tfrac{1}{2}|\Phi_1\rangle_{AC}\otimes\sigma_1|\Psi\rangle_B \\ +\tfrac{1}{2}|\Phi_2\rangle_{AC}\otimes\sigma_2|\Psi\rangle_B+\tfrac{1}{2}|\Phi_3\rangle_{AC}\otimes\sigma_3|\Psi\rangle_B.\end{aligned} \tag{18.7.8}$$

Note that all we have done so far is to rewrite the state of the three particles in a convenient form. The above right-hand side allows us to understand what happens when Alice measures the state of AC in the Bell basis. If she finds

- $|\Phi_0\rangle_{AC}$, the B state becomes $|\Psi\rangle_B$,
- $|\Phi_1\rangle_{AC}$, the B state becomes $\sigma_1|\Psi\rangle_B$,
- $|\Phi_2\rangle_{AC}$, the B state becomes $\sigma_2|\Psi\rangle_B$,
- $|\Phi_3\rangle_{AC}$, the B state becomes $\sigma_3|\Psi\rangle_B$.

If Alice gets $|\Phi_0\rangle_{AC}$, then Bob is in possession of the teleported state and has to do nothing. If Alice gets $|\Phi_i\rangle_{AC}$, Bob's particle is in the state $\sigma_i|\Psi\rangle_B$. Bob applies the ith box, which multiplies his state by σ_i, giving him the desired state $|\Psi\rangle_B$. The teleporting is thus complete.

Note that Alice is left with one of the Bell states $|\Phi_\mu\rangle_{AC}$, which has no information whatsoever about the constants α and β that defined the state to be teleported. Thus, the process did not create a copy of the state. The original state was destroyed in the process of teleportation. This is consistent with the no-cloning principle (section 18.9).

All the computational work above led to the key result (18.7.8), which is neatly summarized as the following identity valid for arbitrary states $|\Psi\rangle$:

$$|\Phi_0\rangle_{AB}\otimes|\Psi\rangle_C = \tfrac{1}{2}\sum_{\mu=0}^{3}|\Phi_\mu\rangle_{AC}\otimes\sigma_\mu|\Psi\rangle_B. \tag{18.7.9}$$

This is an identity for a state of three particles. It expresses the tensor product of an entangled state of the first two particles times a third as a sum of products that involve entangled states of the first and third particle times a state of the second particle.

18.8 EPR and Bell Inequalities

In this section we begin by studying some properties of the singlet state of two particles of spin one-half. We then turn to the claims of Einstein, Podolsky, and Rosen (EPR) concerning entangled states in quantum mechanics. Finally, we discuss the so-called Bell inequalities that would follow if EPR were right. Of course, quantum mechanics violates these inequalities, which experiment indeed shows. EPR were wrong.

We have been talking about the singlet state of two spin one-half particles. This state emerges, for example, in particle decays. The neutral η_0 meson, of rest mass 547 MeV, sometimes decays into a muon and an antimuon of opposite charge:

$$\eta_0 \to \mu^+ + \mu^-. \tag{18.8.1}$$

The meson is a spinless particle that being at rest has zero orbital angular momentum. As a result, it has zero total angular momentum. As it decays, the final state of the two muons must have zero total angular momentum as well. Most often, the state of the two muons has zero orbital angular momentum. In such a situation, conservation of angular momentum requires zero total spin angular momentum. The muon antimuon pair, flying away from each other with zero orbital angular momentum, are in a singlet state. This state takes the form

$$|\Psi\rangle = \tfrac{1}{\sqrt{2}}\big(|+\rangle_1|-\rangle_2 - |-\rangle_1|+\rangle_2\big). \tag{18.8.2}$$

As an angular momentum singlet, this state is rotational invariant (see also problem 18.1). The state is is fact the same for whatever direction $\mathbf{n}$ we use to define a basis of spin states:

$$|\Psi\rangle = \tfrac{1}{\sqrt{2}}\big(|\mathbf{n};+\rangle_1|\mathbf{n};-\rangle_2 - |\mathbf{n};-\rangle_1|\mathbf{n};+\rangle_2\big). \tag{18.8.3}$$

We now ask: In this singlet what is the probability $P(\mathbf{a},\mathbf{b})$ that the first particle is in the state $|\mathbf{a};+\rangle$, and the second particle is in the state $|\mathbf{b};+\rangle$, with $\mathbf{a}$ and $\mathbf{b}$ two arbitrarily chosen unit vectors? To help ourselves, we write the singlet state using the first vector:

$$|\Psi\rangle = \tfrac{1}{\sqrt{2}}\big(|\mathbf{a};+\rangle_1|\mathbf{a};-\rangle_2 - |\mathbf{a};-\rangle_1|\mathbf{a};+\rangle_2\big). \tag{18.8.4}$$

By definition, the probability we want is

$$P(\mathbf{a},\mathbf{b}) = \Big|{}_1\langle\mathbf{a};+|{}_2\langle\mathbf{b};+|\Psi\rangle\Big|^2. \tag{18.8.5}$$

Only the first term in (18.8.4) contributes and we get

$$P(\mathbf{a},\mathbf{b}) = \tfrac{1}{2}\big|\langle\mathbf{b};+|\mathbf{a};-\rangle\big|^2. \tag{18.8.6}$$

We recall that the overlap squared between two spin states is given by the cosine squared of half the angle in between them (example 14.2). Using figure 18.2, we see that the angle between $\mathbf{b}$ and $-\mathbf{a}$ is $\pi-\theta_{ab}$, where θ_{ab} is the angle between $\mathbf{b}$ and $\mathbf{a}$. Therefore,

$$P(\mathbf{a},\mathbf{b}) = \tfrac{1}{2}\cos^2\!\big(\tfrac{1}{2}(\pi-\theta_{ab})\big). \tag{18.8.7}$$

Our final result is therefore

$$\boxed{P(\mathbf{a},\mathbf{b}) = \tfrac{1}{2}\sin^2\!\big(\tfrac{1}{2}\theta_{ab}\big).} \tag{18.8.8}$$

As a simple consistency check, if $\mathbf{b}=-\mathbf{a}$, then $\theta_{ab}=\pi$, and $P(\mathbf{a},-\mathbf{a})=1/2$, which is what we expect. If we measure using orthogonal vectors, like the unit vectors $\hat{\mathbf{x}}$ and $\hat{\mathbf{z}}$, we get

$$P(\hat{\mathbf{z}},\hat{\mathbf{x}}) = \tfrac{1}{2}\sin^2 45^\circ = \tfrac{1}{2}\cdot\tfrac{1}{2} = \tfrac{1}{4}. \tag{18.8.9}$$

This is all we will need to know about singlet states.

The statements by EPR dealt with entangled states and formulated the notion of **local realism**. This is understood as two properties of measurement:

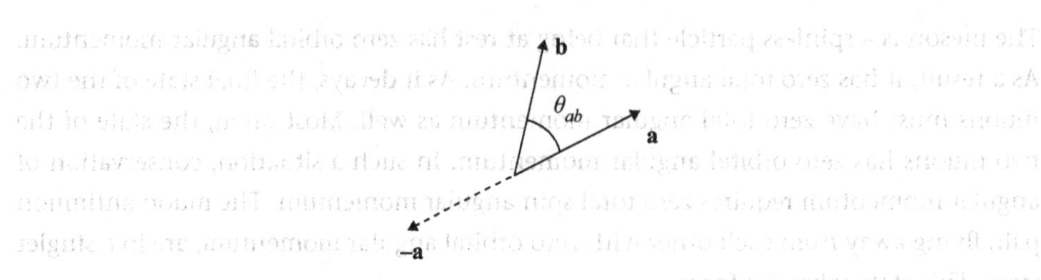

Figure 18.2
Directions associated with the vectors **a** and **b**.

1. The result of a measurement at one point cannot depend on whatever action takes place at a faraway point at the same time.
2. The result of a measurement on a particle in an entangled pair corresponds to some element of reality. If the measurement of an observable gives a value, that value was a definite property of the state before measurement.

Both properties seem eminently reasonable at first thought, but both are violated in quantum mechanics. The violation of the first is disturbing, given our intuition that simultaneous, spatially separated events can't affect each other. But there is something nonlocal about quantum mechanics. The violation of the second is something we have already seen repeatedly in nonentangled contexts. Measurement involves collapse of the wave function; the result is not preordained and does not correspond to an unequivocal property of the system.

Following the logic of EPR, the so-called entangled singlet pairs are just pairs of particles that have definite spins. Moreover, in this logic the results of quantum mechanical measurements are reproduced if our large ensemble of pairs has the following distribution of states:

- In 50% of pairs, particle 1 has spin along $\hat{\mathbf{z}}$, and particle 2 has spin along $-\hat{\mathbf{z}}$.
- In 50% of pairs, particle 1 has spin along $-\hat{\mathbf{z}}$, and particle 2 has spin along $\hat{\mathbf{z}}$.

This would explain the perfect correlations between the spins of the two particles and is consistent, for example, with $P(\hat{\mathbf{z}}, -\hat{\mathbf{z}}) = 1/2$, which we obtained quantum mechanically.

The challenge for the EPR proposal is to keep reproducing the results of more complicated measurements. Suppose each of the two observers can measure spin along two possible axes: the x- and z-axes. They measure in any of these two directions. EPR logic would posit that in any entangled pair each particle has a definite state of spin in these two directions. For example, a particle of type $[\hat{\mathbf{z}}, -\hat{\mathbf{x}}]$ is one that if measured along z always gives $\hbar/2$, while if measured along x gives $-\hbar/2$. This of course is not possible in quantum mechanics: if we want to guarantee that measurement along z gives $\hbar/2$, the state of the particle must be $|+\rangle$, in which case the result of measuring along x cannot be predicted. EPR logic implies that there are states of particles in which noncommuting variables have fixed, predictable values. In this setup the observed quantum mechanical results are matched if our ensemble of pairs has the following properties:

- 25% of pairs have particle 1 in $[\hat{\mathbf{z}}, \hat{\mathbf{x}}]$ and particle 2 in $[-\hat{\mathbf{z}}, -\hat{\mathbf{x}}]$,
- 25% of pairs have particle 1 in $[\hat{\mathbf{z}}, -\hat{\mathbf{x}}]$ and particle 2 in $[-\hat{\mathbf{z}}, \hat{\mathbf{x}}]$,
- 25% of pairs have particle 1 in $[-\hat{\mathbf{z}}, \hat{\mathbf{x}}]$ and particle 2 in $[\hat{\mathbf{z}}, -\hat{\mathbf{x}}]$,
- 25% of pairs have particle 1 in $[-\hat{\mathbf{z}}, -\hat{\mathbf{x}}]$ and particle 2 in $[\hat{\mathbf{z}}, \hat{\mathbf{x}}]$.

First, note some perfect correlations in this EPR-like setup: particles 1 and 2 have opposite spins in each possible direction, so knowing the spin of one in a particular direction tells us the spin of the second. This is, of course, needed to match the properties of the quantum mechanical singlets: if both measure in the same direction, they must get opposite values of the spin. We can ask: What is the probability $P(\hat{\mathbf{z}}, -\hat{\mathbf{z}})$ that particle 1 is along $\hat{\mathbf{z}}$ and particle 2 along $-\hat{\mathbf{z}}$? The first two cases above apply, and thus this probability is $1/2$, consistent with quantum mechanics. We can also ask for $P(\hat{\mathbf{z}}, \hat{\mathbf{x}})$. This time only the second case applies, giving us a probability of $1/4$, as we obtained earlier in (18.8.9). The quantum mechanical answers indeed arise for all questions.

The insight of Bell was to realize that when we can measure in *three* directions the quantum mechanical answers *cannot* be reproduced by suitable ensembles with pairs of particles having their own definite spin states. For this, he showed that such ensembles imply inequalities—the Bell inequalities—that are violated in quantum mechanics. Indeed, suppose each observer can measure along any one of the three vectors $\mathbf{a}, \mathbf{b}, \mathbf{c}$. Each particle is just measured once, along one of these directions. Let us assume that we have a large number N of pairs that, following the EPR logic, contain particles with well-defined spins on these three directions. A particle of type $[\mathbf{a}, -\mathbf{b}, \mathbf{c}]$, for example, would give $\hbar/2$, $-\hbar/2$, and $\hbar/2$ if measured along $\mathbf{a}$, $\mathbf{b}$, or $\mathbf{c}$, respectively. Again, this kind of state does not exist in quantum mechanics. Now, EPR logic would try to give a distribution of pairs of different types that would match quantum mechanical results. We will now show that any distribution will disagree with quantum mechanics. To do this, we keep the values of the populations general:

Populations	Particle 1	Particle 2
N_1	$[\mathbf{a}, \mathbf{b}, \mathbf{c}]$	$[-\mathbf{a}, -\mathbf{b}, -\mathbf{c}]$
N_2	$[\mathbf{a}, \mathbf{b}, -\mathbf{c}]$	$[-\mathbf{a}, -\mathbf{b}, \mathbf{c}]$
N_3	$[\mathbf{a}, -\mathbf{b}, \mathbf{c}]$	$[-\mathbf{a}, \mathbf{b}, -\mathbf{c}]$
N_4	$[\mathbf{a}, -\mathbf{b}, -\mathbf{c}]$	$[-\mathbf{a}, \mathbf{b}, \mathbf{c}]$
N_5	$[-\mathbf{a}, \mathbf{b}, \mathbf{c}]$	$[\mathbf{a}, -\mathbf{b}, -\mathbf{c}]$
N_6	$[-\mathbf{a}, \mathbf{b}, -\mathbf{c}]$	$[\mathbf{a}, -\mathbf{b}, \mathbf{c}]$
N_7	$[-\mathbf{a}, -\mathbf{b}, \mathbf{c}]$	$[\mathbf{a}, \mathbf{b}, -\mathbf{c}]$
N_8	$[-\mathbf{a}, -\mathbf{b}, -\mathbf{c}]$	$[\mathbf{a}, \mathbf{b}, \mathbf{c}]$

As required, all spins are properly correlated in particles 1 and 2. Moreover, $N = \sum_{i=1}^{8} N_i$. To find the probability $P(\mathbf{a}, \mathbf{b})$, for example, we look for the lines in the table that have particle 1 in $\mathbf{a}$ as well as particle 2 in $\mathbf{b}$. The populations that satisfy this are N_3 and N_4,

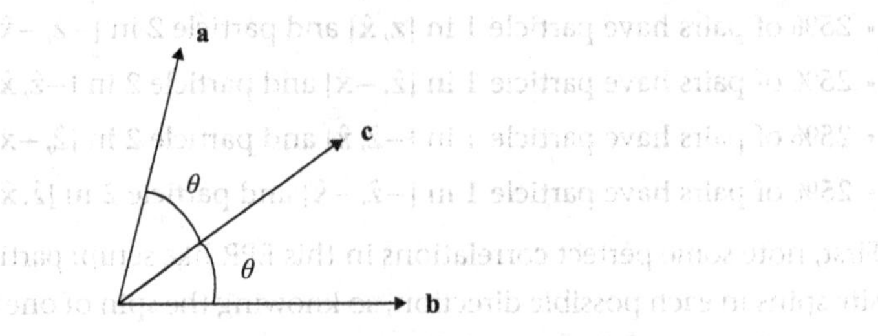

Figure 18.3
A planar configuration for the unit vectors **a**, **b**, and **c**. For $\theta < \frac{\pi}{2}$ the Bell inequality is violated quantum mechanically.

and their sum divided by the total number N of pairs is the desired probability. In this way we record the following probabilities:

$$P(\mathbf{a},\mathbf{b}) = \frac{N_3+N_4}{N}, \quad P(\mathbf{a},\mathbf{c}) = \frac{N_2+N_4}{N}, \quad P(\mathbf{c},\mathbf{b}) = \frac{N_3+N_7}{N}. \tag{18.8.10}$$

Consider now the trivially correct inequality:

$$N_3+N_4 \leq N_2+N_4+ N_3+N_7 \tag{18.8.11}$$

which on account of (18.8.10) implies the **Bell inequality:**

$$\boxed{P(\mathbf{a},\mathbf{b}) \leq P(\mathbf{a},\mathbf{c})+P(\mathbf{c},\mathbf{b}).} \tag{18.8.12}$$

If true quantum mechanically, given (18.8.8) we would have

$$\tfrac{1}{2}\sin^2 \tfrac{1}{2}\theta_{ab} \leq \tfrac{1}{2}\sin^2 \tfrac{1}{2}\theta_{ac} + \tfrac{1}{2}\sin^2 \tfrac{1}{2}\theta_{cb}. \tag{18.8.13}$$

But this is violated for many choices of angles. Take, for example, the planar configuration in figure 18.3:

$$\theta_{ab} = 2\theta, \quad \theta_{ac} = \theta_{cb} = \theta. \tag{18.8.14}$$

For this situation, the inequality becomes

$$\tfrac{1}{2}\sin^2\theta \leq \sin^2 \tfrac{1}{2}\theta. \tag{18.8.15}$$

This fails for sufficiently small θ: $\frac{1}{2}\theta^2 \leq \frac{\theta^2}{4}$ is just plain wrong. In fact, the Bell inequality is violated in quantum mechanics for any $\theta < \frac{\pi}{2}$. Experimental results have confirmed that Bell inequalities are violated, and thus the original claim of local realism by EPR is incorrect.

The arguments of Bell have been extended in several directions, and various examples have provided extra insight into the remarkable properties of entangled quantum states. Some of these directions are considered in the problems. Let us look at some of the ideas:

1. We have spoken about correlations. Let's be more precise about this term. In the spin singlet state, the spins of the two particles are said to be perfectly anticorrelated *when* measured in the same direction. Measuring the spin of one along **n** and reading the

result tells us the spin of the other particle along $\mathbf{n}$: it will be the opposite. There are also *correlation functions*. In a composite system AB with parts A and B and in the state $|\Psi_{AB}\rangle$, the expectation value of $\mathcal{O}_A \otimes \mathcal{O}_B$,

$$\langle \mathcal{O}_A \otimes \mathcal{O}_B \rangle = \langle \Psi_{AB} | \mathcal{O}_A \otimes \mathcal{O}_B | \Psi_{AB} \rangle, \tag{18.8.16}$$

with $\mathcal{O}_A \in \mathcal{L}(\mathcal{H}_A)$ and $\mathcal{O}_B \in \mathcal{L}(\mathcal{H}_B)$ is called the correlation function of $\mathcal{O}_A$ and $\mathcal{O}_B$. If the state of the system is a product state $|\Psi_{AB}\rangle = |\Psi_A\rangle \otimes |\Psi_B\rangle$, then

$$\begin{aligned} \langle \mathcal{O}_A \otimes \mathcal{O}_B \rangle &= \langle \Psi_A | \otimes \langle \Psi_B | \mathcal{O}_A \otimes \mathcal{O}_B | \Psi_A \rangle \otimes | \Psi_B \rangle \\ &= \langle \Psi_A | \mathcal{O}_A | \Psi_A \rangle \langle \Psi_B | \mathcal{O}_B | \Psi_B \rangle \\ &= \langle \mathcal{O}_A \rangle \langle \mathcal{O}_B \rangle, \end{aligned} \tag{18.8.17}$$

showing that the correlator factorizes into the product of expectation values when the state of the full system is not entangled. If the state is entangled, the correlation generically fails to factorize. For example, in the spin singlet state of two spin one-half particles and for unit vectors $\mathbf{a}$ and $\mathbf{b}$ you will show (problem 18.4) that the product of spin operators has the correlation

$$\langle \mathbf{a} \cdot \boldsymbol{\sigma}^{(1)} \otimes \mathbf{b} \cdot \boldsymbol{\sigma}^{(2)} \rangle = -\cos\theta_{ab}, \tag{18.8.18}$$

where θ_{ab} is the angle between the two unit vectors. This right-hand side does not factorize into the product of a function that depends on $\mathbf{a}$ and a function that depends on $\mathbf{b}$, showing the entanglement of the state. The EPR analogs of correlators of this type lead to a version of Bell inequalities (problem 18.4) that is indeed violated by the quantum correlators.

2. A Bell-type inequality obtained by Clauser, Horn, Shimony, and Holt (CHSH) is explored in problem 18.5. They consider an entangled system AB and two operators $\hat{A}_1, \hat{A}_2$ in system A as well as two operators $\hat{B}_1, \hat{B}_2$ in system B. All these operators have eigenvalues ± 1. As it turns out, on the entangled state the expectation value of the operator $\hat{Q}$ defined as

$$\hat{Q} \equiv \hat{A}_1 \otimes \hat{B}_1 - \hat{A}_1 \otimes \hat{B}_2 + \hat{A}_2 \otimes \hat{B}_1 + \hat{A}_2 \otimes \hat{B}_2 \tag{18.8.19}$$

exceeds the value it would take if the $\hat{A}$ and $\hat{B}$ operators were replaced by random variables with deterministic values of ± 1.

3. Bell inequalities can be derived from simple classes of *local hidden variable* theories. A hidden variable is a hypothetical quantity whose values are not accessible to the experimentalist. Let us denote by the generic label λ a set of hidden variables. These hidden variables are presumed to come with a probability distribution. When EPR talk about a particle measured by Alice of type $[\hat{\mathbf{z}}, \hat{\mathbf{x}}]$, it means that Alice measures $+1$ for σ_z and for σ_x. In a theory of hidden variables, one would have two functions $A(z, \lambda)$ and $A(x, \lambda)$, both taking values of ± 1. The first one, $A(z, \lambda)$, is Alice's measured value of spin along z when the hidden variable takes values λ. Similarly, $A(x, \lambda)$ is Alice's measured value of spin along x when the hidden variable takes values λ. For a particle of type $[\hat{\mathbf{z}}, \hat{\mathbf{x}}]$, one must have λ such that both these functions

give +1. Bob has similar functions, and for each entangled pair, the value of λ is the same as Alice's, reflecting the locality of the hidden variable theory. The functions determine uniquely the measured values once we know λ. The analog of quantum expectation values is the averaging over the values of the hidden variables, weighted by the probability distribution. Interestingly, independent of the probability distribution one can derive inequalities that are violated by quantum probabilities. The Bell inequality considered in problem 18.4 can be derived from a hidden variable theory.

4. Bell inequalities, as discussed above, involve probabilities or expectation values of operators. To determine these quantities we require repeated measurements. Some tests of quantum mechanics involve deterministic quantities—namely, the measurement of operators that happen to take definite values on the entangled state. The hidden variable theory analysis is done with a definite value of the hidden variable, and its prediction plainly disagrees with the quantum prediction. A particularly simple and elegant entangled state of three spin one-half particles illustrating this possibility (problem 18.6) was discussed by Greenberg, Horne, and Zeilinger (GHZ):

$$|\Phi\rangle = \frac{1}{\sqrt{2}}\big(|+\rangle|+\rangle|+\rangle - |-\rangle|-\rangle|-\rangle\big). \tag{18.8.20}$$

One considers operators $\hat{O}_i$, with $i = 1, 2, 3$, each acting on the state space of the three particles. The state $|\Phi\rangle$ is an eigenstate of all three $\hat{O}_i$'s, with eigenvalues all equal to +1. The operators are such that $\hat{O}_1\hat{O}_2\hat{O}_3 = -\hat{O}$, with $\hat{O}$ an operator for which $|\Phi\rangle$ must be an eigenstate of eigenvalue −1. In the hidden variable theory, whenever the $\hat{O}_i$ analogs have value +1 so does the analog of $\hat{O}$, giving us a discrepancy with quantum mechanics.

5. If Alice and Bob are far away and each has a particle from an entangled pair, they cannot use this pair to send information to each other. The correlations in the pair allow for no signaling. While we have focused on nonrelativistic quantum mechanics, signaling would be a problem in relativistic quantum mechanics, for it could allow information to be sent faster than light. We will discuss no signaling in detail in section 22.4 (theorem 22.4.2). You can see in problem 18.7 the failure of signaling with a particular strategy. Sharing entangled pairs, however, can somehow help Alice and Bob, who are far apart and cannot communicate with each other, do better in a game they play against Charlie (problems 18.8 and 18.9).

Exercise 18.17. *Alice, Bob, and Charlie are in possession of particles, 1, 2, and 3, respectively, in an entangled state*

$$|\Phi\rangle = \frac{1}{\sqrt{2}}\big(|+\rangle|+\rangle|+\rangle - |-\rangle|-\rangle|-\rangle\big). \tag{18.8.21}$$

Assume Alice measures the spin of particle 1 along the z-direction. Describe the possible states of particles 2 and 3 after her measurement. Are particles 2 and 3 entangled? Assume instead that Alice measures the spin of particle 1 along the x-direction. Describe the possible states of particles 2 and 3 after measurement. Are particles 2 and 3 entangled?

18.9 No-Cloning Property

It is often said that one cannot copy or clone a quantum state. While the statement captures the essence of a key quantum property, the precise version of the *no-cloning* property of quantum mechanics is more nuanced. A quantum cloning machine can be designed to clone a few states in a finite-dimensional state space, but a machine that clones arbitrary states cannot be built.

To understand the result properly, one must define what one means by cloning of a quantum state $|\psi\rangle \in V$, with V some vector space. A quantum state is the state of some quantum system—a particle or an atom, for example. In cloning there is no magical device that, for example, takes an electron in some quantum state and creates another electron in that quantum state. We must start with *two* electrons—call them electron one and electron two. Assume electron one is in some state that we want cloned. The cloning machine must copy the state of electron one into electron two without changing the state of electron one. This is analogous to the way a photocopier works: there is a page with information, the original, and there is a blank page. The machine copies the information of the original onto the blank page without altering the original. This is why the no-cloning property is sometimes referred to as the *no-xeroxing* property. Recall that in teleportation we copied a state, but it was done at the cost of destroying the original state.

For an arbitrary spin one-half state $|\psi\rangle = a_+|+\rangle + a_-|-\rangle$ to be cloned, we would require a second spin one-half particle in some "blank" state $|b\rangle$. The blank state is just some arbitrary but fixed state of the spin, perhaps $|+\rangle$ or $|-\rangle$. As we try cloning different states $|\psi\rangle$, the blank state is kept fixed. The cloning machine must act as follows:

$$\text{Cloning machine:}\ (a_+|+\rangle + a_-|-\rangle)\otimes|b\rangle \ \rightarrow\ e^{i\phi}\,(a_+|+\rangle + a_-|-\rangle)\otimes(a_+|+\rangle + a_-|-\rangle), \tag{18.9.1}$$

with ϕ an arbitrary phase. If perfectly functional, for a fixed blank state $|b\rangle$ the machine must implement this map for all values of a_+ and a_-. The inclusion of an arbitrary phase in the final state will be seen to be immaterial, as it does not help get cloning to work. The above action (18.9.1) with the blank state set equal to $|+\rangle$ is represented in figure 18.4.

Since quantum mechanical evolution is unitary, we assume that the cloning machine is just a unitary operator U acting on the tensor product $V\otimes V$. If, instead, the machine

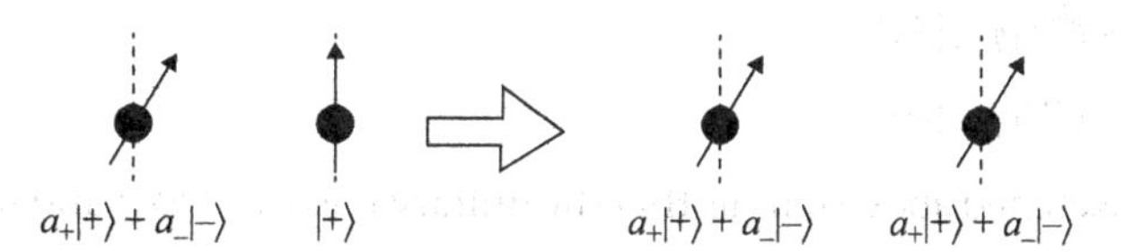

Figure 18.4
Cloning machine, represented by an arrow, copies the state $a_+|+\rangle + a_-|-\rangle$ of the left particle onto the second particle, the blank particle, initially in the state $|+\rangle$.

involved measurements, the output would not be deterministic, and we do not want this. In general, for a normalized state $|\psi\rangle \in V$ to be copied, and a blank state $|b\rangle \in V$ also normalized, we must have

$$U: \; |\psi\rangle \otimes |b\rangle \;\to\; e^{i\phi}|\psi\rangle \otimes |\psi\rangle, \qquad \langle\psi|\psi\rangle = 1, \; \langle b|b\rangle = 1, \tag{18.9.2}$$

with ϕ a phase the could depend on $|\psi\rangle$ and $|b\rangle$. The no-cloning property is summarized by the statement of the following theorem:

Theorem 18.9.1. *For the arbitrary but fixed, normalized state $|b\rangle \in V$, there is no unitary operator $U \in \mathcal{L}(V \otimes V)$ that implements the map (18.9.2) for arbitrary normalized $|\psi\rangle \in V$.*

Corollary. *In a vector space V of dimension n, the maximal number of states that can be cloned by a unitary U is n. These vectors comprise an orthonormal basis of V.*

Proof. Consider the map (18.9.2) for some fixed state $|\psi_1\rangle$ to be copied:

$$U: \; |\psi_1\rangle|b\rangle \to e^{i\phi_1}|\psi_1\rangle|\psi_1\rangle. \tag{18.9.3}$$

There is certainly a unitary $U: V \otimes V \to V \otimes V$ that implements this map. This is clear because the states to the left and to the right of the arrow have the same norm:

$$\begin{aligned} \big\| |\psi_1\rangle|b\rangle \big\|^2 &= \langle\psi_1|\psi_1\rangle\langle b|b\rangle = 1, \\ \Big\| e^{i\phi_1}|\psi_1\rangle|\psi_1\rangle \Big\|^2 &= \langle\psi_1|\psi_1\rangle\langle\psi_1|\psi_1\rangle = 1. \end{aligned} \tag{18.9.4}$$

Indeed, we can explain this more generally, in a way that helps build the rest of the argument. In an arbitrary space W (corresponding to $V \otimes V$ in our case of interest), a unitary operator U can be easily constructed that maps one chosen unit vector $|e_1\rangle$ to some other unit vector $|f_1\rangle$. For this we use Gram-Schmidt to extend $|e_1\rangle$ to an orthonormal basis $|e_i\rangle, i = 1, \ldots, \dim W$, and similarly, we extend $|f_1\rangle$ to an orthonormal basis $|f_i\rangle$, $i = 1, \ldots, \dim W$. The requisite unitary operator U is then $U = \sum_i |f_i\rangle\langle e_i|$. Similarly, given $p \le \dim W$ orthonormal basis vectors $|e_1\rangle, \ldots, |e_p\rangle$ that are mapped, one by one, into p orthonormal basis vectors $|f_1\rangle, \ldots, |f_p\rangle$, there is also a unitary operator that accomplishes this. Again, the operator is constructed by completing the set of vectors into a full orthonormal set.

Applied to the situation in (18.9.3), the above argument shows the existence of a unitary that maps one unit vector in $V \otimes V$, the state to the left of the arrow, to another unit vector in $V \otimes V$, the state to the right of the arrow. Having shown there is a unitary operator U that realizes the map (18.9.3), we ask if the unitary operator can be modified so that it also clones a second state $|\psi_2\rangle$:

$$\begin{aligned} U: \quad & |\psi_1\rangle|b\rangle \to e^{i\phi_1}|\psi_1\rangle|\psi_1\rangle, \\ & |\psi_2\rangle|b\rangle \to e^{i\phi_2}|\psi_2\rangle|\psi_2\rangle. \end{aligned} \tag{18.9.5}$$

The argument sketched above tells us that the unitary U exists if $|\psi_1\rangle$ and $|\psi_2\rangle$ are orthogonal and thus orthonormal vectors in V. This implies that the two $V \otimes V$ states to the left of the arrows are orthonormal and so are the resulting cloned states to the right of the arrows. It is clear now that a unitary U exists that can clone a chosen set of dim V orthonormal states in V. But as we see next, we cannot do better.

For this, consider again the case of two states to be cloned as in (18.9.5), and recall that a unitary operator preserves inner products (see (14.5.21)). If a unitary exists, the inner product of the states to be cloned must equal the inner product of the cloned states:

$$\Big\langle |\psi_1\rangle|b\rangle, |\psi_2\rangle|b\rangle\Big\rangle = \Big\langle e^{i\phi_1}|\psi_1\rangle|\psi_1\rangle, e^{i\phi_2}|\psi_2\rangle|\psi_2\rangle\Big\rangle. \tag{18.9.6}$$

This gives the constraint

$$\langle\psi_1|\psi_2\rangle = e^{i(\phi_2-\phi_1)}(\langle\psi_1|\psi_2\rangle)^2. \tag{18.9.7}$$

One solution of this constraint is $\langle\psi_1|\psi_2\rangle = 0$. We knew this: cloning the two states is possible if the states are orthogonal. The other solution is

$$\langle\psi_1|\psi_2\rangle = e^{i(\phi_1-\phi_2)}. \tag{18.9.8}$$

This condition implies $|\langle\psi_1|\psi_2\rangle|^2 = 1$. But the Schwarz inequality requires $|\langle\psi_1|\psi_2\rangle|^2 \leq \langle\psi_1|\psi_1\rangle\langle\psi_2|\psi_2\rangle = 1$. Since the inequality is saturated, $|\psi_2\rangle$ equals $|\psi_1\rangle$ up to a phase. Then $|\psi_2\rangle$ is not a new state we can clone. This argument implies that if we have a set of clonable states, a *new* state can only be cloned if it is orthogonal to all the clonable states.

Thus, we can begin by picking one state to clone and then add successively orthonormal states that we can clone. Imagine now we have a U that clones dim V orthonormal states, the maximal number of orthonormal states that can be obtained in V. No nontrivial linear combination of these clonable states can be cloned because no such state is orthogonal to all the clonable states. This completes the proof of the no-cloning theorem, as well as that of the corollary. □

Exercise 18.18. *Alice and Bob are far away from each other but share an entangled pair of spin one-half particles in the singlet state. They aim to communicate information by agreeing that Alice will measure her particle along x if she wins the lottery and along z if she loses. Convince yourself that the strategy will not work unless Bob has a quantum cloning machine.*

Problems

Problem 18.1. *Spin states for two particles.*

Consider the entangled state $|\Psi\rangle$ of two spin one-half particles:

$$|\Psi\rangle = \tfrac{1}{\sqrt{2}}\big(|+\rangle\otimes|-\rangle - |-\rangle\otimes|+\rangle\big).$$

We want to show this is a rotationally invariant state. For this consider a spin rotation operator $\hat{R}$, a unitary operator with the property that $\hat{R}|+\rangle = |\mathbf{n};+\rangle$, where $\mathbf{n}$ is some unit vector.

1. Explain why $\hat{R}|-\rangle$ is, up to a phase, equal to $|\mathbf{n};-\rangle$.
2. In the tensor product, the rotation operator is $\hat{R}\otimes\hat{R}$. Show that $(\hat{R}\otimes\hat{R})\,|\Psi\rangle$ is in fact equal to $|\Psi\rangle$ up to a phase. This is the statement of rotational invariance of $|\Psi\rangle$.

Problem 18.2. *Rotating a state of two spin one-half particles (A. Harrow).*

Suppose we are given a state $|\psi\rangle$ of two spin one-half particles:

$$|\psi\rangle \equiv \alpha\,|x;+\rangle\otimes|x;+\rangle \;+\; \beta\,|x;-\rangle\otimes|x;-\rangle, \quad \text{with } \alpha,\beta\in\mathbb{C}.$$

Here $|x;\pm\rangle \equiv \frac{1}{\sqrt{2}}(|+\rangle\pm|-\rangle)$. We would like to transform $|\psi\rangle$ into the state $|\psi'\rangle$ given by

$$|\psi'\rangle \equiv \alpha|+\rangle\otimes|+\rangle \;+\; \beta|-\rangle\otimes|-\rangle.$$

We wish to do so by unitary time evolution: $e^{-i\hat{H}'t'/\hbar}|\psi\rangle = |\psi'\rangle$, with $\hat{H}' = \hat{H}_0\otimes\mathbb{1}+\mathbb{1}\otimes\hat{H}_0$, for some $\hat{H}_0$, time independent, and some time t'. The Hamiltonian $\hat{H}'$ should do this for *any* values of α and β.

1. Substitute the above expression for $\hat{H}'$ into $e^{-i\hat{H}'t'/\hbar}$, and write this operator in the form $e^{-i\hat{H}'t'/\hbar} = e^{i\hat{K}}\otimes e^{i\hat{K}}$. What is $\hat{K}$ in terms of $t'\hat{H}_0/\hbar$?
2. Since $\hat{K}$ determines $\hat{H}'$, we just want to determine $\hat{K}$. The operator $\hat{K}$ is unique if you assume that $e^{i\hat{K}}$ is a rotation operator. Find $\hat{K}$ under this assumption.
3. Calculate explicitly $e^{i\hat{K}}|x;\pm\rangle$. Write your answer in the $\{|+\rangle,|-\rangle\}$ basis.

Problem 18.3. *Entanglement generation (A. Harrow).*

A pair of spin one-half particles interact via the Hamiltonian

$$\hat{H} = \hbar\omega\,\sigma_3\otimes\sigma_3,$$

where ω is a scalar with units of frequency. As usual $|\pm\rangle \equiv |z;\pm\rangle$, and $|x;\pm\rangle \equiv \frac{1}{\sqrt{2}}(|+\rangle\pm|-\rangle)$. For the two-particle states, define the *Z-basis* to be $|1\rangle=|++\rangle$, $|2\rangle=|+-\rangle$, $|3\rangle=|-+\rangle$, $|4\rangle=|--\rangle$, where $|++\rangle\equiv|+\rangle\otimes|+\rangle$, etc.

1. Write down the matrix $\hat{H}$ and the matrix $e^{-i\hat{H}t/\hbar}$ in the Z-basis.
2. Calculate the state $e^{-i\hat{H}t/\hbar}|x;+\rangle\otimes|x;+\rangle$ in the Z-basis. For what values of t is this state *not* entangled?

Problem 18.4. *Bell inequality from correlation functions.*

Let $\mathbf{a}$ and $\mathbf{b}$ denote two unit vectors. Following EPR logic consider the correlation coefficient $C(\mathbf{a},\mathbf{b})$ that takes the average, over an "ensemble" of singlet states of two spin one-half particles, of the product of the measured spin of particle 1 along $\mathbf{a}$ and the measured spin of particle 2 along $\mathbf{b}$:

$$C(\mathbf{a},\mathbf{b}) \equiv \left[\frac{4}{\hbar^2}S_{\mathbf{a}}^{(1)}S_{\mathbf{b}}^{(2)}\right]_{av}.$$

Since any value of measured spin can be only $\pm\hbar/2$, this ensemble average must range between $+1$ and -1. Now consider three directions $\mathbf{a}$, $\mathbf{b}$, and $\mathbf{c}$ and the following quantity $g(\mathbf{a},\mathbf{b},\mathbf{c})$ to be measured:

$$g(\mathbf{a},\mathbf{b},\mathbf{c}) \equiv -\frac{4}{\hbar^2}S_{\mathbf{a}}^{(1)}S_{\mathbf{b}}^{(1)}\left(1-\frac{4}{\hbar^2}S_{\mathbf{b}}^{(1)}S_{\mathbf{c}}^{(1)}\right).$$

Note that all superscripts refer to particle 1. In the sense of EPR, the $S_{\mathbf{n}}^{(i)}$ are not operators but measured values. Thus, for any unit vector $\mathbf{n}$ we have $S_{\mathbf{n}}^{(1)} S_{\mathbf{n}}^{(1)} = \hbar^2/4$, and the anticorrelations of singlets imply that $S_{\mathbf{n}}^{(1)} = -S_{\mathbf{n}}^{(2)}$.

1. Show that

$$[g(\mathbf{a},\mathbf{b},\mathbf{c})]_{av} = C(\mathbf{a},\mathbf{b}) - C(\mathbf{a},\mathbf{c}).$$

2. Using the inequality $\left|[g]_{av}\right| \le [\,|g|]_{av}$ where $|\ldots|$ denote absolute values, show that

$$\left|C(\mathbf{a},\mathbf{b}) - C(\mathbf{a},\mathbf{c})\right| - C(\mathbf{b},\mathbf{c}) \le 1. \tag{1}$$

This is an inequality that must be obeyed by a good theory in the sense of EPR.

3. Turning to quantum mechanics, $S_{\mathbf{a}}^{(1)}$ and $S_{\mathbf{b}}^{(2)}$ become, respectively, the spin operators $\hat{S}_{\mathbf{a}}^{(1)}$ and $\hat{S}_{\mathbf{b}}^{(2)}$ of the first and second particle (recall that $\hat{S}_{\mathbf{n}} \equiv \mathbf{n}\cdot\hat{\mathbf{S}}$). Moreover, $C(\mathbf{a},\mathbf{b})$, now denoted with a hat, becomes the expectation value of their operator product in the singlet state:

$$\hat{C}(\mathbf{a},\mathbf{b}) \equiv \left\langle \frac{4}{\hbar^2}\hat{S}_{\mathbf{a}}^{(1)}\hat{S}_{\mathbf{b}}^{(2)} \right\rangle = \left\langle \mathbf{a}\cdot\boldsymbol{\sigma}^{(1)}\,\mathbf{b}\cdot\boldsymbol{\sigma}^{(2)} \right\rangle.$$

Prove that $\hat{C}(\mathbf{a},\mathbf{b}) = -\cos\theta_{ab}$, where θ_{ab} is the angle between the vectors $\mathbf{a}$ and $\mathbf{b}$.

4. Let $\mathbf{a},\mathbf{b},\mathbf{c}$ be coplanar vectors with $\theta_{ab} = \theta_{bc} = \theta$ and $\theta_{ac} = 2\theta$. Plot the left-hand side of equation (1), with Cs replaced by the hatted versions $\hat{C}$, as a function of θ, and show that the inequality is violated for $\theta \le \pi/2$.

Problem 18.5. *Clauser-Horne-Shimony-Holt (CHSH) inequality.*

Suppose that A_1, A_2, B_1, B_2 are numbers, and each can only take the values ± 1. Define the number Q by the relation

$$Q \equiv A_1B_1 - A_1B_2 + A_2B_1 + A_2B_2.$$

1. Show that the variable Q is bounded as $-a \le Q \le a$, with a a real positive constant. Find the smallest value of a for which the inequality holds.

Define the θ-dependent, Hermitian spin operator $\hat{W}_\theta$ as follows:

$$\hat{W}_\theta \equiv \sigma_x \sin\theta + \sigma_z \cos\theta.$$

2. What are the eigenvalues of $\hat{W}_\theta$? How should θ and θ' be related for $\hat{W}_\theta$ and $\hat{W}_{\theta'}$ to commute?

Let $|\Psi\rangle \equiv \frac{1}{\sqrt{2}}\big(|+\rangle \otimes |-\rangle - |-\rangle \otimes |+\rangle\big)$ denote the singlet state of two spin one-half particles.

3. Determine $\langle\Psi|\hat{W}_\theta \otimes \hat{W}_{\theta'}|\Psi\rangle$, and express your answer as a function of $\theta - \theta'$.

Alice and Bob share an entangled singlet pair $|\Psi\rangle$. Alice is in possession of the first particle, and Bob is in possession of the second particle. Alice has two operators $\hat{A}_1$ and $\hat{A}_2$ she can measure, and Bob has two operators $\hat{B}_1$ and $\hat{B}_2$ he can measure. These operators are

$$\hat{A}_1 = \hat{W}_0, \quad \hat{A}_2 = \hat{W}_{\pi/2},$$

$$\hat{B}_1 = \hat{W}_{\pi/4}, \quad \hat{B}_2 = \hat{W}_{3\pi/4}.$$

The possible values that can be measured for each of these operators is ± 1.

4. Do Alice's operators $\hat{A}_1$ and $\hat{A}_2$ commute? Do Bob's operators $\hat{B}_1$ and $\hat{B}_2$ commute? Consider now the following operator $\hat{Q}$ on the state space of the two particles:

$$\hat{Q} \equiv \hat{A}_1 \otimes \hat{B}_1 - \hat{A}_1 \otimes \hat{B}_2 + \hat{A}_2 \otimes \hat{B}_1 + \hat{A}_2 \otimes \hat{B}_2.$$

Calculate the expectation value $\langle \Psi | \hat{Q} | \Psi \rangle$. The answer should violate the range in (1) for the EPR analog Q. What does this mean?

5. Armed with a very large supply of entangled pairs, how would you have Alice and Bob use them to experimentally find $\langle \Psi | \hat{Q} | \Psi \rangle$?

Problem 18.6. *Deterministic test of hidden variables in the entanglement of three particles.*

Consider a decay process of a particle into a state of three separate but entangled spin one-half particles, the first detected by Alice, the second detected by Bob, and the third detected by Charlie. All three can measure spin in the x and y directions. The decay is characterized by some value λ of hidden variables that determine the measured values without uncertainty. Thus, we have the function $A(x, \lambda) \in \pm 1$ giving the value of the spin of Alice's particle in the x direction and $A(y, \lambda) \in \pm 1$ for the value of the spin of Alice's particle in the y direction. There are similar functions for Bob and Charlie, so we have six functions:

$$A(x, \lambda), \quad A(y, \lambda), \quad B(x, \lambda), \quad B(y, \lambda), \quad C(x, \lambda), \quad C(y, \lambda),$$

all of which have definite values for our fixed λ and can be either $+1$ or -1. Alice, Bob, and Charlie are able to do the measurements so that all six values are determined for this particular decay. Define three products, each involving one measurement along x and two measurements along y:

$$O_1 = A(x, \lambda)\, B(y, \lambda) C(y, \lambda),$$

$$O_2 = A(y, \lambda)\, B(x, \lambda) C(y, \lambda),$$

$$O_3 = A(y, \lambda)\, B(y, \lambda) C(x, \lambda).$$

1. If we find $O_1 = O_2 = O_3 = 1$, explain why we must also find $O \equiv A(x, \lambda) B(x, \lambda) C(x, \lambda) = 1$.

In quantum mechanics σ_x and σ_y measure spin in the x and y directions, respectively. Therefore, we consider the six operators $\sigma_x^A, \sigma_y^A; \sigma_x^B, \sigma_y^B; \sigma_x^C, \sigma_y^C$ and define

$$\hat{O}_1 = \sigma_x^A \otimes \sigma_y^B \otimes \sigma_y^C,$$

$$\hat{O}_2 = \sigma_y^A \otimes \sigma_x^B \otimes \sigma_y^C,$$

$$\hat{O}_3 = \sigma_y^A \otimes \sigma_y^B \otimes \sigma_x^C,$$

$$\hat{O} = \sigma_x^A \otimes \sigma_x^B \otimes \sigma_x^C.$$

Finally, consider the Greenberg, Horn, Zeilinger (GHZ) state $|\Phi\rangle = \frac{1}{\sqrt{2}}(|+\rangle|+\rangle|+\rangle - |-\rangle|-\rangle|-\rangle)$.

2. Show that $|\Phi\rangle$ is an eigenstate of $\hat{O}_1, \hat{O}_2, \hat{O}_3$, and $\hat{O}$. Determine the eigenvalues, and discuss the discrepancy with the results of (1).

3. Relate the product $\hat{O}_1 \hat{O}_2 \hat{O}_3$ to $\hat{O}$. Discuss the result.

Problem 18.7. *Attempting to signal with an entangled pair (A. Guth).*

Alice and Bob share an entangled pair of spin one-half particles in the singlet state. The particles are far apart. Alice has access to particle 1, and Bob has access to particle 2.

1. Alice measures the spin of her particle along the direction $\mathbf{n}$ defined by angles θ, ϕ. What is the probability $P_A(\mathbf{n})$ that she measures spin up along $\mathbf{n}$?
2. Assume we start again with another entangled pair. Bob measures spin along the z-direction and finds it up. What is the probability $P_A(\mathbf{n}|b{=}\uparrow)$ that Alice finds her spin up along $\mathbf{n}$? If, instead, Bob finds the spin down along z, what is the probability $P_A(\mathbf{n}|b{=}\downarrow)$ that Alice finds her spin along $\mathbf{n}$?
3. Bob and Alice have at their disposal a huge amount of entangled pairs. Bob considers sending a signal to Alice. If he does not measure, Alice finds spin up along $\mathbf{n}$ with probability $P_A(\mathbf{n})$, as in (1). But if he measures, she finds two different probabilities, as in (2). The experiment is repeated many times, and Alice knows that either Bob is never measuring or Bob is always measuring. We want to understand if Alice could decide if Bob is or is not measuring. For this purpose find the probability that Alice measures spin up along $\mathbf{n}$ when Bob is always measuring.

Problem 18.8. *Beating the odds using entangled states: part 1.*

Consider two entangled spins in the state $|\Psi\rangle$ corresponding to the first Bell basis state:

$$|\Psi\rangle = \frac{1}{\sqrt{2}}\big(|+\rangle\otimes|+\rangle + |-\rangle\otimes|-\rangle\big). \tag{1}$$

Given a state $|\psi\rangle = \alpha|+\rangle + \beta|-\rangle$, we define the associated "bar" state $|\bar{\psi}\rangle \equiv \alpha^*|+\rangle + \beta^*|-\rangle$.

1. Consider a spin state $|v_a\rangle$ of the first particle and a spin state $|w_b\rangle$ of the second particle. Show that the probability $P(v_a, w_b)$ of getting those states upon measurement on the entangled state is

$$P(v_a, w_b) = \tfrac{1}{2}\big|\langle\bar{v}_a|w_b\rangle\big|^2. \tag{2}$$

Alice and Bob, who are far away from each other and incommunicado, play a game against Charlie. The game uses bits, which are variables that can take two values: zero or one. In a given round of the game, Charlie supplies a bit x to Alice and a bit y to Bob. Alice does not know what bit Bob got, and Bob does not know what bit Alice got. Alice must use the bit x to output a bit $a(x)$, and Bob must use his bit y to output a bit $b(y)$. Alice and Bob win the round against Charlie if

$$a + b \equiv xy \pmod{2}.$$

Thus, for example, if $x{=}y{=}1$ the right-hand side is one, and winning requires $a \neq b$ ($a{=}0, b{=}1$ or $a{=}1, b{=}0$). If either x or y is zero, the right-hand side is zero, and winning works for $a{=}b$ (either both one or both zero). If Charlie supplies bits randomly, a good strategy for winning is for Alice and Bob to output $a{=}b{=}1$ (or zero) for all inputs x, y. The left-hand side is then zero, and this will win for all cases except $x{=}y{=}1$. The probability of winning is then $3/4$.

2. In a classical strategy, Alice chooses a function $a(x)$ for her output bit. She has four choices that can be represented by the functions $\{0, 1, x, 1-x\}$. Explain why. Similarly, Bob's choice of a function $b(y)$ runs over the possibilities $\{0, 1, y, 1-y\}$. For any choice they make,

$$a(x)+b(y) \equiv c_0 + c_1 x + c_2 y \pmod 2,$$

with some constants c_0, c_1, and c_2 that can take values zero or one. Show that there is no choice of these constants for which the right-hand side above equals $xy \pmod 2$, thus giving a strategy that always wins. Explain why this result implies that no (classical) strategy has a winning probability bigger than 3/4.

Problem 18.9. *Beating the odds using entangled states: part 2.*

Alice and Bob, in their game against Charlie, devise a quantum strategy using their shared entangled pair $|\Psi\rangle = \frac{1}{\sqrt{2}}(|+\rangle \otimes |+\rangle + |-\rangle \otimes |-\rangle)$. Their strategy requires fixing four real constants $\alpha_0, \alpha_1, \beta_0$, and β_1, naturally thought of as angles. Alice will make use of the orthonormal basis

$$\begin{aligned} |v_0^x\rangle &\equiv \cos\tfrac{\alpha_x}{2}|+\rangle + \sin\tfrac{\alpha_x}{2}|-\rangle, \\ |v_1^x\rangle &\equiv -\sin\tfrac{\alpha_x}{2}|+\rangle + \cos\tfrac{\alpha_x}{2}|-\rangle. \end{aligned}$$

Since the bit x takes two values, Alice actually has a couple of bases: one for $x=0$, comprising the two states above with α_0, and another for $x=1$, comprising the two states above for α_1. Similarly, Bob will use

$$\begin{aligned} |w_0^y\rangle &\equiv \cos\tfrac{\beta_y}{2}|+\rangle + \sin\tfrac{\beta_y}{2}|-\rangle, \\ |w_1^y\rangle &\equiv -\sin\tfrac{\beta_y}{2}|+\rangle + \cos\tfrac{\beta_y}{2}|-\rangle. \end{aligned}$$

Since y takes two values, Bob also has a couple of bases: one for $y=0$, comprising the two states above with β_0, and another for $y=1$, comprising the two states above for β_1.

The quantum strategy is now as follows. For any given bit x, Alice measures her entangled spin along the basis $(|v_0^x\rangle, |v_1^x\rangle)$. If she finds the spin along the first basis vector, she outputs $a=0$; if she finds the spin along the second basis vector, she outputs $a=1$. For any given bit y, Bob measures his entangled spin along the basis $(|w_0^y\rangle, |w_1^y\rangle)$. If he finds the spin along the first basis vector, he outputs $b=0$; if he finds the spin along the second basis vector, he outputs $b=1$.

1. Show that with the above strategy the probability $P[a=b\,|\,x,y]$ that a is equal to b for input values x, y is given by

$$P[a=b\,|\,x,y] = \cos^2 \tfrac{1}{2}(\alpha_x - \beta_y).$$

2. Find a formula for the probability P of winning in terms of the four unknown angles $\alpha_0, \alpha_1, \beta_0$, and β_1. Show that for $\alpha_0=0, \alpha_1=\pi/2$, and $\beta_0=-\beta_1=\pi/4$, you get

$$P = \cos^2\tfrac{\pi}{8} = \tfrac{1}{2} + \tfrac{1}{2\sqrt{2}} \simeq 0.85355.$$

This gives about a 14% better chance of winning than classically. Note that in the quantum strategy, the outputs a, b for a given x, y are not deterministic.

Problem 18.10. *Swap operator.*

Letting $\sigma_0 \equiv \mathbb{1}$, we define the operator F acting on the tensor product of two spin one-half spaces as follows:

$$F \equiv \frac{1}{2}\sum_{\mu=0}^{3} \sigma_\mu \otimes \sigma_\mu = \frac{1}{2}\Big(\mathbb{1} \otimes \mathbb{1} + \sum_{i=1}^{3} \sigma_i \otimes \sigma_i\Big).$$

Show that this is a swap operator; that is,

$$F\,(|\alpha\rangle \otimes |\beta\rangle) = |\beta\rangle \otimes |\alpha\rangle, \quad \forall\, |\alpha\rangle, |\beta\rangle.$$

Find the matrix representation of F in the ordered basis $\{|+\rangle|+\rangle, |+\rangle|-\rangle, |-\rangle|+\rangle, |-\rangle|-\rangle\}$ in two ways: (i) using the swap property on the basis states and (ii) from the formula above in terms of tensor products of Pauli matrices.

Problem 18.11. *Cloning and entanglement.*

When it clones, a unitary operator takes a nonentangled state to a nonentangled state (why?). Assume that you have a cloning operator for spin one-half states that can clone $|+\rangle$ and $|-\rangle$. Explain why it cannot clone $a_+|+\rangle + a_-|-\rangle$ when both a_+ and a_- are nonzero.

19 Angular Momentum and Central Potentials: Part II

We use index and vector notation to streamline our work on angular momentum and to define scalar and vector operators. Using the algebra of angular momentum operators, we show that multiplets are labeled by j, with $2j$ a nonnegative integer, and contain $2j+1$ states with different values of the z-component of angular momentum. We list the complete set of commuting observables for a Hamiltonian with a three-dimensional central potential. We discuss the free particle solutions, which are used to solve classes of radial problems and to derive Rayleigh's formula: a representation of a plane wave in terms of an infinite sum of spherical waves. We examine the three-dimensional isotropic harmonic oscillator, describing the spectrum in terms of multiplets of angular momentum. Finally, focusing on the hydrogen atom, we discuss the conservation of the classical Runge-Lenz vector and that of its quantum analogue.

19.1 Angular Momentum and Quantum Vector Identities

We had our first exposure to angular momentum in chapter 10. There we learned how to build orbital angular momentum operators using position and momentum operators. Here, we revisit this construction using vector notation and index manipulation. This enables us to streamline computations involving angular momentum as well as position and momentum operators. Moreover, it allows us to define operators that are scalars under rotations and operators that are vectors under rotations.

In order to use index notation, we must call the x-, y-, and z-components of vectors the first, second, and third components. We use indices $i,j,k,\ldots=1,2,3$ that run over three values (context will distinguish between an index i and the imaginary complex number i). Therefore, for position, momentum, and angular momentum operators we have $(\hat{x}_1,\hat{x}_2,\hat{x}_3)$ instead of $(\hat{x},\hat{y},\hat{z})$, $(\hat{p}_1,\hat{p}_2,\hat{p}_3)$ instead of $(\hat{p}_x,\hat{p}_y,\hat{p}_z)$, and $(\hat{L}_1,\hat{L}_2,\hat{L}_3)$ instead of $(\hat{L}_x,\hat{L}_y,\hat{L}_z)$. The triplets of position and momentum operators satisfy commutation relations that are fully summarized by

$$[\hat{x}_i,\hat{p}_j] = i\hbar\,\delta_{ij}, \quad [\hat{x}_i,\hat{x}_j] = 0, \quad [\hat{p}_i,\hat{p}_j] = 0. \tag{19.1.1}$$

The angular momentum operators, inspired by their classical analogues, were defined by

$$\hat{L}_1 \equiv \hat{x}_2\,\hat{p}_3 - \hat{x}_3\,\hat{p}_2, \quad \hat{L}_2 \equiv \hat{x}_3\,\hat{p}_1 - \hat{x}_1\,\hat{p}_3, \quad \hat{L}_3 \equiv \hat{x}_1\,\hat{p}_2 - \hat{x}_2\,\hat{p}_1. \tag{19.1.2}$$

The angular momentum operators are Hermitian since $\hat{x}_i$ and $\hat{p}_i$ are Hermitian, and the products can be reordered without cost:

$$\hat{L}_i^\dagger = \hat{L}_i. \tag{19.1.3}$$

We will write triplets of operators as boldface vectors so that

$$\hat{\mathbf{r}} \equiv (\hat{x}_1, \hat{x}_2, \hat{x}_3), \quad \hat{\mathbf{p}} \equiv (\hat{p}_1, \hat{p}_2, \hat{p}_3), \quad \hat{\mathbf{L}} \equiv (\hat{L}_1, \hat{L}_2, \hat{L}_3). \tag{19.1.4}$$

These are not familiar vectors, as their components are operators. We therefore call them **operator-valued vectors**. Operator-valued vectors are useful whenever we want to use the dot and cross products of three-dimensional space. The identities of vector analysis have analogues for operator-valued vectors. We will develop these identities now. Let us therefore consider operator-valued vectors $\hat{\mathbf{a}}$ and $\hat{\mathbf{b}}$:

$$\hat{\mathbf{a}} = (\hat{a}_1, \hat{a}_2, \hat{a}_3), \quad \hat{\mathbf{b}} = (\hat{b}_1, \hat{b}_2, \hat{b}_3), \tag{19.1.5}$$

and we will assume that the components $\hat{a}_i$ and $\hat{b}_i$ are operators that, in general, fail to commute. The following are our definitions of dot and cross products for our operator-valued vectors:

$$\begin{aligned} \hat{\mathbf{a}} \cdot \hat{\mathbf{b}} &\equiv \hat{a}_i\,\hat{b}_i, \\ (\hat{\mathbf{a}} \times \hat{\mathbf{b}})_i &\equiv \epsilon_{ijk}\,\hat{a}_j\,\hat{b}_k. \end{aligned} \tag{19.1.6}$$

In the second equation, the left-hand side is the ith component of the cross product. Repeated indices are summed over the three possible values 1,2, and 3. The order of the operators on the above right-hand sides cannot be changed; it was chosen to be the same as the order of the operators on the left-hand sides. We also define

$$\hat{\mathbf{a}}^2 \equiv \hat{\mathbf{a}} \cdot \hat{\mathbf{a}}. \tag{19.1.7}$$

Since the operators do not commute, familiar properties of vector analysis do not hold. For example, $\hat{\mathbf{a}} \cdot \hat{\mathbf{b}}$ is not equal to $\hat{\mathbf{b}} \cdot \hat{\mathbf{a}}$. Indeed,

$$\hat{\mathbf{a}} \cdot \hat{\mathbf{b}} = \hat{a}_i\,\hat{b}_i = [\,\hat{a}_i, \hat{b}_i\,] + \hat{b}_i\,\hat{a}_i \tag{19.1.8}$$

so that

$$\boxed{\hat{\mathbf{a}} \cdot \hat{\mathbf{b}} = \hat{\mathbf{b}} \cdot \hat{\mathbf{a}} + [\,\hat{a}_i, \hat{b}_i\,].} \tag{19.1.9}$$

As an application we have

$$\hat{\mathbf{r}} \cdot \hat{\mathbf{p}} = \hat{\mathbf{p}} \cdot \hat{\mathbf{r}} + [\,\hat{x}_i, \hat{p}_i\,]. \tag{19.1.10}$$

The rightmost commutator gives $i\hbar\,\delta_{ii} = 3i\hbar$ so that we have the amusing three-dimensional identity

$$\boxed{\hat{\mathbf{r}} \cdot \hat{\mathbf{p}} = \hat{\mathbf{p}} \cdot \hat{\mathbf{r}} + 3\,i\hbar.} \tag{19.1.11}$$

For cross products we typically have $\hat{\mathbf{a}}\times\hat{\mathbf{b}}\neq-\hat{\mathbf{b}}\times\hat{\mathbf{a}}$. Indeed,

$$\begin{aligned}(\hat{\mathbf{a}}\times\hat{\mathbf{b}})_i &= \epsilon_{ijk}\,\hat{a}_j\,\hat{b}_k = \epsilon_{ijk}\left([\hat{a}_j,\,\hat{b}_k]+\hat{b}_k\,\hat{a}_j\right)\\ &= -\,\epsilon_{ikj}\,\hat{b}_k\,\hat{a}_j+\epsilon_{ijk}\,[\hat{a}_j,\,\hat{b}_k],\end{aligned} \tag{19.1.12}$$

where we flipped the k, j indices in one of the epsilon tensors in order to identify a cross product. Indeed, we now have

$$\boxed{(\hat{\mathbf{a}}\times\hat{\mathbf{b}})_i = -(\hat{\mathbf{b}}\times\hat{\mathbf{a}})_i+\epsilon_{ijk}\,[\hat{a}_j,\,\hat{b}_k].} \tag{19.1.13}$$

The simplest example of the use of this identity is one where we use $\hat{\mathbf{r}}$ and $\hat{\mathbf{p}}$. Certainly,

$$\hat{\mathbf{r}}\times\hat{\mathbf{r}} = 0, \quad \text{and} \quad \hat{\mathbf{p}}\times\hat{\mathbf{p}} = 0, \tag{19.1.14}$$

but more nontrivially,

$$(\hat{\mathbf{r}}\times\hat{\mathbf{p}})_i = -(\hat{\mathbf{p}}\times\hat{\mathbf{r}})_i+\epsilon_{ijk}\,[\hat{x}_j,\,\hat{p}_k]. \tag{19.1.15}$$

The last term vanishes, for it is equal to $i\hbar\,\epsilon_{ijk}\delta_{jk}=0$: the epsilon symbol is antisymmetric in j, k, while the delta is symmetric in j, k, resulting in a zero result. We therefore have, quantum mechanically,

$$\boxed{\hat{\mathbf{r}}\times\hat{\mathbf{p}} = -\,\hat{\mathbf{p}}\times\hat{\mathbf{r}}.} \tag{19.1.16}$$

Thus, $\hat{\mathbf{r}}$ and $\hat{\mathbf{p}}$ can be moved across in the cross product but not in the dot product.

Exercise 19.1. *Prove the following identities for Hermitian conjugation:*

$$\begin{aligned}(\hat{\mathbf{a}}\cdot\hat{\mathbf{b}})^\dagger &= \hat{\mathbf{b}}^\dagger\cdot\hat{\mathbf{a}}^\dagger,\\ (\hat{\mathbf{a}}\times\hat{\mathbf{b}})^\dagger &= -\,\hat{\mathbf{b}}^\dagger\times\hat{\mathbf{a}}^\dagger.\end{aligned} \tag{19.1.17}$$

The angular momentum operators are in fact given by the cross product of the operator-valued vectors $\hat{\mathbf{r}}$ and $\hat{\mathbf{p}}$:

$$\boxed{\hat{\mathbf{L}} = \hat{\mathbf{r}}\times\hat{\mathbf{p}} = -\hat{\mathbf{p}}\times\hat{\mathbf{r}}.} \tag{19.1.18}$$

Given the definition of the product, we have

$$\boxed{\hat{L}_i = \epsilon_{ijk}\,\hat{x}_j\,\hat{p}_k.} \tag{19.1.19}$$

If you evaluate the right-hand side for $i=1,2,3$, you will recover the expressions in (19.1.2). The Hermiticity of the angular momentum operator is verified using (19.1.17), and recalling that $\hat{\mathbf{r}}$ and $\hat{\mathbf{p}}$ are Hermitian,

$$\hat{\mathbf{L}}^\dagger = (\hat{\mathbf{r}}\times\hat{\mathbf{p}})^\dagger = -\hat{\mathbf{p}}^\dagger\times\hat{\mathbf{r}}^\dagger = -\hat{\mathbf{p}}\times\hat{\mathbf{r}} = \hat{\mathbf{L}}. \tag{19.1.20}$$

The use of vector notation implies that, for example,

$$\hat{\mathbf{L}}^2 = \hat{\mathbf{L}}\cdot\hat{\mathbf{L}} = \hat{L}_1\hat{L}_1+\hat{L}_2\hat{L}_2+\hat{L}_3\hat{L}_3 = \hat{L}_i\hat{L}_i. \tag{19.1.21}$$

The classical angular momentum $\vec{L}$ is orthogonal to both $\vec{r}$ and $\vec{p}$, as it is built from the cross product of these two vectors. Happily, these properties also hold for the quantum analogues. Take, for example, the dot product of $\hat{\mathbf{r}}$ with $\hat{\mathbf{L}}$:

$$\hat{\mathbf{r}}\cdot\hat{\mathbf{L}} = \hat{x}_i \hat{L}_i = \hat{x}_i \epsilon_{ijk} \hat{x}_j \hat{p}_k = \epsilon_{ijk} \hat{x}_i \hat{x}_j \hat{p}_k = 0. \tag{19.1.22}$$

The last expression is zero because the $\hat{x}$'s commute and thus form an object symmetric in i, j, while the epsilon symbol is antisymmetric in i, j. Similarly,

$$\hat{\mathbf{p}}\cdot\hat{\mathbf{L}} = \hat{p}_i \hat{L}_i = -\hat{p}_i(\hat{\mathbf{p}}\times\hat{\mathbf{r}})_i = -\hat{p}_i \epsilon_{ijk} \hat{p}_j \hat{x}_k = -\epsilon_{ijk} \hat{p}_i \hat{p}_j \hat{x}_k = 0. \tag{19.1.23}$$

In summary,

$$\boxed{\hat{\mathbf{r}}\cdot\hat{\mathbf{L}} = \hat{\mathbf{p}}\cdot\hat{\mathbf{L}} = 0.} \tag{19.1.24}$$

In manipulating multiple cross products, the following identities are quite useful:

$$\epsilon_{ijk}\,\epsilon_{ipq} = \delta_{jp}\delta_{kq} - \delta_{jq}\delta_{kp} \quad \Rightarrow \quad \epsilon_{ijk}\,\epsilon_{ijq} = 2\delta_{kq}. \tag{19.1.25}$$

The most familiar application involves triple products, which we consider now for the operator case. Taking care not to move operators across each other, we find that

$$\begin{aligned}
[\hat{\mathbf{a}}\times(\hat{\mathbf{b}}\times\hat{\mathbf{c}})]_k &= \epsilon_{kji}\,\hat{a}_j\,(\hat{\mathbf{b}}\times\hat{\mathbf{c}})_i \\
&= \epsilon_{kji}\,\epsilon_{ipq}\,\hat{a}_j\,\hat{b}_p\,\hat{c}_q \\
&= -\,\epsilon_{ijk}\,\epsilon_{ipq}\,\hat{a}_j\,\hat{b}_p\,\hat{c}_q \\
&= -\,(\delta_{jp}\delta_{kq} - \delta_{jq}\delta_{kp})\hat{a}_j\,\hat{b}_p\,\hat{c}_q \\
&= \hat{a}_j\,\hat{b}_k\,\hat{c}_j - \hat{a}_j\,\hat{b}_j\,\hat{c}_k.
\end{aligned} \tag{19.1.26}$$

At this point, there is a dot product in the second term. In the first term, the components of $\hat{\mathbf{a}}$ and $\hat{\mathbf{c}}$ are contracted, but there is an operator $\hat{b}_k$ in between. To have a dot product, the components of $\hat{\mathbf{a}}$ and $\hat{\mathbf{b}}$ must be next to each other. Commuting $\hat{a}_j$ and $\hat{b}_k$, we get

$$\begin{aligned}
[\hat{\mathbf{a}}\times(\hat{\mathbf{b}}\times\hat{\mathbf{c}})]_k &= [\,\hat{a}_j,\,\hat{b}_k\,]\hat{c}_j + \hat{b}_k\,\hat{a}_j\,\hat{c}_j - \hat{a}_j\,\hat{b}_j\,\hat{c}_k \\
&= [\,\hat{a}_j,\,\hat{b}_k\,]\hat{c}_j + \hat{b}_k\,(\hat{\mathbf{a}}\cdot\hat{\mathbf{c}}) - (\hat{\mathbf{a}}\cdot\hat{\mathbf{b}})\,\hat{c}_k.
\end{aligned} \tag{19.1.27}$$

We can write this as

$$\hat{\mathbf{a}}\times(\hat{\mathbf{b}}\times\hat{\mathbf{c}}) = \hat{\mathbf{b}}\,(\hat{\mathbf{a}}\cdot\hat{\mathbf{c}}) - (\hat{\mathbf{a}}\cdot\hat{\mathbf{b}})\,\hat{\mathbf{c}} + [\,\hat{a}_j,\,\hat{\mathbf{b}}\,]\,\hat{c}_j. \tag{19.1.28}$$

The first two terms are all there is in the familiar classical identity; the last term is quantum mechanical. Another familiar relation from classical vector analysis is

$$(\vec{a}\times\vec{b})^2 \equiv (\vec{a}\times\vec{b})\cdot(\vec{a}\times\vec{b}) = \vec{a}^2\,\vec{b}^2 - (\vec{a}\cdot\vec{b})^2. \tag{19.1.29}$$

In deriving this equation, the vector components are assumed to be commuting numbers. If we have operator-valued vectors, additional terms arise.

Exercise 19.2. *Show that*

$$\begin{aligned}
(\hat{\mathbf{a}}\times\hat{\mathbf{b}})^2 = \;& \hat{\mathbf{a}}^2\,\hat{\mathbf{b}}^2 - (\hat{\mathbf{a}}\cdot\hat{\mathbf{b}})^2 \\
& - \hat{a}_j\,[\hat{a}_j, \hat{b}_k]\,\hat{b}_k + \hat{a}_j\,[\hat{a}_k, \hat{b}_k]\,\hat{b}_j - \hat{a}_j\,[\hat{a}_k, \hat{b}_j]\,\hat{b}_k - \hat{a}_j\hat{a}_k\,[\hat{b}_k, \hat{b}_j],
\end{aligned} \tag{19.1.30}$$

and verify that this yields

$$(\hat{\mathbf{a}}\times\hat{\mathbf{b}})^2 = \hat{\mathbf{a}}^2\,\hat{\mathbf{b}}^2-(\hat{\mathbf{a}}\cdot\hat{\mathbf{b}})^2+\gamma\,\hat{\mathbf{a}}\cdot\hat{\mathbf{b}}, \quad \textit{if}\;\; [\hat{a}_i,\hat{b}_j]=\gamma\,\delta_{ij},\;\; \gamma\in\mathbb{C},\;\; [\hat{b}_i,\hat{b}_j]=0. \tag{19.1.31}$$

As an application we calculate $\hat{\mathbf{L}}^2=(\hat{\mathbf{r}}\times\hat{\mathbf{p}})^2$. Equation (19.1.31) can be applied with $\hat{\mathbf{a}}=\hat{\mathbf{r}}$ and $\hat{\mathbf{b}}=\hat{\mathbf{p}}$. Since $[\hat{a}_i,\hat{b}_j]=[\hat{x}_i,\hat{p}_j]=i\hbar\,\delta_{ij}$, we identify $\gamma=i\hbar$ so that

$$\boxed{\hat{\mathbf{L}}^2 = \hat{\mathbf{r}}^2\,\hat{\mathbf{p}}^2-(\hat{\mathbf{r}}\cdot\hat{\mathbf{p}})^2+i\hbar\,\hat{\mathbf{r}}\cdot\hat{\mathbf{p}}.} \tag{19.1.32}$$

Another useful and simple identity is the following:

$$\hat{\mathbf{a}}\cdot(\hat{\mathbf{b}}\times\hat{\mathbf{c}}) = (\hat{\mathbf{a}}\times\hat{\mathbf{b}})\cdot\hat{\mathbf{c}}, \tag{19.1.33}$$

as you should confirm in a one-line computation. In commuting vector analysis, this triple product is known to be cyclically symmetric. Note that in the above, no operator has been moved across another—that's why it holds.

19.2 Properties of Angular Momentum

The orbital angular momentum operators act nicely on the $\hat{x}_i$ and $\hat{p}_i$ operators. The action here is by commutators, and you should verify that

$$\boxed{\begin{aligned}[\hat{L}_i\,,\,\hat{x}_j\,] &= i\hbar\,\epsilon_{ijk}\,\hat{x}_k\,,\\ [\hat{L}_i\,,\,\hat{p}_j\,] &= i\hbar\,\epsilon_{ijk}\,\hat{p}_k\,.\end{aligned}} \tag{19.2.1}$$

We say that these equations mean that $\hat{\mathbf{r}}$ and $\hat{\mathbf{p}}$ are vectors under rotations.

Exercise 19.3. *Prove the commutator relations (19.2.1).*

Exercise 19.4. *Use the above relations and (19.1.13) to show that*

$$\hat{\mathbf{p}}\times\hat{\mathbf{L}} = -\,\hat{\mathbf{L}}\times\hat{\mathbf{p}} \,+\, 2i\hbar\,\hat{\mathbf{p}}. \tag{19.2.2}$$

Hermitization is the process by which we construct a Hermitian operator starting from a non-Hermitian one. Say Ω is not Hermitian. Its Hermitization Ω_h is defined to be

$$\Omega_h \equiv \tfrac{1}{2}(\Omega+\Omega^\dagger). \tag{19.2.3}$$

Exercise 19.5. *Show that the Hermitization of* $\hat{\mathbf{p}}\times\hat{\mathbf{L}}$ *is*

$$(\hat{\mathbf{p}}\times\hat{\mathbf{L}})_h = \tfrac{1}{2}(\hat{\mathbf{p}}\times\hat{\mathbf{L}} \,-\, \hat{\mathbf{L}}\times\hat{\mathbf{p}}) = \hat{\mathbf{p}}\times\hat{\mathbf{L}} \,-\, i\hbar\,\hat{\mathbf{p}}. \tag{19.2.4}$$

We have stated that $\hat{\mathbf{r}}$ and $\hat{\mathbf{p}}$ are vectors under rotations. More generally, we declare that an operator $\hat{\mathbf{u}}=(\hat{u}_1,\hat{u}_2,\hat{u}_3)$ is a **vector under rotations** if

$$[\hat{L}_i\,,\,\hat{u}_j\,] = i\hbar\,\epsilon_{ijk}\,\hat{u}_k\,. \tag{19.2.5}$$

We will also declare that an operator $\hat{Z}$ that commutes with all angular momentum operators is a **scalar under rotations:**

$$[\hat{L}_i\,,\,\hat{Z}\,] = 0\,. \tag{19.2.6}$$

Exercise 19.6. *Assume* $\mathbf{n}$ *is a constant vector, and* $\hat{\mathbf{u}}$ *is a vector under rotations. Use the commutator (19.2.5) to show that*

$$[\mathbf{n}\cdot\hat{\mathbf{L}},\, \hat{\mathbf{u}}] = -i\hbar\, \mathbf{n}\times\hat{\mathbf{u}}. \tag{19.2.7}$$

We first learned in section 10.1 that angular momentum operators generate rotations, meaning they can be used to build unitary operators that rotate states. These unitary rotation operators also rotate operators, as shown in detail in section 14.7 and summarized in (14.7.25) for arbitrary angular momentum operators. When the unitary rotation operators $\hat{R}_{\mathbf{n}}(\alpha)$ act on a vector operator $\hat{\mathbf{u}}$, the vector is rotated by the action of a rotation matrix $\mathcal{R}_{\mathbf{n}}(\alpha)$, as follows:

$$\hat{R}_{\mathbf{n}}^{\dagger}(\alpha)\,\hat{\mathbf{u}}\,\hat{R}_{\mathbf{n}}(\alpha) = \mathcal{R}_{\mathbf{n}}(\alpha)\,\hat{\mathbf{u}}\,, \quad \hat{R}_{\mathbf{n}}(\alpha) = e^{-i\frac{\alpha}{\hbar}\mathbf{n}\cdot\hat{\mathbf{L}}}. \tag{19.2.8}$$

For more details on the notation, see section 14.7.

Exercise 19.7. *Prove the above formula by going over the steps in theorem 14.7.1 that led to the analogous (14.7.25).*

A scalar operator is simply left invariant by the action of the rotation operators $\hat{R}_{\mathbf{n}}(\alpha)$. We will see later in this chapter that the Hamiltonian for a central potential is a scalar operator.

Given two operators $\hat{\mathbf{u}}$ and $\hat{\mathbf{v}}$ that are vectors under rotations, their dot product is a scalar under rotations, and their cross product is a vector under rotations:

$$\boxed{\begin{aligned} [\hat{L}_i, \hat{\mathbf{u}}\cdot\hat{\mathbf{v}}] &= 0, \\ [\hat{L}_i, (\hat{\mathbf{u}}\times\hat{\mathbf{v}})_j] &= i\hbar\,\epsilon_{ijk}\,(\hat{\mathbf{u}}\times\hat{\mathbf{v}})_k. \end{aligned}} \tag{19.2.9}$$

Exercise 19.8. *Prove the above equations.*

A number of useful commutator identities follow from (19.2.9). Most importantly, from the second one, taking $\hat{\mathbf{u}}=\hat{\mathbf{r}}$ and $\hat{\mathbf{v}}=\hat{\mathbf{p}}$, we get

$$[\hat{L}_i, (\hat{\mathbf{r}}\times\hat{\mathbf{p}})_j] = i\hbar\,\epsilon_{ijk}\,(\hat{\mathbf{r}}\times\hat{\mathbf{p}})_k, \tag{19.2.10}$$

which gives a conceptually clear rederivation of the well-known algebra of angular momentum (10.3.12):

$$\boxed{[\hat{L}_i, \hat{L}_j] = i\hbar\,\epsilon_{ijk}\,\hat{L}_k.} \tag{19.2.11}$$

Note that $\hat{\mathbf{L}}$ itself is a vector under rotations. More explicitly, the above commutators read

$$[\hat{L}_x, \hat{L}_y] = i\hbar\hat{L}_z, \quad [\hat{L}_y, \hat{L}_z] = i\hbar\hat{L}_x, \quad [\hat{L}_z, \hat{L}_x] = i\hbar\hat{L}_y. \tag{19.2.12}$$

Note the cyclic nature of these equations: take the first and cycle indices ($x\to y\to z\to x$) once to obtain the second, and cycle again to obtain the third. This is how you can remember these relations by heart! Since the dot product of vector operators is a scalar operator, we have

$$[\hat{L}_i, \hat{\mathbf{r}}^2] = [\hat{L}_i, \hat{\mathbf{p}}^2] = [\hat{L}_i, \hat{\mathbf{r}}\cdot\hat{\mathbf{p}}] = 0, \tag{19.2.13}$$

and, very importantly,

$$[\hat{L}_i, \hat{\mathbf{L}}^2] = 0. \tag{19.2.14}$$

The operator $\hat{\mathbf{L}}^2$ will feature in the complete set of commuting observables for a central potential. An operator, such as $\hat{\mathbf{L}}^2$, that commutes with all $\hat{L}_i$ is called a *Casimir* operator of the algebra of angular momentum. Note that the validity of (19.2.14) just uses the algebra of the $\hat{L}_i$ operators, not their explicit definition in terms of $\hat{\mathbf{r}}$ and $\hat{\mathbf{p}}$. Since the spin operators also satisfy the algebra of angular momentum, we have $[\hat{S}_i, \hat{\mathbf{S}}^2] = 0$. While $\hat{\mathbf{L}}^2$ commutes with the angular momentum operators, it *fails* to commute with the $\hat{x}_i$ operators and with the $\hat{p}_i$ operators.

Exercise 19.9. *Use the algebra of $\hat{\mathbf{L}}$ operators to show that*

$$\hat{\mathbf{L}}\times\hat{\mathbf{L}} = i\hbar\hat{\mathbf{L}}. \tag{19.2.15}$$

This is a very elegant way to express the algebra of angular momentum. You can also check that the three components of this vector equation in fact imply the relations (19.2.12). It follows that (19.2.15) is an equivalent statement of the algebra of angular momentum:

$$\boxed{\hat{\mathbf{L}}\times\hat{\mathbf{L}} = i\hbar\hat{\mathbf{L}} \quad\Longleftrightarrow\quad [\hat{L}_i, \hat{L}_j] = i\hbar\,\epsilon_{ijk}\hat{L}_k.} \tag{19.2.16}$$

More generally, commutation relations of the form

$$[\hat{a}_i, \hat{b}_j] = \epsilon_{ijk}\,\hat{c}_k \tag{19.2.17}$$

admit a natural rewriting in terms of cross products. From (19.1.13),

$$(\hat{\mathbf{a}}\times\hat{\mathbf{b}})_i + (\hat{\mathbf{b}}\times\hat{\mathbf{a}})_i = \epsilon_{ijk}[\hat{a}_j, \hat{b}_k] = \epsilon_{ijk}\epsilon_{jkp}\hat{c}_p = 2\,\hat{c}_i. \tag{19.2.18}$$

This means that

$$\boxed{[\hat{a}_i, \hat{b}_j] = \epsilon_{ijk}\,\hat{c}_k \quad\Longrightarrow\quad \hat{\mathbf{a}}\times\hat{\mathbf{b}} + \hat{\mathbf{b}}\times\hat{\mathbf{a}} = 2\,\hat{\mathbf{c}}.} \tag{19.2.19}$$

The arrow *does not* work in the reverse direction. One finds that $[\hat{a}_i, \hat{b}_j] = \epsilon_{ijk}\,\hat{c}_k + \hat{s}_{ij}$ where $\hat{s}_{ij} = \hat{s}_{ji}$ is arbitrary and undetermined. If the arrow could be reversed, then the relation $\hat{\mathbf{a}}\times\hat{\mathbf{b}} + \hat{\mathbf{b}}\times\hat{\mathbf{a}} = 0$ would incorrectly imply that $\hat{\mathbf{a}}$ and $\hat{\mathbf{b}}$ commute. Indeed, while $\hat{\mathbf{r}}\times\hat{\mathbf{p}} + \hat{\mathbf{p}}\times\hat{\mathbf{r}} = 0$ (see (19.1.18)), the operators $\hat{\mathbf{r}}$ and $\hat{\mathbf{p}}$ don't commute.

Since a vector $\hat{\mathbf{u}}$ under rotations satisfies equation (19.2.5), our result above implies that it also satisfies the vector identity:

$$\hat{\mathbf{u}}\text{ is a vector under rotations} \quad\Longrightarrow\quad \hat{\mathbf{L}}\times\hat{\mathbf{u}} + \hat{\mathbf{u}}\times\hat{\mathbf{L}} = 2\,i\hbar\,\hat{\mathbf{u}}. \tag{19.2.20}$$

In spherical coordinates, the angular momentum operators are combinations of angular derivatives. They have no radial dependence, as is expected for operators that generate rotations. We derived such results in section 10.4. For reference, they are repeated here:

$$\hat{L}_z = \frac{\hbar}{i}\frac{\partial}{\partial\phi}, \quad \hat{L}_\pm \equiv \hat{L}_x \pm i\hat{L}_y = \pm\hbar e^{\pm i\phi}\Big(\frac{\partial}{\partial\theta} \pm i\cot\theta\frac{\partial}{\partial\phi}\Big). \tag{19.2.21}$$

19.3 Multiplets of Angular Momentum

In our first encounter with angular momentum in chapter 10, we learned that there are angular wave functions $Y_{\ell m}(\theta, \phi)$, called spherical harmonics, that are eigenfunctions of the differential operators $\hat{L}_z$ and $\hat{\mathbf{L}}^2$ (see (10.5.32)):

$$\begin{aligned} \hat{\mathbf{L}}^2\, Y_{\ell m} &= \hbar^2\, \ell(\ell+1)\, Y_{\ell m}, \quad && \ell = 0, 1, \ldots \\ \hat{L}_z\, Y_{\ell m} &= \hbar m\, Y_{\ell m}, && m = -\ell, \ldots, \ell. \end{aligned} \tag{19.3.1}$$

For any fixed integer value of ℓ, there are $2\ell+1$ spherical harmonics, as the index m quantifying the z-component of angular momentum varies from $-\ell$ to ℓ in integer steps. These wave functions can be thought of as eigenstates of the operators $\hat{L}_z$ and $\hat{\mathbf{L}}^2$. They form a *multiplet*, a collection of $2\ell+1$ states that have a common value of $\hat{\mathbf{L}}^2$.

In fact, the theory of spin one-half angular momentum supplies two basis states $|\pm\rangle$ that also form a multiplet. For this recall that

$$\hat{S}_z|\pm\rangle = \pm\tfrac{\hbar}{2}|\pm\rangle, \tag{19.3.2}$$

suggesting that the states correspond to $m = \pm\frac{1}{2}$ in the notation we used for orbital angular momentum. But how about the $\hat{\mathbf{S}}^2$ eigenvalue? Recalling that $\hat{S}_i = \frac{\hbar}{2}\sigma_i$, with σ_i the Pauli matrices, we see that

$$\hat{\mathbf{S}}^2 = \hat{S}_1^2 + \hat{S}_2^2 + \hat{S}_3^2 = \tfrac{1}{4}\hbar^2(\sigma_1^2 + \sigma_2^2 + \sigma_3^2). \tag{19.3.3}$$

Since all Pauli matrices square to the 2×2 identity matrix, we have

$$\hat{\mathbf{S}}^2 = \tfrac{3}{4}\hbar^2\mathbb{1}. \tag{19.3.4}$$

It follows that acting on the $|\pm\rangle$ states and writing $\frac{3}{4} = \frac{1}{2}\frac{3}{2}$ we get

$$\begin{aligned} \hat{\mathbf{S}}^2|\pm\rangle &= \hbar^2\tfrac{1}{2}\tfrac{3}{2}|\pm\rangle, \\ \hat{S}_z|\pm\rangle &= \pm\tfrac{\hbar}{2}|\pm\rangle, \end{aligned} \tag{19.3.5}$$

where we copied (19.3.2) on the second line. In analogy to (19.3.1), for spin operators and associated states $|s, m_s\rangle$ we write

$$\begin{aligned} \hat{\mathbf{S}}^2\, |s, m_s\rangle &= \hbar^2\, s(s+1)\, |s, m_s\rangle, \\ \hat{S}_z\, |s, m_s\rangle &= \hbar m_s\, |s, m_s\rangle. \end{aligned} \tag{19.3.6}$$

Comparing with (19.3.5), we see that the states $|\pm\rangle$ correspond to $s = \frac{1}{2}$. Moreover, we deduce that $m_s = \pm\frac{1}{2}$ and identify

$$|\pm\rangle = |s = \tfrac{1}{2}, m_s = \pm\tfrac{1}{2}\rangle. \tag{19.3.7}$$

In fact, the rule we had for orbital multiplets seems to hold here as well: $m_s = -s, \ldots, s$ in integer steps. In this example, $s = \frac{1}{2}$, and there are two states. Note that orbital angular momentum does not allow ℓ to be equal to $\frac{1}{2}$. We do not have spherical harmonics corresponding to fractional values of ℓ. It is perhaps not too surprising that orbital angular momentum and spin angular momentum can have different kinds of eigenstates. After all, the orbital operators are built from position and momenta, while spin operators cannot be built this way.

A general understanding of angular momentum requires learning what multiplets are *allowed*. For this we want to use only the algebra of angular momentum so the results will apply to orbital angular momentum, spin one-half angular momentum, and any other angular momentum that may exist. In order to speak generally of angular momentum operators, we do not call them $\hat{L}_i$ or $\hat{S}_i$ but rather use the new symbol $\hat{J}_i$, with $i = 1, 2, 3$. The operator-valued vector is denoted as $\hat{\mathbf{J}}$.

Any triplet of *Hermitian* operators $\hat{\mathbf{J}} = (\hat{J}_1, \hat{J}_2, \hat{J}_3) = (\hat{J}_x, \hat{J}_y, \hat{J}_z)$ is said to satisfy the algebra of angular momentum if the following commutation relations hold:

$$\text{Algebra of angular momentum:} \quad [\hat{J}_i, \hat{J}_j] = i\hbar\,\epsilon_{ijk}\,\hat{J}_k. \tag{19.3.8}$$

All the properties of the $\hat{L}_i$ operators derived earlier using only the algebra of angular momentum clearly hold for $\hat{J}_i$ operators, which satisfy the same algebra. Thus, on account of (19.2.14) we have

$$[\hat{J}_i, \hat{\mathbf{J}}^2] = 0. \tag{19.3.9}$$

It is useful to begin our work by considering an alternative rewriting of the algebra of angular momentum.

Rewriting the algebra of angular momentum Recall that in the harmonic oscillator the Hermitian operators $\hat{x}$ and $\hat{p}$ could be traded for the non-Hermitian operators $\hat{a}$ and $\hat{a}^\dagger$ that are in fact Hermitian conjugates of each other. We do something analogous here. We will define two non-Hermitian operators $\hat{J}_\pm$ starting from the Hermitian operators $\hat{J}_x$ and $\hat{J}_y$:

$$\begin{aligned} \hat{J}_+ &\equiv \hat{J}_x + i\hat{J}_y, \\ \hat{J}_- &\equiv \hat{J}_x - i\hat{J}_y. \end{aligned} \tag{19.3.10}$$

The two operators are Hermitian conjugates of each other:

$$(\hat{J}_+)^\dagger = \hat{J}_-. \tag{19.3.11}$$

Note that both $\hat{J}_x$ and $\hat{J}_y$ can be solved for in terms of $\hat{J}_+$ and $\hat{J}_-$. We can now compute the algebra of the operators $\hat{J}_+, \hat{J}_-$, and $\hat{J}_z$. This is, again, the algebra of angular momentum, in terms of redefined operators. We begin by computing the product $\hat{J}_+\hat{J}_-$:

$$\hat{J}_+\hat{J}_- = \hat{J}_x^2 + \hat{J}_y^2 - i[\hat{J}_x, \hat{J}_y] = \hat{J}_x^2 + \hat{J}_y^2 + \hbar\hat{J}_z. \tag{19.3.12}$$

Together with the product in the opposite order, we have

$$\begin{aligned} \hat{J}_+\hat{J}_- &= \hat{J}_x^2 + \hat{J}_y^2 + \hbar\hat{J}_z, \\ \hat{J}_-\hat{J}_+ &= \hat{J}_x^2 + \hat{J}_y^2 - \hbar\hat{J}_z, \end{aligned} \tag{19.3.13}$$

which can be summarized as

$$\hat{J}_{\pm}\hat{J}_{\mp} = \hat{J}_x^2 + \hat{J}_y^2 \pm \hbar \hat{J}_z. \tag{19.3.14}$$

From (19.3.13) we can quickly get the commutator:

$$[\hat{J}_+, \hat{J}_-] = 2\hbar \hat{J}_z. \tag{19.3.15}$$

We also compute the commutator of $\hat{J}_{\pm}$ with $\hat{J}_z$:

$$[\hat{J}_z, \hat{J}_+] = [\hat{J}_z, \hat{J}_x] + i[\hat{J}_z, \hat{J}_y] = i\hbar \hat{J}_y + i(-i\hbar \hat{J}_x) = \hbar(\hat{J}_x + i\hat{J}_y) = \hbar \hat{J}_+. \tag{19.3.16}$$

Similarly, $[\hat{J}_z, \hat{J}_-] = -\hbar \hat{J}_-$, and therefore, all in all,

$$[\hat{J}_z, \hat{J}_{\pm}] = \pm \hbar \hat{J}_{\pm}. \tag{19.3.17}$$

This is similar to our harmonic oscillator commutators $[\hat{N}, \hat{a}^\dagger] = \hat{a}^\dagger$ and $[N, \hat{a}] = -\hat{a}$, if we identify $\hat{N}$ with $\hat{J}_z$, $\hat{a}^\dagger$ with $\hat{J}_+$, and $\hat{a}$ with $\hat{J}_-$. For the oscillator they taught us that, acting on states, $\hat{a}^\dagger$ raises the $\hat{N}$ eigenvalue by one unit, while $\hat{a}$ decreases it by one unit. As we will see, $\hat{J}_+$ adds $\hbar$ to the $\hat{J}_z$ eigenvalue, and $\hat{J}_-$ subtracts $\hbar$ from the $\hat{J}_z$ eigenvalue. We collect our results to emphasize that we can take the algebra of angular momentum to be the following:

$$\text{Algebra of angular momentum:} \quad [\hat{J}_+, \hat{J}_-] = 2\hbar \hat{J}_z, \quad [\hat{J}_z, \hat{J}_{\pm}] = \pm \hbar \hat{J}_{\pm}. \tag{19.3.18}$$

If one views the above as *the* definition of the algebra of angular momentum, one must also state that $(\hat{J}_{\pm})^\dagger = \hat{J}_{\mp}$ and that $\hat{J}_z$ is Hermitian. You can then convince yourself that defining $\hat{J}_x$ and $\hat{J}_y$, consistent with (19.3.10), is as follows:

$$\hat{J}_x = \tfrac{1}{2}(\hat{J}_+ + \hat{J}_-), \quad \hat{J}_y = \tfrac{1}{2i}(\hat{J}_+ - \hat{J}_-) \tag{19.3.19}$$

results in Hermitian $\hat{J}_x$ and $\hat{J}_y$. These operators, together with $\hat{J}_z$, satisfy the original angular momentum algebra (19.3.8) on account of (19.3.18).

We also write $\hat{\mathbf{J}}^2$ in a nice way beginning from (19.3.13), which gives two expressions for $\hat{J}_x^2 + \hat{J}_y^2$:

$$\hat{J}_x^2 + \hat{J}_y^2 = \hat{J}_+\hat{J}_- - \hbar \hat{J}_z = \hat{J}_-\hat{J}_+ + \hbar \hat{J}_z. \tag{19.3.20}$$

Adding $\hat{J}_z^2$ to both sides of the equation, we find that

$$\hat{\mathbf{J}}^2 = \hat{J}_+\hat{J}_- + \hat{J}_z^2 - \hbar \hat{J}_z = \hat{J}_-\hat{J}_+ + \hat{J}_z^2 + \hbar \hat{J}_z. \tag{19.3.21}$$

Of course, since $\hat{J}_i$ and $\hat{\mathbf{J}}^2$ commute,

$$[\hat{J}_{\pm}, \hat{\mathbf{J}}^2] = 0. \tag{19.3.22}$$

Multiplets of angular momentum Our analysis will only use the algebra of the operators $\hat{J}_i$ and their Hermiticity to discuss the allowed angular momentum multiplets. Since $\hat{\mathbf{J}}^2$ and $\hat{J}_z$ are Hermitian and commute, they can be simultaneously diagonalized. In fact, there are no more operators in the angular momentum algebra that can be added to this list of simultaneously diagonalizable operators. Our work in section 15.7 shows that the common eigenstates form an orthonormal basis for the vector space the operators act on. We thus introduce eigenstates $|j, m\rangle$, with $j, m \in \mathbb{R}$, where the first label relates to the $\hat{\mathbf{J}}^2$ eigenvalue and the second label to the $\hat{J}_z$ eigenvalue:

$$\begin{aligned} \hat{\mathbf{J}}^2 |j,m\rangle &= \hbar^2 j(j+1)\,|j,m\rangle\,, \\ \hat{J}_z\,|j,m\rangle &= \hbar m\,|j,m\rangle. \end{aligned} \tag{19.3.23}$$

A multiplet of angular momentum is a special collection of linearly independent states, or vectors. It is useful to think of the multiplet as the vector space V generated as the span of these vectors. This vector space V must be invariant under the action of the angular momentum operators: $\hat{J}_i : V \to V$, for $i = 1, 2, 3$. Multiplets, moreover, are *irreducible*, meaning there is no proper subspace of V that is invariant under the action of the angular momentum operators. The set of states that generate a multiplet thus mix completely under the action of the angular momentum operators. In summary, a **multiplet of angular momentum** is a vector space that is irreducible and invariant under the action of the angular momentum operators. In general, if we have a vector space W on which angular momentum operators act, W can be written as a direct sum of multiplets—that is, a direct sum of the corresponding vector spaces.

As we can anticipate from our examples and our previous experience, the states $|j, m\rangle$ that generate a multiplet would be defined by an allowed, *fixed* value of j and a set of associated values of m. We will see below that no operator in the algebra of angular momentum can change the value of j. States with different eigenvalues of Hermitian operators must be orthogonal; therefore, we can assume that the states generating a multiplet are orthogonal:

$$\langle j, m' | j, m\rangle = \delta_{m',m}. \tag{19.3.24}$$

We assume that we do not have to deal with continuous values of j, m that would require delta function normalization (this will be confirmed below). Since j and m are real, the eigenvalues of the Hermitian operators are real, as they have to be.

The value of j in (19.3.23) is constrained by a positivity condition. Indeed, $\hbar^2 j(j+1)$ must be nonnegative:

$$\hbar^2 j(j+1) = \langle j, m|\hat{\mathbf{J}}^2|j,m\rangle = \sum_{i=1}^{3} \langle j, m|\hat{J}_i\hat{J}_i|j,m\rangle = \sum_{i=1}^{3} ||\hat{J}_i|j,m\rangle||^2 \geq 0, \tag{19.3.25}$$

where in the first step we used the eigenvalue definition and orthonormality. Therefore, the condition

$$j(j+1) \geq 0 \tag{19.3.26}$$

is the only a priori condition on the values of j. Since what matters is the eigenvalue of $\hat{\mathbf{J}}^2$, to label the state we can use any of the two j's that give a particular value of $j(j+1)$. As shown in figure 19.1, the positivity of $j(j+1)$ requires $j \geq 0$ or $j \leq -1$. We will use $j \geq 0$:

$$\text{States are labeled as } |j,m\rangle \text{ with } j \geq 0. \tag{19.3.27}$$

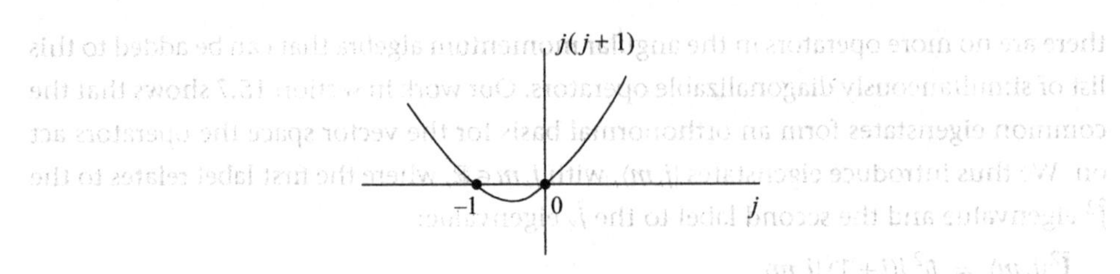

Figure 19.1
Since $j(j+1) \geq 0$, for consistency we can label the states $|j, m\rangle$ using $j \geq 0$.

You should not think there are two different states, with two different j's associated with the eigenvalue $\hbar j(j+1)$. It is just one state, labeled in an unusual way.

Just as we did for the case of the simple harmonic oscillator, we will assume that a state $|j, m\rangle$ of norm one exists and find out what the algebra of angular momentum implies for the existence of additional states and the possible values of j and m that have no inconsistencies. As in the harmonic oscillator case, an inconsistency would be the appearance of negative norm states.

Let us now investigate what the operators $\hat{J}_\pm$ do when acting on $|j, m\rangle$. Since they commute with $\hat{\mathbf{J}}^2$, the operators $\hat{J}_+$ or $\hat{J}_-$ do not change the j value of a state:

$$\hat{\mathbf{J}}^2(\hat{J}_\pm|j,m\rangle) = \hat{J}_\pm\hat{\mathbf{J}}^2|j,m\rangle = \hbar j(j+1)(\hat{J}_\pm|j,m\rangle) \tag{19.3.28}$$

so that we must have

$$\hat{J}_\pm|j,m\rangle \propto |j,m'_\pm\rangle, \quad \text{for some } m'_\pm. \tag{19.3.29}$$

Since $\hat{J}_z$ also commutes with $\hat{\mathbf{J}}^2$, this operator also does not change the value of j, so as mentioned earlier, no operator in the algebra of angular momentum can change the value of j. On the other hand, as anticipated above, the $\hat{J}_\pm$ operators change the value of m:

$$\begin{aligned}\hat{J}_z(\hat{J}_\pm|j,m\rangle) &= ([\hat{J}_z,\hat{J}_\pm]+\hat{J}_\pm\hat{J}_z)|j,m\rangle \\ &= (\pm\hbar\hat{J}_\pm+\hbar m\hat{J}_\pm)|j,m\rangle \\ &= \hbar(m\pm 1)\,\hat{J}_\pm|j,m\rangle,\end{aligned} \tag{19.3.30}$$

from which we learn that

$$\hat{J}_\pm|j,m\rangle = C_\pm(j,m)|j,m\pm 1\rangle, \tag{19.3.31}$$

where $C_\pm(j,m)$ is a constant to be determined. Indeed, $\hat{J}_+$ raises the m eigenvalue by one unit, while $\hat{J}_-$ lowers the m eigenvalue by one unit. To determine $C_\pm(j,m)$, we first take the adjoint of the above equation:

$$\langle j,m|\hat{J}_\mp = \langle j,m\pm 1|C_\pm(j,m)^*, \tag{19.3.32}$$

and then form the overlap

$$\langle j,m|\hat{J}_\mp\hat{J}_\pm|j,m\rangle = |C_\pm(j,m)|^2. \tag{19.3.33}$$

To evaluate the left-hand side, use (19.3.21) in the form $\hat{J}_\mp\hat{J}_\pm = \hat{\mathbf{J}}^2 - \hat{J}_z^2 \mp \hbar\hat{J}_z$:

$$|C_\pm(j,m)|^2 = \langle j,m|(\hat{\mathbf{J}}^2 - \hat{J}_z^2 \mp \hbar\hat{J}_z)|j,m\rangle = \hbar^2 j(j+1) - \hbar^2m^2 \mp \hbar^2 m. \tag{19.3.34}$$

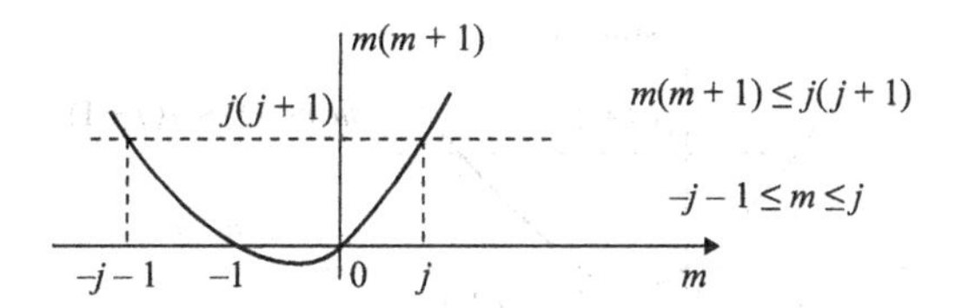

Figure 19.2
Solving the inequality $m(m+1) \leq j(j+1)$.

We thus find that

$$|C_{\pm}(j,m)|^2 = \hbar^2\left(j(j+1) - m(m\pm 1)\right) = ||\hat{J}_{\pm}|j,m\rangle||^2. \tag{19.3.35}$$

Because the norm $||\hat{J}_{\pm}|j,m\rangle||$ is nonnegative, we can take $C_{\pm}(j,m)$ to be real and equal to the positive square root of the middle term:

$$C_{\pm}(j,m) = \hbar\sqrt{j(j+1) - m(m\pm 1)}. \tag{19.3.36}$$

We have thus obtained, in (19.3.31),

$$\hat{J}_{\pm}|j,m\rangle = \hbar\sqrt{j(j+1) - m(m\pm 1)}\;|j,m\pm 1\rangle. \tag{19.3.37}$$

Given a consistent state $|j,m\rangle$, how far can we raise or lower the value of m? Our classical intuition is that $|\hat{J}_z| \leq |\mathbf{J}|$. So we should get something like $|m| \leq \sqrt{j(j+1)}$. We analyze the situation in two steps:

1. For the raised state to be consistent, we must have $||J_+|j,m\rangle||^2 \geq 0$, and therefore,

$$j(j+1) - m(m+1) \geq 0 \quad\Rightarrow\quad m(m+1) \leq j(j+1). \tag{19.3.38}$$

   The solution to this inequality is obtained using figure 19.2:

$$-j-1 \leq m \leq j. \tag{19.3.39}$$

   Since we are raising m, we can focus on the troubles that raising can give because $m \leq j$. Assume $m = j - \beta$ with $0 < \beta < 1$ so that the inequality (19.3.39) is satisfied, and m is less than one unit below j. Then raising once gives us a state with $m' = m + 1 > j$, and since the inequality is now violated, $J_+|j,m'\rangle$ is an inconsistent state. To prevent such inconsistency, the process of raising must terminate: there must be a state on which raising gives no state (the zero state). That happens only if $m = j$, since then $C_+(j,j) = 0$:

$$\hat{J}_+|j,j\rangle = 0. \tag{19.3.40}$$

2. For the lowered state to be consistent, we must have $||\hat{J}_-|j,m\rangle||^2 \geq 0$, and therefore,

$$j(j+1) - m(m-1) \geq 0 \quad\Rightarrow\quad m(m-1) \leq j(j+1). \tag{19.3.41}$$

   The solution to this inequality is obtained using figure 19.3:

$$-j \leq m \leq j+1. \tag{19.3.42}$$

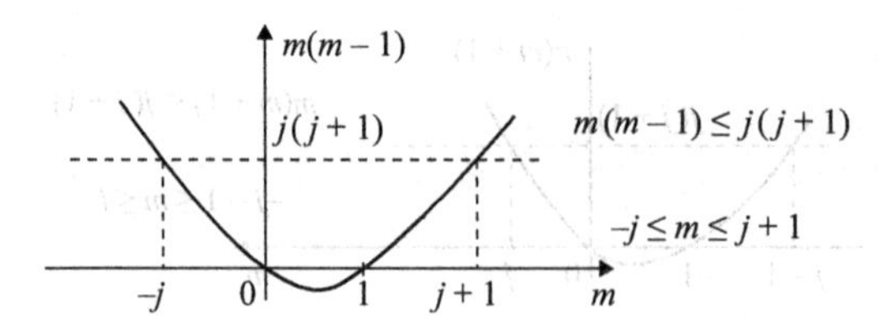

Figure 19.3
Solving the inequality $m(m-1) \leq j(j-1)$.

This time we can focus on $m \geq -j$ and the complications due to lowering. Assume $m = -j + \beta$ with $0 < \beta < 1$ so the constraint (19.3.42) is satisfied, and m is less than one unit above $-j$. Then lowering once gives us a state with $m' = m - 1 < -j$, and since the inequality is now violated, the state $\hat{J}_-|j, m'\rangle$ is an inconsistent state. To prevent such inconsistency, the process of lowering must terminate: there must be a state on which lowering gives no state (the zero state). That happens only if $m = -j$ since then $C_-(j, -j) = 0$:

$$\hat{J}_-|j, -j\rangle = 0. \tag{19.3.43}$$

The above analysis shows that for consistency the starting state $|j, m\rangle$ must have m in the range

$$-j \leq m \leq j, \tag{19.3.44}$$

which is the condition that combines (19.3.39) *and* (19.3.42). Additionally, m must be such that as it is increased by unit steps it reaches j and as it is decreased by unit steps it reaches $-j$. It follows that the distance $2j$ between j and $-j$ must be an integer:

$$\boxed{2j \in \mathbb{Z} \quad \Rightarrow \quad j \in \mathbb{Z}/2, \quad \Rightarrow \quad j = 0, \tfrac{1}{2}, 1, \tfrac{3}{2}, 2, \ldots .} \tag{19.3.45}$$

This is the fundamental quantization of angular momentum. Angular momentum can be integral or half integral. For any allowed value of j, the m values will be $j, j-1, \ldots, -j$. Thus, the multiplet with angular momentum j is a vector space with the following $2j+1$ orthonormal basis states:

$$\text{Multiplet of angular momentum } j: \quad \begin{array}{c} |j, j\rangle, \\ |j, j-1\rangle, \\ \vdots \\ |j, -j+1\rangle, \\ |j, -j\rangle. \end{array} \tag{19.3.46}$$

For $j = 0$ there is just one state, the **singlet** with $m = 0$: $|0, 0\rangle$. The multiplet is then a one-dimensional vector space. For $j = \frac{1}{2}$, we have a **doublet**—that is, two basis states, one with $m = \frac{1}{2}$, the other with $m = -\frac{1}{2}$:

$$\text{Multiplet of angular momentum } j = \tfrac{1}{2}: \quad |\tfrac{1}{2}, \tfrac{1}{2}\rangle, \ |\tfrac{1}{2}, -\tfrac{1}{2}\rangle. \tag{19.3.47}$$

As we discussed below equation (19.3.5), when $\hat{\mathbf{J}}$ is the spin angular momentum $\hat{\mathbf{S}}$ these are basis states of a spin one-half particle, the conventional $|\pm\rangle$ states with $\hat{S}_z = \pm\frac{\hbar}{2}$. The multiplet is the state space $\mathbb{C}^2$ for spin states. For $j=1$ we have three states, a **triplet:**

$$\text{Multiplet of angular momentum } j=1: \quad |1,\ 1\rangle,\ \ |1,\ 0\rangle,\ \ |1,-1\rangle. \tag{19.3.48}$$

If $\hat{\mathbf{J}}$ is orbital angular momentum $\hat{\mathbf{L}}$, these would be the $\ell=1$ spherical harmonics with $m=1,0,-1$ (see (19.3.1)). If $\hat{\mathbf{J}}$ is spin angular momentum $\hat{\mathbf{S}}$, the above states would be the basis vectors on which 3×3 spin matrices act, obeying the algebra of angular momentum. We would be describing a spin one particle. The details of the matrix construction are given below in example 19.1 and explored further in problem 19.2. For $j=3/2$, we have four states:

$$\text{Multiplet of angular momentum } j=\tfrac{3}{2}: \quad |\tfrac{3}{2},\ \tfrac{3}{2}\rangle,\ |\tfrac{3}{2},\ \tfrac{1}{2}\rangle,\ |\tfrac{3}{2},-\tfrac{1}{2}\rangle,\ |\tfrac{3}{2},-\tfrac{3}{2}\rangle. \tag{19.3.49}$$

This cannot arise from orbital angular momentum but does arise from spin angular momentum. It corresponds to spin $\frac{3}{2}$. The above basis states allow a construction of the spin operators as 4×4 matrices.

On any state of a multiplet with angular momentum j, the eigenvalue J^2 of $\hat{\mathbf{J}}^2$ can be used to define the "magnitude" J of the angular momentum. We have

$$J^2 \;=\; \hbar^2 j(j+1) \quad\Rightarrow\quad \tfrac{1}{\hbar}J \;=\; \sqrt{j(j+1)}. \tag{19.3.50}$$

In the limit as j is large: $\frac{1}{\hbar}J = j\sqrt{1+\frac{1}{j}} \simeq j+\frac{1}{2}+\mathcal{O}(1/j)$.

Example 19.1. *Matrix construction of angular momentum operators for $j=1$.*

For a $j=1$ multiplet, we have the three basis states listed in (19.3.48). We order them as follows: $|1\rangle \equiv |1,\ 1\rangle$, $|2\rangle \equiv |1,\ 0\rangle$, $|3\rangle \equiv |1,-1\rangle$. In this basis the operator $\hat{J}_z$ is diagonal, and since $\hat{J}_z|1,m\rangle = \hbar m|1,m\rangle$, we have

$$\hat{J}_z = \hbar\begin{pmatrix} 1 & 0 & 0 \\ 0 & 0 & 0 \\ 0 & 0 & -1 \end{pmatrix}. \tag{19.3.51}$$

The matrices for $\hat{J}_x$ and $\hat{J}_y$ are obtained by first calculating the matrices for $\hat{J}_+$ and $\hat{J}_-$. Equation (19.3.37) with $j=1$ gives $\hat{J}_+|1,m\rangle = \hbar\sqrt{2-m(m+1)}\,|1,m+1\rangle$. This means that

$$\hat{J}_+|1,1\rangle = 0,\ \ \hat{J}_+|1,0\rangle = \hbar\sqrt{2}\,|1,1\rangle,\ \ \hat{J}_+|1,-1\rangle = \hbar\sqrt{2}\,|1,0\rangle, \tag{19.3.52}$$

or equivalently, $\hat{J}_+|1\rangle = 0$, $\hat{J}_+|2\rangle = \hbar\sqrt{2}\,|1\rangle$, and $\hat{J}_+|3\rangle = \hbar\sqrt{2}\,|2\rangle$. This is all we need to build the matrix for $\hat{J}_+$:

$$\hat{J}_+ = \hbar\sqrt{2}\begin{pmatrix} 0 & 1 & 0 \\ 0 & 0 & 1 \\ 0 & 0 & 0 \end{pmatrix}, \quad \hat{J}_- = \hbar\sqrt{2}\begin{pmatrix} 0 & 0 & 0 \\ 1 & 0 & 0 \\ 0 & 1 & 0 \end{pmatrix}, \tag{19.3.53}$$

where the matrix for $\hat{J}_-$ is the Hermitian conjugate of that for $\hat{J}_+$, given that $\hat{J}_- = \hat{J}_+^\dagger$. The matrices for $\hat{J}_x = \frac{1}{2}(\hat{J}_+ + \hat{J}_-)$ and $\hat{J}_y = \frac{1}{2i}(\hat{J}_+ - \hat{J}_-)$ now follow immediately:

$$\hat{J}_x = \frac{\hbar}{\sqrt{2}}\begin{pmatrix} 0 & 1 & 0 \\ 1 & 0 & 1 \\ 0 & 1 & 0 \end{pmatrix}, \quad \hat{J}_y = \frac{\hbar}{\sqrt{2}}\begin{pmatrix} 0 & -i & 0 \\ i & 0 & -i \\ 0 & i & 0 \end{pmatrix}. \tag{19.3.54}$$

The above matrices are Hermitian, as expected for Hermitian operators. It is a good check of our arithmetic that the above matrices satisfy $[\hat{J}_x, \hat{J}_y] = i\hbar\hat{J}_z$ and the other relations in the algebra of angular momentum. The matrices for $\hat{J}_x, \hat{J}_y$, and $\hat{J}_z$ represent the operators as they act on a three-dimensional complex vector space $\mathbb{C}^3$, itself identified as the vector space of the $j = 1$ multiplet. □.

Exercise 19.10. *Show that* $\text{tr}\hat{J}_i = 0$, *for all* $i = 1, 2, 3$, *on any multiplet of angular momentum. Note that for* $\hat{J}_3$ *this follows quickly from the matrix representation.*

Relation to spherical harmonics The states we have found are described abstractly by (19.3.23). If $\hat{\mathbf{J}}$ refers to orbital angular momentum, we would write

$$\begin{aligned}\hat{\mathbf{L}}^2|\ell, m\rangle &= \hbar^2\,\ell(\ell+1)\,|\ell, m\rangle\,,\\ \hat{L}_z\,|\ell, m\rangle &= \hbar m\,|\ell, m\rangle.\end{aligned} \tag{19.3.55}$$

Here the form of the operators is not specified, and the states are not described concretely. We can compare these to the relations (19.3.1)

$$\begin{aligned}\hat{\mathcal{L}}^2\,Y_{\ell m}(\theta,\phi) &= \hbar^2\,\ell(\ell+1)\,Y_{\ell m}(\theta,\phi),\\ \hat{\mathcal{L}}_z\,Y_{\ell m}(\theta,\phi) &= \hbar m\,Y_{\ell m}(\theta,\phi),\end{aligned} \tag{19.3.56}$$

where we have rewritten the operators $\hat{\mathbf{L}}^2$ and $\hat{L}_z$ with calligraphic symbols to emphasize that here they are differential operators that satisfy the algebra of the abstract operators and act on the spherical harmonics. The correspondence between the two sets of relations can be established with the help of some formal notation. In one dimension, $\psi(x) = \langle x|\psi\rangle$ expresses a wave function in terms of a state and position eigenstates. Here, we introduce angular position eigenstates $|\theta\phi\rangle$. To make their definition more concrete, these states satisfy an orthogonality relation analogous to the one-dimensional $\langle x|x'\rangle = \delta(x - x')$:

$$\langle\theta\phi|\theta'\phi'\rangle = \delta(\cos\theta - \cos\theta')\delta(\phi - \phi'). \tag{19.3.57}$$

The delta functions to the right are the natural ones in spherical coordinates. Moreover, the completeness of these position states, in analogy to $\int dx|x\rangle\langle x| = \mathbb{1}$, would read

$$\int d\Omega\,|\theta\phi\rangle\langle\theta\phi| = \mathbb{1}, \qquad \int d\Omega = \int_{-1}^{1} d(\cos\theta)\int_0^{2\pi} d\phi. \tag{19.3.58}$$

The integral is over solid angle. Equipped with angular position states, we now write

$$Y_{\ell m}(\theta,\phi) \equiv \langle\theta\phi|\ell, m\rangle, \tag{19.3.59}$$

identifying the spherical harmonics as the wave functions associated to the $|\ell, m\rangle$ states! To see how this relation is used, we now derive the completeness and orthogonality relations for spherical harmonics from the completeness and orthogonality relations of the $|\ell, m\rangle$ states:

$$\sum_{\ell=0}^{\infty}\sum_{m=-\ell}^{\ell}|\ell, m\rangle\langle\ell, m| = \mathbb{1}, \qquad \langle\ell', m'|\ell, m\rangle = \delta_{\ell',\ell}\delta_{m',m}. \tag{19.3.60}$$

Start with the orthogonality relation, and introduce a complete set of position states:

$$\int d\Omega\, \langle \ell', m'|\theta\phi\rangle\langle\theta\phi|\ell, m\rangle = \delta_{\ell',\ell}\delta_{m',m}. \tag{19.3.61}$$

This is, in fact, the familiar orthogonality property of the spherical harmonics:

$$\int d\Omega\, Y^*_{\ell'm'}(\theta,\phi)\, Y_{\ell m}(\theta,\phi) = \delta_{\ell',\ell}\delta_{m',m}. \tag{19.3.62}$$

Analogously, acting with $\langle\theta'\phi'|$ from the left and $|\theta\phi\rangle$ from the right on the completeness relation (19.3.60) gives us

$$\sum_{\ell=0}^{\infty}\sum_{m=-\ell}^{\ell} \langle\theta'\phi'|\ell, m\rangle\langle\ell, m|\theta\phi\rangle = \delta(\cos\theta - \cos\theta')\delta(\phi-\phi'). \tag{19.3.63}$$

This expresses the completeness of the spherical harmonics. After taking complex conjugates, it reads

$$\sum_{\ell=0}^{\infty}\sum_{m=-\ell}^{\ell} Y^*_{\ell m}(\theta',\phi')Y_{\ell m}(\theta,\phi) = \delta(\cos\theta - \cos\theta')\delta(\phi-\phi'). \tag{19.3.64}$$

We can elaborate on the wave function/state correspondence. For the case of wave functions in one dimension, we have the following relations:

$$\begin{aligned} \hat{p}\,|p\rangle &= p\,|p\rangle, \\ \frac{\hbar}{i}\frac{\partial}{\partial x}\,\langle x|p\rangle &= p\,\langle x|p\rangle, \end{aligned} \tag{19.3.65}$$

where $\frac{\hbar}{i}\frac{\partial}{\partial x}$ is the differential operator that represents momentum in coordinate space. These two relations were used in section 14.10 to demonstrate the interplay between the operator and its differential representation for any general state $|\psi\rangle$:

$$\langle x|\hat{p}|\psi\rangle = \frac{\hbar}{i}\frac{\partial}{\partial x}\,\langle x|\psi\rangle. \tag{19.3.66}$$

Similarly, the two relations

$$\begin{aligned} \hat{\mathbf{L}}^2|\ell, m\rangle &= \hbar^2\,\ell(\ell+1)\,|\ell, m\rangle\,, \\ \hat{\mathcal{L}}^2\, Y_{\ell m}(\theta,\phi) &= \hbar^2\,\ell(\ell+1)Y_{\ell,m}(\theta,\phi) \end{aligned} \tag{19.3.67}$$

can be used to show that for any angular state $|F\rangle$,

$$\langle\theta\phi|\hat{\mathbf{L}}^2|F\rangle = \hat{\mathcal{L}}^2\,\langle\theta\phi|F\rangle. \tag{19.3.68}$$

To do this, start from the left-hand side, and introduce a complete set of angular momentum states:

$$\langle\theta\phi|\hat{\mathbf{L}}^2|F\rangle = \sum_{\ell,m}\langle\theta\phi|\ell, m\rangle\langle\ell, m|\hat{\mathbf{L}}^2|F\rangle = \sum_{\ell,m}\hbar^2\ell(\ell+1)Y_{\ell m}(\theta,\phi)\langle\ell, m|F\rangle, \tag{19.3.69}$$

where $\hat{\mathbf{L}}^2$ acted on the bra using the first equation of (19.3.67). Now we use the second equation there to find

$$\langle\theta\phi|\hat{\mathbf{L}}^2|F\rangle = \sum_{\ell,m}(\hat{\mathcal{L}}^2 Y_{\ell m}(\theta,\phi))\,\langle\ell, m|F\rangle = \hat{\mathcal{L}}^2\sum_{\ell,m} Y_{\ell m}(\theta,\phi)\langle\ell, m|F\rangle, \tag{19.3.70}$$

where the operator $\hat{\mathcal{L}}^2$ can be brought out of the sum since $\langle \ell, m|F\rangle$ has no angular dependence. Rewriting the $Y_{\ell m}$ again in terms of bra-kets, we show that

$$\langle\theta\phi|\hat{\mathbf{L}}^2|F\rangle = \hat{\mathcal{L}}^2 \sum_{\ell,m} \langle\theta\phi|\ell, m\rangle\langle\ell, m|F\rangle = \hat{\mathcal{L}}^2 \langle\theta\phi|F\rangle, \tag{19.3.71}$$

as we wanted. For $|F\rangle = |\ell, m\rangle$ this relation gives $\langle\theta\phi|\hat{\mathbf{L}}^2|\ell, m\rangle = \hat{\mathcal{L}}^2 Y_{\ell m}$. An analogous derivation results in the relation $\langle\theta\phi|\hat{L}_z|F\rangle = \hat{\mathcal{L}}_z\langle\theta\phi|F\rangle$.

Example 19.2. *Spherical harmonics from lowering operators.*
We want to confirm that

$$Y_{\ell 0}(\theta, \phi) = \frac{1}{\sqrt{(2\ell)!}} \left(\frac{\hat{\mathcal{L}}_-}{\hbar}\right)^{\ell} Y_{\ell\ell}(\theta, \phi). \tag{19.3.72}$$

This equation makes precise the relation $Y_{\ell 0} \sim (\hat{\mathcal{L}}_-)^{\ell} Y_{\ell\ell}$, which we expect to hold because each $\hat{\mathcal{L}}_-$ lowers the value of the m quantum number by one unit. The above formula can be used in practice to calculate spherical harmonics. The expression for $Y_{\ell\ell}$ is given in (10.5.35), and the angular form of $\hat{\mathcal{L}}_-$ is in (19.2.21).

To prove (19.3.72) we recall (19.3.37), which for orbital angular momentum reads

$$\hat{L}_-|\ell, m\rangle = \hbar\sqrt{\ell(\ell+1) - m(m-1)}\, |\ell, m-1\rangle\,. \tag{19.3.73}$$

For spherical harmonics, and factorizing the expression inside the square root, we have

$$\frac{\hat{\mathcal{L}}_-}{\hbar} Y_{\ell,m} = \sqrt{(\ell+m)(\ell-m+1)}\, Y_{\ell,m-1}\,. \tag{19.3.74}$$

We thus see that

$$\begin{aligned}
\frac{\hat{\mathcal{L}}_-}{\hbar} Y_{\ell\ell} &= \sqrt{2\ell \cdot 1}\quad Y_{\ell,\ell-1}\,,\\
\frac{\hat{\mathcal{L}}_-}{\hbar} Y_{\ell,\ell-1} &= \sqrt{(2\ell-1)\,2}\; Y_{\ell,\ell-2}\,,\\
\vdots\qquad & \qquad \vdots\\
\frac{\hat{\mathcal{L}}_-}{\hbar} Y_{\ell,1} &= \sqrt{(\ell+1)\ell}\; Y_{\ell,0}.
\end{aligned} \tag{19.3.75}$$

If you look inside the square root prefactors, you will see that they include all integers from 1 to 2ℓ. Therefore,

$$\left(\frac{\hat{\mathcal{L}}_-}{\hbar}\right)^{\ell} Y_{\ell\ell} = \sqrt{(2\ell)!}\, Y_{\ell 0}, \tag{19.3.76}$$

which is indeed the relation we wanted to prove. □

Example 19.3. *Wave functions for p orbitals.*
A p orbital is a state with angular momentum $\ell = 1$. This terminology is commonly used to describe states of electrons in atoms. With three possible values of the m quantum number, $m = -1, 0, 1$, there are three p orbitals. The p_z orbital is defined as the $\hat{L}_z = 0$ state $|1, 0\rangle$. Thus, the angular wave function $\psi_{p_z}(\theta, \phi)$ for this orbital is

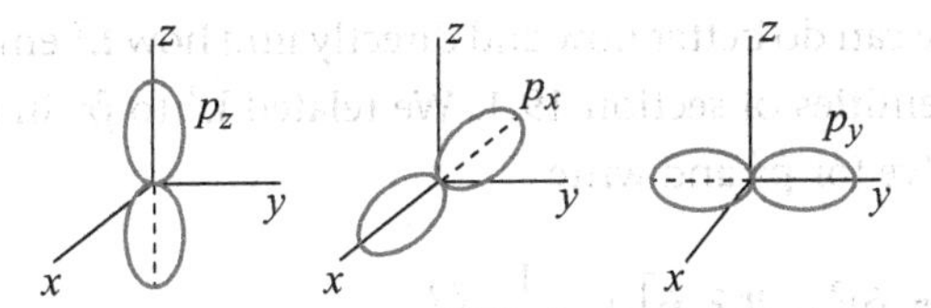

Figure 19.4
From left to right, the angular behavior of the p_z, p_x, and p_y orbitals.

$$\psi_{p_z}(\theta,\phi) = \langle\theta\phi|1,0\rangle = Y_{1,0}(\theta,\phi) = \sqrt{\frac{3}{4\pi}}\cos\theta = \sqrt{\frac{3}{4\pi}}\,\frac{z}{r}, \tag{19.3.77}$$

where we read the spherical harmonic from (10.5.33). The p_x orbital is defined as the $\ell=1$ state with $\hat{L}_x=0$. Using the matrix representation (19.3.54) in the standard $(\hat{\mathbf{L}}^2, \hat{L}_z)$ basis, we see that $\hat{L}_x$ kills the vector $(1,0,-1)$:

$$\hat{L}_x\begin{pmatrix}1\\0\\-1\end{pmatrix} = \frac{\hbar}{\sqrt{2}}\begin{pmatrix}0&1&0\\1&0&1\\0&1&0\end{pmatrix}\begin{pmatrix}1\\0\\-1\end{pmatrix} = 0, \tag{19.3.78}$$

confirming that the normalized state $|\psi_{p_x}\rangle$ with $\hat{L}_x=0$ is

$$|\psi_{p_x}\rangle = \tfrac{1}{\sqrt{2}}(|1,1\rangle - |1,-1\rangle) \;\Rightarrow\; \psi_{p_x} = \tfrac{1}{\sqrt{2}}(Y_{1,1} - Y_{1,-1}). \tag{19.3.79}$$

This quickly gives

$$\begin{aligned}\psi_{p_x} &= -\sqrt{\frac{3}{4\pi}}\sin\theta\cos\phi = -\sqrt{\frac{3}{4\pi}}\,\frac{x}{r},\\ \psi_{p_y} &= -i\sqrt{\frac{3}{4\pi}}\sin\theta\sin\phi = -i\sqrt{\frac{3}{4\pi}}\,\frac{y}{r},\end{aligned} \tag{19.3.80}$$

where we included the p_y orbital that, with $\hat{L}_y=0$, is the state $\frac{1}{\sqrt{2}}(|1,1\rangle + |1,-1\rangle)$. The three orbitals $|\psi_{p_x}\rangle$, $|\psi_{p_y}\rangle$, and $|\psi_{p_z}\rangle$ form an orthonormal basis for the $\ell=1$ multiplet. Note the nice symmetry between their wave functions, made manifest in the plots in figure 19.4. These orbitals are useful in atomic physics because they represent the choices made by electrons in multielectron atoms as they fill $\ell=1$ spaces (example 21.5). □

19.4 Central Potentials and Radial Equation

Angular momentum plays a crucial role in the study of Hamiltonians for a particle that moves in a three-dimensional *central potential*, a potential V that depends only on the distance r from the particle to the origin. The Hamiltonian $\hat{H}$ takes the form:

$$\hat{H} = \frac{\hat{\mathbf{p}}^2}{2m} + V(r). \tag{19.4.1}$$

Although technically r is an operator, for convenience we leave the hat off of r while using a hat for the vector position $\hat{\mathbf{r}}$. The angular momentum operator $\hat{\mathbf{L}}^2$ turns out to be the angular part of the Laplacian operator. This result was obtained in chapter 10 by a long calculation. We can do better now and directly find how $\hat{\mathbf{L}}^2$ enters the Hamiltonian using the vector identities of section 19.1. We related $\hat{\mathbf{L}}^2$ to $\hat{\mathbf{p}}^2$ in (19.1.32). From this equation we can solve for $\hat{\mathbf{p}}^2$ and write

$$\frac{\hat{\mathbf{p}}^2}{2m} = \frac{1}{2m}\frac{1}{r^2}\Big[(\hat{\mathbf{r}}\cdot\hat{\mathbf{p}})^2 - i\hbar\,\hat{\mathbf{r}}\cdot\hat{\mathbf{p}}\Big] + \frac{1}{2mr^2}\hat{\mathbf{L}}^2, \tag{19.4.2}$$

where we use $r^2 = \hat{\mathbf{r}}^2$, a useful notation in coordinate space. The first term on the above right-hand side is purely radial. While the gradient has components along all spherical unit vectors, the radial component is quite simple: $\hat{\mathbf{r}}\cdot\hat{\mathbf{p}} = \frac{\hbar}{i}r\frac{\partial}{\partial r}$. Therefore,

$$(\hat{\mathbf{r}}\cdot\hat{\mathbf{p}})^2 - i\hbar\,\hat{\mathbf{r}}\cdot\hat{\mathbf{p}} = -\hbar^2\left(r\frac{\partial}{\partial r}r\frac{\partial}{\partial r} + r\frac{\partial}{\partial r}\right) = -\hbar^2\left(r^2\frac{\partial^2}{\partial r^2} + 2r\frac{\partial}{\partial r}\right). \tag{19.4.3}$$

It then follows that

$$\frac{1}{r^2}\Big[(\hat{\mathbf{r}}\cdot\hat{\mathbf{p}})^2 - i\hbar\,\hat{\mathbf{r}}\cdot\hat{\mathbf{p}}\Big] = -\hbar^2\left(\frac{\partial^2}{\partial r^2} + \frac{2}{r}\frac{\partial}{\partial r}\right) = -\hbar^2\,\frac{1}{r}\frac{\partial^2}{\partial r^2}\,r, \tag{19.4.4}$$

where the last step is readily checked by explicit expansion. Back to the kinetic term (19.4.2), we can now rewrite the three-dimensional Hamiltonian as

$$\hat{H} = -\frac{\hbar^2}{2m}\frac{1}{r}\frac{\partial^2}{\partial r^2}\,r + \frac{1}{2mr^2}\hat{\mathbf{L}}^2 + V(r). \tag{19.4.5}$$

This is a useful result because the eigenfunctions and eigenvalues of $\hat{\mathbf{L}}^2$ are well understood.

Possibly the most important property of central potential problems is that the angular momentum operators commute with the Hamiltonian:

$$\boxed{\text{Central potential Hamiltonians: } [\hat{L}_i, \hat{H}] = 0.} \tag{19.4.6}$$

We have seen that $\hat{L}_i$ commutes with $\hat{\mathbf{p}}^2$, so it is only needed to show that any $\hat{L}_i$ commutes with $V(r)$. But we have also seen that $\hat{L}_i$ commutes with $\hat{\mathbf{r}}^2 = \hat{\mathbf{r}}\cdot\hat{\mathbf{r}}$, and therefore it commutes with any function of $\mathbf{r}^2$. Since $\mathbf{r}^2 = r^2$, it is clear that $r = \sqrt{\mathbf{r}^2}$ is a function of $\mathbf{r}^2$ and so is any $V(r)$. Thus, $\hat{L}_i$ commutes with $V(r)$ and, as a result, commutes with the central potential Hamiltonian. This is true for all i. This vanishing commutator implies that the $\hat{L}_i$ operators are conserved in central potentials:

$$i\hbar\frac{d}{dt}\langle\hat{L}_i\rangle = \langle[\hat{L}_i, \hat{H}]\rangle = 0. \tag{19.4.7}$$

Since $[\hat{L}_i, \hat{H}] = 0$, if $|\psi\rangle$ is an energy eigenstate with some energy E, $\hat{L}_i|\psi\rangle$ if nonzero is also an energy eigenstate with energy E. We learned in the previous section that states can be organized into angular momentum multiplets characterized by the value of ℓ. All states in a multiplet are connected by the action of the angular momentum operators $\hat{L}_\pm$. It follows that each angular momentum multiplet corresponds to a set of degenerate states! The whole energy spectrum can be organized as multiplets of angular momentum. To summarize:

Claim: The energy spectrum of a central potential Hamiltonian can be described as a collection of angular momentum multiplets. Each multiplet is a set of degenerate eigenstates.

The above statement is consistent with the insight provided by the set of commuting observables of a central potential Hamiltonian. Consider the following list of operators always available in a central potential problem:

$$\begin{aligned}&\hat{H},\ \hat{x}_1,\hat{x}_2,\hat{x}_3,\ \hat{p}_1,\hat{p}_2,\hat{p}_3,\ \hat{L}_1,\hat{L}_2,\hat{L}_3,\ \hat{\mathbf{r}}^2,\ \hat{\mathbf{p}}^2,\ \hat{\mathbf{r}}\cdot\hat{\mathbf{p}},\ \hat{\mathbf{L}}^2,\\&\hat{x}_i\hat{x}_j,\ \hat{p}_i\hat{p}_j,\ \hat{x}_i\hat{p}_j,\ \hat{x}_i\hat{L}_j,\ \hat{p}_i\hat{L}_j,\ \hat{L}_i\hat{L}_j,\ldots,\ (i,j=1,2,3).\end{aligned} \tag{19.4.8}$$

For the time being, we explicitly include all operators up to squares of coordinates, momenta, and angular momenta. There is a bit of redundancy here; some combinations of the $\hat{x}_i\hat{p}_j$ products appear in the angular momentum operators. Commuting observables are required to label states uniquely with their eigenvalues. Since we want to understand the spectrum of the Hamiltonian, one of the labels of states will be the energy, and $\hat{H}$ must be in the list of commuting observables. Because $\hat{H}$ contains the potential $V(r)$, none of the $\hat{p}_i$ operators commutes with the Hamiltonian. Because it contains a $\hat{\mathbf{p}}^2$, none of the $\hat{x}_i$ commutes with the Hamiltonian. Nor will $\hat{\mathbf{r}}^2, \hat{\mathbf{p}}^2, \hat{\mathbf{r}}\cdot\hat{\mathbf{p}}, \hat{x}_i, \hat{x}_j, \hat{p}_i\hat{p}_j$. Moreover, the only combinations of $\hat{x}_i\hat{p}_j$ that commute with $\hat{H}$ are the angular momentum operators. The list, reduced to $\hat{H}$ and the operators that commute with $\hat{H}$, reads

$$\hat{H},\ \hat{L}_1,\hat{L}_2,\hat{L}_3,\ \hat{\mathbf{L}}^2,\hat{L}_i\hat{L}_j\ \ldots. \tag{19.4.9}$$

But all these operators do not commute with each other. From the $\hat{L}_i$ we can only pick at most one, for then the other two necessarily do not commute with the chosen one. Happily, we can also keep $\hat{\mathbf{L}}^2$ because of its Casimir property (19.2.14). Conventionally, everybody chooses $\hat{L}_3=\hat{L}_z$ as the angular momentum component in the set of commuting observables. Thus, we have the following:

$$\boxed{\text{Central potential commuting observables:}\ \ \hat{H},\ \hat{\mathbf{L}}^2,\ \hat{L}_z.} \tag{19.4.10}$$

Remarks:

1. The labels associated to $\hat{H}, \hat{\mathbf{L}}^2$, and $\hat{L}_z$ are E, ℓ, and m, respectively. The eigenvalues of $\hat{H}, \hat{\mathbf{L}}^2$, and $\hat{L}_z$ are $E, \hbar^2\ell(\ell+1)$, and $\hbar m$, respectively.
2. One can clearly add to the list (19.4.10) any product of operators in the list, like $\hat{L}_z\hat{L}_z$. But this would not help labeling states since on any simultaneous eigenstate of the original operators the eigenvalue of any such product operator is redundant information.
3. Although we did not give a complete proof, we claim that there is no extra operator built from the $\hat{x}_i$ and $\hat{p}_i$ that can be added to the list (19.4.10) with an eigenvalue that is not determined by the eigenvalues of the original operators on the list.

4. Since we claimed above that the central potential spectrum can be organized in terms of angular momentum multiplets, given an energy E for which there is a single multiplet of angular momentum ℓ, the $2\ell+1$ states are indeed uniquely specified by the energy and the eigenvalues of $\hat{\mathbf{L}}^2$ and $\hat{L}_z$.
5. We can ask if the above set is complete. More explicitly, are the energy eigenstates of a central potential uniquely labeled by the eigenvalues of the operators in the list? The way this could fail is apparent from the previous remark. If at some fixed energy E there are two or more multiplets with the *same* angular momentum ℓ, our labels do not suffice to identify the states uniquely because they cannot distinguish the multiplets. We will explicitly see that this *cannot happen* for the bound state spectrum of a central potential. If the particle has spin or other degrees of freedom, however, the complete set of commuting observables will include additional operators.

The radial equation This equation was derived in section 10.6. The energy eigenstates $\psi_{E\ell m}$, appropriately labeled by quantities encoding the eigenvalues of the set of commuting observables, take the form

$$\psi_{E\ell m}(r,\theta,\phi) = \frac{u_{E\ell}(r)}{r} Y_{\ell m}(\theta,\phi). \tag{19.4.11}$$

You can look back at the derivation of the radial equation satisfied by $u_{E\ell}(r)$, or better, just redo it using the rewriting of the Hamiltonian in (19.4.5) and starting from $H\psi_{E\ell m} = E\psi_{E\ell m}$. The result is

$$-\frac{\hbar^2}{2m}\frac{d^2 u_{E\ell}}{dr^2} + V_{\text{eff}}(r) u_{E\ell} = E u_{E\ell}, \tag{19.4.12}$$

where the effective potential V_{eff} is the potential $V(r)$ supplemented by a contribution from the angular momentum, a centrifugal barrier preventing the particle from reaching the origin:

$$V_{\text{eff}}(r) \equiv V(r) + \frac{\hbar^2 \ell(\ell+1)}{2mr^2}. \tag{19.4.13}$$

The range of motion is $r \in [0,\infty)$. The state $\psi_{E\ell m}$ is normalized if

$$\int_0^\infty dr\, |u_{E\ell}(r)|^2 = 1, \tag{19.4.14}$$

a rather natural constraint for a function $u_{E\ell}(r)$ that obeys a one-dimensional Schrödinger equation.

When the centrifugal barrier dominates the potential as $r \to 0$, the behavior of the radial solution at the origin is known. In that case we showed that

$$u_{E\ell} \sim c\, r^{\ell+1}, \quad \text{as } r \to 0. \tag{19.4.15}$$

This allows for a constant nonzero wave function at the origin only for $\ell = 0$.

Definite statements about the $r \to \infty$ behavior of the wave function are possible when the potential $V(r)$ vanishes beyond some radius or, at least, decays fast enough as the radius grows without bound:

$$V(r) = 0, \text{ for } r > r_0, \text{ or } \lim_{r\to\infty} rV(r) = 0. \tag{19.4.16}$$

The above assumptions are violated for the $1/r$ potential of the hydrogen atom, so our conclusions below require some modification in that case (see problem 19.4 for details). Under the above assumptions, as $r \to \infty$ we can ignore the effective potential completely, and the radial equation becomes

$$\frac{d^2 u_{E\ell}}{dr^2} = -\frac{2mE}{\hbar^2} u_{E\ell}. \tag{19.4.17}$$

The resulting $r \to \infty$ behavior follows immediately:

$$\begin{aligned} E<0, \quad & u_{E\ell} \sim \exp\Big(-\sqrt{\frac{2m|E|}{\hbar^2}}\, r\Big), \\ E>0, \quad & u_{E\ell} \sim \exp(\pm ikr), \quad k=\sqrt{\frac{2mE}{\hbar^2}}. \end{aligned} \tag{19.4.18}$$

The first behavior for $E<0$ is typical of bound states. For $E>0$ we have a continuous spectrum with degenerate solutions (hence the $\pm$). Having understood the behavior of solutions near $r=0$ and for $r\to\infty$, this allows for qualitative plots of radial solutions.

The discrete spectrum of a central potential is organized as follows. We have energy eigenstates for all values of ℓ. For each value of ℓ, the potential V_{eff} in the radial equation is different. The radial equation must therefore be solved for $\ell = 0, 1, \ldots$. For each *fixed* ℓ, we have a one-dimensional problem. Recalling that one-dimensional potentials have no degeneracies in the bound state spectrum, we conclude that we have no degeneracies in each fixed-ℓ bound state spectrum. We have a set of allowed values of energies that depend on ℓ and are numbered using an integer $n = 1, 2 \ldots$. For each allowed energy $E_{n\ell}$, we have a single radial solution $u_{n\ell}$.

$$\text{Fixed } \ell: \text{ energies } E_{n\ell}, \quad \text{radial function } u_{n\ell}, \quad n=1,2,\ldots. \tag{19.4.19}$$

Of course, each solution $u_{n\ell}$ of the radial equation represents $2\ell+1$ degenerate solutions to the Schrödinger equation, corresponding to the possible values of the $\hat{L}_z/\hbar$ eigenvalue m in the range $(-\ell, \ell)$. Note that n has replaced the label E in the radial solution, and the energies have now been labeled. This is illustrated in figure 19.5, where each solution of the radial equation is shown as a short line atop an ℓ label on the horizontal axis. This is the spectral diagram for the central potential Hamiltonian. Each line of a given ℓ represents an angular momentum multiplet, the $(2\ell+1)$ degenerate basis states obtained with $m = -\ell, \ldots, \ell$. As explained above, the radial equation can't have any degeneracies for any fixed ℓ: for a fixed ℓ, all solutions have different energies. Thus, all the lines on the diagram are single lines, representing single multiplets. Of course, other types of degeneracies of the spectrum can exist: multiplets with different values of ℓ may have the same energy. In other words, the states may match across columns on the figure. This happens, for example, in the hydrogen atom spectrum. Since the effective potential of a general central potential becomes more positive as ℓ is increased,

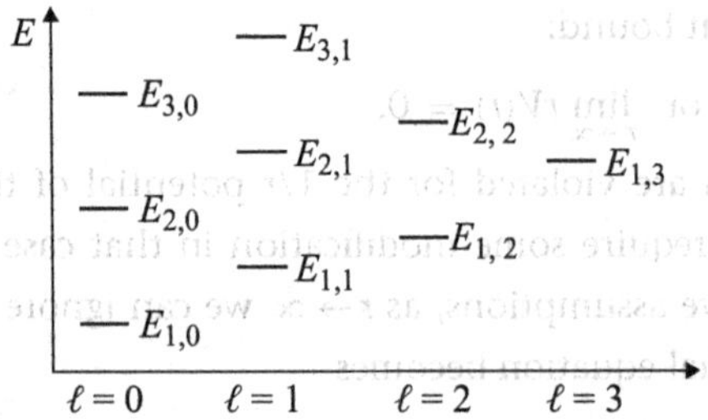

Figure 19.5
The generic discrete spectrum of a central potential Hamiltonian, showing the angular momentum ℓ multiplets and their energies $E_{n\ell}$.

the lowest-energy bound state occurs for $\ell=0$. The energy $E_{1,\ell}$ of the lowest-ℓ multiplet increases as we increase ℓ. These facts are consequences of the variational argument discussed in problem 7.15.

19.5 Free Particle and Spherical Waves

It may sound surprising, but it is useful to find radial solutions that describe a free particle! A free particle moves in the potential $V(r)=0$ that, while trivially so, is a central potential. Here we will investigate in detail the corresponding radial solutions. These solutions can be used to construct, for example, the energy eigenstates for an infinite spherical well. In this case the potential is zero inside the well, and the radial solutions apply. The infinite wall at some fixed radius a simply imposes a boundary condition.

The radial solutions for $V(r)=0$ correspond to positive energy eigenstates. When the time dependence of the wave function is included, they represent spherical waves of definite energy. On the other hand, we have the familiar plane-wave solutions of the $V=0$ theory. One of our goals will be to find a representation of a plane wave as a superposition of spherical waves. This representation is given by the Rayleigh formula, which we will derive in the following section.

We had a first look at the $V(r)=0$ radial equation in example 10.1. The equation for the radial function $u_{E\ell}(r)$ corresponding to $E>0$ eigenstates is

$$-\frac{d^2u_{E\ell}}{dr^2}+\frac{\ell(\ell+1)}{r^2}u_{E\ell}=k^2u_{E\ell},\qquad k\equiv\sqrt{\frac{2mE}{\hbar^2}}.\tag{19.5.1}$$

There should be no quantization of the energy since all energies are possible for a free particle. This is also clear because the energy parameter k can be removed from the above equation by introducing a new radial variable $\rho=kr$:

$$-\frac{d^2u_{E\ell}}{d\rho^2}+\frac{\ell(\ell+1)}{\rho^2}u_{E\ell}=u_{E\ell}.\tag{19.5.2}$$

The energy parameter reappears in the solution when ρ is ultimately expressed in terms of r. The $\ell=0$ solutions of the above equation are simple: $\sin\rho$ and $\cos\rho$. Only the first of these two is consistent with the requisite behavior as $r \to 0$: we need $u_{E0} \sim r$. For $\ell \neq 0$, the solutions are more complicated and are given in terms of functions j_ℓ and n_ℓ:

$$u_{E\ell}(\rho) = A_\ell\, \rho\, j_\ell(\rho) + B_\ell\, \rho\, n_\ell(\rho), \tag{19.5.3}$$

with free real coefficients A_ℓ and B_ℓ since radial solutions can be chosen to be real. More explicitly, dropping an overall factor of k,

$$u_{E\ell}(r) = A_\ell\, r\, j_\ell(kr) + B_\ell\, r\, n_\ell(kr). \tag{19.5.4}$$

In the above, $j_\ell(\rho)$ is the spherical Bessel function, and $n_\ell(\rho)$ is the spherical Neumann function. We constructed these functions with the factorization method; their general explicit form is given in (17.8.32). As $\rho \to 0$, the function j_ℓ is regular and vanishes for $\ell \geq 1$, but n_ℓ diverges for all ℓ:

$$\begin{aligned} j_\ell(\rho) &\simeq \frac{\rho^\ell}{(2\ell+1)!!}, \\ n_\ell(\rho) &\simeq -\frac{(2\ell-1)!!}{\rho^{\ell+1}}. \end{aligned} \tag{19.5.5}$$

Recall that double factorials skip integers: $5!! = 5 \cdot 3 \cdot 1$, and $6!! = 6 \cdot 4 \cdot 2$, for example. As $\rho \to \infty$, we find that

$$\begin{aligned} \rho\, j_\ell(\rho) &\to \ \ \sin\left(\rho - \tfrac{\ell\pi}{2}\right), \\ \rho\, n_\ell(\rho) &\to -\cos\left(\rho - \tfrac{\ell\pi}{2}\right). \end{aligned} \tag{19.5.6}$$

This behavior turns out to be fairly accurate once $\rho > \ell$.

Going back to the $V = 0$ solutions we are trying to write, we now note that near the origin we must have $u_{E\ell} \sim \rho^{\ell+1}$. This means that the solutions are

$$u_{E\ell} = r\, j_\ell(kr). \tag{19.5.7}$$

We do not include a normalization constant because the solutions are not normalizable. This is not surprising. Plane waves representing a free particle are also not normalizable.

Back in (19.4.11), the $V = 0$ energy eigenstates then take the form

$$\psi_{E\ell m}(\mathbf{r}) = j_\ell(kr)\, Y_{\ell m}(\Omega). \tag{19.5.8}$$

The degenerate eigenstates of fixed energy E are labeled with $\ell = 0, 1, \ldots$ and with $m = -\ell, \ldots, \ell$, for each ℓ. The eigenvalues of the complete set of commuting observables (19.4.10) label the states uniquely. Finally, note that the large r behavior in (19.5.6) implies that

$$\psi_{E\ell m}(\mathbf{r}) \simeq \frac{1}{2ik}\left(\frac{e^{ikr - i\frac{\ell\pi}{2}}}{r} - \frac{e^{-ikr + i\frac{\ell\pi}{2}}}{r}\right) Y_{\ell m}(\Omega), \quad r \to \infty. \tag{19.5.9}$$

Recalling that the full wave function has an additional time-dependent factor $e^{-iEt/\hbar}$, we infer that the first term in parentheses represents an outgoing spherical wave, while the second term represents an ingoing spherical wave. This is why the general solutions (19.5.8) are spherical waves. In fact, each fixed-ℓ solution is called a **partial wave**.

Example 19.4. *The infinite spherical well.*
An infinite spherical well of radius a is a potential that forces the particle to be within the sphere $r \leq a$. The potential is zero for $r \leq a$, and it is infinite for $r > a$:

$$V(r) = \begin{cases} 0, & \text{if } r \leq a, \\ \infty, & \text{if } r > a. \end{cases} \tag{19.5.10}$$

For $r \leq a$ the potential vanishes, and the radial solutions (19.5.7) apply:

$$u_{E\ell}(r) = r j_\ell(kr). \tag{19.5.11}$$

The wave functions are now normalizable because the range of r is finite; the solutions must vanish for $r \geq a$. In fact, the quantization of the energy arises because $u_{E\ell}(r)$ must vanish for $r = a$, the location of the hard wall:

$$j_\ell(ka) = 0. \tag{19.5.12}$$

For each ℓ, the condition that ka be a nontrivial zero of j_ℓ selects the possible values of k and thus the possible values of the energy. We need a nontrivial zero because $k = 0$ is not of interest; the solution vanishes. The nontrivial zeroes of j_ℓ are denoted by $z_{n,\ell}$ where

$$z_{n,\ell} \text{ is the } n\text{th zero of } j_\ell : \; j_\ell(z_{n,\ell}) = 0. \tag{19.5.13}$$

Calling $k_{n,\ell}$ the nth wave number for a fixed ℓ, we have the simple condition

$$k_{n,\ell}\, a = z_{n,\ell}, \tag{19.5.14}$$

and the energies are quantized as

$$E_{n,\ell} = \frac{\hbar^2 k_{n,\ell}^2}{2m} = \frac{\hbar^2}{2ma^2}(k_{n,\ell}a)^2 = \frac{\hbar^2}{2ma^2} z_{n,\ell}^2. \tag{19.5.15}$$

Since $\frac{\hbar^2}{2ma^2}$ is the natural energy scale for this potential, it is convenient to define the unit-free scaled energies $\mathcal{E}_{n,\ell}$ by dividing $E_{n,\ell}$ by the natural energy

$$\mathcal{E}_{n,\ell} \equiv \frac{2ma^2}{\hbar^2} E_{n,\ell} = z_{n,\ell}^2. \tag{19.5.16}$$

The unit-free energies are just given by the zeroes squared! The zeroes of $j_0(\rho) = \sin\rho/\rho$ are $z_{n,0} = n\pi$, but the zeroes for $\ell \geq 1$ must be found numerically. For $\ell = 1$, for example, we have $z_{1,1} = 4.4934$, $z_{2,1} = 7.7252$, and $z_{3,1} = 10.904$. These correspond to the unit-free energies $\mathcal{E}_{1,1} = 20.191$, $\mathcal{E}_{2,1} = 59.679$, and $\mathcal{E}_{3,1} = 118.90$.

There are no accidental degeneracies: the zeroes of the various Bessel functions do not coincide. This seems clear from the inspection of many zeroes and perhaps can be proven in general. It follows that energies for different values of ℓ do not coincide. More explicitly, with $\ell \neq \ell'$ we have that $\mathcal{E}_{n,\ell} \neq \mathcal{E}_{n',\ell'}$ for any choices of n and n'. However nice and simple the infinite spherical well is, it does not have enough symmetry to produce a degenerate spectrum. □

19.6 Rayleigh's Formula

Rayleigh's formula expresses a plane wave as a superposition of spherical waves with different values of the orbital angular momentum ℓ. For each value of $\ell = 0, 1, \ldots \infty$, the waves, called **partial waves**, contain both incoming and outgoing components. This formula is essential in three-dimensional scattering theory, where it is the basis for calculating the scattering amplitude in terms of partial-wave phase shifts (section 30.3). For us at this moment, it illuminates the relations between energy eigenstates described using two alternative sets of commuting observables. Moreover, the derivation puts to good use much of what we have learned about spherical harmonics.

Let us begin by recalling the description of a *free* particle in three dimensions. Since the potential vanishes, the Hamiltonian is simply $\hat{H} = \frac{\hat{\mathbf{p}}^2}{2m}$. Solutions of the time-independent Schrödinger equation can be written as plane waves with momentum $\mathbf{p}$ and energy E:

$$\psi(\mathbf{r}) = e^{i\mathbf{k}\cdot\mathbf{r}}, \quad \mathbf{p} = \hbar\mathbf{k}, \quad E = \frac{\hbar^2 k^2}{2m}, \qquad k = |\mathbf{k}|. \tag{19.6.1}$$

The states are infinitely degenerate: for any value of the energy, the magnitude $|\mathbf{k}|$ of $\mathbf{k}$ is fixed, but not its direction. In this description the operators

$$\{H, \hat{p}_x, \hat{p}_y, \hat{p}_z\} \tag{19.6.2}$$

form a complete set of commuting observables. Plane waves are eigenstates of all these operators, and different plane waves are characterized by different eigenvalues of these operators. There is a bit of redundancy here. If the eigenvalues of the momenta are given, the energy eigenvalue is determined. Alternatively, we can fix any solution by giving the value of the energy and the *direction* of the momentum, specified by spherical angles θ, ϕ.

If we work in spherical coordinates, it is natural to consider the complete set of commuting observables $\{\hat{H}, \hat{\mathbf{L}}^2, \hat{L}_z\}$ associated with rotational symmetry. The solutions here are

$$\psi_{E\ell m}(\mathbf{r}) = j_\ell(kr)\, Y_{\ell m}(\Omega), \tag{19.6.3}$$

where the degenerate states of energy E are distinguished by quantum numbers ℓ and m.

Since the spherical wave solutions are complete, it should be possible to write any plane-wave solution as a superposition of spherical wave solutions. In particular for a wave propagating in the z-direction, we must have

$$e^{ikz} = e^{ikr\cos\theta} = \sum_{\ell=0}^{\infty} \sum_{m=-\ell}^{\ell} A_{\ell m} j_\ell(kr) Y_{\ell m}(\Omega), \tag{19.6.4}$$

with calculable coefficients $A_{\ell m}$. On the left-hand side, we have a plane wave with energy $E = \hbar^2 k^2/(2m)$, and all the waves on the right-hand side are waves of the same energy. Since the left-hand side has no ϕ dependence, $m = 0$ is the only allowed value on the right-hand side. Therefore, we find that

$$e^{ikz} = e^{ikr\cos\theta} = \sum_{\ell=0}^{\infty} a_\ell \, j_\ell(kr) Y_{\ell 0}(\Omega) \tag{19.6.5}$$

for some coefficients a_ℓ that must be determined. This takes some effort but is worth doing.

We begin our work by multiplying both sides of equation (19.6.5) by $Y^*_{\ell' 0}$ and integrating over solid angle to get

$$a_{\ell'} j_{\ell'}(kr) = \int d\Omega \, Y^*_{\ell' 0}(\theta) e^{ikr\cos\theta}, \tag{19.6.6}$$

where we used the orthonormality of spherical harmonics. We now relabel $\ell' \to \ell$ and express $Y_{\ell 0}$ in terms of $Y_{\ell\ell}$ using (19.3.72):

$$\begin{aligned} a_\ell j_\ell(kr) &= \frac{1}{\sqrt{(2\ell)!}} \int d\Omega \left[\left(\frac{\hat{L}_-}{\hbar} \right)^\ell Y_{\ell\ell}(\theta, \phi) \right]^* e^{ikr\cos\theta}, \\ &= \frac{1}{\sqrt{(2\ell)!}} \int d\Omega \, Y^*_{\ell\ell}(\theta, \phi) \left(\frac{\hat{L}_+}{\hbar} \right)^\ell e^{ikr\cos\theta} \end{aligned} \tag{19.6.7}$$

because relative to the inner product induced by angular integration, $\hat{L}_+$ is the Hermitian conjugate of $\hat{L}_-$ (see (10.5.38)).

Exercise 19.11. *Recalling the definition $(\psi_1, \psi_2)_a = \int d\Omega \, \psi_1^* \psi$ of the angular inner product, confirm that $\hat{L}_+$ is the Hermitian conjugate of $\hat{L}_-$.*

To evaluate the iterated action of $\hat{L}_+$ on a function of θ, we need the following identity:

$$\left(\frac{\hat{L}_+}{\hbar} \right)^\ell F(\theta) = (-1)^\ell e^{i\ell\phi} (\sin\theta)^\ell \left(\frac{d}{d(\cos\theta)} \right)^\ell F(\theta). \tag{19.6.8}$$

This can be proven using an induction argument, as well as the familiar expression

$$\frac{\hat{L}_+}{\hbar} = e^{i\phi} \left(\frac{\partial}{\partial\theta} + i\cot\theta \frac{\partial}{\partial\phi} \right). \tag{19.6.9}$$

Exercise 19.12. *Prove (19.6.8) by checking it first for $\ell = 1$ and then using an induction argument.*

Using (19.6.8), we quickly find that

$$\left(\frac{\hat{L}_+}{\hbar} \right)^\ell e^{ikr\cos\theta} = (-1)^\ell e^{i\ell\phi} (\sin\theta)^\ell \, (ikr)^\ell e^{ikr\cos\theta}. \tag{19.6.10}$$

We recognize $e^{i\ell\phi}(\sin\theta)^\ell \sim Y_{\ell\ell}(\theta, \phi)$, with the precise relation following from (10.5.35):

$$2^\ell \ell! \sqrt{\frac{4\pi}{(2\ell+1)!}} \; Y_{\ell\ell}(\theta, \phi) = (-1)^\ell e^{i\ell\phi} (\sin\theta)^\ell. \tag{19.6.11}$$

Therefore,

$$\left(\frac{\hat{L}_+}{\hbar}\right)^\ell e^{ikr\cos\theta} = 2^\ell \ell! \sqrt{\frac{4\pi}{(2\ell+1)!}}\, Y_{\ell\ell}(\theta,\phi)\,(ikr)^\ell e^{ikr\cos\theta}. \tag{19.6.12}$$

We can now use this result in (19.6.7), which then gives

$$a_\ell j_\ell(kr) = \frac{2^\ell \ell!}{\sqrt{(2\ell)!}} \sqrt{\frac{4\pi}{(2\ell+1)!}}\,(ikr)^\ell \int d\Omega\, Y^*_{\ell\ell}(\theta,\phi) Y_{\ell\ell}(\theta,\phi) e^{ikr\cos\theta}, \tag{19.6.13}$$

where the radial factor $(ikr)^\ell$ could be taken out of the angular integral. The strategy to calculating a_ℓ is to equate the leading behavior of both sides of the equation for small kr. For the left-hand side, we use (19.5.5). For the right-hand side, we have a factor $(ikr)^\ell$ multiplied by an angular integral. If we imagine Taylor expanding the $e^{ikr\cos\theta}$ factor in the integrand, we find that

$$\begin{aligned}\int d\Omega\, Y^*_{\ell\ell}(\theta,\phi) Y_{\ell\ell}(\theta,\phi) e^{ikr\cos\theta} &= \int d\Omega\, Y^*_{\ell\ell}(\theta,\phi) Y_{\ell\ell}(\theta,\phi)(1+\mathcal{O}(kr\cos\theta)) \\ &= 1+\mathcal{O}(kr),\end{aligned} \tag{19.6.14}$$

using the orthonormality of the spherical harmonics. The integral thus evaluates to one to leading order. It follows now that the leading terms in kr on each side of the equation are

$$a_\ell \frac{(kr)^\ell}{(2\ell+1)!!} = i^\ell (kr)^\ell \frac{2^\ell \ell!}{\sqrt{(2\ell)!}} \sqrt{\frac{4\pi}{(2\ell+1)!}}. \tag{19.6.15}$$

Canceling the common factor of $(kr)^\ell$ and noting that $2^\ell \ell!(2\ell+1)!! = (2\ell+1)!$, we find that

$$a_\ell = i^\ell \sqrt{4\pi}\sqrt{2\ell+1}. \tag{19.6.16}$$

Having determined the a_ℓ coefficients, from (19.6.5) we obtain the *Rayleigh* formula:

$$\boxed{e^{ikz} = \sqrt{4\pi} \sum_{\ell=0}^{\infty} \sqrt{2\ell+1}\; i^\ell j_\ell(kr)\, Y_{\ell 0}(\theta).} \tag{19.6.17}$$

This nontrivial relation expresses a plane wave as a linear superposition of spherical waves, partial waves with all possible values of the angular momentum ℓ. Each partial wave is clearly an exact solution when $V=0$ and so is their sum. Since $j_\ell(kr)$ for large r is a superposition of an incoming and an outgoing wave (see (19.5.9)), each partial wave contains incoming and outgoing spherical waves.

Rayleigh's formula is sometimes written using Legendre polynomials. Recalling the simple relation (10.5.34) between $Y_{\ell 0}(\theta)$ and $P_\ell(\cos\theta)$, we find that

$$e^{ikz} = \sum_{\ell=0}^{\infty} i^\ell\,(2\ell+1)\, j_\ell(kr)\, P_\ell(\cos\theta). \tag{19.6.18}$$

19.7 The Three-Dimensional Isotropic Oscillator

For a particle of mass m in a three-dimensional isotropic harmonic oscillator, the Hamiltonian $\hat{H}$ defining the system takes the form

$$\hat{H} = \frac{\hat{p}_x^2}{2m} + \frac{\hat{p}_y^2}{2m} + \frac{\hat{p}_z^2}{2m} + \tfrac{1}{2}m\omega^2(x^2+y^2+z^2). \tag{19.7.1}$$

Isotropy means that the system does not single out any special direction in space. The isotropy of the above Hamiltonian is a consequence of a single frequency ω appearing in the potential term: the frequencies for the independent x, y, and z oscillations are identical. This Hamiltonian is the sum of three commuting, one-dimensional Hamiltonians,

$$\hat{H} = \hat{H}_x + \hat{H}_y + \hat{H}_z, \tag{19.7.2}$$

where

$$\hat{H}_x = \frac{\hat{p}_x^2}{2m} + \tfrac{1}{2}m\omega^2x^2, \quad \hat{H}_y = \frac{\hat{p}_y^2}{2m} + \tfrac{1}{2}m\omega^2y^2 \quad \hat{H}_z = \frac{\hat{p}_z^2}{2m} + \tfrac{1}{2}m\omega^2z^2. \tag{19.7.3}$$

The Hamiltonian $\hat{H}$ does not describe three particles oscillating in different directions; it describes just one particle. Interestingly, as we will see below, tensor products are relevant here, the state space being the tensor product of three one-dimensional oscillator state spaces. The isotropy of the oscillator is manifest when we note that $x^2+y^2+z^2 = r^2$ and rewrite the Hamiltonian as follows:

$$\hat{H} = \frac{\hat{\mathbf{p}}^2}{2m} + \tfrac{1}{2}m\omega^2r^2. \tag{19.7.4}$$

Both terms are manifestly rotational invariant. The potential term here is a central potential:

$$V(r) = \tfrac{1}{2}m\omega^2r^2. \tag{19.7.5}$$

Therefore, the spectrum of this system consists of multiplets of angular momentum—in fact, an infinite number of them. As we will see, however, the spectrum has degeneracies: multiplets with different values of ℓ but the same energy. These degeneracies are the result of a *hidden symmetry*, a symmetry that is far from obvious in the Hamiltonian. The quantum three-dimensional oscillator is a lot more symmetric than the infinite spherical well, which had no degeneracies whatsoever.

To describe the three-dimensional oscillator, we can use the creation and annihilation operators $\hat{a}_x^\dagger, \hat{a}_y^\dagger, \hat{a}_z^\dagger$ and $\hat{a}_x, \hat{a}_y, \hat{a}_z$ associated with one-dimensional oscillations in the x-, y-, and z-directions. The Hamiltonian then takes the form

$$\hat{H} = \hbar\omega(\hat{N}_x + \hat{N}_y + \hat{N}_z + \tfrac{3}{2}) = \hbar\omega(\hat{N} + \tfrac{3}{2}), \tag{19.7.6}$$

where we define the total number operator $\hat{N}$ as $\hat{N} \equiv \hat{N}_x + \hat{N}_y + \hat{N}_z$. Here, $\hat{N}_x = \hat{a}_x^\dagger\hat{a}_x$, $\hat{N}_y = \hat{a}_y^\dagger\hat{a}_y$, and $\hat{N}_z = \hat{a}_z^\dagger\hat{a}_z$ are the number operators associated with the three directions x, y, and z, respectively.

Tensor products are relevant to the three-dimensional oscillator. We have used tensor products to describe multiparticle states. Tensor products, however, are also relevant to single particles if they have degrees of freedom that live in different spaces or more than one set of attributes, each of which is described by states in some vector space. For the three-dimensional oscillator, the Hamiltonian is the sum of commuting Hamiltonians of one-dimensional oscillators for the x, y, and z directions. Thus, the general states are obtained by tensoring the state spaces $\mathcal{H}_x$, $\mathcal{H}_y$, and $\mathcal{H}_z$ of the three independent oscillators. It is a single particle oscillating, but the description of what it is doing entails saying what it is doing in each of the independent directions. Thus, for the three-dimensional state space $\mathcal{H}_{3D}$ we write

$$\mathcal{H}_{3D} = \mathcal{H}_x \otimes \mathcal{H}_y \otimes \mathcal{H}_z. \tag{19.7.7}$$

In this notation the total Hamiltonian in (19.7.2) is written as

$$\hat{H} = \hat{H}_x \otimes \mathbb{1} \otimes \mathbb{1} + \mathbb{1} \otimes \hat{H}_y \otimes \mathbb{1} + \mathbb{1} \otimes \mathbb{1} \otimes \hat{H}_z. \tag{19.7.8}$$

The vacuum state $|0\rangle$ of the three-dimensional oscillator can be viewed as

$$|0\rangle \equiv |0\rangle_x \otimes |0\rangle_y \otimes |0\rangle_z. \tag{19.7.9}$$

The associated ground state wave function φ is

$$\varphi(x,y,z) = \langle x| \otimes \langle y| \otimes \langle z|\, |0\rangle = \langle x|0\rangle_x \langle y|0\rangle_y \langle z|0\rangle_z = \varphi_0(x)\varphi_0(y)\varphi_0(z), \tag{19.7.10}$$

where φ_0 is the ground state wave function of the one-dimensional oscillator. Recalling the form of the (nonnormalized) basis states for $\mathcal{H}_x, \mathcal{H}_y$, and $\mathcal{H}_z$,

$$\begin{aligned} &\mathcal{H}_x: \ (\hat{a}_x^\dagger)^{n_x}|0\rangle_x, \ n_x = 0, 1, \ldots, \\ &\mathcal{H}_y: \ (\hat{a}_y^\dagger)^{n_y}|0\rangle_y, \ n_y = 0, 1, \ldots, \\ &\mathcal{H}_z: \ (\hat{a}_z^\dagger)^{n_z}|0\rangle_z, \ n_z = 0, 1, \ldots, \end{aligned} \tag{19.7.11}$$

the basis states for the three-dimensional oscillator state space are

$$(\hat{a}_x^\dagger)^{n_x}|0\rangle_x \otimes (\hat{a}_y^\dagger)^{n_y}|0\rangle_y \otimes (\hat{a}_z^\dagger)^{n_z}|0\rangle_z, \quad n_x, n_y, n_z \in \{0, 1, \ldots\}. \tag{19.7.12}$$

This is what we would expect intuitively: we simply pile arbitrary numbers of $\hat{a}_x^\dagger$, $\hat{a}_y^\dagger$, and $\hat{a}_z^\dagger$ operators on the vacuum. For brevity, we write such basis states as

$$(\hat{a}_x^\dagger)^{n_x}(\hat{a}_y^\dagger)^{n_y}(\hat{a}_z^\dagger)^{n_z}|0\rangle. \tag{19.7.13}$$

Each of these states has a wave function that is the product of x-, y-, and z-dependent wave functions. Once we form superpositions of such states, the total wave function can no longer be factored into x-, y-, and z dependent wave functions. The x, y, and z-dependences become "entangled." In that sense these are the analogues of entangled states of three particles.

We are ready to construct the states of the three-dimensional isotropic oscillator. The key property is that the states must organize themselves into representations of angular momentum, the multiplets we explored in section 19.3. Since angular momentum operators commute with the Hamiltonian, angular momentum multiplets represent degenerate states. As we will see, the multiplets that appear at each level can be deduced by simple reasoning.

We already built the ground state, which is a single state with $\hat{N}$ eigenvalue $N=0$. All other states have higher energies, so this state must be, by itself, a representation of angular momentum. It can only be the singlet $\ell=0$. Thus, we have, from (19.7.6),

$$N=0,\ E=\tfrac{3}{2}\hbar\omega,\ \ |0\rangle \longleftrightarrow \ell=0. \tag{19.7.14}$$

The states with $N=1$ have $E=\frac{5}{2}\hbar\omega$ and are

$$\hat{a}_x^\dagger|0\rangle,\ \hat{a}_y^\dagger|0\rangle,\ \hat{a}_z^\dagger|0\rangle. \tag{19.7.15}$$

These three degenerate states fit precisely into an $\ell=1$ multiplet (a triplet). There is, in fact, no other possibility. Any higher-ℓ multiplet has too many states. Moreover, we cannot have three singlets because this would mean a degeneracy in the bound state spectrum of the $\ell=0$ radial Schrödinger equation, which is impossible. The $\ell=0$ ground state and the $\ell=1$ triplet at the first excited level are indicated in figure 19.7.

Let us proceed now with the states at $N=2$ or $E=\frac{7}{2}\hbar\omega$. There are six states:

$$\hat{a}_x^\dagger\hat{a}_x^\dagger|0\rangle,\ \hat{a}_y^\dagger\hat{a}_y^\dagger|0\rangle,\ \hat{a}_z^\dagger\hat{a}_z^\dagger|0\rangle,\ \hat{a}_x^\dagger\hat{a}_y^\dagger|0\rangle,\ \hat{a}_x^\dagger\hat{a}_z^\dagger|0\rangle,\ \hat{a}_y^\dagger\hat{a}_z^\dagger|0\rangle. \tag{19.7.16}$$

To help ourselves in trying to find the angular momentum multiplets, recall that the number of states # for a given ℓ is $2\ell+1$:

ℓ	0	1	2	3	4	5	6	7
#	1	3	5	7	9	11	13	15

We cannot use the triplet twice: this would imply a degeneracy in the spectrum of the $\ell=1$ radial equation. In general, at any fixed energy level any ℓ multiplet that appears can only appear once. Therefore, the only way to get six states is having five from $\ell=2$ and one from $\ell=0$:

$$\text{Six } N=2 \text{ states:}\ \ (\ell=2)\oplus(\ell=0). \tag{19.7.17}$$

We use the direct sum, not the tensor product. The six states define a six-dimensional vector space spanned by five vectors in $\ell=2$ and one vector in $\ell=0$.

Let us continue to determine the pattern. At $N=3$, with $E=\frac{9}{2}\hbar\omega$, we actually have ten states (count them!). It would seem now that there are two options for multiplets:

$$(\ell=3)\oplus(\ell=1)\ \text{ or }\ (\ell=4)\oplus(\ell=0). \tag{19.7.18}$$

We can see that the second option is problematic. If true, an $\ell=3$ multiplet, which has not appeared yet, would not arise at this level. If it appeared eventually, it would do so at a higher energy, and we would have the lowest $\ell=3$ multiplet with higher energy than the lowest $\ell=4$ multiplet. This is not possible, as discussed at the end of section 19.4. You may think perhaps that $\ell=3$ multiplets never appear, avoiding the inconsistency, but this is not true. At any rate we will provide below a rigorous argument that confirms the first option is the one realized. Therefore, the ten degenerate states at $N=3$ consist of the following multiplets:

$$\text{Ten } N=3 \text{ states:}\ \ (\ell=3)\oplus(\ell=1). \tag{19.7.19}$$

Let us do one more level. At $N=4$ we find fifteen states. Instead of writing them out, let us count them without listing them. In fact, we can easily do the general case of

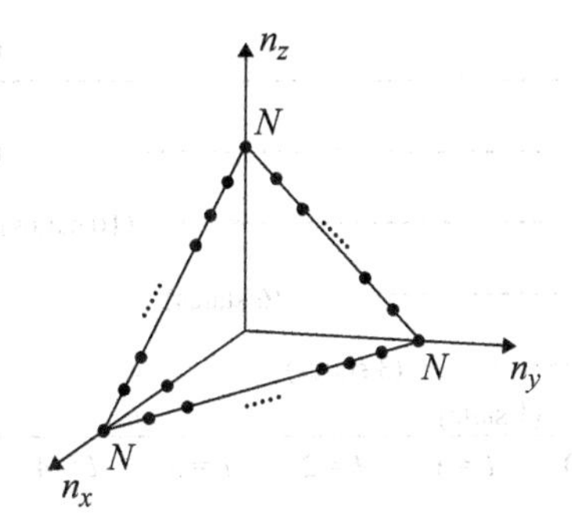

Figure 19.6
Counting the number of degenerate states with total number N in the isotropic harmonic oscillator.

arbitrary integer $N \geq 1$. The states we are looking for are of the form

$$(\hat{a}_x^\dagger)^{n_x}(\hat{a}_y^\dagger)^{n_y}(\hat{a}_z^\dagger)^{n_z}|0\rangle, \quad \text{with } n_x + n_y + n_z = N. \tag{19.7.20}$$

The number of different solutions of $n_x + n_y + n_z = N$, with $n_x, n_y, n_z \geq 0$, is the number of degenerate states at total number N. To visualize this think of $n_x + n_y + n_z = N$ as the equation for a plane in three-dimensional space with axes n_x, n_y, n_z. Since no integer can be negative, we are looking for points with integer coordinates in the region of the plane that lies on the positive octant, as shown in figure 19.6. Starting at one of the three corners, say, $(n_x, n_y, n_z) = (N, 0, 0)$, we have one point, then moving toward the origin we encounter two points, then three, and so on until we find $N + 1$ points on the (n_y, n_z) plane. Thus, the number deg(N) of degenerate states with number N is

$$\deg(N) = 1 + 2 + \cdots + (N+1) = \frac{(N+1)(N+2)}{2}. \tag{19.7.21}$$

Back to the $N = 4$ level, $\deg(4) = 15$. We rule out a single $\ell = 7$ multiplet since the multiplets $\ell = 4, 5, 6$ have not appeared yet. No two of those multiplets can appear simultaneously either, for it would imply that the ground states of two potentials with different ℓ coincide. Therefore, the only one that can appear is the lowest-ℓ multiplet, the $\ell = 4$ multiplet, with nine states. The remaining six states must then appear as $\ell = 2$ plus $\ell = 0$. we then have

$$\text{Fifteen } N = 4 \text{ states: } (\ell = 4) \oplus (\ell = 2) \oplus (\ell = 0). \tag{19.7.22}$$

We can see that ℓ jumps by steps of two, starting from the maximal ℓ. This is in fact the rule. It is quickly confirmed that the 21 states with $N = 5$ ($\deg(5) = 21$), would arise from $(\ell = 5) \oplus (\ell = 3) \oplus (\ell = 1)$. All this is shown in figure 19.7.

Some of the structure of angular momentum multiplets can be seen more explicitly by trading the $\hat{a}_x$ and $\hat{a}_y$ operators for complex linear combinations $\hat{a}_L$ and $\hat{a}_R$:

$$\hat{a}_L = \frac{1}{\sqrt{2}}(\hat{a}_x + i\hat{a}_y), \quad \hat{a}_R = \frac{1}{\sqrt{2}}(\hat{a}_x - i\hat{a}_y). \tag{19.7.23}$$

The $\hat{a}_L^\dagger$ and $\hat{a}_R^\dagger$ operators are obtained by taking the Hermitian conjugate of those above. L and R operators commute with each other while, as expected, $[\hat{a}_L, \hat{a}_L^\dagger] = [\hat{a}_R, \hat{a}_R^\dagger] = 1$.

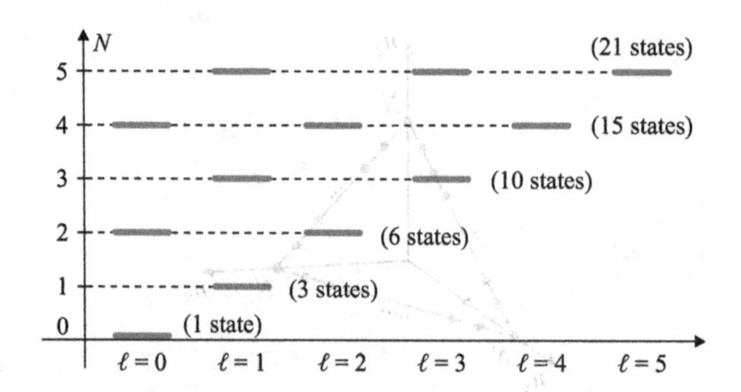

Figure 19.7
Spectral diagram for angular momentum multiplets in the three-dimensional isotropic harmonic oscillator. The energy is $E=\hbar\omega(N+\frac{3}{2})$, with N the total number eigenvalue. At level N the maximal ℓ also equals N.

With number operators $\hat{N}_R=\hat{a}_R^\dagger\hat{a}_R$ and $\hat{N}_L=\hat{a}_L^\dagger\hat{a}_L$, the Hamiltonian reads

$$\hat{H}=\hbar\omega(\hat{N}_R+\hat{N}_L+\hat{N}_z+\tfrac{3}{2}). \tag{19.7.24}$$

More importantly, the z-component $\hat{L}_z$ of angular momentum takes the simple form

$$\hat{L}_z=\hbar(\hat{N}_R-\hat{N}_L). \tag{19.7.25}$$

The above claims are confirmed in problem 19.7. Note that $\hat{a}_z$ carries no z-component of angular momentum. States are built acting with arbitrary numbers of $\hat{a}_L^\dagger$, $\hat{a}_R^\dagger$, and $\hat{a}_z^\dagger$ operators on the vacuum. The $N=1$ states are then presented as

$$\hat{a}_R^\dagger|0\rangle,\ \hat{a}_z^\dagger|0\rangle,\ \hat{a}_L^\dagger|0\rangle. \tag{19.7.26}$$

We see that the first state has $\hat{L}_z$ eigenvalue $L_z=\hbar$, the second $L_z=0$, and the third $L_z=-\hbar$, exactly the three expected values of the $\ell=1$ multiplet previously identified. For number $N=2$, the state with highest $\hat{L}_z$ eigenvalue is $(\hat{a}_R^\dagger)^2|0\rangle$, which has $L_z=2\hbar$. This shows that the highest ℓ multiplet is $\ell=2$. For arbitrary positive integer number N, the state with highest $\hat{L}_z$ eigenvalue is $(\hat{a}_R^\dagger)^N|0\rangle$, which has $L_z=\hbar N$. This shows we must have an $\ell=N$ multiplet. This is in fact what we got before. We can also understand why the top multiplet $\ell=N$ is accompanied by an $\ell=N-2$ multiplet and no $\ell=N-1$ multiplet. Consider the above state with maximal $L_z/\hbar$ equal to N and then the states with one and two units less of $L_z/\hbar$:

$$\begin{aligned}
L_z/\hbar=N: &\quad (\hat{a}_R^\dagger)^N|0\rangle,\\
L_z/\hbar=N-1: &\quad (\hat{a}_R^\dagger)^{N-1}\hat{a}_z^\dagger|0\rangle,\\
L_z/\hbar=N-2: &\quad (\hat{a}_R^\dagger)^{N-2}(\hat{a}_z^\dagger)^2|0\rangle,\ (\hat{a}_R^\dagger)^{N-1}\hat{a}_L^\dagger|0\rangle.
\end{aligned} \tag{19.7.27}$$

While there is only one state with one unit less of $L_z/\hbar$, there are two states with two units less. One linear combination of these two states must belong to the $\ell=N$ multiplet, but the other linear combination must be the top state of an $\ell=N-2$ multiplet! This is the reason for the jump of two units.

For arbitrary N we can see why the total number of states deg(N) in (19.7.21) can be reproduced by ℓ multiplets skipping by two:

$$\begin{aligned} N\,\text{odd}: \ \deg(N) &= \underbrace{1+2}_{\ell=1}+\underbrace{3+4}_{\ell=3}+\underbrace{5+6}_{\ell=5}+\underbrace{7+8}_{\ell=7}+\cdots+\underbrace{N+(N+1)}_{\ell=N}, \\ N\,\text{even}: \ \deg(N) &= \underbrace{1}_{\ell=0}+\underbrace{2+3}_{\ell=2}+\underbrace{4+5}_{\ell=4}+\underbrace{6+7}_{\ell=6}+\cdots+\underbrace{N+(N+1)}_{\ell=N}. \end{aligned} \tag{19.7.28}$$

The degeneracy of the spectrum is "explained" if we identify operators that commute with the Hamiltonian and connect the various ℓ multiplets that appear for a fixed number N. Such operators are symmetries of the theory. Consider, for example,

$$K \equiv \hat{a}_R^\dagger \hat{a}_L. \tag{19.7.29}$$

It is simple to check that K commutes with the Hamiltonian. With more work one can show that acting on the top state of the $\ell = N-2$ multiplet, K gives the top state of the $\ell = N$ multiplet (problem 19.7). This means that these two multiplets are degenerate. A more systematic analysis of the degeneracy would begin by finding all the independent operators that commute with the Hamiltonian. We will not do this here.

The two-dimensional isotropic oscillator also has degeneracies that ask for an explanation. In this case one can show that there are three operators $\hat{J}_i$ that commute with the Hamiltonian and in fact form an algebra of angular momentum (problem 19.8). This is *not* the algebra of orbital angular momentum because there are no such operators for motion in a plane; while there is an $\hat{L}_z = \hat{x}\hat{p}_y - \hat{y}\hat{p}_x$, there is no $\hat{L}_x$ or $\hat{L}_y$ because we have no $(\hat{z}, \hat{p}_z)$ operators. The angular momentum operators $\hat{J}_i$ here generate a *hidden* symmetry, and degenerate states must fall into multiplets of $\hat{\mathbf{J}}$. It turns out that at each energy level of the two-dimensional isotropic oscillator there is a single multiplet of $\hat{\mathbf{J}}$. In fact the full spectrum is precisely the list of *all* possible multiplets $j = 0, \frac{1}{2}, 1, \frac{3}{2}, \ldots$. This is clearly a remarkable system.

19.8 The Runge-Lenz Vector

We studied the hydrogen atom in chapter 11. The Hamiltonian $\hat{H}$ is very simple; it contains a kinetic energy term and a Coulomb potential term $V(r) = -e^2/r$. We calculated the spectrum of this Hamiltonian and found degenerate multiplets. While the energy eigenvalues depend only on a principal quantum number n, for each n there are degenerate multiplets of angular momentum with $\ell = 0, 1, \ldots, n-1$.

The large amount of degeneracy in this spectrum asks for an explanation. The hydrogen Hamiltonian has in fact some hidden symmetry: there is a conserved quantum Runge-Lenz vector operator. In the following we discuss the *classical* Runge-Lenz vector and its conservation. In the end-of-chapter problems, you will learn about the quantum Runge-Lenz operator. In chapter 20 this knowledge will be used to give a fully algebraic derivation of the hydrogen atom spectrum.

Since the following analysis is classical, the vectors are not operators and carry no hats. Consider the energy function for a particle of momentum $\mathbf{p}$ moving in a central potential $V(r)$:

$$E = \frac{\mathbf{p}^2}{2m} + V(r). \tag{19.8.1}$$

The force $\mathbf{F}$ on the particle is given by the negative gradient of the potential:

$$\mathbf{F} = -\nabla V = -V'(r)\frac{\mathbf{r}}{r}, \tag{19.8.2}$$

Here primes denote derivatives with respect to the argument. Newton's equation sets the rate of change of the momentum equal to the force:

$$\frac{d\mathbf{p}}{dt} = -V'(r)\frac{\mathbf{r}}{r}. \tag{19.8.3}$$

Here, $\mathbf{p} = m\dot{\mathbf{r}}$. You should confirm that for motion in a central potential the angular momentum $\mathbf{L} = \mathbf{r} \times \mathbf{p}$ is conserved:

$$\frac{d\mathbf{L}}{dt} = 0. \tag{19.8.4}$$

Let us now calculate the time derivative of $\mathbf{p} \times \mathbf{L}$:

$$\begin{aligned} \frac{d}{dt}(\mathbf{p} \times \mathbf{L}) &= \frac{d\mathbf{p}}{dt} \times \mathbf{L} = -\frac{V'(r)}{r}\mathbf{r} \times (\mathbf{r} \times \mathbf{p}) \\ &= -\frac{mV'(r)}{r}\mathbf{r} \times (\mathbf{r} \times \dot{\mathbf{r}}) = -\frac{mV'(r)}{r}\left[\mathbf{r}(\mathbf{r} \cdot \dot{\mathbf{r}}) - \dot{\mathbf{r}}r^2\right]. \end{aligned} \tag{19.8.5}$$

We now note that

$$\mathbf{r} \cdot \dot{\mathbf{r}} = \frac{1}{2}\frac{d}{dt}(\mathbf{r} \cdot \mathbf{r}) = \frac{1}{2}\frac{d}{dt}r^2 = r\dot{r}. \tag{19.8.6}$$

Using this result, the derivative of $\mathbf{p} \times \mathbf{L}$ becomes

$$\begin{aligned} \frac{d}{dt}(\mathbf{p} \times \mathbf{L}) &= -\frac{mV'(r)}{r}\left[\mathbf{r}r\dot{r} - \dot{\mathbf{r}}r^2\right] = mV'(r)r^2\left[\frac{\dot{\mathbf{r}}}{r} - \frac{\mathbf{r}\dot{r}}{r^2}\right] \\ &= mV'(r)r^2\,\frac{d}{dt}\left(\frac{\mathbf{r}}{r}\right). \end{aligned} \tag{19.8.7}$$

Because of the factor $V'(r)r^2$, the right-hand side fails to be the time derivative of some quantity. But if we focus on potentials for which this factor is a constant, the right-hand side is a time derivative, and we get a conserved quantity. So assume that for some constant γ we have

$$V'(r)\,r^2 = \gamma. \tag{19.8.8}$$

It then follows that

$$\frac{d}{dt}(\mathbf{p} \times \mathbf{L}) = m\gamma\,\frac{d}{dt}\left(\frac{\mathbf{r}}{r}\right) \quad \Rightarrow \quad \frac{d}{dt}\left(\mathbf{p} \times \mathbf{L} - m\gamma\,\frac{\mathbf{r}}{r}\right) = 0. \tag{19.8.9}$$

The complicated vector inside the parentheses is constant in time. The condition (19.8.8) on the potential implies that

$$\frac{dV}{dr} = \frac{\gamma}{r^2} \quad \Rightarrow \quad V(r) = -\frac{\gamma}{r} + c_0. \tag{19.8.10}$$

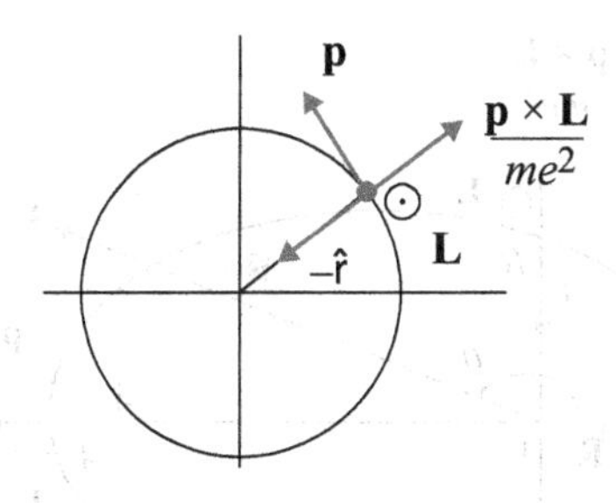

Figure 19.8
The Runge-Lenz vector vanishes for a circular orbit.

This is the most general potential for which we get a conserved vector. For $c_0 = 0$ and $\gamma = e^2$, we have the hydrogen atom potential $V(r) = -e^2/r$. For this case the conservation equation reads

$$\frac{d}{dt}\Big(\mathbf{p}\times\mathbf{L} - me^2\,\frac{\mathbf{r}}{r}\Big) = 0. \tag{19.8.11}$$

Factoring a constant we obtain the unit-free conserved **Runge-Lenz** vector $\mathbf{R}$ associated with the hydrogen atom classical dynamics:

$$\boxed{\mathbf{R} \equiv \frac{1}{me^2}\,\mathbf{p}\times\mathbf{L} - \frac{\mathbf{r}}{r}, \qquad \frac{d\mathbf{R}}{dt} = 0.} \tag{19.8.12}$$

The conservation of the Runge-Lenz vector is a property of inverse-squared central forces. The second term in $\mathbf{R}$ is the inward-directed unit radial vector.

To familiarize ourselves with the Runge-Lenz vector, we first examine its value for a circular orbit, as shown in figure 19.8. With counterclockwise motion, the vector $\mathbf{L}$ points out of the page, and $\mathbf{p}\times\mathbf{L}$ points radially outward. The vector $\mathbf{R}$ is thus a competition between the outward-pointing first term along $\mathbf{p}\times\mathbf{L}$ and the inward-pointing second term along $-\hat{\mathbf{r}}$. If these two terms did not cancel, the result would be a radial vector, outward or inward, but in any case not conserved as it rotates with the particle. This cannot happen, therefore the two terms must cancel. Indeed, for a circular orbit

$$m\frac{v^2}{r} = \frac{e^2}{r^2} \quad\Rightarrow\quad \frac{mv^2 r}{e^2} = 1 \quad\Rightarrow\quad \frac{(mv)(mvr)}{me^2} = 1 \quad\Rightarrow\quad \frac{pL}{me^2} = 1, \tag{19.8.13}$$

which states that in a circular orbit the first term in $\mathbf{R}$ is a unit vector. Since it points outward, it cancels with the second term, and the Runge-Lenz vector vanishes for a circular orbit.

We now argue that for an elliptic orbit the Runge-Lenz vector is not zero. Consider figure 19.9, showing a particle in counterclockwise motion around an elliptic orbit. One focus of the ellipse is at the origin $\mathbf{r} = 0$. At all times the conserved $\mathbf{L}$ points off the page. At the aphelion A, the point farthest away from the focal center, the first term in $\mathbf{R}$ points outward, and the second term point inward. Thus, if $\mathbf{R}$ does not vanish it must

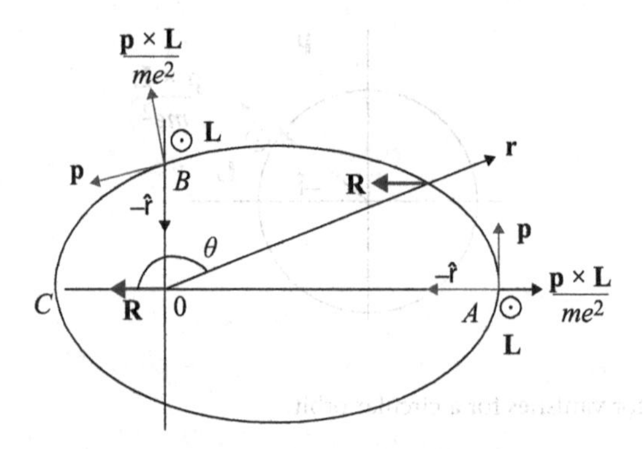

Figure 19.9
In an elliptic orbit, the Runge-Lenz vector is a vector along the major axis of the ellipse and points from the focus to the perihelion *C*.

be a vector along the axis joining the focus and the aphelion, a horizontal vector on the figure. Now consider point *B*, shown directly above the focus of the orbit. Here, **p** is no longer perpendicular to the radial vector, and therefore $\mathbf{p} \times \mathbf{L}$ is no longer radial. As you can see, it points slightly to the left of the vertical. Since **R** is conserved, and we know it is horizontal, its pointing to the left allows us to conclude that in an elliptic orbit **R** is a nonzero vector pointing *from* the focus *to* the perihelion *C*, the point of closest approach in the orbit.

Since **R** vanishes for circular orbits, the length *R* of **R** must measure the deviation of the orbit from circular. In fact, the magnitude *R* of the Runge-Lenz vector is precisely the eccentricity of the orbit. To see this we form the dot product of **R** with the radial vector **r**:

$$\mathbf{r} \cdot \mathbf{R} = \frac{1}{me^2}\mathbf{r} \cdot (\mathbf{p} \times \mathbf{L}) - r. \tag{19.8.14}$$

Referring to figure 19.9, let θ be the angle measured at the origin, increasing clockwise and defined with $\theta = 0$ the direction to the perihelion. The angle between **r** and **R** is then θ and we get

$$rR\cos\theta = \frac{1}{me^2}\mathbf{L} \cdot (\mathbf{r} \times \mathbf{p}) - r = \frac{1}{me^2}L^2 - r. \tag{19.8.15}$$

Collecting terms proportional to *r*,

$$r(1 + R\cos\theta) = \frac{L^2}{me^2} \quad \Rightarrow \quad \boxed{\frac{1}{r} = \frac{me^2}{L^2}(1 + R\cos\theta).} \tag{19.8.16}$$

This is one of the standard presentations of an elliptic orbit, and *R* appears at the place one conventionally has the eccentricity *e*, thus $e = R$. If $R = 0$, the orbit is circular because *r* does not depend on θ. The identification of *R* with *e* follows from the definition

$$e \equiv \frac{r_{\max} - r_{\min}}{r_{\max} + r_{\min}}. \tag{19.8.17}$$

Here r_{min} and r_{max} are, respectively, the minimum and maximum distances to the focus located at the center of force.

Exercise 19.13. *Use equations (19.8.16) and (19.8.17) to confirm that, indeed, $e = R$.*

This analysis thus far has been classical. Quantum mechanically, some things must be changed; happily, not that much! The definition of **R** only has to be changed to guarantee that $\hat{\mathbf{R}}$ is a Hermitian operator. Hermitization gives

$$\hat{\mathbf{R}} \equiv \frac{1}{2me^2}\,(\hat{\mathbf{p}}\times\hat{\mathbf{L}} - \hat{\mathbf{L}}\times\hat{\mathbf{p}}) - \frac{\hat{\mathbf{r}}}{r}. \tag{19.8.18}$$

Here $\hat{\mathbf{r}}$ is the position vector operator (not a unit vector).

Exercise 19.14. *Confirm that $\hat{\mathbf{R}}$ defined above is Hermitian and reduces to **R** when vector operators become classical vectors.*

The quantum mechanical conservation of $\hat{\mathbf{R}}$ is the statement that it commutes with the hydrogen Hamiltonian:

$$[\,\hat{\mathbf{R}}, \hat{H}\,] = 0. \tag{19.8.19}$$

The required calculation (problem 19.10) is the quantum analogue of the above classical calculation that showed that the time derivative of **R** is zero. Moreover, the length squared of the operator $\hat{\mathbf{R}}$ is also of interest. The result (problem 19.11) is

$$\hat{\mathbf{R}}^2 = 1 + \frac{2}{me^4}\,\hat{H}\,(\hat{\mathbf{L}}^2 + \hbar^2). \tag{19.8.20}$$

These facts above will be used in section 20.8 to show that the symmetries generated by $\hat{\mathbf{R}}$ and the angular momentum $\hat{\mathbf{L}}$ determine completely the spectrum of the hydrogen atom.

Problems

Problem 19.1. *Uncertainty in an angular momentum state.*

Calculate the uncertainty of the operator $\hat{L}_x$ on the state $|\ell, m\rangle$.

Problem 19.2. *Measurement of angular momentum for a particle with spin one.*

A quantum particle is known to have total spin angular momentum one; that is, $s = 1$. Use the eigenstates $|s, m\rangle = |1, m\rangle$ of $\hat{\mathbf{S}}^2$ and $\hat{S}_3$ as a basis, ordered as $|1\rangle = |1, 1\rangle$, $|2\rangle = |1, 0\rangle$, $|3\rangle = |1, -1\rangle$. Note that example 19.1 gives you the matrix representations of $\hat{\mathbf{S}}^2$, $\hat{S}_+$, $\hat{S}_-$, $\hat{S}_1$, $\hat{S}_2$, and $\hat{S}_3$.

1. Find expressions for the $\hat{S}_1$ eigenstates $|1, m_1\rangle_1$ with $m_1 = 1, 0, -1$, as superpositions of $\hat{S}_3$ eigenstates.
2. If a particle is in the state $|1, m_1 = 1\rangle_1$ and a measurement is made of the $\hat{S}_3$-component of its angular momentum, what are the possible results and the associated probabilities?

3. A particle is in the state $|1, m_1 = 1\rangle_1$. The $\hat{S}_3$-component of its angular momentum is measured, and the result $m_3 = -1$ is obtained. Immediately afterward, the $\hat{S}_1$-component of angular momentum is measured. Explain what results are obtained and with what probability. Suppose you measured $m_1 = -1$, and now you decide to measure $\hat{S}_3$ again. What are the possible outcomes and with what probability?

Problem 19.3. *Parity operator in three dimensions.*

The goal of this problem is to explore properties of the parity operator Π and to show that it acts on spherical harmonics as follows: $\Pi|\ell, m\rangle = (-1)^{\ell}|\ell, m\rangle$. The parity operator Π is defined by its action on position states: $\Pi|\mathbf{x}\rangle = |-\mathbf{x}\rangle$.

1. Show $\Pi^2 = \mathbb{1}$ and that Π is Hermitian and unitary. What is $\Pi|\mathbf{p}\rangle$?
2. Show that $\Pi\,\hat{x}_i = -\hat{x}_i\,\Pi$, and $\Pi\,\hat{p}_i = -\hat{p}_i\,\Pi$, where $\hat{x}_i$ and $\hat{p}_i$ are position and momentum operators. Conclude that the parity operator commutes with the angular momentum operators: $[\Pi, \hat{L}^i] = 0$.
3. Show that $\Pi|\ell, m\rangle = \lambda_{\ell,m}|\ell, m\rangle$ for some real eigenvalues $\lambda_{\ell,m}$ that satisfy $(\lambda_{\ell,m})^2 = 1$.
4. Use $[\Pi, L_{\pm}] = 0$ to prove that $\lambda_{\ell,m}$ is independent of m: $\Pi|\ell, m\rangle = \lambda_{\ell}|\ell, m\rangle$.
5. Given that $\Pi|\mathbf{x}\rangle = |-\mathbf{x}\rangle$, what does Π give when acting on the position state $|r\theta\phi\rangle$ in spherical coordinates? Use the wave function

$$\langle\theta\phi|\ell, \ell\rangle = Y_{\ell\ell}(\theta, \phi) = N_{\ell}\left(\sin\theta\, e^{i\phi}\right)^{\ell}$$

to show that $\lambda_{\ell} = (-1)^{\ell}$, thus proving the desired result.

Problem 19.4. *Large-distance behavior of the radial equation.*

Consider the radial equation for an $E > 0$ eigenstate $u(r)$, assuming large r so we can ignore the centrifugal barrier:

$$-\frac{\hbar^2}{2m}u'' + V(r)u = E\,u, \quad E = \frac{\hbar^2 k^2}{2m}.$$

To explore the large-r behavior of the solution, we write $u(r) = e^{ikr}g(r)$, where we aim to have $g(r)$ slowly varying for large r. To find what this requires from $V(r)$, show that by ignoring the second derivatives of g the radial equation implies that

$$g(r) = g(r_0)\exp\left(-\frac{im}{\hbar^2 k}\int_{r_0}^{r} V(r')dr'\right)$$

for r_0, a fixed value of r. If the integral converges for $r \to \infty$, then $g(r)$ approaches a constant value at infinity. Find $g(r)$ for the Coulomb potential $V(r) = -e^2/r$, and confirm that the effect of the potential does not die off for any r. Conclude that for $V(r) = c/r^{1+\epsilon}$ with $\epsilon > 0$, $g(r)$ does approach a fixed value at infinity. Such potentials satisfy $\lim_{r\to\infty} rV(r) = 0$.

Problem 19.5. *Qualitative behavior of the radial wave function.*

Consider a particle of mass m moving under the influence of an attractive Yukawa potential $V(r) = -ge^{-\alpha r}/r$, with g and α positive constants.

1. Write the Schrödinger equation for the radial wave function $u(r)$. Define a dimensionless radial variable $x \equiv \alpha r$ and rewrite the radial equation in the form

$$\left[-\frac{d^2}{dx^2} + \frac{\ell(\ell+1)}{x^2} - g'\frac{e^{-x}}{x}\right] u(x) = \lambda u(x).$$

What is the effective potential $V_{\text{eff}}(x)$ for the scaled equation? Relate λ and g' to the original parameters of the problem.

2. Show graphically that the parameter g' can be chosen so the effective potential has a) no bound states or b) many bound states (for any $\ell \neq 0$). Note that it is the interplay between the interaction $V(r)$ and the angular momentum barrier, $\hbar^2\ell(\ell+1)/2mr^2$, that determines the number of bound states.

3. Suppose the parameters are such that there are four distinct bound states with $\ell = 2$. Sketch, as accurately as you can, the wave function of the second, most tightly bound, bound state. You should include the behavior as $x \to 0$, the behavior as $x \to \infty$, the correct number of nodes, and the relative magnitude of the wave function at large, small, and intermediate x.

Problem 19.6. *The finite spherical well.*

A particle of mass m is in a potential $V(r)$ that represents a finite-depth spherical well of radius a:

$$V(r) = \begin{cases} -V_0, & \text{for } r < a, \\ 0, & \text{for } r > a. \end{cases}$$

Here V_0 is a positive constant with units of energy. As usual, we define $z_0^2 = 2ma^2V_0/\hbar^2$.

1. For the potential to have bound states, it should be deep enough. Show that this requires $z_0 \geq \frac{\pi}{2}$. [Hint: barely having a bound state means that the lowest $\ell = 0$ state must have an energy of essentially zero.]

2. Consider the general problem of finding the $\ell = 0$ bound states when V_0 is deep enough to have them. Explain why the eigenstates are in fact the parity-odd eigenstates of a finite square well of width $2a$ and depth V_0, restricted to positive values of the coordinate.

   Assume now that z_0 is a large number. Show that the lowest-energy bound states for low integers n take values

$$E \simeq -V_0 + (n\pi)^2 \frac{\hbar^2}{2ma^2}.$$

   For $n = 1$, find a better approximation for the energy, including the first nontrivial correction that would vanish as $z_0 \to \infty$.

3. Show that in order to have $\ell = 1$ bound states we need $z_0 \geq 2.74371$.

4. Now consider the delta function potential

$$V(r) = -\frac{4\pi}{3}(V_0 L^3)\,\delta(\mathbf{x}),$$

where $V_0 > 0$ is a constant with units of energy, and L is a constant with units of length that is needed for dimensional reasons. It is not easy to solve this problem, so we will regulate it by replacing this potential by a potential $V_a(r)$ of the form

$$V_a(r) = \begin{cases} -V_0\left(\frac{L}{a}\right)^3, & \text{for } r < a, \\ 0, & \text{for } r > a. \end{cases}$$

The regulator parameter a helps represent the delta function. Confirm that $\int d^3x V(r) = \int d^3x V_a(r)$, showing this is a good representation of the delta function as $a \to 0$.

A successful regulation would mean that the bound state energies are independent of the artificial regulator a in the limit as $a \to 0$. Show that this does *not* happen.

One can understand the complications by dimensional analysis. The parameters of this theory are $\hbar$, m, and the quantity $V_0 L^3$ (not V_0 and L separately). Use dimensional analysis to construct the "natural" energy of the bound states. Then argue that this result is absurd!

Problem 19.7. *Degeneracies in the three-dimensional isotropic oscillator.*

For the three-dimensional isotropic oscillator, it is useful to replace the $\hat{a}_x$ and $\hat{a}_y$ operators by operators $\hat{a}_L = \frac{1}{\sqrt{2}}(\hat{a}_x + i\hat{a}_y)$ and $\hat{a}_R = \frac{1}{\sqrt{2}}(\hat{a}_x - i\hat{a}_y)$. For the z-direction, we keep $\hat{a}_z$ and $\hat{a}_z^\dagger$.

1. Show that $\hat{H} = \hbar\omega(\hat{N}_R + \hat{N}_L + \hat{N}_z + \frac{3}{2})$, with $\hat{N}_R = \hat{a}_R^\dagger \hat{a}_R$ and $\hat{N}_L = \hat{a}_L^\dagger \hat{a}_L$. Show that $\hat{L}_z = \hbar(\hat{N}_R - \hat{N}_L)$.
2. Show that $\hat{L}_+ = \sqrt{2}\hbar(\hat{a}_z^\dagger \hat{a}_L - \hat{a}_R^\dagger \hat{a}_z)$. Confirm that $[\hat{L}_+, \hat{L}_-] = 2\hbar\hat{L}_z$, as expected.
3. Find the linear combination $|\psi_{\rm top}\rangle$ of the states with $L_z/\hbar = N-2$ shown in (19.7.27) that represents the highest $\hat{L}_z$ state of the $\ell = N-2$ multiplet. Show that $K = \hat{a}_R^\dagger \hat{a}_L$ acting on $|\psi_{\rm top}\rangle$ gives the top state of the $\ell = N$ multiplet.

Problem 19.8. *A three-dimensional angular momentum in the two-dimensional oscillator.*

As in problems 15.12 and 19.7, we define $\hat{a}_L = \frac{1}{\sqrt{2}}(\hat{a}_x + i\hat{a}_y)$ and $\hat{a}_R = \frac{1}{\sqrt{2}}(\hat{a}_x - i\hat{a}_y)$ as well as $\hat{N}_R = \hat{a}_R^\dagger \hat{a}_R$ and $\hat{N}_L = \hat{a}_L^\dagger \hat{a}_L$. Our subject now is the *two*-dimensional isotropic oscillator, for which one finds

$$\hat{H} = \hbar\omega(\hat{N}_R + \hat{N}_L + 1), \quad \hat{L}_z = \hbar(\hat{N}_R - \hat{N}_L).$$

We are going to show that there is a "hidden" three-dimensional algebra of angular momentum here. The operators are going to be

$$\hat{J}_z = \alpha \hat{L}_z = \alpha\,\hbar(\hat{N}_R - \hat{N}_L), \quad \hat{J}_+ = \beta\,\hbar\,\hat{a}_R^\dagger \hat{a}_L, \quad \hat{J}_- = \beta\,\hbar\,\hat{a}_L^\dagger \hat{a}_R,$$

where α and β are (real) constants to be determined. Note that, as required, $\hat{J}_+^\dagger = \hat{J}_-$.

1. Find α and β such that the operators above obey the algebra of angular momentum.
2. Show that all $\hat{J}_i$'s commute with the Hamiltonian (they are conserved).
3. Associated to this angular momentum are states $|j, m\rangle$ with $J^2 = \hbar^2 j(j+1)$ and $J_z = \hbar m$. Show that $(a_R^\dagger)^n|0\rangle$ is a state $|j, m\rangle$ with $j = m = n/2$.

4. Express $\hat{\mathbf{J}}^2$ in terms of $\hat{N}_L$ and $\hat{N}_R$ and show that eigenstates of $\hat{N}_L$ and $\hat{N}_R$ have angular momentum $j=(N_L+N_R)/2$, with N_L and N_R the eigenvalues of $\hat{N}_L$ and $\hat{N}_R$, respectively.
5. Describe the full spectrum of states of the two-dimensional harmonic oscillator in terms of representations of the hidden angular momentum.
6. Comment why, in retrospect, the constant α could not have been equal to one. This shows that the hidden angular momentum is really well hidden!

Problem 19.9. *A curious factorization of the hydrogen Hamiltonian (R. Jackiw).*

Consider the hydrogen atom Hamiltonian $\hat{H}=\frac{\hat{\mathbf{p}}^2}{2m}-\frac{e^2}{r}$. We will write it as

$$\hat{H}=\gamma+\frac{1}{2m}\sum_{k=1}^{3}\Big(\hat{p}_k+i\beta\frac{\hat{x}_k}{r}\Big)\Big(\hat{p}_k-i\beta\frac{\hat{x}_k}{r}\Big),$$

where $\hat{p}_k$ and $\hat{x}_k$ are, respectively, the Cartesian components of the momentum and position operators, and β and γ are real constants to be adjusted so that the two Hamiltonians are the same.

1. Calculate β and γ in terms of e^2, the Bohr radius a_0, and other constants.
2. Explain why for any state $\langle\hat{H}\rangle\geq\gamma$. Find the wave function of the state that saturates this inequality, and confirm that it is the ground state of hydrogen.

Problem 19.10. *Quantum conservation of the Runge-Lenz vector.*

For the hydrogen atom Hamiltonian $\hat{H}$, the Hermitian, conserved Runge-Lenz operator $\hat{\mathbf{R}}$ is given in (19.8.18). In this problem we want to show that $\hat{\mathbf{R}}$ is conserved: $[\hat{\mathbf{R}},\hat{H}]=0$. To verify this claim, let us follow the steps below.

1. Prove that for an arbitrary radial function $f(r)$

$$[\hat{\mathbf{p}}^2,f(r)\,\hat{\mathbf{r}}]\;=\;\frac{\hbar}{i}\Big(\,(\hat{\mathbf{p}}\cdot\hat{\mathbf{r}})\,\hat{\mathbf{r}}\,\frac{f'(r)}{r}\;+\;\frac{f'(r)}{r}\,\hat{\mathbf{r}}\,(\hat{\mathbf{r}}\cdot\hat{\mathbf{p}})\;+\;\hat{\mathbf{p}}\,f(r)+f(r)\,\hat{\mathbf{p}}\,\Big).\tag{1}$$

   Here $\hat{\mathbf{r}}$ is the position operator. Confirm that the operator in parentheses is Hermitian.
2. Now compute the commutator

$$[\hat{\mathbf{p}}\times\hat{\mathbf{L}}-\hat{\mathbf{L}}\times\hat{\mathbf{p}}\,,\,f(r)]\;=\;\frac{\hbar}{i}\Big(\;\cdots\qquad\cdots\qquad\Big).\tag{2}$$

   Your goal as you compute this is to get a Hermitian operator inside the parentheses, with a structure analogous to that in (1).
3. Attempt to make the problem more general by setting

$$\hat{H}_f\;\equiv\;\frac{\hat{\mathbf{p}}^2}{2m}-f(r),\quad e^2\hat{\mathbf{R}}_f\;\equiv\;\frac{1}{2m}(\hat{\mathbf{p}}\times\hat{\mathbf{L}}-\hat{\mathbf{L}}\times\hat{\mathbf{p}})\;-\;f(r)\,\hat{\mathbf{r}}.$$

   Use the above commutators to show that $[\hat{H}_f,\hat{\mathbf{R}}_f]=0$ if and only if $rf'(r)\;=\;-f(r)$. Verify that the unique solution of this equation is $f(r)=c/r$ with c an arbitrary constant.

Problem 19.11. *Length squared of the quantum Runge-Lenz vector.*

Check that the Runge-Lenz operator $\hat{\mathbf{R}}$, given in (19.8.18), can be alternatively written as

$$\hat{\mathbf{R}} = \frac{1}{me^2}\big(\,\hat{\mathbf{p}}\times\hat{\mathbf{L}} - i\hbar\,\hat{\mathbf{p}}\big) - \frac{\hat{\mathbf{r}}}{r} = \frac{1}{me^2}\big(-\hat{\mathbf{L}}\times\hat{\mathbf{p}} + i\hbar\,\hat{\mathbf{p}}\big) - \frac{\hat{\mathbf{r}}}{r}.$$

In this problem we want to calculate $\hat{\mathbf{R}}^2$:

$$\hat{\mathbf{R}}^2 = \Big(\frac{1}{me^2}(-\hat{\mathbf{L}}\times\hat{\mathbf{p}} + i\hbar\,\hat{\mathbf{p}}) - \frac{\hat{\mathbf{r}}}{r}\Big)\cdot\Big(\frac{1}{me^2}(\,\hat{\mathbf{p}}\times\hat{\mathbf{L}} - i\hbar\,\hat{\mathbf{p}}) - \frac{\hat{\mathbf{r}}}{r}\Big).$$

To do the computation more easily, first prove the three following identities:

$$(\hat{\mathbf{L}}\times\hat{\mathbf{p}})\cdot\hat{\mathbf{r}} = -\hat{\mathbf{L}}^2, \quad i\hbar\Big(\hat{\mathbf{p}}\cdot\frac{\hat{\mathbf{r}}}{r} - \frac{\hat{\mathbf{r}}}{r}\cdot\hat{\mathbf{p}}\Big) = \frac{2\hbar^2}{r}, \quad (\hat{\mathbf{L}}\times\hat{\mathbf{p}})\cdot(\hat{\mathbf{p}}\times\hat{\mathbf{L}}) = -\hat{\mathbf{p}}^2\,\hat{\mathbf{L}}^2.$$

Now show that

$$\hat{\mathbf{R}}^2 = 1 + \frac{2}{me^4}\hat{H}(\hat{\mathbf{L}}^2 + \hbar^2).$$

20 Addition of Angular Momentum

We show that if a space V_1 carries a representation of angular momentum operators $\hat{\mathbf{J}}_1$ and a space V_2 carries a representation of angular momentum operators $\hat{\mathbf{J}}_2$, then the tensor product $V_1 \otimes V_2$ carries a representation of the sum $\hat{\mathbf{J}}_1 + \hat{\mathbf{J}}_2$ of angular momentum operators. We obtain the classic result that the tensor product of two spin one-half vector spaces is the direct sum of a total spin one vector space and a total spin zero vector space. We examine the hyperfine splitting of the ground state of the hydrogen atom and then turn to the spin-orbit interaction, where we discuss complete sets of commuting observables. This leads us to the general study of uncoupled and coupled basis states and the calculation of Clebsch-Gordan coefficients. We derive the selection rule that tells us what representations of the total angular momentum appear in the tensor product $j_1 \otimes j_2$ of two angular momentum representations. We give an algebraic derivation of the spectrum of the hydrogen atom using the conserved orbital angular momentum and Runge-Lenz vector to discover two commuting angular momenta.

20.1 Adding Apples to Oranges?

We are going to be adding angular momenta in a variety of ways. Since angular momenta are operators in quantum mechanics, we are going to be adding angular momentum operators in various ways as well. We may add the *spin* angular momentum $\hat{\mathbf{S}}$ of a particle to its *orbital* angular momentum $\hat{\mathbf{L}}$. Or we may want to add the spin angular momentum $\hat{\mathbf{S}}^{(1)}$ of a particle to the spin angular momentum $\hat{\mathbf{S}}^{(2)}$ of another particle. At first sight we may feel like we are trying to add apples to oranges! For a given particle, spin angular momentum and orbital angular momentum act on different degrees of freedom of the particle. Adding the spins of two different particles also seems unusual if, for example, the particles are far away from each other. Vectors that live at different places are seldom added: you don't typically add the electric field at one point to the electric field at another point because the sum has no obvious interpretation. This is even more severe in general relativity: you *cannot* add vectors that "live" at different points of space-time. To add them you need a procedure to first bring them to a common point. Once they both live at that common point, you can add them.

Despite some differences, however, at an algebraic level all angular momenta are apples (Granny Smith, Red Delicious, McIntosh, Fuji, and so on). Therefore, they can be added, and it is natural to add them. We are not adding apples to oranges; we are adding apples to apples! The physics requires it. For a particle with spin and orbital angular momentum, for example, the sum of these two is the total angular momentum, which is the complete generator of rotations, implementing any rotation on both the spatial and spin degrees of freedom of the particle. Moreover, we will see that in large classes of Hamiltonians, energy eigenstates are eigenstates of a sum of angular momenta. The mathematics allows it: the sum of angular momenta *is* an angular momentum acting in the appropriate *tensor* product. As we will see below, while each angular momentum operator lives on a different vector space, the sum finds a *home* in the tensor product of the vector spaces.

What is an angular momentum? It is a triplet $\hat{J}_i$ of Hermitian operators on some complex vector space V satisfying the commutation relations

$$[\hat{J}_i, \hat{J}_j] = i\hbar\,\epsilon_{ijk}\,\hat{J}_k. \tag{20.1.1}$$

As we have learned, this is a very powerful statement. When coupled with the requirement that no negative norm-squared states exist, it implies that the vector space V on which these operators act can be decomposed into sums of multiplets, finite-dimensional subspaces that carry irreducible representations of angular momentum.

Let us now assume we have two angular momenta:

$$\begin{aligned}&\text{Hermitian operators } \hat{J}_i^{(1)} \text{ on } V_1 \text{ satisfying } [\hat{J}_i^{(1)}, \hat{J}_j^{(1)}] = i\hbar\,\epsilon_{ijk}\,\hat{J}_k^{(1)},\\ &\text{Hermitian operators } \hat{J}_i^{(2)} \text{ on } V_2 \text{ satisfying } [\hat{J}_i^{(2)}, \hat{J}_j^{(2)}] = i\hbar\,\epsilon_{ijk}\,\hat{J}_k^{(2)}.\end{aligned} \tag{20.1.2}$$

Our claim is that the "sum" of angular momenta is an angular momentum in the tensor product space:

$$\boxed{\hat{J}_i \equiv \hat{J}_i^{(1)} \otimes \mathbb{1} + \mathbb{1} \otimes \hat{J}_i^{(2)} \text{ satisfy } [\hat{J}_i, \hat{J}_j] = i\hbar\,\epsilon_{ijk}\,\hat{J}_k \text{ acting on } V_1 \otimes V_2.} \tag{20.1.3}$$

Certainly, the sum operator, as defined above, is an operator on $V_1 \otimes V_2$. It is in fact a Hermitian operator on $V_1 \otimes V_2$. We just need to check that the commutator holds:

$$\begin{aligned}[\hat{J}_i, \hat{J}_j] &= [\hat{J}_i^{(1)} \otimes \mathbb{1} + \mathbb{1} \otimes \hat{J}_i^{(2)}, \hat{J}_j^{(1)} \otimes \mathbb{1} + \mathbb{1} \otimes \hat{J}_j^{(2)}]\\ &= [\hat{J}_i^{(1)} \otimes \mathbb{1}, \hat{J}_j^{(1)} \otimes \mathbb{1}] + [\mathbb{1} \otimes \hat{J}_i^{(2)}, \mathbb{1} \otimes \hat{J}_j^{(2)}],\end{aligned} \tag{20.1.4}$$

since the mixed terms, which represent commutators of the operators in the different spaces, vanish:

$$[\hat{J}_i^{(1)} \otimes \mathbb{1}, \mathbb{1} \otimes \hat{J}_j^{(2)}] = 0, \quad [\mathbb{1} \otimes \hat{J}_i^{(2)}, \hat{J}_j^{(1)} \otimes \mathbb{1}] = 0. \tag{20.1.5}$$

Writing out the commutators, we see that (20.1.4) becomes

$$[\hat{J}_i, \hat{J}_j] = [\hat{J}_i^{(1)}, \hat{J}_j^{(1)}] \otimes \mathbb{1} + \mathbb{1} \otimes [\hat{J}_i^{(2)}, \hat{J}_j^{(2)}]. \tag{20.1.6}$$

We can now use the independent algebras of angular momentum to find

$$\begin{aligned}[\hat{J}_i, \hat{J}_j] &= i\hbar\,\epsilon_{ijk}\hat{J}_k^{(1)} \otimes \mathbb{1} + i\hbar\,\epsilon_{ijk}\,\mathbb{1} \otimes \hat{J}_k^{(2)} \\ &= i\hbar\,\epsilon_{ijk}\left(\hat{J}_k^{(1)} \otimes \mathbb{1} + \mathbb{1} \otimes \hat{J}_k^{(2)}\right) \\ &= i\hbar\,\epsilon_{ijk}\hat{J}_k,\end{aligned} \tag{20.1.7}$$

which is what we set out to prove.

It is important to note that had we added the two angular momenta with some arbitrary coefficients the sum would not have been an angular momentum. Indeed, suppose we use two *nonzero* constants α and β and write

$$\tilde{J}_i \equiv \alpha\hat{J}_j^{(1)} \otimes \mathbb{1} + \beta\,\mathbb{1} \otimes \hat{J}_j^{(2)}. \tag{20.1.8}$$

If the constants are not real, $\tilde{J}_i$ is not Hermitian. The commutator calculation above this time yields

$$[\tilde{J}_i, \tilde{J}_j] = i\hbar\,\epsilon_{ijk}\left(\alpha^2\hat{J}_k^{(1)} \otimes \mathbb{1} + \beta^2\,\mathbb{1} \otimes \hat{J}_k^{(2)}\right). \tag{20.1.9}$$

We have an algebra of angular momentum if the operator in parentheses is $\tilde{J}_k$. This requires $\alpha^2 = \alpha$ and $\beta^2 = \beta$. Since neither α nor β is zero, the only solution is $\alpha = \beta = 1$. This confirms that (20.1.3) is the *unique* way to add two angular momenta to form a new angular momentum.

Since any vector space where angular momentum operators act can be decomposed into the direct sum of irreducible representations of angular momentum (multiplets), the space $V_1 \otimes V_2$ can be decomposed into sums of irreducible representations of the algebra of *total* angular momentum. This property gives us a powerful tool to understand the spectrum of the Hamiltonian in the physical state space $V_1 \otimes V_2$.

20.2 Adding Two Spin One-Half Angular Momenta

We will now consider the simplest and perhaps the most important case of addition of angular momentum. We assume we have two spin one-half particles, and we focus just on spin degrees of freedom. In such a case, as you know, the two-particle system has a four-dimensional state space, described by basis vectors in which particle one and particle two can be either up or down along z. We want to answer this question: What are the possible values of the total angular momentum, and what are the basis states that realize those values? The question is simply one of finding a *new* basis for the *same* state space.

To set up notation, recall that for a spin one-half particle the state space is $\mathbb{C}^2$, spanned by states $|\frac{1}{2}, \frac{1}{2}\rangle$ and $|\frac{1}{2}, -\frac{1}{2}\rangle$, satisfying

$$\begin{aligned}\hat{\mathbf{S}}^2|\tfrac{1}{2}, m_s\rangle &= \tfrac{3}{4}\hbar^2|\tfrac{1}{2}, m_s\rangle, \\ \hat{S}_z|\tfrac{1}{2}, m_s\rangle &= \hbar m_s\,|\tfrac{1}{2}, m_s\rangle.\end{aligned} \tag{20.2.1}$$

Here, $m_s = \pm\frac{1}{2}$. Now consider a system that features two spin one-half particles. For the first particle, we have the triplet of spin operators $\hat{\mathbf{S}}^{(1)}$ acting on a vector space V_1 spanned by

$$|\tfrac{1}{2}, \tfrac{1}{2}\rangle_1, \quad |\tfrac{1}{2}, -\tfrac{1}{2}\rangle_1. \tag{20.2.2}$$

For the second particle, we have the triplet spin operators $\hat{\mathbf{S}}^{(2)}$ acting on the vector space V_2 spanned by

$$|\tfrac{1}{2},\tfrac{1}{2}\rangle_2, \quad |\tfrac{1}{2},-\tfrac{1}{2}\rangle_2. \tag{20.2.3}$$

We now form the total spin operators $\hat{S}_i$:

$$\hat{S}_i \equiv \hat{S}_i^{(1)} \otimes \mathbb{1} + \mathbb{1} \otimes \hat{S}_i^{(2)}, \tag{20.2.4}$$

which, for brevity, we write as

$$\hat{S}_i = \hat{S}_i^{(1)} + \hat{S}_i^{(2)}, \tag{20.2.5}$$

with the understanding that each operator on the right-hand side acts on the appropriate factor in the tensor product. The state space for the dynamics of the two particles must contain the tensor product $V_1 \otimes V_2$; more spaces might be needed if the particles have orbital angular momentum or are moving. As we learned before, $V_1 \otimes V_2$ is a four-dimensional complex vector space spanned by the products of states in (20.2.2) and (20.2.3):

$$|\tfrac{1}{2},\tfrac{1}{2}\rangle_1 \otimes |\tfrac{1}{2},\tfrac{1}{2}\rangle_2, \quad |\tfrac{1}{2},\tfrac{1}{2}\rangle_1 \otimes |\tfrac{1}{2},-\tfrac{1}{2}\rangle_2, \quad |\tfrac{1}{2},-\tfrac{1}{2}\rangle_1 \otimes |\tfrac{1}{2},\tfrac{1}{2}\rangle_2,$$
$$|\tfrac{1}{2},-\tfrac{1}{2}\rangle_1 \otimes |\tfrac{1}{2},-\tfrac{1}{2}\rangle_2. \tag{20.2.6}$$

Since the total spin operators have a well-defined action on these states and the vector space they span, it must be possible to describe the vector space as a sum of subspaces that carry irreducible representations of the total spin angular momentum. The irreducible representations of an angular momentum $\hat{\mathbf{J}}$ are those characterized by the number j that appears in the eigenvalue of $\hat{\mathbf{J}}^2$. A representation is said to be reducible if it is the direct sum of irreducible representations.

We have four basis states, so the possibilities for multiplets of total spin s are as follows:

1. Four singlets ($s=0$).
2. Two doublets ($s=\frac{1}{2}$).
3. One doublet ($s=\frac{1}{2}$) and two singlets ($s=0$).
4. One triplet ($s=1$) and one singlet ($s=0$).
5. One $s=\frac{3}{2}$ multiplet.

Only the last option is an irreducible representation. All others are reducible. It may be instructive at this point if you pause to make sure no other option exists and then to consider which option you think is the correct one!

The main clue is that the states in the tensor product are eigenstates of $\hat{S}_z$, the total z-component of angular momentum. We see by inspection of (20.2.6) that the possible values of $\hat{S}_z/\hbar$ are $+1, 0$, and -1. Since we have a state with $m=1$ and no state with higher m, we must have a triplet $s=1$. Thus, the only option is (4): a triplet and a singlet. This is written as

$$(s=\tfrac{1}{2}) \otimes (s=\tfrac{1}{2}) = (s=1) \oplus (s=0). \tag{20.2.7}$$

Note that on the left-hand side we have the tensor product of the two state spaces, but on the right-hand side we have the *direct sum* of the representations of total spin angular momentum. This is a fundamental result and is written more briefly as

$$\boxed{\tfrac{1}{2}\otimes\tfrac{1}{2} = \mathbf{1}\oplus\mathbf{0}.} \tag{20.2.8}$$

We use bold type for the numbers representing j values to make clear that these represent vector spaces. Note that the $\mathbf{0}$ on the right-hand side is neither a zero vector nor a vanishing vector space. It represents the singlet $s=0$, a vector space with one basis vector, a one-dimensional vector space. Similarly, $\mathbf{1}$ represents the triplet $s=1$. The equality in (20.2.8) is in fact an equality of vector spaces, and therefore the dimensionality of the vector spaces must agree. This follows from the relation $2\times 2=3+1$, where 2×2 is the dimension of the tensor product of two two-dimensional spaces, and $3+1$ is the dimension of the direct sum of a three-dimensional space and a one-dimensional space.

Let us understand the decomposition $\mathbf{1}\oplus\mathbf{0}$ explicitly by organizing the basis states according to the eigenvalue m of $\hat{S}_z/\hbar$, the total z-component of angular momentum:

$$\begin{aligned} m &= \;\;\,1: \quad |\tfrac{1}{2},\tfrac{1}{2}\rangle_1\otimes|\tfrac{1}{2},\tfrac{1}{2}\rangle_2, \\ m &= \;\;\,0: \quad |\tfrac{1}{2},\tfrac{1}{2}\rangle_1\otimes|\tfrac{1}{2},-\tfrac{1}{2}\rangle_2, \quad |\tfrac{1}{2},-\tfrac{1}{2}\rangle_1\otimes|\tfrac{1}{2},\tfrac{1}{2}\rangle_2, \\ m &= -1: \quad |\tfrac{1}{2},-\tfrac{1}{2}\rangle_1\otimes|\tfrac{1}{2},-\tfrac{1}{2}\rangle_2. \end{aligned} \tag{20.2.9}$$

We get two states with $m=0$. This is as it should be. One linear combination of these two states must be the $m=0$ state of the triplet, and another linear combination must be the singlet $s=m=0$. Those two states are in fact entangled states. Denoting by $|s,m\rangle$ the eigenstates of $\hat{\mathbf{S}}^2$ and $\hat{S}_z$ (total spin), we must have a triplet with states

$$\begin{aligned} |1,1\rangle &= |\tfrac{1}{2},\tfrac{1}{2}\rangle_1\otimes|\tfrac{1}{2},\tfrac{1}{2}\rangle_2, \\ |1,0\rangle &= \alpha\,|\tfrac{1}{2},\tfrac{1}{2}\rangle_1\otimes|\tfrac{1}{2},-\tfrac{1}{2}\rangle_2 + \beta\,|\tfrac{1}{2},-\tfrac{1}{2}\rangle_1\otimes|\tfrac{1}{2},\tfrac{1}{2}\rangle_2, \\ |1,-1\rangle &= |\tfrac{1}{2},-\tfrac{1}{2}\rangle_1\otimes|\tfrac{1}{2},-\tfrac{1}{2}\rangle_2, \end{aligned} \tag{20.2.10}$$

for some constants α and β, as well as a singlet

$$|0,0\rangle = \gamma\,|\tfrac{1}{2},\tfrac{1}{2}\rangle_1\otimes|\tfrac{1}{2},-\tfrac{1}{2}\rangle_2 + \delta\,|\tfrac{1}{2},-\tfrac{1}{2}\rangle_1\otimes|\tfrac{1}{2},\tfrac{1}{2}\rangle_2, \tag{20.2.11}$$

for some constants γ and δ. We must determine these four constants.

Let us begin by calculating the $|1,0\rangle$ state in the triplet. To do this, we first recall the general formula

$$\hat{J}_\pm|j,m\rangle = \hbar\sqrt{j(j+1)-m(m\pm 1)}\,|j,m\pm 1\rangle, \tag{20.2.12}$$

which quickly gives us the following preparatory results:

$$\begin{aligned} \hat{J}_-|1,1\rangle &= \hbar\sqrt{2}\,|1,0\rangle, \\ \hat{J}_-\,|\tfrac{1}{2},\tfrac{1}{2}\rangle &= \hbar\sqrt{\tfrac{1}{2}\cdot\tfrac{3}{2}-\tfrac{1}{2}\cdot(-\tfrac{1}{2})}\;|\tfrac{1}{2},-\tfrac{1}{2}\rangle = \hbar\,|\tfrac{1}{2},-\tfrac{1}{2}\rangle, \\ \hat{J}_+|\tfrac{1}{2},-\tfrac{1}{2}\rangle &= \hbar\sqrt{\tfrac{1}{2}\cdot\tfrac{3}{2}-(-\tfrac{1}{2})\cdot(\tfrac{1}{2})}\;|\tfrac{1}{2},-\tfrac{1}{2}\rangle = \hbar\,|\tfrac{1}{2},\tfrac{1}{2}\rangle. \end{aligned} \tag{20.2.13}$$

For our application $\hat{J}$ is total spin $\hat{S}$ so that $\hat{J}_-$ is $\hat{S}_-$. We now apply the lowering operator $\hat{S}_- = \hat{S}_-^{(1)} + \hat{S}_-^{(2)}$ to the top state of the triplet in (20.2.10). We have

$$\hat{S}_-|1,1\rangle = (\hat{S}_-^{(1)}|\tfrac{1}{2},\tfrac{1}{2}\rangle_1)\otimes|\tfrac{1}{2},\tfrac{1}{2}\rangle_2 + |\tfrac{1}{2},\tfrac{1}{2}\rangle_1\otimes(\hat{S}_-^{(2)}|\tfrac{1}{2},\tfrac{1}{2}\rangle_2). \tag{20.2.14}$$

Using the first line of (20.2.13) for the left-hand side and the second line for the right-hand side, we find that

$$\sqrt{2}\hbar\,|1,0\rangle = \hbar|\tfrac{1}{2},-\tfrac{1}{2}\rangle_1\otimes|\tfrac{1}{2},\tfrac{1}{2}\rangle_2 + |\tfrac{1}{2},\tfrac{1}{2}\rangle_1\otimes\hbar|\tfrac{1}{2},-\tfrac{1}{2}\rangle_2. \tag{20.2.15}$$

Canceling the common factors of $\hbar$ and switching the order of the terms, we find that the $|1,0\rangle$ state takes the form

$$|1,0\rangle = \frac{1}{\sqrt{2}}\Big(|\tfrac{1}{2},\tfrac{1}{2}\rangle_1\otimes|\tfrac{1}{2},-\tfrac{1}{2}\rangle_2 + |\tfrac{1}{2},-\tfrac{1}{2}\rangle_1\otimes|\tfrac{1}{2},\tfrac{1}{2}\rangle_2\Big). \tag{20.2.16}$$

Note that the right-hand side, as it should be, is a unit-normalized state. Having found the $m=0$ state of the $s=1$ multiplet, there are a number of ways to find the $m=0$ state of the $s=0$ singlet. One way is orthogonality: the single $|0,0\rangle$ must be orthogonal to the $|1,0\rangle$ state above because these are two states with different eigenvalue s of the Hermitian operator $\hat{\mathbf{S}}^2$. Since the overall sign or phase is irrelevant, we can simply take for the singlet the linear combination of $m=0$ states with a minus sign:

$$|0,0\rangle = \frac{1}{\sqrt{2}}\Big(|\tfrac{1}{2},\tfrac{1}{2}\rangle_1\otimes|\tfrac{1}{2},-\tfrac{1}{2}\rangle_2 - |\tfrac{1}{2},-\tfrac{1}{2}\rangle_1\otimes|\tfrac{1}{2},\tfrac{1}{2}\rangle_2\Big). \tag{20.2.17}$$

You probably remember that we found this state in example 18.3 by searching for a state that is annihilated by the sum of spin angular momentum operators. This is exactly the condition for a singlet.

As an instructive calculation, let us confirm that $\hat{\mathbf{S}}^2$ is zero acting on $|0,0\rangle$. For this it is useful to note that

$$\begin{aligned}\hat{\mathbf{S}}^2 = (\hat{\mathbf{S}}^{(1)} + \hat{\mathbf{S}}^{(2)})^2 &= (\hat{\mathbf{S}}^{(1)})^2 + (\hat{\mathbf{S}}^{(2)})^2 + 2\hat{\mathbf{S}}^{(1)}\cdot\hat{\mathbf{S}}^{(2)} \\ &= (\hat{\mathbf{S}}^{(1)})^2 + (\hat{\mathbf{S}}^{(2)})^2 + \hat{S}_+^{(1)}\hat{S}_-^{(2)} + \hat{S}_-^{(1)}\hat{S}_+^{(2)} + 2\hat{S}_z^{(1)}\hat{S}_z^{(2)},\end{aligned} \tag{20.2.18}$$

where in the second step we used the general result

$$\hat{\mathbf{J}}^{(1)}\cdot\hat{\mathbf{J}}^{(2)} = \tfrac{1}{2}(\hat{J}_+^{(1)}\hat{J}_-^{(2)} + \hat{J}_-^{(1)}\hat{J}_+^{(2)}) + \hat{J}_z^{(1)}\hat{J}_z^{(2)}, \tag{20.2.19}$$

valid for arbitrary angular momenta. Written in explicit tensor notation, it reads

$$\hat{\mathbf{J}}^{(1)}\cdot\hat{\mathbf{J}}^{(2)} \equiv \sum_{i=1}^{3}\hat{J}_i^{(1)}\otimes\hat{J}_i^{(2)} = \tfrac{1}{2}(\hat{J}_+^{(1)}\otimes\hat{J}_-^{(2)} + \hat{J}_-^{(1)}\otimes\hat{J}_+^{(2)}) + \hat{J}_z^{(1)}\otimes\hat{J}_z^{(2)}. \tag{20.2.20}$$

Exercise 20.1. *Prove (20.2.20).*

Back to our calculation, all states have $s_1 = s_2 = \frac{1}{2}$, and therefore $(\hat{\mathbf{S}}^{(1)})^2 = (\hat{\mathbf{S}}^{(2)})^2 = \frac{3}{4}\hbar^2$. We thus have

$$\hat{\mathbf{S}}^2\,|0,0\rangle = \tfrac{3}{2}\hbar^2\,|0,0\rangle + (\hat{S}_+^{(1)}\hat{S}_-^{(2)} + \hat{S}_-^{(1)}\hat{S}_+^{(2)} + 2\hat{S}_z^{(1)}\hat{S}_z^{(2)})|0,0\rangle. \tag{20.2.21}$$

It is simple to see that

$$2\hat{S}_z^{(1)}\hat{S}_z^{(2)}\,|0,0\rangle = 2\,\tfrac{\hbar}{2}\cdot(-\tfrac{\hbar}{2})|0,0\rangle = -\tfrac{1}{2}\hbar^2\,|0,0\rangle \tag{20.2.22}$$

because the singlet is a superposition of tensor states where each has one state up and one state down. Similarly, recalling that

$$\hat{S}_{\pm}|\tfrac{1}{2},\mp\tfrac{1}{2}\rangle = \hbar\,|\tfrac{1}{2},\pm\tfrac{1}{2}\rangle, \tag{20.2.23}$$

we quickly find that

$$(\hat{S}_{+}^{(1)}\hat{S}_{-}^{(2)} + \hat{S}_{-}^{(1)}\hat{S}_{+}^{(2)})|0,0\rangle = -\hbar^2\,|0,0\rangle, \tag{20.2.24}$$

since each of the operators $\hat{S}_{+}^{(1)}\hat{S}_{-}^{(2)}$ and $\hat{S}_{-}^{(1)}\hat{S}_{+}^{(2)}$ kills one term in the singlet, and acting on the other term, it gives $\hbar^2$ times the killed one. Check it! Going back to (20.2.21), we get

$$\hat{\mathbf{S}}^2\,|0,0\rangle = \tfrac{3}{2}\hbar^2\,|0,0\rangle + (-\hbar^2 - \tfrac{1}{2}\hbar^2)|0,0\rangle = 0, \tag{20.2.25}$$

as we wanted to show.

Let us summarize our results. The triplet states and singlet states are given by

$$\begin{aligned}
|1,1\rangle &= |\tfrac{1}{2},\tfrac{1}{2}\rangle_1 \otimes |\tfrac{1}{2},\tfrac{1}{2}\rangle_2,\\
|1,0\rangle &= \tfrac{1}{\sqrt{2}}\Big(|\tfrac{1}{2},\tfrac{1}{2}\rangle_1 \otimes |\tfrac{1}{2},-\tfrac{1}{2}\rangle_2 + |\tfrac{1}{2},-\tfrac{1}{2}\rangle_1 \otimes |\tfrac{1}{2},\tfrac{1}{2}\rangle_2\Big),\\
|1,-1\rangle &= |\tfrac{1}{2},-\tfrac{1}{2}\rangle_1 \otimes |\tfrac{1}{2},-\tfrac{1}{2}\rangle_2.\\
|0,0\rangle &= \tfrac{1}{\sqrt{2}}\Big(|\tfrac{1}{2},\tfrac{1}{2}\rangle_1 \otimes |\tfrac{1}{2},-\tfrac{1}{2}\rangle_2 - |\tfrac{1}{2},-\tfrac{1}{2}\rangle_1 \otimes |\tfrac{1}{2},\tfrac{1}{2}\rangle_2\Big).
\end{aligned} \tag{20.2.26}$$

For briefer notation we replace $|\tfrac{1}{2},\tfrac{1}{2}\rangle \to |\uparrow\rangle$ and $|\tfrac{1}{2},-\tfrac{1}{2}\rangle \to |\downarrow\rangle$:

$$\begin{aligned}
|1,1\rangle &= |\uparrow\rangle_1 \otimes |\uparrow\rangle_2,\\
|1,0\rangle &= \tfrac{1}{\sqrt{2}}\Big(|\uparrow\rangle_1 \otimes |\downarrow\rangle_2 + |\downarrow\rangle_1 \otimes |\uparrow\rangle_2\Big),\\
|1,-1\rangle &= |\downarrow\rangle_1 \otimes |\downarrow\rangle_2.\\
|0,0\rangle &= \tfrac{1}{\sqrt{2}}\Big(|\uparrow\rangle_1 \otimes |\downarrow\rangle_2 - |\downarrow\rangle_1 \otimes |\uparrow\rangle_2\Big).
\end{aligned} \tag{20.2.27}$$

With the understanding that the first arrow refers to the first particle and the second arrow to the second particle, we can finally write all of this quite briefly. With a dashed line separating the triplet and the singlet, we have

$$\boxed{\begin{aligned}
|1,1\rangle &= |\uparrow\uparrow\rangle,\\
|1,0\rangle &= \tfrac{1}{\sqrt{2}}(|\uparrow\downarrow\rangle + |\downarrow\uparrow\rangle),\\
|1,-1\rangle &= |\downarrow\downarrow\rangle.\\
&\text{- - - - - - - - - - - - - - -}\\
|0,0\rangle &= \tfrac{1}{\sqrt{2}}(|\uparrow\downarrow\rangle - |\downarrow\uparrow\rangle).
\end{aligned}} \tag{20.2.28}$$

This decomposition of the tensor product of two spin one-half state spaces is needed to calculate the hyperfine splitting in the hydrogen atom (section 20.4). For this system, the relevant spins are those of the proton and the electron.

20.3 A Primer in Perturbation Theory

The purpose of this section is to familiarize you with two basic results in perturbation theory. The first of these results was actually anticipated in our discussion of the Hellmann-Feynman lemma; the second was not. These results are derived in detail in chapter 25, where they are also extended in a number of directions. Here we just state them so they can be used to discuss the application of addition of angular momentum to the hydrogen atom.

In perturbation theory we have a Hamiltonian $\hat{H}^{(0)}$ that we assume is well understood—that is, we know its eigenstates $|k^{(0)}\rangle$ and eigenvalues $E_k^{(0)}$:

$$\hat{H}^{(0)}|k^{(0)}\rangle = E_k^{(0)}|k^{(0)}\rangle. \tag{20.3.1}$$

The "zero" superscripts indicate that these quantities are all *unperturbed*. We now consider adding a perturbation δH to the original Hamiltonian, giving us a total Hamiltonian $\hat{H}$:

$$\hat{H} = \hat{H}^{(0)} + \delta H. \tag{20.3.2}$$

The perturbation δH is a Hermitian operator. It usually makes the total Hamiltonian very complicated to analyze exactly. In perturbation theory the idea is to find approximate energies and approximate eigenstates of $\hat{H}$ when the perturbation δH is small. It turns out that one must distinguish two cases: it makes a difference if we are looking at a nondegenerate state or at a set of degenerate states. While the treatment of nondegenerate states is easier, for the hydrogen atom the unperturbed energy levels are highly degenerate, so we face a more intricate situation.

Case 1: Nondegenerate energy level Assume the state $|k^{(0)}\rangle$ with energy $E_k^{(0)}$ is not degenerate. In this case the perturbation changes the state and the energy. Calling the energy of the perturbed state E_k, we write

$$E_k = E_k^{(0)} + \delta E_k + \mathcal{O}(\delta H^2). \tag{20.3.3}$$

Here, $E_k^{(0)}$ is the energy of the state absent the perturbation, and δE_k is the correction to the energy to first order in δH. The $\mathcal{O}(\delta H^2)$ term indicates that there are higher-order corrections. It is a very nice fact that finding the energy correction does not require finding how the state changes! Indeed, we find that

$$\delta E_k = \langle k^{(0)}|\delta H|k^{(0)}\rangle. \tag{20.3.4}$$

The correction to the energy is just the expectation value of the perturbation δH on the *unperturbed* state $|k^{(0)}\rangle$. Since the state is assumed known, this is a very simple result. It is sometimes said that this is the most important result in perturbation theory! We actually derived this result in example 7.3, as an application of the Hellman-Feynman lemma.

Case 2: A degenerate energy level Suppose we have a degenerate energy level of energy $E_n^{(0)}$ with N degenerate eigenstates $|n^{(0)};l\rangle$ with $l=1,\ldots,N$, chosen to be orthonormal:

$$\text{Energy } E_n^{(0)} \text{ states: } |n^{(0)};1\rangle,\ldots,|n^{(0)};N\rangle. \tag{20.3.5}$$

To proceed, one calculates the $N\times N$ matrix $[\delta H]$ representing the operator δH in the degenerate subspace. The matrix elements are

$$\delta H_{ij}\equiv\langle n^{(0)};i|\delta H|n^{(0)};j\rangle \tag{20.3.6}$$

and can be calculated explicitly. The $N\times N$ matrix $[\delta H]$ must now be diagonalized to find the N eigenvectors $|\psi_I^{(0)}\rangle$, labeled by I, with their associated eigenvalues δE_{nI}:

$$[\delta H]\,|\psi_I^{(0)}\rangle=\delta E_{nI}\,|\psi_I^{(0)}\rangle,\quad I=1,\ldots,N. \tag{20.3.7}$$

If you find an eigenvector in component form, the associated state is immediately constructed:

$$\text{eigenvector: } \begin{pmatrix} a_{I1}^{(0)} \\ \vdots \\ a_{IN}^{(0)} \end{pmatrix} \quad\Rightarrow\quad |\psi_I^{(0)}\rangle=\sum_k |n^{(0)};k\rangle a_{Ik}^{(0)}. \tag{20.3.8}$$

Before the perturbation is included, the N degenerate states are on the same footing. After the perturbation is included, the zeroth-order approximation to the new energy eigenstates are the $|\psi_I^{(0)}\rangle$. They are zeroth order because, while selected by the perturbation, the coefficients a_I are not proportional to δH. The energy correction for $|\psi_I^{(0)}\rangle$ is in fact the eigenvalue δE_{nI}, and therefore the total energy E_{nI} of the state $|\psi_I^{(0)}\rangle$ is

$$E_{nI}=E_n^{(0)}+\delta E_{nI}+\mathcal{O}(\delta H^2). \tag{20.3.9}$$

In specific situations, it is sometimes possible to choose a basis in the degenerate subspace for which $[\delta H]$ is diagonal. Such a basis is called a *good basis*. The basis vectors are then the zeroth-order energy eigenstates, and the energy corrections are precisely the elements in the diagonal of $[\delta H]$. In the problems we will consider, with two angular momenta, the eigenstates of the *total* angular momenta will provide a good basis.

20.4 Hyperfine Splitting

As our first physical application, we will consider hyperfine splitting. For this example the construction of eigenstates of total angular momentum is simple and was considered earlier. We will discuss here some of the general features of addition of angular momentum, leaving the study of complete sets of observables to our second example, the case of spin-orbit coupling. Since both of these phenomena are considered in the context of the hydrogen atom, let us recall the basics of this system.

We learned in section 11.3 that the hydrogen atom bound state spectrum consists of states with energies E_n, where

$$E_n=-\frac{e^2}{2a_0}\frac{1}{n^2},\quad n=1,2,\ldots,\quad a_0=\frac{\hbar^2}{me^2}. \tag{20.4.1}$$

For each value $n \geq 1$, we have the following ℓ multiplets of orbital angular momentum:

$$\ell = 0, 1, \ldots, n-1. \tag{20.4.2}$$

A simple computation shows that this gives a total of n^2 states (ignoring the degeneracy due to the electron spin). So at each n, we have a degenerate n^2-dimensional space $\mathcal{H}_n$ of energy eigenstates that can be written as the direct sum of angular momentum multiplets:

$$\mathcal{H}_n = (\ell = n-1) \oplus (\ell = n-2) \oplus \cdots \oplus (\ell = 0). \tag{20.4.3}$$

The hydrogen Hamiltonian takes the form

$$\hat{H}^{(0)} = \frac{\hat{\mathbf{p}}^2}{2m} - \frac{e^2}{r}, \tag{20.4.4}$$

where, in the spirit of perturbation theory, we consider this to be the unperturbed Hamiltonian, whose spectrum is well known.

Turning to hyperfine splitting, the simple hydrogen atom Hamiltonian $\hat{H}^{(0)}$ receives a small correction because both the proton and the electron have magnetic dipole moments. The proton magnetic dipole creates a magnetic field. The contribution to the Hamiltonian is the energy of the electron dipole in the magnetic field created by the proton. Recall from section 17.5 that the proton and electron dipole moments are

$$\hat{\boldsymbol{\mu}}_p = \frac{g_p e}{2m_p c}\hat{\mathbf{S}}_p, \quad g_p \simeq 5.59, \qquad \hat{\boldsymbol{\mu}}_e = -\frac{e}{m_e c}\hat{\mathbf{S}}_e. \tag{20.4.5}$$

The extra term ΔH in the Hamiltonian is

$$\delta H = -\hat{\boldsymbol{\mu}}_e \cdot \hat{\mathbf{B}}_p = \frac{e}{m_e c}\hat{\mathbf{S}}_e \cdot \hat{\mathbf{B}}_p, \tag{20.4.6}$$

where $\hat{\mathbf{B}}_p$ is the magnetic field at the electron due to the proton dipole. This magnetic field is an operator (thus the hat) because it is proportional to the proton magnetic moment operator. The magnetic field of a magnetic dipole contains two terms:

$$\hat{\mathbf{B}}_p = \frac{1}{r^3}[3\mathbf{n}(\mathbf{n} \cdot \hat{\boldsymbol{\mu}}_p) - \hat{\boldsymbol{\mu}}_p] + \frac{8\pi}{3}\hat{\boldsymbol{\mu}}_p \delta(\mathbf{r}), \tag{20.4.7}$$

with $\mathbf{n}$ the unit vector pointing from the proton to the electron. The first term is the familiar piece that falls off like $1/r^3$. The second term is a delta function contribution at the position of the dipole, required for the magnetic field of the dipole to have the correct integral over space.

We choose to work out the hyperfine correction for the $n=1, \ell=0$ ground states $|\psi_{100}\rangle$ of the hydrogen atom, a set of four states because we must include the electron and proton spin degrees of freedom. We will now show that for such states the familiar $1/r^3$ term in $\hat{\mathbf{B}}_p$ does *not* contribute an energy correction. For this term, using (20.4.6), the Hamiltonian contribution $\delta H'$ is given by

$$\delta H' \sim \frac{1}{r^3}\hat{\mathbf{S}}_e \cdot [3\mathbf{n}(\mathbf{n} \cdot \hat{\mathbf{S}}_p) - \hat{\mathbf{S}}_p] = \frac{1}{r^3}[3(\mathbf{n} \cdot \hat{\mathbf{S}}_e)(\mathbf{n} \cdot \hat{\mathbf{S}}_p) - \hat{\mathbf{S}}_e \cdot \hat{\mathbf{S}}_p], \tag{20.4.8}$$

recalling that $\hat{\boldsymbol{\mu}}_p \sim \hat{\mathbf{S}}_p$. To assess the contribution of $\delta H'$, we must evaluate its expectation value on the proton-electron ground states. Happily, we do not need to calculate much to see that it vanishes. The expectation value calculation involves both the spatial part and the spin part of the proton-electron states. Focus on the spatial part of the computation, which requires integration over all of space of $\delta H'$, multiplied by the norm squared of the spatial wave function $\langle \mathbf{r}|\psi_{100}\rangle$. Being just a function of r, the wave function does not affect the angular part of the integration. The angular integral is in fact the integral over the direction of the radial unit vector $\mathbf{n}$, and therefore it is proportional to

$$\int d\Omega[3(\mathbf{n}\cdot\hat{\mathbf{S}}_e)(\mathbf{n}\cdot\hat{\mathbf{S}}_p)-\hat{\mathbf{S}}_e\cdot\hat{\mathbf{S}}_p]=3\hat{S}_{e,i}\hat{S}_{p,j}\int d\Omega\, n_in_j-4\pi\,\hat{\mathbf{S}}_e\cdot\hat{\mathbf{S}}_p=0, \tag{20.4.9}$$

where we used $\int d\Omega = 4\pi$, and the last equality follows because in fact

$$\int d\Omega\, n_in_j=\tfrac{4\pi}{3}\,\delta_{ij}. \tag{20.4.10}$$

To show this, note that, by rotational symmetry, $\int d\Omega\, n_1^2=\int d\Omega\, n_2^2=\int d\Omega\, n_3^2$. Since $n_1^2+n_2^2+n_3^2=1$, the sum of these integrals has value 4π, and each integral is equal to $\frac{4\pi}{3}$, consistent with the above claim for $i=j$. When $i\neq j$, the integral vanishes, again, by symmetry considerations. Consider, for example, the integral $\int d\Omega\, n_xn_y$. For each point on the unit sphere with coordinates (n_x, n_y, n_z) and contributing to the integral, there is a point $(-n_x, n_y, n_z)$ that gives exactly a canceling contribution. Thus, the integral vanishes. Having shown that (20.4.10) holds, we have demonstrated that the $1/r^3$ term in the magnetic field does not contribute to the energy correction.

It follows that the relevant δH arises from the second term in the magnetic field (20.4.7), and it is given by

$$\delta H \;=\; \frac{e}{mc}\hat{\mathbf{S}}_e\cdot\frac{8\pi}{3}\,\hat{\boldsymbol{\mu}}_p\,\delta(\mathbf{r}) \;=\; \frac{4\pi}{3}\frac{g_pe^2}{m_pm_ec^2}\,\hat{\mathbf{S}}_e\cdot\hat{\mathbf{S}}_p\,\delta(\mathbf{r}). \tag{20.4.11}$$

In this situation, the Hamiltonian $\hat{H}^{(0)}$ before inclusion of the perturbation is the hydrogen atom Hamiltonian that depends on the spin degrees of freedom of neither the electron nor the proton. Since each particle can be in either of two states, up or down, we have four degenerate energy eigenstates that share a common spatial wave function, the ground state $|\psi_{100}\rangle$, but differ on the spin degrees of freedom:

$$|\psi_{100}\rangle\otimes|\uparrow\uparrow\rangle,\quad |\psi_{100}\rangle\otimes|\uparrow\downarrow\rangle,\quad |\psi_{100}\rangle\otimes|\downarrow\uparrow\rangle,\quad |\psi_{100}\rangle\otimes|\downarrow\downarrow\rangle. \tag{20.4.12}$$

Before we include the perturbation δH, these states are all degenerate with energy equal to the ground state energy of the hydrogen atom. As discussed in case 2 of section 20.3, we can get the perturbed energies if we find new basis states, linear combinations of the above states for which the matrix $[\delta H]$ is diagonal. Let us denote these "good" basis states as

$$|\psi_{100}\rangle|s_i\rangle,\quad i=1,2,3,4, \tag{20.4.13}$$

where the $|s_i\rangle$ are some yet *undetermined* combinations of the spin states for the proton-electron pair. The matrix elements δH_{ij} are given by

$$\begin{aligned}\delta H_{ij} &= \langle\psi_{100}|\langle s_i|\,\delta H\,|\psi_{100}\rangle|s_j\rangle,\\ &= \frac{4\pi}{3}\frac{g_p e^2}{m_p m_e c^2}\int d^3x|\psi_{100}(r)|^2\delta(\mathbf{r})\langle s_i|\hat{\mathbf{S}}_e\cdot\hat{\mathbf{S}}_p|s_j\rangle,\\ &= \frac{4\pi}{3}\frac{g_p e^2}{m_p m_e c^2}|\psi_{100}(0)|^2\langle s_i|\hat{\mathbf{S}}_e\cdot\hat{\mathbf{S}}_p|s_j\rangle,\\ &= \frac{4g_p e^2}{3m_p m_e c^2 a_0^3}\langle s_i|\hat{\mathbf{S}}_e\cdot\hat{\mathbf{S}}_p|s_j\rangle,\end{aligned} \tag{20.4.14}$$

where we recalled that $|\psi_{100}(0)|^2=1/(\pi a_0^3)$. Therefore, we have

$$\delta H_{ij} = \Delta E\,\langle s_i|\,\tfrac{1}{\hbar^2}\hat{\mathbf{S}}_e\cdot\hat{\mathbf{S}}_p\,|s_j\rangle, \tag{20.4.15}$$

where ΔE is the energy scale relevant to hyperfine splitting:

$$\Delta E \equiv \frac{4e^2 g_p}{3m_p m_e c^2}\frac{\hbar^2}{a_0^3}. \tag{20.4.16}$$

If the states $|s_i\rangle$ are chosen properly, the matrix $\langle s_i|\hat{\mathbf{S}}_e\cdot\hat{\mathbf{S}}_p|s_j\rangle$ is diagonal and so is $[\delta H]$.

To make progress, let us relate the operator product $\hat{\mathbf{S}}_e\cdot\hat{\mathbf{S}}_p$ to the total spin angular momentum. This product is a tensor product of operators:

$$\hat{\mathbf{S}}_e\cdot\hat{\mathbf{S}}_p \equiv \hat{S}_{e,1}\otimes\hat{S}_{p,1}+\hat{S}_{e,2}\otimes\hat{S}_{p,2}+\hat{S}_{e,3}\otimes\hat{S}_{p,3} = \sum_i \hat{S}_{e,i}\otimes\hat{S}_{p,i}. \tag{20.4.17}$$

We now define the total spin operator $\hat{\mathbf{S}}\equiv\hat{\mathbf{S}}_e+\hat{\mathbf{S}}_p$, or more precisely,

$$\hat{S}_i \equiv \hat{S}_{e,i}\otimes\mathbb{1}+\mathbb{1}\otimes\hat{S}_{p,i}. \tag{20.4.18}$$

As we have seen, the $\hat{S}_i$ operators satisfy the algebra of angular momentum. It is now a simple computation to expand $\hat{\mathbf{S}}^2$:

$$\begin{aligned}\hat{\mathbf{S}}^2 &= \sum_i \hat{S}_i\hat{S}_i = \sum_i\big(\hat{S}_{e,i}\otimes\mathbb{1}+\mathbb{1}\otimes\hat{S}_{p,i}\big)\big(\hat{S}_{e,i}\otimes\mathbb{1}+\mathbb{1}\otimes\hat{S}_{p,i}\big)\\ &= \sum_i\big(\hat{S}_{e,i}\hat{S}_{e,i}\otimes\mathbb{1}+2\,\hat{S}_{e,i}\otimes\hat{S}_{p,i}+\mathbb{1}\otimes\hat{S}_{p,i}\hat{S}_{p,i}\big)\\ &= \hat{\mathbf{S}}_e^2\otimes\mathbb{1}+2\sum_i\hat{S}_{e,i}\otimes\hat{S}_{p,i}+\mathbb{1}\otimes\hat{\mathbf{S}}_p^2.\end{aligned} \tag{20.4.19}$$

Given our definition of $\hat{\mathbf{S}}_e\cdot\hat{\mathbf{S}}_p$, the above result is in fact

$$\hat{\mathbf{S}}^2 = \hat{\mathbf{S}}_e^2\otimes\mathbb{1}+2\,\hat{\mathbf{S}}_e\cdot\hat{\mathbf{S}}_p+\mathbb{1}\otimes\hat{\mathbf{S}}_p^2. \tag{20.4.20}$$

With a little abuse of notation, one simply writes

$$\hat{\mathbf{S}}^2 = \hat{\mathbf{S}}_e^2+2\,\hat{\mathbf{S}}_e\cdot\hat{\mathbf{S}}_p+\hat{\mathbf{S}}_p^2. \tag{20.4.21}$$

This expression is the "obvious" result for the square of the sum of two commuting operators. We have derived it to ensure you understand how the tensor product works on each term, but next time you should write it directly. We then have the useful rewriting:

$$\boxed{\hat{\mathbf{S}}_e \cdot \hat{\mathbf{S}}_p = \tfrac{1}{2}(\hat{\mathbf{S}}^2 - \hat{\mathbf{S}}_e^2 - \hat{\mathbf{S}}_p^2).} \tag{20.4.22}$$

This shows the relevance of the total spin angular momentum: we can effectively trade $\hat{\mathbf{S}}_e \cdot \hat{\mathbf{S}}_p$ for $\hat{\mathbf{S}}^2$ because the other operators $\hat{\mathbf{S}}_e^2$ and $\hat{\mathbf{S}}_p^2$ on the right-hand side are simply multiples of the identity: $\hat{\mathbf{S}}_e^2 = \hat{\mathbf{S}}_p^2 = \frac{3}{4}\hbar^2 \mathbb{1}$. Given this, the operator whose matrix elements we are considering in (20.4.15) is

$$\frac{1}{\hbar^2}\hat{\mathbf{S}}_e \cdot \hat{\mathbf{S}}_p = \tfrac{1}{2}\big(\tfrac{1}{\hbar^2}\hat{\mathbf{S}}^2 - \tfrac{3}{2}\mathbb{1}\big). \tag{20.4.23}$$

Here $\mathbb{1}$ is the identity matrix in the four-dimensional tensor product space. It follows from this result that

$$\delta H_{ij} = \tfrac{1}{2}\,\Delta E\,\langle s_i|\big(\tfrac{1}{\hbar^2}\hat{\mathbf{S}}^2 - \tfrac{3}{2}\mathbb{1}\big)|s_j\rangle. \tag{20.4.24}$$

This matrix is diagonal when the states $|s_i\rangle$ are eigenstates of the total spin $\hat{\mathbf{S}}$! Those total spin eigenstates thus define a good basis, and the diagonal elements of this matrix are the energy shifts, as we discussed in the previous section. We have seen that the four states in the tensor product of two spin one-half particles combine into a triplet $(s=1)$ and a singlet $(s=0)$. Thus, we take $|s_i\rangle$ with $i=1,2,3$ to be the triplet and $|s_4\rangle$ to be the singlet:

$$|s_1\rangle = |\uparrow\uparrow\rangle,\quad |s_2\rangle = \tfrac{1}{\sqrt{2}}\big(|\uparrow\downarrow\rangle + |\downarrow\uparrow\rangle\big),\quad |s_3\rangle = |\downarrow\downarrow\rangle;\quad |s_4\rangle = \tfrac{1}{\sqrt{2}}\big(|\uparrow\downarrow\rangle - |\downarrow\uparrow\rangle\big). \tag{20.4.25}$$

Any state of the triplet $s=1$ is an $(\frac{1}{\hbar^2}\hat{\mathbf{S}}^2 - \frac{3}{2})$ eigenstate with eigenvalue one-half:

$$\big(\tfrac{1}{\hbar^2}\hat{\mathbf{S}}^2 - \tfrac{3}{2}\big)|s_i\rangle = \big(2 - \tfrac{3}{2}\big)|s_i\rangle = \tfrac{1}{2}|s_i\rangle,\quad i=1,2,3. \tag{20.4.26}$$

Moreover, the singlet $s=0$ is an $(\frac{1}{\hbar^2}\hat{\mathbf{S}}^2 - \frac{3}{2})$ eigenstate with eigenvalue $-\frac{3}{2}$:

$$\big(\tfrac{1}{\hbar^2}\hat{\mathbf{S}}^2 - \tfrac{3}{2}\big)|s_4\rangle = \big(0 - \tfrac{3}{2}\big)|s_4\rangle = -\tfrac{3}{2}|s_4\rangle. \tag{20.4.27}$$

Therefore, the matrix $[\delta H]$ in (20.4.24) takes the form

$$[\delta H] = \begin{pmatrix} \frac{1}{4}\Delta E & 0 & 0 & 0 \\ 0 & \frac{1}{4}\Delta E & 0 & 0 \\ 0 & 0 & \frac{1}{4}\Delta E & 0 \\ 0 & 0 & 0 & -\frac{3}{4}\Delta E \end{pmatrix}. \tag{20.4.28}$$

The three states on the triplet are pushed up $\frac{1}{4}\Delta E$, and the singlet is pushed down $\frac{3}{4}\Delta E$. The total split between the two sets of states is ΔE, which is now identified as the magnitude of the hyperfine splitting of the ground state of the hydrogen atom (figure 20.1).

Let us determine the value of ΔE, as given in (20.4.16):

$$\Delta E = \frac{4g_p e^2}{3m_p m_e c^2}\frac{\hbar^2}{a_0^3} = \frac{4g_p}{3m_p m_e c}\frac{e^2}{\hbar c}\Big(\frac{\hbar}{a_0}\Big)^3. \tag{20.4.29}$$

Recalling that $\alpha = \frac{e^2}{\hbar c}$, we see that

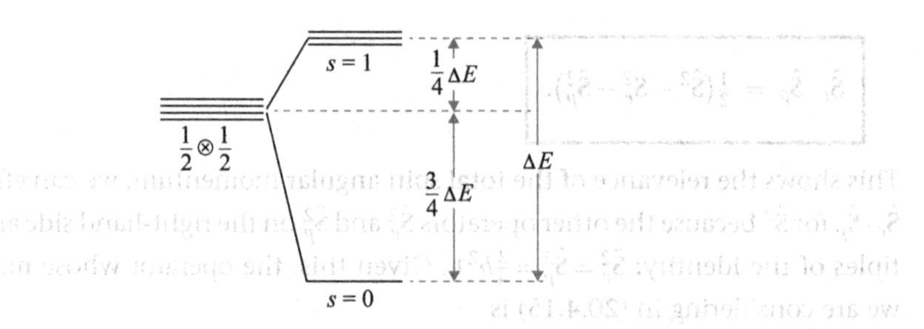

Figure 20.1
The hyperfine splitting of the electron-proton fourfold degenerate ground states $\frac{1}{2}\otimes\frac{1}{2}$ into a total-spin triplet, pushed up, and a total-spin singlet, pushed down.

$$a_0 = \frac{\hbar^2}{m_e e^2} = \frac{\hbar}{m_e c\alpha} \Rightarrow \frac{\hbar}{a_0} = m_e c\,\alpha. \tag{20.4.30}$$

This helps us rewrite ΔE in a clearer way:

$$\Delta E = \frac{4g_p}{3m_p m_e c}\,\alpha^4 m_e^3 c^3 = \frac{4}{3}\,g_p\Big(\frac{m_e}{m_p}\Big)\alpha^4(m_e c^2). \tag{20.4.31}$$

Bohr's energy is of order $\alpha^2(m_ec^2)$—that is, the rest energy of the electron times a suppression factor $\alpha^2 \simeq 1/19{,}000$. The spin-orbit correction, to be studied in section 20.6, is of order $\alpha^4(m_ec^2)$, thus suppressed relative to the Bohr energy by an additional factor of α^2. Finally, hyperfine splitting is suppressed relative to spin-orbit by a factor $\frac{m_e}{m_p}$, of order one over two thousand. We quickly plug in the numbers:

$$\Delta E = \frac{4}{3}\cdot 5.59\cdot\frac{1}{1{,}836}\cdot\Big(\frac{1}{137}\Big)^4\cdot 511{,}000\text{ eV} \simeq 5.88\times 10^{-6}\text{ eV}. \tag{20.4.32}$$

In a transition from the triplet to the singlet, the energy ΔE is carried away by a photon of wavelength λ given by

$$\lambda = \frac{c}{\nu} = \frac{c}{\frac{\Delta E}{h}} = \frac{2\pi\hbar c}{\Delta E} = \frac{2\pi\cdot 197\,\text{MeV}\cdot\text{fm}}{5.88\times 10^{-6}\,\text{eV}} \simeq 21.1\text{ cm}. \tag{20.4.33}$$

The corresponding photon frequency is $\nu = 1{,}420$ MHz. This 21 cm spectral line was first observed by Harold Irving Ewen and Edward Purcell in 1951, who detected the emission of such photons from neutral hydrogen in the Milky Way. The hyperfine splitting is so small that even microwave background photons, of typical wavelength 2 mm, can easily excite the ground state into the triplet. The lifetime of the triplet state, however, is extremely long, about ten million years. As a result, the 21 cm line is extraordinarily sharp and very useful for astronomical Doppler shift measurements. The rotation curve of our galaxy has been calculated using the 21 cm line.

20.5 Computation of $1\otimes\frac{1}{2}$

We already worked out how the tensor product of two $\frac{1}{2}$ multiplets breaks into irreducible representations of the total angular momentum. We found that the result is encapsulated by the relation $\frac{1}{2}\otimes\frac{1}{2} = \mathbf{1}\oplus\mathbf{0}$.

This example is perhaps too simple to illustrate the general ways in which we approach the problem of adding two angular momenta. In this section we will consider two multiplets: a multiplet $j_1=1$ of an angular momentum $\hat{\mathbf{J}}_1$ and a multiplet $j_2=\frac{1}{2}$ of an angular momentum $\hat{\mathbf{J}}_2$. We are going to tensor these two vector spaces to form $1\otimes\frac{1}{2}$ and find out how the tensor product decomposes in multiplets of the total angular momentum $\hat{\mathbf{J}}\equiv\hat{\mathbf{J}}_1+\hat{\mathbf{J}}_2$. This calculation is instructive and, moreover, will be useful in our analysis of spin-orbit coupling in the following section.

Since the $j_1=1$ multiplet is three-dimensional and the $j_2=\frac{1}{2}$ multiplet is two-dimensional, the tensor product has six basis states:

$$|j_1,m_1\rangle\otimes|j_2,m_2\rangle,\quad j_1=1,\quad j_2=\tfrac{1}{2},\tag{20.5.1}$$

which exist for all combinations of $m_1=-1,0,1$, and $m_2=\pm\frac{1}{2}$. We have chosen to write the j_1 multiplet states to the left of the j_2 multiplet states. The states above form what is called the *uncoupled basis* because the products are eigenstates of both $(\hat{\mathbf{J}}_1^2,\hat{J}_{1,z})$ and $(\hat{\mathbf{J}}_2^2,\hat{J}_{2,z})$, the uncoupled angular momenta. Our goal in this section is to determine the *coupled* states, the eigenstates of the total angular momentum operators $\hat{\mathbf{J}}^2$ and $\hat{J}_z$.

We note that even the uncoupled states are eigenstates of $\hat{J}_z=\hat{J}_{1,z}+\hat{J}_{2,z}$:

$$\begin{aligned}\hat{J}_z|j_1,m_1\rangle\otimes|j_2,m_2\rangle &= \hat{J}_{1,z}|j_1,m_1\rangle\otimes|j_2,m_2\rangle+|j_1,m_1\rangle\otimes\hat{J}_{2,z}|j_2,m_2\rangle\\ &= \hbar m_1\,|j_1,m_1\rangle\otimes|j_2,m_2\rangle+\hbar m_2|j_1,m_1\rangle\otimes|j_2,m_2\rangle\\ &= \hbar(m_1+m_2)|j_1,m_1\rangle\otimes|j_2,m_2\rangle.\end{aligned}\tag{20.5.2}$$

Thus, the $\hat{J}_z$ eigenvalue of an uncoupled state is obtained by adding the values of m_1 and m_2 and multiplying by $\hbar$. Our first step in the analysis is to organize the six states by the $\hat{J}_z$ eigenvalue. We quickly find the following table:

$$\begin{aligned}\tfrac{1}{\hbar}J_z &= \tfrac{3}{2}: && |1,1\rangle\otimes|\tfrac{1}{2},\tfrac{1}{2}\rangle,\\ \tfrac{1}{\hbar}J_z &= \tfrac{1}{2}: && |1,0\rangle\otimes|\tfrac{1}{2},\tfrac{1}{2}\rangle\quad |1,1\rangle\otimes|\tfrac{1}{2},-\tfrac{1}{2}\rangle,\\ \tfrac{1}{\hbar}J_z &= -\tfrac{1}{2}: && |1,0\rangle\otimes|\tfrac{1}{2},-\tfrac{1}{2}\rangle\quad |1,-1\rangle\otimes|\tfrac{1}{2},\tfrac{1}{2}\rangle,\\ \tfrac{1}{\hbar}J_z &= -\tfrac{3}{2}: && |1,-1\rangle\otimes|\tfrac{1}{2},-\tfrac{1}{2}\rangle.\end{aligned}\tag{20.5.3}$$

We want to discover the $\hat{\mathbf{J}}^2$ eigenstates. A couple of states are quickly recognized. The top state on this list has $J_z=\frac{3}{2}\hbar$, and since there are no states with higher J_z, it must be the top state of a $j=\frac{3}{2}$ multiplet:

$$|j=\tfrac{3}{2},m=\tfrac{3}{2}\rangle = |1,1\rangle\otimes|\tfrac{1}{2},\tfrac{1}{2}\rangle.\tag{20.5.4}$$

Notice that $\hat{J}_+=\hat{J}_{1,+}+\hat{J}_{2,+}$ kills the state, as it should: $\hat{J}_{1,+}$ kills the first factor, and $\hat{J}_{2,+}$ kills the second factor. The bottom state of this $j=\frac{3}{2}$ multiplet is also recognized; it is the bottom state in the table above:

$$|j=\tfrac{3}{2},m=-\tfrac{3}{2}\rangle = |1,-1\rangle\otimes|\tfrac{1}{2},-\tfrac{1}{2}\rangle.\tag{20.5.5}$$

Since we must get complete multiplets, we must get the four states in $j=\frac{3}{2}$. The state with $J_z=\frac{1}{2}\hbar$ must arise from a linear combination of states on the second line of the table, and the state with $J_z=-\frac{1}{2}\hbar$ must arise from a linear combination of states on the third line of the table. We are then left with two basis states, one with $J_z=\frac{1}{2}\hbar$ and one with $J_z=-\frac{1}{2}\hbar$. These *must* assemble into an additional $j=\frac{1}{2}$ multiplet! Therefore, we write

$$\mathbf{1}\otimes\tfrac{1}{2} = \tfrac{3}{2}\oplus\tfrac{1}{2}. \tag{20.5.6}$$

The left-hand side indicates that we have tensored the vector spaces of $\ell=1$ and $s=\frac{1}{2}$. On the right-hand side, we have the direct sum of $j=\frac{3}{2}$ and $j=\frac{1}{2}$. In terms of numbers of basis states, the above relation holds because $3\times 2=4+2$. To find the $|j=\frac{3}{2},m=\frac{1}{2}\rangle$ state, we will apply $\hat{J}_-$ to equation (20.5.4). Recalling that

$$\hat{J}_\pm|j,m\rangle = \hbar\sqrt{j(j+1)-m(m\pm 1)}\,|j,m\pm 1\rangle, \tag{20.5.7}$$

we see that the left-hand side gives

$$\hat{J}_-|j=\tfrac{3}{2},m=\tfrac{3}{2}\rangle = \hbar\sqrt{\tfrac{3}{2}\cdot\tfrac{5}{2}-\tfrac{3}{2}\cdot\tfrac{1}{2}}\,|j=\tfrac{3}{2},m=\tfrac{1}{2}\rangle = \hbar\sqrt{3}\,|j=\tfrac{3}{2},m=\tfrac{1}{2}\rangle. \tag{20.5.8}$$

On the right-hand side, we see that

$$\begin{aligned}\hat{J}_-|1,1\rangle\otimes|\tfrac{1}{2},\tfrac{1}{2}\rangle &= (\hat{J}_{1,-}\otimes\mathbb{1}+\mathbb{1}\otimes\hat{J}_{2,-})|1,1\rangle\otimes|\tfrac{1}{2},\tfrac{1}{2}\rangle\\ &= \hat{J}_-|1,1\rangle\otimes|\tfrac{1}{2},\tfrac{1}{2}\rangle+|1,1\rangle\otimes\hat{J}_-|\tfrac{1}{2},\tfrac{1}{2}\rangle\\ &= \hbar\sqrt{2}\,|1,0\rangle\otimes|\tfrac{1}{2},\tfrac{1}{2}\rangle+\hbar\,|1,1\rangle\otimes|\tfrac{1}{2},-\tfrac{1}{2}\rangle.\end{aligned} \tag{20.5.9}$$

Equating the results of the two last calculations, we find the desired state:

$$|j=\tfrac{3}{2},m=\tfrac{1}{2}\rangle = \sqrt{\tfrac{2}{3}}\,|1,0\rangle\otimes|\tfrac{1}{2},\tfrac{1}{2}\rangle+\sqrt{\tfrac{1}{3}}\,|1,1\rangle\otimes|\tfrac{1}{2},-\tfrac{1}{2}\rangle. \tag{20.5.10}$$

The state came out normalized, as it should! As expected, this state is a linear combination of the two states with $J_z=\frac{1}{2}\hbar$. The state $|j=\frac{1}{2},m=\frac{1}{2}\rangle$ must be an orthogonal linear combination of the *same* two states because eigenstates of Hermitian operators with different eigenvalues are orthogonal. This allows us to easily write

$$|j=\tfrac{1}{2},m=\tfrac{1}{2}\rangle = -\sqrt{\tfrac{1}{3}}\,|1,0\rangle\otimes|\tfrac{1}{2},\tfrac{1}{2}\rangle+\sqrt{\tfrac{2}{3}}\,|1,1\rangle\otimes|\tfrac{1}{2},-\tfrac{1}{2}\rangle. \tag{20.5.11}$$

The overall sign of this state is arbitrary and a matter of convention. The above state could also have been calculated as follows.

Exercise 20.2. *Consider the ansatz* $|j=\frac{1}{2},m=\frac{1}{2}\rangle=\alpha\,|1,0\rangle\otimes|\frac{1}{2},\frac{1}{2}\rangle+\beta\,|1,1\rangle\otimes|\frac{1}{2},-\frac{1}{2}\rangle$ *and determine the coefficients* α *and* β*, up to normalization, by the condition that the state is annihilated by* $\hat{J}_+$.

We have by now identified four out of the six coupled states (three states in the $j=\frac{3}{2}$ multiplet and one in the $j=\frac{1}{2}$ multiplet). The remaining two are quickly obtained. Writing the $|j,m\rangle$ states without the explicit $j=\ldots$ and $m=\ldots$, the full $j=\frac{3}{2}$ and $j=\frac{1}{2}$ multiplets are given below.

$$
\begin{aligned}
&\mathbf{j}=\tfrac{3}{2}:\\
&|\tfrac{3}{2},\tfrac{3}{2}\rangle = |1,1\rangle\otimes|\tfrac{1}{2},\tfrac{1}{2}\rangle\\
&|\tfrac{3}{2},\tfrac{1}{2}\rangle = \sqrt{\tfrac{2}{3}}\,|1,0\rangle\otimes|\tfrac{1}{2},\tfrac{1}{2}\rangle+\sqrt{\tfrac{1}{3}}\,|1,1\rangle\otimes|\tfrac{1}{2},-\tfrac{1}{2}\rangle\\
&|\tfrac{3}{2},-\tfrac{1}{2}\rangle = \sqrt{\tfrac{2}{3}}\,|1,0\rangle\otimes|\tfrac{1}{2},-\tfrac{1}{2}\rangle+\sqrt{\tfrac{1}{3}}\,|1,-1\rangle\otimes|\tfrac{1}{2},\tfrac{1}{2}\rangle\\
&|\tfrac{3}{2},-\tfrac{3}{2}\rangle = |1,-1\rangle\otimes|\tfrac{1}{2},-\tfrac{1}{2}\rangle\\
&\text{-------------------------------}\\
&\mathbf{j}=\tfrac{1}{2}:\\
&|\tfrac{1}{2},\tfrac{1}{2}\rangle = -\sqrt{\tfrac{1}{3}}\,|1,0\rangle\otimes|\tfrac{1}{2},\tfrac{1}{2}\rangle+\sqrt{\tfrac{2}{3}}\,|1,1\rangle\otimes|\tfrac{1}{2},-\tfrac{1}{2}\rangle\\
&|\tfrac{1}{2},-\tfrac{1}{2}\rangle = \sqrt{\tfrac{1}{3}}\,|1,0\rangle\otimes|\tfrac{1}{2},-\tfrac{1}{2}\rangle-\sqrt{\tfrac{2}{3}}\,|1,-1\rangle\otimes|\tfrac{1}{2},\tfrac{1}{2}\rangle.
\end{aligned}
\tag{20.5.12}
$$

Let us comment on the states we did not derive. The state $|\tfrac{3}{2},-\tfrac{1}{2}\rangle$ can be found by applying $\hat{J}_-$ to $|\tfrac{3}{2},\tfrac{1}{2}\rangle$ or by applying $\hat{J}_+$ to $|\tfrac{3}{2},-\tfrac{3}{2}\rangle$, both of which were determined earlier. Similarly, the state $|\tfrac{1}{2},-\tfrac{1}{2}\rangle$ can be found by applying $\hat{J}_-$ to $|\tfrac{1}{2},\tfrac{1}{2}\rangle$ or by demanding orthogonality to $|\tfrac{3}{2},-\tfrac{1}{2}\rangle$. You can test your knowledge by doing these short computations.

Exercise 20.3. *Are the states in (20.5.12) eigenstates of $\hat{\mathbf{J}}_1^2$? Are they eigenstates of $\hat{\mathbf{J}}_2^2$?*

20.6 Spin-Orbit Coupling

The hydrogen atom Hamiltonian we have used so far is nonrelativistic. It is a good description of the spectrum because an estimate of the typical electron velocity v gives $v/c \simeq \alpha \simeq 1/137$ (problem 11.4). Relativistic corrections are of interest, and they determine the so-called fine-structure of the hydrogen atom, in which some of the degeneracies we have encountered are removed. These corrections can be derived systematically using the relativistic Dirac equation of the electron. Such analysis gives three correction terms to the Hamiltonian: one called, perhaps redundantly, the relativistic correction, a second called spin-orbit, and a third called the Darwin correction. We will study these three in detail in chapter 25. In this section we preview the spin-orbit interaction to further develop our understanding of addition of angular momentum. A complete and systematic analysis of hydrogen will be conducted in chapter 25.

Spin-orbit couplings are a relativistic interaction affecting the particle's spin as it moves inside a potential. While this coupling is relevant in a variety of settings, including the motion of electrons in semiconductors, it is most familiar in atomic physics. Indeed, for hydrogen it refers to a correction to the Hamiltonian that can be interpreted as an interaction of the electron magnetic moment with the magnetic field the electron experiences as it travels around the proton. This magnetic field is itself a relativistic effect: the proton only creates an electric field $\mathbf{E}$, but any particle moving with some velocity $\mathbf{v}$ in an electric field $\mathbf{E}$ will observe a magnetic field $\mathbf{B}$ of the form

$$\mathbf{B} \simeq \mathbf{E} \times \frac{\mathbf{v}}{c} \sim \frac{e}{c} \frac{\mathbf{r} \times \mathbf{v}}{r^3}, \tag{20.6.1}$$

when $|\mathbf{v}| \ll c$. Since $\mathbf{r} \times \mathbf{v} \sim \mathbf{L}$, we conclude that the magnetic field at the electron is proportional to the orbital angular momentum of the electron. Since the magnetic dipole of the electron is proportional to its spin, the extra coupling $-\hat{\boldsymbol{\mu}} \cdot \mathbf{B}$ in the Hamiltonian is proportional to $\hat{\mathbf{S}} \cdot \hat{\mathbf{L}}$. The spin and orbital operators commute, so the order in which they are multiplied is irrelevant. If we kept track of the constant factors, we would get a result that is well known to be off by a factor of two. This error arises because one must implement a further *Thomas correction*, a subtle effect present because the electron rest frame is not inertial. A complete derivation would take us far afield, so here we will just state the resulting correction to the Hamiltonian and explore some of the consequences. One finds that,

$$\boxed{\text{spin-orbit correction:} \quad \delta H = \frac{e^2\hbar^2}{2m^2c^2r^3} \frac{1}{\hbar^2} \hat{\mathbf{L}} \cdot \hat{\mathbf{S}}.} \tag{20.6.2}$$

To estimate the magnitude δE of this correction, we set $\hat{\mathbf{L}} \cdot \hat{\mathbf{S}}/\hbar^2 \sim 1$ and $r \sim a_0$ to find

$$\delta E \sim \frac{e^2\hbar^2}{m^2c^2a_0^3} = \frac{e^8}{\hbar^4c^4} \cdot mc^2 = \alpha^4(mc^2). \tag{20.6.3}$$

Since the ground state energy is of order $\alpha^2(mc^2)$, the fine structure of the hydrogen atom is in effect a factor $\alpha^2 \simeq 1/19{,}000$ smaller than the scale of the energy levels!

Consider the $\hat{\mathbf{L}} \cdot \hat{\mathbf{S}}$ factor in δH. As we discussed for hyperfine splitting, this operator product is a tensor product $\hat{\mathbf{L}} \cdot \hat{\mathbf{S}} = \sum_i L_i \otimes S_i$. This time we define the total angular momentum

$$\hat{\mathbf{J}} \equiv \hat{\mathbf{L}} + \hat{\mathbf{S}}, \tag{20.6.4}$$

or more precisely, $\hat{J}_i \equiv \hat{L}_i \otimes \mathbb{1} + \mathbb{1} \otimes \hat{S}_i$. The computation of $\hat{\mathbf{J}}^2$ is completely analogous to the computation we did earlier and gives $\hat{\mathbf{J}}^2 = \hat{\mathbf{L}}^2 \otimes \mathbb{1} + 2\, \hat{\mathbf{L}} \cdot \hat{\mathbf{S}} + \mathbb{1} \otimes \hat{\mathbf{S}}^2$, which leads to the useful

$$\hat{\mathbf{L}} \cdot \hat{\mathbf{S}} = \tfrac{1}{2}(\hat{\mathbf{J}}^2 - \hat{\mathbf{L}}^2 - \hat{\mathbf{S}}^2). \tag{20.6.5}$$

Complete set of commuting observables (CSCO) Let us now consider the CSCO suitable for the spin-orbit supplemented hydrogen Hamiltonian. For the unperturbed hydrogen atom Hamiltonian $\hat{H}^{(0)}$, the complete set of observables is

$$\{\hat{H}^{(0)}, \hat{\mathbf{L}}^2, \hat{L}_z\}. \tag{20.6.6}$$

Recall that $[\hat{H}^{(0)}, \hat{L}_i] = 0$, so all the operators here commute. Suppose we now consider the spin of the electron but *do not* change the Hamiltonian. In this case the list would be enlarged to

$$\{\hat{H}^{(0)}, \hat{\mathbf{L}}^2, \hat{L}_z, \hat{\mathbf{S}}^2, \hat{S}_z\}. \tag{20.6.7}$$

Since $\hat{H}^{(0)}$ contains no spin operator, $[\hat{H}^{(0)}, \hat{S}_i]=0$ trivially. Moreover, since $\hat{L}_i$'s and $\hat{S}_j$'s commute, all operators on the list commute. Let us now consider the case when the full Hamiltonian $\hat{H}$ includes the spin-orbit correction:

$$\hat{H}=\hat{H}^{(0)}+\delta H, \quad \delta H \sim \frac{1}{r^3}\hat{\mathbf{L}}\cdot\hat{\mathbf{S}}. \tag{20.6.8}$$

With this new Hamiltonian, the list of commuting observables must be rethought. Consider first the list

$$\{\hat{H}, \hat{\mathbf{L}}^2, \hat{\mathbf{S}}^2, \ldots\}, \tag{20.6.9}$$

where the dots indicate additional operators we may be able to add. Notice that so far all is good since both $\hat{\mathbf{L}}^2$ and $\hat{\mathbf{S}}^2$ commute with $\hat{H}^{(0)}$ and with δH and thus with $\hat{H}$. This is because δH is built from $\hat{L}_i$ and $\hat{S}_i$ operators that commute with $\hat{\mathbf{L}}^2$ and $\hat{\mathbf{S}}^2$. δH also contains r, but r commutes with any $\hat{L}_i$ and commutes trivially with any $\hat{S}_i$. Neither $\hat{L}_i$ nor $\hat{S}_i$, for any value of i, can be added to the list above because

$$[\hat{L}_i, \hat{\mathbf{L}}\cdot\hat{\mathbf{S}}] \neq 0, \text{ and } [\hat{S}_i, \hat{\mathbf{L}}\cdot\hat{\mathbf{S}}] \neq 0, \tag{20.6.10}$$

as one can quickly check. Since the separate spin and orbital angular momenta no longer are conserved (they don't commute with the new Hamiltonian), we should consider the total angular momentum $\hat{\mathbf{J}}=\hat{\mathbf{L}}+\hat{\mathbf{S}}$. Can we add $\hat{\mathbf{J}}^2$ to the list? Well,

$$\begin{aligned}
&[\hat{\mathbf{J}}^2, \hat{H}^{(0)}] = 0, \quad \text{since } \hat{H}^{(0)} \text{ commutes with all } \hat{L}_i, \hat{S}_i,\\
&[\hat{\mathbf{J}}^2, \hat{\mathbf{L}}^2] = 0, \quad \text{since } \hat{\mathbf{L}}^2 \text{ commutes with all } \hat{L}_i, \hat{S}_i,\\
&[\hat{\mathbf{J}}^2, \hat{\mathbf{S}}^2] = 0, \quad \text{since } \hat{\mathbf{S}}^2 \text{ commutes with all } \hat{L}_i, \hat{S}_i,\\
&[\hat{\mathbf{J}}^2, \hat{\mathbf{L}}\cdot\hat{\mathbf{S}}] = [\hat{\mathbf{J}}^2, \tfrac{1}{2}(\hat{\mathbf{J}}^2 - \hat{\mathbf{L}}^2 - \hat{\mathbf{S}}^2)] = 0, \text{ using (20.6.5)},\\
&[\hat{\mathbf{J}}^2, f(r)] = 0, \quad \text{since } \hat{L}_i \text{ and } \hat{S}_i \text{ commute with } r.
\end{aligned} \tag{20.6.11}$$

The last two lines imply that $\hat{\mathbf{J}}^2$ commutes with δH, and therefore it commutes with $\hat{H}$. So, yes, we can add $\hat{\mathbf{J}}^2$ to the list. The new list is

$$\{\hat{H}, \hat{\mathbf{L}}^2, \hat{\mathbf{S}}^2, \hat{\mathbf{J}}^2, \ldots\}. \tag{20.6.12}$$

We can try now to add some $\hat{J}_i$. You can see that $\hat{J}_i$ commutes with $\hat{H}^{(0)}, \hat{\mathbf{L}}^2, \hat{\mathbf{S}}^2$, and $\hat{\mathbf{J}}^2$, the last one by the Casimir property. Moreover,

$$[\hat{J}_i, \hat{\mathbf{L}}\cdot\hat{\mathbf{S}}] = [\hat{J}_i, \tfrac{1}{2}(\hat{\mathbf{J}}^2 - \hat{\mathbf{L}}^2 - \hat{\mathbf{S}}^2)] = 0. \tag{20.6.13}$$

Since $[\hat{J}_i, f(r)]=0$ by the same arguments as before, we can add one $\hat{J}_i$ to the list. Note that we have demonstrated that $\hat{J}_i$, for all i, commutes with the new Hamiltonian:

$$[\hat{J}_i, \hat{H}]=0, \tag{20.6.14}$$

confirming that the total angular momentum is conserved in the presence of the spin-orbit correction. Following convention, we add $\hat{J}_z$ to find our final list, which we claim is a CSCO:

$$\boxed{\text{CSCO for spin-orbit coupling: } \{\hat{H}, \hat{\mathbf{L}}^2, \hat{\mathbf{S}}^2, \hat{\mathbf{J}}^2, \hat{J}_z\}.} \tag{20.6.15}$$

We will confirm that all eigenstates can be labeled uniquely by eigenvalues of the above operators.

Computation of the perturbation We will look into the simplest example. Since we want some nonvanishing angular momentum, we will take $\ell=1$. The lowest-energy states with $\ell=1$ have principal quantum number $n=2$. The $\ell=1$ multiplet has three states $|1,m\rangle$, with $m=\pm1$ and $m=0$. Since the spin of the electron can be either up or down, we have a total of *six states*. The states can be written as

$$R_{2,1}(r)\,|1,m\rangle\otimes|\tfrac{1}{2},m_s\rangle,\quad \text{with}\quad m=\pm1,0,\quad m_s=\pm\tfrac{1}{2}. \tag{20.6.16}$$

Here, $|\frac{1}{2},m_s\rangle$ is the spin state, and $|1,m\rangle$ is the angular state, whose explicit coordinate form is $Y_{1,m}(\theta,\phi)=\langle\theta\phi|1,m\rangle$. Finally, $R_{n\ell}(r)$, with $n=2$ and $\ell=1$, denotes the radial wave function, which is common for all six states above. These states are degenerate in the original hydrogen atom Hamiltonian, but the degeneracy will be partially removed by spin-orbit correction to the Hamiltonian. To find the energy corrections, we need to find a basis $|i\rangle$, $i=1,\ldots,6$, formed by linear combinations of the above states such that the matrix $\langle i|\delta H|j\rangle$ is diagonal, in which case the energy shift of $|i\rangle$ is simply $\langle i|\delta H|i\rangle$. The six states in (20.6.16), apart from the $R_{2,1}(r)$ factor, are the tensor product of an $\ell=1$ multiplet and an $s=1/2$ multiplet. They are in fact $\mathbf{1}\otimes\frac{1}{2}$. This makes our work in the previous section useful. Since we know that

$$\mathbf{1}\otimes\tfrac{1}{2}=\tfrac{3}{2}\oplus\tfrac{1}{2}, \tag{20.6.17}$$

the states will break down into multiplets $\frac{3}{2}$ and $\frac{1}{2}$ of the total angular momentum $\hat{\mathbf{J}}$.

The uncoupled states (20.6.16) are *all* eigenstates of $\hat{\mathbf{L}}^2$ and $\hat{\mathbf{S}}^2$ with eigenvalues $\hbar^2\ell(\ell+1)$, with $\ell=1$, and $\hbar^2 s(s+1)$, with $s=1/2$, respectively. Any linear combinations of these states are still eigenstates of $\hat{\mathbf{L}}^2$ and $\hat{\mathbf{S}}^2$. This includes the particular linear combinations that will be eigenstates of $\hat{\mathbf{J}}^2$. Therefore, we see that the basis of $(\hat{\mathbf{J}}^2,\hat{J}_z)$ eigenstates are in fact eigenstates of all the operators in the above CSCO, except for $\hat{H}$. Crucially, as we will see below, $\hat{H}$ is diagonal in the subspace spanned by this basis.

To write the matrix elements clearly, it is convenient to factor out the r dependence of the basis states by writing

$$|\,i\rangle = R_{2,1}(r)|\,\underline{i}\,\rangle,\quad i=1,\ldots,6, \tag{20.6.18}$$

where $|\,\underline{i}\,\rangle$ contains the angular and spin parts of the state. The matrix elements of interest are then written using the expression for δH in (20.6.2). Including only the spatial integral for the r dependence,

$$\begin{aligned}\delta H_{ij}\equiv\langle i|\delta H|j\rangle &= \frac{e^2\hbar^2}{2m^2c^2}\left[\int_0^\infty r^2dr|R_{2,1}(r)|^2\frac{1}{r^3}\right]\frac{1}{\hbar^2}\langle\,\underline{i}\,|\hat{\mathbf{L}}\cdot\hat{\mathbf{S}}\,|\,\underline{j}\,\rangle\\ &= \frac{e^2\hbar^2}{2m^2c^2}\left\langle\frac{1}{r^3}\right\rangle_{2,1}\frac{1}{\hbar^2}\langle\,\underline{i}\,|\hat{\mathbf{L}}\cdot\hat{\mathbf{S}}\,|\,\underline{j}\,\rangle.\end{aligned} \tag{20.6.19}$$

Here the subscripts 2, 1 on the expectation value indicate that one is using the $R_{2,1}$ radial wave function. If the $|\,\underline{i}\,\rangle$ are eigenstates of $\hat{\mathbf{L}}\cdot\hat{\mathbf{S}}$, the matrix $[\delta H]$ will be diagonal, and the diagonal elements are the energy shifts.

All the hard work has already been done. We just have to put the various pieces together. Begin with the simplification of $\hat{\mathbf{L}}\cdot\hat{\mathbf{S}}$ in (20.6.5), given that $\ell=1$ and $s=\frac{1}{2}$:

$$\begin{aligned}\frac{1}{\hbar^2}\hat{\mathbf{L}}\cdot\hat{\mathbf{S}} &= \frac{1}{2\hbar^2}(\hat{\mathbf{J}}^2-\hat{\mathbf{L}}^2-\hat{\mathbf{S}}^2)=\frac{1}{2\hbar^2}(\hat{\mathbf{J}}^2-\hbar^2\cdot 1\cdot 2-\hbar^2\cdot\tfrac{1}{2}\cdot\tfrac{3}{2})\\ &= \tfrac{1}{2}\left(\tfrac{1}{\hbar^2}\hat{\mathbf{J}}^2-\tfrac{11}{4}\right).\end{aligned}\tag{20.6.20}$$

Additionally, the expectation value of $1/r^3$ follows from (11.4.20): $\langle 1/r^3\rangle_{2,1}=1/(24a_0^3)$. From (20.6.19), we now get

$$\delta H_{ij}=\Delta E_0\,\langle \underline{i}|\left(\tfrac{1}{\hbar^2}\hat{\mathbf{J}}^2-\tfrac{11}{4}\right)|\underline{j}\rangle,\ \text{with}\ \Delta E_0\equiv\frac{e^2\hbar^2}{96m^2c^2a_0^3}.\tag{20.6.21}$$

It is now clear that a good basis is one where the first four states are the states $|j,m\rangle$ of the $j=\frac{3}{2}$ multiplet, and the last two states are the states $|j,m\rangle$ of the $j=\frac{1}{2}$ multiplet. In that basis δH_{ij} is diagonal. More explicitly, letting the i,j and $\underline{i},\underline{j}$ labels denote the composite index j,m, we have

$$\begin{aligned}\delta H_{j'm',jm} &= \Delta E_0\,\langle j',m'|\left(\tfrac{1}{\hbar^2}\hat{\mathbf{J}}^2-\tfrac{11}{4}\right)|j,m\rangle\\ &= \Delta E_0\left(j(j+1)-\tfrac{11}{4}\right)\delta_{j'j}\delta_{m'm}.\end{aligned}\tag{20.6.22}$$

The matrix is now manifestly diagonal, and the diagonal elements are the energy shifts. Calling $\Delta E_{j,m}$ the energy shift of the state $|j,m\rangle$, equation (20.6.22) implies that

$$\Delta E_{j,m} = \Delta E_0\left(j(j+1)-\tfrac{11}{4}\right),\tag{20.6.23}$$

and the shifts are m independent. Thus, the six states split into four $j=\frac{3}{2}$ states that shift up together and two $j=\frac{1}{2}$ states that shift down together:

$$\begin{aligned}j=\tfrac{3}{2}:\quad \Delta E_{\frac{3}{2},m} &= \Delta E_0\left(\tfrac{3}{2}\cdot\tfrac{5}{2}-\tfrac{11}{4}\right) = \Delta E_0,\\ j=\tfrac{1}{2}:\quad \Delta E_{\frac{1}{2},m} &= \Delta E_0\left(\tfrac{1}{2}\cdot\tfrac{3}{2}-\tfrac{11}{4}\right) = -2\,\Delta E_0.\end{aligned}\tag{20.6.24}$$

This is the final result for the spin-orbit correction to the $n=2$, $\ell=1$ levels of the hydrogen atom. The value of ΔE_0 is given in (20.6.21). This correction must be combined with other relativistic corrections to obtain the fine structure of hydrogen.

20.7 General Aspects of Addition of Angular Momentum

We have examined the tensor states of two particles with spin one-half and the states of a particle that has both orbital and spin angular momentum and whose state space is a tensor product of $\ell=1$ orbital states and spin states. In both cases we found it useful to form a basis for the tensor product in which states were eigenstates of the total angular momentum. Now we will consider the general situation and discuss some important regularities.

Consider two state spaces $\hat{\mathcal{H}}_1$ and $\hat{\mathcal{H}}_2$, each of which may contain a number of multiplets of angular momenta $\hat{\mathbf{J}}_1$ and $\hat{\mathbf{J}}_2$, respectively:

$$\hat{\mathbf{J}}_1 : \ \hat{\mathcal{H}}_1 = \bigoplus_{j_1} \mathcal{H}_1^{j_1}, \qquad \mathcal{H}_1^{j_1} = \bigoplus_{m_1=-j_1}^{j_1} |j_1, m_1\rangle,$$
$$\hat{\mathbf{J}}_2 : \ \hat{\mathcal{H}}_2 = \bigoplus_{j_2} \mathcal{H}_2^{j_2}, \qquad \mathcal{H}_2^{j_2} = \bigoplus_{m_2=-j_2}^{j_2} |j_2, m_2\rangle. \tag{20.7.1}$$

As indicated above, $\hat{\mathcal{H}}_i$ (with $i=1,2$) is a direct sum of multiplets $\mathcal{H}_i^{j_i}$ with angular momentum j_i, each of which contains the familiar $2j_i+1$ states with different values of m_i. The direct sum symbol is appropriate as the various j_i multiplet vector spaces are put together as orthogonal subspaces inside the big total spaces $\hat{\mathcal{H}}_1$ and $\hat{\mathcal{H}}_2$. The list of j_1 multiplets appearing in $\hat{\mathcal{H}}_1$ and the list of j_2 multiplets appearing in $\hat{\mathcal{H}}_2$ are left arbitrary. The goal is to form the tensor product vector space $\hat{\mathcal{H}}_1 \otimes \hat{\mathcal{H}}_2$ and find a basis of states that are eigenstates of the total angular momentum; more precisely, eigenstates of $\hat{\mathbf{J}}^2$ and $\hat{J}_z$, where

$$\hat{\mathbf{J}} = \hat{\mathbf{J}}_1 \otimes \mathbb{1} + \mathbb{1} \otimes \hat{\mathbf{J}}_2. \tag{20.7.2}$$

The full tensor product is given as a direct sum of the tensor product of the various multiplet subspaces:

$$\hat{\mathcal{H}}_1 \otimes \hat{\mathcal{H}}_2 = \bigoplus_{j_1, j_2} \mathcal{H}_1^{j_1} \otimes \mathcal{H}_2^{j_2}. \tag{20.7.3}$$

It is therefore sufficient for us to understand the structure of the general summand V_{j_1,j_2} in the above relation:

$$V_{j_1,j_2} \equiv \mathcal{H}_1^{j_1} \otimes \mathcal{H}_2^{j_2}, \qquad \dim V_{j_1,j_2} = (2j_1+1)(2j_2+1). \tag{20.7.4}$$

We want to understand the general features of V_{j_1,j_2}. For this purpose we note that we have two natural choices of basis states corresponding to alternative CSCOs.

1. The **uncoupled basis states** for V_{j_1,j_2} follow naturally by the tensor product of $\mathcal{H}_1^{(j_1)}$ basis states and $\mathcal{H}_2^{(j_2)}$ basis states:

$$|j_1, j_2, m_1, m_2\rangle \ \equiv \ |j_1, m_1\rangle \otimes |j_2, m_2\rangle. \tag{20.7.5}$$

   Here m_1 runs over $2j_1+1$ values, and m_2 runs over $2j_2+1$ values. The CSCO$_u$, with subscript u for uncoupled, is given by the following list:

$$\text{CSCO}_u: \ \ \{\hat{\mathbf{J}}_1^2, \hat{\mathbf{J}}_2^2, \hat{J}_{1,z}, \hat{J}_{2,z}\}. \tag{20.7.6}$$

   The states $|j_1, j_2, m_1, m_2\rangle$ are eigenstates of the operators in the list with eigenvalues given, respectively, by

$$\{\, \hbar^2 j_1(j_1+1), \ \hbar^2 j_2(j_2+1), \ \hbar m_1, \ \hbar m_2 \,\}. \tag{20.7.7}$$

2. The **coupled basis states** for V_{j_1,j_2} are eigenstates of the CSCO$_c$, with subscript c for coupled, that contains the following operators:

$$\text{CSCO}_c: \ \ \{\hat{\mathbf{J}}_1^2, \hat{\mathbf{J}}_2^2, \hat{\mathbf{J}}^2, \hat{J}_z\}. \tag{20.7.8}$$

Here $\hat{\mathbf{J}} = \hat{\mathbf{J}}_1 + \hat{\mathbf{J}}_2$. The basis states are

$$|j_1, j_2; j, m\rangle, \tag{20.7.9}$$

and the eigenvalues for the list of operators in the CSCO$_c$ are given by

$$\{\hbar^2 j_1(j_1+1),\ \hbar^2 j_2(j_2+1),\ \hbar^2 j(j+1),\ \hbar m\}. \tag{20.7.10}$$

The states (20.7.9) come in full $\hat{\mathbf{J}}$ multiplets $\mathcal{H}_j$ that contain all allowed m values:

$$\mathcal{H}_j = \bigoplus_{m=-j}^{j} |j_1, j_2; j, m\rangle. \tag{20.7.11}$$

Since we aim to give a basis for the space V_{j_1,j_2}, which has dimension $(2j_1+1)(2j_2+1)$, we will need a collection of $\hat{\mathbf{J}}$ multiplets $\mathcal{H}_j$ with various values of j:

$$V_{j_1,j_2} = \bigoplus_{j \in S(j_1,j_2)} \mathcal{H}_j. \tag{20.7.12}$$

Here $S(j_1, j_2)$ is a list of values that depend on j_1 and j_2. In combining two spin one-half particles, we learned that for $j_1 = j_2 = \frac{1}{2}$ the list is $j \in \{1, 0\}$. In doing spin-orbit coupling, we learned that for $j_1 = 1, j_2 = \frac{1}{2}$ we have $j \in \{\frac{3}{2}, \frac{1}{2}\}$. What will we get for general j_1 and j_2?

Both the coupled and uncoupled bases are orthonormal bases. One natural question is how to relate the two bases. We did this explicitly for the case of $\frac{1}{2} \otimes \frac{1}{2}$ and for $\mathbf{1} \otimes \frac{1}{2}$, in both cases figuring out how to write the coupled states in terms of the uncoupled ones. The question of relating the two sets of basis vectors is the problem of calculating the **Clebsch-Gordan** (CG) coefficients. To write this out clearly, consider the completeness relation for the uncoupled basis of the space V_{j_1,j_2}. With the sums over m_1 and m_2 running over the usual values,

$$\mathbb{1} = \sum_{m_1,m_2} |j_1 j_2; m_1 m_2\rangle\langle j_1 j_2; m_1 m_2|, \quad \text{acting on} \quad V_{j_1,j_2}. \tag{20.7.13}$$

Apply both sides of this equation to the coupled basis state $|j_1, j_2; j, m\rangle$ to find

$$|j_1, j_2; j, m\rangle = \sum_{m_1,m_2} |j_1 j_2; m_1 m_2\rangle \underbrace{\langle j_1 j_2; m_1 m_2 | j_1, j_2; j, m\rangle}_{\text{Clebsch-Gordan coefficients}}. \tag{20.7.14}$$

The overlaps selected above are the CG coefficients. They are numbers that, if known, determine the relations between the two sets of basis vectors. Schematically, a CG coefficient is an overlap $\langle u_i | c_j\rangle$, where u_i denotes an uncoupled basis element, and c_j denotes a coupled basis element. Since the two bases are orthonormal, the transformation from one to the other is produced by a unitary operator U that can be written as

$$U = \sum_k |c_k\rangle\langle u_k| \tag{20.7.15}$$

and manifestly maps $|u_p\rangle$ to $|c_p\rangle$. The matrix elements of this operator,

$$\langle u_i | U | u_j\rangle = \sum_k \langle u_i | c_k\rangle\langle u_k | u_j\rangle = \sum_k \langle u_i | c_k\rangle \delta_{kj} = \langle u_i | c_j\rangle, \tag{20.7.16}$$

are precisely the CG coefficients. In fact, $\langle u_i|U|u_j\rangle = \langle c_i|U|c_j\rangle$ because U is a basis-changing operator.

There are selection rules for CG coefficients that help their computation:

1. We claim that

$$\boxed{\langle j_1 j_2; m_1 m_2 | j_1, j_2; j, m\rangle = 0, \quad \text{for } m \neq m_1 + m_2.} \tag{20.7.17}$$

This is easily proven by inserting the $\hat{J}_z$ operator in between the overlap and setting equal the two possible ways of evaluating the matrix element:

$$\langle j_1 j_2; m_1 m_2 | \hat{J}_z | j_1, j_2; j, m\rangle = \langle j_1 j_2; m_1 m_2 | (\hat{J}_{1,z} + \hat{J}_{2,z}) | j_1, j_2; j, m\rangle. \tag{20.7.18}$$

On the left-hand side, we let the operator act on the ket, and on the right-hand side, we let the operator act on the bra by using its Hermiticity. We then get

$$\hbar m \langle j_1 j_2; m_1 m_2 | j_1, j_2; j, m\rangle = (\hbar m_1 + \hbar m_2) \langle j_1 j_2; m_1 m_2 | j_1, j_2; j, m\rangle. \tag{20.7.19}$$

Collecting terms, we find that

$$\hbar(m - (m_1 + m_2)) \langle j_1 j_2; m_1 m_2 | j_1, j_2; j, m\rangle = 0. \tag{20.7.20}$$

This confirms the claim that the CG coefficient must vanish when $m \neq m_1 + m_2$. Note that when we computed coupled states in terms of uncoupled ones we used this rule: a coupled state with $J_z = \hbar m$ could only be related to uncoupled states in which the $\hat{J}_{1,z}$ and $\hat{J}_{2,z}$ eigenvalues added to $\hbar m$; see, for example, the states in (20.5.12).

2. The second selection rule specifies the values of j for which the coupled state can have a nonvanishing overlap with the uncoupled states $\mathcal{H}_{j_1} \otimes \mathcal{H}_{j_2}$. When writing (20.7.12),

$$V_{j_1, j_2} = \mathcal{H}_1^{j_1} \otimes \mathcal{H}_2^{j_2} = \bigoplus_{j \in S(j_1, j_2)} \mathcal{H}_j, \tag{20.7.21}$$

we asked what values of j appear in the list $S(j_1, j_2)$. Only for those values of j does the CG coefficient $\langle j_1 j_2; m_1 m_2 | j_1, j_2; j, m\rangle$ need not vanish. Additionally, the sum of the dimensions of the $\mathcal{H}_j$'s in the equation above must equal the dimensionality of V_{j_1, j_2}.

We can quickly deduce the largest value of j in the list $S(j_1, j_2)$. The largest value of $\hat{J}_z$ in $\mathcal{H}_1^{j_1} \otimes \mathcal{H}_2^{j_2}$ is the largest value of $\hat{J}_{1,z} + \hat{J}_{2,z}$. The largest value of $\hat{J}_{1,z}$ is $\hbar j_1$, and the largest value of $\hat{J}_{2,z}$ is $\hbar j_2$. Thus, the largest value of $\hat{J}_z$ is $\hbar(j_1 + j_2)$, implying that the largest value of j in the list is $j_1 + j_2$. As it turns out, the smallest value of j in the list is $|j_1 - j_2|$, and the full list includes all integer values of j in between:

$$S(j_1, j_2) = \{j_1 + j_2, j_1 + j_2 - 1, \ldots, |j_1 - j_2|\}. \tag{20.7.22}$$

Note that each value of j appears only once. This result (20.7.21) is usually written by labeling the multiplets with their j value:

$$\boxed{j_1 \otimes j_2 = (j_1 + j_2) \oplus (j_1 + j_2 - 1) \oplus \ldots \oplus |j_1 - j_2|.} \tag{20.7.23}$$

This is an equality between vector spaces showing how to write the vector space on the left-hand side as a direct sum of irreducible, invariant subspaces of $\hat{\mathbf{J}}$. It demonstrates that $j_1 \otimes j_2$ forms a *reducible* representation of the total angular momentum.

There is a basic test of (20.7.23). The dimensionality of $j_1 \otimes j_2$ is $(2j_1+1)(2j_2+1)$, so the dimensionality of the right-hand side must be the same. Since the j_1 and j_2 labels can be exchanged, we can assume that $j_1 \geq j_2$ without loss of generality. The dimensionality of the right-hand side is the sum of the dimensions of each of the representations. The sum is readily done, and the expected value is obtained:

$$\begin{aligned}\sum_{j=j_1-j_2}^{j_1+j_2}(2j+1) &= 2\Big(\sum_{j=0}^{j_1+j_2} j - \sum_{j=0}^{j_1-j_2-1} j\Big) + (j_1+j_2-(j_1-j_2)+1)\\ &= (j_1+j_2)(j_1+j_2+1)-(j_1-j_2-1)(j_1-j_2)+2j_2+1\\ &= (j_1+j_2)^2-(j_1-j_2)^2+2j_1+2j_2+1\\ &= 2j_1(2j_2+1)+2j_2+1=(2j_1+1)(2j_2+1).\end{aligned} \tag{20.7.24}$$

It is possible to explain the selection rule for j by considering explicitly all combinations of states in the j_1 and j_2 multiplets—again, assuming that $j_1 \geq j_2$. Let us use two columns to write all the states:

$$\begin{array}{llll}
[1] & |j_1, j_1\rangle & & \\
[2] & |j_1, j_1-1\rangle & & \\
[3] & |j_1, j_1-2\rangle & & \\
\vdots & \vdots & & \\
[2j_1-2j_2+1] & |j_1, 2j_2-j_1\rangle & |j_2, j_2\rangle & [1] \\
 & |j_1, 2j_2-j_1-1\rangle & |j_2, j_2-1\rangle & [2] \\
\vdots & \vdots & \vdots & \vdots \\
[2j_1+1] & |j_1, -j_1\rangle & |j_2, -j_2\rangle & [2j_2+1].
\end{array} \tag{20.7.25}$$

The data is aligned at the bottom: the lowest m states are on the same line. Since $j_1 \geq j_2$, the higher m states of the j_1 multiplet need not have same-line counterparts in the j_2 multiplet. We have also added number labels to the states. They appear in brackets to the left of the j_1 multiplet and to the right of the j_2 multiplet. Those labels allow us to refer quickly to some particular tensor product. Thus, for example,

$$|j_1, j_1-2\rangle \otimes |j_2, j_2-1\rangle \;\Leftrightarrow\; [3]\times[2]. \tag{20.7.26}$$

We now walk through the construction of figure 20.2, which shows why the selection rule for j holds. The heavy dots are states, and dots on a horizontal line are states with the same value of $J_z = \hbar m$, the total z-component of angular momentum.

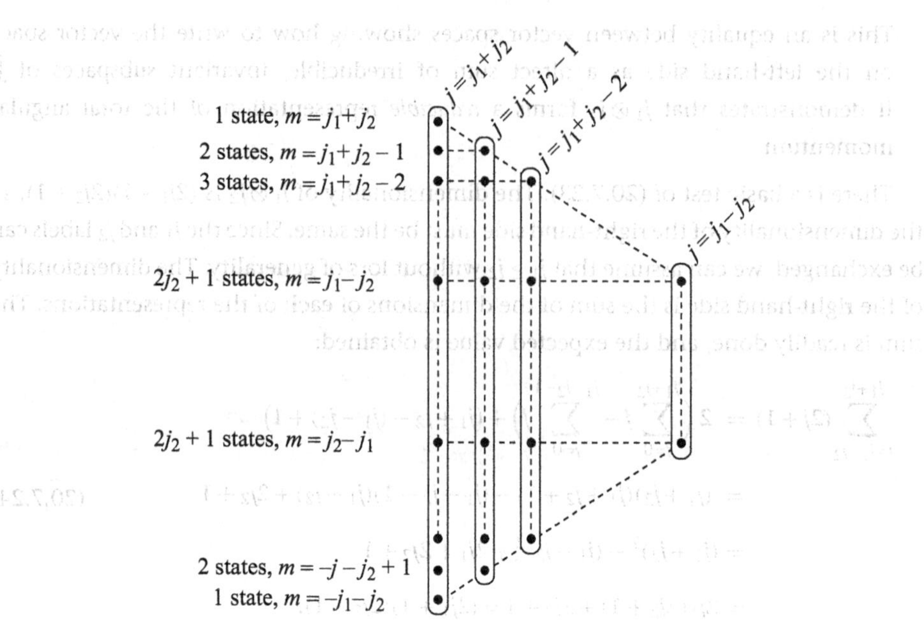

Figure 20.2
Diagram illustrating how $j_1 \otimes j_2$, with $j_1 \geq j_2$, contains total angular momentum representations from $j = j_1 + j_2$ all the way down to $j = j_1 - j_2$. Each heavy dot in the figure is a state. One can view the columns as j multiplets, with j value indicated on the topmost state.

- The first row consists of the single state $[1] \times [1]$, with $m = j_1 + j_2$.
- For the second line, we get two states $[1] \times [2]$ and $[2] \times [1]$, both with $m = j_1 + j_2 - 1$.
- For the third line, we get three states $[1] \times [3]$, $[2] \times [2]$, and $[3] \times [1]$, all with $m = j_1 + j_2 - 2$.
- The number of states on each line will keep growing until we get a maximum of $2j_2 + 1$ states:

$$[1] \times [2j_2 + 1], \; \ldots \; [2j_2 + 1] \times [1], \tag{20.7.27}$$

  all with $m = j_1 - j_2$. At this point we are using all of the states of the j_2 multiplet.
- The number of states on each line now remains the same until we get the set of states

$$[2j_1 - 2j_2 + 1] \times [2j_2 + 1], \; \ldots, \; [2j_1 + 1] \times [1], \tag{20.7.28}$$

  all with $m = j_2 - j_1 = -(j_1 - j_2)$. At this point we are tensoring all the $2j_2 + 1$ matched states at the bottom of the list (20.7.25). Since m values change by one unit as we move from line to line in figure 20.2, there are a total of $j_1 - j_2 - (j_2 - j_1) + 1 = 2(j_1 - j_2) + 1$ lines with the maximum number of states. This implies that the shortest column is the right-most, and it has $2(j_1 - j_2) + 1$ states, with m ranging from $j_1 - j_2$ to $-(j_1 - j_2)$. This corresponds to the $j = j_1 - j_2$ multiplet.
- After this last maximally long horizontal line in figure 20.2, the number of states on each line decreases by one unit as we move from line to line. The last line on the diagram has one state, $[2j_1 + 1] \times [2j_2 + 1]$, with $m = -(j_1 + j_2)$.

- All the columns contain the proper range of m values to form complete j multiplets. The longest column is the first, with $j = j_1 + j_2$. The shortest column was identified before, with $j = j_1 - j_2$. We have columns for all intermediate values of j changing by one unit at a time. This is what we wanted to demonstrate.

This argument shows that there are the right number of states at all values of m to be assembled into the expected multiplets of total angular momentum. In fact, they *have* to assemble in that way, as we explain now. Consider the single state on the first row in the figure, a state with $m = j_1 + j_2$. This m value implies that we have the multiplet $j = j_1 + j_2$. Looking at the second row, we have two states with $m = j_1 + j_2 - 1$. One linear combination must belong to the $j_1 + j_2$ multiplet already identified, but the other linear combination requires a multiplet with $j = j_1 + j_2 - 1$. Looking at the third row, we have three states with $m = j_1 + j_2 - 2$. Two linear combinations belong to the two multiplets already identified, and the third requires a new multiplet with $j = j_1 + j_2 - 2$. Continuing this way, we need a new multiplet with one unit less of j at each step until we get to the first occurrence of the largest row. That row has $2j_2 + 1$ states, all with $m = j_1 - j_2$. We have already identified $2j_2$ multiplets, requiring $2j_2$ linear combinations of the available states. The remaining one linear combination of the states requires one last new multiplet, with $j = j_1 - j_2$. No more multiplets are needed, as the rows no longer grow. We have therefore shown that the expected multiplets appear. Note that if we literally associate the columns in the figure with the multiplets, then each dot will represent some linear combination of the uncoupled basis states listed in the construction.

20.8 Hydrogen Atom and Hidden Symmetry

Our goal here is to give an algebraic derivation of the bound state spectrum of the unperturbed hydrogen atom Hamiltonian $\hat{H}^{(0)}$. The electron spin will play no role here and will be suppressed. For simplicity of notation, we will write the Hamiltonian as $\hat{H}$:

$$\hat{H} = \frac{\hat{\mathbf{p}}^2}{2m} - \frac{e^2}{r}. \tag{20.8.1}$$

For this Hamiltonian, the orbital angular momentum operators $\hat{\mathbf{L}}$ are conserved:

$$[\hat{H}, \hat{\mathbf{L}}] = 0, \qquad \hat{\mathbf{L}} \times \hat{\mathbf{L}} = i\hbar \hat{\mathbf{L}}. \tag{20.8.2}$$

The bound state spectrum of the Hamiltonian consists of states with energies $E_n = -\frac{e^2}{2a_0}\frac{1}{n^2}$ with $n = 1, 2, \ldots$. For each value $n \geq 1$, we have the degenerate n^2-dimensional space $\mathcal{H}_n$ described as a direct sum of ℓ multiplets with values ranging from zero to $n-1$: $\mathcal{H}_n = (\ell = n-1) \oplus (\ell = n-2) \oplus \cdots \oplus (\ell = 0)$.

The algebraic derivation of the spectrum makes use of another conserved vector, the Runge-Lenz vector operator $\hat{\mathbf{R}}$ defined in (19.8.18):

$$\hat{\mathbf{R}} \equiv \frac{1}{2me^2}(\hat{\mathbf{p}} \times \hat{\mathbf{L}} - \hat{\mathbf{L}} \times \hat{\mathbf{p}}) - \frac{\hat{\mathbf{r}}}{r}. \tag{20.8.3}$$

As written this operator is manifestly Hermitian. It is also conserved, as it too commutes with the Hamiltonian (problem 19.10):

$$[\hat{H}, \hat{\mathbf{R}}] = 0. \tag{20.8.4}$$

Two rewritings of $\hat{\mathbf{R}}$ are possible using the identity $\hat{\mathbf{p}} \times \hat{\mathbf{L}} = -\hat{\mathbf{L}} \times \hat{\mathbf{p}} + 2i\hbar\hat{\mathbf{p}}$ obtained in (19.2.2). These are

$$\hat{\mathbf{R}} = \frac{1}{me^2}(\hat{\mathbf{p}} \times \hat{\mathbf{L}} - i\hbar\hat{\mathbf{p}}) - \frac{\hat{\mathbf{r}}}{r} = \frac{1}{me^2}(-\hat{\mathbf{L}} \times \hat{\mathbf{p}} + i\hbar\hat{\mathbf{p}}) - \frac{\hat{\mathbf{r}}}{r}. \tag{20.8.5}$$

If $\hat{\mathbf{R}}$ and $\hat{\mathbf{L}}$ are conserved, all of their scalar products are conserved too since the Hamiltonian would commute with them using the derivation property of commutators. We are already familiar with the Casimir $\hat{\mathbf{L}}^2$ that in fact commutes with all $\hat{L}_i$. Let us consider now $\hat{\mathbf{R}}^2$, which must be a conserved scalar. Its evaluation gives (19.8.20):

$$\hat{\mathbf{R}}^2 = 1 + \frac{2\hat{H}}{me^4}(\hat{\mathbf{L}}^2 + \hbar^2). \tag{20.8.6}$$

Notice that $\hat{H}$, which appears on the above right-hand side, can be moved, if desired, to the right of the parentheses, as it commutes with $\hat{\mathbf{L}}$. The right-hand side is indeed a conserved scalar. We now look into $\hat{\mathbf{R}} \cdot \hat{\mathbf{L}}$, which must also be conserved. Classically, $\hat{\mathbf{R}} \cdot \hat{\mathbf{L}}$ vanishes since $\hat{\mathbf{R}}$ lies on the plane of the orbit (along the major axis of the ellipse), while $\hat{\mathbf{L}}$ is orthogonal to the plane of the orbit. The dot product of the quantum operators also vanishes, as we now confirm. First recall that

$$\hat{\mathbf{r}} \cdot \hat{\mathbf{L}} = 0, \quad \hat{\mathbf{p}} \cdot \hat{\mathbf{L}} = 0. \tag{20.8.7}$$

Using these and the first equality in (20.8.5), we find

$$\hat{\mathbf{R}} \cdot \hat{\mathbf{L}} = \frac{1}{me^2}(\hat{\mathbf{p}} \times \hat{\mathbf{L}}) \cdot \hat{\mathbf{L}}. \tag{20.8.8}$$

But we now notice that

$$(\hat{\mathbf{p}} \times \hat{\mathbf{L}}) \cdot \hat{\mathbf{L}} = \epsilon_{ijk}\hat{p}_j\hat{L}_k\hat{L}_i = \hat{p}_j\,\epsilon_{jki}\hat{L}_k\hat{L}_i = \hat{p}_j(\hat{\mathbf{L}} \times \hat{\mathbf{L}})_j = \hat{\mathbf{p}} \cdot i\hbar\hat{\mathbf{L}} = 0. \tag{20.8.9}$$

As a result, we have shown that

$$\boxed{\hat{\mathbf{R}} \cdot \hat{\mathbf{L}} = 0.} \tag{20.8.10}$$

The Runge-Lenz vector $\hat{\mathbf{R}}$ is a vector under rotations. This we know without any computation since it is built using cross products from $\hat{\mathbf{p}}$ and $\hat{\mathbf{L}}$, which are both vectors under rotations. Therefore we must find that

$$[\hat{L}_i, \hat{R}_j] = i\hbar\,\epsilon_{ijk}\hat{R}_k. \tag{20.8.11}$$

Recall (19.2.20), which states that for any vector $\hat{\mathbf{u}}$ under rotations, $\hat{\mathbf{L}} \times \hat{\mathbf{u}} + \hat{\mathbf{u}} \times \hat{\mathbf{L}} = 2i\hbar\,\hat{\mathbf{u}}$. It follows that

$$\hat{\mathbf{L}} \times \hat{\mathbf{R}} + \hat{\mathbf{R}} \times \hat{\mathbf{L}} = 2i\hbar\,\hat{\mathbf{R}}. \tag{20.8.12}$$

Exercise 20.4. *Show that* $\hat{\mathbf{R}}\cdot\hat{\mathbf{L}} = \hat{\mathbf{L}}\cdot\hat{\mathbf{R}}$. *Thus, both* $\hat{\mathbf{R}}\cdot\hat{\mathbf{L}} = 0$ *and* $\hat{\mathbf{L}}\cdot\hat{\mathbf{R}} = 0$.

In order to understand the commutator of two $\hat{\mathbf{R}}$ operators, we need a simple result: The commutator of two conserved operators is a conserved operator. To prove this consider two conserved operators $\hat{S}_1$ and $\hat{S}_2$:

$$[\hat{S}_1, \hat{H}] = [\hat{S}_2, \hat{H}] = 0. \tag{20.8.13}$$

The Jacobi identity (5.2.9) applied to the operators $\hat{S}_1, \hat{S}_2$, and $\hat{H}$ reads

$$[[\hat{S}_1, \hat{S}_2], \hat{H}] + [[\hat{H}, \hat{S}_1], \hat{S}_2] + [[\hat{S}_2, \hat{H}], \hat{S}_1] = 0. \tag{20.8.14}$$

The second term on the left-hand side vanishes by conservation of $\hat{S}_1$ and the third by conservation of $\hat{S}_2$. It follows that $[[\hat{S}_1, \hat{S}_2], \hat{H}] = 0$, which establishes the conservation of the commutator $[\hat{S}_1, \hat{S}_2]$. This result tells us that the commutator $[\hat{R}_i, \hat{R}_j]$ must be some conserved object. We can focus, equivalently, on the cross product of two $\hat{\mathbf{R}}$s that encodes the commutator. We must have

$$\hat{\mathbf{R}}\times\hat{\mathbf{R}} = (\cdots)\text{ "conserved vector,"} \tag{20.8.15}$$

where the dots represent some conserved scalar. Since $\hat{\mathbf{L}}$ and $\hat{\mathbf{R}}$ are conserved vectors, the possible vectors on the right-hand side are $\hat{\mathbf{L}}, \hat{\mathbf{R}}$, and $\hat{\mathbf{L}}\times\hat{\mathbf{R}}$. To narrow down the options, we examine the behavior of various vectors under the parity transformation $\hat{\mathbf{r}} \to -\hat{\mathbf{r}}$. Under this transformation we must have

$$\hat{\mathbf{p}} \to -\hat{\mathbf{p}}, \quad \hat{\mathbf{L}} \to \hat{\mathbf{L}}, \quad \hat{\mathbf{R}} \to -\hat{\mathbf{R}}. \tag{20.8.16}$$

The first follows because parity must preserve the commutator of $\hat{\mathbf{r}}$ and $\hat{\mathbf{p}}$, the second from $\hat{\mathbf{L}} = \hat{\mathbf{r}}\times\hat{\mathbf{p}}$, and the third from the expression for $\hat{\mathbf{R}}$ in terms of $\hat{\mathbf{r}}, \hat{\mathbf{p}}$, and $\hat{\mathbf{L}}$. Since the left-hand side of (20.8.15) does not change sign under the parity transformation, neither should the right-hand side. Note now that there are no parity-odd conserved scalars; the only candidate, $\mathbf{R}\cdot\mathbf{L}$, vanishes. On the other hand, there are parity-even conserved scalars, such as the Hamiltonian itself. Thus, the conserved vector must be parity even. From our choices $\hat{\mathbf{L}}, \hat{\mathbf{R}}$, and $\hat{\mathbf{L}}\times\hat{\mathbf{R}}$, we can only have $\hat{\mathbf{L}}$. We must therefore have $\hat{\mathbf{R}}\times\hat{\mathbf{R}} = (\cdots)\hat{\mathbf{L}}$, with the expression in parentheses a conserved parity-even scalar. A calculation (problem 20.13) gives

$$\hat{\mathbf{R}}\times\hat{\mathbf{R}} = i\hbar\left(-\frac{2\hat{H}}{me^4}\right)\hat{\mathbf{L}}. \tag{20.8.17}$$

This completes the determination of all commutators relevant to $\hat{\mathbf{L}}$ and $\hat{\mathbf{R}}$.

Now we come to the main point. We will derive algebraically the characterization of the subspaces of degenerate energy eigenstates. For this we will focus on one such subspace $\mathcal{H}_\nu$, at some energy E_ν, where ν is a parameter to be specified below. We will look at our operators *in that subspace*. Indeed, since both $\hat{\mathbf{L}}$ and $\hat{\mathbf{R}}$ are conserved, $\mathcal{H}_\nu$ is invariant under the action of these operators. In our operator relations (20.8.6) and (20.8.17), we are actually allowed to replace $\hat{H}$ by the energy E_ν, given that $\hat{H}$, which commutes with $\hat{\mathbf{L}}$, can be brought to the right, directly in front of the states. We then have

$$\hat{H} \to E_\nu = -\frac{me^4}{2\hbar^2}\frac{1}{\nu^2}, \quad \nu\in\mathbb{R}, \tag{20.8.18}$$

where we have written E_ν in terms of a unit-free, real constant ν, to be determined. The rest of the factors, except for a convenient factor of two, provide the right units. Of course, we know that the correct answer for these energies emerges if ν is a positive integer. This, however, is something we will be able to derive. It follows from the above equation that we can set

$$-\frac{2\hat{H}}{me^4} = \frac{1}{\hbar^2\nu^2}. \tag{20.8.19}$$

We can use this expression to simplify our key relations (20.8.17) and (20.8.6):

$$\begin{aligned}\hat{\mathbf{R}} \times \hat{\mathbf{R}} &= i\hbar\,\frac{1}{\hbar^2\nu^2}\,\hat{\mathbf{L}},\\ \hat{\mathbf{R}}^2 &= 1 - \frac{1}{\hbar^2\nu^2}\,(\hat{\mathbf{L}}^2 + \hbar^2).\end{aligned} \tag{20.8.20}$$

A few further rearrangements give

$$\boxed{\begin{aligned}(\hbar\nu\hat{\mathbf{R}}) \times (\hbar\nu\hat{\mathbf{R}}) &= i\hbar\,\hat{\mathbf{L}},\\ \hat{\mathbf{L}}^2 + \hbar^2\nu^2\hat{\mathbf{R}}^2 &= \hbar^2(\nu^2 - 1).\end{aligned}} \tag{20.8.21}$$

These are clear and simple algebraic relations between our operators. The first one shows that $\hbar\nu\hat{\mathbf{R}}$ has the units of angular momentum and sort of behaves like one, except that the operator to the right is not $\hbar\nu\hat{\mathbf{R}}$ but rather $\hat{\mathbf{L}}$.

Our next step is to show that with the help of $\hat{\mathbf{L}}$ and $\hat{\mathbf{R}}$ we can construct two independent, commuting algebras of angular momentum. Of course, it is clear that $\hat{\mathbf{L}}$ is an algebra of angular momentum. But by using suitable linear combinations of $\hat{\mathbf{L}}$ and $\hat{\mathbf{R}}$, we will obtain two such algebras. Indeed, define $\hat{\mathbf{J}}_1$ and $\hat{\mathbf{J}}_2$ as follows:

$$\begin{aligned}\hat{\mathbf{J}}_1 &\equiv \tfrac{1}{2}(\hat{\mathbf{L}} + \hbar\nu\hat{\mathbf{R}}),\\ \hat{\mathbf{J}}_2 &\equiv \tfrac{1}{2}(\hat{\mathbf{L}} - \hbar\nu\hat{\mathbf{R}}).\end{aligned} \tag{20.8.22}$$

We can also solve for $\hat{\mathbf{L}}$ and $\hbar\nu\hat{\mathbf{R}}$ in terms of $\hat{\mathbf{J}}_1$ and $\hat{\mathbf{J}}_2$:

$$\begin{aligned}\hat{\mathbf{L}} &= \hat{\mathbf{J}}_1 + \hat{\mathbf{J}}_2,\\ \hbar\nu\hat{\mathbf{R}} &= \hat{\mathbf{J}}_1 - \hat{\mathbf{J}}_2.\end{aligned} \tag{20.8.23}$$

It is important to realize that $\hat{\mathbf{L}}$ is nothing but the sum of $\hat{\mathbf{J}}_1$ and $\hat{\mathbf{J}}_2$. It is the total angular momentum.

We now claim that the operators $\hat{\mathbf{J}}_1$ and $\hat{\mathbf{J}}_2$ commute with each other. This is quickly confirmed by direct computation:

$$\begin{aligned}[\hat{J}_{1i}, \hat{J}_{2j}] &= \tfrac{1}{4}\,[\hat{L}_i + \hbar\nu\hat{R}_i,\, \hat{L}_j - \hbar\nu\hat{R}_j]\\ &= \tfrac{1}{4}\big(i\hbar\,\epsilon_{ijk}\hat{L}_k - \hbar\nu[\hat{L}_i, \hat{R}_j] - \hbar\nu[\hat{L}_j, \hat{R}_i] - i\hbar\,\epsilon_{ijk}\hat{L}_k\big) = 0,\end{aligned} \tag{20.8.24}$$

where we note that the first and last terms on the right-hand side cancel each other out, and the second and third terms also cancel each other out using (20.8.11). Now we want to show that $\hat{\mathbf{J}}_1$ and $\hat{\mathbf{J}}_2$ are indeed angular momenta. We check both operators at the same time using the notation $\hat{\mathbf{J}}_\pm$, with $+$ for $\hat{\mathbf{J}}_1$ and $-$ for $\hat{\mathbf{J}}_2$:

$$\begin{aligned}\hat{\mathbf{J}}_\pm \times \hat{\mathbf{J}}_\pm &= \tfrac{1}{4}(\hat{\mathbf{L}} \pm \hbar\nu\hat{\mathbf{R}}) \times (\hat{\mathbf{L}} \pm \hbar\nu\hat{\mathbf{R}})\\ &= \tfrac{1}{4}\big(i\hbar\hat{\mathbf{L}} + i\hbar\hat{\mathbf{L}} \pm (\hat{\mathbf{L}} \times \hbar\nu\hat{\mathbf{R}} + \hbar\nu\hat{\mathbf{R}} \times \hat{\mathbf{L}})\big)\\ &= \tfrac{1}{4}(2i\hbar\hat{\mathbf{L}} \pm 2i\hbar\hbar\nu\hat{\mathbf{R}})\\ &= i\hbar\tfrac{1}{2}(\hat{\mathbf{L}} \pm \hbar\nu\hat{\mathbf{R}}) = i\hbar\hat{\mathbf{J}}_\pm.\end{aligned} \tag{20.8.25}$$

In the first step, we used the first equation in (20.8.21), and in the second step, we used (20.8.12). In summary, we have confirmed that $\hat{\mathbf{J}}_1$ and $\hat{\mathbf{J}}_2$ are indeed two commuting angular momentum operators:

$$\begin{aligned}\hat{\mathbf{J}}_1 \times \hat{\mathbf{J}}_1 &= i\hbar\hat{\mathbf{J}}_1,\\ \hat{\mathbf{J}}_2 \times \hat{\mathbf{J}}_2 &= i\hbar\hat{\mathbf{J}}_2,\\ [\hat{\mathbf{J}}_1, \hat{\mathbf{J}}_2] &= 0.\end{aligned} \tag{20.8.26}$$

The constraint $\hat{\mathbf{R}} \cdot \hat{\mathbf{L}} = 0$ gives us crucial information on the angular momenta. Using (20.8.23) and the commutativity of $\hat{\mathbf{J}}_1$ with $\hat{\mathbf{J}}_2$, we find that

$$(\hat{\mathbf{J}}_1 + \hat{\mathbf{J}}_2) \cdot (\hat{\mathbf{J}}_1 - \hat{\mathbf{J}}_2) = 0 \quad \Rightarrow \quad \hat{\mathbf{J}}_1^2 = \hat{\mathbf{J}}_2^2. \tag{20.8.27}$$

Both angular momenta have the same "magnitude" on the subspace $\mathcal{H}_\nu$ of degenerate energy eigenstates. Let us look at $\hat{\mathbf{J}}_1^2$. Again, using $\hat{\mathbf{R}} \cdot \hat{\mathbf{L}} = 0$ and the second of (20.8.21), we find

$$\hat{\mathbf{J}}_1^2 = \tfrac{1}{4}\left(\hat{\mathbf{L}}^2 + \hbar^2\nu^2\hat{\mathbf{R}}^2\right) = \tfrac{1}{4}\hbar^2(\nu^2 - 1). \tag{20.8.28}$$

Note that the energy parameter ν determines the magnitude of $\hat{\mathbf{J}}_1^2$. This is our "eureka" moment: the quantization of angular momentum is going to imply the quantization of the energy!

Since both $\hat{\mathbf{J}}_1$ and $\hat{\mathbf{J}}_2$ commute with the Hamiltonian, the degenerate subspace $\mathcal{H}_\nu$ must furnish a *simultaneous* representation of both of these angular momenta! All states in the subspace carry the same value of $\hat{\mathbf{J}}_1^2$ and carry the same value of $\hat{\mathbf{J}}_2^2$. So all states are simultaneously in some (irreducible) representation j_1 of $\hat{\mathbf{J}}_1$ and in some (irreducible) representation j_2 of $\hat{\mathbf{J}}_2$. The equality $\hat{\mathbf{J}}_1^2 = \hat{\mathbf{J}}_2^2$, however, implies $j_1 = j_2 \equiv j$. We thus have

$$\hat{\mathbf{J}}_1^2 = \hat{\mathbf{J}}_2^2 = \tfrac{1}{4}\hbar^2(\nu^2 - 1) = \hbar^2 j(j+1). \tag{20.8.29}$$

Since j is an angular momentum, it is quantized: $2j \in \mathbb{Z}$. Solving for ν in terms of j, we now get the quantized energies:

$$\nu^2 = 1 + 4j(j+1) = 4j^2 + 4j + 1 = (2j+1)^2 \quad \Rightarrow \quad \nu = 2j + 1. \tag{20.8.30}$$

Note that as anticipated, the energy is determined by the value of j. This shows that in fact each subspace $\mathcal{H}_\nu$ of degenerate energy eigenstates cannot carry more than one value of j. As j runs over all possible values, ν takes all positive integer values and thus can be indentified with the principal quantum number n:

$$j = 0, \ \tfrac{1}{2}, \ 1, \ \tfrac{3}{2}, \ldots, \qquad (20.8.31)$$
$$n \equiv \nu = 2j+1 = 1, \ 2, \ 3, \ 4, \ldots.$$

We have recovered the quantization of the energy levels in the hydrogen atom!

What is the structure of the degenerate subspace $\mathcal{H}_n$, with $n=2j+1$? We already know that each state must be an eigenstate of $\hat{\mathbf{J}}_1^2$ with eigenvalue $\hbar^2 j(j+1)$ and *at the same time* an eigenstate of $\hat{\mathbf{J}}_2^2$ with the same eigenvalue. Since the two angular momenta are independent, the space can be described as a tensor product of a space that carries the representation j of $\hat{\mathbf{J}}_1$ with a space that carries the representation j of $\hat{\mathbf{J}}_2$. The degenerate subspace must be a $j \otimes j$ vector space. A little elaboration helps make this clear. Consider the set of basis states we are speaking about: the states spanning $\mathcal{H}_{2j+1}$. Since they are all $\hat{\mathbf{J}}_1^2$ eigenstates of eigenvalue $\hbar^2 j(j+1)$, they must form a *collection* of j multiplets. Therefore, the full set of basis states can be organized in the form

$$\big\{ |j, m_1\rangle \otimes |\varphi_1\rangle, \ \ldots, \ |j, m_1\rangle \otimes |\varphi_k\rangle \big\}, \quad -j \leq m_1 \leq j. \qquad (20.8.32)$$

Here k is an integer to be determined, and the $|\varphi_i\rangle$ states, with $i=1,\ldots,k$, are fixed, independent of the value of m_1. In this way, we have k j multiplets of $\hat{\mathbf{J}}_1$, each tensored with a different $|\varphi\rangle$ state. Focus now on the second angular momentum $\hat{\mathbf{J}}_2$. Since all basis states in the space are $\hat{\mathbf{J}}_2^2$ eigenstates and the $\hat{\mathbf{J}}_2$ operators do not act on the $|j, m_1\rangle$ states, each of the $|\varphi\rangle$ states must be a state in a j multiplet of $\hat{\mathbf{J}}_2$. Since a j multiplet has $2j+1$ basis states, the construction of the whole space requires choosing $k=2j+1$ and letting

$$|\varphi_1\rangle = |j, j\rangle, \ \ldots, \ |\varphi_{2j+1}\rangle = |j, -j\rangle. \qquad (20.8.33)$$

In this way, the set of states in (20.8.32) becomes

$$\big\{ |j, m_1\rangle \otimes |j, j\rangle, \ \ldots, \ |j, m_1\rangle \otimes |j, -j\rangle \big\}, \quad -j \leq m_1 \leq j. \qquad (20.8.34)$$

This is in fact the space spanned by

$$\big\{ |j, m_1\rangle \otimes |j, m_2\rangle \big\}, \quad -j \leq m_1, m_2 \leq j, \qquad (20.8.35)$$

immediately recognized as $j \otimes j$. Note that this is the "minimal" solution for the space $\mathcal{H}_{2j+1}$. A direct sum of a number of $j \otimes j$ spaces would also be consistent with the algebraic constraints.

Therefore, assuming minimality, $\mathcal{H}_{n=2j+1}$ is the space $j \otimes j$:

$$\mathcal{H}_{n=2j+1} = j \otimes j, \quad \text{basis states} \quad |j; m_1\rangle \otimes |j; m_2\rangle, \quad -j \leq m_1, m_2 \leq j. \qquad (20.8.36)$$

Since m_1 and m_2 each take $2j+1$ values, the dimension of $\mathcal{H}_{n=2j+1}$ is $(2j+1)^2 = n^2$. This is indeed the expected number of states that we have at this energy level. As we are familiar with, the tensor product breaks into a sum of representations of the *sum* of angular momenta. But the sum here is simply the conventional angular momentum $\hat{\mathbf{J}}_1 + \hat{\mathbf{J}}_2 = \hat{\mathbf{L}}$. Since we know that

$$j \otimes j = 2j \oplus 2j - 1 \oplus \cdots \oplus 0, \qquad (20.8.37)$$

the representations on the right-hand side are the ℓ multiplets that arise. Thus, the degenerate subspace is a direct sum of the ℓ values $(\ell=2j)\oplus(\ell=2j-1)\oplus\cdots\oplus 0$. Recalling that $2j+1=n$, we have obtained

$$\mathcal{H}_n = (\ell=n-1)\oplus(\ell=n-2)\oplus\cdots\oplus(\ell=0). \tag{20.8.38}$$

This is exactly the familiar set of ℓ multiplets at the degenerate subspace labeled by the principal quantum number n. This completes the algebraic derivation of the spectrum.

We should emphasize that the above analysis characterizes the *possible* subspaces $\mathcal{H}_n$ of degenerate energy eigenstates. These subspaces are labeled by the values of j in the infinite list $\{0, \frac{1}{2}, 1, \frac{3}{2}, 2, \frac{5}{2}, \ldots\}$. The algebraic analysis alone *cannot* tell us which values of j in the this list are used by the hydrogen atom. In physics, however, it is often the case that whatever is possible is in fact compulsory. So it is not surprising that all possible values of j actually appear in the hydrogen atom spectrum.

As the simplest nontrivial example of a degenerate subspace, consider $j=\frac{1}{2}$, which gives us $n=2$. We then have $\mathcal{H}_2=\frac{1}{2}\otimes\frac{1}{2}=\mathbf{1}\oplus\mathbf{0}$, where the right-hand side includes the triplet $\ell=1$ and the singlet $\ell=0$. The uncoupled basis states are of the form $|\frac{1}{2},m_1\rangle\otimes|\frac{1}{2},m_2\rangle$, and the four of them can be written briefly as $|\uparrow\uparrow\rangle, |\uparrow\downarrow\rangle, |\downarrow\uparrow\rangle, |\downarrow\downarrow\rangle$. The singlet and triplet are therefore

$$\ell=1:\begin{cases}|1,1\rangle = |\uparrow\uparrow\rangle,\\ |1,0\rangle = \frac{1}{\sqrt{2}}\big(|\uparrow\downarrow\rangle+|\downarrow\uparrow\rangle\big),\\ |1,-1\rangle = |\downarrow\downarrow\rangle,\end{cases}\qquad \ell=0:\quad |0,0\rangle=\frac{1}{\sqrt{2}}\big(|\uparrow\downarrow\rangle-|\downarrow\uparrow\rangle\big). \tag{20.8.39}$$

These are the four states of the $n=2$ energy level. Note that we have built them out of spin one-half states. These are mathematical entities not related in any way to elementary particle spin. The hydrogen atom Hamiltonian we have used assumes the electron and the proton are spinless. This is reminiscent of our discussion of the spectrum of the isotropic two-dimensional oscillator, at the end of section 19.7. There also, the energy levels of the oscillator formed representations of a "hidden" angular momentum and included representations with fractional angular momentum.

Problems

Problem 20.1. *Rotating a tensor product state.*

Consider the following entangled state of two spin one-half particles:

$$|\Psi\rangle=\frac{1}{\sqrt{2}}\big(|\uparrow\downarrow\rangle+|\downarrow\uparrow\rangle\big) = |1,0\rangle.$$

This is the $s=1, m=0$ state of the total spin angular momentum $\hat{\mathbf{S}}$. Calculate the state obtained by rotating $|\Psi\rangle$ by an angle γ around the x-axis. Express your answer in terms of eigenstates $|s,m\rangle$ of the total spin operators $\hat{\mathbf{S}}^2, \hat{S}_z$.

Problem 20.2. *Traces and the splittings of multiplets.*

Consider the space $j_1 \otimes j_2$, where j_1 is a multiplet of an angular momentum $\hat{\mathbf{J}}_1$, and j_2 is a multiplet of angular momentum $\hat{\mathbf{J}}_2$. The two angular momenta commute. Calculate the trace of $\hat{\mathbf{J}}_1 \cdot \hat{\mathbf{J}}_2$ on the space $j_1 \otimes j_2$. Your answer should explain a pattern in the splittings of multiplets for the hyperfine interaction and for spin-orbit coupling.

Problem 20.3. *Hyperfine splitting of the hydrogen ground state in a magnetic field.*

For a magnetic field of magnitude B along the z-direction, the hydrogen atom Hamiltonian relevant to the hyperfine splitting of the ground state has an additional term proportional to the electron spin in the z-direction:

$$\hat{H}' = \epsilon \tfrac{2}{\hbar} \hat{S}_{e,z} + \epsilon' \tfrac{4}{\hbar^2} \hat{\mathbf{S}}_e \cdot \hat{\mathbf{S}}_p.$$

Here, $\hat{\mathbf{S}}_e$ and $\hat{\mathbf{S}}_p$ are the electron and proton spins, respectively. Moreover, ϵ and ϵ' are positive constants with units of energy. In particular, $\epsilon = \mu_B B$.

There are two natural bases in this problem. In the uncoupled basis $|m_e m_p\rangle$, the first entry refers to the electron, and the second entry refers to the proton:

Uncoupled basis: $|1\rangle = |\uparrow\uparrow\rangle$, $|2\rangle = |\uparrow\downarrow\rangle$, $|3\rangle = |\downarrow\uparrow\rangle$, $|4\rangle = |\downarrow\downarrow\rangle$.

There is the coupled basis $|jm\rangle$ of eigenstates of $\hat{\mathbf{J}}^2$ and $\hat{J}_z$, where $\hat{\mathbf{J}} = \hat{\mathbf{S}}_e + \hat{\mathbf{S}}_p$:

Coupled basis: $|1\rangle = |1,1\rangle$, $|2\rangle = |1,0\rangle$, $|3\rangle = |1,-1\rangle$, $|4\rangle = |0,0\rangle$.

1. Find the matrix elements of $\hat{H}'$ in the uncoupled basis. Calculate the energy eigenvalues and the eigenvectors.
2. Find the matrix elements of $\hat{H}'$ in the coupled basis. Calculate the energy eigenvalues and the eigenvectors.
3. Sketch the energy eigenvalues as a function of the magnetic field. Which basis is more suitable for small magnetic fields, and which is more suitable for large magnetic fields?
4. Find the energy eigenvalues and eigenstates correct to first order in the magnetic field B, when this magnetic field is small (the eigenstates need not be normalized).

Problem 20.4. *Particle of spin one-half and $\ell = 1$ in a magnetic field.*

Consider a spin one-half particle in a state with orbital angular momentum $\ell = 1$. Label the eigenstates in the uncoupled basis by $|\ell s m_\ell m_s\rangle$, the eigenvalues of $\hat{\mathbf{L}}^2, \hat{\mathbf{S}}^2, \hat{L}_z$, and $\hat{S}_z$, respectively. Label the states in the coupled basis by $|j, m\rangle$, the eigenvalues of $\hat{\mathbf{J}}^2$ and $\hat{J}_z$, with $\hat{\mathbf{J}}$ the total angular momentum. These states, of course, are also eigenstates of $\hat{\mathbf{L}}^2$ and $\hat{\mathbf{S}}^2$.

1. What is the expectation value of $\hat{L}_z$ in the state with $j = 1/2$, $m = 1/2$? What is the expectation value of $\hat{S}_z$ in this state?
2. Suppose this particle moves in an external magnetic field in the z-direction, $\mathbf{B} = B\,\mathbf{z}$. Assume the particle is an electron, and take $g = 2$. The Hamiltonian describing the

interaction of the electron with the field is

$$\hat{H}_B = \frac{\mu_B}{\hbar}\mathbf{B}\cdot(\hat{\mathbf{L}}+2\hat{\mathbf{S}}).$$

What is $\langle \hat{H}_B \rangle$ in each of the eigenstates $|j, m\rangle$?

3. For the eigenstate $j=1/2$, $m=1/2$, what are the possible values of the magnetic energy, and what are their probabilities?

Problem 20.5. *Addition of angular momentum when tensoring two equal-size multiplets.*

Consider the addition of angular momentum for two particles each of angular momentum j. We write $\hat{\mathbf{J}}=\hat{\mathbf{J}}_1+\hat{\mathbf{J}}_2$ for the total angular momentum in terms of the angular momentum of the first and second particles. As you know, the result is

$$j\otimes j = (2j)\oplus(2j-1)\oplus\ldots\oplus 0.$$

1. Construct the (normalized) states of highest and second-highest J_z for total angular momentum $2j$.
2. Construct the (normalized) states of highest and second-highest J_z for total angular momentum $2j-1$.
3. Consider the states in (1). Are they symmetric, antisymmetric, or neither under the exchange of the two particles? Answer the same question for the states in (2).
4. Do you expect all states in the $2j$ multiplet and all states in the $2j-1$ multiplet to have the same exchange property? Explain.

Problem 20.6. *Coupled basis for* $\mathbf{1}\otimes\mathbf{1}$.

The tensor product $\mathbf{1}\otimes\mathbf{1}$ of two angular momentum multiplets is spanned by uncoupled basis vectors $|1, m_1\rangle\otimes|1, m_2\rangle$, with $m_1, m_2\in\{1, 0, -1\}$. The tensor product breaks into the sum $\mathbf{2}\oplus\mathbf{1}\oplus\mathbf{0}$ of multiplets of total angular momentum. Construct the basis states for those multiplets as a linear superposition of the uncoupled basis vectors.

Problem 20.7. *General addition of* $\hat{\mathbf{L}}$ *and* $\hat{\mathbf{S}}$.

Consider two angular momenta $\hat{\mathbf{L}}$ and $\hat{\mathbf{S}}$ and the states $\ell\otimes\frac{1}{2}$. We define $\hat{\mathbf{J}}=\hat{\mathbf{L}}+\hat{\mathbf{S}}$. Derive a formula for the "coupled" basis states in the $j=\ell+\frac{1}{2}$ multiplet:

$$|j=\ell+\tfrac{1}{2}, m=M+\tfrac{1}{2}\rangle, \qquad -\ell-1\leq M\leq\ell,$$

in terms of suitable superpositions of uncoupled states $|\ell, \frac{1}{2}; m_\ell, m_s\rangle$. Your general formula should reduce to familiar results when $M=\ell$ and $M=-\ell-1$. Since M is arbitrary, the strategy of using lowering or raising operators is not suitable. [Hint: use $\hat{\mathbf{J}}^2$.] Similarly, calculate the states $|j=\ell-\frac{1}{2}, m=M-\frac{1}{2}\rangle$ of the $j=\ell-\frac{1}{2}$ multiplet in the tensor product $(-\ell+1\leq M\leq\ell)$.

Problem 20.8. *Adding three angular momenta.*

Consider three commuting angular momenta $\hat{\mathbf{J}}_1, \hat{\mathbf{J}}_2$, and $\hat{\mathbf{J}}_3$ and the tensor product

$$j_1\otimes j_2\otimes j_3, \quad \text{with} \quad j_1\geq j_2\geq j_3, \qquad j_i\in\mathbb{Z}.$$

As noted above, we restrict ourselves to integer angular momenta. Consider now the decomposition of the above tensor product into multiplets of the total angular momentum $\hat{\mathbf{J}} = \hat{\mathbf{J}}_1 + \hat{\mathbf{J}}_2 + \hat{\mathbf{J}}_3$.

1. What is the highest-value $j_{\max}$ of j in the tensor product?
2. What is the lowest-value $j_{\min}$ of j in the tensor product ? The answer can be put in the form

$$j_{\min} = \begin{cases} \cdots\cdots, & \text{if positive,} \\ 0, & \text{otherwise.} \end{cases}$$

   Determine the function of j_1, j_2, and j_3 that goes where the dots are in the formula above.

3. Assume that $j_1 = j_2 = j_3 = j$, with $j \geq 2$ and $j \in \mathbb{Z}$. We still consider the tensor product $j \otimes j \otimes j$. The following questions refer to representations of the *total* angular momentum contained in this product.
   a. How many times does the singlet $\mathbf{0}$ appear?
   b. How many times does the triplet $\mathbf{1}$ appear?
   c. How many times does the j multiplet appear?

Problem 20.9. *Hamiltonian for three spin-one particles.*

Consider three distinguishable spin-one particles with spin operators $\hat{\mathbf{S}}_1, \hat{\mathbf{S}}_2$, and $\hat{\mathbf{S}}_3$. The spins are placed along a circle, and the interactions are between nearest neighbors. The Hamiltonian takes the form

$$\hat{H} = \frac{\Delta}{\hbar^2}\left(\hat{\mathbf{S}}_1 \cdot \hat{\mathbf{S}}_2 + \hat{\mathbf{S}}_2 \cdot \hat{\mathbf{S}}_3 + \hat{\mathbf{S}}_3 \cdot \hat{\mathbf{S}}_1\right),$$

with $\Delta > 0$ a constant with units of energy. For this problem it is useful to consider the total spin operator $\hat{\mathbf{S}} = \hat{\mathbf{S}}_1 + \hat{\mathbf{S}}_2 + \hat{\mathbf{S}}_3$.

1. What is the dimensionality of the state space of the three combined particles? Determine the energy eigenvalues for $\hat{H}$ and the degeneracies of these eigenvalues.
2. Express the normalized ground state as a superposition of states of the form $|m_1, m_2, m_3\rangle \equiv |1, m_1\rangle \otimes |1, m_2\rangle \otimes |1, m_3\rangle$, where $\hbar m_i$ is the eigenvalue of $\hat{S}_{i,z}$.

Problem 20.10. *Hamiltonian for three spin one-half particles.*

Consider three distinguishable spin one-half particles with spin operators $\hat{\mathbf{S}}_1, \hat{\mathbf{S}}_2$, and $\hat{\mathbf{S}}_3$. The spins are placed along a circle, and the interactions are between nearest neighbors. The Hamiltonian takes the form

$$\hat{H} = \frac{\Delta}{\hbar^2}\left(\hat{\mathbf{S}}_1 \cdot \hat{\mathbf{S}}_2 + \hat{\mathbf{S}}_2 \cdot \hat{\mathbf{S}}_3 + \hat{\mathbf{S}}_3 \cdot \hat{\mathbf{S}}_1\right),$$

with $\Delta > 0$ a constant with units of energy. For this problem it is useful to consider the total spin operator $\hat{\mathbf{S}} \equiv \hat{\mathbf{S}}_1 + \hat{\mathbf{S}}_2 + \hat{\mathbf{S}}_3$.

1. What is the dimensionality of the state space of the three combined particles? Determine the energy eigenvalues for $\hat{H}$ and the degeneracies of these eigenvalues.
2. Give the $\hat{H}$ eigenstate with maximal $\hat{S}_z$. Construct the full $\hat{\mathbf{S}}$ multiplet this state belongs to.
3. Find explicit forms for the ground state multiplets. Your answers will not be unique, showing that $\{\hat{\mathbf{S}}^2, \hat{S}_z\}$ is not a CSCO for this system.
4. Let $\hat{\mathbf{S}}_{ij} \equiv \hat{\mathbf{S}}_i + \hat{\mathbf{S}}_j$. Show that the following are a set of commuting observables:

$$\{\hat{\mathbf{S}}_1^2, \hat{\mathbf{S}}_2^2, \hat{\mathbf{S}}_3^2, \hat{\mathbf{S}}_{12}^2, \hat{\mathbf{S}}^2, \hat{S}_z\}.$$

Prove this is a CSCO by finding an explicit construction of the ground state multiplets with all states specified uniquely by the eigenvalues of the observables.
5. List an alternative CSCO, and determine the associated ground state multiplets. How are the multiplets in the two CSCOs related?
6. Reconsider the excited state multiplet. Are the states in this multiplet $\hat{\mathbf{S}}_{12}^2$ eigenstates? Are they $\hat{\mathbf{S}}_{23}^2$ eigenstates? If yes, what are the eigenvalues? If not, explain why not. Does $\hat{\mathbf{S}}_{12}^2$ commute with $\hat{\mathbf{S}}_{23}^2$?

Problem 20.11. *Isospin and nuclear reactions.*

Heisenberg suggested that the nucleon is an isospin one-half particle, $I = \frac{1}{2}$. The "up" state and the "down" state, with $I_z = \pm\frac{1}{2}$, respectively, are identified as the proton p and the neutron n:

$$|p\rangle = |\tfrac{1}{2}, \tfrac{1}{2}\rangle, \quad |n\rangle = |\tfrac{1}{2}, -\tfrac{1}{2}\rangle.$$

The proton has electric charge $+1$, and the neutron is neutral. The pions π^+, π^0, and π^- are an isospin triplet $I = 1$:

$$|\pi^+\rangle = |1, 1\rangle, \quad |\pi^0\rangle = |1, 0\rangle, \quad |\pi^-\rangle = |1, -1\rangle.$$

The superscript on π denotes the electric charge of the particle. The "Delta" particle is in fact a collection of four particles $(\Delta^{++}, \Delta^+, \Delta^0, \Delta^-)$ that form a multiplet of isospin $I = \frac{3}{2}$:

$$|\Delta^{++}\rangle = |\tfrac{3}{2}, \tfrac{3}{2}\rangle, \quad |\Delta^+\rangle = |\tfrac{3}{2}, \tfrac{1}{2}\rangle \quad |\Delta^0\rangle = |\tfrac{3}{2}, -\tfrac{1}{2}\rangle, \quad |\Delta^-\rangle = |\tfrac{3}{2}, -\tfrac{3}{2}\rangle.$$

The superscripts on Δ denote the electric charge. We will now consider strong interactions of these particles that conserve electric charge and, to a very good approximation, conserve isospin. Isospin conservation can be used to deduce a number of facts, as you will do below.

1. Electric charge conservation allows the Δ^+ to decay in four ways:

1. $\Delta^+ \to pn$. 2. $\Delta^+ \to \pi^0 p$. 3. $\Delta^+ \to \pi^+ n$. 4. $\Delta^+ \to \pi^+ \pi^0$.

Imagine we hold a very large number of Δ^+ particles. Call f_i the fraction of particles using the ith decay. Determine f_i for $i = 1, 2, 3, 4$.

2. The deuteron d is a combination of two nucleons that is an isospin *singlet*. The collision of high-energy deuterons is a reaction R_i of the form

$$R_i: \; d+d \to d+d+X_i.$$

The deuterons scatter off each other but create pions, denoted by the extra term X_i in the reaction R_i. The index i takes two values, for the two possible sets of pions observed: $X_1 = \pi^0\pi^0$, and $X_2 = \pi^+\pi^-$. Call P_i the probability of reaction R_i. Determine P_1/P_2.

Problem 20.12. *Conservation of angular momentum in particle decays (J. Thaler).*

An unstable particle X decays to two daughter particles A and B. In the decay process $X \to AB$, total angular momentum is conserved. In the X rest frame, the total angular momentum $\hat{\mathbf{J}}$ is given just by the spin $\hat{\mathbf{S}}_X$ of particle X. After the decay, the total angular momentum receives three contributions:

$$\hat{\mathbf{J}} = \hat{\mathbf{S}}_A + \hat{\mathbf{S}}_B + \hat{\mathbf{L}},$$

where $\hat{\mathbf{S}}_A$ is the spin of particle A, $\hat{\mathbf{S}}_B$ is the spin of particle B, and $\hat{\mathbf{L}}$ is the relative orbital angular momentum between A and B. Because angular momentum is conserved in this decay, if the initial state is an eigenstate of $\hat{\mathbf{J}}^2$ and $\hat{\mathbf{J}}_z$, then the final state is also an eigenstate with the same eigenvalues.

1. Assume X is spin zero and both A and B are spin one-half (i.e., $s_X = 0$, $s_A = 1/2$, $s_B = 1/2$). What values of the orbital angular momentum ℓ are allowed, consistent with angular momentum conservation?
2. Assume $s_X = 3/2$, $s_A = 1/2$, and $s_B = 1$. Again, what values of the orbital angular momentum ℓ are allowed, consistent with angular momentum conservation?
3. There are certain processes for which a two-body decay is forbidden. Explain why a neutron n *cannot* decay to a proton p and an electron e^- via the decay $n \to pe^-$, even though this decay is consistent with energy and charge conservation. Recall that both the proton and the neutron are spin one-half particles. [Note: the three-body decay $n \to pe^-\bar{\nu}$ is allowed, where $\bar{\nu}$ is a spin one-half antineutrino.]
4. A mystery particle X of unknown spin s_X is polarized such that $m_X = +s_X$. It decays via $X \to AB$, where $s_A = 1/2$, $s_B = 0$, but the relative orbital angular momentum ℓ is also unknown (assume the decay yields a single value of ℓ). After the decay the z-component m_A of the spin of particle A is measured, and the possible results are found with probabilities:

$$P(m_A = \tfrac{1}{2}) = \tfrac{1}{5}, \qquad P(m_A = -\tfrac{1}{2}) = \tfrac{4}{5}.$$

What is s_X and what is ℓ? [Hint: you may want to consider the action of $\hat{\mathbf{J}}_+$ on both the initial and final states.]

Problem 20.13. *Commutator of Runge-Lenz operators (M. Weitzman).*

Proving equation (20.8.17) is equivalent to showing that $[\hat{R}_i, \hat{R}_j] = i\hbar\epsilon_{ijk}\left(-\frac{2\hat{H}}{me^4}\right)\hat{L}_k$. To confirm this relation, due to cyclicity of the basic commutators in the theory, it suffices to show that

$$[\hat{R}_x, \hat{R}_y] = i\hbar\Big(-\frac{2\hat{H}}{me^4}\Big)\,\hat{L}_z. \tag{1}$$

Using the expression for $\hat{\mathbf{R}}$ in the first rewriting of equation (20.8.5), we see that

$$[\hat{R}_x, \hat{R}_y] = \frac{1}{m^2e^4}\Big[\,\hat{p}_y\hat{L}_z - \hat{p}_z\hat{L}_y - i\hbar\hat{p}_x - \frac{\hat{x}}{r}\;\hat{p}_z\hat{L}_x - \hat{p}_x\hat{L}_z - i\hbar\hat{p}_y - \frac{\hat{y}}{r}\,\Big]. \tag{2}$$

1. Imagine expanding the above commutator. Ignoring the commutators that involve $\hat{x}/r$ and $\hat{y}/r$, we get nine commutators. Show that the contributions from the five commutators that involve $i\hbar\hat{p}_x$, or $i\hbar\hat{p}_y$, or both add up to zero.
2. Show that the contributions from the remaining four terms (out of the nine) give $i\hbar(\frac{-2}{me^4})\frac{\hat{\mathbf{p}}^2}{2m}\hat{L}_z$.
3. Evaluate the remaining commutators in equation (2) that involve $\hat{x}/r$ and $\hat{y}/r$. You may find the following identities useful (prove them!):

   $$\Big[\,\hat{p}_i, \frac{\hat{x}_j}{r}\,\Big] = -i\hbar\,\frac{\delta_{ij}}{r} + i\hbar\,\frac{\hat{x}_i\hat{x}_j}{r_3}, \quad \hat{x}\hat{L}_x + \hat{y}\hat{L}_y = -\hat{z}\hat{L}_z,$$

   the second of which follows from $\hat{\mathbf{r}}\cdot\hat{\mathbf{L}} = 0$. Show that together with the contributions computed in (2) you can now confirm the commutator in equation (1).

21 Identical Particles

The description of identical particles in quantum mechanics faces an exchange degeneracy problem: the existence of a number of different candidate states for the description of any state of $N \geq 2$ identical particles. To deal with this problem, we discuss permutation groups and projectors acting on the space $V^{\otimes N}$ relevant to N identical particles, each of which lives in V. The symmetrization postulate gives a satisfactory resolution of the exchange degeneracy conundrum, stating that physical states are either totally symmetric or totally antisymmetric under permutations of the particles. Particles described by symmetric states are called bosons, and particles described by antisymmetric states are called fermions. An efficient description of the states of bosons and fermions uses the language of occupation numbers. We study how the constraints of symmetrization affect the probabilities of finding bosons or fermions in particular states, as compared to those for distinguishable particles. We discuss why exotic statistics are possible for systems in two spatial dimensions.

Two particles are said to be *identical* if no experiment can distinguish them; identical particles are indistinguishable. All electrons, for example, are identical. Electrons are elementary particles, but composite particles can also be identical: the proton is built from quarks and gluons, yet all protons are identical. All hydrogen atoms are identical as well, so we can treat them as identical particles. In general, identical particles share intrinsic properties such as mass, charge, spin, and magnetic moments, for example.

Suppose Alice and Blake each gives Charles an electron. Charles plays with the two electrons for a while, returning one to Alice and one to Blake. There is no way Alice or Blake can tell if they got the same electron they gave Charles. In fact, depending on what he did with the electrons, Charles himself may not know which electron he gave to Alice and which he gave to Blake.

Of course, two identical particles can be in different states, and then we can usually tell them apart. One electron can have spin up, and another can have spin down. Or one can be at rest and the other moving. Any particle has a state space relevant to its description. Identical particles each use a copy of the same state space. Part of our challenge will be to construct the N-particle state space, needed when $N \geq 2$ identical particles interact.

Sometimes the concept of identical particles is only approximate. Nuclear physicists often declare the neutron and the proton to be identical nucleons that are just in different states of an *isospin* quantum number. The small mass difference between these particles (about one-tenth of 1%) helps. It also helps that the nuclear forces are much stronger than the electromagnetic forces, and therefore the electric charge difference between a proton and a neutron leads to minor corrections.

21.1 Identical Particles and Exchange Degeneracy

The notion of identical particles as indistinguishable is also applicable to classical mechanics. Just as in quantum mechanics, in classical mechanics there is a list of intrinsic properties associated with classically identical particles. But there is a major difference. When identical particles come together and interact in classical mechanics, we can keep track of their trajectories and therefore speak unambiguously of the time evolution of *each* particle. We can assign a label to each particle, say, particle one and particle two, and follow them until the end of the experiment so we can tell which is particle one and which is particle two. The results of the experiment are, of course, label independent: they do not depend on which particle was called one and which was called two.

If we have two identical particles, the classical Hamiltonian that describes their time evolution must make this clear: $H(\mathbf{x}_1, \mathbf{p}_1; \mathbf{x}_2, \mathbf{p}_2) = H(\mathbf{x}_2, \mathbf{p}_2; \mathbf{x}_1, \mathbf{p}_1)$. The exchange of $(\mathbf{x}_1, \mathbf{p}_1)$ and $(\mathbf{x}_2, \mathbf{p}_2)$ must leave the Hamiltonian invariant. This will implement the label independence of the physics. If we are given two particles, we label them as if the particles were different and follow them throughout. Had we chosen different labeling, the results would have been equivalent.

In quantum mechanics it is not possible to track the position of a particle. This is clear from the uncertainty principle: if at some time we know the position of a particle exactly, its momentum is completely uncertain, and then we have no idea where the particle will be in the next moment. In particular, we cannot follow identical particles once they overlap and interact.

This is illustrated in the two-photon experiment of Hong, Ou, and Mandel (1987). The experiment involves a single balanced beam splitter of the type discussed in chapter 2. The setup is sketched in figure 21.1. Two identical photons are simultaneously incident on the top and bottom ports of a beam splitter. There are also two detectors, D1 and D2, ready to capture the outgoing photons. If the two photons do not hit simultaneously, we can tell them apart, and given that each has an equal chance of being reflected or transmitted at the beam splitter, we will end 25% of the time with both photons at D1, 25% of the time with both photons at D2, and 50% of the time with a photon in each detector. The same will happen if the photons are polarized differently, and again, we can tell them apart, even if they arrive at the beam splitter

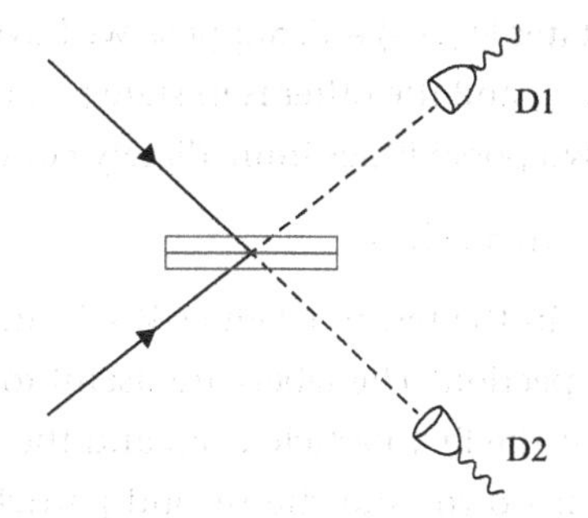

Figure 21.1
Two identical photons hit a balanced beam splitter, one from the top and one from the bottom. An interference hinging on the identical nature of the particles implies that the detectors D1 and D2 never click simultaneously; the two photons both end on D1 or on D2.

simultaneously. The interesting case occurs when photons with the same polarization state arrive at the beam splitter simultaneously. We can imagine two different ways in which each photon ends on a different detector: by both being reflected or both being transmitted at the beam splitter. In fact, these processes, leading to the *same* end result, interfere destructively with each other! We get two photons in the top detector with probability one half, and we get two photons in the bottom detector with probability one half. One never finds simultaneous clicks of the two detectors! Once we develop the framework to deal with identical particles, we will give the computational steps that lead to the above conclusions (example 21.2, section 21.4).

Exchange degeneracy Let us now examine the complication we face when we try to describe states of identical particles in quantum mechanics. As we will see, we encounter an exchange degeneracy problem.

We have learned how states of multiple *distinguishable* particles in quantum mechanics are described using tensor products to combine the various state spaces. Consider, for example, the case of two distinguishable particles, one with state space V and the other with state space W. The state

$$|v\rangle_{(1)} \otimes |w\rangle_{(2)} \in V \otimes W \tag{21.1.1}$$

describes particle one in state $v \in V$ and particle two in state $w \in W$. This is not a general state in the tensor product but suffices for our purposes here. In the above expression, the subscript on a ket denotes the particle label. Labels are conventionally ordered in increasing value as we move to the right. Since the particles are distinguishable, it does not matter which we call particle one and which we call particle two. We can work with $V \otimes W$ or with $W \otimes V$.

Let us consider now the case of two *identical* spin one-half particles. We focus only on the spin degrees of freedom; the spatial degrees of freedom are assumed fixed. There is now a single state space V, the state space of a spin one-half particle. Therefore, we will use the space $V \otimes V$ to describe the states of the two particles. Each particle can be in a superposition of two basis states $|+\rangle, |-\rangle \in V$. Suppose we have prepared the particles so that one of them is in state $|+\rangle$, and the other is in state $|-\rangle$. How are we to describe the state of the two particles? Two possibilities immediately come to mind:

$$|+\rangle_{(1)} \otimes |-\rangle_{(2)} \qquad \text{or} \qquad |-\rangle_{(1)} \otimes |+\rangle_{(2)}. \tag{21.1.2}$$

These are two *different* states in the tensor product $V \otimes V$; in fact, they are orthogonal states, as you can see by inspection! The labels are useful for purposes of description: for the first state, we say that the first particle is up and the second down; for the second state, the first particle is down, and the second particle is up. Which one shall we pick?

One possibility, inspired by classical mechanics, is to pick one state and stick to it. The state you pick should not matter; they are equivalent choices. Perhaps surprisingly, we can show that this is not a viable strategy. It leads to physical ambiguities!

The problem arises as follows. If the two states above are equivalent choices, the principles of quantum mechanics tell us that any superposition of them is also an equivalent choice. We would have to admit that even the states

$$|\psi\rangle = \alpha|+\rangle_{(1)} \otimes |-\rangle_{(2)} + \beta|-\rangle_{(1)} \otimes |+\rangle_{(2)}, \quad |\alpha|^2 + |\beta|^2 = 1 \tag{21.1.3}$$

are equivalent for all possible values of α and β. This ambiguity in the specification of a state of identical particles is called *exchange degeneracy*. It begins with the discrete degeneracy of the basis states in (21.1.2) and leads to the continuous degeneracy of the superposition states above.

To show that there is trouble, we demonstrate that the answer to a physical question can be ambiguous. We ask for the probability p_0 of finding these particles in the state $|\psi_0\rangle$ defined as follows:

$$|\psi_0\rangle = |x; +\rangle_{(1)} \otimes |x; +\rangle_{(2)}. \tag{21.1.4}$$

This state is cleverly chosen to avoid the exchange degeneracy, as both particles are in the same state, with their spins aligned along the positive x-axis. Written in terms of the usual z-based states, we find that

$$\begin{aligned}|\psi_0\rangle &= \tfrac{1}{\sqrt{2}}\big(|+\rangle_{(1)} + |-\rangle_{(1)}\big) \otimes \tfrac{1}{\sqrt{2}}\big(|+\rangle_{(2)} + |-\rangle_{(2)}\big) \\ &= \tfrac{1}{2}\big(|+\rangle_{(1)} \otimes |+\rangle_{(2)} + |+\rangle_{(1)} \otimes |-\rangle_{(2)} + |-\rangle_{(1)} \otimes |+\rangle_{(2)} + |-\rangle_{(1)} \otimes |-\rangle_{(2)}\big).\end{aligned} \tag{21.1.5}$$

The probability p_0 that $|\psi\rangle$ is found in $|\psi_0\rangle$ is determined by the overlap:

$$p_0 = |\langle\psi_0|\psi\rangle|^2 = \left|\tfrac{1}{2}(\alpha + \beta)\right|^2. \tag{21.1.6}$$

There is an ambiguity because if we choose to represent ψ using

$$\alpha = \beta = \tfrac{1}{\sqrt{2}}, \quad \text{then } p_0 = \tfrac{1}{2}, \tag{21.1.7}$$

while if we choose to represent ψ using

$$\alpha = -\beta = \tfrac{1}{\sqrt{2}}, \quad \text{then } p_0 = 0. \tag{21.1.8}$$

This is an unacceptable physical ambiguity following from exchange degeneracy. The chosen values of α and β matter when they should not.

The exchange degeneracy is larger if we have three identical particles. If $|a\rangle, |b\rangle, |c\rangle$ are three possible states of each particle, a state of the three particles with one in $|a\rangle$, one in $|b\rangle$, and one in $|c\rangle$ could be described by any of the six states in the following list:

$$\begin{aligned}
&|a\rangle_{(1)}\otimes|b\rangle_{(2)}\otimes|c\rangle_{(3)}, \qquad &&|a\rangle_{(1)}\otimes|c\rangle_{(2)}\otimes|b\rangle_{(3)},\\
&|b\rangle_{(1)}\otimes|c\rangle_{(2)}\otimes|a\rangle_{(3)}, \qquad &&|b\rangle_{(1)}\otimes|a\rangle_{(2)}\otimes|c\rangle_{(3)},\\
&|c\rangle_{(1)}\otimes|a\rangle_{(2)}\otimes|b\rangle_{(3)}, \qquad &&|c\rangle_{(1)}\otimes|b\rangle_{(2)}\otimes|a\rangle_{(3)}.
\end{aligned}$$

In fact, any linear combination would also do. The problem of the exchange degeneracy will be solved by the symmetrization postulate, which we will state and discuss in section 21.4.

21.2 Permutation Operators

We now build the tools needed to understand explicitly the resolution of the exchange degeneracy problem. We will consider permutation operators and the permutation groups they form. These are relevant when we have identical particles that can be exchanged without physical consequence. These operators act on the tensor-product space:

$$V^{\otimes N} \equiv \underbrace{V\otimes\cdots\otimes V}_{N}, \tag{21.2.1}$$

which is the candidate state space for the description of N identical particles. We will also consider projectors to some important subspaces of $V^{\otimes N}$. We will begin with the case $N=2$ and then turn to larger N, with some special note of the case $N=3$.

In this section we do not yet implement the constraints due to identical particles. The only constraint we are imposing is that each of the particles in the N-particle collection lives in a state space V, thus $V^{\otimes N}$ is the state space relevant to the collection. For the moment it is best to think of the particles as distinguishable.

Two-particle systems Assume that we have two particles, particle one and particle two, each described by the same state space V spanned by orthonormal basis states $|u_i\rangle$, with $i=1,2,\ldots$. Consider the two-particle state in which particle one is in state $|u_i\rangle$, and particle two is in state $|u_j\rangle$, with $u_i\neq u_j$:

$$|u_i\rangle_{(1)}\otimes|u_j\rangle_{(2)} \in V\otimes V. \tag{21.2.2}$$

Note that $|u_j\rangle_{(1)}\otimes|u_i\rangle_{(2)}$ is a different state in $V\otimes V$, the state in which particle one is in $|u_j\rangle$, and particle two is in state $|u_i\rangle$. The states of the particles have been exchanged. Let us define a linear operator $\hat{P}_{21}$ acting on $V\otimes V$ that in fact produces this exchange by permuting the i and j labels:

$$\hat{P}_{21}\Big[|u_i\rangle_{(1)}\otimes|u_j\rangle_{(2)}\Big] \equiv |u_j\rangle_{(1)}\otimes|u_i\rangle_{(2)}, \tag{21.2.3}$$

for all u_i, u_j. This describes the action of $\hat{P}_{21} \in \mathcal{L}(V \otimes V)$ on all the basis vectors of $V \otimes V$ and thus defines the operator completely. We call $\hat{P}_{21}$ a **transposition** operator. It transposes the states of particle one and particle two. Note that the iterated action of $\hat{P}_{21}$ gives the identity operator:

$$\hat{P}_{21} \hat{P}_{21} = \mathbb{1}. \tag{21.2.4}$$

This means that $\hat{P}_{21}$ is its own inverse. We now claim that the operator $\hat{P}_{21}$ is in fact Hermitian:

$$\hat{P}_{21}^{\dagger} = \hat{P}_{21}. \tag{21.2.5}$$

Proof. Recall that an operator $\hat{M}$ is Hermitian if $\langle \hat{M}\alpha, \beta \rangle = \langle \alpha, \hat{M}\beta \rangle$ for all states α, β. Writing the states more briefly, $u_i \otimes u_j \equiv |u_i\rangle_{(1)} \otimes |u_j\rangle_{(2)}$, we have the inner product

$$\big\langle u_k \otimes u_l, u_i \otimes u_j \big\rangle = \delta_{ki}\delta_{lj}. \tag{21.2.6}$$

We can then easily check that

$$\big\langle \hat{P}_{21}\, u_k \otimes u_l, u_i \otimes u_j \big\rangle = \big\langle u_l \otimes u_k, u_i \otimes u_j \big\rangle = \delta_{li}\delta_{kj} \tag{21.2.7}$$

is equal to

$$\big\langle u_k \otimes u_l, \hat{P}_{21}\, u_i \otimes u_j \big\rangle = \big\langle u_k \otimes u_l, u_j \otimes u_i \big\rangle = \delta_{kj}\delta_{li}, \tag{21.2.8}$$

confirming the claimed Hermiticity. Since the transposition operator is Hermitian and it squares to itself, it is a *unitary* operator:

$$\hat{P}_{21}^{\dagger} \hat{P}_{21} = \mathbb{1}. \tag{21.2.9}$$

This suggests that the transposition operator could be used to define a symmetry.

Given a Hermitian operator, such as $\hat{P}_{21}$, it is natural to ask about its eigenvalues and its eigenvectors. Since $\hat{P}_{21}^2 = \mathbb{1}$, any eigenvalue λ of $\hat{P}_{21}$ must satisfy $\lambda^2 = 1$. Thus, $\lambda = \pm 1$ are the only possibilities. The corresponding eigenstates in $V \otimes V$ have names:

If $\hat{P}_{21}|\psi\rangle = \ |\psi\rangle$, then $|\psi\rangle$ is a *symmetric* state,

if $\hat{P}_{21}|\psi\rangle = -|\psi\rangle$, then $|\psi\rangle$ is an *antisymmetric* state.

The state $|\psi\rangle$ with eigenvalue $+1$ is said to be a symmetric state, and the state $|\psi\rangle$ with eigenvalue -1 is said to be an antisymmetric state. They are symmetric and antisymmetric, respectively, under transpositions. The set of symmetric states forms a linear subspace of $V \otimes V$ and so does the set of antisymmetric states. The subspaces are called $\text{Sym}(V \otimes V)$ and $\text{Anti}(V \otimes V)$.

Symmetric and antisymmetric states can be constructed with the help of two Hermitian operators $\hat{S}$ and $\hat{A}$ defined as follows:

$$\hat{S} \equiv \tfrac{1}{2}(\mathbb{1} + \hat{P}_{21}), \qquad \hat{A} \equiv \tfrac{1}{2}(\mathbb{1} - \hat{P}_{21}). \tag{21.2.10}$$

Note that $\hat{S}$ and $\hat{A}$ are such that

$$\hat{P}_{21}\hat{S} = \hat{S} \quad \text{and} \quad \hat{P}_{21}\hat{A} = -\hat{A}. \tag{21.2.11}$$

This claim is quickly checked:

$$\hat{P}_{21}\hat{S} = \tfrac{1}{2}(\hat{P}_{21} + \hat{P}_{21}\hat{P}_{21}) = \tfrac{1}{2}(\hat{P}_{21} + \mathbb{1}) = \hat{S},$$
$$\hat{P}_{21}\hat{A} = \tfrac{1}{2}(\hat{P}_{21} - \hat{P}_{21}\hat{P}_{21}) = \tfrac{1}{2}(\hat{P}_{21} - \mathbb{1}) = -\hat{A}.$$

It now follows that given a generic state $|\psi\rangle \in V \otimes V$ we have a symmetric state $|\psi_S\rangle$ and an antisymmetric state $|\psi_A\rangle$ defined by

$$\begin{aligned} |\psi_S\rangle &\equiv \hat{S}|\psi\rangle \in \mathrm{Sym}(V \otimes V), \\ |\psi_A\rangle &\equiv \hat{A}|\psi\rangle \in \mathrm{Anti}(V \otimes V). \end{aligned} \tag{21.2.12}$$

Indeed, we easily check that

$$\begin{aligned} \hat{P}_{21}|\psi_S\rangle &= \hat{P}_{21}\hat{S}|\psi\rangle = \hat{S}|\psi\rangle = |\psi_S\rangle, \\ \hat{P}_{21}|\psi_A\rangle &= \hat{P}_{21}\hat{A}|\psi\rangle = -\hat{A}|\psi\rangle = -|\psi_A\rangle. \end{aligned} \tag{21.2.13}$$

Because of this, the Hermitian operator $\hat{S}$ is a projector to $\mathrm{Sym}(V \otimes V)$, and the Hermitian operator $\hat{A}$ is a projector to $\mathrm{Anti}(V \otimes V)$. As projectors, they satisfy the relations:

$$\hat{S}\hat{S} = \hat{S}, \quad \hat{A}\hat{A} = \hat{A}, \tag{21.2.14}$$

which you should verify. Because they are also Hermitian operators, both $\hat{S}$ and $\hat{A}$ are in fact *orthogonal* projectors. This means that for each operator its range and its kernel are orthogonal subspaces that in fact add to the full space. The operators $\hat{S}$ and $\hat{A}$ are *complementary* projectors—that is

$$\hat{S} + \hat{A} = \mathbb{1}, \qquad \hat{S}\hat{A} = \hat{A}\hat{S} = 0. \tag{21.2.15}$$

Exercise 21.1. *Show that* □

$$V \otimes V = \mathit{range}\,\hat{S} \oplus \mathit{range}\,\hat{A}, \quad \mathit{and} \quad \mathit{range}\,\hat{S} \perp \mathit{range}\,\hat{A}. \tag{21.2.16}$$

Since $\mathrm{range}\,\hat{S} = \mathrm{Sym}(V \otimes V)$, and $\mathrm{range}\,\hat{A} = \mathrm{Anti}(V \otimes V)$, the full space $V \otimes V$ can be decomposed into the orthogonal subspaces of symmetric and antisymmetric states.

***N*-particle systems** The case of $N = 2$ particles considered above is important and instructive, but it is special. To appreciate the general features of identical particles requires going to $N > 2$. As we go beyond two particles, some facts from group theory become useful.

For $N > 2$, we speak of *permutation* operators that scramble the ordering of the particles. For $N = 2$, we only had a single permutation operator, $\hat{P}_{21}$, which in fact was a transposition. The permutation group of two objects, called the symmetric group S_2, is a two-element group containing the identity operator $\mathbb{1}$ that does not permute anything and the transposition $\hat{P}_{21}$. A set of operators form a group if the product of two operators is an operator in the set, if each operator has an inverse, if the product is associative, and if there is an identity operator in the set. These conditions are clearly satisfied by the set of operators $\{\mathbb{1}, \hat{P}_{21}\}$, defining the group S_2.

In an N-particle system, we can define $N!$ permutation operators because there are $N!$ ways in which we can reorder a set of N objects. These $N!$ operators are the elements of the *symmetric group* S_N. A transposition is a permutation operator in which only two objects in the list are exchanged. Since for $N=2$ there are only two objects, the only nontrivial permutation happened to be a transposition. For $N \geq 3$ we have permutations that are not transpositions.

We will use a special notation to describe permutations in $V^{\otimes N}$. Our permutation operators will be written as $\hat{P}_{i_1,\ldots,i_N}$ where the list $\{i_1,\ldots,i_N\}$ is some reordering of the list $\{1,\ldots,N\}$. The operator $\hat{P}_{12\ldots N}$ will be the identity. For a three-particle system, for example, consider the operator

$$\hat{P}_{ijk}, \tag{21.2.17}$$

with $\{i,j,k\}$ some reordering of the list $\{1,2,3\}$. Since i appears in the first position, we will take this to mean that the state of the ith particle now becomes the state of the first particle. Similarly, the state of the jth particle becomes the state of the second particle, and the state of the kth particle becomes the state of the third particle:

$$\hat{P}_{ijk} \text{ moves the state of } \begin{cases} \text{the } i\text{th particle to the first particle,} \\ \text{the } j\text{th particle to the second particle,} \\ \text{the } k\text{th particle to the third particle.} \end{cases} \tag{21.2.18}$$

For example,

$$\hat{P}_{231}|u_r\rangle_{(1)} \otimes |u_s\rangle_{(2)} \otimes |u_t\rangle_{(3)} = |u_s\rangle_{(1)} \otimes |u_t\rangle_{(2)} \otimes |u_r\rangle_{(3)}. \tag{21.2.19}$$

You should check that the inverse of $\hat{P}_{231}$ is $\hat{P}_{312}$ so that

$$\hat{P}_{231}\hat{P}_{312} = \mathbb{1}. \tag{21.2.20}$$

More formally, we define a *permutation* of N objects by the function α that maps the ordered integers $1,\ldots,N$ into some arbitrary ordering:

$$\alpha : [1,\ldots,N] \rightarrow [\alpha(1),\ \ldots,\ \alpha(N)]. \tag{21.2.21}$$

The permutation operator $\hat{P}_\alpha$ associated to the permutation α is written as

$$\hat{P}_\alpha \equiv \hat{P}_{\alpha(1),\alpha(2),\ldots,\alpha(N)}, \tag{21.2.22}$$

and extending the rule stated for the three-particle case, this operator acts as follows:

$$\boxed{\hat{P}_\alpha |u_1\rangle_{(1)} \otimes \cdots \otimes |u_N\rangle_{(N)} = |u_{\alpha(1)}\rangle_{(1)} \otimes \cdots \otimes |u_{\alpha(N)}\rangle_{(N)}.} \tag{21.2.23}$$

For example, we see that

$$\hat{P}_{3142}|u_1\rangle_{(1)} \otimes |u_2\rangle_{(2)} \otimes |u_3\rangle_{(3)} \otimes |u_4\rangle_{(4)} = |u_3\rangle_{(1)} \otimes |u_1\rangle_{(2)} \otimes |u_4\rangle_{(3)} \otimes |u_2\rangle_{(4)}. \tag{21.2.24}$$

Dropping the subscripts on the states and recalling that they are always ordered from 1 to N in ascending order, we can write

$$\hat{P}_{3142}|a\rangle \otimes |b\rangle \otimes |c\rangle \otimes |d\rangle = |c\rangle \otimes |a\rangle \otimes |d\rangle \otimes |b\rangle. \tag{21.2.25}$$

Table 21.1
$A \cdot B$ matrix for S_3.

$A\backslash B$	$\mathbb{1}$	$\hat{P}_{312}$	$\hat{P}_{231}$	(23)	(12)	(13)
$\mathbb{1}$	$\mathbb{1}$	$\hat{P}_{312}$	$\hat{P}_{231}$	(23)	(12)	(13)
$\hat{P}_{312}$	$\hat{P}_{312}$	$\hat{P}_{231}$	$\mathbb{1}$	(12)	(13)	(23)
$\hat{P}_{231}$	$\hat{P}_{231}$	$\mathbb{1}$	$\hat{P}_{312}$	(13)	(23)	(12)
(23)	(23)	(13)	(12)	$\mathbb{1}$	$\hat{P}_{231}$	$\hat{P}_{312}$
(12)	(12)	(23)	(13)	$\hat{P}_{312}$	$\mathbb{1}$	$\hat{P}_{231}$
(13)	(13)	(12)	(23)	$\hat{P}_{231}$	$\hat{P}_{312}$	$\mathbb{1}$

Omitting the tensor product symbol ⊗ for further streamlining, we would write

$$\hat{P}_{3142}|abcd\rangle = |cadb\rangle. \tag{21.2.26}$$

The $N=3$ permutation operators form the symmetric group S_3 with $3!=6$ elements:

$$\hat{P}_{123}=\mathbb{1}\,,\ \hat{P}_{312}\,,\ \hat{P}_{231}\,,\ \underbrace{\hat{P}_{132}\,,\ \hat{P}_{213}\,,\ \hat{P}_{321}}_{\text{transpositions}}\,. \tag{21.2.27}$$

The last three elements of the list are transpositions. A transposition is a permutation in which the list of labels remains in canonical ascending order except for two labels that are exchanged. For example, $\hat{P}_{132}$ is a transposition in which the states of the second and third particles are exchanged, while the state of the first particle is left unchanged. For transpositions we sometimes use the notation in which we just indicate the two labels that are being transposed. Those two labels could be written in any order without risk of confusion, but we will use ascending order:

$$(23) \equiv \hat{P}_{132}, \quad (12) \equiv \hat{P}_{213}, \quad (13) \equiv \hat{P}_{321}. \tag{21.2.28}$$

The multiplication table for the group S_3 is indicated in table 21.1.

Returning to the case of general N, note that any permutation can be written as a product of transpositions: any set of integers can be arranged into any arbitrary position by successive transpositions. The decomposition of a permutation into a product of transpositions is not unique, but the number of transpositions in the product is in fact unique modulo 2. Hence, we say that every permutation is either *even* or *odd*. A permutation is said to be even if it is the product of an even number of transpositions, and it is said to be odd if it is the product of an odd number of transpositions. An even permutation is said to have even parity, and an odd permutation is said to have odd parity. The identity element is an even permutation.

All transpositions are Hermitian and unitary; the proof we gave for $\hat{P}_{21}$ easily generalizes. Since the product of unitary operators is unitary, *any permutation is a unitary operator.* Although transpositions are also Hermitian, an arbitrary product of them is not Hermitian because the transpositions do not necessarily commute. Thus, general permutations are not Hermitian operators.

Unitarity implies that the Hermitian conjugate of a permutation is its inverse, a permutation of the same parity. This is clear from writing $\hat{P}_\alpha$ as a product of transpositions $\hat{P}_{t_i}$:

$$\hat{P}_\alpha = \hat{P}_{t_1} \dots \hat{P}_{t_k} \quad \rightarrow \quad \hat{P}_\alpha^\dagger = \hat{P}_{t_k}^\dagger \dots \hat{P}_{t_1}^\dagger = \hat{P}_{t_k} \dots \hat{P}_{t_1}, \tag{21.2.29}$$

and therefore, as expected, $\hat{P}_\alpha \hat{P}_\alpha^\dagger = \mathbb{1}$.

It is perhaps surprising that in any symmetric group, the numbers of even and odd permutations are the same.

Claim: For any S_N the number of even permutations is the same as the number of odd permutations.

Proof. Consider the map m_{12} that multiplies any even permutation by the transposition (12) from the left. It is manifest that m_{12} maps even permutations to odd permutations. This map is one to one: if σ, σ' are even, then $(12)\sigma = (12)\sigma'$ implies $\sigma = \sigma'$ by multiplying from the left by (12), which is the inverse of (12). This map is also surjective, or onto: for any odd β, we have $\beta = (12)[(12)\beta]$, where the element in brackets is even. Since m_{12} is one to one and onto, the number of even permutations is equal to the number of odd permutations. □

21.3 Complete Symmetrizer and Antisymmetrizer

Permutation operators do not all commute, so we cannot expect to find a complete *basis* of states that are eigenstates of all permutation operators. It is possible, however, to find *some* states that are simultaneous eigenstates of *all* permutation operators. This is not all that unusual. When two operators fail to commute, there is no complete basis of simultaneous eigenstates, but there can be *some* simultaneous eigenstates.

Consider N particles, each with state space V so that the collection of particles lives in $V^{\otimes N}$. Let $\hat{P}_\alpha$ denote an arbitrary permutation in S_N. We define symmetric states $|\psi_S\rangle$ that are left invariant by the action of all permutations:

$$\text{Symmetric state } |\psi_S\rangle: \quad \hat{P}_\alpha |\psi_S\rangle = |\psi_S\rangle, \quad \forall \alpha. \tag{21.3.1}$$

The symmetric states are eigenstates of all permutation operators with eigenvalue $+1$. The symmetric states cannot form a basis for $V^{\otimes N}$, so they will form a subspace $\text{Sym}^N\, V$. We will later discuss projectors to this subspace.

Let us now define antisymmetric states. We could expect those states to change sign when acted on by permutation operators, but this is not quite right. The identity operator, for example, is a permutation and cannot change the sign of a state. Our intuition is that a transposition must change the sign of an antisymmetric state, but then a permutation built from two transpositions could not change the sign of an antisymmetric state. This intuition is summarized by saying that odd permutations change the sign of an antisymmetric state, but even permutations do not. We thus define antisymmetric states $|\psi_A\rangle$ as follows:

$$\text{Antisymmetric state } |\psi_A\rangle: \quad \hat{P}_\alpha|\psi_A\rangle = \epsilon_\alpha|\psi_A\rangle, \quad \forall\alpha. \tag{21.3.2}$$

Here, ϵ_α is the **parity** of the permutation:

$$\epsilon_\alpha = \begin{cases} +1 & \text{if } \hat{P}_\alpha \text{ is an even permutation,} \\ -1 & \text{if } \hat{P}_\alpha \text{ is an odd permutation.} \end{cases} \tag{21.3.3}$$

The eigenvalue of a permutation acting on an antisymmetric state is the parity of the permutation. The antisymmetric states form a subspace $\text{Anti}^N V \subset V^{\otimes N}$.

Exercise 21.2. *Consider the permutation group S_3 whose elements are permutations $\hat{P}_{ijk}$, with $i \neq j \neq k$ and $i, j, k \in \{1, 2, 3\}$. Check explicitly that ϵ_{ijk} is the parity of $\hat{P}_{ijk}$.*

Our next goal is the construction of projectors $\hat{S}$ and $\hat{A}$ from $V^{\otimes N}$ into $\text{Sym}^N V$ and $\text{Anti}^N V$, respectively. As we will confirm below, the projectors are defined as follows:

$$\hat{S} \equiv \frac{1}{N!}\sum_\alpha \hat{P}_\alpha \quad \text{and} \quad \hat{A} \equiv \frac{1}{N!}\sum_\alpha \epsilon_\alpha \hat{P}_\alpha, \tag{21.3.4}$$

where the sums are over all $N!$ permutations. $\hat{S}$ is called the *symmetrizer*, and $\hat{A}$ is called the *antisymmetrizer*. The first thing to note is that both $\hat{S}$ and $\hat{A}$ are Hermitian operators:

$$\hat{S} = \hat{S}^\dagger \qquad \hat{A} = \hat{A}^\dagger \tag{21.3.5}$$

Hermitian conjugation is a one-to-one invertible map from the set of all permutations to itself. In fact, due to unitarity, Hermitian conjugation maps each permutation to its inverse. In any group the map from elements to their inverses is a one-to-one onto map of the group to itself. Thus, the sum in $\hat{S}$ is simply reordered, making it clear that $\hat{S}$ is not changed. We can introduce a bit of notation to make this clearer. Given an α permutation, we define the $\alpha^\dagger$ permutation via

$$P_{\alpha^\dagger} \equiv P_\alpha^\dagger. \tag{21.3.6}$$

Here $\alpha^\dagger : [\alpha^\dagger(1), \ldots, \alpha^\dagger(N)]$ is the list that makes the above equation hold. Since the set of all $\alpha^\dagger$'s is equal to the set of all α's, it follows that for any function $f(\alpha)$ of α we have

$$\sum_\alpha f(\alpha^\dagger) = \sum_\alpha f(\alpha). \tag{21.3.7}$$

Indeed, both sides of the equation simply compute the sum of the evaluation of f over all permutations. Now consider the Hermitian conjugation of $\hat{S}$. We have

$$\hat{S}^\dagger = \frac{1}{N!}\sum_\alpha \hat{P}_\alpha^\dagger = \frac{1}{N!}\sum_\alpha \hat{P}_{\alpha^\dagger}. \tag{21.3.8}$$

We now use the identity (21.3.7) to find that

$$\hat{S}^\dagger = \frac{1}{N!}\sum_\alpha \hat{P}_\alpha = \hat{S}. \tag{21.3.9}$$

The antisymmetrizer $\hat{A}$ is also unchanged because Hermitian conjugation does not change the parity of a permutation—namely, $\epsilon_{\alpha^\dagger} = \epsilon_\alpha$. As a result,

$$\hat{A}^\dagger = \frac{1}{N!}\sum_\alpha \epsilon_\alpha \hat{P}_{\alpha^\dagger} = \frac{1}{N!}\sum_\alpha \epsilon_{\alpha^\dagger} \hat{P}_{\alpha^\dagger} = \frac{1}{N!}\sum_\alpha \epsilon_\alpha \hat{P}_\alpha = \hat{A}, \tag{21.3.10}$$

using (21.3.7) to pass from the second to the third sum. It is also important to see what happens when $\hat{S}$ or $\hat{A}$ is multiplied by a permutation operator. We claim that

$$\begin{aligned} \hat{P}_{\alpha_0}\hat{S} &= \hat{S}\hat{P}_{\alpha_0} = \hat{S}, \\ \hat{P}_{\alpha_0}\hat{A} &= \hat{A}\hat{P}_{\alpha_0} = \epsilon_{\alpha_0}\hat{A}. \end{aligned} \tag{21.3.11}$$

Proof. Note that $\hat{P}_{\alpha_0}$ acting on the set of all permutations simply rearranges the set: the map from a group to itself induced by multiplication by a group element is one to one and onto. As an example, consider the multiplication table 21.1 for the group S_3. Each row is obtained by multiplying all the elements of the group by one single element from the left. As you can see, each row is simply a rearrangement of the list of elements of the group. More generally, we can introduce a multiplication rule for labels via $\hat{P}_{\alpha_0}\hat{P}_\alpha = \hat{P}_{\alpha_0\cdot\alpha}$. Here $\alpha_0\cdot\alpha$ is the new list of integers that describes the resulting permutation. But again, the point is that $\sum_\alpha f(\alpha_0\cdot\alpha) = \sum_\alpha f(\alpha)$. Therefore, for the action on $\hat{S}$ we see that

$$\hat{P}_{\alpha_0}\hat{S} = \frac{1}{N!}\sum_\alpha \hat{P}_{\alpha_0}\hat{P}_\alpha = \frac{1}{N!}\sum_\alpha \hat{P}_{\alpha_0\cdot\alpha} = \frac{1}{N!}\sum_\alpha \hat{P}_\alpha = \hat{S}. \tag{21.3.12}$$

Analogously, for $\hat{A}$ we have

$$\hat{P}_{\alpha_0}\hat{A} = \frac{1}{N!}\sum_\alpha \epsilon_\alpha \hat{P}_{\alpha_0}\hat{P}_\alpha = \frac{1}{N!}\sum_\alpha \epsilon_\alpha\epsilon_{\alpha_0}\epsilon_{\alpha_0}\hat{P}_{\alpha_0}\hat{P}_\alpha, \tag{21.3.13}$$

where in the last step we introduced the product $\epsilon_{\alpha_0}\epsilon_{\alpha_0} = 1$. Moving one of the ϵ_{α_0} factors outside and noting that $\epsilon_{\alpha_0}\epsilon_\alpha = \epsilon_{\alpha_0\cdot\alpha}$, we get

$$\hat{P}_{\alpha_0}\hat{A} = \epsilon_{\alpha_0}\frac{1}{N!}\sum_\alpha \epsilon_{\alpha_0\cdot\alpha}\hat{P}_{\alpha_0\cdot\alpha} = \epsilon_{\alpha_0}\frac{1}{N!}\sum_\alpha \epsilon_\alpha\hat{P}_\alpha = \epsilon_{\alpha_0}\hat{A}. \tag{21.3.14}$$

This is what we wanted to show. When the permutation operator acts from the right, the proof proceeds in exactly analogous form. □

This preparatory work finally allow us to show that both $\hat{S}$ and $\hat{A}$ are in fact orthogonal projectors:

$$\boxed{\hat{S}^2 = \hat{S}, \quad \hat{A}^2 = \hat{A}, \quad \hat{S}\hat{A} = \hat{A}\hat{S} = 0.} \tag{21.3.15}$$

The verification of these properties is rather straightforward:

$$\hat{S}^2 = \frac{1}{N!}\sum_\alpha \hat{P}_\alpha\hat{S} = \frac{1}{N!}\sum_\alpha \hat{S} = \frac{1}{N!}N!\,\hat{S} = \hat{S}, \tag{21.3.16}$$

$$\hat{A}^2 = \frac{1}{N!}\sum_\alpha \epsilon_\alpha\hat{P}_\alpha\hat{A} = \frac{1}{N!}\sum_\alpha \epsilon_\alpha\epsilon_\alpha\hat{A} = \frac{1}{N!}\sum_\alpha \hat{A} = \frac{1}{N!}N!\,\hat{A} = \hat{A}, \tag{21.3.17}$$

$$\hat{A}\hat{S} = \frac{1}{N!}\sum_\alpha \epsilon_\alpha\hat{P}_\alpha\hat{S} = \frac{1}{N!}\sum_\alpha \epsilon_\alpha\hat{S} = \frac{\hat{S}}{N!}\sum_\alpha \epsilon_\alpha = 0. \tag{21.3.18}$$

Here, $\sum_\alpha \epsilon_\alpha = 0$ follows because there are equal numbers of even and odd permutations.

We now confirm that $\hat{S}$ and $\hat{A}$ project to symmetric and antisymmetric states. This means that for an arbitrary $|\psi\rangle \in V^{\otimes N}$, we find $\hat{S}|\psi\rangle \in \text{Sym}^N V$. Indeed, this is the case because $\hat{S}|\psi\rangle$ satisfies the definition (21.3.1) of a symmetric state:

$$\hat{P}_\alpha \hat{S}|\psi\rangle = \hat{S}|\psi\rangle, \quad \forall \alpha. \tag{21.3.19}$$

Analogously, for an arbitrary $|\psi\rangle \in V^{\otimes N}$ we find $\hat{A}|\psi\rangle \in \text{Anti}^N V$. Indeed, this is the case because $\hat{A}|\psi\rangle$ satisfies the definition (21.3.2) of an antisymmetric state:

$$\hat{P}_\alpha \hat{A}|\psi\rangle = \epsilon_\alpha \hat{A}|\psi\rangle, \quad \forall \alpha. \tag{21.3.20}$$

Hence, as claimed, $\hat{S}$ and $\hat{A}$ are projectors into the symmetric and antisymmetric subspaces:

$$\hat{S}: V^{\otimes N} \to \text{Sym}^N V, \qquad \hat{A}: V^{\otimes N} \to \text{Anti}^N V. \tag{21.3.21}$$

Exercise 21.3. *Explain why an $\hat{S}$ eigenstate of eigenvalue one is a symmetric state. Explain why an $\hat{A}$ eigenstate of eigenvalue one is an antisymmetric state.*

Example 21.1. *Symmetrizer and antisymmetrizer for S_3.*
As we have seen before, the symmetric group S_3 has six permutation operators: $\hat{P}_{123} = \mathbb{1}$, $\hat{P}_{312}$, $\hat{P}_{231}$, $\hat{P}_{132}$, $\hat{P}_{213}$, and $\hat{P}_{321}$, the last three of which are transpositions. In this case the symmetrizer $\hat{S}$ and antisymmetrizer $\hat{A}$ are

$$\begin{aligned} \hat{S} &= \tfrac{1}{6}(\mathbb{1} + \hat{P}_{312} + \hat{P}_{231} + \hat{P}_{132} + \hat{P}_{213} + \hat{P}_{321}), \\ \hat{A} &= \tfrac{1}{6}(\mathbb{1} + \hat{P}_{312} + \hat{P}_{231} - \hat{P}_{132} - \hat{P}_{213} - \hat{P}_{321}). \end{aligned} \tag{21.3.22}$$

While for $N=2$ the operators $\hat{S}$ and $\hat{A}$ add up to the identity, this does not hold for $N=3$:

$$\hat{S} + \hat{A} = \tfrac{1}{3}(\mathbb{1} + \hat{P}_{312} + \hat{P}_{231}) \neq \mathbb{1}. \tag{21.3.23}$$

In fact, for $N > 2$ the direct sum of the purely symmetric and purely antisymmetric subspaces is a proper subspace of $V^{\otimes N}$:

$$\text{Sym}^N V \oplus \text{Anti}^N V \subset V^{\otimes N}, \quad N > 2. \tag{21.3.24}$$

For $N > 2$, the state space $V^{\otimes N}$ is not spanned by purely symmetric and purely antisymmetric states. For a decomposition of $V^{\otimes 3}$ states, see problem 21.6. □

For $N=2$, the permutations group is commutative or abelian—that is, the permutation operators commute. For $N \geq 3$, the permutation group is nonabelian; the permutation operators, as we have seen, do not generally commute. The permutation operators of a permutation group can be represented by matrices. Let m_{α_i} be the matrix associated with $\hat{P}_{\alpha_i}$. These form a representation if whenever $\hat{P}_{\alpha_1}\hat{P}_{\alpha_2} = \hat{P}_{\alpha_3}$ we have $m_{\alpha_1} m_{\alpha_2} = m_{\alpha_3}$. A symmetric state defines a one-dimensional vector space that furnishes a representation of the group elements as 1×1 matrices, all of which are equal to 1. An antisymmetric state also defines a one-dimensional vector space that furnishes a representation of the group elements as 1×1 matrices. Here, associated to $\hat{P}_\alpha$, the matrix is the number ϵ_α, the parity of the permutation. In both cases the matrices, admittedly just numbers, multiply consistent with the way the operators multiply. There are representations with higher-dimensional matrices. The group S_3, for example, has an irreducible representation with 2×2 matrices.

21.4 The Symmetrization Postulate

We now have all the elements needed for the resolution of the exchange degeneracy problem. We will state the *symmetrization postulate*, which should be considered an extra assumption of quantum mechanics. We cannot prove it from the four axioms A1, A2, A3, and A4 discussed in section 16.6. As usual in quantum mechanics, however, it is difficult to see how one could have an alternative statement that works.

Symmetrization postulate:

In a system with N identical particles, the states that are physically realized are not arbitrary states in $V^{\otimes N}$ but rather are totally symmetric (that is, belonging to $\text{Sym}^N V$), in which case the particles are said to be bosons, or they are totally antisymmetric (that is, belonging to $\text{Anti}^N V$), in which case they are said to be fermions.

It is intuitively clear that the exchange degeneracy problem has been solved. If we have two identical particles, one in state $|a\rangle$ and the other in a different state $|b\rangle$, the state of the two particles is neither $|a\rangle \otimes |b\rangle$ nor $|b\rangle \otimes |a\rangle$—the degenerate possibilities—but rather $|a\rangle \otimes |b\rangle \pm |b\rangle \otimes |a\rangle$, up to scale. Interestingly, these options are entangled states. Let us spell out in detail how the symmetrization postulate does the job. For this we must state the exchange degeneracy problem precisely.

Let $|u\rangle \in V^{\otimes N}$ be an arbitrary vector in the tensor product. Associated with $|u\rangle$, we introduce the vector subspace $V_{|u\rangle}$ generated by acting on $|u\rangle$ with all the permutation operators in S_N:

$$V_{|u\rangle} \equiv \text{span}\{\hat{P}_\alpha |u\rangle\,, \forall \alpha\} \subset V^{\otimes N}\,. \tag{21.4.1}$$

Depending on the choice of the state $|u\rangle$, the dimension of $V_{|u\rangle}$ can go from one to $N!$. This dimensionality, if different from one, is the degeneracy due to exchange. The exchange degeneracy problem is the ambiguity we face in selecting a representative for the physical state in $V_{|u\rangle}$. The problem is solved by the symmetrization postulate if we can show the following:

Claim: Up to a multiplicative constant, $V_{|u\rangle}$ contains at most a single state in $\text{Sym}^N V$ and at most a single state in $\text{Anti}^N V$.

Proof. We first show that, up to a multiplicative constant, $V_{|u\rangle}$ contains at most a single ket in $\text{Sym}^N V$. Suppose we have a state $|\psi\rangle \in V_{|u\rangle}$ that is symmetric: $|\psi\rangle \in \text{Sym}^N V$. Since $|\psi\rangle \in V_{|u\rangle}$, we can write it as follows:

$$|\psi\rangle = \sum_\alpha c_\alpha P_\alpha |u\rangle, \tag{21.4.2}$$

with c_α some coefficients. Since $|\psi\rangle \in \text{Sym}^N V$, it is left invariant by the action of $\hat{S}$:

$$|\psi\rangle = \hat{S}|\psi\rangle = \hat{S}\sum_\alpha c_\alpha \hat{P}_\alpha |u\rangle = \sum_\alpha c_\alpha \hat{S}\hat{P}_\alpha |u\rangle = \sum_\alpha c_\alpha \hat{S}|u\rangle = \hat{S}|u\rangle \sum_\alpha c_\alpha. \tag{21.4.3}$$

This shows that any symmetric $|\psi\rangle$ in $V_{|u\rangle}$ must be proportional to $\hat{S}|u\rangle$ and is therefore unique up to a multiplicative constant. The argument is similar for the antisymmetric states. Suppose we have a state $|\psi\rangle \in V_{|u\rangle}$ that is antisymmetric: $|\psi\rangle \in \text{Anti}^N V$. Again, we write it as

$$|\psi\rangle = \sum_\alpha d_\alpha P_\alpha |u\rangle, \tag{21.4.4}$$

with d_α some coefficients. Since $|\psi\rangle \in \text{Anti}^N V$, it is left invariant by the action of $\hat{A}$:

$$|\psi\rangle = \hat{A}|\psi\rangle = \hat{A}\sum_\alpha d_\alpha \hat{P}_\alpha |u\rangle = \sum_\alpha d_\alpha \hat{A}\hat{P}_\alpha |u\rangle = \sum_\alpha \epsilon_\alpha d_\alpha \hat{A}|u\rangle = \hat{A}|u\rangle \sum_\alpha \epsilon_\alpha d_\alpha. \tag{21.4.5}$$

This shows that any $|\psi\rangle$ must be proportional to $\hat{A}|u\rangle$ and is therefore unique up to a multiplicative constant.

Note that the claim is stated as saying that "at most" we get one symmetric state or one antisymmetric state. It may be that for a given $|u\rangle$ we have $\hat{S}|u\rangle = 0$, and no state in $V_{|u\rangle}$ is in $\text{Sym}^N V$. If $|u\rangle$ takes the simple form $|u\rangle = |u_{i_1}\rangle \otimes \cdots \otimes |u_{i_N}\rangle$, the state in $\text{Sym}^N V$ will always exist and be nonzero. The construction will also fail to give a state $V_{|u\rangle}$ that is in $\text{Anti}^N V$ if $\hat{A}|u\rangle = 0$. This special case is discussed below. □

Exercise 21.4. *Assume the state $|u\rangle$ satisfies $\hat{P}_\alpha|u\rangle = -|u\rangle$ for some permutation $\hat{P}_\alpha$. Show that $\hat{S}|u\rangle = 0$.*

The state in $\text{Anti}^N V$ will fail to exist in $V_{|u\rangle}$ if two or more V states appearing in $|u\rangle = |u_{i_1}\rangle \otimes \cdots \otimes |u_{i_N}\rangle$ are the same. This is the content of **Pauli's exclusion principle**, which states that two or more identical fermions cannot be found in the same state. This is easily confirmed. Assume $|u\rangle = |u_{i_1}\rangle \otimes \cdots \otimes |u_{i_N}\rangle$ is such that two fermions, the pth and the qth, are in the same state $|u_k\rangle$. It then follows that the transposition (pq) leaves $|u\rangle$ invariant: $(pq)|u\rangle = |u\rangle$. Acting on this relation with the antisymmetrizer, we find $\hat{A}(pq)|u\rangle = \hat{A}|u\rangle$. Since (pq) is an odd permutation, this gives $-\hat{A}|u\rangle = \hat{A}|u\rangle$, resulting in $\hat{A}|u\rangle = 0$. No state exists.

A related proof of Pauli's exclusion principle can be constructed without reference to a starting state $|u\rangle$. It states that any antisymmetric state $|\psi_A\rangle$ where two fermions are in the same state $|u_k\rangle$ must vanish. To see this, collect all terms $|\psi_{ij}\rangle$ in $|\psi_A\rangle$ such that $|u_k\rangle$ appears in positions i and j with $i < j$. Write $|\psi_A\rangle = |\psi_{ij}\rangle + |\hat{\psi}\rangle$, where $|\hat{\psi}\rangle$ are the other terms in the ket. The transposition (ij) manifestly leaves $|\psi_{ij}\rangle$ invariant: $(ij)|\psi_{ij}\rangle = +|\psi_{ij}\rangle$, and acting on $|\hat{\psi}\rangle$ cannot give terms in $|\psi_{ij}\rangle$, only terms in $|\hat{\psi}\rangle$. Since (ij) fails to change the sign of $|\psi_{ij}\rangle$ but must change the sign of $|\psi_A\rangle$, we must have $|\psi_{ij}\rangle = 0$. Letting i, j run over all possible values $1 \le i < j \le N$, we exhaust all terms in $|\psi_A\rangle$, and we conclude that $|\psi_A\rangle = 0$.

Let us make some additional remarks on the symmetrization postulate.

1. The postulate describes particles that have two different kind of statistics: bosons and fermions. Because of the constraint on their wave functions, the statistical behavior of bosons and fermions is strikingly different.
2. Quantum field theory, together with the constraints of special relativity, can be used to prove the *spin-statistics theorem*, which shows that bosons are particles of integer spins (0, 1, 2, ...), while fermions are particles of half-integer spin (1/2, 3/2, ...).
3. The symmetrization postulate for elementary particles leads to a definite statistic for composite particles. They are either bosons or fermions, themselves obeying the symmetrization postulate. Consider, for example, two hydrogen atoms:

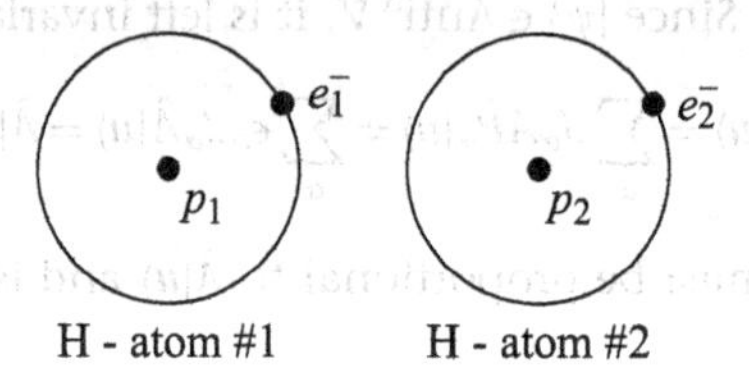

The two-atom system is made by four particles, and its wave function is $\psi(p_1, e_1^-; p_2, e_2^-)$. Here p_1 and e_1^- are the constituent proton and electron of the first atom, and p_2 and e_2^- are the constituent proton and electron of the second atom. Since the two electrons are identical particles of spin one-half, the wave function must be antisymmetric under the exchange $e_1^- \leftrightarrow e_2^-$:

$$\psi(p_1, e_2^-; p_2, e_1^-) = -\psi(p_1, e_1^-; p_2, e_2^-). \tag{21.4.6}$$

Exactly the same argument applies to the protons so that

$$\psi(p_2, e_1^-; p_1, e_2^-) = -\psi(p_1, e_1^-; p_2, e_2^-). \tag{21.4.7}$$

Therefore, under the simultaneous exchange of electrons and protons, we find that

$$\psi(p_2, e_2^-; p_1, e_1^-) = \psi(p_1, e_1^-; p_2, e_2^-). \tag{21.4.8}$$

The exchange in (21.4.8) is an exchange of the two hydrogen atoms! Since the wave function is symmetric under this exchange, the hydrogen atom is a boson!
4. The postulate is in fact a description of the most general statistics of particles that live in three spatial dimensions. A closer look into the physics requires thinking of permutation-induced exchanges as *physical* processes in which the particles actually move in some trajectories to realize the permutations. It turns out on close examination that in three spatial dimensions there are no new possibilities, just bosons or fermions. But when space is two-dimensional, there are new possibilities. Transpositions of particles need not produce only sign factors on the wave function; general phases are in principle possible. Particles with such behavior are called anyons. We will discuss such phases at the end of this section.

A Hamiltonian for identical particles must be such that if we have a state in $\text{Sym}^N V$ or in $\text{Anti}^N V$ at time equal zero, the time evolution induced by the Hamiltonian must guarantee that the state remains in $\text{Sym}^N V$ or in $\text{Anti}^N V$ for all times. To discuss this constraint, we must understand how operators transform under permutations.

Any operator $\hat{B} \in \mathcal{L}(V)$ acting on the single-particle state space V can be used to define operators acting on the tensor product $V^{\otimes N}$ relevant for a system of N identical particles. We define $\hat{B}(k) \in \mathcal{L}(V^{\otimes N})$ as the operator that acts on the kth state in any basis vector of $V^{\otimes N}$. One has, for example, $\hat{B}(1) = \hat{B} \otimes \mathbb{1} \cdots \otimes \mathbb{1}$, $\hat{B}(2) = \mathbb{1} \otimes \hat{B} \otimes \cdots \mathbb{1}$, up to $\hat{B}(N) = \mathbb{1} \otimes \cdots \otimes \mathbb{1} \otimes \hat{B}$.

Permutations act by conjugation on the $\hat{B}(k)$ operators and change the state space they act on. We claim that

$$\hat{P}_\alpha^\dagger \hat{B}(k) \hat{P}_\alpha = B(\alpha(k)). \tag{21.4.9}$$

Such an equation is checked by letting it act on an arbitrary basis state $|u_1\rangle_{(1)} \otimes \cdots \otimes |u_N\rangle_{(N)}$. First imagine that the $\hat{B}$ operator is not present. In that case the effect of $\hat{P}_\alpha^\dagger \hat{P}_\alpha$ is first for $\hat{P}_\alpha$ to scramble the states in $|u_1\rangle_{(1)} \otimes \cdots \otimes |u_N\rangle_{(N)}$ and then for $\hat{P}_\alpha^\dagger = (\hat{P}_\alpha)^{-1}$ to unscramble the states back to their original positions. When $\hat{P}_\alpha^\dagger B(k) \hat{P}_\alpha$ acts on $|u_1\rangle_{(1)} \otimes \cdots \otimes |u_N\rangle_{(N)}$, the operator $\hat{B}(k)$ will act on the state that $\hat{P}_\alpha$ places on the kth position—namely, the state $|u_{\alpha(k)}\rangle$ (recall (21.2.23)). After this, $\hat{P}_\alpha^\dagger$ brings back this state, with $\hat{B}$ acting on it, to its original position as the state of the $\alpha(k)$ particle. Hence, $\hat{B}(k)$ is turned into $\hat{B}(\alpha(k))$ under this conjugation.

Consider now a more general operator $\hat{\Theta}(1, \ldots, N)$, constructed in terms of an arbitrary set of operators acting on the various particles. In here, $1, \ldots, N$ are the labels of the operators, indicating the state space they act on. It is clear from (21.4.9) that we have

$$\hat{P}_\alpha^\dagger \hat{\Theta}(1, \ldots, N) \hat{P}_\alpha = \hat{\Theta}(\alpha(1), \ldots, \alpha(N)), \tag{21.4.10}$$

the arguments on the right-hand side just a reordering of the list $1, \ldots, N$. A Hermitian operator $\hat{M}(1, 2, \ldots, N)$ is said to be a *completely symmetric* observable if for any permutation α we have that

$$\hat{M}(\alpha(1), \ldots, \alpha(N)) = \hat{M}(1, 2, \ldots, N). \tag{21.4.11}$$

For such an operator, equation (21.4.10) gives

$$\hat{P}_\alpha^\dagger \hat{M}(1, 2, \ldots, N) \hat{P}_\alpha = \hat{M}(\alpha(1), \ldots, \alpha(N)) = \hat{M}(1, 2, \ldots, N), \tag{21.4.12}$$

the last equality following from the complete symmetry property of $\hat{M}$. Multiplying this equation by $\hat{P}_\alpha$ from the left, we conclude that

$$[\hat{M}(1, 2, \ldots, N), \hat{P}_\alpha] = 0, \quad \forall \alpha. \tag{21.4.13}$$

A completely symmetric operator commutes with all permutation operators.

For a system of identical particles, the Hamiltonian $\hat{H}(1, \ldots, N)$ must be a completely symmetric observable; this is the physical requirement following from the indistinguishability of the particles. Therefore, $\hat{H}(1, \ldots, N)$ must commute with all permutation operators:

$$[\hat{H}(1, 2, \ldots, N), \hat{P}_\alpha] = 0, \quad \forall \alpha. \tag{21.4.14}$$

This means that all permutation operators $\hat{P}_\alpha$ are conserved, and we have

$$\frac{d}{dt}\langle\hat{P}_\alpha\rangle = \frac{i}{\hbar}\Big\langle[\hat{H}(1,\ldots,N),\hat{P}_\alpha]\Big\rangle = 0. \tag{21.4.15}$$

Since both the symmetrizer $\hat{S}$ and the antisymmetrizer $\hat{A}$ are sums of permutation operators, they, too, are conserved:

$$[\hat{H}(1,2,\ldots,N)\,,\,\hat{S}\,]=0, \quad [\hat{H}(1,2,\ldots,N)\,,\,\hat{A}\,]=0. \tag{21.4.16}$$

It follows that $\hat{S}$ and $\hat{A}$ commute with the unitary operator $\mathcal{U}$ generating time evolution; after all, $\mathcal{U}$ is built solely from the Hamiltonian. As a result, if a state is fully symmetric at time equals zero, meaning an $\hat{S}$ eigenstate with eigenvalue one, it will remain symmetric for all times. Similarly, if a state is totally antisymmetric at $t=0$, meaning an $\hat{A}$ eigenstate with eigenvalue one, it will remain antisymmetric for all times. This is what we wanted to guarantee.

The simplest construction of completely symmetric Hamiltonians is obtained by adding N single-particle Hamiltonians, one for each particle. Calling $\hat{H}_0$ the single-particle Hamiltonian, we consider operators $\hat{H}_0(k)$ that act on the kth particle. The total Hamiltonian $\hat{H}_{\rm tot}$ is defined to the the sum

$$\hat{H}_{\rm tot} = \hat{H}_0(1) + \cdots + \hat{H}_0(N) = \sum_{k=0}^{N}\hat{H}_0(k). \tag{21.4.17}$$

In this Hamiltonian the various particles do not interact. The spectrum of $\hat{H}_{\rm tot}$ follows from that of $\hat{H}_0$ and is specified by occupation numbers for the $\hat{H}_0$ levels. The Hamiltonian $\hat{H}_{\rm tot}$ is invariant under the action of any permutation:

$$\hat{P}_\alpha^\dagger\hat{H}_{\rm tot}\hat{P}_\alpha = \sum_{k=0}^{N}\hat{P}_\alpha^\dagger\hat{H}_0(k)\hat{P}_\alpha = \sum_{k=0}^{N}\hat{H}_0(\alpha(k)) = \hat{H}_{\rm tot}, \tag{21.4.18}$$

where the last equality follows because the set $\alpha(k)$, with $k=1,\ldots,N$, is a permutation of $1,\ldots,N$. This means that

$$[\hat{H}_{\rm tot}\,,\,\hat{P}_\alpha\,]=0, \quad \forall\,\alpha \tag{21.4.19}$$

so that, as required, $\hat{H}_{\rm tot}$ is completely symmetric. The Hamiltonian $\hat{H}_{\rm tot}$ becomes interacting if supplemented with completely symmetric interaction terms. If we had two particles, for example, a potential $V(|\mathbf{x}_1-\mathbf{x}_2|)$ that only depends on the distance between the particles would be symmetric under the exchange of the particles and thus a consistent interaction.

More general phases and statistics We mentioned earlier that options more general than those allowed by the symmetrization postulate are possible. To see this, we must think physically of the action of permutation operators: their exchanges must be realized by moving the particles. In this picture, states are described in terms of particle positions, not the general kets used in the symmetrization postulate. The issue of phases or sign factors under permutations becomes then a physical one: realize the exchange by motion of the particles along some prescribed trajectories, and *determine* how the final wave function is related to the original one. To avoid extraneous effects, the motion is

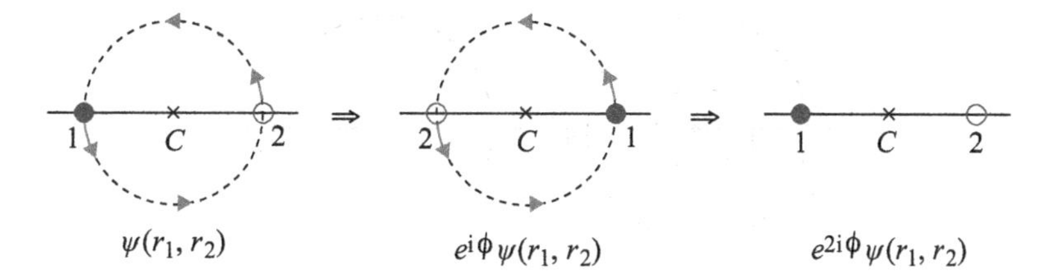

Figure 21.2
Identical particles 1 and 2 are exchanged by an angle π rotation about the midpoint C. A second exchange by a further angle π rotation brings them back to their original position. Below we indicate possible phases acquired by the wave function through these processes.

required to be slow, the process being adiabatic. We will study such processes in detail in chapter 28.

After the physical exchange, the final wave function could differ by a sign factor if the particles are fermions or remain the same if the particles are bosons. By restricting ourselves to adiabatic processes, we aim to ignore any other phase, dynamical or otherwise, that could enter in the process. The sign factor ± 1 in this case is intrinsic to the exchange process and the nature of the particles. But perhaps, more generally, the final wave function could differ from the original by an arbitrary phase determined by the type of particle and the nature of the exchange. We will assume that this phase is *topological*, meaning that continuous deformations of the motion, consistent with the initial and final positions, should not change the phase. In defining continuous deformations, we *do not allow* the particles to be at the same point. This is physically motivated: general particles are composites, and their whole nature could be affected in uncontrolled ways if they coincide.

For simplicity, consider just two identical particles, particle 1 to the left of particle 2, as shown in figure 21.2. Suppose the particles are interchanged by rotating them slowly about a point C located at the midpoint of an imaginary rod joining them. The exchange is realized after a rotation by an angle π. A second exchange, done by a further rotation by an angle π, brings the particles back to their original positions.

This two-step process, we claim, is topologically equivalent to one in which particle 1 is fixed, and particle 2 wraps around particle 1. This means that the two-step process is continuously deformable into a wraparound process. To see this just imagine sliding the fixed point C on the rod toward particle 1. At each stage we rotate the particles about C by an angle of 2π, particle 1 now doing smaller and smaller circles and particle 2 doing larger and larger circles. It is clear that as C reaches particle 1, this particle no longer moves, while particle 2 does a large circle (figure 21.3). Two points should be noted. For each position of C, the initial and final configurations are the same. For each position of C away from the midpoint, the configuration obtained after a rotation by π is no longer an exchange of the particles. This is an unavoidable property of the deformation, as the final wraparound process features no exchange.

Now consider what happens in three spatial dimensions (3D). Here, having one particle wrapping around a fixed particle is continuously deformable into having no motion,

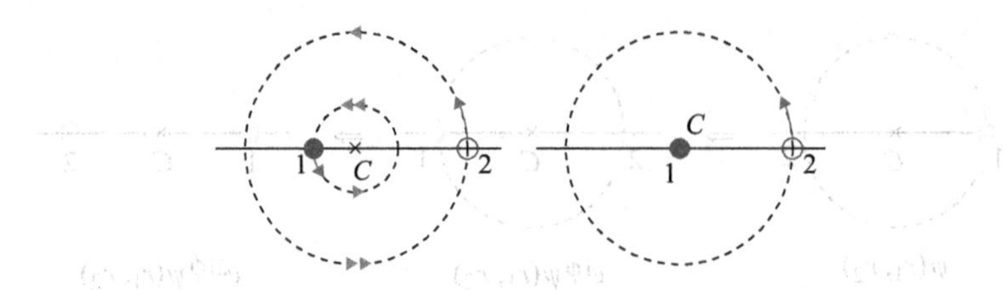

Figure 21.3
Left: Sliding the rotation point C toward particle 1, particle 1 traces a smaller circle, while particle 2 traces a larger circle. *Right*: The end result, as C reaches particle 1, is a static particle 1, with particle 2 wrapping around it.

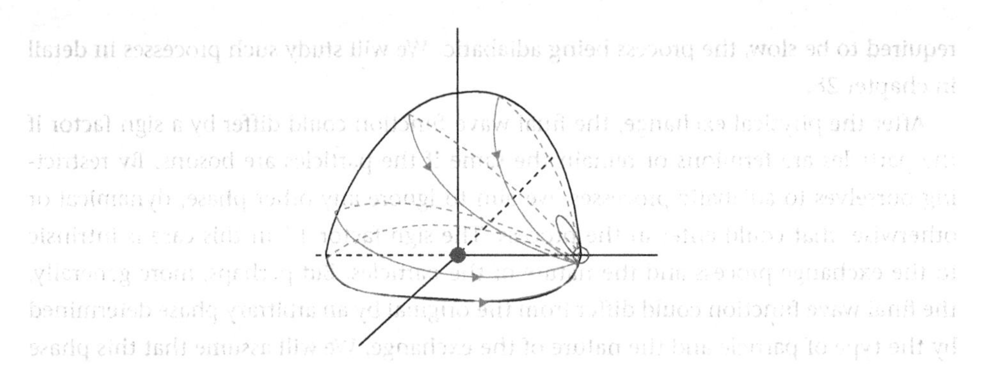

Figure 21.4
In three dimensions the wraparound process on the plane can be deformed continuously into a process with no motion by lifting the path into the third dimension and contracting it without hitting particle 1 at the origin.

as shown in figure 21.4, where the path, originally on the plane, is contracted by sliding it over the surface of a hemisphere. We now ask if after the two-step exchange one could find that the wave function has changed by some finite phase. The answer is no, by our topological assumption above: no phase would be associated to a process in which there is no motion, and the two-step exchange, in 3D, is deformable into a no-motion process. There is no phase after the two-exchange process in 3D; the wave function must return to itself. If that is the case, the single exchange process can only lead to a ± 1 factor. In this way the physical argument reproduces the result of the symmetrization postulate in 3D.

It is now simple to see that if space is two-dimensional, things are different. While the two-step exchange is still equivalent to the wraparound process, the latter is no longer equivalent to no motion. The process cannot be lifted to a third dimension, and contraction of the loop, while keeping the original and final point of the loop at the initial position of particle 2, cannot be realized without bumping into particle 1, which is not permissible. It follows that a phase factor *is* allowed for the two-step exchange:

$$\text{Two-step exchange: } \psi(\mathbf{r}_1, \mathbf{r}_2) \rightarrow e^{2i\phi}\, \psi(\mathbf{r}_1, \mathbf{r}_2). \tag{21.4.20}$$

This means that a single exchange of the particles leads to the following:

$$\text{Single exchange:}\ \psi(\mathbf{r}_1,\mathbf{r}_2) \to e^{i\phi}\,\psi(\mathbf{r}_1,\mathbf{r}_2). \tag{21.4.21}$$

These phases are shown in figure 21.3. The key role played by two dimensions was realized by John Magne Leinaas and Jan Myrheim (1977) and by Frank Wilczek (1982), who called particles with such phases *anyons*, for "any" phase allowed, any statistics.

The arguments above indicate that particles, or at least "quasiparticles" of arbitrary statistics, could logically exist. In fact, on a two-dimensional lattice of spins, with a Hamiltonian defining a *toric-code model*, one finds excited states that are anyons! In the continuum, physicists study interacting fermions moving in two dimensions, subject to strong magnetic fields. Such a system can realize *fractional quantum Hall states*, where quasiparticle excitations are anyons carrying a fraction of the electron charge.

It is worth noting that some of the above analysis is relevant even if we do not have identical particles. While each exchange process is no longer truly an exchange, the particles being distinguishable, the two-exchange process is still equivalent to the wraparound process, and this process is nontrivial in two dimensions. Thus, in two dimensions, even with distinguishable particles, we may obtain a phase after one of the particles wraps around the other. An adiabatic process dealing with that situation is considered in problem 28.5.

21.5 Building Symmetrized States and Probabilities

In this section we begin to explore the building of suitable states of identical particles using the symmetrization postulate. We will focus on fermions, as a canonical construction exists for them. We will then explore how to normalize states of multiple particles and how probabilities are defined consistent with symmetrization. We conclude with an example involving photons that explains the results of the Hong-Ou-Mandel experiment discussed in section 21.1.

Building antisymmetric states Suppose you want to build a three-fermion state starting from the following state $|u\rangle \in V^{\otimes 3}$:

$$|u\rangle = |\omega_1\rangle_{(1)} \otimes |\omega_2\rangle_{(2)} \otimes |\omega_3\rangle_{(3)}. \tag{21.5.1}$$

Of course, you know that the answer is obtained by acting with the antisymmetrizer on $|u\rangle$:

$$\hat{A}|u\rangle = \frac{1}{3!}\sum_{\alpha} \epsilon_\alpha \hat{P}_\alpha\, |\omega_1\rangle_{(1)} \otimes |\omega_2\rangle_{(2)} \otimes |\omega_3\rangle_{(3)}. \tag{21.5.2}$$

We claim that this state can be written as a determinant:

$$\hat{A}|u\rangle = \frac{1}{3!}\begin{vmatrix} |\omega_1\rangle_{(1)} & |\omega_1\rangle_{(2)} & |\omega_1\rangle_{(3)} \\ |\omega_2\rangle_{(1)} & |\omega_2\rangle_{(2)} & |\omega_2\rangle_{(3)} \\ |\omega_3\rangle_{(1)} & |\omega_3\rangle_{(2)} & |\omega_3\rangle_{(3)} \end{vmatrix} = \frac{1}{3!}\epsilon_{ijk}\, |\omega_i\rangle_{(1)} \otimes |\omega_j\rangle_{(2)} \otimes |\omega_k\rangle_{(3)}. \tag{21.5.3}$$

Repeated indices are summed over. You may recall that ϵ_{ijk} is in fact the parity of the permutation $\hat{P}_{ijk}$. In the above matrix of states, the row index labels the different states, and the column index labels the different particles.

Exercise 21.5. *Confirm by direct expansion that the determinant in (21.5.3) gives the action of $\hat{A}$ on $|u\rangle$.*

Exercise 21.6. *Verify that the determinant in (21.5.3) is in fact given by the rightmost expression with the epsilon symbol.*

Now let us do this generally and explain why the so-called **Slater determinant** always supplies the correct answer. Recall first the formula for the determinant of an $N \times N$ matrix B_{ij}:

$$\det B = \sum_{\alpha} \epsilon_{\alpha} B_{\alpha(1),1} B_{\alpha(2),2} \dots B_{\alpha(N),N}. \tag{21.5.4}$$

As before, $\alpha = [\alpha(1), \dots, \alpha(N)]$ is a rearrangement of the ordered list $[1, \dots, N]$, and $P_{\alpha} = P_{\alpha(1)\cdots\alpha(N)}$. This expression is the general version of the formula given in (21.5.3) for the case $N = 3$. Now, let $|\omega\rangle \in V^{\otimes N}$ be a state of the form

$$|\omega\rangle = |\omega_1\rangle_{(1)} \otimes \cdots \otimes |\omega_N\rangle_{(N)}, \tag{21.5.5}$$

with $|\omega_i\rangle \in V$ for $i = 1, \dots, N$, a collection of N normalized states. We aim to construct the N-fermion state $|\psi_{\omega}\rangle$ associated with $|\omega\rangle$. Up to normalization, we write

$$|\psi_{\omega}\rangle = \hat{A}\,|\omega\rangle, \tag{21.5.6}$$

with $\hat{A}$ the antisymmetrizer. To calculate the action of $\hat{A}$, we first note that the action of the permutation $\hat{P}_{\alpha}$ on $|\omega\rangle$ gives

$$\hat{P}_{\alpha}|\omega\rangle = |\omega_{\alpha(1)}\rangle_{(1)} \otimes \cdots \otimes |\omega_{\alpha(N)}\rangle_{(N)}. \tag{21.5.7}$$

The antisymmetrizer action therefore gives

$$|\psi_{\omega}\rangle = \hat{A}|\omega\rangle = \frac{1}{N!}\sum_{\alpha} \epsilon_{\alpha}\hat{P}_{\alpha}|\omega\rangle = \frac{1}{N!}\sum_{\alpha} \epsilon_{\alpha}\,|\omega_{\alpha(1)}\rangle_{(1)} \otimes \cdots \otimes |\omega_{\alpha(N)}\rangle_{(N)}. \tag{21.5.8}$$

Now define a matrix ω_{ij} of kets, with the row index labeling the different states and the column index labeling the various particles:

$$\omega_{ij} \equiv |\omega_i\rangle_{(j)}. \tag{21.5.9}$$

With this definition the above expression for $\hat{A}|\omega\rangle$ becomes

$$|\psi_{\omega}\rangle = \frac{1}{N!}\sum_{\alpha} \epsilon_{\alpha}\, \omega_{\alpha(1),1} \cdots \omega_{\alpha(N),N} = \frac{1}{N!}\det \omega, \tag{21.5.10}$$

on account of the definition (21.5.4) of the determinant. We have thus shown that

$$|\psi_{\omega}\rangle = \hat{A}|\omega\rangle = \frac{1}{N!}\begin{vmatrix} |\omega_1\rangle_{(1)} & \dots & |\omega_1\rangle_{(N)} \\ \vdots & & \vdots \\ |\omega_N\rangle_{(1)} & \dots & |\omega_N\rangle_{(N)} \end{vmatrix}. \tag{21.5.11}$$

The determinant of states above is the Slater determinant. We will consider the normalization of this state below.

We can use the above determinant to construct the position-space wave function for a system of N identical fermions. We simply imagine hitting both sides of equation (21.5.11) with the position-space bra

$$\langle \mathbf{x}_1, \ldots, \mathbf{x}_N| \equiv {}_{(1)}\langle \mathbf{x}_1| \otimes \cdots \otimes_{(N)} \langle \mathbf{x}_N|. \tag{21.5.12}$$

On the left-hand side, we get

$$\psi_\omega(\mathbf{x}_1, \ldots, \mathbf{x}_N) \equiv \langle \mathbf{x}_1, \ldots \mathbf{x}_N| \psi_\omega\rangle. \tag{21.5.13}$$

On the right-hand side, we still get a determinant since we can bring the bras in, attaching $\langle \mathbf{x}_k|$ to the states with particle label k:

$$ {}_{(1)}\langle \mathbf{x}_1| \otimes \cdots \otimes_{(N)} \langle \mathbf{x}_N| \begin{vmatrix} |\omega_1\rangle_{(1)} & \ldots & |\omega_1\rangle_{(N)} \\ \vdots & & \vdots \\ |\omega_N\rangle_{(1)} & \ldots & |\omega_N\rangle_{(N)} \end{vmatrix} = \begin{vmatrix} \langle \mathbf{x}_1|\omega_1\rangle & \ldots & \langle \mathbf{x}_N|\omega_1\rangle \\ \vdots & & \vdots \\ \langle \mathbf{x}_1|\omega_N\rangle & \ldots & \langle \mathbf{x}_N|\omega_N\rangle \end{vmatrix}. \tag{21.5.14}$$

Therefore, the wave function for N fermions in a state defined by the N single-particle wave functions $\omega_1(\mathbf{x}), \ldots, \omega_N(\mathbf{x})$ is

$$\psi_\omega(\mathbf{x}_1, \ldots, \mathbf{x}_N) = \mathcal{N} \begin{vmatrix} \omega_1(\mathbf{x}_1) & \ldots & \omega_1(\mathbf{x}_N) \\ \vdots & & \vdots \\ \omega_N(\mathbf{x}_1) & \ldots & \omega_N(\mathbf{x}_N) \end{vmatrix}, \tag{21.5.15}$$

with $\mathcal{N}$ some normalization constant. The right-hand side is the more familiar form of the Slater determinant, now written with wave functions. Let us note some simple but important properties of wave functions for identical particles. Note that Pauli's exclusion principle is manifest in this formulation of the state of N fermions. If $\omega_i(\mathbf{x}) = \omega_j(\mathbf{x})$ for any $i \neq j$, the determinant vanishes because the ith and jth rows become identical. Moreover, the wave function is antisymmetric under the exchange of any two coordinates $\mathbf{x}_i$ and $\mathbf{x}_j$, for $i \neq j$. This exchange is equivalent to an exchange of columns.

Normalizing wave functions and probabilities We have already defined position-space bras: $\langle \mathbf{x}_1, \ldots, \mathbf{x}_N| = {}_{(1)}\langle \mathbf{x}_1| \otimes \cdots \otimes_{(N)} \langle \mathbf{x}_N|$. The overlap of such states with similarly defined kets is declared to be

$$\langle \mathbf{x}_1, \ldots, \mathbf{x}_N|\mathbf{y}_1, \ldots, \mathbf{y}_N\rangle = \delta(\mathbf{x}_1 - \mathbf{y}_1) \cdots \delta(\mathbf{x}_N - \mathbf{y}_N). \tag{21.5.16}$$

The completeness relation in the N-particle state space takes the form

$$\int d\mathbf{x}_1 \cdots d\mathbf{x}_N \, |\mathbf{x}_1, \ldots, \mathbf{x}_N\rangle\langle \mathbf{x}_1, \ldots, \mathbf{x}_N| = \mathbb{1}. \tag{21.5.17}$$

Each integral here is three-dimensional and runs over all of space. The correctness of this can be verified by letting both sides of the equation act on $|\mathbf{y}_1,\ldots,\mathbf{y}_N\rangle$ and confirming that the left-hand side has no effect.

We use position states to build the position-space wave function $\psi(\mathbf{x}_1,\ldots,\mathbf{x}_N)$ associated to an N-particle ket $|\psi\rangle \in V^{\otimes N}$. Of course, the state $|\psi\rangle$ is either fully symmetric or fully antisymmetric. As stated before, we define

$$\psi(\mathbf{x}_1,\ldots,\mathbf{x}_N) \equiv \langle \mathbf{x}_1,\ldots,\mathbf{x}_N|\psi\rangle = \Big({}_{(1)}\langle \mathbf{x}_1| \otimes \cdots \otimes_{(N)}\langle \mathbf{x}_N|\Big)|\psi\rangle. \tag{21.5.18}$$

We will adopt the normalization condition in which the wave function satisfies

$$\int d\mathbf{x}_1 \cdots d\mathbf{x}_N \big|\psi(\mathbf{x}_1,\ldots,\mathbf{x}_N)\big|^2 = 1. \tag{21.5.19}$$

Note that $\big|\psi(\mathbf{x}_1,\ldots,\mathbf{x}_N)\big|^2$ is a *totally symmetric* function of the arguments—this is true for identical bosons and for identical fermions because any exchange of arguments changes the wave function only by a sign. Equation (21.5.19) determines the probabilistic interpretation of the wave function for identical particles.

Consider, for example, the case of two identical particles and the expression

$$d\mathbf{x}_1 d\mathbf{x}_2 \,|\psi(\mathbf{x}_1,\mathbf{x}_2)|^2. \tag{21.5.20}$$

If the particles are distinguishable, this is the probability of finding the first particle at $\mathbf{x}_1$ within the volume element $d\mathbf{x}_1$ and the second particle at $\mathbf{x}_2$ within the volume element $d\mathbf{x}_2$. Since we cannot tell which is the first particle and which is the second particle, this would not be a clear statement for identical particles. For identical particles, a reasonable question would be: What is the probability of finding one particle at $\mathbf{x}_A$ within $d\mathbf{x}_A$ and the other at $\mathbf{x}_B$ within $d\mathbf{x}_B$? To find the answer, we note that in the context of the double integral

$$\int d\mathbf{x}_1 d\mathbf{x}_2 |\psi(\mathbf{x}_1,\mathbf{x}_2)|^2 = 1, \tag{21.5.21}$$

the configuration we are looking for is realized when $(\mathbf{x}_1,\mathbf{x}_2)=(\mathbf{x}_A,\mathbf{x}_B)$ with corresponding ranges or when $(\mathbf{x}_1,\mathbf{x}_2)=(\mathbf{x}_B,\mathbf{x}_A)$, also with appropriate ranges. Since $|\psi|^2$ does not change under this exchange of arguments, each configuration gives the same contribution, and the desired probability is given as follows:

$$\begin{aligned}&\text{Probability of finding a particle at } \mathbf{x}_A \text{ within } d\mathbf{x}_A\\ &\quad\text{and a particle at } \mathbf{x}_B \text{ within } d\mathbf{x}_B = 2\cdot d\mathbf{x}_A\, d\mathbf{x}_B \big|\psi(\mathbf{x}_A,\mathbf{x}_B)\big|^2.\end{aligned} \tag{21.5.22}$$

More generally, the probability of finding in an N-particle state a particle at each of the positions $(\mathbf{x}_1,\cdots,\mathbf{x}_N)$ with ranges $(d\mathbf{x}_1,\cdots,d\mathbf{x}_N)$ is

$$N!\, d\mathbf{x}_1 \cdots d\mathbf{x}_N \big|\psi(\mathbf{x}_1,\ldots,\mathbf{x}_N)\big|^2. \tag{21.5.23}$$

Note, however, that the probability of finding *all* identical particles at $\mathbf{x}$ within $d\mathbf{x}$ does not have the extra $N!$ factor. We have

$$\underbrace{d\mathbf{x}\cdots d\mathbf{x}}_{N-\text{factors}} \,|\psi(\mathbf{x},\ldots,\mathbf{x})|^2. \tag{21.5.24}$$

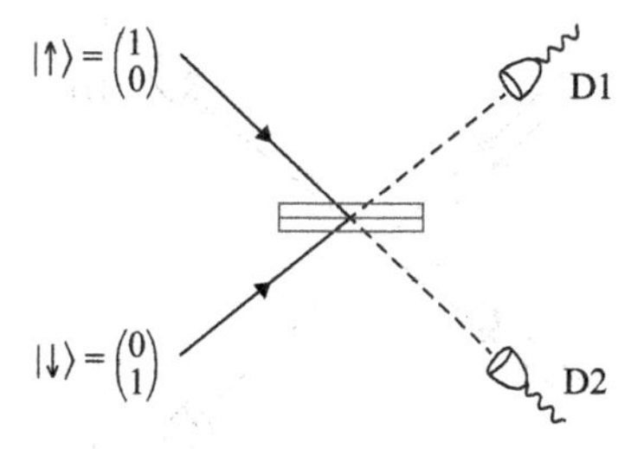

Figure 21.5
Two identical photons hit a balanced beam splitter, one from the top and one from the bottom. The D1 and D2 detectors never click simultaneously; the two photons both end up together on either D1 or D2.

Let us return to address the normalization of the N-fermion wave function (21.5.15). This cannot be done in general unless we know the orthogonality properties of the one-particle wave functions. Assume the $\omega_1(\mathbf{x}), \ldots, \omega_N(\mathbf{x})$ are orthonormal:

$$\int d\mathbf{x}\, \omega_i^*(\mathbf{x})\, \omega_j(\mathbf{x}) = \delta_{ij}, \quad i, j = 1, \ldots, N. \tag{21.5.25}$$

In this case, we achieve proper normalization with

$$\psi(\mathbf{x}_1, \ldots, \mathbf{x}_N) = \frac{1}{\sqrt{N!}} \begin{vmatrix} \omega_1(\mathbf{x}_1) & \ldots & \omega_1(\mathbf{x}_N) \\ \vdots & & \vdots \\ \omega_N(\mathbf{x}_1) & \ldots & \omega_N(\mathbf{x}_N) \end{vmatrix}. \tag{21.5.26}$$

This is clear. The determinant is the sum of $N!$ terms. In the calculation of $\int d\mathbf{x}_1 \cdots d\mathbf{x}_N |\psi|^2$, because of orthonormality each term from the determinant gives a plus-one contribution when it is integrated against its complex conjugate and zero otherwise. The $N!$ terms give a total of $N!$ that is canceled precisely by the square of the determinant prefactor in the above expression. Thus, this ψ satisfies $\int d\mathbf{x}_1 \cdots d\mathbf{x}_N |\psi|^2 = 1$, as required.

Example 21.2. *Hong-Ou-Mandel two-photon experiment.*
It is now possible to derive the results claimed in section 21.1 regarding the experiment in which two identical photons hit a beam splitter simultaneously. The claim was that both photons must end in one detector or the other; that the processes in which one photon ends in each detector interfere destructively. The setup is illustrated in figure 21.5.

The action of a beam splitter with two input ports, top and bottom, can be represented by a 2×2 unitary matrix U of the form

$$U = \frac{1}{\sqrt{2}} \begin{pmatrix} 1 & 1 \\ 1 & -1 \end{pmatrix}, \quad |\uparrow\rangle \equiv \begin{pmatrix} 1 \\ 0 \end{pmatrix}, \quad |\downarrow\rangle \equiv \begin{pmatrix} 0 \\ 1 \end{pmatrix}. \tag{21.5.27}$$

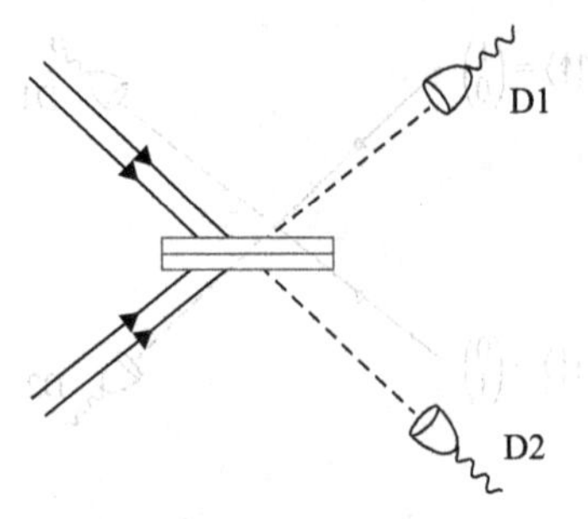

Figure 21.6
Four identical photons hit a balanced beam splitter, two from the top and two from the bottom. What are the probabilities of the various detections? See problem 21.7.

We have also defined the state $|\Uparrow\rangle$, representing a photon incident from the top, and the state $|\Downarrow\rangle$, representing a photon incident from the bottom. To the right of the beam splitter, the state $|\Uparrow\rangle$ is a photon heading to D1, while the state $|\Downarrow\rangle$ is a photon heading to D2.

The incoming state of two photons, one incident from the top and one incident from the bottom, $|\Uparrow\rangle \otimes |\Downarrow\rangle$, must be symmetrized because the particles are identical bosons. The normalized incident state $|\Psi_{\text{inc}}\rangle$ is therefore

$$|\Psi_{\text{inc}}\rangle = \frac{1}{\sqrt{2}}\big(|\Uparrow\rangle_{(1)} \otimes |\Downarrow\rangle_{(2)} + |\Downarrow\rangle_{(1)} \otimes |\Uparrow\rangle_{(2)}\big). \tag{21.5.28}$$

Since each photon experiences the action of the beam splitter, the outgoing state $|\Psi_{\text{out}}\rangle$ is

$$|\Psi_{\text{out}}\rangle = (U \otimes U)|\Psi_{\text{inc}}\rangle = \frac{1}{\sqrt{2}}\big(U|\Uparrow\rangle_{(1)} \otimes U|\Downarrow\rangle_{(2)} + U|\Downarrow\rangle_{(1)} \otimes U|\Uparrow\rangle_{(2)}\big). \tag{21.5.29}$$

The action of U on the basis states follows from (21.5.27), and we get

$$|\Psi_{\text{out}}\rangle = \frac{1}{2\sqrt{2}}\Big(\big(|\Uparrow\rangle + |\Downarrow\rangle\big)_{(1)} \otimes \big(|\Uparrow\rangle - |\Downarrow\rangle\big)_{(2)} + \big(|\Uparrow\rangle - |\Downarrow\rangle\big)_{(1)} \otimes \big(|\Uparrow\rangle + |\Downarrow\rangle\big)_{(2)}\Big). \tag{21.5.30}$$

Expanding out, we see that the mixed terms $|\Uparrow\rangle_{(1)} \otimes |\Downarrow\rangle_{(2)}$ and $|\Downarrow\rangle_{(1)} \otimes |\Uparrow\rangle_{(2)}$ cancel out! These are the terms where one photon ends in D1 and the other ends in D2. The result is

$$|\Psi_{\text{out}}\rangle = \frac{1}{\sqrt{2}}\Big(|\Uparrow\rangle_{(1)} \otimes |\Uparrow\rangle_{(2)} - |\Downarrow\rangle_{(1)} \otimes |\Downarrow\rangle_{(2)}\Big). \tag{21.5.31}$$

The first term gives the amplitude for both photons to end up in D1, and the second term gives the amplitude for both photons to end up in D2. As claimed, we can't get a single photon in each detector.

More photons can be incident on a beam splitter. If we have four identical photons incident simultaneously on a beam splitter, two from the upper port and two from the lower port, we have the situation shown in figure 21.6. As it turns out, this time it is possible to get two photons on each detector, as well as four in either one. But no detector ever finds a single photon, or three photons. You can work this out, as well as the corresponding probabilities in problem 21.7. □

21.6 Particles with Two Sets of Degrees of Freedom

Consider a particle that has both spatial degrees of freedom and spin degrees of freedom. The state space of the particle can be spanned by basis states that take the form

$$|\psi_{\text{spatial}}\rangle \otimes |\psi_{\text{spin}}\rangle. \tag{21.6.1}$$

Let V be the vector space of spatial states, and let W be the vector space of spin states. The general states $|\psi\rangle$ of the particle belong to the tensor product:

$$|\psi\rangle \in V \otimes W. \tag{21.6.2}$$

The space $V \otimes W$ is the state space of the particle. If we had N such particles, the states of the system would live in the Nth tensor product of the single-particle state space:

$$(V \otimes W)^{\otimes N} = \underbrace{(V \otimes W) \cdots (V \otimes W)}_{N \text{times}}. \tag{21.6.3}$$

If you think of basis elements of tensor products as ordered lists of states, it is clear that one can identify the following spaces:

$$(V \otimes W)^{\otimes N} \simeq V^{\otimes N} \otimes W^{\otimes N}. \tag{21.6.4}$$

On the right-hand side, we are listing first the N basis vectors in V and then the N basis vectors in W. For the case of two identical particles, $N = 2$, the above identification reads

$$(V \otimes W)^{\otimes 2} \simeq (V \otimes V) \otimes (W \otimes W). \tag{21.6.5}$$

Now consider a state of two particles and the identification above:

$$(|v_1\rangle_{(1)} \otimes |w_1\rangle_{(1)}) \otimes (|v_2\rangle_{(2)} \otimes |w_2\rangle_{(2)}) \simeq |v_1\rangle_{(1)} \otimes |v_2\rangle_{(2)} \otimes |w_1\rangle_{(1)} \otimes |w_2\rangle_{(2)}. \tag{21.6.6}$$

Let us work with the state on the right. Think of it as a list of four states, and use the language of permutations of four objects. In this case the exchange of the two particles is implemented by $\hat{P}_{2143}$:

$$\hat{P}_{2143}|v_1\rangle_{(1)} \otimes |v_2\rangle_{(2)} \otimes |w_1\rangle_{(1)} \otimes |w_2\rangle_{(2)} = |v_2\rangle_{(1)} \otimes |v_1\rangle_{(2)} \otimes |w_2\rangle_{(1)} \otimes |w_1\rangle_{(2)}. \tag{21.6.7}$$

Indeed, the permutation $\hat{P}_{2143}$ exchanges both the V and W states of the particles. We can now write $\hat{P}_{2143}$ as a product of two permutations:

$$\hat{P}_{2143} = \hat{P}_{2134} \cdot \hat{P}_{1243}, \tag{21.6.8}$$

where $\hat{P}_{2134}$ permutes the V states, leaving the W states alone, and $\hat{P}_{1243}$ permutes the W states, leaving the V states alone. These two permutation operators commute; each squares to the identity, and each has eigenvalues ± 1. They can therefore be diagonalized simultaneously. In fact, they also commute with the product operator $\hat{P}_{2143}$, consistent with the fact that once $\hat{P}_{2134}$ and $\hat{P}_{1243}$ are diagonal, so is $\hat{P}_{2143}$. Denoting by λ with subscripts the eigenvalues of these operators on the space of simultaneous eigenstates, we see that

$$\lambda_{2143} = \lambda_{2134}\, \lambda_{1243}. \tag{21.6.9}$$

The possible values the λ's can take are given in the table below.

λ_{2143}	λ_{2134}	λ_{1243}
1	1	1
1	−1	−1
−1	1	−1
−1	−1	1

Since $\hat{P}_{2143}$ is the operator that exchanges the two particles, the states with $\lambda_{2143}=1$ live in $\text{Sym}^2(V\otimes W)$ and are bosons. Analogously, the states with $\lambda_{2143}=-1$ live in $\text{Anti}^2(V\otimes W)$ and are fermions. Let us begin with the bosons. The value $\lambda_{2143}=1$ is realized in two ways:

$$\lambda_{2134}=\lambda_{1243}=1, \quad \text{or} \quad \lambda_{2134}=\lambda_{1243}=-1. \tag{21.6.10}$$

The states for which $\lambda_{2134}=1$ are in $\text{Sym}^2 V$, and the states for which $\lambda_{1243}=1$ are in $\text{Sym}^2 W$, implying that the states for which $\lambda_{2134}=\lambda_{1243}=1$ are in $\text{Sym}^2 V\otimes\text{Sym}^2 W$. Similarly, the states for which $\lambda_{2134}=-1$ are in $\text{Anti}^2 V$, and the states for which $\lambda_{1243}=-1$ are in $\text{Anti}^2 W$, implying that the states for which $\lambda_{2134}=\lambda_{1243}=-1$ are in $\text{Anti}^2 V\otimes\text{Anti}^2 W$. Since either option yields states in $\text{Sym}^2(V\otimes W)$, we find that

$$\text{Sym}^2(V\otimes W)\simeq(\text{Sym}^2 V\otimes\text{Sym}^2 W)\oplus(\text{Anti}^2 V\otimes\text{Anti}^2 W). \tag{21.6.11}$$

The states with $\lambda_{2143}=-1$ belong to $\text{Anti}^2(V\otimes W)$ and can arise in two ways:

$$\lambda_{2134}=-\lambda_{1243}=1, \quad \text{or} \quad -\lambda_{2134}=\lambda_{1243}=1. \tag{21.6.12}$$

This leads to the decomposition:

$$\text{Anti}^2(V\otimes W)\simeq(\text{Sym}^2 V\otimes\text{Anti}^2 W)\oplus(\text{Anti}^2 V\otimes\text{Sym}^2 W). \tag{21.6.13}$$

In summary, we find

$$\boxed{\begin{aligned}\text{Sym}^2(V\otimes W)&\simeq(\text{Sym}^2 V\otimes\text{Sym}^2 W)\oplus(\text{Anti}^2 V\otimes\text{Anti}^2 W),\\ \text{Anti}^2(V\otimes W)&\simeq(\text{Sym}^2 V\otimes\text{Anti}^2 W)\oplus(\text{Anti}^2 V\otimes\text{Sym}^2 W).\end{aligned}} \tag{21.6.14}$$

This result is simpler than it looks in the above notation. Imagine again a particle with both spatial and spin degrees of freedom. We have shown that a symmetric state of two such particles is realized in either of two ways: combining symmetric spatial states with symmetric spin states or combining antisymmetric spatial states with antisymmetric spin states. An antisymmetric state of two such particles is also realized in either of two ways: combining symmetric spatial states with antisymmetric spin states or combining antisymmetric spatial states with symmetric spin states. The generalization to two particles belonging to $U\otimes V\otimes W$ is simple. For three or more particles, the analysis is more intricate.

Exercise 21.7. *Write the analogue of (21.6.14) for two particles, each living in $U\otimes V\otimes W$.*

21.7 States of Two-Electron Systems

Consider the wave function ψ for two electrons. This wave function depends on the position of each electron and the spin state of each electron. This is, indeed, an example of the situation considered in the previous section, in which each particle lives on a tensor product of position and spin state spaces, and we are dealing with two particles. Since electrons are spin one-half particles, we can denote the spin state by the eigenvalue $\hbar m$ of $\hat{S}_z$. We write m_1 and m_2 for the spin states of the particles and $\mathbf{x}_1$ and $\mathbf{x}_2$ for their positions. The wave function we will consider is of the form

$$\psi(\mathbf{x}_1, m_1; \mathbf{x}_2, m_2) = \phi(\mathbf{x}_1, \mathbf{x}_2) \cdot \chi(m_1, m_2). \tag{21.7.1}$$

More general states can be built by the superposition of states of this form. In the above, assume that χ is a *normalized* spin state. It can be viewed as a superposition of the triplet and the singlet states that arise by combining the two spins. We would like to focus on two important alternatives, denoted as ψ_1 and ψ_2. Since the total wave function must be antisymmetric under the exchange of the electrons, on account of (21.6.14) the alternatives are

$$\begin{aligned} \psi_1(\mathbf{x}_1, m_1; \mathbf{x}_2, m_2) &= \phi_+(\mathbf{x}_1, \mathbf{x}_2) \cdot \chi_{\text{singlet}}(m_1, m_2), \\ \psi_2(\mathbf{x}_1, m_1; \mathbf{x}_2, m_2) &= \phi_-(\mathbf{x}_1, \mathbf{x}_2) \cdot \chi_{\text{triplet}}(m_1, m_2). \end{aligned} \tag{21.7.2}$$

Since the singlet is antisymmetric and the triplet is symmetric, the spatial wave functions must obey the following relations:

$$\phi_\pm(\mathbf{x}_1, \mathbf{x}_2) = \pm\phi_\pm(\mathbf{x}_2, \mathbf{x}_1). \tag{21.7.3}$$

Assuming we do not measure spin and recalling (21.5.22), the probability of finding one electron within $d\mathbf{x}_1$ of $\mathbf{x}_1$ and the other electron within $d\mathbf{x}_2$ of $\mathbf{x}_2$ is given by $2dP$, where

$$dP = d\mathbf{x}_1 d\mathbf{x}_2 \, |\phi(\mathbf{x}_1, \mathbf{x}_2)|^2. \tag{21.7.4}$$

For simplicity, we will assume that the electrons do not interact with each other and that the total Hamiltonian $\hat{H}_{\text{tot}}$ that governs the dynamics is spin independent so that

$$\hat{H}_{\text{tot}} = \hat{H} \otimes \mathbb{1} + \mathbb{1} \otimes \hat{H}, \quad \text{with} \quad \hat{H} = \frac{\hat{\mathbf{p}}^2}{2m} + V(\mathbf{x}). \tag{21.7.5}$$

Here $V(\mathbf{x})$ is the common potential both electrons feel; that could be, for example, the potential created by a few fixed nuclei. The time-independent Schrödinger equation for the spatial part of the wave function is

$$\left[-\frac{\hbar^2}{2m}\nabla^2_{\mathbf{x}_1} + V(\mathbf{x}_1) - \frac{\hbar^2}{2m}\nabla^2_{\mathbf{x}_2} + V(\mathbf{x}_2)\right]\phi(\mathbf{x}_1, \mathbf{x}_2) = E\,\phi(\mathbf{x}_1, \mathbf{x}_2). \tag{21.7.6}$$

This equation is separable, so there is a solution of the form

$$\phi(\mathbf{x}_1, \mathbf{x}_2) = \phi_A(\mathbf{x}_1)\phi_B(\mathbf{x}_2), \tag{21.7.7}$$

where ϕ_A and ϕ_B satisfy

$$\left[-\frac{\hbar^2}{2m}\nabla_{\mathbf{x}}^2+V(\mathbf{x})\right]\phi_A(\mathbf{x})=E_A\phi_A(\mathbf{x}),$$
$$\left[-\frac{\hbar^2}{2m}\nabla_{\mathbf{x}}^2+V(\mathbf{x})\right]\phi_B(\mathbf{x})=E_B\phi_B(\mathbf{x}), \tag{21.7.8}$$

with $E_A+E_B=E$. We can choose the ϕ_A and ϕ_B wave functions to be normalized:

$$\int d\mathbf{x}\,|\phi_A(\mathbf{x})|^2=1, \qquad \int d\mathbf{x}\,|\phi_B(\mathbf{x})|^2=1, \tag{21.7.9}$$

but they need not be orthogonal:

$$\langle\phi_A,\phi_B\rangle = \int d\mathbf{x}\,\phi_A^*(\mathbf{x})\phi_B(\mathbf{x})\equiv\alpha_{AB}\neq 0. \tag{21.7.10}$$

Since states of a Hamiltonian with different energies must be orthogonal, a nonzero α_{AB} requires ϕ_A and ϕ_B to be degenerate states in the spectrum of $\hat{H}$—that is, $E_A=E_B$. A simple example is the case of an electron shared by two separated, fixed protons: one state could have the electron mostly around one of the protons, and the other, degenerate, state would have the electron mostly around the other proton. By Schwarz's inequality,

$$|\langle\phi_A,\phi_B\rangle|^2\leq|\langle\phi_A,\phi_A\rangle||\langle\phi_B,\phi_B\rangle|=1, \tag{21.7.11}$$

showing that $|\alpha_{AB}|^2\leq 1$, or equivalently,

$$|\alpha_{AB}|\leq 1. \tag{21.7.12}$$

We want spatial wave functions with definite exchange symmetry. These are easily obtained from (21.7.7):

$$\phi_\pm(\mathbf{x}_1,\mathbf{x}_2)=\frac{N_\pm}{\sqrt{2}}\big(\phi_A(\mathbf{x}_1)\phi_B(\mathbf{x}_2)\pm\phi_A(\mathbf{x}_2)\phi_B(\mathbf{x}_1)\big), \tag{21.7.13}$$

with $N_\pm$ a real normalization constant that would be equal to one if ϕ_A and ϕ_B were orthogonal. We can calculate $N_\pm$ in terms of the constant α_{AB} introduced above. Writing out the probability density, we find that

$$\begin{aligned} d\mathbf{x}_1 d\mathbf{x}_2\,|\phi_\pm(\mathbf{x}_1,\mathbf{x}_2)|^2 = \tfrac{1}{2}N_\pm^2 d\mathbf{x}_1 d\mathbf{x}_2\Big\{&|\phi_A(\mathbf{x}_1)|^2|\phi_B(\mathbf{x}_2)|^2+|\phi_A(\mathbf{x}_2)|^2|\phi_B(\mathbf{x}_1)|^2\\ &\pm 2\,\mathrm{Re}[\phi_A^*(\mathbf{x}_1)\phi_B^*(\mathbf{x}_2)\phi_A(\mathbf{x}_2)\phi_B(\mathbf{x}_1)]\Big\}. \end{aligned} \tag{21.7.14}$$

The integral of the left-hand side must be one, so we see that

$$\begin{aligned} 1=\tfrac{1}{2}N_\pm^2\int d\mathbf{x}_1 d\mathbf{x}_2\Big\{&|\phi_A(\mathbf{x}_1)|^2|\phi_B(\mathbf{x}_2)|^2+|\phi_A(\mathbf{x}_2)|^2|\phi_B(\mathbf{x}_1)|^2\\ &\pm 2\,\mathrm{Re}[\phi_A^*(\mathbf{x}_1)\phi_B(\mathbf{x}_1)\phi_A(\mathbf{x}_2)\phi_B^*(\mathbf{x}_2)]\Big\}. \end{aligned} \tag{21.7.15}$$

This gives

$$1=\tfrac{1}{2}N_\pm^2(1+1\pm 2|\alpha_{AB}|^2) \quad\Rightarrow\quad N_\pm=\frac{1}{\sqrt{1\pm|\alpha_{AB}|^2}}. \tag{21.7.16}$$

Compared with the case when ϕ_A and ϕ_B are orthogonal, the normalization constant is reduced when the spatial wave function is symmetric and is increased when the spatial wave function is antisymmetric.

Now consider the probability $dP_\pm$ of finding both electrons within $d\mathbf{x}$ of the same position $\mathbf{x}$. From (21.7.14) this probability is given by

$$\begin{aligned} dP_\pm = d\mathbf{x}d\mathbf{x}\,|\phi_\pm(\mathbf{x},\mathbf{x})|^2 &= \tfrac{1}{2}N_\pm^2\,d\mathbf{x}d\mathbf{x}\left\{2|\phi_A(\mathbf{x})|^2|\phi_B(\mathbf{x})|^2 \pm 2|\phi_A(\mathbf{x})|^2|\phi_B(\mathbf{x})|^2\right\} \\ &= N_\pm^2\,d\mathbf{x}d\mathbf{x}\,|\phi_A(\mathbf{x})|^2|\phi_B(\mathbf{x})|^2(1\pm 1). \end{aligned} \tag{21.7.17}$$

Using the value of $N_\pm$ determined above, we thus get

$$dP_+ = \frac{2}{1+|\alpha_{AB}|^2}|\phi_A(\mathbf{x})|^2|\phi_B(\mathbf{x})|^2\,d\mathbf{x}\,d\mathbf{x}, \quad \text{and} \quad dP_- = 0. \tag{21.7.18}$$

Recall that dP_+ is associated with the spin singlet, while dP_- is associated with the spin triplet. Electrons avoid each other when the spatial wave function is antisymmetric—that is, when they are in the triplet state. As we explain now, in the states with a symmetric spatial wave function the probability dP_+ of being at the same point is *enhanced* relative to the case when the particles are distinguishable. For distinguishable particles the normalized wave function is

$$\phi_D(\mathbf{x}_1,\mathbf{x}_2) = \phi_A(\mathbf{x}_1)\phi_B(\mathbf{x}_2). \tag{21.7.19}$$

This is the amplitude for the particle in state A to be at $\mathbf{x}_1$ and the particle in state B to be at $\mathbf{x}_2$. The probability dP_D for the two distinguishable particles to be at the same point is then

$$dP_D = |\phi_A(\mathbf{x})|^2|\phi_B(\mathbf{x})|^2\,d\mathbf{x}\,d\mathbf{x}. \tag{21.7.20}$$

Recalling the value of dP_+ in (21.7.18), we find that

$$dP_+ = \frac{2}{1+|\alpha_{AB}|^2}\,dP_D. \tag{21.7.21}$$

If the one-particle states had been orthogonal, we would have $\alpha_{AB}=0$, and the probability of finding identical particles at the same place is enhanced relative to the distinguishable case by a factor of two. The overlap of the one-particle states reduces this a bit, but since $|\alpha_{AB}|<1$, we always have

$$dP_+ \geq dP_D. \tag{21.7.22}$$

This is what we wanted to show.

More generally, if we look for particles at different points, what can we compare? We can ask for the probability of finding one particle in $\mathbf{x}_1$ and the other in $\mathbf{x}_2$, within ranges $d\mathbf{x}_1$ and $d\mathbf{x}_2$, respectively. For identical particles, this probability $dP_{I,\pm}$, with I for identical, is

$$dP_{I,\pm} = 2\,d\mathbf{x}_1 d\mathbf{x}_2\,|\phi_\pm(\mathbf{x}_1,\mathbf{x}_2)|^2. \tag{21.7.23}$$

For the distinguishable particles, the same quantity is computed by adding two probabilities: the probability that the particle in state A is at $\mathbf{x}_1$, while the one in state B is

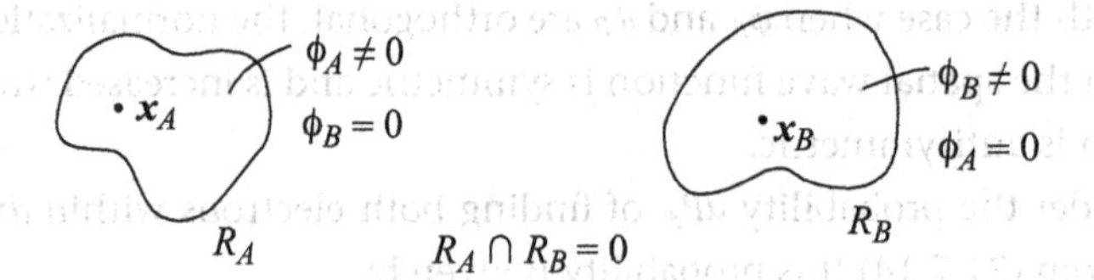

Figure 21.7
Two particles localized in disjoint regions R_A and R_B. The one-particle wave function ϕ_A and ϕ_B have support in R_A and R_B, respectively.

at $\mathbf{x}_2$, *plus* the probability that the particle in state A is at $\mathbf{x}_2$, while the one in state B is at $\mathbf{x}_1$. The ranges, as before, are $d\mathbf{x}_1$ and $d\mathbf{x}_2$. This total probability $dP_{D,\text{tot}}$ is given by

$$dP_{D,\text{tot}} = d\mathbf{x}_1 d\mathbf{x}_2 \left(|\phi_A(\mathbf{x}_1)|^2|\phi_B(\mathbf{x}_2)|^2 + |\phi_A(\mathbf{x}_2)|^2|\phi_B(\mathbf{x}_1)|^2\right). \tag{21.7.24}$$

In fact, for the $+$ case and with $\mathbf{x}_1 = \mathbf{x}_2 = \mathbf{x}$ as well as $d\mathbf{x}_1 = d\mathbf{x}_2 = d\mathbf{x}$, we find that $dP_{I,\pm} = 2dP_\pm$, and $dP_{D,\text{tot}} = 2dP_D$, so we are comparing the right items, and the result in (21.7.21) would follow. More generally, we should have $dP_{I,+} > dP_{D,\text{tot}}$ when $\mathbf{x}_1$ is near $\mathbf{x}_2$, in a region that could be determined in specific examples. Moreover, we also expect $dP_{I,-} < dP_{D,\text{tot}}$, for $\mathbf{x}_1$ near $\mathbf{x}_2$. An explicit computation of the regions where these inequalities hold is explored in problem 21.2.

To conclude, we ask a natural but outrageous question: Why don't we have to symmetrize our wave functions for electrons in the lab with respect to electrons on the moon, or all other electrons in the universe? When we speak of an electron on the moon, we have in mind that the wave function of this electron has no support in our laboratory on Earth, and the wave functions of our electrons in the laboratory have no support on the moon. The question is really one of dealing with identical particles that are localized and have no overlap in their wave functions. We will show that in such a case there is no need for symmetrization or antisymmetrization.

To address this situation, consider the case of two identical particles whose one-particle wave functions are localized and have no overlaps. Assume that the wave function $\phi_A(\mathbf{x})$ is nonzero only in a region R_A, and the wave function $\phi_B(\mathbf{x})$ is nonzero only in a region R_B, with R_A and R_B disjoint: $R_A \cap R_B = 0$ (figure 21.7). It then follows that

$$\alpha_{AB} = \int d^3\mathbf{x}\, \phi_A^*(\mathbf{x})\phi_B(\mathbf{x}) = 0 \quad \Rightarrow \quad N_\pm = 1. \tag{21.7.25}$$

The overlap vanishes because nonzero ϕ_A requires $\mathbf{x} \in R_A$, and nonzero ϕ_B requires $\mathbf{x} \in R_B$, both of which cannot happen for any $\mathbf{x}$. Using our symmetrized wave functions, the probability dP of finding an electron in $d\mathbf{x}_A$ around $\mathbf{x}_A \in R_A$ and another in $d\mathbf{x}_B$ around $\mathbf{x}_B \in R_B$ is

$$dP = d\mathbf{x}_A d\mathbf{x}_B \left(|\phi_\pm(\mathbf{x}_A, \mathbf{x}_B)|^2 + |\phi_\pm(\mathbf{x}_B, \mathbf{x}_A)|^2\right). \tag{21.7.26}$$

Note that using (21.7.13) with $N_\pm = 1$ gives

$$\phi_\pm(\mathbf{x}_A, \mathbf{x}_B) = \frac{1}{\sqrt{2}}\left(\phi_A(\mathbf{x}_A)\phi_B(\mathbf{x}_B) \pm \phi_B(\mathbf{x}_A)\phi_A(\mathbf{x}_B)\right) = \frac{1}{\sqrt{2}}\phi_A(\mathbf{x}_A)\phi_B(\mathbf{x}_B), \tag{21.7.27}$$

and, analogously,

$$\phi_\pm(\mathbf{x}_B, \mathbf{x}_A) = \pm\frac{1}{\sqrt{2}}\,\phi_B(\mathbf{x}_B)\phi_A(\mathbf{x}_A). \tag{21.7.28}$$

Therefore, from (21.7.26) we find that

$$\begin{aligned} dP &= d\mathbf{x}_A d\mathbf{x}_B\big(\tfrac{1}{2}|\phi_A(\mathbf{x}_A)\phi_B(\mathbf{x}_B)|^2 + \tfrac{1}{2}|\phi_B(\mathbf{x}_B)\phi_A(\mathbf{x}_A)|^2\big) \\ &= d\mathbf{x}_A d\mathbf{x}_B|\phi_A(\mathbf{x}_A)|^2\,|\phi_B(\mathbf{x}_B)|^2. \end{aligned} \tag{21.7.29}$$

This is in fact the probability for distinguishable particles. There is no need to symmetrize or antisymmetrize the separate electrons. We can treat them as distinguishable.

21.8 Occupation Numbers

Consider a system of N identical particles, each living in a vector space V. The N-particle quantum system lives in $V^{\otimes N}$. If the particles are bosons, the states lie in $\text{Sym}^N V$; if the particles are fermions, the states lie in $\text{Anti}^N V$. To construct such symmetric and antisymmetric states, we can start with a basis state in $V^{\otimes N}$ and apply the symmetrizer $\hat{S}$ or the antisymmetrizer $\hat{A}$. It turns out, however, that many different basis states in $V^{\otimes N}$ can lead to the *same* state in $\text{Sym}^N V$ and $\text{Anti}^N V$ after the application of the projectors. Moreover, the states in $\text{Sym}^N V$ and $\text{Anti}^N V$ are themselves long lists of superpositions of basis states. There is certainly a need for an economical but complete specification of the states in $\text{Sym}^N V$ and $\text{Anti}^N V$. One defines **occupation numbers** to distinguish basis states in $V^{\otimes N}$ that, after application of $\hat{S}$ or $\hat{A}$, are linearly independent, as well as to give a simple description of the physical states.

To describe occupation numbers concretely, we first introduce the basis states of the single-particle state space V. Let $|u_i\rangle$, with $i = 1, 2, \ldots$, form an orthonormal basis of V:

$$V = \text{span}\{|u_1\rangle, |u_2\rangle, \ldots\}. \tag{21.8.1}$$

For the N-particle system, basis states in $V^{\otimes N}$ take the form

$$|u_{i_1}\rangle_{(1)} \otimes \cdots |u_{i_N}\rangle_{(N)}, \tag{21.8.2}$$

where the subscripts $i_1, \ldots, i_N$ can take *any* values in the list of labels $1, 2, \ldots$ of the V basis states. By applying $\hat{S}$ or $\hat{A}$ to the full set of basis states in $V^{\otimes N}$, we can obtain a spanning set for $\text{Sym}^N V$ and $\text{Anti}^N V$. But the fact is that many different states in $V^{\otimes N}$ can give rise to the same states in $\text{Sym}^N V$ or in $\text{Anti}^N V$.

Given any basis state $|\omega\rangle \in V^{\otimes N}$, we will assign to it a set $\{n_1, n_2, \ldots\}$ of occupation numbers. The set includes an integer $n_i \geq 0$ for each basis vector $|u_i\rangle$ in V:

$$\underset{n_1}{|u_1\rangle}\,,\ \underset{n_2}{|u_2\rangle}\,,\ \ldots\,,\ \underset{n_i}{|u_i\rangle}\,,\ \ldots. \tag{21.8.3}$$

For any $i \geq 1$, the integer $n_i \geq 0$ is the number of times that $|u_i\rangle$ appears in the basis state $|\omega\rangle$. Thus, by inspection of $|\omega\rangle$ we can read all the occupation numbers $n_1, n_2, \ldots$. The full list of occupation numbers is as long as the dimension of V, and it could be infinite. The list of the *nonzero* occupation numbers, however, is finite and has at most

N elements because we are describing states of N particles. Briefly said, the occupation numbers tell us how many particles are in each of the available states furnished by V.

Example 21.3. *Occupation numbers for two particles in a three-state system.*

To illustrate the use of occupation numbers, consider a single-particle state space V with three basis states $|u_1\rangle$, $|u_2\rangle$, and $|u_3\rangle$. Assume we have two particles so that $N=2$. There are nine basis states for $V^{\otimes N} = V \otimes V$. For any basis state, we have three occupation numbers forming a list $\{n_1, n_2, n_3\}$, with $n_i \in 0, 1, 2$ telling us how many particles are in $|u_i\rangle$. Let us enumerate the $V \otimes V$ basis states and write down the corresponding occupation numbers. If both particles are in the same state, we have the following basis vectors:

$$
\begin{aligned}
&|u_1\rangle_{(1)} \otimes |u_1\rangle_{(2)} \;\Rightarrow\; \{2,0,0\},\\
&|u_2\rangle_{(1)} \otimes |u_2\rangle_{(2)} \;\Rightarrow\; \{0,2,0\},\\
&|u_3\rangle_{(1)} \otimes |u_3\rangle_{(2)} \;\Rightarrow\; \{0,0,2\}.
\end{aligned}
\tag{21.8.4}
$$

These basis vectors are automatically symmetric under the transposition (12), and therefore they could be used to represent bosons. If the two particles are in different states, however, we find:

$$
\begin{aligned}
&|u_1\rangle_{(1)} \otimes |u_2\rangle_{(2)},\; |u_2\rangle_{(1)} \otimes |u_1\rangle_{(2)} \;\Rightarrow\; \{1,1,0\},\\
&|u_1\rangle_{(1)} \otimes |u_3\rangle_{(2)},\; |u_3\rangle_{(1)} \otimes |u_1\rangle_{(2)} \;\Rightarrow\; \{1,0,1\},\\
&|u_2\rangle_{(1)} \otimes |u_3\rangle_{(2)},\; |u_3\rangle_{(1)} \otimes |u_2\rangle_{(2)} \;\Rightarrow\; \{0,1,1\}.
\end{aligned}
\tag{21.8.5}
$$

On each line we have two $V \otimes V$ basis states that lead to the same set of occupation numbers. This happens because the two particles are in different states, so there is the familiar exchange degeneracy. If we have bosons, by picking a symmetric superposition on each line, we get three possible states in $\mathrm{Sym}\, V^{\otimes 2}$. If we have fermions, by picking an antisymmetric superposition on each line we get three possible states in $\mathrm{Anti}\, V^{\otimes 2}$. In both cases the states are labeled by the occupation numbers. While there are a total of nine basis states in $V^{\otimes 2}$, there are only six possible sets of occupation numbers. All six can be used to describe states of bosons. Only three can be used to describe states of fermions.

A pictorial representation of occupation numbers makes matters quite clear. For the basis states of V, we use three lines, the lowest line for $|u_1\rangle$, the middle line for $|u_2\rangle$, and the top line for $|u_3\rangle$. A particle in a given state is represented by a small circle at the appropriate level. The six configurations representing the possible occupation numbers are shown in figure 21.8. All six configurations can be used to build states for bosons. Only the last three can be used to build states of fermions. □

Back to the general situation, it should be clear that the basis state $|\omega\rangle \in V^{\otimes N}$ and the state obtained acting with a permutation operator on $|\omega\rangle$ have exactly the same occupation numbers (as illustrated on each line of (21.8.5)). Moreover, two basis states in $V^{\otimes N}$ with the same occupation numbers can be mapped into each other by a permutation operator. It follows that they lead to the same state in $\mathrm{Sym}^N V$ and to the same state

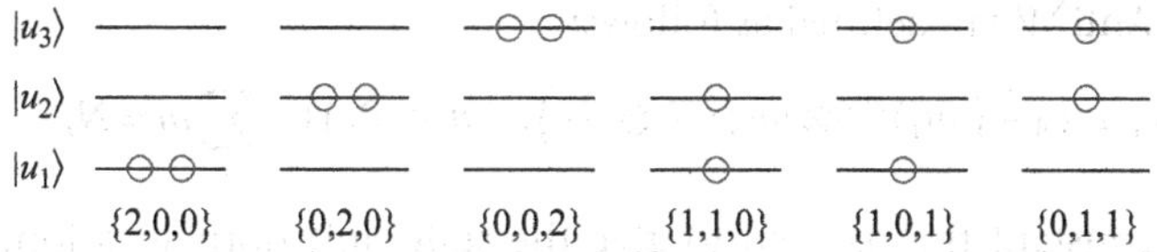

$|u_3\rangle$

$|u_2\rangle$

$|u_1\rangle$

{2,0,0} {0,2,0} {0,0,2} {1,1,0} {1,0,1} {0,1,1}

Figure 21.8
Two particles in a three-state system. The three levels are indicated by horizontal lines, and the particles are shown as small circles. There are six possible sets of occupation numbers.

(up to a sign) in $\text{Anti}^N V$. Two basis states in $V^{\otimes N}$ with different occupation numbers cannot be mapped into each other by a permutation operator. They must lead to different states in $\text{Sym}^N V$ and to different states in $\text{Anti}^N V$, unless they give zero. This shows that we can characterize *uniquely* the states in $\text{Sym}^N V$ and in $\text{Anti}^N V$ with occupation numbers.

Given the occupation numbers of a basis state in $V^{\otimes N}$, the associated basis state in $\text{Sym}^N V$ can be denoted as a ket labeled by the list of all occupation numbers:

$$|n_1, n_2, \ldots\rangle_S \in \text{Sym}^N V, \qquad n_i \geq 0, \quad \sum_i n_i = N. \tag{21.8.6}$$

The subscript S on the state reminds us that we have a state in $\text{Sym}^N V$. Since we have an N-particle state, the occupation numbers have to add up to N. Explicitly, the state is built by the action of the symmetrizer $\hat{S}$ on a basis state constructed from the list of occupation numbers:

$$|n_1, n_2, \ldots\rangle_S \equiv c_S \hat{S}\Big(\underbrace{|u_1\rangle \ldots |u_1\rangle}_{n_1 \text{ times}} \otimes \underbrace{|u_2\rangle \ldots |u_2\rangle}_{n_2 \text{ times}} \otimes \cdots\Big). \tag{21.8.7}$$

Here, c_S is a constant required to give the state unit normalization. More briefly, we can write the above state as

$$|n_1, n_2, \ldots\rangle_S \equiv c_S \hat{S}\Big(|u_1\rangle^{\otimes n_1} \otimes |u_2\rangle^{\otimes n_2} \otimes \cdots\Big), \tag{21.8.8}$$

where $|u_i\rangle^{\otimes n_i}$ is equal to 1 when $n_i = 0$. The states defined by occupation numbers form an orthonormal basis in $\text{Sym}^N V$ with an inner product:

$${}_S\langle n_1', n_2', \ldots | n_1, n_2, \ldots\rangle_S = \delta_{n_1, n_1'} \delta_{n_2, n_2'} \cdots. \tag{21.8.9}$$

The right-hand side is zero unless all occupation numbers agree. The space $\text{Sym}^N V$ relevant to identical bosons is spanned by kets with all possible lists of occupation numbers:

$$\text{Sym}^N V = \text{Span}\ \Big\{|n_1, n_2, \ldots\rangle_S \,\Big|\, \sum_i n_i = N,\ n_i \geq 0,\ \forall i\Big\}. \tag{21.8.10}$$

Here, the index i runs over the list of basis vectors in the single-particle state space V.

The case of fermions differs in one basic way: no occupation number can be larger than one. Any state with an occupation number two or larger is killed by $\hat{A}$ because it cannot be antisymmetrized; this is Pauli's exclusion principle at work. Orthonormal basis states in $\text{Anti}^N V$ are defined as follows:

$$|n_1, n_2, \ldots\rangle_A \equiv c_A \hat{A}\left(|u_1\rangle^{\otimes n_1} \otimes |u_2\rangle^{\otimes n_2} \otimes \cdots\right), \quad n_i \in \{0, 1\}, \quad \sum_i n_i = N, \tag{21.8.11}$$

where c_A is a constant that is used to give the state unit normalization. These states indeed satisfy

$${}_A\langle n_1', n_2', \ldots | n_1, n_2, \ldots\rangle_A = \delta_{n_1, n_1'} \delta_{n_2, n_2'} \cdots. \tag{21.8.12}$$

The space $\text{Anti}^N V$ relevant to identical fermions is spanned by kets with all possible lists of occupation numbers:

$$\text{Anti}^N V = \text{Span}\left\{|n_1, n_2, \ldots\rangle_A \,\Big|\, \sum_i n_i = N,\ n_i \in \{0, 1\}, \forall i\right\}. \tag{21.8.13}$$

Example 21.4. *Number of states of N identical particles in a k-level system.*

Let V be a k-dimensional vector space, visualized as a k-level system:

$$\dim V = k. \tag{21.8.14}$$

Consider N identical particles, each one living on V. We want to determine the dimension of $\text{Sym}^N V$, which is the number of possible N-boson states, as well as the dimension of $\text{Anti}^N V$, which is the number of possible N-fermion states. Note that the dimension of $\text{Anti}^N V$ vanishes unless $k \geq N$—that is, unless the number of levels is equal or larger than the number of particles. Both for bosons and for fermions, we must count the number of possible lists of occupation numbers. We imagine distributing the identical particles on the k levels without constraint for bosons and with the constraint that no more than one particle can be at any given level for fermions.

We can build some intuition considering the cases of low numbers of identical particles. With $N = 1$ (one particle), the number of boson states is k and so is the number of fermion states; in both cases the particle can be at any of the k levels of the system.

When $N = 2$ (two particles), we will have k states with both particles on the same level and $k(k-1)/2$ states with the particles at two different energy levels. Hence, the number of boson states is $\text{Sym}^2 V = k + k(k-1)/2 = k(k+1)/2$. For fermions there are no antisymmetric states with both particles at the same level. Hence, the number of fermion states is $\text{Anti}^2 V = k(k-1)/2$. For two particles on a $k = 3$ (three-level) system, this confirms our count in example 21.3: there are six boson states and three fermion states.

Let us now look at the number of boson states for arbitrary N. We need to count the number of ways N particles can occupy k energy levels. We approach this problem by thinking of the k energy levels as being separated by $k-1$ dividers and representing the N particles as N balls. We now have $N + k - 1$ objects (balls and dividers) that we can place in a line. Each different ordering of these objects is equivalent to a set of occupation numbers. Considering each object as distinguishable, there are $(N + k - 1)!$ possible orderings for those objects. However, the balls are indistinguishable, so we must divide by $N!$, and the dividers are also indistinguishable, so we must divide by $(k-1)!$. We then find that the number of boson states is

$$\dim \mathrm{Sym}^N V = \frac{(N+k-1)!}{N!(k-1)!} = \frac{k}{1} \cdot \frac{k+1}{2} \cdot \frac{k+2}{3} \cdots \frac{k+N-1}{N}. \tag{21.8.15}$$

For the general antisymmetric case, we have k levels and N particles, with $k \geq N$. Each particle must go into a different level. The number of possible assigments is equal to the number of ways we can choose N levels from the k available ones. Thus, we have

$$\dim \mathrm{Anti}^N V = \binom{k}{N} = \frac{k!}{N!(k-N)!}. \tag{21.8.16}$$

This is the number of possible N-particle states of fermions. □

Example 21.5. *States of multielectron atoms.*
The spectrum of bound states of the hydrogen atom is spanned by states specified by quantum numbers (n, ℓ, m, m_s); with $n = 1, 2, \ldots$; $\ell = 0, \ldots; n-1$; $m = -\ell, \ldots; \ell$; and $m_s = \pm\frac{1}{2}$, accounting for the electron spin. A multielectron (neutral) atom has $Z > 1$ electrons and a nucleus with Z protons, as well as some neutrons. It is a very complicated system: each electron feels the Coulomb attraction of the nucleus, but it also feels the Coulomb repulsion from all the other electrons. This repulsion makes matters difficult.

In a very rough approximation, one ignores the electron repulsion, in which case the full atomic Hamiltonian $\hat{H}_{\text{approx}}$ would be the sum of Z independent Hamiltonians, one for each electron:

$$\hat{H}_{\text{approx}} = \sum_{i=1}^{Z} \hat{H}_i, \qquad \hat{H}_i = \frac{\hat{\mathbf{p}}_i^2}{2m} - \frac{Ze^2}{r_i}. \tag{21.8.17}$$

The various Hamiltonians all commute, and each electron operates independently. The total energy is the sum of the energies of all the electrons. The spectrum of each electron is hydrogenic; that is, the quantum numbers are the same as those for hydrogen, but the energies and wave functions differ by the replacement $e^2 \to Ze^2$, which also implies that the Bohr radius changes as $a_0 \to a_0/Z$. This is intuitively clear; with Z protons, the nucleus pulls the electrons with greater strength, and the atom size scale is reduced. Let $\mathcal{H}_1$ denote the state space of a *single* electron in the Z-proton nucleus. It follows that the state space of the approximate Hamiltonian—that is, the state space of the Z mutually noninteracting electrons—is

$$\mathcal{H}_1 \otimes \cdots \otimes \mathcal{H}_1 = (\mathcal{H}_1)^{\otimes Z}. \tag{21.8.18}$$

Since electrons are fermions, no state can be occupied by more than one electron. As a result, Z different states must be occupied and a given state of the atom is specified by a list of Z states with occupation number one. The state of the atom is written by letting the antisymmetrizer $\hat{A}$ act on the tensor basis state that includes the list of all occupied states. It is useful to note that for each (n, ℓ, m) we can have two electrons, both using this spatial wave function, if the spin wave function is antisymmetric. The two electrons must therefore be in the singlet state, thus implementing the requisite antisymmetry under the exchange of the two electrons.

An **orbital** is defined to be the states of fixed quantum numbers (n, ℓ, m). This means that each orbital can have up to two electrons. All the states with a fixed principal quantum number n compose what is called a **shell**. For electron configurations it is customary to use a notation where $\ell=0$ is denoted by the letter s, $\ell=1$ is denoted by p, $\ell=2$ is denoted by d, and so on. Hydrogen has an electron configuration $1s$, with the number in front of the s referring to $n=1$ and s telling us it is an $\ell=0$ state. The next atom, helium, has an electron configuration $(1s)^2$, with the exponent telling us that the number of s electrons is two. In helium, a noble gas, the $n=1$ shell is filled. The next atom, lithium, has three electrons, with configuration $(1s)^2(2s)$, noting that the third electron is an $\ell=0$ electron in the $n=2$ shell. Of course, in this toy model of the multielectron atom all $n=2$ states are degenerate, and it is not clear which one should be used by the third electron. This degeneracy, however, is resolved in real atoms, as we will discuss below, and the above guess is correct. The next atom, beryllium, has four electrons and a configuration $(1s)^2(2s)^2$. By the time we reach neon, with $Z=10$, we fill the six states of the $2p$ orbitals so that the configuration is $(1s)^2(2s)^2(2p)^6$. The $n=2$ shell is filled in neon, which is also a noble gas.

If we now include the Coulomb interaction of the electrons, the orbitals survive but their energies shift. The general question is clear: for a fixed n, what happens to the degeneracy between the ℓ multiplets, with $\ell=0, 1, \ldots, n-1$? The decisive issue here is the possible "shielding" of the nuclear charge. An electron in a state that has a large probability of being found near the nucleus is not shielded from the nuclear charge. It feels the effect of all Z protons and thus will be most bound. Another electron that does not have a large probability of being found near the nucleus is mostly farther away so that other electrons lying near the nucleus shield some of the nuclear charge. Such an electron is less bound. Some care is needed to decide which states are more bound. If we look for the expectation value $\langle r \rangle$ as a guide, one actually finds that it is largest for $\ell=0$ and decreases as ℓ increases (see (11.4.20)). The most probable value of r in the orbitals also follows this behavior. This would suggest that the larger the ℓ value, the more bound the state. This is incorrect, however. It turns out that the key factor is the behavior of the wave function near the origin $r=0$. For a given ℓ, the wave function behaves as $\psi \sim r^\ell$ as r approaches zero. The larger the ℓ, the more suppressed the wave function near the origin. In fact, the $\ell=0$ state is the one with the most support at the origin, the one less shielded, and the one most bound. As ℓ increases, the wave function has less and less support in the neighborhood of the origin, and the state perceives a more shielded nuclear charge, resulting in less binding. In summary, for a given n the degeneracy is broken, with the lowest energy for $\ell=0$ and increasing energy as ℓ increases.

The ordering of the first few energy levels is represented schematically in figure 21.9. As shown on the left, in the $n=2$ shell, the $2s$ orbitals lie below the $2p$ orbitals, which are written as $2p_x, 2p_y$, and $2p_z$ (see example 19.3 for a discussion of these states). Then comes the $3s$ orbital, naturally below the $3p$ orbitals. But then, before one finds the $3d$ orbitals, the $4s$ orbital shows up! The ordering in which electronic states are filled as the atomic number increases follows this diagram. But as we move higher up along the

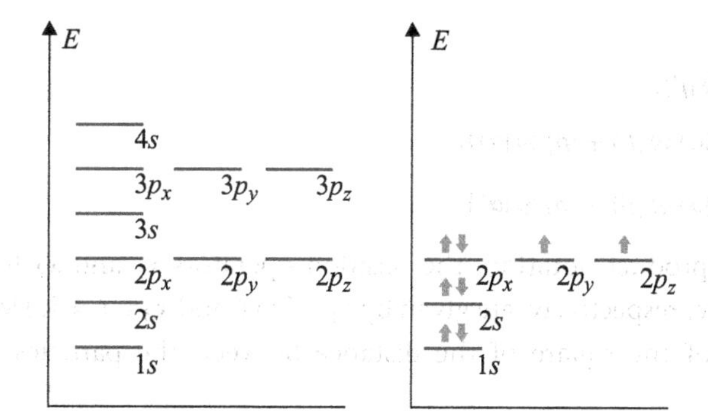

Figure 21.9
Left: A sketch (not to scale) of the ordering of the orbital energies for a multielectron atom. *Right*: For oxygen ($Z=8$), the $2p$ orbitals are not completely filled. The aligned spins in the $2p_y$ and $2p_z$ orbitals are part of what is known as Hund's rule.

periodic table, there is no definite ordering of energy levels. We encounter situations in which an orbital may be filled in an atom, but as we add one electron, the orbital loses one of the electrons. This first happens when going from vanadium ($Z=23$), which has a $(4s)^2$ orbital and $(3d)^3$ orbitals, to chromium ($Z=24$), which has a $(4s)^1$ orbital and $(3d)^5$ orbitals. On the right side of the figure, we show the electronic configuration of the oxygen atom. Its eight electrons fill the $1s$ and $2s$ orbitals completely but the $2p$ orbitals only partially. In fact the $2p$ orbitals $2p_x, 2p_y$, and $2p_z$ are filled sequentially. In the $Z=7$ atom (nitrogen), each $2p$ orbital has one electron.

The silver atom Ag ($Z=47$) played a role in the Stern-Gerlach experiment (section 12.2). Its electronic configuration is the following:

$$\text{Ag:}\ \ (1s)^2(2s)^2(2p)^6\Big|_{\text{Ne}}(3s)^2(3p)^6\Big|_{\text{Ar}}(4s)^2(3d)^{10}(4p)^6\Big|_{\text{Kr}}(4d)^{10}\Big|_{\text{Pd}}(5s)^1, \tag{21.8.19}$$

where we marked with vertical lines the points at which one would have the neon, argon, krypton, and palladium atoms. All orbitals, except for the $5s$, are filled with two electrons in a singlet state. So, effectively, the spin of the $5s$ electron, or more precisely, its associated magnetic dipole moment, is the one measured in the experiment. □

Problems

Problem 21.1. *Exchange "forces."*

As a result of symmetrization or antisymmetrization, bosons are effectively pulled together, while fermions are pushed apart. This effect is often misleadingly attributed to an exchange "force." The effect is real; the force is not.

1. *Spatial wave functions:* Let $|\alpha\rangle$ and $|\beta\rangle$ be two orthogonal single-particle states describing motion in one dimension x, with $\psi_\alpha(x)=\langle x|\alpha\rangle$ and $\psi_\beta(x)=\langle x|\beta\rangle$ the corresponding wave functions. Define the distinguishable (D), symmetric (S), and antisymmetric

(A) states by

$$|\Psi_D\rangle \equiv |\alpha\rangle \otimes |\beta\rangle,$$

$$|\Psi_S\rangle \equiv \frac{1}{\sqrt{2}}\big(|\alpha\rangle \otimes |\beta\rangle + |\beta\rangle \otimes |\alpha\rangle\big),$$

$$|\Psi_A\rangle \equiv \frac{1}{\sqrt{2}}\big(|\alpha\rangle \otimes |\beta\rangle - |\beta\rangle \otimes |\alpha\rangle\big).$$

Using tensor product notation, the position operators $\hat{x}_1$ and $\hat{x}_2$ for the first and second particle, respectively, are given by $\hat{x}_1 \equiv \hat{x} \otimes \mathbb{1}$ and $\hat{x}_2 \equiv \mathbb{1} \otimes \hat{x}$. Define the expectation value of the square of the distance between the particles in the various states:

$$R_X \equiv \langle \Psi_X | (\hat{x}_1 - \hat{x}_2)^2 | \Psi_X \rangle, \quad \text{for } X = D, S, A.$$

Calculate $R_S - R_D$ and $R_A - R_D$, and express them in terms of the matrix elements $x_{\alpha\alpha} = \langle \alpha|\hat{x}|\alpha\rangle$, $x_{\alpha\beta} = \langle \alpha|\hat{x}|\beta\rangle$, $x_{\beta\alpha} = \langle \beta|\hat{x}|\alpha\rangle$, and $x_{\beta\beta} = \langle \beta|\hat{x}|\beta\rangle$. Order R_D, R_S, and R_A from smallest to largest.

2. *Spins:* Now consider two spin s particles and single-particle states $|s, m_a\rangle$ and $|s, m_b\rangle$, with $m_a \neq m_b$. Define the distinguishable, symmetric, and antisymmetric states by

$$|\Psi_D\rangle \equiv |s, m_a\rangle \otimes |s, m_b\rangle,$$

$$|\Psi_S\rangle \equiv \frac{1}{\sqrt{2}}\big(|s, m_a\rangle \otimes |s, m_b\rangle + |s, m_b\rangle \otimes |s, m_a\rangle\big),$$

$$|\Psi_A\rangle \equiv \frac{1}{\sqrt{2}}\big(|s, m_a\rangle \otimes |s, m_b\rangle - |s, m_b\rangle \otimes |s, m_a\rangle\big).$$

Define $R_X \equiv \langle \Psi_X | (\hat{\mathbf{S}}_1 - \hat{\mathbf{S}}_2)^2 | \Psi_X \rangle$, for $X = D, S, A$. Calculate $R_S - R_D$ and $R_A - R_D$, leaving your answer in terms m_a, m_b, s, and the Kronecker deltas δ_{m_a, m_b+1} and δ_{m_a, m_b-1}, as appropriate. Order R_D, R_S, and R_A from smallest to largest.

Problem 21.2. *Probability distributions for distinguishable particles versus identical bosons.*

Consider two noninteracting particles, both of mass m, moving on a line under the influence of a one-dimensional harmonic oscillator potential. We consider two possibilities.

- Distinguishable particles: Particle one, with coordinate x_1, is in the first excited state φ_1 of the oscillator, while particle two, with coordinate x_2, is in the ground state φ_0. Let $\Psi_D(x_1, x_2)$ be the normalized wave function, and let $\mathcal{P}_D(x_1, x_2) = |\Psi_D(x_1, x_2)|^2$ be the probability distribution $\int dx_1 dx_2 \mathcal{P}_D(x_1, x_2) = 1$.
- Identical bosons: One particle is in the φ_0 state, and one is in the φ_1 state. Let $\Psi_B(x_1, x_2)$ be the normalized wave function, and let $\mathcal{P}_B(x_1, x_2) = |\Psi_B(x_1, x_2)|^2$ be the probability distribution $\int dx_1 dx_2 \mathcal{P}_B(x_1, x_2) = 1$.

In both cases, scale x_1, x_2 into unit-free u_1, u_2, with $x_i = L_0 u_i$, with L_0 the oscillator size parameter. Thus, work with functions $\Psi_D(u_1, u_2)$, $\Psi_B(u_1, u_2)$, $\mathcal{P}_D(u_1, u_2)$, and $\mathcal{P}_B(u_1, u_2)$, the latter two integrating to one under $\int du_1 du_2$.

1. Calculate $\Psi_D(u_1, u_2)$ and $\mathcal{P}_D(u_1, u_2)$. Show in the u_1, u_2 plane the regions where $\mathcal{P}_D(u_1, u_2)$ is large.

2. Calculate $\Psi_B(u_1, u_2)$ and $\mathcal{P}_B(u_1, u_2)$. Show in the u_1, u_2 plane the regions where $\mathcal{P}_B(u_1, u_2)$ is appreciable.
3. Since both probability distributions integrate to the same number, it is interesting to compare them. Find the region in the (u_1, u_2) plane where $\mathcal{P}_B(u_1, u_2) \geq \mathcal{P}_D(u_1, u_2)$. You should find that for $u_1 = u_2$ the bosonic distribution is twice the distinguishable one.
4. Consider the two following quantities. The first, for bosons, is the probability that one particle is in u_1, and the other is in u_2, within du_1 and du_2, respectively. The second, for the distinguishable particles in the state above, is the sum of two probabilities: the probability that particle one is in u_1, while particle two is in u_2, plus the probability that particle one is in u_2, while particle two is in u_1 (again, with ranges du_1 and du_2). Find the (u_1, u_2) regions where the first quantity exceeds the second one.

Problem 21.3. *Spin-dependent interaction and Heisenberg Hamiltonian.*

Consider a system of two noninteracting electrons with Hamiltonian $\hat{H}$ written as

$$\hat{H} = \hat{H}_0(\mathbf{r}_1) + \hat{H}_0(\mathbf{r}_2),$$

where $\hat{H}_0$ is a spin-independent Hamiltonian for an electron. Consider two possible one-particle *spatial* states $|\psi_1\rangle$ and $|\psi_2\rangle$, with wave functions $\psi_1(\mathbf{r})$ and $\psi_2(\mathbf{r})$, respectively. Assume, for simplicity, that ψ_1 and ψ_2 are distinct, orthogonal eigenstates of $\hat{H}_0$ of the *same* energy E_0. Moreover, assume that one electron is always in $|\psi_1\rangle$, while the other is always in $|\psi_2\rangle$. Finally, let $\hat{\mathbf{S}}_1$ and $\hat{\mathbf{S}}_2$ denote the spin one-half operators for the first and second electron.

1. Consider the two-electron states $|\Psi_a\rangle$ where the spatial wave function is antisymmetric. What is the dimensionality of this space of states? Let $\hat{\mathbf{S}} = \hat{\mathbf{S}}_1 + \hat{\mathbf{S}}_2$ be the total spin. Construct basis states that are eigenstates of $\hat{\mathbf{S}}^2$ and $\hat{S}_z$.
2. Repeat (1) for the case of two-electron states $|\Psi_s\rangle$ with symmetric spatial wave function.
3. So far all the states enumerated in (1) and (2) have the same energy. We now add to the free Hamiltonian $\hat{H}$ the term

$$\hat{H}' = -J\,\hat{\mathbf{S}}_1 \cdot \hat{\mathbf{S}}_2, \tag{1}$$

   representing the interaction of the magnetic moment of each particle with the magnetic field generated by the other. Here J is a constant with units of energy divided by $\hbar^2$. What are the eigenstates of the system, including the interaction $\hat{H}'$? What is the energy E_a of the states $|\Psi_a\rangle$? What is the energy E_s of the states $|\Psi_s\rangle$?
4. Now suppose we *ignore* the interaction in equation (1), and consider the free Hamiltonian $\hat{H}$ supplemented by $\hat{H}''$ representing the Coulomb repulsion between the two electrons:

$$\hat{H}'' = \frac{e^2}{|\mathbf{r}_1 - \mathbf{r}_2|}. \tag{2}$$

Use perturbation theory to compute the first-order contribution of $\hat{H}''$ to the energy difference $\epsilon \equiv E_a - E_s$ between the antisymmetric and symmetric states [Hint: you may use (20.3.4), which applies because the perturbation $\hat{H}''$ operator is diagonal in the basis of spin states of total angular momentum]. You may leave your answer in terms of e^2 and the integral I defined by

$$I = \iint d^3\mathbf{r}_1 d^3\mathbf{r}_2 \frac{F^*(\mathbf{r}_1)F(\mathbf{r}_2)}{|\mathbf{r}_1 - \mathbf{r}_2|}, \quad F(\mathbf{r}) \equiv \psi_1^*(\mathbf{r})\psi_2(\mathbf{r}).$$

5. Suppose that the system has no spin-spin interaction $\hat{H}'$ but does have the Coulomb repulsion $\hat{H}''$. Argue that such a two-electron system can be described by having no Coulomb interaction but rather an *effective* spin-spin interaction

$$\hat{H}'_{\text{eff}} = -J_{\text{eff}}\, \hat{\mathbf{S}}_1 \cdot \hat{\mathbf{S}}_2.$$

Determine the effective coupling J_{eff} in terms of the integral I and constants e and $\hbar$. From your intuition about the expected "distance" between particles in spatially symmetric or antisymmetric states, what should the sign of J be? [A rigorous proof, not asked for here, requires assessing the sign of I. You may take this as a challenge.]

Problem 21.4. *Three spin one particles.*

If we combine three spin one particles, we get the total spin angular momentum multiplets indicated on the right-hand side of the following equation:

$$\mathbf{1} \otimes \mathbf{1} \otimes \mathbf{1} = \mathbf{3} \oplus \mathbf{2} \oplus \mathbf{2} \oplus \mathbf{1} \oplus \mathbf{1} \oplus \mathbf{1} \oplus \mathbf{0}.$$

We wish to understand how the various multiplets of total angular momentum behave under permutations of the three particles.

1. Show that the total angular momentum operator $\hat{\mathbf{S}} = \hat{\mathbf{S}}_1 + \hat{\mathbf{S}}_2 + \hat{\mathbf{S}}_3$ is invariant under permutations: $\hat{P}_\alpha^\dagger\, \hat{\mathbf{S}}\, \hat{P}_\alpha = \hat{\mathbf{S}}$, for any permutation $\hat{P}_\alpha$ in the group S_3.
2. Show that if the highest S_z state of a multiplet of total angular momentum is totally symmetric or totally antisymmetric, so are the rest of the states in the multiplet.
3. Consider the total spin singlet $\mathbf{0}$. Justify the following claim: Since there is just one singlet, it must be either a symmetric or an antisymmetric state. Which is it?
4. There are two multiplets of total spin two: $\mathbf{2} \oplus \mathbf{2}$. Are they symmetric, antisymmetric, a mix, or neither? What happens when the multiplets are acted on by permutation operators?
5. There are three multiplets of spin one: $\mathbf{1} \oplus \mathbf{1} \oplus \mathbf{1}$. How many antisymmetric multiplets can we get? How many symmetric multiplets can we get? Once we have formed anti-symmetric and symmetric multiplets, how many multiplets without full symmetry do we get?

Problem 21.5. *Three particles on a three-level system (Cohen-Tannoudji).*

Consider a Hamiltonian $\hat{H}_0$ for the spatial degrees of freedom of a particle, with normalized eigenstates $\psi_1(x)$ of energy equal to zero, $\psi_2(x)$ of energy E_0, and $\psi_3(x)$ of energy $2E_0$, with $E_0 > 0$. Now consider three identical particles with Hamiltonian $\hat{H}$ given by

$$\hat{H} = \hat{H}_0(1) + \hat{H}_0(2) + \hat{H}_0(3).$$

Here, $\hat{H}_0(k)$ acts on the kth particle.

1. Assume the three particles are electrons. The energies for the three-particle states conceivably run from 0 to $6E_0$. Find the energy levels of the Hamiltonian $\hat{H}$ and their degeneracies.

   By the Pauli exclusion principle, there is only one state in which the three electrons have their spin up. Write out the normalized coordinate wave function $\phi(x_1, x_2, x_3)$ for this state. Write your answer in terms of the nine quantities $\psi_i(x_j)$, $i, j = 1, 2, 3$.
2. Assume the three particles are spin zero bosons. Find the energy levels of the Hamiltonian $\hat{H}$ and the degeneracies.
3. Assume the three particles have no spin and are distinguishable. Find the energy levels of the Hamiltonian $\hat{H}$ and the degeneracies.

Problem 21.6. *Decomposing the identity operator in the space $V \otimes V \otimes V$.*

For the case of two particles in the state space $V \otimes V$, the symmetrizer $\hat{S}$ and the antisymmetrizer $\hat{A}$ are complementary projectors: $\mathbb{1} = \hat{S} + \hat{A}$. This is not true for the vector space $V \otimes V \otimes V$ of three particles. For this case, in addition to the projectors $\hat{S}$ and $\hat{A}$ we need two additional operators.

1. Consider the operators $\hat{Y}$ and $\hat{Y}'$ defined as follows:

$$\hat{Y} \equiv (\mathbb{1} - (13))(\mathbb{1} + (12)),$$

$$\hat{Y}' \equiv (\mathbb{1} - (12))(\mathbb{1} + (13)).$$

   Find how $\hat{Y}$ and $\hat{Y}'$, together with $\hat{S}$ and $\hat{A}$, can provide a resolution of the identity: $\mathbb{1} = \hat{S} + \hat{A} + \cdots$.
2. Calculate the following operator products:

$$\hat{Y}\hat{Y}, \quad \hat{Y}'\hat{Y}', \quad \hat{Y}\hat{Y}', \quad \hat{Y}'\hat{Y}, \quad \hat{S}\hat{Y}, \quad \hat{S}\hat{Y}', \quad \hat{A}\hat{Y}, \quad \hat{A}\hat{Y}'.$$

   Only the first two require significant computation (use table 21.1 for the group S_3). The answers are either zero or can be written in terms of (some of) $\hat{Y}$, $\hat{Y}'$, $\hat{A}$, and $\hat{S}$.
3. Consider the six-dimensional space $V_{|u\rangle}$ spanned by $|u\rangle = |abc\rangle \in V \otimes V \otimes V$ and the other five states obtained by acting with the permutation group S_3 (assume $a \neq b \neq c$). What are the dimensions of the spaces obtained by acting with $\hat{S}$, $\hat{A}$, $\hat{Y}$, and $\hat{Y}'$ on $V_{|u\rangle}$?

Problem 21.7. *Interference in a four-photon process.*

Refer to example 21.2 (section 21.5) to analyze the situation in which four identical photons hit a balanced beam splitter (figure 21.6). Show that the outgoing state $|\Psi_{\text{out}}\rangle$

is given by

$$|\Psi_{\text{out}}\rangle = \sqrt{\tfrac{3}{8}}\,\big(|\uparrow\uparrow\uparrow\uparrow\rangle + |\downarrow\downarrow\downarrow\downarrow\rangle\big) - \sqrt{\tfrac{1}{24}}\big(|\uparrow\uparrow\downarrow\downarrow\rangle + |\uparrow\downarrow\uparrow\downarrow\rangle + |\uparrow\downarrow\downarrow\uparrow\rangle$$
$$+ |\downarrow\downarrow\uparrow\uparrow\rangle + |\downarrow\uparrow\downarrow\uparrow\rangle + |\downarrow\uparrow\uparrow\downarrow\rangle\big).$$

The four photons in each ket are ordered in standard form; $|\downarrow\uparrow\uparrow\downarrow\rangle$, for example, stands for the tensor product $|\downarrow\rangle_{(1)} \otimes |\uparrow\rangle_{(2)} \otimes |\uparrow\rangle_{(3)} \otimes |\downarrow\rangle_{(4)}$. What are the possible detections and their corresponding probabilities?

Problem 21.8. *Variational approach to the helium atom ground state.*

The helium atom has two electrons, $Z = 2$, and a nucleus with two protons and two neutrons. The Hamiltonian $\hat{H}$ is the sum of two single-electron Hamiltonians $\hat{H}_0$, each electron interacting with the Z-charged nucleus, plus the Coulomb repulsion term:

$$\hat{H} = \hat{H}_0(1) + \hat{H}_0(2) + \frac{e^2}{r_{12}}, \quad \hat{H}_0(i) = \frac{\hat{\mathbf{p}}_i^2}{2m} - \frac{Ze^2}{r_i}, \quad i = 1, 2,$$

where $r_1 = |\mathbf{x}_1|$, $r_2 = |\mathbf{x}_2|$, and $r_{12} = |\mathbf{x}_1 - \mathbf{x}_2|$. Here you will work out a variational estimate for the ground state energy E_{gs}, experimentially known to be $E_{gs,exp} = -78.8$ eV. For this purpose take a trial wave function

$$\psi(\mathbf{x}_1, \mathbf{x}_2) = \sqrt{\frac{Z'^3}{\pi a_0^3}}\, e^{-Z' r_1/a_0} \sqrt{\frac{Z'^3}{\pi a_0^3}}\, e^{-Z' r_2/a_0} = \frac{Z'^3}{\pi a_0^3}\, e^{-Z'(r_1+r_2)/a_0},$$

with $Z' > 0$, a number to be fixed. If $Z' = 2$, the above is the wave function for two noninteracting electrons, each in the $1s$ state of the atom. By introducing a Z' different from $Z = 2$, we are allowing the possible description of shielding of the nuclear charge.

1. Is the spatial wave function $\psi(\mathbf{x}_1, \mathbf{x}_2)$ normalized? What would be the normalized spin wave function consistent with the fermionic nature of the electrons? The spin wave function plays no role in the following.
2. Evaluate the expectation value of $\hat{H}_0(1)$ on the state $\psi(\mathbf{x}_1, \mathbf{x}_2)$. You may use the virial theorem to simplify your work. The answer should be written in terms of Z, Z', e^2, and a_0.
3. Evaluate the expectation value of the Coulomb repulsion term on the state $\psi(\mathbf{x}_1, \mathbf{x}_2)$. This computation is simplified with the help of this nontrivial integral, which you need not prove:

$$\int\!\!\int d^3u_1 d^2u_2 \frac{e^{-u_1-u_2}}{u_{12}} = 20\pi^2,$$

where $u_1 = |\mathbf{u}_1|$, $u_2 = |\mathbf{u}_2|$, and $u_{12} = |\mathbf{u}_1 - \mathbf{u}_2|$, with all u variables unit-free.

4. All in all, you should have the expectation value of $\hat{H}$ on the state $\psi(\mathbf{x}_1, \mathbf{x}_2)$ as

$$\langle \hat{H} \rangle_\psi = \big(2Z'^2 - 4ZZ' + \tfrac{5}{4}Z'\big)\frac{e^2}{2a_0}.$$

Show that the optimum Z' is $Z' = Z - \frac{5}{16} = 1.6875$, which gives an excellent variational estimate of the ground state energy: $E_{gr,var} = -77.5$ eV. The error is less than 2%!

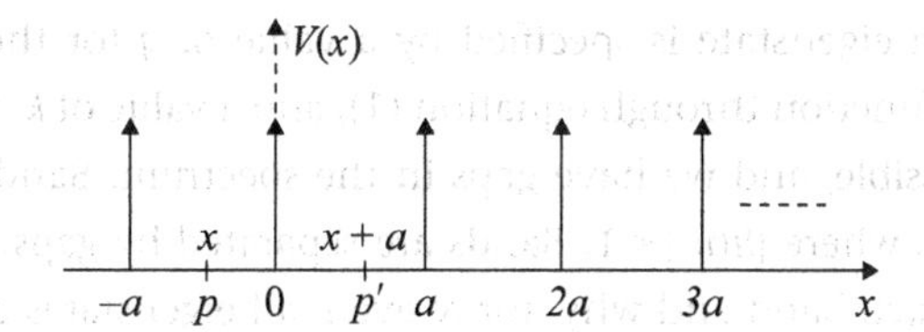

Figure 21.10
Periodic potential of delta functions.

Problem 21.9. *A periodic potential and bands of energy eigenstates.*

A simple model can be used to show that a periodic potential exhibits bands of thinly separated energy eigenstates. The potential satisfies $V(x+a)=V(a)$, with a some fixed distance. The Hamiltonian $\hat{H}$ commutes with the operator $T_a=\exp(-i\hat{p}a/\hbar)$ that implements a translation by a. It follows that we can simultaneously diagonalize both $\hat{H}$ and T_a. Since T_a is unitary, its possible eigenvalues are pure phases, and the energy eigenstates $|\psi\rangle$ will satisfy $T_a|\psi\rangle=e^{-iqa}|\psi\rangle$, introducing a wave number $q\in\mathbb{R}$ to parameterize the phase. Hitting this from the left with $\langle x|$, one finds $\psi(x-a)=e^{-iqa}\psi(x)$ or, equivalently,

$$\psi(x+a)=e^{iqa}\psi(x). \tag{1}$$

To make the model more realistic, we assume that it extends only over an interval of length Na along the x-axis. We impose a periodic boundary condition, making the interval a circle:

$$\psi(x+Na)=\psi(x). \tag{2}$$

1. Show that equations (1) and (2) imply the quantization condition

$$qa=\frac{2\pi n}{N}, \quad \text{with } n=0,1,\ldots,N-1. \tag{3}$$

The range of values for n gives the set of inequivalent T_a eigenvalues.

Let the potential consist of N delta functions along the x-axis. We can place them at positions $x=-a,0,a,\ldots,(N-1)a$, with the first and the last identified. The potential, shown in figure 21.10, is then $V(x)=\alpha\sum_{l=-1}^{N-2}\delta(x-la)$, with $\alpha>0$.

2. Consider the expression for an energy eigenstate $\psi(x)$ with energy $E=\hbar^2k^2/(2m)$, valid in the range $x\in(0,a)$:

$$\psi(x)=A\cos kx+B\sin kx, \quad x\in(0,a). \tag{4}$$

Here, A and B are constant coefficients. Note in the figure the points p and p', separated by a distance a, and the relation $\psi|_p=e^{-iqa}\psi|_{p'}$. Use this relation and equation (4) to write an expression for $\psi(x)$ when $x\in(-a,0)$. Use the continuity of ψ and the proper discontinuity of ψ' at $x=0$ to eliminate the coefficients A and B and prove that

$$\cos qa=\cos u+\gamma\frac{\sin u}{u}\equiv h(u), \tag{5}$$

with $u\equiv ka$ and $\gamma\equiv\frac{m\alpha a}{\hbar^2}$, a unit-free version of the strength α of the potential.

3. A solution for an eigenstate is specified by a value of q (or the integer n), which affects the wave function through equation (1), and a value of k. When $|h(u)| > 1$, no solutions are possible, and we have gaps in the spectrum. Bands of solutions exist in the u intervals where $|h(u)| \leq 1$. Bands are separated by gaps. Explain why there are N states on each band and why, for N even, all eigenstates except two come in degenerate pairs. Consider the case $N = 16$ and $\gamma = 7$. Plot $h(u)$ and the horizontal lines for the allowed values of $\cos qa$. Find the u interval for the first band. How many bands are there with $u < 10$?

Problem 21.10. *Fermi surface for a harmonic trap.*

Imagine populating some potential with noninteracting electrons, where each electron is placed in the lowest-energy state available to it. The energy of the last electron to be added to the potential is known as the Fermi energy E_F. The Fermi surface comprises the set of states of Fermi energy. For free electrons in 3D, the Fermi surface is visualized as a two-dimensional sphere in momentum space.

The concept of a Fermi surface and a Fermi energy extends beyond the free-electron gas model. Consider a noninteracting electron gas in a two-dimensional isotropic harmonic trap—namely, N electrons confined to a potential

$$V(x, y) = \tfrac{1}{2} m_e \omega^2 (x^2 + y^2).$$

The electrons are confined to the (x, y) plane, and we ignore all electron-electron interactions. You may assume that N is very large so that sums can be replaced by suitable integrals. [In statistical physics, this problem would describe a zero-temperature system.]

1. In a harmonic trap, the one-particle energy eigenstates are not labeled by momenta, as in the free gas. Construct the relevant one-particle states and energies in the two-dimensional harmonic trap. If the states are characterized by two integers $n_x, n_y \geq 0$ (with ground state $n_x = n_y = 0$) and spin m_s, what is the energy cost of adding an electron with n_x, n_y, and m_s? What is the degeneracy of the second excited state?
2. The Fermi surface can be represented in the positive quadrant of a plane with axes n_x and n_y. Find a mathematical description of the Fermi surface when the Fermi energy is E_F. Write your equation for the surface as a function $n_y(n_x)$ in terms of n_x, E_F, $\hbar$, and ω.
3. For large N, what is the Fermi energy E_F as a function of the total number N of electrons? What is the total energy E_{tot} of the N-particle ground state? Calculate the chemical potential $\mu = \frac{\partial E_{\text{tot}}}{\partial N}$. How is it related to the Fermi energy?
4. In a harmonic trap, the size of the Fermi "cloud" is the Fermi radius r_F, where r_F^2 is the expectation value $\langle \hat{x}^2 + \hat{y}^2 \rangle$ for an electron at the Fermi surface. Relate r_F to E_F.

III Applications

22 Density Matrix and Decoherence

Incomplete knowledge about a quantum system adds a new layer of randomness to the state of the system, which can no longer be described by a vector in the state space, a pure state. The description requires ensembles that represent mixed states. The essence of ensembles is encoded in the density matrix, a Hermitian, positive semidefinite operator of unit trace. For quantum systems with two interacting subsystems, also called bipartite systems, the state of any subsystem is generally a mixed state which requires a density matrix for its description, even when the total system is in a pure state. The structure of pure, entangled states of a bipartite system is captured by the Schmidt decomposition, which relates it to the density matrices of the subsystems. We look at decoherence, the process in which the pure state of a subsystem becomes mixed, and consider a phenomenological description of the process via the Lindblad equation. We conclude with a look at measurements in quantum mechanics, contrasting the postulates of the Copenhagen interpretation with insights from decoherence and other interpretations.

22.1 Ensembles and Mixed States

We have seen that probabilities play a central and inescapable role in quantum mechanics. This is quite striking, given that in classical physics probabilities arise *only* due to lack of knowledge about the system. If we toss dice and are unable or unwilling to investigate the myriad factors that in principle determine the outcome, we have to consider probabilities for the possible results. In quantum mechanics, however, perfect knowledge still does not do away with probabilities. Consider a state $|\psi\rangle \in V$, with V an N-dimensional complex vector space. Even if the state is known exactly, its properties, as defined by the observables in the theory, are only determined probabilistically. This is not for lack of information, we believe. We have seen that, at least in the simplest setups, local hidden variables carrying information whose absence leads to probabilities are not consistent with experiment. Given the probabilistic interpretation of quantum mechanics, experiments are understood in the framework provided by an *ensemble*: multiple copies of the quantum system, all in the same state $|\psi\rangle$. Measurements performed on each of the elements of the ensemble can be used to confirm the expected probabilities. This is the *intrinsic* randomness of quantum mechanics.

Interestingly, the randomness that arises in classical mechanics due to lack of knowledge, which requires probabilities, has a counterpart in quantum mechanics and adds a *new layer of randomness* to the theory. We will first consider how this new layer also arises in quantum mechanics due to incomplete knowledge. As we will see, in this situation it is useful to consider more general kinds of ensembles. At a more fundamental level, however, we will note that this new randomness arises naturally in quantum mechanics *even* with complete knowledge. This happens in the description of a *subsystem* that happens to be entangled with the rest of the system. This is a good reason to view this new layer of randomness as a general feature of quantum mechanics. As we discuss the issues and complications associated with general ensembles, we will be led to the concept of a density matrix, an *operator* on the state space of the theory that encodes the quantum state of the system and includes this added layer of randomness. A **pure state** is a familiar state $|\psi\rangle \in V$, a vector in the Hilbert space of the theory, a wave function. On the other hand, if we have extra randomness and the state of the quantum system cannot be described by a vector in V, we have a **mixed state.**

To show how lack of knowledge introduces randomness in quantum mechanics, let us reconsider the Stern-Gerlach experiment. In this experiment the beam of silver atoms that emerges from the hot oven is unpolarized: the spin one-half state of the atoms is random. If we denote the spin state of an atom as $|\mathbf{n}\rangle$ with $\mathbf{n}$ a unit vector, the different atoms have vectors $\mathbf{n}$ pointing in random directions. Can we find a quantum state $|\psi\rangle$ whose intrinsic randomness affords a description of the atoms in the beam as an ensemble of $|\psi\rangle$? The answer is clearly no. The general state is

$$|\psi\rangle = a_+|+\rangle + a_-|-\rangle, \quad a_+, a_- \in \mathbb{C}, \tag{22.1.1}$$

with $|\pm\rangle$ the familiar $\hat{S}_z$ eigenstates. The state $|\psi\rangle$, a pure state, is fixed when the coefficients a_+ and a_- are fixed, but this also fixes the direction $\mathbf{n}$ of the spin state. Thus, a state $|\psi\rangle$ as above does not describe states $|\mathbf{n}\rangle$ with random $\mathbf{n}$.

While we will deal with the case of random $\mathbf{n}$ later, let us consider a simpler situation. Assume you have an oven in which 50% of the atoms come out polarized as $|+\rangle$ and the other 50% come out polarized as $|-\rangle$. We can describe the beam by writing the pairs $(p_i, |\psi_i\rangle)$ in which we give the probability p_i of a given atom to be in the quantum state $|\psi_i\rangle$. For the situation we just described, we would write

$$E_z = \left\{ \left(\tfrac{1}{2}, |+\rangle\right),\ \left(\tfrac{1}{2}, |-\rangle\right) \right\}. \tag{22.1.2}$$

We used the label E_z for the *ensemble* of z-polarized states. We say that this ensemble has two *entries*, each entry consisting of a state and its probability. The ensemble here is providing a representation of the mixed state of our system. We can visualize the collection of atoms as a very large ensemble built by joining two equal-size ensembles, one built solely from states that are all $|+\rangle$, and the other built solely from states that are all $|-\rangle$. This is a more general ensemble than one in which all copies of the system are in the same quantum state.

For a general ensemble E associated to a quantum system with state space V, we have a list of states and probabilities:

$$E = \Big\{ (p_1, |\psi_1\rangle), \ldots, (p_n, |\psi_n\rangle) \Big\}, \quad p_1, \ldots, p_n > 0, \quad p_1 + \cdots + p_n = 1. \tag{22.1.3}$$

Here $n \geq 1$ is an integer denoting the number of entries in the ensemble. The ensemble provides a description of a general mixed state. The states $|\psi_a\rangle \in V$ above are all normalized:

$$\langle \psi_a | \psi_a \rangle = 1 \quad \text{for all} \quad a = 1, \ldots, n. \tag{22.1.4}$$

However, they are *not* required to be orthogonal to each other. We can imagine the ensemble E containing a large number M of copies of the system, with $p_a \cdot M$ copies in the state $|\psi_a\rangle$, for each $a = 1, \ldots, n$. The number n need not be related to the dimensionality $\dim V$ of the state space. We can have $n = 1$, in which case $p_1 = 1$, and the ensemble represents a pure state $|\psi_1\rangle$; all elements of the ensemble are in this state. For $n \geq 2$, we have a mixed state. We can also have $n > \dim V$ since the states $|\psi_a\rangle$ are not required to be linearly independent. In fact, nothing goes wrong if $n = \infty$, and the ensemble contains an infinite set of entries.

If $\hat{Q}$ denotes a Hermitian operator we are to measure, its expectation value $\langle \hat{Q} \rangle_E$ in the ensemble E is given by

$$\langle \hat{Q} \rangle_E = \sum_{a=1}^{n} p_a \langle \psi_a | \hat{Q} | \psi_a \rangle = p_1 \langle \psi_1 | \hat{Q} | \psi_1 \rangle + \cdots + p_n \langle \psi_n | \hat{Q} | \psi_n \rangle. \tag{22.1.5}$$

This is clear if we imagine measuring $\hat{Q}$ on the full ensemble E. The expectation value $\langle \psi_a | \hat{Q} | \psi_a \rangle$ of $\hat{Q}$ in the ath subensemble of states $|\psi_a\rangle$ must be weighted by the probability p_a that gives the fraction of all states in E that are in the subensemble. Then we must add the contributions from all values of a. In our example above, where silver atoms emerge as described by the ensemble E_z in (22.1.2), we would see that

$$\langle \hat{Q} \rangle_{E_z} = \tfrac{1}{2} \langle + | \hat{Q} | + \rangle + \tfrac{1}{2} \langle - | \hat{Q} | - \rangle. \tag{22.1.6}$$

Suppose, however, that you are now in possession of an oven that produces 50% of atoms in the state $|x; +\rangle$ and the other 50% in the state $|x; -\rangle$. The ensemble E_x here would be

$$E_x = \Big\{ \big(\tfrac{1}{2}, |x; +\rangle\big), \ \big(\tfrac{1}{2}, |x; -\rangle\big) \Big\}. \tag{22.1.7}$$

The expectation value of $\hat{Q}$ in this ensemble is

$$\langle \hat{Q} \rangle_{E_x} = \tfrac{1}{2} \langle x; + | \hat{Q} | x; + \rangle + \tfrac{1}{2} \langle x; - | \hat{Q} | x; - \rangle. \tag{22.1.8}$$

A curious result emerges if we use $|x; \pm\rangle = \frac{1}{\sqrt{2}}(|+\rangle \pm |-\rangle)$ to rewrite $\langle \hat{Q} \rangle_{E_x}$:

$$\langle \hat{Q} \rangle_{E_x} = \tfrac{1}{4} \big((\langle + | + \langle - |) \hat{Q} (|+\rangle + |-\rangle) \big) + \tfrac{1}{4} \big((\langle + | - \langle - |) \hat{Q} (|+\rangle - |-\rangle) \big). \tag{22.1.9}$$

The off-diagonal matrix elements of $\hat{Q}$ cancel out, and we are left with

$$\langle\hat{Q}\rangle_{E_x} = \tfrac{1}{2}\langle+|\hat{Q}|+\rangle + \tfrac{1}{2}\langle-|\hat{Q}|-\rangle = \langle\hat{Q}\rangle_{E_z}. \tag{22.1.10}$$

The expectation values are identical in the two ensembles E_z and E_x. Since this is true for arbitrary observables, we must conclude that no matter how different the ensembles are they are indistinguishable and thus physically equivalent. Both ensembles in fact represent the same beam coming out of the oven; they represent the same mixed state. With rather different ensembles turning out to be equivalent, we are led to find a better way to represent the mixed quantum state of a particle in the beam. This will be done with density matrices.

Example 22.1. *Unpolarized ensemble.*

The oven in the Stern-Gerlach experiment produces unpolarized silver atoms. We wish to write the expectation value of $\hat{Q}$ in this ensemble and compare it with the result for the E_z ensemble.

In an unpolarized state, the values of $\mathbf{n}$ are uniformly distributed over solid angle. Since the total solid angle is 4π, the probability that the vector $\mathbf{n}$ is within a solid angle $d\Omega$ is $d\Omega/(4\pi)$. The unpolarized ensemble E_{unp} is defined with an infinite number of entries composed by probabilities and states for all possible $d\Omega$:

$$E_{\text{unp}} = \bigcup_{d\Omega}\left(\tfrac{d\Omega}{4\pi}, |\mathbf{n}(\theta,\phi)\rangle \right), \quad |\mathbf{n}(\theta,\phi)\rangle = \cos\tfrac{\theta}{2}|+\rangle + \sin\tfrac{\theta}{2}e^{i\phi}|-\rangle. \tag{22.1.11}$$

The expectation value of any observable $\hat{Q}$ in this ensemble is obtained by integration:

$$\begin{aligned} \langle\hat{Q}\rangle_{E_{\text{unp}}} &= \int \frac{d\Omega}{4\pi}\langle\mathbf{n}(\theta,\phi)|\,\hat{Q}\,|\mathbf{n}(\theta,\phi)\rangle \\ &= \frac{1}{4\pi}\int \sin\theta d\theta d\phi \big(\cos\tfrac{\theta}{2}\langle+| + \sin\tfrac{\theta}{2}e^{-i\phi}\langle-|\big)\hat{Q}\big(\cos\tfrac{\theta}{2}|+\rangle + \sin\tfrac{\theta}{2}e^{i\phi}|-\rangle\big). \end{aligned} \tag{22.1.12}$$

The integral over ϕ kills the off-diagonal matrix elements of $\hat{Q}$, and we find that

$$\langle\hat{Q}\rangle_{E_{\text{unp}}} = \tfrac{1}{2}\int_0^{\pi} \sin\theta d\theta \big(\cos^2\tfrac{\theta}{2}\langle+|\hat{Q}|+\rangle + \sin^2\tfrac{\theta}{2}\langle-|\hat{Q}|-\rangle\big). \tag{22.1.13}$$

Both integrals evaluate to one, and the result is

$$\langle\hat{Q}\rangle_{E_{\text{unp}}} = \tfrac{1}{2}\langle+|\hat{Q}|+\rangle + \tfrac{1}{2}\langle-|\hat{Q}|-\rangle). \tag{22.1.14}$$

This is once more the same expectation value we found in the ensembles E_z and E_x. This shows that the unpolarized ensemble E_{unp} is in fact physically equivalent to the ensembles where half the states are polarized in one direction and the other half in the opposite direction. We checked this for states along z and along x. You should now check it for arbitrary direction. □

Exercise 22.1. *Prove that the ensemble where 50% of the states are $|\mathbf{n};+\rangle$ and the other 50% are $|\mathbf{n};-\rangle$, for arbitrary but fixed unit vector $\mathbf{n}$, is physically equivalent to the unpolarized ensemble.*

An even simpler example of quantum states described by ensembles is provided by a pair of entangled states. Let Alice and Bob each have one of two entangled spin one-half states. The entangled state $|\psi_{AB}\rangle$ they share is the singlet state of total spin equal to zero:

$$|\psi_{AB}\rangle = \frac{1}{\sqrt{2}}\big(|+\rangle_A|-\rangle_B - |-\rangle_A|+\rangle_B\big). \tag{22.1.15}$$

Assume Alice measures the spin of her state along the z-direction. If Alice gets $|+\rangle$, then the state of Bob is $|-\rangle$; if Alice gets $|-\rangle$, the state of Bob is $|+\rangle$. The state of Bob is known if we know the measurement Alice did *and* the result she found. If we do not know the result of her measurement, the situation for Bob is less clear.

Suppose all we know is that Alice measured along the z-direction. What then is the state of Bob's particle? To answer this we can again think in terms of an ensemble in which each element contains the entangled pair $|\psi_{AB}\rangle$. If Alice measures along z, about half of the time she will get $|+\rangle$, and the other half of the time she will get $|-\rangle$. As a consequence, in half of the elements of the ensemble the state of Bob will be $|-\rangle$, and in the other half, the state of Bob will be $|+\rangle$. The state of Bob can be described by the ensemble E_{Bob} that reads

$$E_{\text{Bob}} = \Big\{\big(\tfrac{1}{2}, |+\rangle\big), \big(\tfrac{1}{2}, |-\rangle\big)\Big\}. \tag{22.1.16}$$

Suppose, instead, that Alice decides to measure in an arbitrary direction $\mathbf{n}$. To analyze this, it is convenient to use the rotational invariance of the singlet state to rewrite it as follows:

$$|\psi_{AB}\rangle = \frac{1}{\sqrt{2}}\big(|\mathbf{n};+\rangle_A|\mathbf{n};-\rangle_B - |\mathbf{n};-\rangle_A|\mathbf{n};+\rangle_B\big). \tag{22.1.17}$$

If Alice measures and the result is not known, the state of Bob is again an ensemble. Since the probabilities that she finds $|\mathbf{n};-\rangle$ and $|\mathbf{n};+\rangle$ are the same, the correlations in the entangled state imply that this time the ensemble for Bob's state is

$$E_{\text{Bob}} = \Big\{\big(\tfrac{1}{2}, |\mathbf{n};+\rangle\big), \big(\tfrac{1}{2}, |\mathbf{n};-\rangle\big)\Big\}. \tag{22.1.18}$$

The two ensembles that we get, from Alice measuring along z and along $\mathbf{n}$, are equivalent, as we demonstrated before by considering expectation values. Again, we wish to have a better understanding of why the state of Bob's particle did not depend on the direction Alice used to make her measurement.

Example 22.2. *Absence of a pure-state description of an entangled particle.*

Let us now consider again the same entangled state of two particles, one held by Alice and the other by Bob:

$$|\psi_{AB}\rangle = \frac{1}{\sqrt{2}}\big(|+\rangle_A|-\rangle_B - |-\rangle_A|+\rangle_B\big). \tag{22.1.19}$$

Is there a state $|\psi_A\rangle$ of Alice's particle that summarizes all we know about this particle?

If such a state existed, we would require the expectation value of any observable $\hat{Q}$ in $|\psi_A\rangle$ to be equal to the expectation value of $\hat{Q}\otimes\mathbb{1}$ in $|\psi_{AB}\rangle$:

$$\langle\psi_A|\hat{Q}|\psi_A\rangle = \langle\psi_{AB}|\hat{Q}\otimes\mathbb{1}|\psi_{AB}\rangle? \tag{22.1.20}$$

We will see that there is *no* such state $|\psi_A\rangle$. No pure state can represent Alice's state if it is entangled with another state. To see this we examine the cases when $\hat{Q}$ is σ_x, σ_y, and σ_z. Note that

$$\begin{aligned}
\sigma_x \otimes \mathbb{1}|\psi_{AB}\rangle &= \tfrac{1}{\sqrt{2}}\big(|-\rangle_A|-\rangle_B - |+\rangle_A|+\rangle_B\big),\\
\sigma_y \otimes \mathbb{1}|\psi_{AB}\rangle &= i\tfrac{1}{\sqrt{2}}\big(|-\rangle_A|-\rangle_B + |+\rangle_A|+\rangle_B\big), \qquad (22.1.21)\\
\sigma_z \otimes \mathbb{1}|\psi_{AB}\rangle &= \tfrac{1}{\sqrt{2}}\big(|+\rangle_A|-\rangle_B + |-\rangle_A|+\rangle_B\big).
\end{aligned}$$

It follows quickly that all three expectation values vanish:

$$\langle\psi_{AB}|\sigma_x \otimes \mathbb{1}|\psi_{AB}\rangle = \langle\psi_{AB}|\sigma_y \otimes \mathbb{1}|\psi_{AB}\rangle = \langle\psi_{AB}|\sigma_z \otimes \mathbb{1}|\psi_{AB}\rangle = 0. \qquad (22.1.22)$$

If a pure-state representative $|\psi_A\rangle$ of Alice's particle exists, then it must then satisfy

$$\langle\psi_A|\sigma_x|\psi_A\rangle = \langle\psi_A|\sigma_y|\psi_A\rangle = \langle\psi_A|\sigma_z|\psi_A\rangle = 0. \qquad (22.1.23)$$

There is in fact a simple proof that the state $|\psi_A\rangle$ with vanishing expectation values for σ_x, σ_y, and σ_z does not exist. Any spin state points somewhere, and therefore $|\psi_A\rangle = \gamma|\mathbf{n}\rangle$, for some constant $\gamma \neq 0$ and some direction $\mathbf{n}$. But we also have $\langle\mathbf{n}|\mathbf{n}\cdot\boldsymbol{\sigma}|\mathbf{n}\rangle = 1$. This means $\mathbf{n}\cdot\boldsymbol{\sigma}$ has a nonzero expectation value in $|\psi_A\rangle$, but this is impossible if all three Pauli matrices have zero expectation value in $|\psi_A\rangle$. □

We have thus shown that there is no pure state representing the quantum state of Alice's particle when entangled. How do we describe such a quantum state? By using a density matrix. The description of the state of an entangled particle is given in all generality in section 22.4.

22.2 The Density Matrix

We will now see how the information that defines a general ensemble can be used to construct an interesting operator ρ acting on the Hilbert space V of the theory. The operator $\rho \in \mathcal{L}(V)$ is called a density matrix. Our work begins with the general ensemble of equation (22.1.3):

$$E = \Big\{(p_1, |\psi_1\rangle), \ldots, (p_n, |\psi_n\rangle)\Big\}, \quad p_1, \ldots, p_n > 0, \quad p_1 + \cdots + p_n = 1. \qquad (22.2.1)$$

We recall that for this ensemble the expectation value of any observable $\hat{Q}$ is given by

$$\langle\hat{Q}\rangle_E = \sum_{a=1}^{n} p_a\langle\psi_a|\hat{Q}|\psi_a\rangle. \qquad (22.2.2)$$

We now use the trace identity $\text{tr}\big(|u\rangle\langle w|\big) = \langle w|u\rangle$, derived in (14.9.27), to rewrite the above expectation value in terms of the trace of $\hat{Q}|\psi_a\rangle\langle\psi_a|$:

$$\langle\hat{Q}\rangle_E = \sum_{a=1}^{n} p_a \text{tr}\big(\hat{Q}|\psi_a\rangle\langle\psi_a|\big). \qquad (22.2.3)$$

For any constant p and any matrix A, we have $p\,\text{tr}A = \text{tr}(pA)$, and therefore

$$\langle\hat{Q}\rangle_E = \sum_{a=1}^{n} \text{tr}\big(\hat{Q}p_a|\psi_a\rangle\langle\psi_a|\big) = \text{tr}\Big(\hat{Q}\sum_{a=1}^{n} p_a|\psi_a\rangle\langle\psi_a|\Big), \tag{22.2.4}$$

where we also recalled the linearity property $\sum_i \text{tr}(A_i) = \text{tr}(\sum_i A_i)$. This result shows that *all* the relevant information about the ensemble E is encoded in the operator inside the trace, to the right of $\hat{Q}$. This operator will be called the density matrix operator $\rho_E \in \mathcal{L}(V)$ associated to the ensemble or mixed state E:

$$\text{Density matrix for } E: \quad \rho_E \equiv \sum_{a=1}^{n} p_a|\psi_a\rangle\langle\psi_a|. \tag{22.2.5}$$

Indeed, all the information about E is encoded in ρ_E because all we compute are expectation values of observables, and for them we now find that

$$\langle\hat{Q}\rangle_E = \text{tr}(\hat{Q}\rho_E). \tag{22.2.6}$$

The expectation value of $\hat{Q}$ is obtained by multiplying $\hat{Q}$ by ρ_E and taking the trace.

The density matrix makes immediately clear some of the results we discussed before. For the E_z and E_x ensembles introduced in (22.1.2) and (22.1.7),

$$\begin{aligned}\rho_{E_z} &= \tfrac{1}{2}|+\rangle\langle+| + \tfrac{1}{2}|-\rangle\langle-| = \tfrac{1}{2}\mathbb{1},\\ \rho_{E_x} &= \tfrac{1}{2}|x;+\rangle\langle x;+| + \tfrac{1}{2}|x;-\rangle\langle x;-| = \tfrac{1}{2}\mathbb{1},\end{aligned} \tag{22.2.7}$$

where we simply used two different resolutions of the identity $\mathbb{1}$, one for the $|\pm\rangle$ basis states (*first line*) and one for the $|x;\pm\rangle$ basis states (*second line*). The two density matrices are identical, explaining why all observables in the two ensembles are the same. Recall that these ensembles represented unpolarized beams, which are, arguably, maximally random beams. We now see that for such a random state the density matrix is a multiple of the identity matrix. While a detailed computation would confirm it, the density matrix for the unpolarized ensemble E_{unp} is also $\mathbb{1}/2$. We know this is true because we showed that the expectation values in this ensemble are the same as in the E_z ensemble.

Remarks:

1. The density matrix is a Hermitian operator.

   This is manifest because the p_i are real, and $(|\chi\rangle\langle\eta|)^\dagger = |\eta\rangle\langle\chi|$, implying that $(|\psi\rangle\langle\psi|)^\dagger = |\psi\rangle\langle\psi|$. It follows that ρ can always be diagonalized and has real eigenvalues.

2. The density matrix is a positive semidefinite operator. All its eigenvalues are nonnegative.

   In a complex vector space with inner product $\langle\cdot,\cdot\rangle$, an operator M is said to be positive semidefinite if for any vector v we have $\langle v, Mv\rangle \geq 0$. Using bra-ket notation, the condition that the density matrix ρ be positive semidefinite is

$$\langle\psi|\rho|\psi\rangle \geq 0 \quad \text{for all } |\psi\rangle. \tag{22.2.8}$$

Using the expression for ρ, we have

$$\langle\psi|\rho|\psi\rangle = \sum_{a=1}^{n} p_a \langle\psi|\psi_a\rangle\langle\psi_a|\psi\rangle = \sum_{a=1}^{n} p_a |\langle\psi_a|\psi\rangle|^2 \geq 0, \tag{22.2.9}$$

since we are adding the products of nonnegative numbers. It now follows that the eigenvalues of ρ cannot be negative. Indeed, a matrix M with a negative eigenvalue cannot be positive semidefinite because $\langle v, Mv\rangle < 0$ when v is the corresponding eigenvector.

3. The trace of ρ is equal to one:

$$\text{tr}\,\rho = 1. \tag{22.2.10}$$

This property also follows from direct computation:

$$\text{tr}\,\rho = \text{tr}\Big(\sum_{a=1}^{n} p_a |\psi_a\rangle\langle\psi_a|\Big) = \sum_{a=1}^{n} p_a \text{tr}(|\psi_a\rangle\langle\psi_a|), \tag{22.2.11}$$

by linearity of the trace. It then follows that

$$\text{tr}\,\rho = \sum_{a=1}^{n} p_a \langle\psi_a|\psi_a\rangle = \sum_{a=1}^{n} p_a = 1. \tag{22.2.12}$$

4. The density matrix removes redundancies from the ensemble description of mixed states.

We have seen that different ensembles E and E' sometimes give the same density matrices: $\rho_E = \rho_{E'}$. Since ρ is a Hermitian operator, we can use an orthonormal basis of V to write it as

$$\rho = \sum_{i,j=1}^{N} \rho_{ij} |i\rangle\langle j|, \tag{22.2.13}$$

where the ρ_{ij} are the entries of a Hermitian $N \times N$ matrix, with N the dimension of the vector space V of the theory. Such a matrix is specified by N^2 real numbers. Since the trace of this matrix is one, ρ is specified by $N^2 - 1$ real numbers if there are no additional constraints (in fact there are none!). On the other hand, ensembles can require arbitrarily large amounts of data, especially when the number of entries $(p_a, |\psi_a\rangle)$ is large. Thus, the density matrix removes redundant data in the ensemble description of the mixed state.

5. The phases in the states $|\psi_a\rangle$ of the ensemble are irrelevant to the density matrix.

Since the states $|\psi_a\rangle$ entering the description of the ensemble are normalized, they are only ambiguous up to a phase. That phase cancels out in the density matrix, so they have no physical import. Indeed, if $|\psi_a'\rangle = e^{i\theta_a}|\psi_a\rangle$, we have

$$|\psi_a'\rangle\langle\psi_a'| = e^{i\theta_a}|\psi_a\rangle\,\langle\psi_a|e^{-i\theta_a} = |\psi_a\rangle\langle\psi_a|. \tag{22.2.14}$$

6. With a little abuse of terminology, physicists sometimes speak of the density matrix as the "state" or the "state operator" of the quantum system. The density matrix can describe pure states and mixed states.

Exercise 22.2. *Show that an operator M that is positive semidefinite must be Hermitian.*

If the system is described by a pure state $|\psi\rangle$, the ensemble collapses to one entry, $E=\{(1,|\psi\rangle)\}$, and the associated density matrix is the

$$\text{pure state:} \quad \rho=|\psi\rangle\langle\psi|. \tag{22.2.15}$$

In this case ρ is in fact a *rank-one orthogonal projector* to the subspace of V generated by $|\psi\rangle$. It is an orthogonal projector because ρ is Hermitian and satisfies

$$\rho^2=|\psi\rangle\langle\psi|\psi\rangle\langle\psi|=|\psi\rangle\langle\psi|=\rho. \tag{22.2.16}$$

It is rank one because it has unit trace or, equivalently, because it projects to a one-dimensional subspace of V. Note that given the density matrix ρ of a pure state we can easily recover the state $|\psi\rangle$: we simply let ρ act on any vector in V that is not in the kernel of ρ, the result being a vector along $|\psi\rangle$. Because of the projector property and the trace property of ρ, we see that

$$\operatorname{tr}\rho^2=\operatorname{tr}\rho=1, \quad \text{for a pure state.} \tag{22.2.17}$$

Interestingly, the value of $\operatorname{tr}\rho^2$ allows us to decide if we have a pure state or a mixed state:

Theorem 22.2.1. *$\operatorname{tr}\rho^2\leq 1$, with the inequality saturated only for pure states.*

Proof. We begin by computing the object of interest, the trace of ρ^2:

$$\begin{aligned}\operatorname{tr}\rho^2&=\operatorname{tr}\Big(\sum_{a=1}^{n}p_a|\psi_a\rangle\langle\psi_a|\sum_{b=1}^{n}p_b|\psi_b\rangle\langle\psi_b|\Big)=\operatorname{tr}\Big(\sum_{a,b=1}^{n}p_ap_b|\psi_a\rangle\langle\psi_a|\psi_b\rangle\langle\psi_b|\Big)\\&=\sum_{a,b=1}^{n}p_ap_b\langle\psi_a|\psi_b\rangle\operatorname{tr}|\psi_a\rangle\langle\psi_b|=\sum_{a,b=1}^{n}p_ap_b|\langle\psi_a|\psi_b\rangle|^2.\end{aligned} \tag{22.2.18}$$

Adding and subtracting "one" we have

$$\operatorname{tr}\rho^2=\sum_{a,b=1}^{n}p_ap_b\Big(1-(1-|\langle\psi_a|\psi_b\rangle|^2)\Big)=\sum_{a,b=1}^{n}p_ap_b-\sum_{a,b=1}^{n}p_ap_b(1-|\langle\psi_a|\psi_b\rangle|^2). \tag{22.2.19}$$

The first term on the right-hand side is $\sum_{a,b=1}^{n}p_ap_b=\sum_{a=1}^{n}p_a\sum_{b=1}^{n}p_b=1\cdot 1=1$. Therefore,

$$\operatorname{tr}\rho^2=1-\sum_{a\neq b}p_ap_b(1-|\langle\psi_a|\psi_b\rangle|^2). \tag{22.2.20}$$

The sum over unrestricted a,b running from 1 to n was changed to include only $a\neq b$ terms since the terms with $a=b$ vanish. The Schwarz inequality tells us that the expression in parentheses is always nonnegative because

$$|\langle\psi_a|\psi_b\rangle|^2\leq\langle\psi_a|\psi_a\rangle\cdot\langle\psi_b|\psi_b\rangle=1. \tag{22.2.21}$$

Since, additionally, all $p_a > 0$, equation (22.2.20) implies that

$$\text{tr}\,\rho^2 \leq 1. \tag{22.2.22}$$

This inequality is saturated only if $|\langle\psi_a|\psi_b\rangle|^2 = 1$ for all $a \neq b$ in the ensemble. This saturation of Schwarz's inequality requires $|\psi_a\rangle$ to be parallel to $|\psi_b\rangle$ for all $a \neq b$. Given that the states are normalized, it means *all* the states in the ensemble are the same up to phases. Since phases are immaterial to the density matrix, the ensemble involves only one state, and it describes a pure state. This is what we wanted to prove. □

In summary, the various traces of the density matrix satisfy

$$\boxed{\text{tr}\rho^2 \leq \text{tr}\,\rho = 1.} \tag{22.2.23}$$

The value of $\text{tr}\rho^2$ can be used to characterize quantum states. We define the **purity** $\zeta(\rho)$ of a density matrix ρ using this value:

$$\zeta(\rho) \equiv \text{tr}\,\rho^2. \tag{22.2.24}$$

When $\zeta = 1$, the state is pure. As the value of the purity ζ goes below one, the state becomes mixed. We declare that the lower the value of ζ, the less pure or more mixed the state. Thus, a *maximally mixed* state is a state with the lowest possible value of ζ. We will now show that for a maximally mixed state the density matrix is actually a multiple of the identity matrix. This fixes the density matrix, as the trace must be equal to one. The unpolarized spin density matrix discussed in (22.2.7) is in fact maximally mixed.

Since the density matrix is diagonalizable, we can search for the lowest-purity one by restricting ourselves to diagonal matrices. Writing $N = \dim V$, we write

$$\rho = \text{diag}\,(p_1, \ldots, p_N) \quad \Rightarrow \quad \text{tr}\rho^2 = \sum_{i=1}^{N} p_i^2. \tag{22.2.25}$$

The p_i can be viewed as probabilities in an ensemble $E = \{(p_1, |e_1\rangle), \ldots, (p_N, |e_N\rangle)\}$ for which $\rho = p_1|e_1\rangle\langle e_1| + \cdots + p_N|e_N\rangle\langle e_N|$, with $|e_i\rangle$ orthonormal basis vectors that make ρ diagonal. We want to minimize the sum of squares of probabilities, subject to the trace condition $\sum_{i=1}^{N} p_i = 1$ and the positivity condition $p_i \geq 0$, for all i. The trace condition can be implemented with a Lagrange multiplier λ. We need not worry about the positivity condition because, as it will turn out, the only critical point we find has nonnegative p_i's. We must therefore find the stationary point of the following function L of the p_i's and λ:

$$L(p_1, \ldots, p_N; \lambda) = \sum_{i=1}^{N} p_i^2 - \lambda\Big(-1 + \sum_{i=1}^{N} p_i\Big). \tag{22.2.26}$$

The stationarity conditions are

$$0 = \frac{\partial L}{\partial p_i} = 2p_i - \lambda, \;\; i = 1, \ldots, N, \qquad 0 = \frac{\partial L}{\partial \lambda} = 1 - \sum_{i=1}^{N} p_i. \tag{22.2.27}$$

The first condition sets $p_i = \lambda/2$ for all i, setting all probabilities equal. The second condition fixes λ as it sets the sum of the N probabilities equal to one: $N\lambda/2 = 1$. All in all,

$$p_i = \tfrac{\lambda}{2}, \quad N\tfrac{\lambda}{2} = 1 \quad \Rightarrow \quad p_i = \tfrac{1}{N}, \quad \text{for all } i. \tag{22.2.28}$$

As claimed, the lowest-purity density matrix, which we call $\bar{\rho}$, is a multiple of the identity matrix:

$$\text{Maximally mixed state:} \qquad \bar{\rho} = \frac{1}{N}\,\mathbb{1}. \tag{22.2.29}$$

Here $\mathbb{1}$ is the $N \times N$ identity matrix, with N the dimensionality of the state space. It follows also that the purity of the maximally mixed state is $1/N$:

$$\zeta(\bar{\rho}) = \operatorname{tr} \bar{\rho}^2 = \frac{1}{N^2} \operatorname{tr} \mathbb{1} = \frac{1}{N}. \tag{22.2.30}$$

This is the minimum possible value of the purity. The concept of purity is natural because the purity of a system is conserved under unitary time evolution, as we will see in section 22.3.

Exercise 22.3. *Examine the following two density matrices and determine if the states are pure or mixed. If pure, write the density matrix in the form $|\psi\rangle\langle\psi|$ that makes this manifest.*

$$\rho_1 = \begin{pmatrix} \frac{1}{2} & 0 \\ 0 & \frac{1}{2} \end{pmatrix}, \quad \rho_2 = \begin{pmatrix} \frac{1}{2} & \frac{1}{2} \\ \frac{1}{2} & \frac{1}{2} \end{pmatrix}. \tag{22.2.31}$$

Example 22.3. *Density matrix for spin one-half pure states.*

Here the density matrix is very simple: $|\mathbf{n}\rangle\langle\mathbf{n}|$ for a pure state $|\mathbf{n}\rangle$ pointing along the direction of the unit vector $\mathbf{n}$. Since the density matrix is a Hermitian operator and the set of 2×2 Hermitian matrices is spanned by the identity and the Pauli matrices, we should be able to write

$$|\mathbf{n}\rangle\langle\mathbf{n}| = \tfrac{1}{2} a_0\, \mathbb{1} + \tfrac{1}{2} \sum_{i=1}^{3} a_i \sigma_i, \tag{22.2.32}$$

where the factors of $\frac{1}{2}$ have been included for convenience, and a_0 as well as a_1, a_2, a_3 are real constants. But we have solved this problem before! We did so when constructing the projector $P_{\mathbf{n}} = |\mathbf{n}\rangle\langle\mathbf{n}|$ to the spin state $|\mathbf{n}\rangle$. The answer was given in (14.9.33), which we copy here:

$$\boxed{|\mathbf{n}\rangle\langle\mathbf{n}| = \tfrac{1}{2}(\mathbb{1} + \mathbf{n} \cdot \boldsymbol{\sigma}).} \tag{22.2.33}$$

We will discuss the density matrix for mixed spin one-half states below. □

We have shown that starting from an ensemble, the associated density matrix is a positive semidefinite matrix with unit trace (the condition of Hermiticity follows from positivity, as you showed in exercise 22.2). There are no extra conditions on the density matrix. Indeed, we claim the following:

Theorem 22.2.2. *To any unit trace, positive semidefinite matrix $M \in \mathcal{L}(V)$, we can associate an ensemble for which M is the density matrix.*

Proof. The matrix M, being Hermitian and positive semidefinite, can be diagonalized and will have nonnegative eigenvalues $\lambda_i \geq 0$ with $i = 1, \ldots, N$, with $N = \dim V$. Let us call $|e_i\rangle$ the eigenvector associated with λ_i. We then have

$$M = \sum_{i=1}^{N} \lambda_i |e_i\rangle\langle e_i|, \qquad \sum_{i=1}^{N} \lambda_i = 1. \tag{22.2.34}$$

The second equality arises from the unit trace condition on M. Now consider the ensemble E_M defined by

$$E_M \equiv \{(\lambda_1, |e_1\rangle), \ldots, (\lambda_N, |e_N\rangle)\}, \tag{22.2.35}$$

which is consistent because the λ_i are all nonnegative and add up to one. Note that in this ensemble we allow the value zero for some of the λ_i's; this, of course, means the corresponding terms do not contribute to the density matrix. The density matrix ρ_{E_M} associated with the ensemble E_M is constructed as usual:

$$\rho_{E_M} = \sum_{i=1}^{N} \lambda_i |e_i\rangle\langle e_i| = M, \tag{22.2.36}$$

using (22.2.34) and confirming the claim of the theorem. □

Example 22.4. *Density matrix for general spin one-half states.*
We again begin by writing the density matrix ρ as a general 2×2 Hermitian matrix:

$$\rho = \tfrac{1}{2} a_0 \mathbb{1} + \tfrac{1}{2} \mathbf{a} \cdot \boldsymbol{\sigma}, \quad a_0, a_1, a_2, a_3 \in \mathbb{R}. \tag{22.2.37}$$

As before, $\text{tr}\, \rho = 1$ fixes $a_0 = 1$. On account of the above theorem, the only remaining condition on ρ is that of positivity: none of its eigenvalues can be negative. The eigenvalues of $\mathbf{a} \cdot \boldsymbol{\sigma}$ are $\pm|\mathbf{a}|$, and therefore the eigenvalues of ρ are

$$\tfrac{1}{2}(1 \pm |\mathbf{a}|) \geq 0. \tag{22.2.38}$$

Positivity requires $1 - |\mathbf{a}| \geq 0$ or, equivalently, $|\mathbf{a}| \leq 1$. All in all, the density matrix for a general mixed or pure state is as follows:

$$\boxed{\text{Spin one-half density matrix: } \rho = \tfrac{1}{2}(\mathbb{1} + \mathbf{a} \cdot \boldsymbol{\sigma}), \quad |\mathbf{a}| \leq 1.} \tag{22.2.39}$$

The set of allowed pure and mixed states above is called the *Bloch ball*, a unit ball in the euclidean three-dimensional space $\{a_1, a_2, a_3\}$. When $\mathbf{a}$ is a unit vector, ρ is of the type (22.2.33) and represents a pure state. Thus, the boundary $|\mathbf{a}| = 1$ of the Bloch ball is a two-sphere's worth of pure states. The interior of the Bloch ball represents mixed states.

The center $\mathbf{a}=0$ of the Bloch ball represents the unpolarized state; it is the maximally mixed state. □

Measurement along an orthonormal basis Recall that we can measure a pure state $|\psi\rangle$ along an orthonormal basis $|1\rangle, \ldots, |N\rangle$ of a dimension N vector space V, and the probability $p(i)$ of being in the state $|i\rangle$ is $|\langle i|\psi\rangle|^2$ (see (16.6.9)). After measurement, the state will be in one of the states $|i\rangle$.

This is readily extended to cases in which the measurement is conducted on a mixed-state ensemble (22.1.3). This time the probability $p(i)$ of finding $|i\rangle$ is obtained by weighting, with probability p_a, the probability $|\langle i|\psi_a\rangle|^2$ of finding the state $|\psi_a\rangle$ in $|i\rangle$:

$$p(i)=\sum_{a=1}^{n} p_a|\langle i|\psi_a\rangle|^2=\sum_{a=1}^{n} p_a\langle i|\psi_a\rangle\,\langle\psi_a|i\rangle=\langle i|\sum_{a=1}^{n} p_a|\psi_a\rangle\langle\psi_a|i\rangle. \tag{22.2.40}$$

From this we get the simple expression

$$p(i)=\langle i|\rho|i\rangle. \tag{22.2.41}$$

As it should, this probability depends only on ρ and not on the ensemble that defines ρ. After this measurement the system will be in one of the basis states. If we obtain $|i\rangle$, the density matrix ρ becomes $|i\rangle\langle i|$. This state is in fact the orthogonal projector $M_i \equiv |i\rangle\langle i|$ where

$$M_i^\dagger=M_i, \quad M_iM_i=M_i, \quad \sum_i M_i=\mathbb{1}. \tag{22.2.42}$$

The collection $\{M_i\}$ forms a complete set of orthogonal projectors.

Assume that a measurement along the basis has been performed, but the result is not available to us. What becomes of the density matrix? We know that after measurement we have, for each i, the probability $p(i)$ for the system to be in the state $|i\rangle$. The new ensemble $\tilde{E}$ is therefore

$$\tilde{E}=\Big\{(p(1), |1\rangle), \ldots, (p(N), |N\rangle)\Big\}. \tag{22.2.43}$$

The new, after-measurement density matrix $\tilde{\rho}$ is now easily constructed:

$$\tilde{\rho}=\sum_i p(i)|i\rangle\langle i|=\sum_i |i\rangle\,\langle i|\rho|i\rangle\,\langle i|, \tag{22.2.44}$$

where we used our result for $p(i)$. This can now be rewritten in a suggestive way:

$$\boxed{\tilde{\rho}=\sum_i M_i\,\rho\,M_i.} \tag{22.2.45}$$

This passage from ρ to $\tilde{\rho}$ gives us the effect of measurement along a basis on a quantum system when the result is not available. We quickly check that, as required, the trace of the new density matrix remains equal to one:

$$\text{tr}\,\tilde{\rho}=\sum_i \text{tr}(M_i\,\rho\,M_i)=\sum_i \text{tr}(\rho\,M_iM_i)=\sum_i \text{tr}(\rho\,M_i)=\text{tr}\,(\rho\sum_i M_i)=\text{tr}\rho=1. \tag{22.2.46}$$

In the various steps, we used the cyclicity of the trace as well as the second and third properties in (22.2.42).

22.3 Dynamics of Density Matrices

Since the density matrix describes in all generality the possible quantum states of a system, it is of interest to see how it evolves in time. To this end we need the Schrödinger equation written for a ket $|\psi\rangle$ as well as for a bra $\langle\psi|$:

$$\frac{\partial}{\partial t}|\psi\rangle = -\frac{i}{\hbar}\hat{H}|\psi\rangle, \qquad \frac{\partial}{\partial t}\langle\psi| = \frac{i}{\hbar}\langle\psi|\hat{H}. \tag{22.3.1}$$

Here $\hat{H}$ is the Hamiltonian. We can now compute the rate of change of the projector $|\psi\rangle\langle\psi|$:

$$\frac{\partial}{\partial t}|\psi\rangle\langle\psi| = -\frac{i}{\hbar}\hat{H}|\psi\rangle\langle\psi| + \frac{i}{\hbar}|\psi\rangle\langle\psi|\hat{H} = -\frac{i}{\hbar}\,[\hat{H}, |\psi\rangle\langle\psi|]. \tag{22.3.2}$$

Using the ensemble definition of the density matrix, we then get

$$\begin{aligned}\frac{\partial\rho}{\partial t} &= \frac{\partial}{\partial t}\sum_{a=1}^{n} p_a|\psi_a\rangle\langle\psi_a| = \sum_{a=1}^{n} p_a\frac{\partial}{\partial t}|\psi_a\rangle\langle\psi_a| = -\frac{i}{\hbar}\sum_{a=1}^{n} p_a[\hat{H},\, |\psi_a\rangle\langle\psi_a|] \\ &= -\frac{i}{\hbar}\Big[\hat{H},\, \sum_{a=1}^{n} p_a|\psi_a\rangle\langle\psi_a|\Big].\end{aligned} \tag{22.3.3}$$

This implies the simple result

$$\boxed{i\hbar\frac{\partial\rho}{\partial t} = [\hat{H},\, \rho].} \tag{22.3.4}$$

This equation determines the time evolution of the density matrix of a quantum system. It manifestly preserves the Hermiticity of ρ because it sets its derivative $\frac{\partial\rho}{\partial t}$ equal to a Hermitian operator. Indeed, $\frac{1}{i\hbar}[\hat{H}, \rho]$ is Hermitian because the commutator of Hermitian operators is anti-Hermitian, and the factor of i makes it Hermitian. Moreover, the trace of ρ is unchanged:

$$\frac{d}{dt}\text{tr}\rho = \text{tr}\left(\frac{\partial\rho}{\partial t}\right) = -\frac{i}{\hbar}\text{tr}[\hat{H}, \rho] = 0, \tag{22.3.5}$$

since the trace of a commutator vanishes due to cyclicity. This is automatic in finite-dimensional vector spaces but must be checked carefully when working in infinite-dimensional vector spaces.

Suppose we solve for the time evolution of states by constructing the unitary operator $\mathcal{U}(t)$ that evolves states as follows:

$$|\psi(t)\rangle = \mathcal{U}(t)|\psi(0)\rangle. \tag{22.3.6}$$

It is then clear that the density matrix, which at any time is a sum of terms of the form $|\psi_a(t)\rangle\langle\psi_a(t)|$, evolves as

$$\rho(t) = \mathcal{U}(t)\rho(0)\mathcal{U}^\dagger(t). \tag{22.3.7}$$

This evolution, of course, is consistent with the differential equation (22.3.4) when we recall the differential equation satisfied by the unitary operator (see (16.3.2)). The above expression for $\rho(t)$ makes it manifest that if $\rho(0)$ is positive semidefinite, so is $\rho(t)$ for all times t. Indeed, for any vector v in the state space we see that

$$\langle v|\rho(t)|v\rangle = \langle v|\mathcal{U}\rho(0)\mathcal{U}^\dagger|v\rangle = \langle \mathcal{U}^\dagger v|\rho(0)|\mathcal{U}^\dagger v\rangle \geq 0. \tag{22.3.8}$$

We know from the Schrödinger equation that a pure state $|\psi\rangle$ remains pure under time evolution. This is also visible from the density matrix $\rho(t) = \mathcal{U}(t)|\psi\rangle\langle\psi|\mathcal{U}^\dagger(t)$. In fact, a more general result holds. We can quickly see that the purity $\zeta = \text{tr}\rho^2$ does not change over time:

$$\begin{aligned}\frac{d\zeta}{dt} &= \frac{d}{dt}\text{tr}(\rho\,\rho) = \text{tr}\Big(\frac{d\rho}{dt}\rho + \rho\frac{d\rho}{dt}\Big) = 2\text{tr}\Big(\rho\frac{d\rho}{dt}\Big)\\ &= \tfrac{2}{i\hbar}\text{tr}(\rho[\hat{H},\rho]) = \tfrac{2}{i\hbar}\text{tr}(\rho\hat{H}\rho - \rho\rho\hat{H}) = 0,\end{aligned} \tag{22.3.9}$$

by repeated use of the cyclicity of the trace. Since the purity does not change under unitary time evolution and a pure state has purity equal to one, a pure state will remain pure.

Exercise 22.4. *What we observed for the purity is in fact part of a simple pattern. Show that tr(ρ^n) is conserved under unitary time evolution for n an* arbitrary *positive integer.*

The considerations of time evolution in this section apply to isolated systems. They change in an interesting way when we consider the density matrix of a *subsystem* of an isolated system, as we will begin exploring next.

22.4 Subsystems and Schmidt Decomposition

Let us consider the physics of a quantum system A that is a part, or a subsystem, of a composite system AB. The composite system is isolated from the rest of the world and is defined on a state space $\mathcal{H}_A \otimes \mathcal{H}_B$, with $\mathcal{H}_A$ and $\mathcal{H}_B$ the state spaces for A and B subsystems, respectively. Typically, one has interactions that couple the two systems A and B, and the systems are entangled. The system AB is a **bipartite** system, which just means it is composed of two parts. In example 22.2 we considered a composite system AB in a pure entangled state and demonstrated that there is no pure state that represents the system A. We need a density matrix. This is truly the only option if AB is not in a pure state.

To fix notation, assume the spaces $\mathcal{H}_A$ and $\mathcal{H}_B$ have dimensions d_A and d_B, respectively, and have orthonormal bases given by

$$\begin{aligned}&\dim \mathcal{H}_A = d_A, \quad (e_1^A, \ldots, e_{d_A}^A) \text{ orthonormal basis,}\\ &\dim \mathcal{H}_B = d_B, \quad (e_1^B, \ldots, e_{d_B}^B) \text{ orthonormal basis.}\end{aligned} \tag{22.4.1}$$

Let there be a density matrix ρ_{AB} for the full AB system. We then ask: What is the relevant density matrix ρ_A that can be used to compute the results of measurements on

A? The answer turns out to be quite simple. Since the system B plays no role here, ρ_A, sometimes called the *reduced density matrix*, is obtained by taking the partial trace over $\mathcal{H}_B$ of the full density matrix:

$$\rho_A = \text{tr}_B \rho_{AB} = \sum_k \langle e_k^B | \rho_{AB} | e_k^B \rangle \in \mathcal{L}(\mathcal{H}_A). \tag{22.4.2}$$

This proposal passes a basic consistency check: if ρ_{AB} is a density matrix, so is ρ_A. To see this first note that the trace works out correctly:

$$\text{tr}_A \rho_A = \text{tr}_A \text{tr}_B \rho_{AB} = \text{tr}\, \rho_{AB} = 1, \tag{22.4.3}$$

where tr is the full trace in the tensor product space—see equation (18.4.14). Moreover, ρ_A is a positive semidefinite operator. To prove this we must show that $\langle v_A | \rho_A | v_A \rangle \geq 0$ for any $|v_A\rangle \in \mathcal{H}_A$. This is not complicated:

$$\langle v_A | \rho_A | v_A \rangle = \langle v_A | \sum_k \langle e_k^B | \rho_{AB} | e_k^B \rangle | v_A \rangle = \sum_k \langle v_A | \langle e_k^B | \, \rho_{AB} \, | v_A \rangle | e_k^B \rangle \geq 0, \tag{22.4.4}$$

since every term in the sum is nonnegative because ρ_{AB} is positive semidefinite. Being a positive semidefinite operator of unit trace, ρ_A can represent a density matrix.

The formula (22.4.2) is justified by showing that for an arbitrary operator $\mathcal{O}_A \in \mathcal{L}(\mathcal{H}_A)$ the expectation value obtained using ρ_A equals the expectation value of $\mathcal{O}_A \otimes \mathbb{1}_B$ using the full density matrix ρ_{AB}:

$$\text{tr}_A (\rho_A \mathcal{O}_A) = \text{tr}(\rho_{AB} \mathcal{O}_A \otimes \mathbb{1}_B). \tag{22.4.5}$$

To see this let us write a general density matrix ρ_{AB} as the most general operator on $\mathcal{H}_A \otimes \mathcal{H}_B$. Recalling that $\mathcal{L}(\mathcal{H}_A \otimes \mathcal{H}_B) = \mathcal{L}(\mathcal{H}_A) \otimes \mathcal{L}(\mathcal{H}_B)$, we use basis operators $|e_i^A\rangle\langle e_j^A| \in \mathcal{L}(\mathcal{H}_A)$ and $|e_k^B\rangle\langle e_l^B| \in \mathcal{L}(\mathcal{H}_B)$ to write the most general linear superposition of tensor products:

$$\rho_{AB} = \sum_{i,j,k,l} \rho_{ij,kl} \, |e_i^A\rangle\langle e_j^A| \otimes |e_k^B\rangle\langle e_l^B|. \tag{22.4.6}$$

It follows that

$$\rho_A = \text{tr}_B \rho_{AB} = \sum_{i,j,k,l} \rho_{ij,kl} \, |e_i^A\rangle\langle e_j^A| \otimes \langle e_l^B | e_k^B \rangle = \sum_{i,j,k} \rho_{ij,kk} \, |e_i^A\rangle\langle e_j^A|, \tag{22.4.7}$$

and as a result, the left-hand side of (22.4.5) is

$$\text{tr}_A (\rho_A \, \mathcal{O}_A) = \sum_{i,j,k} \rho_{ij,kk} \, \langle e_j^A | \mathcal{O}_A | e_i^A \rangle. \tag{22.4.8}$$

Similarly, we compute the right-hand side of (22.4.5):

$$\begin{aligned} \text{tr}(\rho_{AB} \, \mathcal{O}_A \otimes \mathbb{1}_B) &= \text{tr}_A \text{tr}_B \sum_{i,j,k,l} \rho_{ij,kl} \, |e_i^A\rangle\langle e_j^A | \mathcal{O}_A \otimes |e_k^B\rangle\langle e_l^B| \\ &= \sum_{i,j,k} \rho_{ij,kk} \, \langle e_j^A | \mathcal{O}_A | e_i^A \rangle, \end{aligned} \tag{22.4.9}$$

making it clear that (22.4.5) holds and thus justifying the claimed formula for the density matrix ρ_A of the subsystem A. Note that nowhere in the proof of (22.4.5) have we used any particular property of the density operator ρ_{AB}. This means that this identity is true for arbitrary operators:

Theorem 22.4.1. *Let $S_{AB} \in \mathcal{L}(\mathcal{H}_A \otimes \mathcal{H}_B)$ be an arbitrary operator, and $S_A = \text{tr}_B S_{AB}$. Then for any $\mathcal{O}_A \in \mathcal{L}(\mathcal{H}_A)$ we find that*

$$\text{tr}_A(S_A \mathcal{O}_A) = \text{tr}(S_{AB} \mathcal{O}_A \otimes \mathbb{1}_B). \tag{22.4.10}$$

Example 22.5. *A pure state of two entangled spins and density matrix of a subsystem.*
Consider the pure-state system AB of two spins examined before:

$$|\psi_{AB}\rangle = \tfrac{1}{\sqrt{2}}\big(|+\rangle_A|-\rangle_B - |-\rangle_A|+\rangle_B\big). \tag{22.4.11}$$

Alice has particle A, and Bob has particle B. We aim to find the density matrix ρ_B for subsystem B, Bob's particle.

The density matrix for the full system is just $\rho_{AB} = |\psi_{AB}\rangle\langle\psi_{AB}|$, which we can write out conveniently as sums of tensor products of operators:

$$\begin{aligned}\rho_{AB} &= \tfrac{1}{\sqrt{2}}\big(|+\rangle_A|-\rangle_B - |-\rangle_A|+\rangle_B\big)\tfrac{1}{\sqrt{2}}\big(\langle+|_A\langle-|_B - \langle-|_A\langle+|_B\big)\\ &= \tfrac{1}{2}\,(|+\rangle\langle+|)_A \otimes (|-\rangle\langle-|)_B\\ &\quad - \tfrac{1}{2}\,(|+\rangle\langle-|)_A \otimes (|-\rangle\langle+|)_B\\ &\quad - \tfrac{1}{2}\,(|-\rangle\langle+|)_A \otimes (|+\rangle\langle-|)_B\\ &\quad + \tfrac{1}{2}\,(|-\rangle\langle-|)_A \otimes (|+\rangle\langle+|)_B.\end{aligned} \tag{22.4.12}$$

We can now take the trace over A to find the density matrix for B. From the four terms displayed on the last right-hand side, the second and third have zero tr_A. The nonvanishing contributions give

$$\rho_B = \text{tr}_A\,\rho_{AB} = \tfrac{1}{2}|-\rangle\langle-| + \tfrac{1}{2}|+\rangle\langle+|. \tag{22.4.13}$$

The state of B is maximally mixed. This was probably expected, as the original pure state $|\psi_{AB}\rangle$ seems as entangled as can be. Another curious fact emerges when we recall that in (22.1.16) we obtained the ensemble that Bob gets when Alice does a measurement of her particle but does not communicate the result. The ensemble is exactly the same as that described by the above ρ_B. Thus, whether or not Alice measures, the state of B is the same, in this case the maximally mixed state. We will understand this remarkable coincidence more generally in the latter part of this section. □

Exercise 22.5. *Consider the following state $|\hat{\psi}_{AB}\rangle$ of two spin one-half particles A and B:*

$$\begin{aligned}|\hat{\psi}_{AB}\rangle &= \tfrac{1}{\sqrt{2}}|+\rangle_A|+\rangle_B + \tfrac{1}{2}|-\rangle_A|+\rangle_B - \tfrac{1}{2}|-\rangle_A|-\rangle_B,\\ &= \tfrac{1}{\sqrt{2}}|+\rangle_A|+\rangle_B + \tfrac{1}{\sqrt{2}}|-\rangle_A|x;-\rangle_B.\end{aligned} \tag{22.4.14}$$

Show that the density matrix ρ_A for particle A is

$$\rho_A = \tfrac{1}{2}|+\rangle\langle+| + \tfrac{1}{2\sqrt{2}}|+\rangle\langle-| + \tfrac{1}{2\sqrt{2}}|-\rangle\langle+| + \tfrac{1}{2}|-\rangle\langle-|. \tag{22.4.15}$$

Exercise 22.6. *Diagonalize the above density matrix ρ_A and show that*

$$\rho_A = \tfrac{1}{2}\big(1+\tfrac{1}{\sqrt{2}}\big)\,|x;+\rangle\langle x;+| + \tfrac{1}{2}\big(1-\tfrac{1}{\sqrt{2}}\big)\,|x;-\rangle\langle x;-|. \tag{22.4.16}$$

Schmidt decomposition The *pure* states $|\psi_{AB}\rangle$ of a bipartite system AB can be written in an insightful way by using as a guide the associated density matrices ρ_A and ρ_B of the subsystems. The result is the *Schmidt decomposition* of the pure state $|\psi_{AB}\rangle$, named in honor of Erhard Schmidt (1876–1959), also known for the Gram-Schmidt procedure that yields orthonormal basis vectors from a set of nonorthonormal ones. The decomposition displays a simple structure—simpler than the general structure allowed from the tensor product:

1. The state $|\psi_{AB}\rangle$ is written in terms of an orthonormal basis $\{|k_A\rangle\}$ of $\mathcal{H}_A$ and an orthonormal basis $\{|k_B\rangle\}$ of $\mathcal{H}_B$ that, respectively, make the reduced density matrices ρ_A and ρ_B diagonal.
2. The decomposition defines an integer r, called the Schmidt index, that characterizes the degree of entanglement of the subsystems A and B.

Suppose we have a bipartite system AB and a pure state $|\Psi_{AB}\rangle$ in which A is entangled with B:

$$|\Psi_{AB}\rangle \in \mathcal{H}_A \otimes \mathcal{H}_B. \tag{22.4.17}$$

The dimensions and basis states for $\mathcal{H}_A$ and $\mathcal{H}_B$ are as in (22.4.1). Assume that we choose to label the systems so that

$$d_A \le d_B. \tag{22.4.18}$$

A state $|\Psi_{AB}\rangle$ is typically written as an expansion over the obvious basis states $|e_i^A\rangle \otimes |e_j^B\rangle$ of $\mathcal{H}_A \otimes \mathcal{H}_B$:

$$|\Psi_{AB}\rangle = \sum_{i=1}^{d_A}\sum_{j=1}^{d_B} \psi_{ij}|e_i^A\rangle \otimes |e_j^B\rangle. \tag{22.4.19}$$

Here the ψ_{ij} are $d_A \cdot d_B$ expansion coefficients. In the Schmidt decomposition, we will do much better than this. Actually, the above expression can be rewritten as

$$|\Psi_{AB}\rangle = \sum_{i=1}^{d_A} |e_i^A\rangle \otimes |\psi_i^B\rangle, \quad \text{with } |\psi_i^B\rangle = \sum_{j=1}^{d_B} \psi_{ij}|e_j^B\rangle. \tag{22.4.20}$$

While this is nicer, not much can be said about the $|\psi_i^B\rangle$ states; in particular, they need not be orthonormal. In the Schmidt decomposition, the sum analogous to the sum over i runs up to an integer r called the Schmidt index, which can be smaller than d_A. Moreover, the states from $\mathcal{H}_B$ as well as those from $\mathcal{H}_A$ are orthonormal.

To derive the Schmidt decomposition, we consider the state $|\psi_{AB}\rangle$ and the associated density matrices:

$$\rho_{AB} = |\Psi_{AB}\rangle\langle\Psi_{AB}|, \quad \rho_A = \text{tr}_B \rho_{AB} = \text{tr}_B\big(|\Psi_{AB}\rangle\langle\Psi_{AB}|\big). \tag{22.4.21}$$

By construction, ρ_A is a Hermitian positive semidefinite $d_A \times d_A$ matrix and can therefore be diagonalized. Let $(p_k, |k_A\rangle)$ with $k = 1, \ldots, d_A$ be the eigenvalues and eigenvectors of ρ_A, with the eigenvectors $|k_A\rangle$ chosen to be an orthonormal basis for $\mathcal{H}_A$ and the eigenvalues p_k nonnegative. We see that the density matrix ρ_A has furnished us with a second orthonormal basis of states for $\mathcal{H}_A$. The density matrix ρ_A can then be written as

$$\rho_A = \sum_{k=1}^{d_A} p_k |k_A\rangle\langle k_A|, \quad \sum_{k=1}^{d_A} p_k = 1. \tag{22.4.22}$$

It may happen, for example, that ρ_A is a pure state, in which case the above sum has just one term, and only one p_k is nonzero. In general, the sum defining ρ_A has $r \le d_A$ terms. Let us assume this and order the list of eigenvectors and eigenvalues so that the first r eigenvalues are nonzero and the rest vanish. Then we will write

$$\rho_A = \sum_{k=1}^{r} p_k |k_A\rangle\langle k_A|, \quad r \le d_A, \text{ and } p_{k>r} = 0. \tag{22.4.23}$$

Let us now consider $|\psi_{AB}\rangle$. Since the $|k_A\rangle$ span $\mathcal{H}_A$, we can write

$$|\psi_{AB}\rangle = \sum_{k=1}^{d_A} |k_A\rangle \otimes |\psi_k^B\rangle, \tag{22.4.24}$$

with $|\psi_k^B\rangle$ some collection of states in $\mathcal{H}_B$. Note that we have at most d_A terms, not the $d_A \cdot d_B$ terms that would arise if we used the basis states of $\mathcal{H}_B$ to expand the $|\psi_k^B\rangle$ states. Forming the density matrix associated with $|\psi_{AB}\rangle$, we find that

$$\rho_{AB} = \sum_{k,\tilde{k}=1}^{d_A} |k_A\rangle \otimes |\psi_k^B\rangle \; \langle \tilde{k}_A| \otimes \langle \psi_{\tilde{k}}^B|. \tag{22.4.25}$$

Taking the trace over B, we now get

$$\rho_A = \text{tr}_B \rho_{AB} = \sum_{k,\tilde{k}=1}^{d_A} |k_A\rangle\langle \tilde{k}_A| \; \langle \psi_{\tilde{k}}^B | \psi_k^B\rangle. \tag{22.4.26}$$

Compare now with our previous expression for ρ_A in (22.4.23), where no state $|k_A\rangle$ with $k > r$ appears. This means that we should reconsider our ansatz for $|\psi_{AB}\rangle$: no state $|k_A\rangle$ with $k > r$ can appear there either. If they did, there would be some nonvanishing terms in ρ_A that are not included in (22.4.23). We therefore rewrite

$$|\psi_{AB}\rangle = \sum_{k=1}^{r} |k_A\rangle \otimes |\psi_k^B\rangle, \tag{22.4.27}$$

which leads to

$$\rho_{AB} = \sum_{k,\tilde{k}=1}^{r} |k_A\rangle \otimes |\psi_k^B\rangle\, \langle \tilde{k}_A| \otimes \langle \psi_{\tilde{k}}^B| \;\Rightarrow\; \rho_A = \sum_{k,\tilde{k}=1}^{r} |k_A\rangle\langle \tilde{k}_A|\, \langle \psi_{\tilde{k}}^B | \psi_k^B\rangle. \tag{22.4.28}$$

Once again comparing with ρ_A in (22.4.23), we see there should be no terms with $k \neq \tilde{k}$. Full agreement then requires that

$$\langle \psi_{\tilde{k}}^B | \psi_k^B\rangle = p_k\, \delta_{k\tilde{k}}, \quad k, \tilde{k} = 1, \ldots, r. \tag{22.4.29}$$

In other words, states $|\psi_k^B\rangle$ with different values of k must be orthogonal. It is therefore useful to introduce normalized versions $|k_B\rangle$ of the states $|\psi_k^B\rangle$ as follows:

$$|k_B\rangle \equiv \frac{|\psi_k^B\rangle}{\sqrt{p_k}}, \quad k = 1, \ldots, r. \tag{22.4.30}$$

These states satisfy

$$\langle k_B | k'_B\rangle = \delta_{k,k'}, \quad k, k' = 1, \ldots, r. \tag{22.4.31}$$

If $r < d_B$, one can define additional orthonormal vectors to have a full basis for $\mathcal{H}_B$. These extra vectors will not feature below.

We have already shown that the pure state $|\psi_{AB}\rangle$ of the bipartite system AB can always be written as a sum of r terms. From (22.4.27) and (22.4.30), we now get the Schmidt decomposition of the pure state $|\psi_{AB}\rangle$:

$$\boxed{|\psi_{AB}\rangle = \sum_{k=1}^{r} \sqrt{p_k}\, |k_A\rangle \otimes |k_B\rangle, \quad r \leq d_A \leq d_B.} \tag{22.4.32}$$

In here,

$$\sum_{k=1}^{r} p_k = 1, \quad p_k > 0, \quad k = 1, \ldots, r, \tag{22.4.33}$$

and the states $|k_A\rangle \in \mathcal{H}_A$ and $|k_B\rangle \in \mathcal{H}_B$, with $k = 1, \ldots, r$, form orthonormal sets:

$$\langle k_A | k'_A\rangle = \delta_{k,k'}, \quad \langle k_B | k'_B\rangle = \delta_{k,k'}. \tag{22.4.34}$$

Despite the similar notation, the $|k_A\rangle$ and $|k_B\rangle$ states have nothing to do with each other; they live in different spaces. The Schmidt decomposition (22.4.32) has the properties we anticipated before. It involves the sum of $r \leq d_A$ terms, each a basis state of $\mathcal{H}_A$ multiplied by some state in $\mathcal{H}_B$. Moreover, the $\mathcal{H}_B$ states $|k_B\rangle$ multiplying the $|k_A\rangle$ basis states also form an orthonormal set. Finally, since the construction is inspired by density matrices, the reduced density matrix ρ_A, and in fact $\rho_B = \text{tr}_A\, \rho_{AB}$ as well, are nicely written in the above language. We already had ρ_A from (22.4.23), and ρ_B follows from a very brief calculation:

$$\boxed{\rho_A = \sum_{k=1}^{r} p_k |k_A\rangle\langle k_A|, \qquad \rho_B = \sum_{k=1}^{r} p_k |k_B\rangle\langle k_B|.} \tag{22.4.35}$$

The $\mathcal{H}_A$ and $\mathcal{H}_B$ basis vectors used in the Schmidt decomposition make the density matrices ρ_A and ρ_B diagonal. Moreover, ρ_A and ρ_B have exactly the same nonzero eigenvalues! This is an important result for any bipartite system in a pure state. Since any pure state of a bipartite system AB has a Schmidt decomposition, the value of r is unambiguously determined. This value is called the **Schmidt number** of the state.

If a state of AB has Schmidt number one, the A and B subsystems are not entangled: the Schmidt decomposition provides a manifest description of the AB state as the tensor product of a state in $\mathcal{H}_A$ and a state in $\mathcal{H}_B$. Moreover, if the Schmidt number r is greater than one, the subsystems are definitely entangled. This is clear because the reduced density matrices ρ_A and ρ_B are mixed (they have $r > 1$ terms), and a state of AB where A and B are not entangled always leads to density matrices ρ_A and ρ_B that represent pure states.

Exercise 22.7. *Prove that for any pure entangled state of AB the purity of ρ_A equals the purity of ρ_B with value:*

$$\zeta(\rho_A) = \zeta(\rho_B) = \sum_{k=1}^{r} p_k^2. \tag{22.4.36}$$

Consider a bipartite system AB where A and B have state spaces of the same dimensionality, and the state of AB is pure. The result of the above exercise shows that when ρ_A is maximally mixed, so is ρ_B. In such a case, we say that A and B are *maximally entangled*.

Example 22.6. *Schmidt decomposition of a state.*

We considered in exercise 22.5 a pure state $|\hat{\psi}_{AB}\rangle$ of a bipartite system AB:

$$|\hat{\psi}_{AB}\rangle = \tfrac{1}{\sqrt{2}}|+\rangle_A|+\rangle_B + \tfrac{1}{2}|-\rangle_A|+\rangle_B - \tfrac{1}{2}|-\rangle_A|-\rangle_B. \tag{22.4.37}$$

We aim to find its Schmidt decomposition. Some of the relevant work was done already. You diagonalized the reduced density matrix ρ_A, finding that

$$\rho_A = \tfrac{1}{2}\big(1 + \tfrac{1}{\sqrt{2}}\big)\,|x;+\rangle\langle x;+| \,+ \tfrac{1}{2}\big(1 - \tfrac{1}{\sqrt{2}}\big)\,|x;-\rangle\langle x;-|. \tag{22.4.38}$$

On account of the relation between ρ_A and the state $|\psi_{AB}\rangle$ it arises from, exemplified in equations (22.4.35) and (22.4.32), we have a simple ansatz for the state $|\hat{\psi}_{AB}\rangle$:

$$|\hat{\psi}_{AB}\rangle = \tfrac{1}{\sqrt{2}}\sqrt{1 + \tfrac{1}{\sqrt{2}}}\,|x;+\rangle_A|1_B\rangle + \tfrac{1}{\sqrt{2}}\sqrt{1 - \tfrac{1}{\sqrt{2}}}\,|x;-\rangle_A|2_B\rangle, \tag{22.4.39}$$

where the states $|1_B\rangle$ and $|2_B\rangle$ are orthonormal states to be determined. Writing the $|\pm\rangle_A$ states in the expression (22.4.37) for $|\hat{\psi}_{AB}\rangle$ in terms of $|x;\pm\rangle_A$, a short calculation gives

$$\begin{aligned} |\hat{\psi}_{AB}\rangle = {} & \tfrac{1}{2}|x;+\rangle_A \otimes \Big(\big(1 + \tfrac{1}{\sqrt{2}}\big)|+\rangle_B - \tfrac{1}{\sqrt{2}}|-\rangle_B\Big) \\ & + \tfrac{1}{2}|x;-\rangle_A \otimes \Big(\big(1 - \tfrac{1}{\sqrt{2}}\big)|+\rangle_B + \tfrac{1}{\sqrt{2}}|-\rangle_B\Big). \end{aligned} \tag{22.4.40}$$

The earlier result (22.4.39) tells us how to rewrite this in a way that orthonormality is manifest. We find that

$$
\begin{aligned}
|\hat{\psi}_{AB}\rangle = \tfrac{1}{\sqrt{2}}\sqrt{1+\tfrac{1}{\sqrt{2}}}\ |x;+\rangle_A \otimes \sqrt{1-\tfrac{1}{\sqrt{2}}}\Big(\big(1+\tfrac{1}{\sqrt{2}}\big)|+\rangle_B - \tfrac{1}{\sqrt{2}}|-\rangle_B\Big) \\
+\tfrac{1}{\sqrt{2}}\sqrt{1-\tfrac{1}{\sqrt{2}}}\ |x;-\rangle_A \otimes \sqrt{1+\tfrac{1}{\sqrt{2}}}\Big(\big(1-\tfrac{1}{\sqrt{2}}\big)|+\rangle_B + \tfrac{1}{\sqrt{2}}|-\rangle_B\Big).
\end{aligned}
\tag{22.4.41}
$$

This is the Schmidt decomposition of $|\hat{\psi}_{AB}\rangle$. You can check that the states to the right of $|x;\pm\rangle_A$ are orthonormal. The Schmidt number is two, and the subsystems A and B are entangled. □

Measurement along a basis in a subsystem We now want to extend to bipartite systems the result $\tilde{\rho} = \sum_i M_i \rho M_i$ (see (22.2.45)), giving the density matrix $\tilde{\rho}$ after measurement along an orthonormal basis $\{|i\rangle\}$ when the result of the measurement is not known, and the original density matrix is ρ. Here, $M_i = |i\rangle\langle i|$.

Consider therefore a bipartite system AB, and imagine that Alice measures the state of A along a basis $\{|i\rangle_A\}$ associated with projectors $M_i^A = |i\rangle_A\,{}_A\langle i|$, satisfying

$$
M_i^{A\dagger} = M_i^A, \quad M_i^A M_i^A = M_i^A, \quad \sum_i M_i^A = \mathbb{1}_A. \tag{22.4.42}
$$

Assume that we start with a density matrix ρ_{AB}, and we do not know the result of Alice's measurement. In analogy to the previous result, the density matrix $\tilde{\rho}_{AB}$ after measurement is

$$
\tilde{\rho}_{AB} = \sum_i (M_i^A \otimes \mathbb{1}_B)\, \rho_{AB}\, (M_i^A \otimes \mathbb{1}_B). \tag{22.4.43}
$$

While clearly very plausible, this claim can be proven explicitly (problem 22.4). With this result we can learn something important about entanglement. We now look for the reduced density matrix $\tilde{\rho}_B$ of B following from $\tilde{\rho}_{AB}$ to see the effect on Bob due to Alice's measurement on A. Can Bob tell that Alice did a measurement? We have already seen in some particular case (example 22.5) that Bob cannot.

We begin our work with the density matrix of B *after* measurement:

$$
\tilde{\rho}_B = \mathrm{tr}_A \tilde{\rho}_{AB} = \mathrm{tr}_A \sum_i (M_i^A \otimes \mathbb{1}_B)\, \rho_{AB}\, (M_i^A \otimes \mathbb{1}_B). \tag{22.4.44}
$$

We wish to compare $\tilde{\rho}_B$ with the density matrix $\rho_B = \mathrm{tr}_A \rho_{AB}$ of B *before* measurement. To analyze this we use a general representation of the original bipartite density matrix in terms of a collection of operators $\mathcal{O}_k^A$ and $\mathcal{O}_k^B$ indexed by some label k:

$$
\rho_{AB} = \sum_k \mathcal{O}_k^A \otimes \mathcal{O}_k^B, \quad \mathcal{O}_k^A \in \mathcal{L}(H_A), \quad \mathcal{O}_k^B \in \mathcal{L}(H_B), \quad \forall k. \tag{22.4.45}
$$

Then we have

$$
\begin{aligned}
\tilde{\rho}_B &= \mathrm{tr}_A \sum_{i,k} (M_i^A \otimes \mathbb{1}_B)\, \mathcal{O}_k^A \otimes \mathcal{O}_k^B\, (M_i^A \otimes \mathbb{1}_B) \\
&= \mathrm{tr}_A \sum_{i,k} M_i^A \mathcal{O}_k^A M_i^A \otimes \mathcal{O}_k^B = \sum_{i,k} \mathrm{tr}_A(M_i^A \mathcal{O}_k^A M_i^A)\, \mathcal{O}_k^B.
\end{aligned}
\tag{22.4.46}
$$

Recalling the cyclicity of the trace and the projector properties of M_i^A listed above, we have $\text{tr}_A(M_i^A \mathcal{O}_k^A M_i^A) = \text{tr}_A(M_i^A \mathcal{O}_k^A)$. Since the sum of M_i^A's over i gives the identity matrix, we get

$$\tilde{\rho}_B = \sum_k \sum_i \text{tr}_A(M_i^A \mathcal{O}_k^A)\, \mathcal{O}_k^B = \sum_k \text{tr}_A(\mathcal{O}_k^A)\, \mathcal{O}_k^B = \text{tr}_A \rho_{AB} = \rho_B. \tag{22.4.47}$$

Alice's measurement, with results unknown to Bob, does *not* change Bob's density matrix. This means that Alice cannot use a measurement to communicate information instantaneously to Bob. Since the particles in entangled pairs can be very far away, this prevents superluminal transfer of information, thus avoiding conflict with special relativity. Note that the result did not depend on using the density matrix of AB. The result holds for an arbitrary operator S_{AB} on AB:

Theorem 22.4.2. No-signaling theorem. *Let $S_{AB} \in \mathcal{L}(\mathcal{H}_A \otimes \mathcal{H}_B)$ be an arbitrary operator and $\tilde{S}_{AB}$ be defined by*

$$\tilde{S}_{AB} = \sum_i (M_i^A \otimes \mathbb{1}_B)\, S_{AB}\, (M_i^A \otimes \mathbb{1}_B), \tag{22.4.48}$$

with M_i^A orthogonal projectors satisfying (22.4.42). Then, $\text{tr}_A \tilde{S}_{AB} = \text{tr}_A S_{AB}$.

If Alice and Bob share an entangled pair of quantum systems, this theorem prevents Alice from sending a message or a signal to Bob instantaneously by performing measurements. Thus the name *no-signaling theorem.*

We can also imagine that Alice, instead of measuring, applies some Hamiltonian to her system, causing some unitary evolution represented by the operator $\mathcal{U}_A$. In this case the evolved density matrix $\hat{\rho}_{AB}$ of the bipartite system whose initial density matrix is ρ_{AB} takes the form

$$\hat{\rho}_{AB} = (\mathcal{U}_A \otimes \mathbb{1}_B)\, \rho_{AB}\, (\mathcal{U}_A^\dagger \otimes \mathbb{1}_B). \tag{22.4.49}$$

It is now simple to show, just as above, that the density matrix for Bob is not affected:

$$\hat{\rho}_B \equiv \text{tr}_A \hat{\rho}_{AB} = \rho_B, \tag{22.4.50}$$

where $\rho_B = \text{tr}_A \rho_{AB}$ is the density matrix of B before Alice subjected her particle to unitary evolution. Alice cannot signal Bob by acting on her system with arbitrary unitary evolution.

Exercise 22.8. *Prove that (22.4.50) holds.*

22.5 Open Systems and Decoherence

We now consider an isolated quantum system AE, which we will view as a bipartite system composed of a subsystem A, the focus of our interest, and a subsystem E called the *environment*. In practice A could be a small quantum system, perhaps the spin state of a single nucleus in an NMR experiment, with E the thousands of nearby spin states that interact with our selected spin. Or, in principle, the AE system could consist of two spin one-half particles, one comprising the A system and the other the E system. We call A an **open system** because it is not isolated; it interacts with another quantum system, the environment E.

Starting from the density matrix ρ_{AE} for a system AE, we can define a reduced density matrix ρ_A for A by tracing over the environment E. We are interested in this reduced density matrix; we do not aim to describe the environment. Our main question is: How does ρ_A evolve in time?

The full AE density matrix represents an isolated system, so it must have unitary time evolution, as in (22.3.7), with $\mathcal{U}$ the evolution operator associated to the complete Hamiltonian of the AE system. The time evolution of ρ_A, however, is *not* unitary. This is easily seen in an example to be discussed in detail below: if we have two spins, one A and one E, and the initial AE state is a pure nonentangled state, the density ρ_A begins as that of a pure state. The interactions, however, entangle the two spins, and as we have seen before, this means that ρ_A becomes a mixed state. Under unitary evolution a pure state remains a pure state (section 22.3), and thus ρ_A does *not* experience unitary evolution.

For the AE system, the open subsystem A can experience **decoherence**, the process where ρ_A begins as a pure state but turns into a mixed state. At time equal zero, A may be in a pure state $|\psi_A\rangle$, a superposition of $\mathcal{H}_A$ basis states. But then, by interacting with the environment E, the state of A becomes mixed. The data defining the original superposition is no longer available on subsystem A, having migrated at least partially into correlations of A with the environment. Decoherence is a problem for a quantum computer: the quantum circuit is expected to be in a pure state during any computation (see example 22.7 below).

In this section we first discuss the time evolution of the reduced density matrix ρ_A of an AE system. We then consider an example illustrating the evolution of the density matrix of a spin one-half particle coupled to another one. We conclude with a description of the density matrices relevant to the microcanonical and macrocanonical ensembles. In the following section, we explore a phenomenological approach to the question of decoherence using an equation proposed by Lindblad to model the time evolution of the density matrix of an open subsystem coupled to a large environment.

Consider the isolated system AE. The operation tr_E, trace over the environment, maps a density matrix ρ_{AE} of AE into a density matrix ρ_A that describes subsytem A:

$$\text{tr}_E\colon \rho_{AE} \to \rho_A. \tag{22.5.1}$$

The time evolution of ρ_A is controlled by this map: we must find the time evolution of ρ_{AE} and then take the environment trace to find the time evolution of ρ_A:

$$\text{tr}_E : \rho_{AE}(t) \to \rho_A(t). \tag{22.5.2}$$

This is the strategy we follow. Assume the Hamiltonian $\hat{H}$ of the full AE system is known so that we also have the unitary operator $\mathcal{U}$ that evolves states of AE. Let $\mathcal{U} = \mathcal{U}(t)$ denote the operator that turns states at $t=0$ into states at t. We then have

$$\rho_{AE}(t) = \mathcal{U}\rho_{AE}(0)\,\mathcal{U}^\dagger. \tag{22.5.3}$$

As a result, the time-dependent density matrix for subsystem A is given by

$$\rho_A(t) = \text{tr}_E\left[\mathcal{U}\rho_{AE}(0)\,\mathcal{U}^\dagger\right]. \tag{22.5.4}$$

Note that the partial trace does not satisfy cyclicity when acting on operators in the full space, and therefore the $\mathcal{U}$ and $\mathcal{U}^\dagger$ cannot be brought together to cancel each other. The above formula satisfies the consistency conditions that make ρ_A into a density matrix. Indeed, recalling that $\text{tr} = \text{tr}_A\,\text{tr}_E$ we show that

$$\begin{aligned} \text{tr}_A\,\rho_A(t) &= \text{tr}_A\,\text{tr}_E[\mathcal{U}\rho_{AE}(0)\mathcal{U}^\dagger] = \text{tr}[\mathcal{U}\rho_{AE}(0)\mathcal{U}^\dagger] \\ &= \text{tr}[\rho_{AE}(0)\mathcal{U}^\dagger\mathcal{U}] = \text{tr}\,\rho_{AE}(0) = 1. \end{aligned} \tag{22.5.5}$$

Exercise 22.9. *Show that $\rho_A(t)$, as defined, is positive semidefinite if $\rho_{AE}(0)$ is.*

A reasonable assumption is to take the state of the environment at the initial time $t=0$ to be pure and equal to some fixed state $|\psi_E\rangle$. We could also take the state of A to be pure, but let us first consider the possibility that it is a density matrix $\rho_A(t_0)$. In this case the initial density matrix of AE is

$$\rho_{AE}(0) = \rho_A(0) \otimes |\psi_E\rangle\langle\psi_E|. \tag{22.5.6}$$

The time evolution equation (22.5.4) then gives

$$\rho_A(t) = \text{tr}_E\left[\mathcal{U}\big(\rho_A(0) \otimes |\psi_E\rangle\langle\psi_E|\big)\mathcal{U}^\dagger\right], \quad \mathcal{U} = \mathcal{U}(t). \tag{22.5.7}$$

This shows how the density matrix of the subsystem evolves in time. Keeping the time t and the state $|\psi_E\rangle$ fixed, the right-hand side defines a linear map from density matrices to density matrices, taking $\rho_A(0)$ to $\rho_A(t)$. If at $t=0$ system A is a pure state $|\phi_A\rangle$, we find that

$$\rho_A(t) = \text{tr}_E\left[\mathcal{U}\big(|\phi_A\rangle\langle\phi_A| \otimes |\psi_E\rangle\langle\psi_E|\big)\mathcal{U}^\dagger\right]. \tag{22.5.8}$$

Example 22.7. *Decoherence of a qubit.*

Consider a spin one-half particle representing a qubit or quantum bit in a quantum computer; a qubit being just a quantum two-state system. The qubit is in the state

$$|\psi_A\rangle = \alpha|\uparrow\rangle + \beta|\downarrow\rangle, \quad |\alpha|^2 + |\beta|^2 = 1. \tag{22.5.9}$$

This qubit is coupled to a state $|0_E\rangle$ of the environment so that the total state $|\psi_{AE}\rangle$ of the system is given by

$$|\psi_{AE}\rangle = |\psi_A\rangle \otimes |0_E\rangle = \big(\alpha|\uparrow\rangle + \beta|\downarrow\rangle\big) \otimes |0_E\rangle = \alpha|\uparrow\rangle \otimes |0_E\rangle + \beta|\downarrow\rangle \otimes |0_E\rangle. \tag{22.5.10}$$

The qubit density matrix ρ_A from this state takes the form given by

$$\rho_A = \text{tr}_E\,\rho_{AE} = \text{tr}_E|\psi_{AE}\rangle\langle\psi_{AE}| = \text{tr}_E|\psi_A\rangle\otimes|0_E\rangle\langle\psi_A|\otimes\langle 0_E|$$
$$= |\psi_A\rangle\langle\psi_A| = \begin{pmatrix}|\alpha|^2 & \alpha\beta^* \\ \beta\alpha^* & |\beta|^2\end{pmatrix}, \tag{22.5.11}$$

using basis vectors $|1\rangle = |\uparrow\rangle, |2\rangle = |\downarrow\rangle$. This is, of course, the density matrix of a pure state. Now assume that an interaction of the qubit with the environment changes the state of the environment for the term where the qubit is down:

$$|\psi_{AE}\rangle = \alpha|\uparrow\rangle\otimes|0_E\rangle + \beta|\downarrow\rangle\otimes|0_E\rangle \quad\rightarrow\quad |\psi'_{AE}\rangle = \alpha|\uparrow\rangle\otimes|0_E\rangle + \beta|\downarrow\rangle\otimes|1_E\rangle, \tag{22.5.12}$$

with $|1_E\rangle$ that other state of the environment. The primed state has the qubit entangled with the environment. Assuming the states $|0_E\rangle$ and $|1_E\rangle$ are normalized and orthogonal to each other, the density matrix ρ'_A is now

$$\rho'_A = \text{tr}_E\,\rho'_{AE} = \text{tr}_E|\psi'_{AE}\rangle\langle\psi'_{AE}|$$
$$= \text{tr}_E\big(\alpha|\uparrow\rangle\otimes|0_E\rangle + \beta|\downarrow\rangle\otimes|1_E\rangle\big)\big(\alpha^*\langle\uparrow|\otimes\langle 0_E| + \beta^*\langle\downarrow|\otimes\langle 1_E|\big) \tag{22.5.13}$$
$$= |\alpha|^2\,|\uparrow\rangle\langle\uparrow| + |\beta|^2\,|\downarrow\rangle\langle\downarrow|.$$

As a matrix, we have

$$\rho'_A = \begin{pmatrix}|\alpha|^2 & 0 \\ 0 & |\beta|^2\end{pmatrix}. \tag{22.5.14}$$

This is a mixed state if $\alpha \neq 0$ and $\beta \neq 0$. Indeed, in this case the purity of the state is less than one:

$$\text{tr}(\rho'_A)^2 = |\alpha|^4 + |\beta|^4 = (|\alpha|^2 + |\beta|^2)^2 - 2|\alpha|^2|\beta|^2 = 1 - 2|\alpha|^2|\beta|^2 < 1. \tag{22.5.15}$$

The qubit experienced decoherence. This is an insidious problem for a quantum computer. You can imagine that we do not want the environment to change the state of a qubit, but this issue can be taken care of by redundancy or error correction. The decoherence problem is more difficult: we do not want the qubit, or the qubits on a circuit, to affect the environment! Preventing such decoherence becomes harder as the circuit becomes larger. It is indeed hard to suppress thermal and other couplings to the environment.

It is worth comparing the density matrices ρ_A and ρ'_A. The former, as you can see, has off-diagonal matrix elements, storing the information about the relative phases of the different components of the wave function. The latter does not. Decoherence reduces or deletes the off-diagonal matrix elements of the density matrix in the basis dictated by the coupling to the environment. □

Example 22.8. *From pure to mixed: two coupled spin one-half particles.*
To illustrate some of the above ideas, we consider two spin one-half particles: particle one and particle two, interacting through an *Ising* Hamiltonian:

$$\hat{H} = -\hbar\omega\,\hat{\sigma}_z^{(1)}\hat{\sigma}_z^{(2)}, \quad \omega > 0. \tag{22.5.16}$$

This interaction tends to align both spins along the z-axis, as such a configuration gives a minimum of the energy. Our focus will be on particle one, and we will treat particle two as the environment. We will determine the time evolution of the reduced density matrix ρ_1 for particle one. The result will display how a pure state of particle one evolves into a mixed state. This evolution is expected because the interaction can turn a nonentangled state of the two particles into an entangled state.

As usual, we assume that at time equal zero the state of the two particles is a pure state. The most general such state takes the form

$$|\psi_{12}(0)\rangle = \tfrac{1}{2}\left(a_+|\uparrow\uparrow\rangle + a_-|\uparrow\downarrow\rangle + b_+|\downarrow\uparrow\rangle + b_-|\downarrow\downarrow\rangle\right). \tag{22.5.17}$$

In here the kets contain two arrows. The first corresponds to the state of particle one and the second to the state of particle two. We will use the same convention for bras. We have introduced a total of four coefficients, $a_\pm, b_\pm$, which are complex constants. The normalization condition reads

$$\tfrac{1}{4}\left(|a_+|^2 + |a_-|^2 + |b_+|^2 + |b_-|^2\right) = 1. \tag{22.5.18}$$

For arbitrary coefficients the two particles are entangled at $t=0$, and the state of particle one is not pure. The density matrix ρ_{12} of the whole system at $t=0$ is

$$\begin{aligned}\rho_{12}(0) = \tfrac{1}{4}&\left(a_+|\uparrow\uparrow\rangle + a_-|\uparrow\downarrow\rangle + b_+|\downarrow\uparrow\rangle + b_-|\downarrow\downarrow\rangle\right)\\ &\cdot\left(a_+^*\langle\uparrow\uparrow| + a_-^*\langle\uparrow\downarrow| + b_+^*\langle\downarrow\uparrow| + b_-^*\langle\downarrow\downarrow|\right).\end{aligned} \tag{22.5.19}$$

The evolution operator $\mathcal{U}(t) = \exp(-i\hat{H}t/\hbar)$ is given by

$$\mathcal{U} = e^{i\omega t\,\hat{\sigma}_z^{(1)}\hat{\sigma}_z^{(2)}}, \quad \mathcal{U}^\dagger = e^{-i\omega t\,\hat{\sigma}_z^{(1)}\hat{\sigma}_z^{(2)}}. \tag{22.5.20}$$

We then have $\rho_{12}(t) = \mathcal{U}\rho_{12}(0)\mathcal{U}^\dagger$, resulting in

$$\begin{aligned}\rho_{12}(t) = \tfrac{1}{4}&\left(a_+e^{i\omega t}|\uparrow\uparrow\rangle + a_-e^{-i\omega t}|\uparrow\downarrow\rangle + b_+e^{-i\omega t}|\downarrow\uparrow\rangle + b_-e^{i\omega t}|\downarrow\downarrow\rangle\right)\\ &\cdot\left(a_+^*e^{-i\omega t}\langle\uparrow\uparrow| + a_-^*e^{i\omega t}\langle\uparrow\downarrow| + b_+^*e^{i\omega t}\langle\downarrow\uparrow| + b_-^*e^{-i\omega t}\langle\downarrow\downarrow|\right).\end{aligned} \tag{22.5.21}$$

Taking the trace over the second state space, we obtain the time-dependent density matrix $\rho_1(t)$ for the first particle:

$$\begin{aligned}\rho_1(t) = \mathrm{tr}_2\,\rho_{12}(t) = \;&\tfrac{1}{4}\left(|a_+|^2 + |a_-|^2\right)|\uparrow\rangle\langle\uparrow|\\ &+\tfrac{1}{4}\left(a_+b_+^*e^{2i\omega t} + a_-b_-^*e^{-2i\omega t}\right)|\uparrow\rangle\langle\downarrow|\\ &+\tfrac{1}{4}\left(a_+^*b_+e^{-2i\omega t} + a_-^*b_-e^{2i\omega t}\right)|\downarrow\rangle\langle\uparrow|\\ &+\tfrac{1}{4}\left(|b_+|^2 + |b_-|^2\right)|\downarrow\rangle\langle\downarrow|.\end{aligned} \tag{22.5.22}$$

This is the general answer and can be applied to any initial condition. Let us consider the case when $a_+ = a_- = b_+ = b_- = 1$, consistent with normalization. Then the initial state is

$$
\begin{aligned}
|\psi_{12}(0)\rangle &= \tfrac{1}{2}\left(|\uparrow\uparrow\rangle + |\uparrow\downarrow\rangle + |\downarrow\uparrow\rangle + |\downarrow\downarrow\rangle\right) \\
&= \tfrac{1}{\sqrt{2}}(|\uparrow\rangle + |\downarrow\rangle) \otimes \tfrac{1}{\sqrt{2}}(|\uparrow\rangle + |\downarrow\rangle) \\
&= |x;+\rangle \otimes |x;+\rangle.
\end{aligned} \tag{22.5.23}
$$

The two particles are not entangled at $t=0$. The density matrix for the first particle can be obtained from (22.5.22):

$$
\rho_1(t) = \tfrac{1}{2}|\uparrow\rangle\langle\uparrow| + \tfrac{1}{2}\cos 2\omega t\,(|\uparrow\rangle\langle\downarrow| + |\downarrow\rangle\langle\uparrow|) + \tfrac{1}{2}|\downarrow\rangle\langle\downarrow|. \tag{22.5.24}
$$

The diagonal terms lead to the required trace, and the off-diagonal terms oscillate. At $t=0$, the density matrix is $\rho_1(0) = |x;+\rangle\langle x;+|$, as expected, since the two particles are not entangled, and this is the density matrix for the pure state $|x;+\rangle$ of the first particle. But as time increases, the states became entangled, and the first particle experiences decoherence: its density matrix becomes that of a mixed state. This is manifest at any time t_* when $\cos 2\omega t_* = 0$, making the off-diagonal terms of ρ_1 vanish. At any such t_*, the state of the first particle is maximally mixed, and the density matrix is proportional to the identity matrix. For arbitrary times it is useful to compute $\text{tr}\rho_1^2$. A short calculation gives

$$
\text{tr}\rho_1^2 = 1 - \tfrac{1}{2}\sin^2 2\omega t \leq 1. \tag{22.5.25}
$$

Since the density matrix represents a pure state if and only if the above inequality is saturated, we see that the state is pure when $\sin 2\omega t = 0$. In this simple system, there is oscillatory behavior. While an initially pure state decoheres, it becomes pure again at a later time. □

A model of decoherence can be built by coupling the spin one-half particle we focus on—call it particle A—to all of the particles in a set of N interacting spin one-half particles. For large N and for random couplings among the N particles, one would expect the dynamics to turn a pure, unentangled state of A into a maximally mixed state. Decoherence would set in. There are other well-studied models of decoherence, such as spin-boson models in which a single spin is coupled to a collection of harmonic oscillators. The Caldeira-Legget model features the coupling of an ordinary particle to a collection of harmonic oscillators. This model has been studied in great detail.

Thermal states In statistical mechanics one usually considers ensembles. We have the microcanonical ensemble and the macrocanonical ensemble. The microcanonical ensemble applies when we have a system with fixed total energy, and the canonical ensemble applies when we have an open system in contact with a large reservoir at some temperature T. We would like to find the relevant density matrices that describe the quantum state of the full system in the first case and the quantum state of the open system in the second case.

For the microcanonical ensemble, we consider the states of the system A with total fixed energy E. These states form a subspace U_E of the state space of A. We call $\Omega(E)$ the dimension of U_E. Since the energy is fixed, we are thinking of the system A as isolated. Let $\Pi(E)$ denote the orthogonal projector to the subspace U_E. Since it projects to an $\Omega(E)$-dimensional subspace, $\Pi(E)$ has rank $\Omega(E)$, and therefore

$$\operatorname{tr}\Pi(E)=\Omega(E). \tag{22.5.26}$$

If the states of energy E are described as $|n\rangle$ with $n=1,\ldots,\Omega(E)$, we can write

$$\Pi(E)=\sum_{n=1}^{\Omega(E)}|n\rangle\langle n|. \tag{22.5.27}$$

The density matrix $\rho(E)$ that we postulate for the microcanonical ensemble is the maximally mixed one, arguably the simplest state given a complete lack of information about the system. In this ensemble all $\Omega(E)$ states of energy E are equally probable. We thus write as the

$$\text{microcanonical ensemble:}\quad \rho(E)=\frac{1}{\Omega(E)}\,\Pi(E). \tag{22.5.28}$$

This operator has trace equal to one, as desired. It is maximally mixed because on the relevant subspace U_E the density matrix is proportional to the identity.

Let us now turn to the canonical ensemble. In this case the system A is an open system in contact with a large environment, the "reservoir" R. The reservoir and the system can exchange energy, and the reservoir is at a temperature T that remains constant. In statistical mechanics the definition of temperature relates it to the rate of change of the number of states of the system as we vary the energy. Calling $\Omega^R(E)$ the number of states of the reservoir with energy E, we have

$$\frac{1}{k_BT}=\frac{d}{dE}\ln\Omega^R(E), \tag{22.5.29}$$

with k_B the Boltzman constant. This relation is easily integrated to find how the number of states of the reservoir depends on energy:

$$\ln\Omega^R(E_1)-\ln\Omega^R(E_0)=\frac{1}{k_BT}(E_1-E_0). \tag{22.5.30}$$

If we let $E_1=E-E_n$ and $E_0=E$, we find that

$$\ln\Omega^R(E-E_n)=\ln\Omega^R(E)-\frac{E_n}{k_BT}. \tag{22.5.31}$$

By exponentiation we finally get an explicit formula for the number of states of the reservoir:

$$\Omega^R(E-E_n)=\Omega^R(E)e^{-\frac{E_n}{k_BT}}. \tag{22.5.32}$$

We want a density matrix for the open subsystem A. The strategy consists of writing a microcanonical density matrix for the full, isolated system AR and then tracing over R to get the density matrix for A. Let us first introduce notation for the energy eigenstates:

$$\begin{aligned} |n\rangle &: \hat{H}^A \text{ eigenstates with energy } E_n, \\ |R_{n,\gamma}\rangle &: \hat{H}^R \text{ eigenstates with energy } E - E_n. \end{aligned} \tag{22.5.33}$$

The label n runs over all energy eigenstates of A. For each value of n, the index γ runs over the set of $\hat{H}^R$ eigenstates with energies $E - E_n$. We need not assume that the energy eigenstates of A are nondegenerate. The systems A and R are, to a good approximation, noninteracting and their energies add. Therefore, the states $|n\rangle \otimes |R_{n,\gamma}\rangle$ of AR have energy E. The operator $|n\rangle\langle n| \otimes |R_{n,\gamma}\rangle\langle R_{n,\gamma}|$ is a projector to the state $|n\rangle \otimes |R_{n,\gamma}\rangle$. As a result, the projector $\Pi^{AR}(E)$ to the states of AR with energy E is

$$\Pi^{AR}(E) = \sum_{n,\gamma} |n\rangle\langle n| \otimes |R_{n,\gamma}\rangle\langle R_{n,\gamma}|. \tag{22.5.34}$$

The trace of this projector is the number of states $\Omega^{AR}(E)$ of AE with energy E:

$$\Omega^{AR}(E) = \text{tr}\, \Pi^{AR}(E). \tag{22.5.35}$$

For the whole system AR at fixed energy E, the density matrix is microcanonical:

$$\rho^{AR}(E) = \frac{1}{\Omega^{AR}(E)} \Pi^{AR}(E). \tag{22.5.36}$$

The density matrix ρ^A is obtained by tracing over the reservoir R:

$$\begin{aligned} \rho^A &= \text{tr}_R\, \rho^{AR}(E) = \frac{1}{\Omega^{AR}(E)} \text{tr}_R \sum_{n,\gamma} |n\rangle\langle n| \otimes |R_{n,\gamma}\rangle\langle R_{n,\gamma}| \\ &= \frac{1}{\Omega^{AR}(E)} \sum_n |n\rangle\langle n| \sum_\gamma \langle R_{n,\gamma}|R_{n,\gamma}\rangle \\ &= \frac{1}{\Omega^{AR}(E)} \sum_n |n\rangle\langle n|\, \Omega^R(E - E_n). \end{aligned} \tag{22.5.37}$$

In the last step, we used $\langle R_{n,\gamma}|R_{n,\gamma}\rangle = 1$ so that the sum over γ, for fixed n, simply counts the number of states of the reservoir at energy $E - E_n$. Now we use (22.5.32) to evaluate $\Omega^R(E - E_n)$:

$$\rho^A = \frac{\Omega^R(E)}{\Omega^{AR}(E)} \sum_n |n\rangle\langle n|\, e^{-\frac{E_n}{k_BT}}. \tag{22.5.38}$$

The sum above is in fact a simple operator. Indeed, recalling that $\hat{H}^A|n\rangle = E_n|n\rangle$ and that $\sum_n |n\rangle\langle n| = \mathbb{1}_A$ in the state space of A, we have

$$e^{-\frac{\hat{H}^A}{k_BT}} = e^{-\frac{\hat{H}^A}{k_BT}} \sum_n |n\rangle\langle n| = \sum_n e^{-\frac{E_n}{k_BT}} |n\rangle\langle n|. \tag{22.5.39}$$

The density matrix ρ^A is then, up to a constant prefactor, equal to the operator $\exp(-\frac{\hat{H}^A}{k_BT})$. It is now possible to write this density matrix using just this operator and its trace:

$$\rho^A = \frac{1}{Z}\exp\Big(-\frac{\hat{H}^A}{k_BT}\Big), \qquad Z \equiv \operatorname{tr}\exp\Big(-\frac{\hat{H}^A}{k_BT}\Big). \tag{22.5.40}$$

Division by the constant factor Z, called the *partition function*, implements the trace condition for ρ^A; the other constants written in (22.5.38) were just unwieldy. The above equation is our final form for the thermal density matrix of an open system in contact with a reservoir at temperature T. The density matrix ρ^A encodes the *Boltzmann distribution*, giving us the probabilities of finding the system A in any of its energy eigenstates. To make that result manifest, it is better to keep the sum over n and write

$$\rho^A = \frac{1}{Z}\sum_n |n\rangle\langle n|\, e^{-\frac{E_n}{k_BT}}. \tag{22.5.41}$$

From the ensemble interpretation of the density matrix, the probability of finding the state $|n\rangle$ is $\frac{1}{Z}\exp(-\frac{E_n}{k_BT})$. The probabilities in this density matrix define the Boltzmann distribution.

22.6 The Lindblad Equation

It is possible to study a variety of models of decoherence by exploring the Schrodinger equation for the system both numerically and analytically. It turns out, however, that for many cases of interest a more phenomenological approach is rather useful. One such approach is provided by the **Lindblad equation**, which is in fact a class of equations for the possible dynamics of the density matrix of an open system. Instead of looking at the microscopic elements of the system and the environment, solving the Schrödinger equation for the wave function, and then exploring the resulting dynamics of the density matrix, one jumps directly to consider possible ways in which the density matrix of an open system could evolve, consistent with positive semidefiniteness and unit trace.

The Lindblad equation for the density matrix $\rho(t)$ generalizes our unitary evolution equation and takes the form

$$\frac{\partial\rho}{\partial t} = \frac{1}{i\hbar}[H,\rho] + \sum_k\Big(L_k\rho L_k^\dagger - \tfrac{1}{2}\{L_k^\dagger L_k,\rho\}\Big), \tag{22.6.1}$$

where the L_k are a set of Lindblad operators, with k an index that can run over an arbitrary set of values. The number of Lindblad operators depends on the model we are trying to build. The Lindblad operators need not be Hermitian. The curly brackets denote the anticommutator: $\{A,B\} = AB + BA$. If all Lindblad operators vanish, we recover the unitary evolution of the density matrix via the Hamiltonian. The right-hand side of the above equation is a Hermitian operator, as it should be in order for ρ to remain Hermitian at all times. In fact, each summand of fixed k is Hermitian, the first term manifestly and the second because it is the anticommutator of Hermitian operators. The Lindblad operators appear quadratically, and operators with different k do not mix.

The prescribed form in which the L_k's enter the right-hand side of the Lindblad equation is constrained by the requirement that the trace of ρ is one for all times. Indeed, we check that

$$\frac{d}{dt}\text{tr}\rho = \text{tr}\frac{\partial\rho}{\partial t} = \sum_k \text{tr}\Big(L_k\rho L_k^\dagger - \tfrac{1}{2}\{L_k^\dagger L_k, \rho\}\Big), \qquad (22.6.2)$$

since the term involving the Hamiltonian $\hat{H}$ has already been checked to be traceless. Expanding out the anticommutator and using the cyclicity of the trace, we find that

$$\frac{d}{dt}\text{tr}\rho = \sum_k \text{tr}\Big(L_k^\dagger L_k\rho - \tfrac{1}{2}L_k^\dagger L_k\rho - \tfrac{1}{2}\rho L_k^\dagger L_k\Big) = 0. \qquad (22.6.3)$$

There is one more consistency check: ρ must remain positive semidefinite under time evolution. This can be confirmed by showing that $\rho(t+dt)$, with infinitesimal $dt > 0$, is positive semidefinite if $\rho(t)$ is (problem 22.5).

The Lindblad equation is by no means a general equation that describes all possible time dependent density matrices. It is a useful approximation for cases when the environment is large, so that the transfer of information from the system of interest occurs rapidly and changes the environment very little. Under these conditions, it is possible to give a derivation of the equation, but we will not attempt to do this here. We will just aim to give insight into the Lindblad equation by studying how it describes a situation of significant experimental interest.

Modeling decoherence in NMR Let us discuss how to model the decoherence behavior of a spin state in a nuclear magnetic resonance (NMR) experiment. Recall the setup discussed in section 17.6. We have a large magnetic field along z, the longitudinal direction. At room temperature (300 K), and even with magnetic fields as large as 10 tesla, the populations of states aligned and anti-aligned with the magnetic field differ by less than 1 in 100,000, but this is enough to put a large number of spins at play. The application of a brief radio-frequency (RF) signal gets the spins to rotate in the x, y plane. With the RF signal off, the spins would continue to rotate in the x, y plane were it not for decoherence effects. Focusing on one particular spin at the instant the RF signal turns off, its state is pure, and the associated density matrix in the z-basis $|\pm\rangle$ is a 2×2 matrix with all entries nonzero. Imagine, for example, that at this instant the state of the spin is along the positive x-axis: $|x; +\rangle$. The density matrix then takes the form

$$\rho = |x; +\rangle\langle x; +| = \begin{pmatrix} \frac{1}{2} & \frac{1}{2} \\ \frac{1}{2} & \frac{1}{2} \end{pmatrix}. \qquad (22.6.4)$$

The off-diagonal elements of the density matrix are sometimes called *coherences* and remain nonzero as long as the spin is a pure state rotating in the (x, y) plane. The name coherences is appropriate; if they vanished the diagonal elements of the density matrix would represent a maximally mixed state.

After a little time, the interaction between the spins has substantial effects. With time constant T_2, called the *transverse* relaxation constant, the rotation of the spin decoheres, and in this process the off-diagonal parts of the density matrix vanish away. We are left with a diagonal density matrix whose components reflect populations. A second time constant T_1, called the *longitudinal* relaxation constant, and usually significantly longer than T_2, controls the timescale for the populations to adjust to the value appropriate for the external magnetic field.

A simple model of the above processes can be obtained by letting the density matrix of a spin one-half particle evolve with three Lindblad operators. For simplicity, we will also assume a zero external magnetic field, implying that $\hat{H}=0$ and that the populations for spin up and spin down will be the same at large times. The three Lindblad operators L_1, L_2, and L_3 are given by

$$\begin{aligned} L_1 &= \alpha|+\rangle\langle-|, \\ L_2 &= \alpha|-\rangle\langle+|, \\ L_3 &= \beta\sigma_z = \beta\,(|+\rangle\langle+| - |-\rangle\langle-|). \end{aligned} \tag{22.6.5}$$

Here α and β are real constants whose signs do not matter because only their squares appear in the Lindblad equation. The operator L_1 generates transitions $|-\rangle \to |+\rangle$, and L_2 generates the reverse transitions. These are population-changing transitions, and thus α must have an impact on the longitudinal relaxation time T_1 governing the evolution of the diagonal terms of the density matrix. The operator L_3 is not associated with transitions; instead, it drives to zero the off-diagonal elements (coherences) of the density matrix. We get a hint of this by realizing that the term $L_3\rho L_3^\dagger$ appearing on the Lindblad equation is proportional to $\sigma_z\rho\sigma_z$, which is the density matrix with the sign of the off-diagonal elements changed. This contribution to the time derivative of ρ will drive the coherences to zero, and therefore β must affect the transverse relaxation constant T_2. Since the Lindblad equation also features a term of the form $\{L_k^\dagger L_k, \rho\}$, it is not easy to anticipate without computation the full effect of the operators. In fact, the constant α also ends up contributing to the transverse relaxation time T_2.

To solve for the density matrix evolution in this model, we write the Lindblad equation as follows (using dots for time derivatives):

$$\dot{\rho} = \sum_{k=1}^{3} R_k, \quad R_k \equiv L_k\rho L_k^\dagger - \tfrac{1}{2}\{L_k^\dagger L_k, \rho\}, \tag{22.6.6}$$

and we evaluate separately the three operators R_1, R_2, and R_3. We begin with R_1:

$$R_1 = \alpha^2|+\rangle\langle-|\rho|-\rangle\langle+| - \tfrac{1}{2}\alpha^2\big\{|-\rangle\langle-|, \rho\big\}. \tag{22.6.7}$$

To evaluate the first term and the anticommutator, it is useful to write out the density matrix:

$$\rho = \begin{pmatrix} \rho_{++} & \rho_{+-} \\ \rho_{-+} & \rho_{--} \end{pmatrix} = \rho_{++}|+\rangle\langle+| + \rho_{+-}|+\rangle\langle-| + \rho_{-+}|-\rangle\langle+| + \rho_{--}|-\rangle\langle-|. \tag{22.6.8}$$

A short calculation then gives

$$R_1 = \alpha^2 \begin{pmatrix} \rho_{--} & -\frac{1}{2}\rho_{+-} \\ -\frac{1}{2}\rho_{-+} & -\rho_{--} \end{pmatrix}. \tag{22.6.9}$$

Since R_1 is a contribution to the time derivative $\dot{\rho}$, we see that $\dot{\rho}_{++} \sim \rho_{--}$, making ρ_{++} increase, and $\dot{\rho}_{--} \sim -\rho_{--}$, making ρ_{--} decay. This is consistent with our expectation that L_1 would generate transitions from $|-\rangle$ to $|+\rangle$. Note that, additionally, R_1 will tend to suppress the off-diagonal terms. The calculation of R_2 is quite analogous and gives

$$R_2 = \alpha^2 \begin{pmatrix} -\rho_{++} & -\frac{1}{2}\rho_{+-} \\ -\frac{1}{2}\rho_{-+} & \rho_{++} \end{pmatrix}. \tag{22.6.10}$$

The calculation of R_3 is even simpler. Since $\sigma_z\sigma_z = 1$, we get

$$R_3 = \beta^2 \sigma_z \rho \sigma_z - \tfrac{1}{2}\beta^2(\rho + \rho) = \beta^2 \begin{pmatrix} 0 & -2\rho_{+-} \\ -2\rho_{-+} & 0 \end{pmatrix}. \tag{22.6.11}$$

The contribution of R_3 to the time derivative of ρ implies $\dot{\rho}_{\pm,\mp} \sim -\rho_{\pm,\mp}$, consistent with suppressing those off-diagonal elements. We can now bring together all our results, and the Lindblad equation defining the time evolution of ρ is

$$\begin{pmatrix} \dot{\rho}_{++} & \dot{\rho}_{+-} \\ \dot{\rho}_{-+} & \dot{\rho}_{--} \end{pmatrix} = \begin{pmatrix} -\alpha^2(\rho_{++} - \rho_{--}) & -(\alpha^2 + 2\beta^2)\rho_{+-} \\ -(\alpha^2 + 2\beta^2)\rho_{-+} & -\alpha^2(\rho_{--} - \rho_{++}) \end{pmatrix}. \tag{22.6.12}$$

The asymptotic $t \to \infty$ values are obtained by setting the derivatives to zero and are $\rho_{\pm\mp} = 0$, and $\rho_{++} = \rho_{--} = \frac{1}{2}$, also using the trace condition. The off-diagonal terms obey the equations

$$\dot{\rho}_{\pm\mp} = -(\alpha^2 + 2\beta^2)\rho_{\pm\mp}, \tag{22.6.13}$$

which are solved in terms of initial $t = 0$ values as follows:

$$\rho_{\pm\mp} = \rho_{\pm\mp}(0)\, e^{-t/T_2}, \quad T_2 = \frac{1}{\alpha^2 + 2\beta^2}. \tag{22.6.14}$$

This gives the transverse relaxation time T_2 for decoherence in terms of the parameters of the model. Subtracting the equations along the diagonal, we find

$$\dot{\rho}_{++} - \dot{\rho}_{--} = -2\alpha^2(\rho_{++} - \rho_{--}). \tag{22.6.15}$$

This is solved by

$$(\rho_{++} - \rho_{--})(t) = (\rho_{++} - \rho_{--})(0)e^{-t/T_1}, \quad T_1 = \frac{1}{2\alpha^2}. \tag{22.6.16}$$

Together with the trace condition $\rho_{++} + \rho_{--} = 1$, valid for all times, we find that

$$\rho_{\pm\pm}(t) = \tfrac{1}{2} + e^{-t/T_1}\big(\rho_{\pm\pm}(0) - \tfrac{1}{2}\big). \tag{22.6.17}$$

Indeed, the diagonal elements of ρ approach one-half, to give a maximally mixed state. They do so with a longitudinal relaxation constant T_1 that, as expected, is just a function of the coefficient α appearing in the first two Lindblad operators. In summary, the two time constants are

$$T_1 = \frac{1}{2\alpha^2}, \quad T_2 = \frac{1}{\alpha^2 + 2\beta^2}. \tag{22.6.18}$$

The first two Lindblad operators, containing α, would have sufficed to give nonzero values to the two time constants. Nevertheless, they would have resulted in $T_1 = \frac{1}{2}T_2 < T_2$, which is experimentally wrong. The third operator, containing β, is needed. The time constants are equal when $2\beta^2 = \alpha^2$. For $2\beta^2 > \alpha^2$, we get $T_1 > T_2$. Typically, T_1 is significantly larger than T_2, and for any fixed α, this can be arranged by taking β sufficiently large.

Exercise 22.10. *Confirm that, given (22.6.18), we always have $T_2 \leq 2T_1$.*

22.7 A Theory of Measurement?

Throughout this textbook we have followed the Copenhagen interpretation of quantum mechanics, largely devised by Niels Bohr and Werner Heisenberg from 1925 to 1927. As we noted while discussing the axioms of quantum mechanics in section 16.6, states evolve unitarily according to the Schrödinger equation, except at measurements. Measurements determine the values of observables, which are Hermitian operators. For any observable $\mathcal{O}$, its eigenvalues are the possible results of measurements. When we measure $\mathcal{O}$, the state collapses instantaneously into an $\mathcal{O}$ eigenstate, and the result of the measurement is the $\mathcal{O}$ eigenvalue of the eigenstate. The various possible results of the measurement appear with probabilities governed by Born's rule. We assume that for any Hermitian operator $\mathcal{O}$ there is some measuring device or apparatus, constructed with suitable ingenuity, that can carry out the measurement. Moreover, the result of each measurement, while not predictable, is one of the eigenvalues, exactly. We also elaborated on the idea of measurement, stating that one can measure along any orthonormal basis set of states in the state space, and one can also do partial measurements on a subsystem that is entangled with the rest of the quantum system.

Some questions arise, however, when one tries to understand the workings of the measurement apparatus and what happens during measurement such that, somehow, unitary evolution fails to hold. These questions have stimulated much work, but surprisingly, no clear answers have emerged. In the "orthodox" Copenhagen interpretation, at least as explained by Bohr and Heisenberg, the measuring devices are classical, and thus classical physics has an inescapable role in quantum mechanics. Landau and Lifshitz (1977), in their quantum mechanics monograph, also emphasize the need for a classical domain where the results of experiments are recorded and analyzed. Some physicists speak of a "Heisenberg cut," a dividing line between a quantum domain and a classical domain.

Nowadays the perspective has changed somewhat. Most physicists agree quantum mechanics should hold for all scales, and therefore there is no "cut" between a microscopic domain where quantum mechanics applies and a macroscopic domain where classical physics applies. We believe that classical physics is what quantum physics looks like, to a good approximation, for wide classes of macroscopic systems. Some large systems, however, can exhibit quantum, and thus nonclassical, behavior. This perspective is supported by experiments in which larger and larger *mesoscopic* systems have been shown to exhibit quantum behavior. We mentioned states of SQUIDS, for example, in section 1.4. Such devices support quantum superpositions of oppositely circulating currents, each with a billion electrons!

We will discuss below a plausible picture of measurement suggested by ideas of decoherence. We will then briefly mention a completely different interpretation of measurement in which there is no collapse of the wave function. This is the many-worlds interpretation of quantum mechanics. Before going into this, let us consider more explicitly how certain devices do measurements.

Measurements and collapse of the wave function In the following we discuss measurements, paying particular attention to the moment when, within the Copenhagen interpretation, the wave function collapses.

A simple example of measurement is provided by the detection of a single photon. This can be done by a photomultiplier tube. A single photon comes into the device and hits a photocathode, a surface with a thin conducting layer. The photon then ejects an electron from a metal surface via the photoelectric effect. The electron is directed by the focusing electrode toward the electron multiplier, a collection of metal plates where further electrons are released in a cascading effect. This is a process of amplification, where the original photon to be detected eventually ends up producing a macroscopic electric current that is easily measured in the domain of classical electromagnetism. The wave function of the photon, originally extending over a possibly large spatial domain, collapses, and the photon is found localized at the photodetector.

A similar principle is used to build a screen where we can detect the position of many incident photons, in this way creating an image. Such a screen can use charge-coupled devices (CCDs). The pixels on the screen are represented by capacitors constructed as doped metal-oxide semiconductors (MOS). These capacitors transform an incoming photon into a macroscopic electric charge at the metal-oxide interface. The data from these charges can then be read out and irreversibly placed in the classical domain, usually in digitized form.

We can also use photons to illustrate measurement along a basis. For this purpose, consider a calcite crystal exhibiting birefringence: the index of refraction depends on the polarization and direction of the light beam. The outputs of a suitably prepared crystal will then be two possible polarization states that we can call an $|H\rangle$-output, for

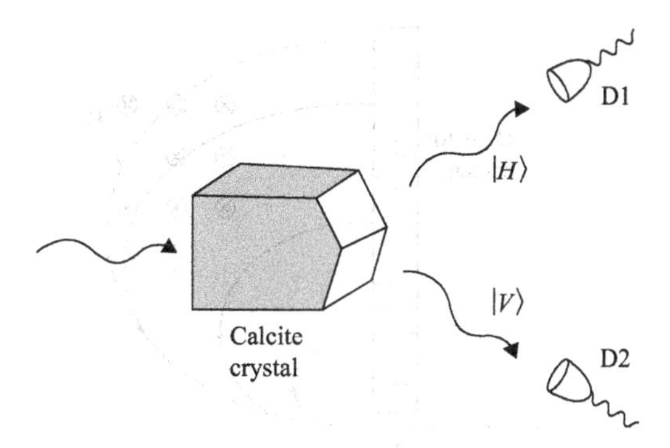

Figure 22.1
Measuring a photon along a basis of polarization states with a birefringent calcite crystal. The incoming photon emerges from the crystal in a superposition of $|H\rangle$ and $|V\rangle$ states, for horizontal and vertical polarizations. The photodetectors complete the measurement, forcing the photon wave function to collapse into a polarization state.

horizontal polarization, and a $|V\rangle$-output, for vertical polarization (figure 22.1). These are the basis states of the photon we are measuring along, with the help of the calcite crystal. The crystal alone *does not* complete the measurement, however. Suppose we send in a photon polarized at 45° relative to the $|H\rangle$ and $|V\rangle$ axes. After hitting the calcite crystal, the photon will have some amplitude to emerge on $|H\rangle$ and some amplitude to emerge on $|V\rangle$. The photon is now in a superposition of the two possible output states, and unitary evolution still holds. The measurement can be completed by placing two photodetectors, one at each of the two outputs. This forces the collapse of the wave function of the photon, and the photon is found in either the H photodetector, thus in the $|H\rangle$ state, or in the V photodetector, in the $|V\rangle$ state.

If you recall the Stern-Gerlach apparatus, the situation is in fact completely analogous: a spin one-half particle subject to a magnetic field gradient along z prepares the state in a superposition of up and down states, $|\uparrow\rangle$ and $|\downarrow\rangle$, appearing at the separate outputs of the apparatus. The measurement is only finalized by having a screen placed at the outputs that records irreversibly the position of the detected particle, telling us which output the particle came from and thus its spin state. An even simpler example, again with photons, is provided by a beam splitter, a device we discussed in the context of the Mach-Zehnder interferometer. For an incident photon, a beam splitter puts the photon in the superposition state of two beams propagating in different directions. Only after we put photodetectors on the paths does the wave function collapse, and the photon is detected in one and only one of the detectors.

We have often talked about measuring momentum. One way to measure the momentum of a charged particle is to place it inside a uniform magnetic field, orthogonal to the velocity of the particle. The particle will then move in a circle whose radius can be determined and used to calculate the value of the momentum. If the particle is in a superposition state of various momenta, the introduction of the magnetic field will

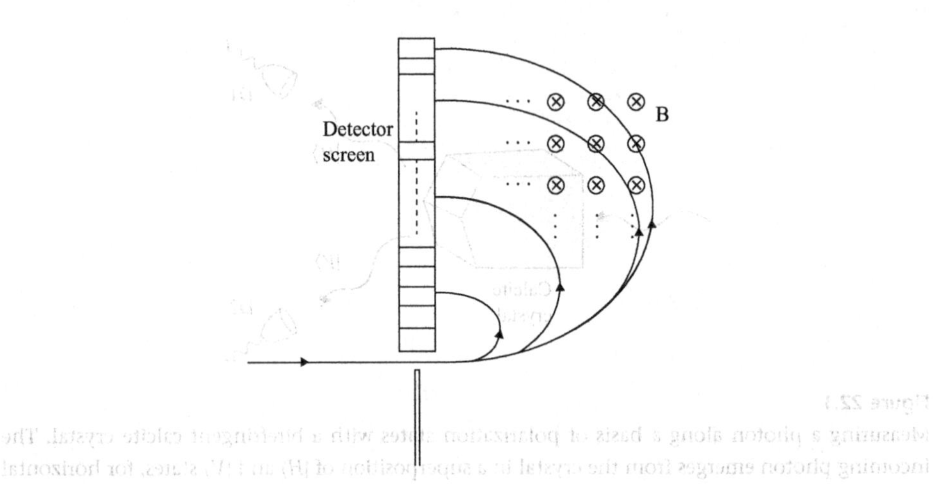

Figure 22.2
Measuring the momentum of a charged particle in a superposition of momentum eigenstates. A magnetic field places the particle in a superposition of circular trajectories of various radii. The detectors at the screen complete the measurement.

simply put the particle in a superposition of circular orbits of various radii. The measurement has not happened yet. If we now put a collection of detectors at locations corresponding to various radii (see figure 22.2), we will force the collapse of the wave function, and the particle will be observed at one of the detectors, again by some amplification effect. This will tell us the radius of the orbit state the particle collapsed into and thus the value of the momentum the particle collapsed into, due to the measurement.

In all of the above measurements, the collapse happens by detection, in which a quantum state is amplified into a macroscopic signal that is well understood classically. The above measurements do not display one aspect of idealized measurements: if we measure an observable $\mathcal{O}$ and we find an eigenstate, an immediate repeated measurement of $\mathcal{O}$ should give the *same* eigenstate. Photon detection destroys the photon, preventing a repeated measurement. When we measure the momentum of a particle, as explained above, the process leaves us without the particle. Indeed, most measurements in quantum chemistry, quantum optics, and elementary particle physics involve single systems perturbed or destroyed by the measurement. More delicate measurements must be designed in order to have wave function collapse while preserving the quantum state. Quantum **nondemolition measurements** are a class of measurements in which the integrity of the quantum state is preserved. They were discussed in the context of gravitational wave detection where small quantum mechanical amplitudes of the oscillation of large objects must be measured over and over with minimal perturbation to the quantum state. The interaction-free measurements we discussed in sections 2.2 and 2.3 are an extreme example. They show that it is possible to gain some information about a system, in this case the presence or absence of a detector, without having any interaction with the system.

Striking examples of quantum nondemolition measurements have been achieved using Rydberg atoms and resonant cavities. The field, pioneered by Serge Haroche (Nobel Prize in Physics, 2012), is called cavity quantum electrodynamics. Highly excited Rydberg atoms are useful because their energy levels are closely spaced, allowing transitions that emit or absorb photons with macroscopic wavelengths. With a properly tuned cavity, one can produce atom-cavity oscillations in which a single atom, in an excited state, emits a photon that spends a little time in the cavity before being reabsorbed by the atom, which goes back to its original excited state to start the oscillation again. The interactions of a moving Rydberg atom with photons within the cavity results in a tiny force felt by the atom. The addition of more photons enhances this force, allowing for a remarkable measurement. It becomes possible to infer the number of photons inside the cavity by measuring the time that an atom with known initial velocity takes to cross the cavity! Before the atom goes in, the number of photons in the cavity is ambiguous, the cavity being in a quantum superposition of states with different numbers of photons. At the moment the experimenter detects the outgoing Rydberg atom, the wave function of the cavity collapses to a state with a definite number of photons. This is an ideal measurement! Any subsequent Rydberg atom sent in to measure the number of photons will find the same value. If the experiment is repeated, starting each time with the original field in the cavity, the statistics of the photon number distribution can be determined.

Premeasurement with a quantum apparatus If quantum mechanics is the theory of the world at all scales, it is natural to try to understand the measurement process as a quantum interaction between a quantum system $\mathcal{S}$ and a quantum apparatus $\mathcal{A}$. In our discussion of measurement above, the measuring apparatus combines a quantum element with an amplification that turns the signal into the classical domain. In order to represent a purely quantum apparatus, we will consider a device with **pointer states**, states of a moving pointer that determine the value of the measurement.

Let the system $\mathcal{S}$ to be measured have an observable $\mathcal{O}_\mathcal{S}$ with eigenvectors $|s_i\rangle$ whose eigenvalues are s_i:

$$\mathcal{O}_\mathcal{S}|s_i\rangle = s_i|s_i\rangle, \quad i = 1, \ldots, n. \tag{22.7.1}$$

Here, we assume that i is an index that runs over a finite set of values. Let us view the apparatus $\mathcal{A}$ as quantum. We then have an observable $\mathcal{O}_\mathcal{A}$ acting on the state space of the apparatus with eigenstates $|a_i\rangle$, where a_i is the eigenvalue of $|a_i\rangle$:

$$\mathcal{O}_\mathcal{A}|a_j\rangle = a_j|a_j\rangle, \quad j = 1, \ldots, m, \quad m \geq n. \tag{22.7.2}$$

The states $|a_j\rangle$ are pointer states that can be identified with the results of a measurement $\mathcal{A}$ is designed to produce. The idea is that we have a correspondence between the eigenstates of $\mathcal{O}_\mathcal{S}$ and those of $\mathcal{O}_\mathcal{A}$:

$$|s_j\rangle \longleftrightarrow |a_j\rangle. \tag{22.7.3}$$

We aim to arrange it so that when the system is in $|s_j\rangle$ and interacts with the apparatus, the pointer state becomes $|a_j\rangle$. The pointer state tells us the value s_j of the system observable. This is why one must assume $m \geq n$: the apparatus must have at least as many states as the system in order to be able to carry out a measurement. Suppose that at $t=0$ the system is in the state

$$|\psi(0)\rangle_S = \sum_{i=1}^{n} c_i|s_i\rangle, \tag{22.7.4}$$

a superposition of $\mathcal{O}_S$ eigenstates. At time equal zero we can imagine the apparatus $\mathcal{A}$ in some definite state $|\varphi(0)\rangle_A$ so that the complete state of the system and apparatus is

$$|\psi(0)\rangle_{SA} = \Big(\sum_{i=1}^{n} c_i|s_i\rangle\Big) \otimes |\varphi(0)\rangle_A. \tag{22.7.5}$$

At $t=0$, an interaction between the system and the apparatus turns on; this interaction is governed by some Hamiltonian H_{SA}. The Hamiltonian is suitably designed such that the *premeasurement* that follows is carried out by time $t=\tau>0$, for some value of τ. The requirement is that at time τ the state of the combined system and apparatus becomes

$$|\psi(\tau)\rangle_{SA} = \sum_{i=1}^{n} c_i e^{\phi_i}|s_i\rangle \otimes |a_i\rangle, \tag{22.7.6}$$

where the ϕ_i are arbitrary phases. The interaction has established the desired entanglement between the system S and the apparatus $\mathcal{A}$. We call this premeasurement because the state of the whole system $S\mathcal{A}$ is still pure but is now a superposition of terms in which the $\mathcal{O}_S$ and $\mathcal{O}_A$ eigenstates are properly correlated. There has been no wave function collapse, and therefore the measurement has not yet happened.

We can quickly illustrate that this kind of correlation is relatively easy to achieve. Assume that both the system S and the apparatus $\mathcal{A}$ are two-level systems, so both can be described in the language of spin states. We will take

$$\mathcal{O}_S = \sigma_z^S, \quad \mathcal{O}_A = \sigma_z^A, \tag{22.7.7}$$

where the superscripts on the Pauli matrices tell us if we are dealing with the system or with the apparatus. Let us also assume that the Hamiltonian $\hat{H}_{SA}$ coupling S to $\mathcal{A}$ is

$$\hat{H}_{SA} = \tfrac{1}{2}\hbar\omega(1+\sigma_z^S) \otimes \sigma_x^A, \tag{22.7.8}$$

with $\omega > 0$ a constant frequency. Let the initial state of the system plus apparatus be

$$|\psi(0)\rangle_{SA} = (c_+|+\rangle_S + c_-|-\rangle_S) \otimes |-\rangle_A. \tag{22.7.9}$$

As indicated, at $t=0$ the system is in a superposition of the two basis states $|\pm\rangle$ along the z-axis, while the apparatus is in a single state. For the apparatus, it is convenient to work with σ_x eigenstates so that the Hamiltonian is diagonal. Using $|-\rangle = \frac{1}{\sqrt{2}}(|x;+\rangle - |x;-\rangle)$

for the first term in the above wave function, we see that

$$|\psi(0)\rangle_{SA} = \frac{1}{\sqrt{2}} c_+ |+\rangle_S \otimes \big(|x;+\rangle_{\mathcal{A}} - |x;-\rangle_{\mathcal{A}}\big) + c_- |-\rangle_S \otimes |-\rangle_{\mathcal{A}}. \tag{22.7.10}$$

The unitary operator $\mathcal{U}$ that implements time evolution is

$$\mathcal{U}(t) = \exp(-i\hat{H}_{SA}t/\hbar) = \exp\Big(-\tfrac{1}{2} i\omega t(1+\sigma_z^S) \otimes \sigma_x^{\mathcal{A}})\Big). \tag{22.7.11}$$

Note that

$$\begin{aligned} \mathcal{U}(t)|-\rangle_S \otimes |-\rangle_{\mathcal{A}} &= |-\rangle_S \otimes |-\rangle_{\mathcal{A}}, \\ \mathcal{U}(t)|+\rangle_S \otimes |x;\pm\rangle_{\mathcal{A}} &= e^{\mp i\omega t}|+\rangle_S \otimes |x;\pm\rangle_{\mathcal{A}}. \end{aligned} \tag{22.7.12}$$

The time-evolved state is therefore

$$|\psi(t)\rangle_{SA} = \frac{1}{\sqrt{2}} c_+ |+\rangle_S \otimes \big(e^{-i\omega t}|x;+\rangle_{\mathcal{A}} - e^{i\omega t}|x;-\rangle_{\mathcal{A}}\big) + c_- |-\rangle_S \otimes |-\rangle_{\mathcal{A}}. \tag{22.7.13}$$

Passing back to the $|\pm\rangle$ basis for the apparatus, a short computation now gives

$$|\psi(t)\rangle_{SA} = -i(\sin\omega t)\, c_+|+\rangle_S \otimes |+\rangle_{\mathcal{A}} + (\cos\omega t) c_+|+\rangle_S \otimes |-\rangle_{\mathcal{A}} + c_-|-\rangle_S \otimes |-\rangle_{\mathcal{A}}. \tag{22.7.14}$$

If we choose a time t_* such that $\omega t_* = \pi/2$, then the second term vanishes, and we get

$$|\psi(t_*)\rangle_{SA} == -i\, c_+|+\rangle_S \otimes |+\rangle_{\mathcal{A}} + c_-|-\rangle_S \otimes |-\rangle_{\mathcal{A}}. \tag{22.7.15}$$

This is of the form (22.7.6); the system and apparatus observables are correlated nicely, as we wanted to demonstrate. This discussion so far follows the early work of von Neumann on measurement. For him the apparatus is a large system, and along the lines we discussed, a correlation between the system $\mathcal{S}$ and the apparatus is established. Von Neumann took the viewpoint that the large quantum system is "classical" and that the value of the apparatus observable can be read out as a classical variable. Still, the mystery of collapse has not been resolved. We get a picture of how a quantum system can interact with a quantum measuring device in a useful way, but the total system is still in a quantum superposition.

Measurement and decoherence It seems clear that if we keep using the unitary evolution of quantum mechanics applied to a system, we are not going to see wave function collapse. A picture that seems plausible in motivating the origin of probabilities, without speaking about collapse, uses the environment and decoherence. Recall that we saw that for an open system time evolution is not unitary. The idea in the picture we describe now is to establish that the density matrix of the open system $\mathcal{SA}$ describes an ensemble with probabilistic interpretation.

In realistic conditions, $\mathcal{S}$ and a macroscopic $\mathcal{A}$ are always in contact with the environment $\mathcal{E}$. Interactions with $\mathcal{E}$ cause further entanglement, and in reality we have a state $|\psi\rangle_{SAE}$ that takes the form

$$|\psi\rangle_{SAE} = \sum_i c_i |s_i\rangle \otimes |a_i\rangle \otimes |\mathcal{E}_i\rangle, \tag{22.7.16}$$

where $|\mathcal{E}_i\rangle$ are states of the environment $\mathcal{E}$. We can now form the density matrix ρ_{SA} for SA by tracing over the environment:

$$\begin{aligned}\rho_{SA} &= \mathrm{tr}_{\mathcal{E}}\, \rho_{SAE} = \mathrm{tr}_{\mathcal{E}}\, |\Psi\rangle_{SAE}\langle\Psi|_{SAE} \\ &= \mathrm{tr}_{\mathcal{E}} \sum_{i,j=1}^{n} c_i c_j^* |s_i\rangle|a_i\rangle|\mathcal{E}_i\rangle\, \langle s_j|\langle a_j|\langle\mathcal{E}_j| \\ &= \sum_{i,j=1}^{n} c_i c_j^* \langle\mathcal{E}_j|\mathcal{E}_i\rangle\, |s_i\rangle|a_i\rangle\langle s_j|\langle a_j|.\end{aligned} \tag{22.7.17}$$

Given that the environment is described by a Hilbert space of extremely high dimensionality, it seems plausible that $\langle\mathcal{E}_j|\mathcal{E}_i\rangle \sim \delta_{ij}$: any two generic states of the environment are likely to have very small overlap. But perhaps something more subtle is going on, and the "pointer" states $|a_i\rangle$ are such that they naturally get entangled with orthogonal states of the environment. At any rate, under the assumption $\langle\mathcal{E}_j|\mathcal{E}_i\rangle = \delta_{ij}$ we see that

$$\rho_{SA} = \sum_{i=1}^{n} |c_i|^2 \cdot |s_i\rangle|a_i\rangle\langle s_i|\langle a_i|. \tag{22.7.18}$$

An observer focused on SA thus observes a density matrix, which can be interpreted as arising from the ensemble

$$E = \left\{(|c_1|^2 |s_1\rangle|a_1\rangle), \ldots, (|c_n|^2 |s_n\rangle|a_n\rangle)\right\}. \tag{22.7.19}$$

In the ensemble there is a probability $|c_i|^2$ of finding the state $|s_i\rangle|a_i\rangle$ in which the pointer state is consistent with the state of the system. One has to claim now that these are the states one observes when looking at SA after the measurement.

This kind of argument also suggests that we can never achieve a superposition of a cat-alive and a cat-dead state, as in the example proposed by Schrödinger:

$$\tfrac{1}{\sqrt{2}}\big(\,|\text{cat-alive}\rangle + |\text{cat-dead}\rangle\,\big). \tag{22.7.20}$$

In such a state, the cat is neither dead nor alive. But certainly there is also an environment, and you could argue that at some time we have the state

$$\tfrac{1}{\sqrt{2}}\big(\,|\text{cat-alive}\rangle + |\text{cat-dead}\rangle\big) \otimes |\mathcal{E}_0\rangle. \tag{22.7.21}$$

But this situation is untenable: the cat, dead or alive, is interacting with the environment, even if contained in a box. A living cat is breathing air. A dead cat is lukewarm and still emits and absorbs blackbody photons. The interactions imply that, almost instantaneously, the above state becomes

$$\tfrac{1}{\sqrt{2}}\big(\,|\text{cat-alive}\rangle \otimes |\mathcal{E}_1\rangle + |\text{cat-dead}\rangle \otimes |\mathcal{E}_2\rangle\big), \tag{22.7.22}$$

with two environment states $|\mathcal{E}_1\rangle$ and $|\mathcal{E}_1\rangle$ that are to great accuracy orthogonal: $\langle\mathcal{E}_1|\mathcal{E}_2\rangle \simeq 0$. As in the discussion above, if we trace over the environment, the result is then a density matrix ρ_{cat} given by

$$\rho_{\text{cat}} = \tfrac{1}{2}|\text{cat-alive}\rangle\langle\text{cat-alive}| + \tfrac{1}{2}|\text{cat-dead}\rangle\langle\text{cat-dead}| \tag{22.7.23}$$

that describes an ensemble of "cat systems" in which half of the cats are alive and half are dead. We cannot observe any subtle superposition.

The above arguments leave a number of open questions. We could get ensembles different from (22.7.19) if the coupling to the environment in (22.7.16) were different; for example, if each environment state were to couple to linear combinations of pointer states. Additionally, the interpretation of the density matrix as an ensemble is not unique: we have seen that an unpolarized spin can be prepared with ensembles in infinitely many different ways. In here, the resulting expression for the density matrix is taken as uniquely selected.

Taking the observer out of the physics discussion is one of the goals of the *decoherent histories* approach to quantum mechanics developed by Murray Gell-Mann and James Hartle, following early work by Robert Griffiths and Roland Omnes. A quasiclassical world is argued to be an emergent feature of long sequences of chance events. Applied to cosmology, it asserts that this quasi-classical world emerges long after earlier times dominated by quantum fluctuations. No classical domain is needed to interpret the results of measurements.

Many worlds? Some physicists feel that the discussion of measurement can be made clearer by adopting various alternative ideas. This is not the place to survey the many proposals, so we will simply make some remarks on one approach that has attracted considerable attention. This is the so-called many worlds interpretation of quantum mechanics. The key idea here is that the wave function *never collapses*. Each time there is a measurement, however, the universe undergoes some splitting, or branching.

More concretely, let us assume that Alice has a spin state $|\psi\rangle$ of the form

$$|\psi\rangle = c_+|+\rangle + c_-|-\rangle, \quad |c_+|^2 + |c_-|^2 = 1, \tag{22.7.24}$$

and she decides to measure the spin along the z-axis. In the Copenhagen interpretation, when she measures the state the wave function collapses. She finds $|+\rangle$ with probability $|c_+|^2$ and $|-\rangle$ with probability $|c_-|^2$. What happens in the many-worlds interpretation? To understand Alice's perspective, however, she must be included in the wave function! Thus, before she performs a measurement on some sunny Monday morning, the relevant state would be

$$|\psi\rangle = \big(c_+|+\rangle + c_-|-\rangle\big) \otimes |\text{Alice ready to measure}\rangle. \tag{22.7.25}$$

In the many-worlds picture, after measurement there is no collapse, but there is entanglement of the form

$$|\psi\rangle = c_+|+\rangle \otimes |\text{Alice sees }+\rangle \; + \; c_-|-\rangle \otimes |\text{Alice sees }-\rangle. \tag{22.7.26}$$

This is the kind of entanglement we found in the von Neumann analysis of measurement, with the state of Alice playing the role of the apparatus! Indeed, the state of Alice is consistently correlated with the state of the spin. The interpretation of the above state is that the universe or world, including Alice and her spin, has split into two branches or two worlds. In one of these branches, the first term in the above superposition is realized: the spin is up, and Alice sees the spin as up. In the other branch or world, the second term is realized: the spin is down, and Alice sees the spin as down. The two copies of Alice are equally real but cannot communicate with each other. As each of the Alices keeps doing quantum measurements, the universe keeps splitting, and more and more copies of Alice are generated. As their experiences differ, after a long time some of the Alices in existence could be quite different. Each one, however, thinks she is *the* Alice that on a sunny Monday morning measured a spin state.

The meaning of probabilities in the many-worlds interpretation is not a priori obvious given that *every* possibility is realized in some world. It has been suggested that they are meant to be "self-locating" probabilities. Imagine the measurement is done, and there are now two Alices and two spins. Before looking at her spin, neither one of them knows which branch of the universe she is in. Then the probabilities assigned to the branches are viewed as the best information the Alices have for guessing their branch before looking. Some heuristic arguments are advanced to argue that those probabilities must be $|c_+|^2$ and $|c_-|^2$ when the wave function is that in (22.7.26).

We have been led to talk about multiple, ever-increasing branches of worlds as well as quantum kets like $|\text{cat-alive}\rangle$, $|\text{cat-dead}\rangle$, and $|\text{Alice sees} \ldots\rangle$. The first seem rather speculative ideas and the latter dubious objects. Common sense suggests that now is a good time to recommend the interested reader follow the current literature for further insight.

Problems

Problem 22.1. *No pure-state description of an entangled particle.*

In example 22.2 we showed that there is no pure-state representative of an entangled particle by arguing that a spin state with zero expectation value for all the Pauli matrices must vanish. Here we prove the same result by a direct computation. Consider the general form the state can take:

$$|\psi_A\rangle = \alpha|+\rangle + \beta|-\rangle, \quad \alpha, \beta \in \mathbb{C}, \quad |\alpha|^2 + |\beta|^2 = 1.$$

Determine what conditions on α and β are imposed by the vanishing of the σ_z, σ_x, and σ_y expectation values in $|\psi_A\rangle$. Show that these conditions cannot be satisfied.

Problem 22.2. *Ensemble with three sets of states (A. Harrow).*

Write down three spin one-half states $|\psi_1\rangle, |\psi_2\rangle, |\psi_3\rangle$ such that if each occurs with probability one-third, the density operator for the ensemble is $\frac{1}{2}\mathbb{1}$—that is, maximally mixed.

Problem 22.3. *Purification of a mixed state.*

Consider a mixed state of a spin one-half system A with density matrix ρ_A given by

$$\rho_A = \tfrac{1}{2}|+\rangle\langle+| - \tfrac{1}{6}|+\rangle\langle-| - \tfrac{1}{6}|-\rangle\langle+| + \tfrac{1}{2}|-\rangle\langle-|. \tag{22.7.27}$$

Purify the state ρ_A; that is, couple A to another spin one-half system B, and write a *pure state* $|\psi_{AB}\rangle$ for the AB system. The state is called a *purification* of ρ_A if ρ_A arises by partial trace over B of the density matrix ρ_{AB}. Explain how much freedom you have in choosing the state of B.

Problem 22.4. *Effect of partial measurement on the density matrix.*

Let AB be a bipartite system with density matrix ρ_{AB}. Alice measures the state of subsystem A along a basis $\{|i\rangle_A\}$ with associated projectors $M_i^A = |i\rangle_{AA}\langle i|$ satisfying

$$M_i^{A\dagger} = M_i^A, \quad M_i^A M_i^A = M_i^A, \quad \sum_i M_i^A = \mathbb{1}_A.$$

Assume we do not know the result of Alice's measurement. Show that the after-measurement density matrix $\tilde{\rho}_{AB}$ is

$$\tilde{\rho}_{AB} = \sum_i (M_i^A \otimes \mathbb{1}_B)\, \rho_{AB}\, (M_i^A \otimes \mathbb{1}_B).$$

This is the result claimed in (22.4.43). [Hint: Attempt a proof similar to that leading to (22.2.45), possibly starting with a diagonal representation of a general ρ_{AB}. Partial measurement was discussed in section 18.6.]

Problem 22.5. *Lindblad equation and positivity of the density matrix.*

Establish that the Lindblad equation (22.6.1) is consistent with ρ remaining positive semidefinite under time evolution. Do this by proving that if $\rho(t)$ is positive semidefinite, so is $\rho(t+dt)$ with $dt > 0$. Begin with writing

$$\langle v|\rho(t+dt)|v\rangle = \langle v|\rho(t)|v\rangle + dt\langle v|\,\dot{\rho}(t)|v\rangle + \mathcal{O}(dt^2),$$

where a dot indicates a time derivative, and $|v\rangle$ is an *arbitrary* state in the state space.

1. The challenge is to show that the *sum* of the two terms on the right-hand side are positive up to terms of order $\mathcal{O}(dt^2)$ for any $|v\rangle$. [Hint: try showing this first without the Lindblad terms, just keeping the commutator involving $\hat{H}$.] The Lindblad terms of the form $\{L_k^\dagger L_k, \rho\}$ can be treated similarly, while the terms $L_k \rho L_k^\dagger$ must be kept separate.
2. A corollary of your proof is the so-called Kraus form:

$$\rho(t+dt) = M_0\, \rho(t)\, M_0^\dagger + \sum_k M_k\, \rho(t)\, M_k^\dagger \; + \mathcal{O}(dt^2),$$

valid with $M_k = \sqrt{dt} L_k$, and $dt > 0$. Did you identify what M_0 is?

Problem 22.6. *Lindblad equation and the evolution of purity.*

Use the Lindblad equation to prove that the purity $\zeta(\rho) = \text{tr}(\rho^2)$ of a density matrix ρ will not increase

$$\frac{d}{dt}\text{tr}(\rho^2) \leq 0$$

if the Lindblad operators L_k satisfy the relation $\sum_k [L_k, L_k^\dagger] = 0$. Your proof may require the following identity, valid for Hermitian M:

$$\mathrm{tr}\Big(MFMF^\dagger - M^2F^\dagger F\Big) = -\tfrac{1}{2}\mathrm{tr}\Big([M,F]^\dagger[M,F] - M^2[F,F^\dagger]\Big), \quad M^\dagger = M.$$

Confirm this identity holds.

Problem 22.7. *Lasers versus light bulbs (A. Harrow).*

1. The state of a laser is represented by a coherent state $|\alpha\rangle$ that in the number basis takes the form

$$|\alpha\rangle = e^{-\frac{|\alpha|^2}{2}} \sum_{n=0}^{\infty} \frac{\alpha^n}{\sqrt{n!}} |n\rangle.$$

   Here, $|n\rangle$ is the state with n photons. In practice, however, we know $|\alpha|$ but do *not* know the phase of α. We model this by setting $\alpha = re^{i\phi}$, where the value of $r \geq 0$ is fixed, and ϕ is a random variable uniformly distributed on the interval $[0, 2\pi]$. Write down the resulting density operator ρ_{laser} in the number basis and as a function of r. What is $\langle \hat{n} \rangle_{\text{laser}}$ as a function of r?

2. An incandescent light bulb produces light that is in a thermal state. Consider only light of a fixed angular frequency ω. Use the Boltzmann distribution to write the density operator for the thermal state ρ_{thermal} at temperature T in the number basis. Express this as a function of the dimensionless quantity $\gamma \equiv \hbar\omega/k_BT$. What is $\langle \hat{n} \rangle_{\text{thermal}}$?

3. The state of a laser and the thermal state cannot be distinguished by observing the average photon number $\langle \hat{n} \rangle$ alone. For this one must measure fluctuations in photon number—that is, $(\Delta n)^2 \equiv \langle \hat{n}^2 \rangle - \langle \hat{n} \rangle^2$. Compute $(\Delta n)^2$ for the laser and for the light bulb. In both cases express your answers in terms of the corresponding value of $\langle \hat{n} \rangle$. Explain how your result can be used to distinguish these two sources of light.

Problem 22.8. *Evolution by random Hamiltonians.*

Consider a pure state $|\psi\rangle$ at time equal zero, evolving under the effect of a time-independent Hamiltonian $\hat{H}(b)$, where b is a continuous random variable with probability distribution $f(b)$ so that $\int f(b)db = 1$. Write down an integral expression for the density matrix $\rho(t)$ that results after time evolution of $|\psi\rangle$ for a time t.

Generalize the above result for the case that all you know is the density matrix $\rho(0)$ of the system at time equal zero. What then is $\rho(t)$?

Problem 22.9. *Gaussian phase error (A. Harrow).*

Consider an electron spin in a state described by the density matrix

$$\rho = \begin{pmatrix} \rho_{++} & \rho_{+-} \\ \rho_{-+} & \rho_{--} \end{pmatrix},$$

where the basis vectors are $|1\rangle = |+\rangle$ and $|2\rangle = |-\rangle$. The electron experiences a magnetic field of magnitude B in the z-direction As a result, the Hamiltonian is $\hat{H}(B) = -\gamma B\hat{S}_z$. Suppose that the field strength B is drawn from a Gaussian distribution with mean 0 and variance σ^2. For this distribution the probability density $f(B)$ is given by

$$f(B) = \frac{1}{\sqrt{2\pi\sigma^2}} e^{-\frac{B^2}{2\sigma^2}}, \quad \int_{-\infty}^{\infty} f(B)dB = 1.$$

Determine the density matrix $\rho(t)$ that results from applying this random Hamiltonian for time t. What is the large time limit of the density matrix?

Problem 22.10. *Spin one-half thermal relaxation and dephasing (A. Harrow).*

Consider a spin one-half particle in a magnetic field undergoing thermal relaxation and dephasing noise. The Hamiltonian is $\hat{H} = -\gamma B\hat{S}_z$, with $\gamma > 0$ and $B > 0$. Associated to this $\hat{H}$ is a thermal density matrix $\rho_{\text{th}} = e^{-\beta\hat{H}}/\text{tr}[e^{-\beta\hat{H}}]$. Additionally, we have positive time constants T_1 associated with the approach to thermal equilibrium and T_2 associated with dephasing. Assume that the state of the system evolves according to

$$\dot{\rho} = -\frac{i}{\hbar}[\hat{H}, \rho] - \frac{1}{T_1}(\rho - \rho_{\text{th}}) - \frac{1}{T_2}\begin{pmatrix} 0 & \rho_{+-} \\ \rho_{-+} & 0 \end{pmatrix}. \tag{1}$$

We parameterize the state of the particle with a time-dependent vector $\mathbf{a}$, recalling that the density matrix of a mixed state takes the form $\rho = \frac{1}{2}(\mathbb{1} + \mathbf{a} \cdot \boldsymbol{\sigma})$ with $|\mathbf{a}| \leq 1$.

1. Calculate the thermal density matrix and show that it is given by

$$\rho_{\text{th}} = \tfrac{1}{2}(\mathbb{1} + \tanh(\tfrac{1}{2}\beta\gamma B\hbar)\sigma_z).$$

2. Show that the evolution equation (1) can be put in the form

$$\frac{d\mathbf{a}}{dt} = \hat{M}\mathbf{a} + \mathbf{b}, \tag{2}$$

with $\hat{M}$ a 3×3 matrix and $\mathbf{b} \in \mathbb{R}^3$. Find $\hat{M}$ and $\mathbf{b}$. You should find that

$$\hat{M} = \begin{pmatrix} \#_1 & \gamma B & 0 \\ -\gamma B & \#_2 & 0 \\ 0 & 0 & \#_3 \end{pmatrix}, \quad \mathbf{b} = \begin{pmatrix} 0 \\ 0 \\ \#_4 \end{pmatrix},$$

where #'s are quantities that you should determine.

3. Solve equation (2) and show that the solution is

$$\mathbf{a}(t) = e^{t\hat{M}}(\mathbf{a}(0) + \hat{M}^{-1}\mathbf{b}) - \hat{M}^{-1}\mathbf{b}.$$

For the explicit solution, you must exponentiate $\hat{M}$, which is not hard because it has a 2×2 block structure. Give your answers for $a_x(t), a_y(t)$ and $a_z(t)$ in terms of their time equal zero values $a_x(0), a_y(0)$, and $a_z(0)$ and functions of $t, T_1, T_2, \gamma, B, \beta$, and $\hbar$.

4. Assume that $T_1 \gg T_2 \gg 1/\gamma B$. Describe the time evolution of the vector $\mathbf{a}(t)$. How does it evolve for small times? What is the steady state at large times?

Problem 22.11. *Decoherence and thermalization (A. Harrow).*

Model an atom as a two-level system with ground state $|g\rangle = |1\rangle$ and excited state $|e\rangle = |2\rangle$. The atom interacts with a photon field, represented by a harmonic oscillator. The interaction Hamiltonian $\hat{H}$ is given by

$$\hat{H} = \hbar\Omega\left(|g\rangle\langle e| \otimes \hat{a}^\dagger + |e\rangle\langle g| \otimes \hat{a}\right),$$

with Ω a constant with units of frequency. The atom state is described by a time-dependent density matrix ρ taking the form

$$\rho = \begin{pmatrix} \rho_{11} & \rho_{12} \\ \rho_{21} & \rho_{22} \end{pmatrix} = \begin{pmatrix} \rho_{gg} & \rho_{ge} \\ \rho_{eg} & \rho_{ee} \end{pmatrix}.$$

This problem will involve the following decoherence process:

- Tensor the photon state $|0\rangle\langle 0|$ to the atom state ρ to form the initial state $\rho \otimes |0\rangle\langle 0|$.
- Let the Hamiltonian $\hat{H}$ evolve the state for time τ.
- Trace over the photon state space.

After this process the density matrix ρ' for the atom takes the form

$$\rho' = \mathrm{tr}_{\text{photon}}\left[e^{-i\hat{H}\tau/\hbar}(\rho \otimes |0\rangle\langle 0|)e^{i\hat{H}\tau/\hbar}\right].$$

1. Compute ρ' to order $O(\tau^2)$; that is, neglecting τ^3 and higher terms. Show that ρ' takes the form

$$\rho' = \rho + \tau^2\Omega^2 \begin{pmatrix} \rho_{ee} & \#_1 \\ \#_2 & -\rho_{ee} \end{pmatrix},$$

where we wrote part of the answer to help you check your calculation. Determine the missing terms.

2. We would like to approximate this process with a continuous-time evolution by taking $\tau \to 0$. In order to obtain a nontrivial answer, however, we have to let Ω diverge as $\tau \to 0$. In fact, as $\tau \to 0$ we will hold $\eta \equiv \Omega^2\tau$ fixed and positive. Derive a differential equation for ρ of the form $\dot{\rho} = L[\rho]$ where $L[\rho]$ is a matrix-valued function of ρ. Solve the differential equation for the components of the density matrix in terms of some assumed initial values at $t = 0$. Determine the steady-state solution.
[Note: The scaling of Ω has a physical explanation: We computed decoherence due to coupling to a single photon mode. But in the timescale τ, the number of modes coupling to the atom is of order $1/\tau$, by the energy-time uncertainty principle. So the more accurate result would have an extra factor of $1/\tau$, which is, in fact, the effect of setting $\Omega^2 = \eta/\tau$.]

3. Now modify the original process so that instead of tensoring a photon field in state $|0\rangle\langle 0|$ at each step we tensor a thermal photon state ρ_{thermal} with inverse temperature β and photon angular frequency ω:

$$\rho_{\text{thermal}} = \sum_{n=0}^{\infty} p_n |n\rangle\langle n|, \quad p_n = \frac{1}{Z} e^{-\beta\hbar\omega n}, \quad Z = \sum_{n=0}^{\infty} e^{-\beta\hbar\omega n} = \frac{1}{1 - e^{-\beta\hbar\omega}}.$$

Repeat the analysis to show that the resulting differential equation for ρ is

$$\dot{\rho} = \eta Z \begin{pmatrix} \rho_{\text{ee}} - \rho_{\text{gg}} e^{-\beta\hbar\omega} & -\frac{1}{2}(1 + e^{-\beta\hbar\omega})\rho_{\text{ge}} \\ -\frac{1}{2}(1 + e^{-\beta\hbar\omega})\rho_{\text{eg}} & -\rho_{\text{ee}} + \rho_{\text{gg}} e^{-\beta\hbar\omega} \end{pmatrix}.$$

Write down the steady state solution, and confirm that the equilibrium state for the atom has been thermalized.

23 Quantum Computation

A qubit, or quantum bit, is a quantum system with two basis states, $|0\rangle$ and $|1\rangle$, that are identified with the possible values 0 and 1 of a classical bit. While the state of n bits is specified by a set of n digits, all of which are either zero or one, the state of n qubits is specified by 2^n complex numbers. Quantum gates are unitary operators acting on the states of multiple qubits and allowing for reversible computation. Quantum circuits implement a kind of parallelism through superposition, leading to computational speedup. We discuss Deutsch's algorithm that computes, with a single evaluation, a property of a function that classically requires two evaluations. We then turn to a search problem in which one looks for the unique input to an oracle function that produces a particular value. When the input can take N values, classical algorithms require $\mathcal{O}(N)$ calls of the oracle. Grover's quantum algorithm finds the unique input with high probability with $\mathcal{O}(\sqrt{N})$ calls.

We saw some aspects of quantum information when we discussed quantum key distribution (chapter 12) as well as when we analyzed teleportation, Bell inequalities, and the no-cloning property of quantum systems (chapter 18). Let us now consider the possibility of using quantum mechanics to perform the computations commonly done with ordinary computers. This is the subject of *quantum computation.* The machines, called quantum computers, would perform computations in a novel way using superposition, entanglement, and interference in quantum systems.

Instead of using the bits of ordinary computers, quantum computers use *qubits* or quantum bits. A qubit is a quantum device. Instead of having two possible values 0 and 1, like the bit, it has the infinite number of states that arise by superposition of two basis states, called $|0\rangle$ and $|1\rangle$. The qubit is simply what we have called a two-state system. Instead of having classical gates, the quantum computer has quantum gates that change the quantum states of qubits by acting on them with unitary transformations.

Plenty of work has been done the last couple of decades to investigate the types of computations that could be done with a quantum computer if one existed. Remarkably, a quantum computer could do certain computations much faster than ordinary computers. The speed advantage is sometimes so striking that, while it has not yet been proven, many scientists believe that no progress in classical computation theory could close the gap.

As of 2019, a fifty-three-qubit quantum computer built by Google was capable of doing in about two hundred seconds a computation that the most powerful supercomputer in the world would take at least three days to complete. The computation was of no particular interest. It was chosen with a simple criterion: it would be doable with a still-primitive quantum computer but very difficult to do with a classical computer. More interesting computations will have to await better hardware for quantum computers.

Although this will not be the focus here, the development of quantum computers raises interesting questions in computer science. In this field the abstract notion of a programmable computer, called a *Turing machine*, was developed by Alan Turing. The *Church-Turing thesis*, named in recognition of Turing's contributions as well as those of Alonzo Church, claims that for any algorithm running on any computer there is an algorithm that accomplishes this task on a Turing machine. In this situation the original algorithm is said to be simulated by the Turing machine. In fact, any calculation on a quantum computer *can* be simulated on a classical computer, thus also on a Turing machine. The Church-Turing thesis is widely believed to be true.

There is, however, a *strong* version of the Church-Turing thesis that introduces the issue of efficiency and reads as follows: Any algorithm running on any computer can be simulated *efficiently* on a Turing machine. If the algorithm is efficient on the original computer, it would have an efficient simulation on a Turing machine. Defining the efficiency of algorithms precisely takes effort, but one roughly considers the number of operations the algorithm takes to solve a problem of "size" N. This size could be, for example, the number of digits of an integer whose prime number decomposition we wished to find. An algorithm in which the number of operations required is *polynomial* in N is said to be "efficient." An algorithm in which the number of operations required is exponential in N is not efficient. Quantum computation is a potential challenge to the strong Church-Turing thesis because a quantum computer can efficiently solve some problems for which there are no known efficient solutions with a classical computer. In establishing the challenge, however, one must take into account any issue that may affect the operation of a quantum computer. Unavoidable thermal and other couplings of the quantum circuit to the environment cause difficulties. A state of the quantum circuit can be altered, which requires quantum error-correction codes. Moreover, the circuit can decohere quickly by simply changing the state of the environment, as discussed in example 22.7. It seems likely that even with these extra complications the computational advantage of quantum computers will persist.

Quantum computers could allow for efficient simulations of quantum systems of many particles. Such systems would be very hard to simulate on a classical computer, a point of view emphasized by Richard Feynman as early as 1981. Much of the current work on quantum computation was stimulated by Peter Shor's discovery in 1994 of an efficient quantum computer algorithm for the factorization of an integer into its prime factors. No efficient algorithm for factoring integers is currently known for classical computers.

23.1 Qubits and Gates

In this section we discuss qubits and the gates that act on them. We will consider gates acting on one qubit and gates acting on two qubits. These will be used to show how to compute functions using circuits built with gates.

A qubit is a two-state quantum system. A clear example is provided by the spin degrees of freedom of a spin one-half particle. When that particle represents a qubit, the two basis states can be taken to be spin up and spin down along z: $|z;\pm\rangle$. Since we are interested in computation, we will use binary code to label the two states:

$$|0\rangle \equiv |z;+\rangle = \begin{pmatrix} 1 \\ 0 \end{pmatrix}, \qquad |1\rangle \equiv |z;-\rangle = \begin{pmatrix} 0 \\ 1 \end{pmatrix}. \tag{23.1.1}$$

Note that $|0\rangle$ is the first column vector (usually called $|1\rangle$) and $|1\rangle$ is the second column vector (usually called $|2\rangle$). This must be kept in mind when using the matrix representation of operators. The general state $|\psi\rangle$ of this qubit is

$$|\psi\rangle = a_0|0\rangle + a_1|1\rangle, \quad a_0, a_1 \in \mathbb{C}, \quad |a_0|^2 + |a_1|^2 = 1. \tag{23.1.2}$$

The $|0\rangle$ and $|1\rangle$ states are the **computational basis states** of the qubit.

A classical bit state is either 0 or 1. The qubit state is, in general, an arbitrary superposition of the states $|0\rangle$ and $|1\rangle$. For a qubit state $|\psi\rangle$, we customarily measure along the basis states $|0\rangle$ and $|1\rangle$. From this measurement we get just one bit of information: we view a $|0\rangle$ result as the bit 0, and we view a $|1\rangle$ result as the bit 1.

If we have two qubits, the quantum state $|\psi\rangle$ that describes a general configuration is

$$|\psi\rangle = a_{00}|0\rangle \otimes |0\rangle + a_{01}|0\rangle \otimes |1\rangle + a_{10}|1\rangle \otimes |0\rangle + a_{11}|1\rangle \otimes |1\rangle, \tag{23.1.3}$$

with $a_{ij} \in \mathbb{C}$, and for normalization, $\sum_{i,j} |a_{ij}|^2 = 1$, with i, j running over the values 0, 1. For brevity we sometimes omit the $\otimes$ and use single kets with two entries, as in $|ij\rangle = |i\rangle \otimes |j\rangle$:

$$|\psi\rangle = a_{00}|00\rangle + a_{01}|01\rangle + a_{10}|10\rangle + a_{11}|11\rangle. \tag{23.1.4}$$

For two qubits we have four computational basis states $|ij\rangle$ with $i, j \in \{0, 1\}$. Each $|ij\rangle$ encodes the two-bit sequence (ij). If measuring the qubits in $|\psi\rangle$ gives the state $|10\rangle$, we obtain the two-bit sequence (10). A classical two-bit device can only be in one of the four two-bit states 00, 01, 10, and 11. The two-qubit system has a wave function in which the four qubit basis states are all present in superposition. Note also that using binary, both the two-bit and two-qubit states encode the digits 0, 1, 2, 3:

$$0 \leftrightarrow |00\rangle, \quad 1 \leftrightarrow |01\rangle, \quad 2 \leftrightarrow |10\rangle, \quad 3 \leftrightarrow |11\rangle. \tag{23.1.5}$$

The two labels in each ket are the binary representation of the integer to the left.

It is simple now to consider a system with n qubits, with n an arbitrary positive integer. The state space of n qubits has dimension N, with

$$N = 2^n = \text{dimension of the } n\text{-qubit state space.} \tag{23.1.6}$$

The associated computational basis states take the form

$$|x_1 \cdots x_n\rangle \equiv |x_1\rangle \otimes \cdots \otimes |x_n\rangle, \tag{23.1.7}$$

with bits x_i equal to either 0 or 1. Again, the sequences $x_1 \cdots x_n$ for all possible values of the bits x_i encode the binary representation of all N integers from 0 to $2^n - 1$. The general n-qubit state is

$$\begin{aligned} |\psi\rangle = \; & a_{0\ldots00}\,|0\cdots 00\rangle \\ & + a_{0\ldots01}\,|0\cdots 01\rangle \\ & + a_{0\ldots10}\,|0\cdots 10\rangle \\ & \;\;\vdots \qquad\;\; \vdots \\ & + a_{1\ldots11}\,|1\cdots 11\rangle. \end{aligned} \tag{23.1.8}$$

The state has 2^n complex coefficients $a_{x_1 \cdots x_n}$ that must be specified to fix the state (up to normalization). This can quickly become a very large number of coefficients. For a fifty-three-qubit circuit (Google), we have a state space of dimension $2^{53} = 9\,007\,199\,254\,740\,992 \simeq 10^{16}$. One can wonder how much memory is needed to store the information that defines the quantum state on an ordinary computer. Assuming conservatively that each complex coefficient requires 8 bytes of memory (1 byte = 8 bits), storing all coefficients requires $2^{53} \cdot 2^3 = 2^{56}$ bytes. A petabyte Pb, at least as defined for IBM computers, is $2^{50} \simeq 10^{15}$ bytes. A petabyte is $2^{10} = 1{,}024 \simeq 1{,}000$ times bigger than a terabyte (Tb), which itself is 2^{10} times bigger than the familiar gigabyte (Gb). Therefore, the quantum state requires for storage 2^{56} bytes $= 2^6$ Pb $= 64$ Pb. This is about a quarter of the full storage capacity of 250 Pb available on the IBM Summit supercomputer at the Oak Ridge National Laboratory in Tennessee.

Single qubit gates Let us now consider gates. A single-qubit gate is simply a unitary operator acting on a qubit. Unitary transformations are reversible and preserve the norm of the qubit state. Recall that the three Pauli operators are unitary because they are both Hermitian and square to the identity. In quantum computation one writes X for σ_1, Y for σ_2, and Z for σ_3. Thus, we have

$$X = \begin{pmatrix} 0 & 1 \\ 1 & 0 \end{pmatrix}, \quad Y = \begin{pmatrix} 0 & -i \\ i & 0 \end{pmatrix}, \quad Z = \begin{pmatrix} 1 & 0 \\ 0 & -1 \end{pmatrix}. \tag{23.1.9}$$

Recalling the column vector expressions (23.1.1) for $|0\rangle$ and $|1\rangle$, we see that

$$X|0\rangle = |1\rangle, \qquad X|1\rangle = |0\rangle, \tag{23.1.10}$$

showing that the X gate is the quantum version of the NOT gate, which turns the 0 bit into the 1 bit, and vice versa. More generally, for a bit x we write its NOT as $\bar{x}$ with

$$\bar{x} = x \oplus 1, \tag{23.1.11}$$

where the symbol $\oplus$ is defined to be addition modulo 2 and should not be confused with the direct sum symbol for vector spaces. We represent the quantum gate X as follows:

$$\text{—}[X;\ x \to \bar{x}]\text{—}\ . \tag{23.1.12}$$

In this diagram the input is to the left, and the output is to the right. The x and $\bar{x}$ inside the box tell us how this gate acts on computational states. With x a bit, the gate gives

$$|x\rangle \ \text{—}[X;\ x \to \bar{x}]\text{—}\ |\bar{x}\rangle. \tag{23.1.13}$$

We thus say $X: |x\rangle \to |\bar{x}\rangle$. Linearity of the gate implies that

$$a_0|0\rangle + a_1|1\rangle \ \text{—}[X]\text{—}\ a_0|1\rangle + a_1|0\rangle, \tag{23.1.14}$$

omitting for brevity the instructions on the box. For the Z gate, we have

$$a_0|0\rangle + a_1|1\rangle \ \text{—}[Z]\text{—}\ a_0|0\rangle + a_1|1\rangle. \tag{23.1.15}$$

Another single-qubit unitary operator that will feature prominently in the rest of the chapter is the **Hadamard gate** H (not to be confused with the Hamiltonian $\hat{H}$). Explicitly,

$$H = \frac{1}{\sqrt{2}}\begin{pmatrix} 1 & 1 \\ 1 & -1 \end{pmatrix}. \tag{23.1.16}$$

The matrix H is, of course, unitary. Its action on the computational states is clear from the matrix representation:

$$H|0\rangle = \tfrac{1}{\sqrt{2}}(|0\rangle + |1\rangle), \quad H|1\rangle = \tfrac{1}{\sqrt{2}}(|0\rangle - |1\rangle). \tag{23.1.17}$$

We represent these results as

$$|0\rangle \ \text{—}[H]\text{—}\ \frac{|0\rangle + |1\rangle}{\sqrt{2}} \qquad |1\rangle \ \text{—}[H]\text{—}\ \frac{|0\rangle - |1\rangle}{\sqrt{2}}. \tag{23.1.18}$$

The general superposition is then processed as follows:

$$a_0|0\rangle + a_1|1\rangle \ \text{—}[H]\text{—}\ a_0\left(\frac{|0\rangle + |1\rangle}{\sqrt{2}}\right) + a_1\left(\frac{|0\rangle - |1\rangle}{\sqrt{2}}\right). \tag{23.1.19}$$

Two-qubit gates Let us now consider some two-qubit gates. A popular one is the *controlled-NOT* gate. The classical version has a control bit and a target bit. If the control bit is 0, the target bit is unchanged. If the control bit is 1, the target bit is reversed—namely, acted upon by NOT. In the classical case, we represent it as follows:

(23.1.20)

where x, y are bits. The top line is the control bit and contains a heavy dot, and the output is the same as the input. The bottom line is the target bit and contains the symbol $\oplus$, and the output $x \oplus y$ implements the NOT operation on y when $x = 1$ and leaves y unchanged when $x = 0$, as desired. The quantum gate is represented in a completely analogous way. The only difference is that we put computational basis states at the inputs and outputs'

$|x\rangle$ —●— $|x\rangle$

$|y\rangle$ —⊕— $|y \oplus x\rangle$ (23.1.21)

In writing, this figure states:

$$\text{Quantum C-NOT}: \ |x\rangle \otimes |y\rangle \rightarrow |x\rangle \otimes |y \oplus x\rangle, \tag{23.1.22}$$

or, in briefer notation,

$$\text{Quantum C-NOT}: \ |x, y\rangle \rightarrow |x, y \oplus x\rangle. \tag{23.1.23}$$

The commas separating the entries improve readability for generic inputs; when dealing with 0s and 1s, we will omit them. The above relation defines the action of the gate on computational basis states. Therefore, by linearity, this defines the action of the gate on arbitrary input states that live in the tensor product of the two qubits.

With the understanding that the heavy dot is on the control line and the $\oplus$ is on the target line, we have the following variant:

(23.1.24)

We now ask: Is the quantum C-NOT gate unitary? It better be. Let us see this by looking explicitly at its matrix representation. For this purpose we use $|x, y\rangle \rightarrow |x, y \oplus x\rangle$ to determine how the basis states transform when acted by the gate:

$$\begin{aligned}
|00\rangle &\rightarrow |00\rangle \\
|01\rangle &\rightarrow |01\rangle \\
|10\rangle &\rightarrow |11\rangle \\
|11\rangle &\rightarrow |10\rangle.
\end{aligned} \tag{23.1.25}$$

The 4 X 4 matrix $U_{\text{C-NOT}}$ representing the action is unambiguous if we order the basis states, which we do using the value of the binary representative: $|00\rangle$, $|01\rangle$, $|10\rangle$, $|11\rangle$:

$$U_{\text{C-NOT}} = \begin{pmatrix} 1 & 0 & 0 & 0 \\ 0 & 1 & 0 & 0 \\ 0 & 0 & 0 & 1 \\ 0 & 0 & 1 & 0 \end{pmatrix}. \tag{23.1.26}$$

The matrix is manifestly Hermitian and clearly squares to itself. It is therefore unitary. Using 2 X 2 blocks, we see that

$$U_{\text{C-NOT}} = \begin{pmatrix} \mathbb{1}_2 & 0 \\ 0 & X \end{pmatrix}, \tag{23.1.27}$$

making it clear that when acting on computational states it leaves the first two states invariant while exchanging the third and fourth states, as indicated in (23.1.25).

It is worth noting that the quantum C-NOT gate can generate entanglement: given a nonentangled input state, the output can be entangled. In that case although we can associate a state to the top input of the gate and a state to the bottom input, we cannot associate a state to each of the separate outputs. For example, for a nonentangled input $(a_0|0\rangle + a_1|1\rangle) \otimes |0\rangle$ we have the following:

$$\text{Quantum C-NOT}: \ (a_0|0\rangle + a_1|1\rangle) \otimes |0\rangle \ \rightarrow \ a_0|0\rangle \otimes |0\rangle + a_1|1\rangle \otimes |1\rangle, \tag{23.1.28}$$

and the output is clearly entangled when a_0 and a_1 are nonvanishing. The representation of this action with a diagram is as follows:

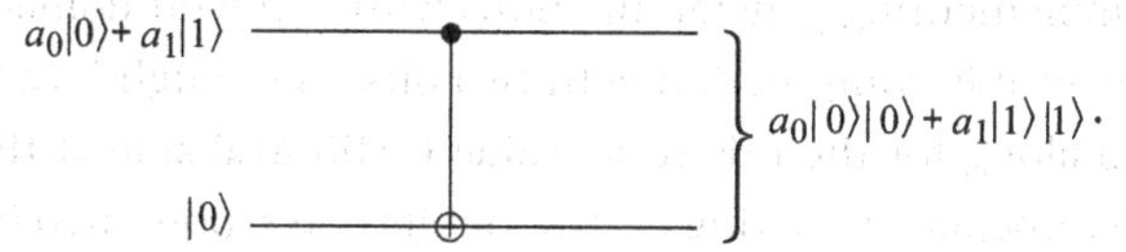

$a_0|0\rangle + a_1|1\rangle$

$a_0|0\rangle|0\rangle + a_1|1\rangle|1\rangle.$

$|0\rangle$

(23.1.29)

Since the output state is entangled, there is no possible association of states to each of the two output lines. Even for the input in (23.1.28), the association of states to the separate lines is ambiguous since the product state is unaffected if the first state is multiplied by a constant and the second is divided by the same constant. When the input itself is entangled, no association of states to the lines is possible.

The quantum C-NOT gate also illustrates the no-cloning theorem of quantum mechanics. If cloning takes

$$|\psi\rangle \otimes |0\rangle \rightarrow |\psi\rangle \otimes |\psi\rangle, \tag{23.1.30}$$

the C-NOT gate certainly takes $|0\rangle \otimes |0\rangle \rightarrow |0\rangle \otimes |0\rangle$ and $|1\rangle \otimes |0\rangle \rightarrow |1\rangle \otimes |1\rangle$, so it can clone both $|0\rangle$ and $|1\rangle$. This is consistent with the discussion in section 18.9, which showed that a cloning machine can be designed to clone a set of mutually orthogonal state vectors. The machine, however, cannot clone linear superpositions. This is clear from (23.1.28): any cloned state $|\psi\rangle \otimes |\psi\rangle$ is nonentangled, while the output of the gate is entangled when a_0 and a_1 are both different from zero.

Exercise 23.1. *Find the output of the iterated circuit below.*

(23.1.31)

The computation of functions Let us consider how we use quantum circuits to compute functions. Suppose we have a function $f(x)$ with a one-bit domain ($x \in \{0,1\}$) and a one-bit range ($f \in \{0,1\}$). How can we construct a circuit that implements this function?

More concretely, we ask: Is there a one-qubit gate that acting on $|x\rangle$ gives us $|f(x)\rangle$? Suppose $f(0)=0$ and $f(1)=1$. Then no programming is needed. The gate is the identity operator. The circuit is just a line with no operator inserted. On the other hand, if $f(0)=1$ and $f(1)=0$, the gate must just do NOT, so it is the X gate.

But how about a function for which $f(0)=f(1)$? This cannot work with the one-qubit realization. The reason is unitarity. Suppose the function is implemented by the unitary operator U_f. Then we would have $U_f|0\rangle = U_f|1\rangle$. Applying the inverse operator $U_f^\dagger$ to these equations, we get $|0\rangle = |1\rangle$, a manifestly false conclusion. This particular function cannot be realized as a one-qubit quantum circuit because the function is not invertible, while unitary operations are always invertible. When $f(0)=f(1)$, the computation done by $f(x)$ is not reversible: knowing the output does not allow us, in principle, to know the input. This is a general lesson in quantum computation: quantum gates implement reversible computations.

The solution to our task of implementing general functions with a one-bit domain and a one-bit range is to do a reversible computation with two bits. The rough idea is clear: the input bit must just go along for the ride so its value is still available at the output. The gate is now a unitary operator U_f acting on two qubits and represented by the following box:

x — U_f — x

y — U_f — $y \oplus f(x)$ (23.1.32)

The action of the box on computational basis states can be written as

$$U_f: \ |x,y\rangle \ \rightarrow \ |x, y \oplus f(x)\rangle. \tag{23.1.33}$$

If we set $|y\rangle = |0\rangle$, we can read the value of the function by measuring the second qubit: $U_f : |x\rangle|0\rangle \rightarrow |x\rangle|f(x)\rangle$. It is necessary to confirm that U_f is unitary. For this we first show that $U_f U_f = \mathbb{1}$. This is quickly confirmed by letting U_f act twice on a computational state:

$$U_f U_f |x,y\rangle = U_f |x, y \oplus f(x)\rangle = |x, y \oplus f(x) \oplus f(x)\rangle = |x,y\rangle, \tag{23.1.34}$$

where we noted that $f(x) \oplus f(x) = 0$ modulo 2. Having proven that U_f squares to one, it remains to show that U_f is Hermitian. For this we write the operator explicitly in bra-ket notation:

$$U_f = \sum_{x,y\in\{0,1\}} |x, y \oplus f(x)\rangle\langle x, y|, \tag{23.1.35}$$

where the sum runs over the four values the pair (x, y) can take. By construction, this operator gives the expected transformations of all computational states. The Hermitian conjugate is obtained by following the familiar rule $(|a\rangle\langle b|)^\dagger = |b\rangle\langle a|$, and gives

$$U_f^\dagger = \sum_{x,y\in\{0,1\}} |x, y\rangle\langle x, y \oplus f(x)|. \tag{23.1.36}$$

To rewrite the sum, it is useful to define primed bits:

$$\begin{aligned} x' &= x, \\ y' &= y \oplus f(x). \end{aligned} \tag{23.1.37}$$

As the pairs (x, y) run over all four possible values, so do (x', y'), for any function f. Indeed, it is clear from inspection that x' runs over the two possible values, and the second equation shows that for any fixed x, so does y'. We can thus replace the sum over $x, y \in \{0, 1\}$ by a sum over $x', y' \in \{0, 1\}$. Note also that the second relation above implies that $y = y' \oplus f(x) = y' \oplus f(x')$. Therefore, we have

$$\begin{aligned} U_f^\dagger &= \sum_{x',y'\in\{0,1\}} |x', y' \oplus f(x')\rangle\langle x', y' \oplus f(x') \oplus f(x')| \\ &= \sum_{x',y'\in\{0,1\}} |x', y' \oplus f(x')\rangle\langle x', y'| = U_f, \end{aligned} \tag{23.1.38}$$

by comparison with the original expression (23.1.35) and showing that U_f is indeed Hermitian and therefore unitary.

Let us now illustrate how we build the four possible functions with a one-bit domain and a one-bit range. What we want is the gate U_f that implements $U_f : |x, y\rangle \to |x, y \oplus f(x)\rangle$. First we do the two noninvertible functions and then the remaining invertible ones:

1. $f(0) = f(1) = 0$. In this case we would have

   $$U_f: \; |x, y\rangle \to |x, y \oplus f(x)\rangle = |x, y\rangle, \tag{23.1.39}$$

   since $f(x) = 0$ for all x. Therefore, this function is implemented by the identity operator $U_f = \mathbb{1}$ on the two-qubit space. The circuit is

   $|x\rangle$ ———————— $|x\rangle$

   $|y\rangle$ ———————— $|y\rangle$ (23.1.40)

2. $f(0) = f(1) = 1$. In this case we would have

   $$U_f: \; |x, y\rangle \to |x, y \oplus f(x)\rangle = |x, y \oplus 1\rangle = |x, \bar{y}\rangle, \tag{23.1.41}$$

since $f(x)=1$ for all x, and $\bar{y}=y\oplus 1$. Therefore, this function is implemented by including the X operator on the second qubit line:

$$|x\rangle \longrightarrow |x\rangle$$
$$|y\rangle \longrightarrow \boxed{X} \longrightarrow |\bar{y}\rangle \tag{23.1.42}$$

3. $f(0)=0, f(1)=1$. In this case $f(x)=x$, so we use this to note that

$$U_f:\ |x,y\rangle \rightarrow |x, y\oplus f(x)\rangle = |x, y\oplus x\rangle, \tag{23.1.43}$$

which is recognized as the CNOT gate:

(23.1.44)

4. $f(0)=1, f(1)=0$. In this case $f(x)=1\oplus x$, so we use this to note that

$$U_f:\ |x,y\rangle \rightarrow |x, y\oplus f(x)\rangle = |x, y\oplus 1\oplus x\rangle = |x, \bar{y}\oplus x\rangle, \tag{23.1.45}$$

which is recognized as the CNOT gate preceded by a NOT gate X acting on the second qubit:

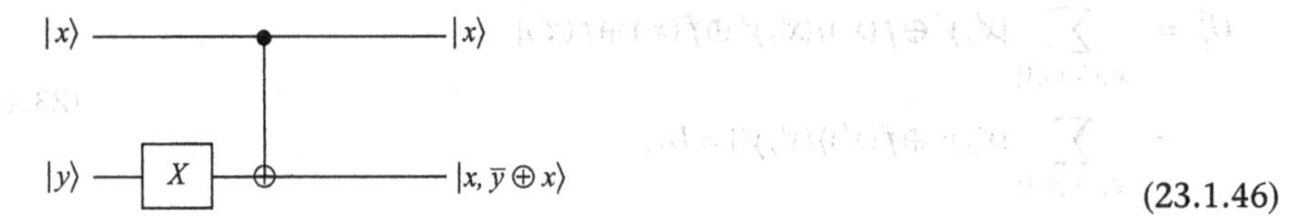

(23.1.46)

This completes the construction of all the functions with a one-bit domain and a one-bit range.

A classical computer gate having a two-bit input and a one-bit output cannot be reversible since there are only two possible outputs (0 and 1) but four different two-bit inputs (00,01,10,11). Thus, the so-called NAND gate, which acting on the pair of bits (x,y) gives $\overline{x\cdot y}$, is not reversible. The NAND gate is universal in classical computation: using only NAND gates, one can construct *any* function on n-bit strings. The Toffoli gate is a reversible gate with a three-bit input and a three-bit output. It can be used to simulate the NAND gate. Therefore, any function on n-bit strings can be computed *reversibly* using Toffoli gates. For some basic facts on the quantum version of the Toffoli gate, see problem 23.1.

For our later purposes, we will assume that, just as we did for a function $f(x)$ with a one-bit domain and a one-bit range, we can construct a gate that evaluates the function

$$f(x_1, \ldots, x_n),$$

with an n-bit domain and a one-bit range using a unitary operator U_f that acts on $n+1$ qubits. On computational states, the first n qubits simply carry the information of the inputs through the box, helping with reversibility. The last qubit, called the oracle qubit, carries the information on the value of the function. We therefore define

$$U_f: \; |x_1, \ldots, x_n, y\rangle \;\rightarrow\; |x_1, \ldots, x_n, y \oplus f(x_1, \ldots, x_n)\rangle. \tag{23.1.47}$$

The diagram for the gate represented by the unitary operator U_f is

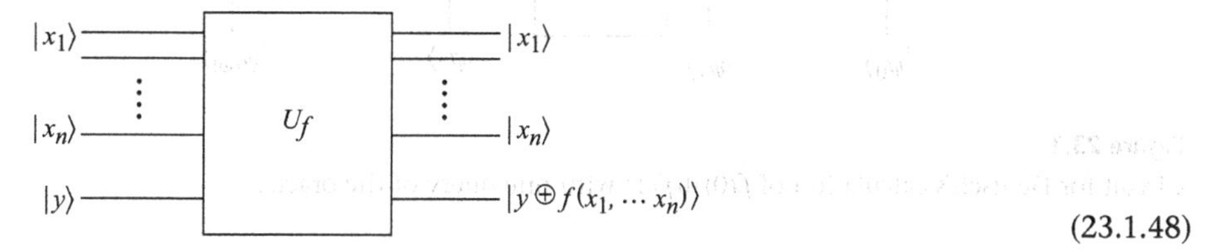

(23.1.48)

The box U_f itself is sometimes called an **oracle**: one queries the oracle with an n-qubit computational state input while setting $y=0$ and out comes the answer for the value of the function.

Exercise 23.2. *Write down an explicit bra-ket representation of the operator U_f, and show the operator is unitary.*

23.2 Deutsch's Computation

Consider a function $f(x)$ with a one-bit domain and a one-bit range. Assume the function is not known. In principle, the oracle that evaluates the function must be queried two times to learn what the function is: first we ask for $f(0)$ and then we ask for $f(1)$. We will illustrate here how a simple quantum circuit (computer) can, with a single query of the oracle, construct a state that contains information about *both* $f(0)$ and $f(1)$. It is not possible to extract the two separate values of the function from the state, but surprisingly, we can extract the value of the (mod 2) sum $f(0) \oplus f(1)$. We get this value from a single query of the oracle, something that would be impossible in the classical situation. Any classical computer would require two queries of the oracle.

The oracle is the two-qubit gate U_f we introduced before, which computes $f(x)$ reversibly and is represented by the following:

$$|x\rangle \;—\; U_f \;—\; |x\rangle, \qquad |y\rangle \;—\; U_f \;—\; |y \oplus f(x)\rangle \tag{23.2.1}$$

Recall that this means that

$$U_f: \; |x\rangle \otimes |y\rangle \;\rightarrow\; |x\rangle \otimes |y \oplus f(x)\rangle. \tag{23.2.2}$$

Using a linear superposition for the first qubit and the state $|0\rangle$ for the second qubit, we have

$$U_f: \; \tfrac{1}{\sqrt{2}}\big(|0\rangle + |1\rangle\big) \otimes |0\rangle \;\rightarrow\; \tfrac{1}{\sqrt{2}}\big(|0\rangle \otimes |f(0)\rangle + |1\rangle \otimes |f(1)\rangle\big). \tag{23.2.3}$$

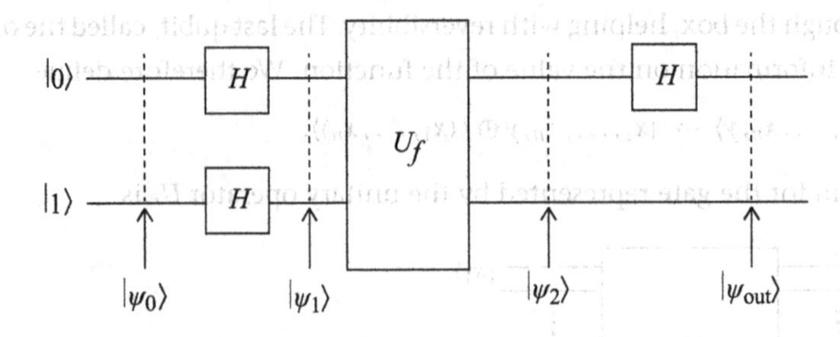

Figure 23.1
Circuit for Deutsch's calculation of $f(0) \oplus f(1)$ with one query of the oracle.

With one single call of the oracle, we have a state in which both $f(0)$ and $f(1)$ appear. Note that the first qubit in the input state is in fact $H|0\rangle$, with H the Hadamard gate (23.1.17). Therefore, the input state above is $H|0\rangle \otimes |0\rangle$.

In 1985 David Deutsch found a way to obtain the value of $f(0) \oplus f(1)$ with a simple circuit that uses two H gates for the input, one at the output and, most importantly, *one* action of U_f. We discuss his construction below. The analysis is simplified by noting that $f(0) \oplus f(1)$ can take two values as follows:

- $f(0) \oplus f(1) = 0$. This requires that

$$f(0) = f(1). \tag{23.2.4}$$

- $f(0) \oplus f(1) = 1$. This requires $f(0) \neq f(1)$, which means that

$$f(0) \oplus 1 = f(1), \quad \text{or equivalently,} \quad f(0) = 1 \oplus f(1). \tag{23.2.5}$$

The circuit for Deutsch's calculation is shown in figure 23.1. Indicated at various positions are the quantum states of the system: $|\psi_0\rangle$ as the input, then $|\psi_1\rangle$, $|\psi_2\rangle$, and finally the output state $|\psi_{\text{out}}\rangle$. We take the input state to be

$$|\psi_0\rangle = |0\rangle \otimes |1\rangle. \tag{23.2.6}$$

After the action of the two Hadamard gates, this becomes

$$\begin{aligned} |\psi_1\rangle &= H \otimes H |\psi_0\rangle = H|0\rangle \otimes H|1\rangle = \tfrac{1}{\sqrt{2}}(|0\rangle + |1\rangle) \otimes \tfrac{1}{\sqrt{2}}(|0\rangle - |1\rangle) \\ &= \tfrac{1}{2}|0\rangle \otimes (|0\rangle - |1\rangle) + \tfrac{1}{2}|1\rangle \otimes (|0\rangle - |1\rangle). \end{aligned} \tag{23.2.7}$$

Now comes the action of U_f to give

$$|\psi_2\rangle = U_f|\psi_1\rangle = \tfrac{1}{2}|0\rangle \otimes \big(|f(0)\rangle - |1 \oplus f(0)\rangle\big) + \tfrac{1}{2}|1\rangle \otimes \big(|f(1)\rangle - |1 \oplus f(1)\rangle\big). \tag{23.2.8}$$

As the circuit indicates, the output state arises as $|\psi_{\text{out}}\rangle = (H \otimes \mathbb{1})|\psi_2\rangle$. We then get

$$\begin{aligned} |\psi_{\text{out}}\rangle = {} & \tfrac{1}{2\sqrt{2}}(|0\rangle + |1\rangle) \otimes (|f(0)\rangle - |1 \oplus f(0)\rangle) \\ & + \tfrac{1}{2\sqrt{2}}(|0\rangle - |1\rangle) \otimes (|f(1)\rangle - |1 \oplus f(1)\rangle). \end{aligned} \tag{23.2.9}$$

We group the terms for which the first qubit is $|0\rangle$ and the terms for which the first qubit is $|1\rangle$:

$$|\psi_{\text{out}}\rangle = \tfrac{1}{2}|0\rangle \otimes \tfrac{1}{\sqrt{2}}\big(|f(0)\rangle - |1\oplus f(0)\rangle + |f(1)\rangle - |1\oplus f(1)\rangle\big)$$
$$+\tfrac{1}{2}|1\rangle \otimes \tfrac{1}{\sqrt{2}}\big(|f(0)\rangle - |1\oplus f(0)\rangle - |f(1)\rangle + |1\oplus f(1)\rangle\big). \tag{23.2.10}$$

We now consider the two possible situations examined in (23.2.4) and (23.2.5). When $f(0)=f(1)$, we see that the state in parentheses on the second line vanishes. When $f(0)\neq f(1)$, the state in parentheses on the first line vanishes. We thus have

$$|\psi_{\text{out}}\rangle = \begin{cases} |0\rangle \otimes \tfrac{1}{\sqrt{2}}\big(|f(0)\rangle - |1\oplus f(0)\rangle\big), & f(0)=f(1), \\ |1\rangle \otimes \tfrac{1}{\sqrt{2}}\big(|f(0)\rangle - |1\oplus f(0)\rangle\big), & f(0)\neq f(1). \end{cases} \tag{23.2.11}$$

The final step is realizing that when $f(0)=f(1)$, we have $|0\rangle = |f(0)\oplus f(1)\rangle$. Moreover, when $f(0)\neq f(1)$, we have $|1\rangle = |f(0)\oplus f(1)\rangle$. As a result, we have obtained the state

$$|\psi_{\text{out}}\rangle = |f(0)\oplus f(1)\rangle \otimes \tfrac{1}{\sqrt{2}}\big(|f(0)\rangle - |1\oplus f(0)\rangle\big). \tag{23.2.12}$$

The second state in the product is never vanishing; in fact

$$|f(0)\rangle - |1\oplus f(0)\rangle = (-1)^{f(0)}(|0\rangle - |1\rangle). \tag{23.2.13}$$

As a result, the output state can be written as

$$|\psi_{\text{out}}\rangle = |f(0)\oplus f(1)\rangle \otimes \frac{(-1)^{f(0)}}{\sqrt{2}}\big(|0\rangle - |1\rangle\big). \tag{23.2.14}$$

If we measure the first qubit in $|\psi_{\text{out}}\rangle$, we indeed read the value of $f(0)\oplus f(1)$. The circuit gave us this value with a single action of U_f, a single query of the oracle. Note that the value of $f(0)$, appearing through an overall sign factor, cannot be read.

23.3 Grover's Algorithm

Consider a search problem over a set with N elements. We label the elements using the integers $0,\ldots,N-1$ and assume $N=2^n$ with $n\geq 1$, to work conveniently in binary. We use the label x to denote these integers, thus x ranges in

$$x=0,1,\ldots,N-1=2^n-1. \tag{23.3.1}$$

The search problem has a function $f(x)$ that can be evaluated with an oracle for all allowed values of x. The function will be assumed to have a one-bit range, so $f(x)$ for any x can be just 0 or 1. Since x can be described in binary by n bits, the function $f(x)$ can be thought of as a function with an n-bit domain and a one-bit range. The search problem aims to find those x's for which $f(x)=1$. If $f(x)=1$, we say x is a solution of the search problem. If $f(x)=0$, x is not a solution of the search problem. We will assume there are M solutions, with $M<N$. Typically, we are interested in $M\ll N$, or sometimes even $M=1$. It takes a lot of trying to find a solution. Nothing about the function $f(x)$ is presumed known to help the search; we simply have an oracle that for an input x gives us $f(x)$. This means that the only way to find solutions is by iterated queries of the oracle over the possible values of x.

If $M = 1$, there is just one x for which $f(x) = 1$. To find it with a classical computer, we expect to have to query the oracle about $N/2$ times on average. We will show that for large N the quantum circuit designed by Lov Grover in 1996 requires only about $\frac{\pi}{4}\sqrt{N} \simeq 0.785\sqrt{N}$ queries of the oracle.

For any integer x in the range from 0 to $N-1$, the binary representation of x is an n-bit sequence:

$$x \longleftrightarrow x_1 \dots x_n, \tag{23.3.2}$$

with all x_i being either zero or one. We will use the notation in which $|x\rangle$ is an n-qubit computational state:

$$|x\rangle \equiv |x_1\rangle \otimes \cdots \otimes |x_n\rangle, \quad f(x) \equiv f(x_1, \dots, x_n). \tag{23.3.3}$$

Recall that f can only take values 0 and 1.

The quantum oracle will be a circuit defining a unitary operator O_f (O for oracle) that evaluates the function f and, as discussed before, takes the following form:

$$|x\rangle \begin{cases} |x_1\rangle \\ \vdots \\ |x_n\rangle \end{cases} \quad |q\rangle \quad \boxed{O_f} \quad \begin{rcases} |x_1\rangle \\ \vdots \\ |x_n\rangle \end{rcases} |x\rangle \quad |q \oplus f(x)\rangle \tag{23.3.4}$$

We write the action of O_f as follows:

$$O_f: \; |x\rangle \otimes |q\rangle \rightarrow |x\rangle \otimes |q \oplus f(x)\rangle, \tag{23.3.5}$$

where $|q\rangle$ is the oracle qubit. A clever choice of $|q\rangle$ is the state $H|1\rangle$:

$$O_f: \; |x\rangle \otimes \tfrac{1}{\sqrt{2}}(|0\rangle - |1\rangle) \rightarrow |x\rangle \otimes \tfrac{1}{\sqrt{2}}(|f(x)\rangle - |1 \oplus f(x)\rangle). \tag{23.3.6}$$

For $f(x) = 0$, the second state in the tensor product is in fact $H|1\rangle$, and for $f(x) = 1$, it is minus that state: $-H|1\rangle$. We thus find the following:

$$O_f: \; |x\rangle \otimes H|1\rangle \rightarrow (-1)^{f(x)}|x\rangle \otimes H|1\rangle. \tag{23.3.7}$$

The action of O_f on such a state simply changes the sign of the state when x is a solution ($f(x) = 1$) and does nothing to the state when x is not a solution ($f(x) = 0$). Since the state of the oracle qubit is not changed by this operation, or by any other operation we will consider below, we will stop writing it out explicitly and just work with the action:

$$O_f: \; |x\rangle \rightarrow (-1)^{f(x)}|x\rangle. \tag{23.3.8}$$

Let us consider the preparation of the initial n-qubit state $|\psi_0\rangle$ that goes into the oracle (accompanied implicitly by the $H|1\rangle$ state that we no longer write). We take $|\psi_0\rangle$ to be given by the equally weighted sum of all possible computational basis states of the n-qubit state space:

$$\begin{aligned} |\psi_0\rangle &= \frac{1}{\sqrt{N}} \sum_{x=0}^{N-1} |x\rangle = \frac{1}{\sqrt{N}}(|0\rangle + |1\rangle + |2\rangle + \cdots + |N-1\rangle) \\ &= \frac{1}{\sqrt{N}}(|0\cdots 0\rangle + |0\cdots 01\rangle + \cdots + |1\cdots 1\rangle). \end{aligned} \tag{23.3.9}$$

In the last line, we used the binary representation of all the integers that appear in the sum. The state is correctly normalized. This state is easily built starting with the ket $|0\cdots 0\rangle = |0\rangle \otimes \cdots \otimes |0\rangle$ and acting on each state with the Hadamard operator. That is,

$$|\psi_0\rangle = (H|0\rangle) \otimes \cdots (H|0\rangle) = \Big(\tfrac{1}{\sqrt{2}}(|0\rangle + |1\rangle)\Big)^{\otimes n}. \tag{23.3.10}$$

Indeed, the normalization factor multiplying the basis kets works out: $(1/\sqrt{2})^n = 1/\sqrt{2^n} = 1/\sqrt{N}$. Moreover, when we expand the tensor product $(|0\rangle + |1\rangle)^{\otimes n}$, it is clear that we get the sum of all possible length-n products of kets $|0\rangle$ and $|1\rangle$. As a result, the input state can be visualized as

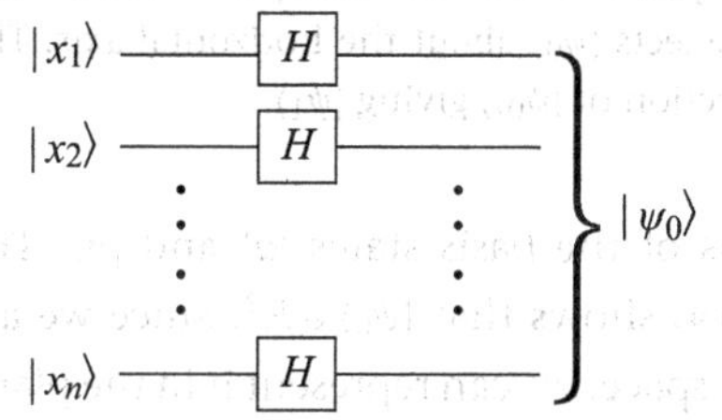

$$\begin{array}{l} |x_1\rangle \;—\boxed{H}— \\ |x_2\rangle \;—\boxed{H}— \\ \quad\vdots \qquad\quad \vdots \\ |x_n\rangle \;—\boxed{H}— \end{array} \Bigg\} \; |\psi_0\rangle \tag{23.3.11}$$

Let us now split the ket $|\psi_0\rangle$ into those kets that are not solutions and those kets that are:

$$|\psi_0\rangle = \frac{1}{\sqrt{N}} \sum_{x=0}^{N-1} |x\rangle = \frac{1}{\sqrt{N}} \sum_{\{x|f(x)=0\}} |x\rangle + \frac{1}{\sqrt{N}} \sum_{\{x|f(x)=1\}} |x\rangle. \tag{23.3.12}$$

In the first sum, we include those kets $|x\rangle$ for which $f(x) = 0$—namely, nonsolutions. In the second sum, we include the rest of the kets, those for which $f(x) = 1$, which are therefore solutions. It is convenient to normalize separately each of these sums. Recalling that we have M solutions and $N - M$ nonsolutions, we trivially rewrite $|\psi_0\rangle$ as follows:

$$|\psi_0\rangle = \sqrt{\frac{N-M}{N}} \Big(\frac{1}{\sqrt{N-M}} \sum_{\{x|f(x)=0\}} |x\rangle \Big) + \sqrt{\frac{M}{N}} \Big(\frac{1}{\sqrt{M}} \sum_{\{x|f(x)=1\}} |x\rangle \Big). \tag{23.3.13}$$

This allows us to write the initial state as the superposition of two vectors $|\alpha\rangle$ and $|\beta\rangle$:

$$|\psi_0\rangle = \sqrt{\frac{N-M}{N}}\, |\alpha\rangle + \sqrt{\frac{M}{N}}\, |\beta\rangle, \tag{23.3.14}$$

where we defined

$$|\alpha\rangle \equiv \frac{1}{\sqrt{N-M}} \sum_{\{x|f(x)=0\}} |x\rangle, \qquad |\beta\rangle \equiv \frac{1}{\sqrt{M}} \sum_{\{x|f(x)=1\}} |x\rangle. \tag{23.3.15}$$

The state $|\alpha\rangle$ contains the kets that are not solutions, while the state $|\beta\rangle$ contains the kets that *are* solutions. Since $\langle x|x'\rangle = \delta_{xx'}$, we note that

$$\langle\alpha|\alpha\rangle = \langle\beta|\beta\rangle = 1, \quad \langle\alpha|\beta\rangle = 0. \tag{23.3.16}$$

The two states $|\alpha\rangle$ and $|\beta\rangle$ are thus normalized and orthogonal. We will be working with the *real* two-dimensional vector space $\mathbb{R}^2$ spanned by these vectors; our analysis

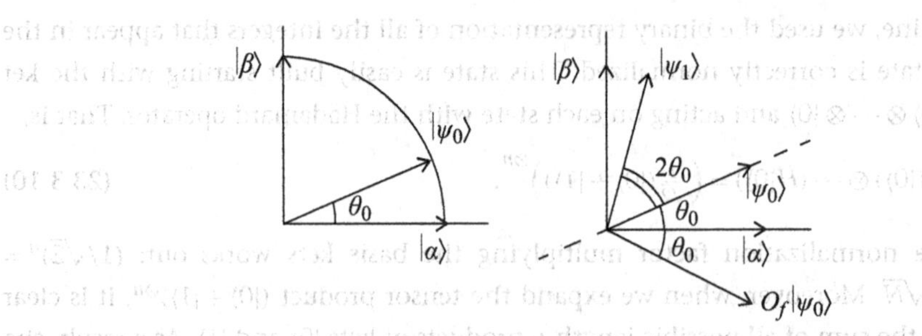

Figure 23.2
Left: The orthogonal unit vectors $|\alpha\rangle$ and $|\beta\rangle$ on the $\mathbb{R}^2$ plane. The initial state $|\psi_0\rangle$ is at an angle θ_0 relative to the horizontal. *Right*: The action of O_f reflects $|\psi_0\rangle$ about the horizontal axis. The subsequent action of R_0 reflects the state about the direction of $|\psi_0\rangle$, giving $|\psi_1\rangle$.

below will only require real linear combinations of the basis states $|\alpha\rangle$ and $|\beta\rangle$. The decomposition of the initial state $|\psi_0\rangle$ in (23.3.14) shows that $|\psi_0\rangle \in \mathbb{R}^2$. Since we are dealing with a real vector subspace of the full state space, we can represent it in the plane with the orthogonal basis vectors $|\alpha\rangle$ and $|\beta\rangle$ lying along the horizontal and vertical directions, respectively (figure 23.2).

The vector $|\psi_0\rangle$ in the $\mathbb{R}^2$ plane is a unit vector at an angle $\theta_0 \in (0, \frac{\pi}{2})$ relative to the horizontal axis (figure 23.2, *left*). The value of θ_0 follows from writing the expression (23.3.14) for $|\psi_0\rangle$ as follows:

$$|\psi_0\rangle = \cos\theta_0|\alpha\rangle + \sin\theta_0|\beta\rangle, \quad \sin\theta_0 = \sqrt{\frac{M}{N}}. \tag{23.3.17}$$

If $M \ll N$, the vector $|\psi_0\rangle$ points very close to the direction of $|\alpha\rangle$ because almost all states are not solutions. The idea for the Grover algorithm now becomes apparent: the quantum computer must act on $|\psi_0\rangle$ and rotate it into the solution ket $|\beta\rangle$. At that point, a measurement of the rotated state would give us one of the solution states! As we will see, this strategy is right except that in general we can only map the initial state very close to the direction of $|\beta\rangle$. This is good enough, in fact.

Let us first see what the oracle operator does to $|\psi_0\rangle$. Recall that acting on $|x\rangle$ the oracle gives $(-1)^{f(x)}|x\rangle$, and therefore, it changes the sign of solutions while doing nothing to states that are not solutions. As a result,

$$O_f|\beta\rangle = -|\beta\rangle, \quad O_f|\alpha\rangle = |\alpha\rangle, \tag{23.3.18}$$

and on the initial state,

$$O_f|\psi_0\rangle = \cos\theta_0|\alpha\rangle - \sin\theta_0|\beta\rangle. \tag{23.3.19}$$

The vector $O_f|\psi_0\rangle$ is the reflection of $|\psi_0\rangle$ about the horizontal axis (figure 23.2, *right*).

We need one more operator, a unitary R_0 that reflects states in $\mathbb{R}^2$ about the direction of $|\psi_0\rangle$. We claim that operator takes the form

$$R_0 \equiv 2|\psi_0\rangle\langle\psi_0| - \mathbb{1}. \tag{23.3.20}$$

R_0 is manifestly Hermitian. Additionally, we quickly verify that it squares to the identity:

$$R_0R_0 = 4|\psi_0\rangle\langle\psi_0|\psi_0\rangle\langle\psi_0| - 4|\psi_0\rangle\langle\psi_0| + \mathbb{1} = \mathbb{1}, \tag{23.3.21}$$

confirming that R_0 is unitary. It is also simple to see that R_0 performs the indicated reflection. Let $|\psi_0^\perp\rangle$ denote a unit vector in $\mathbb{R}^2$ orthogonal to $|\psi_0\rangle$. It is clear that $|\psi_0\rangle$ together with $|\psi_0^\perp\rangle$ form a basis for $\mathbb{R}^2$. Therefore, we have the completeness relation

$$\mathbb{1} = |\psi_0\rangle\langle\psi_0| + |\psi_0^\perp\rangle\langle\psi_0^\perp|,$$

and using this in the expression for R_0, we get

$$R_0 = |\psi_0\rangle\langle\psi_0| - |\psi_0^\perp\rangle\langle\psi_0^\perp|. \tag{23.3.22}$$

This presentation makes it clear that, acting on any vector, R_0 leaves the component along $|\psi_0\rangle$ unchanged but reverses the component along $|\psi_0^\perp\rangle$. This is what a reflection about the direction of $|\psi_0\rangle$ is supposed to do.

The operator R_0 can be realized in terms of simple operations. Indeed, we can write

$$R_0 = H^{\otimes n}(2|0\rangle\langle 0| - \mathbb{1})H^{\otimes n}, \tag{23.3.23}$$

where $|0\rangle = |0\cdots 0\rangle$. This is easily verified, recalling that $|\psi_0\rangle = H^{\otimes n}|0\rangle$. The operator $2|0\rangle\langle 0| - \mathbb{1}$ can be defined and realized as a function that acts on basis states as follows:

$$2|0\rangle\langle 0| - \mathbb{1}: \ |0\rangle \to |0\rangle, \quad |x\rangle \to -|x\rangle \text{ for } x > 0. \tag{23.3.24}$$

We now act with R_0 on the ket $O_f|\psi_0\rangle$ to obtain a state we call $|\psi_1\rangle$:

$$|\psi_1\rangle \equiv R_0 O_f |\psi_0\rangle. \tag{23.3.25}$$

Referring to the right side of figure 23.2, we see that $O_f|\psi_0\rangle$ is at an angle of $2\theta_0$ relative to $|\psi_0\rangle$. Therefore, the reflected state $|\psi_1\rangle$ is also at an angle $2\theta_0$ relative to $|\psi_0\rangle$ but in the opposite direction. Calling θ_1 the angle of $|\psi_1\rangle$ with respect to the horizontal axis, we have

$$\theta_1 = \theta_0 + 2\theta_0 = 3\theta_0. \tag{23.3.26}$$

Define the *Grover operator* G to be the unitary that first acts with O_f and then with R_0:

$$G \equiv R_0 O_f. \tag{23.3.27}$$

We have shown that $|\psi_1\rangle = G|\psi_0\rangle$ where $|\psi_1\rangle$ is obtained by a $2\theta_0$ counterclockwise rotation of $|\psi_0\rangle$. The circuit for the Grover operator is shown in figure 23.3. Note that the three boxes to the right of O_f construct R_0, as given in the product representation (23.3.23).

Geometrically, the Grover operator G first reflects about the horizontal direction (via O_f) and then reflects the result about the direction of $|\psi_0\rangle$ (via R_0). For an arbitrary vector $|\psi\rangle$ in $\mathbb{R}^2$ at an angle γ relative to the horizontal (figure 23.4), the reflection generated by O_f changes the angle to $-\gamma$. The action of R_0 then changes the angle to $-\gamma + 2(\theta_0 + \gamma) = \gamma + 2\theta_0$. It means that the Grover operator is in fact a counterclockwise rotation by an angle $2\theta_0$.

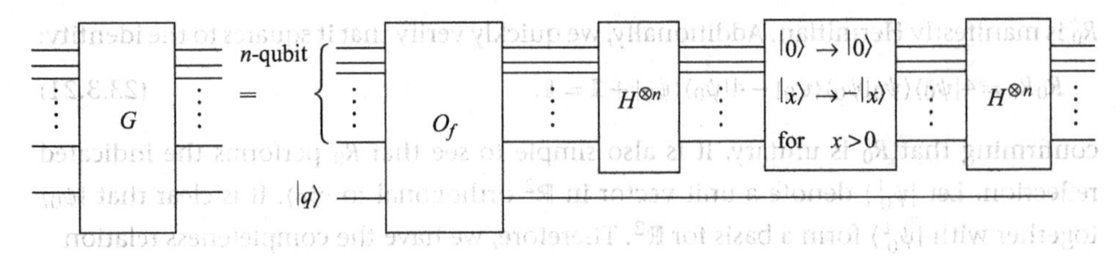

Figure 23.3
The Grover unitary G first evaluates O_f on an n-qubit input, and the result is acted upon R_0, represented by the action of the three boxes to the right of O_f. The oracle qubit is not operated upon by R_0.

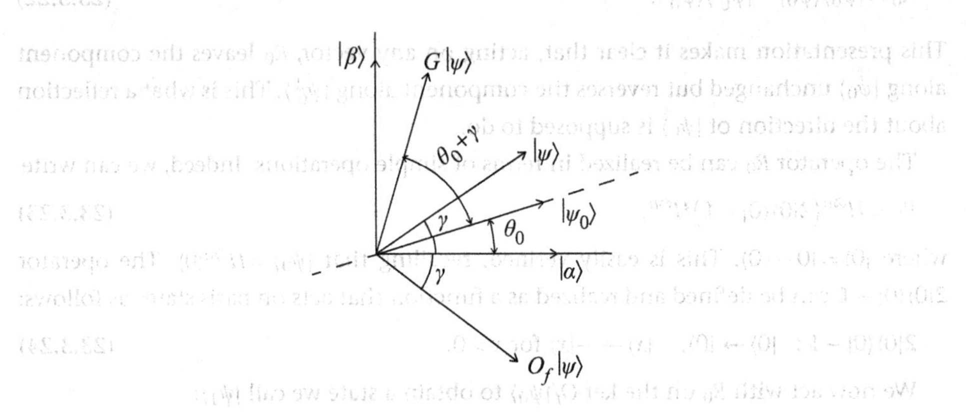

Figure 23.4
The Grover unitary $G = R_0 O_f$ rotates an arbitrary state $|\psi\rangle \in \mathbb{R}^2$ by an angle $2\theta_0$ in the counter-clockwise direction.

A second Grover iteration takes $|\psi_1\rangle$ into $|\psi_2\rangle \equiv G|\psi_1\rangle$, with angle $\theta_2 = \theta_1 + 2\theta_0 = 5\theta_0$. In general, with $|\psi_k\rangle \equiv G^k|\psi_0\rangle$, the angle θ_k of this vector relative to the horizontal axis will be $\theta_k = 2k\theta_0 + \theta_0 = (2k+1)\theta_0$, and therefore,

$$\theta_k = (2k+1)\theta_0 = (2k+1)\arcsin\sqrt{\frac{M}{N}}. \tag{23.3.28}$$

Our aim is to find a value of $k \in \mathbb{Z}^+$ for which θ_k is close to $\pi/2$ so that $|\psi_k\rangle$ is closely aligned with solution ket $|\beta\rangle$. Of course, a situation where θ_k is equal to $\pi/2$ plus a multiple of π also aligns the state properly, but then k is larger and our aim is to find the lowest positive integer k that works. After all, k represents the number of times we query the oracle. Suppose we find a value of k. The probability P_k of success in finding a solution is given by the square of the amplitude for $|\psi_k\rangle$ to be found along $|\beta\rangle$:

$$P_k = |\langle\beta|\psi_k\rangle|^2 = \cos^2(\theta_k - \tfrac{\pi}{2}) = \sin^2\theta_k. \tag{23.3.29}$$

This is a high probability when θ_k is close to $\pi/2$.

Exercise 23.3. *When we measure $|\psi_k\rangle$, we are doing so along the basis computational states $|x\rangle$. Start with the natural expression for the probability P_k of the state to be found in any of*

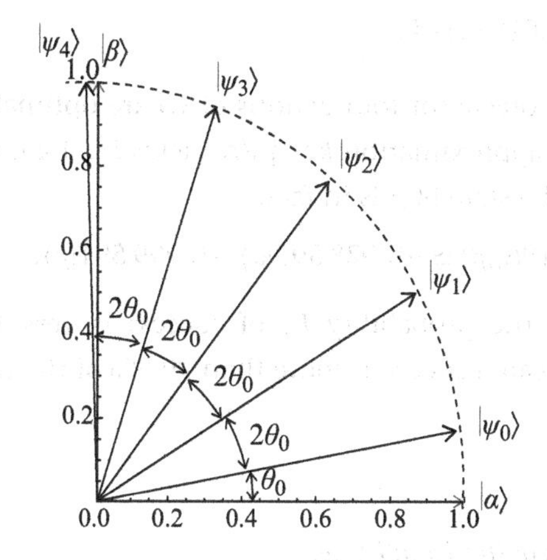

Figure 23.5
A Grover search over $N=32$ integers with one solution ($M=1$). Shown are the initial state $|\psi_0\rangle$ and the iterates $|\psi_1\rangle$, $|\psi_2\rangle$, $|\psi_3\rangle$, and $|\psi_4\rangle$, which exceeds the $\frac{\pi}{2}$ angle of the solution ket by a tiny amount. Four evaluations of the oracle will lead to the solution 99.9% of the time.

the solutions,

$$P_k = \sum_{\{x|f(x)=1\}} |\langle x|\psi_k\rangle|^2, \tag{23.3.30}$$

and show that indeed $P_k = \sin^2\theta_k$.

For $M \ll N$, we use the small angle approximation for arcsine and, taking $\theta_k \simeq \pi/2$, equation (23.3.28) gives

$$\frac{\pi}{2} \simeq (2k+1)\sqrt{\frac{M}{N}} \quad \Rightarrow \quad k \simeq \frac{\pi}{4}\sqrt{\frac{N}{M}}, \quad M \ll N. \tag{23.3.31}$$

This k is of order $\sqrt{N/M}$, which is much lower than the order N/M oracle queries that a classical computer would require. The approximations we made are reflected in the estimate for k not being an integer. For $M=1$, we find $k \simeq \frac{\pi}{4}\sqrt{N}$. In this case a search problem with $N = 2^{30} \simeq 10^9$ would typically require about half a billion queries in a classical computer. With a quantum computer, you would need about $\frac{\pi}{4} \cdot 2^{15} \simeq 26{,}000$ queries—much fewer.

Exercise 23.4. *Show that for* $N=4$ *and* $M=1$ *a single application of the Grover operator on* $|\psi_0\rangle$ *gives* exactly *the solution ket.*

Example 23.1. *Searching over* $N=32$ *elements for a single solution* ($M=1$).
When $N = 32 = 2^5$ and $M=1$, the angle θ_0 that $|\psi_0\rangle$ forms relative to the horizontal is

$$\theta_0 = \sin^{-1}\sqrt{\tfrac{M}{N}} = \sin^{-1}\tfrac{1}{\sqrt{32}} \quad \rightarrow \quad \theta_0 \simeq 0.17777 \simeq 10.18°. \tag{23.3.32}$$

This means that for $k=4$, leading to $\theta_4 = (2\cdot 4+1)\theta_0 = 9\theta_0$, we find $|\psi_4\rangle$ at an angle rather close to $\pi/2$ (figure 23.5):

$$\theta_4 = 9\theta_0 \simeq 91.64° \simeq (1.01821) \cdot \tfrac{\pi}{2}. \tag{23.3.33}$$

Four evaluations of the oracle (or four actions of G) are optimal to find the solution. In this case the large N approximation $k \simeq \frac{\pi}{4}\sqrt{N}$ yields $k \simeq 4.44$, indeed suggesting the value $k = 4$ we found. The state $|\psi_4\rangle$ is in fact,

$$|\psi_4\rangle = \cos 9\theta_0 |\alpha\rangle + \sin 9\theta_0 |\beta\rangle \simeq -0.028\,595|\alpha\rangle + 0.999\,591|\beta\rangle. \tag{23.3.34}$$

When measuring $|\psi_4\rangle$, the probability P_4 of success is very high: $P_4 = (\sin 9\theta_0)^2 \simeq 0.999\,182$. The Grover search succeeds more than 99.9% of the time. □

Problems

Problem 23.1. *Properties of the Toffoli gate.*

The classical Toffoli gate has a three-bit input and a three-bit output. With input bits (x, y, z), the first two are control bits, and the third is the target bit. The control bits emerge unchanged, while the target bit is changed by the (mod 2) addition of the product $x \cdot y$, as shown here:

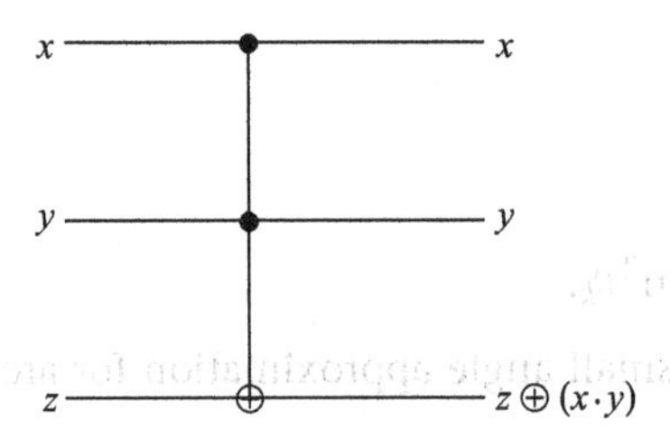

In the quantum Toffoli gate, the action on computational basis states is $|x\rangle|y\rangle|z\rangle \to |x\rangle|y\rangle|z \oplus (x \cdot y)\rangle$.

1. Show that the quantum Toffoli gate is its own inverse.
2. The NAND gate with input bits x, y outputs $\overline{x \cdot y} = 1 \oplus (x \cdot y)$. How do we choose the target bit z in the Toffoli gate so that the input (x, y, z) gives the output $(x, y, \overline{x \cdot y})$?
3. Using the three-qubit basis states $\{|0\rangle|0\rangle|0\rangle, \ldots, |1\rangle|1\rangle|1\rangle\}$, ordered by the binary value of the three-bit sequence, write the 8×8 matrix representative U_{Toff} of the gate. Describe its block structure. Show that U_{Toff} is unitary.
4. Let Z_1, Z_2, Z_3 be the $Z = \sigma_3$ operator acting on the first, second, and third qubit, respectively. Verify the following identities:

$$U_{\text{Toff}} Z_1 U_{\text{Toff}}^\dagger = Z_1,$$
$$U_{\text{Toff}} Z_2 U_{\text{Toff}}^\dagger = Z_2,$$
$$U_{\text{Toff}} Z_3 U_{\text{Toff}}^\dagger = \tfrac{1}{2}(\mathbb{1} + Z_1 + Z_2 - Z_1 Z_2) \otimes Z_3.$$

Problem 23.2. *Three-qubit flip correcting code.*

The Pauli X is the qubit flip operator since $X|0\rangle = |1\rangle$ and $X|1\rangle = |0\rangle$. A flip error changes a qubit $|\psi\rangle$ into $X|\psi\rangle$, and for an unknown $|\psi\rangle$, there is no way to tell if there was an error. To allow for error correction, we encode the qubit $|\psi\rangle$ as a state $|\Psi\rangle$ in a three-qubit state space $\widetilde{\mathcal{H}}$. To do this $|0\rangle$ is coded as $|0\rangle_c \equiv |000\rangle$, and $|1\rangle$ is coded as $|1\rangle_c \equiv |111\rangle$. By linearity, $|\psi\rangle = a|0\rangle + b|\psi\rangle$, with $a, b \in \mathbb{C}$, is coded as $|\Psi\rangle = a|0\rangle_c + b|1\rangle_c = a|000\rangle + b|111\rangle$. The space $\widetilde{\mathcal{H}}$ contains the *code subspace* C defined as the span of $|0\rangle_c$ and $|1\rangle_1$. Here C is a two-dimensional subspace of the eight-dimensional space $\widetilde{\mathcal{H}}$. After coding, any $|\psi\rangle$ becomes a $|\Psi\rangle \in C$.

1. Design a simple three-qubit circuit in which, with first input $|\psi\rangle$ the output is the coded $|\Psi\rangle$. For this, let both the second and third inputs be $|0\rangle$, and use two CNOT gates.
2. Let $P_0 : \widetilde{\mathcal{H}} \to C$ be a projection to the code subspace. Write P_0 in bra-ket notation. Let X_1, X_2, and X_3 denote flip operators acting, respectively, on the first, second, and third qubits in $\widetilde{\mathcal{H}}$ ($X_2 = \mathbb{1} \otimes X \otimes \mathbb{1}$, for example). Let $C_i \equiv X_i C$, with $i = 1, 2, 3$, denote the subspaces obtained by acting with X_i on C. Write out the projectors $P_i : \widetilde{\mathcal{H}} \to C_i$. Demonstrate that the four subspaces in the list C, C_1, C_2, C_3 are all mutually orthogonal, and moreover, $C \oplus C_1 \oplus C_2 \oplus C_3 = \widetilde{\mathcal{H}}$.
3. After coding $|\psi\rangle$ into $|\Psi\rangle \in C$, assume each of the three qubits is sent through an independent copy of a channel where a flip can happen. Carefully explain the following statement: If *at most* one qubit is flipped, we can tell if there is no error or, alternatively, which qubit has been changed by measuring the projectors P_0, P_1, P_2, P_3 on the state. If the ith qubit was changed, the state can then be corrected by an application of X_i.
4. Suppose that for each qubit in a flip noisy channel there is a probability p of a flip and a probability $1 - p$ of no flip. If no error-correcting code is used, there is a probability p of failure. As already shown, the three-qubit code allows the identification of errors when a single qubit flips and fails to identify them for flips of two or three qubits. Show that the probability of failure to correct is $p_{\text{fail}} = 3p^2 - 2p^3$. Show that $p_{\text{fail}} \le p$ for $p \le \frac{1}{2}$, thus demonstrating the advantage of coding for correcting flip errors for p small.
5. To improve the code for the correction of flip errors, consider a five-qubit code where $|0\rangle$ is coded as $|0\rangle_c = |00000\rangle$, and $|1\rangle$ is coded as $|1\rangle_c = |11111\rangle$, states in the thirty-two-dimensional state space of five qubits. The code subspace is, again, the span of $|0\rangle_c$ and $|1\rangle_c$. Explain how this code allows for the identification, and thus correction of flip errors when at most two qubits are flipped. Do this by identifying a collection of sixteen mutually orthogonal subspaces of the five-qubit space whose direct sum builds the five-qubit space. Determine the failure probability p_{fail} in terms of the probability p of flip for each qubit, and show that $p_{\text{fail}} \le p$ for $p \le \frac{1}{2}$. Confirm that for $p = 0.01$ the five-qubit code reduces p_{fail} relative to the three-qubit code by a factor of about thirty!

Problem 23.3. *Fourier transform as a unitary transformation.*

1. In the discrete case, the Fourier transform acts on a sequence $x_0, \ldots, x_{N-1}$ of N numbers and gives another sequence $y_0, \ldots, y_{N-1}$ of N numbers,

$$y_j \equiv \frac{1}{\sqrt{N}} \sum_{k=0}^{N-1} e^{\frac{2\pi i}{N} jk} x_k. \tag{1}$$

To describe this action as a unitary transformation U_F, we introduce orthonormal basis states $|j\rangle$, with $j = 0, \ldots, N-1$, and define the operator by its action on the basis vectors:

$$U_F|j\rangle = \frac{1}{\sqrt{N}} \sum_{k=0}^{N-1} e^{\frac{2\pi i}{N} jk} |k\rangle. \tag{2}$$

We now claim that this definition implements the Fourier transform as follows:

$$U_F \sum_{j=0}^{N-1} x_j |j\rangle = \sum_{j=0}^{N-1} y_j |j\rangle. \tag{3}$$

Prove that, indeed, (1) follows from (2) and (3). Demonstrate that U_F is unitary.

2. In the continuous case, rather than working with dimensional x and p, it is simplest to work with unit-free parameters $u, v, t, \cdots \in \mathbb{R}$. Two functions Ψ and Φ of a real variable are said to be related by Fourier transformation if

$$\Phi(v) = \frac{1}{\sqrt{2\pi}} \int_{-\infty}^{\infty} du\, e^{iuv} \Psi(u). \tag{4}$$

To describe a unitary action, we introduce a continuous basis of states $|t\rangle$, with $t \in \mathbb{R}$ and $\langle t'|t\rangle = \delta(t - t')$, and define

$$U_F|t\rangle \equiv \frac{1}{\sqrt{2\pi}} \int_{-\infty}^{\infty} ds\, e^{ist} |s\rangle. \tag{5}$$

We now claim that this definition implements the Fourier transform as follows:

$$U_F \int_{-\infty}^{\infty} dt\, \Psi(t)|t\rangle = \int_{-\infty}^{\infty} dt\, \Phi(t)|t\rangle. \tag{6}$$

Prove that, indeed, (4) follows from (5) and (6). Demonstrate that U_F is unitary. Note that for a continuous basis and matrix elements $U_{st} \equiv \langle s|U|t\rangle$ the familiar $U^\dagger U = \mathbb{1}$ reads

$$\int_{-\infty}^{\infty} dt\, U_{st}^\dagger U_{ts'} = \delta(s - s').$$

Problem 23.4. *Controlled gates for Fourier transform.*

1. Show that the Hadamard gate H acts on a qubit state $|j_1\rangle$ ($j_1 \in \{0, 1\}$) as follows:

$$H|j_1\rangle = \tfrac{1}{\sqrt{2}}\left(|0\rangle + e^{2\pi i[j_1]}|1\rangle\right), \quad [j_1] \equiv \tfrac{j_1}{2}. \tag{1}$$

2. The single-cubit gate S is defined by the matrix

$$S = \begin{pmatrix} 1 & 0 \\ 0 & i \end{pmatrix}. \tag{2}$$

The S gate is the square root of the Z gate. A controlled-S gate is a two-qubit gate. It acts as $|j_1\rangle|j_2\rangle \to |j_1\rangle S^{j_1}|j_2\rangle$, where S^{j_1} literally means S to the power j_1. The first qubit is the control qubit (denoted by a heavy dot), and the second qubit is the target qubit:

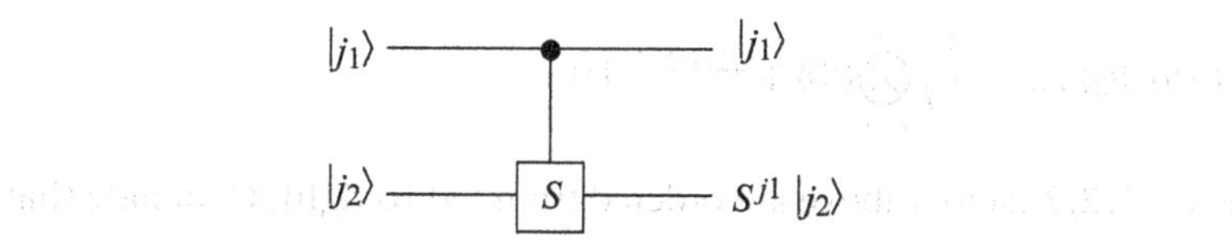

If the control qubit is $|0\rangle$, the target qubit is left unchanged; if the control qubit is $|1\rangle$, the target qubit is acted by S. We write $C_{c,t}S$ for the controlled-S gate, with c specifying the control bit and t specifying the target bit. The above diagram shows the $C_{1,2}S$ gate. We also let H_i denote Hadamard action on the ith qubit. Show that

$$C_{2,1}S \cdot H_1|j_1\rangle|j_2\rangle = \tfrac{1}{\sqrt{2}}\big(|0\rangle + e^{2\pi i[j_1 j_2]}|1\rangle\big)|j_2\rangle, \quad [j_1 j_2] \equiv 2^{-1}j_1 + 2^{-2}j_2. \tag{3}$$

Note that the action on the qubits is represented by the circuit:

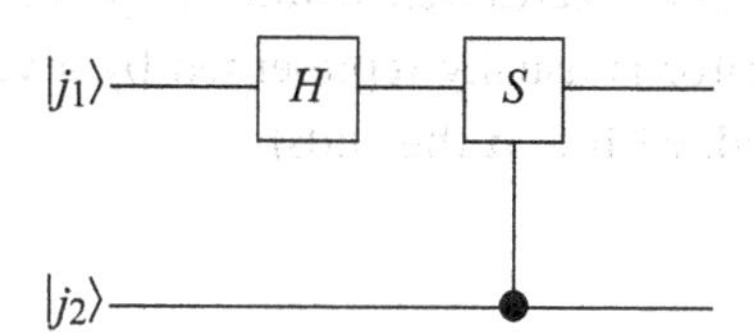

3. The single-qubit gate T is defined by the matrix

$$T = \begin{pmatrix} 1 & 0 \\ 0 & e^{i\pi/4} \end{pmatrix}. \tag{4}$$

One can view the T gate as the square root of the S gate. Show that on a three-qubit state,

$$C_{3,1}T \cdot C_{2,1}S \cdot H_1\,|j_1\rangle|j_2\rangle|j_3\rangle = \tfrac{1}{\sqrt{2}}\big(|0\rangle + e^{2\pi i[j_1 j_2 j_3]}|1\rangle\big)|j_2\rangle|j_3\rangle, \tag{5}$$

with $[j_1 j_2 j_3] \equiv 2^{-1}j_1 + 2^{-2}j_2 + 2^{-3}j_3$. Draw the gate performing the above operations.

Problem 23.5. *Fourier transformations and a circuit realization.*

The two previous problems have laid the groundwork for the explicit construction of a multiqubit gate that performs Fourier transformation. The essence of the procedure is captured by the construction on a three-qubit space—generalization is straightforward. We will do the analysis in two steps.

1. Consider the action of U_F on $|j\rangle$ as defined in equation (2) of problem 23.3, adapted for binary by choosing $N=2^n$:

$$U_F|j\rangle = \frac{1}{2^{n/2}} \sum_{k=0}^{2^n-1} e^{2\pi i 2^{-n}kj}|k\rangle. \tag{1}$$

Specialize to $n=3$ and use $|j\rangle = |j_1\rangle|j_2\rangle|j_3\rangle$ and $|k\rangle = |k_1\rangle|k_2\rangle|k_3\rangle$, with $j=2^2j_1+2j_2+j_3$, $k=2^2k_1+2k_2+k_3$, with $j_1, j_2, j_3, k_1, k_2, k_3 \in \{0,1\}$. Perform the sums over k_1, k_2, and k_3 in equation (1) to show that

$$U_F|j_1\rangle|j_2\rangle|j_3\rangle = \frac{1}{\sqrt{8}} \bigotimes_{\ell=1}^{3} (|0\rangle + e^{2\pi i j 2^{-\ell}}|1\rangle). \tag{2}$$

The $\ell = 1, 2, 3$ factors above are ordered from left to right. Conclude that

$$U_F|j_1\rangle|j_2\rangle|j_3\rangle = \frac{1}{\sqrt{8}}\Big(|0\rangle + e^{2\pi i [j_3]}|1\rangle\Big)\Big(|0\rangle + e^{2\pi i [j_2 j_3]}|1\rangle\Big)\Big(|0\rangle + e^{2\pi i [j_1 j_2 j_3]}|1\rangle\Big). \tag{3}$$

Consistent with definitions in problem 23.4, $[j_1 \ldots j_p] \equiv 2^{-1}j_1 + 2^{-2}j_2 + \cdots + 2^{-p}j_p$.

2. Let $\tilde{S}_{p,q}$ denote the swap operator that exchanges the states in the pth and qth qubits. Show that the state in (3) is

$$\tilde{S}_{1,3} \cdot H_3 \cdot C_{3,2}S \cdot H_2 \cdot C_{3,1}T \cdot C_{2,1}S \cdot H_1\, |j_1\rangle|j_2\rangle|j_3\rangle, \tag{4}$$

where the controlled-S and T gates were defined in problem 23.4. Draw the quantum circuit (the swap operator is usually represented by a vertical line joining the two qubits being exchanged, with $\times$ at the ends).

24 Charged Particles in Electromagnetic Fields

We discuss here charged quantum particles interacting with electromagnetic fields. The fields enter the Schrödinger equation via their associated vector potential $\mathbf{A}$ *and scalar potential* Φ*. Any two sets of potentials related by gauge transformations are equivalent descriptions of an electromagnetic field. The wave function of a charged particle also transforms under gauge transformations, making it possible for the Schrödinger equation to describe the same physics using gauge-related potentials. We find that a uniform magnetic field on a torus must be quantized with the total magnetic flux of an integer multiple of a flux quantum. For a particle moving on a plane under the influence of a constant transverse magnetic field, we find equally spaced Landau levels, each infinitely degenerate. When the particle is confined to a finite area sample, each Landau level has finite degeneracy. We explain how the nonrelativistic Pauli equation for the electron predicts the coupling of the electron spin to external magnetic fields and briefly discuss the relativistic Dirac equation.*

24.1 Electromagnetic Potentials

We have studied in detail how the Schrödinger equation governs the quantum dynamics of a particle moving in some potential $V(\mathbf{x})$. We have not yet studied, however, how arbitrary electromagnetic fields affect the quantum dynamics of charged particles. One could imagine that in some way the electric and magnetic fields, $\mathbf{E}$ and $\mathbf{B}$, would enter the Schrödinger equation. This does not happen; it is the potentials $\mathbf{A}$ and Φ, representing $\mathbf{E}$ and $\mathbf{B}$, that appear in the Schrödinger equation. Because quantum mechanics requires potentials, physicists now believe that the potentials $(\mathbf{A}, \Phi)$ are more fundamental than the fields $(\mathbf{E}, \mathbf{B})$. The original formulation of classical electrodynamics was given in terms of $\mathbf{E}$ and $\mathbf{B}$ fields that obey Maxwell's equations, and the potentials were viewed as mathematical devices introduced for solving these equations. This is no longer the case.

Let us recall how potentials were introduced. From the Maxwell equation

$$\nabla \cdot \mathbf{B} = 0, \tag{24.1.1}$$

one concludes that at least *locally* $\mathbf{B}$ is the curl of a vector field $\mathbf{A}$ called the vector potential:

$$\mathbf{B} = \nabla \times \mathbf{A}. \tag{24.1.2}$$

When we say *locally*, we mean that to represent the magnetic field we may need to use different vector potentials over different regions of space, in a way that will be made clear later. The other source-free Maxwell equation is

$$\nabla \times \mathbf{E} = -\frac{1}{c}\frac{\partial \mathbf{B}}{\partial t}. \tag{24.1.3}$$

Using the representation of $\mathbf{B}$ in terms of $\mathbf{A}$ and commuting spatial and time derivatives, this gives

$$\nabla \times \left(\mathbf{E} + \frac{1}{c}\frac{\partial \mathbf{A}}{\partial t}\right) = 0. \tag{24.1.4}$$

Since a curl-free field is *locally* the gradient of a scalar, we introduce the scalar potential Φ and write

$$\mathbf{E} + \frac{1}{c}\frac{\partial \mathbf{A}}{\partial t} = -\nabla \Phi. \tag{24.1.5}$$

It is therefore possible to write $\mathbf{E}$ in terms of the potentials:

$$\mathbf{E} = -\nabla \Phi - \frac{1}{c}\frac{\partial \mathbf{A}}{\partial t}. \tag{24.1.6}$$

Potentials are not uniquely defined in the sense that they can be altered without changing the resulting $\mathbf{E}$ and $\mathbf{B}$. Suppose we have a scalar function $\Lambda(\mathbf{x}, t)$. You can quickly check that new primed potentials $(\mathbf{A}', \Phi')$ defined from $(\mathbf{A}, \Phi)$ by

$$\begin{aligned} \mathbf{A}' &= \mathbf{A} + \nabla \Lambda, \\ \Phi' &= \Phi - \frac{1}{c}\frac{\partial \Lambda}{\partial t} \end{aligned} \tag{24.1.7}$$

give rise to the same electric and magnetic fields:

$$\mathbf{E}(\mathbf{A}', \Phi') = \mathbf{E}(\mathbf{A}, \Phi), \quad \mathbf{B}(\mathbf{A}', \Phi') = \mathbf{B}(\mathbf{A}, \Phi). \tag{24.1.8}$$

We say that $(\mathbf{A}', \Phi')$ are obtained from $(\mathbf{A}, \Phi)$ by *gauge transformations* with *gauge parameter* Λ. Since quantum mechanics tells us that the potentials are the fundamental quantities, we are faced with defining precisely what electromagnetic fields are. The right definition makes use of gauge transformations:

> *An electromagnetic field configuration is represented by potentials* $(\mathbf{A}, \Phi)$. *Any other set of potentials* $(\mathbf{A}', \Phi')$ *differing from* $(\mathbf{A}, \Phi)$ *by a gauge transformation must be considered an equivalent representation of the electromagnetic field.*

In other words, an electromagnetic field configuration can be imagined as an *equivalence class* of potentials, all of which differ from each other by gauge transformations and with any element of the class a valid representative. This definition implies some surprising possibilities that, in fact, are often relevant in practice.

- Consider two sets of potentials $(\mathbf{A}, \Phi)$ and $(\mathbf{A}', \Phi')$ that give rise to identical fields $(\mathbf{E}, \mathbf{B})$. Suppose is it impossible to find a gauge parameter Λ such that the two sets of potentials are related by a gauge transformation. In this case, and even though they give rise to identical $\mathbf{E}, \mathbf{B}$ fields, the two sets of potentials must be thought of as *inequivalent*. Quantum effects could distinguish them. The Bohm-Aharonov effect is an example where this subtlety plays a role (problem 24.8).
- Consider the fields $(\mathbf{E}, \mathbf{B})$ that satisfy Maxwell's equations. Suppose, however, that there are no potentials $(\mathbf{A}, \Phi)$ from which $(\mathbf{E}, \mathbf{B})$ arise as usual. We must then conclude that $(\mathbf{E}, \mathbf{B})$, despite satisfying Maxwell's equations, are not valid fields. We could not do quantum mechanics with them. We will see an example of this in section 24.4.

These subtleties happen when we have particles that move in spaces with nontrivial topology. This is not hard to arrange. We can constrain particles to move, say, on a circle or on a two-dimensional surface that forms a torus. We will see examples later in this chapter.

24.2 Schrödinger Equation with Electromagnetic Potentials

To find how the Schrödinger equation is modified when a particle has charge q and is coupled to electromagnetic fields, we must determine how the Hamiltonian of the particle should be modified. A bit of the answer is known to you. After all, when discussing the hydrogen atom you included the effect of the electric field created by the proton on the electron. This was done by adding to the Hamiltonian a term $q\Phi(\hat{\mathbf{x}}, t)$, or $q\Phi(\mathbf{x}, t)$ in coordinate space, representing the electrostatic energy of the charge q located at $\mathbf{x}$ when the electric potential at $\mathbf{x}$ is $\Phi(\mathbf{x}, t)$. The way the vector potential enters is less obvious at first sight. One must replace the canonical momentum operator $\hat{\mathbf{p}}$ as follows:

$$\hat{\mathbf{p}} \to \hat{\mathbf{p}} - \frac{q}{c}\mathbf{A}(\hat{\mathbf{x}}, t). \tag{24.2.1}$$

This replacement is sometimes called a **minimal coupling** of the particle to the electromagnetic field. In addition to this minimal coupling, there can be nonminimal couplings, but we will not consider them here as they are not generic and often play no role. Here, $\mathbf{A}(\hat{\mathbf{x}}, t)$ is an operator because it is a function of the position operator. Still, we will not use a hat for $\mathbf{A}$. Despite the above replacement, we do not change the basic commutation relations of position and momenta. The momentum operator $\hat{\mathbf{p}}$ still satisfies

$$[\hat{x}_i, \hat{p}_j] = i\hbar\delta_{ij}, \quad \text{and} \quad \hat{\mathbf{p}} = \frac{\hbar}{i}\nabla \text{ in coordinate space.} \tag{24.2.2}$$

With the above change, the Hamiltonian $\hat{H}$ describing the minimal coupling is therefore

$$\hat{H} = \frac{1}{2m}\Big(\hat{\mathbf{p}} - \frac{q}{c}\mathbf{A}(\hat{\mathbf{x}}, t)\Big)^2 + q\,\Phi(\hat{\mathbf{x}}, t). \tag{24.2.3}$$

In coordinate space we write this Hamiltonian as follows:

$$\hat{H} = \frac{1}{2m}\Big(\frac{\hbar}{i}\nabla - \frac{q}{c}\mathbf{A}(\mathbf{x}, t)\Big)^2 + q\,\Phi(\mathbf{x}, t). \tag{24.2.4}$$

As a result, the Schrödinger equation for a wave function $\Psi(\mathbf{x}, t)$ will read

$$i\hbar\frac{\partial\Psi}{\partial t} = \Big[\frac{1}{2m}\Big(\frac{\hbar}{i}\nabla - \frac{q}{c}\mathbf{A}(\mathbf{x}, t)\Big)^2 + q\,\Phi(\mathbf{x}, t)\Big]\Psi. \tag{24.2.5}$$

The position dependence of the potentials implies that the momentum operator $\hat{\mathbf{p}}$ fails to commute with the potentials. The potentials, however, commute with each other because they have position but no momentum dependence. Getting ahead of ourselves a little bit, we note that in the presence of a vector potential the momentum operator $\hat{\mathbf{p}}$ cannot be identified with $m\hat{\mathbf{v}}$, mass times a "velocity" operator. The first term in the Hamiltonian will still be the kinetic energy, and therefore we will end up identifying $\hat{\mathbf{p}} - \frac{q}{c}\mathbf{A}$ with $m\hat{\mathbf{v}}$:

$$\hat{\mathbf{p}} - \frac{q}{c}\mathbf{A} \equiv m\hat{\mathbf{v}}. \tag{24.2.6}$$

This will be quite clear in the Heisenberg picture. It is reassuring to expand the square in the Hamiltonian to see what is in there. We get:

$$\hat{H} = \frac{\hat{\mathbf{p}}^2}{2m} - \frac{q}{2mc}(\hat{\mathbf{p}}\cdot\mathbf{A} + \mathbf{A}\cdot\hat{\mathbf{p}}) + \frac{q^2}{2mc^2}\mathbf{A}\cdot\mathbf{A} + e\Phi. \tag{24.2.7}$$

Noting that $\hat{\mathbf{p}}$ is a differential operator that in $\hat{\mathbf{p}}\cdot\mathbf{A}$ acts on everything to its right, we find that

$$\hat{\mathbf{p}}\cdot\mathbf{A} = \frac{\hbar}{i}\nabla\cdot\mathbf{A} + \mathbf{A}\cdot\hat{\mathbf{p}}. \tag{24.2.8}$$

Therefore,

$$\hat{H} = \frac{\hat{\mathbf{p}}^2}{2m} - \frac{q}{mc}\mathbf{A}\cdot\hat{\mathbf{p}} + \frac{iq\hbar}{2mc}\nabla\cdot\mathbf{A} + \frac{q^2}{2mc^2}\mathbf{A}\cdot\mathbf{A} + e\Phi. \tag{24.2.9}$$

When possible, the gauge choice $\nabla\cdot\mathbf{A} = 0$ simplifies the Hamiltonian. This gauge choice is called the Coulomb gauge, or radiation gauge.

Given that the Hamiltonian features the potentials, we are forced to address the issue of gauge invariance of the physics. When given the potentials $(\mathbf{A}, \Phi)$, we can use them in the Schrödinger equation, or we can use any other gauge equivalent set $(\mathbf{A}', \Phi')$. We must make sure that the choice of potentials does not change the physics. One could naively ask whether a wave function Ψ that solves the Schrödinger equation with potentials $(\mathbf{A}, \Phi)$ should also be a solution for any other gauge-related potentials $(\mathbf{A}', \Phi')$. But this is too much to ask and does not happen. *We must allow the wave function to change under gauge transformations.* Showing that the physics does not change will take some effort.

Here is the way we state the gauge invariance. If we have a solution Ψ of the Schrödinger equation with potentials $(\mathbf{A}, \Phi)$,

$$i\hbar\frac{\partial\Psi}{\partial t} = \Big[\frac{1}{2m}\Big(\frac{\hbar}{i}\nabla - \frac{q}{c}\mathbf{A}\Big)^2 + q\Phi\Big]\Psi, \tag{24.2.10}$$

then this implies that the following equation also holds:

$$i\hbar\frac{\partial\Psi'}{\partial t} = \Big[\frac{1}{2m}\Big(\frac{\hbar}{i}\nabla - \frac{q}{c}\mathbf{A}'\Big)^2 + q\Phi'\Big]\Psi', \tag{24.2.11}$$

where

$$\mathbf{A}' = \mathbf{A} + \nabla\Lambda, \quad \Phi' = \Phi - \frac{1}{c}\frac{\partial\Lambda}{\partial t}, \quad \text{and} \quad \Psi' = \exp\Big(i\frac{q\Lambda}{\hbar c}\Big)\Psi. \tag{24.2.12}$$

This states that the gauge-transformed wave function Ψ' solves the Schrödinger equation with the gauge-transformed potentials. The gauge transformation multiplies Ψ by a phase factor that depends on the gauge parameter $\Lambda(\mathbf{x}, t)$. More explicitly, we have

$$\Psi'(\mathbf{x}, t) = \exp\Big(i\frac{q\Lambda(\mathbf{x}, t)}{\hbar c}\Big)\Psi(\mathbf{x}, t). \tag{24.2.13}$$

This transformation is a reasonable choice because it does not change the normalization of the wave function. Note also that the charge q of the particle appears in the gauge transformation of the wave function. This phase factor occurs often enough that we give it a name, $U(\Lambda)$, defined by

$$U(\Lambda) \equiv \exp\Big(i\frac{q\Lambda}{\hbar c}\Big), \quad \Psi' = U(\Lambda)\Psi. \tag{24.2.14}$$

Since the wave function is only changed by a phase, the probability density $\rho(\mathbf{x}, t) = \Psi^*(\mathbf{x}, t)\Psi(\mathbf{x}, t)$ does not change under a gauge transformation. This is a consistency check; we certainly expect that the probability of finding a particle at some point and given time does not depend on our gauge choice.

Verifying gauge invariance means showing that equations (24.2.10) and (24.2.11) imply each other when the potentials and the wave functions are related by (24.2.12). This is not hard to do once we establish the following useful identity:

$$\Big(\frac{\hbar}{i}\nabla - \frac{q}{c}\mathbf{A}'\Big)U(\Lambda) = U(\Lambda)\Big(\frac{\hbar}{i}\nabla - \frac{q}{c}\mathbf{A}\Big). \tag{24.2.15}$$

This says that as we move the phase factor U across the factor $\hat{\mathbf{p}} - \frac{q}{c}\mathbf{A}'$, the net effect is to turn $\mathbf{A}'$ into $\mathbf{A}$. To check this we simply expand the left-hand side acting on an arbitrary wave function Ψ:

$$\Big(\frac{\hbar}{i}\nabla - \frac{q}{c}\mathbf{A}'\Big)U(\Lambda)\Psi = \frac{\hbar}{i}(\nabla U)\Psi + U\frac{\hbar}{i}\nabla\Psi - U\frac{q}{c}(\mathbf{A} + \nabla\Lambda)\Psi, \tag{24.2.16}$$

where we used the expression for $\mathbf{A}'$ in terms of $\mathbf{A}$ and Λ and noticed that U commutes with the potentials and the gauge parameter. Note now that

$$\frac{\hbar}{i}\nabla U(\Lambda) = \frac{q}{c}(\nabla\Lambda)\,U(\Lambda), \tag{24.2.17}$$

and therefore back above, we have

$$\Big(\frac{\hbar}{i}\nabla - \frac{q}{c}\mathbf{A}'\Big)U(\Lambda)\Psi = \frac{q}{c}(\nabla\Lambda)\,U(\Lambda)\,\Psi + U(\Lambda)\,\Big(\frac{\hbar}{i}\nabla - \frac{q}{c}\mathbf{A}\Big)\Psi$$
$$- \frac{q}{c}\,U(\Lambda)(\nabla\Lambda)\Psi. \tag{24.2.18}$$

The first and last terms on the right-hand side cancel, and we are left with the right-hand side of (24.2.15) acting on Ψ, thus proving the equation.

Let us now use this identity to show that equation (24.2.11) implies equation (24.2.10). We begin with equation (24.2.11) written as

$$i\hbar\frac{\partial(U\Psi)}{\partial t} = \Big[\frac{1}{2m}\Big(\frac{\hbar}{i}\nabla - \frac{q}{c}\mathbf{A}'\Big)^2 + q\Phi'\Big]U\Psi. \tag{24.2.19}$$

Noting that $i\hbar\partial_t U = -U\frac{q}{c}\partial_t\Lambda$, recalling the gauge transformation of Φ, and using the identity (24.2.15) twice to move U to the left for the first term on the right-hand side, we get

$$-U\frac{q}{c}(\partial_t\Lambda)\Psi + U\,i\hbar\frac{\partial\Psi}{\partial t} = U\Big[\frac{1}{2m}\Big(\frac{\hbar}{i}\nabla - \frac{q}{c}\mathbf{A}\Big)^2 + q\Phi - \frac{q}{c}\partial_t\Lambda\Big]\Psi. \tag{24.2.20}$$

The leftmost and rightmost terms cancel, and the result is the desired Schrödinger equation (24.2.10) multiplied by U from the left. Since U is invertible, this factor can be removed, and we have shown that the equation for Ψ' implies the equation for Ψ. This confirms the statement of gauge invariance.

We now consider the question of observables in this theory. As usual, observables must be Hermitian operators, but this time we want to guarantee that the result of measurements does not depend on the choice of gauge. Not all Hermitian operators will do this, but we will consider a class of operators, called *gauge-covariant operators*, that yield gauge-invariant measurements.

Under a gauge transformation with parameter $\Lambda(\mathbf{x}, t)$, an operator $\mathcal{O}[\Phi, \mathbf{A}]$, written as $\mathcal{O}$ for brevity, goes into a new operator $\mathcal{O}'$ given by $\mathcal{O}[\Phi', \mathbf{A}']$:

$$\text{Gauge transformation:}\;\; \mathcal{O}[\Phi, \mathbf{A}] \;\rightarrow\; \mathcal{O}' = \mathcal{O}[\Phi', \mathbf{A}']. \tag{24.2.21}$$

The prescription is simple: if the operator has $\mathbf{A}$ or Φ dependence, the gauge-transformed operator is the same operator with $\mathbf{A}$ replaced by the gauge-transformed $\mathbf{A}'$ and with Φ replaced by the gauge-transformed Φ'. Operators like $\hat{\mathbf{x}}$ and $\hat{\mathbf{p}}$ that do not depend on the potentials do not change under a gauge transformation.

Physical observables are Hermitian operators $\mathcal{O}$ that are **gauge covariant**—namely, they satisfy the condition

$$\mathcal{O}' = U(\Lambda)\,\mathcal{O}\,U^{-1}(\Lambda), \tag{24.2.22}$$

or more explicitly,

$$\mathcal{O}[\Phi', \mathbf{A}'] = U(\Lambda)\,\mathcal{O}[\Phi, \mathbf{A}]\,U^{-1}(\Lambda). \tag{24.2.23}$$

For a gauge-covariant operator the effect of a gauge transformation is reproduced by left multiplication of the original operator by U and right multiplication by the inverse U^{-1}.

Note that gauge-covariant observables form a real vector space: the product of a gauge-covariant observable and a real number is another gauge-covariant observable (complex numbers would ruin the Hermiticity). Similarly, the sum of gauge-covariant observables is a gauge-covariant observable. In addition, the product of gauge-covariant observables is gauge covariant, as you can quickly confirm.

We now show that gauge-covariant observables give gauge-*invariant* measurements. Suppose $|\Psi\rangle$ is an eigenstate of a gauge-covariant $\mathcal{O}$ with eigenvalue $\lambda_{\mathcal{O}}$:

$$\mathcal{O}|\Psi\rangle = \lambda_{\mathcal{O}}|\Psi\rangle. \tag{24.2.24}$$

We then find that

$$\mathcal{O}'|\Psi'\rangle = U\mathcal{O}U^{-1}U|\Psi\rangle = U\mathcal{O}|\Psi\rangle = U\lambda_{\mathcal{O}}|\Psi\rangle = \lambda_{\mathcal{O}}|\Psi'\rangle, \tag{24.2.25}$$

showing that measurement with the gauge-transformed operator on the gauge-transformed state gives the same result. Similarly, we quickly check that the expectation value of a gauge-covariant observable $\mathcal{O}$ is gauge invariant:

$$\langle\Psi'|\mathcal{O}'|\Psi'\rangle = \langle\Psi|U^{-1}(U\mathcal{O}U^{-1})U|\Psi\rangle = \langle\Psi|\mathcal{O}|\Psi\rangle. \tag{24.2.26}$$

You will consider in problem 24.1 some simple questions about gauge-covariant operators. The position operator $\hat{\mathbf{x}}$, for example, is gauge covariant, but the momentum operator $\hat{\mathbf{p}}$ is not. If you look at equation (24.2.15) closely, you will see that it states that $\hat{\mathbf{p}} - \frac{q}{c}\mathbf{A}$ is gauge covariant. In problem 24.7 you will consider the definition of the probability current $\mathbf{J}$ suitable for this theory.

24.3 Heisenberg Picture

This picture is based on folding the time dependence of states into the operators. Given a Schrödinger operator $\hat{A}_S$, the associated Heisenberg operator $\hat{A}_H(t)$ is given by

$$\hat{A}_H(t) = \mathcal{U}^\dagger(t)\hat{A}_S\mathcal{U}(t), \tag{24.3.1}$$

where $\mathcal{U}(t)$ is the unitary time evolution operator that evolves states as $|\Psi(t)\rangle = \mathcal{U}(t)|\Psi(0)\rangle$. Heisenberg operators satisfy the Heisenberg equations of motion (16.5.7):

$$\frac{d\hat{A}_H(t)}{dt} = \frac{i}{\hbar}[\hat{H}_H(t), \hat{A}_H(t)] + \frac{\partial\hat{A}_H(t)}{\partial t}. \tag{24.3.2}$$

These are often quantum analogues of classical equations of motion. The last term above is needed when the Schrödinger operator $\hat{A}_S$ itself has some explicit time dependence. Another important point is that given a Schrödinger Hamiltonian

$$\hat{H}_S = \hat{H}\big(\hat{\mathbf{x}}, \hat{\mathbf{p}}, \mathbf{A}(\hat{\mathbf{x}}, t), \Phi(\hat{\mathbf{x}}, t)\big), \tag{24.3.3}$$

the Heisenberg version is the same function above but with all Schrödinger operators replaced by their Heisenberg versions:

$$\hat{H}_H = \hat{H}\big(\hat{\mathbf{x}}_H(t), \hat{\mathbf{p}}_H(t), \mathbf{A}(\hat{\mathbf{x}}_H(t), t), \Phi(\hat{\mathbf{x}}_H(t), t)\big). \tag{24.3.4}$$

We stated earlier that $\hat{\mathbf{p}} - \frac{q}{c}\mathbf{A} = m\hat{\mathbf{v}}$, where $\hat{\mathbf{v}}$ is a velocity operator. At the time, this was only a suggestive *definition* of the velocity operator. We will now see that in the Heisenberg picture, where we can define a velocity operator as the rate of change of the

Heisenberg position operator, this is a nontrivial equation. We therefore consider the Heisenberg equation of motion for $\hat{\mathbf{x}}_H$:

$$\frac{d\hat{x}_{H,i}}{dt} = \frac{i}{\hbar}\Big[\frac{1}{2m}(\hat{\mathbf{p}}_H - \frac{q}{c}\mathbf{A}_H)\cdot(\hat{\mathbf{p}}_H - \frac{q}{c}\mathbf{A}_H) + q\Phi_H, \hat{x}_{H,i}\Big], \tag{24.3.5}$$

where $\mathbf{A}_H = \mathbf{A}(\hat{\mathbf{x}}_H, t)$. The term with the scalar potential does not contribute, and the commutator of the kinetic term with $\hat{x}_H$ is evaluated using the derivation property $[AA, B] = A[A, B] + [A, B]A$, with each term on the right-hand side giving the same contribution because, in our case, the $[A, B]$ commutator is a number. We then get

$$\begin{aligned}
\frac{d\hat{x}_{H,i}}{dt} &= \frac{i}{\hbar m}\sum_j(\hat{\mathbf{p}}_H - \frac{q}{c}\mathbf{A}_H)_j\Big[(\hat{\mathbf{p}}_H - \frac{q}{c}\mathbf{A}_H)_j, \hat{x}_{H,i}\Big] \\
&= \frac{i}{\hbar m}\sum_j(\hat{\mathbf{p}}_H - \frac{q}{c}\mathbf{A}_H)_j[\hat{p}_{H,j}, \hat{x}_{H,i}] \\
&= \frac{i}{\hbar m}\sum_j(\hat{\mathbf{p}}_H - \frac{q}{c}\mathbf{A}_H)_j(-i\hbar\delta_{ij}) = \frac{1}{m}(\hat{\mathbf{p}}_H - \frac{q}{c}\mathbf{A}_H)_i.
\end{aligned} \tag{24.3.6}$$

We have therefore shown that

$$\frac{d\,\hat{\mathbf{x}}_H}{dt} = \frac{1}{m}(\hat{\mathbf{p}}_H - \frac{q}{c}\mathbf{A}_H). \tag{24.3.7}$$

The Heisenberg velocity operator $\hat{\mathbf{v}}_H$ is naturally taken to be

$$\hat{\mathbf{v}}_H = \frac{d\,\hat{\mathbf{x}}_H}{dt}. \tag{24.3.8}$$

Therefore, we have the relation

$$\boxed{m\hat{\mathbf{v}}_H = \hat{\mathbf{p}}_H - \frac{q}{c}\mathbf{A}_H,} \tag{24.3.9}$$

confirming the identification of $m\hat{\mathbf{v}}$ with $\hat{\mathbf{p}} - \frac{q}{c}\mathbf{A}$.

The classical picture suggests we should see the emergence of a quantum version of the Lorentz force equation. To find this quantum law, let us consider the time derivative of the velocity operator, multiplied by the mass:

$$m\frac{d\hat{v}_{H,i}}{dt} = \frac{d}{dt}(\hat{\mathbf{p}}_H - \frac{q}{c}\mathbf{A}_H)_i. \tag{24.3.10}$$

The operator $\hat{\mathbf{p}}$ is a time-independent Schrödinger operator, but $\mathbf{A}(\hat{\mathbf{x}}, t)$ is a time-dependent Schrödinger operator, so this time the Heisenberg equation of motion gives

$$m\frac{d\hat{v}_{H,i}}{dt} = \frac{i}{\hbar}\Big[\hat{H}_H, (\hat{\mathbf{p}}_H - \frac{q}{c}\mathbf{A}_H)_i\Big] - \frac{q}{c}\frac{\partial A_{H,i}}{\partial t}. \tag{24.3.11}$$

Here $\mathbf{A}_H = \mathbf{A}(\hat{\mathbf{x}}_H(t), t)$, and we emphasize that the time partial derivative in the last term does not act on the Heisenberg operator $\hat{\mathbf{x}}_H(t)$ (see the discussion below equation (16.5.5)). Additionally, writing the Hamiltonian in terms of the velocity operator, we find

$$m\frac{d\hat{v}_{H,i}}{dt} = \frac{i}{\hbar}\Big[\tfrac{1}{2}m\hat{\mathbf{v}}_H^2 + q\Phi_H,\ \big(\hat{\mathbf{p}}_H - \frac{q}{c}\mathbf{A}_H\big)_i\Big] - \frac{q}{c}\frac{\partial A_{H,i}}{\partial t}. \tag{24.3.12}$$

The commutator involving the scalar potential is readily computed:

$$\Big[q\Phi_H,\ \big(\hat{\mathbf{p}}_H - \frac{q}{c}\mathbf{A}_H\big)_i\Big] = [q\Phi_H,\ \hat{p}_{H,i}] = i\hbar q\,\partial_i\Phi_H. \tag{24.3.13}$$

With this result, equation (24.3.12) becomes

$$m\frac{d\hat{v}_{H,i}}{dt} = \frac{i}{\hbar}\Big[\tfrac{1}{2}m\hat{\mathbf{v}}_H^2,\ m\hat{v}_{H,i}\Big] - q(\nabla\Phi_H)_i - \frac{q}{c}\frac{\partial A_{H,i}}{\partial t}. \tag{24.3.14}$$

We recognize the electric force in the last two terms on the right-hand side:

$$m\frac{d\hat{v}_{H,i}}{dt} = qE_{H,i} + \frac{i}{2\hbar}\Big[m^2\hat{\mathbf{v}}_H^2,\ \hat{v}_{H,i}\Big]. \tag{24.3.15}$$

The second term on the right-hand side will give the magnetic force. To evaluate this commutator, it is convenient to work out a simpler one first:

$$[\,m\hat{v}_{H,i}\,,\ m\hat{v}_{H,j}\,] = \Big[\hat{\mathbf{p}}_{H,i} - \frac{q}{c}\mathbf{A}_{H,i}\,,\ \hat{\mathbf{p}}_{H,j} - \frac{q}{c}\mathbf{A}_{H,j}\Big] = \Big[\hat{p}_i - \frac{q}{c}A_i\,,\ \hat{p}_j - \frac{q}{c}A_j\Big]_H, \tag{24.3.16}$$

recalling that the commutator of Heisenberg operators can be evaluated with the corresponding Schrödinger operators and then turned into a Heisenberg one (see (16.4.7)). We then have

$$\Big[\hat{p}_i - \frac{q}{c}A_i\,,\ \hat{p}_j - \frac{q}{c}A_j\Big] = -\frac{q}{c}\frac{\hbar}{i}\left(\partial_iA_j - \partial_jA_i\right) = \frac{i\hbar q}{c}\left(\partial_iA_j - \partial_jA_i\right). \tag{24.3.17}$$

A little index gymnastics gives us

$$\partial_iA_j - \partial_jA_i = \epsilon_{ijk}\epsilon_{kpq}\partial_pA_q = \epsilon_{ijk}(\nabla\times A)_k = \epsilon_{ijk}B_k, \tag{24.3.18}$$

recalling that $\mathbf{B} = \nabla\times\mathbf{A}$. Therefore,

$$\boxed{\Big[\hat{p}_i - \frac{q}{c}A_i\,,\ \hat{p}_j - \frac{q}{c}A_j\Big] = \frac{i\hbar q}{c}\,\epsilon_{ijk}B_k.} \tag{24.3.19}$$

Going back to (24.3.16), we have found that

$$[\,m\hat{v}_{H,i}\,,\ m\hat{v}_{H,j}\,] = \frac{i\hbar q}{c}\epsilon_{ijk}B_{H,k}. \tag{24.3.20}$$

With this identity you should be able to quickly show that

$$\Big[m^2\hat{\mathbf{v}}_H^2\,,\ \hat{v}_{H,i}\Big] = \frac{i\hbar q}{c}\big(-\hat{\mathbf{v}}_H\times\mathbf{B}_H + \mathbf{B}_H\times\hat{\mathbf{v}}_H\big)_i. \tag{24.3.21}$$

From (24.3.15), we finally get

$$\boxed{m\frac{d\hat{\mathbf{v}}_H}{dt} = q\mathbf{E}_H + \frac{q}{c}\,\frac{1}{2}\big(\hat{\mathbf{v}}_H\times\mathbf{B}_H - \mathbf{B}_H\times\hat{\mathbf{v}}_H\big).} \tag{24.3.22}$$

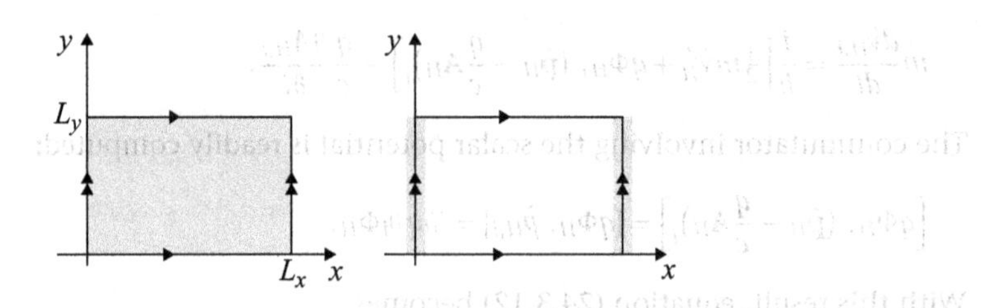

Figure 24.1
Left: A two-dimensional torus as a rectangular region with boundaries identified. *Right*: The candidate gauge potential $A_y(x, y) = B_0 x$ takes different values on the left and right vertical boundaries.

In classical physics **B** and **v** commute, and the terms in parentheses combine to give the familiar $\mathbf{v} \times \mathbf{B}$ contribution to the Lorentz force equation. Here, the antisymmetrization is needed for the right-hand side to be Hermitian. The emergence of a quantum version of the Lorentz force law is clear evidence that we have incorporated electromagnetic fields correctly into quantum mechanics.

24.4 Magnetic Fields on a Torus

A two-dimensional torus T^2 is the space (x, y) subject to two independent identifications that make each coordinate a circle:

$$\begin{aligned}(x, y) &\sim (x + L_x, y), \\ (x, y) &\sim (x, y + L_y).\end{aligned} \tag{24.4.1}$$

One can represent the torus as the rectangular region shown in figure 24.1 with $0 \le x \le L_x$ and $0 \le y \le L_y$ and with identifications on the boundaries, as indicated by the arrows. We will begin by imagining that two out of the three dimensions of space are in fact this torus. This is a world with periodic boundary conditions for everything in the x- and y-directions but with a standard z-direction. Since our world is not of that form, at the end we will discuss what happens when the torus is a surface inside three-dimensional space.

We now consider the possibility that there is a constant magnetic field B_0 in this torus. One can think of B_0 as a constant magnetic field extending forever in the z-direction, as in some kind of infinite solenoid. Certainly, any constant magnetic field, accompanied by a vanishing electric field, is a solution of Maxwell's equations: $\nabla \cdot \mathbf{B} = 0$ and $\nabla \times \mathbf{B} = 0$. We will find, however, that if there exist point particles with charge q, the magnetic field must be quantized. More precisely, the flux of the magnetic field through the torus, $B_0 L_x L_y$, is quantized in integral multiples of a flux quantum $\Phi_0 \propto \hbar c/q$. This is because only for quantized values of the flux can we find a vector potential that gives rise to the magnetic field.

If the magnetic field is B_0 along the z-direction, we must then have

$$B_0 = \partial_x A_y - \partial_y A_x. \tag{24.4.2}$$

We can now attempt to pick a simple solution. We can satisfy the equation by taking

$$A_x(x,y)=0\,,\quad A_y(x,y)=B_0\,x. \tag{24.4.3}$$

We must ask, however, if this configuration of the vector potential is well defined. Because the torus is the rectangle with boundaries identified, the configuration is well defined when the potentials are periodic. Since $A_x=0$ is certainly periodic, we only need to consider A_y. Its periodicity requires that

$$A_y(x,y)=A_y(x+L_x,y)\,,\quad\text{and}\quad A_y(x,y)=A_y(x,y+L_y). \tag{24.4.4}$$

With $A_y=B_0x$, the second relation is satisfied, but the first one is not! In particular, looking at figure 24.1, *right*, we see that A_y on the left vertical line is different from A_y on the right vertical line.

At this point we must reconsider what we *really* need to have a well-defined gauge potential. Periodicity would suffice, but it is not strictly necessary. The freedom allowed by gauge transformations indicates that we only need A_y on the left and right vertical boundary lines of the rectangle to differ by a gauge transformation. If so, they would be equivalent on those two lines and this suffices. There is a general lesson here as well: a configuration of a vector potential on an arbitrary space is one in which we can have different expressions for the vector potential that are valid in various regions, with the condition that at the overlap of those regions the various potentials must differ by a gauge transformation.

Call $A_y^R(y)$ the value of A_y on the right boundary $x=L_x$ of the rectangle:

$$A_y^R(y)=A_y(x=L_x,y)=B_0L_x. \tag{24.4.5}$$

Let us call $A_y^L(y)$ the value of A_y on that same line on the torus but deduced from the left boundary $x=0$ of the rectangle:

$$A_y^L(y)=A_y(x=0,y)\,=\,0. \tag{24.4.6}$$

These two values on that line differ by a gauge transformation induced by a gauge parameter $\Lambda(x,y)$ if, on that line

$$A_y^R(y)=A_y^L(y)+\partial_y\Lambda. \tag{24.4.7}$$

Given the values of A_y^R and A_y^L, we have the equation

$$B_0L_x=0+\partial_y\Lambda(x,y). \tag{24.4.8}$$

One easily writes the following solution:

$$\Lambda(x,y)=B_0L_x\,y+f(x)\,, \tag{24.4.9}$$

with $f(x)$ arbitrary. It is clear that a nontrivial $f(x)$ spoils matters for the A_x potential, which is changed by a gauge parameter with x dependence. The potential A_x was zero everywhere and thus periodic. We wish to keep A_x unchanged, so we set $f(x)=0$, and take

$$\Lambda(x,y)=B_0L_x\,y. \tag{24.4.10}$$

Since this gauge parameter appears to do the job, it would seem that we have managed to construct an acceptable configuration for the vector potential. But we are not really done yet. We must also ask whether the gauge parameter Λ is now well defined on the torus. Since it has no x dependence, it suffices for it to be well defined on the circle $y \sim y + L_y$. But it is clear that Λ is not periodic in this circle.

This obstacle has a solution. All the gauge transformations, the one for the gauge potentials and the one for the wave function, can be written in terms of the phase factor $U(x, y)$ instead of Λ:

$$U(x,y) = \exp\Big(i\frac{q\Lambda}{\hbar c}\Big). \tag{24.4.11}$$

This is manifest for the wave function, since $\Psi' = U\Psi$. As for the potentials, we can write

$$\begin{aligned} \mathbf{A}' &= \mathbf{A} + \nabla\Lambda = \mathbf{A} - \tfrac{i\hbar c}{q}\, U^{-1}\nabla U, \\ \Phi' &= \Phi - \frac{1}{c}\partial_t\Lambda = \Phi + \tfrac{i\hbar}{q}\, U^{-1}\partial_t U. \end{aligned} \tag{24.4.12}$$

So actually, we *do not* need Λ to be periodic; it is U that must be periodic. Indeed, with the value of Λ determined above, we find that

$$U(x,y) = \exp\Big(i\frac{q}{\hbar c}B_0 L_x y\Big), \tag{24.4.13}$$

which is manifestly periodic in x. The condition of y periodicity gives

$$U(x,y) = U(x, y + L_y) \quad \Rightarrow \quad \frac{q}{\hbar c}B_0 L_x L_y = 2\pi n, \quad n \in \mathbb{Z}. \tag{24.4.14}$$

This is the quantization condition:

$$B_0\, L_x L_y = \frac{2\pi\hbar c}{q}\, n. \tag{24.4.15}$$

We call $\Phi_B \equiv B_0 L_x L_y$ the flux of B_0 through the torus. The quantization condition is sometimes written as

$$\Phi_B = \frac{2\pi\hbar c}{q}\, n = \hat{\Phi}_0\, n, \tag{24.4.16}$$

where we introduced the **flux quantum** $\hat{\Phi}_0$:

$$\hat{\Phi}_0 \equiv \frac{2\pi\hbar c}{q}. \tag{24.4.17}$$

This is the smallest nonzero flux allowed. Any magnetic field on the torus must have a flux that is an integer multiple of $\hat{\Phi}_0$. The value of this flux depends on the charge q of the particle that lives on the torus.

Exercise 24.1. *Check that $\hbar c/q$ has units of flux.*

In superconductivity, Cooper pairs of electrons with charge $q = 2e$ are the relevant carriers. Physicists define the magnetic flux quantum Φ_0 (without a hat) by

$$\Phi_0 \equiv \frac{2\pi \hbar c}{2e} = \frac{hc}{2e} \simeq 2.067 \times 10^{-7} \text{ gauss} \cdot \text{cm}^2. \tag{24.4.18}$$

This well-known value of Φ_0 helps us compute $\hat{\Phi}_0$ for arbitrary values of the charge q. We have $\hat{\Phi}_0 = (2e/q)\Phi_0$.

Interestingly, we can view the quantization condition of the flux as a quantization condition on the possible charges that may exist. We do this in the following set of exercises.

Exercise 24.2. *Suppose we know there is a particle with charge $q = e$ and consider the set $S(e)$ of possible fluxes on a torus consistent with this charge. Now assume you discover a new particle with charge ke with $k > 1$ an integer. Is the set $S(e)$ still consistent? What is the maximal set of fluxes consistent with the existence of both particles?*

Exercise 24.3. *Suppose we know there is a particle with charge $q = e$ and consider the set $S(e)$ of possible fluxes on a torus consistent with this charge. Now assume you discover a new particle with charge e/k with $k > 1$ an integer. Is the set $S(e)$ still consistent? What is the maximal set of fluxes consistent with the existence of both particles?*

Exercise 24.4. *Suppose we know there is a particle with charge q and another particle with charge q' such that q/q' is* irrational. *Are there consistent fluxes on a torus?*

Let us conclude by discussing what happens if our two-dimensional torus is, like the surface of a doughnut, a boundaryless surface inside our familiar three-dimensional space. The quantization result is very robust: the magnetic flux through the surface of such a two-dimensional torus is still quantized as in (24.4.16). This time, however, a nonvanishing magnetic flux Φ across the surface of the torus is not consistent with $\nabla \cdot \mathbf{B} = 0$ holding everywhere inside the torus. Indeed, on account of the divergence theorem,

$$\Phi = \int_{\text{torus}} \mathbf{B} \cdot d\mathbf{a} = \int_{\text{vol}} \nabla \cdot \mathbf{B}\, d^3\mathbf{x} = 0. \tag{24.4.19}$$

The only way to have a flux in this case is if magnetic monopoles exist. For a monopole with flux Φ at a point $\mathbf{x}_0$, the Maxwell equation $\nabla \cdot \mathbf{B} = 0$ would become

$$\nabla \cdot \mathbf{B} = \Phi\, \delta^3(\mathbf{x} - \mathbf{x}_0). \tag{24.4.20}$$

If $\mathbf{x}_0$ is inside the torus, the flux through the torus will indeed be Φ. We cannot describe the magnetic field with a vector potential in a neighborhood of $\mathbf{x}_0$, but we can still describe the magnetic field with a potential on the surface of the torus. Since quantization applies, the existence of a magnetic monopole of flux Φ implies, via (24.4.16), the quantization of electric charge:

$$\Phi = \frac{2\pi n \hbar c}{q} \quad \Rightarrow \quad q = \frac{2\pi n \hbar c}{\Phi}. \tag{24.4.21}$$

All charges would have to be multiples of $e_0 \equiv 2\pi\hbar c/\Phi$. In fact, Dirac suggested that the observed quantization of electric charge would be explained by the existence of magnetic monopoles. So far, however, no magnetic monopole has been detected.

24.5 Particles in Uniform Magnetic Field: Landau Levels

We now consider a particle of mass m and charge q moving in a constant uniform magnetic field that can be chosen to point in the z-direction. The classical result is that, up to motion with constant velocity in the z-direction, the charged particle performs circular motion in the (x,y) plane in an orbit that regardless of its radius always has constant angular velocity ω_c, the so-called **cyclotron frequency**. This follows from a simple calculation that equates the Lorentz force to the mass times the centripetal acceleration of circular motion. For a magnetic field of magnitude B, an orbit of radius r, and motion with speed v, we have

$$q\frac{v}{c}B = m\frac{v^2}{r} \quad \rightarrow \quad \frac{v}{r} = \frac{qB}{mc} = \omega_c. \tag{24.5.1}$$

The larger the radius, the larger the velocity and the larger the kinetic energy of the particle.

The problem we want to analyze is that of a quantum charged particle in a uniform magnetic field. We will end up finding Landau levels, energy levels uniformly separated by an energy interval of magnitude Δ, with Δ given by

$$\Delta = \hbar\omega_c = \frac{q\hbar}{mc}B. \tag{24.5.2}$$

If the particles have the whole (x,y) plane to execute their motion, then each Landau level has an infinite degeneracy. If instead the particles only move in a portion of the (x,y) plane, we find a large but *finite* degeneracy for each Landau level.

When the particle in question is an electron, we can also ask whether it is possible to ignore its spin. In general, the answer is no. This is why we will consider the coupling of a spin one-half particle to electromagnetic fields in sections 24.6 and 24.7. We will *derive* the familiar coupling $\hat{H} = -\hat{\mu}\cdot\mathbf{B}$ of the magnetic moment of the particle to the magnetic field, with the magnetic moment related to the spin as expected. For the electron this Zeeman coupling to the magnetic field is given by

$$\hat{H}_Z = \frac{e\hbar}{2m_e c}\,\sigma\cdot\mathbf{B}. \tag{24.5.3}$$

In a constant magnetic field of magnitude B, this Hamiltonian produces energy levels

$$E_{Z\pm} = \pm\frac{e\hbar}{2m_e c}B. \tag{24.5.4}$$

The splitting here is identical to Δ when evaluated for $q=e$ and $m=m_e$. In our analysis below, we will ignore the spin of the electron or particle that is being studied. In some materials, however, the value of Δ is affected by band structure that effectively replaces the electron mass by a much smaller mass in the formula for ω_c. In such cases the Zeeman splitting is a small perturbation.

Let us begin. We consider the motion of a particle of mass m and charge q in the presence of a constant uniform magnetic field of magnitude B pointing in the z-direction.

This magnetic field, in what is called the *Landau gauge*, is represented by the vector potential

$$\mathbf{A} = (-By, 0, 0) \tag{24.5.5}$$

and, of course, zero electric potential Φ. Other choices of gauge are possible and in fact interesting to consider. In the Landau gauge, the vector potential has no x dependence, and this will prove useful soon. With the potentials specified, the Hamiltonian is given by

$$\hat{H} = \frac{1}{2m}\Big(\hat{p}_x - \frac{q}{c}(-By)\Big)^2 + \frac{1}{2m}\hat{p}_y^2 + \frac{1}{2m}\hat{p}_z^2. \tag{24.5.6}$$

We assume motion in the (x, y) plane, so we can ignore the term proportional to $\hat{p}_z$ and take the Hamiltonian to be

$$\hat{H} = \frac{1}{2m}\Big(\hat{p}_x + \frac{qB}{c}y\Big)^2 + \frac{1}{2m}\hat{p}_y^2. \tag{24.5.7}$$

Since there is no x dependence in $\hat{H}$, we have $[\hat{p}_x, \hat{H}] = 0$, and energy eigenstates can be assumed to be $\hat{p}_x$ eigenstates. We will thus assume that the wave function $\psi(x, y)$ takes the form

$$\psi(x, y) = \psi(y)\, e^{ik_x x}, \tag{24.5.8}$$

with k_x an arbitrary wave number. This is a state with $\hat{p}_x$ eigenvalue p_x given by

$$p_x = \hbar k_x. \tag{24.5.9}$$

We restrict the action of $\hat{H}$ to the $\hbar k_x$ eigenspace of $\hat{p}_x$ and call this restriction $\hat{H}_{k_x}$. Since $\hat{p}_x$ commutes with y, the effect of this restriction is to change $\hat{p}_x$ in (24.5.7) to $\hbar k_x$:

$$\hat{H}_{k_x} = \frac{1}{2m}\hat{p}_y^2 + \frac{1}{2m}\Big(\frac{qB}{c}y + \hbar k_x\Big)^2. \tag{24.5.10}$$

This is now recognized as a harmonic oscillator in the y-direction with the equilibrium point shifted:

$$\hat{H}_{k_x} = \frac{1}{2m}\hat{p}_y^2 + \frac{1}{2}m\Big(\frac{qB}{mc}\Big)^2\Big(y - \Big(-\frac{\hbar k_x c}{qB}\Big)\Big)^2. \tag{24.5.11}$$

The frequency of this harmonic oscillator is equal to the cyclotron frequency ω_c, and the minimum of the quadratic potential is at y_0, defined by

$$y_0 = -\frac{\hbar k_x c}{qB}. \tag{24.5.12}$$

The square of the length scale in a harmonic oscillator is $\frac{\hbar}{m\omega}$. In the present setup, with $\omega = \omega_c$, the length scale is called the **magnetic length** ℓ_B:

$$\ell_B^2 = \frac{\hbar}{m\omega_c} = \frac{\hbar c}{qB}. \tag{24.5.13}$$

Note that in terms of the magnetic length, the equilibrium position y_0 reads

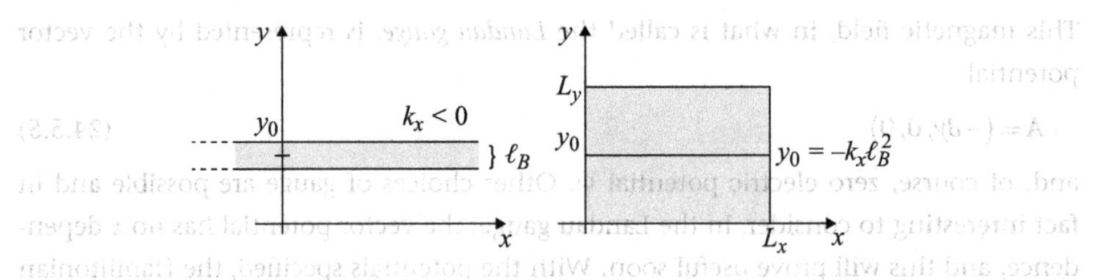

Figure 24.2
Left: The wave function $\psi(x, y)$ is supported in a band of width ℓ_B centered at $y_0 > 0$ if $k_x < 0$. *Right*: A finite rectangular sample. The minimum of the harmonic oscillator is shifted by an amount proportional to k_x. With k_x quantized, the degeneracy of the Landau levels is finite.

$$y_0 = -k_x \ell_B^2. \tag{24.5.14}$$

If we now use for $\psi(y)$ a harmonic oscillator wave function, the energy eigenstate $\psi(x, y)$ in (24.5.8) is centered at some $y_0 > 0$, if $k_x < 0$, with an approximate width ℓ_B, and is fully delocalized in the x-direction (figure 24.2).

If we use the nth state φ_n of the harmonic oscillator ($n = 0, 1, \cdots$) for $\psi(y)$, the wave function takes the form

$$\psi(x, y) = \varphi_n(y - y_0)\, e^{ik_x x}, \tag{24.5.15}$$

or in ket notation,

$$|\psi\rangle = e^{ik_x x}\, |n\rangle_y. \tag{24.5.16}$$

The energy E_{n,k_x} of the states is given by

$$E_{k_x,n} = \hbar\omega_c(\tfrac{1}{2} + n). \tag{24.5.17}$$

Comments:

1. Remarkably, the energy does not depend on the value of k_x. We are used to momenta contributing quadratic terms to the energy. Not here; the value of the momentum k_x ended up defining the value of the shift y_0 of the oscillator. Since k_x can take values in a continuum, we have an infinite degeneracy for each value of $n = 0, 1, \ldots$. The infinitely degenerate eigenstates that exist for each fixed n compose the *Landau levels*. For $n = 0$, we have the first Landau level, for $n = 1$, the second Landau level, and so on. As the value of k_x is changed, the value of y_0 changes. With $k_x \in (-\infty, \infty)$, the strip-like states cover the full plane.
2. Classical intuition suggested that particles would move in circles centered anywhere in the (x, y) plane. The picture we have arrived at seems rather different, with delocalized strip-like states. The x-delocalization is not really required. Since all states with different values of k_x are degenerate, any linear combination of them is still an energy eigenstate in the same Landau level. We can therefore superpose the momentum eigenstates in the x-direction to construct energy eigenstates that are localized in x. Our Landau-gauge solution can be thought of as the coherent superposition

of infinitely many circular orbits centered on the $y = y_0$ line, at all possible values of x.

3. Solutions in different gauges can look *very* different. One can gain some intuition by working with a symmetric gauge $\mathbf{A} = \frac{1}{2}B(-y, x, 0)$ (problem 24.4). Here, we find that circles are the sets of constant $|\psi|^2$ for energy eigenstates. At a fixed Landau level, however, the different eigenstates correspond to circles of different sizes. Therefore, even in this gauge the loci of large $|\psi|^2$ are not the orbits of the classical theory.
4. The classical intuition that particles move in circles centered anywhere in the (x, y) plane can be realized at the operator level. There are operators $\hat{X}_0$, $\hat{Y}_0$ that are constants of the motion and represent the center of the orbit (problem 24.3). In fact, one can also show that an orbit-radius operator squared $\hat{r}_c^2$ is proportional to the Hamiltonian: $\frac{1}{2}m\omega_c^2\hat{r}_c^2 = \hat{H}$, consistent with the idea that the energy is the kinetic energy of the circular motion of a particle of mass m with cyclotron frequency ω_c and radius $\hat{r}_c$. Since classically the energy of the particle grows with the radius of the orbit, Landau levels of different energy are associated with orbits of different size. Semiclassical orbits of arbitrary center and radius are obtained as coherent states that involve the superposition of states in different Landau levels (problem 24.5). One can then explicitly see the time-dependent expectation values of position tracing the expected orbits.

If instead of considering the full (x, y) plane we take a finite rectangular sample with sides L_x and L_y, and $0 \le y \le L_y$, we can count the number of states in any given Landau level (figure 24.2). Finite size in x implies the quantization of k_x via the required periodicity of $e^{ik_x x}$ under $x \to x + L_x$. This gives the relation

$$k_x L_x = 2\pi n_x\,, \quad n_x \in \mathbb{Z}. \tag{24.5.18}$$

In order for the states to be inside the sample we need $0 \le y_0 \le L_y$, of course, in the approximation $\ell_B \ll L_y$. Since $y_0 = -k_x \ell_B^2$, the positivity of y_0 requires $n_x < 0$. The lowest allowed value of n_x, called $-D$, with D a positive integer, is that for which $y_0 = L_y$:

$$L_y = -k_x \ell_B^2 = -\frac{2\pi(-D)}{L_x}\,\ell_B^2. \tag{24.5.19}$$

From this we find that

$$D = \frac{L_x L_y}{\ell_B^2} \cdot \frac{1}{2\pi}. \tag{24.5.20}$$

We are working in the approximation that D is very large. Taken literally, we can choose D to be the largest integer that is less than or equal to the right-hand side. Since D is the number of possible n_x values, it is precisely the degeneracy D of the Landau level—that is, the number of degenerate energy eigenstates allowed by the geometry of the rectangular sample. With area $A = L_x L_y$ and using the formula for ℓ_B, we have

$$D = \frac{A}{\frac{\hbar c}{qB}}\frac{1}{2\pi} = \frac{BA}{\frac{2\pi\hbar c}{q}}. \tag{24.5.21}$$

We recognize this as the ratio of the magnetic flux Φ_B divided by the flux quantum $\hat{\Phi}_0$:

$$D=\frac{\Phi_B}{\hat{\Phi}_0}. \tag{24.5.22}$$

With $\hat{\Phi}_0=4.136\times10^{-7}$ gauss cm^2, following from (24.4.18), a sample of area 1 cm^2 with a 1 gauss magnetic field would have about 2.4 million degenerate states of electrons in each Landau level.

Our computation above gives us an indication of the area A_0 that a state in each Landau level requires. From the relation

$$y_0=-k_x\ell_B^2=-\frac{2\pi n_x}{L_x}\ell_B^2, \tag{24.5.23}$$

we see that a change in n_x by one unit implies a change Δy in y_0, with

$$\Delta y=\frac{2\pi}{L_x}\ell_B^2. \tag{24.5.24}$$

Since the state extends from $x=0$ to $x=L_x$, the area A_0 is

$$A_0=L_x\Delta y=2\pi\ell_B^2. \tag{24.5.25}$$

The area $A_0=2\pi\ell_B^2$ is the area required by each degenerate state in a maximally filled Landau level. It is independent of the Landau level. Being a constant, this area does not correlate with the expected area of the orbit, which grows with the Landau level. It is more like the area required for the smeared center of each orbit, as we will explain below. Our result for A_0 is consistent with the expression for the degeneracy D in (24.5.20): the numerator on the right-hand side is the area A of the sample, and the denominator is the area A_0 occupied by each state.

This value for A_0 also arises when studying the noncommutativity of the operators $\hat{X}_0$ and $\hat{Y}_0$, defined as the x and y quantum coordinates of the center of each orbit (problem 24.3). One finds the commutator $[\hat{X}_0,\hat{Y}_0]=-i\ell_B^2$. The uncertainty inequality applied to these operators gives $\Delta X_0\Delta Y_0\geq\frac{1}{2}\ell_B^2$, just like $[\hat{x},\hat{p}]=i\hbar$ leads to $\Delta x\Delta p\geq\frac{1}{2}\hbar$. For position and momenta, the phase space volume $\Delta x\Delta p$ occupied by a quantum state is $2\pi\hbar$. For our noncommutative coordinates $\hat{X}_0$ and $\hat{Y}_0$, the value of ℓ_B^2 plays the role of $\hbar$. Therefore, the phase space area $\Delta X_0\Delta Y_0$ occupied by a quantum state is $2\pi\ell_B^2$. This is indeed the value of A_0 obtained above.

Exercise 24.5. *Consider momentum states $e^{ipx/\hbar}$ on a segment $x\in[0,L]$, imposing periodic boundary conditions. Show that the phase space volume $L\Delta p$ occupied by each state is $2\pi\hbar$.*

Exercise 24.6. *Given $[\hat{X}_0,\hat{Y}_0]=-i\ell_B^2$, we can think of $\hat{X}_0$ as a position operator with coordinates x_0 and $\hat{Y}_0=i\ell_B^2\frac{\partial}{\partial x_0}$ as a momentum operator. Consider momentum states $e^{-iy_0x_0/\ell_B^2}$ of momentum y_0 living on the interval $x_0\in[0,L]$ and satisfying periodic boundary conditions. Explain why the phase space volume $L\Delta y_0$ occupied by each state is $2\pi\ell_B^2$.*

24.6 The Pauli Equation

New features appear when we couple a spin one-half particle to electromagnetic fields. The nonrelativistic way to do this was discovered by Pauli, and we will discuss how this works below. The *relativistic* coupling of a spin one-half particle to electromagnetic fields was discovered by Dirac, and we will briefly consider that work in the following section.

One way we detect the spin of the electron is through interactions with magnetic fields. The spin of the electron gives rise to a magnetic dipole moment that couples to magnetic fields. The quantum mechanical magnetic moment differs from the naive classical expectation by a factor of two. This factor, called the g-factor of the electron, was discussed in section 17.5. The Pauli equation for the electron predicts this value and helps us understand how the electron couples to external electromagnetic fields.

The magnetic moment operator $\hat{\mu}$ of a particle is related to the spin operator $\hat{\mathbf{S}}$ as follows:

$$\hat{\mu} = g\frac{q}{2mc}\hat{\mathbf{S}}, \tag{24.6.1}$$

for an electron $q=-e$ and $g=2$. Therefore, the electron magnetic moment is given by

$$\hat{\mu} = 2\frac{-e}{2m_e c}\hat{\mathbf{S}} = -2\frac{e\hbar}{2m_e c}\frac{\hat{\mathbf{S}}}{\hbar} = -2\frac{e\hbar}{2m_e c}\tfrac{1}{2}\sigma = -\frac{e\hbar}{2m_e c}\sigma. \tag{24.6.2}$$

For numerical applications we note the value of the *Bohr magneton*:

$$\mu_B = \frac{e\hbar}{2m_e c} \simeq 9.274\times 10^{-21}\,\frac{\text{erg}}{\text{gauss}} = 5.79\times 10^{-9}\,\frac{\text{eV}}{\text{gauss}}. \tag{24.6.3}$$

(For SI units use $\mu_B = \frac{e\hbar}{2m} = 5.79\times 10^{-5}$ eV/T. Here T = tesla = $(10^4/c)$ gauss.) The coupling of an electron to an external magnetic field is therefore represented by a Hamiltonian $\hat{H}_B$ given by

$$\boxed{\hat{H}_B = -\hat{\mu}\cdot\mathbf{B} = \frac{e\hbar}{2m_e c}\sigma\cdot\mathbf{B}.} \tag{24.6.4}$$

Our goal now is to show that this coupling to an *external* magnetic field, and its associated prediction of $g=2$, arises naturally from a nonrelativistic equation for an electron that Pauli invented.

Consider first the Schrödinger equation for a free particle:

$$i\hbar\frac{\partial\Psi}{\partial t} = \frac{\hat{\mathbf{p}}^2}{2m}\Psi. \tag{24.6.5}$$

Since a spin one-half particle has two degrees of freedom, usually assembled into a column vector χ, the expected equation for a free spin one-half particle is

$$i\hbar\frac{\partial\chi}{\partial t} = \frac{\hat{\mathbf{p}}^2}{2m}\chi, \quad \text{with} \quad \chi = \begin{pmatrix}\chi_1\\ \chi_2\end{pmatrix}. \tag{24.6.6}$$

Here χ is called a Pauli spinor. Note there's an implicit 2×2 identity matrix $\mathbb{1}_{2\times 2}$ in the above Hamiltonian:

$$\hat{H} = \frac{\hat{\mathbf{p}}^2}{2m}\mathbb{1}_{2\times 2} = \begin{pmatrix} \frac{\hat{\mathbf{p}}^2}{2m} & 0 \\ 0 & \frac{\hat{\mathbf{p}}^2}{2m} \end{pmatrix}. \quad (24.6.7)$$

If we now use minimal coupling, letting $\hat{\mathbf{p}} \to \hat{\mathbf{p}} - \frac{q}{c}\mathbf{A}$, we do not get the "right" Hamiltonian for a particle with spin coupled to electromagnetic fields. In fact, the coupling $\hat{H}_B$ in (24.6.4) would not arise. It was Pauli's contribution to invent the correct Hamiltonian, valid for low energies. Following Pauli, we first rewrite the above Hamiltonian using Pauli matrices. For this, recall the identity (13.3.26),

$$(\boldsymbol{\sigma} \cdot \mathbf{a})(\boldsymbol{\sigma} \cdot \mathbf{b}) = \mathbf{a} \cdot \mathbf{b}\, \mathbb{1}_{2\times 2} + i\boldsymbol{\sigma} \cdot (\mathbf{a} \times \mathbf{b}), \quad (24.6.8)$$

valid for arbitrary vector *operators* **a** and **b**. Setting both **a** and **b** equal to the momentum operator and recognizing that $\hat{\mathbf{p}} \times \hat{\mathbf{p}} = 0$, we have

$$(\boldsymbol{\sigma} \cdot \hat{\mathbf{p}}) \cdot (\boldsymbol{\sigma} \cdot \hat{\mathbf{p}}) = \hat{\mathbf{p}}^2\, \mathbb{1}_{2\times 2}. \quad (24.6.9)$$

This means that the Hamiltonian (24.6.7) can be rewritten as

$$\hat{H} = \frac{1}{2m}(\boldsymbol{\sigma} \cdot \hat{\mathbf{p}})(\boldsymbol{\sigma} \cdot \hat{\mathbf{p}}). \quad (24.6.10)$$

So far this is just rewriting, with no change in physics. But new things happen when the particle has charge q, and we couple it to external electromagnetic fields. With minimal coupling, we replace the momentum operator as follows:

$$\hat{\mathbf{p}} \to \hat{\boldsymbol{\pi}} \equiv \hat{\mathbf{p}} - \frac{q}{c}\mathbf{A}(\hat{\mathbf{x}}, t). \quad (24.6.11)$$

Moreover, if there is an electromagnetic scalar potential $\Phi(\mathbf{x}, t)$, it contributes an additional term $q\,\Phi(\hat{\mathbf{x}}, t)$ to the Hamiltonian.

With the replacement (24.6.11) applied to the Hamiltonian (24.6.10) and the inclusion of the coupling to the scalar potential, we get the **Pauli Hamiltonian** $\hat{H}_{\text{Pauli}}$:

$$\boxed{\hat{H}_{\text{Pauli}} = \frac{1}{2m}(\boldsymbol{\sigma} \cdot \hat{\boldsymbol{\pi}})(\boldsymbol{\sigma} \cdot \hat{\boldsymbol{\pi}}) + q\,\Phi(\hat{\mathbf{x}}, t).} \quad (24.6.12)$$

This time, using the identity (24.6.8), the second term survives:

$$\hat{H}_{\text{Pauli}} = \frac{1}{2m}\left[(\hat{\boldsymbol{\pi}} \cdot \hat{\boldsymbol{\pi}})\mathbb{1} + i\boldsymbol{\sigma} \cdot (\hat{\boldsymbol{\pi}} \times \hat{\boldsymbol{\pi}})\right] + q\,\Phi(\hat{\mathbf{x}}, t). \quad (24.6.13)$$

We have $\hat{\boldsymbol{\pi}} \times \hat{\boldsymbol{\pi}} \neq \mathbf{0}$ because the various $\hat{\pi}_i$'s do not commute. Note that the replacement (24.6.11) applied to the original Hamiltonian (24.6.7) would not have given us the $\hat{\boldsymbol{\pi}} \times \hat{\boldsymbol{\pi}}$ term. For the evaluation of this term, we first relate it to a commutator:

$$(\hat{\boldsymbol{\pi}} \times \hat{\boldsymbol{\pi}})_k = \epsilon_{ijk}\hat{\pi}_i\hat{\pi}_j = \tfrac{1}{2}\epsilon_{ijk}[\hat{\pi}_i, \hat{\pi}_j]. \quad (24.6.14)$$

The commutator evaluates to

$$[\hat{\pi}_i, \hat{\pi}_j] = \left[\hat{p}_i - \frac{q}{c}A_i, \hat{p}_j - \frac{q}{c}A_j\right] = \frac{i\hbar q}{c}\,\epsilon_{ijk}B_k, \tag{24.6.15}$$

as we showed in (24.3.19). Therefore,

$$(\hat{\pi} \times \hat{\pi})_k = \tfrac{1}{2}\epsilon_{ijk}\frac{i\hbar q}{c}\,\epsilon_{ijp}B_p = \tfrac{1}{2}\,2\delta_{kp}\,\frac{i\hbar q}{c}B_p = \frac{i\hbar q}{c}B_k. \tag{24.6.16}$$

In vector notation,

$$\hat{\pi} \times \hat{\pi} = \frac{i\hbar q}{c}\mathbf{B}. \tag{24.6.17}$$

Back in the Pauli Hamiltonian (24.6.13), leaving identity matrices implicit and setting $q = -e$ and $m = m_e$, we find

$$\begin{aligned} \hat{H}_{\text{Pauli}} &= \frac{1}{2m_e}\left(\hat{\mathbf{p}} + \frac{e}{c}\mathbf{A}\right)^2 + \frac{i}{2m_e}\,\boldsymbol{\sigma}\cdot\frac{i\hbar(-e)}{c}\mathbf{B} - e\Phi(\hat{\mathbf{x}}, t) \\ &= \frac{1}{2m_e}\left(\hat{\mathbf{p}} + \frac{e}{c}\mathbf{A}\right)^2 + \frac{e\hbar}{2m_e c}\boldsymbol{\sigma}\cdot\mathbf{B} - e\Phi(\hat{\mathbf{x}}, t). \end{aligned} \tag{24.6.18}$$

The second term in this expanded Pauli Hamiltonian gives the coupling of the electron spin to the magnetic field and agrees precisely with the expected coupling (24.6.4). We thus see that the Pauli equation predicts the $g = 2$ value in the electron magnetic moment. The electron g-factor has been measured to extraordinary accuracy. It is not exactly two but rather $g = 2.002\,319\,304\,361$. Remarkably, the corrections to the value $g = 2$ can be computed in quantum *field theory* perturbation theory. This agreement between theory and experiment is striking.

There is more interesting information in the Pauli Hamiltonian. The first term in $\hat{H}_{\text{Pauli}}$, when expanded, includes the coupling of a constant external magnetic field to the *orbital* angular momentum of the electron (problem 24.9).

In the hydrogen atom, the spin-orbit coupling arises because the electron is moving in the electric field of the proton. Since the electron is moving relative to the frame where we have a static electric field, the electron also sees a magnetic field $\mathbf{B}$. The spin-orbit coupling is the coupling $-\hat{\boldsymbol{\mu}}\cdot\mathbf{B}$ of that magnetic field to the magnetic dipole moment $\hat{\boldsymbol{\mu}}$ of the electron. This magnetic field is not external, and it has a relativistic origin. Thus, spin-orbit does not arise in the Pauli equation; for this we need the Dirac equation.

24.7 The Dirac Equation

As discovered by Dirac, to give a relativistic description of the electron one has to work with matrices, and the Pauli spinor must be upgraded to a four-component spinor. The analysis begins with the relation between relativistic energies and momenta:

$$E^2 - c^2\mathbf{p}^2 = m^2c^4 \quad \Rightarrow \quad E = \sqrt{c^2\mathbf{p}^2 + m^2c^4}. \tag{24.7.1}$$

This suggests that a relativistic Hamiltonian $\hat{H}$ for a free particle could take the form

$$\hat{H} = \sqrt{c^2\hat{\mathbf{p}}^2 + m^2c^4}, \tag{24.7.2}$$

with the associated Schrödinger equation

$$i\hbar\frac{\partial\Psi}{\partial t} = \sqrt{c^2\hat{\mathbf{p}}^2 + m^2c^4}\,\Psi. \tag{24.7.3}$$

It is not clear, however, how to treat the square root to form a differential equation. Dirac aimed to avoid square roots and find a Hamiltonian linear in momenta. This would be possible if one could write the relativistic energy as the square of a linear function of the momentum:

$$c^2\hat{\mathbf{p}}^2 + m^2c^4 = (c\boldsymbol{\alpha}\cdot\hat{\mathbf{p}} + \beta mc^2)^2 = (c\alpha_1\hat{p}_1 + c\alpha_2\hat{p}_2 + c\alpha_3\hat{p}_3 + \beta mc^2)^2, \tag{24.7.4}$$

with $\alpha_1, \alpha_2, \alpha_3$, and β to be determined. The object inside parentheses on the final expression is the candidate for the energy operator $\hat{H}_{\text{Dirac}}$:

$$\hat{H}_{\text{Dirac}} = \sum_i c\alpha_i\hat{p}_i + \beta mc^2. \tag{24.7.5}$$

Expanding the right-hand side of (24.7.4) and equating coefficients, one finds that the following must hold:

$$\begin{aligned} &\alpha_1^2 = \alpha_2^2 = \alpha_3^2 = \beta^2 = 1\,, \\ &\alpha_i\alpha_j + \alpha_j\alpha_i = \{\alpha_i, \alpha_j\} = 0\,, \quad i \neq j\,, \\ &\alpha_i\beta + \beta\alpha_i = \{\alpha_i, \beta\} = 0. \end{aligned} \tag{24.7.6}$$

Here we are using anticommutators: $\{A, B\} \equiv AB + BA$. The relations on the second and third lines imply that α's and β's can't be numbers. If they were, the equations would imply that $\alpha_i\alpha_j = 0$ for $i \neq j$ and $\beta\alpha_i = 0$ for all i, while the first equation says that none of the α's nor β can be zero.

It turns out that the α's and β can be matrices—in fact, Hermitian matrices so that the energy operator (24.7.5) is Hermitian. The first equation implies that these matrices have eigenvalues that can only be ± 1. The second equation implies that the α's are traceless matrices. Indeed, one readily sees that, for example,

$$\alpha_1\alpha_2 = -\alpha_2\alpha_1 \quad \Rightarrow \quad \alpha_1 = -\alpha_2\alpha_1\alpha_2 \tag{24.7.7}$$

by multiplying the first equation by α_2 from the right. Taking traces of the last equation,

$$\text{tr}\,\alpha_1 = -\text{tr}(\alpha_2\alpha_1\alpha_2) = -\text{tr}(\alpha_2^2\alpha_1) = -\text{tr}\,\alpha_1, \tag{24.7.8}$$

where we used the cyclicity of the trace. This shows $\text{tr}\,\alpha_1 = 0$. An exactly analogous argument shows that all the α's are traceless and so is β. Traceless matrices with eigenvalues ± 1 must be even-dimensional; otherwise, you cannot get zero trace.

We cannot, however, find a solution for the α's and β with 2×2 matrices. We outline the difficulty here; for a complete proof, see problem 24.10. The space of traceless Hermitian matrices is spanned by the three Pauli matrices. The Pauli matrices anticommute with each other and square to the identity, making them good candidates for the three

α's. The problem, however, is that we then cannot find a Hermitian traceless β that anticommutes with all the Pauli matrices: β would have to be a linear combination of the Pauli matrices, and one can quickly convince oneself that none works.

It turns out, however, that there is a solution for the α's and β with 4×4 Hermitian matrices. The α's are constructed from Pauli matrices, used as off-diagonal blocks, and β is built with two identity blocks in the diagonal and with opposite signs:

$$\alpha = \begin{pmatrix} 0 & \sigma \\ \sigma & 0 \end{pmatrix}, \quad \beta = \begin{pmatrix} \mathbb{1} & 0 \\ 0 & -\mathbb{1} \end{pmatrix}. \tag{24.7.9}$$

The Dirac Hamiltonian in (24.7.5) is now written as

$$\hat{H}_{\text{Dirac}} = c\alpha \cdot \hat{\mathbf{p}} + \beta mc^2. \tag{24.7.10}$$

The Dirac equation is

$$i\hbar \frac{\partial \Psi}{\partial t} = \left(c\alpha \cdot \hat{\mathbf{p}} + \beta mc^2 \right) \Psi, \tag{24.7.11}$$

where Ψ is a Dirac spinor, a *four-component* column vector that can be considered to be composed of two two-component vectors. We call those the Pauli spinors χ and η:

$$\Psi = \begin{pmatrix} \chi \\ \eta \end{pmatrix}, \quad \chi = \begin{pmatrix} \chi_1 \\ \chi_2 \end{pmatrix}, \quad \eta = \begin{pmatrix} \eta_1 \\ \eta_2 \end{pmatrix}. \tag{24.7.12}$$

The spinors χ and η appear, respectively, as the top and bottom components of the Dirac spinor Ψ. The Pauli equation arises from the Dirac equation in the low-energy approximation. At low energies one can show that the η spinor is small compared to the χ spinor, and the evolution of the Pauli spinor χ is governed by the Pauli Hamiltonian (problem 24.11).

The coupling of an electron to electromagnetic fields is done as before, letting $\hat{\mathbf{p}} \to \hat{\mathbf{p}} + \frac{e}{c}\mathbf{A}$. The resulting Dirac equation takes the form

$$i\hbar \frac{\partial \Psi}{\partial t} = \left[c\alpha \cdot \left(\hat{\mathbf{p}} + \frac{e}{c}\mathbf{A} \right) + \beta mc^2 + V(r) \right] \Psi, \tag{24.7.13}$$

where the coupling of the electron to the scalar potential $\Phi(r)$ is included via $V(r) = -e\Phi(r)$. The great advantage of the Dirac equation (24.7.13) is that the relativistic corrections to the hydrogen Bohr Hamiltonian can be derived systematically. The first-order corrections to the Bohr Hamiltonian are called fine-structure corrections. We will discuss this subject further in section 25.7.

Problems

Problem 24.1. *Gauge invariance and the Schrödinger equation.*

1. Give brief proofs for the following statements:
   a. The sum of gauge-covariant operators is a gauge-covariant operator.
   b. The product of gauge-covariant operators is a gauge-covariant operator.
   c. The operator $\hat{x}_i$ is gauge covariant.

d. The operator $\hat{p}_i$ is *not* gauge covariant.

e. The velocity operator $\hat{v}_i$ is gauge covariant.

2. Let $\hat{H}$ be the Hamiltonian coupling a charged (spinless) particle to electromagnetic fields. Answer yes or no, with brief explanations:
   a. Is $\hat{H}$ gauge covariant under arbitrary gauge transformations?
   b. Is $\hat{H}$ gauge covariant under gauge transformations with Λ independent of $\mathbf{x}$?
   c. Is $\hat{H}$ gauge covariant under gauge transformations with Λ independent of t?

Problem 24.2. *Classical motion in a magnetic field.*

Consider a particle of mass m and charge q moving in the (x, y) plane along a trajectory $\mathbf{x}(t) = (x(t), y(t))$ in the presence of a constant magnetic field B along the z-direction.

1. Use the Lorentz force law

$$m\frac{d\mathbf{v}}{dt} = \frac{q}{c}\mathbf{v} \times \mathbf{B}$$

to show that the general solution represents circular motion with arbitrary radius r about a center with (constant) coordinates (X_0, Y_0) and with angular speed $|\omega_c|$, where

$$\omega_c = \frac{qB}{mc}.$$

Write the general solution for $x(t)$ and $y(t)$, fixing the origin of time such that at $t=0$ the particle is at $(X_0 + r, Y)$. Write your answers in terms of X_0, Y_0, r, ω_c, and t.

2. Find expressions for X_0 and Y_0 in terms of the (time-dependent) coordinates (x, y) and velocities (v_x, v_y) of the particle, as well as ω_c. Confirm your answer by checking that the differentiation of X_0 and Y_0, so written, gives zero.

Problem 24.3. *General aspects of quantum motion in a magnetic field.*

Consider a particle of mass m and charge q moving in the (x, y) plane in the presence of a magnetic field. Assume the electric field is vanishing. The questions in this problem should be answered without explicitly choosing a gauge.

1. Determine the commutator $[\hat{v}_x, \hat{v}_y]$ of the velocity operators in terms of the z-component B_z of the magnetic field and other constants $(\hbar, m, q, c)$.
2. Let the magnetic field be $B\hat{\mathbf{z}}$, with B constant. Motivated by the analogous classical expressions, we introduce quantum center-of-orbit coordinate operators $\hat{X}_0$ and $\hat{Y}_0$:

$$\hat{X}_0 \equiv \hat{x} + \frac{\hat{v}_y}{\omega_c}, \qquad \hat{Y}_0 \equiv \hat{y} - \frac{\hat{v}_x}{\omega_c}.$$

Are $\hat{X}_0$ and $\hat{Y}_0$ gauge covariant? Are they Hermitian? Are they unitary? Find the commutator $[\hat{X}_0, \hat{Y}_0]$. Give your answer in terms of numerical constants and the magnetic length ℓ_B. The $\hat{X}_0$ and $\hat{Y}_0$ operators are the simplest example in physics of **noncommutative** coordinates!

3. Calculate the commutator of the coordinate operators and the Hamiltonian $\hat{H}$,

$$[\hat{X}_0, \hat{H}] = \dots, \qquad [\hat{Y}_0, \hat{H}] = \dots.$$

Write your answers in terms of $\hat{X}_0$, $\hat{Y}_0$, m, $\hbar$, and ℓ_B. [Hint: It is convenient to write the Hamiltonian in a form $\hat{H} = \frac{1}{2}m(\hat{v}_x^2 + \hat{v}_y^2)$ and first find the commutators between $\hat{X}_0$, $\hat{Y}_0$ and $\hat{v}_x$, $\hat{v}_y$. The answer for the commutators should make physical sense.]

4. Show that the angular momentum relative to the orbit center, $\hat{\Theta}_z \equiv (\hat{x} - \hat{X}_0)m\hat{v}_y - (\hat{y} - \hat{Y}_0)m\hat{v}_x$, is in fact proportional to the Hamiltonian: $\hat{\Theta}_z = -\frac{2}{\omega_c}\hat{H}$.

5. Define the operator $\hat{R}^2$ as the distance square of the orbit center to the origin,

$$\hat{R}^2 \equiv \hat{X}_0^2 + \hat{Y}_0^2.$$

Find the spectrum of $\hat{R}^2$. That is, find the eigenvalues R_n^2 of the operator using an integer $n = 0, 1, \dots$ to enumerate them. (Write your answer in terms of n and ℓ_B.) [Hint: think of $\hat{R}^2$ as a harmonic oscillator Hamiltonian.]

6. Define the *orbit radius* operator $\hat{r}_c$ via the classically inspired relation

$$\hat{r}_c^2 \equiv (\hat{x} - \hat{X}_0)^2 + (\hat{y} - \hat{Y}_0)^2.$$

Does $\hat{r}_c^2$ commute with $\hat{R}^2$? Relate $\hat{r}_c^2$ to the Hamiltonian $\hat{H}$. Find the spectrum of $\hat{r}_c^2$; that is, find the eigenvalues r_m^2 of the operator using an integer $m = 0, 1, \dots$ to enumerate them. (Write your answer in terms of m and ℓ_B.) [Hint: write $\hat{r}_c^2$ in terms of velocities.]

7. The angular momentum operator $\hat{L}_z = \hat{x}\hat{p}_y - \hat{y}\hat{p}_x$ is not gauge covariant. To define a gauge-covariant version $\hat{\mathcal{L}}_z$, we take

$$\hat{\mathcal{L}}_z = \hat{x}\, m\hat{v}_y - \hat{y}\, m\hat{v}_x + \cdots,$$

where the dots are terms you should determine in terms of $\hat{x}$, $\hat{y}$, m, and ω_c using the condition that $\hat{\mathcal{L}}_z$ reduces to the familiar $\hat{L}_z$ in the circular gauge $(A_x, A_y) = \frac{1}{2}B(-y, x)$.

One way to see that the "angular momentum" $\hat{\mathcal{L}}_z$ is a constant of the motion is to relate it to other constants of the motion. Show that $\hat{\mathcal{L}}_z$ is proportional to $\hat{R}^2 - \hat{r}_c^2$:

$$\hat{\mathcal{L}}_z = \beta(\hat{R}^2 - \hat{r}_c^2),$$

where you must find the constant of proportionality β in terms of $\hbar$ and ℓ_B.

Problem 24.4. *Landau levels in the symmetric gauge.*

Consider the Hamiltonian for a charged particle of charge q and mass m moving in a constant magnetic field B in the z-direction. Assume the motion is two-dimensional and restricted to the (x, y) plane, and the magnetic field arises from the vector potential: $\mathbf{A} = \frac{1}{2}B(-y,\, x,\, 0)$.

1. Write the Hamiltonian for this quantum system. Determine the constant α such that

$$\hat{Q} \equiv \alpha\Big(\hat{p}_x + \frac{q\hat{y}B}{2c}\Big), \quad \text{and} \quad \hat{P} \equiv \hat{p}_y - \frac{q\hat{x}B}{2c}$$

form a coordinate/momentum pair. Write the Hamiltonian in terms of $\hat{Q}$ and $\hat{P}$, show that you get a single harmonic oscillator, and read the value of the oscillation frequency ω.

2. Introduce creation and annihilation operators $\hat{a}^\dagger$ and $\hat{a}$ associated with the coordinate/momentum pair $(\hat{Q}, \hat{P})$, and consider the equation for the ground state: $\hat{a}|0\rangle = 0$. Write this condition as a partial differential equation for the ground state wave function $\psi(x, y)$. The equation will take the form

$$\mathcal{O}\Big(\frac{\partial}{\partial x}, \frac{\partial}{\partial y}; x, y\Big)\psi(x, y) = 0, \tag{1}$$

where $\mathcal{O}$ is an operator you must determine (up to an irrelevant multiplicative constant).

3. To understand equation (1) better, consider using complex variables $z \equiv x + iy$ and $\bar{z} = x - iy$ such that

$$\frac{\partial}{\partial z} = \frac{1}{2}\Big(\frac{\partial}{\partial x} - i\frac{\partial}{\partial y}\Big), \quad \frac{\partial}{\partial \bar{z}} = \frac{1}{2}\Big(\frac{\partial}{\partial x} + i\frac{\partial}{\partial y}\Big).$$

Verify that these partials with respect to z and $\bar{z}$ are consistent by letting each of them act on z and $\bar{z}$ (i.e., $\partial_z z = 1$, $\partial_{\bar{z}} z = 0$, etc.). Show that the differential equation (1) can be written as

$$\Big(\frac{\partial}{\partial \bar{z}} + \beta\, z\Big)\psi(z, \bar{z}) = 0, \tag{2}$$

and identify the constant β.

4. Consider a general solution of equation (2) of the form $\exp(-\beta z\bar{z}) f(z, \bar{z})$. What condition do you get on $f(z, \bar{z})$? Describe an infinite set of degenerate solutions $\psi_k(z, \bar{z})$ by using an integer $k = 0, 1, 2, \ldots$ to characterize the possible choices of the function f.

5. Find the value r_k of $r = \sqrt{z\bar{z}}$ for which the probability density associated with $\psi_k(z, \bar{z})$ is maximum *for large* k. With this assumption determine the number N of ground states that would fit in a sample of radius R. Express your answer in terms of R, B, and the quantum flux $\hat{\Phi}_0 = 2\pi\hbar c/q$.

6. Tying a loose end: The original Hamiltonian had a pair of commuting canonical pairs: $(\hat{x}, \hat{p}_x)$ and $(\hat{y}, \hat{p}_y)$. Find the explicit form of the second canonical pair $(\hat{Q}', \hat{P}')$ that commutes with both operators in $(\hat{Q}, \hat{P})$.

Problem 24.5. *Landau levels in the symmetric gauge: building semiclassical circular orbits.*

Consider again the Hamiltonian for a charged particle of charge q and mass m moving in a constant magnetic field B in the z-direction. Assume the motion is two-dimensional and restricted to the (x, y) plane, and the vector potential is $\mathbf{A} = \frac{1}{2}B(-y,\, x,\, 0)$. As usual, let $\omega_c = \frac{qB}{mc}$, and assume it is positive. Also, let $\ell_B = \sqrt{\frac{\hbar}{m\omega_c}}$.

1. Show that the Hamiltonian $\hat{H}$ of the system can be written as

$$\hat{H} = \hat{H}_{xy}(\tfrac{\omega_c}{2}) - \tfrac{\omega_c}{2}\hat{L}_z,$$

where $\hat{H}_{xy}$ is an isotropic two-dimensional harmonic oscillator with mass m and frequency $\omega_c/2$, and $\hat{L}_z$ is the z-component of the angular momentum.

2. For the two-dimensional oscillator part of the Hamiltonian pass to a left-right basis, as in problem 19.8. Thus, take $\hat{a}_L = \frac{1}{\sqrt{2}}(\hat{a}_x + i\hat{a}_y)$ and $\hat{a}_R = \frac{1}{\sqrt{2}}(\hat{a}_x - i\hat{a}_y)$ as well as $\hat{N}_R = \hat{a}_R^\dagger \hat{a}_R$ and $\hat{N}_L = \hat{a}_L^\dagger \hat{a}_L$. One finds that $\hat{L}_z = \hbar(\hat{N}_R - \hat{N}_L)$. Show that

$$\hat{H} = \hbar\omega_c(\hat{N}_L + \tfrac{1}{2}).$$

Write the $\hat{x}, \hat{y}, \hat{p}_x, \hat{p}_y, \hat{v}_x/\omega_c, \hat{v}_y/\omega_c, \hat{X}_0, \hat{Y}_0$ operators in terms of the left-right creation and annihilation operators, ℓ_B, and $\hbar$. Here is a partial answer:

$$\hat{p}_x = \frac{i}{2\sqrt{2}} \frac{\hbar}{\ell_B}(\hat{a}_R^\dagger - \hat{a}_R + \hat{a}_L^\dagger - \hat{a}_L).$$

Note also that $\hat{X}_0$ and $\hat{Y}_0$ depend only on the right-moving oscillators, and $\hat{v}_x/\omega_c$ and $\hat{v}_y/\omega_c$ depend only on the left-moving operators. You can also use $[\hat{X}_0, \hat{Y}_0] = -i\ell_B^2$ as a check of your calculations.

3. Now consider a normalized coherent state $|\alpha_L, \alpha_R\rangle \equiv |\alpha_L\rangle \otimes |\alpha_R\rangle$ built using the left and right operators and satisfying $\hat{a}_L|\alpha_L, \alpha_R\rangle = \alpha_L|\alpha_L, \alpha_R\rangle$, and $\hat{a}_R|\alpha_L, \alpha_R\rangle = \alpha_R|\alpha_L, \alpha_R\rangle$. Here α_L and α_R are complex numbers: $\alpha_L = |\alpha_L|e^{i\phi_L}$, and $\alpha_R = |\alpha_R|e^{i\phi_R}$.
   a. Calculate the time-dependent coherent state $|\alpha_L, \alpha_R; t\rangle$.
   b. Show that the expectation values of $\hat{X}_0$ and $\hat{Y}_0$ are time-independent and given by

$$\begin{aligned}\langle\hat{X}_0\rangle &= \sqrt{2}|\alpha_R|\ell_B\cos\phi_R\,,\\ \langle\hat{Y}_0\rangle &= -\sqrt{2}|\alpha_R|\ell_B\sin\phi_R.\end{aligned}$$

   c. Show that the time-dependent expectation values of $\hat{x} - \hat{X}_0$ and $\hat{y} - \hat{Y}_0$ take the form

$$\begin{aligned}\langle\hat{x} - \hat{X}_0\rangle &= \sqrt{2}|\alpha_L|\ell_B\cos(\phi_L - \omega_c t),\\ \langle\hat{y} - \hat{Y}_0\rangle &= \sqrt{2}|\alpha_L|\ell_B\sin(\phi_L - \omega_c t).\end{aligned}$$

   This result shows that the expectation value of the particle position follows an orbit of radius $\sqrt{2}|\alpha_L|\ell_B$, rotating clockwise with angular frequency ω_c.
   d. Show that $\langle\hat{H}\rangle = \hbar\omega_c(|\alpha_L|^2 + \frac{1}{2})$, and confirm that this is approximately the energy of a particle rotating with angular velocity ω_c on a circle of the radius previously identified.
   e. For large α_L, the uncertainties in the coherent state are small compared with expectation values, supporting the semiclassical picture. Check that in fact $\Delta X_0 = \Delta Y_0 = \ell_B/\sqrt{2}$, thus independent of α_L. Moreover, show that $\Delta H = \hbar\omega_c|\alpha_L|$.

Problem 24.6. *Landau levels in an electric field (M. Metlitski).*

1. Consider a particle of charge q and mass m confined to the (x, y) plane and moving in a uniform magnetic field $\mathbf{B} = B\hat{z}$ *and* a uniform electric field $\mathbf{E} = \mathcal{E}\hat{y}$. The Hamiltonian $\hat{H}$ is

$$\hat{H} = \frac{1}{2m}\left(\hat{\mathbf{p}} - \frac{q}{c}\mathbf{A}\right)^2 - q\mathcal{E}y.$$

Find the energy spectrum and the energy eigenstates. Is the degeneracy of each Landau level broken? [Hint: use the Landau gauge $\mathbf{A} = (-By, 0)$.]

2. Suppose that the magnetic field is as in (1), but now instead of a uniform electric field along the y-direction, we apply an arbitrary potential $V(y)$ that only depends on y:

$$\hat{H} = \frac{1}{2m}\left(\hat{\mathbf{p}} - \frac{q}{c}\mathbf{A}\right)^2 + V(y).$$

Assume that $V(y)$ is a *slowly* varying function of y. Find the approximate energy spectrum to leading order. Does your result agree with the energy spectrum in (1) to leading order in $\mathcal{E}$? [Hints: this part requires no calculations. Does $V(y)$ vary much over the support of each eigenstate in the x, y plane?]

Problem 24.7. *Electromagnetic current density in quantum mechanics.*

The probability flux in the Schrödinger equation can be related to the electromagnetic current density, provided proper attention is paid to the effects of the vector potential. Without electromagnetic fields, the probability current $\mathbf{J}$ is

$$\mathbf{J}(\mathbf{x}, t) = \frac{\hbar}{m}\text{Im}\,[\Psi^*\nabla\Psi] = \frac{1}{m}\text{Re}\Big[\Psi^*\frac{\hbar}{i}\nabla\,\Psi\Big] = \frac{1}{m}\text{Re}\Big[\Psi^*\,(\hat{\mathbf{p}}\,\Psi)\Big],$$

using $\text{Im}(z) = \text{Re}(z/i)$ and noting that $\hat{\mathbf{p}}$ is acting on the wave function to the right, as a differential opertor. Note that $\mathbf{J}(\mathbf{x}, t)$ is just a function of $\mathbf{x}$ and t, built from the wave function.

1. Derive the probability current *in the presence of electric and magnetic fields*. To do so, you can follow the strategy used to derive $\mathbf{J}$ in the absence of electromagnetic fields. Begin with the conservation equation $\partial_t\rho + \nabla\cdot\mathbf{J} = 0$, and use $\rho = \Psi^*\Psi$, as well as the Schrödinger equation for a particle in an electromagnetic field, to determine the current $\mathbf{J}$ that makes the conservation equation work. Your answer should be of the form

$$\mathbf{J}(\mathbf{x}, t) = \frac{\hbar}{m}\text{Im}\,[\Psi^*\nabla\Psi] - \ldots$$

where the missing terms can be written in terms of the vector potential $\mathbf{A}$ as well as Ψ, Ψ^* and the constants q, m, and c. The expression for the current can also be written as follows:

$$\mathbf{J}(\mathbf{x}, t) = \text{Re}\left(\Psi^*(\hat{\mathcal{O}}\,\Psi)\right),$$

where the operator $\mathcal{O}$ just acts on Ψ. Find the operator $\hat{\mathcal{O}}$.

2. Since the particle is charged, there must be a way to go from the probability current density $\mathbf{J}$ to a (quantum) electric current density $\mathbf{j}_e$. Consider the possible relation

$$\mathbf{j}_e \equiv q\mathbf{J}.$$

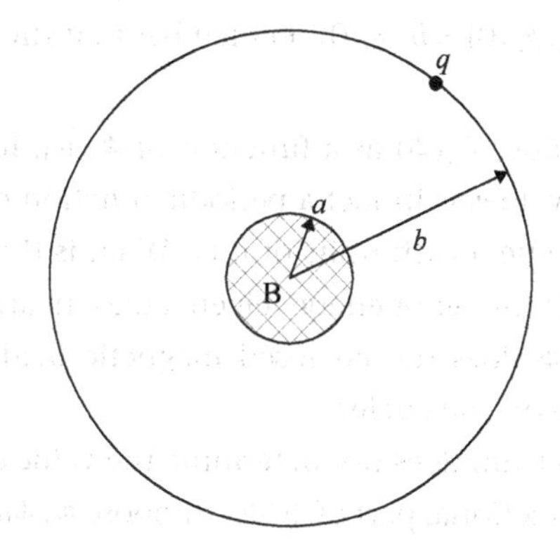

Figure 24.3
Motion of a particle on a circle of radius b. A solenoid of radius a runs perpendicular to the plane of the circle through its center.

Using Q, L, and T for units of charge, length, and time, respectively, what are the units of $\mathbf{j}_e$? Does $\mathbf{j}_e$ have the conventional units of electric current density? How would you define an electric charge density ρ_e to go along with the electric current density $\mathbf{j}_e$ satisfying charge conservation?

Is $\mathbf{J}(\mathbf{x}, t)$ a gauge-*invariant* function of $\mathbf{x}$ and t? In other words, is $\mathbf{J}'$, calculated in terms of $\mathbf{A}'$ and Ψ', identical to $\mathbf{J}$, calculated in terms of $\mathbf{A}$ and Ψ?

Problem 24.8. *The Aharonov-Bohm effect on energy eigenvalues.*

The Aharonov-Bohm effect modifies the energy eigenvalues of certain quantum mechanical systems. As an example, consider a particle of mass m and charge q constrained to move in the (x, y) plane on a circle of radius b, centered at the origin. Along the z-axis, and going through its origin, runs a solenoid of radius a, with $a < b$, carrying a magnetic field $\mathbf{B} = (0, 0, B)$ (figure 24.3). The magnetic field is uniform inside the solenoid and vanishes outside the solenoid. The magnetic flux Φ through the solenoid is $\Phi = \pi a^2 B$. As usual, $\hat{\Phi}_0 = 2\pi\hbar c/q$ is the flux quantum.

1. We want to construct a vector potential $\mathbf{A}$ that describes the magnetic field both inside and outside the solenoid. Using cylindrical coordinates z, r, ϕ, you can find a field configuration with $A_\phi(r) \neq 0$, while $A_r = A_z = 0$. Determine $A_\phi(r)$ both for $r < a$ and $r > a$. Give your answer in terms of the flux Φ, r, and a.
2. Now consider the motion on the $r = b$ circle. For this, we need a wave function $\psi(\phi)$ where ϕ is the angular variable that represents the position of the particle on the circle. Find the Hamiltonian $\hat{H}$ that describes the dynamics of such a wave function, and write it in terms of ∂_ϕ derivatives, the flux Φ and $\hat{\Phi}_0$. [Hint: since the z and r positions of the particle have been fixed externally and do not fluctuate, no z or r derivatives can act on the wave function.]
3. Solve the Schrödinger equation, and find the energy eigenvalues $E_n(\Phi)$ and the normalized energy eigenstates $\psi_n(\phi)$, with $n \in \mathbb{Z}$. (To fix conventions, require $\psi_n(0)$ to be

real and positive, and $\partial_\phi \psi_n(0) = in\psi_n(0)$. Remember that the particle is on a circle of radius b.)

4. Plot the energy eigenvalues $E_n(\Phi)$ as a function of $\Phi/\hat{\Phi}_0$, for a few values of $n \in \mathbb{Z}$. The *set* of energy eigenvalues is in fact a periodic function of Φ: the set is the same for Φ as for $\Phi + \eta$, for some η with units of flux. What is the smallest positive value of η? (Assume $\hat{\Phi}_0 > 0$.) Is the set of energy eigenvalues invariant under $\Phi \to -\Phi$?

   For what values of Φ does the enclosed magnetic field have no effect on the spectrum of the particle on the circle?

   Explain why the spectrum does not determine the value of $\Phi/\hat{\Phi}_0$ but rather gives information about the fractional part of $\Phi/\hat{\Phi}_0$. Suppose $\Phi/\hat{\Phi}_0 = k + f$, where $k \in \mathbb{Z}$ and $f \in [0, 1)$. Can the value of f be determined?

5. Suppose we introduce a defect on the circle at $\phi = 0$, which can trap the particle. There now exist new trapped states in which the wave function of the particle is localized around $\phi = 0$. For simplicity, assume the trapped state wave functions vanish outside an interval $(-\phi_0, \phi_0)$ for some $\phi_0 < \pi$. We want to show that the energy of a trapped state does *not* depend on the existence of the solenoid.

   To do this, find a gauge in which the vector potential vanishes identically in the region where the trapped state wave functions are supported. For concreteness, let's assume $\phi_0 = \pi/2$, so the trap is in the range $-\frac{\pi}{2} \leq \phi \leq \frac{\pi}{2}$. This is a piece of the full range $-\pi \leq \phi \leq \pi$ of the circle, where the points $\phi = \pm\pi$ are identified. Given the original gauge field $A_\phi(r)$ at the bead, we need to find a gauge parameter $\Lambda(\phi)$ such that

   a. A'_ϕ is zero in the interval $(-\frac{\pi}{2}, \frac{\pi}{2})$,
   b. $\oint \mathbf{A}' \cdot d\mathbf{l} = \oint \mathbf{A} \cdot d\mathbf{l}$ over the circle, as this integral is equal to the flux of $\mathbf{B}$.

   What does condition (b) imply for $\Lambda(\phi)$? (We work at $r = b$, so the r dependence of Λ need not be considered.)

   Construct a satisfactory $\Lambda(\phi)$ that is continuous, vanishes at $\phi = \pi$ and $\phi = 0$, and is piecewise linear on three intervals: $(-\pi, -\frac{\pi}{2})$, $(-\frac{\pi}{2}, \frac{\pi}{2})$, and $(\frac{\pi}{2}, \pi)$. Give the formula that describes $\Lambda(\phi)$ on these intervals. (Give your answers in terms of Φ, ϕ and numerical constants.)

   Give the formula that describes the new gauge potential $A'_\phi(\phi)$ on the same three intervals.

   Using this gauge, the vector potential vanishes at the location of the particle, and the Schrödinger equation features no electromagnetic contribution. Clearly, the physics is unaffected by the magnetic field on the solenoid. This argument does not apply to states that are not localized on the circle.

[Moral of problem: Although the particle on the circle is in a region where $\mathbf{B} = 0$, the presence of a nonzero $\mathbf{A}$ affects the energy eigenvalues of states whose wave functions extend over the whole circle. The vector potential does *not* affect the energies of localized states. This is, for the energy spectrum, the counterpart of the statement that the Aharonov-Bohm interference pattern is shifted *if and only if* the relevant paths enclose the solenoid.]

Problem 24.9. *Coupling orbital angular momentum to a magnetic field.*

Consider the kinetic term $\hat{H}_{\rm kin}$ of the Pauli Hamiltonian for an electron:

$$\hat{H}_{\rm kin} = \frac{1}{2m}\left(\hat{\mathbf{p}} + \frac{e}{c}\mathbf{A}\right)^2 .$$

Assume there is a constant, uniform magnetic field of magnitude B along the positive z-direction. That magnetic field can be represented by the vector potential: $(A_x, A_y, A_z) = \frac{1}{2}B(-y, x, 0)$. Expand $\hat{H}_{\rm kin}$ and determine the term $\delta\hat{H}_B$ linear in B. Write this term in terms of angular momentum operators ($\hat{L}_x$, $\hat{L}_y$, or $\hat{L}_z$), B, and other constants.

Problem 24.10. *Matrices for the Dirac equation.*

We claimed that one cannot find matrices $\alpha_1, \alpha_2, \alpha_3$ and β that are traceless and Hermitian, each squaring to the identity and satisfying the additional conditions: $\{\alpha_i, \alpha_j\} = 0$ for $i \neq j$, and $\{\alpha_i, \beta\} = 0$ for $i = 1, 2, 3$. We can easily write an ansatz that constructs traceless, Hermitian matrices:

$$\alpha_i = \mathbf{a}_{(i)} \cdot \boldsymbol{\sigma}\,, \quad i = 1, 2, 3, \quad \beta = \mathbf{b} \cdot \boldsymbol{\sigma}\,,$$

with real vectors $\mathbf{a}_{(1)}, \mathbf{a}_{(2)}, \mathbf{a}_{(2)}$, and $\mathbf{b}$. Formulate the conditions required for the α_i and β matrices in terms of conditions on the four real vectors. Show that there is no possible solution to these conditions.

Problem 24.11. *Large and small spinors in the Dirac equation.*

Consider the Dirac Hamiltonian (24.7.10) in the absence of electromagnetic fields. Let $\psi = \begin{pmatrix}\chi \\ \eta\end{pmatrix}$ be an energy eigenstate of relativistic energy E, where χ and η are two-component spinors. Write the equation $\hat{H}_{\rm Dirac}\psi = E\psi$ in the form $(\hat{H}_{\rm Dirac} - E\mathbb{1})\psi = 0$, using 2×2 blocks. Solve for η in terms of χ *without* having to invert any matrix.

Show that for a nonrelativistic particle $\eta \ll \chi$, meaning that η is obtained from χ by the action of an operator with small eigenvalues. Derive the equation of motion of χ in the nonrelativistic approximation. This must be the Pauli equation for a free spin one-half particle.

25 Time-Independent Perturbation Theory

We consider time-independent Hamiltonians with a known spectrum perturbed by additional time-independent interactions. These perturbations affect nondegenerate states and degenerate ones differently. We first consider nondegenerate states, finding the corrections to the states and their energies. Degenerate states require consideration of a "good basis" in the degenerate subspace. Their treatment differs depending on whether the perturbation lifts the degeneracies to first order or not. We consider both situations, including the possibility that the degeneracies are only lifted to second order in the perturbation. The methods are illustrated with a detailed analysis of the fine structure of hydrogen, a set of corrections to the spectrum suppressed by a factor of order α^2. Placing a hydrogen atom in a weak uniform magnetic field, we find the removal of all degeneracies, with all multiplets splitting into equally spaced levels. This is the weak-field Zeeman effect.

25.1 Time-Independent Perturbations

It often happens that the Hamiltonian of a system differs slightly from a Hamiltonian that is well studied and completely understood. This is a case in which perturbation theory can be useful. Perturbation theory allows us to make statements about the Hamiltonian of the system using what we know about the well-studied Hamiltonian. In this chapter we will consider the situation in which both the well-known Hamiltonian and the Hamiltonian of interest are time independent.

The well-studied Hamiltonian could be that of the simple harmonic oscillator in one, two, or three dimensions. In a diatomic molecule, for example, the potential that controls the vibrations is not exactly quadratic; it has extra terms that make the vibrations slightly anharmonic. In that situation, the extra terms in the potential represent perturbations of the Hamiltonian. The hydrogen atom Hamiltonian is also a well-understood system. If we place the atom inside a weak external magnetic field or electric field, the situation is described by adding some small terms to the hydrogen Hamiltonian. Similarly, the interaction between the magnetic moments of the proton and the electron can be incorporated by modifying the original hydrogen atom Hamiltonian. The interaction between two neutral hydrogen atoms at a distance, leading to the van der Waals force, can be studied in perturbation theory by thinking of the two atoms as electric dipoles.

The Hamiltonian of interest is written as the sum of an understood, original Hamiltonian $\hat{H}^{(0)}$ and a perturbation δH:

$$\hat{H}^{(0)} + \delta H . \tag{25.1.1}$$

Both $\hat{H}^{(0)}$ and δH are time independent. Since $\hat{H}^{(0)}$ is Hermitian and the sum must be a Hermitian Hamiltonian, the perturbation operator δH must also be Hermitian. It is convenient to introduce a unit-free constant $\lambda \in [0, 1]$ and to consider, instead, a λ-dependent Hamiltonian $\hat{H}(\lambda)$ that takes the form

$$\hat{H}(\lambda) = \hat{H}^{(0)} + \lambda\, \delta H . \tag{25.1.2}$$

Here λ allows us to consider a family of Hamiltonians that interpolate from $\hat{H}^{(0)}$, when λ is equal to zero, to the Hamiltonian of interest, when λ is equal to one. In many cases perturbations can be considered for different values of their intensity; think, for example, of an atom in an external magnetic field whose magnitude can be set to any of a continuum of values. We may be particularly interested in fields of magnitude 1 gauss, but knowing what happens when the magnitude is 0.01 gauss could also be useful. We can view λ as the parameter that allows us to adjust the perturbation; the perturbation is said to be *on* when $\lambda \neq 0$. The parameter λ is also useful in organizing the perturbation analysis, as we will see below.

We spoke of a Hamiltonian that differs slightly from $\hat{H}^{(0)}$. In order to use perturbation theory, we need $\lambda \delta H$ to be a small perturbation of the Hamiltonian $\hat{H}^{(0)}$. We will have to deal with the meaning of *small*. At first sight we may imagine that small means that, viewed as matrices, the largest entries in $\lambda \delta H$ are smaller than the largest entries in $\hat{H}^{(0)}$. While this is necessary, more is needed, as we will see soon enough. An additional advantage of using λ is that by taking it to be sufficiently small we can surely make $\lambda \delta H$ small.

We assume that the Hamiltonian $\hat{H}^{(0)}$ is understood—namely, we know the eigenstates and eigenvalues of $\hat{H}^{(0)}$. Our goal is to find the eigenstates and eigenvalues of $\hat{H}(\lambda)$. One may be able to calculate those exactly, but this is rarely possible. Diagonalizing δH is seldom useful to find exact eigenstates since δH and H do not generally commute, and therefore δH eigenstates need not be eigenstates of $\hat{H}(\lambda)$. In perturbation theory, the key assumption is that the eigenvalues and eigenvectors of $\hat{H}(\lambda)$ can be found as series expansions in positive powers of λ. We hope, of course, that there are some values of λ for which the series converges, or at least gives useful information.

In figure 25.1 we illustrate some of the phenomena that can occur in the spectrum of a system with Hamiltonian $\hat{H}(\lambda)$, displaying how the energies of the various states may change as the parameter λ is increased from zero. The two lowest-energy eigenstates of $\hat{H}^{(0)}$ are nondegenerate, and their energies can go up and down as λ varies. Next up in energy, we have two degenerate states of $\hat{H}^{(0)}$, represented by the two heavy dots to the left of the vertical axis. The perturbation splits the two levels, which are

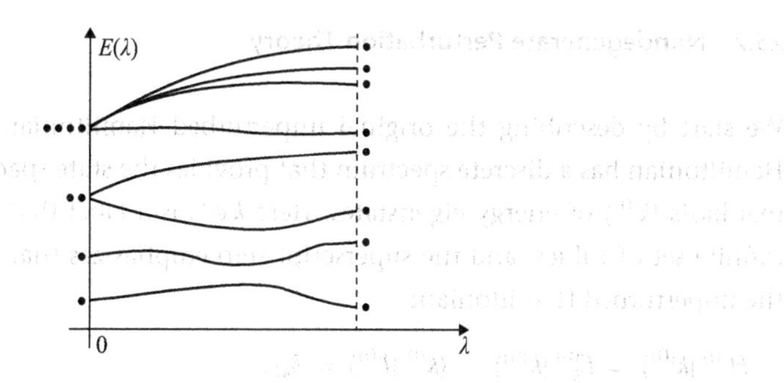

Figure 25.1
The energy eigenvalues of $\hat{H}(\lambda)$ change as λ goes from zero to one. On the $\lambda=0$ vertical axis, the $\hat{H}^{(0)}$ eigenstates are represented by heavy dots. For $\lambda=1$, the dots have shifted.

no longer degenerate when $\lambda \neq 0$. The split is happening to first order in λ, as shown by the different slopes of the two curves that emerge from the $\lambda=0$ energy. In other words, viewed as power series in λ the energies of the two states have different linear terms in λ. The last level shown in the figure represents four degenerate states. The perturbation to first order in λ splits the states into a group of three states and a fourth. To second order in λ, the three states split further. A single Hamiltonian can exhibit behavior like this, with many possible variations. Note that the perturbations do not create new states; they only modify the eigenstates of the $\hat{H}^{(0)}$ theory. In perturbation theory we will be assuming that the state space of $\hat{H}^{(0)}$, called $\mathcal{H}$, is also the state space of $\hat{H}(\lambda)$. The eigenstates of $\hat{H}(\lambda)$ are ultimately going to be expressed as superpositions of $\hat{H}^{(0)}$ eigenstates.

In analyzing the behavior of states and energies as functions of λ, we encounter two different situations:

1. We are following a nondegenerate state, or
2. we are following a collection of degenerate states.

The challenges are quite different and require separate analysis. Clearly, both situations can occur for a single Hamiltonian, depending on the spectrum of $\hat{H}^{(0)}$. To follow a nondegenerate state, we use *nondegenerate perturbation theory*. It does not matter that there are degeneracies at other levels. To follow a set of degenerate states, we use *degenerate perturbation theory*. Since Hamiltonians generally have both nondegenerate and degenerate states, we need to consider both types of perturbation theory. We begin with nondegenerate perturbation theory. Throughout this chapter we are doing time-independent perturbation theory. Not only are $\hat{H}^{(0)}$ and δH time independent, but λ is just a constant. When we consider different values of λ, we do not think of λ as a function of time. If it were, we would be doing time-dependent perturbation theory. That we will do in later chapters. A brief summary of some of the results of this chapter was provided in section 20.3, in order to discuss some applications of addition of angular momentum.

25.2 Nondegenerate Perturbation Theory

We start by describing the original unperturbed Hamiltonian $\hat{H}^{(0)}$. We assume this Hamiltonian has a discrete spectrum that provides the state space $\mathcal{H}$ with an orthonormal basis $|k^{(0)}\rangle$ of energy eigenstates. Here $k \in \mathbb{Z}$ is a label that ranges over a possibly infinite set of values, and the superscript zero emphasizes that this is an eigenstate of the unperturbed Hamiltonian:

$$\hat{H}^{(0)}\,|k^{(0)}\rangle \;=\; E_k^{(0)}\,|k^{(0)}\rangle\,, \quad \langle k^{(0)}|l^{(0)}\rangle \;=\; \delta_{kl}\,. \tag{25.2.1}$$

We will let $k=0$ denote the ground state, and we order the states so that the energies generally increase as the value of the label increases:

$$E_0^{(0)} \;\le E_1^{(0)} \;\le E_2^{(0)} \;\le \cdots\,. \tag{25.2.2}$$

The equal signs are needed because some states may be degenerate.

In this section we focus on a nondegenerate state $|n^{(0)}\rangle$ with fixed n. This means that $|n^{(0)}\rangle$ is a single state that is separated by some finite energy from all the states with more energy and from all the states with less energy. In other words, the following must be part of the sequence of inequalities in (25.2.2):

$$\cdots \le E_{n-1}^{(0)} \;<\; E_n^{(0)} \;<\; E_{n+1}^{(0)} \;\le \cdots\,. \tag{25.2.3}$$

If the chosen nondegenerate state is the ground state, we have $n=0$ and $E_0^{(0)} \;<\; E_1^{(0)}$.

As the perturbation is on by setting $\lambda \neq 0$, the energy eigenstate $|n^{(0)}\rangle$ of $\hat{H}^{(0)}$ will become some energy eigenstate $|n\rangle_\lambda$ of $\hat{H}(\lambda)$ with energy $E_n(\lambda)$:

$$\hat{H}(\lambda)|n\rangle_\lambda \;=\; E_n(\lambda)\,|n\rangle_\lambda\,, \tag{25.2.4}$$

where

$$|n\rangle_{\lambda=0} \;=\; |n^{(0)}\rangle\,, \quad \text{and} \quad E_n(\lambda=0) \;=\; E_n^{(0)}\,. \tag{25.2.5}$$

As we said, the state and its energy are assumed to take the form of a regular power series expansion in λ. To make this clear, consider a function $f(\lambda)$ for which derivatives of all orders exist at $\lambda=0$. In that case we have a Taylor expansion:

$$f(\lambda) = \sum_{n=0}^{\infty} \frac{1}{n!} f^{(n)}(0)\,\lambda^n \;=\; f(0) + f'(0)\,\lambda + \tfrac{1}{2} f''(0)\,\lambda^2 + \tfrac{1}{3!} f'''(0)\,\lambda^3 + \cdots\,. \tag{25.2.6}$$

The expansion is a power series in λ, with coefficients $f(0), f'(0), \cdots$ that are λ independent, being the value of the function and its derivatives at $\lambda=0$.

For our problem we note the values of $|n\rangle_\lambda$ and $E_n(\lambda)$ for $\lambda=0$ in (25.2.5) and postulate the expansion:

$$\begin{aligned} |n\rangle_\lambda \;&=\; |n^{(0)}\rangle + \lambda|n^{(1)}\rangle + \lambda^2|n^{(2)}\rangle + \lambda^3|n^{(3)}\rangle + \cdots\,, \\ E_n(\lambda) \;&=\; E_n^{(0)} \;+ \lambda\, E_n^{(1)} \;+ \lambda^2\, E_n^{(2)} \;+ \lambda^3\, E_n^{(3)} \;+ \cdots\,. \end{aligned} \tag{25.2.7}$$

The superscripts on the states and energies denote the power of λ that accompanies them in the above expressions. The above equations are a natural assumption; they state that the perturbed states and energies, being functions of λ, admit a Taylor expansion around $\lambda = 0$. This also means that the state and its energy vary continuously as a function of λ. Our aim is to calculate the states

$$|n^{(1)}\rangle,\ |n^{(2)}\rangle,\ |n^{(3)}\rangle, \ldots \tag{25.2.8}$$

and the energies

$$E_n^{(1)},\ E_n^{(2)},\ E_n^{(3)},\ \ldots. \tag{25.2.9}$$

All these states and energies are, by definition, λ independent. Here $|n^{(1)}\rangle$ is the leading correction to the state $|n^{(0)}\rangle$ for $\lambda \neq 0$. Similarly, $E_n^{(1)}$ is the leading correction to the energy for $\lambda \neq 0$. Since we assume that the state space $\mathcal{H}$ of $\hat{H}^{(0)}$ is also the state space of $\hat{H}(\lambda)$, the states $|n^{(k)}\rangle$ above will be determined in terms of superpositions of $\hat{H}^{(0)}$ eigenstates. We will *not* impose the requirement that $|n\rangle_\lambda$ is normalized. It suffices that $|n\rangle_\lambda$ is normalizable, which it will be for sufficiently small perturbations.

For $\lambda = 1$, (25.2.7) gives us the state and the energy when $\hat{H} = \hat{H}^{(0)} + \delta H$:

$$\begin{aligned} |n\rangle \equiv |n\rangle_1 &= |n^{(0)}\rangle + |n^{(1)}\rangle + |n^{(2)}\rangle + |n^{(3)}\rangle + \cdots, \\ E_n \equiv E_n(1) &= E_n^{(0)} + E_n^{(1)} + E_n^{(2)} + E_n^{(3)} + \cdots. \end{aligned} \tag{25.2.10}$$

It is now time to do the work! We rewrite the Schrödinger equation (25.2.4) in the form

$$\big(\hat{H}^{(0)} + \lambda\delta H - E_n(\lambda)\big)|n\rangle_\lambda = 0 \tag{25.2.11}$$

and use the λ expansion of $|n\rangle_\lambda$ and $E_n(\lambda)$ in (25.2.7) to find that

$$\begin{gathered} \Big((\hat{H}^{(0)} - E_n^{(0)}) - \lambda(E_n^{(1)} - \delta H) - \lambda^2 E_n^{(2)} - \lambda^3 E_n^{(3)} - \cdots - \lambda^k E_n^{(k)} + \cdots\Big) \\ \Big(|n^{(0)}\rangle + \lambda|n^{(1)}\rangle + \lambda^2|n^{(2)}\rangle + \lambda^3|n^{(3)}\rangle + \cdots + \lambda^k|n^{(k)}\rangle + \cdots\Big) = 0. \end{gathered} \tag{25.2.12}$$

Multiplying out we get a power series in λ whose coefficients are λ-independent states in the state space $\mathcal{H}$. If this is to vanish for all values of λ, every coefficient must be equal to zero. Collecting the coefficients for each power of λ, you should confirm that we get the following set of conditions:

$$\begin{aligned} \lambda^0:&\quad (\hat{H}^{(0)} - E_n^{(0)})\,|n^{(0)}\rangle = 0, \\ \lambda^1:&\quad (\hat{H}^{(0)} - E_n^{(0)})\,|n^{(1)}\rangle = (E_n^{(1)} - \delta H)|n^{(0)}\rangle, \\ \lambda^2:&\quad (\hat{H}^{(0)} - E_n^{(0)})\,|n^{(2)}\rangle = (E_n^{(1)} - \delta H)|n^{(1)}\rangle + E_n^{(2)}\,|n^{(0)}\rangle, \\ \lambda^3:&\quad (\hat{H}^{(0)} - E_n^{(0)})\,|n^{(3)}\rangle = (E_n^{(1)} - \delta H)|n^{(2)}\rangle + E_n^{(2)}\,|n^{(1)}\rangle + E_n^{(3)}\,|n^{(0)}\rangle, \\ &\quad\vdots \qquad\qquad \vdots \qquad\qquad \vdots \\ \lambda^k:&\quad (\hat{H}^{(0)} - E_n^{(0)})\,|n^{(k)}\rangle = (E_n^{(1)} - \delta H)|n^{(k-1)}\rangle + E_n^{(2)}\,|n^{(k-2)}\rangle + \ldots + E_n^{(k)}\,|n^{(0)}\rangle. \end{aligned} \tag{25.2.13}$$

Each equation is the condition that the coefficient multiplying the power of λ indicated to the left vanishes. That power is reflected as the sum of superscripts on each term, counting δH as having superscript one, which is reasonable since δH appears in $\hat{H}(\lambda)$ multiplied by one power of λ. This gives a simple consistency check on our equations. These are equations for the kets $|n^{(1)}\rangle, |n^{(2)}\rangle, \ldots$ as well as for the energy corrections $E_n^{(1)}, E_n^{(2)}, \ldots$. Since λ does not enter into the equations, the kets and energy corrections are manifestly λ independent.

The first equation, corresponding to λ^0, is satisfied by construction. The second equation, corresponding to λ, should allow us to solve for the first correction $|n^{(1)}\rangle$ to the state and the first correction $E_n^{(1)}$ to the energy. Once these are known, the equation corresponding to λ^2 involves only the unknowns $|n^{(2)}\rangle$ and $E_n^{(2)}$ and can be used to determine them. Working recursively, at each stage each equation has only two unknowns: a state correction $|n^{(k)}\rangle$ and an energy correction $E_n^{(k)}$.

A useful choice We now claim that, without loss of generality, we can assume that all the state corrections $|n^{(k)}\rangle$, with $k \geq 1$, contain no vector along the unperturbed state $|n^{(0)}\rangle$. Explicitly,

$$\boxed{0 = \langle n^{(0)}|n^{(1)}\rangle = \langle n^{(0)}|n^{(2)}\rangle = \langle n^{(0)}|n^{(3)}\rangle = \ldots.} \tag{25.2.14}$$

To show this we explain how we can manipulate a solution that does not have this property into one that does. Suppose you have a solution in which the state corrections $|n^{(k)}\rangle$ have components along $|n^{(0)}\rangle$:

$$|n^{(k)}\rangle = |n^{(k)}\rangle' - a_k|n^{(0)}\rangle, \quad k \geq 1, \tag{25.2.15}$$

with some constants a_k and with $|n^{(k)}\rangle'$ orthogonal to $|n^{(0)}\rangle$. Then the solution for the full corrected state is

$$\begin{aligned} |n\rangle_\lambda &= |n^{(0)}\rangle + \lambda\left(|n^{(1)}\rangle' - a_1|n^{(0)}\rangle\right) + \lambda^2\left(|n^{(2)}\rangle' - a_2|n^{(0)}\rangle\right) + \cdots \\ &= (1 - a_1\lambda - a_2\lambda^2 - \cdots)|n^{(0)}\rangle + \lambda|n^{(1)}\rangle' + \lambda^2|n^{(2)}\rangle' + \cdots. \end{aligned} \tag{25.2.16}$$

Since this is an eigenstate of the Hamiltonian $\hat{H}(\lambda)$, it will still be an eigenstate if we change its normalization by dividing it by any function of λ. Dividing by the coefficient of $|n^{(0)}\rangle$, we have the physically identical solution $|n\rangle'_\lambda$ given by

$$|n\rangle'_\lambda = |n^{(0)}\rangle + \frac{1}{(1 - a_1\lambda - a_2\lambda^2 - \cdots)}\left[\lambda|n^{(1)}\rangle' + \lambda^2|n^{(2)}\rangle' + \cdots\right]. \tag{25.2.17}$$

We can Taylor expand the denominator so that we get

$$|n\rangle'_\lambda = |n^{(0)}\rangle + \lambda|n^{(1)}\rangle' + \lambda^2\left(|n^{(2)}\rangle' + a_1|n^{(1)}\rangle'\right) + \cdots. \tag{25.2.18}$$

The explicit expressions do not matter. The key point, actually visible in (25.2.17), is that we have a physically identical solution of the same equation in which the state corrections are all orthogonal to $|n^{(0)}\rangle$. This shows that we can always impose conditions (25.2.14).

Solving the equations Let us finally begin solving equations (25.2.13). For this, note that the Schrodinger equation for the ket $|n^{(0)}\rangle$ implies that for the bra we have

$$\langle n^{(0)}|(\hat{H}^{(0)} - E_n^{(0)}) = 0 . \tag{25.2.19}$$

This means that acting with $\langle n^{(0)}|$ on the left-hand side of any of the equations in (25.2.13) will give zero. Consistency requires that acting with $\langle n^{(0)}|$ on the right-hand side of any of the equations in (25.2.13) also gives zero and, presumably, some interesting information. For the equation arising from order λ, this gives

$$0 = \langle n^{(0)}|(E_n^{(1)} - \delta H)|n^{(0)}\rangle . \tag{25.2.20}$$

Since $|n^{(0)}\rangle$ is normalized and $E_n^{(1)}$ is a number, it follows that

$$\boxed{E_n^{(1)} = \langle n^{(0)}|\delta H|n^{(0)}\rangle.} \tag{25.2.21}$$

This is the *most famous* result in perturbation theory: the first correction to the energy of a nondegenerate energy eigenstate is simply the expectation value of the correction to the Hamiltonian in the *uncorrected* state. You need not know the correction to the state to determine the first correction to the energy! The Hermiticity of δH implies that, as required, the energy correction is a real number.

Using this method, we can find some interesting though not fully explicit formulae for the higher-energy corrections. For the λ^2 equation, acting with $\langle n^{(0)}|$ on the right-hand side gives

$$0 = \langle n^{(0)}|\Big((E_n^{(1)} - \delta H)|n^{(1)}\rangle + E_n^{(2)}\,|n^{(0)}\rangle\Big) . \tag{25.2.22}$$

Recalling our orthogonality assumption, we have $\langle n^{(0)}|n^{(1)}\rangle = 0$, and the term with $E_n^{(1)}$ drops out. We then get

$$E_n^{(2)} = \langle n^{(0)}|\delta H|n^{(1)}\rangle , \tag{25.2.23}$$

which states that the second correction to the energy is determined if we have the first correction $|n^{(1)}\rangle$ to the state. Note that this expression is not explicit enough to make it manifest that $E_n^{(2)}$ is real. This, and the earlier result for the first correction to the energy, has a simple generalization. Acting with $\langle n^{(0)}|$ on the last equation of (25.2.13), we get

$$0 = \langle n^{(0)}|\Big(\,(E_n^{(1)} - \delta H)|n^{(k-1)}\rangle + E_n^{(2)}\,|n^{(k-2)}\rangle + \cdots + E_n^{(k)}\,|n^{(0)}\rangle\Big) . \tag{25.2.24}$$

Using the orthogonality of $|n^{(0)}\rangle$ and all the state corrections, we have

$$0 = -\langle n^{(0)}|\delta H|n^{k-1}\rangle + E_n^{(k)} , \tag{25.2.25}$$

and therefore,

$$\boxed{E_n^{(k)} = \langle n^{(0)}|\,\delta H\,|n^{(k-1)}\rangle .} \tag{25.2.26}$$

At any stage of the recursive solution, the energy at a fixed order is known if the state correction is known to previous order. So it is time to calculate the corrections to the states!

Let us solve for the first correction $|n^{(1)}\rangle$ to the state. Since $|n^{(1)}\rangle$ is some particular superposition of the original energy eigenstates $|k^{(0)}\rangle$, we know $|n^{(1)}\rangle$ if we know its components along all the $|k^{(0)}\rangle$. For this we look at the order λ equation:

$$(\hat{H}^{(0)} - E_n^{(0)})\,|n^{(1)}\rangle \;=\; (E_n^{(1)} - \delta H)|n^{(0)}\rangle\,. \tag{25.2.27}$$

This is a vector equation: the left-hand-side vector is set equal to the right-hand-side vector. As with any vector equation, all of its information can be extracted using a basis set of vectors: forming the inner product of each and every basis vector with both the left-hand side and the right-hand side, we must get equal numbers. We already acted on the above equation with $\langle n^{(0)}|$ to find $E_n^{(1)}$. The remaining information in this equation can be obtained by acting with all the states $\langle k^{(0)}|$, with $k \neq n$:

$$\langle k^{(0)}|(\hat{H}^{(0)} - E_n^{(0)})\,|n^{(1)}\rangle \;=\; \langle k^{(0)}|(E_n^{(1)} - \delta H)|n^{(0)}\rangle\,, \quad k \neq n\,. \tag{25.2.28}$$

On the left-hand side, we can let $\hat{H}^{(0)}$ act on the bra. On the right-hand side, because $k \neq n$ the term with $E_n^{(1)}$ vanishes. We then have

$$(E_k^{(0)} - E_n^{(0)})\,\langle k^{(0)}\,|\,n^{(1)}\rangle \;=\; -\langle k^{(0)}|\delta H|n^{(0)}\rangle\,, \quad k \neq n\,. \tag{25.2.29}$$

To simplify notation we define a shorthand for the matrix elements of δH in the original basis:

$$\delta H_{mn} \;\equiv\; \langle m^{(0)}|\delta H|n^{(0)}\rangle\,. \tag{25.2.30}$$

Note that the Hermiticity of δH implies that

$$\delta H_{nm} = (\delta H_{mn})^*\,. \tag{25.2.31}$$

With this notation, equation (25.2.29) gives

$$\langle k^{(0)}\,|\,n^{(1)}\rangle \;=\; -\frac{\delta H_{kn}}{E_k^{(0)} - E_n^{(0)}}\,, \qquad k \neq n\,. \tag{25.2.32}$$

Since we now know its component along each basis state, $|n^{(1)}\rangle$ has been determined. Indeed, we can use the completeness of the basis to write

$$|n^{(1)}\rangle \;=\; \sum_k |k^{(0)}\rangle\langle k^{(0)}|n^{(1)}\rangle = \sum_{k\neq n} |k^{(0)}\rangle\langle k^{(0)}|n^{(1)}\rangle\,, \tag{25.2.33}$$

since the term with $k = n$ vanishes because of the orthogonality assumption. Using the overlaps (25.2.32), we finally get

$$\boxed{|n^{(1)}\rangle \;=\; -\sum_{k\neq n} \frac{|k^{(0)}\rangle\delta H_{kn}}{E_k^{(0)} - E_n^{(0)}}\,.} \tag{25.2.34}$$

The first correction $|n^{(1)}\rangle$ can have components along all basis states except $|n^{(0)}\rangle$. The component along a state $|k^{(0)}\rangle$ vanishes if the perturbation δH does not couple $|n^{(0)}\rangle$ to $|k^{(0)}\rangle$—namely, if δH_{kn} vanishes. The assumption of nondegeneracy is needed here for the answer to make sense. We are summing over all states $|k^{(0)}\rangle \neq |n^{(0)}\rangle$; if any such state had energy $E_n^{(0)}$, the energy denominator would vanish, causing trouble!

Now that we have the first-order correction to the states, the *second-order* correction to the energy follows from (25.2.23):

$$E_n^{(2)} = \langle n^{(0)}|\delta H|n^{(1)}\rangle = -\sum_{k\neq n} \frac{\langle n^{(0)}|\delta H|k^{(0)}\rangle \delta H_{kn}}{E_k^{(0)} - E_n^{(0)}}. \tag{25.2.35}$$

In the last numerator, we have $\langle n^{(0)}|\delta H|k^{(0)}\rangle = \delta H_{nk} = (\delta H_{kn})^*$, and therefore

$$E_n^{(2)} = -\sum_{k\neq n} \frac{|\delta H_{kn}|^2}{E_k^{(0)} - E_n^{(0)}}. \tag{25.2.36}$$

This is the second-order energy correction. This explicit formula makes the reality of $E_n^{(2)}$ manifest. In summary, going back to (25.2.7) we have that the states and energies for $\hat{H}(\lambda) = \hat{H}^{(0)} + \lambda\delta H$ are given by

$$\boxed{\begin{aligned} |n\rangle_\lambda &= |n^{(0)}\rangle - \lambda \sum_{k\neq n} \frac{\delta H_{kn}}{E_k^{(0)} - E_n^{(0)}} |k^{(0)}\rangle + \mathcal{O}(\lambda^2), \\ E_n(\lambda) &= E_n^{(0)} + \lambda\, \delta H_{nn} - \lambda^2 \sum_{k\neq n} \frac{|\delta H_{kn}|^2}{E_k^{(0)} - E_n^{(0)}} + \mathcal{O}(\lambda^3). \end{aligned}} \tag{25.2.37}$$

Remarks:

1. The energy of the (nondegenerate) ground state to first order in λ *overstates* the exact ground state energy. To see this consider the ground state energy $E_0^{(0)} + \lambda E_0^{(1)}$ to first order in λ. Writing this in terms of expectation values, with $|0^{(0)}\rangle$ denoting the unperturbed ground state, we have

$$\begin{aligned} E_0^{(0)} + \lambda E_0^{(1)} &= \langle 0^{(0)}|\hat{H}^{(0)}|0^{(0)}\rangle + \lambda\langle 0^{(0)}|\delta H|0^{(0)}\rangle \\ &= \langle 0^{(0)}|(\hat{H}^{(0)} + \lambda\delta H)|0^{(0)}\rangle \\ &= \langle 0^{(0)}|\hat{H}(\lambda)|0^{(0)}\rangle. \end{aligned} \tag{25.2.38}$$

By the variational principle, the expectation value of the Hamiltonian on an arbitrary (normalized) state is larger than the ground state energy $E_0(\lambda)$; therefore,

$$E_0^{(0)} + \lambda E_0^{(1)} = \langle 0^{(0)}|H(\lambda)|0^{(0)}\rangle \geq E_0(\lambda), \tag{25.2.39}$$

which is what we wanted to prove. Given this overestimate at first order, the second-order correction to the ground state energy is always negative. Indeed, using (25.2.36) for $n = 0$,

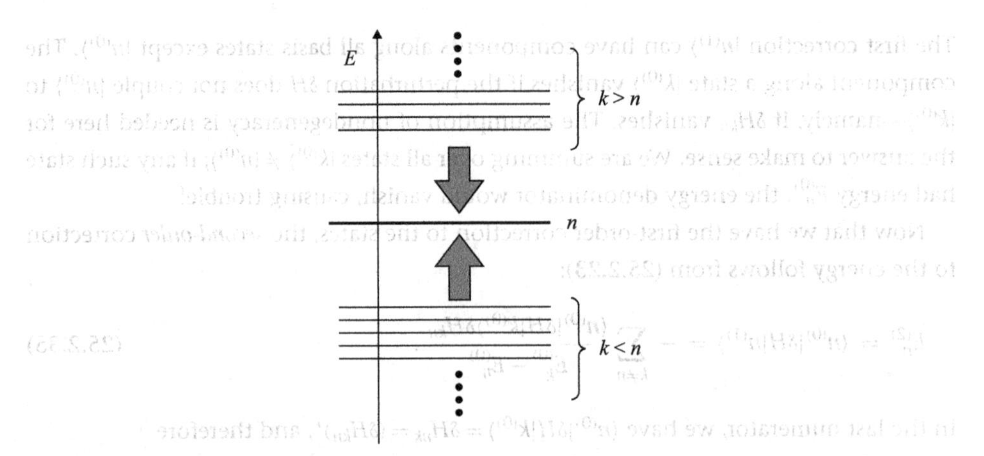

Figure 25.2
The second-order correction to the energy of the state $|n^{(0)}\rangle$ receives negative contributions from the higher-energy states and positive contributions from the lower-energy states. Visualizing states on the energy scale, we have an effective repulsion preventing the state $|n^{(0)}\rangle$ from approaching the neighboring states.

$$E_0^{(2)} = -\lambda^2 \sum_{k\neq 0} \frac{|\delta H_{k0}|^2}{E_k^{(0)} - E_0^{(0)}} < 0, \tag{25.2.40}$$

with each term negative because the unperturbed excited state energies $E_k^{(0)}$ $(k \neq 0)$ exceed the unperturbed ground state energy $E_0^{(0)}$.

2. The second-order correction to the energy of the $|n^{(0)}\rangle$ eigenstate exhibits *level repulsion*: the levels with $k > n$ push the energy of the state down, and the levels with $k < n$ push the energy of the state up. Indeed,

$$E_n^{(2)} = -\lambda^2 \sum_{k\neq n} \frac{|\delta H_{kn}|^2}{E_k^{(0)} - E_n^{(0)}} = -\lambda^2 \sum_{k>n} \frac{|\delta H_{kn}|^2}{E_k^{(0)} - E_n^{(0)}} + \lambda^2 \sum_{k<n} \frac{|\delta H_{kn}|^2}{E_n^{(0)} - E_k^{(0)}}. \tag{25.2.41}$$

The first term on the final right-hand side gives the negative contribution from the higher-energy states, and the second term gives the positive contribution from the lower-energy states (see figure 25.2).

The systematics of solving the equations is now apparent. For each equation the full content is obtained by taking inner products with all states in the state space. Taking the inner product with $\langle n^{(0)}|$ gives us the energy as in (25.2.26). Taking the inner product with $\langle k^{(0)}|$ with all $k \neq n$ gives us the state. It is a good exercise to calculate $|n^{(2)}\rangle$ and $E_n^{(3)}$ (problem 25.3).

Exercise 25.1. *The state $|n\rangle_\lambda$ is not normalized. Use (25.2.37) to calculate to order λ^2 the quantity $Z_n(\lambda)$ defined by*

$$\frac{1}{Z_n(\lambda)} \equiv {}_\lambda\langle n|n\rangle_\lambda. \tag{25.2.42}$$

What is the probability that the state $|n\rangle_\lambda$ will be found along its unperturbed version $|n^{(0)}\rangle$?

Validity of the perturbation expansion What do we mean when we say that $\lambda\delta H$ is small? While we expect that $\lambda\delta H$ must be small compared to the original Hamiltonian $\hat{H}^{(0)}$, it is not a priori clear what this means, as both objects are operators. We will not attempt a general discussion of the issue. Rather, we will consider an example that gives significant insight.

Let $\hat{H}^{(0)}$ be a 2×2 diagonal matrix with nondegenerate eigenvalues:

$$\hat{H}^{(0)} = \begin{pmatrix} E_1^{(0)} & 0 \\ 0 & E_2^{(0)} \end{pmatrix}, \quad E_1^{(0)} \neq E_2^{(0)}. \tag{25.2.43}$$

We assume that the perturbation, called $\lambda \hat{V}$, has only off-diagonal elements so that

$$\hat{H}(\lambda) = \hat{H}^{(0)} + \lambda \hat{V} \equiv \begin{pmatrix} E_1^{(0)} & \lambda V \\ \lambda V^* & E_2^{(0)} \end{pmatrix}, \tag{25.2.44}$$

with V a complex number. In this simple example, there is no need to use perturbation theory since the eigenvalues, E_+ and E_-, can be calculated exactly as functions of λ. You can quickly check that

$$E_\pm(\lambda) = \tfrac{1}{2}(E_1^{(0)} + E_2^{(0)}) \pm \tfrac{1}{2}(E_1^{(0)} - E_2^{(0)}) \sqrt{1 + \left[\frac{\lambda |V|}{\frac{1}{2}(E_1^{(0)} - E_2^{(0)})}\right]^2}. \tag{25.2.45}$$

The perturbative expansion of the energies is obtained by Taylor expansion of the square root in powers of λ. The required expansion can be calculated for a variable z:

$$f(z) \equiv \sqrt{1+z^2} = 1 + \frac{z^2}{2} - \frac{z^4}{8} + \frac{z^6}{16} + \mathcal{O}(z^8). \tag{25.2.46}$$

While we only need the result for z real, the expansion holds for complex z. In fact, the expansion of $f(z)$ around $z=0$ has a radius of convergence equal to one: the series converges for $|z| < 1$ and diverges for $|z| > 1$. This can be understood using ideas of complex analysis. The function $f(z)$ is multivalued because of the square root and exhibits branch points at $z = \pm i$ (figure 25.3). A point p is a branch point of a function if the

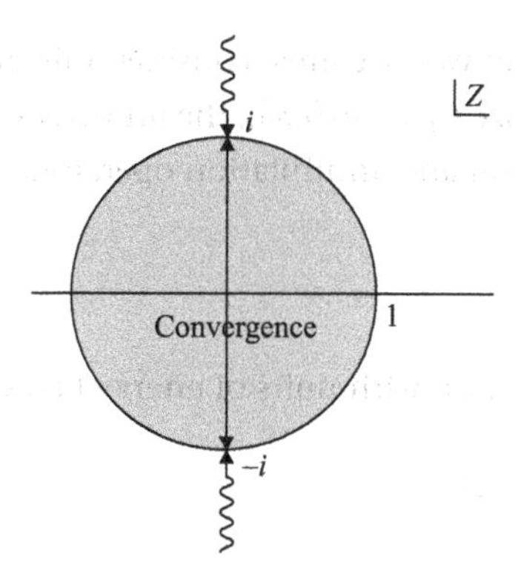

Figure 25.3
The Taylor expansion of the function $f(z) = \sqrt{1+z^2}$ about $z=0$ has a radius of convergence equal to one. The points $z = \pm i$ are branch points.

function is not continuous along arbitrarily small circles around p. Convergence holds for the largest disk centered at $z=0$ that contains no branch points. This is the $|z|<1$ disk.

Applied to our expansion of (25.2.45) about $\lambda=0$, we get convergence only when

$$\frac{|\lambda||V|}{\frac{1}{2}|E_1^{(0)}-E_2^{(0)}|}<1 \;\Rightarrow\; |\lambda V|<\tfrac{1}{2}|E_1^{(0)}-E_2^{(0)}|. \tag{25.2.47}$$

We learn that for convergence the perturbation must be small compared with *energy differences* in $\hat{H}^{(0)}$. It is not sufficient that the magnitude of $\lambda\delta H$ matrix elements be small compared to those in $\hat{H}^{(0)}$; energy differences matter. Indeed, we note in our expressions (25.2.37) that the energy denominators involve energy differences. For small energy differences, the corrections are large, making the perturbation expansion less reliable.

25.3 The Anharmonic Oscillator

Let the unperturbed Hamiltonian $\hat{H}^{(0)}$ be the simple harmonic oscillator Hamiltonian:

$$\hat{H}^{(0)}=\frac{\hat{p}^2}{2m}+\tfrac{1}{2}m\omega^2\hat{x}^2. \tag{25.3.1}$$

In a *classical* harmonic oscillator, the frequency of oscillation is independent of the amplitude of oscillation. In the quantum harmonic oscillator, all energy levels are equally spaced. The frequencies associated with transitions between various levels are therefore integer multiples or harmonics of the frequency associated to a transition between the first excited state and the ground state.

We want to explore the effect on $\hat{H}^{(0)}$ of a perturbation proportional to $\hat{x}^4$. This effectively changes the original quadratic potential by the addition of a quartic term. The system becomes an anharmonic oscillator. In a *classical* anharmonic oscillator, the frequency of oscillation depends on the amplitude of oscillation. In the quantum anharmonic oscillator, the spacing between the energy levels is not uniform, thus the anharmonicity.

To do our analysis in a clear way, we must consider units for the perturbation. Recalling the harmonic oscillator size $L_0=\sqrt{\hbar/(m\omega)}$, the unit-free coordinate $\hat{x}/L_0$ has a simple expression in terms of creation and annihilation operators:

$$\frac{\hat{x}}{L_0}=\frac{1}{\sqrt{2}}(\hat{a}+\hat{a}^\dagger). \tag{25.3.2}$$

It follows that an $\hat{x}^4$ perturbation with units of energy takes the form

$$\delta H=\hbar\omega\frac{\hat{x}^4}{L_0^4}=\tfrac{1}{4}\hbar\omega(\hat{a}+\hat{a}^\dagger)^4. \tag{25.3.3}$$

Using the unit-free parameter $\lambda\in[0,1]$, the perturbed Hamiltonian will therefore be

$$\hat{H}(\lambda) = \hat{H}^{(0)} + \lambda\, \tfrac{1}{4}\, \hbar\omega (\hat{a} + \hat{a}^\dagger)^4 . \tag{25.3.4}$$

We are adding the quartic term with a positive coefficient. This guarantees that the potential remains bounded below. We identify the states $|k^{(0)}\rangle$ of $\hat{H}^{(0)}$ with the number eigenstates $|k\rangle$, $k = 0, 1, \ldots$, of the harmonic oscillator:

$$E_k^{(0)} = \hbar\omega (k + \tfrac{1}{2}), \quad |k^{(0)}\rangle = \frac{(\hat{a}^\dagger)^k}{\sqrt{k!}} |0\rangle . \tag{25.3.5}$$

The Hamiltonian $\hat{H}(\lambda)$ defines an anharmonic oscillator.

First the simplest question: What is the first-order correction $E_0^{(1)}$ to the energy of the ground state? For this, following (25.2.21) we simply calculate the expectation value of the perturbation on the ground state:

$$E_0^{(1)} = \langle 0 | \tfrac{1}{4}\, \hbar\omega (\hat{a} + \hat{a}^\dagger)^4 | 0 \rangle = \tfrac{1}{4}\, \hbar\omega \langle 0 | (\hat{a} + \hat{a}^\dagger)^4 | 0 \rangle = \tfrac{3}{4}\, \hbar\omega, \tag{25.3.6}$$

where we used $\langle 0 | (\hat{a} + \hat{a}^\dagger)^4 | 0 \rangle = 3$, which you should verify. It follows that the corrected energy is

$$E_0(\lambda) = E_0^{(0)} + \lambda E_0^{(1)} + \mathcal{O}(\lambda^2) = \tfrac{1}{2}\hbar\omega + \lambda \tfrac{3}{4}\hbar\omega + \mathcal{O}(\lambda^2) = \tfrac{1}{2}\hbar\omega \left(1 + \tfrac{3}{2}\lambda + \mathcal{O}(\lambda^2)\right). \tag{25.3.7}$$

The energy of the ground state increases with $\lambda > 0$. This is reasonable as the quartic term in the modified potential squeezes the ground state. How about the second-order correction to the ground state energy? For this we use (25.2.36), with $n = 0$:

$$E_0^{(2)} = -\sum_{k \neq 0} \frac{|\delta H_{k0}|^2}{E_k^{(0)} - E_0^{(0)}} . \tag{25.3.8}$$

The sum is over all $k \geq 1$ such that δH_{k0} is nonvanishing. Here,

$$\delta H_{k0} = \tfrac{1}{4}\hbar\omega\, \langle k | (\hat{a} + \hat{a}^\dagger)^4 | 0 \rangle . \tag{25.3.9}$$

We consider $(\hat{a} + \hat{a}^\dagger)^4 |0\rangle$, which up to constants is the result of acting with $\hat{x}^4$ on the ground state wave function. This should give an even wave function. So $(\hat{a} + \hat{a}^\dagger)^4 |0\rangle$ must be a superposition of $|0\rangle$, $|2\rangle$, and $|4\rangle$. We cannot get states with a higher number because there are at most four creation operators acting on the vacuum. A short calculation (do it!) confirms that

$$(\hat{a} + \hat{a}^\dagger)^4 |0\rangle = 3|0\rangle + 6\sqrt{2}\, |2\rangle + \sqrt{4!}\, |4\rangle . \tag{25.3.10}$$

This immediately gives

$$\delta H_{00} = \tfrac{3}{4}\, \hbar\omega, \quad \delta H_{20} = \tfrac{3\sqrt{2}}{2}\, \hbar\omega, \quad \delta H_{40} = \tfrac{\sqrt{6}}{2}\, \hbar\omega, \tag{25.3.11}$$

the first of which we have already determined and is not needed for the second-order computation. Evaluating (25.3.8), we have

$$E_0^{(2)} = -\frac{|\delta H_{20}|^2}{2\hbar\omega} - \frac{|\delta H_{40}|^2}{4\hbar\omega} = -\frac{(\hbar\omega)^2}{2\hbar\omega}\frac{9}{2} - \frac{(\hbar\omega)^2}{4\hbar\omega}\frac{3}{2} = -\left(\tfrac{9}{4} + \tfrac{3}{8}\right)\hbar\omega = -\tfrac{21}{8}\hbar\omega . \tag{25.3.12}$$

Therefore, the corrected ground state energy to quadratic order is

$$E_0^{(0)} + \lambda E_0^{(1)} + \lambda^2 E_0^{(2)} = \tfrac{1}{2}\hbar\omega\left(1 + \tfrac{3}{2}\lambda - \tfrac{21}{4}\lambda^2\right). \tag{25.3.13}$$

The computation can be carried to higher order, as first done by Carl Bender and Tai Tsun Wu (1969):

$$E_0(\lambda) = \tfrac{1}{2}\hbar\omega\left(1 + \tfrac{3}{2}\lambda - \tfrac{21}{4}\lambda^2 + \tfrac{333}{8}\lambda^3 - \tfrac{30885}{64}\lambda^4 + \tfrac{916731}{128}\lambda^5 + \mathcal{O}(\lambda^6)\right). \tag{25.3.14}$$

As it turns out, the coefficients keep growing, and the series does not converge for any nonzero λ; the radius of convergence is actually zero! This does not mean the series is not useful. It is an asymptotic expansion. This means that for a given small value of λ the magnitude of successive terms generally decreases until, at some point, it starts growing again. A good approximation to the desired answer is obtained by including only the part of the sum where the terms are decreasing.

The failure of the series for $E_0(\lambda)$ to converge for any nonzero λ means that there is some nonanalyticity in $E_0(\lambda)$ at $\lambda = 0$. The error in $E_0(\lambda)$ for small λ can still be small because a nonanalytic function can be arbitrarily small: consider, for example, $\exp(-1/\lambda)$, which is nonanalytic at $\lambda = 0$ but vanishingly small as $\lambda \to 0^+$. The precise nature of the nonanalyticity in the anharmonic oscillator is in fact rather intricate, but could have been expected physically. If $E_0(\lambda)$ had a radius of convergence, it would converge for some neighborhood of $\lambda = 0$ in the complex λ plane. In particular, it would converge for some negative λ. But for negative λ, the full potential is unbounded below, and we no longer have bound states. This complication is reflected in the nonanalyticity.

Exercise 25.2. *Calculate the first-order correction $E_n^{(1)}$ to the energy for the state $|n\rangle$. Exhibit the anharmonicity of the oscillator by using this result to find, to first order in λ, the energy separation $\Delta E_n(\lambda) = E_n(\lambda) - E_{n-1}(\lambda)$ between levels.*

Let us now find the first-order correction to the ground state wave function. Using (25.2.34) with $n = 0$, we have

$$|0^{(1)}\rangle = -\sum_{k\neq 0} \frac{\delta H_{k0}}{E_k^{(0)} - E_0^{(0)}}\,|k\rangle. \tag{25.3.15}$$

We then find

$$\begin{aligned} |0^{(1)}\rangle &= -\frac{\delta H_{20}}{2\hbar\omega}|2\rangle - \frac{\delta H_{40}}{4\hbar\omega}|4\rangle = -\tfrac{3}{4}\sqrt{2}|2\rangle - \tfrac{1}{16}\sqrt{4!}|4\rangle \\ &= -\tfrac{3}{4}\,\hat{a}^\dagger\hat{a}^\dagger|0\rangle - \tfrac{1}{16}\,\hat{a}^\dagger\hat{a}^\dagger\hat{a}^\dagger\hat{a}^\dagger|0\rangle. \end{aligned} \tag{25.3.16}$$

This means that to first order, the ground state of the perturbed oscillator is

$$|0\rangle_\lambda = |0\rangle - \lambda\left(\tfrac{3}{4}\,\hat{a}^\dagger\hat{a}^\dagger|0\rangle + \tfrac{1}{16}\,\hat{a}^\dagger\hat{a}^\dagger\hat{a}^\dagger\hat{a}^\dagger|0\rangle\right) + \mathcal{O}(\lambda^2). \tag{25.3.17}$$

As given, the state $|0\rangle_\lambda$ is not normalized (see exercise 25.1).

25.4 Degenerate Perturbation Theory

If the spectrum of $\hat{H}^{(0)}$ has degenerate states, tracking the evolution of those states as λ becomes nonzero presents new challenges. We first show that naive extrapolation of our results for nondegenerate states does not work. We will also be able to appreciate the key complication we face: it is not clear what states to choose for the *zeroth*-order eigenstates. Then we will set up the perturbation theory in detail, working out the energy shifts and energy eigenstates to first order in perturbation theory, under the assumption that the perturbation breaks the degeneracy to first order.

A simple toy model Consider an example with 2×2 matrices. The unperturbed matrix $\hat{H}^{(0)}$ will be set equal to the identity matrix:

$$\hat{H}^{(0)} = \begin{pmatrix} 1 & 0 \\ 0 & 1 \end{pmatrix}. \tag{25.4.1}$$

We have a degeneracy here as the two eigenvalues are identical (and equal to one). The perturbation matrix δH is chosen to be off-diagonal:

$$\delta H = \begin{pmatrix} 0 & 1 \\ 1 & 0 \end{pmatrix}. \tag{25.4.2}$$

We then have

$$\hat{H}(\lambda) = \hat{H}^{(0)} + \lambda \delta H = \begin{pmatrix} 1 & \lambda \\ \lambda & 1 \end{pmatrix}. \tag{25.4.3}$$

Using labels $n = 1, 2$, the unperturbed eigenstates can be taken to be

$$|1^{(0)}\rangle = \begin{pmatrix} 1 \\ 0 \end{pmatrix}, \quad |2^{(0)}\rangle = \begin{pmatrix} 0 \\ 1 \end{pmatrix}, \qquad E_1^{(0)} = E_2^{(0)} = 1, \tag{25.4.4}$$

with the corresponding eigenvalues indicated as well. To first order in λ, the eigenvalues predicted from nondegenerate perturbation theory (25.2.37) are $E_n(\lambda) = E_n^{(0)} + \lambda \delta H_{nn}$. This gives

$$\begin{aligned} E_1(\lambda) &\stackrel{?}{=} E_1^{(0)} + \lambda \delta H_{11} = 1 + \lambda \cdot 0 = 1, \\ E_2(\lambda) &\stackrel{?}{=} E_2^{(0)} + \lambda \delta H_{22} = 1 + \lambda \cdot 0 = 1. \end{aligned} \tag{25.4.5}$$

The eigenvalues are unperturbed to first order in λ since the matrix δH is off-diagonal. These answers, however, are *wrong*. We can compute the exact eigenvalues of $\hat{H}(\lambda)$, and they are $1 \pm \lambda$. There is also a problem with the state corrections. Equation (25.2.37) states that

$$|n\rangle_\lambda = |n^{(0)}\rangle - \lambda \sum_{k \neq n} \frac{\delta H_{kn}}{E_k^{(0)} - E_n^{(0)}} |k^{(0)}\rangle + \mathcal{O}(\lambda^2). \tag{25.4.6}$$

If we attempt to use this formula, we note that $E_1^{(0)} = E_2^{(0)}$ makes the energy denominator zero. Since $\delta H_{12} = \delta H_{21} = 1$, the numerator does not vanish, and we get an infinite result.

What can we do? A direct calculation is possible in this simple example and shows that

$$\hat{H}(\lambda) \text{ has eigenvectors } \quad \frac{1}{\sqrt{2}}\begin{pmatrix}1\\1\end{pmatrix}, \quad \text{with eigenvalue } 1+\lambda\,, \tag{25.4.7}$$

$$\text{and} \quad \frac{1}{\sqrt{2}}\begin{pmatrix}1\\-1\end{pmatrix}, \quad \text{with eigenvalue } 1-\lambda\,. \tag{25.4.8}$$

You may think the eigenvectors jump from those of $\hat{H}^{(0)}$ indicated in (25.4.4) to those of $\hat{H}(\lambda)$ as soon as λ becomes nonzero. Such discontinuity is inconsistent with perturbation theory, where corrections must be small for small λ. Happily, there need not be a discontinuity. The eigenvectors of $\hat{H}^{(0)}$ are in fact *ambiguous*, precisely due to the degeneracy. The eigenvectors of $\hat{H}^{(0)}$ are actually the *span* of the two vectors listed in (25.4.4). The perturbation selected a particular combination of these eigenvectors. This particular combination is the one we should use *even* for $\lambda=0$. The lesson is that we must choose the basis in the degenerate subspace of $\hat{H}^{(0)}$ to get states that vary continuously as λ becomes nonzero. We will call that carefully selected basis the **good basis**, and we will learn how to find it.

Systematic analysis As before, we are looking at the perturbed Hamiltonian

$$\hat{H}(\lambda)=\hat{H}^{(0)}+\lambda\delta H\,, \tag{25.4.9}$$

where $\hat{H}^{(0)}$ has known eigenvectors and eigenvalues. We will focus this time on a degenerate subspace of eigenvectors of dimension $N>1$—that is, a space with N linearly independent eigenstates of the same energy. In the basis of eigenstates, $\hat{H}^{(0)}$ is a diagonal matrix that contains a string of $N>1$ identical entries:

$$\hat{H}^{(0)}=\text{diag}\,\{E_1^{(0)},E_2^{(0)},\ldots,\underbrace{E_n^{(0)},\ldots,E_n^{(0)}}_{N},\ldots\}\,. \tag{25.4.10}$$

In the degenerate subspace, we choose an *arbitrary* collection of N orthonormal eigenstates:

$$|n^{(0)};1\rangle\,,\;\ldots,\;|n^{(0)};N\rangle\,. \tag{25.4.11}$$

Accordingly, we have

$$\begin{aligned}\langle n^{(0)};p\,|n^{(0)};l\rangle &= \delta_{p,l}\,,\\ \hat{H}^{(0)}|n^{(0)};k\rangle &= E_n^{(0)}|n^{(0)};k\rangle\,.\end{aligned} \tag{25.4.12}$$

This set of vectors spans a degenerate subspace of dimension N that we will call $\mathbb{V}_N$:

$$\mathbb{V}_N \equiv \text{span}\{\,|n^{(0)};k\rangle,\;k=1,\ldots,N\}\,. \tag{25.4.13}$$

The total state space of the theory, denoted by $\mathcal{H}$, is written as a direct sum:

$$\mathcal{H}=\mathbb{V}_N\oplus V_\perp\,, \tag{25.4.14}$$

where $V_\perp$ is spanned by those eigenstates of $\hat{H}^{(0)}$ that are not in $\mathbb{V}_N$ and thus are orthogonal to any vector in $\mathbb{V}_N$. We denote by $|p^{(0)}\rangle$ with $p\in\mathbb{Z}$ a basis for $V_\perp$. That basis may include both degenerate and nondegenerate states. Together with the states in $\mathbb{V}_N$, we have an orthonormal basis for the whole state space:

$$\langle p^{(0)}|q^{(0)}\rangle = \delta_{pq}, \qquad \langle p^{(0)}|n^{(0)};k\rangle = 0. \tag{25.4.15}$$

Our notation distinguishes clearly the states in $\mathbb{V}_N$ and those in $V_\perp$: the former have two labels, and the latter have only one.

Since the basis states $|n^{(0)};k\rangle$ of $\mathbb{V}_N$ were chosen arbitrarily, they are, in general, not going to serve as good zeroth-order states. A perturbation δH that breaks the degeneracy will select the good basis. The basis vectors in the good basis are linear combinations of the basis vectors $|n^{(0)};k\rangle$, but we must find those linear combinations. In degenerate perturbation theory, the zeroth-order states are not obvious!

Let us call the good basis vectors $|\psi_I^{(0)}\rangle$ with $I=1,\ldots,N$ labeling the N vectors. We express our ignorance about good basis vectors by stating that we are searching for the right linear combinations:

$$|\psi_I^{(0)}\rangle = \sum_{k=1}^{N} |n^{(0)};k\rangle\, a_{Ik}^{(0)}, \quad I=1,\ldots,N. \tag{25.4.16}$$

For each I, the constants $a_{Ik}^{(0)}$, with $k=1,\ldots,N$, determine the state $|\psi_I^{(0)}\rangle$. The zero superscripts in the notation indicate the objects are of zeroth order. We can view $a_I^{(0)}$ as the column vector whose components $a_{Ik}^{(0)}$ give a representation of $|\psi_I^{(0)}\rangle$ in $\mathbb{V}_N$. Our first goal is to find those vectors $a_I^{(0)}$ and thus the good basis. If the degeneracy is broken by δH, as we will assume in this section, the states $|\psi_I^{(0)}\rangle$ will be automatically orthonormal:

$$\langle \psi_J^{(0)}|\psi_I^{(0)}\rangle = \delta_{IJ}. \tag{25.4.17}$$

This is physically clear. As soon as these states are perturbed, their energies are different, and thus they are necessarily orthonormal. If they are orthonormal for arbitrarily small perturbations, they will remain orthonormal for zero perturbation.

We set up the perturbation theory as usual by expressing the exact eigenstates and their energies as a power series in λ:

$$\begin{aligned} |\psi_I\rangle_\lambda &= |\psi_I^{(0)}\rangle + \lambda|\psi_I^{(1)}\rangle + \lambda^2|\psi_I^{(2)}\rangle + \cdots, \\ E_{nI}(\lambda) &= E_n^{(0)} + \lambda\, E_{nI}^{(1)} + \lambda^2\, E_{nI}^{(2)} + \lambda^3\, E_{nI}^{(3)} + \cdots. \end{aligned} \tag{25.4.18}$$

The energy expansion accounts for the degeneracy: the index I does not appear in the zeroth-order energies. To first order, however, we expect different energies for the different values of I.

Just as we did in the nondegenerate case, we can demand that the order $k\geq 1$ correction $|\psi_I^{(k)}\rangle$ in $|\psi_I\rangle_\lambda$ has no component along the zeroth-order $|\psi_I^{(0)}\rangle$:

$$\langle \psi_I^{(0)}|\psi_I^{(k)}\rangle = 0, \quad k\geq 1. \tag{25.4.19}$$

This does not rule out $|\psi_I^{(k)}\rangle$ having a component in $\mathbb{V}_N$ since the above constraint does not assert that $|\psi_I^{(k)}\rangle$ is orthogonal to the basis vectors $|\psi_J^{(0)}\rangle$, with $J\neq I$.

Introducing the above expansions for states and energies into the Schrödinger equation

$$\hat{H}(\lambda)|\psi_I\rangle_\lambda = E_{nI}(\lambda)|\psi_I\rangle_\lambda \tag{25.4.20}$$

gives the by now familiar equations, of which we list the first three:

$$\begin{aligned}
\lambda^0: &\quad (\hat{H}^{(0)} - E_n^{(0)})\,|\psi_I^{(0)}\rangle = 0,\\
\lambda^1: &\quad (\hat{H}^{(0)} - E_n^{(0)})\,|\psi_I^{(1)}\rangle = (E_{nI}^{(1)} - \delta H)|\psi_I^{(0)}\rangle,\\
\lambda^2: &\quad (\hat{H}^{(0)} - E_n^{(0)})\,|\psi_I^{(2)}\rangle = (E_{nI}^{(1)} - \delta H)|\psi_I^{(1)}\rangle + E_{nI}^{(2)}\,|\psi_I^{(0)}\rangle.
\end{aligned} \tag{25.4.21}$$

The zeroth-order equation is trivially satisfied since the $|\psi_I^{(0)}\rangle$ are linear combinations of states that satisfy the equation. Let us now consider the order λ equation. Acting on both sides of the equation with $\langle n^{(0)}; \ell|$, the left-hand side vanishes, and we find that

$$0 = \langle n^{(0)}; \ell|(E_{nI}^{(1)} - \delta H)|\psi_I^{(0)}\rangle. \tag{25.4.22}$$

Using the expansion of $|\psi_I^{(0)}\rangle$ along the original basis states, we have

$$0 = \sum_{k=1}^{N}\Big(E_{nI}^{(1)} a_{Ik}^{(0)}\delta_{\ell k} - \langle n^{(0)}; \ell|\delta H|n^{(0)}; k\rangle a_{Ik}^{(0)}\Big). \tag{25.4.23}$$

We use shorthand for the matrix elements of δH in the original basis,

$$\delta H_{n\ell,nk} \equiv \langle n^{(0)}; \ell|\delta H|n^{(0)}; k\rangle. \tag{25.4.24}$$

We include the n in the double indices $n\ell$ and nk to make it manifest that we are referring to basis vectors associated with the degenerate subspace. We define the $N \times N$ matrix $[\delta H]$ as the *restriction* of δH to the degenerate subspace, thus

$$[\delta H]_{\ell k} \equiv \delta H_{n\ell,nk}. \tag{25.4.25}$$

Using these definitions and with a little rearrangement, (25.4.23) becomes

$$\sum_{k=1}^{N}\Big([\delta H]_{\ell k} a_{Ik}^{(0)} - E_{nI}^{(1)}\,\delta_{\ell k}\,a_{Ik}^{(0)}\Big) = 0. \tag{25.4.26}$$

The sum over k is implementing matrix multiplication: $[\delta H]$ acting on the vector $a_I^{(0)}$ in the first term and the identity matrix acting on the same vector in the second term. We therefore have

$$\big([\delta H] - E_{nI}^{(1)}\mathbb{1}\big)\,a_I^{(0)} = 0. \tag{25.4.27}$$

This equation states that $a_I^{(0)}$ is the eigenvector of $[\delta H]$ with eigenvalue $E_{nI}^{(1)}$. We have therefore determined the good basis and the leading energy corrections: *the good basis is that composed by the eigenvectors of* $[\delta H]$, *with energy corrections given by the associated eigenvalues.*

This key result can also be appreciated directly in the good basis. Hitting the order λ equation with $\langle\psi_K^{(0)}|$ from the left, the left-hand side vanishes, and we find that

$$0 = \langle\psi_K^{(0)}|(E_{nI}^{(1)} - \delta H)|\psi_I^{(0)}\rangle = E_{nI}^{(1)}\delta_{KI} - \delta H_{KI}\,, \tag{25.4.28}$$

where we recalled the orthonormality of the good basis and defined

$$\delta H_{IJ} \equiv \langle\psi_I^{(0)}|\delta H|\psi_J^{(0)}\rangle\,. \tag{25.4.29}$$

Relabeling $K \to I \to J$, equation (25.4.28) gives

$$\delta H_{IJ} = E_{nJ}^{(1)}\,\delta_{IJ}\,. \tag{25.4.30}$$

This states that in the good basis the perturbation δH is indeed diagonal, with the energy corrections given by the diagonal elements. Indeed, setting $I = J$ in the above equation, we get $E_{nI}^{(1)} = \delta H_{II}$. In summary:

$$\boxed{\text{The basis } |\psi_I^{(0)}\rangle,\ I = 1,\ldots,N \text{ makes } \delta H \text{ diagonal in the space } \mathbb{V}_N.} \tag{25.4.31}$$

Going back to (25.4.18), we see that

$$\boxed{E_{nI}(\lambda) = E_n^{(0)} + \lambda\,\delta H_{II} + \mathcal{O}(\lambda^2)\,.} \tag{25.4.32}$$

The repeated index in δH_{II} is *not* summed over. Here are three important statements about the results we have just obtained.

Remarks:

1. The relation $E_{nI}^{(1)} = \delta H_{II}$ is true *always*, even if the degeneracy is not lifted. The degeneracy is lifted when all eigenvalues of $[\delta H]$ are different:

$$E_{nI}^{(1)} \neq E_{nJ}^{(1)}\,, \quad \text{whenever } I \neq J,\ \ I,J = 1,\ldots,N\,. \tag{25.4.33}$$

   This assumption will be used to compute the corrections to the states. If the degeneracy is lifted, the basis states $|\psi_I^{(0)}\rangle$ that make δH diagonal in $\mathbb{V}_N$ are confirmed to form a good basis. This means they are the basis states in $\mathbb{V}_N$ that get deformed continuously as λ becomes nonzero. If the degeneracy is not lifted to first order, $[\delta H]$ has degeneracies, and the eigenvectors are not uniquely fixed. The determination of the good basis has to be attempted to second order.

2. The perturbation δH is diagonalized in the subspace $\mathbb{V}_N$. The perturbation δH is *not* diagonal on the whole state space; it is only diagonal within $\mathbb{V}_N$. Alternatively, we can see this via the action of δH on the basis states. Introducing a resolution of the identity, we have

$$\begin{aligned}\delta H|\psi_I^{(0)}\rangle &= \sum_J |\psi_J^{(0)}\rangle\,\langle\psi_J^{(0)}|\delta H|\psi_I^{(0)}\rangle + \sum_p |p^{(0)}\rangle\,\langle p^{(0)}|\delta H|\psi_I^{(0)}\rangle\\ &= \sum_J |\psi_J^{(0)}\rangle\,\delta_{JI}E_{nJ}^{(1)} + \sum_p |p^{(0)}\rangle\,\langle p^{(0)}|\delta H|\psi_I^{(0)}\rangle\\ &= E_{nI}^{(1)}|\psi_I^{(0)}\rangle + \sum_p |p^{(0)}\rangle\,\langle p^{(0)}|\delta H|\psi_I^{(0)}\rangle\,.\end{aligned} \tag{25.4.34}$$

This shows that the states $|\psi_I^{(0)}\rangle$ are *almost* δH eigenstates with eigenvalues equal to the first-order energy corrections. The failure is an extra state along $V_\perp$. Of course, the states $|\psi_I^{(0)}\rangle$ are $[\delta H]$ eigenstates.

3. We can sometimes assess without computation that a certain basis in $\mathbb{V}_N$ makes $[\delta H]$ diagonal:

Rule: The matrix $[\delta H]$ is diagonal for a choice of basis in $\mathbb{V}_N$ if for any two different basis vectors there is a Hermitian operator K that commutes with δH for which the two basis vectors are K eigenstates with different eigenvalues.

This rule is quickly established. Assume we are testing a $\mathbb{V}_N$ basis $|n^{(0)};k\rangle$, with $k = 1,\ldots,N$. Take two different basis states: $|n^{(0)};p\rangle$ and $|n^{(0)};q\rangle$, with $p \neq q$. Assume these have different K eigenvalues k_p and k_q, respectively. Since, by assumption, $[\delta H, K] = 0$,

$$0 = \langle n^{(0)};p|[\delta H,K]|n^{(0)};q\rangle = (k_q - k_p)\langle n^{(0)};p|\,\delta H\,|n^{(0)};q\rangle\,. \qquad (25.4.35)$$

Since $k_p \neq k_q$, this implies that the off-diagonal matrix element of δH vanishes. Given that we assume that for any pair of different basis vectors such an operator exists, we know that all off-diagonal matrix elements of δH vanish. In specific applications, a single operator K enables us to show we have a good basis if it has different eigenvalues on each of the basis states. More generally, we need several Hermitian operators $K_1, K_2, \ldots$, each taking care of a subset of all off-diagonal matrix elements.

To find the first-order state corrections as well as second-order energy corrections, we will analyze equations (25.4.21) in two steps:

1. Use the order λ equation to calculate the components of $|\psi_I^{(1)}\rangle$ in $V_\perp$.
2. Form the overlap of the order λ^2 equation with $\langle\psi_K^{(0)}|$ to determine the second-order energy correction $E_{nI}^{(2)}$ *and* the component of $|\psi_I^{(1)}\rangle$ in $\mathbb{V}_N$.

It may be surprising that we have to go to *second* order in λ for the complete determination of the *first*-order correction to the state. The analysis will make it obvious why this is so.

Step 1. Acting on the order λ equation in (25.4.21) with $\langle p^{(0)}|$ gives useful information:

$$(E_p^{(0)} - E_n^{(0)})\langle p^{(0)}|\psi_I^{(1)}\rangle = \langle p^{(0)}|(E_{nI}^{(1)} - \delta H)|\psi_I^{(0)}\rangle = -\langle p^{(0)}|\delta H|\psi_I^{(0)}\rangle, \qquad (25.4.36)$$

using the orthogonality of $V_\perp$ and $\mathbb{V}_N$. Letting

$$\delta H_{pI} \equiv \langle p^{(0)}|\delta H|\psi_I^{(0)}\rangle, \qquad (25.4.37)$$

we then have

$$\langle p^{(0)}|\psi_I^{(1)}\rangle = -\frac{\delta H_{pI}}{E_p^{(0)} - E_n^{(0)}}\,. \qquad (25.4.38)$$

Since the good basis is now assumed known, the matrix elements δH_{pI} are calculable. The state correction, restricted to $V_\perp$, is therefore

$$|\psi_I^{(1)}\rangle\Big|_{V_\perp} = -\sum_p \frac{|p^{(0)}\rangle\delta H_{pI}}{E_p^{(0)} - E_n^{(0)}}\,. \tag{25.4.39}$$

The order λ equation does not give any further information. We just considered the overlap with states in $V_\perp$. Before this, we considered the overlap with general $|\psi_K^{(0)}\rangle$ states, finding that the $|\psi_I^{(0)}\rangle$ provide a basis where δH is diagonal within $\mathbb{V}_N$. The order λ equation does *not* fix the component of the first-order state correction along the degenerate subspace. This completes step 1.

Step 2. Having determined the component of $|\psi_I^{(1)}\rangle$ in $V_\perp$, we now write

$$|\psi_I^{(1)}\rangle = -\sum_p \frac{|p^{(0)}\rangle\delta H_{pI}}{E_p^{(0)} - E_n^{(0)}} + |\psi_I^{(1)}\rangle\Big|_{\mathbb{V}_N}, \tag{25.4.40}$$

where we include explicitly the still undetermined component of $|\psi_I^{(1)}\rangle$ along $\mathbb{V}_N$. We now hit the order λ^2 equation with $\langle\psi_K^{(0)}|$. The left-hand side vanishes, and using the above expression for $|\psi_I^{(1)}\rangle$, we find that

$$\begin{aligned} 0 = &-\langle\psi_K^{(0)}|(E_{nI}^{(1)} - \delta H)\sum_p |p^{(0)}\rangle \frac{\delta H_{pI}}{E_p^{(0)} - E_n^{(0)}} \\ &+ \langle\psi_K^{(0)}|(E_{nI}^{(1)} - \delta H)|\psi_I^{(1)}\rangle\Big|_{\mathbb{V}_N} + E_{nI}^{(2)}\delta_{KI}\,. \end{aligned} \tag{25.4.41}$$

In the first term on the right-hand side, the part proportional to $E_{nI}^{(1)}$ vanishes by orthonormality. On the second line, the term including δH can be simplified because δH is diagonal within $\mathbb{V}_N$. Writing (25.4.34) in bra form, we have

$$\langle\psi_K^{(0)}|\delta H = E_{nK}^{(1)}\langle\psi_K^{(0)}| + \sum_p \langle\psi_K^{(0)}|\delta H|p^{(0)}\rangle\langle p^{(0)}|\,. \tag{25.4.42}$$

The second term drops out for our case of interest:

$$\langle\psi_K^{(0)}|\delta H|\psi_I^{(1)}\rangle\Big|_{\mathbb{V}_N} = E_{nK}^{(1)}\langle\psi_K^{(0)}|\psi_I^{(1)}\rangle\Big|_{\mathbb{V}_N} = E_{nK}^{(1)}\langle\psi_K^{(0)}|\psi_I^{(1)}\rangle\,. \tag{25.4.43}$$

In the last equality, the restriction to $\mathbb{V}_N$ was dropped because it is no longer needed. Returning to equation (25.4.41), we now get:

$$\sum_p \frac{\delta H_{Kp}\delta H_{pI}}{E_p^{(0)} - E_n^{(0)}} + \left(E_{nI}^{(1)} - E_{nK}^{(1)}\right)\langle\psi_K^{(0)}|\psi_I^{(1)}\rangle + E_{nI}^{(2)}\delta_{KI} = 0\,. \tag{25.4.44}$$

This equation will give us the second-order correction to the energy as well as the first-order correction to the state in the degenerate subspace. Setting $K = I$ and using the Hermiticity of δH to conclude that $\delta H_{Ip} = (\delta H_{pI})^*$, the second correction to the energies is found to be

$$E_{nI}^{(2)} = -\sum_p \frac{|\delta H_{pI}|^2}{E_p^{(0)} - E_n^{(0)}}\,. \tag{25.4.45}$$

For $I \neq K$, the equation fixes the component of $|\psi_I^{(1)}\rangle$ along $|\psi_K^{(0)}\rangle$:

$$\sum_p \frac{\delta H_{Kp}\delta H_{pI}}{E_p^{(0)} - E_n^{(0)}} + \left(E_{nI}^{(1)} - E_{nK}^{(1)}\right)\langle\psi_K^{(0)}|\psi_I^{(1)}\rangle = 0, \quad I \neq K. \tag{25.4.46}$$

The assumed nondegeneracy of the first-order corrections is relevant here. We can solve for the overlap $\langle\psi_K^{(0)}|\psi_I^{(1)}\rangle$ *as long as* $E_{nK}^{(1)} \neq E_{nI}^{(1)}$:

$$\langle\psi_K^{(0)}|\psi_I^{(1)}\rangle = -\frac{1}{E_{nI}^{(1)} - E_{nK}^{(1)}} \sum_p \frac{\delta H_{Kp}\delta H_{pI}}{E_p^{(0)} - E_n^{(0)}}, \quad I \neq K. \tag{25.4.47}$$

We thus have

$$|\psi_I^{(1)}\rangle\Big|_{\mathbb{V}_N} = -\sum_{K\neq I} |\psi_K^{(0)}\rangle \frac{1}{E_{nI}^{(1)} - E_{nK}^{(1)}} \sum_p \frac{\delta H_{Kp}\delta H_{pI}}{E_p^{(0)} - E_n^{(0)}}. \tag{25.4.48}$$

It may seem that this first-order correction to the state is higher order: its numerator contains two powers of δH. But this expression also has a denominator $E_{nI}^{(1)} - E_{nK}^{(1)}$ in which each term is of order δH. All in all, the correction to the state is properly first order in δH.

Summarizing our results, we have the following:

Degenerate perturbation theory with degeneracies lifted at $\mathcal{O}(\lambda)$:

$$|\psi_I\rangle_\lambda = |\psi_I^{(0)}\rangle - \lambda\Big(\sum_p \frac{\delta H_{pI}}{E_p^{(0)} - E_n^{(0)}}|p^{(0)}\rangle + \sum_{K\neq I} \frac{|\psi_K^{(0)}\rangle}{E_{nI}^{(1)} - E_{nK}^{(1)}} \sum_p \frac{\delta H_{Kp}\delta H_{pI}}{E_p^{(0)} - E_n^{(0)}}\Big) + \mathcal{O}(\lambda^2),$$

$$E_{nI}(\lambda) = E_n^{(0)} + \lambda\,\delta H_{II} - \lambda^2 \sum_p \frac{|\delta H_{pI}|^2}{E_p^{(0)} - E_n^{(0)}} + \mathcal{O}(\lambda^3), \qquad E_{nI}^{(1)} = \delta H_{II}.$$

(25.4.49)

25.5 Degeneracy Lifted at Second Order

We now investigate the case when the degeneracy is completely unbroken to first order in the perturbation. This actually happens in a variety of examples; it is not an exotic situation. Consider, for example, a particle on a circle $x \sim x + L$, moving freely with zero potential. The ground state is the zero-momentum state of constant wave function. The excited states, however, are doubly degenerate. For each energy they can be represented by momentum eigenstates with opposite values of the quantized momentum. The simplest perturbation, $\delta H \sim \cos(2\pi x/L)$, has zero effect on the energies of the states to first order. The degeneracy of the first excited states, however, is resolved to second order in perturbation theory and can be calculated with the methods discussed below (problem 25.8).

The situation and the setup are similar to the one we just considered: we have a degenerate subspace $\mathbb{V}_N$ of dimension N, and the rest of the space is called $V_\perp$. This time, however, we will assume that the degeneracy of $\hat{H}^{(0)}$ is not broken to first order in δH. This means that in an orthonormal basis of $\mathbb{V}_N$ that makes δH diagonal all the diagonal entries are the same. Therefore, δH restricted to the degenerate subspace is a multiple of the identity. Such a matrix, in fact, remains a multiple of the identity in any basis. This means that in *any* orthonormal $\mathbb{V}_N$ basis $|\psi_I^{(0)}\rangle$ with $I=1,\dots,N$, we have

$$\langle\psi_I^{(0)}|\delta H\,|\psi_J^{(0)}\rangle \;=\; E_n^{(1)}\,\delta_{IJ}\,. \tag{25.5.1}$$

The first-order energy correction is the same and equal to $E_n^{(1)}$ for all basis states in $\mathbb{V}_N$. Compare with (25.4.30), where the energy correction on the right-hand side reads $E_{nJ}^{(1)}$ with the extra subscript J to account for different diagonal values.

Since the degeneracy is not broken to first order, we do not know at this point what the good basis in $\mathbb{V}_N$ is. We will consider here the case when the degeneracy is completely lifted to second order. Once again we write

$$|\psi_I^{(0)}\rangle=\sum_{k=1}^{N}|n^{(0)};k\rangle\, a_{Ik}^{(0)}\,,\quad I=1,\dots,N\,,\qquad \langle\psi_I^{(0)}|\psi_J^{(0)}\rangle=\delta_{IJ}\,. \tag{25.5.2}$$

The good basis states are characterized by the vectors $a_I^{(0)}$. We set up the perturbation theory as usual:

$$\begin{aligned}|\psi_I\rangle_\lambda \;&=\; |\psi_I^{(0)}\rangle+\lambda|\psi_I^{(1)}\rangle+\lambda^2|\psi_I^{(2)}\rangle+\cdots\,,\\ E_{nI}(\lambda) \;&=\; E_n^{(0)}\;+\lambda\,E_n^{(1)}\;+\lambda^2\,E_{nI}^{(2)}\;+\lambda^3\,E_{nI}^{(3)}+\cdots\,.\end{aligned} \tag{25.5.3}$$

Note that in the energy expansion we have accounted for the degeneracy to zeroth and first order: the index I first appears in the second-order corrections to the energy. The Schrödinger equation

$$\hat{H}(\lambda)|\psi_I\rangle_\lambda \;=\; E_{nI}(\lambda)|\psi_I\rangle_\lambda \tag{25.5.4}$$

gives the by now familiar equations, of which we list the first four:

$$\begin{aligned}\lambda^0:&\qquad (\hat{H}^{(0)}-E_n^{(0)})\,|\psi_I^{(0)}\rangle \;=\; 0\,,\\ \lambda^1:&\qquad (\hat{H}^{(0)}-E_n^{(0)})\,|\psi_I^{(1)}\rangle \;=\; (E_n^{(1)}-\delta H)|\psi_I^{(0)}\rangle\,,\\ \lambda^2:&\qquad (\hat{H}^{(0)}-E_n^{(0)})\,|\psi_I^{(2)}\rangle \;=\; (E_n^{(1)}-\delta H)|\psi_I^{(1)}\rangle\;+\;E_{nI}^{(2)}\,|\psi_I^{(0)}\rangle\,,\\ \lambda^3:&\qquad (\hat{H}^{(0)}-E_n^{(0)})\,|\psi_I^{(3)}\rangle \;=\; (E_n^{(1)}-\delta H)|\psi_I^{(2)}\rangle+E_{nI}^{(2)}\,|\psi_I^{(1)}\rangle\;+\;E_{nI}^{(3)}\,|\psi_I^{(0)}\rangle\,.\end{aligned} \tag{25.5.5}$$

The zeroth-order equation is trivially satisfied. For the order λ equation, the overlap with $\langle\psi_K^{(0)}|$ works out automatically, giving us $E_n^{(1)}\delta_{KI}=\delta H_{KI}$, which we already knew from (25.5.1). Acting on the $\mathcal{O}(\lambda)$ equation with $\langle p^{(0)}|$ gives

$$(E_p^{(0)}-E_n^{(0)})\langle p^{(0)}|\psi_I^{(1)}\rangle \;=\; \langle p^{(0)}|(E_n^{(1)}-\delta H)|\psi_I^{(0)}\rangle \;=\; -\langle p^{(0)}|\delta H|\psi_I^{(0)}\rangle\,, \tag{25.5.6}$$

using the orthogonality of $V_\perp$ and $\mathbb{V}_N$. With the familiar notation $\delta H_{pI} \equiv \langle p^{(0)}|\delta H|\psi_I^{(0)}\rangle$, we then have

$$\langle p^{(0)}|\psi_I^{(1)}\rangle = -\frac{\delta H_{pI}}{E_p^{(0)} - E_n^{(0)}}\,. \tag{25.5.7}$$

Since the ket $|\psi_I^{(0)}\rangle$ is still undetermined, it makes sense to write this information about $|\psi_I^{(1)}\rangle$ in terms of the unknown $a_I^{(0)}$ coefficients. We have

$$\delta H_{pI} \equiv \sum_{k=1}^{N} \langle p^{(0)}|\delta H|n^{(0)};k\rangle\, a_{Ik}^{(0)} = \sum_{k=1}^{N} \delta H_{p,nk}\, a_{Ik}^{(0)}\,. \tag{25.5.8}$$

Going back to (25.5.7), we get

$$\langle p^{(0)}|\psi_I^{(1)}\rangle = -\frac{1}{E_p^{(0)} - E_n^{(0)}} \sum_{k=1}^{N} \delta H_{p,nk}\, a_{Ik}^{(0)}\,. \tag{25.5.9}$$

This gives the piece of $|\psi_I^{(1)}\rangle$ in $V_\perp$ in terms of the unknown zeroth-order eigenstates.

We have extracted all the information from the $\mathcal{O}(\lambda)$ equation. We now look at the $\mathcal{O}(\lambda^2)$ equation, which contains the second-order corrections to the energy and therefore should help us determine the zeroth-order good states. We hit that equation with $\langle n^{(0)};\ell|$, and we get

$$0 = \langle n^{(0)};\ell|(E_n^{(1)} - \delta H)|\psi_I^{(1)}\rangle\Big|_{V_\perp} + \langle n^{(0)};\ell|(E_n^{(1)} - \delta H)|\psi_I^{(1)}\rangle\Big|_{\mathbb{V}_N} + E_{nI}^{(2)} a_{I\ell}^{(0)}\,. \tag{25.5.10}$$

Happily, the second term, involving the components of $|\psi_I^{(1)}\rangle$ along $\mathbb{V}_N$, vanishes because of the by now familiar property (25.4.34) adapted to this case. The piece with $E_n^{(1)}$ on the first term also vanishes. We are thus left with

$$0 = -\langle n^{(0)};\ell|\delta H|\psi_I^{(1)}\rangle\Big|_{V_\perp} + E_{nI}^{(2)} a_{I\ell}^{(0)}\,. \tag{25.5.11}$$

After we introduce a resolution of the identity to the immediate right of δH, only the basis states in $V_\perp$ contribute, and we find that

$$0 = -\sum_p \langle n^{(0)};\ell|\delta H|p^{(0)}\rangle\langle p^{(0)}|\psi_I^{(1)}\rangle + E_{nI}^{(2)} a_{I\ell}^{(0)}\,, \tag{25.5.12}$$

where there is no need to copy the $|_{V_\perp}$ restriction anymore. Using the result in (25.5.9) and introducing an extra sum over k on the last term, we now get

$$0 = \sum_p \delta H_{n\ell,p} \frac{1}{E_p^{(0)} - E_n^{(0)}} \sum_{k=1}^{N} \delta H_{p,nk}\, a_{Ik}^{(0)} + \sum_{k=1}^{N} E_{nI}^{(2)} \delta_{\ell k} a_{Ik}^{(0)}\,. \tag{25.5.13}$$

Reordering sums and multiplying by minus one, we get

$$\sum_{k=1}^{N}\left(-\sum_p \frac{\delta H_{n\ell,p}\delta H_{p,nk}}{E_p^{(0)} - E_n^{(0)}} - E_{nI}^{(2)}\delta_{\ell k}\right) a_{Ik}^{(0)} = 0\,. \tag{25.5.14}$$

To better understand this equation, we define the $N \times N$ Hermitian matrix $M^{(2)}$:

$$M^{(2)}_{\ell k} \equiv -\sum_p \frac{\delta H_{n\ell,p}\,\delta H_{p,nk}}{E^{(0)}_p - E^{(0)}_n}\,. \tag{25.5.15}$$

Equation (25.5.14) then becomes

$$\sum_{k=1}^{N}\Big(M^{(2)}_{\ell k} - E^{(2)}_{nI}\,\delta_{\ell k}\Big)a^{(0)}_{Ik} = 0\,. \tag{25.5.16}$$

Since the Kronecker delta is the matrix representation of the identity, we have

$$\Big(M^{(2)} - E^{(2)}_{nI}\mathbb{1}\Big)\,a^{(0)}_I = 0\,. \tag{25.5.17}$$

This is an eigenvalue equation that tells us that the energy corrections $E^{(2)}_{nI}$ are the eigenvalues of $M^{(2)}$, and the vectors $a^{(0)}_I$ are the associated normalized eigenvectors. These determine, via (25.5.2), the orthonormal basis of good zeroth-order states. If δH is known, the Hermitian matrix $M^{(2)}$ is computable and can be diagonalized.

We will leave out the details of the computation of the component of $|\psi^{(1)}_I\rangle$ on the degenerate subspace. This component can be evaluated if the degeneracy is completely broken to quadratic order—namely, if the eigenvalues of $M^{(2)}$ are all different. The computation takes some effort, and one must use the order λ^3 equation.

The results are then summarized by

$$\begin{aligned}
|\psi_I\rangle_\lambda &= |\psi^{(0)}_I\rangle + \lambda\Big(\sum_p |p^{(0)}\rangle \frac{\delta H_{pI}}{E^{(0)}_n - E^{(0)}_p} + \sum_{J\neq I}|\psi^{(0)}_J\rangle\, a^{(1)}_{IJ}\Big) + \mathcal{O}(\lambda^2)\,,\\
E_{In}(\lambda) &= E^{(0)}_n + \lambda\, E^{(1)}_n + \lambda^2 E^{(2)}_{In} + \lambda^3 E^{(3)}_{In} + \cdots + \mathcal{O}(\lambda^3)\,.
\end{aligned} \tag{25.5.18}$$

Here the $a^{(1)}_{IJ}$ are coefficients that determine the component of the first correction to the states along the degenerate subspace. Their value turns out to be

$$a^{(1)}_{IJ} = \frac{1}{E^{(2)}_{nI} - E^{(2)}_{nJ}}\Bigg[\sum_{p,q}\frac{\delta H_{Jp}\,\delta H_{pq}\,\delta H_{qI}}{(E^{(0)}_p - E^{(0)}_n)(E^{(0)}_q - E^{(0)}_n)} - E^{(1)}_n\sum_p \frac{\delta H_{Jp}\,\delta H_{pI}}{(E^{(0)}_p - E^{(0)}_n)^2}\Bigg]. \tag{25.5.19}$$

The third-order corrections to the energy are

$$E^{(3)}_{In} = \sum_{p,q}\frac{\delta H_{Ip}\,\delta H_{pq}\,\delta H_{qI}}{(E^{(0)}_p - E^{(0)}_n)(E^{(0)}_q - E^{(0)}_n)} - E^{(1)}_n\sum_p \frac{|\delta H_{pI}|^2}{(E^{(0)}_p - E^{(0)}_n)^2}\,. \tag{25.5.20}$$

25.6 Review of Hydrogen Atom

We now turn to the hydrogen atom for an instructive and important application of time-independent perturbation theory. We have seen the hydrogen atom Hamiltonian a few times already. We have found the bound state spectrum in more than one way and learned about the remarkably large degeneracy that exists in this spectrum. We will call the hydrogen atom Hamiltonian $\hat{H}^{(0)}$, and it is given by

$$\hat{H}^{(0)} = \frac{\hat{\mathbf{p}}^2}{2m} - \frac{e^2}{r} . \tag{25.6.1}$$

The Hamiltonian $\hat{H}^{(0)}$ is nonrelativistic and spin independent: it must be thought to include the tensor product with the identity operator in the space of the electron spin. The mass m in $\hat{H}^{(0)}$ is the reduced mass of the electron and proton, which we can set equal to the mass of the electron with little error. This Hamiltonian is supplemented with corrections that can be studied in perturbation theory. The study of the corrections to the spectrum is the subject of the rest of this chapter. This spectrum can be calculated to great accuracy, showing the power of quantum mechanics. The hydrogen atom is also an excellent case study, complicated enough to better understand the perturbation theory developed in the previous sections. We begin, however, with some review and comments.

The Bohr radius is the length scale built from $\hbar$, m, and e^2:

$$a_0 \equiv \frac{\hbar^2}{me^2} \simeq 53\,\text{pm}. \tag{25.6.2}$$

The energy levels are enumerated using a *principal* quantum number n, an integer that must be greater than or equal to one:

$$E_n = -\frac{e^2}{2a_0}\frac{1}{n^2}, \quad n = 1, 2, \ldots, \qquad \text{Ry} \equiv \frac{e^2}{2a_0} = \frac{me^4}{2\hbar^2} \simeq 13.6\,\text{eV}. \tag{25.6.3}$$

While the system is nonrelativistic, the energy scale relevant to the bound state spectrum can be better appreciated using the speed of light to write it in terms of the fine-structure constant and the rest energy of the electron. With $\alpha = \frac{e^2}{\hbar c} \simeq \frac{1}{137}$, we see that

$$\text{Ry} = \left(\frac{e^4}{\hbar^2 c^2}\right)\tfrac{1}{2}mc^2 = \alpha^2\,\tfrac{1}{2}mc^2 . \tag{25.6.4}$$

This states that the energy scale of hydrogen bound states is a factor of α^2 smaller than the rest energy of the electron—that is, about 19,000 times smaller. We can thus rewrite the possible energies as:

$$E_n = -\tfrac{1}{2}\,\alpha^2\, mc^2\, \frac{1}{n^2} . \tag{25.6.5}$$

The typical momentum p in the hydrogen atom is

$$p \simeq \frac{\hbar}{a_0} = \frac{me^2}{\hbar} = \frac{e^2}{\hbar c} mc \quad \rightarrow \quad p \simeq \alpha(mc) , \tag{25.6.6}$$

which written as $p \simeq m(\alpha c)$ says that the typical velocity is $v \simeq \alpha c$, which is low enough to ensure that the nonrelativistic approximation is fairly accurate.

We call $\mathcal{H}_n$ the degenerate subspace of bound states with principal quantum number n. The degeneracy is made clear by the relation $n = N + \ell + 1$, where $N \geq 0$ is the degree

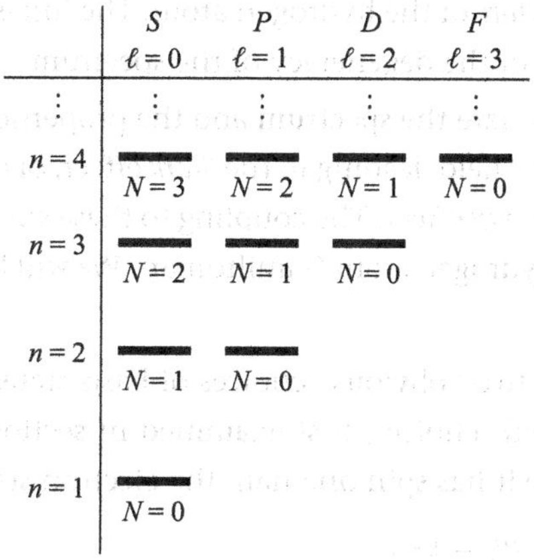

	S $\ell=0$	P $\ell=1$	D $\ell=2$	F $\ell=3$
$\vdots$	$\vdots$	$\vdots$	$\vdots$	$\vdots$
$n=4$	$N=3$	$N=2$	$N=1$	$N=0$
$n=3$	$N=2$	$N=1$	$N=0$	
$n=2$	$N=1$	$N=0$		
$n=1$	$N=0$			

Figure 25.4
The unperturbed spectrum of bound states of the hydrogen atom (not to scale). Each bar is an angular momentum multiplet, doubled to include the two possible spin states of the electron. Here $n=N+\ell+1$.

of a polynomial in r that appears in the wave function (see (11.3.36)), and $\ell \geq 0$ is the angular momentum of the state. For each fixed n, the value of ℓ ranges from zero to $n-1$. For each multiplet $\mathcal{H}_{n,\ell}$ of fixed ℓ, the eigenvalue of $\hat{L}_z$ is $m\hbar$, with m ranging from $-\ell$ up to ℓ. The total state space $\mathcal{H}$ of bound states is therefore given by

$$\mathcal{H}=\bigoplus_{n=1}^{\infty}\mathcal{H}_n\,, \quad \text{with } \mathcal{H}_n=\bigoplus_{\ell=0}^{n-1}\mathcal{H}_{n,\ell}\,. \tag{25.6.7}$$

This gives a total of n^2 states for each value of n (see (11.3.33)). Accounting for the spin degeneracy of the electron, we get $\dim \mathcal{H}_n = 2n^2$. The bound states of hydrogen are shown in figure 25.4, which is *not* drawn to scale. Capital letters are used to denote the various values of the orbital angular momentum ℓ. An S state is a state with $\ell=0$, a P state is a state with $\ell=1$, a D state is a state with $\ell=2$, and an F state is a state with $\ell=3$. Any hydrogen eigenstate is specified by the three quantum numbers n, ℓ, m, and the value m_s of the spin, as we will see below.

Remarks:

1. The degeneracy of the various ℓ multiplets with fixed n is explained by the existence of a conserved quantum Runge-Lenz vector (section 20.8). In the semiclassical picture, the states with various ℓ's correspond to electron orbits of different eccentricity but the same semimajor axis. The $\ell=0$ orbit is the most eccentric, and the $\ell=n-1$ orbit is the least eccentric.
2. For each fixed value of ℓ, the number N increases with the energy. The number N is the number of nodes in the solution of the radial equation.
3. We will supplement $\hat{H}^{(0)}$ with corrections that arise from relativity and from the spin of the electron. This will be the main subject of the following section. It will

determine the *fine structure* of the hydrogen atom. The fine-structure corrections will break some but not all of the degeneracy of the spectrum.

4. In order to better appreciate the spectrum and the properties of the hydrogen atom, one can apply an electric field, leading to the *Stark* effect, or one can apply a magnetic field, leading to the *Zeeman* effect. The coupling to these external fields is represented by extra terms in the hydrogen atom Hamiltonian. We will look at the Zeeman effect in section 25.8.

Let us now review the two "obvious" choices of basis states for the hydrogen atom bound state spectrum. Both choices, first examined in section 20.6, properly include the electron spin. Because it has spin one-half, the electron states are labeled as

$$|s, m_s\rangle\,, \quad \text{with} \quad s = \tfrac{1}{2}\,, \; m_s = \pm\tfrac{1}{2}\,. \tag{25.6.8}$$

In the hydrogen atom, the angular momentum ℓ can take different values, but the spin of the electron is always one-half. As a result, the label s is often omitted, and we usually only record the value of m_s. For hydrogen basis states, we thus have quantum numbers n, ℓ, m_ℓ, *and* m_s. To avoid confusion, we have included the ℓ subscript in m_ℓ, thus stressing that this is the azimuthal quantum number for orbital angular momentum. Since we are not combining the electron spin to its orbital angular momentum, the states form the *uncoupled basis*:

$$\text{Uncoupled basis quantum numbers:} \quad (n, \ell, m_\ell, m_s). \tag{25.6.9}$$

The bound states are completely specified by these quantum numbers. As we let those quantum numbers run over all possible values, we obtain an orthonormal basis of states for the bound state sector of the state space.

It is often useful to use an alternative basis where the states are eigenstates of $\hat{\mathbf{J}}^2$ and $\hat{J}_z$, where $\hat{\mathbf{J}}$ is the total angular momentum, obtained by adding the orbital angular momentum $\hat{\mathbf{L}}$ to the spin angular momentum $\hat{\mathbf{S}}$:

$$\hat{\mathbf{J}} = \hat{\mathbf{L}} + \hat{\mathbf{S}}\,. \tag{25.6.10}$$

The coupled basis is one where states are organized into j multiplets. To find the description in terms of j multiplets, we consider the decomposition of the tensor product $\ell \otimes s$ of a full ℓ multiplet to an $s = 1/2$ multiplet. All states in $\ell \otimes s$ are eigenstates of $\hat{\mathbf{L}}^2$ with the *same* eigenvalue and eigenstates of $\hat{\mathbf{S}}^2$ with the *same* eigenvalue, so ℓ and s are good constant quantum numbers for all j multiplets that arise in the tensor product. Each j multiplet has states labeled by (j, m_j). While the states in j multiplets are no longer $\hat{L}_z$ or $\hat{S}_z$ eigenstates, the ℓ quantum number survives. The s quantum number also survives, but here it is trivial, being always equal to 1/2. Therefore, we have the following:

$$\text{Coupled basis quantum numbers:} \quad (n, \ell, j, m_j)\,. \tag{25.6.11}$$

The (m_ℓ, m_s) quantum numbers of the uncoupled basis have been traded for (j, m_j) quantum numbers, and we have kept the n, ℓ quantum numbers. The coupled states are each linear combinations of uncoupled states that involve different values of m_ℓ and m_s, weighted by Clebsch-Gordan coefficients. At a deeper level, $\hat{\mathbf{J}}$ is relevant because in the relativistic formulation neither $\hat{\mathbf{S}}$ nor $\hat{\mathbf{L}}$ are separately conserved, but their sum $\hat{\mathbf{J}}$ is (problem 25.16).

To find the list of coupled basis states, we must tensor *each* ℓ multiplet in the hydrogen atom spectrum with the spin doublet $\frac{1}{2}$. The rules of addition of angular momentum imply that we find two j multiplets:

$$\ell \otimes \tfrac{1}{2} = (j=\ell+\tfrac{1}{2}) \oplus (j=\ell-\tfrac{1}{2}), \quad \ell \geq 1. \tag{25.6.12}$$

For $\ell=0$, we only obtain a $j=1/2$ multiplet. We use the notation L_j for the coupled multiplets, with $L=S, P, D, F$ for $\ell=0, 1, 2, 3$. The change of basis is summarized by the replacements

$$\ell \otimes \tfrac{1}{2} = L(\ell)_{j=\ell+\frac{1}{2}} \oplus L(\ell)_{j=\ell-\frac{1}{2}} \tag{25.6.13}$$

or, more explicitly, for the first few values of ℓ:

$$0 \otimes \tfrac{1}{2} = S_{\frac{1}{2}}, \quad 1 \otimes \tfrac{1}{2} = P_{\frac{3}{2}} \oplus P_{\frac{1}{2}}, \quad 2 \otimes \tfrac{1}{2} = D_{\frac{5}{2}} \oplus D_{\frac{3}{2}}, \quad 3 \otimes \tfrac{1}{2} = F_{\frac{7}{2}} \oplus F_{\frac{5}{2}}. \tag{25.6.14}$$

Indeed, combined with the electron spin doublet, each $\ell=0$ multiplet gives one $j=\frac{1}{2}$ multiplet, each $\ell=1$ multiplet gives $j=\frac{3}{2}$ and $j=\frac{1}{2}$ multiplets, each $\ell=2$ multiplet gives $j=\frac{5}{2}$ and $j=\frac{3}{2}$ multiplets, and so on. For hydrogen, the principal quantum number is placed ahead to denote the coupled multiplets as follows:

$$\text{Labels for coupled-basis multiplets:} \quad \boxed{nL_j} \tag{25.6.15}$$

Using this labeling of coupled-basis multiplets, the diagram of hydrogen atom bound states is shown in figure 25.5. The total number of states associated with each line is indicated in parentheses right below it. Each degenerate subspace $\mathcal{H}_n$ is the direct sum of some multiplets. For example,

$$\mathcal{H}_2 = 2S_{\frac{1}{2}} \oplus 2P_{\frac{3}{2}} \oplus 2P_{\frac{1}{2}}. \tag{25.6.16}$$

25.7 Fine Structure of Hydrogen

The Bohr Hamiltonian $\hat{H}^{(0)}$ is nonrelativistic, and one can wonder what the correction to the energy due to relativity is. As discussed at the beginning of section 24.7, the relativistic Hamiltonian of a *free* particle with momentum $\hat{\mathbf{p}}$ could read

$$\hat{H} = \sqrt{c^2\hat{\mathbf{p}}^2 + m^2c^4}. \tag{25.7.1}$$

It is not clear how to treat the square root in general. For momenta small compared to mc, however, the Hamiltonian can be formally expanded:

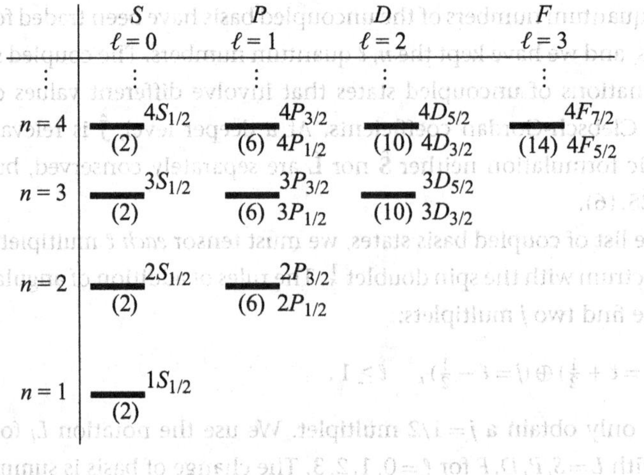

	S $\ell=0$	P $\ell=1$	D $\ell=2$	F $\ell=3$
⋮	⋮	⋮	⋮	⋮
$n=4$	$4S_{1/2}$ (2)	$4P_{3/2}$ (6) $4P_{1/2}$	$4D_{5/2}$ (10) $4D_{3/2}$	$4F_{7/2}$ (14) $4F_{5/2}$
$n=3$	$3S_{1/2}$ (2)	$3P_{3/2}$ (6) $3P_{1/2}$	$3D_{5/2}$ (10) $3D_{3/2}$	
$n=2$	$2S_{1/2}$ (2)	$2P_{3/2}$ (6) $2P_{1/2}$		
$n=1$	$1S_{1/2}$ (2)			

Figure 25.5
The unperturbed spectrum of the hydrogen atom with bound states labeled by coupled basis quantum numbers (*diagram not to scale*). The total number of states for each value of n and ℓ is shown in parentheses under the corresponding bar. The states for each $\ell \neq 0$ include two degenerate j multiplets.

$$
\begin{aligned}
\hat{H} &= mc^2\sqrt{1+\frac{\hat{\mathbf{p}}^2}{m^2c^2}} = mc^2\left[1+\frac{\hat{\mathbf{p}}^2}{2m^2c^2}-\frac{1}{8}\left(\frac{\hat{\mathbf{p}}^2}{m^2c^2}\right)^2+\cdots\right] \\
&= mc^2+\frac{\hat{\mathbf{p}}^2}{2m}-\frac{1}{8}\frac{\hat{\mathbf{p}}^4}{m^3c^2}+\cdots,
\end{aligned}
\tag{25.7.2}
$$

where $\hat{\mathbf{p}}^4 \equiv \hat{\mathbf{p}}^2\hat{\mathbf{p}}^2$. The leading term in the expansion is the constant rest energy and is followed by the familiar nonrelativistic kinetic energy operator. The next term is the first nontrivial relativistic correction. We expect this correction to be present in the more accurate Hamiltonian for the hydrogen atom.

The corrections to the Bohr Hamiltonian $\hat{H}^{(0)}$ can be derived systematically using the Dirac equation, which we discussed briefly in section 24.7. For an electron in the potential of a proton, equation (24.7.13) implies that the Dirac Hamiltonian $\hat{H}_{\text{Dirac}}$ is

$$
\hat{H}_{\text{Dirac}} = c\boldsymbol{\alpha}\cdot\left(\hat{\mathbf{p}}+\frac{e}{c}\mathbf{A}\right)+\beta mc^2-\frac{e^2}{r}. \tag{25.7.3}
$$

Energy eigenstates correspond to solutions of

$$
\hat{H}_{\text{Dirac}}\,\psi = E\,\psi, \quad \psi = \begin{pmatrix}\chi\\ \eta\end{pmatrix}. \tag{25.7.4}
$$

The Dirac spinor ψ is composed of two Pauli spinors χ and η. The corrections to the hydrogen Hamiltonian $\hat{H}^{(0)}$ are obtained by finding the effective Hamiltonian $\hat{H}$ that acts on the Pauli spinor χ. The analysis can be done by setting $\mathbf{A}=0$, since the stationary proton creates no vector potential. The elimination of the η spinor from the dynamics takes some work and will not be discussed here. At the end of the analysis, one finds $\hat{H}\chi = E\chi$, where $\hat{H}$ is the corrected Hamiltonian we are after and takes the form

$$\boxed{\hat{H} = \underbrace{\frac{\hat{\mathbf{p}}^2}{2m} + V}_{\hat{H}^{(0)}} \underbrace{- \frac{\hat{\mathbf{p}}^4}{8m^3c^2}}_{\delta H_{\text{rel.}}} + \underbrace{\frac{1}{2m^2c^2}\frac{1}{r}\frac{dV}{dr}\hat{\mathbf{S}}\cdot\hat{\mathbf{L}}}_{\delta H_{\text{spin-orbit}}} + \underbrace{\frac{\hbar^2}{8m^2c^2}\nabla^2 V}_{\delta H_{\text{Darwin}}}.} \tag{25.7.5}$$

The first correction is the relativistic energy correction anticipated above. The second is the spin-orbit coupling, first examined in section 20.6. The third is the *Darwin* correction, which as we shall see affects only $\ell = 0$ states. The above are the fine-structure corrections to $\hat{H}^{(0)}$.

Recall that the energy scale of $\hat{H}^{(0)}$ eigenstates is $\alpha^2 mc^2$. We will now see that *all the above energy corrections are of order* $\alpha^4 mc^2$—in other words, smaller by a factor of $\alpha^2 \simeq \frac{1}{19{,}000}$ than the zeroth-order energies. This shows that for the hydrogen atom, the role of the unit-free parameter λ of perturbation theory is taken by the fine-structure constant squared: $\lambda \sim \alpha^2$. We cannot adjust the value of α^2 nor take it to zero, but happily, it is rather small.

For the relativistic correction, recalling that $p \simeq \alpha mc$, we indeed have

$$\delta H_{\text{rel.}} = -\frac{\hat{\mathbf{p}}^4}{8m^3c^2} \sim -\alpha^4 mc^2. \tag{25.7.6}$$

For spin-orbit we first rewrite the term using

$$\frac{1}{r}\frac{dV}{dr} = \frac{1}{r}\frac{d}{dr}\left(-\frac{e^2}{r}\right) = \frac{e^2}{r^3} \tag{25.7.7}$$

so that

$$\delta H_{\text{spin-orbit}} = \frac{e^2}{2m^2c^2}\frac{1}{r^3}\hat{\mathbf{S}}\cdot\hat{\mathbf{L}}. \tag{25.7.8}$$

For estimation we set $\hat{\mathbf{S}}\cdot\hat{\mathbf{L}} \sim \hbar^2$, $r \sim a_0$ and recall that $a_0 = \frac{\hbar}{mc}\frac{1}{\alpha}$:

$$\delta H_{\text{spin-orbit}} \sim \frac{e^2}{m^2c^2}\frac{\hbar^2}{a_0^3} = \frac{\alpha\hbar c}{m^2c^2}\frac{\hbar^2}{a_0^3} = \alpha\left(\frac{\hbar}{mca_0}\right)^3 mc^2 = \alpha^4 mc^2. \tag{25.7.9}$$

We can evaluate the Darwin term using $V = -e^2/r$:

$$\delta H_{\text{Darwin}} = -\frac{e^2\hbar^2}{8m^2c^2}\nabla^2\left(\frac{1}{r}\right) = -\frac{e^2\hbar^2}{8m^2c^2}(-4\pi\delta(\mathbf{r})) = \frac{\pi}{2}\frac{e^2\hbar^2}{m^2c^2}\delta(\mathbf{r}). \tag{25.7.10}$$

To estimate this correction, note that, due to the δ function, the integral in the expectation value will introduce a factor $|\psi(0)|^2 \sim a_0^{-3}$. We will therefore have

$$\delta H_{\text{Darwin}} \sim \frac{e^2\hbar^2}{m^2c^2a_0^3} \sim \alpha^4 mc^2, \tag{25.7.11}$$

as this is exactly the same combination of constants that we had for the spin-orbit term.

The fine structure of hydrogen is the spectrum of bound states once one takes into account the corrections indicated in (25.7.5). After the partial simplifications considered above, we have

$$\hat{H} = \underbrace{\frac{\hat{\mathbf{p}}^2}{2m} + V}_{\hat{H}^{(0)}} \underbrace{- \frac{\hat{\mathbf{p}}^4}{8m^3c^2}}_{\delta H_{\text{rel.}}} + \underbrace{\frac{e^2}{2m^2c^2}\frac{\hat{\mathbf{S}}\cdot\hat{\mathbf{L}}}{r^3}}_{\delta H_{\text{spin-orbit}}} + \underbrace{\frac{\pi}{2}\frac{e^2\hbar^2}{m^2c^2}\delta(\mathbf{r})}_{\delta H_{\text{Darwin}}} . \tag{25.7.12}$$

All the terms correcting $\hat{H}^{(0)}$ define the fine-structure Hamiltonian $\delta H_{\text{f.s.}}$ so that $\hat{H} = \hat{H}^{(0)} + \delta H_{\text{f.s.}}$. We will study each of these terms separately and then combine our results. As it will turn out, the coupled basis is the one for which the above sum of perturbations are diagonal. There are additional *smaller* corrections that we will not examine here, such as hyperfine splitting (considered in section 20.4) and the Lamb effect.

Darwin correction Since the Darwin correction has a delta function at the origin, the first-order correction to the energy vanishes unless the wave function is nonzero at the origin. This can only happen for nS states. There is no need to use the apparatus of degenerate perturbation theory even though the nS states are actually a doublet with $\hat{S}_z = \pm\frac{\hbar}{2}$. While these basis states are degenerate, the Darwin perturbation is diagonal in this basis because it commutes with $\hat{S}_z$, which has different values for these states. Thus, there is no need to include the spin in the calculation, and we have

$$E^{(1)}_{n00,\text{Darwin}} = \langle \psi_{n00} | \delta H_{\text{Darwin}} | \psi_{n00} \rangle = \frac{\pi}{2}\frac{e^2\hbar^2}{m^2c^2} |\psi_{n00}(\mathbf{0})|^2 . \tag{25.7.13}$$

As is demonstrated in problem 25.13, the radial equation can be used to determine the value of the normalized nS wave functions at the origin. You will find that

$$|\psi_{n00}(\mathbf{0})|^2 = \frac{1}{\pi n^3 a_0^3} . \tag{25.7.14}$$

As a result,

$$E^{(1)}_{n00,\text{Darwin}} = \frac{e^2\hbar^2}{2m^2c^2}\frac{1}{a_0^3 n^3} = \alpha^4(mc^2)\frac{1}{2n^3} . \tag{25.7.15}$$

This completes the evaluation of the Darwin correction.

The Darwin term in the Hamiltonian, just like the other fine-structure terms, arises from the elimination of one of the two two-component spinors in the Dirac equation. As we will show now, such a term would arise from a nonlocal correction to the potential energy, as *if* the electron had grown from a point to a ball with a radius of the order of its Compton wavelength $\frac{\hbar}{m_e c}$. While a simple estimate of this nonlocal potential energy reproduces the Darwin correction up to a constant factor, one must not conclude that the electron is not a point particle. The coincidence occurs because in a relativistic treatment of the electron its Compton wavelength is relevant; it is the shortest distance an electron can be localized. The Darwin term is a reflection of the *trembling motion* phenomenon, or *zitterbewegung*, discussed in early quantum theory, where using the Heisenberg equations of motion one finds that the position of the Dirac electron shows a rapid oscillation with an amplitude proportional to its Compton wavelength.

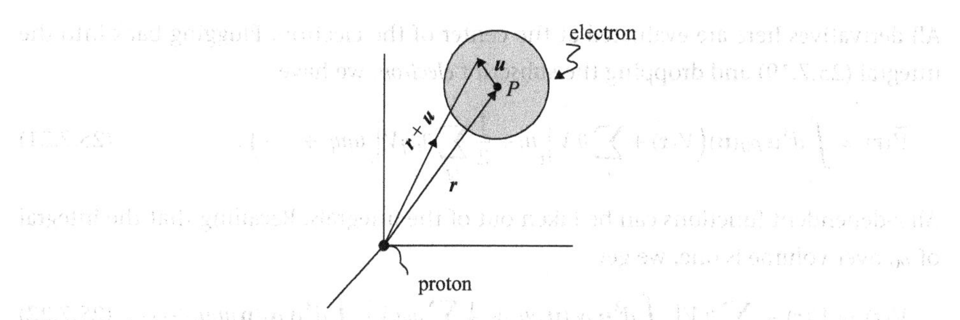

Figure 25.6
A Darwin-type correction to the energy arises if the electron charge is smeared over a region of size comparable to its Compton wavelength. The center of the spherically symmetric electron cloud is at P, and the proton is at the origin. The vector $\mathbf{u}$ is radial from P.

The potential energy $V(\mathbf{r})$ of the electron, as a point particle, is the product of the electron charge $(-e)$ times the electric potential $\Phi(\mathbf{r})$ created by the proton:

$$V(\mathbf{r}) = (-e)\Phi(\mathbf{r}) = (-e)\,\frac{e}{r}\,. \tag{25.7.16}$$

Let us call $\widetilde{V}(\mathbf{r})$ the potential energy when the electron is a charge distribution centered at a point $\mathbf{r}$ with $|\mathbf{r}| = r$ (figure 25.6). This energy is obtained by integration over the electron distribution. Using the vector $\mathbf{u}$ to define position relative to the center P of the electron and letting $\rho(\mathbf{u})$ denote the position-dependent charge density, we see that

$$\widetilde{V}(\mathbf{r}) = \int_{\text{electron}} d^3\mathbf{u}\,\rho(\mathbf{u})\Phi(\mathbf{r}+\mathbf{u})\,, \tag{25.7.17}$$

where, as shown in the figure, $\mathbf{r}+\mathbf{u}$ is the position of the integration point, measured relative to the proton at the origin. It is convenient to write the electron charge density in terms of a normalized function $\rho_0(u)$:

$$\rho(\mathbf{u}) = -e\,\rho_0(\mathbf{u}) \quad \Rightarrow \quad \int_{\text{electron}} d^3\mathbf{u}\,\rho_0(\mathbf{u}) = 1\,, \tag{25.7.18}$$

which guarantees that the integral of ρ over the electron is indeed $(-e)$. Recalling that $-e\Phi(\mathbf{r}+\mathbf{u}) = V(\mathbf{r}+\mathbf{u})$, we now rewrite (25.7.17) as

$$\widetilde{V}(\mathbf{r}) = \int_{\text{electron}} d^3\mathbf{u}\,\rho_0(\mathbf{u})V(\mathbf{r}+\mathbf{u})\,. \tag{25.7.19}$$

This equation has a clear interpretation: the potential energy is obtained as a weighted integral of the potential due to the proton over the extended electron. If the electron charge were perfectly localized, $\rho_0(\mathbf{u}) = \delta(\mathbf{u})$ and $\tilde{V}(\mathbf{r})$ would just be equal to $V(\mathbf{r})$. We will assume that the distribution of charge is spherically symmetric so that $\rho_0(\mathbf{u}) = \rho_0(u)$.

To evaluate (25.7.19), we first do a Taylor expansion of the potential in the integrand about the point $\mathbf{u} = 0$:

$$V(\mathbf{r}+\mathbf{u}) = V(\mathbf{r}) + \sum_i \partial_i V\Big|_{\mathbf{r}} u_i + \frac{1}{2}\sum_{i,j} \partial_i\partial_j V\Big|_{\mathbf{r}} u_i u_j + \cdots. \tag{25.7.20}$$

All derivatives here are evaluated at the center of the electron. Plugging back into the integral (25.7.19) and dropping the subscript *electron*, we have

$$\tilde{V}(\mathbf{r}) = \int d^3\mathbf{u}\, \rho_0(\mathbf{u}) \Big(V(\mathbf{r}) + \sum_i \partial_i V\Big|_{\mathbf{r}} u_i + \frac{1}{2}\sum_{i,j} \partial_i\partial_j V\Big|_{\mathbf{r}} u_i u_j + \cdots \Big). \tag{25.7.21}$$

All $\mathbf{r}$-dependent functions can be taken out of the integrals. Recalling that the integral of ρ_0 over volume is one, we get

$$\tilde{V}(\mathbf{r}) = V(\mathbf{r}) + \sum_i \partial_i V\Big|_{\mathbf{r}} \int d^3\mathbf{u}\, \rho_0(\mathbf{u}) u_i + \tfrac{1}{2}\sum_{i,j} \partial_i\partial_j V\Big|_{\mathbf{r}} \int d^3\mathbf{u}\, \rho_0(\mathbf{u}) u_i u_j + \cdots. \tag{25.7.22}$$

Due to the spherical symmetry of ρ_0, the first integral vanishes: the integrand is odd under the flip $u_i \to -u_i$, and integrals of odd functions over a domain invariant under the flip must vanish. The second takes the form

$$\int d^3\mathbf{u}\, \rho_0(\mathbf{u}) u_i u_j = \tfrac{1}{3}\delta_{ij} \int d^3\mathbf{u}\, \rho_0(\mathbf{u}) u^2. \tag{25.7.23}$$

Indeed, the integral must vanish for $i \neq j$ because the integrand is odd under $u_i \to -u_i$. Moreover, by spherical symmetry, the integral must take equal values for $i = j = 1, 2, 3$. Since $u^2 = u_1^2 + u_2^2 + u_3^2$, the result follows. Therefore, we get

$$\tilde{V}(\mathbf{r}) = V(\mathbf{r}) + \tfrac{1}{6}\sum_i \partial_i\partial_i V\Big|_{\mathbf{r}} \int d^3\mathbf{u}\, \rho_0(\mathbf{u}) u^2 + \cdots = V(\mathbf{r}) + \tfrac{1}{6}\nabla^2 V \int d^3\mathbf{u}\, \rho_0(\mathbf{u}) u^2 + \cdots.$$

The second term represents the correction δV to the potential energy:

$$\delta V = \tfrac{1}{6}\nabla^2 V \int d^3\mathbf{u}\, \rho_0(\mathbf{u}) u^2. \tag{25.7.24}$$

To get an estimate for δV, let us assume that the charge is distributed uniformly over a sphere of radius u_0. This means that $\rho_0(\mathbf{u}) = \frac{3}{4\pi u_0^3}$ for $u < u_0$ and vanishes for $u > u_0$. The integral in (25.7.24) then gives

$$\int d^3\mathbf{u}\, \rho_0(\mathbf{u}) u^2 = \int_0^{u_0} \frac{4\pi u^2 du\, u^2}{\frac{4\pi}{3} u_0^3} = \frac{3}{u_0^3} \int_0^{u_0} u^4 du = \tfrac{3}{5} u_0^2. \tag{25.7.25}$$

Therefore, in (25.7.24) we find that $\delta V = \frac{1}{10} u_0^2 \nabla^2 V$. If we choose the radius u_0 of the charge distribution to be the (reduced) Compton wavelength $\frac{\hbar}{mc}$ of the electron, we get

$$\delta V = \frac{\hbar^2}{10 m^2 c^2} \nabla^2 V. \tag{25.7.26}$$

Comparing with (25.7.5), we see that, up to a small correction (the $\frac{1}{10}$ should be $\frac{1}{8}$), this is the Darwin energy shift. The agreement is surprisingly good for what is, admittedly, a rather heuristic argument.

Relativistic correction We now turn to the relativistic correction from (25.7.12). The energy shifts of the hydrogen states can be analyzed among the degenerate state space $\mathcal{H}_n$ with principal quantum number n. We thus need to consider the matrix elements

$$-\frac{1}{8m^3c^2}\langle\psi_{n\ell'm'_\ell m'_s}|\hat{\mathbf{p}}^2\hat{\mathbf{p}}^2|\psi_{n\ell m_\ell m_s}\rangle\,. \tag{25.7.27}$$

We now explain why the matrix elements are only nonvanishing along the diagonal, meaning the uncoupled basis is a good basis. This is checked using remark 3 in section 25.4. The perturbing operator $\hat{\mathbf{p}}^2\hat{\mathbf{p}}^2$ commutes with $\hat{\mathbf{L}}^2$, with $\hat{L}_z$, and with $\hat{S}_z$. The first operator guarantees that the matrix elements vanish unless $\ell'=\ell$, the second guarantees that the matrix elements vanish unless $m'_\ell=m_\ell$, and the third guarantees, rather trivially, that the matrix elements vanish unless $m'_s=m_s$. The corrections are therefore given by the diagonal elements:

$$E^{(1)}_{n\ell m_\ell m_s;\text{rel}} = -\frac{1}{8m^3c^2}\langle\psi_{n\ell m_\ell m_s}|\hat{\mathbf{p}}^2\hat{\mathbf{p}}^2|\psi_{n\ell m_\ell m_s}\rangle\,. \tag{25.7.28}$$

To evaluate the matrix element, we use the Hermiticity of $\hat{\mathbf{p}}^2$ to move one of the factors into the bra:

$$E^{(1)}_{n\ell m_\ell m_s;\text{rel}} = -\frac{1}{8m^3c^2}\langle\hat{\mathbf{p}}^2\psi_{n\ell m}|\hat{\mathbf{p}}^2\psi_{n\ell m}\rangle\,, \tag{25.7.29}$$

and we eliminate the m_s label by evaluating the trivial expectation value for the spin degrees of freedom. To simplify the evaluation, we use the Schrödinger equation to write

$$\Big(\frac{\hat{\mathbf{p}}^2}{2m}+V\Big)\psi_{n\ell m} = E^{(0)}_n\psi_{n\ell m} \quad\Rightarrow\quad \hat{\mathbf{p}}^2\psi_{n\ell m}=2m(E^{(0)}_n-V)\psi_{n\ell m}\,. \tag{25.7.30}$$

Using this both for the bra and the ket,

$$E^{(1)}_{n\ell m_\ell m_s;\text{rel}} = -\frac{1}{2mc^2}\Big\langle (E^{(0)}_n-V)\psi_{n\ell m}\Big|(E^{(0)}_n-V)\psi_{n\ell m}\Big\rangle\,. \tag{25.7.31}$$

The operator $E^{(0)}_n-V$ is also Hermitian and can be moved from the bra to the ket:

$$\begin{aligned}E^{(1)}_{n\ell m_\ell m_s;\text{rel}} &= -\frac{1}{2mc^2}\Big\langle\psi_{n\ell m}\Big|\big((E^{(0)}_n)^2-2VE^{(0)}_n+V^2\big)\Big|\psi_{n\ell m}\Big\rangle\\ &= -\frac{1}{2mc^2}\Big[(E^{(0)}_n)^2-2E^{(0)}_n\langle V\rangle_{n\ell m}+\langle V^2\rangle_{n\ell m}\Big]\,.\end{aligned} \tag{25.7.32}$$

The problem has been reduced to the computation of the expectation value of $V(r)$ and $V^2(r)$ in the $\psi_{n\ell m}$ state. The expectation value of $V(r)$ is obtained from the virial theorem applied to the Coulomb potential and gives $\langle V\rangle_{n\ell m}=2E^{(0)}_n$ (see (11.4.5)). For $V^2(r)$ we use the expectation value computed using the Hellmann-Feynman lemma and recorded in (11.4.20):

$$\langle V^2\rangle=e^4\Big\langle\frac{1}{r^2}\Big\rangle=e^4\frac{1}{a_0^2n^3\,(\ell+\frac{1}{2})} = \Big(\frac{e^2}{2a_0}\frac{1}{n^2}\Big)^2\frac{4n}{\ell+\frac{1}{2}}=(E^{(0)}_n)^2\cdot\frac{4n}{\ell+\frac{1}{2}}\,. \tag{25.7.33}$$

Going back to (25.7.32), we find that

$$E^{(1)}_{n\ell m_\ell m_s;\text{rel}} = -\frac{(E_n^{(0)})^2}{2mc^2}\left[\frac{4n}{\ell+\frac{1}{2}} - 3\right] = -\tfrac{1}{8}\,\alpha^4\frac{(mc^2)}{n^4}\left[\frac{4n}{\ell+\frac{1}{2}} - 3\right]. \tag{25.7.34}$$

The degeneracy of all ℓ multiplets for a given n has been broken. That degeneracy of $\hat{H}^{(0)}$ was explained by the conserved Runge-Lenz vector. It is clear that in the presence of the relativistic correction the Runge-Lenz vector is no longer conserved. Rotational invariance of the perturbation explains the lack of m_ℓ dependence of the answer (see problem 25.10). Similarly, the manifest spin invariance of the perturbation explains the lack of m_s dependence of the answer.

We have computed the above correction using the uncoupled basis

$$E^{(1)}_{n\ell m_\ell m_s;\text{rel}} = \langle n\ell m_\ell m_s|\delta H_{\text{rel}}|n\ell m_\ell m_s\rangle = \mathcal{E}(n,\ell)\,, \tag{25.7.35}$$

with the definition of $\mathcal{E}$ emphasizing that the matrix elements depend only on n and ℓ. We have already seen that in the full degenerate subspace $\mathcal{H}_n$ the matrix for δH_{rel} is diagonal in the uncoupled basis. But now we see that in each degenerate subspace of fixed n and ℓ,

$$\mathcal{H}_{n,\ell} = \ell \otimes \tfrac{1}{2}\,, \tag{25.7.36}$$

the perturbation δH_{rel} is in fact a multiple of the *identity* matrix since the matrix elements are independent of m_ℓ and m_s, the $\hat{L}_z$ and $\hat{S}_z$ eigenvalues:

$$\langle n\ell m'_\ell m'_s|\delta H_{\text{rel}}|n\ell m_\ell m_s\rangle = \mathcal{E}(n,\ell)\,\delta_{m'_\ell,m_\ell}\delta_{m'_s,m_s}\,. \tag{25.7.37}$$

A matrix equal to a multiple of the identity is invariant under any orthonormal change of basis. For any $\mathcal{H}_{n,\ell}$, the resulting j multiplets,

$$\mathcal{H}_{n,\ell} = (j=\ell+\tfrac{1}{2}) \oplus (j=\ell-\tfrac{1}{2})\,, \tag{25.7.38}$$

provide an alternative orthonormal basis. The invariance of a matrix proportional to the identity implies that, in the basis of $\mathcal{H}_{n,\ell}$ provided by the states of the $j=\ell\pm\frac{1}{2}$ multiplets, we have

$$\langle n\ell j' m'_j|\delta H_{\text{rel}}|n\ell j m_j\rangle = \mathcal{E}(n,\ell)\delta_{j',j}\delta_{m'_j,m_j}\,. \tag{25.7.39}$$

It follows that the energy shifts in the coupled basis are

$$E^{(1)}_{n\ell j m_j,\text{rel}} = \langle n\ell j m_j|\delta H_{\text{rel}}|n\ell j m_j\rangle = \mathcal{E}(n,\ell)\,, \tag{25.7.40}$$

with the same function $\mathcal{E}(n,\ell)$ as in (25.7.35). This is what we wanted to show. The preservation of the matrix elements can also be argued more explicitly. Indeed, any state in the coupled basis is a superposition of orthonormal uncoupled basis states with constant coefficients c_i:

$$|n\ell j m_j\rangle = \sum_i c_i|n\ell m^i_\ell m^i_s\rangle\,, \quad \text{with} \quad \sum_i |c_i|^2 = 1 \tag{25.7.41}$$

because the state on the left-hand side must also have unit norm. Since the matrix elements in the uncoupled basis are diagonal, we get, as claimed,

$$\begin{aligned}\langle n\ell jm_j|\delta H_{\rm rel}|n\ell jm_j\rangle &= \sum_{i,k} c_i^* c_k \langle n\ell m_\ell^i m_s^i|\delta H_{\rm rel}|n\ell m_\ell^k m_s^k\rangle \\ &= \sum_i |c_i|^2 \langle n\ell m_\ell^i m_s^i|\delta H_{\rm rel}|n\ell m_\ell^i m_s^i\rangle \\ &= \sum_i |c_i|^2 \mathcal{E}(n,\ell) = \mathcal{E}(n,\ell)\sum_i |c_i|^2 = \mathcal{E}(n,\ell).\end{aligned} \tag{25.7.42}$$

Spin-orbit coupling The spin-orbit contribution to the Hamiltonian was given in (25.7.12):

$$\delta H_{\text{spin-orbit}} = \frac{e^2}{2m^2c^2}\frac{1}{r^3}\hat{\mathbf{S}}\cdot\hat{\mathbf{L}}. \tag{25.7.43}$$

We gave an extended discussion of the effect of this perturbation on the spectrum in section 20.6. We will be briefer here. Note that the spin-orbit perturbation commutes with $\hat{\mathbf{L}}^2$ because $\hat{\mathbf{L}}^2$ commutes with any $\hat{L}_i$ and any $\hat{S}_i$. Moreover, the perturbation also commutes with $\hat{\mathbf{J}}^2$ and with $\hat{J}_z$ since, in fact, $[\hat{J}_i, \hat{\mathbf{S}}\cdot\hat{\mathbf{L}}]=0$ for any i, which follows quickly from

$$\hat{\mathbf{S}}\cdot\hat{\mathbf{L}} = \tfrac{1}{2}(\hat{\mathbf{J}}^2 - \hat{\mathbf{S}}^2 - \hat{\mathbf{L}}^2). \tag{25.7.44}$$

As a result, $\delta H_{\text{spin-orbit}}$ is diagonal in the level n degenerate subspace if we use the coupled basis $|n\ell jm_j\rangle$. In fact, as we will see, the matrix elements are m_j independent. This, in fact, follows because $\delta H_{\text{spin-orbit}}$ is a scalar under $\hat{\mathbf{J}}$ (problem 25.10). An operator $\hat{A}$ is said to be a *scalar under* $\hat{\mathbf{J}}$ if $[\hat{J}_i, \hat{A}]=0$ for all i.

To compute the matrix elements of $\delta H_{\text{spin-orbit}}$, we use (25.7.44):

$$\begin{aligned}E^{(1)}_{n\ell jm_j;\,\text{spin-orbit}} &= \frac{e^2}{2m^2c^2}\Big\langle n\ell jm_j\Big|\frac{1}{r^3}\hat{\mathbf{S}}\cdot\hat{\mathbf{L}}\Big|n\ell jm_j\Big\rangle \\ &= \frac{e^2}{2m^2c^2}\frac{\hbar^2}{2}\left[j(j+1)-\ell(\ell+1)-\tfrac{3}{4}\right]\Big\langle n\ell jm_j\Big|\frac{1}{r^3}\Big|n\ell jm_j\Big\rangle.\end{aligned} \tag{25.7.45}$$

We need the expectation value of $1/r^3$ in these states. On the uncoupled eigenstates, the result is given in (11.4.20):

$$\Big\langle n\ell m_\ell\Big|\frac{1}{r^3}\Big|n\ell m_\ell\Big\rangle = \frac{1}{n^3 a_0^3\,\ell\left(\ell+\frac{1}{2}\right)(\ell+1)}. \tag{25.7.46}$$

Because of the m_ℓ independence of this expectation value (and its obvious m_s independence), the operator $1/r^3$ is a multiple of the identity matrix in each $\ell\otimes\frac{1}{2}$ multiplet. It follows that it is the same multiple of the identity in the coupled basis description. Therefore,

$$\Big\langle n\ell jm_j\Big|\frac{1}{r^3}\Big|n\ell jm_j\Big\rangle = \frac{1}{n^3 a_0^3\,\ell\left(\ell+\frac{1}{2}\right)(\ell+1)}. \tag{25.7.47}$$

Using this in (25.7.45),

$$E^{(1)}_{n\ell jm_j;\,\text{spin-orbit}} = \frac{e^2\hbar^2}{4m^2c^2}\frac{\left[j(j+1)-\ell(\ell+1)-\frac{3}{4}\right]}{n^3a_0^3\,\ell\left(\ell+\frac{1}{2}\right)(\ell+1)}\,. \tag{25.7.48}$$

Working out the constants in terms of $E_n^{(0)}$ and rest energies, we get

$$E^{(1)}_{n\ell jm_j;\,\text{spin-orbit}} = \frac{(E_n^{(0)})^2}{mc^2}\frac{n\left[j(j+1)-\ell(\ell+1)-\frac{3}{4}\right]}{\ell\left(\ell+\frac{1}{2}\right)(\ell+1)}\,,\quad \ell\neq 0. \tag{25.7.49}$$

Since $\hat{\mathbf{L}}$ vanishes identically acting on any $\ell=0$ state, it is physically reasonable to assume that the spin-orbit correction vanishes for $\ell=0$ states. We find, however, that the limit of the above formula as $\ell\to 0$ is nonzero. To take the limit, we first set $j=\ell+\frac{1}{2}$, since the other possibility, $j=\ell-\frac{1}{2}$, does not apply for $\ell=0$:

$$\begin{aligned} E^{(1)}_{n\ell jm_j;\,\text{spin-orbit}}\Big|_{j=\ell+\frac{1}{2}} &= \frac{(E_n^{(0)})^2}{mc^2}\frac{n\left[(\ell+\frac{1}{2})(\ell+\frac{3}{2})-\ell(\ell+1)-\frac{3}{4}\right]}{\ell\left(\ell+\frac{1}{2}\right)(\ell+1)}\,,\\ &= \frac{(E_n^{(0)})^2}{mc^2}\frac{n}{\left(\ell+\frac{1}{2}\right)(\ell+1)}\,. \end{aligned} \tag{25.7.50}$$

Now taking the limit,

$$\lim_{\ell\to 0} E^{(1)}_{n\ell jm_j;\,\text{spin-orbit}}\Big|_{j=\ell+\frac{1}{2}} = \frac{(E_n^{(0)})^2}{mc^2}(2n) = \alpha^4 mc^2\frac{1}{2n^3}\,. \tag{25.7.51}$$

We see that in this limit the spin-orbit correction is in fact identical to the Darwin shift (25.7.15) of the nS states. This is a bit surprising and will play a technical role below.

Combining results For $\ell\neq 0$ states, we can add the energy shifts from spin-orbit and from the relativistic correction, both expressed as expectation values in the coupled basis. The result, therefore, will give the shifts of the coupled states. Collecting our results (25.7.34) and (25.7.49), we have

$$\begin{aligned} &\left\langle n\ell jm_j\middle|\delta H_{\text{rel}}+\delta H_{\text{spin-orbit}}\middle|n\ell jm_j\right\rangle\\ &\quad= \frac{(E_n^{(0)})^2}{2mc^2}\left\{3-\frac{4n}{\ell+\frac{1}{2}}+\frac{2n\left[j(j+1)-\ell(\ell+1)-\frac{3}{4}\right]}{\ell\left(\ell+\frac{1}{2}\right)(\ell+1)}\right\}\\ &\quad= \frac{(E_n^{(0)})^2}{2mc^2}\left\{3+2n\left[\frac{j(j+1)-3\ell(\ell+1)-\frac{3}{4}}{\ell\left(\ell+\frac{1}{2}\right)(\ell+1)}\right]\right\}. \end{aligned} \tag{25.7.52}$$

These are the fine-structure energy shifts for all states in the spectrum of hydrogen. The states in a coupled multiplet are characterized by ℓ, j, and m_j, and each multiplet as a whole is shifted according to the above formula. The degeneracy within each multiplet is unbroken because the formula has no m_j dependence. This formula, however, hides some additional degeneracies. We uncover those next.

In the above formula, there are two cases to consider for any *fixed* value of j: the multiplet can have $\ell = j - \frac{1}{2}$, or the multiplet can have $\ell = j + \frac{1}{2}$. We will now see something rather surprising. In both of these cases, the shift turns out to be the same, meaning that the shift is in fact ℓ independent! It only depends on j. Call $f(j, \ell)$ the term in brackets above:

$$f(j,\ell) \equiv \frac{j(j+1) - 3\ell(\ell+1) - \frac{3}{4}}{\ell\left(\ell+\frac{1}{2}\right)(\ell+1)}. \tag{25.7.53}$$

The evaluation of this expression in both cases gives the same result:

$$\begin{aligned} f(j,\ell)\Big|_{\ell=j-\frac{1}{2}} &= \frac{j(j+1) - 3(j-\frac{1}{2})(j+\frac{1}{2}) - \frac{3}{4}}{(j-\frac{1}{2})j(j+\frac{1}{2})} = \frac{-2j^2+j}{j\left(j-\frac{1}{2}\right)\left(j+\frac{1}{2}\right)} = -\frac{2}{\left(j+\frac{1}{2}\right)}, \\ f(j,\ell)\Big|_{\ell=j+\frac{1}{2}} &= \frac{j(j+1) - 3(j+\frac{1}{2})(j+\frac{3}{2}) - \frac{3}{4}}{\left(j+\frac{1}{2}\right)(j+1)\left(j+\frac{3}{2}\right)} = \frac{-2j^2-5j-3}{\left(j+\frac{1}{2}\right)(j+1)\left(j+\frac{3}{2}\right)} = -\frac{2}{\left(j+\frac{1}{2}\right)}. \end{aligned} \tag{25.7.54}$$

We can therefore replace in (25.7.52) the result of our evaluation, which we label as fine-structure (f.s.) shifts:

$$E^{(1)}_{n\ell jm_j;\text{f.s.}} = -\frac{(E_n^{(0)})^2}{2mc^2}\left[\frac{4n}{j+\frac{1}{2}} - 3\right] = -\alpha^4(mc^2)\frac{1}{2n^4}\left[\frac{n}{j+\frac{1}{2}} - \frac{3}{4}\right]. \tag{25.7.55}$$

More briefly,

$$\boxed{E^{(1)}_{n\ell jm_j;\text{f.s.}} = -\alpha^4 mc^2 \cdot S_{n,j}\,, \quad \text{with} \quad S_{n,j} \equiv \frac{1}{2n^4}\left[\frac{n}{j+\frac{1}{2}} - \frac{3}{4}\right].} \tag{25.7.56}$$

Remarks:

1. The j dependence and m_j independence of the energy shifts could have been anticipated from the Dirac equation. The rotation generator that commutes with the Dirac Hamiltonian is $\hat{\mathbf{J}} = \hat{\mathbf{L}} + \hat{\mathcal{S}}$, with $\hat{\mathcal{S}}$ the extension of the spin operator to Dirac spinors. The operator $\hat{\mathbf{J}}$ simultaneously rotates position, momenta, and spin states. Neither $\hat{\mathbf{L}}$ nor $\hat{\mathcal{S}}$ are separately conserved (problem 25.16). With $\hat{\mathbf{J}}$ a symmetry, energy eigenstates must fall into j multiplets. Their energies can depend on j but cannot depend on m_j.
2. While a large amount of the degeneracy of $\hat{H}^{(0)}$ has been broken, for fixed quantum number n, multiplets with the same value of j remain degenerate regardless of ℓ.
3. The formula (25.7.56) works for nS states! For these $\ell = 0$ states, we were supposed to add the relativistic correction and the Darwin correction since their spin-orbit correction is zero. But we noticed that the limit $\ell \to 0$ of the spin-orbit correction reproduces the Darwin term. Whether or not this is a meaningful coincidence, it means the sum performed above gives the right answer for $\ell = 0$.

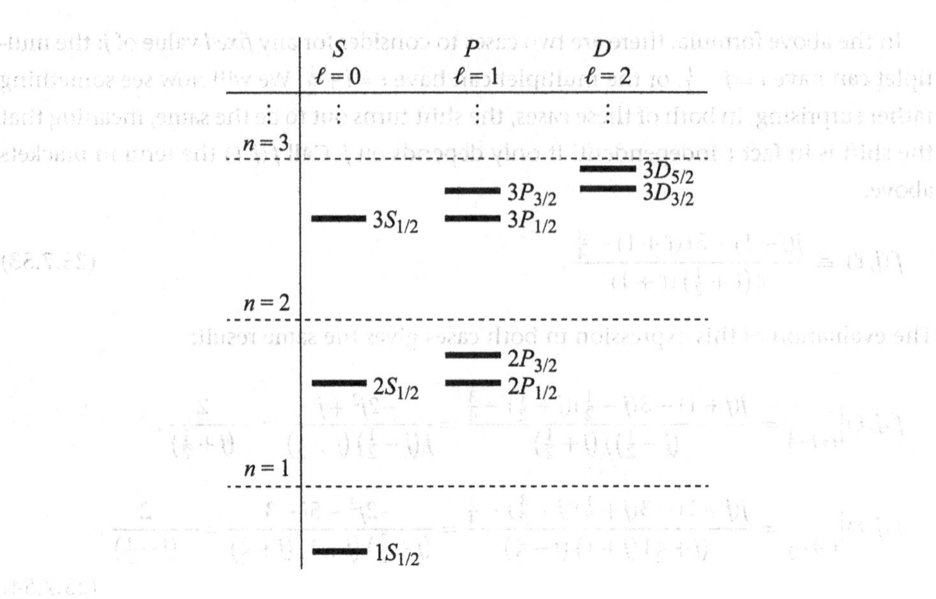

Figure 25.7
A sketch (*not to scale*) of the bound state spectrum of the hydrogen atom, including fine-structure corrections. The corrections depend on n and j only.

4. Since $S_{n,j} > 0$ all energy shifts are down. Indeed,

$$\frac{n}{j+\frac{1}{2}} \geq \frac{n}{j_{\max}+\frac{1}{2}} = \frac{n}{\ell_{\max}+\frac{1}{2}+\frac{1}{2}} = \frac{n}{n} = 1 \quad \Rightarrow \quad \frac{n}{j+\frac{1}{2}} - \frac{3}{4} \geq \frac{1}{4}. \tag{25.7.57}$$

5. For a given fixed n, states with lower values of j get pushed further down. As n increases splittings fall off like n^{-3}.

   For the first three values of n, the values of $S_{n,j}$ are given here:

$$\begin{aligned} &S_{1,\frac{1}{2}} = \tfrac{1}{8}, \\ &S_{2,\frac{1}{2}} = \tfrac{5}{128}, \quad S_{2,\frac{3}{2}} = \tfrac{1}{128}, \\ &S_{3,\frac{1}{2}} = \tfrac{1}{72}, \quad S_{3,\frac{3}{2}} = \tfrac{1}{216}, \quad S_{3,\frac{5}{2}} = \tfrac{1}{648}. \end{aligned} \tag{25.7.58}$$

The energy diagram for states up to $n=3$ is shown in figure 25.7. For the record, the total energy of the hydrogen states is the zeroth value plus the fine-structure correction. Together they give

$$E_{n\ell jm_j} = -\frac{e^2}{2a_0}\frac{1}{n^2}\left[1+\frac{\alpha^2}{n^2}\left(\frac{n}{j+\frac{1}{2}}-\frac{3}{4}\right)\right]. \tag{25.7.59}$$

This is the fine structure of hydrogen! There are, of course, finer corrections. The so-called Lamb shift, for example, breaks the degeneracy between $2S_{1/2}$ and $2P_{1/2}$ and is of order α^5. There is also hyperfine splitting, which arises from the coupling of the magnetic moment of the proton to the magnetic moment of the electron. Such coupling leads to a splitting that is a factor m_e/m_p smaller than fine structure. Additionally, there are additional corrections that are α^2 suppressed relative to fine structure but are seldom of interest.

25.8 Zeeman Effect

In a remarkable experiment conducted in 1896, the Dutch physicist Pieter Zeeman (1865–1943) discovered that atomic spectral lines are split in the presence of an external magnetic field. For this work Zeeman was awarded the Nobel Prize in 1902. The proper understanding of this phenomenon, however, had to wait for quantum mechanics.

The splitting of atomic energy levels by a constant, uniform, external magnetic field, the *Zeeman effect*, has been used as a tool to measure inaccessible magnetic fields. In observing the solar spectrum, a single atomic line, as seen from light emerging away from a sunspot, splits into various equally spaced lines inside the sunspot. From this we have learned that magnetic fields inside a sunspot typically reach three thousand gauss. Sunspots are a bit darker and have lower temperatures than the rest of the solar surface. They can last from hours to months, and their magnetic energy can turn into powerful solar flares.

To study the Zeeman effect, we consider the interaction of an external magnetic field with the total magnetic moment of the electron in the hydrogen atom. The electron has magnetic moment due to its orbital angular momentum and a magnetic moment due to its spin:

$$\hat{\boldsymbol{\mu}}_\ell = -\frac{e}{2mc}\,\hat{\mathbf{L}}, \quad \hat{\boldsymbol{\mu}}_s = -\frac{e}{mc}\,\hat{\mathbf{S}}. \tag{25.8.1}$$

We included the $g=2$ factor in the spin contribution. The Zeeman Hamiltonian δH_{Zeeman} is thus given by

$$\delta H_{\text{Zeeman}} = -(\hat{\boldsymbol{\mu}}_\ell + \hat{\boldsymbol{\mu}}_s)\cdot\mathbf{B} = \frac{e}{2mc}\,(\hat{\mathbf{L}} + 2\,\hat{\mathbf{S}})\cdot\mathbf{B}. \tag{25.8.2}$$

Conventionally, we align the magnetic field with the positive z-axis so that $\mathbf{B} = B\,\mathbf{z}$ and thus get

$$\delta H_{\text{Zeeman}} = \frac{eB}{2mc}\,(\hat{L}_z + 2\hat{S}_z). \tag{25.8.3}$$

When we consider the Zeeman effect on the hydrogen atom, we must not forget the fine-structure corrections $\delta H_{\text{f.s.}}$. The full Hamiltonian to be considered is

$$\hat{H} = \hat{H}^{(0)} + \delta H_{\text{f.s.}} + \delta H_{\text{Zeeman}}. \tag{25.8.4}$$

Recall that in fine structure, there is an internal magnetic field B_{int} associated with spin-orbit coupling: the magnetic field seen by the electron as it goes around the proton. Comparing the external magnetic field B to the internal magnetic field B_{int}, we can consider two different limits:

1. Weak-field Zeeman effect: $B \ll B_{\text{int}}$. In this case the Zeeman effect is small compared with fine-structure effects. Accordingly, the original Hamiltonian $\hat{H}^{(0)}$, *together* with the fine-structure Hamiltonian $\delta H_{\text{f.s.}}$, is thought of as the "known" Hamiltonian $\tilde{H}^{(0)}$, and the Zeeman Hamiltonian is the perturbation:

$$\hat{H} = \underbrace{\hat{H}^{(0)} + \delta H_{\text{f.s.}}}_{\tilde{H}^{(0)}} + \delta H_{\text{Zeeman}} \,. \tag{25.8.5}$$

Of course, the magnetic field must not be so small that other corrections, such as hyperfine splitting, or Lamb shift, overwhelm its effects.

2. Strong-field Zeeman effect: $B \gg B_{\text{int}}$. In this case the Zeeman effect is much larger than fine-structure effects. Accordingly, the original Hamiltonian $\hat{H}^{(0)}$, *together* with the Zeeman Hamiltonian, is thought of as the "known" Hamiltonian $\check{H}^{(0)}$, and the fine-structure Hamiltonian $\delta H_{\text{f.s.}}$ is the perturbation:

$$H = \underbrace{\hat{H}^{(0)} + \delta H_{\text{Zeeman}}}_{\check{H}^{(0)}} + \delta H_{\text{f.s.}} \,. \tag{25.8.6}$$

You may think that $\hat{H}^{(0)} + \delta H_{\text{Zeeman}}$ does not qualify as known, but happily this is a very simple Hamiltonian.

When the Zeeman magnetic field is neither weak nor strong, we must take the sum of the Zeeman and fine-structure Hamiltonians as the perturbation. No simplification is possible, and one must explicitly diagonalize the perturbation.

Weak-field Zeeman effect The approximate eigenstates of $\tilde{H}^{(0)}$ are the coupled states $|n\ell jm_j\rangle$ that exhibit fine-structure corrections and whose energies are a function of n and j, as shown in the fine-structure diagram. Degeneracies in this spectrum occur for fixed j for different values of ℓ, and different values of m_j.

To figure out the effect of the Zeeman interaction on this spectrum, we thus consider the matrix elements:

$$\langle n\ell jm_j|\delta H_{\text{Zeeman}}|n\ell' jm_j'\rangle \,. \tag{25.8.7}$$

Since $\delta H_{\text{Zeeman}} \sim \hat{L}_z + 2\hat{S}_z$, we see that δH_{Zeeman} commutes with $\hat{\mathbf{L}}^2$ and with $\hat{J}_z$. The matrix element thus vanishes unless $\ell' = \ell$ and $m_j' = m_j$, and the Zeeman perturbation is diagonal in the degenerate fine-structure eigenspaces. The energy corrections are therefore

$$E^{(1)}_{n\ell jm_j} = \frac{e\hbar}{2mc} B\, \langle n\ell jm_j|(\hat{L}_z + 2\hat{S}_z)|n\ell jm_j\rangle \frac{1}{\hbar} \,, \tag{25.8.8}$$

where we multiplied and divided by $\hbar$ to make the units of the result manifest. The result of the evaluation of the matrix element will show a remarkable feature: a linear dependence $E^{(1)} \sim \hbar m_j$ on the azimuthal quantum numbers. The states in each j multiplet split into equally separated energy levels, a key signature of the Zeeman effect. We will try to understand this result as a property of matrix elements of vector operators. First, however, note that $\hat{L}_z + 2\hat{S}_z = \hat{J}_z + \hat{S}_z$, and therefore the matrix element of interest in the above equation satisfies

$$\langle n\ell jm_j|(\hat{L}_z + 2\hat{S}_z)|n\ell jm_j\rangle = \hbar m_j + \langle n\ell jm_j|\hat{S}_z|n\ell jm_j\rangle\,. \tag{25.8.9}$$

It follows that we only need to concern ourselves with $\hat{S}_z$ matrix elements.

To calculate these matrix elements, we now digress and talk about vector operators. As we learned in section 19.2, an operator $\hat{\mathbf{V}}$ is said to be a vector operator under an angular momentum $\hat{\mathbf{J}}$ if the following commutator holds for all values of $i, j = 1, 2, 3$:

$$[\hat{J}_i\,,\,\hat{V}_j] = i\hbar\,\epsilon_{ijk}\,\hat{V}_k\,. \tag{25.8.10}$$

It follows from the familiar $\hat{\mathbf{J}}$ commutators that $\hat{\mathbf{J}}$ is a vector operator under $\hat{\mathbf{J}}$. Additionally, if $\hat{\mathbf{V}}$ is a vector operator, it has a simple commutator with $\hat{\mathbf{J}}^2$. One can quickly confirm that (problem 25.18)

$$[\,\hat{\mathbf{J}}^2\,,\hat{\mathbf{V}}] = 2i\hbar\,(\hat{\mathbf{V}}\times\hat{\mathbf{J}} - i\hbar\,\hat{\mathbf{V}})\,. \tag{25.8.11}$$

If $\hat{\mathbf{V}}$ is chosen to be $\hat{\mathbf{J}}$, the left-hand side vanishes by the standard property of $\hat{\mathbf{J}}^2$, and the right-hand side vanishes because $\hat{\mathbf{J}}\times\hat{\mathbf{J}} = i\hbar\hat{\mathbf{J}}$. Finally, by repeated use of the above identities you will show (problem 25.18) that the following formula holds:

$$\frac{1}{(2i\hbar)^2}\Big[\,\hat{\mathbf{J}}^2\,,[\,\hat{\mathbf{J}}^2\,,\hat{\mathbf{V}}]\Big] = (\hat{\mathbf{V}}\cdot\hat{\mathbf{J}})\hat{\mathbf{J}} - \tfrac{1}{2}\,(\hat{\mathbf{J}}^2\,\hat{\mathbf{V}} + \hat{\mathbf{V}}\hat{\mathbf{J}}^2)\,. \tag{25.8.12}$$

Consider $(\hat{\mathbf{J}}^2, \hat{J}_z)$ eigenstates $|k; jm_j\rangle$ where k stands for some other quantum numbers that bear no relation to angular momentum. The matrix elements of the left-hand side of (25.8.12) on such eigenstates are necessarily zero:

$$\langle k'; jm_j'|\Big[\,\hat{\mathbf{J}}^2\,,[\,\hat{\mathbf{J}}^2\,,\hat{\mathbf{V}}]\Big]|k; jm_j\rangle = 0\,, \tag{25.8.13}$$

as can be seen by expanding the outer commutator and noticing that $\hat{\mathbf{J}}^2$ gives the same eigenvalue when acting on the bra and on the ket. Therefore, the matrix elements of the right-hand side are related as follows:

$$\langle k'; jm_j'|(\hat{\mathbf{V}}\cdot\hat{\mathbf{J}})\hat{\mathbf{J}}|k; jm_j\rangle = \hbar^2 j(j+1)\,\langle k'; jm_j'|\hat{\mathbf{V}}|k; jm_j\rangle\,, \tag{25.8.14}$$

which implies that

$$\boxed{\langle k'; jm_j'|\,\hat{\mathbf{V}}\,|k; jm_j\rangle = \frac{\langle k'; jm_j'|\,(\hat{\mathbf{V}}\cdot\hat{\mathbf{J}})\hat{\mathbf{J}}\,|k; jm_j\rangle}{\hbar^2 j(j+1)}\,.} \tag{25.8.15}$$

This is the identity we wanted to establish. Using the less explicit notation $\langle\cdots\rangle$ for the matrix elements, we have found that

$$\langle\hat{\mathbf{V}}\rangle = \frac{\langle(\hat{\mathbf{V}}\cdot\hat{\mathbf{J}})\hat{\mathbf{J}}\rangle}{\langle\hat{\mathbf{J}}^2\rangle}\,. \tag{25.8.16}$$

This is sometimes called the **projection lemma:** the matrix elements of a vector operator $\hat{\mathbf{V}}$ are those of the projection of $\hat{\mathbf{V}}$ onto $\hat{\mathbf{J}}$. Recall that the projection of a vector $\mathbf{v}$ onto the vector $\mathbf{j}$ is $(\mathbf{v}\cdot\mathbf{j})\,\mathbf{j}\,/\mathbf{j}^2$.

Let us now return to the question of interest, the computation of the expectation value of $\hat{S}_z$ in (25.8.9). Since $\hat{\mathbf{S}}$ is a vector operator under $\hat{\mathbf{J}}$, we can use (25.8.15). Specializing to the z-component,

$$\langle n\ell\, jm_j|\hat{S}_z|n\ell\, jm_j\rangle = \frac{\hbar m_j\, \langle n\ell\, jm_j|\hat{\mathbf{S}}\cdot\hat{\mathbf{J}}\,|n\ell\, jm_j\rangle}{\hbar^2 j(j+1)}\,. \tag{25.8.17}$$

We already see the appearance of the predicted $\hbar m_j$ factor. The matrix element in the numerator is still to be calculated, but it will introduce no m_j dependence. In fact, $\hat{\mathbf{S}}\cdot\hat{\mathbf{J}}$ is a scalar operator (it commutes with all $\hat{J}_i$), and therefore it is diagonal in m_j. But even more is true; the expectation value of a scalar operator is in fact independent of m_j (problem 25.10). We will confirm this here by direct computation. Since $\hat{\mathbf{L}}=\hat{\mathbf{J}}-\hat{\mathbf{S}}$, we find that

$$\hat{\mathbf{S}}\cdot\hat{\mathbf{J}} = \tfrac{1}{2}(\hat{\mathbf{J}}^2+\hat{\mathbf{S}}^2-\hat{\mathbf{L}}^2)\,, \tag{25.8.18}$$

and therefore, back in (25.8.17),

$$\langle n\ell\, jm_j|\hat{S}_z|n\ell\, jm_j\rangle = \frac{\hbar m_j}{2j(j+1)}\left(j(j+1)-\ell(\ell+1)+\tfrac{3}{4}\right). \tag{25.8.19}$$

Indeed, no further m_j dependence has appeared. Returning to (25.8.9) we find that

$$\langle n\ell jm_j|(\hat{L}_z+2\hat{S}_z)|n\ell jm_j\rangle = \hbar m_j\left(1+\frac{j(j+1)-\ell(\ell+1)+\frac{3}{4}}{2j(j+1)}\right). \tag{25.8.20}$$

The constant of proportionality in parentheses is called the **Lande g-factor** $g_j(\ell)$:

$$g_j(\ell) \equiv 1+\frac{j(j+1)-\ell(\ell+1)+\frac{3}{4}}{2j(j+1)}\,. \tag{25.8.21}$$

We finally have for the Zeeman energy shifts in (25.8.8)

$$E^{(1)}_{n\ell jm_j} = \frac{e\hbar}{2mc}B\,g_j(\ell)\,m_j = \mu_B B\,g_j(\ell)\,m_j\,. \tag{25.8.22}$$

Here the Bohr magneton $\mu_B = \frac{e\hbar}{2mc} \simeq 5.79\times 10^{-9}$ eV/gauss. This is our final result for the weak-field Zeeman energy corrections to the fine-structure energy levels. The name g-factor for $g_j(\ell)$ is appropriate: just like the g-factor of a free electron, it can be thought to modify the classical value of the magnetic moment. Since all degeneracies within j multiplets are broken and j multiplets with different ℓ values split differently due to the ℓ dependence of $g_j(\ell)$, the weak-field Zeeman effect removes all degeneracies of the hydrogen spectrum!

Strong-field Zeeman effect When the Zeeman perturbation is much larger than the fine-structure corrections, we must take the original hydrogen Hamiltonian together with the Zeeman Hamiltonian to form the "known" Hamiltonian $\check{H}^{(0)}$:

$$\check{H}^{(0)} = \hat{H}^{(0)} + \frac{e}{2mc}(\hat{L}_z+2\hat{S}_z)B\,. \tag{25.8.23}$$

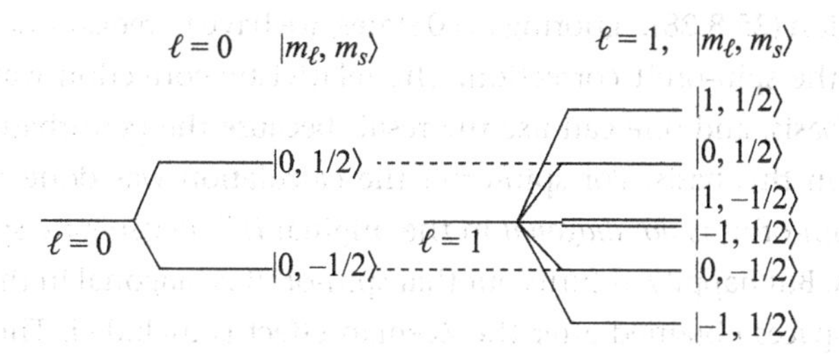

Figure 25.8

Degeneracies remain for the $(\ell=0)\otimes\frac{1}{2}$ and $(\ell=1)\otimes\frac{1}{2}$ multiplets after the inclusion of the Zeeman term in the Bohr Hamiltonian. There are two degenerate states in $1\otimes\frac{1}{2}$, and each of the two states in the $0\otimes\frac{1}{2}$ multiplet is degenerate with a state in the $1\otimes\frac{1}{2}$ multiplet.

Actually, $\check{H}^{(0)}$ is simple because the Zeeman Hamiltonian commutes with the zeroth-order hydrogen Hamiltonian:

$$[\hat{L}_z+2\hat{S}_z\,,\,\hat{H}^{(0)}] = 0\,. \tag{25.8.24}$$

We can thus find eigenstates of both simultaneously. Those are in fact the uncoupled basis states! We see that

$$\begin{aligned}\hat{H}^{(0)}|n\ell m_\ell m_s\rangle &= E_n^{(0)}|n\ell m_\ell m_s\rangle\,,\\(\hat{L}_z+2\hat{S}_z)|n\ell m_\ell m_s\rangle &= \hbar(m_\ell+2m_s)|n\ell m_\ell m_s\rangle\,.\end{aligned} \tag{25.8.25}$$

The uncoupled basis states are therefore the *exact* energy eigenstates of $\check{H}^{(0)}$, and their energies are

$$E_{n\ell m_\ell m_s} = E_n^{(0)}+\mu_B B(m_\ell+2m_s)\,. \tag{25.8.26}$$

While some of the degeneracy of $\hat{H}^{(0)}$ has been removed, some remains. For a fixed principal quantum number n, there are degeneracies among $\ell\otimes\frac{1}{2}$ states as well as degeneracies among such multiplets with $\ell\neq\ell'$. This is illustrated in figure 25.8.

The problem now is to compute the corrections due to $\delta H_{\text{f.s.}}$ on the nondegenerate and on the degenerate subspaces of $\check{H}^{(0)}$. The nondegenerate cases are straightforward, but the degenerate cases could involve diagonalization. We must therefore consider the matrix elements

$$\langle n\ell' m_\ell' m_s'|\,\delta H_{\text{f.s.}}\,|n\ell m_\ell m_s\rangle \tag{25.8.27}$$

with the condition

$$m_\ell'+2m_s' = m_\ell+2m_s \tag{25.8.28}$$

needed for the two states in the matrix element to belong to a degenerate subspace. Since $\hat{\mathbf{L}}^2$ commutes with $\delta H_{\text{f.s.}}$, the matrix elements vanish unless $\ell=\ell'$, and therefore it suffices to consider the matrix elements

$$\langle n\ell m_\ell' m_s'|\,\delta H_{\text{f.s.}}\,|n\ell m_\ell m_s\rangle\,, \tag{25.8.29}$$

still with condition (25.8.28). Ignoring $\ell=0$ states, we have to reexamine the relativistic correction and the spin-orbit correction. The relativistic correction was computed in the uncoupled basis, and one can use the result because the perturbation was shown to be diagonal in this basis. For spin-orbit the calculation was done in the coupled basis because spin-orbit is *not diagonal* in the original $\hat{H}^{(0)}$ degenerate spaces using the uncoupled basis. But happily, it turns out that spin-orbit is diagonal in the more limited degenerate subspaces obtained after the Zeeman effect is included. The details of the strong-field Zeeman effect are discussed in problem 25.17.

Problems

Problem 25.1. *Oscillator with cubic anharmonic perturbation.*

Consider the anharmonic oscillator with Hamiltonian

$$\hat{H}=\frac{\hat{p}^2}{2m}+\tfrac{1}{2}m\omega^2\hat{x}^2+\lambda\sqrt{2}\,\hbar\omega\,\frac{\hat{x}^3}{L_0^3},$$

where $L_0^2=\frac{\hbar}{m\omega}$, and we treat the $\hat{x}^3$ term as a perturbation.

1. Find the shift in the ground state energy to order λ^2.
2. Find the corrected ground state $|0\rangle$ to order λ, writing your answer as a sum of harmonic oscillator states. Include, if required, the appropriate normalization constant that makes the state of unit norm to first order in λ.
3. Find the $\hat{x}$ expectation value $\langle 0|\hat{x}|0\rangle$ in the corrected ground state to leading order in λ.
4. Sketch the potential $V(x)$ as a function of x/L_0 for small $\lambda>0$. Is the state you found in (2) anything like a true energy eigenstate?
5. Establish the anharmonicity of the oscillator by computing $E_n(\lambda)-E_0(\lambda)$, including leading-order corrections in λ.

Problem 25.2. *Perturbation of the three-dimensional isotropic harmonic oscillator.*

The spectrum of the three-dimensional isotropic harmonic oscillator is degenerate. In this problem, we see how a certain perturbation reduces the degeneracy. [This problem does not require the tools of degenerate perturbation theory.]

Consider a system described by the Hamiltonian $\hat{H}=\hat{H}^{(0)}+\lambda\delta H$:

$$\hat{H}^{(0)}=\frac{\hat{\mathbf{p}}^2}{2m}+\tfrac{1}{2}m\omega^2\,\hat{\mathbf{x}}^2\,,\quad \delta H=\omega\hat{L}_2\,.$$

Here, $\hat{\mathbf{x}}=(\hat{x}_1,\hat{x}_2,\hat{x}_3)$, $\hat{\mathbf{p}}=(\hat{p}_1,\hat{p}_2,\hat{p}_3)$, $\hat{L}_2=\hat{x}_3\hat{p}_1-\hat{x}_1\hat{p}_3$ is the component of angular momentum in the y-direction, and λ is unit-free.

1. Set $\lambda=0$ so that $\hat{H}=\hat{H}^{(0)}$. Use creation and annihilation operators in the 1, 2, and 3 directions, with number operators $\hat{N}_1$, $\hat{N}_2$, $\hat{N}_3$, respectively. Denote eigenstates of these number operators by their eigenvalues, as $|n_1,n_2,n_3\rangle$. What is the energy

E_{n_1,n_2,n_3} of the state $|n_1, n_2, n_3\rangle$? How many linearly independent states are there with energy $E = \frac{7}{2}\hbar\omega$?

2. Express δH in terms of creation and annihilation operators. Find the matrix representation of δH in the $E = \frac{5}{2}\hbar\omega$ degenerate subspace spanned by $|1\rangle \equiv |1,0,0\rangle$, $|2\rangle \equiv |0,1,0\rangle$, $|3\rangle \equiv |0,0,1\rangle$.
3. What are the eigenvalues and eigenstates of δH in the $E = \frac{5}{2}\hbar\omega$ degenerate subspace? Give the eigenvalues and eigenstates of $\hat{H} = \hat{H}^{(0)} + \lambda\delta H$ in the degenerate subspace. What is the matrix representation of $\hat{H}^{(0)} + \lambda\delta H$ in the degenerate subspace in this basis of eigenvectors?
4. Calculate $[\hat{H}^{(0)}, \delta H]$. What can be said about $\langle\phi|\delta H|\psi\rangle$ when $|\psi\rangle$ and $|\phi\rangle$ are eigenstates of $\hat{H}^{(0)}$ with *different* energy eigenvalues?

   [The results you will find mean that δH is a "nongeneric" perturbation of $\hat{H}^{(0)}$. We were able to focus on a single degenerate subspace and analyzed $\hat{H} = \hat{H}^{(0)} + \lambda\delta H$, without assuming that λ was small. If δH were "generic," we would have had to assume that λ was small in order to make progress.]

Problem 25.3. *Higher-order corrections in nondegenerate perturbation theory.*

Calculate the second-order state correction $|n^{(2)}\rangle$ and the third-order energy correction $E_n^{(3)}$. The answers can be put in the form

$$|n^{(2)}\rangle = \sum_{m\neq n}\left(\frac{\alpha}{E_m^{(0)} - E_n^{(0)}}\sum_{k\neq n}\frac{\cdots}{E_k^{(0)} - E_n^{(0)}} + \frac{\beta\ \cdots}{(E_m^{(0)} - E_n^{(0)})^2}\right)|m^0\rangle,$$

$$E_n^{(3)} = \sum_{m,k\neq n}\frac{\gamma\cdots}{(E_m^{(0)} - E_n^{(0)})(E_k^{(0)} - E_n^{(0)})} + \delta\cdots\sum_{m\neq n}\frac{\cdots}{(E_m^{(0)} - E_n^{(0)})^2}.$$

In the above, α, β, γ, and δ are numerical constants you must determine, and the dots are expressions built from matrix elements of the perturbation δH.

Problem 25.4. *Energy shift due to finite nuclear size.*

The Coulomb potential of the proton does not hold all the way to the origin. The proton charge is smeared out over a sphere of roughly 10^{-13} cm in radius. This has a small effect on the energy levels of the hydrogen atom. To find out, model the electric charge distribution of the proton as a uniformly charged sphere of radius R. You may ignore fine structure for this problem.

1. Find the electrostatic potential energy $V(r)$ of the electron for all $r > 0$. Your $V(r)$ must be continuous and become $V(r) = -e^2/r$ for $r > R$. [Hint: use Gauss's law $\nabla\cdot\mathbf{E} = 4\pi\rho$ to find the electric field everywhere and then integrate the electric field to find the potential and the potential energy.]
2. Use lowest-order perturbation theory to calculate the shift δE in the energy of the ground state of hydrogen due to this modification of the potential. Assume that R is sufficiently small that the unperturbed wave function varies very little over $0 < r < R$ and can thus be replaced by the value at $r = 0$. Use R and a_0 to express your

answer for $E^{(1)}_{1,0,0}$ as a fraction of the binding energy Ry(= 13.6 eV) of the ground state.

3. Why is this effect most important for states with orbital angular momentum zero? Without calculation, estimate the factor by which this effect is smaller for an $\ell = 1$ state compared to an $\ell = 0$ state.
4. A precise but somewhat controversial measurement of the proton radius came from the Paul Scherrer Institut (PSI) experiment, giving $R = 0.84184(67) \times 10^{-13}$ cm [R. Pohl et al., "The size of the proton," Nature **466**, 213 (2010)]. Explain why the PSI experiment could get such impressive accuracy using muonic hydrogen (a muon-proton bound state) instead of ordinary hydrogen (an electron-proton bound state). [Hint: the muon is 206.8 times heavier than the electron.]

Problem 25.5. *Landau level with an impurity (M. Metlitski).*

Consider a particle of charge q and mass m moving in the (x, y) plane in a uniform magnetic field $\mathbf{B} = B\hat{\mathbf{z}}$. The Hamiltonian is $\hat{H}^{(0)} = \frac{1}{2m}(\hat{\mathbf{p}} - \frac{q}{c}\mathbf{A})^2$. Consider an impurity located at the origin of the sample. We model this as a δ function potential perturbation δH added to $\hat{H}^{(0)}$:

$$\hat{H} = \hat{H}^{(0)} + \lambda\delta H, \quad \delta H = -\hbar\omega_c \ell_B^2\, \delta^2(\mathbf{r})\,.$$

Here $\lambda > 0$ is a dimensionless real parameter, $\omega_c = \frac{qB}{mc}$, and $\ell_B^2 = \frac{\hbar c}{qB}$. Compute the energy shifts of all states in the lowest Landau level to first order in λ. Is the degeneracy of the lowest Landau level (partially) lifted? [Hint: in the symmetric gauge, $\mathbf{A} = \frac{B}{2}(-y, x)$, the normalized wave functions of states in the lowest Landau level are given by

$$\psi_{0,m}(x, y) = \frac{1}{\sqrt{2\pi m!}\,\ell_B}\left(\frac{z}{\ell_B\sqrt{2}}\right)^m \exp\left(-\frac{|z|^2}{4\ell_B^2}\right), \quad m = 0, 1, 2, \ldots,$$

where $z = x + iy$. Note that $0! = 1$.]

Problem 25.6. *Polarizability of a particle on a ring: the methanol molecule.*

Consider a particle of mass m constrained to move in the xy plane on a circular ring of radius a. The only variable of the system is the azimuthal angle ϕ measured with respect to the x-axis. The state of the system is described by a wave function $\psi(\phi)$ that must be periodic and normalized:

$$\psi(\phi + 2\pi) = \psi(\phi), \quad \int_0^{2\pi} |\psi(\phi)|^2 d\phi = 1.$$

The kinetic energy of the particle can be written as

$$\hat{H}^{(0)} = \frac{\hat{L}_z^2}{2ma^2}, \quad \hat{L}_z = \frac{\hbar}{i}\frac{d}{d\phi}.$$

1. Calculate the eigenvalues $E_n^{(0)}$ and eigenfunctions $\psi_n^{(0)}(\phi)$ of $\hat{H}^{(0)}$, with $\psi_n^{(0)}(\phi)$ a normalized $\hat{L}_z$ eigenstate of eigenvalue $\hbar n$. Associate the states $|n^{(0)}\rangle$ with the wave functions $\psi_n^{(0)}$. Which of the energy levels are degenerate?

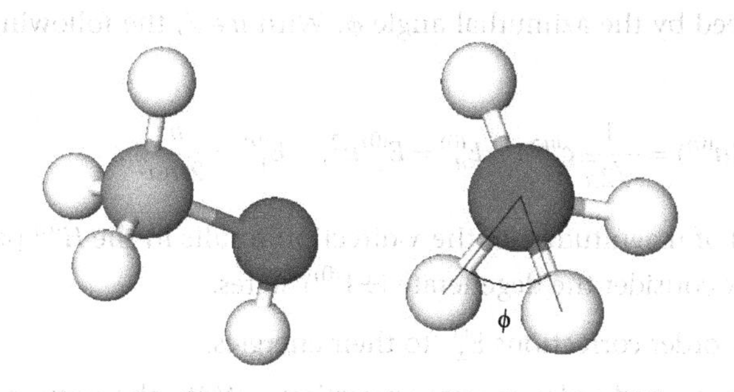

Figure 25.9
Left: A ball-and-stick model of the methanol model. *Right*: The same molecule, seen end on. The angle ϕ indicates the position of the hydrogen attached to the oxygen relative to a hydrogen atom in the stationary CH_3 group.

2. Now assume that the particle has a charge q and is placed in a uniform electric field of magnitude ε in the x-direction. We must therefore add to the Hamiltonian the perturbation

$$\delta H = -q\varepsilon a \cos\phi.$$

Calculate the corrected ground state $|0\rangle$ to first order in ε. Use this state to evaluate the induced electric dipole moment $d = \langle 0|q\hat{x}|0\rangle$ in the x-direction. Determine the *polarizability* constant α relating the dipole moment and the applied field as in $d = \alpha\varepsilon$.

3. Now turn off the electric field of (2), and consider the methanol molecule CH_3OH. We will consider the rotation of the hydroxyl (OH) hydrogen about the straight line joining the oxygen and carbon atoms, as sketched in figure 25.9. To zeroth approximation this rotation is free, and the Hamiltonian $\hat{H}^{(0)}$ describes the rotational kinetic energy (with ma^2 replaced by some effective moment of inertia).

We now take the electrostatic interaction energy between the CH_3 group and the OH group into account as a perturbation. To implement the threefold symmetry of the CH_3 group, we add to $\hat{H}^{(0)}$ a term δH of the form

$$\delta H = \eta\, E_1^{(0)} \cos 3\phi.$$

Here $E_1^{(0)} > 0$ is the first nonzero energy eigenvalue, and $\eta > 0$ is a real unit-free constant. Moreover, $\phi = 0$ corresponds to the OH aligned with some CH vector. Calculate the energy $E_0(\eta)$ of the new ground state to second order in η and the position representation $\psi_0(\phi)$ of this state to first order in η.

Give a physical interpretation of the result. In particular, what are the values of ϕ for which the hydrogen in the OH group is more likely to be found?

Problem 25.7. *Polarizability of a particle on a ring: degenerate states.*

Consider again the situation in the previous problem: a charged particle of charge q and mass m constrained to move in the xy plane on a circular ring of radius a, with the position described by the azimuthal angle ϕ. With $n \in \mathbb{Z}$, the following are the energy eigenstates:

$$\psi_n(\phi)^{(0)} = \langle \phi | n^{(0)} \rangle = \frac{1}{\sqrt{2\pi}} e^{in\phi}, \quad E_n^{(0)} = E_1^{(0)} n^2, \quad E_1^{(0)} = \frac{\hbar^2}{2ma^2}.$$

An electric field of magnitude ε in the x-direction results in the $\hat{H}^{(0)}$ perturbation $\delta H = -q\varepsilon a \cos\phi$. Now consider the degenerate $|\pm 1^{(0)}\rangle$ states.

1. Find the first-order corrections $E_\pm^{(1)}$ to their energies.
2. Calculate the second-order energy corrections. Write the answer in terms of the energy $(q\varepsilon a)^2 / E_1^{(0)}$.
3. What are the wave functions $\psi^{(0)}(\phi)$ for the good basis to zeroth order in the perturbation? How does your result make sense in light of the existence of a conserved "parity" operator?

Problem 25.8. *Particle on a circle and a perturbative potential.*

A free particle of mass m moves on a circle $x \sim x + L$ of circumference L.

1. Determine the normalized momentum eigenstates $\psi_n(x)$, labeled by the integer n running from minus to plus infinity. These happen to be energy eigenstates, so give their energies E_n, and display the degeneracies of the spectrum in a graph.

Now we add a periodic perturbative potential $V(x)$ to the Hamiltonian $\hat{H}^{(0)} = \frac{\hat{p}^2}{2m}$ so that $\hat{H} = \hat{H}^{(0)} + V(\hat{x})$, with

$$V(x) = V_0 \cos\left(\frac{2\pi q x}{L}\right).$$

Here q is a positive integer, and V_0 has units of energy.

2. Compute the matrix element $V_{k\ell} = \langle \psi_k | V(x) | \psi_\ell \rangle$ for arbitrary integers k and ℓ. Leave your answer in terms of Kronecker deltas and constants.
3. Assume q is an odd integer. Describe what happens to the full spectrum to *first order* in perturbation theory, by stating with brief explanations,
   i. which states receive energy corrections and which do not and
   ii. what degeneracies are lifted.
4. Assume q is an even integer so that $q = 2k$ with k a positive integer. As in (3), describe the fate of the spectrum to first order in perturbation theory.
5 Consider $V(x)$ with arbitrary but fixed q. Compute the energy correction to the states ψ_q, ψ_{-q} up to second order. Without performing any calculations could you have anticipated what the "good" basis states are in this subspace?

Problem 25.9. *Three-by-three matrix and degenerate perturbation theory (L. Schiff).*

J. J. Sakurai calls this problem "a challenge for the experts." We will break the original problem in Schiff into several parts. Consider the matrix

$$\begin{pmatrix} E_1^{(0)} & 0 & \epsilon_{13} \\ 0 & E_1^{(0)} & \epsilon_{23} \\ \epsilon_{13}^* & \epsilon_{23}^* & E_3^{(0)} \end{pmatrix} = \begin{pmatrix} E_1^{(0)} & 0 & 0 \\ 0 & E_1^{(0)} & 0 \\ 0 & 0 & E_3^{(0)} \end{pmatrix} + \begin{pmatrix} 0 & 0 & \epsilon_{13} \\ 0 & 0 & \epsilon_{23} \\ \epsilon_{13}^* & \epsilon_{23}^* & 0 \end{pmatrix}.$$

Assume that $E_1^{(0)} > E_3^{(0)}$ and the ϵ_i's are small perturbations. You can think of the above right-hand side as $\hat{H}^{(0)} + \delta H$. The degeneracy of the two (unperturbed) $E_1^{(0)}$ eigenvectors is not lifted to first order in the perturbation. [In this problem $\lambda = 1$, and the ϵ's in δH are perturbation parameters.]

The two eigenstates of $\hat{H}^{(0)}$ with energy $E_1^{(0)}$ will be denoted by

$$|1^{(0)};1\rangle = |1\rangle = \begin{pmatrix} 1 \\ 0 \\ 0 \end{pmatrix}, \quad \text{and} \quad |1^{(0)};2\rangle = |2\rangle = \begin{pmatrix} 0 \\ 1 \\ 0 \end{pmatrix}.$$

The eigenstate of $\hat{H}^{(0)}$ with energy $E_3^{(0)}$ will be denoted by

$$|3^{(0)}\rangle = |3\rangle = \begin{pmatrix} 0 \\ 0 \\ 1 \end{pmatrix}.$$

For convenience, define quantities $\Delta E > 0$ and $\epsilon > 0$ as follows:

$$\Delta E \equiv E_1^{(0)} - E_3^{(0)} > 0, \quad \epsilon^2 \equiv \epsilon_{13}^*\epsilon_{13} + \epsilon_{23}^*\epsilon_{23}.$$

1. Find the energy $E_3(\epsilon)$ of the nondegenerate state to second order and the state $|3\rangle_\epsilon$ itself to first order.
2. Use degenerate perturbation theory to find the good basis eigenvectors to zeroth order and the energy corrections to second order. Call the eigenvectors v_1 and v_2 and their energies E_{v_1} and E_{v_2}, with $E_{v_1} > E_{v_2}$. The eigenvectors to zeroth order are denoted by $v_1^{(0)}$ and $v_2^{(0)}$, and the order-k energy corrections are denoted by $E_{v_1}^{(k)}$ and $E_{v_2}^{(k)}$.
3. Find the first-order correction to the good basis (v_1, v_2) along the $|3^{(0)}\rangle$ state.
4. Calculate the exact eigenvalues, written as (E_+, E_-, E'), with $E_+ > E_-$, and find their expansion to second order in ϵ. Verify that $E' = E_1^{(0)}$ exactly by constructing the eigenvector v'.

   Who do E_+, E_-, and E' correspond to in the list (E_{v_1}, E_{v_2}, E_3)?
5. Expand the exact eigenvectors v_+ and v_-, corresponding to the energies E_+ and E_-, to first order in ϵ. Compare with the perturbative results.

Problem 25.10. *Matrix elements of a scalar operator.*

Recall the angular momentum identities

$$\hat{J}_\pm|j,m\rangle = \hbar\sqrt{j(j+1) - m(m\pm 1)}\,|j,m\pm 1\rangle, \quad \hat{J}_+\hat{J}_- = \hat{\mathbf{J}}^2 - \hat{J}_z^2 + \hbar\hat{J}_z.$$

Consider a scalar operator $\hat{O}$, meaning $[\hat{J}_i, \hat{O}] = 0$, and define $f(j,m) \equiv \langle j,m|\hat{O}|j,m\rangle$.

- Evaluate $\langle j,m|\hat{O}J_+J_-|j,m\rangle$ directly. Assume $m \neq -j$.
- Evaluate $\langle j,m|\hat{O}J_+J_-|j,m\rangle$ by noticing it is equal to $\langle j,m|J_+\hat{O}J_-|j,m\rangle$.

Use your result to show that $f(j,m) = f(j,m-1)$, for $-j < m \leq j$. In words, the expectation value of a scalar operator is independent of the $\hat{J}_z$ eigenvalue of the state.

Problem 25.11. *Stark effect: ground state energy shift.*

When an atom is placed in a uniform external electric field $\mathbf{E}_{\text{ext}}$, the energy levels are shifted, a phenomenon known as the **Stark effect**. Here we analyze the Stark effect for the ground states of hydrogen. Let the electric field point in the *z*-direction so the electrostatic potential of the electron is

$$\delta H_{\text{Stark}} = eE_{\text{ext}}z,$$

where E_{ext} is the magnitude of the electric field. Treat this as a perturbation of the Bohr Hamiltonian, and ignore spin and fine-structure effects.

1. Is the ground state energy $E_{1,0,0}$ affected by δH_{Stark} to first order in perturbation theory?
2. The second-order shift to $E_{1,0,0}$ is challenging to calculate. We aim here to find a bound on this shift. For this, we use two preparatory results. First, calculate γ defined by

$$\gamma \equiv \sum_{\alpha} |\langle\alpha|z|1,0,0\rangle|^2,$$

where α runs over all states of the hydrogen atom, bound or unbound. Write your answer in terms of a_0 and numerical constants. Second, find the largest value of δ for which the following inequality holds for all $\alpha \neq (1,0,0)$:

$$\frac{1}{E^{(0)}_{1,0,0} - E^{(0)}_{\alpha}} \geq \delta .$$

3. Use the previous results to argue that the second-order shift $E^{(2)}_{1,0,0}$ of the ground state energy satisfies

$$E^{(2)}_{1,0,0} \geq -Ca_0^3(E_{\text{ext}})^2,$$

for some unit-free numerical constant C. What is the value of C that you get?

Discussion: The exact calculation of $E^{(2)}_{1,0,0}$ requires summing over bound states and integrating over unbound states. This gives $C = \frac{9}{4}$. In fact, Ca_0^3 is the polarizability of the ground state.

Problem 25.12. *Stark effect: energy shifts for first excited states.*

Once more, for an electric field of magnitude E_{ext} pointing in the *z*-direction the perturbation of the Bohr Hamiltonian $\hat{H}^{(0)}$ is $\delta H_{\text{Stark}} = eE_{\text{ext}}z$. The first excited state of $\hat{H}^{(0)}$ is fourfold degenerate, with $|n,\ell,m\rangle$ states $|2,0,0\rangle$, $|2,1,1\rangle$, $|2,1,0\rangle$ and $|2,1,-1\rangle$. The aim of this problem is to compute the Stark effect shifts due to δH_{Stark}.

1. We first develop a useful lemma. Consider nonnegative integers n_x, n_y, n_z, and a function $f(r)$ of the radial variable $r = \sqrt{x^2 + y^2 + z^2}$. State a simple condition on the integers n_x, n_y, n_z that guarantees the following integral vanishes:

$$I_{n_x,n_y,n_z} \equiv \int dx\,dy\,dz\; x^{n_x} y^{n_y} z^{n_z} f(r).$$

Assume that $f(r)$ is such that the integral is always defined.

2. Using degenerate perturbation theory, determine the first-order shifts of the energy of the four $n=2$ states. Are there degeneracies left? (Ignore spin.)
3. What are the "good" wave functions for the $n=2$ states? Find the expectation value of the electric dipole moment $\hat{\mathbf{p}}_e \equiv -e\hat{\mathbf{r}}$ in each of these "good" states. Are these dipole moments permanent (i.e., do they exist for zero electric field)?
4. Calculate the electric field strength E_0 (in V/cm) at which the magnitude of the Stark energy shift calculated above becomes equal to the fine-structure splitting between the $2S_{1/2}$ and the $2P_{3/2}$ levels.

   Only for $E_{\text{ext}} \gg E_0$ can one ignore fine structure when analyzing the Stark perturbation. A common household electric field strength is about 100 V/cm. Can we ignore fine-structure shifts in the Stark effect analysis at this field strength?

For reference, here you have some possibly relevant, normalized hydrogen wave functions:

$$\langle \mathbf{r}|2,0,0\rangle = \frac{1}{4\sqrt{2\pi}\,a_0^{3/2}}\left(2 - \frac{r}{a_0}\right) e^{-r/2a_0},$$

$$\langle \mathbf{r}|2,1,0\rangle = \frac{1}{4\sqrt{2\pi}\,a_0^{3/2}}\,\frac{z}{a_0}\,e^{-r/2a_0},$$

$$\langle \mathbf{r}|2,1,\pm 1\rangle = \frac{1}{8\sqrt{\pi}\,a_0^{3/2}}\,\frac{x \pm iy}{a_0}\,e^{-r/2a_0}.$$

Problem 25.13. *Wave function at the origin for spherically symmetric eigenstates (J. J. Sakurai).*

Consider a particle in an $\ell = 0$ bound state of a central potential $V(r)$. The wave function $\psi(\mathbf{x})$ can be written as

$$\psi(\mathbf{x}) = \frac{1}{\sqrt{4\pi}}\,\frac{u(r)}{r}.$$

A surprising result relates the value $\psi(\mathbf{0})$ of the normalized wave function at the origin to the expectation value of a derivative of the potential $V(r)$:

$$|\psi(\mathbf{0})|^2 = N(\hbar, m)\left\langle \frac{dV}{dr} \right\rangle.$$

1. Derive such a relation and fix the value of the constant N, which depends on m, $\hbar$, and numerical constants. [Hint: begin with the radial equation for $u(r)$, multiply the equation by $u'(r)$, and integrate the equation from $r=0$ to $r=\infty$.]

2. Use the result to calculate $|\psi_{n00}(0)|^2$ for the $\ell = 0$ states of the hydrogen atom. Verify that you got the right answer for $n = 1$.

Problem 25.14. *Numerical estimate of magnetic fields.*

1. Find the magnitude of an external magnetic field that acting on a free electron produces energy levels that have a separation equal to the splitting between the $3P_{1/2}$ and $3P_{3/2}$ states. Express your answer in tesla (T).
2. Estimate the magnitude of the internal magnetic field *at* the electron in the 3*P* states using the heuristic formula $\mathbf{B} = \frac{e}{mcr^3}\mathbf{L}$ by setting $|\mathbf{L}| = \hbar$ and using the expectation value $\langle 1/r^3 \rangle$ in (11.4.20), valid for hydrogen eigenstates. Express your answer in tesla (T).

Problem 25.15. *Complete sets of commuting observables for the hydrogen atom.*

Consider the Bohr hydrogen Hamiltonian: $\hat{H}^{(0)} = \frac{\hat{\mathbf{p}}^2}{2m} - \frac{e^2}{r}$. A complete set of commuting observables (CSCO) is a set of commuting Hermitian operators whose simultaneous eigenspaces are each one-dimensional and together span the space. Equivalently, the eigenvalues of all the operators in a CSCO uniquely specify the basis states of the theory.

Consider the familiar CSCOs for the bound state spectrum of the hydrogen atom:

- $\{\hat{H}^{(0)}, \hat{\mathbf{L}}^2, \hat{L}_z, \hat{S}_z\}$ form CSCO_1 with eigenbasis $\{|n\ell m_\ell m_s\rangle\}$, and
- $\{\hat{H}^{(0)}, \hat{\mathbf{L}}^2, \hat{\mathbf{J}}^2, \hat{J}_z\}$ form CSCO_2 with eigenbasis $\{|n\ell j m_j\rangle\}$.

For each of the following sets of operators, either explain why they are a CSCO for the bound state spectrum, or explain why they are *not* a CSCO, indicating if they fail to commute or fail to be complete.

1. $\{\hat{H}^{(0)}, \hat{\mathbf{L}}^2, \hat{\mathbf{L}} \cdot \hat{\mathbf{S}}, \hat{J}_z\}$,
2. $\{\hat{H}^{(0)}, \hat{\mathbf{L}}^2, \hat{L}_z, \hat{S}_x\}$,
3. $\{\hat{H}^{(0)}, \hat{\mathbf{L}}^2, \hat{J}_z, \hat{S}_z\}$,
4. $\{\hat{H}^{(0)}, \hat{\mathbf{J}}^2, \hat{J}_z, \hat{S}_z\}$,
5. $\{\hat{H}^{(0)}, \hat{\mathbf{J}}^2, \hat{\mathbf{L}} \cdot \hat{\mathbf{S}}, \hat{J}_z\}$.

Problem 25.16. *Dirac equation and angular momentum.*

The Dirac Hamiltonian for a free particle is given by $\hat{H}_{\text{Dirac}} = c\boldsymbol{\alpha} \cdot \hat{\mathbf{p}} + \beta mc^2$ with the $\boldsymbol{\alpha}$ and β matrices defined in (24.7.9).

1. Calculate the commutator $[\hat{H}_{\text{Dirac}}, \hat{L}_i]$, where $\hat{L}_i$ is the ith component of the (standard) orbital angular momentum operator $\hat{\mathbf{L}}$.
2. The spin operator $\hat{\mathcal{S}}$ in Dirac's theory is given by $\hat{\mathcal{S}} = \frac{\hbar}{2}\boldsymbol{\Sigma} = \frac{\hbar}{2}\begin{pmatrix} \boldsymbol{\sigma} & 0 \\ 0 & \boldsymbol{\sigma} \end{pmatrix}$. Acting on either of the Pauli spinors in the Dirac spinor, $\hat{\mathcal{S}}$ reduces to the familiar spin operator $\frac{\hbar}{2}\boldsymbol{\sigma}$. Confirm that $\hat{\mathcal{S}}$ satisfies the commutation properties of an angular momentum operator. Calculate $[\hat{H}_{\text{Dirac}}, \hat{\mathcal{S}}_i]$, where $\hat{\mathcal{S}}_i$ is the ith component of $\hat{\mathcal{S}}$.

3. Confirm that $\hat{\mathbf{L}}+\hat{\mathcal{S}}$ is an angular momentum operator that is conserved in the Dirac theory.

Problem 25.17. *Strong-field Zeeman effect in hydrogen.*

In the strong-field Zeeman effect, the uncoupled states $|n\ell m_\ell m_s\rangle$ are eigenstates of $\hat{H}^{(0)}+\delta H_{\text{Zeeman}}$, with δH_{Zeeman} the Zeeman Hamiltonian. We take these states to be unperturbed eigenstates, with energies shifted by an amount proportional to $m_\ell+2m_s$ due to the Zeeman Hamiltonian. The perturbation is due to the fine-structure terms. We explore here a shortcut that helps compute these corrections.

We claim that the fine-structure corrections can be viewed as contributing to the Hamiltonian the following term:

$$\delta H_{\text{f.s.}}=-\frac{mc^2\alpha^4}{2n^3}\Big(\frac{1}{\hat{j}+\frac{1}{2}}-\frac{3}{4n}\Big). \tag{1}$$

Here $\hat{j}$ is an operator satisfying $\hat{\mathbf{J}}^2=\hbar^2\hat{j}(\hat{j}+1)$. Indeed, in the coupled basis the operator $\delta\hat{H}_{\text{f.s.}}$ above reproduces the correct fine-structure shifts in (25.7.55).

1. First we note that the shift proportional to m_l+2m_s does not remove all the degeneracies of $\hat{H}^{(0)}$. Display in a diagram those remaining degeneracies for the $n=3$ level. How many degenerate spaces are there, and what are their dimensions?
2. To compute the first-order energy shifts, we assume we can use nondegenerate perturbation theory (as we will justify below) and aim to evaluate

$$I=\Big\langle n\ell m_\ell m_s\Big|\frac{1}{\hat{j}+\frac{1}{2}}\Big|n\ell m_\ell m_s\Big\rangle. \tag{2}$$

If I is known, the expectation value of $\delta H_{\text{f.s.}}$ in the uncoupled basis immediately follows. Use the following strategy to evaluate I:

(a) First compute the expectation value of $\hat{\mathbf{J}}^2$ on the $|n\ell m_\ell m_s\rangle$ state.

(b) Now imagine that we measure $\hat{j}$ in the state $|n\ell m_\ell m_s\rangle$. Use your calculation to find the probabilities P_+ and P_- of the two outcomes $j=\ell+\frac{1}{2}$ and $j=\ell-\frac{1}{2}$, respectively.

(c) Since the two outcomes for $\hat{j}$ define orthogonal states, I is given in terms of these probabilities by

$$I=\frac{P_+}{\ell+1}+\frac{P_-}{\ell}. \tag{3}$$

Use the result for I to go back to equation (1), and including the Zeeman shifts, finally write the full expression for the energies including fine-structure corrections to the strong-field Zeeman effect:

$$E_{n\ell m_\ell m_s}=-\tfrac{1}{2}\,\alpha^2 mc^2\frac{1}{n^2}+\mu_B B(m_\ell+2m_s)+\frac{\alpha^4 mc^2}{2n^3}\big(\cdots\big),$$

where the dots represent a term that you should calculate and that depends on all the quantum numbers of the state.

3. Our computation of the shifts in equation (2) used nondegenerate perturbation theory. Explain carefully and in detail why, despite the remaining degeneracies, the above argument is correct. This can be done by showing that the matrix element of $\delta H_{\text{f.s.}}$

$$\langle n\ell' m'_\ell m'_s | \delta H_{\text{f.s.}} | n\ell m_\ell m_s \rangle$$

between two *degenerate* states of $\hat{H}^{(0)} + \delta H_{\text{Zeeman}}$ vanishes unless $\ell' = \ell$, $m'_\ell = m_\ell$, and $m'_s = m_s$. [Hint: ask yourself which operators in the list $\{\hat{\mathbf{L}}, \hat{\mathbf{L}}^2, \hat{\mathbf{S}}, \hat{\mathbf{S}}^2, \hat{\mathbf{J}}, \hat{\mathbf{J}}^2\}$ commute with $\delta H_{\text{f.s.}}$.]

Problem 25.18. *Identities with vector operators.*

Consider a set of angular momentum operators $\hat{J}_i$, $i = 1, 2, 3$, that define an angular momentum $\hat{\mathbf{J}}$. A set of operators $\hat{W}_i$, with $i = 1, 2, 3$, form a vector operator $\hat{\mathbf{W}}$ under $\hat{\mathbf{J}}$ if $[\hat{J}_i, \hat{W}_j] = i\hbar\, \epsilon_{ijk}\, \hat{W}_k$.

Note that $\hat{\mathbf{J}}$ itself is a vector operator under $\hat{\mathbf{J}}$. An operator is said to be a *scalar* operator under $\hat{\mathbf{J}}$ if it commutes with all $\hat{J}_i$. Recall that if $\hat{\mathbf{U}}$ and $\hat{\mathbf{V}}$ are vector operators under $\hat{\mathbf{J}}$, $\hat{\mathbf{U}} \cdot \hat{\mathbf{V}}$ and $\hat{\mathbf{U}} \times \hat{\mathbf{V}}$ are, respectively, scalar and vector operators under $\hat{\mathbf{J}}$.

1. Show that if $\hat{\mathbf{V}}$ is a vector operator, then

$$[\hat{\mathbf{J}}^2, \hat{\mathbf{V}}] = 2i\hbar\,(\hat{\mathbf{V}} \times \hat{\mathbf{J}} - i\hbar \hat{\mathbf{V}})\,.$$

Check that this formula holds when we choose $\hat{\mathbf{V}} = \hat{\mathbf{J}}$.

2. Show that for a vector operator $\hat{\mathbf{V}}$, the following identity holds:

$$\frac{1}{(2i\hbar)^2}\left[\hat{\mathbf{J}}^2, [\hat{\mathbf{J}}^2, \hat{\mathbf{V}}]\right] = (\hat{\mathbf{V}} \cdot \hat{\mathbf{J}})\hat{\mathbf{J}} - \tfrac{1}{2}(\hat{\mathbf{J}}^2 \hat{\mathbf{V}} + \hat{\mathbf{V}}\hat{\mathbf{J}}^2).$$

26 WKB and Semiclassical Approximation

For slowly varying potentials, the time-independent Schrödinger equation can be solved in the WKB approximation. This approximation, which can be treated formally as an $\hbar \to 0$ limit, is accurate when the local version of the de Broglie wavelength of the particle is small compared to relevant length scales in the problem. The power of the WKB method is enhanced by connection formulae, that allow us to patch solutions across classical turning points where the WKB approximation is not valid. The connection formulae are obtained via Airy functions relevant to the physics of linear potentials. The WKB approximation is used to calculate bound state energies and to estimate tunneling probabilities across potential barriers. For double-well potentials, it allows the calculation of the exponentially suppressed level splitting.

26.1 The Classical Limit

The WKB approximation provides approximate solutions for linear differential equations with coefficients that have slow spatial variation. The acronym WKB stands for Wentzel, Kramers, and Brillouin, who independently discovered the approximation scheme in 1926. In fact, the approximation was discovered earlier, in 1923, by the mathematician Harold Jeffreys. When applied to quantum mechanics, it is called the *semiclassical* approximation, since classical physics then illuminates the main features of the *quantum* wave function.

The de Broglie wavelength λ of a particle can help us assess whether classical physics is relevant to the physical situation. For a particle with momentum p, we have

$$\lambda = \frac{h}{p}. \tag{26.1.1}$$

Classical physics provides useful physical insight when λ is much smaller than the relevant length scale in the system we are investigating. Alternatively, if we take the formal limit $h \to 0$, this will make $\lambda \to 0$, and λ will be smaller than the length scale of the system. Being a constant of nature, we cannot really make $h \to 0$, so taking this limit is a thought experiment in which we imagine worlds where h takes smaller and smaller values, making classical physics more and more applicable. The semiclassical approximation studied here will be applicable if a suitable generalization of the de Broglie wavelength, discussed below, is small and slowly varying. The semiclassical approximation will be set up mathematically by thinking of $\hbar$ as a formally small expansion parameter.

Our discussion in this chapter will focus on one-dimensional problems. Consider, therefore, a particle of mass m and total energy E moving in a potential $V(x)$. In classical physics, $E - V(x)$ is the kinetic energy of the particle at x. This kinetic energy depends on position through $V(x)$. Since kinetic energy is $\frac{p^2}{2m}$, this suggests the definition of the **local momentum** $p(x)$:

$$p^2(x) \equiv 2m(E - V(x)). \tag{26.1.2}$$

The local momentum $p(x)$ is the momentum of the classical particle when it is located at x. With a notion of local momentum, we can define a **local de Broglie wavelength** $\lambda(x)$ by the familiar relation:

$$\boxed{\lambda(x) \equiv \frac{h}{p(x)} = \frac{2\pi\hbar}{p(x)}.} \tag{26.1.3}$$

The time-independent Schrödinger equation

$$-\frac{\hbar^2}{2m}\frac{d^2}{dx^2}\psi(x) = (E - V(x))\psi(x) \tag{26.1.4}$$

can be written nicely in terms of the local momentum squared:

$$-\hbar^2\frac{d^2}{dx^2}\psi = p^2(x)\,\psi. \tag{26.1.5}$$

Using the momentum operator, this equation takes the suggestive form

$$\boxed{\hat{p}^2\,\psi(x) = p^2(x)\,\psi(x).} \tag{26.1.6}$$

This has the flavor of an eigenvalue equation, but it is not one: the momentum operator squared acting on the wave function is not really proportional to the wave function. It equals the wave function multiplied by the position-dependent function $p^2(x)$.

A bit of extra notation is useful. If we are in the classically allowed region $E > V(x)$, then $p^2(x)$ is positive, and we write

$$p^2(x) = 2m(E - V(x)) = \hbar^2 k^2(x), \tag{26.1.7}$$

introducing the local, real wave number $k(x) > 0$. If we are in the classically forbidden region $V(x) > E$, then $p^2(x)$ is negative, and we write

$$-p^2(x) = 2m(V(x) - E) = \hbar^2\kappa^2(x), \tag{26.1.8}$$

introducing the local, real $\kappa(x) > 0$.

The wave functions we use in the WKB approximation are often expressed in polar form. Just like any complex number, z can be written as $re^{i\theta}$, with r and θ the magnitude and phase of z, respectively. We can write the wave function in a similar way:

$$\Psi(\mathbf{x}, t) = \sqrt{\rho(\mathbf{x}, t)}\, \exp\Big(\frac{i}{\hbar}\mathcal{S}(\mathbf{x}, t)\Big). \tag{26.1.9}$$

We are using three-dimensional notation here for generality. By definition, the functions $\rho(\mathbf{x}, t)$ and $\mathcal{S}(\mathbf{x}, t)$ are real. The function ρ is nonnegative, and the function $\mathcal{S}(\mathbf{x}, t)$ as written has units of $\hbar$. The name $\rho(\mathbf{x}, t)$ is well motivated, for it is in fact the probability density:

$$|\Psi(\mathbf{x}, t)|^2 = \rho(\mathbf{x}, t). \tag{26.1.10}$$

Let's compute the probability current $\mathbf{J}$ associated with $\Psi(\mathbf{x}, t)$. For this we begin by taking the gradient of the wave function:

$$\nabla\Psi = \frac{1}{2}\frac{\nabla\rho}{\sqrt{\rho}}\exp\Big(\frac{i\mathcal{S}}{\hbar}\Big) + \frac{i}{\hbar}\nabla\mathcal{S}\,\Psi. \tag{26.1.11}$$

We then form the product

$$\Psi^*\nabla\Psi = \frac{1}{2}\nabla\rho + \frac{i}{\hbar}\rho\,\nabla\mathcal{S}. \tag{26.1.12}$$

The probability current is given by $\mathbf{J} = \frac{\hbar}{m}\,\mathrm{Im}\,(\Psi^*\nabla\Psi)$, and therefore,

$$\boxed{\mathbf{J} = \rho\frac{\nabla\mathcal{S}}{m}.} \tag{26.1.13}$$

This result implies that the probability current $\mathbf{J}$ is perpendicular to the surfaces of constant $\mathcal{S}$, the surfaces of constant phase in the wave function.

In classical physics a fluid with density $\rho(\mathbf{x})$ moving with velocity $\mathbf{v}(\mathbf{x})$ has a current density $\rho\mathbf{v} = \rho\frac{\mathbf{p}}{m}$. Comparing with the above expression for the quantum probability current, we deduce that

$$\mathbf{p}(\mathbf{x}) \simeq \nabla\mathcal{S}. \tag{26.1.14}$$

We use $\simeq$ instead of an equality because this is an association of limited validity. Nevertheless, it holds for the "basic" WKB solutions to be discussed later. This association also holds for a free particle. Consider a free particle with momentum $\mathbf{p}$ and energy E. Its wave function $\Psi(\mathbf{x}, t)$ is

$$\Psi(\mathbf{x}, t) = \exp\Big[\frac{i\mathbf{p}\cdot\mathbf{x}}{\hbar} - \frac{iEt}{\hbar}\Big]. \tag{26.1.15}$$

Here we identify $\mathcal{S} = \mathbf{p}\cdot\mathbf{x} - Et$, and therefore $\nabla\mathcal{S} = \mathbf{p}$. In this case $\nabla\mathcal{S}$ is equal to the momentum eigenvalue, a constant.

26.2 WKB Approximation Scheme

Our aim here is to find approximate solutions for the wave function $\psi(x)$ that solve the time-independent Schrödinger equation in one dimension:

$$-\frac{\hbar^2}{2m}\frac{d^2\psi}{dx^2} + V(x)\psi = E\psi. \tag{26.2.1}$$

Taking the limit $\hbar \to 0$ on this form of the Schrödinger equation is not useful. If we set the first term of the equation to zero, the differential equation turns algebraic. Moreover, it becomes inconsistent as it equates a position-dependent potential $V(x)$ to a constant energy. To find a way to approximate, we must think more physically. In the limit as potentials become constant, the solutions become plane waves. A plane wave looks like

$$\psi \sim \exp\left(\frac{i}{\hbar}px\right), \tag{26.2.2}$$

and it is natural to suspect that the modifications required for slowly varying potentials would modify the *argument* of the exponent. This suggests that a good approximation scheme would parameterize the wave function as the exponential of another quantity.

Earlier we wrote the polar decomposition (26.1.9) of the wave function. We will set the approximation scheme by using a *single complex function* $S(x)$ to represent the time-independent wave function $\psi(x)$. For this, we use a *pure* exponential without any prefactor:

$$\psi(x) = \exp\left(\frac{i}{\hbar}S(x)\right), \quad S(x) \in \mathbb{C}. \tag{26.2.3}$$

As before, S must have units of $\hbar$. This $S(x)$ here is a *complex* number because wave functions are not in general pure phases. The real part of S, divided by $\hbar$, is the phase of the wave function. The imaginary part of $S(x)$ determines the magnitude of the wave function.

Let us plug this into the Schrödinger equation (26.1.5):

$$-\hbar^2\frac{d^2}{dx^2}\left(e^{\frac{i}{\hbar}S(x)}\right) = p^2(x)e^{\frac{i}{\hbar}S(x)}. \tag{26.2.4}$$

Calculating the two derivatives on the left-hand side, we get

$$-\hbar^2\frac{d^2}{dx^2}\left(e^{\frac{i}{\hbar}S(x)}\right) = -\hbar^2\frac{d}{dx}\left(\frac{i}{\hbar}S'(x)e^{\frac{i}{\hbar}S(x)}\right) = -\hbar^2\left(\frac{iS''}{\hbar} - \frac{(S')^2}{\hbar^2}\right)e^{\frac{i}{\hbar}S(x)}. \tag{26.2.5}$$

Going back to the differential equation and canceling the common exponential,

$$-\hbar^2\left(\frac{iS''}{\hbar} - \frac{(S')^2}{\hbar^2}\right) = p^2(x). \tag{26.2.6}$$

With minor rearrangements, we get our final form of the equation for S:

$$\boxed{(S'(x))^2 - i\hbar S''(x) = p^2(x).} \tag{26.2.7}$$

The presence of an explicit i in the equation tells us that the solution for S, as expected, cannot be real. At first sight one may be baffled: we started with the linear Schrödinger equation for ψ and obtained a *nonlinear* equation for S. This was unavoidable because ψ is a nonlinear function of S; linear equations remain linear only under linear changes of variables. As it turns out, the exponential function relating S to ψ ended up giving us the rather tractable nonlinear equation (26.2.7). This equation allows us to set up an approximation scheme in which $\hbar$ is considered small, and thus the term $i\hbar S''(x)$ is small.

We now argue that $\hbar S''(x)$ is in fact small for slowly varying potentials. Indeed, if $V(x) = V_0$ is a constant, then the local momentum $p(x)$ is equal to a constant p_0. Equation (26.2.7) is then solved by taking $S' = p_0$. For this choice $S'' = 0$, and the term $i\hbar S''$ vanishes identically for constant V. It should therefore be small for slowly varying $V(x)$. Intuitively, the term $i\hbar S''$ is small as $\hbar \to 0$, which makes the local de Broglie wavelength go to zero. In that situation, the potential looks constant to the quantum particle.

We will thus take $\hbar$ to be the small parameter in a systematic expansion of $S(x)$:

$$S(x) = S_0(x) + \hbar S_1(x) + \hbar^2 S_2(x) + \mathcal{O}(\hbar^3). \tag{26.2.8}$$

Here S_0, just like S, has units of $\hbar$. The next correction S_1 has no units, and the following S_2 has units of one over $\hbar$. Now plug this expansion into our nonlinear equation (26.2.7):

$$\left(S_0' + \hbar S_1' + \hbar^2 S_2' + \cdots\right)^2 - i\hbar\left(S_0'' + \hbar S_1'' + \hbar^2 S_2'' + \cdots\right) - p^2(x) = 0. \tag{26.2.9}$$

The left-hand side is a power series expansion in $\hbar$. Just as we argued for the parameter λ in perturbation theory, here we want the left-hand side to vanish for all values of $\hbar$. This requires the coefficient of each power of $\hbar$ to vanish. We will work to leading and subleading order only. Collecting the terms independent of $\hbar$ and linear in $\hbar$, we find that

$$(S_0')^2 - p^2(x) + \hbar\left(2S_0'S_1' - iS_0''\right) + \mathcal{O}\left(\hbar^2\right) = 0, \tag{26.2.10}$$

and as a result, we get the following equations:

$$\begin{aligned} (S_0')^2 - p^2(x) &= 0, \\ 2S_0'S_1' - iS_0'' &= 0. \end{aligned} \tag{26.2.11}$$

The first equation is easily solved for S_0:

$$S_0' = \pm p(x) \quad \Rightarrow \quad S_0(x) = \pm \int_{x_0}^{x} p(x')dx', \tag{26.2.12}$$

where x_0 is a constant of integration to be adjusted. The next equation allows us to find S_1, which is in fact imaginary:

$$S_1' = \frac{i}{2}\frac{S_0''}{S_0'} = \frac{i}{2}\frac{(\pm p'(x))}{(\pm p(x))} = \frac{i}{2}\frac{p'}{p}. \tag{26.2.13}$$

This is readily solved to give

$$iS_1(x) = -\tfrac{1}{2}\ln p(x) + C'. \tag{26.2.14}$$

The function $S_0(x)$, which contributes to the phase of the wave function, is given by an integral. The function S_1 is in fact written directly in terms of $p(x)$. Being purely imaginary, S_1 contributes to the magnitude of the wave function. Let us now reconstruct the wave function to this order of approximation:

$$\psi(x) = \exp\left[\frac{i}{\hbar}(S_0 + \hbar S_1 + \mathcal{O}(\hbar^2))\right] \simeq \exp\left[\frac{i}{\hbar}S_0\right]\exp\left[iS_1\right]. \tag{26.2.15}$$

Using our results for S_0 and S_1, we have the approximate solution:

$$\psi(x) = \exp\left[\pm\frac{i}{\hbar}\int_{x_0}^{x} p(x')dx'\right]\exp\left[-\tfrac{1}{2}\ln p(x) + C'\right]. \tag{26.2.16}$$

We used the equal sign in the sense that this *is* the approximate solution. Redefining the multiplicative constant and rearranging, we find that

$$\boxed{\psi(x) = \frac{A}{\sqrt{p(x)}}\exp\left[\pm\frac{i}{\hbar}\int_{x_0}^{x} p(x')dx'\right].} \tag{26.2.17}$$

This is the **basic solution** in the WKB approximation. We do not attempt to normalize this wave function because, in fact, the region of validity of this result is still to be determined.

Remarks:

1. The probability density ρ for a basic solution is given by

$$\rho = \psi^*\psi = \frac{|A|^2}{p(x)} = \frac{|A|^2}{mv(x)}, \tag{26.2.18}$$

where $v(x)$ is the local classical velocity. Note that ρ is large where v is small, as the particle lingers in such regions and is more likely to be found there. This is an intuition we developed long ago (section 7.2) that is justified by this result.

2. The probability current of the basic solution can be determined using the polar decomposition (26.1.9) of the wave function. In this notation, the basic solution gives

$$\mathcal{S}(x) = \int_{x_0}^{x} p(x')dx'. \tag{26.2.19}$$

As anticipated, the gradient of $\mathcal{S}$ in the basic solution is the local momentum. Recalling the result (26.1.13) for the current, we have

$$J = \rho\,\frac{1}{m}\frac{\partial\mathcal{S}}{\partial x} = \frac{|A|^2}{p(x)}\frac{1}{m}p(x) = \frac{|A|^2}{m}. \tag{26.2.20}$$

The fact that the current is a constant should not take us by surprise. An energy eigenstate cannot have a position-dependent current because it would conflict with the current conservation equation $\partial_t \rho + \partial_x J = 0$. Since ρ is time independent, this equation requires $\partial_x J = 0$.

3. We cannot expect the WKB basic solutions to hold near the turning points of a potential. Indeed, near such points the momentum $p(x)$ goes to zero, implying that the de Broglie wavelength goes to infinity. The semiclassical approximation hinges on a small de Broglie wavelength, which makes the potential look slowly varying to the quantum particle. We will therefore have to determine how to relate or connect WKB solutions across turning points.

We can use the basic solution to write general solutions that apply to classically allowed and to classically forbidden regions. On the classically allowed region, where $E - V(x) > 0$, we wrote $p^2(x) = \hbar^2 k^2(x)$, with $k(x) > 0$. The differential equation (26.1.5) becomes the following:

$$\text{Classically allowed region:} \quad \psi''(x) = -k^2(x)\psi(x). \tag{26.2.21}$$

The general solution is a superposition of the two basic solutions (26.2.17), representing waves that propagate in opposite directions. Using $p(x) = \hbar k(x)$, we find that

$$\psi(x) = \frac{A}{\sqrt{k(x)}} \exp\left[-i \int_{x_0}^{x} k(x')dx'\right] + \frac{B}{\sqrt{k(x)}} \exp\left[i \int_{x_0}^{x} k(x')dx'\right]. \tag{26.2.22}$$

The wave with coefficient A moves to the left, while the wave with coefficient B moves to the right. This can be seen by recalling that in a full solution the above energy eigenstate $\psi(x)$ is accompanied by the time-dependent factor $e^{-iEt/\hbar}$. Moreover, the phase associated with the A term becomes more negative as x grows, while the phase associated with the B term grows as x grows. This solution is valid deep in the classically allowed region—that is, far away from turning points where $k(x)$ becomes zero.

On the classically forbidden region, where $E - V(x) < 0$, we wrote $p^2(x) = -\hbar^2\kappa^2(x)$, with $\kappa(x) > 0$. The differential equation (26.1.5) becomes the following:

$$\text{Classically forbidden region:} \quad \psi''(x) = \kappa^2(x)\psi(x). \tag{26.2.23}$$

We can take $p(x) = i\hbar\kappa(x)$ in the basic solutions to find that

$$\psi(x) = \frac{C}{\sqrt{\kappa(x)}} \exp\left[-\int_{x_0}^{x} \kappa(x')dx'\right] + \frac{D}{\sqrt{\kappa(x)}} \exp\left[\int_{x_0}^{x} \kappa(x')dx'\right]. \tag{26.2.24}$$

The argument of the first exponential becomes more negative as x grows. Thus, the first term, with coefficient C, is a decreasing function as x grows. The argument of the second exponential becomes more positive as x grows. Thus, the second term, with coefficient D, is an increasing function as x grows. This solution is valid deep in the classically forbidden region—that is, far away from turning points where $\kappa(x)$ becomes zero.

One can find approximate WKB solutions for general differential equations without considerations of $\hbar, m$, or other dimensionful quantities. A differential equation of the form

$$\frac{d^2y}{dx^2} = h(x)\,y \tag{26.2.25}$$

is tractable in the WKB approximation. Over a region where $h(x) \ll 0$, we compare with (26.2.21) and write solutions analogous to (26.2.22). Over the region where $h(x) \gg 0$, we compare with (26.2.23) and write solutions analogous to (26.2.24). In both cases we must stay away from points where $h(x) = 0$.

Validity of the approximation While the equations of the semiclassical approximation were derived thinking of $\hbar$ as a small expansion parameter, we must understand physically the nature of the approximation. To do so we reconsider the expansion (26.2.10) of the differential equation:

$$(S_0')^2 - p^2(x) + \hbar\left(2S_0'S_1' - iS_0''\right) + \mathcal{O}\left(\hbar^2\right) = 0. \tag{26.2.26}$$

The $\mathcal{O}(\hbar)$ terms in the differential equation must be much smaller in magnitude than the $\mathcal{O}(1)$ terms. At each of these orders, we have two terms that are set equal to each other by the differential equations. It therefore suffices to check that one of the $\mathcal{O}(\hbar)$ terms is much smaller than one of the $\mathcal{O}(1)$ terms. Thus, for example, we must have

$$|\hbar S_0' S_1'| \ll |S_0'|^2. \tag{26.2.27}$$

Canceling one factor of $|S_0'|$ and recalling that $|S_0'| = |p|$, we see that

$$|\hbar S_1'| \ll |p|. \tag{26.2.28}$$

From (26.2.13) we note that $|S_1'| \sim |p'/p|$, and therefore we get

$$\left|\hbar\frac{p'}{p}\right| \ll |p|. \tag{26.2.29}$$

There are two useful ways to think about this relation. First, we write it as follows:

$$\left|\frac{\hbar}{p}\right|\left|\frac{dp}{dx}\right| \ll |p| \quad \Rightarrow \quad \lambda\left|\frac{dp}{dx}\right| \ll |p|, \tag{26.2.30}$$

which tells us that the changes in the local momentum over a distance equal to the de Broglie wavelength are small compared to the momentum. Alternatively, (26.2.29) can also be written in the following form:

$$\left|\hbar\frac{p'}{p^2}\right| \ll 1 \quad \Rightarrow \quad \left|\hbar\frac{d}{dx}\frac{1}{p}\right| \ll 1. \tag{26.2.31}$$

This now means that

$$\boxed{\left|\frac{d\lambda}{dx}\right| \ll 1.} \tag{26.2.32}$$

The de Broglie wavelength must vary slowly. Note the consistency with units, the left-hand side of the inequality being unit-free. More intuitive, perhaps, is the version obtained by multiplying the above by λ:

$$\left|\lambda \frac{d\lambda}{dx}\right| \ll \lambda. \tag{26.2.33}$$

This tells us that the variation of the de Broglie wavelength λ over a distance λ must be much smaller than λ. It is not hard to figure out what the above constraints tell us about the rate of change of the potential. Taking one spatial derivative of the equation $p^2 = 2m(E - V(x))$, we get

$$|pp'| = m\left|\frac{dV}{dx}\right| \quad \Rightarrow \quad \left|\frac{dV}{dx}\right| = \frac{1}{m}|pp'|. \tag{26.2.34}$$

Multiplying by the absolute value of $\lambda = h/p$, we find that

$$\left|\lambda(x)\frac{dV}{dx}\right| = \frac{2\pi\hbar}{m}|p'| \ll \frac{p^2}{m}, \tag{26.2.35}$$

where the last inequality follows from (26.2.29). Hence,

$$\boxed{\left|\lambda(x)\frac{dV}{dx}\right| \ll \frac{p^2(x)}{2m}.} \tag{26.2.36}$$

The change in the potential over a distance equal to the de Broglie wavelength must be much smaller than the kinetic energy. This is the precise meaning of a slowly changing potential in the WKB approximation.

The slow variation conditions needed for the basic WKB solutions to be accurate fail near turning points. This was anticipated earlier, given that at turning points the local momentum becomes zero, and the de Broglie wavelength becomes infinite. Under general conditions and sufficiently near a turning point, the potential $V(x)$ is approximately linear, as shown in figure 26.1. The turning point $x = a$ occurs for energy E, and for x near a we can approximate the difference $V(x) - E$ by a linear function that vanishes for $x = a$:

$$V(x) - E \simeq g(x-a), \quad g > 0, \quad x \sim a. \tag{26.2.37}$$

In the allowed region $x < a$, the local momentum is

$$p^2(x) = 2m(E - V(x)) \simeq 2mg(a-x). \tag{26.2.38}$$

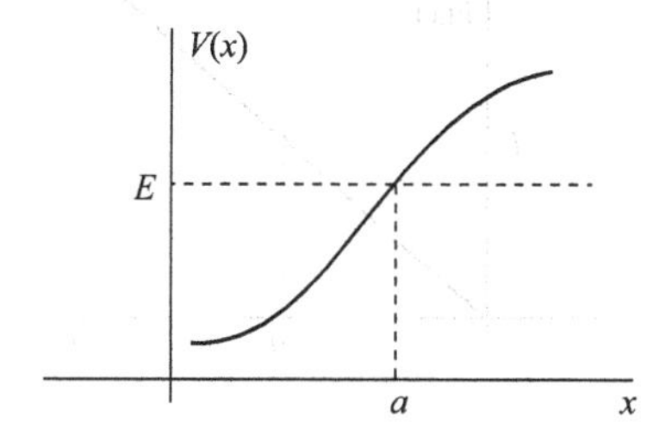

Figure 26.1
A potential $V(x)$ for which a state of energy E has a turning point $x = a$.

As a result, the de Broglie wavelength is given by

$$\lambda(x) = \frac{2\pi\hbar}{p} \simeq \frac{2\pi\hbar}{\sqrt{2mg}\sqrt{a-x}}. \tag{26.2.39}$$

Taking a derivative, we find

$$\left|\frac{d\lambda}{dx}\right| \simeq \frac{\pi\hbar}{\sqrt{2mg}}\frac{1}{(a-x)^{3/2}}. \tag{26.2.40}$$

The right-hand side goes to infinity as $x \to a$. Therefore, the key condition (26.2.32) is violated as we approach turning points. Our basic WKB solutions can be valid only as long as we remain away from turning points. If we have a turning point, such as $x=a$ in the figure, we need a "connection formula" that tells us how WKB solutions, far to the left and far to the right of the turning point, are related when they together form a single solution.

The following two examples consider a linear potential and the WKB solutions valid far away from the turning points. As we noted, potentials look linear near turning points. These examples will therefore be key to the subsequent derivation of the connection formulae.

Example 26.1. *Linear potential and the Airy equation.*

We studied the linear potential in section 6.7, using the Fourier transform to solve the time-independent Schrödinger equation in momentum space. A class of solutions was written in terms of an Airy function, defined by an integral representation. Here, we reconsider the same linear potential,

$$V(x) = gx, \qquad g > 0, \tag{26.2.41}$$

assumed to hold for all x. Our analysis will be in position space. Since the potential is unbounded below, energy eigenstates are not normalizable. The normalizable eigenstates of section 6.7 were those for which the potential included a hard wall at $x=0$. Assume the energy E has a value such that the classical turning point is at $x=a$ (figure 26.2):

$$V(x) - E = g(x-a), \quad a = E/g. \tag{26.2.42}$$

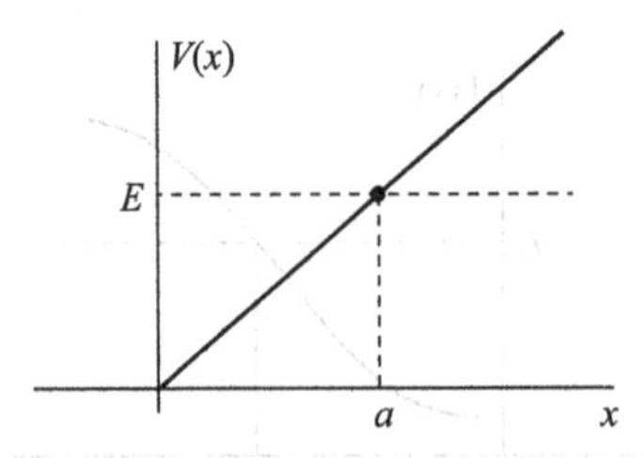

Figure 26.2
The linear potential $V(x)=gx$ and an energy eigenstate with energy E for which the turning point occurs at $x=a$.

The classically allowed region is $x < a$, while the classically forbidden region is $x > a$. The Schrödinger equation becomes

$$-\frac{\hbar^2}{2m}\psi'' + g(x-a)\psi = 0. \tag{26.2.43}$$

To remove the units from this equation, let $x = L\tilde{u}$ with $\tilde{u}$ unit-free and L a quantity with units of length:

$$-\frac{\hbar^2}{2m}\frac{1}{gL^3}\frac{d^2\psi}{d\tilde{u}^2} + \left(\tilde{u} - \frac{a}{L}\right)\psi = 0. \tag{26.2.44}$$

Setting

$$L^3 = \frac{\hbar^2}{2mg} \quad \text{and} \quad u = \tilde{u} - \frac{a}{L} = \frac{1}{L}(x-a), \tag{26.2.45}$$

the differential equation becomes the remarkably simple-looking Airy equation:

$$\frac{d^2\psi}{du^2} = u\psi. \tag{26.2.46}$$

The relevant solution $\psi(u)$ of this differential equation is the Airy function $\mathrm{Ai}(u)$, which represents the energy eigenstates of the linear potential. In fact, using the above relation between u and x, and a and E, we find that

$$\psi(u) = \mathrm{Ai}(u) = \mathrm{Ai}\left(\tilde{u} - \tfrac{a}{L}\right) = \mathrm{Ai}\left(\tfrac{1}{L}\left(x - \tfrac{E}{g}\right)\right). \tag{26.2.47}$$

Given a solution $\mathrm{Ai}(u)$, these relations express it in terms of the physical variables. In the absence of a hard wall, all energies are allowed. □

Example 26.2. *WKB solutions of the Airy equation* $\psi'' = u\psi$.

We reduced the calculation of the energy eigenstates for the linear potential to the problem of solving the Airy differential equation $\psi''(u) = u\psi$. Recalling the discussion of equation (26.2.25), we can write the WKB solutions that hold for $u \gg 1$ and for $u \ll -1$—that is, deep into the forbidden and allowed regions, respectively. The WKB solutions are not expected to hold near the turning point $u = 0$.

For $u \gg 1$, we set $\kappa = u^{1/2}$, and following (26.2.24), we write

$$\psi(u) = \frac{C}{u^{1/4}}\exp\left[-\int_{u_0}^{u}\sqrt{u'}du'\right] + \frac{D}{u^{1/4}}\exp\left[\int_{u_0}^{u}\sqrt{u'}du'\right]. \tag{26.2.48}$$

The lower limit of integration u_0 can be chosen arbitrarily, as long as it is lower than u and, in this case, nonnegative. It can even be chosen differently for each integral. It is convenient to choose $u_0 = 0$, however, since zero is the turning point for the differential equation. Had we chosen $u_0 = 1$, or any other positive constant, the integrals would just differ from their value for $u_0 = 0$ by numerical constants that can be absorbed into the definition of C and D. So setting $u_0 = 0$, we have

$$\psi(u) = \frac{C}{u^{1/4}} \exp\left[-\int_0^u \sqrt{u'}du'\right] + \frac{D}{u^{1/4}} \exp\left[\int_0^u \sqrt{u'}du'\right]. \tag{26.2.49}$$

Doing the integrals, we get

$$\psi(u) = \frac{C}{u^{1/4}} \exp\left[-\frac{2}{3}u^{3/2}\right] + \frac{D}{u^{1/4}} \exp\left[\frac{2}{3}u^{3/2}\right], \quad u \gg 1. \tag{26.2.50}$$

For $u \ll -1$, we set $k = \sqrt{-u} = |u|^{1/2}$, and following (26.2.22), we write

$$\psi(u) = \frac{A}{|u|^{1/4}} \exp\left[i \int_u^0 \sqrt{-u'}\, du'\right] + \frac{B}{|u|^{1/4}} \exp\left[-i \int_u^0 \sqrt{-u'}\, du'\right]. \tag{26.2.51}$$

The limits of integration were set so the lower limit is smaller than the upper limit, again chosen at zero; in doing so, we induced a minus sign in the phases of (26.2.22). Doing the integrals, we get

$$\psi(u) = \frac{A}{|u|^{1/4}} \exp\left[i\frac{2}{3}|u|^{3/2}\right] + \frac{B}{|u|^{1/4}} \exp\left[-i\frac{2}{3}|u|^{3/2}\right], \quad u \ll -1. \tag{26.2.52}$$

Equations (26.2.50) and (26.2.52) give general solutions of $\psi'' = u\psi$ far to the right and far to the left of the origin, respectively. If they represented a *single* solution of the differential equation, there would have to be some relation between the coefficients A, B and C, D. This information would allow us to connect the solutions in the two regions $u \gg 1$ and $u \ll -1$. We will address this question in section 26.5.

Our WKB solutions were supposed to solve the Airy differential equation. It is interesting to see what differential equation they actually solve. Take one of the approximate solutions, called $\psi_a(u)$:

$$\psi_a(u) = \frac{1}{u^{1/4}} \exp\left(-\frac{2}{3}u^{3/2}\right). \tag{26.2.53}$$

By taking two derivatives, you can show that ψ_a satisfies

$$\frac{d^2\psi_a}{du^2} = \left(u + \frac{5}{16}\frac{1}{u^2}\right)\psi_a. \tag{26.2.54}$$

This differs from the Airy equation by a u^{-2} term. As expected, this term becomes negligible for $u \gg 1$. □

Exercise 26.1. *Verify that $\psi_a(u)$ solves the above differential equation.*

26.3 Using Connection Formulae

We have taken preliminary steps in establishing the connection formulae. Before continuing in this direction, however, it is useful to see explicitly what kind of formulae these are and how they can be used in practice. In this section we get a bit ahead of ourselves, stating these formulae and showing how to use them in a concrete example. After doing this, we will resume their derivation in the following section.

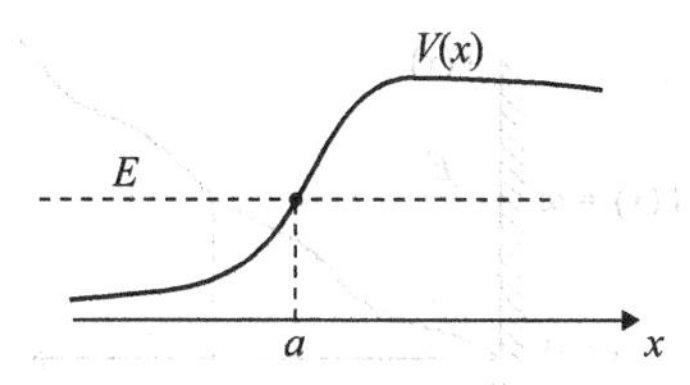

Figure 26.3
A potential $V(x)$ with a turning point at $x=a$. Connection formulae relate WKB solutions far to the left and far to the right of $x=a$.

We consider now the connection formulae for solutions away from a turning point $x=a$ separating a classically allowed region to the left and a classically forbidden region to the right (figure 26.3). The WKB solutions to the right are exponentials that grow or decay, and the WKB solutions to the left are oscillatory functions. They connect via the following relations:

$$\frac{2}{\sqrt{k(x)}}\cos\left(\int_x^a k(x')dx' - \frac{\pi}{4}\right) \Longleftarrow \frac{1}{\sqrt{\kappa(x)}}\exp\left(-\int_a^x \kappa(x')dx'\right), \tag{26.3.1}$$

$$-\frac{1}{\sqrt{k(x)}}\sin\left(\int_x^a k(x')dx' - \frac{\pi}{4}\right) \Longrightarrow \frac{1}{\sqrt{\kappa(x)}}\exp\left(\int_a^x \kappa(x')dx'\right). \tag{26.3.2}$$

The key feature in the above relations is the presence of arrows. In the first relation, the arrow tells us that if the solution is known to be a pure *decaying* exponential to the right of $x=a$, the solution to the left of $x=a$ is accurately determined and given by the phase-shifted cosine function the arrow points to. The second relation states that if the solution to the left of $x=a$ is of the displayed oscillatory type, the *growing part* of the solution to the right of $x=a$ is determined and given by the exponential the arrow points to. The decaying part cannot be reliably determined. As we will elaborate upon later, a connection formula is not to be used in the direction that goes against the arrow. Now we turn to an example that shows how to use these formulae.

Example 26.3. *Quantization condition for a potential with a wall.*
We now attempt to find the quantization condition that governs the energies of bound states in a potential $V(x)$ that includes a hard wall at $x=0$. Assume $V(x)$ increases monotonically and without bound, as illustrated in figure 26.4.

Let E denote the energy of our searched-for eigenstate. Clearly, the energy and the potential $V(x)$ determine the turning point $x=a$. The solution for $x>a$ must be a decaying exponential since the forbidden region extends forever to the right of $x=a$. The wave function for $x>a$ is therefore of the type shown on the right-hand side of the connection formula (26.3.1). This means that we get an accurate representation of the wave function well to the left of $x=a$. Setting the arbitrary normalization constant equal to one, we have

$$\psi(x) = \frac{1}{\sqrt{k(x)}}\cos\left(\int_x^a k(x')dx' - \tfrac{\pi}{4}\right), \quad 0\leq x \ll a. \tag{26.3.3}$$

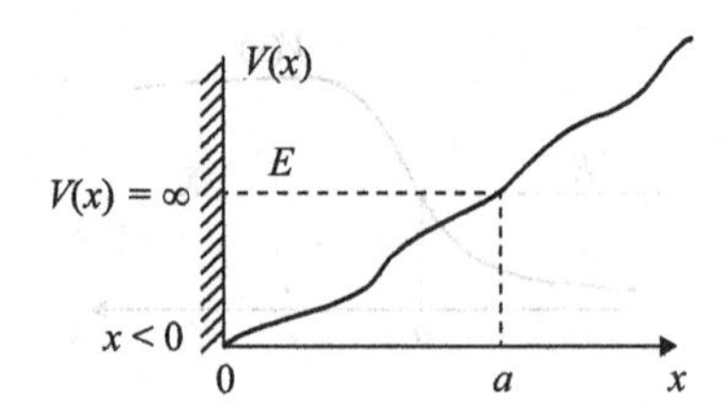

Figure 26.4
A monotonically increasing potential with a hard wall at $x=0$. For an energy eigenstate of energy E, the turning point is at $x=a$.

The wave function must vanish at the hard wall $x=0$. The condition $\psi(0)=0$ requires that

$$\cos\Delta=0, \quad \text{with} \quad \Delta\equiv\int_0^a k(x')dx'-\tfrac{\pi}{4}. \tag{26.3.4}$$

This is satisfied when

$$\int_0^a k(x')dx'-\tfrac{\pi}{4} = \tfrac{\pi}{2}+n\pi, \quad n\in\mathbb{Z}. \tag{26.3.5}$$

The quantization condition is therefore

$$\boxed{\int_0^a k(x')dx'=\left(n+\tfrac{3}{4}\right)\pi, \quad n=0,1,2,\ldots.} \tag{26.3.6}$$

Negative integers are not allowed because the left-hand side is manifestly positive. The above is easy to use in practice. Using the expression for $k(x)$ in terms of E and $V(x)$, we have

$$\int_0^a \sqrt{\frac{2m}{\hbar^2}(E-V(x'))}\,dx'=\left(n+\tfrac{3}{4}\right)\pi, \quad n=0,1,2,\ldots. \tag{26.3.7}$$

In special cases, the turning point position a can be determined explicitly in terms of E, and the integral can be done analytically. More generally, the analysis can always be done numerically, exploring the value of the integral on the left-hand side as a function of E and selecting the energies for which it takes the quantized values from the right-hand side.

Let us rewrite the wave function (26.3.3) using $\int_x^a=\int_0^a-\int_0^x$:

$$\begin{aligned}\psi(x)&=\frac{1}{\sqrt{k(x)}}\cos\left(\int_0^a k(x')dx'-\tfrac{\pi}{4}-\int_0^x k(x')dx'\right)\\&=\frac{1}{\sqrt{k(x)}}\cos\left(\Delta-\int_0^x k(x')dx'\right)\\&=\frac{1}{\sqrt{k(x)}}\sin\Delta\,\sin\left(\int_0^x k(x')dx'\right),\end{aligned} \tag{26.3.8}$$

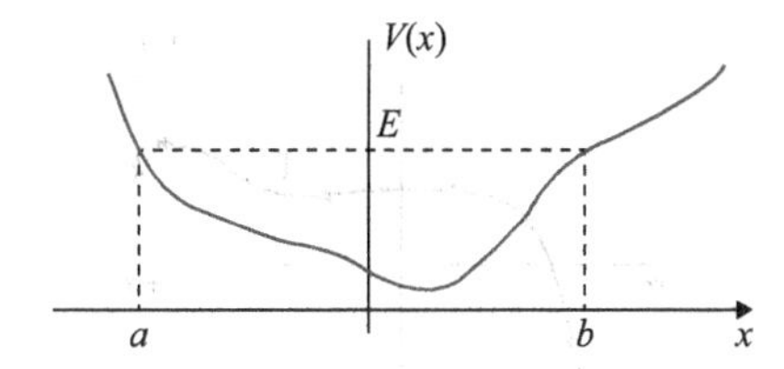

Figure 26.5
A potential for which a particle with energy E encounters turning points at $x=a$ and $x=b$. In this case the WKB quantization condition is given by (26.3.9).

where we expanded the cosine of a sum of angles and recalled that $\cos\Delta=0$. This form makes $\psi(x=0)=0$ manifest. More interestingly, the quantization condition (26.3.6) indicates that the excursion of the phase of $\psi(x)$ from $x=0$ to $x=a$, equal to $(n+\frac{3}{4})\pi$, is a bit higher than $n\pi$ but less than $(n+1)\pi$. Thus, even though the WKB wave function is not reliable all the way to $x=a$, it produces the n nodes the nth excited state must have! □

The quantization condition (26.3.6) applies for a potential with a wall and a single turning point. This result has a simple but important modification for potentials in which states of a given energy have two turning points, a and b, with $a<b$, as shown in figure 26.5. In that case you will be able to show that (problem 26.3)

$$\int_a^b k(x')dx' = (n+\tfrac{1}{2})\pi, \quad n=0,1,2,\ldots. \tag{26.3.9}$$

Note that the offset to the integer n on the right-hand side changed from $\frac{3}{4}$ to $\frac{1}{2}$.

26.4 Airy Functions and Their Expansions

We now revisit and develop the analysis of Airy functions that we began in section 6.7. We will follow a similar approach, based on Fourier transforms but done more generally. Recall that the Schrödinger equation for a linear potential gave a first-order differential equation in momentum space and thus one solution. But the original, coordinate space Schrödinger equation for the linear potential is a second-order differential equation and must have two solutions. How does one obtain the second solution, the second Airy function? Furthermore, we want to have enough control over the solution so that we can derive the asymptotic expansions of the Airy functions, both for large and positive argument as well as for large and negative argument.

Airy functions arise from the study of a differential equation we have already considered—that governing the eigenfunctions of a quantum particle in a linear potential. We showed that those eigenfunctions are determined if we know the solution of equation (26.2.46):

$$\frac{d^2\psi}{du^2} = u\psi. \tag{26.4.1}$$

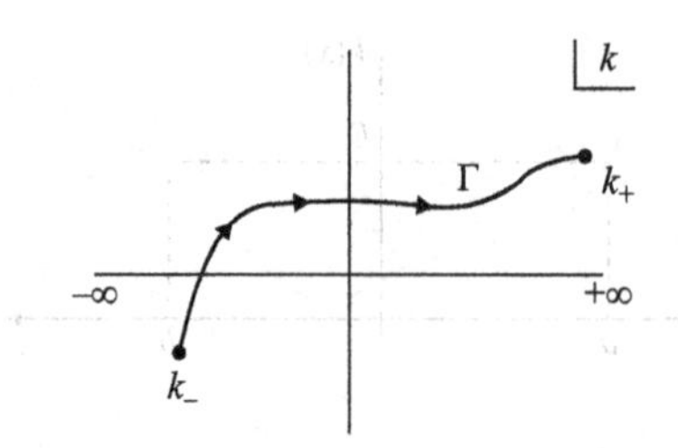

Figure 26.6
A contour Γ in the complex k plane used to construct a solution of (26.4.1). The contour begins at k_- and ends at k_+.

This has an integral solution, obtained by Fourier transformation. We write

$$\psi(u) = \int_{-\infty}^{\infty} \frac{dk}{2\pi}\, \tilde{\psi}(k)\, e^{iku}, \tag{26.4.2}$$

where $\tilde{\psi}(k)$ denotes the Fourier transform, and both k and u are unit-free. For more generality, rather than integrating over the full real line, as we do for the Fourier transform, let us use some oriented contour Γ in the complex k plane, a contour that begins at a point k_- and ends at a point k_+, as shown in figure 26.6. We then write

$$\psi(u) = \int_{\Gamma} \frac{dk}{2\pi}\, \tilde{\psi}(k)\, e^{iku}. \tag{26.4.3}$$

For the moment, the contour Γ is left undefined. This equation is our ansatz for the solution.

The left-hand side of the differential equation (26.4.1) then gives

$$\frac{d^2\psi}{du^2} = \frac{d^2}{du^2} \int_{\Gamma} \frac{dk}{2\pi}\, \tilde{\psi}(k)\, e^{iku} = \int_{\Gamma} \frac{dk}{2\pi}\, (-k^2 \tilde{\psi}(k))\, e^{iku}. \tag{26.4.4}$$

The right-hand side of the same equation results in

$$\begin{aligned} u\psi &= \int_{\Gamma} \frac{dk}{2\pi}\, \tilde{\psi}(k)\, u e^{iku} = \int_{\Gamma} \frac{dk}{2\pi}\, \tilde{\psi}(k) \frac{1}{i}\frac{d}{dk} e^{iku} \\ &= \int_{\Gamma} \frac{dk}{2\pi} \Big[\frac{1}{i}\frac{d}{dk}\Big(\tilde{\psi}(k) e^{iku}\Big) - \frac{1}{i}\frac{d\tilde{\psi}}{dk} e^{iku} \Big] \\ &= \frac{1}{2\pi i}\, \tilde{\psi}(k) e^{iku}\Big|_{k_-}^{k_+} - \int_{\Gamma} \frac{dk}{2\pi} \frac{1}{i} \frac{d\tilde{\psi}}{dk} e^{iku}, \end{aligned} \tag{26.4.5}$$

with k_- and k_+, respectively, the initial and final points in the Γ contour. We can now assemble the whole differential equation:

$$0 = \frac{d^2\psi}{du^2} - u\psi = \int_{\Gamma} \frac{dk}{2\pi} \Big(-k^2 \tilde{\psi}(k) + \frac{1}{i}\frac{d\tilde{\psi}}{dk} \Big) e^{iku} - \frac{1}{2\pi i}\, \tilde{\psi}(k) e^{iku}\Big|_{k_-}^{k_+}. \tag{26.4.6}$$

We have a solution if the integrand in the first term vanishes:

$$-k^2 \tilde{\psi}(k) + \frac{1}{i}\frac{d\tilde{\psi}}{dk} = 0, \tag{26.4.7}$$

and if the boundary contributions from the second term separately vanish,

$$\tilde{\psi}(k_+)e^{ik_+u} = 0, \quad \tilde{\psi}(k_-)e^{ik_-u} = 0. \tag{26.4.8}$$

The attempt to have the boundary contributions cancel each other out encounters difficulties. It would require $\tilde{\psi}(k_+)e^{i(k_+-k_-)u} = \tilde{\psi}(k_-)$, and the only way to make the left-hand side u independent is to set $k_- = k_+$. This would make the contour closed and thus shrinkable to zero size, giving zero $\psi(u)$. The first condition (26.4.7) gives an easily solved *first-order* differential equation:

$$\frac{d\tilde{\psi}}{dk} = ik^2\tilde{\psi}(k) \quad \Rightarrow \quad \psi(k) = e^{ik^3/3}, \tag{26.4.9}$$

where we have set the overall constant multiplying the right-hand side equal to one since we are not attempting to normalize this wave function. With this, our solution is

$$\psi(u) = \int_\Gamma \frac{dk}{2\pi} e^{ik^3/3} e^{iku}, \quad \text{if} \quad e^{ik_+^3/3} e^{ik_+u} = e^{ik_-^3/3} e^{ik_-u} = 0. \tag{26.4.10}$$

For the solution to hold, the boundary terms at the end points of Γ must vanish.

A term $e^{ik^3/3}$ can vanish as $|k| \to \infty$ if k^3 has a positive imaginary part:

$$\operatorname{Im} k^3 > 0. \tag{26.4.11}$$

The intuition is clear: the imaginary part of k^3 can contribute a large negative number in the exponent of $e^{ik^3/3}$. To see this, let us write k in the form $k = |k|e^{i\theta_k}$. The condition $\operatorname{Im} k^3 > 0$ then becomes

$$\operatorname{Im} k^3 = |k|^3 \operatorname{Im}(e^{3i\theta_k}) = |k^3| \sin 3\theta_k > 0. \tag{26.4.12}$$

For any fixed θ_k for which this condition holds, the imaginary part of k^3 will be positive and grow without bounds as $|k| \to \infty$, making $e^{ik^3/3}$ go to zero. The condition requires $\sin 3\theta_k > 0$, and this is satisfied for $3\theta_k \in [0, \pi]$, as well as for $3\theta_k \in [2\pi, 3\pi]$ and for $3\theta_k \in [4\pi, 5\pi]$. The regions with $\operatorname{Im} k^3 > 0$ are therefore

$$0 < \theta_k < \tfrac{\pi}{3}, \quad \tfrac{2\pi}{3} < \theta_k < \pi, \quad \tfrac{4\pi}{3} < \theta_k < \tfrac{5\pi}{3}. \tag{26.4.13}$$

These allowed sectors in the complex k plane are shown shaded in figure 26.7. If k_+ and k_- approach infinity in the shaded sectors, the boundary terms will vanish, and we have a solution of the differential equation. Note that the boundaries of the sectors are lines with $\operatorname{Im} k^3 = 0$. Therefore, having k_+ or k_- go to infinity on those lines is delicate.

If we take Γ to be the contour C_1 running all over the real line (figure 26.7), the result is the Airy function $\operatorname{Ai}(u)$:

$$\operatorname{Ai}(u) = \int_{-\infty}^{\infty} \frac{dk}{2\pi} e^{ik^3/3} e^{iku} = \frac{1}{\pi} \int_0^\infty dk \cos\left(\tfrac{1}{3}k^3 + ku\right). \tag{26.4.14}$$

The integral gives a finite result even though the integrand remains finite as $k \to \infty$. The finite result arises because the integrand oscillates faster and faster, suppressing possible contributions. This delicate convergence is consistent with the choice of $k_\pm$ lying on

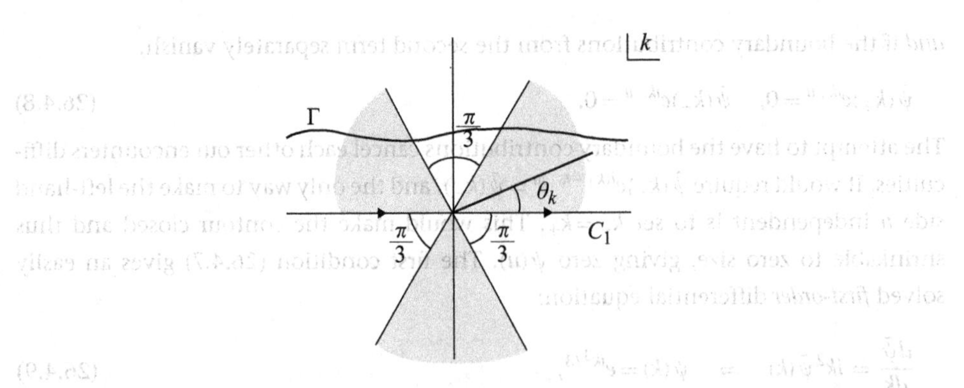

Figure 26.7
If the end points of contour Γ approach infinity within the shaded regions, we have a solution of the differential equation. For the contour C_1 running along the real axis, we get the conventional representation of the Airy function Ai(u), but the integral's convergence is delicate.

the boundary of the allowed sectors. As we will see, Ai(u) is oscillatory for negative u and decays to zero rapidly for positive, increasing u.

To get another solution of the differential equation, we need another contour that *cannot* be deformed into C_1 while keeping the end points within the allowed sectors. The contour C_2 that goes from $-\infty$ to zero, right above the real axis, and then goes down from zero along the negative imaginary axis (figure 26.8) is useful. The second solution of the Airy equation, called Bi(u), is defined so it is oscillatory for negative u, just like Ai(u). But unlike Ai(u), it grows without bound for positive, increasing u. This is obtained by using a sum of contours:

$$\mathrm{Bi}(u) = -i\int_{C_1} \frac{dk}{2\pi} e^{ik^3/3} e^{iku} + 2i\int_{C_2} \frac{dk}{2\pi} e^{ik^3/3} e^{iku}. \tag{26.4.15}$$

Since each contour integral gives a solution, the sum does too. A calculation shows that this contour integral can be rewritten as

$$\mathrm{Bi}(u) = \frac{1}{\pi}\int_0^\infty dk\left(e^{-k^3/3}e^{ku} + \sin\Big(\frac{k^3}{3} + ku\Big)\right). \tag{26.4.16}$$

The relevance of the Ai and Bi functions to our WKB story is through their asymptotic behavior. Some functions have series or asymptotic expansions valid for much of the complex plane. The Taylor expansion of the exponential function e^z, for example, is in fact valid for all z. For the Airy functions, however, we have the Stokes phenomenon: the asymptotic behavior of functions can differ in different regions of the complex plane. For the Airy function Ai(u), the expansion for large negative u is oscillatory and for large positive u is a decaying exponential.

Let us find the asymptotic expansion of Ai(u) for $u \gg 1$ using the integral representation:

$$\mathrm{Ai}(u) = \int_{C_1} \frac{dk}{2\pi} e^{i\left(\frac{k^3}{3} + ku\right)}. \tag{26.4.17}$$

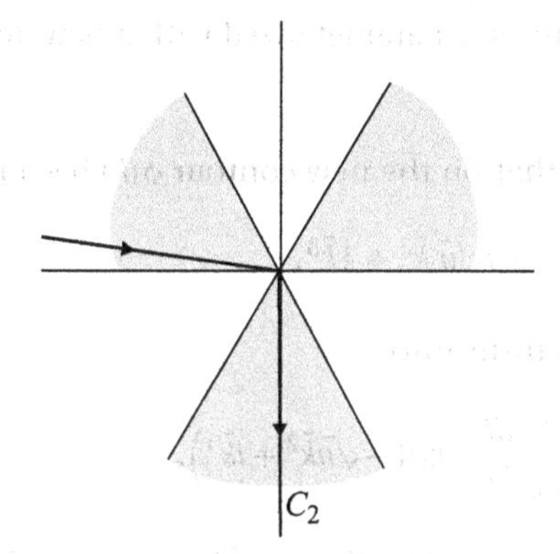

Figure 26.8
The contour C_2 that plays a role in the definition of the Airy function $\mathrm{Bi}(u)$.

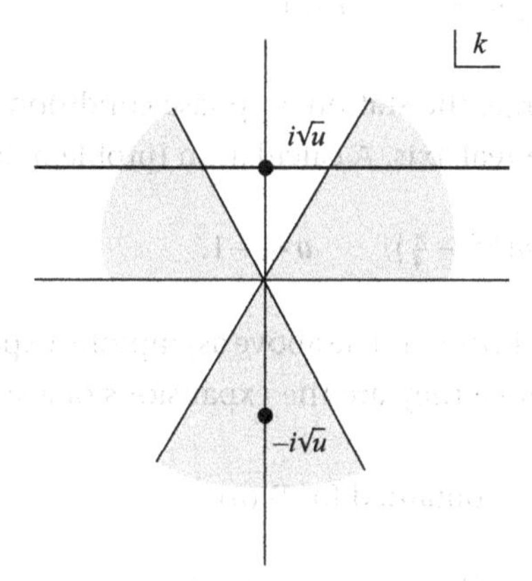

Figure 26.9
The asymptotic expansion of $\mathrm{Ai}(u)$ for $u > 0$ can be obtained by using the stationary phase method, shifting the C_1 contour up to become the line with an imaginary part equal to $i\sqrt{u}$.

The bulk of the contribution to the integral comes from points where the phase is stationary. The phase $\phi(k)$ in the integrand is

$$\phi(k) = \tfrac{1}{3}k^3 + ku. \tag{26.4.18}$$

Then $\phi'(k) = k^2 + u = 0$ defines the points in the k plane with stationary phase. Since $u > 0$, we see that

$$k = \pm i\sqrt{u}. \tag{26.4.19}$$

In the complex plane, we are allowed to deform the contour of integration. It is convenient to move the contour C_1 so that it goes through a stationary phase point. The contour C_1 can only be shifted upward to go through $k = i\sqrt{u}$, as shown in figure 26.9. This shift is possible because the integrand has no poles in the upper half plane and vanishes at the end points as long as we stay in the shaded regions, which we do in this deformation. The new contour is parameterized with a new, real variable $\tilde{k} \in (-\infty, \infty)$:

$$k = i\sqrt{u} + \tilde{k}. \tag{26.4.20}$$

A short calculation shows that on the new contour $\phi(k)$ becomes

$$\phi(k) = \phi(i\sqrt{u} + \tilde{k}) = \tfrac{2}{3} i u^{3/2} + i\sqrt{u}\,\tilde{k}^2 + \tfrac{1}{3}\tilde{k}^3. \tag{26.4.21}$$

Then the integral for $\text{Ai}(u)$ turns into

$$\text{Ai}(u) = \exp\left(-\tfrac{2}{3}u^{3/2}\right) \int_{-\infty}^{\infty} \frac{d\tilde{k}}{2\pi} \exp\left(-\sqrt{u}\tilde{k}^2 + i\tilde{k}^3\right). \tag{26.4.22}$$

For large u, the suppression created by the $(-\sqrt{u}\tilde{k}^2)$ term implies that we can ignore the $i\tilde{k}^3$ term. We then do the Gaussian integral, finding the asymptotic expansion

$$\text{Ai}(u) \simeq \frac{1}{2\sqrt{\pi}} \frac{1}{u^{1/4}} \exp(-\tfrac{2}{3}u^{3/2}), \qquad u \gg 1. \tag{26.4.23}$$

When u is negative and large, the stationary phase condition gives two real roots. The contour can stay along the real axis. A calculation (problem 26.5) then gives

$$\text{Ai}(u) \simeq \frac{1}{\sqrt{\pi}} \frac{1}{|u|^{1/4}} \cos\left(\tfrac{2}{3}|u|^{3/2} - \tfrac{\pi}{4}\right), \qquad u \ll -1. \tag{26.4.24}$$

We have here a connection formula. The above asymptotic expansions, for $u \gg 1$ and for $u \ll -1$, are "connected" since they are the expansions of a single object, the function $\text{Ai}(u)$.

Similar expansions can be obtained for $\text{Bi}(u)$:

$$\begin{aligned} \text{Bi}(u) &\simeq \frac{1}{\sqrt{\pi}} \frac{1}{u^{1/4}} \exp\left(\tfrac{2}{3}u^{3/2}\right), && u \gg 1, \\ \text{Bi}(u) &\simeq -\frac{1}{\sqrt{\pi}} \frac{1}{|u|^{1/4}} \sin\left(\tfrac{2}{3}|u|^{3/2} - \tfrac{\pi}{4}\right), && u \ll -1. \end{aligned} \tag{26.4.25}$$

These relations also represent a connection formula.

26.5 Connection Formulae Derived

With all the preparatory work done, we are ready to establish the connection formulae across turning points. For this purpose consider a general potential $V(x)$, and focus on the region around the turning point $x = a$ (figure 26.10). The potential sufficiently near $x = a$ is approximately linear, and $V(x) - E$ vanishes at $x = a$. We therefore have

$$V(x) - E \simeq g(x - a), \quad g > 0, \tag{26.5.1}$$

with g some constant. We now consider WKB basic solutions ψ_R far to the right (R) of the turning point:

$$\psi_R(x) = \frac{C}{\sqrt{\kappa(x)}} \exp\left(-\int_a^x \kappa(x')dx'\right) + \frac{D}{\sqrt{\kappa(x)}} \exp\left(\int_a^x \kappa(x')dx'\right), \quad x \gg a. \tag{26.5.2}$$

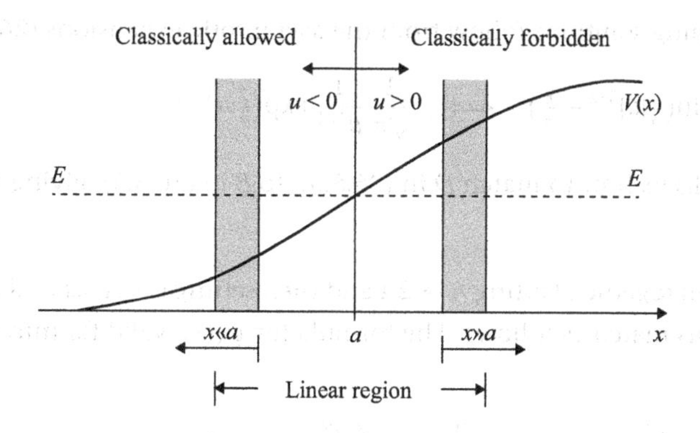

Figure 26.10
A turning point $x=a$ of a potential $V(x)$ that is approximately linear near $x=a$. We need to have regions, *shown shaded*, where $V(x)$ is approximately linear and that are far enough from $x=a$ that the WKB expressions are valid.

This solution was in fact evaluated in section 26.2 under the assumption that the potential is strictly linear. To use this evaluation, we therefore need the existence of a region in which both $x \gg a$ and the potential $V(x)$ is still accurately linear. This region, assumed to exist, is shown shaded in figure 26.10. The result of the evaluation was given in (26.2.50), following a series of redefinitions that used a variable $u=\frac{1}{L}(x-a)$ for which the turning point is at $u=0$:

$$\psi_R(u) = \frac{C}{u^{1/4}} \exp\left[-\frac{2}{3}u^{3/2}\right] + \frac{D}{u^{1/4}} \exp\left[\frac{2}{3}u^{3/2}\right], \quad u \gg 1. \tag{26.5.3}$$

Let us now write the WKB solutions far to the left of $x=a$, this time using sines and cosines instead of exponentials and with a shift by $\pi/4$ that will prove convenient:

$$\psi_L(x) = \frac{A}{\sqrt{k(x)}} \cos\left(\int_x^a k(x')dx' - \frac{\pi}{4}\right) + \frac{B}{\sqrt{k(x)}} \sin\left(\int_x^a k(x')dx' - \frac{\pi}{4}\right), \quad x \ll a. \tag{26.5.4}$$

In section 26.2 we also evaluated the WKB solution (26.2.22) of the linear potential to the left of the barrier. The result was given in (26.2.52), which requires minor modifications now:

$$\psi_L(u) = \frac{A}{|u|^{1/4}} \cos\left[\frac{2}{3}|u|^{3/2} - \frac{\pi}{4}\right] + \frac{B}{|u|^{1/4}} \sin\left[\frac{2}{3}|u|^{3/2} - \frac{\pi}{4}\right], \quad u \ll -1. \tag{26.5.5}$$

In both the evaluation of $\sqrt{\kappa(x)}$ for ψ_R and $\sqrt{k(x)}$ for ψ_L, we dropped the *same* multiplicative constants.

From the Ai(u) asymptotic expansions in (26.4.23) and (26.4.24), both of which arise from the *same* function, we have the matching solutions:

$$\frac{1}{\sqrt{\pi}} \frac{1}{|u|^{1/4}} \cos\left(\tfrac{2}{3}|u|^{3/2} - \tfrac{\pi}{4}\right) \quad \Longleftrightarrow \quad \frac{1}{2\sqrt{\pi}} \frac{1}{u^{1/4}} \exp\left(-\tfrac{2}{3}\, u^{3/2}\right). \tag{26.5.6}$$

This relation tells us how to match C in (26.5.3) to A in (26.5.5), giving us

$$C = \tfrac{1}{2}A. \tag{26.5.7}$$

The Bi(u) matching solutions follow from the asymptotic expansions (26.4.25):

$$-\frac{1}{\sqrt{\pi}}\frac{1}{|u|^{1/4}}\sin\left(\tfrac{2}{3}|u|^{3/2}-\tfrac{\pi}{4}\right) \Longleftrightarrow \frac{1}{\sqrt{\pi}}\frac{1}{u^{1/4}}\exp\left(\tfrac{2}{3}u^{3/2}\right). \tag{26.5.8}$$

This relation tells us how to match D in (26.5.3), to B in (26.5.5), giving us

$$D=-B. \tag{26.5.9}$$

We now put it all together. Letting $A\to 2A$ and then setting $C=A$, as well as $D=-B$, the WKB expressions match as follows. The formula for $\psi_R(x)$, valid far into the forbidden region,

$$\frac{A}{\sqrt{\kappa(x)}}\exp\left(-\int_a^x \kappa(x')dx'\right)-\frac{B}{\sqrt{\kappa(x)}}\exp\left(\int_a^x \kappa(x')dx'\right), \qquad x\gg a, \tag{26.5.10}$$

matches with the formula for $\psi_L(x)$, valid far into the allowed region:

$$\frac{2A}{\sqrt{k(x)}}\cos\left(\int_x^a k(x')dx'-\frac{\pi}{4}\right)+\frac{B}{\sqrt{k(x)}}\sin\left(\int_x^a k(x')dx'-\frac{\pi}{4}\right), \qquad x\ll a. \tag{26.5.11}$$

Up to an important subtlety discussed below, the above are the connection formulae. Note that with $A=1$ and $B=0$, and then with $A=0$ and $B=-1$, these become the relations anticipated in (26.3.1) and (26.3.2), respectively.

The subtlety in question leads to the *arrows* in the connection conditions. In the regions that they are valid, our WKB solutions are only approximate. If we use a WKB solution representing a growing exponential, the other solution representing a decaying exponential cannot be used or trusted. The exact solution, when analyzed, will contain the WKB growing exponential, but it may contain a subleading term that, while smaller than the growing exponential, is still larger than the decaying WKB exponential.

Let us see explicitly how that plays out for the naive relations. Let $B=0$, and take $A=1$. Then we have

$$\frac{2}{\sqrt{k(x)}}\cos\left(\int_x^a k(x')dx'-\frac{\pi}{4}\right) \overset{?}{\Longleftrightarrow} \frac{1}{\sqrt{\kappa(x)}}\exp\left(-\int_a^x \kappa(x')dx'\right). \tag{26.5.12}$$

Suppose that for $x\gg a$ we know we have only a decaying exponential, described by the term to the right. The arrow pointing left tells us, correctly, that we match the object to the left. On the other hand, suppose we know we have the object to the left. There is always the uncertainty that there could be a sine wave with a tiny coefficient B (as in (26.5.11)). In that case, that solution would connect to a growing exponential for $x\gg a$ whose effect could well overwhelm that of the decaying exponential above. Thus, the formula cannot be used from left to right. The correct connection formula is

$$\frac{2}{\sqrt{k(x)}}\cos\left(\int_x^a k(x')dx'-\frac{\pi}{4}\right) \Longleftarrow \frac{1}{\sqrt{\kappa(x)}}\exp\left(-\int_a^x \kappa(x')dx'\right). \tag{26.5.13}$$

We can discuss analogously the second relation. Setting $A=0$ and $B=1$, we have

$$-\frac{1}{\sqrt{k(x)}}\sin\left(\int_x^a k(x')dx'-\frac{\pi}{4}\right) \overset{?}{\Longleftrightarrow} \frac{1}{\sqrt{\kappa(x)}}\exp\left(\int_a^x \kappa(x')dx'\right). \tag{26.5.14}$$

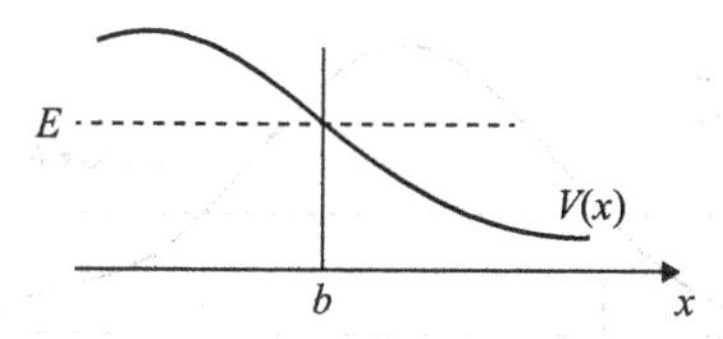

Figure 26.11
A turning point at $x = b$, with the classically forbidden region on the left and the classically allowed region on the right. The relevant connection formulae are (26.5.16) and (26.5.17).

It is clear we cannot go from right to left: with a growing exponential present to the right, a decaying exponential would essentially be invisible while giving on the left a cosine wave comparable to the wave already present. On the other hand, we can go from left to right. A small error on the left, in the form of a cosine wave with small amplitude, would only lead to a decaying exponential to the right, which would be negligible anyway. The correct relation is

$$-\frac{1}{\sqrt{k(x)}}\sin\left(\int_x^a k(x')dx' - \frac{\pi}{4}\right) \Longrightarrow \frac{1}{\sqrt{\kappa(x)}}\exp\left(\int_a^x \kappa(x')dx'\right). \tag{26.5.15}$$

For reference it is useful to give the form of the connection formulae when the linear potential that approximates the region near the turning point slopes downward. This time the turning point is at $x = b$, with the classically allowed region to the right and the classically forbidden region to the left. The connection formulae in this case read

$$\frac{1}{\sqrt{\kappa(x)}}\exp\left(-\int_x^b \kappa(x')dx'\right) \Longrightarrow \frac{2}{\sqrt{k(x)}}\cos\left(\int_b^x k(x')dx' - \frac{\pi}{4}\right), \tag{26.5.16}$$

$$-\frac{1}{\sqrt{\kappa(x)}}\exp\left(\int_x^b \kappa(x')dx'\right) \Longleftarrow \frac{1}{\sqrt{k(x)}}\sin\left(\int_b^x k(x')dx' - \frac{\pi}{4}\right). \tag{26.5.17}$$

The logic behind the arrows is exactly the same as in the case of the upwards-sloping turning point. We can summarize the statement about arrows in the connection formulae in one sentence, valid for all cases:

> We can connect away from a decaying exponential and into a growing exponential.

(26.5.18)

This is relative to a turning point: a decaying exponential is one that decays as we move away from the turning point into the forbidden region, and a growing exponential is one that grows as we move away from the turning point into the forbidden region. As far as factors are concerned, the decaying exponential goes into a cosine wave with twice its "amplitude" and a shift of $\pi/4$. The sine wave, also with a shift of $\pi/4$, goes into a growing exponential with the same "amplitude" but the opposite sign.

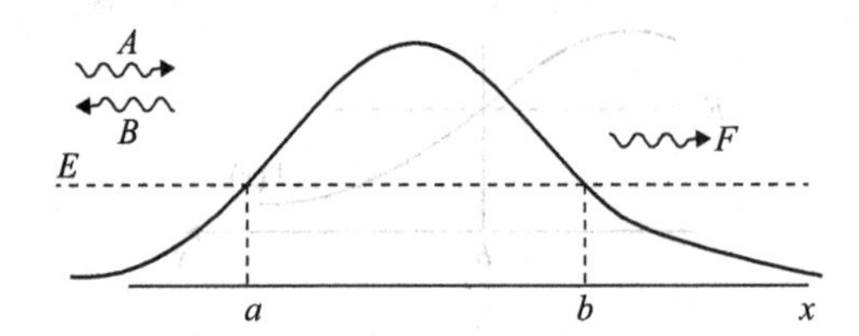

Figure 26.12
Tunneling in the WKB approximation. The energy E is smaller than the height of the potential, and the barrier must be wide and smooth.

26.6 Tunneling through a Barrier

As a useful application of the WKB approximation, we will determine the tunneling probability T for a wave of fixed energy incident on a smoothly varying wide barrier. In order to have tunneling, the energy E of the wave must be smaller than the height of the barrier. In this problem there is an incident wave, a reflected wave, and a transmitted wave. The associated WKB expressions will have amplitudes controlled by constants A, B, and F, respectively. The situation is shown in figure 26.12.

Let us consider a few remarks:

1. We expect the transmission probability T to be small. Little probability flux goes through a wide barrier, and therefore the reflected wave is essentially as large as the incoming wave: $|B| \approx |A|$.
2. For $x \gg b$, there is just an outgoing wave with amplitude controlled by F.
3. Within the barrier the component that decays as x grows is more relevant than the component that grows. It is the only component that can be estimated reliably.

Our strategy is to start from the right, with the transmitted wave. After writing a WKB basic solution for this wave, we work our way into the barrier and then to the left of the barrier. Consider therefore the transmitted right-moving wave $\psi_{\rm tr}$, valid for $x \gg b$:

$$\psi_{\rm tr}(x) = \frac{F}{\sqrt{k(x)}} \exp\left(i \int_b^x k(x')dx' - i\tfrac{\pi}{4}\right), \quad x \gg b. \tag{26.6.1}$$

This is a WKB solution with an extra phase of $\pi/4$ chosen to make the argument of the exponential have the familiar form appearing in the connection formulae. Expanding the exponential, we have

$$\psi_{\rm tr}(x) = \frac{F}{\sqrt{k(x)}} \cos\left(\int_b^x k(x')dx' - \tfrac{\pi}{4}\right) + \frac{iF}{\sqrt{k(x)}} \sin\left(\int_b^x k(x')dx' - \tfrac{\pi}{4}\right), \quad x \gg b. \tag{26.6.2}$$

This is in *standard form*. We can now match the second term to an exponential that grows as we move to the left of $x = b$ by using (26.5.17):

$$\psi_{\rm barr}(x) = -\frac{iF}{\sqrt{\kappa(x)}} \exp\left(\int_x^b \kappa(x')dx'\right), \quad a \ll x \ll b. \tag{26.6.3}$$

The subscript *barr* indicates a solution in the barrier region. If we attempted to match the first term in (26.6.2), we would get an exponential that decays as we move to the left of $x=b$, and it is unreliable given the growing exponential. We can now refer this solution inside the barrier to the turning point $x=a$:

$$\psi_{\text{barr}}(x) = -\frac{iF}{\sqrt{\kappa(x)}} \exp\left(\int_a^b \kappa(x')dx' - \int_a^x \kappa(x')dx'\right), \quad a \ll x \ll b. \tag{26.6.4}$$

Defining a real unit-free constant θ by

$$\theta \equiv \int_a^b \kappa(x')dx', \tag{26.6.5}$$

we see that

$$\psi_{\text{barr}}(x) = -\frac{iFe^{\theta}}{\sqrt{\kappa(x)}} \exp\left(-\int_a^x \kappa(x')dx'\right), \quad a \ll x \ll b. \tag{26.6.6}$$

Since this is a decaying exponential to the right of $x=a$, we can connect it to a solution to the left of $x=a$ using (26.3.1):

$$\psi(x) = -\frac{2iFe^{\theta}}{\sqrt{k(x)}} \cos\left(\int_x^a k(x')dx' - \tfrac{\pi}{4}\right). \tag{26.6.7}$$

This is a superposition of two waves: a wave ψ_{inc} moving to the right and a wave ψ_{ref} moving to the left. The incident part is

$$\psi_{\text{inc}}(x) = -\frac{iFe^{\theta}}{\sqrt{k(x)}} \exp\left(-i\int_x^a k(x')dx' + i\tfrac{\pi}{4}\right). \tag{26.6.8}$$

The sign in front of the integral may seem unusual for a wave moving to the right. It is correct, however, because the argument x appears in the lower limit of integration so that the phase increases, becoming less negative, as x moves to the right. The transmission coefficient T is the ratio of the transmitted probability current over the incident probability current. Given the result in (26.2.20), we find that

$$T = \frac{\text{probability current for } \psi_{\text{tr}}}{\text{probability current for } \psi_{\text{inc}}} = \frac{|F|^2}{|-iFe^{\theta}|^2} = e^{-2\theta}. \tag{26.6.9}$$

This is the well-known exponential suppression of the transmission coefficient. Using the earlier definition of θ, the result is

$$T_{\text{WKB}} = \exp\left(-2\int_a^b \kappa(x')dx'\right). \tag{26.6.10}$$

We added the subscript *WKB* to emphasize that this is the WKB approximation to the exact transmission coefficient. The integral extends in between the two turning points and captures information about the height and the width of the barrier. The integrand and the turning points depend on the energy E of the incident wave. We can display this by using the explicit value of $\kappa(x)$:

$$T_{\rm WKB} = \exp\left(-2\int_a^b \sqrt{\tfrac{2m}{\hbar^2}(V(x') - E)}\, dx'\right). \qquad (26.6.11)$$

The WKB approximation only captures the exponentially decaying part of the transmission coefficient. There are corrections, usually written as a prefactor to the exponential.

Example 26.4. *Transmission coefficient for a rectangular barrier.*

A rectangular barrier is an instructive example where we can easily compute the WKB transmission probability and compare its value with the exact result. The barrier is defined by the potential $V(x)$ given by

$$V(x), = \begin{cases} V_0, & \text{for } |x| < a, \\ 0, & \text{otherwise.} \end{cases} \qquad (26.6.12)$$

Assume the barrier is large. In terms of the familiar unit-free constant z_0 used to characterize square wells, this means that

$$z_0^2 \equiv \frac{2mV_0a^2}{\hbar^2} \gg 1. \qquad (26.6.13)$$

A large barrier means a large z_0 or large V_0a^2. Moreover, we assume E is smaller than V_0; we cannot expect the approximation to work as the energy approaches the top of the barrier.

The WKB estimation, using (26.6.11), is immediate. Since the barrier extends from $x=-a$ to $x=a$ and the potential is constant, we find that

$$T_{\rm WKB} = \exp\left(-2\int_{-a}^{a} \sqrt{\frac{2m}{\hbar^2}(V_0 - E)}\, dx\right) = \exp\left(-\frac{4a}{\hbar}\sqrt{2m(V_0 - E)}\ \right). \qquad (26.6.14)$$

This is the answer in the WKB approximation. In terms of z_0, this reads

$$T_{\rm WKB} = \exp\left(-4z_0\sqrt{1 - \tfrac{E}{V_0}}\ \right). \qquad (26.6.15)$$

The validity of the WKB approximation requires the argument in the exponent to be large. This is why we need $z_0 \gg 1$ and the energy not to approach V_0. When E is very small compared to V_0, the result simplifies further:

$$T_{\rm WKB} \sim \exp(-4z_0), \quad E \ll V_0. \qquad (26.6.16)$$

Since $z_0 \sim a\sqrt{V_0}$, this result is the basis for the general intuition that the exponential suppression of the tunneling probability is proportional to the width of the barrier and the square root of its height.

The exact formula for the tunneling probability of the square barrier (26.6.12) is derived in a number of textbooks. The computation is a bit lengthy and uses the methods in chapter 8. The result is

$$\frac{1}{T} = 1 + \frac{V_0^2}{4E(V_0 - E)}\sinh^2\left(2z_0\sqrt{1 - \tfrac{E}{V_0}}\right) \qquad (26.6.17)$$

This formula will allow us to confirm the WKB exponential suppression and to find the prefactor. Under the conditions $z_0 \gg 1$ and E/V_0 not approaching one, the argument of the sinh function is large. Therefore, this function can be replaced by its growing exponential, that is, $\sinh x \sim \frac{1}{2}e^x$:

$$\frac{1}{T} \simeq 1 + \frac{V_0}{16E(1-\frac{E}{V_0})} \exp\Big(4z_0\sqrt{1-\tfrac{E}{V_0}}\Big). \tag{26.6.18}$$

The additive unit on the right-hand side can be neglected, and we have

$$T \simeq 16\,\tfrac{E}{V_0}\,(1-\tfrac{E}{V_0})\,\exp\Big(-4z_0\sqrt{1-\tfrac{E}{V_0}}\Big). \tag{26.6.19}$$

This is the same exponential suppression as in the WKB result (26.6.15). More interestingly, the prefactor to the exponential gives us the leading correction to the WKB result. In the limit $E \ll V_0$, it modifies (26.6.16), giving us

$$T \simeq 16\,\tfrac{E}{V_0}\,\exp(-4z_0), \quad E \ll V_0. \tag{26.6.20}$$

With large z_0, the prefactor has a small effect on the value of $\ln T$. □

The WKB approximation for the tunneling probability is often used for the estimation of lifetimes. The physical situation is represented in figure 26.13. We have a particle of mass m and energy E localized between the turning points $x=a$ and $x=b$ of the potential. The classical particle cannot escape because of the energy barrier stretching from $x=b$ to $x=c$. The quantum particle, however, can tunnel. Note that the quantum particle, when localized between a and b, is not in an energy eigenstate. This is clear because the wave function for any energy eigenstate would have to be nonzero for $x>c$ and thus nonnormalizable. Moreover, if the particle were in an energy eigenstate, it would remain localized forever. To estimate the lifetime τ of this particle, we use an eclectic mix of tools: a set of classical estimates and the quantum transmission probability.

We first show that the lifetime τ is the inverse of the tunneling probability per unit time w, that is, $\tau = 1/w$. To see this, we examine the function $P(t)$ that represents the probability of having the particle still localized at time t, if it was localized at $t=0$. The lifetime is said to be τ if $P(t)$ has the time dependence $P(t)=e^{-t/\tau}$. To see that an

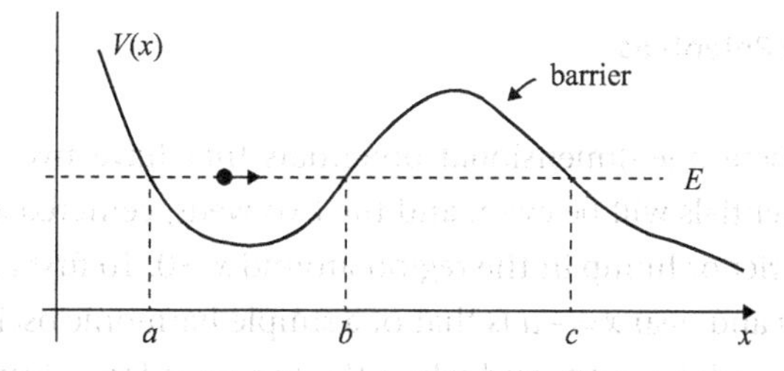

Figure 26.13
Estimating the lifetime of a particle with energy E temporarily localized in $x \in [a,b]$.

exponentially decaying probability arises when the tunneling rate w is constant, note that

$$P(t+dt) = P(t)\cdot(1-w\,dt), \tag{26.6.21}$$

where the factor in parentheses is the probability that the particle did not decay in the time dt. This quickly gives the differential equation, and solution,

$$\frac{dP}{dt} = -w\,P(t) \quad\Rightarrow\quad P(t) = e^{-wt}, \tag{26.6.22}$$

from which we identify, as claimed, that

$$\tau = \frac{1}{w}. \tag{26.6.23}$$

The tunneling rate w, or tunneling probability per unit time, is estimated by first computing the *hit rate* n_{hit}: the number of times the classical particle, bouncing between the turning points a and b, hits the barrier at b per unit time. The hit rate is multiplied by the tunneling probability T to give w:

$$w = n_{\text{hit}}\,T = \frac{1}{\Delta t}\,T. \tag{26.6.24}$$

Here Δt is the time the classical particle takes to go from b to a and back to b:

$$\Delta t = 2\int_a^b \frac{dx}{v(x)} = 2m\int_a^b \frac{dx}{p(x)}. \tag{26.6.25}$$

We can now put all the pieces together to get the lifetime. Using the WKB approximation for T, we find

$$\tau = \frac{\Delta t}{T} \simeq 2m\int_a^b \frac{dx}{p(x)}\cdot\exp\Big(2\int_b^c \kappa(x)dx\Big). \tag{26.6.26}$$

The smaller the tunneling probability, the larger the lifetime of the localized state. Note that the classical prefactor Δt carries the units of the result. We must view the above result as an order-of-magnitude estimate of the lifetime since the WKB approximation for T is itself an order-of-magnitude approximation. Often, the exact value of the lifetime is not that important, and there is plenty of insight to be gained from the dependence of the lifetime on the various parameters of the theory.

26.7 Double-Well Potentials

We will consider here one-dimensional potentials that have two wells; double-well potentials. The potentials will be even, and the two wells, centered at $x = \pm a$, are separated by a large barrier or hump in the region around $x = 0$. To first approximation, the potential near $x = a$ and near $x = -a$ is that of a simple harmonic oscillator. The ground state wave function will be even, and when the barrier is large, intuition tells us that near $x = \pm a$ it will be approximately equal to the ground state wave function of each separate oscillator, thus forming a symmetric combination. The first excited state will then be, approximately, an antisymmetric combination of the separate ground state wave functions. The ground state and the excited state will be exactly symmetric and antisymmetric, respectively, but their description as linear combinations of localized ground states is only approximate. For a large barrier in between the wells, the energies of these states differ by a very small amount—an amount that is in fact supressed exponentially by the effect of the barrier. The goal of this section is compute the splitting of these two levels. We will use the WKB approximation to find this splitting. As the analysis below will make clear, there are a few subtleties along the way. The result will be powerful: we will find not only the exponential suppression of the splitting but also the prefactor multiplying this exponential!

To set the stage for the calculation, we first discuss a couple of double-well potentials $V_1(x)$ and $V_2(x)$. Associated to these potentials are the Hamiltonians $\hat{H}_1$ and $\hat{H}_2$, simply given by

$$\hat{H}_i = -\frac{\hbar^2}{2m}\frac{d^2}{dx^2} + V_i(x), \quad i=1,2. \tag{26.7.1}$$

The potentials are as follows:

$$V_1(x) = \frac{1}{2}m\omega^2(|x|-a)^2, \quad V_2(x) = \frac{m\omega^2}{8a^2}(x^2-a^2)^2. \tag{26.7.2}$$

As expected, both potentials are even and have wells centered at $x=\pm a$ (figure 26.14). For $x \geq 0$, the potential V_1 is just a parabola with its vertex at $x=a$, and for $x<0$, the potential is a parabola with its vertex at $x=-a$. The potential V_1 has a cusp at $x=0$, where the two parabolas meet. Both for $x>0$ and for $x<0$, the potential V_1 is exactly a quadratic potential, with oscillation frequency ω. This potential is shown in figure 26.14. The potential V_2 is smooth, but near $x=\pm a$ it is not an exact quadratic potential. Expanded around $x=a$, for example, the Taylor series terminates at quartic order and takes the following exact form:

$$V_2(x) = \frac{1}{2}m\omega^2(x-a)^2 + \frac{m\omega^2}{2a}(x-a)^3 + \frac{m\omega^2}{8a^2}(x-a)^4. \tag{26.7.3}$$

The first term in $V_2(x)$ makes it clear that, to first approximation, the well at $x=a$ defines an oscillator of frequency ω. This is also clear directly from (26.7.2), letting $x \sim a$. But there is also anharmonicity, as shown by the cubic and quartic terms in V_2. To better appreciate the nature of the potentials, let us introduce the unit-free coordinate u and a unit-free parameter λ that encodes the value of a making use of the oscillator size L_0:

$$x = L_0 u, \quad L_0^2 = \frac{\hbar}{m\omega}, \quad \lambda \equiv \frac{a}{L_0}. \tag{26.7.4}$$

In this language the potentials take the simpler form

$$V_1(x) = \frac{\hbar\omega}{2}(|u|-\lambda)^2, \quad V_2(x) = \frac{\hbar\omega}{8}\frac{1}{\lambda^2}(u^2-\lambda^2)^2. \tag{26.7.5}$$

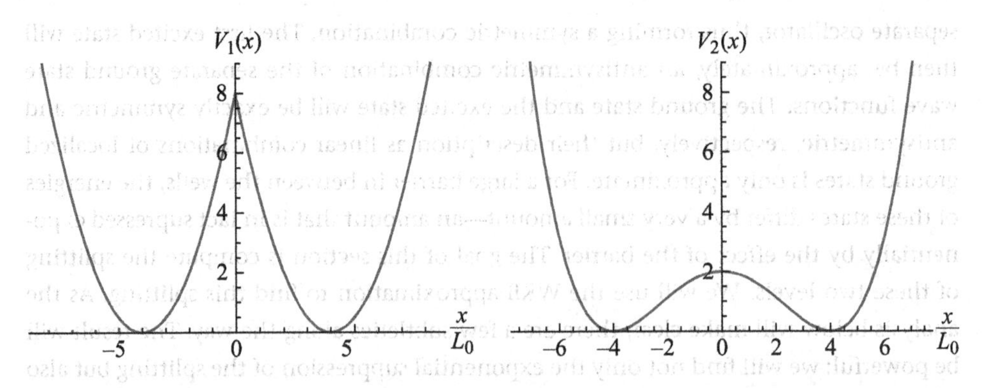

Figure 26.14
Left: The potential $V_1(x)$ is constructed from two parabolas, with vertices at $x = \pm a$ (here $a = 4L_0$), combined with a cusp at $x = 0$. *Right*: The potential $V_2(x)$ is quartic, with wells at $x = \pm a$ (here $a = 4L_0$). Both potentials have the same quadratic behavior near $x = \pm a$.

As λ increases the separation between the wells increases. Not only that, but the height of the barrier, as measured by the value of the potential at $x = 0$, also increases with λ. Indeed, from the previous expressions we immediately see that

$$V_1(0) = \frac{\hbar\omega}{2}\lambda^2, \quad V_2(0) = \frac{\hbar\omega}{8}\lambda^2. \tag{26.7.6}$$

We work in the large λ approximation, where both $V_1(0)$ and $V_2(0)$ are much larger than $\hbar\omega$. Note, however, that for a given value of λ the bump in the "cuspy" potential V_1, as measured by the value of $V_1(0)$, is four times higher than the bump in the quartic potential V_2. For a given λ, we expect the splitting to be smaller for V_1 than for V_2.

Let us consider the potential V_2. Focusing on the well centered at $x = a$, we claim that

$$|x - a| \ll a \;\Rightarrow\; \text{small anharmonic corrections.} \tag{26.7.7}$$

To confirm this, we look at the magnitude of the cubic term in (26.7.3) and see that it is suppressed relative to the quadratic term when the above inequality holds:

$$\frac{m\omega^2}{2a}|x - a|^3 = \frac{m\omega^2}{2a}(x - a)^2\,|x - a| \ll \frac{m\omega^2}{2a}(x - a)^2\, a = \tfrac{1}{2}m\omega^2\,(x - a)^2. \tag{26.7.8}$$

It is also clear that the small anharmonicity condition is also satisfied when $|x - a| \gtrsim L_0$—that is, when x is within a few L_0's away from a. This follows because in the large λ approximation $L_0 \ll a$.

It follows from the above discussion that separately computing the energies of the ground state and the excited state is very difficult. The separation between the two levels is expected to be exponentially suppressed in terms of λ, but corrections due to anharmonicity are perturbative in $1/\lambda$, as can be seen by a quick rewriting of the potential (26.7.3), which gives

$$V_2(u) = \tfrac{1}{2}\hbar\omega(u - \lambda)^2 + \tfrac{1}{2}\hbar\omega\tfrac{1}{\lambda}\,(u - \lambda)^3 + \tfrac{1}{8}\hbar\omega\tfrac{1}{\lambda^2}(u - \lambda)^4. \tag{26.7.9}$$

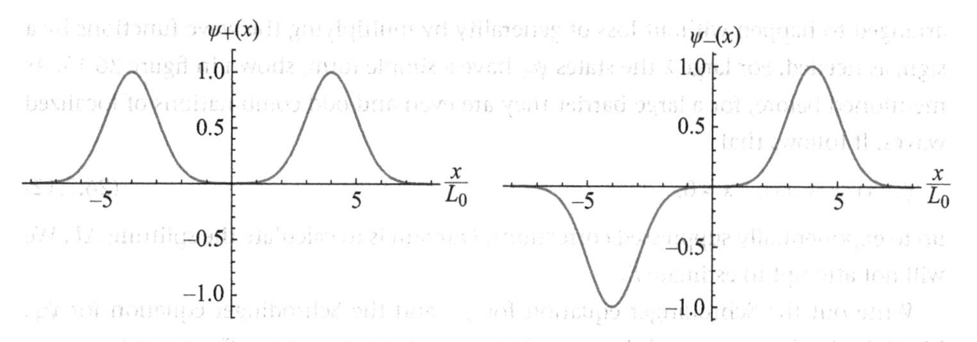

Figure 26.15

Left: The ground state $\psi_+(x)$ of a double well is roughly the even function built by adding oscillator ground states centered at $x=\pm a$ (shown for $a=4L_0$). *Right*: The first excited state $\psi_-(x)$ of a double well is roughly the odd function built by adding to the oscillator ground state centered at $x=a$, *minus* the ground state centered at $x=-a$. This approximation does not reflect the effect of the barrier and is not accurate near $x=0$.

Thus, an accurate calculation of the separate energy levels would require evaluating perturbative corrections to the accuracy of the nonperturbative splittings, something that is generally not doable. We claim, however, that the computation of the splitting is not affected by perturbative corrections. The challenge is to do the computation in a way that this is manifest. Below we follow an approach suggested by Landau and Lifshitz and elaborated by M. Metlitski.

As mentioned at the beginning of the section, the ground state and the first excited state of a double well are even and odd functions, respectively. We call them ψ_+ and ψ_-, with the subscripts reflecting their parity under $x\to -x$. For large barriers, these wave functions are approximated by the sum and the difference of separate ground state wave functions on the wells (figure 26.15). Since these linear combinations do not take the barrier into account, $\psi_\pm$ cannot be expected to be an accurate description of the wave functions near $x=0$. Yes, the wave functions there are very small and so they are in this ansatz, but if we need those small numbers, the error is certainly large. The wave functions are quite accurate, however, near $x=\pm a$.

Our analysis will be completely general, independent of the particular form of the double-well potential, assuming only that the potential is even, the wells are centered at $x=\pm a$, and the oscillation frequency at the wells is ω. At the end, we will specialize to V_1 and then V_2. We begin our work by considering two exact solutions $\psi_\pm(x)$ of a double-well potential:

$$\hat{H}\psi_\pm = E_\pm\,\psi_\pm. \tag{26.7.10}$$

Here, $\psi_\pm(-x)=\pm\psi_\pm(x)$, with ψ_+ and ψ_- the ground state and the first excited state. We will write

$$E_\pm = \bar{E}\mp\tfrac{1}{2}\Delta E, \quad \Delta E>0 \tag{26.7.11}$$

so that the gap between the two states is $\Delta E=E_- - E_+$. The two states $\psi_\pm$ are orthonormal. We will also assume that both wave functions are positive for large x; this can be arranged to happen without loss of generality by multiplying the wave functions by a sign, as needed. For large λ the states $\psi_\pm$ have a simple form, shown in figure 26.15. As mentioned before, for a large barrier they are even and odd combinations of localized waves. It follows that

$$\psi_+(x) \simeq \psi_-(x), \quad x > 0, \tag{26.7.12}$$

up to exponentially suppressed corrections. Our aim is to calculate the splitting ΔE. We will not attempt to estimate $\bar{E}$.

Write out the Schrödinger equation for ψ_- and the Schrödinger equation for ψ_+. Multiply the first by ψ_+ and the second by ψ_- and subtract them. The result is

$$-\frac{\hbar^2}{2m}(\psi_+\psi_-'' - \psi_-\psi_+'') = \Delta E\, \psi_+\psi_-. \tag{26.7.13}$$

Recognizing a total derivative on the left-hand side and integrating from zero to infinity,

$$-\frac{\hbar^2}{2m}\int_0^\infty dx \frac{d}{dx}(\psi_+\psi_-' - \psi_-\psi_+') = \Delta E \int_0^\infty dx\, \psi_+\psi_-. \tag{26.7.14}$$

The wave functions vanish at infinity. Moreover, since $\psi_-(0) = \psi_+'(0) = 0$, the second term on the left-hand side does not contribute. We thus get

$$\frac{\hbar^2}{2m}\psi_+(0)\psi_-'(0) = \Delta E \int_0^\infty dx\, \psi_+\psi_-. \tag{26.7.15}$$

So far no approximations have been made. Both the values of $\psi_+(0)$ and $\psi_-'(0)$ are expected to be exponentially suppressed in λ, making ΔE exponentially suppressed. Near $x = 0$, the even wave function $\psi_+(x)$ is just a very small constant, while the odd wave function $\psi_-(x)$ is linear in x with a very small coefficient. These features could only be gleaned in figure 26.15 with major magnification of the $x = 0$ region.

We now explain that the integral on the right-hand side of (26.7.15) is in fact equal to $\frac{1}{2}$, up to exponential corrections. Such corrections do not matter, since they are exponential corrections to an already exponentially suppressed ΔE. The argument is simple: for $x > 0$, as noted in (26.7.12), $\psi_-(x) \simeq \psi_+(x)$ up to exponential corrections. Thus, $\int_0^\infty \psi_+\psi_- dx \simeq \int_0^\infty \psi_+\psi_+ dx = \frac{1}{2}$ since ψ_+ is even and normalized. It follows that the equation above establishes that

$$\Delta E \simeq \frac{\hbar^2}{m}\psi_+(0)\psi_-'(0). \tag{26.7.16}$$

This is the relation we were looking for.

The next step is to write WKB-type solutions for ψ_+ and ψ_- that will be applicable well within the forbidden region. Since our focus is on single-well ground states, with energy $E = \hbar\omega/2$, the forbidden region extends from $x_1 = a - L_0$ down to $-x_1$. In this estimate we assume, correctly, negligible anharmonicity, and we recall that the classical amplitude of an oscillator in the ground state is L_0. We record the position of the $x > 0$ turning point:

$$x_1 = a - L_0. \tag{26.7.17}$$

To be well within the forbidden region, we need to be a number of L_0's to the left of x_1 as well as a number of L_0's to the right of $-x_1$. We then write the WKB ansatzes:

$$\begin{aligned}\psi_+(x) &\simeq \frac{c_+}{\sqrt{\kappa(x)}}\cosh\Big(\int_0^x \kappa(x')dx'\Big),\\ \psi_-(x) &\simeq \frac{c_-}{\sqrt{\kappa(x)}}\sinh\Big(\int_0^x \kappa(x')dx'\Big),\end{aligned} \tag{26.7.18}$$

with $c_\pm$ constants to be determined. These constants are exponentially suppressed, as they largely determine the values of the wave functions near $x=0$. As usual, we have $\kappa^2 = 2m(V(x)-E)/\hbar^2$, with $E=\hbar\omega/2$, the ground state energy of the oscillator defined by the quadratic term in the wells. The error this induces will be an exponentially small correction to the already suppressed coefficients $c_\pm$. These expressions have the appropriate symmetry under $x\to -x$: the integrals are odd in x, and the hyperbolic functions make ψ_+ even and ψ_- odd. The chosen plus signs for the arguments of the hyperbolic functions guarantee that, as expected, the wave functions grow as x grows away from zero. Note that $\psi_+(0)=c_+/\sqrt{\kappa(0)}$, and $\psi_-'(0)=c_-\sqrt{\kappa(0)}$ so that $\psi_+(0)\psi_-'(0)=c_+c_-$, and equation (26.7.16) becomes

$$\Delta E \simeq \frac{\hbar^2}{m}\, c_+c_-. \tag{26.7.19}$$

We now write solutions valid near $x=\pm a$ and then attempt connection to the above WKB expressions. We claim that

$$\psi_\pm = \tfrac{1}{\sqrt{2}}\big(\varphi_0(x-a)\pm\varphi_0(x+a)\big), \tag{26.7.20}$$

with $\varphi_0(x)=N_0\exp(-\frac{x^2}{2L_0^2})$, the oscillator ground state wave function, and $N_0^2=\frac{1}{\sqrt{\pi}L_0}$. This is properly normalized, up to exponentially suppressed corrections. The above expressions for $\psi_\pm$ are valid as long as anharmonic corrections are small. Focusing on $x>0$, this holds as long as $|x-a|\ll a$, as noted in (26.7.7). On the other hand, the WKB under-the-barrier expressions in (26.7.18) are valid for $|x-a|\gg L_0$ while x is to the left of a. The two conditions,

$$L_0 \ll |x-a| \ll a, \tag{26.7.21}$$

can clearly be satisfied simultaneously because $L_0\ll a$ in the large λ approximation. The region of simultaneous validity of the conditions is shown in figure 26.16.

Matching of the two expressions for $\psi_\pm$ is thus possible. In considering the hyperbolic functions, we must take the leading exponentials in both cosh and sinh. Working with $x>0$, only the ground state centered at $x=a$ is relevant; the ground state centered at $x=-a$ is exponentially supressed. Therefore, the matching of (26.7.18) and (26.7.20) at some point x satisfying conditions (26.7.21) requires that

$$\frac{c_\pm}{2\sqrt{\kappa(x)}}\exp\Big(\int_0^x \kappa(x')dx'\Big) \simeq \frac{1}{\sqrt{2}}\varphi_0(x-a). \tag{26.7.22}$$

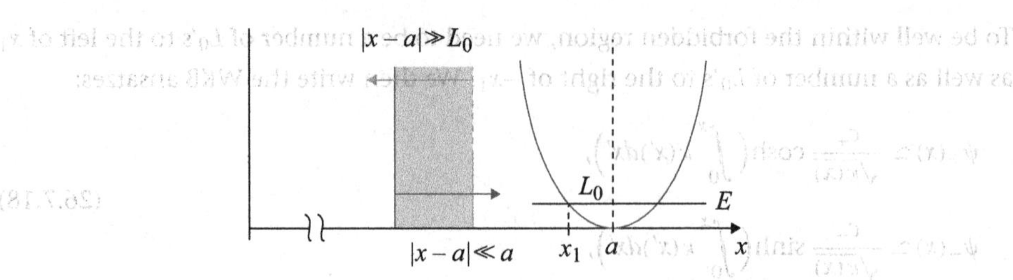

Figure 26.16
The wave function built from single-well oscillator ground states holds accurately when x satisfies $|x-a| \ll a$—that is, when the distance to the well at a is small compared to a (*the region selected by the right-pointing arrow*). The WKB solutions hold when x is well into the forbidden region $x < x_1$ for the energy E. This requires $|x-a| \gg L_0$ (*the region selected by the left-pointing arrow*). The region where both conditions hold is shown shaded.

Using the expression for the oscillator ground state,

$$\frac{c_\pm}{2\sqrt{\kappa(x)}}\exp\Big(\int_0^x \kappa(x')dx'\Big) \simeq \frac{1}{\pi^{1/4}\sqrt{2L_0}}\exp(-\tfrac{1}{2L_0^2}(x-a)^2). \tag{26.7.23}$$

The challenge now is to rewrite the exponential on the left-hand side in order to show that all the x dependence in this equation cancels. This will show that the $c_\pm$ are indeed constants and will determine their values. It is already clear from this expression, however, that $c_+ \simeq c_-$; the constants are the same in the large λ approximation. The strategy is to replace the integral from zero to x with an integral from zero to $x_1 = a - L_0$, the classical turning point for the ground state energy to the left of a (see (26.7.17)), and then to subtract the added piece:

$$\int_0^x \kappa(x')dx' = \int_0^{x_1} \kappa(x')dx' - \int_x^{x_1} \kappa(x')dx'. \tag{26.7.24}$$

This leads to

$$\exp\Big(\int_0^x \kappa(x')dx'\Big) = \exp\Big(\int_0^{x_1} \kappa(x')dx'\Big)\cdot\exp\Big(-\int_x^{x_1} \kappa(x')dx'\Big). \tag{26.7.25}$$

Focus on the right-hand side. The integral in the first exponential is x independent. It is a constant that we do not need to calculate *yet*, but it depends on the potential all the way from x_1 down to $x=0$. The integral in the second exponential needs work. For x' in its region of integration, the potential is just $V(x') \simeq \frac{1}{2}m\omega^2(x'-a)^2$ since this is the region where anharmonic corrections are small. We thus have

$$\begin{aligned}\int_x^{x_1} \kappa(x')dx' &= \int_x^{x_1} dx'\sqrt{\tfrac{2m}{\hbar^2}\big(\tfrac{1}{2}m\omega^2(x'-a)^2 - \tfrac{1}{2}\hbar\omega\big)} \\ &= \int_u^{u_1} du\sqrt{(\lambda-u)^2-1} = \int_1^{\lambda-u} dt\sqrt{t^2-1},\end{aligned} \tag{26.7.26}$$

where we first passed to unit-free variables by setting $x' = L_0 u$ and $x_1 = L_0 u_1$, giving $u_1 = \lambda - 1$, and then changed the integration variable to $t = \lambda - u$. The first inequality in (26.7.21) implies that $a-x \gg L_0$, and therefore $\lambda - u \gg 1$. The final integral above can be done analytically and the result expanded for a large upper limit. Indeed, in general

$$\begin{aligned}\int_1^{\Lambda} dt\sqrt{t^2-1} &= \tfrac{1}{2}\Lambda\sqrt{\Lambda^2-1} - \tfrac{1}{2}\ln[\Lambda+\sqrt{\Lambda^2-1}] \\ &\simeq \tfrac{1}{2}\Lambda^2 - \tfrac{1}{2}\ln\Lambda - \tfrac{1}{4} - \ln\sqrt{2} + \mathcal{O}(\Lambda^{-2}), \quad \Lambda \gg 1.\end{aligned} \tag{26.7.27}$$

Applied to our case of interest, we see that

$$\begin{aligned}\exp\Big(-\int_x^{x_1}\kappa(x')dx'\Big) &\simeq \exp\Big(-\tfrac{1}{2}(\lambda-u)^2 + \tfrac{1}{2}\ln(\lambda-u) + \tfrac{1}{4} + \ln\sqrt{2}\Big) \\ &= e^{1/4}\sqrt{2}\,\sqrt{\lambda-u}\,\exp(-\tfrac{1}{2}(\lambda-u)^2).\end{aligned} \tag{26.7.28}$$

Back in (26.7.23) we have

$$\frac{c_\pm e^{1/4}\sqrt{2}}{2\sqrt{\kappa(x)}}\sqrt{\lambda-u}\,\exp\Big(\int_0^{x_1}\kappa(x')dx'\Big)\,e^{-\frac{1}{2}(\lambda-u)^2} \simeq \frac{e^{-\frac{1}{2}(\lambda-u)^2}}{\pi^{1/4}\sqrt{2L_0}}. \tag{26.7.29}$$

Canceling the common Gaussians and recalling that $\kappa(x) = \frac{1}{L_0}\sqrt{(\lambda-u)^2-1} \simeq \frac{1}{L_0}(\lambda-u)$, we see that all the u dependence cancels, as required. We therefore have

$$\frac{c_\pm e^{1/4}\sqrt{L_0}}{\sqrt{2}}\exp\Big(\int_0^{x_1}\kappa(x')dx'\Big) \simeq \frac{1}{\pi^{1/4}\sqrt{2L_0}}. \tag{26.7.30}$$

Finally, we can solve for constants $c_\pm$:

$$c_\pm \simeq \frac{1}{(\pi\, e)^{1/4}L_0}\exp\Big(-\int_0^{x_1}\kappa(x')dx'\Big). \tag{26.7.31}$$

This result, at last, determines the energy splitting ΔE. Using (26.7.19), we find that

$$\Delta E \simeq \frac{\hbar^2}{m}\frac{1}{(\pi\, e)^{1/2}L_0^2}\exp\Big(-2\int_0^{x_1}\kappa(x')dx'\Big). \tag{26.7.32}$$

Replacing the explicit value of L_0^2 gives the general formula for the splitting ΔE:

$$\boxed{\Delta E \simeq \frac{\hbar\omega}{\sqrt{\pi\, e}}\exp\Big(-2\int_0^{x_1}\kappa(x')dx'\Big).} \tag{26.7.33}$$

The splittings of higher states of the double well can be computed with identical methods, the calculations taking more effort. The result for the lowest states, as given above, is still not that easy to use for a given potential. The evaluation of the integral is a bit delicate; it generally gives a term plus the logarithmic correction. The first gives the exponential suppression, and the second affects the prefactor. Both are important to get the final explicit result for the energy splitting.

Let us carry out this last step for the potential V_1 with partial parabolas and a cusp. Happily, the work was done in equation (26.7.26) because in fact $V(x) = \frac{1}{2}m\omega^2(x-a)^2$ holds exactly for any $x > 0$. We only need to set the lower limit of integration equal to zero; that is, $x=0$ and therefore $u=0$. This can be done all the way to (26.7.28), giving

$$\exp\Big(-2\int_0^{x_1}\kappa(x')dx'\Big)\simeq e^{1/2}\,2\lambda\,e^{-\lambda^2}. \tag{26.7.34}$$

Returning to the general formula (26.7.33), we get

$$\Delta E_1 \simeq \hbar\omega\,\frac{2\lambda}{\sqrt{\pi}}\,e^{-\lambda^2},\quad \lambda=\sqrt{\frac{2V_1(0)}{\hbar\omega}}. \tag{26.7.35}$$

The subscript '1' on ΔE is for V_1. Interestingly, a variational calculation manages to reproduce this result. Moreover, the value of ΔE_1 can be compared with the splittings that one gets by solving the Schrödinger equation numerically with the shooting method. The agreement is remarkably good. These matters are discussed in problem 26.8. Additionally, here the perturbative corrections to the levels go to zero with increasing λ so that for large λ one gets energy levels $\frac{1}{2}\hbar\omega\pm\frac{1}{2}\Delta E_1$.

For the quartic potential $V_2(x)$, the requisite integral is more challenging since this time the potential is not parabolic for $x>0$. This evaluation is done in problem 26.9, and the result for the splitting of the ground states is

$$\Delta E_2 \simeq \hbar\omega\,\frac{4\lambda}{\sqrt{\pi}}\,e^{-\frac{2}{3}\lambda^2},\quad \lambda=\sqrt{\frac{8V_2(0)}{\hbar\omega}}. \tag{26.7.36}$$

As noted before, for V_2 one must take larger λ to achieve a high barrier. Numerical work must be done with high precision. Setting $\lambda=3$, one finds results consistent with the above splittings with an error of about 4.3%. This time, the perturbative anharmonic corrections are gigantic compared to the splittings. Both levels are pushed down.

Problems

Problem 26.1. *The equation satisfied by the approximate WKB solution.*

In trying to solve the equation

$$-\hbar^2\psi''=p^2(x)\psi, \tag{1}$$

we wrote the approximate solution $\psi_a(x)$ given by

$$\psi_a(x)=\frac{1}{\sqrt{p(x)}}\exp\Big(\frac{i}{\hbar}\int^x p(x')dx'\Big).$$

1. Find the *exact* differential equation satisfied by the approximate solution, and show it can be written as

   $$-\hbar^2\psi_a''=\big[p^2(x)+\hbar^2(\cdots)\big]\,\psi_a,$$

   where the dots represent additional terms, not present in (1), involving functions of $p(x)$ and its derivatives. Determine those terms.
2. Consider the extra terms you found above, and explore the condition that each one is much smaller than $p^2(x)$. Express the conditions as two independent constraints involving the local de Broglie wavelength $\lambda(x)$ and its derivatives.

Problem 26.2. *Quantum mechanics of a bouncing ball (Griffiths).*

A particle of mass m can move vertically, bouncing off a table under the influence of a gravitational potential that gives it a constant acceleration g_0.

1. Find the WKB approximation to the allowed energies E_n of the particle with $n=1$, $2, \ldots$. Write your answer in terms of n, m, g_0, and $\hbar$.
2. Compare your WKB results with the energies E_n arising from the linear potential $V(x)=gx$, plus a wall at $x=0$. As given in (6.7.17),

$$E_n = |a_n| \left(\frac{\hbar^2 g^2}{2m}\right)^{1/3},$$

with a_n the nth zero of $\text{Ai}(u)$, with the zeroes ordered by increasing magnitude.

The WKB answer provides an approximation $\tilde{a}_n$ for the zeroes a_n of the Airy function. Write a formula for $\tilde{a}_n$ as a function of n. Find the relative errors e_n given by

$$e_n \equiv \left|\frac{\tilde{a}_n - a_n}{a_n}\right|, \quad \text{for } n=1,2,3.$$

3. Estimate in eV the ground state energy of a neutron on a horizontal surface in the earth's gravitational field. (See V. V. Nesvizhevsky et al., Nature **415**, 297 (2002) and arXiv:hep-ph/0306198 for an experimental measurement of the quantum mechanical ground state energy for neutrons bouncing on a horizontal surface in the earth's gravitational field.)
4. Now imagine dropping a ball of mass one gram from a height of two meters and letting it bounce. Compute the classical energy of the ball. The quantum mechanical state corresponding to a ball following this classical trajectory must be a coherent superposition of energy eigenstates, with mean energy equal to the classical energy. How large is the mean value $\bar{n}$ of the quantum number n in this state?

Problem 26.3. *WKB quantization of the energy with two turning points.*

Show that for the energy eigenstates of a potential with two turning points a and b (figure 26.5), the quantization condition for the energy in the WKB approximation takes the form given in equation (26.3.9):

$$\int_a^b k(x')dx' = \left(n+\tfrac{1}{2}\right)\pi, \quad n=0,1,2,\ldots.$$

Problem 26.4. *Semiclassical approximation of the quartic potential.*

Consider the Schrödinger equation for a particle of mass m moving in the quartic potential $V(x)=\alpha x^4$. Let the bound state energies be $E_0 < E_1 < \cdots$ and define the dimensionless energies $e_n = \frac{E_n}{\gamma}$ where $\gamma \equiv \left(\frac{\hbar^4 \alpha}{m^2}\right)^{1/3}$. Using the shooting method, the first few energies are found to be

$$e_0=0.667986,\ e_1=2.39364,\ e_2=4.69680,\ e_3=7.33573,\ e_4=10.2443,\ e_5=13.3793.$$

In this problem, we will estimate these energies using semiclassical methods.

1. Assume that the turning points are at $-x_0$ and x_0 with $x_0 > 0$. Express the energy E in terms of α and x_0. Set up the WKB quantization condition (26.3.9) to determine the energies E_n in the spectrum, with $n = 0, 1, \ldots$.
2. Let $\tilde{E}_n$ denote the estimate of the nth energy that is obtained from the WKB result, while E_n represents the true energy, as determined by the shooting method. Develop your result in (1) to obtain a formula for $\tilde{e}_n \equiv \tilde{E}_n/\gamma$ in terms of n. The answer should be of the form

$$\tilde{e}_n = \lambda(n+\delta)^{\epsilon},$$

where λ, δ, and ϵ are constants to be determined. You may find the following expression useful:

$$\int_0^1 \sqrt{1-t^4}\, dt = \frac{\sqrt{\pi}\,\Gamma(\frac{1}{4})}{8\Gamma(\frac{7}{4})} \simeq 0.874019.$$

Write down $\tilde{e}_0$, $\tilde{e}_1$, $\tilde{e}_2$, $\tilde{e}_3$, $\tilde{e}_4$, $\tilde{e}_5$ and the relative errors $\left|\frac{\tilde{e}_0-e_0}{e_0}\right|$, $\left|\frac{\tilde{e}_2-e_2}{e_2}\right|$, and $\left|\frac{\tilde{e}_5-e_5}{e_5}\right|$.

Problem 26.5. *Asymptotic expansion of Ai(u) for $u \ll -1$.*

When u is negative and large, the integral representation (26.4.17) of the Airy function Ai(u) can be evaluated in the stationary phase approximation. Verify that the phase of the integrand is stationary at $k = \pm\sqrt{-u} = \pm|u|^{1/2}$. Since the k integration contour passes through the stationary phase points, you can simply add their contributions to find the asymptotic expansion of Ai(u) for $u \ll -1$. Show that, as claimed in section 26.4,

$$\text{Ai}(u) \simeq \frac{1}{\sqrt{\pi}}\frac{1}{|u|^{1/4}}\cos\left(\tfrac{2}{3}|u|^{3/2} - \tfrac{\pi}{4}\right), \quad u \ll -1.$$

Problem 26.6. *Field electron emission in WKB.*

We consider the emission of electrons from a metal induced by an electrostatic field normal to the surface (figure 26.17).

Assume we have a metal for $x < 0$ and a vacuum for $x > 0$. The electrons cannot escape the metal because there is a work function $\Phi > 0$, representing the extra energy required

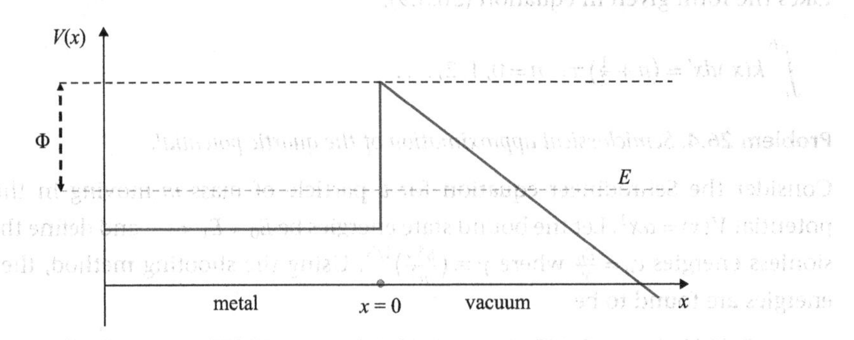

Figure 26.17
Emission of electrons from a metal.

for the electrons to escape. This work function can be viewed as a potential barrier of height Φ relative to the energy E of the electrons. Consider applying an electrostatic field of magnitude $\mathcal{E} > 0$ pointing in the $-x$ direction so that there is an *extra* potential term $V_{\mathcal{E}}(x)$ valid for $x > 0$ and given by

$$V_{\mathcal{E}}(x) = -e\mathcal{E}x.$$

There is no extra potential for $x < 0$. Calculate the WKB tunneling probability T for electrons to escape. Leave your answer in terms of $\Phi, \mathcal{E}, e, m_e, \hbar$. Your result is the *Fowler-Nordheim* formula. Sketch the dependence of T on the field magnitude $\mathcal{E}$.

Problem 26.7. *Tunneling and the Stark effect.*

The Stark effect concerns the physics of an atom in an electric field. Here you will explore the possibility that when placed in an electric field, the electron in an atom can tunnel out, making the atomic bound states unstable. We consider this effect in a simpler one-dimensional analogue.

Consider an electron trapped in a one-dimensional square well of depth V_0 and width d:

$$V(x) = \begin{cases} -V_0, & \text{for } |x| < d/2, \\ 0, & \text{for } |x| \geq d/2. \end{cases}$$

Suppose we turn on a weak constant electric field in the negative x-direction with magnitude $\mathcal{E}$. With this field present, the potential experienced by the electron becomes $V_{\mathcal{E}}(x)$, given by

$$V_{\mathcal{E}}(x) = V(x) - e\mathcal{E}x, \quad (e > 0).$$

Assume throughout this problem that the two following inequalities hold:

$$\text{Inequality A:} \quad \frac{\hbar^2}{2md^2} \ll V_0, \qquad \text{Inequality B:} \quad e\mathcal{E}d \ll \frac{\hbar^2}{2md^2}.$$

It is worth reflecting on what these inequalities mean.

1. Set $\mathcal{E} = 0$ in this part of the problem. Estimate the ground state energy E_1 by recalling that the well is very deep (Inequality A). Is the true ground state energy (still with $\mathcal{E} = 0$) lower or higher than what you've estimated? Explain clearly.
2. Sketch the potential $V_{\mathcal{E}}(x)$ with $\mathcal{E} \neq 0$. Are there bound states in this potential?
3. Use the semiclassical approximation to calculate the tunneling probability T for the ground state. [Hint: use inequalities A *and* B to simplify this part of the problem and to conclude that the particle energy is approximately $-V_0$.] As usual, write the answer in the form $T = \exp(\cdots)$, where the dots indicate an expression that depends on $\hbar, m, V_0, e$, and $\mathcal{E}$.
4. Use classical arguments to convert the tunneling probability T into an estimate of the lifetime τ of the bound state. The answer can be written in the form $\tau = \Delta t/T$. Determine Δt in terms of m, d, $\hbar$, and numerical constants. (Ignore the effect of the perturbation for the purpose of this computation.)

5. Now consider numbers that are characteristic of an atomic system. Calculate the lifetime τ for $V_0 = 20$ eV, $d = 2 \times 10^{-8}$ cm, and an electric field $\mathcal{E} = 7 \times 10^4$ V/cm. Write your answer in the form $\tau \simeq \exp(\beta)$s. Write the age of the universe τ_u in a similar fashion: $\tau_u \simeq \exp(\gamma)$s. Write $\tau = \exp(\delta)\tau_u$ and determine δ.
6. How does the decay rate $w = 1/\tau$ depend on the perturbation $\mathcal{E}$? How do w and its derivatives with respect to $\mathcal{E}$ behave as $\mathcal{E} \to 0^+$? Could w be calculated by treating $\mathcal{E}$ as a perturbation to the Hamiltonian?

Problem 26.8. *Variational analysis of a double-well potential with a cusp (Merzbacher).*

Consider the dynamics of a mass m particle governed by a Hamiltonian $\hat{H}$ that includes the even double-well potential $V_1(x) = \frac{1}{2}m\omega^2(|x| - a)^2$ discussed in section 26.7. As in that section, $\lambda \equiv \frac{a}{L_0}$, $L_0^2 \equiv \frac{\hbar}{m\omega}$. The objective of this problem is to find the splitting between the ground and the first excited states of the double well using a variational method. We use trial wave functions $\psi_\pm$ with $\pm$ denoting the parity of the wave function:

$$\psi_\pm(x) = N_\pm\big(\varphi_0(x-a) \pm \varphi_0(x+a)\big),$$

where $\varphi_0(x) = N_0 \exp(-\frac{1}{2}\frac{x^2}{L_0^2})$, with $N_0^2 = \frac{1}{\sqrt{\pi}L_0}$, is the normalized oscillator ground state.

1. Show that $\psi_\pm$ is normalized if $N_\pm$ satisfies $N_\pm^2 = \frac{1}{2}\frac{1}{1 \pm e^{-\lambda^2}}$.
2. Show that the expectation value $\langle \hat{H} \rangle_\pm$ of the Hamiltonian in $\psi_\pm$ is given by

$$\langle \hat{H} \rangle_\pm = \frac{\alpha_0 \pm \beta_0}{1 \pm e^{-\lambda^2}}, \quad \alpha_0 \equiv \int dx \varphi_0(x-a)\hat{H}\varphi_0(x-a), \quad \beta_0 \equiv \int dx \varphi_0(x+a)\hat{H}\varphi_0(x-a).$$

3. Evaluate the constants α_0 and β_0, showing that

$$\alpha_0 = \hbar\omega\Big[\frac{1}{2} + \frac{2\lambda}{\sqrt{\pi}}\int_{-\infty}^{0} u e^{-(u-\lambda)^2} du\Big], \quad \beta_0 = \hbar\omega\Big[\frac{1}{2} - \frac{\lambda}{\sqrt{\pi}}\Big]e^{-\lambda^2}.$$

You can verify that the integral above can be written as

$$\int_{-\infty}^{0} u e^{-(u-\lambda)^2} du = -\tfrac{1}{2}e^{-\lambda^2} + \lambda\int_{\lambda}^{\infty} dv e^{-v^2}.$$

Derive or look up the expansion of the last integral for large λ and then show that

$$\alpha_0 = \hbar\omega\Big[\frac{1}{2} - \frac{1}{2\sqrt{\pi}}\frac{1}{\lambda}e^{-\lambda^2} + \mathcal{O}\big(\tfrac{e^{-\lambda^2}}{\lambda^3}\big)\Big].$$

[Hint: It can help to rewrite $V(x) = \frac{1}{2}m\omega^2(x-a)^2 + 2m\omega^2 axG(x)$, with $G(x)$ a function that is equal to 1 for $x < 0$ and vanishes for $x > 0$.]

4. Writing $E_\pm = \langle \hat{H} \rangle_\pm$, which are good estimates for the energies of our wave functions, show that to leading order in large λ we have

$$E_\pm = \hbar\omega\Big[\frac{1}{2} \mp \frac{\lambda}{\sqrt{\pi}}e^{-\lambda^2}\Big] \Rightarrow \Delta E = \hbar\omega\frac{2\lambda}{\sqrt{\pi}}e^{-\lambda^2}. \tag{1}$$

This is exactly the same result obtained by the WKB analysis in (26.7.35).

5. If the $t=0$ wave function is $\psi(x;0)=\varphi_0(x-a)$, localized about $x=a$, what is the shortest positive time t_* for which $\psi(x,t_*)=\varphi_0(x+a)$, up to a phase?
6. Use the shooting method to compute the energies $E_\pm$ for various values of λ. For this you can work with $x>0$, where the Schrödinger equation with unit-free coordinate u and unit-free energy $e=E/(\hbar\omega)$ takes the form

$$-\frac{1}{2}\frac{d^2\psi}{du^2}-[e-\tfrac{1}{2}(u-\lambda)^2]\psi=0.$$

Integrate the equation from $u=0$, selecting appropriate initial conditions for even and odd solutions. Let $e_\pm$ be the unit-free energies and $\Delta e=e_- - e_+$. Confirm numerically that for $\lambda=3$ you get $\Delta e=4.1975\times10^{-4}$, which agrees with the predicted value from equation (1) to better than one part in two hundred. Confirm that the average energy of the two levels differs from $\frac{1}{2}$ by just about 3% of the magnitude of the splitting.

Problem 26.9. *The quartic double-well potential: evaluation of the energy splitting.*

1. The goal here is to give an evaluation of the energy splitting ΔE_2 for the potential $V_2(x)$ discussed in section 26.7. The potential is

$$V_2(x)=\frac{m\omega^2}{8a^2}(a^2-x^2)^2=\frac{\hbar\omega}{8}\frac{1}{\lambda^2}(\lambda^2-u^2)^2,$$

using the unit-free coordinate $u=x/L_0$, $\lambda=a/L_0$, and $L_0^2=\frac{\hbar}{m\omega}$. Given the general result (26.7.33), all we need to do is to compute the integral

$$I=\int_0^{x_1}\kappa(x)dx=\frac{1}{\hbar}\int_0^{x_1}dx\sqrt{2m(V_2(x)-E)},\quad E=\tfrac{1}{2}\hbar\omega,\quad x_1=a-L_0.$$

The strategy here is to choose some constant $b>0$ satisfying $L_0\ll b\ll a$ and to use it to split the integral into two pieces, each of which can be evaluated using different approximations. At the end one combines the results, which works consistently if the b dependence drops out (up to highly suppressed corrections). We thus write

$$I=\frac{1}{\hbar}\int_0^{a-b}dx\sqrt{2m(V_2(x)-E)}+\frac{1}{\hbar}\int_{a-b}^{x_1}dx\sqrt{2m(V_2(x)-E)}.$$

For the first integral, one must use the exact potential V_2, but $E\ll V_2$ so one can expand the square root, keeping the first correction:

$$I=\frac{1}{\hbar}\int_0^{a-b}dx\sqrt{2mV_2(x)}-\frac{1}{\hbar}\int_0^{a-b}dx\frac{\sqrt{m}E}{\sqrt{2V_2(x)}}+\frac{1}{\hbar}\int_{a-b}^{x_1}dx\sqrt{2m(V_2(x)-E)}.$$

Here, the first integral can be computed by integrating the exact V_2 in the range $[0,a]$ and then subtracting the integral over $[a-b,a]$, where V_2 can be approximated by the harmonic potential. The second integral also uses the exact potential. The third integral is fully within the region where V_2 is accurately harmonic. Show that at the end you find

$$I \simeq \frac{1}{\hbar}\int_0^a \sqrt{2mV_2(x)} - \frac{1}{2}\ln\left(\frac{2a}{L_0}\right) - \frac{1}{4} - \ln\sqrt{2},$$

up to corrections that vanish by letting $b/a \to 0$ and then $L_0/b \to 0$. Use the above result to show that the expression for ΔE_2 in (26.7.36) holds.

2. Perform a numerical analysis of the potential using the shooting method. With $e = E/(\hbar\omega)$, the unit-free energy, the differential equation is

$$-\frac{1}{2}\frac{d^2\psi}{du^2} - [e - \tfrac{1}{8\lambda^2}(u^2 - \lambda^2)^2]\psi = 0.$$

To work with this, it is useful to do numerics with a working precision of twenty-five digits. Take $\lambda = 6$, and confirm that the energy splitting and the prediction from (26.7.36) differ by an error of about 4.3%. Verify that the splitting is tiny compared with the perturbative corrections to the energies: the two levels are pushed down by an amount about fifteen million times larger!

27 Time-Dependent Perturbation Theory

We consider time-independent Hamiltonians $\hat{H}^{(0)}$ that are modified by time-dependent perturbations $\delta H(t)$. The analysis uses the interaction picture of quantum mechanics, where the time dependence generated by $\hat{H}^{(0)}$ is folded into operators, as in the Heisenberg picture, and the time dependence generated by $\delta H(t)$ is realized in the states, as in the Schrödinger picture. We develop the perturbative series in powers of $\delta H(t)$ and examine in detail first-order transition probabilities for familiar perturbations. For transitions between an initial discrete state and a final state that is part of a continuum, we obtain transition rates summarized by Fermi's golden rule. We also model explicitly and solve the Schrödinger equation for a discrete state decaying into a continuum via a constant perturbation. As applications, we consider the phenomenon of autoionization, hydrogen ionization by radiation, and absorption and stimulated emission rates in two-level atomic systems. We use Einstein's analysis of two-level atoms and blackbody radiation in thermal equilibrium to determine the rate for spontaneous emission.

27.1 Time-Dependent Hamiltonians

Time-dependent Hamiltonians are needed to study a number of physical processes. Spins in time-dependent magnetic fields and atoms subject to electromagnetic radiation are two familiar examples. The Schrödinger equation with time-dependent Hamiltonians is generally harder to solve than the Schrödinger equation with time-independent Hamiltonians. We have seen before that solutions for time-dependent Hamiltonians $\hat{H}(t)$ are sometimes written in terms of a unitary operator $\mathcal{U}(t)$ that can be used to evolve any physical state as a function of time. Alas, there is no simple expression for $\mathcal{U}(t)$, although it can be found as a formal power series in which each term is a nested integral of time-ordered products of the Hamiltonian $\hat{H}(t)$ (section 16.3).

A time-dependent $\hat{H}(t)$ does not have energy eigenstates. Recall that the existence of energy eigenstates is predicated on the factorization of solutions $\Psi(\mathbf{x}, t)$ of the full Schrödinger equation into a space-dependent part $\psi(\mathbf{x})$ and a time-dependent part that turned out to be $e^{-iEt/\hbar}$, with E the energy. Such factorization is only possible when the

Hamiltonian $\hat{H}$ is time independent. The equation for the energy eigenstates $\psi(\mathbf{x})$ is

$$\hat{H}\psi(\mathbf{x}) = E\psi(\mathbf{x}), \tag{27.1.1}$$

and $\Psi(\mathbf{x},t) = \psi(\mathbf{x})e^{-iEt/\hbar}$ solves the full Schrödinger equation $i\hbar\partial_t\Psi = \hat{H}\Psi$. If the Hamiltonian is time dependent, there are no factorized solutions, and this approach does not work. Mimicking (27.1.1), one could try finding some kind of time-dependent instantaneous eigenstates $\psi(\mathbf{x},t)$ satisfying

$$\hat{H}(t)\psi(\mathbf{x},t) = E(t)\psi(\mathbf{x},t), \tag{27.1.2}$$

for each value of t. The term *instantaneous eigenstates* is appropriate since at any time t they are eigenstates of the instantaneous Hamiltonian $\hat{H}(t)$. These are interesting states that will feature in our studies of the adiabatic approximation, but we note here that the $\psi(\mathbf{x},t)$ are *not* solutions of the Schrödinger equation $i\hbar\partial_t\Psi(\mathbf{x},t) = \hat{H}(t)\Psi(\mathbf{x},t)$, as you can quickly check. Moreover, there is no general procedure to construct such solutions $\Psi(\mathbf{x},t)$ starting from $\psi(\mathbf{x},t)$. In summary, for general time-dependent systems we do not have the apparatus of energy eigenstates or anything closely resembling energy eigenstates.

In this chapter we begin the study of time-dependent Hamiltonians in the framework of perturbation theory. In doing so, we will postulate that the time dependence arises as a perturbation. We will therefore assume that, as before, we have a Hamiltonian $\hat{H}^{(0)}$ that is known and is *time independent*. *Known* means we know the energy eigenstates and the energy eigenvalues. The perturbation to the Hamiltonian, denoted as $\delta H(t)$, will be time dependent, and as a result, the full Hamiltonian $\hat{H}(t)$ is also time dependent:

$$\hat{H}(t) = \hat{H}^{(0)} + \delta H(t). \tag{27.1.3}$$

While $\hat{H}^{(0)}$ has a well-defined spectrum, $\hat{H}(t)$ does not. Therefore, our goal is simply to find the time-dependent solutions of the time-dependent Schrödinger equation. Since we are going to focus on time dependence, we will suppress the labels associated with space and other degrees of freedom. We simply say we are trying to find the solution $|\Psi(t)\rangle$ to the Schrödinger equation

$$i\hbar\frac{\partial}{\partial t}|\Psi(t)\rangle = \left(\hat{H}^{(0)} + \delta H(t)\right)|\Psi(t)\rangle. \tag{27.1.4}$$

We have often written time-dependent states as $|\Psi, t\rangle$; here we will use the notation $|\Psi(t)\rangle$.

In many situations the perturbation $\delta H(t)$ vanishes for $t < t_i$, exists for some finite time, and then vanishes for $t > t_f$:

$$\underset{\quad}{\hat{H}^{(0)}} \;\big|_{t_i}\; \hat{H}^{(0)} + \delta H(t) \;\big|_{t_f}\; \hat{H}^{(0)} \longrightarrow t\,. \tag{27.1.5}$$

For $t < t_i$, the system is in an eigenstate of $\hat{H}^{(0)}$, or in a linear combination of $\hat{H}^{(0)}$ eigenstates. When the perturbation $\delta H(t)$ is small or $t_f - t_i$ is a short time, we can use time-dependent perturbation theory to determine the state of the system for $t > t_f$. Both initial and final states are nicely described in terms of eigenstates of $\hat{H}^{(0)}$ since this *is* the Hamiltonian for $t < t_i$ and $t > t_f$. Even while the time-dependent perturbation is on, we can use the eigenstates of $\hat{H}^{(0)}$ to describe and analyze the system. This is because these eigenstates form a complete basis, and therefore any time-dependent state can be written as a superposition of these eigenstates with *time-dependent* coefficients. We will discuss this further in the following section.

Many physical questions can be couched in the language of transitions between eigenstates due to time-dependent perturbations. Assume, for example, that we have a hydrogen atom in its ground state. We turn on an electromagnetic field for some time interval. We can then ask: What are the probabilities of finding the atom in each of the various excited states after the perturbation is turned off? This is a typical question addressed by time-dependent perturbation theory.

27.2 The Interaction Picture

In order to solve efficiently for the state $|\Psi(t)\rangle$, we will introduce the **interaction picture** of quantum mechanics. This picture uses some elements of the Heisenberg picture and some elements of the Schrödinger picture. We will use the known Hamiltonian $\hat{H}^{(0)}$ to define some Heisenberg operators, and the perturbation δH will be used to write a Schrödinger equation.

We begin by recalling some facts from the Heisenberg picture. For *any* Hamiltonian, time dependent or not, one can determine the unitary operator $\mathcal{U}(t)$ that generates time evolution as follows:

$$|\Psi(t)\rangle = \mathcal{U}(t)|\Psi(0)\rangle . \tag{27.2.1}$$

The Heisenberg operator $\hat{A}_H$ associated with a Schrödinger operator $\hat{A}_S$ is obtained by considering a rewriting of expectation values:

$$\langle\Psi(t)|\hat{A}_S|\Psi(t)\rangle = \langle\Psi(0)|\mathcal{U}^\dagger(t)\hat{A}_S\mathcal{U}(t)|\Psi(0)\rangle = \langle\Psi(0)|\hat{A}_H|\Psi(0)\rangle , \tag{27.2.2}$$

where

$$A_H \equiv \mathcal{U}^\dagger(t)\hat{A}_S\mathcal{U}(t) . \tag{27.2.3}$$

This definition applies even for Schrödinger operators with explicit time dependence. Note that the operator $\mathcal{U}^\dagger$ brings states to rest. It does so in the sense that applied to a state at time t it gives us the state at time zero:

$$\mathcal{U}^\dagger(t)|\Psi(t)\rangle = \mathcal{U}^\dagger(t)\mathcal{U}(t)|\Psi(0)\rangle = |\Psi(0)\rangle . \tag{27.2.4}$$

In our problem the known Hamiltonian $\hat{H}^{(0)}$ is time independent, and the associated unitary time evolution operator $\mathcal{U}_0(t)$ takes the simple form

$$\mathcal{U}_0(t) = e^{-i\hat{H}^{(0)}t/\hbar} . \tag{27.2.5}$$

The state $|\Psi(t)\rangle$ in our problem evolves through the agency of $\hat{H}^{(0)}$ and δH. Motivated by (27.2.4) we define the auxiliary ket $|\widetilde{\Psi}(t)\rangle$ as the ket $|\Psi(t)\rangle$ partially brought to rest through $\hat{H}^{(0)}$:

$$|\widetilde{\Psi}(t)\rangle \equiv e^{i\hat{H}^{(0)}t/\hbar}|\Psi(t)\rangle. \tag{27.2.6}$$

We should expect that the Schrödinger equation for $|\widetilde{\Psi}(t)\rangle$ will be simpler, as the above must have taken care of the time dependence generated by $\hat{H}^{(0)}$: if δH were zero, then $|\widetilde{\Psi}(t)\rangle$ would be simply a constant. Of course, if we can determine $|\widetilde{\Psi}(t)\rangle$ we can easily get back the desired state $|\Psi(t)\rangle$ by inverting the above relation to find that

$$|\Psi(t)\rangle = e^{-i\hat{H}^{(0)}t/\hbar}|\widetilde{\Psi}(t)\rangle. \tag{27.2.7}$$

This is intuitive: acting on $|\widetilde{\Psi}(t)\rangle$ with the evolution operator for $\hat{H}^{(0)}$ gives us the full state $|\Psi(t)\rangle$. It is useful to note that the two kets $|\Psi(t)\rangle$ and $|\widetilde{\Psi}(t)\rangle$ agree at $t=0$:

$$|\widetilde{\Psi}(0)\rangle = |\Psi(0)\rangle. \tag{27.2.8}$$

Our objective now is to find the Schrödinger equation for $|\widetilde{\Psi}(t)\rangle$. Taking the time derivative of (27.2.6) and using the original Schrödinger equation (27.1.4), we find that

$$\begin{aligned} i\hbar\frac{d}{dt}|\widetilde{\Psi}(t)\rangle &= -\hat{H}^{(0)}|\widetilde{\Psi}(t)\rangle + e^{i\hat{H}^{(0)}t/\hbar}(\hat{H}^{(0)} + \delta H(t))|\Psi(t)\rangle \\ &= \left[-\hat{H}^{(0)} + e^{i\hat{H}^{(0)}t/\hbar}(\hat{H}^{(0)} + \delta H(t))e^{-i\hat{H}^{(0)}t/\hbar}\right]|\widetilde{\Psi}(t)\rangle \\ &= e^{i\hat{H}^{(0)}t/\hbar}\delta H(t)e^{-i\hat{H}^{(0)}t/\hbar}|\widetilde{\Psi}(t)\rangle, \end{aligned} \tag{27.2.9}$$

where the dependence on $\hat{H}^{(0)}$ canceled out. We have thus found the Schrödinger equation in the interaction picture:

$$i\hbar\frac{d}{dt}|\widetilde{\Psi}(t)\rangle = \widetilde{\delta H}(t)|\widetilde{\Psi}(t)\rangle. \tag{27.2.10}$$

Here, $\widetilde{\delta H}(t)$ is defined by

$$\widetilde{\delta H}(t) \equiv e^{i\hat{H}^{(0)}t/\hbar}\,\delta H(t)\,e^{-i\hat{H}^{(0)}t/\hbar}. \tag{27.2.11}$$

The time evolution of $|\widetilde{\Psi}(t)\rangle$ is generated by $\widetilde{\delta H}(t)$ via a Schrödinger equation. The operator $\widetilde{\delta H}(t)$ is nothing else but the Heisenberg version of δH generated using $\hat{H}^{(0)}$. This is an *interaction picture*, a mixture of Heisenberg's and Schrödinger's picture. While we have some Heisenberg operators relative to $\hat{H}^{(0)}$, there is still a time-dependent state $|\widetilde{\Psi}(t)\rangle$ and a Schrödinger equation for it. If the time-dependent perturbation $\delta H(t)$

vanishes, the state $|\widetilde{\Psi}(t)\rangle$ is time independent and thus equal to its value $|\Psi(0)\rangle$ at $t=0$ (recall (27.2.8)).

So far all of our analysis is exact; we have made no approximations. Let us now consider an example in which the interaction picture helps find an exact solution to the Schrödinger equation.

Example 27.1. *Nuclear magnetic resonance (NMR) with rotating magnetic field.*
The Hamiltonian for a particle with a magnetic moment inside a magnetic field can be written in the form

$$\hat{H}=\boldsymbol{\omega}\cdot\hat{\mathbf{S}}, \tag{27.2.12}$$

where $\hat{\mathbf{S}}$ is the spin operator, and $\boldsymbol{\omega}$ is the Larmor angular velocity, a function of the magnetic field (section 17.5). Let us take the unperturbed Hamiltonian to be

$$\hat{H}^{(0)}=\omega_0 S_z=\tfrac{\hbar}{2}\,\omega_0\sigma_z\,, \tag{27.2.13}$$

corresponding to a magnetic field in the z-direction. For NMR applications the above longitudinal magnetic fields are of order a tesla, and Larmor frequencies $\omega_0\approx 100\,\text{MHz}$ are common. We add to $\hat{H}^{(0)}$ the effect of a magnetic field on the (x,y) plane rotating with the Larmor frequency ω_0 of $\hat{H}^{(0)}$:

$$\delta H(t)=\Omega\left(\hat{S}_x\cos\omega_0 t+\hat{S}_y\sin\omega_0 t\right)\,. \tag{27.2.14}$$

Here Ω is another frequency whose magnitude depends on the strength of the rotating magnetic field. This situation here is a particular case of the general NMR problem considered in section 17.6, where the rotating field frequency can be different from the Larmor frequency of the longitudinal field. Here the system is at resonance. This allows for a simple solution using the interaction picture, as we now see.

By its definition in (27.2.11), the perturbation $\widetilde{\delta H}$ is given by

$$\widetilde{\delta H}(t)=\exp\Big[i\omega_0 t\frac{\sigma_z}{2}\Big]\,\Omega\left(\hat{S}_x\cos\omega_0 t+\hat{S}_y\sin\omega_0 t\right)\exp\Big[-i\omega_0 t\frac{\sigma_z}{2}\Big]\,. \tag{27.2.15}$$

While one can compute this directly, it is possible to show that the right-hand side in fact has zero time derivative (it is a good exercise—do it!). This means that $\widetilde{\delta H}(t)$ is in fact equal to the right-hand side evaluated at $t=0$:

$$\widetilde{\delta H}(t)=\Omega\hat{S}_x\,. \tag{27.2.16}$$

Since $\widetilde{\delta H}(t)$ is a time-independent Hamiltonian, the Schrödinger equation for $|\widetilde{\Psi}\rangle$ is immediately solved:

$$|\widetilde{\Psi}(t)\rangle \;=\; \exp\Big[-i\frac{\widetilde{\delta H}t}{\hbar}\Big]|\widetilde{\Psi}(0)\rangle \;=\; \exp\Big[-i\Omega t\frac{\sigma_x}{2}\Big]|\Psi(0)\rangle. \tag{27.2.17}$$

The complete and *exact* answer for the evolution of the state is therefore

$$\boxed{|\Psi(t)\rangle=\exp\Big[-\frac{i\hat{H}^{(0)}t}{\hbar}\Big]|\widetilde{\Psi}(t)\rangle=\exp\Big[-i\omega_0 t\frac{\sigma_z}{2}\Big]\exp\Big[-i\Omega t\frac{\sigma_x}{2}\Big]|\Psi(0)\rangle.} \tag{27.2.18}$$

The spin, aligned along $\hat{z}$ at $t=0$, will move toward the x, y plane with angular velocity Ω while rotating around the z-axis with angular velocity ω_0. This motion was sketched in figure 17.3. □

Interaction picture as seen in a basis In order to familiarize ourselves further with the interaction picture equations, we look at them using an orthonormal basis. Let $|n\rangle$ be the complete orthonormal basis of states for $\hat{H}^{(0)}$:

$$\hat{H}^{(0)}|n\rangle = E_n|n\rangle . \tag{27.2.19}$$

This time there is no need for (0) superscripts since neither the states nor their energies will be corrected—recall that there are no energy eigenstates in the time-dependent theory. We then write an ansatz for our unknown interaction-picture state:

$$\boxed{|\widetilde{\Psi}(t)\rangle = \sum_n c_n(t)|n\rangle .} \tag{27.2.20}$$

Here the functions $c_n(t)$ are unknown. This expansion is justified since the states $|n\rangle$ form a basis. At any fixed time, a superposition of them describes the state of the system, and as a result, a superposition with time-dependent coefficients can be used to describe the state as it evolves in time due to $\delta\widetilde{H}$. The original wave function is then

$$|\Psi(t)\rangle = e^{-i\hat{H}^{(0)}t/\hbar}|\widetilde{\Psi}(t)\rangle = \sum_n c_n(t)e^{-iE_nt/\hbar}|n\rangle . \tag{27.2.21}$$

The time dependence due to $\hat{H}^{(0)}$ is displayed here. If we had $\delta H=0$, the state $|\widetilde{\Psi}(t)\rangle$ would be a constant, as demanded by the Schrödinger equation (27.2.10), and the $c_n(t)$ would be constants. The solution above would give the expected time evolution of the states under $\hat{H}^{(0)}$.

To see what the Schrödinger equation tells us about the functions $c_n(t)$, we plug the ansatz (27.2.20) into (27.2.10):

$$i\hbar\frac{d}{dt}\sum_m c_m(t)|m\rangle = \delta\widetilde{H}(t)\sum_n c_n(t)|n\rangle . \tag{27.2.22}$$

Using dots for time derivatives and introducing a resolution of the identity on the right-hand side, we find that

$$\begin{aligned}\sum_m i\hbar\,\dot{c}_m(t)|m\rangle &= \sum_m |m\rangle\langle m|\delta\widetilde{H}(t)\sum_n c_n(t)|n\rangle \\ &= \sum_m |m\rangle\sum_n \langle m|\delta\widetilde{H}(t)|n\rangle c_n(t) \\ &= \sum_{m,n}\delta\widetilde{H}_{mn}(t)c_n(t)|m\rangle ,\end{aligned} \tag{27.2.23}$$

where we have used the familiar matrix-element notation

$$\delta\widetilde{H}_{mn}(t) \equiv \langle m|\delta\widetilde{H}(t)|n\rangle . \tag{27.2.24}$$

Equating the coefficients of the basis kets $|m\rangle$ in (27.2.23), we get the equations

$$i\hbar\,\dot{c}_m(t) \;=\; \sum_n \widetilde{\delta H}_{mn}(t)\,c_n(t)\,. \tag{27.2.25}$$

The Schrödinger equation has become an infinite set of coupled first-order differential equations. The matrix elements $\widetilde{\delta H}_{mn}$ can be simplified by passing to nontilde variables:

$$\widetilde{\delta H}_{mn}(t) = \left\langle m\right| e^{i\hat{H}^{(0)}t/\hbar}\delta H(t)\,e^{-i\hat{H}^{(0)}t/\hbar}\left|n\right\rangle = e^{i(E_m-E_n)t/\hbar}\langle m|\delta H(t)|n\rangle\,. \tag{27.2.26}$$

If we define

$$\omega_{mn} \equiv (E_m - E_n)/\hbar\,, \tag{27.2.27}$$

we then see that

$$\widetilde{\delta H}_{mn}(t) \;=\; e^{i\omega_{mn}t}\,\delta H_{mn}(t)\,. \tag{27.2.28}$$

The coupled equations (27.2.25) for the functions $c_m(t)$ then become

$$\boxed{i\hbar\,\dot{c}_m(t) \;=\; \sum_n e^{i\omega_{mn}t}\,\delta H_{mn}(t)c_n(t).} \tag{27.2.29}$$

This is a very explicit form of the evolution equations. If we solve for the $c_m(t)$, the state is then given by (27.2.21):

$$\big|\Psi(t)\big\rangle = \sum_n c_n(t)e^{-iE_nt/\hbar}\,|n\rangle\,. \tag{27.2.30}$$

Note that so far we have not made any approximation, and our equations are exact: they hold for arbitrary $\delta H(t)$, small or large.

Example 27.2. *Transitions in a two-state system.*

Consider a two-state system with basis states $|1\rangle$ and $|2\rangle$, eigenstates of $\hat{H}^{(0)}$ with energies E_1 and E_2, respectively. Call $\omega_{12} \equiv (E_1 - E_2)/\hbar$, and consider an off-diagonal perturbation $\delta H(t)$ defined by a function $f(t)$, possibly complex:

$$\delta H(t) = \begin{pmatrix} 0 & f(t) \\ f^*(t) & 0 \end{pmatrix}. \tag{27.2.31}$$

Assume that the function $f(t)$ is only nonzero for $-T < t < T$, for some fixed $T > 0$. In particular, $f(-\infty) = f(\infty) = 0$. An example of this function is shown in figure 27.1. Take the system to be in the state $|1\rangle$ for $t = -\infty$. We want to know the probability that it is in $|2\rangle$ for $t = +\infty$. We will find the equations that allow us to answer this question.

If the system is in $|1\rangle$ at $t = -\infty$, it will remain in state $|1\rangle$ until $t = -T$, just before the perturbation turns on. This is because $|1\rangle$ is an $\hat{H}^{(0)}$ eigenstate. In fact, up to a constant phase, we have

$$|\Psi(t)\rangle \;=\; e^{-iE_1t/\hbar}|1\rangle\,, \quad \text{for} \quad -\infty < t \leq -T\,. \tag{27.2.32}$$

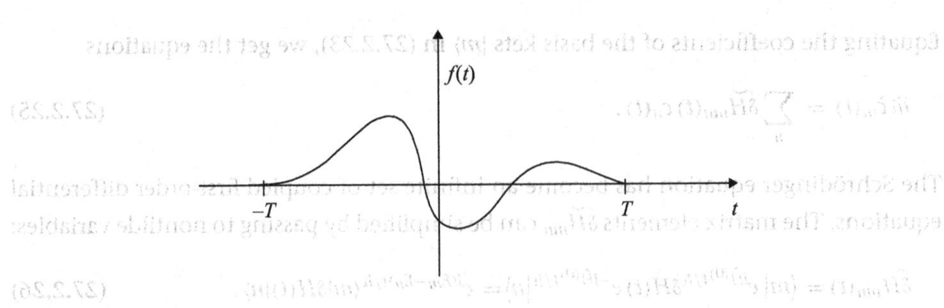

Figure 27.1
A function $f(t)$ that vanishes for $t > |T|$.

The constant phase ambiguity could be fixed if we declared the state to be $|1\rangle$ at some specific time, but working with $t = -\infty$ does not allow us to do so. We wish to know the probability of finding the state in $|2\rangle$ at $t = \infty$, but in fact this probability is the same as the probability of finding the state in $|2\rangle$ at $t = T$. This is because the perturbation does not exist for $t > T$. Indeed, if the state at $t = T$ is

$$|\Psi(T)\rangle = \gamma_1|1\rangle + \gamma_2|2\rangle, \tag{27.2.33}$$

with γ_1 and γ_2 constants, then the state for any time $t > T$ will be

$$|\Psi(t)\rangle = \gamma_1|1\rangle e^{-iE_1(t-T)/\hbar} + \gamma_2|2\rangle e^{-iE_2(t-T)/\hbar}. \tag{27.2.34}$$

The probability $p_2(t)$ of finding the state $|2\rangle$ at time $t > T$ will be

$$p_2(t) = |\langle 2|\Psi(t)\rangle|^2 = \left|\gamma_2 e^{-iE_2t/\hbar}\right|^2 = |\gamma_2|^2, \tag{27.2.35}$$

and, as expected, is time independent. It follows that to find the effect of the perturbation we must just find the state at $t = T$ and determine the constants γ_1 and γ_2.

Following (27.2.30) for a system with two basis states, the unknown state is written as

$$\big|\Psi(t)\big\rangle = c_1(t)e^{-iE_1t/\hbar}|1\rangle + c_2(t)e^{-iE_2t/\hbar}|2\rangle. \tag{27.2.36}$$

The initial condition tells us that the state is $|1\rangle$ at $t = -T$, so we can set

$$c_1(-T) = 1, \quad c_2(-T) = 0. \tag{27.2.37}$$

The differential equations (27.2.29) take the form

$$\begin{aligned} i\hbar\,\dot{c}_1(t) &= e^{i\omega_{12}t}\delta H_{12}(t)c_2(t), \\ i\hbar\,\dot{c}_2(t) &= e^{i\omega_{21}t}\delta H_{21}(t)c_1(t). \end{aligned} \tag{27.2.38}$$

The couplings are off-diagonal because $\delta H_{11} = \delta H_{22} = 0$. Using the form of the δH matrix elements, we find that

$$\begin{aligned} i\hbar\,\dot{c}_1(t) &= e^{i\omega_{12}t}\, f(t)\, c_2(t), \\ i\hbar\,\dot{c}_2(t) &= e^{-i\omega_{12}t} f^*(t)\, c_1(t). \end{aligned} \tag{27.2.39}$$

These equations, together with the initial conditions (27.2.37), determine the solution for all $t > -T$. They are not easily solved analytically for general $f(t)$, but numerical

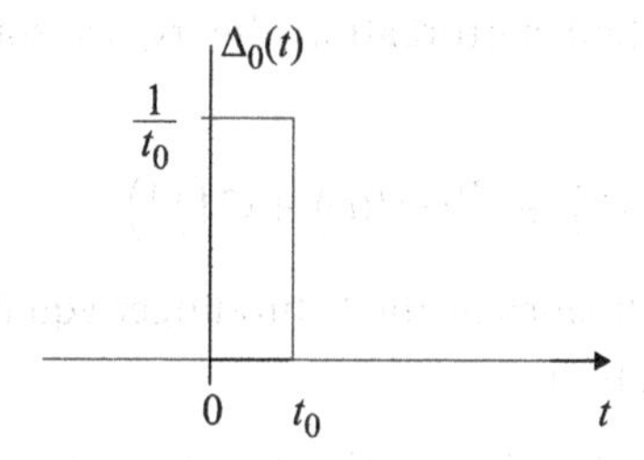

Figure 27.2
A pulse $\Delta_0(t)$ of duration t_0 and height $1/t_0$ (exercise 27.2).

solutions are always possible. Perturbative solutions are also possible, as we will soon see. Once solved, the probability $p_2(\infty)$ of finding the state at $|2\rangle$ in the infinite future is given by $p_2(\infty) = p_2(T) = |c_2(T)|^2$. □

Exercise 27.1. *Show that (27.2.39) implies the following second-order linear differential equation for $c_1(t)$:*

$$\ddot{c}_1(t) - \left(i\omega_{12} + \tfrac{\dot{f}(t)}{f(t)}\right)\dot{c}_1(t) + \tfrac{1}{\hbar^2}f^*(t)f(t)\,c_1(t) = 0\,. \tag{27.2.40}$$

Exercise 27.2. *Consider the two-state system with off-diagonal perturbation of example 27.2, and take $f(t) = \alpha\,\Delta_0(t)$, with α a complex constant, to be a short pulse, nonzero in the interval $t \in [0, t_0]$. The function $\Delta_0(t)$, shown in figure 27.2, takes the value $1/t_0$ for $0 < t < t_0$ and is zero otherwise. Assume that t_0 is chosen sufficiently small such that $\omega_{12}t_0 \ll 1$, and the exponentials in (27.2.39) can be assumed to be equal to one. Assume $c_1(0) = 1$ and $c_2(0) = 0$, and calculate $c_1(t_0)$ and $c_2(t_0)$. Show that the probability that the system is in $|2\rangle$ at time t_0 is* $\sin^2(|\alpha|/\hbar)$.

27.3 Perturbative Solution in the Interaction Picture

Let us now, finally, do perturbation theory. Our aim is to develop a perturbative solution to the interaction-picture Schrödinger equation (27.2.10):

$$i\hbar\frac{d}{dt}|\widetilde{\Psi}(t)\rangle = \widetilde{\delta H}(t)|\widetilde{\Psi}(t)\rangle\,. \tag{27.3.1}$$

In order to set up the perturbative expansion systematically, we include a unit-free small parameter λ multiplying the perturbation δH in the time-dependent Hamiltonian (27.1.3):

$$\hat{H}(t) = \hat{H}^{(0)} + \lambda\delta H(t)\,. \tag{27.3.2}$$

With such a replacement, the interaction picture Schrödinger equation above becomes

$$i\hbar\frac{d}{dt}|\widetilde{\Psi}(t)\rangle = \lambda\widetilde{\delta H}(t)|\widetilde{\Psi}(t)\rangle\,. \tag{27.3.3}$$

As we did in time-independent perturbation theory, we start by expanding $|\widetilde{\Psi}(t)\rangle$ in powers of the parameter λ:

$$|\widetilde{\Psi}(t)\rangle = |\widetilde{\Psi}^{(0)}(t)\rangle + \lambda|\widetilde{\Psi}^{(1)}(t)\rangle + \lambda^2|\widetilde{\Psi}^{(2)}(t)\rangle + \mathcal{O}\left(\lambda^3\right). \tag{27.3.4}$$

We now insert this into both sides of the Schrödinger equation (27.3.3), and using ∂_t for time derivatives, we find that

$$\begin{aligned} i\hbar\partial_t|\widetilde{\Psi}^{(0)}(t)\rangle + \lambda\, i\hbar\partial_t|\widetilde{\Psi}^{(1)}(t)\rangle + \lambda^2\, i\hbar\partial_t|\widetilde{\Psi}^{(2)}(t)\rangle + \lambda^3\, i\hbar\partial_t|\widetilde{\Psi}^{(3)}(t)\rangle + \mathcal{O}\left(\lambda^4\right) \\ = \lambda\,\widetilde{\delta H}|\widetilde{\Psi}^{(0)}(t)\rangle + \lambda^2\,\widetilde{\delta H}\,|\widetilde{\Psi}^{(1)}(t)\rangle + \lambda^3\,\widetilde{\delta H}\,|\widetilde{\Psi}^{(2)}(t)\rangle + \mathcal{O}\left(\lambda^4\right). \end{aligned} \tag{27.3.5}$$

The equality of these two series requires that for each power of λ the coefficients on the left-hand side and on the right-hand side must be equal. This gives

$$\begin{aligned} i\hbar\partial_t|\widetilde{\Psi}^{(0)}(t)\rangle &= 0, \\ i\hbar\partial_t|\widetilde{\Psi}^{(1)}(t)\rangle &= \widetilde{\delta H}|\widetilde{\Psi}^{(0)}(t)\rangle, \\ i\hbar\partial_t|\widetilde{\Psi}^{(2)}(t)\rangle &= \widetilde{\delta H}|\widetilde{\Psi}^{(1)}(t)\rangle, \\ \vdots\quad &= \quad\vdots \\ i\hbar\partial_t|\widetilde{\Psi}^{(n+1)}(t)\rangle &= \widetilde{\delta H}|\widetilde{\Psi}^{(n)}(t)\rangle. \end{aligned} \tag{27.3.6}$$

The origin of the pattern is clear. Since the Schrödinger equation has an explicit λ multiplying the right-hand side, the time derivative of the nth ket is coupled to the $\widetilde{\delta H}$ perturbation acting on the $(n-1)$th ket.

Let us consider the initial condition in detail. We will assume that the state is known at $t=0$—that is, we know $|\Psi(0)\rangle$. We want the state for $t>0$, and of course, we assume that the perturbation is on for $t>0$. Recalling that $|\widetilde{\Psi}(t)\rangle$ and $|\Psi(t)\rangle$ agree at $t=0$, the expansion (27.3.4) evaluated at $t=0$ implies that

$$|\widetilde{\Psi}(0)\rangle = |\Psi(0)\rangle = |\widetilde{\Psi}^{(0)}(0)\rangle + \lambda|\widetilde{\Psi}^{(1)}(0)\rangle + \lambda^2|\widetilde{\Psi}^{(2)}(0)\rangle + \mathcal{O}\left(\lambda^3\right). \tag{27.3.7}$$

This equation must hold for all values of λ. As a result, the coefficient of each power of λ must vanish, and we have

$$\begin{aligned} |\widetilde{\Psi}^{(0)}(0)\rangle &= |\Psi(0)\rangle, \\ |\widetilde{\Psi}^{(n)}(0)\rangle &= 0, \quad n\geq 1. \end{aligned} \tag{27.3.8}$$

These are the relevant initial conditions.

Now consider the first equation in (27.3.6). It states that $|\widetilde{\Psi}^{(0)}(t)\rangle$ is time independent. This is reasonable: if the perturbation vanishes, this is the only equation we get, and we should expect that $|\widetilde{\Psi}(t)\rangle$ is constant. Using the time-independence of $|\widetilde{\Psi}^{(0)}(t)\rangle$ and the initial condition, we find that

$$|\widetilde{\Psi}^{(0)}(t)\rangle = |\widetilde{\Psi}^{(0)}(0)\rangle = |\Psi(0)\rangle, \tag{27.3.9}$$

and so we have solved the first equation completely:

$$\boxed{|\widetilde{\Psi}^{(0)}(t)\rangle = |\Psi(0)\rangle.} \tag{27.3.10}$$

Using this result, the $\mathcal{O}(\lambda)$ equation reads

$$i\hbar\partial_t|\widetilde{\Psi}^{(1)}(t)\rangle = \widetilde{\delta H}(t)|\Psi(0)\rangle. \tag{27.3.11}$$

The solution can be written as an integral:

$$\boxed{|\widetilde{\Psi}^{(1)}(t)\rangle = \int_0^t \frac{\widetilde{\delta H}(t')}{i\hbar}|\Psi(0)\rangle\, dt'.} \tag{27.3.12}$$

By setting the lower limit of integration at $t=0$, we have correctly implemented the initial condition $|\widetilde{\Psi}^{(1)}(0)\rangle = 0$. The next equation, of order λ^2, reads

$$i\hbar\partial_t|\widetilde{\Psi}^{(2)}(t)\rangle = \widetilde{\delta H}|\widetilde{\Psi}^{(1)}(t)\rangle, \tag{27.3.13}$$

and its solution is

$$|\widetilde{\Psi}^{(2)}(t)\rangle = \int_0^t \frac{\widetilde{\delta H}(t')}{i\hbar}|\widetilde{\Psi}^{(1)}(t')\rangle\, dt', \tag{27.3.14}$$

consistent with the initial condition. Using our previous result to write $|\widetilde{\Psi}^{(1)}(t')\rangle$, we now have an iterated integral expression:

$$\boxed{|\widetilde{\Psi}^{(2)}(t)\rangle = \int_0^t \frac{\widetilde{\delta H}(t')}{i\hbar}dt' \int_0^{t'} \frac{\widetilde{\delta H}(t'')}{i\hbar}|\Psi(0)\rangle dt''.} \tag{27.3.15}$$

Here is a graphic representation of the nested integrals for $|\widetilde{\Psi}^{(2)}(t)\rangle$:

$$\begin{array}{cccc} |\widetilde{\Psi}^{(2)}(t)\rangle & \frac{\widetilde{\delta H}(t')}{i\hbar} & \frac{\widetilde{\delta H}(t'')}{i\hbar} & |\Psi(0)\rangle \\ t & t' & t'' & 0 \end{array} \quad . \tag{27.3.16}$$

$$t \geqslant t' \geqslant t'' \geqslant 0$$

The insertions of $\widetilde{\delta H}$ are ordered with later times to the left of earlier ones. It should be clear from this discussion how to write the iterated integral expression for $|\widetilde{\Psi}^{(k)}(t)\rangle$, with $k > 2$. The complete solution, setting $\lambda = 1$, is given by

$$|\Psi(t)\rangle = e^{-i\hat{H}^{(0)}t/\hbar}\Big(|\Psi(0)\rangle + |\widetilde{\Psi}^{(1)}(t)\rangle + |\widetilde{\Psi}^{(2)}(t)\rangle + \cdots\Big). \tag{27.3.17}$$

Let us use perturbation theory to calculate the probability $P_{m\leftarrow n}(t)$ of a transition from $|n\rangle$ at $t=0$ to $|m\rangle$, with $m \neq n$, at time t. By definition,

$$P_{m\leftarrow n}(t) = |\langle m|\Psi(t)\rangle|^2. \tag{27.3.18}$$

Using the above expression for $|\Psi(t)\rangle$ and noting that the phase that arises from the action of the $\hat{H}^{(0)}$ exponential on the bra drops out of the norm squared, we see that

$$P_{m\leftarrow n}(t) = \Big|\langle m|\Big(|\Psi(0)\rangle + |\widetilde{\Psi}^{(1)}(t)\rangle + |\widetilde{\Psi}^{(2)}(t)\rangle + \cdots\Big)\Big|^2. \tag{27.3.19}$$

Since $|\Psi(0)\rangle = |n\rangle$ and $|n\rangle$ is orthogonal to $|m\rangle$, the transition probability simplifies a bit:

$$P_{m\leftarrow n}(t) = \left|\langle m|\widetilde{\Psi}^{(1)}(t)\rangle + \langle m|\widetilde{\Psi}^{(2)}(t)\rangle + \cdots\right|^2 . \tag{27.3.20}$$

To first order in perturbation theory, we keep only the first term in the sum, and using our result for $|\widetilde{\Psi}^{(1)}(t)\rangle$, we find

$$P^{(1)}_{m\leftarrow n}(t) = \left|\langle m|\int_0^t \frac{\delta\widetilde{H}(t')}{i\hbar}|n\rangle\, dt'\right|^2 = \left|\int_0^t \frac{\langle m|\delta\widetilde{H}(t')|n\rangle}{i\hbar}dt'\right|^2 . \tag{27.3.21}$$

The superscript 1 on P reminds us this result was obtained using first-order perturbation theory, even though the expression itself is of order $(\delta H)^2$. Recalling the relation between the matrix elements of $\delta\widetilde{H}$ and those of δH, we finally have our result for the transition probability to first order in perturbation theory:

$$\boxed{P^{(1)}_{m\leftarrow n}(t) = \left|\int_0^t e^{i\omega_{mn}t'}\frac{\delta H_{mn}(t')}{i\hbar}dt'\right|^2 , \quad m \neq n.} \tag{27.3.22}$$

This is a key result and will be very useful in the applications we will be considering.

Exercise 27.3. *Prove that to first order in perturbation theory we have a remarkable equality of transition probabilities:*

$$\boxed{P^{(1)}_{m\leftarrow n}(t) = P^{(1)}_{n\leftarrow m}(t) .} \tag{27.3.23}$$

Perturbative solution in a basis It will also be useful to have our perturbation theory results in terms of the time-dependent coefficients $c_n(t)$ introduced earlier through the expansion (27.2.20):

$$|\widetilde{\Psi}(t)\rangle = \sum_n c_n(t)|n\rangle . \tag{27.3.24}$$

Since $|\Psi(0)\rangle = |\widetilde{\Psi}(0)\rangle$, the initial condition reads

$$|\Psi(0)\rangle = \sum_n c_n(0)|n\rangle = |\widetilde{\Psi}^{(0)}(0)\rangle, \tag{27.3.25}$$

where we also used (27.3.9). In this notation, the $c_n(t)$ functions also have a λ expansion. Writing

$$|\widetilde{\Psi}^{(k)}(t)\rangle = \sum_n c_n^{(k)}(t)|n\rangle , \quad k = 0, 1, 2, \ldots , \tag{27.3.26}$$

the earlier expansion

$$|\widetilde{\Psi}(t)\rangle = |\widetilde{\Psi}^{(0)}(t)\rangle + \lambda|\widetilde{\Psi}^{(1)}(t)\rangle + \lambda^2|\widetilde{\Psi}^{(2)}(t)\rangle + \mathcal{O}\left(\lambda^3\right) , \tag{27.3.27}$$

now gives

$$c_n(t) = c_n^{(0)}(t) + \lambda\, c_n^{(1)}(t) + \lambda^2 c_n^{(2)}(t) + \cdots . \tag{27.3.28}$$

Since $|\widetilde{\Psi}^{(0)}(t)\rangle$ is in fact constant, we find that

$$|\widetilde{\Psi}^{(0)}(t)\rangle = \sum_n c_n^{(0)}(t)|n\rangle = |\Psi(0)\rangle = \sum_n c_n(0)|n\rangle\,, \tag{27.3.29}$$

using (27.3.25). We therefore conclude that

$$c_n^{(0)}(t) = c_n^{(0)}(0) = c_n(0)\,. \tag{27.3.30}$$

This result makes the expansion in (27.3.28) clearer:

$$c_n(t) = c_n(0) + \lambda\, c_n^{(1)}(t) + \lambda^2 c_n^{(2)}(t) + \cdots\,. \tag{27.3.31}$$

The other initial conditions given earlier in (27.3.8) imply that all other $c_n^{(k)}(t)$ functions vanish at time equal zero:

$$c_n^{(k)}(0) = 0\,, \quad k \geq 1\,. \tag{27.3.32}$$

Having obtained the initial conditions, we can use the solution (27.3.12) for $|\widetilde{\Psi}^{(1)}(t)\rangle$ to deduce the value of the first corrections $c_n^{(1)}(t)$:

$$|\widetilde{\Psi}^{(1)}(t)\rangle = \sum_n c_n^{(1)}(t)|n\rangle = \int_0^t \frac{\widetilde{\delta H}(t')}{i\hbar}dt' \sum_n c_n(0)|n\rangle\,. \tag{27.3.33}$$

Acting with the bra $\langle m|$ from the left, we find that

$$c_m^{(1)}(t) = \sum_n \int_0^t \frac{\langle m|\widetilde{\delta H}(t')|n\rangle}{i\hbar} c_n(0)\, dt'\,. \tag{27.3.34}$$

Expressing the matrix element of $\widetilde{\delta H}$ in terms of that of δH, we obtain our final form for the first-order corrections:

$$\boxed{c_m^{(1)}(t) = \sum_n \int_0^t dt'\, e^{i\omega_{mn}t'} \frac{\delta H_{mn}(t')}{i\hbar}\, c_n(0)\,.} \tag{27.3.35}$$

The probability $P_m(t)$ of being found in the state $|m\rangle$ at time t is

$$P_m(t) = |\langle m|\Psi(t)\rangle|^2 = \left|\langle m|\widetilde{\Psi}(t)\rangle\right|^2 = |c_m(t)|^2\,. \tag{27.3.36}$$

Using the perturbative expansion of $c_m(t)$ (with $\lambda = 1$),

$$P_m(t) = \left|c_m(0) + c_m^{(1)}(t) + \mathcal{O}((\delta H)^2)\right|^2\,. \tag{27.3.37}$$

This is true in all generality. When $c_m(0) \neq 0$, we have

$$P_m(t) = \left|c_m(0)\right|^2 + c_m(0)^* c_m^{(1)}(t) + c_m^{(1)}(t)^* c_m(0) + \mathcal{O}(\delta H^2)\,. \tag{27.3.38}$$

Note that the $|c_m^{(1)}(t)|^2$ term cannot be kept to this order of approximation since it is of the same order as contributions that would arise from $c_m^{(2)}(t)$. On the other hand, if $c_m(0) = 0$, we get a stronger result:

$$P_m(t) = \left|c_m^{(1)}(t)\right|^2 + \mathcal{O}((\delta H)^3)\,, \quad c_m(0) = 0\,. \tag{27.3.39}$$

This result is equivalent to (27.3.22), which applies when the initial state is $|n\rangle$ so that the sum in (27.3.35) collapses to a single term.

Example 27.3. *NMR at resonance in perturbation theory.*

In example 27.1 we found the solution to the spin evolution in NMR, when the rotating magnetic field is in resonance with the Larmor frequency associated with the longitudinal magnetic field. Here we treat the rotating magnetic field as a perturbation and obtain the spin-state evolution to first order in the perturbation.

Example 27.1 showed that $\delta\widetilde{H}(t) = \Omega\hat{S}_x$ and gave an exact solution for the state $|\widetilde{\Psi}\rangle$:

$$|\widetilde{\Psi}(t)\rangle = \exp\Big[-\frac{i}{\hbar}\Omega\hat{S}_x t\Big]|\Psi(0)\rangle = \exp\Big[-i\Omega t\frac{\sigma_x}{2}\Big]|\Psi(0)\rangle. \tag{27.3.40}$$

We quickly check that first-order perturbation theory gives a result consistent with this one. Making use of (27.3.12), we get

$$|\widetilde{\Psi}^{(1)}(t)\rangle = \int_0^t \frac{\delta\widetilde{H}(t')}{i\hbar}|\Psi(0)\rangle\, dt' = -\frac{i}{\hbar}\Omega t\hat{S}_x|\Psi(0)\rangle = -i\Omega t\frac{\sigma_x}{2}|\Psi(0)\rangle. \tag{27.3.41}$$

This is indeed the first nontrivial term in the Taylor expansion of the exact answer (27.3.40). This first term, however, is an accurate representation of the answer only for small times $\Omega t \ll 1$. The result to first order in the perturbation is not an accurate description of the system for arbitrarily long times. This complication is generic and will feature in the analysis of Fermi's golden rule. □

27.4 Constant Perturbations

We now use perturbation theory to discuss transitions that occur when the perturbation of the Hamiltonian is in fact time independent. This may sound strange. Since $\hat{H}^{(0)}$ is, by assumption, time independent, if we add a time-independent perturbation, the whole problem is time independent. As such, there should be no transitions among the energy eigenstates of the *full* Hamiltonian. The new eigenstates and energies can be obtained from time-independent perturbation theory. As we will discuss now, the kinds of questions we have in mind are best studied with time-dependent perturbation theory.

Let's first set the notation. In this case the perturbation, called V, is time independent:

$$\hat{H} = \hat{H}^{(0)} + V. \tag{27.4.1}$$

The perturbation can have spatial dependence or spin dependence, but such dependences, if present, are left implicit in the notation. The perturbation V is often present for all times. In the analysis that follows, however, we assume that at some fixed initial time—call it $t=0$—we have an $\hat{H}^{(0)}$ eigenstate. We then incorporate the effect of V to determine the state that we get at time t_0, again described in terms of $\hat{H}^{(0)}$ eigenstates. In practice this is *as if* the perturbation turned on at $t=0$, remained constant, and then turned off at $t=t_0$. An analysis based on *time-independent* perturbation theory would be more cumbersome. We would have to first find the new spectrum, then express the initial $\hat{H}^{(0)}$ eigenstate in terms of the new eigenstates, perform time evolution, and, finally, reexpress the result in terms of $\hat{H}^{(0)}$ eigenstates. Constant perturbations are relevant for the phenomenon of *autoionization*, where an internal electronic transition in an atom is accompanied by the ejection of an electron. The perturbation V for autoionization is the Coulomb interaction between the electrons. It is always present, but we assume we start with an approximate energy eigenstate that does not take V into account and then see what effect V has.

Let us begin. Consider initial states described in terms of $\hat{H}^{(0)}$ eigenstates at $t=0$. Identifying $\delta H(t)=V$ and recalling (27.3.35) for transition amplitudes to first order in perturbation theory, we find that

$$c_m^{(1)}(t)=\sum_n \int_0^t dt' e^{i\omega_{mn}t'} \frac{V_{mn}}{i\hbar} c_n(0)\,. \tag{27.4.2}$$

To represent an initial state $|i\rangle$ at $t=0$, we take $c_n(0)=\delta_{n,i}$. For a transition to a final state $|f\rangle \neq |i\rangle$ at time t_0, we set $m=f$, and we have an integral that is easily performed:

$$c_f^{(1)}(t_0)=\frac{1}{i\hbar}\int_0^{t_0} V_{fi}\, e^{i\omega_{fi}t'}dt' = \frac{V_{fi}}{i\hbar}\frac{e^{i\omega_{fi}t'}}{i\omega_{fi}}\bigg|_0^{t_0}. \tag{27.4.3}$$

Evaluating the limits and simplifying, we get

$$c_f^{(1)}(t_0)=\frac{V_{fi}}{E_f-E_i}\left(1-e^{i\omega_{fi}t_0}\right)=\frac{V_{fi}\, e^{i\omega_{fi}t_0/2}}{E_f-E_i}(-2i)\sin\left(\frac{\omega_{fi}t_0}{2}\right). \tag{27.4.4}$$

Note that $c_f^{(1)}(0)=0$ since initially the state is in $|i\rangle \neq |f\rangle$. The transition probability to go from i at $t=0$ to f at $t=t_0$ is then $|c_f^{(1)}(t_0)|^2$ (see (27.3.39)). Therefore,

$$P_{f\leftarrow i}(t_0)=\frac{|V_{fi}|^2}{\hbar^2}\frac{\sin^2\left(\frac{\omega_{fi}t_0}{2}\right)}{\left(\frac{E_f-E_i}{2\hbar}\right)^2}\,. \tag{27.4.5}$$

This first-order result is expected to be accurate at time t_0 if $P_{f\leftarrow i}(t_0)\ll 1$. Certainly, a large transition probability at first order could not be trusted and would require the examination of higher orders. A useful function features in the above transition amplitude. We will write

$$\boxed{P_{f\leftarrow i}(t_0)=\frac{|V_{fi}|^2}{\hbar^2}F(\omega_{fi};t_0),} \tag{27.4.6}$$

defining the two-argument function $F(\omega;t)$ as follows:

$$F(\omega;t)\equiv\frac{\sin^2\left(\frac{\omega t}{2}\right)}{\left(\frac{\omega}{2}\right)^2}\,. \tag{27.4.7}$$

Note that

$$\lim_{\omega\to 0}F(\omega;t)=t^2\,. \tag{27.4.8}$$

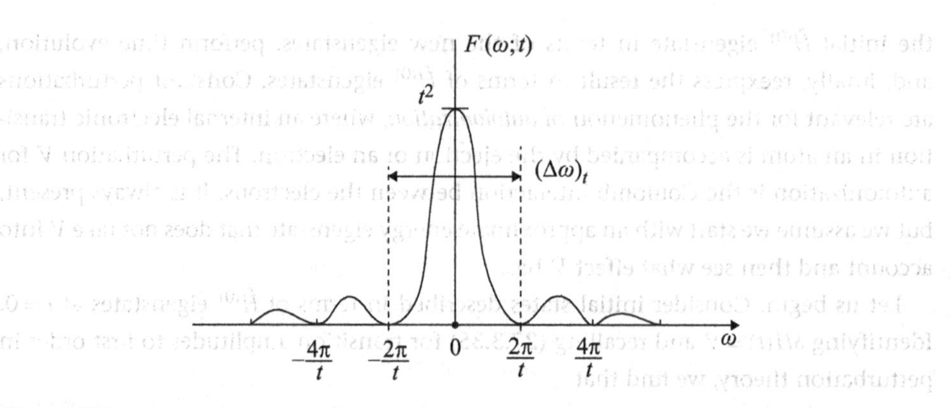

Figure 27.3
Plot of $F(\omega; t)$ as a function of ω for a fixed t. This function enters the transition probability $P_{f\leftarrow i}(t)$ in (27.4.6). As t grows, the width $(\Delta\omega)_t$ of the main lobe decreases like $1/t$, while the peak value for $\omega = 0$ grows like t^2.

A plot of $F(\omega; t)$ as a function of ω at fixed t is shown in figure 27.3. Note the main lobe centered around $\omega = 0$ and the smaller lobes to the sides. The width of the function $F(\omega; t)$ is a time-dependent ω interval that we denote as $(\Delta\omega)_t$ and define as the distance between the two zeroes closest to the origin. It follows that

$$(\Delta\omega)_t \equiv \frac{4\pi}{t}. \tag{27.4.9}$$

The function $F(\omega; t)$ is suppressed for $|\omega| > (\Delta\omega)_t$.

To understand the main features of the transition probability, we examine how it behaves for different values of the final energy E_f. If $E_f \neq E_i$, the transition is said to be energy nonconserving. There is no contradiction here with energy conservation, which states that a time-independent Hamiltonian does not allow for transitions between energy eigenstates or, more generally, that the expectation value of the energy is time independent. The initial $\hat{H}^{(0)}$ eigenstate experiences a potential that turns on, remains constant, and then turns off. This is, effectively, a time-dependent perturbation that can change the $\hat{H}^{(0)}$ energy. The energy is supplied or absorbed by the source that generates the term V. The mechanics of the transition is clear too. The initial state is an $\hat{H}^{(0)}$ eigenstate; it is not an eigenstate of $\hat{H}^{(0)} + V$ but rather a superposition of $\hat{H}^{(0)} + V$ eigenstates. This superposition evolves while V is on. When V turns off, the state at that time can be re-expressed as a superposition of $\hat{H}^{(0)}$ eigenstates. Such a superposition allows a range of values of the energy. If $E_f = E_i$, the transition is said to be energy conserving. Both energy nonconserving and energy-conserving transitions are possible, and let us consider them in turn.

1. $E_f \neq E_i$. The associated transition probability $P_{f\leftarrow i}(t_0)$, given in (27.4.6), is shown in figure 27.4. The probability is periodic in t_0 with period $2\pi/|\omega_{fi}|$. If the amplitude multiplying the sine squared is much less than one—that is,

$$\frac{4|V_{fi}|^2}{(E_f - E_i)^2} \ll 1, \tag{27.4.10}$$

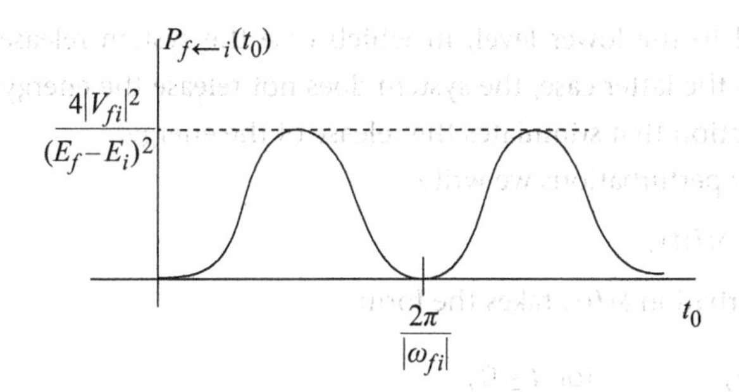

Figure 27.4
The transition probability $P_{f\leftarrow i}$ for constant perturbations, shown as a function of time.

then this first-order transition probability $P_{f\leftarrow i}(t_0)$ is accurate for all times t_0 as it remains small for all times. The amplitude is suppressed as $|E_f - E_i|$ grows, due to the factor in the denominator. This indicates that the larger the energy "violation" the smaller the probability of transition. This is happening because a perturbation that effectively turns on, remains constant, and then turns off is not an efficient supply of energy.

2. $E_f = E_i$. The result for the transition probability in this case is obtained quickly by taking the $\omega_{fi} \to 0$ limit of the probability in (27.4.6). Using (27.4.8), we find that

$$P_{f\leftarrow i}(t_0)\Big|_{E_f=E_i} = \frac{|V_{fi}|^2}{\hbar^2}t_0^2\,. \tag{27.4.11}$$

The probability for an energy-conserving transition grows quadratically in time, without bound. This result, of course, can only be trusted for small enough t_0 such that $P_{f\leftarrow i}(t_0) \ll 1$.

Note that a quadratic growth of $P_{f\leftarrow i}$ is also visible in the energy nonconserving case for very small times t_0. Indeed, (27.4.6) leads again to

$$\lim_{t_0\to 0} P_{f\leftarrow i}(t_0) = \frac{|V_{fi}|^2}{\hbar^2}t_0^2\,, \quad \text{while } E_f \neq E_i. \tag{27.4.12}$$

This behavior can be noted near the origin in figure 27.4. In this case we know how this quadratic growth eventually stops, and the transition probability becomes oscillatory.

27.5 Harmonic Perturbations

Having studied the effect of constant perturbations, we now consider truly time-dependent perturbations. A *harmonic* perturbation is one in which $\delta H(t)$ is periodic in time with some frequency ω. Such a perturbation can efficiently trigger transitions between discrete energy levels separated by an energy approximately equal to $\hbar\omega$. The perturbation can cause a transition from the lower to the higher level, in which case the system absorbs energy from the perturbation. It can also cause a transition from the higher level to the lower level, in which case the system releases energy to the perturbation. In the latter case, the system does not release the energy spontaneously; it is the perturbation that stimulates the release of the energy.

For harmonic perturbations we write

$$\hat{H}(t) = \hat{H}^{(0)} + \delta H(t), \tag{27.5.1}$$

where the perturbation $\delta H(t)$ takes the form

$$\delta H(t) = \begin{cases} 0, & \text{for } t \leq 0, \\ 2H' \cos \omega t, & \text{for } t > 0. \end{cases} \tag{27.5.2}$$

Here, by definition,

$$\omega > 0, \tag{27.5.3}$$

and H' is some time-independent Hamiltonian. The inclusion of an extra factor of two in the relation between δH and H' is convenient. As we will see in the following section, it results in a "golden rule" with the same overall prefactor as in the case of constant perturbations.

We again begin by considering transitions from an initial state $|i\rangle$ with energy E_i to a final state $|f\rangle$ with energy E_f. The transition amplitude follows from (27.3.35):

$$c_f^{(1)}(t_0) = \frac{1}{i\hbar} \int_0^{t_0} dt' \, e^{i\omega_{fi} t'} \delta H_{fi}(t'). \tag{27.5.4}$$

Using the explicit form of $\delta H(t)$, the integral can be evaluated:

$$\begin{aligned} c_f^{(1)}(t_0) &= \frac{1}{i\hbar} \int_0^{t_0} e^{i\omega_{fi} t'} 2H'_{fi} \cos \omega t' dt' \\ &= \frac{H'_{fi}}{i\hbar} \int_0^{t_0} \left(e^{i(\omega_{fi}+\omega)t'} + e^{i(\omega_{fi}-\omega)t'} \right) dt' \\ &= -\frac{H'_{fi}}{\hbar} \left[\frac{e^{i(\omega_{fi}+\omega)t_0} - 1}{\omega_{fi} + \omega} + \frac{e^{i(\omega_{fi}-\omega)t_0} - 1}{\omega_{fi} - \omega} \right]. \end{aligned} \tag{27.5.5}$$

Comments:

1. The amplitude takes the form of a factor multiplying the sum of two terms, each one a fraction. As $t_0 \to 0$, each fraction goes to it_0. For finite t_0, which is our case of interest, each numerator is a complex number of bounded absolute value that oscillates in time from zero up to two. In comparing the two terms, the relevant one is the one with the smallest denominator. This is how we compare any two waves that are superposed: the one with larger amplitude is more relevant, even though at some special times—as it crosses the value of zero, for example—it is smaller than the other wave.
2. The second term is relevant for $\omega_{fi} \simeq \omega$—that is, when $E_f \simeq E_i + \hbar\omega$. Since $\omega > 0$, we have $E_f > E_i$. Energy is transferred from the perturbation to the system, and we have a process of energy "absorption" in which the system moves to a higher-energy state. This is shown on the left in figure 27.5.

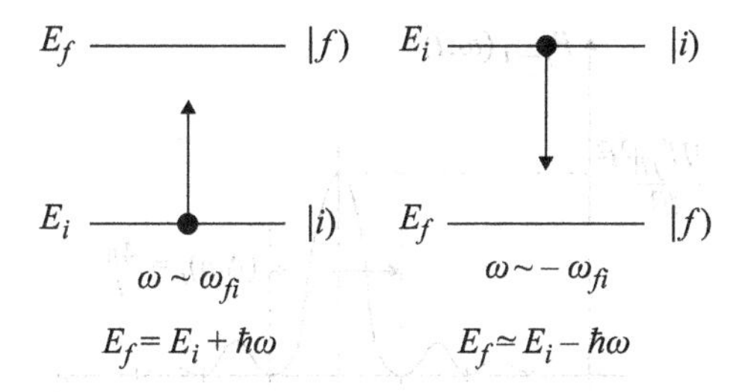

Figure 27.5
Left: Absorption process in which the source supplies the energy for the transition from the lower-energy state $|i\rangle$ into the higher-energy state $|f\rangle$. *Right*: Stimulated emission process in which the source stimulates the system to transition from the higher-energy state $|i\rangle$ into the lower-energy state $|f\rangle$ while releasing energy.

3. The first term is relevant for $\omega_{fi} \simeq -\omega$—that is, when $E_f = E_i - \hbar\omega$. Since $\omega > 0$, we have $E_i > E_f$. Here, the system begins on the higher-energy state E_i, and we have a process of *stimulated emission* in which the source has stimulated the system into a transition that gives away energy $\hbar\omega$. This is shown on the right in figure 27.5.

Both absorption and stimulated emission are of interest. Let us do the calculations for the case of absorption; the answer for the case of stimulated emission will be completely analogous. Since $\omega_{fi} \simeq \omega$, the second term in the last line of (27.5.5) is much more important than the first as long as

$$|\omega - \omega_{fi}| \ll |\omega_{fi}|\,. \tag{27.5.6}$$

Keeping only the second term, we find that

$$c_f^{(1)}(t_0) = -\frac{H'_{fi}}{\hbar}\frac{e^{\frac{i}{2}(\omega_{fi}-\omega)t_0}}{\omega_{fi}-\omega}\,2i\sin\left(\frac{\omega_{fi}-\omega}{2}t_0\right), \tag{27.5.7}$$

and the transition probability is

$$P_{f\leftarrow i}(\omega; t_0) = |c_f^{(1)}(t_0)|^2 = \frac{|H'_{fi}|^2}{\hbar^2}\frac{\sin^2\left(\frac{\omega_{fi}-\omega}{2}t_0\right)}{(\frac{\omega_{fi}-\omega}{2})^2}\,. \tag{27.5.8}$$

We have added to $P_{f\leftarrow i}$ the argument ω to remind us that the transition probability depends on the frequency ω of the perturbation. Note that at this point we are holding both E_i and E_f fixed. The transition probability is exactly the same as that for constant perturbations (see (27.4.5)) with $V \to H'$, and $\omega_{fi} \to \omega_{fi} - \omega$. A sketch of $P_{f\leftarrow i}(\omega; t)$ as a function of ω is shown in figure 27.6.

We can now ask about conditions on the time t_0 for the above transition probability to be valid. There are two observations to be made.

1. Consider again our neglect of the first wave in the transition amplitude, the term peaking for $\omega_{fi} \simeq -\omega$ in (27.5.5). That wave, if included, would contribute to $P_{f\leftarrow i}(\omega; t_0)$ by itself and through interference with the wave we kept. Consider the contribution from its square magnitude. This contribution can be visualized by

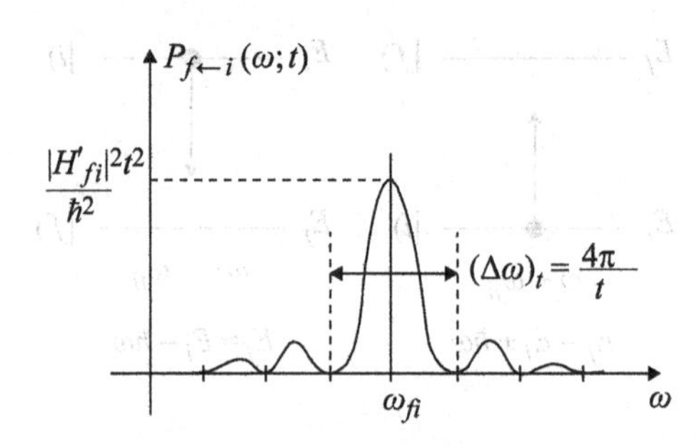

Figure 27.6
The ω dependence of the transition probability $P_{f\leftarrow i}(\omega;t)$ from a state $|i\rangle$ of energy E_i at $t=0$ to a state $|f\rangle$ of energy E_f at time t, under a harmonic perturbation with frequency ω. The probability peaks when $\omega=\omega_{fi}=(E_f-E_i)/\hbar$.

extending the ω-axis in figure 27.6 to include negative values, where we would find a similar bump centered at $-\omega_{fi}$. We need the width $(\Delta\omega)_t$ of the main lobe to be rather small compared to the distance $2|\omega_{fi}|$ between the peaks of the probability distributions:

$$\frac{4\pi}{t_0}\ll 2|\omega_{fi}|\,. \tag{27.5.9}$$

Since $\omega_{fi}\simeq\omega$, we see that

$$t_0\gg\frac{1}{|\omega_{fi}|}\simeq\frac{1}{\omega}\,. \tag{27.5.10}$$

This is reasonable: t_0 must include a number of periods of the wave so that we can identify the perturbation as oscillatory.

2. We also do not want $P_{f\leftarrow i}(\omega;t_0)$ to become too large. To estimate its growth, consider the case of resonance: $\omega=\omega_{fi}$. We then get the condition

$$P_{f\leftarrow i}(\omega_{fi};t_0)=\frac{|H'_{fi}|^2}{\hbar^2}t_0^2\ll 1\,. \tag{27.5.11}$$

This requires that

$$t_0\ll\frac{\hbar}{|H'_{fi}|}\,. \tag{27.5.12}$$

The two conditions we have found for t_0 can be put together as follows:

$$\frac{1}{|\omega_{fi}|}\ll t_0\ll\frac{\hbar}{|H'_{fi}|}\,. \tag{27.5.13}$$

For a suitable range of t_0 to exist, we need to have

$$\frac{1}{|\omega_{fi}|}\ll\frac{\hbar}{|H'_{fi}|}\quad\Rightarrow\quad |H'_{fi}|\ll\hbar|\omega_{fi}|\,. \tag{27.5.14}$$

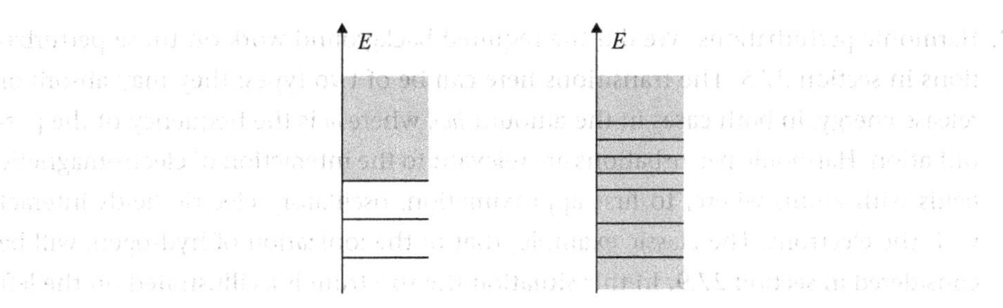

Figure 27.7
Left: A spectrum with discrete states and a separate continuous component. *Right*: A spectrum in which discrete states exist for energies that are in the range of the continuous part of the spectrum.

This is a reasonable constraint: the matrix element of the perturbation is an energy that must be much smaller than the $\hat{H}^{(0)}$ energy separating the two levels.

27.6 Fermi's Golden Rule

Let us now consider transitions in which the initial state of the system is part of a discrete spectrum, but the final state is part of a continuum of states. The ionization of a hydrogen atom is perhaps the most familiar example: the initial state may be one of the discrete bound states of the atom, while the final state includes a free electron, thus an approximate momentum eigenstate that is part of a continuum of nonnormalizable states. That kind of spectrum is shown on the left in Figure 27.7. Another possibility is realized in atoms with more than one electron. As shown on the right in the figure, discrete states (bound states) may appear at energies for which a continuous part of the spectrum also exists. A transition from a discrete state to a state in the continuum can then conserve energy.

While the probability of transition between two discrete states exhibits periodic dependence in time, if the final state is part of a continuum, an integral over final states is needed, and the result will be a transition probability *linear* in time. To such a transition probability, we will be able to associate a constant transition *rate*. The transition rate is given by Fermi's golden rule. We will consider two different cases in full detail:

1. Constant perturbations. We did the required background work on those perturbations in section 27.4. Constant perturbations, as we will see, produce energy-conserving transitions to the continuum. Such transitions can happen when the spectrum is as illustrated on the right in figure 27.7. We will qualitatively discuss autoionization to illustrate some of the main ideas (section 27.7). Moreover, in section 27.8 we will solve the Schrödinger equation for a discrete state decaying into a continuum via a constant perturbation. This gives insight into the long-time behavior of the system and the energy distribution of the decay products.

2. Harmonic perturbations. We did the required background work on those perturbations in section 27.5. The transitions here can be of two types: they may absorb or release energy, in both cases in the amount $\hbar\omega$, where ω is the frequency of the perturbation. Harmonic perturbations are relevant to the interaction of electromagnetic fields with atoms where, to first approximation, oscillatory electric fields interact with the electrons. The classic example, that of the ionization of hydrogen, will be considered in section 27.9. In this situation the spectrum is as illustrated on the left in figure 27.7.

For both types of perturbations above, we already know the time-dependent transition probabilities from some initial state $|i\rangle$ with energy E_i to some final state $|f\rangle$ with energy E_f. We must now learn how to integrate such probabilities over a continuum of final states.

In preparation for this step, we must find a strategy to deal with continuum states. In fact, we need to count the states in the continuuum, so we will replace infinite space by a very large cubic box of side length L, and we will impose periodic boundary conditions on the wave functions. While any reasonable boundary condition should work, periodic boundary conditions are the simplest to implement. If our system is inside the gigantic box, locality of the physics suggests that the effect of the box is negligible and goes to zero as its size goes to infinity. The large box will have almost no effect on the discrete part of the spectrum, but it will replace the continuous spectrum by a discrete spectrum where the separation between the states is infinitesimal and can be made arbitrarily small by making L sufficiently large (see figure 27.8).

If the potential is short range, momentum eigenstates of large energy are a good representation of the continuum. We call L a regulator, as it allows us to deal with infinite quantities such as the number of continuum states. At the end of our calculations, the value of L must drop out. This is a consistency check on the regulation.

The momentum eigenstates $\psi(\mathbf{x})$ take the form

$$\psi(\mathbf{x}) = \frac{1}{\sqrt{L^3}} e^{ik_x x} e^{ik_y y} e^{ik_z z}, \tag{27.6.1}$$

with constant $\mathbf{k} = (k_x, k_y, k_z)$. It is clear that the states are normalized correctly, as

$$\int_{\text{box}} |\psi(\mathbf{x})|^2 d^3x = \frac{1}{L^3} \int_{\text{box}} d^3x = 1. \tag{27.6.2}$$

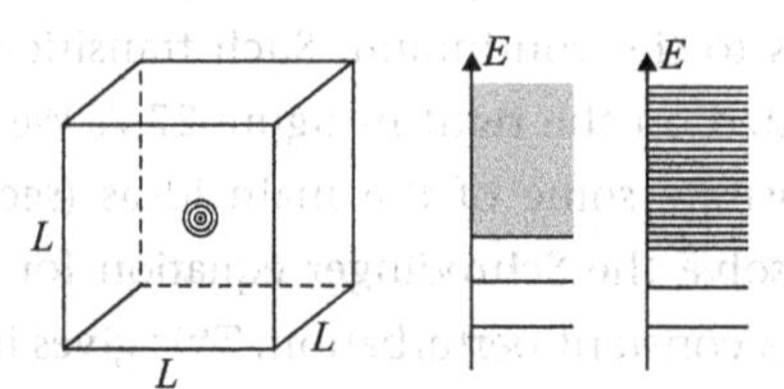

Figure 27.8
To deal with continuum states, we use a cubic box of large size L. The box makes the continuous part of the spectrum discrete, with states very closely packed.

The k's are quantized by the periodicity condition on the wave function:

$$\psi(x+L,y,z) = \psi(x,y+L,z) = \psi(x,y,z+L) = \psi(x,y,z). \tag{27.6.3}$$

The quantization gives

$$\begin{aligned} k_xL=2\pi n_x \quad &\Rightarrow \quad Ldk_x=2\pi dn_x, \\ k_yL=2\pi n_y \quad &\Rightarrow \quad Ldk_y=2\pi dn_y, \\ k_zL=2\pi n_z \quad &\Rightarrow \quad Ldk_z=2\pi dn_z. \end{aligned} \tag{27.6.4}$$

Define ΔN as the total number of states within the little volume element d^3k. It follows from (27.6.4) that we have

$$\Delta N \equiv dn_x\, dn_y\, dn_z = \left(\frac{L}{2\pi}\right)^3 d^3k. \tag{27.6.5}$$

Note that ΔN only depends on d^3k and not on $\mathbf{k}$ itself: the density of states is constant in momentum space. The volume element d^3k can be described in spherical coordinates using the magnitude k of $\mathbf{k}$ and the angles (θ, ϕ) defining the direction of $\mathbf{k}$ (figure 27.9). Therefore,

$$d^3k = k^2dk\, \sin\theta d\theta\, d\phi = k^2dk\, d\Omega. \tag{27.6.6}$$

We now want to express the density of states as a function of energy. For this we take differentials of the relation between the wave number k and the energy E:

$$E=\frac{\hbar^2k^2}{2m} \quad \Rightarrow \quad kdk=\frac{m}{\hbar^2}dE. \tag{27.6.7}$$

Back in (27.6.6), we have that $d^3k = k\frac{m}{\hbar^2}\, dE\, d\Omega$, and hence,

$$\Delta N = \left(\frac{L}{2\pi}\right)^3 k\frac{m}{\hbar^2}\, d\Omega\, dE. \tag{27.6.8}$$

We now introduce some convenient notation, writing ΔN as

$$\Delta N = \rho(E)dE, \tag{27.6.9}$$

where $\rho(E)$ is a *density* of states. More precisely, it is the number of states per unit energy, at around energy E, with momentum pointing within the solid angle $d\Omega$. The last two

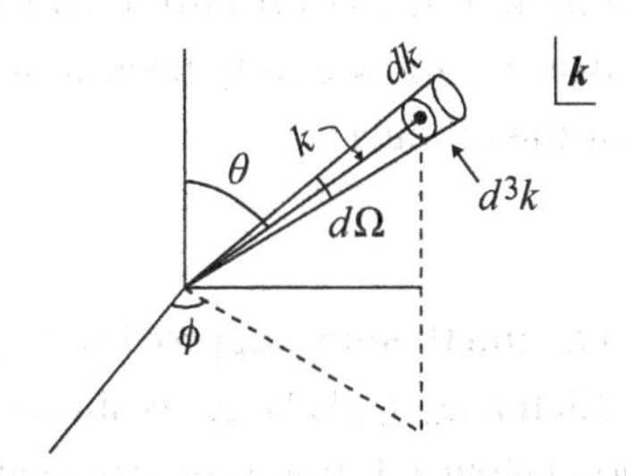

Figure 27.9
A volume element d^3k described in terms of $dk, d\theta$, and $d\phi$, where (θ, ϕ) define the direction of $\mathbf{k}$. The solid-angle element is $d\Omega = \sin\theta d\theta d\phi$.

equations determine this density for us:

$$\rho(E) = \left(\frac{L}{2\pi}\right)^3 \frac{m}{\hbar^2} k\, d\Omega\,. \tag{27.6.10}$$

The utility of $\rho(E)$ is that for a very large box a sum over states can be replaced by an integral as follows:

$$\sum_{\text{states}} \cdots \;\to\; \int \cdots \;\rho(E)dE\,, \tag{27.6.11}$$

where the dots denote an arbitrary function of the momenta $\mathbf{k}$ of the states. We are now ready to perform the integration of the transition amplitudes over final states. We consider the case of constant perturbations first.

Constant perturbations Recall that the Hamiltonian here is $\hat{H}(t) = \hat{H}^{(0)} + V$, with V the constant perturbation. We are not interested in the probability of transition from the initial discrete state to some *particular* state in the continuum. We want the probability that any transition will happen—that is, the total probability that the discrete state will go into a state in the continuum. For this we have to sum the transition probability $P_{f\leftarrow i}$ calculated in (27.4.6) over the now discretized continuum of final states. Remarkably, upon summing, the oscillatory and quadratic behaviors of $P_{f\leftarrow i}$ as a function of time conspire to create a linear behavior!

The sum of transition probabilities over final states is approximated by an integral using our rule (27.6.11). Letting f denote an arbitrary state in the continuum and using the result for the transition amplitude, we find

$$\sum_f P_{f\leftarrow i}(t_0) = \int P_{f\leftarrow i}(t_0)\rho(E_f)dE_f = \int \frac{|V_{fi}|^2}{\hbar^2}\rho(E_f)\,F(\omega_{fi};t_0)\,dE_f\,. \tag{27.6.12}$$

We have not written out explicitly the range of integration because we do not really know how far the continuum extends. In the above integrand $P_{f\leftarrow i}$ peaks for $E_f = E_i$ and is suppressed as $|E_f - E_i|$ becomes large. This is in fact the suppression of $F(\omega_{fi};t_0)$ for large ω_{fi}, this function being large only over a range of order $(\Delta\omega_{fi})_{t_0}$ around $\omega_{fi} = 0$. We therefore expect that the bulk of the contribution to the integral will occur for a narrow range of energies E_f near E_i, and we only need to integrate over some narrow range. Let us assume now that the product

$$K(E_f) \equiv |V_{fi}|^2\rho(E_f) \tag{27.6.13}$$

is a slowly varying function of E_f and therefore approximately constant over the narrow energy interval $(\Delta E)_{t_0}$ over which $F(\omega_{fi}, t_0)$ is large, as shown in figure 27.10. While we will reexamine this assumption below, if it holds, we can evaluate $K(E_f) = |V_{fi}|^2\rho(E_f)$ for E_f set equal to E_i and take it out of the integrand to find that

$$\sum_f P_{f\leftarrow i}(t_0) \simeq \frac{|V_{fi}|^2}{\hbar^2}\rho(E_f = E_i)\cdot I(t_0)\,, \tag{27.6.14}$$

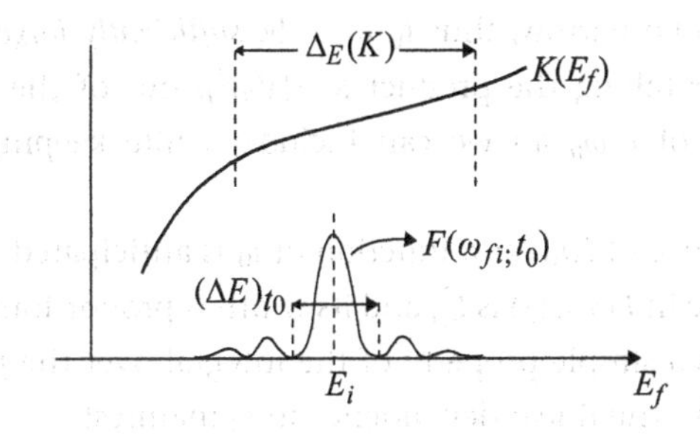

Figure 27.10
Two plots in one, both functions of the energy E_f. The function $F(\omega_{fi};t_0)$ has a main lobe centered at E_i of width $(\Delta E)_{t_0}$. The other function is the product $K(E_F)=|V_{fi}|^2\rho(E_f)$. We indicate the width $\Delta_E(K)$ over which K changes appreciably.

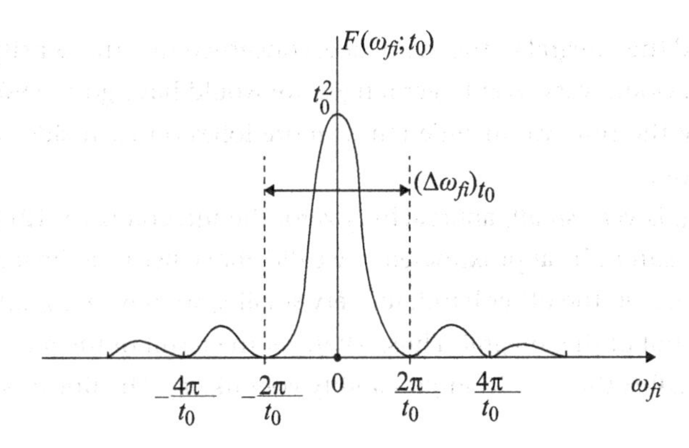

Figure 27.11
Sketch of $F(\omega_{fi};t_0)$ as a function of ω_{fi} for a constant t_0. This function is the integrand in (27.6.15).

with $I(t_0)$ given by

$$I(t_0)\equiv\int F(\omega_{fi};t_0)\,dE_f=\hbar\int F(\omega_{fi};t_0)\,d\omega_{fi}=\hbar\int_{-\infty}^{\infty}\frac{\sin^2(\frac{\omega_{fi}t_0}{2})}{(\frac{\omega_{fi}}{2})^2}\,d\omega_{fi}\,. \tag{27.6.15}$$

Here we noted that $dE_f=\hbar d\omega_{fi}$, given that E_i is taken to be a constant. We also let the integration extend from minus infinity to plus infinity. The plot of $F(\omega_{fi};t_0)$ as a function of ω_{fi} is the same plot as in figure 27.3, with ω replaced by ω_{fi} and t replaced by t_0. We show $F(\omega_{fi};t_0)$ in figure 27.11.

The largest contribution to $I(t_0)$ arises from the main lobe,

$$-\frac{2\pi}{t_0}<\omega_{fi}<\frac{2\pi}{t_0}\,, \tag{27.6.16}$$

whose width is $(\Delta\omega_{fi})_{t_0}$. In terms of the final energy E_f, this corresponds to the range where

$$E_i-\frac{2\pi\hbar}{t_0}<E_f<E_i+\frac{2\pi\hbar}{t_0}\,. \tag{27.6.17}$$

We need this range to be narrow, thus t_0 *must be sufficiently large*. The narrowness is required to justify our taking the product $K=|V_{fi}|^2\rho$ out of the integral. The larger t_0 is, the more lobes of $F(\omega_{fi}, t_0)$ we can include, while keeping the energy range narrow.

The linear dependence of $I(t_0)$ as a function of t_0 is anticipated by noticing that the height of the main lobe in $F(\omega_{fi}; t_0)$ is t_0^2, and its width is proportional to $1/t_0$. The linear dependence, in fact, is a simple property of the integral over the *full* real line. Letting $u=\omega_{fi}t_0/2$, we can make the linear dependence in t_0 manifest:

$$I(t_0) = 2\hbar t_0 \int_{-\infty}^{\infty} \frac{\sin^2 u}{u^2}\, du\,. \tag{27.6.18}$$

The remaining integral evaluates to π, and we get

$$I(t_0) = 2\pi\hbar\, t_0\,. \tag{27.6.19}$$

Had we restricted the integral to the main lobe, concerned that the density of states and matrix elements would vary over larger ranges, we would have gotten 90% of the total contribution. By the time we include ten or more lobes on each side, we are getting 99% of the answer.

Note that if t_0 is very small, approaching zero, the integral (27.6.12) for the transition probability cannot be approximated as we did above because the lobes of $F(\omega_{fi}; t_0)$ become very wide. On the other hand, for very small t_0 we have $F(\omega_{fi}; t_0)\simeq t_0^2$, and this factor then goes out of the integral. Thus, whatever the result of the remaining integration, for very small t_0 the transition probability goes like t_0^2. The linear behavior sets in for larger times.

Having calculated $I(t_0)$, the transition probability (27.6.14) becomes

$$\sum_f P_{f\leftarrow i}(t) \simeq \frac{|V_{fi}|^2}{\hbar^2}\rho(E_f)\, 2\pi\hbar\, t = \frac{2\pi}{\hbar}|V_{fi}|^2\rho(E_f)\, t\,, \tag{27.6.20}$$

where we replaced t_0 by t. While this holds for t when sufficiently large, t cannot be too large because it would make the first-order transition probability large and unreliable. The linear dependence of the transition probability implies we can define a *transition rate* w, or probability of transition per unit time, by dividing the transition probability by t:

$$w \equiv \frac{1}{t}\sum_f P_{f\leftarrow i}(t)\,. \tag{27.6.21}$$

Do not confuse the transition rate w with the similar-looking angular frequency ω, both of which have the same units of inverse time! The result for the transition rate w is Fermi's golden rule for constant perturbations V (and $\hat{H}=\hat{H}^{(0)}+V$):

$$\boxed{\text{Fermi's golden rule:}\quad w = \frac{2\pi}{\hbar}|V_{fi}|^2\rho(E_f)\,,\quad E_f=E_i\,.} \tag{27.6.22}$$

Both the density of states and the matrix element V_{fi} are evaluated at the energy E_i and other observables of the selected final states (spin, momentum, and so on). In this version of the golden rule, the integration over the energy of the final states has been performed. The units are manifestly right: $|V_{fi}|^2$ has units of energy squared, ρ has units of one over energy, and with an $\hbar$ in the denominator, the whole expression has units of one over time, as appropriate for a rate. We can also see that the dependence of w on the size-of-the-box regulator L disappears. The matrix element

$$V_{fi} = \langle f|V|i\rangle \sim L^{-3/2} \tag{27.6.23}$$

because the final state wave function has such dependence (see (27.6.1)). Then the L dependence in $|V_{fi}|^2 \sim L^{-3}$ cancels with the L dependence of the density of states $\rho \sim L^3$, noted in (27.6.10). Finally, note that in terms of the function $K(E)$ defined in (27.6.13) we have

$$w = \frac{2\pi}{\hbar} K(E_i)\,. \tag{27.6.24}$$

Let us summarize the approximations used to derive the golden rule. We have two conditions that must hold simultaneously:

1. We assumed that t_0 is large enough so that the energy range

$$E_i - k\,\frac{2\pi\hbar}{t_0} \;<\; E_f \;<\; E_i + k\,\frac{2\pi\hbar}{t_0}\,, \tag{27.6.25}$$

with k some small integer, is narrow enough that the factor

$$K(E_f) = |V_{fi}|^2 \rho(E_f) \tag{27.6.26}$$

is approximately constant over this range. This allowed us to take this factor out of the integral, making a complete evaluation possible. To state this condition more explicitly, let us call $\Delta_E(K)$, shown in figure 27.10, the energy range over which the change in K is comparable to K. We need the width of the main lobe to be much smaller than this range:

$$\frac{4\pi\hbar}{t_0} \ll \Delta_E(K) \quad\Rightarrow\quad t_0 \gg \frac{\hbar}{\Delta_E(K)}\,. \tag{27.6.27}$$

2. We cannot allow t_0 to be arbitrarily large. Given that $\sum_f P_{f\leftarrow i}(t_0) = wt_0$, the right-hand side must be small for our first-order calculation to be accurate:

$$t_0 \ll \frac{1}{w}\,. \tag{27.6.28}$$

Can the two conditions we imposed above on t_0 be satisfied? Combined together, we have

$$\frac{\hbar}{\Delta_E(K)} \ll t_0 \ll \frac{1}{w}\,. \tag{27.6.29}$$

The conditions can be satisfied if there is a range for t_0 (see figure 27.12).

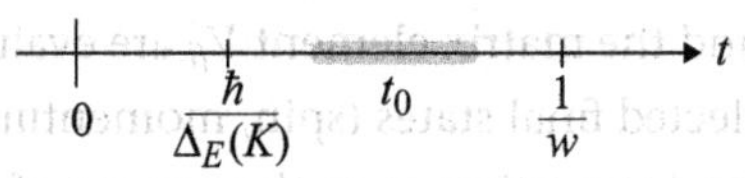

Figure 27.12
The validity of our approximations requires $t_0 \gg \frac{\hbar}{\Delta_E(K)}$ as well as $t_0 \ll \frac{1}{w}$.

A range for t_0 will exist if

$$\frac{\hbar}{\Delta_E(K)} \ll \frac{1}{w} \quad \Rightarrow \quad w \ll \frac{1}{\hbar}\,\Delta_E(K)\,. \tag{27.6.30}$$

Recalling (27.6.24), which gives w in terms of $K(E_i)$, we see that

$$K(E_i) \ll \Delta_E(K)\,. \tag{27.6.31}$$

Note that K has units of energy. The definition of $\Delta_E(K)$ as the energy range over which K changes appreciably from the value at E_i implies that

$$\Delta_E(K)\left|\frac{dK}{dE}\right|_{E_i} \simeq K(E_i)\,. \tag{27.6.32}$$

Going back to the previous inequality, factors of $\Delta_E(K)$ cancel, and we are left with

$$\left|\frac{dK}{dE}\right|_{E_i} \ll 1\,. \tag{27.6.33}$$

In terms of the transition rate $w = 2\pi K/\hbar$, viewed as a function of energy, the above condition simply becomes

$$\left|\frac{d\,\hbar w}{dE}\right|_{E_i} \ll 1\,. \tag{27.6.34}$$

This is our simplest form for the condition of validity of Fermi's golden rule for constant transitions.

Harmonic perturbations The Hamiltonian is now $\hat{H}(t) = \hat{H}^{(0)} + 2H'\cos\omega t$, for $t \geq 0$, and the preparatory work was done in section 27.5. The analysis that follows is completely analogous to the one for constant perturbations, so we can be brief. We write our results for the case of absorption—that is, when $E_f = E_i + \hbar\omega > E_i$. This time we use the transition probability $P_{f\leftarrow i}$ obtained in (27.5.8). Summing over final states, we find

$$\sum_f P_{f\leftarrow i}(\omega; t_0) = \int P_{f\leftarrow i}(\omega; t_0)\,\rho(E_f)dE_f = \int \frac{|H'_{fi}|^2}{\hbar^2}\rho(E_f)\frac{\sin^2\left(\frac{\omega_{fi}-\omega}{2}t_0\right)}{(\frac{\omega_{fi}-\omega}{2})^2}dE_f\,. \tag{27.6.35}$$

This time the main contribution comes from the region where

$$-\frac{2\pi}{t_0} < \omega_{fi} - \omega < \frac{2\pi}{t_0}\,. \tag{27.6.36}$$

In terms of the final energy, this is the band:

$$(E_i + \hbar\omega) - \frac{2\pi\hbar}{t_0} < E_f < (E_i + \hbar\omega) + \frac{2\pi\hbar}{t_0}\,. \tag{27.6.37}$$

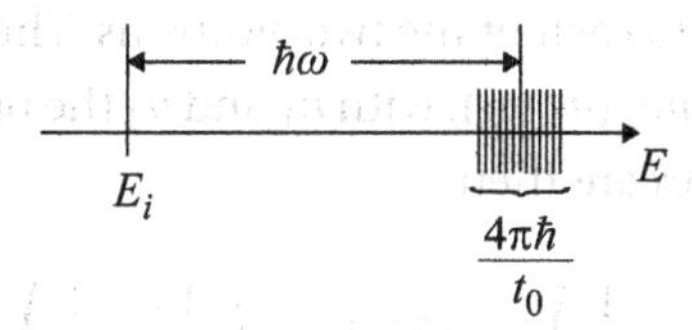

Figure 27.13
In the absorption process, the integral over final states is effectively over a narrow band of states centered at the energy $E_i + \hbar\omega$.

For large t_0, this is the narrow band of states shown in figure 27.13. Assume that over this band $|H'_{fi}|^2\rho$ is constant so that after changing the variable of integration from E_f to ω_{fi} we get

$$\sum_f P_{f\leftarrow i}(\omega; t_0) = \frac{|H'_{fi}|^2}{\hbar}\rho\Big|_{E_i+\hbar\omega}\int_{-\infty}^{\infty}\frac{\sin^2\left(\frac{\omega_{fi}-\omega}{2}t_0\right)}{(\frac{\omega_{fi}-\omega}{2})^2}d\omega_{fi}\,. \tag{27.6.38}$$

The integrand is the same one we had in (27.6.15), only shifted by the constant ω. The value of the integral is therefore the same, $2\pi t_0$, and this gives

$$\sum_f P_{f\leftarrow i}(\omega; t_0) = \frac{2\pi}{\hbar}|H'_{fi}|^2\rho\Big|_{E_i+\hbar\omega} t_0\,. \tag{27.6.39}$$

The transition rate w is finally given by

$$\text{Fermi's golden rule:}\quad w = \frac{2\pi}{\hbar}|H'_{fi}|^2\rho(E_f)\,,\quad E_f = E_i + \hbar\omega\,. \tag{27.6.40}$$

Here $\delta H(t) = 2H'\cos\omega t$. This result is known as Fermi's golden rule for harmonic perturbations. As written, it applies for absorption. For stimulated emission the only change required is in the relation between E_f and E_i, which becomes $E_f = E_i - \hbar\omega$.

The discussion of the validity of (27.6.40) is completely analogous to our previous discussion of constant perturbations, leading to the condition (27.6.34). This condition also applies here, with the proviso that the expression is evaluated at the energy $E_f = E_i + \hbar\omega$:

$$\left|\frac{d\,\hbar w}{dE}\right|_{E_f} \ll 1\,. \tag{27.6.41}$$

27.7 Helium Atom and Autoionization

The helium atom has two protons ($Z = 2$) and two electrons. Let $\hat{H}^{(0)}$ be the Hamiltonian for this system, ignoring the Coulomb repulsion between the electrons:

$$\hat{H}^{(0)} = \frac{\hat{\mathbf{p}}_1^2}{2m} - \frac{Ze^2}{r_1} + \frac{\hat{\mathbf{p}}_2^2}{2m} - \frac{Ze^2}{r_2}\,. \tag{27.7.1}$$

Here, the labels 1 and 2 refer to each of the two electrons. The spectrum of this Hamiltonian consists of *hydrogenic states* (n_1, n_2), with n_1 and n_2 the principal quantum numbers for the electrons. The energies are then

$$E_{n_1,n_2} = -(13.6\,\text{eV})Z^2\left(\frac{1}{n_1^2}+\frac{1}{n_2^2}\right) = -(54.4\,\text{eV})\left(\frac{1}{n_1^2}+\frac{1}{n_2^2}\right). \tag{27.7.2}$$

In this approximation, the helium ground state $(1,1)$ is at $E_{1,1} = -108.8\,\text{eV}$. For the hydrogen atom, we have bound states of negative energy and continuum states of *positive* energy that can be described, to some approximation, as momentum eigenstates of an electron no longer bound to the proton. For the helium atom, since we have two electrons, there are continuum states in the spectrum that have *negative* energy, with one electron bound and the other free.

As we increase the energy from the ground state, continuum states first appear for $n_1 = 1$ in the limit as $n_2 \to \infty$. For $n_1 = 1$ and $n_2 = \infty$, one electron is tightly bound, while the second electron is essentially free and contributes no energy. Thus, a continuum appears for energy $E_{1,\infty} = -54.4\,\text{eV}$. This $(1,\infty)$ continuum extends for all $E \geq -54.4\,\text{eV}$, as the free electron can have arbitrary positive kinetic energy. The level $(2,\infty)$ is the beginning of a second continuum, with $E \geq -13.6\,\text{eV}$. The level $(3,\infty)$ is the beginning of a third continuum, with $E \geq -6.0\,\text{eV}$. In general, the state (n,∞) with $n \geq 1$ marks the beginning of the nth continuum, with $E \geq -54.4/n^2\,\text{eV}$. In each of these continua, one electron is still bound, and the other is free. A diagram showing some discrete states and the $(1,\infty)$, $(2,\infty)$, and $(3,\infty)$ continua is given in figure 27.14.

Self-ionizing energy-conserving transitions can occur because discrete states can find themselves in the middle of a continuum. The state $(2s)^2$, for example, with two electrons on the $2s$ configuration and with energy $E_{2,2} = -27.2$ eV, is in the middle of

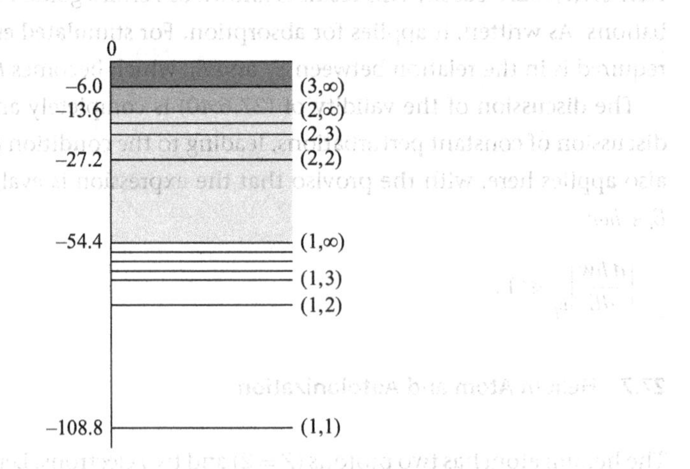

Figure 27.14
Hydrogenic states in helium and continuum states of negative total energy. The energies along the vertical axis are in electron volts, and (n_1, n_2) are the principal quantum numbers of the electrons.

the $(1, \infty)$ continuum. We can view the $(2s)^2$ hydrogenic state as a $t=0$ eigenstate of $\hat{H}^{(0)}$ and treat the Coulomb repulsion as a constant perturbation. We thus have a total Hamiltonian $\hat{H}$ that takes the form

$$\hat{H} = \hat{H}^{(0)} + V, \quad V = \frac{e^2}{|\mathbf{r}_1 - \mathbf{r}_2|}. \tag{27.7.3}$$

The perturbation V produces "autoionizing" transitions to the continuum. We have a transition from the $(2s)^2$ state with energy $E_{2,2} = -27.2\,\text{eV}$ to a state of one bound electron with energy $-54.4\,\text{eV}$ $(n=1)$ and one free electron with kinetic energy of $27.2\,\text{eV}$. That final state is part of a continuum. In fact, a radiative transition from $(2s)^2$ to $(1s)(2s)$, with photoemission, is a lot less likely than the autoionizing transition.

We will not do the quantitative analysis required to determine the lifetime of the $(2s)^2$ state. This would entail using Fermi's golden rule for constant transitions. The main ingredient of the computation would be the calculation of the matrix element of V between the initial state $(2s)^2$ and the final state $1s$-plus-free electron.

In general, autoionization is a process in which an atom or a molecule in an excited state spontaneously emits one of the outer-shell electrons. Autoionizing states are usually short-lived. *Auger transitions* are autoionization processes in which the filling of an inner-shell vacancy is accompanied by the emission of an electron. This effect was discovered by Lise Meitner (1922) and, independently, by Pierre Auger (1923). Our example of the $(2s)^2$ state ejecting one electron while letting the other fill the $1s$ vacancy is an Auger transition. Molecules can have autoionizing Rydberg states in which the little energy needed to remove the Rydberg electron is supplied by a vibrational excitation of the molecule.

27.8 Modeling the Decay of a Discrete State to the Continuum

We want to discuss a model for the decay of a discrete state that lies in the middle of a continuum of states. The discrete state is rendered unstable by a constant perturbation coupling it to the continuum, as in the case of autoionization. The purpose of the following analysis is to go beyond the calculation of the decay rate. We wish now to show explicitly how the probability of remaining in the unstable discrete state decays exponentially, not linearly, as would be naively assumed by taking the decay rate formula to hold for long times. We will also be able to determine the time-dependent probability amplitudes for the final continuum states, showing how they are populated as a result of the decay, as well as the associated probabilities as a function of the energy. This analysis was first performed by Victor Weisskopf and Eugene Wigner (1930).

To model the process, we assume we have a (normalized) discrete state $|i\rangle$ with energy E_i. This is the initial state at $t=0$. In addition, we have a continuum $|\alpha\rangle$ of states with energy E_α. Here α is a continuous variable, and the states are normalized as follows:

$$\langle \alpha | \alpha' \rangle = \delta(\alpha - \alpha'). \tag{27.8.1}$$

This is analogous to the way we normalized position states $|x\rangle$, also labeled by a continuous variable. The state $|i\rangle$ is orthogonal to all states $|\alpha\rangle$. The variable α is correlated with the energy E_α of the states so we can speak of a density of states via

$$d\alpha \equiv \rho(E_\alpha)dE_\alpha\,. \tag{27.8.2}$$

The completeness relation for the states is

$$|i\rangle\langle i| + \int d\alpha\,|\alpha\rangle\langle\alpha| = \mathbb{1}\,. \tag{27.8.3}$$

The constant perturbation V will be assumed to just couple the discrete state to the continuum:

$$V_{i\alpha} = \langle i|V|\alpha\rangle \neq 0\,, \tag{27.8.4}$$

of course with $V_{\alpha i} = V_{i\alpha}^*$. The other matrix elements vanish: $V_{\alpha\alpha'} = 0$ for any two states in the continuum and $V_{ii} = 0$. By assumption, V has no time dependence, thus $V_{i\alpha}$ is also time independent.

Fermi's golden rule for constant perturbations (27.6.22) gives us the transition rate w:

$$w = \frac{2\pi}{\hbar}|V_{\alpha i}|^2\rho(E_i)\,, \tag{27.8.5}$$

where the matrix element $V_{\alpha i}$ is evaluated for α such that $E_\alpha = E_i$; this α value exists since we assumed that E_i lies in the continuum. Just as we did when discussing the conditions for the validity of Fermi's golden rule, it is useful to introduce a function $K(E_\alpha)$ of the energy E_α, allowing a brief rewriting of the transition rate:

$$K(E_\alpha) \equiv |V_{\alpha i}|^2\rho(E_\alpha) \quad \Rightarrow \quad w = \frac{2\pi}{\hbar}K(E_i)\,. \tag{27.8.6}$$

A transition or decay rate w implies that the probability $P_i(t)$ that the state is still $|i\rangle$ at time $t > 0$ is approximately

$$P_i(t) \simeq 1 - wt\,, \quad \text{for} \quad \tfrac{\hbar}{\Delta_E(K)} \ll t \ll \tfrac{1}{w}\,, \tag{27.8.7}$$

where the range comes from (27.6.29). Our first goal is to show that for large t the probability $P_i(t)$ in fact decays exponentially:

$$P_i(t) = e^{-wt}\,. \tag{27.8.8}$$

This also shows that the decay process is irreversible: this probability goes to zero as $t \to \infty$. When we have a two-level system and a perturbation, we get oscillations in the probabilities of being in either state. The continuum, however, makes a difference. In a discretized model of the continuum, we would expect oscillations, but with a period that diverges as the spacing between energy levels goes to zero. Our second goal is to find the energy distribution of the final states.

To solve for the time evolution of the state, we set up the interaction picture wave function as in (27.2.20), modified naturally to include the continuum:

$$|\tilde{\Psi}(t)\rangle = c_i(t)|i\rangle + \int d\alpha\; c_\alpha(t)|\alpha\rangle\,. \tag{27.8.9}$$

Here, $c_i(t)$ and $c_\alpha(t)$ are undetermined amplitudes. Our initial conditions at time equal zero are

$$c_i(0) = 1\,, \quad c_\alpha(0) = 0\,. \tag{27.8.10}$$

Recalling the discretized form (27.2.29) of the Schrödinger equation, $i\hbar\dot{c}_m = \sum_n e^{i\omega_{mn}t}\delta H_{mn}c_n(t)$, we find the coupled equations by using $\delta H = V$ and replacing sums by integrals to account for the continuum:

$$\begin{aligned} i\hbar\,\dot{c}_i(t) &= \int d\alpha\, e^{i(E_i-E_\alpha)t/\hbar}\, V_{i\alpha}\, c_\alpha(t)\,, \\ i\hbar\,\dot{c}_\alpha(t) &= e^{i(E_\alpha-E_i)t/\hbar}\, V_{\alpha i}\, c_i(t)\,. \end{aligned} \tag{27.8.11}$$

Exercise 27.4. *Rederive the above equations starting from the Schrödinger equation, as was done for (27.2.29), and using the completeness relation (27.8.3).*

To make progress here, we will eliminate $c_\alpha(t)$ and find an equation for $c_i(t)$. We use the second equation in (27.8.11) to write an integral expression for $c_\alpha(t)$:

$$c_\alpha(t) = \frac{V_{\alpha i}}{i\hbar}\int_0^t dt' e^{i(E_\alpha-E_i)t'/\hbar}\, c_i(t')\,. \tag{27.8.12}$$

Replacing this result into the equation for $\dot{c}_i$, we find that

$$\dot{c}_i(t) = -\frac{1}{\hbar^2}\int d\alpha\, |V_{\alpha i}|^2 \int_0^t dt'\; e^{i(E_i-E_\alpha)(t-t')/\hbar} c_i(t')\,. \tag{27.8.13}$$

The α integration can be converted to an energy integration using the relations

$$d\alpha |V_{\alpha i}|^2 = dE_\alpha\, \rho(E_\alpha)|V_{\alpha i}|^2 = dE_\alpha\, K(E_\alpha) \tag{27.8.14}$$

and calling the integration variable E_α simply E:

$$\dot{c}_i(t) = -\frac{1}{\hbar^2}\int_0^t dt' c_i(t') \int dE\, K(E)\, e^{i(E_i-E)(t-t')/\hbar}\,. \tag{27.8.15}$$

In writing this equation, we also switched the order of integration. This is an integral/differential equation for $c_i(t)$. Its advantage over the original differential equation is that we can do a physically motivated approximation.

Indeed, consider the integral over E. The function $K(E)$ is slowly varying, as you can see from (27.6.33). The exponential in the integral, however, is a fast-changing function of energy so long as $t - t' \neq 0$. The phase here is in fact a stationary function of the energy E only for $t' = t$. The integral is therefore large only for $t' \simeq t$, and we can therefore replace $c_i(t') \to c_i(t)$. With this replacement, $c_i(t)$ goes out of the integral, and we now have

$$\dot{c}_i(t) = \Big(-\frac{1}{\hbar^2}\int_0^t dt' \int dE\, K(E)\, e^{i(E_i-E)(t-t')/\hbar}\Big)\; c_i(t)\,. \tag{27.8.16}$$

With this major simplification, our goal becomes the evaluation of the function enclosed by large parentheses, which we will call $\mathcal{W}(t)$. We change once more the order of integration and pass from the integration variable t' to $\tau = t - t'$. With these steps we now find that

$$\dot{c}_i(t) = \mathcal{W}(t)\, c_i(t)\,, \quad \text{with} \quad \mathcal{W}(t) = -\frac{1}{\hbar^2}\int dE\, K(E)\int_0^t d\tau\, e^{i(E_i-E)\tau/\hbar}\,. \tag{27.8.17}$$

The key point is that for large time $\mathcal{W}$ is a constant, which we can evaluate. To show this we will simplify the time integral using the following identity:

$$\int_0^\infty d\tau\, e^{iE\tau} = i\mathcal{P}\frac{1}{E} + \pi\delta(E)\,. \tag{27.8.18}$$

The symbol $\mathcal{P}$ here is an instruction to take the **principal value** when doing a *further* integral over E, such as we have in $\mathcal{W}$. Indeed, the above identity is meant to be used inside an integral over E. In that situation the principal value is defined by computing the integral over E, excising a neighborhood $(-\eta, \eta)$ of $E=0$, and then taking the limit as $\eta \to 0^+$. We will prove (27.8.18) at the end of the section. Had the integration on the left-hand side extended from minus to plus infinity, the result would have been $2\pi\delta(E)$ (recall (4.4.5)). For integration from zero to infinity, we get half of that delta function contribution, plus the principal part.

The above identity implies that

$$\lim_{t\to\infty}\int_0^t d\tau\, e^{i(E_i-E)\tau/\hbar} = i\hbar\mathcal{P}\frac{1}{E_i-E} + \hbar\pi\delta(E_i-E)\,. \tag{27.8.19}$$

Indeed, the constant $\hbar$ appears multiplicatively after absorbing it into a redefinition of τ. Additionally, as $t\to\infty$ we get the full integral for which (27.8.18) applies. In practice, we do not really need to take $t\to\infty$; it suffices to take t sufficiently large. How large is a function of the kind of integral the identity (27.8.19) is going to go into. In our case we are integrating against $K(E)$, as indicated in (27.8.17). For accuracy, we need the energy scale associated with $\hbar/t$, responsible for creating the singular features on the right-hand side of (27.8.19), to be much smaller than the scale $\Delta_E(K)$ for appreciable change of K:

$$\frac{\hbar}{t} \ll \Delta_E(K) \quad\Rightarrow\quad t \gg \frac{\hbar}{\Delta_E(K)}\,. \tag{27.8.20}$$

This lower bound on t allows $t \sim 1/w$ as well as $t \gg 1/w$, as you can see from the range displayed in figure (27.12). To compute $\mathcal{W}$ we need to use (27.8.19) inside an integral, as it was meant to be used. For large t we find that

$$\begin{aligned}\mathcal{W}(t) &= -\frac{1}{\hbar}\int dE\, K(E)\Big[i\mathcal{P}\frac{1}{E_i-E} + \pi\delta(E_i-E)\Big]\\ &= -\frac{\pi}{\hbar}K(E_i) - \frac{i}{\hbar}\mathcal{P}\int\frac{dE\, K(E)}{E_i-E}\\ &= -\tfrac{1}{2}w - \tfrac{i}{\hbar}\,\delta E_i\,, \qquad t \gg \tfrac{\hbar}{\Delta_E(K)}\,,\end{aligned} \tag{27.8.21}$$

where we used (27.8.6) and introduced the energy shift δE_i defined by

$$\delta E_i \equiv \mathcal{P}\int\frac{dE\, K(E)}{E_i-E} = -\mathcal{P}\int dE_\alpha\, \rho(E_\alpha)\frac{|V_{\alpha i}|^2}{E_\alpha - E_i}\,, \tag{27.8.22}$$

reverting to the integration variable E_α carrying the continuum label. We now claim that δE_i is in fact the quadratic shift in the energy E_i of the state $|i\rangle$, due to its time-independent coupling V to the continuum. Indeed, compare the above expression for δE_i with the second-order shift (25.2.36) in the case of a fully discrete spectrum:

$$E_n^{(2)} = -\sum_{k\neq n} \frac{|\delta H_{kn}|^2}{E_k^{(0)} - E_n^{(0)}} \,. \tag{27.8.23}$$

We see that $\mathcal{P}\int dE_\alpha\, \rho(E_\alpha)$ is precisely the sum over continuum states, with $\mathcal{P}$ implementing the continuum version of the constraint $k\neq n$. The matrix element squared and the energy differences appear in the same way in (27.8.22) and in (27.8.23). We will see additional confirmation below that $E_i + \delta E_i$ is indeed the approximate energy of the discrete state: it is the energy for which the probability distribution of continuum states peaks. Note that the first-order correction to the energy of the state $|i\rangle$ vanishes because we assumed $V_{ii} = 0$.

Thus, for sufficiently large time the differential equation (27.8.17) for the c_i coefficients becomes

$$\dot{c}_i(t) = -\big(\tfrac{1}{2}w + \tfrac{i}{\hbar}\delta E_i\big)c_i(t)\,. \tag{27.8.24}$$

Consistent with $c_i(0) = 1$, this is solved by

$$c_i(t) = e^{-\frac{w}{2}t}e^{-i\frac{\delta E_i t}{\hbar}}\,, \qquad t \gg \frac{\hbar}{\Delta_E(K)}\,. \tag{27.8.25}$$

It follows that the probability $P_i(t)$ of remaining in the original discrete state for $t \gg \frac{\hbar}{\Delta_E(K)}$ is indeed given by

$$P_i(t) = |c_i(t)|^2 = e^{-wt}\,, \tag{27.8.26}$$

and the lifetime of the state is $\Delta t = 1/w$. Showing this was our first goal.

Having solved for the $c_i(t)$ coefficients, we can now go back and calculate the $c_\alpha(t)$ to see how the amplitudes for continuum states evolve in the decay. Using (27.8.12), we have the simple integral

$$c_\alpha(t) = \frac{V_{\alpha i}}{i\hbar}\int_0^t dt' e^{\left(-\frac{w}{2} + \frac{i}{\hbar}(E_\alpha - E_i - \delta E_i)\right)t'}\,, \tag{27.8.27}$$

which immediately gives

$$c_\alpha(t) \,=\, V_{\alpha i}\, \frac{1 - e^{-\frac{1}{2}wt}e^{i(E_\alpha - E_i - \delta E_i)t/\hbar}}{E_\alpha - E_i - \delta E_i + \frac{i}{2}\hbar w}\,, \qquad t \gg \frac{\hbar}{\Delta_E(K)}\,. \tag{27.8.28}$$

These are the time-dependent amplitudes for the continuum states! It is remarkable that an analytic expression can be found for these amplitudes. For times larger than the lifetime of the discrete state, we can ignore the second term in the numerator and find that

$$c_\alpha(t) \,=\, \frac{V_{\alpha i}}{E_\alpha - E_i - \delta E_i + \frac{i}{2}\hbar w}\,, \qquad t \gg 1/w\,. \tag{27.8.29}$$

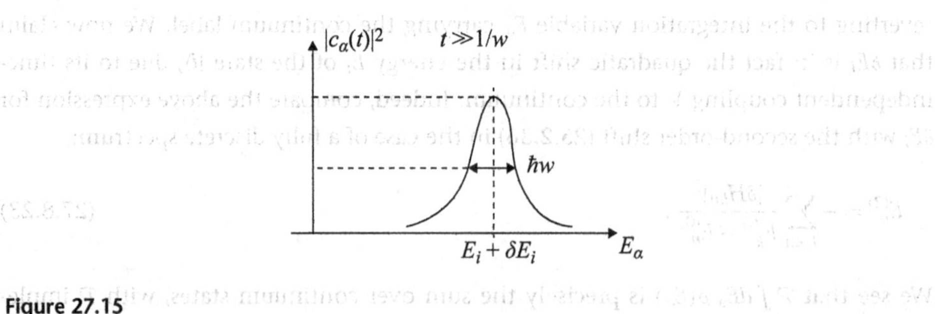

Figure 27.15
The probability distribution $|c_\alpha|^2$ of continuum states $|\alpha\rangle$ populated by the decay of a discrete state $|i\rangle$ with decay rate w and energy $E_i + \delta E_i$.

The associated probability distribution is

$$|c_\alpha(t)|^2 = \frac{|V_{\alpha i}|^2}{[E_\alpha - (E_i + \delta E_i)]^2 + \frac{\hbar^2 w^2}{4}}, \qquad t \gg 1/w. \tag{27.8.30}$$

As a function of the energy E_α of the continuum state, the squared matrix element in the numerator has some slow dependence that we will ignore; it is minimal over an energy range of order $\hbar w$ that displays the features of the distribution. With this assumption, the distribution peaks at $E_i + \delta E_i$, as shown in figure 27.15. Note that this is precisely the corrected energy of the discrete state. Moreover, the width of the distribution is $\hbar w$ since the peak value is reduced by a factor of two for $E_\alpha = E_i + \delta E_i \pm \frac{1}{2}\hbar w$. Note also that the lifetime $\Delta t = 1/w$ of the particle and the energy uncertainty $\Delta E \simeq \hbar w$ of the decay products satisfy the expected uncertainty product $\Delta t \Delta E \simeq \hbar$. This completes our analysis of the decay process. It just remains to prove the identity (27.8.18).

Proving (27.8.18) Recall the singular integral $\int_{-\infty}^{\infty} e^{iEt} dt = 2\pi\delta(E)$, used repeatedly in our discussion of Fourier transforms in section 4.4. The purpose of the following is to find the value of the same integral but with an integration range from zero to infinity: $\int_0^\infty e^{iEt} dt$. Just like the first integral, this is also singular; it will yield a delta function and something else. Since delta functions are not ordinary functions, the above integrals are only well defined if there is a further integration. To evaluate the integral in question, we will introduce a regulator, a small positive constant ϵ, and write

$$\int_0^\infty e^{iEt} dt = \lim_{\epsilon \to 0^+} \int_0^\infty e^{-\epsilon t} e^{iEt} dt = \lim_{\epsilon \to 0^+} \int_0^\infty e^{-(\epsilon - iE)t} dt. \tag{27.8.31}$$

This regulator makes the integral convergent, and for $\epsilon = 0$ we recover the original integral. Evaluating the final integral, we have

$$\int_0^\infty e^{iEt} dt = \lim_{\epsilon \to 0^+} \frac{1}{\epsilon - iE} = i \lim_{\epsilon \to 0^+} \frac{1}{E + i\epsilon} = i \lim_{\epsilon \to 0^+} \frac{E - i\epsilon}{E^2 + \epsilon^2}. \tag{27.8.32}$$

The limit can be taken separately for the real and imaginary parts, as both need to make sense:

$$\int_0^\infty e^{iEt} dt = i \lim_{\epsilon \to 0^+} \frac{E}{E^2 + \epsilon^2} + \lim_{\epsilon \to 0^+} \frac{\epsilon}{E^2 + \epsilon^2}. \tag{27.8.33}$$

The second term on the right-hand side is a representation of a delta function at $E=0$; it is indeed a function of E with width of order ϵ and peak value $1/\epsilon$ at $E=0$. As $\epsilon \to 0$, it becomes narrower with a larger peak, and you can check that the area under the curve equals π, exactly, for any value of $\epsilon > 0$. We therefore see that

$$\int_0^\infty e^{iEt}dt = i \lim_{\epsilon\to 0^+} \frac{E}{E^2+\epsilon^2} + \pi\delta(E). \tag{27.8.34}$$

The remaining limit can also be clarified, leading to the concept of the principal value $\mathcal{P}$ of an integral. The limit is $1/E$, when $E \neq 0$, but such an answer is singular and not well defined for $E=0$, just like the delta function. The limit can be understood, however, when operating within an integral over E. To see this, consider the limit above evaluated after multiplying by a function $f(E)$, continuous at $E=0$, and integrating over E:

$$\lim_{\epsilon\to 0^+} \int_{-\infty}^\infty \frac{E}{E^2+\epsilon^2} f(E)\, dE. \tag{27.8.35}$$

Since the delicate part of the integral is near $E=0$, we split the integration region using a small constant $\eta > 0$ that we can also assume approaches zero:

$$\lim_{\epsilon\to 0^+} \int_{-\infty}^\infty \frac{E}{E^2+\epsilon^2} f(E)dE = \lim_{\epsilon\to 0^+} \lim_{\eta\to 0^+} \left[\int_{-\infty}^{-\eta} + \int_{-\eta}^{\eta} + \int_{\eta}^{\infty} \right] \frac{E}{E^2+\epsilon^2} f(E)\, dE. \tag{27.8.36}$$

The middle integral vanishes as $\eta \to 0$:

$$\lim_{\eta\to 0^+} \int_{-\eta}^{\eta} \frac{E}{E^2+\epsilon^2} f(E)\, dE = f(0) \lim_{\eta\to 0^+} \int_{-\eta}^{\eta} \frac{E}{E^2+\epsilon^2}\, dE = 0. \tag{27.8.37}$$

In the first step, we used the continuity of $f(E)$ at $E=0$, and the final equality follows because the integrand is bounded for any $\epsilon > 0$, and the range of integration goes to zero. For the remaining two integrals, which avoid a neighborhood of $E=0$, we can take the $\epsilon \to 0^+$ limit, getting

$$\lim_{\epsilon\to 0^+} \int_{-\infty}^\infty \frac{E}{E^2+\epsilon^2} f(E)dE = \lim_{\eta\to 0^+} \left[\int_{-\infty}^{-\eta} + \int_{\eta}^{\infty} \right] \frac{1}{E} f(E)\, dE \equiv \mathcal{P} \int_{-\infty}^{\infty} \frac{1}{E} f(E) dE. \tag{27.8.38}$$

The last relation defines the principal value of the integral: it just means excising a symmetric region about $E=0$, computing the integral, and letting the excised region go to zero symmetrically. Using symbolic notation, we remove the integral and the function $f(E)$ and write the previous expression as follows:

$$\lim_{\epsilon\to 0^+} \frac{E}{E^2+\epsilon^2} = \mathcal{P}\frac{1}{E}. \tag{27.8.39}$$

Going back to (27.8.34), we obtain the identity we aimed to prove:

$$\int_0^\infty e^{iEt}dt = i\mathcal{P}\frac{1}{E} + \pi\delta(E). \tag{27.8.40}$$

Exercise 27.5. *Confirm that $\mathcal{P}\int_{-b}^{a} \frac{dx}{x} = \ln\frac{a}{b}$, with $a, b > 0$. Note that this principal value integral vanishes for $a=b$, as should be clear from the graph of $1/x$.*

27.9 Ionization of Hydrogen

We now illustrate the use of Fermi's golden rule by applying it to the calculation of the ionization rate for hydrogen when hit by a linearly polarized plane electromagnetic wave. We assume the hydrogen atom has its electron on the ground state. In this ionization process, at the level of particles, a photon ejects the bound electron, which becomes free.

We will begin by stating our approximations and finding a range of photon energies for which the approximations hold. We then examine the perturbation coupling the electromagnetic wave to the electron. Next is the calculation of the matrix element of the perturbation in between initial and final states. With this result and Fermi's golden rule, we then assemble the ionization rate and examine its dependence on the intensity of the electromagnetic wave and on the energy of the constituent photons.

The kinematics of the ionization process is straightforward. If the electromagnetic field has frequency ω, the incident photons have energy

$$E_\gamma = \hbar\omega\,. \tag{27.9.1}$$

For the ejected electron, its energy E_e and the magnitude k_e of its momentum are given by

$$E_e = \frac{\hbar^2 k_e^2}{2m} = E_\gamma - \text{Ry}\,, \tag{27.9.2}$$

where the last equality follows from energy conservation, assuming that the atom is initially in its ground state. Here, the Rydberg Ry is the magnitude of the ground state energy:

$$2\text{Ry} = \frac{e^2}{a_0} = \frac{\hbar^2}{ma_0^2} = \alpha\,\frac{\hbar c}{a_0}\,, \qquad \text{Ry} \simeq 13.6\ \text{eV}\,. \tag{27.9.3}$$

Let us now discuss two assumptions that simplify the calculation considerably:

1. *Ignoring the spatial dependence of the electromagnetic wave.* This approximation allows us to take the electromagnetic field at the atom to be a uniform, time-dependent field. We can ignore the spatial dependence of the wave if the wavelength λ of the photon is much bigger than the Bohr radius a_0:

$$\frac{\lambda}{a_0} \gg 1\,. \tag{27.9.4}$$

Such a condition puts an *upper* bound on the photon energy since the more energetic the photon, the smaller its wavelength. To bound the energy, we first write

$$\lambda = \frac{2\pi}{k_\gamma} = \frac{2\pi c}{\omega} = \frac{2\pi\hbar c}{\hbar\omega} \tag{27.9.5}$$

and then find that

$$\frac{\lambda}{a_0} = \frac{2\pi}{\hbar\omega}\frac{\hbar c}{a_0} = \frac{4\pi}{\alpha}\frac{\text{Ry}}{\hbar\omega} \simeq 1722\,\frac{\text{Ry}}{\hbar\omega}\,. \tag{27.9.6}$$

The inequality (27.9.4) then gives

$$\hbar\omega \ll 1722\,\text{Ry} \simeq 23\,\text{keV}\,. \tag{27.9.7}$$

Note that this condition, from (27.9.4), can also be written as

$$k_\gamma a_0 \ll 1\,. \tag{27.9.8}$$

2. *Assuming the ejected electron is in a free-particle momentum state.* This requires the ejected electron to be energetic enough to remain mostly unaffected by the Coulomb field. As a result, the photon must be energetic enough, and this constraint provides a *lower* bound for its energy. We thus require that

$$E_e = E_\gamma - \text{Ry} = \hbar\omega - \text{Ry} \gg \text{Ry} \quad \Rightarrow \quad \hbar\omega \gg \text{Ry}\,. \tag{27.9.9}$$

The computation to be discussed below, however, will show that this plane wave condition is surprisingly subtle. Depending on what one calculates, the assumption that the electron state is a plane wave may or may not hold. While it seems intuitively clear that the electron state is a plane wave far away from the proton, it is not clear that closer to the proton this approximation is good, even for a highly energetic electron.

An essentially equivalent condition can be stated using the de Broglie wavelength of the electron. The more energetic the electron, the shorter its de Broglie wavelength λ_e. Energetic electrons will have $\lambda_e \ll a_0$, or equivalently, with $k_e = 2\pi/\lambda_e$,

$$k_e a_0 \gg 1\,. \tag{27.9.10}$$

This is just the reverse of the corresponding photon inequality (27.9.8).

The inequalities (27.9.7) and (27.9.9), put together, demand that

$$\text{Ry} \ll \hbar\omega \ll 1722\,\text{Ry}\,. \tag{27.9.11}$$

If we consider that $1 \ll 10$, the allowed photon energy range is

$$140\,\text{eV} \leq \hbar\omega \leq 2.3\,\text{keV}\,. \tag{27.9.12}$$

Even for the upper limit, the electron is not relativistic: $2.3\,\text{keV} \ll m_e c^2 \simeq 511$ keV. In terms of the photon wavelength λ, the above $\hbar\omega$ range corresponds to $0.54\,\text{nm} \leq \lambda \leq 8.9\,\text{nm}$. These are X-rays. To close the loop, we can determine $k_e a_0$ to see if the inequality $k_e a_0 \gg 1$ holds. Using energy conservation,

$$\frac{\hbar^2 k_e^2}{2m} = \hbar\omega - \text{Ry} = \text{Ry}\left(\frac{\hbar\omega}{\text{Ry}} - 1\right) = \frac{\hbar^2}{2ma_0^2}\left(\frac{\hbar\omega}{\text{Ry}} - 1\right). \tag{27.9.13}$$

Canceling common factors of $\hbar^2/2m$, we find

$$k_e^2 a_0^2 = \frac{\hbar\omega}{\text{Ry}} - 1\,. \tag{27.9.14}$$

On account of (27.9.11), this results in the following range for $(k_e a_0)^2$:

$$1 \ll k_e^2 a_0^2 \ll 1700\,, \tag{27.9.15}$$

clearly consistent with having $k_e a_0 \gg 1$.

The interaction Hamiltonian We can describe the incident electromagnetic plane wave via a nonvanishing vector potential

$$\mathbf{A}(\mathbf{r},t) = -\frac{c}{\omega}\, 2\,\mathbf{E}_0 \cos(\mathbf{k}_\gamma \cdot \mathbf{r} - \omega t)\,, \quad \Phi(\mathbf{r},t) = 0\,. \tag{27.9.16}$$

Here, $\mathbf{r}$ is the position vector, $\mathbf{E}_0$ is a constant vector with units of electric field, and $\mathbf{k}_\gamma$, with magnitude k_γ, points in the direction of propagation. We choose the polarization transverse to the direction of propagation, $\mathbf{k}_\gamma \cdot \mathbf{E}_0 = 0$, and quickly verify that this potential is in the Coulomb gauge $\nabla \cdot \mathbf{A} = 0$. The electric field associated with this vector potential is

$$\mathbf{E} = -\frac{1}{c}\frac{\partial \mathbf{A}}{\partial t} - \nabla\Phi = 2\,\mathbf{E}_0 \sin(\mathbf{k}_\gamma \cdot \mathbf{r} - \omega t)\,. \tag{27.9.17}$$

The magnetic field $\mathbf{B} = \nabla \times \mathbf{A}$ associated with this wave will not play a role here, given our approximations.

To find the coupling of the electron to the electromagnetic field, we look at the kinetic term in the Hamiltonian of a charged electron:

$$\frac{1}{2m}\Big(\hat{\mathbf{p}} + \frac{e}{c}\mathbf{A}\Big)^2 = \frac{\hat{\mathbf{p}}^2}{2m} + \frac{e}{mc}\mathbf{A}\cdot\hat{\mathbf{p}} - \frac{ie\hbar}{2mc}\nabla\cdot\mathbf{A} + \frac{e^2}{2mc^2}\mathbf{A}\cdot\mathbf{A}\,. \tag{27.9.18}$$

Recalling that our vector potential is in the Coulomb gauge and neglecting the term quadratic in $\mathbf{A}$, as it is second order in the strength $|\mathbf{E}_0|$ of the perturbation, we conclude that the coupling of our electron to the electromagnetic wave is given by

$$\delta H = \frac{e}{mc}\mathbf{A}\cdot\hat{\mathbf{p}} = -\frac{e}{m\omega}\, 2\,\mathbf{E}_0 \cdot \hat{\mathbf{p}} \cos(\mathbf{k}_\gamma \cdot \hat{\mathbf{r}} - \omega t)\,. \tag{27.9.19}$$

The "hat" on $\mathbf{r}$ is included to emphasize that, appearing in δH, it is the position operator (not to be confused with a radial unit vector). Since we assumed that the wavelength is much larger than a_0, placing the atom at the origin, we can ignore the $\mathbf{k}_\gamma \cdot \hat{\mathbf{r}}$ term in the argument of the cosine, thus getting

$$\delta H = -\frac{e}{m\omega}\, 2\,\mathbf{E}_0 \cdot \hat{\mathbf{p}} \cos\omega t\,. \tag{27.9.20}$$

In our notation for harmonic perturbations (see (27.5.2)), we write $\delta H = 2H' \cos\omega t$, with H' time independent. We have thus identified H':

$$H' = -\frac{e}{m\omega}\, \mathbf{E}_0 \cdot \hat{\mathbf{p}}\,. \tag{27.9.21}$$

The assumptions above define the *dipole approximation*. Once we take the electric field in (27.9.17) and assume the spatial dependence can be ignored at the atom, we have

$$\mathbf{E} = -2\,\mathbf{E}_0 \sin\omega t\,. \tag{27.9.22}$$

Such an electric field arises as $\mathbf{E} = -\nabla\Phi$ from a potential $\Phi(\mathbf{r}) = -\mathbf{E}\cdot\mathbf{r}$, and zero vector potential. The correction to the Hamiltonian, this time called $\delta\check{H}$, is simply the coupling of the electron charge to Φ:

$$\delta\check{H} = -e\Phi(\hat{\mathbf{r}}) = e\,\mathbf{E}\cdot\hat{\mathbf{r}} = -2\,e\mathbf{E}_0\cdot\hat{\mathbf{r}}\,\sin\omega t\,. \tag{27.9.23}$$

From this we can read the $\check{H}'$ perturbation. The usual time dependence in our conventions is a $\cos\omega t$ that differs by a $\pi/2$ phase from the $\sin\omega t$ above. This corresponds to a time shift that is immaterial in the computation of the norm of the matrix element, which is what we need in Fermi's golden rule. Thus, we read that

$$\check{H}' = -e\,\mathbf{E}_0\cdot\hat{\mathbf{r}}\,. \tag{27.9.24}$$

We claim that these two perturbations H' and $\check{H}'$ give matrix elements that differ by just a phase, which is irrelevant to the computation of the transition rate. This is shown as follows. Begin with the commutator

$$[\hat{H}^{(0)}, \hat{\mathbf{r}}\,] = -\frac{i\hbar}{m}\,\hat{\mathbf{p}}\,. \tag{27.9.25}$$

You can quickly check this, as only the kinetic term $\hat{\mathbf{p}}^2/(2m)$ contributes in $\hat{H}^{(0)}$. Now consider this commutator in between the initial and final energy eigenstates $|i\rangle$ and $|f\rangle$, respectively. We then find that

$$(E_f - E_i)\,\langle f|\,\hat{\mathbf{r}}\,|i\rangle = -\frac{i\hbar}{m}\,\langle f|\,\hat{\mathbf{p}}\,|i\rangle\,. \tag{27.9.26}$$

For any process in which a photon of frequency ω is absorbed, we have $E_f - E_i = \hbar\omega$, and therefore we have the relation

$$\langle f|\,\hat{\mathbf{p}}\,|i\rangle = im\,\omega\langle f|\,\hat{\mathbf{r}}\,|i\rangle\,. \tag{27.9.27}$$

Multiplying by $-\frac{e}{m\omega}\mathbf{E}_0$, we see that

$$-\frac{e}{m\omega}\langle f|\,\mathbf{E}_0\cdot\hat{\mathbf{p}}\,|i\rangle = -ie\langle f|\,\mathbf{E}_0\cdot\hat{\mathbf{r}}\,|i\rangle \;\Rightarrow\; \langle f|H'|i\rangle = i\langle f|\check{H}'|i\rangle\,. \tag{27.9.28}$$

As claimed, the matrix elements are the same up to an irrelevant phase. This claim, of course, assumes that we are using exact energy eigenstates. The problem is that we are going to use approximate final states, plane waves that are not $\hat{H}^{(0)}$ eigenstates. So there is no reason for this equality to hold exactly. Thus, for example, the matrix elements may agree only after we use the approximation $k_e a_0 \gg 1$.

The situation, in fact, is more complicated. Even after the use of such approximations, if we use plane waves for final states, the right-hand side of (27.9.28), involving the matrix element of $\hat{\mathbf{r}}$, takes twice the value of the left-hand side, involving the matrix element of $\hat{\mathbf{p}}$. As it turns out, including corrections to the plane wave changes the value of the right-hand side, bringing it to agreement with the left-hand side. On the left-hand

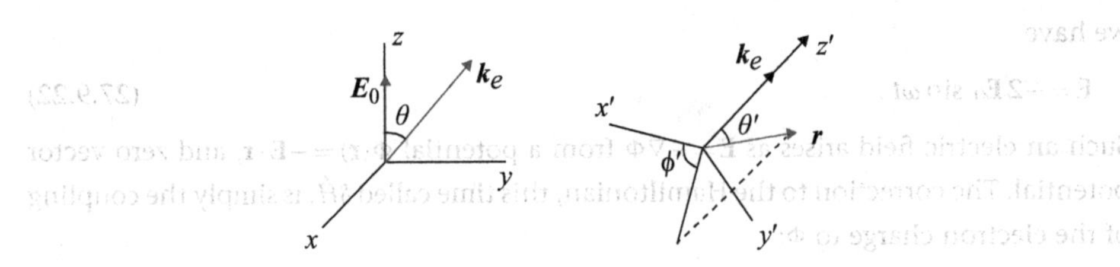

Figure 27.16
Left: An electron ejected at an angle θ relative to the axis of polarization of the incident wave. *Right*: For the integration over $\mathbf{r}$ in (27.9.35), it is convenient to align the electron momentum along a z'-axis and describe $\mathbf{r}$ with spherical coordinates (r', θ', ϕ').

side, the corrections to the plane wave do not change the result. We will use H' to compute the matrix element, essentially the matrix element of the momentum operator $\hat{\mathbf{p}}$. As we said, this matrix element is correctly evaluated using the plane wave assumption for the final states.

Calculating the matrix element Let us take the wave to be polarized along the z-axis so that $\mathbf{E}_0 = E_0\hat{\mathbf{z}}$. We then see that the perturbation H' in (27.9.21) becomes

$$H' = -\frac{eE_0}{m\omega}\,\hat{p}_z\,. \tag{27.9.29}$$

We want to compute the matrix element between an initial state $|i\rangle$ that is the ground state of hydrogen and a final state $|f\rangle$ that is an electron plane wave with momentum $\mathbf{k}_e$:

$$|i\rangle = |\psi_{100}\rangle\,, \quad |f\rangle = |\psi_{\mathbf{k}_e}\rangle\,. \tag{27.9.30}$$

The associated normalized wave functions are

$$\psi_{100}(\mathbf{r}) = \frac{1}{\sqrt{\pi a_0^3}}\,e^{-r/a_0}\,, \quad \psi_{\mathbf{k}_e}(\mathbf{r}) = \frac{1}{L^{3/2}}\,e^{i\mathbf{k}_e\cdot\mathbf{r}}\,, \quad r = |\mathbf{r}|\,. \tag{27.9.31}$$

The only physical angle here is that between the ejected electron momentum $\mathbf{k}_e$ and the electric field polarization, which we have chosen to be along the z-axis. This angle is the conventional latitude angle θ (figure 27.16, *left*). We expect the electron to be ejected maximally along the z-axis.

Since the final state is a momentum eigenstate, we have

$$\hat{p}_z|f\rangle = \hat{p}_z|\psi_{\mathbf{k}_e}\rangle = \hbar(\mathbf{k}_e)_z|\psi_{\mathbf{k}_e}\rangle = \hbar k_e\cos\theta\,|\psi_{\mathbf{k}_e}\rangle\,, \tag{27.9.32}$$

where k_e is the magnitude of $\mathbf{k}_e$. This property and the Hermiticity of $\hat{p}_z$ allow us to simplify considerably the matrix element we are trying to calculate:

$$H'_{fi} \equiv \langle f|H'|i\rangle = -\frac{eE_0}{m\omega}\langle f|\hat{p}_z|i\rangle = -\frac{eE_0}{m\omega}\,\hbar k_e\cos\theta\,\langle f|i\rangle\,. \tag{27.9.33}$$

The overlap between the final and initial states involves an integral:

$$\langle f|i\rangle = \frac{1}{\sqrt{\pi a_0^3 L^3}}\int d^3\mathbf{r}\,e^{-i\mathbf{k}_e\cdot\mathbf{r}}\,e^{-r/a_0}\,. \tag{27.9.34}$$

The integral above has a fixed $\mathbf{k}_e$ vector in an arbitrary direction, and we integrate over all $\mathbf{r}$. We can imagine new (x', y', z') axes with $\mathbf{k}_e$ along z' and the arbitrary $\mathbf{r}$ parameterized by its length r' and the angles θ' and ϕ' (figure 27.16, *right*). In this notation, the integral reads

$$\int d^3\mathbf{r}\, e^{-i\mathbf{k}_e\cdot\mathbf{r}}\, e^{-r/a_0} = 2\pi \int_0^\infty r'^2 dr' \int_{-1}^{1} d(\cos\theta') e^{-ik_e r' \cos\theta} e^{-r'/a_0} = \frac{8\pi a_0^3}{(1+k_e^2 a_0^2)^2}, \tag{27.9.35}$$

where the 2π in the middle expression is from the integral over ϕ'. The final integral takes a little effort but is worth doing—after all, it is the Fourier transform of the ground state wave function.

Exercise 27.6. *Prove the last equality in (27.9.35).*

Returning to the matrix element in (27.9.33), we have found that

$$H'_{fi} = -\frac{eE_0}{m\omega}\,\hbar k_e \cos\theta\, \frac{1}{\sqrt{\pi a_0^3 L^3}}\, \frac{8\pi a_0^3}{(1+k_e^2 a_0^2)^2}\,. \tag{27.9.36}$$

We can rearrange the factors on the right-hand side to write, more clearly,

$$H'_{fi} = -8\sqrt{\pi}\,(eE_0 a_0)\, \frac{\hbar k_e}{m\omega a}\, \frac{a_0^3}{\sqrt{a_0^3 L^3}}\, \frac{1}{(1+k_e^2 a_0^2)^2} \cos\theta\,. \tag{27.9.37}$$

Here the units of the result, energy, are carried by the $eE_0 a$ factor since all other ratios are unit-free. This result is actually less accurate than it looks! In our approximation $k_e^2 a_0^2 \gg 1$, we are not allowed to take seriously the 1 in the $1 + k_e^2 a_0^2$ factors. Moreover, in writing the answer in terms of k_e we can eliminate ω using the relation $\hbar\omega = \hbar^2 k_e^2/(2m)$, which assumes the electron carries the full energy of the photon. As a result of these changes, we see that

$$\begin{aligned} H'_{fi} &= -8\sqrt{\pi}\,(eE_0 a_0)\, \frac{2}{k_e a_0}\, \frac{a_0^3}{\sqrt{a_0^3 L^3}}\, \frac{1}{k_e^4 a_0^4} \cos\theta \\ &= -16\sqrt{\pi}\,(eE_0 a_0)\, \frac{1}{L^{3/2} a_0^{7/2} k_e^5} \cos\theta\,. \end{aligned} \tag{27.9.38}$$

Since the norm squared of H'_{fi} is the required quantity for the ionization rate, we record that

$$|H'_{fi}|^2 = 256\pi\,(eE_0 a_0)^2\, \frac{1}{L^3 a_0^7 k_e^{10}} \cos^2\theta\,. \tag{27.9.39}$$

The transition rate Fermi's golden rule requires the density of electron states, evaluated at the final energy E_e of the electrons. The density of states was given in (27.6.10):

$$\rho(E_e) = \frac{L^3}{8\pi^3}\frac{m}{\hbar^2} k_e\, d\Omega\,. \tag{27.9.40}$$

It follows that using Fermi's golden rule, the rate dw to go into final states in the solid-angle $d\Omega$ is

$$dw = \frac{2\pi}{\hbar}\rho(E_e)|H'_{fi}|^2 = \frac{2\pi}{\hbar}\frac{L^3}{8\pi^3}\frac{m}{\hbar^2}k_e\, d\Omega\, 256\pi(eE_0a_0)^2\frac{\cos^2\theta}{L^3a_0^7k_e^5}\,. \tag{27.9.41}$$

It follows that

$$\boxed{\frac{dw}{d\Omega} = \frac{64}{\pi}\frac{ma_0^2}{\hbar^2}\frac{(eE_0a_0)^2}{\hbar}\frac{1}{(k_ea_0)^9}\cos^2\theta\,.} \tag{27.9.42}$$

$\frac{dw}{d\Omega}$ is the probability of ionization per unit time and per unit solid angle, and $\hbar k_e$ is the momentum of the ejected electron. Note that the units have worked out: $w \sim \frac{1}{[E]}\cdot\frac{[E]^2}{\hbar} \sim \frac{1}{T}$, as expected. In here $2E_0$ is the peak amplitude of the electric field in the wave. The $\cos^2\theta$ implies that, as anticipated, the electron is ejected maximally along the axis of the electric field polarization.

The total ionization probability per unit time (per atom!) is obtained by integration over the solid angle. Using the familiar integral $\int \cos^2\theta\, d\Omega = \frac{1}{3}4\pi$ and recalling that $\frac{\hbar^2}{ma_0^2} = 2\text{Ry}$,

$$w = \int d\Omega\frac{dw}{d\Omega} = \frac{128}{3}\frac{(eE_0a_0)^2}{\hbar\text{Ry}}\frac{1}{(k_ea_0)^9}\,. \tag{27.9.43}$$

Since k_e is a function of the wave's ω, it is reasonable to leave the answer in terms of ω. From (27.9.14), and in the approximation we are using, $ka_0 \simeq \sqrt{\hbar\omega/\text{Ry}}$. This gives

$$w = \frac{128}{3}\frac{(eE_0a_0)^2}{\hbar\text{Ry}}\left(\frac{\text{Ry}}{\hbar\omega}\right)^{9/2}. \tag{27.9.44}$$

For numerical calculations it is convenient to use atomic units, in which the rate takes the form

$$w = \frac{64}{3}\left(\frac{E_p}{E_*}\right)^2\frac{1}{t_*}\left(\frac{\text{Ry}}{\hbar\omega}\right)^{9/2}. \tag{27.9.45}$$

Here $E_p = 2E_0$ is the peak amplitude of the electric field. Moreover, the atomic electric field E_* and the atomic time t_* are given by

$$\begin{aligned} E_* &= \frac{2\text{Ry}}{ea_0} = \frac{e}{a_0^2} = 5.14\times 10^{11}\ \text{V/m}\,, \\ t_* &= \frac{a_0}{\alpha c} = \frac{\hbar}{2\text{Ry}} = 2.42\times 10^{-17}\ \text{s}. \end{aligned} \tag{27.9.46}$$

Note that E_* is the electric field of a proton at a distance a_0 while t_* is the time it takes the electron to travel a distance a_0 at the classical velocity αc in the ground state. A laser intensity I_*, with

$$I_* = 3.55\times 10^{16}\ \text{W/cm}^2\,, \tag{27.9.47}$$

has a peak electric field of magnitude E_*. Since intensities are proportional to peak electric fields squared, we can rewrite w in terms of the intensity I of the incoming wave as

$$w = \frac{64}{3}\frac{I}{I_*}\frac{1}{t_*}\left(\frac{\text{Ry}}{E_\gamma}\right)^{9/2}, \tag{27.9.48}$$

where $E_\gamma = \hbar\omega$ is the photon energy. Setting $E_f \simeq E_\gamma$, you can easily verify now that the validity condition (27.6.41) for Fermi's golden rule gives the following condition on the rate w: $\hbar w \ll E_\gamma$. This states that $w \ll \omega$ and is easily satisfied in usual situations.

Exercise 27.7. *An X-ray laser of wavelength 1 nm has an intensity of 1 MW/cm^2. This laser hits a sample of* 10^{16} *hydrogen atoms for one nanosecond. How many atoms would be ionized? (Answer: about* 376,000.*)*

27.10 Atoms and Light

We have already examined a situation in which an atom interacts with electromagnetic radiation. For the ionization of hydrogen, we used Fermi's golden rule to find the transition rate for the process in which light makes the atom go from its ground state into a state in which the electron is no longer bound to the proton.

We now turn to a new situation in which light also interacts with an atom, or perhaps a molecule. We consider an atom or molecule with two levels, or more accurately, we focus on two levels relevant to our problem. The levels are energy eigenstates for electrons. As a collection of such particles interacts with light, there will be some transitions. Based on our discussion of harmonic perturbations, we can anticipate absorption and stimulated emission processes. But as it turns out, this is not the complete story. There is also a *spontaneous emission* process.

Let us first consider an atom or molecule with two energy levels $|b\rangle$ and $|a\rangle$, with energies E_b and E_a, respectively, and with $E_b > E_a$. We call these level b and level a.

$$\begin{array}{lcl} |b\rangle & \text{———————} & E_b \\ |a\rangle & \text{———————} & E_a. \end{array} \tag{27.10.1}$$

As usual, we define $\omega_{ba} \equiv \frac{1}{\hbar}(E_b - E_a) > 0$. Imagine now shining light into the particles at the resonant frequency ω_{ba}. There are two possible processes depending on the initial state of the particles:

i. The electron is initially in the lower-energy eigenstate $|a\rangle$. We can have an *absorption* process in which a photon gets absorbed, and the electron goes from $|a\rangle$ to $|b\rangle$. In the presence of many photons, we can consider an *absorption rate*.

ii. The electron is initially in the higher-energy eigenstate $|b\rangle$. We can have a *stimulated emission* process in which photons stimulate the electronic transition $|b\rangle \to |a\rangle$ to the ground state with the release of an additional photon. Correspondingly, in the presence of many photons we can consider a *stimulated emission rate*.

These processes are illustrated in the figure below. On the left we show the situation before the process and on the right the situation after the process:

Before After

Absorption $\hbar\omega$ $|b\rangle$ $|a\rangle$ $|b\rangle$ $|a\rangle$

(27.10.2)

Stimulated emission $\hbar\omega$ $|b\rangle$ $|a\rangle$ $|b\rangle$ $|a\rangle$.

In the absorption process, the incoming photon is indeed absorbed as the electron is pushed to the higher level. In the stimulated emission process, the incoming photon just triggers the transition of the electron to the lower level and does so *without* being absorbed. The incoming photon emerges accompanied by a second photon that carries the energy released by the electron transition.

The stimulated emission process is the mechanism that allows the existence of lasers. The name LASER is in fact an acronym for its operating physics: light amplification by stimulated emission of radiation. In this process, a single photon or perhaps a few photons induce transitions that increase the number of photons, stimulating further transitions that in turn increase the number of photons even more and thus the strength of the stimulation effect. The photons emitted by stimulated emission are coherent with the photons that are doing the stimulation, and rather quickly we can have a high-intensity beam of coherent photons bouncing back and forth in the laser.

Lasers use gas molecules with two levels. For molecules with relevant energy levels separated by order 1 eV, statistical physics tells us that at temperatures lower than room temperature essentially all of the molecules will be in the lower-energy state. For the operation of the laser, we need *population inversion*, a large number of molecules in the higher-energy state ready to help generate the coherent laser beam. To generate population inversion, laser designers use molecules with a third level c, called the *pump level*, of energy E_c slightly bigger than E_b. Molecules are pumped with light from level a to level c. A fast transition $|c\rangle \to |b\rangle$ leaves the molecules at the desired place to start laser operation. In the fast transition, the energy released usually goes into a vibrational motion of the molecule.

Einstein's argument Significant insight into the relevant processes and their rates comes from a statistical physics argument developed long ago by Einstein. He considered the constraints imposed by the existence of equilibrium in a system of atoms with two levels interacting with blackbody radiation, all at temperature T. As we go through the argument, we will see the need for a third process, *spontaneous emission*. For Einstein, more than one hundred years ago, the situation was different. He knew about absorption and spontaneous emission. His argument led him to discover stimulated emission.

Consider atoms that can be in states $|a\rangle$ or $|b\rangle$ with $E_b > E_a$. We have a large number of such atoms in a box. Let N_a denote the number of atoms in $|a\rangle$ and N_b the number of atoms in $|b\rangle$. These are the populations.

$$\begin{array}{lll} |b\rangle & \text{———————} & N_b \text{ atoms} \\ |a\rangle & \text{———————} & N_a \text{ atoms}. \end{array} \tag{27.10.3}$$

The atoms are assumed to be in thermal equilibrium with a bath of photons, the whole system at temperature T. Einstein discovered a number of relations that follow from the condition of equilibrium. Three facts enter into his argument:

1. Equilibrium values for the populations mean they do not change in time. Therefore, equilibrium requires vanishing time derivatives:

$$\dot{N}_a = \dot{N}_b = 0. \tag{27.10.4}$$

2. Equilibrium populations are governed by the thermal Boltzmann distribution and therefore obey

$$\frac{N_a}{N_b} = \frac{e^{-\beta E_a}}{e^{-\beta E_b}} = e^{-\beta\hbar\omega_{ab}}, \quad \beta \equiv \frac{1}{k_B T}. \tag{27.10.5}$$

3. For thermal blackbody radiation, the energy $U(\omega)d\omega$ per unit volume in the frequency range $d\omega$ is known to be

$$U(\omega)d\omega = \frac{\hbar}{\pi^2 c^3}\frac{\omega^3 d\omega}{e^{\beta\hbar\omega} - 1}. \tag{27.10.6}$$

In the table below, we have indicated the two obvious processes that must be going on at equilibrium: absorption and stimulated emission.

Process		Rate
Absorption:	$\lvert a\rangle \to \lvert b\rangle$ photon absorbed	$B_{ab}U(\omega_{ba})N_a$
Stimulated emission:	$\lvert b\rangle \to \lvert a\rangle$ photon released	$B_{ba}U(\omega_{ba})N_b$

The rate indicated is the number of transitions per unit time. For absorption, this rate is proportional to the number N_a of atoms in the ground state, capable of absorbing a photon, times $U(\omega_{ba})$, which captures the information about the number of photons available at the right frequency, times a B coefficient B_{ab} that we want to determine. For stimulated emission, the rate is proportional to the number N_b of atoms in the excited state, thus capable of emitting a photon, times $U(\omega_{ba})$ times a B coefficient B_{ba}. The inclusion of $U(\omega_{ba})$ reflects the "stimulated" character of the transition. The B coefficients are assumed to be constants that only depend on the atomic or molecular spectrum.

The question now is: Can we achieve equilibrium with these two processes? We will see that we *cannot*. With the two processes above, the rate of change of the population N_b is given by the absorption rate minus the stimulated emission rate:

$$\dot{N}_b = B_{ab}U(\omega_{ba})N_a - B_{ba}U(\omega_{ba})N_b\,. \tag{27.10.7}$$

At equilibrium $\dot{N}_b = 0$, hence

$$(B_{ab}N_a - B_{ba}N_b)U(\omega_{ba}) = 0 \;\Rightarrow\; N_a\left(B_{ab} - B_{ba}e^{-\beta\hbar\omega_{ba}}\right)U(\omega_{ba}) = 0\,, \tag{27.10.8}$$

using fact (2) above. Since $U(\omega_{ba}) \neq 0$, this is a strange result: equilibrium demands $B_{ab} = B_{ba}e^{-\beta\hbar\omega_{ba}}$. The B coefficients, however, depend on the electronic configurations in the atom and *not* on the temperature T. Thus, this cancellation is not possible for arbitrary temperature unless both B_{ab} and B_{ba} are zero, which is absurd. In conclusion, equilibrium between these two processes is not possible.

What are we missing? Spontaneous emission, a process that can happen in the absence of photons and thus has a rate independent of the number of photons available. We must thus update the table of possible processes to include the following:

Process		Rate
Spontaneous emission:	$\lvert b\rangle \to \lvert a\rangle$	AN_b
	photon released	

The rate is proportional to the number of atoms N_b in the excited state, with a coefficient of proportionality called A. With this extra contribution to $\dot{N}_b$, the equilibrium condition reads

$$\dot{N}_b = B_{ab}U(\omega_{ba})N_a - B_{ba}U(\omega_{ba})N_b - AN_b = 0\,. \tag{27.10.9}$$

Our strategy now is to solve for $U(\omega_{ba})$:

$$A = \left(B_{ab}\frac{N_a}{N_b} - B_{ba}\right)U(\omega_{ba}) \;\Rightarrow\; U(\omega_{ba}) = \frac{A}{B_{ab}}\,\frac{1}{e^{\beta\hbar\omega_{ba}} - \frac{B_{ba}}{B_{ab}}}\,. \tag{27.10.10}$$

Clearly, the equilibrium condition places some important constraints on the photon distribution $U(\omega_{ab})$. From (27.10.6) we know that

$$U(\omega_{ba}) = \frac{\hbar\,\omega_{ba}^3}{\pi^2c^3}\,\frac{1}{e^{\beta\hbar\omega_{ba}} - 1}\,. \tag{27.10.11}$$

Comparing the two expressions for $U(\omega_{ba})$, we deduce that

$$B_{ab} = B_{ba}\,, \qquad \text{and} \qquad \frac{A}{B_{ab}} = \frac{\hbar\omega_{ba}^3}{\pi^2c^3}\,. \tag{27.10.12}$$

These are the key results of Einstein's argument: the absorption and stimulated emission rates per atom are identical, and they are simply related to the spontaneous emission rate A.

In the following section, we will calculate the coefficient B_{ab} from first principles. Calculating A is harder. Happily, thanks to (27.10.12), we can obtain A from B_{ab}. We will not attempt here a first-principles calculation of A. To do such a calculation, one thinks of spontaneous emission as stimulated emission, with the transition stimulated by the vacuum fluctuations of the electromagnetic field. For a given atomic transition, stimulated emission dominates over spontaneous emission at high temperature, when there are more photons around. Spontaneous emission dominates over stimulated emission at very low temperature. These matters are explored in problem 27.8.

27.11 Atom-Light Dipole Interaction

We will now consider how light triggers atomic transitions from one discrete state to another. In particular, we want to derive a formula for the coefficients $B_{ab} = B_{ba}$, introduced by Einstein to describe transition rates in the presence of thermal radiation. To do this we will use time-dependent perturbation theory where the perturbation is the coupling of the time-varying electromagnetic field to the electron. Despite having a transition from one discrete atomic state to another, the derivation and the result will look like those for Fermi's golden rule. The integration in the present case arises because we have thermal radiation that consists of an incoherent superposition of electromagnetic waves with frequency near resonance and with random polarizations. A bonus of our calculation will be a determination of the spontaneous emission rate A. This is the rate in the absence of photon states, and the result, of course, has no trace of the thermal radiation that helps derive it!

We assume that the bound state electron is not relativistic; this is certainly the case in hydrogen. Since magnetic forces are weaker than electric forces by a factor of v/c, we can focus on the electric field $\mathbf{E}$. For optical frequencies $\lambda \sim 6{,}000\,\text{Å}$ and with $a_0 \simeq 0.53\,\text{Å}$, we can also ignore the spatial dependence of the electric field. Thus, the electric field at the atom is just an oscillating vector:

$$\mathbf{E}(t) = E(t)\,\mathbf{n} = 2E_0\mathbf{n}\cos\omega t\,, \tag{27.11.1}$$

where the constant unit vector $\mathbf{n}$ specifies the polarization. As usual in our conventions, the peak value of the electric field is $2E_0$. The coupling of this electric field to the electron is a dipole coupling, as we discussed in detail when considering hydrogen ionization in section 27.9. Associated with this uniform electric field can be defined an electric potential $\Phi(\mathbf{r}) = -\mathbf{r}\cdot\mathbf{E}(t)$, from which $\mathbf{E}(t)$ correctly arises as minus its gradient. For a charged particle with charge q at position $\mathbf{r}$, the coupling δH is therefore

$$\delta H = q\,\Phi(\hat{\mathbf{r}}) = -q\,\hat{\mathbf{r}}\cdot\mathbf{E}(t) = -q\,\hat{\mathbf{r}}\cdot\mathbf{n}\,E(t)\,. \tag{27.11.2}$$

We now define the electric dipole operator $\hat{\boldsymbol{d}}$ by

$$\hat{\boldsymbol{d}} \equiv q\,\hat{\mathbf{r}}\,. \tag{27.11.3}$$

This operator is just the position operator $\hat{\mathbf{r}}$ of the charged particle times its charge, the obvious analogue of the electric dipole in classical electromagnetism. In terms of the dipole operator, the perturbation δH takes the form

$$\delta H = -\hat{\boldsymbol{d}}\cdot\mathbf{n}\,2E_0\cos\omega t = 2(-\hat{\boldsymbol{d}}\cdot\mathbf{n}\,E_0)\cos\omega t\,. \tag{27.11.4}$$

Since we defined harmonic perturbations as $\delta H = 2H' \cos\omega t$ (see (27.5.2)), we deduce that

$$H' = -\hat{\mathbf{d}} \cdot \mathbf{n} E_0 \,. \tag{27.11.5}$$

This is our coupling of the charged particle to the electric field. In a transition from an initial state $|b\rangle$ to a final state $|a\rangle$, the matrix element of H' is

$$H'_{ab} = -\hat{\mathbf{d}}_{ab} \cdot \mathbf{n} E_0 \,, \tag{27.11.6}$$

with $\hat{\mathbf{d}}_{ab} = \langle a|\hat{\mathbf{d}}\,|b\rangle$, since both $\mathbf{n}$ and E_0 are just multiplicative constants. Note that $\hat{\mathbf{d}}_{ab}$ is not an operator; it is a triplet of matrix elements, a vector with components that are complex numbers. This vector appears in H'_{ab} dotted with the vector $\mathbf{n}$.

Recall that to first order in perturbation theory $P_{b\leftarrow a}(t) = P_{a\leftarrow b}(t)$ (27.3.23), so let's consider just the stimulated emission probability $P_{a\leftarrow b}(t)$. We can use equation (27.5.8), even though this equation was obtained for absorption, by replacing $\omega_{fi} \to \omega_{ba} > 0$, thus getting

$$P_{a\leftarrow b}(t) = \frac{|H'_{ab}|^2}{\hbar^2} \frac{\sin^2\left(\frac{\omega_{ba}-\omega}{2}t\right)}{\left(\frac{\omega_{ba}-\omega}{2}\right)^2} = \frac{1}{\hbar^2} E_0^2 \, |\hat{\mathbf{d}}_{ab} \cdot \mathbf{n}|^2 \frac{\sin^2\left(\frac{\omega_{ba}-\omega}{2}t\right)}{\left(\frac{\omega_{ba}-\omega}{2}\right)^2} \,. \tag{27.11.7}$$

The electric field is an *incoherent* superposition of many waves, which we label with an index k, each wave characterized by a frequency ω_k, an amplitude $2E_0(\omega_k)$, and a polarization $\mathbf{n}_{\omega_k}$. Each wave k contributes to the probability of transition with magnitude $P^k_{a\leftarrow b}$ given by

$$P^k_{a\leftarrow b}(t) = \frac{1}{\hbar^2} E_0^2(\omega_k) \, |\hat{\mathbf{d}}_{ab} \cdot \mathbf{n}_{\omega_k}|^2 \frac{\sin^2\left(\frac{\omega_{ba}-\omega_k}{2}t\right)}{\left(\frac{\omega_{ba}-\omega_k}{2}\right)^2} \,. \tag{27.11.8}$$

To make progress, we now discuss the energy density carried by each wave. The energy density u_E in the electric field $E(t) = 2E_0 \cos(\omega t)\mathbf{n}$ is

$$u_E = \frac{|E(t)|^2}{8\pi} = \frac{E_0^2}{2\pi} \cos^2 \omega t \,. \tag{27.11.9}$$

Since this is rapidly oscillating, what matters is the time average $E_0^2/(4\pi)$. In a wave the electric and magnetic energy are the same, and therefore the total energy density u carried by the wave is given by

$$u = \frac{E_0^2}{2\pi} \quad \Rightarrow \quad E_0^2 = 2\pi\, u \,. \tag{27.11.10}$$

While the magnetic field does not couple to the particle in the approximation we are working with, its contribution to the energy cannot be neglected. We can thus write the single mode contribution (27.11.8) to the transition probability using its energy density rather than the field amplitude:

$$P^k_{a\leftarrow b}(t) = \frac{1}{\hbar^2} 2\pi u(\omega_k) \, |\hat{\mathbf{d}}_{ab} \cdot \mathbf{n}_{\omega_k}|^2 \frac{\sin^2\left(\frac{\omega_{ba}-\omega_k}{2}t\right)}{\left(\frac{\omega_{ba}-\omega_k}{2}\right)^2} \,. \tag{27.11.11}$$

The superposition of light is incoherent, so we will add probabilities of transition due to each component of light. Even when we fix a frequency and a field amplitude, we have modes with all polarization directions. If we have, say, one hundred modes with frequency ω_k and energy $u(\omega_k)$, each with different polarization $\mathbf{n}_{\omega_k}$, their contribution is obtained by summing over the one hundred modes but changing the factor $|\hat{\boldsymbol{d}}_{ab} \cdot \mathbf{n}_{\omega_k}|^2$ by its average $\langle|\hat{\boldsymbol{d}}_{ab} \cdot \mathbf{n}|^2\rangle$ over all directions of the vector $\mathbf{n}$. This average is readily calculated. Let $\hat{d}^{\,i}_{ab}$ and n_i denote the Cartesian components of $\hat{\boldsymbol{d}}_{ab}$ and $\mathbf{n}$, respectively. We then have

$$\begin{aligned}\left\langle|\hat{\boldsymbol{d}}_{ab} \cdot \mathbf{n}|^2\right\rangle &= \langle \,|\sum_i \hat{d}^{\,i}_{ab} n_i|^2\rangle = \Big\langle\Big(\sum_i \hat{d}^{\,i}_{ab} n_i\Big)^* \Big(\sum_j \hat{d}^{\,j}_{ab} n_j\Big)\Big\rangle \\ &= \sum_{i,j} (\hat{d}^{\,i}_{ab})^*\, \hat{d}^{\,j}_{ab}\, \langle n_i n_j\rangle\,. \end{aligned} \tag{27.11.12}$$

The average $\langle n_i n_j\rangle$ can be calculated by the integration of $\mathbf{n}$ over solid angle and by dividing by 4π, but it is easier to deduce its value from symmetry considerations. The integral is invariant under reflections about any plane and under rotations. First note that $\langle n_i n_j\rangle = 0$ for $i \neq j$. This follows because a reflection across a plane orthogonal to x_i changes the sign of n_i but not the sign of n_j, changing the sign of the integral. This can only be invariant if the integral is zero to begin with. Moreover, because of rotational symmetry we find that

$$\langle n_1 n_1\rangle = \langle n_2 n_2\rangle = \langle n_3 n_3\rangle\,. \tag{27.11.13}$$

Since

$$\sum_i \langle n_i n_i\rangle = \Big\langle \sum_i n_i n_i\Big\rangle = \langle n^2\rangle = 1\,, \tag{27.11.14}$$

each average in (27.11.13) is equal to $\frac{1}{3}$, and we have $\langle n_i n_j\rangle = \frac{1}{3}\delta_{ij}$. Back in (27.11.12),

$$\left\langle|\hat{\boldsymbol{d}}_{ab} \cdot \mathbf{n}|^2\right\rangle = \tfrac{1}{3}\hat{\boldsymbol{d}}^*_{ab} \cdot \hat{\boldsymbol{d}}_{ab} \equiv \tfrac{1}{3}|\hat{\boldsymbol{d}}_{ab}|^2\,. \tag{27.11.15}$$

The average is proportional to the norm squared of the complex vector $\hat{\boldsymbol{d}}_{ab}$. As we discussed above, we can now write the contribution (27.11.11) of each mode as follows:

$$P^k_{a\leftarrow b}(t) = \frac{2\pi}{3\hbar^2}\, |\hat{\boldsymbol{d}}_{ab}|^2\, u(\omega_k)\, \frac{\sin^2\left(\frac{\omega_{ba}-\omega_k}{2}t\right)}{\left(\frac{\omega_{ba}-\omega_k}{2}\right)^2}\,. \tag{27.11.16}$$

We must now sum over the modes. Consider a little frequency interval $d\omega$ containing a large number of modes ω_k, with the index k in some set $[d\omega]$ that represents all those modes. We then have

$$\sum_{k\in[d\omega]} u(\omega_k) \simeq U(\omega)d\omega\,, \tag{27.11.17}$$

where $U(\omega)d\omega$ is the thermal radiation energy per unit volume in the range $d\omega$ and equals the left-hand side. This relationship still holds if we multiply by a continuous function of ω, leading to the rule needed to turn a sum into an integral:

$$\sum_{k\in[d\omega]} u(\omega_k)f(\omega_k) \simeq U(\omega)f(\omega)d\omega \quad \Rightarrow \quad \sum_k u(\omega_k)f(\omega_k) \simeq \int U(\omega)f(\omega)d\omega\,. \tag{27.11.18}$$

Using this rule, we now find that

$$P_{a\leftarrow b}(t) = \sum_k P^k_{a\leftarrow b}(t) = \frac{2\pi}{3\hbar^2}\,|\hat{\boldsymbol{d}}_{ab}|^2 \int d\omega\, U(\omega)\, \frac{\sin^2\left(\frac{\omega_{ba}-\omega}{2}t\right)}{\left(\frac{\omega_{ba}-\omega}{2}\right)^2}\,. \tag{27.11.19}$$

With the already familiar argument, for sufficiently large t we can take the function $U(\omega)$ out of the integral, evaluated at ω_{ba}, and get

$$P_{a\leftarrow b}(t) = \frac{2\pi}{3\hbar^2}\,|\hat{\boldsymbol{d}}_{ab}|^2\, U(\omega_{ba}) \int d\omega\, \frac{\sin^2\left(\frac{\omega_{ba}-\omega}{2}t\right)}{\left(\frac{\omega_{ba}-\omega}{2}\right)^2}\,. \tag{27.11.20}$$

The integral is the same we had in (27.6.38), and its value is $2\pi t$. Therefore,

$$P_{a\leftarrow b}(t) = \frac{4\pi^2}{3\hbar^2}\,|\hat{\boldsymbol{d}}_{ab}|^2\, U(\omega_{ba})\, t\,. \tag{27.11.21}$$

Finally, the rate $R_{a\leftarrow b}$ associated with the transition is given by

$$\boxed{R_{a\leftarrow b} = \frac{4\pi^2}{3\hbar^2}\,|\hat{\boldsymbol{d}}_{ab}|^2 U(\omega_{ba})\,.} \tag{27.11.22}$$

The rate $R_{a\leftarrow b}$ is the transition probability per unit time for a single atom. In Einstein's argument that rate was written as $B_{ba}U(\omega_{ba})$. We can therefore read the value of the coefficient $B_{ba}(=B_{ab})$:

$$B_{ba} = \frac{4\pi^2}{3\hbar^2}\,|\hat{\boldsymbol{d}}_{ab}|^2\,. \tag{27.11.23}$$

As anticipated, this gives B_{ab} in terms of dipole operator matrix elements between the initial and final states. Moreover, using (27.10.12) we can now find the transition rate A for spontaneous emission:

$$A = \frac{\hbar\omega_{ba}^3}{\pi^2 c^3}\, B_{ba}\,. \tag{27.11.24}$$

Using the value of B_{ba} calculated above, we find that A, the transition probability per unit time for an atom to spontaneously go from $|b\rangle$ to $|a\rangle$, is given by

$$\boxed{A = \frac{4}{3}\frac{\omega_{ba}^3}{\hbar c^3}\,|\hat{\boldsymbol{d}}_{ab}|^2\,.} \tag{27.11.25}$$

This is a very useful result; it allows us to calculate decay rates of excited states of atoms.

If we know the transition rate A for the decay of a state, its lifetime τ is given by $\tau = 1/A$, as explained in section 26.6. In a decay with lifetime τ, by definition, an initial number N_0 of particles at $t=0$ evolves as $N(t) = N_0 e^{-t/\tau}$. If a particle can decay in more than one way, there is a decay rate A_i for each decay mode. With decay rates $A_1, \ldots, A_k$, the rates add up to give the total decay rate A:

$$A = A_1 + \cdots + A_k\,, \quad \text{and } \tau = 1/A\,. \tag{27.11.26}$$

If a state $|a\rangle$ can spontaneously decay into a state $|b\rangle$ and the state $|b\rangle$ can spontaneously decay into a state $|c\rangle$, then we have the processes: $|a\rangle \to |b\rangle \to |c\rangle$. The lifetime of $|a\rangle$ is determined solely by the decay rate from $|a\rangle$ to $|b\rangle$ because once the system is in $|b\rangle$ it is no longer in $|a\rangle$. The decay rate for $|b\rangle \to |c\rangle$ does not affect the lifetime of $|a\rangle$.

27.12 Selection Rules

To compute the decay rate A for an atomic transition, the only quantity that requires an explicit calculation is the dipole matrix element $\hat{\boldsymbol{d}}_{ab}$ connecting the initial and final states (see (27.11.25)). Not only that; if this matrix element vanishes, the decay rate vanishes, and the decay does not happen in the electric-dipole approximation we are considering. It is therefore helpful to be able to tell if $\hat{\boldsymbol{d}}_{ab}$ vanishes due to symmetry considerations. Since $\hat{\boldsymbol{d}} = q\hat{\mathbf{r}}$, we can simply consider the matrix elements $\hat{\mathbf{r}}_{ab}$ of the position operator.

In general we would like to know when the matrix elements

$$\hat{\mathbf{r}}_{12} \equiv \langle \psi_1 | \hat{\mathbf{r}} | \psi_2 \rangle \tag{27.12.1}$$

vanish. Note that $\hat{\mathbf{r}}_{12}$ is a triplet of expectation values: the expectation values of $\hat{x}, \hat{y}$, and $\hat{z}$. One simple criterion is provided by the parity operator P. This is a Hermitian operator $P^\dagger = P$ that squares to one: $PP = 1$. As a result, the eigenvalues of P are ± 1. The parity operator sends $\hat{\mathbf{r}}$ to $-\hat{\mathbf{r}}$, and therefore we have

$$P\hat{\mathbf{r}}P = -\hat{\mathbf{r}}\,. \tag{27.12.2}$$

Assume the states in the expectation value $\hat{\mathbf{r}}_{12}$ are parity eigenstates:

$$P|\psi_1\rangle = \epsilon_1 |\psi_1\rangle\,, \quad P|\psi_2\rangle = \epsilon_2 |\psi_2\rangle\,. \tag{27.12.3}$$

Both ϵ_1 and ϵ_2 can only be one or minus one. Since $PP = 1$, we have

$$\hat{\mathbf{r}}_{12} = \langle \psi_1 | PP\,\hat{\mathbf{r}}\,PP | \psi_2 \rangle = \langle \psi_1 | P^\dagger\, (P\hat{\mathbf{r}}P) P | \psi_2 \rangle = -\epsilon_1 \epsilon_2 \hat{\mathbf{r}}_{12}\,. \tag{27.12.4}$$

As a result,

$$(1 + \epsilon_1 \epsilon_2)\hat{\mathbf{r}}_{12} = 0\,. \tag{27.12.5}$$

This means that

$$\hat{\mathbf{r}}_{12} = 0 \ \text{ unless } \ \epsilon_1 \epsilon_2 = -1\,. \tag{27.12.6}$$

If the states have the same parity, the matrix element must vanish. The states $|\psi_1\rangle$ and $|\psi_2\rangle$ must have opposite parity for the matrix element to have *a chance* not to be zero. Indeed, parity might allow a nonzero matrix element, but another symmetry could well require it to be zero.

Recall that spherical harmonics $Y_{\ell m}$, or the states $|\ell m\rangle$, have parity $\epsilon_{\ell m} = (-1)^\ell$. It then follows that for hydrogen-like states,

$$\langle n\ell m|\hat{\mathbf{r}}|n'\ell' m'\rangle = 0 \text{ unless } (-1)^{\ell+\ell'} = -1, \text{ or equivalently, } \ell + \ell' = \text{odd}. \quad (27.12.7)$$

In particular, $\ell' \neq \ell$ is needed for the matrix element not to be zero. This means that in the dipole approximation, ℓ must change by ± 1, or ± 3, and so on. Thus, $2s$ states of hydrogen cannot decay via the spontaneous emission of a single photon: the only possible decay is to $1s$, and this is forbidden because there is no ℓ change. This does not mean the $2s$ state is completely stable. While lifetimes due to dipole decay are typically of the order of 10^{-10} seconds, the state $2s$ can in fact decay with a lifetime of about one-eighth of a second via the emission of two photons.

The result (27.12.7) is not the strongest constraint involving the angular momentum ℓ. A stronger form arises by considering the nested commutator (25.8.12) that helped us deal with the weak-field Zeeman effect. Using $\hat{\mathbf{L}}$ for $\hat{\mathbf{J}}$ and $\hat{\mathbf{r}}$ for the vector operator $\hat{\mathbf{V}}$, we have

$$\frac{1}{(2i\hbar)^2}\Big[\hat{\mathbf{L}}^2, [\hat{\mathbf{L}}^2, \hat{\mathbf{r}}]\Big] = (\hat{\mathbf{r}}\cdot\hat{\mathbf{L}})\,\hat{\mathbf{L}} - \tfrac{1}{2}\,(\hat{\mathbf{L}}^2\hat{\mathbf{r}} + \hat{\mathbf{r}}\hat{\mathbf{L}}^2)\,. \quad (27.12.8)$$

One can easily check that $\hat{\mathbf{r}}\cdot\hat{\mathbf{L}} = 0$, and therefore we get

$$\Big[\hat{\mathbf{L}}^2, [\hat{\mathbf{L}}^2, \hat{\mathbf{r}}]\Big] = 2\hbar^2\,(\hat{\mathbf{L}}^2\hat{\mathbf{r}} + \hat{\mathbf{r}}\hat{\mathbf{L}}^2)\,. \quad (27.12.9)$$

Hit this equation with $\langle n'\ell' m'|$ from the left and $|n\ell m\rangle$ from the right. One then finds, after canceling factors of $\hbar$,

$$(\ell'(\ell'+1) - \ell(\ell+1))^2 \langle n'\ell' m'|\hat{\mathbf{r}}|n\ell m\rangle = (2\ell(\ell+1) + 2\ell'(\ell'+1))\langle n'\ell' m'|\hat{\mathbf{r}}|n\ell m\rangle\,. \quad (27.12.10)$$

Moving all terms to the left-hand side, we have

$$\Big[(\ell'(\ell'+1) - \ell(\ell+1))^2 - 2\ell(\ell+1) - 2\ell'(\ell'+1)\Big]\langle n'\ell' m'|\hat{\mathbf{r}}|n\ell m\rangle = 0\,. \quad (27.12.11)$$

It takes some ingenuity to factor the expression in brackets, but any algebraic manipulator does it in a flash! We find that

$$((\ell + \ell' + 1)^2 - 1)\,((\ell' - \ell)^2 - 1)\langle n'\ell' m'|\hat{\mathbf{r}}|n\ell m\rangle = 0\,. \quad (27.12.12)$$

The first factor can only vanish if $\ell + \ell' + 1 = \pm 1$. Since $\ell, \ell' \geq 0$, the only possibility is $\ell + \ell' + 1 = 1$, or $\ell = \ell' = 0$. Of course, in this case we already know that the matrix element vanishes anyway, by parity. The second factor vanishes when $\ell' - \ell = \pm 1$, and we thus conclude that

$$\langle n'\ell' m'|\hat{\mathbf{r}}|n\ell m\rangle = 0 \text{ unless } \ell' - \ell = \pm 1\,. \quad (27.12.13)$$

This strengthens the parity result stating that ℓ' and ℓ could differ by an odd number: the odd number must be one.

Further selection rules deal with the azimuthal quantum number. We have, for example,

$$\langle n'\ell'm'|\hat{z}|n\ell m\rangle = 0 \quad \text{unless} \quad m' = m\,. \tag{27.12.14}$$

This is clear because $\hat{L}_z$ commutes with $\hat{z}$. We can also show that matrix elements of $\hat{x}$ and $\hat{y}$ require a change in m quantum number by one unit. Recalling that $[\hat{L}_z, \hat{x}] = i\hbar\hat{y}$ and $[\hat{L}_z, \hat{y}] = -i\hbar\hat{x}$, we get

$$\begin{aligned}\langle n'\ell'm'|[\hat{L}_z, \hat{x}]|n\ell m\rangle &= \hbar(m'-m)\langle n'\ell'm'|\hat{x}|n\ell m\rangle = i\hbar\langle n'\ell'm'|\hat{y}|n\ell m\rangle\,,\\ \langle n'\ell'm'|[\hat{L}_z, \hat{y}]|n\ell m\rangle &= \hbar(m'-m)\langle n'\ell'm'|\hat{y}|n\ell m\rangle = -i\hbar\langle n'\ell'm'|\hat{x}|n\ell m\rangle\,.\end{aligned} \tag{27.12.15}$$

From these we get

$$\begin{aligned}-i(m'-m)\langle n'\ell'm'|\hat{x}|n\ell m\rangle &= \langle n'\ell'm'|\hat{y}|n\ell m\rangle\,,\\ i(m'-m)\langle n'\ell'm'|\hat{y}|n\ell m\rangle &= \langle n'\ell'm'|\hat{x}|n\ell m\rangle\,.\end{aligned} \tag{27.12.16}$$

Replacing the $\hat{x}$ matrix element in the first equation with its value from the second equation, we find that

$$[(m'-m)^2 - 1]\langle n'\ell'm'|\hat{y}|n\ell m\rangle = 0\,, \tag{27.12.17}$$

learning that

$$\langle n'\ell'm'|\hat{y}|n\ell m\rangle = 0 \quad \text{unless} \quad m' - m = \pm 1\,. \tag{27.12.18}$$

Exactly the same happens for $\hat{x}$ matrix elements:

$$\langle n'\ell'm'|\hat{x}|n\ell m\rangle = 0 \quad \text{unless} \quad m' - m = \pm 1\,. \tag{27.12.19}$$

This is the extent of the selection rules. To summarize, defining

$$\Delta\ell \equiv \ell' - \ell\,, \quad \Delta m \equiv m' - m \tag{27.12.20}$$

we see that

$$\boxed{\langle n'\ell'm'|\,\hat{\mathbf{r}}\,|n\ell m\rangle = 0 \quad \text{unless} \quad \Delta\ell = \pm 1 \quad \textit{and} \quad \Delta m = \pm 1, 0\,.} \tag{27.12.21}$$

The quantum number ℓ must change by one unit, and the quantum number m can be conserved or can change by one unit. This means, for example, that any of the $2p$ states of hydrogen can decay to the $1s$ state (see problem 27.9).

Problems

Problem 27.1. *A time-dependent two-state system.*

Consider a two-state system with Hamiltonian $\hat{H}(t)$ taking the form

$$\hat{H}(t) = \begin{pmatrix} E & v(t) \\ v(t) & -E \end{pmatrix},$$

where $v(t)$ is real, and $\int_{-\infty}^{\infty} |v(t)|$ is finite. We will label the states as $|1\rangle = \begin{pmatrix} 1 \\ 0 \end{pmatrix}$, $|2\rangle = \begin{pmatrix} 0 \\ 1 \end{pmatrix}$ and define $\omega_0 \equiv 2E/\hbar$. In both parts of the problem, answers may include integrals involving $v(t)$.

1. Suppose that at $t = -\infty$, the system is in the state $|2\rangle$. Use time-dependent perturbation theory to determine, to lowest order in ν, the probability $P_1(\infty)$ that at $t = \infty$ the system is in the state $|1\rangle$. [Hint: in order to think clearly about initial conditions at $t = -\infty$, it is convenient to imagine that $\nu(t)$ vanishes for times earlier than some large negative time.] Write your answer in the form $P_1(\infty) = \frac{1}{\hbar^2}\big|\cdots\big|^2$.
2. If $E = 0$, the eigenstates of $\hat{H}(t)$ can be chosen to be independent of t. Set $E = 0$, and again assume that the state is $|2\rangle$ at $t = -\infty$. Calculate the exact probability $P_1^{\rm ex}(\infty)$ of a transition to $|1\rangle$ at $t = \infty$. What is the result obtained from first-order time-dependent perturbation theory in this case? What is the condition in which the perturbative result is a good approximation to the exact result?

Problem 27.2. *Expectation values in the interaction picture.*

Consider the familiar time-dependent setup with $\hat{H}(t) = \hat{H}^{(0)} + \delta H(t)$, with $\delta H(t)$ a perturbation. Let the state of the system at $t = 0$ be $|\Psi(0)\rangle$. Consider a time-independent Schrödinger operator $\hat{A}$ and the time-dependent expectation value $\langle \hat{A}\,\rangle(t) \equiv \langle \Psi(t)|\,\hat{A}\,|\Psi(t)\rangle$. This expectation value can be written in the form

$$\langle \hat{A}\,\rangle(t) = \langle \Psi(0)|\hat{A}_H(t)|\Psi(0)\rangle\,,$$

where $\hat{A}_H(t)$ is the fully interacting Heisenberg version of $\hat{A}$. Show that, to first order in perturbation theory,

$$\hat{A}_H(t) = \hat{A}(t) + \int_0^t dt' \Big[\hat{A}(t), \frac{\widetilde{\delta H}(t')}{i\hbar}\Big],$$

where $\hat{A}(t) = e^{i\hat{H}^{(0)}t/\hbar}\hat{A}e^{-i\hat{H}^{(0)}t/\hbar}$ is the $\hat{H}^{(0)}$-based Heisenberg version of $\hat{A}$. *Challenge:* extend the above result to second-order in perturbation theory, writing the additional correction in terms of nested commutators.

Problem 27.3. *Atom and photon.*

Model an atom as a two-level system with a ground state $|g\rangle$, an excited state $|e\rangle$, and energy splitting between these two levels equal to $\hbar\omega_a > 0$. Suppose the atom interacts with an electromagnetic field of frequency ω_p, which we model as a harmonic oscillator. Without interactions, the free Hamiltonian $\hat{H}^{(0)}$, acting on the tensor product of the atom and oscillator state spaces, would be

$$\begin{aligned}\hat{H}^{(0)} &= \tfrac{1}{2}\hbar\omega_a\,(|e\rangle\langle e| - |g\rangle\langle g|)\otimes \mathbb{1} + \hbar\omega_p\,\mathbb{1}\otimes(\hat{a}^\dagger\hat{a} + \tfrac{1}{2})\\ &= \tfrac{1}{2}\hbar\omega_a\,\sigma_z\otimes\mathbb{1} + \hbar\omega_p\,\mathbb{1}\otimes(\hat{a}^\dagger\hat{a} + \tfrac{1}{2})\,,\end{aligned}$$

where we used the Pauli matrix σ_z to describe the atom operator, with the convention $|1\rangle \equiv |e\rangle$, and $|2\rangle \equiv |g\rangle$. It turns out that the electric field strength is proportional to $\hat{a} + \hat{a}^\dagger$, so we can model an atom-photon interaction by

$$\delta H = \alpha\,(|g\rangle\langle e| + |e\rangle\langle g|)\otimes(\hat{a} + \hat{a}^\dagger) = \alpha\,\sigma_x\otimes(\hat{a} + \hat{a}^\dagger)\,,$$

for some constant α.

1. Compute $\widetilde{\delta H}(t) \equiv e^{i\hat{H}_0 t/\hbar}\, \delta H\, e^{-i\hat{H}_0 t/\hbar}$. The answer should be in the form $\widetilde{\delta H}(t) = \alpha\, U_a(t) \otimes U_{\text{em}}(t)$, where $U_a(t)$ is an atomic operator, and $U_{\text{em}}(t)$ is an electromagnetic operator. We demand $U_a(0) = |g\rangle\langle e| + |e\rangle\langle g|$, and $U_{\text{em}}(0) = \hat{a} + \hat{a}^\dagger$.
2. Assume now that $\omega_a = \omega_p \equiv \omega$. Compute the operator

$$\hat{R}(t) \equiv \frac{1}{i\hbar}\int_0^t dt' \widetilde{\delta H}(t') .$$

$\hat{R}(t)$ is the operator that acting on an initial $t=0$ state gives us the first-order correction to the state at time t. Determine the leading terms in $\hat{R}(t)$ for $t \gg 1/\omega$. You should be left with one term that can be interpreted as absorption (which one?) and another that can be interpreted as spontaneous/stimulated emission (which one?).

Problem 27.4. *Gaussian pulse.*

Let $\hat{H}^{(0)}$ be a Hamiltonian with spectrum and energies given by $\hat{H}^{(0)}|n\rangle = E_n|n\rangle$ for $n = 0, 1, 2, \ldots$, with $E_n \leq E_{n'}$, when $n < n'$. Suppose we apply a Gaussian pulse perturbation

$$\delta H(t) = \frac{\exp\left(-\frac{t^2}{2\tau^2}\right)}{\sqrt{2\pi\tau_0^2}}\, \hat{V} ,$$

where $\hat{V}$ is an arbitrary time-independent Hermitian operator with units of energy times time, and $\tau > 0$ and $\tau_0 > 0$ are constants with units of time. The function multiplying $\hat{V}$ is the "pulse." The perturbation is supposed to exist for all times $t \in (-\infty, \infty)$. Also define $\omega_{mn} \equiv \frac{1}{\hbar}(E_m - E_n)$, and $\hat{V}_{mn} \equiv \langle m|\hat{V}|n\rangle$.

1. Assume the system starts in state $|0\rangle$ at time $-\infty$. Use first-order time-dependent perturbation theory to find the probability $P_n(\infty; \tau)$ that our system is in state $|n\rangle$ at time $t = \infty$, with $n \neq 0$. Express your answer in terms of $\hat{V}_{mn}$, ω_{mn}, τ, and constants.
2. Let $\tau_0 = \tau$. What is the limit $\tau \to 0$ of $P_n(\infty; \tau)$, $n \neq 0$? What happens to the pulse in this limit? Now let $\tau_0 \neq \tau$ be a fixed constant. What is the limit $\tau \to \infty$ of $P_n(\infty; \tau)$, $n \neq 0$? What happens to the pulse for very large τ? How is the result for the transition probabilities reasonable?

Problem 27.5. *Vibrational modes of carbon dioxide (CO_2).*

This problem will consider the absorption of infrared radiation by CO_2, a nearly linear molecule. We will treat it as a collection of three point masses at positions x_1, x_2, and x_3, connected by springs of spring constant k. We will make a somewhat less justifiable but qualitatively suitable approximation of setting all atoms to have the same mass m. Thus, the Hamiltonian $\hat{H}^{(0)}$ for the molecule is

$$\hat{H}^{(0)} = \frac{\hat{p}_1^2}{2m} + \frac{\hat{p}_2^2}{2m} + \frac{\hat{p}_3^2}{2m} + \tfrac{1}{2}k(\hat{x}_1 - \hat{x}_2)^2 + \tfrac{1}{2}k(\hat{x}_2 - \hat{x}_3)^2 .$$

We can also write $\hat{H}^{(0)} = \hat{T} + \hat{V}$, where

$$\hat{T} = \frac{\hat{p}_1^2}{2m} + \frac{\hat{p}_2^2}{2m} + \frac{\hat{p}_3^2}{2m}, \quad \hat{V} = \tfrac{1}{2}k(\hat{x}_1 - \hat{x}_2)^2 + \tfrac{1}{2}k(\hat{x}_2 - \hat{x}_3)^2 .$$

All motion is in the x-direction, and $\hat{p}_1$, $\hat{p}_2$, $\hat{p}_3$ and $\hat{x}_1$, $\hat{x}_2$, $\hat{x}_3$ refer, respectively, to the momenta and positions of the three different atoms. We can rewrite $\hat{V}$ in terms of a matrix K as

$$\hat{V} = \hat{\mathbf{x}}^T K \hat{\mathbf{x}} = (\hat{x}_1, \hat{x}_2, \hat{x}_3) K \begin{pmatrix} \hat{x}_1 \\ \hat{x}_2 \\ \hat{x}_3 \end{pmatrix} \quad \text{where} \quad K = \tfrac{1}{2} k \begin{pmatrix} 1 & -1 & 0 \\ -1 & 2 & -1 \\ 0 & -1 & 1 \end{pmatrix}.$$

1. Diagonalize K. That is, find a diagonal matrix Λ (with $\Lambda_{11} \geq \Lambda_{22} \geq \Lambda_{33}$) and a rotation matrix R (i.e., $R^T R = \mathbb{1}$) such that $\Lambda = R^T K R$. The columns of R are the eigenvectors of K and are also called the normal modes. Find the normalized eigenvectors of K, called $\mathbf{u}_1$, $\mathbf{u}_2$, $\mathbf{u}_3$, with $\mathbf{u}_i$ having eigenvalue Λ_{ii}. To fix the phase, require that the first nonvanishing entry of the eigenvectors be positive. One of the eigenvalues of K is zero (so by our convention, $\Lambda_{33} = 0$). What is the physical significance of this?
2. Define a triplet $(\hat{y}_1, \hat{y}_2, \hat{y}_3)$ of normal mode displacement operators $\hat{\mathbf{y}} = R^T \hat{\mathbf{x}}$, or more explicitly,

   $$\hat{y}_i = \sum_{j=1}^{3} R_{ji} \hat{x}_j, \quad \text{for } i = 1, 2, 3.$$

   Write $\hat{V}$ in terms of $\hat{y}_1$, $\hat{y}_2$, $\hat{y}_3$ and the eigenvalues Λ_{11} and Λ_{22}.
3. We wish to define a triplet of normal mode momentum operators $(\hat{\pi}_1, \hat{\pi}_2, \hat{\pi}_3)$ such that they are conjugate to the $\hat{y}$'s: $[\hat{y}_i, \hat{\pi}_j] = i\hbar \delta_{ij}$ for $i, j = 1, 2, 3$. Construct a formula for $\hat{\pi}_i$ in terms of the matrix elements of R and the triplet of momenta $\hat{p}_i$. The formula should take the form $\hat{\pi}_i = \sum_k R_{**} \hat{p}_k$. The formula can also be written in vector/matrix notation. Confirm that your answer makes the above commutator work!
4. Give $\hat{H}^{(0)}$ in terms of the $\hat{\pi}$ and $\hat{\mathbf{y}}$ variables. You should now find that $\hat{H}^{(0)}$ breaks up into three pieces that depend separately on $\hat{\pi}_1$, $\hat{y}_1$, on $\hat{\pi}_2$, $\hat{y}_2$, and on $\hat{\pi}_3$. Show that the first two of these pieces are equivalent to harmonic oscillators, and the third corresponds to a free particle. That is, find frequencies ω_1, ω_2 (in terms of k and m) and operators $\hat{a}_1$, $\hat{a}_2$ (in terms of $\hat{\mathbf{y}}$, $\hat{\pi}$, and other parameters) such that

   $$\hat{H}^{(0)} = \hbar\omega_1 \left(\hat{a}_1^\dagger \hat{a}_1 + \tfrac{1}{2}\right) + \hbar\omega_2 \left(\hat{a}_2^\dagger \hat{a}_2 + \tfrac{1}{2}\right) + \frac{\hat{\pi}_3^2}{2m},$$

   and $\hat{a}_1$, $\hat{a}_2$ satisfy the commutation relations $[\hat{a}_i, \hat{a}_j^\dagger] = \delta_{ij}$ and $[\hat{a}_1, \hat{a}_2] = [\hat{a}_1^\dagger, \hat{a}_2^\dagger] = 0$.

   For the rest of the problem, we will work in the energy eigenbasis of $\hat{H}^{(0)}$. This basis can be written $|n_1, n_2, \pi_3\rangle = |n_1\rangle \otimes |n_2\rangle \otimes |\pi_3\rangle$, where n_1, n_2, and π_3 label eigenstates of $\hat{a}_1^\dagger \hat{a}_1$, $\hat{a}_2^\dagger \hat{a}_2$, and $\hat{\pi}_3$, respectively.
5. We are now ready to add radiation. Unlike the most common gases in the atmosphere (N_2, O_2, and Ar), CO_2 has covalent bonds that are weakly polar: the oxygen atoms attract electrons more strongly than the carbon atom. We model this by assuming that the oxygen atoms each have a charge $-q$, and the carbon atom has a charge $2q$.

The Coulomb interaction is effectively already included in $\hat{H}^{(0)}$, so there is no need to modify the Hamiltonian.

Consider an electric field of magnitude E_0 in the x-direction. The potential is $V=-E_0x$, and in the presence of charges q_i at positions x_i, the contribution to the Hamiltonian is $\delta H=-E_0\sum_i q_i\hat{x}_i=-E_0\,\hat{d}_x$, where $\hat{d}_x$ is the component of the dipole moment in the x-direction. In the low-frequency approximation, an electromagnetic field couples to the molecule through this dipole interaction. For the CO_2 molecule,

$$\hat{d}_x=-q\hat{x}_1+2q\hat{x}_2-q\hat{x}_3=-q(\hat{x}_1-2\hat{x}_2+\hat{x}_3)\,.$$

Write $\hat{d}_x$ in terms of the $\hat{a}_i$ and $\hat{a}_i^\dagger$ operators (as well as m, ω_i, $\hbar$, and q). If an oscillating electric field is applied, which mode of oscillation of the molecule, if any, will contribute to the absorption of light? Let $\hat{A}$ and $\hat{A}^\dagger$ denote the operators that destruct and create a photon, respectively. Write the schematic form of the operator $\mathcal{O}_{CO_2}\otimes\mathcal{O}_{EM}$ that generates the following transition: the absorption of a photon from the electromagnetic field and the molecular transition in which we add one quanta of the first mode of oscillation.

Problem 27.6. *One-dimensional model of ionization.*

Consider an electron in the ground state of a very deep one-dimensional square well:

$$V(x)=\begin{cases}0, & \text{for } x<0,\\ -V_0, & \text{for } 0<x<a,\quad V_0>0,\\ 0, & \text{for } x>a\,.\end{cases}$$

A very deep well means $V_0\gg\frac{\hbar^2}{ma^2}$, or equivalently, $z_0^2\equiv\frac{2ma^2V_0}{\hbar^2}\gg 1$.

An electromagnetic plane wave with electric field $E(t)=2E_0\cos(\omega t)$ parallel to the x-axis acts on the electron. The electron can then escape the well in an "ionization" process.

1. For the final states, use momentum eigenstates *unmodified* by the well. What is the condition on ω for this to be a reasonable approximation? [Write an inequality involving possibly ω, $\hbar$, m, a, V_0 but no additional numerical constants.]
2. Find the relevant density of final states $\rho(E)$ in the continuum. Write your answer in terms of the box regulator L, m, $\hbar$, and the wave number k, including both states with momentum $\hbar k$ and $-\hbar k$. This is possible since the norm $|H'_{fi}|$ of the transition matrix element does not depend on the direction of the momentum.
3. Calculate the norm squared matrix element $|H'_{fi}|^2$ where i denotes the initial (ground) state, and f denotes a final plane wave state of momentum k. You will need the following approximations:
   - Use the *infinite* square-well ground state wave function as the initial state.
   - Assume the energy of the electron on the ground state is $-V_0$.

Your answer, once you use the approximations, will be of the form $|H'_{fi}|^2 = \frac{\gamma}{k^p}$, where γ is a constant built from L, a, e, and E_0, and p is a number.

4. Find the transition rate Γ from the ground state to the continuum of momentum states. Leave the answer in terms of the frequency ω of the plane wave and the other constants of the problem (E_0, e, m, and a).

Problem 27.7. *Spontaneous decay of a charged state in a three-dimensional harmonic oscillator.*

We wish to calculate the lifetime of a particle with charge q and mass m in the first excited states of a three-dimensional isotropic harmonic oscillator of frequency ω. In this oscillator, the energy eigenstates can be labeled as $|n_x, n_y, n_z\rangle$, where n_x, n_y, and n_z are the eigenvalues of the number operators $\hat{N}_x$, $\hat{N}_y$, and $\hat{N}_z$, respectively.

By analogy with the hydrogen atom, we refer to the states $|1,0,0\rangle$, $|0,1,0\rangle$, and $|0,0,1\rangle$ as the $2p$ states, and we call the ground state $|0,0,0\rangle$ the $1s$ state. An alternate basis for the $2p$ states is given by eigenstates of L_z:

$$|m_\ell = \pm 1\rangle = \tfrac{1}{\sqrt{2}}(|1,0,0\rangle \pm i|0,1,0\rangle), \quad |m_\ell = 0\rangle = |0,0,1\rangle .$$

1. Calculate the transition probability per unit time $A(2p, m_\ell \to 1s)$ for the particle to spontaneously make a transition to the ground state while emitting electromagnetic radiation. Show that this decay rate is independent of m_ℓ, and give your formula for $A(2p \to 1s)$ in terms of m, ω, q, and fundamental constants.
2. Thinking of this as a model of hydrogen, let the particle be an electron and set $\hbar\omega = \frac{3}{4}$ Ry (Ry = Rydberg = 13.6 eV). What is the lifetime τ in terms of $\hbar$, Ry, and the fine-structure constant α? What is the lifetime in seconds?

Problem 27.8. *Comparing rates for spontaneous and stimulated emission.*

For transitions with energy difference $\hbar\omega_0$, consider the unit-free ratio r formed by dividing the stimulated emission rate by the spontaneous emission rate, where blackbody radiation at a temperature T is the stimulus:

$$r \equiv \frac{\text{stimulated emission rate}}{\text{spontaneous emission rate}} .$$

1. Calculate the ratio r as a function of ω_0, and $\beta \equiv 1/(kT)$.
2. Consider a single mode of frequency ω_0 of the electromagnetic field associated to a photon of fixed polarization and fixed direction of propagation. Calculate (using statistical physics) the expected number $\bar{n}$ of such photons in the radiation at temperature T as a function of ω_0 and β. Express r in terms of $\bar{n}$. Does the result make sense?
3. Plot the ratio r as a function of $\frac{\hbar\omega_0}{kT}$. At room temperature (300 K), what is the frequency ν_0 (in Hz) for which both rates are the same? Which process dominates for frequencies associated with visible light? Which process dominates at the frequency 10^{10} Hz used in masers?

Problem 27.9. *Decays of 2s and 2p states of hydrogen.*

Calculate the lifetime for each of the four $n = 2$ states of hydrogen. The states, labeled $|n, \ell, m\rangle$, are $|2, 0, 0\rangle$, $|2, 1, 1\rangle$, $|2, 1, 0\rangle$, and $|2, 1, -1\rangle$, with associated wave functions $\psi_{2,0,0}$, $\psi_{2,1,1}$, $\psi_{2,1,0}$, and $\psi_{2,1,-1}$. The lifetimes $\tau_{n,\ell,m}$ can be put in the form

$$\tau_{n,\ell,m} = C_{n,\ell,m} \frac{t_*}{\alpha^3} \qquad \text{with} \quad t_* \equiv \frac{\hbar}{2\text{Ry}} \simeq 2.42 \times 10^{-17}\text{s},$$

where t_* is the atomic time, and the $C_{n,\ell,m}$ are dimensionless numerical constants that you must compute. The following radial wave functions may be useful:

$$R_{10}(r) = \frac{2}{\sqrt{a_0^3}} e^{-r/a_0}, \quad R_{21}(r) = \frac{1}{\sqrt{24a_0^3}} \frac{r}{a_0} e^{-r/2a_0}.$$

Problem 27.10. *Decays of the 3s state of hydrogen (Griffiths).*

An electron in the $n = 3$, $\ell = 0$, $m = 0$ state of hydrogen ($|3, 0, 0\rangle$) decays by a sequence of (electric dipole) transitions to the ground state.

1. What decay routes are open to it? Specify them in the following way:

   $|3, 0, 0\rangle \to |n, \ell, m\rangle \to |n', \ell', m'\rangle \to \ldots \to |1, 0, 0\rangle$.

2. What is the lifetime $\tau_{3,0,0}$ of this state? The result can be put in the form

   $$\tau_{3,0,0} = C_{3,0,0} \frac{t_*}{\alpha^3} \qquad \text{with} \qquad t_* \equiv \frac{\hbar}{2\text{Ry}} \simeq 2.42 \times 10^{-17}\,\text{s},$$

   where you must calculate the dimensionless numerical constant $C_{3,0,0}$.
   The following radial wave functions may be useful:

   $$R_{30}(r) = \frac{2}{\sqrt{27a_0^3}} \left(1 - \frac{2}{3}\frac{r}{a_0} + \frac{2}{27}\frac{r^2}{a_0^2}\right) e^{-r/3a_0}, \quad R_{21}(r) = \frac{1}{\sqrt{24a_0^3}} \frac{r}{a_0} e^{-r/2a_0}.$$

3. If you had a large number of atoms in this state $|3, 0, 0\rangle$, what fraction of them would decay via each route indicated in (1)?

28 Adiabatic Approximation

A quantum system experiences adiabatic evolution if the timescale for changes in the Hamiltonian is much larger than the quantum timescale of the system. In this case time evolution is along instantaneous eigenstates of the Hamiltonian and acquires time-dependent dynamical and geometric phases. If the Hamiltonian varies continuously over an interval $t \in [0, T]$, the probability for a nonadiabatic transition is suppressed by a factor of $1/T^2$. Such transitions can be studied explicitly in a two-state model considered by Landau and Zener. The geometric phase, also called Berry's phase, is observable for time-dependent processes that trace closed loops in the configuration space of the system. The idea of adiabaticity is at the heart of the Born-Oppenheimer approximation used in the study of molecules. We derive the nuclear Hamiltonian by "integration" of the electronic degrees of freedom and examine the key ideas in the context of the simplest molecule, the hydrogen molecular ion.

28.1 Adiabatic Changes and Adiabatic Invariants

Adiabatic processes are ubiquitous in physics. They are processes in which a time dependence is introduced into the physics of a system by letting some parameters that control the dynamics vary slowly in time. Both in classical mechanics and in quantum mechanics, Hamiltonians can be used to define the dynamics. Of course, time-independent Hamiltonians can give time-dependent physics. Two stars going around each other is a system governed by a time-independent Hamiltonian, and certainly, the positions and momenta of the stars have time dependence. In a quantum system described by a time-independent Hamiltonian, any superposition of two nondegenerate energy eigenstates is a time-dependent state with time-dependent expectation values.

We can now imagine some parameters in the Hamiltonian that while usually assumed to be constants, acquire some time dependence. In this case the Hamiltonian becomes time dependent. The change in time of these parameters is said to be *adiabatic* if it is slow compared with the natural timescale of the system. This terminology is used both in classical and in quantum mechanics.

More explicitly, let τ denote a natural timescale of the system. For the two stars orbiting each other, τ can be taken to be the period of the orbit; for the quantum superposition of two energy eigenstates, τ can be taken to be $\hbar$ divided by the energy difference. Moreover, let $\lambda(t)$ be a time-dependent parameter introduced into the Hamiltonian; perhaps the time-dependent mass of the star or, in the quantum system, a magnetic field that controls the energy difference between the two eigenstates. The statement that $\lambda(t)$ changes slowly is the claim that for all times the following inequality holds:

$$\tau\left|\frac{d\lambda}{dt}\right| \ll |\lambda|. \tag{28.1.1}$$

This is saying that the change in λ over a time τ is always very small compared to λ.

Even slowly varying parameters that vary over long times can end up changing by a sizable factor. Let $\Delta\lambda$ denote a fixed, finite change of λ slowly accumulated over a long time T. Holding $\Delta\lambda$ fixed as we vary T means that the larger T is, the slower the change in λ. Let us summarize the various symbols introduced and the key constraint:

$$\begin{aligned}
\tau &: \text{ natural timescale of the system}\\
\lambda(t) &: \text{ slowly varying parameter}\\
\Delta\lambda &: \text{ total change in } \lambda\\
T &: \text{ duration of the adiabatic process that is, } t\in[0,T]\\
\tau\left|\frac{d\lambda}{dt}\right| \ll |\lambda| &: \ \lambda \text{ changes adiabatically.}
\end{aligned} \tag{28.1.2}$$

Often, the natural timescale τ of the evolving system is itself changing in time. We can then take the slowly varying parameter λ to be τ itself, and the above constraint for adiabaticity becomes

$$\left|\frac{d\tau}{dt}\right| \ll 1. \tag{28.1.3}$$

In words this says that the change of τ over a time τ is small compared to τ. This form of the constraint on slow variations is reminiscent of the WKB condition $|\frac{d\lambda}{dx}| \ll 1$ for the position-dependent de Broglie wavelength $\lambda(x)$ (see (26.2.32)).

When the parameter $\lambda(t)$ varies slowly, physical quantities that were formerly conserved can also begin to vary slowly, at least after averaging out rapid oscillations. Just like λ they can also accumulate finite changes over the long time T. Here is where the notion of an **adiabatic invariant** becomes natural. An adiabatic invariant is a quantity I constructed in terms of $\lambda(t)$ and other slowly varying quantities that essentially remains constant. Throughout the time interval $[0, T]$ in which λ changes by $\Delta\lambda$, the adiabatic invariant I changes very little; more precisely, the change ΔI will approach zero as T goes to ∞. Then we have the following definition:

$$I \text{ is an adiabatic invariant if for any } t\in[0,T],\quad |I(t)-I(0)|\to 0 \text{ as } T\to\infty. \tag{28.1.4}$$

Finding an adiabatic invariant usually requires identifying different quantities whose variations are correlated in a slowly varying system.

28.2 From Classical to Quantum Adiabatic Invariants

The simple harmonic oscillator provides a setup to describe concretely adiabatic changes and an interesting example of an adiabatic invariant. In the simple harmonic oscillator, one has two constant parameters: the mass m of the particle and the frequency ω of the oscillations. Let us assume that ω becomes the function of time $\omega(t)$. The classical Hamiltonian H of the oscillator would then be

$$H(x, p, \omega(t)) = \frac{p^2}{2m} + \tfrac{1}{2} m\,\omega^2(t)\,x^2, \tag{28.2.1}$$

where x and p are the position and momentum canonical variables, respectively, and are functions of time. At any instant of time, the value of the Hamiltonian is the value of the energy. The time-varying $\omega(t)$ plays now the role of the time-dependent parameter $\lambda(t)$ discussed above.

Suppose we are going to change ω through a finite amount $\Delta\omega$, comparable to ω during the time interval $t \in [0, T]$ (figure 28.1). The natural timescale τ_ω of the oscillator is the period $2\pi/\omega$. This is still a relevant timescale even when ω changes, as long as it doesn't do so too rapidly:

$$\tau_\omega(t) \equiv \frac{2\pi}{\omega(t)}. \tag{28.2.2}$$

When ω is time dependent, the motion need not be exactly periodic. Following (28.1.1), ω changes adiabatically if the change of ω over the time τ_ω is much smaller than ω:

$$\tau_w \left|\frac{d\omega}{dt}\right| \ll \omega. \tag{28.2.3}$$

Using the definition of τ_ω, this gives

$$\frac{2\pi}{\omega^2}\left|\frac{d\omega}{dt}\right| \ll 1 \quad \rightarrow \quad \left|\frac{d}{dt}\frac{2\pi}{\omega}\right| \ll 1. \tag{28.2.4}$$

This is just the condition

$$\left|\frac{d\tau_\omega}{dt}\right| \ll 1, \tag{28.2.5}$$

just as anticipated in (28.1.3). If satisfied, the change in ω is said to be adiabatic. The condition can also be written as

$$\left|\frac{\dot\omega}{\omega^2}\right| \ll 1. \tag{28.2.6}$$

Consider the total change $\Delta\tau_\omega$ in τ_ω during the time interval $[0, T]$ where changes happen:

$$|\Delta\tau_\omega| = \left|\int d\tau_\omega\right| \leq \int |d\tau_\omega| \ll \int_0^T dt = T, \tag{28.2.7}$$

where we used the relation $|d\tau_\omega| \ll dt$, following from (28.2.5). The above series of relations implies that

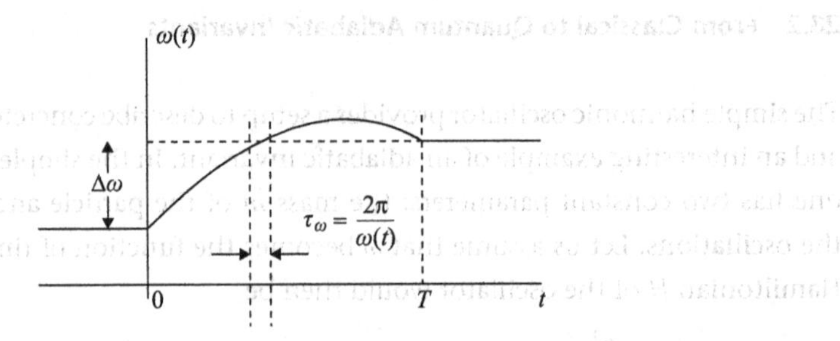

Figure 28.1
A process with a finite change $\Delta\omega$ of ω ocurring in the interval $t \in [0, T]$. The natural timescale of the oscillator is $\tau_\omega = 2\pi/\omega(t)$. In an adiabatic process, $\tau_\omega \ll T$.

$$|\Delta\tau_\omega| \ll T. \tag{28.2.8}$$

For what we call a finite change in τ_ω, we have $|\Delta\tau_\omega| \sim \tau_\omega$, and the above inequality implies that

$$\text{for finite changes in } \tau_\omega: \qquad \tau_\omega \ll T. \tag{28.2.9}$$

The quantities τ_ω and T are shown in figure 28.1.

Having made explicit the condition of adiabaticity, let us now calculate the change of the energy of the oscillator as a function of time. If ω is constant, the energy is constant. But now with a time-dependent $\omega(t)$, the time derivative of H will not be zero. We have

$$\frac{dH}{dt} = \frac{\partial H}{\partial x}\dot{x} + \frac{\partial H}{\partial p}\dot{p} + \frac{\partial H}{\partial t}. \tag{28.2.10}$$

Hamilton's equations of motion (16.2.15) fix the time derivatives of x and p:

$$\dot{x} = \frac{\partial H}{\partial p}, \qquad \dot{p} = -\frac{\partial H}{\partial x}. \tag{28.2.11}$$

These imply that the first two terms on the right-hand side of (28.2.10) vanish, and we have

$$\frac{dH}{dt} = \frac{\partial H}{\partial t} = m\omega\dot{\omega}x^2. \tag{28.2.12}$$

Our claim now is that the ratio $I(t)$ of the energy $H(t)$ and the frequency $\omega(t)$ is in fact an adiabatic invariant:

Claim: $I(t) \equiv \dfrac{H(t)}{\omega(t)}$ is an adiabatic invariant. (28.2.13)

This would entail showing that a finite change in ω realized over a time T leads to a change in I that becomes vanishingly small as $T \to \infty$. We will not try to prove this in generality in this case. We will argue why this is so for a typical situation. In the quantum case, we will do a more complete job, and some of the lessons here will prove useful.

To begin, we compute the time derivative of I:

$$\frac{dI}{dt} = \frac{1}{\omega^2}\Big(\omega\frac{dH}{dt} - H(t)\dot{\omega}\Big) = \frac{1}{\omega^2}\Big(\omega\left(m\omega\dot{\omega}x^2\right) - \Big(\frac{p^2}{2m} + \tfrac{1}{2}m\omega^2x^2\Big)\dot{\omega}\Big) =$$
$$= \frac{\dot{\omega}}{\omega^2}\left(\tfrac{1}{2}m\omega^2x^2 - \frac{p^2}{2m}\right) = \frac{\dot{\omega}}{\omega^2}\left(V(t) - K(t)\right), \tag{28.2.14}$$

where V and K are the potential and kinetic energies of the oscillator, respectively. We want to understand why this right-hand side representing $\dot{I}$ is small. We know from (28.2.6) that $\dot{\omega}/\omega^2$ has small absolute value, but this is not enough. If the factor multiplying $\dot{\omega}/\omega^2$ were approximated as a constant c, the change in I over time T would not be small:

$$I(T) - I(0) = \int_0^T \frac{dI}{dt}dt \sim \int_0^T \frac{\dot{\omega}}{\omega^2}\cdot c = c\Big(\frac{1}{\omega(0)} - \frac{1}{\omega(T)}\Big). \tag{28.2.15}$$

It also would not vanish for large T. As it turns out, the factor $V(t) - K(t)$ is neither small nor slowly varying but in fact does the job. Consider that factor when an oscillator is not time dependent. In that case, we can take x and p to be

$$x = A\sin\omega t \quad \text{and} \quad p = Am\omega\cos\omega t, \tag{28.2.16}$$

where A is the constant amplitude. Therefore,

$$V - K = \tfrac{1}{2}m\omega^2x^2 - \frac{p^2}{2m} = \tfrac{1}{2}m\omega^2A^2\left(\sin^2\omega t - \cos^2\omega t\right) = -\tfrac{1}{2}m\omega^2A^2\cos 2\omega t. \tag{28.2.17}$$

This is a fast variation with zero average value. This last fact proves important. Consider again the change in $I(t)$:

$$I(T) - I(0) = \int_0^T \frac{dI}{dt}dt = \int_0^T \frac{\dot{\omega}}{\omega^2}(t)\left(V(t) - K(t)\right)dt. \tag{28.2.18}$$

It is reasonable to expect that if $\omega(t)$ is slowly varying, then $V(t) - K(t)$ is estimated by (28.2.17) with ω replaced by $\omega(t)$ and A replaced by a slowly varying amplitude $A(t)$:

$$V(t) - K(t) \simeq -\tfrac{1}{2}m\omega^2A(t)^2\cos 2\omega(t)t. \tag{28.2.19}$$

Therefore,

$$I(T) - I(0) \simeq -\tfrac{1}{2}m\int_0^T \dot{\omega}A^2(t)\cos(2\omega(t)t)dt. \tag{28.2.20}$$

The idea is that the integral is in fact small due to the averaging effect of the cosine function. To make that intuition clearer, take the case of a linearly varying ω with total change $\Delta\omega$:

$$\omega(t) = \omega(0) + \frac{t}{T}\Delta\omega \quad \Rightarrow \quad \dot{\omega} = \frac{1}{T}\Delta\omega. \tag{28.2.21}$$

The change in $I(t)$ becomes

$$I(T)-I(0)\simeq-\frac{m\Delta\omega}{2T}\int_0^T A^2(t)\cos(2\omega(t)t)dt. \tag{28.2.22}$$

We want to establish that the integral is not proportional to T—that it is in fact bounded by a constant. Since $A(t)$ is slowly varying, the integral would be essentially zero due to the averaging effect of the cosine. We can estimate the integral as roughly $A^2(t)$ multiplied by the period of the wave, as we take it that the only contribution arises because the oscillations do not fit an integer number of times in T. Thus, we make the estimate that

$$\int_0^T A^2(t)\cos(2\omega(t)t)dt \simeq A^2(t')\frac{2\pi}{2\omega(t')} < C, \tag{28.2.23}$$

where the constant C could be taken to be the largest possible value that $A^2(t')\pi/\omega(t')$ takes on the interval $t'\in[0,T]$. Finally,

$$I(T)-I(0)\simeq-\frac{m\Delta\omega}{2T}C. \tag{28.2.24}$$

This is the desired result. The adiabatic invariant I is really invariant in the limit when the finite change occurs over infinite time—that is, as $T\to\infty$.

The adiabatic invariant E/ω has a nice geometric interpretation in phase space. To see this, consider the classical motion of a constant ω oscillator in phase space (x,p). Periodic oscillations trace the ellipse defined by conservation of the energy E:

$$\frac{p^2}{2m}+\tfrac{1}{2}m\omega^2x^2=E. \tag{28.2.25}$$

The ellipse is shown in figure 28.2. We quickly see that the ellipse intersects the x-axis at $\pm a$ and the p-axis at $\pm b$, with a and b given by

$$a=\sqrt{\frac{2E}{m\omega^2}} \quad\text{and}\quad b=\sqrt{2mE}. \tag{28.2.26}$$

We can calculate the area A of the ellipse now that we have the semimajor and semiminor axes:

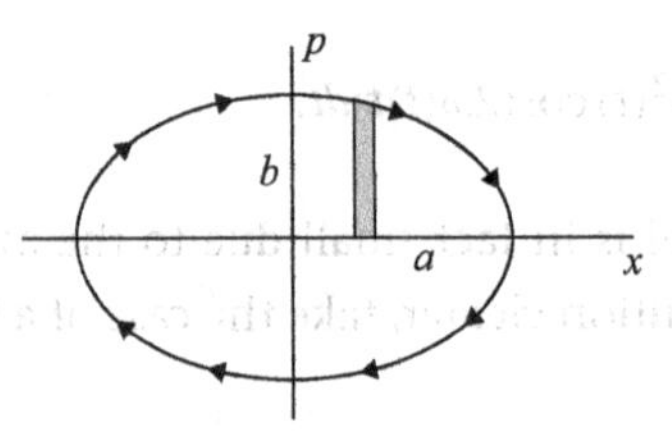

Figure 28.2
The phase-space trajectory for a harmonic oscillator of constant frequency ω and energy E. The ellipse is traced over each period of the oscillator. The constants a and b depend on ω, E, and the mass m of the particle.

$$A = \pi ab = 2\pi \frac{E}{\omega} = 2\pi I, \tag{28.2.27}$$

showing that the area of the ellipse in phase space is the adiabatic invariant! We can actually rewrite the area as a contour integral over the phase-space trajectory:

$$A = \oint p\,dx. \tag{28.2.28}$$

Here we integrate over the *whole* closed trajectory with the orientation shown in figure 28.2. The area above the x-axis is obtained as we integrate from $-a$ to $+a$, left to right, and the magnitude of the area under the x-axis is added as we integrate from $+a$ to $-a$, right to left. In this example, we see that

$$\oint p\,dx = 2\pi I. \tag{28.2.29}$$

More generally, for systems in which the phase-space motion is a closed trajectory, the enclosed area is an adiabatic invariant.

Insights into quantum systems Consider the classical harmonic oscillator adiabatic invariant E/ω, and evaluate it for eigenstates of the quantum harmonic oscillator:

$$\frac{E}{\omega} = \frac{1}{\omega}\hbar\omega\left(n + \tfrac{1}{2}\right) = \hbar\left(n + \tfrac{1}{2}\right). \tag{28.2.30}$$

The adiabatic invariant is the quantum number of the energy eigenstate. This suggests that in quantum mechanics the quantum number does not easily change under adiabatic changes. This is reasonable because quantum numbers are integers, and changes from one integer to another are necessarily discontinuous.

A similar intuition follows more generally from the WKB approximation. Consider a potential with two turning points a, b. In that case the Bohr-Sommerfield quantization condition (26.3.9) reads

$$\frac{1}{\hbar}\int_a^b p(x)dx = \left(n + \tfrac{1}{2}\right)\pi, \quad n = 0, 1, \ldots. \tag{28.2.31}$$

The classical motion here is that of a particle bouncing between the two turning points. In phase space a closed trajectory is obtained when the particle begins at a, goes to b, and then returns to a. For that trajectory we see that

$$\oint p(x)dx = 2\pi\hbar\left(n + \tfrac{1}{2}\right), \quad n = 0, 1, \ldots. \tag{28.2.32}$$

The left-hand side, as noted before, is an adiabatic invariant. Therefore, in the semiclassical approximation the story for the harmonic oscillator holds now for *arbitrary* potential: the adiabatic invariant evaluates to the quantum number, suggesting again that transitions between energy eigenstates are suppressed under adiabatic changes. We will confirm that this is the case.

The result (27.3.22) for transition probabilities at first order in time dependent perturbation theory gives complementary intuition:

$$P_{f\leftarrow i}(t) = \left| \int_0^t e^{i\omega_{fi}t'} \frac{\delta H_{fi}(t')}{i\hbar} dt' \right|^2. \tag{28.2.33}$$

We can mimic adiabatic changes with a constant perturbation in which δH_{fi} is time-independent:

$$P_{f\leftarrow i}(t) = \frac{|\delta H_{fi}|^2}{\hbar^2} \left| \int_0^t e^{i\omega_{fi}t'} dt' \right|^2 = \frac{|\delta H_{fi}|^2}{\hbar^2} \frac{\left| e^{i\omega_{fi}t} - 1 \right|^2}{\omega_{fi}^2}. \tag{28.2.34}$$

If the spectrum is discrete, the transition probability is suppressed by the energy gap squared, as shown by the ω_{fi}^2 factor in the denominator. It is reasonable to expect that for slowly varying perturbations this suppression will remain. So it is difficult in general to change state with constant or slow perturbations, suggesting again that quantum numbers are adiabatic invariants. Efficient transitions between energy levels require oscillatory perturbations at resonance.

28.3 Instantaneous Energy Eigenstates

In a system with a time-dependent Hamiltonian $\hat{H}(t)$, it is possible to find states $|\psi(t)\rangle$ that satisfy the curious equation

$$\hat{H}(t)|\psi(t)\rangle = E(t)|\psi(t)\rangle. \tag{28.3.1}$$

We say this is a curious equation because the similar equation $\hat{H}|\psi\rangle = E|\psi\rangle$ was defined for time independent Hamiltonians and position-dependent but time independent $|\psi\rangle$. The states solving $\hat{H}|\psi\rangle = E|\psi\rangle$ can be promoted to solutions $|\Psi\rangle = e^{-iEt/\hbar}|\psi\rangle$ of the Schrödinger equation.

In equation (28.3.1) the spatial dependence is present but implicit, for brevity. In constrast to energy eigenstates, however, both $|\psi(t)\rangle$ and the energy are time dependent. The solution $|\psi(t)\rangle$ is built by solving the equation $H(t_0)|\psi(t_0)\rangle = E(t_0)|\psi(t_0)\rangle$ for every value of the time t_0 and putting together these solutions into a single solution $|\psi(t)\rangle$ with some $E(t)$. This construction must demand continuity of the state as a function of time because any Hamiltonian can have multiple eigenstates. In other words, while building $|\psi(t)\rangle$, the state at $t + dt$ must be very close to the state at t.

We'll call the state $|\psi(t)\rangle$ defined by (28.3.1) an **instantaneous eigenstate**. The name is appropriate because it is, at any time, an eigenstate of the Hamiltonian at that time. I'd like to emphasize that, in general, an instantaneous eigenstate *is not a solution* of the time-dependent Schrödinger equation. Not even in the time-independent case is $\psi(x)$ a solution of the Schrödinger equation; it must be supplemented by a time-dependent phase to be one. As it turns out, for $|\psi(t)\rangle$ a simple time-dependent phase will not do. In the adiabatic approximation, however, $|\psi(t)\rangle$ can be upgraded to an *approximate* solution to the Schrödinger equation.

Instantaneous eigenstates are less exotic than they may seem at first sight. There is a natural way to generate them. Consider a time-independent Hamiltonian $\hat{H}(R_1,\ldots,R_k)$ that depends on some parameters $R_1,\ldots,R_k$. These parameters could be masses, frequencies, magnetic fields, sizes of wells, parameters in a potential, and so on. We then find the familiar eigenstates

$$\hat{H}(R_1,\ldots,R_k)\,\big|\psi(R_1,\ldots,R_k)\big\rangle = E(R_1,\ldots,R_k)\,\big|\psi(R_1,\ldots,R_k)\big\rangle. \tag{28.3.2}$$

As expected, both the energy and the state depend on the values of the parameters. In fact, they are continuous functions of the parameters. We write this more briefly as

$$\hat{H}(\boldsymbol{R})\,\big|\psi(\boldsymbol{R})\big\rangle = E(\boldsymbol{R})\,\big|\psi(\boldsymbol{R})\big\rangle. \tag{28.3.3}$$

Now imagine the parameters become time dependent in some arbitrary way:

$$\boldsymbol{R} \to \boldsymbol{R}(t). \tag{28.3.4}$$

Since (28.3.3) holds for *arbitrary* values of the parameters, for *any* value of time we find that

$$\hat{H}(\boldsymbol{R}(t))\,\big|\psi(\boldsymbol{R}(t))\big\rangle = E(\boldsymbol{R}(t))\,\big|\psi(\boldsymbol{R}(t))\big\rangle. \tag{28.3.5}$$

The $\big|\psi(\boldsymbol{R}(t))\big\rangle$ are indeed instantaneous eigenstates of the time-dependent Hamiltonian $\hat{H}(\boldsymbol{R}(t))$. This procedure naturally implements the continuity requirement on the instantaneous eigenstates because the original eigenstates $\big|\psi(\boldsymbol{R})\big\rangle$ are continuous functions of the parameters.

We must emphasize that instantaneous eigenstates have a rather important phase ambiguity. This ambiguity originates at the level of the parameter-dependent states. Consider changing the states $\big|\psi(\boldsymbol{R})\big\rangle$ as follows:

$$\big|\psi(\boldsymbol{R})\big\rangle \to e^{i\gamma(\boldsymbol{R})}\,\big|\psi(\boldsymbol{R})\big\rangle, \tag{28.3.6}$$

where $\gamma(\boldsymbol{R})$ is a real function of the parameters. The new states will still satisfy equation (28.3.3), and their normalization has not been altered, so the states are as good as the originals. Similarly, for the time-dependent eigenstates if we let

$$\big|\psi(\boldsymbol{R}(t))\big\rangle \to e^{i\gamma(t;\boldsymbol{R}(t))}\,\big|\psi(\boldsymbol{R}(t))\big\rangle, \tag{28.3.7}$$

equation (28.3.5) will still hold. This phase ambiguity of instantaneous eigenstates will be relevant to the definition of geometric phases.

Example 28.1. *Instantaneous spin one-half states.*

A simple example of instantaneous eigenstates is provided by spin one-half eigenstates. The Hamiltonian for an electron in a magnetic field was obtained in (12.2.17) and reads $\hat{H} = \mu_B \mathbf{B}\cdot\boldsymbol{\sigma}$. For a uniform, time-independent magnetic field $\mathbf{B} = B_0\mathbf{n}$, with $\mathbf{n}$ a unit vector,

$$\hat{H}(B_0,\mathbf{n}) = \mu_B B_0\,\mathbf{n}\cdot\boldsymbol{\sigma}. \tag{28.3.8}$$

Here we have made explicit that B_0 and $\mathbf{n}$ are parameters in the Hamiltonian. The eigenstates are $|\mathbf{n};\pm\rangle$ satisfying $\mathbf{n}\cdot\boldsymbol{\sigma}\,|\mathbf{n};\pm\rangle = \pm|\mathbf{n};\pm\rangle$, and therefore,

$$\hat{H}(B_0,\mathbf{n})|\mathbf{n};\pm\rangle = \pm\mu_B B_0|\mathbf{n};\pm\rangle. \tag{28.3.9}$$

More explicitly, and with $\mathbf{n} = (\sin\theta\cos\phi, \sin\theta\sin\phi, \cos\theta)$, the spin states are (see (12.3.28))

$$|\mathbf{n};+\rangle = \begin{pmatrix} \cos\frac{\theta}{2} \\ \sin\frac{\theta}{2}e^{i\phi} \end{pmatrix}, \qquad |\mathbf{n};-\rangle = \begin{pmatrix} -\sin\frac{\theta}{2}e^{-i\phi} \\ \cos\frac{\theta}{2} \end{pmatrix}. \tag{28.3.10}$$

Now imagine that both the magnitude B_0 and the direction $\mathbf{n}$ of the magnetic field change in time so that $B_0 \to B_0(t)$, and $\mathbf{n} \to \mathbf{n}(t)$. The change of $\mathbf{n}$ can be described explicitly by giving the time dependence $\theta(t)$ and $\phi(t)$ of the spherical angles θ, ϕ that define the direction $\mathbf{n}(t)$. It follows from (28.3.9) that

$$\hat{H}(B_0(t), \mathbf{n}(t))|\mathbf{n}(t);\pm\rangle = \pm\mu_B B_0(t)\,|\mathbf{n}(t);\pm\rangle. \tag{28.3.11}$$

The eigenvalues $\pm\mu_B B_0(t)$ are now time dependent and so are the eigenstates, which become

$$|\mathbf{n}(t);+\rangle = \begin{pmatrix} \cos\frac{\theta(t)}{2} \\ \sin\frac{\theta(t)}{2}e^{i\phi(t)} \end{pmatrix}, \qquad |\mathbf{n}(t);-\rangle = \begin{pmatrix} -\sin\frac{\theta(t)}{2}e^{-i\phi(t)} \\ \cos\frac{\theta(t)}{2} \end{pmatrix}. \tag{28.3.12}$$

The magnitude $B_0(t)$ of the magnetic field appears in the instantaneous eigenvalues but does not appear in the instantaneous eigenstates. □

Building on the instantaneous eigenstates Let us try to understand the relation between the instantaneous eigenstate $|\psi(t)\rangle$ and a solution $|\Psi(t)\rangle$ of the Schrödinger equation

$$i\hbar\partial_t|\Psi(t)\rangle = \hat{H}(t)|\Psi(t)\rangle. \tag{28.3.13}$$

We try an ansatz for $|\Psi(t)\rangle$ in terms of the instantaneous eigenstate:

$$|\Psi(t)\rangle = c(t)\exp\left(\frac{1}{i\hbar}\int_0^t E(t')dt'\right)|\psi(t)\rangle. \tag{28.3.14}$$

Here $c(t)$ is a function of time to be determined, and we have included a time-dependent phase that is a natural generalization of the phase $e^{-iEt/\hbar}$ appropriate for time-independent Hamiltonians. There is, of course, no guarantee that this ansatz will work, but it seems reasonable to try it out.

The left-hand side of the Schrödinger equation then looks like

$$\begin{aligned} i\hbar\partial_t|\Psi(t)\rangle &= i\hbar\dot{c}(t)\exp\left(\frac{1}{i\hbar}\int_0^t E(t')dt'\right)|\psi(t)\rangle \\ &\quad + E(t)|\Psi(t)\rangle + i\hbar c(t)\exp\left(\frac{1}{i\hbar}\int_0^t E(t')dt'\right)|\dot{\psi}(t)\rangle. \end{aligned} \tag{28.3.15}$$

For the right-hand side, using the instantaneous eigenstate equation, we have

$$\hat{H}(t)|\Psi(t)\rangle = c(t)\exp\left(\frac{1}{i\hbar}\int_0^t E(t')dt'\right)\hat{H}(t)|\psi(t)\rangle = E(t)|\Psi(t)\rangle. \tag{28.3.16}$$

Equating the two sides, we get

$$\dot{c}(t)\exp\left(\frac{1}{i\hbar}\int_0^t E(t')dt'\right)|\psi(t)\rangle + c(t)\exp\left(\frac{1}{i\hbar}\int_0^t E(t')dt'\right)|\dot{\psi}(t)\rangle = 0. \tag{28.3.17}$$

Canceling the two exponentials, we find that

$$\dot{c}(t)\,|\psi(t)\rangle \;=\; -c(t)\,|\dot{\psi}(t)\rangle. \tag{28.3.18}$$

Multiply by $\langle\psi(t)|$ to get a differential equation for $c(t)$:

$$\dot{c}(t) \;=\; -c(t)\,\langle\psi(t)|\dot{\psi}(t)\rangle, \tag{28.3.19}$$

which, happily, we can solve. Letting $c(0)=1$, we find that the solution is

$$c(t) = \exp\left(-\int_0^t \langle\psi(t')|\dot{\psi}(t')\rangle dt'\right). \tag{28.3.20}$$

The above exponential is a phase because the bracket in the integrand is actually purely imaginary. Indeed, taking a time derivative of $\langle\psi(t)|\psi(t)\rangle = 1$, we have

$$\langle\dot{\psi}(t)|\psi(t)\rangle + \langle\psi(t)|\dot{\psi}(t)\rangle = 0 \quad\Rightarrow\quad \langle\psi(t)|\dot{\psi}(t)\rangle^* + \langle\psi(t)|\dot{\psi}(t)\rangle = 0, \tag{28.3.21}$$

showing that the real part of $\langle\dot{\psi}(t)|\psi(t)\rangle$ vanishes, and thus this quantity is purely imaginary. To emphasize this fact, we write

$$c(t) = \exp\left(i\int_0^t i\langle\psi(t')|\dot{\psi}(t')\rangle dt'\right). \tag{28.3.22}$$

Having apparently solved for $c(t)$, we now return to our ansatz (28.3.14). We get

$$\boxed{|\Psi(t)\rangle \;\simeq\; c(0)\exp\left(i\int_0^t i\langle\psi(t')|\dot{\psi}(t')\rangle dt'\right)\exp\left(\frac{1}{i\hbar}\int_0^t E(t')dt'\right)|\psi(t)\rangle.} \tag{28.3.23}$$

But there is a mistake in this analysis. We really did not solve the Schrödinger equation! That's why we put a $\simeq$ instead of an equality.

The equation we had to solve, (28.3.18), is a vector equation, and forming the inner product with $\langle\psi(t)|$ gives a necessary condition for the solution but not a sufficient one. We must check the equation forming the overlap with a full basis set of states. Indeed, since $|\psi(t)\rangle$ is known, the equation can only have a solution if the two vectors $|\dot{\psi}(t)\rangle$ and $|\psi(t)\rangle$ are parallel. This does not happen in general. So we really did *not* solve equation (28.3.18). The conclusion is that ultimately the ansatz in (28.3.14) is *not good enough*. Still, (28.3.23) is a fairly accurate solution if the Hamiltonian is slowly varying, as the system tends to remain in suitably improved instantaneous eigenstates. We will show this in the next section.

The approximate solution (28.3.23) is written compactly with the help of the following definitions:

$$\theta(t) \;\equiv\; -\frac{1}{\hbar}\int_0^t E(t')dt', \qquad \nu(t) \;\equiv\; i\langle\psi(t)|\dot{\psi}(t)\rangle, \quad \gamma(t) \;\equiv\; \int_0^t \nu(t')dt'. \tag{28.3.24}$$

Here, $\theta(t)$, $\nu(t)$, and $\gamma(t)$ are all real. The state reads

$$|\Psi(t)\rangle \simeq c(0)\, e^{i\gamma(t)} e^{i\theta(t)} |\psi(t)\rangle. \tag{28.3.25}$$

We call $\theta(t)$ the *dynamical* phase and $\gamma(t)$ the *geometric* phase.

28.4 Quantum Adiabatic Theorem

Let us begin by stating precisely the content of the adiabatic theorem in quantum mechanics. For this, consider a family of instantaneous eigenstates:

$$\hat{H}(t)|\psi_n(t)\rangle = E_n(t)|\psi_n(t)\rangle, \tag{28.4.1}$$

with $E_1(t) < E_2(t) < \ldots$ so that there are no degeneracies in this spectrum at any time. We will also assume, for simplicity, that the spectrum of $\hat{H}$ is only discrete and that the number of energy eigenstates is finite.

Theorem 28.4.1. Adiabatic theorem. *Let $\hat{H}(t)$ be a continuously varying Hamiltonian for $0 \leq t \leq T$. Let the state of the system at $t=0$ be one of the instantaneous eigenstates: $|\Psi(0)\rangle = |\psi_m(0)\rangle$ for some m. Then at any time $t \in [0, T]$ we have $|\psi(t)\rangle \simeq |\psi_m(t)\rangle$ up to a calculable phase. The amplitude to transition to any other instantaneous eigenstate is of order $1/T$. This implies that the probability of remaining in the instantaneous eigenstate is one up to corrections of order $1/T^2$.*

We will prove the adiabatic theorem by carefully finding bounds on the terms in the Schrödinger equation that induce nonadiabatic transitions. To do this we need some preparatory work. We will examine the Schrödinger equation using a basis of instantaneous eigenstates. We will obtain the general expression for the dressing of the instantaneous eigenstates and identify the terms that can produce nonadiabatic transitions—that is, transitions between different instantaneous eigenstates.

We begin by setting an expansion for our normalized, time-dependent state $|\Psi(t)\rangle$:

$$|\Psi(t)\rangle = \sum_n c_n(t)|\psi_n(t)\rangle. \tag{28.4.2}$$

Since the instantaneous eigenstates are normalized, all c_n satisfy $|c_n(t)| \leq 1$. The Schrödinger equation then gives

$$i\hbar \sum_n \left(\dot{c}_n|\psi_n(t)\rangle + c_n|\dot{\psi}_n(t)\rangle\right) = \sum_n c_n(t)E_n(t)|\psi_n(t)\rangle. \tag{28.4.3}$$

Acting with $\langle\psi_k(t)|$ from the left gives

$$i\hbar\dot{c}_k = E_k c_k - i\hbar \sum_n \langle\psi_k|\dot{\psi}_n\rangle c_n. \tag{28.4.4}$$

From the sum we separate out the term $n=k$:

$$i\hbar\dot{c}_k = \left(E_k - i\hbar\langle\psi_k|\dot{\psi}_k\rangle\right)c_k - i\hbar \sum_{n\neq k} \langle\psi_k|\dot{\psi}_n\rangle c_n. \tag{28.4.5}$$

It is clear that the terms in the sum are those that couple the $n \neq k$ eigenstates to the k eigenstate. This is the term that produces transitions. Absent that term, if at $t=0$

we have $c_k(0)=0$, then $c_k(t)=0$ for all times, and no transition to the state $|\psi_k(t)\rangle$ happens.

It is instructive to relate $\langle\psi_k|\dot{\psi}_n\rangle$ to a matrix element of the *time derivative* of $\hat{H}(t)$. We start with the defining equation:

$$\hat{H}(t)|\psi_n(t)\rangle = E_n(t)|\psi_n(t)\rangle, \tag{28.4.6}$$

and take a time derivative

$$\frac{d\hat{H}(t)}{dt}|\psi_n(t)\rangle + \hat{H}(t)|\dot{\psi}_n(t)\rangle = \frac{dE_n(t)}{dt}|\psi_n(t)\rangle + E_n(t)|\dot{\psi}_n(t)\rangle. \tag{28.4.7}$$

Now multiply by $\langle\psi_k(t)|$ from the left, with $k\neq n$:

$$\left\langle\psi_k(t)\middle|\frac{d\hat{H}}{dt}\middle|\psi_n(t)\right\rangle + E_k(t)\langle\psi_k(t)|\dot{\psi}_n(t)\rangle = E_n(t)\langle\psi_k(t)|\dot{\psi}_n(t)\rangle. \tag{28.4.8}$$

Hence,

$$\langle\psi_k(t)|\dot{\psi}_n(t)\rangle = \frac{\left\langle\psi_k(t)\middle|\frac{d\hat{H}}{dt}\middle|\psi_n(t)\right\rangle}{E_n(t)-E_k(t)} \equiv \frac{[\frac{d\hat{H}}{dt}]_{kn}}{E_n-E_k}, \quad k\neq n. \tag{28.4.9}$$

We can plug (28.4.9) back into (28.4.4) to get

$$i\hbar\dot{c}_k = \Big(E_k - i\hbar\langle\psi_k|\dot{\psi}_k\rangle\Big)c_k - i\hbar\sum_{n\neq k}\frac{[\frac{d\hat{H}}{dt}]_{kn}}{E_n-E_k}c_n. \tag{28.4.10}$$

The last term, as we said above, induces transitions. It is not so simple to see that those transition amplitudes are suppressed by factors of $1/T$. We can see that matrix elements of $\frac{d\hat{H}}{dt}$ must go like $1/T$ because the same finite change in the Hamiltonian occurs for any value of T. On the other hand, the effect of this term must be integrated over a time T, so it is not clear that the suppression survives. We will do a detailed analysis below showing that there *is* suppression.

If we ignore the transition-causing terms, we would simply have

$$i\hbar\dot{c}_k = \Big(E_k - i\hbar\langle\psi_k|\dot{\psi}_k\rangle\Big)c_k. \tag{28.4.11}$$

This is easily integrated:

$$\begin{aligned} c_k(t) &= c_k(0)\exp\left[\frac{1}{i\hbar}\int_0^t \big(E_k(t') - i\hbar\langle\psi_k|\dot{\psi}_k\rangle\big)\,dt'\right] \\ &= c_k(0)\exp\left(-\frac{i}{\hbar}\int_0^t E_k(t')dt'\right)\exp\left(i\int_0^t i\langle\psi_k|\dot{\psi}_k\rangle dt'\right). \end{aligned} \tag{28.4.12}$$

Therefore, we have

$$c_k(t) = c_k(0)e^{i\theta_k(t)}e^{i\gamma_k(t)}, \tag{28.4.13}$$

where, as before,

$$\theta_k(t) = -\frac{1}{\hbar}\int_0^t E_k(t')dt', \quad \gamma_k(t) = \int_0^t \nu_k(t')dt', \quad \nu_k(t) = i\langle\psi_k(t)|\dot{\psi}_k(t)\rangle. \tag{28.4.14}$$

This result is valid for all k. In any specific situation, all we need to know to determine the approximate time evolution are the initial values of the c_k's at $t=0$.

We can now go back to the statement of the adiabatic theorem. We are told that $|\Psi(0)\rangle = |\psi_m(0)\rangle$. On account of the expansion (28.4.2), we have $|\Psi(t)\rangle = c_m(t)|\psi_m(t)\rangle$, with $c_m(0)=1$. Thus, using (28.4.13) with $k=m$, we get

$$|\Psi(t)\rangle \simeq e^{i\theta_m(t)}e^{i\gamma_m(t)}|\psi_m(t)\rangle. \tag{28.4.15}$$

Bounding the error We will now prove the adiabatic theorem by showing that transition amplitudes to other instantaneous eigenstates are of order $1/T$. For this we must first simplify further the differential equations satisfied by the $c_k(t)$ coefficients. Let us therefore reconsider the exact equation of motion, last given in (28.4.5):

$$i\hbar\,\dot{c}_k(t) = \Big(E_k - i\hbar\langle\psi_k|\dot{\psi}_k\rangle\Big)c_k(t) - i\hbar\sum_{n\neq k}\langle\psi_k|\dot{\psi}_n\rangle c_n(t). \tag{28.4.16}$$

This equation can be simplified, without any loss of generality, by choosing the instantaneous eigenstates cleverly. We can in fact assume that the instantaneous energy eigenstates $|\psi_i(t)\rangle$ satisfy

$$\langle\psi_i(t)|\dot{\psi}_i(t)\rangle \;=\; 0, \quad \forall i. \tag{28.4.17}$$

If an instantaneous eigenstate $|\psi_i(t)\rangle$ does not satisfy this condition, the redefined state $|\psi_i'(t)\rangle = e^{i\gamma_i(t)}|\psi_i(t)\rangle$, with $\gamma_i(t)$ the geometric phase, will satisfy the condition.

Exercise 28.1. *Verify that* $\langle\psi_i'(t)|\dot{\psi}_i'(t)\rangle = 0$.

So we assume that suitably defined instantaneous eigenstates satisfy (28.4.17), and therefore equation (28.4.16) becomes

$$i\hbar\,\dot{c}_k(t) = E_k c_k(t) - i\hbar\sum_{n\neq k}\langle\psi_k|\dot{\psi}_n\rangle c_n(t). \tag{28.4.18}$$

To simplify the equation further, we introduce new coefficients $\check{c}_k(t)$ that differ from $c_k(t)$ by the geometric phase:

$$c_k(t) = e^{i\theta_k(t)}\check{c}_k(t). \tag{28.4.19}$$

These coefficients agree at $t=0$. Introducing this ansatz into (28.4.18), we now see that

$$\dot{\check{c}}_k(t) = -\sum_{n\neq k}\langle\psi_k|\dot{\psi}_n\rangle e^{i\theta_{nk}(t)}\check{c}_n(t), \qquad \theta_{nk}(t) \equiv \theta_n(t) - \theta_k(t). \tag{28.4.20}$$

We can integrate this equation from time zero to time t:

$$\check{c}_k(t) - \check{c}_k(0) = -\sum_{n\neq k}\int_0^t dt'\,\langle\psi_k|\dot{\psi}_n\rangle(t')e^{i\theta_{nk}(t')}\check{c}_n(t'). \tag{28.4.21}$$

We must now work to show that the right-hand side is bounded. As assumed, the time-dependent process occurs in the interval $t\in[0,T]$. It is therefore convenient to introduce the unit-free time s defined as

$$s \equiv \frac{t}{T}, \quad s \in [0, 1]. \tag{28.4.22}$$

We will use $s \in [0, 1]$ as the main variable to bring into the open the dependence of various quantities on T. We will define some functions of s, indicated by a tilde, to distinguish them from the functions of t they originate from. Given a function $A(t)$, we define the function $\tilde{A}(s)$ by

$$\tilde{A}(s) \equiv A(sT). \tag{28.4.23}$$

This is reasonable: a tilde function at s evaluates to the value of the original function at the associated time $t = sT$. We will use this definition for $\hat{H}(t), E_k(t)$, and $|\psi_k(t)\rangle$:

$$\tilde{H}(s) \equiv \hat{H}(sT), \quad \tilde{E}_k(s) \equiv E_k(sT), \qquad |\tilde{\psi}_k(s)\rangle \equiv |\psi_k(sT)\rangle. \tag{28.4.24}$$

It follows, for example, that the defining relation $\hat{H}(t)|\psi_k(t)\rangle = E_k(t)\,|\psi_k(t)\rangle$ for instantaneous eigenstates becomes

$$\tilde{H}(s)\,|\tilde{\psi}_k(s)\rangle = \tilde{E}_k(s)\,|\tilde{\psi}_k(s)\rangle, \tag{28.4.25}$$

as you should verify. Other quantities involving derivatives or integrals over time can acquire explicit factors of T. For example, consider the dynamical phase, and change variables to $s' = t'/T$ in the integral that defines it:

$$\theta_k(t) = -\frac{1}{\hbar}\int_0^t dt' E_k(t') = -\frac{T}{\hbar}\int_0^{t/T} ds' E_k(s'T) = -\frac{T}{\hbar}\int_0^s ds' \tilde{E}_k(s'), \tag{28.4.26}$$

where $s = t/T$. Defining

$$\Theta_k(s) \equiv -\frac{1}{\hbar}\int_0^s ds' \tilde{E}_k(s'), \tag{28.4.27}$$

we have learned that

$$\theta_k(t) = T\,\Theta_k(s), \qquad s = t/T. \tag{28.4.28}$$

Similarly, for the inner product $\langle\psi_k|\dot{\psi}_n\rangle$ we write

$$\langle\psi_k(t)|\dot{\psi}_n(t)\rangle = \frac{1}{T}\big\langle\tilde{\psi}_k(s)\big|\frac{d}{ds}\big|\tilde{\psi}_n(s)\big\rangle \equiv \frac{1}{T}F_{kn}(s), \quad s = t/T, \tag{28.4.29}$$

where we introduced the function $F_{kn}(s)$. We also note that, in analogy with (28.4.9), we have

$$F_{kn}(s) = \big\langle\tilde{\psi}_k(s)\big|\frac{d}{ds}\big|\tilde{\psi}_n(s)\big\rangle = \frac{1}{\tilde{E}_n(s) - \tilde{E}_k(s)}\big\langle\tilde{\psi}_k(s)\big|\frac{d\tilde{H}}{ds}\big|\tilde{\psi}_n(s)\big\rangle. \tag{28.4.30}$$

This shows that $F_{nk}(s)$ is a bounded function of $s \in [0, 1]$ as long as the states are nondegenerate so that the denominator in the last expression is never zero. Since we assumed that $\tilde{H}(s)$ is a continuous function, its derivative is an operator with finite matrix elements. Note that even from the defining expression we could argue F_{kn} is bounded because instantaneous eigenstates are continuous functions as long as they are not degenerate.

With the relations obtained above, equation (28.4.21) becomes

$$\check{c}_k(t) - \check{c}_k(0) = -\sum_{n\neq k}\int_0^t dt' \frac{F_{kn}(s')}{T} e^{iT\Theta_{nk}(s')}\check{c}_n(t'), \quad s' = t'/T, \tag{28.4.31}$$

where we defined

$$\Theta_{nk}(s) \equiv \Theta_n(s) - \Theta_k(s). \tag{28.4.32}$$

In order to express this result clearly in terms of s and T, we now define coefficients c_k with two arguments:

$$c_k(s, T) = \check{c}_k(t = sT), \quad \text{such that} \quad c_k(0, T) = \check{c}_k(0) = c_k(0). \tag{28.4.33}$$

With these variables and using s' for the integration variable, the integral (28.4.31) becomes

$$c_k(s, T) - c_k(0) = -\sum_{n\neq k}\int_0^s ds' F_{kn}(s') e^{iT\Theta_{nk}(s')} c_n(s', T). \tag{28.4.34}$$

As desired, t has disappeared from the equation. For later use we note that the s derivative of the above equation is given by

$$\frac{d}{ds}c_k(s, T) = -\sum_{n\neq k} F_{kn}(s) e^{iT\Theta_{nk}(s)} c_n(s, T). \tag{28.4.35}$$

Now, we can see that

$$e^{iT\Theta_{nk}(s)} = \frac{i\hbar}{T}\frac{1}{\tilde{E}_{nk}(s)}\frac{d}{ds}e^{iT\Theta_{nk}(s)}, \quad \text{where} \quad \tilde{E}_{nk}(s) = \tilde{E}_n(s) - \tilde{E}_k(s). \tag{28.4.36}$$

This is verified by taking the derivative on the right-hand side and using $\frac{d}{ds}\Theta_k = -\tilde{E}_k/\hbar$. Going back to the integral (28.4.34), we get

$$c_k(s, T) - c_k(0) = -\frac{i\hbar}{T}\sum_{n\neq k}\int_0^s ds' \frac{F_{kn}(s')c_n(s', T)}{\tilde{E}_{nk}(s')}\frac{d}{ds'}e^{iT\Theta_{nk}(s')}. \tag{28.4.37}$$

Integrating by parts, we find that

$$\begin{aligned} c_k(s, T) - c_k(0) = -\frac{i\hbar}{T}\sum_{n\neq k}\Bigg[&\frac{F_{kn}(s')c_n(s', T)}{\tilde{E}_{nk}(s')}e^{iT\Theta_{nk}(s')}\Bigg|_0^s \\ &-\int_0^s ds' \frac{d}{ds'}\Big(\frac{F_{kn}(s')c_n(s', T)}{\tilde{E}_{nk}(s')}\Big)e^{iT\Theta_{nk}(s')}\Bigg]. \end{aligned} \tag{28.4.38}$$

The whole right-hand side has a prefactor $1/T$. We can now argue that the first term inside the brackets appearing on the first line is bounded by a number independent of T. Indeed, as discussed right below equation (28.4.30), F_{kn} is bounded and continuous. Moreover, $|c_n(s', T)| \leq 1$, $|e^{iT\Theta_{nk}(s')}| = 1$, and $1/\tilde{E}_{nk}(s')$ is bounded. Each term in the sum is thus bounded by a number independent of T, and so is the sum that, by assumption, involves a finite number of terms. We therefore write

$$c_k(s,T) - c_k(0) = \mathcal{O}(\tfrac{1}{T}) + \frac{i\hbar}{T}\sum_{n\neq k}\int_0^s ds'\,\frac{d}{ds'}\Big(\frac{F_{kn}(s')c_n(s',T)}{\tilde{E}_{nk}(s')}\Big)e^{iT\Theta_{nk}(s')}. \tag{28.4.39}$$

To see that the remaining terms are bounded by a constant independent of T, we have to take the derivative:

$$\begin{aligned}\frac{d}{ds'}\Big(\frac{F_{kn}(s')c_n(s',T)}{\tilde{E}_{nk}(s')}\Big) &= \frac{d}{ds'}\bigg(\frac{F_{kn}(s')}{\tilde{E}_{nk}(s')}\bigg)c_n(s',T) \\ &\quad - \sum_{m\neq n}\frac{F_{kn}(s')F_{nm}(s')}{\tilde{E}_{nk}(s')}e^{iT\Theta_{mn}(s')}c_m(s',T),\end{aligned} \tag{28.4.40}$$

where we used (28.4.35) to evaluate the derivative of the $c_n(s',T)$ amplitude. Each term on the above right-hand side is bounded by a constant that is T independent. This fact implies that the derivative in the integral (28.4.39) is bounded independently of T and so is the integral. All in all we find that

$$c_k(s,T) - c_k(0) = \mathcal{O}(\tfrac{1}{T}). \tag{28.4.41}$$

This is what we wanted to prove. Indeed, the $c_k(s,T)$ are the coefficients of the expansion of the state at time $t = sT$ in terms of the redefined instantaneous eigenstates we introduced along the proof. If $c_k(0) = 0$, the state is not in the kth state at time equal zero. Then the result $c_k(s,T) = \mathcal{O}(1/T)$ shows that the transition probability to the kth instantaneous eigenstate during the adiabatic process is $\mathcal{O}(1/T^2)$. This concludes the proof of the adiabatic theorem. □

28.5 Landau-Zener Transitions

The adiabatic theorem tells us that *nonadiabatic* transitions, those between different instantaneous eigenstates, are suppressed. An idealized two-state system considered by Landau and Zener can be used to examine nonadiabatic transitions and calculate their probability. A word on terminology: the term *adiabatic transition*, widely used, is essentially an oxymoron because it refers to a system following an instantaneous eigenstate. There is a transition only in the sense that at the end of the process the final state is rather different from the initial state!

Zener considered electronic configurations of a molecule with fixed nuclei separated by a distance R. He examined two eigenstates of electrons $\psi_1(x;R)$ and $\psi_2(x;R)$ with energies $E_1(R) > E_2(R)$ for all values of R. A sketch of the energies as a function of the separation R is shown in figure 28.3. As a thought experiment, one can imagine a situation in which for $R < R_0$, with R_0 some special distance, the ground state represents a polar molecule—one with a permanent dipole moment—while the excited state is nonpolar. As $R > R_0$, their roles are traded: the ground state is nonpolar, while the excited state is polar. Note that the eigenstates are nondegenerate for all values of R; the curves

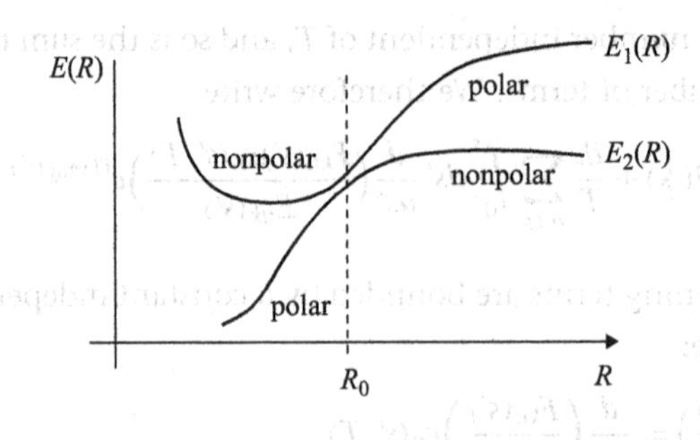

Figure 28.3
The energies $E_1(R) > E_2(R)$ of electronic configurations in a molecule with nuclei separated a distance R. As R crosses the value R_0, the properties of the molecules change.

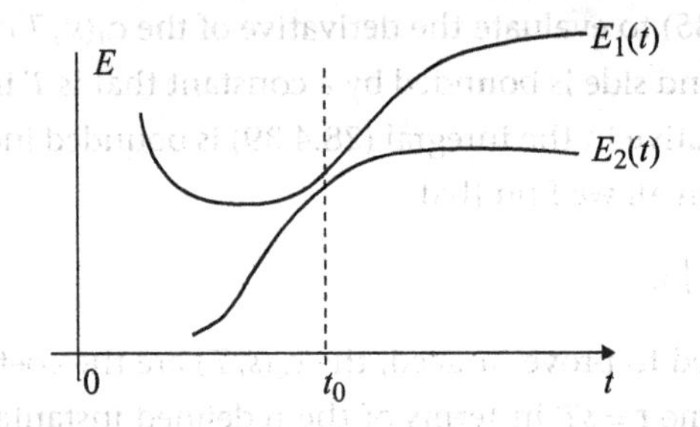

Figure 28.4
The time-dependent counterpart of figure 28.3. This is the result of the system changing the nuclear separation R with constant velocity $\dot{R}$.

representing the energies do not cross, and we have what is usually called an *avoided crossing*.

The energy eigenstates satisfy

$$\hat{H}(R)\,\psi_i(x;R) = E_i(R)\,\psi_i(x;R), \quad i = 1, 2. \tag{28.5.1}$$

If R changes very slowly, a molecule in ψ_1 remains in ψ_1, and a molecule in ψ_2 will remain in ψ_2. In this case we have adiabatic transitions: a molecule in ψ_1 changes from nonpolar to polar, and a molecule in ψ_2 changes from polar to nonpolar. On the other hand, if R changes quickly and the gap between the states at R_0 is small, we can have a nonadiabatic transition: starting with ψ_1 for $R < R_0$, for example, the end result would be ψ_2 for $R > R_0$.

If we think of R as a function of time, we have a time-dependent Hamiltonian $\hat{H}(R(t))$ with instantaneous energy eigenstates $\psi_i(x;R(t))$ of energies $E_i(R(t))$. For constant velocity $\dot{R}$, the plot of the time-dependent energies would be precisely that of figure 28.3, with a different horizontal axis representing time (figure 28.4).

Why are avoided crossings typical? The answer is that a crossing generally requires fine-tuning. A two-state system illustrates the point. With states $|1\rangle$ and $|2\rangle$, the general Hamiltonian $\hat{H}$ is of the form

$$\hat{H} = b_0\mathbf{1} + \mathbf{b}\cdot\boldsymbol{\sigma}, \quad b_0, b_1, b_2, b_3 \in \mathbb{R}. \tag{28.5.2}$$

This is a system with four real parameters. The eigenstates have energies

$$E_{\pm} = b_0 \pm |\mathbf{b}| \quad \Rightarrow \quad E_+ - E_- = 2\,|\mathbf{b}|. \tag{28.5.3}$$

For the levels to cross, we need $|\mathbf{b}| = 0$ and thus $\mathbf{b} = 0$. This requires fine-tuning *three* out of the four parameters. So it will not happen in general. Symmetries can make level crossing less fine-tuned. Consider, for example, a two-level system where $|1\rangle$ and $|2\rangle$ are states with angular momentum $\ell = 0$ and $\ell = 1$, respectively. Assume the Hamiltonian $\hat{H}$ is rotational invariant so that $[\hat{H}, \hat{\mathbf{L}}^2] = 0$. Since $|1\rangle$ and $|2\rangle$ have different $\hat{\mathbf{L}}^2$ eigenvalues, this implies that $\langle 1|\hat{H}|2\rangle = \langle 2|\hat{H}|1\rangle = 0$. In terms of the parameterization of $\hat{H}$ in (28.5.2), we get $b_1 = b_2 = 0$. So this time the setting of a single parameter, b_3, equal to zero will lead to degenerate levels.

Let us now consider the Landau-Zener setup. This is, once more, a two-level system with a time-dependent Hamiltonian. The time dependence is in the diagonal matrix elements, while the off-diagonal matrix elements are time independent. To appreciate the subtleties of the problem, it is convenient to begin with the off-diagonal elements set equal to zero. As usual we work in the conventions where the basis states are

$$|1\rangle = \begin{pmatrix} 1 \\ 0 \end{pmatrix} \text{ (spin up)}, \qquad |2\rangle = \begin{pmatrix} 0 \\ 1 \end{pmatrix} \text{ (spin down)}. \tag{28.5.4}$$

The Hamiltonian $\hat{H}(t)$ will take the form

$$\hat{H}(t) = \begin{pmatrix} \frac{1}{2}\alpha t & 0 \\ 0 & -\frac{1}{2}\alpha t \end{pmatrix}, \quad \text{with} \quad \alpha > 0. \tag{28.5.5}$$

Since $\hat{H}(t)$ is diagonal, the time-independent states $|1\rangle$ and $|2\rangle$ are in fact instantaneous eigenstates with energies $E_1(t)$ and $E_2(t)$ given by

$$E_1(t) = \tfrac{1}{2}\alpha t, \quad E_2(t) = -\tfrac{1}{2}\alpha t. \tag{28.5.6}$$

It is not hard to use the instantaneous eigenstates to build exact solutions of the Schrödinger equation. Take, for example, $|\psi_1(t)\rangle = f_1(t)|1\rangle$. The Schrödinger equation requires

$$i\hbar\partial_t|\psi_1(t)\rangle = \hat{H}|\psi_1(t)\rangle \quad \Rightarrow \quad i\hbar(\partial_t f_1)|1\rangle = E_1(t)\, f_1(t)|1\rangle. \tag{28.5.7}$$

This means that up to a constant multiplicative factor, f_1 is the dynamical phase:

$$f_1(t) = \exp\left(\frac{1}{i\hbar}\int_0^t E_1(t')dt'\right) = \exp\left(-\frac{i\alpha t^2}{4\hbar}\right). \tag{28.5.8}$$

All in all, we have the two solutions:

$$\begin{aligned} |\psi_1(t)\rangle &= \exp\left(\frac{1}{i\hbar}\int_0^t E_1(t')dt'\right)|1\rangle = \exp\left(-\frac{i\alpha t^2}{4\hbar}\right)|1\rangle, \\ |\psi_2(t)\rangle &= \exp\left(\frac{1}{i\hbar}\int_0^t E_2(t')dt'\right)|2\rangle = \exp\left(+\frac{i\alpha t^2}{4\hbar}\right)|2\rangle. \end{aligned} \tag{28.5.9}$$

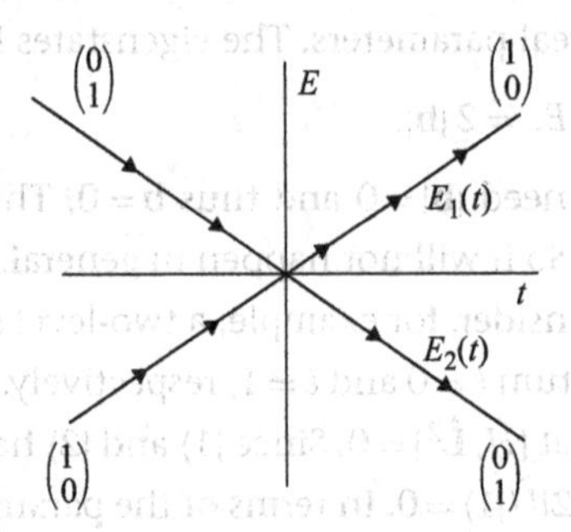

Figure 28.5
The exact time-dependent solutions of the Schrödinger equation for the Hamiltonian in (28.5.5). The sloping-down energy corresponds to a spin-down state, while the sloping-up energy corresponds to a spin-up state. At $t=0$, the states are degenerate.

The time-dependent energies of the state are shown in figure 28.5. The energy of the spin-up state $|\psi_1(t)\rangle$ increases linearly with time, and the energy of the spin-down state $|\psi_2(t)\rangle$ decreases linearly in time. Since the off-diagonal elements of $H(t)$ vanish, we get level crossing when the diagonal elements vanish, which happens at $t=0$. The states are degenerate at the crossing, but they ignore each other. The up state remains up forever, and the down state remains down forever.

Now we consider the complete problem by including in the Hamiltonian constant off-diagonal terms:

$$\hat{H}(t)=\begin{pmatrix}\frac{1}{2}\alpha t & H_{12}\\ H_{12}^{*} & -\frac{1}{2}\alpha t\end{pmatrix}, \quad \text{with} \quad \alpha>0. \tag{28.5.10}$$

Here H_{12} is a constant, possibly complex. At $t=0$, the energy eigenvalues are $\pm|H_{12}|$. We can write the Hamiltonian in terms of Pauli matrices:

$$\hat{H}=\tfrac{1}{2}\alpha t\,\sigma_3+\mathrm{Re}(H_{12})\sigma_1-\mathrm{Im}(H_{12})\sigma_2, \tag{28.5.11}$$

and the energies are given by

$$E_{\pm}(t)=\pm\sqrt{|H_{12}|^2+\tfrac{1}{4}\alpha^2t^2}. \tag{28.5.12}$$

The time-dependent energies are sketched in figure 28.6. Note that we now have an avoided crossing. The states never become degenerate. The upper branch, with positive energies throughout, corresponds to an instantaneous eigenstate $|\psi_+(t)\rangle$. The lower branch, with negative energies throughout, corresponds to an instantaneous eigenstate $|\psi_-(t)\rangle$. Neither one, of course, is an exact solution of the Schrödinger equation, but they should be approximate solutions if the process is adiabatic. We can identify the form of the instantaneous eigenstates for large $|t|$ because the effect of the off-diagonal terms becomes negligible, and the states must coincide with the states determined for vanishing off-diagonal terms. Comparing with figure 28.5, we see that

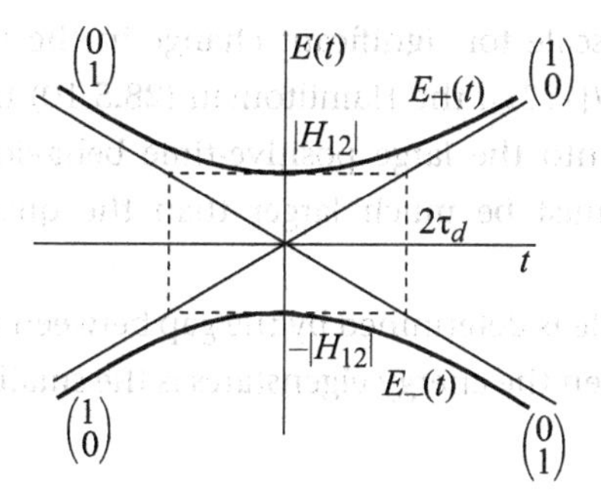

Figure 28.6
The energies $E_+(t)$ and $E_-(t)$, with $E_+(t) > E_-(t)$ plotted as a function of time. The instantaneous eigenstate $|\psi_+(t)\rangle$ has energy $E_+(t)$, and the instantaneous eigenstate $|\psi_-(t)\rangle$ has energy $E_-(t)$. This diagram also shows the definition of the time scale τ_d, called the duration.

$$\begin{aligned}|\psi_+(t)\rangle &\to \begin{pmatrix}0\\1\end{pmatrix} \text{ as } t\to-\infty, \text{ and } |\psi_+(t)\rangle \to \begin{pmatrix}1\\0\end{pmatrix} \text{ as } t\to\infty,\\ |\psi_-(t)\rangle &\to \begin{pmatrix}1\\0\end{pmatrix} \text{ as } t\to-\infty, \text{ and } |\psi_-(t)\rangle \to \begin{pmatrix}0\\1\end{pmatrix}, \text{ as } t\to\infty.\end{aligned} \tag{28.5.13}$$

If the instantaneous eigenstates are accurate solutions, we have adiabatic transitions in the sense that as time goes from minus to plus infinity the instantaneous eigenstates evolve from spin down to spin up (for $|\psi_+\rangle$) and from spin up to spin down (for $|\psi_-\rangle$). These are adiabatic changes. There is also the possibility of nonadiabatic transitions $|\psi_\pm\rangle \to |\psi_\mp\rangle$ as time goes from minus to plus infinity. In this case the system jumps across instantaneous eigenstates, resulting, curiously, in the spin not changing direction.

The instantaneous energy eigenstates are

$$|\psi_\pm(t)\rangle = |\mathbf{n}(t);\pm\rangle \quad \text{with energies} \quad E_\pm(t), \tag{28.5.14}$$

where the unit vector $\mathbf{n}(t)$ is the time-dependent spin arrow. This vector can be calculated in terms of α, H_{12}, and t.

In terms of the original Zener diagram in figure 28.3, we now see that an adiabatic transition changes the molecule from polar to nonpolar and vice versa. A nonadiabatic transition in fact preserves the polar properties of the molecule.

Exercise 28.2. *Assume H_{12} is real, and determine explicitly the time-dependent unit vector $\mathbf{n}(t)$. Where do the states $|\psi_\pm(t)\rangle$ point to at time equal zero?*

Let us now understand the timescales relevant for the transition. Consider the *duration* τ_d defined by letting $2\tau_d$ be the time required for the diagonal entries $E_1(t)$ and $E_2(t)$ in $\hat{H}(t)$ to change by a magnitude $|H_{12}|$ (see figure 28.6). Recalling that $\alpha > 0$,

$$|E_1| = \tfrac{1}{2}\alpha t \;\Rightarrow\; |H_{12}| = \tfrac{1}{2}\alpha(2\tau_d) \;\Rightarrow\; \tau_d = \frac{|H_{12}|}{\alpha}. \tag{28.5.15}$$

We view τ_d as the time scale for significant change in the Hamiltonian. This is reasonable; over the times $|t| < 2\tau_d$, the Hamiltonian (28.5.10) transitions from the large negative-time behavior into the large positive-time behavior. For the process to be adiabatic, the time τ_d must be much larger than the quantum time scale of the system.

The quantum time scale is determined by the gap between the energy levels. At time equal zero, the gap between the energy eigenstates is the smallest, and the Hamiltonian takes the form

$$\hat{H}(0) = \begin{pmatrix} 0 & H_{12} \\ H_{12}^* & 0 \end{pmatrix}. \qquad (28.5.16)$$

Transitions between the states $|1\rangle$ and $|2\rangle$ in this Hamiltonian happen with the *Rabi frequency*

$$\omega_{12} \equiv \frac{|H_{12}|}{\hbar}. \qquad (28.5.17)$$

The quantum time scale is $T_{12} = 2\pi/\omega_{12}$. This is in fact the maximum value of the quantum time scale for the process since ω_{12} is the smallest frequency associated with the energy gap between the energy eigenstates.

The evolution is adiabatic when the largest value T_{12} of the quantum time scale is much smaller than the duration τ_d:

$$T_{12} \ll \tau_d \quad \Rightarrow \quad \frac{2\pi}{\omega_{12}} \ll \tau_d \quad \Rightarrow \quad \boxed{\omega_{12}\tau_d \gg 1.} \qquad (28.5.18)$$

In terms of the H_{12} and α variables of the problem,

$$\text{the process is adiabatic if} \quad \frac{|H_{12}|^2}{\hbar\alpha} \gg 1. \qquad (28.5.19)$$

In order to calculate the probability of nonadiabatic transitions, one must aim to find exact solutions of the time-dependent problem. Given the Hamiltonian, one writes an ansatz for the solution

$$|\psi(t)\rangle = A(t) \exp\left(-\frac{i}{\hbar}\int_0^t E_1(t')dt'\right)|1\rangle + B(t)\exp\left(-\frac{i}{\hbar}\int_0^t E_2(t')dt'\right)|2\rangle. \qquad (28.5.20)$$

Here $A(t)$ and $B(t)$ are functions of time to be determined, satisfying the normalization condition $|A|^2 + |B|^2 = 1$. Note that the kets $|1\rangle$ and $|2\rangle$ represent what is sometimes called a *diabatic* basis. If we start with $A \neq 0$ and $B = 0$ at large negative times, the amplitude A at large positive times is the amplitude for a nonadiabatic transition, while the amplitude B at large positive times is the amplitude for an adiabatic transition. This should be clear from figure 28.6.

In problem 28.2 you will use the Schrödinger equation to derive a set of linear, first-order, coupled differential equations for $A(t)$ and $B(t)$. You will use these equations to calculate the probability of a nonadiabatic transition in the limit as $|H_{12}|$ is very small.

This result is consistent with the *exact* result for the nonadiabatic transition probability $P_{\text{non-ad}}$:

$$P_{\text{non-ad}} = \exp\left(-2\pi\omega_{12}\tau_d\right) = \exp\left(-2\pi\frac{|H_{12}|^2}{\hbar\alpha}\right). \tag{28.5.21}$$

If the process is adiabatic, condition (28.5.19) implies that the probability for a non-adiabatic transition goes to zero. It is not hard to give a heuristic derivation of (28.5.21), as you will do in problem 28.2.

28.6 Berry's Phase

In our statement of the adiabatic theorem, we emphasized that for Hamiltonians varying continuously in the time interval $t \in [0, T]$ and for a state in an instantaneous eigenstate $|\psi_n(t)\rangle$ at $t=0$, the state at any time $t \in [0, T]$ is rather accurately equal to $|\psi_n(t)\rangle$ up to a calculable phase. The probability that the system transitions to another energy eigenstate is suppressed by a factor of order $1/T^2$. Now we want to focus on the calculable phase, which we already determined in (28.4.15):

$$|\Psi(t)\rangle \simeq e^{i\theta_n(t)} e^{i\gamma_n(t)} |\psi_n(t)\rangle. \tag{28.6.1}$$

The dynamical phase $\theta_n(t)$ exists even if the Hamiltonian and the eigenstates are time independent. The geometric phase $\gamma_n(t)$, however, vanishes in this case. This phase is given by

$$\gamma_n(t) = \int_0^t \nu_n(t')dt', \quad \nu_n(t) = i\langle\psi_n(t)|\dot{\psi}_n(t)\rangle. \tag{28.6.2}$$

It clearly vanishes if the instantaneous eigenstates are time independent. In this section we explain why this phase is geometric.

Let the time dependence of $\hat{H}$ be expressed in terms of a set of N parameters that are time dependent. We will refer to these parameters as coordinates in some *configuration* space:

$$\boldsymbol{R} = (R_1, \ldots, R_N). \tag{28.6.3}$$

Assume that we have eigenstates for all values of the coordinates:

$$\hat{H}(\boldsymbol{R})|\psi_n(\boldsymbol{R})\rangle = E(\boldsymbol{R})|\psi_n(\boldsymbol{R})\rangle. \tag{28.6.4}$$

Then, for time-dependent coordinates

$$\boldsymbol{R}(t) = (R_1(t), \ldots, R_N(t)), \tag{28.6.5}$$

we have instantaneous eigenstates:

$$\hat{H}(\boldsymbol{R}(t))\big|\psi_n(\boldsymbol{R}(t))\big\rangle = E(\boldsymbol{R}(t))\big|\psi_n(\boldsymbol{R}(t))\big\rangle. \tag{28.6.6}$$

At each instant of time, a point in configuration space determines the values of all parameters. The time evolution of the Hamiltonian can be thought of as a path in the configuration space, a path parameterized by time (figure 28.7).

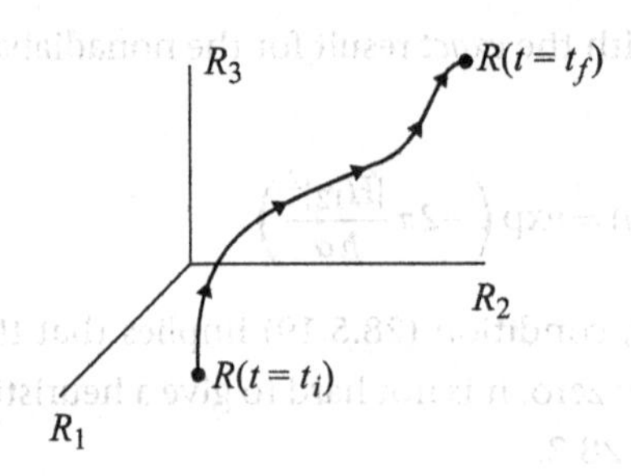

Figure 28.7
A time-dependent adiabatic process can be visualized as motion along a curve in the configuration space for the Hamiltonian.

To evaluate the geometric phase, we start by computing its integrand $\nu_n(t)$:

$$\nu_n(t) = i\langle\psi_n(\boldsymbol{R}(t))|\frac{d}{dt}|\psi_n(\boldsymbol{R}(t))\rangle. \tag{28.6.7}$$

To find the time derivative of the ket, we use the chain rule:

$$\frac{d}{dt}|\psi_n(\boldsymbol{R}(t))\rangle = \sum_{i=1}^{N}\frac{\partial}{\partial R_i}|\psi_n(\boldsymbol{R}(t))\rangle\frac{dR_i}{dt} = \nabla_{\boldsymbol{R}}|\psi_n(\boldsymbol{R}(t))\rangle\cdot\frac{d\boldsymbol{R}(t)}{dt}. \tag{28.6.8}$$

On the rightmost expression, we use a dot product and a gradient $\nabla_{\boldsymbol{R}}$ in the configuration space. Returning to the evaluation of $\nu_n(t)$, we see that

$$\nu_n(t) = i\langle\psi_n(\boldsymbol{R}(t))|\nabla_{\boldsymbol{R}}|\psi_n(\boldsymbol{R}(t))\rangle\cdot\frac{d\boldsymbol{R}(t)}{dt}. \tag{28.6.9}$$

The geometric phase $\gamma_n(t_i, t_f)$ accumulated from some initial time t_i up to some final time t_f is therefore

$$\gamma_n(t_i, t_f) \equiv \int_{t_i}^{t_f}\nu_n(t)dt = \int_{t_i}^{t_f} i\langle\psi_n(\boldsymbol{R}(t))|\nabla_{\boldsymbol{R}}|\psi_n(\boldsymbol{R}(t))\rangle\cdot\frac{d\boldsymbol{R}(t)}{dt}dt. \tag{28.6.10}$$

The crucial fact now is that the dt factors cancel: $\frac{d\boldsymbol{R}}{dt}dt = d\boldsymbol{R}$, and the integral can then be viewed as one in the configuration space along the path Γ_{if} going from $\boldsymbol{R}_i$ to $\boldsymbol{R}_f$ and giving a phase $\gamma_n(\Gamma_{if})$:

$$\gamma_n(\Gamma_{if}) = \int_{\boldsymbol{R}_i}^{\boldsymbol{R}_f} i\langle\psi_n(\boldsymbol{R})|\nabla_{\boldsymbol{R}}|\psi_n(\boldsymbol{R})\rangle\cdot d\boldsymbol{R}. \tag{28.6.11}$$

The integral depends on the path Γ_{if} in configuration space but does not depend on time! The phase is thus geometric: it does not depend on the parameterization of the path by the time parameter. Whether the transition from $\boldsymbol{R}_i$ to $\boldsymbol{R}_f$ along Γ_{if} takes a nanosecond or an hour, the phase $\gamma_n(\Gamma_{if})$ accumulated is the same. The geometric phase is known as **Berry's phase.** Recalling that for a function $f(\mathbf{u})$ of several variables u_i one has $df = \sum_i \frac{\partial f}{\partial_i}du_i$, we recognize that

$$\nabla_{\boldsymbol{R}}|\psi_n(\boldsymbol{R})\rangle\cdot d\boldsymbol{R} = d\,|\psi_n(\boldsymbol{R})\rangle. \tag{28.6.12}$$

This allows a compact rewriting of Berry's phase:

$$\boxed{\gamma_n(\Gamma_{if}) = \int_{\Gamma_{if}} i\langle\psi_n(\boldsymbol{R})|\,d\,|\psi_n(\boldsymbol{R})\rangle.} \tag{28.6.13}$$

The quantity $i\langle\psi_n(\boldsymbol{R})|\nabla_{\boldsymbol{R}}|\psi_n(\boldsymbol{R})\rangle$ is an N-component vector defined everywhere in the configuration space. It is the **Berry connection** $\mathrm{A}_n(\boldsymbol{R})$ associated with the instantaneous eigenstate $|\psi_n(t)\rangle$:

$$\mathrm{A}_n(\boldsymbol{R}) \equiv i\langle\psi_n(\boldsymbol{R})|\nabla_{\boldsymbol{R}}|\psi_n(\boldsymbol{R})\rangle. \tag{28.6.14}$$

In this way we can rewrite the Berry phase naturally:

$$\gamma_n(\Gamma_{if}) = \int_{\Gamma_{if}} \mathrm{A}_n(\boldsymbol{R})\cdot d\boldsymbol{R}. \tag{28.6.15}$$

This integral is similar to those in electrodynamics involving the integral of the vector potential along a path. Like any connection, $\mathrm{A}_n(\boldsymbol{R})$ has gauge transformations. The gauge transformations arise from redefinitions of the instantaneous eigenstates by phases reflecting the ambiguity of these states. Let us consider such a redefinition into "tilde" states:

$$|\psi_n(\boldsymbol{R})\rangle \;\rightarrow\; |\widetilde{\psi}_n(\boldsymbol{R})\rangle \equiv e^{-i\beta(\boldsymbol{R})}|\psi_n(\boldsymbol{R})\rangle, \tag{28.6.16}$$

where $\beta(\boldsymbol{R})$ is an arbitrary real function. Let us see what happens then to the connection:

$$\begin{aligned}
\widetilde{\mathrm{A}}_n(\boldsymbol{R}) &= i\langle\widetilde{\psi}_n(\boldsymbol{R})|\nabla_{\boldsymbol{R}}|\widetilde{\psi}_n(\boldsymbol{R})\rangle \\
&= i\langle\psi_n(\boldsymbol{R})|e^{i\beta(\boldsymbol{R})}\nabla_{\boldsymbol{R}}e^{-i\beta(\boldsymbol{R})}|\psi_n(\boldsymbol{R})\rangle \\
&= i\,(-i\nabla_{\boldsymbol{R}}\beta(\boldsymbol{R}))\,\langle\psi_n(\boldsymbol{R})|\psi_n(\boldsymbol{R})\rangle \,+\, \mathrm{A}_n(\boldsymbol{R}).
\end{aligned} \tag{28.6.17}$$

Since the eigenstates are normalized, we have found that

$$\widetilde{\mathrm{A}}_n(\boldsymbol{R}) = \mathrm{A}_n(\boldsymbol{R}) + \nabla_{\boldsymbol{R}}\beta(\boldsymbol{R}), \tag{28.6.18}$$

just like the gauge transformations of the vector potential in electrodynamics ($\mathrm{A}\rightarrow \mathrm{A}' = \mathrm{A} + \nabla\Lambda$). What happens to Berry's phase under these gauge transformations? We compute

$$\begin{aligned}
\widetilde{\gamma}_n(\Gamma_{if}) &= \int_{\Gamma_{if}} \widetilde{\mathrm{A}}_n(\boldsymbol{R})\cdot d\boldsymbol{R} = \int_{\Gamma_{if}} \mathrm{A}_n(\boldsymbol{R})\cdot d\boldsymbol{R} + \int_{\Gamma_{if}} \nabla_{\boldsymbol{R}}\beta(\boldsymbol{R})\cdot d\boldsymbol{R} \\
&= \gamma_n(\Gamma_{if}) + \int_{\Gamma_{if}} d\beta(\boldsymbol{R}).
\end{aligned} \tag{28.6.19}$$

The second term is an integral of a total derivative and thus picks the values of the integrand at the final and initial points of the path:

$$\widetilde{\gamma}_n(\Gamma_{if}) = \gamma_n(\Gamma_{if}) + \beta(\boldsymbol{R}_f) - \beta(\boldsymbol{R}_i). \tag{28.6.20}$$

This shows that the geometric phase associated to an open path in configuration space (a path that begins and ends at different points) is ambiguous. The phase can be changed by redefinitions of the instantaneous eigenstates. By choosing the function $\beta(\boldsymbol{R})$ suitably, one could even make the geometric phase vanish *all along* the open path. On the other hand, if the path is closed, $\boldsymbol{R}_f = \boldsymbol{R}_i$ and then $\beta(\boldsymbol{R}_f) = \beta(\boldsymbol{R}_i)$, resulting in a gauge-invariant phase. The geometric phase is unambiguously defined for closed paths Γ in parameter space. This quantity is an observable:

$$\gamma_n(\Gamma) \text{ is gauge invariant if } \Gamma \text{ is a closed path.} \tag{28.6.21}$$

Let us make some comments regarding basic properties of Berry's phase.

1. If the instantaneous eigenstates $|\psi_n(t)\rangle$ are real, Berry's phase vanishes. Indeed, we showed in (28.3.21) that $\langle\psi_n(t)|\dot{\psi}_n(t)\rangle$ is purely imaginary. But if ψ_n is real, this overlap cannot produce a complex number, so it can only be zero. The argument we used then can be made very explicit by including the definition of the overlap:

$$\begin{aligned}\nu_n &= i\langle\psi_n(t)|\dot{\psi}_n(t)\rangle = i\int d\mathbf{x}\,\psi_n^*(t,\mathbf{x})\frac{d}{dt}\psi_n(t,\mathbf{x})\\ &= i\int d\mathbf{x}\,\psi_n(t,\mathbf{x})\frac{d}{dt}\psi_n(t,\mathbf{x}) = \frac{i}{2}\int d\mathbf{x}\,\frac{d}{dt}\Big(\psi_n(t,\mathbf{x})\Big)^2 =\\ &= \frac{i}{2}\frac{d}{dt}\int d\mathbf{x}\,|\psi_n(t,\mathbf{x})|^2 = 0,\end{aligned} \tag{28.6.22}$$

since the wave function is normalized.

2. If the configuration space is one-dimensional, Berry's phase vanishes for any closed loop Γ. Indeed, if there is just one coordinate R in the configuration space, any closed loop Γ stretches from some R_i to some R_f and then back from R_f to R_i. Since the Berry connection $A_n(R)$ is just a one-component object, we then get

$$\oint_\Gamma A_n(R)dR = \int_{R_i}^{R_f} A_n(R)dR + \int_{R_f}^{R_i} A_n(R)dR = 0, \tag{28.6.23}$$

since the two contributions cancel each other out.

3. In three-dimensional space a vector potential, or connection $\mathbf{A}$, has an associated magnetic field $\mathbf{B}$, sometimes called the *curvature* of the connection. There is a perfect analogy for Berry connections in three-dimensional *configuration* spaces. Associated with the Berry connection $\mathbf{A}_n$ is **Berry's curvature $\boldsymbol{D}_n$**. To see this, consider Berry's phase for a closed loop Γ in a three-dimensional configuration space. By the curl theorem, we have

$$\gamma_n(\Gamma) = \oint_\Gamma \mathbf{A}_n(\boldsymbol{R})\cdot d\boldsymbol{R} = \iint_S\Big(\nabla_{\boldsymbol{R}}\times\mathbf{A}_n\Big)\cdot d\mathbf{S} \equiv \iint_S \boldsymbol{D}_n\cdot d\mathbf{S}, \tag{28.6.24}$$

where S is the surface with boundary Γ, and $d\mathbf{S}$ is the area element with the orientation defined by the orientation of Γ. In here we define

$$\boldsymbol{D}_n(\boldsymbol{R}) \equiv \nabla_{\boldsymbol{R}}\times\mathbf{A}_n(\boldsymbol{R}). \tag{28.6.25}$$

Berry's curvature is gauge invariant for the same reason ordinary magnetic fields are:

$$\widetilde{\boldsymbol{D}}_n = \nabla_{\boldsymbol{R}}\times\widetilde{\mathbf{A}}_n = \nabla_{\boldsymbol{R}}\times\Big(\mathbf{A}_n + \nabla_{\boldsymbol{R}}\beta\Big) = \nabla_{\boldsymbol{R}}\times\mathbf{A}_n + \nabla_{\boldsymbol{R}}\times\nabla_{\boldsymbol{R}}\beta = \boldsymbol{D}_n. \tag{28.6.26}$$

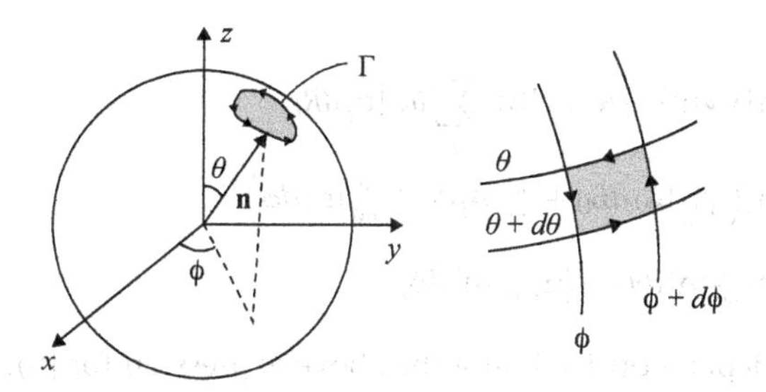

Figure 28.8
Left: The direction of a magnetic field is determined by the unit vector $\mathbf{n}$, specified by spherical angles (θ, ϕ). The tip of $\mathbf{n}$ traces a closed loop Γ on the unit sphere. *Right*: An infinitesimal rectangle in (θ, ϕ) space.

While the phase of a *single* state is not observable, Berry's phase is observable in a setting with more than one state. Consider a Hamiltonian and build a state that at $t = 0$ is in a superposition of two instantaneous eigenstates. Suppose the Hamiltonian is time dependent and traces a loop in configuration space. Each of the two eigenstates will evolve, and each will have a Berry phase. At the end of the evolution, the *difference* between the phases of the two states, as usual, can be observed. Berry's phase has turned out to be a rather useful concept and features in quantum magnetism, spin one-half chains, and topological insulators. It also appears in the treatment of molecules, as we will see in the following section.

Example 28.2. *Berry phase for an electron in a slowly varying magnetic field.*
We considered in example 28.1 an electron in a uniform magnetic field $\mathbf{B} = B_0\mathbf{n}$. The Hamiltonian is $\hat{H} = \mu_B B_0 \mathbf{n} \cdot \boldsymbol{\sigma}$. We then let the magnitude B_0 and the direction $\mathbf{n}$ of the magnetic field change in time so that $B_0 \to B_0(t)$ and $\mathbf{n} \to \mathbf{n}(t)$. The change of $\mathbf{n}$ can be described explicitly by giving the time dependence $\theta(t)$ and $\phi(t)$ of the spherical angles (θ, ϕ) that define the direction $\mathbf{n}$ (figure 28.8). The configuration space is described with three parameters:

$$(R_1, R_2, R_3) = (B_0, \theta, \phi). \tag{28.6.27}$$

Let's assume the magnetic field changes in a way that it traces a closed loop Γ in configuration space. This means the tip of the unit vector $\mathbf{n}$ traces a closed path on the unit sphere, and the magnitude B_0 of the magnetic field changes but returns to its initial value. We want to compute Berry's phase for this loop. Since we have two instantaneous eigenstates $|\mathbf{n}; \pm\rangle$, we have two phases. Let us consider, for definiteness, the phase γ_+ for the eigenstate $|\mathbf{n}\rangle \equiv |\mathbf{n}; +\rangle$:

$$|\mathbf{n}\rangle = \begin{pmatrix} \cos\frac{\theta}{2} \\ \sin\frac{\theta}{2} e^{i\phi} \end{pmatrix}. \tag{28.6.28}$$

Using equations (28.6.15) and (28.6.14), we have

$$\gamma_+(\Gamma) = \int_\Gamma \mathbf{A}_+(\boldsymbol{R}) \cdot d\boldsymbol{R}, \tag{28.6.29}$$

where

$$\begin{aligned} \mathbf{A}_{+}(\boldsymbol{R}) \cdot d\boldsymbol{R} &= i\langle \mathbf{n}|\nabla_{\boldsymbol{R}}|\mathbf{n}\rangle \cdot d\boldsymbol{R} = i\langle \mathbf{n}|\sum_i \partial_{R_i}|\mathbf{n}\rangle\, dR_i \\ &= i\langle \mathbf{n}|\Big(\tfrac{\partial}{\partial B_0}|\mathbf{n}\rangle\, dB_0 + \tfrac{\partial}{\partial \theta}|\mathbf{n}\rangle\, d\theta + \tfrac{\partial}{\partial \phi}|\mathbf{n}\rangle\, d\phi\Big) \\ &= i\langle \mathbf{n}|\tfrac{\partial}{\partial \theta}|\mathbf{n}\rangle\, d\theta + i\langle \mathbf{n}|\tfrac{\partial}{\partial \phi}|\mathbf{n}\rangle\, d\phi, \end{aligned} \tag{28.6.30}$$

since $|\mathbf{n}\rangle$ does not depend on B_0. Using the above expression for $|\mathbf{n}\rangle$, we find

$$\langle \mathbf{n}|\tfrac{\partial}{\partial \theta}|\mathbf{n}\rangle = (\cos\tfrac{\theta}{2}, \sin\tfrac{\theta}{2}e^{-i\phi}) \begin{pmatrix} -\frac{1}{2}\sin\frac{\theta}{2} \\ \frac{1}{2}\cos\frac{\theta}{2}e^{i\phi} \end{pmatrix} = 0,$$

$$\langle \mathbf{n}|\tfrac{\partial}{\partial \phi}|\mathbf{n}\rangle = (\cos\tfrac{\theta}{2}, \sin\tfrac{\theta}{2}e^{-i\phi}) \begin{pmatrix} 0 \\ i\sin\frac{\theta}{2}e^{i\phi} \end{pmatrix} = i\sin^2\tfrac{\theta}{2}. \tag{28.6.31}$$

Therefore,

$$\mathbf{A}_{+}(\boldsymbol{R}) \cdot d\boldsymbol{R} = -\sin^2\tfrac{\theta}{2}\, d\phi = -\tfrac{1}{2}(1-\cos\theta)\, d\phi. \tag{28.6.32}$$

Berry's phase thus takes the form

$$\gamma_{+}(\Gamma) = -\tfrac{1}{2}\int_{\Gamma}(1-\cos\theta)d\phi. \tag{28.6.33}$$

Here Γ refers to the curve traced by the tip of the unit vector $\mathbf{n}$ in the adiabatic process. Let S_Γ denote the surface on the unit sphere whose boundary is Γ: $\partial S_\Gamma = \Gamma$. We will now confirm that the above integral actually computes the solid angle spanned by S_Γ at the origin, with an additional factor of minus one-half.

A simple test case is that of a closed curve of constant $\theta = \theta_0$, with $\phi \in [0, 2\pi]$. The associated solid angle is $\Omega(\theta_0) = 2\pi(1-\cos\theta_0)$. On the other hand, the integral (28.6.33) gives $\gamma_{+} = -\frac{1}{2}(2\pi)(1-\cos\theta_0) = -\frac{1}{2}\Omega(\theta_0)$, providing evidence for the claim that, in general,

$$\gamma_{+}(\Gamma) = -\tfrac{1}{2}\,\Omega(S_\Gamma). \tag{28.6.34}$$

This is an elegant geometric result. To prove it, we show that the integral of $(1-\cos\theta)d\phi$ over a small loop is indeed the area element. For this consider the tiny rectangle $d\Gamma$ with ranges $d\theta$ and $d\phi$ about (θ, ϕ), as shown on the right in figure 28.8. Since the integrand only involves $d\phi$, we get

$$\begin{aligned} \oint_{d\Gamma}(1-\cos\theta)d\phi &= (1-\cos(\theta+d\theta))d\phi - (1-\cos\theta)d\phi \\ &= (\cos\theta - \cos(\theta+d\theta))d\phi \\ &= \sin\theta\, d\theta\, d\phi = dA = d\Omega, \end{aligned} \tag{28.6.35}$$

where we recognize the area element on the unit sphere, equal to the solid-angle element. Since one can represent the surface S_Γ with boundary Γ as the collection of infinitely many infinitesimal rectangular elements, this shows that

$$\int_\Gamma (1-\cos\theta)d\phi = \Omega(S_\Gamma), \tag{28.6.36}$$

and on account of (28.6.33) confirms our claim in (28.6.34). The result satisfies a simple consistency check. Consider a loop defined by $\theta = \pi - \epsilon$, for positive $\epsilon \to 0$, and $\phi \in [0, 2\pi]$. This is a very tiny loop around the south pole of the sphere. The solid angle in this case is just short of 4π and $\gamma_+ \simeq -2\pi$. Such a Berry phase is equivalent to zero phase, which is reasonable because as $\epsilon \to 0$ the direction of the magnetic field is simply not changing! □

28.7 Born-Oppenheimer Approximation

Molecules are much harder to treat than atoms. Atoms are hard because even though the potential created by the nucleus is spherically symmetric, the Coulomb interactions between the electrons break the spherical symmetry. In molecules, even ignoring Coulomb repulsion between electrons, the potential created by nuclei that are spatially separated is not spherically symmetric.

To a good approximation, one can view a molecule as a system in which the nuclei are in classical equilibrium with well-localized positions while the electrons move around in the Coulomb potential created by the nuclei. The electron cloud helps generate the attraction that compensates for the Coulomb repulsion of the nuclei. This approximation is reasonable since typically $m/M \simeq 10^{-4}$, where m is the electron mass and M the nuclear mass. In this picture, slow nuclear vibrations adiabatically deform the electronic states.

In order to make estimates, consider a molecule of size a so that

$$p_{\text{electron}} \sim \frac{\hbar}{a} \quad \text{and} \quad E_{\text{electron}} \sim \frac{\hbar^2}{ma^2}. \tag{28.7.1}$$

The positively charged nuclei repel each other, but the electrons in between create an effective attraction that, at equilibrium, cancels the repulsive forces. There will be nuclear vibrations around the equilibrium configuration. Consider the nuclear oscillations governed by the nuclear Hamiltonian $\hat{H}_N$:

$$\hat{H}_N = \frac{\hat{p}^2}{2M} + \tfrac{1}{2}kx^2. \tag{28.7.2}$$

The restoring force is determined by k, and it is due to the electron system, with no reference to the mass M. Since k has units of energy over length squared, we must find that

$$k \sim \frac{\hbar^2}{ma^4}. \tag{28.7.3}$$

But $k = M\omega^2$, with ω the frequency of nuclear oscillations. Therefore,

$$M\omega^2 \sim \frac{\hbar^2}{ma^4} \quad \Rightarrow \quad \omega^2 \sim \frac{m}{M}\frac{\hbar^2}{m^2a^4} \quad \Rightarrow \quad \hbar\omega \sim \sqrt{\frac{m}{M}}\frac{\hbar^2}{ma^2}. \tag{28.7.4}$$

We thus find that the nuclear vibrational energies $E_{\text{vibration}}$ are related to the energy E_{electron} of the electronic states as follows:

$$E_{\text{vibration}} \sim \sqrt{\frac{m}{M}}\, E_{\text{electron}}\,. \tag{28.7.5}$$

There are also rotations of the molecule, and their energy E_{rotation} is even smaller, as rotations involve no distortion of the molecule. Let L denote the angular momentum of the molecule and I its moment of inertia. We estimate $L^2 \sim \hbar^2$ and $I \sim Ma^2$ so that

$$E_{\text{rotation}} \simeq \frac{L^2}{2I} \sim \frac{\hbar^2}{Ma^2} \sim \frac{m}{M}\frac{\hbar^2}{ma^2} \quad \Rightarrow \quad E_{\text{rotation}} \sim \frac{m}{M}E_{\text{electron}}. \tag{28.7.6}$$

Therefore, we have the following hierarchy of energies:

$$E_{\text{electron}} : E_{\text{vibration}} : E_{\text{rotation}} = 1 : \sqrt{\frac{m}{M}} : \frac{m}{M}. \tag{28.7.7}$$

To begin our detailed work on the Born-Oppenheimer approximation, consider a molecule with N nuclei and n electrons. The Hamiltonian takes the form

$$\hat{H} = \sum_{\alpha=1}^{N} \frac{\hat{\mathbf{P}}_\alpha^2}{2M_\alpha} + V_{NN}(\boldsymbol{R}) + \sum_{i=1}^{n} \frac{\hat{\mathbf{p}}_i^2}{2m} + V_{eN}(\boldsymbol{R}, \mathbf{r}) + V_{ee}(\mathbf{r}), \tag{28.7.8}$$

where $\alpha = 1, \ldots, N$ are labels for the nuclei, $i = 1, \ldots, n$ are labels for the electrons, and

M_α : nuclear masses,

$\hat{\mathbf{P}}_\alpha$, $\boldsymbol{R}_\alpha$: nuclei canonical variables, $\hat{\mathbf{P}}_\alpha = \frac{\hbar}{i}\nabla_{\boldsymbol{R}_\alpha}$,

$\hat{\mathbf{p}}_i$, $\mathbf{r}_i$: electron canonical variables,

$\boldsymbol{R} \equiv (\boldsymbol{R}_1, \ldots, \boldsymbol{R}_N)$, positions for the N nuclei,

$\mathbf{r} \equiv (\mathbf{r}_1, \ldots, \mathbf{r}_n)$, positions for the n electrons,

$V_{NN}(\boldsymbol{R})$: nuclei-nuclei interactions,

$V_{ee}(\mathbf{r})$: electron-electron interactions,

$V_{eN}(\boldsymbol{R}, \mathbf{r})$: electron-nuclei interactions.

The position-space wave function $\psi(\boldsymbol{R}, \mathbf{r})$ for the molecule is a function of all the nuclear positions and all the electron positions.

In the limit when $M_\alpha/m \to \infty$, the nuclear skeleton may be considered fixed, making the positions $\boldsymbol{R}$ fixed. The electrons then move under the effect of the nuclear potential $V_{eN}(\boldsymbol{R}, \mathbf{r})$ and the electron-electron Coulomb repulsion $V_{ee}(\mathbf{r})$. The relevant Hamiltonian $\hat{H}_e$ for the electrons is then

$$\hat{H}_e(\hat{\mathbf{p}}, \mathbf{r}; \boldsymbol{R}) = \sum_{i=1}^{n} \frac{\hat{\mathbf{p}}_i^2}{2m} + V_{eN}(\boldsymbol{R}, \mathbf{r}) + V_{ee}(\mathbf{r}). \tag{28.7.9}$$

This is a different Hamiltonian each time we change the positions $\boldsymbol{R}$ of the nuclei. The associated Schrödinger equation for the electrons is

$$\left[-\frac{\hbar^2}{2m}\sum_{i=1}^{n}\nabla^2_{\mathbf{r}_i} + V_{eN}(\boldsymbol{R},\mathbf{r}) + V_{ee}(\mathbf{r})\right]\phi^{(i)}_{\boldsymbol{R}}(\mathbf{r}) = E^{(i)}_e(\boldsymbol{R})\phi^{(i)}_{\boldsymbol{R}}(\mathbf{r}). \qquad (28.7.10)$$

The wave function for the electrons, as expected, is a function of the positions $\mathbf{r}$ of all the electrons, which appear as the argument of the wave function. Since the wave function depends on the nuclear positions, this dependence is included as a subscript. Finally, the superscript i labels the various wave functions that may appear as solutions of this equation. The associated energies $E^{(i)}_e(\boldsymbol{R})$ depend on the nuclear positions and the label i. If we calculated all the $\phi^{(i)}_{\boldsymbol{R}}(\mathbf{r})$, we would have a full basis of electronic configurations, and we could write an ansatz for the full wave function of the molecule:

$$\psi(\boldsymbol{R},\mathbf{r}) = \sum_i \eta^{(i)}(\boldsymbol{R})\phi^{(i)}_{\boldsymbol{R}}(\mathbf{r}), \qquad (28.7.11)$$

where the $\eta^{(i)}$ are unknown functions of $\boldsymbol{R}$. Substitution into the full Schrödinger equation

$$\hat{H}\psi(\boldsymbol{R},\mathbf{r}) = E\psi(\boldsymbol{R},\mathbf{r}) \qquad (28.7.12)$$

gives an infinite set of coupled differential equations for $\eta^{(i)}(\boldsymbol{R})$. This is too difficult to work with, so we try to make do with a simple product:

$$\psi(\boldsymbol{R},\mathbf{r}) = \eta(\boldsymbol{R})\phi_{\boldsymbol{R}}(\mathbf{r}), \qquad (28.7.13)$$

where we would generally use for $\phi_{\boldsymbol{R}}(\mathbf{r})$ the ground state wave function for the electrons in the frozen nuclear potential. If we know this wave function for all $\boldsymbol{R}$, as we now assume, we also know the value $E_e(\boldsymbol{R})$ of the associated energy, as defined by (28.7.10).

We will do a variational analysis. For this we will compute the expectation value $\langle\psi|\hat{H}|\psi\rangle$ of the full Hamiltonian using $\psi(\boldsymbol{R},\mathbf{r}) = \eta(\boldsymbol{R})\phi_{\boldsymbol{R}}(\mathbf{r})$. The idea is to utilize the known $\phi_{\boldsymbol{R}}(\mathbf{r})$ to integrate the electronic dependence in the expectation value so that we are left with an effective Hamiltonian for the nuclear degrees of freedom. Indeed, consider the calculation of $\langle\psi|\hat{H}|\psi\rangle$:

$$\langle\hat{H}\rangle = \int d\boldsymbol{R}\, d\mathbf{r}\, \psi^*(\boldsymbol{R},\mathbf{r})\hat{H}\,\psi(\boldsymbol{R},\mathbf{r}) = \int d\boldsymbol{R}\, d\mathbf{r}\, \eta^*(\boldsymbol{R})\phi^*_{\boldsymbol{R}}(\mathbf{r})\hat{H}\,\eta(\boldsymbol{R})\phi_{\boldsymbol{R}}(\mathbf{r}). \qquad (28.7.14)$$

Since we assume $\phi_{\boldsymbol{R}}(\mathbf{r})$ is known, we can, at least in principle, perform the integral over $d\mathbf{r}$ and be left with an integral over just $d\boldsymbol{R}$, involving $\eta(\boldsymbol{R})$ but no longer the electron degrees of freedom. The resulting integral would be of the form

$$\langle\hat{H}\rangle = \int d\boldsymbol{R}\, \eta^*(\boldsymbol{R})\hat{H}_{\text{eff}}\,\eta(\boldsymbol{R}), \qquad (28.7.15)$$

and the operator $\hat{H}_{\text{eff}}$ remaining in the integral and defining a variational principle for the nuclear wave function $\eta(\boldsymbol{R})$ can be viewed as the effective Hamiltonian for the nuclear degrees of freedom. As we will see now, the strategy can indeed be carried out in practice! The resulting nuclear dynamics will have a gauge invariance inherited from the product structure of the wave function $\psi(\boldsymbol{R}, \mathbf{r})$. It will also feature Berry connections!

We begin the calculation by rewriting the original Hamiltonian as a sum of two Hamiltonians:

$$\hat{H} = \hat{H}_N + \hat{H}_e, \qquad \hat{H}_N = \sum_{\alpha=1}^{N} \frac{\hat{\mathbf{P}}_\alpha^2}{2M_\alpha} + V_{NN}(\boldsymbol{R}). \tag{28.7.16}$$

As a warm-up we calculate the expectation value of $\hat{H}_e$:

$$\begin{aligned}\langle \hat{H}_e \rangle &= \int d\boldsymbol{R}\, d\mathbf{r}\, \eta^*(\boldsymbol{R})\phi_{\boldsymbol{R}}^*(\mathbf{r})\hat{H}_e\, \eta(\boldsymbol{R})\phi_{\boldsymbol{R}}(\mathbf{r}) = \int d\boldsymbol{R}\, \eta^*(\boldsymbol{R}) \left[\int d\mathbf{r}\, \phi_{\boldsymbol{R}}^*(\mathbf{r})\hat{H}_e\phi_{\boldsymbol{R}}(\mathbf{r}) \right] \eta(\boldsymbol{R}) \\ &= \int d\boldsymbol{R}\, \eta^*(\boldsymbol{R}) E_e(\boldsymbol{R}) \eta(\boldsymbol{R}),\end{aligned} \tag{28.7.17}$$

since $\hat{H}_e\phi_{\boldsymbol{R}}(\mathbf{r}) = E_e(\boldsymbol{R})\phi_{\boldsymbol{R}}(\mathbf{r})$, and $\phi_{\boldsymbol{R}}(\mathbf{r})$ is normalized. This term has contributed to the effective nuclear Hamiltonian the value of the $\boldsymbol{R}$-dependent electron energy $E_e(\boldsymbol{R})$. Now consider the nuclear kinetic term:

$$\left\langle \frac{\hat{\mathbf{P}}_\alpha^2}{2M_\alpha} \right\rangle = -\frac{\hbar^2}{2M_\alpha} \int d\boldsymbol{R}\, d\mathbf{r}\, \eta^*(\boldsymbol{R})\phi_{\boldsymbol{R}}^*(\mathbf{r})\ \nabla_{\boldsymbol{R}_\alpha}\cdot \nabla_{\boldsymbol{R}_\alpha} (\eta(\boldsymbol{R})\phi_{\boldsymbol{R}}(\mathbf{r})) = -\frac{\hbar^2}{2M_\alpha} I. \tag{28.7.18}$$

To manipulate the integral I above efficiently, we write it with streamlined notation:

$$\begin{aligned} I &= \int d\boldsymbol{R}\, d\mathbf{r}\, \eta^*\phi^*\ \nabla\cdot\nabla\,(\eta\phi) = \int d\boldsymbol{R}\, d\mathbf{r}\, \eta^*\phi^*((\nabla^2\eta)\phi + 2\nabla\eta\cdot\nabla\phi + \eta\nabla^2\phi) \\ &= \int d\boldsymbol{R}\, \eta^*\Big[\nabla^2\eta + \int d\mathbf{r}\,(2\nabla\eta\cdot(\phi^*\nabla\phi) + \eta\phi^*\nabla^2\phi)\Big] \\ &= \int d\boldsymbol{R}\, \eta^*\Big[\nabla^2\eta + 2\nabla\eta\cdot\int d\mathbf{r}\,\phi^*\nabla\phi + \eta\int d\mathbf{r}\,\phi^*\nabla^2\phi\Big]. \end{aligned} \tag{28.7.19}$$

It is now convenient to define $\mathbf{A} = i\hbar\int d\mathbf{r}\phi^*\nabla\phi$, or more explicitly,

$$\boxed{\mathbf{A}_\alpha(\boldsymbol{R}) \equiv i\hbar \int d\mathbf{r}\, \phi_{\boldsymbol{R}}^*(\mathbf{r})\nabla_{\boldsymbol{R}_\alpha}\phi_{\boldsymbol{R}}(\mathbf{r}) = -\int d\mathbf{r}\, \phi_{\boldsymbol{R}}^*(\mathbf{r})\hat{\mathbf{P}}_\alpha\phi_{\boldsymbol{R}}(\mathbf{r}).} \tag{28.7.20}$$

This is a Berry connection! Because of the index α, we have a full Berry connection for each nucleus. The connection $\mathbf{A}_\alpha$ has three components, as $\boldsymbol{R}_\alpha$ is a position space vector. The Berry connection arises from the electronic configuration, from the dependence of the electron wave function $\phi_{\boldsymbol{R}}(\mathbf{r})$ on $\boldsymbol{R}$.

With this definition, we continue the evaluation of I, still in simplified notation:

$$\begin{aligned} I &= \int d\boldsymbol{R}\, \eta^*\Big[\nabla^2\eta + \frac{2}{i\hbar}\nabla\eta\cdot A + \eta\int d\mathbf{r}\,\big(\nabla\cdot(\phi^*\nabla\phi) - \nabla\phi^*\nabla\phi\big)\Big] \\ &= \int d\boldsymbol{R}\, \eta^*\Big[\Big(\nabla^2 + \frac{2}{i\hbar}A\cdot\nabla\Big)\eta + \eta\frac{1}{i\hbar}\nabla\cdot A - \eta\int d\mathbf{r}\,\nabla\phi^*\nabla\phi\Big] \\ &= \int d\boldsymbol{R}\, \eta^*\Big[\nabla^2 - \frac{2i}{\hbar}A\cdot\nabla - \frac{i}{\hbar}(\nabla\cdot A) - \int d\mathbf{r}\,\nabla\phi^*\nabla\phi\Big]\eta. \end{aligned} \tag{28.7.21}$$

We can complete the square using the first three terms:

$$I = \int d\boldsymbol{R}\,\eta^*\Big[\big(\nabla - \frac{i}{\hbar}A\big)^2 + \frac{1}{\hbar^2}A^2 - \int d\mathbf{r}\,\nabla\phi^*\nabla\phi\Big]\eta$$
$$= -\frac{1}{\hbar^2}\int d\boldsymbol{R}\,\eta^*\Big[(\hat{P}-A)^2 - A^2 + \hbar^2\int d\mathbf{r}\,\nabla\phi^*\nabla\phi\Big]\eta. \tag{28.7.22}$$

With all the labels back in, (28.7.18) becomes

$$\left\langle\frac{\hat{\mathbf{P}}_\alpha^2}{2M_\alpha}\right\rangle = \int d\boldsymbol{R}\,\eta^*(\boldsymbol{R})\left[\frac{(\hat{\mathbf{P}}_\alpha - \mathbf{A}_\alpha)^2}{2M_\alpha} - \frac{\mathbf{A}_\alpha^2}{2M_\alpha} + \frac{\hbar^2}{2M_\alpha}\int d\mathbf{r}\,\big|\nabla_{\boldsymbol{R}_\alpha}\phi_{\boldsymbol{R}}\big|^2\right]\eta(\boldsymbol{R}). \tag{28.7.23}$$

A useful simplification follows from the following identity:

$$\frac{1}{\hbar^2}\int d\mathbf{r}\,\big|(\hat{\mathbf{P}}_\alpha + \mathbf{A}_\alpha)\phi_{\boldsymbol{R}}\big|^2 = \int d\mathbf{r}\,\big|(\nabla_{\boldsymbol{R}_\alpha} + \tfrac{i}{\hbar}\mathbf{A}_\alpha)\phi_{\boldsymbol{R}}\big|^2 = \int d\mathbf{r}\,\big|\nabla_{\boldsymbol{R}_\alpha}\phi_{\boldsymbol{R}}\big|^2 - \frac{1}{\hbar^2}\mathbf{A}_\alpha^2, \tag{28.7.24}$$

which you can check by expanding the left-hand side and noting that the mixed terms contribute to the $\mathbf{A}_\alpha^2$ term. This means that we can rewrite (28.7.23) as follows:

$$\left\langle\frac{\hat{\mathbf{P}}_\alpha^2}{2M_\alpha}\right\rangle = \int d\boldsymbol{R}\,\eta^*(\boldsymbol{R})\left[\frac{(\hat{\mathbf{P}}_\alpha - \mathbf{A}_\alpha)^2}{2M_\alpha} + \frac{1}{2M_\alpha}\int d\mathbf{r}\,\big|(\hat{\mathbf{P}}_\alpha + \mathbf{A}_\alpha)\phi_{\boldsymbol{R}}\big|^2\right]\eta(\boldsymbol{R}). \tag{28.7.25}$$

An important check can be performed now. The theory has a funny gauge invariance. Note that, from (28.7.13),

$$\psi(\boldsymbol{R},\mathbf{r}) = \eta(\boldsymbol{R})\phi_{\boldsymbol{R}}(\mathbf{r}) = \big(e^{i\beta(\boldsymbol{R})}\eta(\boldsymbol{R})\big)\big(e^{-i\beta(\boldsymbol{R})}\phi_{\boldsymbol{R}}(\mathbf{r})\big), \tag{28.7.26}$$

and therefore the wave function $\psi(\boldsymbol{R},\mathbf{r})$ is invariant under the gauge transformations

$$\eta(\boldsymbol{R}) \to \eta'(\boldsymbol{R}) = e^{i\beta(\boldsymbol{R})}\eta(\boldsymbol{R}), \qquad \phi_{\boldsymbol{R}}(\mathbf{r}) \to \phi'_{\boldsymbol{R}}(\mathbf{r}) = e^{-i\beta(\boldsymbol{R})}\phi_{\boldsymbol{R}}(\mathbf{r}). \tag{28.7.27}$$

It is an important check on our algebra that (28.7.25) be invariant under those changes. Note that the Berry connection indeed transforms. You should check that

$$\mathbf{A}'_\alpha = \mathbf{A}_\alpha + \hbar\,\nabla_{\boldsymbol{R}_\alpha}\beta. \tag{28.7.28}$$

This is the familiar transformation we could have expected. Moreover, you can also check that, as we had for electromagnetic couplings,

$$(\hat{\mathbf{P}}_\alpha - \mathbf{A}'_\alpha)\eta'(\boldsymbol{R}) = e^{i\beta(\boldsymbol{R})}(\hat{\mathbf{P}}_\alpha - \mathbf{A}_\alpha)\eta(\boldsymbol{R}), \tag{28.7.29}$$

explaining the gauge invariance of the first term in (28.7.25). Similarly, one can also verify that

$$(\hat{\mathbf{P}}_\alpha + \mathbf{A}'_\alpha)\phi'_{\boldsymbol{R}}(\mathbf{r}) = e^{-i\beta(\boldsymbol{R})}(\hat{\mathbf{P}}_\alpha + \mathbf{A}_\alpha)\phi_{\boldsymbol{R}}(\mathbf{r}), \tag{28.7.30}$$

which explains the gauge invariance of the second term in (28.7.25). The last two equations show that η and ϕ have opposite charges in their coupling to the Berry connection; recall that the charge q enters the minimal coupling in the form $(\hat{\mathbf{p}} - \frac{q}{c}\mathbf{A})$. This is consistent with the opposite signs in the exponents of the transformations (28.7.27).

Let us now finish the computation of the effective Hamiltonian. Since the nuclear-nuclear potential $V_{NN}(\mathbf{R})$ does not act on the electronic wave functions, we can see that

$$\langle V_{NN}(\boldsymbol{R})\rangle = \int d\boldsymbol{R}\, \eta^*(\boldsymbol{R}) V_{NN}(\boldsymbol{R})\eta(\boldsymbol{R}), \tag{28.7.31}$$

and thus contributes precisely $V_{NN}(\boldsymbol{R})$ to the effective nuclear Hamiltonian. Together with our results for the nuclear kinetic terms and the electronic Hamiltonian, we now get

$$\hat{H}_{\text{eff}} = \sum_\alpha \frac{(\hat{\mathbf{P}}_\alpha - \mathbf{A}_\alpha)^2}{2M_\alpha} + \sum_\alpha \frac{1}{2M_\alpha}\int d\boldsymbol{R}\, \left|(\hat{\mathbf{P}}_\alpha + \mathbf{A}_\alpha)\phi_{\boldsymbol{R}}\right|^2 + V_{NN}(\boldsymbol{R}) + E_e(\boldsymbol{R}). \tag{28.7.32}$$

We can separate the effective nuclear Hamiltonian $\hat{H}_{\text{eff}}$ into kinetic and potential terms:

$$\hat{H}_{\text{eff}} = \sum_\alpha \frac{(\hat{\mathbf{P}}_\alpha - \mathbf{A}_\alpha)^2}{2M_\alpha} + U(\boldsymbol{R}), \tag{28.7.33}$$

where the nuclear effective potential $U(\boldsymbol{R})$ is

$$\boxed{U(\boldsymbol{R}) \equiv V_{NN}(\boldsymbol{R}) + E_e(\boldsymbol{R}) + \sum_\alpha \frac{\hbar^2}{2M_\alpha}\int d\mathbf{r}\, \left|(\nabla_{\boldsymbol{R}_\alpha} + \tfrac{i}{\hbar}\mathbf{A}_\alpha)\phi_{\boldsymbol{R}}\right|^2.} \tag{28.7.34}$$

Since $\hat{\mathbf{P}}_\alpha - \mathbf{A}_\alpha = \frac{\hbar}{i}\left(\nabla_{\boldsymbol{R}_\alpha} - \frac{i}{\hbar}\mathbf{A}_\alpha\right)$, the Schrödinger equation for nuclear motion is

$$\left[-\sum_\alpha \frac{\hbar^2}{2M_\alpha}(\nabla_{\boldsymbol{R}_\alpha} - \tfrac{i}{\hbar}\mathbf{A}_\alpha)^2 + U(\boldsymbol{R})\right]\eta(\boldsymbol{R}) = E\,\eta(\boldsymbol{R}). \tag{28.7.35}$$

The energy E here is the expectation value of the full Hamiltonian $\hat{H}$, as given in (28.7.15). This completes the general analysis of the problem. Despite our adiabatic approximation, and the ensuing simplifications, the effective nuclear Hamiltonian is fairly complicated.

If the electronic wave functions can be chosen to be real, the Berry connections vanish, and the effective nuclear Hamiltonian in (28.7.32) becomes

$$H_{\text{eff}} = \sum_\alpha \frac{\hat{\mathbf{P}}_\alpha^2}{2M_\alpha} + \sum_\alpha \frac{\hbar^2}{2M_\alpha}\int d\mathbf{r}\, \left|\nabla_{\boldsymbol{R}_\alpha}\phi_{\boldsymbol{R}}\right|^2 + V_{NN}(\boldsymbol{R}) + E_e(\boldsymbol{R}). \tag{28.7.36}$$

To first approximation one often ignores the second term in the above Hamiltonian. This gives the lowest-order Born-Oppenheimer effective nuclear Hamiltonian,

$$\hat{H}_{\text{eff}} \simeq -\sum_\alpha \frac{\hbar^2}{2M_\alpha}\nabla^2_{\boldsymbol{R}_\alpha} + V_{NN}(\boldsymbol{R}) + E_e(\boldsymbol{R}), \tag{28.7.37}$$

which we will apply to a simple molecule in the next section.

28.8 The Hydrogen Molecule Ion

Hydrogen gas is composed of hydrogen H_2 molecules. The hydrogen molecule has two protons and two electrons. The hydrogen molecule ion H_2^+ is what we obtain when the H_2 molecule loses one electron. This ion is still a molecule, but this time we have two protons held together by *one* electron in a stable configuration: we have a bound state. This is a remarkable situation that could not be achieved in classical electromagnetic theory. Suppose we had two positively charged particles along the real x-axis at positions $x = \pm x_0$, with $x_0 > 0$ and one negatively charged particle at $x = 0$ (figure 28.9, *left*). The electrostatic energy $U(x_0)$ here is a function of x_0, and assuming all charges have magnitude e, is given by

$$U(x_0) = \frac{e^2}{2x_0} - \frac{e^2}{x_0} - \frac{e^2}{x_0} = -\frac{3}{2}\frac{e^2}{x_0}. \tag{28.8.1}$$

Note that as a function of x_0, the energy $U(x_0)$ has no stationary point. The energy can be lowered by decreasing the value of x_0 all the way to 0. The situation is rather different if we imagine the negative charge at the middle spread uniformly over a ring of radius R (figure 28.9, *right*). In this case the potential energy $U_R(x_0)$ is given by

$$U_R(x_0) = \frac{e^2}{2x_0} - \frac{2e^2}{\sqrt{R^2 + x_0^2}}. \tag{28.8.2}$$

For fixed R the function $U_R(x_0)$ has a global minimum. The minimum happens when $4x_0^3 = (R^2 + x_0^2)^{3/2}$ so that $x_0 = R/\sqrt{4^{2/3} - 1} \simeq 0.811R$. The configuration is stable under changes of x_0, but it is not *fully* stable; full stability is not possible in classical electrostatics. Still, the result suggests that spreading out the negative electric charge should help in creating a bound state. This is, of course, what we naturally get in quantum mechanics: an electronic cloud. In the Born-Oppenheimer approximation, our first step is to determine the electronic cloud in the background of two fixed protons.

There are constraints on the bound state energy E_B of the H_2^+ ion. The energy E_B must be lower than the energy we get after we dissociate the ion in any possible way. If we remove the electron from the ion, the two remaining protons do not form a bound state, so we just get full dissociation and a state of zero energy, telling us that $E_B < 0$, which is not new information. On the other hand, if we remove a proton from the ion, we get a hydrogen atom, formed by the electron and the other proton. This is a bound

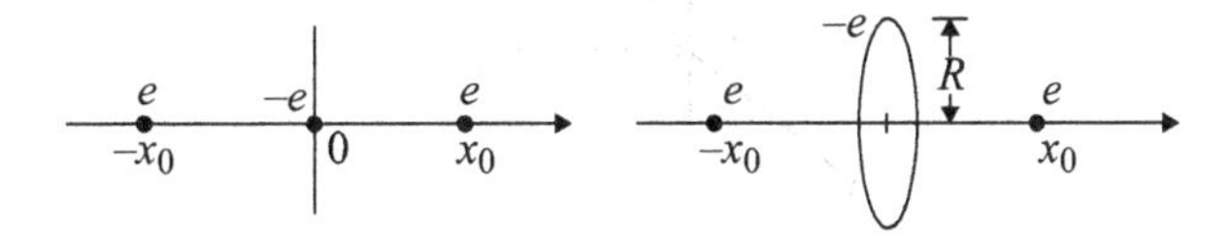

Figure 28.9
Left: Two positive charges at $x = \pm x_0$ and a negative charge at the origin. *Right*: Two positive charges at $x = \pm x_0$ and a negative charge uniformly spread over a circle of radius R centered at the origin.

state of energy −Ry. Therefore,

$$E_B < -\text{Ry} = -13.6\,\text{eV}. \tag{28.8.3}$$

In investigating H_2^+, we would like to determine E_B as well as the size of the molecule, defined as the distance R between the two protons. Additionally, we can ask for the vibration frequencies f of the molecule. Here is what is known experimentally:

$$E_B \simeq -16.4\,\text{eV}, \quad R \simeq 2.0\,a_0, \quad f \simeq 6.6 \times 10^{13}\,\text{Hz}. \tag{28.8.4}$$

Here a_0 is the Bohr radius. The dissociation energy is then $-13.6\,\text{eV} - (-16.4\,\text{eV}) = 2.8\,\text{eV}$.

We would like to use the theory we have developed to estimate those values, so let us consider the Hamiltonian for the electron in the Born-Oppenheimer approximation, as given in (28.7.9). With just one electron, the Hamiltonian takes the form

$$\hat{H}_e(\hat{\mathbf{p}}, \mathbf{r}; \mathbf{R}) = \frac{\hat{\mathbf{p}}^2}{2m} + V_{eN}(\mathbf{R}, \mathbf{r}). \tag{28.8.5}$$

Let R denote the separation between the protons. Proton number one is located at the origin, and proton number two is located at $z = R$, as shown in figure 28.10. Let $r = r_1$ denote the distance from the electron to proton one (at the origin) and r_2 denote the distance from the electron to proton two. The electron Hamiltonian is then

$$\hat{H}_e(r, R) = -\frac{\hbar^2}{2m}\nabla^2 - e^2\left(\frac{1}{r_1} + \frac{1}{r_2}\right). \tag{28.8.6}$$

Here $\nabla^2 = \nabla_{\mathbf{r}}^2$ is relative to the position $\mathbf{r}$ of the electron, specified in spherical coordinates by (r, θ, ϕ). From the geometry of the configuration, we see that

$$r_1 = r, \quad r_2 = \sqrt{R^2 + r^2 - 2rR\cos\theta}. \tag{28.8.7}$$

There is no dependence on the angle ϕ as the proton configuration is axially symmetric.

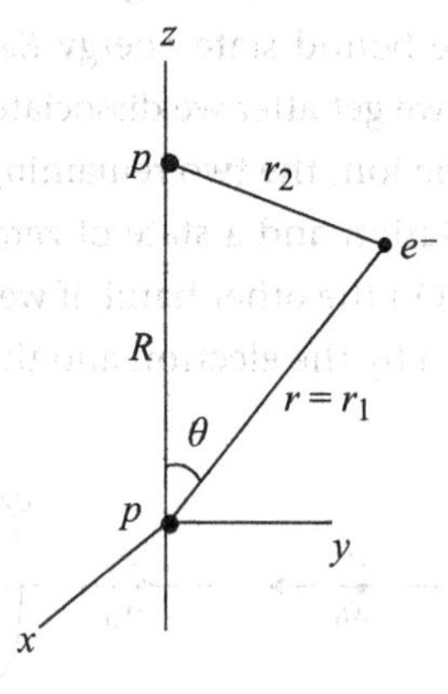

Figure 28.10
The H_2^+ ion, with proton 1 at the origin and proton 2 a distance R away along the z-axis. The electron coordinates r and θ define the distances r_1 and r_2.

We will use the variational method to get upper bounds for the ground state energy of $\hat{H}_e$, for arbitrary but fixed R. The simplest possible ansatz is the *LCAO ansatz*, for linear combination of atomic orbitals. We take the electron wave function $\tilde{\psi}$ to be a sum of a wave function where it is attached to one proton, as if in a hydrogen ground state, and a wave function where it is similarly attached to the other proton:

$$\tilde{\psi} = \tilde{N}\,(\tilde{\psi}_0(r_1) + \tilde{\psi}_0(r_2)), \qquad \tilde{\psi}_0(r) \equiv \frac{1}{\sqrt{\pi a_0^3}} e^{-\frac{r}{a_0}}, \tag{28.8.8}$$

with $\tilde{\psi}_0(r)$ a normalized wave function and $\tilde{N}$ a normalization constant that can be assumed to be real. How good is this ansatz? It is excellent when the protons are far apart, for in this case we would expect the lowest-energy ground state to be a superposition of hydrogen atom wave functions around each proton. It is like the ground state of a particle on two largely separated square wells. On the other hand, as $R \to 0$ the wave function should approach that of an electron in a $Z = 2$ atom, thus $\psi \sim e^{-Zr/a_0}$ with $Z = 2$. This does not happen; we get a $Z = 1$ wave function. So the ansatz is not very good for small R, which suggests it will not lead to a good estimate of the separation R.

While we will explicitly discuss below the consequences of the ansatz (28.8.8) below, it will save us time to consider a more general wave function for the electron. Following Finkelstein and Horowitz (1928), we include an additional constant Z that controls the size of the orbitals:

$$\psi = N\,(\psi_0(r_1) + \psi_0(r_2)), \qquad \psi_0(r) \equiv \sqrt{\frac{Z^3}{\pi a_0^3}}\; e^{-Z\frac{r}{a_0}}, \tag{28.8.9}$$

with N a real normalization constant. We will let the variational approach fix the value of Z. If $Z > 1$ at the optimum, this wave function will be more accurate as $R \to 0$ and a little less accurate for $R \to \infty$. The effect of $Z > 1$ is to contract the original hydrogen orbitals, so this can be expected to help with finding the right size of the molecule.

To begin our calculation, we must normalize the wave function:

$$1 = \int |\psi|^2 = N^2\Big(1 + 1 + 2\int d\mathbf{r}\, \psi_0(r_1)\psi_0(r_2)\Big) = 2N^2(1 + I), \tag{28.8.10}$$

where I is the integral

$$I \equiv \int d\mathbf{r}\, \psi_0(r_1)\psi_0(r_2). \tag{28.8.11}$$

If I is determined, so is the normalization constant N. The integral I takes some effort to evaluate (problem 28.6), but one gets

$$I = \big(1 + Zx + \tfrac{1}{3}Z^2x^2\big)e^{-Zx}, \quad x \equiv \tfrac{R}{a_0}. \tag{28.8.12}$$

In the variational method, the electronic energy $E_e(R)$ is approximately identified with the upper bound provided by the expectation value $\langle \hat{H}_e \rangle$ of the electron Hamiltonian in the above wave function ψ. This energy is one contribution to the nuclear effective potential U_{eff}; the other comes from the proton-proton electrostatic energy. Indeed, from the last two terms in (28.7.37) we find that

$$U_{\text{eff}} = V_{NN}(R) + E_e(R) \simeq \frac{e^2}{R} + \langle \hat{H}_e \rangle = \frac{1}{x}\,(2\text{Ry}) + \langle \hat{H}_e \rangle. \tag{28.8.13}$$

The remaining challenge is the computation of the expectation value of $\hat{H}_e$ on the state ψ. We can do this by first noticing the equations satisfied by our candidate wave functions:

$$\begin{aligned} \Big(-\frac{\hbar^2}{2m}\nabla^2 - \frac{Ze^2}{r_1}\Big)\psi_0(r_1) &= -Z^2\,\text{Ry}\,\psi_0(r_1), \\ \Big(-\frac{\hbar^2}{2m}\nabla^2 - \frac{Ze^2}{r_2}\Big)\psi_0(r_2) &= -Z^2\,\text{Ry}\,\psi_0(r_2). \end{aligned} \tag{28.8.14}$$

It is clear why the first equation holds: $\psi_0(r)$ is simply the ground state of an electron in the potential of a nucleus with Z protons, and since $r_1 = r$, the equation is in standard form. For the second equation, note that $\mathbf{r} = \boldsymbol{R} + \mathbf{r}_2$, where $\boldsymbol{R}$ is the vector going from the proton at the origin to the proton at the z-axis at $z = R$. It thus follows that $\nabla^2 = \nabla^2_{\mathbf{r}_2}$, and the second equation is also in standard form. Using the form of the electron Hamiltonian $\hat{H}_e$ in (28.8.6), we now have

$$\hat{H}_e\psi = N\Big[\Big(-\frac{\hbar^2}{2m}\nabla^2 - \frac{e^2}{r_1}\Big)\psi_0(r_1) + \Big(-\frac{\hbar^2}{2m}\nabla^2 - \frac{e^2}{r_2}\Big)\psi_0(r_2) - e^2\Big(\frac{\psi_0(r_1)}{r_2} + \frac{\psi_0(r_2)}{r_1}\Big)\Big].$$

Using the equations of motion in (28.8.14), the above can be put in the following form:

$$\hat{H}_e\psi = -Z^2\,\text{Ry}\,\psi + N\,e^2(Z-1)\Big(\frac{\psi_0(r_1)}{r_1} + \frac{\psi_0(r_2)}{r_2}\Big) - Ne^2\Big(\frac{\psi_0(r_1)}{r_2} + \frac{\psi_0(r_2)}{r_1}\Big). \tag{28.8.15}$$

We form the inner product by multiplying by ψ from the left and integrating over all of space. With a little algebra, we find that

$$\begin{aligned} \langle \hat{H}_e \rangle &= -Z^2\,\text{Ry} + 2N^2e^2(Z-1)\int \frac{d\mathbf{r}}{r_1}\psi_0(r_1)\psi_0(r_1) \\ &\quad + 2N^2e^2(Z-2)\int \frac{d\mathbf{r}}{r_1}\psi_0(r_1)\psi_0(r_2) - 2N^2e^2\int \frac{d\mathbf{r}}{r_2}\psi_0(r_1)\psi_0(r_1). \end{aligned} \tag{28.8.16}$$

The integrals above can be evaluated, and the results are

$$\begin{aligned} W &\equiv a_0\int \frac{d\mathbf{r}}{r_1}\psi_0(r_1)\psi_0(r_1) = Z, \\ D &\equiv a_0\int \frac{d\mathbf{r}}{r_2}\psi_0(r_1)\psi_0(r_1) = \frac{1}{x} - \Big(Z + \frac{1}{x}\Big)e^{-2Zx}, \\ X &\equiv a_0\int \frac{d\mathbf{r}}{r_1}\psi_0(r_1)\psi_0(r_2) = Z(1+Zx)e^{-Zx}. \end{aligned} \tag{28.8.17}$$

In terms of W, D, and X and using the value of N^2, the $\hat{H}_e$ expectation value can be written as

$$\langle \hat{H}_e \rangle = \Big[-\frac{Z^2}{2} + \frac{1}{1+I}\big((Z-1)W + (Z-2)X - D\big)\Big]\,(2\text{Ry}). \tag{28.8.18}$$

We now have all the ingredients ready. Back in U_{eff}, as given in (28.8.13), and using the evaluations of the integrals, we finally have

$$U_{\text{eff}}(x;Z) = \left[\tfrac{1}{x} - \tfrac{1}{2}Z^2 + \frac{Z^2 - Z - \frac{1}{x} + Z(Z-2)(1+Zx)e^{-Zx} + (Z+\frac{1}{x})e^{-2Zx}}{1 + e^{-Zx}(1+Zx+\frac{1}{3}Z^2x^2)}\right](2\text{Ry}). \tag{28.8.19}$$

This is the effective nuclear potential in terms of the separation parameter $x = R/a_0$ and the variational parameter Z. For each value of Z, the value x_* of x at the minimum of U_{eff} is the predicted nuclear separation, and the bound state energy E_B is $U_{\text{eff}}(x_*; Z)$, the potential evaluated at the minimum.

As a preliminary, we can examine the effective potential for the original ansatz (28.8.8) corresponding to $Z = 1$. The minimum of the potential $U_{\text{eff}}(x; 1)$ is readily obtained numerically. One finds $x_* \simeq 2.4928$, and thus

$$R \simeq 2.4928\, a_0, \quad E_B \simeq -1.1297\,\text{Ry} = -15.36\,\text{eV}. \tag{28.8.20}$$

This corresponds to a dissociation energy of 1.763 eV. Comparing with the quoted results in (28.8.4), we have roughly a 20% error in the molecular size and an almost 40% error in the dissociation energy, which is not very good.

Significant improvement can be obtained by now letting Z vary. We thus minimize $U_{\text{eff}}(x; Z)$ over *both* Z and x. The minimum occurs for $Z = Z_*$ and $x = x_*$, with values

$$Z_* \simeq 1.2380, \quad x_* = 2.0033, \tag{28.8.21}$$

resulting in

$$E_B = U_{\text{eff}}(x_*; Z_*) \simeq -1.17301\,\text{Ry} = -15.953\,\text{eV}. \tag{28.8.22}$$

This implies a dissociation energy of 2.353 eV. Note that $Z_* = 1.2380$ was the variational compromise between $Z = 1$ at $R \to \infty$, and $Z = 2$ for $R \to 0$. Strikingly, $R = 2.0033\, a_0$ is pretty much the exact nuclear separation! The energy also improved substantially: the error in the dissociation energy is now only about 16%. The effective potential $U_{\text{eff}}(x; Z_*)$ is shown in figure 28.11. A contour plot of the wave function ψ is shown on the left in figure 28.12. The plot is in a two-dimensional plane containing both protons, placed symmetrically about the origin. While the two orbitals we add to form ψ are each spherically symmetric about their own proton, the sets of constant $|\psi|^2$ are not circles about the protons. The vibrational frequencies for the nuclei can be estimated by exploring the effective potential near the minimum x_*, with $Z = Z_*$ (problem 28.6). One finds $f \simeq 8.15 \times 10^{13}$ Hz, about 23% larger than the value given in (28.8.4). This is a reasonable but not striking agreement.

An insightful modification of the ansatz (28.8.9) was suggested by Guillemin and Zener in 1929. The new wave function allows for polarization of the orbitals by introducing a second constant Z' supplementing Z:

$$\psi = N\left(e^{-(Zr_1 + Z'r_2)/a_0} + e^{-(Zr_2 + Z'r_1)/a_0}\right). \tag{28.8.23}$$

Here we still expect $Z > 1$ and some small $Z' > 0$. Examine the first wave function, which is an orbital around the first proton when $Z > Z'$. We have

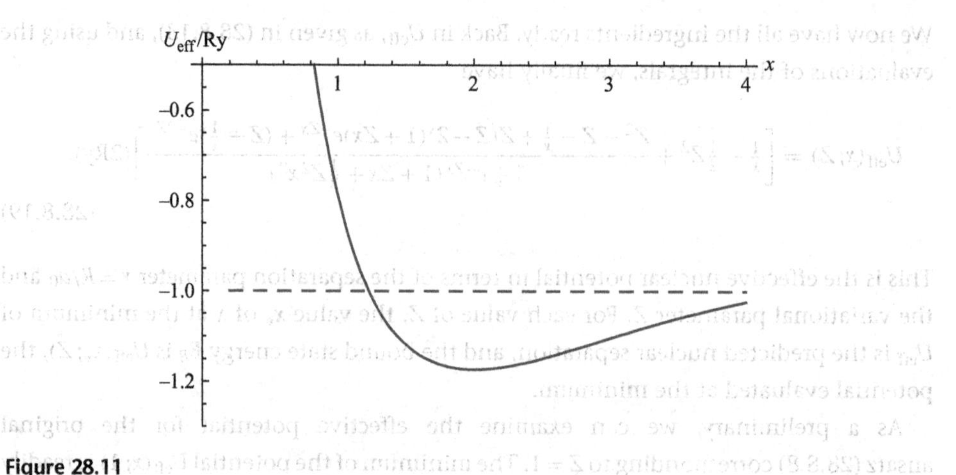

Figure 28.11
The nuclear effective potential $U_{\text{eff}}(x; Z=1.2380)$. The minimum is at $x = 2.003$, with energy $-1.173\,\text{Ry} = -15.953\,\text{eV}$. The horizontal dashed line defines the energy at which the molecule dissociates into a hydrogen atom and a free proton.

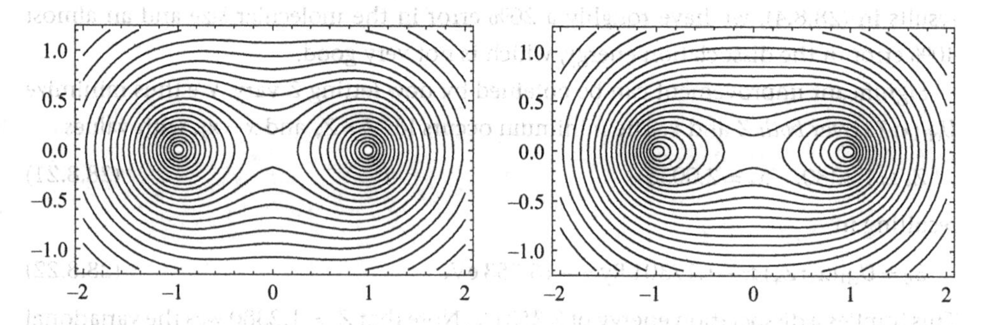

Figure 28.12
Contour plots for the electronic wave functions; contours are the lines of constant $|\psi|^2$. *Left*: Contour plot for the Finkelstein-Horowitz ansatz (28.8.9) of spherically symmetric orbitals with $Z = 1.2380$. *Right*: Contour plot for the Guillemin-Zener ansatz of asymmetric orbitals with $Z = 1.13$ and $Z' = 0.23$.

$$e^{-(Zr_1 + Z'r_2)/a_0} = e^{-Zr_1/a_0} \cdot e^{-Z'r_2/a_0}. \tag{28.8.24}$$

The first exponential on the right-hand side defines a spherically symmetric wave function about the first proton. However, the second exponential, with $Z' > 0$, has the effect of pulling the orbital toward the second proton. Indeed, the second exponential increases the suppression of the wave function as one moves *away* from the second proton. This is exactly what we would expect physically: an electronic cloud around a proton would be partially pulled by electrostatic attraction toward the other proton.

We will not attempt to do the variational calculation following from ansatz (28.8.23). It is laborious and best done using parabolic coordinates. The answers, however, are known. One finds that the optimum is achieved for

$$Z \simeq 1.13, \quad Z' \simeq 0.23. \tag{28.8.25}$$

For these values one finds $R \simeq 2.0\,a_0$, and $E_B \simeq -16.4\,\text{eV}$, both agreeing with the measured values to the stated accuracy. A contour plot of the wave function for the above asymmetric orbitals is shown on the right in figure 28.12. Comparing to the wave function associated with symmetric orbitals shown on the left, we see how the contours are indeed pulled toward the center.

Problems

Problem 28.1. *Adiabatic spin rotation.*

Consider a spin one-half particle at rest with a magnetic dipole moment $\hat{\mu} = 2\mu_B \hat{\mathbf{S}}/\hbar$, where $\hat{\mathbf{S}}$ is the spin operator, and $\mu_B > 0$ is the Bohr magneton. Assume its spin is free to rotate in response to a time-dependent magnetic field. The Hamiltonian $\hat{H}$ of the spin in a magnetic field $\mathbf{B}$ is

$$\hat{H} = -\frac{2\mu_B}{\hbar}\,\hat{\mathbf{S}} \cdot \mathbf{B}(t).$$

In the ground state, the spin will tend to align with $\mathbf{B}$.

We will start at time $t = -T$ with $T > 0$. At this time there is a large magnetic field mostly in the $-\hat{\mathbf{z}}$-direction that we will slowly decrease to zero and then increase in the opposite direction until time $t = T$. We will assume that, throughout, there is a constant small field in the (x, y) plane. The magnetic field is then

$$\mathbf{B}(t) = (B_x, B_y, \gamma t) \quad \text{for } -T \leq t \leq T.$$

Here $\gamma > 0$ is the rate at which the magnetic field in the z-direction grows, and we assume that

$$\gamma T \gg \sqrt{B_x^2 + B_y^2} \equiv \delta B \tag{1}$$

so that for $t = \pm T$, the magnetic field is mostly along the z-axis. Denote the exact, instantaneous ground state at time t by $|\psi_g(t)\rangle$ and the exact, instantaneous excited state at time t by $|\psi_e(t)\rangle$. These correspond, respectively, to spins that are either aligned or anti aligned with the exact magnetic field at time t. Suppose that at time $t = -T$, the spin is in the state $|\psi_g(-T)\rangle$.

1. Use the adiabatic theorem to argue that the particle finishes, up to a phase, in the state $|\psi_g(T)\rangle$ with nearly unit probability, as long as $\delta B > 0$. Find the inequality that must be satisfied by the rate γ of magnetic field growth for the adiabatic theorem to apply. Write the inequality in the form $\gamma \ll \cdots$ where the quantity represented by the dots is built from μ_B, $\hbar$, and δB (do not include additional pure numerical constants).
2. Explain why equation (1) implies $|\psi_g(-T)\rangle \simeq |-\rangle$ and $|\psi_g(T)\rangle \simeq |+\rangle$, where $\sigma_z|\pm\rangle = \pm|\pm\rangle$. In this case, the adiabatic process in (1) will convert the state largely along $|-\rangle$ to a state largely along $|+\rangle$. Use equation (1) and the adiabatic condition you found in part (1) to find $T \gg \ldots$, where the unknown quantity does not depend on γ. Write your answer in terms of δB, μ_B, and $\hbar$, with no additional numerical constants.

3. Instead of a time-varying magnetic field, consider a *spatially* varying magnetic field. We can use the adiabatic theorem to understand the magnetic traps used by atomic physicists to trap very cold gases of *spin-polarized* atoms. Classically, the force on a dipole $\boldsymbol{\mu}$ from a magnetic field gradient is

$$\mathbf{F} = (\boldsymbol{\mu} \cdot \nabla)\mathbf{B}(\mathbf{x}).$$

One can design magnetic field gradients such that an atom that has its spin antiparallel to the local magnetic field $\mathbf{B}(\mathbf{x})$ experiences a force toward the center of the trap. Those atoms with spins parallel to $\mathbf{B}(\mathbf{x})$ feel a force that expels them from the trap. This way one can trap atoms of just one polarization state. But this raises a question: Since $\mathbf{B}(\mathbf{x})$ varies in space, how can we ensure that the thermal movements of the atoms within the trap do not flip the spins that are antialigned with the local magnetic field?

Assume that the atoms have mass m and temperature T. Let B denote the magnitude of the magnetic field and $|\nabla B|$ denote the magnitude of its gradient. State an inequality in terms of m, kT, $\hbar$, μ_B, B, and $|\nabla B|$ that must be satisfied for the adiabatic condition to be valid. [Hint: since B varies in space, the thermal motion of the atoms implies that the spins experience time-dependent magnetic fields.] Write your answer in the form $|\nabla B| \ll \cdots$ or $|\nabla B| \gg \cdots$ as appropriate (do not include pure numerical constants).

Since magnetic *gradients* exert the trapping forces, one might think there would be no problem if at one point in the trap the magnetic field vanishes or is very very small. Is this problematic?

Problem 28.2. *Landau-Zener transitions.*

Consider the time-dependent Hamiltonian (28.5.10) for the Landau-Zener analysis of section 28.5. The natural ansatz for a time-dependent solution was given in (28.5.20):

$$|\Psi(t)\rangle = A(t)\exp\left(-\frac{i}{\hbar}\int_0^t E_1(t')dt'\right)|1\rangle + B(t)\exp\left(-\frac{i}{\hbar}\int_0^t E_2(t')dt'\right)|2\rangle.$$

Here $A(t)$ and $B(t)$ are functions of time to be determined. For convenience, define the energy difference $E_{12}(t)$ and the phase difference $\theta_{12}(t)$ as follows:

$$E_{12}(t) \equiv E_1(t) - E_2(t) = \alpha t, \quad \theta_{12}(t) \equiv \frac{1}{\hbar}\int_0^t dt' E_{12}(t') = \frac{\alpha t^2}{2\hbar}.$$

1. Use the Schrödinger equation to derive the set of linear, first-order, coupled differential equations satisfied by $A(t)$ and $B(t)$:

$$\begin{aligned}\dot{A} &= -\tfrac{i}{\hbar}H_{12}B(t)e^{i\theta_{12}(t)},\\ \dot{B} &= -\tfrac{i}{\hbar}H_{12}^{*}A(t)e^{-i\theta_{12}(t)}.\end{aligned}$$

Derive a second-order linear differential equation for $B(t)$, and write it in the form $\ddot{B} + \cdots = 0$.

2. Assume that the system is in state $|2\rangle$ at $t=-\infty$:

$$A(-\infty)=0, \qquad B(-\infty)=1.$$

Verify that constant A and constant B is a solution for $H_{12}=0$. We can think of this as a solution to zeroth order in the small perturbation H_{12}.

Integrate the equation for $A(t)$, keeping terms to first order in H_{12}, and determine $A(\infty)$ in terms of H_{12}, α, $\hbar$, and numerical constants. To this approximation, give the probability P_2 that the system remains in $|2\rangle$ at $t=\infty$.

3. Reconsider the second-order differential equation for $B(t)$ derived in (1). Construct a unit-free time parameter τ in terms of t, α, and $|H_{12}|$ (and no extra numerical constants), and show that the differential equation for $B(\tau)$ becomes

$$\frac{d^2B}{d\tau^2}+i\gamma\,\tau\frac{dB}{d\tau}+\gamma^2B=0, \tag{1}$$

where γ is a constant that you should determine and can be written in terms of H_{12}, H_{12}^*, $\hbar$, and α. Solve this equation numerically for $\gamma=0.1$, integrating your solution in the interval $\tau\in(-25,25)$ with initial conditions $B=1$ and $A=0$ at $\tau=-25$. Plot the value of $|B(\tau)|^2$, which represents the probability for the system to have made a nonadiabatic transition by time τ. What is the value of $|B(25)|^2$ that follows from your program? Compare this with the prediction that the probability for a nonadiabatic transition is $\exp(-2\pi\gamma)$.

4. We now argue heuristically that if $B(\tau=-\infty)=1$, then $B(\tau=\infty)=\exp(-\gamma\pi)$, implying a nonadiabatic transition probability $\exp(-2\pi\gamma)$, as claimed. For this, work on the complex τ plane, and imagine finding the evolution of B along a very large circle centered at the origin. Thus, take $\tau=Re^{i\theta}$ with fixed $R\gg 1$ and the angle θ denoting the position on the circle. Show that on this circle, equation (1) becomes $\partial_\theta B=-\gamma B$, in the limit $R\to\infty$. The condition $B=1$ in the infinite past implies $B(\theta=\pi)=1$. Given this, $B(\theta)=\exp(-\gamma(\theta-\pi))$. Finding B in the infinite future involves deciding if this future corresponds to $\theta=0$ or $\theta=2\pi$. A full derivation would provide a first-principles answer to this question. Confirm that the desired result arises from identifying $\tau=+\infty$ with $\theta=2\pi$.

Problem 28.3. *Which phase?*

In the adiabatic theorem, an instantaneous eigenstate $|\psi(t)\rangle$ of energy $E(t)$ satisfies the equation $\hat{H}(t)|\psi(t)\rangle=E(t)|\psi(t)\rangle$. The eigenstate is not unique. This is clear because multiplying $|\psi(t)\rangle$ by an arbitrary *time-dependent* phase still gives an instantaneous eigenstate with the same energy.

Suppose that Alice has an instantaneous eigenstate $|\psi^A(t)\rangle$ with energy $E(t)$, while Bob has an instantaneous eigenstate $|\psi^B(t)\rangle$ with the same energy. Assume that their states agree at time $t=0$ so that $|\psi^A(0)\rangle=|\psi^B(0)\rangle$. At later times their solutions differ by a time-dependent phase $\alpha(t)$ so that

$$|\psi^A(t)\rangle=e^{i\alpha(t)}|\psi^B(t)\rangle, \qquad \text{with} \quad \alpha(0)=0.$$

Will this lead Alice and Bob to get different predictions from the adiabatic theorem? More concretely, suppose that at time $t=0$, a system is in state

$$|\Psi(t=0)\rangle = |\psi^A(0)\rangle = |\psi^B(0)\rangle.$$

Suppose that for times $0 \le t \le T$, the Hamiltonian changes adiabatically. Both Alice and Bob predict the state at time T using the adiabatic theorem and their own definitions of the instantaneous eigenstates. Since in the adiabatic theorem we care about phases, equality of their answers means equality including phases. Recall that in the adiabatic theorem we have $|\Psi(t)\rangle = e^{i\theta(t)}e^{i\gamma(t)}|\psi(t)\rangle$, for $0 \le t \le T$. We will use superscripts A, B for quantities computed by Alice and Bob, respectively.

1. How is $\theta^A(t)$ related to $\theta^B(t)$?
2. How is $\gamma^A(t)$ related to $\gamma^B(t)$?
3. How is $|\Psi^A(t)\rangle$ related to $|\Psi^B(t)\rangle$, for $0 \le t \le T$?

Problem 28.4. *Landau levels in an electric field, again (M. Metlitski).*

Consider a particle with charge q and mass m moving in two dimensions in a uniform magnetic field $\mathbf{B} = B\hat{\mathbf{z}}$. The Hamiltonian is

$$\hat{H} = \frac{1}{2m}\left(\hat{\mathbf{p}} - \frac{q}{c}\mathbf{A}\right)^2.$$

We choose the Landau gauge $\mathbf{A} = (-By, 0)$ and take the x-direction to be periodic of length L, $x \sim x + L$.

We next imagine applying to the system a small time-dependent electric field $E(t)$ along the x-direction. We choose to keep the scalar potential $\Phi = 0$ and take the vector potential to be $\mathbf{A} = (-By + \delta A_x(t), 0)$, with $\delta A_x(t)$ a slowly increasing function of time t, independent of x and y. [The case of a time-independent electric field was considered in problem 24.6.]

1. Compute the electric field corresponding to the gauge potential given above. Verify that the magnetic field is unaffected by $\delta A_x(t)$.
2. Since $\mathbf{A}$ is now a function of time, so is the Hamiltonian $\hat{H}(t)$. Find the instantaneous eigenstates and instantaneous energies of $\hat{H}(t)$. Choose the instantaneous eigenstates to be simultaneously eigenstates of $\hat{p}_x$. Do the instantaneous energies depend on δA_x? Do the eigenstates depend on δA_x?
3. Imagine that at $t=0$ the system is in one of the eigenstates you found in the previous part. Apply the adiabatic theorem to write the wave function of the system at a later time t, up to dynamical and geometric phases that you need not compute. In which direction is the particle moving? What is the expectation value of the velocity of the particle? [Hint: you may confirm your intuition using Ehrenfest's theorem.] How is this motion related to the direction of the electric field? This is the Hall effect.
4. When $\delta A_x(t)$ is a slowly varying function of t, the adiabatic approximation is applicable despite the degeneracy of instantaneous eigenstates. Why? [Hint: $\hat{p}_x$ commutes with the Hamiltonian for all t.]

Problem 28.5. *The Aharonov-Bohm effect and Berry's phase (M. Metlitski).*

Consider a particle of charge q and mass m moving on a ring of radius b with magnetic flux Φ through the ring. The Hamiltonian $\hat{H}_0$ is given by

$$\hat{H}_0 = \frac{\hbar^2}{2mb^2}\big(-i\partial_\phi - \tfrac{\Phi}{\Phi_0}\big)^2,$$

where ϕ is the angle around the circle, and $\Phi_0 = \frac{2\pi\hbar c}{q}$. Here we work in the formalism where the wave function ψ is periodic: $\psi(\phi+2\pi) = \psi(\phi)$. Suppose we add a "trapping" potential $V(\phi)$ to the problem so that the resulting Hamiltonian $\hat{H}$ is

$$\hat{H} = \frac{\hbar^2}{2mb^2}\big(-i\partial_\phi - \tfrac{\Phi}{\Phi_0}\big)^2 + V(\phi).$$

Since we are working on the circle, V must be periodic: $V(\phi+2\pi) = V(\phi)$. We assume that $V(\phi)$ is nonzero and attractive near $\phi = 0$.

Suppose the ground state wave function of $\hat{H}$ is $\chi(\phi)$ with energy E. In general, χ and E will depend on the flux Φ. We further assume that χ is localized near $\phi = 0$ over a short length scale. As you saw in problem 24.8, in this case the energy E is *independent* of the flux Φ.

1. Apply the Hellmann-Feynman lemma to $\hat{H}$ in order to relate $\frac{dE}{d\Phi}$ to integrals of χ and $\partial_\phi\chi$. What identity follows from the Φ independence of E?

Next we begin to slowly move the center of the trap. Thus, we get a family of Hamiltonians $\hat{H}_{\phi_0}$ labeled by the position ϕ_0 of the trap center:

$$\hat{H}_{\phi_0} = \frac{\hbar^2}{2mb^2}\big(-i\partial_\phi - \tfrac{\Phi}{\Phi_0}\big)^2 + V(\phi-\phi_0).$$

2. What is the ground state wave function $\psi_{\phi_0}(\phi)$ and energy E_{ϕ_0} of $\hat{H}_{\phi_0}$ expressed in terms of χ and E? As we sweep ϕ_0 from 0 to 2π, the Hamiltonian comes back to its starting point: $\hat{H}_{\phi_0=2\pi} = \hat{H}_{\phi_0=0}$. Is the wave function $\psi_{\phi_0}(\phi)$ the same for $\phi_0 = 2\pi$ and $\phi_0 = 0$?
3. Calculate the Berry connection $\mathcal{A}(\phi_0) = i\langle\psi_{\phi_0}|\partial_{\phi_0}\psi_{\phi_0}\rangle$, with your answer expressed in terms of an integral of χ and $\partial_\phi\chi$. Show that $\mathcal{A}(\phi_0)$ actually does not depend on ϕ_0. What is the Berry phase γ accumulated as ϕ_0 sweeps from 0 to 2π?
4. Use the identity derived in (1) to find an expression for the Berry phase γ.

The Berry phase γ can be thought of as the geometric phase accumulated by moving the particle around around the circle adiabatically.

Problem 28.6. *Calculations in the hydrogen molecule ion.*

1. Show that the integral I used in the variational calculation is given as claimed in (28.8.12):

$$I = \frac{Z^3}{\pi a_0^3}\int d\mathbf{r}\, e^{-Z(r_1+r_2)/a_0} = \big(1 + Zx + \tfrac{1}{3}Z^2x^2\big)e^{-Zx}, \quad x \equiv \tfrac{R}{a_0}.$$

Note that $r_1 = r$, and $r_2^2 = r^2 + R^2 - 2rR\cos\theta$ (figure 28.10). Use spherical coordinates and trade the integration over θ for an integration over r_2. Then, passing to unit-free coordinates $u = Zr/a_0$ and $w = Zr_2/a_0$, show that

$$I = \frac{2}{Zx}\int_0^\infty du\, u\, e^{-u} \int_{|u-Zx|}^{u+Zx} dw\, w e^{-w}.$$

Do the integral over w first and then the integral over u, splitting the region of integration appropriately to deal with the appearance of $|u - Zx|$ on the integrand.

2. To determine the nuclear vibrational frequencies in the hydrogen ion molecule, the effective potential (28.8.19) can be expanded to find its behavior for x near the minimum x_*, obtained with $Z = Z_*$. The result is

$$U_{\text{eff}}(x; Z_*) = \big(-0.58651 + \beta(x - x_*)^2 + \cdots\big)(2\text{Ry}), \quad \beta \simeq 0.070464, \quad x = \frac{R}{a_0}.$$

The above quadratic term must be compared with the canonical quadratic term $\frac{1}{2}k(x_1 - x_2)^2$ in an oscillator to read the value of k and, using the appropriate reduced mass, the value of ω. Show that

$$\hbar\omega = \sqrt{\beta}\sqrt{\frac{m_e}{m_p}}\, 4\,\text{Ry}.$$

Confirm that this gives a frequency $f \simeq 8.15 \times 10^{13}$ Hz.

29 Scattering in One Dimension

Energy eigenstates for finite-range potentials on the half line are fully characterized by an energy-dependent phase shift that also determines the time delay experienced by moving wave packets. A surprising result, Levinson's theorem, relates the number of bound states of a potential to the excursion of the phase shift as the energy goes from zero to infinity. Resonances are characterized by rapid growth of the phase shift, long time delays, and a scattering probability peak governed by the Breit-Wigner distribution. Resonances are best thought of as poles in the scattering amplitude that are found right below the real axis in the complex k plane.

29.1 Scattering on the Half Line

Physicists learn a lot from scattering experiments. In scattering, incident particles are forced to deviate from motion with constant momentum due to interactions with other particles or some medium. Rutherford learned about the structure of the atom by scattering alpha particles off thin layers of gold. We have *elastic* scattering if particles do not change type. Elastic scattering typically requires low energies. At high energies, scattering becomes quite complicated due to the creation of new particles. One then has *inelastic* scattering.

The scattering of a particle off a fixed target can often be studied by representing the effect of the target as a potential $V(\mathbf{x})$ that affects the motion of the particle. Even in the case of particle *collisions*, it is generally possible to use center-of-mass coordinates to view the problem as one in which a particle scatters off a potential.

We begin our study of scattering by considering scattering in one dimension. It is the simplest setting in which we can learn many of the relevant concepts without the technical complications that higher dimensions entail. Scattering is interesting and nontrivial in one dimension. Moreover, our results here will be useful when we turn to scattering in three spatial dimensions in chapter 30.

We will consider elastic scattering in the general potential shown in figure 29.1. This potential is given by

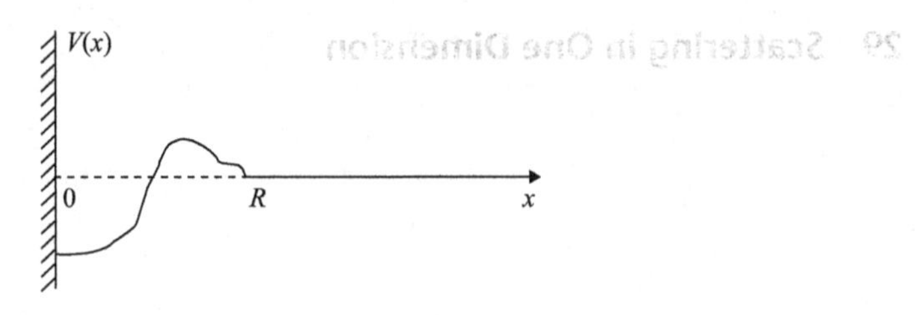

Figure 29.1
A potential of range R. The potential vanishes for $x > R$ and is infinite for $x \leq 0$.

$$V(x) = \begin{cases} \mathcal{V}(x), & 0 < x < R, \\ 0, & x > R, \\ \infty, & x < 0. \end{cases} \tag{29.1.1}$$

We call this a *finite range* potential because the nontrivial part $\mathcal{V}(x)$ of the potential does not extend beyond a distance R from the origin. The distance R is called the range of the potential. Moreover, we have an infinite potential wall at $x = 0$. Thus, all the physics happens for $x > 0$, and incoming waves from $x = \infty$ will eventually be reflected back, giving the physicist information about the potential. This is in fact the purpose of scattering experiments: we learn about the potentials acting on the particles by sending particles in and looking at the outgoing states. The restriction to $x > 0$ has a perfect analogue for scattering in three dimensions: in spherical coordinates we have $r > 0$. The present x, constrained to be nonnegative, can represent a radial coordinate. We will in fact see that in three-dimensional scattering, when the angular momentum is zero, scattering is described as in this chapter, with x replaced by r.

Consider first the case where $\mathcal{V}(x) = 0$ so that

$$V(x) = \begin{cases} 0, & x > 0, \\ \infty, & x < 0, \end{cases} \tag{29.1.2}$$

as shown in figure 29.2. We have a free particle, except for the wall at $x = 0$. The energy eigenstates can be constructed using a linear combination of momentum eigenstates $e^{\pm ikx}$, where

$$k^2 = \frac{2mE}{\hbar^2}. \tag{29.1.3}$$

The energy eigenstate, called $\phi(x)$ for the case of zero $\mathcal{V}$, has an incoming wave e^{-ikx} and an outgoing wave e^{ikx} combined in such a way that $\phi(0) = 0$, as required by the presence of the wall:

$$\phi(x) \sim e^{ikx} - e^{-ikx}. \tag{29.1.4}$$

We divide this solution by $2i$ to find

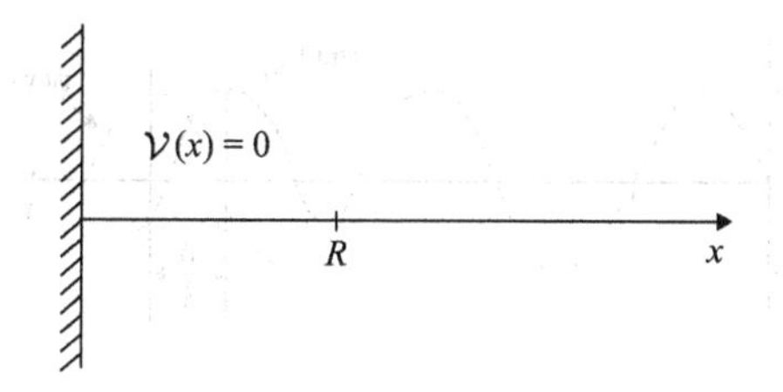

Figure 29.2
The zero potential and its energy eigenstates $\phi(x)=\sin kx$ are needed to compare with the $\mathcal{V}(x)\neq 0$ problem.

$$\boxed{\phi(x)=-\frac{e^{-ikx}}{2i}+\frac{e^{ikx}}{2i}=\sin kx.} \tag{29.1.5}$$

The first term to the right of the first equal sign is a wave incoming from $x=\infty$, and the second term is an outgoing reflected wave. Both carry the same amount of probability current, but in opposite directions.

Consider now the case $\mathcal{V}(x)\neq 0$. Now that the potential is nontrivial, we denote the energy eigenstate by $\psi(x)$. The potential always acts over the finite range $0<x<R$, and we will eventually be interested in computing the energy eigenstate $\psi(x)$ in this region. For the moment, however, let us consider $\psi(x)$ in the region $x>R$ where the potential vanishes and the solution has to be simple. We will take the incoming wave to be the same one we had for the zero-potential solution $\phi(x)$:

$$\text{Incoming wave:}\qquad -\frac{e^{-ikx}}{2i},\qquad x>R. \tag{29.1.6}$$

The outgoing wave to be added to the above requires an e^{ikx} to have the same energy as the incoming wave solution. We now claim that the most general solution includes a constant phase factor:

$$\text{Outgoing wave:}\qquad e^{2i\delta}\,\frac{e^{ikx}}{2i},\quad \delta\in\mathbb{R},\qquad x>R. \tag{29.1.7}$$

We note that the phase δ cannot be a function of x: the free Schrödinger equation for $x>R$ only allows phases linear in x. But, if present, that would change the value of the momentum associated with the outgoing wave, which we have already argued must be k. Furthermore, δ cannot be complex because then we would fail to have equality between incident and reflected probability currents. Indeed, this condition requires that the norm squared of the coefficients multiplying the exponentials $e^{\pm ikx}$ be the same.

While it does not have coordinate dependence, the phase δ is a real function $\delta(E)$ that depends on the energy E of the eigenstate and, of course, on the potential. Since the energy is fixed by k, we can also view δ as a function $\delta(k)$ of the wave number. Introduced in the form $e^{2i\delta}$, a natural range of δ is from zero to π, but it will be easier

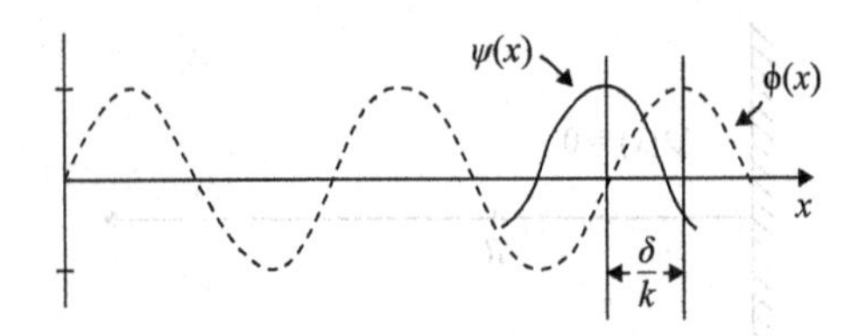

Figure 29.3
The solution $\sin kx = \phi(x)$ for zero potential is shown as a dashed line. For $x > R$, the solution $\sin(kx+\delta) \sim \psi(x)$ is shown as a continuous line. Compared to $\phi(x)$, the eigenstate $\psi(x)$ is shifted toward the origin a distance δ/k.

to let it range through all the real numbers in order to achieve a δ that is a *continuous* function of the energy. Assembling together the incident and reflected components of the $x > R$ solution, we get

$$\boxed{\psi(x) = \frac{1}{2i}\left(-e^{-ikx} + e^{ikx+2i\delta}\right) = e^{i\delta}\sin(kx+\delta), \quad \text{for} \quad x > R.} \tag{29.1.8}$$

This is called the **canonical solution** for $x > R$. In general, given a potential $\mathcal{V}(x)$ our aim is to compute $\delta(E)$. This requires also studying the Schrödinger equation in the region $0 \le x \le R$. If $\delta = 0$, the solution $\psi(x)$ becomes $\phi(x)$. We call δ the **phase shift**, as it is certainly an extra phase in the argument of the sine function. There is an ambiguity in the definition of δ: δ and $\delta \pm \pi$ give exactly the same $\psi(x)$:

$$\delta \sim \delta \pm \pi. \tag{29.1.9}$$

Often, as mentioned above, the ambiguity is partially fixed by insisting that the phase shift be a continuous function of the energy. This is particularly important in the context of Levinson's theorem (section 29.3).

Let us compare the solutions $\sin kx$ and $\sin(kx+\delta)$, representing the waves $\phi(x)$ and $\psi(x)$ for $x > R$, up to an overall phase. A value of $\sin kx$ that we find for $x = x_0$ will be obtained for the $\sin(kx+\delta)$ wave for $kx + \delta = kx_0$, so that

$$x = x_0 - \frac{\delta}{k}. \tag{29.1.10}$$

Therefore, for small $\delta > 0$, the wave $\sin(kx+\delta) \sim \psi(x)$ is *pulled in* a distance δ/k relative to the free $\phi(x)$. We conclude that for $\delta > 0$ the potential is exerting attraction. For small $\delta < 0$, the wave $\psi(x)$ is pushed out a distance $|\delta|/k$ relative to $\phi(x)$, and the potential is exerting repulsion.

We define the *scattered wave* $\psi_s(x)$ as the extra wave in the solution $\psi(x)$ that would vanish for the case of zero potential, that is,

$$\psi(x) = \phi(x) + \psi_s(x), \qquad x > R. \tag{29.1.11}$$

Since we required ψ to have the same incident wave as ϕ, the wave ψ_s must be outgoing. Using our expressions for ψ and ϕ, we indeed find an outgoing wave,

$$\psi_s(x) = \psi(x) - \phi(x) = -\frac{e^{-ikx}}{2i} + \frac{e^{ikx+2i\delta}}{2i} + \frac{e^{-ikx}}{2i} - \frac{e^{ikx}}{2i}$$
$$= \frac{e^{ikx+2i\delta}}{2i} - \frac{e^{ikx}}{2i} = e^{i\delta}\Big(\frac{e^{i\delta}}{2i} - \frac{e^{-i\delta}}{2i}\Big)e^{ikx}. \tag{29.1.12}$$

Our result for the scattered wave is therefore

$$\psi_s(x) = e^{i\delta}\sin\delta\, e^{ikx} = A_s e^{ikx}, \quad \text{with } A_s \equiv e^{i\delta}\sin\delta. \tag{29.1.13}$$

A_s is called the *scattering amplitude*, as befits the amplitude for the scattered wave. While we can't normalize our states yet (for that we need wave packets), the probability of scattering is captured by the norm squared of A_s:

$$|A_s|^2 = \sin^2\delta. \tag{29.1.14}$$

Note that $|\psi_s(x)|^2 = \sin^2\delta$, too, with no x dependence.

Exercise 29.1. *Consider scattering with wave number k in a potential where the hard wall at $x=0$ has been moved to $x=x_0>0$, and the potential vanishes for all $x>x_0$. Determine the phase shift $\delta(k)$. Fix the ambiguity of δ by making the phase shift zero when $x_0=0$.*

29.2 Time Delay

Assume that we send a wave packet into a finite range potential. We can arrange matters so that given the velocity of the packet for $x>R$, the packet would arrive at $x=0$ at $t=0$ in the absence of a potential. For negative times, the peak x_p of the packet would move as

$$x_p = -v_0 t, \tag{29.2.1}$$

as long as $x_p > R$, since we do not know how the packet moves within the range of the potential. For the reflected wave packet, we would find, in general,

$$x_p = v_0(t - t_0), \tag{29.2.2}$$

again only valid for $x_p > R$. If $t_0 > 0$, we interpret this as a *time delay*. Indeed, if there were no potential, the incoming wave packet would reach the origin at zero time and would be reflected at zero time, the reflection obeying $x_p = v_0 t$ for $t>0$. The packet described by (29.2.2) is moving as if it emerged from $x=0$ at time $t=t_0$. If $t_0<0$, we have a negative time delay or, more properly, a time *advance*. Our goal is to compute t_0 for general wave packets and arbitrary potentials.

We will show that the energy dependence of the phase shift $\delta(E)$ determines the time delay of a reflected wave packet. Indeed, we claim that the delay is given by

$$\Delta t = 2\hbar \frac{d\delta}{dE}\Big|_{E=E_0} = 2\hbar\,\delta'(E_0), \tag{29.2.3}$$

where the derivative of δ with respect to energy is evaluated for the central energy E_0 of the superposition that builds the wave packet. In the last expression, the prime denotes the derivative with respect to the argument E of $\delta(E)$. This formula, you may recall, is completely analogous to the delay formula (8.2.18) we derived for the reflected wave when a low-energy wave packet hits a step potential. If $\Delta t < 0$ in (29.2.3), then the particle spends less time near $x=0$, either because the potential is attractive and the particle speeds up or because the potential is repulsive, and the particle bounces before reaching $x=0$. If $\Delta t > 0$, then the particle spends more time near $x=0$, typically because it slows down or gets temporarily trapped in the potential. Both less and more time are meant here in comparison to the doings of the wave packet in the absence of a potential.

We write the incident wave in the form

$$\Psi_{\text{inc}}(x,t) = \int_0^{\infty} dk\, g(k) e^{-ikx} e^{-iE(k)t/\hbar}, \quad x > R, \tag{29.2.4}$$

where $g(k)$ is a real function narrowly peaked around $k=k_0$, for which the energy is $E_0 = E(k_0)$. We write the associated reflected wave by noticing that the above is a superposition of waves as in (29.1.6), and the reflected wave must be the associated superposition of waves as in (29.1.7). We must therefore change the sign of the momentum in the exponent, change the overall sign of the amplitude, and multiply by the phase $e^{2i\delta(k)}$:

$$\Psi_{\text{ref}}(x,t) = -\int_0^{\infty} dk\, g(k) e^{ikx} e^{2i\delta(k)} e^{-iE(k)t/\hbar}, \quad x > R. \tag{29.2.5}$$

We now use the stationary phase approximation to determine the motion of the wave packet peak. We must have a stationary phase when $k=k_0$:

$$\frac{d}{dk}\left(kx + 2\delta(k) - \frac{E(k)t}{\hbar}\right)\bigg|_{k_0} = 0. \tag{29.2.6}$$

This gives

$$x + 2\frac{d\delta}{dk}\bigg|_{k_0} - \frac{dE}{dk}\bigg|_{k_0}\frac{t}{\hbar} = 0. \tag{29.2.7}$$

Using the chain rule, we find that

$$x + 2\hbar\frac{d\delta}{dE}\bigg|_{E(k_0)}\frac{1}{\hbar}\frac{dE}{dk}\bigg|_{k_0} - \frac{1}{\hbar}\frac{dE}{dk}\bigg|_{k_0} t = 0. \tag{29.2.8}$$

Evaluating the derivative dE/dk gives

$$\frac{1}{\hbar}\frac{dE}{dk}\bigg|_{k_0} = \frac{\hbar k_0}{m} \equiv v_0, \tag{29.2.9}$$

where we have called v_0 the group velocity of the wave packet. Going back to (29.2.8) and grouping terms gives

$$x + v_0\left(2\hbar\frac{d\delta}{dE}\bigg|_{E(k_0)} - t\right) = 0. \tag{29.2.10}$$

From this we finally find that the peak x_p of the reflected packet moves as

$$x_p = v_0\left(t - 2\hbar\frac{d\delta}{dE}\bigg|_{E(k_0)}\right). \tag{29.2.11}$$

If there had been no phase shift δ (because $\mathcal{V}(x)=0$), there would be no time delay, and the peak of the reflected wave packet would follow the line $x_p = v_0 t$. Therefore, the delay Δt, as claimed, is given by (29.2.3): $\Delta t = 2\hbar\,\delta'(E_0)$, where $E_0 = E(k_0)$.

In order to have an intuitive grasp for the magnitude of the time delay, we compare it to a natural timescale in the problem: the time the wave packet takes to travel the range R of the potential. For this, we first rewrite our result for Δt as follows:

$$\Delta t = 2\hbar \frac{dk}{dE}\frac{d\delta}{dk} = \frac{2}{\left(\frac{1}{\hbar}\frac{dE}{dk}\right)}\frac{d\delta}{dk} = \frac{2}{v_0}\frac{d\delta}{dk}, \tag{29.2.12}$$

where derivatives are evaluated at k_0. As a result, we have

$$\frac{d\delta}{dk} = \frac{\Delta t}{2} v_0. \tag{29.2.13}$$

Multiplying this by $\frac{1}{R}$, we have

$$\boxed{\frac{1}{R}\frac{d\delta}{dk} = \frac{\Delta t}{\left(\frac{2R}{v_0}\right)} = \frac{\text{delay}}{\text{free transit time}}.} \tag{29.2.14}$$

The left-hand side is unit-free, and the right-hand side is the ratio of the time delay to the time the free particle would take to travel in and out of the range R.

Example 29.1. *Phase shift in an attractive square well.*

In order to illustrate the concepts we have introduced and in order to learn how to compute phase shifts in a concrete situation, we examine scattering off an attractive potential:

$$V(x) = \begin{cases} -V_0, & \text{for } \; 0 < x < a, \\ 0, & \text{for } \; x > a, \\ \infty, & \text{for } \; x < 0. \end{cases} \tag{29.2.15}$$

The potential is shown in figure 29.4, with the energy $E > 0$ of the energy eigenstate indicated with a dashed line. The first step is to write an ansatz for the solution. For $x > a$, the solution must be taken equal to the canonical solution (29.1.8), which includes the unknown phase shift. We thus write

$$\psi(x) = \begin{cases} e^{i\delta}\sin(kx+\delta), & x > a, \\ A\sin(k'x), & 0 < x < a. \end{cases} \tag{29.2.16}$$

The expression for $x < a$ follows from $\psi(0) = 0$ and the potential being zero in this region. The value of the constant A is also unknown. The constants k and k' are given by

$$k^2 = \frac{2mE}{\hbar^2}, \qquad k'^2 = \frac{2m(E+V_0)}{\hbar^2}. \tag{29.2.17}$$

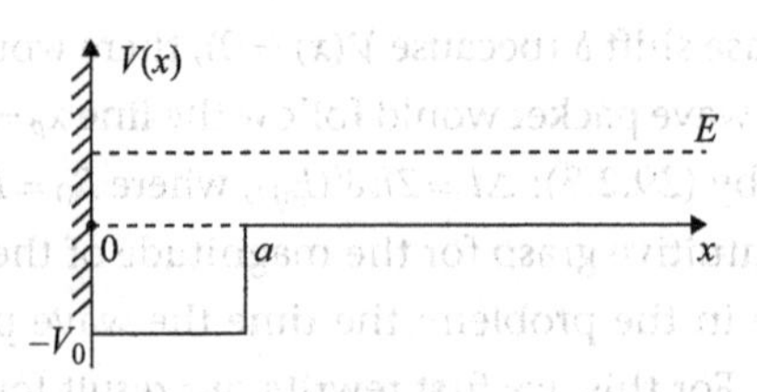

Figure 29.4
An attractive potential. We look for the energy eigenstates for all $E > 0$.

Matching ψ and ψ' at $x = a$, we find the conditions that will give us the unknown phase shift δ:

$$\begin{aligned} A\sin(k'a) &= e^{i\delta}\sin(ka+\delta), \\ k'A\cos(k'a) &= ke^{i\delta}\cos(ka+\delta). \end{aligned} \tag{29.2.18}$$

Dividing the latter equation by the former,

$$k\cot(ka+\delta) = k'\cot k'a. \tag{29.2.19}$$

We now make use of the identity

$$\cot(A+B) = \frac{\cot A\cot B - 1}{\cot A + \cot B}, \tag{29.2.20}$$

from which we find that

$$\frac{k'}{k}\cot k'a = \cot(ka+\delta) = \frac{\cot ka\cot\delta - 1}{\cot ka + \cot\delta}. \tag{29.2.21}$$

Solving this equation for $\cot\delta$ and inverting to find $\tan\delta$, we get

$$\tan\delta = \frac{1 - \frac{k'}{k}\cot k'a\tan ka}{\tan ka + \frac{k'}{k}\cot k'a}. \tag{29.2.22}$$

While this is a complicated formula to analyze directly, we can plot the phase shift as a function of $u \equiv ka$ for fixed values of V_0. As usual, we characterize the well by the unit-free constant z_0, and we can write $v \equiv k'a$ in terms of z_0 and u:

$$z_0^2 = \frac{2mV_0a^2}{\hbar^2}, \quad v^2 = z_0^2 + u^2. \tag{29.2.23}$$

Once δ is known, we can use the top equation in (29.2.18) to determine $|A|$:

$$|A| = \left|\frac{\sin(u+\delta)}{\sin v}\right|. \tag{29.2.24}$$

Figure 29.5 shows the phase factor δ, the quantity $\sin^2\delta$, the time delay $\frac{d\delta}{du} = \frac{1}{a}\frac{d\delta}{dk}$, and the amplitude $|A|$ inside the well, all as functions of $u = ka$, for $z_0 = 2.0$. Let us discuss the main features of the plot.

1. The phase δ is zero for zero energy and reaches $(-\pi)$ for infinite energy. The excursion of the phase is thus π, and as we will see in the next section, this is the case because

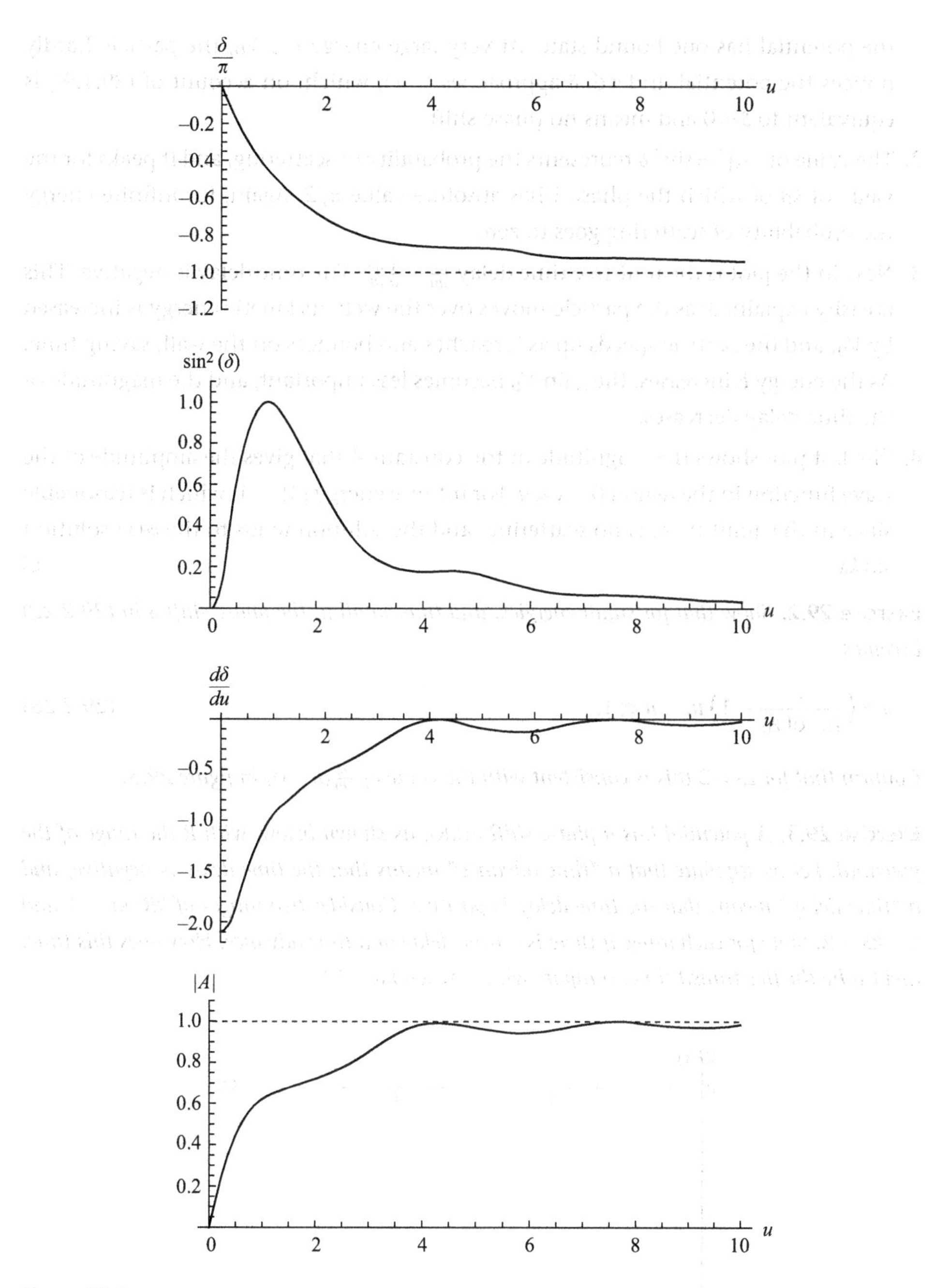

Figure 29.5
Various quantities plotted as functions of $u = ka$, for $z_0 = 2.0$. *Top*: The phase shift, going from zero to $-\pi$. *Second from top*: The scattering amplitude $|A_s|^2 = \sin^2 \delta$. *Third from the top*: The delay relative to the free transit time. *Bottom*: The norm $|A|$ of the amplitude of the wave inside the well.

the potential has one bound state. At very large energy $E \gg V_0$, the particle hardly notices the potential. Indeed, δ approaches $(-\pi)$, which, on account of (29.1.9), is equivalent to $\delta = 0$ and means no phase shift.

2. The value of $|A_s|^2 = \sin^2 \delta$ represents the probability of scattering, and it peaks for the value of ka of which the phase δ has absolute value $\pi/2$. Again, for infinite energy the probability of scattering goes to zero.
3. Next in the plot is the unit-free time delay $\frac{d\delta}{du} = \frac{1}{a}\frac{d\delta}{dk}$. The time delay is negative. This is easily explained: as the particle moves over the well, its kinetic energy is increased by V_0, and the particle speeds up as it reaches and bounces off the wall, saving time. As the energy E increases, the gain V_0 becomes less important, and the magnitude of the time delay decreases.
4. The last plot shows the magnitude of the constant A that gives the amplitude of the wave function in the region $0 < x < a$. For infinite energy $|A| \to 1$, which is reasonable since in this limit there is no scattering, and the solution tends to the $\phi(x)$ solution $\sin kx$. □

Exercise 29.2. *Show that for small energies, and thus small u, the phase shift δ in (29.2.22) becomes*

$$\delta \simeq \Big(\frac{1}{z_0 \cot z_0} - 1\Big)\, u, \quad u \ll 1. \tag{29.2.25}$$

Confirm that for $z_0 = 2$ this is consistent with the value of $\frac{d\delta}{du}(u=0)$ in figure 29.5.

Exercise 29.3. *A potential has a phase shift δ(kR), as shown below, with R the range of the potential. Let us stipulate that a "time advance" means that the time delay is negative, and a "time delay" means that the time delay is positive. Consider two ranges of kR: kR < 1 and 1 < kR < 2. State for each range if there is a time delay or a time advance. How does this time, divided by the free transit time, compare with π when kR < 1?*

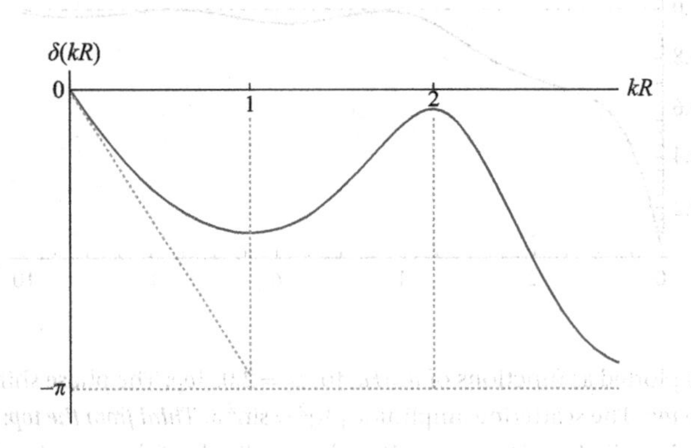

Exercise 29.4. *Consider a situation in which the scattering energy E is slightly above a constant positive potential V_0 in the region $0 < x < R$. As usual, there is a wall at $x = 0$, and the potential vanishes for $x > R$. Use semiclassical reasoning to determine if there should be a time delay or a time advance.*

29.3 Levinson's Theorem

Levinson's theorem relates the number N_b of bound states of a given potential to the excursion of the phase shift $\delta(E)$ as the energy goes from zero to infinity:

$$\boxed{N_b = \frac{1}{\pi}\left(\delta(0) - \delta(\infty)\right).} \tag{29.3.1}$$

This is a rather surprising result because the phase shift is defined for nonnormalizable energy eigenstates with $E > 0$. Somehow, however, the phase shifts encode information about the $E < 0$ bound states of the potential!

To prove this result, consider an arbitrary potential $V(x)$ of range R, with a wall at $x = 0$. This potential, shown on the left in figure 29.6, has a number of bound states, all of which are nondegenerate and can be enumerated. There is also a set of positive energy eigenstates: the scattering states that, belonging to a continuum, cannot be enumerated.

Our proof requires the ability to enumerate states, so we will introduce a second infinite wall, placed at $x = L$ for large L. This will change the spectrum, of course, but as L becomes larger and larger, the changes will become smaller and smaller. We think of L as a *regulator* for the potential that discretizes the spectrum and thus enables us to enumerate the states. It does so because with two infinite walls, the potential becomes a wide infinite well, and all energy eigenstates are bound states. There are an infinite number of bound states, some with negative energy that are regulated versions of the original bound states and some with positive energy that are regulated versions of the original continuum states. With the two walls, there are no degeneracies and no continuum; all the states can be enumerated. The potential with the regulator wall is shown on the right in figure 29.6.

The key to the proof will be to compare the counting of states in the regulated $V \neq 0$ potential to the counting of states in the $V = 0$ potential, also regulated with a second wall at $x = L$. We will be careful never to speak of, or compare, the *total* number of states in those potentials, as this number is infinite. Nor will we speak about the total number of positive energy states, which is also infinite for both potentials.

Consider first the regulated $V = 0$ potential and its eigenstates, which are all *positive* energy. These correspond to the wave function $\phi(x) = \sin kx$, with the second wall requiring $\phi(x = L) = 0$. We thus have

$$kL = n\pi, \quad \text{with} \quad n = 1, 2, \ldots. \tag{29.3.2}$$

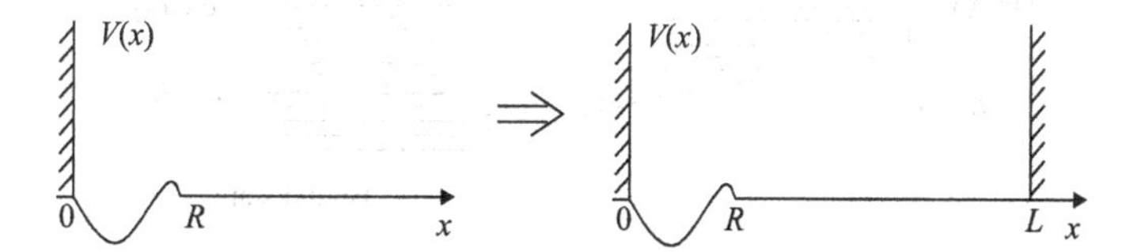

Figure 29.6
Left: An arbitrary one-dimensional potential $V(x)$ of range R. *Right*: The same potential with a regulator wall placed at $x = L$, with $L \gg R$.

The values of k are now quantized. Let dk be an infinitesimal interval in k space and dn the number of states in dk when $V=0$. Thus,

$$dk\,L = dn\,\pi \quad \Rightarrow \quad dn = \frac{L}{\pi}\,dk. \tag{29.3.3}$$

When $V(x) \neq 0$, the *positive energy* solutions are known for $x > R$ and take the form

$$\psi(x) = e^{i\delta} \sin(kx + \delta). \tag{29.3.4}$$

The boundary condition $\psi(L) = 0$ implies a quantization condition:

$$kL + \delta(k) = n'\pi, \tag{29.3.5}$$

with n' an integer. We can again differentiate to determine the number of positive energy states dn' in the interval dk, with $V \neq 0$:

$$dkL + \frac{d\delta}{dk}\,dk = dn'\pi \quad \rightarrow \quad dn' = \frac{L}{\pi}dk + \frac{1}{\pi}\left(\frac{d\delta}{dk}\right)dk. \tag{29.3.6}$$

The number of positive energy solutions *lost* in the interval dk as we turn on the potential V is given by $dn - dn'$, which can be evaluated using (29.3.3) and (29.3.6):

$$dn - dn' = -\frac{1}{\pi}\left(\frac{d\delta}{dk}dk\right). \tag{29.3.7}$$

The *total* number of positive energy solutions lost as the potential V is turned on is then given by integrating the above over the full range of k:

$$\begin{aligned}\text{Number of positive energy solutions lost as } V \text{ turns on} &= \int_0^\infty \left(-\frac{1}{\pi}\frac{d\delta}{dk}\right)dk \\ &= -\frac{1}{\pi}\left(\delta(\infty) - \delta(0)\right).\end{aligned} \tag{29.3.8}$$

Although we lose a number of positive energy solutions as the potential V is turned on, states do not disappear. Imagine changing the potential continuously from zero until it becomes V. As the potential changes, we can track each energy eigenstate without difficulty because there are never degeneracies (see figure 29.7). No state can disappear, and no state can be created. If we lose some positive energy states, those states must now appear as negative energy states, or bound states! Thus, the number N_b

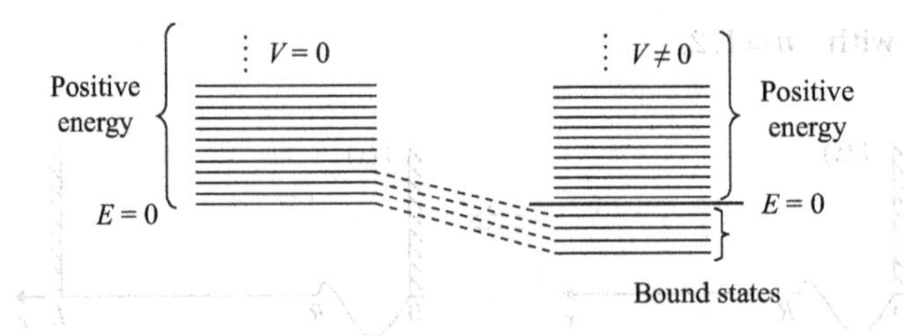

Figure 29.7
The positive energy states of the regulated $V=0$ system shift as the potential is turned on, and some can become bound states.

of bound states in the $V \neq 0$ potential must equal the number of positive energy states lost by turning on V. Using (29.3.8), this implies that

$$N_b = \frac{1}{\pi}(\delta(0) - \delta(\infty)). \tag{29.3.9}$$

This is what we wanted to prove! Note that the phase difference $\delta(0) - \delta(\infty)$ is well defined and unambiguous. We can use the ambiguity $\delta \sim \delta \pm \pi$ to fix the value of $\delta(0)$, but once we do this, we must simply allow $\delta(E)$ to be a continuous function of E. This will fix unambiguously the value of $\delta(\infty)$, making $\delta(0) - \delta(\infty)$ unambiguous.

Exercise 29.5. *We have shown (see (29.3.7)) that as the potential is turned on the change in the density of states per unit wave number at a given value of k is*

$$\frac{dn'}{dk} - \frac{dn}{dk} = \frac{1}{\pi}\frac{d\delta}{dk}. \tag{29.3.10}$$

Write the right-hand side as a factor, multiplying the time delay Δt determined earlier. At a wave number k for which a packet would experience a positive time delay, are the states with the potential V more or less closely packed than with zero potential?

29.4 Resonances

We have calculated the time delay $\Delta t = 2\hbar\delta'(E)$ associated with the reflected wave packet that emerges from a range-R potential. If the time delay is negative, the reflected wave packet emerges ahead of time. If the time delay is positive, the reflected wave packet lingers in the potential well.

We now ask: Can we get an arbitrarily large *negative* time delay? The answer is no. A very large negative time delay would be a violation of causality. It would mean that the incoming packet is reflected even before it reaches $x = R$, which is impossible. In fact, the largest negative time delay would be realized (at least classically) if we had perfect reflection when the incoming packet hits $x = R$. If this happens, the time delay would be $-\frac{2R}{v_0}$, where v_0 is the velocity of the packet. Indeed, $\frac{2R}{v_0}$ is the time saved by the packet that did not have to go in and out of the range. Thus, we expect that the

$$\text{time delay} = 2\hbar\frac{d\delta}{dE} \geq -\frac{2R}{v_0}. \tag{29.4.1}$$

This can be simplified a bit by using k derivatives:

$$2\hbar\frac{d\delta}{dE} = 2\hbar\frac{1}{\frac{dE}{dk}}\frac{d\delta}{dk} = \frac{2}{v_0}\frac{d\delta}{dk} \geq -\frac{2R}{v_0}, \tag{29.4.2}$$

which then gives the constraint

$$\frac{d\delta}{dk} \geq -R. \tag{29.4.3}$$

The argument was not rigorous, but the result is rather accurate, receiving corrections that vanish for packets of large energy.

We now ask about the other possibility: Can we get an arbitrarily large *positive* time delay? The answer is yes. This can happen if the wave packet gets trapped for a long time in the region $0 < x < R$ of the potential; in that case we have a resonance. The wave packet behaves like a bound state in that it gets localized in the potential, at least temporarily. We say we have a resonance at k_0 if the following all happen at once:

i. The eigenstate with energy $E(k_0)$ exhibits a large time delay as a result of a phase shift that grows very rapidly with energy.
ii. The rapid excursion of the phase shift δ includes the value $\pi/2 \pmod{\pi}$, resulting in a narrowly peaked, large scattering amplitude $|A_s| = |\sin\delta| = 1$ for $k \simeq k_0$.
iii. The amplitude for the wave function within the range of the potential peaks at $k \simeq k_0$.

In a resonance, the three signals above—a peak time delay, δ going through $\pi/2 \pmod{\pi}$, and peak amplitude within the well—do not occur for exactly the same value of k_0, although the differences are tiny. Sometimes, we recognize what we call a *quasi-resonance*, a value of k for which we have a modest peak of the time delay as well as a modest peak of the amplitude within the range of the potential. In particular, the phase shift may not cross $\pi/2 \pmod{\pi}$.

These ideas will be illustrated in an example below. In the following section, we will give a useful, approximate model of resonances. We will also discuss a mathematically precise definition of a resonance. This definition gives useful physical insight and makes no fundamental distinction between resonances and quasi-resonances.

For our example, we consider the usual one-dimensional wall potential $V(x) = \infty$ for $x \leq 0$, supplemented by a repulsive delta function at $x = a > 0$:

$$V(x) = g\,\delta(x-a), \quad g > 0, \quad x > 0. \tag{29.4.4}$$

We scatter particles with mass m and energy $E > 0$ off this potential (figure 29.8). We also define

$$k^2 = \frac{2mE}{\hbar^2} \quad \text{and} \quad \lambda \equiv \frac{mag}{\hbar^2},$$

with λ a unit-free measure of the strength g of the delta function.

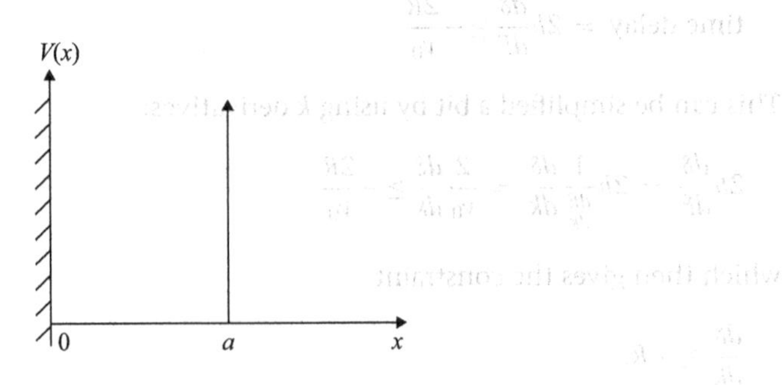

Figure 29.8
A wall plus a delta function create a region $x \in [0, a]$ in which resonances can form.

In order to work out this model, we must solve the Schrödinger equation for the energy eigenstates and calculate the phase shift $\delta(k)$. An ansatz for the energy eigenstates is easily written:

$$\psi(x) = \begin{cases} A \sin kx, & 0 \le x \le a, \\ e^{i\delta} \sin(kx+\delta), & x > a. \end{cases} \tag{29.4.5}$$

The continuity of the wave function at $x=a$ and the discontinuity of the derivative at $x=a$ impose the conditions

$$\begin{aligned} A \sin ka &= e^{i\delta} \sin(ka+\delta), \\ \frac{2mg}{\hbar^2} A \sin ka + kA \cos ka &= ke^{i\delta} \cos(ka+\delta), \end{aligned} \tag{29.4.6}$$

as you should verify. Dividing the second equation by the first, dividing the result by k, and using the definition of λ, we find that

$$\frac{2\lambda}{ka} + \cot ka = \cot(ka+\delta). \tag{29.4.7}$$

A little work is needed to put this answer in a nicer form:

Exercise 29.6. *Use the identity (29.2.20) on the right-hand side of (29.4.7), and solve for* $\tan\delta$, *showing that*

$$\tan\delta = -\frac{(\sin ka)^2}{\sin ka \cos ka + \frac{ka}{2\lambda}}. \tag{29.4.8}$$

With δ determined, we find the k-dependent normalization constant A using the continuity condition in (29.4.5):

$$|A(k)| = \left|\frac{\sin(ka+\delta)}{\sin ka}\right|. \tag{29.4.9}$$

Let us consider the behavior of the phase shift for low energies: $ka \ll 1$. In this case $\sin ka \simeq ka$, and $\cos ka \simeq 1$, giving

$$\tan\delta \simeq -\frac{ka}{1+\frac{1}{2\lambda}}, \quad ka \ll 1. \tag{29.4.10}$$

As $\lambda \to 0$, we have $\tan\delta \to 0$. This makes sense: for $\lambda \to 0$, the delta function disappears, and there is no phase shift. As $\lambda \to \infty$, we see that $\tan\delta \to -ka$. This also makes sense: as $\lambda \to \infty$, the delta function becomes infinitely strong, and it behaves like a wall at $x=a$, forcing the wave function to vanish there. Using the second line in (29.4.5), this requires $\sin(ka+\delta)=0$, which indeed implies $\delta = -ka$, up to shifts by multiples of π. The wall explanation holds in fact for arbitrary energy—that is, arbitrary ka. From (29.4.8) we see that as $\lambda \to \infty$, we get $\tan\delta \to -\tan ka$. This again implies that $\delta = -ka$.

The quickest way to see the appearance of resonances is to do some numerical work. For this we set $\lambda = 5$ and define $u \equiv ka$. In figure 29.9 we plot, for $u \in [0, 12]$, the phase

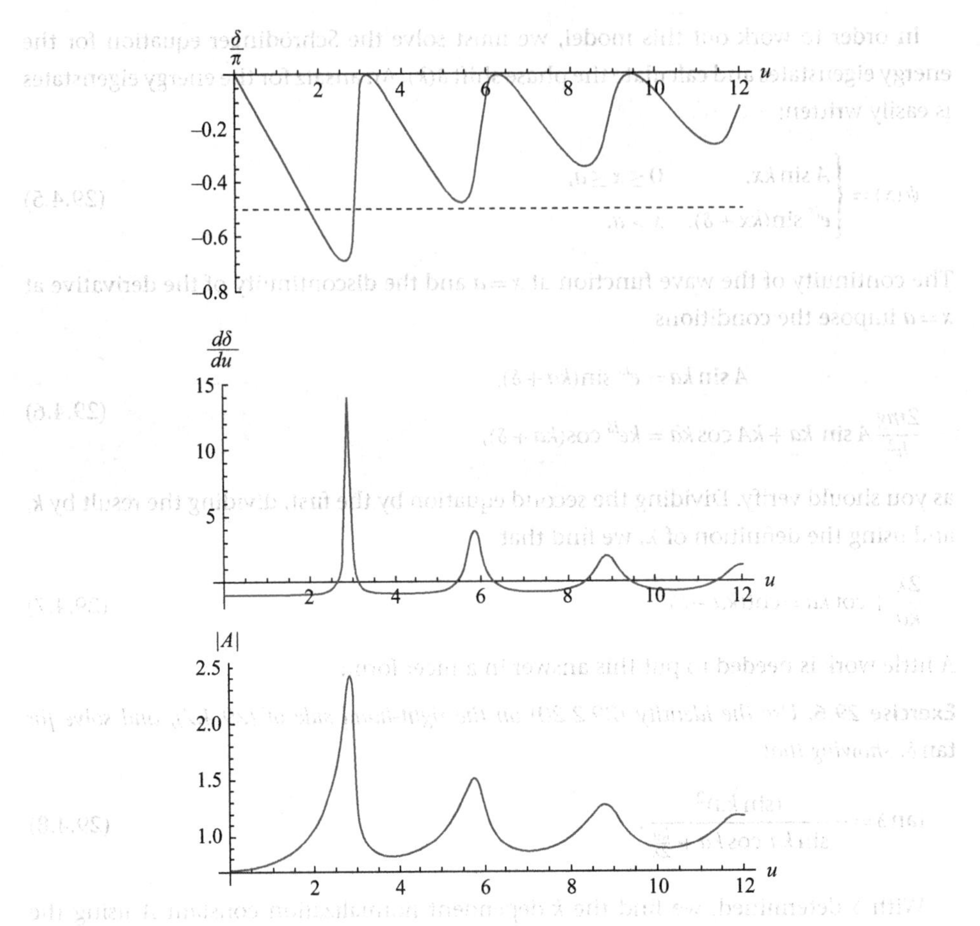

Figure 29.9
Top: The phase shift $\delta(u)$, divided by π, with $u = ka$ and $\lambda = 5$. Resonances occur at the places where the phase shift increases rapidly, with an excursion that includes the value of $\pi/2 \pmod{\pi}$. *Middle*: The unit-free time delay $\frac{d\delta}{du}$ exhibits peaks at various energies. *Bottom*: The value of $|A|$ controlling the scale of the wave function in the region $x \in [0, a]$.

shift $\delta(u)$, the unit-free time delay $d\delta/du$, and the magnitude $|A|$ of the constant that controls the amplitude of the wave function in the region between the wall and the delta function.

Consider the top figure showing $\delta(u)$. For small u, δ decreases linearly, a sign of a negative time delay, as the low-energy waves reflect off the $x = a$ delta function. As δ crosses $-\pi/2$, there is no resonance, even though $|A_s|^2 = \sin^2\delta = 1$. Indeed, since the slope of the phase shift is negative, the bottom figure shows a small time advance. Moreover, in the bottom figure we see no bump in the amplitude $|A|$ for this energy. As the energy increases, the phase shift reaches a minimum value and then begins growing with a large derivative, thus giving a resonance.

A resonance occurs at $u \simeq 2.82$, as defined by the peak in $|A|$, of magnitude $|A| \simeq 2.43$. Around this value of u, the excursion of the phase shift includes the value $-\pi/2$. Finally, we have a large time delay, and thus all the conditions for a resonance are satisfied:

$$\frac{d\delta}{du} = \frac{1}{a}\frac{d\delta}{dk} \simeq 14. \tag{29.4.11}$$

Since the range R of this potential is a, the unit-free time delay (29.2.14) is about 14, and this is the delay divided by the free transit time to go in and out. In the classical picture, we imagine the particle comes in from infinity, gets trapped between the wall and the delta function, bounces about fourteen times, and then escapes back toward infinity.

We also see quasi-resonances. The first occurs at $u \simeq 5.75$, with an amplitude $|A| \simeq 1.51$. The second occurs at $u \simeq 8.79$, with an amplitude $|A| \simeq 1.28$. For both of these, the rapid excursion of the phase shift δ does not include the point $-\pi/2$.

The plot of $\delta(u)$ is consistent with Levinson's theorem. This potential doesn't have any bound states: all energy eigenstates have positive energy and thus are not damped as $x \to \infty$. Consistent with this we see that the phase difference $\delta(0) - \delta(\infty)$ vanishes. The vanishing of $\delta(u)$ as $u \to \infty$ is strongly suggested by figure 29.9, but it is also clear from equation (29.4.8).

29.5 Modeling Resonances

It is good to be able to recognize a resonance by examining various plots of phase shifts, time delays, and wave amplitudes. We want to do better, however, and distill the general features of resonances. Additionally, it would be nice to find a concise definition of resonances, or an equation that defines them.

As a first step, we model the behavior of the phase shift near resonance. Recalling that a resonance requires δ to cross the value $\pi/2 \pmod{\pi}$ and using the equivalence $\delta \sim \delta \pm \pi$, we can choose to have δ vary from nearly zero to nearly π. We can achieve this with the following simple function:

$$\tan\delta = \frac{\beta}{\alpha - k}, \qquad \text{with} \quad \beta > 0, \qquad \alpha > 0. \tag{29.5.1}$$

Here α and β are positive constants with the units of k. To see what the phase δ does as a function of k, we first plot $\beta/(\alpha - k)$ in figure 29.10. Note that this function varies quickly in the region $k \in (\alpha - \beta, \alpha + \beta)$. The variation of the associated phase δ is shown below. To have a sharp increase in the phase, we must have $\beta \ll \alpha$.

Two relatively short calculations give us further insight:

$$\left.\frac{d\delta}{dk}\right|_{k=\alpha} = \frac{1}{\beta}, \qquad |\psi_s|^2 = \sin^2\delta = \frac{\beta^2}{\beta^2 + (k-\alpha)^2}. \tag{29.5.2}$$

The first informs us that the smaller β is the larger the resulting time delay. The second, which uses (29.1.13), gives the normsquared of the scattering amplitude as a function of k, with a peak at $k = \alpha$. Recall that $|\psi_s|^2 = |A_s|^2$ has no spatial dependence. The equation for $|\psi_s|^2$ is most famously expressed in terms of the energy. For this define E_α to be the energy for $k = \alpha$, and note that

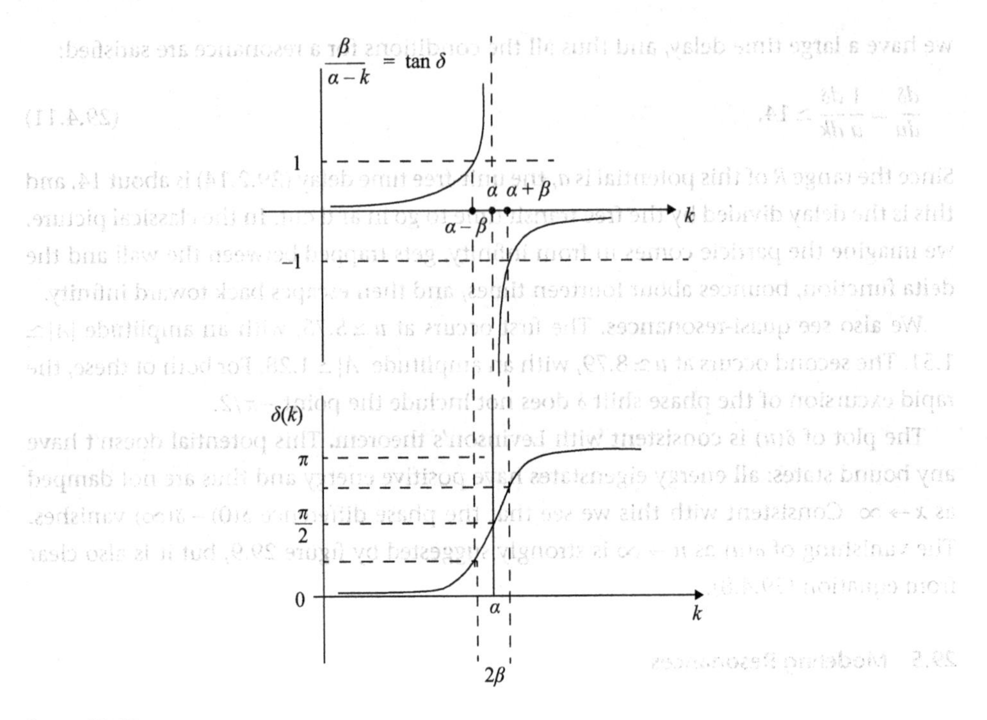

Figure 29.10
The constant β must be small compared to α to obtain a sharp variation. A resonance, as shown here, requires δ increasing rapidly with energy in order to have a large time delay.

$$E - E_\alpha = \frac{\hbar^2}{2m}(k^2 - \alpha^2) = \frac{\hbar^2}{2m}(k+\alpha)(k-\alpha) \simeq \frac{\hbar^2}{2m}(2\alpha)(k-\alpha), \tag{29.5.3}$$

when working with $k \approx \alpha$. It thus follows that

$$(k-\alpha)^2 \simeq \frac{m^2}{\hbar^4\alpha^2}(E-E_\alpha)^2, \tag{29.5.4}$$

and therefore,

$$|\psi_s|^2 \simeq \frac{\beta^2}{\beta^2 + \frac{m^2}{\hbar^4\alpha^2}(E-E_\alpha)^2} = \frac{\frac{1}{4}\Gamma^2}{(E-E_\alpha)^2 + \frac{1}{4}\Gamma^2}, \tag{29.5.5}$$

where we have defined the constant Γ with units of energy:

$$\tfrac{1}{4}\Gamma^2 = \frac{\hbar^4\beta^2\alpha^2}{m^2} \quad \Rightarrow \quad \Gamma = \frac{2\alpha\beta\hbar^2}{m}. \tag{29.5.6}$$

The energy dependence of the squared scattering amplitude $|\psi_s|^2$ near resonance follows the *Breit-Wigner distribution*:

$$\boxed{|\psi_s|^2 \simeq \frac{\frac{1}{4}\Gamma^2}{(E-E_\alpha)^2 + \frac{1}{4}\Gamma^2}.} \tag{29.5.7}$$

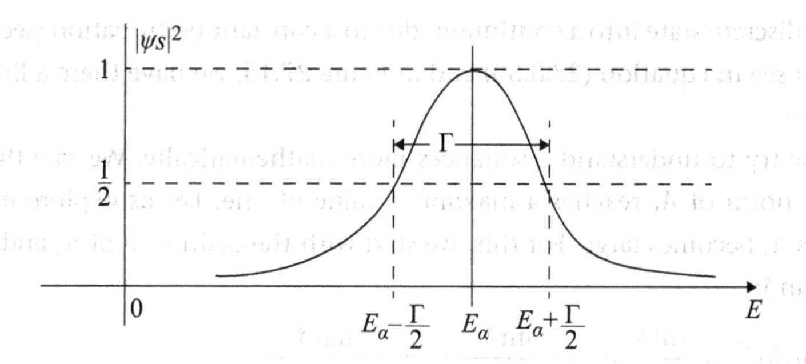

Figure 29.11
The Breit-Wigner distribution. Γ is the width of the distribution at half maximum.

The distribution is shown in figure 29.11. The peak value for $|\psi_S|^2$ is attained for $E = E_\alpha$ and is one. We call Γ the *width at half maximum* because the value of $|\psi_S|^2$ at $E = E_\alpha \pm \frac{1}{2}\Gamma$ is one-half. Small Γ corresponds to a narrow width, or a narrow resonance.

To better understand the significance of Γ, we define the associated time τ, the **lifetime** of the resonance:

$$\tau \equiv \frac{\hbar}{\Gamma} = \frac{m}{2\alpha\beta\hbar}. \tag{29.5.8}$$

As you probably would expect, the lifetime is closely related to the time delay for a wave packet of mean energy equal to the resonant energy. Indeed, we can evaluate the time delay Δt for $k = \alpha$ to get

$$\Delta t = 2\hbar\frac{d\delta}{dE} = 2\hbar\frac{dk}{dE}\bigg|_{k=\alpha}\frac{d\delta}{dk}\bigg|_{k=\alpha} = \frac{2\hbar}{\left(\frac{\hbar^2\alpha}{m}\right)}\left(\frac{1}{\beta}\right) = \frac{2m}{\alpha\beta\hbar} = 4\tau, \tag{29.5.9}$$

where we made use of the first equation in (29.5.2). We therefore conclude that the lifetime and the time delay are the same quantity, up to a factor of four:

$$\tau = \frac{\hbar}{\Gamma} = \tfrac{1}{4}\Delta t. \tag{29.5.10}$$

Unstable particles are sometimes called resonances. The Higgs boson, discovered in 2012, is an unstable particle with mass 125 GeV. It can decay into two photons, or into two taus, or into a $b\bar{b}$ pair, among several possibilities. The width Γ associated with the particle is 4.07 MeV (±4%). Its lifetime τ is about 1.62×10^{-22} seconds! You can confirm this number by working out the following exercise.

Exercise 29.7. *From $\tau = \hbar/\Gamma$, find the pure number N in the expression*

$$\tau = \frac{N}{(\Gamma/\text{MeV})}s, \tag{29.5.11}$$

which allows a quick calculation of τ in seconds when we know Γ in MeV.

The Breit-Wigner distribution also governs the energy dependence of the decay products of an unstable particle or resonance. You may recall our calculation showing the decay of a discrete state into a continuum due to a constant perturbation (section 27.8). As you can see in equation (27.8.30) and in figure 27.15, we have there a Breit-Wigner distribution.

We now try to understand resonances more mathematically. We saw that at resonance the norm of A_s reaches a maximum value of one. Let us explore under what conditions A_s becomes large. For this, we start with the definition of A_s and write it in terms of $\tan\delta$:

$$A_s = \sin\delta\, e^{i\delta} = \frac{\sin\delta}{e^{-i\delta}} = \frac{\sin\delta}{\cos\delta - i\sin\delta} = \frac{\tan\delta}{1 - i\tan\delta}. \tag{29.5.12}$$

At resonance $\delta = \pi/2$, and $A_s = i$, using the first equality. On the other hand, while we usually think of δ as a real number, the final expression above indicates that A_s becomes infinite when

$$\tan\delta = -i. \tag{29.5.13}$$

Given that $\tan iz = i\tanh z$, we deduce that the above condition is satisfied for $\delta \to -i\infty$, a rather strange result. At any rate, A_s becomes infinite, or has a pole, at $\tan\delta = -i$. We will see that the large value $|A_s| = 1$ at resonance can be viewed as the "shadow" of the infinite value A_s reaching nearby in the *complex k plane*. Since real values of k give real values of δ, the only way to get complex values of δ is to explore complex values of k.

Indeed, we can see how A_s behaves near resonance by inserting the near-resonance behavior (29.5.1) of δ into the last expression for A_s in (29.5.12):

$$A_s(k) = \frac{\frac{\beta}{\alpha - k}}{1 - i\frac{\beta}{\alpha - k}} = \frac{\beta}{(\alpha - i\beta) - k}. \tag{29.5.14}$$

At the resonant wave number $k = \alpha$, we get $A_s = i$, as expected. If we now think of the wave number k as a complex variable, we see that A_s has a pole at $k = k_* = \alpha - i\beta$. The real part of k_*, equal to α, encodes the resonant energy. The imaginary part of k_*, equal to β, together with α, encodes the lifetime τ (see (29.5.9)). For small β the resonance is a pole near the real axis, as shown in figure 29.12. The smaller β is, the sharper the resonance and the longer the lifetime. As we can now understand, the value of $|A_s|$ on the real line becomes large for $k = \alpha$ because it is actually infinite a little below the axis.

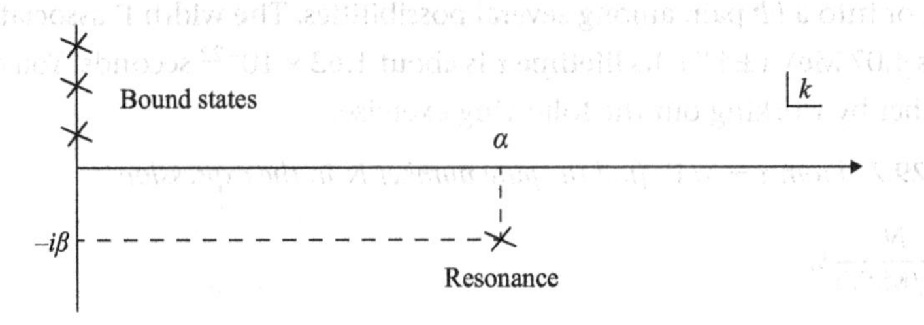

Figure 29.12
In the complex k plane, resonances are poles of the scattering amplitude A_s located slightly below the real axis, here at $k = \alpha - i\beta$. Bound states appear as poles on the positive imaginary axis.

The lesson in all of this is that we can indeed take (29.5.13) seriously and look for resonances by solving for the *complex* k values for which $\tan\delta = -i$:

$$\boxed{\text{Resonance condition: } \tan\delta(k) = -i.} \tag{29.5.15}$$

The real part of those k's give us the resonant energies. The imaginary part then fixes for us the width of the resonances, or their lifetimes.

The idea of a complex k plane is very powerful. Suppose we consider purely imaginary k values of the form $k = i\kappa$, with $\kappa > 0$. Then the energy takes the form

$$E = -\frac{\hbar^2\kappa^2}{2m} < 0, \tag{29.5.16}$$

which is suitable for bound states. Indeed, one can show (problem 29.7) that bound states appear as poles of A_s along the *positive* imaginary axis, as in figure 29.11. The complex k plane has room to fit scattering states, resonances, and bound states!

Example 29.2. *Complex k for the resonance in the delta function model.*
Let us now see how to use the condition $\tan\delta = -i$ to identify the resonance in the delta function model considered in the previous section. Writing $u = ka$, the expression for $\tan\delta$ in (29.4.8) becomes

$$\tan\delta = -\frac{(\sin u)^2}{\sin u\,\cos u + \frac{u}{2\lambda}} = \frac{\cos 2u - 1}{\frac{u}{\lambda} + \sin 2u}. \tag{29.5.17}$$

Just as in the example considered in figure 29.9, we set $\lambda = 5$, and thus the condition for a resonance becomes

$$\frac{\cos 2u - 1}{\frac{u}{5} + \sin 2u} + i = 0. \tag{29.5.18}$$

In the previous analysis, we found a resonance for $u \simeq 2.82$, so we suspect that the complex value of u satisfying the resonance condition must have a real part close to that value. The above equation can be directly solved in an algebraic manipulator. One indeed finds a solution near the expected position:

$$u = u_{\text{re}} + iu_{\text{im}} \simeq 2.87758 - 0.066511\,i. \tag{29.5.19}$$

This result allows us to compute the lifetime τ of the resonance. We note that the units $(a, m, \hbar)$ of the setup determine a timescale $\tau_0 = ma^2/\hbar$, and therefore τ must be proportional to τ_0 up to constants we want to determine. Recall also that $u = ka = (\alpha - i\beta)a = u_{\text{re}} + iu_{\text{im}}$. We thus have, starting from (29.5.9),

$$\tau = \frac{m}{2\alpha\beta\hbar} = -\frac{ma^2}{2u_{\text{re}}u_{\text{im}}\hbar} = -\frac{1}{2u_{\text{re}}u_{\text{im}}}\,\tau_0. \tag{29.5.20}$$

For the resonance in question, the values in (29.5.19) result in $\tau \simeq 2.61\,\tau_0$. The time delay Δt is four times this value: $\Delta t \simeq 10.45\,\tau_0$. □

Problems

Problem 29.1. *Testing Levinson's theorem in an example.*

We explore further scattering off the potential in figure 29.4: $V(x) = -V_0$ for $0 < x < a$, $V(x) = 0$ for $x > a$, and $V(x) = \infty$ for $x < 0$. The phase shift δ was given in (29.2.22). We recall the definitions $k^2 = 2mE/\hbar^2$, $k'^2 = 2m(E + V_0)/\hbar^2$, and $z_0^2 = 2mV_0a^2/\hbar^2$.

1. As $E \to 0$, we have $ka \to 0$. What happens to $k'a$? Find the limit of $\tan\delta$ as $E \to 0$ (answer in terms of z_0).
2. What is the limit of $\tan\delta$ as $E \to \infty$?
3. Let $u \equiv ka$, and write $\tan\delta$ as a function of u and z_0.
4. Numerically compute the phase shift in order to plot $\delta(u; z_0)$ over a range of u for a given z_0. Use Levinson's theorem and your plots to find the number of bound states for $z_0 = 2, 5, 9$. You will have to change the range of u to see all appropriate features; for $z_0 = 5$, for example, $u \in [0, 50]$ is appropriate.

Problem 29.2. *Scattering off a step and a wall.*

Consider the potential $V(x)$ given by

$$V(x) = \begin{cases} V_0, & 0 < x < a, \quad V_0 > 0, \\ 0, & x > a, \\ \infty, & x \leq 0. \end{cases}$$

Our goal is to calculate the phase shift δ as a function of $k \equiv \sqrt{\frac{2mE}{\hbar^2}}$.

1. Compute $\delta(k)$ for $E(k) > V_0$. You may want to do this, for practice, starting from the beginning. Alternatively, you could try to just modify the answer for a *well* of depth V_0. Give your answer in the form $\tan\delta = \ldots$, and write the right-hand side in terms of k and k', where k' is the wave number for $x < a$: $k'^2 \equiv 2m(E - V_0)/\hbar^2$.
2. Compute $\delta(k)$ for $E(k) < V_0$. Again, you can do this from the beginning if you'd like the practice. Alternatively, you can try to employ an analytic continuation of the result for (1). Give your answer in the form $\tan\delta = \ldots$, and write the right-hand side in terms of k and κ where $\kappa^2 \equiv 2m(V_0 - E)/\hbar^2$.
3. Define $z_0^2 \equiv 2mV_0a^2/\hbar^2$ and $u \equiv ka$. When $E(k) > V_0$, what is the relationship between k and k'? When $E(k) < V_0$, what is the relationship between k and κ? When $E = V_0$, what is the value of k?
4. Let $z_0 = 5$, and do a numerical plot of $\delta(u)$ with $u = ka$ varying from zero up to some large number. Based on the plot you've generated, does the phase shift satisfy Levinson's theorem? Using your formula for $\delta(u)$, what is the phase shift when $E = 2V_0$? Make sure it agrees with your plot! Give your answer to five significant figures.

Problem 29.3. *Features of the Breit-Wigner distribution.*

1. The phase shift δ for a resonance was modeled by writing $\tan\delta = \frac{\beta}{\alpha - k}$, leading to a $|\psi_s|^2$ that obeys an approximate Breit-Wigner distribution. Calculate the ratio β/α in terms of the parameters Γ and E_α of the distribution. Confirm that small β/α means a sharp resonance.
2. Calculate the position of the inflection points in the Breit-Wigner distribution (answer in terms of E_α and Γ). Are the inflection points within the width at half maximum?

Problem 29.4. *Resonance in an attractive potential and a barrier.*

To get a resonance, it helps to have an attractive potential and a barrier:

$$V(x) = \begin{cases} \infty, & \text{for } x \leq 0, \\ -V_0, & \text{for } 0 < x < a, \\ V_1, & \text{for } a < x < 2a, \\ 0\,, & \text{for } x > 2a. \end{cases}$$

The potential, with $V_0, V_1 > 0$, is shown in figure 29.13. To search for resonances, we explore energies in the range 0 to V_1.

Given the three relevant regions in the potential, we define

$$k'^2 = \frac{2m(E+V_0)}{\hbar^2}, \qquad \kappa^2 = \frac{2m(V_1 - E)}{\hbar^2}, \qquad k^2 = \frac{2mE}{\hbar^2}$$

and write the following ansatz for the energy eigenstate:

$$\psi(x) = \begin{cases} A\sin(k'x), & 0 < x < a, \\ A\sin(k'a)\cosh\kappa(x-a) + B\sinh\kappa(x-a), & a < x < 2a, \\ e^{i\delta}\sin(kx+\delta), & x > 2a, \end{cases}$$

which implements continuity at $x = a$. Show that after implementing the remaining boundary conditions, one finds that

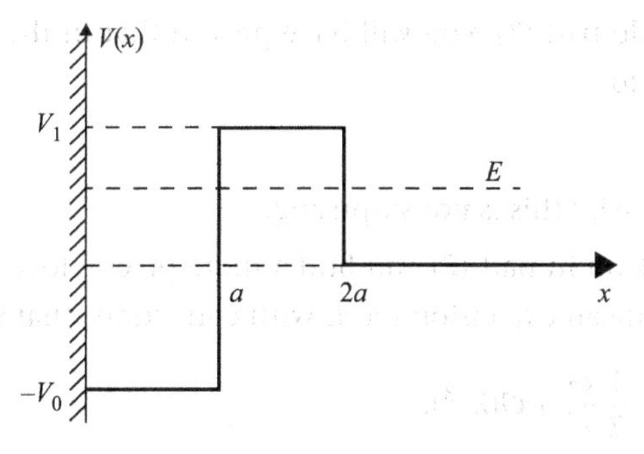

Figure 29.13
Problem 29.4: We search for resonances with energy E in the range $(0, V_1)$.

$$\tan(2ka+\delta)=\frac{ka}{\kappa a}\cdot\frac{\sin k'a\cosh\kappa a+\frac{k'}{\kappa}\cos k'a\sinh\kappa a}{\sin k'a\sinh\kappa a+\frac{k'}{\kappa}\cos k'a\cosh\kappa a}. \quad (1)$$

This expression is fairly intricate, so it is best to do numerical work. For this we define $z_0^2=2mV_0a^2/\hbar^2$, and $z_1^2=2mV_1a^2/\hbar^2$. The energy E can then be written as $E=eV_1$, with $e=(u/z_1)^2$. These allow us to express both $k'a$ and κa as functions of e, z_0, and z_1. At this point (1) can be used to determine δ as a function of e and the constants z_0, z_1. In the following, work with $z_0^2=1$ and $z_1^2=6$. Construct three plots, as functions of the unit-free energy $e\in[0,1]$, showing δ, $\sin^2\delta$, and the value $|A|$ controlling the amplitude of the wave function in the region $x\in[0,a]$. Confirm that the phase exhibits resonant behavior, crossing the $-\pi/2$ value for $E\simeq 0.624V_1$, and that the peak value of the amplitude inside the well is $|A|\simeq 4.41$.

Problem 29.5. *Resonant poles of the delta function potential.*

In this problem we investigate further the resonances of the delta function plus the wall potential studied in section 29.4.

1. For a resonance at $k=\alpha$, the unit-free delay $\mathcal{T}$ is given by

$$\mathcal{T}\equiv\frac{\text{time delay}}{\text{free transit time}}=\frac{1}{a}\frac{d\delta}{dk}\Big|_{k=\alpha},$$

where a is the range of the potential. With the behavior of the phase shift near resonance modeled as $\tan\delta=\beta/(\alpha-k)$, determine $\mathcal{T}$ in terms of some or all of the constants α, β, and a.

Recall that the phase shift can be put in the form (29.5.17):

$$\tan\delta=\frac{\cos 2u-1}{\frac{u}{\lambda}+\sin 2u},\quad u=ka, \quad (1)$$

with $\lambda=mag/\hbar^2>0$. Resonances occur for complex values of u such that $\tan\delta=-i$. We are interested in the first resonance u_*, the one with the lowest real part, when λ is large. We will write

$$u_*=k_*a=(\alpha-i\beta)a,\quad \alpha,\beta\in\mathbb{R}. \quad (2)$$

2. By the time you do part (3), you will have proven that in the limit $\lambda\to\infty$ the first resonance occurs for

$$u_*=k_*a=\pi.$$

Explain *physically* why this is not surprising.

3. To confirm the claim in part (2) and find a more precise location of the resonance for large λ, we write an expansion for u_* with corrections that vanish when λ is big:

$$u_*=\pi+\frac{1}{2}\frac{c_1}{\lambda}+\frac{1}{2}\frac{c_2}{\lambda^2}+\mathcal{O}(\lambda^{-3}). \quad (3)$$

Here c_1 and c_2 are unknown numbers, possibly complex. Consider the equation for resonance, and use the above expression for u_* to evaluate all terms in $\tan\delta$ to order $1/\lambda^2$ accuracy.

i. Determine the values of c_1 and c_2.

ii. Give the values of αa and βa (see equation (2)).

iii. What is the time delay $\mathcal{T}$ in terms of λ and numerical constants? Is the (true) resonant wave number for large λ a little above or below π/a?

Problem 29.6. *Bound states and phase shift for an attractive delta function potential.*

Consider a potential with a hard wall at $x=0$ and an attractive delta function for $x=a$:

$$V(x)=\begin{cases}-g\delta(x-a), & \text{for} \quad x>0,\\ \infty, & \text{for} \quad x<0.\end{cases}$$

Here $g>0$, and we note the unit-free combination $\lambda=\frac{mag}{\hbar^2}\geq 0$ that represents the effective strength of the potential.

1. Write an ansatz for a bound state with energy $E<0$. Define $\xi=\kappa a$, and $\kappa^2=-2mE/\hbar^2$. Find the equation that relates ξ and λ. Show that it takes the form

$$\frac{\xi}{1-e^{-\#\xi}}=\lambda,$$

with # a (positive) constant you must determine. Verify that the above left-hand side is a monotonically increasing function of ξ. What is the minimum value of λ needed to have a bound state?

2. If we now consider scattering states, a calculation (almost identical to that for the positive-sign delta function) gives the following expression for the phase shift associated with this potential:

$$\tan\delta = -\frac{\sin^2 ka}{\sin ka\cos ka-\frac{ka}{2\lambda}}.$$

What is the approximate value of $\tan\delta$ for very small ka? How does $\tan\delta$ change as λ crosses the critical value needed to have a bound state? When does one get a delayed reflection, and when does one get an advanced reflection?

3. For $\lambda=1$, figure 29.14 shows $\tan\delta$ as a function of $ka\in[0,14]$. Use this plot to sketch δ as a function of ka. Explain how your answer is consistent with Levinson's theorem. Do you see resonances?

Problem 29.7. *Bound states as poles on the complex k plane along the imaginary axis.*

Consider a finite-range, one-dimensional potential $V(x)$ that is infinite for $x<0$ and zero for $x>R$. The canonical wave solution for $x>R$ is given by $\psi(x)=e^{i\delta}\sin(kx+\delta)$, where $E=\hbar^2k^2/2m$, and δ is a function of k. If $V(x)=0$ for $x>0$, then the wave function will be exactly $\phi(x)=\sin kx$. The *scattered* wave function $\psi_s(x)$ is the difference $\psi_s(x)=\psi(x)-\phi(x)=A_se^{ikx}$, where $A_s=e^{i\delta}\sin\delta$ is the scattering amplitude. We now investigate how bound states appear as poles of A_s along the imaginary axis.

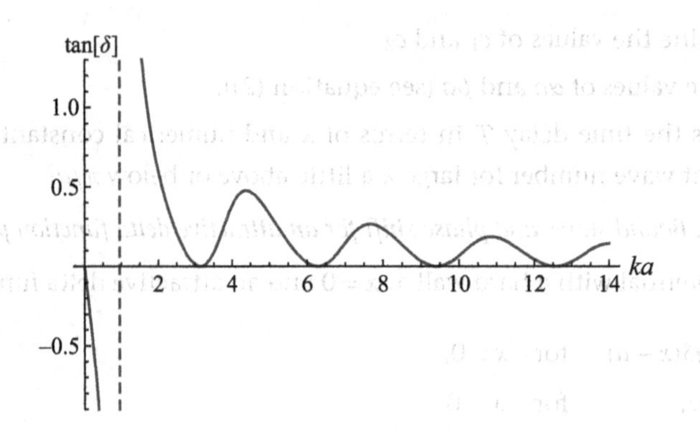

Figure 29.14
Problem 29.6: Plot of $\tan\delta$ as a function of ka.

1. If $A_s \to \infty$ at some complex k, which of the following two quantities becomes large at that value of k: $e^{2i\delta}$ or $e^{-2i\delta}$? Write $\psi(x)$ in terms of complex exponentials in the following form:

$$\psi(x) = \frac{e^{2i\delta}}{2i}(\cdots - \cdots).$$

2. To see how bound states are represented by A_s poles on the imaginary k-axis, set $k = \pm i\kappa$ with $\kappa > 0$. Both signs correspond to an energy

$$E = \frac{\hbar^2 k^2}{2m} = -\frac{\hbar^2 \kappa^2}{2m} < 0,$$

which is what would be required for a bound state to occur. We would like to understand which sign option works. To investigate this, rewrite a scaled version of the wave function $\psi(x)$ in terms of κ using $k = \pm i\kappa$ to obtain two options:

$$2ie^{-2i\delta}\psi_{\pm}(x) = \cdots - \cdots.$$

3. Assume that $A_s \to \infty$, and examine the scaled wave functions $2ie^{-2i\delta}\psi_{\pm}(x)$. Identify which of the wave functions is appropriate for the description of bound states. Are bound states poles of A_s in the positive or negative imaginary axis of the complex k plane?

30 Scattering in Three Dimensions

We consider the elastic scattering of particles interacting via a translationally invariant potential. Working in the center-of-mass frame, this requires the construction of positive-energy eigenstates with asymptotic behavior consistent with incoming particles moving along a fixed direction and then scattered radially by the target. The main observables in scattering are differential and total cross sections, which quantify the ability of the target to scatter the incoming particles. For central potentials, the cross sections are determined in terms of phase shifts in partial waves labeled by the value of the orbital angular momentum. The time-independent Schrödinger equation is recast as an integral equation that naturally implements the asymptotic behavior of the energy eigenstates. This equation leads to the Born approximation to calculate scattering, which is accurate for weak potentials and for general potentials at high energies.

In high-energy physics experiments, a beam of particles hits a target composed of particles. By detecting the by-products of the collision, one aims to study and understand the interactions that occur during the collision. Collisions of particles can be rather intricate. The particles involved may be elementary, as is the case when electrons and positrons collide. The particles can also be composite, as is the case when protons collide with protons, because a proton is not an elementary particle; it is composed of quarks and gluons. Some examples of collisions are

$$\begin{aligned} p+p &\rightarrow p+p+\pi^0, \\ p+p &\rightarrow p+n+\pi^+, \\ e^+ + e^- &\rightarrow \mu^+ + \mu^-. \end{aligned}$$

As you can see, the final set of particles may not be the same as the initial set of particles. Moreover, identical sets of incoming particles can give rise to different outgoing particles, as is illustrated by the first two lines. Particles can be created or destroyed, but of course, not everything is possible. Outcomes are restricted by conservation laws. Charge conservation is familiar, and all the above processes conserve electric charge. Less familiar, but still true as far as we know, are baryon number and lepton number conservation. The proton and the neutron have baryon number one, while the pions

π^+, π^-, π^0 have baryon number zero. Thus, the first two collisions conserve baryon number. The electron e^- and the positron e^+ have opposite lepton numbers; so do the muon μ^- and the antimuon μ^+. The last reaction conserves lepton number. The proton, neutron, and pions have zero lepton number, and the electron and the muon and their antiparticles have zero baryon number. Thus, the first two reactions trivially conserve lepton number, and the last one trivially conserves baryon number.

We have *scattering* when the particles in the initial and final state are the same:

$$a+b \rightarrow a+b.$$

The scattering is *elastic* if none of the internal states of the particles change in the collision. The famous Frank-Hertz experiment involved inelastic scattering; collisions of electrons with mercury atoms in which the mercury atoms were excited. This experiment revealed the existence of energy levels in atoms.

Our focus here will be on the *elastic scattering* of particles. For simplicity, we will assume that the particles have *no spin*. Moreover, we will work in the *nonrelativistic* approximation. We will also assume that the interaction potential $V(\mathbf{r}_1, \mathbf{r}_2)$ is translational invariant:

$$V(\mathbf{r}_1, \mathbf{r}_2) = V(\mathbf{r}_1 - \mathbf{r}_2). \tag{30.0.1}$$

As shown when we considered the hydrogen atom, it follows that in the center-of-mass frame the problem reduces to the scattering of a single particle of reduced mass m off a potential $V(\mathbf{r})$. Finally, we will work with energy eigenstates and will not attempt to justify our construction using wave packets. This, however, would not take much effort.

30.1 Energy Eigenstates for Scattering

We are interested in positive-energy eigenstates of a Hamiltonian $\hat{H}$ of the form

$$\hat{H} = \frac{\hat{\mathbf{p}}^2}{2m} + V(\mathbf{r}). \tag{30.1.1}$$

Here the mass m is the reduced mass. The potential $V(\mathbf{r})$ will be taken to be of finite range: it vanishes for $r > a$ for some constant a. If not of finite range, the potential should vanish as $r \to \infty$ faster than $1/r$. This does not include the Coulomb potential, which is in fact a long-range potential. The analysis of this potential requires some care; much but not all of what we will discuss applies in that case.

The energy $E \geq 0$ of the $\hat{H}$ eigenstates is the energy of a particle far away from the potential, where it is effectively free. We can therefore identify the energy as the kinetic energy of a free particle. We will thus write

$$E = \frac{\hbar^2 k^2}{2m} \geq 0, \tag{30.1.2}$$

and we can think of k as the wave number of the particle far away from the potential. Writing the energy eigenstates as

$$\Psi(\mathbf{r}, t) = \psi(\mathbf{r}) e^{-iEt/\hbar}, \tag{30.1.3}$$

the time-independent Schrödinger equation reads

$$\left[-\frac{\hbar^2}{2m}\nabla^2+V(\mathbf{r})\right]\psi(\mathbf{r})=E\,\psi(\mathbf{r}). \tag{30.1.4}$$

With our writing of the energy E in terms of k, we find that

$$\left[-\frac{\hbar^2}{2m}\left(\nabla^2+k^2\right)+V(\mathbf{r})\right]\psi(\mathbf{r})=0. \tag{30.1.5}$$

This equation has infinitely many solutions. Not all of them represent scattering. In order to search for scattering solutions, we must input some physical ideas. We must use our intuition about the nature of the wave function far away from the potential.

In order to motivate our setup, let us recall the case of scattering on the real line $x\in(-\infty,\infty)$. Assume the potential has range a so that it vanishes for $|x|>a$. If we had an incident wave packet from minus infinity, we would expect a reflected wave packet and a transmitted wave packet. Thus, for energy eigenstates we expect three waves to exist for $|x|>a$: incident, reflected, and transmitted. The reflected and transmitted waves can be thought of as scattered waves, arising from the incident one. The energy eigenstates relevant for scattering will have an incoming wave Ae^{ikx} as well as a reflected wave Be^{-ikx} for $x<-a$ so that

$$\psi(x)\simeq Ae^{ikx}+Be^{-ikx},\qquad \text{for } x<-a. \tag{30.1.6}$$

The eigenstate must also have a transmitted wave for $x>a$:

$$\psi(x)\simeq Ce^{ikx},\qquad \text{for } x>a. \tag{30.1.7}$$

The waves are shown schematically in figure 30.1. Note that the waves defining the energy eigenstate are set up for $|x|>a$. The energy eigenstate is such that B and C are determined in terms of A, once we solve the Schrödinger equation in the region $|x|<a$.

Let us now return to our case of interest, the solutions of (30.1.5). For any fixed energy E, we expect an infinite degeneracy of energy eigenstates. This is obvious when $V(\mathbf{r})\equiv 0$. In that case $e^{i\mathbf{k}\cdot\mathbf{x}}$ is a solution for any $\mathbf{k}$ such that $\mathbf{k}\cdot\mathbf{k}=k^2$. Assume now that the potential $V(\mathbf{r})$ is of finite range a and that the incident wave is moving in the z-direction toward larger z. The wave function, written as $\varphi(\mathbf{r})$, will look like

$$\varphi(\mathbf{r}) = e^{ikz}. \tag{30.1.8}$$

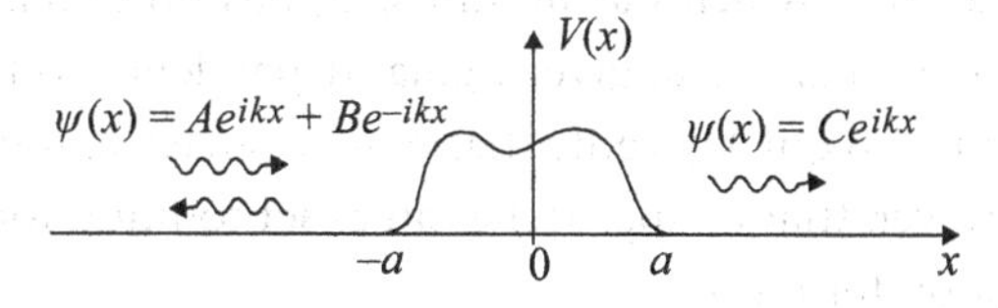

Figure 30.1
The ansatz for an energy eigenstate relevant to scattering in one dimension is written for the region $|x|>a$.

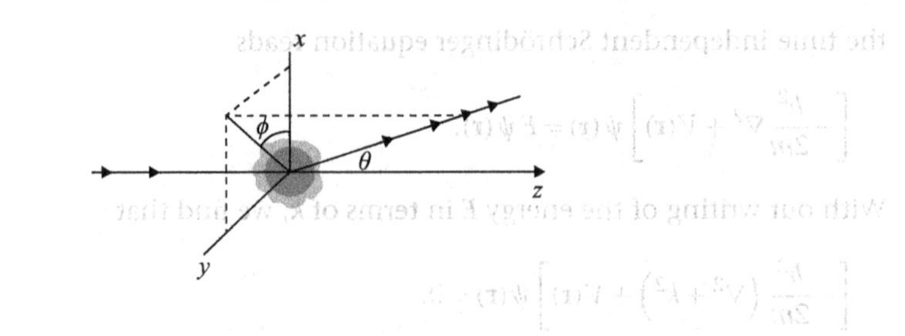

Figure 30.2
Particles incident from the negative z-direction will scatter in all directions defined by the spherical angles θ and ϕ.

This wave function cannot be normalized, and thus we will take it to have no units. Since $V(\mathbf{r})$ is of range a, the wave $\varphi(\mathbf{r})$ satisfies (30.1.5) for any $r > a$. For $r < a$, however, it does not satisfy the equation; $\varphi(\mathbf{r})$ is a solution only where the potential vanishes.

Given an incident wave, we will also have a scattered wave. We should expect the scattered wave to propagate radially out. Could it be a spherical wave of the form $\psi(\mathbf{r}) = e^{ikr}$? It cannot, because

$$\left(\nabla^2 + k^2\right) e^{ikr} \neq 0 \quad \text{for } r \neq 0. \tag{30.1.9}$$

On the other hand, $\psi(\mathbf{r}) = e^{ikr}/r$ would work:

$$\left(\nabla^2 + k^2\right) \frac{e^{ikr}}{r} = 0, \quad \text{for } r \neq 0. \tag{30.1.10}$$

This is consistent with the radial equation having a solution $u(r) = e^{ikr}$ in the region where the potential vanishes. Recall that the full radial solution takes the form $u(r)/r$. Note that $\psi(\mathbf{r}) = e^{ikr}/r$ fails to be a solution for $r = 0$, as the Laplacian acting on it contains a delta function. In our scattering setup, $\psi(\mathbf{r}) = e^{ikr}/r$ is only a solution for $r > a$.

Can the scattered wave therefore be e^{ikr}/r for $r > a$? Yes, but this is not general enough, as there could be some angular dependence. Since the particles approach the target from a fixed direction, defined to be the z-direction, the amplitude of the scattered wave would be expected to depend on the angles θ and ϕ indicated in figure 30.2. Hence, our ansatz for the scattered wave is

$$\psi_s(\mathbf{r}) = f_k(\theta, \phi) \frac{e^{ikr}}{r}. \tag{30.1.11}$$

Since ψ_s will be added to φ, which has no units, ψ_s has no units. It follows that the function $f_k(\theta, \phi)$ introduced above to give angular dependence to the wave amplitude has units of length. One can quickly verify that ψ_s *is not* an exact solution of (30.1.5) for $r > a$. We will see, however, that ψ_s approaches an exact solution for $r \to \infty$. As a result, ψ_s is an accurate solution for $r \gg a$.

Both the incident and the scattered wave must be present in the energy eigenstate. Note that the wave number k is the same in both waves. This is a condition for elastic scattering, implemented here naturally because we have a single energy eigenstate. With

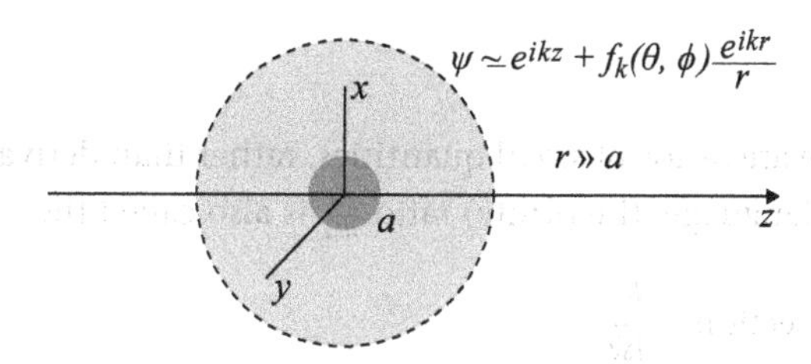

Figure 30.3
The ansatz for an energy eigenstate relevant to scattering in three dimensions is written for the region $r \gg a$, where it holds to a good approximation.

the two waves combined, we have the asymptotic behavior of the energy eigenstate:

$$\text{Asymptotic behavior:} \quad \psi(\mathbf{r}) = \varphi(\mathbf{r}) + \psi_s(\mathbf{r}) \simeq e^{ikz} + f_k(\theta, \phi)\frac{e^{ikr}}{r}, \quad r \gg a. \tag{30.1.12}$$

As indicated, this expression only holds far away from the scattering center (figure 30.3). We physically expect $f_k(\theta, \phi)$ to be determined by $V(\mathbf{r})$, but that will require analyzing the eigenvalue equation in the region where the potential is nonzero. We call $f_k(\theta, \phi)$ the **scattering amplitude**. Before we try to calculate f_k, we will relate it to observable quantities.

30.2 Cross Sections from Scattering Amplitudes

We now relate $f_k(\theta, \phi)$ to the cross section. The key concept is what we call the *differential cross section* $d\sigma$ associated to a small solid angle $d\Omega$. The definition is natural: $d\sigma$ is the area that would remove from the incident beam the particles scattered into the solid angle $d\Omega$. The solid angle $d\Omega$ is defined as a narrow range of θ and ϕ around some values (θ, ϕ) that define the direction of observation. If we include time, we would say that $d\sigma$ is the area that would remove particles from the incident beam at a rate that equals the rate of particles scattered into the solid angle $d\Omega$:

$$\begin{bmatrix}\text{Number of particles scattered per unit time} \\ \text{into solid angle } d\Omega \text{ about } (\theta,\phi)\end{bmatrix} = d\sigma \left[\text{incident flux of particles} = \frac{\#\text{ particles}}{\text{area} \cdot \text{time}}\right]. \tag{30.2.1}$$

Call $\frac{dN^{sc}}{dt}$ the rate at which particles are scattered into the solid angle $d\Omega$. This is exactly the above left-hand side. Call $\frac{dN^{inc}}{dAdt}$ the incident flux of particles. This is exactly the factor in brackets on the above right-hand side. We thus have

$$\frac{dN^{sc}}{dt} = d\sigma\,\frac{dN^{inc}}{dAdt}. \tag{30.2.2}$$

As defined, $\frac{dN^{sc}}{dt}$ is still infinitesimal because it depends on the size of the tiny solid angle $d\Omega$. Also, $d\sigma$ in the equation is infinitesimal. Dividing by $d\Omega$, the above equation therefore takes the form

$$\frac{dN^{sc}}{dtd\Omega} = \frac{d\sigma}{d\Omega}\frac{dN^{inc}}{dAdt}. \tag{30.2.3}$$

The quantities above are ratios of small quantities, rather than derivatives of functions. By a slight abuse of language, the (finite) ratio $\frac{d\sigma}{d\Omega}$ is also called the

$$\text{differential cross section: } \frac{d\sigma}{d\Omega}. \tag{30.2.4}$$

We can now see that this quantity is directly measurable by an experimentalist who uses detectors with known angular aperture relative to the target to determine the left-hand side of (30.2.3) and knows the properties of the incident beam that determine the second factor on the right-hand side.

Let us now go back to (30.2.1), written equivalently as

$$d\sigma = \frac{\begin{bmatrix}\text{Number of particles scattered per unit time}\\ \text{into solid angle } d\Omega \text{ about } (\theta,\phi)\end{bmatrix}}{\left[\text{Flux of incident particles} = \frac{\#\text{ particles}}{\text{area}\cdot\text{time}}\right]}. \tag{30.2.5}$$

Let us calculate the numerator and the denominator. First, we'll do the denominator, which is the probability current of the incoming wave e^{ikz}:

$$\text{Incident flux} = \frac{\hbar}{m}\text{Im}[e^{-ikz}\nabla e^{ikz}] = \frac{\hbar k}{m}\hat{\mathbf{z}}. \tag{30.2.6}$$

To calculate the numerator, we must calculate the flux of the probability current from the scattered wave through the area element $\mathbf{da}$ given by

$$\mathbf{da} = r^2 d\Omega\,\hat{\mathbf{r}}. \tag{30.2.7}$$

Here, r is large, and $\hat{\mathbf{r}}$ is the unit radial vector. To calculate the current $\mathbf{J}_s$ associated with the scattered wave, we first look at the gradient of ψ_s, as given in (30.1.11):

$$\nabla\psi_s = f_k\left(ik\frac{e^{ikr}}{r} - \frac{e^{ikr}}{r^2}\right)\hat{\mathbf{r}} + \frac{1}{r}\frac{\partial f_k}{\partial\theta}\frac{e^{ikr}}{r}\hat{\boldsymbol{\theta}} + \frac{1}{r\sin\theta}\frac{\partial f_k}{\partial\phi}\frac{e^{ikr}}{r}\hat{\boldsymbol{\phi}}. \tag{30.2.8}$$

The current $\mathbf{J}_s = \frac{\hbar}{m}\text{Im}\left[\psi_s^*\nabla\psi_s\right]$ is then

$$\mathbf{J}_s = \frac{\hbar k}{m}\frac{|f_k|^2}{r^2}\hat{\mathbf{r}} + \frac{\hbar}{mr^3}\text{Im}\left(f_k^*\frac{\partial f_k}{\partial\theta}\right)\hat{\boldsymbol{\theta}} + \frac{\hbar}{mr^3\sin\theta}\text{Im}\left(f_k^*\frac{\partial f_k}{\partial\phi}\right)\hat{\boldsymbol{\phi}}. \tag{30.2.9}$$

As $r\to\infty$, the components of the current in the θ and ϕ direction fall off much faster than the radial component, which dominates for large r. The flux, which represents the numerator in (30.2.5), is therefore

$$\mathbf{J}_s\cdot\mathbf{da} = \frac{\hbar k}{m}|f_k(\theta,\phi)|^2\,d\Omega. \tag{30.2.10}$$

Having computed the numerator and denominator of the formula for $d\sigma$, we find that

$$d\sigma = \frac{\frac{\hbar k}{m}|f_k(\theta,\phi)|^2\,d\Omega}{\frac{\hbar k}{m}}. \tag{30.2.11}$$

Hence,

$$\boxed{\frac{\mathrm{d}\sigma}{\mathrm{d}\Omega} = \left|f_k(\theta, \phi)\right|^2.} \tag{30.2.12}$$

This is what we wanted to establish: the scattering amplitude determines the differential cross section. The total cross section σ is obtained by integrating the differential cross section over the full solid angle:

$$\sigma = \int \mathrm{d}\sigma = \int \frac{\mathrm{d}\sigma}{\mathrm{d}\Omega}\mathrm{d}\Omega = \int \left|f_k(\theta, \phi)\right|^2 \mathrm{d}\Omega. \tag{30.2.13}$$

The total cross section represents the area that removes from the incoming beam a flux of particles equal to the flux of the scattered particles.

30.3 Scattering Amplitude in Terms of Phase Shifts

We now consider the calculation of the scattering amplitude for the case of a central potential $V(\mathbf{r}) = V(r)$. While the potential is invariant under rotations, the scattering amplitude does not have full rotational symmetry: the direction z of the incoming wave partially breaks this symmetry. We still have axial symmetry however: the system is invariant under rotations about the z-axis. As a result, the scattering amplitude f_k is no longer a function of the azimuthal angle ϕ and is only a function of θ.

We will calculate the scattering amplitude $f_k(\theta)$ in terms of quantities δ_ℓ called *phase shifts*. The physical picture is one in which the scattering takes place on each of the "partial waves" that, by superposition, represent the full wave function. Each partial wave is a wave of different (quantized) angular momentum ℓ and has an incoming and an outgoing component. For each partial wave, one has a phase shift δ_ℓ relating the incoming and outgoing components of the wave.

As a way of motivation, let us recall the use of phase shifts in one-dimensional scattering (section 29.1). Here the physics happens in $x > 0$, and there is a hard wall at $x = 0$. If the potential vanishes, $V = 0$, the solution $\varphi(x)$ is a sine function and can be written in terms of an incoming wave and an outgoing wave:

$$\varphi(x) = \sin kx = \frac{1}{2i}\Big(e^{ikx} - \underbrace{e^{-ikx}}_{\text{incoming}}\Big), \qquad \text{solution for } V = 0. \tag{30.3.1}$$

This solution is valid for the whole real line. If we now have a nonzero potential $V(x)$ of finite range R, the *exact solution* is hard to find for $x < R$, but for $x > R$ it must take a simple form:

$$\psi(x) = \frac{1}{2i}\Big(e^{ikx}e^{2i\delta_k} - \underbrace{e^{-ikx}}_{\text{incoming}}\Big), \qquad \text{for } x > R. \tag{30.3.2}$$

We have chosen the incoming wave to be the same as for the case $V = 0$, and the remarkable fact is that the outgoing wave can only differ from the $V = 0$ outgoing wave by a phase factor $e^{2i\delta_k}$, with δ_k a *phase shift* that depends on k and the potential $V(x)$. This is the only possibility consistent with having a solution of the Schrödinger equation and conservation of probability: $\psi(x)$ must have a vanishing probability current. The scattering wave ψ_s is an outgoing wave. Since we chose the same incoming waves for the $V=0$ and the $V\neq 0$ solutions, we can write

$$\psi(x)=\varphi(x)+\psi_s(x)\,,\qquad x>R. \tag{30.3.3}$$

A short computation shows that $\psi_s(x)=e^{ikx}e^{i\delta}\sin\delta$.

Let us now return to three-dimensional scattering. Our ansatz (30.1.12), valid at a large distance from the potential, is of a similar form:

$$\psi(\mathbf{r})\simeq e^{ikz}+f_k(\theta)\frac{e^{ikr}}{r}\,,\quad r\gg a. \tag{30.3.4}$$

Just as in the one-dimensional case (30.3.3), the first term on the right-hand side is a $V=0$ solution, and the second term is the outgoing scattered wave. It follows that the incoming waves on the left-hand side must be equal to the incoming waves in e^{ikz}. To express this concretely, we need an expansion of e^{ikz} in spherical waves. This result, called the Rayleigh formula, was derived in section 19.6:

$$e^{ikz}=\sqrt{4\pi}\sum_{\ell=0}^{\infty}\sqrt{2\ell+1}\,i^\ell\,Y_{\ell 0}(\theta)j_\ell(kr)\,. \tag{30.3.5}$$

The plane wave is built as a linear superposition of spherical waves with *all* possible values of the angular momentum! Each ℓ contribution is a *partial wave*. Each partial wave is an exact solution when $V=0$. We can see the spherical ingoing and outgoing waves in each partial wave by expanding the above right-hand side for large r. The asymptotics (19.5.6) of the Bessel function gives $j_\ell(kr)\to\frac{1}{kr}\sin\left(kr-\frac{\ell\pi}{2}\right)$ for large r, and therefore we have

$$e^{ikz}\simeq\frac{\sqrt{4\pi}}{k}\sum_{\ell=0}^{\infty}\sqrt{2\ell+1}\,i^\ell\,Y_{\ell 0}(\theta)\frac{1}{2i}\Big[\underbrace{\frac{e^{i\left(kr-\frac{\ell\pi}{2}\right)}}{r}}_{\text{outgoing}}-\underbrace{\frac{e^{-i\left(kr-\frac{\ell\pi}{2}\right)}}{r}}_{\text{incoming}}\Big],\quad r\to\infty. \tag{30.3.6}$$

We see explicitly the outgoing and incoming spherical waves in each partial wave.

We said that the incoming waves on the left-hand side $\psi(\mathbf{r})$ of (30.3.4) must be equal to the incoming waves in e^{ikz}. We now claim this means that

$$\psi(\mathbf{r})=\frac{\sqrt{4\pi}}{k}\sum_{\ell=0}^{\infty}\sqrt{2\ell+1}\,i^\ell\,Y_{\ell 0}(\theta)\frac{1}{2i}\Big[\underbrace{\frac{e^{i\left(kr-\frac{\ell\pi}{2}\right)}e^{2i\delta_\ell}}{r}}_{\text{outgoing}}-\underbrace{\frac{e^{-i\left(kr-\frac{\ell\pi}{2}\right)}}{r}}_{\text{incoming}}\Big],\quad r\to\infty. \tag{30.3.7}$$

As required, the incoming partial waves in $\psi(\mathbf{r})$ are set equal to the incoming partial waves in e^{ikz}. In writing the outgoing partial waves in $\psi(\mathbf{r})$, we have used the one-dimensional intuition that the amplitude of these waves can only differ from that of the ingoing waves by a phase factor $e^{2i\delta_\ell}$, which can be k dependent. This implements conservation of probability for each partial wave. We will see below (in (30.3.17) and the explanation that follows) that the general solution of the wave equation far away confirms that this ansatz is completely general.

Consider (30.3.4) again and do the partial wave expansion of the right-hand side for large r:

$$\psi(\mathbf{r}) = \frac{\sqrt{4\pi}}{k}\sum_{\ell=0}^{\infty}\sqrt{2\ell+1}\, i^{\ell}\, Y_{\ell 0}(\theta)\frac{1}{2i}\Big[\underbrace{\frac{e^{i\left(kr-\frac{\ell\pi}{2}\right)}}{r}}_{\text{outgoing}} - \underbrace{\frac{e^{-i\left(kr-\frac{\ell\pi}{2}\right)}}{r}}_{\text{incoming}}\Big] + f_k(\theta)\frac{e^{ikr}}{r}. \tag{30.3.8}$$

Setting the right-hand sides of the last two equations equal, we can solve for the scattering amplitude f_k in terms of the phase shifts. The incoming partial waves are identical and thus cancel out. Collecting the outgoing partial waves on one side of the equation, we find that

$$\frac{\sqrt{4\pi}}{k}\sum_{\ell=0}^{\infty}\sqrt{2\ell+1}\, i^{\ell}\, Y_{\ell 0}(\theta)\frac{1}{2i}\left(e^{2i\delta_\ell}-1\right)\frac{e^{ikr}e^{-\frac{i\ell\pi}{2}}}{r} = f_k(\theta)\frac{e^{ikr}}{r}. \tag{30.3.9}$$

Canceling the common r-dependent factors and noticing that

$$\frac{1}{2i}\left(e^{2i\delta_\ell}-1\right) = e^{i\delta_\ell}\sin\delta_\ell, \quad e^{-\frac{i\ell\pi}{2}} = (-i)^{\ell}, \quad \text{and} \quad i^{\ell}(-i)^{\ell} = 1, \tag{30.3.10}$$

we find the result:

$$\boxed{f_k(\theta) = \frac{\sqrt{4\pi}}{k}\sum_{\ell=0}^{\infty}\sqrt{2\ell+1}\, Y_{\ell 0}(\theta)e^{i\delta_\ell}\sin\delta_\ell.} \tag{30.3.11}$$

We can see from this formula that phase shifts are only defined modulo π:

$$\delta_\ell \sim \delta_\ell + n\pi\,, \quad n\in\mathbb{Z}. \tag{30.3.12}$$

Indeed, it is clear that letting $\delta_\ell \to \delta_\ell + \pi$ does not change the scattering amplitude above; it induces two changes of sign that cancel out. More to the point, δ_ℓ was introduced through the phase factor $e^{2i\delta_\ell}$ in the outgoing partial waves, and such a factor is manifestly invariant under changes of δ_ℓ by multiples of π.

Before continuing further, let us confirm the generality of the ansatz for $\psi(\mathbf{r})$. For this consider that for $r > a$ the wave function $\psi(\mathbf{r})$ must satisfy the $V = 0$ wave equation and therefore, restricted to a fixed ℓ, must take the form that follows from the general radial solution considered in (19.5.4):

$$\psi(\mathbf{x})\Big|_{\ell} = \left(A_\ell j_\ell(kr) + B_\ell n_\ell(kr)\right) Y_{\ell 0}(\theta)\,, \qquad r > a. \tag{30.3.13}$$

As mentioned in the discussion of the radial equation, the radial solution can be taken to be real. Therefore, without loss of generality, we can take A_ℓ and B_ℓ to be real coefficients. Next we expand $\psi(\mathbf{x})|_\ell$ for large kr using (19.5.6):

$$\psi(\mathbf{x})\Big|_{\ell} \simeq \left[\frac{A_\ell}{kr}\sin\left(kr-\tfrac{\ell\pi}{2}\right) - \frac{B_\ell}{kr}\cos\left(kr-\tfrac{\ell\pi}{2}\right)\right] Y_{\ell 0}(\theta). \tag{30.3.14}$$

Let us define a phase δ_ℓ as follows:

$$\tan\delta_\ell \equiv -\frac{B_\ell}{A_\ell}. \tag{30.3.15}$$

Now we can set $A_\ell = \gamma\cos\delta_\ell$ and $B_\ell = -\gamma\sin\delta_\ell$, with γ some constant. This gives

$$\begin{aligned}\psi(\mathbf{x})\Big|_\ell &\simeq \frac{\gamma}{kr}\left[\cos\delta_\ell\,\sin\left(kr-\tfrac{\ell\pi}{2}\right)+\sin\delta_\ell\cos\left(kr-\tfrac{\ell\pi}{2}\right)\right]Y_{\ell 0}(\theta)\\ &\simeq \frac{\gamma}{kr}\sin\left(kr-\tfrac{\ell\pi}{2}+\delta_\ell\right)Y_{\ell 0}(\theta),\end{aligned} \tag{30.3.16}$$

where we used the addition formula for trigonometric functions. Expanding the sine function in terms of exponentials, we find that

$$\psi(\mathbf{x})\Big|_\ell \simeq e^{-i\delta_\ell}\,\frac{\gamma}{k}\,\frac{1}{2i}\left[\frac{e^{i\left(kr-\frac{\ell\pi}{2}\right)}e^{2i\delta_\ell}}{r}-\frac{e^{-i\left(kr-\frac{\ell\pi}{2}\right)}}{r}\right]Y_{\ell 0}(\theta). \tag{30.3.17}$$

This confirms that in a general solution the partial wave ℓ comprises an incoming wave and an outgoing wave whose amplitudes only differ by a phase. This justifies our expansion of $\psi(\mathbf{r})$ in (30.3.7). Moreover, we have now identified the phase shift in terms of the coefficients A_ℓ and B_ℓ of the spherical Bessel functions in the exact solution (30.3.13) away from the potential. That relation is $\tan\delta_\ell = -B_\ell/A_\ell$.

With our expression for the scattering amplitude in terms of phase shifts, we can rewrite the differential and total cross sections more explicitly. We had

$$d\sigma = |f_k(\theta)|^2 d\Omega, \tag{30.3.18}$$

and this differential cross section exhibits a θ dependence that can be worked out explicitly if the phase shifts are known. The various partial waves interfere here—there are terms involving products of different partial waves. The full cross section has no angular dependence as it arises from integration over all angles. The integral can be performed explicitly:

$$\begin{aligned}\sigma &= \int |f_k(\theta)|^2\, d\Omega = \int f_k^*(\theta) f_k(\theta) d\Omega\\ &= \frac{4\pi}{k^2}\sum_{\ell,\ell'}\sqrt{2\ell+1}\sqrt{2\ell'+1}\,e^{-i\delta_\ell}\sin\delta_\ell\, e^{i\delta_{\ell'}}\sin\delta_{\ell'}\int d\Omega\, Y_{\ell 0}^*(\Omega)Y_{\ell',0}(\Omega).\end{aligned}$$

Because the spherical harmonics are orthonormal, the above angular integral equals $\delta_{\ell\ell'}$, and we find that

$$\boxed{\sigma = \frac{4\pi}{k^2}\sum_{\ell=0}^{\infty}(2\ell+1)\sin^2\delta_\ell.} \tag{30.3.19}$$

Each partial wave contributes *separately* to the total cross section, and there is no longer any interference. This allows us to define partial-wave cross sections σ_ℓ as the contributions to the cross section from each partial wave:

$$\sigma = \sum_{\ell=0}^{\infty} \sigma_\ell \,, \quad \text{with} \quad \sigma_\ell = \frac{4\pi}{k^2}(2\ell+1)\sin^2\delta_\ell \,. \tag{30.3.20}$$

Since $\sin^2\delta_\ell \le 1$ for any ℓ, the partial-wave cross sections are bounded above:

$$\sigma_\ell \le \frac{4\pi}{k^2}(2\ell+1). \tag{30.3.21}$$

This is sometimes called a *unitarity bound*. Note that the partial-wave cross sections attain the maximum possible value for phase shifts $\delta_\ell = \pi/2$.

Something interesting happens when we consider the value of the scattering amplitude $f_k(\theta)$ in the forward direction $\theta = 0$. Recalling that spherical harmonics with $m = 0$ are in fact Legendre polynomials (see (10.5.34)) and that $P_\ell(1) = 1$ (see (10.5.25)), we see that

$$Y_{\ell 0}(\theta) = \sqrt{\frac{2\ell+1}{4\pi}}P_\ell(\cos\theta) \implies Y_{\ell 0}(\theta = 0) = \sqrt{\frac{2\ell+1}{4\pi}}. \tag{30.3.22}$$

Therefore,

$$f_k(\theta = 0) = \frac{\sqrt{4\pi}}{k}\sum_{\ell=0}^{\infty}\sqrt{2\ell+1}\sqrt{\frac{2\ell+1}{4\pi}}e^{i\delta_\ell}\sin\delta_\ell = \frac{1}{k}\sum_{\ell=0}^{\infty}(2\ell+1)e^{i\delta_\ell}\sin\delta_\ell. \tag{30.3.23}$$

This seems to be closely related to the full cross section σ in (30.3.19). Up to a constant of proportionality, the only difference is that the forward amplitude has $e^{i\delta_\ell}\sin\delta_\ell$, and the cross section has $\sin^2\delta_\ell$. Taking the imaginary part of the forward amplitude corrects for this discrepancy:

$$\text{Im}(f_k(0)) = \frac{1}{k}\sum_{\ell=0}^{\infty}(2\ell+1)\sin^2\delta_\ell = \frac{1}{k}\frac{k^2}{4\pi}\sigma. \tag{30.3.24}$$

We have found a remarkable relation between the total elastic cross section and the imaginary part of the *forward* ($\theta = 0$) scattering amplitude. This relation is known as the

$$\textbf{optical theorem:} \quad \sigma = \frac{4\pi}{k}\text{Im}(f_k(0)). \tag{30.3.25}$$

The formalism of partial waves is so powerful that it gave us this result by simple algebraic manipulations. More insight into the optical theorem can be obtained by a careful analysis of the probability current. Imagine a large sphere surrounding the target. Certainly, the incoming wave e^{ikz} has zero probability flux: whatever goes into the large sphere also goes out. The scattered wave, however, has a finite outgoing probability flux, proportional to the cross section. How can that be consistent with probability conservation? The answer is that the probability current also includes a contribution from the interference of the incoming wave with the scattered wave. A detailed calculation (problem 30.2) shows that this interference happens specifically in the forward direction and in fact supplies *into* the sphere exactly the same probability flux that the scattered wave is moving outward. Thus, the scattering amplitude in the forward direction controls the value of the cross section. One can think of the optical theorem in terms of a "shadow" created by the potential in the forward direction, just as shining light on an object casts a shadow. The particles removed to create a shadow are the scattered ones.

30.4 Computation of Phase Shifts

Any spherically symmetric potential can be analyzed using phase shifts. If we know the phase shifts, we know the differential and total scattering cross sections. It is therefore important to be able to compute and identify the phase shifts for a given potential. We now discuss how to do this and then turn to some examples.

We already showed a way to identify phase shifts in the previous section. When solving for the wave function beyond the range of the potential, we consider solutions $\psi(\mathbf{r})$ that restricted to a single partial wave take the form

$$\psi(\mathbf{r})\Big|_{\ell} = \big(A_\ell j_\ell(kr) + B_\ell n_\ell(kr)\big)\, Y_{\ell 0}(\theta)\,, \qquad r > a. \tag{30.4.1}$$

A nonzero B_ℓ is the signal of a nonvanishing potential because this solution becomes singular as we attempt to extend it to $r=0$. We showed in the previous section that for such an expansion,

$$\tan\delta_\ell \equiv -\frac{B_\ell}{A_\ell}. \tag{30.4.2}$$

Another way to identify the phase shifts directly uses the asymptotic behavior of (30.4.1) for very large r. This was calculated in (30.3.16):

$$\psi(\mathbf{x})\Big|_{\ell} \sim \frac{1}{kr}\sin\left(kr - \tfrac{\ell\pi}{2} + \delta_\ell\right) Y_{\ell,0}(\theta)\,, \quad r\to\infty. \tag{30.4.3}$$

The strategy here aims to find the large r behavior of the solution and to compare it with the above form to read the value of δ_ℓ.

The case of $\ell = 0$ merits special attention. Using the exact form of $j_0(kr)$ and $n_0(kr)$ given in (17.8.33), the solution (30.4.1) becomes, up to a constant that we call C',

$$\psi(\mathbf{r})\Big|_{\ell=0} = \frac{C'}{r}\left(A_0\sin kr - B_0\cos kr\right)\,, \quad r > a. \tag{30.4.4}$$

This is exact for $r > a$. From $\tan\delta_0 = -B_0/A_0$, we find that

$$\psi(\mathbf{r})\Big|_{\ell=0} = \frac{C}{r}\left(\cos\delta_0\sin kr + \sin\delta_0\cos kr\right)\,, \quad r > a, \tag{30.4.5}$$

with C yet another constant. This means that

$$\psi(\mathbf{r})\Big|_{\ell=0} = \frac{C}{r}\sin(kr + \delta_0)\,, \quad r > a. \tag{30.4.6}$$

As a result, the exact ansatz for the $\ell = 0$ radial wave function $u_0(r)$ outside the range of the potential is

$$u_0(r) = C\sin(kr + \delta_0), \quad r > a. \tag{30.4.7}$$

This is very useful in explicit computations where this solution must be matched at $r = a$ to a solution that holds for $r < a$. Note that this is in fact the canonical solution for the wave $\psi(x)$ outside the range of the potential for the one-dimensional scattering problem in the half line $x > 0$ (see equation (29.1.8)). This is the claim we made in chapter 29 that three-dimensional scattering with $\ell = 0$ can be viewed as scattering on the half line.

Example 30.1. *Scattering off a hard sphere.*

We wish to determine the phase shifts and total cross section for the scattering of waves off a hard sphere. The potential for a hard sphere of radius a is infinite for $r \le a$ and zero for $r > a$:

$$V(r) = \begin{cases} \infty, & r \le a \\ 0, & r > a. \end{cases} \tag{30.4.8}$$

As a warm-up consider the $\ell = 0$ phase shift. In this case the radial wave function for $r \ge a$ must be

$$u_0(r) = C\sin(kr + \delta_0). \tag{30.4.9}$$

Since the wave function must vanish at $r = a$, we find

$$\delta_0 = -ka. \tag{30.4.10}$$

The associated partial-wave cross section σ_0 is therefore

$$\sigma_0 = \frac{4\pi}{k^2}(\sin ka)^2 = 4\pi a^2 \cdot \Big(\frac{\sin ka}{ka}\Big)^2. \tag{30.4.11}$$

Since $|\frac{\sin x}{x}| \le 1$, we see that σ_0 is bounded as $\sigma_0 \le 4\pi a^2$. In fact, σ_0 reaches the upper bound as the energy of the waves goes to zero: $ka \to 0$. This is the long-wavelength limit:

$$\lim_{k\to 0} \sigma_0 = 4\pi a^2. \tag{30.4.12}$$

We will confirm below that at low energies the higher partial-wave cross sections $\sigma_{\ell>0}$ give much suppressed contributions. So σ_0 above is in fact the complete cross section at low energy:

$$\lim_{k\to 0} \sigma = 4\pi a^2. \tag{30.4.13}$$

This is perhaps surprisingly large: the cross section σ is given by the area of the sphere, as opposed to the transverse area πa^2 that the sphere presents to the beam! It is as if the waves, having arbitrarily large wavelength, sense the whole sphere rather than its transverse area. Note also that the differential cross section associated with $\ell = 0$ has no angular dependence because $f_k(\theta)$ restricted to $\ell = 0$ has no angular dependence:

$$f_k(\theta)\big|_{\ell=0} = \frac{\sqrt{4\pi}}{k} Y_{00} e^{i\delta_0} \sin\delta_0 = \frac{1}{k} e^{i\delta_0} \sin\delta_0. \tag{30.4.14}$$

Since the $\ell = 0$ partial wave dominates, scattering is therefore isotropic in the low-energy limit. This has an intuitive explanation as well. In the low-energy limit, the wavelength of the incoming particle is becoming very large. Therefore, the phase of the wave is approximately constant all over the extent of the spherically symmetric target. All sense of directionality due to the phase gradient in the z-direction is lost. The scattering is consequently isotropic.

The calculation of the higher-ℓ phase shifts is actually rather straightforward, although the answers are in terms of spherical Bessel functions. Since the potential vanishes for $r > a$, the complete solution $\psi(\mathbf{r})$ can be written as

$$\psi(\mathbf{r}) = \sum_{\ell=0}^{\infty} (A_\ell j_\ell(kr) + B_\ell n_\ell(kr)) P_\ell(\cos\theta). \tag{30.4.15}$$

The wave function must vanish at $r = a$; therefore,

$$\sum_{\ell=0}^{\infty} (A_\ell j_\ell(ka) + B_\ell n_\ell(ka)) P_\ell(\cos\theta) = 0. \tag{30.4.16}$$

Since the Legendre polynomials $P_\ell(\cos\theta)$ are complete functions in the physical range $-1 \le \cos\theta \le 1$, each coefficient of $P_\ell(\cos\theta)$ in the above series must vanish:

$$A_\ell j_\ell(ka) + B_\ell n_\ell(ka) = 0, \quad \forall \ell. \tag{30.4.17}$$

Since this equation determines the ratio of the coefficients A_ℓ and B_ℓ for all ℓ, we have found the phase shifts. Recalling that $\tan\delta_\ell = -\frac{B_\ell}{A_\ell}$, we have that, for all ℓ,

$$\boxed{\tan\delta_\ell = \frac{j_\ell(ka)}{n_\ell(ka)}.} \tag{30.4.18}$$

We easily compute the cross section σ that, recalling (30.3.19), requires the value of $\sin^2\delta_\ell$:

$$\sin^2\delta_\ell = \frac{\tan^2\delta_\ell}{1 + \tan^2\delta_\ell} = \frac{j_\ell^2(ka)}{j_\ell^2(ka) + n_\ell^2(ka)}. \tag{30.4.19}$$

Then the cross section is

$$\sigma = \frac{4\pi}{k^2} \sum_{\ell=0}^{\infty} (2\ell+1) \sin^2\delta_\ell = \frac{4\pi}{k^2} \sum_{\ell=0}^{\infty} (2\ell+1) \frac{j_\ell^2(ka)}{j_\ell^2(ka) + n_\ell^2(ka)}. \tag{30.4.20}$$

Let us reconsider the low-energy limit. Using the small-argument expansions (19.5.5) of the spherical Bessel functions, we get that

$$\tan\delta_\ell = \frac{j_\ell(ka)}{n_\ell(ka)} \simeq -\frac{(ka)^{2\ell+1}}{(2\ell+1)!!(2\ell-1)!!}, \quad ka \ll 1. \tag{30.4.21}$$

Recall that $(2k+1)!! = (2k+1)(2k-1)\cdots 3\cdot 1$. Since the right-hand side is very small for any ℓ, the phase shifts themselves are given by

$$\delta_\ell \simeq -\frac{(ka)^{2\ell+1}}{(2\ell+1)!!(2\ell-1)!!}, \quad ka \ll 1. \tag{30.4.22}$$

The low-energy behavior of the phase shifts is rather generic. We expect that

$$\delta_\ell \sim (ka)^{2\ell+1} \tag{30.4.23}$$

for potentials of range a in the limit $ka \ll 1$. Exceptions happen if the potential admits bound states with vanishingly small bound-state energies (problem 30.4). In this case, low-energy waves experience resonance behavior as they interact with the bound state and the phase shift $\delta_0 \to \frac{\pi}{2}$. If regular behavior holds, however, low-energy scattering is rather accurately computed by using a few partial waves starting with $\ell = 0$.

High-energy scattering off the hard sphere is also of interest. In this case, $ka \gg 1$, and we can use the large-argument limit of the Bessel functions to find that

$$\tan\delta_\ell = \frac{j_\ell(ka)}{n_\ell(ka)} \simeq -\tan\left(ka - \tfrac{\ell\pi}{2}\right) \quad \to \quad \delta_\ell \simeq -ka + \tfrac{\ell\pi}{2}. \tag{30.4.24}$$

One should note a limitation of this result: this only holds for $\ell < ka$. Indeed, the Bessel functions $j_\ell(\rho)$ and $n_\ell(\rho)$ settle into trigonometric behavior only for $\ell < \rho$. For very high ℓ—namely, $\ell \gg \rho$, the small-argument behavior is applicable, and the phase shifts can be seen to go to zero. It follows that the high-energy cross section is evaluated by summing the partial waves for $\ell \leq ka$. A rough estimate confirms the well-known result that at high energies,

$$\sigma \to 2\pi a^2, \qquad ka \gg 1. \tag{30.4.25}$$

One might have expected the classical answer πa^2, as the de Broglie wavelength of the particles is much smaller than the size of the sphere. The explanation of this result is a bit intricate, based on the idea that due to diffraction there is actually no shadow. The extra area πa^2 of the missing shadow must be added to the classical cross section πa^2. □

Exercise 30.1. *Confirm that in the low-energy limit $ka \ll 1$ the hard-sphere cross section is given by*

$$\sigma \simeq \frac{4\pi}{k^2}\sum_{\ell=0}^{\infty}\frac{1}{2\ell+1}\left[\frac{2^\ell \ell!}{(2\ell)!}\right]^4 (ka)^{4\ell+2}. \tag{30.4.26}$$

General computation of the phase shift To compute the $\ell \neq 0$ phase shifts of an arbitrary potential with range a, you need a strategy. You must find a radial solution $R_\ell(r)$ that applies for $r \leq a$. That solution, of course, will depend on the potential and, unless the potential is too singular at the origin, will satisfy the boundary condition $R_\ell(r) \sim r^\ell$ as $r \to 0$. Let us assume you have found $R_\ell(r)$. This must be matched to the general solution $A_\ell j_\ell(kr) + B_\ell n_\ell(kr)$ that holds for $r > a$ where $V = 0$ (see figure 30.4).

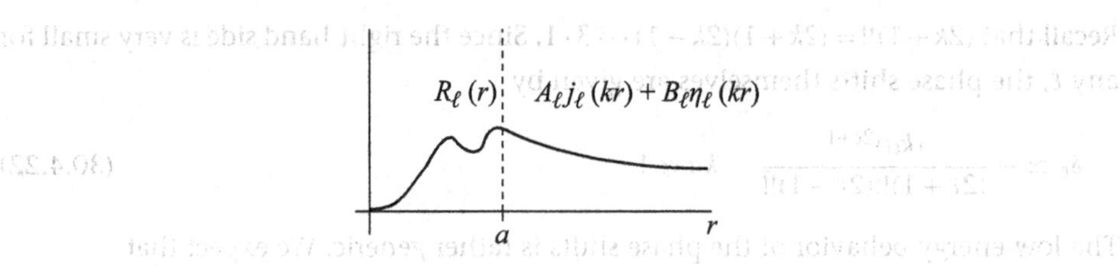

Figure 30.4
The radial solution $R_\ell(r)$, valid for $r < a$, must match at $r = a$ to the general solution $A_\ell j_\ell(kr) + B_\ell n_\ell(kr)$, valid for $r > a$.

At $r = a$, we must match the functions and their derivatives:

$$\begin{aligned} R_\ell(a) &= A_\ell j_\ell(ka) + B_\ell n_\ell(ka), \\ aR'_\ell(a) &= ka\big(A_\ell j'_\ell(ka) + B_\ell n'_\ell(ka)\big). \end{aligned} \tag{30.4.27}$$

Let's form the ratio and define β_ℓ as the logarithmic derivative of the radial solution:

$$\beta_\ell \equiv \frac{aR'_\ell(a)}{R_\ell(a)} = ka\frac{A_\ell j'_\ell(ka) + B_\ell n'_\ell(ka)}{A_\ell j_\ell(ka) + B_\ell n_\ell(ka)} = ka\frac{j'_\ell(ka) + \frac{B_\ell}{A_\ell}n'_\ell(ka)}{j_\ell(ka) + \frac{B_\ell}{A_\ell}n_\ell(ka)}. \tag{30.4.28}$$

This gives

$$\beta_\ell = ka\,\frac{j'_\ell(ka) - \tan\delta_\ell\, n'_\ell(ka)}{j_\ell(ka) - \tan\delta_\ell\, n_\ell(ka)}. \tag{30.4.29}$$

A little algebra allows us to solve for $\tan\delta_\ell$, our objective in this calculation:

$$\tan\delta_\ell = \frac{j_\ell(ka) - \frac{ka}{\beta_\ell} j'_\ell(ka)}{n_\ell(ka) - \frac{ka}{\beta_\ell} n'_\ell(ka)}. \tag{30.4.30}$$

If $\beta_\ell \to \infty$, which happens if $R_\ell(a) = 0$, we recover the phase shifts of a hard sphere. This is as expected, for $R_\ell(a) = 0$ is the hard-sphere condition on the solution. For large β_ℓ, the phase shifts are small deformations of the hard-sphere phase shifts. Of course, if $\beta_\ell \to 0$ (meaning $R'_\ell(a) \to 0$), the phase shifts can change dramatically from those of the hard sphere.

Phase shifts are particularly useful when a few of them dominate the cross section. This happens when $ka < 1$ with a the range of the potential. At fixed k this can happen for short-range potentials, and at fixed range this can happen for low energies. More generally, suppose $ka \sim \ell_0$ for some positive integer ℓ_0. Then partials waves with $\ell \leq \ell_0$ give the largest contribution to scattering. We will now make a few remarks explaining how this comes about from the semiclassical perspective:

1. Assume a particle is incident upon a target with impact parameter b (figure 30.5). We can estimate b in terms of the magnitude L of the angular momentum relative to the origin and the incident momentum $\hbar k$. We have $L = b\,\hbar k$, and in the quantum setting, we write $L \sim \hbar\ell$, as is appropriate given the eigenvalue relation $L^2 = \hbar^2\ell(\ell + 1)$.

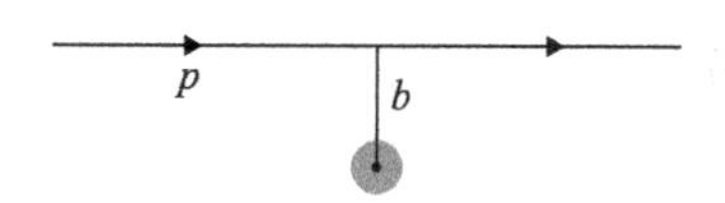

Figure 30.5
A particle of momentum $p = \hbar k$ incident upon a target with impact parameter b.

Therefore,

$$L \simeq \hbar\ell \simeq b\hbar k \quad \Rightarrow \quad b \simeq \frac{\ell}{k}. \tag{30.4.31}$$

Naturally, waves with larger ℓ have a larger impact parameter. For a potential of finite range a, we have no classical scattering when $b > a$. Thus, quantum mechanically, we expect

$$b = \frac{\ell}{k} > a \quad \Rightarrow \quad \text{Expect little scattering for } \ell > ka. \tag{30.4.32}$$

The partial waves with $\ell \leq ka$ give the largest contribution to the cross section.

2. The behavior of the radial wave function $u_\ell \sim r j_\ell(kr)$ for the free ℓth partial wave confirms our intuition about impact parameters. We now show that this wave has negligible value for $kr < \ell$—that is, for $r < \ell/k = b$. This is what we would expect for a wave that has impact parameter b.

   To see this, recall that $r j_\ell(kr)$ is a solution of the $V = 0$ radial equation with angular momentum ℓ and energy $\hbar^2 k^2/(2m)$. With vanishing V, the effective radial potential consists of the centrifugal barrier arising from the angular momentum. This barrier prevents the wave from reaching the origin. If r_* is the position of the turning point for classical motion, the wave should be small in the classically forbidden region $r < r_*$. Setting the effective potential equal to the energy,

$$\frac{\hbar^2 \ell(\ell+1)}{2mr^2} = \frac{\hbar^2 k^2}{2m}, \tag{30.4.33}$$

   we find that the turning point r_* is at

$$k^2 r_*^2 = \ell(\ell+1) \quad \Rightarrow \quad kr_* \simeq \ell \quad \Rightarrow \quad r_* \simeq \frac{\ell}{k} = b. \tag{30.4.34}$$

   Thus, $r j_\ell(kr)$ is rather small for $r < b$, as we aimed to show. See figure 30.6 for a graphic demonstration of this when $\ell = 10$.

3. A semiclassical argument allows estimation of the partial cross sections. We estimate σ_ℓ as the area of an annulus whose inner radius is the impact parameter ℓ/k for angular momentum ℓ and whose outer radius is the impact parameter $(\ell+1)/k$ for angular momentum $\ell + 1$. With Δb denoting the width of the annulus, we therefore have

$$\sigma_\ell \simeq 2\pi \frac{\ell}{k} \Delta b = 2\pi \frac{\ell}{k}\Big(\frac{\ell+1}{k} - \frac{\ell}{k}\Big) = \frac{2\pi}{k^2}\ell. \tag{30.4.35}$$

   This result is of the right order of magnitude (see (30.3.21)) but roughly four times smaller than the maximum value σ_ℓ can take in the quantum theory.

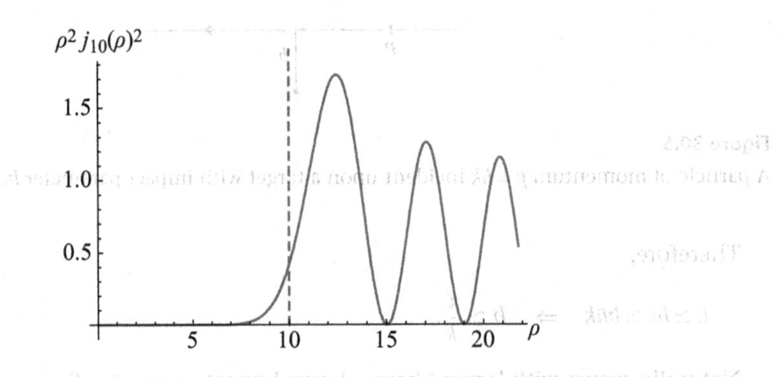

Figure 30.6
Plot of $(\rho j_{10}(\rho))^2$. The function is small for $\rho < 10$.

30.5 Integral Equation for Scattering

Not all potentials are spherically symmetric, and therefore we cannot use our partial-wave analysis for scattering off an arbitrary potential. By rewriting the Schrödinger equation as an integral equation, however, we can make some useful approximations that allow us to deal with these potentials when the particles have high energies or, for arbitrary energies, when the potentials are weak. It is also quite striking that the integral reformulation of the Schrödinger equation naturally incorporates the desired asymptotic behavior (30.1.12) of our energy eigenstates:

$$\psi(\mathbf{r}) \simeq e^{ikz} + f_k(\theta,\phi)\frac{e^{ikr}}{r}, \quad r \gg a. \tag{30.5.1}$$

Let us first derive the integral equation reformulation, and then in the following section, we will consider the rather important *Born approximation.*

Once again, we begin with the time-independent Schrödinger equation:

$$\left[-\frac{\hbar^2}{2m}\nabla^2 + V(\mathbf{r})\right]\psi(\mathbf{r}) = E\psi(\mathbf{r}). \tag{30.5.2}$$

We set the energy equal to that of a plane wave of wave number k and rewrite the potential in terms of a rescaled version $U(\mathbf{r})$ that simplifies the units:

$$E = \frac{\hbar^2 k^2}{2m} \quad \text{and} \quad V(\mathbf{r}) = \frac{\hbar^2}{2m}U(\mathbf{r}). \tag{30.5.3}$$

The Schrödinger equation then becomes

$$\left(\nabla^2 + k^2\right)\psi(\mathbf{r}) = U(\mathbf{r})\psi(\mathbf{r}). \tag{30.5.4}$$

This is the equation we must solve. Assume we have found a Green's function $G(\mathbf{r}-\mathbf{r}')$ for the operator $\nabla^2 + k^2$—that is, a solution of the following differential equation:

$$\left(\nabla^2 + k^2\right)G(\mathbf{r}-\mathbf{r}') = \delta^{(3)}(\mathbf{r}-\mathbf{r}'). \tag{30.5.5}$$

Note that the Laplacian ∇^2 acts on $\mathbf{r}$, not $\mathbf{r}'$. Also let $\psi_0(\mathbf{r})$ denote an arbitrary solution of the homogeneous equation

$$\left(\nabla^2 + k^2\right)\psi_0(\mathbf{r}) = 0. \tag{30.5.6}$$

Then we claim that any solution of the following integral equation

$$\boxed{\psi(\mathbf{r}) = \psi_0(\mathbf{r}) + \int d^3\mathbf{r}' G(\mathbf{r}-\mathbf{r}')U(\mathbf{r}')\psi(\mathbf{r}')} \tag{30.5.7}$$

is a solution of the Schrödinger equation (30.5.4). Indeed, we can show this directly:

$$\begin{aligned}\left(\nabla + k^2\right)\psi(\mathbf{r}) &= \left(\nabla + k^2\right)\int d^3\mathbf{r}' G(\mathbf{r}-\mathbf{r}')U(\mathbf{r}')\psi(\mathbf{r}') \\ &= \int d^3\mathbf{r}'[\left(\nabla + k^2\right)G(\mathbf{r}-\mathbf{r}')]U(\mathbf{r}')\psi(\mathbf{r}') \\ &= \int d^3\mathbf{r}'\delta^{(3)}(\mathbf{r}-\mathbf{r}')U(\mathbf{r}')\psi(\mathbf{r}') = U(\mathbf{r})\psi(\mathbf{r}).\end{aligned} \tag{30.5.8}$$

Conversely, any solution $\psi(\mathbf{r})$ of the Schrödinger equation is a solution of the integral equation for some $\psi_0(\mathbf{r})$ satisfying the homogeneous equation. This is verified by first checking that

$$(\nabla^2 + k^2)\left(\psi(\mathbf{r}) - \int d^3\mathbf{r}' G(\mathbf{r}-\mathbf{r}')U(\mathbf{r}')\psi(\mathbf{r}')\right) = 0, \tag{30.5.9}$$

as you can almost see by inspection. This implies that the object in parentheses is some $\psi_0(\mathbf{r})$ satisfying the homogeneous equation. Therefore,

$$\psi(\mathbf{r}) - \int d^3\mathbf{r}' G(\mathbf{r}-\mathbf{r}')U(\mathbf{r}')\psi(\mathbf{r}') = \psi_0(\mathbf{r}). \tag{30.5.10}$$

This is in fact the integral equation with the homogeneous solution ψ_0.

Let us now address the determination of the Green's function. It is simplest to consider equation (30.5.5) with $\mathbf{r}' = 0$. This entails no loss of generality: by translational invariance of the equation, $G(\mathbf{r}-\mathbf{r}')$ is just $G(\mathbf{r})$ with $\mathbf{r}$ replaced by $\mathbf{r}-\mathbf{r}'$. Therefore, we consider

$$\left(\nabla^2 + k^2\right)G(\mathbf{r}) = \delta^{(3)}(\mathbf{r}). \tag{30.5.11}$$

It is rather simple to find good candidates for the Green's function. Note that away from the origin the operator $\nabla^2 + k^2$ must kill G. We found in (30.1.10) that

$$\left(\nabla^2 + k^2\right)\frac{e^{\pm ikr}}{r} = 0, \qquad \text{for } r \neq 0. \tag{30.5.12}$$

We actually stated this result only for the plus sign, but the minus sign works as well; it corresponds to changing the sign of k. We now ask whether there is a delta function on the left-hand side of the above equation. As $r \to 0$, the function $e^{\pm ikr}/r$ goes into $1/r$, and we recall that

$$\nabla^2 \frac{1}{r} = -4\pi\delta(\mathbf{r}). \tag{30.5.13}$$

This makes it extremely plausible that

$$\boxed{G_\pm(\mathbf{r}) = -\frac{1}{4\pi}\frac{e^{\pm ikr}}{r}} \tag{30.5.14}$$

solves the desired equation

$$(\nabla^2 + k^2)G_\pm(\mathbf{r}) = \delta^{(3)}(\mathbf{r}). \tag{30.5.15}$$

To be sure, you may follow the steps indicated in the following exercise.

Exercise 30.2. *Check that*

$$\nabla^2 e^{\pm ikr} = \left(-k^2 \pm \frac{2ik}{r}\right) e^{\pm ikr}, \tag{30.5.16}$$

and use the identity $\nabla^2(fg) = (\nabla^2 f)g + 2\nabla f \cdot \nabla g + f\nabla^2 g$ *to verify the claim above.*

The Green's function G_+ is called a "retarded" Green's function. Indeed, if we appended the usual time dependence $e^{-iEt/\hbar}$, the function G_+ would represent outgoing waves. The Green's function G_- is called an "advanced" Green's function. With the standard time dependence, it represents incoming waves.

We now come to an important point: How do we choose G and the homogeneous solution $\psi_0(\mathbf{r})$? For this, we think concretely of the scattering problem that we are trying to solve. In it, we have outgoing waves emerging from the region where the potential is nonzero. We also have an incoming plane wave e^{ikz} that is a solution of the homogeneous equation: $(\nabla^2 + k^2)e^{ikz} = 0$. We therefore choose

$$\psi_0(\mathbf{r}) = e^{ikz} \quad \text{and} \quad G = G_+. \tag{30.5.17}$$

With these choices the integral equation (30.5.7) takes the form

$$\psi(\mathbf{r}) = e^{ikz} + \int d^3\mathbf{r}' G_+(\mathbf{r}-\mathbf{r}')U(\mathbf{r}')\psi(\mathbf{r}'), \tag{30.5.18}$$

where our choice of Green's function is

$$G_+(\mathbf{r}-\mathbf{r}') = -\frac{1}{4\pi}\frac{e^{ik|\mathbf{r}-\mathbf{r}'|}}{|\mathbf{r}-\mathbf{r}'|}. \tag{30.5.19}$$

We now want to prove that this choice implements our asymptotic expansion for the energy eigenstates. For that we need to make some approximations in the Green's function that are valid when the observation point $\mathbf{r}$ is far away from the region where the potential is nonzero, the region of integration for $\mathbf{r}'$ in our integral equation. Being far away therefore means

$$r \gg r', \quad \text{with} \quad r = |\mathbf{r}|, \quad r' = |\mathbf{r}'|. \tag{30.5.20}$$

This is illustrated in figure 30.7.

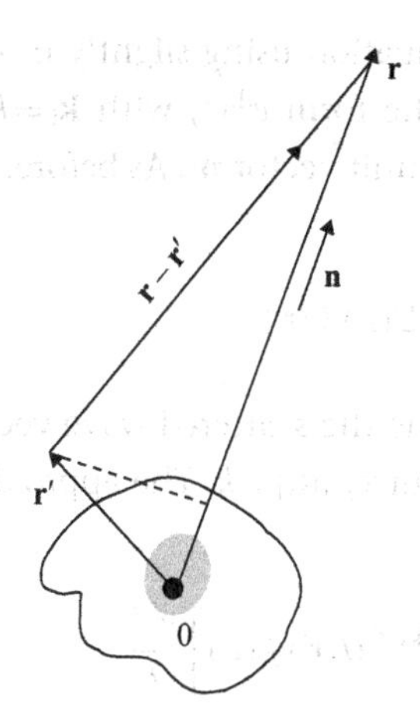

Figure 30.7
Distances relevant to approximations in the Green's function $G_+(\mathbf{r}-\mathbf{r}')$.

We need to evaluate the distance $|\mathbf{r}-\mathbf{r}'|$ in the approximation $r \gg r'$. Including a first correction, one finds that

$$|\mathbf{r}-\mathbf{r}'| \simeq r - \mathbf{n}\cdot\mathbf{r}', \quad \text{with} \quad \mathbf{n} \equiv \frac{\mathbf{r}}{r}. \tag{30.5.21}$$

We can approximate $|\mathbf{r}-\mathbf{r}'| \simeq r$ in the denominator of the Green's function, but for the more sensitive phase in the exponent, we need the first correction above (and further corrections if k is arbitrarily large). Using these approximations, we see that

$$G_+(\mathbf{r}-\mathbf{r}') \simeq -\frac{1}{4\pi r} e^{ikr} e^{-ik\mathbf{n}\cdot\mathbf{r}'}. \tag{30.5.22}$$

With this expression for G_+, the integral equation (30.5.18) becomes

$$\psi(\mathbf{r}) = e^{ikz} + \left[-\frac{1}{4\pi}\int d^3\mathbf{r}' e^{-ik\mathbf{n}\cdot\mathbf{r}'} U(\mathbf{r}')\psi(\mathbf{r}')\right]\frac{e^{ikr}}{r}. \tag{30.5.23}$$

Here, we have kept in the integral only objects that depend on $\mathbf{r}'$. The factor e^{ikr}/r to the right of the integral is the outgoing wave of our asymptotic ansatz. Since $\mathbf{r}'$ is integrated over, the expression in brackets is a function of the unit vector $\mathbf{n}$ in the direction of the observation point $\mathbf{r}$. Thus, this expression can be identified with the scattering amplitude $f_k(\theta,\phi)$ in the asymptotic ansatz:

$$f_k(\theta,\phi) = -\frac{1}{4\pi}\int d^3\mathbf{r}' e^{-ik\mathbf{n}\cdot\mathbf{r}'} U(\mathbf{r}')\psi(\mathbf{r}'). \tag{30.5.24}$$

This shows that the integral equation, through the choice of G_+ and ψ_0, incorporates the asymptotic conditions. Of course, this does not yet determine $f_k(\theta,\phi)$, as the undetermined wave function $\psi(\mathbf{r})$ still appears under the integral. We have not solved the integral equation; we have only shown that it properly represents the expected asymptotic behavior.

Let us rewrite the integral equation using slightly more general notation. The incident wave will be written in the form $e^{i\mathbf{k}_i\cdot\mathbf{r}}$, with $\mathbf{k}_i = k\mathbf{n}_i$ the incident wave vector pointing in the direction of the unit vector $\mathbf{n}_i$. As before, $|\mathbf{k}_i| = k$. The integral equation then reads

$$\psi(\mathbf{r}) = e^{i\mathbf{k}_i\cdot\mathbf{r}} + \int d^3\mathbf{r}' G_+(\mathbf{r}-\mathbf{r}')U(\mathbf{r}')\psi(\mathbf{r}'). \tag{30.5.25}$$

For the outgoing wave, we define the scattered wave vector $\mathbf{k}_s \equiv k\mathbf{n}$ in the direction $\mathbf{n}$ of the observation point. Of course, $|\mathbf{k}_s| = k$. The approximate expression we obtained for $\psi(\mathbf{r})$ in (30.5.23) becomes

$$\psi(\mathbf{r}) = e^{i\mathbf{k}_i\cdot\mathbf{r}} + \left[-\frac{1}{4\pi}\int d^3\mathbf{r}' e^{-i\mathbf{k}_s\cdot\mathbf{r}'}U(\mathbf{r}')\psi(\mathbf{r}')\right]\frac{e^{ikr}}{r}. \tag{30.5.26}$$

Note that in here we have

$$\mathbf{k}_i = k\mathbf{n}_i, \quad \mathbf{k}_s \equiv k\mathbf{n}, \quad |\mathbf{k}_i| = |\mathbf{k}_s| = k. \tag{30.5.27}$$

30.6 The Born Approximation

The Born approximation is a strategy to obtain accurate solutions of the integral equation (30.5.25):

$$\psi(\mathbf{r}) = e^{i\mathbf{k}_i\cdot\mathbf{r}} + \int d^3\mathbf{r}' G_+(\mathbf{r}-\mathbf{r}')U(\mathbf{r}')\psi(\mathbf{r}'). \tag{30.6.1}$$

As we will discuss further below, the Born approximation is valid when the potential is weak or, alternatively, when the energy of the particles is large. In the Born approximation, the unknown $\psi(\mathbf{r}')$ appearing under the integral on the right-hand side of the equation is recursively replaced by its very own expression!

To see how this works, we first rewrite the above equation letting $\mathbf{r} \to \mathbf{r}'$ and using a new $\mathbf{r}''$ for the variable of integration:

$$\psi(\mathbf{r}') = e^{i\mathbf{k}_i\cdot\mathbf{r}'} + \int d^3\mathbf{r}'' G_+(\mathbf{r}'-\mathbf{r}'')U(\mathbf{r}'')\psi(\mathbf{r}''). \tag{30.6.2}$$

Now plug (30.6.2) under the integral in (30.6.1) to get the following:

$$\begin{aligned}\psi(\mathbf{r}) &= e^{i\mathbf{k}_i\cdot\mathbf{r}} + \int d^3\mathbf{r}' G_+(\mathbf{r}-\mathbf{r}')U(\mathbf{r}')e^{i\mathbf{k}_i\cdot\mathbf{r}'} \\ &\quad + \int d^3\mathbf{r}' G_+(\mathbf{r}-\mathbf{r}')U(\mathbf{r}')\int d^3\mathbf{r}'' G_+(\mathbf{r}'-\mathbf{r}'')U(\mathbf{r}'')\psi(\mathbf{r}'').\end{aligned} \tag{30.6.3}$$

This is still an exact equation. We can repeat the procedure one more time to replace the $\psi(\mathbf{r}'')$ under the last integral. We find that

$$\psi(\mathbf{r}) = e^{i\mathbf{k}_i\cdot\mathbf{r}} + \int d^3\mathbf{r}' G_+(\mathbf{r}-\mathbf{r}')\, U(\mathbf{r}')e^{i\mathbf{k}_i\cdot\mathbf{r}'}$$

$$+\int d^3\mathbf{r}' G_+(\mathbf{r}-\mathbf{r}')U(\mathbf{r}')\int d^3\mathbf{r}'' G_+(\mathbf{r}'-\mathbf{r}'')\,U(\mathbf{r}'')e^{i\mathbf{k}_i\cdot\mathbf{r}''} \tag{30.6.4}$$

$$+\int d^3\mathbf{r}' G_+(\mathbf{r}-\mathbf{r}')U(\mathbf{r}')\int d^3\mathbf{r}'' G_+(\mathbf{r}'-\mathbf{r}'')U(\mathbf{r}'')\int d^3\mathbf{r}''' G_+(\mathbf{r}''-\mathbf{r}''')U(\mathbf{r}''')\psi(\mathbf{r}''').$$

This is still exact. The pattern is clear: by iterating this procedure we can form an infinite series, the Born series, which schematically looks as follows:

$$\psi = e^{i\mathbf{k}_i\cdot\mathbf{r}} + \int G\,U e^{i\mathbf{k}_i\mathbf{r}} + \int G\,U \int G\,U e^{i\mathbf{k}_i\mathbf{r}} + \int G\,U \int G\,U \int G\,U e^{i\mathbf{k}_i\mathbf{r}} + \cdots . \tag{30.6.5}$$

The approximation in which we keep the first integral in this iterative series and set all others to zero is called the *first Born approximation* and yields a wave function ψ_{B}:

$$\psi_{\mathrm{B}}(\mathbf{r}) = e^{i\mathbf{k}_i\cdot\mathbf{r}} + \int d^3\mathbf{r}' G_+(\mathbf{r}-\mathbf{r}')U(\mathbf{r}')e^{i\mathbf{k}_i\cdot\mathbf{r}'}. \tag{30.6.6}$$

Note that, effectively, we just replaced $\psi(\mathbf{r}')$ inside the integral by the incoming wave $e^{i\mathbf{k}_i\cdot\mathbf{r}'}$. Approximating the Green's function as we did before for observation points far away, we use (30.5.26) to find that in the first Born approximation

$$\psi_{\mathrm{B}}(\mathbf{r}) = e^{i\mathbf{k}_i\cdot\mathbf{r}} + \left[-\frac{1}{4\pi}\int d^3\mathbf{r}' e^{-i\mathbf{k}_s\cdot\mathbf{r}'}U(\mathbf{r}')e^{i\mathbf{k}_i\cdot\mathbf{r}'}\right]\frac{e^{ikr}}{r}. \tag{30.6.7}$$

We have thus found the scattering amplitude f_k^{B} in the Born approximation:

$$\boxed{f_k^{\mathrm{B}}(\theta,\phi) = -\frac{1}{4\pi}\int d^3\mathbf{r}\, e^{-i\boldsymbol{K}\cdot\mathbf{r}}\,U(\mathbf{r}), \qquad \boldsymbol{K} \equiv \mathbf{k}_s - \mathbf{k}_i.} \tag{30.6.8}$$

Note that we relabeled the integration variable $\mathbf{r}' \to \mathbf{r}$, and we defined the *wave-vector transfer* $\boldsymbol{K}$. The vector $\hbar\boldsymbol{K}$ is the momentum that must be added to the incident momentum $\hbar\mathbf{k}_i = \hbar k\mathbf{n}_i$ to get the scattered momentum $\hbar\mathbf{k}_s = \hbar k\mathbf{n}$. We will call θ the angle between $\mathbf{k}_i$ and $\mathbf{k}_s$. This is in fact the spherical angle θ if we choose axes such that $\mathbf{k}_i$ is along the positive z-axis, and $\mathbf{n}$ is specified by (θ,ϕ). Since both $\mathbf{k}_i$ and $\mathbf{k}_s$ have magnitude k (figure 30.8), we find that

$$K = |\boldsymbol{K}| = 2k\sin\tfrac{\theta}{2}. \tag{30.6.9}$$

The magnitude of $\boldsymbol{K}$ encodes the angle θ, and the direction of $\boldsymbol{K}$ encodes the azimuthal angle ϕ. Equation (30.6.8) tells us that in the Born approximation the scattering amplitude is (up to constants) the Fourier transform of the potential $U(\mathbf{r})$ evaluated at the wave number $\boldsymbol{K}$.

If we have a central potential $V(\mathbf{r}) = V(r)$, we can simplify the expression for the Born scattering amplitude further by performing the radial integration. Using the formula for U in terms of V, we find that

$$f_k^{\mathrm{B}}(\theta) = -\frac{1}{4\pi}\frac{2m}{\hbar^2}\int d^3\mathbf{r}\, e^{-i\boldsymbol{K}\cdot\mathbf{r}}\,V(\mathbf{r}). \tag{30.6.10}$$

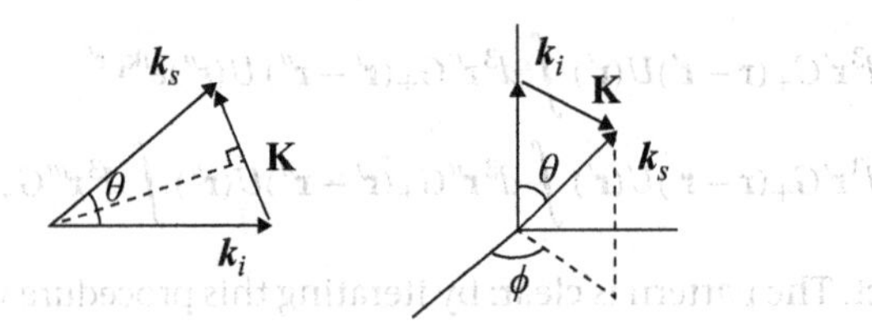

Figure 30.8
Left: The wave vector transfer $\boldsymbol{K}$ has magnitude $K = 2k\sin\frac{\theta}{2}$. *Right*: Choosing $\mathbf{k}_i$ along the z-axis, the wave vector $\boldsymbol{K}$ determines the angles (θ, ϕ).

By spherical symmetry it is clear that this integral just depends on the magnitude K of $\boldsymbol{K}$, not on its direction. This is why the scattering amplitude only depends on θ. To do the integral, think of $\boldsymbol{K}$ as fixed, and let Θ be the angle that the integration variable $\mathbf{r}$ forms with $\boldsymbol{K}$. We then have the following:

$$\begin{aligned} f_k^{\mathrm{B}}(\theta) &= -\frac{m}{2\pi\hbar^2}\int_0^\infty 2\pi r^2 V(r)\,dr \int_{-1}^{1} d\cos\Theta\, e^{-iKr\cos\Theta} \\ &= -\frac{m}{\hbar^2}\int_0^\infty dr\, r^2 V(r)\frac{e^{-iKr}-e^{iKr}}{-iKr} = -\frac{m}{\hbar^2}\int_0^\infty dr\, r^2 V(r)\,\frac{2\sin(Kr)}{Kr}. \end{aligned} \tag{30.6.11}$$

We therefore have the Born approximation for a

$$\text{central potential:}\quad f_k^{\mathrm{B}}(\theta) = -\frac{2m}{\hbar^2 K}\int_0^\infty dr\, rV(r)\sin(Kr), \qquad K = 2k\sin\tfrac{\theta}{2}. \tag{30.6.12}$$

Example 30.2. *Scattering amplitude for the Yukawa potential.*
The Yukawa potential is a central potential featuring a Coulomb potential screened by an exponential that falls off at large r:

$$V(\mathbf{r}) = V(r) = \beta\,\frac{e^{-\mu r}}{r}, \qquad \mu > 0. \tag{30.6.13}$$

Here μ is the parameter that controls the screening, and β is a constant. If $\beta = Z_1 Z_2 e^2$, the Yukawa potential is a good approximation to the potential felt by a *nucleus* of change $Z_1 e$ scattering off an *atom* with atomic number Z_2. Indeed, for small r the nucleus feels the full effect of the atomic nucleus, but for r larger than the atom size, the electron cloud shields the atomic nucleus. This requires choosing $\mu \sim 1/R$, with R the atom size.

The scattering amplitude in the Born approximation follows immediately from (30.6.12). The integral is easily done, and we find that

$$f_k^{\mathrm{B}}(\theta) = -\frac{2m\beta}{\hbar^2 K}\int_0^\infty dr\, e^{-\mu r}\sin(Kr) = -\frac{2m\beta}{\hbar^2(\mu^2+K^2)}. \tag{30.6.14}$$

Here, $K = 2k\sin\frac{\theta}{2}$. Letting $\mu \to 0$, one finds the scattering amplitude for the Coulomb potential, which happens to give a cross section identical to that of classical Rutherford scattering (see problem 30.6). □

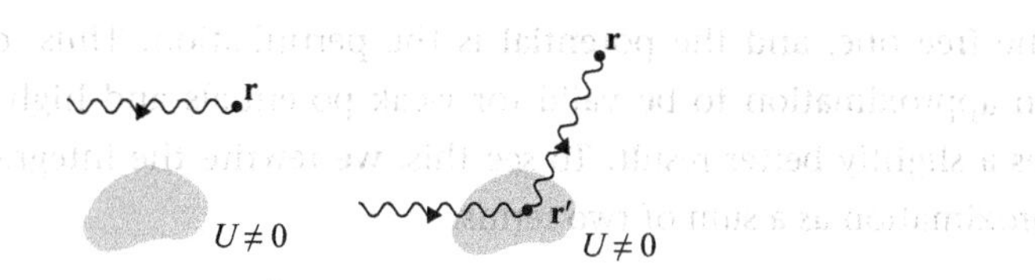

Figure 30.9
Pictorial view of the first Born approximation. The wave at $\mathbf{r}$ is the sum of the free incident wave at $\mathbf{r}$, plus waves coming from secondary sources at points $\mathbf{r}'$ in the region of nonzero potential, induced by the interaction of the incident wave with the potential.

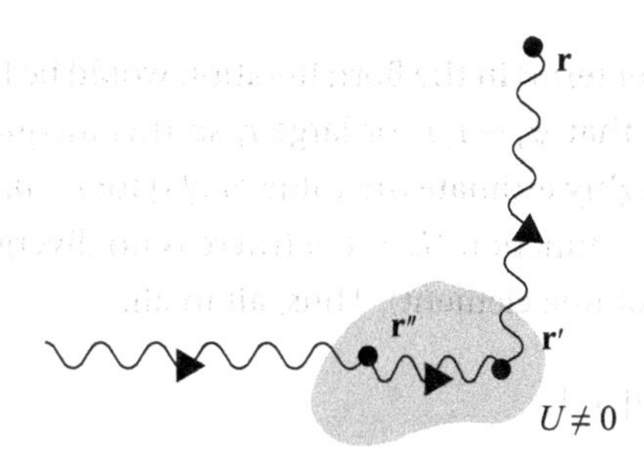

Figure 30.10
Pictorial view of second-order Born term. The incident wave scatters at $\mathbf{r}''$, and the generated wave propagates to $\mathbf{r}'$, where it scatters again, sending a wave all the way to $\mathbf{r}$.

It is possible to give an intuitive graphic representation of the Born series

$$\begin{aligned}\psi(\mathbf{r}) = e^{i\mathbf{k}_i\cdot\mathbf{r}} &+ \int d^3\mathbf{r}' G_+(\mathbf{r}-\mathbf{r}')\, U(\mathbf{r}') e^{i\mathbf{k}_i\cdot\mathbf{r}'} \\ &+ \int d^3\mathbf{r}' G_+(\mathbf{r}-\mathbf{r}') U(\mathbf{r}') \int d^3\mathbf{r}'' G_+(\mathbf{r}'-\mathbf{r}'')\, U(\mathbf{r}'') e^{i\mathbf{k}_i\cdot\mathbf{r}''} + \cdots .\end{aligned} \tag{30.6.15}$$

In the first Born approximation (first line in the above equation, right-hand side), two waves reach the observation point $\mathbf{r}$, as shown in figure 30.9. The first is the direct incident wave. The second is a secondary wave originating at the scattering "material" at a point $\mathbf{r}'$ and carried all the way to $\mathbf{r}$ by the Green's function G_+. The amplitude of the *source* at $\mathbf{r}'$ is given by the product $e^{i\mathbf{k}_i\cdot\mathbf{r}'}U(\mathbf{r}')$ of the incident wave times the "density" $U(\mathbf{r}')$ of scattering material.

In the *second* Born approximation, we include the term on the second line of equation (30.6.15), with two appearances of the Green's function and two appearances of the potential. The graphic representation of this term is shown in figure 30.10. The secondary wave now takes two steps: the incident wave hits scattering material at $\mathbf{r}''$, from which a wave emerges that is propagated by G_+ and hits scattering material at $\mathbf{r}'$, from which a second wave emerges that is propagated by G_+ all the way to $\mathbf{r}$. Higher terms in the Born approximation have an analogous representation.

Let us now briefly discuss the validity of the Born approximation. Note that the Born series is a kind of perturbation theory where the unperturbed Hamiltonian is the free one, and the potential is the perturbation. Thus, one would expect the Born approximation to be valid for weak potentials and high energy. A closer look gives a slightly better result. To see this, we rewrite the integral form (30.6.6) of the approximation as a sum of two terms:

$$\psi_B(\mathbf{r}) = \psi_0(\mathbf{r}) + \psi_1(\mathbf{r}), \quad \psi_0(\mathbf{r}) = e^{i\mathbf{k}_i\cdot\mathbf{r}}, \quad \psi_1 = \int d^3\mathbf{r}' G_+(\mathbf{r}-\mathbf{r}')U(\mathbf{r}')e^{i\mathbf{k}_i\cdot\mathbf{r}'}. \qquad (30.6.16)$$

The Born approximation is good if the term ψ_1 arising from the integral is small compared to the incident wave ψ_0 at all points:

$$|\psi_1| \ll |\psi_0| = 1. \qquad (30.6.17)$$

If this did not hold, higher terms in the Born iteration would be too important to ignore. We have already shown that $\psi_1 \sim 1/r$ for large r, so this inequality will hold for large enough r. Let us now roughly estimate the value of $|\psi_1|$ for $r \sim a$. We assume the integral $\int d^3\mathbf{r}' \sim a^3$ and the Green's function $G_+ \sim 1/a$ (there is no divergence in the integration at $\mathbf{r}' = \mathbf{r}$ because of the volume element). Thus, all in all,

$$|\psi_1| \sim a^3 \frac{1}{a} |U| = a^2 |U| \ll 1, \qquad (30.6.18)$$

with U a typical value of the scaled potential. In terms of V, the required inequality is

$$\frac{2m|V|a^2}{\hbar^2} \ll 1, \quad \text{or} \quad |V| \ll \frac{\hbar^2}{ma^2}. \qquad (30.6.19)$$

This is the condition for a weak potential. In fact, if the potential were constant and negative, this would be the condition that the (three-dimensional) potential has no bound states.

The above condition does not depend on the energy of the incoming particles. For large energy ($ka \gg 1$), however, the condition can be weakened. One now notices that the phase in G_+ will oscillate very quickly, thus cutting off the integral that defines $|\psi_1|$. We will not go into the details, but Landau and Lifschitz do and show that in this case one only needs

$$|V| \ll \frac{\hbar^2}{ma^2} \cdot ka. \qquad (30.6.20)$$

For sufficiently large energy, the Born approximation will always be valid.

Problems

Problem 30.1. *Partial waves.*

The scattering amplitude $f_k(\theta)$ for particles of mass m and momentum $\hbar k$ in a certain potential is given by

$$f_k(\theta) = \frac{1}{k}\left(\frac{\Gamma k}{k_0 - k - ik\Gamma} + 3e^{2i\beta k^3}\sin(2\beta k^3)\cos\theta\right),$$

where Γ, k_0, and β are constants that characterize the potential.

1. What partial waves ℓ are active, and what are the corresponding phase shifts δ_ℓ? Write your answers in terms of k, k_0, Γ, and β. Do they have the typical behavior $\delta_\ell \sim k^{2\ell+1}$ as $k \to 0$?
2. What is the differential cross section $d\sigma/d\Omega$ for general values of k?
3. What are the partial-wave cross sections σ_ℓ?
4. Find the total cross section σ for arbitrary k and the imaginary part of the forward scattering amplitude. How are these two quantities related?

Problem 30.2. *Optical theorem from current conservation.*

We complete here the analysis that began in problem 3.9. With a wave function

$$\Psi(\mathbf{x}) = \Psi_1(\mathbf{x}) + \Psi_2(\mathbf{x}), \quad \text{with} \quad \Psi_1(\mathbf{x}) = e^{ikz}, \quad \text{and} \quad \Psi_2(\mathbf{x}) = \frac{f_k(\theta)}{r} e^{ikr},$$

the total probability current can be written as the sum of three contributions: $\mathbf{J} = \mathbf{J}_1 + \mathbf{J}_2 + \mathbf{J}_{12}$. The optical theorem follows from the condition that the flux of $\mathbf{J}$ across a large sphere of radius R centered at the target is zero. In problem 3.9 you calculated the flux of $\mathbf{J}_1$ and the flux of $\mathbf{J}_2$. To complete the derivation of the optical theorem, we calculate here the flux Φ_{12} of $\mathbf{J}_{12}$:

$$\Phi_{12} \equiv \int R^2 d\Omega \, \hat{\mathbf{r}} \cdot \mathbf{J}_{12}.$$

1. Using the result for $\hat{\mathbf{r}} \cdot \mathbf{J}_{12}$ from problem 3.9, confirm that

$$\Phi_{12} \simeq \frac{\pi \hbar k R}{m} \int_0^\pi d\theta \, \sin\theta (1 + \cos\theta) \left[f_k(\theta) e^{ikR(1-\cos\theta)} + f_k^*(\theta) e^{-ikR(1-\cos\theta)} \right].$$

   With large kR, the integral over θ is evaluated in the stationary phase approximation. Show that the stationary phase condition requires $\sin\theta = 0$ and that from the two possibilities, $\theta = 0$ and $\theta = \pi$, only the first one contributes due to the $(1 + \cos\theta)$ factor in the integrand. Thus,

$$\Phi_{12} \simeq 2\pi \frac{\hbar}{m} kR \int_0^\pi d\theta \, \sin\theta \left[f_k(0) e^{ikR(1-\cos\theta)} + f_k^*(0) e^{-ikR(1-\cos\theta)} \right].$$

2. Consider the integrals $I_\pm(kR)$ required to complete the evaluation of the above flux:

$$I_\pm(kR) \equiv \int_0^\pi d\theta \, \sin\theta \, e^{\pm ikR(1-\cos\theta)}.$$

   The contribution to the integral happens for small θ, and therefore you can use the small-angle approximation in the integrand while, at the same time, extending the integration region to infinity. Verify that with $u \equiv \theta^2/2$, one gets

$$I_\pm = \int_0^\infty du \, e^{\pm ikRu} = \pm \frac{i}{kR}.$$

   Argue that the last equality follows by adding by hand a convergence factor $e^{-\epsilon u}$ to the integral, with $\epsilon > 0$, doing the integral, and then taking the limit as $\epsilon \to 0$.

3. Complete the evaluation of the flux Φ_{12}, and using the results for Φ_1 and Φ_2, derive the optical theorem:

$$\sigma = \frac{4\pi}{k}\mathrm{Im}f_k(0).$$

Problem 30.3. *Spherical well. Part 1: bound states at threshold.*

Consider a particle of mass m in a spherical well of radius R and of depth parameterized by a constant γ with units of one over length:

$$V(\mathbf{r}) = V(r) = \begin{cases} -\frac{\hbar^2}{2m}\gamma^2, & r \leq R, \\ 0, & r > R. \end{cases} \tag{1}$$

In this problem, we work only with the spherical $\ell = 0$ modes.

1. Consider bound states of this potential, setting up the equations that determine their energies $E < 0$. Use the definitions

$$\kappa^2 = \frac{2m|E|}{\hbar^2}, \quad \beta^2 \equiv \gamma^2 - \kappa^2 > 0, \quad x \equiv \beta R, \quad x_0 \equiv \gamma R.$$

Why is $\gamma^2 > \kappa^2$? The equation that determines the bound states takes the form $\tan x = \ldots$, where the right-hand side is a function of x and x_0 that you should determine. This equation is actually satisfied by $x = 0$. Is this a physical solution of the Schrödinger equation?

A bound state with $E = 0^-$ that is arbitrarily close to zero, but still negative, is called a *threshold* bound state. Which values of γR correspond to threshold bound states? Write your answer for γR in terms of an integer $n \geq 0$. For each such value of n, how many bound states $N(n)$ does the potential support? Include the threshold bound state in the count.

2. Consider the case when $\gamma R = \frac{\pi}{2} + \Delta$, with $\Delta \ll 1$ a small constant. Find the threshold bound state energy E_B to leading order in Δ. Write your answer in terms of Δ, R, $\hbar$, m, and numerical constants.

3. Another way to think about threshold bound states is to explore the solutions for zero energy $E = 0$. Determine the radial wave function $u(r)$ for general values of γ. The solutions cannot be normalized, so fix the normalization by requiring that $u \simeq \gamma r$ for $r \to 0$. Write the solution both for $r < R$ and for $r > R$. What happens for $\gamma R = (n + \frac{1}{2})\pi$?

Problem 30.4. *Spherical well. Part 2: scattering solutions.*

This is a continuation of the previous problem, requiring some of the results obtained there. The potential is unchanged, still that in equation (1). This time, however, we consider scattering—in particular, s-wave ($\ell = 0$) scattering of an incoming plane wave with momentum $\hbar k$. The relevant phase shift is δ_0, and the cross section is the partial cross section, $\sigma_0(k) = \frac{4\pi}{k^2}\sin^2\delta_0$.

1. Calculate the phase shift δ_0. Leave your answer in terms of k, R, and the real, positive constant q with units of momentum, defined via the relation

$$q^2 \equiv k^2 + \gamma^2.$$

2. Find the scattering length a_s, defined by $a_s \equiv -\lim_{k\to 0} \frac{\delta_0}{k}$, assuming that $\tan \gamma R$ is finite.

   Your answer should be in terms of γ and R. How does a_s enter into the cross section σ_0 as $k \to 0$? Plot a_s/R as a function of γR. Your plot should have many zeros. For these values of γR, we have $\sigma_0 = 0$, and there is no s-wave scattering. This is the Ramsauer-Townsend effect. Find numerically the smallest positive value of γR for which $a_s = 0$.

   For what values of γR do you get $a_s = \infty$? Describe these values using an integer $n \geq 0$. What happens to δ_0 and σ_0 at these points?

3. Given your answers in (2), you might worry that partial-wave unitarity is violated at low k when a_s is infinite. Working with values of γR that make a_s infinite, compute $\sigma_0(k)$ at small k. How does the value of σ_0 relate to the bound from partial-wave unitarity?

4. The values of γR for which $a_s \to \infty$ actually coincide with those values for which there is a bound state at threshold, as you calculated in the previous problem. Suppose γR is slightly larger than $\pi/2$, so there is a threshold bound state with energy E_B, with E_B negative. Show that for incoming waves of small positive energy E,

$$\sigma_0 \approx \frac{c}{E + |E_B|},$$

   for some constant c that you must find. This result shows that low-energy scattering can be used to detect low-lying bound states. [Hint: assume $\gamma R = \frac{\pi}{2} + \Delta$, with $\Delta \ll 1$ a small positive constant. In the previous problem you found E_B as a function of Δ. Now you must also find δ_0 as a function of Δ.]

Problem 30.5. *Scattering from a δ shell.*

Consider s-wave ($\ell = 0$) scattering of particles of mass m and momentum $\hbar k$ from the potential $V(r)$, whose strength is quantified by a positive unit-free parameter λ:

$$V(r) = \lambda \frac{\hbar^2}{2mR}\, \delta(r - R), \quad \lambda > 0.$$

1. Let u denote the radial solution. By comparing $u'(r)/u(r)$ just inside and just outside the $r = R$ shell, find a formula for $\delta_0(k)$. Write your answer in terms of λ, k, and R.

   When $\lambda \to \infty$, with finite $\cot(kR)$, what does δ_0 become? How does it compare with the phase shift for a hard sphere? What does the cross section σ_0 become?

2. Find the scattering length $a_s \equiv -\lim_{k\to 0} \frac{\delta_0}{k}$ for arbitrary λ (answer in terms of λ and R).

3. Assume $\lambda \gg 1$. Plot $\delta_0(k)$ and $\sin^2(\delta_0(k))$ as functions of kR (take $\lambda = 50$, for example). [Hint: to ensure that $\delta_0(k)$ is continuous, try plotting the function mod π. You may also need to request high resolution to capture all of the features.]

You should observe that as kR approaches $n\pi$ with n a positive integer, $\delta_0(k)$ increases very rapidly by $\sim\pi$. To analyze the large-λ behavior of δ_0 near those points, where in fact $\cot(kR)$ is *not* finite, we have to carefully consider the expression for δ_0. You will find it useful to parameterize the region kR near $n\pi$ using a variable y defined by

$$kR = n\pi - \frac{n\pi}{\lambda} + \frac{n\pi}{\lambda^2} + \left(\frac{n\pi}{\lambda}\right)^2 y.$$

This parameterization was chosen so that δ_0 is just a function of y, and you will see its logic while solving the problem. Find $\delta_0(y)$ (mod π), neglecting terms $O(\lambda^{-1})$.

Assume y ranges from -10 to 10. Using the approximate expression for $\delta_0(y)$ found above, what is the change $\Delta\delta_0/\pi$ in the phase shift in units of π? What is the associated change $\Delta(kR)/\pi$ when $\lambda = 50$, and $kR \sim \pi$ for $n = 1$? What about for $n = 8$? What is the significance of the values $kR = n\pi$? (This question is closely related to the analysis in problem 29.5).

Problem 30.6. *Born approximation, Yukawa and Coulomb potentials, and a computation.*

Consider the scattering of particles of mass m and energy E off a Yukawa potential

$$V(r) = \beta\frac{e^{-\mu r}}{r},$$

where β and $\mu \geq 0$ are constants. Example 30.2 gave the scattering amplitude for this potential in the first Born approximation.

1. Write $\frac{d\sigma}{d\Omega}$ in terms of β, μ, E, θ, m, and $\hbar$. Calculate the associated total cross section σ. The answer can be put in the form $\sigma = \frac{\pi\beta^2}{4E^2}(\cdots)$, where the unknown expression can be constructed in terms of the constant $\delta^2 \equiv \frac{\hbar^2\mu^2}{8mE}$.

   Take $\beta = Q_1Q_2$ and $\mu = 0$ so that we are in the case of a Coulomb potential. Confirm that the differential cross section computed above reproduces the classical Rutherford result:

$$\frac{d\sigma}{d\Omega} = \frac{Q_1^2Q_2^2}{16E^2}\frac{1}{\sin^4(\theta/2)}.$$

2. Differential cross sections $\frac{d\sigma}{d\Omega}$ are what physicists actually use to calculate the rate at which scattered particles will enter their detectors. The number of particles scattered into a solid angle $d\Omega$ per second by a *single scatterer* is given by

$$\frac{d^2N_*^{sc}}{dt\,d\Omega} = \frac{d\sigma}{d\Omega} \times \frac{d^2N^{inc}}{dt\,dA},$$

   where the $*$ is for single scatterer, and $\frac{d^2N^{inc}}{dtdA}$ is the incident flux per unit area transverse to the beam and per unit time (see equation (30.2.3)).

Consider a uniform beam of $\frac{dN^{inc}}{dt}$ particles per second with a cross-sectional area A. This beam strikes a target with n scattering sites per unit volume and thickness T. Give an expression for the number of particles scattered into a detector with angular size $d\Omega$ per unit time:

$$\frac{d^2N^{out}}{dt\, d\Omega} = \cdots .$$

Write your answer in terms of (a subset) of: $\frac{dN^{inc}}{dt}$, n, T, A, $\frac{d\sigma}{d\Omega}$, and $\frac{d^2N^{inc}}{dt\, dA}$. Would the result change if the beam was not uniform across its cross-sectional area A? Explain.

3. Consider a beam of alpha particles ($Q_1 = 2e$) with kinetic energy 8 MeV scattering from a gold foil. Suppose that the beam corresponds to a current of 1 nA (nanoampere). [It is conventional to use MKS (meter-kilogram-second) units for beam currents. One nanoampere is 10^{-9} amperes, meaning 10^{-9} coulombs of charge per second. Each alpha particle has charge $2e$, where $e = 1.602 \times 10^{-19}$ coulombs.] Suppose the gold foil is 1 micron thick. You may assume the alpha particles scatter only off nuclei, not off electrons. You may also assume that each alpha particle scatters only once. You will need to look up the density of gold and the nuclear charge of gold (Q_2). What is the rate $\frac{dN}{dt}$ of alpha particles per second that you expect to be scattered into a detector that occupies a cone of angular extent $d\theta = d\phi = 10^{-2}$ radians, centered at $\theta = \pi/2$?

Problem 30.7. *Born approximation for one-dimensional problems (Griffiths).*

Consider the one-dimensional Schrödinger equation for a particle of mass m moving in a potential $V(x)$. For convenience, define the rescaled potential function $U(x)$, and set the energy E of the particle equal to that of a plane wave with momentum $\hbar k$:

$$V(x) = \frac{\hbar^2}{2m}\, U(x)\,, \qquad E = \frac{\hbar^2 k^2}{2m}.$$

1. Find the explicit form of the Green's function $G(x)$ that will allow you to write the following integral form of the Schrödinger equation for the wave function $\psi(x)$:

$$\psi(x) = \psi_0(x) + \int_{-\infty}^{\infty} dx' G(x - x') U(x') \psi(x'),$$

where $\psi_0(x)$ is a solution with zero potential. Use an outgoing-type Green's function, in analogy to our three-dimensional case. Write $G(x)$ in terms of x, k, and constants.

2. Consider one-dimensional scattering on the open line with a potential $V(x)$ that is nonzero only for $x \in [-x_0, x_0]$ for some $x_0 > 0$. Consider the above integral equation, setting $\psi_0(x) = Ae^{ikx}$, representing the incident wave from the left. Show that to first order in the Born approximation, the reflection coefficient R for this potential takes the form

$$R \simeq C(k, m, \hbar) \left| \int_{-x_0}^{x_0} dx' e^{2ikx'} V(x') \right|^2 .$$

Here $C(k, m, \hbar)$ is a constant that you must determine.

3. Evaluate the above expression for R when the potential is an attractive delta function:

$$V(x) = -\alpha \delta(x) , \qquad \alpha > 0.$$

Write your answer for R in terms of the particle energy E, m, α, and $\hbar$. Compare with the exact reflection coefficient $R = \left(1 + \frac{2\hbar^2 E}{m\alpha^2}\right)^{-1}$, which you could derive if you wished. Do the answers agree exactly? Do they agree to leading order in α?

References

A large number of quantum mechanics textbooks are available today. Some are suitable for undergraduate courses; some are clearly graduate-level. I include below a list of books that I am familiar with and that, in some way or another, have influenced me while writing this book. No doubt other physicists would have made somewhat different selections. This is not a comprehensive list, and perhaps it does not include some of your favorites.

At an undergraduate level, we have the following textbooks, ordered alphabetically:

C. Cohen-Tannoudji, B. Diu, and F. Laloë, *Volume 1: Concepts, Tools, and Applications*; *Volume 2: Angular Momentum, Spin, and Approximation Methods*, Second Edition, Wiley-VCH (2019).

D. J. Griffiths and D. F. Schroeter, *Introduction to Quantum Mechanics*, Third Edition, Cambridge University Press (2018).

H. C. Ohanian, *Principles of Quantum Mechanics*, Prentice Hall (1989).

B. Schumacher and M. Westmoreland, *Quantum Processes, Systems, and Information*, Cambridge University Press (2010).

R. Shankar, *Principles of Quantum Mechanics*, Second Edition, Plenum Press (1994).

J. S. Townsend, *A Modern Approach to Quantum Mechanics*, Second Edition, University Science Books (2012).

Out of these textbooks, Cohen-Tannoudji et al. has a good amount of advanced material, especially in the second volume. At a graduate level, but also containing some material suitable for advanced undergraduates, we have:

L. Ballentine, *Quantum Mechanics, a Modern Development*, Second Edition, World Scientific (2015).

G. Baym, *Lectures on Quantum Mechanics*, Lecture Notes and Supplements in Physics (1969).

H. A. Bethe and R. Jackiw, *Intermediate Quantum Mechanics*, Third Edition, Frontiers of Physics (1997).

P. A. M. Dirac, *The Principles of Quantum Mechanics*, Fourth Edition, Clarendon Press (1982).

L. D. Landau and E. M. Lifshitz, *Quantum Mechanics (Non-relativistic Theory)*, Third Edition, Butterworth-Heinemann, Elsevier Science (1977).

E. Merzbacher, *Quantum Mechanics*, Third Edition, Wiley (1997).

A. Messiah, *Quantum Mechanics*, Second Edition, Dunod (1995), Dover Edition (2014).

M. A. Nielsen and I. L. Chuang, *Quantum Computation and Quantum Information*, Cambridge University Press (2010).

J. J. Sakurai and J. Napolitano, *Modern Quantum Mechanics*, Second Edition, Cambridge University Press (2017).

L. Schiff, *Quantum Mechanics*, Third Edition, McGraw Hill (1968).

S. Weinberg, *Lectures on Quantum Mechanics*, Second Edition, Cambridge University Press (2015).

Also available on the web are fairly complete and useful lecture notes by a number of authors:

- Robert G. Littlejohn (http://bohr.physics.berkeley.edu/classes/221/1819/221.html)
- John Preskill (http://theory.caltech.edu/~preskill/ph219/)
- David Tong (http://www.damtp.cam.ac.uk/user/tong/teaching.html)

Let us now discuss references that helped shape the various chapters in this book or that can provide some extra reading. We do not reference foundational papers that have shaped the historical development of quantum mechanics. Thus, we largely do not cite papers by Einstein, Schrödinger, Heisenberg, Dirac, and others. The interested reader may consult other sources for such references; Weinberg's textbook, for example, is particularly careful in citing original work. Moreover, we do not attempt to find references for what has become familiar material. This means, for example, no references for the harmonic oscillator, or square well potentials, or Heisenberg operators, among many others. Chapters presenting well-known material may have no references. In citing the above books, I will simply use the author's name. For the additional references below, I include the year of publication.

Chapter 1. Some of the discussion here has been influenced by Dirac's foundational book. Oscillations between clockwise and counterclockwise states of currents in SQUIDs, verifying the picture of superposition, have been experimentally observed in Friedman et al. (2000). These matters are described in the enjoyable short book by Gerry and Bruno (2013). M. Headrick suggested discussing how quantum mechanics deals with the classical impossibility of atoms.

Chapter 2. Interaction-free measurements are nicely exemplified by the work in Elitzur and Vaidman (1993). For early experiments demonstrating this effect, see Kwiat et al. (1995). Our presentation was influenced by the discussion of interaction-free measurements in Schumacher and Westmoreland. Thanks to V. Vuletic for clarifying the origin of the peak at the incoming wavelength λ_i in Compton scattering (figure 2.15).

For some discussion of de Broglie wavelength and Galilean symmetry, see Merzbacher and Weinberg.

Chapter 3. The discussion of free-particle wave functions in section 3.1 is based on Ohanian, chapter 2. See also Weinberg for insight into the thinking that led to the Schrödinger equation and the early attempts to interpret the wave function. For a history of the early days of quantum mechanics, see Bloch (1976).

Chapter 7. The heuristic discussion of the node theorem in section 7.4 is based on Moriconi (2007).

Chapter 8. The treatment of wave packets in the step potential (section 8.2) was motivated by the lucid discussion of this topic in Baym, chapter 4.

Chapter 11. Rydberg atoms are remarkable systems with a host of important applications. See the clear discussion in Kleppner et al. (1981), and the more recent exposition by Kleppner (2014), discussing the use of these atoms in modern applications and in cavity quantum electrodynamics.

Chapter 12. For the remarkable story of the Stern-Gerlach experiment, see Friedrich and Herschbach (2003). See also the story of the discovery of electron spin by Pais (1989). Quantum key distribution and related aspects of quantum cryptography are discussed qualitatively in Gerry and Bruno (2013).

Chapters 13 and 14. The discussion of linear algebra in these two chapters has been heavily influenced by the excellent book by Axler (2015), with a title that lives up to its promise. The reader may consult this book to supplement some of the brief arguments presented here. We largely followed Axler in the definitions of vector spaces, subspaces, direct sums, linear operators, matrix representations, operator inverses, eigenvalues and eigenvectors, in chapter 13. Chapter 14 also follows Axler in setting inner products and defining orthonormal bases, adjoints, and orthogonal projectors, as well as Hermitian and unitary operators.

Chapter 16. The proof in section 16.1 that unitary time evolution leads to a Schrödinger equation is based on an argument in Dirac. Our presentation of the axioms of quantum mechanics (section 16.6) has been influenced by the discussion in Nielsen and Chuang and that in Preskill's lecture notes but differs slightly in that we view their composite system axiom as a somewhat less general "postulate." Moreover, we also include the symmetrization postulate.

Chapter 17. This chapter includes section 17.8 explaining the factorization, or supersymmetric, method. Thanks to R. Jaffe for suggesting the inclusion of this material. There are numerous applications of this method that we could not cover, but they are discussed in Cooper, Khare, and Sukhatme (1995).

Chapter 19. The calculations that lead to Rayleigh's formula in section 19.6 follow the lines of Cohen-Tannoudji et al.

Chapter 20. The algebraic solution of the hydrogen atom has its roots in the work of Pauli, using the conservation of a second vector to find the spectrum of the Kepler problem. For early work on the use of group theory for the hydrogen atom, see Bander and Itzykson (1966).

Chapter 21. Our treatment of exchange degeneracy, permutation operators, and symmetrizers follows that of Cohen-Tannoudji et al. The two-photon experiment is presented in Hong, Ou, and Mandel (1987); see also Gerry and Bruno (2013). Thanks are due to M. Metlitski for explanations on possible statistics of particles in two-dimensional space. For early work, see Leinaas and Myrheim (1977) and Wilczek (1982). Interesting material on this subject can be found in Kitaev (2003), Nayak et al. (2008), and Wen (2004).

Chapter 22. In developing the subject of density matrices, I've benefited from the lecture notes of A. Harrow (unpublished), Preskill, and Headrick (2019). The reader can find additional material in Schumacher and Westmoreland and Nielsen and Chuang. For insightful elaborations on measurement, see the book by Auletta, Fortunato, and Parisi (2009). A readable account of cavity electrodynamics using Rydberg atoms for quantum nondemolition measurements can be found in Haroche and Raimond (1993). For more technical aspects of the field, see the review by Walther et al. (2006). For the decoherent histories approach to quantum mechanics, see Gell-Mann and Hartle (2007). For decoherence and measurement, see Zurek (2014). A defense of the many worlds interpretation of quantum mechanics can be found in Carroll (2019).

Chapter 23. The classic reference for quantum computation is Nielsen and Chuang.

Chapter 24. Instructive material on particles in electromagnetic fields and on the physics of Landau levels can be found in Ballentine and in Cohen-Tannoudji et al. For a clear introduction to the Dirac equation, see Shankar.

Chapter 25. The anharmonic oscillator, our example for nondegenerate perturbation theory, was studied in great depth by Bender and Wu (1969). Their results in equation (2.12) must all be multiplied by a factor of 2 to fit our conventions, as they take $A_0 = 1/2$. Our analysis of degenerate perturbation theory is more detailed and goes further than that of most textbooks. The derivation of the fine-structure Hamiltonian for the hydrogen atom from the Dirac equation is given in Shankar. Our calculations of matrix elements for the Zeeman effect use some simple forms of the Wigner-Eckart theorem, which is discussed in all generality in most graduate-level textbooks, such as Merzbacher and Weinberg, for example.

Chapter 26. The Airy functions, used in the WKB connection formulae, are discussed in a practical and useful way by Meissen (2013). Our WKB analysis of level spitting in double-well potentials begins along the lines of Landau and Lifshitz, section 50, but I have relied on notes by M. Metlitski (unpublished) for the full discussion. Such splittings are also computed using methods of quantum field theory; see, for example, Coleman (1979) and Kleinert (2009), section 17.7.

Chapter 27. Our general discussion of time-dependent perturbation theory has been influenced by Cohen Tannoudji et al. and Merzbacher. The presentation of the work of Weisskopf and Wigner (1930) on the decay of a discrete state coupled to a continuum follows Cohen Tannoudji et al. The discussion of hydrogen ionization faced a subtlety in which two apparently equivalent computations give answers that differ by a factor of two. M. Metlitski (unpublished) clarified that one computation is seriously affected by the plane wave approximation for the ejected electron, while the other one is not. Our exposition of selection rules for dipole transitions largely follows Griffiths.

Chapter 28. For perspectives on the adiabatic theorem and references to the original works, see Avron and Elgart (1998). The proof of the quantum adiabatic theorem in section 28.4 follows Messiah, an appendix of Farhi et al. (2008), and notes by M. Metlitski. We followed Merzbacher in the discussion of the Born-Oppenheimer approximation while noting the emergent gauge invariance.

Chapter 29. The presentation of scattering on the half line, the derivation of Levinson's theorem, and some of the discussion of resonances owe plenty to Ohanian, chapter 11.

Chapter 30. For additional discussion of three-dimensional potential scattering theory, see, for example, Merzbacher and Weinberg. Our exposition did not include the important case of the Coulomb potential nor the useful eikonal approximation. The conditions for the validity of the Born approximation are discussed in Landau and Lifshitz, section 45.

Additional References

G. Auletta, M. Fortunato, and G. Parisi, *Quantum Mechanics*, Cambridge University Press (2009).

J. E. Avron and A. Elgart, "An adiabatic theorem without a gap condition" (1998), arXiv:math-ph/9810004

S. Axler, *Linear Algebra Done Right*, Third Edition, Springer (2015).

M. Bander and C. Itzykson, "Group theory and the hydrogen atom," Rev. Mod. Phys. **38**, 330–345 (1966), doi:10.1103/RevModPhys.38.330

C. M. Bender and T. T. Wu, "Anharmonic oscillator," Phys. Rev. **184**, 1231–1260 (1969).

F. Bloch, "Heisenberg and the early days of quantum mechanics," Phys. Today **29**, 12, 23 (1976), doi:10.1063/1.3024633

S. Carroll, *Something Deeply Hidden: Quantum Worlds and the Emergence of Spacetime*, Dutton (2019).

S. Coleman, "The uses of instantons," Subnucl. Ser. **15**, 805 (1979), Contribution to the 15th Erice School of Subnuclear Physics, https://lib-extopc.kek.jp/preprints/PDF/1978/7805/7805043.pdf

F. Cooper, A. Khare, and U. Sukhatme, "Supersymmetry and quantum mechanics," Phys. Rept. **251**, 267–385 (1995), doi:10.1016/0370-1573(94)00080-M, arXiv:hep-th/9405029 [hep-th]

A. Elitzur and L. Vaidman, "Quantum mechanical interaction-free measurements," Found. Phys. **23**, 987 (1993).

E. Farhi, J. Goldstone, S. Gutmann, and D. Nagaj, "How to make the quantum adiabatic algorithm fail," Int. J. Quantum Inf. **6** (3): 503–516 (2008), arXiv:quant-ph/0512159

B. N. Finkelstein and G. E. Horowitz, *Z. Physik*, **48**, 118 (1928).

J. Friedman, V. Patel, W. Chen, S. K. Tolpygo, and J. E. Lukens, "Quantum superposition of distinct macroscopic states," Nature **406**, 43–46 (2000).

B. Friedrich and D. Herschbach, "Stern and Gerlach: How a bad cigar helped reorient atomic physics," Phys. Today **56**, 12, 53 (2003), doi:10.1063/1.1650229

M. Gell-Mann and J. B. Hartle, "Quasiclassical coarse graining and thermodynamic entropy," Phys. Rev. A **76**: 022104 (2007), arXiv:quant-ph/0609190

C. C. Gerry and K. M. Bruno, *The Quantum Divide*, Oxford University Press (2013).

S. Haroche and J-M. Raimond, "Cavity quantum electrodynamics," Scientific American, April 1993.

M. Headrick, "Lectures on entanglement entropy in field theory and holography," arXiv:1907.08126 [hep-th]

C. K. Hong, Z. Y. Ou, and L. Mandel, "Measurement of subpicosecond time intervals between two photons by interference," Phys. Rev. Lett. **59** (18): 2044 (1987).

A. Yu. Kitaev, "Fault tolerant quantum computation by anyons," Ann. Phys. **303**, 2 (2003), quant-ph/9707021

H. Kleinert, *Path Integrals*, Fifth Edition, World Scientific (2009). http://users.physik.fu-berlin.de/~kleinert/b5/psfiles/pthic17.pdf

D. Kleppner. "Rydberg atoms" (2014), https://www.youtube.com/watch?v=e0IWPEhmMho

D. Kleppner, M. G. Littman, and M. L. Zimmerman, "Highly excited atoms," Scientific American **244**, 130 (May 1981).

P. Kwiat, H. Weinfurter, T. Herzog, A. Zeilinger, and M. Kasevich, "Interaction-free measurement," Phys. Rev. Lett. **74** (24): 4763 (1995).

J. M. Leinaas and J. Myrheim, "On the theory of identical particles," Nuovo Cimento B. **37** (1): 1–23 (1977).

E. Meissen, "Integral asymptotics and the WKB approximation," (2013), https://www.yumpu.com/en/document/read/51724902/integral-asymptotics-and-the-wkb-approximation-

M. Moriconi, "Nodes of wavefunctions," Am. J. Phys. **75**, 284 (2007).

C. Nayak, S. H. Simon, A. Stern, M. Freedman, and S. Das Sarma, "Non-abelian anyons and topological quantum computation," Rev. Mod. Phys. **80**, 1083–1159 (2008), arXiv:0707.1889 [cond-mat.str-el]

A. Pais, "George Uhlenbeck and the discovery of electron spin," Phys. Today **42**, 12, 34 (1989), doi:10.1063/1.881186

H. Walther, B. Varcoe, B-G. Englert, and T. Becker, "Cavity quantum electrodynamics," Rep. Prog. Phys. **69**, 1325 (2006).

V. Weisskopf and E. Wigner, "Berechnung der natürlichen linienbreite auf grund der diracschen lichttheorie," Z. Phys. **63**, 54–73 (1930), doi:10.1007/BF01336786

X. G. Wen, *Quantum Field Theory of Many-Body Systems*, Oxford Graduate Texts (2004).

F. Wilczek, "Quantum mechanics of fractional-spin particles," Phys. Rev. Lett. **49** (14): 957 (1982).

W. H. Zurek, "Quantum Darwinism, classical reality, and the randomness of quantum jumps," Phys. Today **67**, 44–50 (2014), arXiv:1412.5206[quant-ph]

Index

Useful Formulae

constants: $\hbar c \simeq 197.3$ MeV·fm, $\dfrac{e^2}{\hbar c} \simeq \frac{1}{137}$, $m_e c^2 \simeq 0.511$ MeV, $m_p c^2 \simeq 938$ MeV.

relativity: $p = \gamma\, mv$, $E = \gamma mc^2$, $E^2 = p^2c^2 + m^2c^4$, $\gamma = \dfrac{1}{\sqrt{1-\beta^2}}$, $\beta = \dfrac{v}{c}$.

de Broglie: $\lambda = \dfrac{h}{p}$, Compton: $\lambda_C = \dfrac{h}{mc}$.

$$\hat{p} = \frac{\hbar}{i}\frac{\partial}{\partial x}, \quad [\hat{x}, \hat{p}\,] = i\hbar, \quad \hat{\mathbf{p}} = \frac{\hbar}{i}\nabla, \quad [\hat{x}_i, \hat{p}_j\,] = i\hbar\,\delta_{ij}, \quad [\hat{p}_i, f(\hat{\mathbf{x}})] = \frac{\hbar}{i}\frac{\partial f}{\partial x_i}.$$

Schrödinger equation: $i\hbar \dfrac{\partial \Psi}{\partial t}(\mathbf{x}, t) = \Big(-\dfrac{\hbar^2}{2m}\nabla^2 + V(\mathbf{x}, t)\Big)\Psi(\mathbf{x}, t)$,

$$\frac{\partial}{\partial t}\rho(\mathbf{x}, t) + \nabla\cdot \mathbf{J}(\mathbf{x}, t) = 0, \quad \rho(\mathbf{x}, t) = |\Psi(\mathbf{x}, t)|^2, \quad \mathbf{J}(\mathbf{x}, t) = \frac{\hbar}{m}\mathrm{Im}\,[\Psi^*\nabla\Psi]\,.$$

Fourier transforms:

$$\Psi(x) = \frac{1}{\sqrt{2\pi}}\int dk\,\Phi(k)e^{ikx}, \quad \Phi(k) = \frac{1}{\sqrt{2\pi}}\int dx\,\Psi(x)e^{-ikx}, \quad \int dx\,|\Psi(x)|^2 = \int dk\,|\Phi(k)|^2.$$

$$\frac{1}{2\pi}\int_{-\infty}^{\infty} e^{ikx}dx = \delta(k), \quad \frac{1}{(2\pi)^3}\int e^{i\mathbf{k}\cdot\mathbf{x}}\,d^3x = \delta^{(3)}(\mathbf{k}).$$

$$\int_{-\infty}^{\infty} dx \exp\Big(-ax^2 + bx\Big) = \sqrt{\frac{\pi}{a}}\,\exp\Big(\frac{b^2}{4a}\Big), \quad \text{when } \mathrm{Re}(a) > 0.$$

expectation value: $\langle \hat{Q}\rangle \equiv \langle \Psi, \hat{Q}\Psi\rangle$, $i\hbar \dfrac{d}{dt}\langle \hat{Q}\rangle = \big\langle [\hat{Q}, H]\big\rangle$, $\hat{Q}$ time independent.

uncertainty: $\Delta A \equiv \|(\hat{A} - \langle \hat{A}\rangle \mathbb{1})\Psi\|$, $(\Delta A)^2 = \langle \hat{A}^2\rangle - \langle \hat{A}\rangle^2 \geq 0$.

$\Delta A\,\Delta B \geq \big|\langle \Psi|\frac{1}{2i}[\hat{A}, \hat{B}]|\Psi\rangle\big|$, saturation: $(\hat{B} - \langle \hat{B}\rangle\mathbb{1})|\Psi\rangle = i\lambda(\hat{A} - \langle \hat{A}\rangle\mathbb{1})|\Psi\rangle$, $\lambda \in \mathbb{R}$.

$\Delta x\,\Delta p \geq \frac{\hbar}{2}$. For $\psi \sim \exp(-\frac{1}{4}\frac{x^2}{\Delta^2})$, $\Delta x = \Delta$ and $\Delta p = \frac{\hbar}{2\Delta}$.

$\Delta H \Delta t \geq \frac{\hbar}{2}$, $\Delta t \equiv \dfrac{\Delta Q}{\big|\frac{d\langle Q\rangle}{dt}\big|}$.

stationary state: $\Psi(\mathbf{x},t)=\psi(\mathbf{x})e^{-iEt/\hbar}$, $\quad -\frac{\hbar^2}{2m}\nabla^2\psi(\mathbf{x})+V(\mathbf{x})\psi(\mathbf{x})=E\,\psi(\mathbf{x}).$

variational principle: $E_{gs}\le\langle\psi,\hat{H}\psi\rangle$, for all normalized ψ.

Hellmann-Feynman: $\hat{H}(\lambda)\psi_n(\lambda)=E_n(\lambda)\psi_n(\lambda)\;\rightarrow\;\frac{dE_n(\lambda)}{d\lambda}=\Big\langle\psi_n(\lambda),\frac{d\hat{H}(\lambda)}{d\lambda}\psi_n(\lambda)\Big\rangle.$

virial theorem: one dimension: $\Big\langle\frac{\hat{p}^2}{2m}\Big\rangle=\frac{1}{2}\Big\langle\hat{x}\frac{dV}{d\hat{x}}\Big\rangle$, central potential: $\Big\langle\frac{\hat{\mathbf{p}}^2}{2m}\Big\rangle=\frac{1}{2}\Big\langle r\frac{\partial V}{\partial r}\Big\rangle.$

commutator identities:

$$[A,BC]=[A,B]C+B[A,C],$$

$$e^ABe^{-A}=e^{\mathrm{ad}_A}B = B+[A,B]+\tfrac{1}{2}[A,[A,B]]+\tfrac{1}{3!}[A,[A,[A,B]]]+\cdots,$$

$$e^ABe^{-A}=B+[A,B],\quad \text{if}\quad [A,[A,B]]=0,$$

$$[B,e^A]=[B,A]e^A,\quad \text{if}\quad [A,[A,B]]=0,$$

$$e^{A+B}=e^Ae^Be^{-\frac{1}{2}[A,B]}=e^Be^Ae^{\frac{1}{2}[A,B]}\quad \text{if } [[A,B],A]=[[A,B],B]=0.$$

matrix exponential: $e^{iM\theta}=\mathbb{1}\cos\theta+iM\sin\theta$, if $M^2=\mathbb{1}$.

index manipulation: $\mathbf{a}\cdot\mathbf{b}=a_ib_i$, $\delta_{ij}B_j=B_i$, $\delta_{ii}=3$, $(\mathbf{a}\times\mathbf{b})_i=\epsilon_{ijk}a_jb_k$, $\epsilon_{123}=1$.

$\epsilon_{ijk}\epsilon_{ipq}=\delta_{jp}\delta_{kq}-\delta_{jq}\delta_{kp}$, $\epsilon_{ijk}\epsilon_{ijq}=2\delta_{kq}$.

linear algebra: matrix representation in the basis $(v_1,\dots,v_n)$: $Tv_i=\sum_k T_{ki}v_k$.

basis change: $u_k=\sum_j A_{jk}v_j$, $T(\{u\})=A^{-1}T(\{v\})A$.

inner product: $\langle v,v\rangle\ge 0$, for all v, $\langle v,v\rangle=0$ if and only if $v=0$, $\langle u,v\rangle=\langle v,u\rangle^*$.

Schwarz inequality: $|\langle u,v\rangle|\le\|u\|\,\|v\|$.

Complex vector space: $\langle v,Tv\rangle=0,\ \forall v\in V\ \rightarrow\ T=0$.

adjoint: $\langle u,Tv\rangle=\langle T^\dagger u,v\rangle$, $(T^\dagger)^\dagger=T$, $(ST)^\dagger=T^\dagger S^\dagger$.

Hermitian operator T: $T^\dagger=T$, unitary operator U: $U^\dagger U=UU^\dagger=\mathbb{1}$.

orthogonal projector: $P_U:V\to U$, $P_UP_U=P_U$, $P_U^\dagger=P_U$, $V=\mathrm{range}\,P_U\oplus\mathrm{null}P_U$.

bra-kets: $\langle u|v\rangle\equiv\langle u,v\rangle$, $|Tv\rangle\equiv T|v\rangle$, $\langle Tv|=\langle v|T^\dagger$, $\langle u|T^\dagger v\rangle=\langle Tu|v\rangle$.

$|\alpha_1v_1+\alpha_2v_2\rangle=\alpha_1|v_1\rangle+\alpha_2|v_2\rangle$, $\langle\alpha_1v_1+\alpha_2v_2|=\alpha_1^*\langle v_1|+\alpha_2^*\langle v_2|$.

$T=|u\rangle\langle w|\ \rightarrow\ T^\dagger=|w\rangle\langle u|$, $\mathrm{tr}\,T=\langle w|u\rangle$.

orthonormal basis $|i\rangle$: $\langle i|j\rangle=\delta_{ij}$, $\mathbb{1}=\sum_i|i\rangle\langle i|$.

$T_{ij}=\langle i|T|j\rangle\ \leftrightarrow\ T=\sum_{i,j}T_{ij}\,|i\rangle\langle j|$, $\langle i|T^\dagger|j\rangle=\langle j|T|i\rangle^*$, $(T^\dagger)_{ij}=(T_{ji})^*$.

$[M,M^\dagger]=0\ \leftrightarrow\ M$ is unitarily diagonalizable.

spectral theorem: $T^\dagger = T \;\rightarrow\; T = \sum_k \lambda_k P_k, \quad P_k^\dagger = P_k, \quad P_k P_l = \delta_{kl} P_l, \quad \sum_k P_k = \mathbb{1}.$

Schrödinger picture: $|\Psi, t\rangle = \mathcal{U}(t, t_0)|\Psi, t_0\rangle, \quad \hat{H}(t) = i\hbar \dfrac{\partial \mathcal{U}(t, t_0)}{\partial t}\mathcal{U}(t_0, t),$

$\mathcal{U}(t, 0) = \exp\left(\dfrac{-i\hat{H}t}{\hbar}\right), \quad$ for $\hat{H}$ time independent.

Heisenberg picture: $\hat{A}_H(t) \equiv \mathcal{U}^\dagger(t, 0)\hat{A}_S\,\mathcal{U}(t, 0), \quad i\hbar \dfrac{d\hat{A}_H(t)}{dt} = [\hat{A}_H(t), \hat{H}_H(t)] + i\hbar \dfrac{\partial \hat{A}_H(t)}{\partial t}.$

harmonic oscillator: $\hat{H} = \dfrac{1}{2m}\hat{p}^2 + \dfrac{1}{2}m\omega^2\hat{x}^2 = \hbar\omega\,(\hat{N} + \tfrac{1}{2}), \quad \hat{N} = \hat{a}^\dagger\hat{a},$

$$\hat{a} = \frac{1}{\sqrt{2}L_0}\left(\hat{x} + \frac{i\hat{p}}{m\omega}\right), \quad \hat{a}^\dagger = \frac{1}{\sqrt{2}L_0}\left(\hat{x} - \frac{i\hat{p}}{m\omega}\right), \quad L_0^2 = \frac{\hbar}{m\omega},$$

$$\hat{x} = \frac{L_0}{\sqrt{2}}(\hat{a} + \hat{a}^\dagger), \quad \hat{p} = \frac{i}{\sqrt{2}}\frac{\hbar}{L_0}(\hat{a}^\dagger - \hat{a}),$$

$$[\hat{a}, \hat{a}^\dagger] = 1, \quad [\hat{N}, \hat{a}] = -\hat{a}, \quad [\hat{N}, \hat{a}^\dagger] = \hat{a}^\dagger,$$

$$\hat{a}\varphi_0 = 0, \quad \varphi_0(x) = N_0 \exp\left(-\tfrac{1}{2}\tfrac{x^2}{L_0^2}\right), \quad N_0^2 = \frac{1}{\sqrt{\pi}L_0},$$

$$\varphi_n = \frac{1}{\sqrt{n!}}(a^\dagger)^n\varphi_0, \quad \hat{H}\varphi_n = \hbar\omega\big(n + \tfrac{1}{2}\big)\varphi_n, \quad \hat{N}\varphi_n = n\varphi_n, \quad \langle\varphi_m, \varphi_n\rangle = \delta_{mn},$$

$$\hat{a}^\dagger\varphi_n = \sqrt{n+1}\,\varphi_{n+1}, \quad \hat{a}\,\varphi_n = \sqrt{n}\,\varphi_{n-1}.$$

$$\hat{x}_H(t) = \hat{x}\,\cos\omega t + \frac{\hat{p}}{m\omega}\sin\omega t,$$

$$\hat{p}_H(t) = \hat{p}\,\cos\omega t - m\omega\,\hat{x}\sin\omega t.$$

coherent states: $|\alpha\rangle \equiv e^{\alpha a^\dagger - \alpha^* a}|0\rangle, \quad \alpha = \frac{\langle\hat{x}\rangle}{\sqrt{2}L_0} + i\,\frac{\langle\hat{p}\rangle L_0}{\sqrt{2}\hbar} \in \mathbb{C},$

$$\hat{a}|\alpha\rangle = \alpha|\alpha\rangle, \quad |\alpha\rangle = e^{-\frac{1}{2}|\alpha|^2}e^{\alpha\,a^\dagger}|0\rangle, \quad |\alpha, t\rangle = e^{-i\omega t/2}|e^{-i\omega t}\alpha\rangle.$$

orbital angular momentum: $\hat{L}_x = \hat{y}\hat{p}_z - \hat{z}\hat{p}_y, \quad \hat{L}_y = \hat{z}\hat{p}_x - \hat{x}\hat{p}_z, \quad \hat{L}_z = \hat{x}\hat{p}_y - \hat{y}\hat{p}_x,$

$$[\hat{L}_x, \hat{L}_y] = i\hbar\hat{L}_z, \quad [\hat{L}_y, \hat{L}_z] = i\hbar\hat{L}_x, \quad [\hat{L}_z, \hat{L}_x] = i\hbar\hat{L}_y,$$

$$\hat{\mathbf{L}}^2 \equiv \hat{L}_x\hat{L}_x + \hat{L}_y\hat{L}_y + \hat{L}_z\hat{L}_z, \quad [\hat{\mathbf{L}}^2, \hat{L}_i] = 0.$$

$$\nabla^2 = \frac{1}{r}\frac{\partial^2}{\partial r^2}r - \frac{1}{r^2}\frac{\hat{\mathbf{L}}^2}{\hbar^2} \qquad \hat{L}^2 = -\hbar^2\left(\frac{\partial^2}{\partial\theta^2} + \cot\theta\frac{\partial}{\partial\theta} + \frac{1}{\sin^2\theta}\frac{\partial^2}{\partial\phi^2}\right),$$

$$\hat{L}_z = \frac{\hbar}{i}\frac{\partial}{\partial\phi}, \quad \hat{L}_\pm = \hbar e^{\pm i\phi}\Big(\pm\frac{\partial}{\partial\theta} + i\cot\theta\frac{\partial}{\partial\phi}\Big).$$

$$Y_{\ell m}(\theta, \phi) = \langle\theta\phi|\ell m\rangle = Y_{\ell m}(\Omega) \equiv \mathcal{N}_{\ell m}P_\ell^m(\cos\theta)e^{im\phi},$$

$$\hat{L}_z Y_{\ell m} = \hbar m\,Y_{\ell m}, \quad \hat{\mathbf{L}}^2\,Y_{\ell m} = \hbar^2\,\ell(\ell+1)\,Y_{\ell m},$$

$$\int d\Omega\, Y^*_{\ell'm'}(\Omega)\, Y_{\ell m}(\Omega) = \delta_{\ell'\ell}\delta_{m'm}, \quad \sum_{\ell=0}^{\infty}\sum_{m=-\ell}^{\ell} Y^*_{\ell m}(\Omega')Y_{\ell m}(\Omega) = \delta(\cos\theta - \cos\theta')\delta(\phi - \phi').$$

$$Y_{0,0}(\theta,\phi) = \frac{1}{\sqrt{4\pi}}, \quad Y_{1,\pm 1}(\theta,\phi) = \mp\sqrt{\frac{3}{8\pi}}\sin\theta\, e^{\pm i\phi}, \quad Y_{1,0}(\theta,\phi) = \sqrt{\frac{3}{4\pi}}\cos\theta.$$

central potentials: $\psi(r,\theta,\phi) = \dfrac{u(r)}{r} Y_{\ell m}(\theta,\phi)$,

$$\Big(-\frac{\hbar^2}{2m}\frac{d^2}{dr^2} + V(r) + \frac{\hbar^2\,\ell(\ell+1)}{2mr^2}\Big)u(r) = Eu(r), \quad u(r) \sim r^{\ell+1}, \quad \text{as } r \to 0.$$

spin one-half: $\hat{H} = -\hat{\mu}\cdot\mathbf{B}$, $\hat{\mu} = g\,\dfrac{e\hbar}{2mc}\,\dfrac{1}{\hbar}\hat{\mathbf{S}}$, $\mu_B = \dfrac{e\hbar}{2m_e c}$, $\hat{\mu}_e = -2\,\mu_B\dfrac{\hat{\mathbf{S}}}{\hbar}$.

$$|1\rangle \equiv |z;+\rangle = |+\rangle = \begin{pmatrix}1\\0\end{pmatrix}, \quad |2\rangle \equiv |z;-\rangle = |-\rangle = \begin{pmatrix}0\\1\end{pmatrix},$$

$$\hat{S}_i = \tfrac{\hbar}{2}\sigma_i, \quad \sigma_x = \begin{pmatrix}0 & 1\\ 1 & 0\end{pmatrix}, \quad \sigma_y = \begin{pmatrix}0 & -i\\ i & 0\end{pmatrix}, \quad \sigma_z = \begin{pmatrix}1 & 0\\ 0 & -1\end{pmatrix}.$$

$$[\sigma_i,\sigma_j] = 2i\epsilon_{ijk}\sigma_k, \quad [\hat{S}_i,\hat{S}_j] = i\hbar\,\epsilon_{ijk}\hat{S}_k,$$

$$\sigma_i\sigma_j = \delta_{ij}\mathbb{1} + i\epsilon_{ijk}\sigma_k, \quad (\sigma\cdot\mathbf{a})(\sigma\cdot\mathbf{b}) = \mathbf{a}\cdot\mathbf{b}\,\mathbb{1} + i\sigma\cdot(\mathbf{a}\times\mathbf{b}),$$

$$e^{i\,\mathbf{a}\cdot\sigma} = \mathbb{1}\cos a + i\sigma\cdot\hat{\mathbf{a}}\sin a, \quad \mathbf{a} = \hat{\mathbf{a}}\,a,\ a = |\mathbf{a}|.$$

$$\hat{S}_{\mathbf{n}} \equiv \mathbf{n}\cdot\hat{\mathbf{S}} = \tfrac{\hbar}{2}\,\mathbf{n}\cdot\sigma, \quad \hat{S}_{\mathbf{n}}\,|\mathbf{n};\pm\rangle = \pm\tfrac{\hbar}{2}\,|\mathbf{n};\pm\rangle, \quad |\mathbf{n}\rangle \equiv |\mathbf{n};+\rangle, \quad \langle\mathbf{n}|\hat{\mathbf{S}}|\mathbf{n}\rangle = \tfrac{\hbar}{2}\mathbf{n},$$

$$|\mathbf{n};+\rangle = \cos\tfrac{\theta}{2}|+\rangle + \sin\tfrac{\theta}{2}e^{i\phi}|-\rangle,$$

$$|\mathbf{n};-\rangle = -\sin\tfrac{\theta}{2}\,e^{-i\phi}|+\rangle + \cos\tfrac{\theta}{2}|-\rangle.$$

$$\hat{R}_{\mathbf{n}}(\alpha) \equiv \exp(-\tfrac{i}{\hbar}\alpha\hat{S}_{\mathbf{n}}), \quad \hat{R}_{\mathbf{n}}(\alpha)|\mathbf{n}'\rangle = |\mathbf{n}''\rangle, \quad \text{with} \quad \mathbf{n}'' = \mathcal{R}_{\mathbf{n}}(\alpha)\,\mathbf{n}',$$

$$\hat{R}^\dagger_{\mathbf{n}}(\alpha)\,\hat{\mathbf{S}}\,\hat{R}_{\mathbf{n}}(\alpha) = \mathcal{R}_{\mathbf{n}}(\alpha)\,\hat{\mathbf{S}}.$$

spin precession: $\hat{H} = \omega_L\cdot\hat{\mathbf{S}}$, $\omega_L = -\gamma\mathbf{B}$, $\hat{\mu} = \gamma\hat{\mathbf{S}}$.

general angular momentum: $[\hat{J}_i,\hat{J}_j] = i\hbar\,\epsilon_{ijk}\hat{J}_k \Leftrightarrow \hat{\mathbf{J}}\times\hat{\mathbf{J}} = i\hbar\hat{\mathbf{J}}$, $[\hat{\mathbf{J}}^2,\hat{J}_i] = 0$.

$$\hat{J}_\pm = \hat{J}_x \pm i\hat{J}_y, \quad (\hat{J}_\pm)^\dagger = \hat{J}_\mp \quad \hat{J}_x = \tfrac{1}{2}(\hat{J}_+ + \hat{J}_-), \quad \hat{J}_y = \tfrac{1}{2i}(\hat{J}_+ - \hat{J}_-),$$

$$[\hat{J}_z,\hat{J}_\pm] = \pm\hbar\hat{J}_\pm, \quad [\hat{J}_+,\hat{J}_-] = 2\hbar\hat{J}_z, \quad [\hat{\mathbf{J}}^2,\hat{J}_\pm] = 0,$$

$$\hat{\mathbf{J}}^2 = \hat{J}_+\hat{J}_- + \hat{J}_z^2 - \hbar\hat{J}_z = \hat{J}_-\hat{J}_+ + \hat{J}_z^2 + \hbar\hat{J}_z.$$

$$\hat{\mathbf{J}}^2|jm\rangle = \hbar^2\,j(j+1)|jm\rangle, \quad \hat{J}_z|jm\rangle = \hbar m|jm\rangle, \quad m = -j,\ldots,j,$$

$$\hat{J}_\pm|jm\rangle = \hbar\sqrt{j(j+1) - m(m\pm 1)}\ |j, m\pm 1\rangle.$$

$$\hat{R}^\dagger_{\mathbf{n}}(\alpha)\,\hat{\mathbf{J}}\,\hat{R}_{\mathbf{n}}(\alpha) = \mathcal{R}_{\mathbf{n}}(\alpha)\,\hat{\mathbf{J}}, \quad \hat{R}_{\mathbf{n}}(\alpha) = e^{-i\frac{\alpha}{\hbar}\mathbf{n}\cdot\hat{\mathbf{J}}}.$$

addition of angular momentum: $\hat{\mathbf{J}} = \hat{\mathbf{J}}_1 + \hat{\mathbf{J}}_2$.

uncoupled basis: $|j_1 j_2; m_1 m_2\rangle$ CSCO: $\{\hat{\mathbf{J}}_1^2, \hat{\mathbf{J}}_2^2, \hat{J}_{1z}, \hat{J}_{2z}\}$.

coupled basis: $|j_1 j_2; jm\rangle$ CSCO: $\{\hat{\mathbf{J}}_1^2, \hat{\mathbf{J}}_2^2, \hat{\mathbf{J}}^2, \hat{J}_z\}$.

$j_1 \otimes j_2 = (j_1 + j_2) \oplus (j_1 + j_2 - 1) \oplus \ldots \oplus |j_1 - j_2|$.

$$|j_1 j_2; jm\rangle = \sum_{m_1+m_2=m} |j_1 j_2; m_1 m_2\rangle \underbrace{\langle j_1 j_2; m_1 m_2 | j_1 j_2; jm\rangle}_{\text{Clebsch-Gordan coefficient}}.$$

$$\hat{\mathbf{J}}_1 \cdot \hat{\mathbf{J}}_2 = \tfrac{1}{2}(\hat{J}_{1+}\hat{J}_{2-} + \hat{J}_{1-}\hat{J}_{2+}) + \hat{J}_{1z}\hat{J}_{2z} = \tfrac{1}{2}(\hat{\mathbf{J}}^2 - \hat{\mathbf{J}}_1^2 - \hat{\mathbf{J}}_2^2).$$

hydrogen atom: $\hat{H} = \dfrac{\hat{\mathbf{p}}^2}{2m} - \dfrac{Ze^2}{r}$, $Z = 1$ for hydrogen.

$$E_n = -\frac{Z^2 e^2}{2a_0}\frac{1}{n^2}, \quad a_0 = \frac{\hbar^2}{me^2} \simeq 52.9\,\text{pm}, \quad \frac{e^2}{2a_0} = \tfrac{1}{2}mc^2\alpha^2 \equiv \text{Ry} \simeq 13.6\,\text{eV}.$$

$$\psi_{n\ell m}(\mathbf{x}) = \mathcal{N}\left(\tfrac{r}{a_0}\right)^{\ell}\left(\text{polynomial in } \tfrac{r}{a_0} \text{ of degree } n-(\ell+1)\right) e^{-\frac{Zr}{na_0}}\, Y_{\ell m}(\theta,\phi),$$

$$\psi_{100}(r,\theta,\phi) = \sqrt{\frac{Z^3}{\pi a_0^3}}\, e^{-Zr/a_0}.$$

$$Z=1: \quad \langle r\rangle = \tfrac{1}{2}a_0(3n^2 - \ell(\ell+1)), \quad \left\langle \frac{1}{r}\right\rangle = \frac{1}{a_0 n^2},$$

$$\left\langle \frac{1}{r^2}\right\rangle = \frac{1}{a_0^2 n^3(\ell+\frac{1}{2})}, \quad \left\langle \frac{1}{r^3}\right\rangle = \frac{1}{a_0^3 n^3 \ell\,(\ell+\frac{1}{2})\,(\ell+1)}.$$

fine structure: $E_{n\ell j m_j} = -\dfrac{e^2}{2a_0}\dfrac{1}{n^2}\left[1 + \dfrac{\alpha^2}{n^2}\left(\dfrac{n}{j+\frac{1}{2}} - \dfrac{3}{4}\right)\right]$.

density matrix: $E = \Big\{ (p_1, |\psi_1\rangle), \ldots, (p_n, |\psi_n\rangle)\Big\}$, $p_1, \ldots, p_n > 0$, $p_1 + \cdots + p_n = 1$.

$$\rho_E \equiv \sum_{a=1}^{n} p_a |\psi_a\rangle\langle\psi_a|, \quad \langle\hat{Q}\rangle_E = \text{tr}(\hat{Q}\rho_E).$$

General ρ is positive semidefinite, and $\text{tr}\,\rho = 1$. Pure state $\leftrightarrow$ $\text{tr}\,\rho^2 = 1$.

spin one-half density matrix: $\rho = \tfrac{1}{2}(\mathbb{1} + \mathbf{a}\cdot\boldsymbol{\sigma})$, $|\mathbf{a}| \leq 1$.

time evolution: $i\hbar \dfrac{\partial \rho}{\partial t} = [\hat{H}, \rho]$.

Schmidt decomposition: $|\psi_{AB}\rangle = \displaystyle\sum_{k=1}^{r} \sqrt{p_k}\, |k_A\rangle \otimes |k_B\rangle$, $r \leq d_A \leq d_B$,

$$\rho_A = \sum_{k=1}^{r} p_k |k_A\rangle\langle k_A|, \quad \rho_B = \sum_{k=1}^{r} p_k |k_B\rangle\langle k_B|, \quad \langle k_A|k'_A\rangle = \delta_{k,k'}, \quad \langle k_B|k'_B\rangle = \delta_{k,k'}.$$

Lindblad equation: $\dfrac{\partial\rho}{\partial t} = \dfrac{1}{i\hbar}[H,\rho] + \displaystyle\sum_k \Big(L_k \rho L_k^\dagger - \tfrac{1}{2}\{L_k^\dagger L_k, \rho\}\Big)$.

electromagnetic couplings: $\hat{H} = \frac{1}{2m}\left(\hat{\mathbf{p}} - \frac{q}{c}\mathbf{A}(\hat{\mathbf{x}}, t)\right)^2 + q\,\Phi(\hat{\mathbf{x}}, t)$ (no spin).

gauge transformations: $\mathbf{A}' = \mathbf{A} + \nabla\Lambda, \quad \Phi' = \Phi - \frac{1}{c}\frac{\partial\Lambda}{\partial t}, \quad \Psi' = \exp\left(i\frac{q\Lambda}{\hbar c}\right)\Psi.$

Pauli Hamiltonian (electron): $\hat{H}_{\text{Pauli}} = \frac{1}{2m_e}\left(\hat{\mathbf{p}} + \frac{e}{c}\mathbf{A}\right)^2 + \frac{e\hbar}{2m_e c}\boldsymbol{\sigma}\cdot\mathbf{B} - e\Phi(\hat{\mathbf{x}}, t).$

time-independent perturbation theory:

nondegenerate: $|n\rangle_\lambda = |n^{(0)}\rangle - \lambda\sum_{k\neq n}\frac{\delta H_{kn}}{E_k^{(0)} - E_n^{(0)}}|k^{(0)}\rangle + \mathcal{O}(\lambda^2),$

$$E_n(\lambda) = E_n^{(0)} + \lambda\,\delta H_{nn} - \lambda^2\sum_{k\neq n}\frac{|\delta H_{kn}|^2}{E_k^{(0)} - E_n^{(0)}} + \mathcal{O}(\lambda^3).$$

degeneracy lifted at $\mathcal{O}(\lambda)$: good basis: $\delta H_{IJ} \equiv \langle\psi_I^{(0)}|\delta H|\psi_J^{(0)}\rangle = E_{nJ}^{(1)}\,\delta_{IJ},$

$$|\psi_I\rangle_\lambda = |\psi_I^{(0)}\rangle - \lambda\Big(\sum_p\frac{\delta H_{pI}}{E_p^{(0)} - E_n^{(0)}}|p^{(0)}\rangle + \sum_{K\neq I}\frac{|\psi_K^{(0)}\rangle}{E_{nI}^{(1)} - E_{nK}^{(1)}}\sum_p\frac{\delta H_{Kp}\delta H_{pI}}{E_p^{(0)} - E_n^{(0)}}\Big) + \mathcal{O}(\lambda^2),$$

$$E_{nI}(\lambda) = E_n^{(0)} + \lambda\,\delta H_{II} - \lambda^2\sum_p\frac{|\delta H_{pI}|^2}{E_p^{(0)} - E_n^{(0)}} + \mathcal{O}(\lambda^3).$$

WKB quantization with a, b turning points or hard walls:

$$\int_a^b k(x')dx' = (n+\beta)\pi, \quad n = 0, 1, 2, \cdots, \quad \beta = 1 - \tfrac{1}{4}(\#\text{ of turning points}).$$

time-dependent perturbations: $\hat{H}(t) = \hat{H}^{(0)} + \delta H(t), \quad |\widetilde{\Psi}(t)\rangle \equiv e^{i\hat{H}^{(0)}t/\hbar}|\Psi(t)\rangle,$

$$i\hbar\frac{d}{dt}|\widetilde{\Psi}(t)\rangle = \widetilde{\delta H}(t)|\widetilde{\Psi}(t)\rangle, \quad \widetilde{\delta H}(t) \equiv e^{i\hat{H}^{(0)}t/\hbar}\,\delta H(t)\,e^{-i\hat{H}^{(0)}t/\hbar},$$

$$|\widetilde{\Psi}(t)\rangle = |\Psi(0)\rangle + \int_0^t\frac{\widetilde{\delta H}(t')}{i\hbar}|\Psi(0)\rangle\,dt' + \mathcal{O}(\delta H^2).$$

Fermi's golden rule: $\hat{H} = \hat{H}^{(0)} + V, \quad w = \frac{2\pi}{\hbar}|V_{fi}|^2\rho(E_f), \quad E_f = E_i,$

$\hat{H} = \hat{H}^{(0)} + 2H'\cos\omega t, \quad w = \frac{2\pi}{\hbar}|H'_{fi}|^2\rho(E_f), \quad E_f = E_i \pm \hbar\omega.$

adiabatic approximation: $|\Psi(t)\rangle \simeq e^{i\theta_k(t)}e^{i\gamma_k(t)}|\psi_k(t)\rangle, \quad \hat{H}(t)|\psi_k(t)\rangle = E_k(t)|\psi_k(t)\rangle,$

$$\theta_k(t) = -\frac{1}{\hbar}\int_0^t E_k(t')dt', \quad \gamma_k(t) = \int_0^t \nu_k(t')dt', \quad \nu_k(t) = i\langle\psi_k(t)|\dot{\psi}_k(t)\rangle.$$

Berry's phase: $\gamma_n(\Gamma_{if}) = \int_{\Gamma_{if}}\mathbf{A}_n(\boldsymbol{R})\cdot d\boldsymbol{R}, \quad \mathbf{A}_n(\boldsymbol{R}) \equiv i\langle\psi_n(\boldsymbol{R})|\nabla_{\boldsymbol{R}}|\psi_n(\boldsymbol{R})\rangle.$

scattering on the half line: $\psi(x) = e^{i\delta}\sin(kx+\delta), \quad \psi_s(x) = e^{i\delta}\sin\delta\, e^{ikx}, \quad x > R.$

time delay: $\Delta t = 2\hbar\,\delta'(E_0).$

Levinson's theorem: $N_b = \frac{1}{\pi}\left(\delta(0) - \delta(\infty)\right).$

resonance, Breit-Wigner: $|\psi_s|^2 \simeq \dfrac{\frac{1}{4}\Gamma^2}{(E-E_\alpha)^2 + \frac{1}{4}\Gamma^2}.$

scattering in 3D: $\psi(\mathbf{r}) = \varphi(\mathbf{r}) + \psi_s(\mathbf{r}) \simeq e^{ikz} + f_k(\theta,\phi)\dfrac{e^{ikr}}{r}, \quad r \gg a.$

$$\frac{d\sigma}{d\Omega} = \left|f_k(\theta,\phi)\right|^2, \quad \sigma = \int \left|f_k(\theta,\phi)\right|^2 d\Omega.$$

Rayleigh: $e^{ikz} = \sqrt{4\pi}\displaystyle\sum_{\ell=0}^{\infty}\sqrt{2\ell+1}\, i^\ell\, Y_{\ell 0}(\theta) j_\ell(kr).$

phase shifts: $f_k(\theta) = \dfrac{\sqrt{4\pi}}{k}\displaystyle\sum_{\ell=0}^{\infty}\sqrt{2\ell+1}\, Y_{\ell 0}(\theta) e^{i\delta_\ell}\sin\delta_\ell, \quad \sigma = \dfrac{4\pi}{k^2}\displaystyle\sum_{\ell=0}^{\infty}(2\ell+1)\sin^2\delta_\ell.$

$$\psi(\mathbf{r})\Big|_\ell = \left(A_\ell j_\ell(kr) + B_\ell n_\ell(kr)\right) Y_{\ell 0}(\theta), \quad r > a, \quad \tan\delta_\ell \equiv -\frac{B_\ell}{A_\ell}.$$